AutoCAD 2007

INSTRUCTOR

A Student Guide to Complete Coverage of AutoCAD's Commands and Features

The McGraw-Hill Graphics Series

Providing you with the highest quality textbooks that meet your changing needs requires feedback, improvement, and revision. The team of authors and McGraw-Hill are committed to this effort. We invite you to become part of our team by offering your wishes, suggestions, and comments for future editions and new products and texts.

Please mail or fax your comments to: James A. Leach
c/o McGraw-Hill Higher Education
Engineering Division
111 Huntington Ave., 6th floor
Boston, MA 02199
FAX: 617-867-9849

Titles in the McGraw-Hill Graphics Series

Consulting Editor
Gary R. Bertoline, Purdue University

Bertoline, *Graphic Drawing Workbook*

Bertoline, *Introduction to Graphics Communications for Engineers*

Bertoline/Wiebe, *Fundamentals of Graphics Communication*

Bertoline/Wiebe, *Technical Graphics Communication*

Duff/Maxson, *The Complete Technical Illustrator*

French/Vierck/Foster, *Engineering Drawing and Graphic Technology*

Howard/Musto, *Introduction to Solid Modeling using SolidWorks*

Jensen/Helsel/Short, *Engineering Drawing and Design Student Edition*

Kelley, *Pro/Engineer Assistant*

Kelley, *Pro/Engineer-Wildfire*

Leach, *AutoCAD Companion*

Leach, *AutoCAD Instructor*

Leake, *Autodesk Inventor*

Lieu, *Graphics Interactive CD-ROM*

Tickoo, *Mechanical Desktop Instructor*

AutoCAD® 2007

2007

INSTRUCTOR

A Student Guide to
Complete Coverage of AutoCAD's
Commands and Features

James A. Leach

University of Louisville

 **Higher Education**

Boston   Burr Ridge, IL   Dubuque, IA   New York   San Francisco   St. Louis
Bangkok   Bogotá   Caracas   Kuala Lumpur   Lisbon   London   Madrid   Mexico City
Milan   Montreal   New Delhi   Santiago   Seoul   Singapore   Sydney   Taipei   Toronto

The McGraw-Hill Companies

Higher Education

AUTOCAD ® 2007 INSTRUCTOR: A STUDENT GUIDE TO COMPLETE COVERAGE OF AUTOCAD'S COMMANDS AND FEATURES

Published by McGraw-Hill, a business unit of The McGraw-Hill Companies, Inc., 1221 Avenue of the Americas, New York, NY 10020. Copyright © 2007 by The McGraw-Hill Companies, Inc. All rights reserved. No part of this publication may be reproduced or distributed in any form or by any means, or stored in a database or retrieval system, without the prior written consent of The McGraw-Hill Companies, Inc., including, but not limited to, in any network or other electronic storage or transmission, or broadcast for distance learning.

Some ancillaries, including electronic and print components, may not be available to customers outside the United States.

This book is printed on acid-free paper.

1 2 3 4 5 6 7 8 9 0 CCI/CCI 0 9 8 7 6

ISBN-13   978-0-07-352262-3
ISBN-10   0-07-352262-7

Publisher: *Suzanne Jeans*
Senior Sponsoring Editor: *Bill Stenquist*
Developmental Editors: *Kathleen L. White/Larry Goldberg*
Executive Marketing Manager: *Michael Weitz*
Project Manager: *Joyce Watters*
Lead Production Supervisor: *Sandy Ludovissy*
Associate Media Producer: *Christina Nelson*
Designer: *John Joran*
Compositor: *Laura Hunter/Visual Q*
Typeface: *10.5/12 Palatino*
Printer: *Courier/Kendalville*

Library of Congress Control Number: 2006928662

This book is dedicated to my best pal, Sam, a black lab. He has been beside me since I wrote the first book, thirteen years ago. It looks like this book will be his last.

DEDICATION

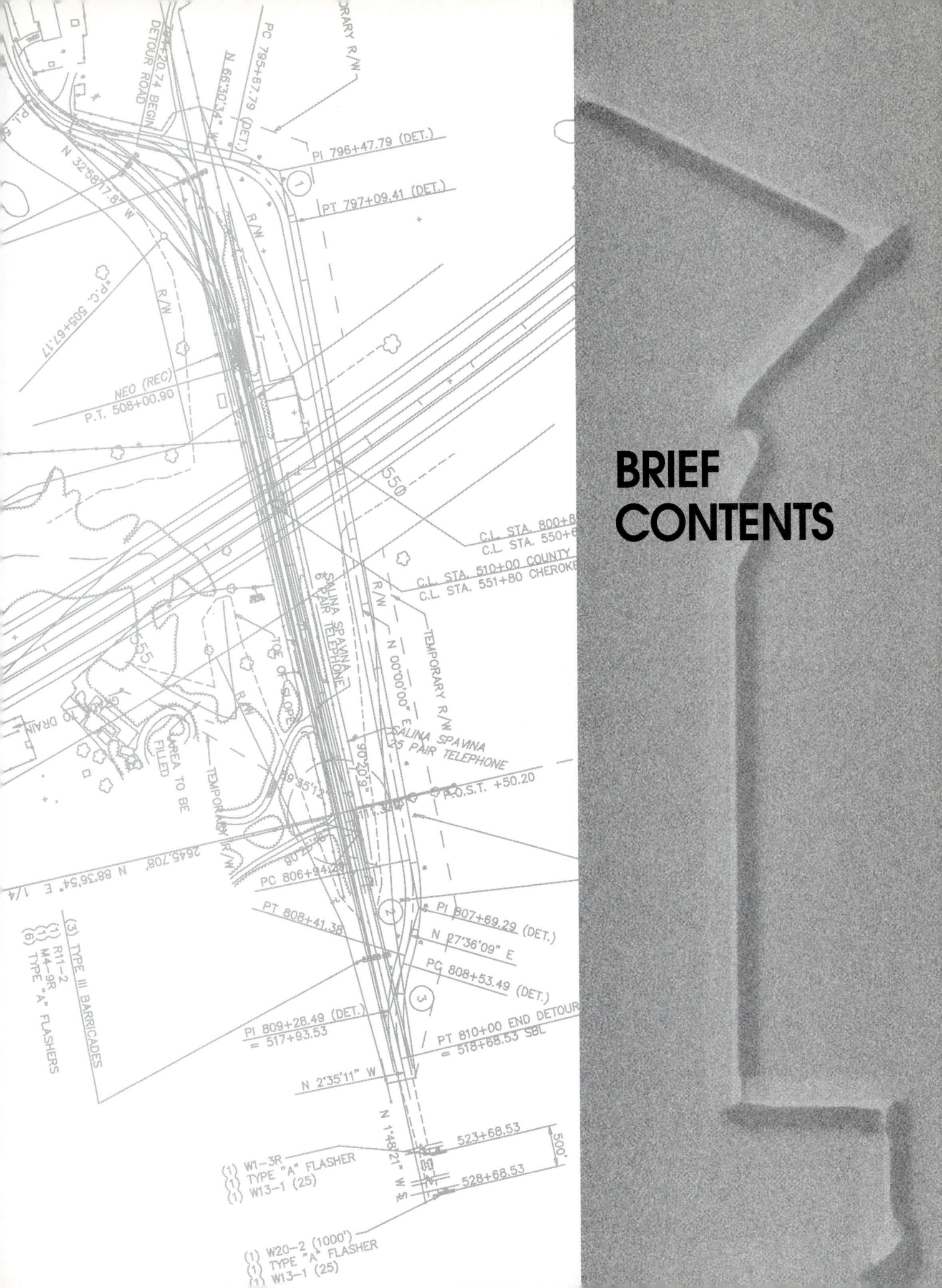

BRIEF
CONTENTS

BRIEF CONTENTS

GUIDED TOUR xxiii
PREFACE xxix
INTRODUCTION xxxv
1.   GETTING STARTED 1
2.   WORKING WITH FILES 25
3.   DRAW COMMAND CONCEPTS 45
4.   SELECTION SETS 67
5.   HELPFUL COMMANDS 81
6.   BASIC DRAWING SETUP 91
7.   OBJECT SNAP AND OBJECT SNAP TRACKING 113
8.   DRAW COMMANDS I 141
9.   MODIFY COMMANDS I 159
10.   VIEWING COMMANDS 199
11.   LAYERS AND OBJECT PROPERTIES 221
12.   ADVANCED DRAWING SETUP 261
13.   LAYOUTS AND VIEWPORTS 275
14.   PRINTING AND PLOTTING 311
15.   DRAW COMMANDS II 341
16.   MODIFY COMMANDS II 371
17.   INQUIRY COMMANDS 407
18.   TEXT AND TABLES 417
19.   GRIP EDITING 465
20.   ADVANCED SELECTION SETS 479
21.   BLOCKS, DESIGNCENTER, AND TOOL PALETTES 501
22.   BLOCK ATTRIBUTES 555
23.   INTERNET TOOLS AND COLLABORATION 587
24.   MULTIVIEW DRAWING 621
25.   PICTORIAL DRAWINGS 641
26.   SECTION VIEWS 655
27.   AUXILIARY VIEWS 681
28.   DIMENSIONING 695
29.   DIMENSION STYLES AND VARIABLES 739
30.   XREFERENCES 799
31.   OBJECT LINKING AND EMBEDDING (OLE) 843
32.   RASTER IMAGES AND VECTOR FILES 865
33.   ADVANCED LAYOUTS AND PLOTTING 887
34.   SHEET SETS 933
35.   3D BASICS, NAVIGATION, AND VISUAL STYLES 961
36.   USER COORDINATE SYSTEMS 1023
37.   WIREFRAME MODELING 1041
38.   SOLID MODELING CONSTRUCTION 1063
39.   SOLID MODEL EDITING 1117
40.   SURFACE MODELING 1177
41.   RENDERING AND ANIMATION 1205
42.   CREATING 2D DRAWINGS FROM 3D MODELS 1269
43.   MISCELLANEOUS COMMANDS AND FEATURES 1207
44.   MISCELLANEOUS CUSTOMIZATION 1293
APPENDICES 1343
INDEX 1447

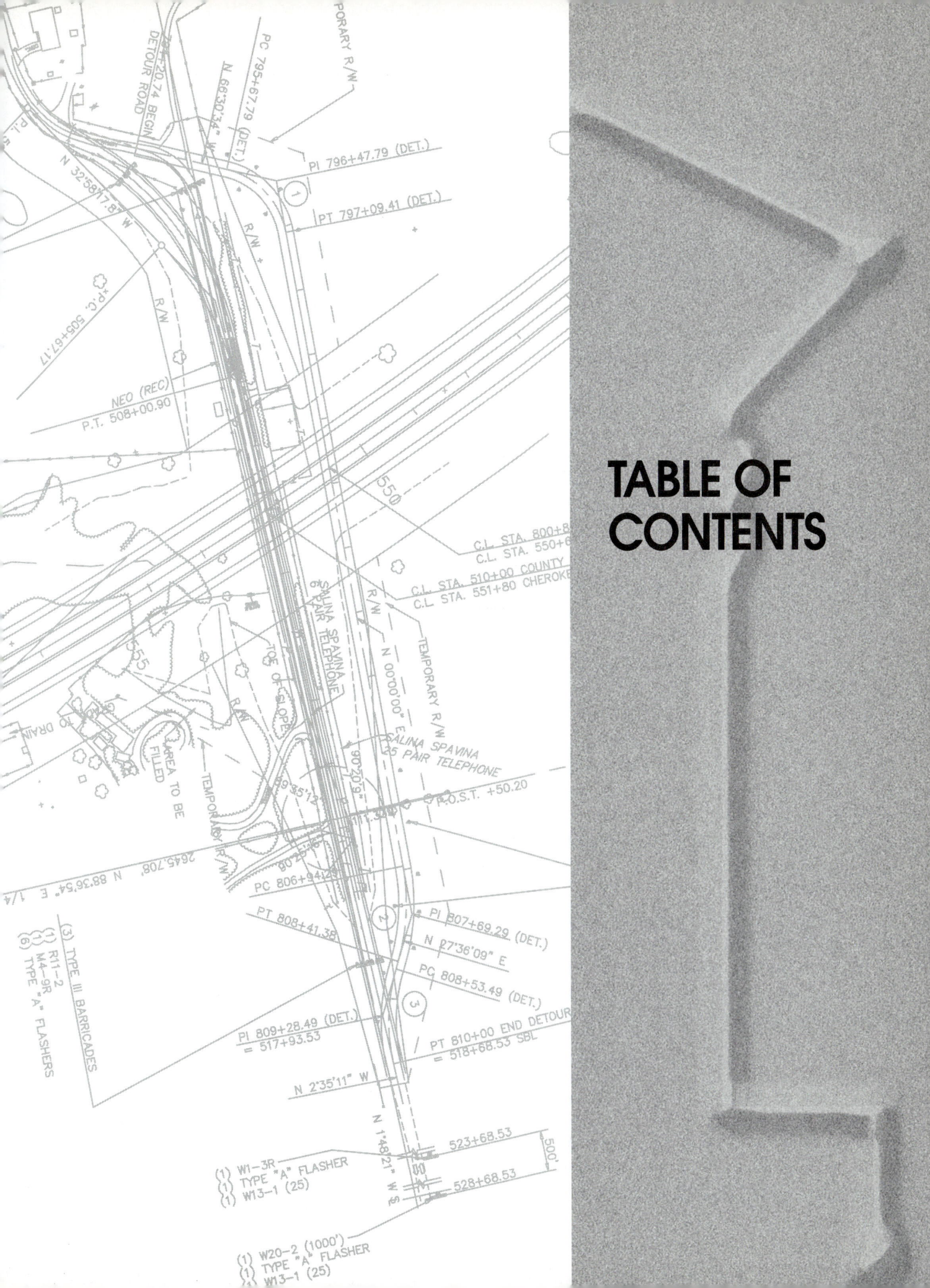

TABLE OF CONTENTS

TABLE OF CONTENTS

GUIDED TOUR xxiii
PREFACE xxix
INTRODUCTION xxxv

CHAPTER 1.
    GETTING STARTED 1
    CONCEPTS 2
        Coordinate Systems 2
        The CAD Database 3
        Angles in AutoCAD 3
        Draw True Size 3
        Plot to Scale 4
    STARTING AutoCAD 5
    THE AutoCAD DRAWING EDITOR 5
        The *Startup* Dialog Box 5
        If AutoCAD Starts in 3D Mode 6
        Graphics Area 6
        Command Line 7
        Palettes 7
        Toolbars .7
        Pull-down Menus 8
        Screen (Side) Menu 8
        Dialog Boxes 9
        Status Bar 10
        Coordinate Display (Coords) 10
        Digitizing Tablet Menu 10
    COMMAND ENTRY 12
        Methods for Entering Commands 12
        Keyboard Entry Features 12
        Using the "Command Tables" in This
            Book to Locate a Particular
            Command 12
        Copy 12
        Shortcut Menus 13
        Mouse and Digitizing Puck Buttons . .14
        Function Keys 14
        Control Key Sequences
            (Accelerator Keys) 15
        Special Key Functions 15
        Drawing Aids 15
        AutoCAD Text Window (F2) 18
        Model Tab and *Layout* Tabs 18
    COMMAND ENTRY METHODS
        PRACTICE 19
    CUSTOMIZING THE AutoCAD
        SCREEN .21
        Toolbars .21
        Clean Screen 22

        Communication Center 22
        Tray Settings 22
        Command Line 23
    CHAPTER EXERCISES 24

CHAPTER 2.
    WORKING WITH FILES 25
    AutoCAD DRAWING FILES 26
        Naming Drawing Files 26
        Beginning and Saving an AutoCAD
            Drawing 26
        Accessing File Commands 26
        File Navigation Dialog Box Functions .27
        Windows Right-Click Shortcut Menus 28
    AutoCAD FILE COMMANDS 29
        New 29
        The *Create New Drawing* Dialog Box . .30
        Qnew 31
        Open 32
        Save 34
        Qsave .34
        Saveas .34
        Close 36
        Closeall 36
        Exit 36
        Quit 37
        Partialopen 37
        Partiaload 38
        Recover 39
        Dwgprops 39
        Savetime 41
        AutoCAD Backup Files 42
        Drawing Recovery 42
        AutoCAD Drawing File Management 42
        Using Other File Formats 43
    CHAPTER EXERCISES 43

CHAPTER 3.
    DRAW COMMAND CONCEPTS 45
    AutoCAD OBJECTS 46
    LOCATING THE DRAW COMMANDS .46
    COORDINATE ENTRY 47
        Coordinate Formats and Types 47
        Coordinate Formats 47
        Coordinate Types 48
        Typical Combinations 48
        Three Methods of Coordinate Input . .48
        Mouse Input 48
        Dynamic Input 49
        Command Line Input 49

PRACTICE USING THE THREE
 COORDINATE INPUT METHODS . . .50
  Line 50
  Circle 50
  Using Mouse Input to Draw Lines
   and Circles 50
  Using Dynamic Input to Draw Lines
   and Circles 52
  Using Command Line Input to Draw
   Lines and Circles 54
  Direct Distance Entry and Angle
   Override 56
  Dynamic Input Settings 56
POLAR TRACKING AND POLAR
 SNAP 57
  Polar Tracking 57
  Polar Tracking with Direct Distance
   Entry 59
  Polar Tracking Override 59
  Polar Snap 59
DRAWING LINES USING POLAR
 TRACKING 60
  Using Polar Tracking and Grid Snap . .60
  Using Polar Tracking and Dynamic
   Input 62
  Using Polar Tracking, Polar Snap, and
   Dynamic Input 63
 CHAPTER EXERCISES 64

CHAPTER 4.
 SELECTION SETS 67
 MODIFY COMMAND CONCEPTS 68
 SELECTION SETS 68
  Selection Set Options 69
  Select 73
 NOUN/VERB SYNTAX 73
 SELECTION SETS PRACTICE 73
  Using *Erase* 74
  Erase 74
  Using *Move* 75
  Move 75
  Using *Copy* 76
  Copy 76
  Noun/Verb Command Syntax
   Practice 77
 CHAPTER EXERCISES 78

CHAPTER 5.
 HELPFUL COMMANDS 81
 CONCEPTS 82
  Help 82

  Assist 85
  Oops 86
  U 86
  Undo 87
  Redo 88
  Mredo 88
  Regen 89
 CHAPTER EXERCISES 89

CHAPTER 6.
 BASIC DRAWING SETUP 91
 STEPS FOR BASIC DRAWING SETUP . .92
 STARTUP OPTIONS 92
 DRAWING SETUP OPTIONS 93
  Start from Scratch 93
  Template 94
  Table of *Start from Scratch Imperial*
   Settings (ACAD.DWT) 94
  Table of *Start from Scratch Metric*
   Settings (ACADISO.DWT) 95
  Wizard 95
 SETUP COMMANDS 97
  Units 97
  Keyboard Input of *Units* Values 99
  Limits 99
  Snap 101
  Grid 103
  Dsettings 104
  Using *Snap* and *Grid* 105
 INTRODUCTION TO LAYOUTS
  AND PRINTING 105
  Model Tab and *Layout* Tabs 106
  Why Set Up Layouts Before
   Drawing? .107
  Printing and Plotting 108
  Setting Layout Options and Plot
   Options 108
 CHAPTER EXERCISES 110

CHAPTER 7.
 **OBJECT SNAP AND OBJECT SNAP
  TRACKING** .113
 CAD ACCURACY 114
 OBJECT SNAP 114
 OBJECT SNAP MODES 115
  Acquisition Object Snap Modes 118
  Object Snap Modifiers 120
 OSNAP SINGLE POINT SELECTION . .121
  Object Snaps (Single Point) 121
  Temporary Override Keys for
   Object Snap 123

OSNAP RUNNING MODE 123
   Object Snaps (Running) 124
   Running Object Snap Modes 124
   Running Object Snap Toggles 124
   Object Snap Cycling 125
   Object Snap Options 125
OBJECT SNAP TRACKING 127
   To Use Object Snap Tracking 128
   Object Snap Tracking with Polar
      Tracking 129
   Object Snap Tracking Settings 130
OSNAP APPLICATIONS 130
OSNAP PRACTICE 130
   Single Point Selection Mode 130
   Running Object Snap Mode 132
   Object Snap Tracking 133
CHAPTER EXERCISES 134

CHAPTER 8.
   DRAW COMMANDS I 141
CONCEPTS 142
   Draw Commands—Simple and
      Complex 142
   Draw Command Access 142
   Coordinate Entry Input Methods 143
COMMANDS 143
   Line 143
   Circle 144
   Arc 146
   Use *Arcs* or *Circles*? 150
   Point 151
   Ddptype 151
   Pline 152
   Pline Arc Segments 154
CHAPTER EXERCISES 154

CHAPTER 9.
   MODIFY COMMANDS I 159
CONCEPTS 160
COMMANDS 161
   Erase 161
   Move 162
   Rotate 163
   Scale 165
   Stretch 166
   Lengthen 167
   Trim 169
   Extend 171
   Trim and *Extend* Shift-Select Option . .172
   Break 173
   Join 174

   Copy 175
   Mirror 177
   Offset 178
   Array 180
   Fillet 183
   Chamfer 186
CHAPTER EXERCISES 188

CHAPTER 10.
   VIEWING COMMANDS 199
CONCEPTS 200
ZOOM and *PAN* with the Mouse Wheel 201
   Zoom and *Pan* 201
   Zoom with the Mouse Wheel 201
   Pan with the Mouse Wheel or
      Third Button 202
COMMANDS 202
   Zoom 202
   Realtime (Rtzoom) 203
   Pan 206
   Zoom and *Pan* with *Undo* and *Redo* . .208
   Scroll Bars 208
   View 208
   Dsviewer 210
   Viewres 211
   Ucsicon 212
   Syswindows 213
   Vports 214
CHAPTER EXERCISES 216

CHAPTER 11.
   LAYERS AND OBJECT
   PROPERTIES 221
CONCEPTS 222
   Assigning Colors, Linetypes, and
      Lineweights 222
   Object Properties 223
   Plot Styles 223
LAYERS AND LAYER PROPERTIES
   CONTROLS .224
   Layer Control Drop-Down List 224
   Layer 224
   Layer Properties Manager Layer List . .224
   Color, Linetype, Lineweight, and
      Other Properties 227
   Layer Filters 228
   Layer States Manager 231
   Displaying an Object's Properties
      and Visibility Settings 233
   Laymcur 233
   Layerp 234

OBJECT-SPECIFIC PROPERTIES
 CONTROLS 234
   Linetype 234
   Linetype Control Drop-Down List . . .236
   Lweight 236
   Lineweight Control Drop-Down List 237
   Color 238
   Color Control Drop-Down List 238
   True Color and *Color Books* 239
CONTROLLING LINETYPE SCALE . . .240
   Ltscale 240
   Celtscale 241
CHANGING OBJECT PROPERTIES 242
   Object Properties Toolbar 242
   Properties 243
   Matchprop .244
LAYER TOOLS 245
   Laymch 245
   Laycur 246
   Layiso 246
   Layuniso 247
   Copytolayer 247
   Laywalk 248
   Layfrz 251
   Laythw 252
   Layoff 252
   Layon 252
   Laylck 253
   Layulk 253
   Layvpi 253
   Laymrg 254
   CHAPTER EXERCISES 255

CHAPTER 12.
 ADVANCED DRAWING SETUP 261
CONCEPTS 262
STEPS FOR DRAWING SETUP 262
USING AND CREATING TEMPLATE
 DRAWINGS .266
   Using Template Drawings 266
   Table of AutoCAD-Supplied
    Template 267
   Creating Template Drawings 271
   Additional Advanced Drawing
    Setup Concepts 272
   CHAPTER EXERCISES 272

CHAPTER 13.
 LAYOUTS AND VIEWPORTS 275
CONCEPTS 276
   Paper Space and Model Space 276

Layouts 276
Viewports 277
Layouts and Viewports Example 279
Guidelines for Using Layouts and
 Viewports .282
CREATING AND SETTING UP
 LAYOUTS 283
   Layoutwizard 283
   Layout 285
   Inserting Layouts with AutoCAD
    DesignCenter 287
   Setting Up the Layout 287
   Options 288
   Pagesetup 289
   Psetupin 290
USING VIEWPORTS IN PAPER SPACE 291
   Vports 291
   Pspace 294
   Mspace 295
   Model 295
   Scaling the Display of Viewport
    Geometry 295
   Locking Viewport Geometry 297
   Linetype Scale in Viewports—
    PSLTSCALE 297
   Vpmax 298
   Vpmin 299
   Scalelistedit 299
   Advanced Applications of Paper
    Space Viewports 300
   CHAPTER EXERCISES 301

CHAPTER 14.
 PRINTING AND PLOTTING 311
CONCEPTS 312
TYPICAL STEPS TO PLOTTING 312
PRINTING, PLOTTING, AND SAVING
 PAGE SETUPS 313
   Plot 313
   -Plot 318
   Viewplotdetails 318
   Pagesetup 319
   Should You Use the *Plot* Dialog Box
    or the *Page Setup* Manager? 321
   Preview 322
PLOTTING TO SCALE 322
   Standard Paper Sizes 323
   Calculating the Drawing Scale
    Factor 323
   Guidelines for Plotting the *Model* Tab
    to Scale 324

Guidelines for Plotting *Layouts*
to Scale 324
TABLES OF *LIMITS* SETTINGS 325
Mechanical Table of *Limits* Settings
for ANSI Sheet Sizes 326
Architectural Table of *Limits* Settings
for Architectural Sheet Sizes 327
Metric Table of *Limits* Settings for
Metric (ISO) Sheet Sizes 328
Metric Table of *Limits* Settings for
ANSI Sheet Sizes 329
Civil Table of *Limits* Settings for
ANSI Sheet Sizes 330
Examples for Plotting to Scale 331
CONFIGURING PLOTTERS AND
PRINTERS 332
Plotting Components 332
Plottermanager 332
Add-A-Plotter Wizard 333
Plotter Configuration Editor 334
.PCP, .PC2, and .PC3 Files 335
Pcinwizard 336
CHAPTER EXERCISES 337

CHAPTER 15.
DRAW COMMANDS II 341
CONCEPTS 342
COMMANDS .342
Xline 342
Ray 344
Polygon 345
Rectang 346
Donut 347
Spline 348
Ellispe 349
Divide 351
Measure 353
Mline 353
Mlstyle 355
Multiline Style Dialog Box 356
New Multiline Style Dialog Box 356
Multiline Styles Example 358
Editing Multilines 358
Sketch 359
The *SKPOLY* Variable 360
Solid 361
Boundary 361
Region 362
Wipeout 364
Revcloud 364
CHAPTER EXERCISES 365

CHAPTER 16.
MODIFY COMMANDS II 371
CONCEPTS 372
COMMANDS .372
Considerations for Changing Basic
Properties of Objects 373
Properties 374
Dblclkedit 377
Matchprop 378
Object Properties Toolbar 379
Chprop 380
Change 380
Explode 381
Align 382
Pedit 383
Vertex Editing 386
Splinedit 389
Editing *Spline* <u>Data Points</u> 389
Editing *Spline* <u>Control Points</u> 391
Mledit 392
Boolean Commands 396
Union 397
Subtract 397
Intersect 398
CHAPTER EXERCISES 399

CHAPTER 17.
INQUIRY COMMANDS 407
CONCEPTS 408
COMMANDS .408
Status 408
List 409
Dblist 410
Area 410
Dist 412
Id 412
Time 413
Setvar 413
Massprop 414
CHAPTER EXERCISES 415

CHAPTER 18.
TEXT AND TABLES 417
CONCEPTS 418
TEXT CREATION COMMANDS 420
Text 420
Mtext 423
Text Formatting Toolbar and
Command Line Options 424
Options Toolbar and *Options* Menu . .426
Bullets and Lists 430

Creating Stacked Text 431
Calculating Text Height for Scaled
    Drawings 431
Spacetrans 432
TEXT STYLES, FONTS, EXTERNAL
    TEXT, AND EXTERNAL EDITORS . .433
Style 433
Importing External Text into
    AutoCAD .436
Using an External Text Editor for
    Multiline Text 437
Format Codes for Creating *Multiline*
    Text in External Text Editors 437
EDITING TEXT 438
Ddedit 439
Mtedit 439
Properties 439
Spell 440
Find 441
Find/Replace in the *Text Formatting*
    Editor 443
Scaletext 443
Justifytext 445
Qtext 446
Textfill 447
Substituting Fonts 447
TEXT FIELDS AND TABLES 448
Text Fields 448
Field 449
Updatefield 450
Creating and Editing Tables 450
Table 451
Tablestyle 452
Tabledit 453
Other Table Editing Methods 454
Text Fields and Calculations
    in Tables 456
CHAPTER EXERCISES 458

CHAPTER 19.
GRIP EDITING 465
CONCEPTS 466
GRIPS FEATURES 466
Activating Grips on Objects 467
Cold and Hot Grips 467
Grip Editing Options 468
Shift to Turn on *ORTHO* 470
Auxiliary Grid 471
Editing Dimensions 471
GUIDELINES FOR USING GRIPS 472
GRIPS SETTINGS 472

Point Specification Precedence 473
More Grips 474
CHAPTER EXERCISES 474

CHAPTER 20.
ADVANCED SELECTION SETS 479
CONCEPTS 480
SELECTION SET VARIABLES 480
Changing the Settings 480
Options 481
PICKFIRST 481
PICKADD 482
PICKAUTO 483
PICKDRAG 484
OBJECT SELECTION FILTERS 484
Qselect 485
Filter 487
OBJECT GROUPS 490
Group 491
Creating a Named Group 491
Changing Properties
    of a Group .493
Groups Examples 495
CHAPTER EXERCISES 497

CHAPTER 21.
BLOCKS, DESIGNCENTER, AND
TOOL PALETTES 501
CONCEPTS 502
COMMANDS .504
Block 504
Block Color, Linetype, and *Lineweight*
    Settings 505
Insert 506
Minsert 507
Explode 507
-Insert with * .508
Xplode 508
Wblock 509
Base 512
Redefining Blocks and the
    Block Editor 512
Bedit 513
Blocks Search Path 515
Purge 515
Rename 517
DESIGNCENTER 518
Adcenter 518
DesignCenter Options 519
To Open a Drawing from
    DesignCenter 524

To *Insert* a *Block* Using
DesignCenter 525
TOOL PALETTES 527
Toolpalettes 527
Creating and Managing Tool
Palettes 529
DYNAMIC BLOCKS 531
Parameters and Actions 532
Steps for Creating Dynamic Blocks . .533
Assigning Parameters to Dynamic
Blocks 534
Associating Actions with
Parameters 538
Visibility States 544
Lookup Actions and Lookup Tables .546
CHAPTER EXERCISES 548

CHAPTER 22.
BLOCK ATTRIBUTES 555
CONCEPTS 556
CREATING ATTRIBUTES 556
Steps for Creating and Inserting
Block Attributes 556
Attdef 557
Saving the Attributed Block 560
Inserting Attributed Blocks 561
The *ATTDIA* Variable 561
The *ATTREQ* Variable 562
DISPLAYING AND EDITING
ATTRIBUTES 562
Attdisp 562
Attedit 563
Multiple Attedit 563
Eattedit 564
Properties 565
Find 565
-Attedit 566
Global and Individual
Editing Example 568
REDEFINING ATTRIBUTES 570
Battman 570
Attredef 573
Refedit, Properties, and
Refclose 574
The Block Editor 574
Grip Editing Attribute Location 575
Attsync 575
EXTRACTING ATTRIBUTES 576
Eattext 576
Attext 580
CHAPTER EXERCISES 583

CHAPTER 23.
INTERNET TOOLS AND
COLLABORATION 587
CONCEPTS 588
Browser 589
Inetlocation .589
Communication Center 590
Communication Center Dialog Box 591
Configuration Settings Dialog Box 591
USING HYPERLINKS 592
Hyperlink 592
Hyperlinkbase 594
Selecturl 594
CREATING EMAIL AND WEB PAGES .594
Publishtoweb 594
Etransmit 597
Transmittal Setups 598
Modify Transmittal Setup
Dialog Box 599
CREATING .DWF FILES 600
Raster Files and Vector Files 601
Creating .DWF Files with ePlot 602
ePlot Options 602
Creating .DWF Files with *Publish* . . .604
Publish 604
AutoCAD's *Sheet Set Manager* 607
VIEWING .DWF FILES 608
Autodesk DWF Viewer 608
DWF Underlay in AutoCAD 610
Dwfattach 610
Dwfclip 612
To Detach a DWF Underlay 613
Viewing URLs (Hyperlinks) in a
.DWF Image 613
MARKUP SET MANAGER 613
The Markup Tool 614
The Markup Process 614
Markups in AutoCAD 615
Opendwfmarkup 615
Markup 615
TRANSFERRING AutoCAD FILES
OVER THE INTERNET 616
Opening and Saving .DWF and
.DWG Files via the Internet 616
Using i-drop to Drag-and-Drop
Drawing Files into AutoCAD 617
CHAPTER EXERCISES 617

CHAPTER 24.
MULTIVIEW DRAWING 621
CONCEPTS 622

PROJECTION AND ALIGNMENT
  OF VIEWS 622
  Using *ORTHO* and *OSNAP* to Draw
    Projection Lines 622
  Using *Polar Tracking* to Draw
    Projection Lines 623
  Using *Object Snap Tracking* to Draw
    Projection Lines Aligned with
    OSNAP Points 624
  Using *Xline* and *Ray* for
    Construction Lines 626
  Using *Offset* for Construction of
    Views 626
  Realignment of Views Using *Polar
    Snap* and *Polar Tracking* 627
USING CONSTRUCTION LAYERS 627
USING LINETYPES 627
  Drawing Hidden and Center Lines . .629
  Managing Linetypes, Lineweights,
    and Colors 630
CREATING FILLETS, ROUNDS, AND
  RUNOUTS 631
GUIDELINES FOR CREATING A TYP-
  ICAL THREE-VIEW DRAWING 633
CHAPTER EXERCISES 637

CHAPTER 25.
  PICTORIAL DRAWINGS 641
CONCEPTS 642
  Types of Pictorial Drawings 642
  Pictorial Drawings Are 2D
    Drawings .643
ISOMETRIC DRAWING IN
  AutoCAD 643
  Snap 644
  Isometric Ellipses 646
  Ellipse 646
  Creating an Isometric Drawing 648
  Dimensioning Isometric Drawings
    in AutoCAD 650
OBLIQUE DRAWING IN AutoCAD 650
CHAPTER EXERCISES 652

CHAPTER 26.
  SECTION VIEWS 655
CONCEPTS 656
DEFINING HATCH PATTERNS AND
  HATCH BOUNDARIES 656
  Step for Creating a Section View
    Using the Hatch Command 657
  Hatch 658

Hatch and Gradient Dialog Box—
  Hatch Tab 659
  Selecting the Hatch Area 662
  Expanded Dialog Box Options 664
  Gradient Tab 666
  -Hatch Command Line Options 667
  Toolpalettes 669
  Creating Tool Palettes 670
EDITING HATCH PATTERNS AND
  BOUNDARIES 671
  Hatchedit 671
  Using Grips with Hatch Patterns 672
  Using *Trim* with Hatch Patterns 672
  Object Selection Features of Hatch
    Patterns .672
  FILLMODE for Solid *Hatch* Fills 673
  Draworder .674
  Texttofront .675
DRAWING CUTTING PLANE
  LINES 675
CHAPTER EXERCISES 677

CHAPTER 27.
  AUXILIARY VIEWS 681
CONCEPTS 682
CONSTRUCTING AN AUXILIARY
  VIEW 682
  Setting Up the Principal Views 682
  Using *Snap Rotate* and *ORTHO* 683
  Rotating *SNAP* Back to the Original
    Position 684
  Using *Polar Tracking* 686
  Using the *Offset* Command 687
  Offset 688
  Using the *Xline* and *Ray* Commands .689
  Xline 689
  Ray 689
  Constructing Full Auxiliary Views . .690
CHAPTER EXERCISES 691

CHAPTER 28.
  DIMENSIONING 695
CONCEPTS 696
DIMENSION DRAWING
  COMMANDS 697
  Dimlinear 697
  Dimaligned .700
  Dimbaseline .701
  Dimcontinue 702
  Dimdiameter 703
  Dimradius 704

Dimjog 705
Dimarc 705
Dimcenter 706
Dimangular 707
Leader 708
Qleader 710
Annotation Tab 711
Leader Line & Arrow Tab 712
Attachment Tab 712
Dimordinate 713
Qdim 715
Tolerance 718
Editing Feature Control Frames 720
Basic Dimensions 720
GDT-Related Dimension Variables . .720
Geometric Dimensioning and
    Tolerancing Example 721
EDITING DIMENSIONS 725
    Grip Editing Dimensions 725
    Dimension Right-Click Menu 726
    Exploding Associative Dimensions . .727
    Dimension Editing Commands 728
    Dimtedit 728
    Dimedit 729
    Properties 730
    Customizing Dimensioning Text 730
    Associative Dimensions 731
    Dimreassociate 733
    Dimdisassociate 734
    Dimregen 735
DIMENSIONING VARIABLES
    INTRODUCTION 735
CHAPTER EXERCISES 736

CHAPTER 29.
DIMENSION STYLES AND
    VARIABLES 739
CONCEPTS 740
    Dimension Variables 740
    Dimension Styles 740
    Dimension Style Families 741
    Dimension Style Overrides 742
DIMENSION STYLES 743
    Dimstyle and *Ddim* 743
    -Dimstyle 746
DIMENSION VARIABLES 748
    Changing Dimension Variables
        Using the Dialog Box Method 748
    Lines Tab 749
    Dimension Lines Section (*Lines* Tab) . .749
    Extension Line Section (*Lines* Tab) 750

Symbols and Arrows Tab 752
Arrowheads Section (*Symbols
    and Arrows* Tab) 752
Center Marks Section (*Symbols
    and Arrows* Tab) 753
Text Tab 753
Text Appearance Section (*Text* Tab) 754
Text Placement Section (*Text* Tab) 755
Text Alignment Section (*Text* Tab) 756
Fit Tab .756
Scale for Dimension Feature
    Section (*Fit* Tab) 756
Fit Options Section (*Fit* Tab) 757
Text Placement Section (*Fit* Tab) 758
Fine Tuning Section (*Fit* Tab) 759
Primary Units Tab 760
Linear Dimensions Section
    (*Primary Units* Tab) 760
Measurement Scale Section
    (*Primary Units* Tab) 762
Zero Suppression Section
    (*Primary Units* Tab) 762
Angular Dimensions Section
    (*Primary Units* Tab) 763
Alternate Units Tab 763
Alternate Units Section (*Alternate
    Units* Tab) 764
Zero Suppression Section (*Alternate
    Units* Tab) 764
Placement Section (*Alternate
    Units* Tab) 765
Tolerances Tab 765
Tolerance Format Section (*Tolerances
    Tab*) .765
Alternate Unit Tolerance Section
    (*Tolerances* Tab) 766
Changing Dimension Variables
    Using Command Line Format 767
DIM...(Variable Name) 767
Associative Dimensions 767
Dimension Variables Table 767
Using *Setup Wizards* and Template
    Drawings 779
MODIFYING EXISTING
    DIMENSIONS 780
Modifying a Dimension Style 780
Properties 781
Creating Dimension Style
    Overrides and Using *Update* 781
Update .781
Dimoverride 782

Matchprop 784
Dimension Right-Click Menu 784
GUIDELINES FOR DIMENSIONING
    IN AutoCAD 785
Dimensioning a Single Drawing 785
Creating Dimension Styles
    as Part of Template Drawings 787
Optional Method for Fixed
    Dimension Text Height in
    Template Drawings 787
DIMENSIONING IN PAPER SPACE
    LAYOUTS 788
When to Dimension in Paper Space ..789
Procedure for Dimensioning in
    Paper Space 790
DIMENSIONING ISOMETRIC
    DRAWINGS 791
CHAPTER EXERCISES 793

CHAPTER 30.
XREFERENCES 799
CONCEPTS 800
An *Xref* Example 804
CREATING AND USING *XREFS* 806
Externalreferences or *Xref* 806
Dependent Objects and Names 811
Xattach 814
Xbind 815
Adcenter 816
VISRETAIN 817
Xclip 818
XCLIPFRAME 820
Demand Loading 820
INDEXCTL 821
XLOADCTL 822
XLOADPATH 823
EDITING *XREFS* AND *BLOCKS* 824
Xopen 825
In-Place *Xref* and *Block* Editing 826
Refedit 826
Using *-Refedit* at the Command Line .828
Refset 829
Refclose 829
XEDIT 830
XFADECTL 831
XREFNOTIFY 831
MANAGING *XREF* DRAWINGS 832
PROJECTNAME 832
Reference Manager 833
XREFCTL 835
CHAPTER EXERCISES 836

CHAPTER 31.
**OBJECT LINKING AND EMBEDDING
    (OLE)** 843
CONCEPTS 844
Cut and Paste Between AutoCAD
    Drawings 844
Cut and Paste Between AutoCAD
    and Other Windows
    Applications (OLE) 845
The Windows Clipboard 847
OLE RELATED COMMANDS 848
Copyclip 849
Copybase 850
Copylink 850
Cutclip 851
Pasteclip 851
Pasteorig 853
Pasteblock 853
Pastespec 854
Insertobj 856
Pasteashyperlink 857
MANAGING OLE OBJECTS 858
Olelinks 858
Olescale 859
Olepen 859
OLE Right-Click Shortcut Menu 859
Other AutoCAD Commands with
    OLE Objects 860
OLE-RELATED SYSTEM VARIABLES ..861
MSOLESCALE 861
OLEHIDE 861
OLEQUALITY 861
OLESTARTUP 861
CHAPTER EXERCISES 862

CHAPTER 32.
**RASTER IMAGES AND VECTOR
    FILES** 865
CONCEPTS 866
Raster Files versus Vector Files 866
Attaching Raster Images and
    Importing Vector Files 867
Exporting Raster and Vector Files ...867
Related Commands 867
ATTACHING RASTER IMAGES 868
Externalreferences 868
Imageattach 871
Imageframe 872
Imageadjust 872
Imageclip 873
Imagequality 874

Transparency 874
Properties 875
Draworder 876
Saveimg 876
Bmpout 876
Jpgout .877
Pngout .877
Tifout .877
IMPORTING AND EXPORTING
    VECTOR FILES 877
Import .878
3Dsin 878
Acisin 879
Acisout 880
Stlout 880
Dxfin 880
Dxfout .880
Dxbin 881
Psout 881
Export .882
Creating PDF Files 883
CHAPTER EXERCISES 884

CHAPTER 33.
    ADVANCED LAYOUTS AND
        PLOTTING 887
CONCEPTS .888
ADVANCED LAYOUTS 888
Layer Visibility for Viewports 889
Layer 890
Vplayer .892
Linetype Scale in Viewports 895
PSLTSCALE 895
Dimensioning in Paper Space 897
The "Reverse Method" for
    Calculating Drawing Scale Factor . .900
Steps for Calculating the Drawing
    Scale Factor by the "Reverse
    Method" 900
Applications for Multiple
    Viewports 901
Using Two Layouts to Plot the Same
    Geometry at Different Scales 902
Using Two Viewports to Display the
    Same Geometry at Different
    Scales .904
Using One Layout to Plot Xref
    Drawing in Multiple Viewports
    at Different Scales 905
Considerations for Using Multiple
    Viewports and Layouts 907

PLOT STYLE TABLES AND PLOT
    STYLES .907
What is the Difference between a
    Plot Style and a Plot Style Table? . .908
Color-Dependent and Named Plot
    Style Tables 910
Attaching a Plot Style Table to a
    Layout .912
Stylesmanager 912
Creating a Plot Style Table 913
Editing Plot Styles 914
Assigning Named Plot Styles to
    Objects 915
Plotstyle .917
Named Plot Style Table Example 918
Convertpstyles 920
Convertctb 921
PLOT STAMPING 921
Plotstamp 921
CHAPTER EXERCISES 923

CHAPTER 34.
    SHEET SETS 933
SHEET SET CONCEPTS 934
Basic Principles 934
SHEET SET COMMANDS 936
Opensheetset 936
The Sheet Set Manager 936
Sheetset .936
To Create a Sheet, Import Layout
    as Sheet 940
To Create a Sheet, New Sheet
    (From Template) 941
To Place a View of a Drawing
    on a Sheet 942
To View and Edit Sheet Set
    Properties 943
To Organize Sheets, Subsets,
    and Views 944
To Place View Label and Callout
    Blocks .945
To Rename and Renumber Sheets
    and Views 946
To Create a Sheet List Table 947
Plotting and Publishing Sheet Sets . . .948
To Transmit and Archive Sheet Sets . .950
Newsheetset 952
An Example Sheet Set 953
Existing Drawings 954
To Create a "Blank" .DST File 955
CHAPTER EXERCISES 955

CHAPTER 35.
3D BASICS, NAVIGATION, AND
    VISUAL STYLE 961
THE AutoCAD 3D INTERFACE 962
    The *Dashboard* .963
TYPES OF 3D MODELS 964
    Wireframe Models 964
    Surface Models 965
    Solid Models .966
3D COORDINATE ENTRY 968
    3D Coordinate Entry Formats 968
COORDINATE SYSTEMS 973
    The World Coordinate System
        (WCS) and WCS Icon 974
    User Coordinate Systems (UCS)
        and Icons 974
    The Right-Hand Rule 975
    Ucsicon 976
3D NAVIGATION AND VIEW
    COMMANDS .978
    3D Navigation Control Panel 979
    Pan 979
    Zoom 979
    3D Orbit Commands 980
    3Dorbit 980
    3Dforbit 982
    3Dcorbit 983
    PERSPECTIVE 983
    3Ddistance 984
    View Presets 985
    View 987
    Adjust the Amount of Perspective . . .989
    3Dclip 990
    Camera 991
    Editing Camera Settings 993
    3Dswivel 994
    Vpoint 994
    Ddvpoint 996
    Plan 996
    Grid 997
    Vports 998
VISUAL STYLES 999
    Vscurrent 1000
    Default Visual Styles 1000
    Visual Styles Properties 1002
    Face Settings .1003
    Environment Settings 1004
    Edge Settings .1005
    Visual Styles Manager 1008
    Adaptive Degradation and
        Performance Tuning 1011

    Shade, SHADEDIF, SHADEDGE 1013
    Render 1013
    Printing 3D Models 1014
CHAPTER EXERCISES 1015

CHAPTER 36.
USER COORDINATE SYSTEMS 1023
CONCEPTS 1024
CREATING AND MANAGING
    USER COORDINATE SYSTEMS . . .1024
    Ucs 1025
    Dducsp 1031
    Ucsman 1032
UCS SYSTEM VARIABLES 1033
    UCSBASE 1034
    UCSVIEW 1034
    UCSORTHO 1034
    UCSFOLLOW 1036
    UCSVP 1036
    UCSAXISANG 1038
USER COORDINATE SYSTEMS AND
    3D MODELING 1038
    Dynamic User Coordinate
        Systems 1038
    The Progression for Learning
        3D Modeling 1038
CHAPTER EXERCISES 1038

CHAPTER 37.
WIREFRAME MODELING 1041
CONCEPTS 1042
WIREFRAME MODELING TUTORIAL 1042
    Hints 1052
COMMANDS .1056
    3Dpoly 1057
    Helix 1058
    Xedges 1059
CHAPTER EXERCISES 1059

CHAPTER 38.
SOLID MODELING
    CONSTRUCTION 1063
CONCEPTS 1064
    Solid Model Construction Process . .1064
    Solid Primitives Commands 1064
    Dynamic User Coordinate Systems .1065
    Commands to Move, Rotate,
        and Mirror Solids 1065
    Boolean Operations 1067
SOLID PRIMITIVES COMMANDS . . .1067
    Box 1068

Cone 1069
Cylinder 1071
Wedge 1072
Sphere 1073
Torus 1074
Pyramid 1074
Polysolid 1076
Extrude 1078
Revolve 1080
Sweep 1081
Loft 1083
DYNAMIC USER COORDINATE
SYSTEMS 1085
COMMANDS FOR MOVING SOLIDS .1087
Move 1087
3Dalign 1088
3Dmove 1090
3Drotate 1092
Mirror3D 1093
3Darray 1096
BOOLEAN OPERATION
COMMANDS 1098
Union 1098
Subtract 1099
Intersect 1100
Chamfer 1102
Fillet 1103
DESIGN EFFICIENCY 1104
CHAPTER EXERCISES 1105

CHAPTER 39.
ADVANCED SOLIDS FEATURES . . .1117
CONCEPTS 1118
EDITING PRIMITIVES USING
THE PROPERTIES PALETTE 1118
Editing Properties of Individual
Primitives 1118
Composite Solid Subobject
Selection 1120
Editing Properties of Composite
Solids 1121
EDITING SOLIDS USING GRIPS 1121
Grip Editing Primitive Solids 1122
Move and Rotate Grip Tools 1127
Grip Editing Composite Solids 1129
Subobjects and Selection 1130
Grip Editing Faces, Edges,
and Vertices 1131
Selecting Overlapping Subobjects . .1135
History Settings 1135
SOLID EDITING COMMANDS 1137

Solidedit 1138
Face Options 1139
Edge Options 1146
Body Options 1147
Slice 1150
Presspull 1151
Sectionplane 1154
Interfere 1157
Imprint 1159
USING SURFACES WITH SOLIDS 1160
Creating Surfaces Composite
with Solids 1160
Planesurf 1161
Slicing Solids with Surfaces 1162
Creating Solids from Surfaces 1164
Thicken 1164
USING SOLIDS DATA 1165
Massprop 1165
Stlout 1167
Acisin 1167
Acisout 1167
CHAPTER EXERCISES 1168

CHAPTER 40.
SURFACE MODELING 1177
CONCEPTS 1178
New Surfaces and Legacy
Surfaces 1178
NON-MESHED SURFACES
WITH STRAIGHT EDGES 1179
3Dface 1179
Edge 1181
Applications of 3Dface 1181
Editing 3Dfaces 1182
CREATING MESHED SURFACES 1182
3Dmesh 1183
CREATING GEOMETRICALLY
DEFINED MESHES 1184
Controlling the Mesh Density 1184
Edgesurf 1184
Rulesurf 1185
Tabsurf 1187
Revsurf 1187
EDITING POLYGON MESHES 1188
Pedit with Polygon Meshes 1189
Variables Affecting Polygon Meshes 1190
SURFTYPE 1190
SPLFRAME 1191
Editing Polygon Meshes with Grips 1191
Editing Polygon Meshes with
Explode 1192

CONSTRUCTION TECHNIQUES
FOR POLYGON MESHES AND
3DFACES 1192
Surface Model Example 1193
Limitations of Polygon Meshes
and *3Dface* 1197
REGION MODELING
APPLICATIONS 1197
USING *THICKNESS* AND
ELEVATION .1198
Thickness 1198
Elevation 1198
Changing *Thickness* and *Elevation* . . .1200
Properties 1200
CHAPTER EXERCISES 1201

CHAPTER 41.
RENDERING AND ANIMATION . . .1205
CONCEPTS 1206
Creating a Default Rendering 1207
Typical Steps for Using Render 1207
Color Systems 1208
DEFINING A BACKGROUND 1210
Background Dialog Box 1211
Renderenvironment 1214
CREATING LIGHTS IN A MODEL 1215
Default Lighting 1216
User-Defined Lighting 1216
Light 1218
Creating a *Point Light* 1220
Creating a *Spotlight* 1221
Creating a *Distant Light* 1221
Using *Sunlight* 1221
Sunproperties 1221
Geographiclocation 1223
Selecting and Modifying Lights 1224
Light Glyphs 1224
Lightlist 1225
Modifying a Light 1225
Convertoldlights 1226
Global *Brightness* and *Contrast* 1226
USING *MATERIALS* IN THE MODEL .1226
Materials Tool Palettes 1227
Materials 1228
Materialattach 1230
CREATING AND MODIFYING
EXISTING MATERIALS 1230
Convertoldmaterials 1235
Creating Wood and Marble
Materials .1235
IMAGE MAPPING 1236

Adjust Bitmap Dialog Box 1238
Removing and Disabling an
Image Map 1240
Materialmap .1241
RENDERING AND SPECIFYING
PREFERENCES 1242
Render 1242
Renderwin 1243
Rendercrop 1244
Rref 1244
Renderpresets 1247
SAVING, PRINTING, AND VIEWING
RENDERINGS 1251
Saveimg 1251
Saving a Rendered Image to a File . .1252
Printing a Rendered Image with
Shade Plot .1252
Miscellaneous Rendering Tips 1253
CREATING AN ANIMATION 1254
Steps for Creating an Animation
Using the *3D Navigation Control
Panel* 1255
*Controlling Animation from the 3D
Navigation Control Panel* 1256
3Dwalk 1257
3Dfly 1258
*Walk and Fly Navigation
Mappings* .1258
Position Locator 1259
Walkflysettings 1260
Creating an Animation
Using *Anipath* 1260
Anipath 1261
Saving an Animation to a File 1263
Playing Back an Animation 1263
CHAPTER EXERCISES 1264

CHAPTER 42.
**CREATING 2D DRAWINGS FROM
3D MODELS** 1269
CONCEPTS 1270
USING *MVSETUP* FOR STANDARD
ENGINEERING DRAWINGS 1271
Mvsetup 1271
Typical Steps for Using the *Standard
Engineering* Option of *Mvsetup* . . .1271
Mvsetup Example 1272
Creating Hidden Lines for a 3D
Wireframe Shown with *Mvsetup* .1275
Creating Dimensions for a 3D
Model Shown with *Mvsetup* 1275

USING *SOLVIEW* AND *SOLDRAW*
TO CREATE MULTIVIEW
DRAWINGS 1276
  Typical Steps for Using *Solview* and
    Soldraw 1276
  Solview 1277
  Soldraw 1278
  Solview and *Soldraw* Example 1278
USING *SOLPROF* AND FLATSHOT
WITH SOLID MODELS 1283
  Solprof 1283
  Solprof 3D Wireframe Example 1285
  Flatshot 1286
  Flatshot versus *Solprof* 1287
CHAPTER EXERCISES 1288

CHAPTER 43.
  MISCELLANEOUS COMMANDS
    AND FEATURES 1293
CONCEPTS 1294
MISCELLANEOUS FEATURES 1294
  Using Wildcards 1294
  Rename 1299
  Purge 1299
  Regenauto 1299
  Fill 1299
  Reinit 1300
  Multiple 1300
  Audit 1301
  FILEDIA 1302
  ATTDIA 1302
USING SLIDES AND SCRIPTS 1302
  Mslide 1303
  Vslide 1304
  Creating and Using Scripts 1304
  Creating a Slide Show 1305
  Delay 1305
  Rscript 1306
  Resume 1306
  Script 1307
  Quickcalc 1308
THE *OPTIONS* DIALOG BOX 1312
  Options 1312
  Files Tab 1312
  Display Tab 1315
  Open and Save Tab 1315
  Plot and Publish Tab 1317
  System Tab 1317
  User Preferences Tab 1319

  Drafting Tab .1321
  3D Modeling Tab 1321
  Selection Tab .1323
  Profiles Tab 1323
CHAPTER EXERCISES 1325

CHAPTER 44.
  MISCELLANEOUS
    CUSTOMIZATION 1329
CONCEPTS 1330
INTRODUCTION TO THE CUI 1330
  CUI or *Toolbar* 1330
  Customizing an Existing
    Toolbar 1331
CREATING CUSTOM LINETYPES 1332
  Creating Simple Linetypes 1333
  Creating Complex Linetypes 1334
CUSTOMIZING THE ACAD.PGP
  FILE 1334
  Specifying External Commands 1335
  Defining Command Aliases 1336
CREATING SCRIPT FILES 1337
  Scripts for Drawing Setups 1338
  Special Script Commands 1339
  Script 1339
  Delay 1339
  Rscript 1339
  Resume 1339
  Logfileon/Logfileoff 1339
MORE CUSTOMIZATION 1341
ADDITIONAL CHAPTERS
  AND RESOURCES 1341

APPENDICES 1343

  APPENDIX A. SYSTEM VARIABLES .1343
  APPENDIX B. AutoCAD COMMAND
    ALIAS LIST SORTED BY
    COMMAND 1413
  APPENDIX C. AutoCAD COMMAND
    ALIAS LIST SORTED BY ALIAS . . .1417
  APPENDIX D. BUTTONS AND
    SPECIAL KEYS 1421
  APPENDIX E. COMMAND TABLE
    INDEX 1425

INDEX 1447

CHAPTER EXERCISE INDEX 1469

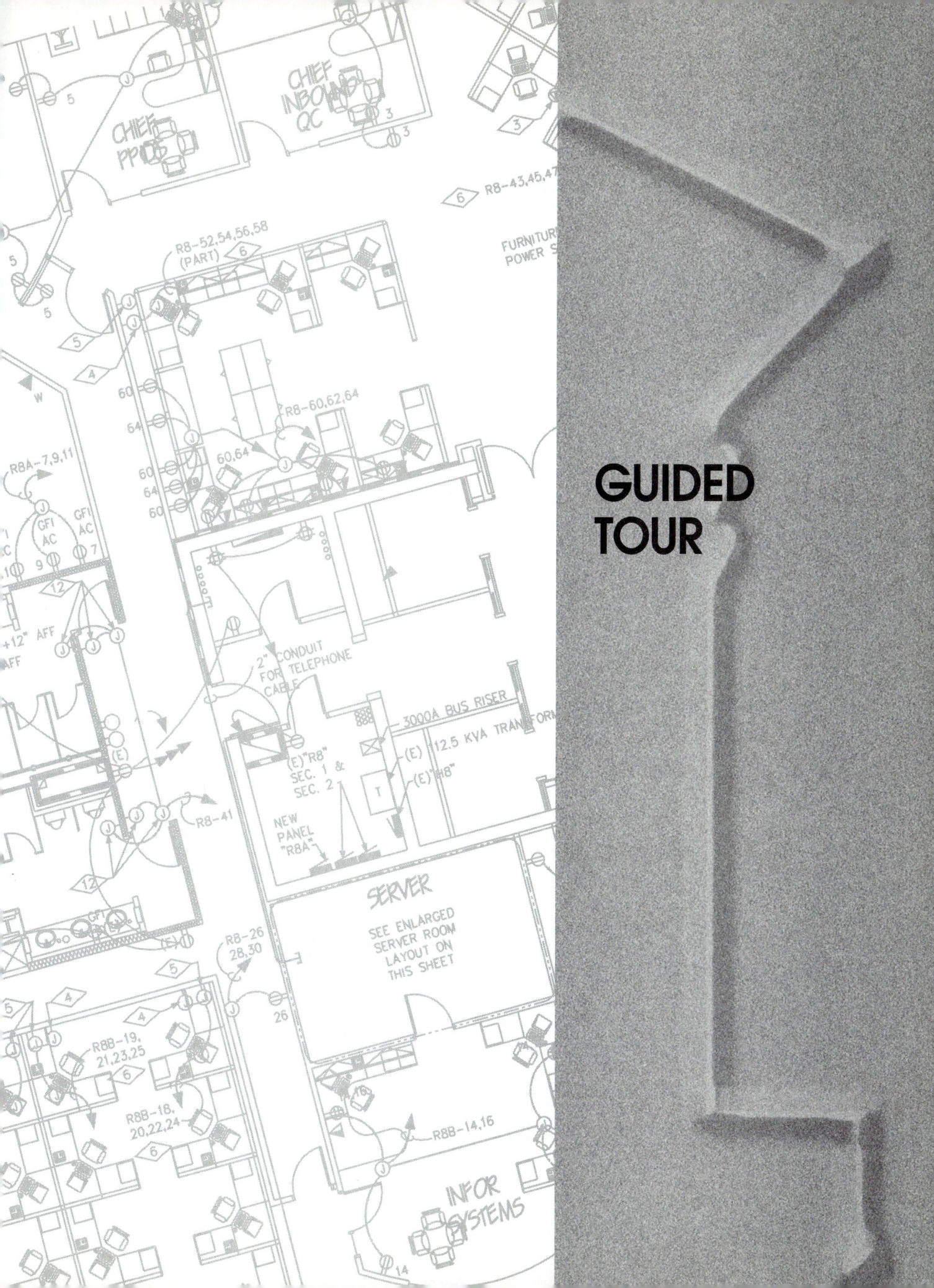

GUIDED TOUR

GUIDED TOUR

Welcome to AutoCAD 2007 Instructor. Here are some features you will find in this book to help you learn AutoCAD 2007.

Chapter Objectives

Each chapter opens with a list of new concepts you can expect to learn. Having objectives in mind helps you focus on the important ideas as you move through each chapter.

C H A P T E R

9

MODIFY COMMANDS I

CHAPTER OBJECTIVES

After completing this chapter you should:

1. be able to *Erase* objects;
2. be able to *Move* objects from a base point to a second point;
3. know how to *Rotate* objects about a base point;
4. be able to enlarge or reduce objects with *Scale*;
5. be able to *Stretch* objects and change the length of *Lines* and *Arcs* with *Lengthen*;
6. be able to *Trim* away parts of objects at cutting edges and *Extend* objects to boundary edges;
7. know how to use the four *Break* options;
8. be able to *Copy* objects and make *Mirror* images of selected objects;
9. know how to create parallel copies of objects with *Offset*;
10. be able to make *rectangular* and *Polar Arrays* of objects;
11. be able to create a *Fillet* and a *Chamfer* between two objects.

---

74   Chapter 4

Begin a *New* drawing to complete the selection set practice exercises. If the *Startup* dialog box appears, select *Start from Scratch*, choose *Imperial* as the default setting, and click the *OK* button. The *Erase*, *Move*, and *Copy* commands can be activated by any one of the methods shown in the command tables that follow.

Using *Erase*

Erase is the simplest editing command. *Erase* removes objects from the drawing. The only action required is the selection of objects to be erased.

Erase

Pull-down Menu	Command (Type)	Alias (Type)	Short-cut	Screen (side) Menu	Tablet Menu
Modify *Erase*	*Erase*	*E*	(Edit Mode) *Erase*	*MODIFY1* *Erase*	*V,14*

1. Draw several *Lines* and *Circles*. Practice using the object selection options with the *Erase* command. The following sequence uses the pickbox, *Window*, and *Crossing Window*.

STEPS	COMMAND PROMPT	PERFORM ACTION	COMMENTS
1.	Command:	type *E* and press **Spacebar**	*E* is the alias for *Erase*, Spacebar can be used like Enter
2.	ERASE Select objects:	use pickbox to select one or two objects	objects are highlighted
3.	Select objects:	type *W* to use a *Window*, then select more objects	objects are highlighted
4.	Select objects:	type *C* to use a *Crossing Window*, then select objects	objects are highlighted
5.	Select objects:	press **Enter**	objects are erased

2. Draw several more *Lines* and *Circles*. Practice using the *Erase* command with the *AUto Window* and *AUto Crossing Window* options as indicated below.

STEPS	COMMAND PROMPT	PERFORM ACTION	COMMENTS
1.	Command:	select the *Modify* pull-down, then *Erase*	
2.	ERASE Select objects:	use pickbox to PICK an open area, drag *Window* to the right to select objects	objects inside *Window* are highlighted
3.	Select objects:	PICK an open area, drag *Crossing Window* to the left to select objects	objects inside and crossing through *Window* are highlighted
4.	Select objects:	press **Enter**	objects are erased

Practice Tables

Step-by-step practice exercises in the early chapters help you systematically tackle fundamental skills.

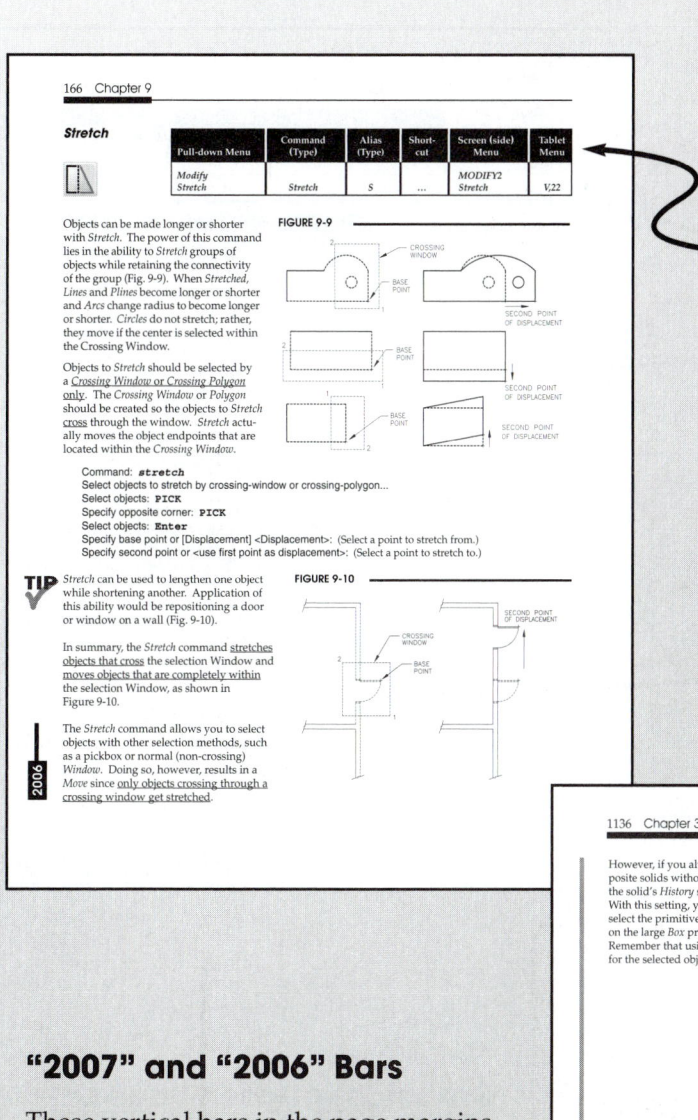

Command Tables

Whenever an AutoCAD command is introduced in the book, this table indicates all the methods you can use to call up the command.

"2007" and "2006" Bars

These vertical bars in the page margins allow you to quickly identify the commands and options that are new in AutoCAD 2007 and in AutoCAD 2006. This feature is especially helpful for readers updating from either AutoCAD 2006 or 2005.

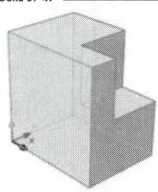

"TIP" Icons

"TIP" icons point out time-saving tricks and tips that otherwise might be discovered only after much experience with AutoCAD.

Offset

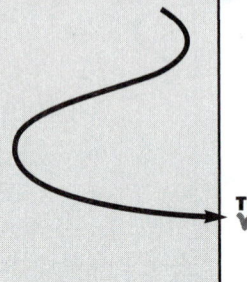

	Pull-down Menu	Command (Type)	Alias (Type)	Short-cut	Screen (side) Menu	Tablet Menu
	Modify Offset	Offset	O	...	MODIFY1 Offset	V,17

Invoke the *Offset* command and specify a distance. The first distance is arbitrary. Specify an appropriate value between the front view inclined plane and the nearest edge of the auxiliary view (20 for the example). *Offset* the new *Line* at a distance of 50 (for the example) or PICK two points (equal to the depth of the view).

TIP Note that the *Offset* lines have lengths equal to the original and therefore require no additional editing (Fig. 27-12).

FIGURE 27-12

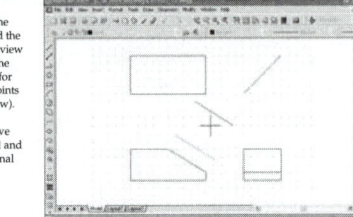

Next, two *Lines* would be drawn between *Endpoints* of the existing offset lines to complete the rectangle. *Offset* could be used again to construct additional lines to facilitate the construction of the two circles in the partial auxiliary view (Fig 27-13).

From this point forward, the construction process would be similar to the example given previously (Figs. 27-5 through 27-8). Even though *Offset* does not require that the *SNAP* be *Rotated*, the complete construction of the auxiliary view could be simplified by using the rotated *SNAP* and *GRID* in conjunction with *Offset*.

FIGURE 27-13

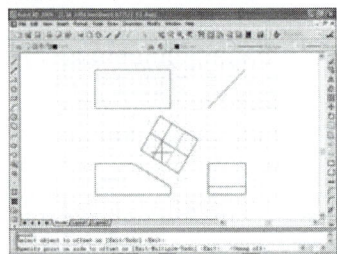

Overlay

An *Xref Overlay* is similar to an *Attached Xref* with one main difference—an *Overlay* cannot be nested. In other words, if you *Attach* a drawing that (itself) has an *Overlay*, the *Overlay* does not appear in your drawing. On the other hand, if the first drawing is *Attached*, it appears when the parent drawing is *Attached* to another drawing (Fig. 30-18).

FIGURE 30-18

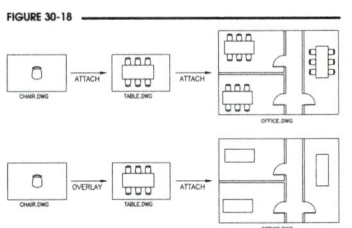

To produce an *Overlay* using the *Xref Manager*, you must first select the *Attach* button to produce the *Select File* dialog box. After selecting the desired file to overlay, press the *Overlay* button in the *External Reference* dialog box that appears (Fig. 30-19). If you type the *-Xref* command, simply use the *Overlay* option and the standard *Select File* dialog box appears.

FIGURE 30-19

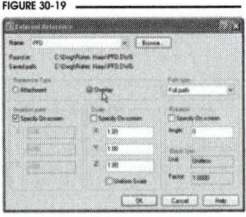

The *Overlay* option prevents "circular" *Xrefs* from appearing by preventing unwanted nested *Xrefs*. This is helpful in a networking environment where many drawings *Xref* other drawings. For example, assume drawing B has drawing A as an *Overlay*. As you work on drawing C, you *Xref* and view only drawing B without drawing B's overlays—namely drawing A. This occurs because drawing A is an *Overlay* to B, not *Attached*. If all drawings are *Overlays*, no nesting occurs (Fig. 30-20).

FIGURE 30-20

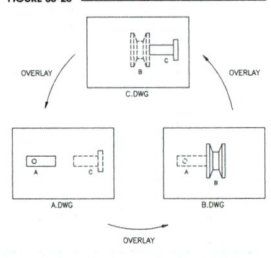

Graphically Driven Page Layout

Approximately 2000 illustrations are given in the book. Explanatory text is located directly next to the related figure.

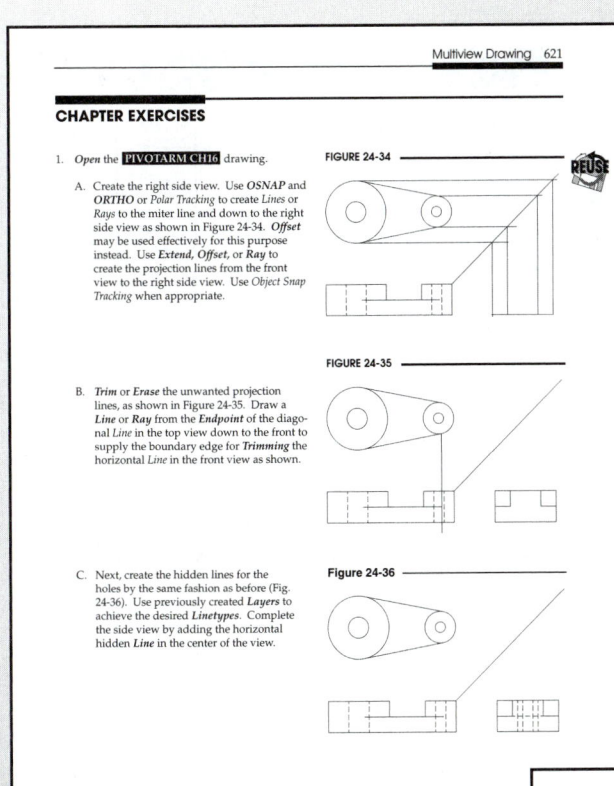

CHAPTER EXERCISES

1. *Open* the **PIVOTARM CH16** drawing.

FIGURE 24-34

A. Create the right side view. Use *OSNAP* and *ORTHO* or *Polar Tracking* to create *Lines* or *Rays* to the miter line and down to the right side view as shown in Figure 24-34. *Offset* may be used effectively for this purpose instead. Use *Extend*, *Offset*, or *Ray* to create the projection lines from the front view to the right side view. Use *Object Snap Tracking* when appropriate.

FIGURE 24-35

B. *Trim* or *Erase* the unwanted projection lines, as shown in Figure 24-35. Draw a *Line* or *Ray* from the *Endpoint* of the diagonal *Line* in the top view down to the front to supply the boundary edge for *Trimming* the horizontal *Line* in the front view as shown.

Figure 24-36

C. Next, create the hidden lines for the holes by the same fashion as before (Fig. 24-36). Use previously created *Layers* to achieve the desired *Linetypes*. Complete the side view by adding the horizontal hidden *Line* in the center of the view.

Chapter Exercises

Exercises help you try out the commands discussed in each chapter. Exercises begin with simple, step-by-step use of commands and progress to more advanced drawings based on synthesis of earlier ideas.

Multi-chapter "Reuse" Exercises

Exercise drawings that are used again in other chapters are designated with this "Reuse" icon. The associated drawing file names are printed in a reversed font. Using these multi-chapter exercises maximizes the student's efforts, creates connections between concepts, and makes the most of the natural pedagogical progression of the material.

5. *Text, Mtext*

Open the **CBRACKET** drawing from Chapter 9 Exercises. Using *romans.shx* font, use *Text* to place the part name and METRIC annotation (Fig. 18-71). Use a *Height* of 5 and 4, respectively, and the *Center Justification* option. For the notes, use *Mtext* to create the boundary as shown. Use the default *Justify* method (*TL*) and a *Height* of 3. Use *Ddedit*, *Mtedit*, or *Properties* if necessary. *SaveAs* CBRACTXT.

FIGURE 18-71

6. *Style*

Create two new styles for each of your template drawings: **ASHEET**, **BSHEET**, and **C-D-SHEET**. Use the *romans.shx* style with the default options for engineering applications or *CityBlueprint* (.TTF) for architectural applications. Next, design a style of your choosing to use for larger text as in title blocks or large notes.

7. *Import Text, Ddedit, Properties*

Use a text editor such as Windows Wordpad or Notepad to create a text file containing words similar to "Temperatures were recorded at Sanderson Field by the National Weather Service." Then *Open* the **TEMPGRPH** drawing and use the *Import Text...* option of *Mtext* to bring the text into the drawing as a note in the graph as shown in Figure 18-72. Use *Ddedit* to edit the text if desired or use *Properties* to change the text style or height.

FIGURE 18-72

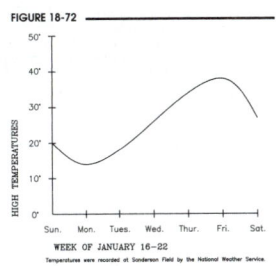

8. *Create a Title Block*

A. Begin a *New* drawing and assign the name **TBLOCK**. Create the title block as shown in Figure 18-73 or design your own, allowing space for eight text entries. The dimensions are set for an A size sheet. Draw on *Layer 0*. Use a *Pline* with .02 *width* for the boundary and *Lines* for the interior divisions. (No *Lines* are needed on the right side and bottom because the title block will fit against the border lines.)

FIGURE 18-73

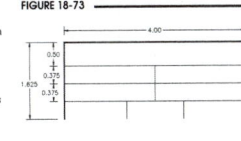

Reference Material

Useful for self study or classroom use, tabbed pages help you quickly locate the Command Table Index, Shortcut Keys, Dimension Variables, System Variables, Tables of Limits Settings, Template Drawings and more.

McGraw-Hill's ARIS—Assessment, Review, and Instruction System

A vast amount of material is available from **www.mhhe.com/leach** for both students and instructors.

- A course management system for creating assignments and maintaining course communication
- Test or review questions for each chapter with automatic grading
- Architecture-specific exercises for each chapter
- Mechanical engineering-specific exercises for each chapter
- Civil engineering-specific exercises for each chapter
- Chapter 45, Customize User Interface
- Chapter 46, CAD Management
- Chapter 47, Express Tools
- Solution drawings for chapter exercises (instructors only)

1330   Appendix E

Command Name (type)	Button	Pull-down Menu	Alias (type)	Short Cut	Screen (side) Menu	Tablet Menu	Chapter in this Text
DIMALIGNED		Dimension Aligned	DAL	...	DIMNSION Aligned	W,4	28
DIMANGULAR		Dimension Angular	DAN	...	DIMNSION Angular	X,3	28
DIMARC		Dimension Arc Length	DAR	...	...	...	28
DIMBASELINE		Dimension Baseline	DBA	...	DIMNSION Baseline	...	28
DIMCENTER		Dimension Center Mark	DCE	...	DIMNSION Center	X,2	28
DIMCONTINUE		Dimension Continue	DCO	...	DIMNSION Continue	...	28
DIMDIAMETER		Dimension Diameter	DDI	...	DIMNSION Diameter	X,4	28
DIMDISASSOCIATE		...	DDA	...	...	...	28
DIMEDIT		Dimension Oblique	DED	...	DIMNSION Dimedit	Y,1	28, 29
DIMJOG		Dimension Jogged	DJO	...	...	...	28
DIMLINEAR		Dimension Linear	DLI	...	DIMNSION Linear	W,5	28
DIMORDINATE		Dimension Ordinate	DOR	...	DIMNSION Ordinate	W,3	28
DIMOVERRIDE		Dimension Override	DOV	...	...	Y,4	29
DIMRADIUS		Dimension Radius	DRA	...	DIMNSION Radius	X,5	28
DIMREASSOCIATE		Dimension Reassociate Dimensions	DRE	...	...	...	28
DIMREGEN		...	...	...	...	...	28
DIMSTYLE		Dimension Style...	D, DST	...	DIMNSION Ddim	Y,5	29
-DIMSTYLE		Dimension Update	...	...	DIMNSION Dimstyle	Y,3	29
DIMTEDIT		Dimension Align Text >	DIMTED	...	DIMNSION Dimtedit	Y,2	28, 29
DIST		Tools Inquiry > Distance	DI	...	TOOLS 1 Dist	T,8	17
DIVIDE		Draw Point > Divide	DIV	...	DRAW 2 Divide	V,13	15
DONUT		Draw Donut	DO	...	DRAW 1 Donut	K,9	15
DRAWINGRECOVERY		Drawing Utilities Recovery	DRM	...	...	...	2

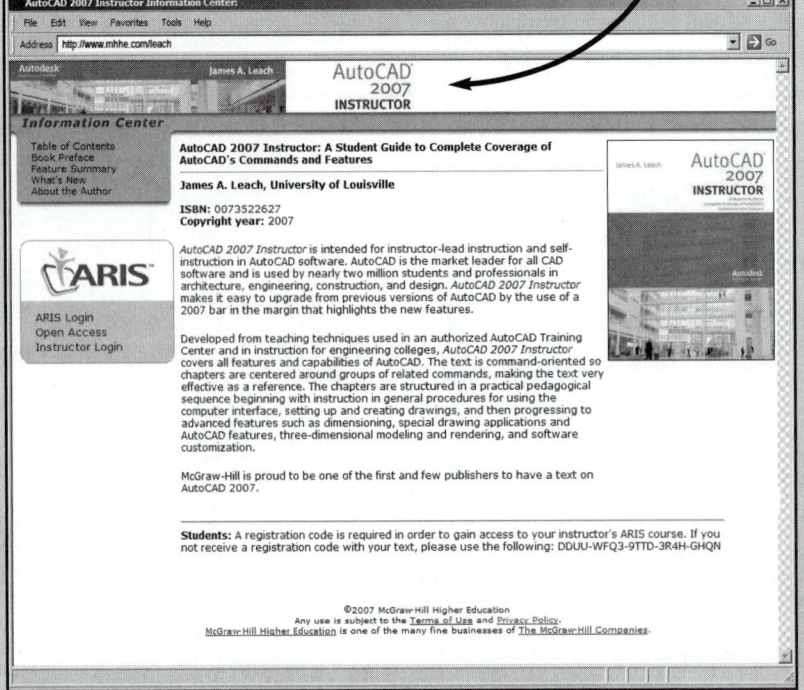

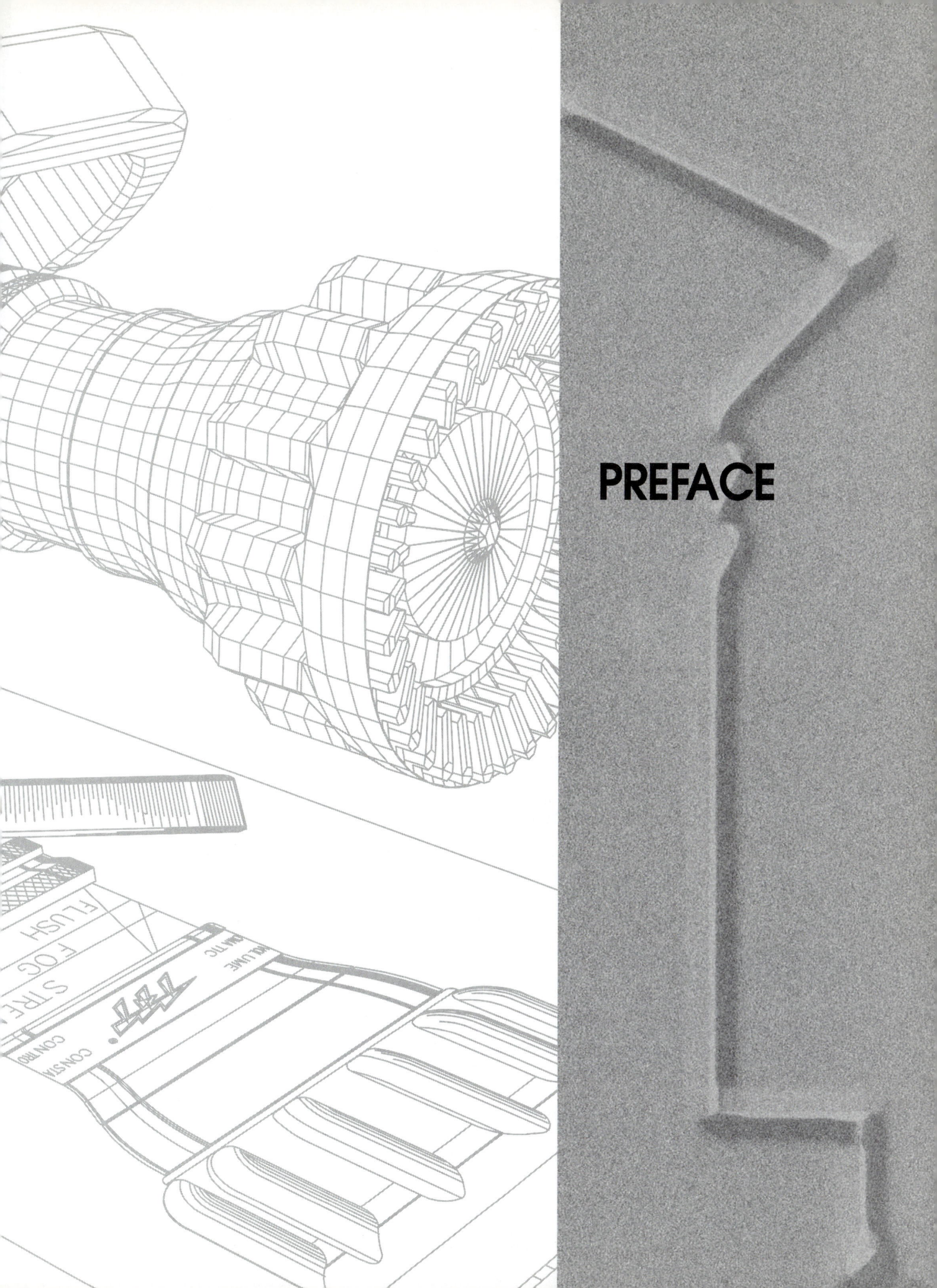

PREFACE

ABOUT THIS BOOK

This Book Is Your AutoCAD 2007 Instructor

The objective of *AutoCAD 2007 Instructor* is to provide you the best possible printed medium for learning AutoCAD, whether you are a professional or student learning AutoCAD on your own or whether you are attending an instructor-led course.

Complete Coverage

AutoCAD 2007 Instructor is written to instruct you in the full range of AutoCAD 2007 features. All commands, system variables, and features within AutoCAD are covered. This text can be used for a two-, three-, or four-course sequence.

Graphically Oriented

Because *AutoCAD 2007 Instructor* discusses concepts that are graphical by nature, many illustrations (approximately 2000) are used to communicate the concepts, commands, and applications.

Easy Update from AutoCAD 2005 and AutoCAD 2006

AutoCAD 2007 Instructor is helpful if you are already an AutoCAD user but are updating from AutoCAD 2005 or 2006. All new commands, concepts, features, and variables are denoted on the edges of the pages by a vertical "2006" bar (denoting an update since 2005) or a "2007" bar (denoting an update since 2006).

Pedagogical Progression

AutoCAD 2007 Instructor is presented in a pedagogical format by delivering the fundamental concepts first, then moving toward the more advanced and specialized features of AutoCAD. The book begins with small pieces of information explained in a simple form and then builds on that experience to deliver more complex ideas, requiring a synthesis of earlier concepts. The chapter exercises follow the same progression, beginning with a simple tutorial approach and ending with more challenging problems requiring a synthesis of earlier exercises.

Multi-chapter "Reuse" Exercises

About half of the Chapter Exercises are used again later in subsequent chapters. This concept emphasizes the natural pedagogical progression of the text, creates connections between concepts, and maximizes the student's efforts. Reuse Exercises are denoted with a "Reuse" (diskette) icon and the related drawing file names are printed in a reversed font.

Important "Tips"

Tips, reminders, notes, and cautions are given in the book and denoted by a "TIP" marker in the margin. This feature helps you identify and remember important concepts, commands, procedures, and tricks used by professionals that would otherwise be discovered only after much experience.

Valuable Reference Guide

AutoCAD 2007 Instructor is structured to be used as a reference guide to AutoCAD. Several important tables, lists, and variable settings are "tabbed" on the edge of the page for easy access. Every command throughout the book is given with a "command table" listing the possible methods of invoking the command. A complete index gives an alphabetical listing of all AutoCAD commands, command options, system variables, and concepts discussed.

For Professionals and Students in Diverse Areas

AutoCAD 2007 Instructor is written for professionals and students in the fields of engineering, architecture, design, construction, manufacturing, and any other field that has a use for AutoCAD. Applications and examples from many fields are given throughout the text. The applications and examples are not intended to have an inclination toward one particular field. Instead, applications to a particular field are used when they best explain an idea or use of a command.

McGraw-Hill's ARIS—Assessment Review and Instruction System

Please visit our web page at **www.mhhe.com/leach** to access a complete homework and course management system for AutoCAD 2007 Instructor. Ancillary materials are available for reading or download and can be assigned using ARIS. Questions for each chapter (true-false, multiple choice, and written answer) are available for homework, review, and testing. Over 400 drawing problems specifically for architectural, mechanical engineering, and civil/electrical applications are available. Solutions for drawing problems and questions can be downloaded by requesting a password from your McGraw-Hill representative.

Additional Chapters

Chapter 45, Customize User Interface, Chapter 46, CAD Management, and Chapter 47, Express Tools, are available free at **www.mhhe.com/leach**.

Have Fun

I predict you will have a positive experience learning AutoCAD. Although learning AutoCAD is not a trivial endeavor, you will have fun learning this exciting technology. In fact, I predict that more than once in your learning experience you will say to yourself, "Sweet!" (or something to that effect).

James A. Leach

ABOUT THE AUTHOR

James A. Leach (B.I.D., M.Ed.) is Professor of Engineering Graphics at the University of Louisville. He began teaching AutoCAD at Auburn University early in 1984 using Version 1.4, the first version of AutoCAD to operate on IBM personal computers. Jim is currently Director of the AutoCAD Training Center established at the University of Louisville in 1985, one of the first fifteen centers to be authorized by Autodesk.

In his 30 years of teaching Engineering Graphics and AutoCAD courses, Jim has published numerous journal and magazine articles, drawing workbooks, and textbooks about Autodesk and engineering graphics instruction. He has designed CAD facilities and written AutoCAD-related course materials for Auburn University, University of Louisville, the AutoCAD Training Center at the University of Louisville, and several two-year and community colleges. Jim is the author of 15 AutoCAD textbooks published by Richard D. Irwin and McGraw-Hill.

CONTRIBUTING AUTHORS

Steven H. Baldock is an engineer at a consulting firm in Louisville and offers CAD consulting as well. Steve is an Autodesk Certified Instructor and teaches several courses at the University of Louisville AutoCAD Training Center. He has 18 years experience using AutoCAD in architectural, civil, and structural design applications. Steve has degrees in engineering, computer science, and mathematics from the University of Louisville and the University of Kentucky. Steve Baldock prepared material for several sections of *AutoCAD 2007 Instructor*, such as xrefs, groups, text, multiline drawing and editing, and dimensioning. Steve also created several hundred figures used in *AutoCAD 2007 Instructor*.

Michael E. Beall is the owner of AutoCAD® Trainer Guy, LLC in Shelbyville, Kentucky, near Louisville. Most recently, Mr. Beall authored *AutoCAD 2006: Transitioning from AutoCAD 2002 to AutoCAD 2006* (Autodesk Official Training Courseware) published by Autodesk, Inc. For New Riders Publishing, he served as the Development Editor for *Inside AutoCAD 2000*, co-authored *AutoCAD 14 Fundamentals*, and was a contributing author to *Inside AutoCAD 14*. Other contributions include co-authoring *AutoCAD Release 13 for Beginners* and *Inside AutoCAD LT for Windows 95*. Since 1982 Michael has presented CAD training seminars in nearly a dozen countries and more than 30 states. He is an Autodesk Certified Instructor (ACI) at the University of Louisville AutoCAD Training Center where he received Autodesk's "Instructional Quality Award" and "Instructional Quality Award in Industry Training." Michael assisted with several topics in *AutoCAD 2007 Instructor* and continues to be a source of information about AutoCAD nuances.

Bruce Duffy has been an AutoCAD trainer and consultant for 21 years. He has aided many Fortune 500 companies with training, curriculum development, and productivity assessment/training. Bruce worked as an instructor at the University of Louisville and the Authorized Autodesk Training Center for seven years. Since 2002, Bruce has been working with Florida International University in Miami, Florida, as an Instructional Design and Online Learning consultant. He still consults and provides AutoCAD training for national and international clients. Bruce Duffy contributed much of the material for Chapters 31 and 32 of *AutoCAD 2007 Instructor*.

Patrick McCuistion, Ph.D., has taught in the Department of Industrial Technology at Ohio University since 1989. Dr. McCuistion taught three years at Texas A&M University and previously worked in various engineering design, drafting, and checking positions at several manufacturing industries. He has provided instruction in geometric dimensioning and tolerancing to many industry, military, and educational institutions and has written one book, several articles and given many presentations on the topic. Dr. McCuistion is an active chairman or member of the following ANSI/ASME subcommittees, Y14.2 Line Conventions & Lettering, Y14.3 Multiview and Section View Drawings, Y14.5 Dimensioning & Tolerancing, Y14.8 Castings, Forging & Molded Part Drawings, Y14.5.2 Geometric Dimensioning Certification, Y14.35 Revisions, Y14.36 Surface Texture, Y14.43 Functional Gages and the Y14 Main committee. Dr. McCuistion contributed the material on geometric dimensioning and tolerancing for *AutoCAD 2007 Instructor*.

Lee Ambrosius is an Autodesk Authorized Author and Developer and the owner of HyperPics, LLC, an AutoCAD consulting company. He has been a user of AutoCAD since AutoCAD Release 12 and of 3D Studio for over 10 years. Lee consults on a variety of topics ranging from customizing AutoCAD to programming with AutoLISP, ObjectARX, and other languages. Lee is also a contributing author to AutoCAD Users Group International's AUGI World and develops technical documents such as white papers for Autodesk. Lee contributed much of the material for Chapter 41, Rendering and Animation, Chapter 45, Customize User interface, and Chapter 46, CAD Standards.

Tom Heffner is presently the Coordinator and AutoCAD Instructor at Winnipeg Technical College, Continuing Education since 1989. Tom prepares the curriculum and teaches the Continuing Education AutoCAD Program to Level I, II, and III (3D) students. Over the past 30 years Tom has worked for several engineering companies in Canada, spending the last 19 years with The City of Winnipeg, Public Works Department, Engineering Division as Supervisor of Drafting and Graphic Services. Tom was instrumental in introducing and training City of Winnipeg Engineering employees AutoCAD starting with Version 2.6 up to the current Autodesk release. Tom has been a contributing author for McGraw-Hill for *AutoCAD 2007 Instructor* and *AutoCAD 2006 Companion* chapter exercises and chapter test/review questions.

ACKNOWLEDGMENTS

I want to thank all of the contributing authors for their assistance in writing *AutoCAD 2007 Instructor*. Without their help, this text could not have been as application-specific nor could it have been completed in the short time frame.

I am very grateful to Gary Bertoline for his foresight in conceiving the McGraw-Hill Graphics Series and for including my efforts in it.

I would like to give thanks to the excellent editorial and production group at McGraw-Hill who gave their talents and support during this project, especially Suzanne Jeans, Katie White, and Larry Goldberg.

A special thanks goes to Laura Hunter for the page layout work of *AutoCAD 2007 Instructor*. She was instrumental in fulfilling my objective of providing the most direct and readable format for conveying concepts. I also want to thank Matt Hunter for his precise and timely proofing and for preparing the Table of Contents and Index for *AutoCAD 2007 Instructor*.

Michael Beall prepared the solution drawings for the Chapter Exercises that are available on the www.mhhe.com/leach Web site. David Morgan and Brianne Duffy deserve credit for preparation of the contents of the "command tables" used throughout this text.

My thanks go to the following colleagues who provided detailed reviews of this text and made many helpful suggestions: Tracy Arras, *Oregon State University*; Nii Attoh-Okine, *University of Delaware*; Danny Benton, *Boise State University*; Jeffrey Connor, *Virginia Polytechnic Institute and State University*; Steven Dyer, *Richland Community College*; William E. Eichinger, *University of Iowa*; Patrick Ferrell, *College of Southern Idaho*; Ishmail Fidan, *Tennessee Technological University*; Mark Hagen, *Portland Community College*; Kevin W. Hall, *Western Illinois University*; Anant Honkan, *Georgia Perimeter College*; James R. King, *The Community and Technical College at West Virginia University Institute of Technology*; Carol M. Lamb, *Youngstown State University*; Douglas Larson, *Jefferson Community and Technical College*; Amlan Mukherjee, *Michigan Technological University*; Bisi Oluyemi, *Morehouse College*; Jeffrey M. Otey, *Texas A&M University*; Paul Palazolo, *University of Memphis*; Minh Quoc Pham, *Houston Community College*; Bruce Savage, *McNeese State University*; Randy Stach, *North Dakota State College of Science*; Hillary Smith Tanner, *University of Georgia*; Mary Tolle, *South Dakota State University*; Richard Youchak, *University of Pittsburgh at Johnstown*.

I also want to thank all of the readers that have contacted me with comments and suggestions on specific sections of this text. Your comments help me improve this book, assist me in developing new ideas, and keep me abreast of ways AutoCAD and this text are used in industrial and educational settings.

I also acknowledge: my colleague and friend Robert A. Matthews for his support of this project and for doing his job well; Charles Grantham of Contemporary Publishing Company of Raleigh, Inc., for generosity and consultation; and Speed School of Engineering Dean's Office for support and encouragement.

Special thanks, once again, to my wife, Donna, for the many hours of copy editing required to produce this and the other texts.

TRADEMARK AND COPYRIGHT ACKNOWLEDGMENTS

The object used in the Wireframe Modeling Tutorial, Chapter 37 appears courtesy of James H. Earle, *Graphics for Engineers, Third Edition*, (pg. 120), ©1992 by Addison-Wesley Publishing Company, Inc. Reprinted by permission of the publisher.

The following drawings used for Chapter Exercises appear courtesy of James A. Leach, *Problems in Engineering Graphics Fundamentals, Series A* and *Series B,* ©1984 and 1985 by Contemporary Publishing Company of Raleigh, Inc.: Gasket A, Gasket B, Pulley, Holder, Angle Brace, Saddle, V-Block, Bar Guide, Cam Shaft, Bearing, Cylinder, Support Bracket, Corner Brace, and Adjustable Mount. Reprinted or redrawn by permission of the publisher.

The following are registered trademarks of Autodesk, Inc., in the USA and/or other countries: 3D Studio, 3D Studio MAX, 3D Studio VIZ, 3ds max, ActiveShapes, ActiveShapes (logo), Actrix, ADI, AEC-X, ATC, AUGI, AutoCAD, AutoCAD LT, Autodesk, Autodesk Design Review, Autodesk Envision, Autodesk Inventor, Autodesk Map, Autodesk MapGuide, Autodesk RealDWG, Autodesk Streamline, Autodesk WalkThrough, Autodesk World, AutoLISP, AutoSketch, backdraft, Biped, bringing information down to earth, Buzzsaw, CAD Overlay, Character Studio, Cinepak, Cinepak (logo), cleaner, Codec Central, combustion, Design Your World, Design Your World (logo), EditDV, Education by Design, gmax, Heidi, HOOPS, Hyperwire, i-drop, IntroDV, lustre, Mechanical Desktop, ObjectARX, Physique, Powered with Autodesk Technology (logo), ProjectPoint, RadioRay, Reactor, Revit, VISION*, Visual, Visual Construction, Visual Drainage, Visual Hydro, Visual Landscape, Visual Roads, Visual Survey, Visual Toolbox, Visual Tugboat, Visual LISP, Volo, WHIP!, and WHIP! (logo).

The following are trademarks of Autodesk, Inc., in the USA and/or other countries: AutoCAD Learning Assistance, AutoCAD LT Learning Assistance, AutoCAD Simulator, AutoCAD SQL Extension, AutoCAD SQL Interface, AutoSnap, AutoTrack, Built with ObjectARX (logo), burn, Buzzsaw.com, CAiCE, Cinestream, Civil 3D, cleaner central, ClearScale, Colour Warper, Content Explorer, Dancing Baby (image), DesignCenter, Design Doctor, Designer's Toolkit, DesignKids, DesignProf, DesignServer, Design Web Format, DWF, DWFit, DWG Linking, DXF, Extending the Design Team, GDX Driver, gmax (logo), gmax ready (logo),Heads-up Design, jobnet, mass, ObjectDBX, onscreen onair online, Plasma, PolarSnap, Productstream, Real-time Roto, Render Queue, Visual Bridge, Visual Syllabus, and Where Design Connects.

All other brand names, product names, or trademarks belong to their respective holders.

LEGEND

The following special treatment of characters and fonts in the textual content is intended to assist you in translating the meaning of words or sentences in *AutoCAD 2007 Instructor.*

<u>Underline</u>	Emphasis of a word or an idea.
Helvetica font	An AutoCAD prompt appearing on the <u>screen</u> at the command line or in a text window.
Italic (Upper and Lower)	An AutoCAD command, option, menu, toolbar, or dialog box name.
UPPER CASE	A file name.
UPPER CASE ITALIC	An AutoCAD system variable or a drawing aid (*OSNAP, SNAP, GRID, ORTHO*).

Anything in **Bold** represents user input:

Bold	What you should <u>type</u> or press on the keyboard.
Bold Italic	An AutoCAD <u>command</u> that you should type or <u>menu item</u> that you should select.
BOLD UPPER CASE	A <u>file name</u> that you should type.
BOLD UPPER CASE ITALIC	A <u>system variable</u> that you should type.
PICK	Move the cursor to the indicated position on the screen and press the <u>select</u> button (button #1 or left mouse button).

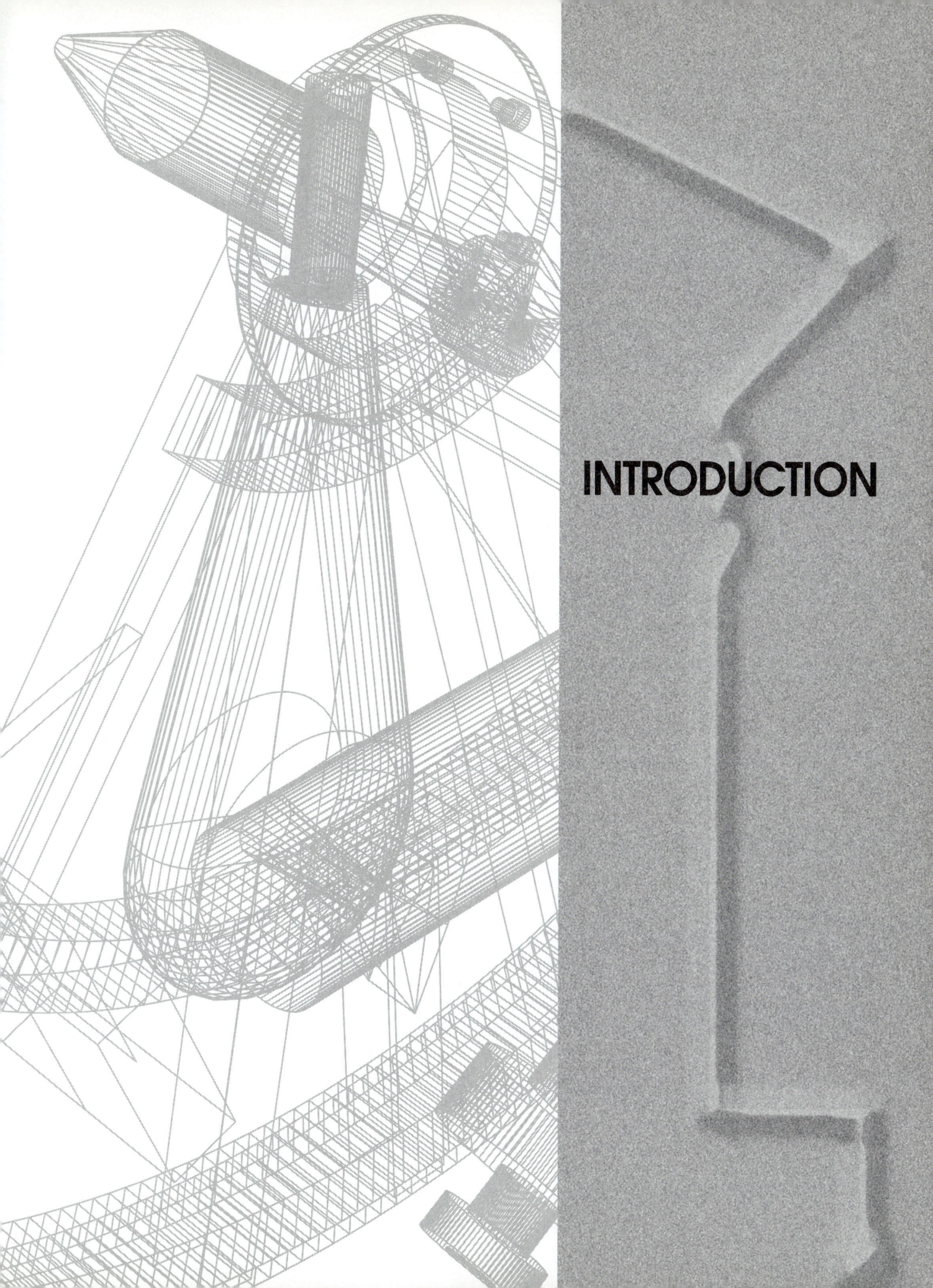

INTRODUCTION

WHAT IS CAD?

CAD is an acronym for Computer-Aided Design. CAD allows you to accomplish design and drafting activities using a computer. A CAD software package, such as AutoCAD, enables you to create designs and generate drawings to document those designs.

Design is a broad field involving the process of making an idea into a real product or system. The design process requires repeated refinement of an idea or ideas until a solution results—a manufactured product or constructed system. Traditionally, design involves the use of sketches, drawings, renderings, two-dimensional (2D) and three-dimensional (3D) models, prototypes, testing, analysis, and documentation. Drafting is generally known as the production of drawings that are used to document a design for manufacturing or construction or to archive the design.

CAD is a <u>tool</u> that can be used for design and drafting activities. CAD can be used to make "rough" idea drawings, although it is more suited to creating accurate finished drawings and renderings. CAD can be used to create a 2D or 3D computer model of the product or system for further analysis and testing by other computer programs. In addition, CAD can be used to supply manufacturing equipment such as lathes, mills, laser cutters, or rapid prototyping equipment with numerical data to manufacture a product. CAD is also used to create the 2D documentation drawings for communicating and archiving the design.

The tangible result of CAD activity is usually a drawing generated by a plotter or printer but can be a rendering of a model or numerical data for use with another software package or manufacturing device. Regardless of the purpose for using CAD, the resulting drawing or model is stored in a CAD file. The file consists of numeric data in binary form usually saved to a magnetic or optical device such as a diskette, hard drive, tape, or CD.

WHY SHOULD YOU USE CAD?

Although there are other methods used for design and drafting activities, CAD offers the following advantages over other methods in many cases:

1. Accuracy
2. Productivity for repetitive operations
3. Sharing the CAD file with other software programs

Accuracy

Since CAD technology is based on computer technology, it offers great accuracy. When you draw with a CAD system, the graphical elements, such as lines, arcs, and circles, are stored in the CAD file as numeric data. CAD systems store that numeric data with great precision. For example, AutoCAD stores values with fourteen significant digits. The value 1, for example, is stored in scientific notation as the equivalent of 1.0000000000000. This precision provides you with the ability to create designs and drawings that are 100% accurate for almost every case.

Productivity for Repetitive Operations

It may be faster to create a simple "rough" drawing, such as a sketch by hand (pencil and paper), than it would by using a CAD system. However, for larger and more complex drawings, particularly those involving similar shapes or repetitive operations, CAD methods are very efficient. Any kind of shape or operation accomplished with the CAD system can be easily duplicated since it is stored in a CAD file. In short, it may take some time to set up the first drawing and create some of the initial geometry, but any of the existing geometry or drawing setups can be easily duplicated in the current drawing or for new drawings.

Likewise, making changes to a CAD file (known as editing) is generally much faster than creating the original geometry. Since all the graphical elements in a CAD drawing are stored, only the affected components of the design or drawing need to be altered, and the drawing can be plotted or printed again or converted to other formats.

As CAD and the associated technology advance and software becomes more interconnected, more productive developments are available. For example, it is possible to make a change to a 3D model that automatically causes a related change in the linked 2D engineering drawing. One of the main advantages of these technological advances is productivity.

Sharing the CAD File with Other Software Programs

Of course, CAD is not the only form of industrial activity that is making technological advances. Most industries use computer software to increase capability and productivity. Since software is written using digital information and may be written for the same or similar computer operating systems, it is possible and desirable to make software programs with the ability to share data or even interconnect, possibly appearing simultaneously on one screen.

For example, word processing programs can generate text that can be imported into a drawing file, or a drawing can be created and imported into a text file as an illustration. (This book is a result of that capability.) A drawing created with a CAD system such as AutoCAD can be exported to a finite element analysis program that can read the computer model and compute and analyze stresses. CAD files can be dynamically "linked" to spreadsheets or databases in such a way that changing a value in a spreadsheet or text in a database can automatically make the related change in the drawing, or vice versa.

Another advance in CAD technology is the automatic creation and interconnectivity of a 2D drawing and a 3D model in one CAD file. With this tool, you can design a 3D model and have the 2D drawings automatically generated. The resulting set has bi-directional associativity; that is, a change in either the 2D drawings or the 3D model is automatically updated in the other.

With the introduction of the new Web technologies, designers and related professionals can more easily collaborate by viewing and transferring drawings over the Internet. CAD drawings can contain Internet links to other drawings, text information, or other related Web sites. Multiple CAD users can even share a single CAD session from remote locations over the Internet.

CAD, however, may not be the best tool for every design related activity. For example, CAD may help develop ideas but probably won't replace the idea sketch, at least not with present technology. A 3D CAD model can save much time and expense for some analysis and testing but cannot replace the "feel" of an actual model, at least not until virtual reality technology is developed and refined.

With everything considered, CAD offers many opportunities for increased accuracy, productivity, and interconnectivity. Considering the speed at which this technology is advancing, many more opportunities are rapidly obtainable. However, we need to start with the basics. Beginning by learning to create an AutoCAD drawing is a good start.

WHY USE AutoCAD?

CAD systems are available for a number of computer platforms: laptops, personal computers (PCs), workstations, and mainframes. AutoCAD, offered to the public in late 1982, was one of the first PC-based CAD software products. Since that time, it has grown to be the world leader in market share for all CAD products. Autodesk, the manufacturer of AutoCAD, is the world's leading supplier of PC design software and multimedia tools. At the time of this writing, Autodesk is one of the largest software producers in the world and has several million customers in more than 150 countries.

Learning AutoCAD offers a number of advantages to you. Since AutoCAD is the most widely used CAD software, using it gives you the highest probability of being able to share CAD files and related data and information with others.

As a student, learning AutoCAD, as opposed to learning another CAD software product, gives you a higher probability of using your skills in industry. Likewise, there are more employers who use AutoCAD than any other single CAD system. In addition, learning AutoCAD as a first CAD system gives you a good foundation for learning other CAD packages because many concepts and commands introduced by AutoCAD are utilized by other systems. In some cases, AutoCAD features become industry standards. The .DXF file format, for example, was introduced by Autodesk and has become an industry standard for CAD file conversion between systems.

As a professional, using AutoCAD gives you the highest possibility that you can share CAD files and related data with your colleagues, vendors, and clients. Compatibility of hardware and software is an important issue in industry. Maintaining compatible hardware and software allows you the highest probability for sharing data and information with others as well as offering you flexibility in experimenting with and utilizing the latest technological advancements. AutoCAD provides you with great compatibility in the CAD domain.

This introduction is not intended as a selling point but to remind you of the importance and potential of the task you are about to undertake. If you are a professional or a student, you have most likely already made up your mind that you want to learn to use AutoCAD as a design or drafting tool. If you have made up your mind, then you can accomplish anything. Let's begin.

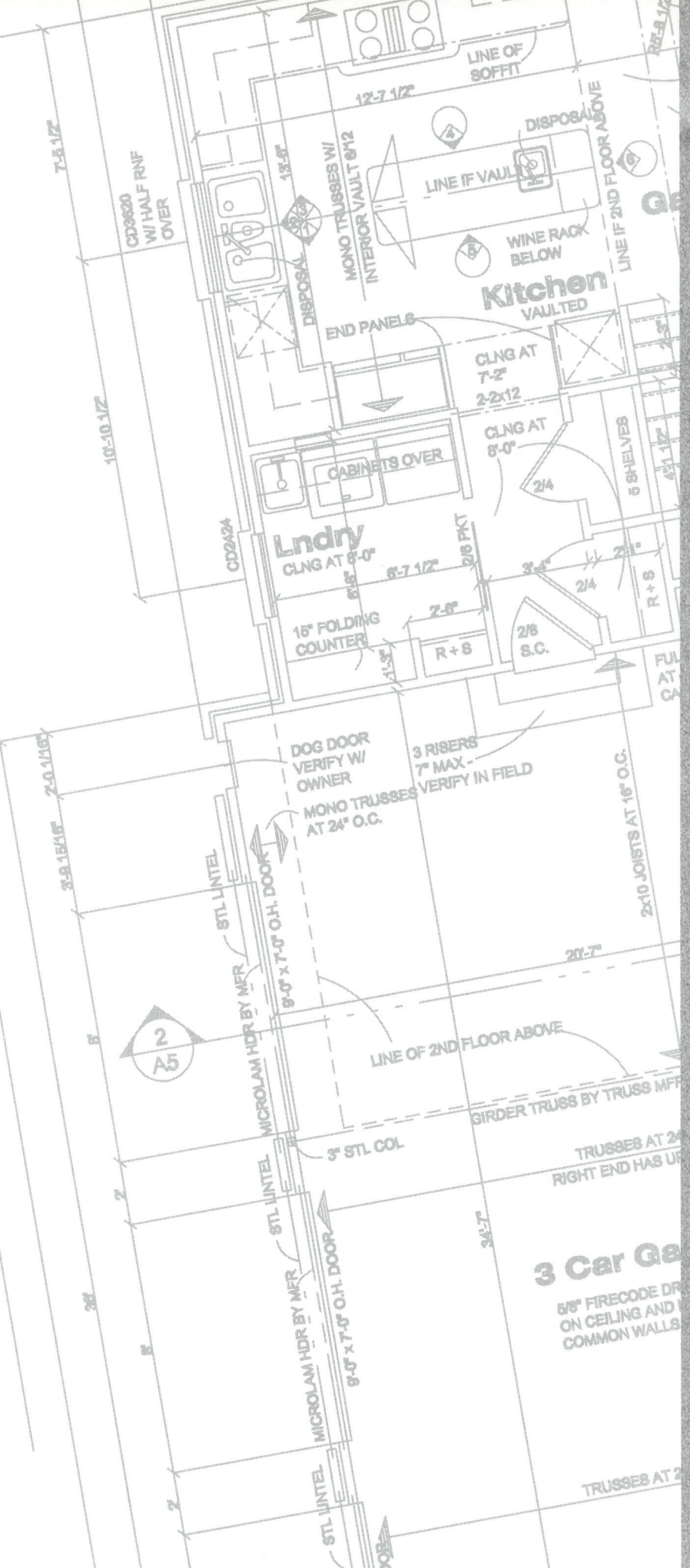

GETTING STARTED

CHAPTER OBJECTIVES

After completing this chapter you should:

1. understand how the X, Y, Z coordinate system is used to define the location of drawing elements in digital format in a CAD drawing file;

2. understand why you should create drawings full size in the actual units with CAD;

3. be able to start AutoCAD to begin drawing;

4. recognize the areas of the AutoCAD Drawing Editor and know the function of each;

5. be able to use the five methods of entering commands;

6. be able to turn on and off the *SNAP, GRID, ORTHO, POLAR,* and *DYN* drawing aids;

7. know how to customize the AutoCAD for Windows screen to your preferences.

CONCEPTS

Coordinate Systems

Any location in a drawing, such as the endpoint of a line, can be described in X, Y, and Z coordinate values (Cartesian coordinates). If a line is drawn on a sheet of paper, for example, its endpoints can be charted by giving the distance over and up from the lower-left corner of the sheet (Fig. 1-1).

These distances, or values, can be expressed as X and Y coordinates; X is the horizontal distance from the lower-left corner (origin) and Y is the vertical distance from that origin. In a three-dimensional coordinate system, the third dimension, Z, is measured from the origin in a direction perpendicular to the plane defined by X and Y.

Two-dimensional (2D) and three-dimensional (3D) CAD systems use coordinate values to define the location of drawing elements such as lines and circles (called underline objects in AutoCAD).

In a 2D drawing, a line is defined by the X and Y coordinate values for its two endpoints (Fig. 1-2).

In a 3D drawing, a line can be created and defined by specifying X, Y, and Z coordinate values (Fig. 1-3). Coordinate values are always expressed by the X value first separated by a comma, then Y, then Z.

FIGURE 1-1

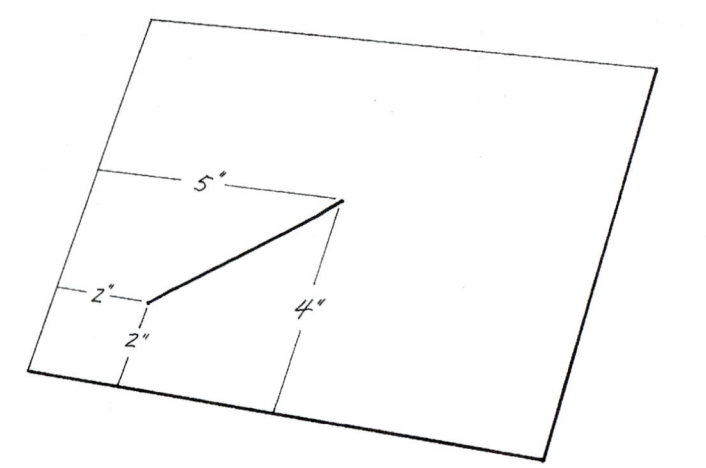

FIGURE 1-2

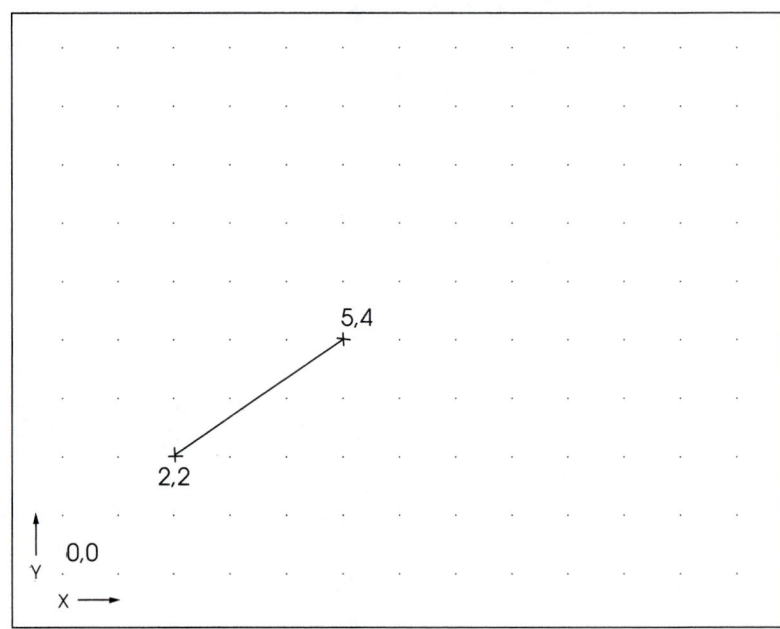

FIGURE 1-3

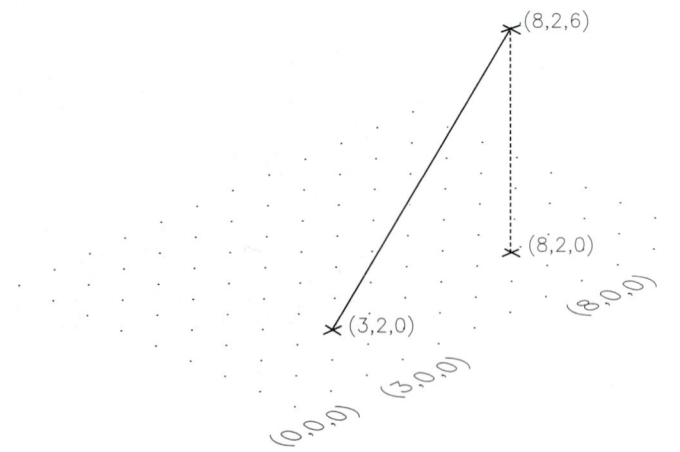

The CAD Database

A CAD (Computer-Aided Design) file, which is the electronically stored version of the drawing, keeps data in binary digital form. These digits describe coordinate values for all of the endpoints, center points, radii, vertices, etc. for all the objects composing the drawing, along with another code that describes the kinds of objects (line, circle, arc, ellipse, etc.). Figure 1-4 shows part of an AutoCAD DXF (Drawing Interchange Format) file giving numeric data defining lines and other objects. Knowing that a CAD system stores drawings by keeping coordinate data helps you understand the input that is required to create objects and how to translate the meaning of prompts on the screen.

FIGURE 1-4

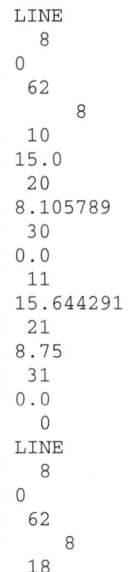

```
LINE
  8
0
  62
        8
  10
15.0
  20
8.105789
  30
0.0
  11
15.644291
  21
8.75
  31
0.0
  0
LINE
  8
0
  62
        8
  18
```

Angles in AutoCAD

Angles in AutoCAD are measured in a <u>counterclockwise direction</u>. Angle 0 is positioned in a positive X direction, that is, horizontally from left to right. Therefore, 90 degrees is in a positive Y direction, or straight up; 180 degrees is in a negative X direction, or to the left; and 270 degrees is in a negative Y direction, or straight down (Fig. 1-5).

The position and direction of measuring angles in AutoCAD can be changed; however, the defaults listed here are used in most cases.

FIGURE 1-5

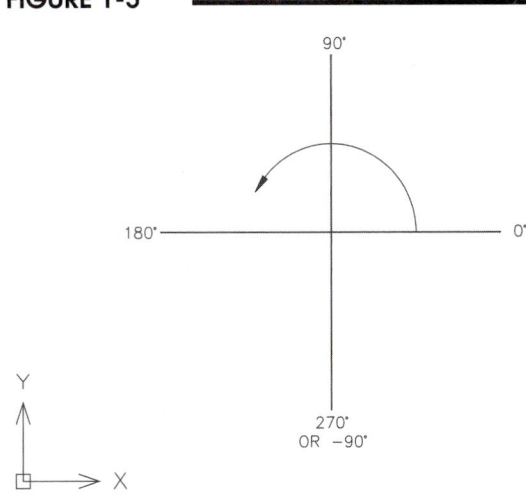

Draw True Size

When creating a drawing with pencil and paper tools, you must first determine a scale to use so the drawing will be proportional to the actual object and will fit on the sheet (Fig. 1-6). However, when creating a drawing on a CAD system, there is no fixed size drawing area. The number of drawing units that appear on the screen is variable and is assigned to fit the application.

The CAD drawing is not scaled until it is physically transferred to a fixed size sheet of paper by plotter or printer.

FIGURE 1-6

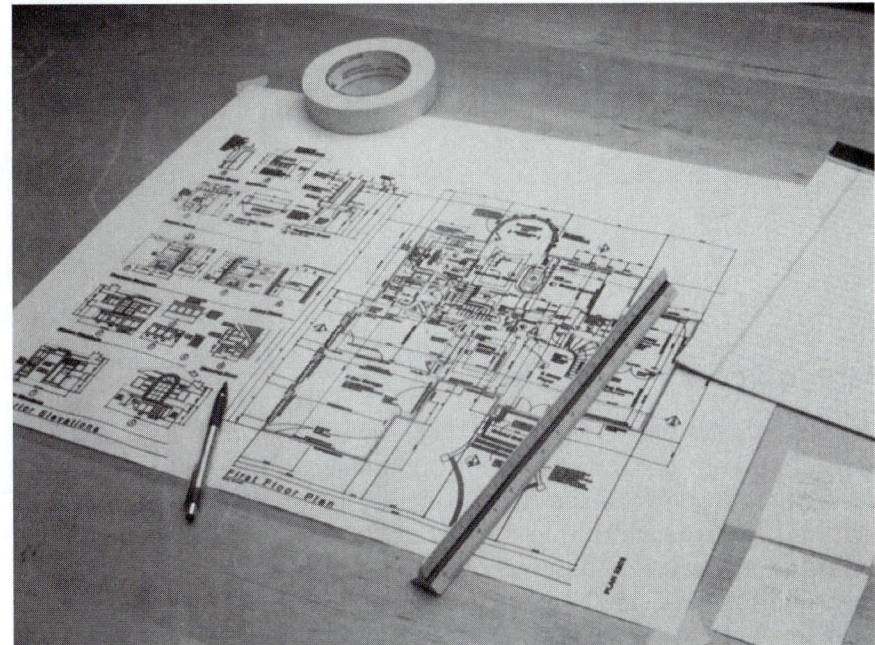

TIP

The rule for creating CAD drawings is that the drawing should be created <u>true size</u> using real-world units. The user specifies what units are to be used (architectural, engineering, etc.) and then specifies what size drawing area is needed (in X and Y values) to draw the necessary geometry (Fig. 1-7).

FIGURE 1-7

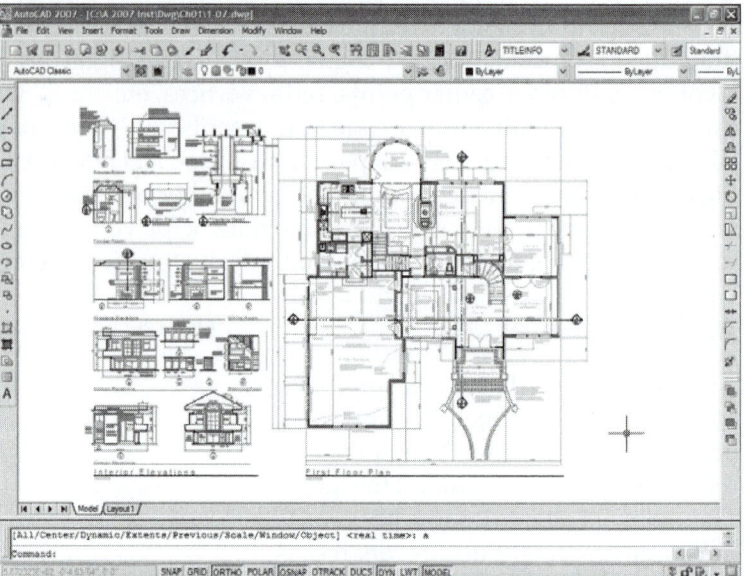

Whatever the specified size of the drawing area, it can be displayed on the screen in its entirety (Fig. 1-7) or as only a portion of the drawing area (Fig. 1-8).

FIGURE 1-8

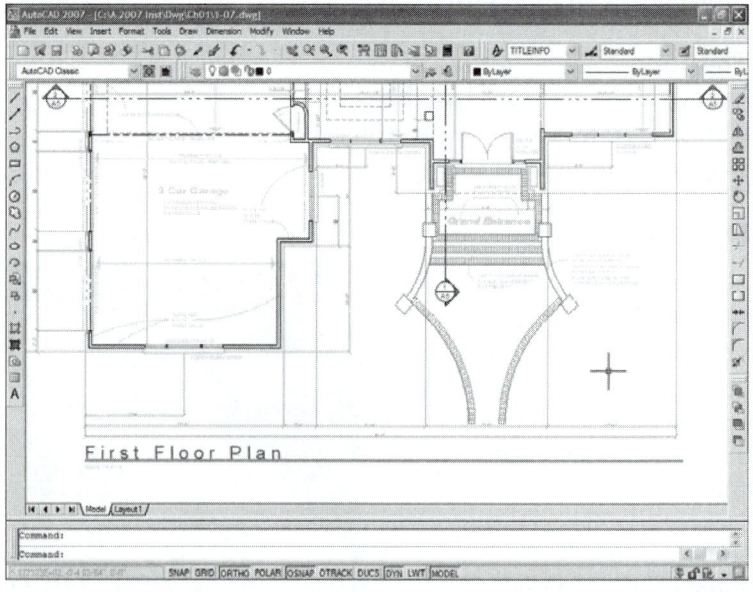

Plot to Scale

As long as a drawing exists as a CAD file or is visible on the screen, it is considered a virtual, full-sized object. Only when the CAD drawing is transferred to paper by a plotter or printer is it converted (usually reduced) to a size that will fit on a sheet. A CAD drawing can be automatically scaled to fit on the sheet regardless of sheet size; however, this action results in a plotted drawing that is not to an accepted scale (not to a regular proportion of the real object). Usually it is desirable to plot a drawing so that the resulting drawing is a proportion of the actual object size. The scale to enter as the plot scale (Fig. 1-9) is simply the proportion of the <u>plotted drawing</u> size to the <u>actual object</u>.

FIGURE 1-9

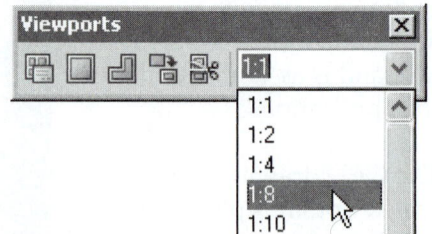

STARTING AutoCAD

FIGURE 1-10

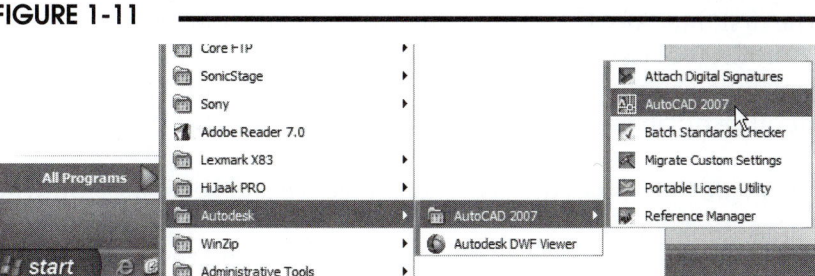

Assuming that AutoCAD has been installed and configured properly for your system, you are ready to begin using AutoCAD.

To start AutoCAD 2007, locate the "AutoCAD 2007" shortcut icon on the desktop (Fig. 1-10). Double-clicking on the icon launches AutoCAD 2007.

If you cannot locate the AutoCAD 2007 shortcut icon on the desktop, select the "Start" button, highlight "Programs" and search for "Autodesk." From the list that appears select "AutoCAD 2007" (Fig. 1-11).

FIGURE 1-11

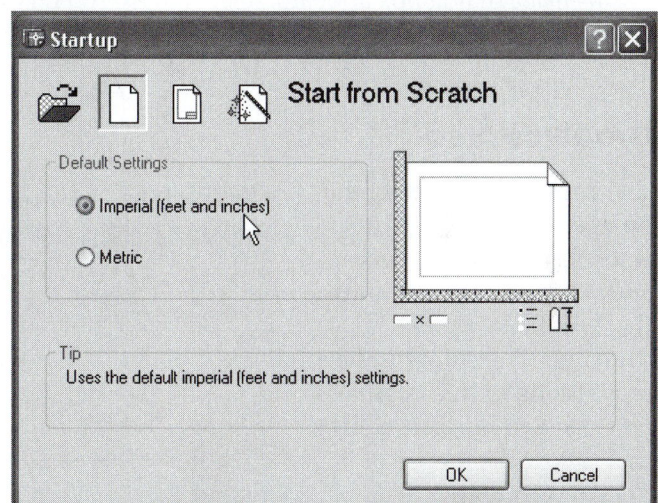

THE AutoCAD DRAWING EDITOR

FIGURE 1-12

The *Startup* Dialog Box

When you start AutoCAD 2007, you are pre-sented with the *Startup* dialog box (Fig. 1-12) or a "blank" drawing screen, depending on how your system has been configured. By default, a blank drawing screen appears. You can turn on the *Startup* dialog box in the *Options* dialog box, *System* tab (or by typing in the *STARTUP* system variable and changing the setting to 1).

This *Startup* dialog box also allows you to begin a new drawing or open an existing drawing, including providing access to drawing tem-plates and setup *Wizards*.

If you want to <u>start AutoCAD with the default English or metric settings</u> using the *Startup* dialog box, you can use one of the following options:

1. Select the *Start from Scratch* button in the *Startup* dialog box. Text appears displaying a choice of *Imperial (feet and inches)* or *Metric*. Make your selection, then press *OK*.

2. Select the *Template* option in the *Startup* dialog box, and then select the *Acad.dwt* (inches) or *Acadiso.dwt* (metric) template drawing.

3. Select the *Cancel* button in the *Startup* dialog box.

All three of these methods begin a new drawing using AutoCAD's default settings for use with inch or metric units. The inch settings (*Acad.dwt*) use a 12" x 9" drawing area (called *Limits*), and the metric set-tings (*Acadiso.dwt*) use a 420mm x 297mm drawing area. The *MEASUREINIT* system variable setting (0=inch, 1=metric) controls which of the two formats is used if you cancel the dialog boxes or if you use the startup *Wizards*.

NOTE: If your session starts with no dialog box and a template other than *Acad.dwt* or *Acadiso.dwt* appears, a specific template was specified by the *Qnew* setting in the *Options* dialog box (see "QNEW" in Chapter 2). The *Startup* dialog box and the *Create New Drawing* dialog boxes are explained in more detail in Chapters 2, 6, and 12. For the examples and exercises in Chapters 1–5, use the first method above to begin a drawing.

If AutoCAD Starts in 3D Mode

The first time AutoCAD is launched after installation, the user chooses to begin working in the 2D (AutoCAD Classic) environment or in the 3D environment. This option is typically suppressed for subsequent openings. Therefore, it is possible that AutoCAD may have already been set up in your particular office or laboratory to begin working in 3D. In that case, follow the instructions below.

To initiate the AutoCAD classic (2D) interface and begin a 2D drawing:

1. Using the *Workspaces* toolbar (upper-left, second row of tools), select *AutoCAD Classic* from the drop-down list.

2. Use the *New* command from the *File* pull-down menu to begin a new drawing. When the *Select Template* dialog box appears, select the ACAD.DWT (for English) or ACADISO.DWT (for metric) template.

Graphics Area

The large central area of the screen is the <u>Graphics area</u>. It displays the lines, circles, and other objects you draw that will make up the drawing. The <u>cursor</u> is the intersection of the crosshairs (vertical and horizontal lines that follow the mouse or puck movements). The default size of the graphics area for English settings is 12 units (X or horizontal) by 9 units (Y or vertical). This usable drawing area (12 x 9) is called the drawing *Limits* and can be changed to any size to fit the application. As you move the cursor, you will notice the numbers in the <u>Coordinate display</u> change (Fig. 1-13, bottom left).

FIGURE 1-13

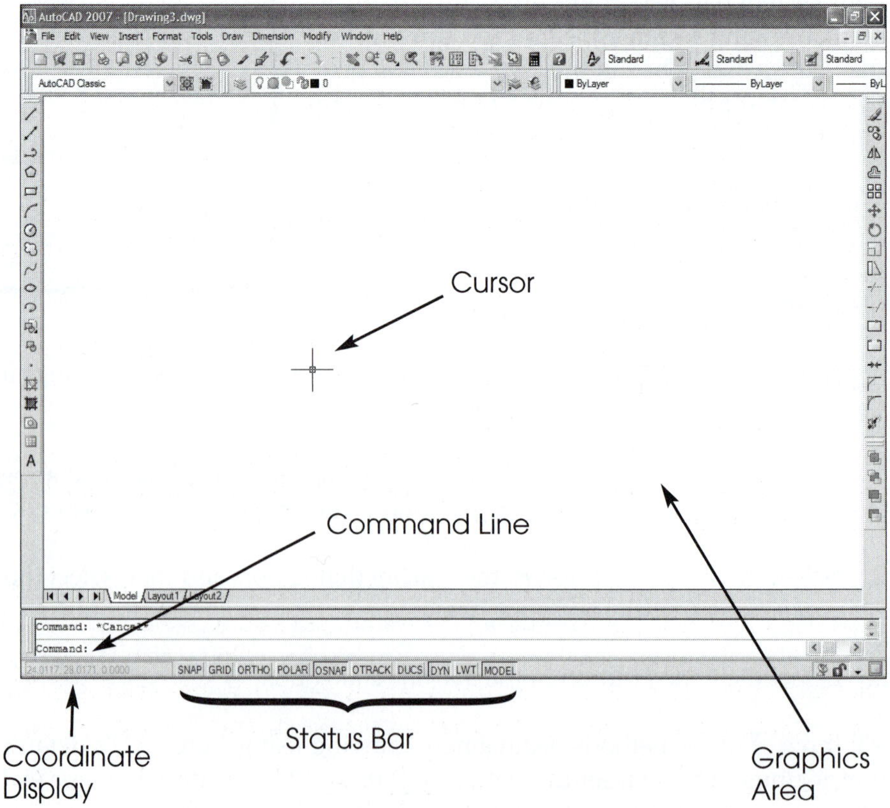

Command Line

The <u>Command line</u> consists of the three text lines at the bottom of the screen (by default) and is the most important area other than the drawing itself (see Fig. 1-13). Any command that is entered or any prompt that AutoCAD issues appears here. The Command line is always visible and gives the current state of drawing activity. You should develop the habit of glancing at the Command line while you work in AutoCAD. The Command line can be set to display any number of lines and/or moved to another location (see "Customizing the AutoCAD Screen" later in this chapter).

Palettes

Several palettes are available in AutoCAD, and one or more palettes may appear on your screen when you start AutoCAD, such as the *Sheet Set Manager* (see Figure 1-14), the *Properties* palette, or the *Tool Palette*. Each palette serves a particular function as explained in later chapters. You can close the palettes by clicking on the "X" in the upper corner of the palette.

Toolbars

AutoCAD provides a variety of <u>tool-bars</u> (Fig. 1-14). Each toolbar contains a number of icon buttons (tools) that can be PICKed to invoke commands for drawing or editing objects (lines, arcs, circles, etc.) or for managing files and other functions.

The <u>Standard toolbar</u> is the row of icons nearest the top of the screen. The Standard toolbar contains many standard icons used in other Windows applications (like *New, Open, Save, Print, Cut, Paste,* etc.) and other icons for AutoCAD-specific functions (like *Zoom* and *Pan*). The <u>Object Properties toolbar</u>, located beneath the standard toolbar, is used for managing properties of objects, such as *Layers* and *Linetypes*. The <u>Draw and Modify toolbars</u> also appear (by default) when you first use AutoCAD. As shown in Figure

FIGURE 1-14

Standard Toolbar   Object Properties Toolbar

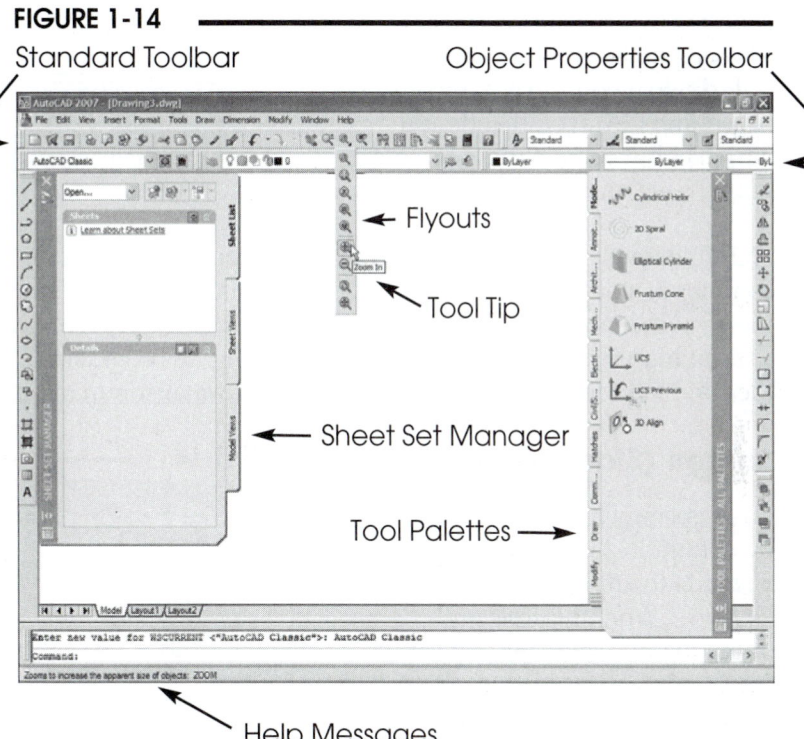

Flyouts

Tool Tip

Sheet Set Manager

Tool Palettes →

Help Messages

1-14, the Draw and Modify toolbars (left and right side) and the Standard and Object Properties toolbars (top) are <u>docked</u>. Many other toolbars are available and can be made to float, resize, or dock (see "Customizing the AutoCAD Screen" later in this chapter).

If you place the pointer on any icon and wait momentarily, a <u>Tool Tip</u> and a <u>Help message</u> appear. Tool Tips pop out by the pointer and give the command name (see Fig. 1-14). The Help message appears at the bottom of the screen, giving a short description of the function. <u>Flyouts</u> are groups of related icons that pop out in a row or column when one of the group is selected. PICKing any icon that has a small black triangle in its lower-right corner causes the related icons to fly out.

Pull-down Menus

The pull-down menu bar is at the top of the screen just under the title bar (Fig. 1-15). Selecting any of the words in the menu bar activates, or pulls down, the respective menu. Selecting a word appearing with an arrow activates a cascading menu with other options. Selecting a word with ellipsis (. . .) activates a dialog box (see "Dialog Boxes"). Words in the pull-down menus are not necessarily the same as the formal command names used when typing commands. Menus can be canceled by pressing Escape or PICKing in the graphics area. The pull-down menus do not contain all of the AutoCAD commands and variables but contain the most commonly used ones.

FIGURE 1-15

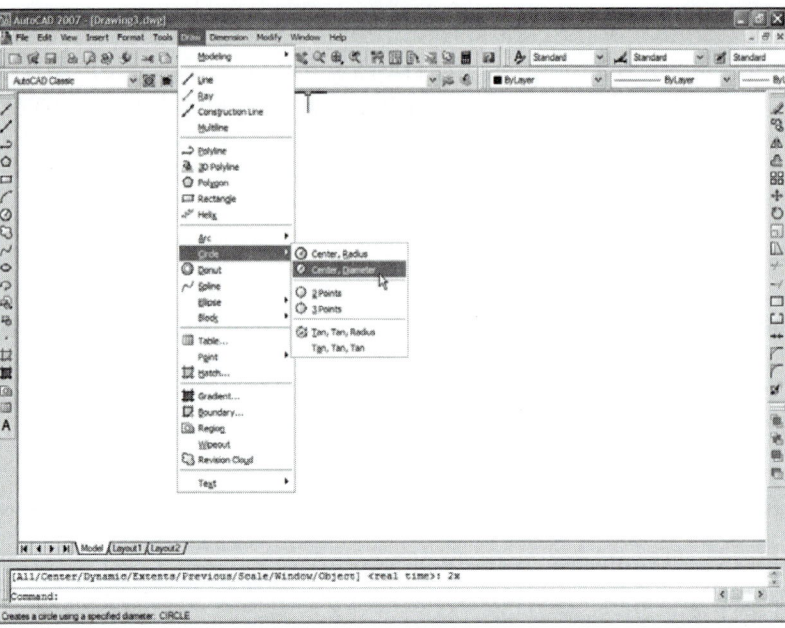

NOTE: Because the words that appear in the pull-down menus are not always the same as the formal commands you would type to invoke a command, AutoCAD can be confusing to learn. The Help message that appears just below the Command line (when a menu is pulled down or the pointer rests on an icon button) can be instrumental in avoiding this confusion. The Help message gives a description of the command followed by colon (:), then the formal command name (see Fig. 1-14 and Fig. 1-15).

Screen (Side) Menu

The screen menu does not appear by default in AutoCAD, but can be made to appear by selecting *Options…* from the *Tools* pull-down menu. This selection causes the *Options* dialog box to appear (Fig. 1-16). Select the *Display* tab and check the *Display screen menu* option. Keep in mind that this menu is not needed if you prefer to use the icon buttons, pull-down menus, or keyboard to enter commands.

FIGURE 1-16

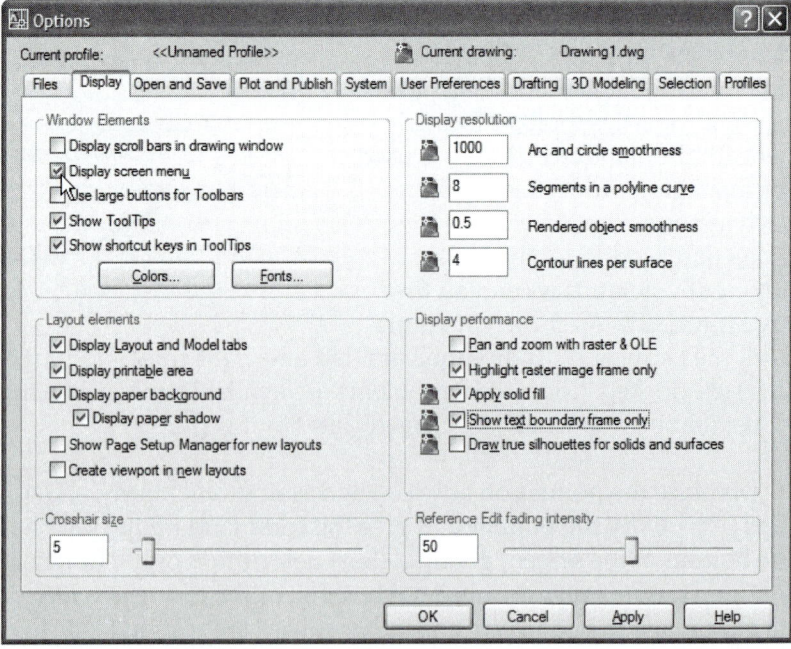

Not all of the AutoCAD commands can be accessed through the menu system located at the right side of the screen. The screen menu has a tree structure.

Menus and commands are accessed by branching out from the root or top-level menu (Fig. 1-17). Commands (in uppercase and lowercase) or menus (capital letters only) are selected by moving the cursor to the desired position until the word is highlighted and then pressing the PICK button. The root menu (top level) is accessed from any level of the structure by selecting the word "AutoCAD" at the top of any menu.

The commands in these menus are generally the formal command names that can also be typed at the keyboard. (Some command names are truncated in the screen menu because only eight characters can be displayed.)

If commands are invoked by typing, pull-downs, or toolbars, the screen menu automatically changes to the current command.

FIGURE 1-17

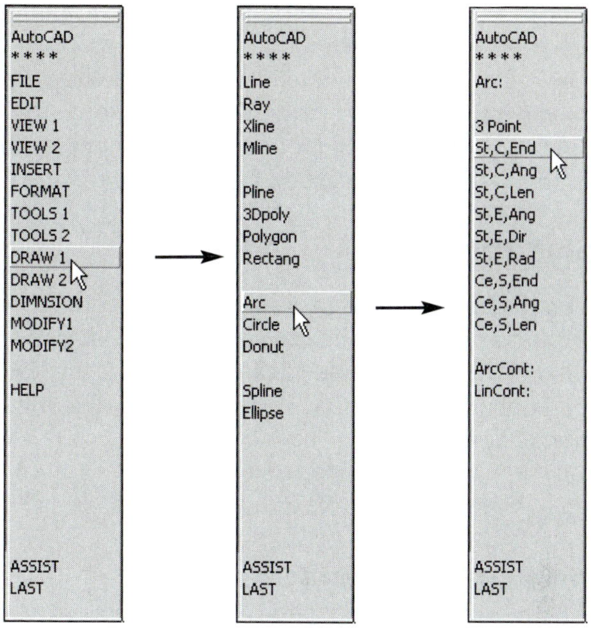

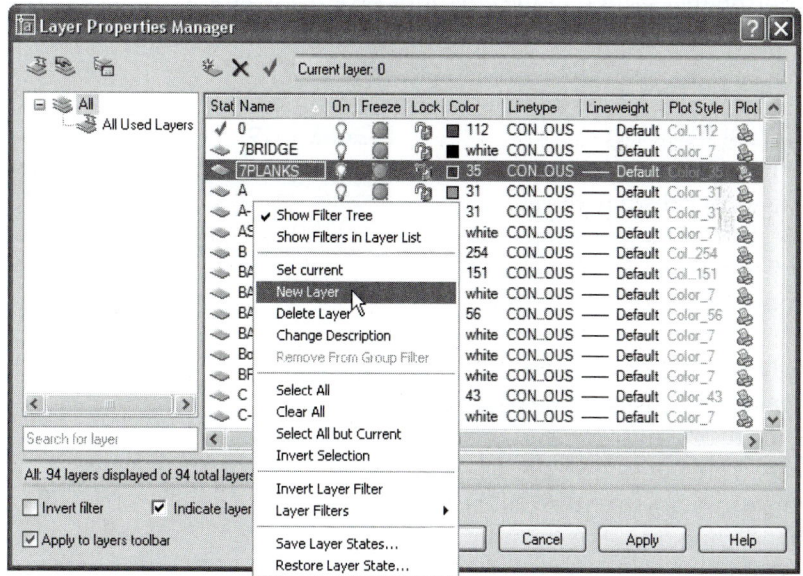

Dialog Boxes

Dialog boxes provide an interface for controlling complex commands or a group of related commands. Depending on the command, the dialog boxes allow you to select among multiple options and sometimes give a preview of the effect of selections. The *Layer Properties Manager* dialog box (Fig. 1-18) gives complete control of layer colors, linetypes, and visibility.

FIGURE 1-18

Dialog boxes can be invoked by typing a command, selecting an icon button, or PICKing from the menus. For example, typing *Layer*, PICKing the *Layers* button, or selecting *Layer* from the *Format* pull-down menu causes the *Layer Properties Manager* dialog box to appear. In the pull-down menu, all commands that invoke a dialog box end with ellipsis points (…).

The basic elements of a dialog box are listed on the next page.

Button	Resembles a push button and triggers some type of action.
Edit box	Allows typing or editing of a single line of text.
Image tile	A button that displays a graphical image.
List box	A list of text strings from which one or more can be selected.
Drop-down list	A text string that drops down to display a list of selections.
Radio button	A group of buttons, only one of which can be turned on at a time.
Checkbox	A checkbox for turning a feature on or off (displays a check mark when on).

Status Bar

The Status bar is a set of informative words or symbols that gives the status of the drawing aids. The Status bar appears at the very bottom of the screen (see Fig. 1-13). The following drawing aids can be toggled on or off by single-clicking (pressing the left mouse button once) on the desired word or by using Function keys or Ctrl key sequences. The following drawing aids are explained in this or in following chapters:

SNAP, GRID, ORTHO, POLAR, OSNAP, OTRACK, DUCS, DYN, LWT, MODEL

Coordinate Display (Coords)

The <u>Coordinate display</u> is located in the lower-left corner of the AutoCAD screen (Fig. 1-19). The Coordinate display (*Coords*) displays the current position of the cursor in one of two possible formats explained below. This display can be very helpful when you draw because it can give the X, Y, and Z coordinate position of the cursor or give the cursor's distance and angle from the last point established. The format of *Coords* is controlled by single-clicking on the *Coords* display (numbers) at the bottom left of the screen. *Coords* can also be toggled off.

Cursor tracking

When *Coords* is in this position, the values display the <u>current location of the cursor</u> in absolute Cartesian (X, Y, and Z) coordinates (see Fig. 1-19).

Relative polar display

This display is possible only if a draw or edit command is in use. The values give the distance and angle of the "rubberband" line from the last point established (not shown).

FIGURE 1-19

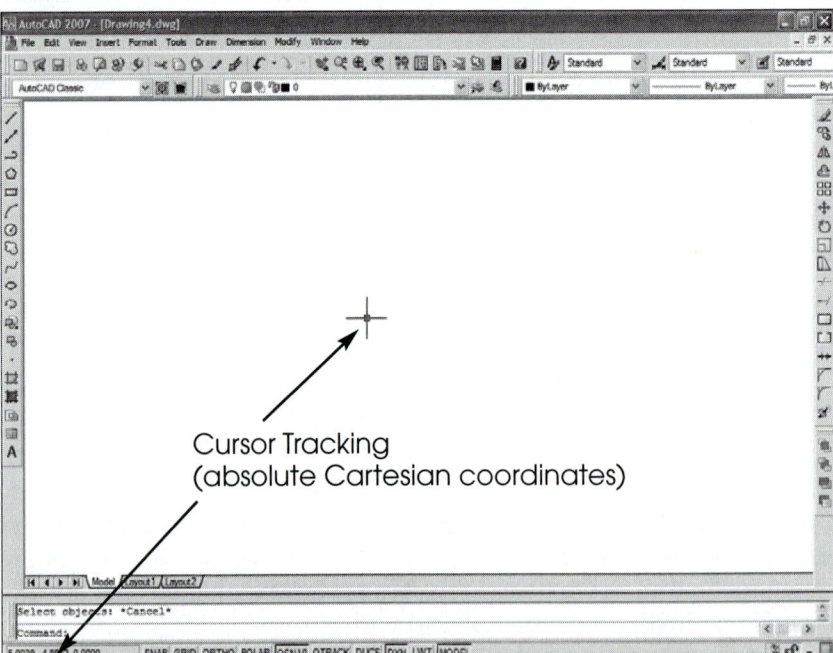

Cursor Tracking
(absolute Cartesian coordinates)

Digitizing Tablet Menu

If you have a digitizing tablet, the AutoCAD commands are available by making the desired selection from the AutoCAD digitizing tablet menu (Fig. 1-20, on the next page). The open area located slightly to the right of center is called the Screen Pointing area. Locating the digitizing puck there makes the cursor appear on the screen. Locating the puck at any other location allows you to select a command.

The icons on the tablet menu are identical to the toolbar icons. Similar to the format of the toolbars, commands are located in groups such as Draw, Edit, View, and Dimension. The tablet menu has the columns numbered along the top and the rows lettered along the left side. The location of each command by column and row is given in the "command tables" in this book. For example, the *Line* command can be found at *10,J* (Fig. 1-20).

FIGURE 1-20

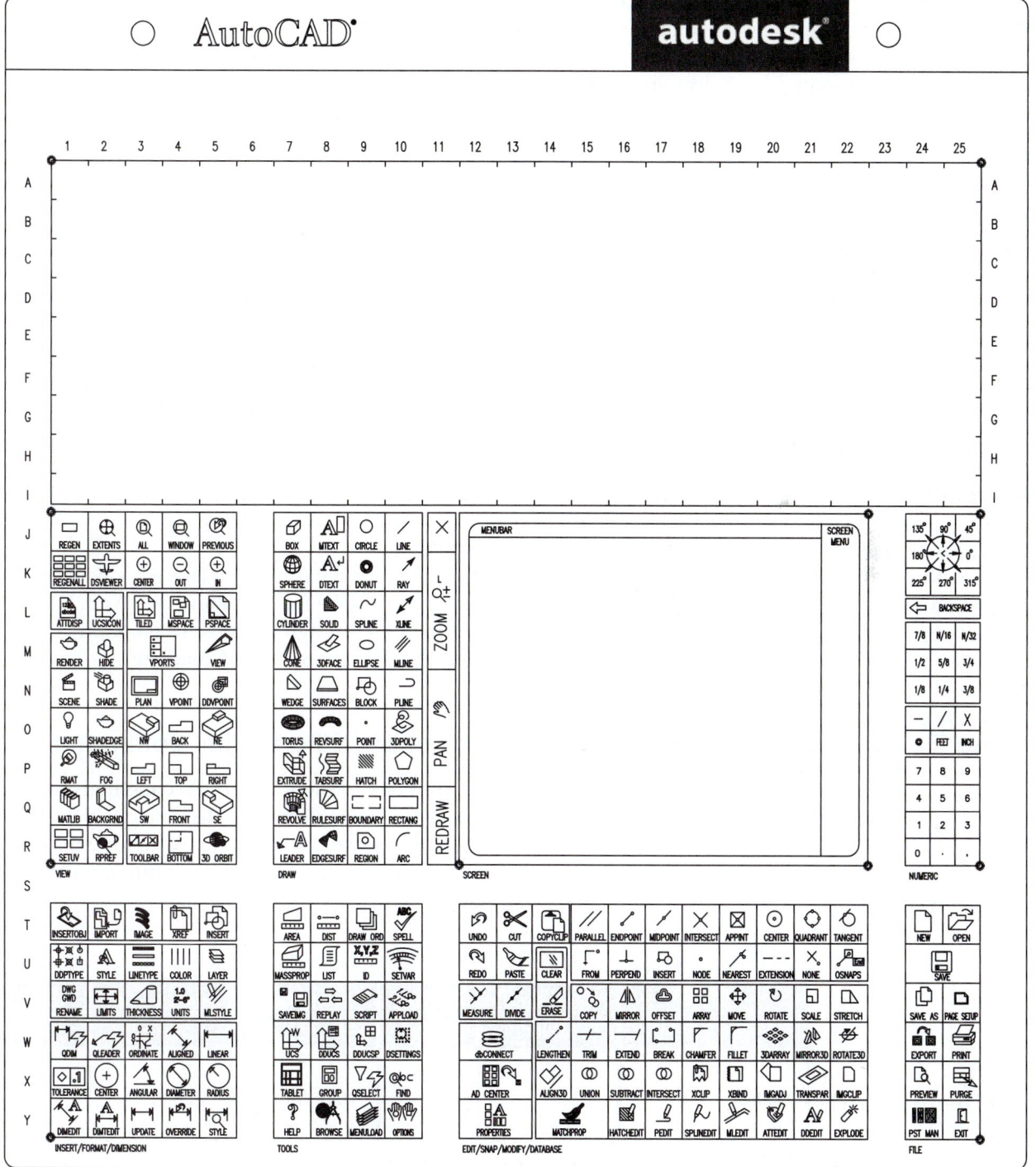

COMMAND ENTRY

Methods for Entering Commands

There are five possible methods for entering commands in AutoCAD depending on your *Options* setting (for the screen menu) and availability of a digitizing tablet. Generally, <u>any one</u> of the five methods can be used to invoke a particular command depending on your system configuration.

1. **Toolbars**   Select the command or dialog box by PICKing an icon (tool) from a toolbar.
2. **Pull-down menu**   Select the command or dialog box from a pull-down menu.
3. **Screen (side) menu**   Select the command or dialog box from the screen (side) menu.
4. **Keyboard**   Type the command name, command alias, or accelerator (Ctrl) keys at the keyboard.
5. **Tablet menu**   Select the command from the digitizing tablet menu (if available).

All five methods of entering commands accomplish the same goals; however, one method may offer a slightly different option or advantage to another depending on the command used.

All menus, including the digitizing tablet, can be customized by editing the ACAD.CUI file, and command aliases can be changed or added in the ACAD.PGP file so that command entry can be designed to your preference (see Chapter 44, Basic Customization, and Chapter 45, Menu Customization).

Keyboard Entry Features

Using the keyboard to type in commands offers several features:

Command name   Type in the full command name.
Command alias   Type the one- or two-letter shortcut for the command (see App. B and C).
Accelerator keys   Type a Ctrl key plus another key to invoke the command (see App. D).
Up and down arrows   Use the up and down arrows to cycle through the most recent input.
Tab key   Type in a few letters and press the Tab key to "AutoComplete" the command.

Using the "Command Tables" in This Book to Locate a Particular Command

Command tables, like the one below, are used throughout this book to show the possible methods for entering a particular command. The table shows the icon used in the toolbars and digitizing tablet, gives the selections to make for the pull-down and screen (side) menus, gives the correct spelling for entering commands and command aliases at the keyboard, and gives the command location (column, row) on the digitizing tablet menu. This example uses the *Copy* command.

Copy

Pull-down Menu	Command (Type)	Alias (Type)	Short-cut	Screen (side) Menu	Tablet Menu
Modify *Copy*	*Copy*	*CO* or *CP*	(Edit Mode) *Copy Selection*	*MODIFY1* *Copy*	*V,15*

Shortcut Menus

AutoCAD makes use of shortcut menus that are activated by pressing the right mouse button (sometimes called right-click menus). Shortcut menus give quick access to command options. There are many shortcut menus to list since they are based on the active command or dialog box. The menus fall into five basic categories listed here.

Default Menu

The default menu appears when you right-click in the drawing area and no command is in progress. Using this menu, you can repeat the last command, select from recent input (like using the up and down arrows), use the Windows Cut, Copy, Paste functions, and select from other viewing and utility commands (Fig. 1-21).

FIGURE 1-21

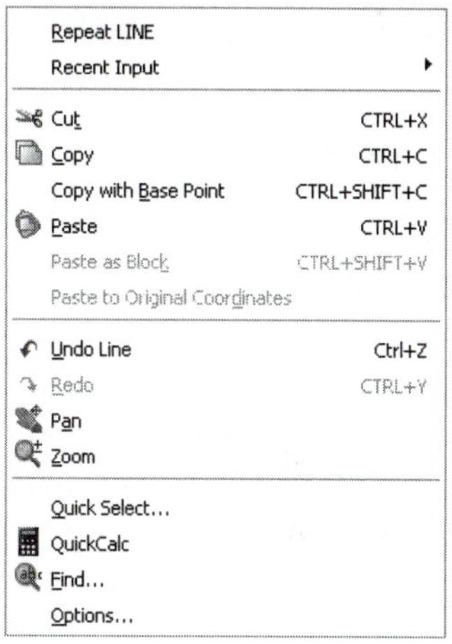

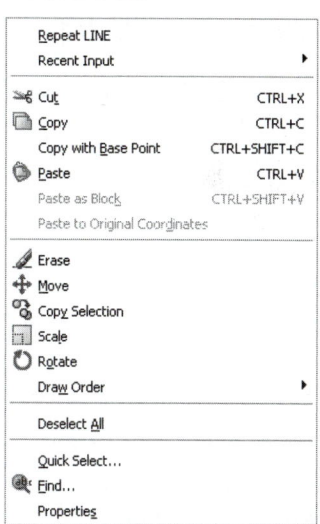

Edit-Mode Menu

This menu appears when you right-click when <u>objects have been selected</u> but no command is in progress. Note that several of AutoCAD's *Modify* commands are available on the menu such as *Erase*, *Move*, *Copy*, *Scale*, and *Rotate* (Fig. 1-22).

NOTE: Edit mode shortcut menus do not appear if the *PICKFIRST* system variable is set to 0 (see "*PICKFIRST*" in Chapter 20).

FIGURE 1-22

Command-Mode Menu

These menus appear when you right-click <u>when a command is in progress</u>. This menu changes since the options are specific to the command (Fig. 1-23).

FIGURE 1-23

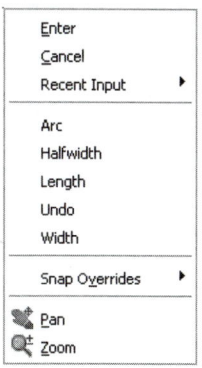

Dialog-Mode Menu

When the pointer is in a dialog box or tab and you right-click, this menu appears. The options on this menu can change based on the current dialog box (Fig. 1-24).

FIGURE 1-24

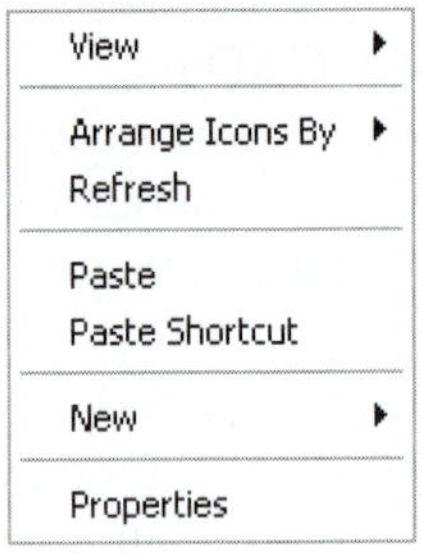

Other Menus

There are other menus that can be invoked. For example, this menu appears if you right-click in the Command line area (Fig. 1-25).

Because there are so many shortcut menus, don't be too concerned about learning these until you have had some experience. The best advice at this time is just remember to experiment by right-clicking often to display the possible options.

FIGURE 1-25

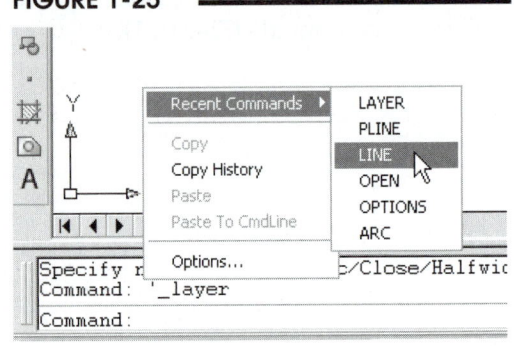

Mouse and Digitizing Puck Buttons

Depending on the type of mouse or digitizing puck used for cursor control, a different number of buttons is available. In any case, the buttons perform the following tasks:

#1 (left button)	PICK	Used to select commands or pick locations on the screen.
#2 (right button)	Enter or shortcut menu	Depending on the status of the drawing or command, this button either performs the same function as the enter key or produces a shortcut menu.
#3 (middle button or wheel)	*Pan*	If you press and drag, you can pan the drawing about on the screen.
	Zoom	If you turn the wheel, you can zoom in and out centered on the location of the cursor.

Function Keys

Several function keys are usable with AutoCAD. They offer a quick method of turning on or off (toggling) drawing aids.

F1	*Help*	Opens a help window providing written explanations on commands and variables.
F2	*Flipscreen*	Activates a text window showing the previous command line activity (command history).
F3	*Osnap Toggle*	If Running Osnaps are set, toggling this tile temporarily turns the Running Osnaps off so that a point can be picked without using Osnaps. If no Running Osnaps are set, F3 produces the *Osnap Settings* dialog box (discussed in Chapter 7).
F4	*Tablet*	Turns the *TABMODE* variable on or off. If *TABMODE* is on, the digitizing tablet can be used to digitize an existing paper drawing into AutoCAD.

F5	*Isoplane*	When using an *Isometric* style *SNAP* and *GRID* setting, toggles the cursor (with *ORTHO* on) to draw on one of three isometric planes.
F6	*DUCS*	Turns *Dynamic UCS* on or off for 3D modeling (see Chapters 36 and 38).
F7	*GRID*	Turns the *GRID* on or off (see "Drawing Aids").
F8	*ORTHO*	Turns *ORTHO* on or off (see "Drawing Aids").
F9	*SNAP*	Turns *SNAP* on or off (see "Drawing Aids").
F10	*POLAR*	Turns *POLAR* on or off (see "Drawing Aids").
F11	*Osnap Tracking*	Turns Object Snap Tracking on or off.
F12	*DYN*	Turns Dynamic Input on or off (see "Drawing Aids").

Control Key Sequences (Accelerator Keys)

Several control key sequences (holding down the Ctrl key or Alt key and pressing another key simultaneously) invoke regular AutoCAD commands or produce special functions (see App. D).

Special Key Functions

Esc	The Escape key cancels a command, menu, or dialog box or interrupts processing of plotting or hatching.
Space bar	In AutoCAD, the space bar performs the same action as the Enter key. Only when you are entering text into a drawing does the space bar create a space.
Enter	If Enter or the space bar is pressed when no command is in use (the open Command: prompt is visible), the last command used is invoked again.

Drawing Aids

This section gives a brief introduction to AutoCAD's Drawing Aids. For a full explanation of the related commands and options, see Chapters 3 and 6.

SNAP (F9 or Status Bar)

SNAP has two modes in AutoCAD: Grid Snap and Polar Snap. Only one of the two modes can be active at one time. Grid Snap is a function that forces the cursor to "snap" to regular intervals (.5 units is the default English setting), which aids in creating geometry accurate to interval lengths. You can use the *Snap* command or the *Drafting Settings* dialog box to specify any value for the Grid Snap increment. Figure 1-26 displays a Snap setting of .25 (note the values in the coordinate display in the lower-left corner).

FIGURE 1-26

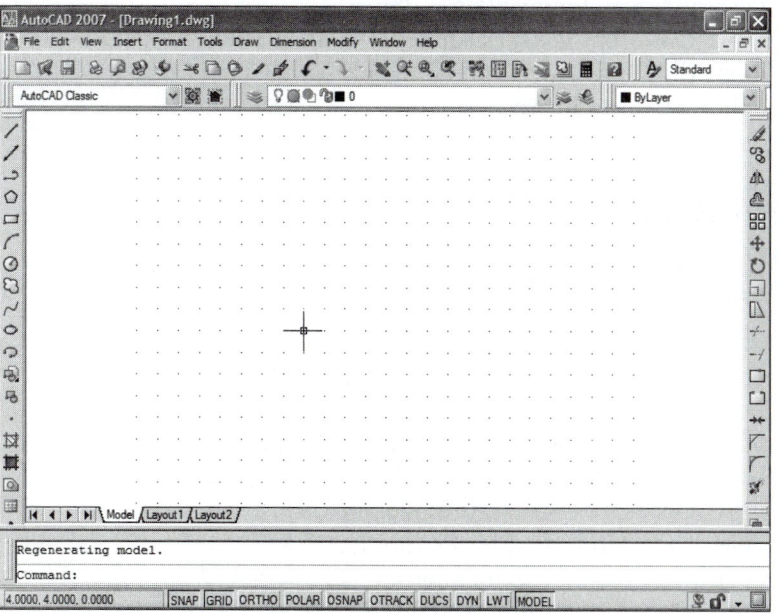

The other mode of Snap is Polar Snap. Polar Snap forces the cursor to snap to regular intervals along angular lines. Polar Snap is functional only when *POLAR* is also toggled on since it works in conjunction with *POLAR*. The Polar Snap interval uses the Grid Snap setting by default but can be changed to any value using the *Snap* command or the *Drafting Settings* dialog box. Polar Snap is discussed in detail in Chapter 3.

Since you can have only one *SNAP* mode on at a time (Grid Snap or Polar Snap), you can select which of the two is on by right-clicking on the word "*SNAP*" on the Status bar (Fig. 1-27). You can also access the *Drafting Settings* dialog box by selecting *Settings…* from the menu.

FIGURE 1-27

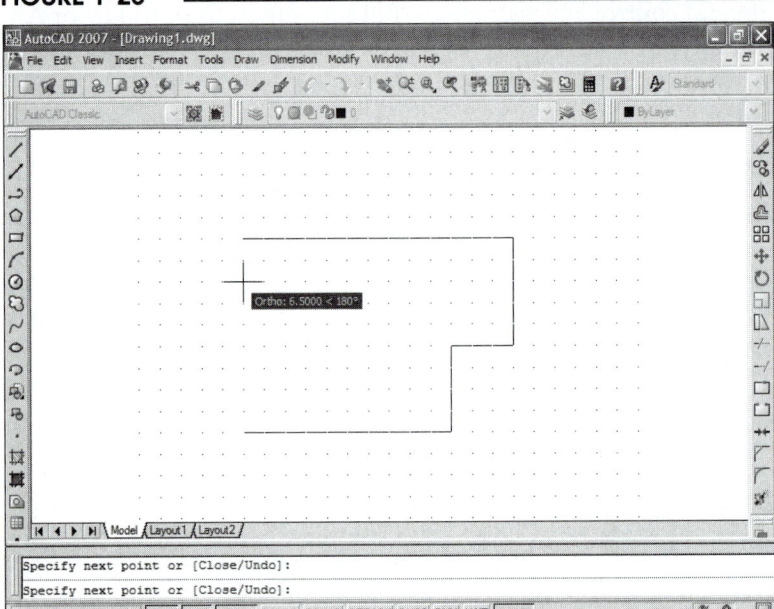

GRID (F7 or Status Bar)

A drawing aid called *GRID* can be used to give a visual reference of units of length. The *GRID* default value for English settings is .5 units. The *Grid* command or *Drawing Aids* dialog box allows you to change the interval to any value. The *GRID* is not part of the geometry and is not plotted. Figure 1-28 displays a *GRID* of .5. *SNAP* and *GRID* are independent functions—they can be turned on or off independently. However, you can force the *GRID* to have the same interval as *SNAP* by entering a *GRID* value of 0 or you can use a proportion of *SNAP* by entering a *GRID* value followed by an "X".

FIGURE 1-28

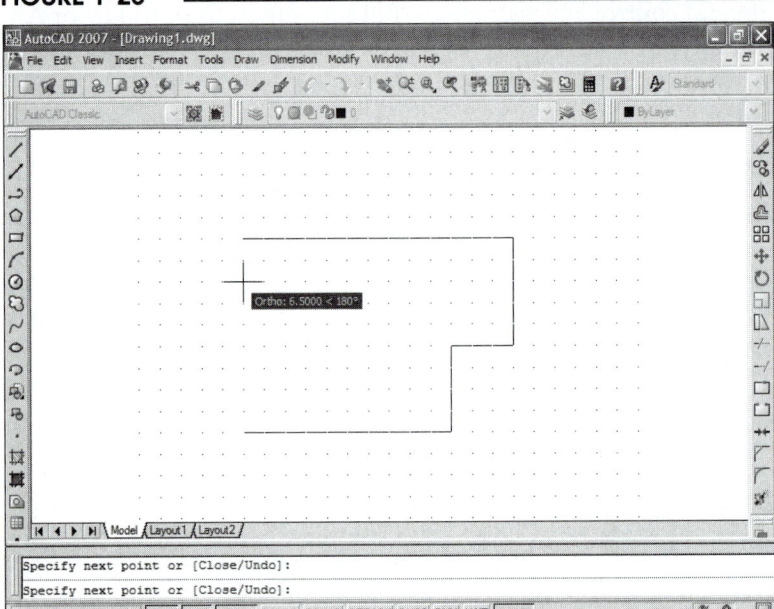

ORTHO (F8 or Status Bar)

If *ORTHO* is on, lines are forced to an orthogonal alignment (horizontal or vertical) when drawing (Fig. 1-28). *ORTHO* is often helpful since so many drawings are composed mainly of horizontal and vertical lines. *ORTHO* can be turned only on or off.

POLAR (F10 or Status Bar)

POLAR makes it easy to draw lines at regular angular increments, such as 30, 45, or 90 degrees. Using the F10 key or *POLAR* button toggles Polar Tracking on or off. When Polar Tracking is on, a polar tracking vector (a faint dotted line) appears when the rubber band line approaches the desired angular increment as shown in Figure 1-29. By default, *POLAR* is set to 90 degrees, but can be set to any angular increment by right-clicking on the *POLAR* button and selecting *Settings*.

FIGURE 1-29

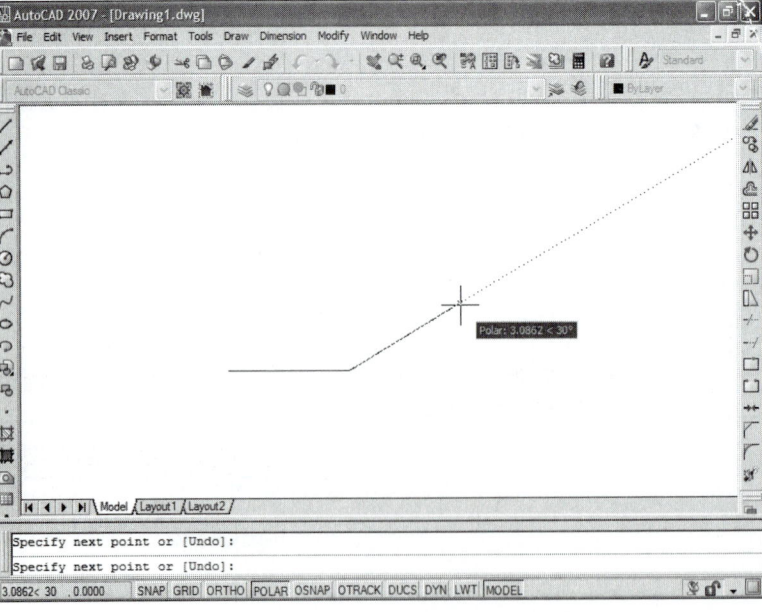

OSNAP (F3 or Status Bar)

The *OSNAP* toggle is discussed in detail in Chapter 7.

OTRACK (F11 or Status Bar)

The *OTRACK* toggle is also discussed in Chapter 7.

DYN (F12 or Status Bar)

DYN, or Dynamic Input, is a feature that helps you visualize and specify coordinate values and angular values when drawing lines, arcs, circles, etc. *DYN* may display absolute Cartesian coordinates (X and Y values) or relative polar coordinates (distance and angle) depending on the current command prompt and the settings you prefer. Dynamic Input is explained further in Chapter 3 (Fig. 1-30).

FIGURE 1-30

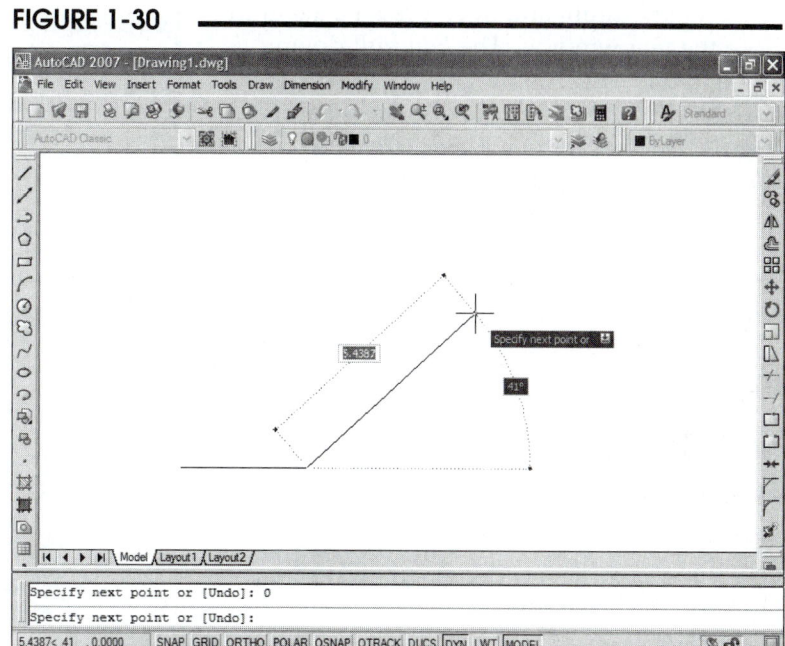

You can toggle Dynamic Input on or off by using the *DYN* button or the F12 key. You can also change the format of the display during input by right-clicking on the *DYN* button and selecting *Settings* from the menu. For example, you can set Dynamic Input to display a "tool tip" that gives the current coordinate location of the cursor (cursor tracking mode) when no commands are in use.

LWT

This feature is discussed in Chapter 11.

MODEL

See Chapter 13 for an explanation of the *MODEL/PAPER* toggle.

Status Bar Control

Right-clicking in the Status bar produces the shortcut menu shown in Figure 1-31. This menu gives you control of which of these drawing aids appear in the Status bar. Clearing a check mark removes the associated drawing aid from the Status bar. It is recommended for most new AutoCAD users to keep all the drawing aids visible on the Status bar.

FIGURE 1-31

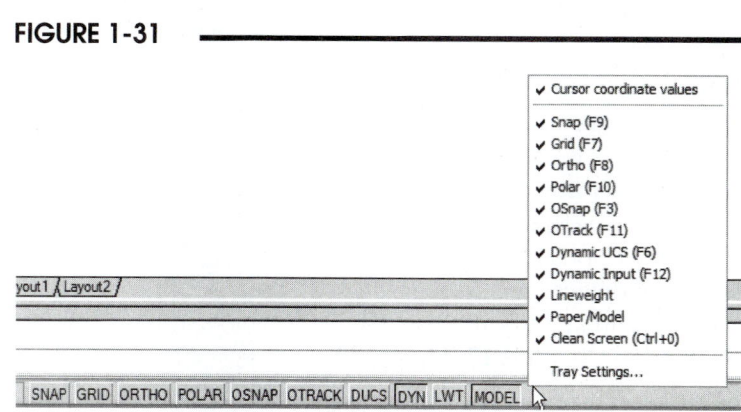

AutoCAD Text Window (F2)

Pressing the F2 key activates the *AutoCAD Text Window*, sometimes called the Command History. Here you can see the text activity that occurred at the Command line—kind of an "expanded" Command line. Press F2 again to close the text window. The *Edit* pull-down menu in this text window provides several options. If you highlight text in the window (Fig. 1-32), you can then *Paste to Cmdline* (Command line), *Copy* it to another program such as a word processor, *Copy History* (entire command history) to another program, or *Paste* text into the window. The *Options* choice invokes the *Options* dialog box (discussed later).

Model Tab and *Layout* Tabs

Model Tab

When you create a new drawing, the *Model* tab is the current tab (see Fig. 1-33, bottom left). This area is also known as <u>model space</u>. In this area you <u>should create the geometry representing the subject of your drawing</u>, such as a floor plan, a mechanical part, or an electrical schematic. Dimensions are usually created and attached to your objects in model space.

Layout Tabs

When you are finished with your drawing, you can plot it directly from the *Model* tab or switch to a *Layout* tab (Fig. 1-34). Layout tabs, sometimes known as paper space, represent sheets of paper that you plot on. You must use several commands to set up the layout to display the geometry and set all the plotting options such as scale, paper size, plot device, and so on. Plotting and layouts are discussed in detail in Chapters 13, 14, and 33.

FIGURE 1-32

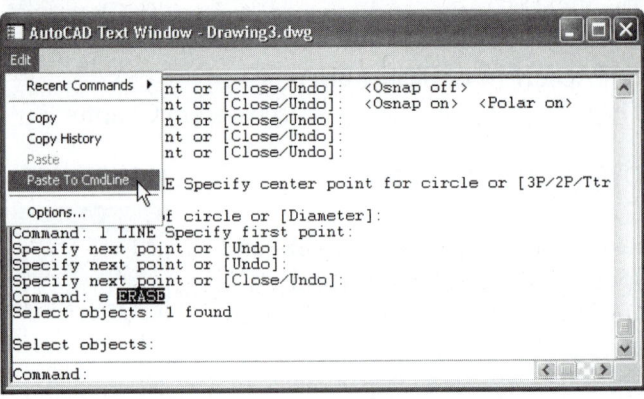

FIGURE 1-33

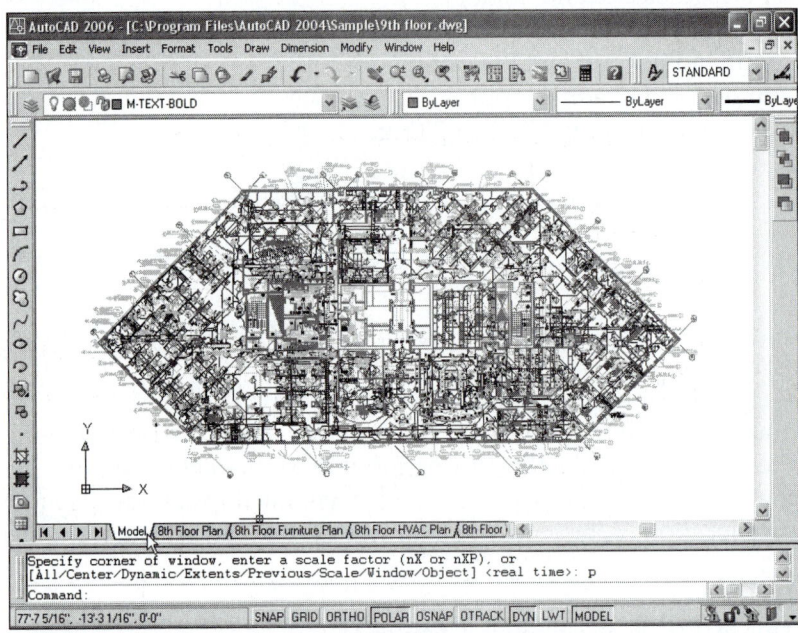

FIGURE 1-34

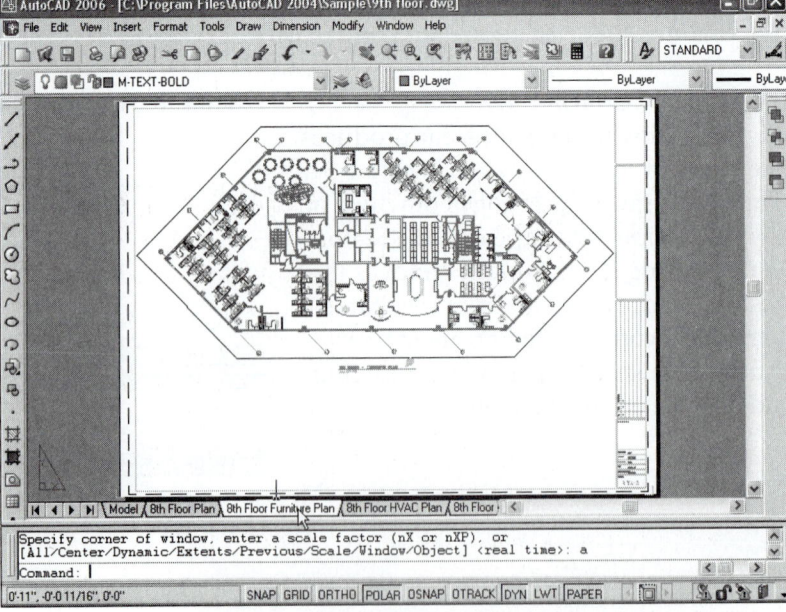

COMMAND ENTRY METHODS PRACTICE

Start AutoCAD. If the *Startup* dialog box appears, select *Start from Scratch*, choose *Imperial* as the default setting, and click the *OK* button. Invoke the *Line* command using each of the command entry methods as follows.

1.   Type the command

STEPS	COMMAND PROMPT	PERFORM ACTION	COMMENTS
1.		press **Escape** if another command is in use	
2.	Command:	type *Line* and press **Enter**	
3.	LINE Specify first point:	**PICK** any point	a "rubberband" line appears
4.	Specify next point or [Undo]:	**PICK** any point	another "rubberband" line appears
5.	Specify next point or [Undo]:	press **Enter**	to complete command

2.   Type the command alias

STEPS	COMMAND PROMPT	PERFORM ACTION	COMMENTS
1.		press **Escape** if another command is in use	
2.	Command:	type *L* and press **Enter**	
3.	LINE Specify first point:	**PICK** any point	a "rubberband" line appears
4.	Specify next point or [Undo]:	**PICK** any point	another "rubberband" line appears
5.	Specify next point or [Undo]:	press **Enter**	to complete command

3.   Pull-down menu

STEPS	COMMAND PROMPT	PERFORM ACTION	COMMENTS
1.	Command:	select the *Draw* menu from the menu bar on top	
2.	Command:	select *Line*	menu disappears
3.	LINE Specify first point:	**PICK** any point	a "rubberband" line appears
4.	Specify next point or [Undo]:	**PICK** any point	another "rubberband" line appears
5.	Specify next point or [Undo]:	press **Enter**	to complete command

4.   **Screen menu** (if activated on your setup)

STEPS	COMMAND PROMPT	PERFORM ACTION	COMMENTS
1.	Command:	select **AutoCAD** from screen menu on the side	only if menu is not at root level
2.	Command:	select *DRAW1* from root screen menu	menu changes to *DRAW 1*
3.	Command:	select *Line*	
4.	LINE Specify first point:	**PICK** any point	a "rubberband" line appears
5.	Specify next point or [Undo]:	**PICK** any point	another "rubberband" line appears
6.	Specify next point or [Undo]:	press **Enter**	to complete command

5.   **Toolbars**

STEPS	COMMAND PROMPT	PERFORM ACTION	COMMENTS
1.	Command:	select the *Line* icon from the Draw toolbar (on the left side of the screen)	the *Line* tool should be located at the top of the toolbar
2.	LINE Specify first point:	**PICK** any point	a "rubberband" line appears
3.	Specify next point or [Undo]:	**PICK** any point	another "rubberband" line appears
4.	Specify next point or [Undo]:	press **Enter**	to complete command

6.   **Digitizing tablet menu** (if available)

STEPS	COMMAND PROMPT	PERFORM ACTION	COMMENTS
1.	Command:	select *LINE* on the digitizing tablet menu	located at *10,J*
2.	LINE Specify first point:	**PICK** any point	a "rubberband" line appears
3.	Specify next point or [Undo]:	**PICK** any point	another "rubberband" line appears
4.	Specify next point or [Undo]:	press **Enter**	to complete command

When you are finished practicing, use the *Files* pull-down menu and select *Exit* to exit AutoCAD. You do not have to "Save Changes."

CUSTOMIZING THE AutoCAD SCREEN

Toolbars

Many toolbars are available, each with a group of related commands for specialized functions. For example, when you are ready to dimension a drawing, you can activate the Dimension toolbar for efficiency. Right-clicking on any toolbar displays a list of all toolbars (Fig. 1-35). Selecting any toolbar name makes that toolbar appear on the screen. You can also select *Toolbars...* from the *View* pull-down menu or type the command *Toolbar* to display the *Customize User Interface* dialog box. Toolbars can be removed from the screen by clicking once on the "X" symbol in the upper right of a floating toolbar.

FIGURE 1-35

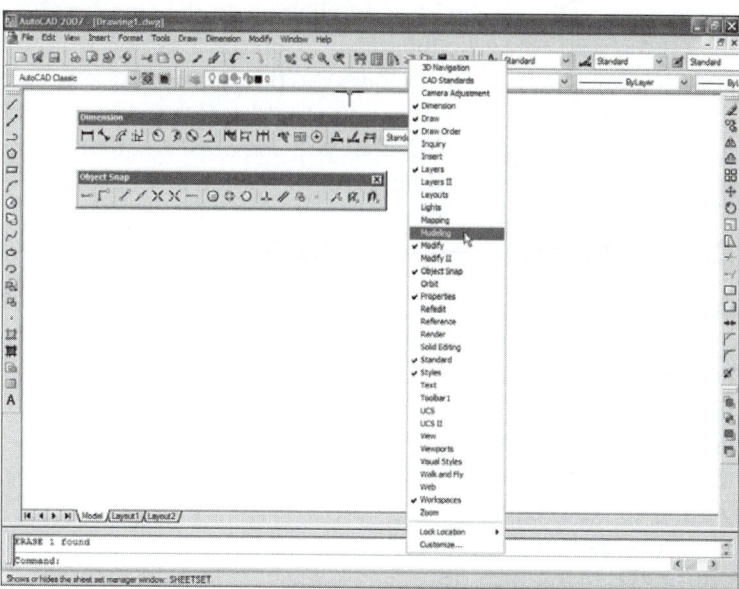

By default, the Object Properties and Standard toolbars (top) and the Draw and Modify toolbars (side) are docked, whereas toolbars that are newly activated are floating (see Fig. 1-35). A floating toolbar can be easily moved to any location on the screen if it obstructs an important area of a drawing. Placing the pointer in the title background allows you to move the toolbar by holding down the left button and dragging it to a new location (Fig. 1-36). Floating toolbars can also be resized by placing the pointer on the narrow border until a two-way arrow appears, then dragging left, right, up, or down (Fig. 1-37).

FIGURE 1-36

FIGURE 1-37

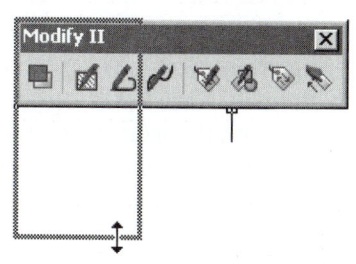

A floating toolbar can be docked against any border (right, left, top, bottom) by dragging it to the desired location (Fig. 1-38). Several toolbars can be stacked in a docked position. By the same method, docked toolbars can be dragged back onto the graphics area. Although the Object Properties and Standard toolbars can be moved onto the graphics area or docked on another border, it is wise to keep these toolbars in their standard position. The Object Properties toolbar only displays its full options when located in a horizontal position.

FIGURE 1-38

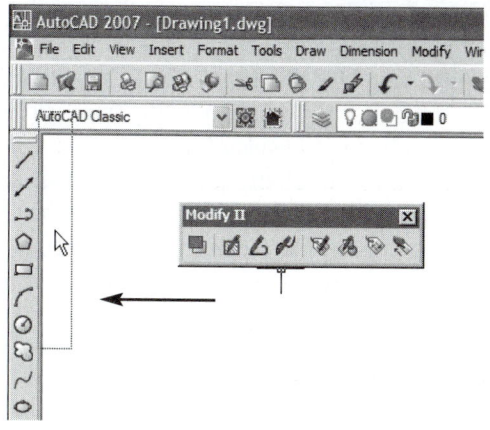

You can lock the placement of toolbars or palettes to prevent accidental movement on the screen by right-clicking on the small padlock icon at the lower-right corner of the screen. Make a selection from the menu that appears (Fig. 1-39). New toolbars can be created and existing toolbars can be customized. Typically, you would create toolbars to include groups of related commands that you use most frequently or need for special activities (see Chapter 44 and Chapter 45).

FIGURE 1-39

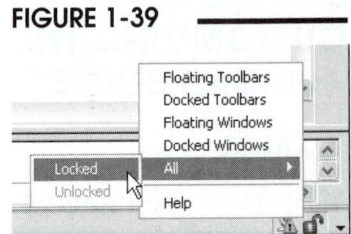

Clean Screen

FIGURE 1-40

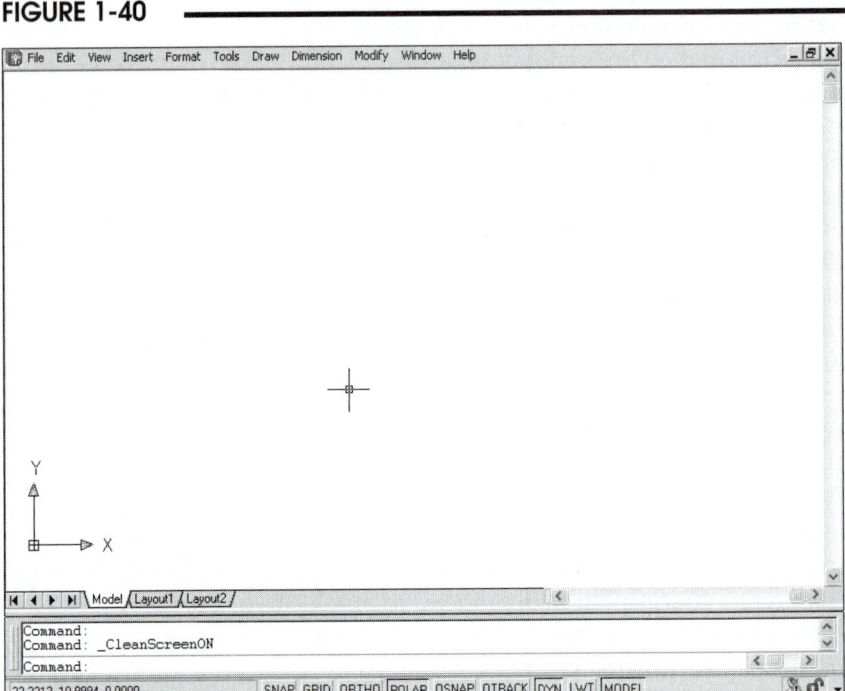

The *Cleanscreenon* and *Cleanscreenoff* commands can be used to change the AutoCAD window from a normal window showing all toolbars to a window with <u>no</u> toolbars (Fig. 1-40). Notice that the Command line and pull-down menus are still visible. This is useful if you want to maximize the drawing area to view a drawing or to make a presentation. Instead of typing these commands or using the *View* pull-down menu, a shortcut keystroke of Ctrl+0 (zero) toggles between the two windows.

Communication Center

The Communication Center feature provides up-to-date product information, software updates, and marketing announcements. When new information or software updates are available, an information bubble appears in the lower-right corner of the Status Bar. If the Communication Center icon is not visible on your system, the icon may be disabled (see "Tray Settings"), or this feature may not have been configured during the installation process. (For more information on Communication Center, see Chapter 23, Internet Tools and Collaboration.)

Tray Settings

FIGURE 1-41

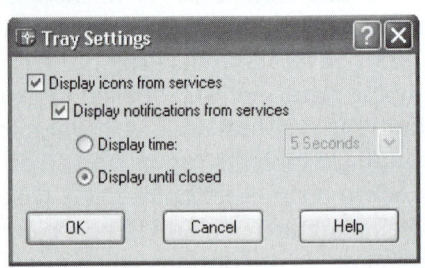

Communication Center and other services are available through icons in the system tray (lower-left corner of the AutoCAD window). Use the *Traysettings* command or right-click in the Status bar area to produce the *Tray Settings* dialog box (Fig. 1-41). If you chose to *Display Icons from Services*, icons appear (for enabled services) such as Communication Center or Digital Signature Validation. If *Display Notifications from Services* is checked, you will receive alert messages, such as when Communication Center updates are available, digital signatures are invalid, or externally referenced drawings need reloading.

Command Line

The command line, normally located near the bottom of the screen, can be resized to display more or fewer lines of text. Moving the pointer to the border between the graphics screen and the command line until two-way arrows appear allows you to slide the border up or down. A <u>minimum of two lines</u> of text is recommended. The Command window can also be moved to any location and can be floating or docked. Point to the two vertical bars on the left of the Command window, then drag the window to the desired location. If the Command window is floating, you can specify transparency by right-clicking on the title bar, selecting *Transparency* from the menu, and setting the **Transparency Level** in the *Transparency* dialog box (Fig. 1-42).

FIGURE 1-42

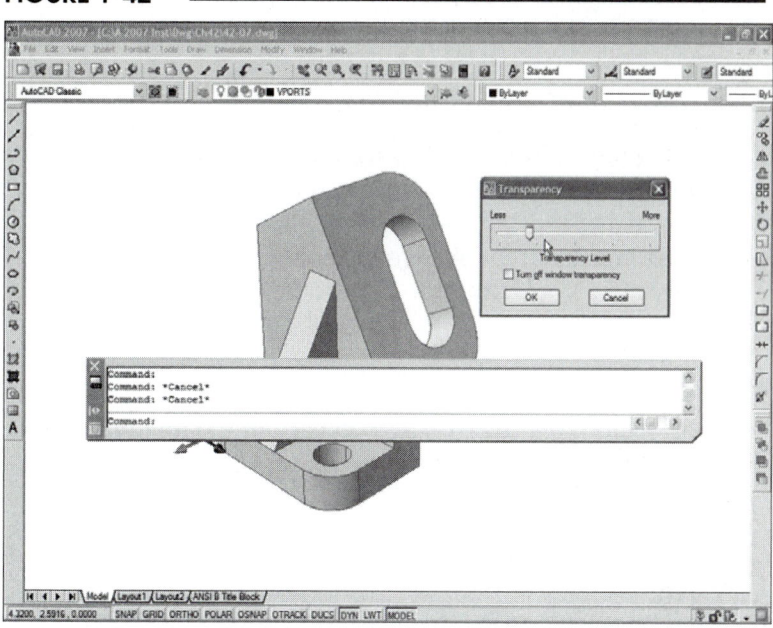

Options

Fonts, colors, and other features of the AutoCAD drawing editor can be customized to your liking by using the *Options* command and selecting the *Display* tab. Typing *Options* or selecting *Options…* from the *Tools* pull-down menu or default shortcut menu activates the dialog box shown in Figure 1-43. The screen menu can also be activated through this dialog box (second checkbox in the dialog box). Selecting the *Color…* tile provides a dialog box for customizing the screen colors.

FIGURE 1-43

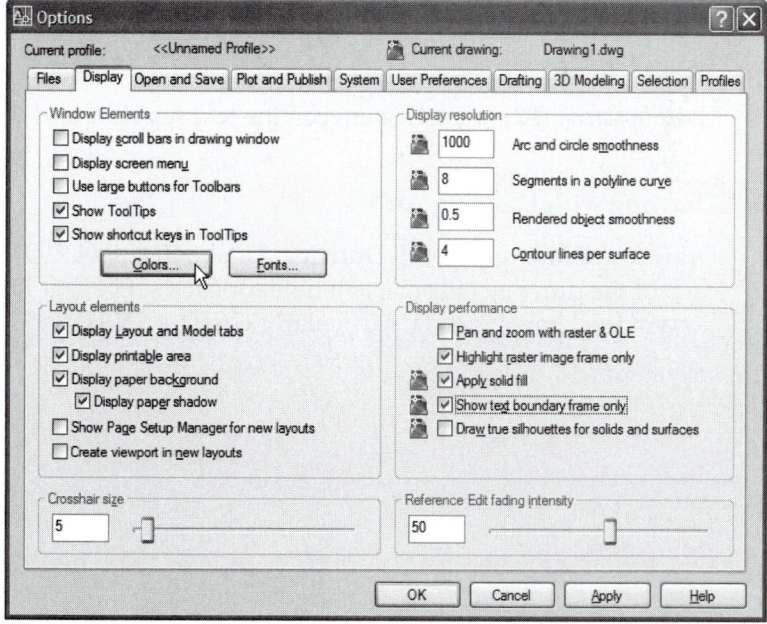

All changes made to the Windows screen by any of the options discussed in this section are automatically saved for the next drawing session. The changes are saved as the current profile. However, if you are working in a school laboratory, it is likely that the computer systems are set up to present the same screen defaults each time you start AutoCAD.

CHAPTER EXERCISES

1. **Starting and Exiting AutoCAD**

   Start AutoCAD by double-clicking the "AutoCAD 2007" shortcut icon or selecting "AutoCAD 2006" from the Programs menu. If the *Startup* dialog box appears, select *Start from Scratch*, choose *Imperial* as the default setting, and click the *OK* button. Draw a *Line*. Exit AutoCAD by selecting the *Exit* option from the *Files* pull-down menu. Answer *No* to the "Save changes to Drawing1.dwg?" prompt. Repeat these steps until you are confident with the procedure.

2. **Using Drawing Aids**

   Start AutoCAD. Turn on and off each of the following modes:

   SNAP, GRID, ORTHO, POLAR, DYN

3. **Understanding Coordinates**

   Begin drawing a **Line** by PICKing a "Specify first point:". Toggle *DYN* on and then toggle *Coords* to display each of the <u>three</u> formats. PICK several other points at the "Specify next point or [Undo]:" prompt. Pay particular attention to the coordinate values displayed for each point and visualize the relationship between that point and coordinate 0,0 (absolute Cartesian value) or the last point established and the distance and angle (relative polar value). Finish the command by pressing Enter.

4. **Using *Flipscreen***

   Use *Flipscreen* (**F2**) to toggle between the text window and the graphics screen.

5. **Drawing with Drawing Aids**

   Draw four **Lines** using <u>each</u> Drawing Aid: **GRID, SNAP, ORTHO, POLAR.** Toggle on and off each of the drawing aids one at a time for each set of four *Lines*. Next, draw **Lines** using combinations of the Drawing Aids, particularly **GRID + SNAP** and **GRID + SNAP + POLAR.**

2

WORKING WITH FILES

CHAPTER OBJECTIVES

After completing this chapter you should:

1. be able to name drawing files;

2. be able use file-related dialog boxes;

3. be able to use the Windows right-click shortcut menus in file dialog boxes;

4. be able to create *New* drawings;

5. be able to *Open* and *Close* existing drawings;

6. be able to *Save* drawings;

7. be able to use *SaveAs* to save a drawing under a different name, path, and/or format;

8. be able to use *Partialopen* and *Partialload* to open part of a drawing;

9. be able to practice good file management techniques.

AutoCAD DRAWING FILES

Naming Drawing Files

What is a drawing file? A CAD drawing file is the electronically stored data form of a drawing. The computer's hard disk is the principal magnetic storage device used for saving and restoring CAD drawing files. Diskettes, compact disks (CDs), networks, and the Internet are used to transport files from one computer to another, as in the case of transferring CAD files among clients, consultants, or vendors in industry. The AutoCAD commands used for saving drawings to and restoring drawings from files are explained in this chapter.

An AutoCAD drawing file has a name that you assign and a file extension of ".DWG." An example of an AutoCAD drawing file is:

PART-024.DWG

file name   extension

The file name you assign must be compliant with the Windows file name conventions; that is, it can have a <u>maximum</u> of 256 alphanumeric characters. File names and directory names (folders) can be in UPPER-CASE, Title Case, or lowercase letters. Characters such as _ - $ # ( ) ^ and spaces can be used in names, but other characters such as \ / : * ? < > | are not allowed. AutoCAD automatically appends the extension of .DWG to all AutoCAD-created drawing files.

NOTE: The chapter exercises and other examples in this book generally list file names in UPPERCASE letters for easy recognition. The file names and directory (folder) names on your system may appear as UPPERCASE, Title Case, or lowercase.

Beginning and Saving an AutoCAD Drawing

When you start AutoCAD, the drawing editor appears and allows you to begin drawing even before using any file commands.   As you draw, you should develop the habit of saving the drawing periodically (about every 15 or 20 minutes) using *Save*. *Save* stores the drawing in its most current state to disk.

The typical drawing session would involve using *New* to begin a new drawing or using *Open* to open an existing drawing. Alternately, the *Startup* dialog box that may appear when starting AutoCAD would be used to begin a new drawing or open an existing one. *Save* would be used periodically, and *Exit* would be used for the final save and to end the session.

FIGURE 2-1

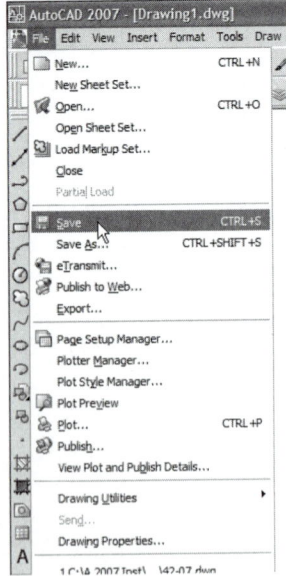

Accessing File Commands

Proper use of the file-related commands covered in this chapter allows you to manage your AutoCAD drawing files in a safe and efficient manner. Although the file-related commands can be invoked by several methods, they are easily accessible via the first pull-down menu option, *File* (Fig. 2-1). Most of the selections from this pull-down menu invoke dialog boxes for selection or specification of file names.

The Standard toolbar at the top of the AutoCAD screen has tools (icon buttons) for *New, Open,* and *Save.* File commands can also be entered at the keyboard by typing the formal command name, the command alias, or using the Ctrl keys sequences. File commands and related dialog boxes are also available from the *File* screen menu, or by selection from the digitizing tablet menu.

File Navigation Dialog Box Functions

There are many dialog boxes appearing in AutoCAD that help you manage files. All of these dialog boxes operate in a similar manner. A few guidelines will help you use them. The *Save Drawing As* dialog box (Fig. 2-2) is used as an example.

- The top of the box gives the title describing the action to be performed. It is <u>very important</u> to glance at the title before acting, especially when saving or deleting files.
- The desired file can be selected by PICKing it, then PICKing *OK*. Double-clicking on the selection accomplishes the same action. File names can also be typed in the *File name:* edit box.
- Every file name has an extension (three letters following the period) called the <u>type.</u> File types can be selected from the *Files of Type:* section of the dialog boxes, or the desired file extension can be entered in the *File name:* edit box.
- The current folder (directory) is listed in the drop-down box near the top of the dialog box. You can select another folder (directory) or drive by using the drop-down list displaying the current path, by selecting the *Back* arrow, or by selecting the *Up One Level* icon to the right of the list. (Rest your pointer on an icon momentarily to make the tool tip appear.)

FIGURE 2-2

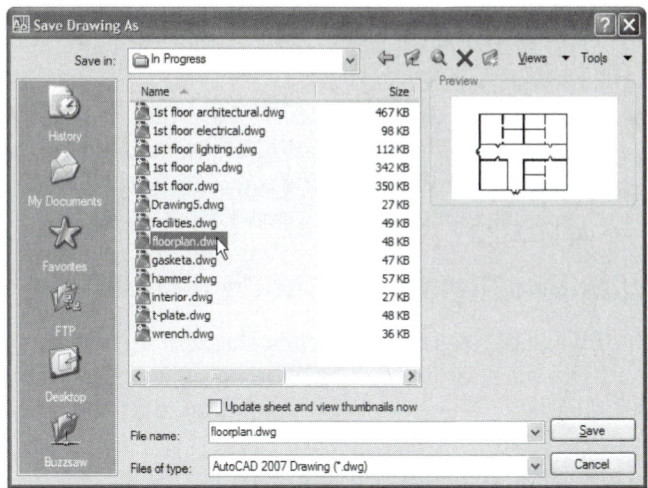

FIGURE 2-3

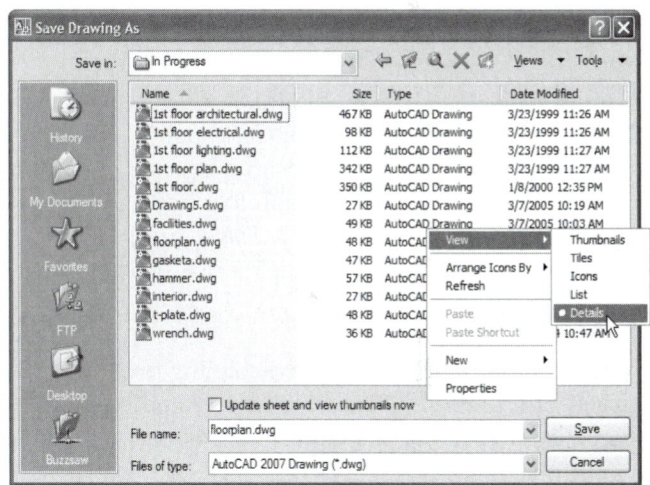

- Selecting one of the folders displayed on the left side of the file dialog boxes allows you to navigate to the following locations. The files or locations found appear in the list in the central area.
  History: Lists a history of files you have opened from most to least recent.
  My Documents: Lists all files and folders saved in the My Documents folder.
  Favorites: Lists all files and folders saved in the Favorites folder.
  Desktop: Shows files, folders, and shortcuts on your desktop.
  FTP: Allows you to browse through the FTP sites you have saved.
  Buzzsaw: Connects you to Buzzsaw—an Autodesk Web site for collaboration of files and information.
- You can use the *Search the Web* icon to display the *Browse the Web* dialog box (not shown). The default site is http://www.autodesk.com/.
- Any highlighted file(s) can be deleted using the "X" (*Delete*) button.
- A new folder (subdirectory) can be created within the current folder (directory) by selecting the *New Folder* icon.
- The *View* drop-down list allows you to toggle the listing of files to a *List* or to show *Details*. The *List* option displays only file names (see Fig. 2-2), whereas the *Details* option gives file-related information such as file size, file type, time, and date last modified (see Fig. 2-3). The detailed list can also be sorted alphabetically by name, by file size, alphabetically by file type, or chronologically by time and date. Do this by clicking the *Name, Type, Size,* or *Date Modified* tiles immediately above the list. Double-clicking on one of the tiles reverses the order of the list.

You can resize the width of a column (*Name, Size, Type*, etc.) by moving the pointer to the "crack" between two columns, then sliding the double arrows that appear in either direction. You can also use the *Preview* option to show a preview of the highlighted file (see Fig. 2-2) or to disable the preview (see Fig. 2-3).

- Use the *Tools* drop-down list to add files or folders to the *Favorites* or *FTP* folder. *Places* adds files or folders to the folder listing on the left side of the dialog box. *Options* allows you to specify formats for saving DXF files, and *Security Options* allows you to apply a password or Digital Signature to the file (see "*Saveas*").

Windows Right-Click Shortcut Menus

AutoCAD utilizes the right-click shortcut menus that operate with the Windows operating systems. Activate the shortcut menus by pressing the right mouse button (right-clicking) <u>inside the file list area</u> of a dialog box. There are two menus that give you additional file management capabilities.

Right-Click, No Files Highlighted

When no files are highlighted, right-clicking produces a menu for file list display and file management options (Fig. 2-3). The menu choices are as follows:

View	Shows *Thumbnails* (see Fig. 2-4), *Tiles* (large icons), *Icons* (small icons), *List* (see Fig. 2-3), or *Details* (see Fig. 2-3).
Arrange Icons By	Sorts by file *Name*, file *Size*, extension *Type*, or date *Modified*.
Paste	If the *Copy* or *Cut* function was previously used (see "Right-Click, File Highlighted," Fig. 2-4), *Paste* can be used to place the copied file into the displayed folder.
Paste Shortcut	Use this option to place a *Shortcut* (to open a file) into the current folder.
New	Creates a new *Folder, Shortcut,* or document.
Properties	Displays a dialog box listing properties of the current folder.

Right-Click, File Highlighted

When a file is highlighted, right-clicking produces a menu with options for the selected file (Fig. 2-4). The menu choices are as follows:

FIGURE 2-4

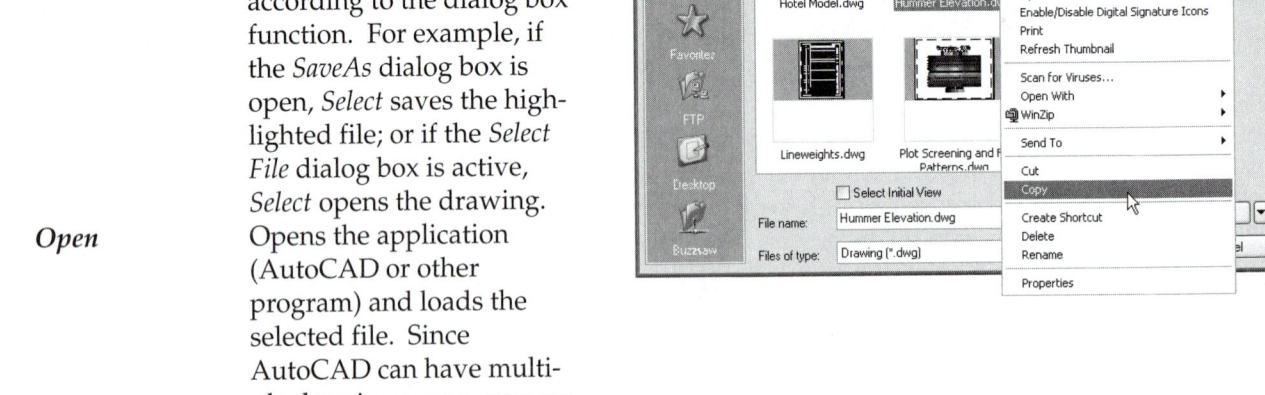

Select	Processes the file (like selecting the *OK* button) according to the dialog box function. For example, if the *SaveAs* dialog box is open, *Select* saves the highlighted file; or if the *Select File* dialog box is active, *Select* opens the drawing.
Open	Opens the application (AutoCAD or other program) and loads the selected file. Since AutoCAD can have multiple drawings open, you can select several drawings to open with this feature.
Print	Sends the selected file to the configured system printer.
Send To	Copies the selected file to the selected device. <u>This is an easy way to copy a file from your hard drive to a diskette in A: drive.</u>

Cut	In conjunction with *Paste,* allows you to move a file from one location to another.
Copy	In conjunction with *Paste,* allows you to copy the selected file to another location. You can copy the file to the same folder, but Windows renames the file to "Copy of . . .".
Create Shortcut	Use this option to create a *Shortcut* (to open the highlighted file) in the current folder. The shortcut can be moved to another location using drag and drop.
Delete	Sends the selected file to the Recycle Bin.
Rename	Allows you to rename the selected file. Move your cursor to the highlighted file name, click near the letters you want to change, then type or use the backspace, delete, space, or arrow keys.
Properties	Displays a dialog box listing properties of the selected file.

Specific features of other dialog boxes are explained in the following sections.

AutoCAD FILE COMMANDS

When you start AutoCAD, one of three situations occur based on how the *Options* are set on your system:

1. The *Startup* dialog box appears.
2. No dialog box appears and the session starts with the ACAD.DWT or the ACADISO.DWT template (determined by the *MEASUREINIT* system variable setting, 0 or 1, respectively).
3. No dialog box appears and the session starts with a template specified by the *Qnew* setting in the *Options* dialog box (see "QNEW").

To control the appearance of the *Startup* dialog box, use the *Options* command to produce the *Options* dialog box, select the *System* tab, and make the desired selection from the *Startup* drop-down list (Fig. 2-5).

FIGURE 2-5

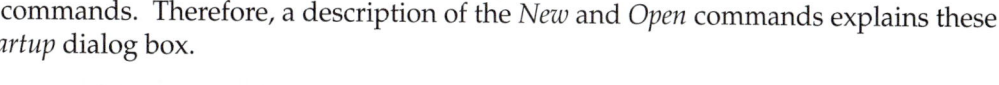

The *Startup* dialog box allows you to create new drawings and open existing drawings. These options are the same as using the *New* and *Open* commands. Therefore, a description of the *New* and *Open* commands explains these functions in the *Startup* dialog box.

New

Pull-down Menu	Command (Type)	Alias (Type)	Short-cut	Screen (side) Menu	Tablet Menu
File *New...*	*New*	*...*	*Ctrl+N*	*FILE* *New*	*T,24*

The *New* command begins a new drawing. The new drawing can be a completely "blank" drawing or it can be based on a template that may already have a title block and some additional desired settings. Based on the settings for your system, the *New* command produces one of two methods to start a new drawing:

1. The *Create New Drawing* dialog box appears (identical to the *Startup* dialog box).
2. The *Select Template* dialog box appears.

Control these two actions by the setting in the *System* tab of the *Options* dialog box (see Fig. 2-5). If you select *Show Startup dialog box*, the *Create New Drawing* dialog box appears when you use *New*. If you select *Do not show a startup dialog*, the *Select Template* dialog box appears. These two dialog boxes are explained next.

The *Create New Drawing* Dialog Box

In the *Create New Drawing* dialog box (Fig. 2-6), the
options for creating a new drawing are the same as in the
Startup dialog box (shown previously in Fig. 1-12): *Start
from Scratch*, *Use a Template*, and *Use a Wizard*.

Start from Scratch

In the *Create New Drawing* dialog box select the *Start from
Scratch* option (Fig. 2-6). Next, choose from either the
Imperial (feet and inches) or *Metric* default settings.
Selecting *Imperial* uses the default ACAD.DWT template
drawing with a drawing area (called *Limits*) of 12 x 9
units. Choosing *Metric* uses the ACADISO.DWT template drawing which has *Limits* settings of 420 x 297. For
more details on these templates, see "Table of AutoCAD-
Supplied Template Drawing Settings" in Chapter 12.

FIGURE 2-6

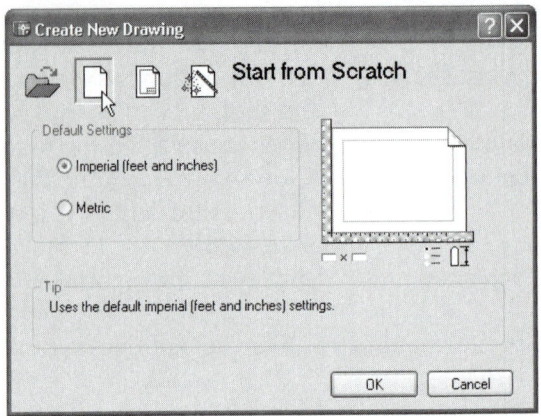

Use a Wizard

In the *Create New Drawing* dialog box select the *Use a
Wizard* option (Fig. 2-7). Next, choose between a *Quick
Setup Wizard* and an *Advanced Setup Wizard*. The *Quick
Setup Wizard* prompts you to select the type of drawing
Units you want to use and to specify the drawing *Area*
(*Limits*). The *Advanced Setup* offers options for units,
angular direction and measurement, and area. The
Quick Setup Wizard and the *Advanced Setup Wizard* are
discussed in Chapter 6, Basic Drawing Setup.

FIGURE 2-7

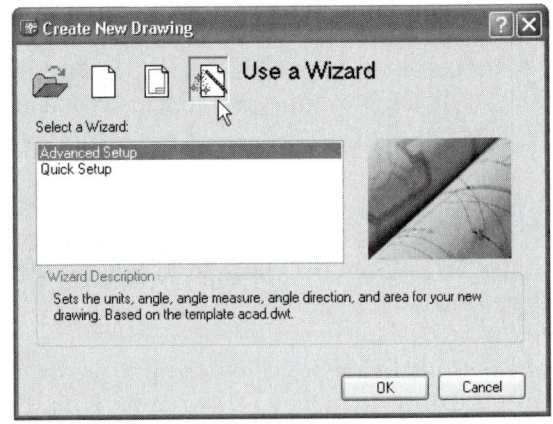

Use a Template

Use this option if you want to create a new drawing
based on an existing template drawing (Fig. 2-8). A template drawing is one that may have some of the setup
steps performed but contains no geometry (graphical
objects).

Selecting the ACAD.DWT template begins a new
drawing using the Imperial default settings with a
drawing area of 12 x 9 units (identical to using the *Start
from Scratch, Imperial* option). Selecting the
ACADISO.DWT template begins a new drawing using
the metric settings having a drawing area of 420 x 279
units (identical to using the *Start from Scratch, Metric*
option). Details of these and other template drawings
are discussed in Chapter 12, Advanced Drawing Setup.

FIGURE 2-8

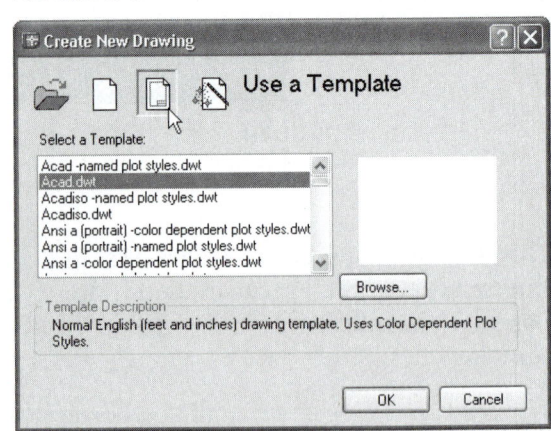

The *Select Template* Dialog Box

The *Select Template* dialog box (Fig. 2-9) appears when you use the *New* command and your system is set to *Do not show a startup dialog box* (see previous Fig. 2-5). The *Select Template* dialog box has the identical outcome as using the *Use a Template* option in the *Create New Drawing* dialog box (see Fig. 2-8); that is, you can select from all AutoCAD-supplied templates (see previous discussion). See Chapter 12, Advanced Drawing Setup, for complete details of all template drawings.

NOTE: For the purposes of learning AutoCAD starting with the basic principles and commands discussed in this text, it is helpful to use the <u>Imperial (inch) settings</u> when begin-

FIGURE 2-9

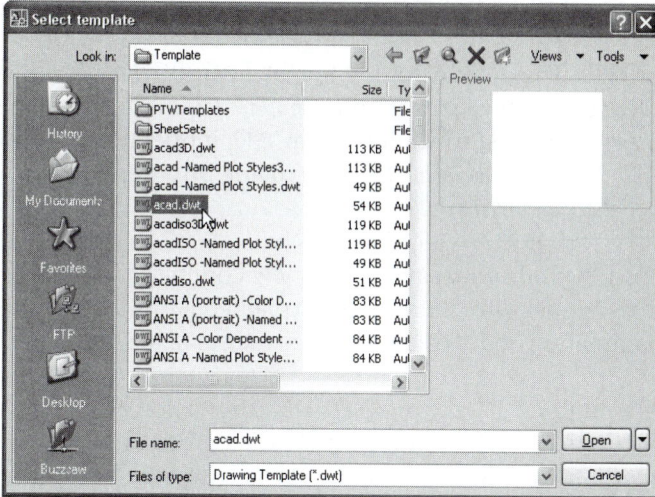

ning a new drawing. Until you read Chapters 6 and 12, you can begin new drawings for completing the exercises and practicing the examples in Chapters 1–5 by any of the following methods.

Select *Imperial (feet and inches)* in the *Start from Scratch* option.
Select the ACAD.DWT template drawing using any method then perform a *Zoom All*.
Cancel the *Startup* dialog when AutoCAD starts (the default setting of 0 for the *MEASURINIT* system variable uses the ACAD.DWT).

Qnew

Pull-down Menu	Command (Type)	Alias (Type)	Short-cut	Screen (side) Menu	Tablet Menu
...	*Qnew*	...	...	...	...

Qnew (Quick New) is intended to immediately begin a new drawing using the template file specified in the *Files* tab of the *Options* dialog box. <u>If the proper settings are made</u>, *Qnew* is faster to use than the *New* command since it does not force the *Create New Drawing* or the *Select Template* dialog box to open, as the *New* command does.

To set *Qnew* to operate quickly when opening a drawing, you must first make two settings (if you are using AutoCAD in a school or office setting, these settings may have already been made):

1. Specify a default template to use when *Qnew* is invoked. Open the *Options* dialog box by selecting *Options* from the *Tools* pull-down menu. Select the

FIGURE 2-10

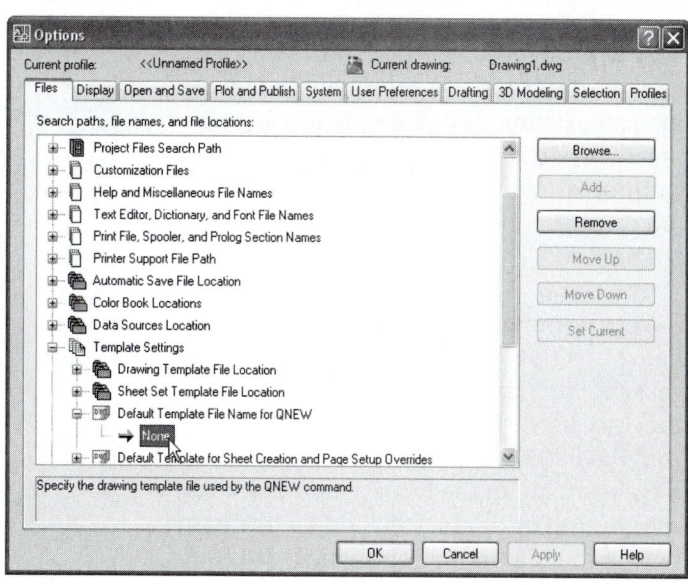

Files tab and expand the *Default Template File Name for QNEW* line (Fig. 2-10). The default setting is *None*. Highlight *None*, then pick the *Browse* button to locate the template file you want to use.

2.   Set your system to *Do not show a startup dialog*. This is accomplished in the *System* tab of the *Options* dialog box (see previous Fig. 2-5).

NOTE: The default setting for the *Default Template File Name for QNEW* line (in the *Options* dialog box) is "None." That means if you use *Qnew* without specifying a template drawing, as will be the case for all users upon first installing AutoCAD, it operates similarly to the *New* command by opening the *Create New Drawing* or *Select Template* dialog box, depending on your setting for *Show Startup dialog box* or *Do not show a startup dialog*.

Since Autodesk has made this new command relatively complex to understand and control, the three possibilities and controls are explained simply here. Based on the settings for your system, the *Qnew* command produces one of three methods to start a new drawing:

Method 1.   The *Create New Drawing* dialog box appears.
Method 2.   The *Select Template* dialog box appears.
Method 3.   A new drawing starts with the *Qnew* template specified in the *Options* dialog box.

Control the previous actions by the following settings:

Method 1.   Select *Show Startup dialog box* in the *System* tab of the *Options* dialog box.
Method 2.   Select *Do not show a startup dialog* in the *System* tab of the *Options* dialog box and ensure the *Drawing Template File Name* is set to "None" in the *Files* tab of the *Options* dialog box.
Method 3.   Select *Do not show a startup dialog* in the *System* tab of the *Options* dialog box and set the *Drawing Template File Name* in the *Files* tab of the *Options* dialog box to the desired template file.

Open

Pull-down Menu	Command (Type)	Alias (Type)	Short-cut	Screen (side) Menu	Tablet Menu
File Open...	Open	...	Ctrl+O	FILE Open	T,25

Use *Open* to select an existing drawing to be loaded into AutoCAD. Normally you would open an existing drawing (one that is completed or partially completed) so you can continue drawing or to make a print or plot. You can *Open* multiple drawings at one time in AutoCAD.

The *Open* command produces the *Select File* dialog box (Fig. 2-11). In this dialog box you can select any drawing from the current directory list. PICKing a drawing name from the list displays a small bitmap image of the drawing in the *Preview* tile. Select the *Open* button or double-click on the file name to open the highlighted drawing. You could instead type the file name (and path) of the desired drawing in the *File name:* edit box and press Enter, but a preview of the typed entry will not appear.

FIGURE 2-11

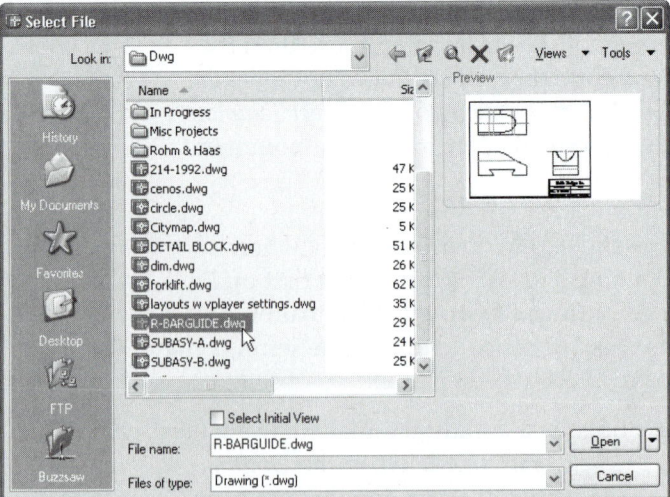

You can open multiple drawings at one time by holding down the Shift key to select a range (all files between and including the two selected) or holding down the Ctrl key to select multiple drawings not in a range.

Using the *Open* drop-down list (lower right) you can select from these options:

Open	Opens a drawing file and allows you to edit the file.
Open Read-Only	Opens a drawing file for viewing, but you cannot edit the file.
Partial Open	Allows you to open only a part of a drawing for editing. See "*Partialopen.*"
Partial Open Read-Only	Allows you to open only a part of a drawing for viewing, but you cannot edit the drawing. See "*Partialopen.*"

Other options in this dialog box are similar to the *Save Drawing As* dialog box described earlier in "File Navigation Dialog Box Functions." For example, to locate a drawing in another folder (directory) or on another drive on your computer or network, select the drop-down list on top of the dialog box next to *Look in:*, or use the *Up one level* button. Remember that you can also display the file details (size, type, modified) by toggling the *Details* option from the *Views* drop-down list. You can open a .DWG (drawing), .DWS (drawing standards), .DXF (drawing interchange format), or .DWT (drawing template) file by selecting from the *Files of type:* drop-down list.

Selecting the *Find...* option from the *Tools* drop-down list in the *Select File* dialog box (upper right) invokes the *Find* dialog box (Fig. 2-12). The dialog box offers two criteria to locate files: *Name & Location* and *Date Modified.*

Using the *Name & Location* tab, you can search for .DWG, .DWS, .DWT, or .DXF files in any directory, including subdirectories if desired. In the *Named:* edit box, enter a drawing name to find. Wildcards can be used (see Chapter 43, Miscellaneous Commands and Features, for valid wildcards in AutoCAD).

The *Date Modified* tab of the *Find* dialog box enables you to search for files meeting specific date criteria that you specify (Fig. 2-13). This feature helps you search for files *between* two dates, *during the previous months,* or *during the previous days,* based on when the files were last saved.

After specifying the search criteria in either tab, select the *Find Now* button. All file names that are found matching the criteria are displayed in the list at the bottom of the dialog box. Double-clicking on the name or using the *OK* button passes the name to the *File Name* section of the *Select File* dialog box.

NOTE: It is considered <u>poor practice to *Open* a drawing from a diskette in A: drive</u>. Normally, drawings should be copied to a directory on a fixed (hard) drive, then opened from that directory. Opening a drawing from a diskette and performing saves to a diskette are much slower and less reliable than operating from a hard drive. In addition, some temporary files may be written to the drive or directory from which the file is opened. See other "NOTE"s at the end of the discussion on *Save* and *Saveas.*

FIGURE 2-12

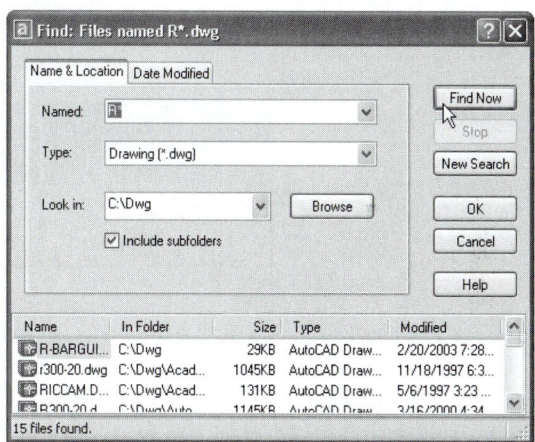

FIGURE 2-13

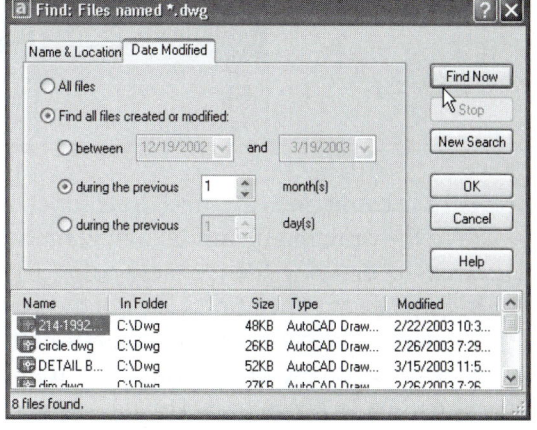

Save

Pull-down Menu	Command (Type)	Alias (Type)	Short-cut	Screen (side) Menu	Tablet Menu
File *Save*	*Save*	...	*Ctrl+S*	...	*U,24-U,25*

The *Save* command is intended to be used periodically during a drawing session (every 15 to 20 minutes is recommended). When *Save* is selected from a menu, the current version of the drawing is saved to disk without interruption of the drawing session. The first time an unnamed drawing is saved, the *Save Drawing As* dialog box (see Fig. 2-2 and Fig. 2-3) appears, which prompts you for a drawing name. Typically, however, the drawing already has an assigned name, in which case *Save* actually performs a *Qsave* (quick save). A *Qsave* gives no prompts or options, nor does it display a dialog box.

When the file has an assigned name, using *Save* by selecting the command from a menu or icon button automatically performs a quick save (*Qsave*). Therefore, *Qsave* automatically saves the drawing in the same drive and directory from which it was opened or where it was first saved. In contrast, typing *Save* always produces the *Save Drawing As* dialog box, where you can enter a new name and/or path to save the drawing or press Enter to keep the same name and path (see *SaveAs*).

NOTE: If you want to save a drawing directly to a diskette in A:, first *Save* the drawing to the hard drive, then use the *Send To* option in the right-click menu in the *Save*, *Save Drawing As*, or *Select File* dialog box (see "Windows Right-Click Shortcut Menus" and Fig. 2-4).

Qsave

Pull-down Menu	Command (Type)	Alias (Type)	Short-cut	Screen (side) Menu	Tablet Menu
...	*Qsave*	...	*Ctrl+S*	*FILE* *Qsave*	...

Qsave (quick save) is normally invoked automatically when *Save* is used (see *Save*) but can also be typed. *Qsave* saves the drawing under the previously assigned file name. No dialog boxes appear nor are any other inputs required. This is the same as using *Save* (from a menu), assuming the drawing name has been assigned. However, if the drawing has not been named when *Qsave* is invoked, the *Save Drawing As* dialog box appears.

Saveas

Pull-down Menu	Command (Type)	Alias (Type)	Short-cut	Screen (side) Menu	Tablet Menu
File *Save As...*	*Saveas*	...	...	*FILE* *Saveas*	*V,24*

The *SaveAs* command can fulfill four functions:
1. Save the drawing file under a new name if desired.
2. Save the drawing file to a new path (drive and directory location) if desired.
3. In the case of either 1 or 2, assign the new file name and/or path to the current drawing (change the name and path of the current drawing).
4. Save the drawing in a format other than the default AutoCAD 2007 format or as a different file type (.DWT, .DWS, or .DXF).

Therefore, assuming a name has previously been assigned, *SaveAs* allows you to save the current drawing under a different name and/or path; but, beware, *SaveAs* sets the current drawing name and/or path to the last one entered. This dialog box is shown in Figures 2-2 and 2-3.

SaveAs can be a benefit when creating two similar drawings. A typical scenario follows. A design engineer wants to make two similar but slightly different design drawings. During construction of the first drawing, the engineer periodically saves under the name DESIGN1 using *Save*. The first drawing is then completed and *Saved*. Instead of starting a *New* drawing, *SaveAs* is used to save the current drawing under the name DESIGN2. *SaveAs* also resets the current drawing name to DESIGN2. The designer then has two separate but identical drawing files on disk which can be further edited to complete the specialized differences. The engineer continues to work on the current drawing DESIGN2.

NOTE: If you want to save the drawing to a diskette in A: drive, do not use *SaveAs*. Invoking *SaveAs* by any method resets the drawing name and path to whatever is entered in the *Save Drawing As* dialog box, so entering A:NAME would set A: as the current drive. This could cause problems because of the speed and reliability of a diskette as opposed to a hard drive. Instead, you should save the drawing to the hard drive (usually C:), then close the drawing (by using *Close*). Next, use a right-click shortcut menu to copy the drawing file to A:. (See "Windows Right-Click Shortcut Menus" and Fig. 2-4.)

You can save an AutoCAD 2007 drawing in several formats other than the default format (*AutoCAD 2007 Drawing *.dwg*). Use the drop-down list at the bottom of the *Save* (or *Save Drawing As*) dialog box (Fig. 2-14) to save the current drawing as an earlier version drawing (.DWG) of AutoCAD or LT, a template file (.DWT), or a .DWS or a .DXF file. In AutoCAD 2007, a drawing can be saved to an earlier release and later *Opened* in AutoCAD 2007 and all new features are retained during the "round trip."

FIGURE 2-14

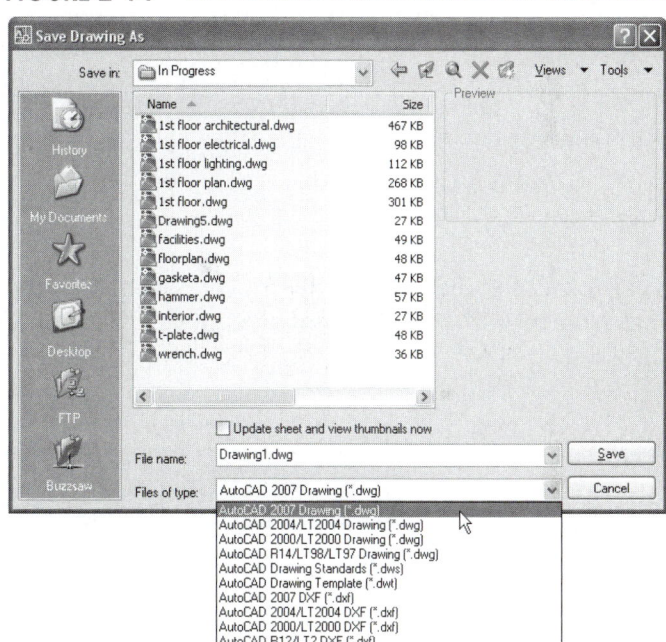

Passwords
An added security feature allows you to save your drawings with password protection. If a password is assigned, the drawing can be opened again in AutoCAD <u>only</u> if you enter the correct password when using the *Open* command. Beware: <u>If you lose or forget the password, the drawing cannot be opened or recovered in any way!</u>

To assign a password during the *Save* or *SaveAs* command, select the *Tools* drop-down menu in the *Save Drawing As* dialog box, then select *Security Options…* (Fig. 2-15) to produce the *Security Options* dialog box. Passwords are not case sensitive.

FIGURE 2-15

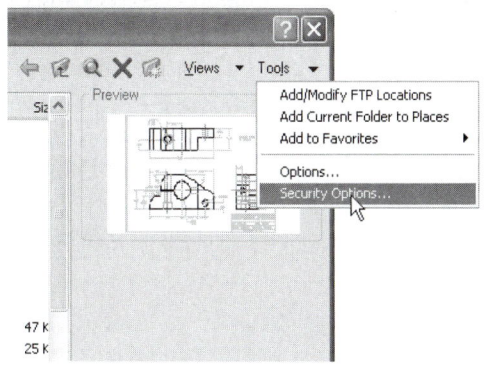

Digital Signatures
You can also attach digital signatures to drawings using the *Security Options* dialog box. If a drawing has a digital signature assigned, it is designated as an original signed and dated drawing. Users are assured that the drawing is original and has not been changed in any way. If such a drawing is later modified, the digital signature is then invalidated, and anyone opening the drawing is notified that the drawing has been changed and the digital signature is invalid.

Close

Pull-down Menu	Command (Type)	Alias (Type)	Short-cut	Screen (side) Menu	Tablet Menu
File Close	Close	...	...	...	...

Use the *Close* command to <u>close the current drawing</u>. Because AutoCAD allows you to have several drawings *Open*, *Close* gives you control to close one drawing while leaving others open. If the drawing has been changed but not saved, AutoCAD prompts you to save or discard the changes. In this case, a warning box appears (Fig. 2-16). *Yes* causes AutoCAD to *Save* then close the drawing; *No* closes the drawing without saving; and *Cancel* aborts the close operation so the drawing stays open.

FIGURE 2-16

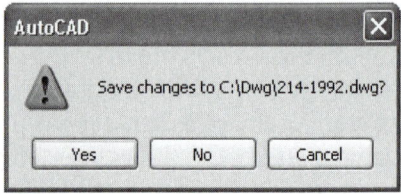

Closeall

Pull-down Menu	Command (Type)	Alias (Type)	Short-cut	Screen (side) Menu	Tablet Menu
Window Closeall	Closeall	...	...	...	...

The *Closeall* command closes all drawings currently open in your AutoCAD session. If any of the drawings have been changed but not saved, you are prompted to save or discard the changes for each drawing.

Exit

Pull-down Menu	Command (Type	Alias (Type)	Short-cut	Screen (side) Menu	Tablet Menu
File Exit	Exit	...	Ctrl+Q	...	Y,25

This is the simplest method to use when you want to exit AutoCAD. If any changes have been made to the drawings since the last *Save*, *Exit* invokes a warning box asking if you want to *Save changes to . . .?* before ending AutoCAD (see Fig. 2-16).

An alternative to using the *Exit* option is to use the standard Windows methods for exiting an application. The two options are (1) select the **"X"** in the extreme upper-right corner of the AutoCAD window, or (2) select the AutoCAD logo in the extreme upper-left corner of the AutoCAD window. Selecting the logo in the upper-left corner produces a pull-down menu allowing you to *Minimize*, *Maximize*, etc., or to *Close* the window. Using this option is the same as using *Exit*.

Quit

Pull-down Menu	Command (Type)	Alias (Type)	Short-cut	Screen (side) Menu	Tablet Menu
...	Quit	EXIT	...	FILE Quit	...

The *Quit* command accomplishes the same action as *Exit*. *Quit* discontinues (exits) the AutoCAD session and usually produces the dialog box shown in Figure 2-16 if changes have been made since the last *Save*.

Partialopen

Pull-down Menu	Command (Type)	Alias (Type)	Short-cut	Screen (side) Menu	Tablet Menu
File Open Partial Open	Partialopen	...	...	...	...

This feature in AutoCAD speeds up and simplifies loading, viewing, and working with large drawings. With versions previous to AutoCAD 2000, you could only *Open* the entire contents of a drawing (all layers and all geometry) in order to work with the drawing. (The earlier *Xref* feature allowed you to view partial drawings, but you could not edit the geometry.) Since AutoCAD 2000, you can select which views and layers, and therefore which geometry, you want to load. This feature frees up system memory, speeds up regenerations, and prevents having to view and manipulate unneeded layers and geometry.

There are two ways you can use *Partialopen*: entering the command at the Command line or using the *Open* command to access the *Partial Open* dialog box interface. Generally you should use the *Open* command to access the *Select File* dialog box, then the *Partial Open* dialog box. Use the *Open* drop-down list in the lower-right corner of the *Select File* dialog box to access *Partial Open* (Fig. 2-17). The Command line version of *Partialopen* is explained at the end of this section.

FIGURE 2-17

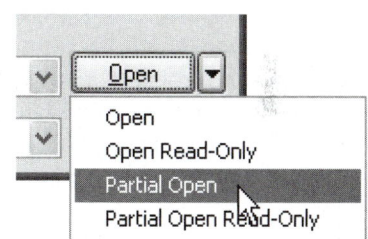

Choosing the *Partial Open* option from the *Select File* dialog box produces the *Partial Open* dialog box (Fig. 2-18). The dialog box allows you to select a *View* (if a named *View* has previously been saved in the drawing) and select which layers you want to open. (See Chapter 10 and Chapter 11 for information on views and layers.) You can select a range of layers by holding down Shift when selecting a layer at both ends of the range.

FIGURE 2-18

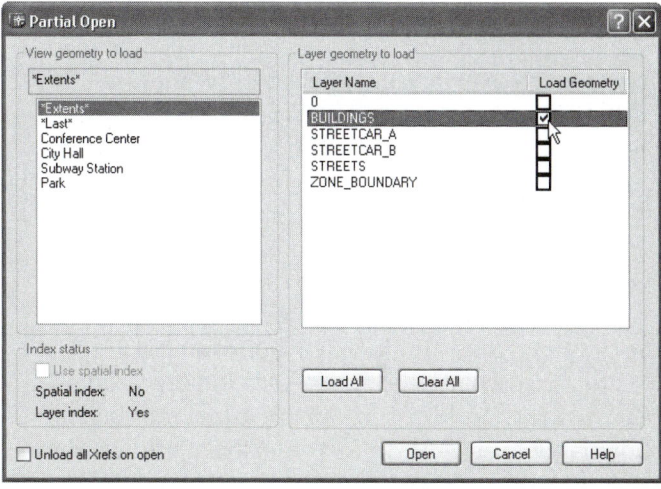

When a drawing is partially open, only the geometry on the selected layers is visible and can be edited. Although all of the layer names will appear in any list of layers in the drawing, <u>only the geometry on the selected layers is actually loaded</u>. In addition, all other named objects in the drawing (blocks, dimension styles, layouts, linetypes, text styles, UCSs, views, and viewport configurations) are loaded into AutoCAD.

For example, assume a drawing of the downtown area was partially opened. Selecting only the BUILDINGS layer (see Fig. 2-18) may yield a display as shown in Figure 2-19 revealing only the objects on the BUILDINGS layer.

FIGURE 2-19

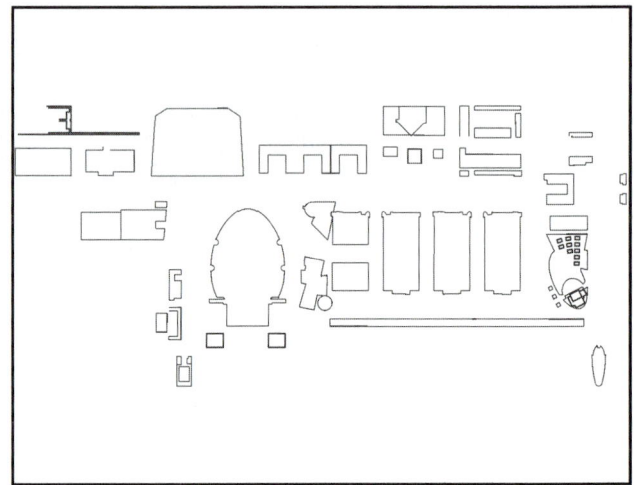

You can also use Spatial and Layer indexing (see Fig. 2-18, lower-left corner) if the drawing's *INDEXCTL* variable is set to allow "demand loading." This feature speeds loading and saves system memory for large drawings. Also, any attached *Xref* drawings can be unloaded if desired (see Chapter 30, Xreferences).

If you choose to enter the *Partialopen* command at the Command line, a text interface appears instead of the dialog box.

> Command: **partialopen**
> Enter name of drawing to open: (Enter drawing name or enter the tilde symbol [~] to display the *Select File* dialog box, then select the *Partial Open* button)
> Enter view to load or [?] <*Extents**>: (Enter view name or press Enter to load the extents)
> Enter layers to load or [?]<none>: (Enter desired layer names or an asterisk [*] to load all layers)
> Unload all Xrefs on open? [Yes/No] <N>: (Entering "Y" loads all attached Xrefs)
> Opening an AutoCAD 2007 format file.
> Regenerating model.
> AutoCAD menu utilities loaded.

Partialload

Pull-down Menu	Command (Type)	Alias (Type)	Short-cut	Screen (side) Menu	Tablet Menu
File *Partial Load*	*Partialload*	…	…	…	…

You must have a partially opened drawing for *Partialload* to operate (see *Partialopen* described previously). *Partialload* loads additional geometry into a partially opened drawing.

Accessing this command by a menu or by typing *Partialload* produces the *Partial Load* dialog. Here you select the additional layers or another area of the drawing you want to open. For example, Figure 2-20 displays the *Partial Load* dialog box displaying *Views* and *Layers* from the same drawing shown in Figure 2-19. In this case, the STREETS layer is loaded to reveal the additional geometry

FIGURE 2-20

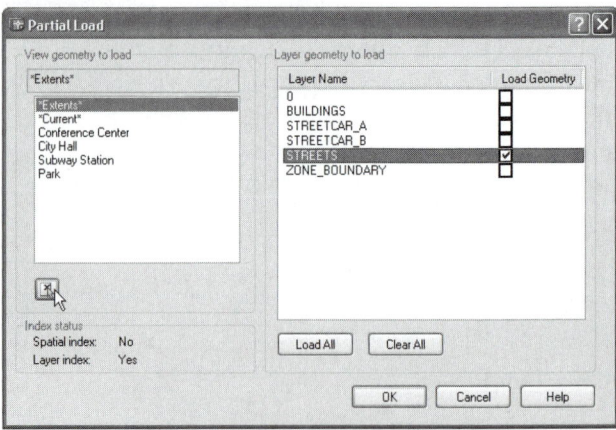

(see Fig. 2-21). Any information that is loaded into the file using *Partiaload* cannot be unloaded, not even with *Undo*. The *Partial Load* dialog box is essentially the same as the *Partial Open* dialog box (see Fig. 2-18) with the exclusion of the *Index* and *Xref* options in the lower-left corner.

Notice the *Pick a Window* button in the lower-left corner of the dialog box. This feature allows you to return to the drawing and define an area (window) of the drawing to view. All geometry within and crossing the windowed area for the defined layers appears in the drawing. For example, the *Pick a Window* feature was used to select the right side of the drawing shown in Figure 2-21. Notice how the STREETS layer is dis-

FIGURE 2-21

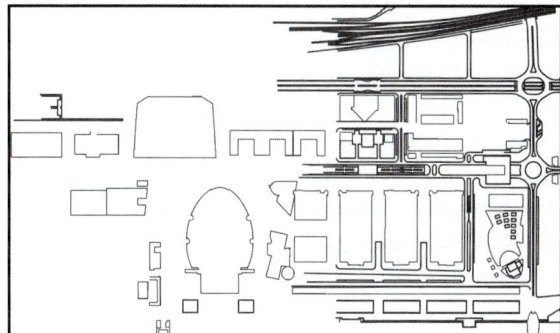

played only on the right side of the drawing. This feature allows you to view two or more areas of the same drawing, each displaying a different set of layers.

If you enter *-Partiaload* at the Command line (with the hyphen prefix), the text interface appears at the Command line.

Recover

Pull-down Menu	Command (Type)	Alias (Type)	Short-cut	Screen (side) Menu	Tablet Menu
File Drawing Utilities > Recover...	Recover	...	...	FILE Recover	...

This option is used only in the case that *Open* does not operate on a particular drawing because of damage to the file. Damage to a drawing file can occur from improper exiting of AutoCAD (such as power failure) or from damage to a diskette. *Recover* can usually reassemble the file to a usable state and reload the drawing. The *Recover* command produces the *Select File* dialog box. Select the drawing to recover. A recovery log is displayed in a text window reporting results of the recover operation.

In the event of a software crash or power outage, you can use the Drawing Recovery Manager to recover backup files. See "Drawing Recovery."

Dwgprops

Pull-down Menu	Command (Type)	Alias (Type)	Short-cut	Screen (side) Menu	Tablet Menu
File Drawing Properties	Dwgprops	...	...	...	...

Invoking *Dwgprops* by any method produces the *Drawing Properties* dialog box (Fig. 2-22, on the next page). This dialog box allows you to input details for better management and tracking of drawings such as title, author, subject, and other keywords. The information stored in the *Drawing Properties* dialog box should be specified by you while other information, such as time and date of last save, is read from the system. You must *Open* the drawing to input data into the user fields and *Save* the drawing to save the properties you enter.

The drawing properties can be read from Windows Explorer or My Computer by right-clicking a .DWG file and selecting *Properties*. Using an AutoCAD feature called DesignCenter, unopened drawings can be located by finding keywords stored in the drawing properties (see Chapter 21). The four tabs in the *Drawing Properties* dialog box are explained here.

General Tab

This tab (Fig. 2-22) displays the drawing type, location, size, and other information. All the information here is supplied by the operating system; therefore, all fields are read-only.

FIGURE 2-22

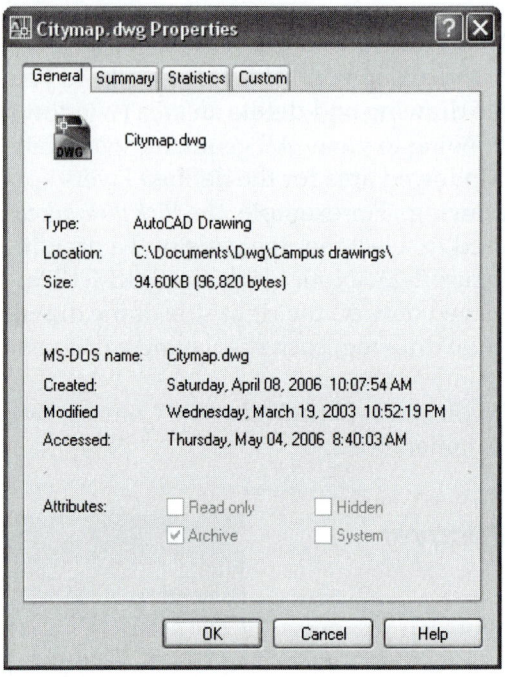

Summary Tab

This tab is used to input information you want to use to track the drawing (Fig. 2-23). Data you enter is viewable through AutoCAD's DesignCenter and Windows Explorer *Properties*. You can enter the drawing title, subject, author, keywords, comments, and a hyperlink base. The *Hyperlink Base* is a base address for all relative links you insert within the drawing using the *Hyperlink* command (see Chapter 19, Internet Tools). You can enter a URL (an Internet address) or a path to a folder on a network drive in the *Hyperlink Base* field.

FIGURE 2-23

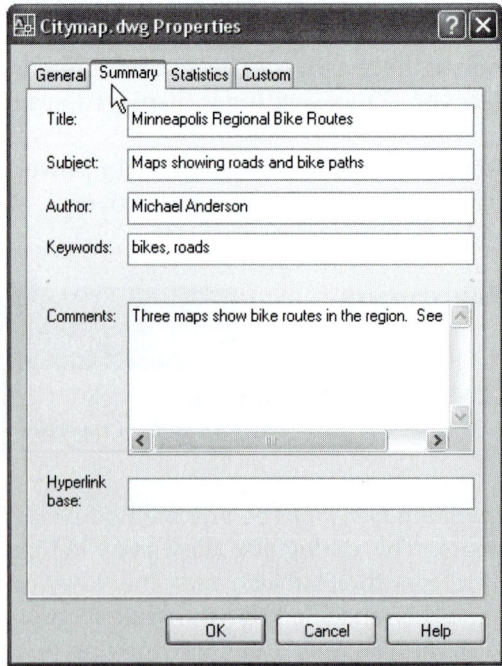

Statistics Tab

This tab shows read-only data such as file size and the dates that files were created and last modified (Fig. 2-24). You can search for all files created at a certain time or modified on a certain date. The *Created*, *Modified*, and *Total Editing Time* values are stored in the *TDCREATE*, *TDUP-DATE*, and *TDINDWG* read-only system variables.

Custom Tab

The custom fields can also be used in searches to help locate the drawing using the *Find* dialog box of AutoCAD DesignCenter (not shown). AutoCAD also provides access to these properties using its native programming language, AutoLISP. Enter any data relevant to the drawing that may be useful in a search.

FIGURE 2-24

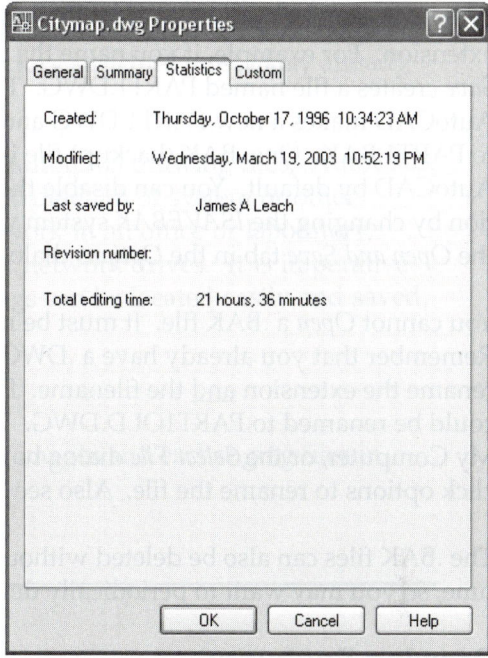

Savetime

Pull-down Menu	Command (Type)	Alias (Type)	Short-cut	Screen (side) Menu	Tablet Menu
Tools *Options...* *Open and Save* *File Safety* *Precautions*	*Savetime*	*...*	*...*	*TOOLS 2* *Options* *Open and Save* *File Safety* *Precautions*	*...*

SAVETIME is actually a system variable that controls AutoCAD's temporary save feature. AutoCAD automatically saves a temporary drawing file for you at time intervals that you specify. The default time interval is 10 (minutes). A value of 0 disables this feature. The value for *SAVETIME* can also be set in the *Open and Save* tab of the *Options* dialog box (see Fig. 2-25, on the next page).

When automatic saving occurs, the drawing is saved under the current name with a randomly generated suffix and a .SV\$ extension is assigned. For example, if the Electrical.DWG drawing were automatically saved, the assigned file might be Electrical_1_1_9912.SV\$. In this way the current drawing name is not overwritten. The path (location) of the automatically saved file can be specified using the *SAVE-FILEPATH* system variable or the *Files* tab of the *Options* dialog box.

NOTE: <u>The automatically saved file is only a temporary file</u>. That is, once you save or close the drawing, the temporary save file is automatically deleted. Therefore, this feature is useful when you experience an improper shutdown of AutoCAD (such as a power outage). In this case, the temporary file can be retrieved and renamed so all of your work is not lost. See "Drawing Recovery."

AutoCAD OBJECTS

The smallest component of a drawing in AutoCAD is called an <u>object</u> (sometimes referred to as an entity). An example of an object is a *Line,* an *Arc,* or a *Circle* (Fig. 3-1). A rectangle created with the *Line* command would contain four objects.

Draw commands <u>create</u> objects. The draw command names are the same as the object names.

Simple objects are *Point, Line, Arc,* and *Circle.*

Complex objects are shapes such as *Polygon, Rectangle, Ellipse, Polyline,* and *Spline* which are created with one command (Fig. 3-2). Even though they appear to have several segments, they are <u>treated</u> by AutoCAD as one object.

FIGURE 3-1

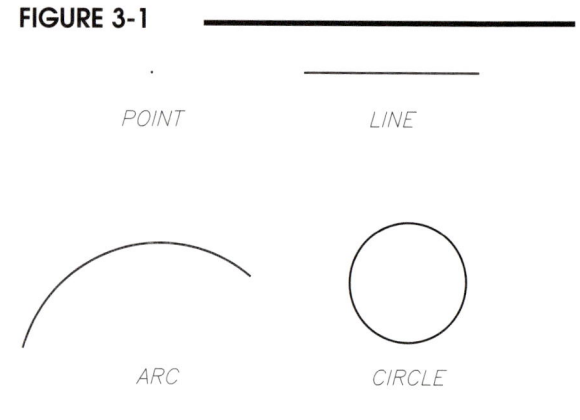

POINT   LINE

ARC   CIRCLE

FIGURE 3-2

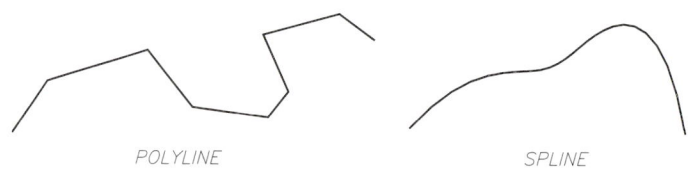

POLYGON   RECTANGLE   ELLIPSE

POLYLINE   SPLINE

It is not always apparent whether a shape is composed of one or more objects. However, if you "hover over" an object with the cursor, an object is "highlighted," or shown in a broken line pattern (Fig. 3-3). This highlighting reveals whether the shape is composed of one or several objects.

FIGURE 3-3

ONE *POLYLINE* SELECTED

ONE OF FOUR *LINES* SELECTED

LOCATING THE DRAW COMMANDS

To invoke *Draw* commands, any of the five command entry methods can be used depending on your computer setup.

1. **Toolbars**   Select the icon from the *Draw* toolbar.
2. **Pull-down menu**   Select the command from the *Draw* pull-down menu.
3. **Screen menu**   Select the command from the *Draw* screen menu.
4. **Keyboard**   Type the command name, command alias, or accelerator keys at the keyboard (a command alias is a one- or two-letter shortcut).
5. **Tablet menu**   Select the icon from the digitizing tablet menu (if available).

For example, a draw command can be activated by PICKing its icon button from the *Draw* toolbar (Fig. 3-4).

FIGURE 3-4

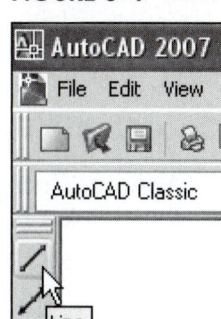

The *Draw* pull-down menu can also be used to select draw commands (Fig. 3-5). Options for a command are found on cascading menus.

FIGURE 3-5

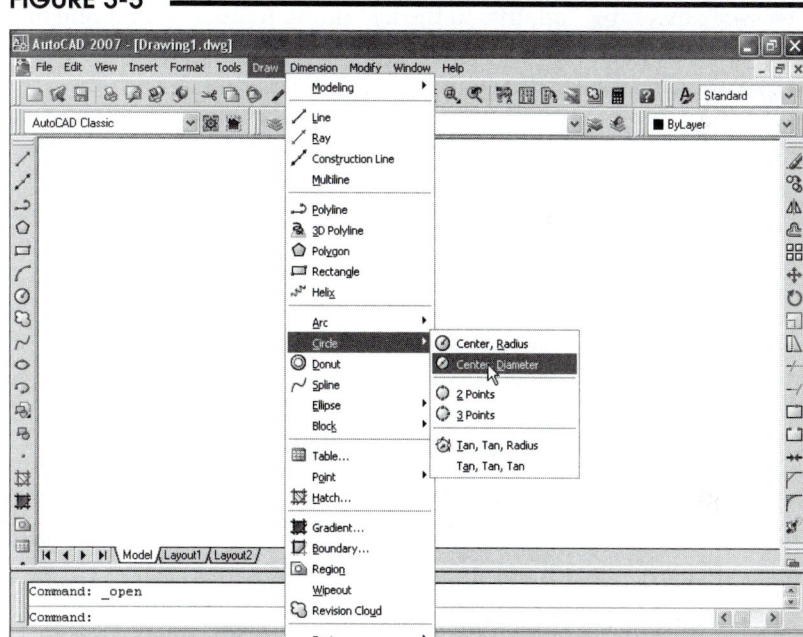

COORDINATE ENTRY

All drawing commands prompt you to specify points, or locations, in the drawing. For example, the *Line* command prompts you to give the "Specify first point:" and "Specify next point:," expecting you to specify locations for the first and second endpoints of the line. After you specify those points, AutoCAD stores the coordinate values to define the line. A two-dimensional line in AutoCAD is defined and stored in the database as two sets of X, Y, and Z values (with Z values of 0), one for each endpoint.

Coordinate Formats and Types

There are many ways you can specify coordinates, or tell AutoCAD the location of points, when you draw or edit objects. You can also use different types and formats for the coordinates that you enter.

Coordinate Formats

1.	Cartesian Format	3,7	Specifies an X and Y value.
2.	Polar Format	6<45	Specifies a distance and angle, where 6 is the distance value, the "less than" symbol indicates "angle of," and 45 is the angular value in degrees.

The Cartesian format is useful when AutoCAD prompts you for a point and you know the exact X,Y coordinates of the desired location. The Polar format is useful when you know the specific length and angle of a line or other object you are drawing.

Coordinate Types

1. Absolute Coordinates	3,7 or 6<45	Values are relative to 0,0 (the origin).
2. Relative Coordinates	@3,7 or @6<45	Values are relative to the "last point," where the "@" (at) symbol is interpreted by AutoCAD as the last point specified.

Absolute coordinates are typically used to specify the "first point" of a line or other object since you are concerned with the location of the point with respect to the origin (0,0) in the drawing. Relative coordinates are typically used for the "next point" of a line or other object since the location of the next point is generally given <u>relative</u> to the first point specified.   For example, the "next point" might be given as "@6<45", meaning "relative to the last point, 6 units at an angle of 45 degrees."

Typical Combinations

Although you can use either coordinate type in either format, the typical combinations are:

Absolute Cartesian coordinates (such as 3,7 for the "first point").
Relative Polar coordinates (such as @6<45 for the "next point").
Relative Cartesian coordinates (such as @3,7 for the "next point").

Even though you can specify coordinates in any type or format using any of the following three methods, AutoCAD stores the data in the drawing file in absolute Cartesian coordinate format.

Three Methods of Coordinate Input

While drawing or editing lines, circles, and other objects, you have the choice to use the following three methods to tell AutoCAD the location of points. Even though you may be simply picking points on the screen, technically you are entering coordinates.

1. Mouse Input	PICK (press the left <u>mouse</u> button to specify a point on the screen).
2. Dynamic Input	When *DYN* is on, you can <u>key values into the edit boxes</u> near the cursor and press Tab to change boxes.
3. Command Line Input	You can <u>key values</u> in any format <u>at the Command prompt</u>.

Mouse Input

Use mouse input to specify coordinates by simply moving the cursor and pressing the left mouse button to pick points. You can use the mouse to input points at any time regardless of the Status Bar settings; however, you should always use a drawing aid with mouse input, such as *SNAP* and *GRID* or *OSNAP*. With *SNAP* and *GRID* on, you can watch the coordinate display in the lower-left corner of the screen to locate points (see Fig. 3-6).

FIGURE 3-6

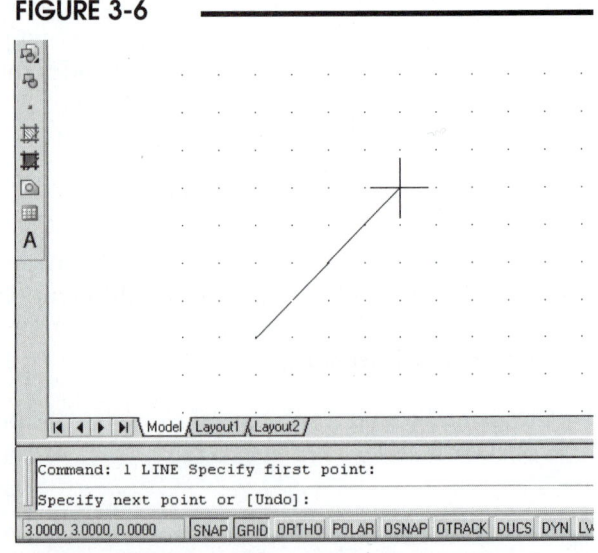

For example, to draw a *Line* from 1,1 to 4,5, first turn on *SNAP* and *GRID*. Then invoke the *Line* command and at the "Specify first point:" prompt, watch the coordinate display while moving the cursor to location 1,1, then press the left mouse button to "pick" that point. Finally, at the "Specify next point:" prompt, locate position 4,5 and pick (see Fig. 3-6).

Dynamic Input

Turn on Dynamic Input by depressing the *DYN* button on the Status Bar. You can then enter exact values into the edit boxes that appear in the drawing area instead of pressing the left mouse button. Press the Tab key to change edit boxes for input, then press Enter to complete the entries. After pressing Enter, the values appear on the Command line. <u>By default, Dynamic Input is in an absolute Cartesian format for the "first point" and a relative polar format when entering the "next point"</u> (see Fig. 3-7). With Dynamic Input, you do not use the "@" (at) symbol to specify relative coordinates. You can change Dynamic Input to Cartesian or relative format if desired. The prompt line near the cursor (for example "Specify next point or") can also be turned off (see "Dynamic Input Settings").

FIGURE 3-7

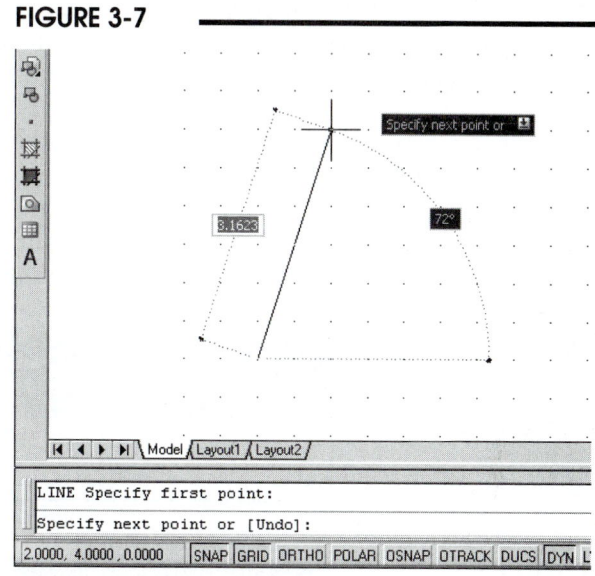

For example, using Dynamic Input to draw a *Line* starting at 1,1 and having a length of 5 units and an angle of 53 degrees, first invoke the *Line* command. Then, at the "Specify first point:" prompt, key "1" into the first edit box, press Tab, then key "1" into the second edit box, then Enter. Finally, at the "Specify next point:" prompt, key "5" into the first edit box, press Tab, then key "53" into the second edit box, then Enter. For some applications you may prefer instead to key a value into only one edit box, press Enter, and specify the other value (length or angle) using mouse input. See also "Direct Distance Entry and Angle Override."

Command Line Input

You can enter coordinate values in any type or format at the Command line. If you enter relative coordinates, you must key in the "@" (at) symbol before Cartesian or polar values to make them relative (see Fig. 3-8). The location of the cursor on the screen has no effect on entering coordinates during Command line input. See also "Direct Distance Entry and Angle Override."

NOTE: To use Command line input, *DYN* can be <u>on or off</u>. If *DYN* is off, values are entered directly at the Command line. However, if *DYN* is on, the values entered appear in the Dynamic Input edit boxes but do not actually appear on the Command line until after pressing Enter.

FIGURE 3-8

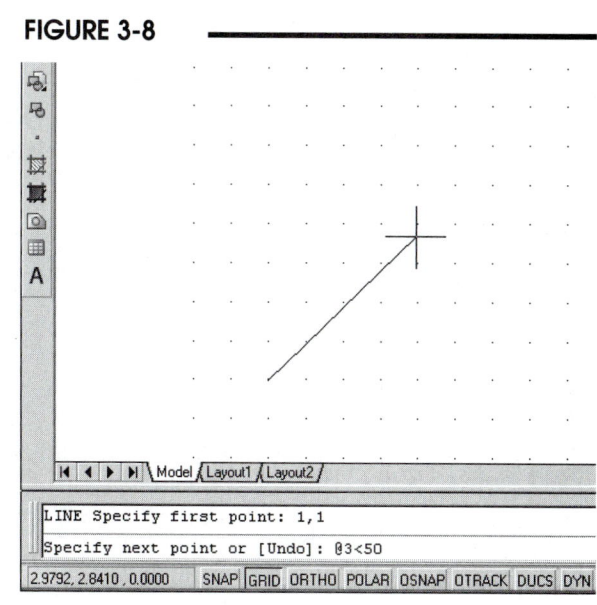

PRACTICE USING THE THREE COORDINATE INPUT METHODS

Because AutoCAD provides three methods to specify coordinates and multiple formats for coordinates, it allows you to be very efficient in your specific application drawing; however, it will take some practice to understand the input methods and formats. After completing these practice exercises, you will have a basic understanding of the methods. Later you will develop a preference for which input methods and which formats are most appropriate for your applications.

For these exercises, begin a new drawing. Do this by selecting the *New* command from the *File* pull-down menu. If the *Select Template* dialog box appears, select the *ACAD.DWT* template, then select *Zoom All* from the *View* pull-down menu. If the *Create New Drawing* dialog box appears, select *Start from Scratch*, choose *Imperial* as the default setting, and click the *OK* button. The *Line* command and the *Circle* command can be activated by any one of the methods shown in the command tables below.

Line

Pull-down Menu	Command (Type)	Alias (Type)	Short-cut	Screen (side) Menu	Tablet Menu
Draw *Line*	*Line*	L	...	*DRAW 1* *Line*	J,10

Circle

Pull-down Menu	Command (Type)	Alias (Type)	Short-cut	Screen (side) Menu	Tablet Menu
Draw *Circle >*	*Circle*	C	...	*DRAW 1* *Circle*	J,9

Using Mouse Input to Draw Lines and Circles

You can use the mouse to pick points <u>at any time</u>, no matter if *DYN* or any other drawing aids appearing on the Status Bar are on or off. However, for clarity it is best to use the settings given below for this exercise.

Status Bar settings (bottom of AutoCAD screen) to begin this exercise: *SNAP*, *GRID*, and *MODEL* should be on (the button appears depressed). All other settings (*ORTHO, POLAR, OSNAP, OTRACK, DUCS, DYN,* and *LWT* should be off (buttons appear protruding) (see Fig. 3-9). Also ensure *Grid Snap* is on, not *Polar Snap*, by right-clicking on the word *SNAP* and selecting *Grid Snap on*. Remember that AutoCAD stores coordinate values to an accuracy of 14 places, so <u>it is important to use *SNAP* and *GRID* or some other appropriate drawing aid with mouse input</u> to assist you in picking exact points.

In the following 4 exercises you will draw three *Lines* and one *Circle* in specific locations as shown in Figure 3-10. Begin by following the steps on the next page to construct the first horizontal *Line*.

FIGURE 3-9

FIGURE 3-10

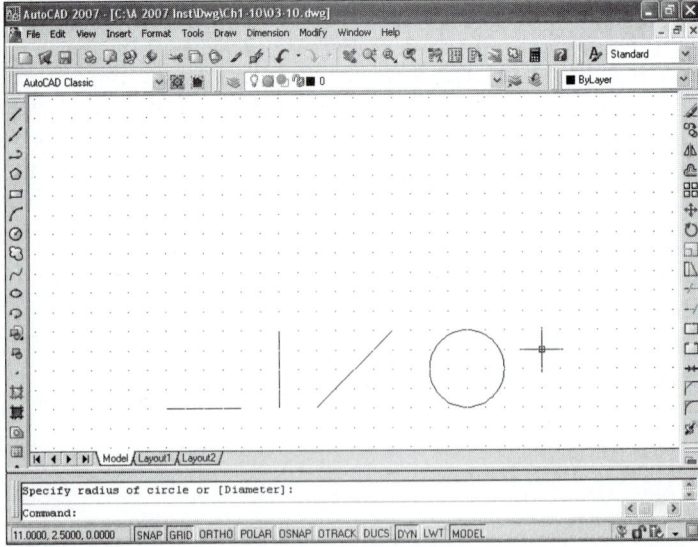

1. Draw a horizontal *Line* of 2 units length starting at 1,1.

Steps	Command Prompt	Perform Action	Comments
1.	Command:	select or type *Line*	use any method
2.	LINE Specify first point:	**PICK** location 1,1	watch coordinate display, lower left
3.	Specify next point or [Undo]:	**PICK** location 3,1	watch coordinate display
4.	Specify next point or [Undo]:	press **ENTER**	completes command

Check your work with the first *Line* as shown in Figure 3-10. Then proceed to construct the other two *Lines* and the *Circle*.

2. Draw a vertical *Line* of 2 units length starting at 4,1.

Steps	Command Prompt	Perform Action	Comments
1.	Command:	select or type *Line*	use any method
2.	LINE Specify first point:	**PICK** location 4,1	watch coordinate display, lower left
3.	Specify next point or [Undo]:	**PICK** location 4,3	watch coordinate display
4.	Specify next point or [Undo]:	press **ENTER**	completes command

3. Draw a diagonal *Line* of 2 units length starting at 5,1.

Steps	Command Prompt	Perform Action	Comments
1.	Command:	select or type *Line*	use any method
2.	LINE Specify first point:	**PICK** location 5,1	watch coordinate display, lower left
3.	Specify next point or [Undo]:	**PICK** location 7,3	watch coordinate display
4.	Specify next point or [Undo]:	press **ENTER**	completes command

4. **Draw a *Circle* of 1 unit radius with the center at 9,2.**

Steps	Command Prompt	Perform Action	Comments
1.	Command:	select or type *Circle*	use any method
2.	CIRCLE Specify center point...	PICK location 9,2	watch coordinate display
3.	Specify radius of circle...	PICK location 10,2	press F6 to change coordinate display to Cartesian coordinates if necessary; completes command

Check your work with the objects shown in Figure 3-10. Remember that when you use mouse input you should use *SNAP* and *GRID*, *OSNAP*, or other appropriate drawing aids to ensure you pick exact points.

Using Dynamic Input to Draw Lines and Circles

Turn on Dynamic Input by depressing the *DYN* button on the Status Bar (bottom of AutoCAD screen). For this exercise, *DYN* and *MODEL* should be on (depressed) and all other settings (*SNAP*, *GRID*, *ORTHO*, *POLAR*, *OSNAP*, *OTRACK*, *DUCS*, and *LWT*) should be off (buttons appear protruding).

Using the default setting for Dynamic Input, the "first point" specified is absolute Cartesian format and the "next point:" is relative polar format.

In the following 4 exercises you will draw another set of three *Lines* and one *Circle* located directly above the previous set as shown in Figure 3-11.

FIGURE 3-11

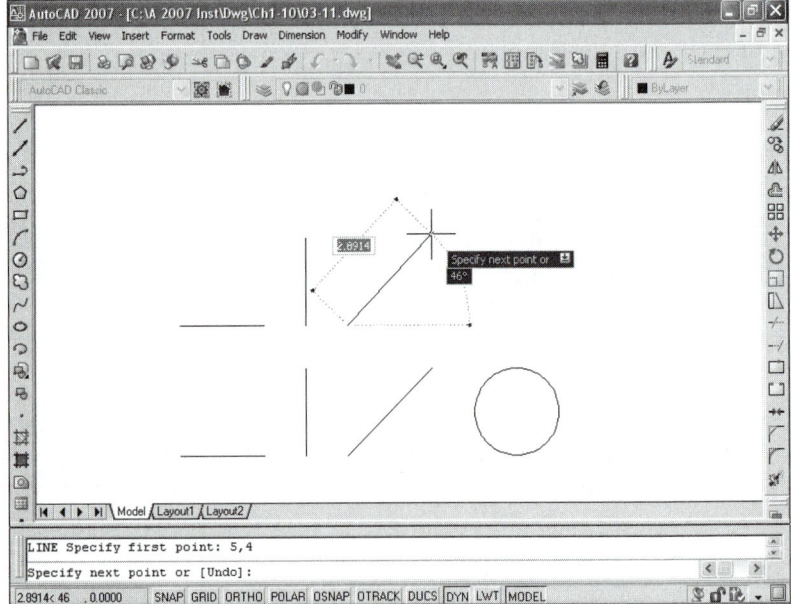

5. Draw a horizontal *Line* of 2 units length starting at 1,4.

Steps	Command Prompt	Perform Action	Comments
1.	Command:	select or type *Line*	use any method
2.	LINE Specify first point:	Enter **1** into first edit box, then press **Tab**	this is the X value for Cartesian format
3.		Enter **4** into second edit box, the press **Enter**	this is the Y value; establishes first endpoint
4.	Specify next point or [Undo]:	Enter **2** into first edit box, then press **Tab**	this is distance value for relative polar format
5.		Enter **0** into second edit box, then press **Enter**	this is angular value; establishes second endpoint
6.	Specify next point or [Undo]:	press **Enter**	completes command

6. Draw a vertical *Line* of 2 units length starting at 4,4.

Steps	Command Prompt	Perform Action	Comments
1.	Command:	select or type *Line*	use any method
2.	LINE Specify first point:	Enter **4** into first edit box, then press **Tab**	this is the X value for Cartesian format
3.		Enter **4** into second edit box, the press **Enter**	this is the Y value; establishes first endpoint
4.	Specify next point or [Undo]:	Enter **2** into first edit box, then press **Tab**	this is distance value for relative polar format
5.		Enter **90** into second edit box, then press **Enter**	this is angular value; establishes second endpoint
6.	Specify next point or [Undo]:	press **Enter**	completes command

7. Draw a diagonal *Line* of 2.8284 units length starting at 5,4.

Steps	Command Prompt	Perform Action	Comments
1.	Command:	select or type *Line*	use any method
2.	LINE Specify first point:	Enter **5** into first edit box, then press **Tab**	this is the X value for Cartesian format
3.		Enter **4** into second edit box, the press **Enter**	this is the Y value; establishes first endpoint
4.	Specify next point or [Undo]:	Enter **2.8284** into first edit box, then press **Tab**	this is distance value for relative polar format

(continued)

2006

Steps	Command Prompt	Perform Action	Comments
5.		Enter **45** into second edit box, then press **Enter**	this is angular value; establishes second endpoint
6.	Specify next point or [Undo]:	press **Enter**	completes command

See Figure 3-11 for an illustration of this exercise.

8. Draw a *Circle* of 1 unit radius with the center at 9,5.

Steps	Command Prompt	Perform Action	Comments
1.	Command:	select or type *Circle*	use any method
2.	CIRCLE Specify first point...	Enter **9** into first edit box, then press **Tab**	this is the X value for Cartesian format
3.		Enter **5** into second edit box, then press **Enter**	this is the Y value; establishes circle center
4.	Specify radius of circle...	Enter **1** into edit box, then press **Enter**	this is distance value; completes command

Using Command Line Input to Draw Lines and Circles

You can use Command line input when *DYN* is on or off; however, for clarity in this exercise it is best to toggle *DYN* off. Although the settings for the other drawing aids do not affect Command line input, in this exercise it is best to set only *MODEL* on (the button appears depressed) and all other settings should be off (buttons appear protruding) (Fig. 3-12).

With Command line input you can key in coordinates at the Command line in any type or format. When using relative format, you must key in the "@" (at) symbol.

FIGURE 3-12

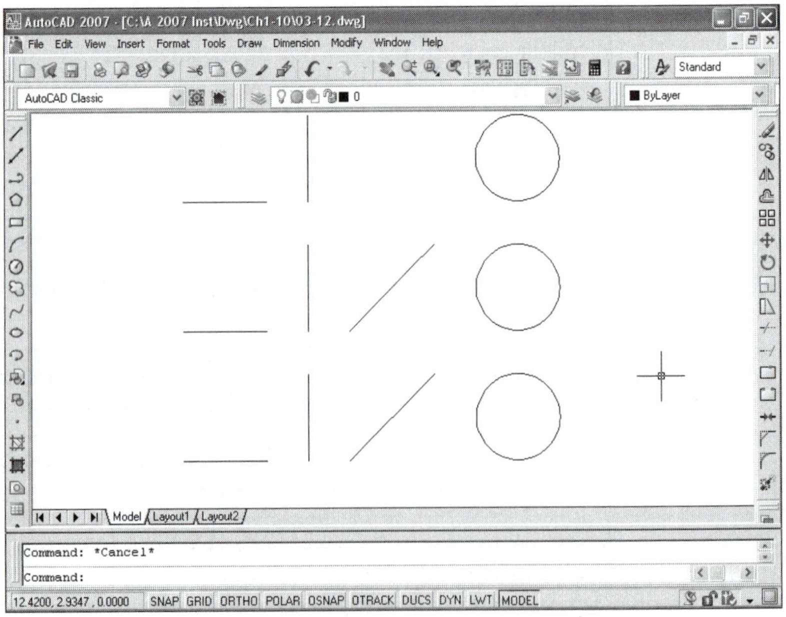

9. Draw a horizontal *Line* of 2 units length starting at 1,7. Use absolute Cartesian coordinates.

Steps	Command Prompt	Perform Action	Comments
1.	Command:	select or type *Line*	use any method
2.	LINE Specify first point:	type **1,7** and press **Enter**	establishes first endpoint
3.	Specify next point or [Undo]:	type **3,7** and press **Enter**	establishes second endpoint
4.	Specify next point or [Undo]:	press **Enter**	completes command

10. Draw a vertical *Line* of 2 units length. Use relative Cartesian coordinates.

Steps	Command Prompt	Perform Action	Comments
1.	Command:	select or type *Line*	use any method
2.	LINE Specify first point:	type **@1,0** and press **Enter**	@ means "last point"
3.	Specify next point or [Undo]:	type **@0,2** and press **Enter**	establishes second endpoint
4.	Specify next point or [Undo]:	press **Enter**	completes command

11. Draw a diagonal *Line* of 2.8284 units length. Use absolute Cartesian and relative polar coordinates.

Steps	Command Prompt	Perform Action	Comments
1.	Command:	select or type *Line*	use any method
2.	LINE Specify first point:	type **5,7** and press **Enter**	establishes first endpoint
3.	Specify next point or [Undo]:	type **@2.8284<45** and press **Enter**	< means "angle of"; establishes second endpoint
4.	Specify next point or [Undo]:	press **Enter**	completes command

12. Draw a *Circle* of 1 unit radius with the center at 9,8. Use absolute Cartesian coordinates.

Steps	Command Prompt	Perform Action	Comments
1.	Command:	select or type *Circle*	use any method
2.	CIRCLE Specify center point...	type **9,8**	establishes center
3.	Specify radius of circle...	type **1**	

Direct Distance Entry and Angle Override

When prompted for a "next point," instead of using relative polar values preceded by the "@" (at) symbol to specify a distance and an angle, you can key in a distance-only value (called Direct Distance Entry) or key in an angle-only value (Angle Override). With these features, you can key in one value (either distance or angle) and specify the other value using mouse input. This capability represents an older, less graphic version of Dynamic Input. Both Direct Distance Entry and Angle Override can be used with Command Line Input and with Dynamic Input.

Direct Distance Entry

Direct Distance Entry allows you to directly key in a distance value. Use Direct Distance Entry in response to the "next point:" prompt. With Direct Distance Entry, you point the mouse so the "rubber-band" line indicates the angle of the desired line, then simply key in the desired distance value (length) at the Command line in response to the "next point:" prompt. For example, if you wanted to draw a horizontal line of 7.5 units, you would first turn on *POLAR*, point the cursor in a horizontal direction, key in "7.5," and press Enter. Direct Distance Entry can be used with *DYN* on or off.

Direct Distance Entry is best used in conjunction with Polar Tracking (turn on *POLAR* on the Status Bar). Polar Tracking assists you in creating a "rubberband" line that is horizontal, vertical, or at a regular angular increment you specify such as 30 or 45 degrees. Indicating the angle of the line with the mouse without using *POLAR* or *ORTHO* will not likely result in a precise angle. See "Polar Tracking and Polar Snap."

Angle Override

If you want to specify that a line is drawn at a specific angle, but specify the line length using mouse input, use Angle Override. At the "next point" prompt, key in the "<" (less than) symbol and an angular value, such as "<30", then press Enter. AutoCAD responds with "Angle Override: 30." The rubberband line is then constrained to 30 degrees but allows you to pick a point to specify the length. Angle Override can be used with Command Line Input and with Dynamic Input.

Dynamic Input Settings

The default setting for Dynamic Input is absolute Cartesian coordinate input for the "first point" and relative polar coordinates for the "second point." To change these settings, right-click on the word *DYN* on the Status Bar and select *Settings…*. This action produces the *Dynamic Input* tab of the *Drafting Settings* dialog box (Fig. 3-13). You can also select *Drafting Settings* from the *Tools* pull-down menu or type the command alias *DS* to produce this dialog box.

The *Pointer Input* section allows you to change settings for the type and format of coordinates that are used during Dynamic Input. The *Dimension Input* section controls the appearance of the dimension lines that are shown around the edit boxes during the "next point" specification.

FIGURE 3-13

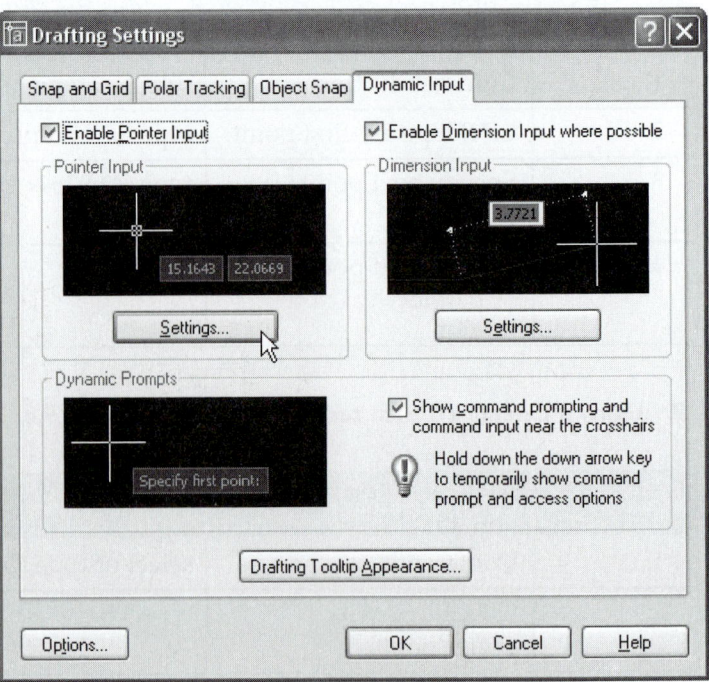

Select *Settings...* in the *Pointer Input* section to produce the *Polar Input Settings* dialog box (Fig. 3-14). Note that *Polar format* and *Relative coordinates* are the default settings for the "next point" specification. You cannot change the format for the "first point" specification, which is always absolute Cartesian format.

FIGURE 3-14

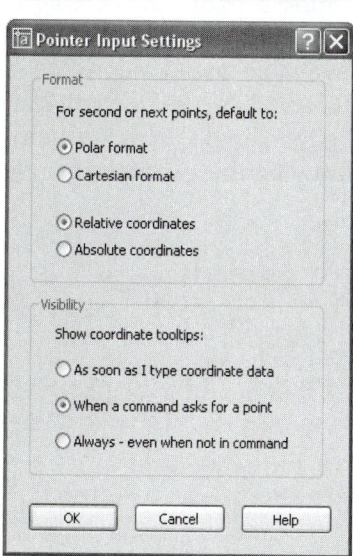

POLAR TRACKING AND POLAR SNAP

Remember that all draw commands prompt you to specify locations, or coordinates, in the drawing. You can indicate these coordinates interactively (using the cursor), entering values (at the keyboard), or using a combination of both. Features in AutoCAD that provide an easy method for specifying coordinate locations interactively are Polar Tracking and Polar Snap. Another feature, Object Snap Tracking, is discussed in Chapter 7. These features help you draw objects at specific angles and in specific relationships to other objects.

Polar Tracking (*POLAR*) helps the rubberband line snap to angular increments such as 45, 30, or 15 degrees. Polar Snap can be used in conjunction to make the rubberband line snap to incremental lengths such as .5 or 1.0. You can toggle these features on and off with the *SNAP* and *POLAR* buttons on the Status Bar or by toggling F9 and F10, respectively. First, you should specify the settings in the *Drafting Settings* dialog box.

Technically, <u>Polar Tracking and Polar Snap would be considered a variation of the Mouse Input method</u> since the settings are determined and specified beforehand, then points on the screen are PICKed using the cursor.

To practice with Polar Tracking and Polar Snap, you should turn *DYN, OSNAP,* and *OTRACK* off (these are on by default when you install AutoCAD). Since some of these features generate alignment vectors, it can be difficult to determine which of the features is operating when they are all on.

Polar Tracking

Polar Tracking (*POLAR*) simplifies drawing *Lines* or performing other operations such as *Move* or *Copy* at specific angle increments. For example, you can specify an increment angle of 30 degrees. Then when Polar Tracking is on, the rubberband line "snaps" to 30-degree increments when the cursor is in close proximity (within the *Aperture* box) to a specified angle. A dotted "tracking" line is displayed and a "tracking tip" appears at the cursor giving the current distance and angle (Fig. 3-15). In this case, it is simple to draw lines at 0, 30, 60, 90, or 120-degree angles and so on.

FIGURE 3-15

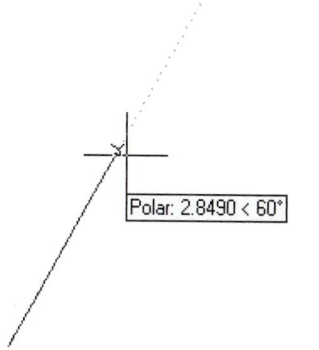

Figure 3-16 displays possible positions for Polar Tracking <u>when a 30-degree angle is specified</u>. Available angle options are 90, 45, 30, 22.5, 18, 15, 10, and 5 degrees, or you can specify any user-defined angle. This is a tremendous aid for drawing lines at typical angles.

FIGURE 3-16

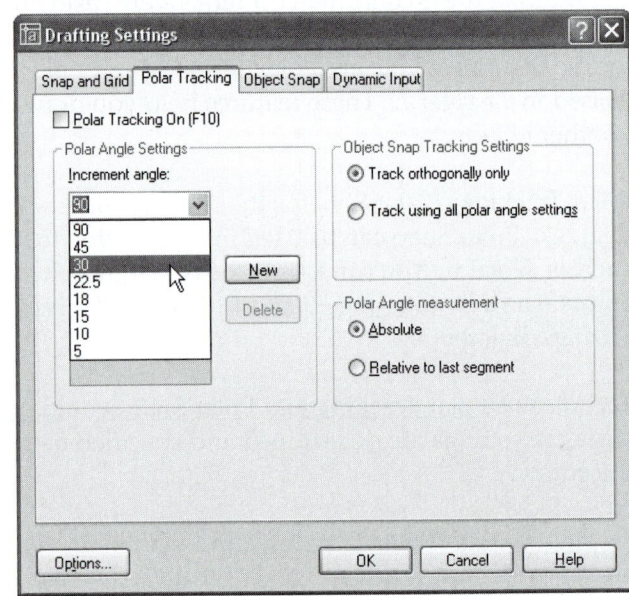

Polar: 1.7705 < 30°

 You can access the settings using the *Polar Tracking* tab of the *Drafting Settings* dialog box (Fig. 3-17). Invoke the dialog box by the following methods:

1. Enter *Dsettings* or *DS* at the Command line, then select the *Polar Tracking* tab.

2. Right-click on the word *POLAR* on the Status Bar, then choose *Settings...* from the menu.

3. Select *Drafting Settings* from the *Tools* pull-down menu, then select the *Polar Tracking* tab.

FIGURE 3-17

In this dialog box, you can select the *Increment Angle* from the drop-down list (as shown) or specify *Additional Angles*. Use the *New* button to create user-defined angles. Highlighting a user-defined angle and selecting the *Delete* button removes the angle from the list.

The *Object Snap Tracking Settings* are used only with Object Snap (discussed in Chapter 7). When the *Polar Angle Measurement* is set to *Absolute*, the angle reported on the "tracking tip" is an absolute angle (relative to angle 0 of the current coordinate system).

Polar Tracking with Direct Distance Entry

With normal Polar Tracking (Polar Snap is off), the line "snaps" to the set angles but can be drawn to any length (distance). This option is particularly useful in conjunction with Direct Distance Entry since you specify the angular increment in the *Polar Tracking* tab of the *Drafting Settings* dialog box, but indicate the distance by <u>entering values at the Command line</u>. In other words, when drawing a *Line*, move the cursor so it "snaps" to the desired angle, then enter the desired distance at the keyboard (Fig. 3-18). See previous discussion, "Direct Distance Entry and Angle Override."

FIGURE 3-18

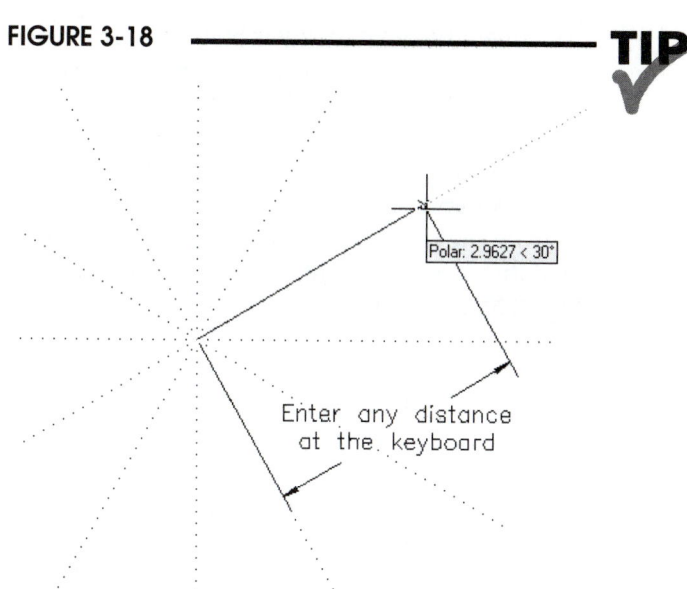

Polar Tracking Override

When Polar Tracking is on, you can override the previously specified angle increment and draw to another specific angle. The new angle is valid only for one point specification. To enter a polar override angle, enter the left angle bracket (<) and an angle value whenever a command asks you to specify a point. The following command prompt sequence shows a 12-degree override entered during a *Line* command.

```
Command: line
Specify first point: PICK
Specify next point or [Undo]: <12
Angle Override:  12
Specify next point or [Undo]: PICK
```

Polar Tracking Override can be used with Command Line Input or Dynamic Input.

Polar Snap

<u>Polar Tracking</u> makes the rubberband line "snap" to <u>angular increments</u>, whereas <u>Polar Snap</u> makes the rubberband line "snap" to <u>distance increments</u> that you specify. For example, setting the distance increment to 2 allows you to draw lines at intervals of 2, 4, 6, 8, and so on. Therefore, Polar Tracking with Polar Snap allows you to draw at specific angular <u>and</u> distance increments (Fig. 3-19). Using Polar Tracking with Polar Snap off, as described in the previous section, allows you to draw at specified angles but not at any specific distance intervals.

FIGURE 3-19

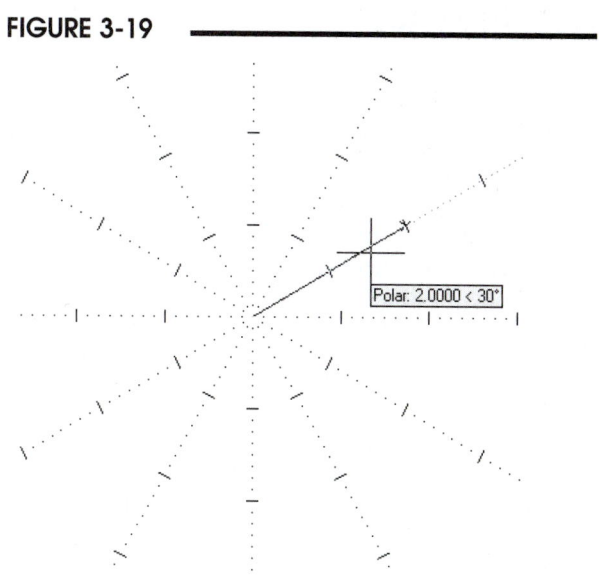

Only one type of Snap, Polar Snap, or Grid Snap, can be used at any one time. Toggle either *Polar Snap* or *Grid Snap* on by setting the radio button in the *Snap and Grid* tab of the *Drafting Settings* dialog box (Fig. 3-20, lower left). Alternately, right-click on the word *SNAP* at the Status Bar and select either *Polar Snap On* or *Grid Snap On* from the shortcut menu (see Fig. 3-21). Set the Polar Snap distance increment in the *Polar Distance* edit box (Fig. 3-20, left-center).

FIGURE 3-20

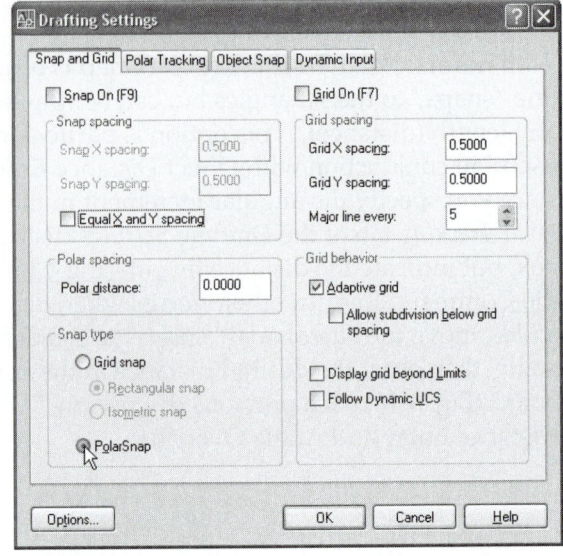

To utilize Polar Snap, both *SNAP* (F9) and *POLAR* (F10) must be turned on, Polar Snap must be on, and a Polar Distance value must be specified in the *Drafting Settings* dialog box. As a check, settings should be as shown in Figure 3-21 (*SNAP* and *POLAR* are recessed), with the addition of some Polar Distance setting in the dialog box. *SNAP* turns on or off whichever of the two snap types is active, Grid Snap or Polar Snap. For example, if Polar Snap is the active snap type, toggling F9 turns only Polar Snap off and on.

FIGURE 3-21

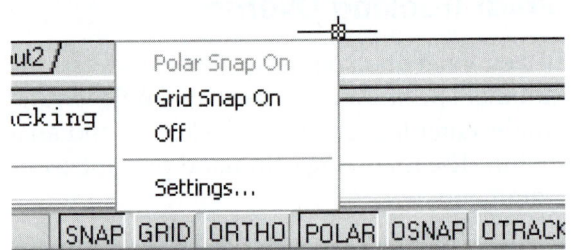

DRAWING LINES USING POLAR TRACKING

For these exercises you will use combinations of Polar Tracking with Polar Snap, Grid Snap, and Dynamic Input. Also, Object Snap and Object Snap Tracking should be turned off. Turn off Object Snap and Object Snap Tracking by toggling the *OSNAP* and *OTRACK* buttons on the Status Bar to the protruding position or using F3 and F11 to do so.

Using Polar Tracking and Grid Snap

Draw the shape in Figure 3-22 using Polar Tracking in conjunction with Grid Snap. Follow the steps below.

1. Begin a *New* drawing. Select **Start from Scratch** and the **Imperial** default settings.

2. Toggle on Grid Snap (rather than Polar Snap) by right-clicking on the word **SNAP** on the Status Bar, then selecting **Grid Snap On** from the menu (see Fig. 3-6).

FIGURE 3-22

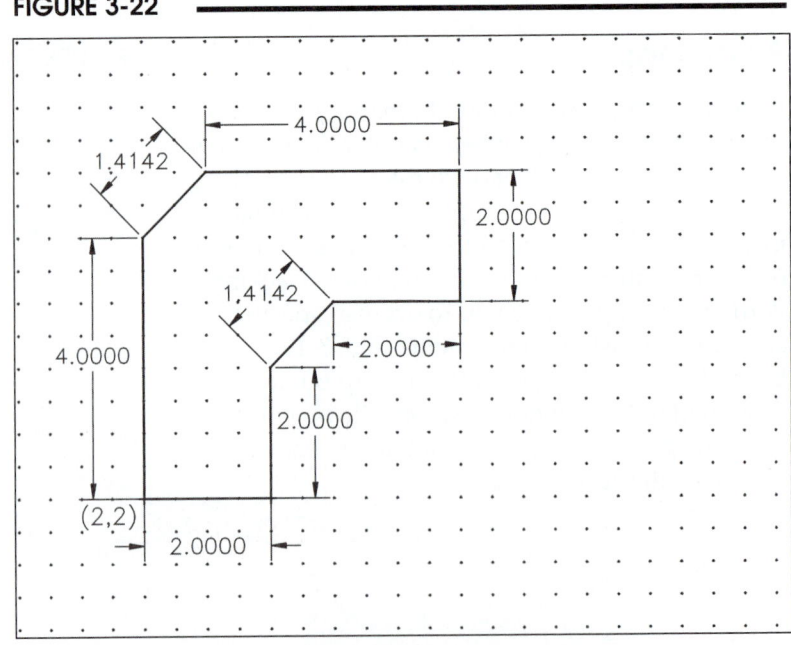

3. Turn on Polar Tracking and make the appropriate settings. Do this by right-clicking on the word *POLAR* on the Status Bar and selecting *Settings* from the menu (Fig. 3-23). In the *Polar Tracking* tab of the *Drafting Settings* dialog box that appears, set the *Increment Angle* to **45.0** (see Fig. 3-17). Select the *OK* button.

FIGURE 3-23

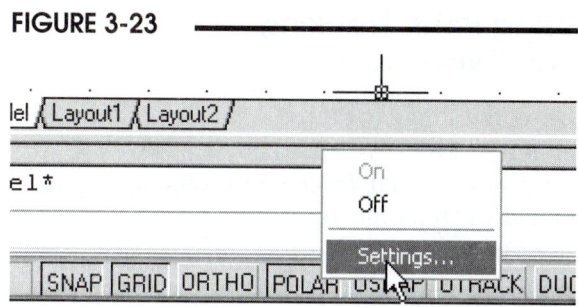

4. Make these other settings by single-clicking the words on the Status Bar or using the Function keys. The Status Bar indicates the on/off status of these drawing aids such that the button in a recessed position means <u>on</u> and a protruding button means <u>off</u> (Fig. 3-24). The command prompt also indicates the on or off status when they are changed.

FIGURE 3-24

SNAP (F9) is on
GRID (F7) is on
ORTHO (F8) is off
POLAR (F10) is on
OSNAP (F3) is off
OTRACK (F11) is off
DUCS (F6) is off
DYN (F12) is off

The resulting Status Bar should look like that in Figure 3-24.

5. Use the *Line* command. Starting at location 2.000, 2.000, 0.000 (watch the *COORDS* display), begin drawing the shape shown in Figure 3-22. The tracking tip should indicate the current length and angle of the lines as you draw (Fig. 3-25).

6. Complete the shape. Compare your drawing to that in Figure 3-22. Use *Save* and name the drawing **PTRACK-GRIDSNAP**. *Close* the drawing.

FIGURE 3-25

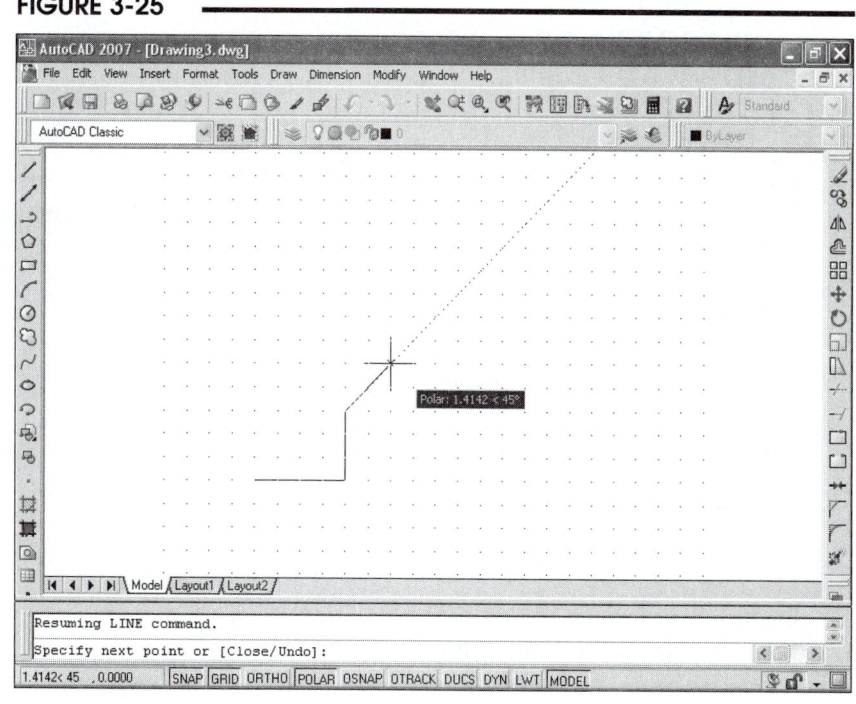

Using Polar Tracking and Dynamic Input

Draw the shape in Figure 3-26. Since the lines are at regular angles but irregular lengths, use Polar Tracking in conjunction with Dynamic Input. Follow the steps below.

1. Begin a *New* drawing. Select *Start From Scratch* and the *Imperial* default settings.

2. Toggle off *SNAP* on the Status Bar. No Snap type is used since distances will be entered at the keyboard.

3. Turn on Polar Tracking and make the appropriate settings. Do this by right-clicking on the word *POLAR* on the Status Bar and selecting *Settings* from the menu. In the *Polar Tracking* tab of the *Drafting Settings* dialog box, ensure the *Increment Angle* is set to **30.0**, and **Polar Tracking** is **On**. Select the **OK** button.

4. Ensure these other settings are correct by clicking the words on the Status Bar or using the Function keys.

> *SNAP* (F9) is off
> *GRID* (F7) is off
> *ORTHO* (F8) is off
> *POLAR* (F10) is on
> *OSNAP* (F3) is off
> *OTRACK* (F11) is off
> *DUCS* (F6) is off
> *DYN* (F12) is on

5. Use the *Line* command. Starting at location 2.000, 2.000, 0.000 (enter absolute coordinates of **2,2**), begin drawing the shape shown in Figure 3-29. Use Polar Tracking by moving the mouse in the desired direction (Fig. 3-27). When the angle edit box indicates the correct angle for each line, enter the distance for each line (**2.3**) into the distance edit box and press Enter.

6. Complete the shape. Compare your drawing to that in Figure 3-29. Use *Save* and name the drawing **PTRACK-DYN**. *Close* the drawing and *Exit* AutoCAD.

FIGURE 3-26

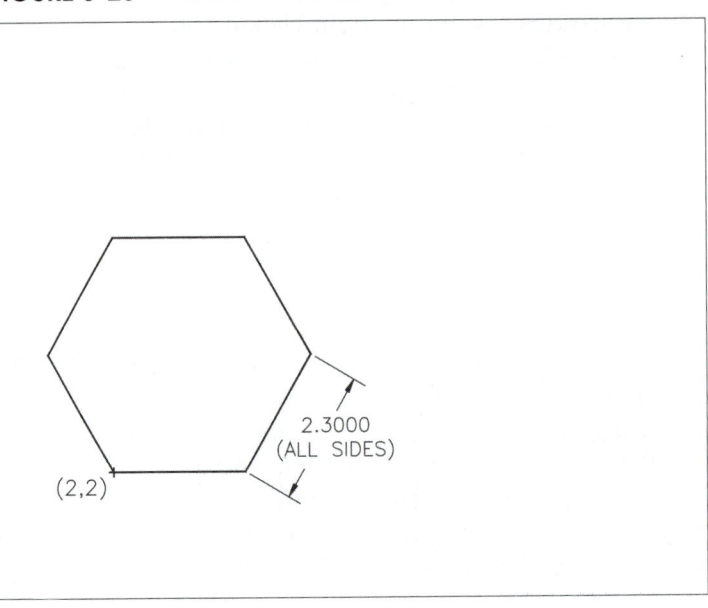

FIGURE 3-27

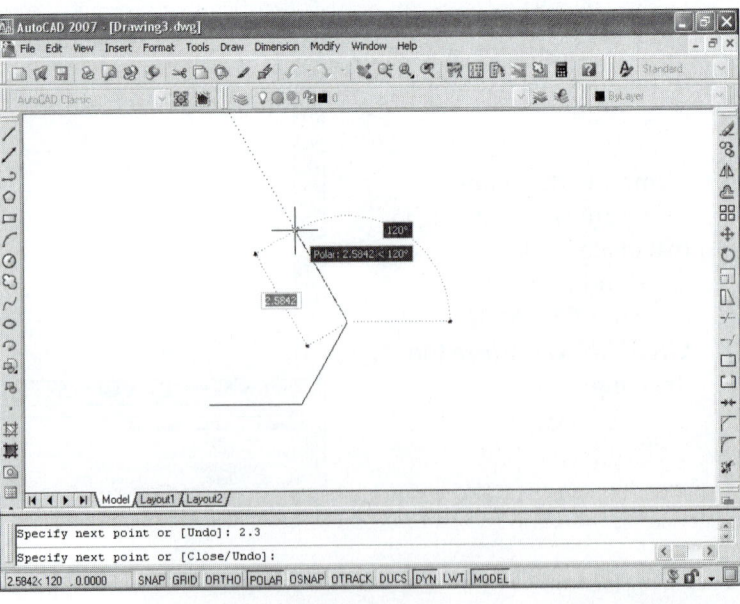

Using Polar Tracking, Polar Snap, and Dynamic Input

Draw the shape in Figure 3-28. Since the lines are regular interval lengths, use Polar Tracking in conjunction with Polar Snap. Follow the steps below.

1. Begin a *New* drawing. Select *Start From Scratch* and the *Imperial* default settings.

2. Toggle on Polar Snap (rather than Grid Snap) by right-clicking on the word *SNAP* on the Status Bar, then selecting *Polar Snap On* from the menu.

3. Make the appropriate Polar Snap settings. Do this by right-clicking on the word *SNAP* on the Status Bar and selecting *Settings* from the menu. In the *Snap and Grid* tab of the *Drafting Settings* dialog box that appears, set the *Polar Distance* to **1.0000** (see Fig. 3-29). Next, access the *Polar Tracking* tab and ensure the *Increment Angle* is set to **45.0**. Select the *OK* button.

4. Ensure these other settings are correct by clicking the words on the Status Bar or using the Function keys.

    SNAP (F9) is on
    GRID (F7) is on
    ORTHO (F8) is off
    POLAR (F10) is on
    OSNAP (F3) is off
    OTRACK (F11) is off
    DUCS (F6) is off
    DYN (F12) is on

5. Use the *Line* command. Starting at location 3.000, 1.000, 0.000 (enter absolute coordinates of **3,1**), begin drawing the shape shown in Figure 3-28. The edit boxes should indicate the current length and angle of the lines as you draw (Fig. 3-30).

6. Complete the shape. Compare your drawing to that in Figure 3-28. Use *Save* and name the drawing **PTRACK-POLARSNAP**. *Close* the drawing.

FIGURE 3-28

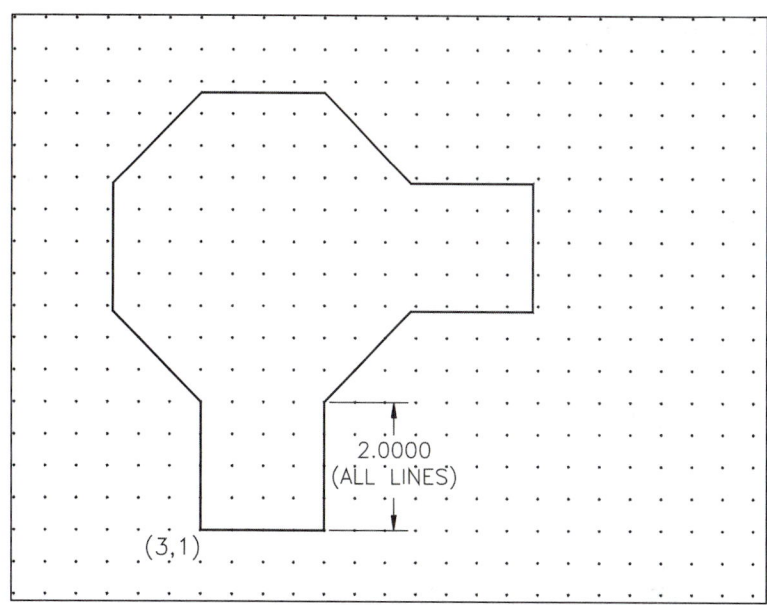

FIGURE 3-29

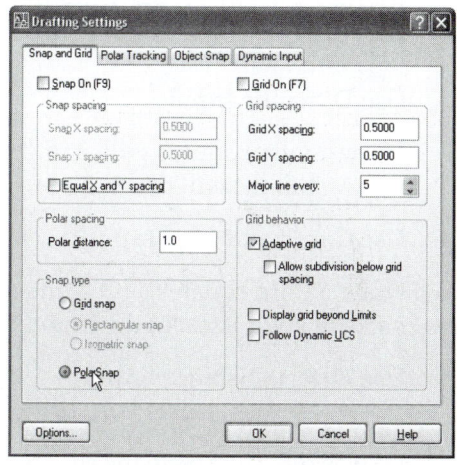

FIGURE 3-30

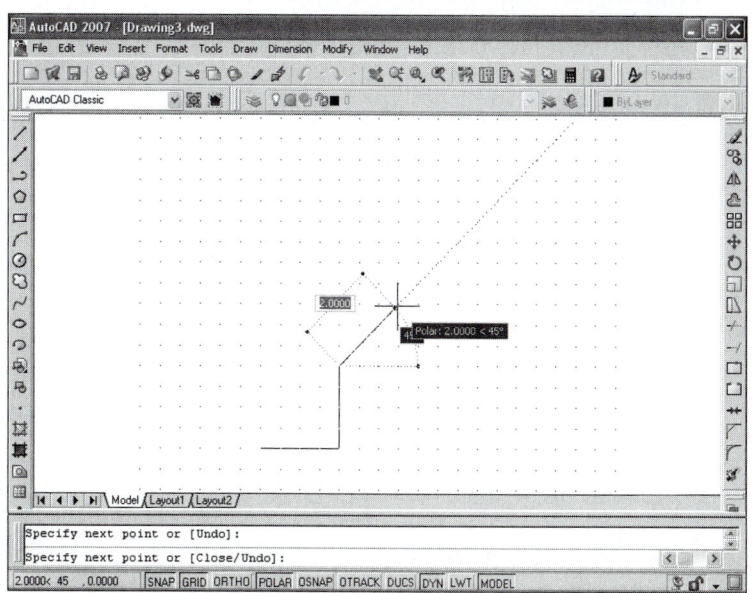

CHAPTER EXERCISES

1. **Start a *New* Drawing**

    Start a *New* drawing. Select the
    Start from Scratch option and
    select **Imperial** settings or use the
    ACAD.DWT template. Next, use
    the *Save* command and save the
    drawing as **CH3EX1.** Remember
    to *Save* often as you complete the
    following exercises. The com-
    pleted exercise should look like
    Figure 3-31.

FIGURE 3-31

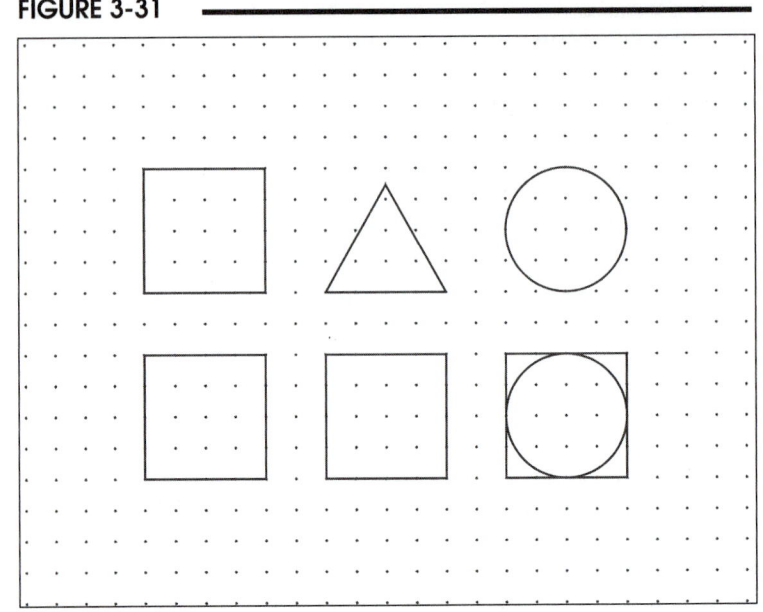

2. **Use Mouse Input**

    Draw a square with sides of 2 unit
    length. Locate the lower-left
    corner of the square at **2,2.** Use the
    Line command with mouse input.
    (HINT: Turn on *SNAP*, *GRID*, and
    POLAR.)

3. **Use Dynamic Input**

    Draw another square with 2 unit sides using the *Line* command. Turn on *DYN*. Enter the correct
    values into the edit boxes.

4. **Use Command Line Input**

    Draw a third square (with 2 unit sides) using the *Line* command. Turn off *DYN*. Enter coordi-
    nates in any format at the Command line.

5. **Use Direct Distance Entry**

    Draw a fourth square (with 2 unit sides) beginning at the lower-left corner of **2,5.** Ensure that
    DYN is off. Complete the square drawing *Lines* using direct distance coordinate entry. Don't
    forget to turn on *ORTHO* or *POLAR*.

6. **Use Dynamic Input and Polar Tracking**

    Draw an equilateral triangle with sides of 2 units. Turn on *POLAR* and *DYN*. Make the appro-
    priate angular increment settings in *Drafting Settings* dialog box. Begin by locating the lower-left
    corner at **5,5.** (HINT: An equilateral triangle has interior angles of 60 degrees.)

7. **Use Mouse Input**

    Turn on *SNAP* and *GRID*. Use mouse input to draw a *Circle* with a 1 unit <u>radius</u>. Locate the
    center at **9,6.**

8. **Use any method**

    Draw another *Circle* with a 2 unit <u>diameter</u>. Using any method, locate the center 3 units below
    the previous *Circle*.

9. **Save your drawing**

    Use *Save*. Compare your results with Figure 3-31. When you are finished, *Close* the drawing.

10. In the next series of steps, you will create the Stamped Plate shown in Figure 3-32. Begin a *New* drawing. Select the *Start from Scratch* option and select *Imperial* settings. Next, use the *Save* command and assign the name **CH3EX2.** Remember to *Save* often as you work.

FIGURE 3-32

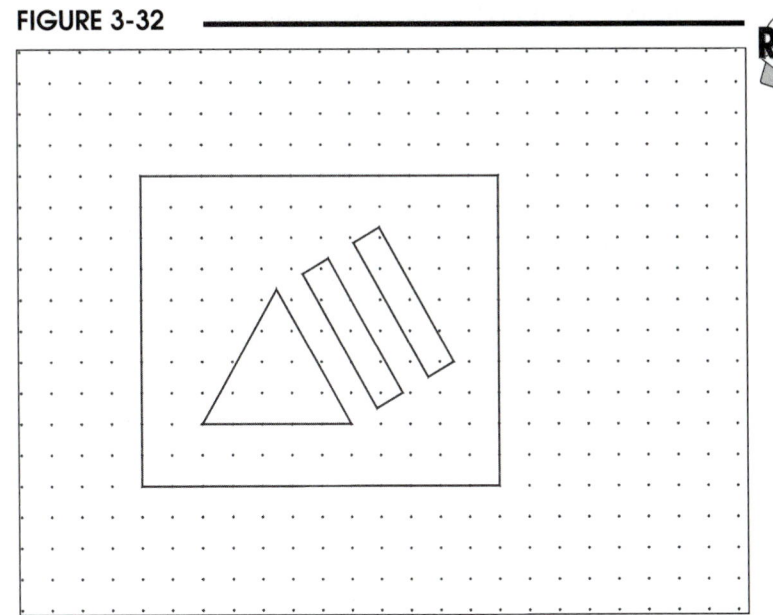

11. First, you will create the equilateral triangle as shown in Figure 3-33. All sides are equal and all angles are equal. To create the shape easily, you should use Polar Tracking with Dynamic Input. Access the *Polar Tracking* tab of the *Drafting Settings* dialog box and set the *Increment Angle* to **30**. (HINT: Right-click on the word *POLAR* and select *Settings* from the shortcut menu.) On the Status Bar, make sure *POLAR* and *DYN* are on (appear recessed), but not *SNAP, ORTHO, OSNAP, OTRACK* or *DUCS. GRID* is optional.

FIGURE 3-33

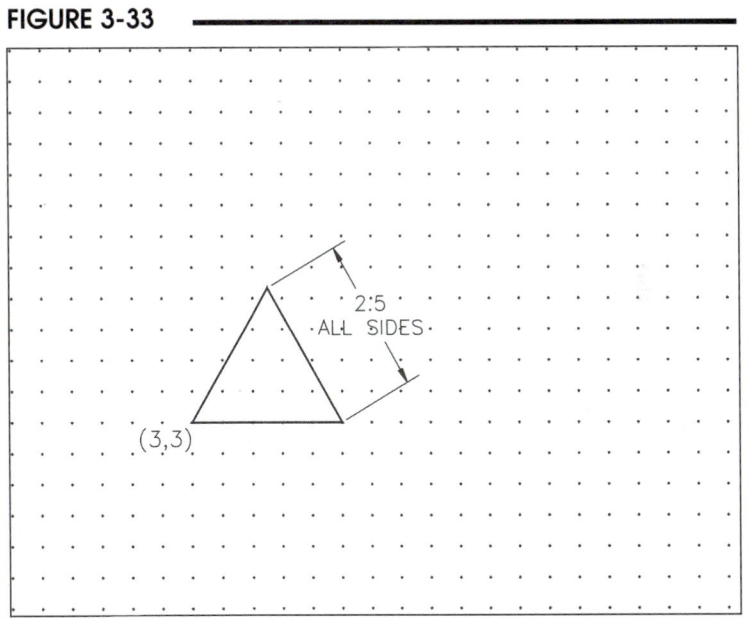

12. Use the *Line* command to create the equilateral triangle. Start at position **3,3** by entering absolute coordinates. All sides should be **2.5** units, so enter the distance values at the keyboard as you position the mouse in the desired direction for each line. The drawing at this point should look like that in Figure 3-33 (not including the notation). *Save* the drawing but do not *Close* it.

13. Next, you should create the outside shape (Fig. 3-34). Since the dimensions are all at even unit intervals, use Grid Snap to create the rectangle. (HINT: Right-click on the word *SNAP* and select *Grid Snap On*.) At the Status Bar, make sure that *SNAP* and *POLAR* are on. Use the *Line* command to create the rectangular shape starting at point **2,2** (enter absolute coordinates).

FIGURE 3-34

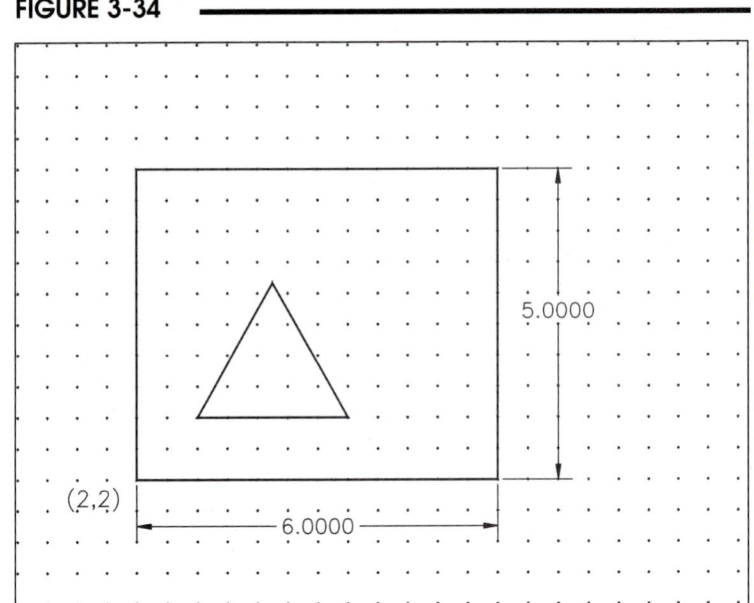

14. The two inside shapes are most easily created using Polar Tracking in combination with Polar Snap. Access the *Grid and Snap* tab of the *Drafting Settings* dialog box. (HINT: Right-click on the word *SNAP* and select *Settings*.) Set the *Snap Type* to *Polar Snap* and set the *Polar Distance* to **.5**. At the Status Bar, make sure that *SNAP, POLAR,* and *DYN* are on. Next, use the *Line* command and create the two shapes as shown in Figure 3-35. Specify the starting positions for each shape (**5.93,3.25** and **6.80,3.75**) by entering absolute coordinates.

FIGURE 3-35

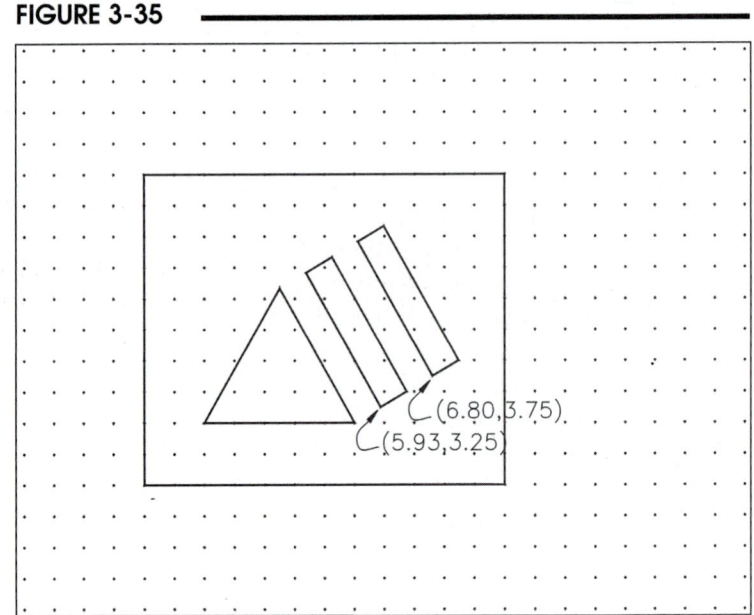

15. When you are finished, *Save* the drawing and *Exit* AutoCAD.

4

SELECTION SETS

CHAPTER OBJECTIVES

After completing this chapter you should:

1. know that *Modify* commands and many other commands require you to select objects;

2. be able to create a selection set using each of the specification methods;

3. be able to *Erase* objects from the drawing;

4. be able to *Move* and *Copy* objects from one location to another;

5. understand Noun/Verb and Verb/Noun order of command syntax.

MODIFY COMMAND CONCEPTS

Draw commands create objects. Modify commands <u>change existing objects</u> or <u>use existing objects to create new ones</u>. Examples are *Copy* an existing *Circle, Move* a *Line,* or *Erase* a *Circle.*

Since all of the Modify commands use or modify <u>existing</u> objects, you must first select the objects that you want to act on. The process of selecting the objects you want to use is called building a <u>selection set</u>. For example, if you want to *Copy, Erase,* or *Move* several objects in the drawing, you must first select the set of objects that you want to act on.

Remember that any of the five command entry methods (depending on your setup) can be used to invoke Modify commands.

1. **Toolbars**   *Modify* or *ModifyII* toolbar.
2. **Pull-down menu**   *Modify* pull-down menu.
3. **Screen menu**   *MODIFY1* or *MODIFY2* screen menus.
4. **Keyboard**   Type the command name <u>or</u> command alias.
5. **Tablet menu**   Select the command icon.

FIGURE 4-1

All of the Modify commands will be discussed in detail later, but for now we will focus on how to build selection sets.

Two *Circles* and five *Lines,* as shown in Figure 4-1, will be used to illustrate the selection process in this chapter. In Figure 4-1, no objects are selected.

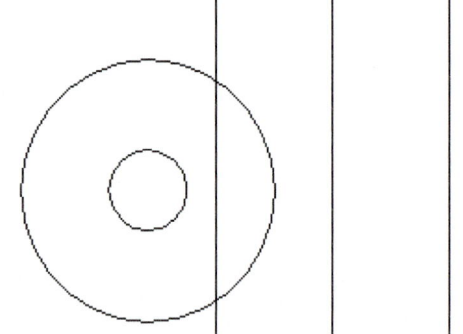

SELECTION SETS

Selection Preview
When you pass the crosshairs or pickbox over an object such as a *Line* or *Circle,* the object becomes thick and dashed (broken) as shown by the vertical *Line* in Figure 4-2. This feature, called selection preview, or sometimes "rollover highlighting," assists you by indicating the objects you are about to select before you actually PICK them. Selection preview may occur both during a modify command at the "Select objects:" prompt (when only the pickbox appears) and also when no commands are in use (when the "crosshairs" appear). This feature, turned on with the default settings, can be turned off or changed if desired.

FIGURE 4-2

Highlighting
<u>After</u> you PICK an object using the pickbox or any of the methods explained next, the objects become "highlighted," or displayed in a dashed linetype, as shown by the large *Circle* in Figure 4-2. Highlighted (selected) objects become part of the <u>selection set</u>.

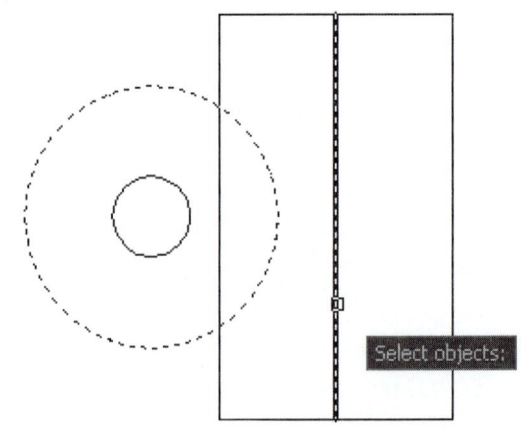

Verb/Noun Selection

There are two times you can select objects. You can select objects before using a command, called Noun/Verb (object/command) order or you can select objects during a Modify command when the "Select objects:" prompt appears, called Verb/Noun (command/object) order. It is recommended that you begin learning the selection options by using the Modify command first (command/object), then select objects during the "Select objects:" prompt. (See "Noun/Verb Syntax" for more information.)

Grips

When you select objects before using a command (Noun/Verb selection), blue squares called Grips may appear on the selected objects. Grips offer advanced editing features that are discussed in Chapter 19. When you begin AutoCAD, Grips are most likely turned on by default. It is recommended to turn off Grips until you learn the basic selection options and Modify commands. Turn Grips off by typing in *Grips* at the Command line and changing the setting to *0* (zero).

Selection

No matter which of the five methods you use to invoke a Modify command, you must specify a selection set during the command operation. For example, examine the command syntax that would appear at the Command line when the *Erase* command is used.

        Command: **erase**
        Select objects:

The "Select objects:" prompt is your cue to use any of several methods to PICK the objects to erase. As a matter of fact, most Modify commands begin with the same "Select objects:" prompt (unless you selected immediately before invoking the command). When the "Select objects:" prompt appears, the "crosshairs" disappear and only a small, square pickbox appears at the cursor.

Use the pickbox to select one object at a time. Once an object is selected, it becomes highlighted. The Command line also verifies the selection by responding with "1 found." You can select as many additional objects as you want, and you can use many selection methods as described on the next few pages. Therefore, object selection is a cumulative process in that selected objects remain highlighted while you pick additional objects. AutoCAD continually responds with the "Select objects:" prompt, so press Enter to indicate when you are finished selecting objects and are ready to proceed with the command.

Selection Set Options

The pickbox is the default option, which can automatically be changed to a window or crossing window by PICKing in an open area (PICKing no objects). Since the pickbox, window, and crossing window (sometimes known as the *AUto* option) are the default selection methods, no action is needed on your part to activate these methods. The other methods can be invoked by typing the capitalized letters shown in the option names following.

All the possible selection set options are listed if you enter a "?" (question mark symbol) at the "Select objects:" prompt.

TIP

        Command: **move**
        Select objects: **?**
        Invalid selection
        Expects a point or
        Window/Last/Crossing/BOX/ALL/Fence/WPolygon/CPolygon/Add/Remove/Multiple/Previous/
          Undo/AUto/Single
        Select objects:

Use any option by typing the indicated uppercase letters and pressing Enter.

pickbox

This default option is used for selecting <u>one object</u> at a time. Locate the pickbox so that an object crosses through it and PICK (Fig. 4-3). You do not have to type or select anything to use this option.

FIGURE 4-3

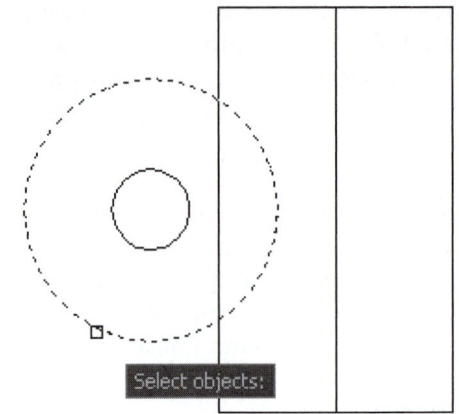

window (*AUto* option)

Only objects <u>completely within</u> the window are selected, not objects that cross through or are outside the window (Fig. 4-4). The window border is a <u>solid linetype</u> rectangular box filled with a transparent blue color.

To use this option, you do not have to type or select anything from a menu. Instead, when the pickbox appears, position it in an open area (so that no objects cross through it); then PICK to start a window. If you <u>drag to the right</u>, a window is created. Next, PICK the other diagonal corner.

FIGURE 4-4

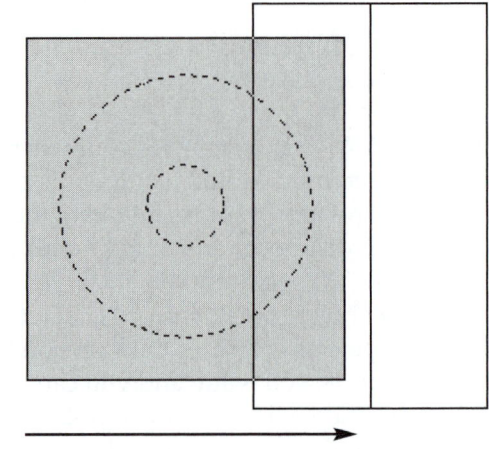

Window

You can also type the letter *W* at the "Select objects:" prompt to create a selection window as shown in Figure 4-4. In this case, no pickbox appears, and you can PICK the diagonal corners of the window by dragging in either direction.

crossing window (*AUto* option)

Note that all objects <u>within and crossing through</u> a crossing window are selected (compare Fig. 4-5 with Fig. 4-4). The crossing window is displayed as a <u>broken linetype</u> rectangular box filled with a transparent green color.

This option is similar to using a window in that when the pickbox appears you can position it in an open area and PICK to start the window. However, <u>drag to the left</u> to form a crossing window. Dragging to the right creates a window instead.

FIGURE 4-5

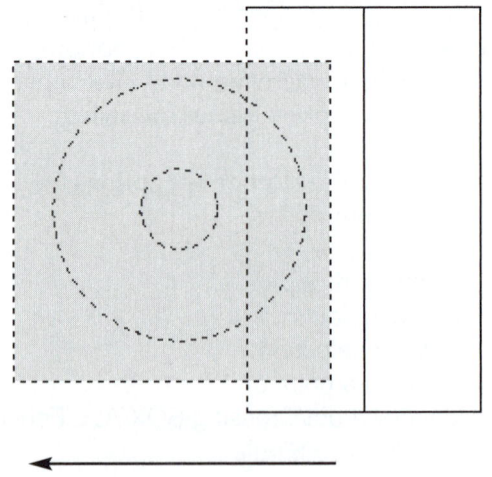

Crossing Window

You can also type the letters *CW* at the "Select objects:" prompt to create a crossing window as shown in Figure 4-5. With this option, no pickbox appears, and you can PICK the diagonal corners by dragging in either direction to form the crossing window.

Window Polygon

Type the letters **WP** at the "Select objects:" prompt to use this option. Note that the *Window Polygon* operates like a *Window* in that only objects <u>completely within</u> the window are selected, but the box can be *any* irregular polygonal shape (Fig. 4-6). You can pick <u>any number of corners to form any shape</u> rather than picking just two diagonal corners to form a regular rectangular *Window*.

FIGURE 4-6

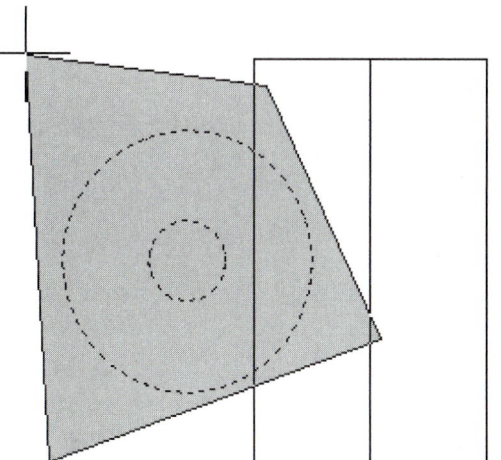

Crossing Polygon

Type the letters **CP** at the "Select objects:" prompt to use this option. The *Crossing Polygon* operates like a *Crossing Window* in that all objects <u>within and crossing through</u> the *Crossing Polygon* are selected; however, the polygon can have any number of corners (Fig. 4-7).

FIGURE 4-7

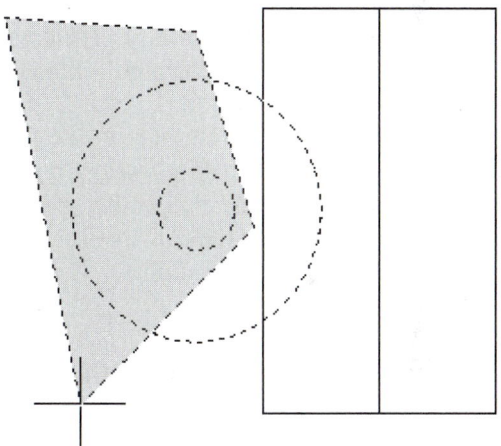

Fence

To invoke this option, type the letter **F** at the "Select objects:" prompt. A *Fence* operates like a <u>crossing line</u>. Any <u>objects crossing</u> the *Fence* are selected. The *Fence* can have any number of segments (Fig. 4-8).

FIGURE 4-8

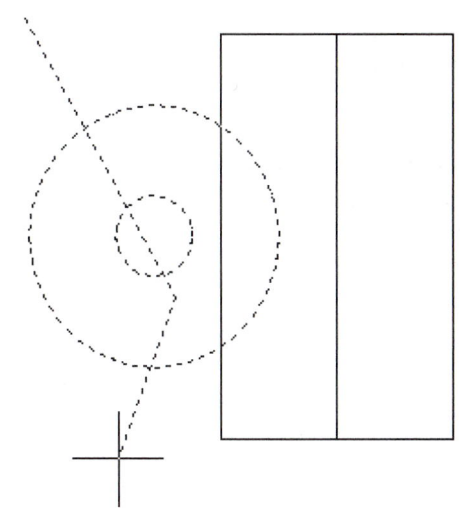

Last

This option automatically finds and selects <u>only</u> the last object <u>created</u>. *Last* does not find the last object modified (with *Move, Stretch,* etc.).

Previous

Previous finds and selects the <u>previous selection set</u>, that is, whatever was selected during the previous command (except after *Erase*). This option allows you to use several editing commands on the same set of objects without having to respecify the set.

ALL

This option selects <u>all objects</u> in the drawing except those on *Frozen* or *Locked* layers (*Layers* are covered in Chapter 11).

BOX

This option is equivalent to the *AUto* window/crossing window option <u>without</u> the pickbox. PICKing diagonal corners from left to right produces a window and PICKing diagonal corners from right to left produces a crossing window (see *AUto*).

Multiple

The *Multiple* option allows selection of objects with the pickbox only; however, the selected objects are not highlighted. Use this method to select very complex objects to save computing time required to change the objects' display to highlighted.

SIngle

This option allows only a <u>single selection</u> using one of the *AUto* methods (pickbox, window, or crossing window), then automatically continues with the command. Therefore, you can select multiple objects (if the window or crossing window is used), but only one selection is allowed. You do not have to press Enter after the selection is made.

Undo

Use *Undo* to cancel the selection of the object(s) most recently added to the selection set.

Remove

Selecting this option causes AutoCAD to <u>switch</u> to the "Remove objects:" mode. Any selection options used from this time on remove objects from the highlighted set.

Add

The *Add* option switches back to the default "Select objects:" mode so additional objects can be added to the selection set.

Shift + Left Button

Holding down the **Shift** key and pressing the left mouse button simultaneously <u>removes</u> objects selected from the highlighted set. This method is generally quicker than, but performs the same action as, *Remove*. The advantage here is that the *Add* mode is in effect unless Shift is held down.

Group

The *Group* option selects groups of objects that were previously specified using the *Group* command. Groups are selection sets to which you can assign a name (see Chapter 20).

Select

Pull-down Menu	Command (Type)	Alias (Type)	Short-cut	Screen (side) Menu	Tablet Menu
...	*Select*	...	...	...	...

The *Select* command can be used to PICK objects to be saved in the selection set buffer for subsequent use with the *Previous* option. Any of the selection methods can be used to PICK the objects.

```
Command: select
Select objects: PICK (use any selection option)
Select objects: Enter (completes the selection process)
Command:
```

The selected objects become unhighlighted when you complete the command by pressing Enter. The objects become highlighted again and are used as the selection set if you use the *Previous* selection option in the <u>next</u> editing command.

NOUN/VERB SYNTAX

An object is the <u>noun</u> and a command is the <u>verb</u>. Noun/Verb syntax order means to pick objects (nouns) first, then use an editing command (verb) second. If you select objects first (at the open Command: prompt) and then immediately choose a Modify command, AutoCAD recognizes the selection set and passes through the "Select objects:" prompt to the next step in the command.

Verb/Noun means to invoke a command and then select objects within the command. For example, if the *Erase* command (verb) is invoked first, AutoCAD then issues the prompt to "Select objects:"; therefore, objects (nouns) are PICKed second. In the previous examples, and with much older versions of AutoCAD, only Verb/Noun syntax order was used.

You can use <u>either</u> order you want (Noun/Verb or Verb/Noun) and AutoCAD automatically understands. If objects are selected first, the selection set is passed to the next editing command used, but if no objects are selected first, the editing command automatically prompts you to "Select objects:".

If you use Noun/Verb order, you are limited to using only the *AUto* options for object selection (pickbox, *Window*, and *Crossing Window*). You can only use the other options (e.g., *Crossing Polygon, Fence, Previous*) if you invoke the desired Modify command first, then select objects when the "Select objects:" prompt appears.

The *PICKFIRST* variable (a very descriptive name) enables Noun/Verb syntax. The default setting is 1 (*On*). If *PICKFIRST* is set to 0 (*Off*), Noun/Verb syntax is disabled and the selection set can be specified only <u>within</u> the editing commands (Verb/Noun).

Setting *PICKFIRST* to 1 provides two options: Noun/Verb and Verb/Noun. You can use <u>either</u> order you want. If objects are selected first, the selection set is passed to the next editing command, but if no objects are selected first, the editing command prompts you to select objects. See Chapter 20 for a complete explanation of *PICKFIRST* and advanced selection set features.

SELECTION SETS PRACTICE

NOTE: While learning and practicing with the editing commands, it is suggested that *GRIPS* be turned off. This can be accomplished by typing in **GRIPS** and setting the *GRIPS* variable to a value of **0**. The AutoCAD default has *GRIPS* on (set to 1). *GRIPS* are covered in Chapter 19.

None

None turns off the *U* and *Undo* commands. If you do use *Undo* while it is turned off, only the *All*, *None*, and *One* options are displayed at the command line.

One

This option limits *Undo* to a single operation, similar to using the *U* command.

Combine

Combine controls whether multiple, consecutive zoom and pan commands are combined as a single operation for undo and redo operations (on by default).

Redo

Pull-down Menu	Command (Type)	Alias (Type)	Short-cut	Screen (side) Menu	Tablet Menu
Edit *Redo*	*Redo*	...	*Crtl+Y* or (Default Menu) *Redo*	*EDIT* *Redo*	*U,12*

The *Redo* command undoes an *Undo*. *Redo* must be used as the <u>next</u> command after the *Undo*. The result of *Redo* is as if *Undo* was never used. *Redo* automatically reverses the action of the previous *Undo*, even when multiple commands were undone. Remember, *U* or *Undo* can be used at any time, but *Redo* has no effect unless used immediately after *U* or *Undo*.

Mredo

Pull-down Menu	Command (Type)	Alias (Type)	Short-cut	Screen (side) Menu	Tablet Menu
...	*Mredo*	...	...	...	...

The *Redo* command reverses the total action of the last *Undo* command. In contrast, *Mredo* (multiple redo) allows you to select how many of the commands, in reverse chronological order, you want to redo. *Mredo* is useful only when the *Undo* command or drop-down list was just used (as the last command) and multiple commands were undone. Then you can immediately use *Mredo* to reverse the action of any number of the previous *Undos*. This feature is available (the icon is enabled) only if *Undo* was just used.

You can achieve a *Mredo* using the *Redo* drop-down list to select from the list of previously undone commands to reverse the action of the *Undo* (Fig. 5-11). Remember, the list includes only commands that were undone during the previous *Undo* command.

FIGURE 5-11

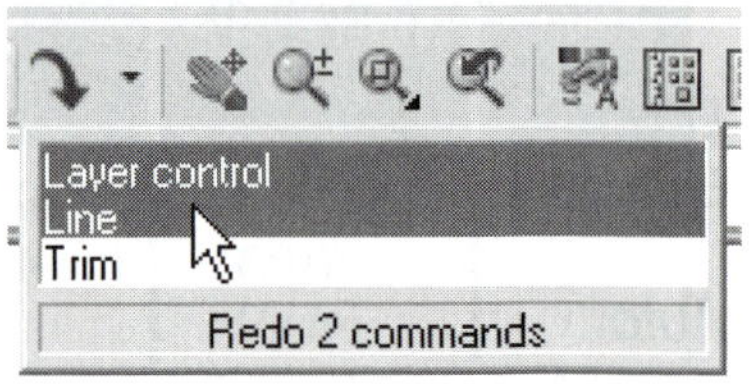

Typing *Mredo* yields the following prompt:

```
Command: mredo
Enter number of actions or [All/Last]:
```

Enter a value for the number of commands you want to redo. *All* will redo all of the undos (all the undos shown in the *Redo* drop-down list), the same as using the *Redo* command. The *Last* option generally has the same result as the *All* option; however, when the *U* command was used multiple times in succession, the *Last* option undoes only the last *U*.

Regen

Pull-down Menu	Command (Type)	Alias (Type)	Short-cut	Screen (side) Menu	Tablet Menu
View *Regen*	*Regen*	*RE*	...	*VIEW 1* *Regen*	*J,1*

The *Regen* command reads the database and redisplays the drawing accordingly. A *Regen* is caused by some commands automatically. Occasionally, the *Regen* command is required to update the drawing to display the latest changes made to some system variables. *Regenall* is used to regenerate all viewports when several viewports are being used.

CHAPTER EXERCISES

Start AutoCAD. If the *Startup* dialog box appears, select **Start from Scratch**, choose **Imperial** as the default setting, and click the **OK** button. Alternately, select the **ACAD.DWT** template. Complete the following exercises.

1. *Help*

   Use **Help** by any method to find information on the following commands. Use the *Contents* tab and select the *Command Reference* to locate information on each command. (Use *Help* at the open Command: prompt, not during a command in use.) Read the text screen for each command.

   New, Open, Save, SaveAs
   Oops, Undo, U

2. **Context-sensitive** *Help*

   Invoke each of the commands listed below. When you see the first prompt in each command, enter **'Help** or **'?** (transparently) or select **Help** from the menus or Standard toolbar. Read the explanation for each prompt. Select a hypertext item in each screen (underlined).

   Line, Arc, Circle, Point

3. *Info Palette*

   If the *Info Palette* is not currently visible on your screen, invoke it by using the *Info Palette* command from the *Help* pull-down menu. Next, use the following commands and take note of how the *Info Palette* contents change to describe each command as you use it. If you need help on these commands, use the information given.

   Line, Circle, Move, and *Copy*

   Next, use the *Auto-hide* function to leave only the *Info Palette* title bar on the screen. Use the following commands.

   Line, Circle

   Bring your cursor to the *Info Palette* title bar (simply hover the pointer over the bar) to make the palette contents appear. The contents should describe the current command or last command you used.

4. *Oops*

   Draw 3 vertical **Lines**. **Erase** one line; then use **Oops** to restore it. Next, **Erase** two *Lines*, each with a separate use of the *Erase* command. Use **Oops**. Only the last *Line* is restored. **Erase** the remaining two *Lines*, but select both with a *Window*. Now use **Oops** to restore both *Lines* (since they were *Erased* at the same time).

5. **Delayed** *Oops*

   Oops can be used at any time, not only immediately after the *Erase*. Draw several horizontal **Lines** near the bottom of the screen. Draw a **Circle** on the **Lines**. Then **Erase** the *Circle*. Use **Move**, select the *Lines* with a *Window*, and displace the *Lines* to another location above. Now use **Oops** to make the *Circle* reappear.

6. *U*

   Press the letter **U** (make sure no other commands are in use). The *Circle* should disappear (*U* undoes the last command—*Oops*). Do this repeatedly to *Undo* one command at a time until the *Circle* and *Lines* are in their original position (when you first created them).

7. *Undo*

   Use the **Undo** command and select the **Back** option. Answer **Yes** to the warning message. This action should *Undo* everything.

   Draw a vertical **Line**. Next, draw a square with four **Line** segments (all drawn in the same *Line* command). Finally, draw a second vertical **Line**. **Erase** the first *Line*.

   Now <u>type</u> **Undo** and enter a value of **3**. You should have only one (the first) *Line* remaining. *Undo* reversed the following three commands:

Erase	The first vertical *Line* was unerased.
Line	The second vertical *Line* was removed.
Line	The four *Lines* comprising the square were removed.

8. *Multiple Undo*

   Draw two *Circles* and two more *Lines* so your drawing shows a total of two *Circles* and three *Lines*. Now use the *Multiple Undo* drop-down list. The list should contain *Line, Line, Circle, Circle, Line*. Highlight the first two lines and the two circles in the list, then left-click. Only one *Line* should remain in your drawing.

9. *Multiple Redo*

   Select the *Multiple Redo* drop-down list. The list should contain *Circle, Circle, Line, Line*. Highlight the first two *Circles* in the list, then left-click. The two *Circles* should reappear in your drawing, now showing a total of two *Circles* and one *Line*.

10. *Exit* AutoCAD and answer *No* to "Save Changes to...?"

6

BASIC
DRAWING SETUP

CHAPTER OBJECTIVES

After completing this chapter you should:

1. know the basic steps for setting up a drawing;

2. know how to use the *Start from Scratch* option and the *Quick Setup* and *Advanced Setup* wizards that appear in the *Startup* and *Create New Drawing* dialog boxes;

3. know how to use the *Select template* dialog box;

4. be able to specify the desired *Units*, *Angles* format, and *Precision* for the drawing;

5. be able to specify the drawing *Limits*;

6. know how to specify the *Snap* and *Grid* increments;

7. understand the basic function and use of a *Layout*.

STEPS FOR BASIC DRAWING SETUP

Assuming the general configuration (dimensions and proportions) of the geometry to be created is known, the following steps are suggested for setting up a drawing:

1. Determine and set the *Units* that are to be used.
2. Determine and set the drawing *Limits;* then *Zoom All.*
3. Set an appropriate *Snap* type and increment.
4. Set an appropriate *Grid* value to be used.

These additional steps for drawing setup are discussed also in Chapter 12, Advanced Drawing Setup.

5. Change the *LTSCALE* value based on the new *Limits.*
6. Create the desired *Layers* and assign appropriate *linetype* and *color* settings.
7. Create desired *Text Styles* (optional).
8. Create desired *Dimension Styles* (optional).
9. Activate a *Layout* tab, set it up for the plot or print device and paper size, and create a viewport (if not already existing).
10. Create or insert a title block and border in the layout.

When you start AutoCAD for the first time or use the *New* command to begin a new drawing, the *Startup, Create New Drawing,* or *Select Template* dialog box appears. These tools make available three options for setting up a new drawing. The three options represent three levels of automation/preparation for drawing setup. The *Start from Scratch* and *Use a Wizard* options are described in this chapter. *Use a Template* and creating template drawings are discussed in Chapter 12, Advanced Drawing Setup.

The *Start from Scratch* option requires you to step through each of the individual commands listed above to set up a drawing to your specifications. The *Use a Wizard* option provides two wizards, the *Quick Setup* and the *Advanced Setup* wizard. These two wizards lead you through the first two steps listed above.

STARTUP OPTIONS

When you start AutoCAD or use the *New* command, you are presented with one of three options, depending on how your system has been configured:

1. the *Startup* or *Create New Drawing* dialog box appears;
2. the *Select Template* dialog box appears; or
3. a new drawing starts with the *Qnew* template specified in the *Options* dialog box.

FIGURE 6-1

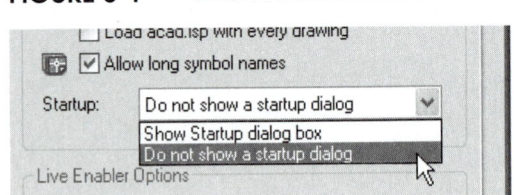

To control which of these actions occurs, make the following settings:

For method 1. Select *Show Startup dialog box* in the *System* tab of the *Options* dialog box (Fig. 6-1).

For method 2. Select *Do not show a startup dialog* in the *System* tab of the *Options* dialog box and ensure the *Drawing Template File Name* is set to "None" in the *Files* tab of the *Options* dialog box.

For method 3. Select *Do not show a startup dialog* in the *System* tab of the *Options* dialog box and set the *Drawing Template File Name for Qnew* in the *Files* tab of the *Options* dialog box to the desired template file.

See "AutoCAD File Commands," "*New,*" and "*Qnew*" in Chapter 2 for more information on these settings.

DRAWING SETUP OPTIONS

There are three general options to set up a new drawing based on which of the three startup options you have configured. They are:

1. *Start from Scratch*;
2. *Use a Template*; and
3. *Use a Wizard*.

All three methods, *Start From Scratch, Use a Template*, and *Use a Wizard*, are available in the *Startup* or *Create New Drawing* dialog boxes (Fig. 6-2). The *Select Template* dialog (Fig. 6-3) is essentially the same as the *Use a Template* option in the *Startup* or *Create New Drawing* dialog box.

FIGURE 6-2

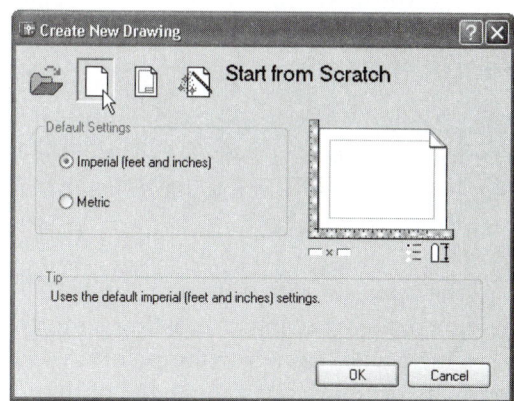

However, no matter which of the three methods you select, you can accomplish essentially the same setup to begin a drawing. Despite all the "bells and whistles" AutoCAD supplies, all of these startup options and drawing setup options generally start the drawing session with one of two setups, either the ACAD.DWT (inch) or the ACADISO.DWT (metric) drawing template. In fact, using *Start from Scratch* or selecting the defaults in either setup *Wizard* from any source, or selecting either of these templates by any method produces the same result—beginning with the ACAD.DWT (inch) or the ACADISO.DWT (metric) drawing template! The template used by the *Wizards* is determined by the *MEASUREINIT* system variable setting (0 for ACAD. DWT, 1 for ACADISO.DWT).

Start from Scratch

The *Start from Scratch* option is available in the *Startup* or *Create New Drawing* dialog box (see Fig. 6-2). Use this option to begin a drawing with basic drawing settings, then determine your own system variable settings using the individual setup commands such as *Units, Limits, Snap,* and *Grid*.

Imperial (feet and inches)
Use the *Imperial (feet and inches)* option if you want to begin with the traditional inch-based AutoCAD default drawing settings, such as *Limits* settings of 12 x 9. This option normally causes AutoCAD to use the ACAD.DWT template drawing. See the "Table of *Start from Scratch, Imperial* Settings (ACAD.DWT)." The same setup can be accomplished by any of the following methods.

Select the *Start from Scratch, Imperial* option in the *Startup* or *Create New Drawing* dialog box.
Select the ACAD.DWT by any *Template* method.
Configure your system for *Do not show a startup dialog*, then set the *Default Template File Name for Qnew* in the *Files* tab of the *Options* dialog box to ACAD.DWT.
Set the *MEASUREINIT* system variable to 0, then cancel the *Startup* dialog box that appears when AutoCAD starts.

Metric
Use the *Metric* option for setting up a drawing for use with metric units. This option causes AutoCAD to use the ACADISO.DWT template drawing. The drawing has *Limits* settings of 420 x 279, equal to a metric A3 sheet measured in mm. See the "Table of *Start from Scratch, Metric* Settings (ACADISO.DWT)." The same setup can be accomplished by any of the following methods.

Select the *Start from Scratch, Metric* option in the *Startup* or *Create New Drawing* dialog box.
Select the ACADISO.DWT by any *Template* method.

Configure your system for *Do not show a startup dialog*, then set the *Default Template File Name for Qnew* in the *Files* tab of the *Options* dialog box to ACADISO.DWT.

Set the *MEASUREINIT* system variable to 1, then cancel the *Startup* dialog box that appears when AutoCAD starts.

Template

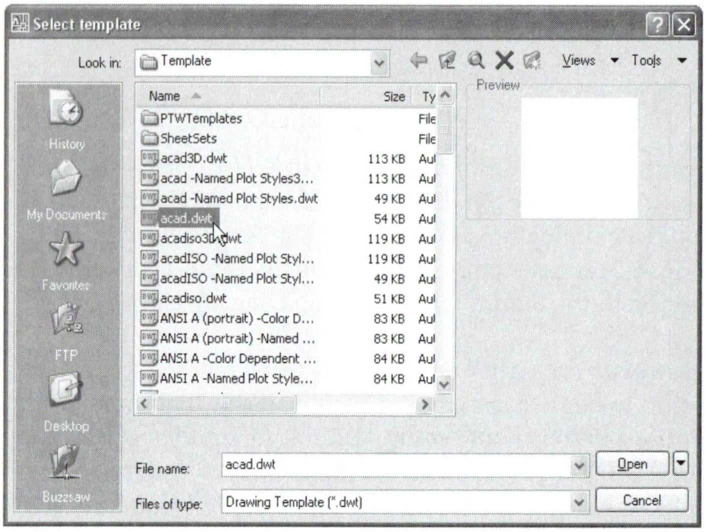

The *Template* option is available in the *Startup* or *Create New Drawing* dialog box (see Fig. 6-2) and from the *Select Template* dialog box (see Fig. 6-3).

Use the *Template* option if you want to begin a drawing using an existing template drawing (.DWT) as a starting point. A template drawing can have many of the drawing setup steps performed but contains no geometry. Several templates are provided by AutoCAD, including the default inch (*Imperial*) template ACAD.DWT and the default metric template ACADISO.DWT. See the "Table of *Start from Scratch, Imperial* Settings (ACAD.DWT)" and the "Table of *Start from Scratch, Metric* Settings (ACADISO.DWT)." More information on this option, including creating template drawings and using templates provided by AutoCAD, is given in Chapter 12, Advanced Drawing Setup.

FIGURE 6-3

NOTE: After selecting the ACAD.DWT template drawing, it is recommended that you perform a *Zoom* with the *All* option. Do this by selecting *Zoom, All* from the *View* pull-down menu or type *Z*, press Enter, then type *A* and Enter. This action ensures that the drawing displays all of the drawing *Limits*.

The following two tables list the AutoCAD settings for the two template drawings, ACAD.DWT and ACADISO.DWT. Many of these settings, such as *Units*, *Limits*, *Snap*, and *Grid*, are explained in the following sections.

Table of *Start from Scratch Imperial* Settings (ACAD.DWT)

Related Command	Description	System Variable	Default Setting
Units	linear units	LUNITS	2 (decimal)
Limits	drawing area	LIMMAX	12.0000,9.0000
Snap	snap increment	SNAPUNIT	.5000, .5000
Grid	grid increment	GRIDUNIT	.5000, .5000
LTSCALE	linetype scale	LTSCALE	1.0000
DIMSCALE	dimension scale	DIMSCALE	1.0000
Text, Mtext	text height	TEXTSIZE	.2000
Hatch	hatch pattern scale	HPSCALE	1.0000

Table of *Start from Scratch Metric* Settings (ACADISO.DWT)

Related Command	Description	System Variable	Default Setting
Units	linear units	LUNITS	2 (decimal)
Limits	drawing area	LIMMAX	420.0000, 297.0000
Snap	snap increment	SNAPUNIT	10.0000, 10.0000
Grid	grid increment	GRIDUNIT	10.0000, 10.0000
LTSCALE	linetype scale	LTSCALE	1.0000
DIMSCALE	dimension scale	DIMSCALE	1.0000
Text, Mtext	text height	TEXTSIZE	2.5000
Hatch	hatch pattern scale	HPSCALE	1.0000

The metric drawing setup is intended to be used with ISO linetypes and ISO hatch patterns, which are pre-scaled for these *Limits,* hence the *LTSCALE* and hatch pattern scale of 1. The individual dimensioning variables for arrow size, dimension text size, gaps, and extensions, etc. are changed so the dimensions are drawn correctly with a *DIMSCALE* of 1 with the ISO-25 dimension style.

Wizard

Selecting the *Use a Wizard* option in the *Startup* or *Create New Drawing* dialog box (see Fig. 6-2) gives a choice of using the *Quick Setup* or *Advanced Setup* wizard. The *Advanced Setup* wizard is an expanded version of the *Quick Setup* wizard.

The wizards use default settings based on either the ACAD.DWT or the ACADISO.DWT template. The template used by the wizards is determined by the *MEASUREINIT* system variable setting for your system (0 = ACAD.DWT, English template, and 1 = ACADISO.DWT, metric template). To change the setting, type *MEASUREINIT* at the Command prompt.

Quick Setup Wizard
The *Quick Setup* wizard automates only the first two steps listed under "Steps for Basic Drawing Setup" on the chapter's first page. Those functions, simply stated, are:

1. *Units*
2. *Limits*

Choosing the *Quick Setup* wizard invokes the *QuickSetup* dialog box (Fig. 6-4). There are two steps, *Units* and *Area.*

FIGURE 6-4

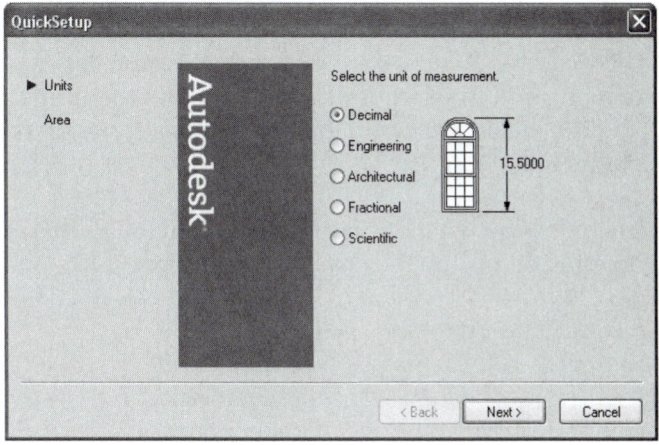

Units

Press the desired radio button to display the units you want to use for the drawing. The options are:

Decimal   Use generic decimal units with a precision 0.0000.
Engineering   Use feet and decimal inches with a precision of 0.0000.
Architectural Use feet and fractional inches with a precision of 1/16 inch.
Fractional   Use generic fractional units with a precision of 1/16 units.
Scientific   Use generic decimal units showing a precision of 0.0000.

Use *Architectural* or *Engineering* units if you want to specify coordinate input using feet values with the apostrophe (') symbol. If you want to set additional parameters for units such as precision or system of angular measurement, use the *Units* or *-Units* command. Keep in mind the setting you select in this step changes only the display of units in the coordinate display area of the Status Bar (*Coords*) and in some dialog boxes, but not necessarily for the dimension text format. Select the *Next* button after specifying *Units*.

Area

Enter two values that constitute the X and Y measurements of the area you want to work in (Fig. 6-5). These values set the *Limits* of the drawing. The first edit box labeled *Width* specifies the X value for *Limits*. The X value is usually the longer of the two measurements for your drawing area and represents the distance across the screen in the X direction or along the long axis of a sheet of paper. The second edit box labeled *Length* specifies the Y value for *Limits* (the Y distance of the drawing area). (Beware: The terms "Width" and "Length" are misleading since "length" is normally the longer of two dimensions. When setting AutoCAD *Limits*, the Y measurement is generally the shorter of the two measurements.) The two values together specify the upper right corner of the drawing *Limits*. When you finish the two steps, you should use *Zoom All* to display the entire *Limits* area in the screen.

FIGURE 6-5

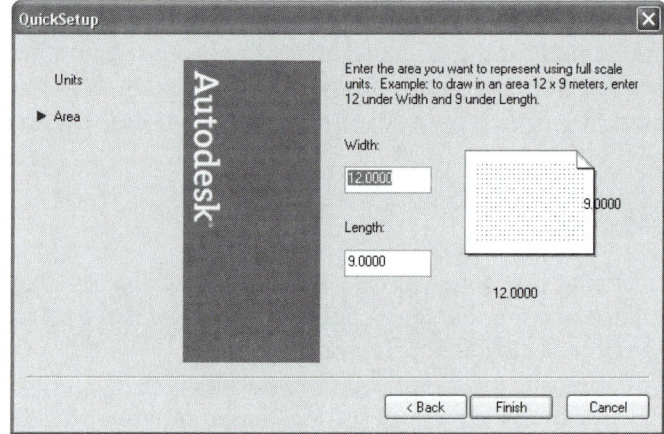

Advanced Setup Wizard

The *Advanced Setup* wizard performs the same tasks as the *Quick Setup* wizard with the addition of allowing you to select units precision and other units options (normally available in the *Units* dialog box). Selecting *Advanced Setup* produces a series of dialog boxes (not all shown). There are five steps involved in the series.

The following list indicates the "steps" in the *Advanced Setup* wizard and the related "Steps for Basic Drawing Setup" on the chapter's first page.

1. *Units* and *Precision*   *Units* command
2. *Angle*   *Units* command
3. *Angle Measure*   *Units* command
4. *Angle Direction*   *Units* command
5. *Area*   *Limits* command

Units

You can select the units of measurement for the drawing as well as the unit's *Precision* in this first step (Fig. 6-6). These are the same options available in the *Drawing Units* dialog box (see "*Units*," this chapter). This is similar to the first step in the *Quick Setup* wizard but with the addition of *Precision*.

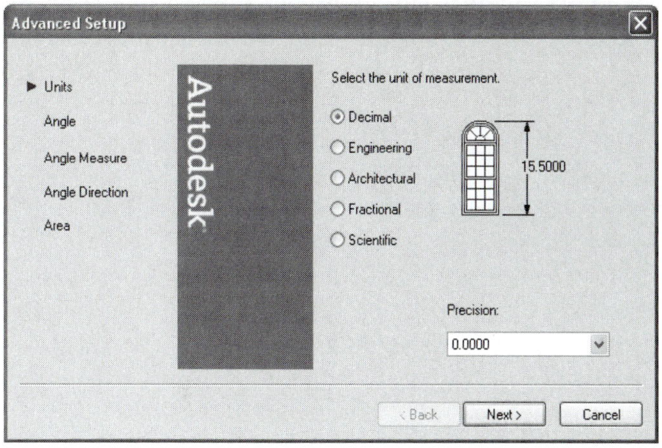

Similar to using the *Units* or *-Units* command, your choices in this and the next three dialog boxes determine the display of units for the coordinate display area of the Status Bar (*Coords*) and in dialog boxes. If you want to use feet units for coordinate input, select *Architectural* or *Engineering*.

FIGURE 6-6

Angle

This step (not shown) provides for your input of the desired system of angular measurement. Select the drop-down list to select the angular *Precision*.

Angle Measure

This step sets the direction for angle 0. East (X positive) is the AutoCAD default. This has the same function as selecting the *Angle 0 Direction* in the *Drawing Units* dialog box.

Angle Direction

Select *Clockwise* if you want to change the AutoCAD default setting for <u>measuring</u> angles. This setting (identical to the *Drawing Units* dialog box option) affects the direction of positive and negative angles in commands such as *Rotate*, *Array Polar*, and dimension commands that measure angles, but does not affect the direction *Arcs* are drawn (always counterclockwise).

Area

Enter values to define the upper right corner for the *Limits* of the drawing. The *Width* refers to the X *Limits* component and the *Length* refers to the Y component. Generally, the *Width* edit box contains the larger of the two values unless you want to set up a vertically oriented drawing area. If you plan to print or plot to a standard scale, your input for *Area* should be based on the intended plot scale and sheet size. See the "Tables of *Limits* Settings" in Chapter 14 for appropriate values to use.

SETUP COMMANDS

If you want to set up a drawing using individual commands instead of the *Quick Setup* or *Advanced Setup* wizard, use the commands given in this section. The *Quick Setup* and *Advanced Setup* wizards use only the first two commands discussed in this section, *Units* and *Limits*.

Units

Pull-down Menu	Command (Type)	Alias (Type)	Short-cut	Screen (side) Menu	Tablet Menu
Format Units...	Units or -Units	UN or -UN	...	FORMAT Units	V,4

The *Units* command allows you to specify the type and precision of linear and angular units as well as the direction and orientation of angles to be used in the drawing. The current setting of *Units* determines the display of values by the coordinates display (*Coords*) and in some dialog boxes.

You can select *Units ...* from the *Format* pull-down or type *Units* (or command alias *UN*) to invoke the *Drawing Units* dialog box (Fig. 6-7). Type *-Units* (or *-UN*) to produce a text screen (Fig. 6-8).

FIGURE 6-7

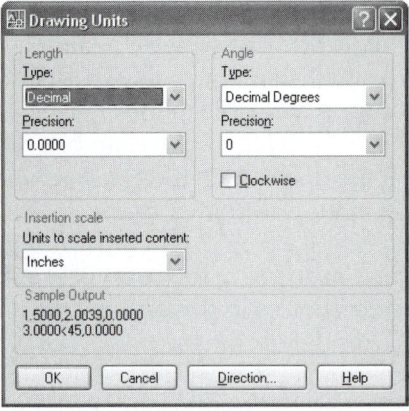

The linear and angular units options are displayed in the dialog box format (Fig. 6-7) and in Command line format (Fig. 6-8). The choices for both linear and angular *Units* are shown in the figures.

FIGURE 6-8

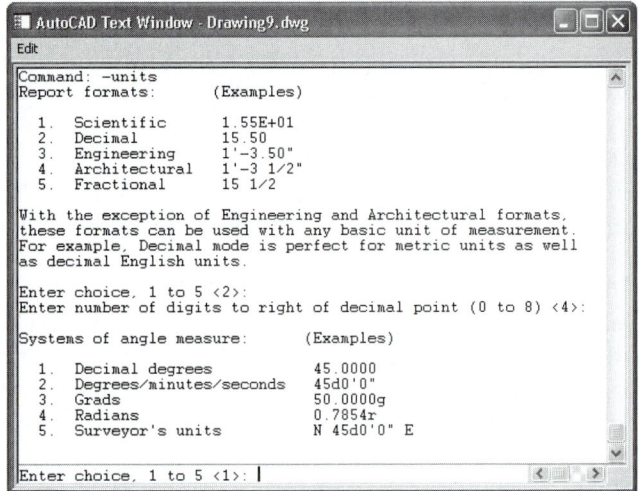

Units	Format	
1. Scientific	1.55E + 01	Generic decimal units with an exponent.
2. Decimal	15.50	Generic decimal usually used for applications in metric or decimal inches.
3. Engineering	1'-3.50"	Explicit feet and decimal inches with notation, one unit equals one inch.
4. Architectural	1'-3 1/2"	Explicit feet and fractional inches with notation, one unit equals one inch.
5. Fractional	15 1/2	Generic fractional units.

Precision

When setting *Units,* you should also set the precision. *Precision* is the number of places to the right of the decimal or the denominator of the smallest fraction to display. The precision is set by making the desired selection from the *Precision* pop-up list in the *Drawing Units* dialog box (Fig. 6-7) or by keying in the desired selection in Command line format.

Precision controls only the <u>display</u> of *COORDS*. The <u>actual precision</u> of the drawing database is always the same in AutoCAD, that is, 14 significant digits.

Angles

You can specify a format other than the default (decimal degrees) for expression of angles. Format options for angular display and examples of each are shown in Figure 6-7 (dialog box format).

The orientation of angle 0 can be changed from the default position (east or X positive) to other options by selecting the *Direction* tile in the *Drawing Units* dialog box. This produces the *Direction Control* dialog box (Fig. 6-9). Alternately, the *-Units* command can be typed to select these options in Command line format.

FIGURE 6-9

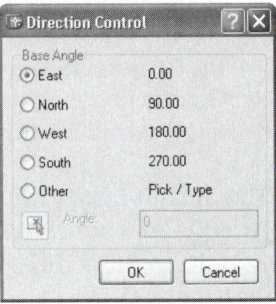

The direction of angular measurement can be changed from its default of counterclockwise to clockwise. The direction of angular measurement affects the direction of positive and negative angles in commands such as *Array Polar, Rotate* and dimension commands that <u>measure</u> angular values but does not change the direction *Arcs* are <u>drawn</u>, which is always counterclockwise.

Insertion Scale

This section in the *Drawing Units* dialog box specifies the units to use when you drag-and-drop *Blocks* from AutoCAD DesignCenter or Tool palettes into the current drawing. There are many choices including *Unitless, Inches, Feet, Millimeters*, and so on. If you intend to insert *Blocks* into the drawing (you are currently setting units for) using drag-and-drop, select a unit from the list that matches the units of the *Blocks* to insert. If you are not sure, select *Unitless*. See Chapter 21, Blocks, DesignCenter, and Tool Palettes for more information on this subject.

Keyboard Input of *Units* Values

When AutoCAD prompts for a point or a distance, you can respond by entering values at the keyboard. The values can be in <u>any format</u>—integer, decimal, fractional, or scientific—regardless of the format of *Units* selected.

You can <u>type in explicit feet and inch values only if *Architectural* or *Engineering*</u> units have been specified as the drawing units. For this reason, specifying *Units* is the first step in setting up a drawing.

Type in explicit feet or inch values by using the apostrophe (') symbol after values representing feet and the quote (") symbol after values representing inches. If no symbol is used, the values are understood by AutoCAD to be <u>inches</u>.

Feet and inches input <u>cannot</u> contain a blank, so a hyphen (-) must be typed between inches and fractions. For example, with *Architectural* units, key in *6'2-1/2"*, which reads "six feet, two and one-half inches." The standard engineering and architectural format for dimensioning, however, places the hyphen between feet and inches (as displayed by the default setting for the *Coords* display).

The *UNITMODE* variable set to **1** changes the display of *Coords* to remind you of the correct format for <u>input</u> of feet and inches (with the hyphen between inches and fractions) rather than displaying the standard format for feet and inch notation (standard format, *UNITMODE* of **0**, is the default setting). If options other than *Architectural* or *Engineering* are used, values are read as generic units.

Limits

Pull-down Menu	Command (Type)	Alias (Type)	Short-cut	Screen (side) Menu	Tablet Menu
Format *Drawing Limits*	*Limits*	...	...	*FORMAT* *Limits*	*V,2*

The *Limits* command allows you to set the size of the drawing area by specifying the lower-left and upper-right corners in X,Y coordinate values.

Command: *limits*
Reset Model space limits
Specify lower left corner or [ON/OFF] <0.0000,0.0000>: **X,Y** or **Enter** (Enter an X,Y value or accept the 0,0 default—normally use 0,0 as lower-left corner.)
Specify upper right corner <12.0000,9.0000>: **X,Y** (Enter new values to change upper-right corner to allow adequate drawing area.)

The default *Limits* values in the ACAD.DWT are 12 and 9; that is, 12 units in the X direction and 9 units in the Y direction (Fig. 6-10). Starting a drawing by the following methods results in *Limits* of 12 x 9:

> Selecting the ACAD.DWT template drawing.
> Selecting the *Imperial* defaults in the *Start from Scratch* option.

FIGURE 6-10

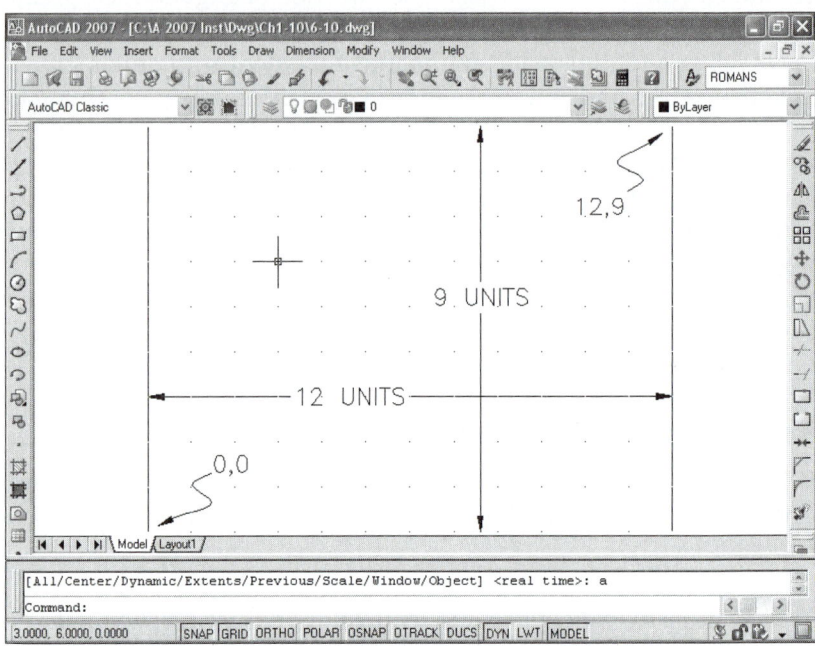

With the default settings for the ACAD.DWT, the *Grid* is displayed only over the *Limits*. The grid display can be controlled using the *Limits* option of *Grid* (see "*Grid*"). The units are generic decimal units that can be used to represent inches, feet, millimeters, miles, or whatever is appropriate for the intended drawing. Typically, however, decimal units are used to represent inches or millimeters. If the default units are used to represent inches, the default drawing size would be 12 x 9 inches.

Remember that when a CAD system is used to create a drawing, the geometry should be drawn <u>full size</u> by specifying dimensions of objects in <u>real-world units</u>. A completed CAD drawing or model is virtually an exact dimensional replica of the actual object. Scaling of the drawing occurs only when plotting or printing the file to an actual fixed-size sheet of paper.

Before beginning to create an AutoCAD drawing, determine the size of the drawing area needed for the intended geometry. After setting *Units,* appropriate *Limits* should be set in order to draw the object or geometry to the <u>real-world size in the actual units</u>. There are no practical maximum or minimum settings for *Limits*.

The X,Y values you enter as *Limits* are understood by AutoCAD as values in the units specified by the *Units* command. For example, if you previously specified *Architectural units,* then the values entered are understood as inches unless the notation for feet (') is given (**240,180** or **20',15'** would define the same coordinate). Remember, you can type in explicit feet and inch values only if *Architectural* or *Engineering* units have been specified as the drawing units.

If you are planning to plot the drawing to scale, *Limits* should be set to a proportion of the <u>sheet size</u> you plan to plot on. For example, setting limits to 22 x 17 (2 times 11 by 8.5) would allow enough room for drawing an object about 20" x 15" and allow plotting at 1/2 size on the 11" x 8.5" sheet. Simply stated, <u>set *Limits* to a proportion of the paper.</u>

Limits also defines the display area when a *Zoom All* is used. *Zoom All* forces the full display of the *Limits*. *Zoom All* can be invoked by typing *Z* (command alias) then *A* for the *All* option.

Changing *Limits* does <u>not</u> automatically change the display. As a general rule, you should make a habit of invoking a *Zoom All* <u>immediately following</u> a change in *Limits* to display the area defined by the new limits (Fig. 6-11).

If you are already experimenting with drawing in different *Linetypes*, a change in *Limits*

FIGURE 6-11

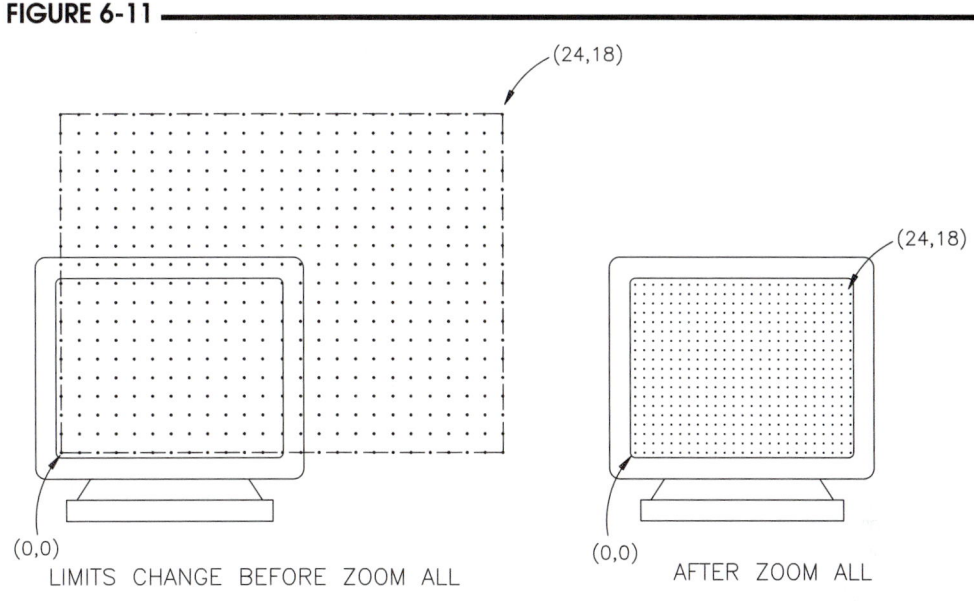

LIMITS CHANGE BEFORE ZOOM ALL

AFTER ZOOM ALL

and *Zoom All* affects the display of the hidden and dashed lines. The *LTSCALE* variable controls the spacing of non-continuous lines. The *LTSCALE* is often changed proportionally with changes in *Limits*.

ON/OFF

If the *ON* option of *Limits* is used, limits checking is activated. Limits checking prevents you from drawing objects outside of the limits by issuing an outside-limits error. This is similar to drawing "off the paper." Limits checking is *OFF* by default.

Snap

Pull-down Menu	Command (Type)	Alias (Type)	Short-cut	Screen (side) Menu	Tablet Menu
Tools *Drafting Settings...* *Snap and Grid*	*Snap*	*SN*	*F9 or* *Ctrl+B*	*TOOLS 2* *Grid* *Snap and Grid*	*W,10*

Snap in AutoCAD has two possible types, <u>*Grid Snap*</u> and <u>*Polar Snap*</u>. When you are setting up the drawing, you should set the desired *Snap type* and increment. You may decide to use both types of Snap, so set both increments initially. You can have only <u>one of the two *Snap types* active at one time</u>. (See Chapter 3, Draw Command Concepts, for further explanation and practice using both Snap types.)

<u>*Grid Snap*</u> forces the cursor to preset positions on the screen, similar to the *Grid*. *Grid Snap* is like an invisible grid that the cursor "snaps" to. This function can be of assistance if you are drawing *Lines* and other objects to set positions on the drawing, such as to every .5 unit. The default value for *Grid Snap* is .5, but it can be changed to any value.

<u>*Polar Snap*</u> forces the cursor to move in set intervals from the previously designated point (such as the "first point" selected during the *Line* command) to the next point. *Polar Snap* operates for cursor movement at any previously set *Polar Tracking* angle, whereas *Grid Snap* (since it is rectangular) forces regular intervals only in horizontal or vertical movements. *Polar Tracking* must also be on (*POLAR* or F10) for *Polar Snap* to operate.

Set the desired snap type and increments using the *Snap* command or the *Drafting Settings* dialog box. In the *Drafting Settings* dialog box, select the *Snap and Grid* tab (Fig. 6-12).

To set the <u>*Polar Snap*</u> increment, first select *Polar Snap* in the *Snap Type* section (lower left). This action causes the *Polar Spacing* section to be enabled. Next, enter the desired *Polar Distance* in the edit box.

To set the <u>*Grid Snap*</u> increment, first select *Grid Snap* in the *Snap Type* section. This action causes the *Snap* section to be enabled. Next, enter the desired *Snap X Spacing* value in the edit box. If you want a non-square snap grid, remove the check by *Equal X and Y spacing*, then enter the desired values in the *Snap X spacing* and *Snap Y spacing* edit boxes.

FIGURE 6-12

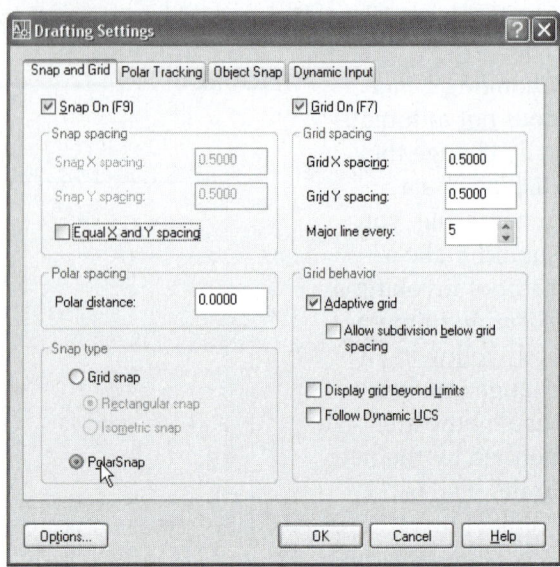

The Command line format of *Snap* is as follows:

      Command: **snap**
      Specify snap spacing or [ON/OFF/Aspect/Style/Type] <0.5000>:

Value
Entering a value at the Command line prompt sets the <u>*Grid Snap* increment only</u>. You must use the *Drafting Settings* dialog box to set the *Polar Snap* distance.

ON/OFF
Selecting *ON* or *OFF* accomplishes the same action as toggling the F9 key or selecting the word *SNAP* on the Status Bar.

Aspect
The *Aspect* option allows specification of unequal X and Y spacing for the snap grid. This is identical to entering unequal *Snap X Spacing* and *Snap Y Spacing* in the dialog box.

Rotate
Although the *Rotate* option is not listed, the *Grid Snap* can be rotated about any point and set to any angle. When Snap has been rotated, the *GRID, ORTHO,* and the "crosshairs" automatically follow this alignment. This action facilitates creating objects oriented at the specified angle, for example, creating an auxiliary view of drawing part or a floor plan at an angle. To do this, use the *Rotate* option in Command line format (see Chapter 27, Auxiliary Views).

Style
The *Style* option allows switching between a *Standard* snap pattern (square or rectangular) and an *Isometric* snap pattern. If using the dialog box, toggle *Isometric* snap. When *Snap Style* or *Rotate* (*Angle*) is changed, the *Grid* automatically aligns with it (see Chapter 25, Pictorial Drawings).

Type
This option switches between *Grid Snap* and *Polar Snap*. Remember, you can have only one of the two snap types active at one time.

Once your snap type and increment(s) are set, you can begin drawing using either snap type or no snap at all. While drawing, you can right-click on the word *SNAP* on the Status Bar to invoke the shortcut menu shown in Figure 6-13. Here you can toggle between *Grid Snap* and *Polar Snap* or turn both off (*POLAR* must also be on to use *Polar Snap*). Left-clicking on the word *SNAP* or pressing F9 toggles on or off whichever snap type is current.

FIGURE 6-13

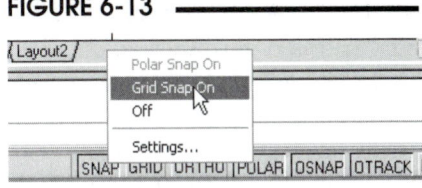

The process of setting the *Snap Type, Grid Snap Spacing, Polar Snap Spacing,* and *Polar Tracking Angle* is usually done during the initial stages of drawing setup, although these settings can be changed at any time. The *Grid Snap Spacing* value is stored in the *SNAPUNIT* system variable and saved in the <u>drawing file</u>. The other settings (*Snap Type, Polar Snap Spacing,* and *Polar Tracking Angle*) are saved in the <u>system registry</u> (as the *SNAPTYPE, POLARDIST,* and *POLARANG* system variables) so that the settings remain in affect for any drawing until changed.

Grid

Pull-down Menu	Command (Type)	Alias (Type)	Short-cut	Screen (side) Menu	Tablet Menu
Tools *Drafting Settings...* *Snap and Grid*	*Grid*	...	*F7 or* *Ctrl+G*	*TOOLS 2* *Grid* *Snap and Grid*	*W,10*

Grid is visible on the screen, whereas *Grid Snap* is invisible. *Grid* is only a <u>visible</u> display of some regular interval. *Grid* and *Grid Snap* can be <u>independent</u> of each other. In other words, each can have separate spacing settings and the active state of each (*ON, OFF*) can be controlled independently. The *Grid* <u>follows</u> the *Snap* if *Snap* is rotated or changed to *Isometric Style*. Although the *Grid* spacing can be different than that of *Snap,* it can also be forced to follow *Snap* by using the *Snap* option. The default *Grid* setting is **0.5.**

The *Grid* <u>cannot</u> be plotted. It is <u>not</u> comprised of *Point* objects and therefore is not part of the current drawing. *Grid* is only a visual aid.

Grid can be accessed by Command line format (shown below) or set via the *Drafting Settings* dialog box (Fig. 6-12). The dialog box can be invoked by menu selection or by typing *Dsettings* or *DS*.

NOTE: When you draw in 3D (normally by beginning with a 3D template drawing, ACAD3D.DWT or ACADISO3D.DWT), <u>lines</u> rather than dots are used to designate the *Grid*.

Command: **grid**
Specify grid spacing(X) or [ON/OFF/Snap/Major/aDaptive/Limits/Follow/Aspect] <0.5000>:

ON/OFF
The *ON* and *OFF* options simply make the *Grid* visible or not (like toggling the F7 key, pressing Ctrl+G, or clicking *GRID* on the Status bar).

Grid Spacing
If you supply a value in Command line format or in the *Grid X spacing* and *Grid Y spacing* edit boxes, the *Grid* is displayed at the specified spacing regardless of the *Snap* spacing. If you key in an *X* as a suffix to the value (for example, 2X), the *Grid* is displayed at that value times the current *Snap* spacing.

Limits (Display grid beyond Limits)

This option determines whether or not the *Grid* is displayed beyond the area set by the *Limits* command. For example, if you chose to display the grid beyond the *Limits*, the grid pattern is displayed to the edge of the drawing area no matter how far you *Zoom* in or out. If you chose not to display the *Grid* beyond the *Limits*, when *Zoomed* out you can "see" the *Limits* by the area filled with the grid pattern. Regardless of this setting, *Zoom All* displays the area defined by the *Limits*.

Adaptive grid

When the grid is <u>not</u> adaptive, dots (or lines used instead of dots for 3D drawings) are displayed at the value specified for *Grid spacing* regardless of how far you *Zoom* in or out. With this setting, at some point the grid becomes too small to display if you *Zoom* out far enough.

If *Adaptive* grid is turned on, the grid adapts to changes in the display due to *Zooming*. For example, if you *Zoom* out, the *Grid* dots may be displayed every 2.5 units instead of every .5 units (the frequency of these grid dots or lines is determined by the setting of the *Major* grid lines). This *Adaptive grid* setting is stored in the *GRIDDISPLAY* system variable.

Allow subdivision below grid spacing

If *Adaptive* grid is turned on, but this option is off, the grid adapts only when you *Zoom* <u>out</u> by increasing the *Grid* spacing. If *Allow subdivision below grid spacing* is turned on, the grid adapts by adding more grid dots (or lines) when you *Zoom* in. For example, if you *Zoom* in, the *Grid* dots may be displayed every .1 units instead of every .5 units. However, keep in mind that even though the grid dots are more closely spaced, the *Snap* setting does not change. Therefore, the closely spaced dots do not improve the *Snap* accuracy.

Major lines

This value specifies the frequency of grid dots (or lines) that appear if *Adaptive grid* is turned on. For example, if *Grid spacing* is set to 1 and the *Major lines* setting is 2, grid dots (or lines) would appear at intervals of 2, 4, 8, and so on, as you continually *Zoom* out. Since *Adaptive grid* requires a *Regen* (regeneration) to display the new settings, a lower *Major lines* value requires *Regens* more often than a higher value. This setting is stored in the *GRIDMAJOR* system variable.

Follow

This option is used only for drawing in 3D. If this option is on, the grid pattern (lines for 3D drawings) automatically follow the XY plane of the Dynamic UCS. This setting is stored in the *GRIDDISPLAY* system variable.

Aspect

The *Aspect* option of *Grid* allows different X and Y spacing (causing a rectangular rather than a square *Grid*).

Dsettings

Pull-down Menu	Command (Type)	Alias (Type)	Short-cut	Screen (side) Menu	Tablet Menu
Tools Drafting Settings...	Dsettings	DS	Status Bar (right-click) Settings...	TOOLS 2 Osnap... (Grid or Polar)	W,10

You can access controls to *Snap* and *Grid* features using the *Dsettings* command. *Dsettings* produces the *Drafting Settings* dialog box described earlier (see Fig. 6-12). The three tabs in the dialog box are *Snap and Grid*, *Polar Tracking*, and *Object Snap*. Use the *Snap and Grid* tab to control settings for *Snap* and *Grid* as previously described. The *Polar Tracking* and *Object Snap* tabs are explained in Chapter 7.

Midpc

Using *Snap* and *Grid*

Using *Snap* and *Grid* for drawing is a personal preference and should be used whenever appropriate for the drawing. Using *Snap* and *Grid* can be beneficial for some drawings, but may not be useful for others.

Generally, *Snap* and *Grid* can be useful in cases where many of the lines to be drawn or other measurements used are at some regular interval, such as 1 mm or 1/2". Typically, small mechanical drawings and some simple architectural drawings may fall into this category. On the other hand, if you anticipate that few of the measurements in the drawing will be at regular interval lengths, *Snap* and *Grid* may be of little value. Drawings such as civil engineering drawings involving site plans would fall into this category.

Neare

Also, *Grid* and *Snap* are useful only in cases where the interval is set to a large enough value relative to the screen size (or *Limits*) to facilitate seeing and PICKing points easily. For example, with *Limits* of 12 x 9 it would be relatively easy to see a *Grid* set to .5 and to PICK points at *Snap* intervals of .125. However, in cases where the *Limits* cover a large area and the desired increments are relatively small, *Snap* may not be of much usefulness. For example, an architectural drawing may have *Limits* that represent hundreds of feet, and although all measurements are to be drawn at 1/4" intervals, it would be almost impossible to interactively PICK points at such a small increment relative to the overall drawing size.

Neares
POLA.
point f
Howe
only) i
struct
angle)

INTRODUCTION TO LAYOUTS AND PRINTING

The last several steps listed in the "Steps for Basic Drawing Setup" on this chapter's first page are also listed below.

Node

5. Change the *LTSCALE* value based on the drawing *Limits*.
6. Create the desired *Layers* and assign appropriate *linetype* and *color* settings.
7. Create desired *Text Styles*.
8. Create desired *Dimension Styles*.
9. Activate a *Layout* tab, set it up for the plot or print device and paper size, and create a viewport (if not already created).
10. Create or insert a title block and border in the layout.

Part of the process of setting the *LTSCALE*, creating *Text*, and creating *Dimension Styles*, (steps 5, 7, and 8) is related to the size of the drawing. Steps 9 and 10 prepare the drawing for making a print or plot. Although steps 9 and 10 are often performed after the drawing is complete and just before making a print or plot, it is wise to consider these steps in the drawing setup process. Because you want hidden line dashes, text, dimensions, hatch patterns, etc. to have the correct size in the finished print or plot, it would be sensible to consider the paper size of the print or plot and the drawing scale before you create text, dimensions, and hatch patterns in the drawing. In this way, you can more accurately set the necessary sizes and system variables before you draw, or as you draw, instead of changing multiple settings upon completion of the drawing.

Perpe

To put it simply, to determine the size for linetypes (*LTSCALE*), text, and dimensions, use the proportion of the drawing to the paper size. In other words, if the size of the drawing area (*Limits*) is 22 x 17 and the paper size you will print on is 11 x 8.5, then the drawing is 2 times the size of the paper. Therefore, set the *LTSCALE* to 2 and create the text and dimensions twice as large as you want them to appear on the print.

This chapter gives an introduction to layouts and creating paper space viewports. Advanced features of drawing setup, layouts, viewports, and plotting are discussed in Chapter 12, Advanced Drawing Setup, Chapter 13, Layouts and Viewports, and Chapter 14, Printing and Plotting.

Quadrant

The *Quadrant* option snaps to the 0, 90, 180, or 270 degree quadrant of a *Circle* (Fig. 7-13). PICK <u>nearest</u> to the desired *Quadrant*.

FIGURE 7-13

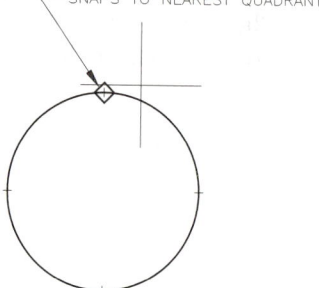

SNAPS TO NEAREST QUADRANT

Tangent

This option calculates and snaps to a tangent point of an *Arc* or *Circle* (Fig. 7-14). PICK the *Arc* or *Circle* as near as possible to the expected *Tangent* point.

FIGURE 7-14

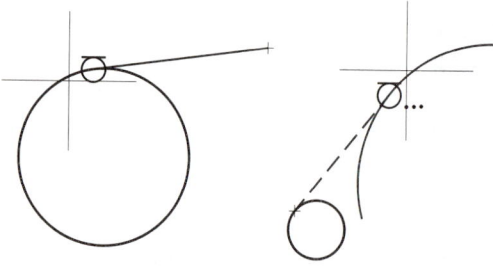

SNAPS TANGENT TO

Apparent intersection

Use this option when you are working with a 3D drawing and want to snap to a point in space where two objects appear to intersect (from your viewpoint) but do not actually physically intersect.

Acquisition Object Snap Modes

Three Object Snap modes require an "acquisition" step: *Parallel, Extension*, and *Temporary Tracking*. When you use these Object Snap modes a dotted line appears, called an <u>alignment vector</u>, that indicates a vector along which the selected point will lie. These modes require an additional step—you must "acquire" (select) a point or object. For example, using the *Parallel* mode, you must "acquire" an object to be parallel to. The process used to "acquire" objects is explained here and is similar to that used for Object Snap Tracking (see "Object Snap Tracking" later in this chapter).

To Acquire an Object

To acquire an object to use for an *Extension* or *Parallel* Object Snap mode, move the cursor over the desired object and pause briefly, <u>but do not pick the object</u>. A small plus sign (+) is displayed when AutoCAD acquires the object (Fig. 7-15). A dotted-line "alignment vector" appears as you move the cursor into a parallel or extension position (Fig. 7-16). <u>You can acquire multiple objects to generate multiple vectors</u>.

To clear an acquired object (in case you decide not to use that object), move the cursor back over the acquisition marker until the plus sign (+) disappears. Acquired points also clear automatically when another command is issued.

FIGURE 7-15

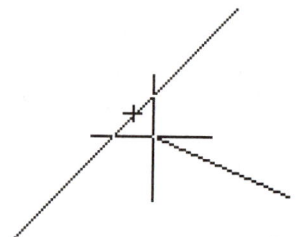

Parallel

The *Parallel* Osnap option snaps the rubberband line into a parallel relationship with any acquired object (see "To Acquire an Object," earlier this section).

Start a *Line* by picking a start point. When AutoCAD prompts to "Specify next point or [Undo]:," acquire an object to use as the parallel source (a line to draw parallel to). Next, move the cursor to within a reasonably parallel position with the acquired line. The current rubberband line snaps into an exact parallel position (Fig. 7-16). A dotted-line parallel alignment vector appears as well as a tool tip indicating the current *Line* length and angle. The parallel Osnap symbol appears on the source parallel line. Pick to specify the current *Line* length. Keep in mind multiple lines can be acquired, giving you several parallel options. The acquired source objects lose their acquisition markers when each *Line* segment is completed.

Consider the use of *Parallel* Osnap with other commands such as *Move* or *Copy*. Figure 7-17 illustrates using *Move* with a *Circle* in a direction *Parallel* to the acquired *Line*.

Extension

The *Extension* Osnap option snaps the rubberband line so that it intersects with an extension of any acquired object (see "To Acquire an Object," earlier this section).

For example, assume you use the *Line* command, then the *Extension* Osnap mode. When another existing object is acquired, the current *Line* segment intersects with an extension of the acquired object. Figure 7-18 depicts drawing a *Line* segment (upper left) to an *Extension* of an acquired *Line* (lower right). Notice the acquisition marker (plus symbol) on the acquired object.

Consider drawing a *Line* to an *Extension* of other objects, such as shown in Figure 7-19. Here a *Line* is drawn to the *Extension* of an *Arc*.

FIGURE 7-16

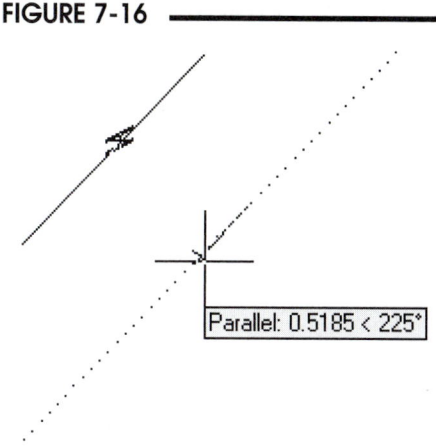

FIGURE 7-17

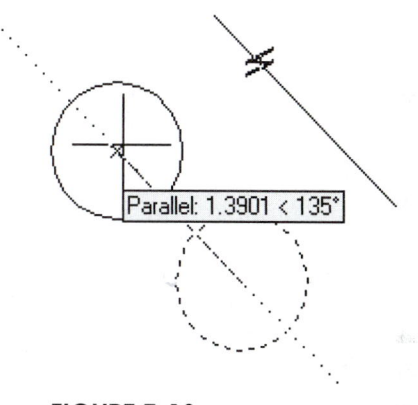

FIGURE 7-18

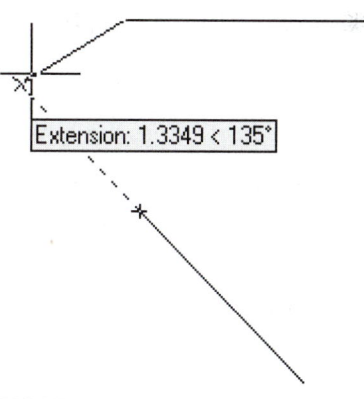

FIGURE 7-19

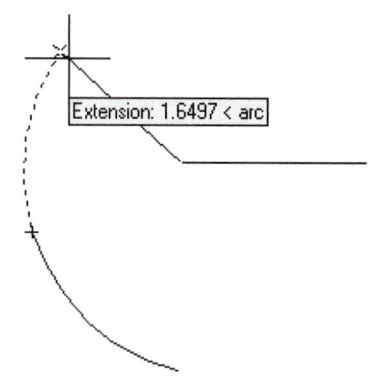

If you set *Extension* as a <u>Running Osnap mode</u>, each newly created *Line* segment automatically becomes acquired; therefore, you can draw an extension of the previous segment (at the same angle) easily (Fig. 7-20). (See "Osnap Running Mode" later in this chapter.)

FIGURE 7-20

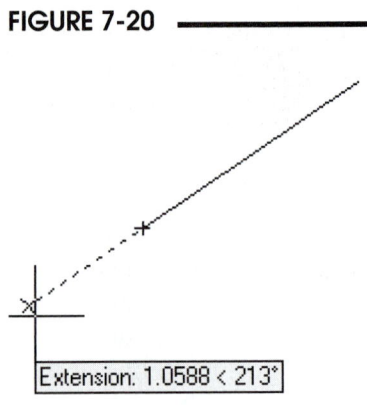

Object Snap Modifiers

Three Object Snap modes can be thought of as Object Snap modifiers—that is, they are intended to be used with other Object Snap options. They are *Temporary Tracking* or *TT* (also requires an acquisition point), *From*, and *Mid Between Two Points* or *M2p*.

Temporary Tracking (TT)

Temporary Tracking sets up a temporary Polar Tracking point. This option allows you to "track" in a polar direction from any point you select. <u>Tracking</u> is the process of moving from a selected point in a preset angular direction. The preset angles are those specified in the *Increment Angle* section in the *Polar Tracking* tab of the *Drafting Settings* dialog box (see "Polar Tracking" in Chapter 3, Fig. 3-17).

For example, if you use the *Line* command, then use the *Temporary Tracking* button or type **TT** at the "Specify first point:" prompt, you can select a point anywhere and that point becomes the temporary tracking point. The cursor moves in a preset angle from that point (Fig. 7-21). The next point selected along the alignment path becomes the *Line's* first point. <u>*POLAR* does not have to be on</u> to use *TT*. Note that the temporary tracking point is indicated by an acquisition marker, so it appears similar to any other acquired point.

FIGURE 7-21

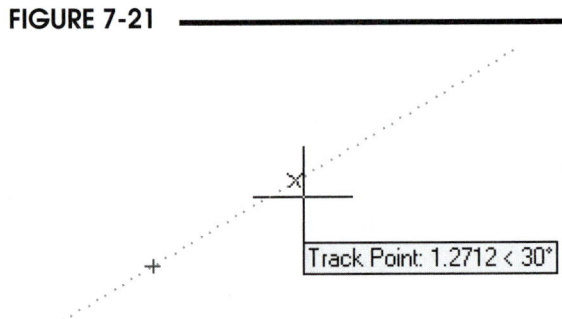

Temporary Tracking should be used <u>in conjunction with another Osnap mode</u> (use *TT*, then another *Osnap* mode to specify the point). As an example, assume you wanted to begin drawing a *Line* directly above the corner of an existing rectangular shape. To do this, use the *Line* command and at the "Specify first point:" prompt invoke *TT*. Then use *Endpoint* to acquire the desired corner of the rectangle. A temporary tracking alignment vector appears from the corner as indicated in Figure 7-22. To specify an exact distance, enter a value (Direct Distance Entry with *DYN* off). The resulting point is the point that satisfies the "Specify first point:" prompt for the *Line*.

FIGURE 7-22

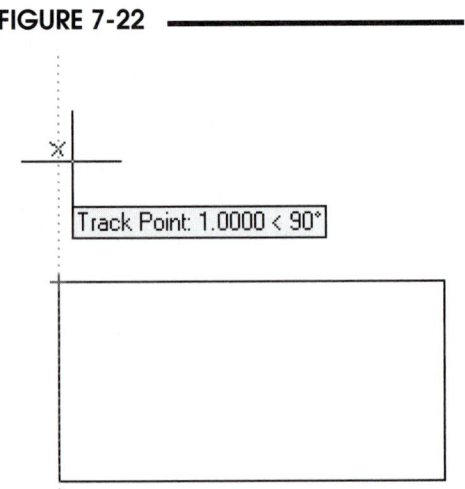

Normally *Temporary Tracking* is used when only one point is required for the operation or possibly for the "first point" of a *Line* or other object. Instead of using *Temporary Tracking* for the "next point," use Object Snap Tracking (see "Object Snap Tracking" later in this chapter).

From

The *From* option is similar to the *Temporary Tracking* mode; however, instead of using a tracking vector to place the desired point, you specify a distance using relative Cartesian or relative polar coordinates. Like *Temporary Tracking*, you use another *Osnap* mode to specify the point to track from, except with the *From* option no tracking vector appears. There are two steps: first, select a "base point" (use another *Osnap* mode to select this point), then specify an "Offset" (enter relative Cartesian or relative polar coordinates).

Like *Temporary Tracking*, the From mode is used for the "first point" or when only one point is required. Use Object Snap Tracking for the "next point" specification rather than *From*.

Mid Between 2 Points (M2p)

M2p finds the midpoint between two points. *M2p* is intended to find a point <u>not necessarily on an object</u>—the point can be anywhere between the two selected points. Other *Osnap* modes should be used to designate the desired two points.

For example, assume you wanted to draw a *Circle* in the middle of a doorway of a floor plan (Fig. 7-23). Begin the *Circle* command, then at the "Specify center point for circle:" prompt, enter "m2p." Select the two points on each side of the doorway as shown. AutoCAD locates the midpoint between and uses it as the center of the *Circle*.

FIGURE 7-23

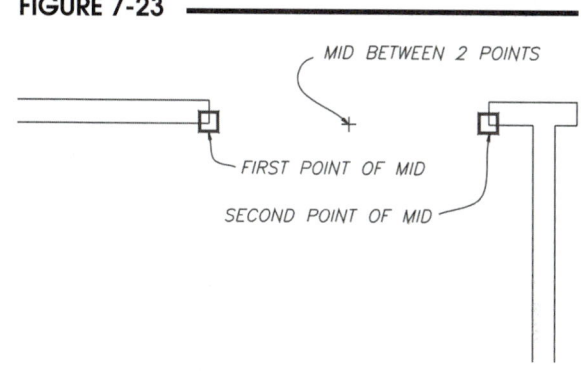

```
Command: circle
Specify center point for circle or [3P/2P/Ttr (tan
   tan radius)]: m2p
First point of mid: PICK
Second point of mid: PICK
Specify radius of circle or [Diameter]:
```

M2p runs only as a single-point mode and should be typed or selected from the *OSNAP* shortcut menu (Shift+right-click). *M2p* is not available as a "running *OSNAP* mode" option (not found in the *Object Snap* tab of the *Drafting Settings* dialog box), but is intended to be used with running *OSNAPs*.

OSNAP SINGLE POINT SELECTION

Object Snaps (Single Point)	Pull-down Menu	Command (Type)	Alias (Type)	Short-cut	Screen (side) Menu	Tablet Menu	Cursor Menu (Shift+button 2)
	...	End, Mid, etc. (first three letters)	...	...	****(asterisks)	T,15 - U,22	Endpoint Midpoint, etc.

There are many methods for invoking *OSNAP* modes for single point selection, as shown in the command table. If you prefer to type, enter only the <u>first three letters</u> of the *OSNAP* mode at the "Specify first point: " prompt, "Specify next point or [Undo]:" prompt, or <u>any time</u> <u>AutoCAD prompts for a point</u>.

If desired, an *Object Snap* toolbar (Fig. 7-24) can be activated to float or dock on the screen by right-clicking on any icon button and selecting *Object Snap* from the list of toolbars.

FIGURE 7-24

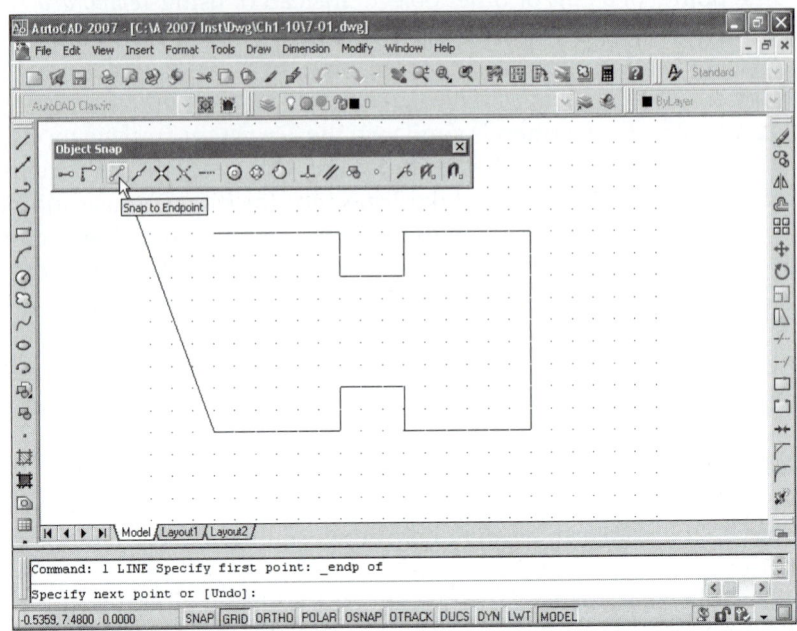

In addition to these options, a special menu called the <u>cursor</u> <u>menu</u> can be used. The cursor menu pops up at the <u>current loca-</u> <u>tion</u> of the cursor and replaces the cursor when invoked (Fig. 7-25). This menu is activated by pressing **Shift+#2** (hold down the Shift key while clicking the right mouse button).

(If you have a wheel mouse, you can also change the *MBUTTON-PAN* system variable to 0. This action allows you to press the wheel to invoke the *Osnap* cursor menu, but disables your ability to *Pan* by holding down the wheel.

FIGURE 7-25

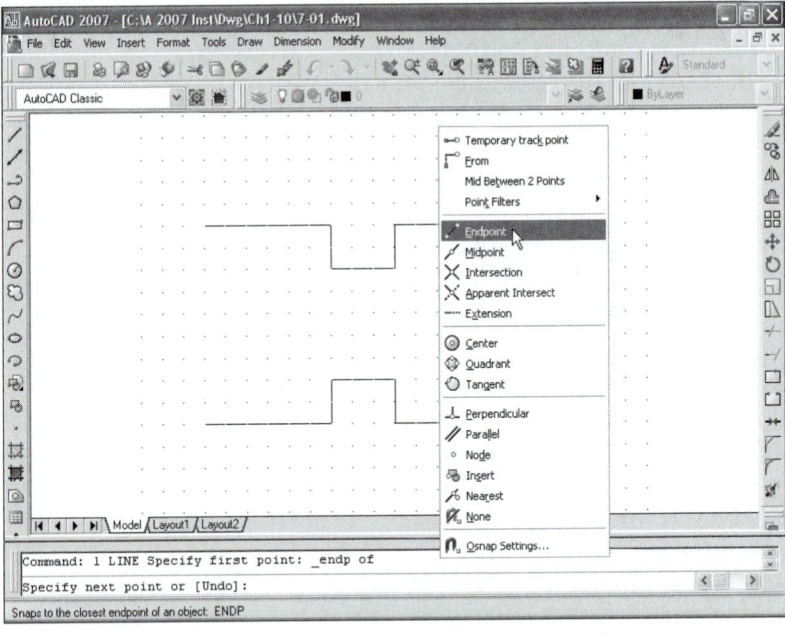

<u>With any of these methods, *OSNAP* modes are active only for selection of a single point</u>. If you want to *OSNAP* to another point, you must select an *OSNAP* mode again for the second point. With this method, the desired *OSNAP* mode is selected <u>transparently</u> (invoked during another command operation) immediately before selecting a point when prompted. In other words, whenever you are prompted for a point (for example, the "Specify first point:" prompt of the *Line* command), select or type an *OSNAP* option. Then PICK near the desired object feature (endpoint, center, etc.) with the cursor. AutoCAD snaps to the feature of the object and uses it for the point specification. Using *OSNAP* in this way allows the *OSNAP* mode to operate only for that <u>single point selection</u>.

For example, when using *OSNAP* during the *Line* command, the Command line reads as shown:

```
Command: _line Specify first point: endp of (PICK)
Specify next point or [Undo]: endp of (PICK)
Specify next point or [Undo]: Enter
Command:
```

Temporary Override Keys for Object Snap

Temporary Override keys allow you to access some AutoCAD drawing aids only while the key sequences are held down. (The keys are called "overrides" because they temporarily override the current settings that appear on the Status Bar without toggling the drawing aid on or off.) For example, the *Endpoint Osnap* mode can be used during a *Line* command if you hold down the "Shift" key and the letter "E" key while specifying the "first point" or "next point." Using one of the Temporary Override Keys shown below is identical to selecting one of these *Osnaps* from the *Object Snap* toolbar or from the Shift+right-click shortcut menu in that the *Osnap* mode is active only for one pick.

Only three *Osnap* modes are programmed as Temporary Override Keys: *Endpoint*, *Midpoint*, and *Center*. For each of these modes two key sequences are available—one on the right side of the keyboard and one on the left. For example, to specify an *Endpoint* for the "next point" of a *Line*, hold down the Shift+E sequence (left side of the keyboard) or the Shift+P sequence (right side of the keyboard). When you then move the cursor near an endpoint, the square *Endpoint* Snap Marker appears so you can pick the endpoint. When you release the keys, the *Endpoint* mode is no longer available.

Left Side	Right Side	Function
Shift+E	Shift+P	*Endpoint*
Shift+V	Shift+M	*Midpoint*
Shift+C	Shift+, (comma)	*Center*

Temporary Overrides can be used only if the *TEMPOVERRIDES* system variable is set to 1. See Appendix D for the complete list of default Temporary Override Keys. See Chapter 45 for information on creating other Temporary Override Keys.

2006

OSNAP RUNNING MODE

A more effective method for using Object Snap is called "Running Object Snap" because one or more *OSNAP* modes (*Endpoint, Center, Midpoint*, etc.) can be turned on and kept running indefinitely. Turn on Running Object Snap by selecting the *OSNAP* button on the Status Bar (Fig. 7-26) or by pressing the F3 key or Ctrl+F key sequence.

FIGURE 7-26

RTHO POLAR OSNAP OTRACK DU(

This method can obviously be more productive because you do not have to continually invoke an *OSNAP* mode each time you need to use one. For example, suppose you have several *Endpoints* to connect. It would be most efficient to turn on the running *Endpoint OSNAP* mode and leave it running during the multiple selections. This is faster than continually selecting the *OSNAP* mode each time before you PICK.

TIP

You normally want <u>several *OSNAP* modes running at the same time</u>. A common practice is to turn on the *Endpoint, Center, Midpoint* modes simultaneously. In that way, if you move your cursor near a *Circle*, the *Center* Marker appears; if you move the cursor near the end or the middle of a *Line*, the *Endpoint*, or the *Midpoint* mode Markers appear.

Object Snaps (Running)

Pull-down Menu	Command (Type)	Alias (Type)	Short-cut	Screen (side) Menu	Tablet Menu	Cursor Menu (Shift+button 2)
Tools *Drafting Settings...* *Object Snap*	*Osnap* or *-Osnap*	*OS* or *-OS*	...	*TOOLS 2* *Osnap...*	*U,22*	*Osnap* *Settings...*

Running Object Snap Modes

All features of Running Object Snaps are controlled by the *Drafting Settings* dialog box (Fig. 7-27). This dialog box can be invoked by the following methods (see the previous Command Table):

1. Type the *OSNAP* command.
2. Type *OS*, the command alias.
3. Select *Drafting Settings...* from the *Tools* pull-down menu.
4. Select *Osnap Settings...* from the bottom of the cursor menu (Shift + right button).
5. Select the *Object Snap Settings* icon button from the *Object Snap* toolbar.
6. Right-click on the words *OSNAP* or *OTRACK* on the Status Bar, then select *Settings...* from the shortcut menu that appears (see Fig. 7-28).

FIGURE 7-27

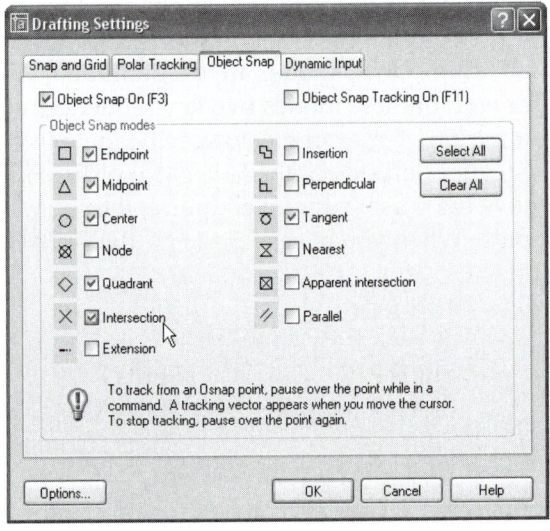

Use the *Object Snap* tab to select the desired Object Snap settings. Try using three or four commonly used modes together, such as *Endpoint, Midpoint, Center,* and *Intersection*. The Snap Markers indicate which one of the modes would be used as you move the cursor near different object features. Using similar modes simultaneously, such as *Center, Quadrant,* and *Tangent,* can sometimes lead to difficulties since it requires the cursor to be placed almost in an exact snap spot, or in some cases it may not find one of the modes (*Quadrant* overrides *Tangent*). In these cases, the Tab key can be used to cycle through the options (see "Object Snap Cycling").

FIGURE 7-28

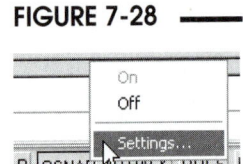

Running Object Snap Toggles

Another feature that makes Running Object Snap effective is Osnap Toggle. If you need to PICK a point without using *OSNAP*, use the Osnap Toggle to temporarily override (turn off) the modes. With Running Osnaps temporarily off, you can PICK any point without AutoCAD forcing your selection to an *Endpoint, Center,* or *Midpoint*, etc. When you toggle Running Object Snap on again, AutoCAD remembers which modes were previously set.

The following methods can be used to toggle Running Osnaps on and off:

1. click the word **OSNAP** that appears on the Status Bar (at the bottom of the screen)
2. press **F3**
3. press **Ctrl+F**

None

 The *None OSNAP* option is a Running Osnap <u>override effective for only one PICK</u>. None is similar to the Running Object Snap Toggle, except it is effective for one PICK, then Running Osnaps automatically come back on without having to use a toggle. If you have *OSNAP* modes running but want to deactivate them for a <u>single</u> PICK, use *None* in response to "Specify first point:" or other point selection prompt. In other words, using *None* <u>during a draw or edit command</u> overrides any Running Osnaps for that single point selection. *None* can be typed at the command prompt (when prompted for a point) or can be selected from the bottom of the Object Snap toolbar.

Object Snap Cycling

In cases when you have multiple Running Osnaps set, and it is difficult to get the desired snap Marker to appear, you can use the <u>Tab</u> key to cycle through the possible *OSNAP* modes for the highlighted object. In other words, pressing the Tab key makes AutoCAD highlight the object nearest the cursor, then cycles through the possible snap Markers (for running modes that are set) that affect the object.

For example, when the *Center* and *Quadrant* modes are both set as Running Osnaps, moving the cursor near a *Circle* makes the *Quadrant* Marker appear but not the *Center* Marker. In this case, pressing the Tab key highlights the *Circle*, then cycles through the four *Quadrant* and one *Center* snap candidates.

Object Snap Options

FIGURE 7-29

You can set other options for using Object Snap in the *Drafting* tab of the *Options* dialog box (Fig. 7-29). Here, more advanced preferences are set such as size and color for the markers and Aperture, and the display of tracking vectors and tool tips. Access the *Options* dialog box by typing *Options*, selecting the *Options* button from the *Drafting Settings* dialog box, or right-clicking in the drawing area while no commands are in progress, then selecting *Options...* from the shortcut menu. The possible settings are explained next.

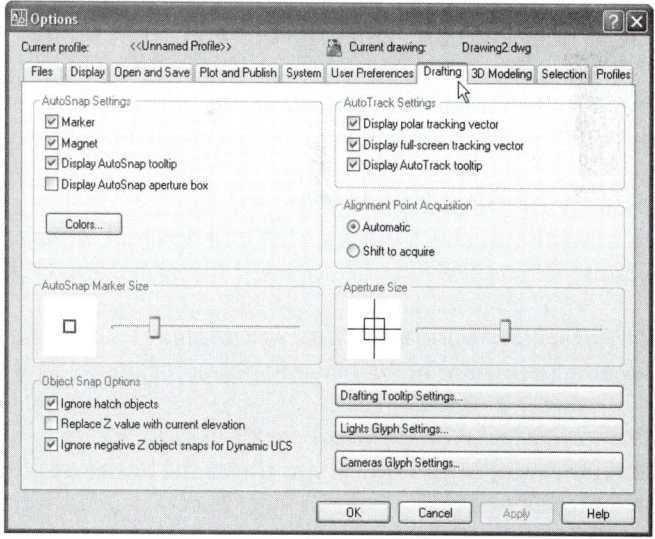

Marker
When this box is checked, a Snap Marker appears when the cursor is moved near an object feature. Each *OSNAP* mode (*Endpoint, Center, Midpoint,* etc.) has a unique Marker. Normally this box should be checked, especially when using Running Object Snap, because the Markers help confirm when and which *OSNAP* mode is in effect.

Magnet
The Magnet feature causes the cursor to "lock" onto the object feature (*Endpoint, Center, Midpoint,* etc.) when the cursor is within the confines of the Marker. The Magnet helps confirm the exact location that will be snapped to for the subsequent PICK.

Display AutoSnap tooltip
The Snap Tip (similar to a Tool Tip) is helpful for beginners because it gives a verbal indication of the *OSNAP* mode in effect (*Endpoint, Center, Midpoint,* etc.) when the cursor is near the object feature. Experienced users may want to turn off the Snap Tips once the Marker symbols are learned.

Display AutoSnap Aperture box
This checkbox controls the visibility of the Aperture. Whether the Aperture is visible or not, its size determines how close the cursor must be to an object before a Marker appears and the displayed *OSNAP* mode takes effect. (See *Aperture size* in the *Running Osnap* dialog box.)

AutoSnap Marker size
This control determines the size of the Markers. Remember that if the Magnet is on, the cursor locks to the object feature when the cursor is within the confines of the Magnet. Therefore, the larger the Marker, the greater the Magnet effective area, and the smaller the Marker, the smaller the Magnet effective area.

Colors
Use this button to select a color for the Markers. You may want to change Marker color if you change the color of the Drawing Editor background. For example, if the background is changed to white, you may want to change the Marker color to dark blue instead of yellow.

Display polar tracking vector
Removing the check from this box disables the display of the alignment vector that appears only when you use Polar Tracking. This option does not affect the display of Object Snap alignment vectors. You can also change the setting of the *TRACKPATH* system variable to 2.

Display full-screen tracking vector
When the check is removed from this box, tracking vectors that appear are displayed from the acquired point to the cursor only, not across the entire screen. You can also change the setting of the *TRACKPATH* system variable to 0 (full-screen vectors) or 1 (short vectors).

Display AutoTracking tooltip
Removing this check disables the tool tips that appear for Polar Tracking, single point selection Object Snaps such as *Temporary Tracking* or *Extension,* and Object Snap Tracking.

Automatic (Alignment Point Acquisition)
Resting the cursor on an existing object for one second automatically acquires that point.

Shift to acquire (Alignment Point Acquisition)
With this option you must hold down the Shift key when resting the cursor on an object to acquire. This is sometimes helpful when you use Object Snap Tracking (discussed later in this chapter) since many unintended points can be automatically acquired.

Aperture size
Notice that the *Aperture Size* can be adjusted through this dialog box (or through the use of the *Aperture* command). Whether the Aperture is visible or not (see Fig. 7-29, *Display AutoSnap aperture box*), its size determines how close the cursor must be to an object before the Marker appears and the displayed *OSNAP* mode takes effect. In other words, the smaller the Aperture, the closer the cursor must be to the object before a Marker appears and the *OSNAP* mode has an effect, and the larger the Aperture, the farther the cursor can be from the object while still having an effect. The Aperture default size is 10 pixels square. The Aperture size influences *Single Point Osnap* selection as well as *Running Osnaps.*

Ignore hatch objects

The default setting is to *Ignore Hatch Objects* for *OSNAPs*. In other words, with this setting you cannot *OSNAP* to a hatch object—the hatch objects are "invisible" to *OSNAP* modes. If this box is unchecked, an *Endpoint* option, for example, would find endpoints for all hatch pattern lines. You can also use the *OSOPTIONS* system variable. (See Chapter 26, Section Views, for information on *Hatch*.)

Replace Z value with current elevation

Normally (with this box unchecked) you can *OSNAP* to any point in 3D space and AutoCAD finds the X, Y, and Z values of the selected point. When this feature is checked and you *OSNAP* to a point in 3D space, AutoCAD uses the selected X and Y coordinates but substitutes the Z value with whatever value you set as the *Elevation*. You could instead use the *OSNAPZ* system variable to turn this feature on and off. (See "*Elevation*" in Chapter 40.)

Ignore Negative Z Object Snaps for Dynamic UCS

Intended for drawing in 3D, this option specifies that object snaps ignore geometry with negative Z values during use of a Dynamic UCS (*OSOPTIONS* system variable).

OBJECT SNAP TRACKING

A feature called Object Snap Tracking helps you draw objects at specific angles and in specific relationships to other objects. Object Snap Tracking works in conjunction with object snaps and displays temporary alignment paths called "tracking vectors" that help you create objects aligned at precise angular positions relative to other objects. You can toggle Object Snap Tracking on and off with the *OTRACK* button on the Status Bar or by toggling F11.

To practice with Object Snap Tracking, try turning the *Extension* and *Parallel* Osnap options off. Since these two new options require point acquisition and generate alignment vectors, it can be difficult to determine which of these features is operating when they are all on.

Object Snap Tracking is similar to Polar Tracking in that it displays and snaps to alignment vectors, but the alignment vectors are generated from <u>other existing objects</u>, not from the current object. These other objects are acquired by Osnapping to them. Once a point (*Endpoint, Midpoint*, etc.) is acquired, alignment vectors generate from them in proximity to the cursor location. This process allows you to construct geometry that has orthogonal or angular relationships to other existing objects.

Using Object Snap Tracking is essentially the same as using the *Temporary Tracking* Osnap option, then using another Osnap mode to acquire the tracking point. (See previous Fig. 7-22 and related explanation to refresh your memory.) Object Snap Tracking can be used with either Single Point or Running Osnap mode.

For example, Figure 7-30 displays an alignment vector generated from an acquired *Endpoint* (top of the left *Line*). The current *Line* can then be drawn to a point that is horizontally aligned with the acquired *Endpoint*. Note that in Figure 7-30 and the related figures following, the Osnap Marker (*Endpoint* marker in this case) is anchored to the acquired point. The alignment vector always rotates about, and passes through, the acquired point (at the *Endpoint* marker). The current *Line* being constructed is at the cursor location.

FIGURE 7-30

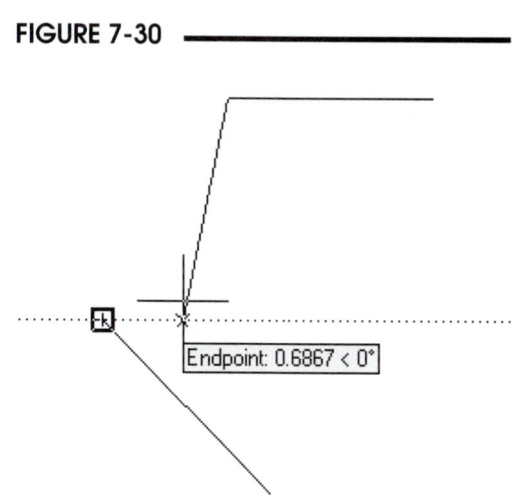

Endpoint: 0.6867 < 0°

Moving the cursor to another location causes a different alignment vector to appear; this one displays vertical alignment with the same *Endpoint* (Fig. 7-31).

FIGURE 7-31

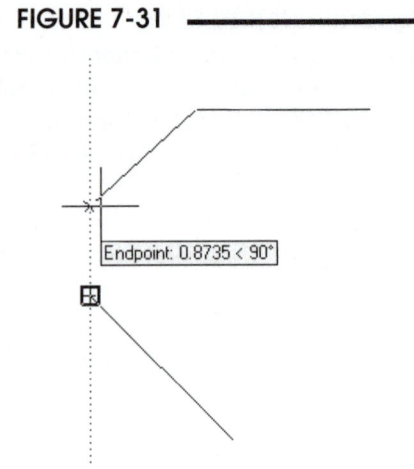

Endpoint: 0.8735 < 90°

In addition, moving the cursor around an acquired point causes an <u>array of alignment vectors to appear based on the current angular increment</u> set in the *Polar Tracking* tab of the *Drafting Settings* dialog box (see Chapter 3, Fig. 3-17, and related discussion). In Figure 7-32, alignment vectors are generated from the acquired point in 30-degree increments.

FIGURE 7-32

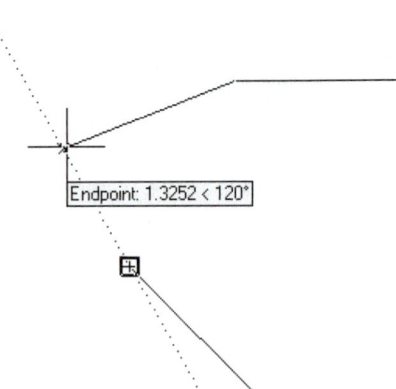

Endpoint: 1.3252 < 120°

To acquire a point to use for Object Snap Tracking (when AutoCAD prompts to specify a point), move the cursor over the object point and pause briefly when the Osnap Marker appears, but <u>do not pick the point</u>. A small plus sign (+) is displayed when AutoCAD acquires the point (Fig. 7-33). The alignment vector appears as you move the cursor away from the acquired point. <u>You can acquire multiple points to generate multiple alignment vectors</u>.

FIGURE 7-33

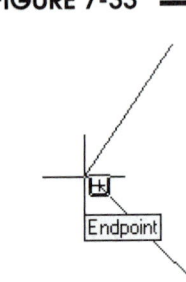

Endpoint

In Figure 7-33, the *Endpoint* object snap is on. Start a *Line* by picking its start point, move the cursor over another line's endpoint to acquire it, and then move the cursor along the horizontal, vertical, or polar alignment vector that appears (not shown in Fig. 7-33) to locate the endpoint you want for the line you are drawing.

To clear an acquired point (in case you decide not to use an alignment vector from that point), move the cursor back over the point's acquisition marker until the plus sign (+) disappears. Acquired points also clear automatically when another command is issued.

To Use Object Snap Tracking:

1. Turn on both Object Snap and Object Snap Tracking (press F3 and F11 or toggle *OSNAP* and *OTRACK* on the Status Bar). You can use *OSNAP* single point selection when prompted to specify a point instead of using Running Osnap.

2. Start a *Draw* or *Modify* command that prompts you to specify a point.

3. Move the cursor over an Object Snap point to temporarily acquire it. <u>Do not PICK the point</u> but only pause over the point briefly to acquire it.

4. Move the cursor away from the acquired point until the desired vertical, horizontal, or polar align-
    ment vector appears, then PICK the desired location for the line along the alignment vector.

Remember, you must set an Object Snap (single or running) before you can track from an object's snap
point. Object Snap and Object Snap Tracking must both be turned on to use Object Snap Tracking. Polar
tracking can also be turned on, but it is not necessary.

Object Snap Tracking with Polar Tracking

For some cases you may want to use Object Snap Tracking in
conjunction with Polar Tracking. This combination allows
you to track from the last point specified (on the current *Line*
or other operation) as well as to connect to an alignment
vector from an existing object.

Figure 7-34 displays both Polar Tracking and Object Snap
Tracking in use. The current *Line* (dashed vertical line) is Polar
Tracking along a vertical vector from the last point specified (on
the horizontal line above). The new *Line's* endpoint falls on the
horizontal alignment vector (Object Snap Tracking) acquired
from the *Endpoint* of the existing diagonal *Line*. Note that the
tool tip displays the Polar Tracking angle ("Polar: <270") and the
Object Snap Tracking mode and angle ("Endpoint: <0").

If the cursor is moved from the previous location (in the
previous figure), additional Polar Tracking options and
Object Snap Tracking options appear. For example, in
Figure 7-35, the current *Line* endpoint is tracking at 210
degrees (Polar Tracking) and falls on a vertical alignment
vector from the *Endpoint* of the diagonal line (Object Snap
Tracking).

FIGURE 7-34

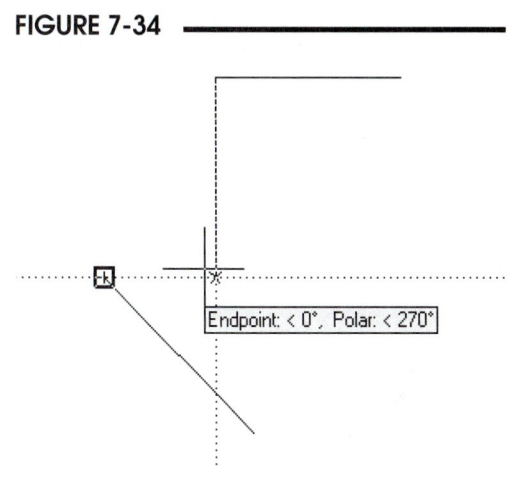

FIGURE 7-35

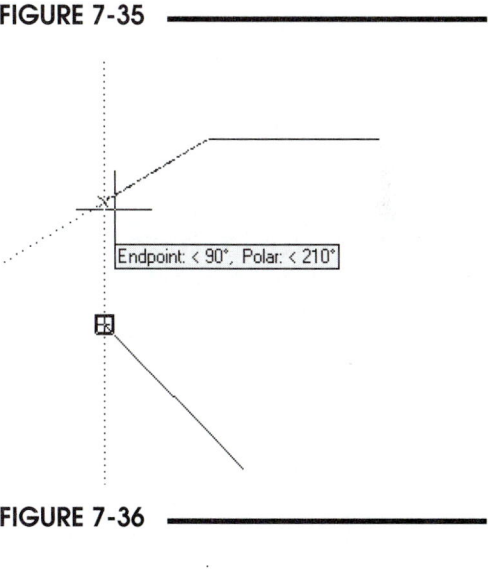

You can accomplish the same capabilities available with the
combination of Polar Tracking and Object Snap Tracking by
using only Object Snap Tracking and multiple acquired
points. For example, Figure 7-36 illustrates the same situation as in Figure 7-34 but only Object Snap tracking is on
(Polar Tracking is off). Notice that two *Endpoints* have been
acquired to generate the desired vertical and horizontal
alignment vectors.

FIGURE 7-36

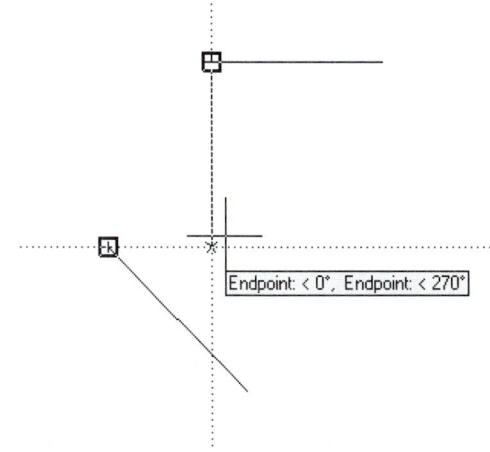

Object Snap Tracking Settings

You can set the *Object Snap Tracking Settings* in the *Drafting Settings* dialog box (Fig. 7-37). The only options are to *Track orthogonally only* or to *Track using all polar angle settings*. The Object Snap Tracking vectors are determined by the *Increment angle* and *Additional angles* set for Polar Tracking (left side of dialog box). If you want to track using these angles, select *Track using all polar angle settings*.

FIGURE 7-37 ──────────

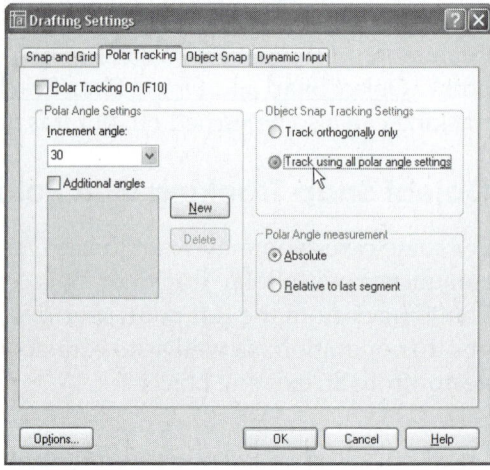

OSNAP APPLICATIONS

OSNAP can be used any time AutoCAD prompts you for a point. This means that you can invoke an *OSNAP* mode during any draw or modify command as well as during many other commands. *OSNAP* provides you with the potential to create 100% accurate drawings with AutoCAD. Take advantage of this feature whenever it will improve your drawing precision. Remember, any time you are prompted for a point, use *OSNAP* if it can improve your accuracy.

OSNAP PRACTICE

Single Point Selection Mode

1. Turn off *SNAP, POLAR, OSNAP, OTRACK, DUCS,* and *DYN*. Draw two (approximately) vertical *Lines*. Follow these steps to draw another *Line* between *Endpoint*s.

STEPS	COMMAND PROMPT	PERFORM ACTION	COMMENTS
1.	Command:	select *Line* by any method	
2.	LINE Specify first point:	type *END* and press **Enter** (or Spacebar)	
3.	endp of	move the cursor near the end of one of the *Lines*	endpoint Marker appears (square box at *Line* end), "Endpoint" Snap Tip may appear
4.		**PICK** while the Marker is visible	rubberband line appears
5.	Specify next point or [Undo]:	type *END* and press **Enter** (or Spacebar)	
6.	endp of	move the cursor near the end of the second *Line*	endpoint Marker appears (square box at *Line* end), "Endpoint" Snap Tip may appear
7.		**PICK** while the Marker is visible	*Line* is created between endpoints
8.	Specify next point or [Undo]:	press **Enter**	completes command

2. Draw two *Circles*. Follow these steps to draw a *Line* between the *Centers*.

STEPS	COMMAND PROMPT	PERFORM ACTION	COMMENTS
1.	Command:	select *Line* by any method	
2.	LINE Specify first point:	invoke the cursor menu (press **Shift+#2** button) and select *Center*	
3.	cen of	move the cursor near a *Circle* object (or where you think the center is)	the Snap Marker appears (small circle at center), "Center" Snap Tip may appear
4.		**PICK** while the Marker is visible	rubberband line appears
5.	Specify next point or [Undo]:	invoke the cursor menu (press **Shift+#2** button) and select *Center*	
6.	cen of	move the cursor near the second *Circle* object (or where you think the center is)	the Snap Marker appears (small circle at center), "Center" Snap Tip may appear
7.		**PICK** the other *Circle* while the Marker is visible	*Line* is created between *Circle* centers
8.	Specify next point or [Undo]:	press **Enter**	completes command

3. *Erase* the *Line* only from the previous exercise. Draw another **Line** anywhere, but <u>not</u> attached to the *Circles*. Follow the steps to *Move* the *Line* endpoint to the *Circle* center.

STEPS	COMMAND PROMPT	PERFORM ACTION	COMMENTS
1.	Command:	select *Move* by any method	
2.	MOVE Select objects:	**PICK** the *Line*	the *Line* becomes highlighted
3.	Select objects:	press **Enter**	completes selection set
4.	Specify base point or displacement:	Select the ***Endpoint*** icon button from the *Object Snap* toolbar	
5.	endp of	move the cursor near a *Line* endpoint	Snap Marker appears (square box), "Endpoint" Snap Tip may appear
6.		**PICK** while Marker is visible	endpoint becomes the "handle" for *Move*
7.	Specify second point of displacement:	select *Center* from *Object Snap* toolbar	
8.	cen of	move the cursor near a *Circle*	Snap Marker appears (small circle), "Center" Snap Tip may appear
9.		**PICK** the *Circle* object while Marker is visible	selected *Line* is moved to *Circle* center

Running Object Snap Mode

4. Draw several *Lines* and *Circles* at random. To draw several *Lines* to *Endpoints* and *Tangent* to the *Circles*, follow these steps.

STEPS	COMMAND PROMPT	PERFORM ACTION	COMMENTS
1.	Command:	turn on *OSNAP* at the Status Bar, then type *OSNAP* or *OS*	*Drafting Settings* dialog box appears
2.		select *Endpoint* and *Tangent* then press *OK*	turns on the Running Osnap modes
3.	Command:	invoke the *Line* command	use any method
4.	LINE Specify first point:	move the cursor <u>near</u> one *Line* endpoint	Snap Marker appears (square box), "Endpoint" Snap Tip may appear
5.		**PICK** while Marker is visible	rubberband line appears, connected to endpoint
6.	Specify next point or [Undo]:	move cursor near endpoint of another *Line*	Snap Marker appears (square box), "Endpoint" Snap Tip may appear
7.		**PICK** while Marker is visible	a *Line* is created between endpoints
8.	Specify next point or [Undo]:	move the cursor near a *Circle* object	Snap Marker appears (small circle with tangent line segment), "Tangent" Snap Tip may appear
9.		**PICK** *Circle* while Marker is visible	a *Line* is created *Tangent* to the *Circle*
10.	Specify next point or [Undo]:	press **Enter**	ends *Line* command
11.	Command:	invoke the *Line* command	use any method
12.	LINE Specify first point:	move cursor near a *Circle*	Snap Marker appears (small circle with tangent line segment), "Deferred Tangent" Snap Tip may appear
13.		**PICK** *Circle* while Marker is visible	rubberband line appears
14.	Specify next point or [Undo]:	move the cursor near a second *Circle* object	Snap Marker appears (small circle with tangent line segment), "Deferred Tangent" Snap Tip may appear
15.		**PICK** *Circle* while Marker is visible	a *Line* is created tangent to the two *Circles*
16.	Specify next point or [Undo]:	click the word *OSNAP* on the Status Bar, press **F3** or**Ctrl+F**	temporarily toggles Running Osnaps off

(Continued)

17.		PICK a point near a *Line* end	*Line* is created, but does not snap to *Line* endpoint
18.	Specify next point or [Undo]:	click the word *OSNAP* on the Status Bar, press **F3** or **Ctrl+F**	toggles Running Osnaps back on
19.		PICK near a *Circle* (when Marker is visible	*Line* is created that snaps tangent to *Circle*
20.	Specify next point or [Undo]:	**Enter**	ends *Line* command
21.	Command:	invoke the cursor menu (press **Shift+#2**) and select *Osnap Settings*…	*Osnap Settings* dialog box appears
22.		Press the *Clear all* button, then *OK*	running *OSNAPS* are turned off

Object Snap Tracking

5. Begin a *New* drawing. Create an "L" shape by drawing one vertical and one horizontal *Line*, each 5 units long.   To draw several *Lines* that track from *Endpoints* and *Midpoints*, follow these steps.

STEPS	COMMAND PROMPT	PERFORM ACTION	COMMENTS
1.	Command:	type *OSNAP* or *OS*	*Drafting Settings* dialog box appears, *Object Snap* tab
2.		select *Endpoint* and *Midpoint* options	turns on the Running Osnap modes
3.		select *Polar Tracking* tab and set *Increment Angle* to **45**, then select *OK*	sets tracking angle for Object Snap Tracking
4.		toggle <u>on</u> *OSNAP* and *OTRACK* and toggle <u>off</u> *SNAP*, *ORTHO*, and *POLAR*	select the words on the Status Bar (recessed is on and protruding is off)
5.	Command:	invoke the *Line* command	
6.	LINE Specify first point:	move the cursor to the right horizontal *Line* endpoint and rest the cursor until the point is acquired	Snap Marker appears (square box) and acquisition marker (plus) appears
7.		PICK while Marker is visible	rubberband line appears connected to endpoint and tracking vectors appear at 45 and 90 positions
8.	Specify next point	move the cursor to endpoint of vertical *Line* and rest the cursor until the point is acquired but <u>do</u> not PICK the point	Snap Marker appears (square box) and acquisition marker (plus) appears

(Continued)

9.		move the cursor so two 90-degree alignment vectors appear (forming a square), then **PICK**	*Line* is drawn from first endpoint at 90-degree alignment to both endpoints
10.	Specify next point:	move cursor near top endpoint of vertical *Line* and PICK the point	*Line* is drawn from last endpoint at 90-degree alignment to endpoint of vertical *Line* to form a square
11.	Specify next point:	press **Enter**	ends *Line* command
12.	Command:	invoke the **Line** command	use any method
13.	LINE Specify next point:	move the cursor to the vertical *Line* endpoint (top) and rest the cursor until the point is acquired	Snap Marker appears (square box) and acquisition marker (plus) appears
14.		**PICK** *Line* while Marker is visible	rubberband line appears connected to endpoint and tracking vectors appear at 45 and 90 positions
15.	Specify next point:	move cursor to midpoint of original (bottom) horizontal *Line* and rest the cursor until the point is acquired but <u>do not</u> PICK the point	Snap Marker appears (triangle) and acquisition marker (plus) appears
16.		move the cursor to the square's center so vertical alignment vector from horizontal line and 45-degree vector from vertical line appears, then **PICK**	diagonal *Line* is drawn above midpoint and aligned (45 degrees) to vertical line endpoint (to center of square)
17.	Specify next point:	press **Enter**	ends *Line* command

CHAPTER EXERCISES

1. *OSNAP* **Single Point Selection**

   Open the CH6EX1A drawing and begin constructing the sheet metal part. Each unit in the drawing represents one inch.

   A. Create four *Circles*. All *Circles* have a radius of **1.685**. The *Circles'* centers are located at **5,5**, **5,13**, **19,5**, and **19,13**.

   B. Draw four *Lines*. The *Lines* (highlighted in Fig. 7-38) should be drawn on the outside of the *Circles* by using the **Quadrant** *OSNAP* mode as shown for each *Line* endpoint.

FIGURE 7-38

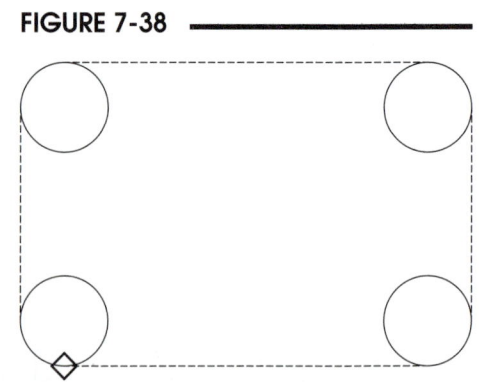

C. Draw two *Lines* from the *Center* of the existing *Circles* to form two diagonals as shown in Figure 7-39.

D. At the *Intersection* of the diagonals create a *Circle* with a **3** unit radius.

FIGURE 7-39

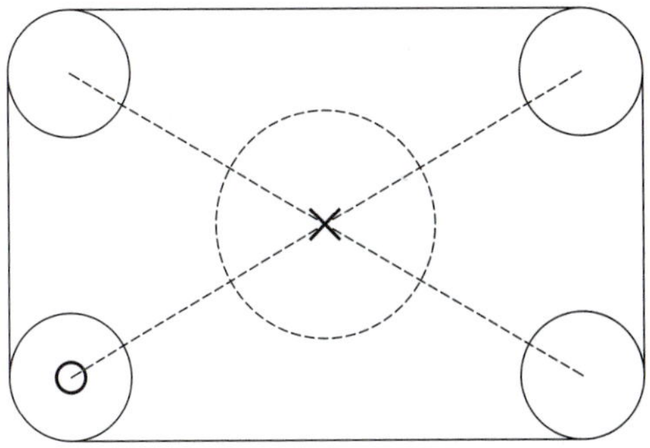

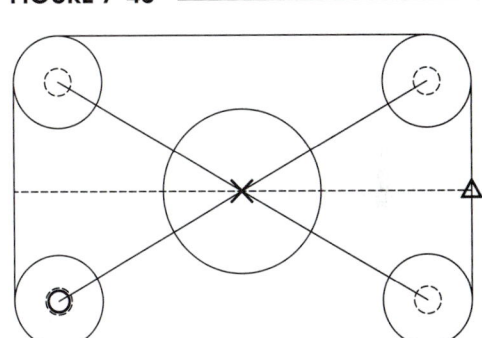

E. Draw two *Lines*, each from the *Intersection* of the diagonals to the *Midpoint* of the vertical *Lines* on each side. Finally, construct four new *Circles* with a radius of **.25,** each at the *Center* of the existing ones (Fig. 7-40).

F. *SaveAs* **CH7EX1.**

FIGURE 7-40

2. *OSNAP* **Single Point Selection**

A multiview drawing of a mechanical part is to be constructed using the **A-METRIC** drawing. All dimensions are in millimeters, so each unit equals one millimeter.

A. *Open* **A-METRIC** from Chapter 6 Exercises. Draw a *Line* from **60,140** to **140,140**. Create two *Circles* with the centers at the *Endpoints* of the *Line*, one *Circle* having a <u>diameter</u> of **60** and the second *Circle* having a diameter of **30**. Draw two *Lines Tangent* to the *Circles* as shown in Figure 7-41. *SaveAs* **PIVOTARM CH7.**

FIGURE 7-41

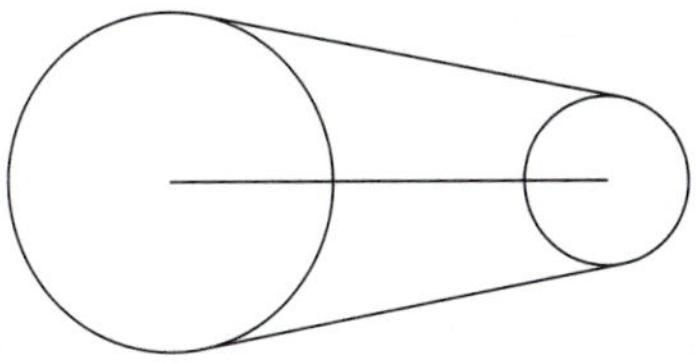

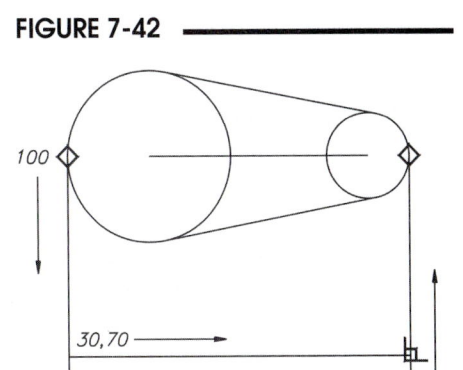

B. Draw a vertical *Line* down from the far left *Quadrant* of the *Circle* on the left. Specify relative polar coordinates using any input method to make the *Line* 100 units at 270 degrees. Draw a horizontal *Line* **125** units from the last *Endpoint* using relative polar or direct distance entry coordinates. Draw another *Line* between that *Endpoint* and the *Quadrant* of the *Circle* on the right. Finally, draw a horizontal *Line* from point **30,70** and *Perpendicular* to the vertical *Line* on the right (Fig. 7-42).

FIGURE 7-42

C. Draw two vertical *Lines* from the
    Intersections of the horizontal *Line*
    and *Circles* and **Perpendicular** to the
    Line at the bottom. Next, draw two
    Circles concentric to the previous
    two and with diameters of **20** and **10**
    as shown in Figure 7-43.

FIGURE 7-43

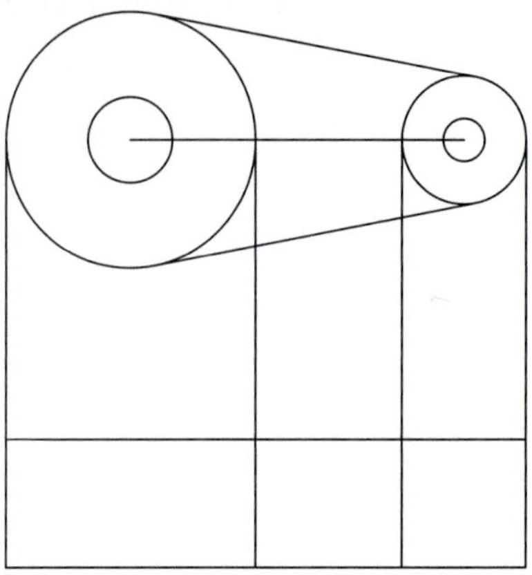

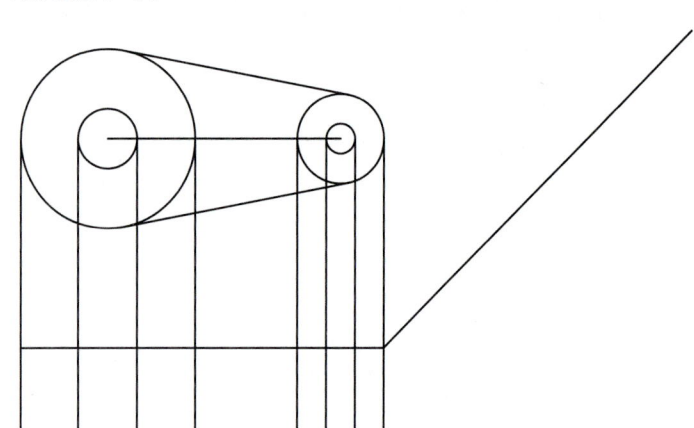

D. Draw four more vertical *Lines* as
    shown in Figure 7-44. Each *Line* is
    drawn from the new *Circles'*
    Quadrant and **Perpendicular** to the
    bottom line. Next, draw a miter
    Line from the **Intersection** of the
    corner shown to **@150<45**. **Save** the
    drawing for completion at a later
    time as another chapter exercise.

FIGURE 7-44

3. **Running Osnap**

    Create a cross-sectional view of a door header composed of two 2 x 6
    wooden boards and a piece of 1/2″ plywood. (The dimensions of a
    2 x 6 are actually 1-1/2″ x 5-3/8″.)

    A. Begin a *New* drawing and assign the name **HEADER**. Draw four
        vertical lines as shown in Figure 7-45.

    B. Use the *OSNAP* command or select **Drafting Settings...** from the
        Tools pull-down menu and turn on the *Endpoint* and
        Intersection modes.

FIGURE 7-45

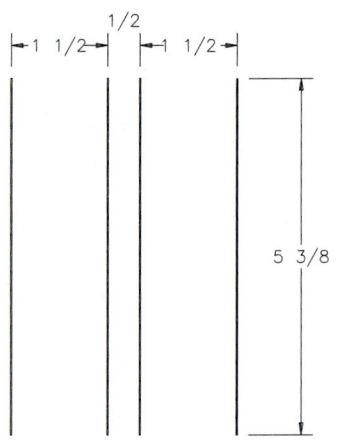

C. Draw the remaining lines as shown in Figure 7-46 to complete the header cross-section. *Save* the drawing.

FIGURE 7-46 ——

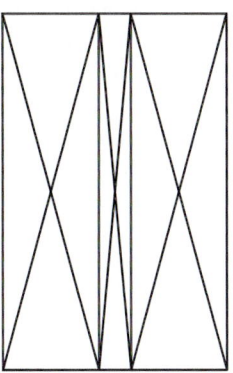

4. **Running Object Snap and *OSNAP* toggle**

Assume you are commissioned by the local parks and recreation department to provide a layout drawing for a major league sized baseball field. Lay out the location of bases, infield, and out-field as follows.

A. *Open* the ███ BALL FIELD CH6.DWG ███ that you set up in Chapter 6. Make sure *Limits* are set to 512',384', *Snap* is set to 10', and *Grid* is set to 20'. Ensure *SNAP* and *GRID* are on. Use *SaveAs* to save and rename the drawing to **BALL FIELD CH7**.

B. Begin drawing the baseball diamond by using the *Line* command and using Dynamic Input. **PICK** the "Specify first point:" at **20',20'** (watch *Coords*). Draw the foul line to first base by turning on *POLAR*, move the cursor to the right (along the X direction) and enter a value of **90'** (don't forget the apostrophe to indicate feet). At the "Specify next point or [Undo]:" prompt, continue by drawing a vertical *Line* of **90'**. Continue drawing a square with **90'** between bases (Fig. 7-47).

FIGURE 7-47 ——————

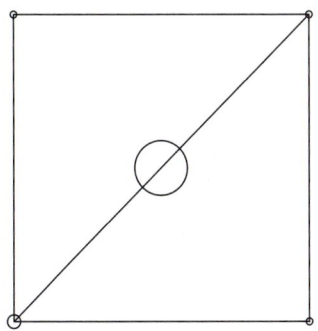

C. Invoke the *Drafting Settings* dialog box by any method. Turn on the *Endpoint, Midpoint,* and *Center* object snaps.

D. Use the *Circle* command to create a "base" at the lower-right corner of the square. At the "Specify center point for circle or [3P/2P/Ttr]:" prompt, **PICK** the *Line Endpoint* at the lower-right corner of the square. Enter a *Radius* of **1'** (don't forget the apostrophe). Since Running Object Snaps are on, AutoCAD should display the Marker (square box) at each of the *Line Endpoints* as you move the cursor near; therefore, you can easily "snap" the center of the bases (*Circles*) to the corners of the square. Draw *Circles* of the same *Radius* at second base (upper-right corner), and third base (upper-left corner of the square). Create home plate with a *Circle* of a **2'** *Radius* by the same method (see Fig. 7-47).

E. Draw the pitcher's mound by first drawing a *Line* between home plate and second base. **PICK** the "Specify first point:" at home plate (*Center* or *Endpoint*), then at the "Specify next point or [Undo]:" prompt, PICK the *Center* or the *Endpoint* at second base. Construct a *Circle* of **8'** *Radius* at the *Midpoint* of the newly constructed diagonal line to represent the pitcher's mound (see Fig. 7-47).

F. *Erase* the diagonal *Line* between home plate and second base. Draw the foul lines from first and third base to the outfield. For the first base foul line, construct a *Line* with the "Specify first point:" at the *Endpoint* of the existing first base line or *Center* of the base. Move the cursor (with *POLAR* on) to the right (X positive) and enter a value of **240'** (don't forget the apostrophe). Press **Enter** to complete the *Line* command. Draw the third base foul line at the same length (in the positive Y direction) by the same method (see Fig. 7-48).

FIGURE 7-48 ————

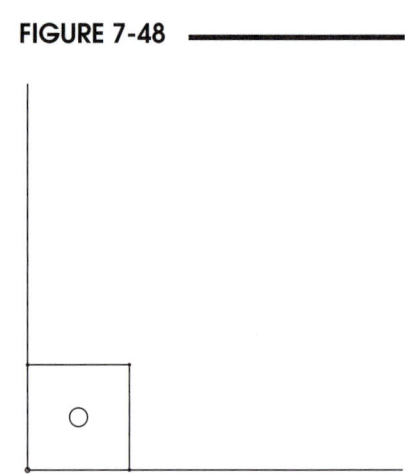

G. Draw the home run fence by using the *Arc* command from the *Draw* pull-down menu. Select the *Start, Center, End* method. **PICK** the end of the first base line (*Endpoint* object snap) for the *Start* of the *Arc* (Fig. 7-49, point 1), **PICK** the pitcher's mound (*Center* object snap) for the *Center* of the *Arc* (point 2), and the end of the third base line (*Endpoint* object snap) for the *End* of the *Arc* (point 3). Don't worry about *POLAR* in this case because *OSNAP* overrides *POLAR*.

FIGURE 7-49 ————

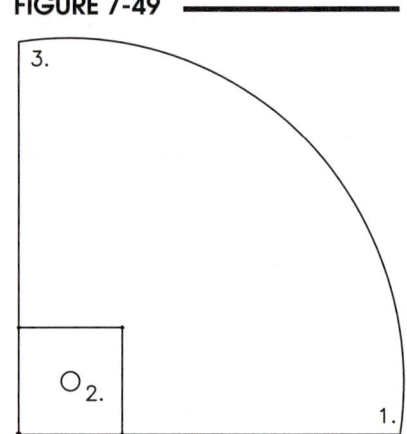

H. Now turn off *POLAR*. Construct an *Arc* to represent the end of the infield. Select the *Arc Start, Center, End* method from the *Draw* pull-down menu. For the *Start* point of the *Arc,* toggle Running Osnaps <u>off</u> by pressing **F3, Ctrl+F,** or clicking the word *OSNAP* on the Status Bar and **PICK** location **140', 20', 0'** on the first base line (watch *Coords* and ensure *SNAP* and *GRID* are on). (See Fig. 7-50, point 1.) Next, toggle Running Osnaps back on, and **PICK** the *Center* of the pitcher's mound as the *Center* of the *Arc* (point 2). Third, toggle Running Osnaps off again and **PICK** the *End* point of the *Arc* on the third base line (point 3).

FIGURE 7-50 ————

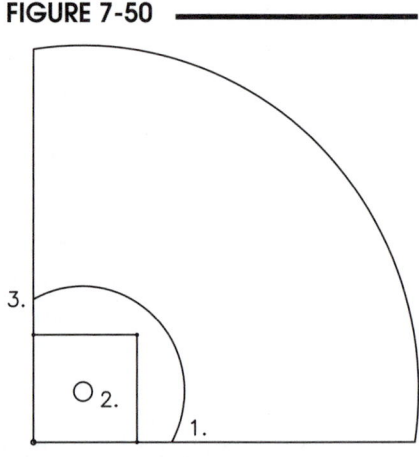

I. Lastly, the pitcher's mound should be moved to the correct distance from home plate. Type *M* (the command alias for *Move*) or select *Move* from the *Modify* pull-down menu. When prompted to "Select objects:" **PICK** the *Circle* representing pitcher's mound. At the "Specify base point or displacement:" prompt, toggle Running Osnaps <u>on</u> and **PICK** the *Center* of the mound. At the "Specify second point of displacement:" prompt, enter **@3'2"<225**. This should reposition the pitcher's mound to the regulation distance from home plate (60'-6"). Compare your drawing to Figure 7-50. *Save* the drawing (as BALL FIELD CH7).

J. Activate a *Layout* tab. Assuming your system is configured for a printer using an 11 x 8.5 sheet and is configured to automatically create a viewport (see "Setting Layout Options and Plot Options" in Chapter 6), a viewport should appear. (If the *Page Setup Manager* automatically appears, select *Close*.) If a viewport appears, proceed to step L. If no viewport appears, complete step K.

K. To make a viewport, type **-Vports** at the Command prompt (don't forget the hyphen) and press **Enter**. Accept the default (*Fit*) option by pressing **Enter**. A viewport appears to fit the printable area.

L. Double-click inside the viewport, then type **Z** for *Zoom* and press **Enter**, and type **A** for *All* and press **Enter**. Activate the *Viewports* drop-down list by right-clicking on any icon button and selecting *Viewports* from the list of available toolbars that appears. Select **1/64"=1'** from the list. Make a print of the drawing on an 11 x 8.5 inch sheet.

M. Activate the *Model* tab. *Save* the drawing as **BALL FIELD CH7.**

5. **Polar Tracking and Object Snap Tracking**

A. In this exercise, you will create a table base using Polar Tracking and Object Snap Tracking to locate points for construction of holes and other geometry. Begin a *New* drawing and use the **ACAD.DWT** template. *Save* the drawing and assign the name **TABLE-BASE.** Set the drawing limits to **48** x **32**.

B. Invoke the *Drafting Settings* dialog box. In the *Snap and Grid* tab, set *Polar Distance* to **1.00** and make *Polar Snap* current. In the *Polar Tracking* tab, set the *Increment Angle* to **45**. In the *Object Snap* tab, turn on the *Endpoint, Midpoint,* and *Center Osnap* options. Select *OK*. On the Status Bar ensure *SNAP, POLAR, OSNAP,* and *OTRACK* are on.

C. Use a *Line* and create the three line segments representing the first leg as shown in Figure 7-51. Begin at the indicated location. Use Polar Tracking and Polar Snap to assist drawing the diagonal lines segments. (Do not create the dimensions in your drawing.)

FIGURE 7-51 ───────────

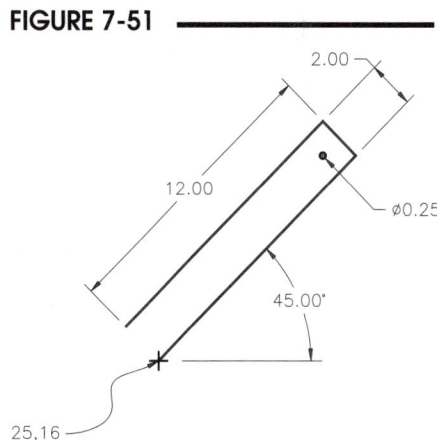

D. Place the first drill hole at the end of the leg using *Circle* with the *Center, Radius* option. Use Object Snap Tracking to indicate the center for the hole (*Circle*). Track vertically and horizontally from the indicated corners of the leg in Figure 7-52. Use *Endpoint Osnap* to snap to the indicated corners. (See Fig. 7-51 for hole dimension.)

FIGURE 7-52 ───────────

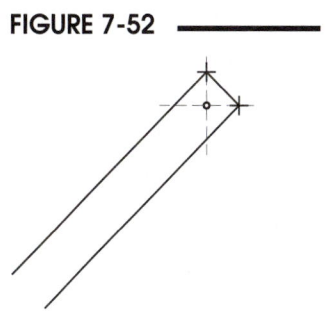

E. Create the other 3 legs in a similar fashion. You can track from the *Endpoints* on the bottom of the existing leg to identify the "first point" of the next *Line* segment as shown in Figure 7-53.

FIGURE 7-53

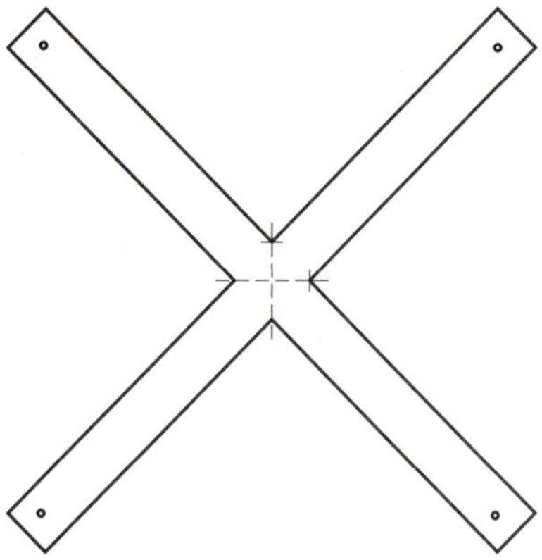

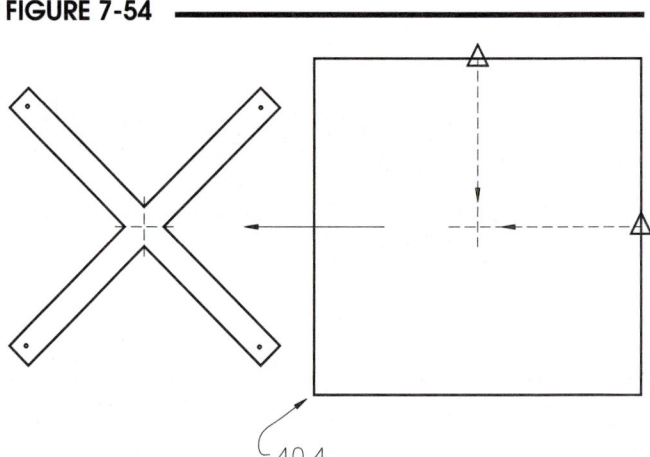

F. Use *Line* to create a 24" x 24" square table top on the right side of the drawing as shown in Figure 7-54. Use Object Snap Tracking and Polar Tracking with the *Move* command to move the square's center point to the center point of the 4 legs (HINT: At the "Specify base point or displacement:" prompt, Osnap Track to the square's *Midpoints*. At the "Specify second point of displacement:" prompt, use *Endpoint* Osnaps to locate the legs' center.

FIGURE 7-54

40,4

G. Create a smaller square (*Line*) in the center of the table which will act as a support plate for the legs. Track from the *Midpoint* of the legs to create the lines as indicated (highlighted) in Figure 7-55.

FIGURE 7-55

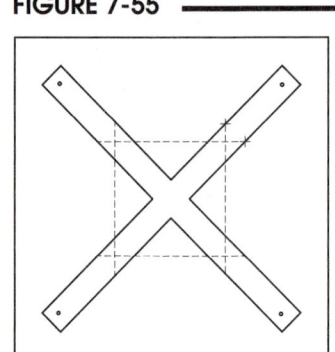

H. *Save* the drawing. You will finish the table base in Chapter 9.

DRAW
COMMANDS I

CHAPTER OBJECTIVES

After completing this chapter you should:

1. know where to locate and how to invoke the draw commands;

2. be able to draw *Lines*;

3. be able to draw *Circles* by each of the five options;

4. be able to draw *Arcs* by each of the eleven options;

5. be able to create *Point* objects and specify the *Point Style*;

6. be able to create *Plines* with width and combined of line and arc segments.

CONCEPTS

Draw Commands—Simple and Complex

Draw commands create objects. An object is the smallest component of a drawing. The draw commands listed immediately below create simple objects and are discussed in this chapter. Simple objects appear as one entity.

FIGURE 8-1

> *Line*, *Circle*, *Arc*, and *Point*

Other draw commands create more complex shapes. Complex shapes appear to be composed of several components, but each shape is usually one object. An example of an object that is one entity but usually appears as several segments is listed below and is also covered in this chapter:

> *Pline*

Other draw commands discussed in Chapter 15 (listed below) are a combination of simple and complex shapes:

> *Xline*, *Ray*, *Polygon*, *Rectangle*, *Donut*, *Spline*, *Ellipse*, *Divide*, *Mline*, *Measure*, *Sketch*, *Solid*, *Region*, and *Boundary*

Draw Command Access

As a review from Chapter 3, Draw Command Concepts, remember that any of the five methods can be used to access the draw commands: *Draw* toolbar (Fig. 8-1), *Draw* pull-down menu (Fig. 8-2), *DRAW1* and *DRAW2* screen menus, keyboard entry of the command or alias, and digitizing tablet icons.

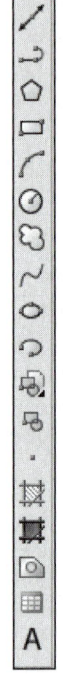

FIGURE 8-2

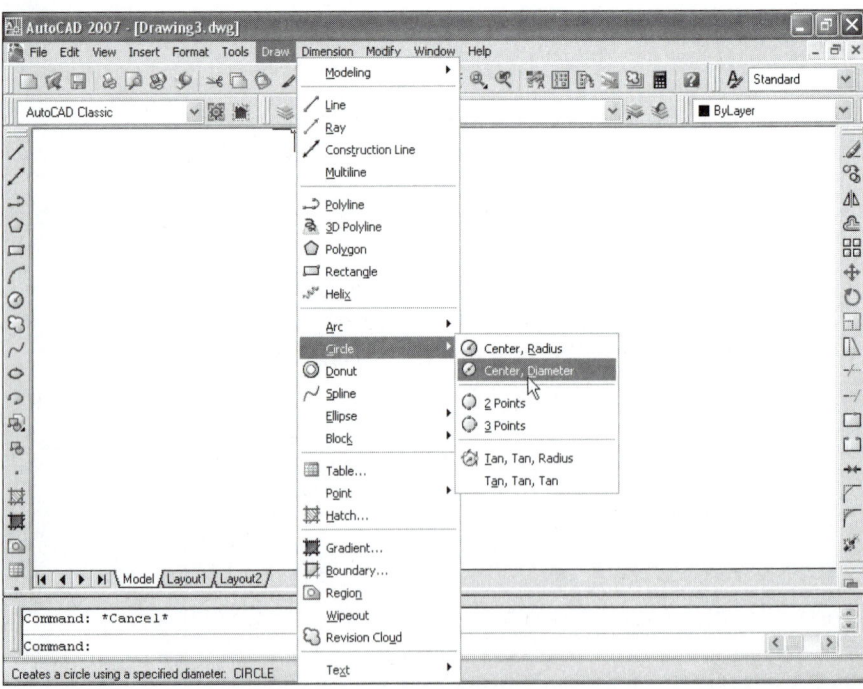

Coordinate Entry Input Methods

When creating objects with draw commands, AutoCAD always prompts you to indicate points (such as endpoints, centers, radii) to describe the size and location of the objects to be drawn. An example you are familiar with is the *Line* command, where AutoCAD prompts for the "Specify first point:". Indication of these points, called <u>coordinate entry</u> can be accomplished by three input methods:

1. Mouse Input   PICK (press the left <u>mouse</u> button to specify a point on the screen).
2. Dynamic Input   When *DYN* is on, you can <u>key values into the edit boxes</u> near the cursor and press Tab to change boxes.
3. Command Line Input   You can <u>key values</u> in any format <u>at the Command prompt</u>.

Any of these methods can be used <u>whenever</u> AutoCAD prompts you to specify points. (For practice with these methods, see Chapter 3, Draw Command Concepts.)

Drawing Aids

Also keep in mind that you can specify points interactively using the following AutoCAD features individually or in combination: Grid Snap, Polar Snap, Ortho, Polar Tracking, Object Snap, and Object Snap Tracking. Use these drawing tools <u>whenever</u> AutoCAD prompts you to select points. (See Chapters 3 and 7 for details on these tools.)

COMMANDS

Line

Pull-down Menu	Command (Type)	Alias (Type)	Short-cut	Screen (side) Menu	Tablet Menu
Draw *Line*	*Line*	L	...	*DRAW 1* *Line*	*J,10*

This is the fundamental drawing command. The *Line* command creates straight line segments; each segment is an object. One or several line segments can be drawn with the *Line* command.

Command: **Line**
Specify first point: **PICK** or (**coordinates**) (A point can be designated by interactively selecting with the input device or by entering coordinates. Use any of the drawing tools to assist with interactive entry.)
Specify next point or [Undo]: **PICK** or (**coordinates**) (Again, device input or keyboard input can be used.)
Specify next point or [Undo]: **PICK** or (**coordinates**)
Specify next point or [Close/Undo]: **PICK** or (**coordinates**) or **C**
Specify next point or [Close/Undo]: press **Enter** to finish command
Command:

FIGURE 8-3

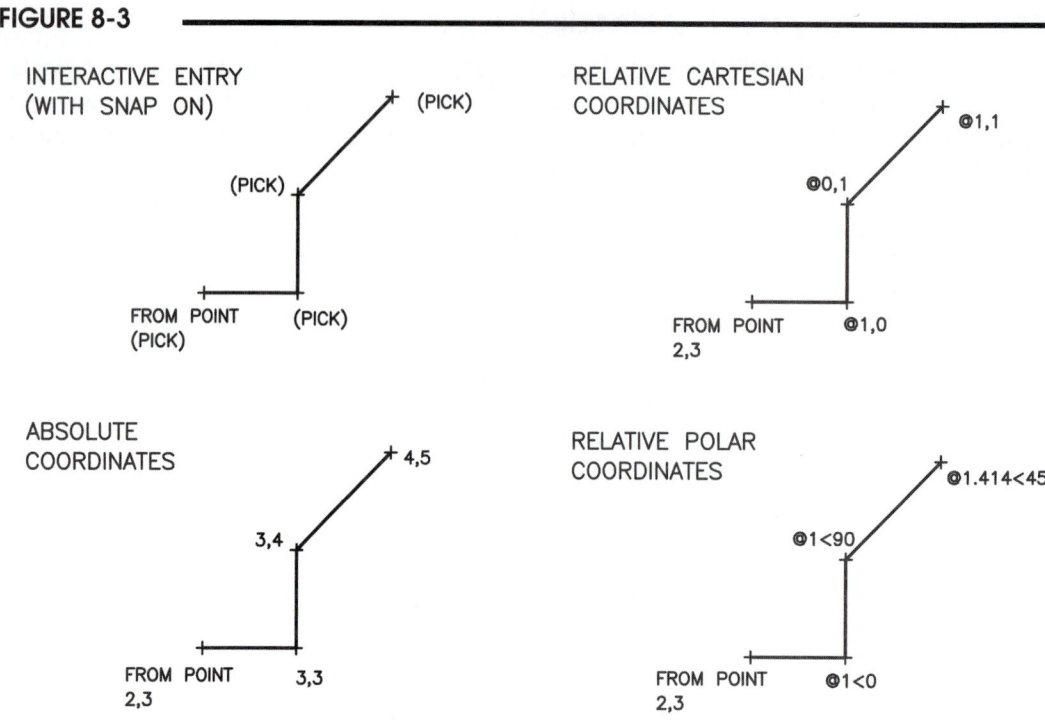

INTERACTIVE ENTRY (WITH SNAP ON) — (PICK), (PICK), FROM POINT (PICK), (PICK)

RELATIVE CARTESIAN COORDINATES — @1,1, @0,1, @1,0, FROM POINT 2,3

ABSOLUTE COORDINATES — 4,5, 3,4, FROM POINT 2,3, 3,3

RELATIVE POLAR COORDINATES — @1.414<45, @1<90, @1<0, FROM POINT 2,3

Figure 8-3 shows four examples of creating the same *Line* segments using different formats of coordinate entry. Refer to Chapter 3, Draw Command Concepts, for examples of drawing vertical, horizontal, and inclined lines using the formats for coordinate entry.

Circle

Pull-down Menu	Command (Type)	Alias (Type)	Short-cut	Screen (side) Menu	Tablet Menu
Draw *Circle >*	*Circle*	*C*	...	*DRAW 1* *Circle*	*J,9*

The *Circle* command creates one object. Depending on the option selected, you can provide two or three points to define a *Circle*. As with all commands, the Command line prompt displays the possible options:

Command: **Circle**
Specify center point for circle or [3P/2P/Ttr (tan tan radius)]: **PICK** or (**coordinates**) or (**option**).
(PICKing or entering coordinates designates the center point for the circle. You can enter 3P, 2P, or T for another option.)

As with most commands, the default and other options are displayed on the Command line. The default option always appears first. The other options can be invoked by typing the numbers and/or uppercase letter(s) that appear in brackets [ ].

All of the options for creating *Circles* are available from the *Draw* pull-down menu and from the *DRAW 1* screen (side) menu. The tool (icon button) for only the *Center, Radius* method is included in the *Draw* toolbar. Although the *Circle* options are not available on the *Draw* toolbar, you can use the *Circle* button, then right-click for a shortcut menu showing all the options.

The options, or methods, for drawing *Circles* are listed below. Each figure gives several possibilities for each option, with and without *OSNAPs*.

Center, Radius

Specify a center point, then a radius (Fig. 8-4).

The *Radius* (or *Diameter*) can be specified by entering values or by indicating a length inter-actively (PICK two points to specify a length when prompted). As always, points can be specified by PICKing or entering coordinates. Watch *Coords* for coordinate or distance (polar format) display. Grid Snap, Polar Snap, Polar Tracking, Object Snap, and Object Snap Tracking can be used for interactive point specification.

FIGURE 8-4

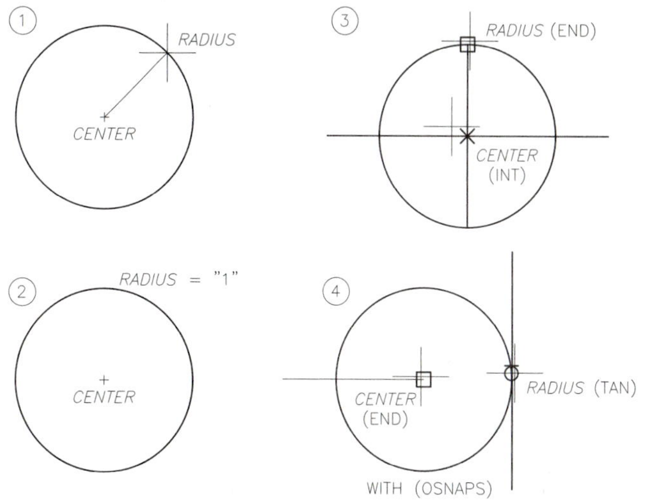

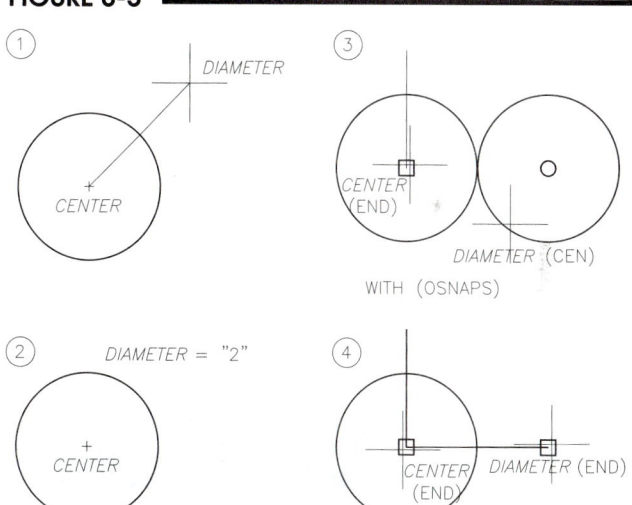

Center, Diameter

Specify the center point, then the diameter (Fig. 8-5).

FIGURE 8-5

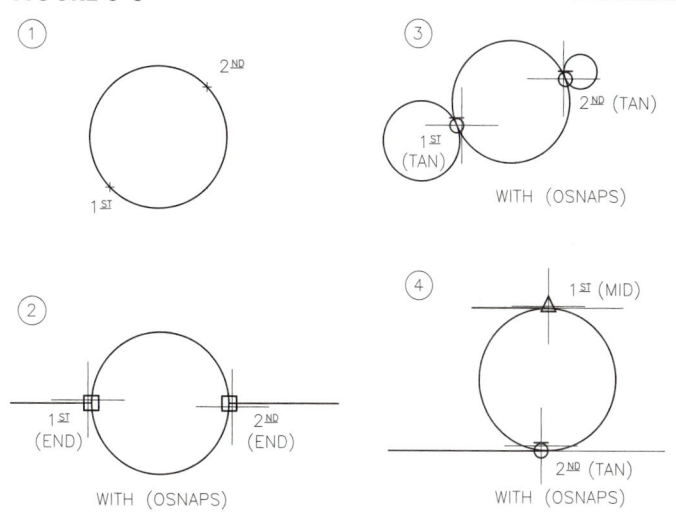

2 Points

The two points specify the location and diameter.

The *Tangent OSNAPs* can be used when selecting points with the *2 Point* and *3 Point* options, as shown in Figures 8-6 and 8-7.

FIGURE 8-6

Start, Center, Length

Length means length of chord. The length of chord is between the start and the other point specified (Fig. 8-12). A negative chord length can be entered to generate an *Arc* of 180+ degrees.

FIGURE 8-12

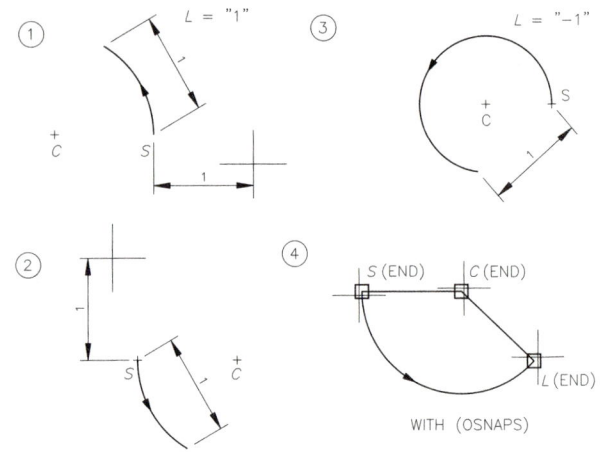

Start, End, Angle

The included angle is between the sides from the center to the endpoints (Fig. 8-13). Negative angles generate clockwise *Arcs*.

FIGURE 8-13

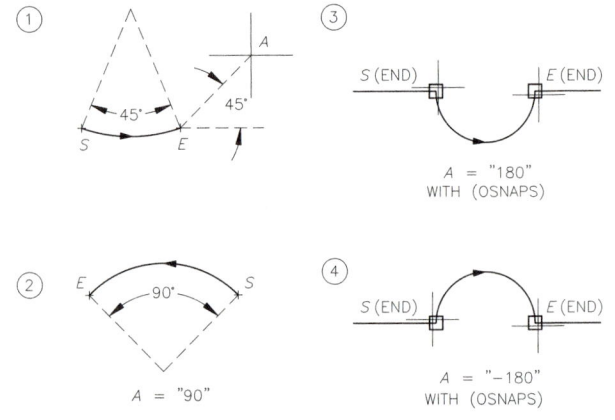

Start, End, Radius

The radius can be PICKed or entered as a value (Fig. 8-14). A negative radius value generates an *Arc* of 180+ degrees.

FIGURE 8-14

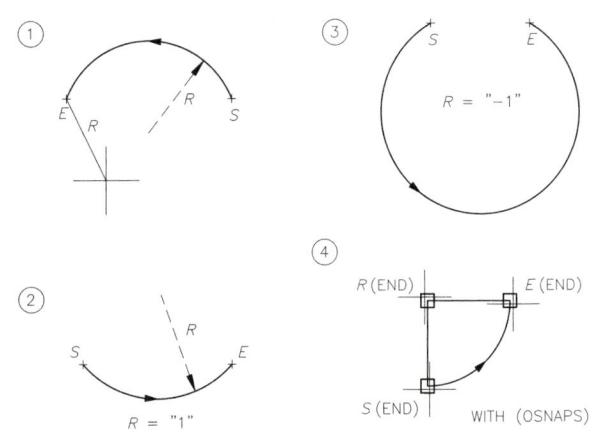

Start, End, Direction

The direction is tangent to the start point (Fig. 8-15).

FIGURE 8-15

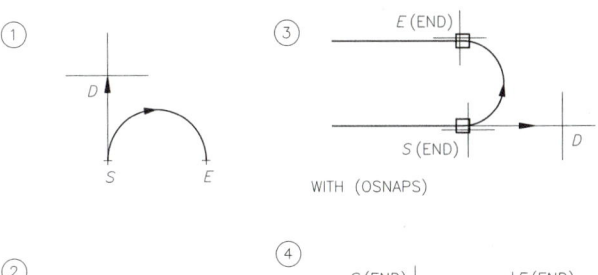

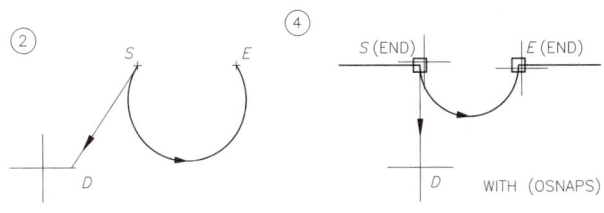

Center, Start, End

This option is like *Start, Center, End* but in a different order (Fig. 8-16).

FIGURE 8-16

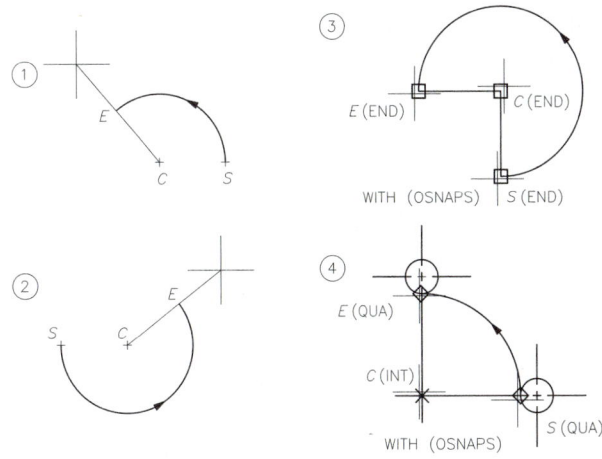

Center, Start, Angle

This option is like *Start, Center, Angle* but in a different order (Fig. 8-17).

FIGURE 8-17

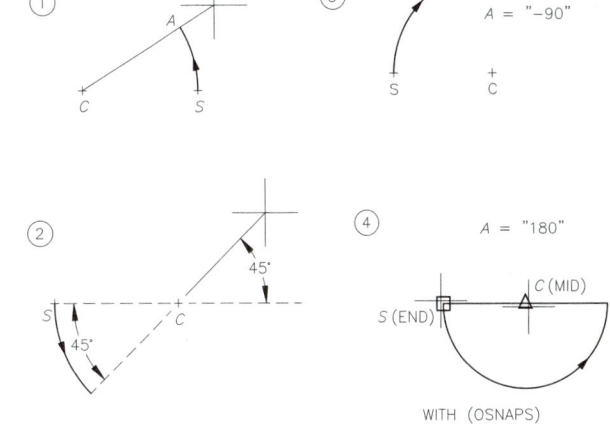

Center, Start, Length

This is similar to the *Start, Center, Length* option but in a different order (Fig. 8-18). *Length* means length of chord.

FIGURE 8-18

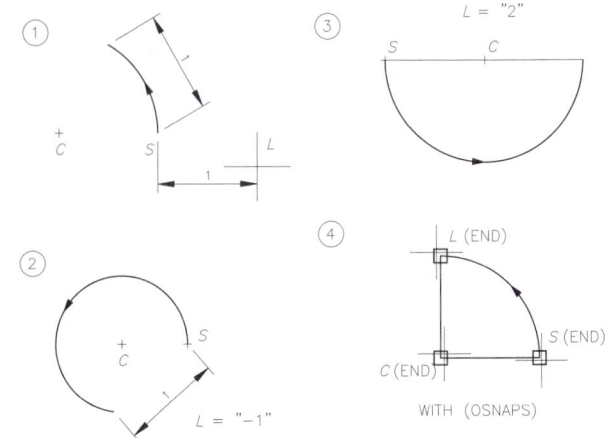

Continue

The new *Arc* continues from and is tangent to the last point (Fig. 8-19). The only other point required is the endpoint of the *Arc*. This method allows drawing *Arcs* tangent to the preceding *Line* or *Arc*.

FIGURE 8-19

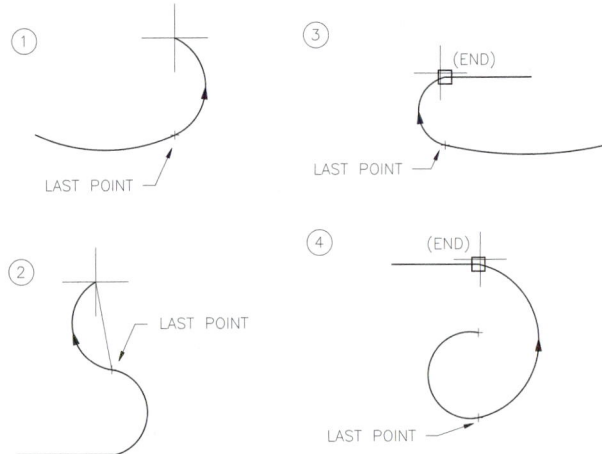

Arcs are always created in a <u>counterclockwise</u> direction. This fact must be taken into consideration when using any method <u>except</u> the *3-Point*, the *Start, End, Direction,* and the *Continue* options. The direction is explicitly specified with *Start, End, Direction,* and *Continue* methods, and direction is irrelevant for *3-Point* method.

As usual, points can be specified by PICKing or entering coordinates. Watch *Coords* to display coordinate values or distances. Grid Snap, Polar Snap, Polar Tracking, Object Snap, and Object Snap Tracking can be used when PICKing. The *Endpoint, Intersection, Center, Midpoint,* and *Quadrant OSNAP* options can be used with great effectiveness. The *Tangent OSNAP* option <u>cannot</u> be used effectively with most of the *Arc* options. The *Radius, Direction, Length,* and *Angle* specifications can be given by entering values or by PICKing with or without *OSNAPs*.

Use *Arcs* or *Circles?*

Although there are sufficient options for drawing *Arcs*, <u>usually it is easier to use the *Circle* command</u> followed by *Trim* to achieve the desired arc. Creating a *Circle* is generally an easier operation than using *Arc* because the counterclockwise direction does not have to be considered. The unwanted portion of the circle can be *Trimmed* at the *Intersection* of or *Tangent* to the connecting objects using *OSNAP*. The *Fillet* command can also be used instead of the *Arc* command to add a fillet (arc) between two existing objects (see Chapter 9, Modify Commands I).

Point

		Command (Type)	Alias (Type)	Short-cut	Screen (side) Menu	Tablet Menu
	Pull-down Menu					
	Draw *Point >* *Single Point* or *Multiple Point*	*Point*	*PO*	...	*DRAW 2* *Point*	*O,9*

A *Point* is an object that has no dimension; it only has location. A *Point* is specified by giving only one coordinate value or by PICKing a location on the screen.

Figure 8-20 compares *Points* to *Line* and *Circle* objects.

> Command: **point**
> Current point modes: PDMODE=0 PDSIZE=0.0000
> Specify a point: **PICK** or (**coordinates**)
> (Select a location for the *Point* object.)

Points are useful in construction of drawings to locate points of reference for subsequent construction or locational verification. The *Node OSNAP* option is used to snap to *Point* objects.

FIGURE 8-20

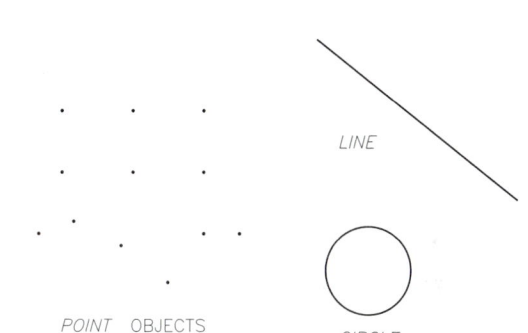

LINE

POINT   OBJECTS

CIRCLE

Points are drawing objects and therefore appear in prints and plots. The default "style" for points is a tiny dot. The *Point Style* dialog box can be used to define the format you choose for *Point* objects (the *Point* type [PDMODE] and size [PDSIZE]).

The *Draw* pull-down menu offers the *Single Point* and the *Multiple Point* options. The *Single Point* option creates one *Point,* then returns to the command prompt. This option is the same as using the *Point* command by any other method. Selecting *Multiple Point* continues the *Point* command until you press the Escape key.

Ddptype

		Command (Type)	Alias (Type)	Short-cut	Screen (side) Menu	Tablet Menu
	Pull-down Menu					
	Format *Point Style...*	*Ddptype*	...	...	*DRAW 2* *Point* *Ddptype:*	*U,1*

The *Point Style* dialog box (Fig. 8-21) is available only through the methods listed in the command table above. This dialog box allows you to define the format for the display of *Point* objects. The selected style is applied immediately to all newly created *Point* objects. The *Point Style* controls the format of *Points* for printing and plotting as well as for the computer display.

You can set the *Point Size* in *Absolute* units or *Relative to Screen* (default option). The *Relative to Screen* option keeps the *Points* the same size on the display when you *Zoom* in and out, whereas setting *Point Size* in *Absolute Units* gives you control over the size of *Points* for prints and plots. The *Point Size* is stored in the *PDSIZE* system variable. The selected *Point Style* is stored in the *PDMODE* variable.

FIGURE 8-21

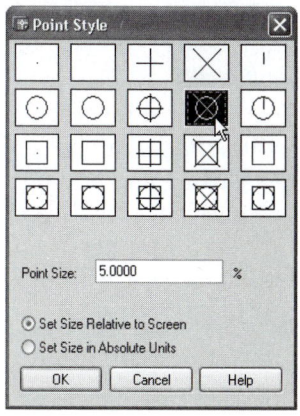

Pline

	Pull-down Menu	Command (Type)	Alias (Type)	Short-cut	Screen (side) Menu	Tablet Menu
	Draw Polyline	Pline	PL	...	DRAW 1 Pline	N,10

A *Pline* (or *Polyline*) has special features that make this object more versatile than a *Line*. Three features are most noticeable when first using *Plines*:

1. A *Pline* can have a specified *width*, whereas a *Line* has no width.
2. Several *Pline* segments created with one *Pline* command are treated by AutoCAD as <u>one</u> object, whereas individual line segments created with one use of the *Line* command are individual objects.
3. A *Pline* can contain arc segments.

Figure 8-22 illustrates *Pline* versus *Line* and *Arc* comparisons.

The *Pline* command begins with the same prompt as *Line*; however, <u>after</u> the "start point:" is established, the *Pline* options are accessible.

FIGURE 8-22

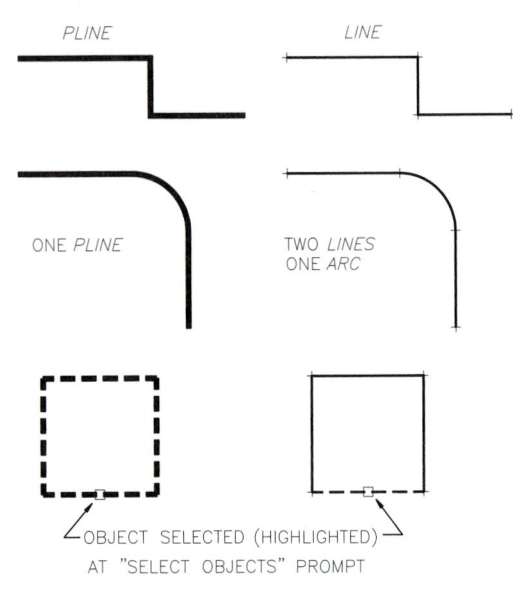

```
Command: Pline
Specify start point: PICK or (coordinates)
Current line-width is 0.0000
Specify next point or [Arc/Close/Halfwidth/Length/Undo/Width]:
```

Similar to the other drawing commands, you can invoke the *Pline* command by any method, then right-click for a shortcut menu at any time during the command to show other possible options at that point.

The options and descriptions follow.

Width
You can use this option to specify starting and ending widths. Width is measured perpendicular to the centerline of the *Pline* segment (Fig. 8-23). *Plines* can be tapered by specifying different starting and ending widths. See NOTE at the end of this section.

Halfwidth
This option allows specifying half of the *Pline* width. *Plines* can be tapered by specifying different starting and ending widths (Fig. 8-23).

FIGURE 8-23

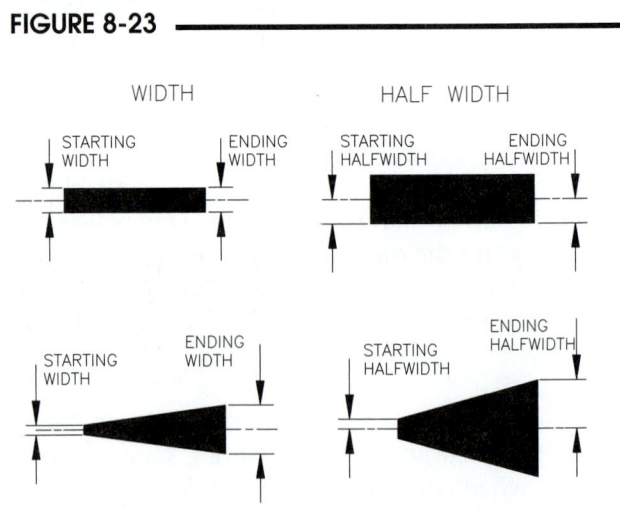

Arc

This option (by default) creates an arc segment in a manner similar to the *Arc Continue* method (Fig. 8-24). Any of several other methods are possible (see *"Pline Arc* Segments").

FIGURE 8-24

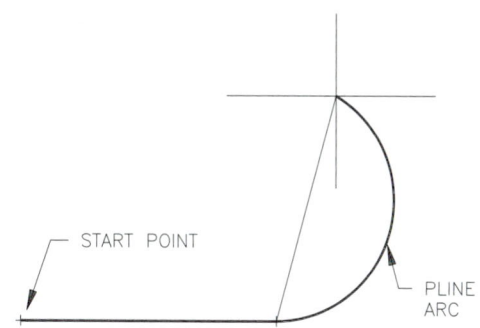

Close

The *Close* option creates the closing segment connecting the first and last points specified with the current *Pline* command as shown in Figure 8-25.

This option can also be used to close a group of connected *Pline* segments into one continuous *Pline*. (A *Pline* closed by PICKing points has a specific start and endpoint.) A *Pline Closed* by this method has special properties if you use *Pedit* for *Pline* editing or if you use the *Fillet* command with the *Pline* option (see *Fillet* in Chapter 9 and *Pedit* in Chapter 16).

FIGURE 8-25

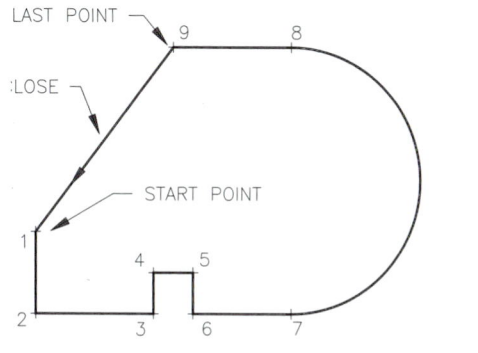

Length

Length draws a *Pline* segment at the same angle as and connected to the previous segment and uses a length that you specify. If the previous segment was an arc, *Length* makes the current segment tangent to the ending direction (Fig. 8-26).

FIGURE 8-26

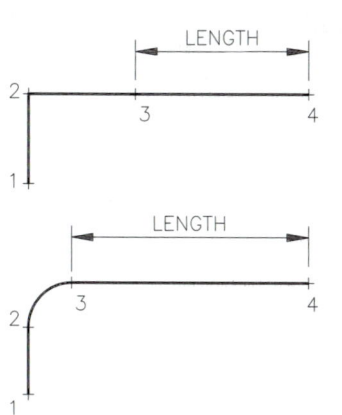

Undo

Use this option to *Undo* the last *Pline* segment. It can be used repeatedly to undo multiple segments.

NOTE: If you change the *Width* of a *Pline*, be sure to respond to <u>both prompts</u> for width ("Specify starting width:" and "Specify ending width:") before you draw the first *Pline* segment. It is easy to hastily PICK the endpoint of the *Pline* segment after specifying the "Specify starting width:" instead of responding with a value (or Enter) for the "Specify ending width:". In this case (if you PICK at the "Specify ending width:" prompt), AutoCAD understands the line length that you interactively specified to be the ending width you want for the next line segment (you can PICK two points in response to the "Specify starting width:" or "Specify ending width:" prompts).

```
Command: Pline
Specify start point: PICK
Current line-width is 0.0000
Specify next point or [Arc/Close/Halfwidth/Length/Undo/Width]: w
Specify starting width <0.0000>: .2
Specify ending width <0.2000>: Enter a value or press Enter—do not PICK the "next point."
```

Pline Arc Segments

When the *Arc* option of *Pline* is selected, the prompt changes to provide the various methods for construction of arcs:

Specify endpoint of arc or [Angle/CEnter/CLose/Direction/Halfwidth/Line/Radius/Second pt/Undo/Width]:

Angle
You can draw an arc segment by specifying the included angle (a negative value indicates a clockwise direction for arc generation).

CEnter
This option allows you to specify a specific center point for the arc segment.

CLose
This option closes the *Pline* group with an arc segment.

Direction
Direction allows you to specify an explicit starting direction rather than using the ending direction of the previous segment as a default.

Line
This switches back to the line options of the *Pline* command.

Radius
You can specify an arc radius using this option.

Second pt
Using this option allows specification of a 3-point arc.

Because a shape created with one *Pline* command is <u>one object</u>, manipulation of the shape is generally easier than with several objects. For some applications, one *Pline* shape can have advantages over shapes composed of several objects (see "*Offset*," Chapter 9). Editing *Plines* is accomplished by using the *Pedit* command. As an alternative, *Plines* can be "broken" back down into individual objects with *Explode*.

Drawing and editing *Plines* can be somewhat involved. As an alternative, you can draw a shape as you would normally with *Line, Circle, Arc, Trim*, etc., and then <u>convert</u> the shape to one *Pline* object using *Pedit*. (See Chapter 16 for details on converting *Lines* and *Arcs* to *Plines*.)

CHAPTER EXERCISES

Create a *New* drawing. Use the *ACAD.DWT* template or select **Start from Scratch** and select **Imperial** settings. Set *Polar Snap On* and set the *Polar Distance* to **.25**. Turn on *SNAP, POLAR, OSNAP, OTRACK*, and *DYN*. *Save* the drawing as **CH8EX.** For each of the following problems, *Open* **CH8EX**, complete one problem, then use *SaveAs* to give the drawing a new name.

1. **Open CH8EX**. Create the geometry shown in Figure 8-27. Start the first *Circle* center at point **4,4.5** as shown. Do not copy the dimensions. *SaveAs* **LINK** (HINT: Locate and draw the two small *Circles* first. Use *Arc, Start, Center, End* or *Center, Start, End* for the rounded ends.)

FIGURE 8-27

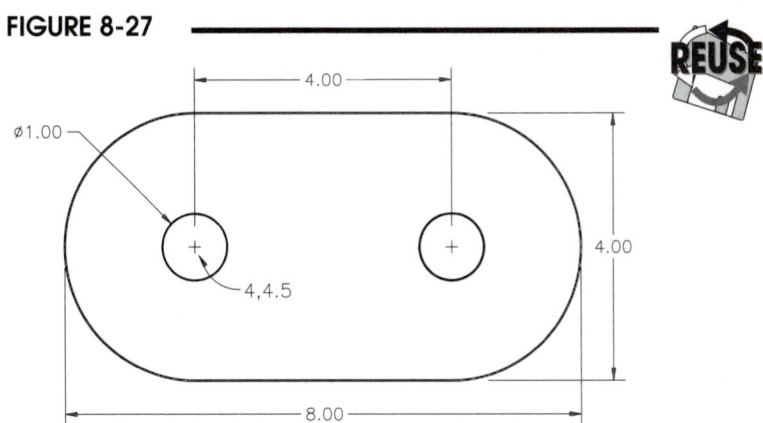

2. **Open CH8EX**. Create the geometry as shown in Figure 8-28. Do not copy the dimensions. Assume symmetry about the vertical axis. *SaveAs* **SLOTPLATE CH8.**

FIGURE 8-28

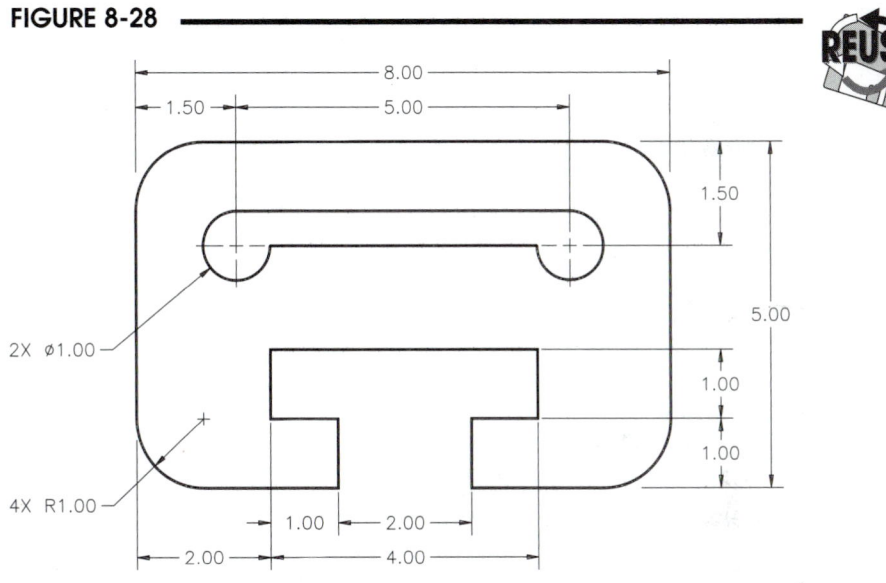

3. **Open CH8EX**. Create the shapes shown in Figure 8-29. Do not copy the dimensions. *SaveAs* **CH8EX3.**

Draw the *Lines* at the bottom first, starting at coordinate **1,3**. Then create *Point* objects at **5,7**, **5.4,7**, **5.8,7**, etc. Change the *Point Style* to an X and *Regen*. Use the *NODe* OSNAP mode to draw the inclined *Lines*. Create the *Arc* on top with the *Start, End, Direction* option.

FIGURE 8-29

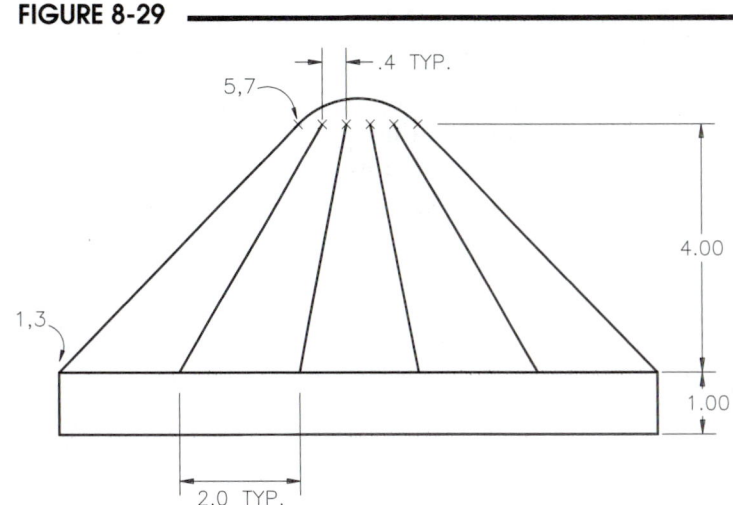

4. **Open CH8EX**. Create the shape shown in Figure 8-30. Draw the two horizontal *Lines* and the vertical *Line* first by specifying the endpoints as given. Then create the *Circle* and *Arcs*. *SaveAs* **CH8EX4**.

   HINT: Use the *Circle 2P* method with *Endpoint OSNAP*s. The two upper *Arcs* can be drawn by the *Start, End, Radius* method.

FIGURE 8-30 —————————

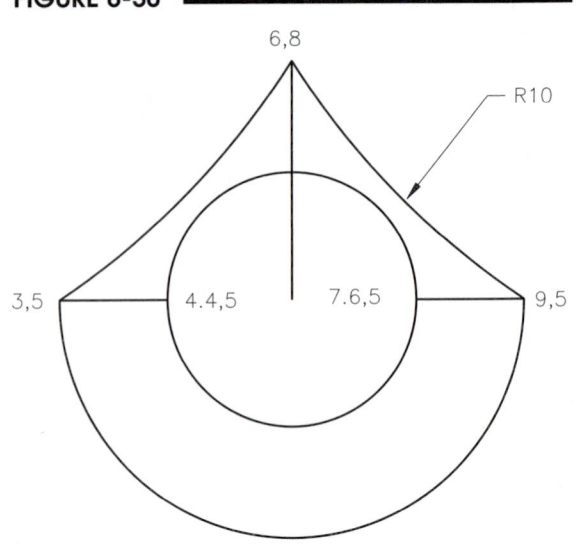

5. **Open CH8EX**. Draw the shape shown in Figure 8-31. Assume symmetry along a vertical axis. Start by drawing the two horizontal *Lines* at the base.

   Next, construct the side *Arcs* by the *Start, Center, Angle* method (you can specify a negative angle). The small *Arc* can be drawn by the *3P* method. Use *OSNAP*s when needed (especially for the horizontal *Line* on top and the *Line* along the vertical axis). *SaveAs* **CH8EX5**.

FIGURE 8-31 —————————

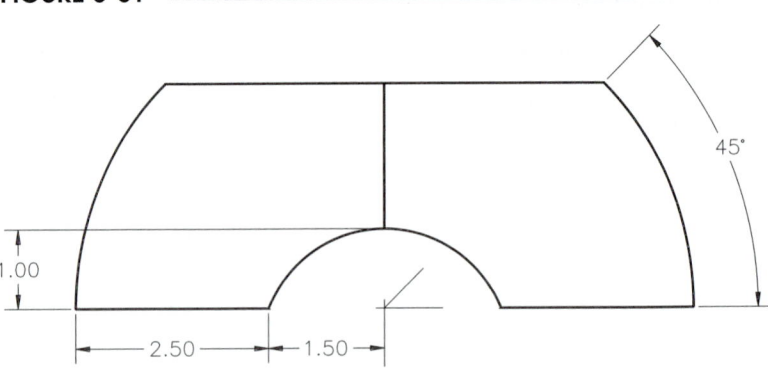

6. **Open CH8EX**. Complete the geometry in Figure 8-32. Use the coordinates to establish the *Lines*. Draw the *Circles* using the *Tangent, Tangent, Radius* method. *SaveAs* **CH8EX6**.

FIGURE 8-32 —————————

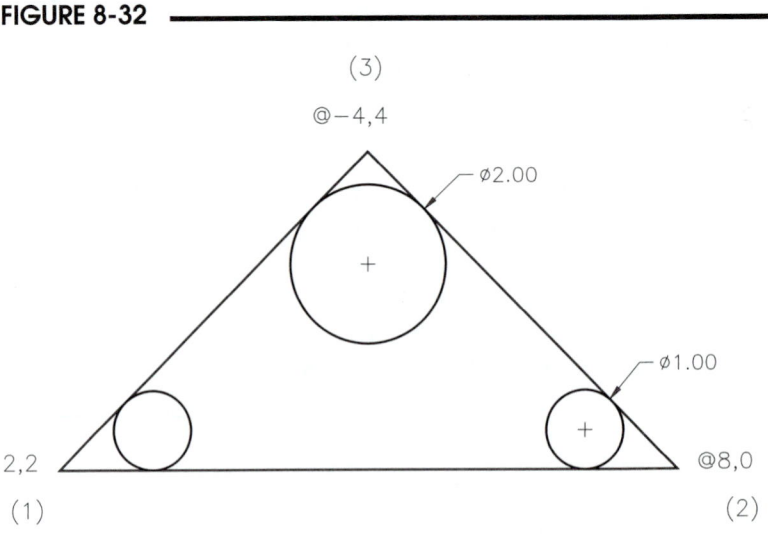

7. **Retaining Wall**

FIGURE 8-33

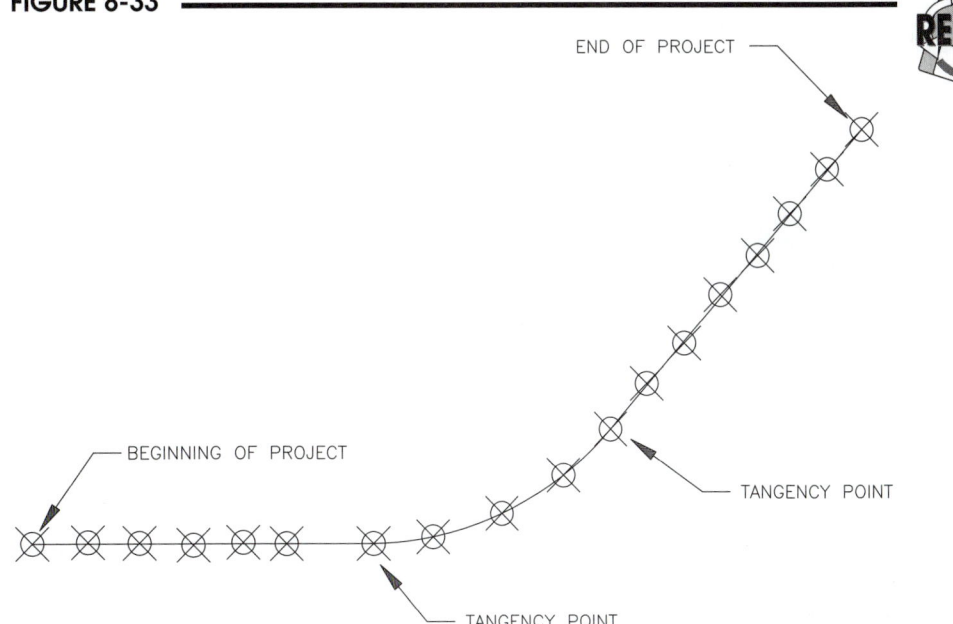

A contractor plans to stake out the edge of a retaining wall located at the bottom of a hill. The following table lists coordinate values based on a survey at the site. Use the drawing you created in Chapter 6 Exercises named **CH6EX2**. *Save* the drawing as **RET-WALL.**

Place **Points** at each coordinate value in order to create a set of data points. Determine the location of one **Arc** and two **Lines** representing the centerline of the retaining wall edge. The centerline of the retaining wall should match the data points as accurately as possible (Fig. 8-33). The data points given in the following table are in inch units.

Pt #	X	Y	
1.	60.0000	108.0000	
2.	81.5118	108.5120	
3.	101.4870	108.2560	
4.	122.2305	107.4880	
5.	141.4375	108.5120	
6.	158.1949	108.0000	
7.	192.0000	108.0000	Recommended tangency point
8.	215.3914	110.5185	
9.	242.3905	119.2603	
10.	266.5006	133.6341	
11.	285.1499	151.0284	Recommended tangency point
12.	299.5069	168.1924	
13.	314.2060	183.4111	
14.	328.3813	201.7784	
15.	343.0811	216.4723	
16.	355.6808	232.2157	
17.	370.3805	249.0087	
18.	384.0000	264.0000	

(Hint: Change the *Point Style* to an easily visible format.)

8. *Pline*

Create the shape shown in Figure 8-34. Draw the outside shape with <u>one continuous</u> *Pline* (with 0.00 width). When finished, *SaveAs*

PLINE1.

FIGURE 8-34

FIGURE 8-34

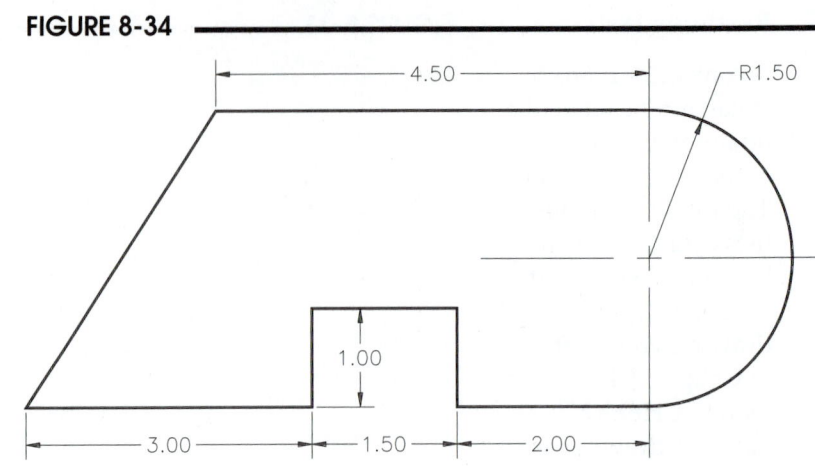

CHAPTER

MODIFY
COMMANDS I

CHAPTER OBJECTIVES

After completing this chapter you should:

1. be able to *Erase* objects;

2. be able to *Move* objects from a base point to a second point;

3. know how to *Rotate* objects about a base point;

4. be able to enlarge or reduce objects with *Scale*;

5. be able to *Stretch* objects and change the length of *Lines* and *Arcs* with *Lengthen*;

6. be able to *Trim* away parts of objects at cutting edges and *Extend* objects to boundary edges;

7. know how to use the four *Break* options;

8. be able to *Copy* objects and make *Mirror* images of selected objects;

9. know how to create parallel copies of objects with *Offset*;

10. be able to make *rectangular* and *Polar Arrays* of objects;

11. be able to create a *Fillet* and a *Chamfer* between two objects.

CONCEPTS

Draw commands are used to create new objects. *Modify* commands are used to change existing objects or to use existing objects to create new and similar objects. The *Modify* commands covered first in this chapter (listed below) only <u>change existing objects</u>.

FIGURE 9-1

    Erase, Move, Rotate, Scale, Stretch, Lengthen, Trim, Extend, and *Break*

The *Modify* commands covered near the end of this chapter (listed below) <u>use existing objects to create new and similar objects</u>. For example, the *Copy* command prompts you to select an object (or set of objects), then creates an identical object (or set).

    Copy, Mirror, Offset, Array, Fillet, and *Chamfer*

Modify commands can be invoked by any of the five command entry methods: toolbar icons, pull-down menus, screen menus, keyboard entry, and digitizing tablet menu. The *Modify* toolbar is visible and docked to the right side of the screen by default (Fig. 9-1).

Several commands are found in the *Edit* pull-down menu such as *Cut, Copy,* and *Paste*. Although these command names appear similar (or the same in the case of *Copy*) to some of those in the *Modify* menu, these AutoCAD command names are actually *Cutclip, Copyclip,* and *Pasteclip*. These commands are for OLE operations (cutting and pasting) between two different AutoCAD drawings or other software applications. See Chapter 31, Object Linking and Embedding, for more information.

The *Modify* pull-down menu (Fig. 9-2) contains commands that change existing geometry and that use existing geometry to create new but similar objects. All of the *Modify* commands are contained in this single pull-down menu.

FIGURE 9-2

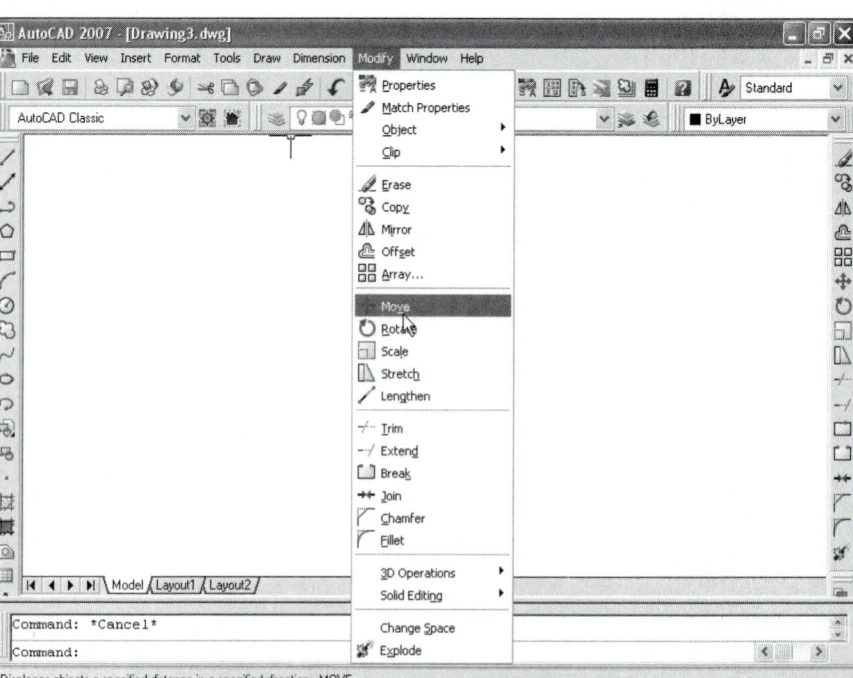

The *MODIFY1* and *MODIFY2* screen menus branch to the individual commands and options (Fig. 9-3).

Since all *Modify* commands affect or use existing geometry, the first step in using most *Modify* commands is to construct a selection set (see Chapter 4). This can be done by one of two methods:

1. Invoking the desired command and then creating the selection set in response to the "Select Objects:" prompt (Verb/Noun syntax order) using any of the select object options;

2. Selecting the desired set of objects with the pickbox or *Auto Window* or *Crossing Window* <u>before</u> invoking the edit command (Noun/Verb syntax order).

FIGURE 9-3

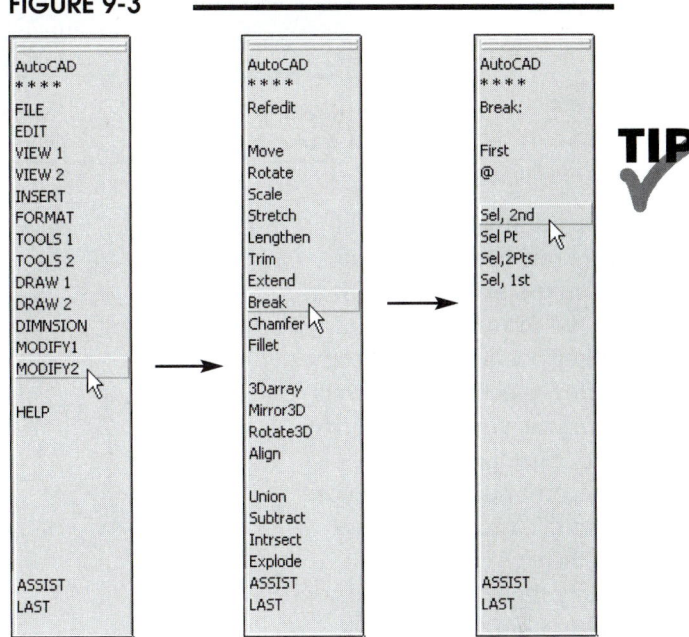

The first method allows use of any of the selection options (*Last, All, WPolygon, Fence,* etc.), while the latter method allows <u>only</u> the use of the pickbox and *Auto Window* and *Crossing Window*.

COMMANDS

Erase

Pull-down Menu	Command (Type)	Alias (Type)	Short-cut	Screen (side) Menu	Tablet Menu
Modify *Erase*	*Erase*	E	(Edit Mode) *Erase*	*MODIFY1* *Erase*	V,14

The *Erase* command deletes the objects you select from the drawing. Any of the object selection methods can be used to highlight the objects to *Erase*. The only other required action is for you to press *Enter* to cause the erase to take effect.

    Command: **erase**
    Select Objects: **PICK** (Use any object selection method.)
    Select Objects: **PICK** (Continue to select desired objects.)
    Select Objects: **Enter** (Confirms the object selection process and causes *Erase* to take effect.)
    Command:

If objects are erased accidentally, *U* can be used immediately following the mistake to undo one step, or *Oops* can be used to bring back into the drawing whatever was *Erased* the last time *Erase* was used (see Chapter 5). If only part of an object should be erased, use *Trim* or *Break*.

Move

Pull-down Menu	Command (Type)	Alias (Type)	Short-cut	Screen (side) Menu	Tablet Menu
Modify *Move*	*Move*	*M*	*(Edit Mode)* *Move*	*MODIFY2* *Move*	*V,19*

Move allows you to relocate one or more objects from the existing position in the drawing to any other position you specify. After selecting the objects to *Move*, you must specify the "base point" and "second point of displacement." You can use any of the five coordinate entry methods to specify these points. Examples are shown in Figure 9-4.

FIGURE 9-4

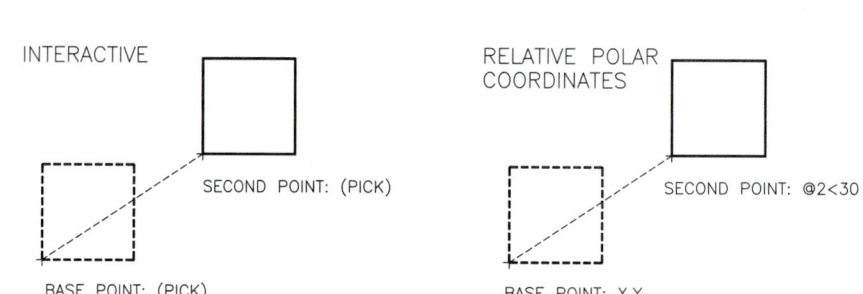

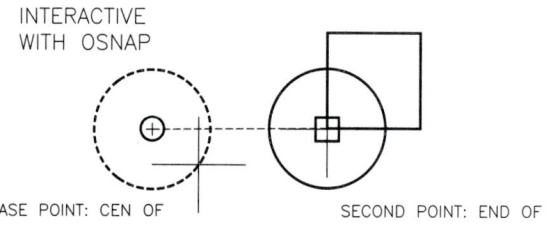

Command: **move**
Select Objects: **PICK**   (Use any of the object selection methods.)
Select Objects: **PICK** (Continue to select other desired objects.)
Select Objects: **Enter** (Press Enter to indicate selection of objects is complete.)
Specify base point or displacement: **PICK** or (**coordinates**) (This is the point to <u>move from</u>. Select a point to use as a "handle." An *Endpoint* or *Center*, etc., can be used.)
Specify second point of displacement or <use first point as displacement>: **PICK** or (**coordinates**)
(This is the point to <u>move to</u>. *OSNAP*s can also be used here.)
Command:

Keep in mind that *OSNAP*s can be used when PICKing any point. It is often helpful to toggle *ORTHO* or *POLAR ON* to force the *Move* in a horizontal or vertical direction.

If you know a specific distance and an angle that the set of objects should be moved, Polar Tracking or relative polar coordinates can be used. In the following sequence, relative polar coordinates are used to move objects 2 units in a 30 degree direction (see Fig. 9-4, relative polar coordinates).

Command: **move**
Select Objects: **PICK**
Select Objects: **PICK**
Select Objects: **Enter**
Specify base point or displacement: **X,Y (coordinates)**
Specify second point of displacement or <use first point as displacement>: **@2<30**
Command:

Dynamic Input and direct distance entry can be used effectively with *Move*. For example, assume you wanted to move the right side view of a multiview drawing 20 units to the right (Fig. 9-5). First, invoke the *Move* command and select all of the objects comprising the right side view in response to the "Select Objects:" prompt. The command sequence is as follows:

FIGURE 9-5

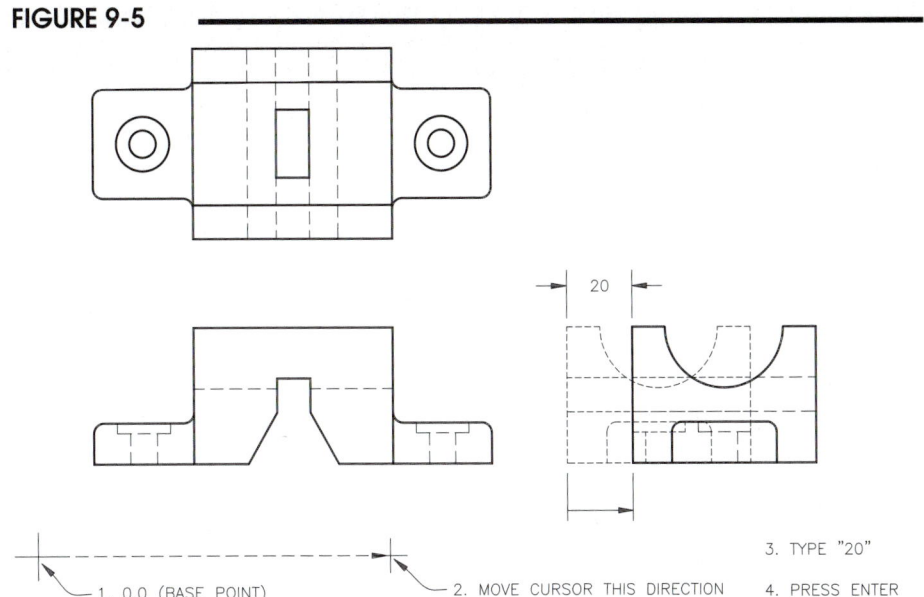

1. 0,0 (BASE POINT)
2. MOVE CURSOR THIS DIRECTION
3. TYPE "20"
4. PRESS ENTER

Command: **move**
Select Objects: **PICK**
Select Objects: **Enter**
Specify base point or displacement: **0,0** (or any value)
Specify second point of displacement or <use first point as displacement>: **20** (with *ORTHO* or *POLAR* on, move cursor to right), then press **Enter**

In response to the "Specify base point or displacement:" prompt, PICK a point or enter any coordinate pairs or single value. At the "Specify second point of displacement :" prompt, move the cursor to the right any distance (using Polar Tracking or *ORTHO* on), then type "20" and press Enter. The value specifies the distance, and the cursor location from the last point indicates the direction of movement.

In the previous example, note that <u>any value</u> can be entered in response to the "Specify base point or displacement:" prompt. If a single value is entered ("3," for example), AutoCAD recognizes it as direct distance entry. The point designated is 3 units from the last PICK point in the direction specified by wherever the cursor is at the time of entry.

Rotate

Pull-down Menu	Command (Type)	Alias (Type)	Short-cut	Screen (side) Menu	Tablet Menu
Modify *Rotate*	*Rotate*	RO	(Edit Mode) *Rotate*	MODIFY2 *Rotate*	V,20

Selected objects can be rotated to any position with this command. After selecting objects to *Rotate,* you select a "base point" (a point to rotate about) then specify an angle for rotation. AutoCAD rotates the selected objects by the increment specified from the original position.

For example, specifying a value of **45** would *Rotate* the selected objects 45 degrees counterclockwise from their current position; a value of **-45** would *Rotate* the objects 45 degrees in a clockwise direction (Fig. 9-6).

FIGURE 9-6

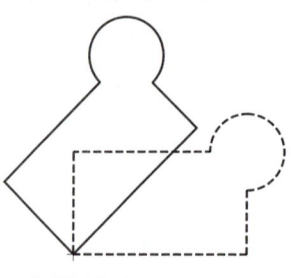

COORDINATE ENTRY

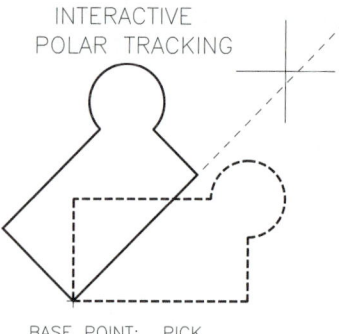

INTERACTIVE POLAR TRACKING

BASE POINT: X,Y
ANGLE: 45

BASE POINT: PICK
ANGLE: TRACK (45)

Command: *rotate*
Current positive angle in UCS: ANGDIR=counterclockwise ANGBASE=0
Select objects: **PICK**
Select objects: **PICK**
Select objects: **Enter** (Indicates completion of object selection.)
Specify base point: **PICK** or (**coordinates**) (Select a point to rotate about.)
Specify rotation angle or [Copy/Reference]: **PICK** or (**coordinates**) (Interactively rotate the set or enter a value for the number of degrees to rotate the object set.)
Command:

The base point is often selected interactively with *OSNAPs*. When specifying an angle for rotation, you can enter an angular value, use Polar Tracking, or turn on *ORTHO* for 90-degree rotation. Note the status of the two related system variables is given: *ANGDIR* (counterclockwise or clockwise rotation) and *ANGBASE* (base angle used for rotation).

The *Copy* option allows you to make a copy of the selected objects—that is, the original set of objects remains in its original orientation while you create rotated copies. After selecting a "base point," type "C" or right-click and select *Copy* at the following prompt.

Specify rotation angle or [Copy/Reference] <0>: *C*

Next, enter a rotation angle value or PICK a point. Using this option would result in, using Figure 9-6 for example, a drawing with both sets of objects: the original set (shown with dashed lines) and the new set.

TIP   The *Reference* option can be used to specify a vector as the original angle before rotation (Fig. 9-7). This vector can be indicated interactively (*OSNAPs* can be used) or entered as an angle using keyboard entry. Angular values that you enter in response to the "New angle:" prompt are understood by AutoCAD as <u>absolute</u> angles for the *Reference* option only.

Specify rotation angle or [Reference]: **R** (Indicates the *Reference* option.)
Specify the reference angle <0>: **PICK** or (**value**) (PICK the first point of the vector.)
Specify second point: **PICK** (Indicates the second point of the vector.)
Specify the new angle: (**value**) (Indicates the new angle.)
Command:

FIGURE 9-7

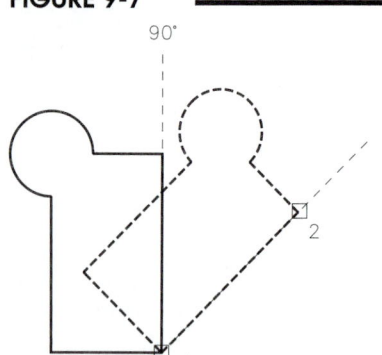

BASE POINT: PICK 1, 2
NEW ANGLE: 90

Scale

Pull-down Menu	Command (Type)	Alias (Type)	Short-cut	Screen (side) Menu	Tablet Menu
Modify Scale	Scale	SC	(Edit Mode) Scale	MODIFY2 Scale	V,21

The *Scale* command is used to increase or decrease the size of objects in a drawing. The *Scale* command does not normally have any relation to plotting a drawing to scale.

After selecting objects to *Scale*, AutoCAD prompts you to select a "Base point:", which is the <u>stationary point</u>. You can then scale the size of the selected objects interactively or enter a scale factor (Fig. 9-8).

FIGURE 9-8

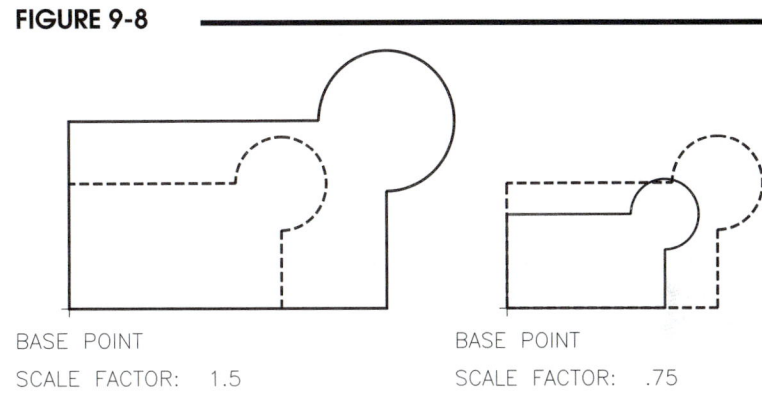

BASE POINT

SCALE FACTOR: 1.5

BASE POINT

SCALE FACTOR: .75

Using interactive input, you are presented with a rubberband line connected to the base point. Making the rubberband line longer or shorter than 1 unit increases or decreases the scale of the selected objects by that proportion; for example, pulling the rubberband line to two units length increases the scale by a factor of two.

    Command: **scale**
    Select objects: **PICK** or (**coordinates**) (Select the objects to scale.)
    Select objects: **Enter** (Indicates completion of the object selection.)
    Specify base point: **PICK** or (**coordinates**) (Select the stationary point.)
    Specify scale factor or [Copy/Reference]: **PICK** or (**value**) or **R** (Interactively scale the set of objects or enter a value for the scale factor.)
    Command:

The *Copy* option allows you to make a copy of the selected objects—that is, the original set of objects remains in its original orientation while you create scaled copies. Type "C" or right-click and select *Copy* in response to "Specify rotation angle or [Copy/Reference] <0>:". Similar to the *Copy* option of the *Rotate* command, using this option would result in two sets of objects: the original selection set and the new set.

It may be desirable in some cases to use the **Reference** option to specify a value or two points to use as the reference length. This length can be indicated interactively (*OSNAPs* can be used) or entered as a value. This length is used for the subsequent reference length that the rubberband uses when interactively scaling. For example, if the reference distance is 2, then the "rubberband" line must be stretched to a length greater than 2 to increase the scale of the selected objects.

Scale normally should <u>not</u> be used to change the scale of an entire drawing in order to plot on a specific size sheet. CAD drawings should be created <u>full size</u> in <u>actual units</u>.

TIP

Stretch

	Pull-down Menu	Command (Type)	Alias (Type)	Short-cut	Screen (side) Menu	Tablet Menu
	Modify *Stretch*	*Stretch*	*S*	...	*MODIFY2* *Stretch*	*V,22*

Objects can be made longer or shorter with *Stretch*. The power of this command lies in the ability to *Stretch* groups of objects while retaining the connectivity of the group (Fig. 9-9). When *Stretched*, *Lines* and *Plines* become longer or shorter and *Arcs* change radius to become longer or shorter. *Circles* do not stretch; rather, they move if the center is selected within the Crossing Window.

Objects to *Stretch* should be selected by a <u>Crossing Window or Crossing Polygon only</u>. The *Crossing Window* or *Polygon* should be created so the objects to *Stretch* <u>cross</u> through the window. *Stretch* actually moves the object endpoints that are located within the *Crossing Window*.

FIGURE 9-9

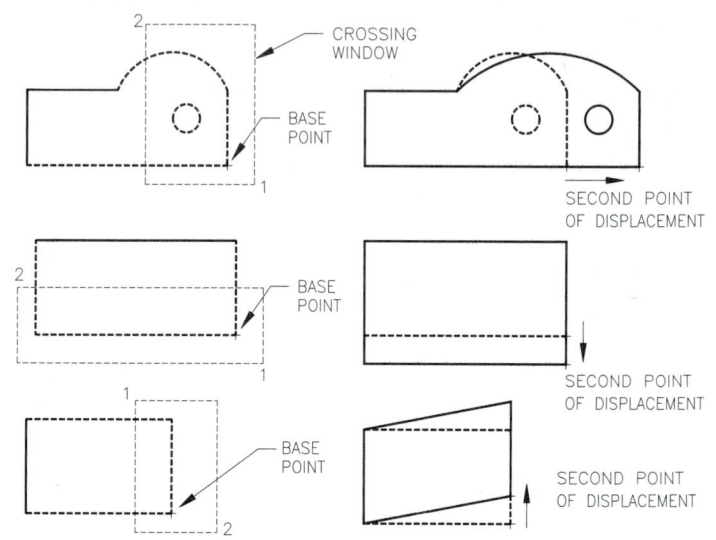

```
Command: stretch
Select objects to stretch by crossing-window or crossing-polygon...
Select objects: PICK
Specify opposite corner: PICK
Select objects: Enter
Specify base point or [Displacement] <Displacement>: (Select a point to stretch from.)
Specify second point or <use first point as displacement>: (Select a point to stretch to.)
```

Stretch can be used to lengthen one object while shortening another. Application of this ability would be repositioning a door or window on a wall (Fig. 9-10).

In summary, the *Stretch* command <u>stretches objects that cross</u> the selection Window and <u>moves objects that are completely within</u> the selection Window, as shown in Figure 9-10.

The *Stretch* command allows you to select objects with other selection methods, such as a pickbox or normal (non-crossing) *Window*. Doing so, however, results in a *Move* since <u>only objects crossing through a crossing window get stretched</u>.

FIGURE 9-10

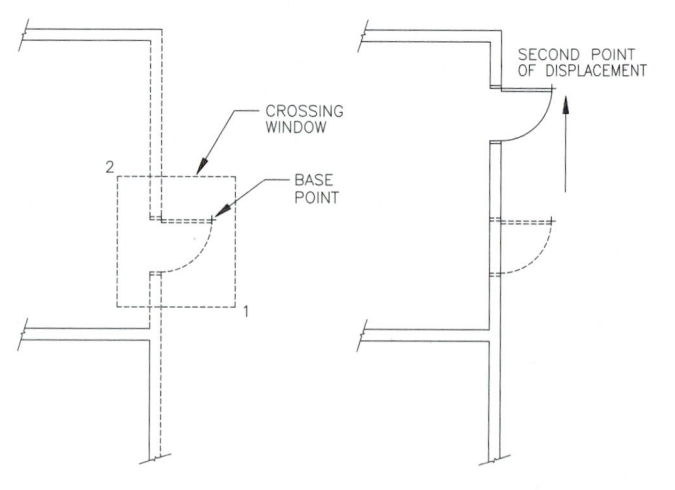

2006

Lengthen

Pull-down Menu	Command (Type)	Alias (Type)	Short-cut	Screen (side) Menu	Tablet Menu
Modify Lengthen	Lengthen	LEN	...	MODIFY2 Lengthen	W,14

Lengthen changes the length (longer or shorter) of linear objects and arcs. No additional objects are required (as with *Trim* and *Extend*) to make the change in length. Many methods are provided as displayed in the command prompt.

Command: **lengthen**
Select an object or [DElta/Percent/Total/DYnamic]:

Select an object
Selecting an object causes AutoCAD to report the current length of that object. If an *Arc* is selected, the included angle is also given.

DElta

Using this option returns the prompt shown next. You can change the current length of an object (including an arc) by any increment that you specify. Entering a positive value increases the length by that amount, while a negative value decreases the current length. The end of the object that you select changes while the other end retains its current endpoint (Fig. 9-11).

Enter delta length or [Angle] <0.0000>:
(**value**) or **a**

FIGURE 9-11

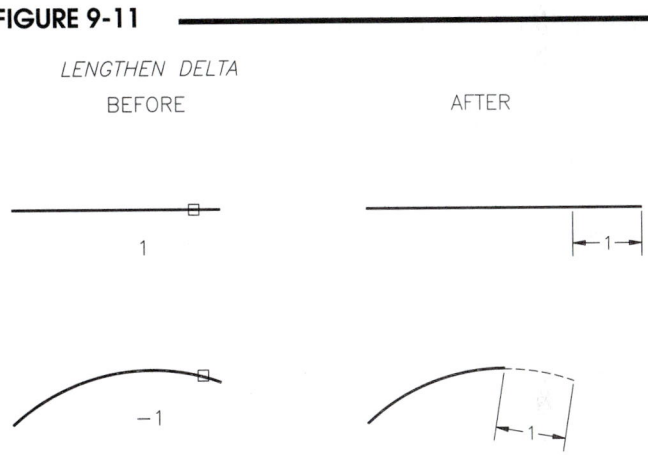

The *Angle* option allows you to change the <u>included angle</u> of an arc (the length along the curvature of the arc can be changed with the *Delta* option). Enter a positive or negative value (degrees) to add or subtract to the current included angle, then select the end of the object to change (Fig. 9-12).

FIGURE 9-12

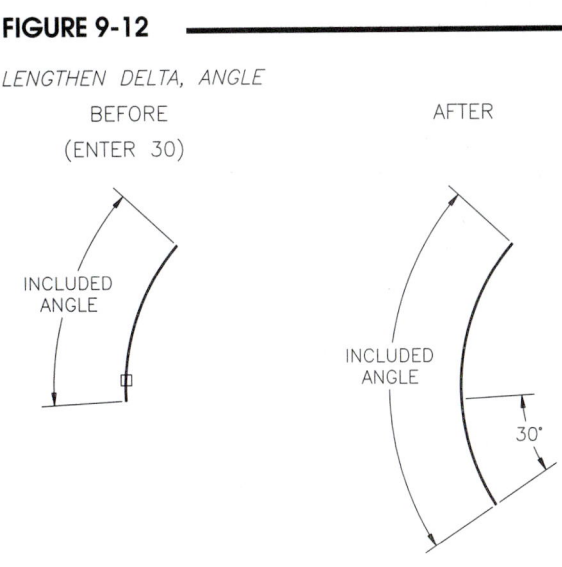

Percent

Use this option if you want to change the length by a percentage of the current total length (Fig. 9-13). For arcs, the percentage applied affects the length and the included angle equally, so there is no *Angle* option. A value of greater than 100 increases the current length, and a value of less than 100 decreases the current length. Negative values are not allowed. The end of the object that you select changes.

FIGURE 9-13

LENGTHEN PERCENT
BEFORE (100%)   AFTER

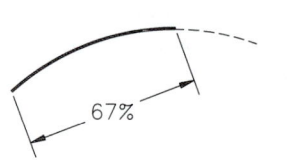

ENTER 133

ENTER 67

Total

This option lets you specify a value for the new total length (Fig. 9-14). Simply enter the value and select the end of the object to change. The angle option is used to change the total included angle of a selected arc.

Specify total length or [Angle] <1.0000)>: (**value**) or **A**

FIGURE 9-14

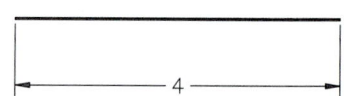

LENGTHEN TOTAL
BEFORE (3 UNITS)   AFTER

ENTER 4

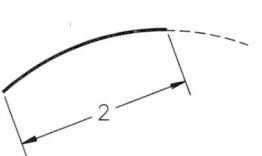

ENTER 2

DYnamic

This option allows you to change the length of an object by dynamic dragging (Fig. 9-15). Select the end of the object that you want to change. Object Snaps and Polar Tracking can be used.

FIGURE 9-15

LENGTHEN DYNAMIC
BEFORE   AFTER

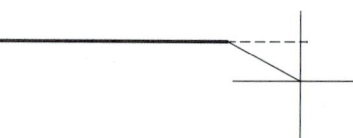

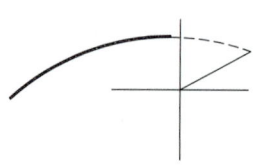

Trim

Pull-down Menu	Command (Type)	Alias (Type)	Short-cut	Screen (side) Menu	Tablet Menu
Modify *Trim*	*Trim*	*TR*	...	*MODIFY2* *Trim*	*W,15*

The *Trim* command allows you to trim (shorten) the end of an object back to the intersection of another object (Fig. 9-16). The middle section of an object can also be *Trimmed* between two inter-secting objects. There are two steps to this command: first, PICK one or more "cutting edges" (existing objects); then PICK the object or objects to *Trim* (portion to remove). The cutting edges are highlighted after selection. Cutting edges themselves can be trimmed if they intersect other cutting edges, but lose their highlight when trimmed.

FIGURE 9-16

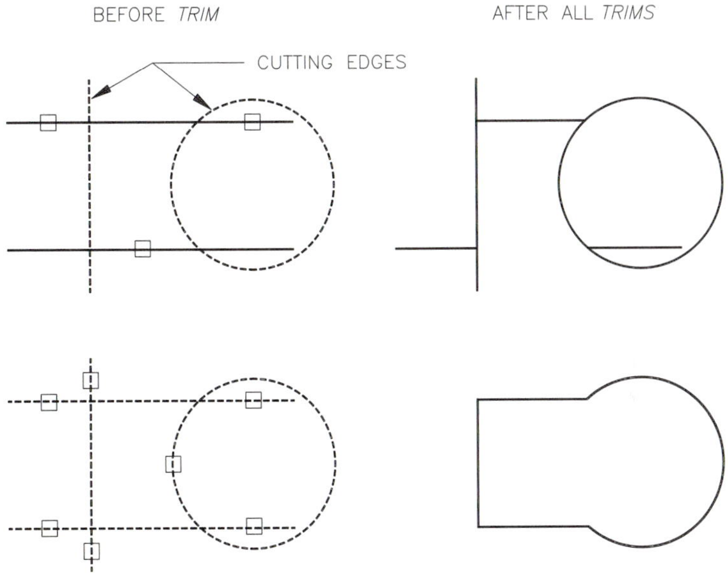

BEFORE *TRIM*   AFTER ALL *TRIMS*

CUTTING EDGES

> Command: **Trim**
> Current settings: Projection=UCS, Edge=None
> Select cutting edges ...
> Select objects or <select all>: **PICK** or press **Enter** to use all objects as cutting edges
> Select objects: **Enter**
> Select object to trim or shift-select to extend or [Fence/Crossing/Project/Edge/eRase/Undo]: **PICK** (Select the part of an object to trim away.)
> Select object to trim or shift-select to extend or [Fence/Crossing/Project/Edge/eRase/Undo]: **PICK**
> Select object to trim or shift-select to extend or [Fence/Crossing/Project/Edge/eRase/Undo]: **Enter**

Select all

If you need to select many objects as <u>cutting edges</u>, you can press Enter at the "Select objects or <select all>:" prompt. In this case, all objects in the drawing become cutting edges; however, the objects are not highlighted.

Fence

The *Fence* and *Crossing Window* options are available for selecting the <u>objects you want to trim</u>. Rather than selecting each object that you want to trim individually with the pickbox, the *Fence* option allows you to select multiple objects using a cross-ing line. The *Fence* can have several line segments (see "Fence" in Chapter 3). This option is especially useful when you have many objects to trim away that are in close proximity to each other such as shown in Figure 9-17.

FIGURE 9-17

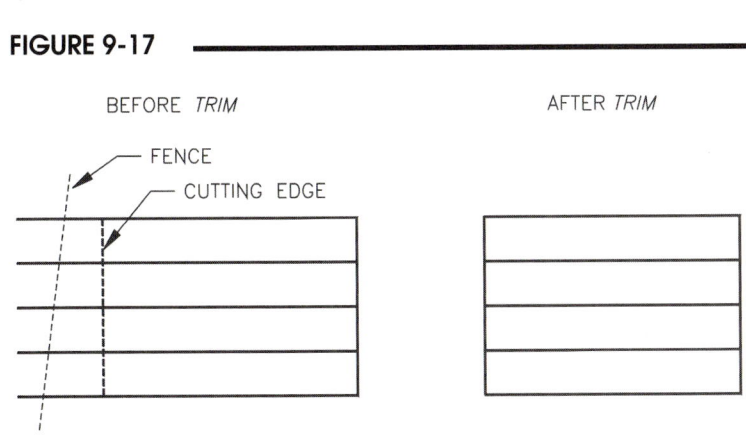

BEFORE *TRIM*   AFTER *TRIM*

FENCE

CUTTING EDGE

Select, Second

This method (the default method) has two steps: select the object to break; then select the second point of the break. The first point used to select the object is <u>also</u> the first point of the break (see Fig. 9-23, Fig. 9-24, Fig. 9-25).

> Command: **break**
> Select object: **PICK** (This is the first point of the break.)
> Specify second break point or [First point]: **PICK** (This is the second point of the break.)
> Command:

Select, 2 Points

This method uses the first selection only to indicate the <u>object</u> to *Break*. You then specify the point that is the first point of the *Break,* and the next point specified is the second point of the *Break*. This option can be used with *OSNAP Intersection* to achieve the same results as *Trim*. The command sequence is as follows:

> Command: **break**
> Select object: **PICK** (This selects only the object to break.)
> Specify second break point or [First point]: **f** (Indicates respecification of the first point.)
> Specify first break point: **PICK** (This is the first point of the break.)
> Specify second break point: **PICK** (This is the second point of the break.)
> Command:

Break at Point

This option creates a break with <u>no</u> space; however, you can select the object you want to *Break* first and the point of the *Break* next (Fig. 9-26).

> Command: **break**
> Select object: **PICK** (This selects only the object to break.)
> Specify second break point or [First point]: **f** (Indicates respecification of the first point.)
> Specify first break point: **PICK** (This is the first point of the break.)
> Specify second break point: **@** (The second point of the break is the last point.)

FIGURE 9-26

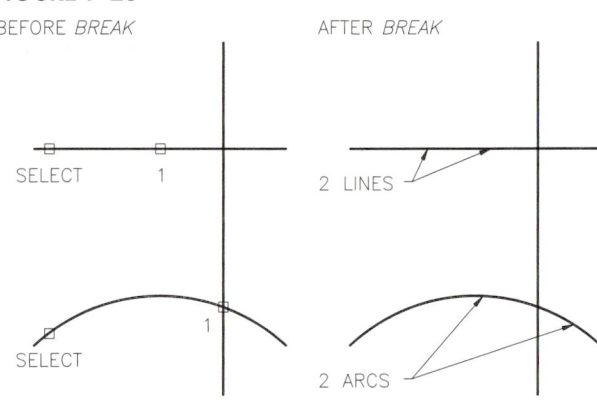

Join

Pull-down Menu	Command (Type)	Alias (Type)	Short-cut	Screen (side) Menu	Tablet Menu
Modify Join	Join	J	...	...	...

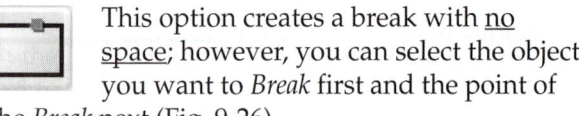

The *Join* command allows you to combine two of the same objects (for example, two collinear *Lines*) into one object (for example, one *Line*). You can also use *Join* to *Close* an *Arc* to form a complete *Circle* or close an elliptical arc to form a complete *Ellipse*. You can *Join* the following objects:

> *Arcs, Elliptical* arcs, *Lines, Polylines,* and *Splines*

Objects to be joined must be located in the same plane. Sample prompts and restrictions for each type of object are given on the next page.

Lines

The *Lines* to join must be collinear, but they can have gaps between them (Fig. 9-27).

    Command: *Join*
    Select source object: **PICK**
    Select lines to join to source: **PICK**
    Select lines to join to source: **Enter**
    1 line joined to source
    Command:

FIGURE 9-27

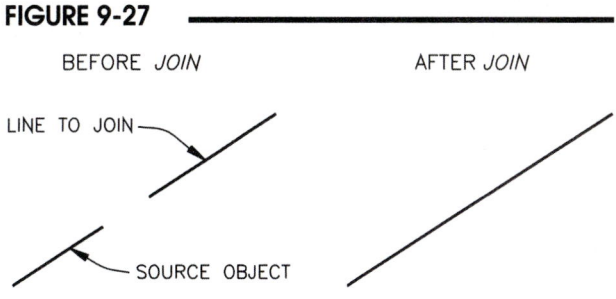

Arc

The *Arc* objects must lie on the same imaginary circle but can have gaps between them. The *Close* option converts the source arc into a *Circle* (Fig. 9-28).

    Command: *Join*
    Select source object: **PICK**
    Select arcs to join to source or [cLose]: *1*
    Arc converted to a circle.
    Command:

FIGURE 9-28

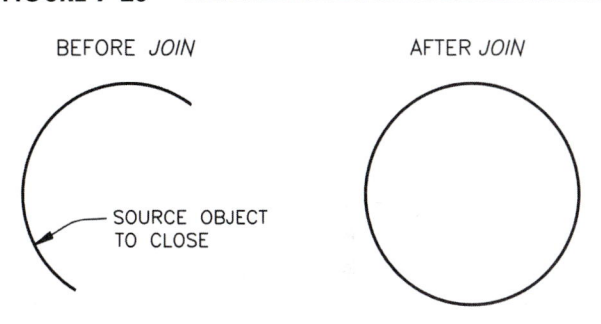

Polyline

Objects to *Join* to *Plines* can be *Plines*, *Lines*, or *Arcs*. The objects cannot have gaps between them and must lie on the UCS XY plane or on a plane parallel to it. If *Arcs* and *Lines* are joined to the *Pline*, they are converted to one *Pline*.

Elliptical Arc

The elliptical arcs must lie on the same ellipse shape but can have gaps between them. The *Close* option closes the source elliptical arc into a complete *Ellipse*.

Spline

The *Spline* objects must lie in the same plane, and the end points to join must be connected.

Copy

Pull-down Menu	Command (Type)	Alias (Type)	Short-cut	Screen (side) Menu	Tablet Menu
Modify *Copy*	*Copy*	*CO* or *CP*	**(Edit Mode)** *Copy Selection*	*MODIFY1* *Copy*	*V,15*

Copy creates a duplicate set of the selected objects and allows placement of those copies. The *Copy* operation is like the *Move* command, except with *Copy* the original set of objects remains in its original location. You specify a "base point:" (point to copy <u>from</u>) and a "specify second point of displacement or <use first point as displacement>:" (point to copy <u>to</u>). See Figure 9-29, on the next page. The command syntax for *Copy* is as follows.

Command: *Copy*
Select objects: **PICK**
Select objects: **Enter**
Specify base point or [Displacement] <Displacement>: **PICK** (This is the point to copy from. Select a point, usually on an object, as a "handle.")
Specify second point or <use first point as displacement>: **PICK** (This is the point to copy to. *OSNAPs* or coordinate entry should be used.)
Specify second point or [Exit/Undo] <Exit>: **Enter**

FIGURE 9-29

In many applications it is desirable to use *OSNAP* options to PICK the "base point:" and the "second point of displacement (Fig. 9-30).

FIGURE 9-30

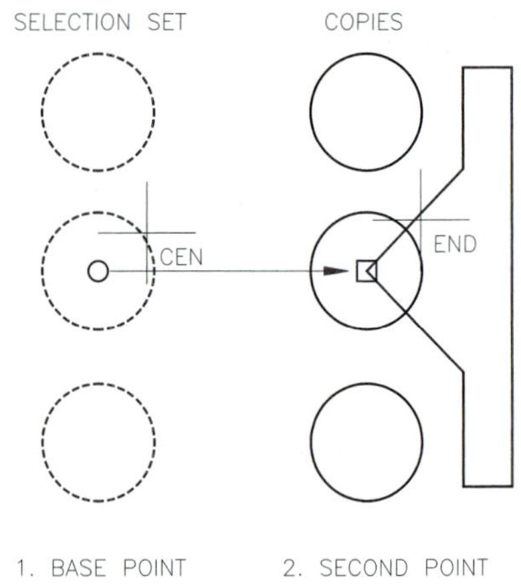

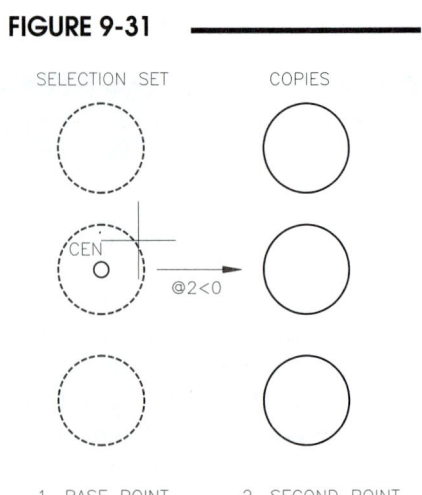

Alternately, you can PICK the "base point:" and use Polar Tracking or enter relative Cartesian coordinates, relative polar coordinates, or direct distance entry to specify the "second point of displacement:" (Fig. 9-31).

FIGURE 9-31

SELECTION SET   COPIES

CEN

@2<0

1. BASE POINT   2. SECOND POINT

The *Copy* command is in a <u>multiple copy mode by default</u>. You do not have to specify if you want multiple copies (as you did with previous versions of AutoCAD). For multiple copies, simply continue to specify a "second point" to create additional copies of the selected objects. For example, in Figure 9-32, only one hole (*Circle*) is drawn, then the *Circle* is copied three times to the *Intersections* to create the other holes.

FIGURE 9-32

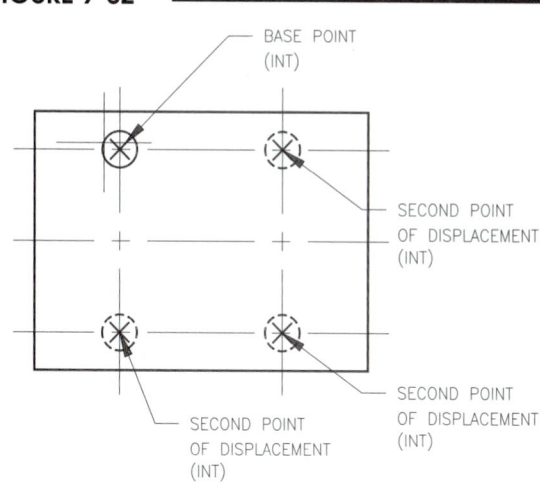

Command: *Copy*
Select objects: **PICK**
Select objects: **Enter**
Specify base point or [Displacement]
<Displacement>: **PICK** (This is the point to copy from.)
Specify second point or <use first point as displacement>: **PICK** (This creates the first copy.)
Specify second point or <use first point as displacement>: **PICK** (This creates a second copy.)
Specify second point or <use first point as displacement>: **PICK** (This creates a third copy.)
Specify second point or [Exit/Undo] <Exit>: **Enter** (completes the command)
Command:

Mirror

Pull-down Menu	Command (Type)	Alias (Type)	Short-cut	Screen (side) Menu	Tablet Menu
Modify *Mirror*	*Mirror*	*MI*	...	*MODIFY1* *Mirror*	*V,16*

This command creates a mirror image of selected existing objects. You can retain or delete the original objects ("source objects"). After selecting objects, you create two points specifying a "rubberband line," or "mirror line," about which to *Mirror*.

The length of the mirror line is unimportant since it represents a vector or axis (Fig. 9-33).

FIGURE 9-33

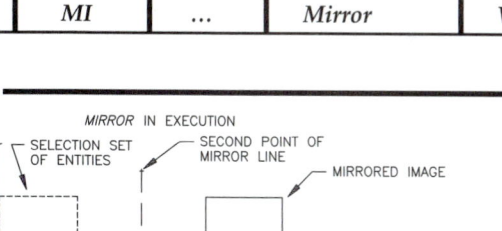

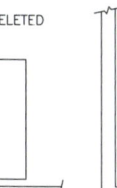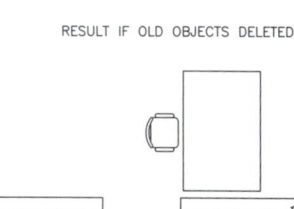

Command: *mirror*
Select objects: **PICK** (Select object or group of objects to mirror.)
Select objects: **Enter** (Press Enter to indicate completion of object selection.)
Specify first point of mirror line: **PICK** or (**coordinates**) (Draw first endpoint of line to represent mirror axis by PICKing or entering coordinates.)
Specify second point of mirror line: **PICK** or (**coordinates**) (Select second point of mirror line by PICKing or entering coordinates.)
Erase source objects? [Yes/No] <N>: **Enter** or **Y** (Press Enter to yield both sets of objects or enter Y to keep only the mirrored set.)
Command:

If you want to *Mirror* only in a vertical or horizontal direction, toggle *ORTHO* or *POLAR On* before selecting the "second point of mirror line."

Mirror can be used to draw the other half of a symmetrical object, thus saving some drawing time (Fig. 9-34).

FIGURE 9-34

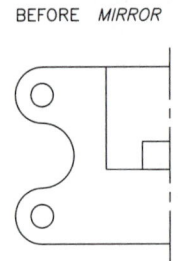
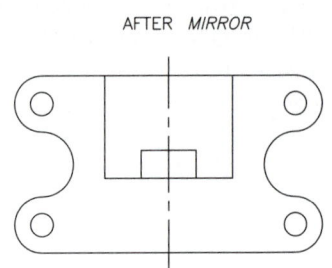

BEFORE   *MIRROR*   AFTER   *MIRROR*

You can control whether <u>text</u> is mirrored by using the *MIRRTEXT* variable. Type *MIRRTEXT* at the Command prompt and change the value to 1 if you want the text to be reflected; otherwise, the default setting of 0 <u>does not</u> mirror text along with other selected objects (Fig. 9-35). Dimensions are <u>not</u> affected by the *MIRRTEXT* variable and therefore are not reflected (that is, if the dimensions are associative). (See Chapter 28 for details on dimensions.)

TIP

FIGURE 9-35

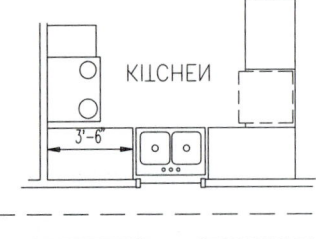
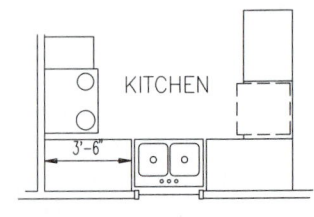
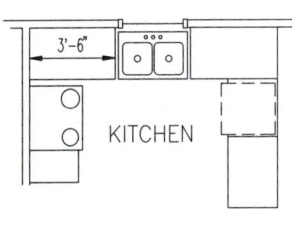
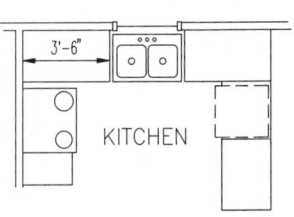

MIRRTEXT = 1   MIRRTEXT = 0

Offset

Pull-down Menu	Command (Type)	Alias (Type)	Short-cut	Screen (side) Menu	Tablet Menu
Modify *Offset*	*Offset*	*O*	...	*MODIFY1* *Offset*	V,17

Offset creates a <u>parallel copy</u> of selected objects. Selected objects can be *Lines*, *Arcs*, *Circles*, *Plines*, or other objects. *Offset* is a very useful command that can increase productivity greatly, particularly with *Plines*.

Depending on the object selected, the resulting *Offset* is drawn differently (Fig. 9-36). *Offset* creates a parallel copy of a *Line* equal in length and perpendicular to the original. *Arcs* and *Circles* have a concentric *Offset*. *Offsetting* closed *Plines* or *Splines* results in a complete parallel shape.

FIGURE 9-36

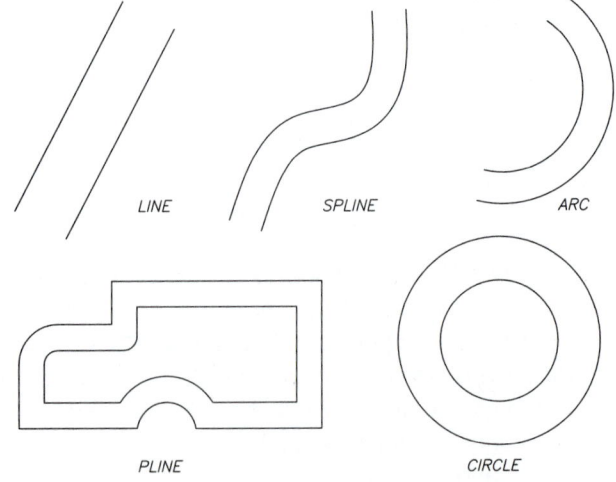

LINE   SPLINE   ARC

PLINE   CIRCLE

Distance

The default option (*distance*) prompts you to specify a distance between the original object and the newly offset object (Fig. 9-37, left).

FIGURE 9-37

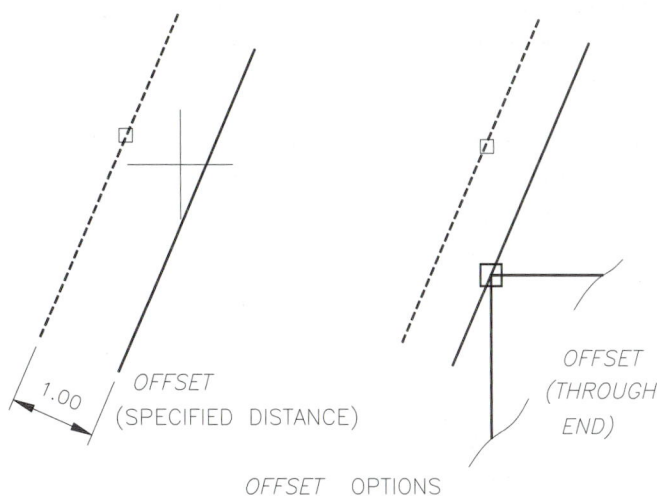

```
Command: offset
Current settings: Erase source=No  Layer=Source  OFFSETGAPTYPE=0
Specify offset distance or [Through/Erase/Layer] <1.0000>: (value) or PICK (Specify an offset distance
by entering a value or picking two points.)
Select object to offset or [Exit/Undo] <Exit>: PICK  (Select the object to offset.)
Specify point on side to offset or [Exit/Multiple/Undo] <Exit>: PICK (Specify which side of original object
to create the offset object.)
Select object to offset or [Exit/Undo] <Exit>: Enter
Command:
```

Through

Use the *Through* option to offset an object so it passes through a specified point (Fig. 9-37, right).

```
Command: offset
Current settings: Erase source=No  Layer=Source  OFFSETGAPTYPE=0
Specify offset distance or [Through/Erase/Layer] <1.0000>: t
Select object to offset or [Exit/Undo] <Exit>: PICK
Specify through point or [Exit/Multiple/Undo] <Exit>: PICK  (Select a point for the object to be offset
through.  Normally, an OSNAP mode is used to specify the point.)
```

Erase

Note that the default *Erase source* setting is *No*. With this setting, the original object to offset remains in the drawing. Changing the *Erase* setting to *Yes* causes the selected (source) object to be erased automatically when the offset is created.

```
Specify offset distance or [Through/Erase/Layer] <1.0000>: e
Erase source object after offsetting? [Yes/No] <No>: y
```

Layer

You can use this setting to create an offset object on a different layer than the original layer of the source object. First, using the *Layer* command (see Chapter 11) change the current layer to the desired layer, next change the *Layer* setting of the *Offset* command to *Current*, then create the offset.

```
Specify offset distance or [Through/Erase/Layer] <1.0000>: 1
Enter layer option for offset objects [Current/Source] <Source>: c
```

Notice the power of using *Offset* with closed *Pline* or *Spline* shapes (Fig. 9-38). Because a *Pline* or a *Spline* is one object, *Offset* creates one complete smaller or larger "parallel" object. Any closed shape composed of *Lines* and *Arcs* can be converted to one *Pline* object (see "*Pedit*," Chapter 16).

FIGURE 9-38

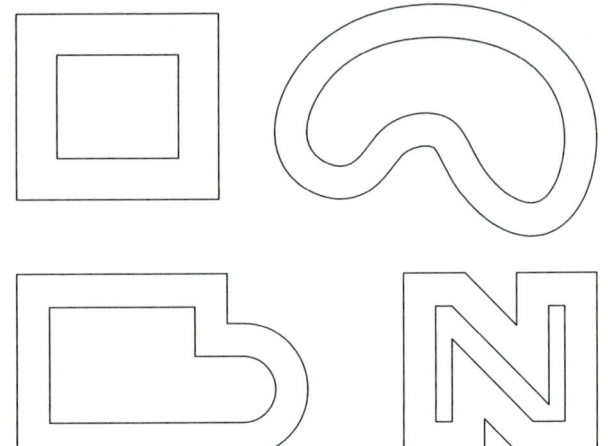

OFFSETGAPTYPE

When you create an offset of a "closed" *Pline* shape with sharp corners, such as the rectangle shown in Figure 9-38, the corners of the new offset object are also sharp (with the default setting of 0). The *OFFSETGAPTYPE* system variable controls how the corners (potential gaps) between the line segments are treated when closed *Plines* are offset (Fig. 9-39).

FIGURE 9-39

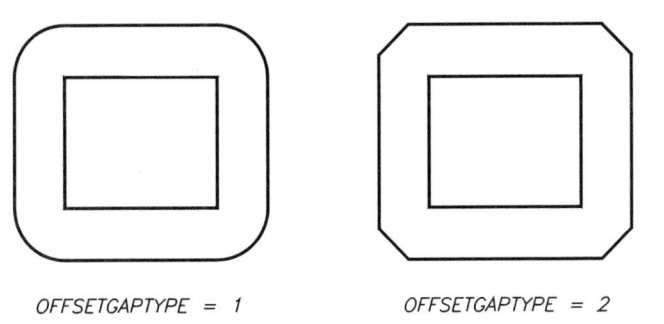

OFFSETGAPTYPE = 1   OFFSETGAPTYPE = 2

0   Fills the gaps by extending the polyline segments
1   Fills the gaps with filleted arc segments (the radius of each arc segment is equal to the offset distance)
2   Fills the gaps with chamfered line segments (the perpendicular distance to each chamfer is equal to the offset distance)

Array

Pull-down Menu	Command (Type)	Alias (Type)	Short-cut	Screen (side) Menu	Tablet Menu
Modify *Array*	*Array*	*AR*	...	*MODIFY1* *Array*	*V,18*

The *Array* command creates either a *Rectangular* or a *Polar* (circular) pattern of existing objects that you select. The pattern could be created from a single object or from a group of objects. *Array* copies a duplicate set of objects for each "item" in the array.

The *Array* command produces a dialog box to enter the desired array parameters and to preview the pattern (Fig. 9-40). The first step is to select the type of array you want (at the top of the box): *Rectangular Array* or *Polar Array*.

FIGURE 9-40

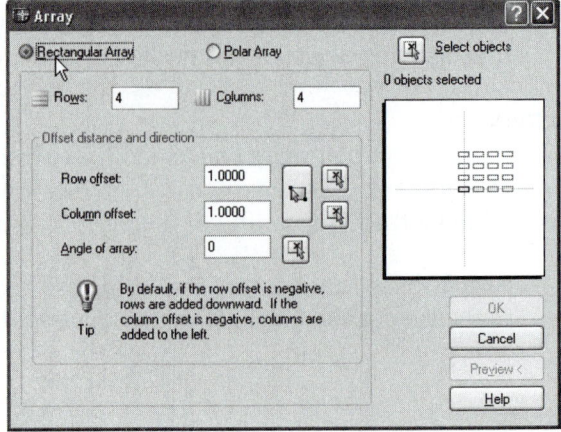

Rectangular Array

A rectangular array is a pattern of objects generated into rows and columns. Selecting a *Rectangular Array* produces the central area of the box and preview image as shown in Figure 9-40. Follow these steps to specify the pattern for the *Rectangular Array*.

FIGURE 9-41

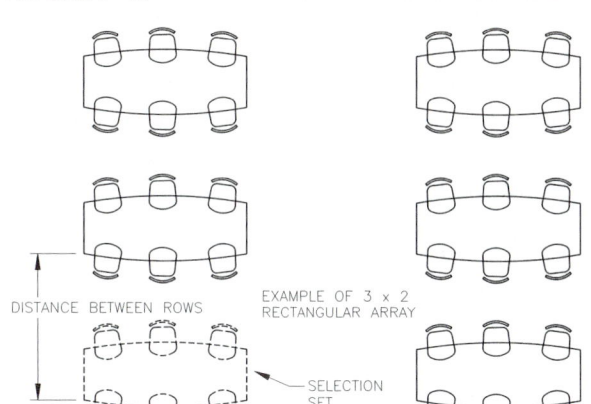

1. Pick the *Select objects* button in the upper-right corner of the dialog box to select the objects you want to array.
2. Enter values in the *Rows:* and *Columns:* edit boxes to indicate how many rows and columns you want. The image area indicates the number of rows and columns you request, but does not indicate the shape of the objects you selected.
3. Enter values in the *Row offset:* and *Column offset:* edit boxes to indicate the distance between the rows and columns (see Fig. 9-41). Enter positive values to create an array to the right (+X) and upward (+Y) from the selected set, or enter negative values to create the array in the –X and –Y directions. You can also use the small select buttons to the far right of these edit boxes to interactively PICK two points for the *Row offset* and *Column offset*. Alternately, select the large button to the immediate right of the *Row offset* and *Column offset* edit boxes to use the "unit cell" method. With this method, you select diagonal corners of a window to specify both directions and both distances for the array (see Fig. 9-42).

FIGURE 9-42

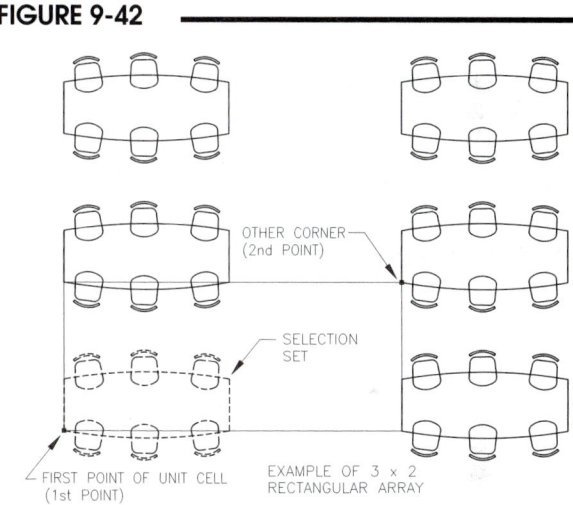

4. If you want to create an array at an angle (instead of generating the rows vertically and columns horizontally), enter a value in the *Angle of array* edit box or use the select button to PICK two points to specify the angle of the array.
5. Select the *Preview* button to temporarily close the dialog box and return to the drawing where you can view the array and choose to *Accept* or *Modify* the array. Selecting *Modify* returns you to the *Array* dialog box.

Polar Array

The *Polar Array* option generates a circular pattern of the selection set. Selecting the *Polar Array* button produces the central area of the box and preview image similar to that shown in Figure 9-43. Typical steps are as follows.

FIGURE 9-43

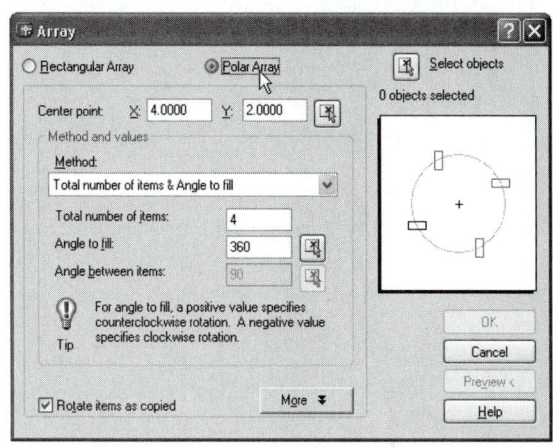

1. Pick the *Select objects* button in the upper-right corner of the dialog box to select the objects you want to array.
2. Enter values for the *Center point* of the circular pattern in the X and Y edit boxes, or use the select button to PICK a center point interactively.

3. Select the *Method* from the drop-down list. The three possible methods are:
   Total number of items & Angle to fill
   Total number of items & Angle between items
   Angle to fill & Angle between items
   Whichever option you choose enables the two applicable edit boxes below.
4. You must supply two of the three possible parameters following to specify the configuration of the array based on your selection in the *Method* drop-down list.
   Total number of items (the total number <u>includes</u> the selected object set)
   Angle to fill
   Angle between items
   Arrays are generated counterclockwise by default. To produce a clockwise array, enter a negative value in *Angle to fill*; or enter a negative value in *Angle to fill*, then switch to the *Total number of items & Angle between items*. The image area indicates the configuration of the array according to the options you selected and values you specified in the *Method and values* section. The image area does not indicate the shape of the object(s) you selected to array.
5. If you want the set of objects to be rotated but remain in the same orientation, remove the check in the *Rotate items as copied* box.
6. Select the *Preview* button to temporarily close the dialog box and return to the drawing where you can view the array and chose to *Accept* or *Modify* the array. Selecting *Modify* returns you to the *Array* dialog box.

Figure 9-44 illustrates a *Polar Array* created using 8 as the *Total number of items*, 360 as the *Angle to fill*, and selecting *Rotate items as copied*.

FIGURE 9-44

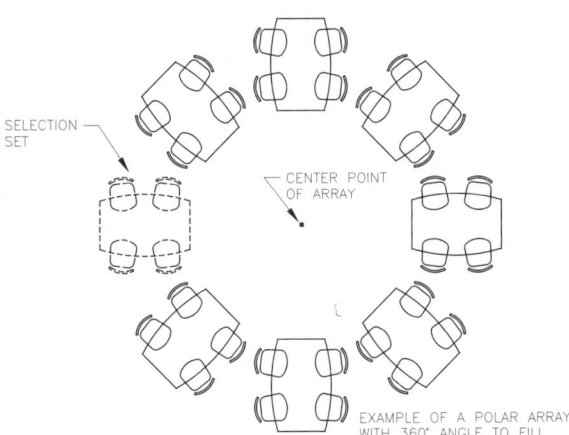

Figure 9-45 illustrates a *Polar Array* created using 5 as the *Total number of items*, 180 as the *Angle to fill*, and selecting *Rotate items as copied*.

FIGURE 9-45

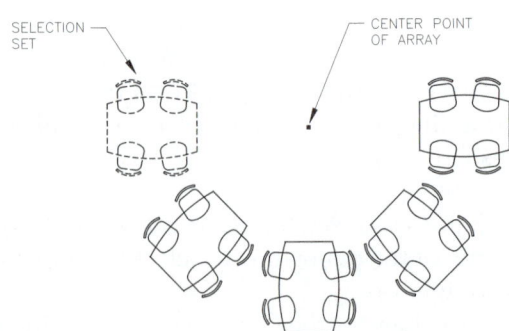

Occasionally an unexpected pattern may result if you remove the check from *Rotate items as copied*. In such a case, the objects may not seem to rotate about the specified center, as shown in Figure 9-46. The reason is that AutoCAD must select a single base point from the set of objects to use for the array such as the end of a line. This base point is not necessarily at the center of the set of objects.

FIGURE 9-46

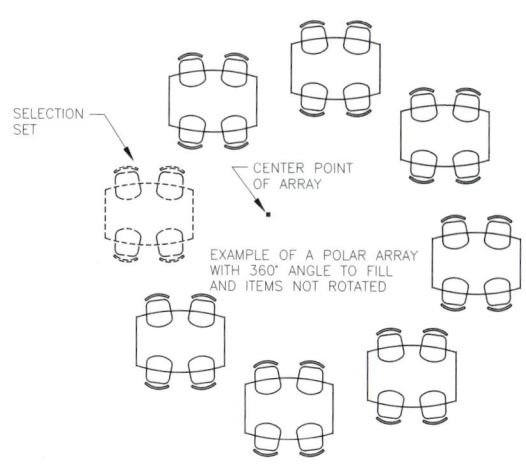

SELECTION SET

CENTER POINT OF ARRAY

EXAMPLE OF A POLAR ARRAY WITH 360° ANGLE TO FILL AND ITEMS NOT ROTATED

The solution to this problem is to use the *More* button near the bottom of the *Array* dialog box to open the lower section of the dialog box shown in Figure 9-47. In this area, remove the check for *Set to object's default*, then specify a *Base point* at the center of the selected set of objects by either entering values or PICKing a point interactively.

FIGURE 9-47

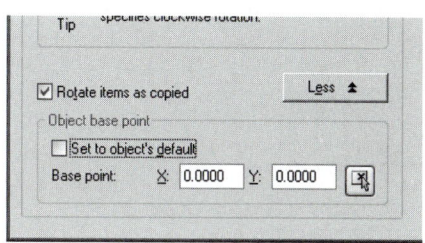

As an alternative to using the *Array* command to produce the *Array* dialog box, you can type *-Array* to use the Command line version. After selecting objects, enter the letter "r" or "p" to produce a *Rectangular Array* or *Polar Array* as shown in the following sequence.

```
Command: -array
Select objects: PICK
Select objects: Enter
Enter the type of array [Rectangular/Polar] <R>: r
Enter the number of rows (—-) <1>: (value)
Enter the number of columns (|||) <1> (value)
Enter the distance between rows or specify unit cell (—-): PICK  or (value)
Specify opposite corner: PICK
Command:
```

Fillet

Pull-down Menu	Command (Type)	Alias (Type)	Short-cut	Screen (side) Menu	Tablet Menu
Modify *Fillet*	*Fillet*	*F*	...	*MODIFY2* *Fillet*	*W,19*

The *Fillet* command automatically rounds a sharp corner (intersection of two *Lines, Arcs, Circles,* or *Pline* vertices) with a radius (Fig. 9-48). You specify only the radius and select the objects' ends to be *Filleted*. The objects to fillet do <u>not</u> have to completely intersect or can overlap. You can specify whether or not the objects are automatically extended or trimmed as necessary (see Fig. 9-49).

FIGURE 9-48

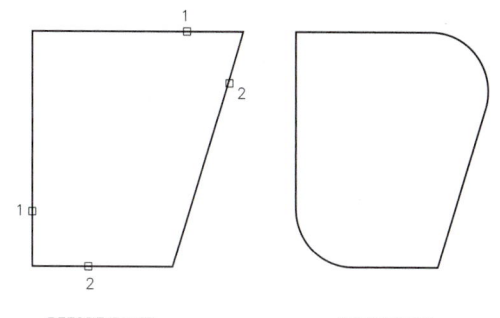

BEFORE *FILLET*   AFTER *FILLET*

Command: **Fillet**
Current settings: Mode = TRIM, Radius = 0.5000
Select first object or [Undo/Polyline/Radius/Trim/Multiple]: **PICK** (Select one *Line, Arc,* or *Circle* near the point where the fillet should be created.)
Select second object or shift-select to apply corner: **PICK** (Select the second object near the desired fillet location.)
Command:

The fillet is created at the corner selected.

Treatment of *Arcs* and *Circles* with *Fillet* is shown in Figure 9-49. Note that the objects to *Fillet* do not have to intersect or can overlap.

FIGURE 9-49

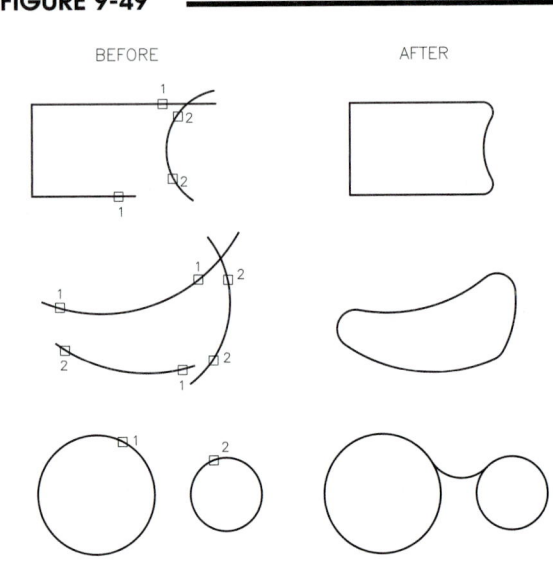

Radius
Use the *Radius* option to specify the desired radius for the fillets. *Fillet* uses the specified radius value for all new fillets until the value is changed.

 If parallel objects are selected, the *Fillet* is automatically created to the correct radius (Fig. 9-50). Therefore, parallel objects can be filleted at any time without specifying a radius value.

FIGURE 9-50

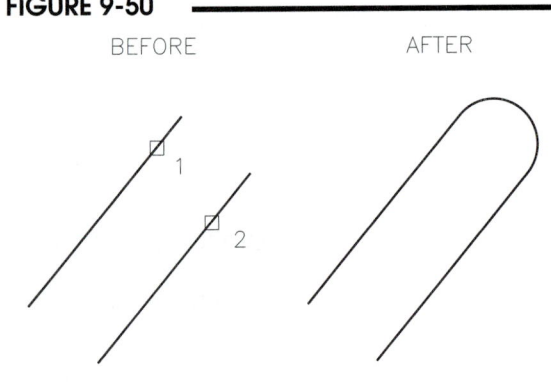

Polyline
Fillets can be created on *Polylines* in the same manner as two *Line* objects. Use *Fillet* as you normally would for *Lines*. However, if you want a fillet equal to the specified radius to be added to each vertex of the *Pline* (except the endpoint vertices), use the *Polyline* option; then select anywhere on the *Pline* (Fig. 9-51).

FIGURE 9-51

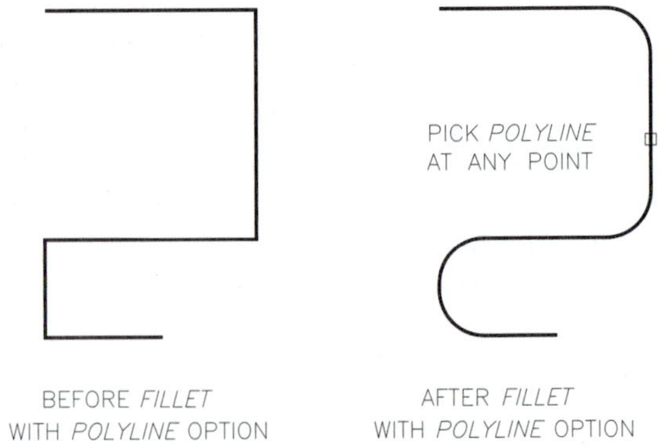

Closed Plines created by the *Close* option of *Pedit* react differently with *Fillet* than *Plines* connected by PICKing matching endpoints. Figure 9-52 illustrates the effect of *Fillet Polyline* on a *Closed Pline* and on a connected *Pline*.

FIGURE 9-52

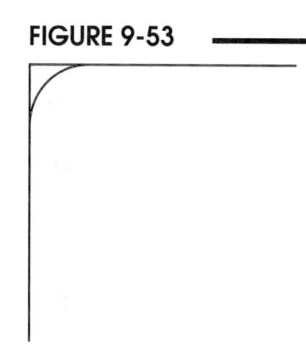

PLINE CONNECTED WITH PICK
AFTER FILLET

CLOSED PLINE
AFTER FILLET

Trim/Notrim

In the previous figures, the objects are shown having been filleted in the *Trim* mode—that is, with automatically trimmed or extended objects to meet the end of the new fillet radius. The *Notrim* mode creates the fillet <u>without</u> any extending or trimming of the involved objects (Fig. 9-53). Note that the command prompt indicates the current mode (as well as the current radius) when *Fillet* is invoked.

FIGURE 9-53

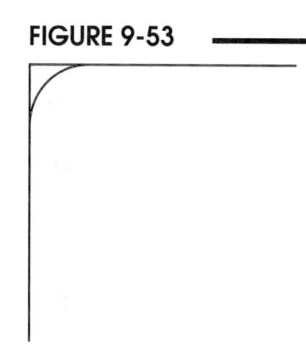

FILLET   NOTRIM

> Command: **Fillet**
> Current settings: Mode = TRIM, Radius = 0.5000
> Select first object or [Undo/Polyline/Radius/Trim/Multiple]: **T** (Invokes the Trim/No trim option)
> Enter Trim mode option [Trim/No trim] <Trim>: **n** (Sets the option to No trim)
> Select first object or [Undo/Polyline/Radius/Trim/Multiple]:

The *Notrim* mode is helpful in situations similar to that in Figure 9-54. In this case, if the intersecting *Lines* were trimmed when the first fillet was created, the longer *Line* (highlighted) would have to be redrawn in order to create the second fillet.

FIGURE 9-54

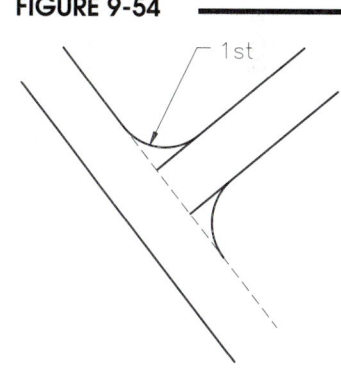

1st

Changing the *Trim/Notrim* option sets the *TRIMMODE* system variable to 0 (*Notrim*) or 1 (*Trim*). This variable controls trimming for both *Fillet* and *Chamfer*, so if you set *Fillet* to *Notrim*, *Chamfer* is also set to *Notrim*.

FILLET   NOTRIM

Multiple

Normally the *Fillet* command allows you to create one fillet, then the command ends. Use the *Multiple* option if you want to create several filleted corners. For example, if you wanted to *Fillet* the four corners of a rectangular shape you could use the *Multiple* option instead of using the *Fillet* command four separate times.

Undo

The *Undo* option operates only in conjunction with the *Multiple* option. *Undo* will remove the most recent of multiple fillets.

2006

Shift-select

If you want to create a sharp corner instead of a rounded (radiused) corner, you can use the Shift-select option. After selecting the first line, hold down the Shift key and select the second line. Because *Fillet* can automatically trim and extend as needed, a sharp corner is created (Fig. 9-55). This action is identical to setting a *Radius* of 0 and selecting two lines normally. The *Trim* mode setting does not affect using Shift-select to create sharp corners.

FIGURE 9-55

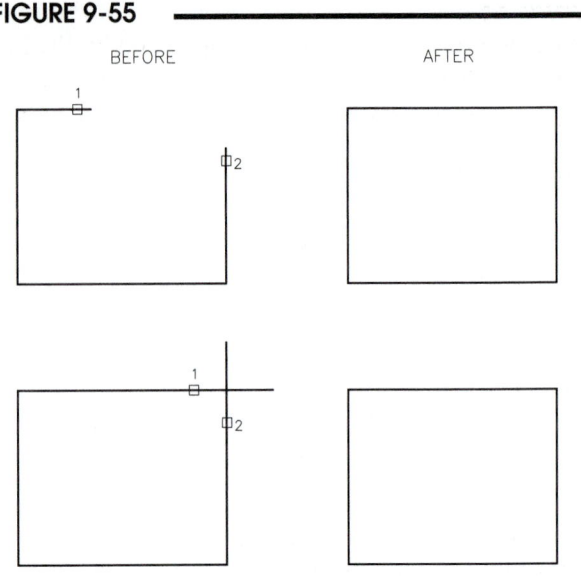

Command: *Fillet*
Current settings: Mode = TRIM, Radius = 0.5000
Select first object or [Undo/Polyline/Radius/Trim/Multiple]: **PICK** (Select the first line.)
Select second object or shift-select to apply corner: **Shift** and **PICK** (Hold down the Shift key and PICK the second line.)

Chamfer

Pull-down Menu	Command (Type)	Alias (Type)	Short-cut	Screen (side) Menu	Tablet Menu
Modify *Chamfer*	*Chamfer*	*CHA*	...	*MODIFY2* *Chamfer*	*W,18*

Chamfering is a manufacturing process used to replace a sharp corner with an angled surface. In AutoCAD, *Chamfer* is commonly used to change the intersection of two *Lines* or *Plines* by adding an angled line. The *Chamfer* command is similar to *Fillet*, but rather than rounding with a radius or "fillet," an angled line is automatically drawn at the distances (from the existing corner) that you specify.

Chamfers can be created by two methods: *Distance* (specify two distances) or *Angle* (specify a distance and an angle). The current method and the previously specified values are displayed at the Command prompt along with the options:

Command: *chamfer*
(TRIM mode) Current chamfer Dist1 = 0.0000, Dist2 = 0.0000
Select first line or [Undo/Polyline/Distance/ Angle/Trim/mEthod/Multiple]:

mEthod

Use this option to indicate which of the two methods you want to use: *Distance* (specify 2 distances) or *Angle* (specify a distance and an angle).

Distance

The *Distance* option is used to specify the two values applied <u>when the *Distance Method* is used</u> to create the chamfer. The values indicate the distances from the corner (intersection of two lines) to each chamfer endpoint (Fig. 9-56).

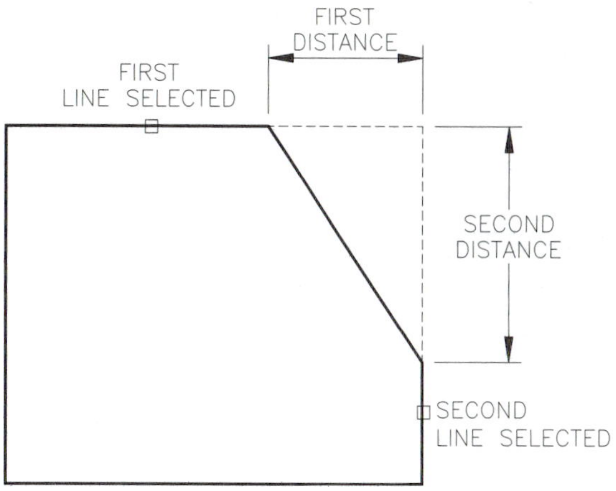

> Command: **chamfer**
> (TRIM mode) Current chamfer Dist1 = 0.0000, Dist2 = 0.0000
> Select first line or [Undo/Polyline/ Distance/Angle/Trim/mEthod/Multiple]: **d**
> Specify first chamfer distance <0.0000>: **(value)**
> (Enter a value for the distance from the existing corner to the endpoint of the chamfer on the first line.)
> Specify second chamfer distance <0.7500>: **Enter** or **(value)** (Press Enter to use the same value as the first, or enter another value.)
> Select first line or [Undo/Polyline/Distance/Angle/Trim/mEthod/Multiple]:

FIGURE 9-56 —————

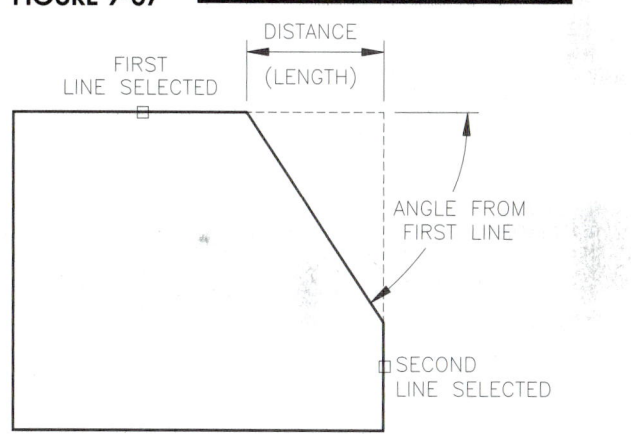

Angle

The *Angle* option allows you to specify the values that are <u>used for the *Angle Method*</u>. The values specify a distance <u>along the first line</u> and an <u>angle from the first line</u> (Fig. 9-57).

> Select first line or [Undo/Polyline/Distance/ Angle/Trim/mEthod/Multiple]: **a**
> Specify chamfer length on the first line <0.0000>: **(value)**
> Specify chamfer angle from the first line <0>: **(value)**

FIGURE 9-57 —————

Trim/Notrim

The two lines selected for chamfering do not have to intersect but can overlap or not connect. The *Trim* setting automatically trims or extends the lines selected for the chamfer (Fig. 9-58), while the *Notrim* setting adds the chamfer without changing the length of the selected lines. These two options are the same as those in the *Fillet* command (see "*Fillet*"). Changing these options sets the *TRIMMODE* system variable to 0 (*Notrim*) or 1 (*Trim*). The variable controls trimming for both *Fillet* and *Chamfer*.

FIGURE 9-58 —————

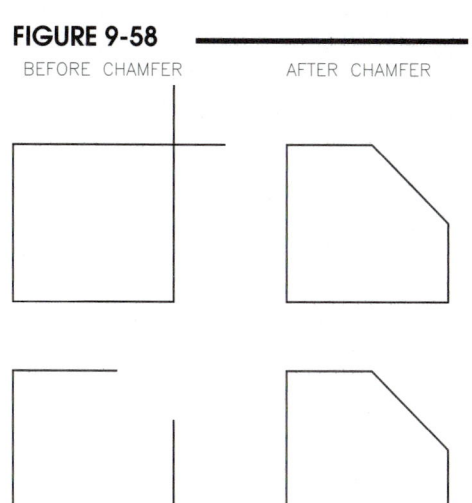

CHAMFER WILL
TRIM OR EXTEND

Polyline

The *Polyline* option of *Chamfer* creates chamfers on all vertices of a *Pline*. All vertices of the *Pline* are chamfered with the supplied distances (Fig. 9-59). The first end of the *Pline* that was <u>drawn</u> takes the first distance. Use *Chamfer* without this option if you want to chamfer only one corner of a *Pline*. This is similar to the same option of *Fillet* (see "Fillet").

FIGURE 9-59

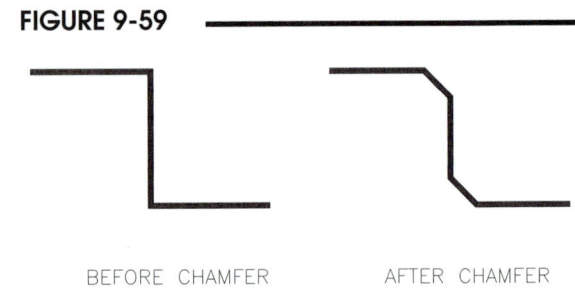

BEFORE CHAMFER   AFTER CHAMFER

A POLYLINE CHAMFER EXECUTES
A CHAMFER AT EACH VERTEX

Multiple

Normally the *Chamfer* command ends after you create one chamfer, similar to the operation of the *Fillet* command. Instead of using the *Chamfer* command several times to create several chamfered corners, use the *Multiple* option.

Undo

The *Undo* option operates only in conjunction with the *Multiple* option. *Undo* will remove the most recent of multiple chamfers.

Shift-select

Similar to the same feature of the *Fillet* command, use the Shift-select option if you want to create a sharp corner instead of a chamfered corner. After selecting the first line to chamfer, hold down the Shift key and select the second line.

CHAPTER EXERCISES

1. *Move*

    Begin a *New* drawing and create the geometry in Figure 9-60A using *Lines* and *Circles*. If desired, set *Polar Distance* to **.25** to make drawing easy and accurate.

    For practice, <u>turn *SNAP OFF*</u> (**F9**). Use the *Move* command to move the *Circles* and *Lines* into the positions shown in Fig. 9-60B. *OSNAPs* are required to *Move* the geometry accurately (since *SNAP* is off). *Save* the drawing as **MOVE1.**

FIGURE 9-60

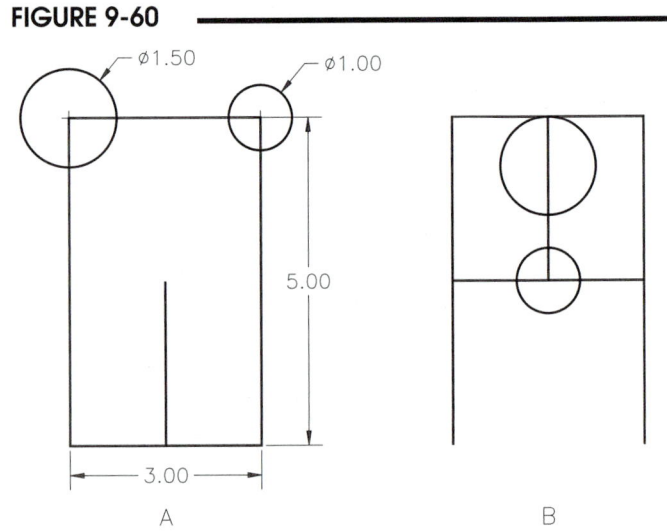

A   B

2. *Rotate*

Begin a *New* drawing and create the geometry in Figure 9-61A.

Rotate the shape into position shown in step B. *SaveAs* **ROTATE1**.

Use the *Reference* option to *Rotate* the box to align with the diagonal *Line* as shown in C. *SaveAs* **ROTATE2**.

FIGURE 9-61

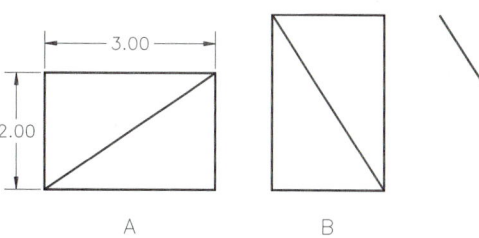

3. *Scale*

Open **ROTATE1** to again use the shape shown in Figure 9-62B and *SaveAs* **SCALE1**. *Scale* the shape by a factor of **1.5**.

Open **ROTATE2** to again use the shape shown in C. Use the *Reference* option of *Scale* to increase the scale of the three other *Lines* to equal the length of the original diagonal *Line* as shown. (HINT: *OSNAPs* are required to specify the *Reference length* and *New length*.) *SaveAs* **SCALE2**.

FIGURE 9-62

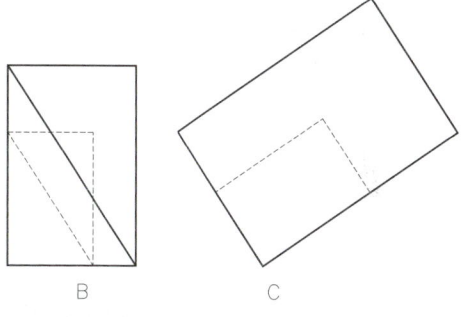

4. *Stretch*

A design change has been requested. *Open* the **SLOTPLATE CH8** drawing and make the following changes.

A. The top of the plate (including the slot) must be moved upward. This design change will add 1" to the total height of the Slot Plate. Use *Stretch* to accomplish the change, as shown in Figure 9-63. Draw the *Crossing Window* as shown.

FIGURE 9-63

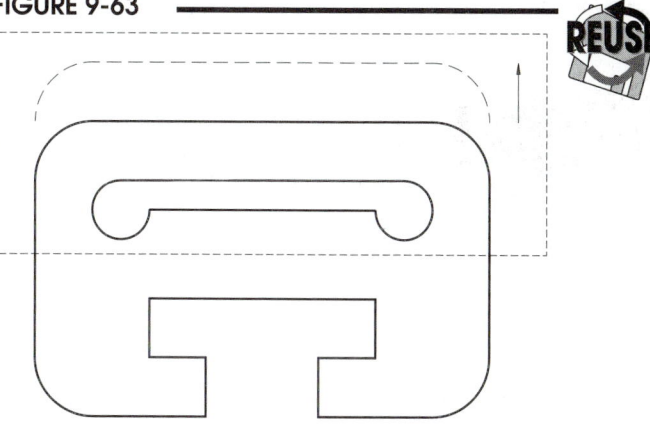

B. The notch at the bottom of the plate must be adjusted slightly by relocating it .50 units to the right, as shown in Figure 9-64. Draw the *Crossing Window* as shown. Use *SaveAs* to reassign the name to **SLOTPLATE 2.**

FIGURE 9-64

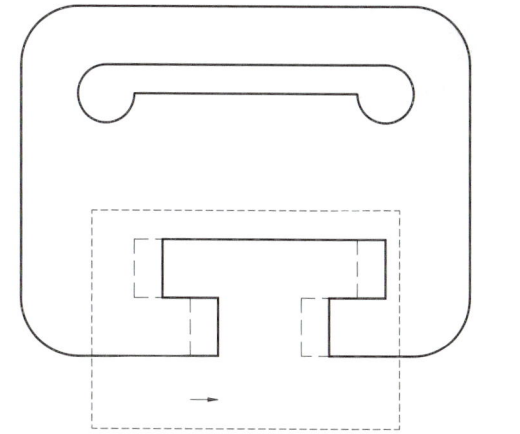

5. *Trim*

   A. Create the shape shown in Figure 9-65A. *SaveAs* **TRIM-EX**.

   B. Use *Trim* to alter the shape as shown in B. *SaveAs* **TRIM1**.

   C. *Open* **TRIM-EX** to create the shapes shown in C and D using *Trim*. *SaveAs* **TRIM2** and **TRIM3**.

6. *Extend*

   Open each of the drawings created as solutions for Figure 9-65 (**TRIM1, TRIM2,** and **TRIM3**). Use *Extend* to return each of the drawings to the original form shown in Figure 9-65A. *SaveAs* **EXTEND1, EXTEND2,** and **EXTEND3**.

7. *Break*

   A. Create the shape shown in Figure 9-66A. *SaveAs* **BREAK-EX**.

   B. Use *Break* to make the two breaks as shown in B. *SaveAs* **BREAK1**. (HINT: You may have to use *OSNAPs* to create the breaks at the *Intersections* or *Quadrants* as shown.)

   C. Open **BREAK-EX** each time to create the shapes shown in C and D with the *Break* command. *SaveAs* **BREAK2** and **BREAK3**.

FIGURE 9-65

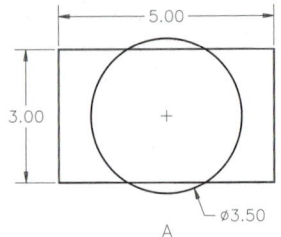
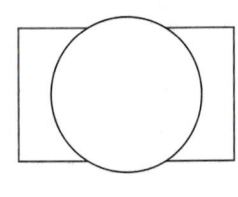
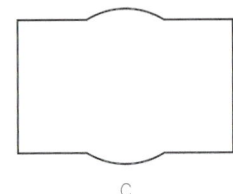
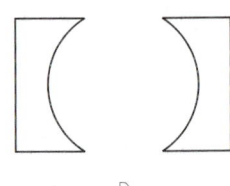

FIGURE 9-66

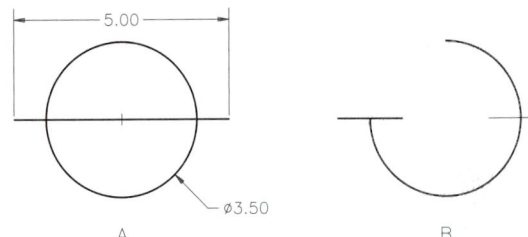

8. **Complete the Table Base**

   A. Open the **TABLE-BASE** drawing you created in the Chapter 7 Exercises. The last step in the previous exercise was the creation of the square at the center of the table base. Now, use *Trim* to edit the legs and the support plate to look like Figure 9-67.

FIGURE 9-67

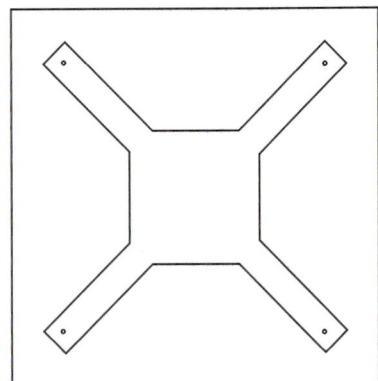

B. Use *Circle* with the *Center, Radius* option to create a 4″ diameter circle to represent the welded support for the center post for this table. Finally, add the other 4 drill holes (.25″ diameter) on the legs at the intersection of the vertical and horizontal lines representing the edges of the base. Use **Polar Tracking** or *Intersection* OSNAP for this step. The table base should appear as shown in Figure 9-68. *Save* the drawing.

FIGURE 9-68

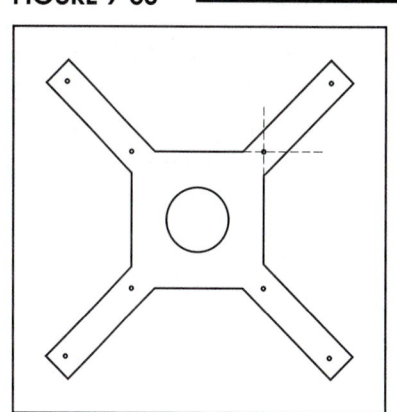

9. *Lengthen*

Five of your friends went to the horse races, each with $5.00 to bet. Construct a simple bar graph (similar to Fig. 9-69) to illustrate how your friends' wealth compared at the beginning of the day.

Modify the graph with *Lengthen* to report the results of their winnings and losses at the end of the day. The reports were as follows: friend 1 made $1.33 while friend 2 lost $2.40; friend 3 reported a 150% increase and friend 4 brought home 75% of the money; friend 5 ended the day with $7.80 and friend 6 came home with $5.60. Use the *Lengthen* command with the appropriate options to change the line lengths accordingly.

Who won the most? Who lost the most? Who was closest to even? Enhance the graph by adding width to the bars and other improvements as you wish (similar to Fig. 9-70). *SaveAs* **FRIENDS GRAPH.**

FIGURE 9-69

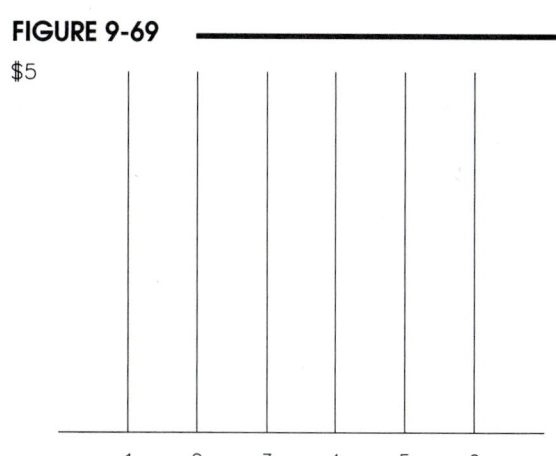

FIGURE 9-70

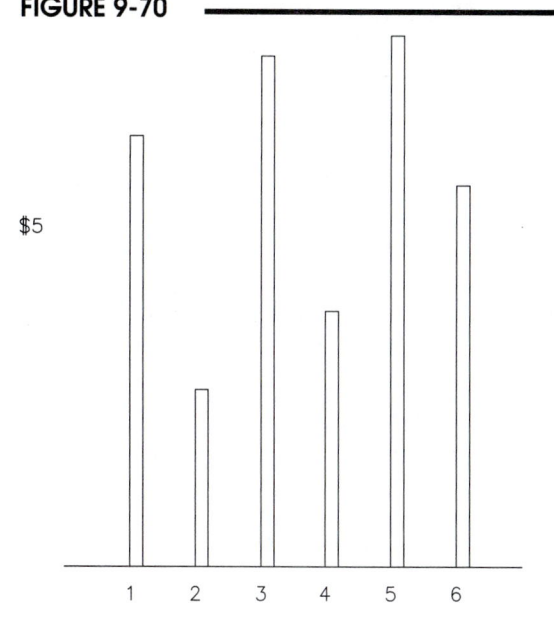

10. *Trim, Extend*

A. *Open* the **PIVOTARM CH7** drawing from the Chapter 7 Exercises. Use *Trim* to remove the upper sections of the vertical *Lines* connecting the top and front views. Compare your work to Figure 9-71.

FIGURE 9-71

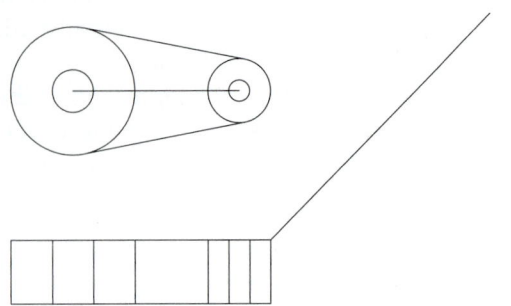

B. Next, draw a horizontal *Line* in the front view between the *Midpoints* of the vertical *Line* on each end of the view, as shown in Figure 9-72. *Erase* the horizontal *Line* in the top view between the *Circle* centers.

FIGURE 9-72

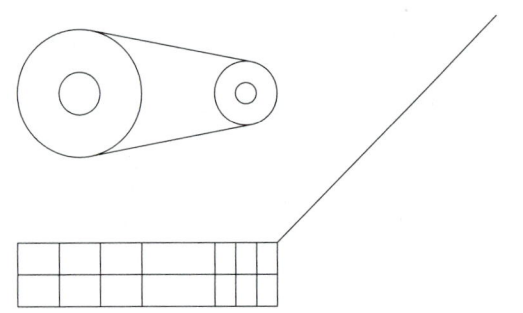

C. Draw vertical *Lines* from the *Endpoints* of the inclined *Line* (side of the object) in the top view down to the bottom *Line* in the front view, as shown in Figure 9-73. Use the two vertical lines as *Cutting edges* for *Trimming* the *Line* in the middle of the front view, as shown highlighted.

FIGURE 9-73

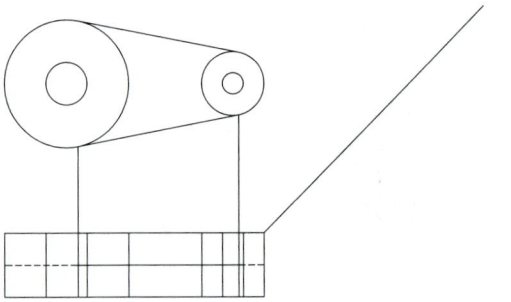

D. Finally, *Erase* the vertical lines used for *Cutting edges* and then use *Trim* to achieve the object as shown in Figure 9-74. Use *SaveAs* and name the drawing **PIVOTARM CH9.**

FIGURE 9-74

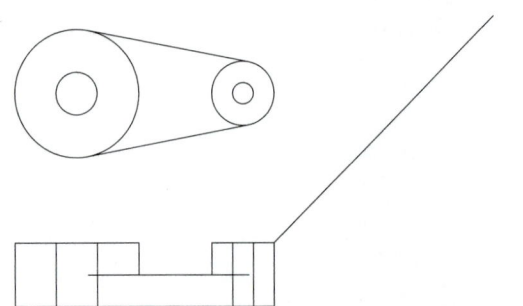

11. **GASKETA**

Begin a *New* drawing. Set *Limits* to **0,0** and **8,6**. Set *SNAP* and *GRID* values appropriately. Create the Gasket as shown in Figure 9-75. Construct only the gasket shape, not the dimensions or centerlines. (HINT: Locate and draw the four .5" diameter *Circles*, then create the concentric .5" radius arcs as full *Circles*, then *Trim*. Make use of *OSNAPs* and *Trim* whenever applicable.) *Save* the drawing and assign the name **GASKETA.**

FIGURE 9-75

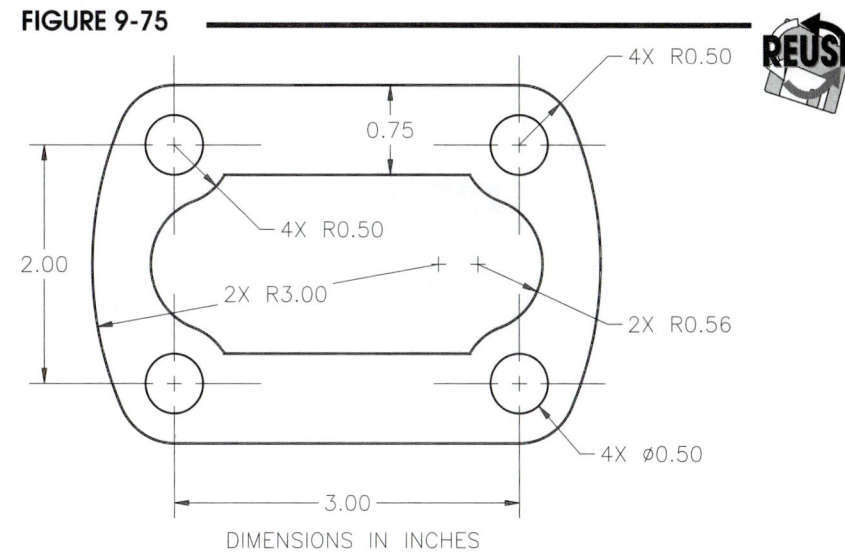

DIMENSIONS IN INCHES

12. **Chemical Process Flow Diagram**

Begin a *New* drawing and re-create the chemical process flow diagram shown in Figure 9-76. Use *Line, Circle, Arc, Trim, Extend, Scale, Break,* and other commands you feel necessary to complete the diagram. Because this is diagrammatic and will not be used for manufacturing, dimensions are not critical, but try to construct the shapes with proportional accuracy. *Save* the drawing as **FLOWDIAG.**

FIGURE 9-76

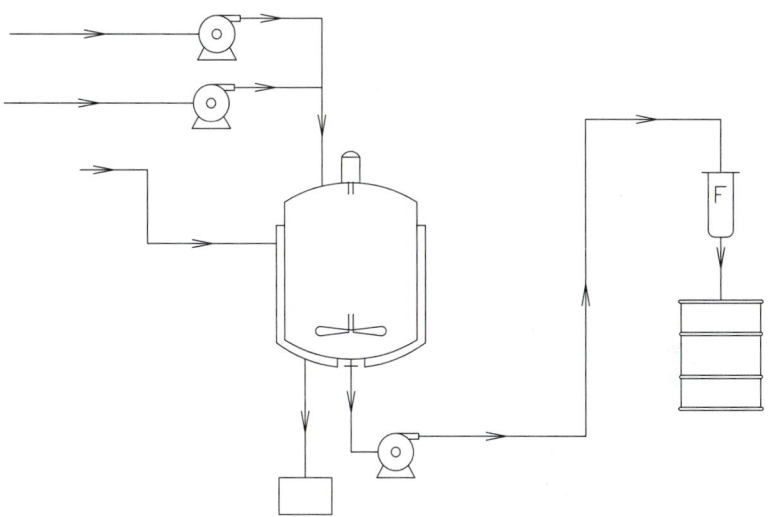

13. *Copy*

Begin a *New* drawing. Set the *Limits* to **24,18**. *Set Polar Snap* to **.5** and set your desired running *OSNAPs*. Turn On *SNAP, GRID, OSNAP,* and *POLAR*. Create the sheet metal part composed of four *Lines* and one **Circle** as shown in Figure 9-77. The lower-left corner of the part is at coordinate 2,3.

FIGURE 9-77

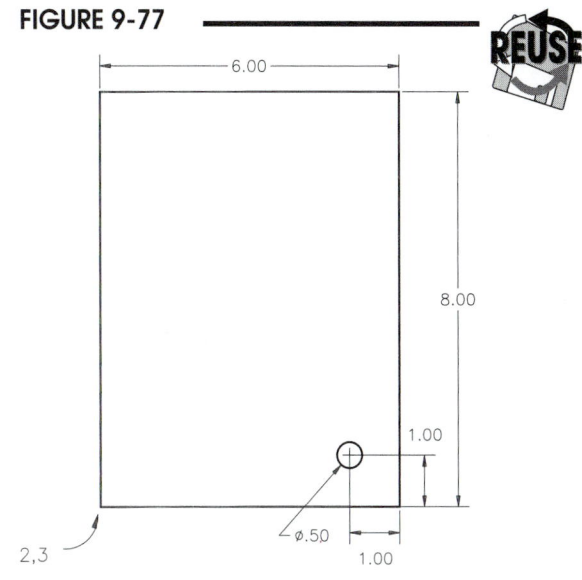

Use *Copy* to create two copies of the rectangle and hole in a side-by-side fashion as shown in Figure 9-78. Allow 2 units between the sheet metal layouts. *SaveAs* **PLATES.**

FIGURE 9-78

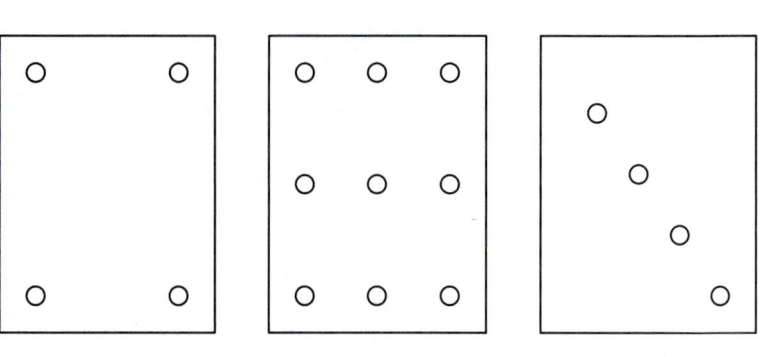

A   B   C

A. Use *Copy* to create the additional 3 holes equally spaced near the other corners of the part as shown in Figure 9-78A. *Save* the drawing.

B. Use *Copy* to create the hole configuration as shown in Figure 9-78B. *Save*.

C. Use *Copy* to create the hole placements as shown in C. Each hole <u>center</u> is **2** units at **125** degrees from the previous one (use relative polar coordinates or set an appropriate *Polar Angle*). *Save* the drawing.

14. *Mirror*

A manufacturing cell is displayed in Figure 9-79. The view is from above, showing a robot <u>centered</u> in a work station. The production line requires 4 cells. Begin by starting a *New* drawing, setting *Units* to *Engineering* and *Limits* to **40′ x 30′**. It may be helpful to set *Polar Snap* to **6″**. Draw one cell to the dimensions indicated. Begin at the indicated coordinates of the lower-left corner of the cell. *SaveAs* **MANFCELL.**

FIGURE 9-79

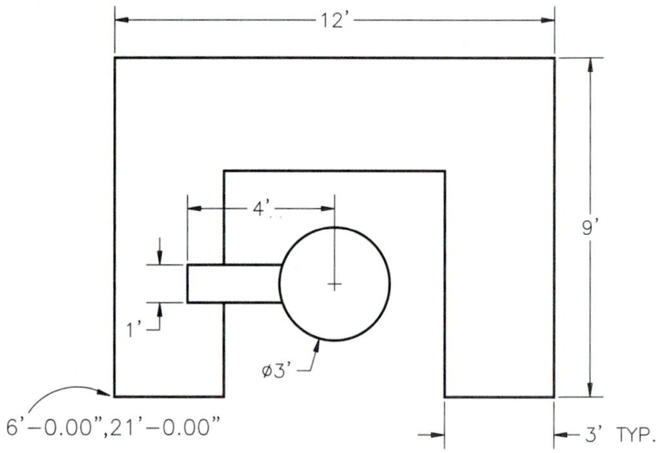

Use *Mirror* to create the other three manufacturing cells as shown in Figure 9-80. Ensure that there is sufficient space between the cells as indicated. Draw the two horizontal *Lines* representing the walkway as shown. *Save*.

FIGURE 9-80

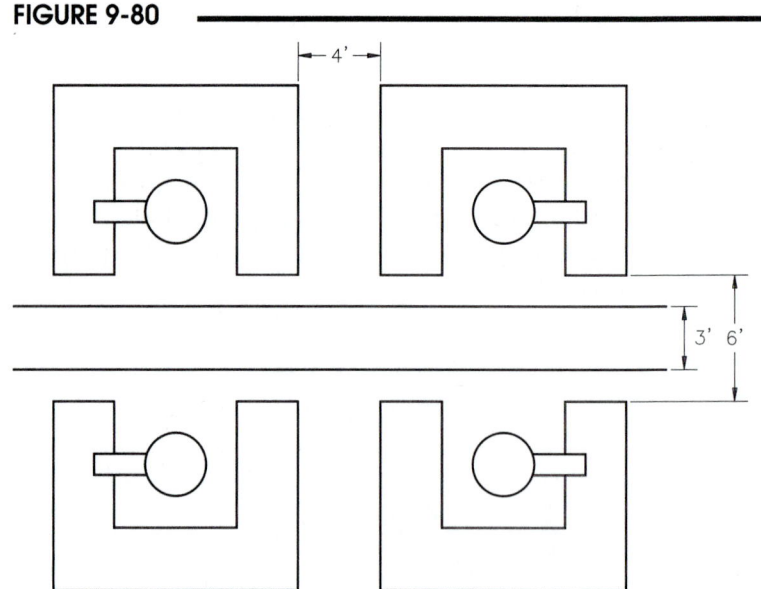

15. *Offset*

Create the electrical schematic as shown in Figure 9-81. Draw *Lines* and *Plines*, then *Offset* as needed. Use *Point* objects with an appropriate (circular) *Point Style* at each of the connections as shown. Check the *Set Size to Absolute Units* radio button (in the *Point Style* dialog box) and find an appropriate size for your drawing. Because this is a schematic, you can create the symbols by approximating the dimensions. Omit the text. *Save* the drawing as **SCHEMATIC1.** Create a *Plot* and check *Scaled to Fit*.

FIGURE 9-81

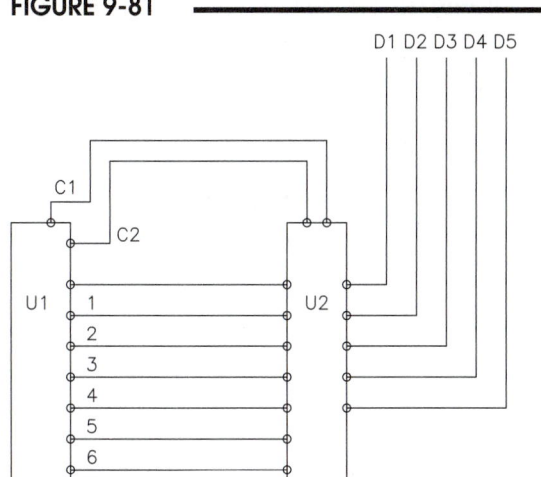

16. *Array, Polar*

Begin a *New* drawing. Create the starting geometry for a Flange Plate as shown in Figure 9-82. *SaveAs* **ARRAY.**

FIGURE 9-82

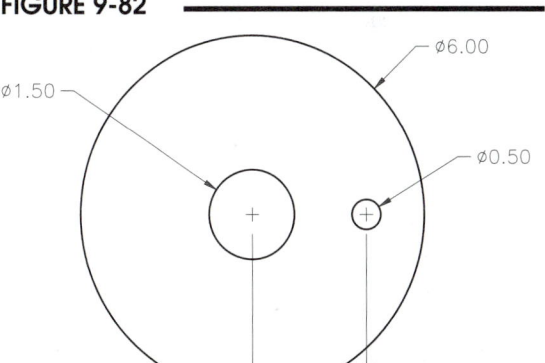

A. Create the *Polar Array* as shown in Figure 9-83A. *SaveAs* **ARRAY1.**

B. *Open* **ARRAY**. Create the *Polar Array* as shown in Figure 9-83B. *SaveAs* **ARRAY2.** (HINT: Use a negative angle to generate the *Array* in a clockwise direction.)

FIGURE 9-83

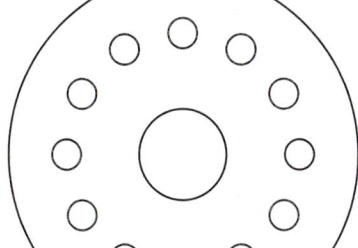

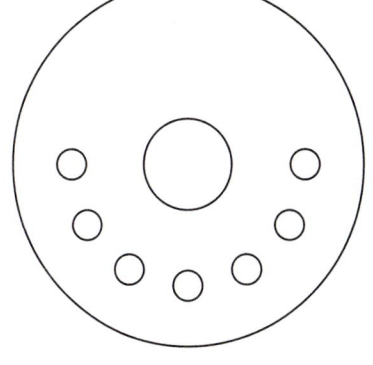

A

B

17. *Array, Rectangular*

Begin a *New* drawing. Select *Start from Scratch*, *Imperial* defaults. Use *Save* and assign the name **LIBRARY DESKS**. Create the *Array* of study carrels (desks) for the library as shown in Figure 9-84. The room size is 36' x 27' (set *Units* and *Limits* accordingly). Each carrel is **30" x 42"**. Design your own chair. Draw the first carrel (highlighted) at the indicated coordinates. Create the *Rectangular Array* so that the carrels touch side to side and allow a 6' aisle for walking between carrels (not including chairs).

FIGURE 9-84

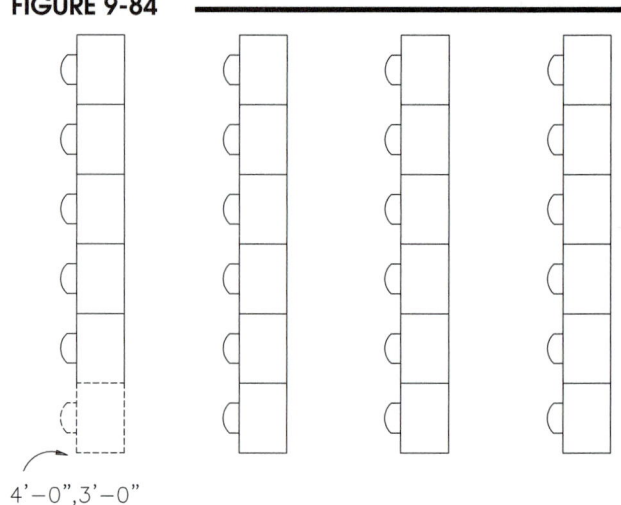

4'–0",3'–0"

18. *Array, Rectangular*

Create the bolt head with one thread as shown in Figure 9-85A. Create the small line segment at the end (0.40 in length) as a *Pline* with a *Width* of .02. *Array* the first thread (both crest and root lines, as indicated in B) to create the schematic thread representation. There is **1** row and **10** columns with **.2** units between each. Add the *Lines* around the outside to complete the fastener. *Erase* the *Pline* at the small end of the fastener (Fig. 9-85B) and replace it with a *Line*. *Save* as **BOLT.**

FIGURE 9-85

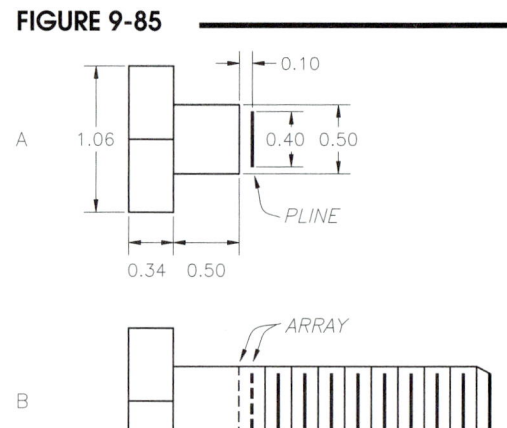

19. *Fillet*

Create the "T" Plate shown in Figure 9-86. Use *Fillet* to create all the fillets and rounds as the last step. When finished, *Save* the drawing as **T-PLATE** and make a plot.

FIGURE 9-86

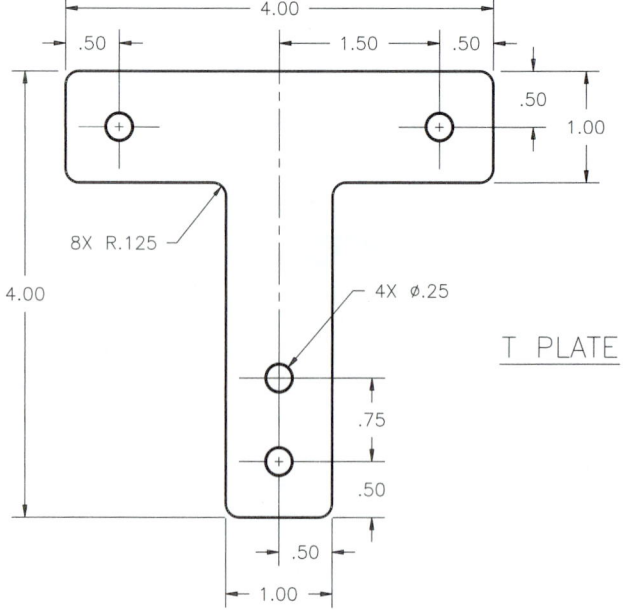

20. *Copy, Fillet*

Create the Gasket shown in Figure 9-87. Use the *Start from Scratch, Imperial* settings when you begin. Include center-lines in your drawing. *Save* the drawing as **GASKETB.**

FIGURE 9-87

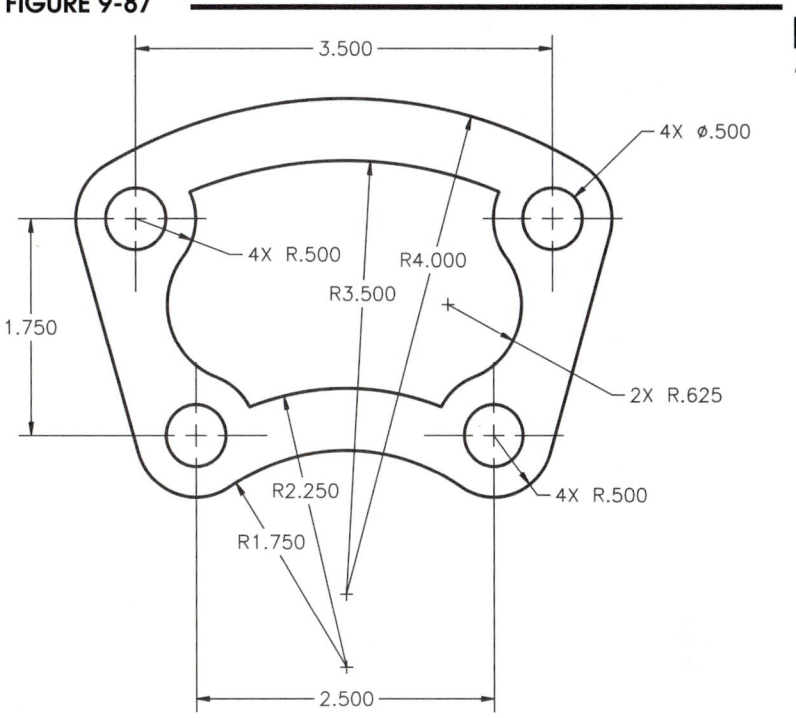

21. *Array*

Create the perforated plate as shown in Figure 9-88. *Save* the drawing as **PERFPLAT**. (HINT: For the *Rectangular Array* in the center, create 100 holes, then *Erase* four holes, one at each corner for a total of 96. The two circular hole patterns contain a total of 40 holes each.)

FIGURE 9-88

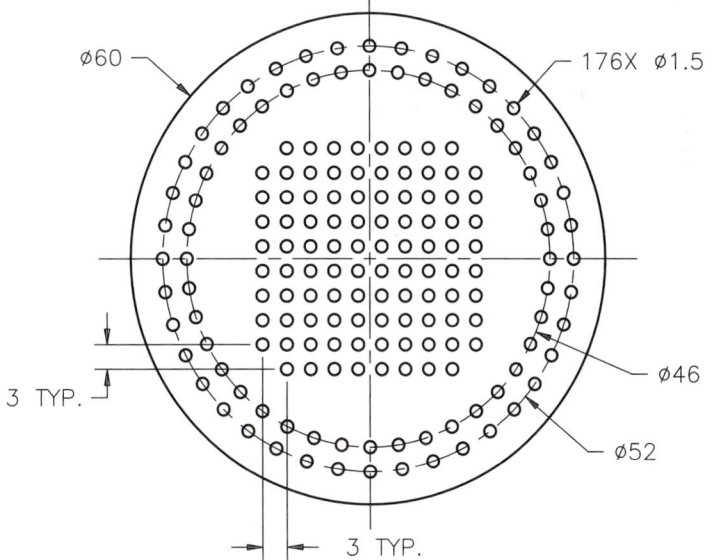

22. *Chamfer*

Begin a *New* drawing and select *Start from Scratch, Imperial* defaults. Set the *Limits* to **279,216**. Next, type *Zoom* and use the *All* option. Set *Snap* to **2** and *Grid* to **10**. Create the Catch Bracket shown in Figure 9-89. Draw the shape with all vertical and horizontal *Lines*, then use *Chamfer* to create the six chamfers. *Save* the drawing as **CBRACKET** and create a plot.

FIGURE 9-89

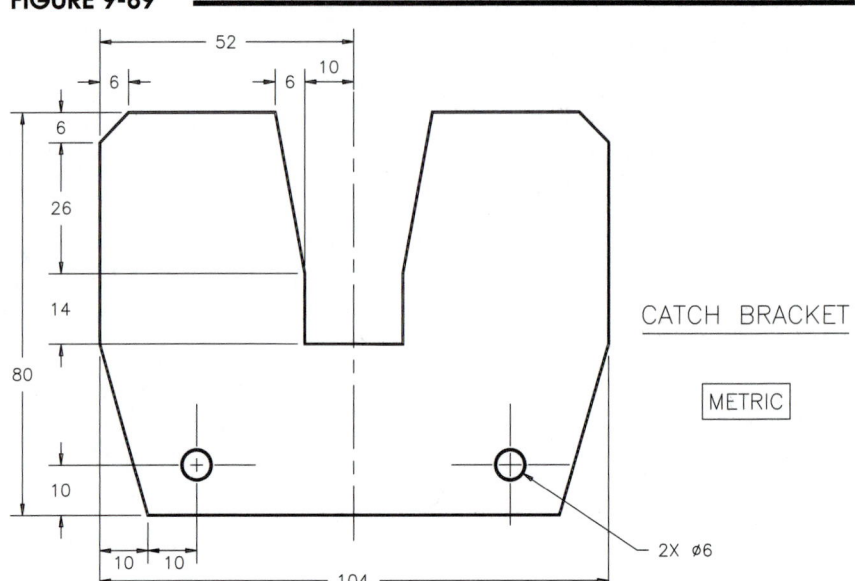

CATCH BRACKET

METRIC

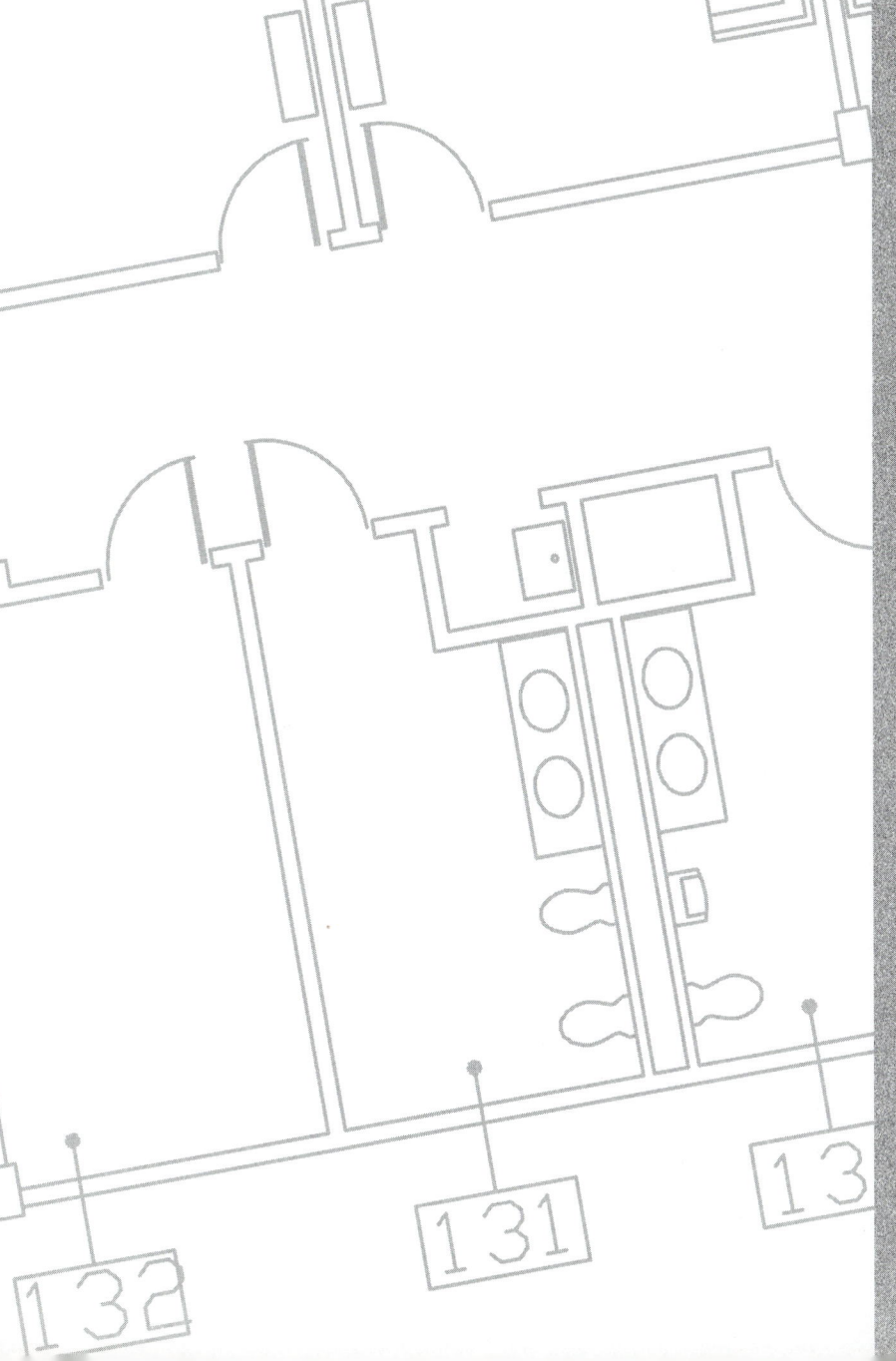

VIEWING COMMANDS

CHAPTER OBJECTIVES

After completing this chapter you should:

1. understand the relationship between the drawing objects and the display of those objects;

2. be able to use all of the *Zoom* options to view areas of a drawing;

3. be able to *Pan* the display about your screen;

4. be able to save and restore *Views*;

5. know how to use *Viewres* to change the display resolution for curved shapes;

6. be able to use *Aerial View* to *Pan* and *Zoom*;

7. be able to turn on and off the *UCS Icon*;

8. be able to create, save, and restore model space viewports using the *Vports* command.

CONCEPTS

The accepted CAD practice is to draw full size using actual units. Since the drawing is a virtual dimensional replica of the actual object, a drawing could represent a vast area (several hundred feet or even miles) or a small area (only millimeters). The drawing is created full size with the actual units, but it can be displayed at any size on the screen. Consider also that CAD systems provide for a very high degree of dimensional precision, which permits the generation of drawings with great detail and accuracy.

Combining those two CAD capabilities (great precision and drawings representing various areas), a method is needed to view different and detailed segments of the overall drawing area. In AutoCAD the commands that facilitate viewing different areas of a drawing are *Zoom*, *Pan*, and *View*.

The viewing commands are found in the *View* pull-down menu (Fig. 10-1).

FIGURE 10-1

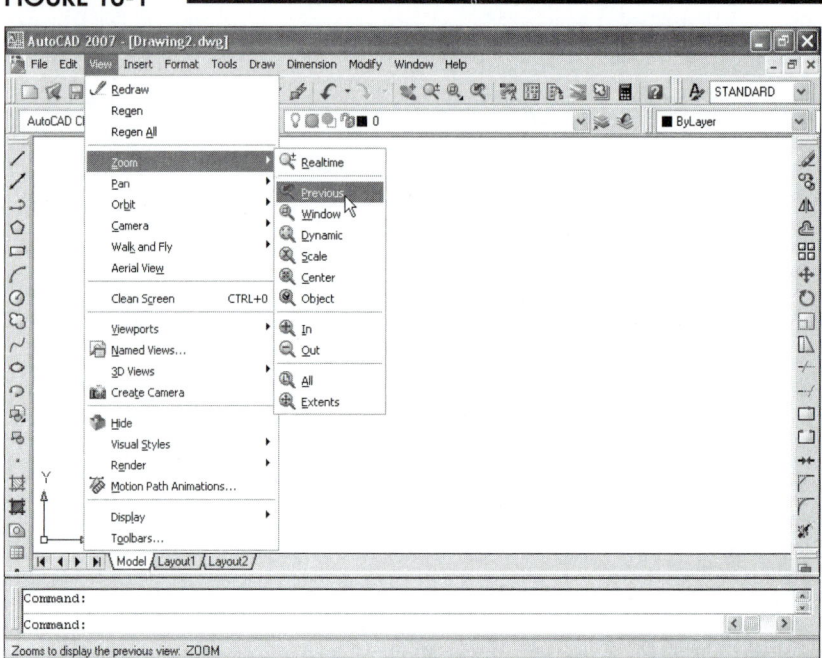

The Standard toolbar contains a group of tools (icon buttons) for the viewing commands located near the right end of the toolbar (Fig. 10-2). The *Realtime* options of *Pan* and *Zoom*, and *Zoom Previous* have icons permanently displayed on the toolbar, whereas the other *Zoom* options are located on flyouts.

Like other commands, the viewing commands can also be invoked by typing the command or command alias, using the screen menu (*VIEW1*), or using the digitizing tablet menu (if available).

The commands discussed in this chapter are:

> *Zoom*, *Pan*, *View*, *Dsviewer*, *Viewres*, *Ucsicon*, *Syswindows*, and *Vports*

The *Realtime* options are the default options for both the *Pan* and *Zoom* commands. *Realtime* options of *Pan* and *Zoom* allow you to <u>interactively</u> change the drawing display by moving the cursor in the drawing area.

FIGURE 10-2

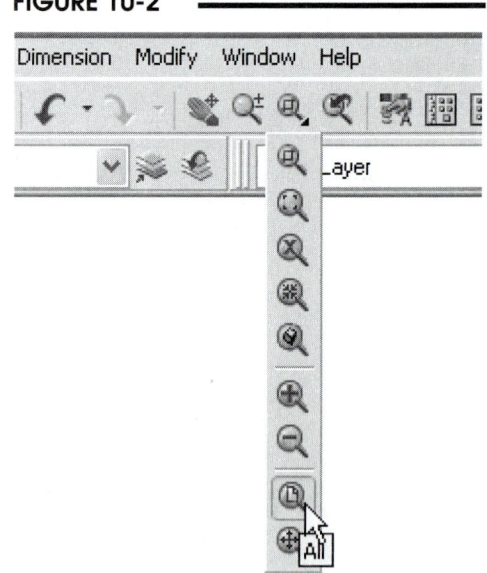

ZOOM AND *PAN* WITH THE MOUSE WHEEL

Zoom and *Pan*

To zoom in means to magnify a small segment of a drawing and to zoom out means to display a larger area of the drawing. <u>Zooming does not change the size of the drawing objects; zooming changes only the display of those objects</u>—the area that is displayed on the screen. All objects in the drawing have the same dimensions before and after zooming. Only your display of the objects changes.

To pan means to move the display area slightly without changing the size of the current view window. Using the pan function, you "drag" the drawing across the screen to display an area outside of the current view in the Drawing Editor.

Although there are many ways to change the area of the drawing you want to view using either the *Zoom* or *Pan* commands, the fastest and easiest method for simple zoom and pan operations is to use the mouse wheel (the small wheel between the two mouse buttons) if you have one. Using the mouse wheel to zoom or pan does not require you to invoke the *Zoom* or *Pan* commands. Additionally, the mouse wheel zoom and pan functions are transparent, meaning that you can use this method while another command is in use. So, if you have a mouse wheel, you can zoom and pan at any time without using any commands.

FIGURE 10-3

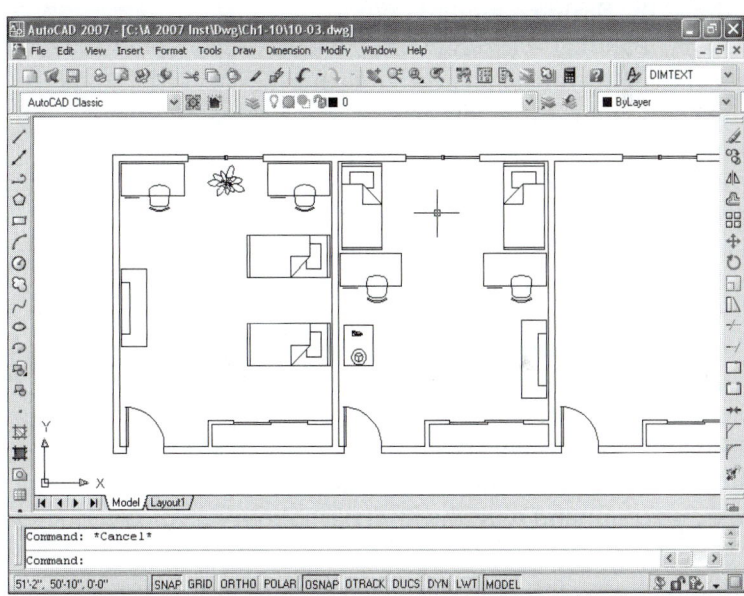

FIGURE 10-4

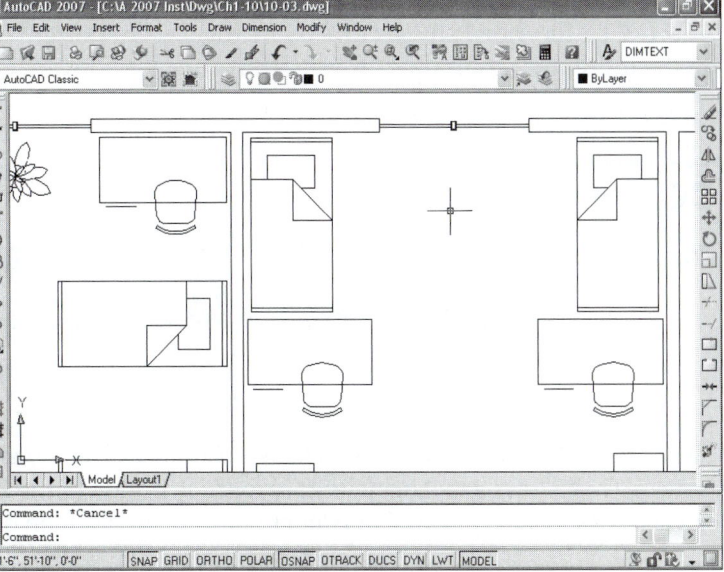

Zoom with the Mouse Wheel

If you have a mouse wheel, you can zoom by simply turning the wheel. Turn the wheel forward to zoom <u>in</u> and turn the wheel backward to zoom <u>out</u>. <u>The current location of the cursor is the center for zooming</u>. In other words, if you want to zoom in to an area, simply locate your cursor on the spot and turn the wheel forward. This type of zooming can be done transparently (during another command operation). You can also pan using the wheel (see "*Pan* with the Mouse Wheel or Third Button").

For example, Figure 10-3 displays a floor plan of a dorm room. Notice the location of the cursor in the upper-right corner of the drawing area. By turning the mouse wheel forward, the display changes by zooming in (enlarging) to the area designated by the cursor location. The resulting display (after turning the mouse wheel forward) is shown in Figure 10-4.

Using the same method, you could zoom out, or change the display from Figure 10-4 to Figure 10-3, by turning the wheel backward. When zooming out, the cursor location also controls the center for zooming.

TIP The amount of zooming in or out that occurs with each turn of the mouse wheel is controlled by the *ZOOMFACTOR* system variable. The default value is 60. Increasing this value causes a greater degree of zooming with each increment of the mouse wheel's forward or backward movement; whereas, decreasing the value results in smaller changes in zooming with the wheel movement. The setting for *ZOOMFACTOR* can be between 3 and 100 and is stored with the individual computer (in the registry).

Pan with the Mouse Wheel or Third Button

If you have a mouse wheel or third mouse button, you can pan by <u>holding down</u> the wheel or button so the "hand" appears (Fig. 10-5), then move the hand cursor in any direction to "drag" the drawing around on the screen. Panning is typically used when you have previously zoomed in to an area but then want to view a different area that is slightly out of the current viewing area (off the screen). Panning with the mouse wheel or third button, similar to zooming with the mouse wheel, does not require you to invoke any commands. In addition, the pan feature is transparent, so you can pan with the wheel or third button while another command is in operation. Using the mouse wheel or third button to pan is essentially the same as using the *Pan* command with the *Realtime* option, except you do not have to invoke the *Pan* command.

FIGURE 10-5

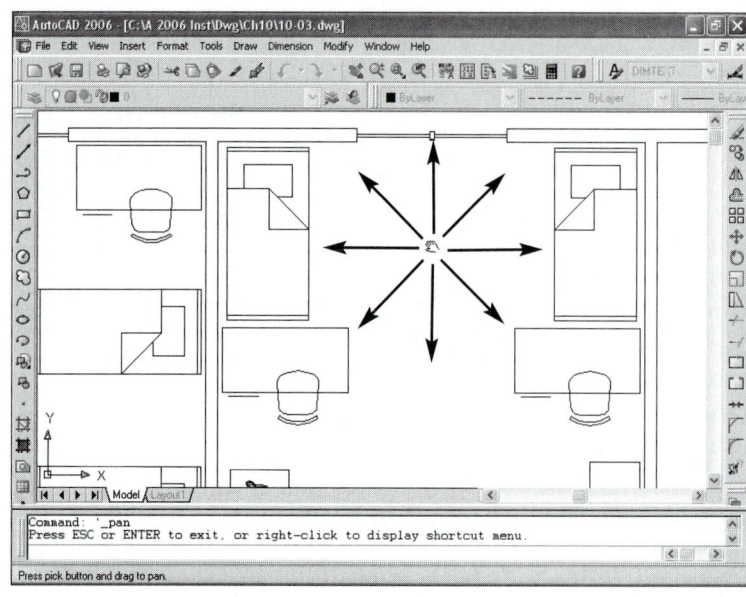

Your ability to pan with the wheel or third button is based on the *MBUTTONPAN* system variable setting. If *MBUTTONPAN* is set to 1, then pressing the wheel or third button activates the *Realtime Pan* feature. If *MBUTTONPAN* is set to 0, pressing the wheel or button triggers the action defined in the ACAD.CUI file, normally set to activate the *Osnap* shortcut menu. The *MBUTTONPAN* setting is saved on the individual computer (in the registry).

COMMANDS

Zoom

Pull-down Menu	Command (Type)	Alias (Type)	Short-cut	Screen (side) Menu	Tablet Menu
View *Zoom >*	*Zoom*	Z	(Default Menu) *Zoom*	*VIEW 1* *Zoom*	*J2 - J,5 or K,3 - K,5*

The *Zoom* options described next can be typed or selected by any method. Unlike most commands, AutoCAD provides a tool (icon button) for each <u>option</u> of the *Zoom* command. If you type *Zoom*, the options can be invoked by typing the first letter of the desired option or pressing Enter for the *Realtime* option.

The Command prompt appears as follows.

Command: **zoom**
Specify corner of window, enter a scale factor (nX or nXP), or
[All/Center/Dynamic/Extents/Previous/Scale/Window/Object] <real time>:

Realtime (Rtzoom)

Realtime is the default option of
Zoom. If you type *Zoom*, just
press Enter to activate the
Realtime option. Alternately, you can
type *Rtzoom* to invoke this option
directly.

With the *Realtime* option you can interactively zoom in and out with <u>vertical</u>
<u>cursor motion</u>. When you activate the
Realtime option, the cursor changes to a
magnifying glass with a plus (+) and a
minus (-) symbol. Move the cursor to
any location on the drawing area and
hold down the PICK (left) mouse button.
Move the cursor up to zoom in and down
to zoom out (Fig. 10-6). Horizontal
movement has no effect. You can zoom
in or out repetitively. Pressing Esc or
Enter exits *Realtime Zoom* and returns to
the Command prompt.

FIGURE 10-6

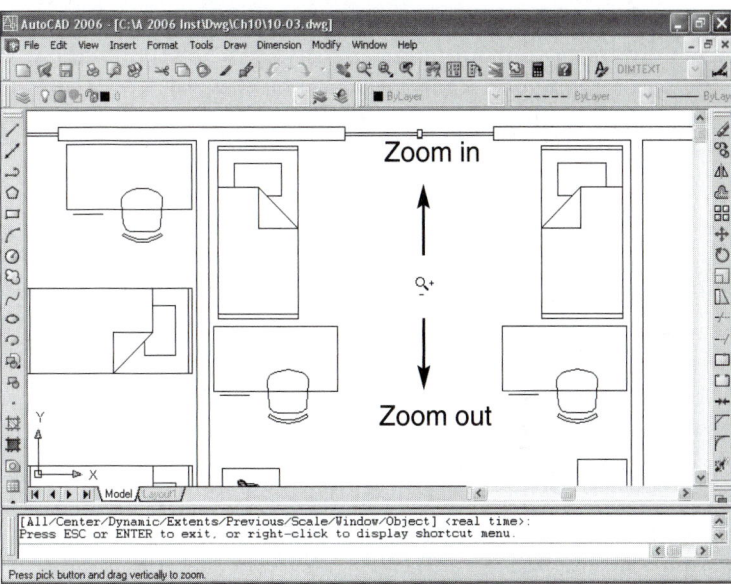

The current drawing window is used to determine the zooming factor. Moving the cursor half the
window height (from the center to the top or bottom) zooms in or out to a zoom factor of 100%. Starting
at the bottom or top allows zooming in or out (respectively) at a factor of 200%.

If you have zoomed in or out repeatedly, you can reach a zoom limit. In this case the plus (+) or
minus (-) symbol disappears when you press the left mouse button. To zoom further with *Rtzoom*, first
type *Regen*.

Pressing the right mouse button (or button #2 on digitizing pucks) produces a small
cursor menu with other viewing options (Fig. 10-7). This same menu and its options
can also be displayed by right-clicking during *Realtime Pan*. The options of the cursor
menu are described below.

FIGURE 10-7

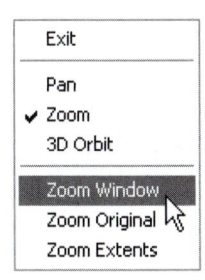

Exit
Select *Exit* to exit *Realtime Pan* or *Zoom* and return to the Command prompt. The
Escape or Enter key can be used to accomplish the same action.

Pan
This selection switches to the *Realtime* option of *Pan*. A check mark appears here if you are currently
using *Realtime Pan*. See "*Pan*" next in this chapter.

Zoom
A check mark appears here if you are currently using *Realtime Zoom*. If you are using the *Realtime* option
of *Pan*, check this option to switch to *Realtime Zoom*.

Zoom Window

Select this option if you want to display an area to zoom in to by specifying a *Window*. This feature operates differently but accomplishes the same action as the *Window* option of *Zoom*. With this option, window selection is made with one PICK. In other words, PICK for the first corner, hold down the left button and drag, then release to establish the other corner. With the *Window* option of the *Zoom* command, PICK once for each of two corners. See "*Window.*"

Zoom Original

Selecting this option automatically displays the area of the drawing that appeared on the screen immediately before using the *Realtime* option of *Zoom* or *Pan*. Using this option successively has no effect on the display. This feature is different from the *Previous* option of the *Zoom* command, in which ten successive previous views can be displayed.

Zoom Extents

Use this option to display the entire drawing in its largest possible form in the drawing area. This option is identical to the *Extents* option of the *Zoom* command. See "*Extents.*"

Window

To *Zoom* with a *Window* is to draw a rectangular window around the desired viewing area. You PICK a first and a second corner (diagonally) to form the rectangle. The windowed area is magnified to fill the screen (Fig. 10-8). It is suggested that you draw the window with a 4 x 3 (approximate) proportion to match the screen proportion. If you type *Zoom* or *Z*, *Window* is an <u>automatic</u> option so you can begin selecting the first corner of the window after issuing the *Zoom* command <u>without</u> indicating the *Window* option as a separate step.

FIGURE 10-8

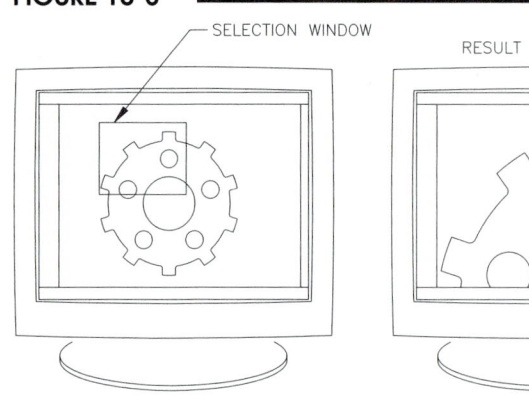

All

This option displays <u>all of the objects</u> in the drawing <u>and all of the *Limits*</u>. In Figure 10-9, notice the effects of *Zoom All*, based on the drawing objects and the drawing *Limits*.

Extents

This option results in the largest possible display of <u>all of the objects</u>, <u>disregarding</u> the *Limits*. (*Zoom All* includes the *Limits*.) See Figure 10-9.

FIGURE 10-9

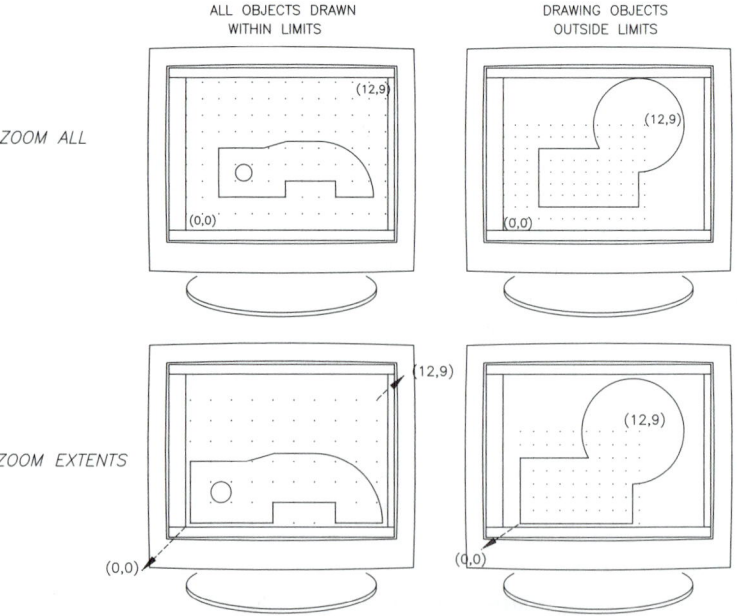

Zoom to Object

You can think of this option as "zoom to extents of object." *Zoom to Object* allows you to select any object, such as a *Line*, *Circle*, or *Pline*, then AutoCAD automatically zooms to display the entire object as large as possible on the screen.

Scale (X/XP)

This option allows you to enter a scale factor for the desired display. The value that is entered can be relative to the full view (*Limits*) or to the current display (Fig. 10-10). A value of **1**, **2**, or **.5** causes a display that is 1, 2, or .5 times the size of the *Limits*, centered on the current display. A value of **1X**, **2X**, or **.5X** yields a display 1, 2, or .5 times the size of the current display. If you are using paper space and viewing model space, a value of **1XP**, **2XP**, or **.5XP** yields a model space display scaled 1, 2, or .5 times paper space units.

If you type *Zoom* or *Z*, you can enter a scale factor at the *Zoom* command prompt without having to type the letter *S*.

FIGURE 10-10

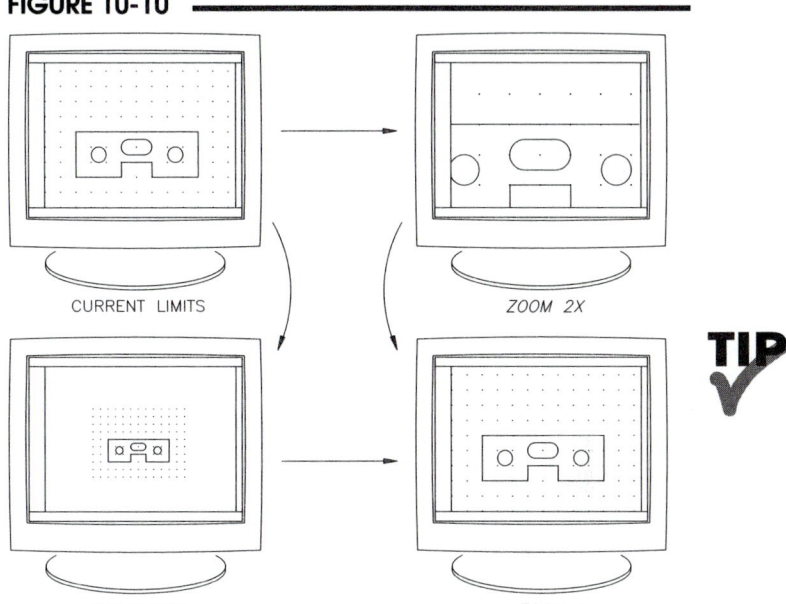

CURRENT LIMITS   ZOOM 2X

ZOOM 0.5X   ZOOM 1

In

Zoom In magnifies the current display by a factor of 2X (2 times the current display). Using this option is the same as entering a *Zoom Scale* of 2X.

Out

Zoom Out makes the current display smaller by a factor of .5X (.5 times the current display). Using this option is the same as entering a *Zoom Scale* of .5X.

Center

First, specify a location as the center of the zoomed area; then specify either a *Magnification factor* (see *Scale X/XP*), a *Height* value for the resulting display, or PICK two points forming a vertical to indicate the height for the resulting display (Fig. 10-11).

FIGURE 10-11

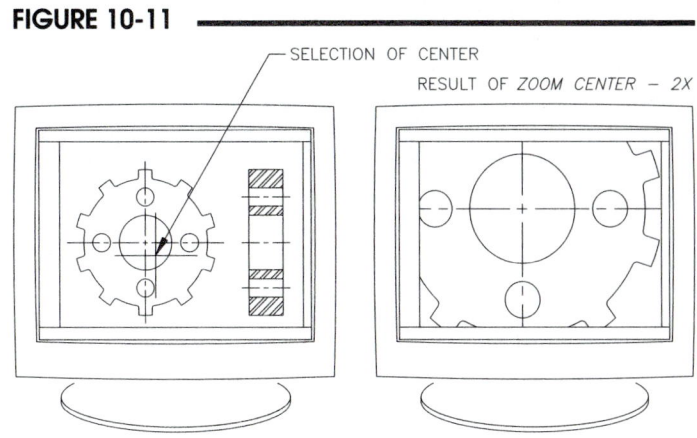

SELECTION OF CENTER
RESULT OF *ZOOM CENTER – 2X*

Previous

Selecting this option automatically changes to the previous display. AutoCAD saves the previous ten displays changed by *Zoom*, *Pan*, and *View*. You can successively change back through the previous ten displays with this option.

Dynamic

With this option you can change the display from one windowed area in a drawing to another without using *Zoom All* (to see the entire drawing) as an intermediate step. *Zoom Dynamic* causes the screen to display the drawing *Extents* bounded by a box. The current view or window is bounded by a box in a broken-line pattern. The view box is the box with an "X" in the center which can be moved to the desired location (Fig. 10-12). The desired location is selected by pressing **Enter** (not the PICK button as you might expect).

Pressing the left button allows you to resize the view box (displaying an arrow instead of the "X") to the desired size. Move the mouse or puck left and right to increase and decrease the window size. Press the left button again to set the size and make the "X" reappear.

FIGURE 10-12

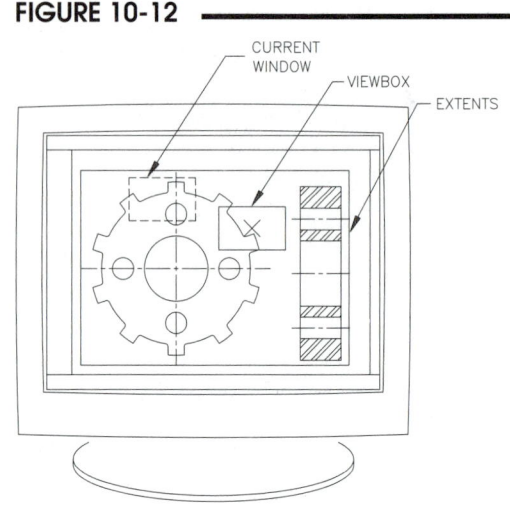

CURRENT WINDOW
VIEWBOX
EXTENTS

Zoom Is Transparent

Zoom is a <u>transparent</u> command, meaning it can be invoked while another command is in operation. You can, for example, begin the *Line* command and at the "Specify next point or [Undo]:" prompt *'Zoom* with a *Window* to better display the area for selecting the endpoint. This transparent feature is automatically entered if *Zoom* is invoked by the screen, pull-down, or tablet menus or toolbar icons, but if typed it must be prefixed by the apostrophe (') symbol, e.g., "Specify next point or [Undo]: *'Zoom*." If *Zoom* has been invoked transparently, the **>>** symbols appear at the command prompt before the listed options as follows:

```
Command: line
Specify first point: 'zoom
>>Specify corner of window, enter a scale factor (nX or nXP), or
[All/Center/Dynamic/Extents/Previous/Scale/Window/Object] <real time>:
```

Pan

Pull-down Menu	Command (Type)	Alias (Type)	Short-cut	Screen (side) Menu	Tablet Menu
View *Pan*	*Pan* or -*Pan*	*P* or -*P*	(Default Menu) *Pan*	*VIEW 1* *Pan*	*N,10-P,11*

Using the *Pan* command by typing or selecting from the tool, screen, or digitizing tablet menu produces the *Realtime* version of *Pan*. The *Point* option is available from the pull-down menu or by typing -*Pan*. The other "automatic" *Pan* options (*Left*, *Right*, *Up*, *Down*) can only be selected from the pull-down menu.

The *Pan* command is useful if you want to move (pan) the display area slightly without changing the <u>size</u> of the current view window. With *Pan*, you "drag" the drawing across the screen to display an area outside of the current view in the drawing area.

Realtime (RTPAN)

Realtime is the default option of *Pan*. It is invoked by selecting *Pan* by any method, including typing *Pan* or *Rtpan*. *Realtime Pan* is essentially the same as using the mouse wheel or third button to pan. That is, *Realtime Pan* allows you to interactively pan by "pulling" the drawing across the screen with the cursor motion. After activating the command, the cursor changes to a hand cursor. Move the hand to any location on the drawing, then hold the PICK (left) mouse button down and drag the drawing around on the screen to achieve the desired view (see Fig. 10-5). When you release the mouse button, panning is discontinued. You can move the hand to another location and pan again without exiting the command.

You must press Escape or Enter or use *Exit* from the pop-up cursor menu to exit *Realtime Pan* and return to the Command prompt. The following Command prompt appears during *Realtime Pan*.

> Command: **pan**
> Press Esc or Enter to exit, or right-click to display shortcut menu.

If you press the right mouse button (#2 button), a small cursor menu pops up. This is the same menu that appears during *Realtime Zoom* (see Fig. 10-7). This menu provides options for you to *Exit*, *Zoom* or *Pan* (realtime), *Zoom Window*, *Zoom Original*, or *Zoom Extents*. See "*Zoom*" earlier in this chapter.

Point

The *Point* option is available only through the pull-down menu or by typing "*-Pan*" (some commands can be typed with a hyphen [-] prefix to invoke the command line version of the command without dialog boxes, etc.). Using this option produces the following Command prompt.

> Command: **-pan**
> Specify base point or displacement: **PICK**
> or (**value**)
> Specify second point: **PICK** or
> (**value**)
> Command:

FIGURE 10-13

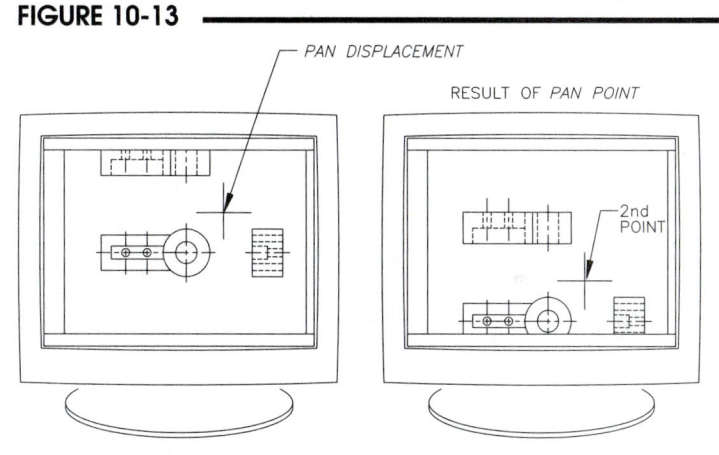

The "base point:" can be thought of as a "handle," or point to move <u>from</u>, and the "second point:" as the new location to move <u>to</u> (Fig. 10-13). You can PICK each of these points with the cursor.

You can also enter coordinate values rather than interactively PICKing points with the cursor. The command syntax is as follows:

> Command: **-pan**
> Specify base point or displacement: **0,-2**
> Specify second point: **Enter**
> Command:

Entering coordinate values allows you to *Pan* to a location outside of the current display. If you use the interactive method (PICK points), you can *Pan* only within the range of whatever is visible on the screen.

The following *Pan* options are available from the *View* pull-down menu only.

Left

Automatically pans to the left, equal to about 1/8 of the current drawing area width.

Right

Pans to the right—similar to but opposite of *Left*.

Up

Automatically pans up, equal to about 1/8 of the current drawing area height.

Down

Pans down—similar to but opposite of *Up*.

Pan Is Transparent

Pan, like *Zoom*, can be used as a transparent command. The transparent feature is automatically entered if *Pan* is invoked by the screen, pull-down, or tablet menus or icons. However, if you <u>type</u> *Pan* during another command operation, it must be prefixed by the apostrophe (') symbol.

```
Command: line
Specify first point: 'pan
>>Press ESC or ENTER to exit, or right-click to display shortcut menu. (Pan, then) Enter
Resuming LINE command.
Specify next point or [Undo]:
```

Zoom and *Pan* with *Undo* and *Redo*

By default, consecutive uses of *Zoom* and *Pan* are treated as one command for *Undo* and *Redo* operations. For example, if you *Zoomed* three times consecutively but then wanted to return to the original display, you would have to *Undo* only once. You can change this setting (*Combine zoom and pan commands*) in the lower-right corner of the *User Preference* tab of the *Options* dialog box.

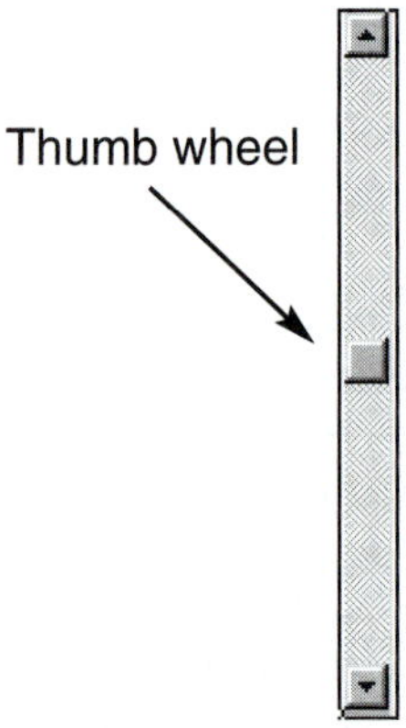

Scroll Bars

You can also use the horizontal and vertical scroll bars directly below and to the right of the graphics area to pan the drawing (Fig. 10-14). These scroll bars pan the drawing in one of three ways: (1) click on the arrows at the ends of the scroll bar, (2) move the thumb wheel, or (3) click inside the scroll bar. The scroll bars can be turned on or off by using the *Display* tab in the *Options* dialog box.

FIGURE 10-14

Thumb wheel

View

Pull-down Menu	Command (Type)	Alias (Type)	Short-cut	Screen (side) Menu	Tablet Menu
View *Named Views...*	*View* or *-View*	*V* or *-V*	...	VIEW 1 *Ddview*	*M,5*

The *View* command provides a dialog box called the *View Manager* for you to create *New* views of a specified display window and restore them (make *Current*) at a later time. For typical applications you would use *View, New* to save that display under an assigned name. Later in the drawing session, the named *View* can be made *Current* any number of times. This method is preferred to continually *Zooming* in and out to display several of the same areas repeatedly. Making a named *View* the *Current* view requires no regeneration time.

2006

Using *View* invokes the *View Manager* (Fig. 10-15) and *-View* produces the Command line format. The options of the *View* command are described below.

The left side of the *View Manager* lists the saved views contained in the current drawing. The central area gives several input fields that allow you to change *Camera* and *Target* settings, *Perspective*, *Field of view*, and *Clipping* plane settings. These features are used almost exclusively for 3D views (see Chapter 35, 3D Basics, Navigation, and Visual Styles).

FIGURE 10-15

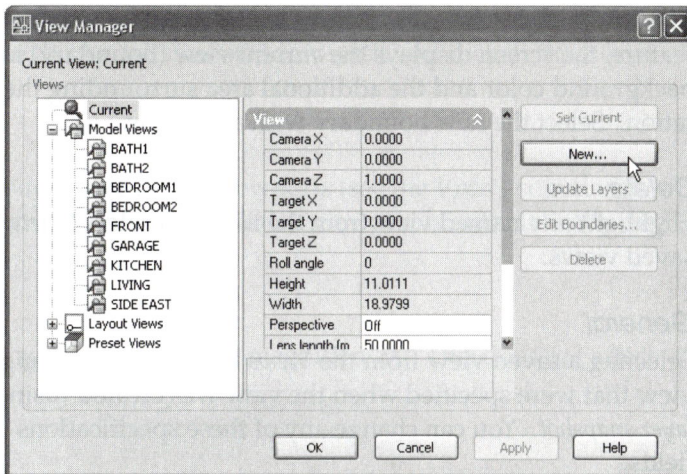

New

Using the *New* button in the *View Manager* box produces the *New View* dialog box (Fig. 10-16) that you should use to create and specify options for a new named view. Enter a *View Name* and (optional) *View Category*, a description field (such as "Elevation" or "Plan").

The *Boundary* section allows you to specify the area of the drawing you want to display when the view is restored. If already zoomed into the desired view area, select *Current Display*. Otherwise, select *Define window* or press the *Define View Window* button to return to the drawing to specify a window.

Under *Settings* you can choose to *Save layer snapshot with view* so that when the view is later restored the layer visibility settings return to those when the view was defined or later updated using the *Update Layers* button. In this way, when the view is restored, AutoCAD resets the layer visibility back to when the view was created no matter what the current visibility settings may be at the time you restore the view. (Layers are discussed in detail in Chapter 11.) The *UCS*, *Live section*, *Visual Style* and *Background* options are primarily for 3D applications (see Chapters 35, 36, 39, and 41).

FIGURE 10-16

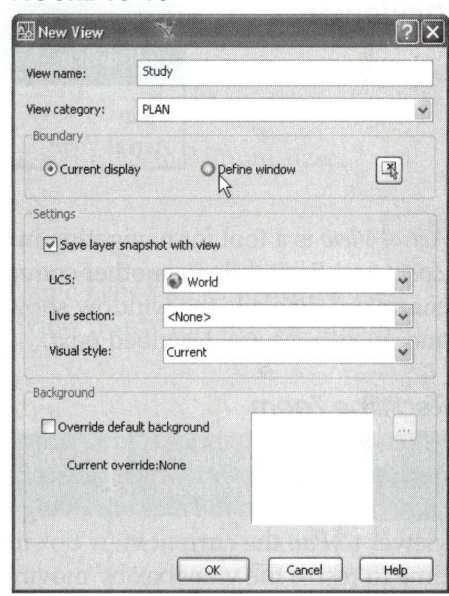

2007

Set Current

To *Restore* a named view, double-click on the name from the list or highlight the name and select *Set Current*, then press *OK*. This action makes the selected view current in the drawing area. If the view you saved was in another *Layout tab* or *Model tab*, then using *Set Current* will return to the correct layout and display the desired view.

Update Layers

If the layer settings were stored when the view was created (the *Store Current Settings with View* option was used) and you want to reset the layer visibility for the view, use this option. Pressing this button causes the layer visibility information saved with the selected named view to update to the current settings (match the layer visibility in the current model space or layout viewport).

You can type -*Vports* to use the Command line version. The format is as follows:

    Command: **-vports**
    Enter an option [Save/Restore/Delete/Join/SIngle/?/2/3/4] <3>:

Save
Allows you to assign a name and save the current viewport configuration. The configuration can be *Restored* at a later time.

Restore
Redisplays a previously *Saved* viewport configuration.

Delete
Deletes a named viewport configuration.

Join
This option allows you to combine (join) two adjacent viewports. The viewports to join must share a common edge the full length of each viewport. For example, if four equal viewports were displayed, two adjacent viewports could be joined to produce a total of three viewports. You must select a *dominant* viewport. The *dominant* viewport determines the display to be used for the new viewport.

SIngle
Changes back to a single screen display using the current viewport's display.

?
Displays the identification numbers and screen positions of named (saved) and active viewport configurations. The screen positions are relative to the lower-left corner of the screen (0,0) and the upper-right corner (1,1).

2, 3, 4
Use these options to create 2, 3, or 4 viewports. You can choose the configuration. The possibilities are illustrated if you use the *Viewports* dialog box.

CHAPTER EXERCISES

1. *Zoom Extents, Realtime, Window, Previous, Center*

   Open the sample drawing supplied with AutoCAD called **DB_SAMP.DWG**. If your system has the standard installation of AutoCAD, the drawing is located in the **C:\Program Files\AutoCAD 2007\Sample** directory. The drawing shows an office building layout (Fig. 10-28). Use *SaveAs* and rename the drawing to **ZOOM TEST** and locate it in your working folder (directory). (When you view sample drawings, it is a good idea to copy them to another name so you do not accidentally change the original drawings.)

FIGURE 10-28

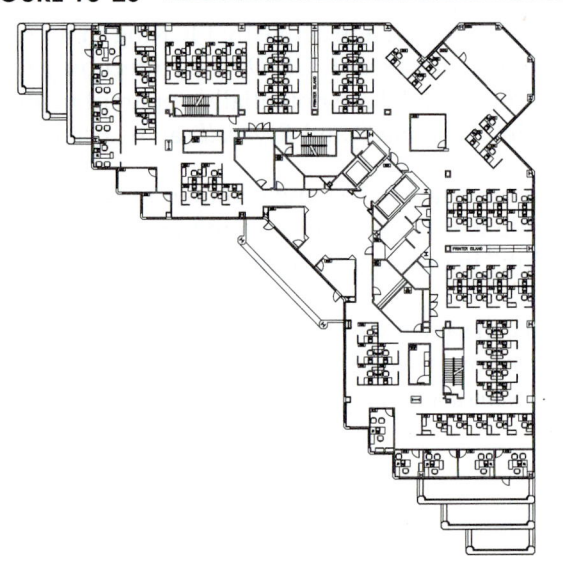

A. First, use *Zoom Extents* to display the office drawing as large as possible on your screen. Next, use *Zoom Realtime* and zoom in to the center of the layout (the center of your screen). You should be able to see rooms 6002 and 6198 located against the exterior wall (Fig. 10-29). Right-click and use *Zoom Extents* from the cursor pop-up menu to display the entire layout again. Press **Esc**, **Enter**, or right-click and select *Exit* to exit *Realtime*.

FIGURE 10-29

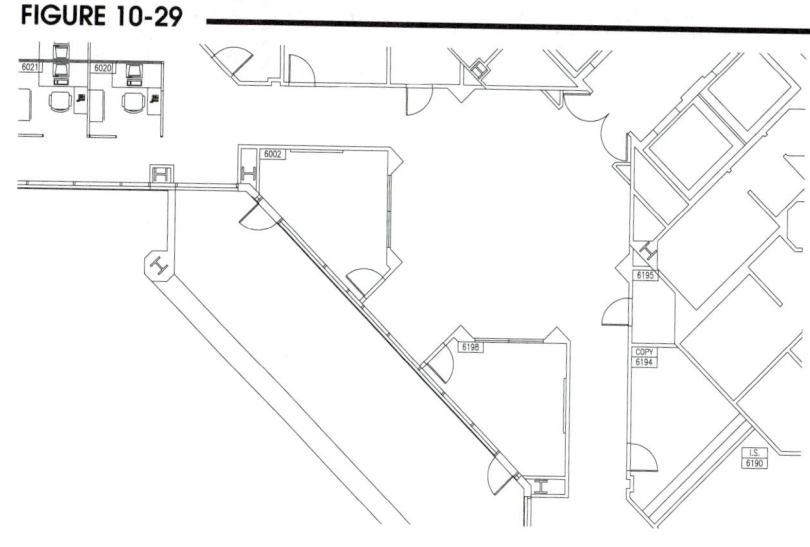

B. Use *Zoom Window* to closely examine room 6050 at the top left of the building. Whose office is it? What items on the desk are magenta in color? What item is cyan (light blue)?

  computer+keyboard   *telephone*   *Laura Ginsburg*

C. Use *Zoom Center*. **PICK** the telephone on the desk next to the computer as the center. Specify a height of **24** (2'). You should see a display showing only the telephone. How many keys are on the telephone's number pad (*Point* objects)? *12*

D. Next, use *Zoom Previous* repeatedly until you see the same display of rooms 6002 and 6198 as before (Fig. 10-29). Use *Zoom Previous* repeatedly until you see the office drawing *Extents*.

2. *Pan Realtime*

A. Using the same drawing as exercise 1 (ZOOM TEST.DWG, originally DB_SAMP.DWG), use *Zoom Extents* to ensure you can view the entire drawing. Next, *Zoom* with a *Window* to an area about 1/4 the size of the building.

B. Invoke the *Real Time* option of *Pan*. *Pan* about the drawing to find a coffee room. Can you find another coffee room? How many coffee rooms are there? *1*

3. *Pan* and *Zoom* **with the Mouse Wheel**

In the following exercise, *Zoom* and *Pan* using the wheel on your mouse (turn the wheel to *Zoom* and press and drag the wheel to *Pan*). If you do not have this capability, use *Realtime Pan* and *Zoom* in concert to examine specific details of the office layout.

A. Find your new office. It is room 6100. (HINT: It is centrally located in the building.) Your assistant is in the office just to the right. What is your assistant's name and room number?

  Jamie Alda Rm 6105

B. Although you have a nice office, there are some disadvantages. How far is it from your office door to the nearest coffee room (nearest corner of the coffee room)? HINT: Use the *Dist* command with *Endpoint OSNAP* to select the two nearest doors. *106'*

C. *Pan* and/or *Zoom* to find the copy room 6006. How far is it to the copy room (direct distance, door to door)?

  76'-9 27/64"

D. Perform a **Zoom Extents**. Use the wheel to zoom in to Kathy Ragerie's office (room 6150) in the lower-right corner of the building. Remember to locate the cursor at the spot in the drawing that you want to zoom in to. If you need to pan, hold down the wheel and "drag" the drawing across the screen until you can view all of room 6150 clearly. How many chairs are in Kathy's office? Finally, use **Zoom Extents** to size the drawing to the screen.

4. *Pan Point*

A. **Zoom** in to your new office (room 6100) so that you can see the entire room and room number. Assume you were listening to music on your computer with the door open and calculated it could be heard about 100' away. Naturally, you are concerned about not bothering the company CEO who has an office in room 6048. Use **Pan Point** to see if the CEO's office is within listening range. (HINT: enter **100'<0** at the "Specify base point or displacement:" prompt.) Should you turn down the system?

B. Do not **Exit** the **ZOOM TEST** drawing.

5. *Syswindows*

A. **Open** the **TABLET** drawing from the **Sample** folder (C:\Program Files\AutoCAD 2007\Sample). Use *SaveAs*, name the new drawing **TABLET TEST**, and locate it in your working folder.

B. You should now have two drawings open. Use the **Window** pull-down menu and select each of the options: *Cascade*, *Tile Horizontally*, and *Tile Vertically*. Which of these options would you generally use to display two drawings so they are optimally sized and arranged? Experiment with *Realtime Pan* and *Zoom* in each window.

C. Next, maximize one of the drawings by clicking the **Maximize** button in the drawing window's upper right corner (the center of the three small buttons). The drawing should fill the entire screen and the other drawing should disappear. Next, toggle between drawings by holding down the **Ctrl** key and pressing the **Tab** key a couple of times.

D. Make the **TABLET TEST** drawing the current drawing.

6. *Viewres, Zoom Dynamic*

A. Make sure the **TABLET TEST** drawing is current. Use **Zoom** with the **Window** option (or you can try *Realtime Zoom*). Zoom in to the first several command icons located in the middle left section of the tablet menu. Your display should reveal icons for *Zoom* and other viewing commands. **Pan** or **Zoom** in or out until you see the *Zoom* icons clearly.

B. Notice how the *Zoom* command icons are shown as polygons instead of circles? Use the **Viewres** command and change the value to **5000**. Do the icons now appear as circles? Use **Zoom** *Realtime* again. Does the new setting change the speed of zooming?

C. Use **Zoom Dynamic**, then try moving the view box immediately. Can you move the box before the drawing is completely regenerated? Change the view box size so it is approximately equal to the size of one icon. Zoom in to the command in the lower-right corner of the tablet menu. What is the command? What is the command in the lower-left corner? For a drawing of this complexity and file size, which option of *Zoom* is faster and easier for your system—*Realtime* or *Dynamic*?

D. If you wish, you can **Exit** the **TABLET TEST** drawing.

7. *View*

   A. Make the **ZOOM TEST** drawing that you used in previous exercises the current drawing. *Zoom* in to the office that you will be moving into (room 6100). Make sure you can see the entire office. Use the *View* command and create a *New* view named **6100**. Next, *Zoom* or *Pan* to room 6048. *Save* the display as a *View* named **6048**.

   B. Use the *View* dialog box again and make view **6100** *Current*.

   C. In order for the CEO to access information on your new computer, a cable must be stretched from your office to room 6048. Find out what length of cable is needed to connect the upper left corner of office 6100 to the upper right corner of office 6048. (HINT: Use *Dist* to determine the distance. Type the *'-View* command <u>transparently</u> [prefix with an apostrophe and hyphen] during the *Dist* command and use *Running Endpoint OSNAP* to select the two indicated office corners. The V*iew* dialog box <u>is not transparent</u>.) What length of cable is needed? Do not *Exit* the drawing.

8. *Aerial View*

   A. It has been decided to connect several other computers to your office computer. Using a similar technique as in the previous exercise (using *Distance*), determine cable lengths to other locations. Rather than saving a *View* for each location, activate *Aerial View*.

   B. Use the *Global* option in *Aerial View* to display the entire office. Next, click inside the viewer and make the view box about the size of your office and right-click to zoom in to it (inside *Aerial View*). Your office should appear in the main drawing area.

   C. Ensure that *Endpoint* is checked in the *Running Osnap* dialog box. Activate *Dist* and **PICK** the lower-left corner of your office (in the main graphics display). Next, in *Aerial View*, select a *Zoom Window* box around the room 6105 (just to the right of your office). What length of cable is needed to connect to the nearest left corner of the room?

   D. Use the same technique to connect the lower corner of office space 6092 to your office (upper-right corner), the lower corner of room 6093 to your office. What lengths of cable are needed?

   E. If you want, *Exit* the **ZOOM TEST** drawing for the next exercise.

   NOTE: If you view and experiment with other sample drawings, remember to immediately use *Saveas* to create new drawings so the originals are not accidentally changed.

9. *Zoom All, Extents*

   A. Begin a *New* drawing. Turn on the *SNAP* (**F9**) and *GRID* (**F7**). Draw two *Circles*, each with a **1.5** unit *radius*. The *Circle* centers are at **3,5** and at **5,5**. See Figure 10-30.

   B. Use *Zoom All*. Does the display change? Now use *Zoom Extents*. What happens? Now use *Zoom All* again. Which option <u>always</u> shows all of the *Limits*?

FIGURE 10-30

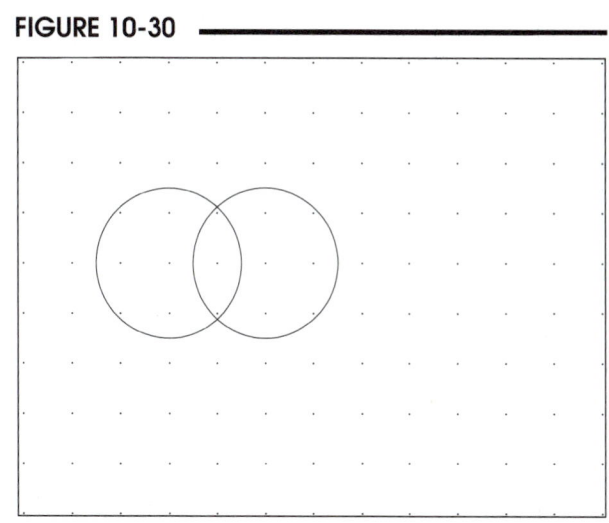

C. Draw a *Circle* with the center at **10,10** and with a *radius* of **5**. Now use *Zoom All*. Notice the GRID appears only on the area defined by the *Limits*. Can you move the cursor to 0,0? Now use *Zoom Extents*. What happens? Can you move the cursor to 0,0?

D. *Erase* the large *Circle*. Use *Zoom All*. Can you move the cursor to 0,0? Use *Zoom Extents*. Can you find point 0,0?

E. *Exit* the drawing and discard changes.

10. *Vports*

A. *Open* the **ZOOM TEST** drawing again. Perform a *Zoom Extents*. Invoke the *Viewports* dialog box by any method. When the dialog box appears, choose the *Three: Right* option. The resulting viewport configuration should appear as Figure 10-25, shown earlier in the chapter.

B. Using *Zoom* and *Pan* in the individual viewports, produce a display with an overall view on the top and two detailed views as shown in Figure 10-27.

C. Type the *Vports* command. Use the *New* option to save the viewport configuration as **3R**.

D. Click in the bottom-left viewport to make it the current viewport. Use *Vports* again and change the display to a *Single* screen. Now use *Realtime Zoom* and *Pan* to locate and zoom to room 6100.

E. Use the *Viewports* dialog box again. Select the *Named Viewports* tab and select **3R** as the viewport configuration to restore. Ensure the **3R** viewport configuration was saved as expected, including the detail views you prepared.

F. Invoke a *Single* viewport again. Finally, use the *Vports* dialog box to create another viewport configuration of your choosing. *Save* the viewport configuration and assign an appropriate name. *Save* the **ZOOM TEST** drawing.

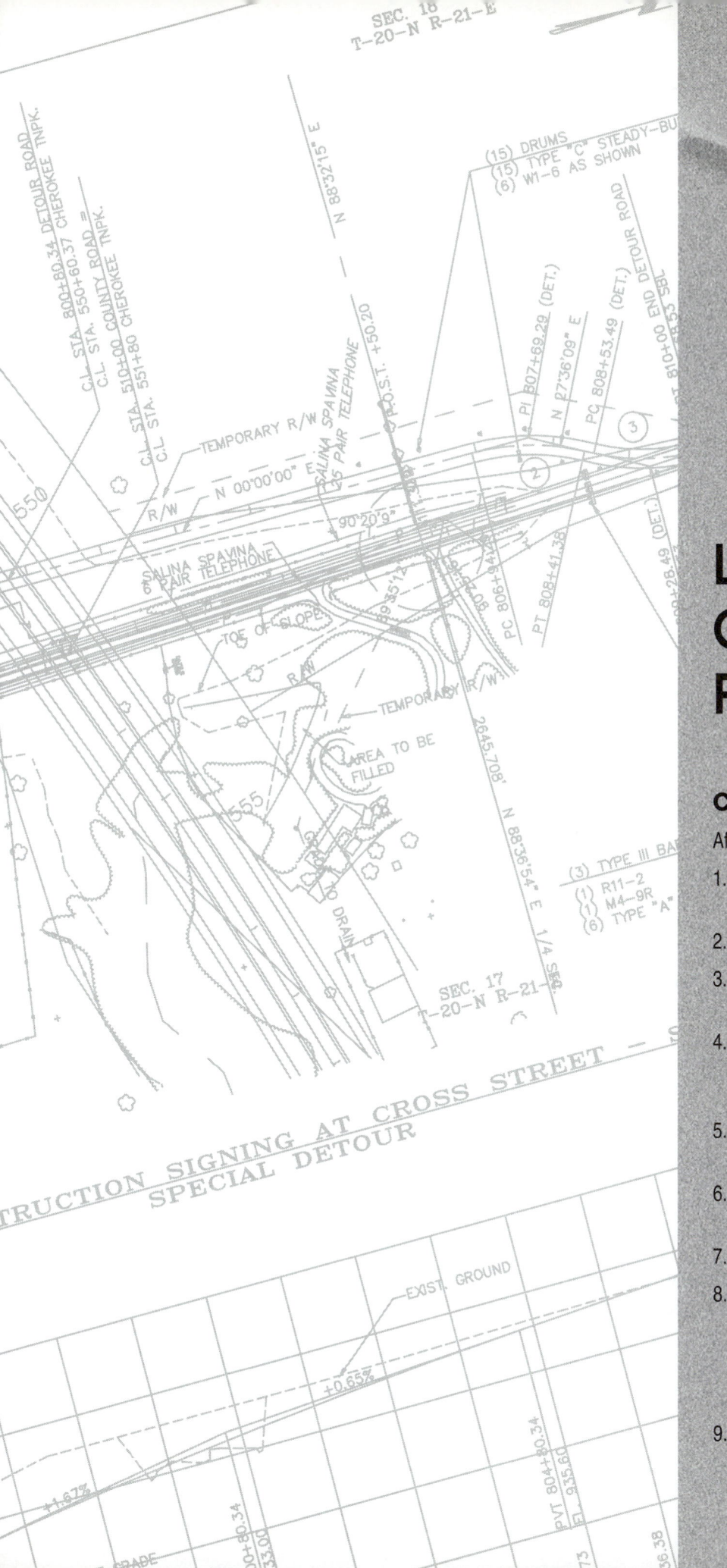

LAYERS AND OBJECT PROPERTIES

CHAPTER OBJECTIVES

After completing this chapter you should:

1. understand the strategy of grouping related geometry with *Layers*;

2. be able to create *Layers*;

3. be able to assign *Color, Linetype,* and *Lineweight* to *Layers*;

4. be able to control a layer's properties and visibility settings (*On, Off, Freeze, Thaw, Lock, Unlock*);

5. be able to manipulate groups of layers by creating *Group Filters* and *Properties Filters*;

6. be able to set *LTSCALE* to adjust the scale of linetypes globally;

7. understand the concept of object properties;

8. be able to change an object's properties (layer, color, linetype, linetype scale, and lineweight) with the *Object Properties* toolbar, the *Properties* palette, and with *Match Properties*;

9. be able to use the *Layer Tools* included with AutoCAD.

CONCEPTS

In a CAD drawing, layers are used to group related objects in a drawing. Objects (*Lines*, *Circles*, *Arcs*, etc.) that are created to describe one component, function, or process of a drawing are perceived as related information and, therefore, are typically drawn on one layer. A single CAD drawing is generally composed of several components and, therefore, several layers. Use of layers provides you with a method to control visible features of the components of a drawing. For each layer, you can control its color on the screen, the linetype and lineweight it will be displayed with, and its visibility setting (on or off). You can also control if the layer is plotted or not.

FIGURE 11-1

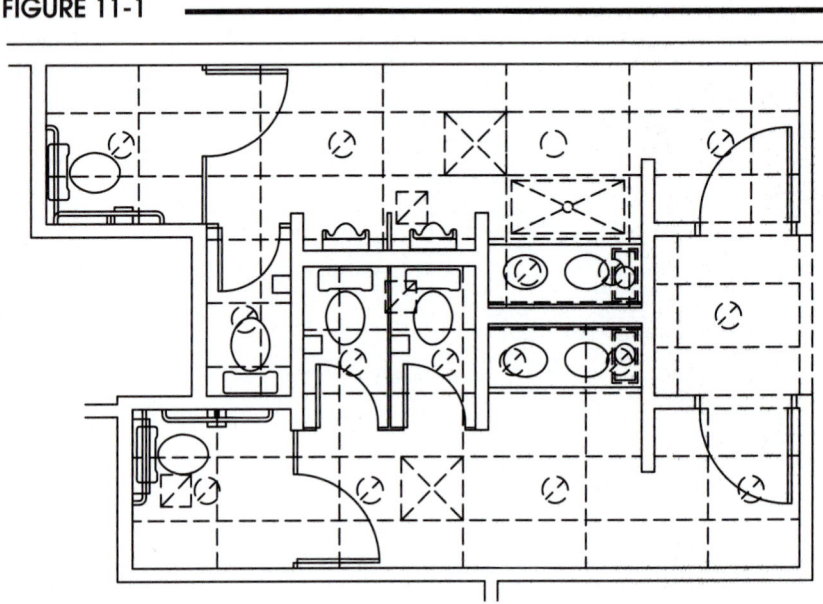

Layers in a CAD drawing can be compared to clear overlay sheets on a manual drawing. For example, in a CAD architectural drawing, the floor plan can be drawn on one layer, electrical layout on another, plumbing on a third layer, and HVAC (heating, ventilating, and air conditioning) on a fourth layer (Fig. 11-1). Each layer of a CAD drawing can be assigned a different color, linetype, lineweight, and visibility setting similar to the way clear overlay sheets on a manual drawing can be used. Layers can be temporarily turned *Off* or *On* to simplify drawing and editing, like overlaying or removing the clear sheets. For example, in the architectural CAD drawing, only the floor plan layer can be made visible while creating the electrical layout, but can later be cross-referenced with the HVAC layout by turning its layer on. Layers can also be made "non-plottable." Final plots can be made of specific layers for the subcontractors and one plot of all layers for the general contractor by controlling the layers' *Plot/No Plot* icon before plotting.

AutoCAD allows you to create a practically unlimited number of layers. You should assign a name to each layer when you create it. The layer names should be descriptive of the information on the layer.

Assigning Colors, Linetypes, and Lineweights

There are two strategies for assigning colors, linetypes, and lineweights in a drawing: assign these properties to layers or assign them to objects.

Assign colors, linetypes, and lineweights to layers

For most drawings, layers are assigned a color, linetype, and lineweight so that all objects drawn on a single layer have the same color, linetype, and lineweight. Assigning colors, linetypes, and lineweights to layers is called *ByLayer* color, linetype, and lineweight setting. Using the *ByLayer* method makes it visually apparent which objects are related (on the same layer). All objects on the same layer have the same linetype and color.

Assign colors, linetypes, and lineweights to individual objects

Alternately, you can assign colors, linetypes, and lineweights to specific objects, overriding the layer's color, linetype, and lineweight setting. This method is fast and easy for small drawings and works well when layering schemes are not used or for particular applications. However, using this method makes it difficult to see which layers the objects are located on.

Object Properties

Another way to describe the assignment of color, linetype, and lineweight properties is to consider the concept of Object Properties. Each object has properties such as a layer, a color, a linetype, and a lineweight. The color, linetype, and lineweight for each object can be designated as *ByLayer* or as a specific color, linetype, or lineweight. Object properties for the two drawing strategies (schemes) are described as follows.

Object Properties

	1. *ByLayer* Drawing Scheme	2. Object-Specific Drawing Scheme
Layer Assignment	Layer name descriptive of geometry on the layer	Layer name descriptive of geometry on the layer
Color Assignment	*ByLayer*	*Red, Green, Blue, Yellow,* or other specific color setting
Linetype Assignment	*ByLayer*	*Continuous, Hidden, Center,* or other specific linetype setting
Lineweight Assignment	*ByLayer*	0.05 mm, 0.15 mm, 0.010", or other specific lineweight setting

It is recommended that beginners use only one method for assigning colors, linetypes, and lineweights. After gaining some experience, it may be desirable to combine the two methods only for specific applications. Usually, the *ByLayer* method is learned first, and the object-specific color, linetype, and lineweight assignment method is used only when layers are not needed or when complex applications are needed. (The *ByBlock* color, linetype, and lineweight assignment has special applications for *Blocks* and is discussed in Chapter 21.)

Plot Styles

Plot style is an object property that controls how an object appears in a plot. A plot style can be assigned to any object, but can also be assigned to a layer (*ByLayer*). A plot style can control such appearances as (plotted) color, lineweight, screen percentage, line end styles, line join styles, and pen number. A plot style is actually another object property, just like color, linetype, and lineweight, except that assigning it is optional. You should assign plot styles only if you want the plot to have features other than the way the drawing appears on the screen.

There are two types of plot styles: color-dependent and named plot styles. The color-dependent plot style is the simpler of the two because it is based on color. Since colors are usually assigned to layers, each layer's color on the screen controls how the drawing is plotted. Each screen color can be assigned to use a separate pen, screening percentage, line end style, and so on. In this way, all objects that appear in one color will be plotted with the same appearance. On the other hand, named plot styles are not based on color, so they can be assigned to any object. With named plot styles, objects that are drawn in one color can be plotted with different appearances. Plot styles are discussed in detail in Chapter 33, Advanced Layouts and Plotting.

LAYERS AND LAYER PROPERTIES CONTROLS

Layer Control Drop-Down List

The Object Properties toolbar contains a drop-down list for making layer control quick and easy (Fig. 11-2). The window normally displays the current layer's name, visibility setting, and properties. When you pull down the list, all the layers (unless otherwise specified in the *Named Layer Filters* dialog box) and their settings are displayed. Selecting any layer <u>name</u> makes it current. Clicking on any of the visibility/properties icons changes the layers' setting as described in the following section. Several layers can be changed in one "drop." You cannot change a layer's color, linetype, or lineweight, nor can you create new layers using this drop-down list.

FIGURE 11-2

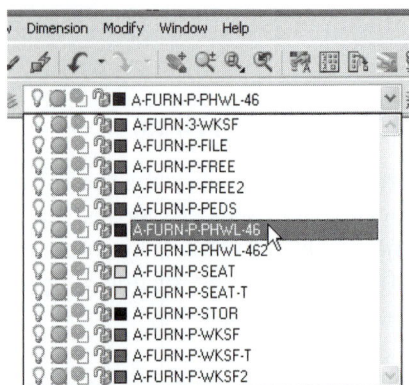

Layer

Pull-down Menu	Command (Type)	Alias (Type)	Short-cut	Screen (side) Menu	Tablet Menu
Format *Layer*	*Layer* or *-Layer*	*LA* or *-LA*	...	*FORMAT* *Layer*	*U,5*

FIGURE 11-3

The way to gain complete layer control is through the *Layer Properties Manager* (Fig. 11-3). The *Layer Properties Manager* is invoked by using the icon button (shown above), typing the *Layer* command or *LA* command alias, or selecting *Layer* from the *Format* pull-down or screen menu.

As an alternative to the *Layer Properties Manager*, the *Layer* command can be used in Command line format by typing *Layer* with a "-" (hyphen) prefix. The Command line format shows most of the options available through the *Layer Properties Manager*.

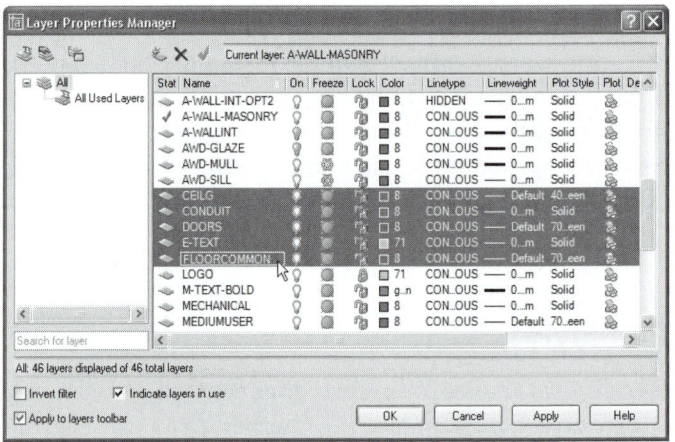

```
Command: -LAYER
Current layer: "0"
Enter an option
[?/Make/Set/New/ON/OFF/Color/Ltype/LWeight/Plot/PStyle/Freeze/Thaw/LOck/Unlock/stAte]:
```

Layer Properties Manager Layer List

The right-central area of the *Layer Properties Manager* (see Fig. 11-3) is the <u>layer list</u> that displays the list of layers in the drawing. Here you can control the current visibility settings and the properties assigned to each layer. This list can display all the layers in the drawing or a subset of layers if a layer filter is applied. Layer filters are created and applied in the <u>Filter Tree View</u> on the left side of the *Layer Properties Manager* (see "Layer Filters" later in this chapter). For new drawings such as those created from the ACAD.DWT template, only one layer may exist in the drawing—layer "0" (zero). New layers can be created by selecting the *New Layer* button or the *New Layer* option of the right-click menu (see *New Layer*).

All properties and visibility settings of layers can be controlled by highlighting the layer name and then selecting one of the icons for the layer such as the light bulb icon (*On, Off*), sun/snowflake icon (*Thaw/Freeze*), padlock icon (*Lock/Unlock*), *Color* tile, *Linetype, Lineweight,* or *Plot/No plot* icon.

Multiple layers can be highlighted (see Fig. 11-3) by holding down the Ctrl key while PICKing (to highlight one at a time) or holding down the Shift key while PICKing (to select a range of layers between and including two selected names). Right-clicking in the list area displays a cursor menu allowing you to *Select All* or *Clear All* names in the list and other options (see Fig. 11-5). You can rename a layer by clicking twice <u>slowly</u> on the name (this is the same as PICKing an already highlighted name).

The column widths can be changed by moving the pointer to the "crack" between column headings until double arrows appear (Fig. 11-4). You can also resize the entire dialog box by placing your pointer on the extreme border or corner until double arrows appear, then clicking and dragging.

FIGURE 11-4

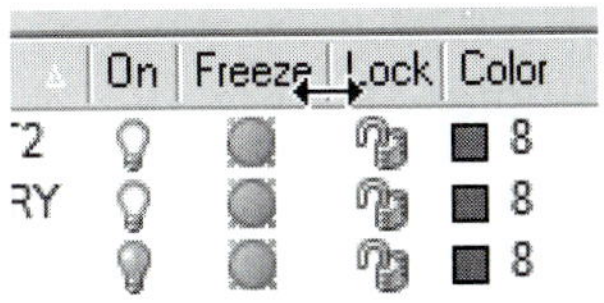

△	On	Freeze	Lock	Color
²2	💡	◯	🔒	■ 8
ꓤY	💡	◯	🔒	■ 8
	💡	◯	🔒	■ 8

A particularly useful feature is the ability to sort the layers in the list by any one of the headings (*Name, On, Freeze, Linetype,* etc.) by clicking on the heading tile. For example, you can sort the list of names in alphabetical order (or reverse order) by clicking once (or twice) on the *Name* heading above the list of names. Or you may want to sort the *Frozen* and *Thawed* layers or sort layers by *Color* by clicking on the column heading.

Right-clicking in the layer list of the *Layer Properties Manager* produces a shortcut menu (Fig. 11-5). You have a choice of selecting icon buttons or selecting shortcut menu options for most of the layer controls available in the *Layer Properties Manager,* as noted in the following sections.

FIGURE 11-5

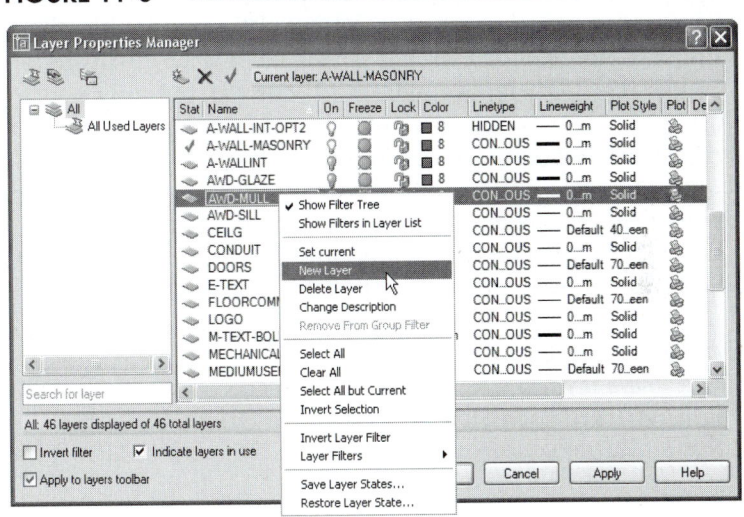

New Layer

The *New Layer* option allows you to make new layers. There is only one layer in the AutoCAD templates (ACAD.DWT and ACADISO.DWT) as they are provided to you "out of the box." That layer is Layer 0. Layer 0 is a part of every AutoCAD drawing because it cannot be deleted. You can, however, change the *Color, Linetype,* and *Lineweight* of Layer 0 from the defaults (*Continuous* linetype, *Default* lineweight, and color #7 *White*). Layer 0 is generally used as a construction layer or for geometry not intended to be included in the final draft of the drawing. Layer 0 has special properties when creating *Blocks* (see Chapter 21).

You should create layers for each group of related objects and assign appropriate layer names for that geometry. You can use up to 256 characters (including spaces) for layer names. There is practically no limit to the number of layers that can be created (although 32,767 has been found to be the actual limit).

A new layer named "Layer1" (or other number) then appears in the list with the default color (*White*), linetype (*Continuous*), and lineweight (*Default*). If an existing layer name is highlighted when you make a new layer, the highlighted layer's properties (*color, linetype, lineweight,* and *plot style*) are used as a template for the new layer. The new layer name initially appears in the rename mode so you can immediately assign a more appropriate and descriptive name for the layer. You can create many new names quickly by typing (renaming) the first layer name, then typing a comma before other names. A comma forces a new "blank" layer name to appear. Colors, linetypes, and lineweights should be assigned as the next step.

If you want to create layers by typing, the *New* and *Make* options of the -*Layer* command can be used. *New* allows creation of one or more new layers. *Make* allows creation of one layer (at a time) and sets it as the current layer.

Delete

 The *Delete Layer* option allows you to delete layers. <u>Only layers with no geometry can be deleted</u>. You cannot delete a layer that has objects on it, nor can you delete Layer 0, the current layer, or layers that are part of externally referenced (*Xref*) drawings. If you attempt to *Delete* such a layer, accidentally or intentionally, a warning appears.

Set Current

 To *Set* a layer as the *Current* layer is to make it the active drawing layer. Any objects created with draw commands are created on the *Current* layer. You can, however, edit objects on any layer, but draw only on the current layer. Therefore, if you want to draw on the FLOORPLAN layer (for example), use the *Set Current* option or double-click on the layer name. If you want to draw with a certain *Color* or *Linetype*, set the layer with the desired *Color, Linetype,* and *Lineweight* as the *Current* layer. Any layer can be made current, but only <u>one layer at a time</u> can be current.

Status and *Indicate Layers in Use*

 Status is not an option; however, the *Status* column indicates whether or not the related layer contains objects. The *Status* icon is a lighter shade of gray if no objects exist on the layer. The *Indicate Layers in Use* option below the layer list <u>must be checked</u> for the *Status* icon to indicate if objects do not exist, otherwise all *Status* icons are dark gray.

On, Off

 If a layer is *On*, it is visible. Objects on visible layers can be edited or plotted. Layers that are *Off* are not visible. Objects on layers that are *Off* will not plot and cannot be edited (unless the *ALL* selection option is used, such as *Erase, All*). It is not advisable to turn the current layer *Off*.

Freeze, Thaw

 Freeze and *Thaw* override *On* and *Off*. *Freeze* is a more protected state than *Off*. Like being *Off*, a frozen layer is not visible, nor can its objects be edited or plotted. Objects on a frozen layer cannot be accidentally *Erased* with the *ALL* option. *Freezing* also prevents the layer from being considered when *Regen*s occur. *Freezing* unused layers speeds up computing time when working with large and complex drawings. *Thawing* reverses the *Freezing* state. Layers can be *Thawed* and also turned *Off*. Frozen layers are not visible even though the light bulb icon is on.

Lock, Unlock

 Layers that are *Locked* are protected from being edited but are still visible and can be plotted. *Locking* a layer prevents its objects from being changed even though they are visible. Objects on *Locked* layers cannot be selected with the *ALL* selection option (such as *Erase, All*). Layers can be *Locked* and *Off*.

Color, Linetype, Lineweight, and Other Properties

Layers have properties of *Color, Linetype,* and *Lineweight* such that (generally) an object that is drawn on, or changed to, a specific layer assumes the layer's linetype and color. Using this scheme (*ByLayer*) enhances your ability to see what geometry is related by layer. It is also possible, however, to assign specific color, linetype, and lineweight to objects which will override the layer's color, linetype, and lineweight (see "*Color, Linetype,* and *Lineweight* Commands" and "Changing Object Properties").

Color

■ green   Selecting one of the small color boxes in the list area of the *Layer Properties Manager* causes the *Select Color* dialog box to pop up (Fig. 11-6). The desired color can then be selected or the name or color number (called the ACI—AutoCAD Color Index) can be typed in the edit box. This action retroactively changes the color assigned to a layer. Since the color setting is assigned to the layer, all objects on the layer that have the *ByLayer* setting change to the new layer color. Objects with specific color assigned (not *ByLayer*) are not affected. Alternately, the *Color* option of the -*Layer* command (hyphen prefix) can be typed to enter the color name or ACI number.

The *Select Color* dialog box also contains tabs for the *Index Color* (ACI), *True Color*, and *Color Books*. See "True Color and Color Books" later in this chapter for information.

FIGURE 11-6

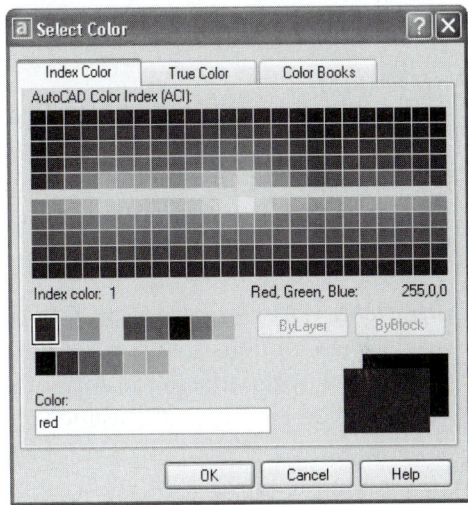

Linetype

To set a layer's linetype, select the *Linetype* (word such as *Continuous* or *Hidden*) in the layer list, which in turn invokes the *Select Linetype* dialog box (Fig. 11-7). Select the desired linetype from the list.

Alternately, the *Linetype* option of the -*Layer* command (hyphen prefix) can be typed. Similar to changing a layer's color, all objects on the layer with *ByLayer* linetype assignment are retroactively displayed in the selected layer linetype while non-*ByLayer* objects remain unchanged.

FIGURE 11-7

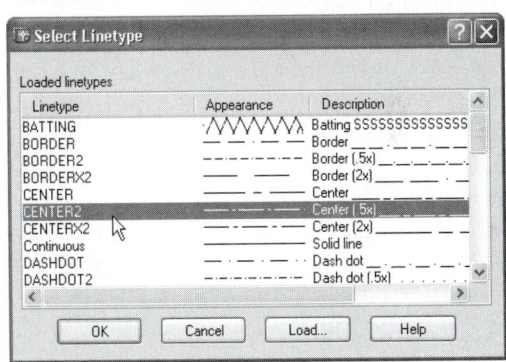

The ACAD.DWT and ACADISO.DWT template drawings as supplied by Autodesk have only one linetype available (*Continuous*). Before you can use other linetypes in a drawing, you must load the linetypes by selecting the *Load* tile (see the *Linetype* command) or by using a template drawing that has the desired linetypes already loaded.

Lineweight

The lineweight for a layer can be set by selecting the *Lineweight* (word such as *Default* or *0.20 mm, 0.50 mm, 0.010"* or *0.020"*, etc.) in the layer list. This action produces the *Lineweight* dialog box (Fig. 11-8). Select the desired lineweight from the list. Alternately, the *Lineweight* option of the -*Layer* command (hyphen prefix) can be typed.

FIGURE 11-8

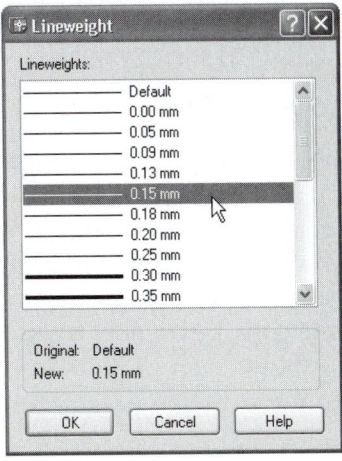

Like the *Color* and *Linetype* properties, all objects on the layer with *ByLayer* lineweight assignment are retroactively displayed in the lineweight assigned to the layer while non-*ByLayer* objects remain unchanged. You can set the units for lineweights (mm or inches) using the *Lineweight* command.

Plot Style

This column of the *Layer Properties Manager* designates the plot style <u>assigned to the layer</u>. If the drawing has a color-dependent Plot Style Table attached, this section is disabled since the plot styles are automatically assigned to colors. If a named Plot Style Table is attached to the drawing you can select this section to assign plot styles from the table to layers. Plot styles are discussed in Chapter 33, Advanced Layouts and Plotting.

Plot/No Plot

You can prevent a layer from plotting by clicking the printer icon so a red line appears over the printer symbol. There are actually three methods for preventing a layer from appearing on the plot: *Freeze* it, turn it *Off*, or change the *Plot/ No Plot* icon. The *Plot/ No Plot* icon is generally preferred since the other two options prevent the layer from appearing in the drawing (screen) as well as in the plot.

Current Viewport Freeze, New Viewport Freeze

These options are used and are displayed in the *Layer Properties Manager* only when paper space viewports exist in the current layout tab. Using these options, you can control what geometry (layers) appears in specific viewports. See Chapters 13 and 33 for more information on these options.

Description

You can add an optional *Description* for each layer that appears in the *Layer Properties Manager*. Select *Change Description* from the right-click menu and add descriptive information about the objects that reside on the layer. For example, layer A-WALL-INT-OPT2 might have the description "Design option 2 for interior walls."

Search for Layer

This edit box below the Filter Tree View allows you to use wildcards to search for matching names in a long list of layers. For example, entering "A-WALL*" would cause only layers beginning with "A-WALL" to appear in the layer list. Clicking in the edit box makes the "*" (asterisk) wildcard automatically appear. <u>Do not press Enter</u> after typing in your entry. (See Chapter 43 for valid wildcards.)

Apply and OK

When you are finished making all the desired changes to the layers in the *Layer Properties Manager*, press the *OK* button (at the bottom of the dialog box) to apply the new settings and dismiss the dialog box. You can also use the *Apply* button to immediately apply any layer settings you have made to the drawing without dismissing the *Layer Properties Manager*.

Layer Filters

When working on drawings with a large number of layers in the *Layer Properties Manager* or in the Layer Control drop-down list, it is time consuming to scroll through a long list of layer names to change visibility and properties settings. A "layer filter" allows you to reduce the list to a subset of the entire list of layers in a drawing. You can create multiple layer filters, and therefore, multiple "groups" of layers, based on object properties, visibility settings, or objects on the layers. For example, you could make filters to display only layers that are *Thawed*, only layers that are red in color, or only layers that contain furniture.

The Filter Tree View area on the left side of the *Layer Properties Manager* allows you to create layer filters (Fig. 11-9). You can even create hierarchical filters, or groups within groups. The groups of layers (filters) that you create are listed in a hierarchical format under the *All* designator. You can create two types of filters: *Group Filters* and *Properties Filters*. If any *Xrefs* are used by the drawing, AutoCAD automatically creates a filter for each *Xref* drawing and displays the list under the *Xref* heading (see Fig. 11-9).

When working with drawings containing a large number of layers, using filters can improve your productivity since filters allow you to reduce the number of names in the layer list. You can click on any group of layers in the Filter Tree View to display its list of layers in the right side of the dialog box. For example, Figure 11-10 displays the list of layers in the "Furniture" group. If *Apply to layer toolbar* is checked, only this list, not the list of all layers, appears in the Layer Control drop-down list. Checking *Invert filter* causes the list to display <u>all but</u> the filter list. For example, in Figure 11-10, if *Invert filter* was checked, all the layers in the drawing <u>except</u> those beginning with A_FURN would be displayed in the list.

Additionally, you can control visibility and lock settings for an entire group. Right-click on a group (filter) name in the Filter Tree View, then select the *Visibility* option, *On, Off, Thaw, Freeze, Lock* or *Unlock* from the shortcut menu (see Fig. 11-9). You can also *Freeze* or *Thaw* the layers for the current viewport (if applicable). A handy *Isolate Group* option is available. Using this feature freezes all layers except those in the group and displays only the group layers in the drawing. However, beware of this feature since no "unisolate" option is available and *Layerp* (Layer Previous) will not undo the layer isolation (see *Layerp*).

Creating a *Group Filter*

Create a group filter by selecting the *New Group Filter* button in the upper-left corner of the *Layer Properties Manager* or by right-clicking on the *All* group and selecting *New Group Filter* from the shortcut menu (see Fig. 11-9). This action causes a new entry named *Group Filter 1* to appear in the filter list under *All*. Assign an appropriate descriptive name for the filter you intend to create, such as "Walls" or "Text." You can then populate the list of layers for the group using two methods. First, you can <u>drag-and-drop layer names</u> from the drawing layer list on the right into the group (layers added to the group are not moved from the drawing layer list, but copied). Second, you can highlight the group name, right-click, and pick *Select Layers* from the shortcut menu (Fig. 11-11). This action returns you to the drawing so you can <u>pick objects on the layers</u> you want to include in the group.

FIGURE 11-9

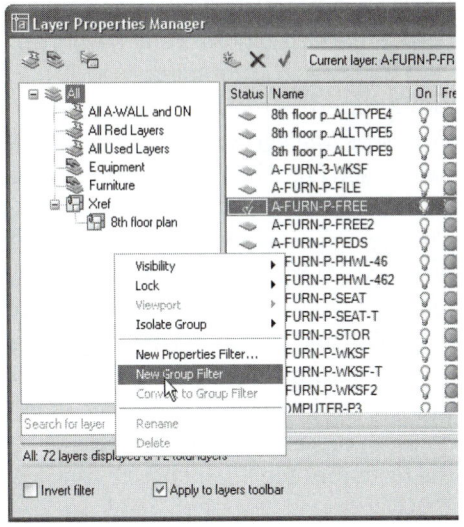

FIGURE 11-10

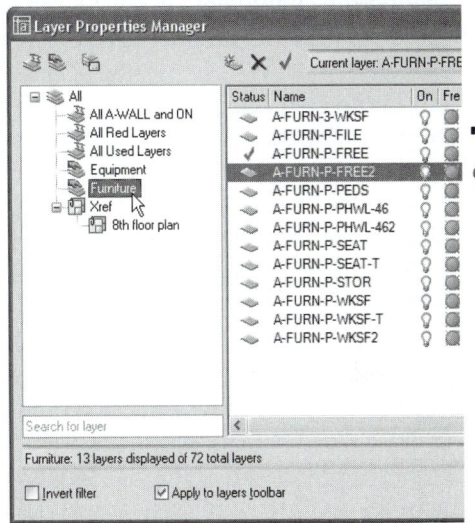

FIGURE 11-11

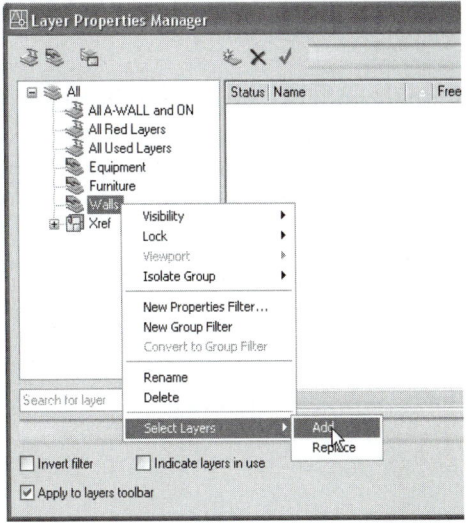

Creating a *Properties Filter*

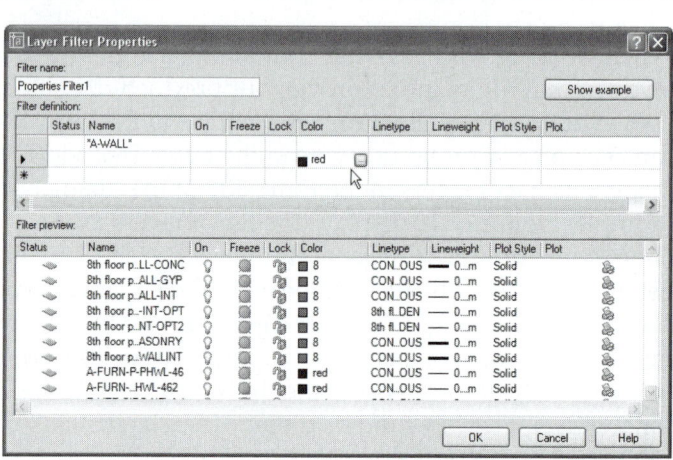

A properties filter is a group of layers that contain the properties that match the properties you specify in the *Filter Definition*. A property could be a *Color, Linetype, Lineweight, On/Off* status, *Freeze/Thaw* status, *Lock/Unlock* status, or specific characters contained in the layer name. Create a new properties filter by selecting the *New Properties Filter* button in the upper-left corner of the *Layer Properties Manager* or by right-clicking on the *All* group and selecting *New Properties Filter* from the shortcut menu (see Fig. 11-9). This action invokes the *Layer Filter Properties* dialog box (Fig. 11-12).

FIGURE 11-12

The layers you want to include in the group must match the criteria you specify in the *Filter definition* area. Multiple properties can be specified, and the layers matching these properties are automatically included in the list and immediately appear in the *Filter preview* list below (you cannot directly alter the contents or the format of the *Filter preview* list).

For example, assume you want to create a property filter to include all layers that include the characters "A-WALL" in the name and all layers that are *Red* in color. In the first line, enter "*A_WALL*" in the *Name* column and in the next line specify *Red* in the color column. The matching layers then appear in the list below.

Based on the filter you want to create, it is important whether you specify the properties on one line or on separate lines in the *Filter definition* box. Specifying properties on <u>separate lines</u> allows <u>any layers that match any property</u> to be included in the list. Specifying properties all on <u>one line</u> requires that <u>only layers matching all of the properties</u> are included in the list. For example, the list in Figure 11-12 includes all layers that have A-WALL in the name <u>or</u> layers that are *Red*; whereas, if the two properties were listed on one line, only layers that include A-WALL <u>and</u> are *Red* would be included.

<u>Properties filters are dynamic, whereas group filters are static</u>. For example, if a property filter definition includes an *On* visibility status, all layers that are currently *On* match this definition and therefore are automatically included in the list. If any layer's *On/Off* visibility is then changed, it would cause the layer to move into or out of the group. In contrast, group filters are based solely on the specific layers that are chosen to be in the group, so changing the group is possible only by "manually" moving a layer into or out of the group.

Property filters can be converted to group filters by right-clicking on the filter name and selecting *Convert to Group Filter* from the shortcut menu. Only layers that <u>currently</u> match the properties are included in the new group filter. Keep in mind that the new group is static unless you "manually" change its members. Group filters cannot be converted to property filters.

Consider also that you may want to create subgroups, or filters "nested" within filters, therefore appearing in a hierarchy in the Filter Tree View. Group filters can be created within other group filters. You can likewise create a property filter within another property filter. You can also create a property filter within a group filter, but you cannot create a group filter within a property filter due to the flexibility of the parent property filter.

Layer States Manager

The *Layer States Manager* is available only through the *Layer Properties Manager* by selecting the icon button or pressing Alt+S when the *Layer Properties Manager* is open. The *Layer States Manager* is the tool that allows you to save and restore "layer states."

A layer state is a combination of layer <u>settings</u> (*On/Off, Freeze/Thaw*, etc.) that can be saved and later restored. Layer states differ from layer filters in that <u>filters</u> simply allow you to "save" short <u>lists</u> of layers, whereas layer <u>states</u> allow you to save different visibility and properties <u>settings</u> for use at a later time. A layer state saves all the visibility and properties settings for all the layers in the drawing. Saving multiple layer states allows you to quickly reset particular *On/Off, Freeze/Thaw*, etc. combinations to improve your efficiency working with particular groups of layers.

For example, you may require that only a certain combination of layers is *On* for editing a particular aspect of a drawing—say, the floor plan, the HVAC layout, and the HVAC dimensions—while a different set of layers is *On* for working with another aspect of the drawing—say, the floor plan, plumbing layout, and fixtures. Rather than manipulating the list of layers and making the appropriate *On/Off* settings <u>each time</u> you work with the floor plan, the HVAC layout, and the HVAC dimensions and <u>each time</u> you work with floor plan, plumbing layout, and fixtures, you can Save these settings as layer states to be *Restored* when needed. A layer state also stores the *Current* layer, so when you *Restore* a layer setting, you are ready to begin drawing on the appropriate layer.

For example, the preceding scenario (saving the HVAC layers settings) would be accomplished by first using the *Layer Properties Manager* as you would normally to make the desired layer visibility settings, that is, turn *On* the floor plan layer, the HVAC layer, and the HVAC dimensions layer, and turn *Off* all other layers. Next, invoke the *Layer States Manager* (Fig. 11-13) to assign a name and save the settings. Then, at any time later, no matter what the current layer settings are, you could use the *Layer States Manager* to *Restore* the visibility settings specified in the HVAC layer state.

FIGURE 11-13

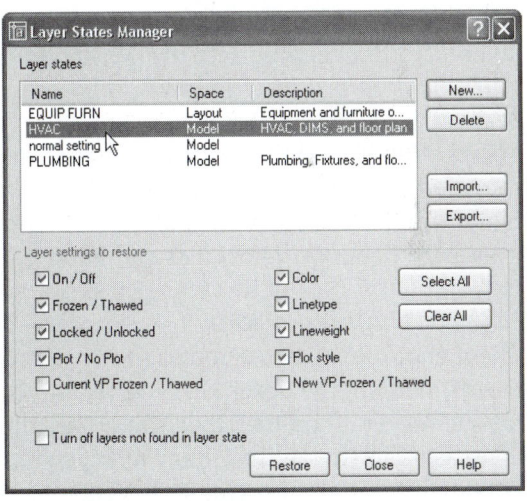

To Save a Layer State:
The procedure for saving a layer state is as follows:

1. Use the *Layer Properties Manager* as you would normally to make the desired settings (*On/Off, Freeze/Thaw*, etc.) for all layers.
2. Invoke the *Layer States Manager* (icon button or Alt+S) from within the *Layer Properties Manager* (see Fig. 11-13).
3. Use the *New* option to produce the *New Layer State to Save* dialog box (Fig. 11-14). Here assign an appropriate name and optional description for the current layer settings configuration, then select *OK* to close this dialog box.
4. *Close* the *Layer States Manager*.
5. Press *OK* or *Apply* in the *Layer Properties Manager*.

FIGURE 11-14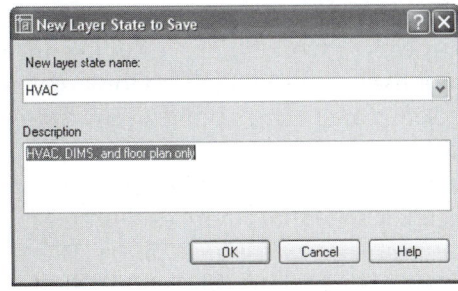

NOTE: <u>You must use *Apply* or *OK* in the *Layer Properties Manager* to save a new layer state.</u>

NOTE: It is highly suggested that you first create a "Normal Settings" layer state when the layer settings are in the normal positions and before you begin creating other new layer states. In this way, you can easily restore the layer settings to their normal positions, no matter what changes are made subsequently. No "*Unrestore*" option exists, and for some cases the *LayerP* command will not re-establish the layer settings back to those previous to restoring a layer state.

Options in the *Layer State Manager* are described here.

New

This option produces the *New Layer State to Save* dialog box (see Fig. 11-14). The *Description* can contain any characters including spaces; however, the layer state name can contain only alpha-numeric characters and spaces.

Delete

Delete a layer state by highlighting the name from the list then pressing *Delete*.

Restore

Highlight a name from the *Layer States* list, then use this option to restore the selected saved layer state. AutoCAD automatically sets the current drawing's global *On, Off, Freeze, Thaw,* etc. settings to those saved under the assigned name.

Export

Once you have created a layer state, you can save it to a file (with a .LAS extension). Each named layer state must be saved as a separate .LAS file; you cannot save the entire set of layer states to one file. A layer state can be imported into other drawings with the *Import* option.

Import

You can import a previously saved layer state from a .LAS file. Exporting and importing layer states is intended to be used primarily when you have several drawings with the same set of layer names (as when similar drawings are created from the same template drawing). Layer states can be saved in one drawing, then *Exported* to files and *Imported* to other similar drawings.

Layer Settings to Restore

These checkboxes allow you to specify which settings you want to save with the layer state. In other words, with all the boxes checked you save all the settings as they appear in the *Layer Properties Manager* for each layer. If on the other hand, only the *On/Off* box is checked, only the *On* or *Off* setting of each  layer as it is listed in the *Layer Properties Manager* is saved when the layer state is created. Therefore, if you are interested in saving only visibility settings (*On/Off, Freeze/Thaw*) for layer groups, it is recommended to uncheck the other boxes.

Although the visibility settings such as *On/Off* and *Freeze/Thaw* are likely to be changed often during the process of working on a drawing, the other properties of layers such as *Color, Linetype, Lineweight,* etc. are not likely to be changed. However, creating layer states using these checkboxes (on the right) can be useful for specific applications. Since properties of layer such as color, lineweight, and plot style usually affect printing and plotting, you can create and restore layer states to prepare for different plot setups. Or, you may prefer a certain color and lineweight "look" on screen for drawing and editing, then restore a layer state that complies with the company standard before saving the drawing or for creating a print or plot. For example, to prepare for printing on a black and white printer, you could create and restore a layer state that quickly sets all layer colors to black. These checkboxes are also helpful if you intend to *Export* a layer state from one drawing, then *Import* it to another. In this case, you can clone one drawing's color, linetype, lineweight, and/or plot characteristics to another drawing, assuming each drawing has the same layer names.

Turn off layers not found in layer state
This checkbox may be useful if you imported a layer state (.LAS file) into your drawing and the current drawing contained more layers than were named in the .LAS file.

Layer States with *Blocks and Xrefs*
Layer states saved with a drawing that is inserted into your current drawing as a *Block* are added to the current drawing. Saved layer states of *Xrefed* drawings are not accessible from the current drawing. (See Chapters 21 and 30 for information on *Blocks* and *Xrefs*.)

Displaying an Object's Properties and Visibility Settings

When the Layer Control drop-down list in the Object Properties toolbar is in the normal position (not "dropped down"), it can be used to display an <u>object's</u> layer properties and visibility settings. Do this by selecting an object (with the pickbox, window, or crossing window) when no commands are in use.

When an object is selected, the Object Properties toolbar displays the layer, color, linetype, lineweight, and plot style <u>of the selected object</u> (Fig. 11-15). The Layer Control box displays the selected object's layer name and layer settings rather than that of the current layer. The Color Control, Linetype Control, Lineweight Control, and Plot Style Control boxes in the Object Properties toolbar (just to the right) also temporarily reflect the color, linetype, lineweight, and plot style properties of the selected object. Pressing the Escape key causes the list boxes to display the current layer name and settings again, as would normally be displayed. If more than one object is selected, and the objects are on different layers or have different color, linetype, and lineweight properties, the list boxes go blank until the objects become unhighlighted or a command is used.

These sections of the Object Properties toolbar have another important feature that allows you to change a highlighted object's (or set of objects) properties. See "Changing Object Properties" near the end of this chapter.

FIGURE 11-15

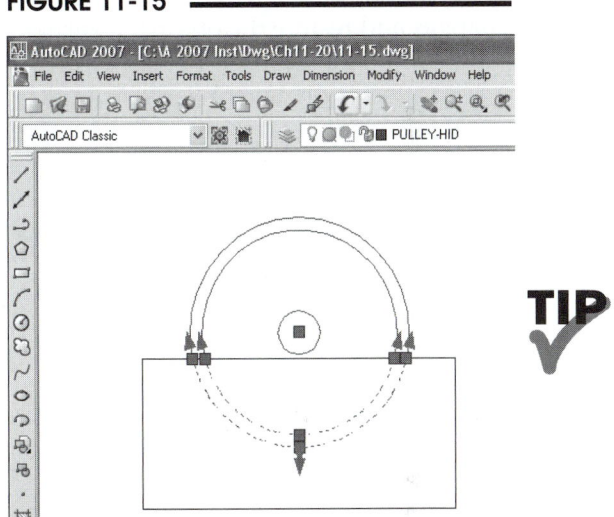

TIP

Laymcur

Pull-down Menu	Command (Type)	Alias (Type)	Short-cut	Screen (side) Menu	Tablet Menu
Format, Layer Tools > Make Object's Layer Current	*Laymcur*	...	...	...	...

This productive feature can be used to make a desired layer current simply by selecting <u>any object on the layer</u>. This option is generally faster than using the *Layer Properties Manager* or *Layer* drop-down list to set a layer current. The *Make Object's Layer Current* feature is particularly useful when you want to draw objects on the same layer as other objects you see, but are not sure of the layer name. TIP

There are two steps in the procedure to make an object's layer current: (1) select the icon and (2) select any object on the desired layer. The selected object's layer becomes current and the new current layer name immediately appears in the Layer Control list box in the Object Properties toolbar.

Layerp

Pull-down Menu	Command (Type)	Alias (Type)	Short-cut	Screen (side) Menu	Tablet Menu
Format *Lauer Tools >* *Layer Previous*	*Layerp*	...	...	...	...

Layerp (*Layer Previous*) is an *Undo* command only for layers. In other words, when you use *Layerp*, all the changes you made the last time you used the *Layer Properties Manager*, the *Layer Control* drop-down list, or the *-Layer* command are undone. Using *Layerp* does not affect any other activities that occurred to the drawing since the last layer settings, such as creating or editing geometry or viewing controls like *Pan* or *Zoom*.

Use *Layerp* to return the drawing to the previous layer settings. For example, if you froze several layers and changed some of the geometry in a drawing, but now want to thaw those frozen layers again without affecting the geometry changes, use *Layerp*. Or, if you changed the color and linetype properties of several layers but later decide you prefer the previous property settings, use *Layerp* to undo the changes and restore the original layer settings.

Layerp affects only layer-related activities; however, *Layerp* does not undo the following changes:
Renamed layers: If you rename a layer and change its properties, *Layerp* restores the original properties but not the original layer name.
Deleted layers: If you delete or purge a layer, using *Layerp* does not restore it.
New layers: If you create a new layer in a drawing, using *Layerp* does not remove it.

OBJECT-SPECIFIC PROPERTIES CONTROLS

The commands in this section are used to control the *color*, *linetype*, and *lineweight* properties of individual objects. This method of object property assignment is used only in special cases—when you want the objects' *color*, *linetype*, and *lineweight* to <u>override those properties assigned to the layers</u> on which the objects reside. Using this method makes it difficult to see which objects are on which layers. In most cases, the *ByLayer* property is assigned instead to individual objects so the objects assume the properties of their layers.

Linetype

Pull-down Menu	Command (Type)	Alias (Type)	Short-cut	Screen (side) Menu	Tablet Menu
Format *Linetype...*	*Linetype* or *-Linetype*	*LT* or *-LT*	...	*FORMAT* *Linetype*	*U,3*

Invoking this command presents the *Linetype Manager* (Fig. 11-16). Even though this looks similar to the dialog box used for assigning linetypes to layers (shown earlier in Fig. 11-7), beware!

FIGURE 11-16

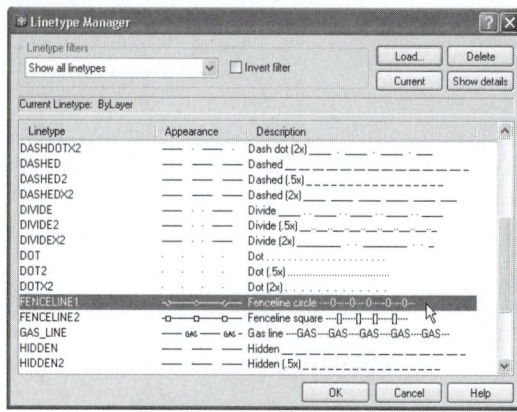

When linetypes are selected and made *Current* using the *Linetype Manager*, they are <u>assigned to objects</u>—not to layers. That is, selecting a linetype by this manner (making it *Current*) causes all objects <u>from that time on</u> to be drawn using that linetype, regardless of the layer that they are on (unless the *ByLayer* type is selected). In contrast, selecting linetypes during the *Layer* command (using the *Layer Properties Manager*) results in assignment of linetypes to <u>layers</u>.

To draw an object in a specific linetype, simply select the linetype from the *Linetype Manager* list, select *Current* (or double-click the linetype), and select the *OK* tile. That linetype stays in effect (all objects are drawn with that linetype) until another linetype is selected. If you want to draw objects in the <u>layer's</u> assigned linetype, select *ByLayer*. *ByBlock* linetype assignment is discussed in Chapter 21.

Load

The *Linetype Manager* also lets you *Load* linetypes that are not already in the drawing. Just select the *Load* button to view and select from the list of available linetypes in the *Load or Reload Linetypes* dialog box (Fig. 11-17). The linetypes are loaded from the ACAD.LIN or ACADISO.LIN file. If you want to load all linetypes into the drawing, right-click to display a small cursor menu, then choose *Select All* to highlight all linetypes (see Fig. 11-17). You can also select multiple linetypes by holding down the Ctrl key or select a range by holding down the Shift key. The loaded linetypes are then available for assignment from both the *Linetype Manager* and the *Layer Properties Manager*.

FIGURE 11-17

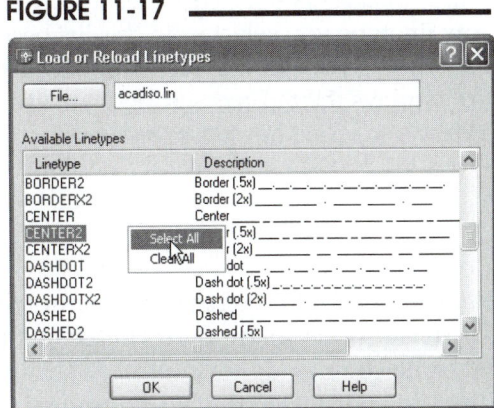

AutoCAD supplies numerous linetypes that can be viewed in the *Load or Reload Linetypes* dialog box (Fig. 11-17). The ACAD_ISO*n*W100 linetypes are intended to be used with metric drawings. For the ACAD_ISO*n*W100 linetypes, *Limits* should be set to metric sheet sizes to accommodate the relatively large linetype spacing. Avoid using both ACAD_ISO*n*W100 linetypes and non-ISO linetypes in one drawing due to the difficulty managing the two sets of linetype scales.

Delete

The *Delete* button is useful for deleting any unused linetypes from the drawing. Unused linetypes are those which have been loaded into the drawing but have not been assigned to layers or to objects. Freeing the drawing of unused linetypes can reduce file size slightly.

The *Linetype Manager* has a *Linetype filters* drop-down list and *Details/Hide Details* button. If details are visible, you can change the *Name* or *Description*, or set the *Global Scale Factor* (see "LTSCALE") or *Current Object Scale* (see "CELTSCALE").

Typing -*Linetype* (use the hyphen prefix) produces the command line format of *Linetype*. Although this format of the command does not offer the capabilities of the dialog box version, you can assign linetypes <u>to objects</u> or assign the *ByLayer* linetype setting.

```
Command: -linetype
Current line type: "ByLayer"
Enter an option [?/Create/Load/Set]:
```

You can list (*?*), *Load*, *Set*, or *Create* linetypes with the typed form of the command. Type a question mark (*?*) to display the list of available linetypes. Type *L* to *Load* any linetypes. The *Set* option of *Linetype* accomplishes the same action as selecting a linetype from the list in the dialog box. That is, you can set a specific linetype for all subsequent objects regardless of the layer's linetype setting, or you can use *Set* to assign the new objects to use the layer's (*ByLayer*) linetype.

You can create your own custom linetypes with the *Create* option or by using a text editor. Both simple linetypes (line, dash, and dot combinations) and complex linetypes (including text or other shapes) are possible. See Chapter 44, Miscellaneous Customization, for more information on creating linetypes.

Linetype Control Drop-Down List

The Object Properties toolbar contains a drop-down list for selecting linetypes (Fig. 11-18). Although this appears to be quick and easy, you can <u>only assign linetypes to objects</u> by this method unless *ByLayer* is selected to use the layers' assigned linetypes. Any linetype you select from this list becomes the current object linetype. If you want to select linetypes for layers, <u>make sure this list displays the *ByLayer* setting</u>, then use the *Layer Properties Manager* to select linetypes for layers.

FIGURE 11-18

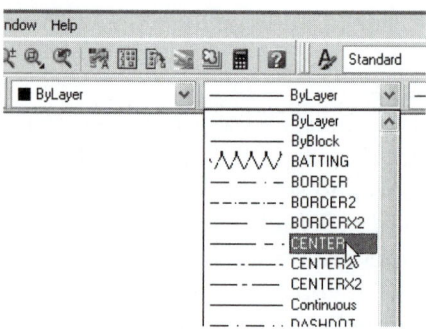

Keep in mind that the Linetype Control drop-down list as well as the Layer Control, Color Control, and Lineweight Control drop-down lists display the current settings for selected objects (objects selected when no commands are in use). When linetypes are assigned to layers rather than objects, the layer drop-down list reports a *ByLayer* setting.

Lweight

Pull-down Menu	Command (Type)	Alias (Type)	Short-cut	Screen (side) Menu	Tablet Menu
Format *Lineweight...*	*Lweight*	*LW*	...	...	...

The *Lweight* command (short for lineweight) produces the *Lineweight Settings* dialog box (Fig. 11-19). This dialog box is the lineweight <u>equivalent</u> of the *Linetype Manager*—that is, this dialog box assigns lineweights <u>to objects, not layers</u> (unless the *ByLayer* or *Default* lineweight is selected). See the previous discussion under "*Linetype*." The *ByBlock* setting is discussed in Chapter 21.

FIGURE 11-19

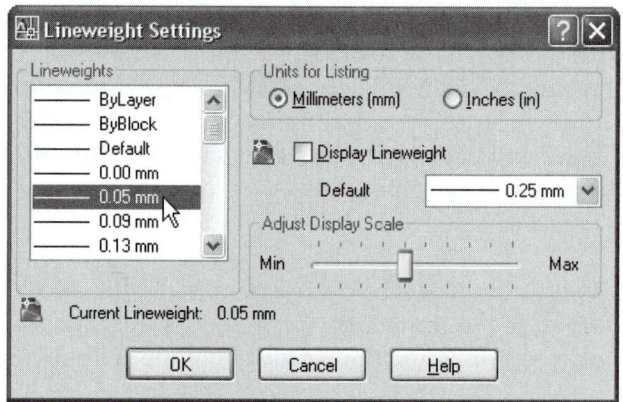

Any lineweight selected in the *Lineweight Settings* dialog box automatically becomes current (without having to select a *Current* button or double-click). The current lineweight is assigned to all subsequently drawn objects. That is, selecting a lineweight by this manner (making it current) causes all objects <u>from that time on</u> to be drawn using that lineweight, regardless of the layer that they are on (unless *ByLayer* or *Default* is selected). In contrast, selecting lineweights during the *Layer* command (using the *Layer Properties Manager*) results in assignment of lineweights to <u>layers</u>.

As a reminder, this type of drawing method (object-specific property assignment) can be difficult for beginning drawings and for complex drawings. If you want to draw objects in the <u>layer's</u> assigned lineweight, select *ByLayer*. If you want all layers to have the same lineweight, select *Default*.

Units for Listing

You can choose *Millimeters (mm)* or *Inches (in)* as the units for the lineweight list. The values reported in the list (0.05 mm, 0.15 mm, 0.010", 0.015", etc.) indicate the thickness of the selected line. The setting is stored in the *LWUNITS* system variable.

Display Lineweight

Checking this option causes the assigned lineweight thickness to be displayed anywhere in the drawing—in model space (the *Model* tab), for model space geometry that appears in paper space viewports (viewports in a *Layout* tab), and in paper space (in a *Layout* tab, but not in a viewport).

When lineweights are displayed in model space (the *Model* tab), the line thickness is relative to the screen size, so zooming in or out does not make the lines appear wider or narrower. However, when lineweights are displayed in paper space (a *Layout* tab), the line thickness is absolute, so zooming changes the lineweight appearance.

FIGURE 11-20

The *Display Lineweight* checkbox has the same function as toggling the word *LWT* on the Status Bar (Fig. 11-20). The *LWT* toggle is somewhat more accessible to use. The status of the lineweight display for the drawing is stored in the *LWDIS-PLAY* system variable. Using either the *Display Lineweight* checkbox or the *LWT* toggle changes the setting of the *LWDIS-PLAY* variable, and vice versa.

Default

Although the *Current* lineweight for new drawings is *ByLayer*, when new layers are created, all layers have the *Default* lineweight. This feature (*Default* rather than *ByLayer*) is exclusive for lineweights. Assigning the *Default* setting ensures that all layers (or objects) with this setting have the same lineweight. The advantage is that the lineweight can be changed globally for all layers (or objects) that have the *Default* setting.

Selecting a new lineweight in the *Default* drop-down list of the *Lineweight Settings* dialog box changes the lineweight thickness for existing layers (or objects) with a *Default* setting and assigns that setting for new layers (or objects) that are subsequently created. In contrast, to change the lineweight for existing layers with a *ByLayer* setting, each layer's lineweight must be changed individually. The *Default* lineweight setting is stored in the *LWDEFAULT* system variable.

Adjust Display Scale

This adjustment affects only the display of lineweights in the *Lineweight Settings* dialog box and the Lineweight Control drop-down list.

Typing *-Lweight* produces the Command line format of *Lweight*. This command line version only allows you to set the current lineweight for new objects and does not offer the capabilities of the dialog box version. Typing *-Lweight* (use the hyphen prefix) produces the following prompt:

```
Command: -lweight
Current lineweight: ByLayer
Enter default lineweight for new objects or [?]:
```

FIGURE 11-21

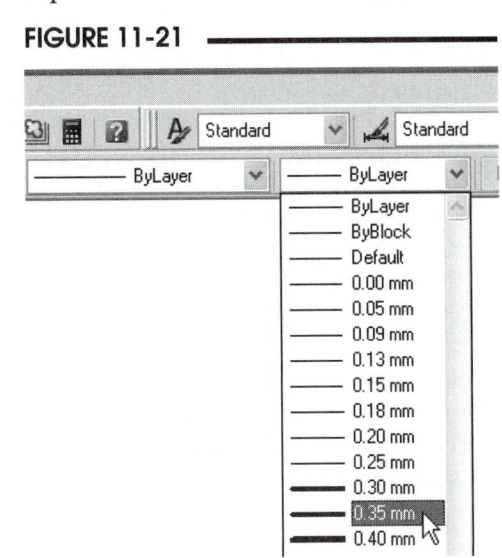

Lineweight Control Drop-Down List

The Object Properties toolbar contains a drop-down list for selecting lineweights (Fig. 11-21). Similar to the function of the Linetype Control drop-down list, you can only assign lineweights to objects by this method unless *ByLayer* is selected to use the layers' assigned lineweights. Any lineweight you select from this list becomes the current object lineweight. If you want to select lineweights for layers, make sure this list displays the *ByLayer* setting, then use the *Layer Properties Manager* to select lineweights for layers.

Remember that the Lineweight Control drop-down list as well as the Layer Control, Color Control, and Linetype Control drop-down lists display the current settings for selected objects (objects selected when no commands are in use). When lineweights are assigned to layers rather than objects, the layer drop-down list reports a *ByLayer* setting.

Color

Pull-down Menu	Command (Type)	Alias (Type)	Short-cut	Screen (side) Menu	Tablet Menu
Format Color...	*Color*	*COL*	...	*FORMAT Color*	*U,4*

Similar to linetypes and lineweights, colors can be assigned to layers or to objects. Using the *Color* command assigns a color for all newly created <u>objects</u>, regardless of the layer's color designation (unless the *ByLayer* color is selected). This color setting <u>overrides</u> the layer color for any newly created objects so that all new objects are drawn with the specified color no matter what layer they are on. This type of color designation prohibits your ability to see which objects are on which layers by their color; however, for some applications object color setting may be desirable. Use the *Layer Properties Manager* to set colors for layers.

Invoking this command by the menus or by typing *Color* presents the *Select Color* dialog box shown in Figure 11-22. This is essentially the same dialog box used for assigning colors to layers; however, the *ByLayer* and *ByBlock* tiles are accessible. (The buttons are grayed-out when this dialog is invoked from the *Layer Properties Manager* because, in that case, any setting is a *ByLayer* setting.) *ByBlock* color assignment is discussed in Chapter 21.

FIGURE 11-22

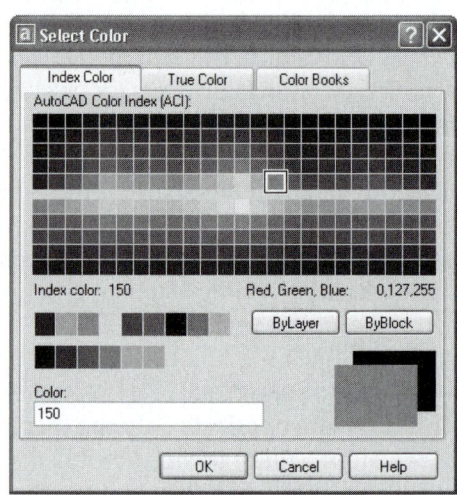

If you type -*Color* (include the hyphen prefix), the Command line format is displayed as follows:

    Command: **-color**
    Enter default object color [Truecolor/COlorbook] <BYLAYER>:

The current color (whether *ByLayer* or specific object color) is saved in the *CECOLOR* (Current Entity Color) variable. The current color can be set by changing the value in the *CECOLOR* variable directly at the command prompt (by typing *CECOLOR*), by using the *Color* command (*Select Color* dialog box), or by using the Color Control drop-down list in the Object Properties toolbar. The *CECOLOR* variable accepts a string value (such as "red" or "bylayer") or accepts the ACI number (0 through 255). Valid values for true colors are a string of integers each from 1 to 255 separated by commas and preceded by RGB. A true color setting is entered as follows: RGB:000,000,000.

Color Control Drop-Down List

When an object-specific color has been set, the top item in the Object Properties toolbar Color Control drop-down list displays the current color (Fig. 11-23). Beware—using this list to select a color assigns an <u>object-specific color unless *ByLayer* is selected</u>. Any color you select from this list becomes the current object color. Choosing *Select Other...* from the bottom of the list invokes the *Select Color* dialog box (see Fig. 11-22). If you want to assign colors to layers, make sure this list displays the *ByLayer* setting, then use the *Layer Properties Manager* to select colors for layers.

FIGURE 11-23

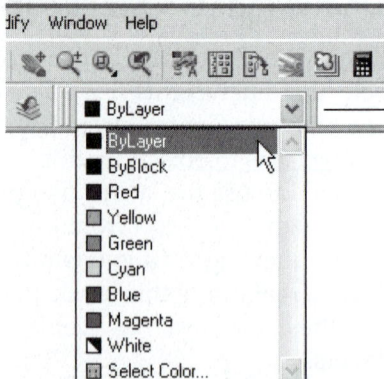

The *Color Control* drop-down list, as well as the others in the Object Properties toolbar, displays the current settings for selected objects (when no commands are in use).

True Color and Color Books

AutoCAD offers support for true color and industry-standard color books. Three tabs are available in the *Select Color* dialog box—*Index Color*, *True Color*, and *Color Books*. The *Index Color* tab, discussed earlier in the chapter (see Fig. 11-22), allows you to specify color using the 256-color ACI (AutoCAD Color Index).

True Color

The *True Color* tab in the *Select Color* dialog box provides a tool for you to specify colors by the *HLS* or the *RGB* color systems. Select the *Color Model* drop-down list to select which system you prefer (Fig. 11-24, upper-right corner). Watch the color swatch in the lower-right corner to dynamically display the color you specify.

HLS

This system specifies the color by *Hue, Luminance,* and *Saturation* (Fig. 11-24). *Hue* controls the pure color (red, yellow, blue), *Luminance* controls the color "value" (white to black—100 to 0), and *Saturation* controls the purity of the color (mix of *Hue* and *Luminance*—100 is pure color, 0 is no color). Use the pointer to specify *Hue* and *Saturation* and the slider for *Luminance* or enter values in the edit boxes.

RGB

This system determines the color by the amount of red, green, and blue components (Fig. 11-25). Each of these colors is specified on a scale of 0 (no color) to 255 (maximum color). You can use the sliders or enter in the RGB values in the edit box.

Color Books

The *Color Books* tab in the *Select Color* dialog box allows you to specify colors using the DIC, Pantone, or RAL commercial color systems. These systems are used largely in the architectural and interior design industries as a standard for specifying colors for paint and interior furnishings. These standard colors have been traditionally supplied to professionals in books displaying color samples for design and color matching.

Select the desired DIC, Pantone, or RAL "book" from the *Color book* drop-down list (Fig. 11-26). Use the slider and up/down arrows to display the color samples. Specify a particular color by clicking on it so the color appears in the color swatch (lower-right corner) and the index number appears in the *Color* edit box. If you know the desired index number, you can enter it directly in the edit box.

FIGURE 11-24

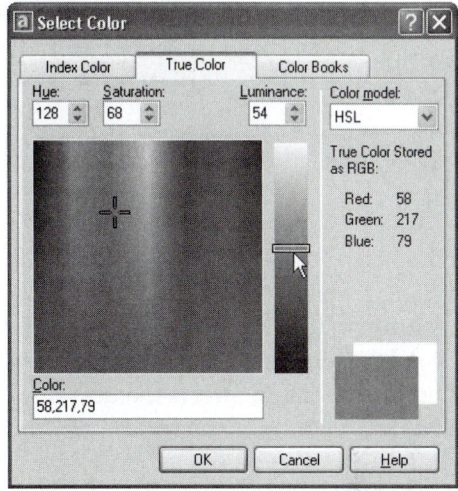

FIGURE 11-25

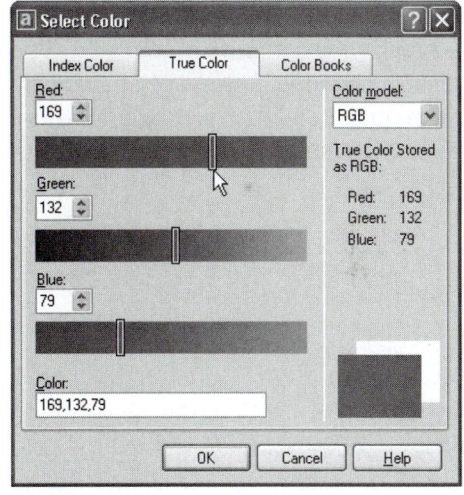

FIGURE 11-26

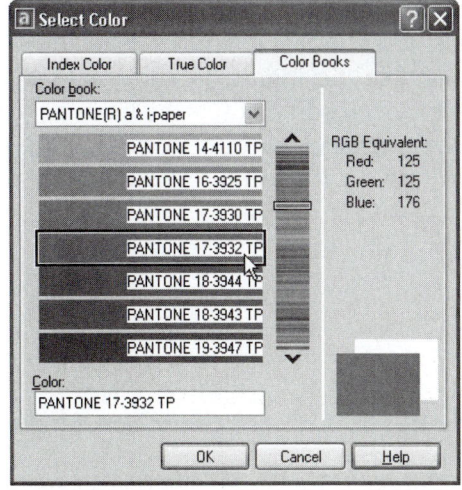

CONTROLLING LINETYPE SCALE

LTSCALE

Pull-down Menu	Command (Type)	Alias (Type)	Short-cut	Screen (side) Menu	Tablet Menu
Format *Linetype...* *Show details >>* *Global scale* *factor*	*Ltscale*	*LTS*	...	*FORMAT* *Linetype* *Show details >>* *Global scale* *factor*	...

Hidden, dashed, dotted, and other linetypes that have spaces are called <u>non-continuous</u> linetypes. When drawing objects that have non-continuous linetypes (either *ByLayer* or object-specific linetype designations), the linetype's dashes or dots are automatically created and spaced. The *LTSCALE* (Linetype Scale) system variable controls the length and spacing of the dashes and/or dots. The value that is specified for *LTSCALE* affects the drawing <u>globally and retroactively</u>. That is, all existing non-continuous lines in the drawing as well as new lines are affected by *LTSCALE*. You can therefore adjust the drawing's linetype scale for all lines at any time with this one command.

TIP

If you choose to make the dashes of non-continuous lines smaller and closer together, reduce *LTSCALE*; if you desire larger dashes, increase *LTSCALE*. The *Hidden* linetype is shown in Figure 11-27 at various *LTSCALE* settings. Any positive value can be specified.

FIGURE 11-27

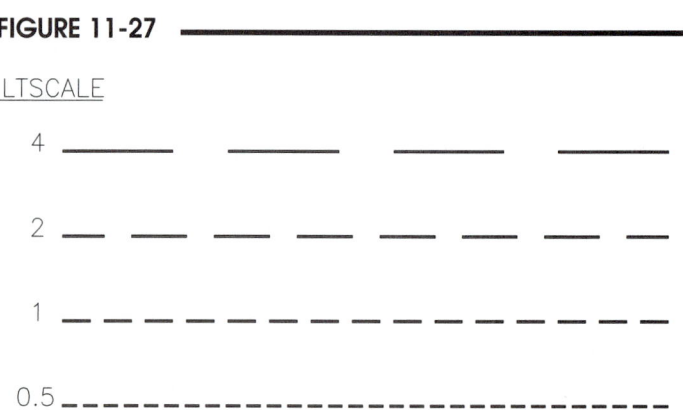

LTSCALE can be set in the *Details* section of the *Linetype Manager* or in Command line format. In the *Linetype Manager*, select the *Details* button to allow access to the *Global scale factor* edit box (Fig. 11-28). Changing the value in this edit box sets the *LTSCALE* variable. Changing the value in the *Current object scale* edit box sets the linetype scale for the current object only (see "*CELTSCALE*"), but does not affect the global linetype scale (*LTSCALE*).

FIGURE 11-28

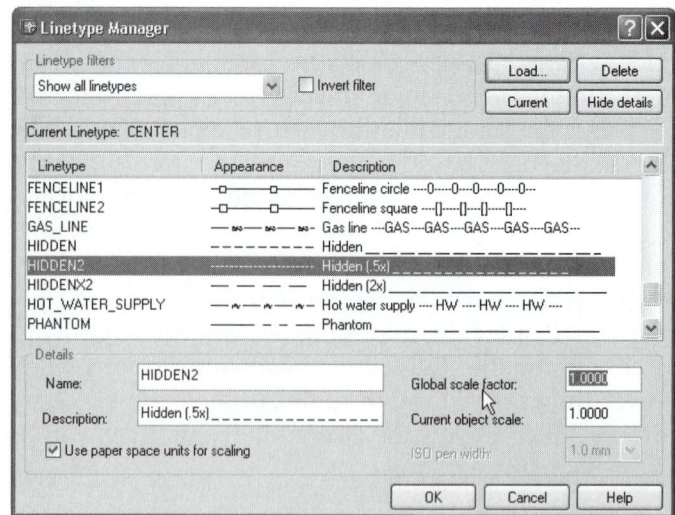

LTSCALE can also be used in the Command line format.

Command: **ltscale**
Enter new linetype scale factor <1.0000>: (**value**) (Enter any positive value.)

The *LTSCALE* for the default template drawing (ACAD.DWT) is 1. This value represents an appropriate *LTSCALE* for objects drawn within the default *Limits* of 12 x 9.

As a general rule, you should change the *LTSCALE* <u>proportionally</u> when *Limits* are changed (more specifically, when *Limits* are changed to other than the intended plot sheet size). For example, if you increase the drawing area defined by *Limits* by a factor of 2 (to 24 x 18) from the default (12 x 9), you might also change *LTSCALE* proportionally to a value of 2. Since *LTSCALE* is retroactive, it can be changed at a later time or repeatedly adjusted to display the desired spacing of linetypes.

If you load the *Hidden* linetype, it displays dashes (when plotted 1:1) of 1/4" with a *LTSCALE* of 1. The ANSI and ISO standards state that hidden lines should be displayed on drawings with dashes of approximately 1/8" or 3mm. To accomplish this, you can use the *Hidden* linetype and change *LTSCALE* to .5. It is recommended, however, that you <u>use the *Hidden2* linetype</u> that has dashes of 1/8" when plotted 1:1. Using this strategy, *LTSCALE* remains at a value of 1 to create the standard linetype sizes for *Limits* of 12 x 9 and can easily be changed in proportion to the *Limits*. (For more information on *LTSCALE*, see Chapter 12 and Chapter 13.)

The ACAD_ISO*n*W100 linetypes are intended to be used with metric drawings. Changing *Limits* to metric sheet sizes automatically displays these linetypes with appropriate linetype spacing. For these linetypes only, *LTSCALE* is changed automatically to the value selected in the *ISO Pen Width* box of the *Linetype* tab (see Fig. 11-28). Using both ACAD_ISO*n*W100 linetypes and other linetypes in one drawing is discouraged due to the difficulty managing two sets of linetype scales.

Even though you have some control over the size of spacing for non-continuous lines, you have almost <u>no control</u> over the <u>placement</u> of the dashes for non-continuous lines. For example, the short dashes of center lines <u>cannot</u> always be controlled to intersect at the centers of a series of circles. The spacing can only be adjusted globally (all lines in the drawing) to reach a compromise. You can also adjust individual objects' linetype scale (*CELTSCALE*) to achieve the desired effect. See "*CELTSCALE*" (next) and Chapter 24, Multiview Drawing, for further discussion and suggestions on this subject.

CELTSCALE

CELTSCALE stands for Current Entity Linetype Scale. *CELTSCALE* is actually a system variable for linetype scale stored with each object—an object property. This setting changes the <u>object-specific linetype scale proportional</u> to the *LTSCALE*. The *LTSCALE* value is global and retroactive, whereas *CELTSCALE* sets the linetype scale for all <u>newly created</u> objects and is <u>not retroactive</u>. *CELTSCALE* is object-specific. Using *CELTSCALE* to set an object linetype scale is similar to setting an object color, linetype, and lineweight in that the properties are <u>assigned to the specific object</u>.

For example, if you wanted all non-continuous lines (dashes and spaces) in the drawing to be two times the default size, set *LTSCALE* to 2 (*LTSCALE* is global and retroactive). If you then wanted only two or three lines to have smaller spacing, change *CELTSCALE* to .5, draw the new lines, and then change *CELTSCALE* back to 1.

Matchprop

Pull-down Menu	Command (Type)	Alias (Type)	Short-cut	Screen (side) Menu	Tablet Menu
Modify *Match* *Properties*	*Matchprop*	*MA*	...	*MODIFY1* *Matchprp*	Y,14 and Y,15

Matchprop is used to "paint" the properties of one object to another. The process is simple. After invoking the command, select the object that has the desired properties (source object), then select the object you want to "paint" the properties to (destination object). The command prompt is as follows.

```
Command: matchprop
Select source object: PICK
Current active settings:  Color Layer Ltype Ltscale Lineweight Thickness
     PlotStyle Dim Text Hatch Polyline Viewport Table Material Shadow display
Select destination object(s) or [Settings]: PICK
Select destination object(s) or [Settings]: Enter
Command:
```

Only one "source object" can be selected, but its properties can be painted to several "destination objects." The "destination object(s)" assume all of the properties of the "source object" (listed as "Current active settings").

Use the *Settings* option to control which of several possible properties and other settings are "painted" to the destination objects. At the "Select destination object(s) or [Settings]:" prompt, type *S* to display the *Property Settings* dialog box (Fig. 11-31). In the dialog box, designate which of the *Basic Properties* or *Special Properties* are to be painted to the "destination objects." The following *Basic Properties* correspond to properties discussed in this chapter.

FIGURE 11-31

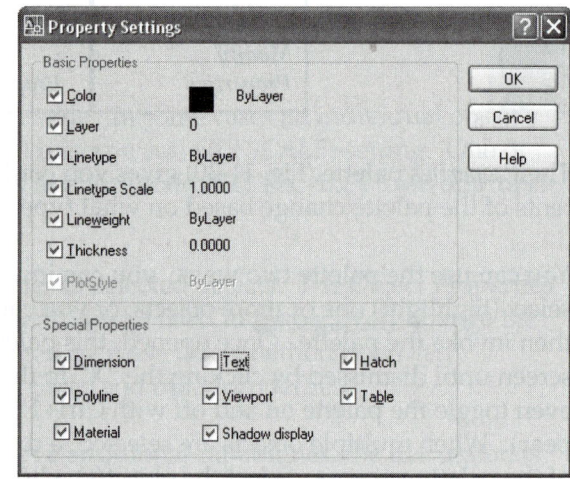

Color	Paints the object-specific or *ByLayer* color.
Layer	Moves selected objects to Source Object layer.
Linetype	Paints the object-specific or *ByLayer* linetype.
Lineweight	Paints the object-specific or *ByLayer* lineweight.
Linetype Scale	Changes the individual object's linetype scale (*CELTSCALE*), not global (*LTSCALE*).

The *Special Properties* of the *Property Settings* dialog box are discussed in Chapter 16, Modify Commands II. Also see Chapter 16 for a full explanation of the *Properties* palette (*Properties*).

LAYER TOOLS

Several very productive and popular layer-related commands were accessible in previous AutoCAD releases only if you installed the Express Tools. This set of useful commands has been included in the AutoCAD 2007 core product and can be found in the *Layer Tools* cascading menu (Fig. 11-32).

FIGURE 11-32

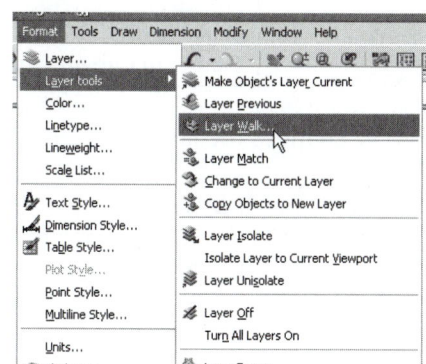

Laymch

Pull-down Menu	Command (Type)	Alias (Type)	Short-cut	Screen (side) Menu	Tablet Menu
Format Layer Tools > Layer Match	*Laymch*	...	...	...	...

Laymch allows you to change the layer of selected objects to the layer of the selected destination object. This command performs the same function as the *Matchprop* command (see Chapter 12), but only with respect to layers (*Matchprop* can change other properties such as color, linetype, or linetype scale).

Using *Laymch* produces the following prompt:

    Command: laymch
    Select objects to be changed:
    Select objects: PICK
    Select objects: PICK
    Select objects: Enter
    2 found.
    Select object on destination layer or [Name]: PICK
    2 objects changed to layer 0.
    Command:

You can use the *Name* option to select the desired layer to change the objects to. The *Change to Layer* dialog box appears (Fig. 11-33). Simply select the destination layer for the selected objects.

FIGURE 11-33

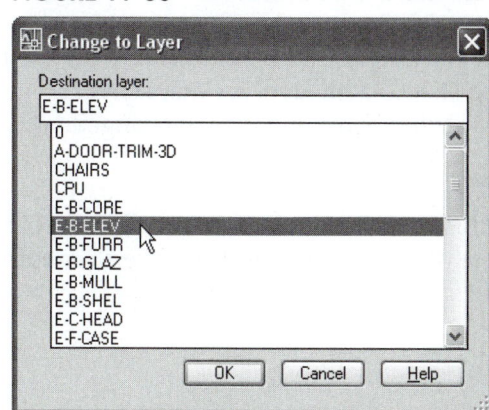

Laycur

Pull-down Menu	Command (Type)	Alias (Type)	Short-cut	Screen (side) Menu	Tablet Menu
Format *Layer Tools >* *Change to Current Layer*	*Laycur*	...	...	...	...

Laycur changes the layer of selected objects to the current layer. The following prompt is issued:

    Command: **laycur**
    Select objects to be changed to the current layer:
    Select objects: **PICK**
    Select objects: **PICK**
    Select objects: **Enter**
    2 found.
    2 objects changed to layer TEXT (the current layer).
    Command:

Layiso

Pull-down Menu	Command (Type)	Alias (Type)	Short-cut	Screen (side) Menu	Tablet Menu
Format *Layer Tools >* *Layer Isolate*	*Layiso*	...	...	...	...

Use this express command if you want to turn *Off* <u>all layers except</u> the layer(s) of the selected object(s). Several layers can be selected to be "isolated." If more than one layer is isolated, the last one selected becomes the current layer. The following prompt is used:

    Command: **layiso**
    Current setting: Viewports=Vpfreeze
    Select objects on the layer(s) to be isolated or [Settings]: **PICK**
    Select objects on the layer(s) to be isolated or [Settings]: **Enter**
    Layer FURNITURE has been isolated.
    Command:

The *Settings* option allows you to specify how layers are frozen in viewports. The setting you choose persists from session to session.

        In paper space viewport use [Vpfreeze/Off] <Vpfreeze>:

Vpfreeze
If you are using this command in paper space, *Vpfreeze* isolates layers in the current viewport. Objects on the isolated layer are displayed and all other layers in the current viewport are frozen. Other viewports in the drawing are unchanged.

Off
This option isolates layers in all viewports—model space and paper space. Objects on the isolated layer are displayed and all other layers are turned off in viewports and in model space.

Layuniso

Pull-down Menu	Command (Type)	Alias (Type)	Short-cut	Screen (side) Menu	Tablet Menu
Format Layer Tools > Layer Unisolate	Layuniso	...	...	...	...

Layuniso simply thaws all layers that were previously frozen with *Layiso*. All layers are restored to the state they were in just before you used *Layiso*. If *Layiso* was not used, *Layuniso* does not restore any layers.

Command: **layuniso**
Layers isolated by LAYISO command have been restored.

Copytolayer

Pull-down Menu	Command (Type)	Alias (Type)	Short-cut	Screen (side) Menu	Tablet Menu
Format Layer Tools > Copy Objects to New Layer	Copytolayer...	...	...	...	...

When you need two different layers that contain similar (or duplicate) objects, such as "Existing Doors" and "Demo Doors," you can use this express feature to select existing object(s), then choose the destination layer for copied objects from a dialog box. The destination layer must exist prior to launching this command.

Command: **copytolayer**
Select objects to copy: **PICK**
Select objects to copy: **Enter**
Select object on destination layer or [Name] <Name>: ***n***
1 object(s) copied and placed on layer "CHAIRS".
Specify base point or [Displacement/eXit] <eXit>: **PICK**
Specify second point of displacement or <use first point as displacement>: **PICK**
Command:

When asked for the "destination layer," you can PICK an object on the desired layer or use the *Name* option to select the desired layer to copy the objects to. The *Name* option produces the *Copy to Layer* dialog box (Fig. 11-34) where you select the destination layer for the copied objects.

Also note that you can select a new location for the copied objects when prompted for a "base point" and "second point of displacement." In some cases, such as copying objects from an "Existing Doors" layer to a "Demo Doors" layer, you can press Enter to copy the new objects to the same coordinates as the existing objects.

FIGURE 11-34

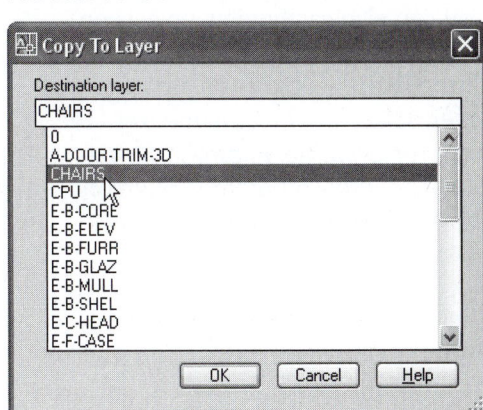

Laywalk

	Pull-down Menu	Command (Type)	Alias (Type)	Short-cut	Screen (side) Menu	Tablet Menu
	Format Layer Tools > Layer Walk	Laywalk	...	...	...	...

The main feature of this utility enables you to view the objects on any layer(s) in a drawing even if the layers are *Off* or *Frozen*. *Laywalk* enables you to view objects on any specific layer or any set of layers, one layer at a time if desired. *Laywalk* is best utilized to examine the layer contents of unfamiliar drawings, such as those created by an outside source. Although *Laywalk* dynamically changes the *Freeze/Thaw* or *Off/On* state of any layers you select to view, you can choose not to keep the changes made using *Laywalk* and restore the original layer visibility settings when you exit this utility. The *LayerWalk* dialog box provides several options and a shortcut menu with additional features to manipulate the visibility of layers. You also have access to the *Layer Manager* utility from the shortcut menu, enabling you to manipulate and save layer states.

NOTE: Since *Laywalk* shows the contents of each layer, you may want to use *Zoom Extents* just prior to launching *Laywalk* with an unfamiliar drawing.

Invoking the *Laywalk* command by any method produces the *LayerWalk* dialog box. When the *LayerWalk* dialog box first appears, layers that are visible in the drawing are highlighted in the list. For example in Figure 11-35, the *LayerWalk* dialog box list indicates (highlights) the names of all visible layers in the drawing, but the A-FURN-* layers are not visible in the drawing and are not highlighted in the list.

NOTE: The *LayerWalk* dialog box is sizable, so enlarge the dialog box if needed to see more of the layer names in the list.

You can also easily view objects on each layer individually, one layer at a time. To view objects on a particular layer, select any layer name from the list. For example, Figure 11-36 displays only the geometry drawn on the Xref layer "8th floor plan | A-WALL-INT."

To "walk" through the layers, highlight the first layer name from the list, then use the down arrow key on your keyboard to go through the list and display the geometry on each layer in the drawing, one layer at a time.

FIGURE 11-35

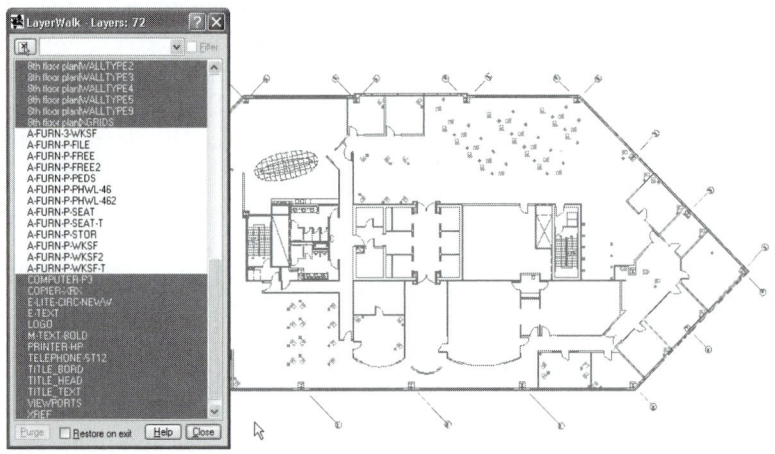

FIGURE 11-36

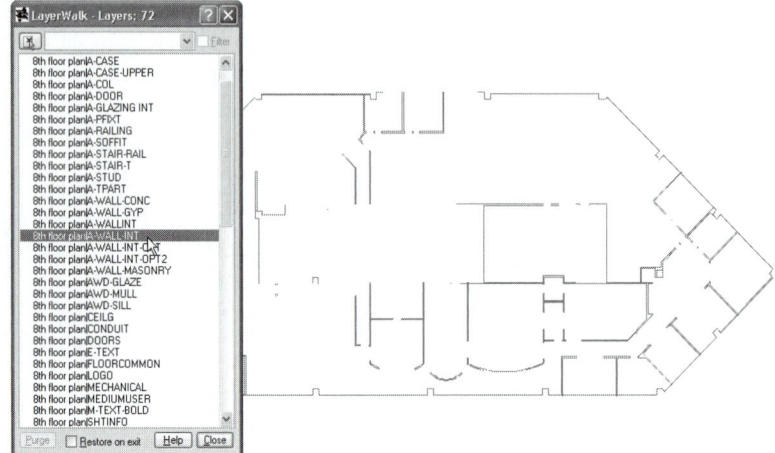

You can also select more than one layer by holding down the left button and dragging your cursor across the layers to view (or use Ctrl to select single layers or Shift to select a range).

Layer name edit box and *Filter* check box

As an alternative, you can enter individual layer names in the edit box at the top of the *LayerWalk* dialog box to display that layer. Wildcards can be used. For example entering "A-FURN-*" in the edit box would display those layers in the drawing whose names began with "A-FURN-" (Fig. 11-37).

FIGURE 11-37

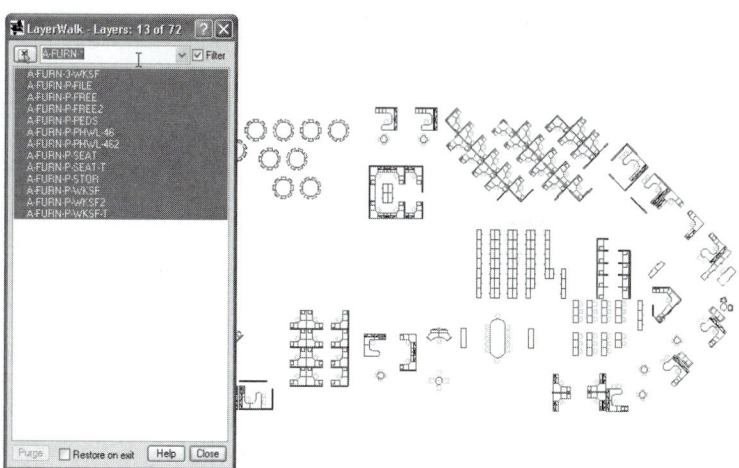

Use a filter to reduce the number of layers appearing in the list and in the drawing. Create a filter by entering layer names with a common prefix, then an asterisk, such as "A-FURN-*" (see Figure 11-37), or enter layer names of an Xref drawing, such as "8ᵗʰ floor plan*." This action creates a temporary filter for the list. When the *Filter* box is checked, only those layers listed in the edit box appear in the list and in the drawing. With a filter in effect, you can still select individual names from the list to view the geometry, etc. Uncheck the *Filter* box to make all layers in the drawing appear in the list. A filter remains in effect until cleared by unchecking the box. Remember that filters are <u>temporary</u> (useful only during the current *LayerWalk* session) unless saved. See "Shortcut Menu," *Save Filter,* to save filters for use in subsequent *LayerWalk* sessions.

Select Objects button

This button (upper-left corner of the dialog box) enables you to select one or multiple objects <u>in the drawing</u>, then returns to the *LayerWalk* dialog box with the objects layer(s) highlighted. This is an excellent way to examine an unfamiliar drawing and find out which layers specific objects were drawn on, especially when selected objects are on more than one layer.

Purge

Purge <u>deletes layers from the drawing</u> (even if *Restore on Exit* is checked). The *Purge* button is enabled only if there are one or more unreferenced layers (layers containing no objects) in the drawing and those layers are highlighted in the list. Use the *Select Unreferenced* option in the shortcut menu to locate and highlight any unused layers, then the *Purge* button to delete those layer names from the drawing and from the layer list.

Restore on Exit check box

Restore on Exit means to <u>disregard</u> any changes made to the layer status using *LayerWalk* and restore the drawing to the visibility state before using *LayerWalk*. Use this option if you want to examine the geometry on specific layers of a drawing but then return to work on the drawing without saving any of those layer visibility changes. In other words, if *Restore on Exit* is checked when you *Close* the *LayerWalk* dialog box, AutoCAD returns to the layers displayed prior to using *LayerWalk*. If you want to keep the layer display attained using *LayerWalk*, but return to work on the drawing, make sure *Restore on Exit* is <u>not</u> checked before you close the dialog box.

2007

Shortcut menu

Right-click in the *LayerWalk* dialog to display the shortcut menu (Fig. 11-38).

FIGURE 11-38

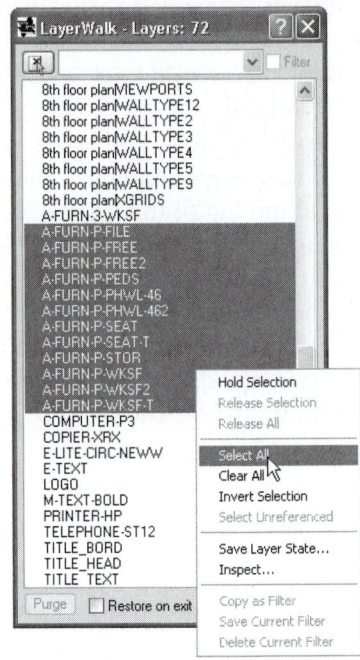

Hold Selection, Release Selection, Release All
If you want to turn on the display of a set of layers, then select other layers from the list to examine the geometry in the drawing, use *Hold Selection*. For example, in an architectural setting you could highlight the floorplan layer names, use *Hold Selection* to keep those layers visible, then select other layer names to display that geometry within the context of the floorplan. When you "hold" a list of layers, high-lighting is removed from the names, so "held" selections appear in the list with an asterisk before the name. You can "hold" multiple selections. Use *Release Selection* to remove one selection from "hold" status and use *Release All* to remove multiple layer selections.

Select All, Clear All
Use *Select All* to select all names in the list or *Clear All* to clear the list. Use *LayerWalk's Select All* feature to quickly turn on all layers in a drawing, even if the layers selected are *Frozen*!

Invert Selection
The *Invert Selection* option highlights (and displays in the drawing) all layers from the list previously not highlighted and clears (and does not display in the drawing) those names previously shown.

Select Unreferenced
The *Select Unreferenced* feature highlights any unreferenced layers in the drawing (layers that contain no geometry). Using *Select Unreferenced* is a quick way to locate unused layers so you can then use the *Purge* button to remove those layer names from the <u>drawing</u>.

Save Layer State
The *Save Layer State* feature enables you to give a name to the highlighted layer group for future use (Fig. 11-39). A *Layer State* is a particular layer visibility setting saved with a name. For example, if you use *LayerWalk* to specify a set of layers for viewing, but expect to view this specific layer set again some-time in the future, *Save* the layer state. In this way you can *Restore* the layer state by name in the future rather than having to select that specific set of layers one by one in the *LayerWalk* dialog box. Named layer states are available in the *Layer Properties Manager*. See previous discussion, "*Layer Properties Manager*," for more information.

FIGURE 11-39

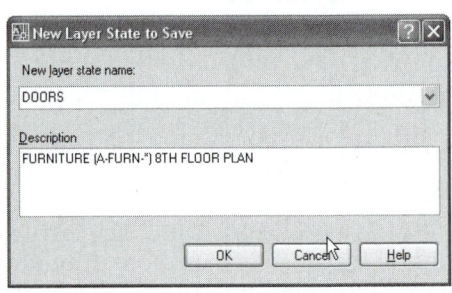

Inspect
Select *Inspect* to open the *Inspect* dialog box (Fig. 11-40). This utility displays the number of layers in the drawing, the number of layers selected (if any), and the number of objects on the selected layers.

FIGURE 11-40

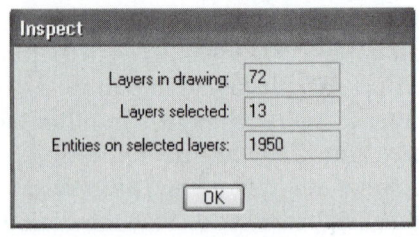

Copy Filter, Save Current Filter, Delete Current Filter
Use the *Copy Filter, Save Current Filter, Delete Current Filter* func-tions to capture and discard layer filter names. Remember that filters are <u>temporary</u> (useful only during the current *LayerWalk* session) unless saved (see previous discussion, "Layer name edit box and *Filter* checkbox").

To save a filter for use in subsequent *LayerWalk* sessions (after closing the *LayerWalk* dialog box), create a filter by entering a layer name string in the edit box (see previous Fig. 11-37), then select *Save Current Filter* from the shortcut menu. Or, use *Copy Filter* if a single layer name has been highlighted and you want to save the name as a filter. The named filter then appears in the drop-down list whenever *LayerWalk* is used again for the drawing. The filter is saved regardless of the setting for *Restore on Exit*. The *LayerWalk* filters are not available as named layer filters from the *Layer Properties Manager*.

Layfrz

Pull-down Menu	Command (Type)	Alias (Type)	Short-cut	Screen (side) Menu	Tablet Menu
Format *Layer Tools >* *Layer Freeze*	*Layfrz*	...	...	...	...

This utility *Freezes* the layer(s) of the selected object(s). More than one object (layer) can be selected for *Freezing* with this command, but you select only one object at a time, then the prompt repeats. You cannot *Freeze* the current layer. AutoCAD gives the following prompt:

```
Command: layfrz
Current settings: Viewports=Vpfreeze, Block nesting level=Block
Select an object on the layer to be frozen or [Settings/Undo]: PICK
Layer "CHAIRS" has been frozen.
Select an object on the layer to be frozen or [Settings/Undo]: PICK
Layer "E-F-DOOR" has been frozen.
Select an object on the layer to be frozen or [Settings/Undo]: Enter
Command:
```

You can use *Undo* to undo the last layer (object) selected. Type *S* for *Settings* to produce the following prompt:

```
Enter setting type for [Viewports/Block selection]:
```

Viewports
Using this option you have the choice to freeze layers only for the current viewport (*Viewports*) or freeze the layer globally for the drawing (*Freeze*).

Block selection
This option allows you to control which layers related to the block are frozen when you select a block: *None*, *Block*, or *Entity*.

None
If a nested *Block* or an *Xref* is selected, freezes the parent layer containing that *Block* or *Xref*.

Block
If a selected object is nested in a *Block*, freezes only the layer of that *Block*. If a selected object is nested in an *Xref*, freezes only the layer of the object.

Entity
Freezes the layers of selected individual objects within the *Block* or *Xref* even if they are nested.

Laythw

Pull-down Menu	Command (Type)	Alias (Type)	Short-cut	Screen (side) Menu	Tablet Menu
Format Layer Tools > Thaw All Layers	*Laythw*	...	...	...	...

Use this command to <u>*Thaw* all layers</u> in the drawing. Layers that are *Frozen* <u>and</u> *Off* are only *Thawed* by this command so they still remain invisible until turned *On*:

    Command: **laythw**
    All layers have been thawed.
    Command:

Layoff

Pull-down Menu	Command (Type)	Alias (Type)	Short-cut	Screen (side) Menu	Tablet Menu
Format Layer Tools > Layer Off	*Layoff*	...	...	...	...

Layoff is used to turn *Off* the layer(s) of the selected object(s). Multiple objects can be selected (one at a time) to turn off their layers. The *Options* are the same as those for *Layfrz*:

    Command: **layoff**
    Current settings: Viewports=Vpfreeze, Block nesting level=Block
    Select an object on the layer to be turned off or [Settings/Undo]: **PICK**
    Layer "E-F-DOOR" has been turned off.
    Select an object on the layer to be turned off or [Settings/Undo]: **Enter**
    Command:

You can use *Undo* to undo the last layer (object) selected. The *Settings* option produces the following prompt:

    Enter setting type for [Viewports/Block selection]:

Viewports

Using this option you have the choice to turn off the selected layers only for the current viewport (*Viewport*) or turn off selected layers globally for the drawing (*Off*).

Block selection

This option allows you to control which layers related to the block are frozen when you select a block: *None*, *Block*, or *Entity*. These options are identical to those described for *Layfrz* (see "*Layfrz*").

Layon

Pull-down Menu	Command (Type)	Alias (Type)	Short-cut	Screen (side) Menu	Tablet Menu
Format Layer Tools > Turn All Layers On	*Layon*	...	...	...	...

This command is a quick way to turn <u>all layers on</u>. Layers that are *Frozen* are not affected; only layers that are *Off* can be turned *On* with this command:

Command: **layon**
Warning: layer 0 is frozen. Will not display until thawed.
All layers have been turned on.
Command:

Laylck

Pull-down Menu	Command (Type)	Alias (Type)	Short-cut	Screen (side) Menu	Tablet Menu
Format Layer Tools > Layer Lock	Laylck	...	...	...	...

Laylck locks the layers of selected objects. There are no options for this command. If objects are selected that are part of an *Xref* or *Block,* the layer that was current when the *Xref* or *Block* was *Attached* or *Inserted* is locked.

Command: **laylck**
Select an object on the layer to be locked: **PICK**
Layer TEXT has been locked.

Layulk

Pull-down Menu	Command (Type)	Alias (Type)	Short-cut	Screen (side) Menu	Tablet Menu
Format Layer Tools > Layer Unlock	Layulk	...	...	...	...

Layulk has the opposite function of *Laylck*; that is, it unlocks the layers of selected objects. If the selected objects are part of an *Xref* or *Block,* the layer that was current when the *Xref* or *Block* was *Attached* or *Inserted* is unlocked. There are no options for this command.

Command: **layulk**
Select an object on the layer to be unlocked: **PICK**
Layer TEXT has been unlocked.
Command:

Layvpi

Pull-down Menu	Command (Type)	Alias (Type)	Short-cut	Screen (side) Menu	Tablet Menu
Format Layer Tools > Isolate Layer to Current Viewport	Layvpi	...	...	...	...

This feature is intended to be used to freeze layers within viewports. *Layvpi* freezes specific layers in all viewports <u>except</u> the current one. You identify the layer(s) you want to freeze by selecting any object on the desired layer. Make sure you select the objects from within the layout in which you do <u>not</u> want the layer(s) frozen.

Command: **layvpi**
Select an object on the layer to be Isolated in viewport or [Settings/Undo]: **PICK**
Layer A-FURN-P-SEAT has been frozen in all viewports but the current one.

When you select an object, its layer is immediately frozen in all other viewports. The prompt repeats so you can continually select objects (on layers to freeze). If you type *S* for *Settings,* the following prompt appears:

Select an option [Layouts/Block Selection]:

The *Block Selection* option here is the same as that found in *Layfrz* which allows you to determine the level of freezing for nested *Blocks* or *Xrefs* (see previous discussion under *"Layfrz"*).

Normally using *Layvpi* affects viewports only in the current layout. If you want to have layers frozen for all layouts, enter *L* for *Layouts* at the previous prompt to choose from the following:

Isolate layers in all viewports except current for [All layouts/Current layout] <Current layout>:

The *All layouts* option specifies that the selected layer is frozen in all layouts. In other words, with this option, selecting an object on a layer in a viewport not only freezes that layer for all other viewports in that layout, but it also freezes the layer in all other viewports in <u>all the layouts</u> in that drawing.

Laymrg

Pull-down Menu	Command (Type)	Alias (Type)	Short-cut	Screen (side) Menu	Tablet Menu
Format *Layer Tools >* *Layer Merge*	*Laymrg*	...	...	...	...

Laymrg combines selected layers with any other existing layer. Objects on the layers to merge are moved to the target layer, then the merged layers are deleted. For example, merging layers CHAIRS and FILE_CABINETS to layer FURNITURE would move the related objects to layer FURNITURE and delete layers CHAIRS and FILE_CABINETS.

```
Command: laymrg
Select object on layer to merge or [Name]: PICK
Selected layers: CHAIRS.
Select object on layer to merge or [Name/Undo]: PICK
Selected layers: CHAIRS,FILE_CABINETS.
Select object on layer to merge or [Name/Undo]: Enter
Select object on target layer or [Name]: PICK
******** WARNING ********
You are about to merge 2 layers into layer "FURNITURE".
Do you wish to continue? [Yes/No] <No>: y
Deleting layer "CHAIRS".
Deleting layer "FILE_CABINETS".
2 layers deleted.
Command:
```

At the "Select object on layer to merge or [Name]:" prompt you can select an object from the drawing or you can enter *N* for the *Name* option. This option produces the *Merge Layers* dialog box in which you can select the desired layer by name (Fig. 11-41).

FIGURE 11-41

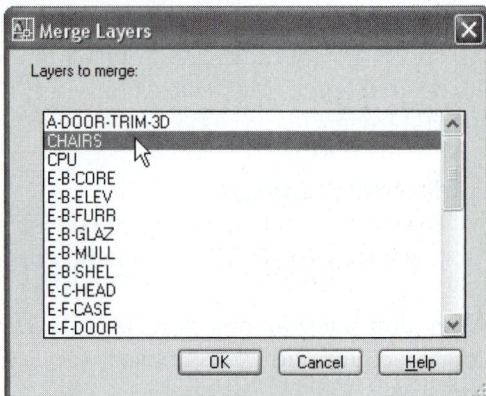

Laydel

Pull-down Menu	Command (Type)	Alias (Type)	Short-cut	Screen (side) Menu	Tablet Menu
Format Layer Tools > Layer Delete	Laydel	...	...	...	...

Laydel deletes all objects on selected layers, then deletes the layers (names). For example, selecting layer E-B-CORE would delete layer E-B-CORE and all objects on that layer.

Command: **laydel**
Select object on layer to delete or [Name]: **PICK**
Selected layers: E-B-CORE.
Select object on layer to delete or [Name/Undo]: **Enter**
******** WARNING ********
You are about to delete layer "E-B-CORE" from this drawing.
Do you wish to continue? [Yes/No] <No>: **y**
Deleting layer "E-B-CORE".
1 layer deleted.
Command:

At the "Select object on layer to delete or [Name]:" prompt you can select an object from the drawing or you can enter *N* for the *Name* option. The *Name* option produces the *Delete Layers* dialog box where you can select the desired layer by name. The dialog box is essentially the same as that used for the *Laymrg* command (see Fig. 11-41).

2007

CHAPTER EXERCISES

1. *Layer Properties Manager,* **Layer Control drop-down list,** *LayerP,* **and** *Make Object's Layer Current*

    Open the **DB_SAMP.DWG** sample drawing located in the C:\Program Files\Acad2007\Sample\ directory. Next, use *Saveas* to save the drawing <u>in your working directory</u> as **DB_SAMP2**.

    A. Invoke the *Layer* command in command line format by typing *-LAYER* or *-LA*. Use the *?* option to yield the list of layers in the drawing. Notice that all the layers have the same linetype except one. Which layer has a different linetype and what is the linetype?

    B. Use the Layer Control drop-down list to indicate the layers of selected items. On the upper-left side of the drawing, **PICK** one or two cyan (light blue) lines. What layer are these lines on? Press the **Esc** key twice to cancel the selection. Next, **PICK** the gray staircase just above the center of the floor plan. What layer is the staircase on? Press the **Esc** key twice to cancel the selection.

    C. Invoke the *Layer Properties Manager* by selecting the icon button or using the *Format* pull-down menu. Turn *Off* the **CPU** layer (click the light bulb icon). Select *OK* to return to the drawing and check the new setting. Are the computers displayed? Next, use the **Layer Control** drop-down list to turn the **CPU** layer back *On*.

D. Use the *Layer Properties Manager* and determine if any layers are *Frozen*. Which layers are *Frozen*? Next, right-click for the shortcut menu and select the *Select All* option. *Freeze* all layers. Which layer will not *Freeze*? Select *OK* in the warning dialog box, then *Apply*. Notice the light bulb icon indicates the layers are still *On* although the layers do not appear in the drawing (*Freeze* overrides *On*). Use the same procedure to select all layers and *Thaw* them. Then *Freeze* the two layers that were previously frozen.

E. Use the *Layer Properties Manager* again and select *All Used Layers* in the Filter Tree View area. Are there any layers in the drawing that are unused (examine the text just below the Filter Tree View)? How many layers are in the drawing? Select the *All* group in the Filter Tree View.

F. Now we want to *Freeze* all the *Cyan* layers. Using the *Layer Properties Manager* again, sort the list by color. **PICK** the first layer <u>name</u> in the list of the four cyan ones. Now hold down the **Shift** key and **PICK** the last name in the list. All four layers should be selected. *Freeze* all four by selecting any one of the sun/snowflake icons. Select *OK* to return to the drawing to examine the changes. Next, use the *LayerP* command to undo the last action in the *Layer Properties Manager*.

G. Assuming you wanted to work only with the Electrical layers (those beginning with "E-"). You can create a group filter to display only those layers. First, activate the *Layer Properties Manager*, then select the *New Group Filter button* above the Filter Tree View. Name the filter "**Electrical**." Next, select the *All* group to make the list of all layers appear. Then enter "**E?***" (E, hyphen, asterisk) in the *Search for Layer* edit box below so only layers beginning with "E-" appear in the list. Select all those layers from the list, then drag and drop them into the new group. Next, in the Filter Tree View select the *All* group, then select the *Electrical* group again to ensure the group was created correctly. Check the *Apply to Layers Toolbar* checkbox, then press *OK* to close the *Layer Properties Manager*. Now examine the Layer Control drop-down list. Does it display only layer names beginning with "E-"? Return to the *Layer Properties Manager* and remove the check from the *Apply to Layers Toolbar* check-box. Finally, *Freeze* all layers <u>not</u> beginning with "E-" by checking the *Invert Filter* checkbox, then selecting and freezing all the layers that appear in the list. Press *OK* to see only the Electrical layers in the drawing. Use *LayerP* to undo the last action. Save the drawing.

H. In the *Layer Properties Manager*, remove the check in the *Invert Filter* checkbox. Then ensure all layers are *Thawed*. Sort the list alphabetically by *Name*. *Freeze* all layers beginning with **C** and **E**. Assuming you were working with only the thawed layers, it would be convenient to display <u>only</u> those names in the drop-down list. Make a new Property Filter by selecting the *New Property Filter* button. In the *Layer Filter Properties* dialog box that appears, assign the new filter name "**Thawed**," then press the Tab key. In the *Filter Definition* area, select the "sun" icon in the *Freeze* column. Select *OK* and then examine the layer list. Select both the *All* layer filter and the *Thawed* filter to display the related names in the layer list. Next, *Freeze* the **GRID-08** and **GRIDLN** layers. Note that the list of names in the Thawed filter changes since property filters are dynamic. Finally, *Thaw* all layers and return to the drawing. **Save** the drawing.

I. Next, you will "move" objects from one layer to another using the Object Properties toolbar. Make layer **0** (zero) the current layer by using the Layer Control drop-down list and selecting the name. Then use the *Layer Properties Manager*, right-click to *Select All* names. *Freeze* all layers (except the current layer). Then *Thaw* layer **CHAIRS**. Select *OK* and return to the drawing to examine the chairs (cyan in color). Use a crossing window to select all of the objects in the drawing (the small blue Grips may appear). Next, list the layers using the **Layer Control** drop-down list and **PICK** layer **0**. Press **Esc** twice to disable the Grips. The objects should be black and on layer 0. You actually "moved" the objects to layer 0. Finally, type *U* to undo the last action so the chairs return to layer CHAIRS (cyan). Also, *Thaw* all layers.

J. Let's assume you want to draw some additional objects on the GRIDLN layer and the E-F-TERR layer. Select the *Make Object's Layer Current* icon button (just to the right of the Layer Control drop-down list). At the "Select object whose layer will become current" prompt, select one of the horizontal or vertical grid lines (gray color). The **GRIDLN** layer name should appear in the Layer Control box as the current layer. Draw a *Line* and it should be on layer GRIDLN and gray in color. Next, use *Make Object's Layer Current* again and select one of the cyan lines at the upper-left corner of the drawing. The **E-F-TERR** layer should appear as the current layer. Draw another *Line*. It should be on layer E-F-TERR and cyan in color. Keep in mind that this method of setting a layer current works especially well when you do not know the layer names or know what objects are on which layers.

K. *Close* the drawing. Do not save the changes.

2. *Layer Properties Manager* **and** *Linetype Manager*

A. Begin a *New* drawing and use *Save* to assign the name **CH11EX2**. Set *Limits* to **11** x **8.5**, then *Zoom All*. Use the *Linetype Manager* to list the loaded linetypes. Are any linetypes already loaded? **PICK** the *Load* button then right-click to *Select All* linetypes. Select *OK*, then close the *Linetype Manager*.

B. Use the *Layer Properties Manager* to create 3 new layers named **OBJ**, **HID**, and **CEN**. Assign the following colors, linetypes, and lineweights to the layers by clicking on the *Color, Linetype,* and *Lineweight* columns.

OBJ	*Red*	*Continuous*	*Default*
HID	*Yellow*	*Hidden2*	*Default*
CEN	*Green*	*Center2*	*Default*

C. Make the **OBJ** layer *Current* and PICK the *OK* tile. Verify the current layer by looking at the *Layer Control* drop-down list, then draw the visible object *Lines* and the *Circle* only (not the dimensions) as shown in Figure 11-42.

FIGURE 11-42

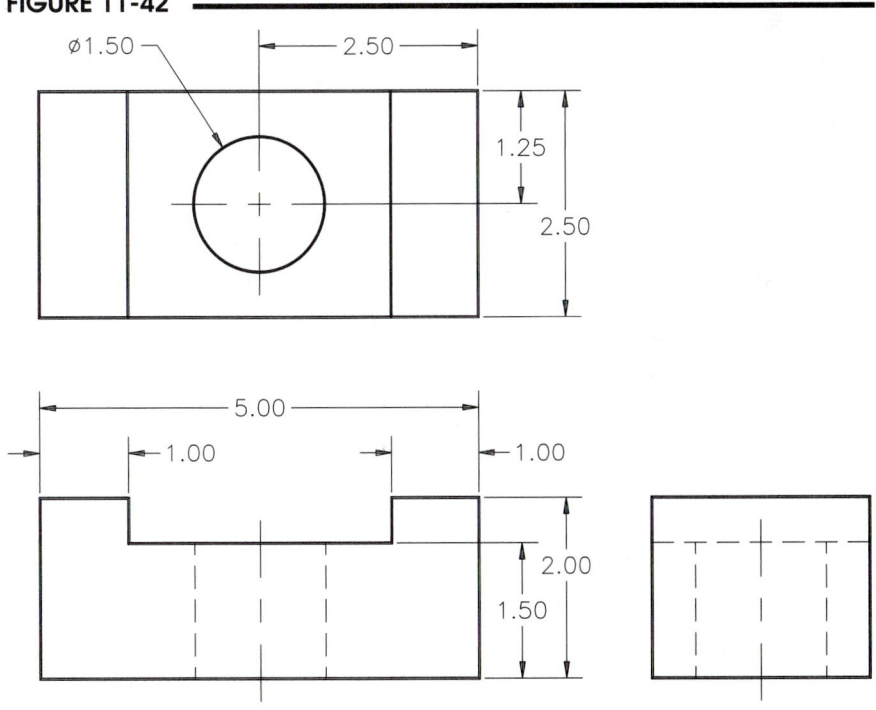

D. When you are finished drawing the visible object lines, create the necessary hidden lines by making layer **HID** the *Current* layer, then drawing *Lines*. Notice that you only specify the *Line* endpoints as usual and AutoCAD creates the dashes.

E. Next, create the center lines for the holes by making layer **CEN** the *Current* layer and drawing *Lines*. Make sure the center lines extend <u>slightly beyond</u> the *Circle* and beyond the horizontal *Lines* defining the hole. *Save* the drawing.

F. Now create a *New* layer named **BORDER** and draw a border and title block of your design on that layer. The final drawing should appear as that in Figure 11-43.

G. Open the *Linetype Manager* again and right-click to *Select All*, then select the *Delete* button. All unused linetypes should be removed from the drawing.

H. *Save* the drawing, then *Close* the drawing.

FIGURE 11-43

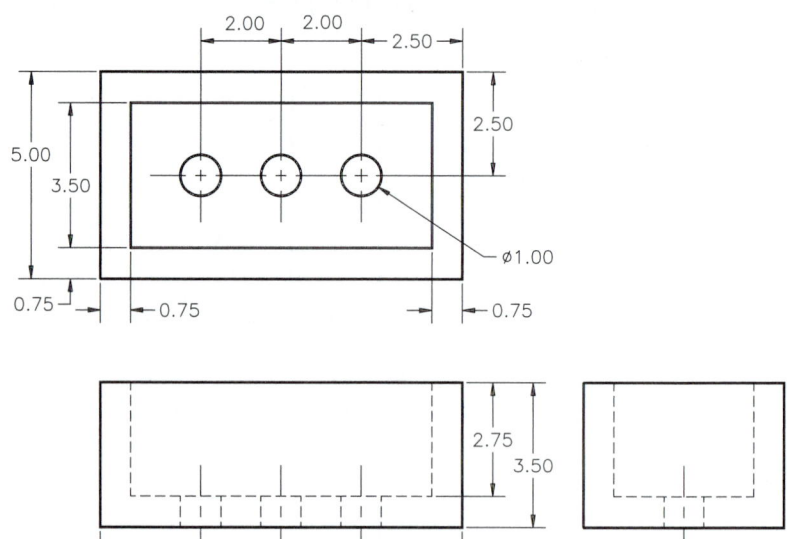

3. *LTSCALE, Properties* **palette,** *Matchprop*

A. Begin a *New* drawing and use *Save* to assign the name **CH11EX3.** Create the same four *New* layers that you made for the previous exercise (**OBJ, HID, CEN,** and **BORDER**) and assign the same linetypes, lineweights, and colors. Set the *Limits* equal to a "C" size sheet, **22 x 17,** then *Zoom All*.

B. Create the part shown in Figure 11-44. Draw on the appropriate layers to achieve the desired linetypes.

FIGURE 11-44

C. Notice that the *Hidden* and *Center* linetype dashes are very small. Use *LTSCALE* to adjust the scale of the non-continuous lines. Since you changed the *Limits* by a factor of slightly less than 2, try using **2** as the *LTSCALE* factor. Notice that all the lines are affected (globally) and that the new *LTSCALE* is retroactive for existing lines as well as for new lines. If the dashes appear too long or short because of the small hidden line segments, *LTSCALE* can be adjusted.

Remember that you cannot control where the multiple <u>short</u> centerline dashes appear for one line segment. Try to reach a compromise between the hidden and center line dashes by adjusting the *LTSCALE*.

D. The six short vertical hidden lines in the front view and two in the side view should be adjusted to a smaller linetype scale. Invoke the *Properties* palette by selecting **Properties** from the **Modify** pull-down menu or selecting the **Properties** icon button. Select <u>only the two short lines in the side view</u>. Use the palette to retroactively adjust the two individual objects' **Linetype Scale** to **0.6000**.

E. When the new *Linetype Scale* (actually, the *CELTSCALE*) for the two selected lines is adjusted correctly, use *Matchprop* to "paint" the *Linetype Scale* to the several other short lines in the front view. Select **Match Properties** from the **Modify** pull-down menu or select the "paint brush" icon. When prompted for the "source object," select one of the two short vertical lines in the side view. When prompted for the "destination object(s)," select <u>each</u> of the short lines in the front view. This action should "paint" the *Linetype Scale* to the lines in the front view.

F. When you have the desired linetype scales, **Exit** AutoCAD and **Save Changes**. Keep in mind that the drawing size may appear differently on the screen than it will on a print or plot. Both the plot scale and the paper size should be considered when you set *LTSCALE*. This topic will be discussed further in Chapter 12, Advanced Drawing Setup.

3D HOUSE.DWG, Courtesy of Autodesk, Inc.

ADVANCED DRAWING SETUP

CHAPTER OBJECTIVES

After completing this chapter you should:

1. know the steps for setting up a drawing;

2. be able to determine an appropriate Limits setting for the drawing;

3. be able to calculate and apply the "drawing scale factor";

4. be able to access existing and create new template drawings;

5. know what setup steps can be considered for creating template drawings.

CONCEPTS

When you begin a drawing, there are several steps that are typically performed in preparation for creating geometry, such as setting *Units, Limits,* and creating *Layers* with *linetypes, lineweights,* and *colors.* Some of these basic concepts were discussed in Chapters 6 and 11. This chapter discusses setting *Limits* for correct plotting as well as other procedures, such as layer creation and variables settings, that help prepare a drawing for geometry creation.

To correctly set up a drawing for printing or plotting to scale, <u>any two</u> of the following three variables must be known. The third can be determined from the other two.

> Drawing *Limits*
> Print or plot sheet (paper) size
> Print or plot scale

The method given in this text for setting up a drawing (Chapters 6 and 12) is intended for plotting from both the *Model* tab (model space) and from *Layout* tabs (paper space). The method applies when you plot from the *Model* tab or plot with <u>one viewport</u> in a *Layout* tab. If multiple viewports are created in one layout, the same general steps would be taken; however, the plot scale might vary for each viewport based on the size and number of viewports. (See Chapters 13, 14, and 33 for more information.)

 Rather than performing the steps for drawing setup each time you begin, you can use "template drawings." Template drawings have many of the setup steps performed but contain no geometry. AutoCAD provides several template drawings and you can create your own template drawings. A typical engineering, architectural, design, or construction office generally produces drawings that are similar in format and can benefit from creation and use of individualized template drawings. The drawing similarities may be subject of the drawing (geometry), plot scale, sheet size, layering schemes, dimensioning styles, and/or text styles. Template drawing creation and use are discussed in this chapter.

STEPS FOR DRAWING SETUP

Assuming that you have in mind the general dimensions and proportions of the drawing you want to create, and the drawing will involve using layers, dimensions, and text, the following steps are suggested for setting up a drawing:

1. Determine and set the *Units* that are to be used.
2. Determine and set the drawing *Limits;* then *Zoom All.*
3. Set an appropriate *Snap Type* (*Polar Snap* or *Grid Snap*), *Snap* spacing, and *Polar* spacing if useful.
4. Set an appropriate *Grid* value if useful.
5. Change the *LTSCALE* value based on the new *Limits.* Set *PSLTSCALE* to 0.
6. Create the desired *Layers* and assign appropriate *linetype, lineweight,* and *color* settings.
7. Create desired *Text Styles* (optional, discussed in Chapter 18).
8. Create desired *Dimension Styles* (optional, discussed in Chapter 29).
9. Activate a *Layout* tab, set it up for the plot or print device and paper size, and create a viewport (if not already existing).
10. Create or insert a title block and border in the layout.

Each of the steps for drawing setup is explained in detail here.

1. **Set *Units***
    This task is accomplished by using the *Units* command or the *Quick Setup* or *Advanced Setup* wizard. Set the linear units and precision desired. Set angular units and precision if needed. (See Chapter 6 for details on the *Units* command and setting *Units* using the wizards.)

2. Set *Limits*

Before beginning to create an AutoCAD drawing, determine the size of the drawing area needed for the intended geometry. Using the actual *Units*, appropriate *Limits* should be set in order to draw the object or geometry to the <u>real-world size</u>. *Limits* are set with the *Limits* command by specifying the lower-left and upper-right corners of the drawing area. Always *Zoom All* after changing *Limits*. *Limits* can also be set using the *Quick Setup* or *Advanced Setup* wizard. (See Chapter 6 for details on the *Limits* command and the setup wizards.)

If you are planning to plot the drawing to scale, *Limits* should be set to a proportion of the sheet size you plan to plot on. For example, if the sheet size is 11" x 8.5", set *Limits* to 11 x 8.5 if you want to plot full size (1"=1"). Setting *Limits* to 22 x 17 (2 times 11 x 8.5) provides 2 times the drawing area and allows plotting at 1/2 size (1/2"=1") on the 11" x 8.5" sheet. Simply stated, set *Limits* to a <u>proportion</u> of the paper size.

Setting *Limits* to the <u>paper size</u> allows plotting at 1=1 scale. Setting *Limits* to a <u>proportion</u> of the sheet  size allows plotting at the <u>reciprocal</u> of that proportion. For example, setting *Limits* to 2 times an 11" x 8.5" sheet allows you to plot 1/2 size on that sheet. Or setting *Limits* to 4 times an 11" x 8.5" sheet allows you to plot 1/4 size on that sheet. (Standard paper sizes are given in Chapter 14.)

Even if you plan to use a layout tab (paper space) and create one viewport, setting model space *Limits* to a proportion of the paper size is recommended. In this way you can easily calculate the <u>viewport scale</u> (the proportion of paper space units to model space units); therefore, the model space geometry can be plotted to a standard scale.

If you plan to plot from a layout, *Limits* in <u>paper space</u> should be set to the actual paper size; however, setting the paper space *Limits* values is automatically done for you when you select a *Plot Device* and *Paper Size* from the *Page Setup* or *Plot* dialog box (see Chapter 13, Layouts and Viewports, and Chapter 14, Printing and Plotting).

To set model space *Limits*, you can use the individual commands for setting *Units* and *Limits* (*Units* command and *Limits* command) or use the *Quick Setup* wizard or *Advanced Setup* wizard.

Drawing Scale Factor

<u>The proportion of the *Limits* to the intended print or plot sheet size is the "drawing scale factor."</u> This factor can be used as a general scale factor for other size-related drawing variables such as *LTSCALE,* dimension *Overall Scale* (*DIMSCALE*), and *Hatch* pattern scale. The drawing scale factor can also be used to determine the viewport scales.

Most size-related AutoCAD drawing variables are set to 1 by default. This means that variables (such as *LTSCALE*) that control sizing and spacing of objects are set appropriately for creating a drawing plotted full size (1=1).

FIGURE 12-1

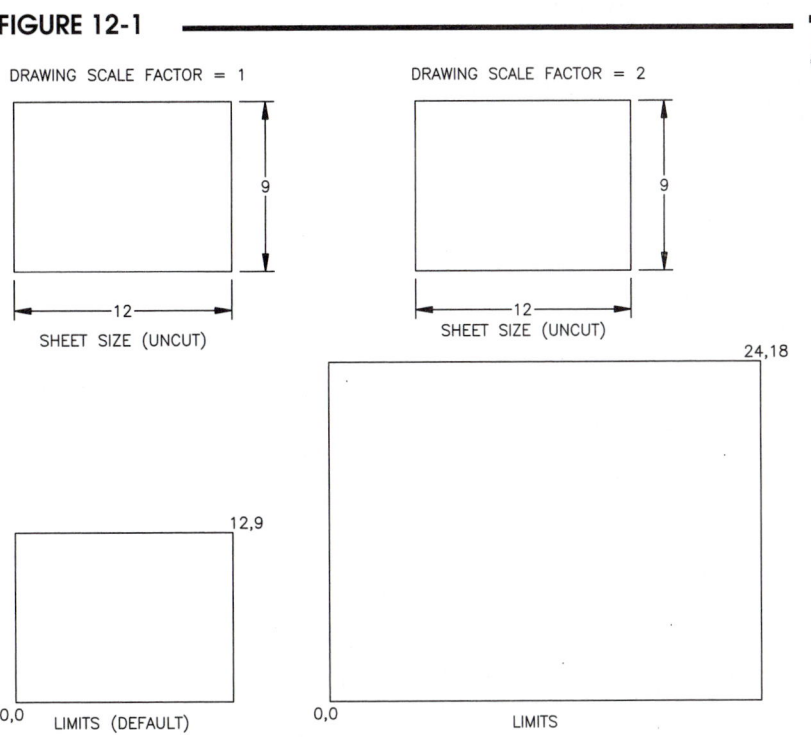

Therefore, when you set *Limits* to the sheet size and print or plot at 1=1, the sizing and spacing of linetypes and other variable-controlled objects are correct. When *Limits* are changed to some proportion of the sheet size, the size-related variables should also be changed proportionally. For example, if you intend to plot on a 12 x 9 sheet and the default *Limits* (12 x 9) are changed by a factor of 2 (to 24 x 18), then 2 becomes the drawing scale factor. Then, as a general rule, the values of variables such as *LTSCALE, DIMSCALE (Overall Scale),* and other scales should be multiplied by a factor of 2 (Fig. 12-1, on the previous page). When you then print the 24 x 18 area onto a 12" x 9" sheet, the plot scale would be 1/2, and all sized features would appear correct.

Limits should be set to the paper size or to a proportion of the paper size used for plotting or printing. In many cases, "cut" paper sizes are used based on the 11" x 8.5" module (as opposed to the 12" x 9" module called "uncut sizes"). Assume you plan to print on an 11" x 8.5" sheet. Setting *Limits* to the sheet size provides plotting or printing at 1=1 scale. Changing the *Limits* to a proportion of the sheet size <u>by some multiplier</u> makes that value the drawing scale factor. The <u>reciprocal</u> of the drawing scale factor is the scale for plotting on that sheet (Fig. 12-2).

FIGURE 12-2

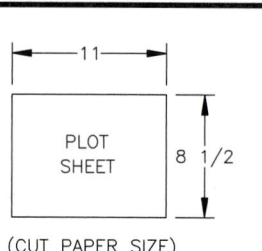

PLOT SHEET

(CUT PAPER SIZE)

LIMITS	SCALE FACTOR	PLOT SCALE	
11 x 8.5	1	1=1	(FULL SIZE)
22 x 17	2	1=2	(HALF SIZE)
44 x 34	4	1=4	(1/4 SIZE)
110 x 85	10	1=10	(1/10 SIZE)

Since the drawing scale factor (DSF) is the proportion of *Limits* to the sheet size, you can use this formula:

$$DSF = \frac{Limits}{Sheet\ size}$$

Because plot/print scale is the reciprocal of the drawing scale factor:

$$Plot\ scale = \frac{1}{DSF}$$

The term "drawing scale factor" is not a variable or command that can be found in AutoCAD or its official documentation. This concept has been developed by AutoCAD users to set system variables appropriately for printing and plotting drawings to standard scales.

3. **Set *Snap***

   Use the *Snap* command or *Snap and Grid* tab of the *Drafting Settings* dialog box to set the *Snap* type (*Grid Snap* and/or *Polar Snap*) and appropriate *Snap Spacing* and/or *Polar Spacing* values. The *Snap Spacing* and *Polar Spacing* values are dependent on the <u>interactive</u> drawing accuracy that is desired.

   The accuracy of the drawing and the size of the *Limits* should be considered when setting the *Snap Spacing* and *Polar Spacing* values. On one hand, to achieve accuracy and detail, you want the values to be the smallest dimensional increments that would commonly be used in the drawing. On the other hand, depending on the size of the *Limits*, the *Snap* value should be large enough to make interactive selection (PICKing) fast and easy. As a starting point for determining appropriate *Snap Spacing* and *Polar Spacing* values, the default values can be multiplied by the "drawing scale factor."

   If you are creating a template drawing (.DWT file), it is of no help to set the *Snap* type (*Grid Snap* or *Polar Snap*), *Polar Spacing* value, or *Polar Tracking* angles since these values are stored in the system registry, not in the drawing. The *Grid Snap Spacing* value, however, is stored in the drawing file.

4. **Set *Grid***

    The *Grid* value setting is usually set equal to or proportionally larger than that of the *Grid Snap* value. *Grid* should be set to a proportion that is <u>easily visible</u> and to some value representing a regular increment (such as .5, 1, 2, 5, 10, or 20). Setting the value to a proportion of *Grid Snap* gives visual indication of the *Grid Snap* increment. For example, a *Grid* value of 1X, 2X, or 5X would give visual display of every 1, 2, or 5 *Grid Snap* increments, respectively. If you are not using *Snap* because only extremely small or irregular interval lengths are needed, you may want to turn *Grid* off. (See Chapter 6 for details on the *Grid* command.)

5. **Set the *LTSCALE* and *PSLTSCALE***

    A change in *Limits* (then *Zoom All*) affects the display of non-continuous (hidden, dashed, dotted, etc.) lines. The *LTSCALE* variable controls the spacing of non-continuous lines. The drawing scale factor can be used to determine the *LTSCALE* setting. For example, if the default *Limits* (based on sheet size) have been changed by a factor of 2, the drawing scale factor=2; so set the *LTSCALE* to 2. In other words, if the DSF=2 and plot scale=1/2, you would want to double the linetype dashes to appear the correct size in the plot, so *LTSCALE*=2. If you prefer a *LTSCALE* setting of other than 1 for a drawing in the default *Limits,* multiply that value by the drawing scale factor.

    American National Standards Institute (ANSI) requires that <u>hidden lines be plotted with 1/8″</u> <u>dashes</u>. An AutoCAD drawing plotted full size (1 unit=1″) with the default *LTSCALE* value of 1 plots the *Hidden* linetype with 1/4″ dashes. Therefore, it is recommended for English drawings that you <u>use the *Hidden2, Center2,* and other *2 linetypes</u> with dash lengths .5 times as long so an *LTSCALE* of 1 produces the ANSI standard dash lengths. Multiply your *LTSCALE* value by the drawing scale factor when plotting to scale. If individual lines need to be adjusted for linetype scale, use *Properties* when the drawing is nearly complete. (See Chapter 11 for details on the *LTSCALE* and *CELTSCALE* variables.)

    Assuming you are planning to create only <u>one viewport</u> in a *Layout* tab, set the *PSLTSCALE* value (paper space *LTSCALE*) to 0. AutoCAD's default setting is 1. A setting of 0 forces the linetype scale to appear the same in *Layout* tabs as it does in the *Model* tab. To set the *PSLTSCALE* value, similar to setting the *LTSCALE*, simply type *PSLTSCALE* at the Command prompt. (See Chapter 13 for details on *PSLTSCALE*.)

6. **Create *Layers*; Assign *Linetypes*, *Lineweights*, and *Colors***

    Use the *Layer Properties Manager* to create the layers that you anticipate needing. Multiple layers can be created with the *New* option. You can type in several new names separated by commas. Assign a descriptive name for each layer, indicating its type of geometry, part name, or function. Include a *linetype* designator in the layer name if appropriate; for example, PART1-H and PART1-V indicate hidden and visible line layers for PART1.

    Once the *Layers* have been created, assign a *Color, Linetype,* and *Lineweight* to each layer. *Colors* can be used to give visual relationships to parts of the drawing. Geometry that is intended to be plotted with different pens (pen size or color) or printed with different plotted appearances should be drawn in different screen colors (especially when color-dependent plot styles are used). Use the *Linetype* command to load the desired linetypes. You should also select *Lineweights* for each layer (if desired) using the *Layer Properties Manager*. (See Chapter 11 for details on creating layers and assigning linetypes, lineweights, and colors.)

7. **Create Text Styles**

    AutoCAD has only one text style as part of the standard template drawings (ACAD.DWT and ACADISO.DWT) and one or two text styles for most other template drawings. If you desire other *Text Styles,* they are created using the *Style* command or the *Text Style…* option from the *Format* pull-down menu. (See Chapter 18, Text and Tables.) If you desire engineering standard text, create a *Text Style* using the ROMANS.SHX or .TTF font file.

8. **Create Dimension Styles**

    If you plan to dimension your drawing, Dimension Styles can be created at this point; however, they are generally created during the dimensioning process. Dimension Styles are names given to groups of dimension variable settings. (See Chapter 29 for information on creating Dimension Styles.)

    Although you do not have to create Dimension Styles until you are ready to dimension the geometry, it is helpful to create Dimension Styles as part of a template drawing. If you produce similar drawings repeatedly, your dimensioning techniques are probably similar. Much time can be saved by using a template drawing with previously created Dimension Styles. Several of the AutoCAD-supplied template drawings have prepared Dimension Styles.

9. **Activate a *Layout* Tab, Set the Plot Device, and Create a Viewport**

    This step can be done just before making a print or plot, or you can complete this and the next step before creating your drawing geometry. Completing these steps early allows you to see how your printed drawing might appear and how much area is occupied by the titleblock and border.

    Activate a *Layout* tab. If your *Options* are set as explained in Chapter 6, Basic Drawing Setup, the layout is automatically set up for your default print/plot device and an appropriate-sized viewport is created. Otherwise, use the *Page Setup Manager* (*Pagesetup* command) to set the desired print/plot device and paper size. Then use the *Vports* command to create one or more viewports. This topic is discussed briefly in Chapter 6, Basic Drawing Setup, and is discussed in more detail in Chapter 13, Layouts and Viewports.

10. **Create a Titleblock and Border**

    For 2D drawings, it is helpful to insert a titleblock and border early in the drawing process. This action gives a visual drawing boundary as well as reserves the space needed by the titleblock.

    Rather than create a new titleblock and border for each new drawing, a common practice is to use the *Insert* command to insert an existing titleblock and border as a *Block*.

    You can draw or *Insert* a titleblock and border in the *Model* tab (model space) or into a layout (paper space), depending on how you want to plot the drawing. If you plan to plot from the *Model* tab, multiply the actual size of the titleblock and border by the drawing scale factor to determine their sizes for the drawing. If you want to plot from a layout, draw or insert the titleblock and border at the actual size (1:1) since a layout represents the plot sheet. (See Chapters 21 and 22 for information on *Block* creation and *Insertion*.)

USING AND CREATING TEMPLATE DRAWINGS

Instead of going through the steps for setup <u>each time</u> you begin a new drawing, you can create one or more template drawings or use one of the AutoCAD-supplied template drawings. AutoCAD template drawings are saved as a .DWT file format. A template drawing has the initial setup steps (*Units, Limits, Layers, Linetypes, Lineweights, Colors*, etc.) completed and saved, but no geometry has been created yet. For some templates, *Layout* tabs have been set up with titleblocks, viewports, and plot settings. Template drawings are used as a <u>template</u> or <u>starting point</u> each time you begin a new drawing. AutoCAD actually makes a copy of the template you select to begin the drawing. The creation and use of template drawings can save hours of preparation.

Using Template Drawings

To use a template drawing, select the *Template* option in the *Startup* or *Create New Drawing* dialog box (Fig. 12-3, on the next page) or in the *Select Template* dialog box. This option allows you to select the desired template (.DWT) drawing from a list including all templates in the Template folder.

Any template drawings that you create using the *SaveAs* command (as described in the next section) appear in the list of available templates. Select the *Browse…* option to browse other folders.

Several template drawings are supplied with AutoCAD (see the selections in Fig. 12-3) and are described in the following table. Generally, the template drawing files include the titleblocks. In some cases, Dimension Styles have been created but other changes have not been made to the template drawings such as layer setups or loading linetypes.

FIGURE 12-3

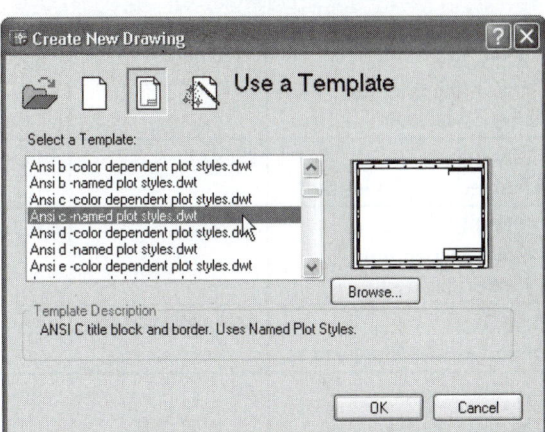

Table of AutoCAD-Supplied Template

Template	Intended Sheet Size	Limits (Model)	Title Block	Orientation	Layers	Dimen. Styles	Plot Style
ACAD - Named Plot Styles.dwt	---	12,9	No	Landscape	0	Standard	Named
Acad.dwt	---	12,9	No	Landscape	0	Standard	Color Dependent
ACAD - Named Plot Styles3d.dwt	---	12,9	No	Landscape	0	Standard	Named
ACAD3d.dwt	---	12,9	No	Landscape	0	Standard	Color Dependent
ACADISO - Named Plot Styles.dwt	---	420,297	No	Landscape	0	Yes[10]	Named
Acadiso.dwt	---	420,297	No	Landscape	0	Yes[10]	Color Dependent
ACADISO - Named Plot Styles3d.dwt	---	420,297	No	Landscape	0	Yes[10]	Named
ACADISO3d.dwt	---	420,297	No	Landscape	0	Yes[10]	Named
ANSI A (portrait) - Color Dependent Plot Styles.dwt	8.5″,11″	12,9	Yes	Portrait	Yes[1]	Standard	Color Dependent
ANSI A (portrait) - Named Plot Styles.dwt	8.5″,11″	12,9	Yes	Portrait	Yes[1]	Standard	Named
ANSI A - Color Dependent Plot Styles.dwt	11″,8.5″	12,9	Yes	Landscape	Yes[1]	Standard	Color Dependent
ANSI A - Named Plot Styles.dwt	11″,8.5″	12,9	Yes	Landscape	Yes[1]	Standard	Named
ANSI B - Color Dependent Plot Styles.dwt	17″x11″	12,9	Yes	Landscape	Yes[1]	Standard	Color Dependent
ANSI B - Named Plot Styles.dwt	17″x11″	12,9	Yes	Landscape	Yes[1]	Standard	Named
ANSI C - Color Dependent Plot Styles.dwt	22″x17″	12,9	Yes	Landscape	Yes[1]	Standard	Color Dependent
ANSI C - Named Plot Styles.dwt	22″x17″	12,9	Yes	Landscape	Yes[1]	Standard	Named

Template	Intended Sheet Size	Limits (MS)	Title Block	Orientation	Layers	Dimen. Styles	Plot Style
ANSI D - Color Dependent Plot Styles.dwt	22"x34"	12,9	Yes	Landscape	Yes[1]	Standard	Color Dependent
ANSI D - Named Plot Styles.dwt	22"x34"	12,9	Yes	Landscape	Yes[1]	Standard	Named
ANSI E - Color Dependent Plot Styles.dwt	34"x44"	12,9	Yes	Landscape	Yes[1]	Standard	Color Dependent
ANSI E - Named Plot Styles.dwt	34"x44"	12,9	Yes	Landscape	Yes[1]	Standard	Named
ANSI Layout templates.dwt	---	12,9	Yes	---	Yes[1]	Standard	Color Dependent
Architectural, English units -Color Dependent Plot Styles.dwt	24"x36"	1'-0"x9"	Yes	Landscape	Yes[1]	Standard	Color Dependent
Architectural, English units -Named Plot Styles.dwt	24"x36"	1'-0"x9"	Yes	Landscape	Yes[1]	Standard	Named
DIN A0 - Color Dependent Plot Styles.dwt	1189mmx841mm	420x297	Yes	Landscape	Yes[2]	Yes[11]	Color Dependent
DIN A0 - Named Plot Styles.dwt	1189mmx841mm	420x297	Yes	Landscape	Yes[2]	Yes[11]	Named
DIN A1 - Color Dependent Plot Styles.dwt	841mmx594mm	420x297	Yes	Landscape	Yes[2]	Yes[11]	Color Dependent
DIN A1 - Named Plot Styles.dwt	841mmx594mm	420x297	Yes	Landscape	Yes[2]	Yes[11]	Named
DIN A2 -Color Dependent Plot Styles.dwt	594mmx420mm	420x297	Yes	Landscape	Yes[2]	Yes[11]	Color Dependent
DIN A2 - Named Plot Styles.dwt	594mmx420mm	420x297	Yes	Landscape	Yes[2]	Yes[11]	Named
DIN A3 -Color Dependent Plot Styles.dwt	420mmx297mm	420x297	Yes	Landscape	Yes[2]	Yes[11]	Color Dependent
DIN A3 -Named Plot Styles.dwt	420mmx297mm	420x297	Yes	Landscape	Yes[2]	Yes[11]	Named
DIN A4 -Color Dependent Plot Styles.dwt	210mmx297mm	420x297	Yes	Portrait	Yes[2]	Yes[11]	Color Dependent
DIN A4 -Named Plot Styles.dwt	210mmx297mm	420x297	Yes	Portrait	Yes[2]	Yes[11]	Named
Gb –Color Dependent Plot Styles.dwt	---	256.8069 x 149.3068	No	Landscape	0	Yes[13]	Color Dependent
Gb –Named Plot Styles.dwt	---	420x297	No	Landscape	0	Yes[13]	Named
Gb_a0 –Color Dependent Plot Styles.dwt	1306mmx924mm	420x 297	Yes	Landscape	Yes[6]	Yes[13]	Color Dependent
Gb_a0 –Named Plot Styles.dwt	1306mmx924mm	1154x821	Yes	Landscape	Yes[6]	Yes[13]	Named
Gb_a1 –Color Dependent Plot Styles.dwt	924mmx653mm	806x574	Yes	Landscape	Yes[6]	Yes[13]	Color Dependent

Template	Intended Sheet Size	Limits (MS)	Title Block	Orientation	Layers	Dimen. Styles	Plot Style
Gb_a1 –Named Plot Styles.dwt	924mmx653mm	420x297	Yes	Landscape	Yes[6]	Yes[13]	Named
Gb_a2 –Color Dependent Plot Styles.dwt	653mmx462mm	420x297	Yes	Landscape	Yes[6]	Yes[14]	Color Dependent
Gb_a2 –Named Plot Styles.dwt	653mmx462mm	559x400	Yes	Landscape	Yes[6]	Yes[13]	Named
Gb_a3 –Color Dependent Plot Styles.dwt	462mmx326mm	390x287	Yes	Landscape	Yes[6]	Yes[13]	Color Dependent
Gb_a3 –Named Plot Styles.dwt	462mmx326mm	420x297	Yes	Landscape	Yes[6]	Yes[13]	Named
Gb_a4 –Color Dependent Plot Styles.dwt	435mmx307mm	420x297	Yes	Portrait	Yes[6]	Yes[13]	Color Dependent
Gb_a4 –Named Plot Styles.dwt	435mmx307mm	180x287	Yes	Portrait	Yes[6]	Yes[13]	Named
Generic 24in x 32in Title Block -Color Dependent Plot Styles.dwt	36"x24"	12x9	Yes	Landscape	Yes[1]	Standard	Color Dependent
Generic 24in x 32in Title Block -Named Plot Styles.dwt	36"x24"	12x9	Yes	Landscape	Yes[1]	Standard	Named
ISO A0 -Color Dependent Plot Styles.dwt	1189mmx841mm	420x297	Yes	Landscape	Yes[3]	Yes[10]	Color Dependent
ISO A0 -Named Plot Styles.dwt	1189mmx841mm	420x297	Yes	Landscape	Yes[4]	Yes[10]	Named
ISO A1 -Color Dependent Plot Styles.dwt	841mmx594mm	420x297	Yes	Landscape	Yes[4]	Yes[10]	Color Dependent
ISO A1 -Named Plot Styles.dwt	841mmx594mm	420x297	Yes	Landscape	Yes[4]	Yes[10]	Named
ISO A2 -Color Dependent Plot Styles.dwt	594mmx420mm	420x297	Yes	Landscape	Yes[4]	Yes[10]	Color Dependent
ISO A2 -Named Plot Styles.dwt	594mmx420mm	420x297	Yes	Landscape	Yes[4]	Yes[10]	Named
ISO A3 -Color Dependent Plot Styles.dwt	420mmx297mm	420x297	Yes	Landscape	Yes[4]	Yes[10]	Color Dependent
ISO A3 -Named Plot Styles.dwt	420mmx297mm	420x297	Yes	Landscape	Yes[4]	Yes[10]	Named
ISO A4 -Color Dependent Plot Styles.dwt	210mmx297mm	420x297	Yes	Portrait	Yes[4]	Yes[10]	Color Dependent
ISO A4 -Named Plot Styles.dwt	210mmx297mm	420x297	Yes	Portrait	Yes[4]	Yes[10]	Named
JIS A0 -Color Dependent Plot Styles.dwt	1189mmx841mm	420x297	Yes	Landscape	Yes[5]	Yes[12]	Color Dependent
JIS A0 -Named Plot Styles.dwt	1189mmx841mm	420x297	Yes	Landscape	Yes[5]	Yes[12]	Named
JIS A1 -Color Dependent Plot Styles.dwt	841mmx594mm	420x297	Yes	Landscape	Yes[5]	Yes[12]	Color Dependent
JIS A1 -Named Plot Styles.dwt	841mmx594mm	420x297	Yes	Landscape	Yes[5]	Yes[12]	Named
JIS A2 -Color Dependent Plot Styles.dwt	594mmx420mm	420x297	Yes	Landscape	Yes[5]	Yes[12]	Color Dependent

Template	Intended Sheet Size	Limits (MS)	Title Block	Orientation	Layers	Dimen. Styles	Plot Style
JIS A2 -Named Plot Styles.dwt	594mmx420mm	420x297	Yes	Landscape	Yes[5]	Yes[12]	Named
JIS A3 -Color Dependent Plot Styles.dwt	420mmx297mm	420x297	Yes	Landscape	Yes[5]	Yes[10]	Color Dependent
JIS A3 -Named Plot Styles.dwt	420mmx297mm	420x297	Yes	Landscape	Yes[5]	Yes[12]	Named
JIS A4 (landscape) - Color Dependent Plot Styles.dwt	297mmx210mm	420x297	Yes	Landscape	Yes[5]	Yes[12]	Color Dependent
JIS A4 (landscape) - Named Plot Styles.dwt	297mmx210mm	420x297	Yes	Landscape	Yes[5]	Yes[10]	Named
JIS A4 (portrait) -Color Dependent Plot Styles.dwt	210mmx297mm	420x297	Yes	Portrait	Yes[5]	Yes[12]	Color Dependent
JIS A4 (portrait) - Named Plot Styles.dwt	210mmx297mm	420x297	Yes	Portrait	Yes[5]	Yes[12]	Named
Metric Layout templates.dwt	---	420x297	Yes	---	Yes[7]	Yes[10]	Color Dependent
Tutorial-iarch.dwt	3'x2'	60'x40'	Yes	Landscape	Yes[8]	Standard	Named
Tutorial-imfg	34"x22"	12x9	Yes	Landscape	Yes[9]	Standard	Named
Tutorial-march.dwt	841x594	20000mm x12000mm	Yes	Landscape	Yes[8]	Yes[10]	Named
Tutorial-mmfg	841x594	420mm x297mm	Yes	Landscape	Yes[9]	Yes[10]	Named

[1] Layers : 0, Title_Block and Viewport.

[2] Layers : 0, 025, 035, 05, 07.

[3] Layers : 0, FRAME 025, FRAME 050, FRAME 070, Title Block, and Viewport.

[4] Layers : 0, FRAME 025, FRAME 050, FRAME 070, FRAME 200, Title Block, and Viewport.

[5] Layers : 0, Object, Title Block, and Viewport.

[6] Layers : \M+5CDBC\M+5BFF2_\M+5B1EA\M+5CCE2\M+5C0B8\M+5CAF4\M+5D0D4
     \M+5CDBC\M+5BFF2_\M+5B1EA\M+5CCE2\M+5C0B8\M+5CEC4\M+5D7D6
     \M+5CDBC\M+5BFF2_\M+5BDC7\M+5CFDF
     \M+5CDBC\M+5BFF2_\M+5C4DA\M+5BFF2\M+5CFDF
     \M+5CDBC\M+5BFF2_\M+5CAD3\M+5BFDA
     \M+5CDBC\M+5BFF2_\M+5CDE2\M+5BFF2\M+5CFDF, 0.

[7] Layers: FRAME 025, FRAME 050, FRAME 070, FRAME 200, Object, RAHMEN025, RAHMEN035, RAHMEN05, RAHMEN07, Title block, Viewport.

[8] 0, A-Doors, A-Electrical, A-Plumbing, A-Walls, A-Windows, Defpoints, Dimensions, Elevation, No Print, Reference, Section AA, Text, Title Block, Viewport.

[9] 0, Defpoints, Dimensions, Model-Detail, Model-Front, Model-Right Side, Model-Top, No Print, Reference, Text, Title Block, Viewport.

[10] Dimension Styles: ISO-25.

[11] Dimension Styles: DIN.

[12] Dimension Styles: JIS.

[13] Dimension Styles: GB-5

[14] Dimension Styles: GB-35

NOTE: The initial *PSLTSCALE* value for all templates is 1. (See Chapter 13 for details on *PSLTSCALE*.)

See the Tables of *Limits* Settings in Chapter 14 for standard plotting scales and intended sheet sizes for the *Limits* settings in these templates.

Creating Template Drawings

To make a template drawing, begin a *New* drawing using a default template drawing (ACAD.DWT) or other template drawing (*.DWT), then make the initial drawing setups. Alternately, *Open* an existing drawing that is set up as you need (layers created, linetypes loaded, *Limits* and other settings for plotting or printing to scale, etc.), and then *Erase* all the geometry. Next, use *SaveAs* to save the drawing under a different descriptive name with a .DWT file extension. Do this by selecting the *AutoCAD Drawing Template File (*.dwt)* option in the *Save Drawing As* dialog box (Fig. 12-4).

AutoCAD automatically (by default) saves the drawing in the folder where other template (*DWT) files are found. Use the Save in: drop-down list at the top of the Save Drawing As dialog box to specify another location for saving the .DWT file.

FIGURE 12-4

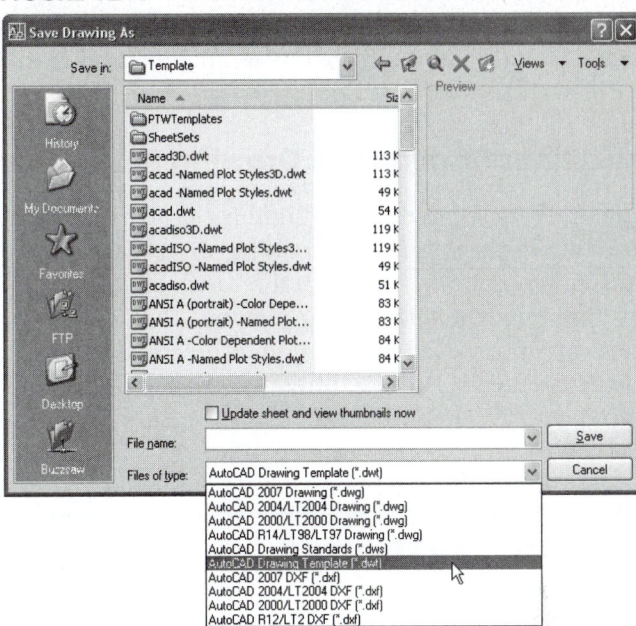

Location of the Template folder aligns with the Microsoft Windows 2000 and XP conventions. That is, the templates are stored by default (for a stand-alone installation) in a folder multiple levels under the user's profile (C:\Documents and Settings\user's name\Local Settings\Application Data\AutoCAD 2007\... and so on). To find the complete path, use the *Options* dialog box, *Files* tab, and expand *Drawing Template Settings*.

After you assign the file name in the *Save Drawing As* dialog box, you can enter a description of the template drawing in the *Template Description* dialog box that appears (Fig. 12-5). The description is saved with the template drawing and appears in the *Startup* and *Create New Drawing* dialog boxes when you highlight the template drawing name (see Fig. 12-3).

FIGURE 12-5

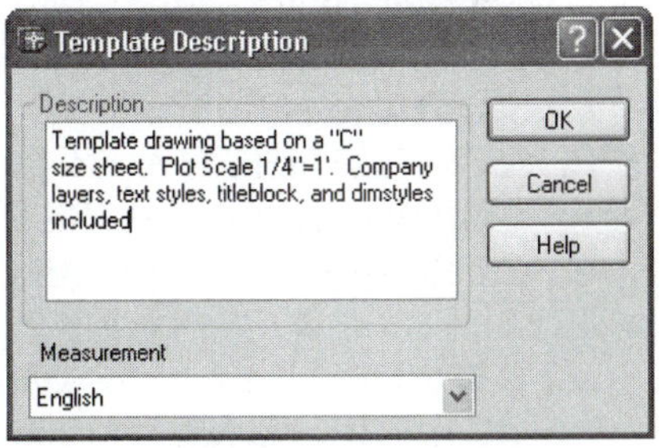

Multiple template drawings can be created, each for a particular situation. For example, you may want to create several templates, each having the setup steps completed but with different *Limits* and with the intention to plot or print each in a different scale or on a different size sheet. Another possibility is to create templates with different layout and layering schemes. There are many possibilities, but the specific settings used depend on your applications (geometry), typical scales, or your plot/print devices.

Typical drawing steps that can be considered for developing template drawings are listed below:

Set *Units*
Set *Limits*
Set *Snap*
Set *Grid*
Set *LTSCALE* and *PSLTSCALE*
Create *Layers* with color and linetypes assigned
Create *Text Styles* (see Chapter 18)
Create Dimension Styles (see Chapter 29)
Create *Layout* tabs, each with titleblock, border and plot settings saved (plot device, paper size, plot scale, etc.). (See Chapter 13, Layouts and Viewports.)

You can create template drawings for different sheet sizes and different plot scales. If you plot from the *Model* tab, a template drawing should be created for each case (scale and sheet size). When you learn to create *Layouts*, you can save one template drawing that contains several layouts, each layout for a different sheet size and/or scale.

Additional Advanced Drawing Setup Concepts

The drawing setup described in this chapter is appropriate for learning the basic concepts of setting up a drawing in preparation for creating layouts and viewports and plotting to scale. In Chapter 13, Layouts and Viewports, you will learn the basics of creating layouts and viewports—particularly for drawings that use one large viewport in a layout. In Chapter 33, Advanced Layouts and Plotting, other advanced ideas are introduced including the "reverse method" for drawing setup, which is a strategy where the *Limits* are not set and the drawing scale factor is not calculated initially, but size-related settings are based on the viewport scale.

CHAPTER EXERCISES

For the first three exercises, you will set up several drawings that will be used in other chapters for creating geometry. Follow the typical steps for setting up a drawing given in this chapter.

1. **Create a Metric Template Drawing**
   You are to create a template drawing for use with metric units. The drawing will be printed full size on an "A" size sheet (not on an A4 metric sheet), and the dimensions are in millimeters. Start a *New* drawing, select *Start from Scratch*, and select the <u>*Imperial*</u> default settings. Follow these steps.

   A. Set *Units* to *Decimal* and *Precision* to **0.00**.
   B. Set *Limits* to a metric sheet size, **279.4 x 215.9**; then *Zoom All* (scale factor is approximately 25).
   C. Set *Snap Spacing* (for *Grid Snap*) to **1**. (*Polar Tracking* and *Polar Snap* settings do not have to be preset since they are saved in the system registry.)
   D. Set *Grid* to **10**.
   E. Set *LTSCALE* to **25** (drawing scale factor of 25.4). Set *PSLTSCALE* to **0**.
   F. *Load* the *Center2* and *Hidden2 Linetypes*.
   G. Create the following *Layers* and assign the *Colors*, *Linetypes*, and *Lineweights* as shown:

OBJECT	*red*	*continuous*	*0.40 mm*
CONSTR	*white*	*continuous*	*0.25 mm*
CENTER	*green*	*center2*	*0.25 mm*
HIDDEN	*yellow*	*hidden2*	*0.25 mm*
DIM	*cyan*	*continuous*	*0.25 mm*
TITLE	*white*	*continuous*	*0.25 mm*
VPORTS	*white*	*continuous*	*0.25 mm*

   H. Use the *SaveAs* command. In the dialog box under *Save as type:*, select *AutoCAD Drawing Template File (.dwt)*. Assign the name **A-METRIC.** Enter **Metric A Size Drawing** in the *Template Description* dialog box.

2. *Quick Setup* **wizard**
   A drawing of a mechanical part is to be made and plotted full size (1"=1") on an 11" x 8.5" sheet. Use the *Quick Setup* wizard to assist with the setup.

A. Begin a *New* drawing. When the *Startup* or *Create New Drawing* dialog box appears, select *Wizard*. Select *Quick Setup*.

B. In the *Quick Setup* dialog box, select *Decimal* in the *Units* step.

C. Press *Next* to proceed to the *Area* step. Enter **11** and **8.5** in the two edit boxes so 11 appears as the *Width* (X direction) and 8.5 appears as the *Length* (in the Y direction).

D. Select the *Finish* button.

E. Check to ensure the *Limits* are set to 11,8.5 and *Grid* is set to .5.

F. Change the *Snap* to **.250**.

G. Set *LTSCALE* to **1**. Set *PSLTSCALE* to **0**.

H. *Load* the *Center2* and *Hidden2 Linetypes*.

I. Create the following *Layers* and assign the given *Colors, Linetypes,* and *Lineweights*:

OBJECT	red	continuous	0.40 mm
CONSTR	white	continuous	0.25 mm
CENTER	green	center2	0.25 mm
HIDDEN	yellow	hidden2	0.25 mm
DIM	cyan	continuous	0.25 mm
TITLE	white	continuous	0.25 mm
VPORTS	white	continuous	0.25 mm

J. Save the drawing as **BARGUIDE** for use in another chapter exercise.

3. *Quick Setup* **wizard**

A floor plan of an apartment has been requested. Dimensions are in feet and inches. The drawing will be plotted at 1/2"=1' scale on an 24" x 18" sheet.

A. Begin a *New* drawing and select the *Quick Setup* wizard.

B. Select *Architectural* in the *Units* step.

C. In the *Area* step, set *Width* and *Length* to **48'** x **36'** (576" x 432"), respectively (use the apostrophe symbol for feet). Select *Finish*.

D. Set the *Snap* value to **1** (inch).

E. Set the *Grid* value to **12** (inches).

F. Set the *LTSCALE* to **12**. Set *PSLTSCALE* to **0**.

G. Create the following *Layers* and assign the *Colors* and *Lineweights*:

FLOORPLN	red	0.016"
CONSTR	white	0.010"
TEXT	green	0.010"
TITLE	yellow	0.010"
DIM	cyan	0.010"
VPORTS	white	0.010"

H. *Save* the drawing and assign the name **APARTMENT** to be used later.

4. *Quick Setup* **wizard**

A drawing is to be created and printed in 1/4"=1" scale on an A size (11" x 8.5") sheet. The *Quick Setup* wizard can be used to automate the setup.

A. Begin a *New* drawing. Select *Quick Setup* wizard.
B. In the *Quick Setup* dialog box, select *Decimal* in the *Units* step.
C. Press *Next* to proceed to the *Area* step. Enter **44** and **34** in the two edit boxes so 44 appears in the *Width* edit box (X direction) and 34 appears in the *Length* edit box (Y direction).
D. Select the *Finish* button.
E. Check to ensure the *Limits* are set to 44,34 and *Snap* and *Grid* are set to 1.
F. Save the drawing as **CH12EX4**.

5. **Create a Template Drawing**
In this exercise, you will create a "generic" template drawing that can be used at a later time. Creating templates now will save you time later when you begin new drawings.

A. Create a template drawing for use with decimal dimensions and using standard paper "A" size format. Begin a *New* drawing and select *Start from Scratch*. Select the *Imperial* default settings.
B. Set *Units* to *Decimal* and *Precision* to **0.00**.
C. Set *Limits* to **11 x 8.5**.
D. Set *Snap* to **.25**.
E. Set *Grid* to **1**.
F. Set *LTSCALE* to **1**. Set *PSLTSCALE* to **0**.
G. Create the following *Layers*, assign the *Linetypes* and *Lineweights* as shown, and assign your choice of *Colors*. Create any other layers you think you may need or any assigned by your instructor.

OBJECT	*continuous*	*0.016"*
CONSTR	*continuous*	*0.010"*
TEXT	*continuous*	*0.010"*
TITLE	*continuous*	*0.010"*
VPORTS	*continuous*	*0.010"*
DIM	*continuous*	*0.010"*
HIDDEN	*hidden2*	*0.010"*
CENTER	*center2*	*0.010"*
DASHED	*dashed2*	*0.010"*

H. Use *SaveAs* and name the template drawing **ASHEET.** Make sure you select *AutoCAD Drawing Template File (*.dwt)* from the *Save as Type:* drop-down list in the *Save Drawing As* dialog box. Also, from the *Save In:* drop-down list on top, select your working directory as the location to save the template.

13

LAYOUTS AND VIEWPORTS

CHAPTER OBJECTIVES

After completing this chapter you should:

1. know the difference between paper space and model space and between tiled viewports and paper space viewports;

2. know that the purpose of a layout is to prepare model space geometry for plotting;

3. know the "Guidelines for Using Layouts and Viewports";

4. be able to create layouts using the *Layout Wizard* and the *Layout* command;

5. know how to set up a layout for plotting using the *Page Setup Manager* and the *Page Setup* dialog box;

6. be able to use the options of *Vports* and *-Vports* to create and control viewports in paper space;

7. know how to scale the model geometry displayed in paper space viewports.

CONCEPTS

You are already somewhat familiar with model space (the *Model* tab) and paper space (*Layout* tabs). Model space (the *Model* tab) is used for construction of geometry, whereas paper space (*Layout* tabs) is used for preparing to print or plot the model geometry. In order to view the model geometry in a layout, one or more viewports must be created. This can be done automatically by AutoCAD when you activate a *Layout* tab for the first time, or you can use the *Vports* command to create viewports in a layout. An introduction to these concepts is given in Chapter 6, Basic Drawing Setup, "Introduction to Layouts and Printing." However, this chapter (Chapter 13) gives a full explanation of paper space, layouts, and creating paper space viewports. Advanced features of layouts and plotting are discussed in Chapter 33 and using layouts for 3D applications is discussed in Chapter 42.

Paper Space and Model Space

The two drawing spaces that AutoCAD provides are model space and paper space. Model space is activated by selecting the *Model* tab and paper space is activated by selecting the *Layout1*, *Layout2*, or other layout tab. When you start AutoCAD and begin a drawing, model space is active by default. Objects that represent the subject of the drawing (model geometry) are normally drawn in model space. Dimensioning is traditionally performed in model space because it is associative—directly associated to the model geometry. The model geometry is usually completed before using paper space. It is possible to create dimensions in paper space that are attached to objects in model space; however, this practice is recommended only for certain applications. (See "Dimensioning in Paper Space Layouts" in Chapter 29.)

Paper space represents the <u>paper that you plot or print on</u>. When you enter paper space (by activating a *Layout* tab) for the first time, you may see only a blank "sheet," unless your system is configured to automatically generate a viewport as explained in Chapter 6. Normally the only geometry you would create in paper space is the title block, drawing border, and possibly some other annotation (text). In order to see any model geometry from paper space, viewports must be created (like cutting rectangular holes) so you can "see" into model space. Viewports are created either automatically by AutoCAD or by using the *Vports* command. Any number or size of rectangular or non-rectangular viewports can be created in paper space. Since there is only <u>one model space</u> in a drawing, you see the same model space geometry in each viewport initially. You can, however, control the size and area of the geometry displayed and which layers are *Frozen* and *Thawed* <u>in each viewport</u>. Since paper space represents the actual paper used for plotting, you should plot from paper space at a scale of 1:1.

The *TILEMODE* system variable controls which space is active—paper space (*TILEMODE*=0) or model space (*TILEMODE*=1). *TILEMODE* is automatically set when you activate the *Model* tab or a *Layout* tab.

Layouts

By default there is a *Layout1* tab and *Layout2* tab. You can produce any configuration of layouts by creating new layouts or copying, renaming, deleting, or moving existing layouts. A shortcut menu is available (right-click while your pointer is on a layout tab) that allows you to *Rename* the selected layout as well as create a *New Layout* from scratch or *From Template*, *Delete*, *Move or Copy* a layout, *Select all Layouts*, or set up for plotting (Fig. 13-1).

The primary function of a layout is to prepare the model geometry for plotting. A layout simulates the sheet of paper you will plot on and allows you to specify plotting parameters and see the changes in settings that you make, similar to a plot preview. Plot specifications are made using the *Page Setup* dialog box to select such parameters as plot device, paper size and orientation, scale, and Plot Style Tables to attach.

Since you can have multiple layouts in AutoCAD, you can set up multiple plot schemes. For example, assume you have a drawing (in model space) of an office

FIGURE 13-1

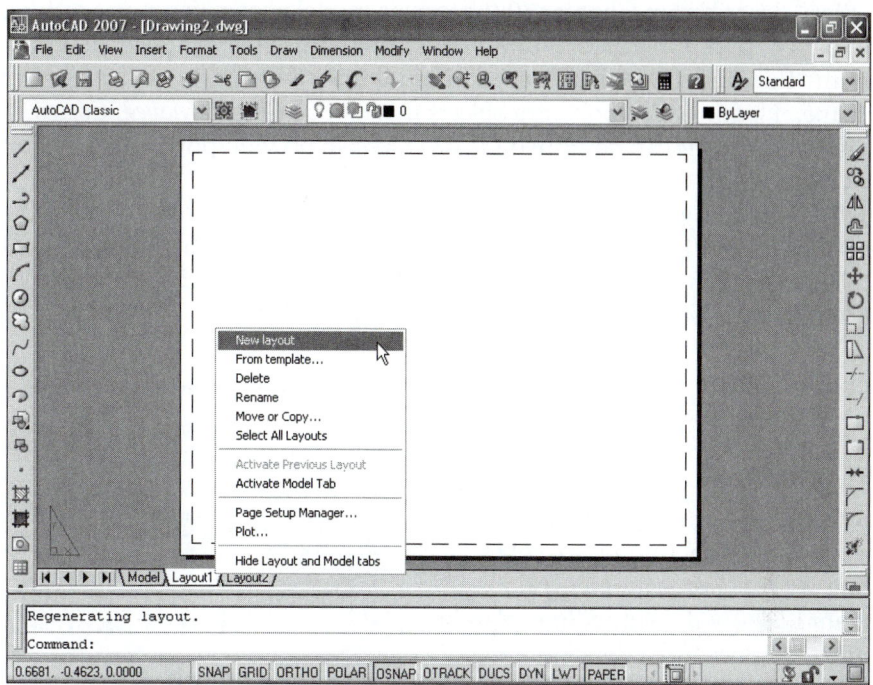

complex. You can set up one layout to plot the office floor plan on a "C" size sheet with an electrostatic plotter in 1/4"=1' scale and set up a second layout to plot the same floor plan on a "A" size sheet with a laser jet printer at 1/8"=1' scale. Since the plotting setup is saved with each layout, you can plot any layout again without additional setup.

NOTE: If the *Model* and *Layout* tabs are not visible at the bottom left of the drawing area, you can make them appear by right-clicking on the layout icon just to the right of the Status bar, then select *Display Layout and Model Tabs* (Fig. 13-2).

FIGURE 13-2

Viewports

When you activate a layout tab (enter paper space) for the first time, you may see a "blank sheet." However, if *Create Viewport in New Layouts* is checked in the *Display* tab of the *Options* dialog box (see Fig. 13-3, lower-left corner), you may see a viewport that is automatically created. If no viewport exists, you can create one or more with the *Vports* command. A viewport in paper space is a window into model space. Without a viewport, no model geometry is visible in a layout.

FIGURE 13-3

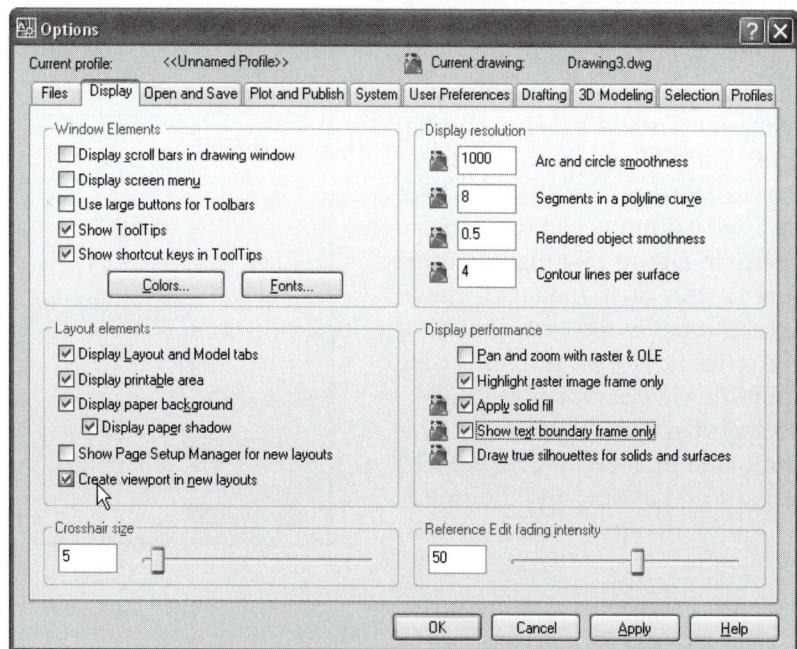

There are two types of viewports in AutoCAD, (1) viewports in model space that divide the screen like "tiles" described in Chapter 10 (known as model space viewports or tiled viewports) and (2) viewports in paper space described in this chapter (known as paper space viewports or floating viewports). The term "floating viewports" is used because these viewports can be moved or resized and can take on any shape. <u>Floating viewports are objects</u> that can be *Erased*, *Moved*, *Copied*, stretched with grips, or the viewport's layer can be turned *Off* or *Frozen* so no viewport "border" appears in the plot.

The command generally used to create viewports, *Vports*, can create either tiled viewports or floating viewports, but only one type at a time depending on the active space—*Model* tab or *Layout* tab. When the *Model* tab is active (*TILEMODE=1*), *Vports* creates tiled viewports. When a *Layout* tab is active (*TILE-MODE=0*), *Vports* creates paper space viewports.

Typically there is only one viewport per layout; however, you can create multiple viewports in one layout. This capability gives you the flexibility to display two or more areas of the same model (model space) in one layout. For example, you may want to show the entire floor plan in one viewport and a detail (enlarged view) in a second viewport, both in the same layout (Fig. 13-4).

FIGURE 13-4

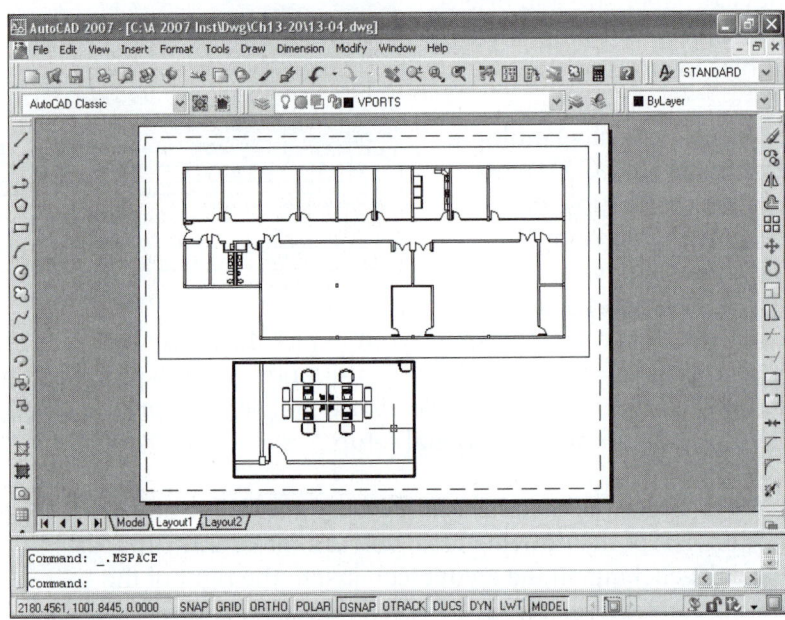

You can control the layer visibility in each viewport. In other words, you can control which layers are visible in which viewports. Do this by using the *Current VP Freeze* button and *New VP Freeze* button in the *Layer Properties Manager* (Fig. 13-5, last two columns). For example, notice in Figure 13-4 that the furniture layer is on in the detail viewport, but not in the overall view of the office floor plan above. (Using multiple viewports and setting viewport-specific layer visibility is discussed in detail in Chapter 33, Advanced Layouts and Plotting.)

FIGURE 13-5

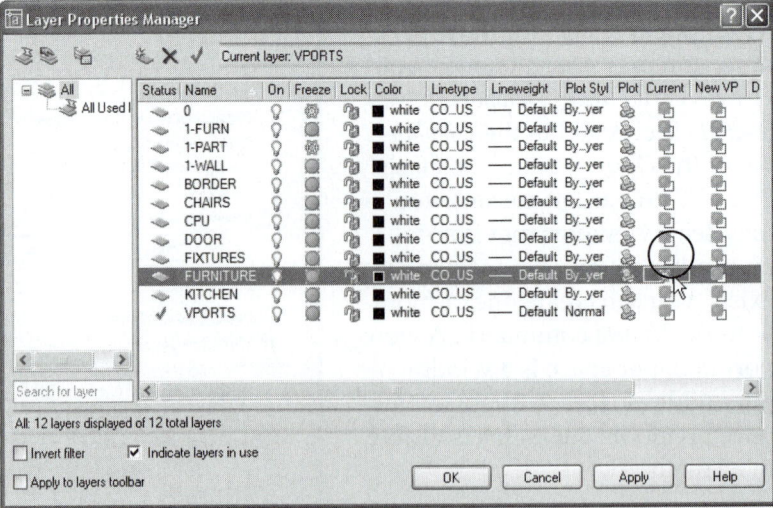

You can also scale the display of the geometry that appears in a viewport. You can set the viewport scale by using the *Viewports* toolbar drop-down list (Fig. 13-6), using the *Properties* palette, or entering a *Zoom XP* factor. This action sets the proportion (or scale) of paper space units to model space units.

FIGURE 13-6

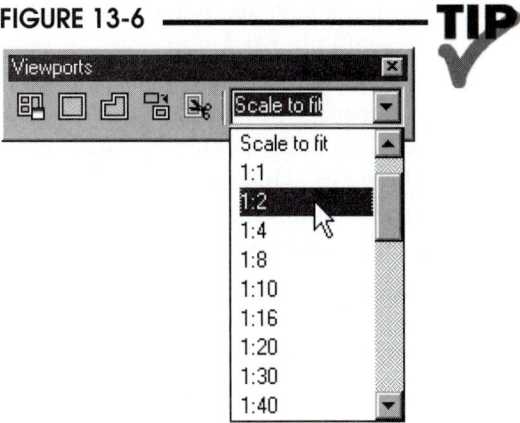

Since a layout represents the plotted sheet, you normally plot the layout at 1:1 scale. However, the geometry in each viewport is scaled to achieve the scale for the drawing objects. For example, you could have two or more viewports in one layout, each displaying the geometry at different scales, as is the case for Figure 13-7. Or, you can set up one layout to display model geometry in one scale and set up a similar layout to display the same geometry at a different scale.

FIGURE 13-7

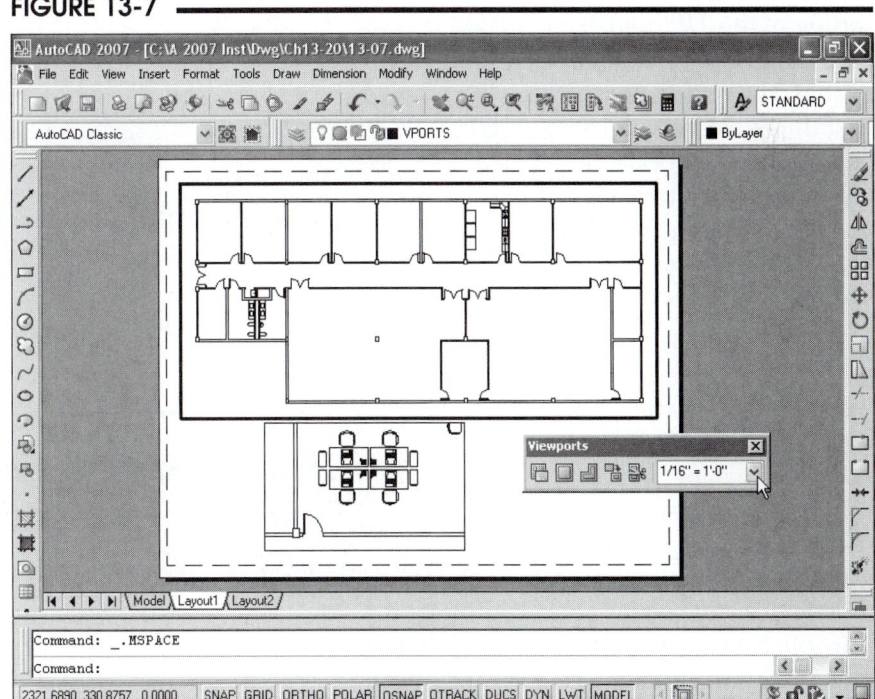

An important concept to remember when using layouts is that the *Plot Scale* that you select in the *Page Setup* and *Plot* dialog boxes should almost always be 1:1. This is because the layout is the actual paper size. Therefore, the geometry displayed in the viewports must be scaled to the desired plot scale (the reciprocal of the "drawing scale factor," which can be set by using the *Viewports* toolbar, the *Properties* palette for the viewport, or a *Zoom XP* factor.

Layouts and Viewports Example

Consider this brief example to explain the basics of using layouts and viewports. In order to keep this example simple, only one viewport is created in the layout to set up a drawing for plotting. This example assumes that no automatic viewports or page setups are created. Your system may be configured to automatically create viewports when a *Layout* tab is activated (as recommended in Chapter 6 to simplify viewport creation). Disabling automatic setups is accomplished by ensuring *Show Page Setup Manager for New Layouts* and *Create Viewport in New Layouts* is unchecked in the *Display* tab in the *Options* dialog box.

First, the part geometry is created in the *Model* tab as usual (Fig. 13-8). Associative dimensions are also normally created in the *Model* tab. This step is the same method that you would have normally used to create a drawing.

When the part geometry is complete, enable paper space by double-clicking a *Layout* tab. AutoCAD automatically changes the setting of the *TILEMODE* variable to 0. When you enable paper space for the first time in a drawing, a "blank sheet" appears (assuming no automatic viewports or page setups are created as a result of settings in the *Options* dialog box).

FIGURE 13-8

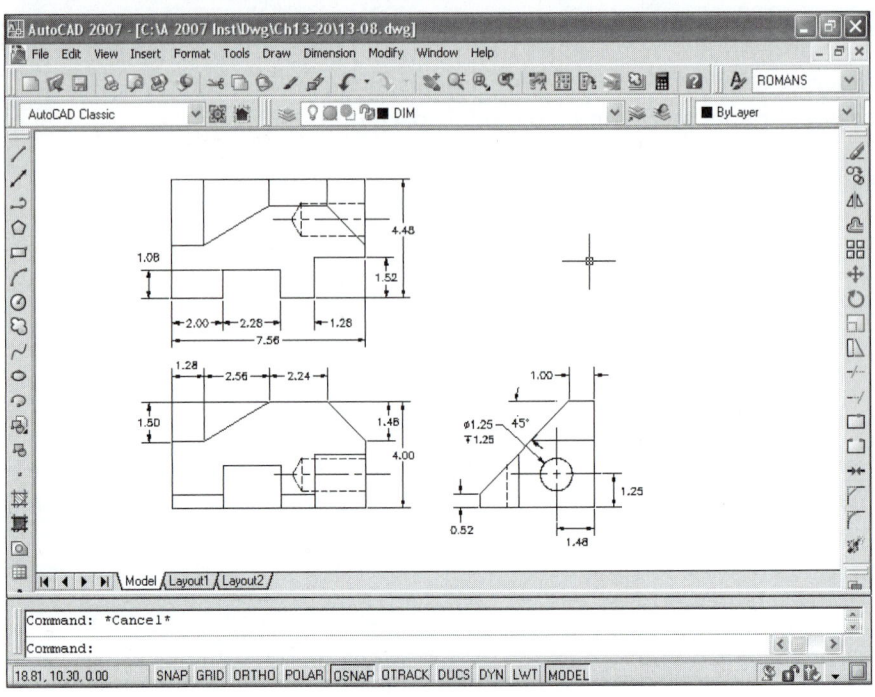

By activating a *Layout* tab, you are automatically switched to paper space. When that occurs, the paper space *Limits* are set automatically based on the selected plot device and sheet size previously selected for plotting. (The default plot device is set in the *Plotting* tab of the *Options* dialog box.) Objects such as a title block and border should be created in the paper space layout (Fig. 13-9). Normally, only objects that are annotations for the drawing (title blocks, tables, border, company logo, etc.) are drawn in the layout.

FIGURE 13-9

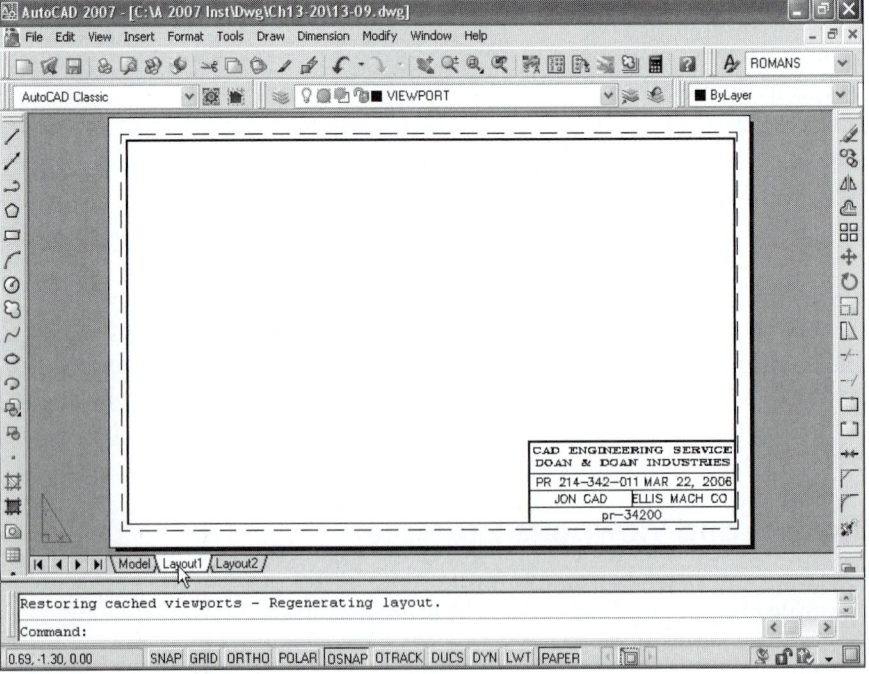

Next, create a viewport in the "paper" with the *Vports* command in order to "look" into model space. All of the model geometry in the drawing initially appears in the viewport. In this example, the *Vports* command is selected from the *Viewports* toolbar. At this point, there is no specific scale relation between paper space units and the size of the geometry in the viewport (Fig. 13-10).

You can control what part of model space geometry you see in a viewport and control the scale of model space to paper space. Two methods are used to control the geometry that is visible in a particular viewport.

FIGURE 13-10

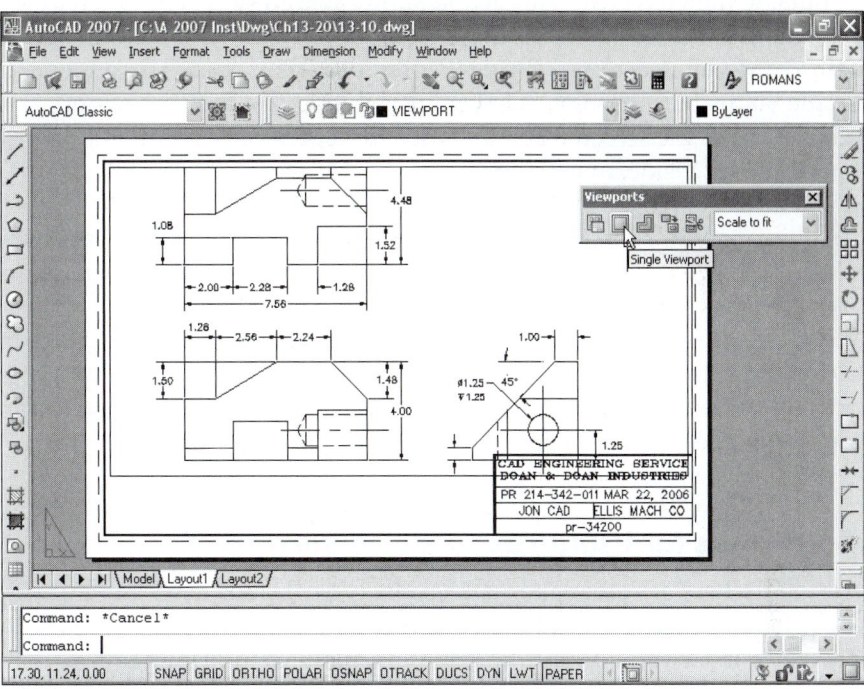

1. **Display commands and Viewport Scale**

    The *Zoom, Pan, View, 3Dorbit* and other display commands allow you to specify what area of model space you want to see in a viewport. In addition, the scale of the model geometry in the viewport can be set by using the *Viewports* toolbar, the *Properties* palette for the viewport, or a *Zoom XP* factor.

2. **Viewport-specific layer visibility control**

    You can control what <u>layers</u> are visible in specific viewports. This function is often used for displaying different model space geometry in separate viewports. You can control which layers appear in which viewports by using the *Layer Properties Manager*. (See Chapter 33 for information on viewport-specific layer visibility control.)

For example, the *Viewports* toolbar could be used to scale the model geometry in a viewport (Fig. 13-11). In this example, double-click inside the viewport and select *1:2* to scale the model geometry to 1/2 size (model space units equal 1/2 times paper space units).

FIGURE 13-11

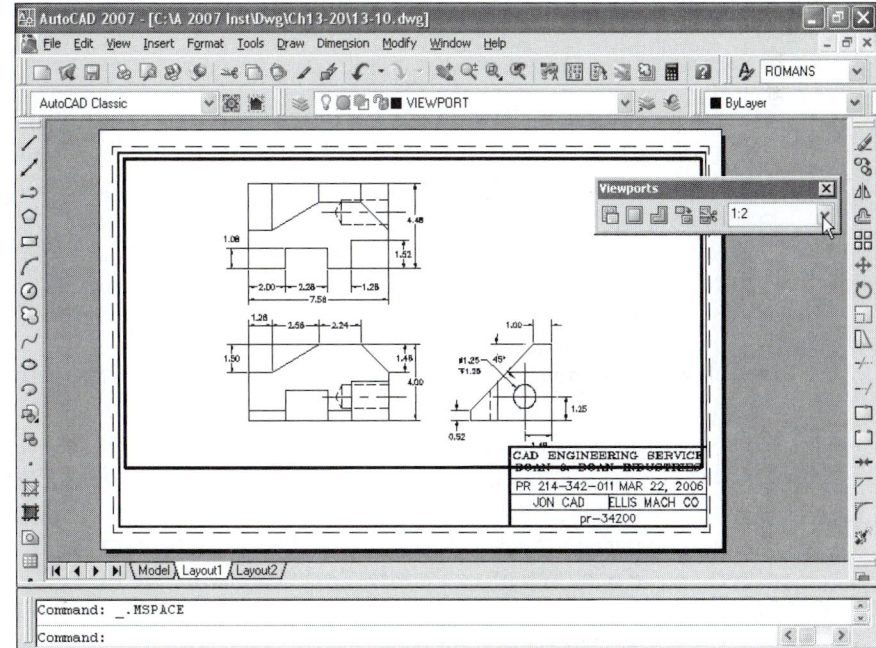

While in a layout with a viewport created to display model space geometry, you can double-click inside or outside the viewport. This action allows you to switch between model space (inside a viewport) and paper space (outside a viewport) so you can draw or edit in either space. You could instead use the *Mspace* (Model Space) and *Pspace* (Paper Space) commands or single-click the words *MODEL* or *PAPER* on the Status Bar to switch between inside and outside the viewport. The "crosshairs" can move completely across the screen when in paper space, but only within the viewport when in model space (model space inside a viewport).

An object <u>cannot be in both spaces</u>. You can, however, draw in paper space and <u>OSNAP to objects in model space</u>. Commands that are used will affect the objects or display of the current space.

When drawing is completed, activate paper space and plot at 1:1 since the layout size (paper space *Limits*) is set to the <u>actual paper size</u>. Typically, the *Page Setup Manager* or the *Plot* dialog box is used to prepare the layout for the plot. The plot scale to use for the layout is almost always 1:1 since the layout represents the plotted sheet and the model geometry is already scaled.

 ## Guidelines for Using Layouts and Viewports

Although there are other alternatives, the typical steps used to set up a layout and viewports for plotting are listed here. (These guidelines assume that no automatic viewports or page setups are created, as described in Chapter 6, "Introduction to Layouts and Printing.")

1. Create the part geometry in model space. Associative dimensions are typically created in model space.

2. Click on a *Layout* tab. When this is done, AutoCAD automatically sets up the size of the layout (*Limits*) for the default plot device, paper size, and orientation. Use the *Page Setup Manager* and *Page Setup* dialog box to ensure the correct device, paper size, and orientation are selected. If not, make the desired selections. (If the layout does not automatically reflect the desired settings, check the *Plotting* tab of the *Options* dialog box to ensure that *Use the Plot Device Paper Size* is checked.)

3. Set up a title block and border as indicated.

    A. Make a layer named BORDER or TITLE and *Set* that layer as current.
    B. Draw, *Insert*, or *Xref* a border and a title block.

4. Make viewports in paper space.

    A. Make a layer named VIEWPORT (or other descriptive name) and set it as the *Current* layer (it can be turned *Off* later if you do not want the viewport objects to appear in the plot).
    B. Use the *Vports* command to make viewports. Each viewport contains a view of model space geometry.

5. Control the display of model space geometry in each viewport. Complete these steps for each viewport.

    A. Double-click inside the viewport or use the *MODEL/PAPER* toggle to "go into" model space (in the viewport). Move the cursor and PICK to activate the desired viewport if several viewports exist.
    B. Use the *Viewport Scale Control* drop-down list, the *Properties* palette for the viewport, or *Zoom XP* to scale the display of the paper space units to model units. This action dictates the plot scale for the model space geometry. The viewport scale is the same as the plot scale factor that would otherwise be used (reciprocal of the "drawing scale factor").
    C. Control the layer visibility for each viewport by using the icons in the *Layer Properties Manager*.

6. Plot from paper space at a scale of 1:1.

    A. Use the *MODEL/PAPER* toggle or double-click outside the viewport to switch to paper space.
    B. If desired, turn *Off* the VIEWPORT layer so the viewport borders do not plot.
    C. Invoke the *Plot* dialog box or the *Page Setup* dialog box to set the plot scale and other options. Normally, set the plot scale for the layout to *1:1* since the layout is set to the actual paper size. The paper space geometry will plot full size, and the resulting model space geometry will plot to the <u>scale</u> selected in the *Viewport Scale* drop-down list, *Properties* palette, or by the designated *Zoom XP* factor.
    D. Make a *Full Preview* to check that all settings are correct. Make changes if necessary. Finally, make the plot.

CREATING AND SETTING UP LAYOUTS

Listed below are several ways to create layouts.

Shortcut menu   Right-click on a *Layout* tab and select *New Layout*, *From Template*, or *Move or Copy* (see Fig. 13-1).
Layout command   Type or select from the *Insert* pull-down menu. Use the *New Layout*, *Layout from Template*, or *Create Layout Wizard* option.
Layoutwizard command Type or select from the *Insert* pull-down menu.

There are several steps involved in correctly setting up a layout, as listed previously in the "Guidelines for Using Layouts and Viewports." The steps involve specifying a plot device, setting paper size and orientation, setting up a new *Layout* tab, and creating the viewports and scaling the model geometry. One of the best methods for accomplishing all these tasks is to use the *Layout Wizard*.

Layoutwizard

Pull-down Menu	Command (Type)	Alias (Type)	Short-cut	Screen (side) Menu	Tablet Menu
Insert *Layout >* *Create Layout Wizard*	*Layoutwizard*	...	...	...	...

The *Layoutwizard* command produces a wizard to automatically lead you through eight steps to correctly set up a layout tab for plotting. The steps are essentially the same as those listed earlier in the "Guidelines for Using Layouts and Viewports."

Begin
The first step is intended for you to enter a name for the layout (Fig. 13-12). The default name is whatever layout number is next in the sequence, for example, *Layout3*. Enter any name that has not yet been used in this drawing. You can change the name later if you want using the *Rename* option of the *Layout* command.

FIGURE 13-12

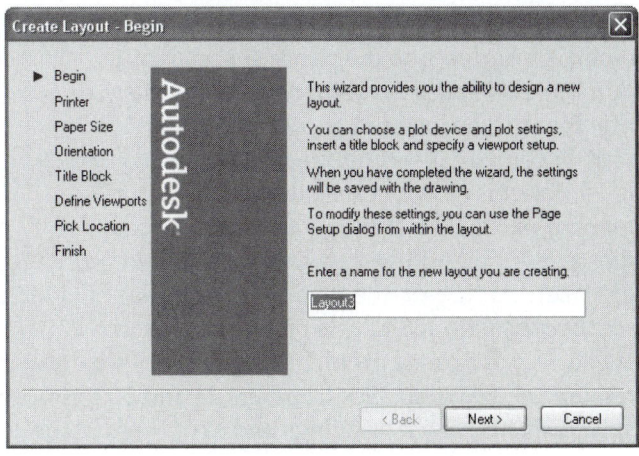

Printer

Because the *Limits* in paper space are automatically set based on the size and orientation of the selected paper, choosing a print or plot device is essential at this step (Fig. 13-13). The list includes all previously configured printer and plotter devices. If you intend to use a different device, it may be wise to *Cancel*, use *Plotter Manager* to configure the new device, then begin *Layoutwizard* again (see Chapter 14, "Configuring Plotters and Printers").

FIGURE 13-13

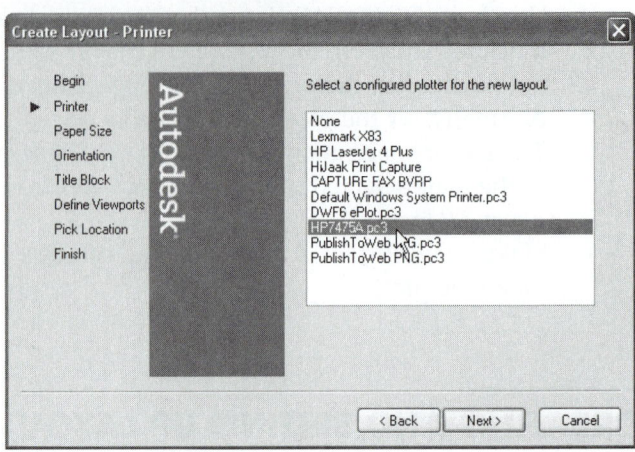

Paper Size

Select the paper size you expect to use for the layout (Fig. 13-14). Make sure you also select the units for the paper (*Drawing Units*) since the layout size (*Limits* in paper space) is automatically set in the selected units.

Orientation

Select either *Portrait* or *Landscape* (not shown). *Landscape* plots horizontal lines in the drawing along the long axis of the paper, whereas *Portrait* plots horizontal lines in the drawing along the short axis of the paper.

FIGURE 13-14

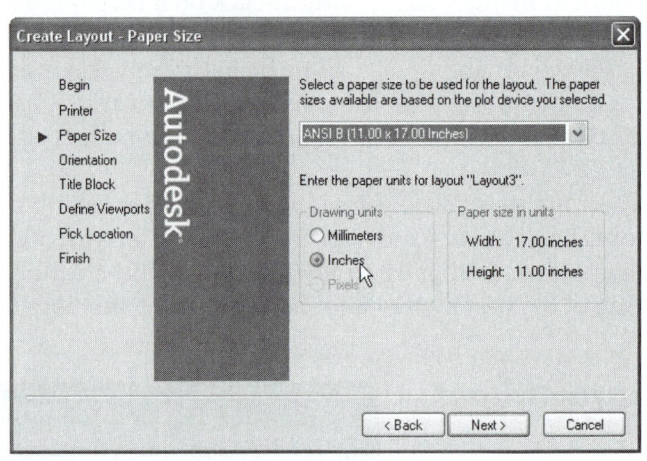

Title Block

In this step, you can select *None* or select from several AutoCAD-supplied ANSI, DIN, ISO and JIS standard title blocks (Fig. 13-15). See "Table of AutoCAD-Supplied Template Drawing Settings" in Chapter 12 for exact sizes and specifications. The selected title block is inserted into the paper space layout. You can also specify whether you want the title block to be inserted as a *Block* or as an *Xref*. Normally inserting the title block as a *Block* object is safer because it becomes a permanent part of the drawing although it occupies a small bit of drawing space (increased file size). An *Xrefed* title block saves on file size because the actual title block is only referenced, but can cause problems if you send the drawing to a client who does not have access to the same referenced drawing. (See Chapters 21 and 30 for more information on these subjects.) You can also create your own title block drawings and save them in the Template folder so they will appear in this list.

FIGURE 13-15

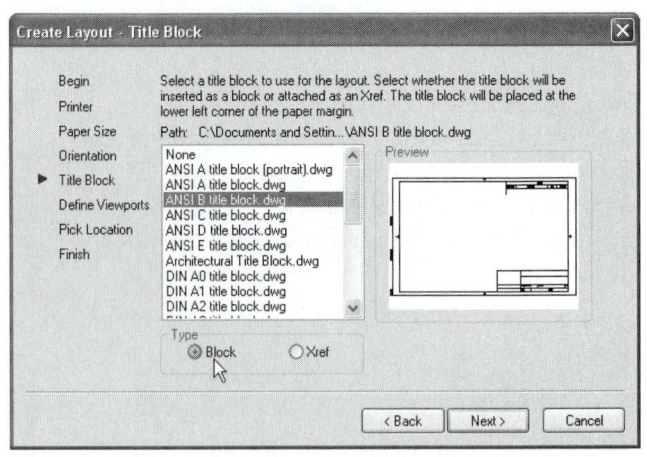

Define Viewports

This step has two parts (Fig. 13-16). First select the type and number of viewports to set up. Use *None* if you want to set up viewports at a later time or want to create a non-rectangular viewport. Normally, use *Single* if you need only one rectangular viewport. Use *Std. 3D Engineering Views* if you have a 3D model. The *Array* option creates multiple viewports with the number of *Rows* and *Columns* you enter in the edit boxes below. Second, determine the *Viewport Scale*. This setting determines the scale that the model space geometry will be displayed in the viewport (ratio of paper space units to model space units).

Pick Location

In this step, pick *Select Location* to temporarily exit the wizard and return to the layout to pick diagonal corners for the viewport(s) to fit within (Fig. 13-17). Normally, you do not want the corners of the viewport(s) to overlap the title block or border. You can bypass this step to have AutoCAD automatically draw the viewport border at the extents of the printable area.

Finish

This is not really a step, rather a confirmation (not shown). In all other steps, you can use *Back* to go backward in the process to change some of the specifications. In this step, *Back* is disabled, so you can only select *Finish* or *Cancel*.

FIGURE 13-16

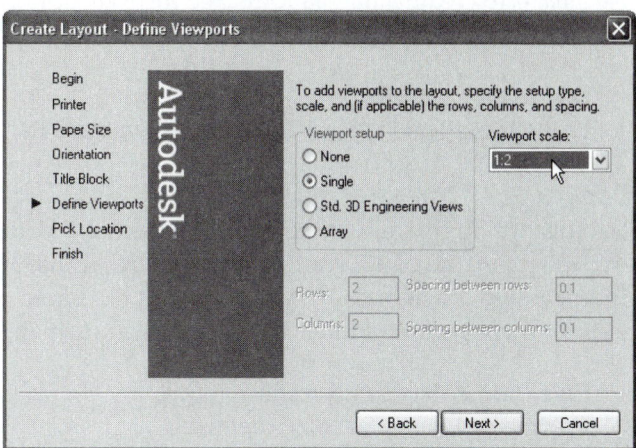

FIGURE 13-17

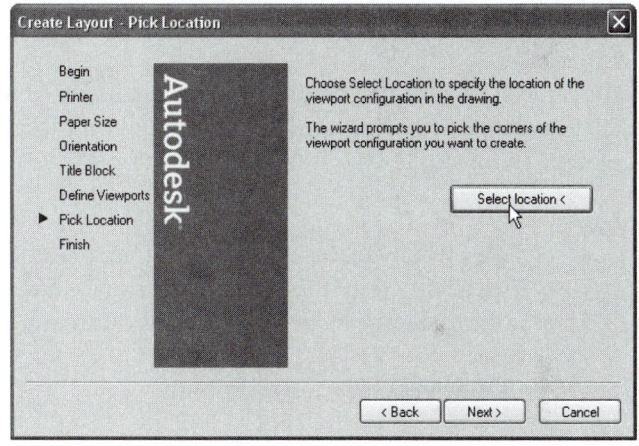

At this time, the new layout is activated so you can see the setup you specified. If you want to make changes to the plot device, paper size, orientation, or viewport scale, you can do so by using the *Page Setup* dialog box. You can also change the viewport sizes or configuration using *Vports* or grips (explained later in this chapter).

Layout

Pull-down Menu	Command (Type)	Alias (Type)	Short-cut	Screen (side) Menu	Tablet Menu
Insert *Layout >* *New Layout*	*Layout*	*LO* or *-LO*	...	...	...

If you prefer to create a layout without using the *Layout Wizard*, use the *Layout* command by typing it or selecting it from the *Insert* pull-down menu or *Layout* toolbar. Using the wizard leads you through all the steps required to create a new layout including selecting plot device, paper size and orientation, and creating viewports. Use the *Layout* command to create viewports "manually" (if your system is not set to automatically create viewports). You will then have to use *Vports* to create viewports in the layout.

It is a good idea to ensure the correct default plot device is specified in the *Options* dialog box before you use *Layout* to create a new layout. See the *New* option for more information. Besides creating a new layout, using a template, and copying an existing layout, you can rename, save, and delete layout.

    Command: **layout**
    Enter layout option [Copy/Delete/New/Template/Rename/SAveas/Set/?] <set>:

Copy

Use this option if you want to copy an existing layout including its plot specifications. If you do not provide the name of a layout to copy, AutoCAD assumes you want to copy the active layout tab. If you do not assign a new name, AutoCAD uses the name of the copied layout (*Floor Plan*, for example) and adds an incremental number in parentheses, such as *Floor Plan (2).*

    Enter layout to copy <current>:
    Enter layout name for copy <default>:

Delete

Use this option to *Delete* a layout. If you do not enter a name, the most current layout is deleted by default.

    Enter name of layout to delete <current>:

Wildcard characters can be used when specifying layout names to delete. You can select several *Layout* tabs to delete by holding down Shift while you pick. If you select all layouts to delete, all the layouts are deleted, and a single layout tab remains named *Layout1*. The *Model* tab cannot be deleted.

New

 This option creates a new layout tab. AutoCAD creates a new layout from scratch based on settings defined by the default plot device and paper size (in the *Options* dialog box). If you choose not to assign a new name, the layout name is automatically generated (*Layout3*, for example).

    Enter new layout name <Layout#>:

Keep in mind that when the new layout is created, its size (*Limits*) and shape are determined by the *Use as Default Output Device* setting specified in the *Plot and Publish* tab of the *Options* dialog. Therefore, it is a good idea to ensure the correct plot device is specified before you use the *New* option. You can, however, change the layout at a later time using the *Page Setup* dialog box, assuming that *Use Plot Device Paper Size* is checked in the *Plot and Publish* tab of the *Options* dialog box (see *Options* in this chapter).

Template

 You can create a new template based on an existing layout in a template file (.DWT) or drawing file (.DWG) with this option. The layout, viewports, associated layers, and all the paper space geometry in the layout (from the specified template or drawing file) are inserted into the current drawing. No dimension styles or other objects are imported. The standard file navigation dialog box (not shown) is displayed for you to select a file to use as a template. Next, the *Insert Layout* dialog box appears for you to select the desired layout from the drawing (Fig. 13-18). See "Table of AutoCAD-Supplied Template Drawings Settings" in Chapter 12.

FIGURE 13-18

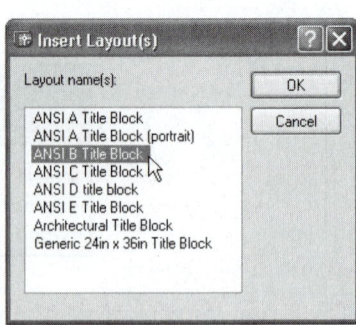

Rename

Use *Rename* to change the name of an existing layout. The last current layout is used as the default for the layout to rename.

    Enter layout to rename <current>:
    Enter new layout name:

Layout names can contain up to 255 characters and are not case sensitive. Only the first 32 characters are displayed in the tab.

Save

This option creates a .DWT file. The file contains all of the (paper space) geometry in the layout and all of the plot settings for the layout, but no model space geometry. All *Block* definitions appearing in the layout, such as a titleblock, are also saved to the .DWT file, but not unused *Block* definitions. All layouts are stored in the template folder as defined in the *Options* dialog box. The last current layout is used as the default for the layout to save.

    Enter layout to save to template <current>:

The standard file selection dialog box is displayed in which you can specify a file name for the .DWT file. When you later create new layout from a template (using the *Template* option of *Layout*), all the saved information (plot settings, layout geometry and *Block* definitions) are imported into the new layout.

Set

This option simply makes a layout current. This option has identical results as picking a layout tab.

? (List Layouts)

Use this option to list all the layouts when you have turned off the layout tabs in the *Display* tab of the *Options* dialog box. Otherwise, layout tabs are visible at the bottom of the drawing area.

Inserting Layouts with AutoCAD DesignCenter

A feature in AutoCAD, called DesignCenter, allows you to insert content from any drawing into another drawing. Content that can be inserted includes drawings, *Blocks*, *Dimstyles*, *Layers*, *Layouts*, *Linetypes*, *Textstyles*, *Xrefs*, raster images, and URLs (web site addresses). If you have created a layout in a drawing and want to insert it into the current drawing, you can use the *Template* option of *Layout* (as previously explained) or use DesignCenter to drag and drop the layout name into the drawing. With DesignCenter, only layouts from .DWG files (not .DWT files) can be imported. See Chapter 21 Blocks, DesignCenter, and Tool Palettes for more information.

Setting Up the Layout

When you do not use the *Layout Wizard*, the steps involved in setting up the layout must be done individually. In cases when you need flexibility, such as creating non-rectangular viewports or when some factors are not yet known, it is desirable to take each step individually. These steps are listed in "Guidelines for Using Layouts and Viewports."

Options

Pull-down Menu	Command (Type)	Alias (Type)	Short-cut	Screen (side) Menu	Tablet Menu
Tools Options...	Options	OP	(Default) Options...	TOOLS2 Options	Y,10

When you create *New* layouts using the *Layout* command, the size and shape of the layout is determined by the selected plot device, paper size, and paper size. Default settings are made in the *Plot and Publish* tab of the *Options* dialog box (Fig. 13-19). Although most settings concerned with plotting a layout can later be changed using the *Page Setup* dialog box, it may be efficient to make the settings in the *Options* dialog box <u>before</u> creating layouts from scratch. Four settings in this dialog box affect layouts.

FIGURE 13-19

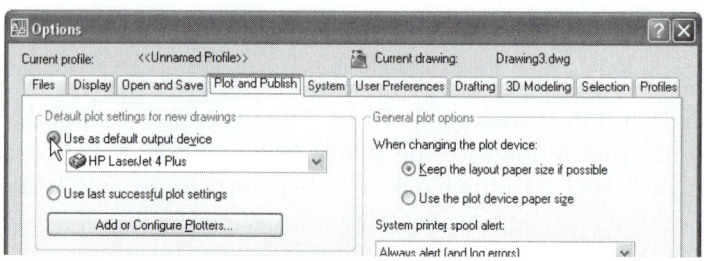

Default Plot Settings for New Drawings
Your choice in this section determines what plot device is used to automatically set the paper size (paper space *Limits*) when new layouts are created. Your setting here applies to <u>new drawings and the current drawing</u>. (Paper size information for each device is stored either in the plotter configuration file [.PC3] or in the default system settings if the output device is a system printer.)

Use as Default Output Device
The selected device is used to determine the paper size (paper space *Limits*) that are set automatically when a new layout is created in the current drawing and for new drawings. This drop-down list contains all configured plot or print devices (any plotter configuration files [PC3] and any system printers that are configured in the system).

Use Last Successful Plot Settings
This button uses the plotting settings (device and paper size) according to those of the last successful plot that was made on the system (a plot must be made from the layout for this option to be valid) and applies them to new layouts that are created. This setting affects new layouts created for the current drawing as well as new drawings. This option overrides the output device appearing in the *Use as Default Output Device* drop-down list.

General Plot Options
These options apply only when plot devices are changed for existing layouts <u>and</u> the layout has a viewport created. Plot devices for layouts can be changed using the *Page Setup* or *Plot* dialog boxes. When you change the plot device for an existing layout, the paper size (*Limits* setting) of the layout is determined by your choice in this section.

Keep the Layout Paper Size if Possible
If this button is pressed, AutoCAD attempts to use the paper size initially specified for the layout if you decide later to change the plot device for the layout. This option is useful if you have two devices that can use the same size paper as specified in the layout, so you can select either device to use without affecting the layout. However, if the selected output device cannot plot to the previously specified paper size, AutoCAD displays a warning message and uses the paper size specified by the plot device you select in the *Page Setup* or *Plot* dialog box. If no viewports have been created in the layout, you can change the plot device and automatically reset the layout limits without consequence. This button sets the *PAPERUPDATE* system variable to 0 (layout paper size is not updated).

Use the Plot Device Paper Size
As the alternative to the previous button, this one sets the paper size (*Limits*) for an existing layout to the paper size of whichever device is selected for that layout in the *Page Setup* or *Plot* dialog box. This option sets *PAPERUPDATE* to 1.

Pagesetup

Pull-down Menu	Command (Type)	Alias (Type)	Short-cut	Screen (side) Menu	Tablet Menu
File Page Setup Manager...	Pagesetup	...	...	...	V,25

Plot specifications that you make and save for a layout are called "page setups." There are two separate dialog boxes that are used for saving a page setup. Using the *Pagesetup* command produces the *Page Setup Manager* (Fig. 13-20) which is a "front end" for the *Page Setup* dialog box. The *Page Setup* dialog box (see Fig. 13-21) is used to select the plot specifications for the layout such as plot device, paper size, plot scale, etc. All settings made in the *Page Setup Manager* and related dialog box are saved with the layout. Therefore, the plot settings are prepared when you are ready to make a plot or print.

If you have created a layout using the *Layout* command and the desired settings were made in the *Options* dialog box as explained in the previous section, the next step in setting up a layout is using the *Page Setup Manager* and the related *Page Setup* dialog box. If you used the *Layout Wizard*, you can optionally use the *Page Setup Manager* and related dialog box to alter some of those settings.

To save a page setup for a layout, first activate the desired layout tab, then use *Pagesetup* to produce the *Page Setup Manager*. The active layout tab's name appears highlighted in the *Page setups* list. Next, select the *Modify* option as shown in Figure 13-20 to produce the *Page Setup* dialog box.

When the *Page Setup* dialog box appears (Fig. 13-21), specify the following plotting options:

1. Select a plotter or printer in the *Printer Plotter* section.

FIGURE 13-20

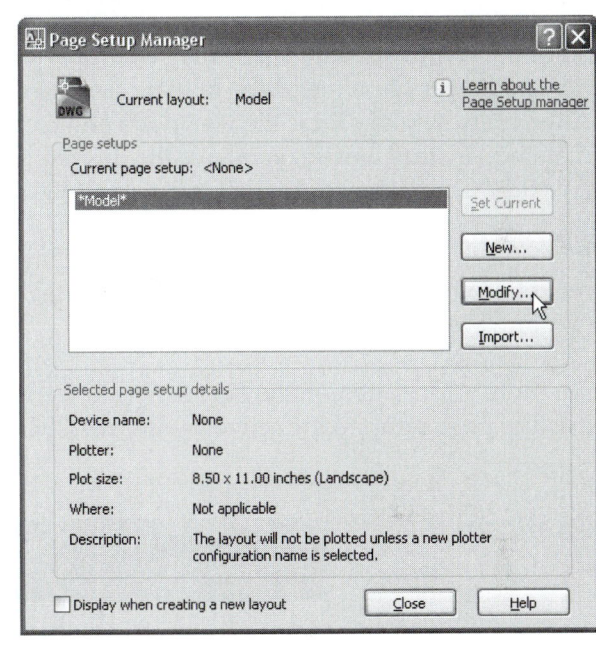

FIGURE 13-21

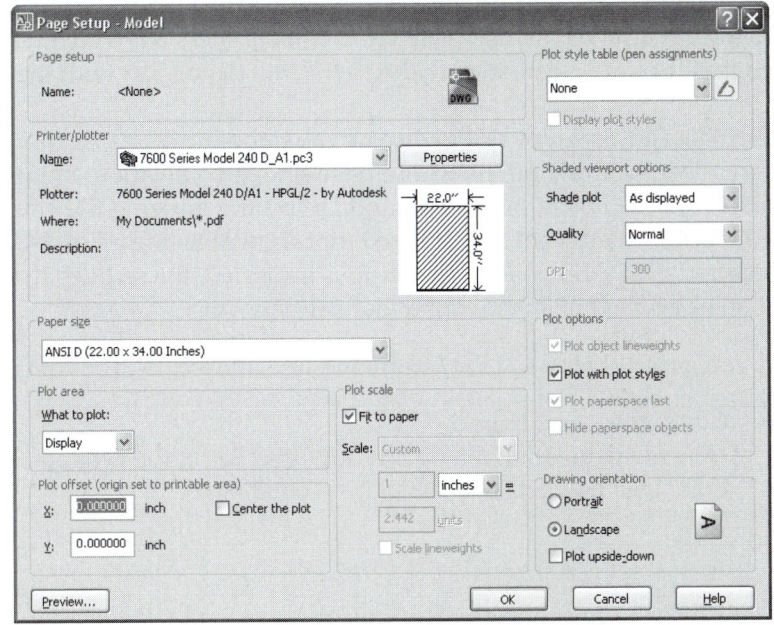

2. Select the desired *Paper size.*

3. Ensure the *Plot scale* is set to 1:1.

4. Select any other desired options such as *Drawing orientation, Plot options*, etc.

5. Select the *OK* button, then *Close* the *Page Setup Manager* to save the settings for the layout.

Remember that the size of the layout (paper space *Limits*) is automatically set based on *Paper Size* and *Drawing Orientation* you select here. Depending on your selection of *Keep the Layout Size if Possible* or *Use the Plot Device Paper Size* in the *Plotting* tab of the *Options* dialog box (see previous section), the layout size and shape may or may not change to the new plot device settings. You can, however, select a new plot device, paper size, or orientation <u>after creating the layout, but before creating viewports</u>.

Since the layout size and shape may change with a new device, viewports that were previously created may no longer fit on the page. In that case, the viewports must be changed to a new size or shape. Alternately, you can *Erase* the viewports and create new ones. Either of these choices is not recommended. Instead, ensure you have the desired plot device and paper settings <u>before creating viewports</u>. (See the "Guidelines for Using Layouts and Viewports.")

Plot Scale is almost always set to *1:1* since the layout represents the sheet you will be plotting on. The model geometry that appears in the layout (in a viewport) will be scaled by setting a viewport scale (see "Scaling Viewport Geometry").

For more information on the *Page Setup Manager* and the *Page Setup* dialog box, such as how to apply an existing page setup to the active layout, see Chapter 14.

Psetupin

Pull-down Menu	Command (Type)	Alias (Type)	Short-cut	Screen (side) Menu	Tablet Menu
...	*Psetupin* or *-Psetupin*	...	...	...	...

Psetupin imports a user-defined page setup into a new drawing layout. This feature provides the ability to import a saved, named page setup from another drawing into the current drawing. User-defined page setups are created using the *Page Setup* dialog box with the *Add* option (see previous section).

Psetupin causes AutoCAD to display the standard file navigation dialog box (not shown) in which you can select the drawing file whose page setups you want to import. The selected .DWG file must contain a user-defined page setup. After you select the drawing file you want to use, the *Import User Defined Page Setups* dialog box is displayed (not shown) listing all user-defined page setups contained in the selected .DWG file. When the setup is imported, the settings that are saved in the named page setup can then be applied to any layout in the new drawing.

If you enter *-Psetupin* at the Command line, the following prompts are displayed.

```
Command: -psetupin
```
(The file navigation dialog box appears. Select the desired .DWG file.)
```
Enter user defined page setup(s) to import or [?]:
```

USING VIEWPORTS IN PAPER SPACE

If you use the *Layout Wizard*, viewports can be created during that process. If you do not use the wizard, viewports should be created for a layout only after specifying the plot device, paper size, and orientation in the *Page Setup* dialog box or through the default plot options set in the *Options* dialog box. You can create viewports with the *Vport* or *-Vport* commands. The Command line version (*-Vports*) produces several options not available in the dialog box, including options to create non-rectangular viewports.

Vports

Pull-down Menu	Command (Type)	Alias (Type)	Short-cut	Screen (side) Menu	Tablet Menu
View *Viewports >*	*Vports* or *-Vports*	...	...	VIEW 1 *Vports*	M,3 and M,4

The *Vports* command produces the *Viewports* dialog box. Here you can select from several configurations to create rectangular viewports as shown in Figure 13-22. Making a selection in the *Standard Viewports* list on the left of the dialog box in turn displays the selected *Preview* on the right.

After making the desired selection from the dialog box, AutoCAD issues the following prompt.

    Command: **vports**
    Specify first corner or [Fit] <Fit>:

You can pick two diagonal corners to specify the area for the viewports to fill. Generally, pick just inside the titleblock and border if one has been inserted. If you use the *Fit* option, AutoCAD automatically fills the printable area with the specified number of viewports.

The *Setup: 3D* and *Change View to:* options are intended for use with 3D models. (See Chapter 42 for more information on these options.)

You may notice that the *Viewports* dialog box is the same dialog box that appears when you use *Vports* while in the *Model* tab. However, when *Vports* dialog is used in the *Model* tab, <u>tiled</u> viewports are created, whereas when the *Vports* is used in a layout (paper space), paper space viewports are created.

FIGURE 13-22

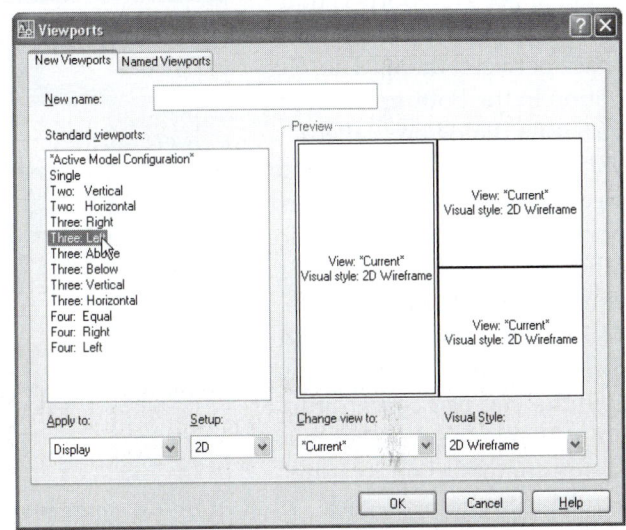

FIGURE 13-23

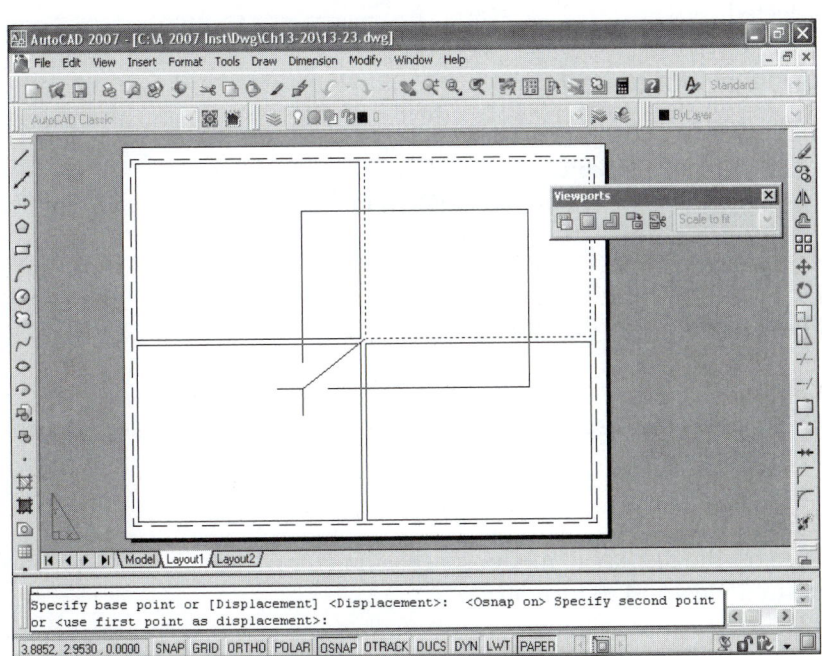

Paper space viewports differ from tiled viewports. AutoCAD treats paper space viewports created with *Vports* as <u>objects</u>. Like other objects, paper space viewports can be affected by most editing commands. For example, you could use *Vports* to create one viewport, then use *Copy* or *Array* to create other viewports. You could edit the size of viewports with *Stretch* or *Scale*. You can use Grips to change paper space viewports. Additionally, you can use *Move* (Fig. 13-23, previous page) to relocate the position of viewports. Delete a viewport using *Erase*. To edit a paper space viewport, you must be in paper space to PICK the viewport objects (borders).

Paper Space viewports can also overlap. This feature makes it possible for geometry appearing in different viewports to occupy the same area on the screen or on a plot.

NOTE: Avoid creating one viewport completely within another's border because visibility and selection problems may result.

Another difference between tiled (*Model* tab) viewports and paper space viewports is that viewports in layouts can have space between them. The option in the bottom left of the *Viewports* dialog box called *Viewport Spacing* allows you to create space between the viewports (if you choose any option other than *Single*) by entering a value in the edit box. Figure 13-24 illustrates a layout with three viewports after entering .5 in the *Viewport Spacing* edit box. Also note several of the -*Vports* command options are available in the right-click shortcut menu that appears when a viewport object is selected.

FIGURE 13-24

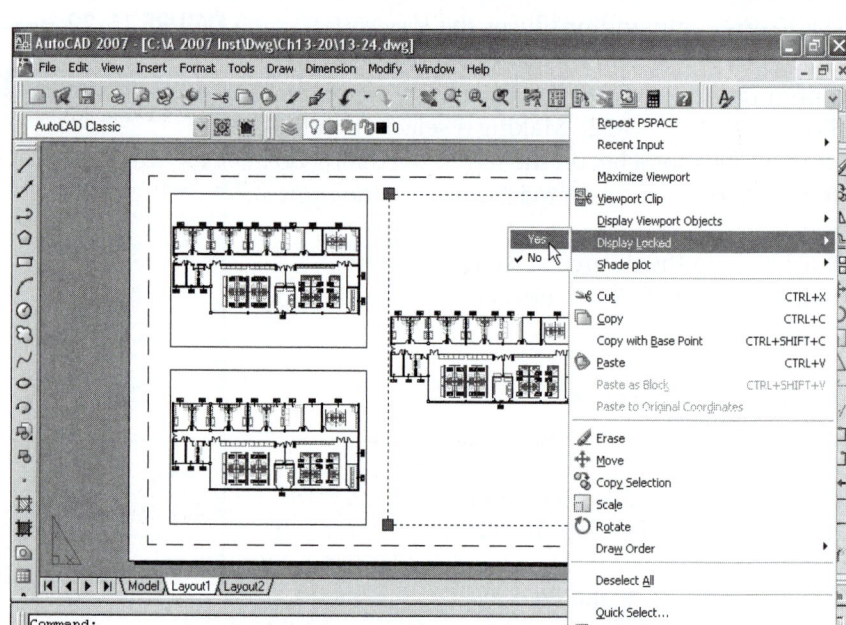

Using the -*Vports* command (with a hyphen prefix) invokes the Command line version and produces the following options.

```
Command: -vports
Specify corner of viewport or [ON/OFF/Fit/Shadeplot/Lock/Object/Polygonal/Restore/2/3/4] <Fit>:
```

The default option (Specify corner of viewport) allows you to pick two diagonal corners for one viewport. The other options are as follows. Remember to select the viewport objects (borders) while in paper space when AutoCAD prompts you to select a viewport.

ON

On turns on the display of model space geometry in the selected viewport. Select the desired viewport to turn on.

OFF

Turn off the display of model space geometry in the selected viewport with this option. Select the desired viewport to turn off.

Fit

Fit creates a new viewport and fits it to the size of the printable area. The new viewport becomes the current viewport.

Shadeplot

If you are displaying a 3D surface or solid model, *Shadeplot* allows you to select how you want to print the current viewport: *As Displayed, Wireframe, Hidden, Visual Styles,* or *Rendered.* (See Chapter 35, 3D Basics, Navigation, and Visual Styles, for more information.)

Lock

Use this option to <u>lock the scale</u> of the current viewport. Turn viewport locking *On* or *Off.* With paper space viewports, you set the scale of each viewport individually using *Zoom* with an *XP* scale factor or by using the *Viewports* toolbar scale drop-down list (see "Scaling Viewport Geometry"). Double-clicking inside a viewport and then using *Zoom* normally changes the scale of a viewport. If *Lock* is used on the viewport, the specified scale of the viewport cannot be changed (unless viewport locking is turned *Off*). If you attempt to *Zoom* inside a locked viewport, the display of the entire layout (paper space) is zoomed.

Object

You can use *Object* to convert a closed *Polyline, Ellipse, Spline, Region,* or *Circle* into a viewport (border). The polyline you specify must be closed and contain at least three vertices. It can be self-intersecting and can contain arcs as well as line segments. Figure 13-25 displays a closed *Pline* converted to a viewport.

FIGURE 13-25

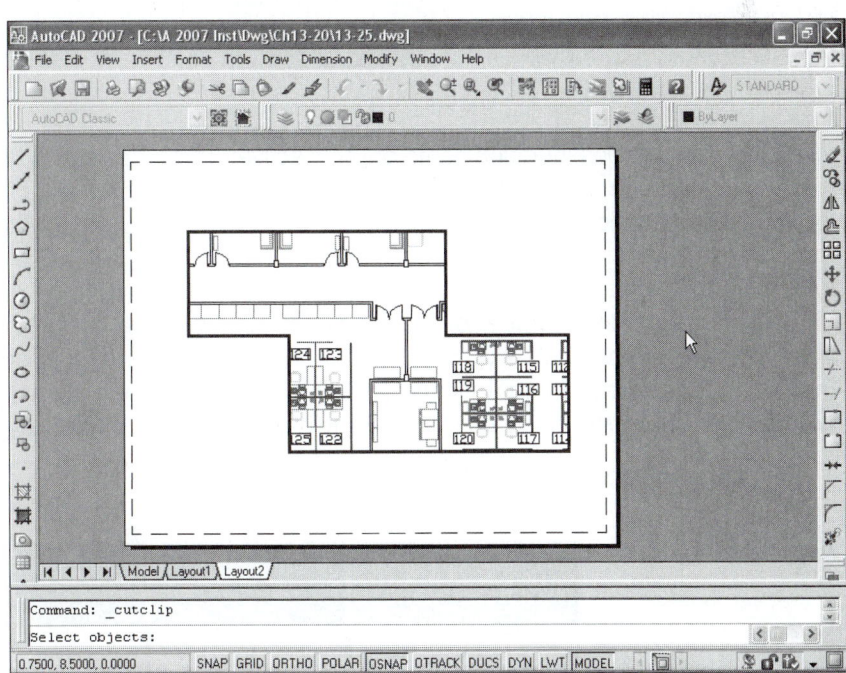

Polygonal

Use this option to create an irregularly shaped viewport defined by specifying points. You can define the viewport by straight line or arc segments.

```
Command: -vports
Specify corner of viewport or [ON/OFF/Fit/Shadeplot/Lock/Object/Polygonal/Restore/2/3/4] <Fit>: p
Specify start point: PICK
Specify next point or [Arc/Close/Length/Undo]: PICK
Specify next point or [Arc/Close/Length/Undo]: PICK
Specify next point or [Arc/Close/Length/Undo]: c
Regenerating model.
```

If you choose the *Arc* option, the following prompts allow all arc creation options, identical to those found in the full *Arc* command.

```
Enter an arc boundary option [Angle/CEnter/CLose/Direction/Line/Radius/Second pt/Undo/Endpoint of arc]
<Endpoint>:
```

Figure 13-26 displays a viewport created with the *Polygonal* option of *-Vports*. Notice the arc segment. Keep in mind both the *Object* and *Polygonal* options can be used to create non-rectangular and curved viewports.

FIGURE 13-26

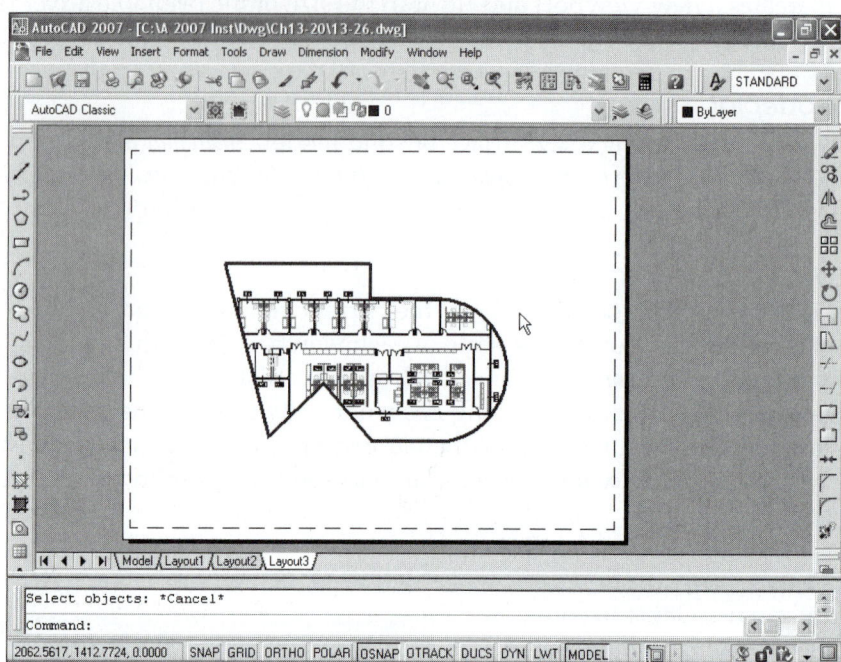

Restore

Use this option to restore a previously created named viewport configuration.

> Enter viewport configuration name or [?] <*Active>: Enter named viewport configuration.
> Specify first corner or [Fit] <Fit>: **PICK** or **F**
> Regenerating model.

2/3/4

These options create a number of viewports within a rectangular area that you specify. The 2 option allows you to arrange 2 viewports either *Vertically* or *Horizontally*. The 4 option automatically divides the specified area into 4 equal viewports. The possible configurations using the 3 option are displayed if you use the *Viewports* dialog box. Using the 3 option yields the following prompt:

> Horizontal/Vertical/Above/Below/Left/<Right>:

Pspace

Pull-down Menu	Command (Type)	Alias (Type)	Short-cut	Screen (side) Menu	Tablet Menu
...	*Pspace*	PS	...	*VIEW 1* *Pspace*	L,5

The *Pspace* command switches from model space (inside a viewport) to paper space (outside a viewport). The cursor is displayed in paper space and can move <u>across the entire screen</u> when in paper space, while the cursor appears only inside the current viewport if the *Mspace* command is used. This command accomplishes the same action as double-clicking outside the viewport in a layout or using the *MODEL/PAPER* toggle on the Status Bar.

If you type this command while in the *Model* tab, the following message appears:

> Command: **pspace**
> PSPACE
> ** Command not allowed in Model Tab **

Mspace

Pull-down Menu	Command (Type)	Alias (Type)	Short-cut	Screen (side) Menu	Tablet Menu
...	Mspace	MS	...	VIEW Mspace	L,4

The *Mspace* command switches from paper space to <u>model space inside a viewport</u>, similar to double-clicking inside a viewport. If several viewports exist in the layout when you use the command, you are switched to the last active viewport. The cursor appears only in that viewport. The current viewport displays a heavy border. You can switch to another viewport (make another current) by PICKing in it. To use *Mspace*, there must be at least one viewport in the layout, and it must be *On* (not turned *Off* by the *-Vports* command) or AutoCAD issues a message and cancels the command.

Model

Pull-down Menu	Command (Typo)	Alias (Type)	Short-cut	Screen (side) Menu	Tablet Menu
...	Model	...	...	...	...

The *Model* command accomplishes the same action as selecting the *Model* tab; that is, *Model* makes the *Model* tab active. If you issue this command while you are in the *Model* tab, nothing happens.

Scaling the Display of Viewport Geometry

Paper space (layout) objects such as title block and border should correspond to the paper at a 1:1 scale. A paper space layout is intended to represent the print or plot sheet. *Limits* in a <u>layout</u> are automatically set to the exact paper size, and the finished drawing is plotted from the layout to a scale of 1:1. The model space geometry, however, should be true scale in real-world units, and *Limits* in <u>model space</u> are generally set to accommodate that geometry.

When model space geometry appears in a viewport in a layout, the size of the <u>displayed</u> geometry can be controlled so that it appears and plots in the correct scale. There are three ways to scale the display of model space geometry in a paper space viewport. You can (1) use the *Viewport Scale Control* drop-down list, (2) use the *Properties* palette, or (3) use the *Zoom* command and enter an *XP* factor. Once you set the scale for the viewport, use the *Lock* option of *-Vports* or the *Properties* palette to lock the scale of the viewport so it is not accidentally changed when you *Zoom* or *Pan* inside the viewport.

When you use layouts, the <u>*Plot Scale* specified in the *Plot* dialog box and *Page Setup* dialog box</u> dictates the scale for the layout, which <u>should normally be set to 1:1</u>. Therefore, <u>the viewport scale actually determines the plot scale of the geometry in a finished plot</u>. Set the viewport scale to the reciprocal of the "drawing scale factor," as explained in Chapter 12, Advanced Drawing Setup.

Viewport Scale Control Drop-Down List
The *Viewport Scale Control* drop-down list is located in the *Viewports* toolbar (see Fig. 13-27). Invoke the *Viewports* toolbar using the *Toolbars...* option at the bottom of the *View* pull-down menu.

To set the viewport scale (scale of the model geometry in the viewport), first make the viewport current by double-clicking in it or typing *Mspace*. When the crosshairs appear in the viewport and the viewport border is highlighted (a wide border appears), select the desired scale from the list.

The *Viewport Scale Control* drop-down list provides only standard scales for decimal, metric, and architectural scales. You can select from the list or enter a value. Values can be entered as a decimal, a fraction (proper or improper), a proportion using a colon symbol (:), or an equation using an equal symbol (=). Feet (') and inch (") unit symbols can also be entered. For example, if you wanted to scale the viewport geometry at 1/2"=1', you could enter any of the following, as well as other, values.

FIGURE 13-27

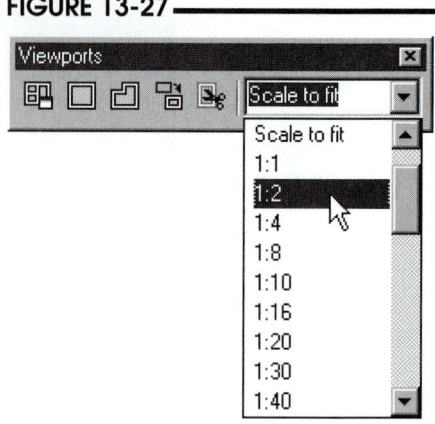

```
1/2"=1'
.5"=1'
1/2=12
1/24
1:24
1=24
.0417
```

In addition, you can edit the contents of the scale list by adding, deleting, or rearranging the entries. See *Scalelistedit* near the end of this chapter.

Remember that the viewport scale is actually the desired plot scale for the geometry in the viewport, which is the reciprocal of the "drawing scale factor" (see Chapter 12).

Properties

The *Properties* palette can also be used to specify the scale for the display of model space geometry in a viewport. Do this by first double-clicking in paper space or issuing the *Pspace* command, then invoking *Properties*. When (or before) the dialog box appears, select the viewport object (border). Ensure the word "Viewport" appears in the top of the box (Fig. 13-28).

FIGURE 13-28

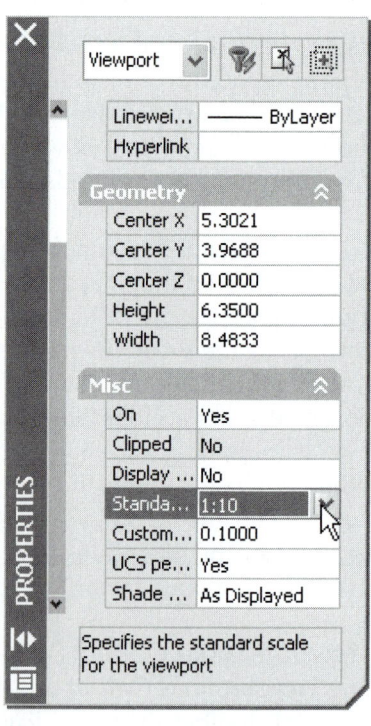

Under *Standard Scale*, select the desired scale from the drop-down list. This method offers only standard scales. Alternately, you can enter any values in the *Custom Scale* box. (See *Properties*, Chapter 16.) Similar to using the *Viewport Scale Control* edit box, you can enter values as a decimal or fraction and use a colon symbol or an equal symbol. Feet and inch unit symbols can also be entered. You can also *Lock* the viewport scale using the *Display Locked* option in the palette. (See *Properties*, Chapter 16.)

ZOOM XP Factors

To scale the display of model space geometry relative to paper space (in a viewport), you can also use the *XP* option of the *Zoom* command. *XP* means "times paper space." Thus, model space geometry is *Zoomed* to some factor "times paper space."

Since paper space is set to the actual size of the paper, the *Zoom XP* factor that should be used for a viewport is equivalent to the plot scale that would otherwise be used for plotting that geometry in model space. The *Zoom XP* factor is the reciprocal of the "drawing scale factor." *Zoom XP* only while you are "in" the desired model space viewport. Fractions or decimals are accepted.

For example, if the model space geometry would normally be plotted at 1/2"=1" or 1:2, the *Zoom* factor would be .5*XP* or 1/2*XP*. If a drawing would normally be plotted at 1/4"=1', the *Zoom* factor would be 1/48*XP*. Other examples are given in the following list.

1:5	*Zoom .2XP* or 1/5 *XP*
1:10	*Zoom .1XP* or 1/10*XP*
1:20	*Zoom .05XP* or 1/20*XP*
1/2"=1"	*Zoom 1/2XP*
3/8"=1"	*Zoom 3/8XP*
1/4"=1"	*Zoom 1/4XP*
1/8"=1"	*Zoom 1/8XP*
3"=1'	*Zoom 1/4XP*
1"=1'	*Zoom 1/12XP*
3/4'=1'	*Zoom 1/16XP*
1/2"=1'	*Zoom 1/24XP*
3/8"=1'	*Zoom 1/32XP*
1/4"=1'	*Zoom 1/48XP*
1/8"=1'	*Zoom 1/96XP*

Refer to the "Tables of Limits Settings," Chapter 14, for other plot scale factors.

Locking Viewport Geometry

Once the viewport scale has been set to yield the correct plot scale, you should lock the scale of the viewport so it is not accidentally changed when you *Zoom* or *Pan* inside the viewport. If viewport lock is on and you use *Zoom* or *Pan* when the viewport is active, AutoCAD automatically switches to paper space (*Pspace*) for zooming and panning.

You can lock the viewport scale by the following methods.

1. Use the *Lock* option of the *-Vports* command (see *-Vports*).
2. Use the *Display Locked* setting in the *Properties* palette (see Fig. 13-28).
3. Use the *Display Locked* option in the right-click shortcut edit menu that appears when a viewport object is selected (see Fig. 13-24).

Linetype Scale in Viewports—*PSLTSCALE*

The *LTSCALE* (linetype scale) setting controls how hidden, center, dashed, and other non-continuous lines appear in the *Model* tab. *LTSCALE* can be set to any value. The *PSLTSCALE* (paper space linetype scale) setting determines if those lines appear and plot the same in *Layout* tabs as they do in the *Model* tab. *PSLTSCALE* can be set to 1 or 0 (on or off).

If *PSLTSCALE* is set to 0, the non-continuous line scaling that appears in *Layout* tabs (and in viewports) is the same as that in the *Model* tab. In this way, the linetype spacing always looks the same relative to model space units, no matter whether you view it from the *Model* tab or from a viewport in a *Layout*. For example, if in the *Model* tab a particular center line shows one short dash and two long dashes, it will look the same when viewed in a viewport in a layout—one short dash and two long dashes. When *PSLTSCALE* is 0, linetype scaling is controlled exclusively by the *LTSCALE* setting. A *PSLTSCALE* setting of 0 is recommended for most drawings unless they contain multiple viewports or layouts.

If *PSLTSCALE* is set to 1 (the default setting for all AutoCAD template drawings), the linetype scale for non-continuous lines in viewports is automatically changed relative to the viewport scale; therefore, the scale of those lines in *Layout* tabs can appear differently than they do in the *Model* tab. <u>A *PSLTSCALE* setting of 1 is recommended for drawings that contain multiple viewports or layouts, especially when they are at different scales</u>. In such a situation, the line dashes would appear the same size in different viewports relative to paper space, even though the viewport scales were different. Technically speaking, if *PSLTSCALE* is set to 1, *LTSCALE* controls linetype scaling globally for the drawing, but lines that appear in viewports are automatically scaled to the *LTSCALE* times the viewport scale.

Until you gain more experience with layouts and viewports, and in particular creating multiple viewports, a *PSLTSCALE* setting of 0 is the simplest strategy. This setting also agrees with the strategy used in this text discussed earlier (in Chapter 6 and Chapter 12) for drawing setup. You can change the default *PSLTSCALE* setting of 1 to 0 by typing *PSLTSCALE* at the Command prompt.

Chapter 33, Advanced Layouts and Plotting, gives a more detailed discussion of *PSLTSCALE*, creating multiple viewports and layouts, and other drawing setup strategies.

Vpmax

Pull-down Menu	Command (Type)	Alias (Type)	Short-cut	Screen (side) Menu	Tablet Menu
...	*Vpmax*	...	...	...	...

You can invoke the *Vpmax* command by typing or using the button located just to the right of the Status Bar <u>only when a Layout tab is active</u> (Fig. 13-29). When a *Layout* tab is active, *Vpmax* maximizes the viewport to enable you to draw and edit inside a viewport (in model space). *Vpmax* is intended to be used in conjunction with *Vpmin* (see *Vpmin* next). This enhanced method is used instead of using the *Paper/Model* toggle on the Status Bar when you to want toggle between paper space and model space as well as maximize your drawing area.

FIGURE 13-29

For example, assume you are working in a layout (such as in Fig. 13-30) but realize that you need to edit some geometry inside a viewport. Use the *Vpmax* command to best prepare for making the edits in model space. *Vpmax* accomplishes two actions: 1) it activates model space inside the viewport, and 2) it removes all paper space area from the screen to make drawing and editing in model space easier.

FIGURE 13-30

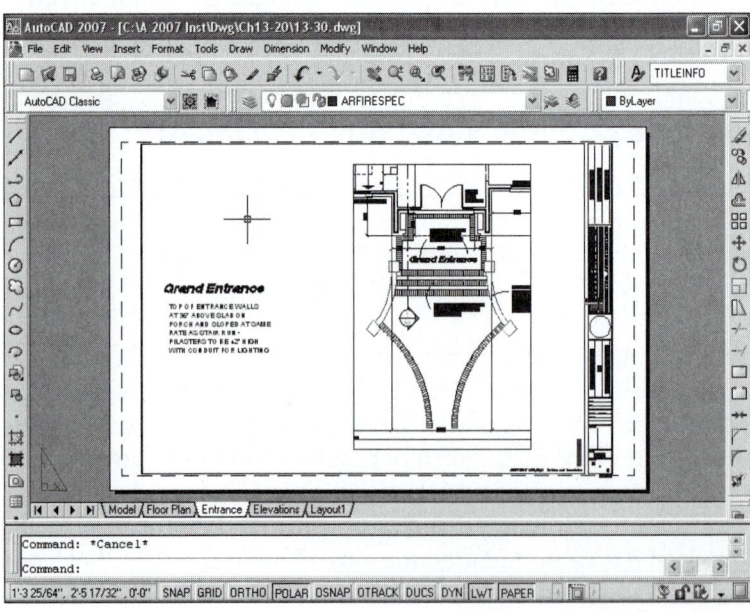

Figure 13-31 shows our example after using *Vpmax*. Model space (inside the viewport) is active and the drawing area has been maximized in the Drawing Editor. Note that *Vpmax* does not *Zoom*, it only displays more area of model space. The size of the model space geometry is the same as before using *Vpmax*. You can, however, use *Zoom* after using *Vpmax* to enlarge the model geometry display. Using *Vpmin* restores the original layout display, even if you used *Zoom* after *Vpmax*.

FIGURE 13-31

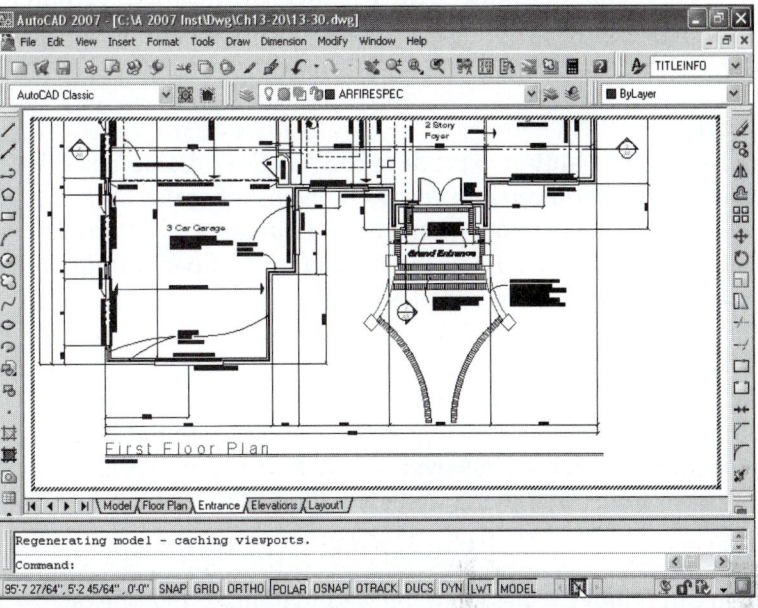

Vpmin

Pull-down Menu	Command (Type)	Alias (Type)	Short-cut	Screen (side) Menu	Tablet Menu
...	*Vpmin*	...	...	...	...

The Vpmin button appears at the same location as the *Vpmax* button, just to the right of the Status Bar (Fig. 13-32). *Vpmin* would be used after finishing the needed changes to the drawing to restore the original layout display (to that before using *Vpmax*). For the example used previously, using *Vpmin* would restore the display to that shown in Figure 13-30.

FIGURE 13-32

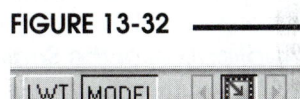

Scalelistedit

Pull-down Menu	Command (Type)	Alias (Type)	Short-cut	Screen (side) Menu	Tablet Menu
...	*Scalelistedit*	...	...	...	...

To set the scale for the display of geometry in a viewport, use the *Viewport* toolbar scale drop-down list as described in "Scaling the Display of Viewport Geometry" and illustrated in Figure 13-27. You can use the *Scalelistedit* command to control the contents of the list of scales. *Scalelistedit* produces the *Edit Scale List* dialog box (Fig. 13-33). You can add new scales, edit existing scales, rearrange the list, and delete unused scales. For example, if you were working in an industry where you typically used only a few scales, you may want to use *Scalelistedit* to move those choices to the top of the list. You can restore the original (default) list at any time.

FIGURE 13-33

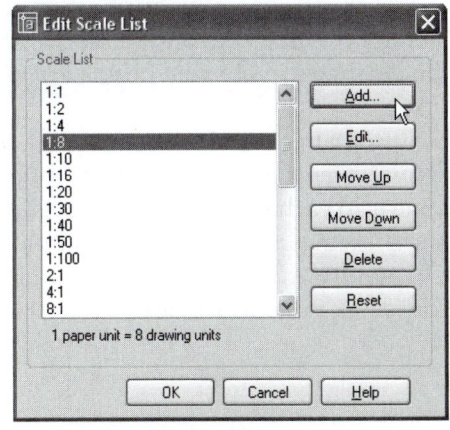

Move Up and Move Down

You can highlight a scale in the list, then use the *Move Up* and *Move Down* buttons to rearrange the list. Only one scale at a time can be moved using these buttons-selecting multiple scales disables the *Move Up* and *Move Down* options.

Add

Select the *Add* button to produce the *Add Scale* dialog box (Fig. 13-34) where you can add entries to the list of scales. The *Name appearing in scale list* can contain numbers or letters, as shown in the figure. Enter the desired ratio of paper units to drawing units in the edit boxes below.

Edit

Selecting the *Edit* button in the *Edit Scale List* dialog box (see Fig. 13-33) produces a dialog box (not shown) similar to the *Add Scale* dialog box; however, you can change only the ratio of paper units to drawing units, not the scale name.

FIGURE 13-34

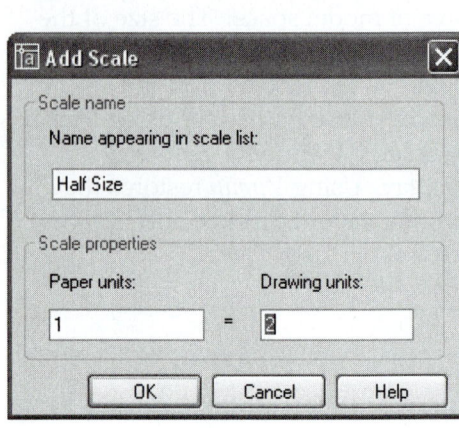

Delete

You can remove scales from the list by highlighting one or more scales (select multiple scales by holding down the Ctrl or Shift keys while selecting), then pressing Delete.

Reset

The *Reset* option removes all the customized scales and restores the default list of scales. In this case you will <u>lose all the scales you added, moved, or otherwise edited</u>. A warning appears requiring your confirmation to reset the list to the defaults.

Application of the Scale List

Although this chapter has been primarily concerned with the scale list that appears in the *Viewport* toolbar drop-down list, editing the scale list using the *Scalelistedit* command affects all the scale lists that appear in AutoCAD. The *Scalelistedit* command affects the following lists.

Viewport toolbar drop-down list
Plot dialog box
Page Setup dialog box
Properties palette
Layout Wizard
Sheet Set Manager

Advanced Applications of Paper Space Viewports

In this chapter the basic applications of layouts and viewports are discussed. The basic procedures include creating a layout, setting up the layout for printing or plotting, creating viewports, and scaling the geometry in the viewports to plot to scale.

Layouts can be used for many other advanced applications such as setting up multiple viewports to display views of one drawing at different scales or for different plot devices, controlling what layers appear in each of several viewports, applying Plot Style Tables, displaying multiple drawings (*Xrefs*) in one drawing, and displaying and plotting several views of a 3D model. These topics are discussed in Chapter 33, Advanced Layouts and Plotting; Chapter 30, Xreferences; and Chapter 42, Creating 2D Drawings from 3D Models.

CHAPTER EXERCISES

1. *Pagesetup, Vports*

A. *Open* the **A-METRIC.DWT** template drawing that you created in the Chapter 12 Exercises. Activate a *Layout* tab. If your *Options* are set as explained in Chapter 6, the layout is automatically set up for your default print/plot device. Otherwise, use the *Page Setup Manager* and select the *Modify* button. Then use the *Page Setup* dialog box to set the desired print/plot device and to select an 11 x 8.5 paper size.

B. If a viewport already exists in the layout, *Erase* it. Make **VPORTS** the *Current* layer. Then use the *Vports* command and select a *Single* viewport and accept the default (*Fit*) option to create one viewport at the maximum size for the printable area. Ensure that *PSLTSCALE* is set to 0 by typing it at the Command prompt. Finally, make the *Model* tab active. *Save* and *Close* the drawing.

2. *Pagesetup, Vports*

A. *Open* the **BARGUIDE** drawing that you set up in the Chapter 12 Exercises. Activate a *Layout* tab. If your *Options* are set as explained in Chapter 6, the layout is automatically set up for your default print/plot device. Otherwise, use the *Page Setup Manager* and select the *Modify* button. Then use the *Page Setup* dialog box that appears to set the desired print/plot device and to select an **11 x 8.5** paper size.

B. If a viewport already exists in the layout, *Erase* it. Make **VPORTS** the *Current* layer. Then use the *Vports* command and select a *Single* viewport and accept the default (*Fit*) option to create one viewport at the maximum size for the printable area. Type *PSLTSCALE* and set the value to **0**. Finally, make the *Model* tab active. *Save* and *Close* the drawing.

3. **Create a Layout and Viewport for the ASHEET Template Drawing**

A. *Open* the **ASHEET.DWT** template drawing you created in the Chapter 12 Exercises. Activate a *Layout* tab. Unless already set up, use *Pagesetup* to set the desired print/plot device and to select an **11 x 8.5** paper size.

B. If a viewport already exists in the layout, *Erase* it. Make **VPORTS** the *Current* layer. Then use the *Vports* command and select a *Single* viewport and accept the default (*Fit*) option to create one viewport at the maximum size for the printable area. Ensure that *PSLTSCALE* is set to 0. Finally, make the *Model* tab active. *Save* and *Close* the ASHEET template drawing.

4. **Create a New BSHEET Template Drawing**
    Complete this exercise if your system is configured with a "B" size printer or plotter.

    A. Using the template drawing in the previous exercise, create a template for a standard engineering "B" size sheet. First, use the *New* command and select the *Template* option from the *Create New Drawing* dialog box. Select the *Browse* button. When the *Select a Template File* dialog box appears, select the **ASHEET.DWT.** When the drawing opens, set the model space *Limits* to **17** x **11** and *Zoom All*.

    B. Activate the *Layout 1* tab. *Erase* the existing viewport. Activate the *Page Setup Manager* and select the *Modify* button. In the *Page Setup* dialog box select a print or plot device that can use a "B" size sheet (17 x 11). In the *Paper Size* drop-down list, select *ANSI B (11 x 17 inches)*. Close the dialog box and the *Page Setup Manager*.

    C. Make **VPORTS** the *Current* layer. Then use the *Vports* command and select a *Single* viewport and accept the default (*Fit*) option to create one viewport at the maximum size for the printable area. Finally, make the *Model* tab active. All other settings and layers are okay as they are. Use *Saveas* and save the new drawing as a template (.**DWT**) drawing in your working directory. Assign the name **BSHEET.**

5. **Create Multiple Layouts for Plotting on C and D Size Sheets**
    Complete this exercise if your system is configured with a "C" and "D" sized plotter.

    A. Using the ASHEET.DWT template drawing from an earlier exercise, create a template for standard engineering "C" and "D" size sheets. First, use the *New* command and select the *Template* option from the *Create New Drawing* dialog box. Select the *Browse* button. When the *Select a Template File* dialog box appears, select the **ASHEET.DWT.** When the drawing opens, set the model space *Limits* to **34** x **22** and *Zoom All*.

    B. Activate the *Layout1* tab. *Erase* the existing viewport. Activate the *Page Setup Manager* and select the *Modify* button. In the *Page Setup* dialog box select a plot device that can use a "C" size sheet (22 x 17). In the *Paper Size* drop-down list, select *ANSI C (22 x 17 inches)*. Close the dialog box and the *Page Setup Manager*.

    C. Make **VPORTS** the *Current* layer. Then use the *Vports* command and select a *Single* viewport and use the *Fit* option to create one viewport at the maximum size for the printable area. Right-click on the *Layout 1* tab and select *Rename* from the shortcut menu. Rename the layout to "**C Sheet**."

    D. Activate the *Layout2* tab. *Erase* the existing viewport if one exists. Activate the *Page Setup Manager,* select the *Modify* button, then select a plot device that can use a "D" size sheet (34 x 22). (This can be the same device you selected in step B, as long as it supports a D size sheet.) In the *Paper Size* drop-down list, select *ANSI D (34 x 22 inches)*. Close this dialog box and the *Page Setup Manager*.

    E. Make **VPORTS** the *Current* layer. Then use the *Vports* command and select *Single* to create one viewport. *Rename* the layout "**D Sheet**."

    F. Finally, make the *Model* tab active. All other settings and layers are okay as they are. Use *Saveas* and save the new drawing as a template (.**DWT**) drawing in your working directory. Assign the name **C-D-SHEET.**

6. *Layout Wizard*

In this exercise, you will open an existing drawing and use the *Layout Wizard* to set up a layout for plotting.

A. *Open* the **DB_SAMP** drawing that is located in the AutoCAD 2007\Sample folder. Use the *Saveas* command and assign the name **VP_SAMP** and specify your working folder as the location to save.

B. Create a new layer named **VPORTS** and make it the *Current* layer. Invoke the *Layout Wizard*. In the *Begin* step, enter a name for the new layout, such as **Layout 3**. Select *Next*.

C. In the *Printer* step, the list of available printers in your lab or office appears. The default printer should be highlighted. Choose any device that you know will operate for your lab or office and select *Next*. If you are unsure of which devices are usable, keep the default selection and select *Next*.

D. For *Paper Size*, select the largest paper size supported by your device. Ensure **Inches** is selected unless you are using a standard metric sheet. Select *Next*.

E. Select **Landscape** for the *Orientation*.

F. Select a *Title Block* that matches the paper size you specified. If you are unsure of which to choose, refer to the Layout Templates table in the chapter. Insert the title block as a **Block**. Select *Next*.

G. In the *Define Viewports* step, select a **Single** viewport. Under *Viewport Scale*, select **Scaled to Fit**. Proceed to the next step.

H. Bypass the next step in the wizard, *Pick Location*, by selecting **Next**. Also, when the *Finish* step appears, select **Next**. This action causes AutoCAD to create a viewport that fits the extents of the printable area. The resulting layout should look similar to that shown in Figure 13-35. As you can see, the model space geometry extends beyond the title block. Your layout may appear slightly different depending on the sheet size and title block you selected. *Save* the drawing.

FIGURE 13-35

I. Use the *Mspace* command or double-click inside the viewport to activate model space. To specify a standard scale for the geometry in the viewport, invoke the *Viewports* toolbar. In the *Viewport Scale Control* drop-down list, select a scale (such as *1/64"=1'*) until the model geometry appears completely within the title block border (Fig. 13-36).

FIGURE 13-36

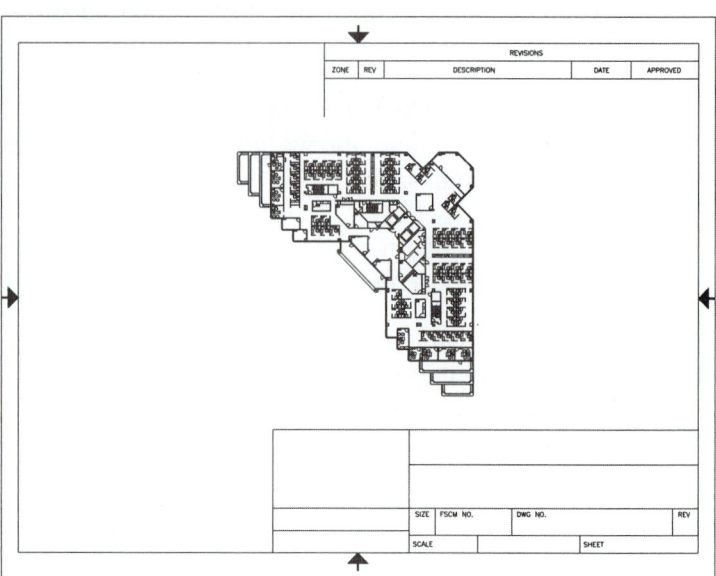

J. *Freeze* layer **VPORTS** so the viewport border does not appear. *Save* the drawing. With the layout active, access the *Plot* dialog box and ensure the *Plot Scale* is set to **1:1**. Select *Plot*.

7. **Page Setup, Vports, and Zoom XP**

In this exercise, you will use an existing metric drawing, access *Layout1*, draw a border and title block, create one viewport, and use *Zoom XP* to scale the model geometry to paper space. Finally, you will *Plot* the drawing to scale on an "A" size sheet.

A. *Open* the **CBRACKET** drawing you worked on in Chapter 9 Exercises. If you have already created a title block and border, *Erase* them.

B. Invoke the *Options* dialog box and access the *Display* tab. In the *Layout Elements* section (lower left), ensure that *Show Page Setup Manager for New Layouts* is checked and *Create Viewport in New Layouts* is not checked.

C. Select the *Layout1* tab. The *Page Setup Manager* should appear. Select the *Modify* button. In the *Page Setup* dialog box select a print or plot device that will allow you to print or plot on an "A" size sheet (11 x 8.5). In the *Paper Size* drop-down list, select the correct sheet size, then select *Landscape* orientation. Select *OK* in the *Page Setup* dialog box and *Close* the *Page Setup Manager*.

D. Set layer **TITLE** *Current*. Draw a title block and border (in paper space). HINT: Use *Pline* with a *width* of .02, and draw the border with a .5 unit margin within the edge of the *Limits* (border size of 10 x 7.5). Provide spaces in the title block for your school or company name, your name, part name, date, and scale as shown in the following figure. Use *Saveas* and assign the name **CBRACKET-PS**.

E. Create layer **VPORTS** and make it *Current*. Use the *Vports* command to create a *Single* viewport. Pick diagonal corners for the viewport at **.5,1.5** and **9.5,7**. The CBRACKET drawing should appear in the viewport at no particular scale, similar to Figure 13-37.

FIGURE 13-37

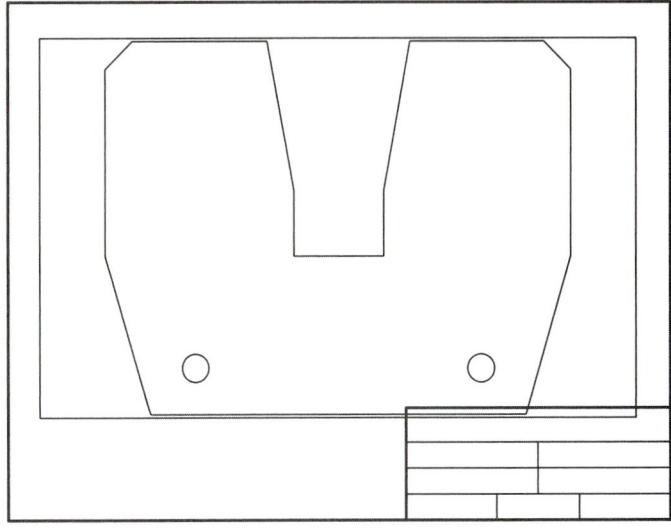

F. Next, use the *Mspace* command, the *MODEL/PAPER* toggle on the Status Bar, or double-click inside the viewport to bring the cursor into model space. Use *Zoom* with a *.03937XP* factor to scale the model space geometry to paper space. (The conversion from inches to millimeters is 25.4 and from millimeters to inches is .03937. Model space units are millimeters and paper space units are inches; therefore, enter a *Zoom XP* factor of .03937.)

G. Finally, activate paper space by using the *Pspace* command, the *MODEL/PAPER* toggle, or double-clicking outside the viewport. Use the *Layer Properties Manager* to make layer **VPORTS** *non-plottable*. Your completed drawing should look like that in Figure 13-38. *Save* the drawing. *Plot* the drawing from the layout at **1:1** scale.

FIGURE 13-38

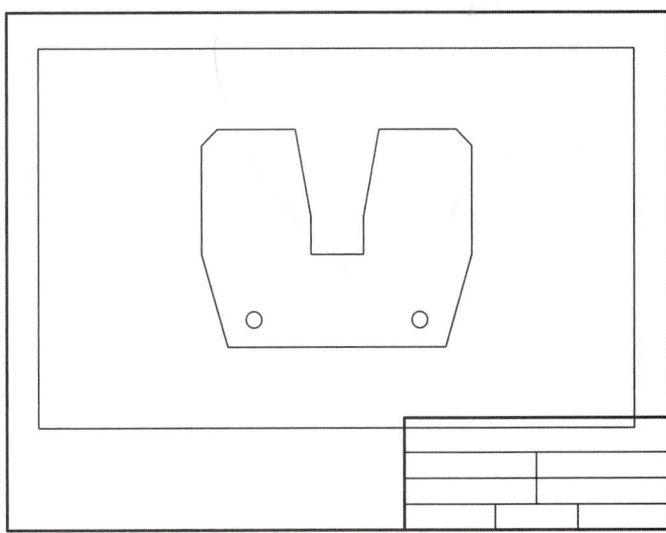

8. **Using a Layout Template**

This exercise gives you experience using a layout template already set up with a title block and border. You will draw the hammer (see Fig. 13-39) and use a layout template to plot on a B size sheet. It is necessary to have a plot or print device configured for your system that can use a "B" size sheet.

A. Start AutoCAD. In the *Options* dialog box, select the *System* tab. In the *General Options* section, select *Show Startup Dialog Box*.

B. Now access the *Display* tab in the *Options* dialog box. In the *Layout Elements* section (lower left), remove the check for both *Show Page Setup Manager for New Layouts* and *Create Viewport in New Layouts*. Select *OK*.

C. Begin a *New* drawing. In the *Create New Drawing* dialog box that appears, select *Use a Wizard* and select the *Advanced Setup Wizard*. In the first step, *Units*, select *Decimal* units and a *Precision* of **0.00**. Accept the defaults for the next three steps. In the *Area* step, enter a *Width* of **17** and a *Length* of **11**.

D. Right-click on an existing layout tab and use the *From Template* option from the shortcut menu. Select the *ANSI B -Color Dependent Plot Styles* template drawing (.DWT) in the *Select Template From File* dialog box, then select the *ANSI B Title Block* in the *Insert Layout(s)* dialog box. Access the new tab. Note that the template loads a layout with a title block and border in paper space and has one viewport created. Access the *Page Setup Manager* box and select the matching plot device and sheet size for the selected ANSI B title block.

E. Pick the *Model* tab. Set up appropriate *Snap* and *Grid* (for the *Limits* of 17 x 11) if desired. Also set *LTSCALE* and any running *Osnaps* you want to use. Type *PSLTSCALE* and set the value to **0**.

F. Create the following layers and assign linetypes and line weights. Assign colors of your choice.

    GEOMETRY   *Continuous*
    CENTER   *Center2*

Notice the *Title Block* and *Viewport* layers already exist as part of the template drawing.

G. Draw the hammer according to the dimensions given in Figure 13-39. *Save the drawing as* **HAMMER.**

FIGURE 13-39

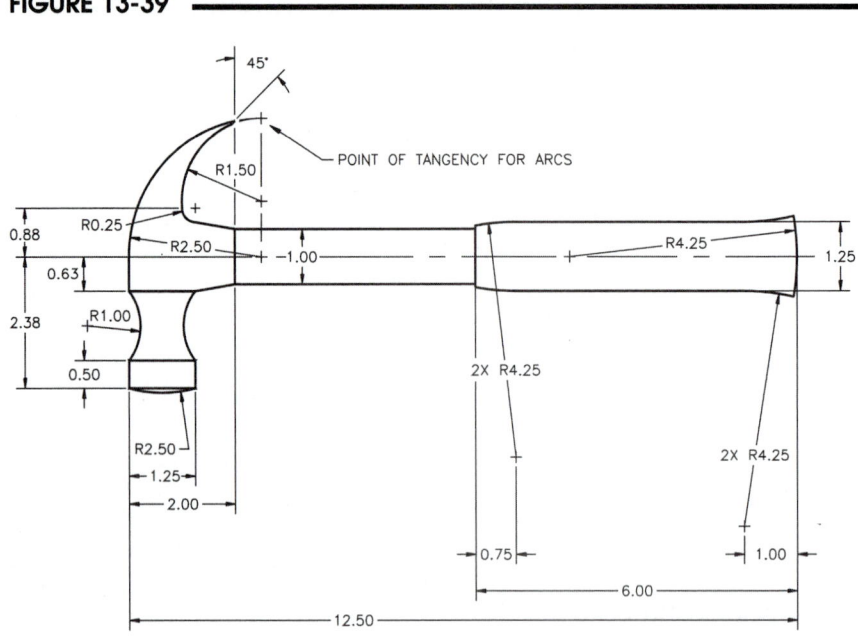

H. When you are finished with the drawing, select the *ANSI B Title Block* tab. Now, scale the hammer to plot to a standard scale. You can use either the *Viewports* toolbar or the *Properties* palette to set the viewport scale to **1:1**. Alternately, use *Zoom 1XP* to correctly size model space units to paper space units.

I. Your completed drawing should look like that in Figure 13-40. Use the *Layer Properties Manager* to make layer **Viewports** *non-plottable. Save* the drawing. Make a *Plot* from the layout at a scale of **1:1**.

FIGURE 13-40

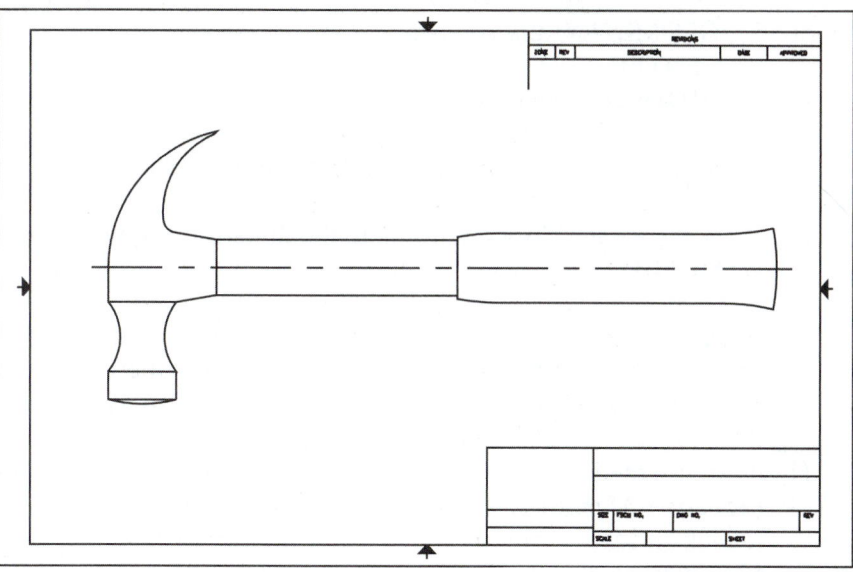

9. **Using a Template Drawing**

This exercise gives you experience using a template drawing that includes a title block and border. You will construct the wedge block in Figure 13-41 and plot on a B size sheet. It is necessary to have a plot or print device configured for your system that can use a "B" size sheet.

FIGURE 13-41

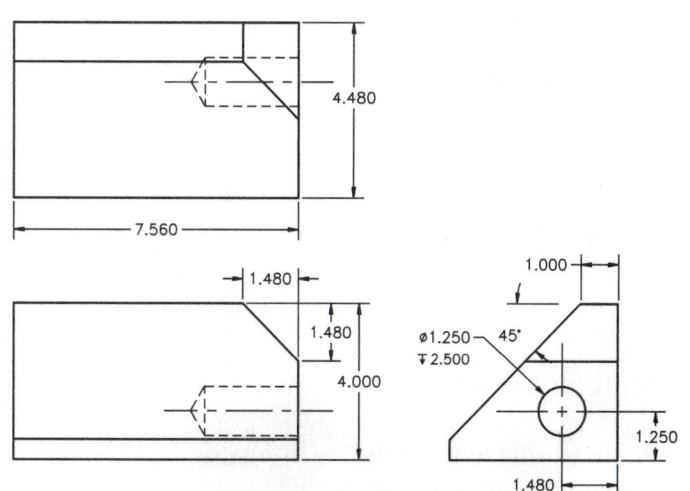

A. Complete steps 8.A. and 8.B. from the previous exercise.

B. Begin a *New* drawing. In the *Create New Drawing* dialog box that appears, select *Use a Template*. Select *ANSI B -Color Dependent Plot Styles.Dwt.*

C. Select the *ANSI B Title Block* tab. Then invoke the *Page Setup Manager* box and select a matching plot or print device and sheet size for the ANSI B title block.

D. Pick the *Model* tab. Set up model space with *Limits* of **17** x **11**. Set up appropriate *Snap* and *Grid* if desired. Also set *LTSCALE* and any running *Osnaps* you want to use. Type *PSLTSCALE* and set the value to **0**.

E. Create the following layers and assign linetypes and line weights. Assign colors of your choice.

GEOMETRY *Continuous*
CENTER *Center2*
HIDDEN *Hidden2*

Notice the *Title Block* and *Viewport* layers already exist as part of the template drawing.

F. Draw the wedge block according to the dimensions given (Fig. 13-41). *Save* the drawing as **WEDGEBK**.

G. When you are finished with the views, select the *ANSI B Title Block* tab. To scale the model space geometry to paper space, you can use either the *Viewports* toolbar or the *Properties* palette to set the viewport scale to **1:2.** You may also need to use *Pan* to center the geometry.

H. Next, use the *Layer Properties Manager* to make layer **Viewports** *non-plot-table*. Your completed drawing should look like that in Figure 13-42. *Save* the drawing. From paper space, plot the drawing on a B size sheet and enter a scale of **1:1**

FIGURE 13-42

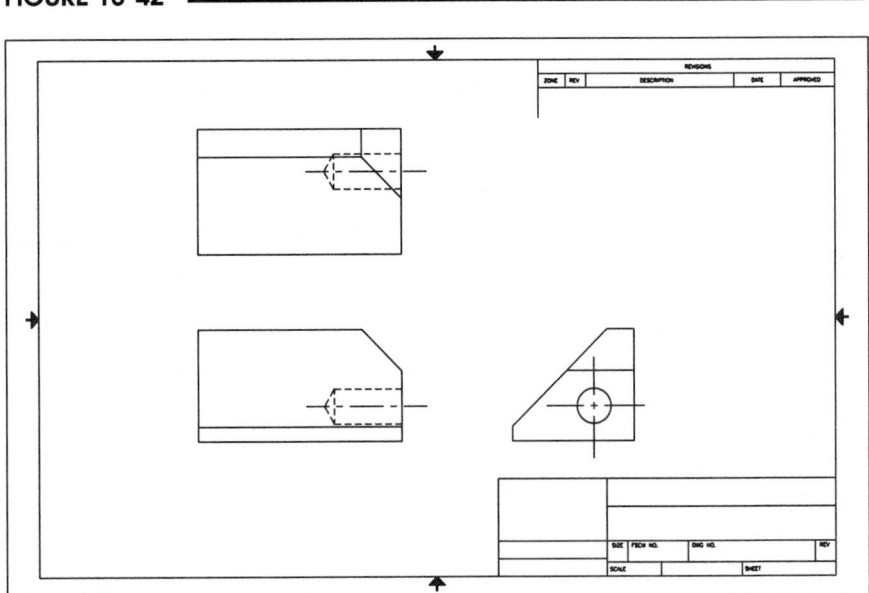

10. **Multiple Layouts**

Assume a client requests that a copy of the HAMMER drawing be faxed to him immediately; however, your fax machine cannot accommodate the B size sheet that contains the plotted drawing. In this exercise you will create a second layout using a layout template to plot the same model geometry on an A size sheet.

A. Open the drawing you created in a previous exercise named **HAMMER**. Access the *Display* tab in the *Options* dialog box. In the *Layout Elements* section (lower left), remove the check for both *Show Page Setup Manager for New Layouts* and *Create Viewport in New Layouts*. Select *OK*.

B. Right-click on an existing layout tab and use the *From Template* option from the short-cut menu. Select the *ANSI A -Color Dependent Plot Styles* template drawing (.DWT) to import. When the layout is imported, click the new *ANSI A Title Block* tab. Note that the template loads a layout with a title block and border in paper space and has one viewport created. Access the *Page Setup Manager* and select the matching print or plot device and sheet size for the selected ANSI A title block.

C. Double-click inside the viewport or use *Mspace* or the *MODEL/PAPER* toggle. Now scale the hammer to plot to a standard scale by using *Zoom 3/4XP* or enter **.75** in the appropriate edit box in the *Viewports* toolbar or *Properties* palette to correctly scale the model space geometry. Use *Pan* in the viewport to center the geometry within the viewport.

D. Your new layout should look like that in Figure 13-43. Access the *ANSI B Title Block* tab you created earlier. Note that you now have two layouts, each layout has settings saved to plot the same geometry using different plot devices and at different scales.

FIGURE 13-43

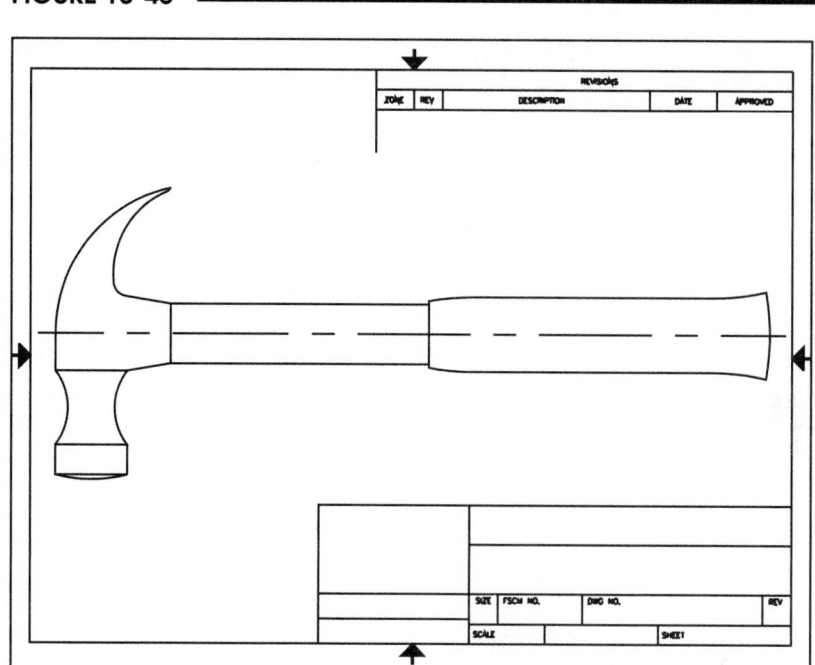

E. Access the new layout tab. *Save* the drawing. Make a *Plot* from the layout at a scale of **1:1** and send the fax.

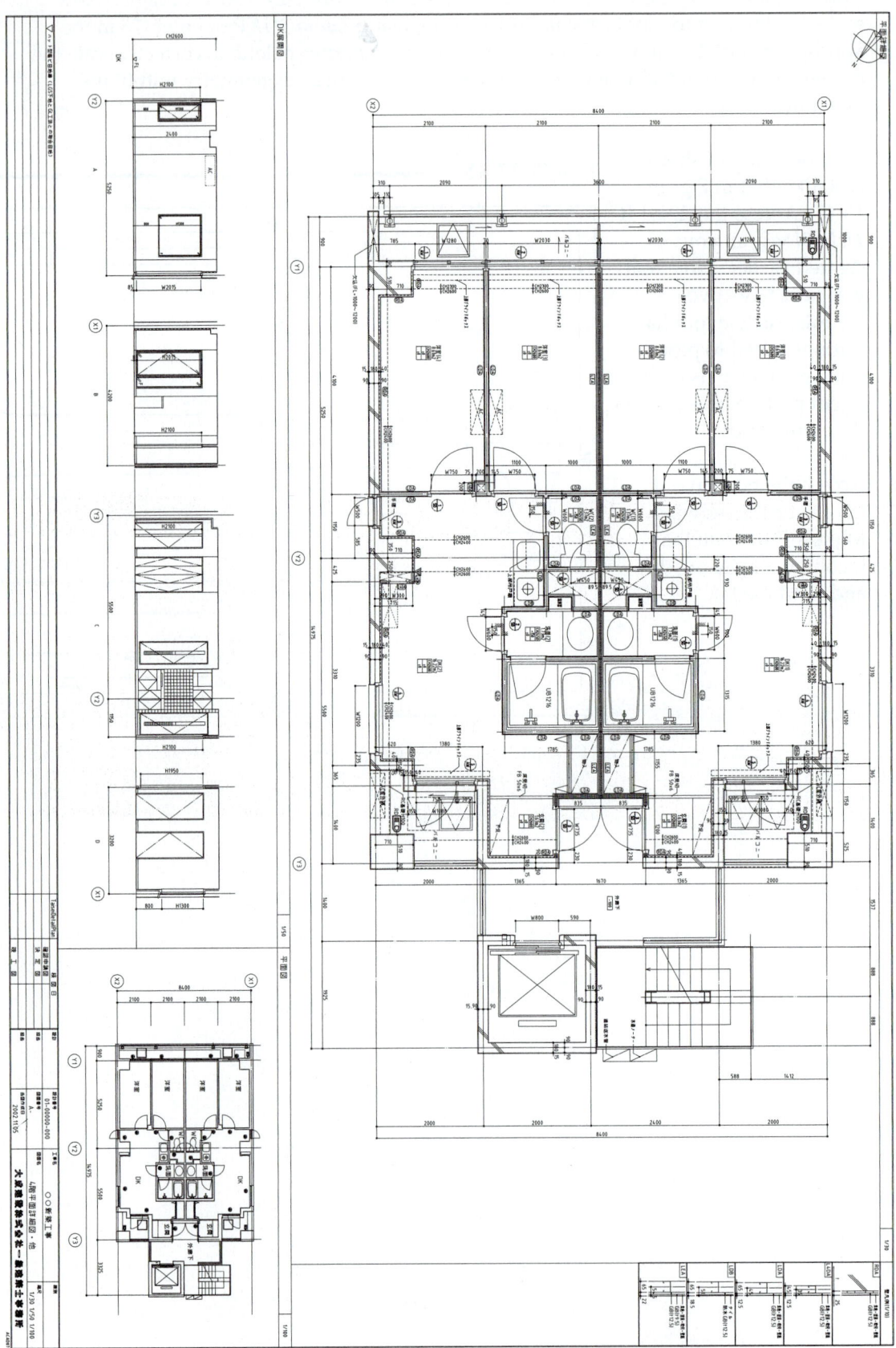

PRINTING AND PLOTTING

CHAPTER OBJECTIVES

After completing this chapter you should:

1. know the typical steps for printing or plotting;

2. be able to invoke and use the *Plot* dialog box and *Page Setup Manager*;

3. be able to select from available plotting devices and set the paper size and orientation;

4. be able to specify what area of the drawing you want to plot;

5. be able to preview the plot before creating a plotted drawing;

6. be able to specify a scale for plotting a drawing;

7. know how to set up a drawing for plotting to a standard scale on a standard size sheet;

8. be able to use the Tables of *Limits* Settings to determine *Limits*, scale, and paper size settings;

9. know how to configure plot and print devices in AutoCAD.

CONCEPTS

Several concepts, procedures, and tools in AutoCAD relate to printing and plotting. Many of these topics have already been discussed, such as setting up a drawing to draw true size, creating layouts and viewports, specifying the viewport scale, and specifying the plot or print options. This chapter explains the typical steps to plotting from either the *Model* tab or *Layout* tabs, using the *Plot* dialog box and the *Page Setup Manager*, configuring print and plot devices, and plotting to scale. Other advanced concepts of plotting, such as using plot styles and plot style tables, plot stamping, and advanced features of layouts, are explained in Chapter 33, Advanced Layouts and Plotting.

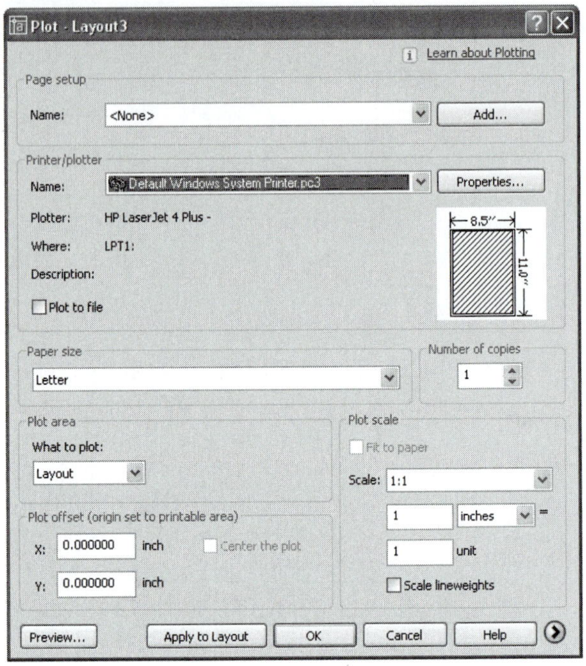

In AutoCAD, the term "plotting" can refer to plotting on a plot device (such as a pen plotter or electrostatic plotter) or printing with a printer (such as a laser jet printer). The *Plot* command is used to create a plot or print by producing the *Plot* dialog box. You can plot only by using the *Plot* command (Fig. 14-1).

FIGURE 14-1

Before making a plot, you specify the parameters of how the drawing will appear on the plot, such as what type of printer or plotter you want to use, the paper size and orientation, what area of the drawing to plot, plot scale, and other options. You can save the plot specifications for each layout. These saved plot specifications are called "page setups." In this way, each layout can have its own page setup—set up to plot or print with a specific device, paper size, orientation, scale, and so on. Therefore, one drawing can be plotted and printed in multiple ways.

Although you can create a plot only by using the *Plot* dialog box, two methods can be used to create and save the plot specifications for a layout:

1. Use the *Plot* dialog box to specify the plot settings and select the *Apply to Layout* button.
2. Use the *Page Setup Manager* and related *Page Setup* dialog box.

These two methods are almost identical since the *Plot* dialog box and the *Page Setup* dialog box have almost the same features and functions except that you cannot plot from the *Page Setup* dialog box. Therefore, for most applications, it is suggested that you use the *Plot* dialog box for saving the page setups and for plotting. This chapter explains the features and distinctions of both methods.

TYPICAL STEPS TO PLOTTING

Assuming your CAD system and plotting devices have been properly configured, the typical basic steps to plotting the *Model* tab or a *Layout* tab are listed below.

1. Use *Save* to ensure the drawing has been saved in its most recent form before plotting (just in case some problem arises while plotting).

2. Make sure the plotter or printer is turned on, has paper (and pens for some devices) loaded, and is ready to accept the plot information from the computer.

3. Invoke the *Plot* dialog box.

4. Select the intended plot device from the *Printer/Plotter* drop-down list. The list includes all devices currently configured for your system.

5. Ensure the desired *Drawing Orientation* and *Paper Size* are selected.

6. Determine and select the desired *Plot Area* for the drawing: *Layout, Limits, Extents, Display, Window,* or *View.*

7. Select the desired scale from the *Scale* drop-down list or enter a *Custom* scale. If no standard scale is needed, select *Scaled to Fit.* If you are plotting from a *Layout,* normally set the scale to 1:1.

8. If necessary, specify a *Plot Offset* or *Center the Plot* on the sheet.

9. If necessary, specify the *Plot Options,* such as plotting with lineweights or plot styles.

10. Always preview the plot to ensure the drawing will plot as you expect. Select a *Preview* to view the drawing objects as they will plot. If the preview does not display the plot as you intend, make the appropriate changes. Otherwise, needless time and media could be wasted.

11. If all settings are acceptable and you want to save the page setup (plot specifications) with the layout (recommended), select the *Apply to Layout* button.

12. If you want to make the plot at this time, select the *OK* button. If you do not want to plot, select *Cancel* and the settings will be saved (if you used *Apply to Layout*).

PRINTING, PLOTTING, AND SAVING PAGE SETUPS

Plot

Pull-down Menu	Command (Type)	Alias (Type)	Short-cut	Screen (side) Menu	Tablet Menu
File *Plot...*	*Plot* or *-Plot*	*PRINT*	*Ctrl+P*	*FILE* *Plot*	*W,25*

Using *Plot* invokes the *Plot* dialog box (see Fig. 14-2 on the next page). This dialog box has a collapsed and expanded mode controlled by the arrow button in the lower-right corner. The *Plot* dialog box allows you to set plotting parameters such as plotting device, paper size, orientation, and scale. Settings made to plotting options can be <u>saved for the *Model* tab and each *Layout* tab</u> in the drawing so they do not have to be re-entered the next time you want to make a plot.

Page Setup

The *Name* drop-down list at the top left corner of the *Plot* dialog box gives the page setups saved in the drawing. You can select a previously saved page setup from the list to apply those settings to the current layout. Although the default page setup is listed as *<None>,* it will be listed as the current tab name (such as *Layout1*) once you save it. You can also assign a specific name using the *Add* button. The list can contain saved page setups (such as *Layout1* or *Layout2*) and "named" (assigned) setups (such as *Setup1, Setup2,* or other assigned names). You can also *Import* page setups from other drawings that have assigned names.

Add

This option produces the *Add Page Setup* dialog box (not shown) where you can assign a name for the current page setup. You can accept the default assigned name (*Setup1*) or enter a name of your choice. You must assign a name if you want to *Import* the page setup into another drawing.

Printer/plotter

Many devices (printers and plotters) can be configured for use with AutoCAD. Once configured, those devices appear in the *Printer/plotter* list. For example, you may configure an "A" size plotter, a "D" size plotter, and a laser printer. Any one of those devices could then be used to plot the current drawing.

FIGURE 14-2 ————————

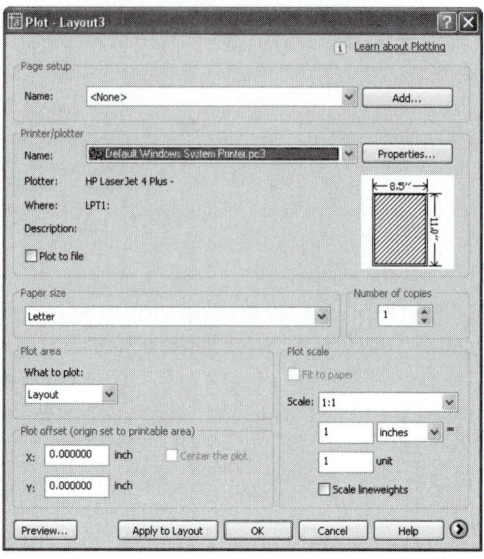
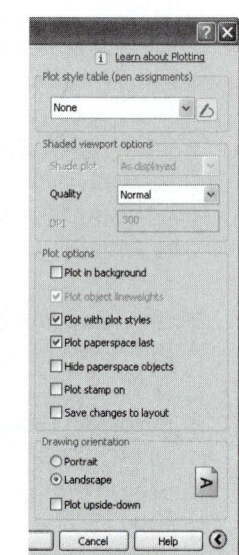

When you install AutoCAD, it automatically configures itself to use the Windows system print devices, as well as a *DWF6 ePlot* and two *PublishToWeb* devices. Additional devices must be configured within AutoCAD by using the *Plotter Manager* (see "Configuring Plotters and Printers" near the end of this chapter).

Name:
Select the desired device from the list. The list is composed of all previously configured devices for your system.

Properties...
This button invokes the *Plotter Configuration Editor*. In the editor, you can specify a variety of characteristics for the specific plot device shown in the *Name:* drop-down list. The *Plotter Configuration Editor* is also accessible through the *Plotter Manager* (see "*Plotter Configuration Editor*" near the end of this chapter).

Plot to File
Choosing this option writes the plot to a file instead of making a plot. This action generally creates a .PLT file type; however, the format of the file (for example, PCL or HP/GL language) depends on the device brand and model that is configured. When you use the *OK* button to create the plot, the *Browse for Plot File* dialog box (not shown) appears for you to specify the name and location for the plot file. The plot file can be printed or plotted later <u>without</u> AutoCAD, assuming the correct interpreter for the device is available.

Paper Size and Number of Copies

The options that appear in this list depend on the selected printer/plotter device. Normally, several paper sizes are available. Select the desired *Paper Size* from the drop-down list, then select the *Number of Copies* you want to create.

Plot Area

In the *What to Plot:* list section you can select what aspect of the drawing you want to plot, described next.

Limits or Layout
This option changes depending on whether you invoke the *Plot* dialog box while the *Model* tab or a *Layout* tab is current. If the model tab is current, this selection plots the area defined by the *Limits* command. If a *Layout* tab is current, the layout is plotted as defined by the selected paper size.

Extents

Plotting *Extents* is similar to *Zoom Extents*. This option plots the entire drawing (all objects), disregarding the *Limits*. Use the *Zoom Extents* command in the drawing to make sure the extents have been updated before using this plot option.

Display

This option plots the current display on the screen. If using viewports, it plots the current viewport display.

View

With this option you can plot a named view previously saved using the *View* command.

Window

Window allows you to plot any portion of the drawing. You must specify the window by picking or supplying coordinates for the lower-left and upper-right corners.

Plot Offset

Plotting and printing devices cannot plot or print to the edge of the paper. Therefore, the lower left corner of your drawing will not plot at exactly the lower left corner of the paper. You can choose to reposition, or offset, the drawing on the paper by entering positive or negative values in the X and Y edit boxes. Home position for plotters is the <u>lower-left</u> corner (landscape orientation) and for printers it is the <u>upper-left</u> corner of the paper (portrait orientation). If plotting from the *Model* tab, you can choose to *Center the Plot* on the paper.

Plot Scale

You can select a standard scale from the drop-down list or enter a custom scale.

Fit to Paper

The *Fit to Paper* option is disabled if a layout tab is active. If the *Model* tab is active, the *Fit to Paper* option is checked and the *Scale* section is disabled. The drawing is automatically sized to fit on the sheet based on the specified area to plot (*Display*, *Limits*, *Extents*, etc.). Changing the *inches* or *mm* specification changes only the units of measure for other areas of the *Plot* dialog box.

Scale

This section provides the scale drop-down list for you to select a specific scale for the drawing to be plotted. These options are the same as those appearing in all other scale drop-down lists (such as the *Viewport* toolbar and *Properties* palette). The entries can be edited using the *Scalelistedit* command (see Chapter 13). First ensure the correct paper units are selected (*inches* or *mm*), then select the desired scale from the list.

If you are plotting from a *Layout* tab, select 1:1 from the Scale drop-down list since you almost always want to plot paper space at 1:1 scale. (To set the scale for the viewport geometry in a layout, use the *Viewport* toolbar, *Zoom XP* factor, or *Properties* palette as explained in Chapter 13.) If you are plotting the *Model* tab, select an option directly from the *Scale* drop-down list to specify the desired plot scale for the drawing. In either case, the plot scale for the drawing is equal to the reciprocal of the drawing scale factor and can be determined by referencing the Tables of *Limits* Settings. (See "Plotting to Scale.")

Custom Edit Boxes

Instead of selecting from the drop-down list, you can enter the desired ratio in the edit boxes. Decimals or fractions can be entered. The first edit box represents paper units and the second represents drawing units. For example, to prepare for a plot of 3/8 size (3/8"=1"), the following ratios can be entered: 3=8, 3/8=1, or .375=1 (paper units =drawing units). For guidelines on plotting a drawing to scale, see "Plotting to Scale" in this chapter.

Scale Lineweights
This option scales the lineweights for plots. When checked, lineweights are scaled proportionally with the plot scale, so a 1mm lineweight, for example, would plot at .5mm when the drawing is plotted at 1:2 scale. Normally (no check in this box) lineweights are plotted at their absolute value (0.010", 0.016", 0.25 mm, etc.) at any scale the drawing is plotted.

The following options are available from the expanded mode of the *Plot* dialog box (Fig. 14-3).

FIGURE 14-3

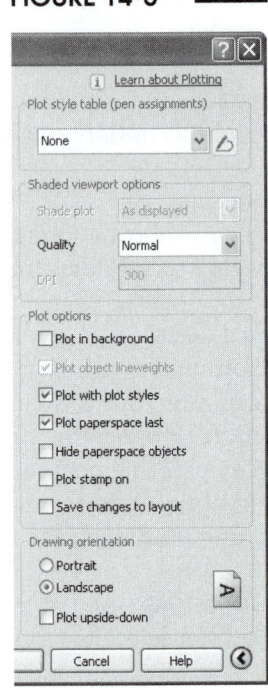

Plot Style Table (pen assignments)
If desired, select a plot style table to use with your plot device. A plot style table can contain several plot styles. Each plot style can assign specific line characteristics (lineweight, screening, dithering, end and joint types, fill patterns, etc.). Normally, a plot style table can be assigned to the *Model* tab or to *Layout* tabs. The plot style tables appearing in the list are managed by the *Plot Style Manager.* Selecting the *New* option invokes the *Add Plot Style Table* wizard. (Plot styles are explained in Chapter 33.)

Shaded Viewport Options
These options are <u>useful only if you are printing from the *Model* tab</u> and you are printing a 3D surface or solid model. These options <u>do not affect the display of 3D objects in viewports</u> of a *Layout* tab. You should have a laser jet, inkjet, or electrostatic type printer configured as the plot device (one that can print solid areas) since a pen plotter only creates line plots.

Shade Plot
Use these options if you are printing from the *Model* tab. You can shade the 3D object on the screen using any Visual Style, then select *As Displayed* in the *Shade Plot* section to make a shaded print, or you can select any option from the drop-down list for printing the drawing no matter which display is visible on screen.

Quality and DPI
You can also select from *Draft* (wireframe), *Preview* (shaded at 150 dpi), *Normal* (shaded at 300 dpi), *Presentation* (shaded at 600 dpi), or *Maximum* (for your printer) to specify the resolution quality of the print. Select *Custom* if you want to supply a different value in the *DPI* (dots per inch) edit box.

NOTE: The label for this section is misleading since it cannot be used to control printing for viewports. If you are printing from a *Layout* tab and want to print shaded images of 3D objects in viewports, you must control *Shadeplot* for each viewport using the *-Vports* command or the *Properties* palette. (See "Printing Shaded 3D Models" in Chapter 35.)

Plot Options
These checkboxes control the following options.

Plot in Background
When this box is not checked, the *Plot Job Progress* window appears while the drawing is processed to the print or plot device so you must wait some time (depending on the complexity of the drawing) before continuing to work. Checking this option allows you to return to the drawing immediately while plotting continues in the background. Plotting in the background generally requires more time to produce the plot.

NOTE: The *Plot in Background* setting is valid only for the current plot and is <u>not saved in the page setup</u>. The default setting for this box (checked or unchecked) is controlled by the *Background processing options* section in the *Plot and Publish* tab of the *Options* dialog box (see Fig. 14-7).

Plot Object Lineweights
Checking this option plots the lineweights as they are assigned in the drawing. Lineweights can be assigned to layers or to objects. Checking *Plot with Plot Styles* disables this option.

Plot with Plot Styles
You can use plot styles to override lineweights assigned in the drawing. You must have a plot style table attached (see *Plot Device Tab, plot style table*) for this option to have effect.

Plot Paperspace Last
Checking this option plots paper space geometry last. Normally, paper space geometry is plotted before model space geometry.

Hide Paperspace Objects
This option is useful <u>only if you have 3D surface or solid objects in paper space</u> and you want to remove hidden lines (edges obscured from view). 3D objects are almost always created in model space. In that case, use the *Shade Plot* options (see *Shaded Viewport Options*) if you are printing from the *Model* tab. If you are printing from a *Layout* tab, you must control *Shadeplot* for each viewport using the *-Vports* command or the *Properties* palette. (See "Printing Shaded 3D Models" in Chapter 35.)

Plot Stamp On
You have the capability of adding an informational text "stamp" on each plot. For example, the drawing name, date, and time of plot can be added to each drawing that you plot for tracking and verification purposes. To specify the plot stamp, use the *Plotstamp* command. Select this option to apply the stamp you specified to the plotted drawing. (See "Plotstamp" in Chapter 33.)

Save Changes to Layout
Check this box if you want settings you make (the page setup) to be saved with the drawing for the selected *Model* tab or *Layout* tab. Selecting this option and making a plot has the same effect as pressing the *Apply to Layout* button. If you want to save changes without plotting, use the *Apply to Layout* button.

Drawing Orientation
Landscape is the typical horizontal orientation (similar to the drawing area orientation) and *Portrait* is a vertical orientation. If plotting from a layout tab, this setting affects the orientation of the paper; therefore, set this option before creating your viewport arrangement and title block to ensure they match the paper orientation. If plotting from the *Model* tab, *Portrait* positions horizontal lines in the drawing along the short axis of the paper.

FIGURE 14-4 ━━━━━━━━

Preview
Selecting the *Preview* button displays a simulated plot sheet to ensure the drawing will be plotted as you expect (Fig. 14-4). This feature is helpful particularly when plot style tables are attached since the resulting plots may appear differently than the *Model* tab or *Layout* tabs display. The *Zoom* feature is on by default so you can check specific areas of the drawing before making the plot. Right-clicking produces a shortcut menu, allowing you to use *Pan* and other *Zoom* options. Changing the display during a *Preview* <u>does not</u> change the area of the drawing to be plotted or printed.

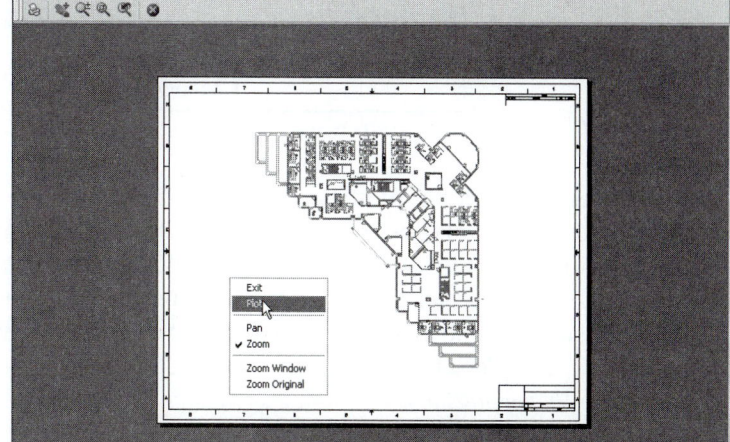

Apply to Layout

Using this button saves the current *Plot* dialog box settings to the active layout as a page setup. If no changes have been made to the default settings, this button is disabled. If this button is disabled and you want to save the current settings, assign a specific page setup name using the *Add* button. If you do not want to plot after saving the page setup, then use *Cancel*.

NOTE: Using this option (in the *Plot* dialog box) accomplishes essentially the same results as using the *Page Setup Manager* and *Page Setup* dialog box to save a page setup.

OK

Press *OK* to create the print or plot as assigned by the current settings in the *Plot* dialog box. The settings are not saved with the layout as a page setup unless the *Save changes to layout* box is checked or you used the *Apply to Layout* option.

Cancel

Cancel the plot and the current settings with this option. If you used the *Apply to Layout* option, *Cancel* will not undo the saved settings.

-Plot

Entering *-Plot* at the Command line invokes the Command line version of *Plot*. This is an alternate method for creating a plot if you want to use the current settings or if you enter a page setup name to define the plot settings.

```
Command:  -plot
Detailed plot configuration? [Yes/No] <No>:
Enter a layout name or [?] <ANSI E Title Block>:
Enter a page setup name <>:
Enter an output device name or [?] <HP 7586B.pc3>:
Write the plot to a file [Yes/No] <N>:
Save changes to page setup [Yes/No]? <N>
Proceed with plot [Yes/No] <Y>:
Effective plotting area:  33.54 wide by 42.60 high
Plotting viewport 2.
Command:
```

Viewplotdetails

	Pull-down Menu	Command (Type)	Alias (Type)	Short-cut	Screen (side) Menu	Tablet Menu
	...	*Viewplotdetails*	...	...	...	...

Once you begin a printing, plotting, or publishing job in an AutoCAD session, a "plotter" icon and an alert bubble appear (default setting) in the lower-right corner of the AutoCAD window (Fig. 14-5). Select the "X" in the upper-right corner of the bubble to dismiss it. However, to view the plot details, you can select *Click to view plot and publish details*, right-click on the plotter icon, or type *Viewplotdetails*.

FIGURE 14-5

Any one of the three actions described above produces the *Plot and Publish Details* dialog box (Fig. 14-6). Complete details about the plot or print job, including any errors encountered, are displayed on the dialog box.

FIGURE 14-6

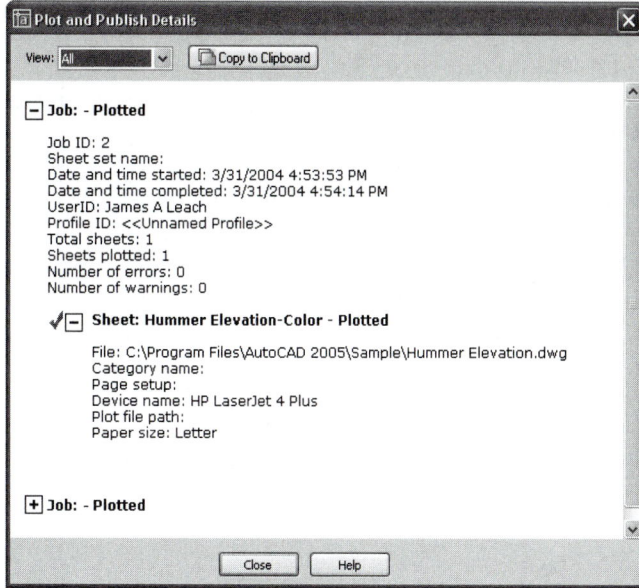

In addition to the plot details, AutoCAD (by default) creates a plot log named *PlotandPublishLog.CSV*. This is a spread sheet format file listing details about the plots such as start time and date, completion time and date, and final status. You can choose to create one continuous or multiple plot logs. The default settings are shown in the lower-left corner of the *Plot and Publish* tab of the *Options* dialog box (Fig. 14-7). You can find or change the location of the plot log file using the *Files* tab of the *Options* dialog box (not shown).

FIGURE 14-7

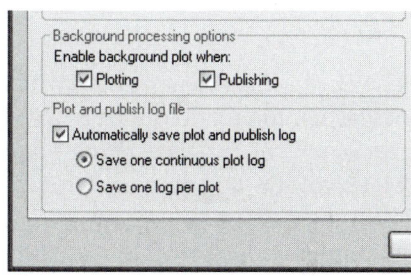

Note that this section of the *Plot and Publish* tab of the *Options* dialog box also contains the control for *Background processing options*. See the *Plot* command, *Plot in Background* option in this chapter for more information.

Pagesetup

Pull-down Menu	Command (Type)	Alias (Type)	Short-cut	Screen (side) Menu	Tablet Menu
File *Page Setup...*	*Pagesetup*	...	...	...	*V,25*

As you already know, AutoCAD allows you to set and save plotting specifications, called page setups, for each layout. In this way, one drawing can contain several layouts, each layout can display the same or different aspects of the drawing, and each layout can be set up to print or plot with the same or different plot devices, paper sizes, scales, plot styles, and so on. Oddly enough, you can create and manage page setups using either the *Plot* dialog box or the *Page Setup Manager* and related *Page Setup* dialog box. The main difference is that <u>you cannot plot from the *Page Setup Manager* or *Page Setup* dialog box</u>. See the following discussion, "Should you use the *Plot* dialog box or the *Page Setup Manager*?"

The *Pagesetup* command produces the *Page Setup Manager* (Fig. 14-8). Produce the *Page Setup Manager* by any method shown in the command table above or right-click on a *Layout* or *Model* tab and select *Page Setup Manager*. To cause the *Page Setup Manager* to appear automatically when you create a new layout, check the box in the lower-left corner of this dialog box or check *Show Page Setup Manager for new layouts* in the *Display* tab of the *Options* dialog box.

FIGURE 14-8

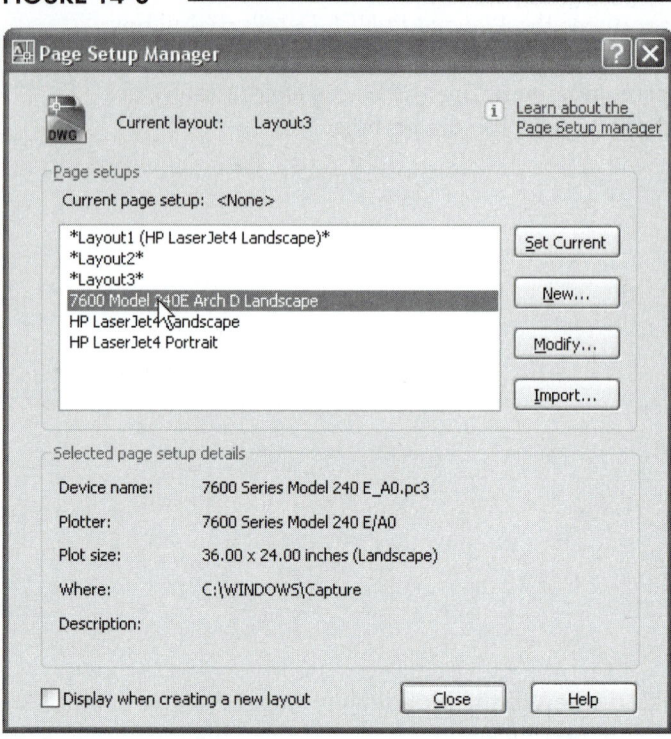

The *Page Setup Manager* has two basic functions: 1) apply existing page setups appearing in the *Page Setups* list to any layout or to a sheet set, and 2) provide access to the *Page Setup* dialog box where you can *Modify* or *Import* existing page setups or create *New* page setups. The *Page Setup* dialog box (see Fig. 14-10) allows you to specify and save the individual settings such as plot device, paper size, and so on.

To apply a page setup to a layout, first activate the desired layout tab, then use *Pagesetup* to produce the *Page Setup Manager*. The *Current layout* note at the top of the manager displays the name of the current layout.

Page Setups List

This area lists the saved page setups that are available in the drawing (or sheet set). There are two types of page setups: those that are <u>unnamed</u> but applied to layouts, such as *Layout1*, and page setups that have an assigned name, or <u>named</u> page setups, such as *HP LaserJet4 Landscape*. "Applied" or "attached" page setups are denoted by asterisks.

By default, every initialized layout (initialized by clicking the layout tab) has a default unnamed page setup automatically applied to it, such as *Layout1*. Layouts that have a named page setup applied to them are enclosed in asterisks with the named page setup in parentheses, such as *Layout 2 (HP LaserJet4 Landscape)*. Therefore, the *Page Setups* list can contain 1) unnamed applied page setups such as *Layout1*, 2) named but unapplied page setups such as *HP LaserJet4 Landscape*, and 3) named page setups that have been applied to a layout such as *Layout 2 (HP LaserJet4 Landscape)*.

If the *Page Setup Manager* is opened from the *Sheet Set Manager*, only named page setups in the page setup overrides file (a drawing template [.DWT] file) that have *Plot Area* set to *Layout* or *Extents* are listed. (See Chapter 34, Sheet Sets.)

Set Current

To <u>apply</u> or attach a page setup from the list to the current layout, highlight the desired page setup, then use *Set Current*. You can select any named page setup, such as *HP LaserJet4 Landscape*, to apply to the current layout. You can also select and apply another layout's unnamed page setup, such as **Layout3**; however, doing so does <u>not</u> change the name of the current layout's page setup. Once you assign a page setup to the current layout, you cannot "un-apply" or "un-attach" it. You can only assign another named page setup. *Set Current* is disabled for sheet sets or when you select the current page setup from the list.

New

This button produces the *New Page Setup* dialog box (Fig. 14-9) where you can assign a name for the new page setup. Assign a descriptive name (the default name being "Setup1"). You can select from the existing page setups in the drawing to "start with" as a template. Selecting the *OK* button produces the *Page Setup* dialog box where you specify the individual plotting options. <u>Creating a *New* page setup does not automatically apply the new page setup to the current layout—you must use *Set Current* to do so.</u>

FIGURE 14-9

As you can see, the *Page Setup* dialog box (Fig. 14-10) is essentially identical to the *Plot* dialog box, except that you cannot create a plot. A few other small differences are explained in the next section (see "Should You Use the *Plot* Dialog Box or *Page Setup Manager*?").

FIGURE 14-10

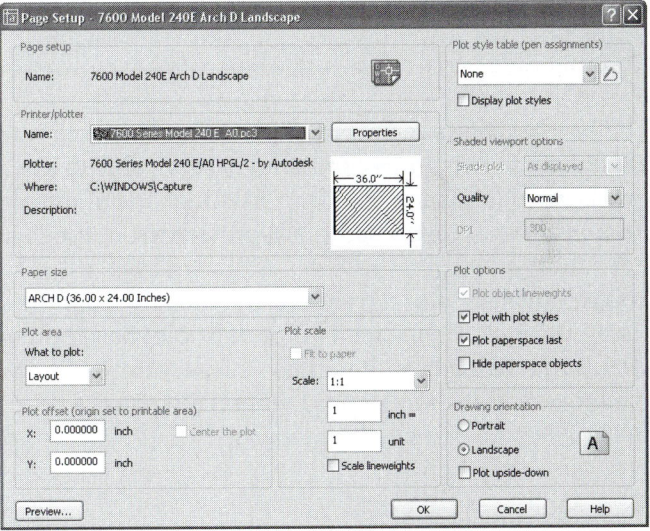

Modify

This button in the *Page Setup Manager* (see previous Fig. 14-8) produces the *Page Setup* dialog box where you can edit the existing settings for the selected page setup. Select <u>any</u> page set up in the *Page Setup* list to modify, including the current layout's page setup.

Import

Selecting this button invokes a standard file selection dialog box (not shown) from which you can import one or more page setups. You can select from any .DWG, .DWT, or .DXF file. Next, the *Import Page Setups* dialog box appears (not shown) for you to select the individual page setups in that specific drawing to import. The imported page setups then appear in the *Page Setup Manager* list, but are not yet applied to any layout.

Should You Use the *Plot* Dialog Box or the *Page Setup Manager*?

As you probably noticed, the *Page Setup Manager* and related *Page Setup* dialog box are similar and have essentially the same functions as the *Plot* dialog box, including managing page setups, with a few exceptions. The main differences are noted below.

1. You cannot plot from the *Page Setup Manager* or *Page Setup* dialog box.
2. The *Plot* dialog box includes all features in one dialog box compared to two for the *Page Setup Manager* and *Page Setup* dialog box.

3. Only the *Plot* dialog box allows you to set options to *Plot to a file*, *Plot in the background*, and turn *Plot Stamp on*.
4. You can modify any page setup from the *Page Setup Manager*, whereas you can modify only the current layout's page setup in the *Plot* dialog box.

So it has probably occurred to you, "Why do I need the *Page Setup Manager*?" Autodesk designed the *Page Setup Manager* interface primarily for use with sheet sets. A "sheet set" is a group of sheets related to one project, and therefore all "sheets" are plotted with the same plot settings (see Chapter 34, Sheet Sets). With the *Page Setup Manager*, you can apply one page setup to the entire sheet set. Without the *Page Setup Manager* and only the *Plot* dialog, you would have to open every sheet in the set (therefore every drawing file) and apply the same page setup to each layout in each drawing. Also, the desired page setup may have to be imported into each sheet (.DWG) individually before applying it.

 So which method should you use? If you want to accomplish typical plotting and printing tasks, use the *Plot* dialog box for simplicity. You can also create, import, and assign page setups easily using the top section of the *Plot* dialog box. However, if you want to manage a large group of page setups, including making modifications to them, use the *Page Setup Manager*. Also use the *Page Setup Manager* when working with sheet sets.

Preview

Pull-down Menu	Command (Type)	Alias (Type)	Short-cut	Screen (side) Menu	Tablet Menu
File *Plot Preview*	*Preview*	**PRE**	...	...	**X,24**

The *Preview* command accomplishes the same action as selecting *Preview* in the *Plot* dialog box. The advantage to using this command is being able to see a full print preview directly without having to invoke the *Plot* dialog box first. All functions of this preview (right-click for shortcut menu to *Pan* and *Zoom*, etc.) operate the same as a preview from the *Plot* dialog box including one important option, *Plot*.

 You can print or plot during *Preview* directly from the shortcut menu (Fig. 14-11). Using the *Plot* option creates a print or plot based on your settings in the *Plot* or *Page Setup* dialog box. If you use this feature, ensure you first set the plotting parameters in the *Plot* or *Page Setup* dialog box.

FIGURE 14-11 ——

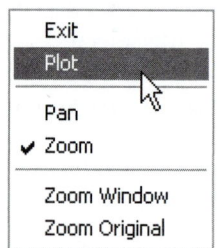

PLOTTING TO SCALE

When you create a <u>manual</u> drawing, a scale is determined before you can begin drawing. The scale is determined by the proportion between the size of the object on the paper and the actual size of the object. You then complete the drawing in that scale so the actual object is proportionally reduced or enlarged to fit on the paper.

With a <u>CAD</u> drawing you are not restricted to drawing on a sheet of paper, so the geometry can be created full size. Set *Limits* to provide an appropriate amount of drawing space; then the geometry can be drawn using the <u>actual dimensions</u> of the object. The resulting drawing on the CAD system is a virtual full-size replica of the actual object. Not until the CAD drawing is <u>plotted</u> on a fixed size sheet of paper, however, is it <u>scaled</u> to fit on the sheet.

You can specify the plot scale in one of two ways. If you are plotting a *Layout* tab, specify the scale of the geometry that appears in the viewport by using the *Viewport* toolbar, the *Properties* palette, or the *Zoom XP* factor, and then select 1:1 in the *Plot* or *Page Setup* dialog box. If you plot the *Model* tab, select the desired scale directly in the *Plot* or *Page Setup* dialog box. In either case, the value or ratio that you specify is the proportion of paper space units to model space units—the same as the proportion of the paper size to the *Limits*. This ratio is also equal to the reciprocal of the "drawing scale factor," as explained in Chapters 12 and 13. Exceptions to this rule are cases in which multiple viewports are used.

For example, if you want to draw an object 15″ long and plot it on an 11″ x 8.5″ sheet, you can set the model space *Limits* to 22 x 17 (2 x the sheet size). The value of **2** is then the drawing scale factor, and the plot scale is **1:2** (or **1/2,** the reciprocal of the drawing scale factor). If, in another case, the calculated drawing scale factor is **4**, then **1:4** (or **1/4**) would be the plot scale to achieve a drawing plotted at 1/4 actual size. In order to calculate *Limits* and drawing scale factors, it is helpful to know the standard paper sizes.

Standard Paper Sizes

Size	Engineering (″) (ANSI)	Architectural (″)
A	8.5 x 11	9 x 12
B	11 x 17	12 x 18
C	17 x 22	18 x 24
D	22 x 34	24 x 36
E	34 x 44	36 x 48

Size	Metric (mm) (ISO)
A4	210 x 297
A3	297 x 420
A2	420 x 594
A1	594 x 841
A0	841 x 1189

Calculating the Drawing Scale Factor

Assuming you are planning to plot from the *Model* tab or from a *Layout* tab using one viewport that fills almost all of the printable area, use the following method to determine the "drawing scale factor" and, therefore, the plot or print scale to specify.

1. In order to provide adequate space to create the drawing geometry full size, it is recommended to set the model space *Limits*, then *Zoom All*. *Limits* should be set to the intended sheet size used for plotting times a factor, if necessary, that provides enough area for drawing. This factor, <u>the proportion of the *Limits* to the sheet size, is the "drawing scale factor."</u>

$$DSF = \frac{Limits}{sheet\ size}$$

You should also use a proportion that will yield a standard drawing scale (1/2″=1″, 1/8″=1′, 1:50, etc.), instead of a scale that is not a standard (1/3″=1″, 3/5″=1′, 1:23, etc.). See the "Tables of *Limits* Settings" for common standard drawing scales.

2. The "drawing scale factor" is used as the value (at least a starting point) for changing all size-related variables (*LTSCALE*, *Overall Scale* for dimensions, text height, *Hatch* pattern scale, etc.). The <u>reciprocal of the drawing scale factor is the plot scale</u> to use for plotting or printing the drawing.

$$Plot\ Scale = \frac{1}{DSF}$$

3. If you using millimeter dimensions, set *Units* to *Decimal* and use metric values for the sheet size. In this way, the reciprocal of the drawing scale factor is the plot scale, the same as feet and inch drawings. However, multiply the drawing's scale factor by 25.4 (25.4mm = 1") to determine the factor for changing all size-related variables (*LTSCALE*, *Overall Scale* for dimensions, etc.).

$$DSF(mm) = \frac{Limits}{sheet\ size} \times 25.4$$

If drawing or inserting a border on the sheet, its maximum size cannot exceed the printable area. The printable area is displayed in a layout by the dashed line border just inside the "paper" edge. Since plotters or printers do not draw or print all the way to the edge of the paper, the border should not be drawn outside of the printable area. Generally, approximately 1/4" to 1/2" (6mm to 12mm) offset from each edge of the paper (margin) is required. When plotting a *Layout*, the printable area is denoted by a dashed border around the sheet. If plotting the *Model* tab, you should multiply the margin (1/2 of the difference between the *Paper Size* and printable area) by the "drawing scale factor" to determine the margin size in model space drawing units.

Guidelines for Plotting the *Model* Tab to Scale

Even though model space *Limits* can be changed and plot scale can be reset at <u>any time</u> in the drawing process, it is helpful to begin the process during the initial drawing setup (also see Chapter 12, Advanced Drawing Setup).

1. Set *Units* (*Decimal, Architectural, Engineering*, etc.) and *Precision* to be used in the drawing.
2. Set model space *Limits*. Set the *Limits* values to a proportion of the desired sheet size that provides enough area for drawing geometry as described in "Calculating the Drawing Scale Factor." The resulting value is the drawing scale factor.
3. Create the drawing geometry in the *Model* tab. Use the DSF as the initial value (or multiplier) for linetype scale, text size, dimension scale, hatch scale, etc.
4. Use the DSF to determine the scale factor to create, *Insert* or *Xref* the title block and border in model space, if one is needed.
5. Access the *Plot* dialog box. Select the desired scale in the *Scale* drop-down list or enter a value in the *Custom* edit boxes. The value to enter is 1/DSF. (Values are also given in the "Tables of *Limits* Settings.") Make a plot *Preview*, then make the plot or make the needed adjustments.

Guidelines for Plotting *Layouts* to Scale

Assuming you are using one viewport that fills almost all of the printable area, plot from the *Layout* tab using this method.

1. Set *Units* (*Decimal, Architectural, Engineering*, etc.) and *Precision* to be used in the drawing.
2. Set model space *Limits* and create the geometry in the *Model* tab. Use the DSF to determine values for linetype scale, text size, dimension scale, hatch scale, etc. Complete the drawing geometry in model space.
3. Click a *Layout* tab to begin setting up the layout. You can also use the *Layout Wizard* to create a new layout and complete steps 2 through 5.
4. Use *Page Setup* to select the desired *Plot Device* and *Paper Size*.
5. Create, *Insert*, or *Xref* the title block and border in the layout. If you are using one of AutoCAD's template drawings or another template, the title block and border may already exist.
6. Create a viewport using *Vports*.
7. Click inside the viewport to set the viewport scale. You can set the scale using the *Viewports* toolbar, the *Properties* palette, or use *Zoom* and enter an *XP* factor. Use 1/DSF as the viewport scale.
8. Activate the *Plot* dialog box and ensure the scale in the *Scale* drop-down list is 1:1. Make a plot *Preview*, then make the plot or make the needed adjustments.

If you want to print the same drawing using different devices and different scales, you can create multiple layouts. With each layout you can save the specific plot settings. Beware—a drawing printed at different scales may require changing or setting different values for linetype scale, text size, dimension variables, hatch scales, etc. for each layout. See Chapter 33 for more information on this topic.

Simplifying the process of plotting to scale can be accomplished by preparing template drawings. One method is to create a template for each sheet size that is used in the lab or office. In this way, the CAD operator begins the session by selecting the template drawing representing the sheet size and then multiplies <u>those</u> *Limits* by some factor to achieve the desired *Limits* and drawing scale factor. Another method is to create multiple layouts in the template drawing, one for each device and sheet size that you have available. In this way, a template drawing can be selected with the final layouts, title blocks, plot devices, and sheet sizes already specified, then you need only set the model space *Limits* and plot scale(s).

TABLES OF *LIMITS* SETTINGS

Rather than making calculations of *Limits*, drawing scale factor, and plot scale for each drawing, the Tables of *Limits* Settings on the following pages can be used to make calculating easier. There are five tables, one for each of the following applications:

> Mechanical Engineering
> Architectural
> Metric (using ISO standard metric sheets)
> Metric (using ANSI standard engineering sheets)
> Civil Engineering

To use the tables correctly, you must know <u>any two</u> of the following three variables. The third variable can be determined from the other two.

> Approximate drawing <u>*Limits*</u>
> Desired print or plot <u>paper size</u>
> Desired print or plot <u>scale</u>

1. If you know the *Scale* and **paper size**:
    Assuming you know the scale you want to use, look along the top row to find the desired <u>scale</u> that you eventually want to print or plot. Find the desired <u>paper size</u> by looking down the left column. The intersection of the row and column yields the model space <u>*Limits*</u> settings to use to achieve the desired plot scale.

2. If you know the approximate model space **Limits** and **paper size**:
    Calculate how much space (minimum *Limits*) you require to create the drawing actual size. Look down the left column of the appropriate table to find the <u>paper size</u> you want to use for plotting. Look along that row to find the next larger <u>*Limits*</u> settings than your required area. Use these values to set the drawing *Limits*. The <u>scale</u> to use for the plot is located on top of that column.

3. If you know the *Scale* and approximate model space **Limits**:
    Calculate how much space (minimum *Limits*) you need to create the drawing objects actual size. Look along the top row of the appropriate table to find the desired <u>scale</u> that you eventually want to plot or print. Look down that column to find the next larger <u>*Limits*</u> than your required minimum area. Set the model space *Limits* to these values. Look to the left end of that row to give the <u>paper size</u> you need to print or plot the *Limits* in the desired scale.

NOTE: Common scales are given in the Tables of *Limits* Settings. If you need to create a print or plot in a scale that is not listed in the tables, you may be able to use the tables as a guide by finding the nearest table and standard scale to match your needs, then calculating <u>proportional</u> settings.

MECHANICAL TABLE OF *LIMITS* SETTINGS
For ANSI Sheet Sizes
(X axis x Y axis)

Paper Size (Inches)	Drawing Scale Factor 1 / Scale 1"=1" / Proportion 1:1	1.33 / 3/4"=1" / 3:4	2 / 1/2"=1" / 1:2	2.67 / 3/8"=1" / 3:8	4 / 1/4"=1" / 1:4	5.33 / 3/16"=1" / 3:16	8 / 1/8"=1" / 1:8
A 11 x 8.5 In.	11.0 x 8.5	14.7 x 11.3	22.0 x 17.0	29.3 x 22.7	44.0 x 34.0	58.7 x 45.3	88.0 x 68.0
B 17 x 11 In.	17.0 x 11.0	22.7 x 14.7	34.0 x 22.0	45.3 x 29.3	68.0 x 44.0	90.7 x 58.7	136.0 x 88.0
C 22 x 17 In.	22.0 x 17.0	29.3 x 22.7	44.0 x 34.0	58.7 x 45.3	88.0 x 68.0	117.0 x 90.7	176.0 x 136.0
D 34 x 22 In.	34.0 x 22.0	45.3 x 29.3	68.0 x 44.0	90.7 x 58.7	136.0 x 88.0	181.3 x 117.3	272.0 x 176.0
E 44 x 34 In.	44.0 x 34.0	58.7 x 45.3	88.0 x 68.0	117.3 x 90.7	176.0 x 136.0	235.7 x 181.3	352.0 x 272.0

ARCHITECTURAL TABLE OF *LIMITS* SETTINGS
For Architectural Sheet Sizes
(X axis x Y axis)

Paper Size (Inches)		Drawing Scale Factor 12 / Scale 1"=1' / Proportion 1:12	16 / 3/4"=1' / 1:16	24 / 1/2"=1' / 1:24	32 / 3/8"=1' / 1:32	48 / 1/4"=1' / 1:48	64 / 3/16"=1' / 1:64	96 / 1/8"=1' / 1:96
A 12 x 9	Ft	12 x 9	16 x 12	24 x 18	32 x 24	48 x 36	64 x 48	96 x 72
	In.	144 x 108	192 x 144	288 x 216	384 x 288	576 x 432	768 x 576	1152 x 864
B 18 x 12	Ft	18 x 12	24 x 16	36 x 24	48 x 32	64 x 48	96 x 64	128 x 96
	In.	216 x 144	288 x 192	432 x 288	576 x 384	768 x 576	1152 x 768	1536 x 1152
C 24 x 18	Ft	24 x 18	32 x 24	48 x 36	64 x 48	96 x 72	128 x 96	192 x 144
	In.	288 x 216	384 x 288	576 x 432	768 x 576	1152 x 864	1536 x 1152	2304 x 1728
D 36 x 24	Ft	36 x 24	48 x 32	72 x 48	96 x 64	144 x 96	192 x 128	288 x 192
	In.	432 x 288	576 x 384	864 x 576	1152 x 768	1728 x 1152	2304 x 1536	3456 x 2304
E 48 x 36	Ft	48 x 36	64 x 48	96 x 72	128 x 96	192 x 144	256 x 192	384 x 288
	In.	576 x 432	768 x 576	1152 x 864	1536 x 1152	2304 x 1728	3072 x 2304	4608 x 3456

METRIC TABLE OF *LIMITS* SETTINGS
For Metric (ISO) Sheet Sizes
(X axis x Y axis)

Paper Size (mm)	Drawing Scale Factor: 25.4 / Scale 1:1 / Proportion 1:1	50.8 / 1:2 / 1:2	127 / 1:5 / 1:5	254 / 1:10 / 1:10	508 / 1:20 / 1:20	1270 / 1:50 / 1:50	2540 / 1:100 / 1:100
A4 297 x 210 mm	297 x 210	594 x 420	1485 x 1050	2970 x 2100	5940 x 4200	14,850 x 10,500	29,700 x 21,000
m	.297 x .210	.594 x .420	1.485 x 1.050	2.97 x 2.10	5.94 x 4.20	14.85 x 10.50	29.70 x 21.00
A3 420 x 297 mm	420 x 297	840 x 594	2100 x 1485	4200 x 2970	8400 x 5940	21,000 x 14,850	42,000 x 29,700
m	.420 x .297	.840 x .594	2.100 x 1.485	4.20 x 2.97	8.40 x 5.94	21.00 x 14.85	42.00 x 29.70
A2 594 x 420 mm	594 x 420	1188 x 840	2970 x 2100	5940 x 4200	11,880 x 8400	29,700 x 21,000	59,400 x 42,000
m	.594 x .420	1.188 x .840	2.97 x 2.10	5.94 x 4.20	11.88 x 8.40	29.70 x 21.00	59.40 x 42.00
A1 841 x 594 mm	841 x 594	1682 x 1188	4205 x 2970	8410 x 5940	16,820 x 11,880	42,050 x 29,700	84,100 x 59,400
m	.841 x .594	1.682 x 1.188	4.205 x 2.970	8.41 x 5.94	16.82 x 11.88	42.05 x 29.70	84.10 x 59.40
A0 1189 x 841 mm	1189 x 841	2378 x 1682	5945 x 4205	11,890 x 8410	23,780 x 16,820	59,450 x 42,050	118,900 x 84,100
m	1.189 x .841	2.378 x 1.682	5.945 x 4.205	11.89 x 8.41	23.78 x 16.82	59.45 x 42.05	118.90 x 84.10

METRIC TABLE OF *LIMITS* SETTINGS
For ANSI Sheet Sizes
(X axis x Y axis)

Paper Size (mm)		Drawing Scale Factor 25.4 Scale 1:1 Proportion 1:1	50.8 1:2 1:2	127 1:5 1:5	254 1:10 1:10	508 1:20 1:20	1270 1:50 1:50	2540 1:100 1:100
A 279.4 x 215.9	mm	279.4 x 215.9	558.8 x 431.8	1397 x 1079.5	2794 x 2159	5588 x 4318	13,970 x 10,795	27,940 x 21,590
	m	0.2794 x 0.2159	0.5588 x 0.4318	1.397 x 1.0795	2.794 x 2.159	5.588 x 4.318	13.97 x 10.795	27.94 x 21.59
B 431.8 x 279.4	mm	431.8 x 279.4	863.6 x 558.8	2159 x 1397	4318 x 2794	8636 x 5588	21,590 x 13,970	43,180 x 27,940
	m	0.4318 x 0.2794	0.8636 x 0.5588	2.159 x 1.397	4.318 x 2.794	8.636 x 5.588	21.59 x 13.97	43.18 x 27.94
C 558.8 x 431.8	mm	558.8 x 431.8	1117.6 x 863.6	2794 x 2159	5588 x 4318	11,176 x 8636	27,940 x 21,590	55,880 x 43,180
	m	0.5588 x 0.4318	1.1176 x 0.8636	2.794 x 2.159	5.588 x 4.318	11.176 x 86.36	27.94 x 21.59	55.88 x 43.18
D 863.6 x 558.8	mm	863.6 x 558.8	1727.2 x 1117.6	4318 x 2794	8636 x 5588	17,272 x 11,176	43,180 x 27,940	86,360 x 55,880
	m	0.8636 x 0.5588	1.7272 x 1.1176	4.318 x 2.794	8.636 x 5.588	17.272 x 11.176	43.18 x 27.94	86.36 x 55.88
E 1117.6 x 863.6	mm	1117.6 x 863.6	2235.2 x 1727.2	5588 x 4318	11,176 x 8636	22,352 x 17,272	55,880 x 43,180	111,760 x 86,360
	m	1.1176 x 0.8636	2.2352 x 1.7272	5.588 x 4.318	11.176 x 8.636	22.352 x 17.272	55.88 x 43.18	111.76 x 86.36

CIVIL TABLE OF *LIMITS* SETTINGS
For ANSI Sheet Sizes
(X axis x Y axis)

Paper Size (Inches)		Drawing Scale Factor 120, Scale Tab 1"=10', Proportion 1:120	240, 1"=20', 1:240	360, 1"=30', 1:360	480, 1"=40', 1:480	600, 1"=50', 1:600
A. 11 x 8.5	In.	1320 x 1020	2640 x 2040	3960 x 3060	5280 x 4080	6600 x 5100
	Ft	110 x 85	220 x 170	330 x 255	440 x 340	550 x 425
B. 17 x 11	In.	2040 x 1320	4080 x 2640	6120 x 3960	8160 x 5280	10,200 x 6600
	Ft	170 x 110	340 x 220	510 x 330	680 x 440	850 x 550
C. 22 x 17	In.	2640 x 2040	5280 x 4080	7920 x 6120	10,560 x 8160	13,200 x 10,200
	Ft	220 x 170	440 x 340	660 x 510	880 x 680	1100 x 850
D. 34 x 22	In.	4080 x 2640	8160 x 5280	12,240 x 7920	16,320 x 10,560	20,400 x 13,200
	Ft	340 x 220	680 x 440	1020 x 660	1360 x 880	1700 x 1100
E. 44 x 34	In.	5280 x 4080	10,560 x 8160	15,840 x 12,240	21,120 x 16,320	26,400 x 20,400
	Ft	440 x 340	880 x 680	1320 x 1020	1760 x 1360	2200 x 1700

Examples for Plotting to Scale

Following are several hypothetical examples of drawings that can be created using the "Guidelines for Plotting to Scale." As you read the examples, try to follow the logic and check the "Tables of *Limits* Settings."

A. A one-view drawing of a mechanical part that is 40" in length is to be drawn requiring an area of approximately 40" x 30", and the drawing is to be plotted on an 8.5" x 11" sheet. The template drawing *Limits* are preset to 0,0 and 11,8.5. (The ACAD.DWT template *Limits* are set to 0,0 and 12,9, which represents an uncut sheet size. Template *Limits* of 11 x 8.5 are more practical in this case.) The expected plot scale is 1/4"=1". The following steps are used to calculate the new *Limits*.

1. *Units* are set to *decimal*. Each unit represents 1.00 inches.
2. Multiplying the *Limits* of 11 x 8.5 by a factor of **4**, the new *Limits* should be set to **44 x 34**, allowing adequate space for the drawing. The "drawing scale factor" is **4**.
3. All size-related variable default values (*LTSCALE, Overall Scale* for dimensions, etc.) are multiplied by **4**. The plot scale (for *Layout* tabs, entered in the *Viewport* toolbar, etc., or for the *Model* tab, entered in the *Plot* or *Page Setup* dialog box) is **1:4** to achieve a plotted drawing of 1/4"=1".

B. A floorplan of a residence will occupy 60' x 40'. The drawing is to be plotted on a "D" size architectural sheet (24" x 36"). No prepared template drawing exists, so the ACAD.DWT template drawing (12 x 9) *Limits* are used. The expected plot scale is 1/2"=1'.

1. *Units* are set to *Architectural*. Each unit represents 1".
2. The floorplan size is converted to inches (60' x 40' = 720" x 480"). The sheet size of 36" x 24" (or 3' x 2') is multiplied by **24** to arrive at *Limits* of **864" x 576"** (72' x 48'), allowing adequate area for the floor plan. The *Limits* are changed to those values.
3. All default values of size-related variables (*LTSCALE, Overall Scale* for dimensions, etc.) are multiplied by **24**, the drawing scale factor. The plot scale (for *Layout* tabs, entered in the *Viewport* toolbar, etc., or for the *Model* tab, entered in the *Plot* or *Page Setup* dialog box) is **1/2"=1'** (or 1/24, which is the reciprocal of the drawing scale factor).

C. A roadway cloverleaf is to be laid out to fit in an acquired plot of land measuring 1500' x 1000'. The drawing will be plotted on "D" size engineering sheet (34" x 22"). A template drawing with *Limits* set equal to the sheet size is used. The expected plot scale is 1"=50'.

1. *Units* are set to *Engineering* (feet and decimal inches). Each unit represents 1.00".
2. The sheet size of 34" x 22" (or 2.833' x 1.833') is multiplied by **600** to arrive at *Limits* of **20400" x 13200"** (1700' x 1100'), allowing enough drawing area for the site. The *Limits* are changed to **1700' x 1100'**.
3. All default values of size-related variables (*LTSCALE, Overall Scale* for dimensions, etc.) are multiplied by **600**, the drawing scale factor. The plot scale (for *Layout* tabs, entered in the *Viewport* toolbar, etc., or for the *Model* tab, entered in the *Plot* or *Page Setup* dialog box) is **1/600** (*inches = drawing units*) to achieve a drawing of 1"=50' scale.

NOTE: Many civil engineering firms use one unit in AutoCAD to represent one foot. This simplifies the problem of using decimal feet (10 parts/foot rather than 12 parts/foot); however, problems occur if architectural layouts are inserted or otherwise combined with civil work.

D. Three views of a small machine part are to be drawn and dimensioned in millimeters. An area of 480mm x 360mm is needed for the views and dimensions. The part is to be plotted on an A4 size sheet. The expected plot scale is 1:2.

1. *Units* are set to *decimal*. Each unit represents 1.00 millimeters.
2. Setting the *Limits* exactly to the sheet size (297 x 210) would not allow enough area for the drawing. The sheet size is multiplied by a factor of **2** to yield **Limits** of **594 x 420**, providing the necessary 480 x 360 area.
3. The plot scale (for *Layout* tabs, entered in the *Viewport* toolbar, etc., or for the *Model* tab, entered in the *Plot* or *Page Setup* dialog box) is **1:2** scale. Since this drawing is metric, the drawing scale factor for changing all size-related variable default values (**LTSCALE, Overall Scale** for dimensions, etc.) is multiplied by 25.4, or 2 x 25.4 = approximately **50**.

CONFIGURING PLOTTERS AND PRINTERS

Plotting Components

When a drawing is plotted, three main components are referenced to develop the finished plot.

1. The plot and page settings you set up in the *Plot* dialog box or *Page Setup* dialog box (saved in the .DWG file)
2. The plotter configuration information you specify in the *Add-A-Plotter Wizard* or *Plotter Configuration Editor* (saved in a .PC3 file)
3. A plot style table you create using the *Add-A-Plot Style Table Wizard* or *New* option in the *Plot* dialog box (saved in a .CTB or .STB file)

The three components are stored separately so they can be applied in different combinations. For example, any plot style table can be used with any plot device, or any page setup can use any plot style table.

The *Plot* and *Page Setup* dialog boxes were explained earlier in this chapter. The following section explains how to configure plotters and create .PC3 files using the *Plotter Manager, Add-A-Plotter Wizard*, and *Plotter Configuration Editor*. Plot style tables (.CTB and .STB files) are discussed in Chapter 33.

If you are using AutoCAD in a lab or office, the devices used for making plots and prints are most likely already configured for your use. However, no matter if you are a student or a professional, you may be required to configure a new device or change configuration settings on existing devices. This section explains those possibilities. In order to accomplish either of the tasks mentioned here, use the *Plottermanager* command to invoke the AutoCAD *Plotter Manager*.

Plotter-manager

Pull-down Menu	Command (Type)	Alias (Type)	Short-cut	Screen (side) Menu	Tablet Menu
File *Plotter Manager...*	*Plottermanager*	...	...	...	...

The AutoCAD *Plotter Manager* can be invoked by several methods as shown in the command table above. In addition to those methods, you can select the *Add or Configure Plotters* button in the *Plotting* tab of the *Options* dialog box.

In any case, the *Plotter Manager* is a separate system window which can run as an independent application to AutoCAD (Fig. 14-12). By default, the window displays icons of all plot devices configured specifically for AutoCAD. The default Windows configured device is included in this group. (When you install AutoCAD, the Windows configured system printer is automatically configured if you have a previously installed Windows printer.) You can add or configure a plotter with the *Plotter Manager*.

FIGURE 14-12

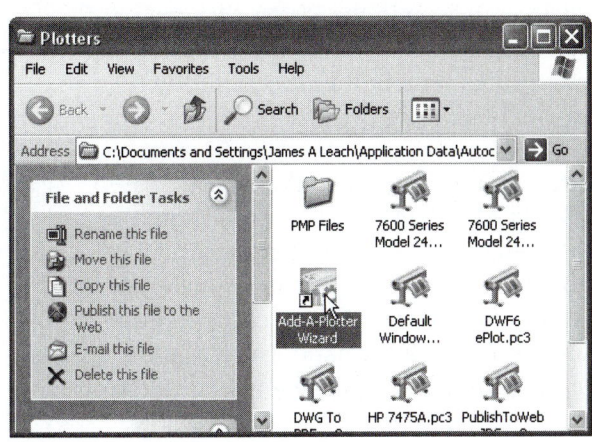

The display of the *Plotter Manager* can also be changed to list *Details* instead of *Icons*, as shown in Figure 14-13 (use the *View* pull-down menu and select *Details*). Note that the defined plotters are actually saved as .PC3 files. If you want to add a new plotter for use with AutoCAD, select the *Add-A-Plotter Wizard*. To change the settings on an existing plotter using the *Plotter Configuration Editor*, double-click on its icon or file name.

FIGURE 14-13

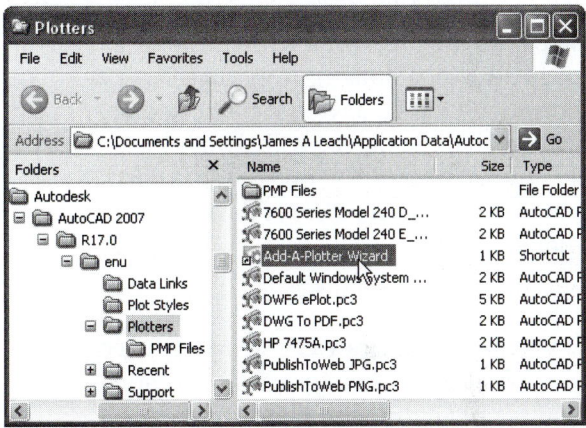

Add-A-Plotter Wizard

The *Add-A-Plotter Wizard* makes the process of configuring a new device very simple. Several choices are available (Fig. 14-14). You can configure a new device to be connected directly to and managed by your system or to be shared and managed by a network server. You can also reconfigure printers that were previously set up as Windows system printers. In this case, icons for the printers will appear in the *Plotter Manager* and you can access and control configuration settings to use specifically for AutoCAD.

FIGURE 14-14

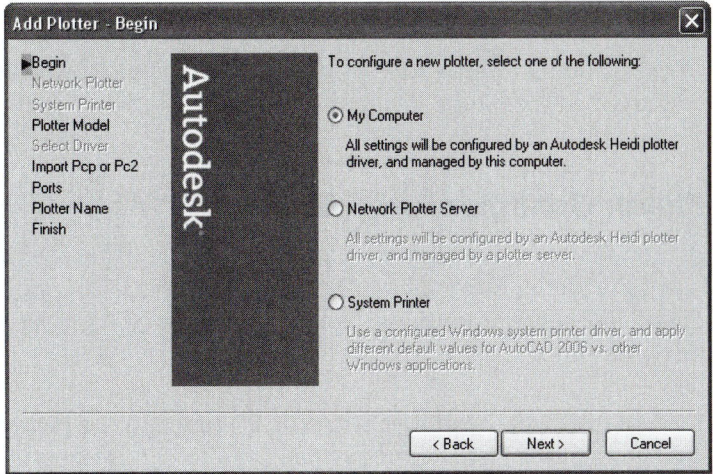

Start the *Add-A-Plotter Wizard* by double-clicking its icon in the *Plotter Manager* or selecting *Add Plotter...* from the *Wizards* cascading menu in the *Tools* pull-down menu.

After you have selected the device (*Plotter Model*) to configure (not shown), you can choose to import information about the plotter if it was previously configured for use with an earlier release of AutoCAD. Information such as paper size, optimization, port names, etc. were previously stored in .PCP and .PC2 files and can be converted to the .PC3 format (Fig. 14-15).

If you selected *My Computer* in the *Begin* step, you are presented with the *Ports* page next (not shown). You can select the port to which the device is connected or specify that the device will always *Plot to a File* or *Autospool*. If you specify a port, you can later select *Plot to File* in the *Plot* dialog box.

If you are using a device previously used as a Windows system printer, make sure you supply a different name for the device than the name Windows uses. This action ensures the new name appears in the *Plot* and *Page Setup* dialog boxes (Fig. 14-16).

FIGURE 14-15

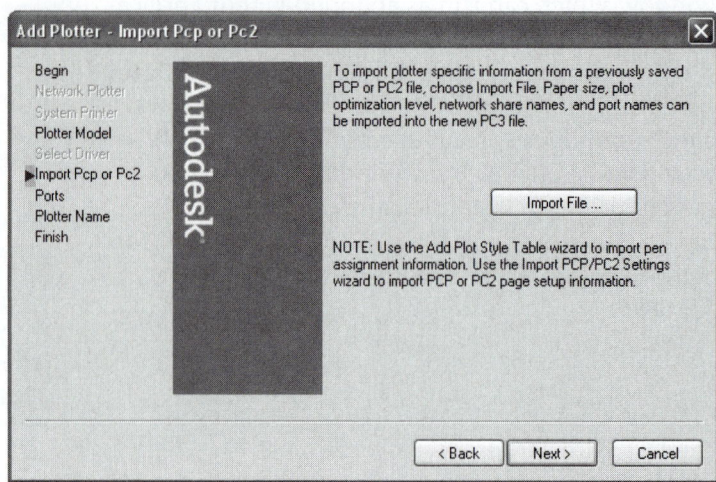

FIGURE 14-16

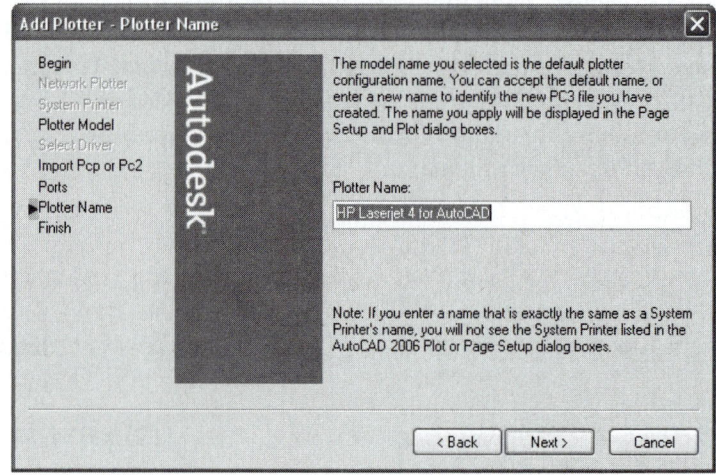

Once the new or previously used Windows system device has been configured for AutoCAD, its icon is displayed in the *Plotter Manager* along with any other configured devices.

Plotter Configuration Editor

After you create a configured plotter (.PC3) file using the *Add-A-Plotter Wizard*, you can edit the file using the *Plotter Configuration Editor*. The *Plotter Configuration Editor* allows you to specify a number of the plotter's output settings. You can start the *Plotter Configuration Editor* by one of the following methods.

1. Double-click the plotter's icon or file name in the *Plotter Manager*.
2. Choose *Edit Plotter Configuration* from the *Finish* step of the *Add-A-Plotter Wizard*.
3. Select *Properties* from the *Page Setup* dialog box or *Plot* dialog box.
4. Double-click a .PC3 file from Windows Explorer.

The *Plotter Configuration Editor* contains three tabs. The *General* tab contains basic information about the plotter. Only the *Description* area is not read-only and can be edited. Information entered in the *Description* is saved with the .PC3 file.

The *Ports* tab allows you to specify ports to connect with the device. If you are configuring a local, non-Windows system plotter, you can respecify the port to which the device is connected. You can choose a serial (LPT), parallel (COM), or network port. If you are using a serial port, the settings within AutoCAD must match the plotter settings. For additional technical information, check AutoCAD *Help, Driver and Peripheral Guide*.

The *Device and Document Settings* tab contains the plotting options. These choices are different depending on your configured plotting device. For example, when you configure a pen plotter you have the option to modify the pen characteristics (Fig. 14-17). The *Device and Document Settings* tab can contain some or all of the following general option areas.

FIGURE 14-17

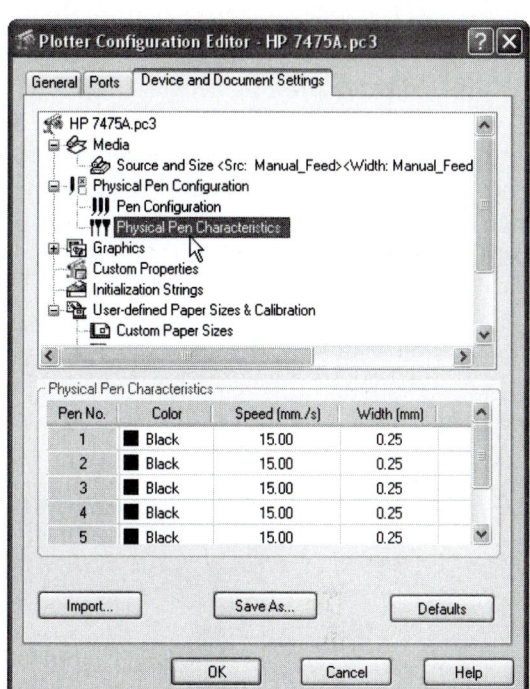

Media	Specifies a paper source, size, type, and destination.
Physical Pen Configuration	Specifies settings for pen plotters.
Graphics	Specifies settings for printing vector and raster graphics and True Type fonts.
Custom Properties	Displays settings related to the device driver.
Initialization Strings	Sets pre-initialization, post-initialization, and termination printer strings.
User-Defined Paper Sizes and Calibration	Attaches a plot model parameter (.PMP) file to the .PC3 file, calibrates plotter, and adds, deletes, or revises custom paper sizes.

The related .PC3 file contains these same categories of settings. Double-click any of the categories to view and change the specific settings.

You can also filter paper sizes that you do not use so they do not appear as a choice in the *Page Setup* and *Plot* dialog boxes. Expand the *User-defined Paper Sizes & Calibration* section, then select *Filter Paper Sizes*.

Because there are so many possible options for all the possible plotting devices it is not appropriate to include the information in this text. Additional technical information on this subject can be found in AutoCAD *Help, Driver and Peripheral Guide*.

.PCP, .PC2, and .PC3 Files

Plotting information for pre-AutoCAD 2000 releases was stored in .PCP and .PC2 files. These file types cannot be used in AutoCAD 2007, but the information in these files can be converted for use with AutoCAD 2007.

.PCP files, used with Release 12 and earlier versions of AutoCAD, contain information about plot settings, specifically pen settings. These files are device-independent, so one .PCP file could be used to define pen settings for many plotters. Although these were optional, pen assignments could be saved to a .PCP file and later imported rather than specifying the settings again when a different device was configured.

In Releases 13 and 14, .PC2 files were used. These files contain pen assignment information as well as the device (plotter or printer) configuration information. In this way, when a .PC2 file was loaded, it automatically installed and configured the device and made it the current plotter. This feature was helpful for batch plotting because different plot devices could be operated while unattended.

AutoCAD 2000 through 2007 store plotter configuration information in .PC3 files. .PC3 files store the complete settings for particular devices, including media, paper sizes, printable area, and pen information. .PC3 files are drawing independent, portable, and sharable, so they can be used for any drawing or by any AutoCAD session.

.PCP and .PC2 can be imported and converted to the .PC3 format using the *Pcinwizard*. Therefore, information you have already saved about your devices can be automatically updated to use with AutoCAD 2007.

Pcinwizard

Pull-down Menu	Command (Type)	Alias (Type)	Short-cut	Screen (side) Menu	Tablet Menu
Tools Wizards > Import Plot Settings...	Pcinwizard	...	...	...	...

Pcinwizard displays the *Import PCP or PC2 Plot Settings* wizard (Fig. 14-18). The wizard imports and updates plotter information previously stored in .PCP and .PC2 files (for Release 14 and earlier releases of AutoCAD) to the AutoCAD 2006 format, .PC3. Information that can be imported from .PCP or .PC2 files includes plot area, rotation, plot offset, plot optimization, plot to file, paper size, plot scale, and pen mapping.

FIGURE 14-18

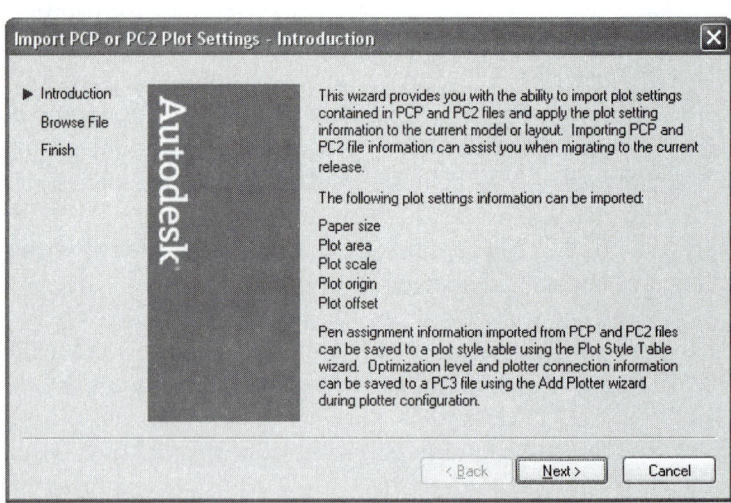

The wizard prompts you for the name of the .PCP or .PC2 configuration file from which you want to import settings. You can view and modify the plot settings prior to importing them.

Like all the wizards, explanatory information is given in all steps. Upon completion, plot settings such as paper size, plot area, plot scale, plot origin, and plot offset are stored for each device and are ready for use.

CHAPTER EXERCISES

1. **Print GASKETA from the *Model* Tab**

   A. *Open* the **GASKETA** drawing that you created in Chapter 9 Exercises. Activate the *Model* tab. What are the model space *Limits*?

   B. From the *Model* tab, *Plot* the drawing *Extents* on an 11" x 8.5" sheet. Select the *Fit to paper* option from the *Plot Scale* section in the *Plot* dialog box.

   C. Next, *Plot* the drawing from the *Model* tab again. Plot the *Limits* on the same size sheet. *Scale to Fit*.

   D. Now, *Plot* the drawing *Limits* as before but plot the drawing at **1:1**. Measure the drawing and compare the accuracy with the dimensions given in the exercise in Chapter 9.

   E. Compare the three plots. What are the differences and why did they occur? (When you finish, there is no need to *Save* the changes.)

2. **Print GASKETA from a *Layout* Tab in Two Scales**

   A. Using the **GASKETA** drawing again, activate the *Layout1* tab. If a viewport exists in the layout, *Erase* it. Activate the *Plot* or *Page Setup* dialog box and configure the layout to use a printer to print on a *Letter* (or "A" size) sheet.

   B. Next, make layer **0** *Current*. Use *Vports* to create a *Single* viewport using the *Fit* option. Double-click inside the viewport and *Zoom All* to see all of the model space *Limits*. Next, set the viewport scale to **1:1** using any method (*Zoom XP*, the *Viewport* toolbar, or the *Properties* palette). You may need to *Pan* to center the geometry, but do <u>not</u> *Zoom*.

   C. (Optional) Create a title block and border for the drawing in paper space.

   D. Activate the *Plot* dialog box. Ensure the *Scale* edit box is set at **1:1**. Plot the *Layout*. When complete, measure the plot for accuracy by comparing with the dimensions given for the exercise in Chapter 9. *Save* the drawing.

   E. Reset the viewport scale to **2:1** using any method (*Zoom XP*, the *Viewport* toolbar, or the *Properties* palette). Activate the *Plot* dialog box and make a second plot at the new scale. *Close* the drawing, but <u>do not save</u> it.

3. **Print the MANFCELL Drawing in Two Standard Scales**

   A. *Open* the **MANFCELL** drawing from Chapter 9. Check to make sure that the *Limits* settings are at **0,0** and **40',30'**.

   B. Activate the *Layout1* tab. If a viewport exists, *Erase* it. Use *Pagesetup* to select a printer to print on a letter ("A") sheet size. Next, create a *Single* viewport with the *Fit* option on layer **0**.

Bisect

This option draws the *Xline* at an angle between two
selected points. First, select the angle vertex, then
two points to define the angle (Fig. 15-5).

> Command: **xline**
> Specify a point or [Hor/Ver/Ang/Bisect/Offset]: **b**
> Specify angle vertex point: **PICK**
> Specify angle start point: **PICK**
> Specify angle end point: **PICK**
> Specify angle end point: **Enter**
> Command:

FIGURE 15-5

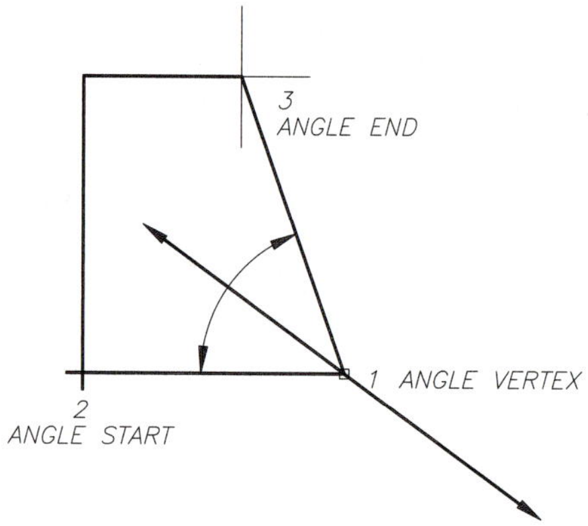

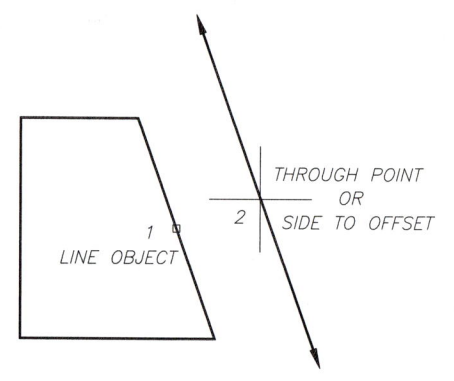

Offset

Offset creates an *Xline* parallel to another line. This
option operates similarly to the *Offset* command. You
can (1) specify a *Distance* from the selected line or (2)
PICK a point to create the *Xline Through*. With the
Distance option, enter the distance value, select a line
(*Line, Xline, Ray,* or *Pline*), and specify on which side
to create the offset *Xline* (Fig. 15-6).

> Command: **xline**
> Specify a point or [Hor/Ver/Ang/Bisect/Offset]: **o**
> Specify offset distance or [Through] <1.0000>:
> (**value**) or **PICK**
> Select a line object: **PICK**
> Specify side to offset: **PICK**
> Select a line object: **Enter**
> Command:

FIGURE 15-6

Using the *Through* option, select a line (*Line, Xline, Ray,* or *Pline*); then specify a point for the *Xline* to pass
through. In each case, the anchor point of the *Xline* is the "root." (See "*Offset,*" Chapter 9.)

Ray

Pull-down Menu	Command (Type)	Alias (Type)	Short-cut	Screen (side) Menu	Tablet Menu
Draw Ray	Ray	...	...	DRAW 1 Ray	K,10

A *Ray* is also a construction line (see *Xline*), but it
extends to infinity in only <u>one direction</u> and has one
"anchored" endpoint. Like an *Xline*, a *Ray* extends
past the drawing area but does not affect the drawing
Limits or *Extents*. The construction process for a *Ray* is
simpler than for an *Xline*, only requiring you to establish a "start point" (endpoint) and a "through point"
(Fig. 15-7). Multiple *Rays* can be created in one
command.

FIGURE 15-7

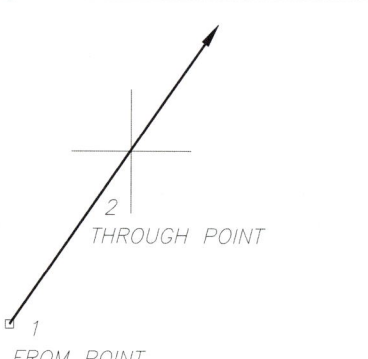

Command: **ray**
Specify start point: **PICK** or **(coordinates)**
Specify through point: **PICK** or **(coordinates)**
Specify through point: **PICK** or **(coordinates)**
Specify through point: **Enter**
Command:

Rays are especially helpful for construction of geometry about a central reference point or for construction of angular features. In each case, the geometry is usually constructed in only one direction from the center or vertex. If horizontal or vertical *Rays* are needed, just toggle on *ORTHO*. To draw a *Ray* at a specific angle, use relative polar coordinates or Polar Tracking. *Endpoint* and other appropriate *Osnaps* can be used with *Rays*. A *Ray* has one *Endpoint* but no *Midpoint*.

Polygon

Pull-down Menu	Command (Type)	Alias (Type)	Short-cut	Screen (side) Menu	Tablet Menu
Draw *Polygon*	*Polygon*	*POL*	...	*DRAW 1* *Polygon*	*P,10*

The *Polygon* command creates a regular polygon (all angles are equal and all sides have equal length). A *Polygon* object appears to be several individual objects but, like a *Pline,* is actually <u>one</u> object. In fact, AutoCAD uses *Pline* to create a *Polygon.* There are two basic options for creating *Polygons:* you can specify an *Edge* (length of one side) or specify the size of an imaginary circle for the *Polygon* to *Inscribe* or *Circumscribe.*

Inscribe/Circumscribe
The command sequence for this default method follows:

Command: **polygon**
Enter number of sides <4>: **(value)**
Specify center of polygon or [Edge]: **PICK** or **(coordinates)**
Enter an option [Inscribed in circle/Circumscribed about circle] <I>: **I** or **C**
Specify radius of circle: **PICK** or **(coordinates)**
Command:

The orientation of the *Polygon* and the imaginary circle are shown in Figure 15-8. Note that the *Inscribed* option allows control of one-half of the distance <u>across the corners</u>, and the *circumscribed* option allows control of one-half of the distance <u>across the flats</u>.

Using *ORTHO ON* with specification of the *radius of circle* forces the *Polygon* to a 90 degree orientation.

Edge
The *Edge* option only requires you to indicate the number of sides desired and to specify the two endpoints of one edge (Fig. 15-8).

FIGURE 15-8

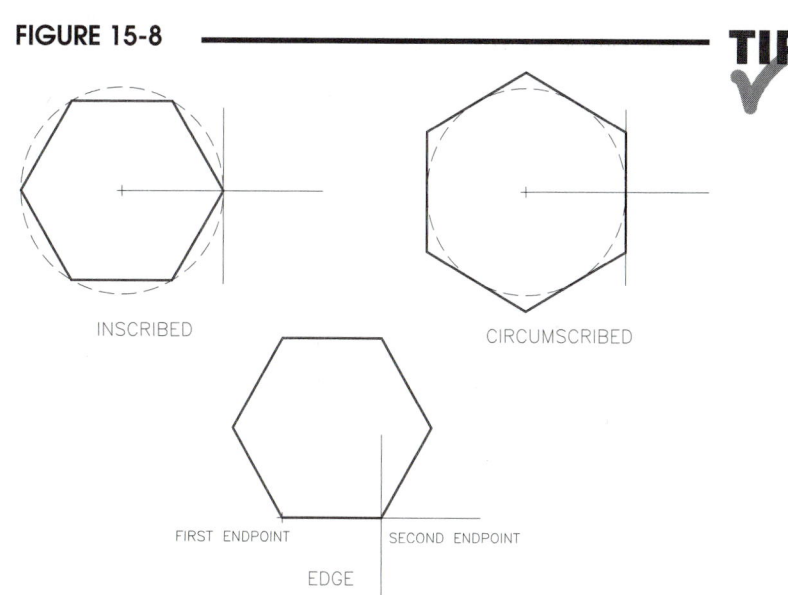

INSCRIBED

CIRCUMSCRIBED

FIRST ENDPOINT   SECOND ENDPOINT

EDGE

```
Command: polygon
Enter number of sides <4>: (value)
Specify center of polygon or [Edge]: e
Specify first endpoint of edge: PICK  or  (coordinates)
Specify second endpoint of edge: PICK  or  (coordinates)
Command:
```

Because *Polygons* are created as *Plines*, *Pedit* can be used to change the line width or edit the shape in some way (see "*Pedit*," Chapter 16). *Polygons* can also be *Exploded* into individual objects similar to the way other *Plines* can be broken down into component objects (see "*Explode*," Chapter 16).

Rectang

	Pull-down Menu	Command (Type)	ALIAS (Type)	Short-cut	Screen (side) Menu	Tablet Menu
	Draw Rectangle	Rectang	REC	...	DRAW 1 Rectang	Q,10

The *Rectang* command only requires the specification of two diagonal corners for construction of a rectangle, identical to making a selection window (Fig. 15-9). The corners can be PICKed or dimensions can be entered. The resulting <u>rectangle is actually a *Pline* object</u>. Therefore, a <u>rectangle is one object</u>, not four separate objects.

```
Specify first corner point or
[Chamfer/Elevation/Fillet/Thickness/Width]:
PICK, (coordinates) or option
Specify other corner point or [Area/Dimensions/Rotation]:
PICK, (coordinates) or option
```

FIGURE 15-9

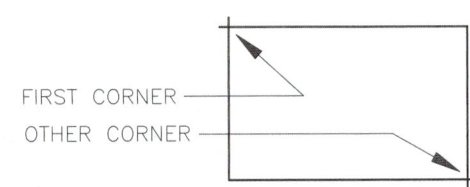

FIRST CORNER
OTHER CORNER

Several options of the *Rectangle* command affect the shape as if it were a *Pline* (see *Pline* and *Pedit*). For example, a *Rectangle* can have *width*:

```
Specify first corner point or
[Chamfer/Elevation/Fillet/Thickness/Width]: w
Specify line width for rectangles <0.0000>: (value)
```

WIDTH=.05

The *Fillet* and *Chamfer* options allow you to specify values for the
fillet radius or chamfer distances:

```
Specify first corner point or
[Chamfer/Elevation/Fillet/Thickness/Width]: f
Specify fillet radius for rectangles <0.0000>: (value)
```

FILLET
RADIUS=.5

```
Specify first corner point or
[Chamfer/Elevation/Fillet/Thickness/Width]: c
Specify first chamfer distance for rectangles <0.0000>:
(value)
Specify second chamfer distance for rectangles <0.5000>:
(value)
```

CHAMFER
1ST DISTANCE=.5
2ND DISTANCE=.5

Thickness and *Elevation* are 3D properties. (See Chapter 40, Surface Modeling, for more information.)

After specifying the first corner of the rectangle, you can enter the other corner or use these options (Fig. 15-10). The *Dimensions* option prompts for the length and width rather than picking two diagonal corners.

FIGURE 15-10

LENGTH=4
WIDTH=2.5

> Specify other corner point or [Area/Dimensions/Rotation]: **d**
> Specify length for rectangles <4.0000>:**(value)**
> Specify width for rectangles <2.5000>:**(value)**

The *Area* option prompts for an area value and either the length or width to use for calculating the area.

AREA=10
LENGTH=4

> Enter area of rectangle in current units <100.0000>:
> **(value)**
> Calculate rectangle dimensions based on [Length/Width]
> <Length>: **1**
> Enter rectangle length <10.0000>:**(value)**

The *Rotation* option allows you to specify a rotational value for the orientation of the *Rectangle*.

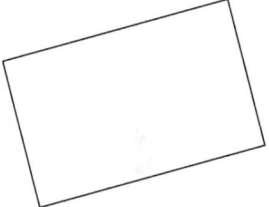

ROTATION=15

> Specify other corner point or [Area/Dimensions/Rotation]: **r**
> Specify rotation angle or [Pick points] <0>: **(value)**

Donut

Pull-down Menu	Command (Type)	Alias (Type)	Short-cut	Screen (side) Menu	Tablet Menu
Draw *Donut*	*Donut*	*DO*	...	*DRAW 1* *Donut*	*K,9*

A *Donut* is a circle with width (Fig. 15-11). Invoking the command allows changing the inside and outside diameters and creating multiple *Donuts*:

FIGURE 15-11

> Command: **donut**
> Specify inside diameter of donut
> <0.5000>: **(value)**
> Specify outside diameter of donut
> <1.0000>: **(value)**
> Specify center of donut or <exit>:
> **PICK**
> Specify center of donut or <exit>:
> **Enter**

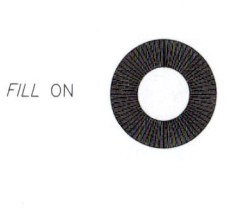

FILL ON

INSIDE DIA = 0.5   INSIDE DIA = 0   INSIDE DIA = 1.8
OUTSIDE DIA = 1   OUTSIDE DIA = 1   OUTSIDE DIA = 2

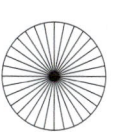
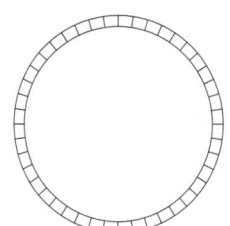

FILL OFF

*Donut*s are actually solid filled circular *Pline*s with width. The solid fill for *Donuts*, *Plines*, and other "solid" objects can be turned off with the *Fill* command or *FILLMODE* system variable.

Spline

	Pull-down Menu	Command (Type)	Alias (Type)	Short-cut	Screen (side) Menu	Tablet Menu
	Draw Spline	Spline	SPL	...	DRAW 1 Spline	L,9

The *Spline* command creates a NURBS (non-uniform rational bezier spline) curve. The non-uniform feature allows irregular spacing of selected points to achieve sharp corners, for example. A *Spline* can also be used to create regular (rational) shapes such as arcs, circles, and ellipses. Irregular shapes can be combined with regular curves, all in one spline curve definition.

Spline is the newer and more functional version of a *Spline*-fit *Pline* (see "*Pedit*," Chapter 16). The main difference between a *Spline*-fit *Pline* and a *Spline* is that a *Spline* curve passes through the points selected, while the points selected for construction of a *Spline*-fit *Pline* only have a "pull" on the curve. Therefore, *Splines* are more suited to accurate design because the curve passes exactly through the points used to define the curve (data points) (Fig. 15-12).

FIGURE 15-12

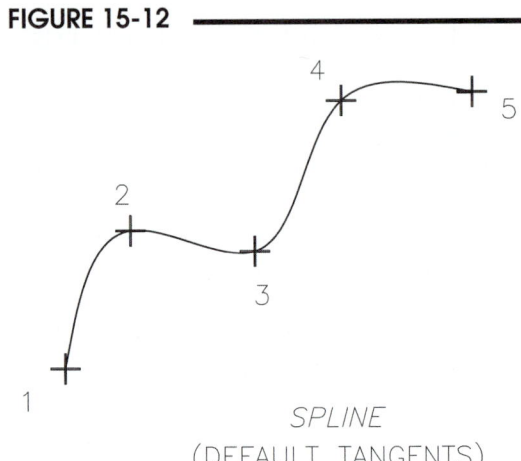

SPLINE
(DEFAULT TANGENTS)

The construction process involves specifying points that the curve will pass through and determining tangent directions for the two ends (for non-closed *Splines*) (Fig. 15-13).

FIGURE 15-13

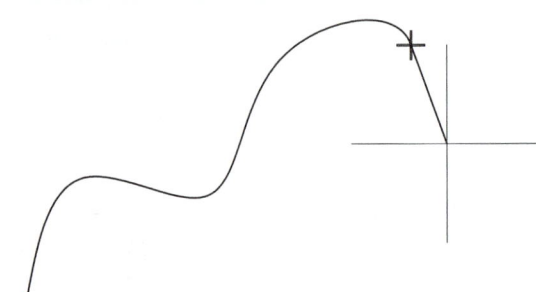

END TANGENT

The *Close* option allows creation of closed *Splines* (these can be regular curves if the selected points are symmetrically arranged) (Fig. 15-14).

FIGURE 15-14

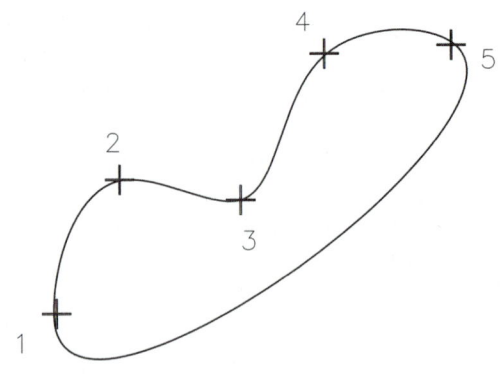

CLOSED SPLINE

Command: **spline**
Specify first point or [Object]: **PICK** or (**coordinates**)
Specify next point: **PICK** or (**coordinates**)
Specify next point or [Close/Fit tolerance] <start tangent>: **PICK** or (**coordinates**)
Specify next point or [Close/Fit tolerance] <start tangent>: **PICK** or (**coordinates**)
Specify next point or [Close/Fit tolerance] <start tangent>: **Enter**
Specify start tangent: **PICK** or **Enter** (Select direction for tangent or Enter for default)
Specify end tangent: **PICK** or **Enter** (Select direction for tangent or Enter for default)
Command:

The *Object* option allows you to convert *Spline*-fit *Plines* into NURBS *Splines*. Only *Spline*-fit *Plines* can be converted (see "*Pedit*," Chapter 16).

Command: **spline**
Specify first point or [Object]: **o**
Select objects to convert to splines ...
Select objects: **PICK**
Select objects: **Enter**
Command:

A *Fit Tolerance* applied to the *Spline* "loosens" the fit of the curve. A tolerance of 0 (default) causes the *Spline* to pass exactly through the data points. Entering a positive value allows the curve to fall away from the points to form a smoother curve (Fig. 15-15).

FIGURE 15-15

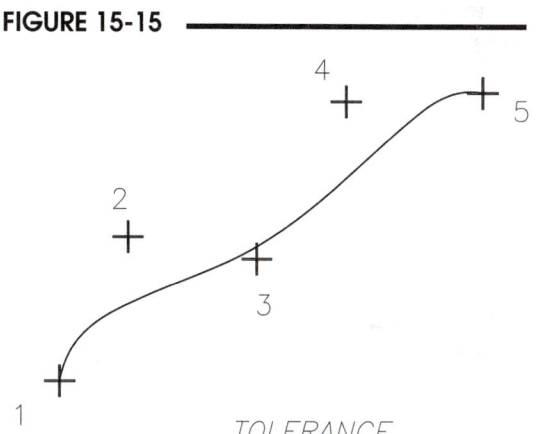

TOLERANCE

Specify next point or [Close/Fit tolerance] <start tangent>: **f**
Specify fit tolerance <0.0000>: (Enter a positive value)

Ellipse

Pull-down Menu	Command (Type)	Alias (Type)	Short-cut	Screen (side) Menu	Tablet Menu
Draw Ellipse	Ellipse	EL	...	DRAW 1 Ellipse	M,9

An *Ellipse* is one object. AutoCAD *Ellipses* are (by default) NURBS curves (see *Spline*). There are three methods of creating *Ellipses* in AutoCAD: (1) specify one <u>axis</u> and the <u>end</u> of the second, (2) specify the <u>center</u> and the ends of each axis, and (3) create an elliptical <u>arc</u>. Each option also permits supplying a rotation angle rather than the second axis length.

Command: **ellipse**
Specify axis endpoint of ellipse or [Arc/Center]: **PICK** or (**coordinates**)
(This is the first endpoint of either the major or minor axis.)
Specify other endpoint of axis: **PICK** or (**coordinates**)
(Select a point for the other endpoint of the first axis.)
Specify distance to other axis or [Rotation]: **PICK** or (**coordinates**)
(This is the distance measured perpendicularly from the established axis.)
Command:

Axis End

This default option requires PICKing three points as indicated in the command sequence above (Fig. 15-16).

FIGURE 15-16

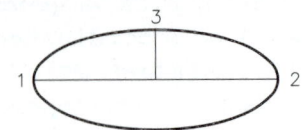
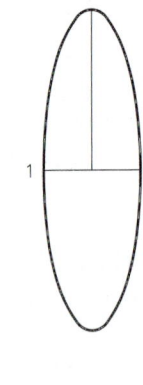

Rotation

If the *Rotation* option is used with the *Axis End* method, the following syntax is used:

    Specify distance to other axis
    or [Rotation]: **r**
    Specify rotation around major
    axis: **PICK** or (**value**)

The specified angle is the number of degrees the shape is rotated <u>from the circular position</u> (Fig. 15-17).

FIGURE 15-17

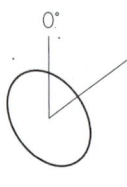
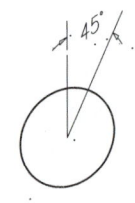
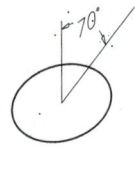

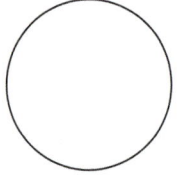
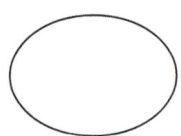
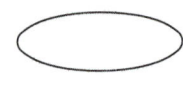

ROTATION = 0   ROTATION = 45   ROTATION = 70

Center

With many practical applications, the center point of the ellipse is known, and therefore the *Center* option should be used (Fig. 15-18):

FIGURE 15-18

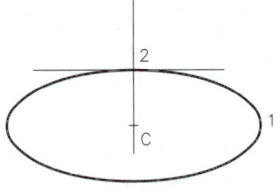
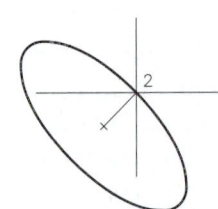
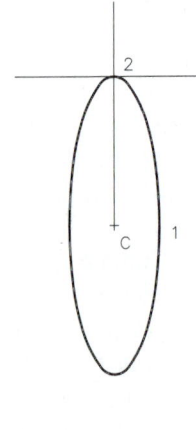

Command: **ellipse**
Specify axis endpoint of ellipse or [Arc/Center]: **c**
Specify center of ellipse: **PICK** or (**coordinates**)
Specify endpoint of axis: **PICK** or (**coordinates**)
(Select a point for the other endpoint of the first axis.)
Specify distance to other axis or [Rotation]: **PICK** or (**coordinates**) (This is the distance measured perpendicularly from the established axis.)
Command:

The *Rotation* option appears and can be invoked after specifying the *Center* and first *Axis endpoint*.

Arc

Use this option to construct an elliptical arc (partial ellipse). The procedure is identical to the *Center* option with the addition of specifying the start- and endpoints for the arc (Fig. 15-19):

FIGURE 15-19

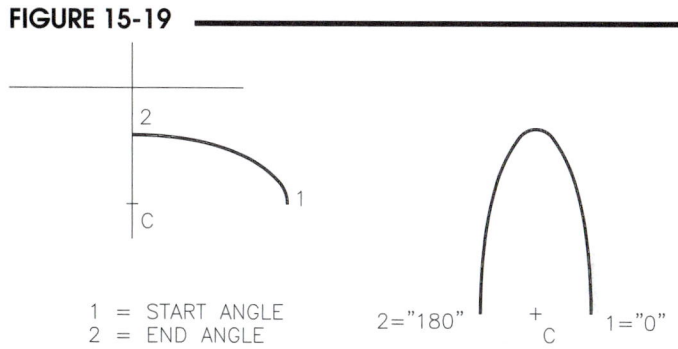

1 = START ANGLE
2 = END ANGLE

2="180"   1="0"

Command: **ellipse**
Specify axis endpoint of ellipse or [Arc/Center]: **a**
Specify axis endpoint of elliptical arc or [Center]: **PICK** or (**coordinates**)
Specify other endpoint of axis: **PICK** or (**coordinates**)
Specify distance to other axis or [Rotation]: **PICK** or (**coordinates**)
Specify start angle or [Parameter]: **PICK** or (**angular value**)
Specify end angle or [Parameter/Included angle]: **PICK** or (**angular value**)
Command:

The *Parameter* option allows you to specify the start point and endpoint for the elliptical arc. The parameters are based on the parametric vector equation: $p(u)=c+a*\cos(u)+b*\sin(u)$, where c is the center of the ellipse and a and b are the major and minor axes.

Divide

Pull-down Menu	Command (Type)	Alias (Type)	Short-cut	Screen (side) Menu	Tablet Menu
Draw *Point >* *Divide*	*Divide*	*DIV*	...	DRAW 2 *Divide*	*V,13*

The *Divide* and *Measure* commands add *Point* objects to existing objects. Both commands are found in the *Draw* pull-down menu under *Point* because they create *Point* objects.

The *Divide* command finds equal intervals along an object such as a *Line, Pline, Spline,* or *Arc* and adds a *Point* object at each interval. The object being divided is <u>not</u> actually broken into parts—it remains as <u>one</u> object. *Point* objects are automatically added to display the "divisions."

The point objects that are added to the object can be used for subsequent construction by allowing you to *OSNAP* to equally spaced intervals (*Nodes*).

The command sequence for the *Divide* command is as follows:

    Command: **divide**
    Select object to divide: **PICK** (Only one object can be selected.)
    Enter the number of segments or [Block]: (**value**)
    Command:

Point objects are added to divide the object selected into the desired number of parts. Therefore, there is <u>one less</u> *Point* added than the number of segments specified.

After using the *Divide* command, the *Point* objects may not be visible unless the point style is changed with the *Point Style* dialog box (*Format* pull-down menu) or by changing the *PDMODE* variable by command line format (see Chapter 8). Figure 15-20 shows *Points* displayed using a *PDMODE* of 3.

FIGURE 15-20

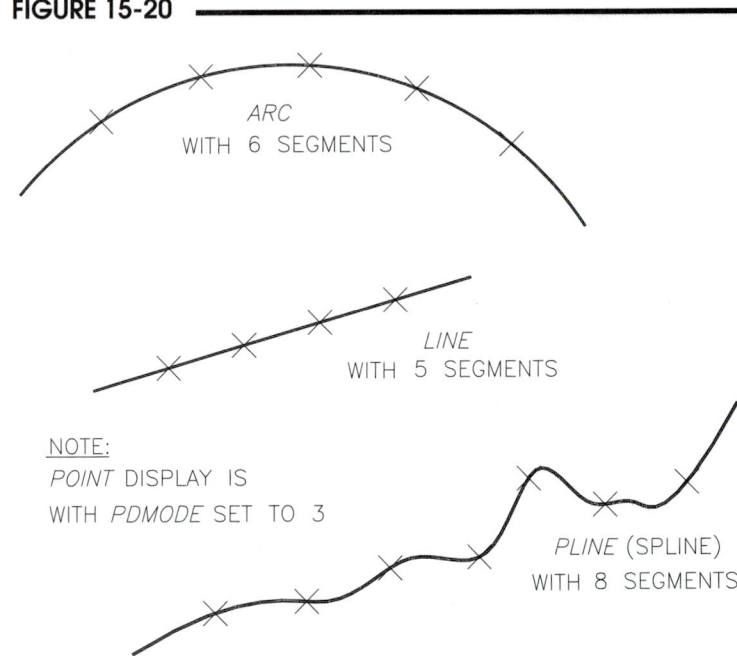

You can request that *Block*s be inserted rather than *Point* objects along equal divisions of the selected object. Figure 15-21 displays a generic rectangular-shaped block inserted with *Divide*, both aligned and not aligned with a *Line*, *Arc*, and *Pline*. In order to insert a *Block* using the *Divide* command, the name of an <u>existing</u> *Block* must be given. (See Chapter 21.)

FIGURE 15-21

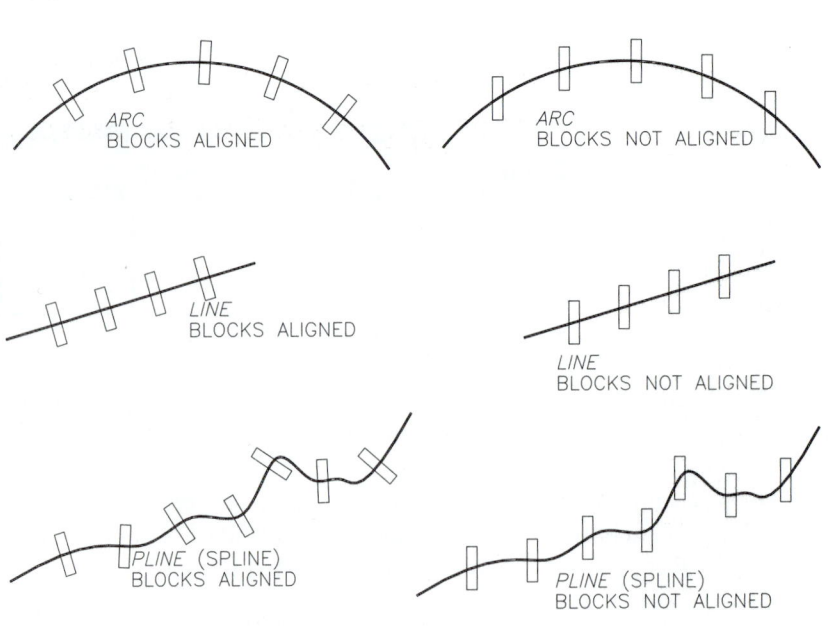

Measure

Pull-down Menu	Command (Type)	Alias (Type)	Short-cut	Screen (side) Menu	Tablet Menu
Draw Point > Measure	*Measure*	*ME*	...	DRAW 2 *Measure*	*V,12*

The *Measure* command is similar to the *Divide* command in that *Point* objects (or *Blocks*) are inserted along the selected object. The *Measure* command, however, allows you to designate the <u>length</u> of segments rather than the <u>number</u> of segments as with the *Divide* command.

    Command: **measure**
    Select object to measure: **PICK** (Only one object can be selected.)
    Specify length of segment or [Block]: (**value**) (Enter a value for length of one segment.)
    Command:

Point objects are added to the selected object at the designated intervals (lengths). One *Point* is added for each interval <u>beginning at the end nearest</u> the end used for object selection (Fig. 15-22). The intervals are of equal length except possibly the last segment, which is whatever length is remaining.

You can request that *Blocks* be inserted rather than *Point* objects at the designated intervals of the selected object. Inserting *Blocks* with *Measure* requires that an <u>existing</u> *Block* be used.

FIGURE 15-22

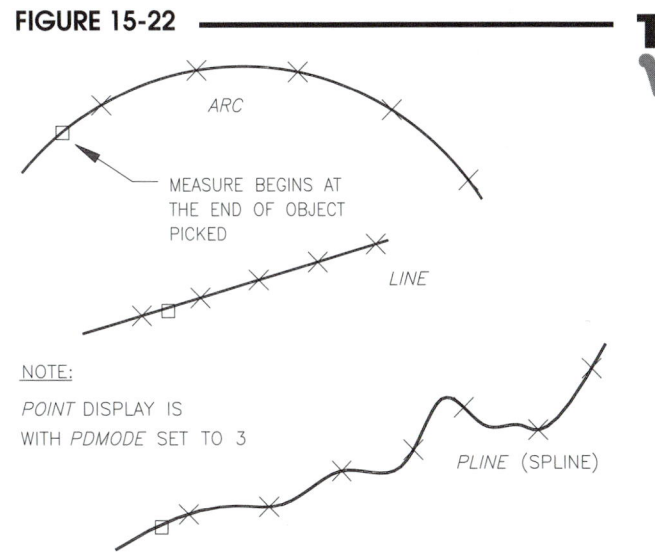

ARC

MEASURE BEGINS AT THE END OF OBJECT PICKED

LINE

<u>NOTE:</u>

POINT DISPLAY IS WITH *PDMODE* SET TO 3

PLINE (SPLINE)

TIP

Mline

Pull-down Menu	Command (Type)	Alias (Type)	Short-cut	Screen (side) Menu	Tablet Menu
Draw Multiline	*Mline*	*ML*	...	DRAW 1 *Mline*	*M,10*

Mline is short for multiline. A multiline is a set of <u>parallel lines</u> that behave as <u>one object</u>. The individual lines in the set are called line <u>elements</u> and are defined to your specifications. The set can contain up to 16 individual lines, each having its own offset, linetype, or color.

Mlines are useful for creating drawings composed of many parallel lines. An architectural floor plan, for example, contains many parallel lines representing walls. Instead of constructing the parallel lines individually (symbolizing the wall thickness), only <u>one</u> *Mline* need be drawn to depict the wall. The thickness and other details of the wall are dictated by the definition of the *Mline*.

<u>Mline</u> is the command to <u>draw</u> multilines. To <u>define</u> the individual line elements of a logical set (offset, linetype, color), use the <u>*Mlstyle*</u> command. An *Mlstyle* (multiline style) is a particular combination of defined line elements. Each *Mlstyle* can be saved in an external file with a .MLN file type. In this way, multilines can be used in different drawings without having to redefine the set. See "*Mlstyle*" next.

The *Mline* command operates similar to the *Line* command, asking you to specify points:

```
Command: mline
Current settings: Justification = Top, Scale = 1.00, Style = STANDARD
Specify start point or [Justification/Scale/STyle]: PICK or (coordinates)
Specify next point: PICK or (coordinates)
Specify next point or [Undo]: PICK or (coordinates)
Specify next point or [Close/Undo]: PICK or (coordinates)
Specify next point or [Close/Undo]: Enter
Command:
```

Mline provides the following three options for drawing multilines.

Justification

Justification determines how the multiline is drawn between the points you specify. The choices are *Top, Zero,* or *Bottom* (Fig. 15-23). *Top* aligns the top of the set of line elements along the points you PICK as you draw. A *Zero* setting uses the center of the set of line elements as the PICK point, and *Bottom* aligns the bottom line element between PICK points.

NOTE: The *Justification* methods are defined for drawing from <u>left to right</u> (positive X direction). In other words, using the *Top* justification, the top of the multiline (as it is defined in the *Mlstyle* dialog box) is between PICK points when you draw from left to right. PICKing points from right to left draws the *Mline* upside down.

FIGURE 15-23

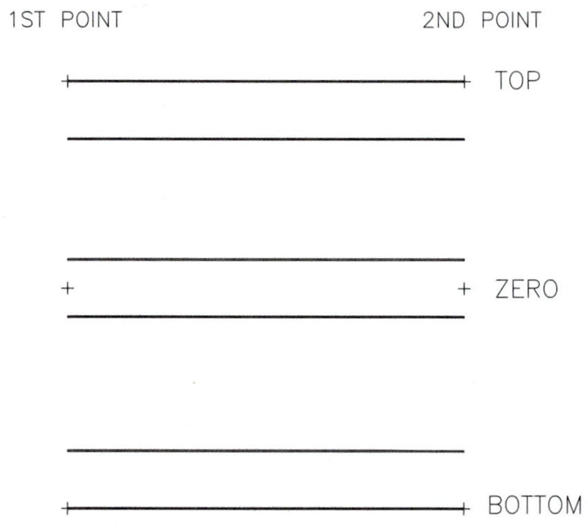

1ST POINT   2ND POINT

TOP

ZERO

BOTTOM

Scale

The *Scale* option controls the overall <u>width</u> of the multi-line. The value you specify for the scale factor increases or decreases the defined multiline width proportionately. In Figure 15-24, a scale factor of 2.0 provides a multiple line pattern that is twice as wide as the *Mlstyle* definition; a scale of 3.0 provides a pattern three times the width.

FIGURE 15-24

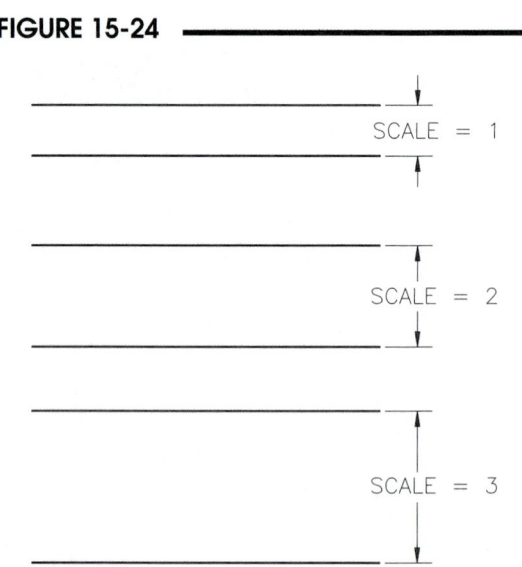

SCALE = 1

SCALE = 2

SCALE = 3

You can also enter negative scale values or a value of 0. A negative scale value flips the order of the multiple line pattern (Fig. 15-25). A value of 0 collapses the multiline into a single line. The utility of the *Scale* option allows you to vary a multiline set proportionally without the need for a complete new multiline definition.

FIGURE 15-25

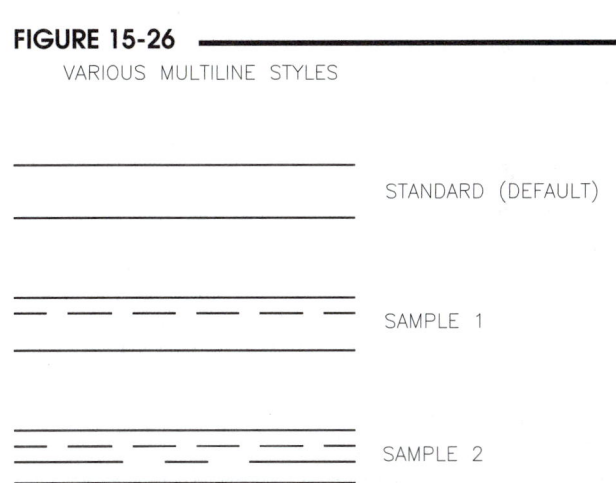

SCALE = 1

SCALE = −1

SCALE = 0

Style

The *Style* option enables you to load different multiline styles provided they have been previously created. Some examples of multiline styles are shown in Figure 15-26. Only one multiline *Style* called STANDARD is provided by AutoCAD. The STANDARD multiline style has two parallel lines defined at 1 unit width.

Using the *Style* option of the *Mline* command provides only the Command line format for setting a current style. Instead, you can use the *Mlstyle* dialog box to select a current multiline style by a more visible method.

FIGURE 15-26

VARIOUS MULTILINE STYLES

STANDARD (DEFAULT)

SAMPLE 1

SAMPLE 2

Mlstyle

Pull-down Menu	Command (Type)	Alias (Type)	Short-cut	Screen (side) Menu	Tablet Menu
Format Multiline Style...	Mlstyle	...	...	DRAW 1 Mline Mlstyle:	V,5

The *Mline* command draws multilines, whereas the *Mlstyle* controls the configuration of the multilines that are drawn. Use the *Mlstyle* command to <u>create new</u> multiline definitions or to <u>load, select, and edit</u> previously created multiline styles. *Mlstyle* is <u>not a draw command</u> and is therefore found not in the *Draw* menus but in the *Format* pull-down and screen menus.

Using the *Mlstyle* command by any method invokes the *Multiline Styles* dialog box (see Fig. 15-27). This and the related dialog boxes are explained next.

FIGURE 15-27

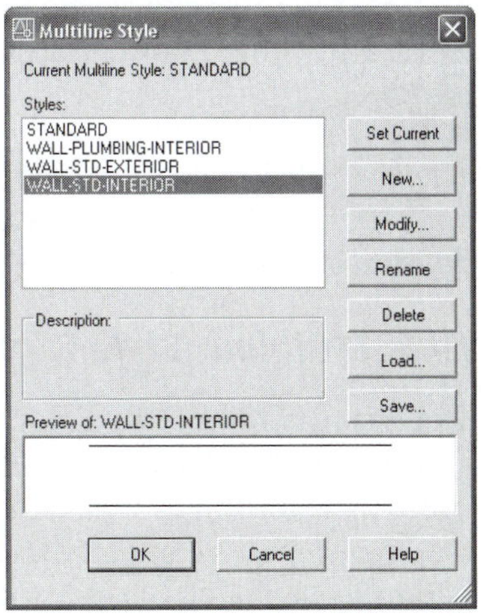

Multiline Style **Dialog Box**

Styles List

This list displays the names of the currently loaded multiline styles. The image at the bottom of the dialog box provides a representation of the highlighted multiline style definition so you can display each of the loaded multiline styles by selecting the names in the list. The *Style* option of the *Mline* command serves the same purpose, but operates only in command line format and requires you to know the specific *Mlstyle* name and definition.

Set Current

Highlight a multiline style from the list, then select *Set Current* to make it the current style. Select the *OK* button to return to the drawing to draw with the current style.

New

Select *New* to create a new multiline style and produce the *Create New Multiline Style* dialog box (not shown). You can select any loaded style to *Start with* as a template. Short names are acceptable since a *Description* of the multiline configuration can be entered in the dialog box that appears next (see "*New Multiline Style* Dialog Box").

Modify

Select any multiline style from the list to *Modify*. The *Modify Multiline* Style dialog box appears, which is essentially the same as the *New Multiline Style* dialog box (see "*New Multiline Style* Dialog Box").

Rename and Delete

You can highlight any name from the list to *Rename*. The *Delete* tile allows you to delete the highlighted multiline style. A confirmation dialog box appears.

Save

Multiline styles that you create are automatically saved with the drawing file. However, using *Save* allows you to save the highlighted multiline style definition to an external (.MLN) file so you can use that multiline style in other drawings without having to recreate the style. The *Save Multiline File* dialog box appears where you can specify the .MLN file name and location. Although multiple multiline styles can be saved to one .MLN file, only one style at a time can be saved. The default multiline definition file is ACAD.MLN, which contains only the STANDARD multiline style. This file can be appended with additional style definitions or a new .MLN file can be created. A .MLN file can contain more than one multiline style.

FIGURE 15-28

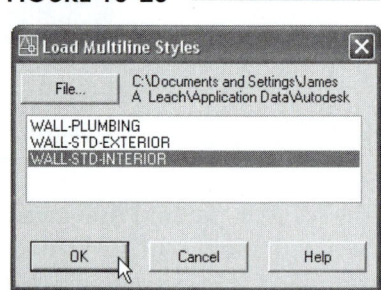

NOTE: You must choose the *OK* button in the *Multiline Style* dialog box to save any new multiline style definitions to the current drawing file.

Load

The *Load* tile invokes the *Load Multiline Styles* dialog box where you can select one existing multiline style definition from an .MLN file (Fig. 15-28).

New Multiline Style **Dialog Box**

The *New Multiline Style* dialog box (Fig. 15-29) and the identical *Modify Multiline Style* dialog box allow you to specify the multiline style definition—the properties of the line elements contained in the style. Options specified in these dialog boxes are immediately displayed in the *Multiline Style* dialog box preview.

FIGURE 15-29

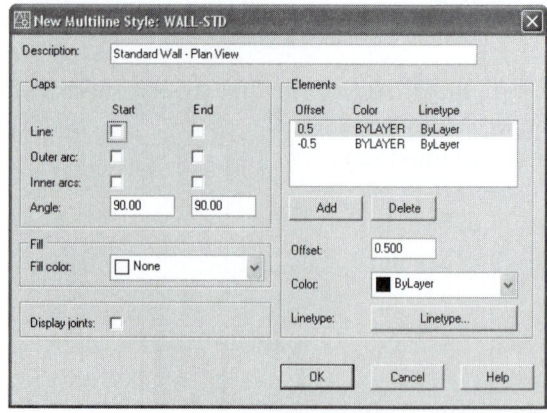

Description

The *Description* edit box allows you to provide an optional description for a new or existing multiline style definition. The description can contain up to 256 characters.

Caps

Use this section to specify how the ends of the multi-lines are "capped." A cap can be set for the *Start* and/or *End* segments of a multiline series (Fig. 15-30).

The *Line* checkboxes create straight line caps across either or both of the *Start* or *End* segments of the multiline. The *Line* caps are drawn across all line elements (Fig. 15-30).

The *Outer Arcs* option creates an arc connecting and tangent to the two outermost line elements. You can control both the *Start* and *End* segments.

Selecting *Inner Arcs* creates an arc between interior line segments. This option works with four or more line elements. In the case of an uneven number of line elements (five or more), the middle line is not capped.

FIGURE 15-30 ───────────

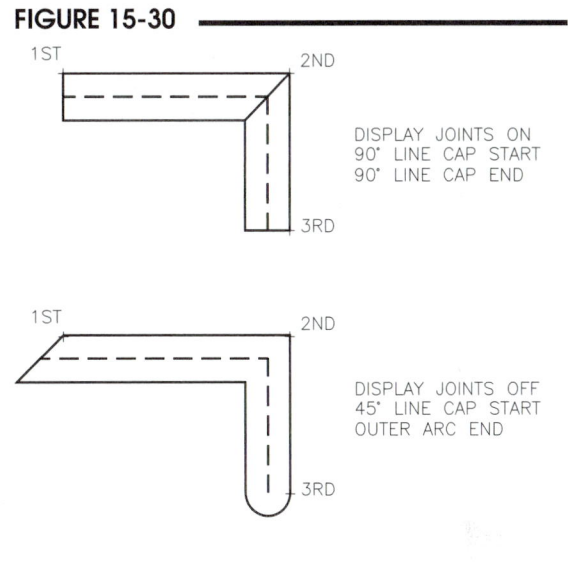

The *Angle* option controls the angle of start and end lines or arcs. Valid values range from 10 to 170 degrees.

Fill

Selecting a *Fill Color* produces a multiline with color filled between the outermost lines. Although the image preview does not display the multiline fill, the color fill appears in the drawing.

Display Joints

When several segments of an *Mline* are drawn, AutoCAD automatically miters the "joints" so the segments connect with sharp, even corners. The "joint" is the bisector of the angle created by two adjoining segments. The *Display Joints* checkbox toggles on and off the display of this miter line running across all the line elements.

Elements

The *Elements* list displays the current settings for the multiline style. Each line entry (element) in a style has an offset, color, and linetype associated with it. A multiline definition can have as many as 16 individual line elements defining it. Select any element in the list before changing its properties with the options below.

Add

New line elements are added to the list by selecting the *Add* button; then the *Offset*, *Color*, and *Linetype* properties can be assigned. The new element appears in the list with an offset of 0, color set to *Bylayer*, and linetype set to *Bylayer*.

Delete

The highlighted line element in the *Elements* list is deleted by selecting the *Delete* button.

Offset

Enter a value in the *Offset* edit box for each line element. The offset value is the distance measured perpendicular to the theoretical zero (center) position along the axis of the *Mline*.

Color

Select any color from the drop-down list or choose *Select Color* to produce the standard *Select Color* dialog box. Any color selected is assigned to the highlighted line element in the list.

Linetype

Use this button to assign a linetype to the highlighted line element. The standard *Select Linetype* dialog box appears. Linetypes not in the current drawing file can be *Loaded*.

Multiline Styles **Example**

Because the process of creating and editing multiline styles is somewhat involved, here is an example that you can follow to create a new multiline style. The specifications for two multilines are given in Figure 15-31. Follow these steps to create SAMPLE1.

1. Invoke the *Multiline Styles* dialog box by any menu or icon or by typing *Mlstyle*.

2. Select the *New* button and enter the name SAMPLE1 in the *Name* edit box. Select *OK* to invoke the *New Multiline Style* dialog box.

3. Select the *Add* button twice. Four line entries should appear in the *Elements* list box. The two new entries have a 0.0 offset.

4. Highlight one of the new entries and change its offset by entering 0.2 in the *Offset* edit box. Next, change the linetype by selecting the *Linetype* button and choosing the *Center* linetype. (If the *Center* linetype is not listed, you should *Load* it.)

5. Highlight the other new entry and change its *Offset* to -0.3. Next, change its *Linetype* to *Hidden*.

6. Select the *OK* button to return to the *Multiline Styles* dialog box. Highlight the new SAMPLE1 style from the list and select *Set Current*. Last, choose the *OK* button to save the multiline style to the drawing.

7. Use *Mline* to draw several segments and to test your new multiline style.

8. Use *Mlstyle* again and select *Modify* to create Line end caps and a *Color Fill* to the style.

Now try to create SAMPLE2 multiline style on your own (see Fig. 15-31).

Editing Multilines

Because a multiline is treated as one object and its complex configuration is defined by the multiline style, a special set of editing functions is designed for *Mlines*. The *Mledit* (multiline edit) command provides the editing functions for existing multilines and is discussed in Chapter 16. Only a few other traditional editing commands such as *Trim* and *Extend* can be used for editing functions. Your usefulness for *Mlines* is limited until you can edit them with Mledit.

FIGURE 15-31

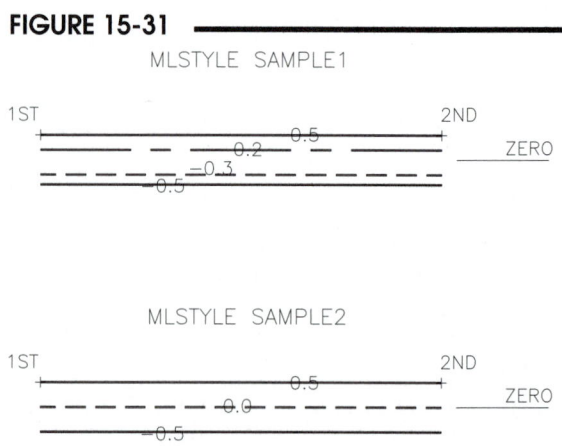

Sketch

Pull-down Menu	Command (Type)	Alias (Type)	Short-cut	Screen (side) Menu	Tablet Menu
...	Sketch	...	...	...	...

FIGURE 15-32

The *Sketch* command is unlike other draw commands. *Sketch* quickly creates many short line segments (individual objects) by following the motions of the cursor. *Sketch* is used to give the appearance of a freehand "sketched" line, as used in Figure 15-32 for the tree and bushes. You do not specify individual endpoints, but rather draw a freehand line by placing the pen down, moving the cursor, and then picking up the pen. This action creates a large number of short *Line* segments. *Sketch* line segments are temporary (displayed in a different color) until *Recorded* or until *eXiting Sketch*.

CAUTION: *Sketch* can increase the drawing file size greatly due to the relatively large number of line segment endpoints required to define the *Sketch* line.

```
Command: sketch
Record increment <0.1000>: (value) or Enter
```
(Enter a value to specify the segment increment length or press Enter to accept the default increment.)
```
Sketch. Pen eXit Quit Record Erase Connect.
```
(**letter**) (Enter "p" or press button #1 to put the pen down and begin drawing or enter another letter for another option. After drawing a sketch line, enter "p" or press button #1 again to pick up the pen.)

FIGURE 15-33

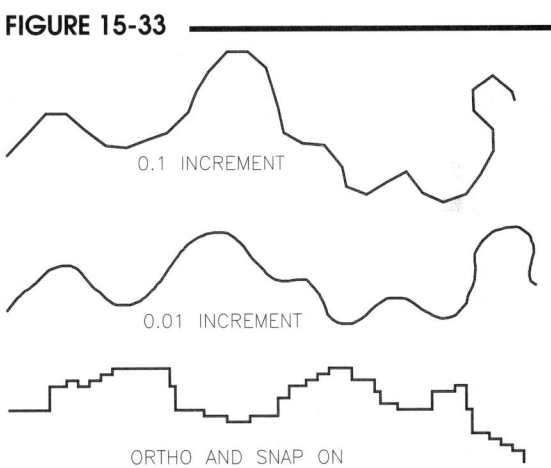

0.1 INCREMENT

0.01 INCREMENT

ORTHO AND SNAP ON

It is important to specify an *increment* length for the short line segments that are created. This *increment* controls the "resolution" of the *Sketch* line (Fig. 15-33).

Too large of an *increment* makes the straight line segments apparent, while too small of an *increment* unnecessarily increases file size. The default *increment* is 0.1 (appropriate for default *Limits* of 12 x 9) and should be changed proportionally with a change in *Limits*. Generally, multiply the default *increment* of 0.1 times the drawing scale factor.

Another important rule to consider whenever using *Sketch* is to <u>turn *SNAP Off* and *ORTHO Off*</u>, unless a "stair-step" effect is desired (see Fig. 15-33). *Polar Snap* and *Polar Tracking* do not affect *Sketch* lines.

Sketch Options
Options of the *Sketch* command can be activated either by entering the corresponding letter(s) shown in uppercase at the *Sketch* command prompt or by pressing the desired mouse or puck button. For example, putting the pen up or down can be done by entering *P* at the Command line or by pressing button #1. The options of *Sketch* are as follows.

Pen	Lifts and lowers the pen. Position the cursor at the desired location to begin the line. Lower the pen and draw. Raise the pen when finished with the line.
Record	Records all temporary lines sketched so far without changing the pen position. After recording, the lines cannot be *Erased* with the *Erase* option of *Sketch* (although the normal *Erase* command can be used).
eXit	Records all temporary lines entered and returns to the Command: prompt.
Quit	Discards all temporary lines and returns to the Command: prompt.
Erase	Allows selective erasing of temporary lines (before recording). To erase, move backward from last sketch line toward the first. Press "p" to indicate the end of an erased area. This method works easily for relatively straight sections. To erase complex sketch lines, *eXit* and use the normal *Erase* command with window, crossing, or pickbox object selection.
Connect	Allows connection to the end of the <u>last</u> temporary sketch line (before recording). Move the cursor to the last sketch line and the pen is automatically lowered.
(period)	Draws a straight line (using *Line*) from the last sketched line to the cursor. After adding the straight lines, the pen returns to the up position.

Several options are illustrated in Figure 15-34. The *Pen* option lifts the pen up and down. A *period* (.) causes a straight line segment to be drawn from the last segment to the cursor location. *Erase* is accomplished by entering *E* and making a reverse motion.

As an alternative to the *Erase* option of *Sketch*, the *Erase* command can be used to erase all or part of the *Sketch* lines. Using a *Window* or *Crossing Window* is suggested to make selection of all the objects easier.

FIGURE 15-34

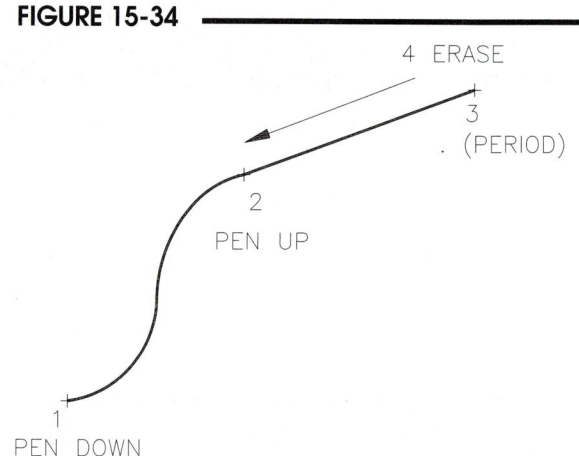

The *SKPOLY* Variable

The *SKPOLY* system variable controls whether AutoCAD creates connected *Sketch* line segments as one *Pline* (one object) or as multiple *Line* segments (separate objects). *SKPOLY* affects <u>newly created</u> *Sketch* lines only (Fig. 15-35).

FIGURE 15-35

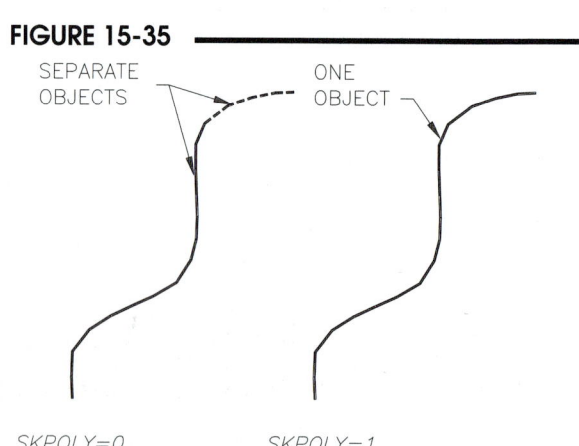

SKPOLY=0	This setting generates *Sketch* segments as individual *Line* objects. This is the <u>default</u> setting.
SKPOLY=1	This setting generates connected *Sketch* segments as <u>one *Pline* object</u>.

Using editing commands with *Sketch* lines can normally be tedious (when *SKPOLY* is set to the default value of 0). In this case, editing *Sketch* lines usually requires *Zooming* in since the line segments are relatively small. However, changing *SKPOLY* to 1 simplifies operations such as using *Erase* or *Trim*. For example, *Sketch* lines are sometimes used to draw break lines (Fig. 15-36) or to represent broken sections of mechanical parts, in which case use of *Trim* is helpful. If you expect to use *Trim* or other editing commands with *Sketch* lines, change *SKPOLY* to 1 before creating the *Sketch* lines.

FIGURE 15-36

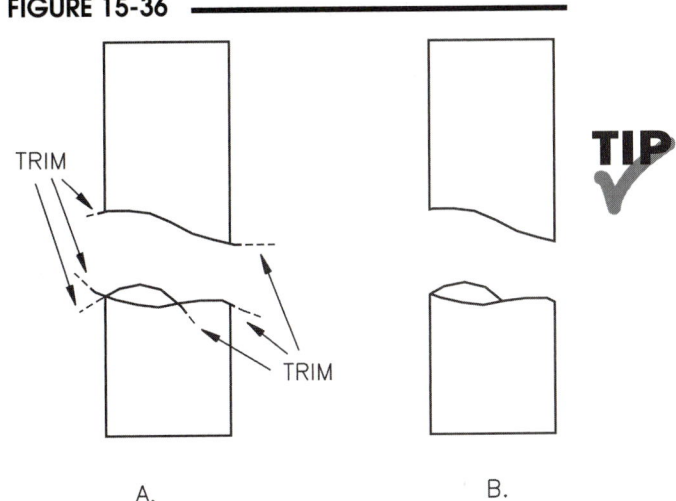

TRIM

TRIM

A.   B.

Solid

Pull-down Menu	Command (Type)	Alias (Type)	Short-cut	Screen (side) Menu	Tablet Menu
Draw Surfaces > 2D Solid		Solid	SO	...	DRAW 2 SURFACES Solid:	L,8

The *Solid* command creates a 2D shape that is <u>filled solid</u> with the current color (object or *ByLayer*). The shape is <u>not</u> a 3D solid. Each 2D solid has either <u>three or four straight edges</u>. The construction process is relatively simple, but the order of PICKing corners is critical. In order to construct a square or rectangle, draw a bow-tie configuration (Fig. 15-37). Several *Solids* can be connected to produce more complex shapes.

FIGURE 15-37

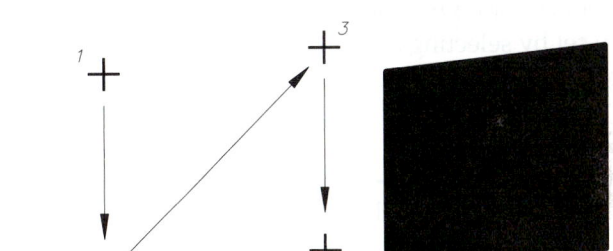

SELECTION ORDER   RESULTING SOLID

A *Solid* object is useful when you need shapes filled with solid color; for example, three-sided *Solids* can be used for large arrowheads. Keep in mind that solid filled areas may cause problems for some output devices such as plotters with felt pens and thin paper. As an alternative to using *Solid*, enclosed areas can be filled with hatch patterns or solid color (see "*Hatch*," Chapter 26).

Boundary

Pull-down Menu	Command (Type)	Alias (Type)	Short-cut	Screen (side) Menu	Tablet Menu
Draw Boundary...	Boundary or -Boundary	BO or -BO	...	DRAW 2 Boundary	Q,9

Boundary finds and draws a boundary from a group of connected or overlapping shapes forming an enclosed area. The shapes can be any AutoCAD objects and can be in any configuration, as long as they form a totally enclosed area. *Boundary* creates either a *Polyline* or *Region* object forming the boundary shape (see *Region* next). The resulting *Boundary* does not affect the existing geometry in any way. *Boundary* finds and includes internal closed areas (called "islands") such as circles (holes) and includes them as part of the *Boundary*.

Wipeout

	Command (Type)	Alias (Type)	Short-cut	Screen (side) Menu	Tablet Menu
Pull-down Menu					
Draw Wipeout	Wipeout	...	...	...	...

Wipeout creates an "opaque" object that is used to "hide" other objects. The *Wipeout* command creates a closed *Pline* or uses an existing closed *Pline*. All objects behind the *Wipeout* become "invisible." For example, assume a *Pline* shape was created so as to cross other objects. You could use *Wipeout* to make the shape appear to be "on top of" the other objects (Fig. 15-42, left).

FIGURE 15-42

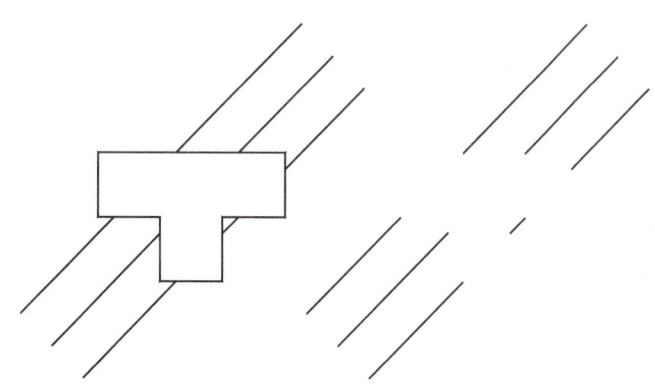

```
Command: wipeout
Specify first point or [Frames/Polyline]
<Polyline>: PICK
Specify next point: PICK
Specify next point or [Undo]: PICK
Specify next point or [Close/Undo]: Enter
Command:
```

You can use the *Frames* option to make the *Wipeout* object disappear. Doing so would result in a display similar to Figure 15-42, right.

```
Specify first point or [Frames/Polyline] <Polyline>: f
Enter mode [ON/OFF] <ON>: off
```

If you have an existing *Pline* and want to use it to create a *Wipeout* of the same shape, use the *Polyline* option.

```
Command: wipeout
Specify first point or [Frames/Polyline] <Polyline>: p
Select a closed polyline: PICK
Erase polyline? [Yes/No] <No>:
```

Selecting *No* to the "Erase polyline?" prompt results in two objects that occupy the same space—the existing *Pline* and the new *Wipeout*. To create an "invisible" *Wipeout* you must specify the *Erase Polyline* option when you create the *Wipeout*, then use the *Frame Off* option to make the *Wipeout* object disappear.

Revcloud

	Command (Type)	ALIAS (Type)	Short-cut	Screen (side) Menu	Tablet Menu
Pull-down Menu					
Draw Revision Cloud	Revcloud	...	...	...	...

Use this command to draw a revision cloud on the current layer. Revision clouds are the customary method of indicating an area of a drawing (usually an architectural application) that contains a revision to the original design. Revision clouds should be created on a separate layer so they can be controlled for plotting.

Revcloud is an automated routine that simplifies drawing revision clouds. All you have to do is pick a point and move the cursor in a circular direction (Fig. 15-43). You do not have to "pick" an endpoint. The cloud automatically closes when your cursor approaches the start point and the command ends. A *Revcloud* is a series of *Pline* arc segments.

FIGURE 15-43

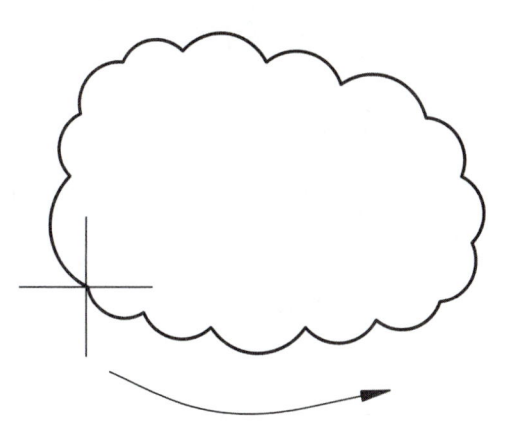

```
Command: revcloud
Minimum arc length: 0.5000   Maximum arc length:
0.5000  Style: Normal
Specify start point or [Arc length/Object] <Object>:
Guide crosshairs along cloud path...
Revision cloud finished.
Command:
```

Arc length

Use the *Arc length* option to specify a value in drawing units for the arc segments. Entering different values for the minimum and maximum arc lengths draws *Revclouds* with different size arcs, similar to that shown in Figure 15-43.

Object

You can convert some closed objects into *Revclouds* using the *Object* option. *Rectangles*, *Polygons*, *Circles*, and *Ellipses* can be converted to *Revclouds*.

Style

Use this option to specify the line style that the *Revcloud* is drawn with, either *Normal* or *Calligraphy*. The *Normal* option creates arcs with a constant line width of 0, such as shown in Figure 15-43. The *Calligraphy* option creates a "stroke" similar to a calligraphy pen. With this option each arc is drawn starting with a thin lineweight and ending with a thick lineweight.

CHAPTER EXERCISES

1. *Polygon*

    Open the **PLINE1** drawing you created in Chapter 8 Exercises. Use *Polygon* to construct the polygon located at the *Center* of the existing arc, as shown in Figure 15-44. When finished, *SaveAs* **POLYGON1.**

FIGURE 15-44

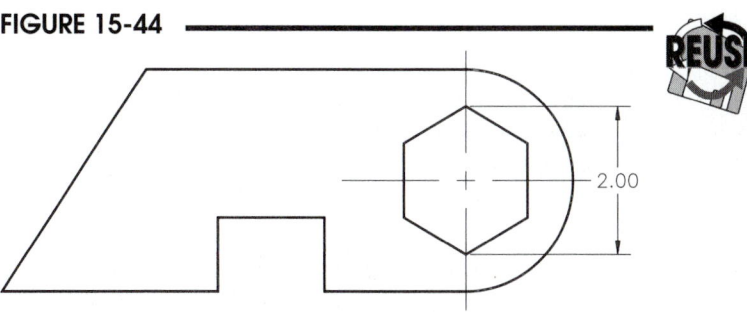

2. *Xline, Ellipse, Rectangle*

    Open the **APARTMENT** drawing that you created in Chapter 12. Draw the floor plan of the efficiency apartment on layer FLOORPLAN as shown in Figure 15-45. Use *SaveAs* to assign the name **EFF-APT.** Use *Xline* and *Line* to construct the floor plan with 8" width for the exterior walls and 5" for interior walls. Use the *Ellipse* and *Rectangle* commands to design and construct the kitchen sink, tub, wash basin, and toilet. *Save* the drawing but do not exit AutoCAD. Continue to Exercise 3.

FIGURE 15-45

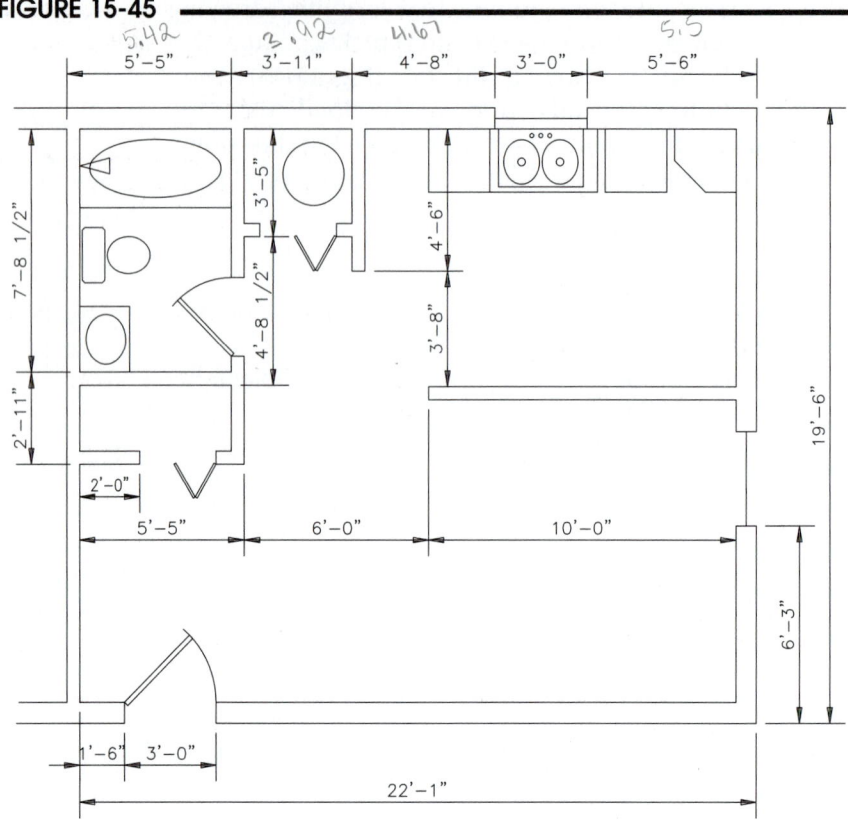

3. *Polygon, Sketch*

    Create a plant for the efficiency apartment as shown in Figure 15-46. Locate the plant near the entry. The plant is in a hexagonal pot (use *Polygon*) measuring 18" across the corners. Use *Sketch* to create 2 leaves as shown in Figure A. Create a *Polar Array* to develop the other leaves similar to Figure B. *Save* the **EFF-APT** drawing.

FIGURE 15-46

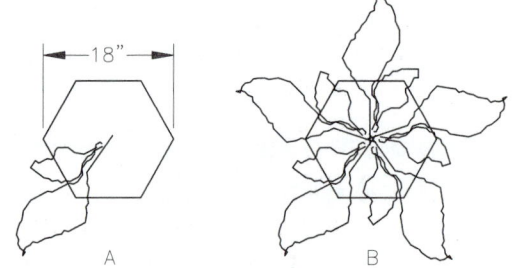

4. *Pline, Page Setup, Vports,* and *Viewport Scale*
    This exercise gives you experience creating a border and plotting the efficiency apartment to scale. To complete the exercise as instructed, you should have a device configured to plot on a "C" size sheet.

    A. Invoke the *Options* dialog box and access the *Display* tab. In the *Layout Elements* section (lower left), ensure that *Show Page Setup Manager for New Layouts* and *Create Viewport in New Layouts* are not checked.

    B. Select the *Layout1* tab. Activate the *Page Setup Manager*. Select *Modify*, then select a plot device from the list that will allow you to plot on an architectural "C" size sheet (**24" x 18"**). Next, select the correct sheet size. Pick *OK* to finish the setup.

C. Make layer **TITLE** the *Current* layer. Draw a border (in paper space) using a *Pline* with a *width* of **.04**, and draw the border with a **.75** unit margin within the edge of the *Limits* (border size of 22.5 x 16.5). *Save* the drawing as **EFF-APT-PS.**

D. Make layer **VPORTS** *Current*. Use *Vports* to create a *Single* viewport. Pick diagonal corners for the viewport at **1,1** and **23,17**. The apartment drawing should appear in the viewport at no particular scale. Use any method to set the viewport scale to *1/2"=1'*.

FIGURE 15-47

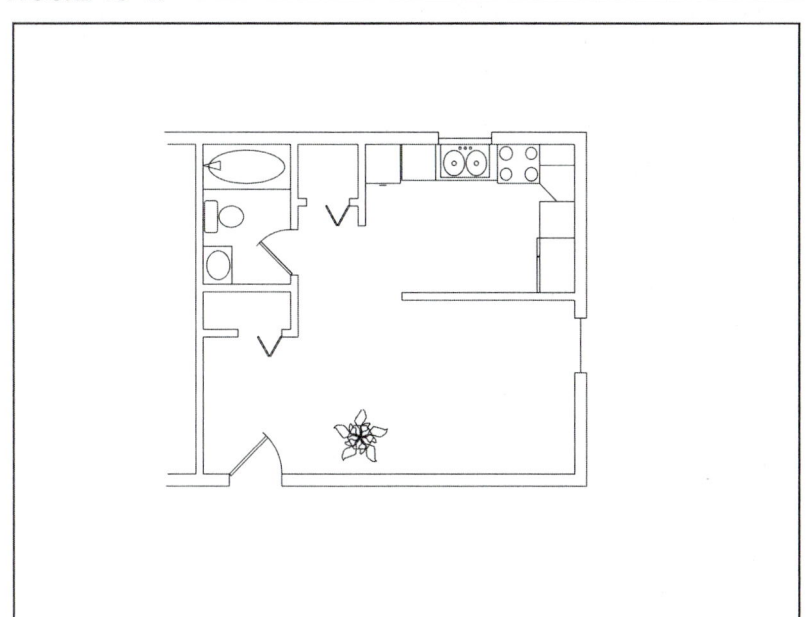

E. Next, *Freeze* layer **VPORTS**. Your completed drawing should look like that in Figure 15-47. *Save* the drawing. *Plot* the drawing from the layout at **1:1** scale.

5. *Donut*

Open **SCHEMATIC1** drawing you created in Chapter 9 Exercises. Use the *Point Style* dialog box (*Format* pull-down menu) to change the style of the *Point* objects to dots instead of small circles. Turn on the *Node Running Osnap* mode. Use the *Donut* command and set the *Inside diameter* and the *Outside diameter* to an appropriate value for your drawing such that the *Inside diameter* is 1/2 the value of the *Outside diameter*. Create a *Donut* at each existing **Node**. *SaveAs* **SCHEMATIC1B**. Your finished drawing should look like that in Figure 15-48.

FIGURE 15-48

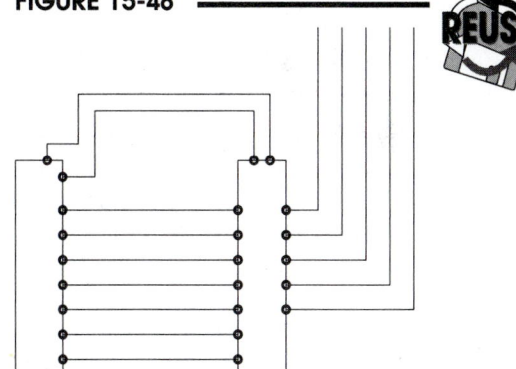

6. *Ellipse, Polygon*

Begin a *New* drawing using the template drawing that you created in Chapter 12 named **A-METRIC** Use *SaveAs* and assign the name **WRENCH.**

FIGURE 15-49

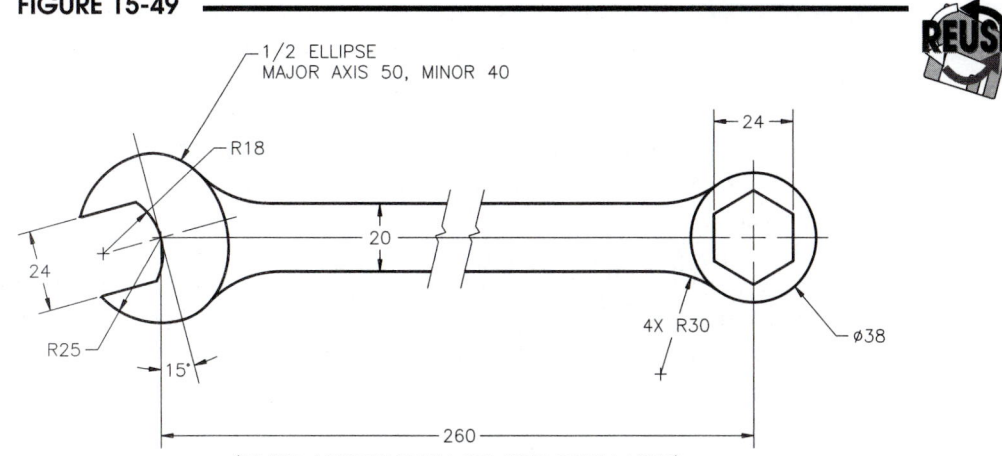

1/2 ELLIPSE
MAJOR AXIS 50, MINOR 40

R18

24

20

24

R25

15°

4X R30

ø38

260

(DRAWN APPROXIMATELY 180 WITH BREAK LINES)

METRIC

Complete the construction of the wrench shown in Figure 15-49. Center the drawing within the *Limits*. Use *Line*, *Circle*, *Arc*, *Ellipse*, and *Polygon* to create the shape. For the "break lines," create one connected series of jagged *Lines*, then *Copy* for the matching set. Utilize *Trim*, *Rotate*, and other edit commands where necessary. HINT: Draw the wrench head in an orthogonal position; then rotate the entire shape 15 degrees. Notice the head is composed of ½ of a *Circle* (radius=25) and ½ of an *Ellipse*. *Save* the drawing when completed. Activate a *Layout* tab, configure a print or plot device using *Pagesetup*, make one *Viewport*, and *Plot* the drawing at a standard scale.

7. *Xline, Spline*

Create the handle shown in Figure 15-50. Construct multiple *Horizontal* and *Vertical* *Xlines* at the given distances on a separate layer (shown as dashed lines in Figure 15-50). Construct two *Splines* that make up the handle sides by PICKing the points indicated at the *Xline* intersections. Note the *End Tangent* for one *Spline*. Connect the *Splines* at each end with horizontal *Lines*. *Freeze* the construction (*Xline*) layer. *Save* the drawing as **HANDLE.**

FIGURE 15-50

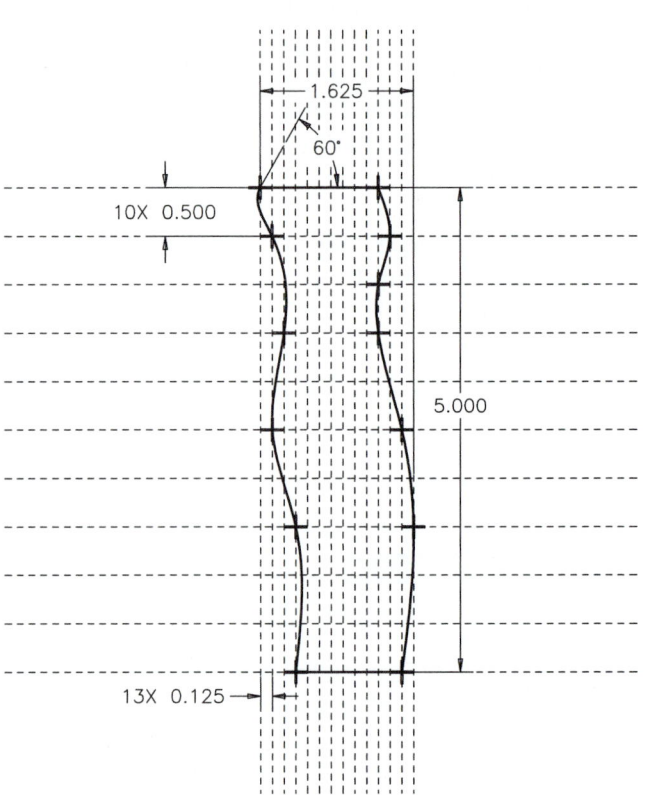

8. *Divide, Measure*

Use the **ASHEET** template drawing, use *SaveAs*, and assign the name **BILLMATL.** Create the table in Figure 15-51 to be used as a bill of materials. Draw the bottom *Line* (as dimensioned) and a vertical *Line*. Use *Divide* along the bottom *Line* and *Measure* along the vertical *Line* to locate *Points* as desired. Create *Offsets Through* the *Points* using *Node OSNAP*. (*ORTHO* and *Trim* may be of help.)

FIGURE 15-51

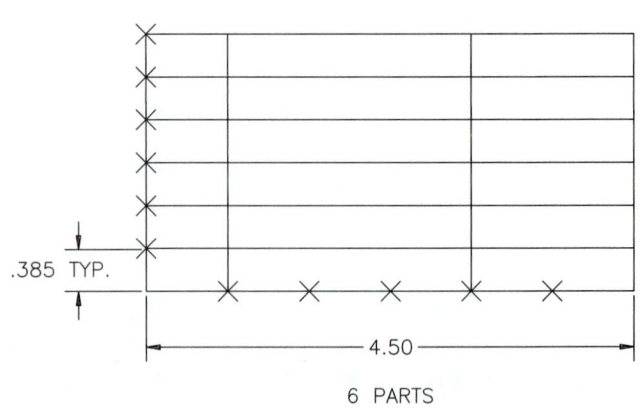

9. *Mline, Mlstyle, Sketch*

FIGURE 15-52

Using the drawing that you created in Chapter 8, add the concrete walls, island, and curbs (Fig. 15-52). Use *Mlstyle* to create a *New* multiline style named **WALL**. The new style should have two line elements with *Offsets* of **1** and **-1** and *Start* and *End Caps* at **90** degrees. Next, use *Mline* with the *Bottom Justification* and draw one *Mline* along the existing retaining wall centerline. Use *Endpoint* to snap to the ends of the existing 2 *Lines* and *Arc*. Construct a second *Mline* above using the following coordinates:

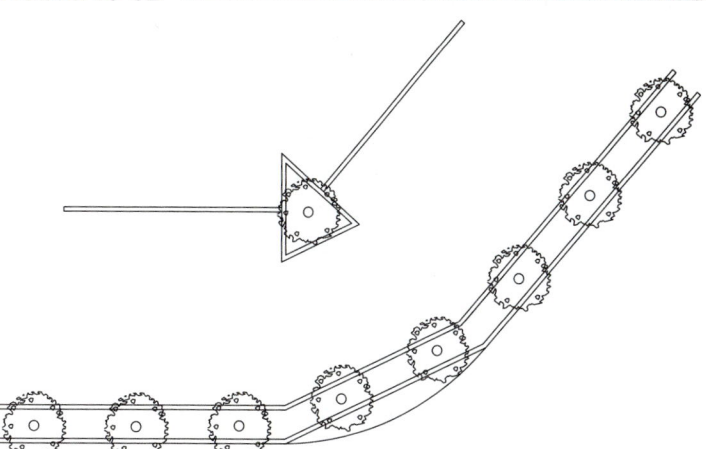

> 60,123
> 192,123
> 274,161
> 372.7,274

Next, construct the *Closed* triangular island with an *Mline* using *Top Justification*. The coordinates are:

> 192,192
> 222.58,206.26
> 192,233

For the parking curbs, draw two more *Mlines*; each *Mline* snaps to the *Midpoint* of the island and is **100′** long. The curb on the right has an angle of **49** degrees (use polar coordinate entry). Finally, create trees similar to the ones shown using *Sketch*. Remember to set the *SKPOLY* variable. Create one tree and *Copy* it to the other locations. Spacing between trees is approximately **48′**. *Save* as **RET-WAL2.**

10. *Boundary*

Open the **GASKETA** drawing that you created in Chapter 9 Exercises. Use the *Boundary* command to create a boundary of the <u>outside shape only</u> (no islands). Use *Move Last* to displace the new shape to the right of the existing gasket. Use *SaveAs* and rename the drawing to **GASKET-AREA.** Keep this drawing to determine the *Area* of the shape after reading Chapter 17.

11. *Region*

Create a gear, using *Regions*. Begin a *New* drawing, *Start from Scratch* with the *English* defaults settings. Use *Save* and assign the name **GEAR-REGION.**

CONCEPTS

This chapter examines commands that are similar to, but generally more advanced and powerful than, those discussed in Chapter 9, Modify Commands I. None of the commands in this chapter create new duplicate objects from existing objects but instead modify the properties of the objects or convert objects from one type to another. Several commands are used to modify specific types of objects such as *Pedit* (modifies *Plines*), *Splinedit* (modifies *Splines*), *Mledit* (modifies *Mlines*), and *Union, Subtract,* and *Intersect* (modify *Regions*). Only one command in this chapter, *Align,* does not modify object properties but combines *Move* and *Rotate* into one operation. Several of the commands and features discussed in this chapter were mentioned in Chapter 11 (*Properties, MatchProp,* and *Object Properties* toolbar) but are explained completely here.

Only some commands in this chapter that modify general object properties have icon buttons that appear in the AutoCAD Drawing Editor by default. For example, you can access *Properties* and *Matchprop* from the *Object Properties* toolbar and *Explode* by using its icon from the *Modify* toolbar.

Other commands that modify specific objects such as *Pedit*, *Mledit*, and *Splinedit* appear in the *Modify II* toolbar (Fig. 16-1). The Boolean operators (*Union, Subtract,* and *Intersect*) appear in the *Solids Editing* toolbar. Activate a toolbar by right-clicking on any tool (icon button) and selecting from the list.

FIGURE 16-1

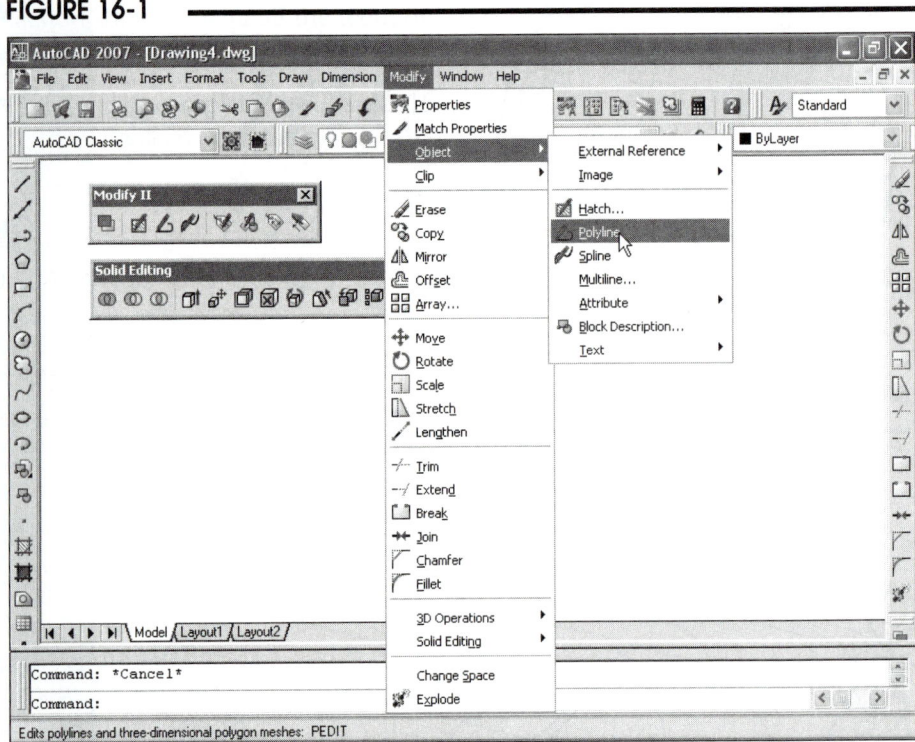

COMMANDS

You can use several methods to change properties of an object or of several objects. If you want to change an object's layer, color, or linetype only, the quickest method is by using the *Object Properties* toolbar (see "*Object Properties*" toolbar). If you want to change many properties of one or more objects (including layer, color, linetype, linetype scale, text style, dimension style, or hatch style) to match the properties of another existing object, use *Matchprop*. If you want to change any property (including coordinate data) for any object, use *Properties*.

You can double-click on any object (assuming the *Dblclkedit* command is set to *On*) to produce the *Properties* palette or a more specific editing tool, such as the *Hatch Edit* dialog box or the *Multiline Text Editor*, depending on the type of object you select.

Considerations for Changing Basic Properties of Objects

The *Properties* palette, the *Object Properties* toolbar, and the *Matchprop*, *Chprop*, and *Change* commands can be used to change the basic properties of objects. Here are several basic properties that can be changed and considerations when doing so.

Layer

By changing an object's *Layer*, the selected object is effectively <u>moved</u> to the designated layer. In doing so, if the object's *Color, Linetype,* and *Lineweight* are set to *BYLAYER*, the object assumes the color, line-type, and lineweight of the layer to which it is moved.

Color

It may be desirable in some cases to assign explicit *Color* to an object or objects independent of the layer on which they are drawn. The properties editing commands allow changing the color of an existing object from one object color to another, or from *BYLAYER* assignment to an object color. <u>An object drawn with an object-specific color can also be changed to *BYLAYER* with this option.</u>

Linetype

An object can assume the *Linetype* assigned *BYLAYER* or can be assigned an object-specific *Linetype*. The *Linetype* option is used to change an object's *Linetype* to that of its layer (*BYLAYER*) or to any object-specific linetype that has been loaded into the current drawing. <u>An object drawn with an object-specific linetype can also be changed to *BYLAYER* with this option.</u>

Linetype Scale

When an individual object is selected, the <u>object's individual linetype scale</u> can be changed with this option, but <u>not the global</u> linetype scale (*LTSCALE*). <u>This is the recommended method to alter an individual object's linetype scale.</u> First, draw all the objects in the current global *LTSCALE*. One of the properties editing commands could then be used with this option to retroactively adjust the linetype scale of <u>specific</u> objects. The result would be similar to setting the *CELTSCALE* before drawing the specific objects. Using this method to adjust a specific object's linetype scale <u>does not reset</u> the global *LTSCALE* or *CELTSCALE* variables.

Thickness

An object's *Thickness* can be changed by this option. *Thickness* is a three-dimensional quality (Z dimension) assigned to a two-dimensional object (see Chapter 40, Surface Modeling).

Lineweight

An object can assume the *Lineweight* assigned *ByLayer* or can be assigned an object-specific *Linetype*. The *Lineweight* option is used to change an object's *Lineweight* to that of its layer (*ByLayer*) or to any object-specific linetype that has been loaded into the current drawing. <u>An object drawn with an object-specific linetype can also be changed to *ByLayer* with this option.</u>

Plotstyle

Use this option to change an object's *Plotstyle*. Plot styles assigned as *ByLayer* or to individual objects can change the way the objects appear in plots, such as having certain screen patterns, line end joints, plotted colors, plotted lineweights, and so on. (See Chapter 33 for more information on plot styles.)

The *Geometry* group lists and allows changing any values that control the selected object's geometry. For example, to change the diameter of a *Circle*, highlight the property and change the value in the right side of the palette (Fig. 16-6).

FIGURE 16-6

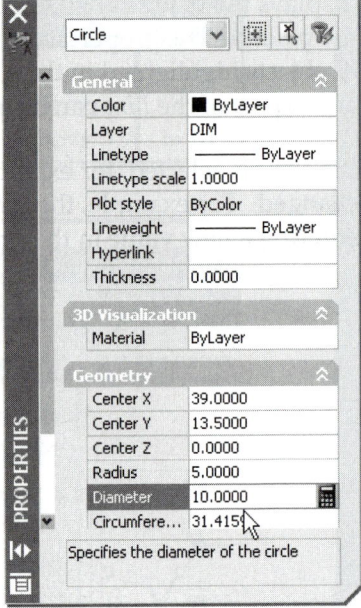

The *3D Visualization* group lists only *Material* by default for most objects. The *Bylayer*, *Byblock*, and *Global* properties determine how materials are attached. This feature is used primarily for rendering 3D objects (see Chapter 41).

Quick Select

The *Quick Select* button appears near the upper-right corner of the *Properties* palette (Fig. 16-7). Selecting this button produces the *Quick Select* dialog box in which a selection set can be constructed based on criteria you choose from the dialog box. (See Chapter 20, Advanced Selection Sets, for information on *Quick Select*.)

FIGURE 16-7

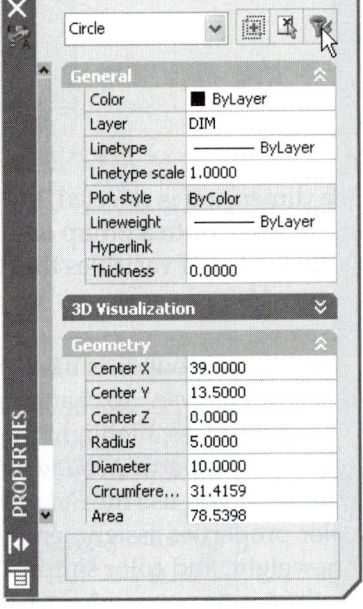

Select Objects

Normally you can select objects by the typical methods any time the *Properties* palette is open. When objects are selected, AutoCAD searches its database and presents the properties of each object, one at a time, in the *Properties* palette. Instead, you can use the *Select Objects* button to save time posting information on each object individually to the window. Using *Select objects*, the properties of the total set of selected objects are not posted to the *Properties* palette until you press Enter.

Toggle Value of Pickadd Sysvar

 This button toggles the *PICKADD* system variable on or off (1 or 0). This variable affects object selection at all times, not only for use with the *Properties* palette. *PICKADD* determines whether objects you PICK are added to the current selection set (normal setting, *PICKADD*=1) or replace the previous object or set (*PICKADD*=0). Changing the *PICKADD* variable to 0 works well with the *Properties* palette because selecting an object replaces the previous set of properties appearing in the palette with the new object's properties.

The button position of "+" (plus symbol) indicates a current setting of 1, or on, for *PICKADD* (PICKed objects are added). Confusing as it appears, a button position of "1" indicates a <u>current setting of 0</u>, or off, for *PICKADD* (PICKed objects replace the previous ones).

NOTE: Since the *PICKADD* setting affects object selection anytime, <u>make sure you set this variable back to your desired setting (usually on, or "+") before dismissing the *Properties* palette</u>. (See Chapter 20, Advanced Selection Sets, for information on *PICKADD*.)

Dblclkedit

Pull-down Menu	Command (Tpye)	Alias (Type)	Short-cut	Screen (side) Menu	Tablet Menu
...	*Dblclkedit*	...	...	...	...

You can double-click on an object to produce the *Properties* palette or other similar dialog box to edit the object. The *Dblclkedit* command (double-click edit) controls whether double-clicking an object produces a dialog box.

Command: **dblclkedit**
Enter double-click editing mode [ON/OFF] <ON>:

If double-click editing is turned on, the *Properties* palette or other dialog box is displayed when an object is double-clicked. When you double-click most objects, the *Properties* palette is displayed.

However, double-clicking some types of objects displays editing tools that are specific to the type of object. For example, double-clicking a line of *Text* produces the *Text Edit* dialog box. The object types (and resulting editing tool) <u>that do not produce the *Properties* palette</u> when the object is double-clicked are listed below. These objects and the related editing tools are discussed fully in upcoming chapters of this text.

Attribute	Displays the *Edit Attribute Definition* dialog box (*Ddedit*).
Attribute within a block	Displays the *Enhanced Attribute Editor* (*Eattedit*).
Block	Displays the *Reference Edit* dialog box (*Refedit*).
Hatch	Displays the *Hatch Edit* dialog box (*Hatchedit*).
Leader text	Displays the *Multiline Text Editor* dialog box (*Ddedit*).
Mline	Displays the *Multiline Edit Tools* dialog box (*Mledit*).
Mtext	Displays the *Multiline Text Editor* dialog box (*Ddedit*).
Text	Displays the *Edit Text* dialog box (*Ddedit*).
Xref	Displays the *Reference Edit* dialog box (*Refedit*).

Dblclkedit is a command, not a system variable.

NOTE: The *PICKFIRST* system variable must be on (set to 1) for the *Properties* palette or other editing tool to appear when an object is double-clicked. (See Chapter 20, Advanced Selection Sets, for information on the *PICKFIRST* system variable.)

Matchprop

Pull-down Menu	Command (Type)	Alias (Type)	Short-cut	Screen (side) Menu	Tablet Menu
Modify *Match Properties*	*Matchprop*	*MA*	...	*MODIFY1* *Matchprp*	Y,14 and Y,15

Matchprop is explained briefly in Chapter 11 but is explained again in this chapter with the full details of the *Special Properties* palette.

Matchprop is used to "paint" the properties of one object to another. Simply invoke the command, select the object that has the desired properties ("source object"), then select the object you want to "paint" the properties to ("destination object"). The Command prompt is as follows:

```
Command: matchprop
Select source object: PICK
Current active settings:  Color Layer Ltype LTSCALE Lineweight Thickness PlotStyle Text Dim Hatch
    Polyline Viewport
Select destination object(s) or [Settings]: PICK
Select destination object(s) or [Settings]: PICK
Select destination object(s) or [Settings]: Enter
Command:
```

You can select several destination objects. The destination object(s) assume all of the "Current active settings" of the source object.

The *Property Settings* dialog box can be used to set which of the *Basic Properties* palette and *Special Properties* palette are to be painted to the destination objects (Fig. 16-8). At the "Select destination object(s) or [Settings]:" prompt, type *S* to display the dialog box. You can control the following *Basic Properties* palette. Only the checked properties are painted to the destination objects.

FIGURE 16-8

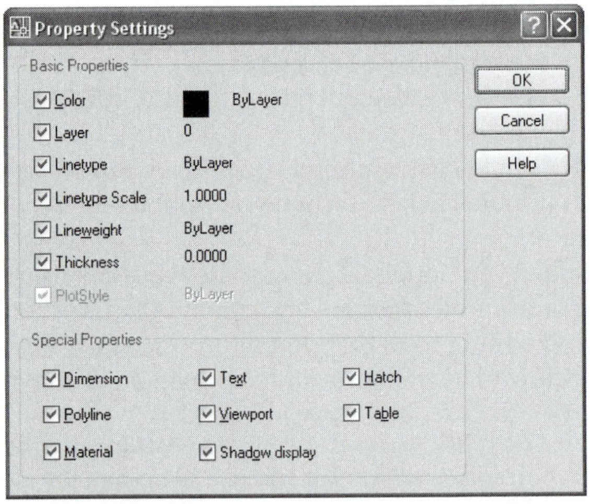

Color	This option paints the object-specific or *ByLayer* color.
Layer	Move selected objects to the source object layer with this option.
Linetype	Paint the object-specific or *ByLayer* linetype of the source object to the destination object.
Linetype Scale	This option changes the individual object's linetype scale (*CELTSCALE*) not global linetype scale (*LTSCALE*).
Thickness	Thickness is a 3-dimensional quality (see Chapter 40).
Lineweight	You can paint the object-specific or *ByLayer* lineweight of the source object to the destination object.
PlotStyle	The plot style of the source object is painted to the destination object when this setting is checked. (See Chapter 33 for information on plot styles.)

The *Special Properties* palette section allows you to specify features of dimensions, text, hatch patterns, viewports, and polylines to match, as explained next.

Dimension	This setting paints the *Dimension Style*. A *Dimension Style* defines the appearance of a dimension such as text style, size of arrows and text. (See Chapter 29.)
Text	This setting paints the source object's text *Style*. The text *Style* defines the text font and many other parameters that affect the appearance of the text. (See Chapter 18.)
Hatch	Checking this box paints the hatch properties such as *Pattern, Angle, Scale,* and other characteristics. (See Chapter 26.)
Polyline	If you use *Matchprop* with *Plines*, the *Width* and *Linetype Generation* properties in addition to basic object properties of the first *Pline* are painted to the second. If the source polyline has variable *Width*, it is not transferred.
Viewport	If you match properties from one paper space viewport to another, the following properties are changed in addition to the basic object properties: *On/Off, Display Locking,* standard or custom *Scale, Shadeplot, Snap, Grid,* and *UCSicon* settings.
Table	Use this option to change the table style of the destination object to that of the source object. (See Chapter 18.)
Material	This setting changes the material applied to the destination object. (See Chapter 41.)
Shadow Display	A 3D object can cast shadows, receive shadows, or both, or it can ignore shadows. This property changes the shadow display. (See Chapter 41.)

2007

Object Properties Toolbar

The five drop-down lists in the Object Properties toolbar (when not "dropped down") generally show the <u>current</u> layer, color, linetype, lineweight, and plot style. However, if an object or set of objects is selected, the information in these boxes changes to display the current <u>object's</u> settings. You can change an object's settings by picking an object (when no commands are in use), then dropping down any of the three lists and selecting a different layer, linetype, or color, etc.

Make sure you select (highlight) an object when no commands are in use. Use the pickbox (that appears on the cursor), *Window*, or *Crossing Window* to select the desired object or set of objects. The entries in the five boxes (Layer Control, Color Control, Linetype Control, etc.) then change to display the settings for the <u>selected object or objects</u>. If several objects are selected that have different properties, the boxes display no information (go "blank").

Next, use any of the drop-down lists to make another selection (Fig. 16-9). The highlighted object's properties are changed to those selected in the lists. Press the Escape key to complete the process.

Remember that in most cases color, linetype, lineweight, and plot style settings are assigned *ByLayer*. In this type of drawing scheme, to change the linetype or color properties of an object, you would change only the object's <u>layer</u> (see Fig. 16-9). If you are using this type of drawing scheme, refrain from using the Color Control and Linetype Control drop-down lists for changing properties.

FIGURE 16-9

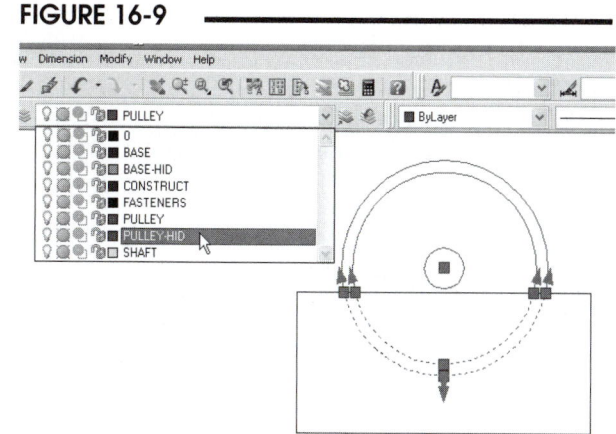

This method for changing object properties is about as quick and easy as *Matchprop*; however, only the layer, linetype, color, lineweight, or plot style can be changed with the *Object Properties* toolbar. The *Object Properties* toolbar method works well if you do not have other existing objects to match or if you want to change <u>only</u> layer, linetype, color, lineweight, and plot style <u>without</u> matching text, dimension, and hatch styles, and viewport or polyline properties.

Chprop

Pull-down Menu	Command (Type)	Alias (Type)	Short-cut	Screen (side) Menu	Tablet Menu
...	Chprop	...	...	...	...

Chprop allows you to change basic properties of one or more objects using Command line format. If you want to change properties of several objects, pick several objects or select with a window or crossing window at the "Select objects:" prompt.

    Command: *chprop*
    Select objects: **PICK**
    Select objects: **Enter**
    Enter property to change [Color/LAyer/LType/ltScale/LWeight/Thickness]:

Change

Pull-down Menu	Command (Type)	Alias (Type)	Short-cut	Screen (side) Menu	Tablet Menu
...	Change	-CH	...	...	...

The *Change* command allows changing three options: *Points*, *Properties*, or *Text*.

Point
This option allows changing the endpoint of an object or endpoints of several objects to one new position:

    Command: *change*
    Select objects: **PICK**
    Select objects: **Enter**
    Specify change point or [Properties]: **PICK** (Select a point to establish as new endpoint of all objects. *OSNAP*s can be used.)

The endpoint(s) of the selected object(s) <u>nearest</u> the new point selected at the "Specify change point or [Properties]:" prompt is changed to the new point (Fig. 16-10).

Properties
These options are discussed previously. The *Elevation* property (a 3D property) of an object can also be changed only with *Change* (see Chapter 40 for information on *Elevation*).

    Specify change point or [Properties]: *p*
    Enter property to change
    [Color/Elev/LAyer/LType/ltScale/
        LWeight/Thickness]:

FIGURE 16-10

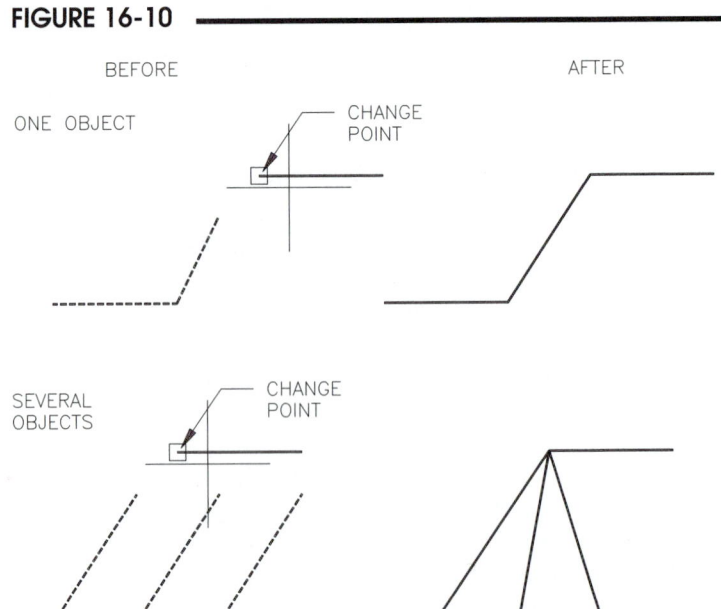

Text

Although the word *"Text"* does <u>not</u> appear as an option, the *Change* command recognizes text created with *Text* if selected. <u>*Change* does not change *Mtext*</u>. (See Chapter 18 for information on *Text* and *Mtext*.)

You can change the following characteristics of *Text* objects:

Text insertion point
Text style
Text height
Text rotation angle
Textual content

To change text, use the following command syntax:

Command: **change**
Select objects: **PICK** (Select one or several lines of text)
Select objects: **Enter**
Specify change point or [Properties]: **Enter**
Specify new text insertion point <no change>: **PICK** or **Enter**
Enter new text style <Standard>: (**text style name**) or **Enter**
Specify new height <0.2000>: (**value**) or **Enter**
Specify new rotation angle <0>: (**value**) or **Enter**
Enter new text <text>: (**new text**) or **Enter** (Enter the complete new line of text)

Explode

Pull-down Menu	Command (Type)	Alias (Type)	Short-cut	Screen (side) Menu	Tablet Menu
Modify *Explode*	*Explode*	*X*	...	*MODIFY2* *Explode*	*Y,22*

Many graphical shapes can be created in AutoCAD that are made of several elements but are treated as one object, such as *Plines, Polygons, Blocks, Hatch* patterns, and dimensions. The *Explode* command provides you with a means of breaking down or "exploding" the complex shape from one object into its many component segments (Fig. 16-11). Generally, *Explode* is used to allow subsequent editing of one or more of the component objects of a *Pline, Polygon,* or *Block,* etc., which would otherwise be impossible while the complex shape is considered one object.

FIGURE 16-11

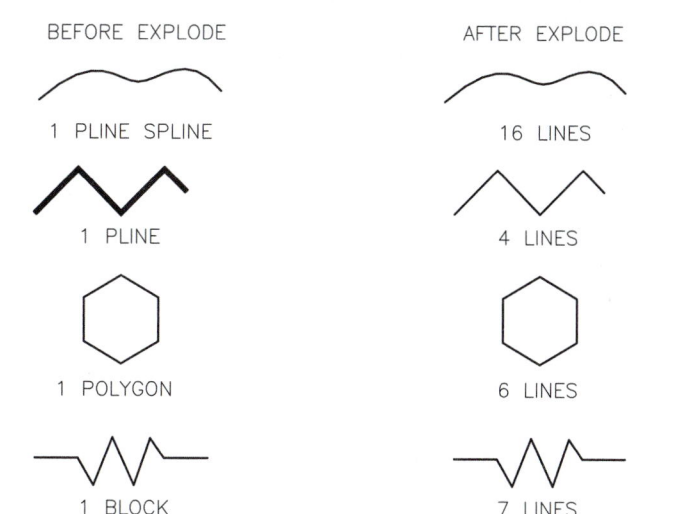

The *Explode* command has no options and is simple to use. You only need to select the objects to *Explode*.

    Command: **explode**
    Select Objects: **PICK** (Select one or more *Plines*, *Blocks*, etc.)
    Select Objects: **Enter** (Indicates selection of objects is complete.)

TIP When *Plines*, *Polygons*, *Blocks*, or hatch patterns are *Exploded*, they are transformed into *Line*, *Arc*, and *Circle* objects. Beware, *Plines* having *width* lose their width information when *Exploded* since *Line*, *Arc*, and *Circle* objects cannot have width. *Exploding* objects can have other consequences such as losing "associativity" of dimensions and hatch objects and increasing file sizes by *Exploding Blocks*.

Align

Pull-down Menu	Command (Type)	Alias (Type)	Short-cut	Screen (side) Menu	Tablet Menu
Modify *3D Operations >* *Align*	*Align*	*AL*	...	*MODIFY2* *Align*	*X,14*

Align provides a means of aligning one shape (a simple object, group of objects, *Pline*, *Boundary*, *Region*, *Block*, or a 3D object) with another shape. *Align* provides a complex motion, usually a combined translation (like *Move*) and rotation (like *Rotate*), in one command.

The alignment is accomplished by connecting source points (on the shape to be moved) to destination points (on the stationary shape). You should use *OSNAP* modes to select the source and destination points to assure accurate alignment. Either a 2D or 3D alignment can be accomplished with this command. The command syntax for alignment in a <u>2D alignment</u> is as follows:

    Command: **align**
    Select objects: **PICK**
    Select objects: **Enter**
    Specify first source point: **PICK** (with *Osnap*)
    Specify first destination point: **PICK** (with *Osnap*)
    Specify second source point: **PICK** (with *Osnap*)
    Specify second destination point: **PICK** (with *Osnap*)
    Specify third source point or <continue>: **Enter**
    Scale objects based on alignment points? [Yes/No] <N>: **Enter** (or *Y* to scale the source object to match destination object)
    Command:

This command performs a translation and a rotation in one motion if needed to align the points as designated (Fig. 16-12).

First, the first source point is connected to (actually touches) the first destination point (causing a translation). Next, the vector defined by the first and second source points is aligned with the vector defined by the first and second destination points (causing rotation).

FIGURE 16-12

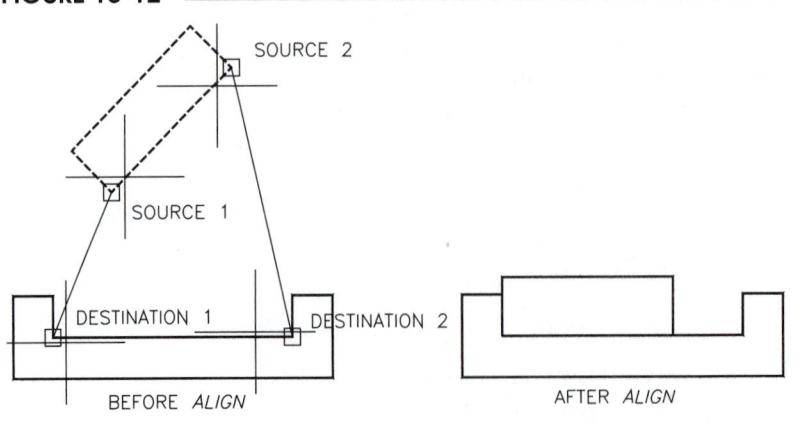

If no third destination point is given (needed only for a 3D alignment), a 2D alignment is assumed and performed on the basis of the two sets of points.

Note that you can scale the source object based on the distance between the source points and the destination points. If you answer "Y" to the "Scale objects based on alignment points?" prompt, the source object is enlarged or reduced so the distance between its alignment points matches that of the destination points.

Pedit

Pull-down Menu	Command (Type)	Alias (Type)	Short-cut	Screen (side) Menu	Tablet Menu
Modify Object > Polyline	Pedit	PE	...	MODIFY1 Pedit	Y,17

This command provides numerous options for editing *Polylines* (*Plines*). As an alternative, *Properties* can be used to change many of the *Pline's* features in dialog box form (see Fig. 16-4).

The list of options below emphasizes the great flexibility possible with *Polylines*. The first step after invoking *Pedit* is to select the *Pline* to edit.

```
Command: pedit
Select polyline or [Multiple]: PICK
Enter an option [Close/Join/Width/Edit vertex/Fit/Spline/Decurve/Ltype gen/Undo]:
```

Multiple

The *Multiple* option allows multiple *Plines* to be edited simultaneously. The selected *Plines* can be totally separate objects and do not have to be connected in any way. Once the *Multiple* option is invoked and the objects are selected, any *Pline* option, such as *Close, Open, Join, Width, Fit, Spline, Decurve,* or *Ltype gen,* operates on the selected *Plines.*

```
Command: Pedit
Select polyline or [Multiple]: m
Select objects: PICK
Select objects: PICK
Select objects: Enter
Enter an option [Close/Open/Join/Edit vertex/Width/Fit/Spline/Decurve/Ltype gen/Undo]:
```

For example, you can change the width of all *Plines* in a drawing simultaneously using the *Multiple* option, selecting all *Plines,* and using the *Width* option.

Close

Close connects the last segment with the first segment of an existing "open" *Pline,* resulting in a "closed" *Pline* (Fig. 16-13). A closed *Pline* is one continuous object having no specific start or endpoint, as opposed to one closed by PICKing points. A *Closed Pline* reacts differently to the *Spline* option and to some commands such as *Fillet, Pline* option (see "Fillet," Chapter 9).

Open

Open removes the closing segment if the *Close* option was used previously (Fig. 16-13).

FIGURE 16-13

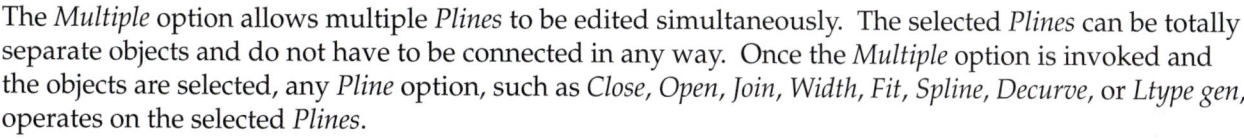

OPEN   CLOSE

Join

The *Join* option combines two or more objects into one *Pline*. The line segments <u>do not have to meet exactly</u> for *Join* to work. You must first use the *Multiple* option, then *Join*.

    Command: **Pedit**
    Select polyline or [Multiple]: **m**
    Select objects: **PICK**
    Select objects: **Enter**
    Enter an option [Close/Open/Join/Width/Fit/Spline/Decurve/Ltype gen/Undo]: **j**
    Join Type = Extend
    Enter fuzz distance or [Jointype] <0.0000>: **.5**
    1 segments added to polyline
    Enter an option [Close/Open/Join/Width/Fit/Spline/Decurve/Ltype gen/Undo]:

If the ends of the line segments do not touch but are within a distance that you can set, called the *fuzz distance*, the ends can be joined by *Pedit*. *Pedit* handles this automatically by either extending and trimming the line segments or by adding a new line segment based on your setting for *Jointype*.

    Select objects: **PICK**
    Enter an option [Close/Open/Join/Width/Fit/Spline/Decurve/Ltype gen/Undo]: **j**
    Join Type = Extend
    Enter fuzz distance or [Jointype] <2.0000>: **j**
    Enter join type [Extend/Add/Both] <Extend>:

Extend
This option causes AutoCAD to join the selected polylines by extending or trimming the segments to the nearest endpoints (see Fig. 16-14).

Add
Use this option to add a straight segment between the nearest endpoints (see Fig. 16-14).

Both
If you use this option, the selected polylines are joined by extending or trimming if possible. If not, as in the case of near parallel lines when an extension would be outside the fuzz distance, a straight segment is added between the nearest endpoints.

FIGURE 16-14

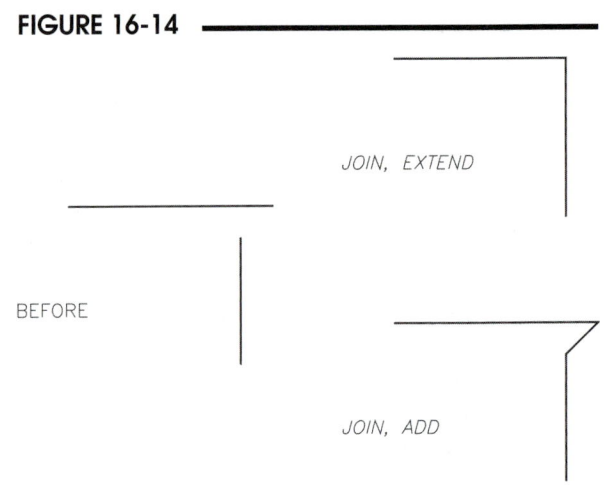

JOIN, EXTEND

BEFORE

JOIN, ADD

If the properties of several objects being joined into a polyline differ, the resulting polyline inherits the properties of the first object you select.

Width
Width allows specification of a uniform width for *Pline* segments (Fig. 16-15). Non-uniform width can be specified with the *Edit vertex* option.

Edit vertex
This option is covered in the next section.

FIGURE 16-15

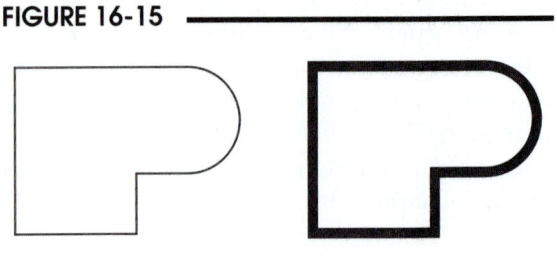

BEFORE WIDTH   AFTER WIDTH

Fit

This option converts the *Pline* from straight line segments to arcs. The curve consists of two arcs for each pair of vertices. The resulting curve can be radical if the original *Pline* consists of sharp angles. The resulting curve passes <u>through all</u> vertices (Fig. 16-16).

FIGURE 16-16

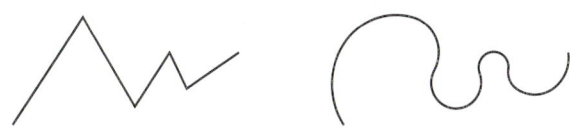

BEFORE FIT   AFTER FIT

Spline

This option converts the *Pline* to a B-spline (Bezier spline) (Fig. 16-17). The *Pline* vertices act as "control points" affecting the shape of the curve. The resulting curve passes through <u>only the end</u> vertices. A *Spline*-fit *Pline* is <u>not the same as a spline curve created with the Spline</u> command. This option produces a less versatile version of the newer *Spline* object.

Decurve

Decurve removes the *Spline* or *Fit* curve and returns the *Pline* to its original straight line segments state (Fig. 16-17).

When you use the *Spline* option of *Pedit*, the amount of "pull" can be affected by setting the *SPLINETYPE* system variable to either 5 or 6 <u>before</u> using the *Spline* option. *SPLINETYPE* applies either a quadratic (5=more pull) or cubic (6=less pull) B-spline function (Fig. 16-18).

FIGURE 16-17

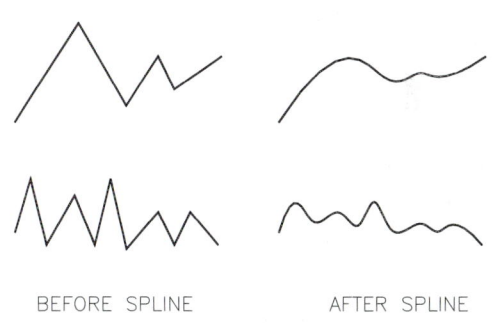

BEFORE SPLINE   AFTER SPLINE
AFTER DECURVE   BEFORE DECURVE

FIGURE 16-18

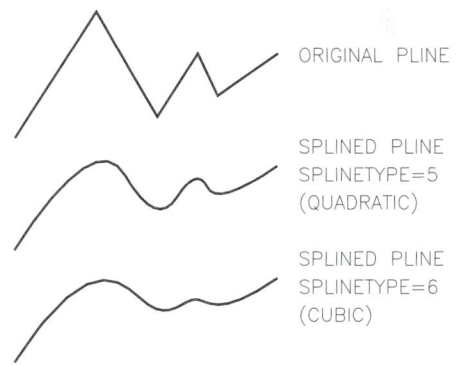

ORIGINAL PLINE

SPLINED PLINE
SPLINETYPE=5
(QUADRATIC)

SPLINED PLINE
SPLINETYPE=6
(CUBIC)

The *SPLINESEGS* system variable controls the number of line segments created when the *Spline* option is used. The variable should be set <u>before</u> using the option to any value (8=default): the higher the value, the more line segments. The actual number of segments in the resulting curve depends on the original number of *Pline* vertices and the value of the *SPLINETYPE* variable (Fig. 16-19).

FIGURE 16-19

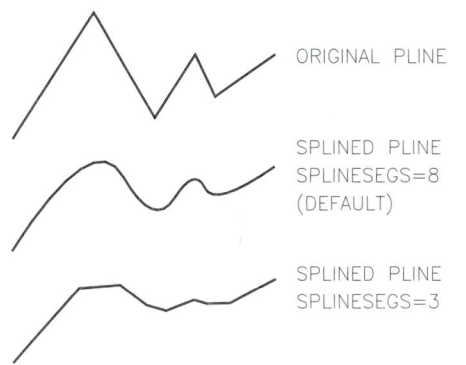

ORIGINAL PLINE

SPLINED PLINE
SPLINESEGS=8
(DEFAULT)

SPLINED PLINE
SPLINESEGS=3

Changing the *SPLFRAME* variable to 1 causes the *Pline* frame (the original straight segments) to be displayed for *Splined* or *Fit Plines*. *Regen* must be used after changing the variable to display the original *Pline* "frame" (Fig. 16-20).

FIGURE 16-20

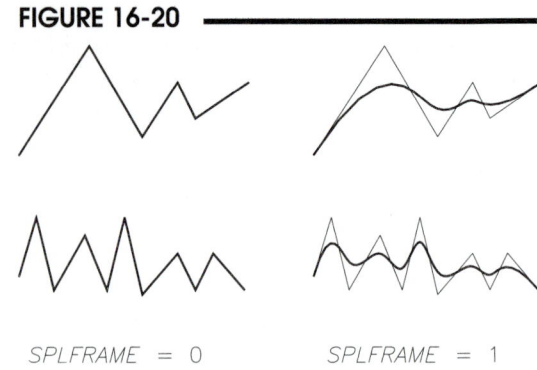

SPLFRAME = 0   SPLFRAME = 1

Ltype gen

This setting controls the generation of non-continuous linetypes for *Plines*. If *Off*, non-continuous linetype dashes start and stop at each vertex, as if the *Pline* segments were individual *Line* segments. For dashed linetypes, each line segment begins and ends with a full dashed segment (Fig. 16-21). If *On*, linetypes are drawn in a consistent pattern, disregarding vertices. In this case, it is possible for a vertex to have a space rather than a dash. Using the *Ltype gen* option retroactively changes *Plines* that have already been drawn. *Ltype gen* affects objects composed of *Plines* such as *Polygons*, *Rectangles*, and *Boundaries*.

Similarly, the *PLINEGEN* system variable controls how new non-continuous linetypes are drawn for *Plines*. A setting of 1 creates a consistent linetype pattern, disregarding vertices (like *Ltype gen On*). A *PLINEGEN* setting of 0 creates linetypes stopping and starting at each vertex (like *Ltype gen Off*). However, *PLINEGEN* is not retroactive—it affects only new *Plines*.

FIGURE 16-21

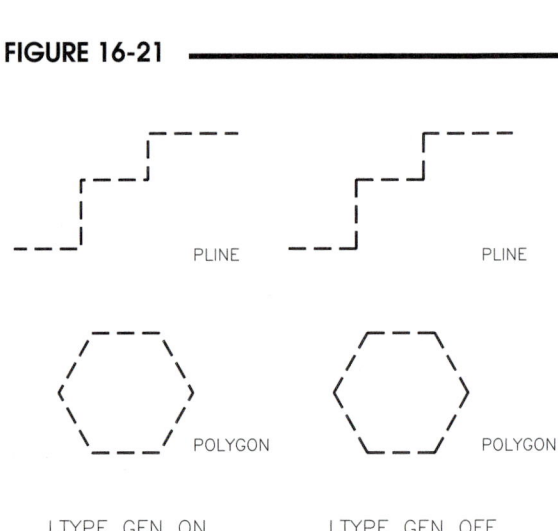

PLINE   PLINE

POLYGON   POLYGON

LTYPE GEN ON   LTYPE GEN OFF

Undo

Undo reverses the most recent *Pedit* operation.

eXit

This option exits the *Pedit* options, keeps the changes, and returns to the Command prompt.

Vertex Editing

Upon selecting the *Edit Vertex* option from the *Pedit* options list, the group of suboptions is displayed on the screen menu and Command line:

```
Command: pedit
Select polyline or [Multiple]: PICK
Enter an option [Close/Join/Width/Edit vertex/Fit/Spline/Decurve/Ltype gen/Undo]: e
Enter a vertex editing option
[Next/Previous/Break/Insert/Move/Regen/Straighten/Tangent/Width/eXit] <N>:
```

Next

AutoCAD places an **X** marker at the first endpoint of the *Pline*. The *Next* and *Previous* options allow you to sequence the marker to the desired vertex (Fig. 16-22).

Previous

See *Next* above.

FIGURE 16-22

VERTEX MARKER

PREVIOUS AND NEXT OPTIONS WILL MOVE THE
MARKER TO THE DESIRED VERTEX

Break

This selection causes a break between the marked vertex and another vertex you then select using the *Next* or *Previous* option (Fig. 16-23).

Enter an option [Next/Previous/Go/eXit] <N>:

Selecting *Go* causes the break. An endpoint vertex cannot be selected.

FIGURE 16-23

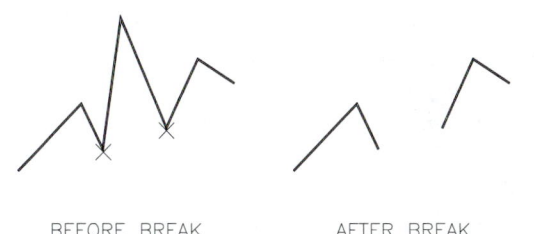

BEFORE BREAK   AFTER BREAK

Insert

Insert allows you to insert a new vertex at any location <u>after</u> the vertex that is marked with the **X** (Fig. 16-24). Place the marker before the intended new vertex, use *Insert*, then PICK the new vertex location.

FIGURE 16-24

BEFORE INSERT   AFTER INSERT

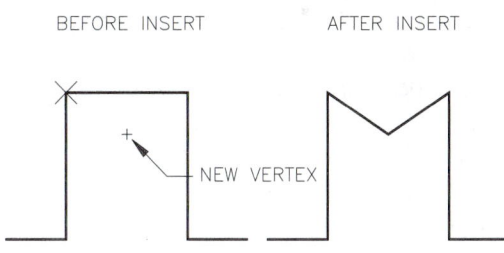

NEW VERTEX

Move

You are prompted to indicate a new location to *Move* the marked vertex (Fig. 16-25).

Regen

In older releases of AutoCAD, *Regen* should be used after the *Width* option to display the new changes.

FIGURE 16-25

BEFORE MOVE   AFTER MOVE

NEW LOCATION

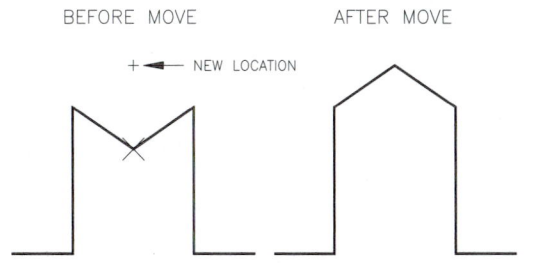

Straighten

You can *Straighten* the *Pline* segments between the current marker and the other marker that you then place by one of these options:

> Enter an option [Next/Previous/Go/eXit] <N>:

Selecting *Go* causes the straightening to occur (Fig. 16-26).

Tangent

Tangent allows you to specify the direction of tangency of the current vertex for use with curve *Fitting*.

Width

This option allows changing the *Width* of the *Pline* segment immediately following the marker, thus achieving a specific width for one segment of the *Pline* (Fig. 16-27). *Width* can be specified with different starting and ending values.

eXit

This option exits from vertex editing, saves changes, and returns to the main *Pedit* prompt.

FIGURE 16-26

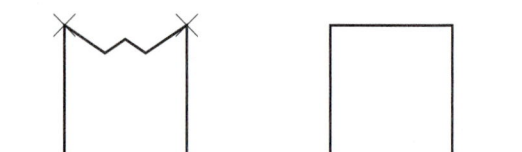

BEFORE STRAIGHTEN   AFTER STRAIGHTEN

FIGURE 16-27

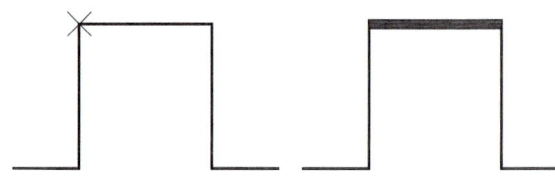

BEFORE WIDTH   AFTER WIDTH

Grips

Plines can also be edited easily using Grips (see Chapter 19). A Grip appears on each vertex of the *Pline*. Editing *Plines* with Grips is sometimes easier than using *Pedit* because Grips are more direct and less dependent on the command interface.

Converting *Lines* and *Arcs* to *Plines*

A very important and productive feature of *Pedit* is the ability to convert *Lines* and *Arcs* to *Plines* and closed *Pline* shapes. Potential uses of this option are converting a series of connected *Lines* and *Arcs* to a closed *Pline* for subsequent use with *Offset* or for inquiry of the area (*Area* command) or length (*List* command) of a single shape. The only requirement for conversion of *Lines* and *Arcs* to *Plines* is that the selected objects must have <u>exact</u> matching endpoints.

To accomplish the conversion of objects to *Plines*, simply select a *Line* or *Arc* object and request to turn it into one:

```
Command: pedit
Select polyline or [Multiple]: PICK  (Select only one Line or Arc)
Object selected is not a polyline
Do you want to turn it into one? <Y> Enter
Enter an option [Close/Join/Width/Edit vertex/Fit/Spline/Decurve/Ltype gen/Undo]: j (Use the Join option)
Select objects: PICK
Select objects: Enter
1 segments added to polyline
Enter an option [Close/Join/Width/Edit vertex/Fit/Spline/Decurve/Ltype gen/Undo]: Enter
Command:
```

The resulting conversion is a one *Polyline* shape.

Splinedit

	Pull-down Menu	Command (Type)	Alias (Type)	Short-cut	Screen (side) Menu	Tablet Menu
	Modify Object > Splinedit	Splinedit	SPE	...	MODIFY1 Splinedt	Y,18

Splinedit is an extremely powerful command for changing the configuration of existing *Splines*. You can use multiple methods to change *Splines*. All of the *Splinedit* methods fall under two sets of options.

The two groups of options that AutoCAD uses to edit *Splines* are based on two sets of points: data points and control points. Data points are the points that were specified when the *Spline* was created—the points that the *Spline* actually passes through (Fig. 16-28).

FIGURE 16-28

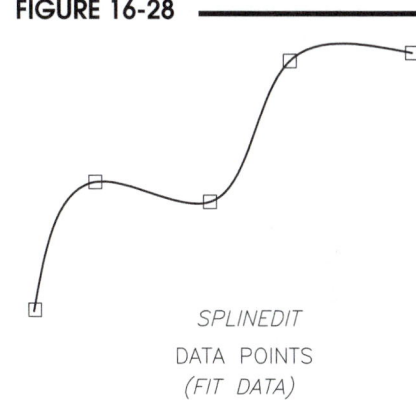

SPLINEDIT
DATA POINTS
(FIT DATA)

Control points are other points outside of the path of the *Spline* that only have a "pull" effect on the curve (Fig. 16-29).

FIGURE 16-29

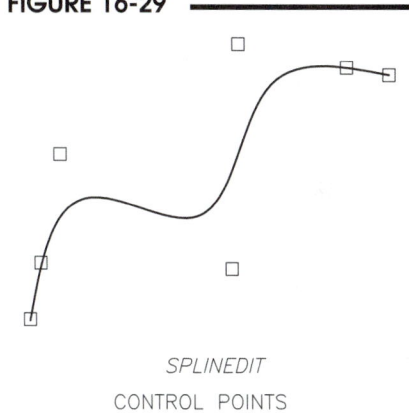

SPLINEDIT
CONTROL POINTS

Editing *Spline* Data Points

The command prompt displays several levels of options. The top level of options uses the control points method for editing. Select *Fit Data* to use data points for editing. The *Fit Data* methods are recommended for most applications. Since the curve passes directly through the data points, these options offer direct control of the curve path.

```
Command: splinedit
Select spline: PICK
Enter an option [Fit data/Close/Move vertex/Refine/rEverse/Undo]: f
Enter a fit data option [Add/Close/Delete/Move/Purge/Tangents/toLerance/eXit] <eXit>:
```

Add

You can add points to the *Spline*. The *Spline* curve changes to pass through the new points. First, PICK an existing point on the curve. That point and the next one in sequence (in the order of creation) become highlighted. The new point will change the curve between those two highlighted data points (Fig. 16-30).

> Specify control point: **PICK** (Select an existing point on the curve before the intended new point.)
> Specify new point: **PICK** (PICK a new point location between the two marked points.)

FIGURE 16-30

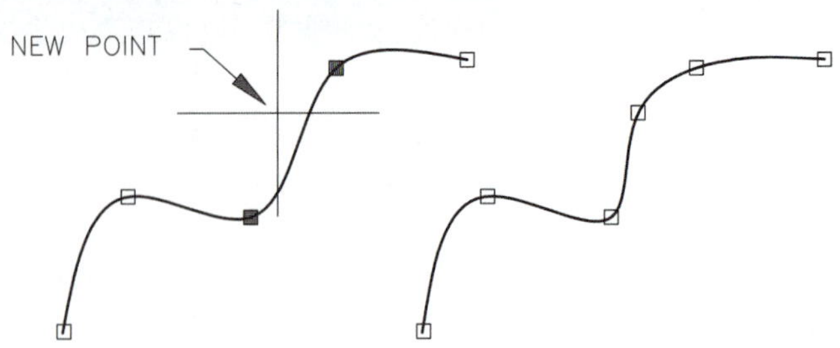

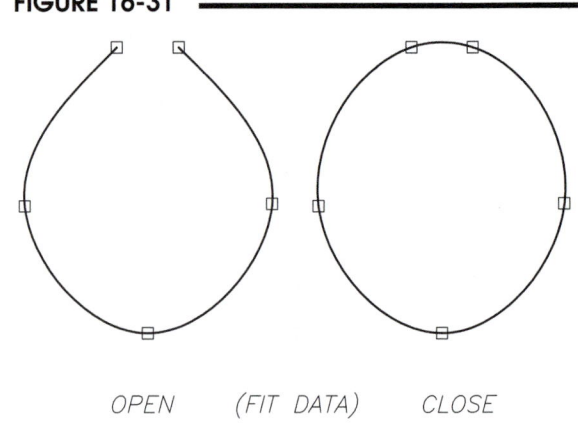

SPLINEDIT   FIT DATA   ADD

Close/Open

The *Close* option appears only if the existing curve is open, and the *Open* prompt appears only if the curve is closed. Selecting either option automatically forces the opposite change. *Close* causes the two ends to become tangent, forming one smooth curve (Fig. 16-31). This <u>tangent continuity</u> is characteristic of *Closed Splines* only. *Splines* that have matching endpoints do not have tangent continuity unless the *Close* option of *Spline* or *Splinedit* is used.

FIGURE 16-31

OPEN   (FIT DATA)   CLOSE

Move

You can move any data point to a new location with this option (Fig. 16-32). The beginning endpoint (in the order of creation) becomes highlighted. Type *N* for next or *S* to select the desired data point to move; then PICK the new location.

> Specify new location or [Next/Previous/Select point/eXit] <N>:

FIGURE 16-32

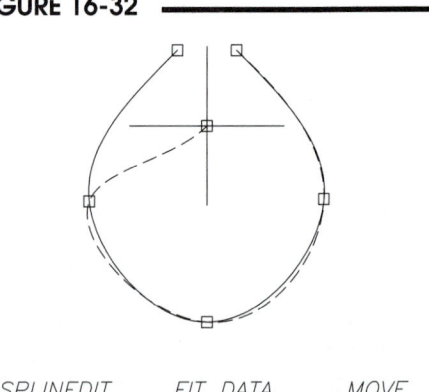

Purge

Purge <u>deletes all data points</u> and renders the *Fit Data* set of options unusable. You are returned to the control point options (top level). To reinstate the points, use *Undo*.

SPLINEDIT   FIT DATA   MOVE

Tangents

You can change the directions for the start and endpoint tangents with this option. This action gives the same control that exists with the "Enter start tangent" and "Enter end tangent" prompts of the *Spline* command used when the curves were created (see Fig. 15-13, *Spline, End Tangent*).

toLerance

Use *toLerance* to specify a value, or tolerance, for the curve to "fall" away from the data points. Specifying a tolerance causes the curve to smooth out, or fall, from the data points. The higher the value, the more the curve "loosens." The *toLerance* option of *Splinedit* is identical to the *Fit Tolerance* option available with *Spline* (see Fig. 15-15, *Spline, Tolerance*).

Editing *Spline* **Control Points**

Use the top level of command options (except *Fit Data*) to edit the *Spline's* control points. These options are similar to those used for editing the data points; however, the results are different since the curve does not pass through the control points.

Command: **splinedit**
Select spline: **PICK**
Enter an option [Fit data/Close/Move vertex/Refine/rEverse/Undo]:

Fit Data
Discussed previously.

Close/Open
These options operate similar to the *Fit Data* equivalents; however, the resulting curve falls away from the control points (Fig. 16-33; see also Fig. 16-31, *Fit Data, Close*).

FIGURE 16-33

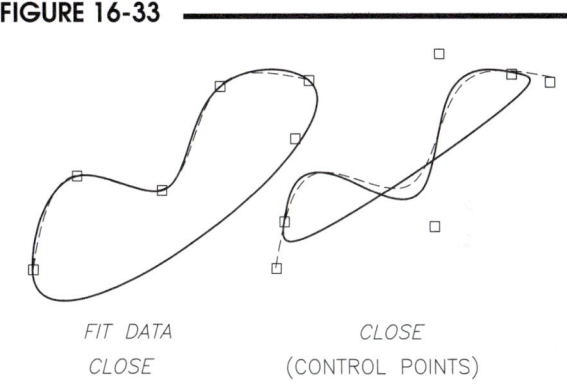

FIT DATA   CLOSE
CLOSE   (CONTROL POINTS)

Move Vertex
Move Vertex allows you to move the location of any control points. This is the control points' equivalent to the *Move* option of *Fit Data* (see Fig. 16-32, *Fit Data, Move*). The method of selecting points (*Next/Previous/Select point/eXit/*) is the same as that used for other options.

Refine
Selecting the *Refine* option reveals another level of options.

Enter a refine option [Add control point/Elevate order/Weight/eXit] <eXit>:

Add control points is the control points' equivalent to *Fit Data Add* (see Fig. 16-30). At the "Select a point on the Spline" prompt, simply PICK a point near the desired location for the new point to appear. Once the *Refine* option has been used, the *Fit Data* options are no longer available.

Elevate order allows you to <u>increase the number of control points</u> uniformly along the length of the *Spline*. Enter a value from *n* to 26, where *n* is the current number of points + one. Once a *Spline* is elevated, it cannot be reduced.

Weight is an option that you use to assign a value to the <u>amount of "pull"</u> that a <u>specific control point</u> has on the *Spline* curve. The higher the value, the more "pull," and the closer the curve moves toward the control point. The typical method of selecting points (*Next/Previous/Select point/eXit/*) is used.

rEverse
The *rEverse* option reverses the direction of the *Spline*. The first endpoint (when created) then becomes the last endpoint. Reversing the direction may be helpful for selection during the *Move* option.

Grips
Splines can also be edited easily using *Grips* (see Chapter 19). The *Grip* points that appear on the *Spline* are identical to the *Fit Data* points. Editing with *Grips* is a bit more direct and less dependent on the command interface.

Mledit

Pull-down Menu	Command (Type)	Alias (Type)	Short-cut	Screen (side) Menu	Tablet Menu
Modify Object > Multiline...	Mledit or -Mledit	...	...	MODIFY1 Mledit	Y,19

The *Mledit* command provides tools for editing multilines created with the *Mline* command. Invoking the *Mledit* command by any method produces the *Multiline Edit Tools* dialog box (Fig. 16-34). Selecting an option closes the dialog box allowing you to select the desired multiline segments for the editing action.

Mledit should be used for most *Mline* modifications. Although some typical editing commands such as *Trim*, *Extend*, *Stretch*, and Grips operate with *Mlines*, the *Break*, *Fillet*, and *Chamfer* commands do not operate with *Mlines* unless the *Mlines* are *Exploded*, which is normally not desired. (See "Other Editing Possibilities for *Mlines*" at the end of this section.)

FIGURE 16-34

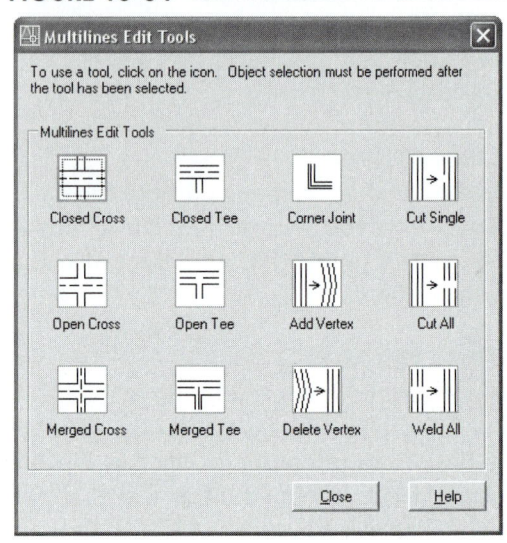

Command: **mledit** (The *Multiline Edit Tools* dialog box appears. Select the desired option.)
Select first mline: **PICK**
Select second mline: **PICK**
Select first mline or [Undo]: **PICK** or **Enter**
Command:

The *Multiline Edit Tools* dialog box is organized in columns as follows:

Intersection—Cross	Intersection—Tee	Corner, Vertices	Lines
Closed Cross	Closed Tee	Corner Joint	Cut Single
Open Cross	Open Tee	Vertex Add	Cut All
Merged Cross	Merged Tee	Vertex Delete	Weld

Closed Cross

Use this option to trim one of two intersecting *Mlines*. The first *Mline* PICKed is trimmed to the outer edges of the second. The second *Mline* is "closed." All line elements in the first multiline are trimmed (Fig. 16-35).

FIGURE 16-35

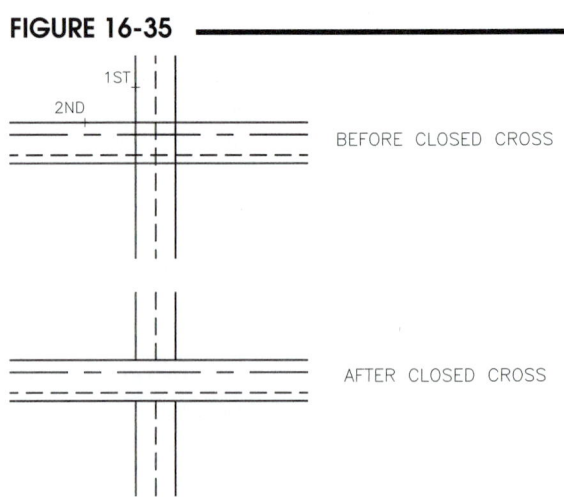

Open Cross

This option <u>trims both</u> intersecting *Mlines*. Both *Mlines* are "open." <u>All</u> line elements of the <u>first</u> *Mline* PICKed are trimmed to the outer edges of the second. Only the outer line elements of the second multiline are trimmed, while the inner lines continue through the intersection (Fig. 16-36).

FIGURE 16-36 ───────

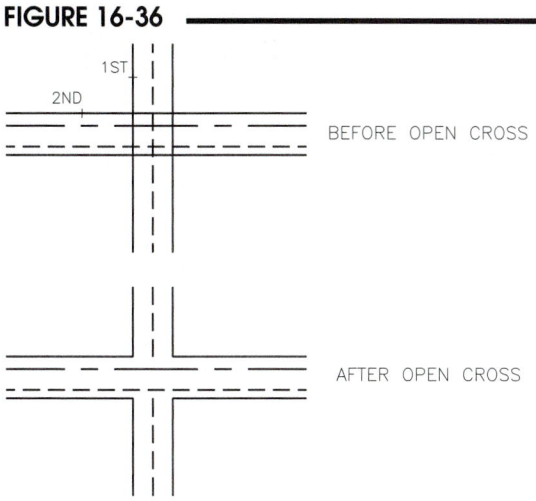

BEFORE OPEN CROSS

AFTER OPEN CROSS

Merged Cross

With this option, the <u>outer</u> line elements of <u>both</u> intersecting *Mlines* are trimmed and the <u>inner line elements merge</u>. The inner line elements merge at the second *Mline's* next set (Fig. 16-37). A full merge (both inner lines continue through) occurs only if the second *Mline* PICKed has no or only one inner line element.

FIGURE 16-37 ───────

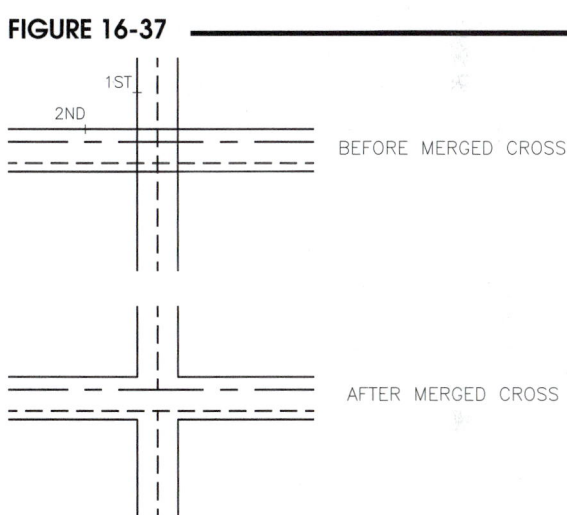

BEFORE MERGED CROSS

AFTER MERGED CROSS

Closed Tee

As indicated by the image tile, a "T" intersection is created rather than a "crossing" intersection. The <u>first</u> *Mline* PICKed is <u>trimmed</u> to the <u>nearest</u> (to the PICK point) outer edge of the second. Only the side of the first *Mline* nearest the PICK point remains. The second *Mline* is not affected (Fig. 16-38).

FIGURE 16-38 ───────

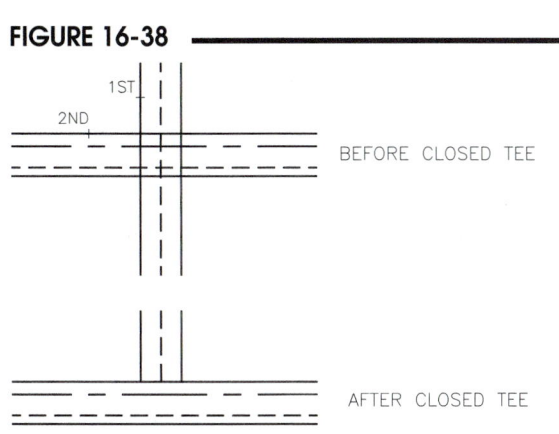

BEFORE CLOSED TEE

AFTER CLOSED TEE

Open Tee

With this "T" intersection, both outer edges of two intersecting *Mlines* are "open." All line elements of the <u>first</u> *Mline* PICKed are <u>trimmed</u> at the outer edges of the second. Only the outer line element of the second is trimmed (Fig. 16-39).

FIGURE 16-39

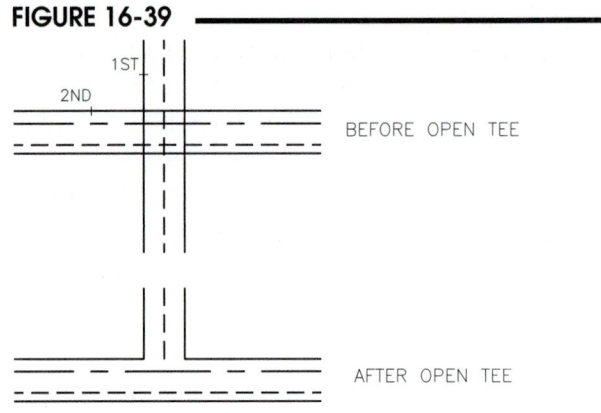

BEFORE OPEN TEE

AFTER OPEN TEE

Merged Tee

This "T" option allows the <u>inner line elements</u> of both intersecting *Mlines* to *merge*. The merge occurs at the second *Mline's* first inner element. Only the side of the first line nearest the PICK point remains (Fig. 16-40).

FIGURE 16-40

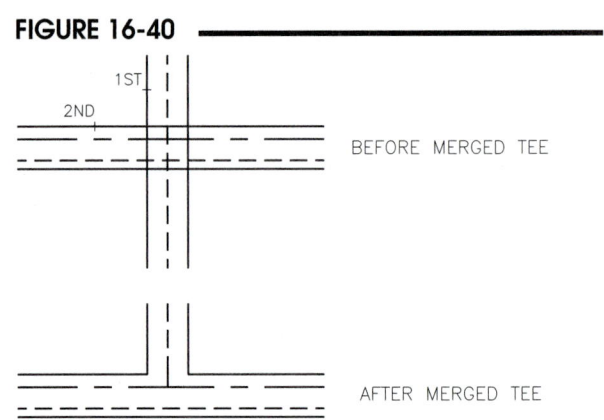

BEFORE MERGED TEE

AFTER MERGED TEE

Corner Joint

This option <u>trims both</u> *Mlines* to create a <u>corner</u>. Only the PICKed sides of the *Mlines* remain and the extending portions of both (if any) are trimmed. All of the inner line elements merge (Fig. 16-41).

FIGURE 16-41

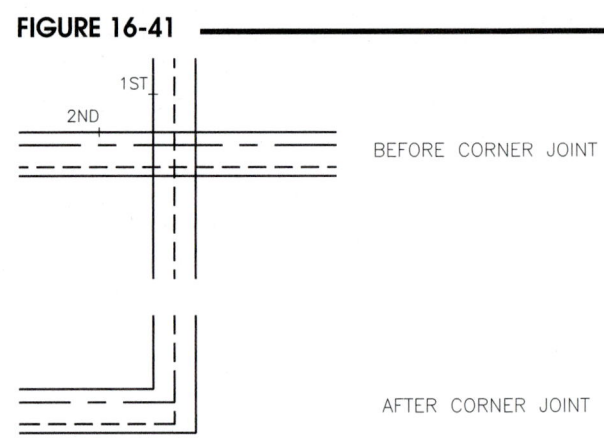

BEFORE CORNER JOINT

AFTER CORNER JOINT

Add Vertex

If you want to add a new corner (vertex) to an existing *Mline*, use this feature. A new vertex is added <u>where you PICK</u> the *Mline*. However, it is not readily apparent that a new vertex exists, nor does the *Mline* visibly change in any way. You must <u>further edit</u> the *Mline* with *Stretch* or Grips (see Chapter 19) to create a new "corner" (Fig. 16-42).

FIGURE 16-42

BEFORE ADD VERTEX

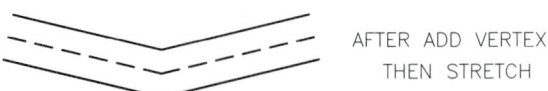

AFTER ADD VERTEX
THEN STRETCH

Delete Vertex

Use this feature to remove a corner (vertex) of an *Mline*. The vertex <u>nearest</u> the location you PICK is deleted. The resulting *Mline* contains only one straight segment between the adjacent two vertices. Unlike *Add Vertex,* the *Mline* immediately changes and no further editing is needed to see the effect (Fig. 16-43).

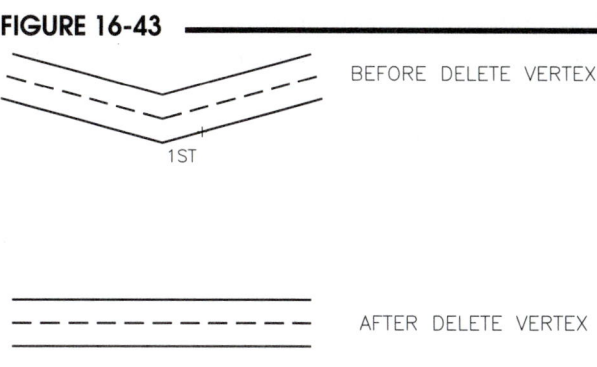

FIGURE 16-43

BEFORE DELETE VERTEX

1ST

AFTER DELETE VERTEX

Cut Single

The *Cut* options are used to break (cut) a space in one *Mline*. *Cut Single* <u>breaks any line element</u> that is selected. Similar to *Break 2Points,* the break occurs <u>between the two PICK points</u> (Fig. 16-44). The break points can be on either side of a vertex.

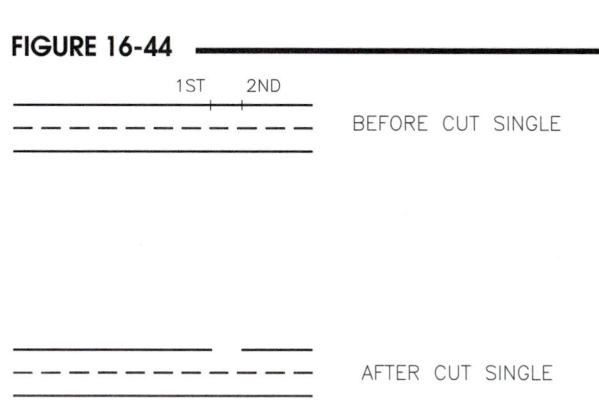

FIGURE 16-44

1ST   2ND

BEFORE CUT SINGLE

AFTER CUT SINGLE

Cut All

This *Cut* option <u>breaks all line elements</u> of the selected *Mline*. Any line elements can be selected. The line elements are cut at the PICK points in a direction perpendicular to the axis of the *Mline* (Fig. 16-45). NOTE: Although the *Mline* appears to be cut into two separate *Mlines*, it <u>remains one single object</u>. Using *Stretch* or Grips to "move" the *Mline* causes the cut to close again.

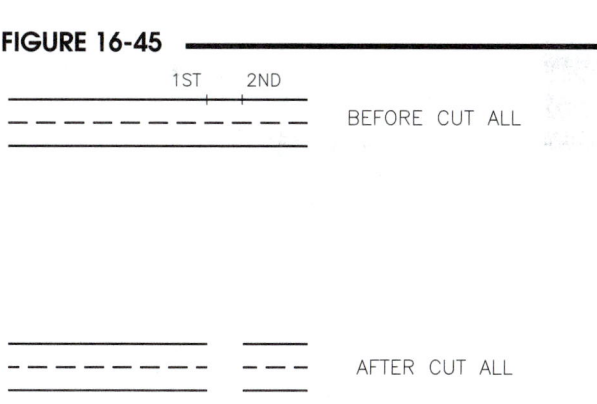

FIGURE 16-45

1ST   2ND

BEFORE CUT ALL

AFTER CUT ALL

Weld

This option <u>reverses the action of a *Cut*</u>. The *Mline* is restored to its original continuous configuration. PICK on both sides of the "break" (Fig. 16-46).

FIGURE 16-46

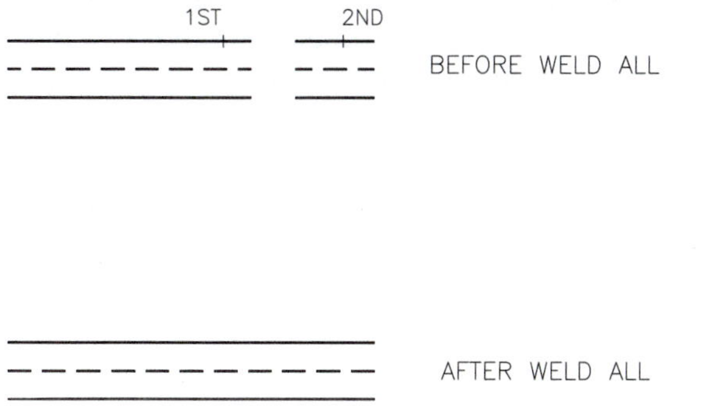

1ST   2ND

BEFORE WELD ALL

AFTER WELD ALL

Like many other dialog box-based commands, *Mledit* can also be used in command line format by using the hyphen (-) symbol as a prefix to the command. Key in "*-MLEDIT*" to force the command line interface. The following prompt appears:

> Command: *-mledit*
> Enter mline editing option [CC/OC/MC/CT/OT/MT/CJ/AV/DV/CS/CA/WA]:

Other Editing Possibilities for *Mlines*

Trim, Extend, Stretch, and Grips operate correctly with *Mlines. Stretch* and Grips can be used to change endpoints and move vertices; however, doing so may affect intersections with other *Mlines.*

The *Trim* and *Extend* commands can be used very effectively with *Mlines.* For the "cutting edges" and "boundary edges," you can select *Mlines* as well as other objects (*Lines, Plines, Arcs,* etc.). If you select an *Mline* as a "cutting edge" or "boundary edge," you can then select from a *Closed, Open,* or *Merged* intersection, producing results <u>identical to using the *Closed Tee, Open Tee,* or *Merged Tee* options of *Mledit.*</u>

> Command: *TRIM*
> Current settings: Projection=UCS, Edge=None
> Select cutting edges ...
> Select objects or <select all>: **PICK**
> Select objects: **Enter**
> Select object to trim or shift-select to extend or [Fence/Crossing/Project/Edge/eRase/Undo]: **PICK** (select *Mline* object)
> Enter mline junction option [Closed/Open/Merged] <Open>:

You may experience situations where *Mledit, Trim, Extend, Stretch,* or Grips cannot handle your desired editing request. It is possible to *Explode* the *Mline,* which converts each line element to an individual object, allowing you to use other editing options. <u>However, this practice is not recommended because once an *Mline* is *Exploded,* it cannot be converted back to an *Mline,* nor can *Mledit* be used.</u>

Boolean Commands

Region combines one or several objects forming a closed shape into one object, a *Region.* The appearance of the object(s) does not change after the conversion, even though the resulting shape is one object (see "*Region,*" Chapter 15).

Although the *Region* appears to be no different than a closed *Pline,* it is more powerful because several *Regions* can be combined to form complex shapes (called "composite *Regions*") using the three Boolean operations explained next. As an example, a set of *Regions* (converted *Circles*) can be combined to form the sprocket with only <u>one</u> *Subtract* operation, as shown in Chapter 15, Figure 15-41.

The Boolean operators, *Union, Subtract,* and *Intersect,* can be used with *Regions* as well as solids. Any number of these commands can be used with *Regions* to form complex geometry. To illustrate each of the Boolean commands, consider the shapes shown in Figure 16-47. The *Circle* and the closed *Pline* are <u>first</u> converted to *Regions;* then *Union, Subtract,* or *Intersection* can be used.

FIGURE 16-47

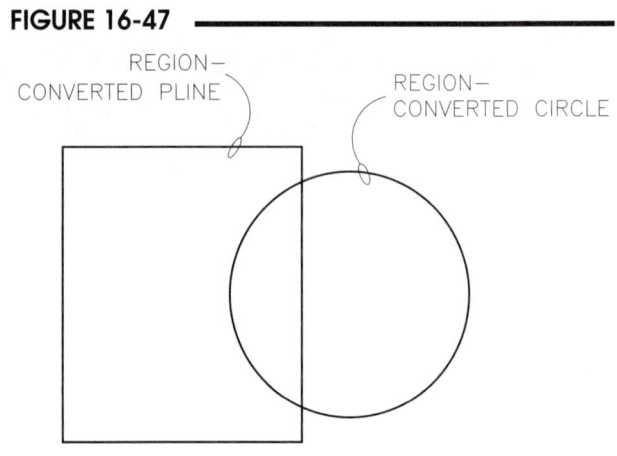

REGION— CONVERTED PLINE

REGION— CONVERTED CIRCLE

Union

Pull-down Menu	Command (Type)	Alias (Type)	Short-cut	Screen (side) Menu	Tablet Menu
Modify Solid Editing > Union	Union	UNI	...	MODIFY2 Union	X,15

Union combines <u>two or more</u> *Regions* (or solids) into <u>one</u> *Region* (or solid). The resulting composite *Region* has the encompassing perimeter and area of the original *Regions*. Invoking *Union* causes AutoCAD to prompt you to select objects. You can select only existing *Regions* (or solids).

```
Command: union
Select Objects: PICK (Region)
Select Objects: PICK (Region)
Select Objects: Enter
Command:
```

The selected *Regions* are combined into one composite *Region* (Fig. 16-48). Any number of Boolean operations can be performed on the *Region*(s).

FIGURE 16-48

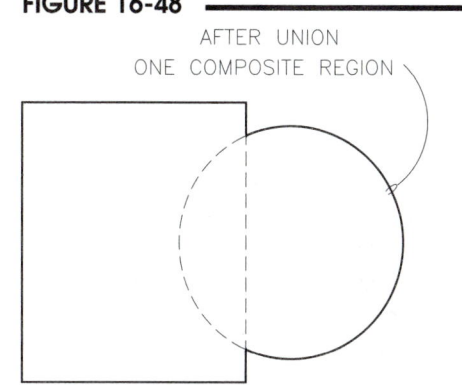

AFTER UNION
ONE COMPOSITE REGION

Several *Regions* can be selected in response to the "Select objects:" prompt. For example, a composite *Region* such as that in Figure 16-49 can be created with one *Union*.

Two or more *Regions* can be *Unioned* even if they do not overlap. They are simply combined into one object although they still appear as two.

FIGURE 16-49

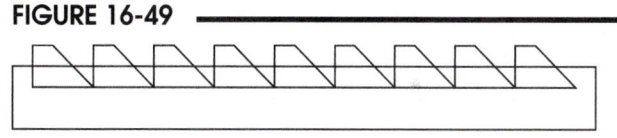

BEFORE— 10 REGIONS

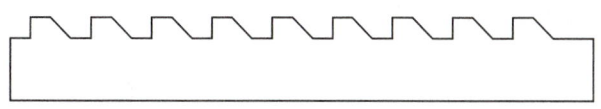

AFTER UNION— ONE REGION

Subtract

Pull-down Menu	Command (Type)	Alias (Type)	Short-cut	Screen (side) Menu	Tablet Menu
Modify Solid Editing > Subtract	Subtract	SU	...	MODIFY2 Subtract	X,16

Subtract enables you to remove one *Region* (or set of *Regions*) from another. The *Regions* must be created before using *Subtract* (or another Boolean operator). *Subtract* also works with solids (as do the other Boolean operations).

There are two steps to *Subtract*. First, you are prompted to select the *Region* or set of *Regions* to "subtract from" (those that you wish to <u>keep</u>), then to select the *Regions* "to subtract" (those you want to <u>remove</u>). The resulting shape is one composite *Region* comprising the perimeter of the first set minus the second (sometimes called "difference").

```
Command: subtract
Select solids and regions to subtract from...
Select Objects: PICK
Select Objects: Enter
Select solids and regions to subtract...
Select Objects: PICK
Select Objects: Enter
Command:
```

Consider the two shapes previously shown in Figure 16-47. The resulting *Region* shown in Figure 16-50 is the result of *Subtracting* the circular *Region* from the rectangular one.

 TIP Keep in mind that <u>multiple</u> *Regions* can be selected as the set to keep or as the set to remove. For example, the sprocket illustrated previously (Fig. 15-41) was created by subtracting several circular *Regions* in one operation. Another example is the removal of material to create holes or slots in sheet metal (discussed in Chapter 40, Surface Modeling).

FIGURE 16-50 ————————————

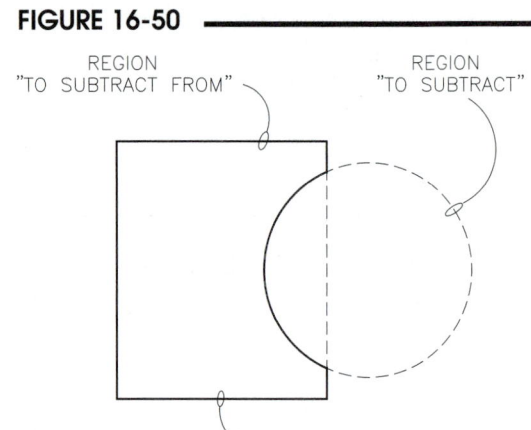

REGION "TO SUBTRACT FROM"

REGION "TO SUBTRACT"

RESULTING REGION

Intersect

Pull-down Menu	Command (Type)	Alias (Type)	Short-cut	Screen (side) Menu	Tablet Menu
Modify *Solid Editing >* *Intersect*	*Intersect*	*IN*	...	*MODIFY2* *Intrsect*	*X,17*

Intersect is the Boolean operator that finds the common area from two or more *Regions*.

Consider the rectangular and circular *Regions* previously shown (Fig. 16-47). Using the *Intersect* command and selecting both shapes results in a *Region* comprising only that area that is shared by both shapes (Fig. 16-51):

```
Command: intersect
Select Objects: PICK
Select Objects: PICK
Select Objects: Enter
Command:
```

FIGURE 16-51 ————————————

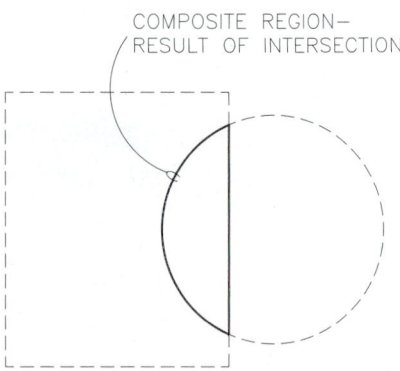

COMPOSITE REGION— RESULT OF INTERSECTION

If more than two *Regions* are selected, the resulting *Intersection* is composed of only the common area from all shapes (Fig. 16-52). If <u>all</u> of the shapes selected do not overlap, a <u>null</u> *Region* is created (all shapes disappear because no area is common to <u>all</u>).

Intersect is a powerful operation when used with solid modeling. Techniques for saving time using Boolean operations are discussed in Chapter 38, Solid Modeling Construction.

FIGURE 16-52

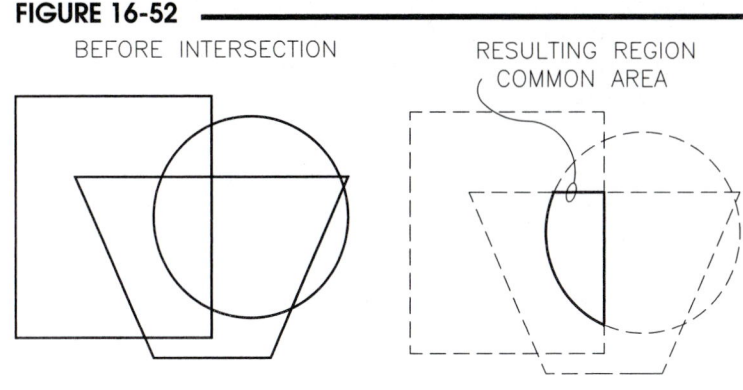

BEFORE INTERSECTION   RESULTING REGION COMMON AREA

CHAPTER EXERCISES

1. *Chprop or Properties*

    Open the **PIVOTARM CH9** drawing that you worked on in Chapter 9 Exercises. *Load* the *Hidden2* and *Center2 Linetypes*. Make two *New Layers* named **HID** and **CEN** and assign the matching linetypes and yellow and green colors, respectively. Check the *Limits* of the drawing; then calculate and set an appropriate *LTSCALE*. Use *Chprop* or *Properties* palette to change the *Lines* representing the holes in the front view to the **HID** layer as shown in Figure 16-53. Use *SaveAs* and name the drawing **PIVOTARM CH16.**

FIGURE 16-53

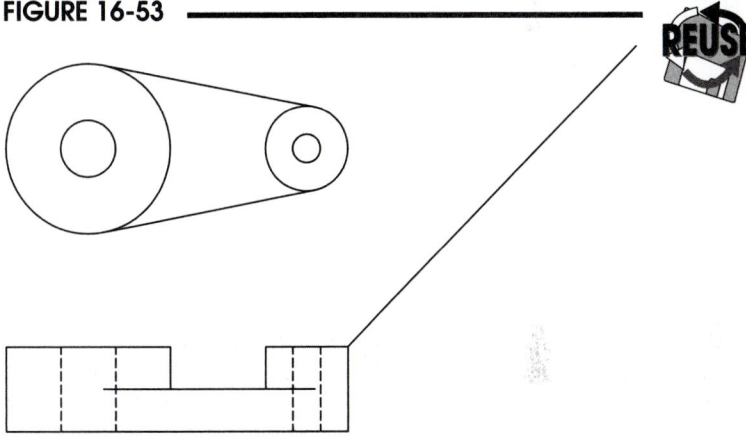

2. *Change*

    Open **CH8EX3.** *Erase* the *Arc* at the top of the object. *Erase* the *Points* with a window. Invoke the *Change* command. When prompted to *Select objects*, **PICK** all of the inclined *Lines* near the top. When prompted to "Specify change point," enter coordinate **6,8**. The object should appear as that in Figure 16-54. Use *SaveAs* and assign the name **CH16EX2.**

FIGURE 16-54

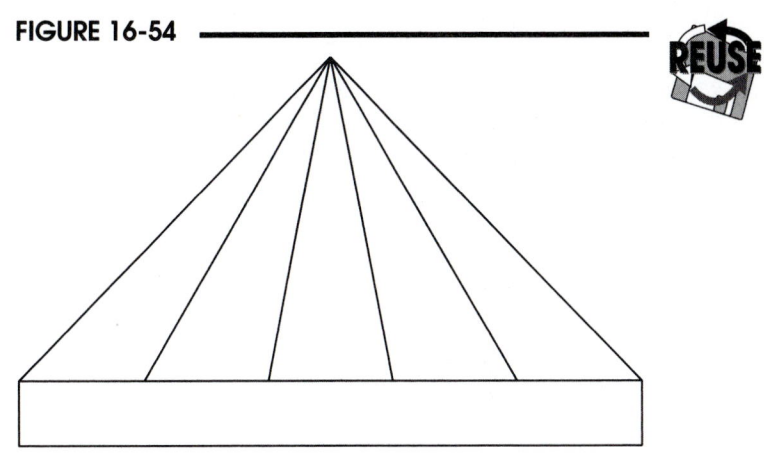

3. ***Properties* Palette**

A design change is required for the bolt holes in **GASKETA** (from Chapter 9 Exercises). *Open* **GASKETA** and invoke the *Properties* palette. Change each of the four bolt holes to *.375* diameter (Fig. 16-55). *Save* the drawing.

FIGURE 16-55

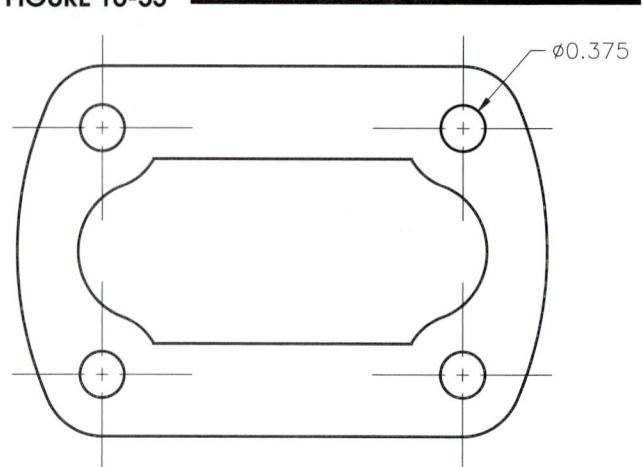

ø0.375

4. *Align*

Open the **PLATES** drawing you created in Chapter 9. The three plates are to be stamped at one time on a single sheet of stock measuring 15″ x 12″. Place the three plates together to achieve optimum nesting on the sheet stock.

A. Use *Align* to move the plate in the center (with 9 holes). Select the *First source* and *destination points* (1S, 1D) and *Second source* and *destination points* (2S, 2D) as shown in Figure 16-56.

FIGURE 16-56

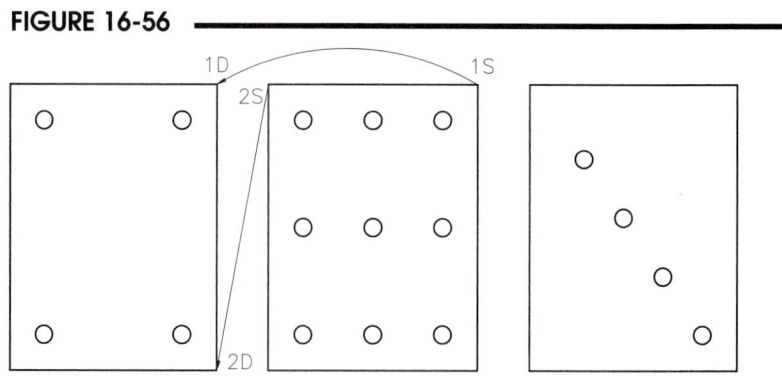

B. After the first alignment is complete, use *Align* to move the plate on the right (with 4 diagonal holes). The *source* and *destination points* are indicated in Figure 16-57.

FIGURE 16-57

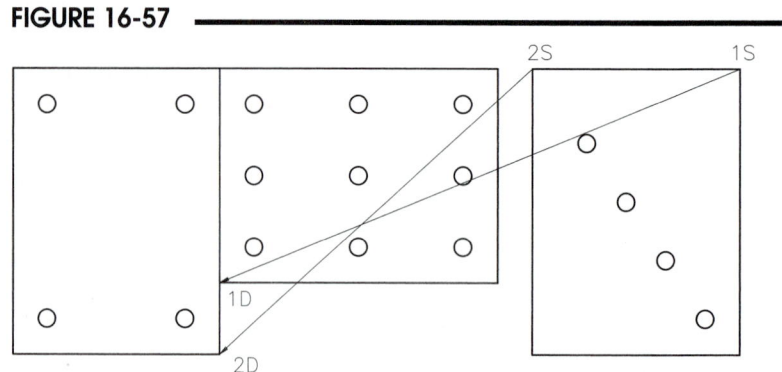

C. Finally, draw the sheet stock outline (15″ x 12″) using *Line* as shown in Figure 16-58. The plates are ready for production. Use *SaveAs* and assign the name **PLATENEST.**

FIGURE 16-58

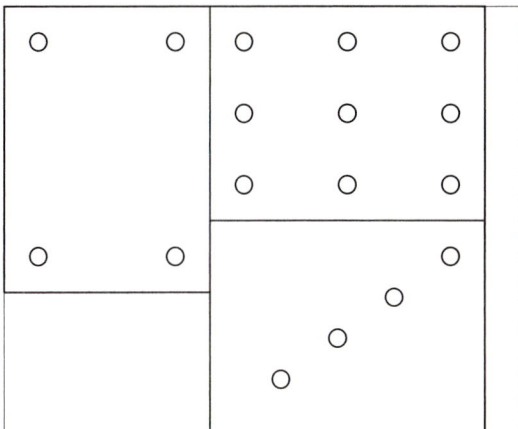

5. *Explode*

Open the **POLYGON1** drawing that you completed in Chapter 15 Exercises. You can quickly create the five-sided shape shown (continuous lines) in Figure 16-59 by *Exploding* the *Polygon*. First, *Explode* the *Polygon* and *Erase* the two *Lines* (highlighted). Draw the bottom *Line* from the two open *Endpoints*. Do not exit the drawing.

FIGURE 16-59

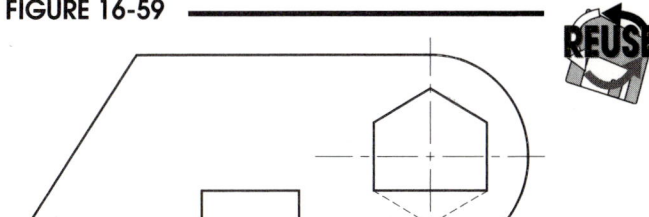

6. *Pedit*

Use *Pedit* with the *Edit vertex* options to alter the shape as shown in Figure 16-60. For the bottom notch, use *Straighten*. For the top notch, use *Insert*. Use *SaveAs* and change the name to **PEDIT1.**

FIGURE 16-60

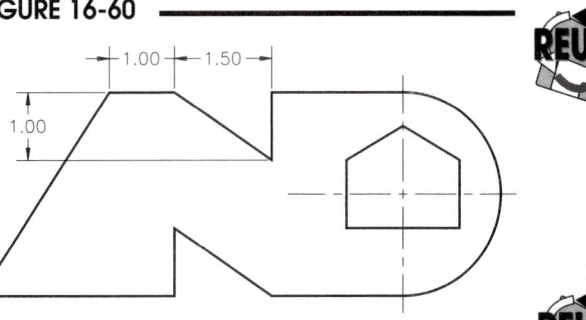

7. *Pline, Pedit*

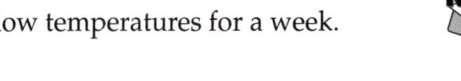

A. Create a line graph as shown in Figure 16-61 to illustrate the low temperatures for a week. The temperatures are as follows:

X axis	Y axis
Sunday	20
Monday	14
Tuesday	18
Wednesday	26
Thursday	34
Friday	38
Saturday	27

FIGURE 16-61

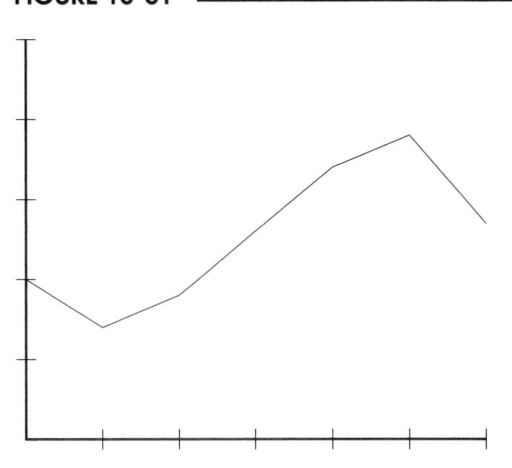

Use equal intervals along each axis. Use a *Pline* for the graph line. *Save* the drawing as **TEMP-A**. (You will label the graph at a later time.)

CONCEPTS

AutoCAD provides several commands that allow you to find out information about the current drawing status and specific objects in a
drawing. These commands as a group are known as "*Inquiry* commands" and are grouped together in the menu systems. The *Inquiry*
commands are located in the *Inquiry* toolbar (Fig. 17-1). You can also use
the *Tools* pull-down menu to access the *Inquiry* commands (Fig. 17-2).

FIGURE 17-1

Using *Inquiry* commands, you
can find out such information
as the amount of time spent
in the current drawing, the
distance between two points,
the area of a closed shape, the
database listing of properties
for specific objects (coordinates of endpoints, lengths,
angles, etc.), and current settings for system variables as
well as other information.
The *Inquiry* commands are:

> *Status, List, Dblist, Area,*
> *Dist, ID, Time,* and *Setvar*

FIGURE 17-2

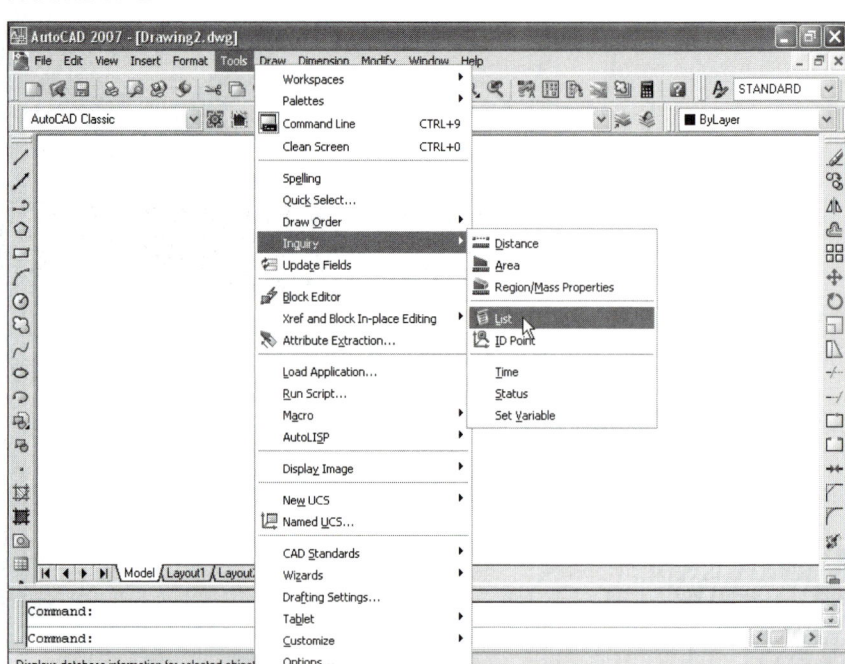

COMMANDS

Status

Pull-down Menu	Command (Type)	Alias (Type)	Short-cut	Screen (side) Menu	Tablet Menu
Tools *Inquiry* > *Status*	*Status*	...	...	*TOOLS 1* *Status*	...

The *Status* command gives many pieces of information related to the current drawing. Typing or
PICKing the command from the icon or one of the menus causes a text screen to appear similar to that
shown in Figure 17-3, on the next page. The information items are:

> Total number of objects in the current drawing.
> Paper space limits: values set by the *Limits* command in Paper Space. (*Limits* in Paper Space are set
>   by selecting a *Paper size* in the *Page Setup* or *Plot* dialog box.)
> Paper space uses: area used by the objects (drawing extents) in Paper Space.
> Model space limits: values set by the *Limits* command in Model Space.
> Model space uses: area used by the objects (drawing extents) in Model Space.

Display shows: current display or windowed area.

Insertion basepoint: point specified by the *Base* command or default (0,0).

Snap resolution: value specified by the *Snap* command.

Grid spacing: value specified by the *Grid* command.

Current space: Paper Space or Model Space

Current layer: name.

Current color: current color assignment

Current linetype: current linetype assignment

Current lineweight: current lineweight assignment.

Current plot style: current plot style assignment.

Current elevation, thickness: 3D properties—current height above the XY plane and Z dimension.

On or off status: *FILL, GRID, ORTHO, QTEXT, SNAP, TABLET.*

Object Snap Modes: current *Running OSNAP* modes.

Free dwg disk: space on the current drawing hard disk drive.

Free temp disk: space on the current temporary files hard disk drive.

Free physical memory: amount of free RAM (total RAM).

Free swap file space: amount of free swap file space (total allocated swap file).

FIGURE 17-3

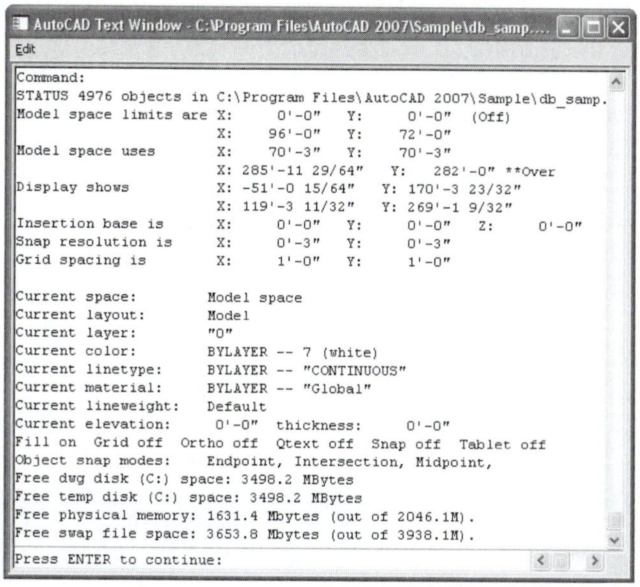

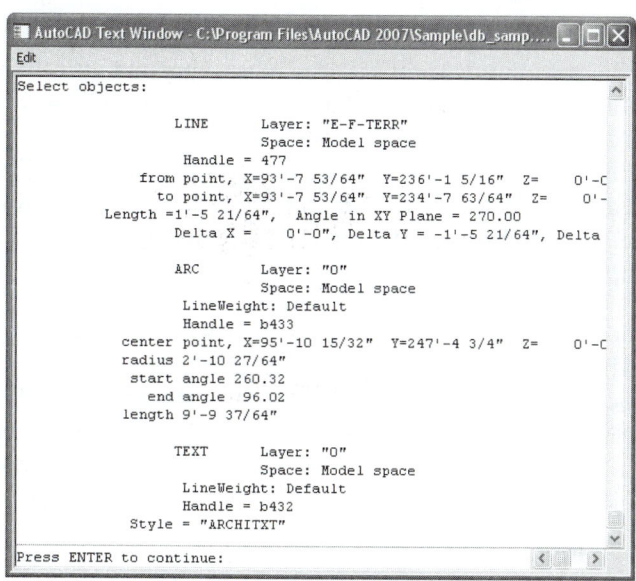

List

Pull-down Menu	Command (Type)	Alias (Type)	Short-cut	Screen (side) Menu	Tablet Menu
Tools *Inquiry >* *List*	*List*	*LS* or *LI*	...	TOOLS 1 List	U,8

The *List* command displays the database list of information in text window format for one or more specified objects. The information displayed depends on the <u>type</u> of object selected. Invoking the *List* command causes a prompt for you to select objects. AutoCAD then displays the list for the selected objects (see Fig. 17-4, right and Fig. 17-5, on the next page). A *List* of a *Line* and an *Arc* is given in Figure 17-4.

For a *Line*, coordinates for the endpoints, line length and angle, current layer, and other information are given.

For an *Arc*, the center coordinate, radius, start and end angles, and length are given.

FIGURE 17-4

The *List* for a *Pline* is shown in Figure 17-5. The location of each vertex is given, as well as the length and perimeter of the entire *Pline*.

Plines are created and listed as *Lwpolylines*, or "lightweight polylines." The data common to all vertices are stored only once, and only the coordinate data are stored for each vertex. Because this data structure saves file space, the *Plines* are known as lightweight *Plines*.

FIGURE 17-5

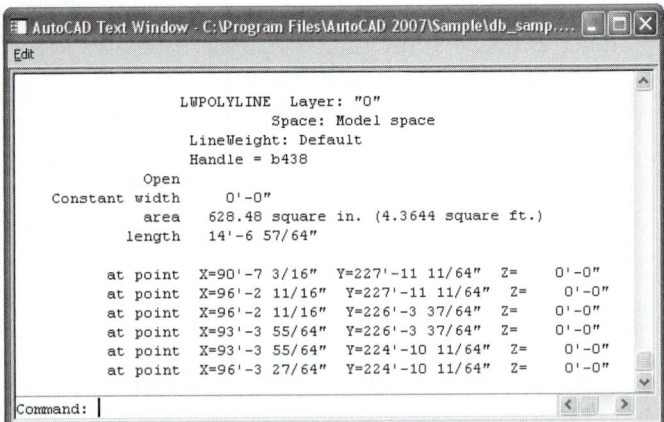

Pull-down Menu	Command (Type)	Alias (Type)	Short-cut	Screen (side) Menu	Tablet Menu
...	*Dblist*	...	...	...	...

Dblist

The *Dblist* command is similar to the *List* command in that it displays the database listing of objects; however, *Dblist* gives information for <u>every</u> object in the current drawing! This command is generally used when you desire to copy the list to a file or when only a few objects are in the drawing. If you use this command in a complex drawing, be prepared to page through many screens of information. Press Escape to cancel *Dblist* and return to the Command: prompt. Press F2 to open and close the text window.

Area

Pull-down Menu	Command (Type)	Alias (Type)	Short-cut	Screen (side) Menu	Tablet Menu
Tools *Inquiry >* *Area*	*Area*	*AA*	...	*TOOLS 1* *Area*	*T,7*

The *Area* command is helpful for many applications. With this command AutoCAD calculates the area and the perimeter of any enclosed shape in a matter of milliseconds. You specify the area (shape) to consider for calculation by PICKing the *Object* (if it is a closed *Pline, Polygon, Circle, Boundary, Region* or other closed object) or by PICKing points (corners of the outline) to define the shape. The options are given below.

Specify first corner point

The command sequence for specifying the area by PICKing points is shown below. This method should be used only for shapes with <u>straight</u> sides. An example of the *Point* method (PICKing points to define the area) is shown in Figure 17-6.

FIGURE 17-6

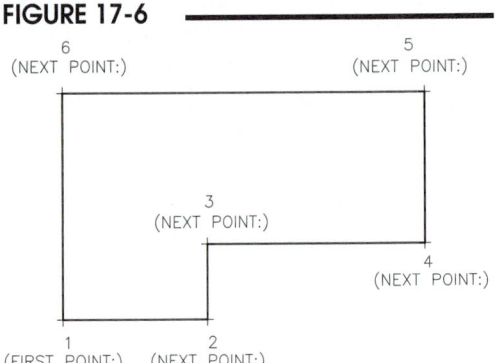

Command: **area**
Specify first corner point or [Object/Add/Subtract]: **PICK** (Locate the first corner to define the shape.)
Specify next corner point or press ENTER for total: **PICK** (Locate the second corner on the shape.)
Specify next corner point or press ENTER for total: **PICK** (Locate the next corner.)
Specify next corner point or press ENTER for total: **PICK** (Continue selecting points until all corners have been defined.)
Specify next corner point or press ENTER for total: **Enter**
Area = *nn.nnn* perimeter = *nn.nnn*
Command:

Object

If the shape for which you want to find the area and perimeter is a *Circle, Polygon, Rectangle, Ellipse, Boundary, Region,* or closed *Pline,* the *Object* option of the *Area* command can be used. Select the shape with one PICK (since all of these shapes are considered as one object).

The ability to find the area of a closed *Pline, Region,* or *Boundary* is extremely helpful. Remember that any closed shape, even if it includes *Arcs* and other curves, can be converted to a closed *Pline* with the *Pedit* command (as long as there are no gaps or overlaps) or can be used with the *Boundary* command. This method provides you with the ability to easily calculate the area of any shape, curved or straight. In short, convert the shape to a closed *Pline, Region,* or *Boundary* and find the *Area* with the *Object* option.

Add, Subtract

Add and *Subtract* provide you with the means to find the area of a closed shape that has islands, or negative spaces. For example, you may be required to find the surface area of a sheet of material that has several punched holes. In this case, the area of the holes is subtracted from the area defined by the perimeter shape. The *Add* and *Subtract* options are used specifically for that purpose. The following command sequence displays the process of calculating an area and subtracting the area occupied by the holes.

Command: **area**
Specify first corner point or [Object/Add/Subtract]: **a** (Use the *Add* option to begin a running total.)
Specify first corner point or [Object/Subtract]: **o** (Use the *Object* option to select the outside shape.)
(ADD mode) Select objects: **PICK** (Select the closed object.)
Area = 13.31, Perimeter = 14.39
Total area = 13.31

(ADD mode) Select objects: **Enter** (Completion of *Add* mode.)
Specify first corner point or [Object/Subtract]: **s** (Switch to *Subtract* mode.)
Specify first corner point or [Object/Add]: **o** (Use *Object* mode.)
(SUBTRACT mode) Select objects: **PICK** (Select the first *Circle* to subtract.)
Area = 0.69, Length = 2.95
Total area = 12.62

(SUBTRACT mode) Select objects: **PICK** (Select the second *Circle* to subtract.)
Area = 0.69, Length = 2.95
Total area = 11.93

(SUBTRACT mode) Select objects: **Enter** (Completion of *Subtract* mode.)
Specify first corner point or [Object/Add]: **Enter** (Completion of *Area* command.)
Command:

Make sure that you press Enter between the *Add* and *Subtract* modes.

An example of the last command sequence used to find the area of a shape minus the holes is shown in Figure 17-7. Notice that the object selected in the first step is a closed *Pline* shape, including an *Arc*.

FIGURE 17-7

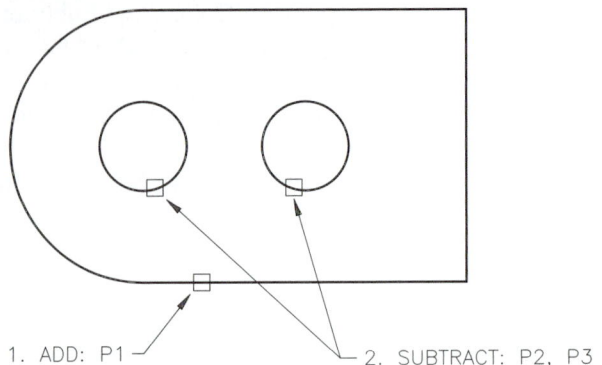

1. ADD: P1   2. SUBTRACT: P2, P3

Dist

Pull-down Menu	Command (Type)	Alias (Type)	Short-cut	Screen (side) Menu	Tablet Menu
Tools Inquiry > Distance	Dist	DI	...	TOOLS 1 Dist	T,8

The *Dist* command reports the distance between any two points you specify. *OSNAP*s can be used to snap to the existing points. This command is helpful in many engineering or architectural applications, such as finding the clearance between two mechanical parts, finding the distance between columns in a building, or finding the size of an opening in a part or doorway. The command is easy to use.

```
Command: dist
Specify first point: PICK  (Use Osnaps if needed.)
Specify second point: PICK  (Use Osnaps if needed.)
Distance = 3.63,  Angle in XY Plane = 165,  Angle from XY Plane = 0
Delta X = -3.50,  Delta Y = 0.97,   Delta Z = 0.00
Command:
```

AutoCAD reports the absolute and relative distances as well as the angle of the line between the points.

Id

Pull-down Menu	Command (Type)	Alias (Type)	Short-cut	Screen (side) Menu	Tablet Menu
Tools Inquiry > ID Point	Id	...	...	TOOLS 1 ID	U,9

The *ID* command reports the coordinate value of any point you select with the cursor. If you require the location associated with a specific object, an *OSNAP* mode (*Endpoint, Midpoint, Center,* etc.) can be used.

```
Command: id
Specify point: PICK  (Use Osnaps if needed.)
X = 7.63    Y = 6.25    Z = 0.00
Command:
```

NOTE: *ID* also sets AutoCAD's "last point." The last point can be referenced in commands by using the @ (at) symbol with relative rectangular or relative polar coordinates.

Time

Pull-down Menu	Command (Type)	Alias (Type)	Short-cut	Screen (side) Menu	Tablet Menu
Tools *Inquiry >* *Time*	*Time*	...	...	*TOOLS 1* *Time*	...

This command is useful for keeping track of the time spent in the current drawing session or total time spent on a particular drawing. Knowing how much time is spent on a drawing can be useful in an office situation for bidding or billing jobs.

The *Time* command reports the information shown in Figure 17-8. The *Total editing time* is automatically kept, starting from when the drawing was first created until the current time. Plotting and printing time is not included in this total, nor is the time spent in a session when changes are discarded.

FIGURE 17-8

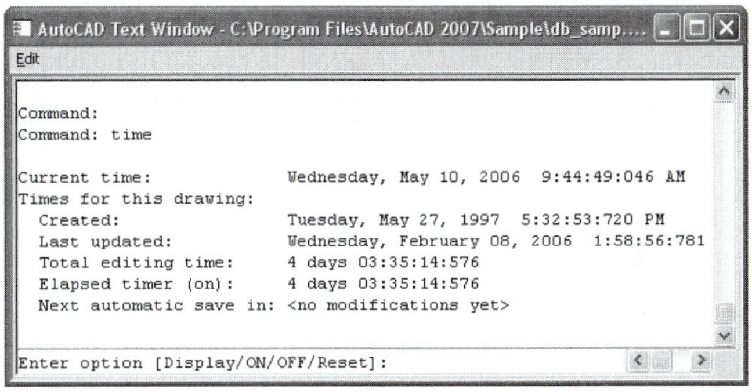

Display

The *Display* option causes *Time* to repeat the display with the updated times.

ON/OFF/Reset

The *Elapsed timer* is a separate compilation of time controlled by the user. The *Elapsed timer* can be turned *ON* or *OFF* or can be *Reset*.

Time also reports when the next automatic save will be made. The time interval of the Automatic Save feature is controlled by the *SAVETIME* system variable. To set the interval between automatic saves, type *SAVETIME* at the Command line and specify a value for time (in minutes). See also *SAVETIME* in Chapter 2.

Setvar

Pull-down Menu	Command (Type)	Alias (Type)	Short-cut	Screen (side) Menu	Tablet Menu
Tools *Inquiry >* *Set Variable*	*Setvar*	*SET*	...	*TOOLS 1* *Setvar*	*U,10*

The settings (values or on/off status) that you make for many commands, such as *Limits, Grid, Snap, Running Osnaps, Fillet* values, *Pline* width, etc. are saved in system variables. There are over 400 system variables. The variables store the settings that are used to create and edit the drawing. *Setvar* ("set variable") gives you access to the system variables. (See Appendix A for a complete list of the system variables, including an explanation, default setting, and possible settings for each.)

Setvar allows you to perform two functions: (1) change the setting for any system variable and (2) display the current setting for one or all system variables. To change a setting for a system variable using *Setvar*, just use the command and enter the name of the variable. For example, the following syntax lists values for the *GRID* setting and current *Fillet* radius:

Command: **setvar**
Enter variable name or [?]: **gridunit**
Enter new value for GRIDUNIT <1.00,1.00>:

Command: **setvar**
Enter variable name or [?]: **filletrad**
Enter new value for FILLETRAD <0.50>:

With recent releases of AutoCAD, _Setvar_ is not needed to set system variables. You can enter the variable name directly at the Command prompt without using _Setvar_ first.

Command: **filletrad**
Enter new value for FILLETRAD <0.50>:

TIP To list the current settings for all system variables, use _Setvar_ with the _?_ (question mark) option. The complete list of system variables is given in a text window with the current setting for each variable (Fig. 17-9).

FIGURE 17-9

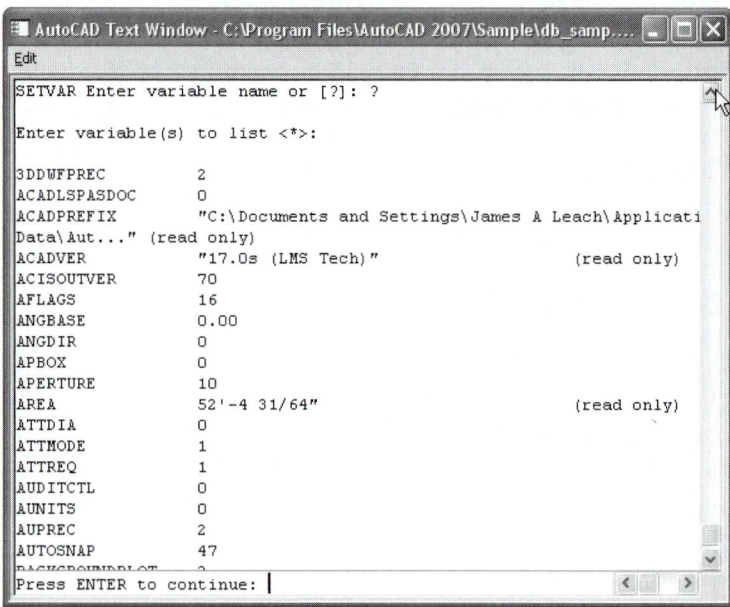

You can also use _Help_ to list the system variables with a short explanation for each. In the _Help Topics_ dialog box, select the _Contents_ tab, select _Command Reference_, then _System Variables_.

Massprop

The _Massprop_ command is used to give mass and volumetric properties for AutoCAD solids. See Chapter 39, Solid Model Editing, for use of _Massprop_.

CHAPTER EXERCISES

1. *List*

   A. *Open* the **PLATENEST** drawing from Chapter 16 Exercises. Assume that a laser will be used to cut the plates and holes from the stock, and you must program the coordinates. Use the *List* command to give information on the *Line*s and *Circle*s for the one plate with four holes in a diagonal orientation. Determine and write down the coordinates for the 4 corners of the plate and the centers of the 4 holes.

   Corners (16,5) (8,5) (16,-1) (8,-1)
   Circles (10.1, 6) (11.7, 1.7) (13.4, 2.6) (15,4)

   B. *Open* the **EFF-APT** drawing from Chapter 15 Exercises. Use *List* to determine the area of the inside of the tub. If the tub were filled with 10″ of water, what would be the volume of water in the tub?

   Area = 874.15 sq in × 10 in = [8741.5 in³]

2. *Area*

   A. *Open* the **EFF-APT** drawing. The entry room is to be carpeted at a cost of $12.50 per square yard. Use the *Area* command (with the PICK points option) to determine the cost for carpeting the room, not including the closet.

   27441.25 sq in × .00088 sq yds / sq in = 21.953 sq yds 190.5642 sq ft × 12.5 = $274.4125

   B. *Open* the **PLATENEST** drawing from Chapter 16 Exercises. Use *SaveAs* to create a file named **PLATE-AREA**. Using the *Area* command, calculate the wasted material (the two pieces of stock remaining after the 3 plates have been cut or stamped). HINT: Use *Boundary* to create objects from the waste areas for determining the *Area*.

   A = 24 + 12 = 36 in

   C. Create a *Boundary* (with islands) of the plate with four holes arranged diagonally. *Move* the new boundary objects **10** units to the right. Using the *Add* and *Subtract* options of *Area*, calculate the surface area for painting 100 pieces, both sides. (Remember to press Enter between the *Add* and *Subtract* operations.) *Save* the drawing.

   48 × 2 × 100 = 96000

3. *Dist*

   A. *Open* the **EFF-APT** drawing. Use the *Dist* command to determine the best location for installing a wall-mounted telephone in the apartment. Where should the telephone be located in order to provide the most equal access from all corners of the apartment? What is the farthest distance that you would have to walk to answer the phone?

   wall by entrance to kitchen 17′

   B. Using the *Dist* command, determine what length of pipe would be required to connect the kitchen sink drain (use the center of the far sink) to the tub drain (assume the drain is at the far end of the tub). Calculate only the direct distance (under the floor).

   11′-6 5/8″

4. *ID*

   Open the **PLATENEST** drawing once again. You have now been assigned to program the laser to cut the plate with 4 holes in the corners. Use the *ID* command (with *OSNAP*s) to determine the coordinates for the 4 corners and the hole centers.

   Corners: (8,3) (8,11) (2,11) (2,3)
   Circles: (7,4) (3,4) (3,10) (7,10)

5. *Time*

Using the *Time* command, what is the total amount of editing time you spent with the **PLATENEST** drawing? How much time have you spent in this session? How much time until the next automatic save?

3:57:59:078

6. *Setvar*

Using the **PLATENEST** drawing again, use *Setvar* to list system variables. What are the current settings for *Fillet* radius (**FILLETRAD**), the last point used (**LASTPOINT**), *Linetype Scale* (**LTSCALE**), automatic file save interval (**SAVETIME**), and text *Style* (**TEXTSTYLE**).

1   0   10   <16, 2.5, 0)   "Standard"

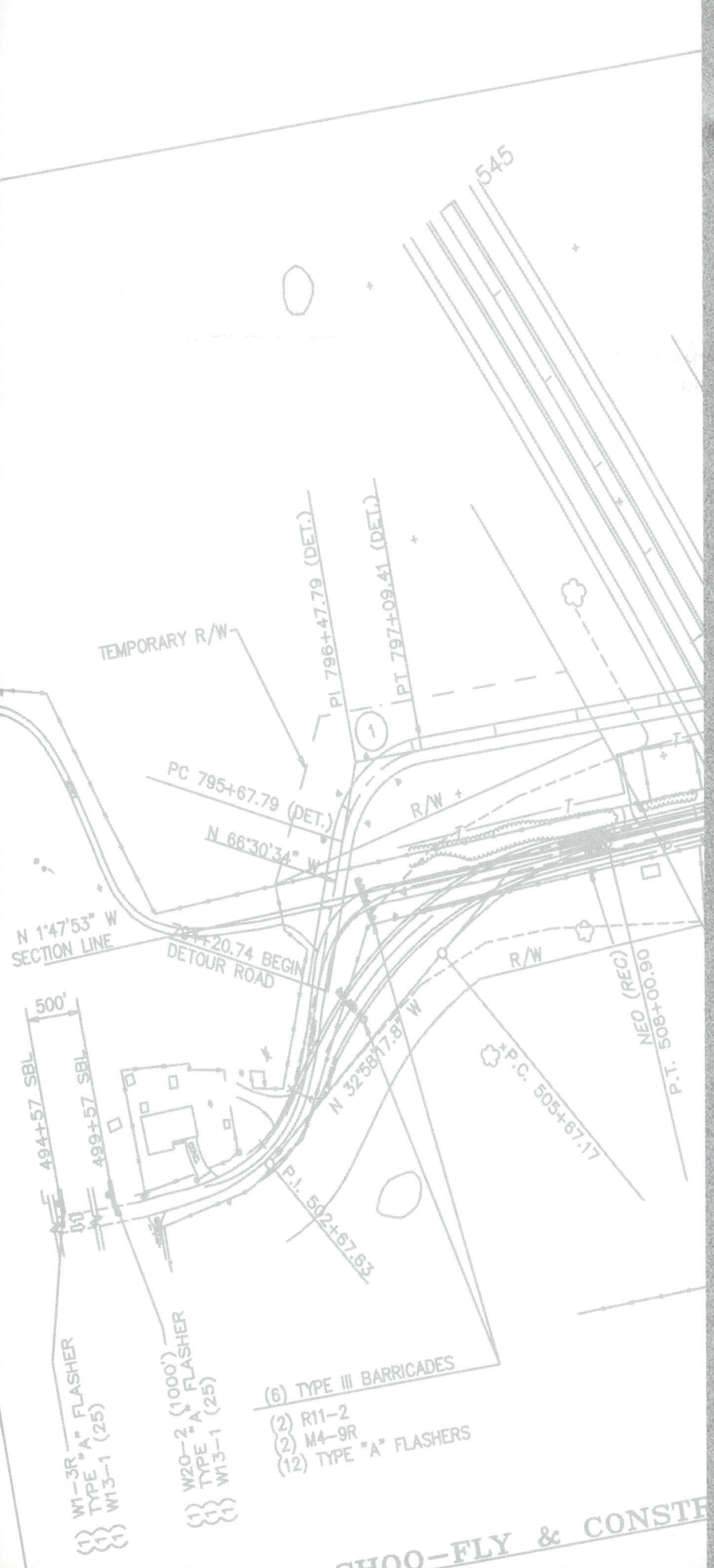

TEXT AND TABLES

CHAPTER OBJECTIVES

After completing this chapter you should:

1. be able to create lines of text in a drawing using *Text;*

2. be able to create and format paragraph text using *Mtext;*

3. know the options available for creating and editing text using the *Text Formatting* Editor;

4. be able to create text styles with the *Style* command;

5. know how to edit text using *Ddebit, Mtedit,* and *Properties;*

6. be able to use *Spell* to check spelling and be able to *Find and Replace* text in a drawing;

7. be able to create and update text *Fields;*

8. be able to create *Tables* that contain text, fields, and calculations.

CONCEPTS

The *Text* and *Mtext* commands provide you with a means of creating text in an AutoCAD drawing. "Text" in CAD drawings usually refers to sentences, words, or notes created from alphabetical or numerical characters that appear in the drawing. The numeric values that are part of specific dimensions are generally <u>not</u> considered "text," since dimensional values are a component of the dimension created automatically with the use of dimensioning commands.

Text in technical drawings is typically in the form of notes concerning information or descriptions of the objects contained in the drawing. For example, an architectural drawing might have written descriptions of rooms or spaces, special instructions for construction, or notes concerning materials or furnishings (Fig. 18-1). An engineering drawing may contain, in addition to the dimensions, manufacturing notes, bill of materials, schedules, or tables (Fig. 18-2). Technical illustrations may contain part numbers or assembly notes. Title blocks also contain text.

FIGURE 18-1

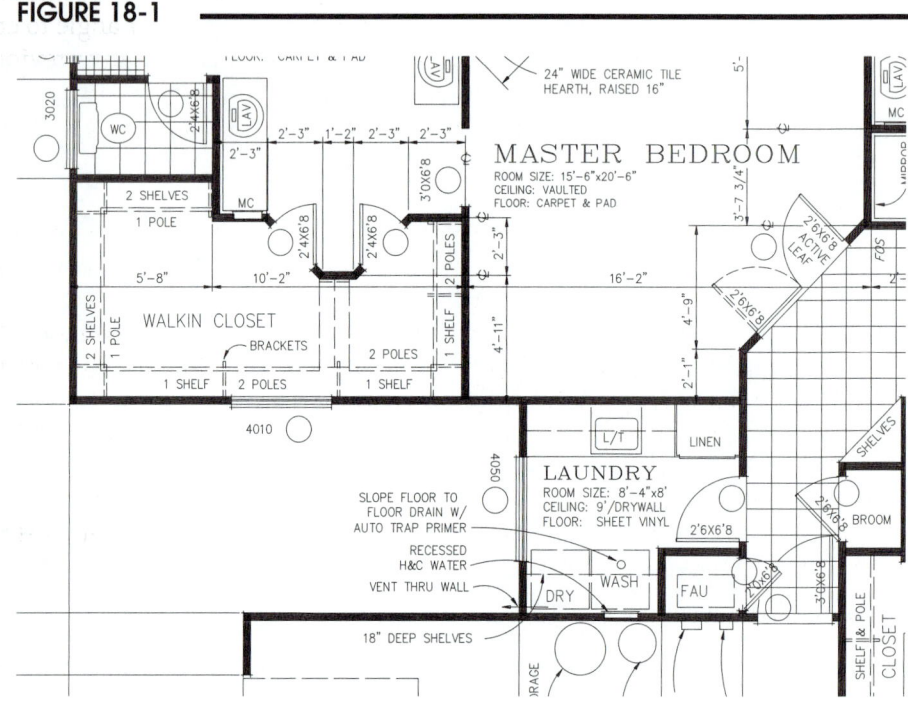

A line of text or paragraph of text in an AutoCAD drawing is treated as an object, just like a *Line* or a *Circle*. Each text object can be *Erased*, *Moved*, *Rotated*, or otherwise edited as any other graphical object. The letters themselves can be changed individually with special text editing commands. A spell checker is available by using the *Spell* command. Since text is treated as a graphical element, the use of many lines of text in a drawing can slow regeneration time and increase plotting time significantly.

FIGURE 18-2

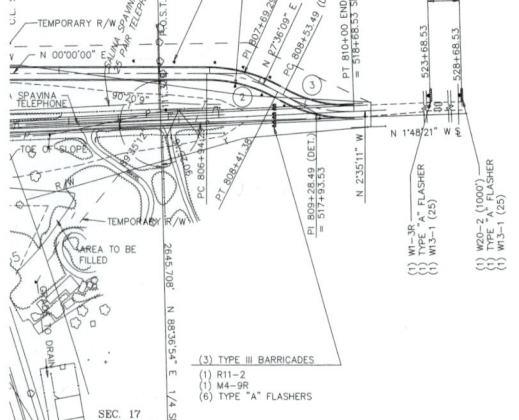

ROUTE	POINT	STATION	COORDINATES	
			NORTH	EAST
₵ BASE LINE	P.I.	502+67.63	452555.90	2855805.56
₵ BASE LINE	P.C.	505+67.16	453207.20	2855642.55
₵ BASE LINE	P.I.	506+86.85	453307.61	2855577.41
₵ BASE LINE	P.T.	508+00.90	453427.17	2855572.01
₵ BASE LINE	P.T.	510+00.00	453626.07	2855563.02
₵ BASE LINE	P.T.	516+50.00	454275.41	2855533.69
₵ BASE LINE	P.O.T.	519+00.00	454264.13	2855783.44
DETOUR @ STA 551+80				
₵ DETOUR	P.O.T.	794+20.74	453044.48	2855659.62
₵ DETOUR	P.C.	795+67.79	453103.09	2855524.75
₵ DETOUR	P.I.	796+47.79	453134.98	2855451.38
₵ DETOUR	P.T.	797+09.41	453214.98	2855451.38
₵ DETOUR	P.C.	806+94.29	454199.87	2855451.38
₵ DETOUR	P.I.	807+69.29	454274.87	2855451.38
₵ DETOUR	P.T.	808+41.38	454341.30	2855486.12
₵ DETOUR	P.C.	808+53.49	454352.03	2855491.73
₵ DETOUR	P.I.	809+28.49	454418.50	2855526.48
₵ DETOUR	P.C.	810+00.00	454493.42	2855523.09

The *Text* and *Mtext* commands perform basically the same function; they create text in a drawing. *Text* creates a single line of text but allows entry of multiple lines with one use of the command. *Mtext* is the more sophisticated method of text entry. With *Mtext* (multiline text) you can create a paragraph of text that "wraps" within a text boundary (rectangle) that you specify. You can also use *Mtext* to import text content from an external file. An *Mtext* paragraph is treated as one AutoCAD object. The *Dtext* command, available in previous releases, has been converted to the *Text* command.

The form or shape of the individual letters is partly determined by the text *Style*. Create a style with the *Style* command, select any Windows standard TrueType font (.TTF) or AutoCAD-supplied font file (.TTF or .SHX shape file), then specify other parameters such as width or angle to create a style to suit your needs. When you use the *Text* or *Mtext* commands, you can select any previously created *Style* and apply one of many options for text justification used for aligning multiple lines of text, such as right, center, or left justified. Only one simple text style, called *Standard*, has been created as part of the ACAD.DWT and ACADISO.DWT template drawings.

Once the text has been placed in the drawing, many options are available for changing, or "editing," the text. For *Text* objects you can edit the text "in place" (directly in the drawing), and for *Mtext* objects you can use the "text editor" you used to create the text, complete with all the formatting options. Options for editing text include changing the actual content of the text, changing the text style, scale, or justification method, spell checking, and using a find and replace tool. You can also use the *Properties* palette to change any aspect of the text. When text is printed you can specify that the text be unfilled (outline text) or that each line be printed as a rectangular block instead of individual letters.

Additional features described in this chapter allow you to create lettered, numbered, or bulleted lists. You can create text fields that contain strings of text that are variable and display information based on current data. You can even create tables and control the format and content of the tables. Tables can be set up to perform calculations.

Commands related to creating or editing text and tables in an AutoCAD drawing include:

Text	Places individual lines of text in a drawing.
Mtext	Places text in paragraph form (with word-wrap) within a text boundary and allows many methods of formatting the appearance of the text using the *Text Formatting* Editor.
Style	Creates text styles for use with any of the text creation commands. You can select from font files and specify other parameters to design the appearance of the letters.
Spell	Checks spelling of existing text in a drawing.
Find	Used to find or replace a text string globally in the drawing.
Ddedit	Allows you to edit *Text* objects in place.
Mtedit	Invokes the *Text Formatting* Editor for editing any aspect of individual *Mtext* characters or the entire *Mtext* paragraph(s).
Spacetrans	Calculates the size of text in a model space viewport in paper space units.
Qtext, Textfill	Changes the appearance of text to speed up regenerations and/or plots.
Scaletext	Allows you to change the text size without changing the insertion point.
Justifytext	Allows you to change the text justification method without moving the text.
Field	Creates variable text that changes based on current data in the drawing or system.
Table	Creates a table based on a table style that contains rows and columns of text.
Tablestyle	Specifies the properties of a table such as text style, text alignment, and border line properties.
Tabledit	Edits text content in a table.

TEXT CREATION COMMANDS

FIGURE 18-3

FIGURE 18-4

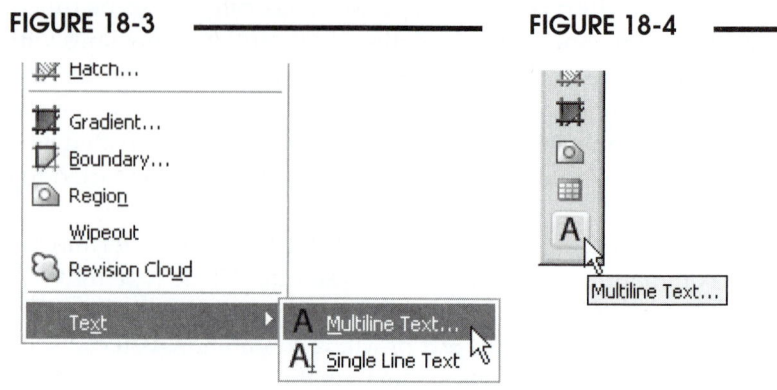

The commands for creating text are formally named *Text* and *Mtext* (these are the commands used for typing). The *Draw* pull-down and screen (side) menus provide access to the two commonly used text commands, *Multiline Text...* (*Mtext*) and *Single-Line Text* (*Text*) (Fig. 18-3). Only the *Mtext* command has an icon button (by default) near the bottom of the *Draw* toolbar (Fig. 18-4).

Text

Pull-down Menu	Command (Type)	Alias (Type)	Short-cut	Screen (side) Menu	Tablet Menu
Draw *Text >* *Single Line Text...*	*Text*	*DT*	...	*DRAW 2* *Dtext*	*K,8*

Text lets you insert single lines of text into an AutoCAD drawing. *Text* displays each character in the drawing as it is typed. You can enter multiple lines of text without exiting the *Text* command. The lines of text do not wrap. The options are presented below:

```
Command: text
Current text style: "Standard" Text height: 0.2000
Specify start point of text or [Justify/Style]:
```

Start Point

The *Start point* for a line of text is the <u>left end</u> of the baseline for the text (Fig. 18-5). *Height* is the distance from the baseline to the top of upper case letters. Additional lines of text are automatically spaced below and left justified. The *rotation angle* is the angle of the baseline (Fig. 18-6).

The command sequence for this option is:

```
Command: text
Current text style: "Standard" Text height: 0.2000
Specify start point of text or [Justify/Style]: PICK or (coordinates)
Specify height <0.2000>: Enter or (value)
Specify rotation angle of text <0>: Enter or (value)
Enter text: (Type the desired line of text and press Enter)
(Press Enter once to add an additional line of text. Press Enter
twice to complete the command.)
Command:
```

FIGURE 18-5

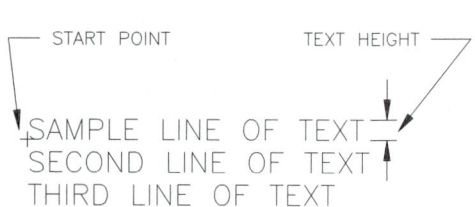

FIGURE 18-6

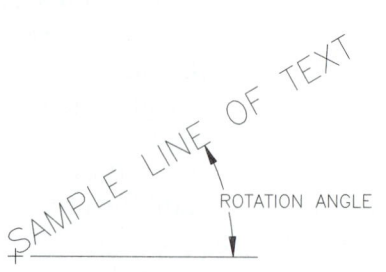

Justify

If you want to use one of the justification methods, invoking this option displays the choices at the prompt:

```
Command: text
Current text style: "Standard" Text height: 0.2000
Specify start point of text or [Justify/Style]: J (Justify option)
Enter an option [Align/Fit/Center/Middle/Right/TL/TC/TR/ML/MC/MR/BL/BC/BR]:
```

After specifying a justification option, you can enter the desired text in response to the "Enter text:" prompt. The text is justified <u>as you type</u>.

Align

Aligns the line of text between the two points specified (P1, P2). The text height is adjusted automatically (Fig. 18-7).

Fit

Fits (compresses or extends) the line of text between the two points specified (P1, P2). The text height does not change (Fig. 18-7).

Center

Centers the baseline of the first line of text at the specified point. Additional lines of text are centered below the first (Fig. 18-8).

Middle

Centers the first line of text both vertically and horizontally about the specified point. Additional lines of text are centered below it (Fig. 18-8).

Right

Creates text that is right justified from the specified point (Fig. 18-8).

TL

Top Left. Places the text in the drawing so the top line (of the first line of text) is at the point specified and additional lines of text are left justified below the point. The top line is defined by the upper case and tall lower case letters (Fig. 18-9).

TC

Top Center. Places the text so the top line of text is at the point specified and the line(s) of text are centered below the point (Fig. 18-9).

TR

Top Right. Places the text so the top right corner of the text is at the point specified and additional lines of text are right justified below that point (Fig. 18-9).

FIGURE 18-7

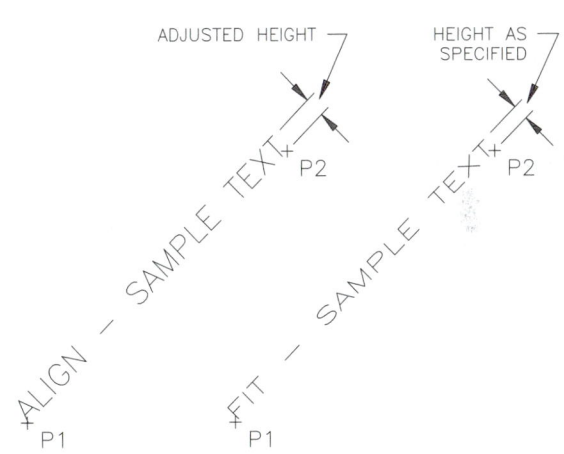

FIGURE 18-8

LEFT JUSTIFIED TEXT
(START POINT)
SAMPLE

RIGHT JUSTIFIED TEXT
(R OPTION)
SAMPLE

CENTER JUSTIFIED TEXT
(C OPTION)
SAMPLE

MIDDLE JUSTIFIED TEXT
(M OPTION)
SAMPLE

FIGURE 18-9

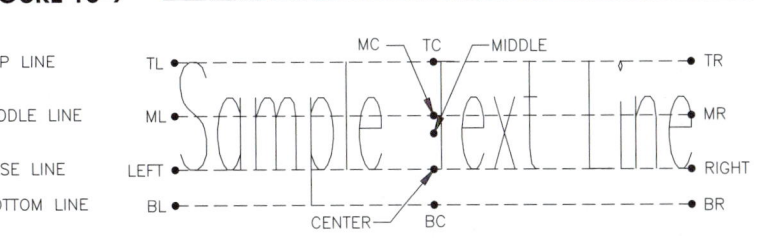

ML

Middle Left. Places text so it is left justified and the middle line of the first line of text aligns with the point specified. The middle line is half way between the top line and the baseline, not considering the bottom (extender) line (Fig. 18-9).

MC

Middle Center. Centers the first line of text both vertically and horizontally about the midpoint of the middle line. Additional lines of text are centered below that point (Fig. 18-9).

MR

Middle Right. Justifies the first line of text at the right end of the middle line. Additional lines of text are right justified (Fig. 18-9).

BL

Bottom Left. Attaches the bottom (extender) line of the first line of text to the specified point. The bottom line is determined by the lowest point of lower case extended letters such as y, p, q, j, and g. If only upper-case letters are used, the letters appear to be located above the specified point. Additional lines of text are left justified (Fig. 18-9).

BC

Bottom Center. Centers the first line of text horizontally about the bottom (extender) line (Fig. 18-9).

BR

Bottom Right. Aligns the bottom (extender) line of the first line of text at the specified point. Additional lines of text are right justified (Fig. 18-9).

NOTE: Because there is a separate baseline and bottom (extender) line, the *MC* and *Middle* points do not coincide and the *BL, BC, BR* and *Left, Center, Right* options differ. Also, because of this feature, when all uppercase letters are used, they ride above the bottom line. This can be helpful for placing text in a table because selecting a horizontal *Line* object for text alignment with *BL, BC,* or *BR* options automatically spaces the text visibly above the *Line*.

Style   (option of *Text* or *Mtext*)

The *style* <u>option</u> of the *Text* or *Mtext* command allows you to select from the <u>existing</u> text styles that have been previously created as part of the current drawing. The style selected from the list becomes the current style and is used when placing text with *Text* or *Mtext*. Alternately, <u>before you use *Text* or *Mtext*</u>, you can select the current text style from the *Text Style Control* drop-down list (just above the right end of the Object Properties toolbar) to make it the current text style.

Since only one text *style, Standard,* is available in the traditional (inch) template drawing (ACAD.DWT) and the metric template drawing (ACADISO.DWT), other styles must be created before the *style* option of *Text* is of any use. Various text styles are created with the *Style* <u>command</u> (this topic is discussed later).

Use the *Style* option of *Text* or *Mtext* to list existing styles and specifications. An example listing is shown below:

```
Command: text
Current text style: "Standard" Text height: 0.2000
Specify start point of text or [Justify/Style]: S  (Style option)
Enter style name or [?] <STANDARD>: ?  (list option)
Enter text style(s) to list <*>: Enter
Text styles:
```

Style name: "ROMANS"   Font files: romans
  Height: 1.50 Width factor: 1.00 Obliquing angle: 0.000
  Generation: Normal

Style name: "STANDARD"   Font files: txt
  Height: 0.00 Width factor: 1.00 Obliquing angle: 0.000
  Generation: Normal

Current text style: "STANDARD"
Text height: 0.30
Specify start point of text or [Justify/Style]:

Mtext

Pull-down Menu	Command (Type)	Alias (Type)	Short-cut	Screen (side) Menu	Tablet Menu
Draw *Text >* *Multiline Text...*	*Mtext* or *-Mtext*	*T, -T* or *MT*	...	*DRAW 2* *Mtext*	*J,8*

Multiline Text (*Mtext*) has more editing options than other text commands. You can apply underlining, color, bold, italic, font, and height changes to individual characters or words within a paragraph or multiple paragraphs of text.

Mtext allows you to create paragraph text defined by a text boundary. The text boundary is a reference rectangle that specifies the paragraph width. The *Mtext* object that you create can be a line, one paragraph, or several paragraphs. AutoCAD references *Mtext* (created with one use of the *Mtext* command) as one object, regardless of the amount of text supplied. Like *Text*, several justification methods are possible.

    Command: **mtext**
    Current text style: "Standard" Text height: 0.2000
    Specify first corner: **PICK**
    Specify opposite corner or [Height/Justify/Line spacing/Rotation/Style/Width]: **PICK** or (**option**)

You can PICK two corners to invoke the *Text Formatting* Editor, or enter the first letter of one of these options: *Height, Justify, Rotation, Style, Line Spacing*, or *Width*. Other options can also be accessed within the *Text Formatting* Editor.

Using the default option the *Mtext* command, you supply a "first corner" and "opposite corner" to define the diagonal corners of the text boundary (like a window). Although this boundary confines the text on two or three sides, one or two arrows indicate the direction text flows if it "spills" out of the boundary (Fig. 18-10). (See "*Justification.*")

FIGURE 18-10

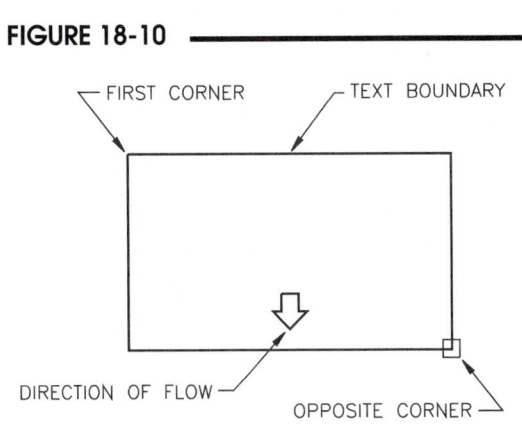

After you PICK the two points defining the text boundary, the *Text Formatting* Editor appears ready for you to enter the desired text (Fig. 18-11). The text wraps based on the width you defined for the text boundary. Select the *OK* button to have the text entered into the drawing.

FIGURE 18-11

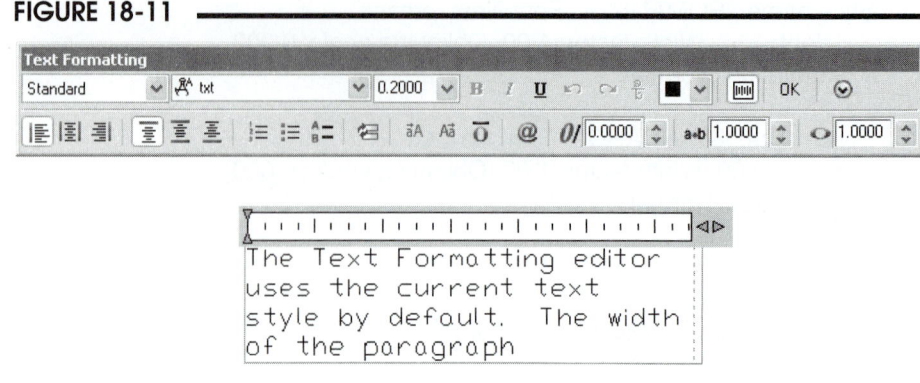

Four ways you can format and edit text using *Mtext* are: 1) the command line options, 2) the *Toolbar* (top toolbar in Fig. 18-11), 3) the *Options* toolbar (second row toolbar in Fig. 18-11), and *Options* menu (same as the right-click shortcut menu).

Text Formatting Toolbar and Command Line Options

Style
This option is the first (far left) drop-down list on the *Text Formatting* toolbar. The box (not dropped down) displays the current text *Style* (*Standard* for most template drawings), and the list displays all existing text styles in the drawing. Select the style you want to use (make current) from this list. Alternately, before you use *Text or Mtext*, you can select the current text style from the *Text Style Control* drop-down list (just above the right end of the Object Properties toolbar) to make it the current text style. (See the *Style* command for information on creating new text styles.)

Font
Choose from any font in the drop-down list. The list includes all fonts registered on your system. Use this feature to change the appearance of selected words in the paragraph. Your selection here overrides the font originally assigned to the text *Style* for the selected or newly created words or paragraphs. Even though you can change the font for the entire *Mtext* object (paragraph) using this feature, it is recommended that you set the paragraph to the desired *Style* (in the first drop-down list), rather than change the characters in the *Style* to a different font for the current *Mtext* object. See following NOTE.

Text Color
Select individual text, then use this drop-down list from the toolbar to select a color for the selected text. This selection overrides the layer color. See following NOTE.

NOTE: The *Font* and *Color* options in the *Text Formatting* toolbar override the properties of the text *Style* and layer color. For example, changing the font in the *Font* list overrides the text style's font used for the paragraph so it is possible to have one font used for the paragraph and others for individual characters within the paragraph. This is analogous to object-specific color and linetype assignment in that you can have a layer containing objects with different linetypes and colors than the linetype and color assigned to the layer. To avoid confusion, it is recommended for global formatting to create text *Styles* to be used for the paragraphs, layer color to determine color for the paragraphs, then if needed, use the *Font* and *Color* options in the *Text Formatting* toolbar to change fonts and colors for selected text rather than for the entire paragraph.

Height
Select from the list or enter a new value for the height to be used globally for the paragraph or for selected words or letters. If *Height* was defined when the selected text *Style* was created for the drawing, this option overrides the *Style's Height*.

Bold, Italic, Underline

Select (highlight) the desired letters or words then PICK the desired button. Only authentic TrueType fonts (not the AutoCAD-supplied .SHX equivalents) can be bolded or italicized.

Undo, Redo

Use *Undo* to reverse the action of the last formatting option used. You can *Redo* the last *Undo*, but only immediately after the *Undo*.

Stack/Unstack

This option is a toggle to *Stack* or *Unstack* specific text (see "Creating Stacked Text"). Highlight the fraction or text, then use this option to stack or unstack the fraction or text.

Ruler

Use the *Ruler* button or *Show Ruler* option from the shortcut menu to make the ruler on top of the text area to appear (Fig. 18-12).

FIGURE 18-12

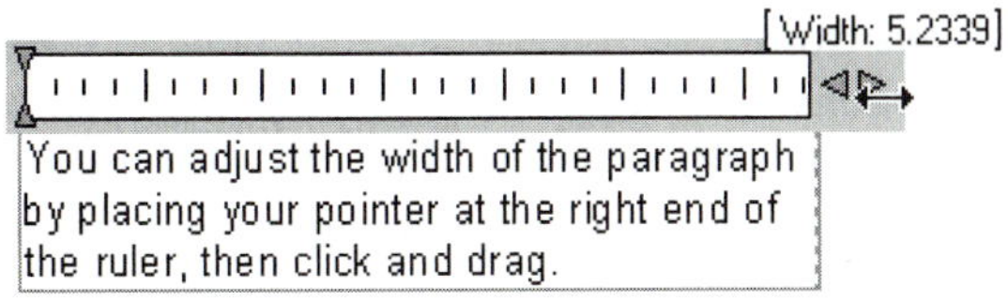

[Width: 5.2339]

You can adjust the width of the paragraph by placing your pointer at the right end of the ruler, then click and drag.

Width

Even though this option does not appear on the *Text Formatting* toolbar, you can interactively change the width of the existing text boundary. Do this by placing your pointer at the right border of the ruler above the text so the double arrows appear, then adjust the paragraph width (Fig. 18-12). The lines of text will automatically "wrap" to adjust to the new text boundary. You can also use the *Width* option from the Command line to set the width to a specific value. Also see "*Justification*" for changing the width of the text boundary with Grips.

Rotation

The *Rotation* option specifies the rotation angle of the entire *Mtext* object (paragraph) including the text boundary. This option is available only by command line; however, the Grips Rotate option can be used after the *Mtext* object has been created. Using the Command line method, you can see the rotated text boundary as you specify the corners (Fig. 18-13).

FIGURE 18-13

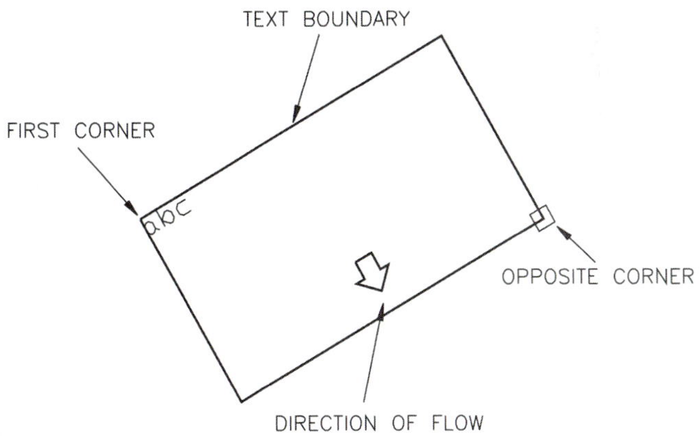

TEXT BOUNDARY

FIRST CORNER

OPPOSITE CORNER

DIRECTION OF FLOW

Line Spacing

The *Line Spacing* option is available only by Command line. Line spacing sets the spacing between lines for the paragraph. The spacing increment is the vertical distance between the bottom (or baseline) of one line of text and the bottom of the next line of text.

```
Specify opposite corner or [Height/Justify/Line spacing/Rotation/Style/Width]: L
Enter line spacing type [At least/Exactly] <At least>:
```

Select *At Least* (the default setting) if you have different size characters in the paragraph. This option automatically adds space between lines based on the height of the largest character in the line so the spacing can vary depending on the size of the characters.

Bullets and Lists

You can create a variety of lettered, numbered, or bulleted lists using these options in the *Options* menu or using the three buttons (*Numbering, Bullets,* and *Upper Case Letters*) on the *Options* toolbar. See "Bullets and Lists" later in this chapter.

Background Mask

FIGURE 18-20

In some cases, it is necessary to create text "on top of" other objects, such as a hatch pattern (Fig. 18-20). With this option, you can specify that the *Mtext* object you create has a background, or mask, to "cover" the objects "behind" the text. The resulting background is similar to that created by a *Wipeout*, only for text.

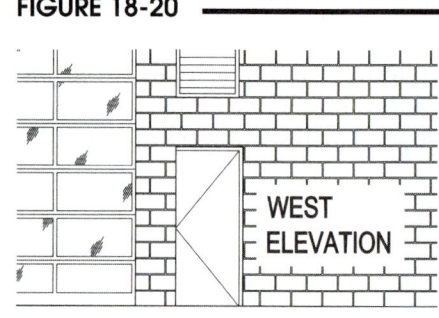

To create a background mask for an *Mtext* object, select *Background Mask* from the *Options* menu or right-click short-cut menu. This produces the *Background Mask* dialog box (Fig. 18-21) where you specify the *Fill color* for the mask and the *Border offset factor* (distance the mask boundary is offset from the actual text inserted). Selecting the *Use drawing background color* option under *Fill Color* forces AutoCAD to use the screen color as the background, therefore producing a "clear" mask.

FIGURE 18-21

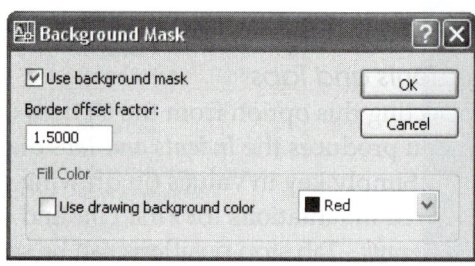

Find and Replace

See "Editing Text" later in this chapter.

Select All

Picking this option automatically selects all the text contained in the *Mtext* object (paragraph or paragraphs).

Change Case

First select the desired text, then select either *UPPERCASE* or *lowercase* from the *Options* toolbar or *Options* menu. All letters are changed; therefore, you cannot have mixed upper and lower case, such as Title Case, unless you change letters individually.

Overline

This option is available in the *Options* toolbar only. Highlight the desired text, then select this button to create a line over the text.

AutoCAPS

This option affects only newly typed and imported text. When checked, *AutoCAPS* turns on Caps Lock on your keyboard so only uppercase letters appear as you type. If needed, use *Change Case* to convert the text back to lowercase.

Remove Formatting

Select the desired text, then select this option to remove bold, italic, or underline formatting.

Combine Paragraphs

If you have more than one paragraph in one *Mtext* object, you can select all the text in the paragraphs, then use this option. The selected paragraphs are converted into a single paragraph and each paragraph return is replaced with a space.

Stack/Unstack

Stack/Unstack appears in the shortcut menu if you highlight only stacked text, such as a fraction, then right-click to invoke the menu. This option is a toggle so that stacked characters become unstacked and vice versa. See "Creating Stacked Text."

Stack Properties

This option appears in the *Options* menu or right-click shortcut menu if you first highlight only stacked text such as a fraction. The *Stack Properties* dialog box appears (Fig. 18-22). See also "Creating Stacked Text."

The *Text* section of the *Stack Properties* dialog box allows you to change the text characters (numbers or letters) contained in the stacked set. Although using a slash (/), pound (#), or carat (^) symbol to specify a stack normally creates horizontal, diagonal, and tolerance format, respectively, that format can be changed with the *Style* drop-down list in this dialog box. Use the *Position* option to align the fraction with the *Top, Center,* or *Bottom* of the other normal characters in the line of text. You can also specify the *Text size* for the stack (which should generally be smaller than normal characters since the stacked set occupies more vertical space).

FIGURE 18-22

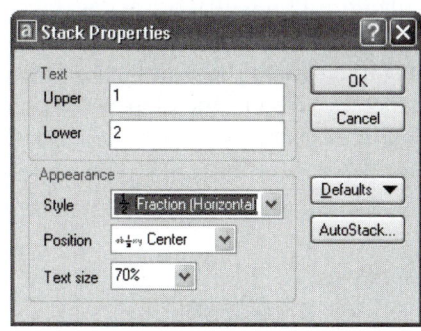

Insert Field

See "Text Fields" later in this chapter.

Import Text

See "Importing External Text into AutoCAD."

Symbol

Common symbols (*Degrees, Plus/Minus, Diameter*) can be inserted as well as many mathematical and other drawing symbols (*Almost Equal* through *Cubed*). Selecting *Other...* produces a character map to select symbols from (See *Other Symbols* next).

Other Symbols

The steps for inserting symbols from the Character Map dialog box (Fig. 18-23) are as follows:

1. Highlight the symbol.
2. Double-click or pick *Select* so the item appears in the *Characters to copy:* edit box.
3. Select the *Copy* button to copy the item(s) to the Windows Clipboard.
4. *Close* the dialog box.
5. In the *Text Formatting* Editor, move the cursor to the desired location to insert the symbol.
6. Finally, right-click and select *Paste* from the menu.

FIGURE 18-23

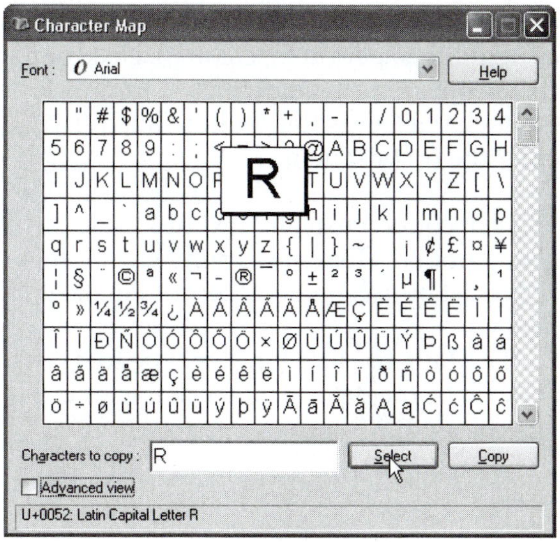

Character Set

This option (available only from the *Options* menu or right-click shortcut menu) displays a menu of code pages. If *Big Fonts* have been used for text styles in the drawing, you can select a text string, then select a code page from the menu to apply it to the text (see "*Big Fonts*").

Oblique Angle

This option, available only from the *Options* toolbar, allows you to apply an angle (incline) to text. Enter the desired angle (in degrees) in the edit box. Apply an *Oblique Angle* to newly created text or existing text.

Tracking

This parameter, also known as kerning, determines the spacing between letters. This does not affect the width of the individual letters. The acceptable range is .75 through 4.0, with 1.0 representing normal spacing. This option is available only from the *Options* toolbar.

Width Factor

Use this option from the *Options* toolbar to increase the width of the individual letters. Apply this characteristic to newly created text or existing text.

Bullets and Lists

FIGURE 18-24

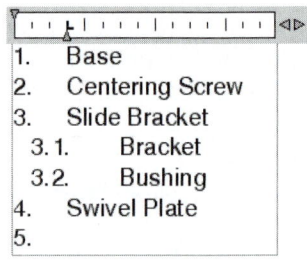

You can use the *Text Formatting* editor or *Options* menu to create a list with numbers, letters, bullets, or other characters as separators (Fig. 18-24). Items in the list are automatically indented based on the tab stops on the ruler. You can add or delete an item or move an item up or down a level and the list numbering automatically adjusts. You can create sub-levels in the list and also remove and reapply list formatting. To turn automatic list creation off completely, remove the check from *Numbering On* in the *Options* menu or right-click shortcut menu.

You can create lists in two ways: create the list as you type or apply list formatting to selected text. To create a list as you type, first select the *Numbers, Bullets*, or *Letters* option from the *Options* toolbar, *Options* menu or right-click shortcut, then begin typing the first item in the list. Pressing Enter adds the next number, letter, or bullet. Press Enter twice to end the list. To create a sub-list (a lower-level list), press the Tab key before the first item in the sub-list. Press Enter to start the next item at the same level, or press Shift+Tab to move the item up a level.

If *Auto-list* is checked in the *Options* or right-click menus, you can type any letter or number, then a period or another delimiter character, then Tab to create a list. The following characters can be used as delimiters: period (.), colon (:), close parenthesis ()), close angle bracket (>), close square bracket (]), and close curly bracket (}). You can also create a bulleted list (without letters or numbers) using other characters as the "bullet," such as pound (#), asterisk (*), or hyphen (-), then press Tab and enter the first list item.

You can convert existing text to a list by selecting the paragraph, then select the *Numbers, Bullets*, or *Letters* option. To remove numbers, bullets, or letters from a list, highlight the list, then select the active list button on the toolbar to make it inactive.

NOTE: Bulleted lists for TrueType fonts use the bullet character that is included in the font. For SHX fonts, the default bullet character is a plus sign.

Creating Stacked Text

"Stacked" text is a combination of two elements of text, one above the other, usually separated by a horizontal line. The most common type of stacked text is a fraction, although letters or words can also be stacked. If you want to create stacked text (numbers or letters), enter a slash (/), pound (#), or carat (^) as a separator between the upper and lower lines of text. Entering a slash, pound, or carat in the *Text Formatting* Editor automatically invokes the *AutoStack Properties* dialog box (Fig. 18-25). Although the slash, pound, and carat each specify a particular stacked format, any of these formats can be changed using the *Properties* option in the *Mtext* shortcut menu (see "*Properties*" in "*Mtext* Right-Click Shortcut Menu").

FIGURE 18-25

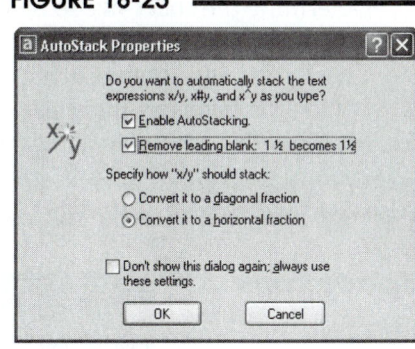

Enable AutoStacking
When this box is checked, fractions are automatically stacked as you type. Anytime you enter numeric characters before and after the carat, slash, or pound character, the numbers are automatically stacked. Each of these characters causes the following style of stacking:

slash (/)   causes a vertical stack separated by a horizontal line
pound (#)   causes a diagonal stack separated by a diagonal line
carat (^)   causes a vertical stack without a horizontal separator line

Remove Leading Blank
This option removes blanks between a whole number and a fraction. For example, you would normally type one and one-half with a space (1 1/2) which would convert to a whole number plus a space before the fraction.

Convert it to a Diagonal Fraction and Convert it to a Horizontal Fraction
These options convert the slash character to a diagonal or horizontal fraction when AutoStack is on. Whether AutoStack is on or off, the pound character is always converted to a diagonal fraction, and the carat character is always converted to a tolerance format.

Don't Show This Dialog Again; Always Use These Settings
Place a check in this box to suppresses display of the *AutoStack Properties* dialog box using the current property settings for all stacked text. When this option is cleared, the *AutoStack Properties* dialog box is automatically displayed if you type two numbers separated by a slash, carat, or pound sign followed by a space or nonnumeric character.

Calculating Text Height for Scaled Drawings

To achieve a specific height in a drawing intended to be plotted to scale, multiply the desired text height for the plot by the drawing scale factor. (See Chapter 14, Printing and Plotting.) For example, if the drawing scale factor is 48 and the desired text height on the plotted drawing is 1/8", enter **6** (1/8 x 48) in response to the "Height:" prompt of the *Text* or *Mtext* command.

If you know the plot scale (for example, 1/4"= 1') but not the drawing scale factor (DSF), calculate the reciprocal of the plot scale to determine the DSF, then multiply the intended text height for the plot by the DSF, and enter the value in response to the "Height:" prompt. For example, if the plot scale is 1/4"=1', then the DSF = 48 (reciprocal of 1/48).

If *Limits* have already been set and you do not know the drawing scale factor, use the following steps to calculate a text height to enter in response to the "Height:" prompt to achieve a specific plotted text height.
1. Determine the sheet size to be used for plotting (for example, 36" x 24").
2. Decide on the text height for the finished plot (for example, .125").
3. Check the *Limits* of the current drawing (for example, 144' x 96' or 1728" x 1152").
4. Divide the *Limits* by the sheet size to determine the drawing scale factor (1728"/36" = 48).
5. Multiply the desired text height by the drawing scale factor (.125 x 48 = 6).

Also see "*Spacetrans*" next.

Spacetrans

Pull-down Menu	Command (Type)	Alias (Type)	Short-cut	Screen (side) Menu	Tablet Menu
...	*Spacetrans*	...	...	...	...

The *Spacetrans* command is used to convert a distance in one space (model space or paper space) to the equivalent distance in the other space. *Spacetrans* is intended to be used transparently (within a command) to pass the translated value to the current command. For example, using *Spacetrans* while drawing a *Line* in paper space allows you to draw the line length to an equivalent model space distance.

 Spacetrans is especially helpful for creating text objects in a model space viewport that print at a specific size in paper space. Instead of having to calculate the text height in model space units times the drawing scale factor (text height x DSF) as described previously in "Calculating Text Height for Scaled Drawings," using *Spacetrans* allows you to enter only the desired plotted text height (assuming you are plotting from paper space at 1:1). Using *Spacetrans* in model space (inside a viewport) allows you to create model space text in paper space units.

As an example, assume you are creating *Text* inside a viewport (in model space) and want the text height to be equivalent to .125 units (1/8") in paper space. You would use a procedure similar to that shown below—that is, invoking *Spacetrans* transparently during the *Text* command to convert a paper space value (.125) to model space units and pass the value to the *Text* command.

```
Command: text
Current text style: "romans" Text height: 0.2000
Specify start point of text or [Justify/Style]: PICK
Specify height <0.2000>: 'spacetrans
>>Specify paper space distance <1.0000>: .125
Resuming TEXT command.
Specify height <0.2000>: 0.500000000000000
Specify rotation angle of text <0>: Enter
Enter text:
```

When you are first prompted to "Specify height," invoke *Spacetrans* transparently. Enter the desired height in paper space units at the "Specify paper space distance" prompt. *Spacetrans* then converts the .125 units to the equivalent distance in model space (0.500000000000000) and passes the value to the second "Specify height" prompt, thus creating text in model space to print at .125 in paper space.

Technically, *Spacetrans* computes the entered paper space units times the reciprocal of the viewport scale (or viewport zoom factor). In the previous example, the viewport scale is 1:4; therefore, .125 x 4 = .5.

Spacetrans cannot be used in the *Model* tab—you must be in a layout with at least one viewport created, and you must be in the desired space when *Spacetrans* is invoked. If you are in paper space when *Spacetrans* is invoked and there is more than one viewport, you are prompted to "Select a viewport."

TEXT STYLES, FONTS, EXTERNAL TEXT, AND EXTERNAL EDITORS

Style

Pull-down Menu	Command (Type)	Alias (Type)	Short-cut	Screen (side) Menu	Tablet Menu
Format *Text Style...*	*Style* or *-Style*	*ST*	...	*FORMAT* *Style:*	*U,2*

Text styles can be created by using the *Style* command. A text *Style* is created by selecting a font file as a foundation and then specifying several other parameters to define the configuration of the letters.

Using the *Style* command invokes the *Text Style* dialog box (Fig. 18-26). All options for creating and modifying text styles are accessible from this device. Selecting the font file is the initial step in creating a style. The font file selected then becomes a foundation for "designing" the new style based on your choices for the other parameters (*Effects*).

FIGURE 18-26 ────────────

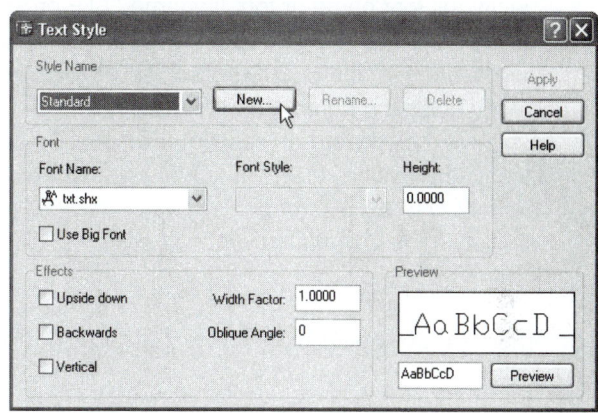

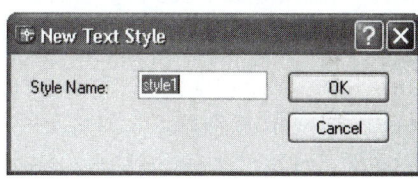

Recommended steps for creating text *Styles* are as follows:

1. Use the *New* button to create a new text *Style*. This button opens the *New Text Style* dialog box (Fig. 18-27). By default, AutoCAD automatically assigns the name Style*n*, where *n* is a number that starts at 1. Enter a descriptive name into the edit box. Style names can be up to 256 characters long and can contain letters, numbers, and the special characters dollar sign ($), underscore (_), and hyphen (-).

FIGURE 18-27 ────────────

2. Select a font to use for the style from the *Font Name* drop-down list (see Fig. 18-26). Authentic TrueType font files (.TTF) as well as AutoCAD equivalents of the TrueType font files (.SHX) are available.
3. Select parameters in the *Effects* section of the dialog box (Fig. 18-26, lower left). Changes to the *Upside down, Backwards, Width Factor,* or *Oblique Angle* are displayed immediately in the *Preview* tile.
4. You can enter specific words or characters in the edit box in the lower-right corner, then press *Preview* to view the characters in the new style.
5. Select *Apply* to save the changes you made in the *Effects* section to the new style.
6. Create other new *Styles* using the same procedure listed in steps 1 through 5.
7. Select *Close*. The new (or last created) text style that is created automatically becomes the current style inserted when *Text* or *Mtext* is used.

You can modify existing styles using this dialog box by selecting the existing *Style* from the list, making the changes, then selecting *Apply* to save the changes to the existing *Style*.

Rename
Select an existing *Style* from the drop-down list, then select *Rename*. Enter a new name in the edit box.

Delete

Select an existing *Style* from the drop-down list, then select *Delete*. You cannot delete the current *Style* or *Styles* that have been used for creating text in the drawing.

Alternately, you can enter *-Style* (use the hyphen prefix) at the command prompt to display the Command line format of *Style* as shown below:

```
Command: -style
Enter name of text style or [?] <Standard>: name (Enter new style name.)
New style.
Specify full font name or font filename (TTF or SHX) <txt>: name (Enter desired font file.)
Specify height of text <0.0000>: Enter or (value)
Specify width factor <1.0000>: Enter or (value)
Specify obliquing angle <0>: Enter or (value)
Display text backwards? [Yes/No] <N>: Enter or Y
Display text upside-down? [Yes/No] <N>: Enter or Y
Vertical? <N> Enter or Y
"(new style)" is now the current text style.
Command:
```

The options that appear in both the *Text Style* dialog box and in the Command line format are described in detail next.

Height <0.000>

The height should be 0.000 if you want to be prompted again for height each time the *Text* command is used. In this way, the height is variable for the style each time you create text in the drawing. If you want the height to be constant, enter a value other than 0. Then, *Text* will not prompt you for a height since it has already been specified. The *Mtext* command, however, allows you to change the height in the *Text Formatting* Editor, even if you specified a height when you created the text style. A specific height assignment with the *Style* command also overrides the *DIMTXT* setting (see Chapter 29).

Width factor <1.000>

A *width factor* of 1 keeps the characters proportioned normally. A value less than 1 compresses the width of the text (horizontal dimension) proportionally; a value of greater than 1 extends the text proportionally.

Obliquing angle <0>

An angle of 0 keeps the font file as vertical characters. Entering an angle of 15, for example, would slant the text forward from the existing position, or entering a negative angle would cause a back-slant on a vertically oriented font (Fig. 18-28).

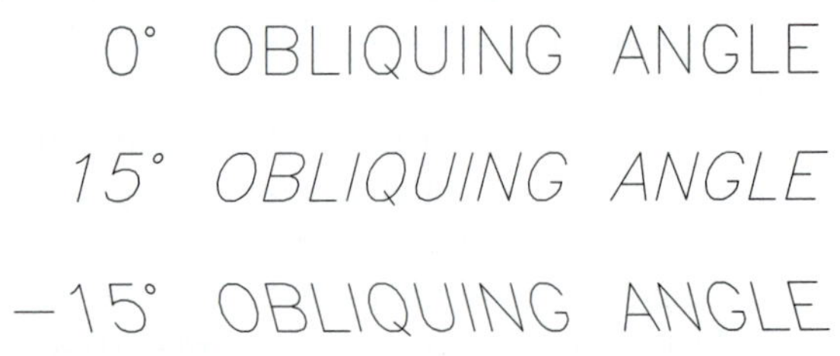

Backwards

Backwards characters can be helpful for special applications, such as those in the printing industry.

Upside-down

Each letter is created upside-down in the order as typed (Fig. 18-29). This is different than entering a rotation angle of 180 in the *Text* command. (Turn this book 180 degrees to read the figure.)

FIGURE 18-28 ────────────────

0° OBLIQUING ANGLE

15° OBLIQUING ANGLE

−15° OBLIQUING ANGLE

FIGURE 18-29 ────────────────

UPSIDE DOWN TEXT

TEXT ROTATED 180°

Vertical

Vertical letters are shown in Figure 18-30. The normal rotation angle for vertical text when using *Text* or *Mtext* is 270. Only .SHX fonts can be used for this option. *Vertical* text does not display in the *Preview* image tile.

FIGURE 18-30 —

Since specification of a font file is an initial step in creating a style, it seems logical that different styles could be created using <u>one</u> font file but changing the other parameters. It is possible, and in many cases desirable, to do so. For example, a common practice is to create styles that reference the same font file but have different obliquing angles (often used for lettering on isometric planes). This can be done by using the *Style* command to assign the same font file for each style, but assign different

FIGURE 18-31 ———————————————————

FONT FILE	STYLE NAME (EXAMPLE)	RESULTING TEXT (BASED ON OTHER PARAMETERS)
ROMAN SIMPLEX (ROMANS.SHX)	ROMANS-VERT	ABCDEFG 1234
	ROMANS-ITAL	*ABCDEFG 1234*
ROMAN DUPLEX (ROMAND.SHX)	ROMAND-EXT	A B C 1 2 3 4
	ROMAND-COMP	ABCDEFG 1234

parameters (obliquing angle, width factor, etc.) and a unique name to each style. The relationship between fonts, styles, and resulting text is shown in Figure 18-31.

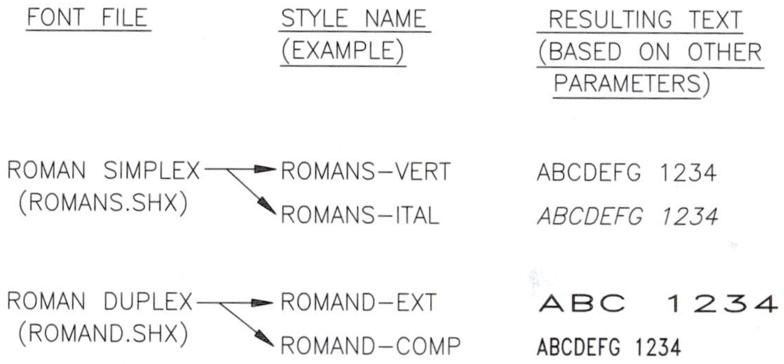

.SHX Fonts

The .SHX fonts are fonts created especially for AutoCAD drawings by Autodesk. They are generally smaller files and are efficient to use for AutoCAD drawings where file size and regeneration time is critical. These fonts, however, are composed of single line segments and are not solid-filled such as most TrueType fonts (use *Zoom* to closely examine and compare .SHX vs. TrueType fonts).

Because the *Text Formatting* Editor is a Windows-compliant dialog box, it can display only fonts that are recognized by Windows. Since AutoCAD .SHX fonts are not recognized by Windows, AutoCAD supplies a TrueType equivalent when you select an .SHX or any other non-TrueType font for display in the *Text Formatting* Editor only. When you then use a text creation command, AutoCAD creates the text in the drawing with the true .SHX font.

All AutoCAD .SHX shape fonts are Unicode fonts. A Unicode font can contain 65,535 characters with shapes for many languages (see *Big Fonts*). Unicode fonts contain many more characters than are shown on your keyboard in your system; therefore, to use a character not directly available from the keyboard, you can enter the escape sequence *U+nnnn* where *nnnn* represents the Unicode hexadecimal value for the character (see "Special Text Characters").

Big Fonts

Big Fonts are used for characters not found in the English language. Text files for some alphabets contain thousands of non-ASCII characters. For these applications, AutoCAD provides a special type of shape definition known as a *Big Font* file that contains many other characters.

When you check *Use Big Font*, the *Font Style* box changes to a *Big Font* box. You can select *Use Big Font* only when using .SHX fonts. Set a *Style* to use both regular and *Big Font* files for other than English languages. Otherwise, for most applications, select only the regular .SHX font files.

TrueType Fonts

AutoCAD uses the Windows operating system directly to display TrueType text on the screen for most applications. However, when text in AutoCAD has been transformed (mirrored, upside-down, backward, oblique, has a width factor not equal to 1, or is in an orientation that is not co-planar with the screen), AutoCAD must draw the TrueType text. Text that has been transformed might appear slightly more bold in some circumstances, especially at lower resolutions. This difference is only in the screen display of the font and does not affect the plotted drawing.

For TrueType fonts only, the value specified for text *Height* might not represent the actual height of uppercase letters. The height specified represents the height of a capital letter plus an "ascent" area reserved for accent marks and other marks used in non-English languages. TrueType fonts also have a "descent" area for portions of characters with extenders (such as the characters y, j, p, g, and q).

Proxy Fonts

For third-party or custom .SHX fonts that have no TrueType equivalent, AutoCAD supplies up to eight different TrueType fonts called proxy fonts. Proxy fonts appear different in the *Text Formatting* Editor from the font they represent to indicate that they are substitutions for the fonts used in the drawing.

Importing External Text into AutoCAD

You can import ASCII text files created in other text editors or word processors into an AutoCAD drawing. There are three methods: use the *Import Text* option in the *Text Formatting* Editor, use *Copy* and *Paste* (using the Windows Clipboard), or "drag and drop" a file icon from the Windows Explorer.

Importing text in ASCII or RTF files can save you drawing time. For example, you can create a text file of standard notes that you include in many drawings or use information prepared for a report or other document. Instead of entering (typing) this information in the drawing, you can import the file. The imported text becomes an AutoCAD text object, which you can edit as if you created it in AutoCAD. Imported text retains its original formatting properties from the text editor in which it was created.

Importing with the Text Formatting Editor

To import text into the *Text Formatting* Editor, first use the *Mtext* command. PICK two points to define the text boundary as usual. When the *Text Formatting* Editor appears, right-click and select *Import Text* from the shortcut menu. The *Select File* dialog box appears for you to locate and select the desired file. The file to import must be in an ASCII or RTF format. (An ASCII text file often has a .TXT file extension.) The selected text appears in the editor and can be manipulated just as any text you entered.

You can also *Copy* text (to the Windows Clipboard) from any Windows document and *Paste* it into AutoCAD. The best method to use with *Copy* and *Paste* is to use the *Text Formatting* Editor. Open the editor, then use the Alt+Tab key combination to switch to the other document. Highlight the desired text and select *Copy* from the *Edit* menu or right-click menu. Switch back to the *Text Formatting* Editor, right-click for the pop-up menu, and select *Paste*. The imported text can be edited like any other AutoCAD *Mtext* object.

Using Drag and Drop to Import Text Files

You can also use the drag-and-drop feature to insert ASCII text into a drawing. Open the Windows Explorer (file manager) and size the window so AutoCAD is also visible on the screen. Drag the file name or icon and drop it into an AutoCAD drawing. The pasted text uses the formats and fonts defined by the current AutoCAD text style. The width of the lines or paragraph is determined by line breaks and carriage returns in the original document. You can drag only files with a file extension of .TXT into an AutoCAD drawing.

Using an External Text Editor for *Multiline Text*

You can specify a different text editor to use instead of the "internal" *Text Formatting* Editor. For example, you may prefer the Windows Notepad or other editor, although these editors do not offer the text formatting capabilities available in the *Text Formatting* Editor.

To specify an external text editor, you can use the *Files* tab of the *Options* dialog box from the *Tools* pull-down menu (Fig. 18-32) or use the *MTEXTED* system variable. Enter "internal" to use the internal AutoCAD *Text Formatting* Editor.

FIGURE 18-32

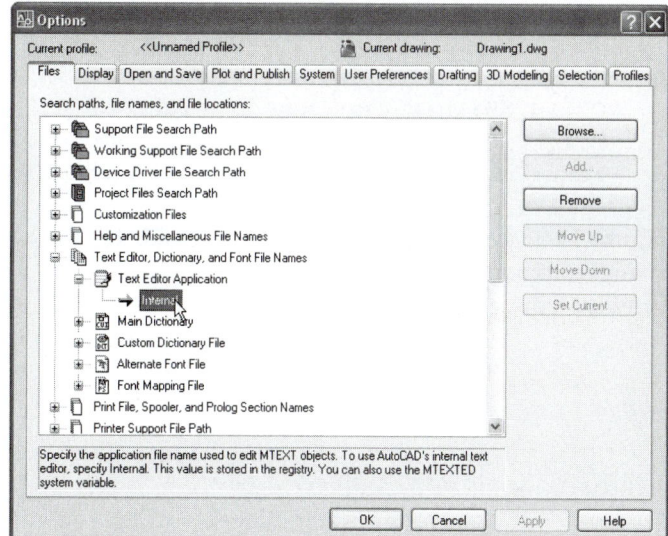

If you use an external text editor (other than the AutoCAD internal editor), you must enter the formatting codes as you enter the text using the text editor. Text features that can be controlled are underline, overline, height, color, spacing, stacked text, and other options. The following table lists the codes and the resulting text.

Format Codes for Creating *Multiline* Text in External Text Editors

Format code	Purpose	Type this. . .	To produce this
\O. . .\o	Turns overline on and off	You have \Omany\o choices	You have $\frac{1}{2}$many choices
\L. . .\l	Turns underline on and off	You have \Lmany\l choices	You have <u>many</u> choices
\~	Inserts a nonbreaking space	Keep these\~words together	Keep these words together
\\	Inserts a backslash	slash\\backslash	slash\ backslash
\{. . .\}	Inserts an opening and closing brace	The \{bracketed\} word	The {bracketed} word
\Cvalue;	Changes to the specified color	Change \C2;these colors	Change these colors
\Ffilename;	Changes to the specified font file	Change \Farial;these fonts	Change these fonts
\Hvalue;	Changes to the specified text height	Change \H2;these sizes	Change these sizes
\S. . .^. . .	Stacks the subsequent text at the \ or ^ symbol	1.000\S+0.010^0.000;	$1.000^{+0.010}_{-0.000}$
\Tvalue;	Adjusts the space between characters, from .75 to 4 times	\T2;TRACKING	T R A C K I N G
\Qangle;	Changes obliquing angle	\Q20;OBLIQUE	*OBLIQUE*
\Wvalue;	Change width factor to produce wide text	\W2;Wide	WIDE
\A	Sets the alignment value to 0 (bottom), 1 (center), or 2 (top)	\A1;Center\S1/2	Center
\P	Ends paragraph	First paragraph\PSecond paragraph	First paragraph Second paragraph

For example, you may enter the following text in the Windows Notepad editor using the *Mtext* command.

```
\C1;This sample line of text is red in color.\P
\C7;\H.25;This text has a height of 0.25.\P
\H.2;\Fromanc;This line is an example of romanc.shx.\P
\Ftxt;This line contains stacked fractions such as \S1/2
```

Figure 18-33 displays the text as it might appear in the AutoCAD drawing editor.

FIGURE 18-33 ━━━━━━━━━━━━━━━

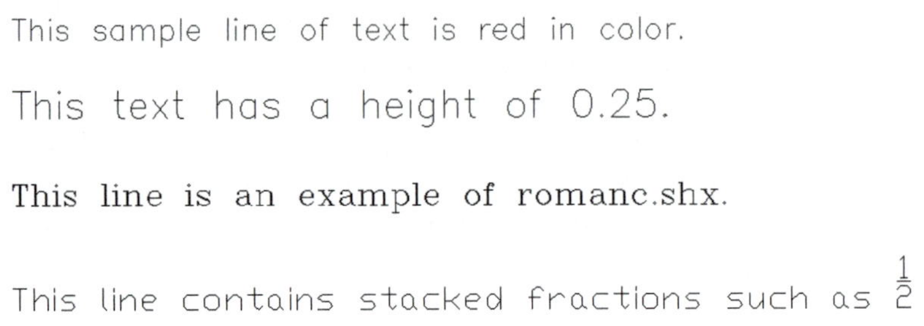

This sample line of text is red in color.

This text has a height of 0.25.

This line is an example of romanc.shx.

This line contains stacked fractions such as $\frac{1}{2}$

Special Text Characters

Special characters that are often used in drawings can be entered easily in the internal *Text Formatting* Editor by selecting *Symbol* from the shortcut menu (see Fig. 18-14). In an external editor or with the *Text* command, you can code the text by using the "%%" symbols or by entering the Unicode values for high-ASCII (characters above the 128 range). The %% symbols can be used with the *Text* command, but the Unicode values must be used for creating *Mtext* with an external text editor. The following codes are typed in response to the "Enter text:" prompt in the *Text* command or entered in the external text editor.

External Editor	Text	Result	Description
\U+2205	%%c	ø	diameter (metric)
\U+00b0	%%d	°	degrees
	%%o		overscored text
	%%u	____	underscored text
\U+00b1	%%p		plus or minus
	%%*nnn*	varies	ASCII text character number
\U+*nnnn*		varies	Unicode text hexadecimal value

For example, entering "**Chamfer 45%%d**" at the "Enter text:" prompt of the *Text* command draws the following text: **Chamfer 45°**.

EDITING TEXT

Although several methods are available to edit text, probably the simplest method to use is to double-click on a *Text* or *Mtext* object (assuming *Dblclkedit* is set to *On*). Doing so invokes the appropriate editing method for the specific text. If you double-click on a *Text* object, you can edit the text "in place." No real text editor appears—the text becomes reversed (in color) and you can change the text itself (see *"Ddedit"*). If you double-click on an *Mtext* object, the *Text Formatting* Editor appears (see *"Mtedit"*).

You can also edit *Text* and *Mtext* objects using the *Properties* palette. The *Properties* palette is useful for editing all properties of *Text* objects; however, to gain the most control over editing *Mtext* objects, double-click, use *Ddedit*, or use *Mtedit*.

Ddedit

	Pull-down Menu	Command (Type)	Alias (Type)	Short-cut	Screen (side) Menu	Tablet Menu
	Modify Object > Text > Edit...	Ddedit	ED	...	MODIFY1 Ddedit	Y,21

Ddedit can be used for editing either *Text* or *Mtext* objects. The following prompt is issued:

Command: **ddedit**
Select an annotation object or [Undo]:

If you select a *Text* object, the text becomes "reversed" and you can edit the contents of the text directly in the drawing (Fig. 18-34). Only the textual content can be changed. If you want to change other properties of the *Text* object such as *Style, Justification, Height, Rotation,* etc., use the *Properties* palette.

FIGURE 18-34 ─────────────

INSIDE DIAMETER TO BE HONED AFTER HEAT TREAT

Mtedit

	Pull-down Menu	Command (Type)	Alias (Type)	Short-cut	Screen (side) Menu	Tablet Menu
	...	Mtedit	...	...	...	...

If you select an *Mtext* object, the *Text Formatting* Editor appears (see previous Fig. 8-14). In the *Text Formatting* Editor, you can edit the textual contents as well as any properties of the *Mtext* object. (See also "*Mtedit.*")

The *Mtedit* command is designed specifically for editing *Mtext* objects by producing the *Text Formatting* Editor. The following prompt appears.

Command: **mtedit**
Select an MTEXT object:

The *Text Formatting* Editor (see previous Fig. 18-14) allows you to change any aspect of the text content and all other properties of the *Mtext* objects. (You can also produce the *Text Formatting* Editor by double-clicking on an *Mtext* object.) *Mtedit* cannot be used to edit *Text* objects.

NOTE: If you select *Mtext* that is too small or too large to read and edit properly, you can *Zoom* with the mouse wheel inside the edit boundary to change the display of the text and text editing boundary.

Properties

	Pull-down Menu	Command (Type)	Alias (Type)	Short-cut	Screen (side) Menu	Tablet Menu
	Modify Properties...	Properties	PR or CH	(Default Menu) Find or Ctrl + 1	MODIFY1 Modify	Y,14

The *Properties* command invokes the *Properties* palette. As described in Chapter 16, the palette that appears is specific to the type of object that is PICKed—kind of a "smart" properties palette.

Remember that you can select objects and then invoke the palette, or you can keep the *Properties* palette on the screen and select objects in the drawing whose properties you want to change. (See Chapter 16 for more information on the *Properties* palette.)

If a line of *Text* is selected, the palette displays properties as shown in Figure 18-35. The *Properties* palette allows you to change almost any properties associated with the text, including the *Contents* (text), *Style*, *Justification*, *Height*, *Rotation*, *Width,* and *Obliquing*. Simply locate the property in the left column and change the value in the right column.

FIGURE 18-35

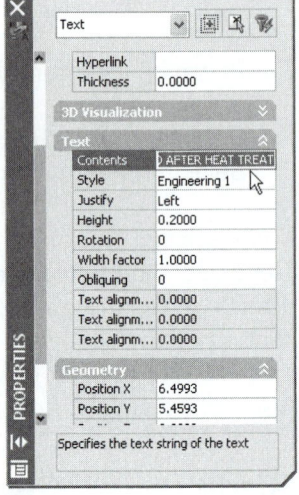

If you select an *Mtext* paragraph, the *Properties* palette appears allowing you to change the properties of the object. However, if you select the *Contents* section, a small button on the right appears (Fig. 18-36). Pressing this button invokes the *Text Formatting* Editor with the *Mtext* contents and allows you to edit the text with the full capabilities of the editor.

FIGURE 18-36

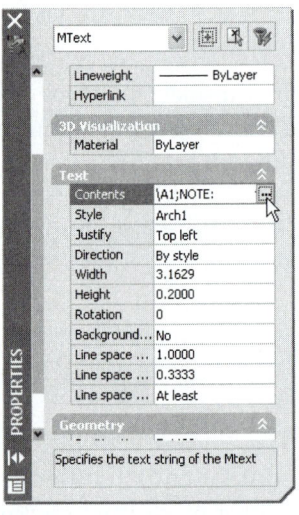

Spell

Pull-down Menu	Command (Type)	Alias (Type)	Short-cut	Screen (side) Menu	Tablet Menu
Tools Spelling	Spell	SP	...	TOOLS 1 Spell	T,10

AutoCAD has an internal spell checker that can be used to spell check and correct existing text in a drawing after using *Text* or *Mtext*. You can also select *Block* attributes and *Blocks* containing text to spell check. The *Check Spelling* dialog box (see Fig. 18-37) has options to *Ignore* the current word or *Change* to the suggested word. The *Ignore All* and *Change All* options treat every occurrence of the highlighted word.

AutoCAD matches the words in the drawing to the words in the current dictionary. If the speller indicates that a word is misspelled but it is a proper name or an acronym you use often, it can be added to a custom dictionary. Choose *Add* if you want to leave a word unchanged but add it to the current custom dictionary.

Selecting the *Change Dictionaries…* tile produces the dialog box shown in Figure 18-38. You can select from other *Main Dictionaries* that are provided with your version of AutoCAD. The current main dictionary can also be changed in the *Files* tab of the *Options* dialog box (see Fig. 18-32), and the name is stored in the *DCTMAIN* system variable.

The AutoCAD default custom dictionary is SAMPLE.CUS. If you use the *Add* function (in the *Check Spelling* dialog box), the selected word is added to the current custom dictionary. You can also create a custom dictionary "on the fly" by entering any name in the *Custom dictionary* edit box (Fig. 18-38). A file extension of .CUS should be used with the name (although other extensions will work). A custom dictionary name must be specified before you can add words. Words can be added by entering the desired word in the *Custom dictionary words* edit box or by using the *Add* tile in the *Check Spelling* dialog box. Custom dictionaries can be changed during a spell check. The custom dictionary can also be changed in the *Files* tab of the *Options* dialog box (see Fig. 18-32), and its name is stored in the *DCTCUST* system variable.

FIGURE 18-37

FIGURE 18-38

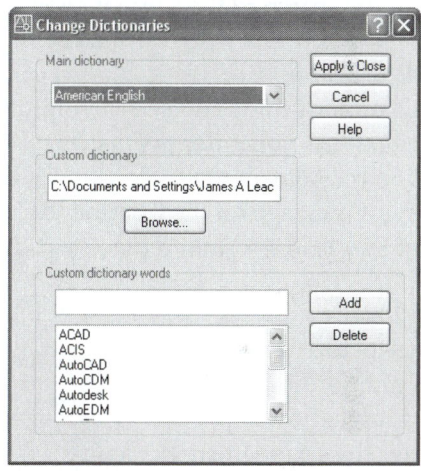

Find

Pull-down Menu	Command (Type)	Alias (Type)	Short-cut	Screen (side) Menu	Tablet Menu
Edit *Find...*	*Find*	...	(Default menu) *Find...*	...	*X,10*

Find is used to find or replace text strings <u>globally</u> in a drawing. It can find and replace any text in a drawing, whether it is *Text* or *Mtext*. *Find* produces the *Find and Replace* dialog box (see Fig. 18-39).

The *Find* command searches a drawing globally and operates with all types of text, including dimensions, tables, and *Block* attributes. In contrast, the *Find/Replace* option in the *Text Formatting* Editor right-click menu (see "*Find/Replace* in the *Text Formatting* Editor" next) operates only for *Mtext* and for only one *Mtext* object at a time.

FIGURE 18-39

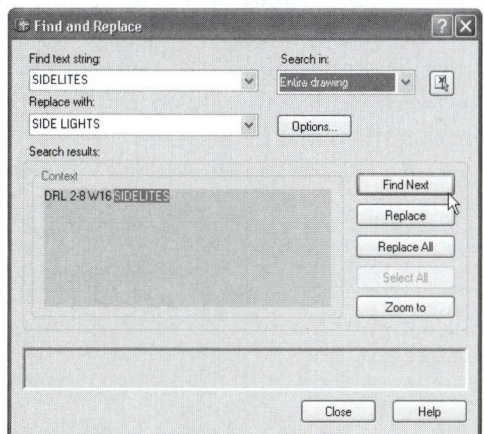

Find Text String, Find/Find Next
To find text only (not replace), enter the desired string in the *Find Text String* edit box and press *Find* (Fig. 18-39). When matching text is found, it appears in the context of the sentence or paragraph in the *Search Results* area. When one instance of the text string is located, the *Find* button changes to *Find Next*.

Replace With, Replace
You can find any text string in a drawing, or you can search for text and replace it with other text. If you want to search for a text string and replace it with another, enter the desired text strings in the *Find Text String* and *Replace With* edit boxes, then press the *Find/Find Next* button. You can verify and replace each instance of the text string found by alternately using *Find Next* and *Replace*, or you can select *Replace All* to globally replace all without verification. The status area confirms the replacements and indicates the number of replacements that were made.

Select Objects
The small button in the upper-right corner of the dialog box allows you to select objects with a pickbox or window to determine your selection set for the search. Press Enter when objects have been selected to return to the dialog box. Once a selection set has been specified, you can search the *Entire drawing* or limit the search to the *Current selection* by choosing these options in the *Search in*: drop-down list.

Select All
This options finds and selects (highlights) all objects in the current selection set containing instances of the text that you enter in *Find Text String*. This option is available only when you use *Select Objects* and set *Search In* to *Current Selection*. When you choose *Select All*, the dialog box closes and AutoCAD displays a message on the Command line indicating the number of objects that it found and selected. Note that *Select All* does not replace text; AutoCAD ignores any text in *Replace With*.

Zoom To

This feature is extremely helpful for finding text in a large drawing. Enter the desired text string to search for in the *Find Text String* edit box, select *Find/Find Next* button, then use *Zoom to*. AutoCAD automatically locates the desired text string and zooms in to display the text (Fig. 18-40). Although AutoCAD searches model space and all layouts defined for the drawing, you can zoom only to text in the current *Model* or *Layout* tab.

FIGURE 18-40

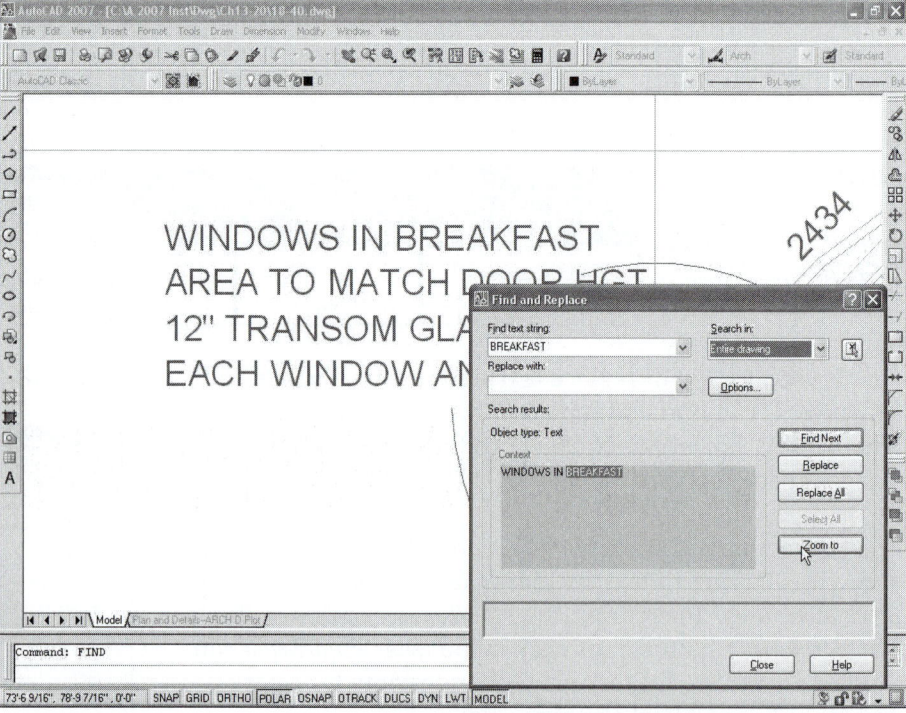

Options

Selecting the *Options* button in the *Find and Replace* dialog box produces the *Find and Replace Options* dialog box (Fig. 18-41). Note that the *Find* feature can locate and replace *Block Attribute Values*, *Dimension Annotation Text*, *Text (Mtext, Text)*, *Table Text*, *Hyperlink Description*, and *Hyperlinks*. Removing any check disables the search for that text type.

FIGURE 18-41

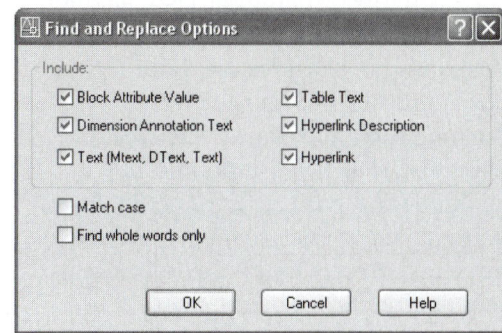

If it is important that the search exactly match the case (upper and lowercase letters) you entered in the *Find Text String* and *Replace With* edit boxes, check the *Match Case* box. If you want to search for complete words only, check the *Find whole words only* box. For example, without a check in either of these boxes, entering "door" in the *Find Text String* edit box would yield a find of "Doors" in the *Search Results* area. With a check in either *Match Case* or *Find whole words only*, a search for "door" would not find "Doors."

Find/Replace in the *Text Formatting* Editor

Another version of the *Find* command is the *Find and Replace* option available from the *Text Formatting* Editor shortcut menu (see previous Fig. 18-14). With this feature you can <u>find and replace text only within the *Text Formatting* Editor</u>. *Find and Replace* operates <u>only for *Mtext* objects</u>, not for other text objects.

Selecting *Find and Replace* from the *Text Formatting* Editor short-cut menu produces the *Replace* dialog box (Fig. 18-42). The options here operate similarly to the same options in the *Find and Replace* dialog box of the *Find* command (see previous explanation of "*Find*").

FIGURE 18-42

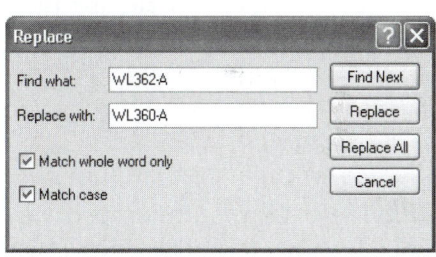

This feature is useful primarily if you are currently using the *Text Formatting* Editor or if you have located a specific *Mtext* object you want to deal with. In most cases, however, the *Find* command is preferred since it provides more features and options for finding and replacing text for all types of AutoCAD text objects.

Scaletext

Pull-down Menu	Command (Type)	Alias (Type)	Short-cut	Screen (side) Menu	Tablet Menu
Modify *Object >* *Text >* *Scale*	*Scaletext*	...	...	...	...

You cannot effectively scale multiple lines of text (*Text* objects) or multiple paragraphs of text (*Mtext* objects) using the normal *Scale* command since all text objects become scaled with relation to only one base point; therefore, individual lines of text lose their original insertion point.

TIP   The text editing command *Scaletext* allows you to scale multiple text objects using this one command, and each line or paragraph is scaled relative to its individual justification point.

For example, consider the three *Text* objects shown in Figure 18-43. Assume each line of text was created using the *Start point* option (left-justified) and with a height of .20. If you wanted to change the height of all three text objects to .25, the following sequence could be used.

FIGURE 18-43 ───────────

SCALETEXT scales each text object individually. Each line is scaled relative to the justification point. These sample lines are created by DTEXT.

```
Command: scaletext
Select objects: PICK
Select objects: PICK
Select objects: PICK
Select objects: Enter
Enter a base point option for scaling
[Existing/Left/Center/Middle/Right/TL/TC/TR/ML/MC/MR/BL/BC/BR] <Existing>: Enter
Specify new height or [Match object/Scale factor] <0.2000>: .25
Command:
```

The resulting three lines of scaled text appear in Figure 18-44. Note that each line increased in height, which is evident since the space between lines has been reduced. Note also that each line of text has retained its insertion point; therefore, the original justification is unchanged.

FIGURE 18-44 ───────────

SCALETEXT scales each text object individually.
Each line is scaled relative to the justification point.
These sample lines are created by DTEXT.

Scaletext also allows you to specify a new justification point. The justification option you select is applied to each text object individually. For example, selecting the *MC* option would scale each line of text in relation to <u>its own</u> middle center point.

```
Enter a base point option for scaling
[Existing/Left/Center/Middle/Right/TL/TC/TR/ML/MC/MR/BL/BC/BR] <Existing>: mc
Specify new height or [Match object/Scale factor] <0.2000>: .25
```

Scale factor

This option allows you to specify a relative scale factor rather than a specific height. Entering a value of 1.5, for example, would scale each line of text 1.5 times the current size. This option is particularly helpful when the drawing contains lines of text at different heights and you want to change all lines proportionally.

```
Specify new height or [Match object/Scale factor] <0.2500>: s
Specify scale factor or [Reference] <0.2000>: 1.5
```

Match object

You can select an existing text object that you want the text to match. After selecting the matching text, its height is listed and the text (that was selected in the first step) is scaled to the listed height.

```
Specify new height or [Match object/Scale factor] <0.2500>: m
Select a text object with the desired height: PICK
Height=0.2400
```

Justifytext

Pull-down Menu	Command (Type)	Alias (Type)	Short-cut	Screen (side) Menu	Tablet Menu
Modify *Object >* *Text >* *Justify*	*Justifytext*	...	...	...	...

Justifytext changes the justification point of selected text objects without changing the text locations. Technically, this command relocates the insertion point (sometimes called attachment point) for the text object (*Mtext* objects, *Text* objects, leader text objects, and block attributes) and then justifies the text to the new insertion point.

Consider the two *Text* objects and the single *Mtext* object shown in Figure 18-45. In each case, the default justification methods were used when creating the text, resulting in left-justified text and insertion points as shown by the small "blip" (the "blip" is not created as part of the text—it is shown here only for illustration).

FIGURE 18-45

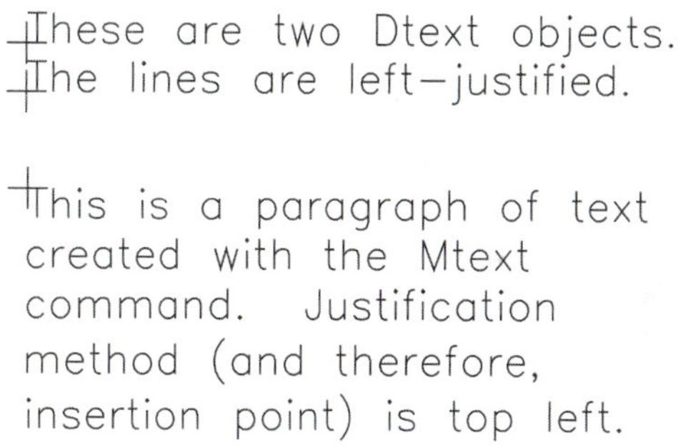

These are two Dtext objects.
The lines are left—justified.

This is a paragraph of text created with the Mtext command. Justification method (and therefore, insertion point) is top left.

Using previously available methods for changing the text justification yielded different results based on the command used or type of text object selected. For example, using the *Properties* palette to change the *Justify* field and selecting the *Bottom Right* option results in the changes shown in Figure 18-46. Note that for the *Text* objects, the insertion point is not changed, but the text objects are moved to justify according to the original insertion point. On the other hand, the insertion point for the *Mtext* paragraph is changed from the original top-left insertion point (shown as dashed in the figure) to the bottom right and the text is justified to the new insertion point.

FIGURE 18-46

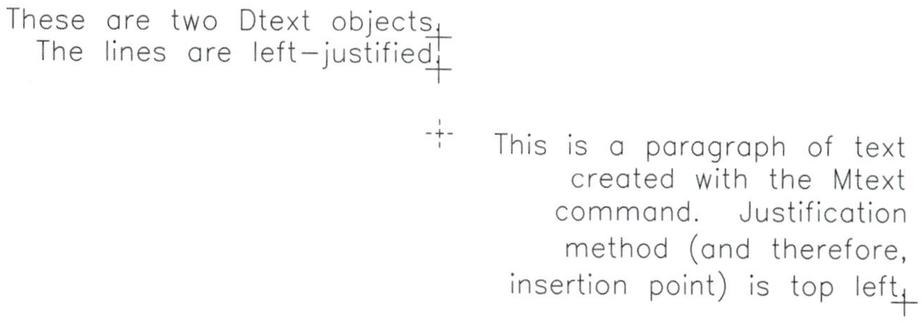

These are two Dtext objects.
The lines are left—justified.

This is a paragraph of text created with the Mtext command. Justification method (and therefore, insertion point) is top left.

To avoid this problem, use the *Justifytext* command to change multiple text objects to produce similar results for all cases. For example, assume you used the *Justifytext* command to change the three text objects shown previously in Figure 18-45 to a new justification point of *Bottom right*.

Command: **justifytext**
Select objects: **PICK**
Select objects: **PICK**
Select objects: **PICK**
Select objects: **Enter**
Enter a justification option
[Left/Align/Fit/Center/Middle/Right/TL/TC/TR/ML/MC/
MR/BL/BC/BR] <TL>: **br**
Command:

The resulting changes are shown in Figure 18-47. Compare these changes to the original text objects in Figure 18-45. Notice that the *Justifytext* command keeps the text in the same location, although the insertion points have been changed. Each text object has been justified in relation to its own new insertion point.

FIGURE 18-47

These are two Dtext objects
The lines are left—justified

This is a paragraph of text
created with the Mtext
command. Justification
method (and therefore,
insertion point) is top left

Qtext

Pull-down Menu	Command (Type)	Alias (Type)	Short-cut	Screen (side) Menu	Tablet Menu
...	*Qtext*	...	...	...	...

Qtext (quick text) allows you to display a line of text <u>as a box</u> in order to speed up drawing and plotting times. Because text objects are treated as graphical elements, a drawing with much text can be relatively slower to regenerate and take considerably more time plotting than the same drawing with little or no text. When *Qtext* is turned *ON* and the drawing is regenerated, each text line is displayed as a rectangular box (Fig. 18-48). Each box displayed represents one line of text and is approximately equal in size to the associated line of text.

FIGURE 18-48

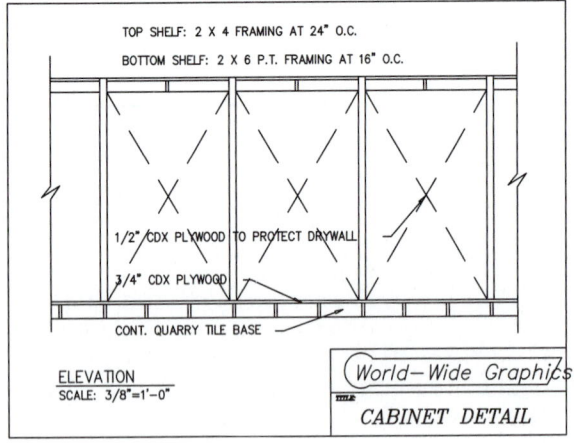

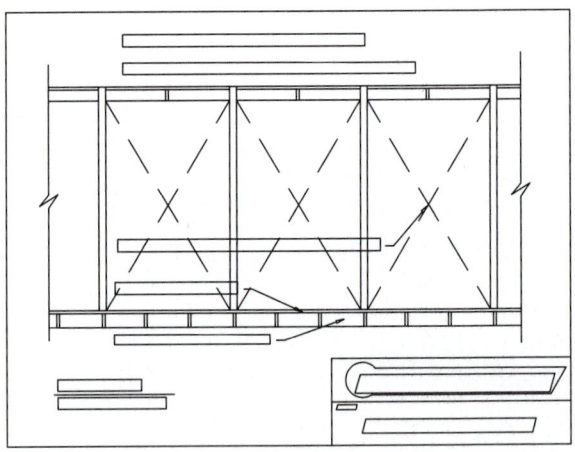

QTEXT OFF   QTEXT ON

TIP For drawings with considerable amounts of text, *Qtext ON* noticeably reduces regeneration time. For check plots (plots made during the drawing or design process used for checking progress), the drawing can be plotted with *Qtext ON*, requiring considerably less plotting time. *Qtext* is then turned *OFF* and the drawing must be *Regenerated* to make the final plot. When *Qtext* is turned *ON*, the text remains in a readable state until a *Regen* is invoked or caused. When *Qtext* is turned *OFF*, the drawing must be regenerated to read the text again.

Textfill

Pull-down Menu	Command (Type)	Alias (Type)	Short-cut	Screen (side) Menu	Tablet Menu
...	*Textfill*	...	...	...	...

The *TEXTFILL* variable controls the display of TrueType fonts for <u>printing and plotting only</u>. *TEXTFILL* does not control the display of fonts for the screen—fonts always appear filled in the Drawing Editor. If *TEXTFILL* is set to 1 (on), these fonts print and plot with solid-filled characters (Fig. 18-49). If *TEXTFILL* is set to 0 (off), the fonts print and plot as outlined text. The variable controls text display globally and retroactively.

FIGURE 18-49

TEXTFILL ON
TEXTFILL OFF

Substituting Fonts

Alternate Fonts

Fonts in your drawing are specified by the font file referenced by the current text *Style* and by individual font formats specified for sections of *Mtext*. Text font files are not stored in the drawing; instead, text objects reference the font files located on the current computer system. Occasionally, a drawing is loaded that references a font file not available on your system. This occurrence is more common with the ability to create text styles using TrueType fonts.

By default, AutoCAD substitutes the SIMPLEX.SHX for unreferenced fonts. In other words, when AutoCAD loads a drawing that contains a font not on your system, the SIMPLEX.SHX font is automatically substituted. You can use the *FONTALT* system variable or the *Files* tab of the *Options* dialog box (Fig. 18-50) to specify other font files that you want to use as an alternate font whenever a drawing with unreferenced fonts is loaded. Typically, .SHX fonts are used as alternates.

FIGURE 18-50

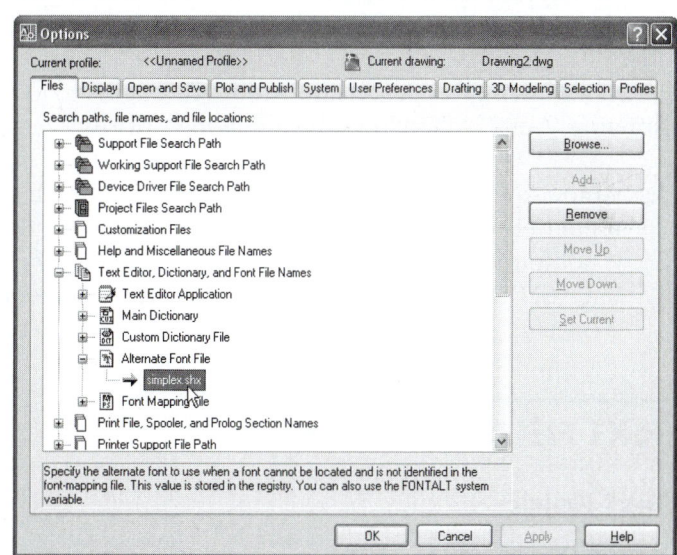

Font Mapping

You can also set up a <u>font mapping table</u> listing several fonts and substitutes to be used whenever a text object is encountered that references one of the fonts. The current table name is stored in the *FONTMAP* system variable (the default is ACAD.FMP). A font mapping table is a plain ASCII file with a .FMP file extension and contains one font mapping per line. Each line contains the base font followed by a semicolon and the substitute font.

You can use font mapping to simplify problems that may occur when exchanging drawings with clients. You can also use a font mapping table to substitute fast-drawing .SHX files while drawing and for test plots, then use another table to substitute back the more complex fonts for the final draft. For example, you might want to substitute the ROMANS.SHX font for the Times TrueType font and the TXT.SHX font for the Arial TrueType font. (The second font in each line is the new font that you want to appear in the drawing.)

```
times;romans.shx
arial;txt.shx
```

AutoCAD provides a sample font mapping table called ACAD.FMP. The table has 16 substitutions that are used to map the Release 13 PostScript fonts to the newer TrueType fonts. The ACAD.FMP contents are shown here.

```
cibt;CITYB___.TTF
cobt;COUNB___.TTF
eur;EURR_____.TTF
euro;EURRO___.TTF
par;PANROMAN.TTF
rom;ROMANTIC.TTF
romb;ROMAB___.TTF
romi;ROMAI___.TTF
sas;SANSS___.TTF
sasb;SANSSB___.TTF
sasbo;SANSSBO_.TTF
saso;SANSSO___.TTF
suf;SUPEF___.TTF
te;TECHNIC_.TTF
teb;TECHB___.TTF
tel;TECHL___.TTF
```

You can edit this table with an ASCII text editor or create a new font mapping table. You can specify a new font mapping file in the *Files* tab of the *Options* dialog box (see Fig. 18-50) or use the *FONTMAP* system variable.

A font mapping table <u>forces</u> the listed substitutions, while an alternate font substitutes the specified font for <u>any</u> font but <u>only</u> when an unreferenced font is encountered.

When a drawing is opened and AutoCAD begins to locate the fonts referenced by text *Styles* in the drawing, AutoCAD follows the following priority:

1.　AutoCAD uses the *FONTMAP* value(s) if defined.
2.　If not defined, AutoCAD uses the font defined in the text *Style*.
3.　If not found, Windows substitutes a similar font (if .TTF) or uses the *FONTALT* specified (if .SHX).
4.　If an .SHX is not found, it prompts you for a new font.

TEXT FIELDS AND TABLES

Text Fields

You can insert a "text field" as part of a *Text* or *Mtext* object. A text field is a string of text that is variable and displays information based on current data. For example, you might insert text strings in a title block that display the drawing creation date, the current date, the file name, and sheet number. You may also use text fields to display the properties of any object, such as area, color, layer, length, linetype, etc. Other possibilities include using fields to list current system variables settings, plot settings, and settings for sheet sets (see Chapter 34). The available text fields are predefined in AutoCAD. In other words, you would use a field to display information about a drawing or about objects in a drawing that you might expect to change. Whenever the drawing is updated and then regenerated, saved, or opened, the field automatically displays the current information. Therefore, using fields prevents you from having to make "manual" changes to text as the drawing changes.

Field

Pull-down Menu	Command (Type)	Alias (Type)	Short-cut	Screen (side) Menu	Tablet Menu
Insert, Field...	Field	...	...	...	...

Field is both a command and an option of the *Text* and *Mtext* commands. You can use the *Field* command directly by any method shown in the command table above, you can select *Insert Field...* from the shortcut menu of *Text* (right-click during the "Enter text:" prompt), or you can select *Insert Field...* from the shortcut menu in the *Text Formatting* Editor. You can also insert a text field in a *Table* (see "Creating and Editing Tables" next).

Using the *Field* command or selecting *Insert Field* from a shortcut menu produces the *Field* dialog box (Fig. 18-51). Select the desired information you want displayed from the *Field Names* list, then select the *Format* or other options on the right (based on the selected *Field Name*). This action defines a variable visible in the *Field Expression:* box below. The actual text displayed in the drawing is the current information that satisfies the variable that is taken from the drawing or from your computer system. For example, the *Date, Filesize,* and *Login* variables read and display information from your computer system. The *Title, Author, Subject, Keywords, Hyperlinkbase,* and *Comments* variables read and display information as previously specified in the drawing properties (*Dwgprops*). Other options, such as *Createdate, Filename,* sheet information, plot information, are taken from information contained in the drawing. Selecting some options, such as *Object, NamedObject, Hyperlink, SystemVariable,* and *DieselExpression,* provide edit boxes on the right side of the *Field* dialog box for you to select or otherwise specify the information.

FIGURE 18-51

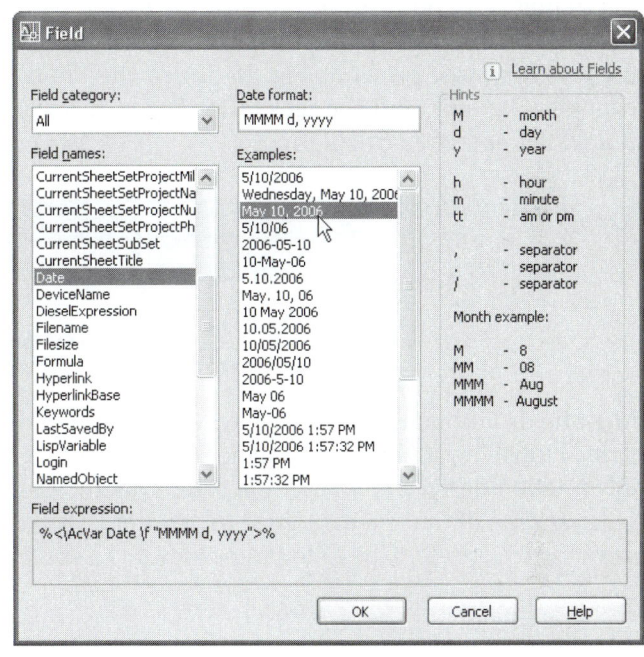

If you want to create a field as a single, independent text string, use the *Field* <u>command</u>. Using the *Field* command, once the desired settings are selected from the *Field* dialog box, an *Mtext* (field) object is placed in your drawing at the location you PICK. The text is entered with the current *text style* and *text height* (previously set using *Mtext* or *Style*).

If you want to insert a field as part of a sentence or paragraph (Fig. 18-52) or use other than the current text style and height, use the *Field* <u>option</u> of the *Text* or *Mtext* command to insert the field. Note that a field object ("222.2 SQ. FT.") is normally displayed with a gray background. Use the *FIELDDISPLAY* system variable to control whether fields are displayed with a gray background (*FIELDDISPLAY*=1) or no background (*FIELDDISPLAY*=0).

FIGURE 18-52

Living Room Area = 222.2 SQ. FT.

One example of the power of using *Fields* is to display an object property. As an example, assume you want to insert a text field that displays the area of a closed *Pline* (the *Pline* might represent a floor plan or the outline of a mechanical part). In the *Field* dialog box (Fig. 18-53), first select *Object* from the *Field names*: list. The right side of the dialog box changes to provide appropriate options depending on your choice of field names. Second, select the *Select Objects* button (see pointer in top-center of figure) to return to the drawing and select the desired *Pline* object. Third, select the object *Property:* and *Format:*. Finally, selecting the *OK* button dismisses the dialog box and prompts you for the location of the desired text. The area of the *Pline* is displayed in the text field (see previous Fig. 18-52). If the Pline changes, the text field changes (after *Regen* or *Save*) to reflect the new area calculation.

FIGURE 18-53

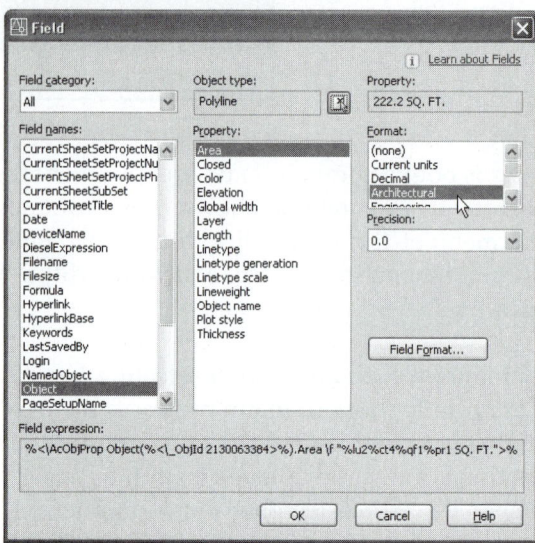

Updatefield

Pull-down Menu	Command (Type)	Alias (Type)	Short-cut	Screen (side) Menu	Tablet Menu
Tools *Update Fields...*	*Updatefield*	...	...	...	...

Most fields automatically update when the related drawing features change and you then use *Regen* or *Save*. However, some fields may not change when you use *Regen* or *Save*, such as fields that display information derived from the computer system. Use the *Updatefield* command to force an update and regeneration of the text contained in <u>any</u> field. You can also enter the *All* option to force all fields in the drawing to update. The following prompt is issued:

```
Command: updatefield
Select objects: PICK
Select objects: Enter
1 field(s) found
1 field(s) updated
Command:
```

Creating and Editing Tables

Many drawings contain tables such as Bills of Materials, Revision Notes, Sheet Lists, Parts Lists, and so on. In early releases of AutoCAD, tables were created with some difficulty by drawing *Lines* and inserting the text into cells using *Text*. Changing the shapes or sizes of the tables or changing the individual text objects or their properties was tedious.

With the current AutoCAD release you can use the *Table* command to create *Table* objects. The *Table* objects are variable in that the number and size of rows and columns is flexible. Text or blocks can be easily inserted by right-clicking within the cells and can be easily re-justified for the entire table. You can specify that particular cells contain title headings, column headings, and sub-headings. Properties such as text style, height, color, and fill can be assigned to cells. Border properties can be assigned such as lineweight, color, and grid visibility. You can specify the table direction as up or down. You can even include cells that make calculations in your tables.

Another important feature of tables is they are based on Table Styles, much the same way as text and dimensions are based on Text Styles and Dimension Styles. In other words, you can use the *Tablestyle* command (*Table Style* dialog box) to quickly create a table based on an existing Table Style having specific settings and properties. Similarly, you can create new styles and modify or delete existing styles. For example, you could create one table style named Revision Notes, another named Parts List, and another named Sheet List. Table styles ensure you are using standard fonts, text heights, colors, lineweights, etc. In this way, an office or studio can create and save common Table Styles as part of a template drawing to facilitate quick and easy table creation based on company practices or standards.

Once a table has been created in a drawing, it is easy to edit the table and its contents. You can double-click on a cell or use the *Tabledit* command to change the text entries. Using grips, you can click and drag the size of rows and columns. With the *Properties* palette, you change the table style, number of rows and columns, and other properties. You can also highlight the table or cells and access right-click short-cut menus with options specifically for the table and its contents.

Table

Pull-down Menu	Command (Type)	Alias (Type)	Short-cut	Screen (side) Menu	Tablet Menu
Draw *Table...*	*Table*	*TB*	...	...	...

The *Table* command produces the *Insert Table* dialog box (Fig. 18-54). Here you can select from existing Table Styles, specify the column and row settings, and specify how you want to insert the table.

FIGURE 18-54

Table Style Settings
First select the Table Style you want to use from the *Table Style Name* drop-down list (upper-left corner). The table style defines the text, cell, and border properties of the table, such as text height, color, border lineweight, and so on. If this is your first table or you are using the ACAD.DWT or other AutoCAD-supplied template, only the *Standard* style exists. To create a new style, use the button to the right of the drop-down list to produce the *Table Style* dialog box. (See "*Tablestyle*" next.)

Insertion Behavior
After selecting the table style, use this section to specify how you want to insert the table. The *Specify insertion point* option allows you to insert the table in the drawing using its upper-right corner, and the table size is based on the options you specify below in the *Column & Row Settings* section. The *Specify window* option allows you to dynamically determine the size of the table during insertion based on the window you specify.

Column & Row Settings
The available options in the *Column & Row Settings* section are based on which of the *Insertion Behavior* options you select. If you chose to insert the table by the *Specify insertion point* option above (see Fig. 18-54), you can then specify the number and width of columns and number and height of rows <u>before</u> insertion. However, if you had selected the *Specify window* option above, you can specify only one column option and one row option since the other parameters are determined dynamically <u>during</u> insertion.

In other words, if you had selected the *Specify window* option above, then specified the number of columns and rows in this section, the resulting width and height of the cells would be determined dynamically based on the window size you pick during insertion. No matter which of these options you select, you can easily edit the table later to include more or fewer rows or columns and change the size of the rows and columns.

Creating the Table Text

Once you have specified the option in the *Insert Table* dialog box, selecting *Close* returns you to the drawing and produces the "Specify insertion point:" prompt. After insertion, the table appears with the first cell highlighted and the *Text Formatting* (*Mtext*) editor above (Fig. 18-55). Additionally, the columns and rows appear with temporary labels for columns (A, B, C, etc.) and rows (1, 2, 3, etc.). (The column

FIGURE 18-55

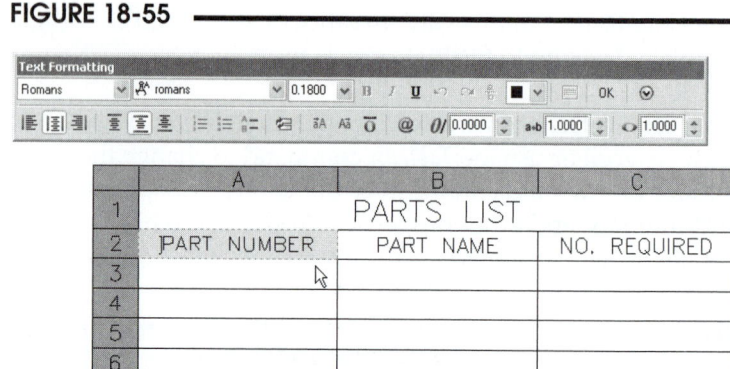

and row labels do not appear in the completed table.) Simply enter the desired text into each cell. To move around the cells use the arrow keys, or use the Tab to move across cells and the Enter key to move down the cells. You can click inside any cell to enter and edit the text. You can also insert *Blocks* or *Fields* into cells (see "*Tabledit*").

Editing the Table Text
See "*Tabledit*."

Tablestyle

Pull-down Menu	Command (Type)	Alias (Type)	Short-cut	Screen (side) Menu	Tablet Menu
Format *Tablestyle...*	*Tablestyle*	*TS*	...	...	...

Table Styles are similar to dimension styles in that the table style defines the appearance of the table like a dimension style defines the appearance of a dimension. Use the *Table Style* dialog box to specify the properties of a table, such as text style, alignment, height, color, fill color, border line color, and lineweight. Similar also to the relationship between dimensions and their dimension styles, an existing table in a drawing is automatically updated whenever its style is changed.

The *Table Style* dialog box (Fig. 18-56) can be produced from within the *Insert Table* dialog box or by using the *Tablestyle* command. This dialog box allows you to modify an existing style, make new styles, delete unused styles, or set a style as the current style.

FIGURE 18-56

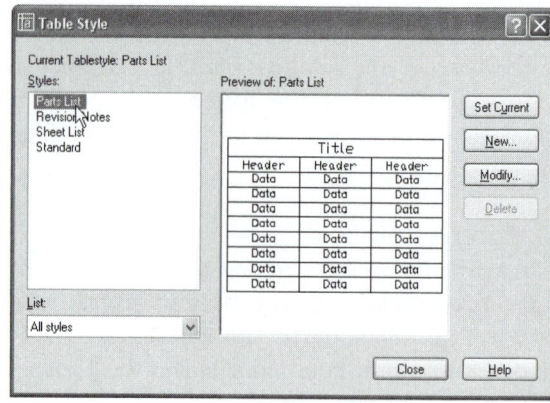

Set Current

Select a style from the *Styles:* list, then select *Set Current*. The current table style is useful only when using the *Table* command. The selected current style appears as the default in the *Insert Table* dialog box (see "*Table*").

New...

Use this option to create a new table style. This button produces the *New Table Style* dialog box (Fig. 18-57). Normally you would first specify the *Table direction* (right side) and *Cell Margins* for the table. The image tile in the upper right displays the settings made in any area of the dialog box.

On the left side of the dialog box, the three tabs, *Data, Column Heads,* and *Title,* each have identical options but apply only to the data, column heads, and title cells within a table. The *Cell Properties* section allows changing all aspects of the <u>text</u> including a fill (background) color. The alignment options are similar to the text commands' justification methods. The *Border Properties* section defines the appearance of the <u>grid lines</u> around and within the table. You can specify the grid's lineweight and color as well as specify the presence or absence of the grid lines themselves using the five buttons in this section.

FIGURE 18-57

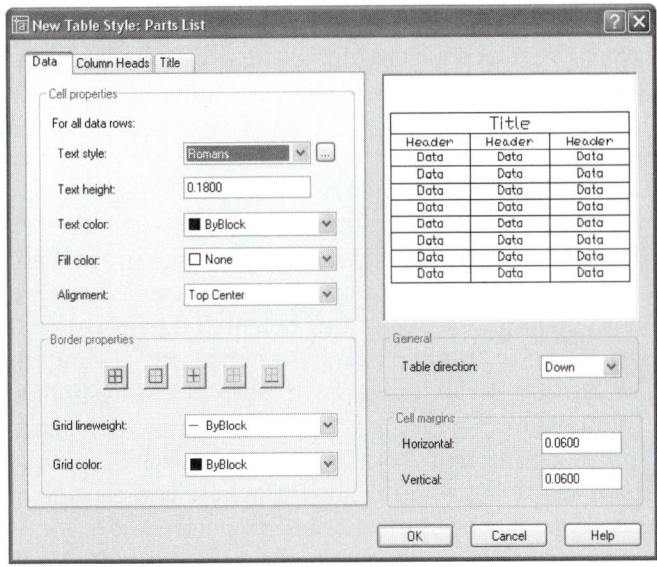

Modify...

The *Modify Table Style* dialog box (not shown) is identical to the *New Table Style* dialog box except for the name. The options for modifying an existing table style are the same as those for creating a new style. However, there is one important feature here: <u>when a table style is modified, all existing tables in the drawing automatically update to display the new style settings</u>. For example, suppose you have a table in the drawing but want to change the alignment of the cells from middle center to left center. Using the *Modify Table Style* dialog box to change the alignment for the style automatically updates the alignment in the existing table and any other tables created with that style.

Tabledit

Pull-down Menu	Command (Type)	Alias (Type)	Short-cut	Screen (side) Menu	Tablet Menu
...	*Tabledit*	*TB*	...	...	...

The *Tabledit* command is used to edit text in a table similar to the way *Ddedit* is used to edit *Mtext*—that is, doing so produces the *Text Formatting* Editor (see previous Fig. 18-55). As an easy alternative to invoking *Tabledit*, you can simply double-click inside any cell (assuming *Dblclkedit* is on) to produce the *Text Formatting* Editor. If you invoke the *Tabledit* command otherwise, you are prompted to pick a cell, then the editor appears.

Command: **tabledit**
Pick a table cell: **PICK**

The options available for editing the text are the same as those used to create the text for the table. Several options are available on the toolbars, the *Options* menu, and the right-click shortcut menu (see "Mtext").

Other Table Editing Methods

Highlight, Right-click Shortcut Menus

Two right-click shortcut menus are available depending on whether you first highlight an individual cell, highlight two or more cells, or highlight the entire table.

Shortcut Menu, Individual Cell Highlighted

The shortcut menu shown in Figure 18-58 displays options when an individual cell is highlighted (select the cell, then right-click). Most of these options allow you to change properties of only the highlighted cell. For example, you can use *Cell Alignment* to re-justify the text, use *Match Cell* to match the properties of the highlighted cell to another, *Insert Block* or *Insert Field* into the cell, *Edit Cell Text, Delete Cell Contents*, or *Remove All Property Overrides* (see "Editing Tables with the *Properties* Palette"). Selecting *Cell Borders...* from this shortcut menu produces the *Cell Borders Properties* dialog box (not shown) where you can specify which grid lines appear and set the border lineweight and color. The column and row options in this menu affect the individual cell or the row/column adjacent to the cell. For example, you can *Insert Columns* to the *Left* or *Right* of the highlighted cell and *Insert Rows Above* or *Below* the cell. Select *Properties* to produce the *Properties* palette (see "Editing Tables with the *Properties* Palette").

FIGURE 18-58

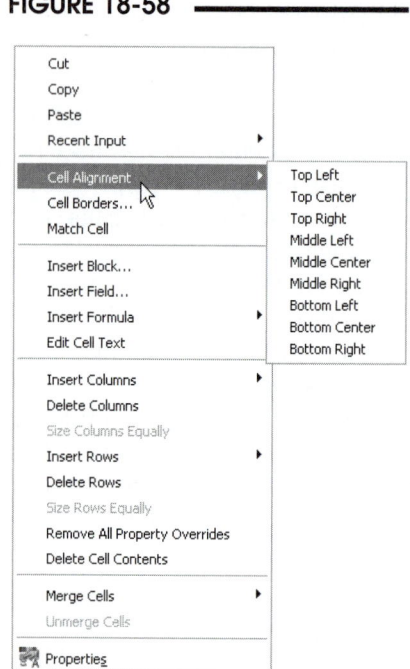

Shortcut Menu, Two or More Cells Highlighted

Highlight multiple cells by holding down the Shift key when selecting. Right-clicking produces the same menu as shown in Figure 18-58, but with all the options enabled. You can *Size Columns Equally* or *Size Rows Equally* for the highlighted cells only. You can also *Merge Cells* where the first cell (left to right) determines the contents of the newly merged cell.

Shortcut Menu, Table Highlighted

If you select the table on its edge gridline, the entire table becomes highlighted rather than an individual cell. Right-clicking produces the menu shown in Figure 18-59. Here you can also *Size Columns Equally* and *Size Rows Equally*; however, these options affect all cells in the table. In this case, the overall size of the table remains constant but the individual cells equalize. You can also *Export* tables to a comma-separated file (.CSV) for use in Microsoft Excel or other applications (see "Importing and Exporting Tables"). The *Table Indicator Color* option allows you to change the row (1, 2, 3) and column (A, B, C) labels (see previous Fig. 18-55).

FIGURE 18-59

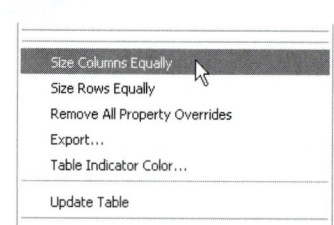

Remove All Property Overrides

Using this option from either table shortcut menu generally removes any table-wide properties that were changed using the *Properties* palette (see "Editing Tables with the *Properties* Palette") and sets the properties back to those settings defined by the table style. These properties, such as the *Cell Properties, Border Properties,* and *Cell Margins,* are defined in the *New Table Style* and *Modify Table Style* dialog boxes (see previous Fig. 18-57). However, some properties such as table width or height (even though they may have been set in the *Properties* palette) or properties changed in the *Text Formatting* Editor may not revert back to the settings defined by the style when this option is selected.

Editing Tables with Grips
If Grips are on, you can select individual cells or select the entire table and dynamically adjust the table size (see Chapter 19). Selecting an individual cell, then a hot grip, you can STRETCH the size of the column or row of the selected cell. With the entire table selected, use any column grip to STRETCH the column width (Fig. 18-60). Holding down the Ctrl key while STRETCHing a column keeps the overall table width constant. Use the upper corner grip to STRETCH all column widths and use the lower corner grip to STRETCH both rows and columns.

FIGURE 18-60

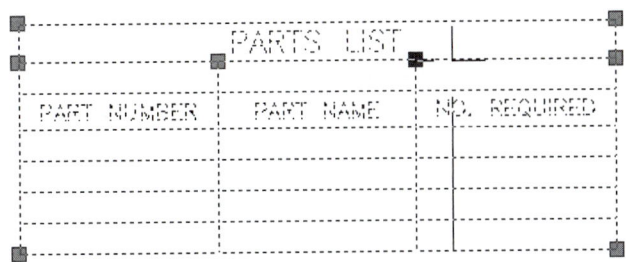

Editing Tables with the Properties Palette
Double-clicking on a cell produces the *Text Formatting* (*Mtext*) editor, while double-clicking on the table edge produces the *Properties* Palette (Fig. 18-61). In the *Table* category you cannot change the number of columns or rows, but you can change a table from one table style to another. You can also change the overall width, height, and the vertical and horizontal cell margins.

FIGURE 18-61

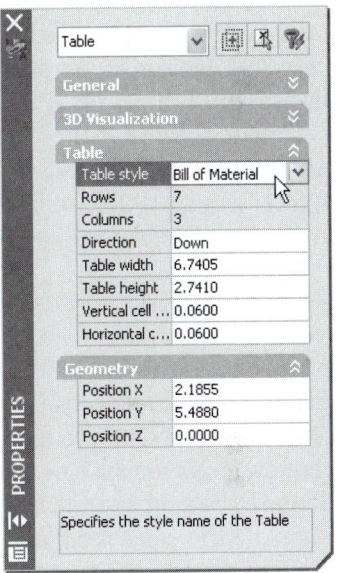

You can also select an individual cell, right-click to invoke the shortcut menu, then select *Properties* to change the properties of the individual highlighted cell using the *Properties* palette (Fig. 18-62). You can alter the *Cell* properties and the *Content* of the cell. Changing the *Cell width* or *Cell height* changes the entire row or column of the cell.

Importing and Exporting Tables
Using the OLE (object linking and embedding) features in AutoCAD, you can paste a Microsoft Excel spreadsheet into your drawing. The pasted object is converted to a native AutoCAD object—a table. The resulting table has all the features of a normal table including the ability to set and modify the table properties as well as the text content.

FIGURE 18-62

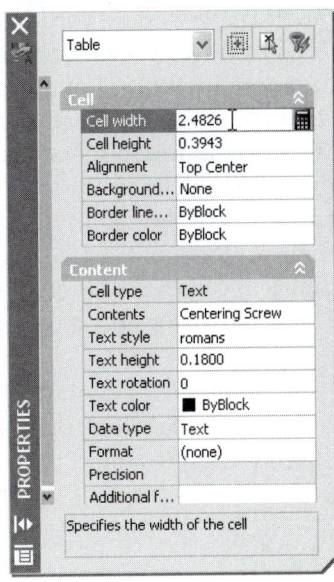

To copy a Microsoft Excel spreadsheet and paste it into AutoCAD, first open the Excel document and highlight the desired cells. Select Copy from the Edit menu in Excel. Next, return to AutoCAD (leaving the Excel application running) and select Paste Special from the Edit menu in AutoCAD. In the Paste Special dialog box select AutoCAD Entities (Fig. 18-63).

FIGURE 18-63

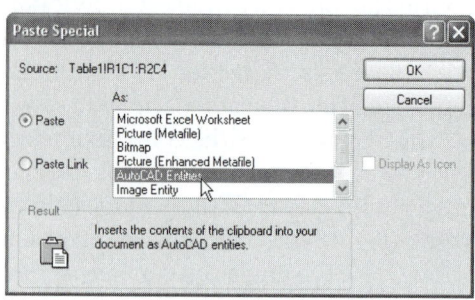

You can also export an AutoCAD table to a comma-separated file (.CSV) for use in Microsoft Excel or other applications. Do this by selecting the entire table and right-clicking, then select *Export* from the shortcut menu (see previous Fig. 18-59). The resulting file can be opened directly into Microsoft Excel.

Text Fields and Calculations in Tables

Traditionally, external spreadsheets have been used to collect information from AutoCAD drawings and to perform calculations with the data. By combining the use of tables and the power of text fields, you can create tables to collect information and perform simple calculations <u>directly in an AutoCAD drawing</u>. This is accomplished by including *Fields* in table cells and composing mathematical expressions within the *Fields*.

Arithmetic calculations that are provided are *Sum, Average,* and *Count.* You can also enter an = (equals) symbol into any table cell, then enter your own formula. You can read values from other cells by specifying the cell labels (such as "B4") or selecting a cell range, then include these cell values in your formulas.

As an example, suppose you want to create a table in a drawing of a house to calculate room square footage and the cost to carpet the rooms (Fig. 18-64). The cells indicated with a gray background contain *Fields* for making the calculations. In this table, cells B4 through B7 calculate the areas for *Plines* that are drawn around each room (see the example in Fig. 18-52). Cells D4 through D7 convert the price in square feet from the price per square yard in cell D2. Cells F4 through F7 calculate the price per room by multiplying the values in columns B and D. Finally, row 8 contains two summation values. Therefore, this table will automatically update when any of the rooms designs are changed (area of the *Plines*) or when a different carpet cost is entered in cell D2. The general steps for creating such a table are:

FIGURE 18-64

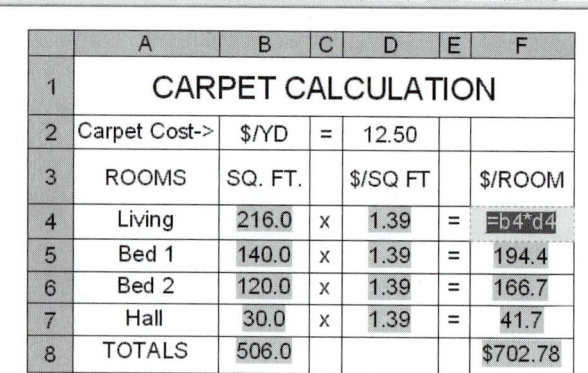

	A	B	C	D	E	F
1	CARPET CALCULATION					
2	Carpet Cost->	$/YD	=	12.50		
3	ROOMS	SQ. FT.		$/SQ FT		$/ROOM
4	Living	216.0	x	1.39	=	=b4*d4
5	Bed 1	140.0	x	1.39	=	194.4
6	Bed 2	120.0	x	1.39	=	166.7
7	Hall	30.0	x	1.39	=	41.7
8	TOTALS	506.0				$702.78

1. Use the *Table* command to create the basic format (rows and columns) of the table.
2. Use the *Tablestyle* command to set the cell properties (text style, height, etc.) and border properties (lineweight, color).
3. Enter in the heading and other static text.
4. Create *Fields* in the desired cells.

Fields in table cells can contain:

1. formulas that you enter directly into the cell by first entering an "=" (equals) symbol,
2. formulas that you create using either the *Field* dialog box or by selecting directly from the cell short-cut menu, or
3. any of the other field contents available using the *Field* dialog box (see "*Field*" earlier in the chapter).

To create *Fields* in cells containing formulas, you can double-click in the cell to produce the *Text Formatting* Editor, or highlight a table cell and right-click to produce a shortcut menu with the options shown in Figure 18-65.

FIGURE 18-65

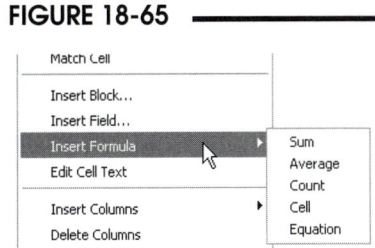

Entering Formulas Directly Into Cells

Produce the *Text Formatting* Editor by either double-clicking in a cell or by highlighting a cell, then selecting *Edit Cell Text* from the shortcut menu (see Fig. 18-65). Enter an "=" (equals) symbol, then enter the desired arithmetic expression. Typical operators are accepted:

plus (+), minus (-), times (*), divided by (/), exponent (^), parentheses ()

In addition, the following functions are accepted:

sum, average, count

You can also read numerical values from other cells by entering the cell labels (such as "B4" or "D2"). You can specify a range of cells to read using a ":" (colon) symbol (such as "A4:D4").

For example, to sum the contents of cells A1 through A25, enter the following formula: "=sum(a1:a25)." As another example, in Figure 18-64 the highlighted cell (F4) displays the formula ("=b4*d4") for calculating the cost of carpeting the living room. The formula multiplies the values for cell B4 (square footage) times D4 (price).

Entering Formulas Using the *Field* Dialog Box

Highlight a single cell (not the entire table), right-click, then select *Insert Field* from the cell shortcut menu. This action produces the *Field* dialog box (Fig. 18-66) with the *Formula* option pre-selected. You can enter a formula directly into the *Formula* edit box, or select the *Average, Sum,* or *Count* functions above. Select the *Cell* button to return to the drawing to specify a cell to enter into the *Formula*. *Evaluate* displays the calculation in the *Preview* box.

FIGURE 18-66

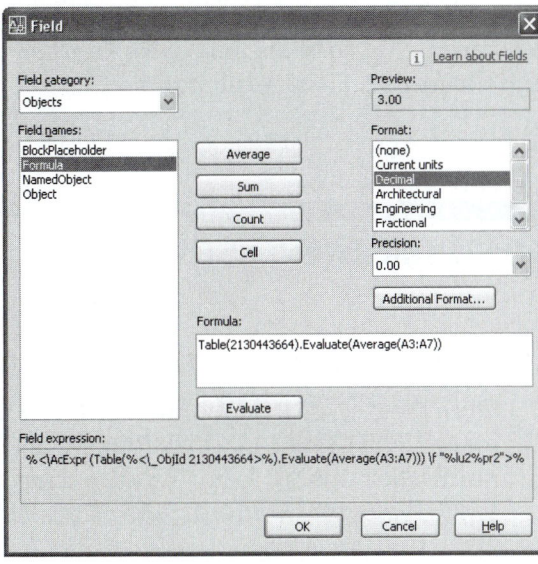

Additional Format
The *Additional Format* dialog box provides options for units conversion (Fig. 18-67). If you need to convert the *Current value* to some other unit of length, volume, etc., enter the desired value into the *Conversion factor* input field. You can add a *Prefix* or a *Suffix* (for example, see the "$" [dollar] symbol in cell F8 in Fig. 18-64). Also choose from number separators or zero suppression.

FIGURE 18-67

Entering Formulas Using Cell Shortcut Menu
Highlight a single cell, then right-click and select *Insert Formula* from the shortcut menu (see Fig. 18-65). You can then select from the standard functions, *Sum, Average,* etc. When prompted to "Select first corner of table range:", select the desired cells using a selection window. AutoCAD automatically enters the formula into the cell and produces the *Text Formatting* Editor so you can further edit the formula if desired.

Entering Other (Non-formula) Fields Using the *Field* Dialog Box
Highlight a single cell, right-click, then select *Insert Field* from the cell shortcut menu. This action produces the *Field* dialog box (Fig. 18-66) with the *Formula* option selected. In the *Field Category* drop-down list, select *All*. Specify any field for the cell. For example in Figure 18-64, the square footage for each room is calculated (column B) by entering *Object* as the field name and specifying that the *Area* for a *Polyline* is displayed in the cell (also see the example in Fig. 18-52).

Text Attributes

When you want text to be entered into a drawing and associated with some geometry, such as a label for a symbol or number for a part, *Block Attributes* can be used. *Attributes* are text objects attached to *Blocks*. When the *Blocks* are *Inserted* into a drawing, the text *Attributes* are also inserted; however, the content of the text can be entered at the time of the insertion. See Chapter 21, Blocks, DesignCenter, and Tool Palettes, and Chapter 22, Block Attributes.

CHAPTER EXERCISES

1. *Text*

    Open the **EFF-APT-PS** drawing. Make *Layer* TEXT *Current* and use *Text* to label the three rooms: **KITCHEN, LIVING ROOM,** and **BATH.** Use the *Standard style* and the *Start point* justification option. When prompted for the *Height:*, enter a value to yield letters of 3/16" on a 1/4"=1' plot (3/16 x the drawing scale factor = text height). *Save* the drawing.

2. *Style, Text, Ddedit*

Open the drawing of the temperature graph you created as **TEMP-D.** Use *Style* to create two styles named **ROMANS** and **ROMANC** based on the *romans.shx* and *romanc.shx* font files (accept all defaults). Use *Text* with the *Center* justification option to label the days of the week and the temperatures (and degree symbols) with the **ROMANS** style as shown in Figure 18-68. Use a *Height* of **1.6**. Label the axes as shown using the **ROMANC** style. Use *Ddedit* for editing any mistakes. Use *SaveAs* and name the drawing **TEMPGRPH.**

FIGURE 18-68

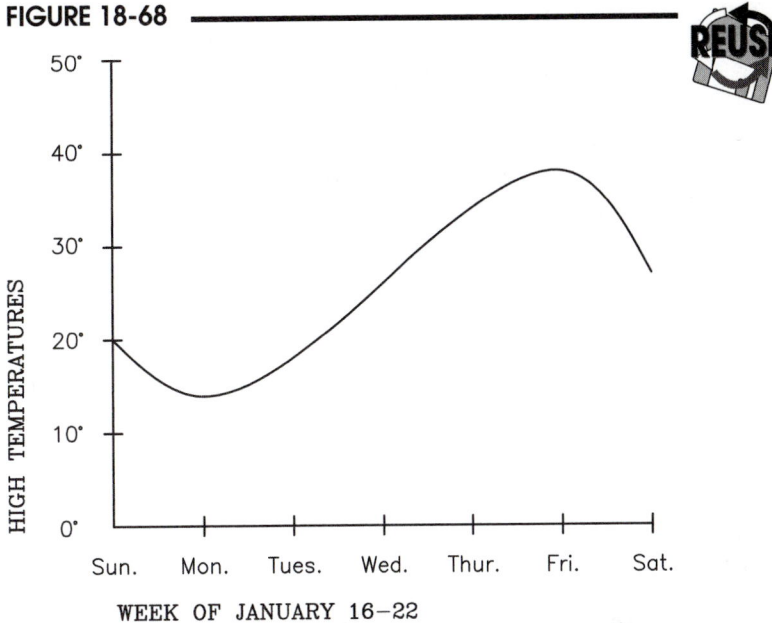

3. *Style, Text*

Open the **BILLMATL** drawing created in the Chapter 15 Exercises. Use *Style* to create a new style using the *romans.shx* font. Use whatever justification methods you need to align the text information (not the titles) as shown in Figure 18-69. Next, type the *Style* command to create a new style that you name as **ROMANS-ITAL.** Use the *romans.shx* font file and specify a **15** degree *obliquing angle*. Use this style for the **NO.**, **PART NAME**, and **MATERIAL.** *SaveAs* **BILLMAT2.**

FIGURE 18-69

NO.	PART NAME	MATERIAL
1	Base	Cast Iron
2	Centering Screw	N/A
3	Slide Bracket	Mild Steel
4	Swivel Plate	Mild Steel
5	Top Plate	Cast Iron

4. *Edit Text*

Open the **EFF-APT-PS** drawing. Create a new style named **ARCH1** using the *CityBlueprint* (.TTF) font file. Next, invoke the *Properties* command. Use this dialog box to modify the text style of each of the existing room names to the new style as shown in Figure 18-70. *SaveAs* **EFF-APT2.**

FIGURE 18-70

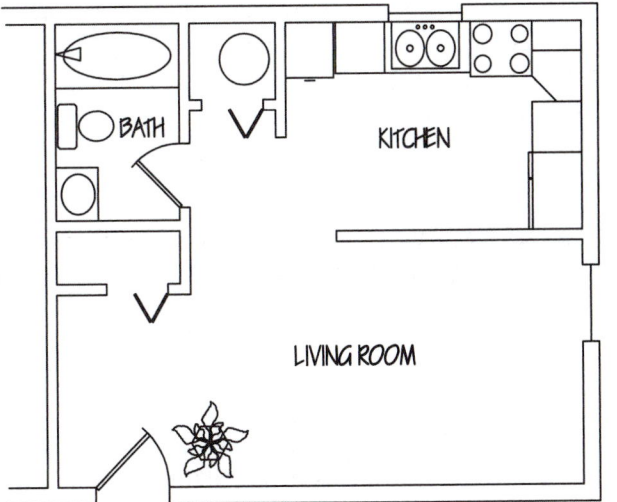

5. *Text, Mtext*

Open the **CBRACKET** drawing from Chapter 9 Exercises. Using *romans.shx* font, use *Text* to place the part name and METRIC annotation (Fig. 18-71). Use a *Height* of **5** and **4**, respectively, and the *Center Justification* option. For the notes, use *Mtext* to create the boundary as shown. Use the default *Justify* method (*TL*)and a *Height* of **3**. Use *Ddedit, Mtedit,* or *Properties* if necessary. *SaveAs* **CBRACTXT.**

FIGURE 18-71

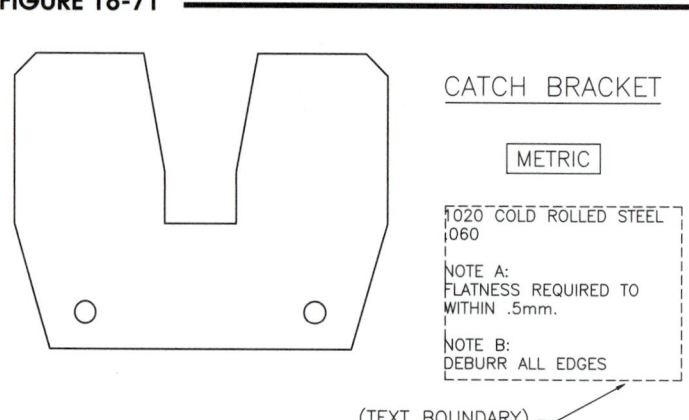

CATCH BRACKET

METRIC

1020 COLD ROLLED STEEL
.060

NOTE A:
FLATNESS REQUIRED TO
WITHIN .5mm.

NOTE B:
DEBURR ALL EDGES

(TEXT BOUNDARY)

6. *Style*

Create two new styles for each of your template drawings: **ASHEET**, **BSHEET**, and **C-D-SHEET.** Use the *romans.shx* style with the default options for engineering applications or *CityBlueprint* (.TTF) for architectural applications. Next, design a style of your choosing to use for larger text as in title blocks or large notes.

7. *Import Text, Ddedit, Properties*

Use a text editor such as Windows Wordpad or Notepad to create a text file containing words similar to "Temperatures were recorded at Sanderson Field by the National Weather Service." Then *Open* the **TEMPGRPH** drawing and use the *Import Text…* option of *Mtext* to bring the text into the drawing as a note in the graph as shown in Figure 18-72. Use *Ddedit* to edit the text if desired or use *Properties* to change the text style or height.

FIGURE 18-72

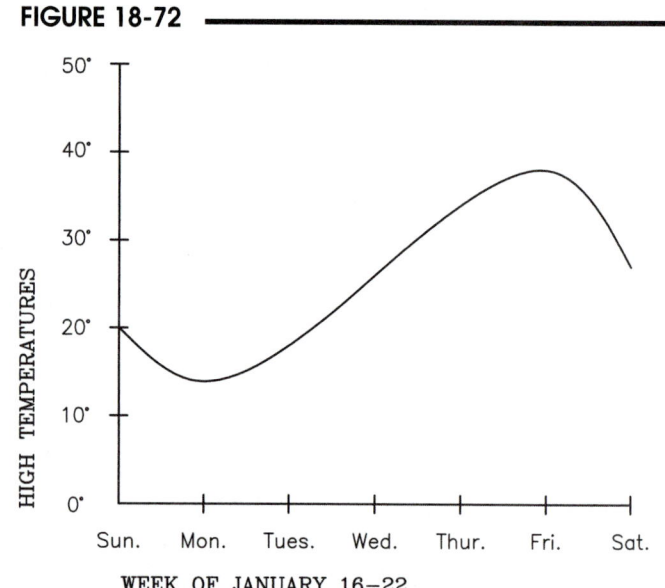

WEEK OF JANUARY 16–22

Temperatures were recorded at Sanderson Field by the National Weather Service.

8. *Create a Title Block*

A. Begin a *New* drawing and assign the name **TBLOCK.** Create the title block as shown in Figure 18-73 or design your own, allowing space for eight text entries. The dimensions are set for an A size sheet. Draw on *Layer* **0**. Use a *Pline* with **.02** *width* for the boundary and *Lines* for the interior divisions. (No *Lines* are needed on the right side and bottom because the title block will fit against the border lines.)

FIGURE 18-73

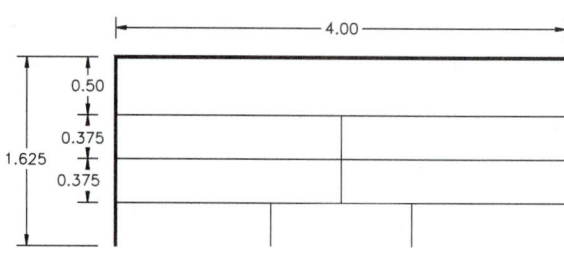

B. Create two text *Styles* using *romans.shx* and *romanc.shx* font files. Insert text similar to that shown in Figure 18-74. Examples of the fields to create are:

> Company or School Name
> Part Name or Project Title
> Scale
> Designer Name
> Checker or Instructor Name
> Completion Date
> Check or Grade Date
> Project or Part Number

Choose text items relevant to your school or office. *Save* the drawing.

FIGURE 18-74 ━━━━━━━━━━

CADD Design Company		
Adjustable Mount		1/2"=1"
Des.— B.R. Smith		Chk.—JRS
1/1/2007	1/3/2007	42B—ADJM

9. *Mtext*

Open the **STORE ROOM** drawing that you created in Chapter 16 Exercises. Use *Mtext* to create text paragraphs giving specifications as shown in Figure 18-75. Format the text below as shown in the figure. Use *CityBlueprint* (.TTF) as the base font file and specify a base *Height* of **3.5″**. Use a *Color* of your choice and *CountryBlueprint* (.TTF) font to emphasize the first line of the Room paragraph. All paragraphs use the *TC Justify* methods except the Contractor Notes paragraph, which is *TL*. Save the drawing as **STORE ROOM2.**

Room: <u>STORAGE ROOM</u>
    11′-2″ x 10′-2″
    Cedar Lined - 2 Walls

Doors: 2 - 2268 DOORS
    Fire Type A
    Andermax

Windows: 2 - 2640 CASE-MENT WINDOWS
    Triple Pane Argon
        Filled
    Andermax

Notes: <u>Contractor Notes:</u>

    Contractor to verify
    all dimensions in
    field. Fill door and
    window roughouts
    after door and
    window placement.

FIGURE 18-75 ━━━━━━━━━━

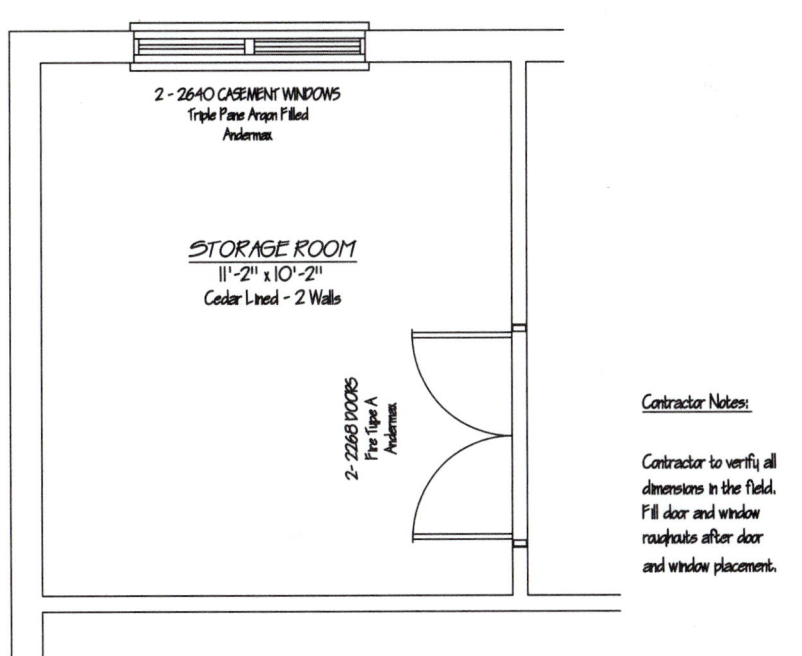

10. *Tablestyle, Table*

Begin a *New* drawing and assign the name **MATERIAL TABLE**. Use the *Tablestyle* command to create a *New* table style named **Bill of Material**. In the *Data, Column Heads* and *Title* tabs, assign a new *Text style* using the **Romans.shx** font file. Accept the default *Text height* and other parameters. Next, use the *Table* command to specify a table with *3 Columns* and *5 Data rows*. Select *Specify window* under *Insertion behavior*. Select *OK*, then PICK the two corners for the table ensuring that you stretch the table wide enough to allow appropriate space in the cells for the intended contents. Enter the text into the cells as shown in Figure 18-76. If you need to make adjustments, highlight the table on an outside edge and use Grips to stretch the table. Also, you could right-click while the table is highlighted and select *Size Columns Equally* or *Size Rows Equally*. *Save* the drawing.

FIGURE 18-76

BILL OF MATERIAL		
NUMBER	PART NAME	MATERIAL
1	Base	Cast Iron
2	Centering Screw	N/A
3	Slide Bracket	Mild Steel
4	Swivel Plate	Mild Steel
5	Top Plate	Cast Iron

11. *Fields*

A. *Open* the **EFT-APT2** drawing you worked on in Exercise 4. Use *SaveAs* and assign the name **EFF-APT-AREA**. Make a new *Layer* called *Area* and make it the *Current* layer. In the floor plan, draw a *Pline* around the Living Room using *Osnap* to connect to the interior corners of the room as shown in Figure 18-77 by the heavy outline. Make sure you *Close* the *Pline*. (You do not have to assign a *Width* for your *Pline*.) Next, draw a second *Pline* around the Kitchen using the same method, as shown in the figure. Finally, draw a third *Pline* around the other space (bath, closets, and hall) using *Osnap* to connect to the interior walls and two other *Plines* as shown in the figure.

FIGURE 18-77

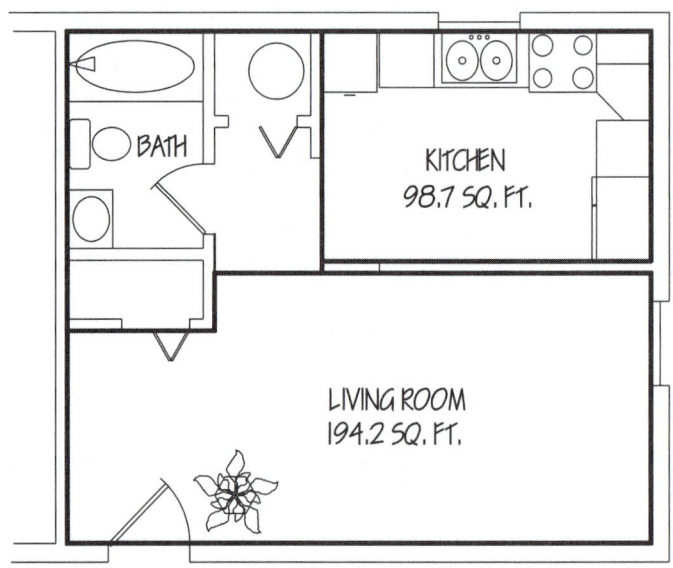

B. Make the text layer *Current*. Use *Mtext* to place text directly below the "LIVING ROOM" text. When the *Text Formatting* Editor appears, select *Insert Field* from the *Options* menu or the right-click menu. In the *Field* dialog box, select *Object* from the *Field names* list and then use the *Select Object* button to return to the drawing and select the *Pline* around the Living Room. Select *Area* in the *Property* list. Set the other *Format* options to display text as shown in Figure 18-77. Next, follow the same procedure to create a *Field* to display the area of the Kitchen. Do not create a field for the other area. *Freeze* the *Area* layer. *Save* the drawing.

12. *Fields* and Calculations in a *Table*

A. If not already open, *Open* the **EFF-APT-AREA** drawing. *Thaw* the *Area* layer. Create a new *Layer* named **Table** and make it *Current*.

B. Use *Tablestyle* to create a *New* table style and assign the name **Area Table**. In the *Data* tab, set the *Text Style* to *Cityblueprint* and the *Height* to *6"*. For the *Column Heads* tab, set *Text Style* to *Cityblueprint* and *Height* to *8"*. In the *Title* tab, set *Text Style* to *Cityblueprint* and *Height* to *9"*. Make **Area Table** the *Current* table style, then close the dialog box.

C. Use the *Table* command to create a table similar to that shown in Figure 18-78. In the *Insert Table* dialog box, choose the following options: *Specify Insertion Point, 2 Columns, 4 Rows, Column Width* of *5'*, then select *OK* to insert the table just to the right of the floor plan. Create the static text as shown in the figure but do not create text for the four cells listing the square footage calculations. If needed, make any adjustments to the table.

FIGURE 18-78 ————————————

Efficiency Apartment Area	
ROOMS	AREA
Living Room	194.2 SQ. FT.
Kitchen	98.7 SQ. FT.
Other	91.2 SQ. FT.
TOTAL	384.0 SQ. FT.

D. To create the area fields for the Living Room, highlight the appropriate cell, then use the *Options* menu or right-click menu and select *Insert Field*. In the *Field* dialog box select the following options: *Object* from *Field Names* list, *Select Objects* button to select the *Pline* around the Living Room, *Area* from *Property* list, *Architectural* from *Format* list. Select *OK* to close the dialog box and create the field displaying the area for the Living Room. Follow the same procedure in the two cells below for the "Kitchen" area and "Other" area.

E. For the last cell (lower-right corner of the table), use *Insert Field*, select *Formula* from *Names* list, *Sum* from the formula choices, and *Decimal* from *Format* list. When prompted to select the corners of the data range, select the three cells displaying the area calculations. When the calculation appears in the last cell, make any adjustments to the table as needed. (Your area calculations may vary slightly from those shown in Figure 18-78 based on where your *Plines* are drawn.) Finally, *Freeze* the **Area** layer. *Save* the drawing.

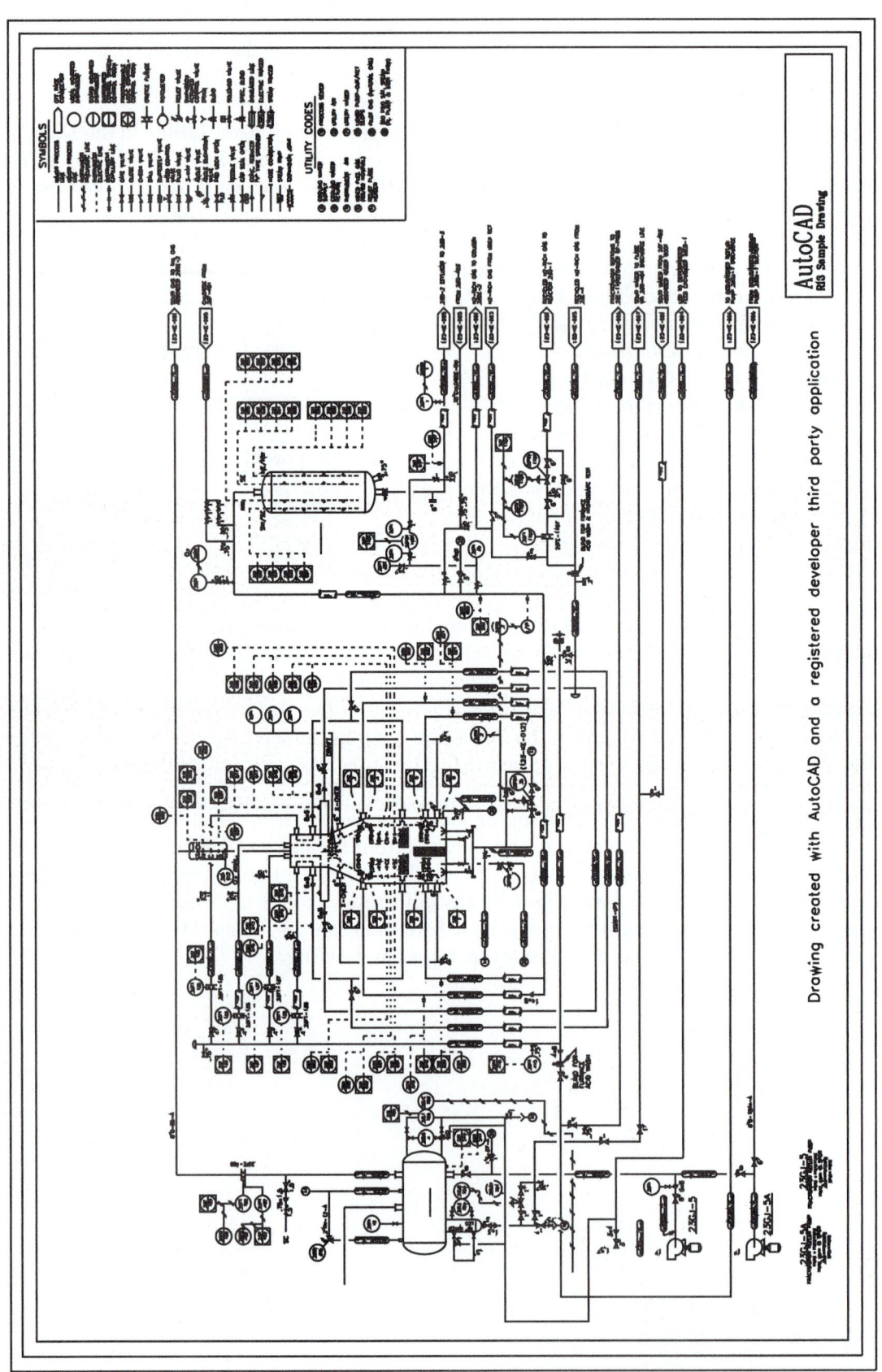

PNID.DWG, Courtesy of Autodesk, Inc.

GRIP EDITING

CHAPTER OBJECTIVES

After completing this chapter you should:

1. be able to use the *GRIPS* variable to enable or disable object grips;

2. be able to activate the grips on any object;

3. be able to make an object's grips hot or cold;

4. be able to use each of the grip editing options, namely, STRETCH, MOVE, ROTATE, SCALE, and MIRROR;

5. be able to use the copy and Base suboptions;

6. be able to use the auxiliary grid that is automatically created when the Copy suboption is used;

7. be able to change grip variable settings using the *Options* dialog box or the Command line format.

CONCEPTS

Grips provide an alternative method of editing AutoCAD objects. The object grips are available for use by setting the *GRIPS* variable to **1**. Object grips are small squares appearing on selected objects at endpoints, midpoints, or centers, etc. The object grips are activated (made visible) by **PICK**ing objects with the cursor pickbox only <u>when no commands are in use</u> (at the open Command prompt). Grips are like small, magnetic *OSNAPs* (*Endpoint*, *Midpoint*, *Center*, *Quadrant*, etc.) that can be used for snapping one object to another, for example. If the cursor is moved within the small square, it is automatically "snapped" to the grip. Grips can replace the use of *OSNAP* for many applications. The grip option allows you to STRETCH, MOVE, ROTATE, SCALE, MIRROR, or COPY objects without invoking the normal editing commands or *OSNAPs*.

As an example, the endpoint of a *Line* could be "snapped" to the quadrant of a *Circle* (shown in Fig. 19-1) by the following steps:

1. Activate the grips by selecting both objects. Selection is done when no commands are in use (during the open Command prompt).
2. Select the grip at the endpoint of the *Line* (1). The grip turns **hot** (red).
3. The ** STRETCH ** option appears in place of the Command prompt.
4. STRETCH the *Line* to the quadrant grip on the *Circle* (2). **PICK** when the cursor "snaps" to the grip.
5. The *Line* and the *Circle* should then have connecting points. The Command prompt reappears. Press Escape to cancel (deactivate) the grips.

FIGURE 19-1

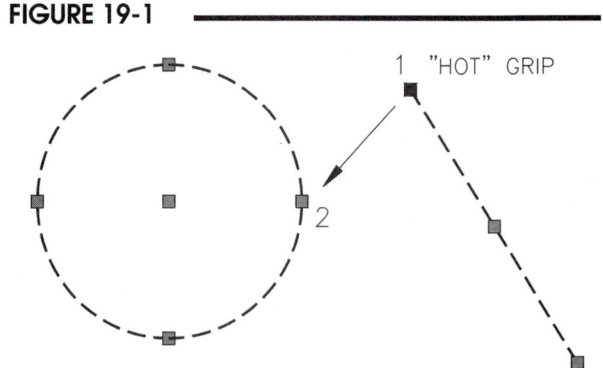

GRIPS FEATURES

Grips are enabled or disabled by changing the setting of the system variable, *GRIPS*. A setting of 1 enables or turns *ON GRIPS* and a setting of **0** disables or turns *OFF GRIPS*. This variable can be typed at the Command prompt, or Grips can be invoked from the *Selection* tab of the *Options* dialog box, or by typing *Ddgrips* (Fig. 19-2). Using the dialog box, toggle *Enable Grips* to turn *GRIPS ON*. The default setting in AutoCAD for the *GRIPS* variable is **1** (*ON*). (See "Grip Settings" near the end of the chapter for explanations of the other options in the *Grips* section of the *Options* dialog box.)

FIGURE 19-2

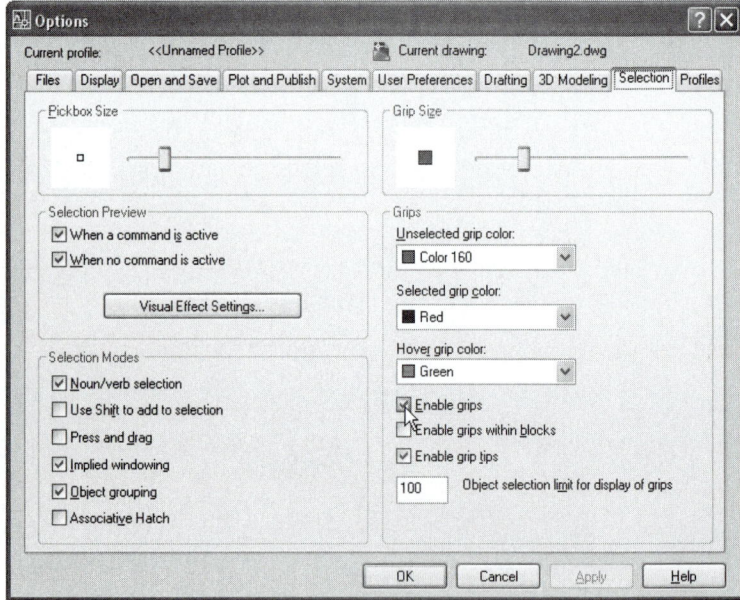

The *GRIPS* variable is saved in the user profile rather than in the current drawing as with most other system variables. Variables saved in the system registry are effective for any drawing session for that <u>particular user</u>, no matter which drawing is current. The reasoning is that grip-related variables (and selection set-related variables) are a matter of personal preference and therefore should remain constant for a particular user's CAD station.

When *GRIPS* have been enabled, a small pickbox (5 pixels is the default size) appears at the center of the cursor crosshairs. (The pickbox also appears if the *PICKFIRST* system variable is set to **1**.) This pickbox operates in the same manner as the pickbox appearing during the "Select objects:" prompt. <u>Only</u> the pickbox, *AUTO window*, or *AUTO crossing window* methods can be used for selecting objects to activate the grips. (These three options are the only options available for Noun/Verb object selection as well.)

Activating Grips on Objects

The grips on objects are activated by selecting desired objects with the cursor pickbox, window, or crossing window. This action is done <u>when no commands are in use</u> (at the open Command prompt). When an object has been selected, two things happen: the grips appear and the object is highlighted. The grips are the small blue (default color) boxes appearing at the endpoints, midpoint, center, quadrants, vertices, insertion point, or other locations depending on the object type (Fig. 19-3). Highlighting indicates that the object is included in the selection set.

FIGURE 19-3

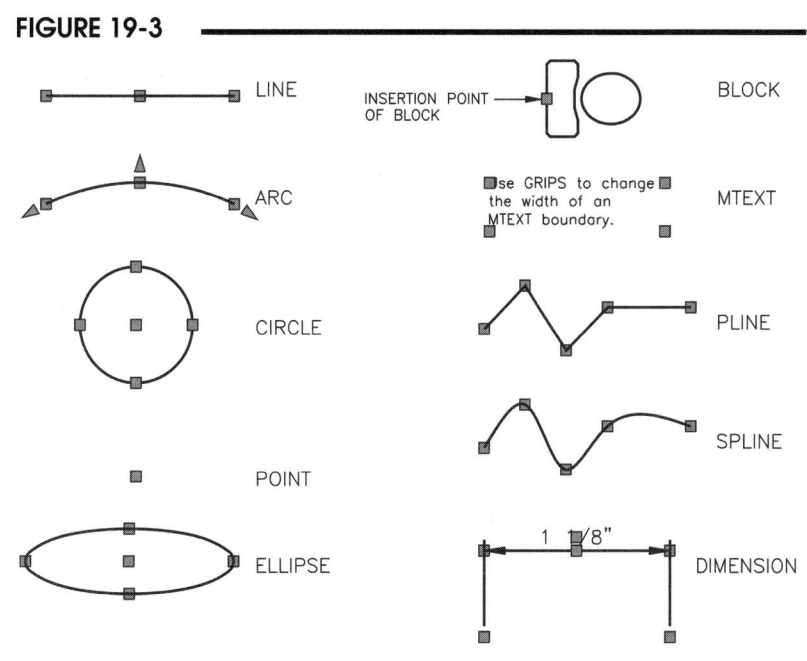

Cold and Hot Grips

Grips can have only two states: **cold** and **hot**. When *Grips* are turned on, and an object is selected (when no commands are in use), the object's grips become **cold**—its grips are displayed in blue (default color) and the object is highlighted.

One or more grips can then be made **hot**. A hot grip is created by "hovering" the cursor over a cold grip until the grip changes to a light green (default) color, then you PICK the grip so it changes to red (default color). The intermittent green grip is called a "hover" grip.

A **hot** grip is red (by default) and its object is almost always highlighted. Any grip can be changed to **hot** by selecting the grip itself. A <u>**hot** grip is the default base point</u> used for the editing action such as MOVE, ROTATE, SCALE, or MIRROR, or is the stretch point for STRETCH. When a **hot** grip exists, a new series of prompts appear in place of the Command prompt that displays the various grip editing options.

The grip editing options are also available from a right-click cursor menu (Fig. 19-4). <u>A grip must be changed to</u> **hot** <u>before the editing options appear</u>. Two or more grips can be made **hot** simultaneously by pressing Shift while selecting <u>each</u> grip.

FIGURE 19-4

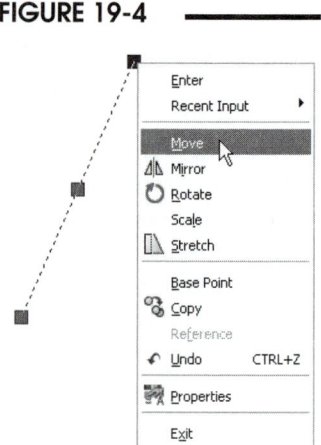

If you have made a *Grip* **hot** and want to deactivate it, possibly to make another *Grip* **hot** instead, press Escape once. This returns the grip to a **cold** state. In effect, this is an undo only for the **hot** grip. Pressing Escape again cancels all *Grips*. Pressing Escape demotes hot grips or cancels all grips:

Grip State	Press Escape once	Press Escape twice
only **cold**	grips are deactivated	
hot and **cold**	**hot** demoted to **cold**	grips are deactivated

NOTE: Beware that there are <u>two right-click (shortcut) menus available when grips are activated</u>. The Grip menu (see previous Fig. 19-4) is available only when a grip is <u>hot</u> (red). If you make any grip hot, then right-click, the <u>Grip menu</u> appears. The Grip menu contains grip editing options. However, if grips are <u>cold</u> and you right-click, the <u>Edit Mode menu</u> appears (Fig. 19-5). This menu appears any time one or more objects are selected, no commands are active, and you right-click (see Chapter 1, Edit Mode Menu). Although the commands displayed in the two menus appear to be the same, they are not. The Edit Mode menu contains the full commands. For example, *Move* in Figure 19-4 is the MOVE grip option, whereas *Move* in Figure 19-5 is the *Move* command.

FIGURE 19-5

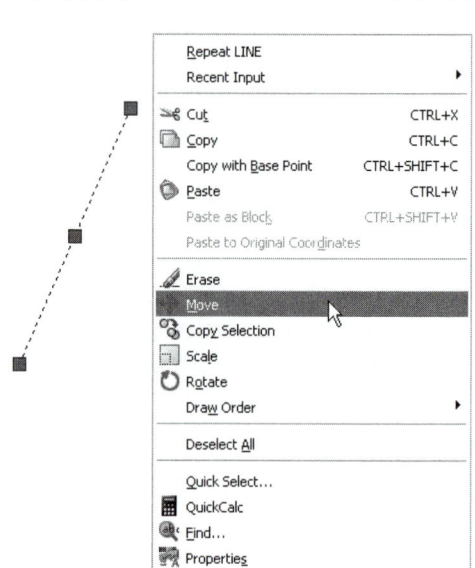

NOTE: When *PICKFIRST* is set to 0, the Edit-mode shortcut menus do not appear—the Default menu appears instead.

Grip Editing Options

When a **hot grip** has been activated, the grip editing options are available. The Command prompt is replaced by the STRETCH, MOVE, ROTATE, SCALE, or MIRROR grip editing options. You can sequentially cycle through the options by pressing the Space bar or Enter key. The editing options are displayed in Figures 19-6 through 19-10 on the next page.

Alternately, you can <u>right-click when a grip is</u> **hot** to activate the grip menu (see Fig. 19-4). This menu has the same options available in Command line format with the addition of *Reference* and *Properties*. The options are described in the following figures.

** STRETCH **
Specify stretch point or
[Basepoint/Copy/Undo/eXit]:

Note that for *Arc* objects only, grips appear at
the endpoints, midpoints, and center. In addition, arrowshaped grips appear at the endpoints and midpoint (Fig. 19-6). These arrow
grips allow you to change the radius (midpoint
arrow) or extend the arc's length (endpoint
arrows) in the STRETCH mode only.

FIGURE 19-6

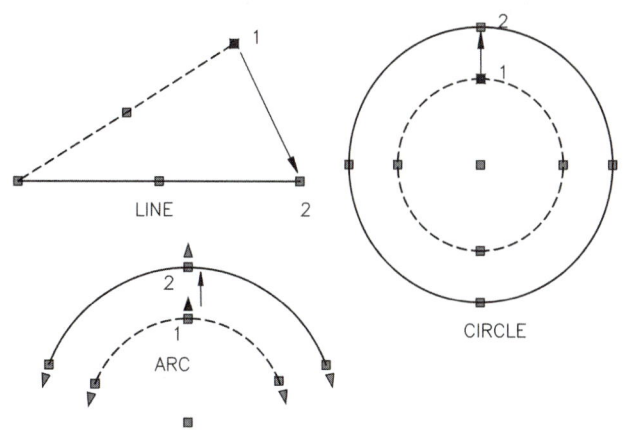

** MOVE **
Specify move point or
[Base point/Copy/Undo/eXit]:
(Fig. 19-7)

FIGURE 19-7

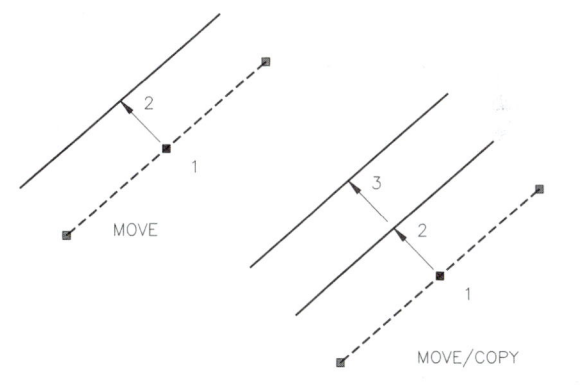

** ROTATE **
Specify rotation angle or
[Base point/Copy/Undo/Reference/eXit]:
(Fig. 19-8)

FIGURE 19-8

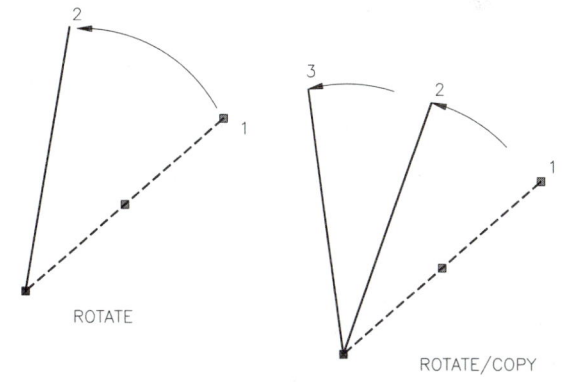

** SCALE **
Specify scale factor or
[Base point/Copy/Undo/Reference/eXit]:
(Fig. 19-9)

FIGURE 19-9

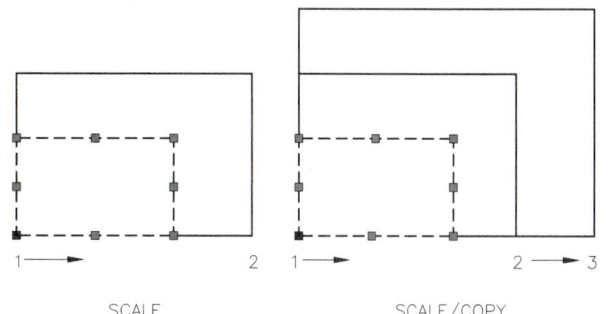

** MIRROR **
Specify second point or
[Base point/Copy/Undo/eXit]:
(Fig. 19-10)

FIGURE 19-10

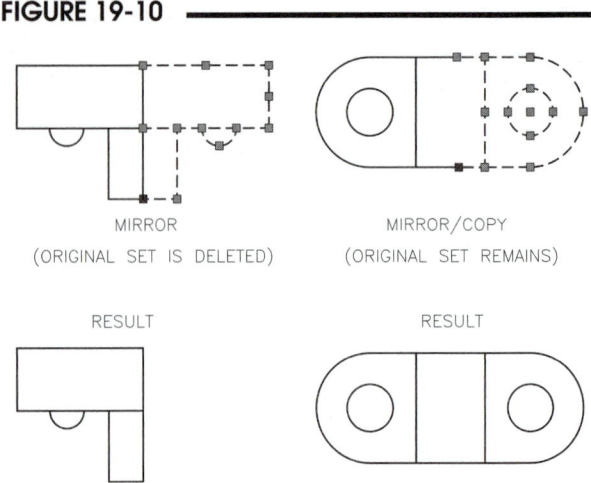

The *Grip* options are easy to understand and use. Each option operates like the full AutoCAD command by the same name. Generally, the editing option used (except for STRETCH) affects all highlighted objects. The **hot** grip is the base point for each operation. The suboptions, Base and Copy, are explained next.

NOTE: The STRETCH option differs from other options in that STRETCH affects only the object that is attached to the **hot** grip, rather than affecting all highlighted (**cold**) objects.

Base

The Base suboption appears with all of the main grip editing options (STRETCH, MOVE, etc.). Base allows using any other grip as the base point instead of the **hot** grip. Type the letter *B* or select from the right-click cursor menu to invoke this suboption.

Copy

Copy is a suboption of every main choice. Activating this suboption by typing the letter *C* or selecting from the right-click cursor menu invokes a Multiple copy mode, such that whatever set of objects is STRETCHed, MOVEd, ROTATEd, etc., becomes the first of an unlimited number of copies (see the previous five figures). The Multiple mode remains active until exiting back to the Command prompt.

Undo

The Undo option, invoked by typing the letter *U* or selecting from the right-click cursor menu will undo the last Copy or the last Base point selection. Undo functions only after a Base or Copy operation.

Reference

This option operates similarly to the reference option of the *Scale* and *Rotate* commands. Use *Reference* to enter or PICK a new reference length (SCALE) or angle (ROTATE). (See "*Scale* and *Rotate*," Chapter 9.) With grips, *Reference* is only enabled when the SCALE or ROTATE options are active.

Properties... (Right-Click Menu Only)

Selecting this option from the right-click grip menu (see Fig. 19-4) activates the *Properties* palette. All highlighted objects become subjects of the palette. Any property of the selected objects can be changed with the palette. (See "*Properties*," Chapter 16.)

Shift to Turn on *ORTHO*

Using the Copy option allows you to make multiple copies, such as MOVE/Copy or ROTATE/Copy. If you hold down the Shift key as you make copies, *ORTHO* is turned on as long as Shift is depressed.

Auxiliary Grid

An auxiliary grid is <u>automatically</u> established on creating the first Copy (Fig. 19-11). The grid is activated by pressing Ctrl while placing the subsequent copies. The subsequent copies are then "snapped" to the grid in the same manner that *SNAP* functions. The spacing of this auxiliary grid is determined by the location of the first Copy, that is, the X and Y intervals between the base point and the second point.

FIGURE 19-11

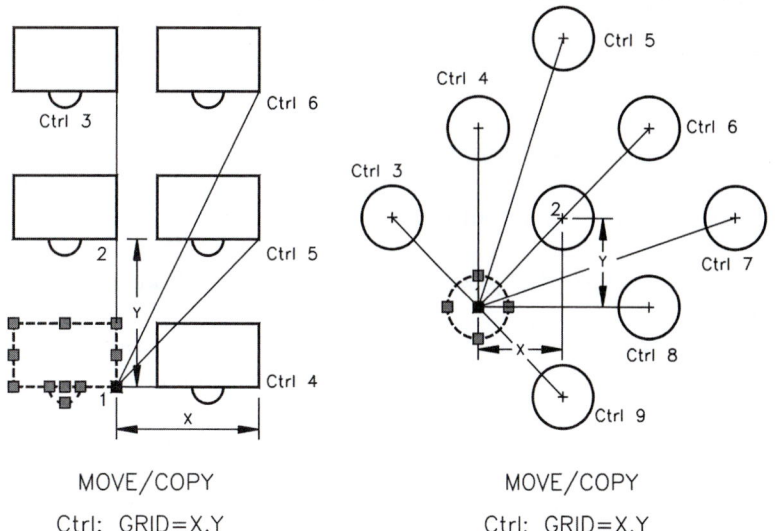

MOVE/COPY
Ctrl: GRID=X,Y

MOVE/COPY
Ctrl: GRID=X,Y

For example, a "polar array" can be simulated with grips by using ROTATE with the Copy suboption (Fig. 19-12). The "array" can be constructed by making one Copy, then using the auxiliary grid to achieve equal angular spacing. The steps for creating a "polar array" are as follows:

1. Select the object(s) to array.

2. Select a grip on the set of objects to use as the center of the array. Cycle to the ROTATE option by pressing Enter or selecting from the right-click cursor menu. Next, invoke the Copy suboption.

3. Make the first copy at any desired location.

4. After making the first copy, activate the auxiliary angular grid by holding down Ctrl while making the other copies.

5. Cancel the grips or select another command from the menus.

FIGURE 19-12

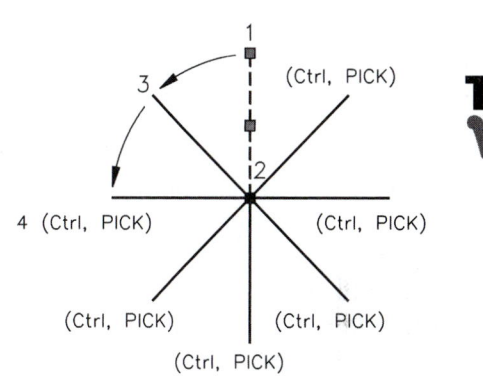

ROTATE/COPY
Ctrl — ANGLES EQUAL

Editing Dimensions

One of the most effective applications of grips is as a dimension editor. Because grips exist at the dimension's extension line origins, arrowhead endpoints, and dimensional value, a dimension can be changed in many ways and still retain its associativity (Fig. 19-13). See Chapter 28, Dimensioning, for further information about dimensions, associativity, and editing dimensions with *Grips*.

FIGURE 19-13

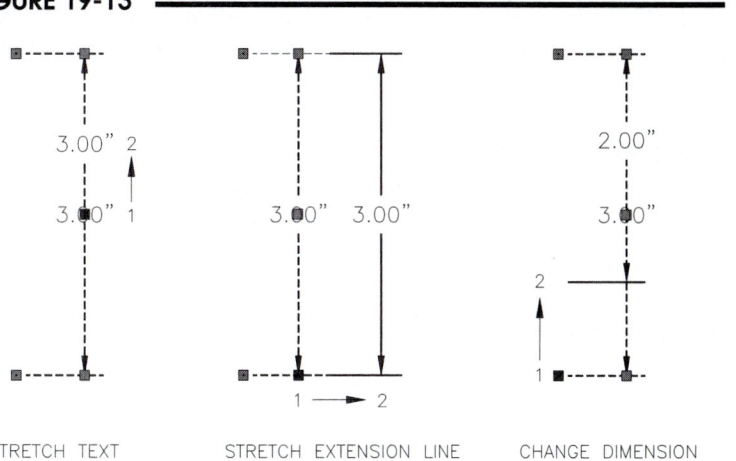

STRETCH TEXT   STRETCH EXTENSION LINE   CHANGE DIMENSION

GUIDELINES FOR USING GRIPS

Although there are many ways to use grips based on the existing objects and the desired application, a general set of guidelines for using grips is given here:

1.   Create the objects to edit.

2.   Select objects to make **cold** grips appear.

3.   Select the desired **hot** grip(s). The *Grip* options should appear in place of the Command line.

4.   Press Space or Enter to cycle to the desired editing option (STRETCH, MOVE, ROTATE, SCALE, MIRROR) or select from the right-click cursor menu.

5.   Select the desired suboptions, if any. If the *Copy* suboption is needed or the base point needs to be re-specified, do so at this time. *Base* or *Copy* can be selected in either order.

6.   Make the desired STRETCH, MOVE, ROTATE, SCALE, or MIRROR.

7.   Cancel the grips by pressing Escape or selecting a command from a menu.

GRIPS SETTINGS

FIGURE 19-14

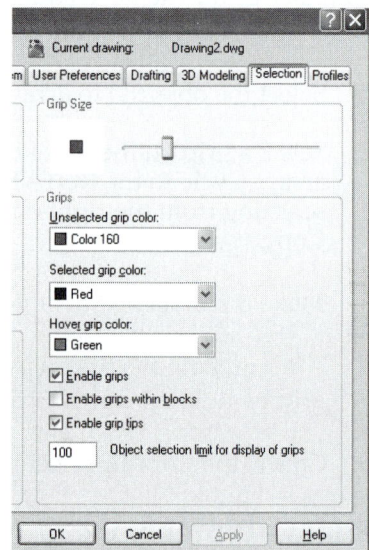

Several settings are available in the *Selection* tab of the *Options* dialog box (Fig. 19-14) that control the way grips appear or operate. The settings can also be changed by typing in the related variable name at the Command prompt.

Grip Size
Use the slider bar to interactively increase or decrease the size of the grip "box." The *GRIPSIZE* variable could alternately be used. The default size is 5 pixels square (*GRIPSIZE*=5).

Unselected Grip Color
This setting enables you to change the color of **cold** grips. Select the desired color from the *Unselected Grip Color* drop-down list. PICKing the *Select Color...* tile produces the *Select Color* dialog box (identical to that used with other color settings). The default setting is blue (ACI number 5). Alternately, the *GRIPCOLOR* variable can be typed and any ACI number from 0 to 255 can be entered.

Selected Grip Color
The color of hot grips can be specified with this option. Select the desired color from the *Selected Grip Color* drop-down list. The *Select Color...* tile produces the *Select Color* dialog box. You can also type *GRIPHOT* and enter any ACI number to make the change. The default color is red (1).

Hover Grip Color
This setting specifies the color of grips when you "hover" your cursor over a cold (blue) grip, usually just before you PICK it to make it a hot grip. The drop-down list operates identically to the other grip color options.

Enable Grips

If a check appears in the checkbox, *Grips* are enabled for the workstation. A check sets the *GRIPS* variable to 1 (on). Removing the check disables grips and sets *GRIPS=0* (off). The default setting is on.

Enable Grips within Blocks

When no check appears in this box, only one grip at the block's insertion point is visible (Fig. 19-15). This allows you to work with a block as one object. The related variable is *GRIPBLOCK*, with a setting of 0=off. This is the default setting (disabled).

FIGURE 19-15

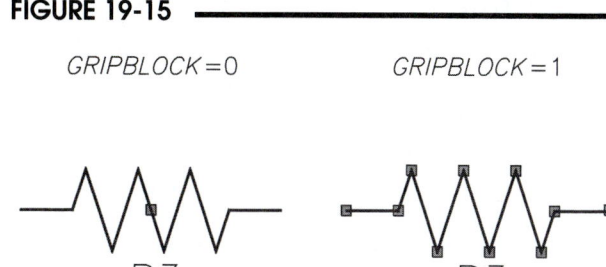

GRIPBLOCK = 0   *GRIPBLOCK = 1*

When the box is checked, *GRIPBLOCK* is set to 1. All grips on all objects contained in the block are visible and functional (Fig. 19-15). This allows you to use any grip on any object within the block. This does <u>not</u> mean that the block is <u>*Exploded*</u>—the block retains its single-object integrity and individual entities in the block cannot be edited independently. This feature permits you to use the grips only on each of the block's components.

Notice in Figure 19-15 how the insertion point is not accessible when *GRIPBLOCK* is set to 1. Notice also that there are two grips on each of the *Normal* attributes and only one on the *Invisible* attribute and that these grips do not change with the two *GRIPBLOCK* settings. Making one of these grips hot allows you to "move" the attribute to any location and the attribute still retains its association with the block.

Enable Grip Tips

This option has no effect on standard AutoCAD objects. This option enables the grip tips only when custom (third-party) objects supporting grip tips are included in the drawing. (Alternately, set *GRIP-TIPS=1*.)

Object Selection Limit for Display of Grips

This value sets the maximum number of selected objects that can display grips. When the initial selection set includes more than the specified number of objects, grips are not displayed. The valid range is 1 to 32,767. (The default setting is 100 for the *GRIPOBJLIMIT* system variable.)

NOTE: All *Grip*-related variable settings are saved in the user profile rather than in the current drawing file. This generally means that changes in these variables remain with the computer, not the drawing being used.

Point Specification Precedence

When you select a point with the input device, AutoCAD uses a point specification precedence to determine which location on the drawing to find. The hierarchy is listed here:

1. Object Snap (*OSNAP*)
2. Explicit coordinate entry (absolute, relative, or polar coordinates)
3. *ORTHO*
4. Point filters (XY filters)
5. *Grips* auxiliary grid (rectangular and circular)
6. *Grips* (on objects)
7. *SNAP* (F9 grid snap)
8. Digitizing a point location

Practically, this means that *OSNAP* has priority over any other point selection mode. As far as *Grips* are concerned, *ORTHO* overrides *Grips*, so turn off *ORTHO* if you want to snap to *Grips*. Although *Grips* override *SNAP* (F9), it is suggested that *SNAP be turned Off* while using *Grips* to simplify PICKing.

More Grips

Grips have a wide range of editing potential. As presented in this chapter, most of the grip-editing options are alternatives to using the *Stretch*, *Move*, *Rotate*, *Scale*, *Mirror*, and *Copy* commands. However, grips also provide editing methods for particular objects that are not available by any other means. Editing dimension objects is one example that is discussed in Chapters 28 and 29. Another very powerful application of grips is editing solid models. These techniques are discussed in Chapters 38, 39, and 40.

CHAPTER EXERCISES

FIGURE 19-16

1. *Open* the **TEMP-D** drawing from Chapter 16 Exercises. Use the grips on the existing *Spline* to generate a new graph displaying the temperatures for the following week. **PICK** the *Spline* to activate grips (Fig. 19-16). Use the **STRETCH** option to stretch the first data point grip to the value indicated below for Sunday's temperature. **Cancel** the grips; then repeat the steps for each data point on the graph. *Save* the drawing as **TEMP-F**.

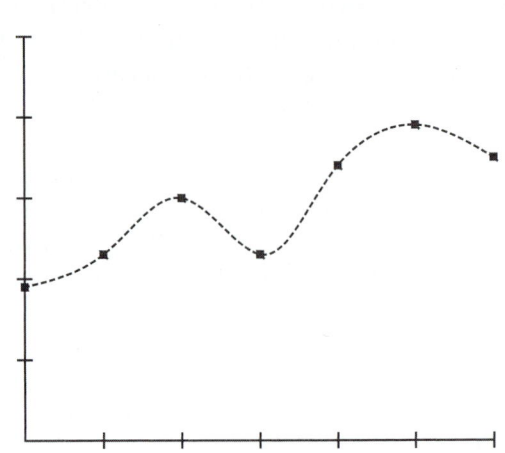

X axis	Y axis
Sunday	29
Monday	32
Tuesday	40
Wednesday	33
Thursday	44
Friday	49
Saturday	45

2. *Open* **CH16EX2** drawing. Activate **grips** on the *Line* to the far right. Make the top grip **hot**. Use the **STRETCH** option to stretch the top grip to the right to create a vertical *Line*. **Cancel** the grips.

FIGURE 19-17

Next, activate the **grips** for all the other *Lines* (not including the vertical one on the right); as shown in Figure 19-17. **STRETCH** the top of all inclined *Lines* to the top of the vertical *Line* by making the common top grips **hot**, then *OSNAPing* to the top **ENDpoint** of the vertical *Line*. *Save* the drawing as **CH19EX2**.

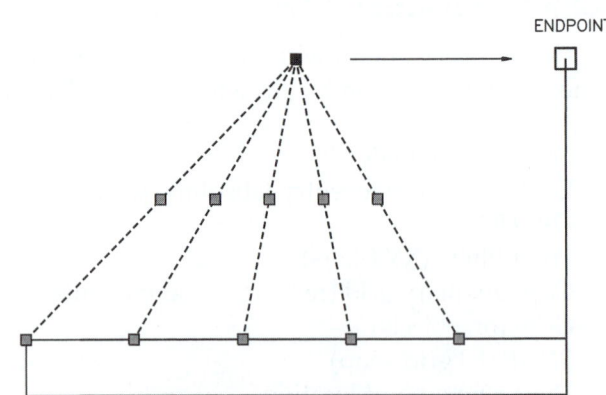
ENDPOINT

3. This exercise involves all the options of grip editing to create a Space Plate. Begin a *New* drawing or use the **ASHEET** *template* and use *SaveAs* to assign the name **SPACEPLT**.

A. Set the *Snap* value to **.125**. Set **Polar Snap** to **.125** and turn on **Polar Tracking**. Draw the geometry shown in Figure 19-18 using the *Line* and *Circle* commands.

FIGURE 19-18

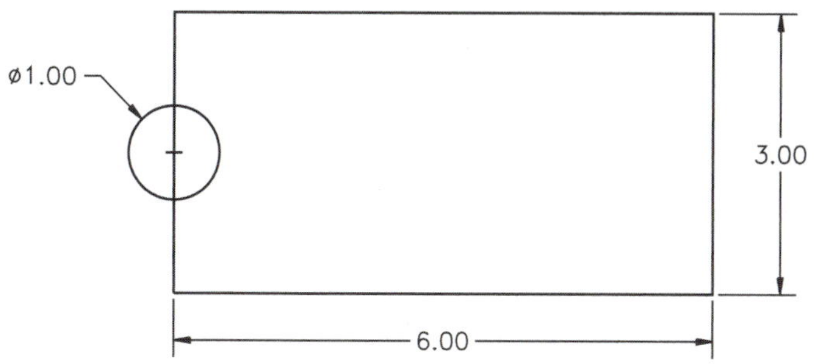

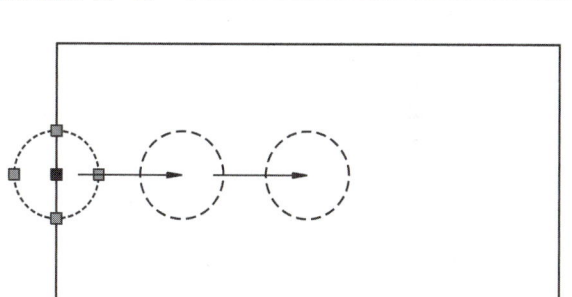

B. Activate the **grips** on the *Circle*. Make the center grip **hot**. Cycle to the **MOVE** option. Enter *C* for the **Copy** option. You should then get the prompt for **MOVE (multiple)**. Make two copies as shown in Figure 19-19. The new *Circles* should be spaced evenly, with the one at the far right in the center of the rectangle. If the spacing is off, use **grips** with the **STRETCH** option to make the correction.

FIGURE 19-19

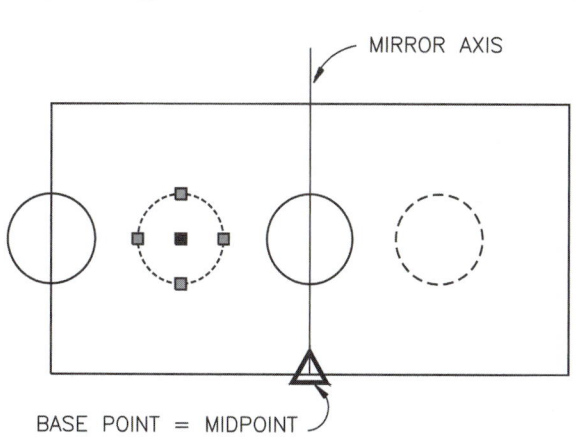

C. Activate the **grips** on the number 2 *Circle* (from the left). Make the center *Circle* grip **hot** and cycle to the **MIRROR** option. Enter *C* for the **Copy** option. (You'll see the **MIRROR (multiple)** prompt.) Then enter *B* to specify a new base point as indicated in Figure 19-20. Press Shift to turn *On ORTHO* and specify the mirror axis as shown to create the new *Circle* (shown in Fig. 19-20 in hidden linetype).

FIGURE 19-20

MIRROR AXIS

BASE POINT = MIDPOINT

D. Use *Trim* to trim away the outer half of the *Circle* and the interior portion of the vertical *Line* on the left side of the Space Plate. Activate the **grips** on the two vertical *Lines* and the new *Arc*. Make the common grip **hot** (on the *Arc* and *Line* as shown in Fig. 19-21) and **STRETCH** it downward .5 units. (Note how you can affect multiple objects by selecting a common grip.) **STRETCH** the upper end of the *Arc* upward .5 units.

FIGURE 19-21

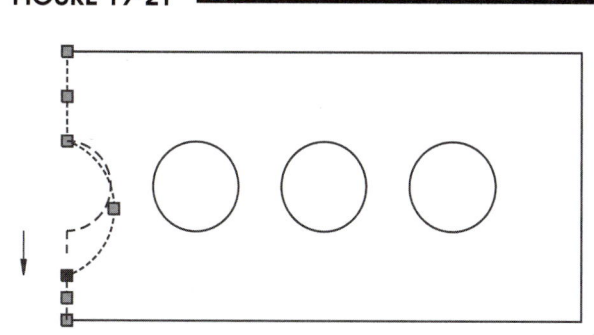

E. *Erase* the *Line* on the right side of the Space Plate. Use the same method that you used in step C to **MIRROR** the *Lines* and *Arc* to the right side of the plate (as shown in Fig. 19-22).

(REMINDER: After you select the **hot** grip, use the **Copy** option <u>and</u> the **Base** point option. Use Shift to turn on *ORTHO*.)

FIGURE 19-22

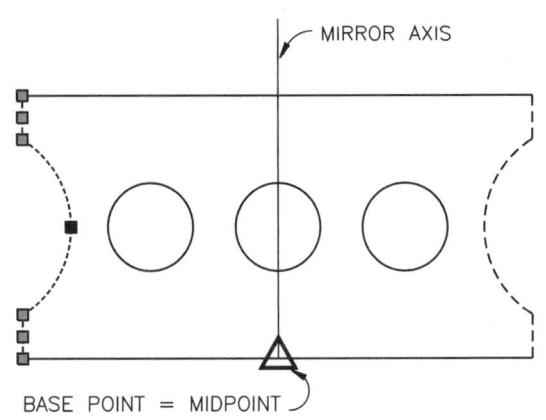

F. In this step, you will **STRETCH** the top edge upward one unit and the bottom edge downward one unit by selecting <u>multiple</u> **hot** grips.

Select the desired horizontal <u>and</u> attached vertical *Lines*. Hold down Shift while selecting <u>each</u> of the endpoint grips, as shown in Figure 19-23. Although they appear **hot**, you must select one of the two again to activate the **STRETCH** option.

FIGURE 19-23

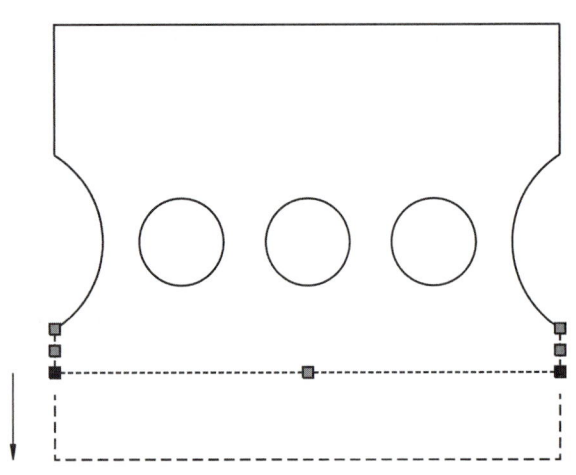

G. In this step, two more *Circles* are created by the **ROTATE** option (see Fig. 19-24). Select the two outside *Circles* to make the grips appear. PICK the center grip to make it **hot**. Cycle to the **ROTATE** option. Enter *C* for the **Copy** option. Next, enter *B* for the **Base** option and *OSNAP* to the *CENter* as shown. Press Shift to turn on *ORTHO* and create the new *Circles*.

FIGURE 19-24

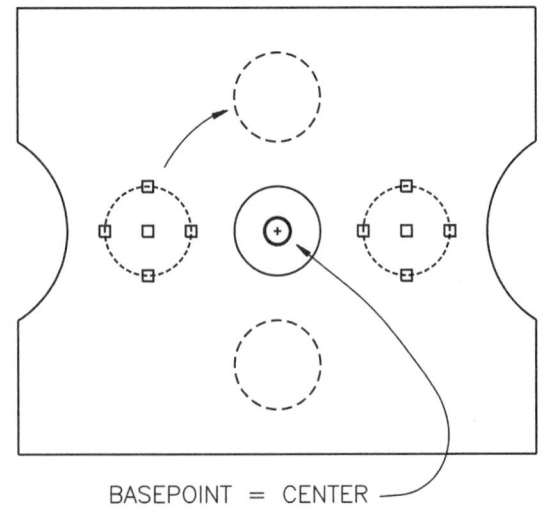

H. Select the center *Circle* and make the
    center grip **hot** (Fig. 19-25). Cycle to the
    SCALE option. The scale factor is **1.5**.
    Since the *Circle* is a 1 unit diameter, it can be
    interactively scaled (watch the *Coords*
    display), or you can enter the value. The
    drawing is complete. *Save* the drawing as
    SPACEPLT.

FIGURE 19-25

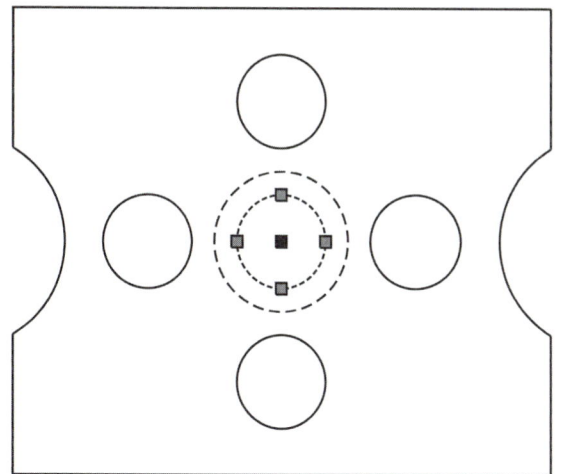

4. *Open* the **STORE ROOM2** drawing from Chapter 18 Exercises. Use the grips to **STRETCH**,
    MOVE, or **ROTATE** each text paragraph to achieve the results shown in Figure 19-26. *SaveAs*
    STORE ROOM3.

FIGURE 19-26

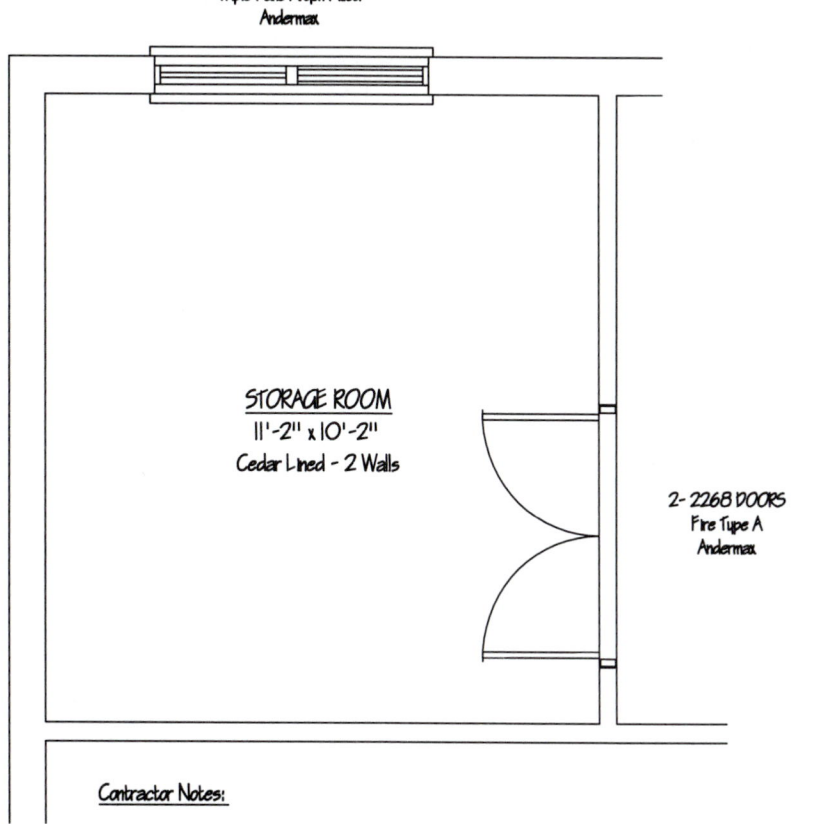

5.   *Open* the **T-PLATE** drawing that you created in Chapter 9. An order has arrived for a modified version of the part. The new plate requires two new holes along the top, a 1″ increase in the height, and a .5″ increase from the vertical center to the hole on the left (Fig. 19-27). Use grips to **STRETCH** and **Copy** the necessary components of the existing part. Use *SaveAs* to rename the part to **TPLATEB**.

FIGURE 19-27

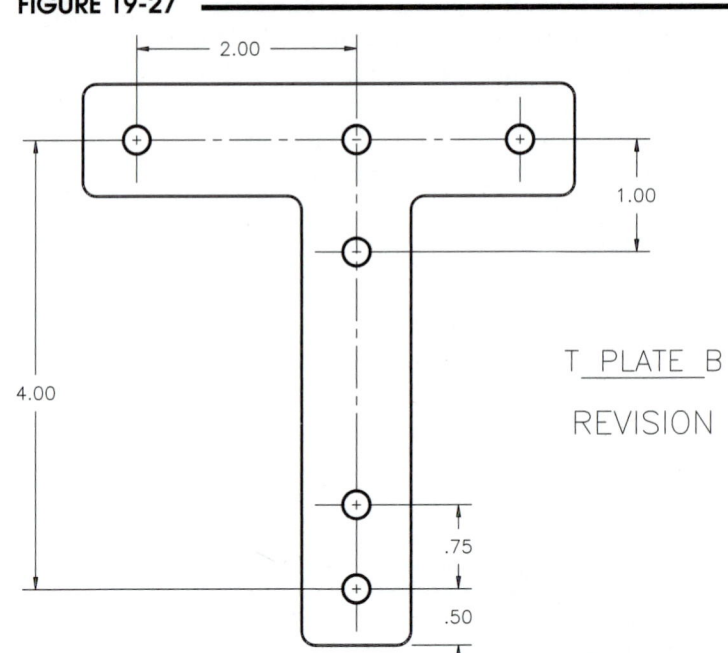

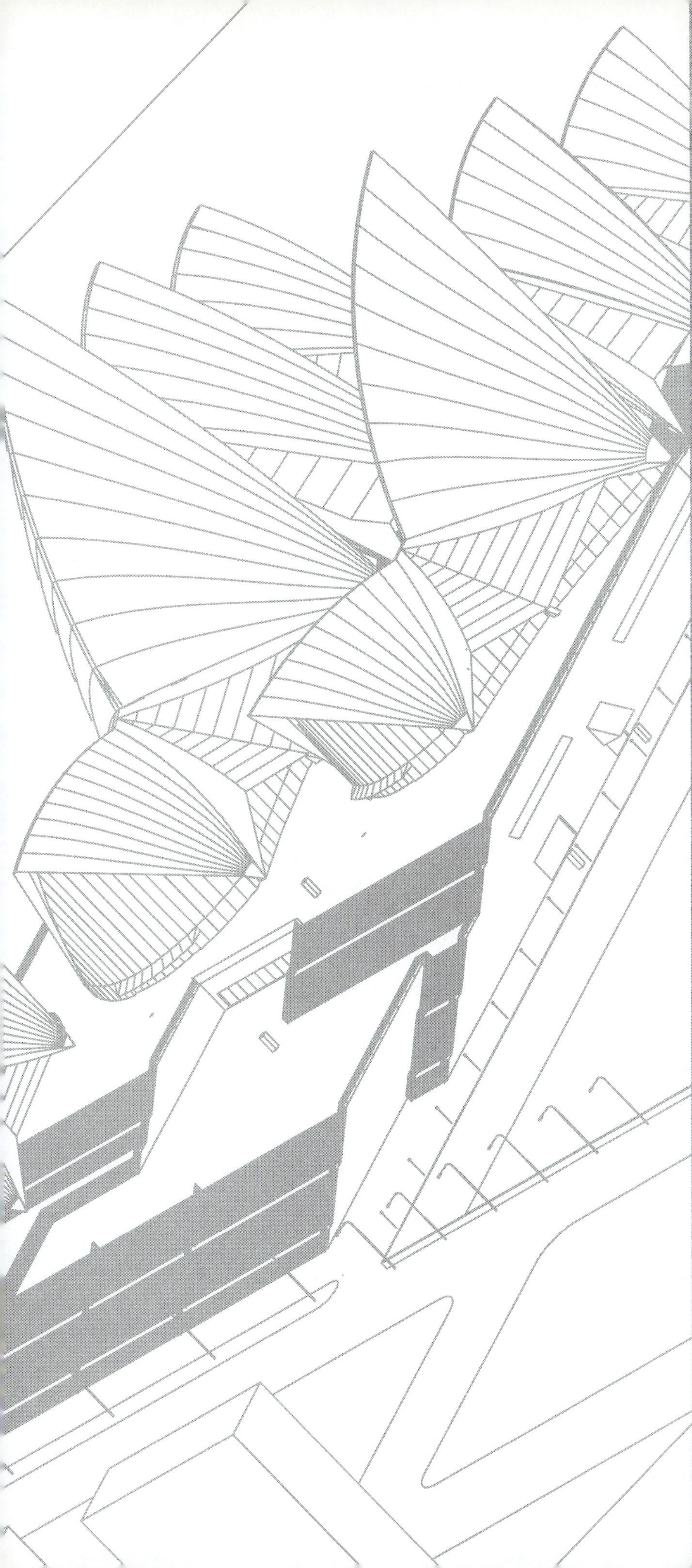

20

ADVANCED SELECTION SETS

CHAPTER OBJECTIVES

After completing this chapter you should:

1. be able to control the way objects are selected using the following dialog box options (or variables): *Noun/Verb Selection* (*PICKFIRST*), *Use Shift to Add to Selection* (*PICKADD*), *Implied Windowing* (*PICKAUTO*), and *Press and Drag* (*PICKDRAG*);

2. be able to use *Qselect* to specify criteria for creating a selection set to use with an editing command;

3. be able to use *Object Selection Filters* in a complex drawing to find a selection set for use with an editing command;

4. be able to create named selection sets (Object Groups) with the *Group* command and use options in the *Object Grouping* dialog box.

CONCEPTS

When you "Select Objects:", you determine which objects in the drawing are affected by the subsequent editing action by specifying a <u>selection set</u>. You can select the set of objects in several ways. The fundamental methods of object selection (such as PICKing with the pickbox, a *Window*, a *Crossing Window*, etc.) are explained in Chapter 4, Selection Sets. This chapter deals with advanced methods of specifying selection sets and the variables that control your preferences for how selection methods operate. Specifically, this chapter explains:

> Selection Set Variables
> Object Filters
> Object Groups

SELECTION SET VARIABLES

Four variables allow you to customize the way you select objects. Keep in mind that selecting objects occurs <u>only for editing</u> commands; therefore, the variables discussed in this chapter affect how you select objects when you <u>edit</u> AutoCAD objects. The variable names and the related action are briefly explained here:

Variable Name	Default Setting	Related Action
PICKFIRST	1	Enables and disables Noun/Verb command syntax. Noun/Verb means *PICK* objects (noun) *FIRST* and then use the edit command (verb).
PICKADD	1	Controls whether objects are *ADD*ed to the selection set when *PICK*ed or replace the selection set when picked. Also controls whether the Shift key + #1 button combination removes or adds selected objects to the selection set.
PICKAUTO	1	Enables or disables the *PICK*box *AUTO*matic window/crossing window feature for object selection.
PICKDRAG	0	Enables or disables single *PICK* window *DRAG*ging. When *PICKDRAG* is set to 1, you start the window by pressing the *PICK* button, then draw the window by holding the button down and *DRAG*ging to specify the diagonal corner, and close the window by releasing the button. In other words, windowing is done with one PICK and one release rather than with two PICKs.

Like many system variables, the selection set variables listed above hold an integer value of either 1 or 0. The 1 designates a setting of *ON* and 0 designates a setting of *OFF*. System variables that hold an integer value of either 1 or 0 "toggle" a feature *ON* or *OFF*.

Changing the Settings

Settings for the selection set variables can be changed by any of two ways:

1. You can type the variable name at the Command: prompt (just like an AutoCAD command) to change the setting.

2. Use the *Selection* tab of the *Options* dialog box (Fig. 20-1) to change the four selection set variables listed above. The four variables can be changed in this dialog box, but the syntax in the dialog box is <u>not</u> the same as the variable name. A check in the checkbox by each choice does <u>not</u> necessarily mean a setting of *ON* for the related variable.

When you change any of these four variables, the setting is recorded for the user profile, rather than in the current drawing file as with most variable settings. In this way, the change (which is generally the personal preference of the operator) is established at the workstation, not the drawing file.

Options

Pull-down Menu	Command (Type)	Alias (Type)	Short-cut	Screen (side) Menu	Tablet Menu
Tools Options...	Options	OP	(Default menu) Options...	TOOLS2 Options	Y,10

The *Selection* tab of the *Options* dialog box (Fig. 20-1) can be used to change the four selection set variables listed previously, namely, *PICKFIRST*, *PICKADD*, *PICKAUTO*, and *PICKDRAG*. Although the checkboxes on the left side of the *Selection* tab do not indicate the formal variable names, they are intended to be more descriptive.

The checkbox names and the respective variables are listed here:

Noun/Verb Selection	*PICKFIRST*
Use Shift to Add to Selection	*PICKADD*
Press and Drag	*PICKDRAG*
Implied Windowing	*PICKAUTO*

These selection set variables are explained in the sections next; each section is designated by the formal variable name.

FIGURE 20-1

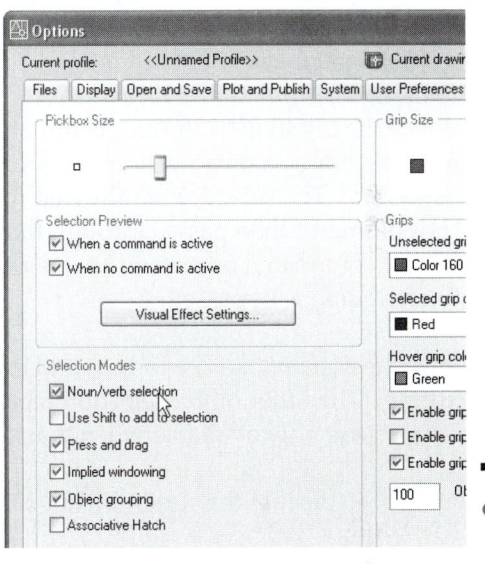

NOTE: It is advisable to turn off Grips when you are experimenting with these variables to avoid confusion. Turn off Grips by removing the check from *Enable Grips* on the right side of the dialog box or by typing *GRIPS* at the command prompt and changing the setting to 0 (off).

PICKFIRST

NOTE: If the *GRIPS* variable is set to 1, object grips are enabled. Object grips do not hinder your ability to use *PICKFIRST*, but they may distract your attention from the current topic. Setting the *GRIPS* variable to 0 disables object grips. You can type *GRIPS* at the command prompt to change the setting. See Chapter 19 for a full discussion of object grips.

Changing the setting of *PICKFIRST* is easily accomplished in command line mode (typing). However, the *Selection* tab of the *Options* dialog box can also be used to toggle the setting. The choice is titled *Noun/Verb Selection*, and a check appearing in the box means that *PICKFIRST* is set to 1 (*ON*).

The *PICKFIRST* system variable enables or disables the ability to select objects <u>before</u> using a command. If *PICKFIRST* is set to 1, or *ON*, you can select objects at the Command: prompt <u>before</u> a command is used. *PICKFIRST* means that you PICK the objects FIRST and then invoke the desired command. *PICKFIRST* set to 1 makes the small pickbox appear at the cursor when you are not using a command (at the open Command: prompt). *PICKFIRST* set to 1 enables you to select objects with the pickbox, *AUto* window, or *Crossing Window* methods when a command is not in use.

This order of editing is called "Noun/Verb"; the objects are the nouns and the command is the verb. Noun/Verb editing is preferred by some users because you can decide what objects need to be changed, then decide how (what command) you want to change them. Noun/Verb editing allows AutoCAD to operate like some other CAD systems; that is, the objects are PICKed FIRST and then the command is chosen.

The command syntax for Noun/Verb editing is given next using the *Move* command as an example:

> Command: **PICK** (Use the cursor pickbox or auto window/crossing window to select objects.)
> Command: **PICK** (Continue selecting desired objects.)
> Command: **Move** (Enter the desired command and AutoCAD responds with the number of objects selected.) 2 found
> Specify base point or displacement: **PICK** or (**coordinates**)
> Specify second point of displacement, or <use first point as displacement>: **PICK** or (**coordinates**)
> Command:

Notice that as soon as the edit command is invoked, AutoCAD reports the number of objects found and uses these as the selection set to act on. You do not get a chance to "Select objects:" within the command. The selection set PICKed immediately before the command is used for the editing action. The command then passes through the "Select objects:" step to the next prompt in the sequence. All editing commands operate the same as with Verb/Noun syntax order with the exception that the "Select objects:" step is bypassed.

 Only the pickbox, auto window, and crossing window can be used for object selection with Noun/Verb editing. The other object selection methods (*ALL, Last, Previous, Fence, Window Polygon*, and *Crossing Polygon*) are only available when you are presented with the "Select objects:" prompt.

NOTE: To disable the cursor pickbox, *PICKFIRST* and *GRIPS* must be OFF.

NOTE: When *PICKFIRST* is set to 0, the Edit-mode shortcut menus do not appear (see "Shortcut Menus," Chapter 1).

PICKADD

The *PICKADD* variable controls whether objects are ADDed to the selection set when they are PICKed or whether selected objects replace the last selection set. This variable is *ON* (set to 1) by default. Most AutoCAD operators work in this mode.

Until you reached this chapter, it is probable that all PICKing you did was with *PICKADD* set to 1. In other words, every time you selected an object it was added to the selection set. In this way, the selection set is cumulative; that is, each object PICKed is added to the current set. This mode also allows you to use multiple selection methods to build the set. You can PICK with the pickbox, then with a window, then with any other method to continue selecting objects. The "Select objects:" process can only be ended by pressing Enter.

With the default option (when *PICKADD* is set to 1 or *ON*), the Shift+#1 key combination allows you to deselect, or remove, objects from the current selection set. This has the same result as using the *Remove* option. Deselecting is helpful if you accidentally select objects or if it is easier in some situations to select *ALL* and then deselect (Shift+#1) a few objects.

When the *PICKADD* variable is set to 0 (*OFF*), objects that you select replace the last selection set. Let's say you select five objects and they become highlighted. If you then select two other objects with a window, they would become highlighted and the other five objects automatically become deselected and unhighlighted.

The two new objects would replace the last five to define the new selection set. Figure 20-2 illustrates a similar scenario.

The *PICKADD* variable also controls whether the Shift+#1 button combination removes from or adds to the selection set. When *PICKADD* is set to 1 (*ON*), the Shift+#1 combination deselects (removes) objects from the selection set. When *PICKADD* is set to 0 (*OFF*), the Shift+#1 combination toggles objects in or out of the selection set, depending on the object's current state. In other words (when *PICKADD* is *OFF*), if the object is included in the set (highlighted), Shift+#1 deselects (unhighlights) it, or if the object is not in the selection set, Shift+#1 adds it.

FIGURE 20-2

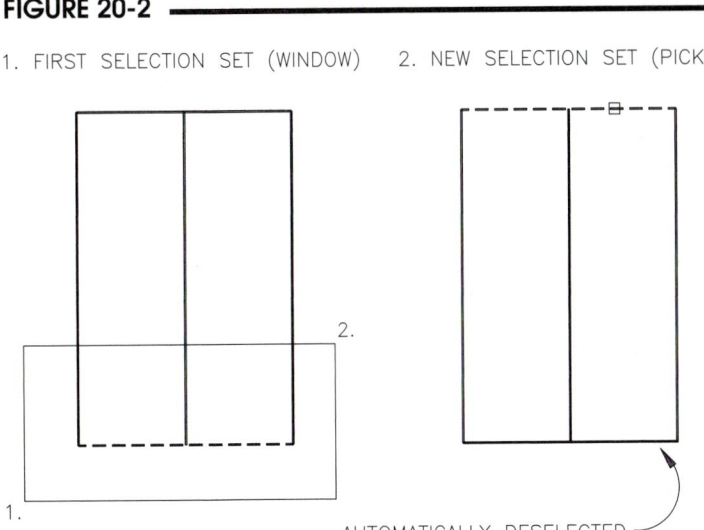

1. FIRST SELECTION SET (WINDOW)   2. NEW SELECTION SET (PICK)

AUTOMATICALLY DESELECTED

You can change the variable by typing in *PICKADD* at the Command prompt. Changing the *PICKADD* variable in the *Options* dialog box is accomplished by making the desired choice in the checkbox. Changing the setting in this way is confusing because a check by *Use Shift to Add* means that *PICKADD* is *OFF*! Normally, a check means the related variable is *ON*. To avoid confusion, use only one method (either typing or dialog box) to change the setting until you are familiar with it.

You can also change the *PICKADD* variable using the *Toggle value of PICKADD Sysvar* button in the upper-right corner of the *Properties* palette. The symbol on this button changes (" + " or "1") to reflect the current setting for *PICKADD*; however, the symbols are confusing since changing *PICKADD* to 0 (*OFF*) changes the button to display a "1."

It may occur to you that you cannot imagine a practical application for using *PICKADD OFF*. It does make sense to operate AutoCAD for most applications with *PICKADD ON*. However, it can be useful to turn *PICKADD OFF* when using the *Properties* palette (so selecting one set of objects replaces the previous set in the *Properties* palette). For more information on *Properties*, see Chapter 16, Modify Commands II.

PICKAUTO

The *PICKAUTO* variable controls automatic windowing when the "Select objects:" prompt appears. Automatic windowing (*Implied Windowing*) is the feature that starts the first corner of a window or crossing window when you PICK in an open area. Once the first corner is established, if the cursor is moved to the right, a window is created, and if the cursor is moved to the left, a crossing window is started. See Chapter 4 for details of this feature.

When *PICKAUTO* is set to 1 (or *Implied Windowing* is checked in the dialog box, Fig. 20-1), the automatic window/crossing is available whenever you select objects. When *PICKAUTO* is set to 0 (or no check appears by *Implied Windowing*), the automatic windowing feature is disabled. However, the *PICKAUTO* variable is overridden if *GRIPS* are *ON* or if the *PICKFIRST* variable is *ON*. In either of these cases, the <u>cursor</u> pickbox and auto windowing are enabled so that objects can be selected at the open Command: prompt.

Auto windowing is a helpful feature and can be used to increase your drawing efficiency. The default setting of *PICKAUTO* (*ON*) is the typical setting for AutoCAD users.

PICKDRAG

This variable controls the method of drawing a selection window. *PICKDRAG* set to 1 or *ON* allows you to draw the window or crossing window by clicking at the first corner, holding down the button, dragging to the other corner, and releasing the button at the other corner (Fig. 20-3). With *PICKDRAG ON* you can specify diagonal corners of the window with one press and one release rather than with two clicks. GUIs (graphical user interfaces) of some other software packages use this method of mouse control.

FIGURE 20-3

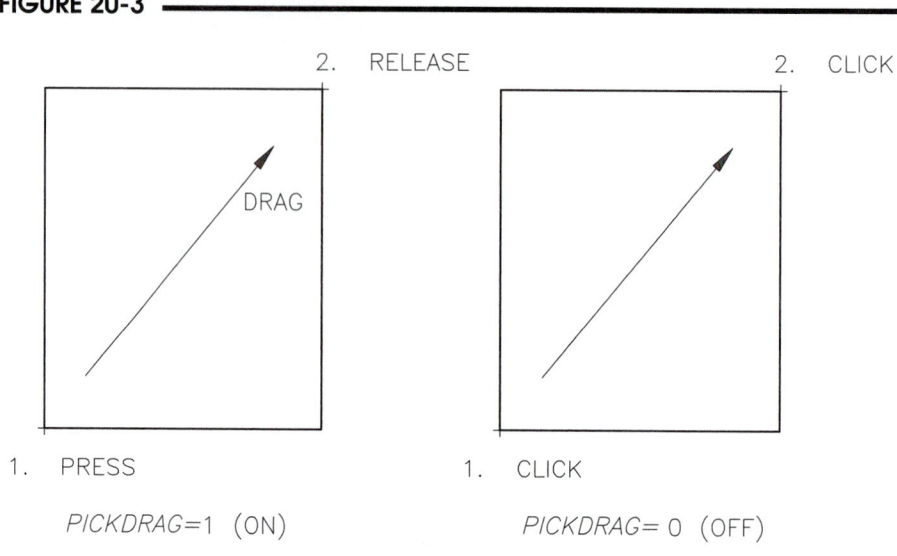

The default setting of *PICKDRAG* is 0 or *OFF*. This method allows you to create a window by clicking at one corner and again at the other corner. Most AutoCAD users are accustomed to this style, which accounts for the default setting of *PICKDRAG* to 0 (*OFF*).

OBJECT SELECTION FILTERS

Large and complex AutoCAD drawings may require advanced methods of specifying a selection set. For example, consider working with a drawing of a manufacturing plant layout and having to select all of the doors on all floors, or having to select all metal washers of less than 1/2″ diameter in a complex mechanical assembly drawing. Rather than spend several minutes PICKing objects, it may be more efficient to use one of the object filters that AutoCAD provides.

There are two selection filters provided in AutoCAD: *Qselect* (*Quick Select* dialog box) and *Filter* (*Object Selection Filter* dialog box). *Qselect* is simpler to use than *Filter*.

Both object selection filters allow you to specify a criteria based on object type and/or object property. In other words, you specify what type of object (*Line, Arc, Mtext,* etc.) and/or what property (*Layer, Color, Linetype,* etc.) and AutoCAD searches (applies the filter to) the drawing to find objects that match the criteria. When the objects are found, they can be used for some editing action, such as *Move, Copy, Erase,* or change properties.

The *Quick Select* dialog can be invoked by typing *Qselect* or is accessible through other dialog boxes, such as the *Properties* palette. *Qselect* should be used immediately before an editing command to locate the desired objects you want to edit. *Qselect* is <u>not</u> transparent (cannot be used during another command, except when it is accessible from other dialog boxes).

The *Object Selection Filter* dialog box can be invoked by typing *Filter*. *Filter* is more advanced than *Qselect* and can be used to specify a more complex filter criteria. Unlike *Qselect*, *Filter* is transparent, so it can be used during another command by typing *'Filter* (with an apostrophe prefix).

Qselect

	Pull-down Menu	Command (Type)	Alias (Type)	Short-cut	Screen (side) Menu	Tablet Menu
	Tools Quick Select...	Qselect	...	(Default Menu) Quick Select...	...	X,9

With *Qselect*, you create a selection set of all objects that either match or do not match the object type and object property criteria that you specify. You can search the entire drawing or an existing selection set. If you use *Qselect* with a partially opened drawing, objects that are not loaded are not considered in the search. When the *OK* button is pressed in the *Quick Select* dialog box, AutoCAD finds and highlights the objects matching the criteria. The objects become the current selection set so any editing command used immediately after *Qselect* (by Noun/Verb selection) automatically finds the matching selection set.

Qselect is the simpler of the two available object selection filters. Although you cannot name and save selection sets with *Qselect*, you can search and find objects by *Object type* and *Properties* much easier than using the *Object Selection Filters* dialog box (explained in detail in the following section).

Qselect can be invoked by several methods listed in the command table above, including choosing *Quick Select* from the shortcut menu that appears when you right-click in the drawing area. The *Qselect* button also appears in the *Properties* palette (see Fig. 16-7). *Qselect* produces the *Quick Select* dialog box (Fig. 20-4).

To use the *Quick Select* dialog box, follow these steps:

1. Invoke the *Quick Select* dialog box by any method described above.

2. Specify the *Entire Drawing* to search or use the *Select Object* button to specify a smaller search area.

3. If you want to find one *Object Type*, select it from the drop-down list; otherwise *Multiple* will be used.

4. Select the desired *Object Property* (only one) to search for (unless you need only to search by *Object Type*).

5. Specify the *Operator* and *Value*.

6. Specify if you want to *Include in New Selection Set* (search for the object type and property you specified) or *Exclude from New Selection Set* (search for everything but the object type and property you specified).

7. Press *OK*. AutoCAD highlights all objects that match the criteria.

8. Use an editing command to act on the highlighted objects (*Move*, *Erase*, change properties, etc.)

FIGURE 20-4 ─────────────

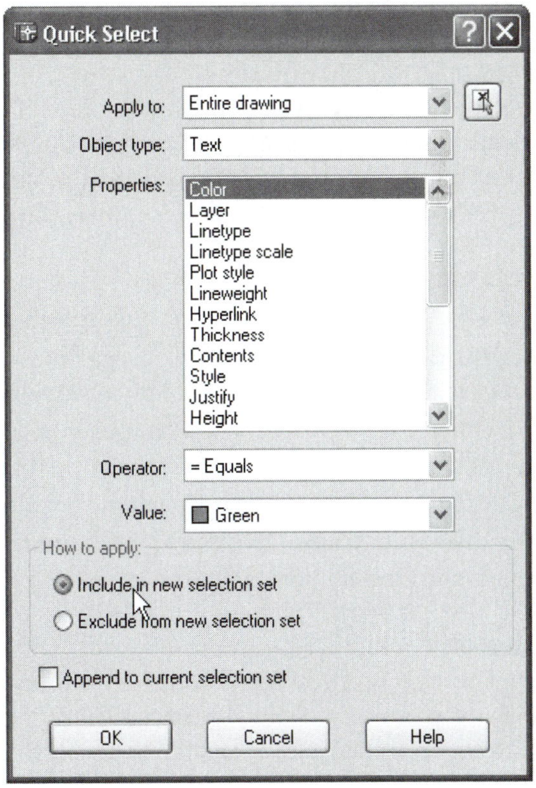

For example, assume the criteria shown in Figure 20-4 were applied to a process flow diagram, AutoCAD would find and highlight all *Text* objects in the drawing whose *Color Equals Green* (Fig. 20-5). Note the Grips (small squares) that appear on the highlighted objects when Grips are enabled.

FIGURE 20-5

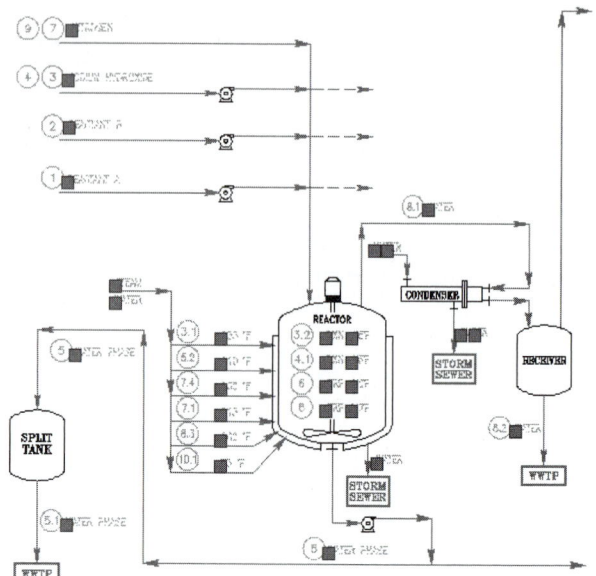

The specific areas of the dialog box are explained as follows.

Apply to
Specify whether you want to apply the filtering criteria to the entire drawing or the current selection set. If there is no current selection set when you invoke *Qselect*, *Entire Drawing* is the default value. Use the *Select Objects* button to return to the drawing and create a selection set.

Select Objects
The *Select Objects* button (upper right corner) temporarily returns to the drawing so you can select the objects on which you want to perform the *Quick Select* operation. AutoCAD uses the new selection set as the area to search for objects matching the specified criteria.

Object Type
Specifies the object type you want to search for (filter). The default is *Multiple*; therefore, if no previous selection set exists, the *Object Type* list includes all and only object types available in the <u>current drawing</u>. For example, if only *Lines* and *Circles* have been created in the drawing, only *Line* and *Circle* object types appear in the list. If a selection set exists, the list includes only the object types of the selected objects.

Properties
This field is specific to the *Object Type* selected in the list above. The *Properties* list <u>includes all searchable properties for the selected object type</u>. For some object types, the list of properties can be very long; for example, dimension objects contain a property field for each dimensioning variable. This feature gives you considerable power. For example, you could use this tool to highlight all *Linear Dimensions* that have a specific type of *Tolerance Display* or find all *Text* that matches a specific *Height* value. The sort order for the properties list (alphabetic or categorized) is based on the current sort order in the *Object Properties* palette (see "*Properties*" in Chapter 16). The property you select determines the options available in the *Operator* and *Value* boxes.

Operator
The *Operator* that you select controls if the search *Equals*, is *Not Equal To*, *Greater Than*, or *Less Than* the text string or numerical value in the *Value* field. Options in the *Operator* field vary depending on the *Property*. For example, *Greater Than* and *Less Than* are not available for some properties.

Value
Possibilities in this field depend on the selected object and property. If known values for the selected property are stored in the drawing, the *Value* box becomes a drop-down list from which you can choose an available value. For example, if *Color* is selected in the *Properties* list, *ByLayer*, *ByBlock*, *Red*, *Yellow*, and so on appear in the list. If no choices appear in the *Value* field, input a text string or numerical value.

How To Apply

Your choice here specifies whether you want the new selection set to include or exclude objects that match the specified filtering criteria. Choose *Include in new selection set* if you want to create a new selection set composed only of objects that <u>match</u> the filtering criteria. Choose *Exclude from new selection set* to create a new selection set composed of all objects that <u>do not match</u> the filtering criteria.

OK

After specifying the criteria for the objects to search in the fields described above, select *OK*. All objects matching (or not matching) the criteria set in the *Quick Select* dialog box become highlighted in the current drawing. These objects are treated as if you had selected them by the pickbox, window, or other selection method. Assuming Noun/Verb selection is enabled (*PICKFIRST=1*), the next editing command issued automatically finds the *Qselect* objects.

Append to Current Selection Set

Although you can only select one specific object type and one property at a time, you can use *Qselect* <u>repeatedly</u> to search for multiple object types and properties. Do this by specifying one set of criteria, then select *OK*. AutoCAD highlights all objects matching that criteria. Next, <u>while the objects are highlighted</u>, right-click to select *Qselect* again. Specify a second set of criteria and include *Append to current selection set*. When you press *OK*, AutoCAD highlights objects matching the new criteria and "adds" them to the original highlighted set.

Filter

Pull-down Menu	Command (Type)	Alias (Type)	Short-cut	Screen (side) Menu	Tablet Menu
...	*Filter*	*FI*	...	*ASSIST* *Filters*	...

The *Object Selection Filters* dialog box is a more advanced version of *Qselect*. Here you can specify multiple selection criteria and save a criteria set under a name that you assign for use at a later time. You can also invoke *Filter* during a command (transparently) to select objects for use with the current command. Using the *Filter* command by any method produces the *Object Selection Filters* dialog box. Figure 20-6 displays the dialog box when it first appears. The *Select Filter* cluster near the left center is where you specify the search criteria. After you designate the filters and values, use *Add to List* to cause your choices to appear in the large area at the top of the dialog box. All filters appearing at the top of the box are applied to the drawing when *Apply* is selected.

FIGURE 20-6 ⸺⸺⸺⸺

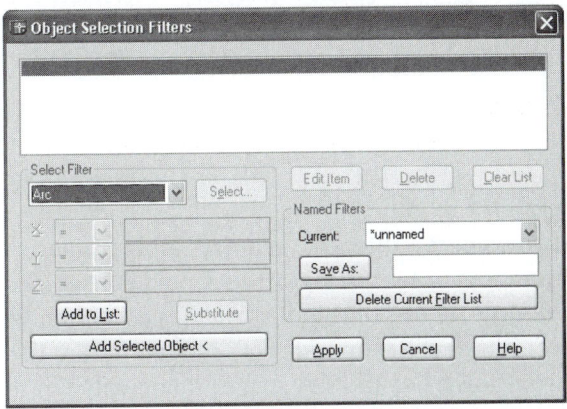

Select Filter (Drop-Down List)

Activating the drop-down list reveals the possible selection criteria (filters) that you can use, such as *Arc* or *Layer*. You can select one or more of the selection filters shown in the following list. You can also specify a set of values that applies to each filter, for example, *Arcs* with a *radius* of less than 1.00" or *Layers* that begin with "AR."

Arc	Attribute Position	Block Position
Arc Center	Attribute Tag	Block Rotation
Arc Radius	Block	Body
Attribute	Block Name	Circle

Circle Center	Line End	Solid Body
Circle Radius	Line Start	Spline
Color	Linetype	Text
Dimension	Linetype Scale	Text Height
Dimension Style	Multiline	Text Position
DWF Underlay	Multiline Style	Text Rotation
Elevation	Normal Vector	Text Style Name
Ellipse	Point	Text Value
Ellipse Center	Point Position	Thickness
Hatch	Polyline	Tolerance
Hatch Pattern Name	Ray	Trace
Helix	Region	Viewport
Image	Section Plane	Viewport Center
Image Position	Shape	Xdata
Layer	Shape Name	Xline
Leader	Shape Position	3dface
Line	Solid	

The list also provides a series of typical database grouping operators which include AND, OR, XOR, and NOT. For example, you may want to select all *Text* in a drawing but *NOT* the *Text* in a specific text style.

Select

If you select a filter that has multiple values, such as *Layer, Block Name,* and *Text Style Name,* the *Select* button is activated and enables you to select a value from the list (Fig. 20-7). If the *Select* button is grayed out, no existing values are available.

FIGURE 20-7

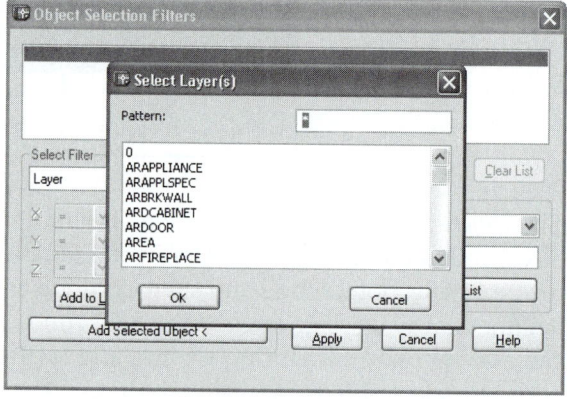

X

If you select a filter that requires alphanumeric input (numbers or text strings), the *X* field is used. For example, if you selected *Arc Radius* as a filter, you would select the operator (=, <, >, etc.) from the first drop-down list and key in a numeric value in the edit box (Fig. 20-8). Other examples requiring numeric input are *Circle Radius, Text Rotation,* and *Text Height.*

If you select a filter that requires a text string, the string is entered in the *X* field edit box. Examples are *Attribute Tag* and *Text Value.*

The *X* field is also used (in conjunction with the *Y* and *Z* fields) for filters that require coordinate input in an *X,Y,Z* format. Example filters include *Arc Center, Block Position,* and *Line Start.*

FIGURE 20-8

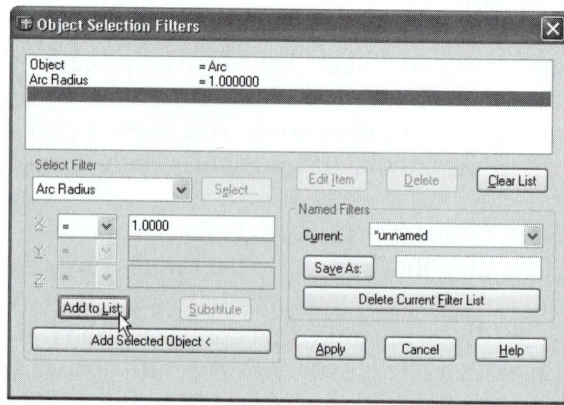

Y,Z

These fields are used for coordinate entry of the Y and Z coordinates when needed as one of the criteria for the filter (see "X" on the previous page). Additionally, you may use standard database relational operators such as equal, less than, or greater than, to more effectively specify the desired value.

Add to List

Use the *Add to List* button after selecting a filter to force the new selection to appear in the list at the top of the dialog box.

Substitute

After choosing a value from the *Select* listing or entering values for X, Y, and Z, *Substitute* replaces the value of the <u>highlighted</u> item in the list (at the top of the dialog box) with the new choice or value.

Add Selected Object

Use this button to select an object from the drawing that you want to include in the filter list.

Edit Item

Once an item is added to the list, you must use *Edit Item* to change it. First, highlight the item in the filter list (on top); then select *Edit Item*. The filter and values then appear in the *Select Filter* cluster ready for you to specify new values in the X,Y,Z fields or you can select a new filter by using the *Select* button.

Delete

This button simply deletes the highlighted item in the filter list.

Clear List

Use *Clear List* to delete all filters in the existing filter list.

Named Filters

The *Named Filters* cluster contains features that enable you to save the current filter configuration to a file for future use with the current drawing. This eliminates the need to rebuild a selection filter used previously.

Current

By default, the current filter list of criteria is *unnamed*, in much the same fashion as the default dimension style. Once a filter list has been saved by name, it appears as a selectable object filter configuration in the drop-down list of named filters.

Save As

Once object filter(s) have been specified and added to the list (on top), you can save the list by name. First, enter a name for the list; then PICK the *Save As* button. The list is automatically set as the current object filter configuration and can be recalled in the future for use in the drawing.

Delete Current Filter List

Deletes the named filter configuration displayed in the *Current* field.

Apply

The *Apply* button applies the current filter list to the current AutoCAD drawing. Using *Apply* causes the dialog box to disappear and the "Select objects:" prompt to appear. <u>You must specify a selection set (by any method) to be tested against the filter criteria</u>. If you want the filters to apply to the entire drawing (except *Locked* and *Frozen* layers), enter **ALL**. If you only need to search a smaller area of the drawing, PICK objects or use a window or other selection method.

The same procedure is used to combine several objects into a group for each furniture item. The complete suite of groups may appear as shown in the dialog box shown in Figure 20-17. Toggling on *Include Unnamed* forces the unnamed group to appear in the list. The unnamed group was automatically created when the CHAIR group was copied.

Using the named groups, completing the layout is simplified. Each group (CHAIR, DESK, etc.) is selected and *Copied* to complete the room layout.

FIGURE 20-17

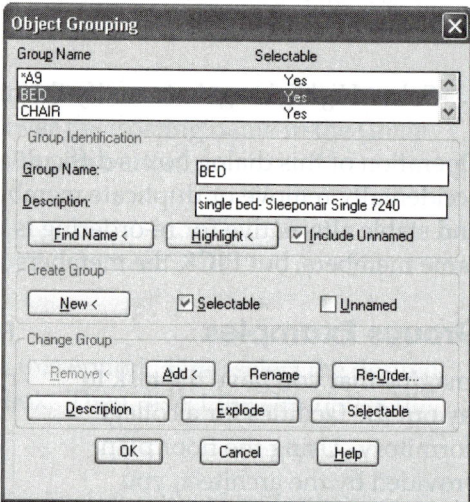

The next step in the drawing is to lay out several other rooms. Assuming that other rooms will have the same arrangement of furniture, another group could be made of the entire room layout. (Objects can be members of more than one group.) A group called ROOM is created in the same manner as before, and all objects that are part of furniture items are selected to be members of the group. If the ROOM group is made selectable, all items can be highlighted with one PICK and copied to the other rooms (Fig. 20-18).

FIGURE 20-18

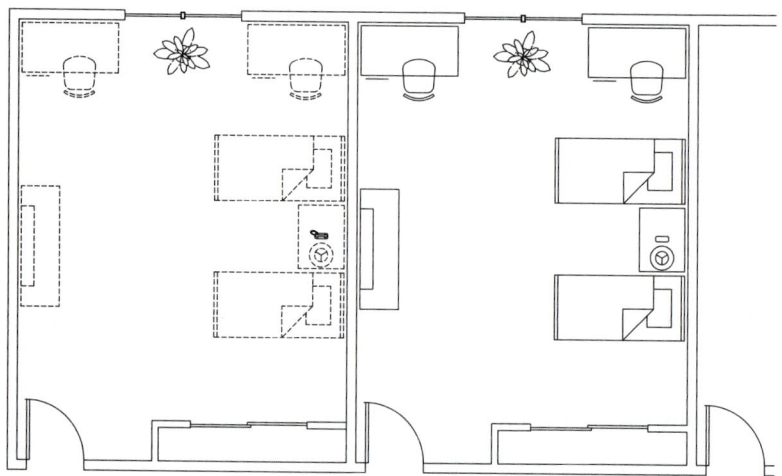

After submitting the layout to the client for review, a request is made for you to present alternative room layouts. Going back to the drawing, you discover that each time you select an individual furniture item to manipulate, the entire ROOM group is highlighted. To remedy the situation, the ROOM group is changed to nonselectable with the *GROUP* command. Now the individual furniture items can be selected since the original groups (CHAIR, DESK, BED, etc.) are still *Selectable* (Fig. 20-19).

FIGURE 20-19

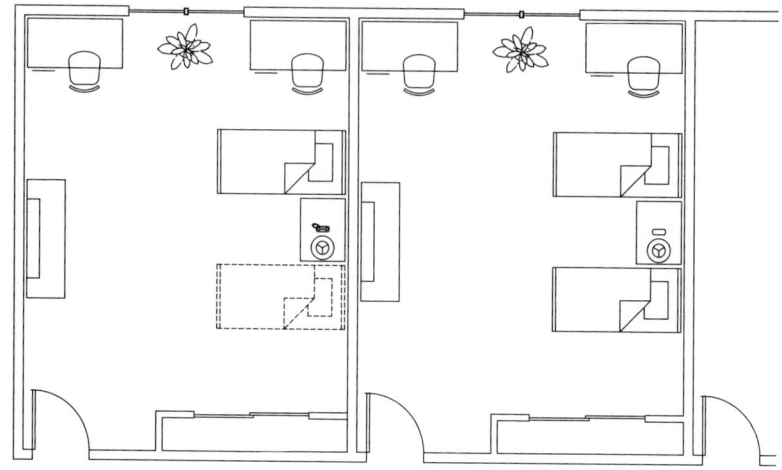

Another, more typical application is the use of *Groups* in conjunction with *Blocks*. A block is a set of objects combined into one object with the *Block* command. A block always behaves as <u>one object</u>. In our dorm room layout, each furniture item could be *Inserted* as a previously defined block. The entire set of blocks (bed, chair, desk, etc.) could be combined into a group called ROOM. In this case, the ROOM group would be set to *Selectable*. This would allow you to select <u>all</u> furniture items as a group by PICKing only one. However, if you wanted to select only <u>one</u> furniture item (Block), use Ctrl+H to change *PICKSTYLE* to 0. In this way, individual members of any group (blocks in this case) can be PICKed without having to change the *Selectable* status. See Chapter 21 for more information on *Blocks*.

CHAPTER EXERCISES

1. *PICKFIRST*

   Check to ensure that **PICKFIRST** is set to **1** (*ON*). Also make sure *GRIPS* are set to 0 (*OFF*) by typing **GRIPS** at the command prompt or invoking the **Selection** tab of the **Options** dialog box. For each of the following editing commands you use, make sure you PICK the objects FIRST, then invoke the editing command.

   A. *Open* the PLATES drawing that you created as an exercise in Chapter 9 (not the PLATNEST drawing). *Erase* 3 of the 4 holes from the plate on the left, leaving only the hole at the lower-left corner (**PICK** the *Circles*; then invoke *Erase*). Create a *Rectangular Array* with **7** rows and **5** columns and **1** unit between each hole (PICK the *Circle* before invoking *Array*). The new plate should have 35 holes as shown in Figure 20-20, plate A.

FIGURE 20-20

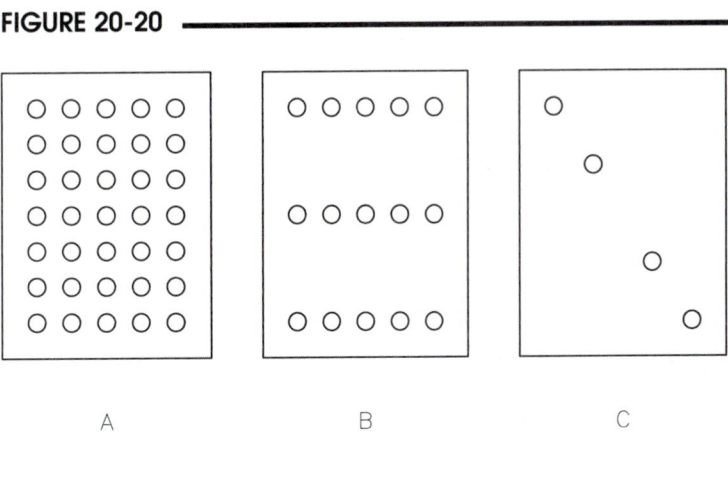

A   B   C

   B. For the center plate, **PICK** the 3 holes on the <u>vertical</u> center; then invoke *Copy*. Create the additional 2 sets of 3 holes on each side of the center column. Your new plate should look like that in Figure 20-20, plate B. Use *SaveAs* to assign a new name, **PLATES2**.

   C. For the last plate (on the right), **PICK** the two top holes and *Move* them upward. Use the **Center** of the top hole as the "Specify base point or displacement, or [Multiple]:" and specify a "Specify second point of displacement or <use first point as displacement>:" as **1** unit from the top and left edges. Compare your results to Figure 20-20, plate C. *Save* the drawing (as **PLATES2**).

Remember that you can leave the *PICKFIRST* setting *ON* always. This means that you can use both Noun/Verb or Verb/Noun editing at any time. You always have the choice of whether you want to PICK objects first or use the command first.

2. **PICKADD**

FIGURE 20-21

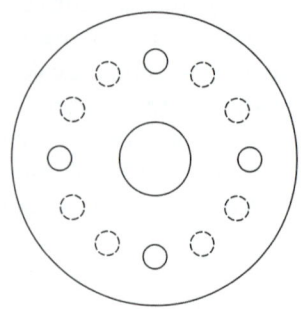

Change *PICKADD* to **0** (*OFF*). Remember that if you want to PICK objects and add them to the existing highlighted selection set, hold down the **SHIFT** key; then **PICK**.

A. *Open* the **ARRAY1** drawing from Chapter 9. (The settings for *GRIPS* and *PICKFIRST* have not changed by *Opening* a new drawing.) *Erase* the holes (highlighted *Circles* shown in Figure 20-21) leaving only 4 holes. Use either Noun/Verb or Verb/Noun editing. *SaveAs* **FLANGE1**.

B. *Open* drawing **ARRAY2.** Select all the holes and *Rotate* them **180** degrees to achieve the arrangement shown in Figure 20-22. Any selection method may be used. Try several selection methods to see how *PICKADD* reacts. (The *Fence* option can be used to select all holes without having to use the **Shift** key.) *SaveAs* **FLANGE2**. Change the *PICKADD* setting back to **1** (*ON*).

FIGURE 20-22

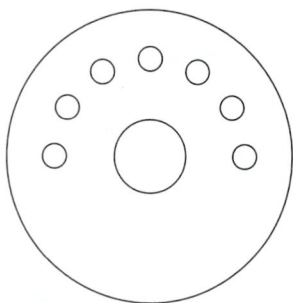

3. **PICKDRAG**

FIGURE 20-23

Open the **CH11EX3** drawing. Set *PICK-DRAG* to **1** (*ON*). Use *Stretch* to change the front and side views to achieve the new base thickness as indicated in Figure 20-23. (Remember to hold down the PICK button when dragging the mouse or puck.) When finished, *SaveAs* **HOLDR-CUP**. Change the *PICKDRAG* setting back to **0** (*OFF*).

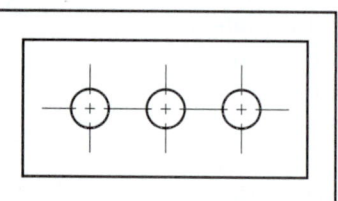

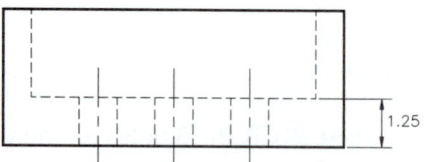

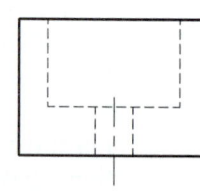

1.25

4. **Qselect**

This exercise involves using object selection filters to change the properties of objects in the drawing. *Open* the **DB_SAMP.DWG** from the AutoCAD 2006\Sample folder.

A. Activate the *Model* tab. Notice that all the doors in the drawing are green in color. Using *List*, verify that any door is on the E-F-Door layer. ✓

B. Invoke *Qselect* by any method. Ensure that *Entire Drawing* appears in the *Apply to*: section. In the *Object Type* section, select *Block Reference* from the drop-down list. In *Properties*, *Operator* and *Value*, select *Layer*, *Equals*, and **E-F-DOOR**, respectively. Ensure *Include in new selection set* is checked and *Append to current selection set* is not checked. Then select the *OK* button. All the doors should be highlighted. The command line reports "39 items selected." ✓

C. With the doors highlighted, use the Layer Control drop-down list and select layer **0**. All the doors are now "moved" to layer 0 and therefore change to black or white (the *ByLayer* color assignment) depending on your background color. Press **Escape** to clear the highlighted items and grips (if enabled).

FIGURE 20-24

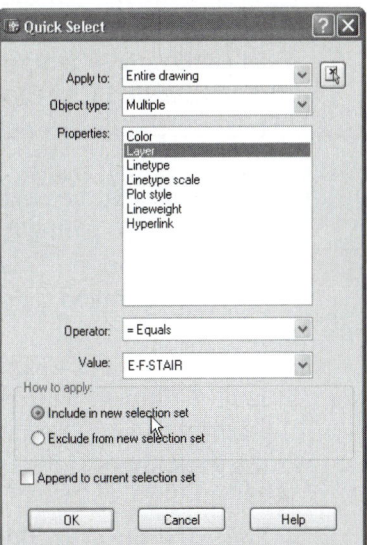

D. Invoke the *Properties* palette (not shown). Use the **Quick Select** button to produce the *Quick Select* dialog box. In the dialog box (Fig. 20-24), ensure that **Entire Drawing** appears in the *Apply to*: section. In the *Object Type* section, select **Multiple** from the drop-down list. In *Properties*, *Operator* and *Value*, select **Layer, Equals,** and **E-F-STAIR**, respectively. Ensure **Include in new selection set** is checked and *Append to current selection set* is not checked. Then select the **OK** button. The stairs should be highlighted and the command line reports "217 items selected." ✓

E. In the *Properties* palette, select the **Color** property and select **Green**. Click in the drawing area and press **Escape** to deselect the stair objects. The stairs should now be green.

F. **Close** the drawing and <u>do not save</u> the changes. ✓

Ch20ex4

5. *Filter*

FIGURE 20-25

In this exercise, you will use the *Filter* command for several operations.

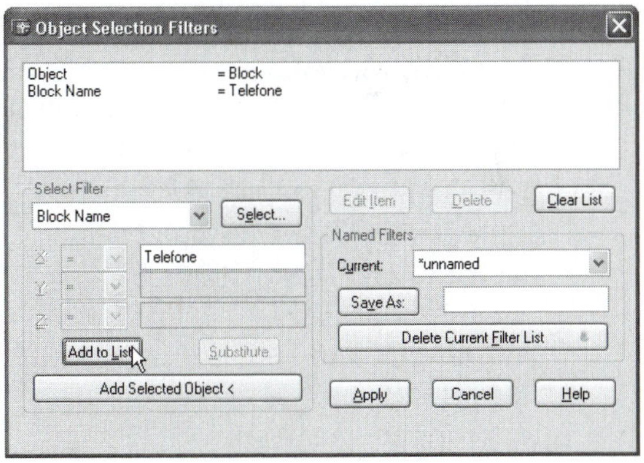

A. **Open** the **EXPO 98 maps.DWG** (download this drawing from the www.mhhe.com/leach Web site). Select the *Model* tab. The map indicates information about services such as telephone locations, transportation, restaurants, and so on.

B. Use the *Object Selection Filter* dialog box to count the number of telephones. Invoke *Filter*. When the dialog box appears, select **Block Name** from the *Select Filter* section drop-down list. Then, click on the **Select** button to select from the list of Blocks in the drawing. Select **TELEFONE** from the bottom of the list, then pick **OK**. In the *Object Selection Filter* dialog box, select **Add to List**. Two entries in the filter list above should appear (Fig. 20-25).

C. Select the **Apply** button. The dialog box disappears and the command line prompts to "Select objects:". Enter **All** so the filter is applied to the entire drawing. AutoCAD reports the number of (all) objects found and the number that were filtered out (not telephones). At the "Select objects:" prompt, press **Enter**. Although you have exited the *Filter* command, the objects remain highlighted. 75

D. Use the *Select* command to do the arithmetic. Type **Select** at the command prompt. AutoCAD reports "75 found."

E. Next, you will change the cafeteria symbols to layer 0. Do this by typing *Change* at the command line. At the "Select objects:" prompt, enter *'Filter* (with the apostrophe prefix). When the dialog box appears, highlight the existing items (telephone blocks) from the filter list and select *Delete* to remove them. Next, go through the process (as before) to create a filter for *Block Names* called **CAFETERIA**. Make sure you *Add to List*. Finally, select *Apply* to exit the dialog box. At the "Select objects:" prompt, type *All* so AutoCAD searches the entire drawing.

F. Press **Enter** at the next "Select objects:" prompt. Remember, you are still in the *Change* command, so AutoCAD prompts to "Specify change point or [Properties]:". Type *P* for the *Properties* option. Enter *LA* for *Layer* and specify layer **0** as the new layer name. Complete the *Change* command by pressing **Enter**.

G. Since the CAFETERIA objects have been moved to layer 0, use the *Layer* command to make layer **0** *Current*, then right-click to *Select All* and *Freeze* all layers. AutoCAD reports that the current layer (0) cannot be frozen. Pick *OK* to exit the *Layer Properties Manager*. The cafeteria symbols should be visible.

H. *Close* the drawing, but do not save the changes.

6. **Object Groups**

FIGURE 20-26

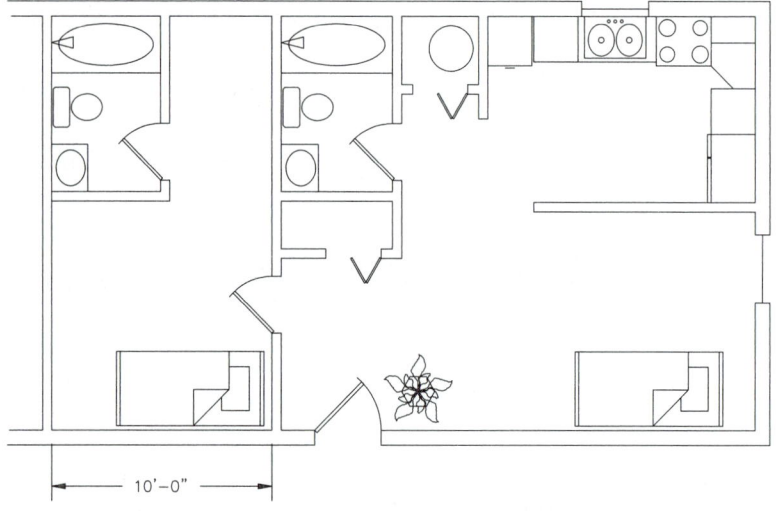

A. *Open* the **EFF-APT2** drawing that you last worked on in Chapter 18 exercises. Create a *Group* for each of the three bathroom fixtures. *Name* one group **WASHBASIN**, give it a *Description*, and include the *Ellipse* and the surrounding *Rectangle* as its members. Make two more groups named **WCLOSET** and **BATHTUB** and select the appropriate members. Next, combine the three fixture groups into one *Group* named **BATHROOM** and give it a description. Then, draw a bed of your own design and combine its members into a *Group* named **SINGLEBED**. Make all five groups *Selectable*. Use *SaveAs* to save and rename the drawing to **2BR-APT.**

B. A small bedroom and bathroom are to be added to the apartment, but a 10' maximum interior span is allowed. One possible design is shown in Figure 20-26. Other more efficient designs are possible. Draw the new walls for the addition, but design the bathroom and bedroom door locations to your personal specifications. Use *Copy* or *Mirror* where appropriate. Make the necessary *Trims* and other edits to complete the floor plan.

C. Next, *Copy* the bathroom fixture groups to the new bathroom. If your design is similar to that in the figure, the BATHROOM group can be copied as a whole. If you need to *Copy* only individual fixtures, take the appropriate action so you don't select the entire BATHROOM group. *Copy* or *Mirror* the **SINGLEBED**. Once completed, activate the *Object Grouping* dialog box and toggle on *Include Unnamed*. Can you explain what happened? Assign a *Description* to each of the unnamed groups and *Save* the drawing.

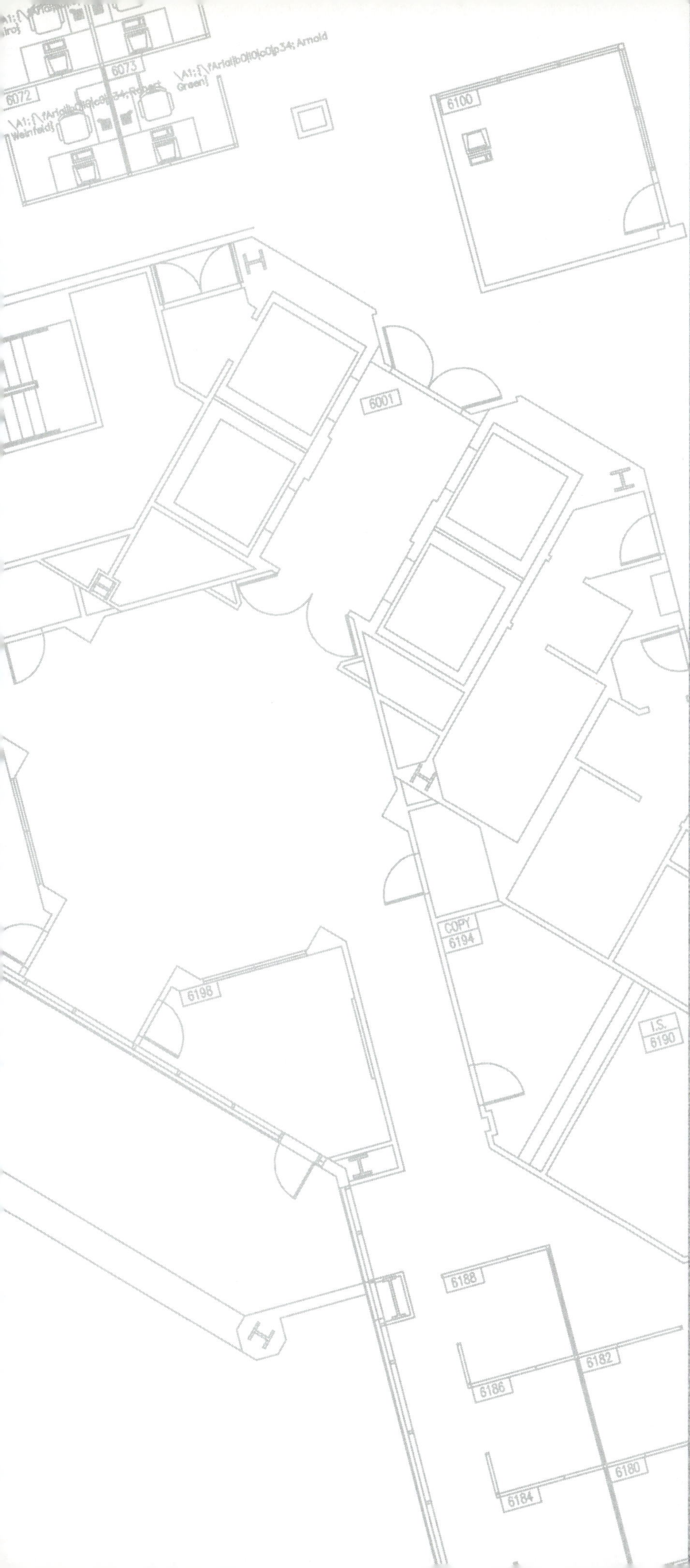

BLOCKS, DESIGN-CENTER, AND TOOL PALETTES

CHAPTER OBJECTIVES

After completing this chapter you should:

1. be able to use the *Block* command to transform a group of objects into one object that is stored in the current drawing's block definition table;

2. be able to use the *Insert* and *Minsert* commands to bring *Blocks* into drawings;

3. know that *color, linetype* and *lineweight* of *Blocks* are based on conditions when the *Block* is made;

4. be able to convert *Blocks* to individual objects with *Explode*;

5. be able to use *Wblock* to prepare .DWG files for insertion into other drawings;

6. be able to redefine and globally change previously inserted *Blocks*;

7. be able to use DesignCenter™ and Tool Palettes to drag and drop *Blocks* into the current drawing;

8. be able to create dynamic blocks that change based on assigned parameters and associated actions.

CONCEPTS

A *Block* is a <u>group</u> of objects that are combined into <u>one</u> object with the *Block* command. The typical application for *Blocks* is in the use of symbols. Many drawings contain symbols, such as doors and windows for architectural drawings, capacitors and resistors for electrical schematics, or pumps and valves for piping and instrumentation drawings. In AutoCAD, symbols are created first by constructing the desired geometry with objects like *Line*, *Arc*, and *Circle*, then transforming the set of objects comprising the symbol into a *Block*. A description of the objects comprising the *Block* is then stored in the drawing's "block definition table." The *Blocks* can then each be *Inserted* into a drawing many times and treated as a single object. Text can be attached to *Blocks* (called *Attributes*) and the text can be modified for each *Block* when inserted.

Figure 21-1 compares a shape composed of a set of objects and the same shape after it has been made into a *Block* and *Inserted* back into the drawing. Notice that the original set of objects is selected (highlighted) individually for editing, whereas, the *Block* is only one object.

FIGURE 21-1 ───────────

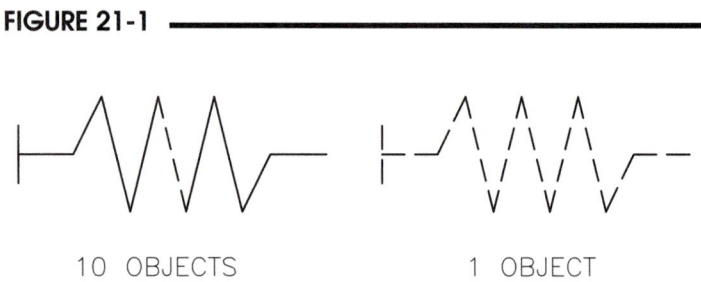

10 OBJECTS   1 OBJECT

Since an inserted *Block* is one object, it uses less file space than a set of objects that is copied with *Copy*. The *Copy* command creates a duplicate set of objects, so that if the original symbol were created with 10 objects, 3 copies would yield a total of 40 objects. If instead the original set of 10 were made into a *Block* and then *Inserted* 3 times, the total objects would be 13 (the original 10 + 3).

Upon *Inserting* a *Block*, its scale can be changed and rotational orientation specified without having to use the *Scale* or *Rotate* commands (Fig. 21-2). If a design change is desired in the *Blocks* that have already been *Inserted*, the original *Block* can be redefined and the previously inserted *Blocks* are automatically updated. *Blocks* can be made to have explicit *Linetype*, *Lineweight* and *Color*, regardless of the layer they are inserted onto, or they can be made to assume the *Color*, *Linetype*, and *Lineweight* of the layer onto which they are *Inserted*.

FIGURE 21-2 ───────────

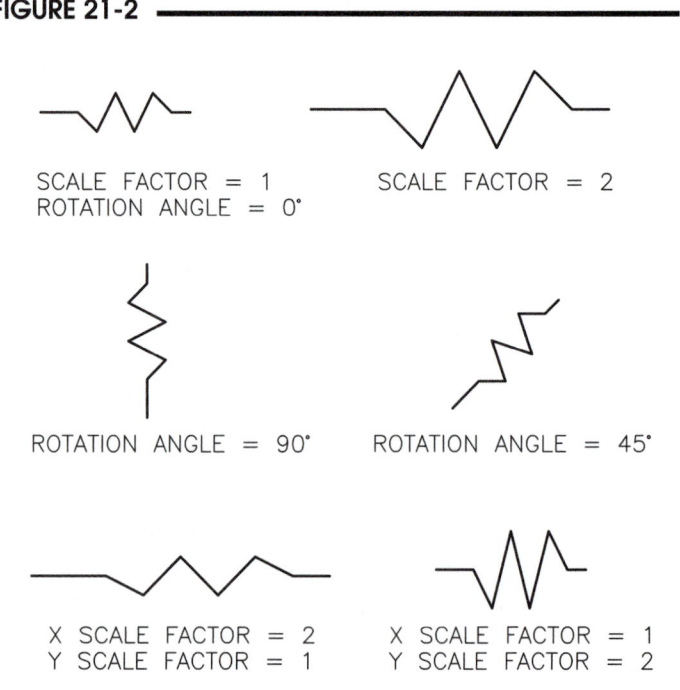

SCALE FACTOR = 1   SCALE FACTOR = 2
ROTATION ANGLE = 0°

ROTATION ANGLE = 90°   ROTATION ANGLE = 45°

X SCALE FACTOR = 2   X SCALE FACTOR = 1
Y SCALE FACTOR = 1   Y SCALE FACTOR = 2

Blocks can be nested; that is, one *Block* can reference another *Block*. Practically, this means that the definition of *Block* "C" can contain *Block* "A" so that when *Block* "C" is inserted, *Block* "A" is also inserted as part of *Block* "C" (Fig. 21-3).

FIGURE 21-3

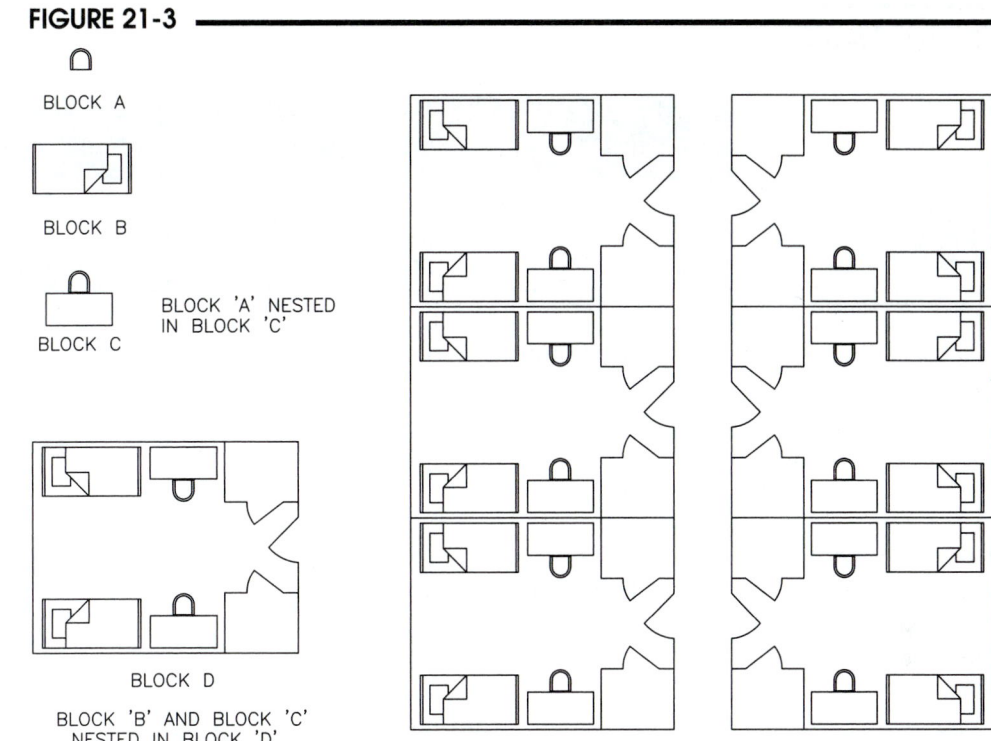

BLOCK A

BLOCK B

BLOCK 'A' NESTED IN BLOCK 'C'

BLOCK C

BLOCK D

BLOCK 'B' AND BLOCK 'C' NESTED IN BLOCK 'D'

Blocks created within the current drawing can be copied to disk as complete and separate drawing files (.DWG file) by using the *Wblock* command (Write Block). This action allows you to *Insert* the *Blocks* into other drawings.

Commands related to using *Blocks* are:

Block	Creates a *Block* from individual objects
Insert	Inserts a *Block* into a drawing
Minsert	Permits a multiple insert in a rectangular pattern
Explode	Breaks a *Block* into its original set of multiple objects
Wblock	Writes an existing *Block* or a set of objects to a file on disk
Base	Allows specification of an insertion base point
Purge	Deletes uninserted *Blocks* from the block definition table
Rename	Allows renaming *Blocks*
Bedit	Invokes the Block Editor for editing blocks

The following additional tools, discussed in this chapter, provide powerful features for creating, inserting, and editing blocks.

Tool palettes	Tool palettes allow you to drag-and-drop *Blocks* and *Hatch* patterns into a drawing. This process is similar, but easier, than using *Insert*. You can easily add blocks to create your own tool palettes. Tool palettes are discussed also in Chapters 26 and 45.
DesignCenter	This tool allows you to drag-and-drop *Blocks* from other drawings into the current drawing. You can easily locate content from other drawings such as *Dimension Styles*, *Layers*, *Linetypes*, *Text Styles*, and *Xref* drawings to drag and drop into the current drawing. DesignCenter is discussed also in Chapter 30.
Dynamic Blocks	Dynamic blocks are created using the Block Editor. Dynamic blocks can be dynamically changed after being inserted into the drawing based on predefined parameters and actions.

2006

COMMANDS

Block

Pull-down Menu	Command (Type)	Alias (Type)	Short-cut	Screen (side) Menu	Tablet Menu
Draw *Block >* *Make...*	*Block,* or *-Block*	*B* or *-B*	...	DRAW 2 *Bmake*	*N,9*

Selecting the icon button, using the pull-down or screen menu, or typing *Block* or *Bmake*, produces the *Block Definition* dialog box shown in Figure 21-4. This dialog box provides the same functions as using the *-Block* command (a hyphen prefix produces the Command line equivalent).

To make a *Block*, first create the *Lines, Circles, Arcs*, or other objects comprising the shapes to be combined into the *Block*. Next, use the *Block* command to transform the objects into one object—a *Block*.

In the *Block Definition* dialog box, enter the desired *Block* name in the *Name* edit box. Then use the *Select Objects* button (top center) to return to the drawing temporarily to select the objects you wish to comprise the *Block*. After selection of objects, the dialog box reappears. Use the *Pick Insertion Base Point* button (in the *Base Point* section of the dialog box) if you want to use a point other than the default 0,0,0 as the "insertion point" when the *Block* is later inserted. Usually select a point in the corner or center of the set of objects as the base point. When you select *OK*, the new *Block* is defined and stored in the drawing's block definition table awaiting future insertions.

FIGURE 21-4

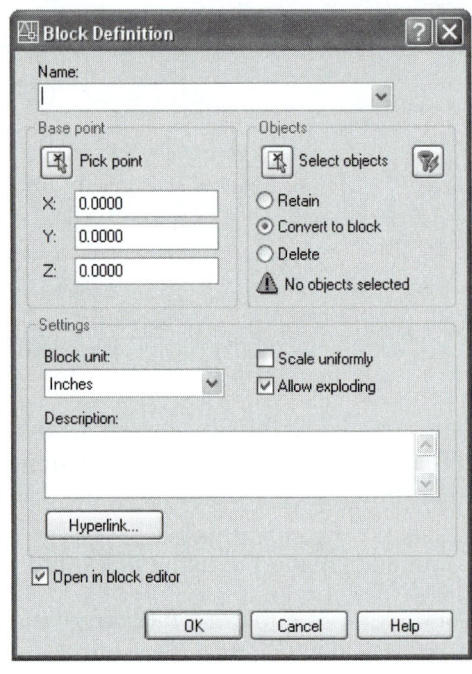

If *Delete* is selected in the *Objects* section of the dialog box, the original set of "template" objects comprising the *Block* disappear even though the definition of the *Block* remains in the table. Checking *Retain* forces AutoCAD to retain the original objects (similar to using *Oops* after the *Block* command), or selecting *Convert to Block* keeps the original set of objects visible in the drawing but transforms them into a *Block*.

The *Block Unit* drop-down option and the *Description* edit box are used to specify how *Blocks* are described when DesignCenter is used to drag and drop the *Blocks* into a drawing instead of using the *Insert* command. The *Block Units* options are described later in this chapter. If the *Scale uniformly* checkbox is not checked, you have the option to scale the block non-uniformly when it is *Inserted* into the drawing. If the *Allow exploding* box is checked, blocks can be "broken down" into the individual objects used to create the block (*Line, Arc, Circle*, etc.) using the *Explode* command. Use the *Hyperlink* button to produce the *Insert Hyperlink* dialog box for attaching a hyperlink to a *Block*. If *Open in block editor* is checked, the Block Editor is opened when you press the *OK* button. Normally, remove this check if you are making a block and do not want to edit it further. The *Names* drop-down list (at the top) is used to select existing *Blocks* if you want to redefine a *Block* (see "Redefining Blocks and the Block Editor" later in this chapter).

If you prefer to type, use -*Block* to produce the Command line equivalent of the *Block Definition* dialog box. The command syntax is as follows:

Command: **-Block**
Block name (or ?): (**name**) (Enter a descriptive name for the *Block* up to 255 characters.)
Insertion base point: **PICK** or (**coordinates**) (Select a point to be used later for insertion.)
Select objects: **PICK**
Select objects: **PICK** (Continue selecting all desired objects.)
Select objects: **Enter**

The *Block* then <u>disappears</u> as it is stored in the current drawing's "block definition table." The *Oops* command can be used to restore the original set of "template" objects (they reappear), but the definition of the *Block* remains in the table. Using the *?* option of the *Block* command lists the *Blocks* stored in the block definition table.

Block Color, Linetype, and Lineweight Settings

The <u>color</u>, <u>linetype</u>, and <u>lineweight</u> of the *Block* are determined by one of the following settings when the *Block* is created:

1.   When a *Block* is inserted, it is drawn on its original layer with its original *color, linetype* and *lineweight* (when the objects were <u>created</u>), regardless of the layer or *color, linetype* and *lineweight* settings that are current when the *Block* is inserted (unless conditions 2 or 3 exist).

2.   If a *Block* is created on <u>Layer 0</u> (Layer 0 is current when the original objects comprising the *Block* are created), then the *Block* assumes the *color, linetype* and *lineweight* of any layer that is current when it is inserted (Fig. 21-5).

FIGURE 21-5

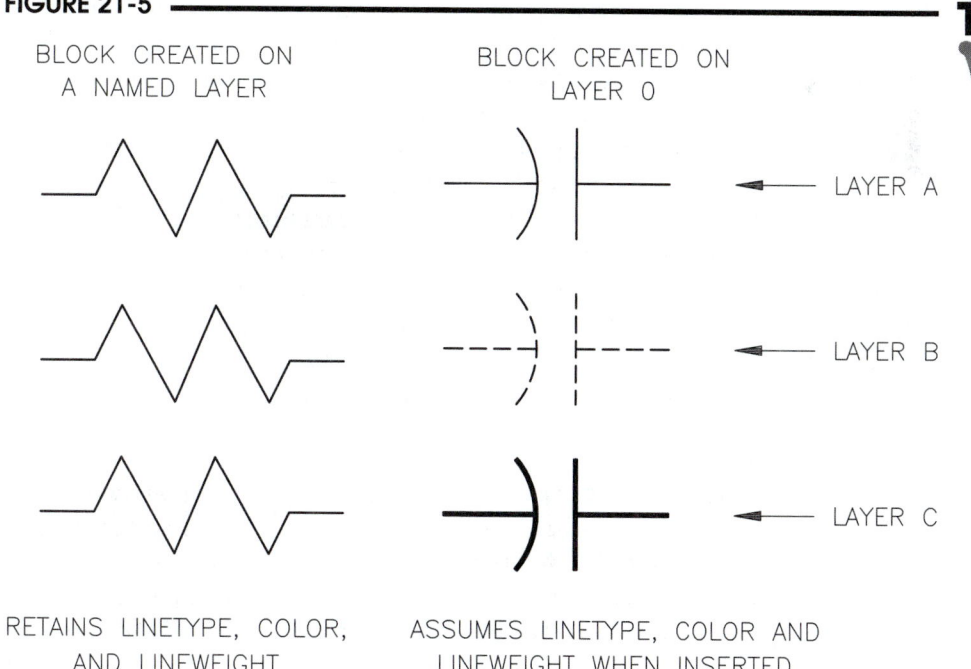

BLOCK CREATED ON
A NAMED LAYER

BLOCK CREATED ON
LAYER 0

LAYER A

LAYER B

LAYER C

RETAINS LINETYPE, COLOR,
AND LINEWEIGHT

ASSUMES LINETYPE, COLOR AND
LINEWEIGHT WHEN INSERTED

3.   If the *Block* is created with the special *BYBLOCK color, linetype* and *lineweight* setting, the *Block* is inserted with the *color, linetype* and *lineweight* settings that are <u>current during insertion</u>, whether the *BYLAYER* or explicit object *color, linetype* and *lineweight* settings are current.

Insert

	Pull-down Menu	Command (Type)	Alias (Type)	Short-cut	Screen (side) Menu	Tablet Menu
	Insert Block...	Insert or -Insert	I or -I	...	INSERT Ddinsert	T,5

Once the *Block* has been created, it is inserted back into the drawing at the desired location(s) with the *Insert* command. The *Insert* command produces the *Insert* dialog box (Fig. 21-6) which allows you to select which *Block* to insert and to specify the *Insertion Point, Scale,* and *Rotation*, either interactively (*On-screen*) or by specifying values.

First, select the *Block* you want to insert. All *Blocks* located in the drawing's block definition table are listed in the *Name* drop-down list. Next, determine the parameters for *Insertion Point, Scale,* and *Rotation*. You can enter values in the edit boxes if you have specific parameters in mind or check *Specify On-screen* to interactively supply the parameters. For example, with the settings shown in Figure 21-6, AutoCAD would allow you to preview the *Block* as you dragged it about the screen and picked the *Insertion Point*. You would not be prompted for a *Scale* or *Rotation* angle since they are specified in the dialog box as 1.0000 and 0 degrees, respectively. Entering any other values in the *Scale* or *Rotation* edit boxes causes AutoCAD to preview the *Block* at the specified scale and rotation angle as you drag it about the drawing to pick the insertion point. Remember that *Osnaps* can be used when specifying the parameters interactively. Check *Uniform Scale* to ensure the X, Y, and Z values are scaled proportionally. *Explode* can also be toggled, which would insert the *Block* as multiple objects (see "*Explode*").

FIGURE 21-6

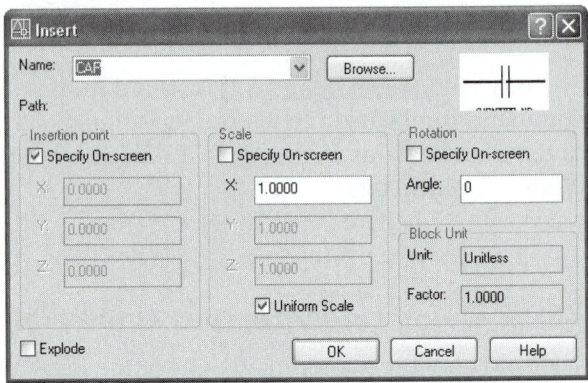

Inserting Other Drawings as *Blocks*

Selecting the *Browse* tile in the *Insert* dialog box produces the *Select Drawing* dialog box (Fig. 21-7). Here you can select <u>any</u> drawing (.DWG file) for insertion. When one drawing is *Inserted* into another, the entire drawing comes into the current drawing as a *Block*, or as <u>one object</u>. If you want to edit individual objects in the inserted drawing, you must *Explode* the object.

If you prefer the Command line equivalent, type -*Insert*.

FIGURE 21-7

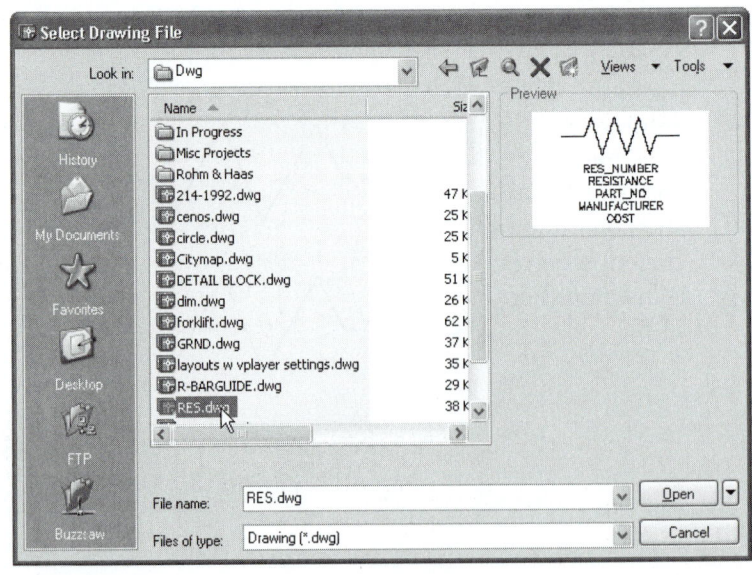

```
Command: -insert
Enter block name or [?] : name
(Type the name of an existing block
or .DWG file to insert.)
Specify insertion point or
[Basepoint/Scale/X/Y/Z/Rotate/PScale/PX/PY/PZ/PRotate]: PICK or option
Enter X scale factor, specify opposite corner, or [Corner/XYZ] <1>: value, PICK or option
Enter Y scale factor <use X scale factor>: value or PICK
Specify rotation angle <0>: value or PICK
```

Sometimes it is desirable to see the *Block* in the intended scale factor or rotation angle before you choose the insertion point. Presets ("PScale/PC/PY/PZ/PRotate") allow you to specify a rotation angle or scale factor <u>before</u> you dynamically drag the *Block* to pick the insertion point. (Normally, you would have to select the insertion point before the prompts for scale factor and rotation angle appear.)

Minsert

Pull-down Menu	Command (Type)	Alias (Type)	Short-cut	Screen (side) Menu	Tablet Menu
...	*Minsert*	...	...	...	...

This command allows a <u>multiple insert</u> in a rectangular pattern (Fig. 21-8). *Minsert* is actually a combination of the *Insert* and the *Array Rectangular* commands. The *Blocks* inserted with *Minsert* are associated (the group is treated as one object) and cannot be edited independently (unless *Exploded*).

Examining the command syntax yields the similarity to a *Rectangular Array*.

FIGURE 21-8

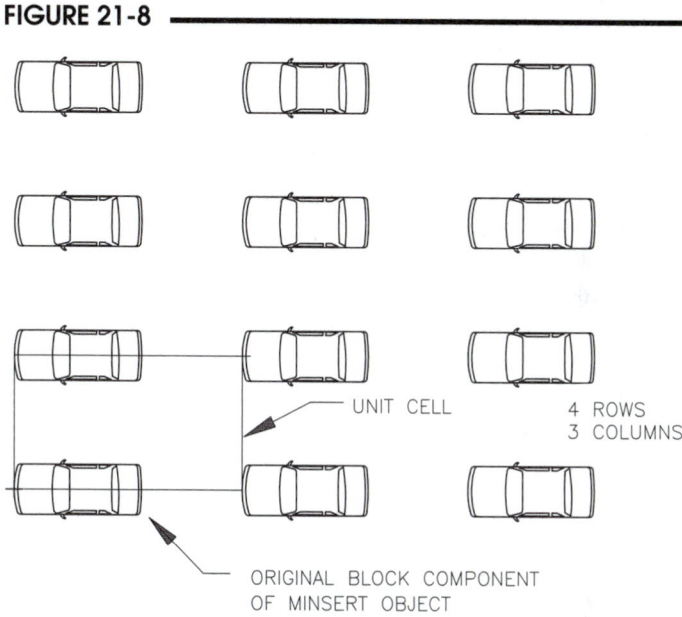

UNIT CELL   4 ROWS 3 COLUMNS

ORIGINAL BLOCK COMPONENT OF MINSERT OBJECT

Command: **Minsert**
Enter block name [or ?] <current>: **name**
Specify insertion point or [Basepoint/Scale/X/Y/Z/Rotate/PScale/PX/PY/PZ/PRotate]: (**value**), **PICK** or option
Enter X scale factor, specify opposite corner, or [Corner/XYZ] <1>: (**value**) or **Enter**
Enter Y scale factor <use X scale factor>: (**value**) or **Enter**
Specify rotation angle <0>: (**value**) or **Enter**
Enter number of rows (—-) <1>: (**value**)
Enter number of columns (|||) <1>: (**value**)
Enter distance between rows or specify unit cell (—-): (**value**) or **PICK** (Value specifies Y distance from *Block* corner to *Block* corner; PICK allows drawing a unit cell rectangle.)
Distance between columns: (**value**) or **PICK** (Specifies X distance between *Block* corners.)
Command:

Explode

Pull-down Menu	Command (Type)	Alias (Type)	Short-cut	Screen (side) Menu	Tablet Menu
Modify *Explode*	*Explode*	*X*	...	*MODIFY2* *Explode*	*Y,22*

Explode breaks a <u>previously</u> inserted *Block* back into its original set of objects (Fig. 21-9, on the next page), which allows you to edit individual objects comprising the shape. If *Allow exploding* was unchecked in the *Block Definition* dialog box when the *Block* was created, the *Block* cannot be *Exploded*. *Blocks* that have been *Minserted* cannot be *Exploded*.

Command: *explode*
Select objects: **PICK**
Select objects: **Enter**
Command:

FIGURE 21-9

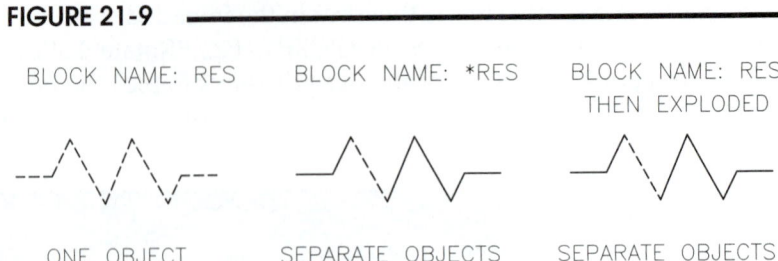

BLOCK NAME: RES   BLOCK NAME: *RES   BLOCK NAME: RES
                                                THEN EXPLODED

ONE OBJECT   SEPARATE OBJECTS   SEPARATE OBJECTS

-Insert with *

Using the *-Insert* command with the asterisk (*) allows you to insert a *Block,* not as one object, but as the original set of objects comprising the *Block.* In this way, you can edit individual objects in the *Block,* otherwise impossible if the *Block* is only one object (Fig. 21-9).

The normal *-Insert* command is used (type *-Insert* with a hyphen); however, when the desired <u>Block name</u> is entered, it is prefaced by the asterisk (*) symbol:

Command: **-insert**
Enter Block name (or ?): ***** (**name**) (Type the * symbol, then the name of an existing block or .DWG
    file to insert.)
Specify insertion point for block:
Specify scale factor for XYZ axes:
Specify rotation angle <0>:
Command:

This action accomplishes the same goal as using the *Insert* dialog box; then *Explode.*

Xplode

Pull-down Menu	Command (Type)	Alias (Type)	Short-cut	Screen (side) Menu	Tablet Menu
...	*Xplode*	XP	...	...	...

When you *Insert* a *Block* into a drawing, all layers, linetypes, and colors contained in the *Block* are also inserted into the parent drawing if they do not already exist. If you use *Explode* to break down the *Block* into its component entities, those entities retain their native properties of layer, linetype, and color. For example, you may insert a *Block* that also inserts its layer named FLOORPLAN. If that *Block* is *Exploded,* its objects remain on layer FLOORPLAN. You may instead want to *Explode* the *Block* and have the objects reside on a layer existing in the current drawing, AR-WALL, for example.

The *Xplode* command is an expanded version of *Explode.* The *Xplode* command allows you to specify new properties for the objects that are exploded. When you use *Xplode,* you can also specify the new *Layer, Linetype, Color, Lineweight,* or choose to *Explode* the *Block* normally.

Command: **xplode**
Select objects to XPlode.
Select objects: **PICK**
Select objects: **Enter**
Enter an option [All/Color/LAyer/LType/LWeight/Inherit from parent block/Explode] <Explode>:

The options are as follows.

Color

Use this option to specify a color for the exploded objects. Entering *ByLayer* causes the component objects to inherit the color of the exploded object's layer.

```
Enter an option
[All/Color/LAyer/LType/LWeight/Inherit from parent block/Explode] <Explode>: c
New color [Truecolor/COlorbook] <BYLAYER>:
```

Layer

With this option you can specify the layer of the component objects after you explode them. The default option is to inherit the current layer rather than the layer of the exploded object.

```
Enter an option [All/Color/LAyer/LType/LWeight/Inherit from parent block/Explode] <Explode>: la
Enter new layer name for exploded objects <0>:
```

Linetype

You can enter the name of any linetype that is loaded in the drawing. The exploded objects assume the specified linetype. Entering *ByLayer* causes the component objects to inherit the linetype of the exploded object's layer. Entering *ByBlock* causes the component objects to inherit the object-specific linetype of the exploded object.

```
Enter an option [All/Color/LAyer/LType/LWeight/Inherit from parent block/Explode] <Explode>: lt
Enter new linetype name for exploded objects <BYLAYER>:
```

Lineweight

Use this option to specify the lineweight of the objects when they are exploded. The value you enter specifies the lineweight in inches or millimeters, whichever is set in the *Lineweight Settings* dialog box.

```
Enter an option
[All/Color/LAyer/LType/LWeight/Inherit from parent block/Explode] <Explode>: lw
Enter new lineweight <ByLayer>:
```

Inherit from parent block

This option sets the color, linetype, lineweight, and layer of the exploded objects to that of the *Block* if the *Block* was created using *ByBlock* color, linetype, and lineweight and the objects were drawn on layer 0.

All

Use this option to specify a layer, color, linetype, and lineweight for the exploded objects.

Wblock

Pull-down Menu	Command (Type)	Alias (Type)	Short-cut	Screen (side) Menu	Tablet Menu
File *Export...*	*Wblock*	*W*	...	FILE *Export*	*W,24*

The *Wblock* command writes a *Block* out to disk as a separate and complete drawing (.DWG) file. The *Block* used for writing to disk can exist in the current drawing's *Block* definition table or can be created by the *Wblock* command. Remember that the *Insert* command inserts *Blocks* (from the current drawing's block definition table) or finds and accepts .DWG files and treats them as *Blocks* upon insertion.

There are two ways to create a *Wblock* from the current drawing, (1) using an existing *Block* and (2) using a set of objects not previously defined in the current drawing as a *Block*. If you are using an existing *Block*, a copy of the *Block* is essentially transformed by the *Wblock* command to create a complete AutoCAD drawing (.DWG) file. The original block definition remains in the current drawing's block definition table. In this way, *Blocks* that were originally intended for insertion into the current drawing can be easily inserted into other drawings.

Figure 21-10 illustrates the relationship among a *Block*, the current drawing, and a *WBlock*. In the figure, SCHEM1.DWG contains several *Blocks*. The RES block is written out to a .DWG file using *Wblock* and named RESISTOR. RESISTOR is then *Inserted* into the SCHEM2 drawing.

FIGURE 21-10 ──

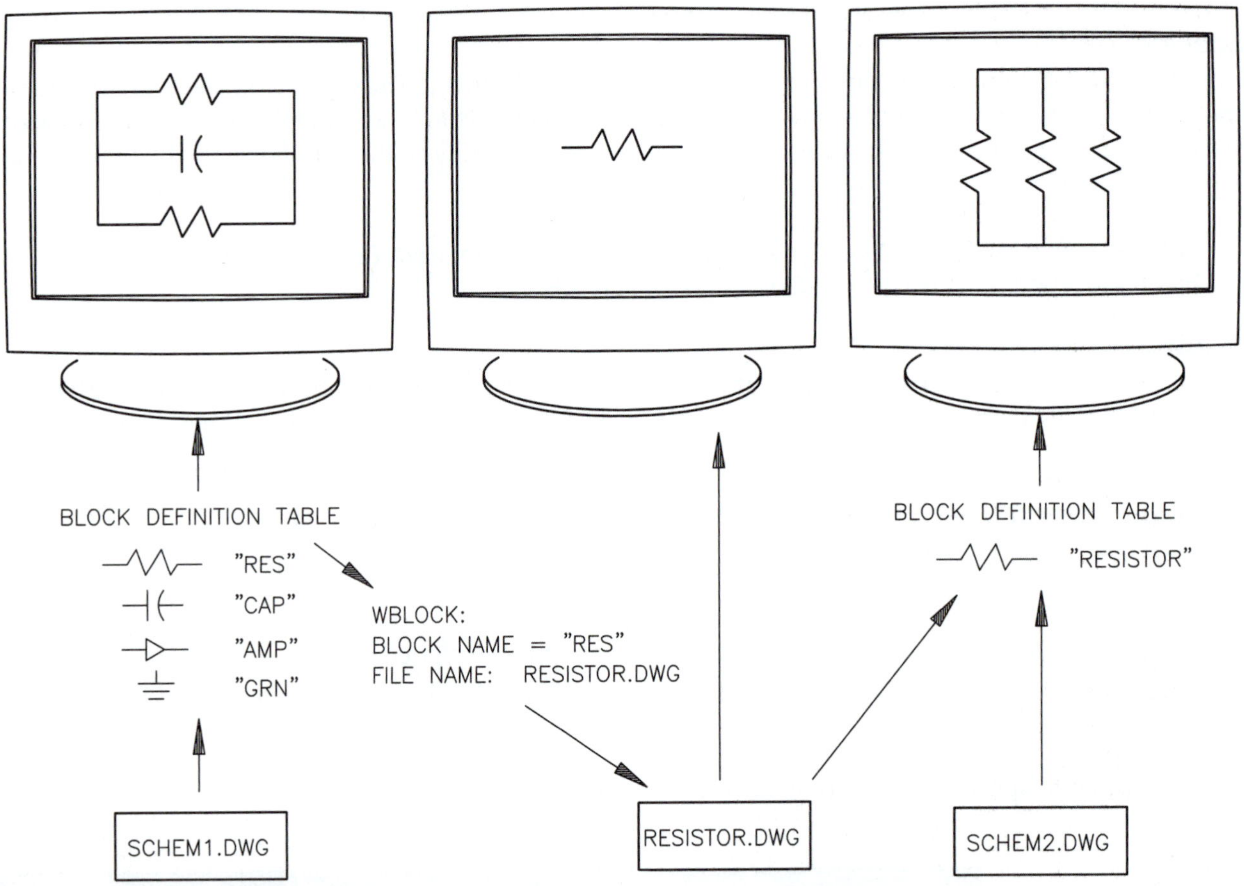

BLOCK DEFINITION TABLE

—W— "RES"

—||— "CAP"

—▷— "AMP"

⏚ "GRN"

WBLOCK:
BLOCK NAME = "RES"
FILE NAME: RESISTOR.DWG

BLOCK DEFINITION TABLE

—W— "RESISTOR"

SCHEM1.DWG

RESISTOR.DWG

SCHEM2.DWG

TIP If you want to transform a set of objects into a *Block* to be used in other drawings but not in the current one, you can use *Wblock* to transform (a copy of) the objects in the current drawing into a separate .DWG file. This action does not create a *Block* in the current drawing. As an alternative, if you want to create symbols specifically to be inserted into other drawings, each symbol could be created initially as a separate .DWG file.

The *Wblock* command produces the *Write Block* dialog box (Fig. 21-11). You should notice similarities to the *Block Definition* dialog box. Under *Source*, select *Block* if you want to write out an existing *Block* and select the *Block* name from the list, or select *Objects* if you want to transform a set of objects (not a previously defined *Block*) into a separate .DWG file.

FIGURE 21-11

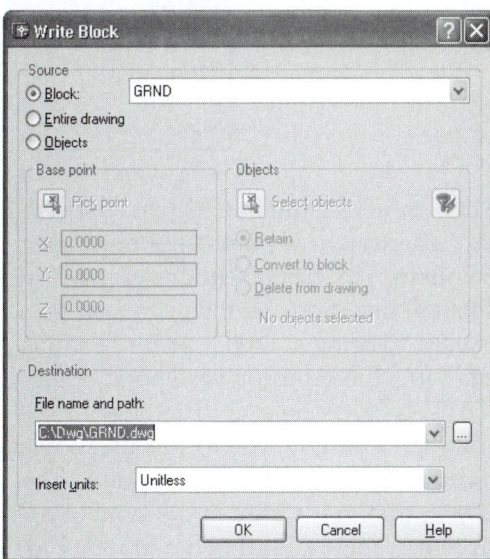

The *Base Point* section allows you to specify a base point to use upon insertion of the *Block*. Enter coordinate values or use the *Specify Insert Base Point* button to pick a location in the drawing. The *Objects* section allows you to specify how you want to treat selected objects if you create a new .DWG file from objects (not from an existing *Block*). You can *Retain* the objects in their current state, *Convert to Block*, or *Delete from Drawing*.

The *Destination* section defines the desired *File Name and Path*, and *Insert Units*. Your choice for *Insert Units* is applicable only when you drag and drop a *Block*, as with DesignCenter. See "DesignCenter" later in this chapter for an explanation of this subject.

If you prefer the Command line equivalent to the *Write Block* dialog box, type *-Wblock* and follow the prompt sequence shown here to create *Wblocks* (.DWG files) <u>from existing *Blocks*</u>,

> Command: **-wblock**
> (At this point, the *Create Drawing File* dialog box appears, prompting you to supply a name for the .DWG file to be created. Typically, a new descriptive name would be typed in the edit box rather than selecting from the existing names.)
> Enter name of existing block or [= (block=output file)/* (whole drawing)] <define new drawing>:
> (Enter the name of the desired existing *Block*. If the file name given in the previous step is the same as the existing *Block* name, an "=" symbol can be entered, or enter an asterisk to write out the entire drawing.)
> Command:

A copy of the existing *Block* is then created in the selected directory as a *Wblock* (.DWG file).

An alternative method for creating a *Wblock* is using the *Export Data* dialog box accessed from the *File* pull-down menu. Make sure you select *Block (*.DWG)* as the type of file to export (bottom of the dialog box). You can create a *Wblock* (.DWG file) from an existing *Block* or from objects that you select from the screen. After specifying the .DWG name you want to create and the dialog box disappears, you are presented with the same command syntax as shown previously for the *-Wblock* command.

When *Wblocks* are *Inserted*, the *Color*, *Linetype*, and *Lineweight* settings of the *Wblock* are determined by the settings current when the original objects comprising the *Wblock* were created. The three possible settings are the same as those for *Blocks* (see "Block," *Color*, *Linetype*, and *Lineweight* Settings).

When a *Wblock* is *Inserted*, its parent (original) layer is also inserted into the current drawing. *Freezing* <u>either</u> the parent layer or the layer that was current during the insertion causes the *Wblock* to be frozen.

Base

Pull-down Menu	Command (Type)	Alias (Type)	Short-cut	Screen (side) Menu	Tablet Menu
Draw Block > Base	Base	...	...	DRAW 2 Base	...

The *Base* command allows you to specify an "insertion base point" (see the *Block* command) in the current drawing for subsequent insertions. If the *Insert* command is used to bring a .DWG file into another drawing, the insertion base point of the .DWG is 0,0 by default. The *Base* command permits you to specify another location as the insertion base point. The *Base* command is used in the symbol drawing, that is, used in the drawing <u>to be inserted</u>. For example, while creating separate symbol drawings (.DWGs) for subsequent insertion into other drawings, the *Base* command is used to specify an appropriate point on the symbol geometry for the *Insert* command to use as a "handle" other than point 0,0 (Fig. 21-12).

FIGURE 21-12

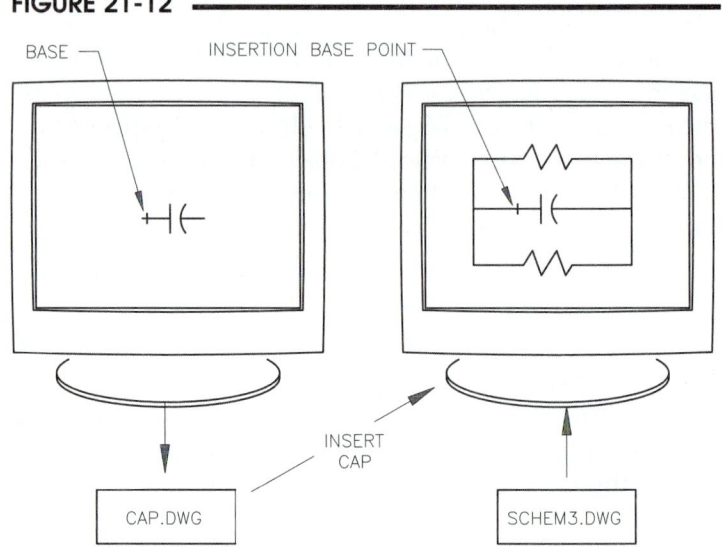

Redefining Blocks and the Block Editor

If you want to change the configuration of a *Block*, even after it has been inserted, it can be accomplished by redefining the *Block*. When you redefine a block (change a block and save it), all of the previous *Block* insertions in the drawing are automatically and globally updated (Fig. 21-13). For each *Block* insertion in the drawing, AutoCAD stores two fundamental pieces of information–the insertion point and the *Block* name. The actual block definition is stored in the block definition table. Redefining the *Block* involves changing that definition.

FIGURE 21-13

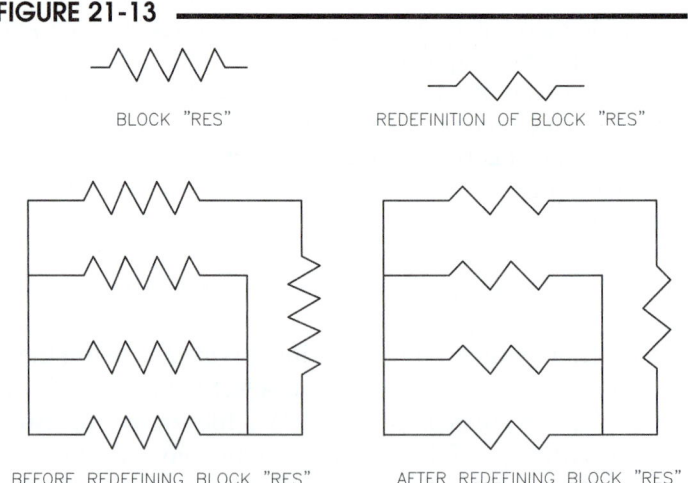

You can use three methods to redefine a *Block*: 1) use *Refedit* to change and save the block, 2) redraw the block geometry, then use the *Block* command to save the new geometry under the old block name, and 3) use the Block Editor to change the block, then *Save* the block from the editor.

To Redefine a Block Using *Refedit*

The *Refedit* command "opens" the *Block* for editing, then you make the necessary changes, and finally use *Refclose* to "close" the editing session and save the *Block*. This process has the same result as redefining the *Block*. This method is probably the least efficient method. See "*Refedit*" in Chapter 30, Xreferences.

To Redefine a Block Using the *Block* Command

To redefine a *Block* by this method, you must first draw the new geometry or change the original "template" set of objects. Alternately, you can *Explode* one insertion of the block in the Drawing Editor, then make the desired changes. (The original unexploded block cannot be included in the new *Block* because a block definition cannot reference itself.) Next, use the *Block* command and select the new or changed geometry. The old block is redefined with the new geometry as long as the original block name is used.

To Redefine a Block Using the Block Editor

To use the Block Editor, double-click on a block or use the *Bedit* command. See "*Bedit*" next.

Bedit

Pull-down Menu	Command (Type)	Alias (Type)	Short-cut	Screen (side) Menu	Tablet Menu
Tools *Block Editor*	*Bedit*	*BE*	...	...	...

To edit and redefine existing blocks, or to create dynamic blocks, use the Block Editor. Produce the Block Editor by using the *Bedit* command by any method shown in the command table above. Alternately, double-click on any block (without attributes) in the drawing (assuming *Dblclkedit* is on) to automatically open the Block Editor.

Before the Block Editor appears, the *Edit Block Definition* dialog box appears for you to select the desired block name from the *Block to create or edit* list (Fig. 21-14).

Once the desired block is selected in the *Edit Block Definition* dialog box, the Block Editor appears in place of the Drawing Editor (Fig. 21-15). The Block Editor has a light colored background by default. The Block Editor is similar to the Drawing Editor in that you can draw and edit geometry as you would normally. Most AutoCAD commands operate in the Block Editor, although a few commands, such as those related to plotting, publishing, layouts, views, and viewports, are not allowed.

The selected block geometry also appears in the drawing editor. However, instead of the block appearing as one object, it appears in its original "exploded" state—that is, the individual *Lines, Arcs, Circles,* or other objects that make up the block can be changed.

FIGURE 21-14

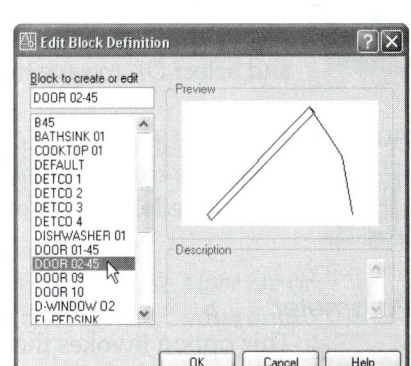

FIGURE 21-15

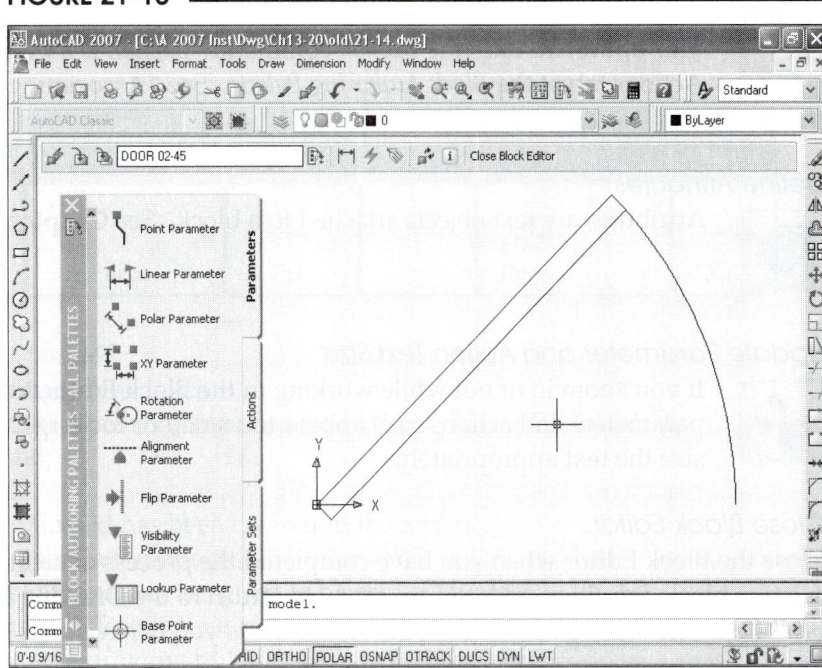

Since these named objects occupy a small amount of space in the drawing, using *Purge* to delete unused named objects can reduce the file size. This may be helpful when drawings are created using template drawings that contain many unused named objects or when other drawings are *Inserted* that contain many unused *Layers* or *Blocks*, etc.

 Purge is especially useful for getting rid of unused *Blocks* because unused *Blocks* can occupy a huge amount of file space compared to other named objects (*Blocks* are the only geometry-based named objects). Although some other named objects can be deleted by other methods (such as selecting *Delete* in the *Text Style* dialog box), *Purge* is the only method for deleting *Blocks*.

The *Purge* command produces the *Purge* dialog box (Fig. 21-17). Here you can view all named objects in the drawing, but you can purge only those items that are not used in the current drawing.

FIGURE 21-17

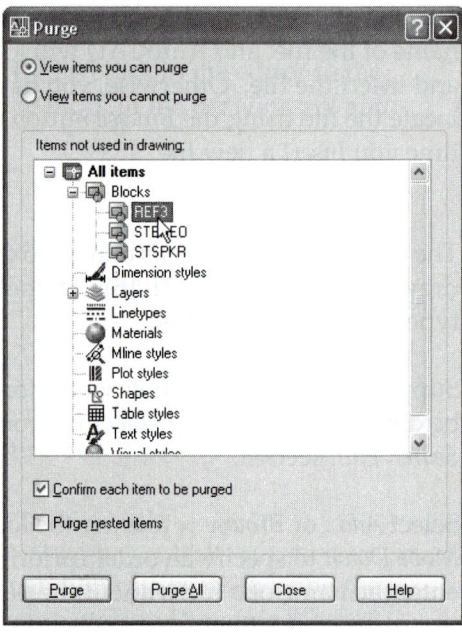

View items you can purge
Generally this option is the default choice since you can remove unused named objects from your drawing only if this option is checked.

 ### View items you cannot purge
This option is useful if you want to view all the named objects contained in the drawing. You can also get an explanation appearing at the bottom of the dialog box as to why a selected item cannot be removed.

Purge
Use the *Purge* button to remove only the items you select from the list. First, select the items, then select *Purge*. You can select more than one item using the Ctrl or Shift keys while selecting multiple items or a range, respectively.

Purge All
Use this button to remove all unused named objects from the drawing. You do not have to first select items from the list. To be safe, you should also select *Confirm each item to be purged* when using this option.

Confirm each item to be purged
If the *Confirm each item to be purged* box is checked, AutoCAD asks for confirmation before removing each object. If this box is not checked and you *Purge All*, all unused named objects in the drawing are removed at once without notification.

Purge nested items
You can remove unused nested *Blocks* and other nested named objects with this option. This option is particularly helpful if you have *Inserted* other drawings or *Xrefs* into the current drawing since all the named items contained in the inserted drawings are also inserted.

A Command line version can be invoked by typing -*Purge*.

```
Command: -purge
Enter type of unused objects to purge
[Blocks/Dimstyles/LAyers/LTypes/Materials/Plotstyles/SHapes/textSTyles/Mlinestyles/Tablestyles/Visual
Styles/Regapps/All]: b
Enter name(s) to purge <*>: Enter or (name)
Verify each name to be purged? [Yes/No] <Y>:
```

Rename

Pull-down Menu	Command (Type)	Alias (Type)	Short-cut	Screen (side) Menu	Tablet Menu
Format *Rename...*	*Rename* or *-Rename*	*REN* or *-REN*	...	*FORMAT* *Rename*	*V,1*

This utility command allows you to rename a *Block* or <u>any named object</u> that is part of the current drawing. *Rename* allows you to rename the named objects listed here:

Blocks, Dimension Styles, Layers, Linetypes, Materials, Plot Styles, Table Styles, Text Styles, User Coordinate Systems, Views, and *Viewport* configurations.

If you prefer to use dialog boxes, you can type *Rename* or select *Rename...* from the *Format* pull-down menu to access the dialog box shown in Figure 21-18.

You can select from the *Named Objects* list to display the related objects existing in the drawing. Then select or type the old name so it appears in the *Old Name:* edit box. Specify the new name in the *Rename To:* edit box. <u>You must then PICK *the Rename To:* tile</u> and the new names will appear in the list. Then select the *OK* tile to confirm.

FIGURE 21-18

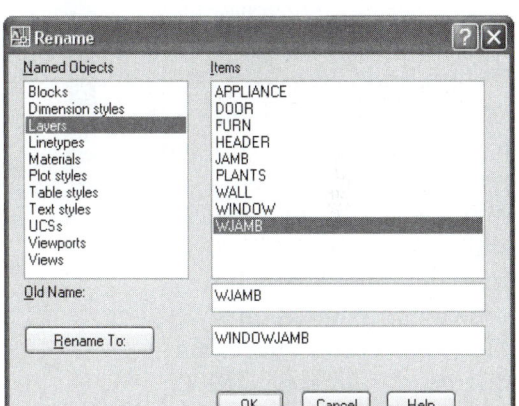

The *Rename* dialog box can be used with wildcard characters (see Chapter 43) to specify a list of named objects for renaming. For example, if you desired to rename all the "DIM-*" layers so that the letters "DIM" were replaced with only the letter "D", the following sequence would be used.

1. In the *Old Name* edit box, enter "DIM-*" and press Enter. All of the layers beginning with "DIM-" are highlighted.

2. In the *Rename To:* edit box, enter "D-*." Next, the *Rename To:* tile must be PICKed. Finally, PICK the *OK* tile to confirm the change.

The *-Rename* command (with a hyphen prefix) can be used to rename objects one at a time. For example, the command sequence might be as follows:

```
Command: -rename
Enter object type to rename
[Block/Dimstyle/LAyer/LType/Material/Plotstyle/textStyle/Tablestyle/Ucs/VIew/VPort]: b
Enter old block name: chair7
Enter new block name: desk-chair
Command:
```

Wildcard characters are not allowed with the *-Rename* command but are allowed with the dialog box.

DESIGNCENTER

The DesignCenter window (Fig. 21-19) allows you to navigate, find, and preview a variety of content, including *Blocks*, located anywhere accessible to your workstation, then allows you to open or insert the content using drag-and-drop. "Content" that can be viewed and managed includes other drawings, *Blocks*, *Dimstyles*, *Layers*, *Layouts*, *Linetypes*, *Table Styles*, *Textstyles*, *Xrefs*, raster images, and URLs (Web site addresses). In addition, if you have multiple drawings open, you can streamline your drawing process by copying and pasting content, such as layer definitions, between drawings. (Dimension Styles, Xrefs, and Raster Images are discussed in Chapters 29, 30, 32, respectively.)

FIGURE 21-19

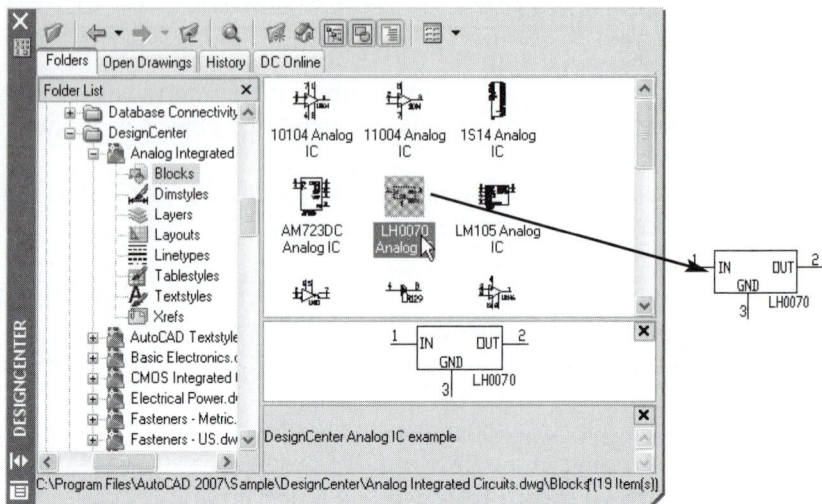

Adcenter

Pull-down Menu	Command (Type)	Alias (Type)	Short-cut	Screen (side) Menu	Tablet Menu
Tools Palettes > DesignCenter...	Adcenter	ADC	Ctrl + 2	...	...

Accessing *Adcenter* by any method produces the DesignCenter palette. There are two sections to the window. The left side is called the Tree View and displays a Windows Explorer-type hierarchical directory (folder) structure of the local system (Fig. 21-20). The right side is called the Content Area and displays lists, icons, or thumbnail sketches of the content selected in the Tree View. The Content Area can display *Blocks*, *Dimstyles*, *Layers*, *Layouts*, *Linetypes*, *Table Styles*, *Textstyles*, *Xrefs*, and a variety of other content.

Generally, the Content Area is used to drag and drop the icons or thumbnails from the content area into the current drawing (see Fig. 21-19). However, you can also streamline a variety of tasks such as those listed here:

- Browse sources of drawing content including open drawings, other drawings, raster images, content within the drawings (*Blocks*, *Dimstyles*, *Layers*, and so on), content on network drives, or content on a Web page.
- Insert, attach, or copy and paste the content (drawings, images, *Blocks*, *Layers*, etc.) into the current drawing.

- Create shortcuts to drawings, folders, and Internet locations that you access frequently.

- Use a special search engine to find drawing content on your computer or network drives. You can specify criteria for the search based on key words, names of *Blocks, Dimstyles, Layers,* etc., or the date a drawing was last saved. Once you have found the content, you can load it into DesignCenter or drag it into the current drawing.

- Open drawings by dragging a drawing (.DWG) file from the Content Area into the drawing area.

These features of DesignCenter are explained in the following descriptions of the DesignCenter toolbar icons.

FIGURE 21-20

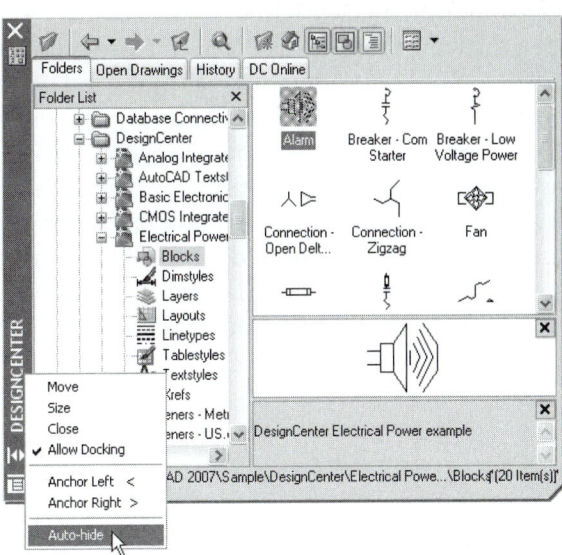

Resizing, Docking, and Hiding the DesignCenter Palette

You can change the width and height of DesignCenter by resting your pointer on one of the borders (not the title bar) until a double arrow appears, then dragging to the desired size. Alternately, rest your pointer on the lower corner (where the bevel appears) to resize both height and width. You can also move the bar between the Content Area and the Tree View area. Dock DesignCenter by clicking the title bar, then dragging it to either edge of the drawing window. You can hide the DesignCenter palette by using the *Auto-hide* option (click on the *Properties* button at the bottom of the title bar to produce the *Properties* menu as shown in Fig. 21-20). When *Auto-hide* is on, the palette is normally hidden (only the title bar is visible) and the palette appears when you bring the pointer to the title bar.

DesignCenter Options

In addition to selecting the icons from the toolbar, you can right-click the Palette background to produce the shortcut menu and choose the desired option (Fig. 21-21).

Folders Tab

Folders is the default display for the Tree View side of DesignCenter. This choice displays a hierarchical structure of the desktop (local workstation). Because this arrangement is similar to Windows Explorer, you can navigate and locate content anywhere accessible to your system, including network drives. Figure 21-21 displays a typical hierarchical structure on the Tree View side. For example, you may want to import layer definitions (including color and linetype information) from a drawing into the current drawing by dragging and dropping.

FIGURE 21-21

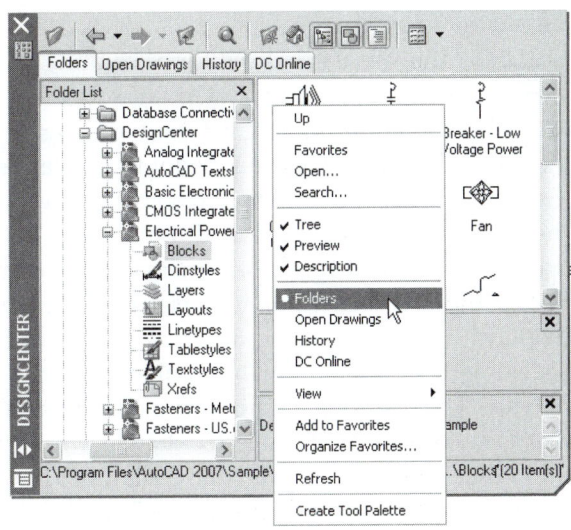

Open Drawings Tab

This option changes the Tree View to display all open drawings (Fig. 21-22). This feature is helpful when you have several drawings open and want to locate content from one drawing and import it into the current drawing. As shown in Figure 21-22, you may want to locate *Block* definitions from one drawing and *Insert* them into <u>another</u> drawing. Ensure you make the desired "target" drawing current in the drawing area before you drag and drop content from the Content Area.

FIGURE 21-22

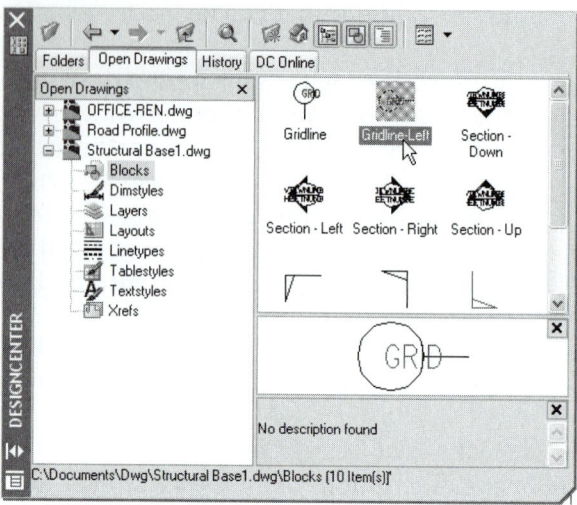

History Tab

This option displays a history (chronological list) of the last 20 file locations accessed through DesignCenter (Fig. 21-23). The purpose of this feature is simply to locate the file and load it into the Content Area. Load the file into the Content Area by double-clicking on it.

FIGURE 21-23

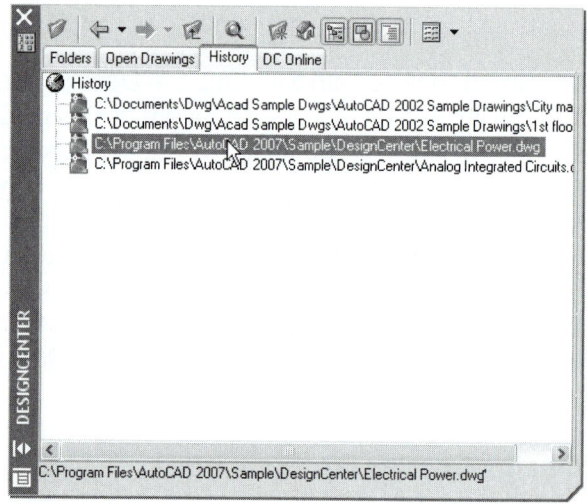

Tree View Toggle

 Tree View is helpful for navigating your system for content. Once the desired folder or drawing is found and highlighted in Tree View, you may want to toggle Tree View off so only the Content Area is displayed with the desired content. The desired content may be drawings, *Blocks*, images, or a variety of other content. For example, consider previous Figure 21-22 which displays Tree View on. Figure 21-24 illustrates Tree View toggled off and only the Content Area displayed. The resulting configuration displays only the *Blocks* contained in the selected drawing. Keep in mind that you can also change the Views of the Content Area to display *Large Icons*, *Small Icons*, a *List*, or *Details* (see "Views").

FIGURE 21-24

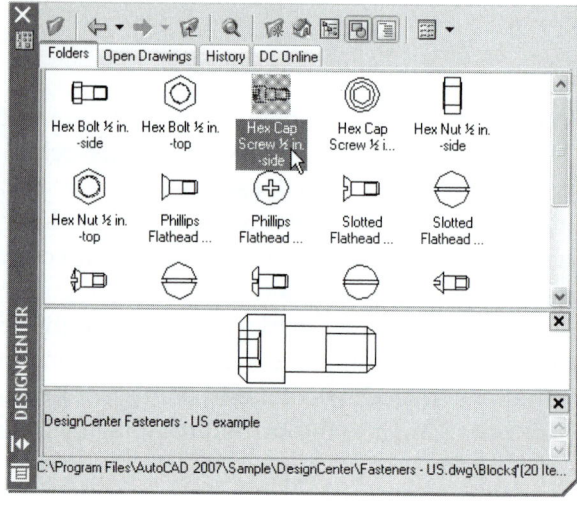

Favorites

This button displays the contents of the AutoCAD Favorites folder in the Content Area. The Tree View section displays the highlighted folder in the Desktop view.

You can add folders and files to *Favorites* by highlighting an item in Tree View or the Content Area, right-clicking on it, and selecting *Add to Favorites* from the shortcut menu (Fig. 21-25).

FIGURE 21-25

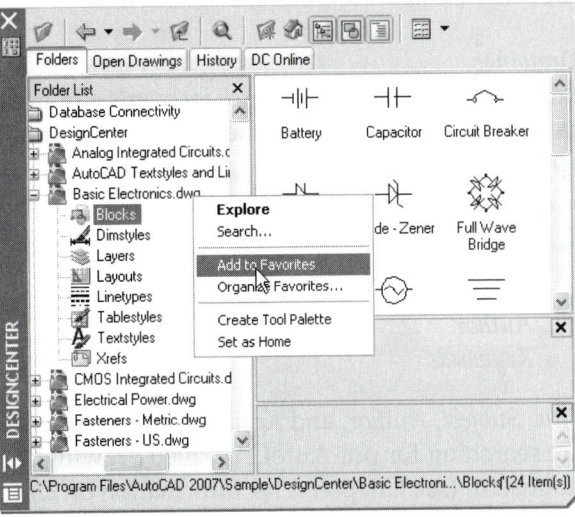

Load

Displays the *Load* dialog box (not shown), in which you can load the Content Area with content from anywhere accessible from your system. The *Load* dialog box is identical to the *Select File* dialog box (see "*Open*," Chapter 2, Working with Files). After selecting a file, DesignCenter automatically finds the file in Tree View and loads its content (*Blocks*, *Layers*, *Dimstyles*, etc.) into the Content Area. You can also load files into the Content Area using Windows Explorer (see "Loading the Content Area with Windows Explorer" at the end of this section).

Search

This button invokes the *Search* dialog box (Fig. 21-26), in which you can specify search criteria to locate drawing files, *Blocks*, *Layers*, *Dimstyles*, and other content within drawings. <u>Once the desired content is found in the list at the bottom of the dialog box, double-click on it to load it into the Content Area</u>.

FIGURE 21-26

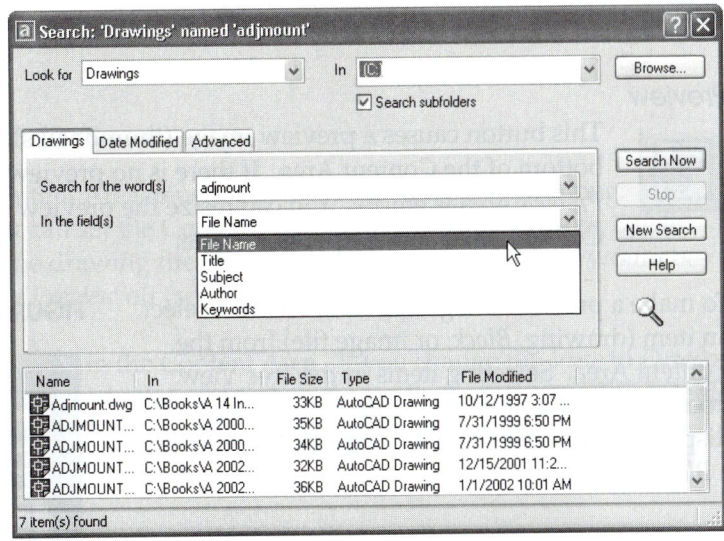

This feature is extremely powerful and easy to use. The list of possible items to search for is displayed in the *Look For* drop-down list and includes the following choices:

Blocks, Dimstyles, Drawings, Drawings and Blocks, Hatch Patterns, Hatch Pattern Files, Layers, Layouts, Linetypes, Tablestyles, Textstyles, Xrefs

Using the *Tool Palette* window, you can preset the *Block* insertion properties or hatch pattern properties of any tool on a tool palette. For example, you can change the insertion scale of a *Block* or the angle of a hatch pattern. Do this by right-clicking on a tool and selecting *Properties...* (see "Tool *Properties*").

You can also change the scale and rotation angle after inserting the *Block* using Grips (see Chapter 19, Grip Editing).

Palette *Properties*

The options and settings for tool palettes are accessible from shortcut menus in different areas on the *Tool Palettes* window. Right-clicking inside the palette produces the menu shown in Figure 21-35. You can also click on the *Properties* button (bottom of the vertical title bar) to produce essentially the same menu with the addition of *Move, Size,* and *Close* options. Palette properties are saved with your AutoCAD profile.

FIGURE 21-35

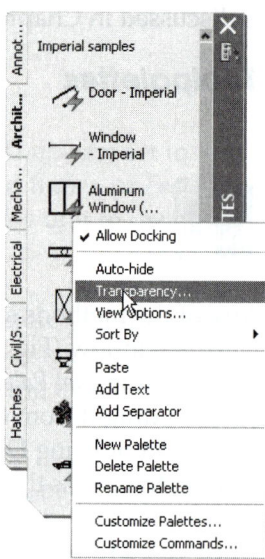

Move, Size, Close, Allow Docking, Auto-hide
These options are identical to the same features in DesignCenter. See "Resizing, Docking, and Hiding the DesignCenter Palette" earlier in this chapter.

Transparency...
Use this option to make the *Tool Palettes* window transparent so you can "see through" the window to the drawing underneath. Set the desired level in the *Transparency* dialog box that appears (Fig. 21-36).

NOTE: In order to use transparency, you must select *Software Acceleration* in the *Options* dialog box. Select the *System* tab, then *Properties...* under *Current 3D Graphics Display*. Transparency is not available if you are using the Microsoft Windows NT operating system.

FIGURE 21-36

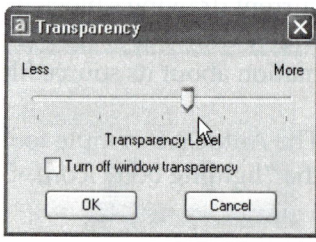

View Options...
This option produces the *View Options* dialog box (Fig. 21-37). You can specify the *Image size* and the *View style*. The *List View* (shown in Fig. 21-35) places the tools with text in a vertical column regardless of the palette width, whereas the *Icon with text* option is best to use when you increase the window width to create multiple columns.

Add Text, Add Separator, New Palette, Delete Palette, Rename Palette, and *Customize*
See "Creating and Managing Tool Palettes" later in this section.

FIGURE 21-37

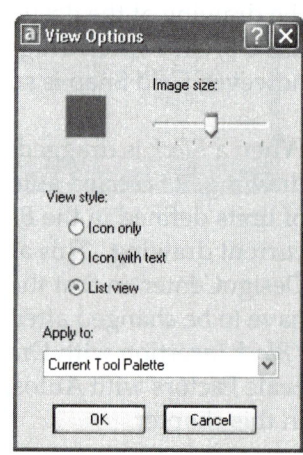

Tool *Properties*

If you right-click on a tool (*Block* or hatch pattern), a shortcut menu appears (Fig. 21-38) allowing you to change properties <u>for that specific tool</u>. You can *Cut* any tool and *Paste* it to another palette or *Copy* a tool to another palette. Selecting *Delete* removes it from that palette only. If the source block file has been changed, you can *Update tool image*. Select *Block Editor* to edit the current block in the Block Editor.

FIGURE 21-38

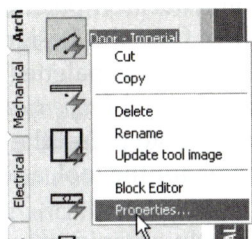

You can change the properties (such as scale and rotation angle) for any specific tool. Do this by right-clicking on the tool and selecting *Properties...* from the shortcut menu (see Fig. 21-38) to produce the *Tool Properties* dialog box (Fig. 21-39).

The *Tool Properties* dialog box can have three categories of properties—*Insert* (or *Pattern*) properties, *General* properties, and *Custom* properties. You can control object-specific properties such as *Scale* and *Rotation* angle in the *Insert* list.

Use the *General* properties section <u>only</u> if you want to override the current drawing's property settings such as *Layer*, *Color*, and *Linetype*. (This action is sometimes referred to as setting a tool property <u>override</u>.) When a property override is set, special conditions may exist. For example, if a specified layer does not exist in the drawing, that layer is created automatically when the *Block* or hatch is inserted. If a *Block* or hatch is inserted on a layer that is turned off or frozen, the *Block* or hatch is created on the current layer instead. *Custom* properties are used for dynamic blocks.

FIGURE 21-39

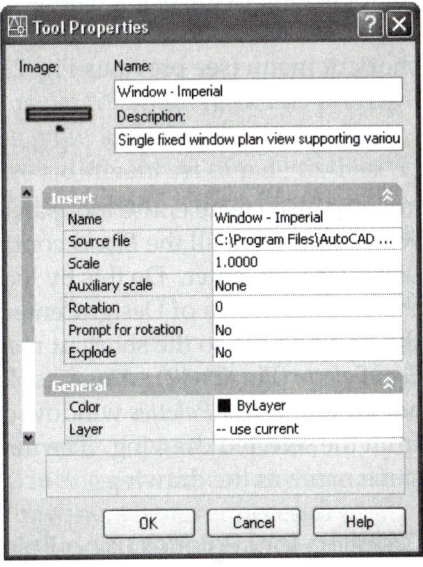

Creating and Managing Tool Palettes

Creating Tool Palettes using DesignCenter

DesignCenter provides a simple method for creating new tool palettes. Begin by opening both the DesignCenter window and the *Tool Palettes* window. Locate the desired *Blocks* or hatch patterns you want to insert into a new (not yet created) palette and ensure they appear in the DesignCenter Content Area. Next, right-click on any tool you want to move into the new palette so a shortcut menu appears (Fig. 21-40, center). Select *Create Tool Palette* from the menu. A new palette appears with the selected *Block* shown at the top of the palette. Near the new *Block*, a small edit box (not shown) appears for you to enter in the desired tool palette name. After doing so, you are presented with the new palette containing the selected *Block* (as shown in Fig. 21-40, right side).

FIGURE 21-40

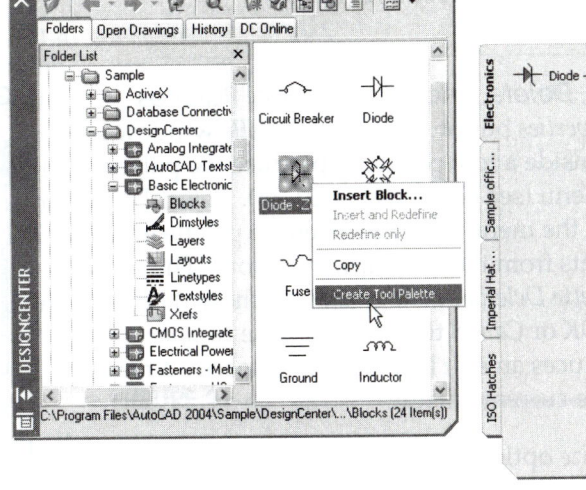

XY Parameter

Similar in performance to a *Polar Parameter*, the *XY Parameter* allows the specified geometry to be changed in any direction, but in separate X and Y distances from the base point you specify. In the drawing, you can change the X and Y distances in the *Properties* palette or by using the key grips. In the Block Editor, the parameter displays as a horizontal and vertical pair of "Distances" that share a common base point (Fig. 21-54). *Move, Scale, Stretch,* or *Array Actions* can be associated with this parameter.

FIGURE 21-54

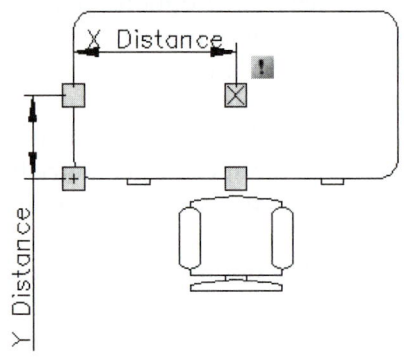

Rotation Parameter

Apply this parameter to a block or part of a block that you expect to rotate in the drawing. The selected objects rotate about the base point you define (designated by the "X" marker), and the angle shown in the *Properties* palette (when you later select the block in a drawing) is measured relative to the base "Angle" you specify for the parameter (Fig. 21-55). In the Block Editor, the parameter displays as a circle. Only a *Rotate Action* can be associated with this parameter. See "*Rotate Action.*"

FIGURE 21-55

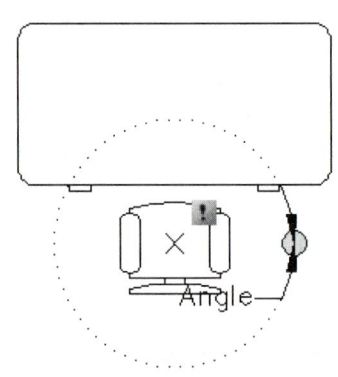

Alignment Parameter

An *Alignment Parameter* is defined by an X,Y location and an angle. When you assign this parameter, select a "base point" on the block to designate as the key point grip, then specify the alignment direction to align with a prominent object of the block geometry. In our chair-desk example, the base point appears as the "X" marker with the attached grip and the alignment direction appears as the dotted line (Fig. 21-56). If needed, you can define an "alignment type" as *Perpendicular* (default) or *Tangent.* An alignment parameter always applies to the entire block and <u>needs no action associated with it.</u>

FIGURE 21-56

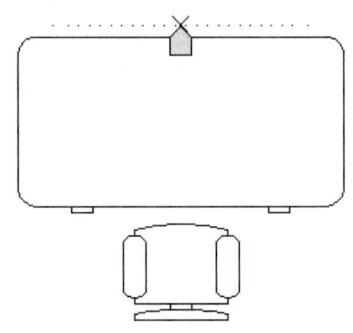

When the block is inserted into the drawing, the alignment parameter allows the block to automatically rotate around the key point so that it aligns with any other object that it touches. For example, the chair-desk block automatically aligns with any wall at any angle when the block is inserted (Fig. 21-57).

FIGURE 21-57

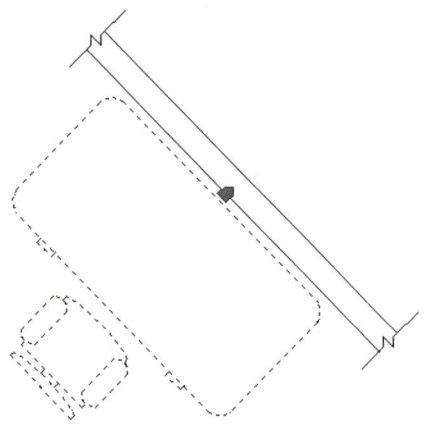

These two met
the *Baction* cor
displays only
parameter. Fo
prompt. You
Block Editor.

    Command
    Select par
    Enter acti

Remember th
etc. any norm
fact that you
that can be af
ters, the selec
actions to par
the steps are:

1.   Associate
2.   For some
3.   Select the
     drawing.
     part of th
4.   Dependi
5.   Indicate
     does not
     in the dr

Move Actio

You can asso
Parameter. F
select the de
block is edit
block geome
21-60 displa
Parameter. C
be associated

Once the bl
on the block
and the *Mo*
allows you
chair and th
you can cha
in the *Prope*
block.

Flip Parameter

This parameter is designed to flip blocks on a "reflection line," similar to mirroring an object. You specify two points to designate the reflection line. This first point of the reflection line is the default location for the flip parameter grip. In the chair-desk block example, the grip appears on the left as the arrow-shaped grip and the reflection line is the top edge of the desk denoted by the dotted line (Fig. 21-58). If a *Flip Action* is then attached to a *Flip Parameter*, you click the arrow grip to flip the block in the drawing. The Properties palette displays the *Flip State* as *Flipped* or *Not Flipped*. See "*Flip Action.*"

FIGURE 21-58

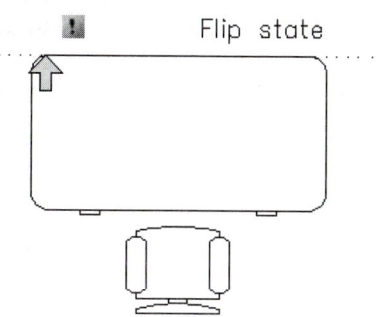

Visibility Parameter

Use this parameter to control the visibility of objects in the block. You control which objects are visible for different "visibility states" that you name. Therefore, changing a visibility state in the drawing changes the part of the block that is visible. In a drawing, you click the key grip to display a list of named visibility states available, then select the desired state. A *Visibility Parameter* always applies to the entire block and <u>needs no action associated with it</u>. See "Visibility States" after the section on applying actions.

Lookup Parameter

This parameter defines a custom property such that other parameters are linked to values in a table you define. The parameter is associated with a *Lookup Action* that defines the lookup table. When the block is inserted into the drawing, you can click the key grip to display a list of available properties, so selecting an option forces the block to change based on values in the table. See "Lookup Actions and Lookup Tables" for more information.

Base Point Parameter

This parameter allows you to define an additional base point for the dynamic block reference relative to the geometry in the block. This parameter cannot be associated with any actions, but it can belong to an action's selection set. Therefore, use this parameter if you need additional base points to use with multiple actions. In the Block Editor, this parameter displays as a circle with crosshairs (not shown).

NOTE: Since a *Base point Parameter* needs no associated actions, you can use the *Base point Parameter* with any normal (non-dynamic) block to reset its base point for insertion.

Parameter Custom Properties

When you assign a parameter using either the *Parameters* tab of the *Block Authoring Palettes* or the *Bparameter* command, the first prompt you receive at the command line is similar to the following prompt, but varies depending on the parameter:

    Specify start point or [Name/Label/Chain/Description/Base/Palette/Value set]:

When you select the dynamic block reference in a drawing, these properties (such as *Name, Label,* etc.) and possibly other properties are shown in the *Properties* palette under *Misc* or *Custom*. Use these options at the command prompt to specify other than the default properties.

NOTE: Although these options appear in the prompts when you first assign parameters, in many cases it is easier to assign the custom properties using the *Properties* palette <u>after the block has been created</u>. This process is accomplished using the Block Editor to edit an existing dynamic block, then opening the *Properties* palette <u>in the Block Editor</u>.

Name/L
Each pa
"Linear
mended
the *Prog*
tive lab
applyin
from "I
that ap

Base
This op
point c
edited
cated t
remair
from tl

Chain
It is pc
than o
can tri
either

Palette
By de:
in a d:

Value
This p
how t
optioi
Rotate
For tl
speci
edited

Ass

An a
a pai
ertie:
and
ter, a
cases
than

To a:
Fig.
ousl

Scale Action

A *Scale Action* can be associated with a *Linear, Polar,* or *XY Parameter*. This action allows a set of objects to be scaled relative to a base point when the block is edited in the drawing. First you are prompted to specify which objects in the block are to be scaled. For example, Figure 21-62 displays a *Scale Action* associated with a *Polar Parameter*. Only the desk is specified as the selection set for the action—the chair is not selected. The base point is denoted by an "X" marker.

FIGURE 21-62

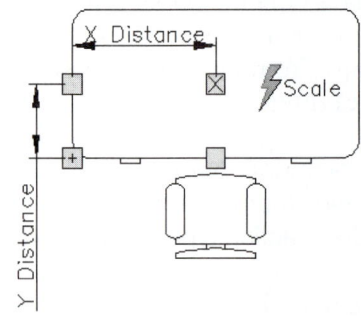

Once the chair-desk block is inserted into the drawing, the key grip is dragged to scale the desk proportionally in the X and Y distances (Fig. 21-63). The chair is not changed since it was not included in the selection set associated with the action.

FIGURE 21-63

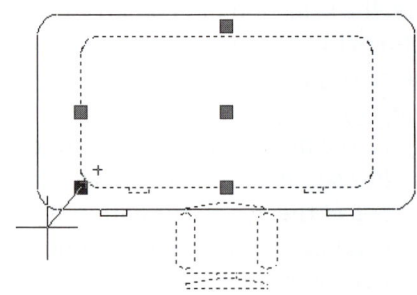

When you associate a *Scale Action* with a *Linear* or *Polar Parameter,* the following prompt is displayed:

> Specify action location or [Base type]:

By default, a *Dependent* base point is used, in which case the geometry is scaled relative to the base point of the parameter. An *Independent* base point is one that you specify. An independent base point is shown in the Block Editor as an "X" marker.

If you select an *XY Parameter* to associate with the *Scale Action*, the following prompt is displayed:

> Specify action location or [Base type/XY]:

Here you can specify whether the (scale) distance is applied to the X distance only, Y distance only, or XY distance from the base point.

Stretch Action

You can associate a stretch action with a *Point, Linear, Polar,* or *XY Parameters*. In a dynamic block reference, a stretch action causes objects to move and stretch a specified distance in a specified location. First you are prompted to select the parameter to associate, then you "Specify parameter point to associate with action" (select the grip to designate as the key grip).

Next you specify a "stretch frame," then specify a selection set. These two steps appear to be simple but require special attention. The frame must cross the objects you want to stretch and include the key grip; however, the frame only <u>partially</u> determines which objects stretch and move (objects entirely within the frame are moved, objects that cross the frame are stretched). However, <u>objects you want to stretch and move must also be included in the selection set</u>. In our example, the stretch frame and the selection set include the same objects, as shown (Fig. 21-64). The chair is not included with the selection set for either action.

FIGURE 21-64

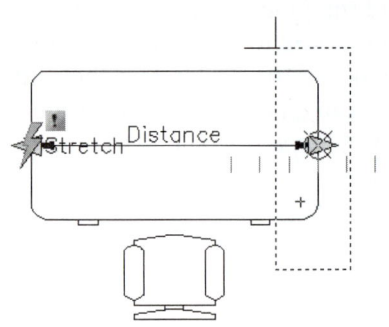

Often two *Stretch Actions* are associated with one *Linear Parameter*. In our example, one *Stretch Action* has been associated with a *Linear Parameter* and appears on the left side of the desk (see Fig. 21-64). A second *Stretch Action* is in process, with the stretch frame being applied to the right side of the desk.

The result, when the block is edited in the drawing, the desk can be stretched by moving either of the key grips to stretch only the associated side of the desk (Fig 21-65). The chair is not affected. In addition, the *Linear Parameter* has a parameter value set to 6"; therefore, the desk can be stretched in 6" increments only. Accomplish this using the *Value set* option when assigning the parameter, or use the *Properties* palette at any time in the Block Editor to add the values.

FIGURE 21-65

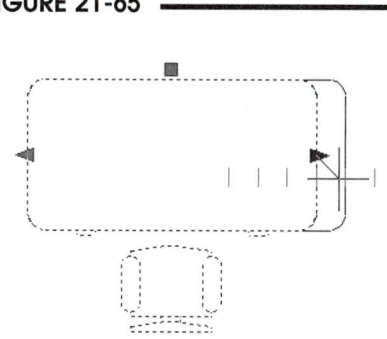

Polar Stretch Action

You can only associate a *Polar Stretch Action* with a *Polar Parameter*. In a block reference, a *Polar Stretch Action* rotates, and/or moves, and/or stretches objects a specified angle and distance when the key point on the associated polar parameter is changed using either a grip or the *Properties* palette. In other words, with this action you can specify that some objects in the block stretch while others move while still others rotate only. This control is accomplished by 1) associating the *Polar Stretch Action* with a *Polar Parameter*, 2) specifying a stretch "frame," 3) specifying a selection set for the action, and 4) specifying a selection set for rotating only. This is a fairly involved process. The resulting action operates by these rules:

    The stretch frame determines only which objects stretch and it must include the key grip.
    The selection set determines which objects move and rotate.
    Objects to stretch must 1) cross the frame and 2) be in the selection set.
    Objects selected to rotate only do just that.

To clear up any misconceptions acquired by experimentation, reading the prompts, or reading the AutoCAD *Help*, these points are offered.

    Objects entirely within the frame only do nothing.
    The frame does not determine what objects move and rotate.
    The frame does not determine where the stretch occurs, only the objects that get stretched.
    Objects not in the selection set do nothing, even if included in the frame.
    A selection set crossing window does not determine where the stretch occurs.

When associating the *Polar Stretch Action* to the *Polar Parameter* in our chair-desk example (Fig. 21-66), note that the stretch frame (vertical crossing window) includes the key grip and indicates only the two lines crossing the frame that get stretched. The object selection set (large crossing window) includes the left side of the desk and left desk handle (these objects get stretched and/or moved). The objects selected to rotate only are the chair and right end of the desk.

FIGURE 21-66

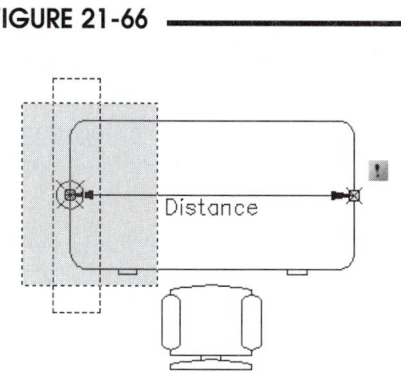

When the block is edited in the drawing, the left end of the desk stretches and rotates, and the chair and the right end of the desk rotate only (Fig. 21-67). Note that the left desk handle moves and rotates since it was included in the selection set, even though it was not included in the stretch frame. The right handle does nothing since it was not included in the selection set or the rotate-only set.

FIGURE 21-67

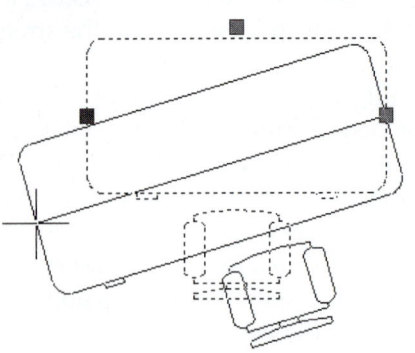

Rotate Action

You can associate a *Rotate Action* only with a *Rotation Parameter*. Although you could rotate any normal block about its insertion grip, this action allows you to rotate only selected objects in the block. For the chair-desk block example, you could apply the rotation parameter to the chair only (Fig. 21-68).

FIGURE 21-68

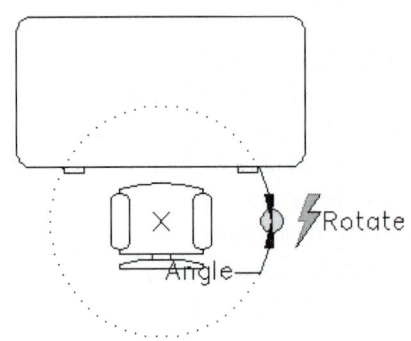

To complete this procedure, simply associate the *Rotate Action* to the matching parameter, then specify the objects within the block you want to rotate. Once the block has been inserted into the drawing, select the block to display its grips, then rotate the objects associated with the *Rotate Action* using the key grip (Fig. 21-69). Alternately, the associated objects can be rotated using the *Properties* palette.

FIGURE 21-69

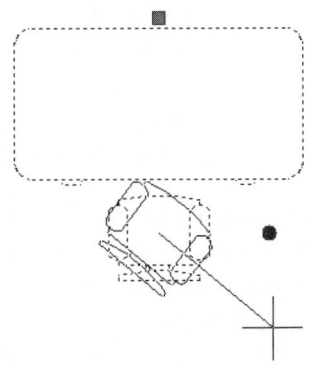

For particular applications, you may want to associate multiple *Rotate Actions* with one *Rotation Parameter*, therefore allowing you to rotate multiple objects in the block individually. To accomplish this, rather than using the base point of the associated *Rotation Parameter*, set the *Base Type* to *Independent*, then specify a new base point location for the rotate action in the proximity of the unique set of objects to rotate. An independent base point is shown in the Block Editor as an "X" marker.

Flip Action

You can associate a *Flip Action* only with a *Flip Parameter*. Flip is similar to the action of the *Mirror* command. With a *Flip Action*, you can flip a dynamic block reference about a specified axis called a reflection line. Since the reflection line is specified previously when the *Flip Parameter* is assigned (Fig. 21-70), this procedure requires only that you select the geometry to be affected by the action.

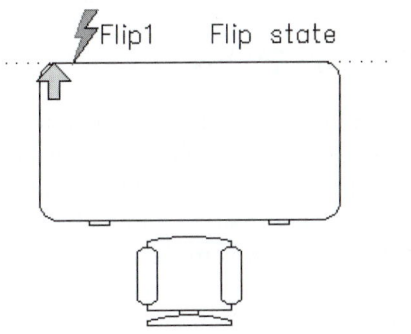

FIGURE 21-70

Once the block has been inserted into the drawing, you can select the block, then click the arrow grip to flip the geometry. You can also use the *Properties* palette to set the *Flip State* to *Flipped* or *Not Flipped*. Although you can specify that only some objects are affected by the action, in the chair-desk example the entire block is flipped (Fig. 21-71).

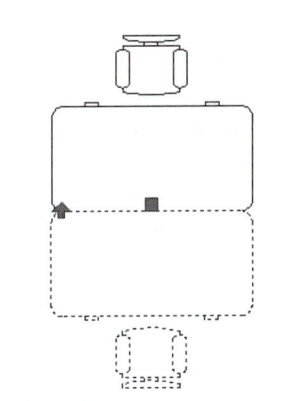

FIGURE 21-71

Array Action

An *Array Action* can be associated with a *Linear, Polar,* or *XY Parameter*. An *Array Action* causes its associated objects to create a rectangular array pattern when the associated parameter is edited using the grips or using the *Properties* palette.

In our chair-desk example, the *Array Action* is associated with an *XY Parameter* (Fig. 21-72). As with most actions, first associate an *Array Action* with a parameter, then select the objects within the block that you want to be affected by the action. Finally, specify a distance between rows and columns by entering values or by using a unit cell (picking two diagonal corners).

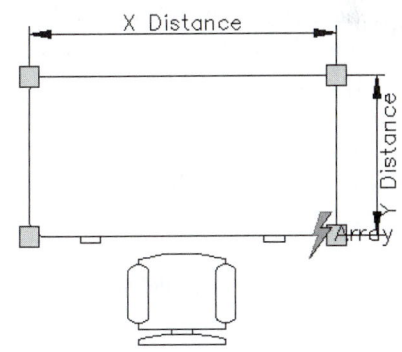

FIGURE 21-72

When you later edit the inserted block in the drawing, simply stretch the key grip to create an array. If the *Array Action* is associated with either a *Polar Parameter* or an *XY Parameter*, like our chair-desk example, stretching the grip diagonally creates a rectangular array (Fig. 21-73). Alternately, you could change the parameter values (*X Distance* and *Y Distance*, in this case) in the *Properties* palette to create the array. Stretching either horizontally only or vertically only would create a single row or single column. However, when you associate an array action with a *Linear Parameter*, you can create only a linear array (one column or row, depending on the parameter direction).

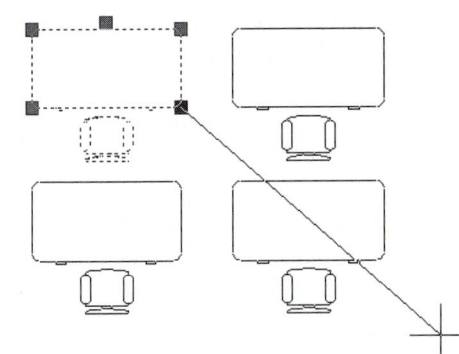

FIGURE 21-73

Lookup Action
See "Lookup Actions and Lookup Tables" after the "Visibility States" section.

Visibility States

FIGURE 21-74

"Visibility states" are simply names you assign to display different parts of a block. Changing the visibility state of a block in the drawing (selecting a name from the list) turns on or off (makes visible or invisible) the assigned geometry for that state. For example, a dynamic block currently display-ing one standard fastener could contain additional geometry that described other similar fasteners (Fig 21-74). Selecting the key grip and then selecting another choice from the list of "visibility states" would display other geometry currently invisible—other fasteners in this case.

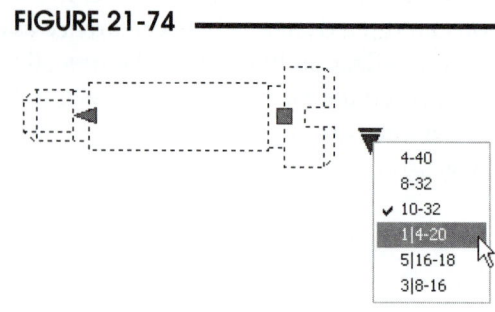

Use the Block Editor to create blocks with visibility states. When the block geometry is created, assign a *Visibility Parameter* to the entire block. Once you have assigned this parameter, the visibility tools on the right end of the Block Editor toolbar are enabled (Fig. 21-75).

FIGURE 21-75

Alternately, use the *Bvstate, Bvhide,* and *Bvmode* commands in the Block Editor. Use these tools to assign names for your visibility states and to select which parts of the block you want to be visible or invisible for each state, then save the block. Remember that a *Visibility Parameter* always applies to the entire block and needs no action associated with it.

Once you have inserted a block with visibility states into the drawing, you can click on the visibility grip to get a list of the visibility states assigned to the block, then select the desired state to display that part of the block. Changing a visibility state in the drawing changes the part of the block that is visible.

A typical application for visibility states is using multiple blocks that are similar but combining them into one block with multiple visibility states. Each state displays one of the original blocks. For example, assume you had three blocks: a small desk chair, a large desk chair, and a conference chair. Rather than keep the three chairs as separate blocks in the library and insert each one individually, you can create only one dynamic block with visibility states that contains each of the chairs, but displays only one at a time. In this way you insert one dynamic block, then click on the key grip to display the desired chair.

Steps for Defining Visibility States

The steps for creating a dynamic block with visibility states are listed below. Although there are other variations of this method, a simple procedure is listed here:

1. Open the Block Editor.
2. Create the desired geometry containing all possible variations of the block displays. Alternately, insert the desired blocks to combine into one dynamic block.
3. Assign a *Visibility Parameter* to the geometry. The visibility tools become enabled.
4. Use the *Bvstate* command or *Manage Visibility States* button to produce the *Visibility States* dialog box, then select the *New* button.
5. Assign a name for each desired visibility state where each name is a descriptor of the associated geometry, such as fastener1, fastener2, or chair1, chair2, etc. Also select the *Show all existing geometry in new state* option. Select OK in the *New Visibility States* dialog box and select *OK* in the *Visibility States* dialog box.

6. In the Block Editor, note that one new visibility state name is displayed in the drop-down list. Use the *Bvhide* command or *Make Invisible* button, then select all of the geometry that is not intended to be part of that visibility state.
7. Highlight another visibility state name from the list, then use the same method to make invisible all the geometry that is not part of the current state. Repeat this for each visibility state (each set of geometry).
8. Use the *Save Block As* button to assign a name for the new dynamic block.
9. *Close* the Block Editor and *Insert* the new dynamic block into your drawing to test it.

For example, assume you have three blocks: a small desk chair, a large desk chair, and a conference chair, and you want to combine these into one dynamic block. First, open the Block Editor. Insert each of the three chair blocks. Ensure the chairs are all located at the same insertion point (Fig. 21-76). Assign a *Visibility Parameter*. The location of the parameter icon determines the location of the key grip used to display the visibility states when the block is later edited in the drawing. Assigning the parameter also enables the visibility toolbar (see previous Fig. 21-75).

FIGURE 21-76

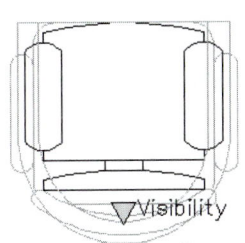

Next, use the *Manage Visibility States* button (or *Bvstate* command) to produce the *Visibility States* dialog box (Fig. 21-77). Select the *New* option to produce the *New Visibility States* dialog box (Fig 21-78). Assign three new visibility state names for the chair blocks, such as "Desk-Small," "Desk-Large," and "Conference." Select *Show all existing objects in new state* in this dialog box, since you can use the buttons on the toolbar to specify visibility at a later time. Select the *OK* button to create the new states and return to the *Visibility States* dialog box. Select *OK* in the *Visibility States* dialog box to return to the Block Editor.

FIGURE 21-77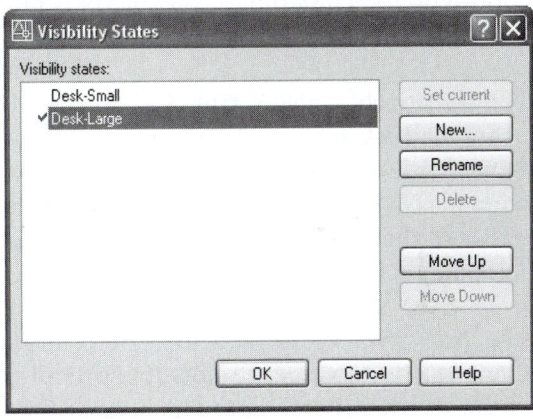

Note that the first visibility state name, "Desk-Small," for example, appears in the visibility states drop-down list. Note that all of the geometry included in the block is displayed by default, since the *Show all existing objects in new state* option was used. Use the *Make Invisible* button or *Bvhide* command, then select all of the geometry that should not be part of this chair (not visible for this state). Next, select another chair name (visibility state name) from the list to display all of its geometry. Again, all geometry included in the block is displayed, so use the *Make Invisible* button or *Bvhide* command, then select all of the geometry that should not be part of this chair. Repeat this procedure for the other visibility states. Use *Save Block As* and assign a name for the new dynamic block. Close the Block Editor.

FIGURE 21-78

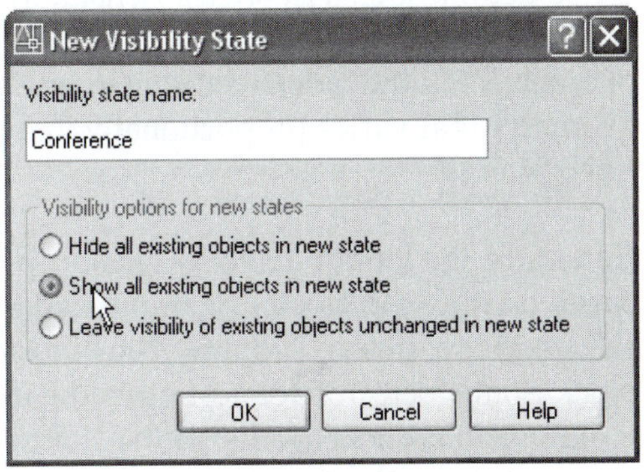

Finally, Insert the new dynamic block into a drawing. Selecting the
block displays the key (down arrow-shaped) grip. Select the grip to
display the visibility state names (Fig. 21-79). Selecting each name
should display only the appropriate chair.

FIGURE 21-79

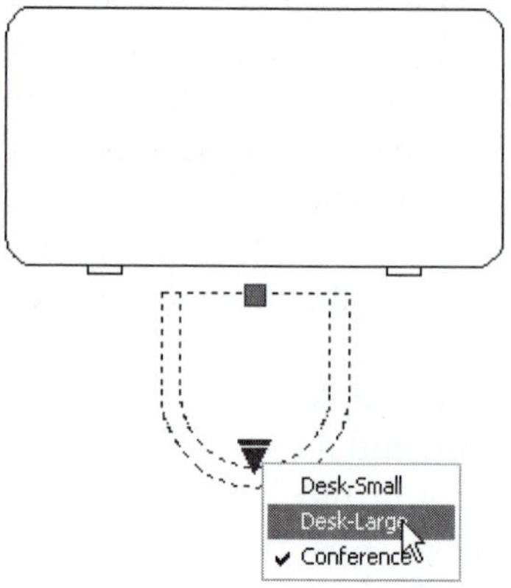

Lookup Actions and Lookup Tables

You can associate a lookup table to a dynamic block to control the geometry in the block. For example,
as an alternative to using visibility states to display different sets of geometry, you can use a lookup table
to control the display of one set of geometry to give different variations. Using a group of mechanical
fasteners as an example, you can use visibility states to display different size fasteners such that each visibility state controls which parts of the geometry are displayed (which fastener is displayed).
Alternately, you could create a dynamic block with only one set of geometry, then use a lookup table to
control how each part of the geometry is proportionally stretched to display different variations, therefore different fasteners.

You can attach a lookup table to a dynamic block
using the *Lookup Parameter* and matching *Lookup
Action*. A *Lookup Action* can only be associated
with a *Lookup Parameter*. When you add a *Lookup
Action* to the block definition, the *Property Lookup
Table* dialog box is displayed (Fig. 21-80).

FIGURE 21-80

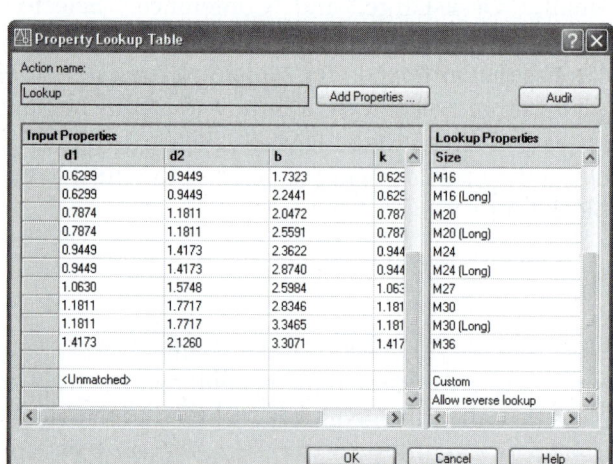

The Property Lookup Table dialog box allows you
to input a set of values to store for specific aspects
of the dynamic block geometry. Each <u>column</u> of
Input Properties <u>is assigned to a parameter</u>, such as
a *Linear Parameter*, and each parameter has an associated action, such as a *Stretch Action*. In this case,
each column in the table controls a specific length
of an object in the block resulting in one set of
geometry that varies proportionally in a predefined manner.

Generally, the lookup table assigns property values to the dynamic block
based on how the block is manipulated in a drawing. If a *Reverse Lookup* is
selected for a lookup property, the block displays a lookup grip when the
block is later selected for editing in the drawing (Fig. 21-81). When the
lookup grip is clicked, a list of the *Lookup Properties* is displayed. Selecting
an option from that list changes the display of the dynamic block based on
the values specified in the *Input Properties*.

FIGURE 21-81

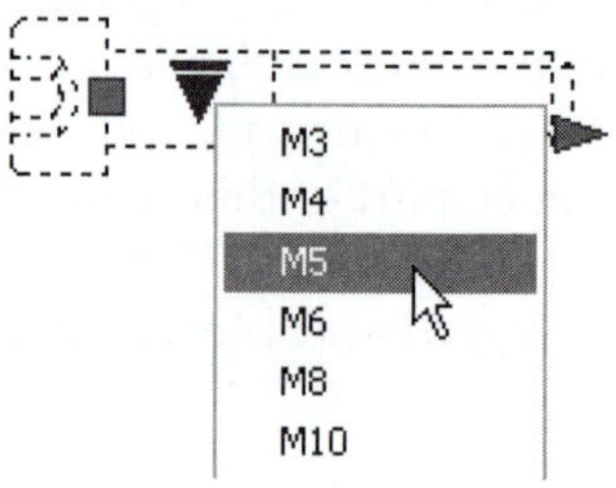

The process of assigning a column of *Input Properties* is simple. This action is accomplished by giving the columns the same name as the parameter names. As a matter of fact, this happens automatically when you use the *Add properties* button in the *Property Lookup Table* to add *Input Properties*. Doing so produces the *Add Parameter Properties* dialog box already populated with the current block's parameter names (Fig. 21-82). Select the *Add input properties* option and *OK* button to add that parameter's property name to the *Input Properties* as a column head in the *Property Lookup Table*. Any values subsequently entered into the column automatically associate with the parameter.

FIGURE 21-82

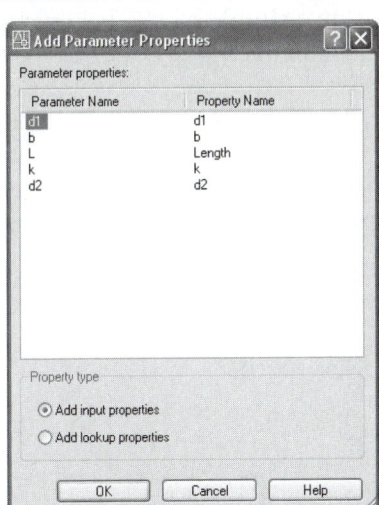

Figure 21-83 displays the matching dynamic block of a fastener. Compare the parameter names ("d1," "d2," "b," and "k") with the entries automatically populated in the *Add Parameter Properties* dialog box and the subsequent column heads in the *Property Lookup Table* dialog box.

FIGURE 21-83

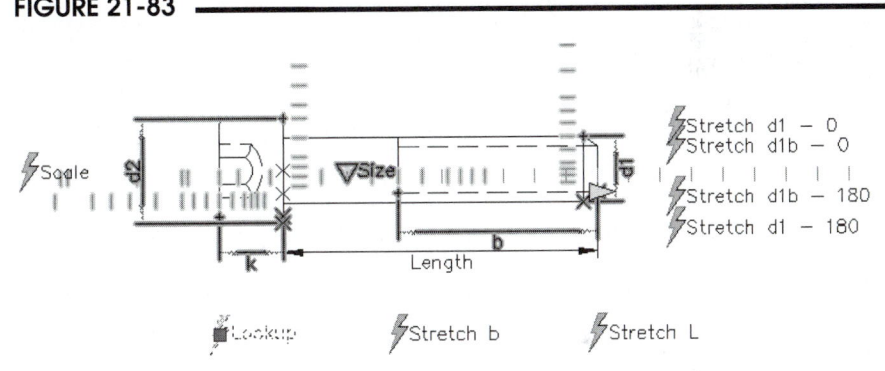

Steps for Creating Lookup Tables
Although other methods exist for creating lookup tables, below are steps for a simple method of creating a dynamic block containing a lookup table:

1. Open the Block Editor and create the geometry for the block or insert an existing block.

2. Assign the parameters (such as *Linear Parameter*, for example), then associate the related actions (such as *Stretch Action*). Give the parameters and actions specific labels, such as "Distance D1" and "Stretch D1."

3. Assign a *Lookup Parameter* to the block and associate a *Lookup Action*.

4. When the *Property Lookup Table* dialog box appears, select *Add Properties*.

5. The *Add Parameter Properties* dialog box is populated with all of the existing parameter labels in the block. Select the desired property labels, select the *Add input properties* option, and select *OK*.

6. In the *Property Lookup Table*, the *Input Properties* column heads contain the parameter labels. Key in the desired values for each cell. Next, key in the desired *Lookup Properties* name for each set of values in *Input Properties*.

7. If you want to change the block later in the drawing using a drop-down list when you select the block, select *Allow reverse lookup* at the bottom of the *Lookup Properties* column. Select OK.

8. Save the block, *Close* the Block Editor, and test your new dynamic block.

You should begin the process of creating lookup tables by using a simple block. For example, assume you create a block in the Block Editor of a simple rectangle (Fig. 21-84). Assign two *Linear Parameters* and associate two *Stretch Actions* such that the block could be dynamically stretched in two directions. Assign labels for the parameters (such as "Width" and "Height") and labels for the actions (such as "Stretch Width" and "Stretch Height").

FIGURE 21-84

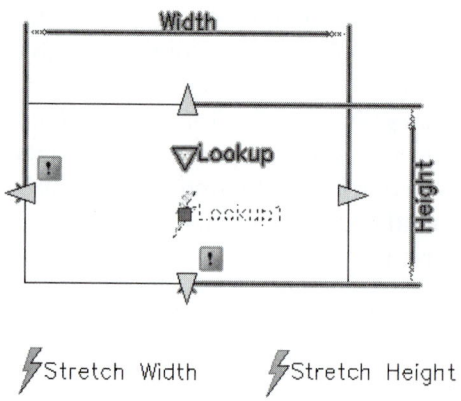

Next, assign a Lookup Parameter and a *Lookup Action*. The *Property Lookup Table* dialog box appears automatically. Select the *New* button to produce the *Add Parameters Properties* dialog box (not shown). This process (see previous "Steps for Creating Lookup Tables") allows you to populate the *Input Properties* in the *Property Lookup Table* (Fig. 21-85). (Compare the column heads with the parameter labels for the block.) Next, enter the desired names for the Lookup *Properties*. Select *OK*.

FIGURE 21-85

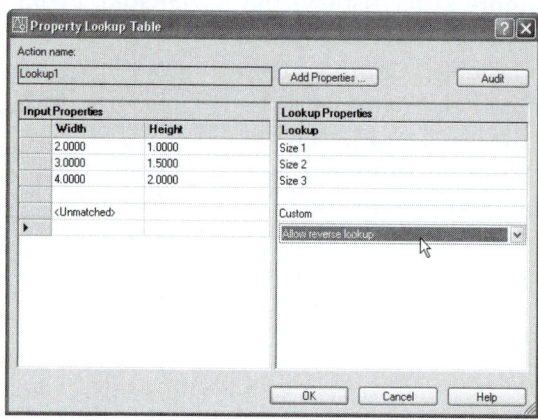

Save the new dynamic block and *Close* the Block Editor. When the new block is *Inserted* into the drawing, you can select the block for editing. Doing so displays the lookup grip (Fig. 21-86). Click on the grip to display the list of lookup properties (as assigned in the *Lookup Properties* section of the table). Selecting any choice from the list changes the block to match the values assigned to the related parameters in the block.

FIGURE 21-86

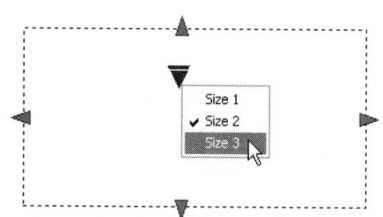

CHAPTER EXERCISES

1. *Block, Insert*

   In the next several exercises, you will create an office floor plan, then create pieces of furniture as *Blocks* and *Insert* them into the office. All of the block-related commands are used.

   A. Start a *New* drawing. Select *Start from Scratch* and use the *Imperial* defaults. Use *Save* and assign the name **OFF-ATT.** Set up the drawing as follows:

1. *Units*   *Architectural*   **1/2″ Precision**
2. *Limits*   **48' x 36'**   (1/4″=1' scale on an A size sheet), drawing scale factor = 48
3. *Snap, Grid*   **3**
4. *Grid*   **12**
5. *Layers*   **FLOORPLAN**   *continuous*   *.014*   **colors of your choice**
                    FURNITURE   *continuous*   *.060*
                    ELEC-HDWR   *continuous*   *.060*
                    ELEC-LINES   *hidden 2*   *.060*
                    DIM-FLOOR   *continuous*   *.060*
                    DIM-ELEC   *continuous*   *.060*
                    TEXT   *continuous*   *.060*
                    TITLE   *continuous*   *.060*
6. *Text Style*   *CityBlueprint*   *CityBlueprint (TrueType font)*
7. *Ltscale*   **48**

B. Create the floor plan shown in Figure 21-87. Center the geometry in the *Limits*. Draw on layer **FLOORPLAN**. Use any method you want for construction (e.g., *Line*, *Pline*, *Xline*, *Mline*, *Offset*).

FIGURE 21-87

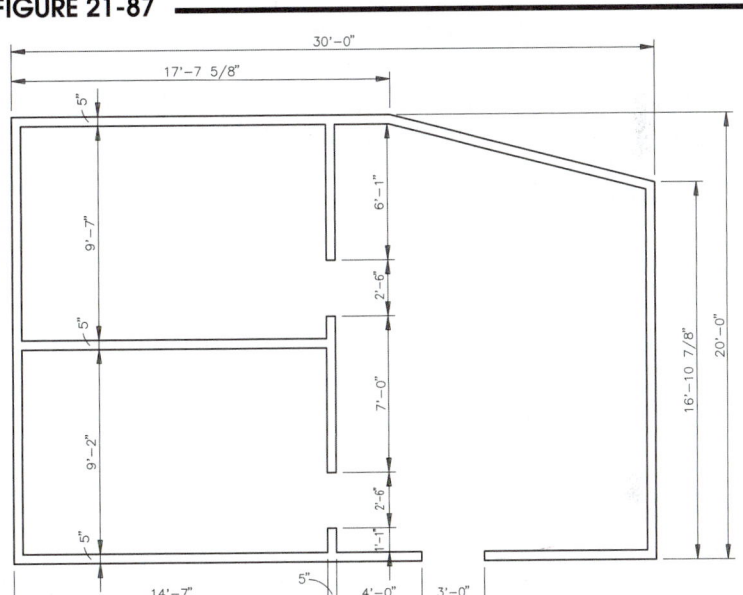

C. Create the furniture shown in Figure 21-88. Draw on layer **FURNITURE**. Locate the pieces anywhere for now. Do <u>not</u> make each piece a *Block*. *Save* the drawing as **OFF-ATT**.

Now make each piece a *Block*. Use the *name* as indicated and the *insertion base point* as shown by the "blip." Next, use the *Block* command again but only to check the list of *Blocks*. Use *SaveAs* and rename the drawing **OFFICE.**

FIGURE 21-88

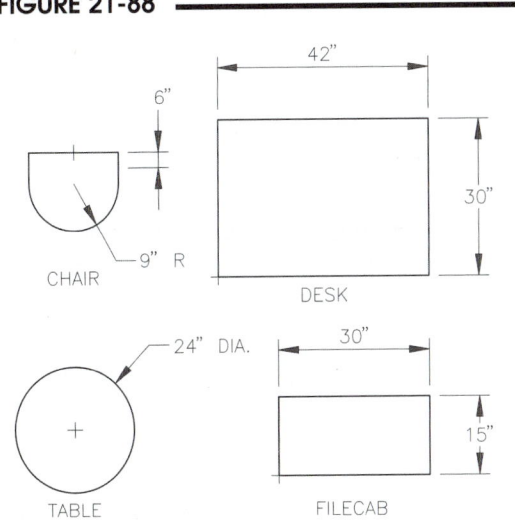

D. Use *Insert* to insert the furniture into the drawing, as shown in Figure 21-89. You may use your own arrangement for the furniture, but *Insert* the same number of each piece as shown. *Save* the drawing.

FIGURE 21-89

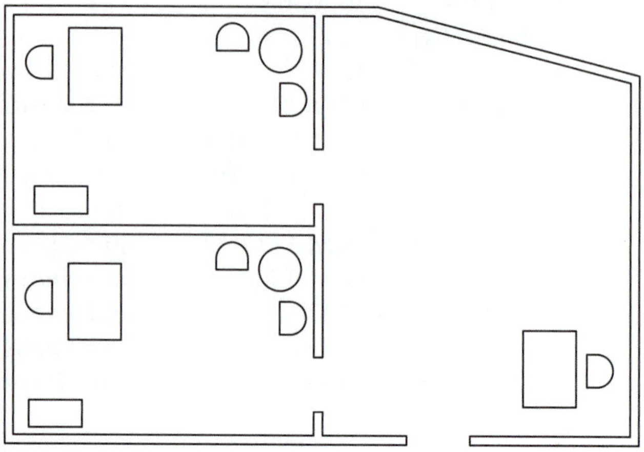

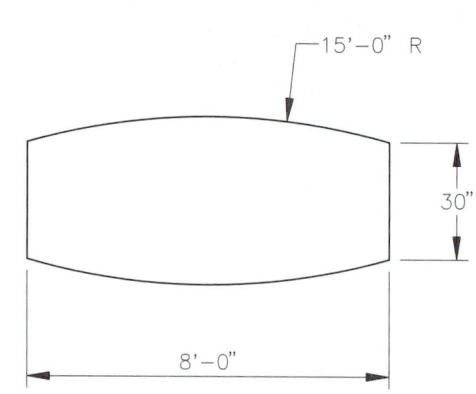

2. **Creating a .DWG file for** *Insertion, Base*

FIGURE 21-90

Begin a *New* drawing. Assign the name **CONFTABL**. Create the table as shown in Figure 21-90 on *Layer* **0**. Since this drawing is intended for insertion into the **OFFICE** drawing, use the *Base* command to assign an insertion base point at the lower-left corner of the table.

When you are finished, *Save* the drawing.

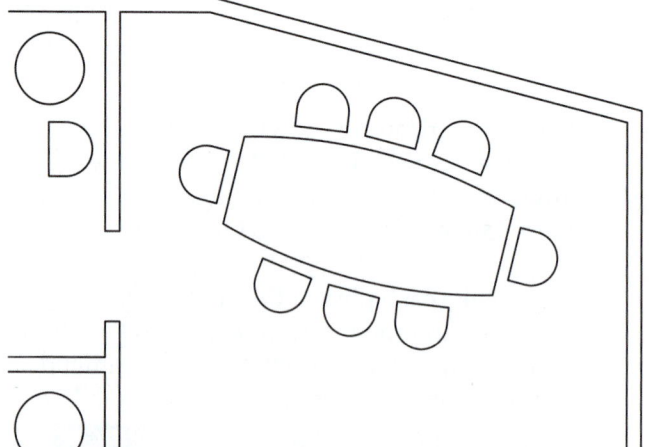

3. *Insert, Explode, Divide*

FIGURE 21-91

A. *Open* the **OFFICE** drawing. Ensure that layer **FURNITURE** is current. Use **DesignCenter** to bring the **CONFTABL** drawing in as a *Block* in the placement shown in Figure 21-91.

Notice that the CONFTABL assumes the linetype and color of the current layer, since it was created on layer **0**.

B. *Explode* the CONFTABL. The *Exploded* CONFTABL returns to *Layer* **0**, so use the *Properties* palette to change it back to *Layer* **FURNITURE**. Then use the *Divide* command (with the *Block* option) to insert the **CHAIR** block as shown in Figure 21-91. Also *Insert* a **CHAIR** at each end of the table. *Save* the drawing.

4. *Wblock, BYBLOCK setting*

FIGURE 21-92

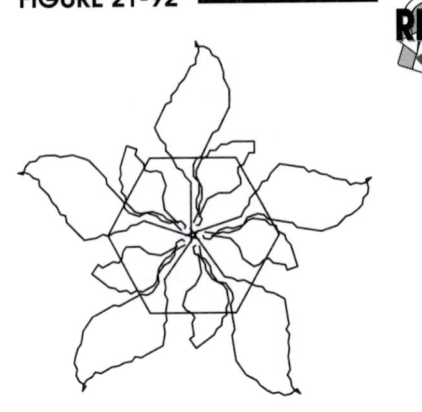

PLANT

A. *Open* the **EFF-APT** drawing you worked on in Chapter 15 Exercises. Use the *Wblock* command to transform the plant into a .DWG file (Fig. 21-92). Use the name **PLANT** and specify the *Insertion base point* at the center. Do not save the EFF-APT drawing. *Close* **EFF-APT**.

B. *Open* the **OFFICE** drawing and use **DesignCenter** to *Insert* the **PLANT** into one of the three rooms. The plant probably appears in a different color than the current layer. Why? Check the *Layer* listing to see if any new layers came in with the PLANT block. *Erase* the PLANT block.

C. *Open* the **PLANT** drawing. Change the *Color, Linetype*, and *Lineweight* setting of the plant objects to *BYBLOCK*. *Save* the drawing.

D. *Open* the **OFFICE** drawing again and *Insert* the **PLANT** onto the **FURNITURE** layer. It should appear now in the current layer's *color, linetype*, and *lineweight*. *Insert* a **PLANT** into each of the 3 rooms. *Save* the drawing.

5. **Redefining a *Block***

FIGURE 21-93

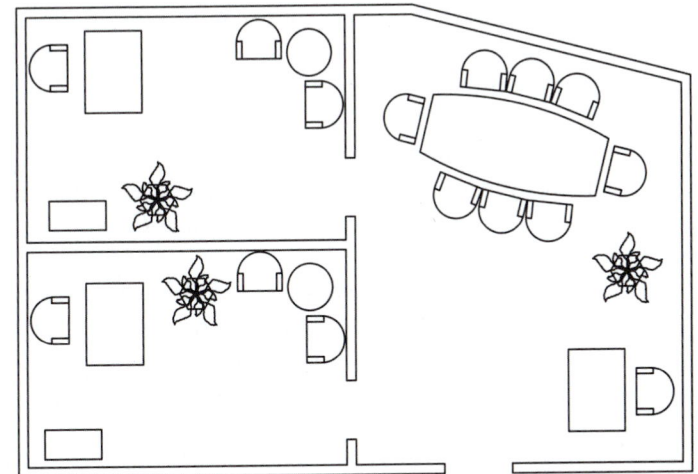

After a successful meeting, the client accepts the proposed office design with one small change. The client requests a slightly larger chair than that specified. Open the **Block Editor** and select the **CHAIR** block to edit. Redesign the chair to your specifications. **Save** the CHAIR block and *Close* the Block Editor. All previous insertions of the CHAIR should reflect the design change. *Save* the drawing. Your design should look similar to that shown in Figure 21-93. *Plot* to a standard scale based on your plotting capabilities.

6. *Rename*

Open the **OFFICE** drawing. Enter *Rename* (or select from the *Format* pull-down menu) to produce the *Rename* dialog box. Rename the *Blocks* as follows:

New Name	Old Name
DSK	**DESK**
CHR	**CHAIR**
TBL	**TABLE**
FLC	**FILECAB**

Next, use the *Rename* dialog box again to rename the *Layers* as indicated below:

New Name | Old Name
FLOORPLN-DIM | **DIM-FLOOR**
ELEC-DIM | **DIM-ELEC**

Use the *Rename* dialog box again to change the *Text Style* name as follows:

New Name | Old Name
ARCH-FONT | **CITYBLUEPRINT**

Use *SaveAs* to save and rename the drawing to **OFFICE-REN**.

7. *Purge*

Using the Windows Explorer, check the file size of **OFFICE-REN.DWG**. Now use *Purge* to remove any unreferenced named objects. *Exit AutoCAD* and *Save Changes*. Check the file size again. Is the file size slightly smaller?

8. **Use DesignCenter to create a Tool Palette**

    A. *Close* all open drawings and begin a *New* drawing. Next, open *DesignCenter* and use the *Folders* tab to locate the **OFFICE** drawing in the Tree View area. Expand the plus (+) symbol next to OFFICE.DWG, then click on the word *Blocks* just below the drawing name. All of the *Blocks* contained in the drawing should appear in the Content Area.

    B. Use the *Toolpalettes* command or press **Ctrl+3** to produce the Tool Palettes window. Now, back in DesignCenter, locate the OFFICE drawing in the Tree View area and right-click on the drawing name. From the shortcut menu select *Create Tool Palette*.

    C. This action should have automatically created a new tool palette named OFFICE and populated it with all the *Blocks* contained in the drawing. Examine the new tool palette. Right-click on any tool and examine its *Properties*. Now you have a new tool palette ready to use for inserting the office furniture *Blocks* into any other drawings.

    D. Use *DesignCenter* again to locate the **HOME-SPACE PLANNER** drawing in the AutoCAD 2006 /Sample/DesignCenter folder. In the Tree View area, right-click on the drawing name, then select *Create Tool Palette* from the shortcut menu. A new palette should appear in the *Tool Palettes* window fully populated with furniture.

9. **Using Tool Palettes**

    In this exercise you will change some of the blocks in the OFFICE-REN drawing using blocks from the *Tool Palettes* window.

    A. *Open* the **OFFICE-REN** drawing again if not already open. Assume as a result of the meeting with the client, new desks and chairs are requested for the receptionist and the two offices. If not already available, produce the Tool Palettes window using the *Toolpalettes* command or pressing **Ctrl+3**. Select the new *Home-Space Planner* palette (showing the home furniture).

B. Right-click on the *Chair-Desk* tool and select *Properties*. In the *Tool Properties* dialog box that appears, note that the scale is 1.00 and the layer is set to use current. Check the settings for the *Desk - 30 x 60 in*. With these settings, the new furniture should be usable without changes.

C. Make **FURNITURE** the current layer. *Erase* the three desks and the three desk chairs. Drag and drop three chairs and three desks from the palette window and place them near the desired locations in the office. Use *Rotate* and *Move* to position the new *Blocks* into suitable locations and orientations. Your drawing should look similar to that in Figure 21-94. Use *SaveAs* and name the drawing **OFFICE-REV.**

FIGURE 21-94

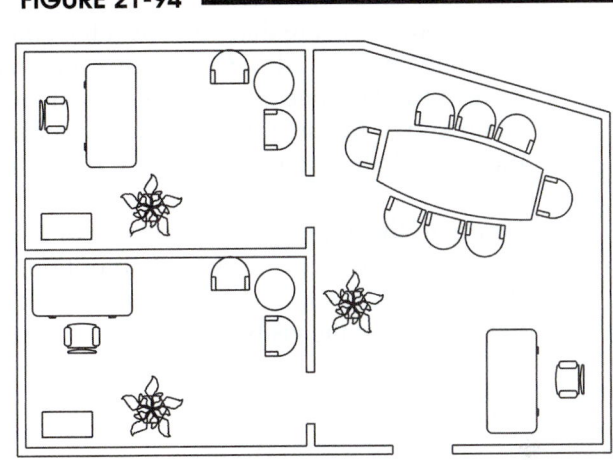

10. **Creating Dynamic Blocks**

A. *Open* the **OFFICE-REN** drawing again (not OFFICE-REV) if not already open. Open the **Block Editor** and select the **DSK** block to edit. Open the *Block Authoring Palettes* if not already open.

B. Assign a *Linear Parameter* to the desk by attaching the "Start point" and the "Endpoint" by *Osnapping* to the *Midpoints* of each side. Next, associate a *Stretch Action* to the linear parameter. Specify a crossing window on the left side of the desk as the "stretch frame" and again at the "Specify objects to stretch" prompt as shown in Figure 21-95. Assign another *Linear Parameter* and a matching *Stretch Action* to the right side of the desk.

FIGURE 21-95

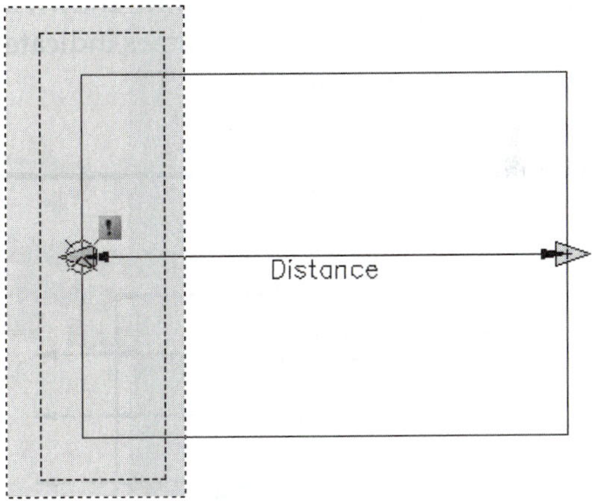

Distance

C. Return to the drawing and select any **DSK** block in the drawing. Note that the stretch grips appear. Do not stretch the desk yet. Instead, activate the *Properties* palette. Ensure the desk is highlighted in the drawing, then examine the *Custom* properties for the block. Only the *Distance* value is listed.

D. Open the **DSK** block again in the **Block Editor**. Open the *Properties* palette in the Block Editor. Select the *Linear Parameter*, then examine the *Custom* properties. In the *Value Set* section, change the *Distance type* to *Increment*. Next, change the *Dist increment* to **6″**. Now change the *Dist minimum* to **3'** (feet) and the *Dist maximum* to **5'**. *Save* the block and close the editor.

E. Return to the drawing. Highlight a desk to dynamically stretch its length. Change several desks, such as making the receptionist's desk shorter and one of the office desks longer.

F. Open the *Tool Palettes* window and drag and drop a *Chair-Desk* block from the *Home-Space Planner* palette into the drawing at any location. Erase the block from the drawing (although it will remain in the block definition table).

Default Attribute Value

This is the <u>default text</u> that appears with the block when it is *Inserted*. Supply the typical or expected words or numbers. If you are using the *Attribute Definition* dialog box, you can enter a text *Field*. In this way, the attribute can automatically change based on changes in the *Field*. (See Chapter 18.)

Justify

Any of the typical text justification methods can be selected using this option.

Style

This option allows you to select from existing text *Styles* in the drawing. Otherwise, the attribute is drawn in the current *Style*.

As an example, assume that you are using the *Attdef* command to make attributes attached to a resistor symbol to be *Blocked* (Fig. 22-2). The scenario may appear as follows or as presented in the *Attribute Definition* dialog box in Figure 22-1:

```
Command: -attdef
Current attribute modes: Invisible=N  Constant=N  Verify=N  Preset=N  Lock position=Y
Enter an option to change [Invisible/Constant/Verify/Preset/Lock position] <done>: Enter
Enter attribute tag name: RES_NUMBER
Enter attribute prompt: Enter Resistor Number:
Enter default attribute value: R1
Current text style: "ROMANS" Text height: 0.2000
Specify start point of text or [Justify/Style]: J
Enter an option [Align/Fit/Center/Middle/Right/TL/TC/TR/ML/MC/MR/BL/BC/BR]: C
Specify center point of text: PICK
Specify height <0.2000>: .125
Specify rotation angle of text <0>: Enter
```

The attribute would then appear at the selected location, as shown in Figure 22-2. Keep in mind that the *Lines* comprising the resistor were created before creating attributes. Also remember that the resistor and the attribute are not yet *Blocked*. The *Attribute Tag* <u>only</u> is displayed until the *Block* command is used.

Two other attributes could be created and positioned beneath the first one by pressing **Enter** when prompted for the "<Start point>:" (Command line format) or selecting the "*Align below previous attribute definition*" checkbox (dialog box format). The command syntax may read like this:

```
Command: -attdef
Current attribute modes: Invisible=N  Constant=N  Verify=N  Preset=N  Lock position=Y
Enter an option to change [Invisible/Constant/Verify/Preset/Lock position] <done>: Enter
Enter attribute tag name: RESISTANCE
Enter attribute prompt: Enter Resistor Number:
Enter default attribute value: 0 ohms
Current text style: "ROMANS" Text height: 0.1250
Specify start point of text or [Justify/Style]: Enter

Command: -attdef
Current attribute modes: Invisible=N  Constant=N  Verify=N  Preset=N  Lock position=Y
Enter an option to change [Invisible/Constant/Verify/Preset/Lock position] <done>: Enter
Enter attribute tag name: PART_NO
Enter attribute prompt: Enter Part Number:
Enter default attribute value: 0-0000
Current text style: "ROMANS" Text height: 0.1250
Specify start point of text or [Justify/Style]: Enter
```

The resulting unblocked symbol and three attributes appear as shown in Figure 22-3. Only the _Attribute Tags_ are displayed at this point. The _Block_ command has not yet been used.

FIGURE 22-3

RES_NUMBER
RESISTANCE
PART_NO

The _Attributes Modes_ define the appearance or the action required when you _Insert_ the attributes. The options are as follows.

Invisible

An _Invisible_ attribute is not displayed in the drawing after insertion. This option can be used to prevent unnecessary information from cluttering the drawing and slowing regeneration time. The _Attdisp_ command can be used later to make these attributes visible.

Constant

This option gives the attribute a fixed value for all insertions of the block. In other words, the attribute always has the same text and cannot be changed. You are not prompted for the value upon insertion.

Verify

When attributes are entered in Command line format, this option forces you to verify that the attribute is correct when the block is inserted. This option has no effect when attributes are entered in dialog box format (when _ATTDIA_=1).

Preset

This option prevents you from having to enter a value for the attribute during insertion if you are using the Command line format. In this case, you are not prompted for the attribute value during insertion. Instead, the default value is used for the block, although it can be changed later with the _Attedit_ command.

Lock Position

This option locks the position of the attribute within the block definition. If not locked, the attribute can be moved with normal grips once the block is inserted into the drawing. In a dynamic block, an attribute's position must be locked if you intend to include it in an action's selection set.

The _Attribute Modes_ are toggles (Yes or No). In Command line format, the options are toggled by entering the appropriate letter, which reverses its current position (Y or N). For example, to create an _Invisible_ attribute, type **"I"** at the prompt:

```
Command: -attdef
Current attribute modes: Invisible=N Constant=N Verify=N Preset=N Lock position=Y
Enter an option to change [Invisible/Constant/Verify/Preset/Lock position] <done>: I
Current attribute modes: Invisible=Y Constant=N Verify=N Preset=N Lock position=Y
Enter an option to change [Invisible/Constant/Verify/Preset/Lock position] <done>: Enter
```

Note that the second prompt (fourth line) reflects the new option for the previous request.

If _DYN_ is on and set to the default options, select the desired option from the _DYN_ menu (Fig. 22-4). If _DYN_ is off, you can right-click to produce a shortcut menu with the same options.

FIGURE 22-4

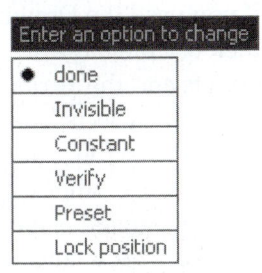

2006

Alternately, attribute modes could be defined using the *Attribute Definition* dialog box (see Fig. 22-5). The *Mode* cluster of checkboxes (upper-left corner) makes setting attribute modes simpler than in Command line format. Note that the *Lock position in block* option is at the lower-left corner of the dialog box.

Using our previous example, two additional attributes can be created with specific *Invisible* and *Verify* attribute modes. Either the dialog box or Command line format can be used. The following command syntax shows the creation of an *Invisible* attribute:

```
Command: -attdef
Current attribute modes: Invisible=N  Constant=N  Verify=N  Preset=N
Enter an option to change [Invisible/Constant/Verify/Preset] <done>: I
Current attribute modes: Invisible=Y  Constant=N  Verify=N  Preset=N
Enter an option to change [Invisible/Constant/Verify/Preset] <done>: Enter
Enter attribute tag name: MANUFACTURER
Enter attribute prompt: Enter Manufacturer:
Enter default attribute value: Enter
Current text style:  "ROMANS" Text height:  0.125000
Specify start point of text or [Justify/Style]: Enter
Command:
```

The first attribute has been defined. Next, a second attribute is defined having *Invisible* and *Verify* parameters. Figure 22-5 shows the correct entries in the *Attribute Definition* dialog box for creation of this attribute. *Tags* for *Invisible* attributes appear as any other attribute tags. The *Values*, however, are invisible when the block is inserted.

Saving the Attributed Block

As indicated earlier, you can create the block geometry and the attributes using either the Drawing Editor or the Block Editor. The next step in the process of creating an attributed block is to save the *Block*. This step is different depending on the method used to create the geometry and attributes.

FIGURE 22-5

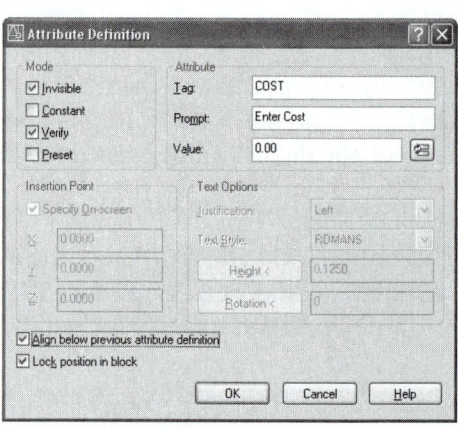

Using the Block Editor

If you used the Block Editor for the previous steps, saving the attributed block is simple. Select the *Save Block Definition* button or use the *Bsave* command to complete the process of creating the attributed *Block*. However, using the Block Editor you have no control of the order in which the attributes appear in the *Enter Attributes* dialog box or at the command prompt—you are prompted to enter the attribute values in the order they were created. In other words, when you use the *Insert* command to insert the attributed block into the drawing, prompts to enter the attribute text appear in the order the attributes were created.

Using the Drawing Editor

If you used the Drawing Editor for the previous steps, you must then use the *Block* or *-Block* command to transform the drawing objects and the attributes into a *Block*. When you are prompted to "Select objects:" during the *Block* command, select the attributes in the order you want their prompts to appear during insertion. In other words, the first attribute selected during the *Block* command is the first attribute prompted for editing during the *Insert* command. PICK the objects with the pickbox <u>one at a time</u> in the desired order. Figure 22-6 shows this sequence.

FIGURE 22-6

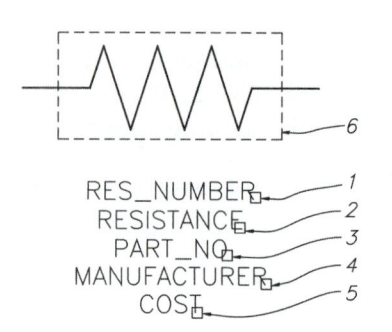

RES_NUMBER — 1
RESISTANCE — 2
PART_NO — 3
MANUFACTURER — 4
COST — 5

If you select the *Convert to block* option in the *Block Definition* dialog box, then press the *OK* button, you are prompted to enter the attribute text immediately. These attributes are for the original template objects that are being converted to a *Block*.

Inserting **Attributed Blocks**

FIGURE 22-7

Use the *Insert* command to bring attributed blocks into your drawing as you would to bring any block into your drawing. When the *Insert* dialog box appears (not shown), make the desired settings and select *OK*. Next, specify the insertion point (where you want to place the block in your drawing). At this point, the *Edit Attributes* dialog box appears (Fig. 22-7). Enter the desired attribute values for each block insertion.

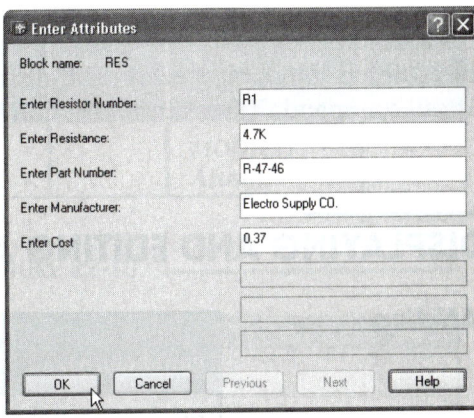

Notice that the prompts appearing in the dialog box are those that you specified as the "attribute prompt" with the *Attdef* command. The order of the prompts matches the order of selection during the *Block* command.

The appearance of the *Enter Attributes* dialog box is controlled by the *ATTDIA* system variable (see "The *ATTDIA* Variable"). If *ATTDIA* is set to 0, the dialog box does not appear and you are prompted to enter the attribute text at the Command line. For example, if *ATTDIA* is set to 0, the following prompt appears during the *Insert* command. Note the repeated prompt (*Verify*) for the COST attribute.

```
Command: insert
Specify insertion point or [Basepoint/Scale/X/Y/Z/Rotate/PScale/PX/PY/PZ/PRotate]: PICK
Enter attribute values
Enter Resistor Number: <0>: R1
Enter Resistance: <0 ohms>: 4.7K
Enter Part Number: <0-0000>: R4746
Enter Manufacturer: Electro Supply Co.
Enter Cost <0.00>: .37
Verify attribute values
Enter Cost <.37>: Enter
Command:
```

Entering the attribute values as indicated by either method above yields the block insertion shown in Figure 22-8. Notice the absence of the last two attribute values since they are *Invisible*.

FIGURE 22-8

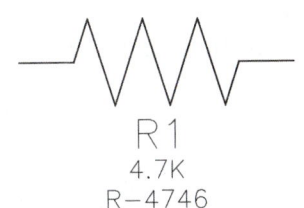

The *ATTDIA* Variable

When you use the *Insert* command to insert a *Block* containing attributes, you are normally prompted to enter the desired text for the attributes. The *ATTDIA* variable controls how you are prompted. If *ATTDIA* is set to 1, the *Enter Attributes* dialog box appears (see previous Fig. 22-7). If *ATTDIA* is set to 0, the dialog box does not appear and you are prompted to enter the attribute text at the Command line.

Using the *Find* command produces the *Find and Replace* dialog box (Fig. 22-15). To use *Find* to edit attributes in a drawing, you must ensure that *Find* looks for block attributes when the search is made. To do this, use the *Options* button in the *Find and Replace* dialog box.

FIGURE 22-15

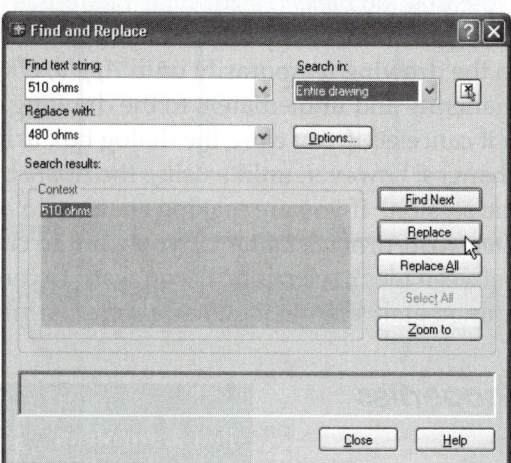

In the *Find and Replace Options* dialog box that appears, select *Block Attribute Value* to be included in the search (Fig. 22-16). Next, select *OK* to return to the main dialog box and to proceed with the search.

Specify the attribute value to find in the *Find text string* edit box (see Fig. 22-15). Enter the replacement attribute value in the *Replace with* edit box. Select the *Find* button to begin the search, then use *Replace* and *Find Next*, or *Replace All* as needed. Refer to Chapter 18 for details on these and other features of the *Find and Replace* dialog box, such as the *Zoom to* feature.

FIGURE 22-16

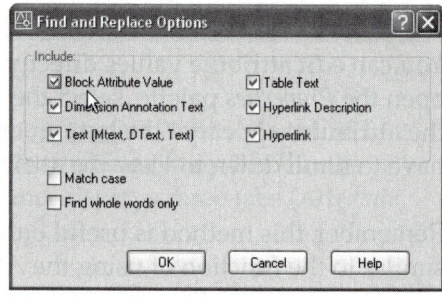

-Attedit

Pull-down Menu	Command (Type)	Alias (Type)	Short-cut	Screen (side) Menu	Tablet Menu
Modify Object > Attribute > Global	*-Attedit*	*-ATE*	...	*MODIFY1 Attedit*	...

-Attedit (Attribute Edit) is an older Command line attribute editing tool that has several features similar to those included in the newer dialog box-based commands. *-Attedit* is similar to *Attedit* and *Eattedit* in that you can edit the values of selected attributes in a drawing. *-Attedit* also has options similar to *Eattedit* that allow you to change the attribute text options and some properties. You can also use *-Attedit* in a manner similar to *Find* to globally edit attribute values in a drawing.

> Command: **-attedit**
> Edit attributes one at a time? [Yes/No] <Y>:

The following prompts depend on your response to the first prompt:

N (No)

Indicates that you want global editing. (In other words, all attributes are selected.) Attributes can then be further selected by block name, tag, or value. <u>Only attribute values can be edited</u> using this global mode.

Y (Yes)

Allows you to PICK each attribute you want to edit. You can further filter (restrict) the selected set by block name, tag, and value. You can then <u>edit any property or placement</u> of the attribute.

No matter which option you choose, the following prompts appear, which allow you to filter, or further restrict, the set of attributes for editing. You can specify the selection set to include only specific block names, tags, or values:

```
Enter block name specification <*>: Enter desired block name(s)
Enter attribute tag specification <*>: Enter desired tag(s)
Enter attribute value specification <*>: Enter desired value(s)
```

During this portion of the selection set specification, sets of names, tags, or values can be entered only to filter the set you choose. Commas and wildcard characters can be used (see Chapter 43 for information on using wildcards). Pressing Enter accepts the default option of all (*) selected.

Attribute values are <u>case sensitive</u> (upper- and lowercase letters must be specified exactly). If you have entered null values (for example, if *ATTREQ* was set to 0 during block insertion), you can select all null values to be included in the selection set by using the backslash symbol (\).

The subsequent prompts depend on your previous choice of global or individual editing.

No (Global Editing—Editing Values Only)

```
Command: -attedit
Edit attributes one at a time? [Yes/No] <Y>: n
Performing global editing of attribute values.
Edit only attributes visible on screen? [Yes/No] <Y>: (option)
Enter block name specification <*>: Enter or (desired block name)
Enter attribute tag specification <*>: Enter or (desired tag)
Enter attribute value specification <*>: Enter or (desired value)
Select Attributes: PICK
Select Attributes: PICK
Select Attributes: Enter
2 attributes selected.
Enter string to change: (Enter any existing string)
Enter new string: (Enter any new string)
Command:
```

The "only attributes visible on the screen" refers to those visible (Y) or those through the entire drawing (N). Answering "N" to this prompt and Enter to the next three would find all attributes in the drawing.

AutoCAD searches for the "string to change" and replaces all occurrences of the string with the new string. No change is made if the string is not found. Remember that only attribute values can be edited with the global mode. This feature has the same capability as using *Find*, described previously.

Yes (Individual Editing—Editing Any Property)

After filtering for the block name, attribute tag, and value, AutoCAD prompts:

```
Select Attributes: PICK
Select Attributes: PICK (continue selection)
Select Attributes: Enter
n attributes selected.
Enter an option [Value/Position/Height/Angle/Style/Layer/Color/Next] <N>:
```

A marker appears at one of the attributes. Use the *Next* option to move the marker to the attribute you wish to edit. You can then select any other option. After editing the marked attribute, the change is displayed immediately and you can position the marker at another attribute to edit.

You can change the attribute's *Value, Position, Height, Angle, Style, Layer*, or *Color* by selecting the appropriate option. These options are Command line versions of the *Text Options* and *Properties* tabs of the *Enhanced Attribute Editor* (see "*Eattedit*").

Global and Individual Editing Example

Assume that the RES block and a CAP block were inserted several times to create the partial schematic shown in Figure 22-17. The *Attdisp* command is set to *On* to display all the attributes, including the last two for each block, which are normally *Invisible*.

FIGURE 22-17

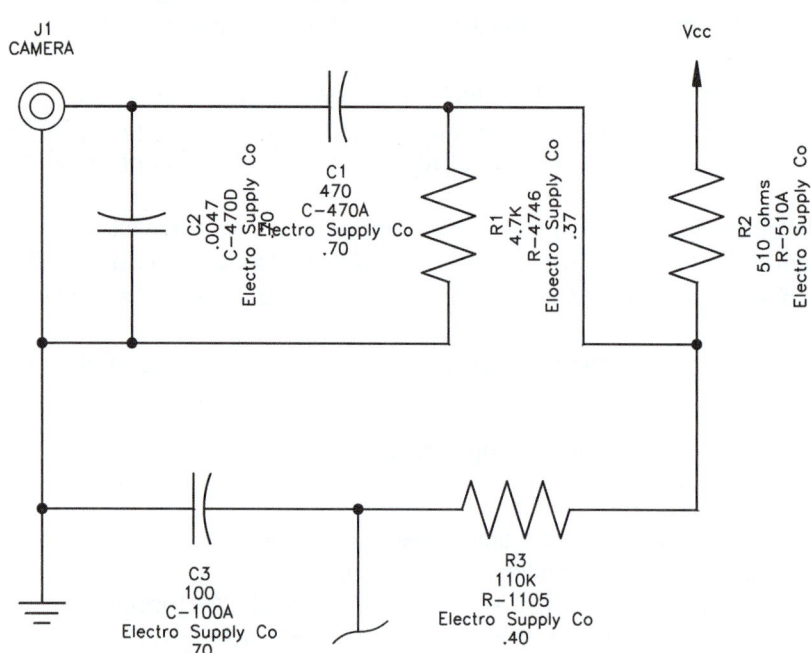

-*Attedit* is used to edit the attributes of RES and CAP. If you wish to change the value of one or several attributes, the global editing mode of -*Attedit* would generally be used, since values only can be edited in global mode. For this example, assuming that the supplier changed, the MANUFACTURER attribute for all blocks in the drawing is edited. The command syntax is as follows:

```
Command: -attedit
Edit attributes one at a time? [Yes/No] <Y>: n
Performing global editing of attribute values.
Edit only attributes visible on screen? [Yes/No] <Y>: Enter
Enter block name specification <*>: Enter
Enter attribute tag specification <*>: Manufacturer
Enter attribute value specification <*>: Enter
Select Attributes: PICK (Select specifically the 6 MANUFACTURER attributes)
Select Attributes: Enter
6 attributes selected.
Enter string to change: Electro Supply Co
Enter new string: Sparks R Us
Command:
```

All instances of the "Electro Supply Co" redisplay with the new string. Note that resistor R1 did not change because the string did not match (due to the misspelling, Fig. 22-18). *Attedit* or *Eattedit* can be used to edit the value of that attribute individually.

The individual editing mode of -*Attedit* or *Eattedit* can be used to change the <u>height</u> of text for the RES_NUMBER and CAP_NUMBER attribute for both the RES and CAP blocks. The command syntax is as follows:

FIGURE 22-18

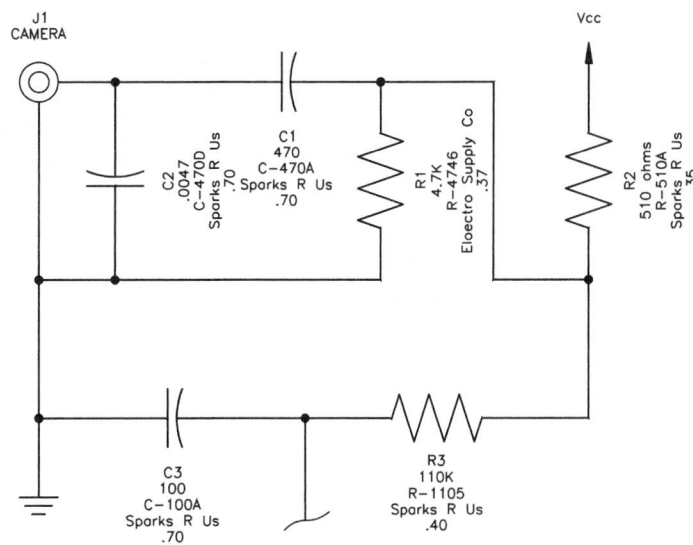

Command: **-attedit**
Edit attributes one at a time? [Yes/No] <Y>: **Enter**
Enter block name specification <*>: **Enter**
Enter attribute tag specification <*>: **Enter**
Enter attribute value specification <*>: **Enter**
Select Attributes: **PICK**
Select Attributes: **PICK** (Continue selecting all 6 attributes)
Select Attributes: **Enter**
6 attributes selected.
Enter an option [Value/Position/Height/Angle/Style/Layer/Color/Next] <N>: **H**
Specify new height <0.1250>: **.2**
Enter an option [Value/Position/Height/Angle/Style/Layer/Color/Next] <N>: **N**
Enter an option [Value/Position/Height/Angle/Style/Layer/Color/Next] <N>: **H**
Specify new height <0.1250>: **.2**
     etc.

The result of the new attribute text height for CAP and RES blocks is shown in Figure 22-19. *Attdisp* has been changed to *Normal* (*Invisible* attributes do not display).

FIGURE 22-19

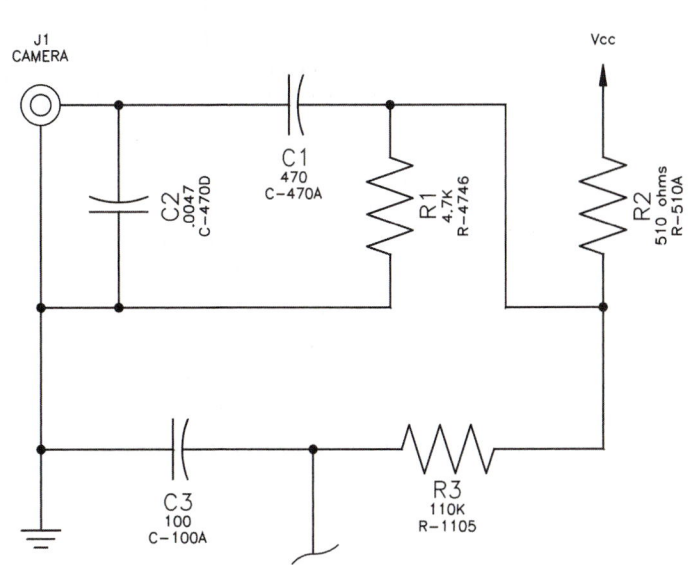

The block definition is updated, and all instances of the block automatically and globally change to display the new changes. All attributes that were changed reflect the default values. *Attedit* can be used later to make the desired corrections to the individual attributes.

Figure 22-26 displays one of the RES blocks after redefining. Notice that the fourth attribute is invisible as it was before it was *Exploded*. For this attribute, the mode was not changed. The new last attribute, however, was changed to a normal attribute mode and therefore appears after using *Attredef*.

FIGURE 22-26

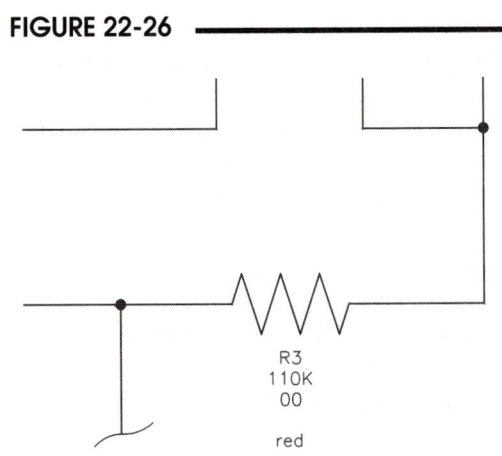

Refedit, Properties, and Refclose

Another method of redefining blocks is to use *Refedit*, the *Properties* palette, and *Refclose*. This technique is also known as "in-place editing" because the references (*Blocks* or *Xrefs*) can be edited after being inserted or attached. However, the *Block Attribute Manager* makes the *Refedit* method seem relatively involved and limited; therefore, this method for the purpose of redefining block attributes is explained briefly in this chapter only for reference. Refer to Chapter 30, Xreferences, for a full explanation of *Refedit* and other related commands and variables for use with *Blocks* and *Xrefs*.

To redefine one block attribute at a time using *Refedit*, *Properties*, and *Refclose*, follow these steps.

1. Invoke the *Properties* palette.
2. Invoke *Refedit*. Select the <u>Block</u> (not the attributes) containing the attributes you want to redefine.
3. In the *Reference Edit* dialog box that appears, select the *Display attribute definitions for editing* checkbox. Select *OK*.
4. At the "Select nested objects" prompt, select the *Block* again (not the attributes).
5. Use the *Properties* palette to change the *Tag* or *Prompt* property. When finished, press Escape twice to deselect the *Block*.
6. Use the *Refclose* command and *Save* option to save the changes and redefine the attributes. All insertions of the *Block* are redefined.

When *Blocks* are redefined using this method, the block definition is changed in the current drawing. However, when *Xrefs* are changed using *Refedit*, the original externally referenced drawing can be changed from within the current drawing (see Chapter 30, Xreferences).

The Block Editor

You can use the Block Editor to change attributed blocks; however, changes to attributes are not automatically updated for existing blocks in the drawing until you use the *Attsync* command (see "Attsync"). You can use three commands in the Block Editor to redefine blocks: *Ddedit*, *Properties*, or *Attdef*. First open the Block Editor and select the desired block to edit. Next use one of the three commands to edit the attributes. When you are finished in the Block Editor, save the block when you close the editor. The three options are explained next.

Ddedit

Ddedit, the dialog box for editing text, recognizes the text as attributes and allows you to change the attribute *Tags*, *Prompts*, and *Default Values*. Figure 22-27 illustrates using the *Edit Attribute Definition* dialog box for changing the attribute *Prompt*.

FIGURE 22-27

FIGURE 22-27

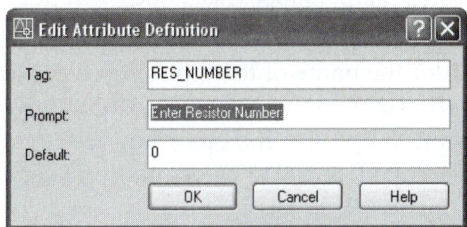

Properties

Optionally, the *Properties* palette can be invoked. Not only does *Properties* recognize the text as attributes, but the *Tag, Prompt,* default *Value,* and attribute modes can be modified. Figure 22-28 displays this dialog box for the same operation. Notice all of the possible changes that can be made.

FIGURE 22-28

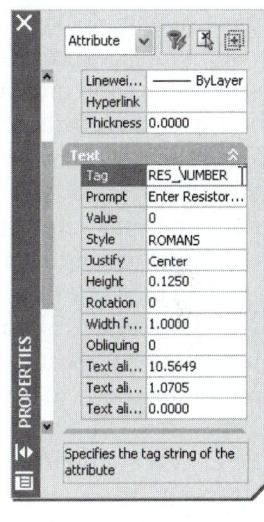

Attdef

Instead of using *Ddedit* or *Properties* to change <u>existing</u> attributes, new attributes can be created "from scratch" with *Attdef*.

Grip Editing Attribute Location

Grips can be used to "move" the location of the attributes after an attributed *Block* has been inserted. To do this (with Grips on), PICK the *Block* so grips appear at each attribute and at the block insertion point. Next, PICK the attribute you want to move so its grip becomes "hot." The attribute can be moved to any location and still retain its association with the block. (See Chapter 19, Grip Editing.)

Attsync

Pull-down Menu	Command (Type)	Alias (Type)	Short-cut	Screen (side) Menu	Tablet Menu
...	*Attsync*	...	...	...	...

The *Attsync* command synchronizes the selected block attributes in the drawing to change according to the definition as it exists in the block definition table. *Attsync* does not affect individual attribute values, only changes made to the properties of the inserted attributes that are different from what is established in the definition table for the block. *Attsync* is usually not needed if you used the *Block Attribute Manager* or *Attredef* to redefine block attributes. This command may be helpful in the following cases.

Use *Attsync* if you used the Block Editor or *Block* command to redefine a block containing attributes. Redefining a block with the Block Editor or the *Block* command does not update attributes in blocks inserted in the drawing; these methods update only the drawing geometry. For example, you may have opened an attributed block in the Block Editor, deleted an attribute, then saved the *Block* to redefine it. Using *Attsync* would be necessary to update all the existing blocks in the drawing to the new definition.

Use *Attsync* if you edited the format of the attributes by some method and want to change them back to the original definition. For example, if you previously used the *Enhanced Attribute Editor* to change specific properties of an attribute in the drawing such as text height or color, using *Attsync* would change the properties back to the original settings as established in the block's definition. *Attsync* does not affect any changes (text or numeric content) made to the attribute values themselves.

When *Attsync* is invoked, you are prompted for the names of blocks you want to update with the current attributes defined for the blocks.

Command: **attsync**
Enter an option [?/Name/Select] <Select>:

Enter the name of the block you want to update. Entering a question mark (?) displays a list of all block definitions in the drawing. Alternately, you can press Enter to select the block whose attributes you want to update. If a block you specify does not contain attributes or does not exist, an error message is displayed, and you are prompted to specify another block.

EXTRACTING ATTRIBUTES

By extracting block and attribute information from AutoCAD drawings, you can create a report of the existing blocks and related text information contained in the drawings. In other words, you can select a list of blocks and attributes in a drawing, then extract all the contained information into a table. You can choose to convert this text and/or numeric information into an AutoCAD *Table* or to an external file in spreadsheet format. The resulting table can be used as a report indicating a variety of data such as the name, number, location, layer, and scale factor of the blocks, as well as any attribute information such as prices, manufacturers, sizes, dates, or any other attribute data that may be defined in the block inser-tions. Because of this attribute extract feature, AutoCAD drawings contain the intelligence to compile or calculate bills of materials, supplier information, employee records, quotations, billing totals, mainte-nance schedules, catalogs, and so on. Since an AutoCAD *Table* generated from attribute data is linked to the attributes, the data content of the table is updated when changes are made to the attributes. You can use two methods to extract attributes: the *Attribute Extraction* wizard and the *Attext* command.

TIP The *Eattext* command produces the *Attribute Extraction* wizard. This wizard makes the process of extracting attributes far simpler and more visual than the earlier methods in AutoCAD. The wizard leads you through the steps of selecting drawings, selecting *Blocks* and/or *Xrefs*, selecting the desired attribute information, selecting the output format, and saving templates for future use. The *Attribute Extraction* wizard provides you with capabilities not available by other means.

Alternately, you can use the *Attext* command to produce the older *Attribute Extraction* dialog box or type *-Attext* to invoke the Command line equivalent. Using this method is far more complex than the newer *Attribute Extraction* wizard because template files must be created externally to AutoCAD and the chances of error are great.

Eattext

Pull-down Menu	Command (Type)	Alias (Type)	Short-cut	Screen (side) Menu	Tablet Menu
Tools *Attribute Extraction...*	*Eattext*	...	...	...	...

To produce the *Attribute Extraction* wizard, use the *Eattext* command or any option shown in the command table. The *Attribute Extraction* wizard leads you through the six steps to extract block attribute (text and numeric) information from a drawing into a table. The resulting table can be inserted into the drawing as an AutoCAD *Table* or generated as an external spreadsheet format file.

The *Attribute Extraction* wizard offers a number of features that are not available by other means. With the *Attribute Extraction* wizard, you can extract attribute data from the current drawing or from multiple drawings and sheet sets. You can also extract attribute data from nested *Blocks* and *Xrefs*. In addition to the ability to choose the type of table to create, you can view and edit the organization of the data before creating the table.

The six "pages" (steps) required in the *Attribute Extraction* wizard are explained here. Use the *Next* button to proceed after each page and the *Back* button to go back at any time and change an option.

Begin (Page 1 of 6)

In the first page (Fig. 22-29), decide if you want to create a table from scratch or use a template. If this is your first time to use the *Attribute Extraction* wizard, select *Create table or external file from scratch*. If you have used the *Attribute Extraction* wizard previously and have saved a template file in .BLK format, or one exists for your project or company, you can select *Use template*, then use the browse button to locate and select the file. Template files contain specifications indicating which blocks and attributes are to be extracted and the format of the output table. The *Attribute Extraction* wizard cannot use template files in .TXT format created for early versions of AutoCAD.

FIGURE 22-29

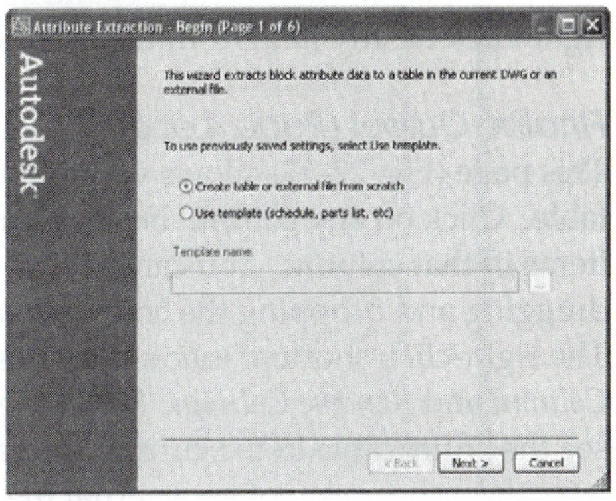

Select Drawings (Page 2 of 6)

In this second page you are prompted to specify the "data source" you want to extract attribute and block information from (Fig. 22-30). The *Select objects* option allows you to select only specific *Blocks* from the current drawing. Use the *Current drawing* option to have all blocks in the current drawing considered for extraction. Use the *Select drawings/sheet sets* option to locate and specify multiple drawings or sheet sets containing the blocks you want to extract information from. The list of drawings you selected appears in the *Drawing files* list. In the (next) *Select Attributes* page, you can exclude any of the blocks you specified in this page if desired.

FIGURE 22-30

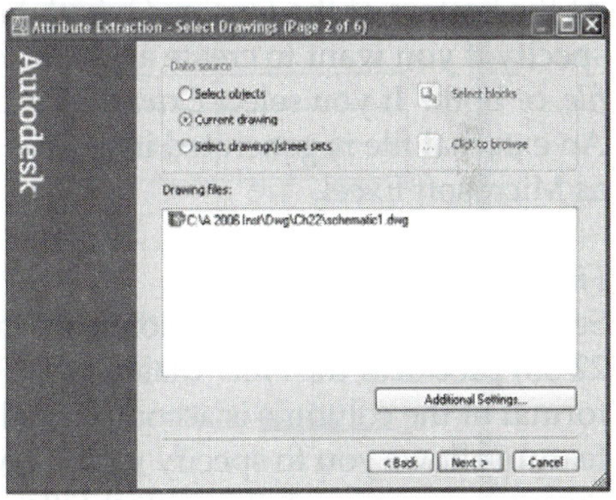

Additional Settings

Selecting this button in the second page produces the dialog box shown in Figure 22-31. The data sources you selected previously may contain nested blocks (blocks within blocks) or attached *Xref* drawings that contain blocks. Your choices here determine if you want to include blocks from these sources in the table. For most applications, remove checks from *Include Xrefs in block counts* and *Include blocks from entire drawing* so any attached *Xrefs* or any blocks in layouts (such as title blocks) are not included. If these sources are not contained in the selected drawing(s) from the previous step, your selection here is irrelevant.

FIGURE 22-31

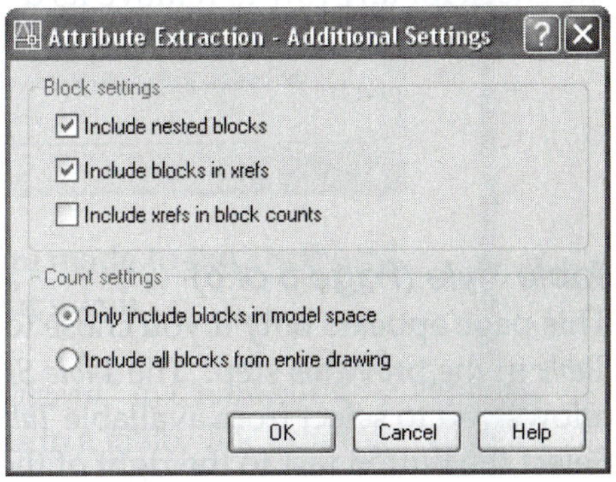

Select Attributes (Page 3 of 6)

The *Select Attributes* page (Fig. 22-32) demonstrates the power and flexibility of the *Attribute Extraction* wizard. Here you include or exclude which blocks and which attributes for each block you want to use in the table. The *Blocks* list on the left contains all the blocks included from your previous selections. You can *Exclude blocks without attributes* or select individual blocks from the list to include or exclude. For each checked block in the *Blocks* list, the related attributes for that block are listed in the *Properties for checked blocks* list on the right.

FIGURE 22-32

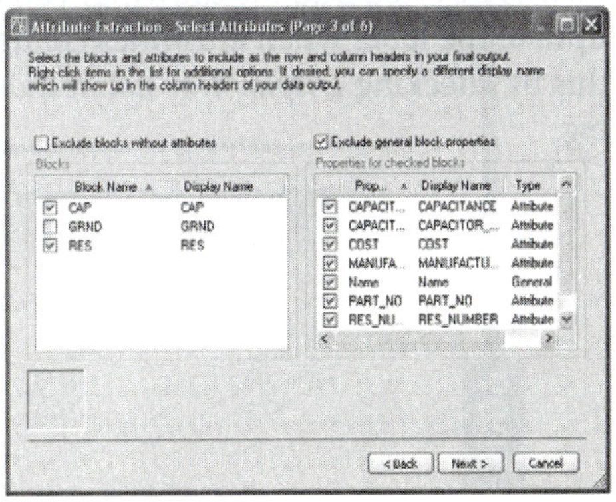

Attext

Pull-down Menu	Command (Type)	Alias (Type)	Short-cut	Screen (side) Menu	Tablet Menu
...	*Attext* or *-Attext*	...	...	...	...

Before AutoCAD 2002, the *Attext* command was used to extract attributes. The *Attext* command produces the *Attribute Extraction* dialog box shown in Figure 22-39. This method is more complex and allows attribute extraction from the current drawing only; however, you can create an output file in three formats using this method. Before invoking *Attext*, you must create the template file "manually" in a text editor or word processor. The steps are given below.

Steps for Creating Output Files Using Attext

1. In AutoCAD, *Insert* all desired *Blocks* and related attributes into the drawing. *Save* the drawing.

2. With a word processor or text editor, create a template file. The template file specifies what information should be included in the output file and how the output file should be structured. (A template file is not required if you specify a *DXF* format for the output file.)

3. In AutoCAD, use the *Attext* or *-Attext* command to specify the name of the template file to be used and to create the output file.

4. Examine the output file in the text editor or word processor or import the output file into a database or spreadsheet program.

Creating the Template File (Step 2)

The template file is created in a text editor (such as Windows Notepad) or word processor (such as Microsoft Word) in non-document mode. The template file must be in straight ASCII form (no internal word processing codes). The template file must be given a .TXT file extension.

Each line of the template file specifies a column of information, or field, to be written in the output file, including the name of the column, the width of the column (number of characters or numerals), and the numerical precision. Each *Block* that matches a line in the template file creates a column in the output file.

The possible fields that AutoCAD allows you to specify and the format that you must use for each are shown below. Use any of the fields you want when creating the template file. The first two columns below are included in the template file (not the comments in the third column):

```
BL:LEVEL       Nwwwddd      (Block nesting level)
BL:NAME        Cwww000      (Block name)
BL:X           Nwwwddd      (X coordinate of Block insertion point)
BL:Y           Nwwwddd      (Y coordinate of Block insertion point)
BL:Z           Nwwwddd      (Z coordinate of Block insertion point)
BL:NUMBER      Nwwwddd      (Block counter; same for all members of MINSERT)
BL:HANDLE      Cwww000      (Block handle; same for all members of MINSERT)
BL:LAYER       Cwww000      (Block insertion layer name)
BL:ORIENT      Nwwwddd      (Block rotation angle)
BL:XSCALE      Nwwwddd      (X scale factor of Block)
BL:YSCALE      Nwwwddd      (Y scale factor of Block)
BL:ZSCALE      Nwwwddd      (Z scale factor of Block)
BL:XEXTRUDE    Nwwwddd      (X component of Block's extrusion direction)
BL:YEXTRUDE    Nwwwddd      (Y component of Block's extrusion direction)
```

```
BL:ZEXTRUDE       Nwwwddd        (Z component of Block's extrusion direction)
BL:SPACE          Cwww000        (Space between fields)
Attribute Tag     Cwww000        (Attribute tag, character)
Attribute Tag     Nwwwddd        (Attribute tag, numeric)
```

All items in the first column (in Courier font) must be spelled exactly as shown, if used. Use <u>spaces</u> rather than tabs or indents to separate the second column entries. In the second column, the first letter must be N or C, representing a numerical or character field. The three "w"s indicate digits that specify the field width (number of characters or numbers). The last three digits (0 or d) specify the number of decimal places you desire for the field (0 indicates that no decimals can be specified for that field).

The template file can specify any or all of the possible fields in any order, but should be listed in the order you wish the extract file to display. Each template file must include <u>at least one</u> attribute field. If a block contains any of the specified attributes, it is listed in the extract file; otherwise it is skipped. If a block contains some, but not all, of the specified attributes, the blank fields are filled with spaces or zeros.

A sample template file for the schematic drawing example in Figure 22-19 is shown here:

```
BL:NAME           C004000
BL:X              N005002
BL:Y              N005002
BL:SPACE          C002000
RES_NUMBER        C004000
RESISTANCE        C009000
CAP_NUMBER        C004000
CAPACITANCE       C006000
PART_NO           C007000
MANUFACTURER      C012000
COST              N005002
```

Creating the Extract File (Step 3)

Once you have created a template file, you can use the *-Attext* command (for Command line format) or invoke the *Attribute Extraction* dialog box by the *Attext* command. The Command line format syntax is as follows:

```
Command: -attext
Enter extraction type or enable object selection [Cdf/Sdf/Dxf/Objects] <C>: Enter or (option)
```

Instead of using the *-Attext* command, you may prefer to use *Attext*, which invokes the *Attribute Extraction* dialog box (Fig. 22-40). This dialog box serves the same functions as the *Attext* command.

You can use the *Select Objects* option (or the *Objects* option in Command line format) if you want to PICK only certain block attributes to be included in the extract file.

File Format specifies the structure of the extract file. You can specify either a *SDF*, *CDF*, or *DXF* format for the extract file.

FIGURE 22-40

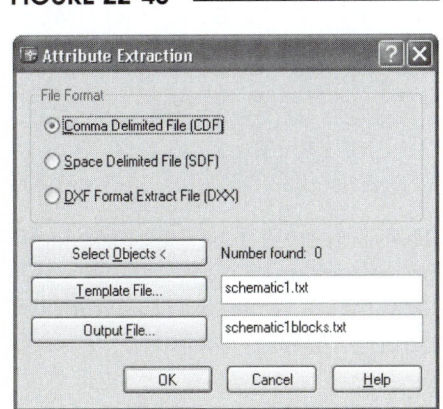

SDF

SDF (Space Delimited File) uses spaces to separate the fields. The *SDF* format is more readable because it appears in columnar format. Numerical fields are right justified, whereas character fields are left justified. Therefore, it may be necessary to include a space field (`BL:SPACE` or `BL:DUMMY`) after a numeric field that would otherwise be followed immediately by a character field. This method is used in the example template file shown previously.

CDF

CDF (Comma Delimited File) uses a character that you specify to separate the fields of the extract file. The default character for *CDF* is a comma (,). Some database packages require a *CDF* format for files to be imported.

DXF

The *DXF* format is a variation of the standard AutoCAD *DXF* (Data Interchange File format); however, it contains only block and attribute information. This format contains more information than the *SDF* and *CDF* files and is generally harder to interpret. *DXF* is a standard format and therefore does not request a template file. (Selecting *Export* from the *File* pull-down menu, then *DXX Extract,* allows you to create a *DXF* format extract file only. The *Attext* command must be typed if you want an *SDF* or *CDF* format extract file.)

After specifying the desired format, you must specify the *Template File* (name of previously created template file) and *Output File* (name of the extract file to create). The default extract file name is the same as the drawing name but with the .TXT extension.

Finally, AutoCAD then creates the extract file based on the specified parameters, displaying a message similar to the following:

```
6 records in extract file.
Command:
```

If you receive an error message, check the format of your template file. An error can occur if, for example, you have a `BL:NAME` field with a width of 10 characters, but a block in the drawing has a name 12 characters long.

Below is the sample extract file created using the electrical schematic drawing and the previous example template file.

Sample Extract File

```
RES   7.25 5.75  R1   4.7                    R-4746 Sparks R Us  0.37
RES  10.00 5.75  R2   510  ohms              R-510A Sparks R Us  0.35
CAP   6.00 7.00              C1  470   C-470A Sparks R Us  0.70
CAP   3.75 5.75              C2  .0047 C-470D Sparks R Us  0.70
CAP   4.50 2.75              C3  100   C-100A Sparks R Us  0.70
RES   8.00 2.75  R3   110K                   R-1105 Sparks R Us  0.40
```

This extract file lists the block name, X and Y location of the block resistor number, resistance value, capacitor number, capacitance value, manufacturer, and cost. Compare the resulting extract file above with the matching template file shown previously. Remember that each line in the template file creates a column in the extract file.

CHAPTER EXERCISES

1. **Create an Attributed Title Block**

    A. Open the **TBLOCK** drawing that you created in Chapter 18, Exercise 8. Because we need to create block attributes, *Erase* the existing text so only the lines remain. (Alternately, you can begin a *New* drawing and follow the instructions for Chapter 18, Exercise 8A. Do not complete 8B.)

    B. *SaveAs* **TBLOCKAT.** Check the drawing to ensure that 2 text *Styles* exist using *romans* and *romanc* font files (if not, create the two *Styles*). Next, define attributes similar to those described below using *Attdef*. Substitute your personalized *Tags*, *Prompts*, and *Values* where appropriate or as assigned (such as your name in place of B. R. Smith):

Tag	Prompt	Value	Mode
COMPANY		CADD Design Co.	const
PROJ_TITLE	Enter Project Title	PROJ-	
SCALE	Enter Scale	1"=1"	
DES_NAME		Des.-B.R.Smith	const
CHK_NAME	Enter Checker Name	Ch.-	
COMP_DATE	Enter Completion Date	1/1/05	
CHK_DATE	Enter Check Date	1/1/05	
PROJ_NO	Enter Project Number	PROJ	verify

The completed attributes should appear similar to those shown in Figure 22-41. *Save* the drawing.

C. Use the *Base* command to assign the *Insertion base point*. **PICK** the lower-right corner (shown in Figure 22-41 by the "blip") as the *Insertion base point*.

Since the entire TBLOCKAT drawing (.DWG file) will be inserted as a block, you do not have to use the *Block* command to define the attributed block. *Save* the drawing (as **TBLOCKAT**).

FIGURE 22-41

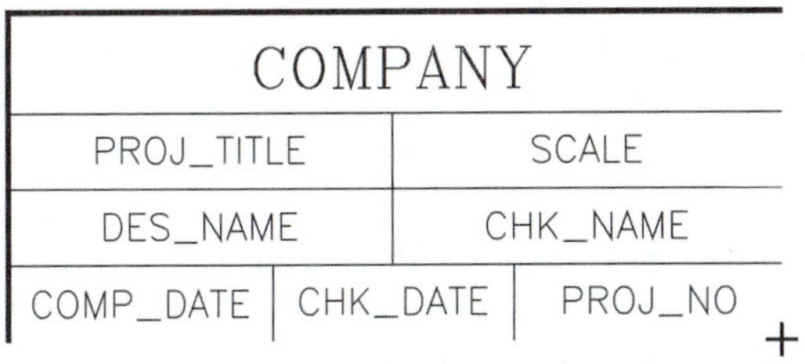

D. Begin a *New* drawing and use the **ASHEET** drawing as a *template*. Set the **TITLE** layer *current* and draw a border with a *Pline* of .02 *width*. Set the *ATTDIA* variable to **1** (*On*). Use the *Insert* command to bring the **TBLOCKAT** drawing in as a block. The *Enter Attributes* dialog box should appear. Enter the attributes to complete the title block similar to that shown in Figure 22-42. Do not *Save* the drawing.

FIGURE 22-42

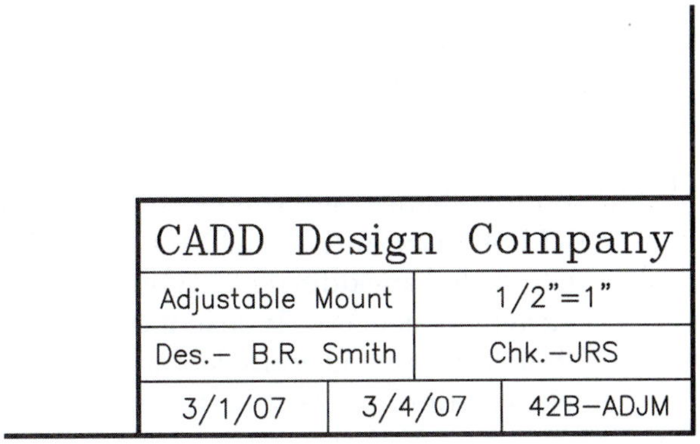

2. **Create Attributed Blocks for the OFF-ATT Drawing**

FIGURE 22-43

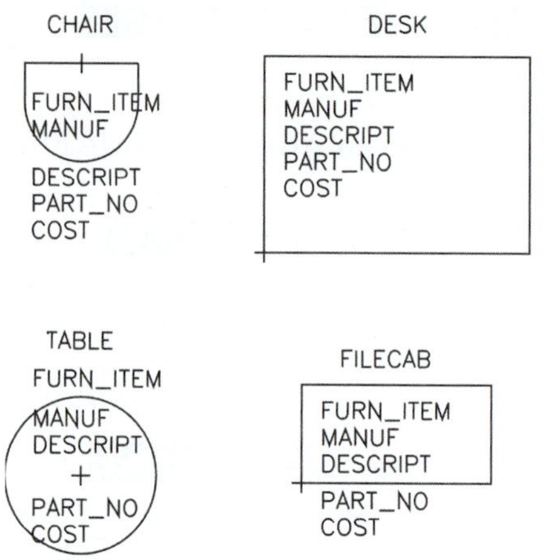

CHAIR   DESK

CHAIR:
FURN_ITEM
MANUF
DESCRIPT
PART_NO
COST

DESK:
FURN_ITEM
MANUF
DESCRIPT
PART_NO
COST

TABLE   FILECAB

TABLE:
FURN_ITEM
MANUF
DESCRIPT
+
PART_NO
COST

FILECAB:
FURN_ITEM
MANUF
DESCRIPT
PART_NO
COST

A. *Open* the **OFF-ATT** drawing (not OFFICE-REN) that you prepared in Chapter 21 Exercises. The drawing should have the office floor plan completed and the geometry drawn for the furniture blocks. The furniture has not yet been transformed to *Blocks*. You are ready to create attributes for the furniture, as shown in Figure 22-43.

Include the information below for the proposed blocks. Use the same *Tag*, *Prompt*, and *Mode* for each proposed block. The *Values* are different, depending on the furniture item.

Tag	Prompt	Mode
FURN_ITEM	Enter furniture item	
MANUF	Enter manufacturer	
DESCRIPT	Enter model or size	
PART_NO	Enter part number	Invisible
COST	Enter dealer cost	Invisible

Values

CHAIR	DESK	TABLE	FILE CABINET
WELLTON	KERSEY	KERSEY	STEELMAN
SWIVEL	50" x 30"	ROUND 2'	28 x 3
C-143	KER-29	KER-13	3-28-L
139.99	449.99	142.50	229.99

B. Next, use the **Block** command to make each attributed block Use the block *names* shown (Fig. 22-43). Select the attributes <u>in the order</u> you want them to appear upon insertion; then select the geometry. Use the **Insertion base points** as shown. *Save* the drawing.

C. Set the *ATTDIA* variable to **0** (*Off*). **Insert** the blocks into the office and accept the default values. Create an arrangement similar to that shown in Figure 22-44.

FIGURE 22-44

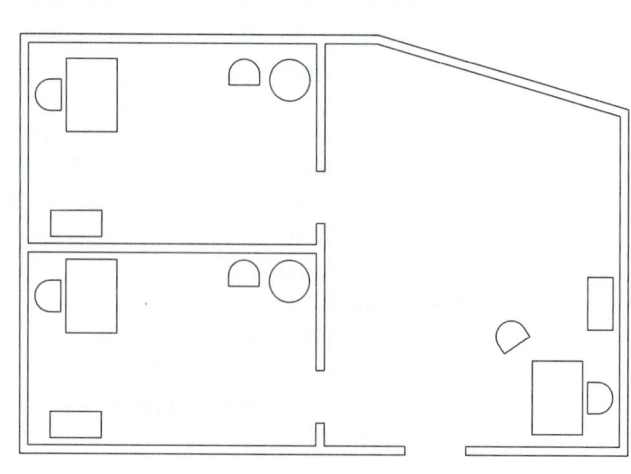

D. Use the *ATTDISP* variable and change the setting to *Off*. Do the attributes disappear?

E. A price change has been reported for all KERSEY furniture items. Your buyer has negotiated an additional $20 off the current dealer cost. Type in the **Multiple Attedit** command and select each block one at a time to view the attributes.

Make the COST changes as necessary. Use *SaveAs* and assign the new name **OFF-ATT2**.

F. Change the *ATTDISP* setting back to *Normal*. It would be helpful to increase the *Height* of the FURN_ITEM attribute (CHAIR, TABLE, etc.). Use the *Enhanced Attribute Editor* to change the **FUNR_ITEM** tag for each block. You will have to select each block individually, then use the *Text Options* tab to change the height to **3"** for each block. If everything looks correct, *Save* the drawing.

3. **Extract Attributes**

A. A cost analysis of the office design is required to check against the $7,000 budget allocated for furnishings. This analysis can be accomplished by extracting the attribute data into an **AutoCAD** *Table*. Use the *Attribute Extraction* wizard to extract the block attributes. In the *Select Attributes* step, include all the text attributes in the table as shown below. To force the table to display each insertion of each furniture item as shown below, remove the check for *Exclude general block properties*, then check the *X Position* and *Y Position* attributes. Then in the *Finalize Output* step, hide the *X Position* and *Y Position* columns. In the *Table Style* step, select *Romans* font for the table. Set the *Text height* for the data, column heads, and title to **4"**, **5"**, and **6"**, respectively. Set the *Cell Margins* to **1"**. Also, ensure you check the *Display tray notification when data needs refreshing* option. In the *Finalize Output* step, opt to *Save a template* in case you have to recreate the table. When finished, insert the table into the drawing. The table should appear similar to that below (Fig. 22-45).

B. Total the COST column to acquire the total furnishings cost so far for the job. You may want to create a new row for the table, then add a *Field* containing a *Formula* to *Sum* the cost values. (However, when you make "manual" changes to the table, this row will be lost when the table is updated.) If you plan to purchase the $1800 conference table and 8 more chairs, how much money will be left in the budget to purchase plants and wall decorations?

FIGURE 22-45

FURNISHINGS DATA					
Quantity	FURN_ITEM	DESCRIPT	MANUF	PART_NO	COST
1	CHAIR	SWIVEL	WELLTON	C-143	139.99
1	CHAIR	SWIVEL	WELLTON	C-143	139.99
1	CHAIR	SWIVEL	WELLTON	C-143	139.99
1	CHAIR	SWIVEL	WELLTON	C-143	139.99
1	CHAIR	SWIVEL	WELLTON	C-143	139.99
1	CHAIR	SWIVEL	WELLTON	C-143	139.99
1	DESK	50"x30"	KERSEY	KER-29	429.95
1	DESK	50"x30"	KERSEY	KER-29	429.95
1	DESK	50"x30"	KERSEY	KER-29	429.95
1	FILE CABINET	28x3	STEELMAN	3-28-L	229.99
1	FILE CABINET	28x3	STEELMAN	3-28-L	229.99
1	FILE CABINET	28x3	STEELMAN	3-28-L	229.99
1	TABLE	ROUND 2'	KERSEY	KER-13	122.50
1	TABLE	ROUND 2'	KERSEY	KER-13	122.50

C. After negotiating with the furniture supplier for additional chairs, a 10% discount is offered on all the chairs. Calculate the cost per chair with the discount. Next, use the *Multiple* command modifier with the *Attedit* command to change the **COST** attribute for the inserted chairs to the discounted value. (Alternately, double-click on each chair to changes its COST attribute value.) *Update* the Furnishings Data table.

D. With the savings, you decide to purchase a more expensive desk chair for the two offices. Open the **Block Editor** with the **CHAIR** block. Add arms to the chair. Also, redefine the following attributes using the *Properties* palette in the Block Editor to the following:

FURN_ITEM	CHAIR-A
DESCRIPT	ARM CHAIR
PART_NO	C-242
COST	159.99

E. In the Block Editor, use the *Save Block As* option to assign the name **CHAIR-A**. Close the editor. Finally, *Erase* the two office desk chairs and *Insert* two new arm chairs. *Update* the table. Are the new arm chair blocks listed in the table? Since the original table setup did not include this new block, you must create a new table. *Erase* the existing table. Use the *Attribute Extraction* wizard again, but in the *Begin* step, select the *Use a template* option and locate the template you saved previously. Accept all the previous settings. Note that the *Select Attributes* page lists the new CHAIR-A block, but you must check it to include its attributes in the new table. Complete the wizard and insert the new table into the drawing. Does the new table reflect the new changes? *Save* the drawing.

4. Create the electrical schematic illustrated in Figure 22-46. Use *Blocks* for the symbols. The text associated with the symbols should be created as attributes. Use the following block names and tags:

Block Names: RES
               CAP
               GRD
               AMP

Attribute Tags: PART_NUM
                  MANUF_NUM
                  RESISTANCE
                  CAPACITANCE

(Only the RES and CAP blocks have values for resistance or capacitance.) When you are finished with the drawing, create an external .CVS file reporting information for all attribute tags (coordinate data is not needed). Save the drawing as **SCHEM2**.

FIGURE 22-46 ————————————————————————

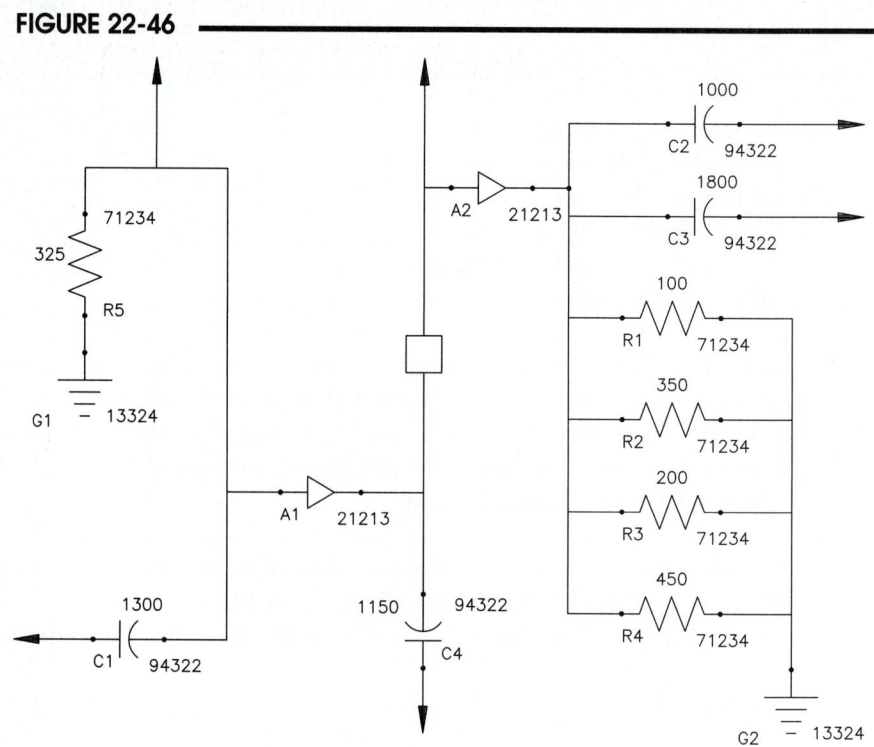

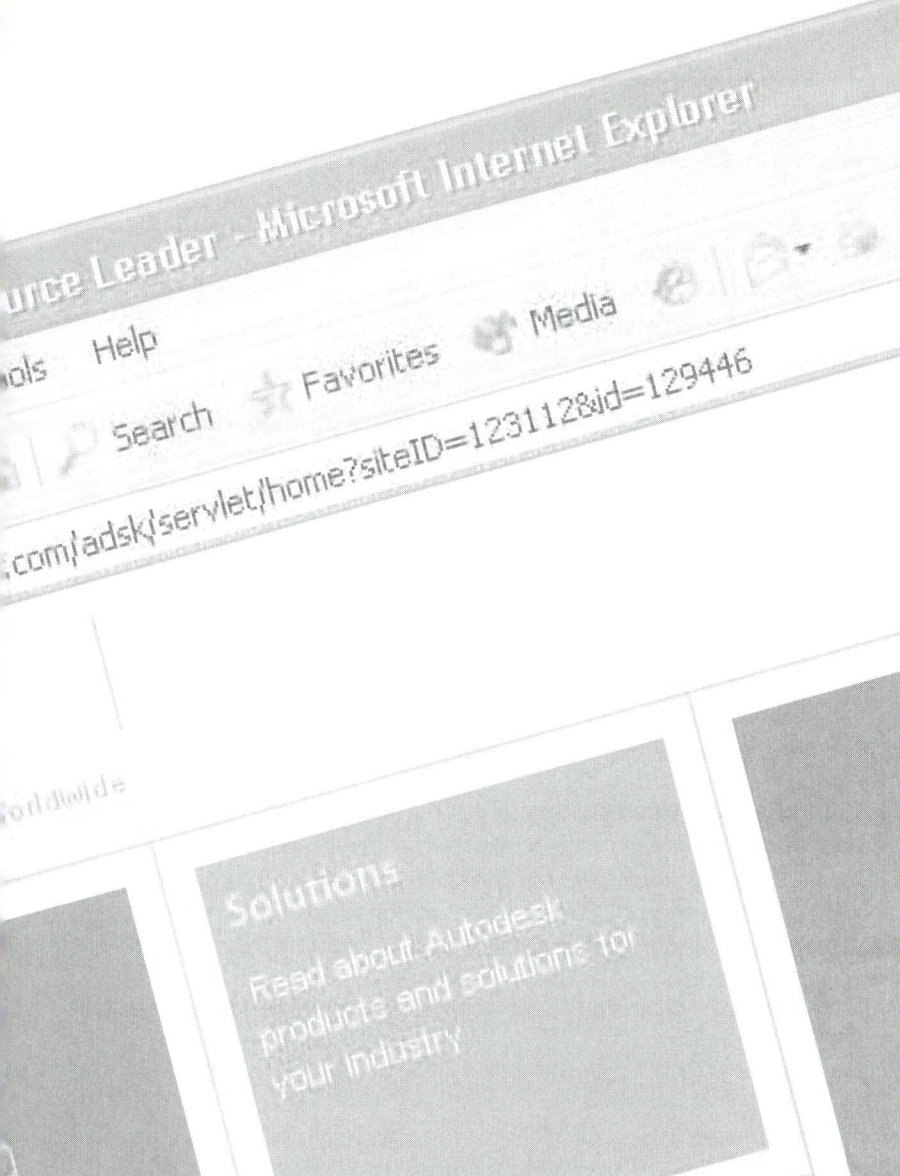

INTERNET TOOLS AND COLLABORATION

CHAPTER OBJECTIVES

After completing this chapter you should:

1. be able to use the *Browser* command to launch your Web browser from within AutoCAD;

2. be able to use *Hyperlink* to link your drawings to other drawings, text documents, or spreadsheets on your computer, your network, or the Internet;

3. be able to use *Publishtoweb* to create Web pages displaying AutoCAD drawings;

4. know how to attach drawings to your email messages using *Etransmit*;

5. know how to use the *DWF6 ePlot* device driver to create .DWF files;

6. be able to use *Publish* to create drawing sets in .DWF format;

7. be able to use AutoCAD and the Autodesk DWF Viewer to view .DWF files.

CONCEPTS

Through recent years Autodesk has created features to allow AutoCAD users to fully utilize the latest Internet-related technological advances to enhance their productivity. Through the recent releases of AutoCAD, multiple tools have been added to enable you to collaborate and share your designs with colleagues and clients using Internet technologies.

AutoCAD includes features that allow you to be more connected to Web technology by enabling capabilities such as publishing Web pages that include drawings, emailing drawings, linking other remotely located drawings and documents to your drawing, creating single drawings or complete drawing sets that are viewable in standard Web browsers, and allowing automatic updating to new features through the Autodesk Web site. The commands explained in this chapter are listed below.

Browser	Launch your Web browser from within AutoCAD with the *Browser* command so you can have instant access to the Internet while in your drawing environment.
Communication Center	This new feature keeps you up to date by notifying you when new features or updates for AutoCAD are available through the Internet.
Hyperlink	Use the *Hyperlink* command to create pointers in your AutoCAD drawings that provide jumps to other files you want to associate to the current drawing. The other files can be located on your local computer or network or can link to Web pages on the Internet. In this way your AutoCAD drawings can "contain" additional graphical, numerical, or textual information.
PublishToWeb	With this powerful tool you can generate Web pages complete with embedded AutoCAD drawing images in .JPG, .PNG, and .DWF format. This tool is automatic so no experience with .HTML is required.
eTransmit	*eTransmit* opens your email system and allows you to include an AutoCAD drawing (and any associated files) as an email attachment.
Publish	This feature allows you to create drawing sets that are readable, but not editable, from Autodesk DWF Viewer or Microsoft Internet Explorer. The drawings are created in .DWF format.
Plot Drivers	You can create a Drawing Web Format (.DWF) file of the current drawing using the *DWF6 ePlot* device from the *Plot* dialog box. A .DWF file is a compact, vector-based file that is viewable without AutoCAD and easily transportable over the Internet. You can also create raster files of the current drawing using the *PublishToWeb PNG* or *PublishToWeb JPG* device.
Autodesk DWF Viewer	This viewer can be used to view .DWF files. The Autodesk DWF Viewer is free and downloadable, and can be installed by anyone—without a license for AutoCAD.
Dwfattach	Use this command to insert a DWF image directly into AutoCAD. You cannot alter the DWF image in AutoCAD, but you can *Osnap* to the geometry and trace over it like an "underlay."
Markup	This command produces the *Markup Set Manager*. This utility allows you to read .DWF files that have been "marked up" in Autodesk DWF Composer with design revisions that are needed in the original design drawings.

2007

Browser

Pull-down Menu	Command (Type)	Alias (Type)	Short-cut	Screen (side) Menu	Tablet Menu
...	*Browser*	...	...	...	...

AutoCAD includes this command specifically to launch your Web browser from within AutoCAD. An icon button on the *Web* toolbar (Fig. 23-1) and in the file navigation dialog boxes is available to invoke the *Browser* command.

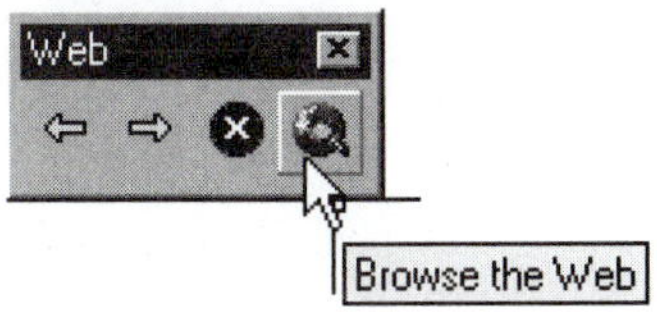

The information about your browser (location and name of the executable file) is stored in the system registry. Therefore, when you use *Browser*, AutoCAD automatically locates and launches your current (default) Web browser from within AutoCAD. This action makes it possible to view any Web site or select any viewable files located on your computer. If you have Autodesk DWF Viewer, you can view .DWF files after you have created them, view other .DWF files on the Internet, and drag-and-drop related .DWG files from the Internet directly into your AutoCAD session.

When you use *Browser*, your registered Web browser is launched. By default (if you have not specified a different file or URL), the Autodesk home page is displayed (Fig. 23-2). You can locate another file or URL by entering it in the *Address* edit box of the browser.

FIGURE 23-1

FIGURE 23-2

Inetlocation

Pull-down Menu	Command (Type)	Alias (Type)	Short-cut	Screen (side) Menu	Tablet Menu
Tool *Options...* *Files*	*Inetlocation*	...	...	*TOOLS 2* *Options...* *Files*	...

The initial Web address that is located when *Browser* is used can be specified by using the *INETLOCATION* system variable or by using the *Options* dialog box. The default *INETLOCATION* is:

www.autodesk.com

You can specify any other viewable file or Internet address. Using the Command line method, the following prompt is issued:

```
Command: inetlocation
Enter new value for INETLOCATION <"http://www.autodesk.com">: (Enter desired URL)
Command:
```

If you prefer to use the *Options* dialog box (Fig. 23-3), expand the *Help and Miscellaneous File Names* section. Expand *Default Internet Location*, highlight the address, and select the *Remove* button. Then enter the desired new address.

FIGURE 23-3

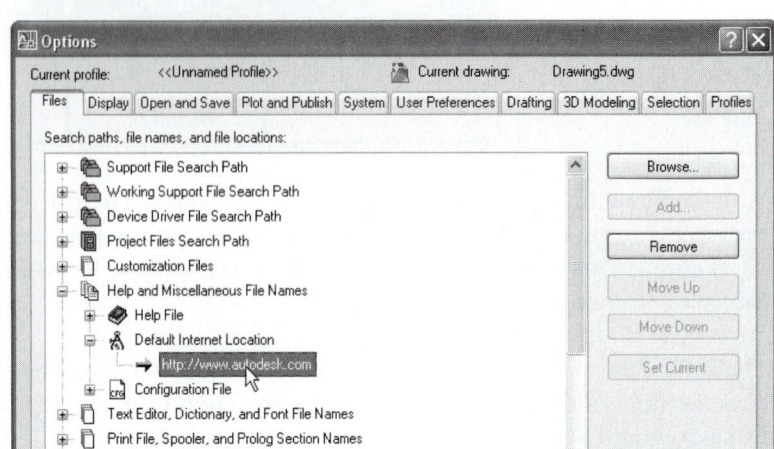

Communication Center

Communication Center is not a command, rather it is a feature that can be installed with AutoCAD. Communication Center may or may not be installed on your system depending on options that were selected during installment. (See "Tray Settings" in Chapter 1 for information on enabling the Communication Center icon.) Communication Center is an interactive feature that connects your computer to the Autodesk Web site. Autodesk monitors information about your computer and AutoCAD installation and notifies you when new information and updates are available.

You must be connected to the Internet for Autodesk to deliver content and information. Each time Communication Center is connected, it sends information from your computer to Autodesk so that the correct resources can be returned to you. All information is sent to Autodesk anonymously to maintain your privacy. The following information is sent to Autodesk:

Product Name	the name of the product in which you are using Communication Center
Product Release Number	the version of the product
Product Language	the language version of your product
Country	the country that was specified in the Communication Center settings

Autodesk compiles statistics using the information sent from Communications Center to monitor how AutoCAD is being used and if it needs updating. Autodesk states that it will maintain your information in accordance with Autodesk's published privacy policy. The privacy policy is available at:

http://www.autodesk.com/privacy

If Communication Center is available on your computer, a small "satellite dish" icon appears in the extreme lower-right corner of the AutoCAD window. Right-click on the icon to produce the *Communication Center* dialog box. Left-click on the icon or select *Settings* from the *Communication Center* dialog box to produce the *Configurations Settings* dialog box.

Communication Center Dialog Box

The *Communication Center* dialog box (Fig. 23-4) indicates whether your product is up to date and may give you a variety of other information depending on your configuration settings. Click on any of the items in the list to view the content. If you have not been connected to the Internet for some time, or you specified to update information *On Demand*, you can choose to *Refresh Content*. Selecting *Settings* produces the *Configuration Settings* dialog box.

Configurations Settings Dialog Box

Settings Tab

In the *Settings* tab (Fig. 23-5), specify the country you are in and how often you want to check for new content (*Daily, Weekly, Monthly,* or *On Demand*). Checking *Enable Balloon Notification for new announcements* causes a balloon message to appear in the lower-right corner of the AutoCAD window when new content is available.

Channels Tab

The *Channels* tab (Fig. 23-6) allows you to indicate what information should be sent to you. The following four content areas are available.

Live Update Maintenance Patches
You can receive notifications whenever new maintenance patches are released from Autodesk.

Subscription Info and Extension Announcements
This option provides announcements and subscription program news if you are an Autodesk subscription member.

Articles and Tips
Check this box to be notified when new articles and tips are available on Autodesk Web sites (see Fig. 23-4).

Product Support Information
You can get breaking news from the Product Support team at Autodesk with this feature. Also stay informed about Autodesk company news and product announcements.

Controlling Communication Center on a Network

If you have a network installation of AutoCAD, you can use the CAD Manager Control utility to turn Communication Center on and off. For example, if you want to prevent Communication Center from sending information to Autodesk, you can turn it off. The utility can be installed from the AutoCAD CD under the *Network Deployment* option.

FIGURE 23-4

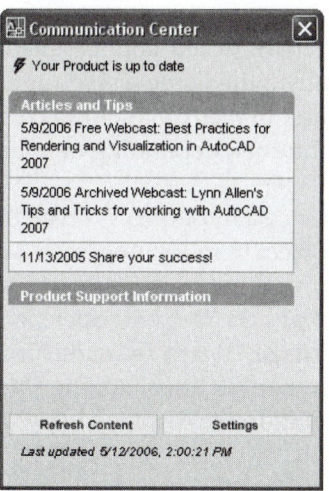

FIGURE 23-5

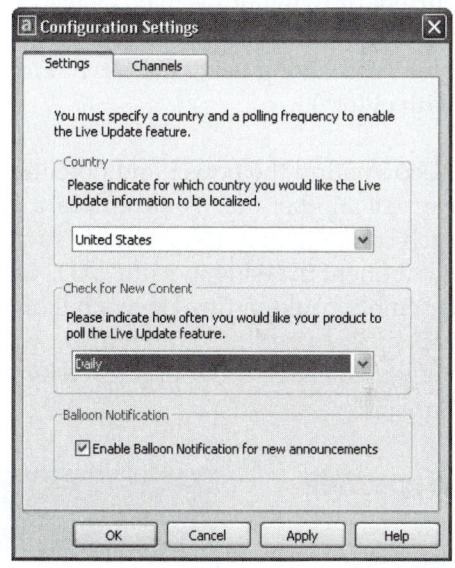

FIGURE 23-6

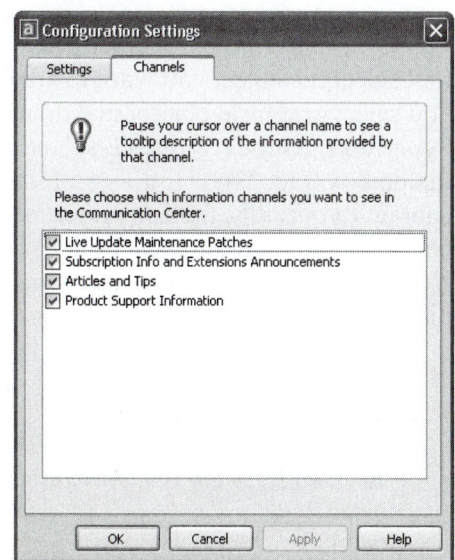

USING HYPERLINKS

The *Hyperlink* command allows you to attach a link to any AutoCAD graphical object. The link is actually a file name or Web address (URL) associated with the object. Therefore, *Hyperlink* can be used to link to files on the local computer or network (without Internet access) or to connect to files over the Internet, depending on what information is entered as the link.

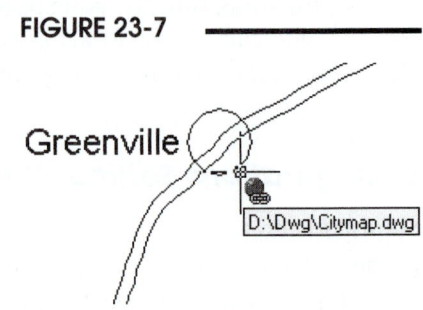

Once a hyperlink is created, you can press Ctrl and left-click on the object to follow the link. The link automatically opens the related program (word processor, spreadsheet, or CAD program) and displays the file, or it opens your Web browser and displays the page or image. When the cursor passes over an AutoCAD object that has a hyperlink attached, the *Hyperlink* symbol and the related link information are displayed (Fig. 23-7).

A *Hyperlink* can be used to create links within two types of AutoCAD files: (1) to create links in the current AutoCAD drawing (.DWG) and (2) to create links for a .DWF image that is created from the current drawing but is to be viewed from a Web browser. To create links to be used in .DWF images, use the *Hyperlink* command as you would normally to attach links to objects in the current drawing, then plot the drawing using the *DWF ePlot* device to create the related .DWF file (see "Creating .DWF Files with ePlot").

Keep in mind the power and potential usefulness of hyperlinks in AutoCAD drawings. Any hyperlink can call another file. For example, a drawing of a mechanical assembly could contain hyperlinks for each component of the assembly such that following the link for each component could open the related component drawing in AutoCAD. Or, as another example, opening another hyperlink for the entire assembly could in turn open a bill of materials in a word processing or spreadsheet program. The same capability hyperlinks added to an AutoCAD drawing can be contained in a .DWF image of the drawing viewed in a Web browser if those hyperlinks call other images or HTML files normally viewable in a browser.

Hyperlink

Pull-down Menu	Command (Type)	Alias (Type)	Short-cut	Screen (side) Menu	Tablet Menu
Insert Hyperlink...	*Hyperlink*	...	*Ctrl+K*	...	...

When the *Hyperlink* command is invoked, you must first select the object you want to attach the link to. If you select graphical objects that do not already contain hyperlinks, AutoCAD displays the *Insert Hyperlink* dialog box (Fig. 23-8). If the graphical objects already contain hyperlinks, the *Edit Hyperlink* dialog box is displayed, which has the same options as the *Insert Hyperlink* dialog box. The Command prompt and the dialog box options are explained next.

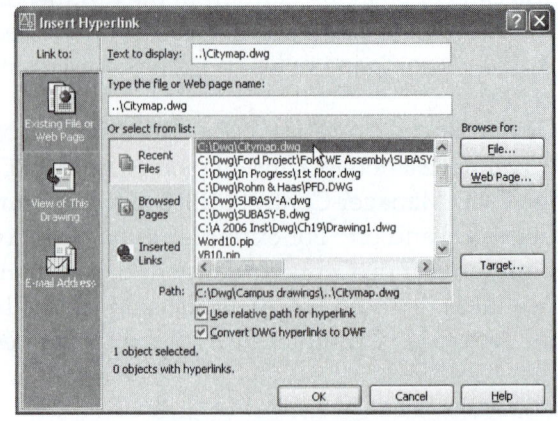

Command: **hyperlink**
Select objects: **PICK**
Select objects: **Enter**
(The *Insert Hyperlink* or *Edit Hyperlink* dialog box appears.)
Command:

When the *Insert Hyperlink* dialog box appears, first select (from the buttons on the left side of the dialog box) what kind of link you want to attach to the selected object—an *Existing File or Web Page*, a view of the current drawing (by selecting *View of this Drawing*), or an *Email Address*.

Existing File or Web Page

Selecting the *Existing File or Web Page* button allows you to type the information or select from *Recent Files*, *Browsed Pages*, or *Inserted Links*. In each case, files, pages, and URLs you recently used are displayed. There are also buttons to select another *File* or *Web Page* that is not displayed in the list. Selecting the *Target* button opens the *Select Place in Document* dialog box where you can select a named *View* or layout in a drawing to link to (not the current drawing). The text you enter in the *Text to display:* edit box is the text that appears in the drawing when the cursor passes over the object containing the hyperlink (see Fig. 23-4).

The *Use Relative Path for Hyperlink* checkbox toggles the use of a relative path for the current drawing. If this option is selected, the full path to the linked file is not stored in the drawing with the hyperlink. AutoCAD sets the relative path to the value specified by the *HYPERLINKBASE* system variable or, if this variable is not set, the current drawing path. If the *Convert DWG hyperlinks to DWF* option is checked and the hyperlink is another drawing, the hyperlinked drawing will be converted to a DWF file hyperlink if you publish or plot the current drawing to a DWF file (see "Creating .DWF Files").

The *Remove Link* button appears only in the *Edit Hyperlink* dialog box. You can use this option to remove a previously attached hyperlink from the selected object.

View of This Drawing

If you want a specific view or layout of the current drawing to be associated to the selected object as a hyperlink, you can use this button to select from a list of *Views* that are contained in the current drawing (Fig. 23-9). In other words, this feature allows you to select an object in the current drawing and attach a hyperlink that causes a jump to a different layout or saved view of the drawing. AutoCAD displays that portion of the drawing (view or layout) when the hyperlink is opened.

Email Address

This feature allows you to attach a hyperlink to an object so that when this hyperlink is opened, it causes the computer's default email system to open ready to type a new message. The email address contained in the link (appearing in the *E-mail address:* edit box of the *Insert Hyperlink* dialog box, Fig. 23-10) is automatically entered in the "Send to:" box of your new email message and the subject text in the link (contained in the *Subject:* edit box) is automatically entered in the "Subject:" box of your email message. You simply type the message and select *Send* in your email program. Use the *Text to display:* edit box to supply the text that appears in the drawing when the cursor passes over the object containing the hyperlink.

FIGURE 23-9

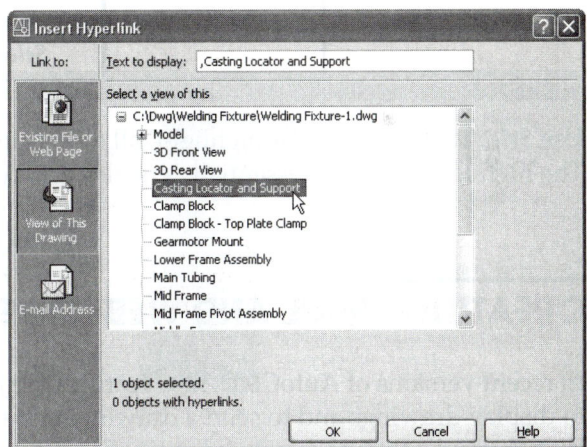

FIGURE 23-10

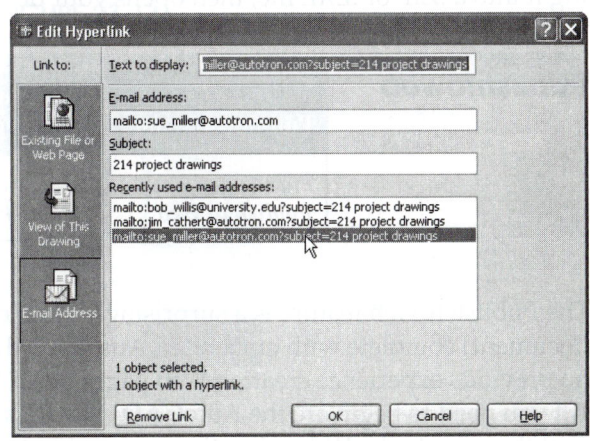

Files Tree Tab

The *Files Tree* tab lists the files to be included in the transmittal set (Fig. 23-17). When you first access the tab, all files associated with the current drawing (such as related Xrefs, drawing standards files, plot styles, plotter configurations, and fonts) are listed in a hierarchical format and have a check appearing in the checkbox before each file name in the list. Remove a check if you do not want the related file included in the set to transmit. For example, if you want to send a drawing that contains Xrefs, but you do not want to include the Xrefs in the transmittal set, remove the checks from all the Xrefed drawings. You can also right-click in the file list to display a shortcut menu from which you can *Check All* or *Uncheck All* and *Expand All* or *Collapse All*. Use the *Add File* button to add additional files to the set. Files that are referenced by URLs (hyperlinks) do not appear in the list and are not included in the transmittal set.

Files Table Tab

The list displays the same files as the Files Tree tab, but in a different format. In this list, files are given in alphabetical order by default but can be sorted by *Filename, Type, Size, Version,* or *Date* in normal or reverse order by clicking on the desired column heading.

Add File

Use this button to open the standard file selection dialog box where you can select additional files to include in the transmittal set.

Transmittal Setups

See *"Transmittal Setups* Dialog Box."

View Report

This button displays the report (Fig. 23-18) that is automatically generated and included with the transmittal set. You can add additional notes in the report by entering the information in the edit box at the bottom of the *Create Transmittal* dialog box. The information automatically generated by AutoCAD explains what steps must be taken by the recipient for the transmittal set to work properly. For example, if AutoCAD detects .PC3 (printer configuration) files or .CTB/.STB (plot style table) files in one of the transmittal drawings, the report instructs the recipient where to copy these files so AutoCAD can detect the files when the transmittal drawing is opened.

FIGURE 23-18

Save As

This button (in the *View Transmittal Report* dialog box) opens the *Save report file* as dialog box where you can specify a location to save a report (.TXT) file. A report file is always generated and included with all transmittal sets; however, you can save an additional copy of the report file for archival purposes using this option.

Transmittal Setups

A transmittal setup is a saved set of parameters that determine how the transmittal package is created. Selecting the *Transmittal Setups* button in the *Create Transmittal* dialog box produces the *Transmittal Setups* dialog box (not shown) as a "front end" for the *Modify Transmittal Setup* dialog box (see Fig. 23-19).

Transmittal Setups Dialog Box

This dialog box (not shown) allows you to select the setup name you want to use for the transmittal and also acts as an intermediate step to create *New* setups and *Modify* existing setups. If this is your first setup, select the *New* button to create a new setup (using *Standard* as a template) or select *Modify* to change the *Standard* setup. In either case, your choice produces the *Modify Transmittal Setup* dialog box where you can select the settings for the current setup. Any new setups are saved for future use and stored for each user on the computer. You can also select existing setups from the list to *Rename* or *Delete*.

Modify Transmittal Setup Dialog Box

Transmittal Package Type

Here you can select from *Folder (set of files), Self-executable (*.exe), or Zip (*.zip)*. The Folder option creates a new folder or uses an existing folder and creates the uncompressed transmittal files. The *Self-executable (*.exe)* option creates one self-executable file that the user can double-click to decompress and restore the original files. The *Zip (*.zip)* option creates a transmittal set of files as a compressed .ZIP file. The user needs a decompression utility such as the shareware application WinZip to decompress and restore the files.

FIGURE 23-19

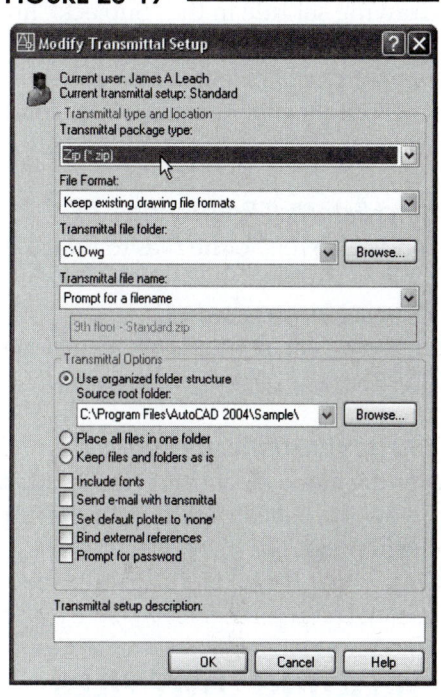

File Format

You can *Keep the existing file formats* or select from the drop-down list to convert all the drawings to *AutoCAD 2004/LT 2004 drawing format* or *AutoCAD 2000/LT 2000 drawing format.*

Transmittal File Folder

Use this list to select the location on your computer or network where the transmittal set is to be created. You can use the *Browse* button to specify a new location.

Transmittal File Name

Here you specify how the file name is specified when you create the transmittal package. Selecting *Prompt for a file name* allows you to enter the name of the transmittal package. *Increment file name if necessary* uses a logical default file name, and if the file name already exists, a number is added to the end. *Overwrite if necessary* uses a logical default file name, and if the file name already exists, the existing file is automatically overwritten.

Use Organized Folder Structure

When the transmittal package is created, this option duplicates the folder structure for the files being transmitted. The root folder is the top level folder within a hierarchical folder tree. Relative paths remain unchanged and absolute paths within the root folder tree are converted to relative paths. Other folders for fonts, plot configuration files, and sheet sets are created if necessary.

Place all Files in One Folder

When the transmittal package is created, all files are installed to a single, specified target folder.

Keep Files and Folders As Is

Use this option to preserve the folder structure of all files in the transmittal package. This option is good for decompression and installation on another similar system.

Include Fonts

If you remove this check, all the current drawing's associated font files (.TXT and .SHX) are removed from the list and the font files are not included in the transmittal set. Generally, most users already have the associated .TXT and .SHX fonts on their systems since they are included with AutoCAD. Because TrueType fonts are proprietary, they are never included with the transmittal set. However, if any required TrueType fonts are not registered on the computer that receives the transmittal set, the font specified by the *FONTALT* system variable is substituted.

Send E-mail with Transmittal

Check this box to launch your default email program when the transmittal set is created. When the program opens, AutoCAD automatically attaches the transmittal set to the email message and enters the drawing set title in the "Subject:" line of your note.

Set Default Plotter to 'None'

This option changes the printer/plotter setting in the transmittal package to *None* since your local printer/plotter settings are usually not relevant to the recipient.

Bind External References

Binds all Xrefs to the files to which they were attached (see Chapter 30 for information on binding Xrefs).

Prompt for password

Here you can set a password that is required to open the files. Make sure you notify the recipient of the files that the password is required.

Transmittal Setup Description

If you enter a description for the transmittal setup, it is displayed in the *Transmittal Setups* dialog box below the list of transmittal file setups. You can select any transmittal setup in the list to display its description.

CREATING .DWF FILES

CAD files contain very precise and detailed geometric information as well as an enormous amount of data beyond the drawing geometry that you see, such as coordinate data, named dependent objects, system variable settings, user preferences, and so on. Most CAD files are also proprietary—viewable only using the software product that created the CAD file. In today's collaborative environment, however, it is necessary for many people who are involved with various aspects of the design process to view drawings with some precision, but not edit the drawings or have access to the other data contained in the drawing file. An example of implementing this idea is AutoCAD's *Publish to Web* capability, which facilitates posting drawing images on a Web page in .JPG, .DWF, or .PNG format, so anyone with a Web browser can view the drawings.

Autodesk's .DWF file format is also intended for this purpose—viewing a detailed drawing image without AutoCAD and without the ability to access unneeded data. Since the data is compressed, the file is easily transportable over the Internet. Although .JPG and .PNG files can be viewed easily over the Internet, they are raster files and cannot contain much detail. Therefore, Autodesk developed the Design Web Format, or .DWF, which is a vector-based file that can contain incredible geometric detail and additional information, and it is highly compressed for easy transport over the Internet. The .DWF files cannot be viewed using a standard Web browser—they require AutoCAD or Autodesk DWF Viewer. Autodesk DWF Viewer is free and downloadable from the www.autodesk.com Web site and it is included with AutoCAD.

.DWF files are a good way to share AutoCAD drawing files with colleagues who don't have AutoCAD. Because the Autodesk DWF Viewer is free and the interface is easy to use, even individuals with no CAD knowledge can easily view and navigate a .DWF file. You cannot change the .DWF image from within the viewer, although you can pan, zoom, print, manipulate layers and views, and activate hyperlinks. (See "Viewing .DWF Files.")

Raster Files and Vector Files

The common World Wide Web graphic standard uses a raster image format such as .GIF or .JPG. These files can be highly compressed, making the information displayed in the images quickly transportable over data transmission lines and devices used for the Internet. However, because raster images are pixel-based, or composed of tiny "dots," viewing detail is not possible—zooming in only enlarges the "dots." Figure 23-20 displays a raster image which has been enlarged.

Raster images, such as a "screen capture" for example, are defined in the data file as a map of pixels—rows and columns of picture elements either in black and white or in color. Raster images are sometimes referred to as "bitmaps," since the file retains the map of each dot (placement within the rows and columns) with its color information.

FIGURE 23-20

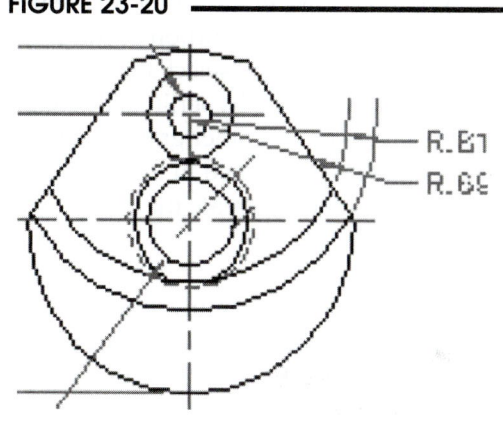

Examples of raster image file formats (file name extensions) are: .BMP, .CLP, .DIB, .GIF, .ICO, .IFF, .JPG, .MAC, .MSP, .NIF, .PBM, .PCX, .PSD, .RAS, .SGI, .TGA, .TIF, .XBM, .XPM, .XWD.

Raster images are of little use to professionals who require drawings with detailed information. Instead, the engineering, architecture, design, and construction industries use vector images to retain data with great precision and detail. CAD programs, which are the primary graphics creation and storage media for these industries, generate <u>vector</u> data. Figure 23-21 illustrates the same vector image (a CAD drawing) in a "zoomed in" display.

FIGURE 23-21

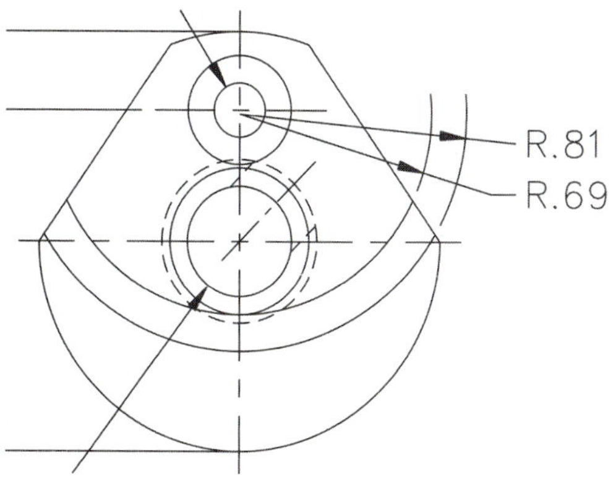

Because a vector file defines a line, for example, as a vector between two endpoints, only one dimension is retained—length. Normally, "zooming in" to a detailed section of the drawing does not create a second dimension (thickness or width) for the line. The thickness of the displayed line on a computer monitor is always one pixel, the size of which is determined by the display device.

NOTE: When creating a .DWF file, you can choose to display the drawings in .DWF format with line weight to give a representation of the printed image. In this case, zooming does change line thickness.

Examples of vector file formats include: .CDR, .CGM, .CMX, .DGN, .DWG, .DWF, .DXF, .GEM, .IGS, .MCS, .P10, .PCL, .PGL, .PIC, .PLT, .PRT, .WRL.

A .DWF image is a compressed <u>vector</u> file whose geometry can be viewed in detail by zooming in and out. Before the .DWF standard was introduced, the World Wide Web provided inadequate usefulness for the engineering, architecture, design, and construction industries. Raster files are unable to retain the necessary detailed information, while most data-rich vector files generated by CAD programs are too large to be "viewable" over the Web.

Preview

Shows a preview of the selected layout from the list.

Add Sheets

This option produces the standard *Select Drawings* dialog box. Select the drawing files you want to include in the .DWF file to be created. See also "*Model tab and Layout tabs.*"

Remove Sheets

Highlight any layout and select this option to remove the layout from the list.

Move Sheet Up and Move Sheet Down

These buttons allow you to specify the order of the drawing sheets in the *Sheets to Publish* list. The list order dictates the order of sheets <u>when they are viewed</u> (in the list of Autodesk DWF Viewer).

Load Sheet List

If you saved a list of drawings previously using *Save Sheet List*, you can load the same list again for publishing or modifying. You can load .DSD (drawing sheet description) or .BP3 (batch plot) files.

Save Sheet List

This choice saves the current list of drawings and layouts in the *Sheets to Publish* as a .DSD file.

Plot Stamp Settings

<u>Only if the sheets are printed or plotted</u> instead of published to a .DWF file (see *Plotter named in page setup*), you can specify that each sheet contains a plot stamp. A plot stamp may contain a date, time, drawing name, etc. Using this option produces the *Plot Stamp* dialog box. (See "Plot Stamping" in Chapter 33.)

Include Plot Stamp

This box must be checked for plot stamps (as specified in the *Plot Stamp Settings*) to be applied to printed or plotted sheets and only when *Plotter named in page setups* is also checked.

Plotter named in page setup

<u>If you select this option, *Publish* does not create a .DWF file.</u> Instead, when you select *Publish*, each sheet is printed, plotted, or plotted to a file as specified by the device named in its *Page Setup* column. You can change the page setup by clicking the page setup name in the *Sheets to Publish* list and selecting another page setup from the list or importing a named page setup. Selecting this radio button also enables the *Number of Copies* option and reverse order button.

DWF file

Select this radio button to specify that one .DWF file containing all the sheets in the *Sheets to Publish* is created when you use the *Publish* button to create the .DWF file (also see "Publish Options").

Model tab and Layout tabs

When drawings are added to the list using *Add Sheets*, you can specify whether or not the *Model* layout or layout tabs are automatically included in the list of *Sheets to Publish*. These options have no effect on drawings already in the list.

Show Details

Use this button to expand the *Publish* dialog box to give *Selected sheet information* and *Selected page setup information*. Highlight any sheet from the list to display its information below.

Publish

Select *Publish* only when you are ready to begin the publishing operation (all other options are specified). *Publish* creates a .DWF file or plots to a device or file depending on your choice of *Plotter named in page setup* or *DWF file*. If you are publishing to a .DWF file, specify the desired .*DWF file name* and location in the *DWF file name* dialog box that subsequently appears.

Publish Options

Select this button to produce the *Publish Options* dialog box (Fig. 23-26). Here you specify options when .DWF files are created using *Publish*.

Default output location
When you *Publish*, this setting determines the default path for the .DWF files or when the layouts are plotted to files.

FIGURE 23-26

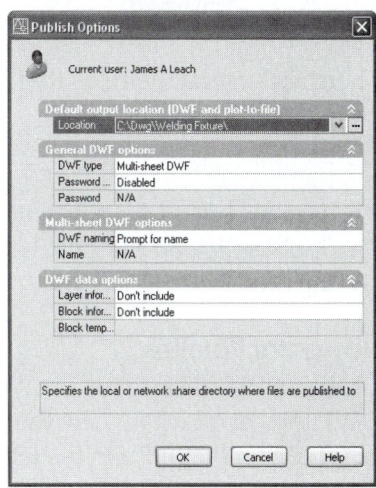

General DWF options
If you have several layouts in the *Sheets to Publish* list, the *Single sheet DWF* option will produce one .DWF file for each layout. Selecting *Multi-sheet DWF* will produce one .DWF fie containing all layouts. You can specify that a password is automatically applied to all created DWF files (choose *Specify password*, then enter the password in the *Password* edit box) or that you are prompted for a password when each .DWF file is created (choose *Prompt for password*). .DWF passwords are case sensitive and can be made up of letters, numbers, punctuation, or non-ASCII characters. If you lose or forget the password, it cannot be recovered.

Multi-sheet DWF options
You can choose to be prompted for a name when each .DWF file is created or you can specify a name in the *Name* edit box below.

DWF data options
You can include *Layer information* and *Block information* in the .DWF files if desired. If layer information is included, you can control the *On/Off* visibility of the layers in the .DWF image when the resulting .DWF is viewed in the viewer, similar to controlling layer visibility in AutoCAD (see "Viewing .DWF Files"). If this box is cleared, the resulting .DWF in the viewer offers no layer listing or visibility control. If you choose to include *Block information* in the .DWF files, the block property and attribute information will be visible in the published .DWF files (see Chapters 21 and 22).

AutoCAD's *Sheet Set Manager*

Another AutoCAD tool, the *Sheet Set Manager* can be used to create a group of drawings related to one project, called a sheet set. You can use *Sheet Set Manager* to *Publish* a sheet set as a multi-drawing .DWF file. See Chapter 34, Sheet Sets.

VIEWING .DWF FILES

Once .DWF files are created from AutoCAD drawings, you can view the .DWF files two ways:

1. You can use Autodesk DWF Viewer to view single- or multi-sheet .DWF files. Autodesk DWF Viewer comes with AutoCAD and can be downloaded free from www.autodesk.com for anyone who does not have AutoCAD.
2. You can view .DWF files directly in AutoCAD. This is accomplished by inserting an individual .DWF sheet as a *DWF Underlay*.

Autodesk DWF Viewer

Autodesk DWF Viewer comes with AutoCAD, so everyone with AutoCAD can view DWF images directly with Autodesk DWF Viewer. Alternately, Autodesk DWF Viewer can be downloaded free from www.autodesk.com and installed by anyone—<u>without AutoCAD</u>. Keep in mind that AutoCAD DWF Viewer is used to <u>view the images only</u>; you cannot draw in Autodesk DWF Viewer or make any changes to a .DWF file.

There are several methods that can be used to open .DWF files for viewing:

1. use the *File* pull-down menu, then use *Open* in Autodesk DWF Viewer to locate and open a .DWF file;
2. drag-and-drop the .DWF file from Windows Explorer or My Computer into Autodesk DWF Viewer;
3. double-click on the .DWF file in Windows Explorer or My Computer to launch DWF Viewer to view the file.

Navigator Palettes

Once a .DWF file is loaded in Autodesk DWF Viewer, you can view the individual sheets of the file. On the left side of the viewer are several "navigator palettes" that expand to view and control options within the palettes. For example, the top palette, labeled *Contents*, expands to provide thumbnail images of each sheet (Fig. 23-27). Click on any thumbnail image to view the related sheet.

FIGURE 23-27

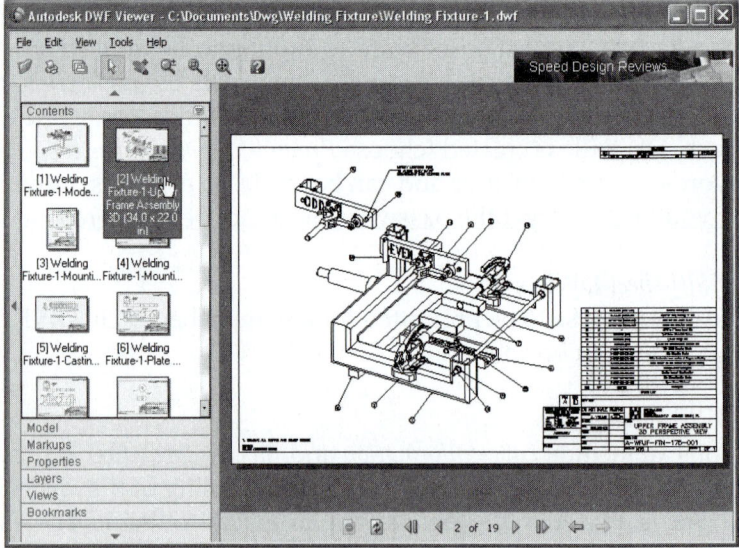

Model
The *Model* palette is used for .DWF files created from 3D solid model drawings. The palette contains a tree view of the 3D model's objects and subobjects.

Markups
The *Markups* palette is used to locate and view a markup. A markup is created in Autodesk DWF Composer and can be viewed in DWF Viewer or AutoCAD (see "Markup Set Manager").

Properties
The *Properties* palette allows you to view the properties of a selected sheet, markup, or object. Click on the *Palette Options* button next to the word *Properties* and select *Sheet Properties*, *Markup Properties*, or *Object Properties*. The properties for the selected item display in the *Properties* palette.

Layers

The *Layers* palette lists all layers for each sheet. You can control which layers are visible in the DWF sheet by toggling the "light bulb" icon on or off. For example, Figure 23-28 displays a sheet with the DIM layer off, so the sheet's dimensions are no longer visible.

Views

Similar to an AutoCAD drawing, the related .DWF file can contain the same views. Use the *Views* palette to list views contained in the file. For 3D models, the standard views (top, front, right, etc.) are listed, which you can click to display.

FIGURE 23-28

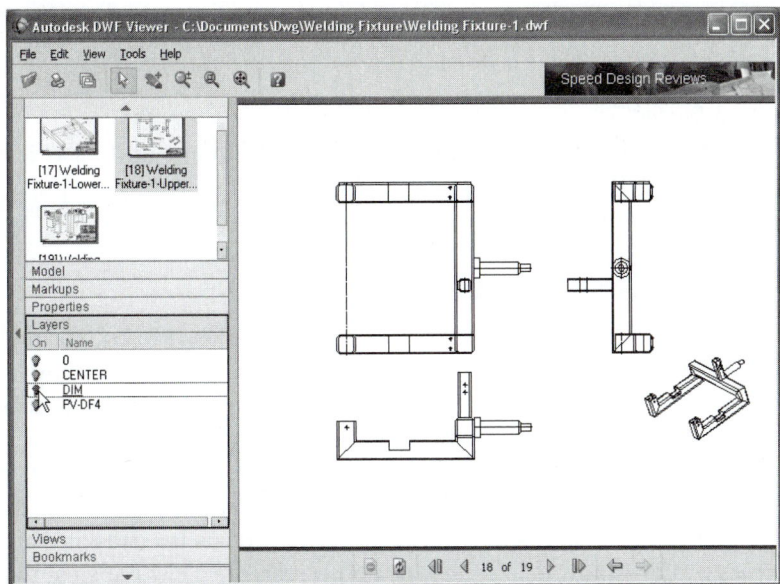

Bookmarks

If the original drawings were created in Autodesk Revit®, the *Bookmarks* tab lists locations identified by the drawing's author. Clicking a bookmark displays the bookmarked sheet.

Shortcut Menu, Pull-down Menus

Right-clicking in the viewing area produces a shortcut menu (Fig. 23-29). This menu provides the same options available from both the *View* pull-down menu and *Zoom* and *Pan* buttons in the upper left of the viewer.

Pan

This option puts you in real-time *Pan* mode and the pointer becomes a "hand" icon. Press the left mouse button to drag the image around on the screen in a manner identical to the *Realtime* option of *Pan* in AutoCAD.

FIGURE 23-29

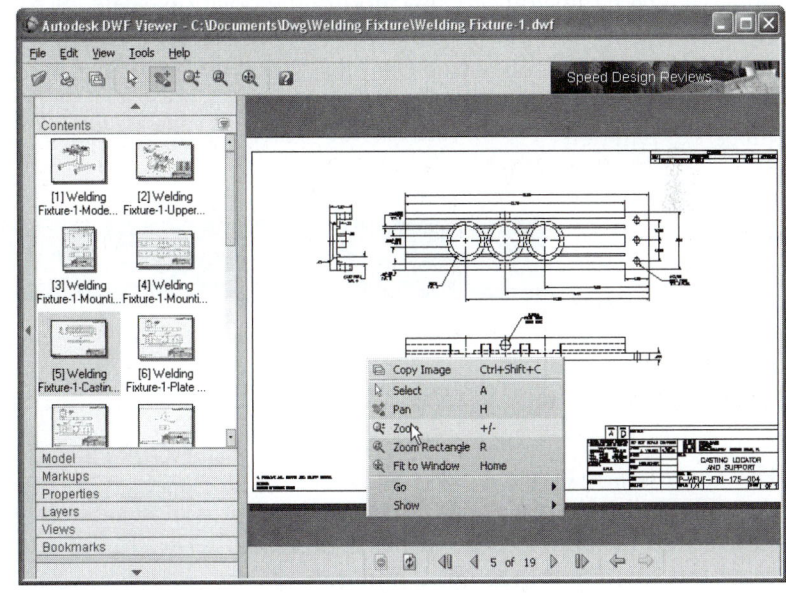

Zoom Options

With *Zoom* capabilities of a DWF image, genuine detail can be viewed, similar to capabilities in the original CAD program with the associated CAD drawing. By default, you can *Zoom* at any time (without selecting a menu option) by turning the mouse wheel if you have a wheel mouse. The *Zoom Rectangle* and *Fit in Window* options operate identically to the *Window* and *All* options, respectively, in AutoCAD (see Chapter 10, Viewing Commands).

Show

A cascading menu appears with toggles for you to show *Hyperlinks* (if hyperlinks were included in the drawing), *Markups* (if the .DWF files were marked using DWF Composer), and to show the *Paper Background* in the viewer.

Print

You can print the DWF image using this option. Your system's *Print* dialog box is displayed with all printers configured for your computer.

Copy Image

Use this option if you want to copy the DWF image and *Paste* it into AutoCAD as an OLE object. (See Chapter 32, Raster Images and Vector Files.)

DWF Underlay in AutoCAD

One of the primary purposes of using .DWF files is to collaborate with clients or colleagues concerning particular designs and details contained in AutoCAD drawings. Using a DWF image, you can pass drawing information to clients or colleagues electronically without risking accidental changes being made to the original drawing file. Using Autodesk DWF Viewer, the drawings can only be viewed. Alternately, you can insert .DWF files directly into AutoCAD as a *DWF Underlay*. This feature allows you not only to "see" the DWF images, but the image can be treated as an "underlay" for tracing. Therefore, you can use the DWF image as a basis for drawing and to *Osnap* directly to objects in the image. For example, a colleague could recreate a similar design based on an existing design you submitted in .DWF format. Another possibility in the collaboration process is to use Autodesk DWF Composer to make design changes by creating "markups" directly in the .DWF file (see "Markup Set Manager").

Once a .DWF file is attached to your drawing as an underlay, the geometry remains visible and available with *Osnaps*. Although you specify the insertion point, scale factor, and rotation angle at the time you attach a .DWF file, you can also modify the position, scale, or rotation at any time. Because an attached .DWF underlay acts like an attached raster image, you can move, scale, or rotate, the image as you can any other attached image (see Chapter 32, Raster Images and Vector Files). You can also affect alterations after you select a DWF underlay by choosing commands from the right-click menu or from the *Properties* palette.

Dwfattach

Pull-down Menu	Command (Type)	Alias (Type)	Short-cut	Screen (side) Menu	Tablet Menu
Insert DWF Underlay	*Dwfattach*	...	...	...	...

Use *Dwfattach* to attach a .DWF file to an AutoCAD drawing. *Dwfattach* lets you locate the desired file and specify the insertion point, scale and rotation for the DWF underlay. Keep in mind that one .DWF file can include multiple sheets, so each sheet that you select is considered a separate underlay.

When you invoke *Dwfattach*, the standard *Select DWF File* navigation dialog box appears for you to select the desired parent .DWF file (not shown). Once selected, the *Attach DWF Underlay* dialog appears (Fig. 23-30).

FIGURE 23-30

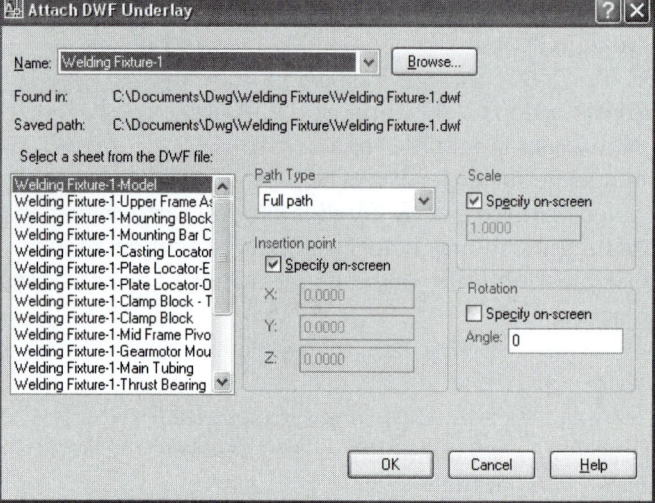

Name

This section identifies the .DWF file you previously selected to attach. The drop-down lists previously attached DWF underlays. To add another copy of a DWF underlay that is already attached, select the DWF image name from the list. Use the *Browse* button to open the *Select DWF File* dialog box.

Select a sheet from the DWF file

This list displays all of the sheets that are found in the .DWF file. Select the desired sheet from the list to attach to the drawing as a underlay. If the .DWF file contains only a single sheet, that sheet is listed. If the .DWF file contains multiple sheets, only a single sheet at a time can be selected for attachment. 3D sheets are not listed.

Path Type

This list allows you to specify one of three types of folder path information to save with an attached DWF underlay: an absolute path, a relative path, or no path. Similar to attaching *Xrefs* or other images, these options are important depending on whether the drawing file or attached image file is later relocated to another folder or computer.

The *Full Path* option saves the entire (absolute) path with the .DWF file and is used when the location of the related files is not expected to change. Use the *Relative Path* option when you expect to relocate the related files together to another drive but using the same directory structure. The *No Path* option saves only the .DWF file name; therefore, this option should be used only when the .DWF file is located in the folder with the current drawing file.

Insertion Point

Identical to inserting *Blocks*, this section specifies the point in the drawing you want to insert the selected .DWF file. Select *Specify on-screen* to pick a point in the drawing, or remove the check and input coordinates.

Scale

Specifies the scale factor for the selected DWF underlay. Select *Specify On-Screen* to pick a point in the drawing, or clear the checkbox to enter a value for the scale factor.

Rotation

This section specifies the rotation angle of the selected DWF underlay using the same methods as the *Scale* and *Insertion point* sections.

Once the image is inserted into the drawing, you can use it as an "underlay" by tracing directly over the objects or simply using the image as a reference for checking an associated drawing. If you want to *Osnap* to objects in the image, *DWF Object Snap* must be turned on. Check this using the right-click shortcut menu.

Shortcut Menu

To produce the shortcut menu (Fig. 23-31), first select the DWF image so it becomes highlighted and the insertion point grip appears (assuming *Grips* are on).

FIGURE 23-31

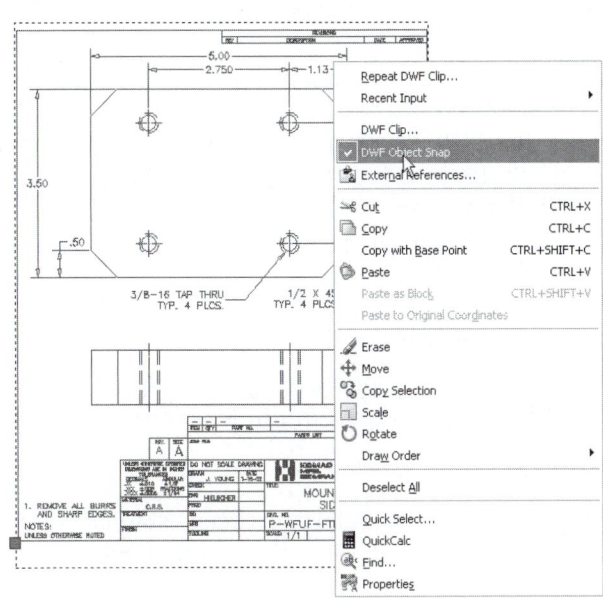

DWF Object Snap

This option must be checked for you to be able to *Osnap* to objects within the image. This setting is saved in the *DWFOSNAP* system variable.

DWF Clip

This option invokes the *Dwfclip* command (see "*Dwfclip*"). With this command you can create a new boundary for the DWF image. For example, you may want to show only a small area of the entire image, therefore "clipping" the original boundary.

Erase, Move, Copy, Scale, Rotate

These options affect the entire DWF image—you cannot affect individual objects within the image. Effectively, these options allow you to change the original insertion point, scale, and rotation angle.

Dwfclip

Pull-down Menu	Command (Type)	Alias (Type)	Short-cut	Screen (side) Menu	Tablet Menu
...	*Dwfclip*	...	...	...	...

For some applications it is not desirable to show the entire DWF image in the drawing. For example, an image may display a title block and border included on the sheet, but you may want to display only the design geometry rather than the entire sheet. Use the *Dwfclip* command to create a new, smaller, boundary.

For example, examine previous Figure 23-31 which displays the Mounting Block Side Clamp sheet. Using *Dwfclip*, a smaller boundary could be created to display only the profile view, clipping out the title block, border, and flat view (Fig. 23-32).

FIGURE 23-32

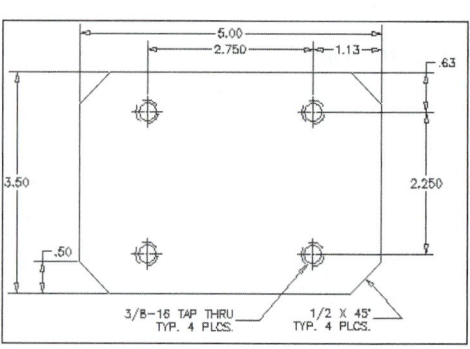

To create a new clipping boundary, follow the command sequence below.

    Command: **dwfclip**
    Select DWF to clip: **PICK**
    Enter DWF clipping option [ON/OFF/Delete/New boundary] <New boundary>: **Enter**
    Enter clipping type [Polygonal/Rectangular] <Rectangular>: **Enter**
    Specify first corner point: **PICK**
    Specify opposite corner point: **PICK**
    Command:

New Boundary

Using the default option, you can create a *Rectangular* boundary by picking two points or use the *Polygonal* option to pick multiple points in any pattern to define the clipping boundary.

ON/OFF

Once a clipping boundary has been established, use the *On* and *Off* options to turn the display of the boundary on and off. In the *Off* setting, the original boundary is displayed (see Fig. 23-31).

Delete

If you want to delete a clipping boundary rather than turn it off, use this option.

To Detach a DWF Underlay

To sever the link between the AutoCAD drawing and the DWF image, follow the steps below. For more information on external references, see Chapter 30, Xreferences.

1. From the *Insert* pull-down menu, select *External References*.

2. In the *External References* palette that appears, locate the *File References* section and select the DWF underlay you want to detach.

3. Right-click on the DWF underlay name and select *Detach* from the shortcut menu.

The DWF underlay is no longer linked to the drawing file. All instances of the underlay are removed from the drawing.

Viewing URLs (Hyperlinks) in a .DWF Image

When viewing a .DWF image in Autodesk DWF Viewer or in AutoCAD, passing the mouse pointer over a URL-linked object displays a small hand image. Clicking on the linked object automatically links to the new URL, which causes the text and/or images to display in Microsoft Internet Explorer or Autodesk DWF Viewer, depending on the file format (.HTM, .DWF, .JPG, etc.). For example, clicking on a door, window, or fixture in an architectural layout could, in turn, display a table of text information listing price, manufacturer, specifications, etc. Or, clicking on one component of a mechanical assembly could, in turn, display a detailed drawing of the selected part. It would also be possible to view a map, select a building, and see written information about the business or link to the Web site of the business.

NOTE: Depending on the file format, clicking on a URL-linked object in Autodesk DWF Viewer may display the linked object in Microsoft Internet Explorer or another program, not in Autodesk DWF Viewer.

To create URLs in .DWF images, *Open* the original drawing in AutoCAD and use the *Hyperlink* command to specify a URL. Next, use the *Plot* command and select an ePlot device to create the related .DWF file with hyperlinks included (see "Using Hyperlinks" and "Creating .DWF Files with ePlot").

MARKUP SET MANAGER

A tool in AutoCAD, called the *Markup Set Manager*, is used for viewing markups in .DWF files so changes to the corresponding original drawing files can be made. This tool is part of a suite of Autodesk tools that can be used by engineers, designers, and architects to collaborate with other professionals using the enhanced .DWF file publishing, viewing, and markup capabilities. Using AutoCAD and a few other Autodesk products, the drawing authors can easily *Publish* single- or multi-sheet .DWF files for all others to view. Next, using a markup process, clients and colleagues can suggest design changes within the .DWF files and publish those markups. Finally, the markups can be viewed by the original authors in AutoCAD using the *Markup Set Manager*, and changes to the original set of drawings (.DWG files) can be made.

2007

The Markup Tools

The following tools from Autodesk can be used in the markup process.

1. AutoCAD: Used to create the original drawings and publish .DWF files. At the other end of the process, markups are viewed using the *Markup Set Manager* so changes to the original drawings can be made.

2. Autodesk DWF Viewer: Used only for viewing and printing .DWF files.

3. Autodesk DWF Composer: A more powerful .DWF-viewing tool from Autodesk. This tool is not included with AutoCAD but can be purchased as a separate application that allows you to read and mark up .DWF files. In this way, project team members can view .DWF files, measure distances, control layers and views, and "red line" drawings by creating marked up drawings with comments, sketches, dimensions, and text. Autodesk DWF Composer also allows you to reorganize sheets in a sheet set and republish the .DWFs with markups.

4. Autodesk DWFwriter: For those without AutoCAD, this free printer driver can be downloaded from the Autodesk site, installed on any Windows system, and set as a system printer. Therefore, DWFwriter enables .DWF files to be created from CAD applications other than AutoCAD.

The Markup Process

The markup process is a way for collaborators involved in a project to help make improvements to the design. First, the designer creates drawings in AutoCAD and publishes them as a sheet set, creating a multi-sheet .DWF file.

Next, the project coordinator, client, or other member of the project team checks the drawings and creates the markups using Autodesk DWF Composer. "Markups" are areas designated with corrections and suggested changes to the drawings in the set. The markups, also called "redlines," generally are made using a *Revcloud* tool and are accompanied by a note indicating the nature of the needed change (Fig. 23-33).

FIGURE 23-33

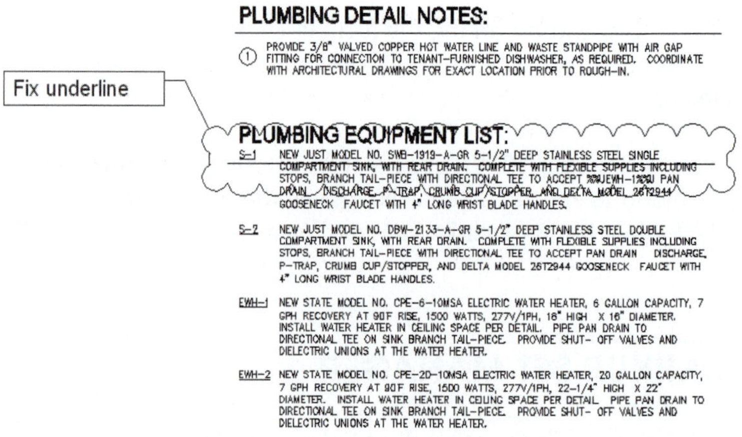

Markups to .DWF files are created only in Autodesk DWF Composer— AutoCAD and Autodesk DWF Viewer cannot create markups for .DWF files. These markups are saved and then published as new .DWF files (called a markup set) from Autodesk DWF Composer. The advantage here is that only the .DWF files, not the .DWG files, contain the markups.

Thirdly, changes to the actual drawings (.DWG) files in AutoCAD are completed by the drawing author with the help of the *Markup Set Manager*. Although .DWF files can be viewed and "redlined" in Autodesk DWF Composer and markups can only be viewed in Autodesk DWF Viewer, changes to the original .DWG files can be made only in AutoCAD.

Markups in AutoCAD

Once markups have been made in Autodesk DWF Composer, they can be easily viewed in AutoCAD, and the changes can be made to the corresponding .DWG files. The *Markup Set Manager* in AutoCAD facilitates this process. You can open a markup set using the *Opendwfmarkup* command or by opening through the *Markup Set Manager*.

Opendwf-markup

Pull-down Menu	Command (Type)	Alias (Type)	Short-cut	Screen (side) Menu	Tablet Menu
File *Load Markup Set...*	*Opendwfmarkup*	...	...	...	...

The *Opendwfmarkup* command simply produces a standard file navigation dialog box. Use it to locate and open any .DWF markup set (.DWF) file. Opening the .DWF file automatically invokes the *Markup Set Manager* displaying the contents of the opened markup set.

Markup

Pull-down Menu	Command (Type)	Alias (Type)	Short-cut	Screen (side) Menu	Tablet Menu
Tools *Palettes >* *Markup Set* *Manager*	*Markup*	...	*Ctrl+7*	...	...

Use any of the options shown in the previous command table (or use the *Markup* command) to produce the *Markup Set Manager*. The *Markup Set Manager* appears in a palette format (Fig. 23-34). Use the drop-down list options at the top of the *Markup Set Manager* to open a markup set (.DWF file). Doing so displays the markup set list in the *Markups* area of the manager. Each sheet is shown in a hierarchical fashion under the .DWF set name. Any markups created for the sheets are displayed by expanding each sheet in the list (clicking the plus symbol next to the sheet name). The *Details* section below gives information about each markup. The *Preview* pane (not shown) displays only the entire sheet, not the marked up area.

FIGURE 23-34

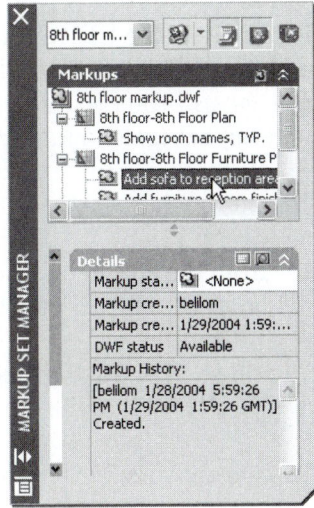

You can view the individual markups in the AutoCAD Drawing Editor by double-clicking on the markup name in the list (Fig. 23-35). This is an extremely useful feature. This action opens the indicated .DWF file and automatically zooms to the area of the markup. Additionally, the corresponding .DWG file is opened. You can toggle between viewing the .DWF and the .DWG files using the two buttons in the upper-right corner of the manager titled *View DWG Geometry (Alt+3)* and *View DWF Geometry (Alt+4)*. In this way, you can see the section of the .DWF marked for change, then toggle to the drawing to actually make the change. You can also use the *View Redline Geometry (Alt+2)* button to toggle the display of the markup in both the .DWF and in the actual corresponding drawing file.

FIGURE 23-35

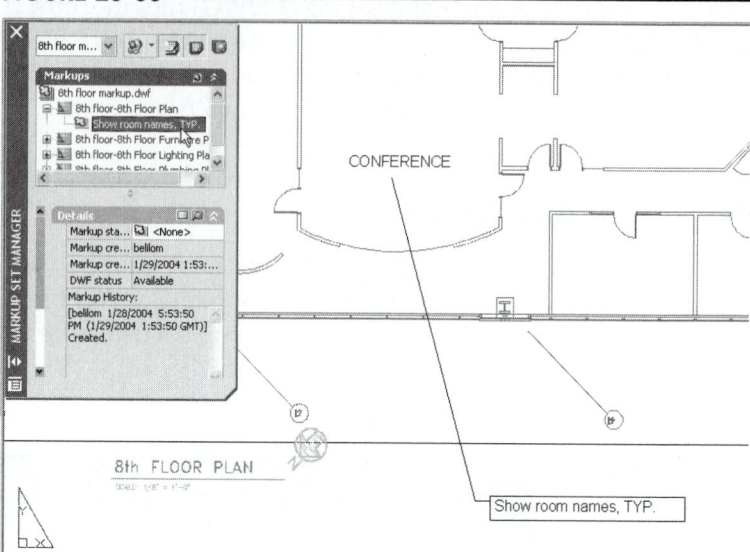

When changes to the original drawing files are made, use the *Markup Sheet Manager* to republish the .DWF files. You can choose to publish only the drawings that were changed or the entire set.

TRANSFERRING AutoCAD FILES OVER THE INTERNET

Opening and Saving .DWF and .DWG Files via the Internet

You can use AutoCAD to open and save files from the Internet using the typical *Open* and *Saveas* commands. If you use *Open* to open a drawing from a remote Internet site, the drawing file that you specify is downloaded to your computer and opened in the AutoCAD drawing area. You can then edit the drawing and save it, either locally or back to any Internet or intranet location assuming you have sufficient access privileges.

If you know the URL to the file you want to open or save to, you can enter it directly in the *Select File* dialog box that appears when you use *Open* or the *Save Drawing As* dialog box that appears when you use *Saveas*. You can also use the *Search the Web* button to produce the *Browse the Web* dialog box. Here you can navigate to find the Internet location for the file.

To open an AutoCAD file from the Internet using the *Select File* dialog box:

1. from the *File* menu, choose *Open*; then
2. enter the URL to the file in *File Name* and select *Open*.

You must enter the transfer protocol (for example, http:// or ftp://) and the extension (for example, .DWG or .DWT) of the file you want to open.

To save an AutoCAD file to an Internet location using the *Save Drawing As* dialog box:

1. from the *File* menu, select *Save As*;
2. enter the URL to the file in *File Name*; then

3. select a file format from the *Save As Type* list, and then choose *Save*.

In step 2 you must enter the File Transfer Protocol (ftp://) and the extension (for example, .DWG or .DWT) of the file you want to save. You can only save AutoCAD files to an Internet location using FTP.

Using i-drop to Drag-and-Drop Drawing Files into AutoCAD

Autodesk provides a utility called "i-drop" to allow users to drag and drop AutoCAD drawings from a Web page directly into AutoCAD. One component of this technology is the i-drop Indicator that modifies your browser behavior so that you can initiate the drag-and-drop action. If you have your own Web page, you can activate the i-drop capability so visitors to your page can drag and drop drawing files into an AutoCAD session. For information about using i-drop, downloading the i-drop Indicator, and creating an i-drop handle on your Web site, see the i-drop documentation on the Autodesk Web site at www.autodesk.com/idrop. If you use *PublishToWeb* in AutoCAD to create your Web pages, select the *Enable i-drop* option to make use of this capability (see *PublishToWeb*).

CHAPTER EXERCISES

1. *Browser, Inetlocation,* **Autodesk DWF Viewer**

    A. Start AutoCAD if not already running. Type in the *Browser* command. Note that the default URL is http://www.autodesk.com. Press **Enter** to display your system's default browser and the Autodesk Web site.

    B. On the Autodesk site, locate **Products**, then click to follow the link. On the Products page, locate Autodesk DWF Viewer and go to that page. If Autodesk DWF Viewer is not installed on your computer, select the **Download Autodesk DWF Viewer** button and follow the instructions.

    C. Once the Autodesk DWF Viewer is installed, start AutoCAD again if not already started. Enter *Inetlocation* at the command prompt. Change the default location to your favorite news source such as **http://www.usatoday.com.** Next, use *Browser* again, press **Enter** to accept the new URL, and validate that the new site appears in your Internet browser.

2. *Publish to Web*

    A. Start AutoCAD if not already running. *Save* the current drawing. (It is recommended to always save the current drawing before using *PublishToWeb*.)

    B. Invoke *Publishtoweb*. Follow the necessary steps in the wizard to create a new Web page named **Project Drawings** and locate it in your working directory. Write an appropriate *description* when prompted. Specify *JPEG* image type and *Small* image size. For the template, specify *Array plus Summary*. Select a *Theme* of your choice. In the *Select Drawings* step, specify four of your favorite drawings that you completed in the previous Chapter Exercises. Proceed to generate the images and *Preview* them. If the Web page needs improving, go *Back* and make the changes, or use *Publishtoweb* again and select *Edit an Existing Web Page* to generate new images. Finally, *Post* the Web page to the **Project Drawings** folder.

C. Start your browser, locate and open the **ACWEBPUBLISH. HTM** file to view your Web page. Your new Web page should appear in the browser and look similar to Figure 23-36, depending on the drawings you selected to display. Click on any image to produce a larger view of the drawing. If you like, you can improve the page as you see fit by changing the drawings in AutoCAD and regenerating the Web page again with *Publishtoweb*.

FIGURE 23-36

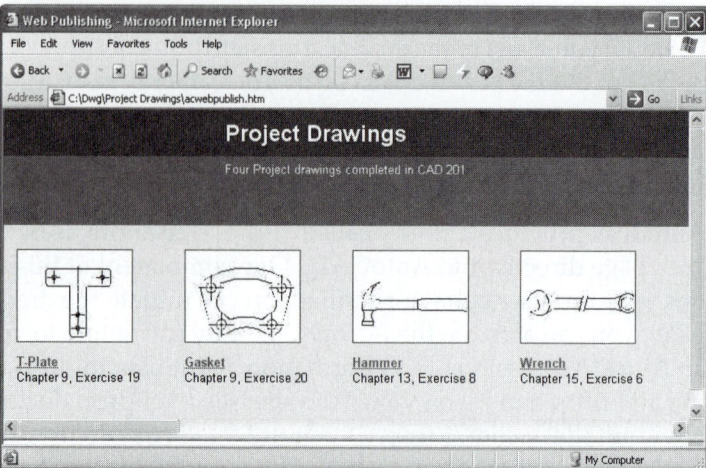

3. *Etransmit*

A. Start AutoCAD if not already running. *Open* a drawing that you want to transmit to a friend. If you make any changes, *Save* the current drawing.

B. Invoke *eTransmit*. The *Create Transmittal* dialog box should appear. In the notes section, enter "Here is the drawing that I would like for you to review," or similar message. Select *Transmittal Setups*, then *Modify* the *Standard* setup. Specify a *Self-extracting executable (*exe)* as the *Transmittal package type*. Use the *Browse* button to specify your working directory as the *Transmittal file folder*. Select *Place all files in one folder*, then clear the other options below. Do <u>not</u> select *OK* yet.

C. If you have an email system available on the system, select *Send e-mail with transmittal*. If not, clear the check for this option. Select *OK*.

D. Depending on your selection in step C, your email system may open with the .EXE file as an attachment. If you chose not to have AutoCAD open your email program, copy the .EXE file to disk, and attach it to an email message at a later time. In either case, send an email to a friend and ask him or her to extract the file and view the drawing in AutoCAD. Request a confirmation and a reply email reporting on the success of the transmittal.

4. **Use *ePlot* to create .DWF files**

A. In AutoCAD, *Open* the **T-PLATE** drawing from Chapter 9 Exercises. Activate the *Model* tab. Use the *Plot* dialog box to select the *DWF6 ePlot.pc3* device. Select *OK* to make the .DWF file. Accept the default name (T-PLATE-MODEL) and ensure your working directory is specified as the destination for the file.

B. Start Autodesk DWF Viewer. Use the *File* pull-down menu, then the *Open* command, and locate the **T-PLATE-MODEL.DWF** to verify the .DWF was created correctly.

C. In AutoCAD, *Open* the **GASKETA** drawing from Chapter 9 Exercises. Follow step 1. to make a **.DWF** file. Use Autodesk DWF Viewer to view it.

D. Create another **.DWF** file similar to the previous two, but this time activate a *Layout* tab that was set up previously with a viewport. If no layouts were set up, create one viewport but make sure the viewport border is slightly smaller than the printable area (dashed line), then make the .DWF file of the layout.

E. View the new .DWF image with Autodesk DWF Viewer to verify it was created correctly. The image should appear similar to the Model tab .DWFs, except the viewport border should be visible.

5. **Use *Publish* to create a .DWF file**

A. In AutoCAD, use ***Publish***. When the *Publish* dialog box appears, select ***Add Sheets***. Select four drawings of your choice (for example, **WRENCH**, **GASKETA**, **HAMMER**, and **T-PLATE**). In the *Sheets to publish* list, specify the layouts you want to publish. Select ***DWF file*** in the ***Publish to*** section below. Invoke the ***Publish Options*** and specify a ***Multi-sheet DWF***, then enter the desired ***DWF file name*** and location, such as **Exercise Drawings**.

B. Select ***Publish*** in the *Publish* dialog box. When publishing is complete, open Autodesk DWF Viewer and view the multi-sheet .DWF file.

C. Your new multi-sheet .DWF file should appear in the Autodesk DWF Viewer. Use the *Thumbnail* or *List* area on the left of the viewer to inspect each of the drawings, similar to that in Figure 23-37. When you are finished, close the viewer.

FIGURE 23-37

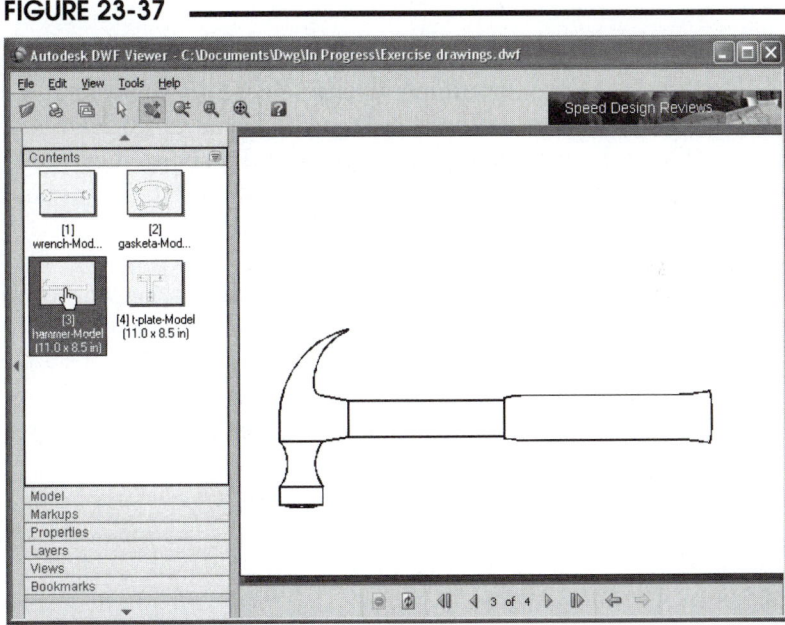

6. **Insert *Hyperlinks***

A. In AutoCAD, ***Open*** the **GASKETA** drawing. Activate the *Model* tab. Use *SaveAs* to save the drawing as **GASKETA2**. Close the drawing.

B. Open the **T-PLATE** drawing. Activate the *Model* tab. Use *SaveAs* to save the drawing as **T-PLATE2**. *Close* the drawing.

C. *Open* the **GASKET2** drawing. Use *Text* to place text at the bottom of the drawing. Enter "**Click here to view the T-PLATE2 drawing**." The text should appear similar to that shown in Figure 23-38. Create a *Hyperlink* for the entire line of text. Associate the **T-PLATE2.DWG** as the file to link. *Save* and *Close* the drawing.

D. *Open* the **T-PLATE2** drawing. Use *Text* to place text at the bottom of the drawing, stating "**Click here to view the GASKETA2 drawing**." Create a *Hyperlink* for the text that links to the **GASKETA2.DWG**. Save and *Close* the drawing.

E. *Open* either drawing. Hover over the text until the link information appears, then hold down the Ctrl key and left-click to follow the link. You should be able to do the same in both drawings to "toggle" between the two drawings.

FIGURE 23-38

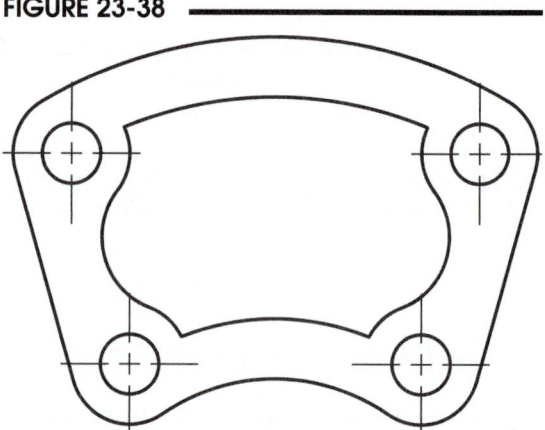

Click here to view the T-PLATE2 drawing

7. **Create .DWF files with *Hyperlinks***

A. *Open* the **GASKETA2** drawing. Activate the *Model* tab. Select the hyperlink, right-click to produce the shortcut menu, and select *Edit Hyperlink*. Edit the hyperlink to link to a .DWF, not the .DWG; for example, change C:\DWG\GASKETA2.DWG to **C:\DWG\GASKETA2.DWF**. Open the *Plot* dialog box to create a **.DWF** file. Use the *DWF6 ePlot.pc3* device. Make sure you specify **GASKETA2** for the name of the .DWF file, <u>not</u> GASKETA2-Model. Select *OK* to produce the .DWF file. *Save* and *Close* the drawing.

B. *Open* the **T-PLATE2** drawing. Follow the same procedure as in step A to create a .DWF file for the T-PLATE2 drawing. Make sure you convert the hyperlink extension to .DWF and save the .DWG with the correct name. *Save* and *Close* the drawing.

C. Use Autodesk DWF Viewer to view the .DWF images and test the hyperlinks. Make any necessary corrections.

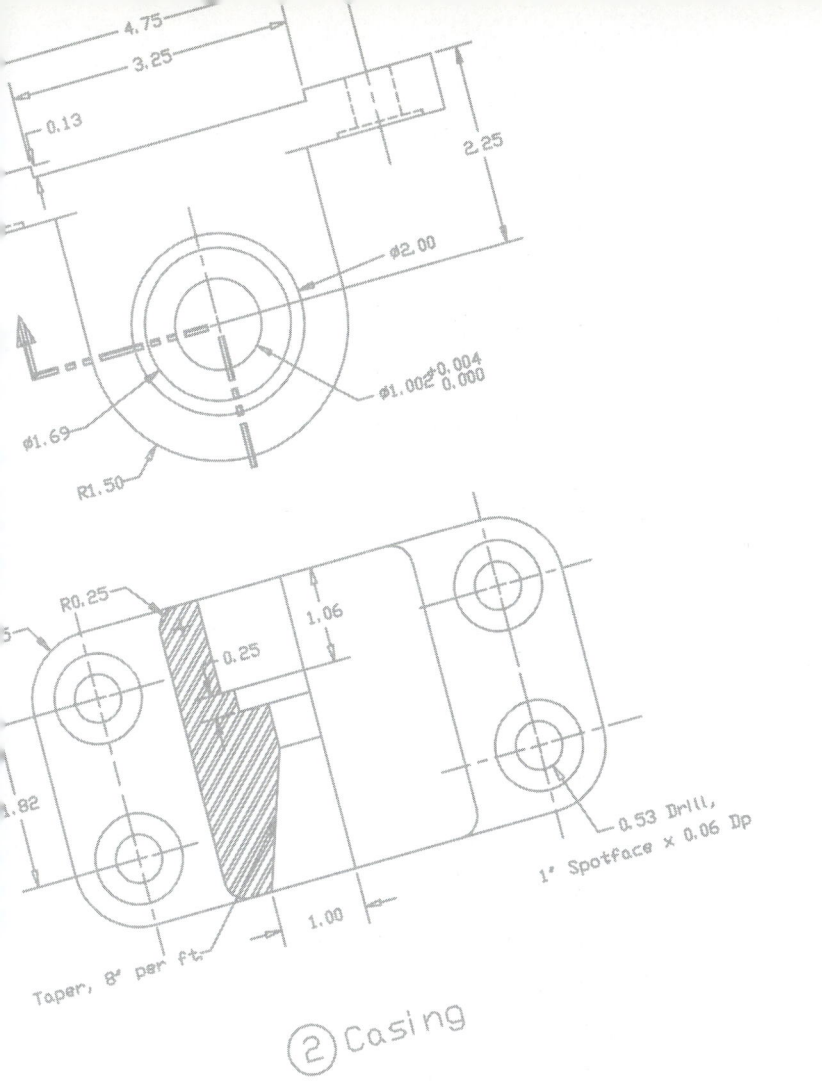

0.13
3.25
4.75
2.25
⌀2.00
⌀1.00$^{+0.004}_{0.000}$
⌀1.69
R1.50

R0.25
0.25
1.06
.82
0.53 Drill,
1" Spotface x 0.06 Dp
1.00
Taper, 8° per ft.

② Casing

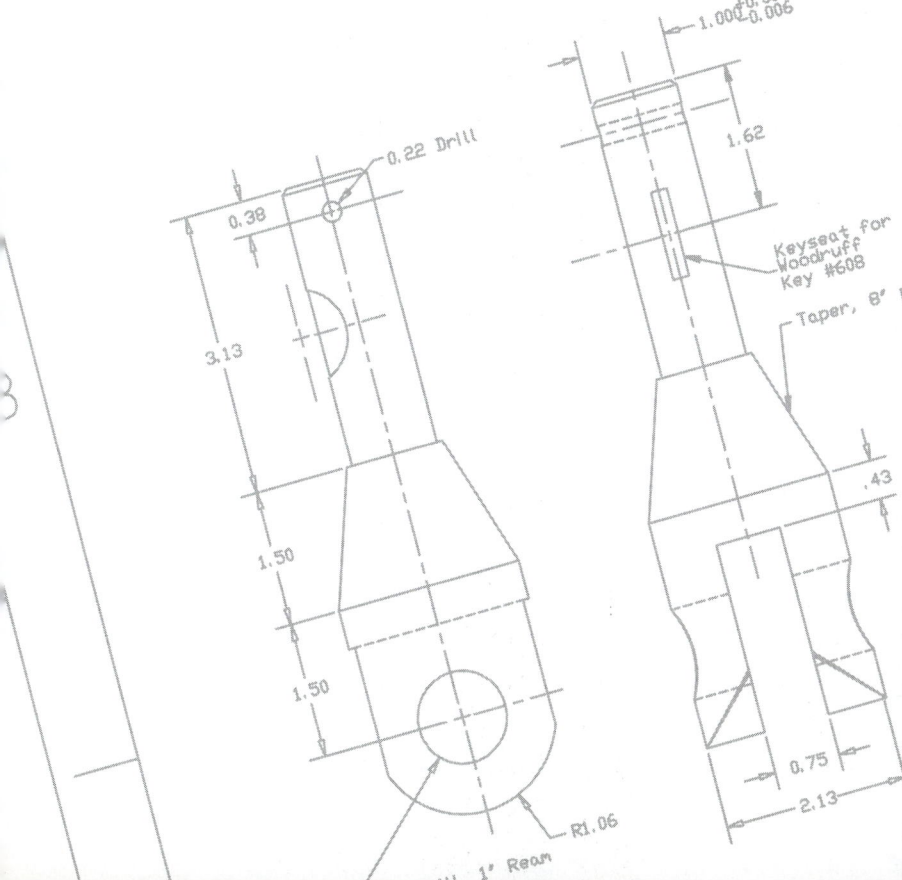

1.00$^{+0.003}_{-0.006}$
0.22 Drill
1.62
0.38
Keyseat for
Woodruff
Key #608
Taper, 8° p
3.13
.43
1.50
1.50
0.75
R1.06
1" Ream
2.13

MULTIVIEW DRAWING

CHAPTER OBJECTIVES

After completing this chapter you should:

1. be able to draw projection lines using *ORTHO* and *SNAP*, and *Polar Tracking*;

2. be able to use *Xline* and *Ray* to create construction lines;

3. know how to use *Offset* for construction of views;

4. be able to use *Object Snap Tracking* for alignment of lines and views;

5. be able to use construction layers for managing construction lines and notes;

6. be able to use linetypes, lineweights, and layers to draw and manage ANSI standards;

7. know how to create fillets, rounds, and runouts;

8. know the typical guidelines for creating a three-view multiview drawing.

CONCEPTS

Multiview drawings are used to represent 3D objects on 2D media. The standards and conventions related to multiview drawings have been developed over years of using and optimizing a system of representing real objects on paper. Now that our technology has developed to a point that we can create 3D models, some of the methods we use to generate multiview drawings have changed, but the standards and conventions have been retained so that we can continue to have a universally understood method of communication.

This chapter illustrates methods of creating 2D multiview drawings with AutoCAD (without a 3D model) while complying with industry standards. (Creating 2D drawings from 3D models is addressed in Chapter 42.) Many techniques can be used to construct multiview drawings with AutoCAD because of its versatility. The methods shown in this chapter are the more common methods because they are derived from traditional manual techniques. Other methods are possible.

PROJECTION AND ALIGNMENT OF VIEWS

Projection theory and the conventions of multiview drawing dictate that the views be aligned with each other and oriented in a particular relationship. AutoCAD has particular features, such as *SNAP*, *ORTHO*, construction lines (*Xline, Ray*), Object Snap, Polar Tracking and Object Snap Tracking, that can be used effectively for facilitating projection and alignment of views.

Using *ORTHO* and *OSNAP* to Draw Projection Lines

ORTHO (F8) can be used effectively in concert with *OSNAP* to draw projection lines during construction of multiview drawings. For example, drawing a Line interactively with *ORTHO ON* forces the *Line* to be drawn in either a horizontal or vertical direction.

Figure 24-1 simulates this feature while drawing projection *Lines* from the top view over to a 45 degree miter line (intended for transfer of dimensions to the side view). The "first point:" of the *Line* originated from the *Endpoint* of the *Line* on the top view.

FIGURE 24-1

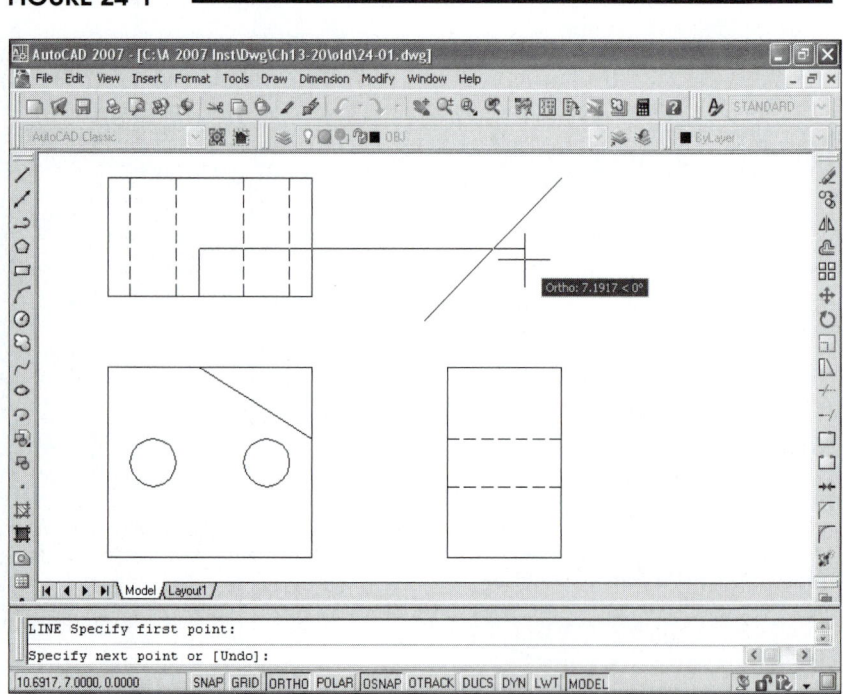

Figure 24-2 illustrates the next step. The vertical projection *Line* is drawn from the *Intersection* of the 45 degree line and the last projection line. *ORTHO* forces the *Line* to the correct vertical alignment with the side view.

Remember that any draw <u>or</u> edit command that requires PICKing is a candidate for *ORTHO* and/or *OSNAP*.

NOTE: *OSNAP* overrides *ORTHO*. If *ORTHO* is *ON* and you are using an *OSNAP* mode to PICK the "next point:" of a *Line*, the *OSNAP* mode has priority; and, therefore, the construction may not result in an orthogonal *Line*.

FIGURE 24-2

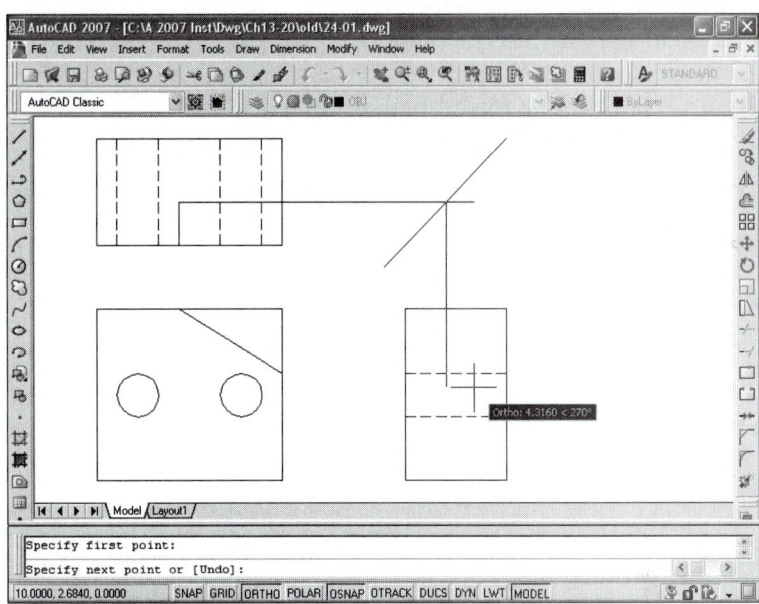

Using *Polar Tracking* to Draw Projection Lines

Similar to using *ORTHO* and *OSNAP*, you can use *Polar Tracking* and *OSNAP* options to draw projection lines at 90-degree increments. Using the previous example, *Polar Tracking* is used to project dimensions between the top and side views through the 45-degree miter line.

To use *Polar Tracking*, set the desired *Polar Angle Settings* in the *Polar Tracking* tab of the *Drafting Settings* dialog box. Ensure *POLAR* appears recessed on the Status Bar. You can also set a *Polar Snap* increment or a *Grid Snap* increment to use with *Polar Tracking* (see Chapter 3 for more information on these settings).

For example, you could use the *Endpoint OSNAP* option to snap to the *Line* in the top view, then *Polar Tracking* forces the line to a previously set polar increment (0 degrees in this case).

Note that you can use *OSNAPs* in conjunction with *Polar Tracking*, as shown in Figure 24-3, such that the current horizontal line *OSNAPs* to its *Intersection* with the 45-degree miter line. (*OSNAPs* <u>cannot</u> be used effectively with *ORTHO*, since *OSNAPs* override *ORTHO*.)

FIGURE 24-3

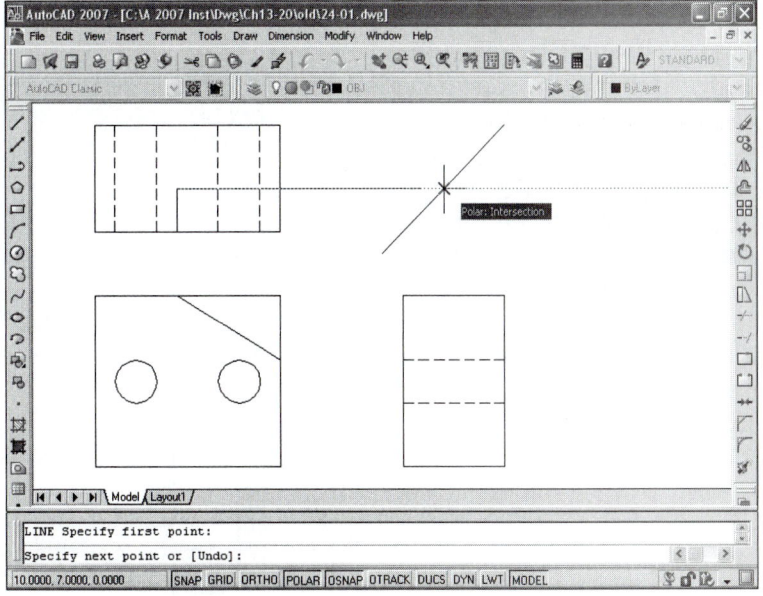

In the following step, use *Intersection OSNAP* option to snap to the intersection of the previous *Line* and the 45-degree miter line (Fig. 24-4). Again *Polar Tracking* forces the *Line* to a vertical position (270 degrees).

FIGURE 24-4

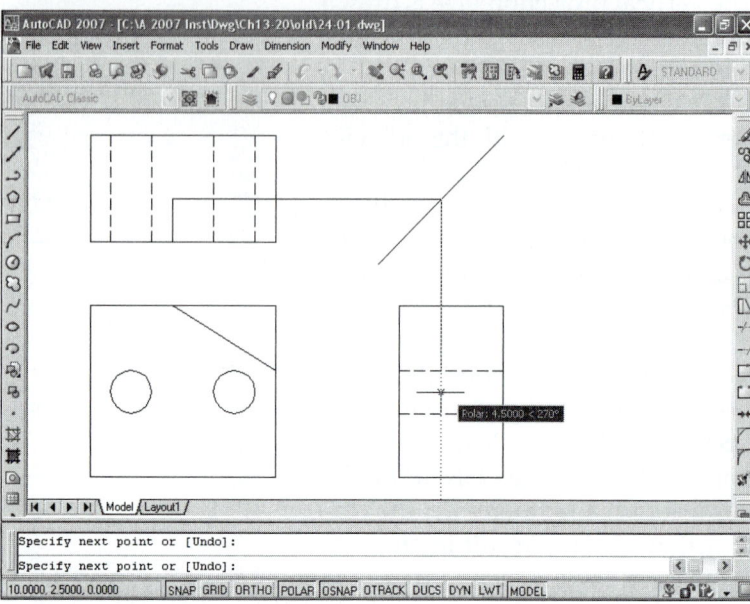

Using *Object Snap Tracking* to Draw Projection Lines Aligned with *OSNAP* Points

This feature, although the most complex, provides the greatest amount of assistance in constructing multiview drawings. The advantage is the availability of the features described in the previous method (*Polar Tracking* and *OSNAP* for alignment of vertical and horizontal *Lines*) in addition to the creation of *Lines* and other objects that align with *OSNAP* points (*Endpoint*, *Intersection*, *Midpoint*, etc.) of other objects.

To use *Object Snap Tracking*, first set the desired running *OSNAP* options (*Endpoint*, *Intersection*, etc.). Next, toggle on *OTRACK* at the Status Bar. Use *Polar Tracking* in conjunction with *Object Snap Tracking* by setting polar angle increments (see previous discussion) and toggling on *POLAR* on the Status Bar.

For example, to begin the *Line* in Figure 24-5, first "acquire" the *Endpoint* of the *Line* shown in the front view. Note that the "first point" of the new *Line* (in the top view) aligns vertically with the acquired *Endpoint*. At this point in time, the short vertical *Line* for the top view could be drawn from the intersection shown. *Object Snap Tracking* prevents having to draw the vertical projection line from the front up to the top view.

FIGURE 24-5

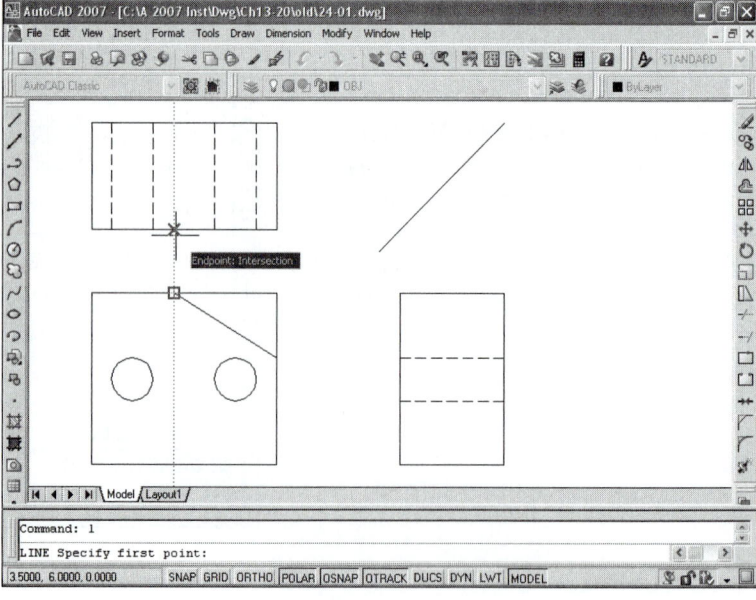

After constructing the two needed *Lines* for the top view, a tracking vector aligns with the acquired *Endpoint* of the indicated line in the top view and the *Intersection* of the 45-degree miter line (Fig. 24-6). Note that a <u>horizontal projection line is not needed</u> since *Object Snap Tracking* ensures the "first point" aligns with the appropriate point from the top view.

FIGURE 24-6

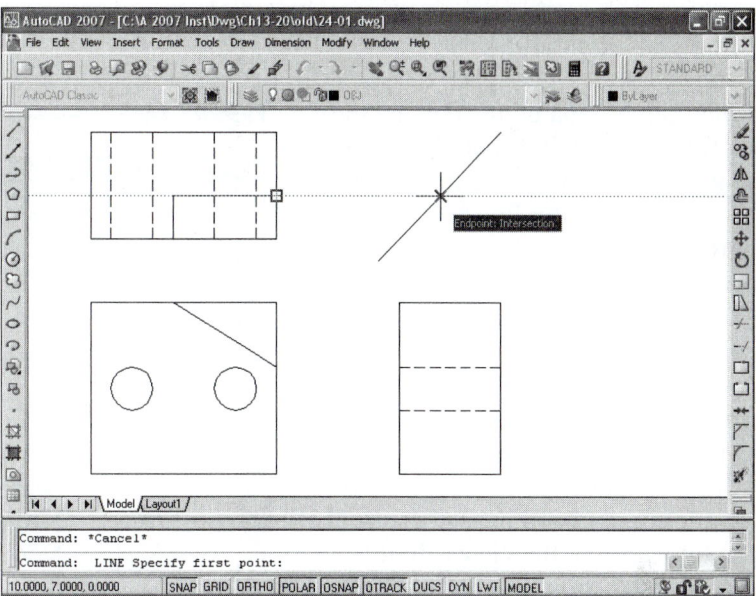

The next step is to draw the short horizontal visible line in the side view (Fig. 24-7). Coming from the "first point" (on the 45-degree miter line found in the previous step), draw the vertical projection line down to the appropriate point in the side view that "tracks" with the related *Endpoint* in the front view (see horizontal tracking vector). Since this *Line* is constructed to the correct endpoint, <u>*Trimming* is unnecessary here</u>, but it would be needed in the previous two methods.

FIGURE 24-7

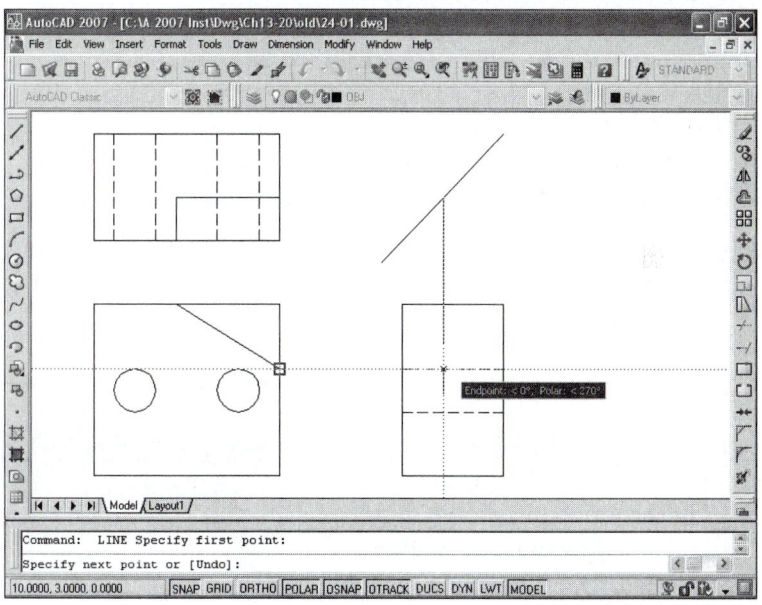

These AutoCAD features (*Object Snap Tracking* in conjunction with *Polar Tracking*) are probably the most helpful features for construction of multiview drawing since the introduction of AutoCAD in 1982.

Using *Xline* and *Ray* for Construction Lines

Another strategy for constructing multiview drawings is to make use of the AutoCAD construction line commands *Xline* and *Ray*.

Xlines can be created to "box in" the views and ensure proper alignment (Fig. 24-8). *Ray* is suited for creating the 45 degree miter line for projection between the top and side views. The advantage to using this method is that horizontal and vertical *Xlines* can be created quickly.

FIGURE 24-8

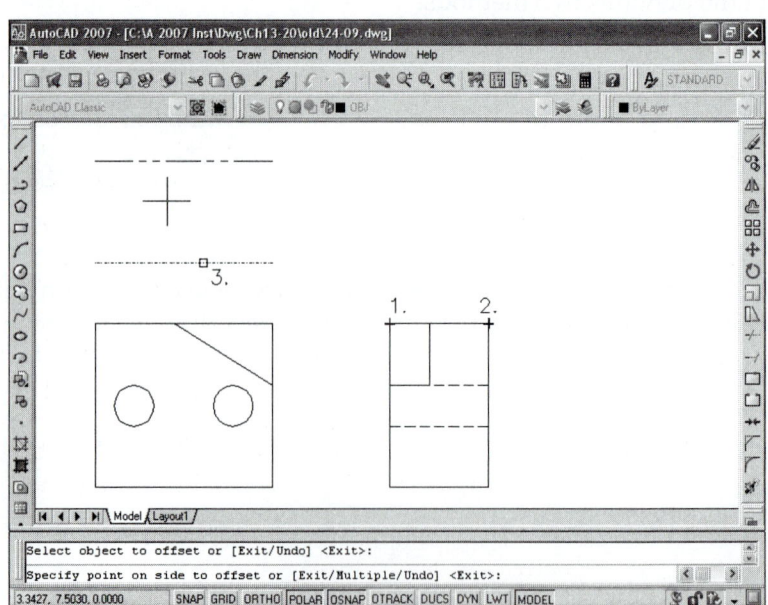

These lines can be *Trimmed* to become part of the finished view. Alternately, other lines could be drawn "on top of" the construction lines to create the final lines of the views. In this case, <u>construction lines should be drawn on a separate layer</u> so that the layer can be frozen before plotting. If you intend to *Trim* the construction lines so that they become part of the final geometry, draw them originally on the view layers.

TIP

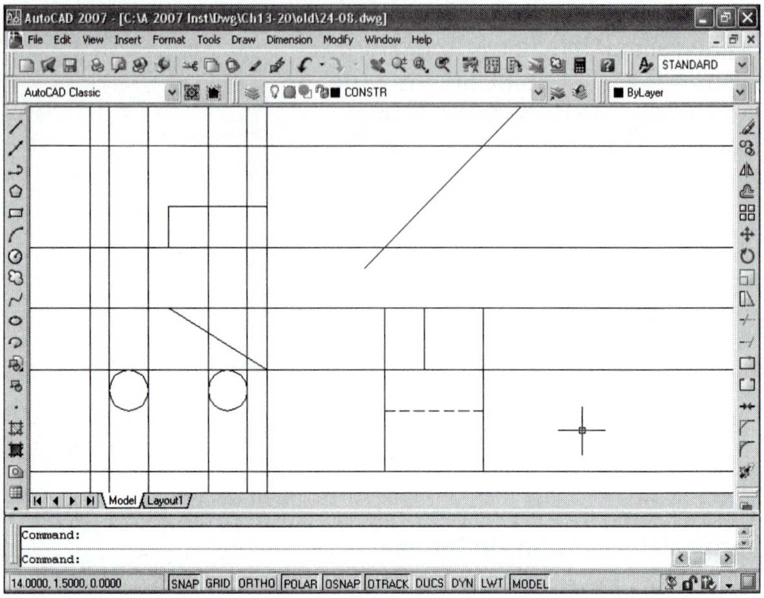

Using *Offset* for Construction of Views

An alternative to using the traditional miter line method for construction of a view by projection, the *Offset* command can be used to transfer distances from one view and to construct another. The *Distance* option of *Offset* provides this alternative.

For example, assume that the side view was completed and you need to construct a top view (Fig. 24-9). First, create a horizontal line as the inner edge of the top view (shown highlighted) by *Offset* or other method. To create the outer edge of the top view (shown in phantom linetype), use *Offset* and PICK points (1) and (2) to specify the *distance*. Select the existing line (3) as the *"Object to Offset:"*, then PICK the *"side to offset"* at the current cursor position.

FIGURE 24-9

Realignment of Views Using *Polar Snap* and *Polar Tracking*

Another application of *Polar Snap* and *Polar Tracking* is the use of *Move* to change the location of an entire view while retaining its orthogonal alignment.

For example, assume that the views of a multiview (Fig. 24-10) are complete and ready for dimensioning; however, there is not enough room between the front and side views. You can invoke the *Move* command, select the entire view with a window or other option, and "slide" the entire view outward. *Polar Tracking* ensures proper orthogonal alignment. *Polar Snap* forces the movement to a regular increment so the coordinate points of the geometry retain a relationship to the original points (for example, moving exactly 1 unit).

FIGURE 24-10

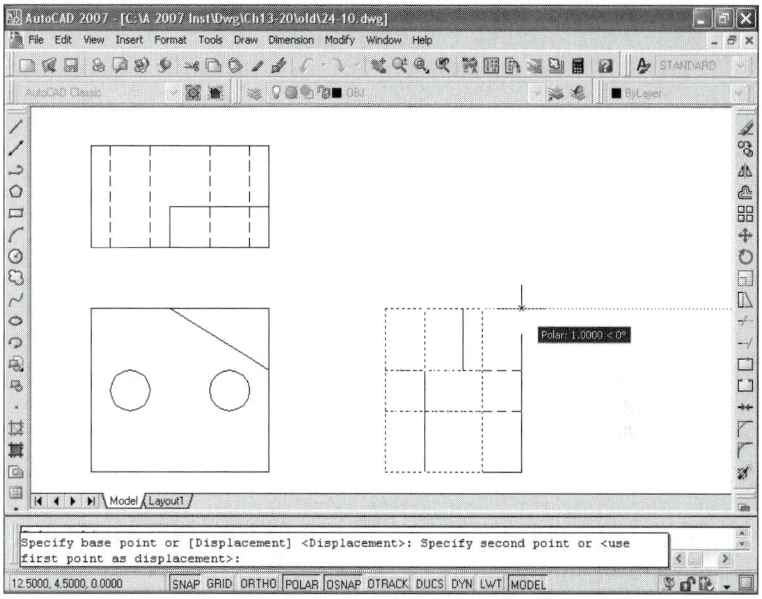

An alternative to moving a view using mouse input is using Dynamic Input. Turn *DYN* on, then key in the desired distance value to move in the distance edit box, and enter an angular value in the angle edit box. Another alternative is moving the view using Direct Distance Entry. With this method, indicate the direction to move with the rubberband line (assuming *POLAR* is on), then key in the desired distance at the Command prompt.

USING CONSTRUCTION LAYERS

The use of layers for isolating construction lines can make drawing and editing faster and easier. The creation of multiview drawings can involve construction lines, reference points, or notes that are not intended for the final plot. Rather than *Erasing* these construction lines, points, or notes before plotting, they can be created on a separate layer and turned *Off* or made *Frozen* before running the final plot. If design changes are required, as they often are, the construction layers can be turned *On*, rather than having to recreate the construction.

There are two strategies for creating construction objects on separate layers:

1. <u>Use Layer 0</u> for construction lines, reference points, and notes. This method can be used for fast, simple drawings.

2. <u>Create</u> a <u>new</u> layer for construction lines, reference points, and notes. Use this method for more complex drawings or drawings involving use of *Blocks* on Layer 0.

For example, consider the drawing during construction in Figure 24-11. A separate layer has been created for the construction lines, notes, and reference points.

FIGURE 24-11

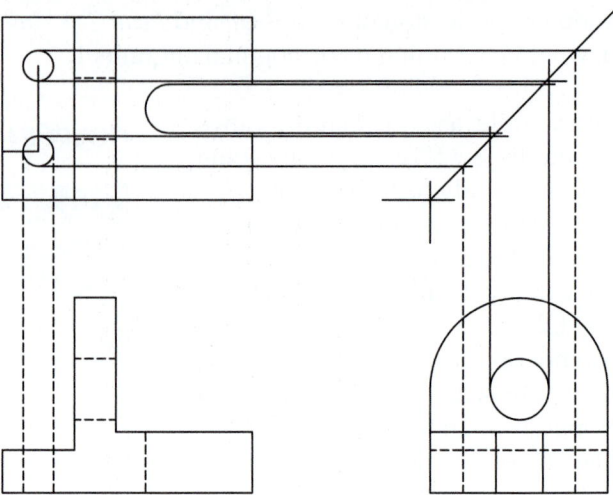

In Figure 24-12, the drawing is ready for making a print. Notice that the construction layer is *Frozen*.

FIGURE 24-12

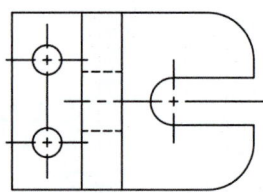

If you are printing the *Limits*, the construction layer should be *Frozen*, rather than turned *Off*, since *Zoom All* and *Zoom Extents* are affected by geometry on layers turned *Off* and not *Frozen* (unless the objects are *Xlines* or *Rays*).

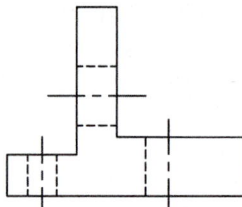

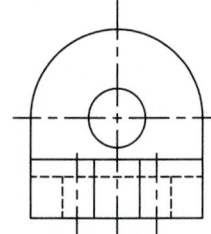

USING LINETYPES

Different types of lines are used to represent different features of a multiview drawing. Linetypes in AutoCAD are accessed by the *Linetype* command or through the *Layer Properties Manager*. *Linetypes* can be changed retroactively by the *Properties* palette. *Linetypes* can be assigned to individual objects specifically or to layers (*ByLayer*). See Chapter 11 for a full discussion on this topic.

AutoCAD complies with the ANSI and ISO standards for linetypes. The principal AutoCAD linetypes used in multiview drawings and the associated names are shown in Figure 24-13.

Many other linetypes are provided in AutoCAD. Refer to Chapter 11 for the full list and illustration of the linetypes.

FIGURE 24-13

CONTINUOUS	————————————	DARK, WIDE
HIDDEN	— — — — — — — —	MEDIUM
CENTER	—— — —— — ——	MEDIUM
PHANTOM	—— — — ——	VARIES
DASHED	— — — — — —	VARIES

Other standard lines are created by AutoCAD automatically. For example, dimension lines can be auto-matically drawn when using dimensioning commands (Chapter 28), and section lines can be automati-cally drawn when using the *Hatch* command (Chapter 26).

Objects in AutoCAD can have lineweight. This is accomplished by using the *Lineweight* command or by assigning *Lineweight* in the *Layer Properties Manager* (see Chapter 11). Additionally, lineweights can be assigned by using plot styles or assigning plot device lineweights or pen thickness.

Drawing Hidden and Center Lines

FIGURE 24-14

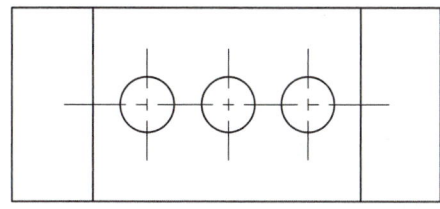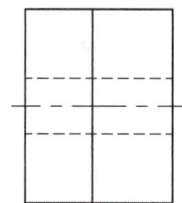

Figure 24-14 illustrates a typical applica-tion of AutoCAD *Hidden* and *Center* line-types. Notice that the horizontal center line in the front view does not automati-cally locate the short dashes correctly, and the hidden lines in the right side view incorrectly intersect the center vertical line.

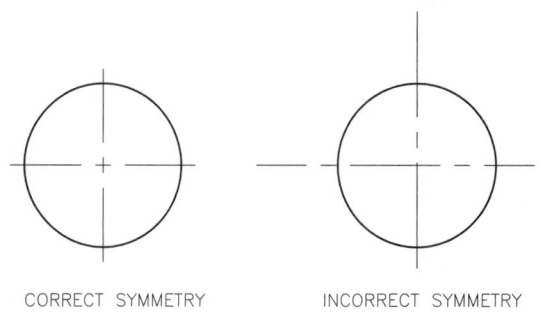

Although AutoCAD supplies ANSI standard linetypes, the application of those linetypes does not always follow ANSI standards. For example, you do <u>not</u> have control over the placement of the individual dashes of center lines and hidden lines. (You have control of only the endpoints of the lines and the *Ltscale*.) Therefore, the short dashes of center lines may not cross exactly at the circle centers, or the dashes of hidden lines may not always intersect as desired.

You do, however, have control of the endpoints of the lines. Draw lines with the *Center* linetype such that the endpoints are symmetric about the circle or group of circles. This action assures that the short dash occurs at the center of the circle (if an odd number of dashes are generated). Figure 24-15 illustrates correct and incorrect technique.

FIGURE 24-15

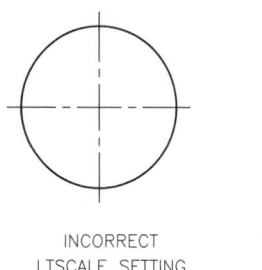

CORRECT SYMMETRY   INCORRECT SYMMETRY

You can also control the relative size of non-continuous linetypes with the *Ltscale* variable. <u>In some cases</u>, the variable can be adjusted to achieve the desired results.

For example, Figure 24-16 demonstrates the use of *Ltscale* to adjust the center line dashes to the correct spacing. Remember that *Ltscale* adjusts linetypes <u>globally</u> (all linetypes across the drawing).

FIGURE 24-16

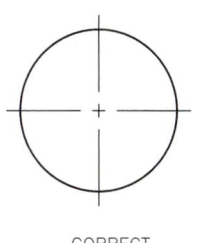

INCORRECT   CORRECT
LTSCALE SETTING   LTSCALE SETTING

When the *Ltscale* has been optimally adjusted for the drawing <u>globally</u>, use the *Properties* palette to adjust the linetype scale of <u>individual objects</u>. In this way, the drawing lines can originally be created to the global linetype scale without regard to the *Celtscale*. The finished drawing linetype scale can be adjusted with *Ltscale* globally; then <u>only those objects</u> that need further adjusting can be fine-tuned retroactively with *Properties*.

TIP

The *Dimcenter* command (a dimensioning command) can be used to draw center lines automatically with correct symmetry and spacing (see Chapter 28, Dimensioning).

ANSI Standards require that multiview drawings are created with object lines having a dark lineweight, and hidden, center, dimension, and other reference lines created in a medium lineweight (see Figure 24-13).

You can assign *Lineweight* to objects using the *Lineweight* command or assign *Lineweight* to layers in the *Layer Properties Manager*. As described previously in Chapter 11, *Lineweight* can be assigned to individual objects or to layers (*ByLayer*). Use the *Lineweight Settings* dialog box to assign *Lineweight* property to objects. Use the *Layer Properties Manager* to assign *Lineweight* to layers. Generally, the *ByLayer Lineweight* assignment is preferred, similar to the *ByLayer* method of assigning *Color* and *Linetype*.

Additionally, lineweights can be assigned by using plot styles or assigning plot device lineweights or pen thickness (see Chapter 33 for information on plot styles).

Managing Linetypes, Lineweights, and Colors

There are two strategies for assigning linetypes, lineweights, and colors: *ByLayer* and object-specific assignment. In either case, thoughtful layer utilization for linetypes will make your drawings more flexible and efficient.

BYLAYER Linetypes, Lineweights, and Colors

The *ByLayer* linetype, lineweight, and color settings are recommended when you want the most control over linetype visibility and plotting. This is accomplished by creating layers with the *Layer Properties Manager* and assigning *Linetype, Lineweight,* and *Color* for each layer. After you assign *Linetypes* to specific layers, you simply set the layer (with the desired linetype) as the *Current* layer and draw on that layer in order to draw objects in a specific linetype.

A template drawing for creating typical multiview drawings could be set up with the layer and linetype assignments similar to that shown in Figure 24-17. Each layer has its own *Color, Linetype,* and *Lineweight* setting, and the layer name indicates the type of lines used (hidden, center, object, etc.). This strategy is useful for simple, generic applications.

FIGURE 24-17

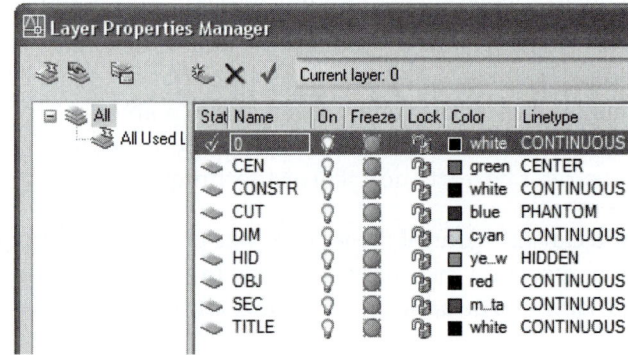

TIP

Using the *ByLayer* strategy (*ByLayer Linetype, Lineweight,* and *Color* assignment) gives you flexibility. You can control the linetype visibility by controlling the layer visibility (show only object layers or hidden line layers). You can also <u>retroactively</u> change the *Linetype, Lineweight,* and *Color* of an existing object by changing the object's *Layer* property with the *Properties* palette or *Matchprop*. Objects changed to a different layer assume the *Color, Linetype,* and *Lineweight* of the new layer.

Another strategy for multiview drawings involving several parts, such as an assembly, is to create layers for each linetype <u>specific to each part</u>, as shown in Figure 24-18. With this strategy, each part has the complete set of linetypes, but only <u>one color per part</u> in order to distinguish the part from others in the display.

Related layer groups could be organized and managed in the *Layer Properties Manager* by creating *Group* filters and *Property* filters (Fig. 24-18). For example, *Group* filters could be created for the "Base," "Fasten," and "Mount" groups. *Property* filters could be created based on *Linetypes* for "Hidden," "Center," and "Object" lines.

Object *Linetypes*, *Lineweights*, and *Colors*

Although this method can be complex, object-specific *Linetype, Lineweight,* and *Color* assignment can also be managed by skillful utilization of layers. One method is to create one layer for each part or each group of related geometry (Fig. 24-19). The *Color, Lineweight,* and *Linetype* settings should be left to the defaults. Then object-specific linetype settings can be assigned (for hidden, center, visible, etc.) using the *Linetype, Lineweight,* and *Color* commands.

For assemblies, all lines related to one part would be drawn on one layer. Remember that you can draw everything with one linetype and color setting, then use *Properties* to retroactively set the desired color and linetype for each set of objects. Visibility of parts can be controlled by *Freezing* or *Thawing* part layers. You cannot isolate and control visibility of linetypes or colors by this method.

FIGURE 24-18

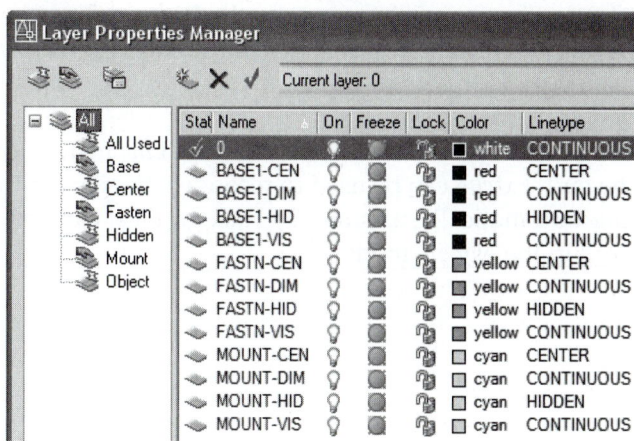

FIGURE 24-19

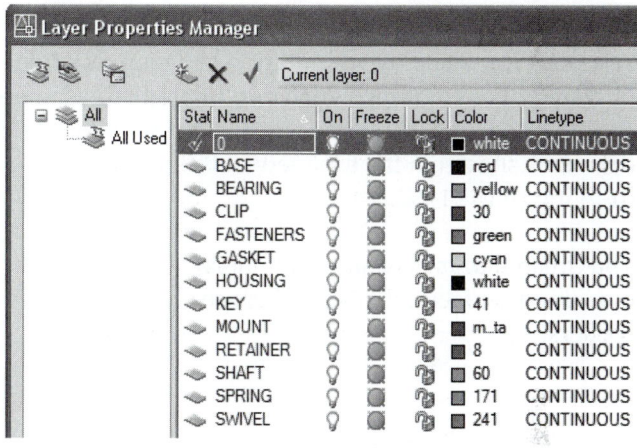

CREATING FILLETS, ROUNDS, AND RUNOUTS

Many mechanical metal or plastic parts manufactured from a molding process have slightly rounded corners. The otherwise sharp corners are rounded because of limitations in the molding process or for safety. A convex corner is called a <u>round</u> and a concave corner is called a <u>fillet</u>. These fillets and rounds are created easily in AutoCAD by using the *Fillet* command.

The example in Figure 24-20 shows a multiview drawing of a part with sharp corners before the fillets and rounds are drawn.

FIGURE 24-20

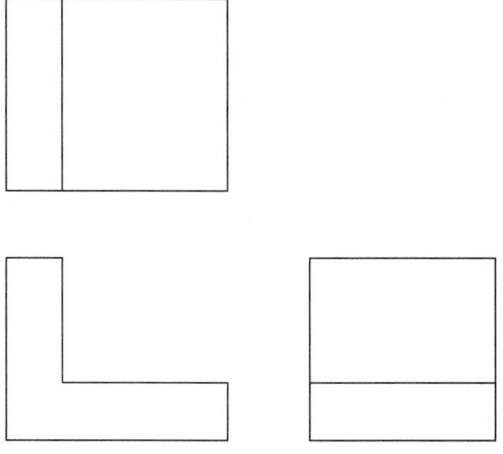

The corners are rounded using the *Fillet* command (Fig. 24-21). First, use *Fillet* to specify the *Radius*. Once the *Radius* is specified, just select the desired lines to *Fillet* near the end to round.

If the *Fillet* is in the middle portion of a *Line* instead of the end, *Extend* can be used to reconnect the part of the *Line* automatically trimmed by *Fillet*, or *Fillet* can be used in the *Notrim* mode.

FIGURE 24-21

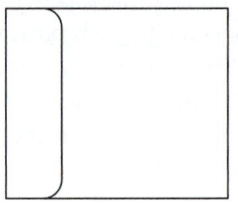

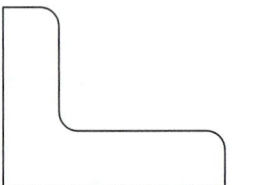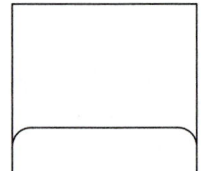

A <u>runout is a visual representation</u> of a complex fillet or round. For example, when two filleted edges intersect at less than a 90 degree angle, a runout should be drawn as shown in the top view of the multiview drawing (Fig. 24-22).

The finish marks (V-shaped symbols) indicate machined surfaces. Finished surfaces have sharp corners.

FIGURE 24-22

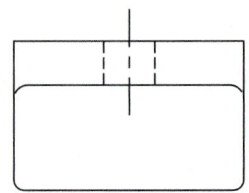

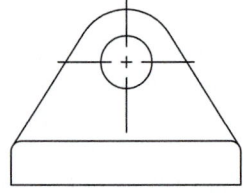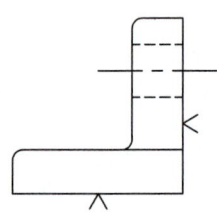

A close-up of the runouts is shown in Figure 24-23. No AutoCAD command is provided for this specific function. The *3point* option of the *Arc* command can be used to create the runouts, although other options can be used. Alternately, the *Circle TTR* option can be used with *Trim* to achieve the desired effect. As a general rule, use the same radius or slightly larger than that given for the fillets and rounds, but draw it less than 90 degrees.

FIGURE 24-23

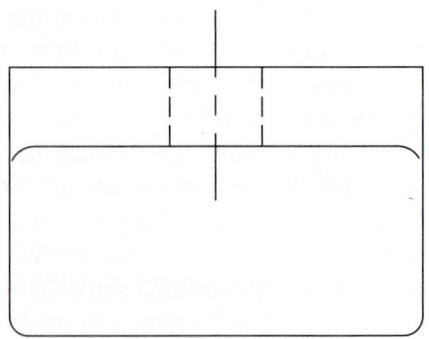

GUIDELINES FOR CREATING A TYPICAL THREE-VIEW DRAWING

Following are some guidelines for creating the three-view drawing in Figure 24-24. This object is used only as an example. The steps or particular construction procedure may vary, depending on the specific object drawn. Dimensions are shown in the figure so you can create the multiview drawing as an exercise.

FIGURE 24-24

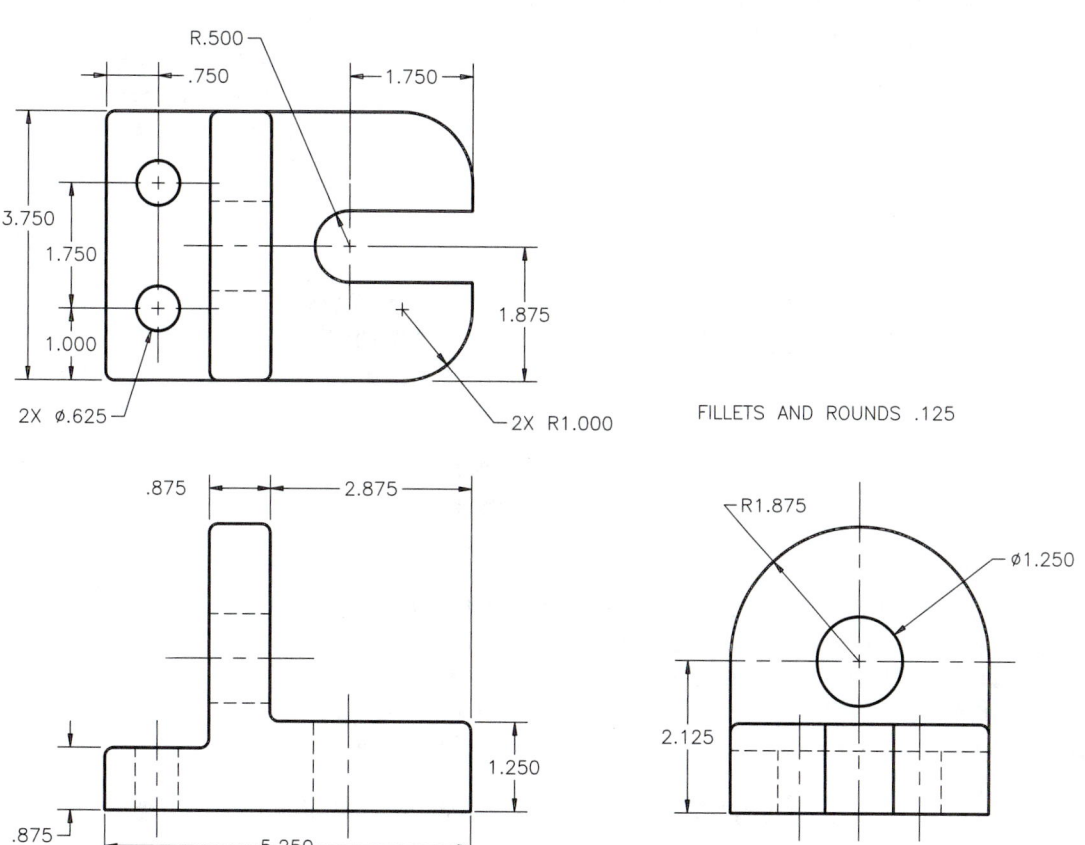

1. Drawing Setup

   Units are set to *Decimal* with 3 places of *Precision*. *Limits* of 22 x 17 are set to allow enough drawing space for both views. The finished drawing can be plotted on a B size sheet at a scale of 1"=1" or on an A size sheet at a scale of 1/2"=1". *Snap* is set to an increment of .125. *Grid* is set to an increment of .5. Although it is recommend that you begin drawing the views using *Grid Snap*, a *Polar Snap* increment of .125 is set, and *Polar Tracking* angles are set. Turn *POLAR* on. Set the desired running *OSNAPs*, such as *Endpoint*, *Midpoint*, *Intersection*, *Quadrant*, and *Center*. Turn on *OTRACK*. *Ltscale* is not changed from the default of 1. Layers are created (OBJ, HID, CEN, DIM, BORDER, and CONSTR) with appropriate *Linetypes*, *Lineweights*, and *Colors* assigned.

2. An outline of each view is "blocked in" by drawing the appropriate *Lines* and *Circles* on the OBJ layer similar to that shown in Figure 24-25. Ensure that *SNAP* is *ON*. *ORTHO* or *Polar Tracking* should be turned *ON* when appropriate. Use the cursor to ensure that the views align horizontally and vertically. Note that the top edge of the front view was determined by projecting from the *Circle* in the right side view.

FIGURE 24-25 —————————

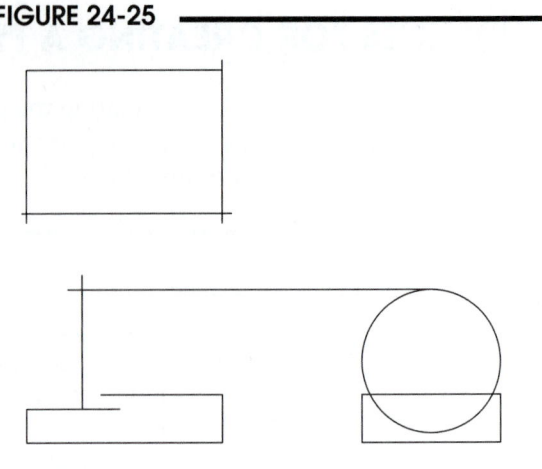

Another method for construction of a multi-view drawing is shown in Figure 24-26. This method uses the *Xline* and *Ray* commands to create construction and projection lines. Here all the views are blocked in and some of the object lines have been formed. The construction lines should be kept on a separate layer, except in the case where *Xlines* and *Rays* can be trimmed and converted to the object lines. (The following illustrations do not display this method because of the difficulty in seeing which are object and construction lines.)

FIGURE 24-26 —————————

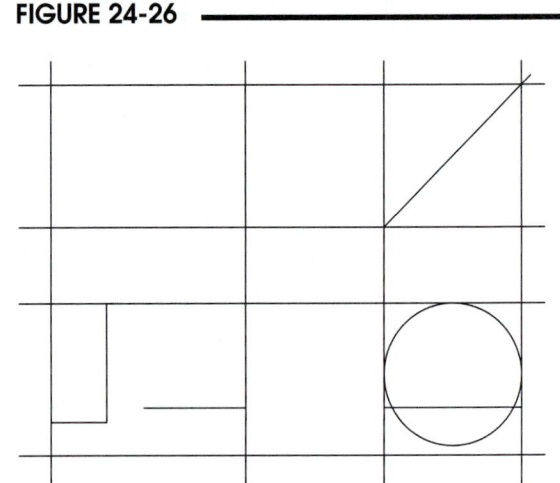

3. This drawing requires some projection between the top and side views (Fig. 24-27). The CONSTR layer is set as *Current*. Two *Lines* are drawn from the inside edges of the two views (using *OSNAP* and *ORTHO* for alignment). A 45 degree miter line is constructed for the projection lines to "make the turn." A *Ray* is suited for this purpose.

FIGURE 24-27 —————————

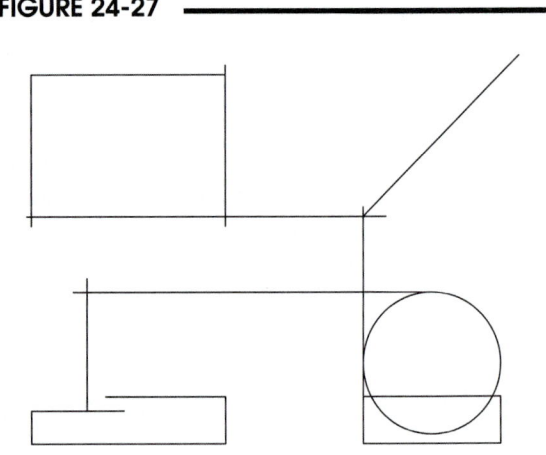

4. Details are added to the front and top views (Fig. 24-28). The projection line from the side view to the front view (previous figure) is *Trimmed*. A *Circle* representing the hole is drawn in the side view and projected up and over to the top view and to the front view. The object lines are drawn on layer OBJ and some projection lines are drawn on layer CONSTR. The horizontal projection lines from the 45 degree miter line are drawn on layer HID awaiting *Trimming*. Alternately, those two projection lines could be drawn on layer CONSTR and changed to the appropriate layer with *Properties* after *Trimming*. Keep in mind that *Object Snap Tracking* can be used here to ensure proper alignment with object features and to prevent having to actually draw some construction lines.

FIGURE 24-28

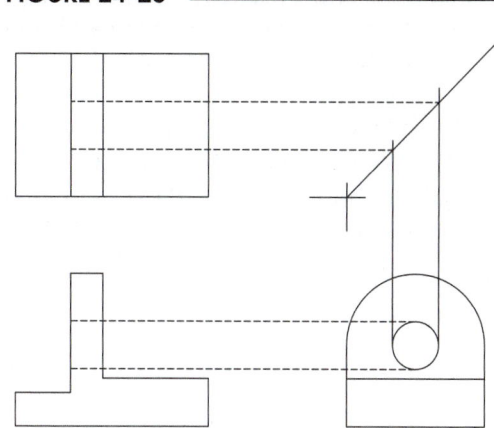

5. The hidden lines used for projection to the top view and front view (previous figure) are *Trimmed* (Fig. 24-29). The slot is created in the top view with a *Circle* and projected to the side and front views. It is usually faster and easier to draw round object features in their circular view first, then project to the other views. Make sure you use the correct layers (OBJ, CONSTR, HID) for the appropriate features. If you do not, *Properties* or *Matchprop* can be used retroactively.

FIGURE 24-29

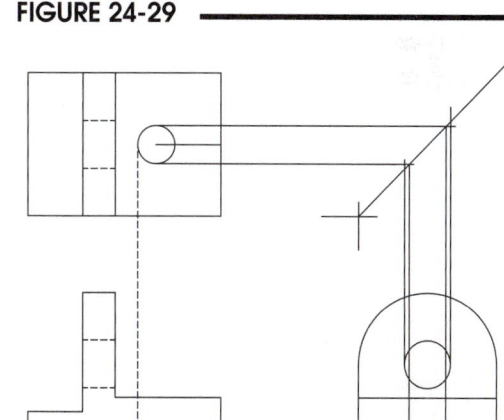

6. The lines shown in the previous figure as projection lines or construction lines for the slot are *Trimmed* or *Erased* (Fig. 24-30). The holes in the top view are drawn on layer OBJ and projected to the other views. The projection lines and hidden lines are drawn on their respective layers.

FIGURE 24-30

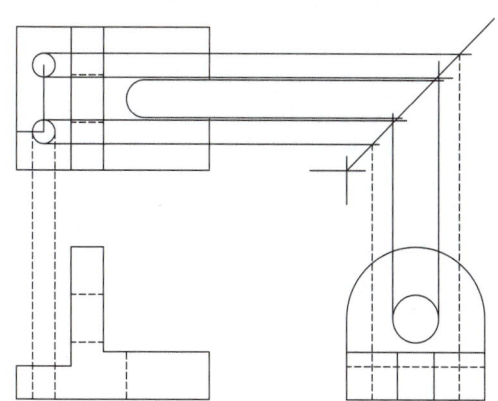

7. *Trim* the appropriate hidden lines (Fig. 24-31). *Freeze* layer CONSTR. On layer OBJ, use *Fillet* to create the rounded corners in the top view. Draw the correct center lines for the holes on layer CEN. The value for *Ltscale* should be adjusted to achieve the optimum center line spacing. The *Properties* palette can be used to adjust individual object line-type scale.

FIGURE 24-31

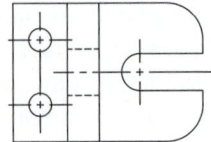
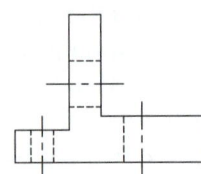
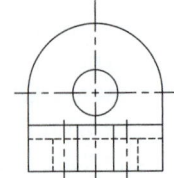

8. Fillets and rounds are added using the *Fillet* command (Fig. 24-32). The runouts are created by drawing a *3point Arc* and *Trimming* or *Extending* the *Line* ends as necessary. Use *Zoom* for this detail work.

FIGURE 24-32

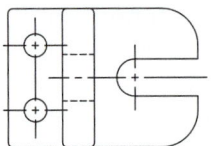
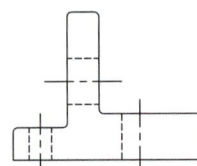
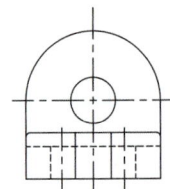

9. Activate a *Layout* tab, configure a print or plot device using *Pagesetup*, make one *Viewport*, and set the *Viewport scale* to a standard scale (Fig. 24-33). Add a border and a title block using *Pline*. Include the part name, company, draftsperson, scale, date, and drawing file name in the title block. The drawing is ready for dimensioning and manufacturing notes.

FIGURE 24-33

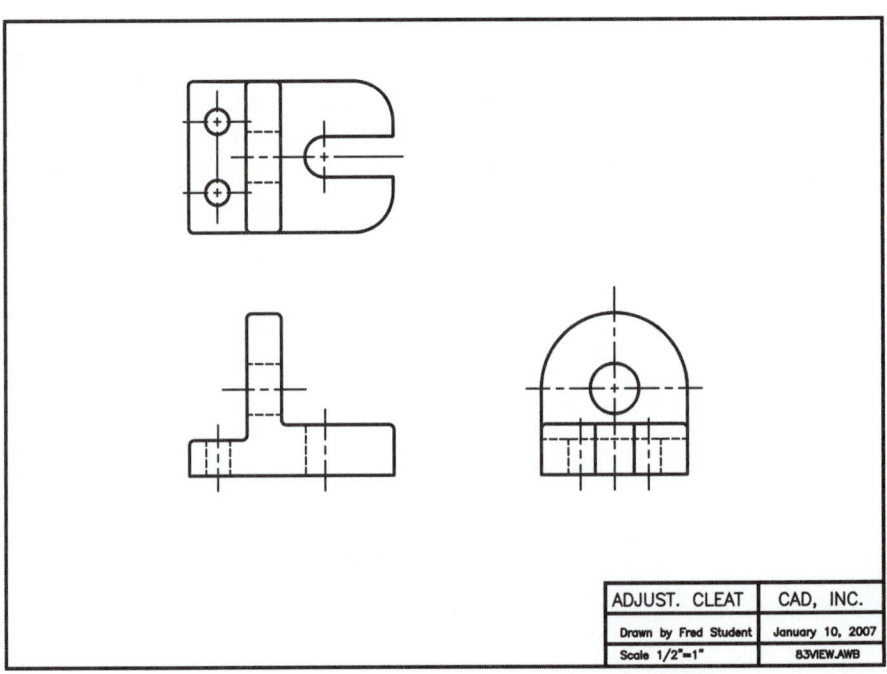

ADJUST. CLEAT	CAD, INC.
Drawn by Fred Student	January 10, 2007
Scale 1/2"=1"	83VIEW.AWB

CHAPTER EXERCISES

1. *Open* the PIVOTARM CH16 drawing.

    A. Create the right side view. Use *OSNAP* and *ORTHO* or *Polar Tracking* to create *Lines* or *Rays* to the miter line and down to the right side view as shown in Figure 24-34. *Offset* may be used effectively for this purpose instead. Use *Extend, Offset,* or *Ray* to create the projection lines from the front view to the right side view. Use *Object Snap Tracking* when appropriate.

FIGURE 24-34

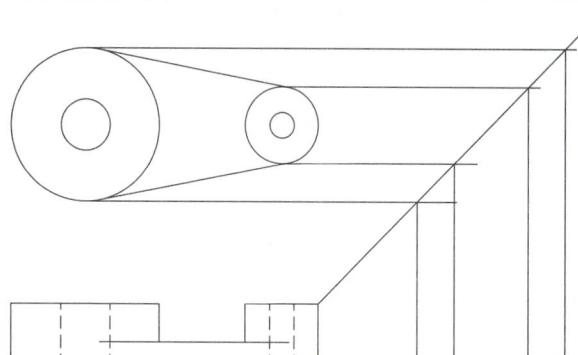

    B. *Trim* or *Erase* the unwanted projection lines, as shown in Figure 24-35. Draw a *Line* or *Ray* from the *Endpoint* of the diagonal *Line* in the top view down to the front to supply the boundary edge for *Trimming* the horizontal *Line* in the front view as shown.

FIGURE 24-35

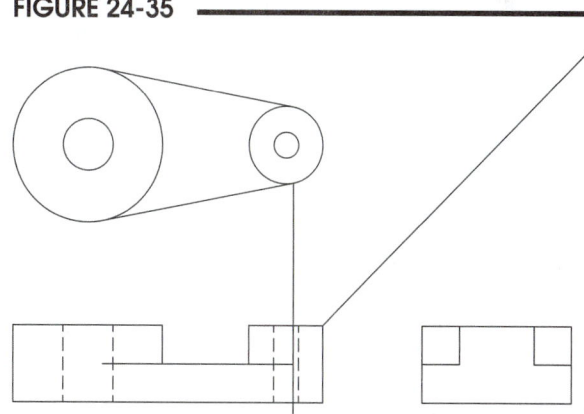

    C. Next, create the hidden lines for the holes by the same fashion as before (Fig. 24-36). Use previously created *Layers* to achieve the desired *Linetypes*. Complete the side view by adding the horizontal hidden *Line* in the center of the view.

FIGURE 24-36

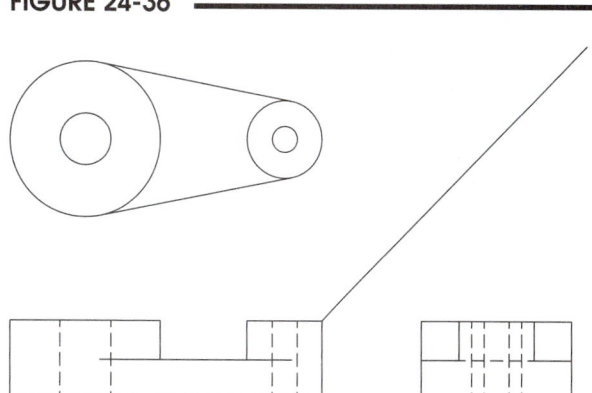

D. Another hole for a set screw must be added to the small end of the Pivot Arm. Construct a *Circle* of **4mm** diameter with its center located **8mm** from the top edge in the side view as shown in Figure 24-37. Project the set screw hole to the other views.

FIGURE 24-37

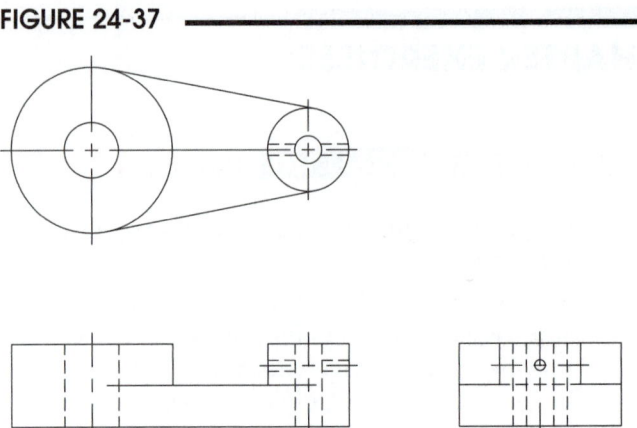

E. Make new layers **CONSTR, OBJ**, and **TITLE** and change objects to the appropriate layers with *Properties*. *Freeze* layer **CONSTR**. Add centerlines on the **CEN** layer as shown in Figure 24-37. Change the *Ltscale* to **18**. Activate a *Layout* tab, configure a print or plot device using *Pagesetup*, make one *Viewport*, and set the *Viewport scale* to a standard scale. To complete the PIVOTARM drawing, draw a *Pline* border (*width .02* x scale factor) in the layout and *Insert* the **TBLOCKAT** drawing that you created in Chapter 22 Exercises. *SaveAs* **PIVOTARM CH24.**

For exercises 2 through 5, construct and plot the multiview drawings as instructed. Use an appropriate *template* drawing for each exercise unless instructed otherwise. Use conventional practices for *layers* and *linetypes*. Draw a *Pline* border with the correct *width* and *Insert* your **TBLOCK** or **TBLOCKAT** drawing.

2. Make a two-view multiview drawing of the Clip (Fig. 24-38). *Plot* the drawing full size (**1=1**).

Use the **ASHEET** template drawing to achieve the desired plot scale. *Save* the drawing as **CLIP.**

FIGURE 24-38

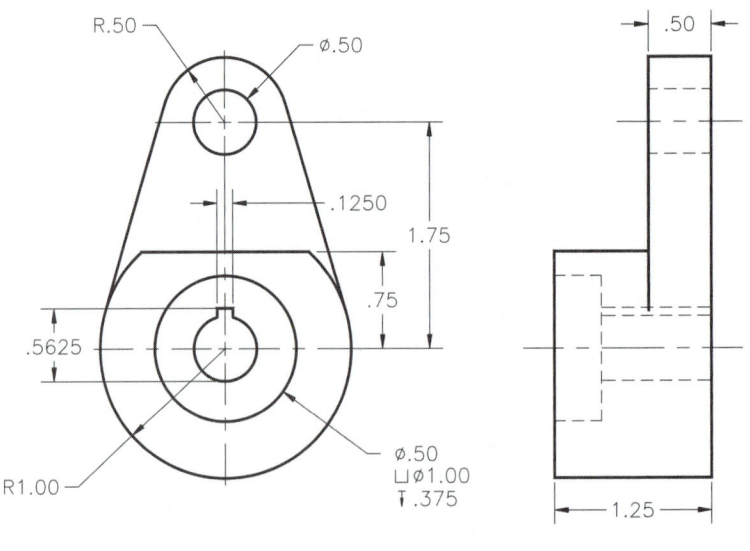

CLIP

3. Make a three-view multiview drawing of the Bar Guide and *Plot* it **1=1** (Fig. 24-39). Use the **BARGUIDE** drawing you set up in Chapter 12 Exercises. Note that a partial *Ellipse* will appear in one of the views.

FIGURE 24-39

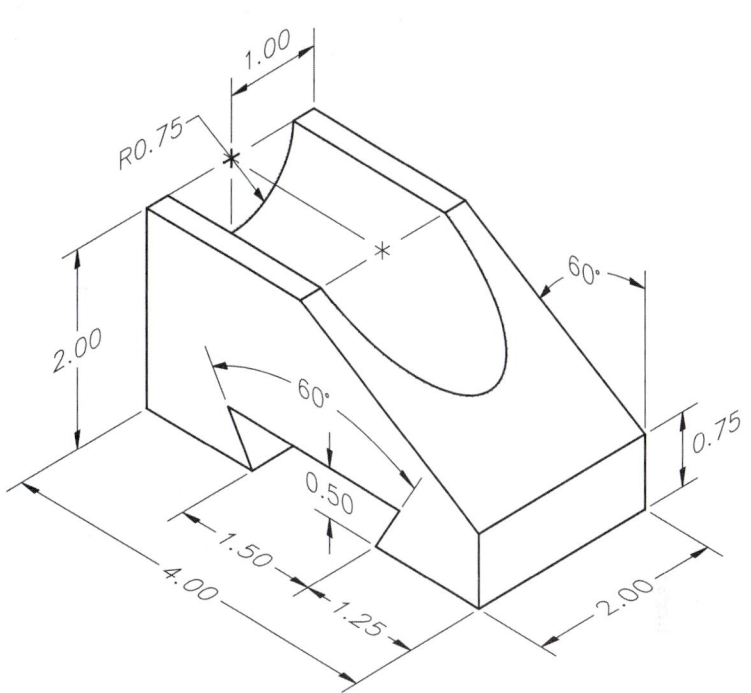

4. Construct a multiview drawing of the Saddle (Fig. 24-40). Three views are needed. The channel along the bottom of the part intersects with the saddle on top to create a slotted hole visible in the top view. *Plot* the drawing at **1=1**.

Save as **SADDLE.**

FIGURE 24-40

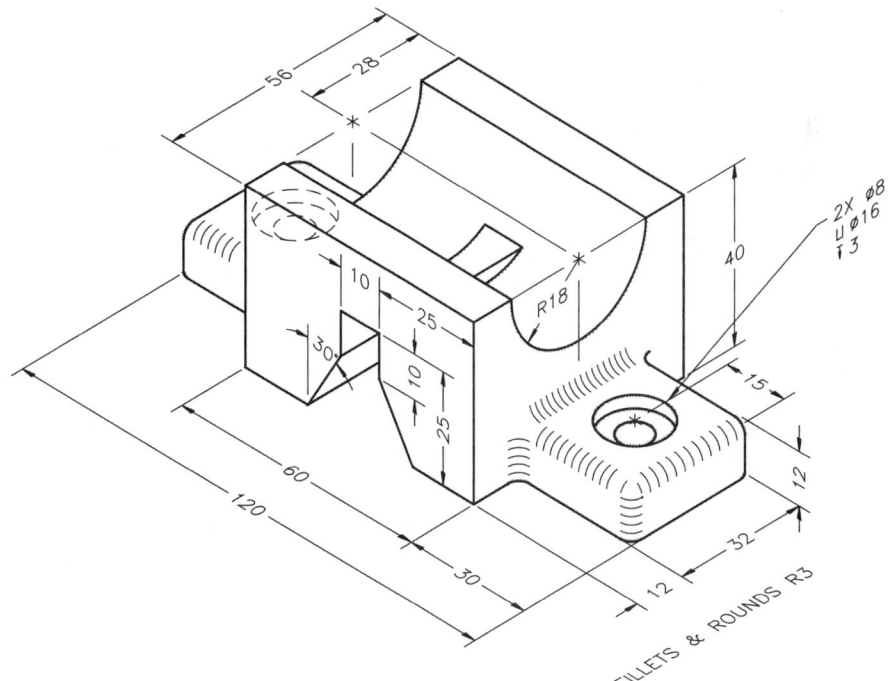

5.   Draw a multiview of the Adjustable Mount shown in Figure 24-41. Determine an appropriate template drawing to use and scale for plotting based on your plotter capabilities. *Plot* the drawing to an accepted scale. *Save* the drawing as **ADJMOUNT.**

FIGURE 24-41

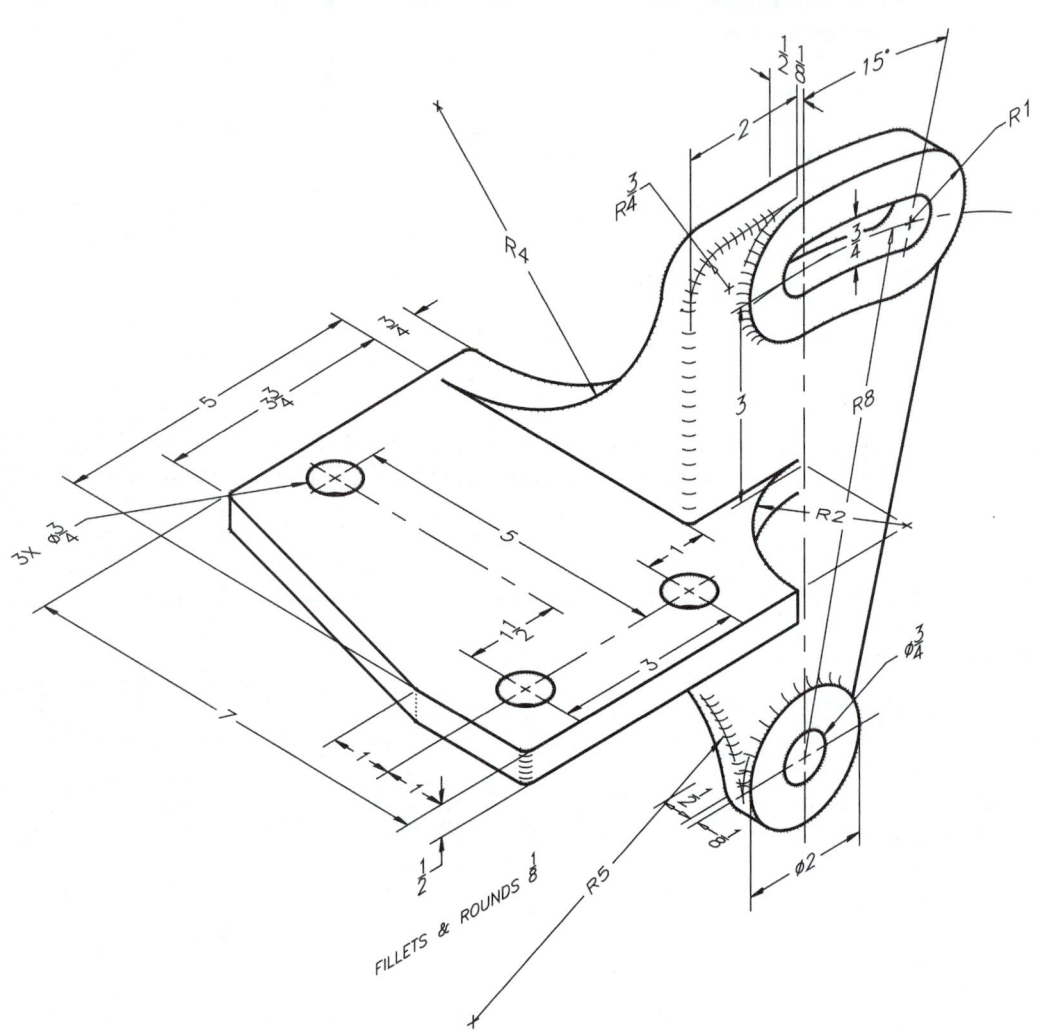

25

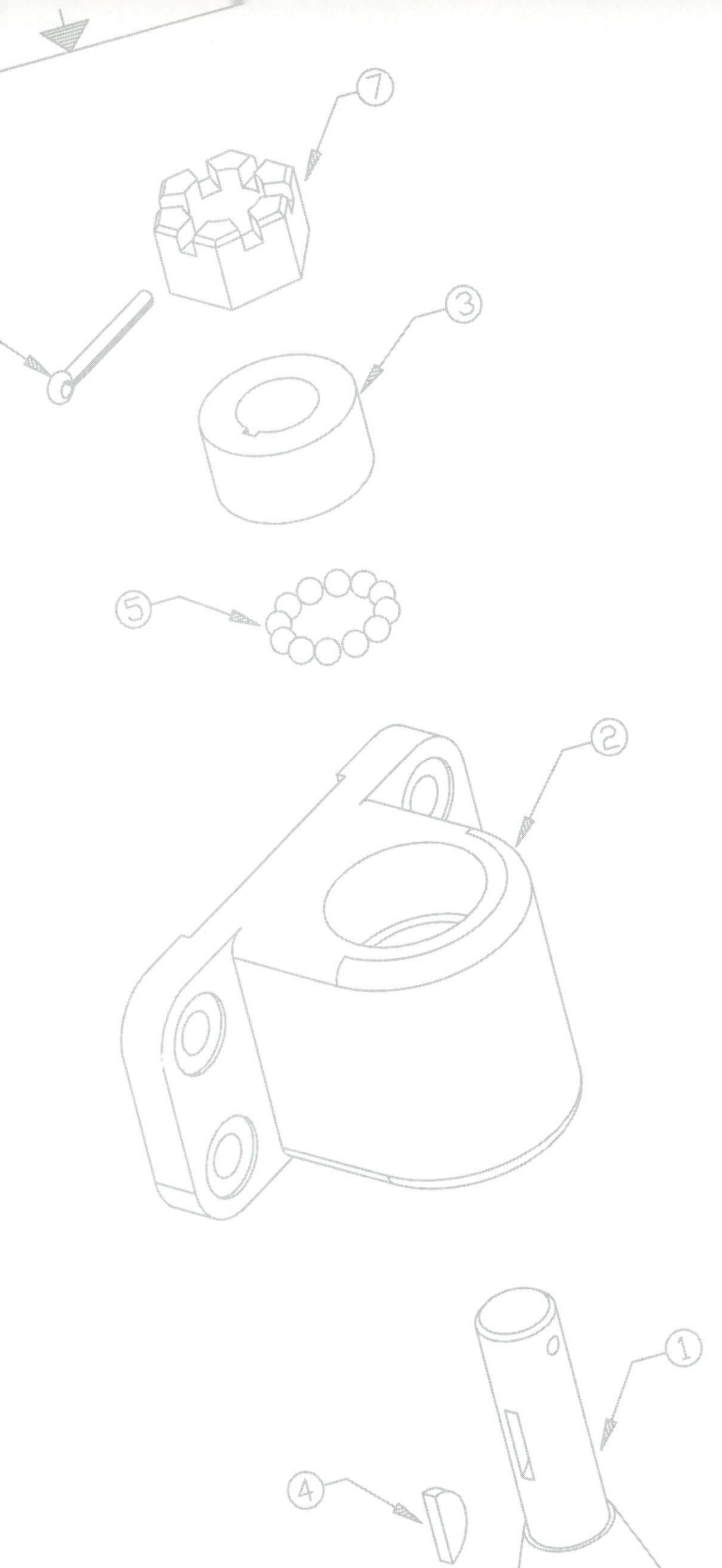

PICTORIAL DRAWINGS

CHAPTER OBJECTIVES

After completing this chapter you should:

1. be able to activate the *Isometric Style* of *Snap* for creating isometric drawings;

2. know how to draw on the three isometric planes by toggling *Isoplane* using Ctrl+E;

3. be able to create isometric ellipses with the *Isocircle* option of *Ellipse*;

4. be able to construct an isometric drawing in AutoCAD;

5. be able to create Oblique Cavalier and Cabinet drawings in AutoCAD.

CONCEPTS

Isometric drawings and oblique drawings are pictorial drawings. Pictorial drawings show three principal faces of the object in one view. A pictorial drawing is a drawing of a 3D object as if you were positioned to see (typically) some of the front, some of the top, and some of the side of the object. All three dimensions of the object (width, height, and depth) are visible in a pictorial drawing.

Multiview drawings differ from pictorial drawings because a multiview only shows two dimensions in each view, so two or more views are needed to see all three dimensions of the object. A pictorial drawing shows all dimensions in the one view. Pictorial drawings depict the object similar to the way you are accustomed to viewing objects in everyday life, that is, seeing all three dimensions. Figure 25-1 and Figure 25-2 show the same object in multiview and in pictorial representation, respectively. Notice that multiview drawings use hidden lines to indicate features that are normally obstructed from view, whereas <u>hidden lines are normally omitted</u> in isometric drawings (unless certain hidden features must be indicated for a particular function or purpose).

Types of Pictorial Drawings

Pictorial drawings are classified as follows:

1. Axonometric drawings
    a. Isometric drawings
    b. Dimetric drawings
    c. Trimetric drawings

2. Oblique drawings

Axonometric drawings are characterized by how the angle of the edges or axes (axon-) are measured (-metric) with respect to each other.

Isometric drawings are drawn so that each of the axes have equal angular measurement. ("Isometric" means equal measurement.) The isometric axes are always drawn at 120 degree increments (Fig. 25-3). All rectilinear lines on the object (representing horizontal and vertical edges—not inclined or oblique) are drawn on the isometric axes.

A 3D object seen "in isometric" is thought of as being oriented so that each of three perpendicular faces (such as the top, front, and side) are seen equally. In other words, the angles formed between the line of sight and each of the principal faces are equal.

FIGURE 25-1

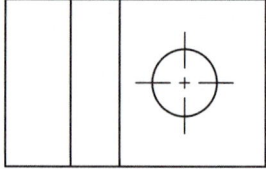
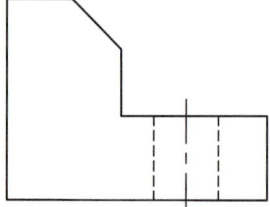
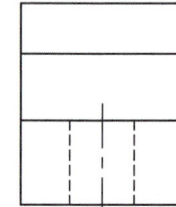

FIGURE 25-2

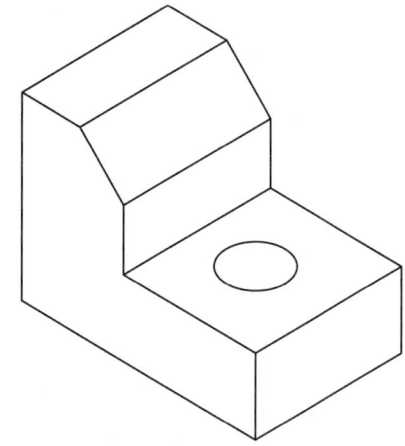

FIGURE 25-3

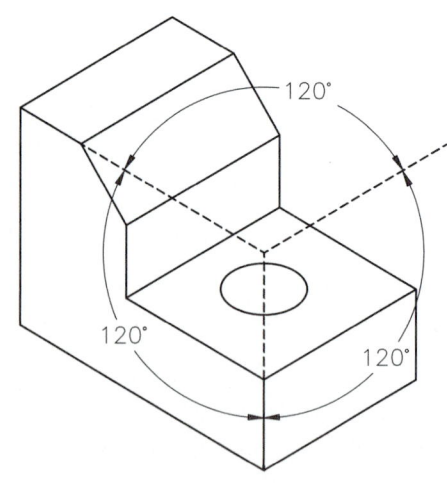

Dimetric drawings are constructed so that the angle between any two of the three axes is equal. There are many possibilities for dimetric axes. A common orientation for dimetric drawings is shown in Figure 25-4. For 3D objects seen from a dimetric viewpoint, the angles formed between the line of sight and each of two principal faces are equal.

FIGURE 25-4

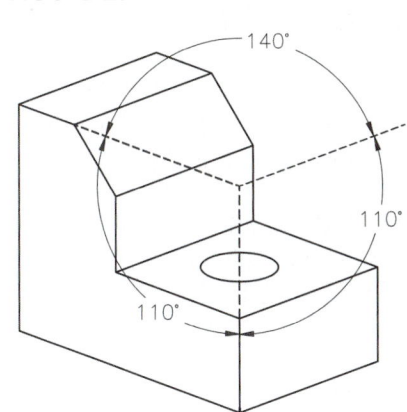

Trimetric drawings have three unequal angles between the axes. Numerous possibilities exist. A common orientation for trimetric drawings is shown in Figure 25-5.

FIGURE 25-5

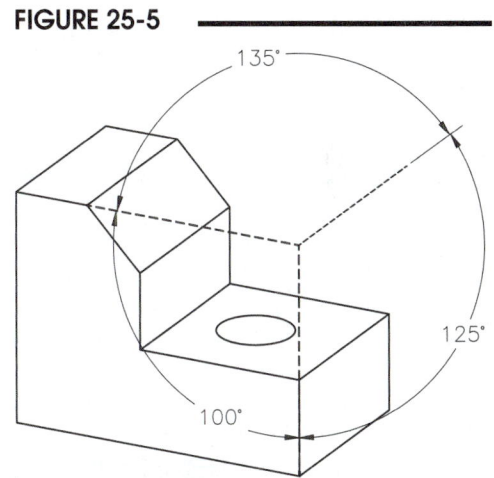

Oblique drawings are characterized by a vertical axis and horizontal axis for the two dimensions of the front face and a third (receding) axis of either 30, 45, or 60 degrees (Fig. 25-6). Oblique drawings depict the true size and shape of the front face, but add the depth to what would otherwise be a typical 2D view. This technique simplifies construction of drawings for objects that have contours in the profile view (front face) but relatively few features along the depth. Viewing a 3D object from an oblique viewpoint is not possible.

This chapter will explain the construction of isometric and oblique drawings in AutoCAD.

FIGURE 25-6

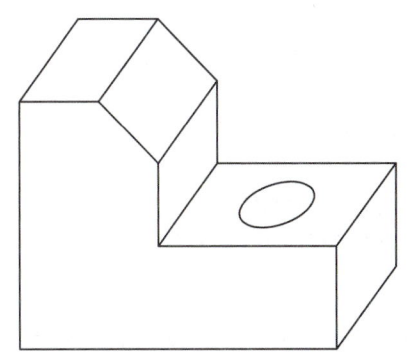

Pictorial Drawings Are 2D Drawings

Isometric, dimetric, trimetric, and oblique drawings are 2D drawings, whether created with AutoCAD or otherwise. Pictorial drawing was invented before the existence of CAD and therefore was intended to simulate a 3D object on a 2D plane (the plane of the paper). If AutoCAD is used to create the pictorial, the geometry lies on a 2D plane—the XY plane. All coordinates defining objects have X and Y values with a Z value of 0. When the *Isometric* style of *Snap* is activated, an isometrically structured *SNAP* and *GRID* appear on the XY plane.

Figure 25-7 illustrates the 2D nature of an isometric drawing created in AutoCAD. Isometric lines are created on the XY plane. The *Isometric SNAP* and *GRID* are also on the 2D plane. (The *3Dorbit* command was used to give other than a *Plan* view of the drawing in this figure.)

Although pictorial drawings are based on the theory of projecting 3D objects onto 2D planes, it is physically possible to achieve an axonometric (isometric, dimetric, or trimetric) viewpoint of a 3D object using a 3D CAD system. In AutoCAD, the *Vpoint* and *3Dorbit* commands can be used to specify the observer's position in 3D space with respect to a 3D model. Chapter 35 discusses the specific commands and values needed to attain axonometric viewpoints of a 3D model.

FIGURE 25-7

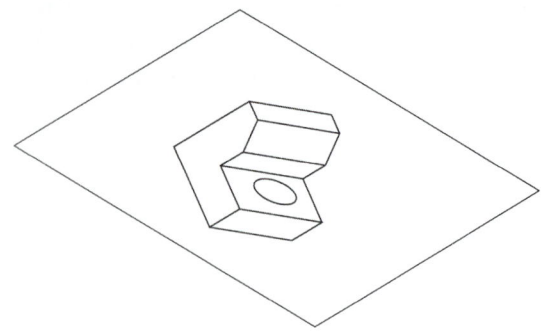

ISOMETRIC DRAWING IN AutoCAD

AutoCAD provides the capability to construct isometric drawings. An isometric *SNAP* and *GRID* are available, as well as a utility for creation of isometrically correct ellipses. Isometric lines are created with the *Line* command. There are no special options of *Line* for isometric drawing, but isometric *SNAP* and *GRID* can be used to force *Lines* to an isometric orientation. Begin creating an isometric drawing in AutoCAD by activating the *Isometric Style* option of the *Snap* command. This action can be done using any of the options listed in the following Command table.

Snap

Pull-down Menu	Command (Type)	Alias (Type)	Short-cut	Screen (side) Menu	Tablet Menu
Tools *Drafting Settings...* *Snap and Grid*	*Snap*	*SN*	*F9 or Ctrl+B*	*TOOLS 2* *Grid*	*W,10*

FIGURE 25-8

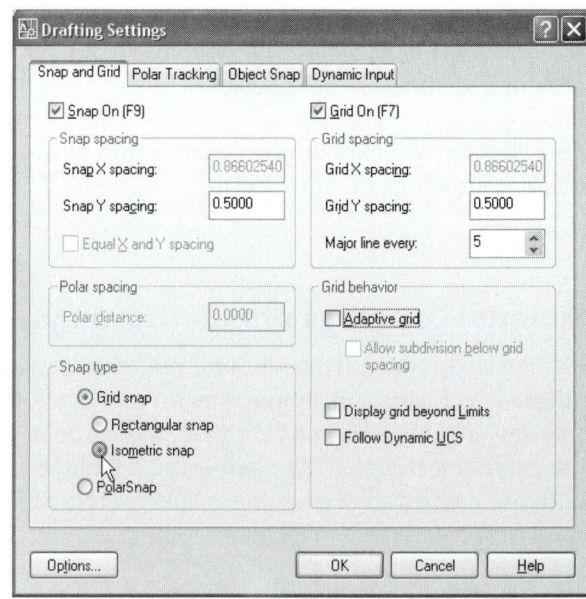

Command: **snap**
Specify snap spacing or
[ON/OFF/Aspect/Rotate/Style/Type] <0.5000>: **s**
Enter snap grid style [Standard/Isometric] <S>: **i**
Specify vertical spacing <0.5000>: **Enter**
Command:

Alternately, toggling the indicated checkbox in the lower-left corner of the *Drafting Settings* dialog box activates the *Isometric Snap* and *Grid Snap* (Fig. 25-8).

NOTE: For AutoCAD 2007 or newer drawings, toggle *Adaptive Grid* off (Fig. 25-8, right). Otherwise, the *GRID* will not display at the set increment when the *Isometric snap* style is current.

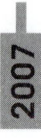

Figure 25-9 illustrates the effect of setting the *Isometric SNAP* and *GRID*. Notice the new orientation of the cursor.

FIGURE 25-9

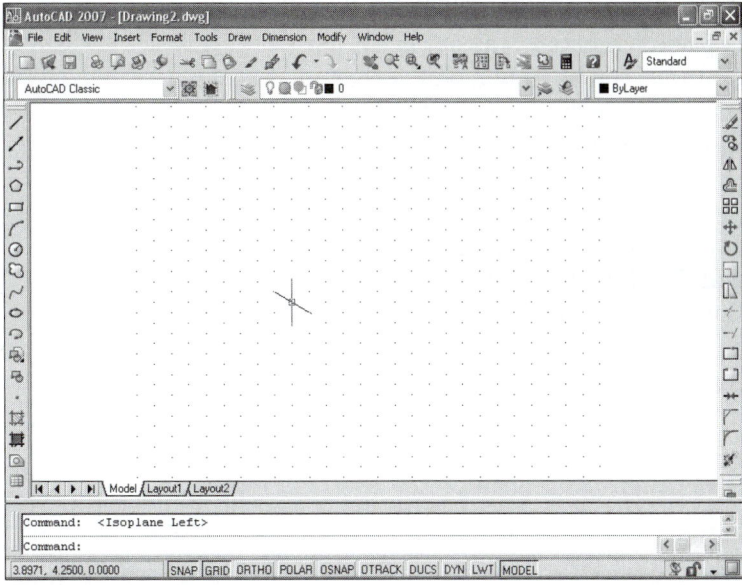

Using <u>Ctrl+E</u> (pressing the Ctrl key and the letter "E" simultaneously) toggles the cursor to one of three possible *Isoplanes* (AutoCAD's term for the three faces of the isometric pictorial). If *ORTHO* is *ON*, only <u>isometric</u> lines are drawn; that is, you can only draw *Lines* aligned with the isometric axes. *Lines* can be drawn on only two axes for each isoplane. Ctrl+E allows drawing on the two axes aligned with another face of the object. *ORTHO* is *OFF* in order to draw inclined or oblique lines (not on the isometric axis). The functions of *GRID* (F7) and *SNAP* (F9) remain unchanged.

With *SNAP ON*, toggle *Coords* (F6) several times and examine the readout as you move the cursor. The <u>Cartesian coordinate format is of no particular assistance</u> while drawing in isometric because of the configuration of the *GRID*. The <u>relative polar</u> format, however, is <u>very helpful</u>. Use relative polar format for *Coords* while drawing in isometric (Fig. 25-10).

Alternately, use *Polar Tracking* instead of *ORTHO*. Set the *Polar Angle Settings* to 30 degrees. The advantage of using *Polar Tracking* is that the current line length is given on the polar tracking tip (see Fig. 25-10). The disadvantage is that it is possible to draw non-isoplane lines accidentally. Only one setting at a time can be used in AutoCAD—*Polar Tracking* or *ORTHO*. (The remainder of the figures illustrate the use of *ORTHO*.)

FIGURE 25-10

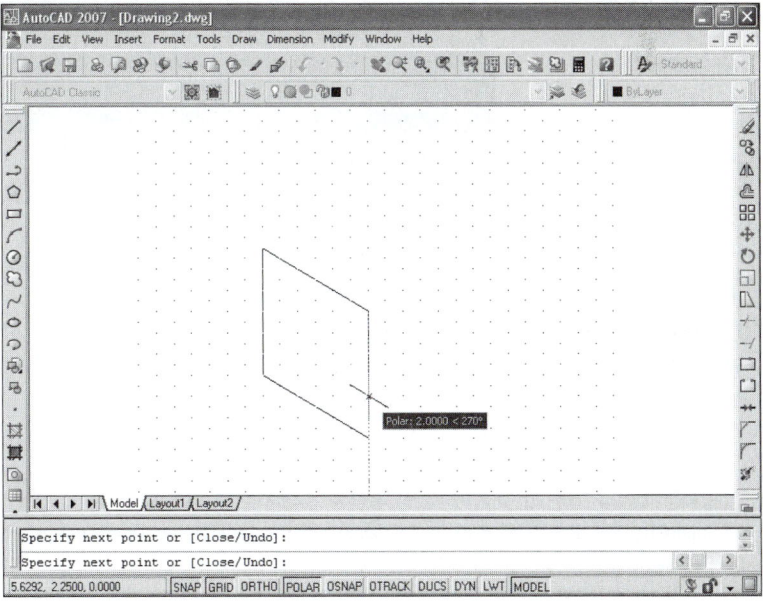

The effects of changing the *Isoplane* are shown in the following figures. Press Ctrl+E to change *Isoplane*.

With *ORTHO ON*, drawing a *Line* is limited to the two axes of the current *Isoplane*. Only one side of a cube, for example, can be drawn on the current *Isoplane*. Watch *Coords* (in a polar format) or turn on *DYN* to give the length of the current *Line* as you draw.

Toggling Ctrl+E switches the cursor and the effect of *ORTHO* to another *Isoplane*. One other side of a cube can be constructed on this *Isoplane* (Fig. 25-11).

FIGURE 25-11

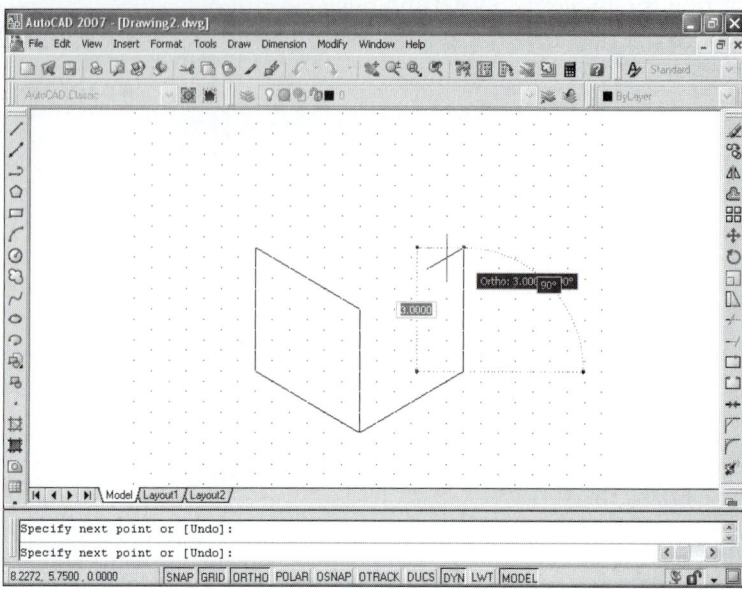

Direct Distance Entry can be of great help when drawing isometric lines. Use Ctrl+E and *ORTHO* to force the *Line* to the correct orientation, then enter the desired distance value at the Command line.

Isometric Ellipses

Isometric ellipses are easily drawn in AutoCAD by using the *Isocircle* option of the *Ellipse* command. This option appears <u>only</u> when the isometric *SNAP* is *ON*.

Ellipse

Pull-down Menu	Command (Type)	Alias (Type)	Short-cut	Screen (side) Menu	Tablet Menu
Draw *Ellipse*	*Ellipse*	*EL*	...	*DRAW 1* *Ellipse*	*M,9*

Although the *Isocircle* option does not appear in the pull-down or digitizing tablet menus, it can be invoked as an option of the *Ellipse* command. The *Isocircle* option of *Ellipse* appears only when the *Snap Type* is set to *Isometric*. You <u>must type "I"</u> to use the *Isocircle* option. The command syntax is as follows:

    Command: **ellipse**
    Specify axis endpoint of ellipse or [Arc/Center/Isocircle]: **i**
    Specify center of isocircle: **PICK** or **(coordinates)**
    Specify radius of isocircle or [Diameter]: **PICK** or **(coordinates)**
    Command:

After selecting the center point of the *Isocircle*, the isometrically correct ellipse appears on the screen on the current *Isoplane*. Use Ctrl+E to toggle the ellipse to the correct orientation. When defining the radius interactively, use *ORTHO* to force the rubberband line to an isometric axis (Fig. 25-12, next page).

Since isometric angles are equal, all isometric ellipses have the same proportion (major to minor axis). The only differences in isometric ellipses are the size and the orientation (*Isoplane*).

FIGURE 25-12

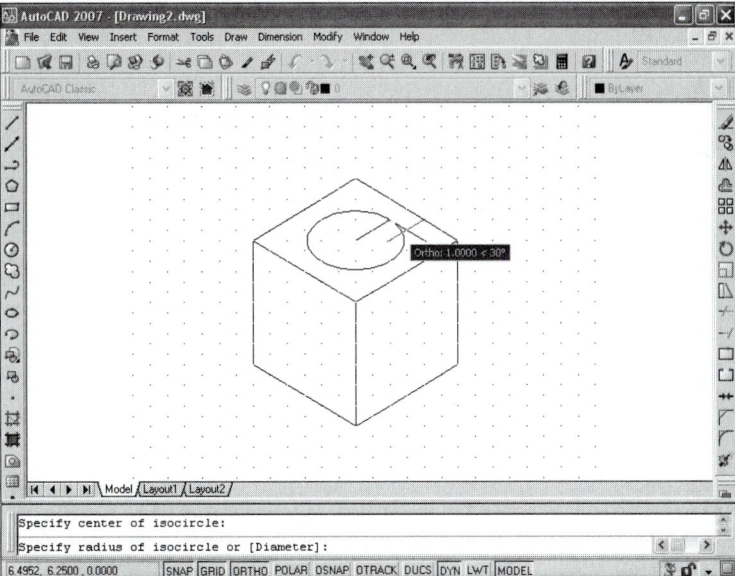

Figure 25-13 shows three ellipses correctly oriented on their respective faces. Use Ctrl+E to toggle the correct *Isoplane* orientation: *Isoplane Top*, *Isoplane Left*, or *Isoplane Right*.

FIGURE 25-13

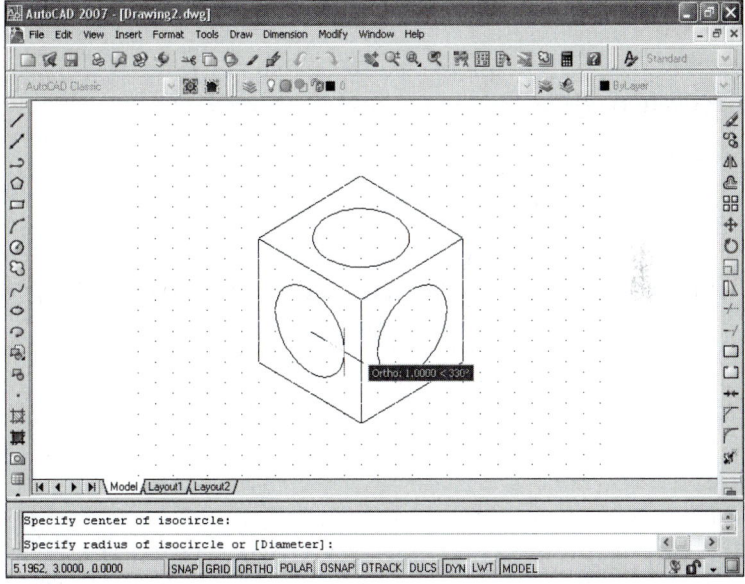

When defining the radius or diameter of an ellipse, it should always be measured in an isometric direction. In other words, an isometric ellipse is always measured on the two isometric axes (or center lines) parallel with the plane of the ellipse.

If you define the radius or diameter interactively, use *ORTHO ON*. If you enter a value, AutoCAD automatically applies the value to the correct isometric axes.

Creating an Isometric Drawing

In this exercise, the object in Figure 25-14 is drawn in isometric.

The initial steps to create an isometric drawing begin with the typical setup (see Chapter 6, Drawing Setup):

1. Set the desired *Units*.

2. Set appropriate *Limits*.

3. Set the *Isometric Style* of *Snap* and specify an appropriate value for spacing. Toggle *Adaptive grid* off.

FIGURE 25-14

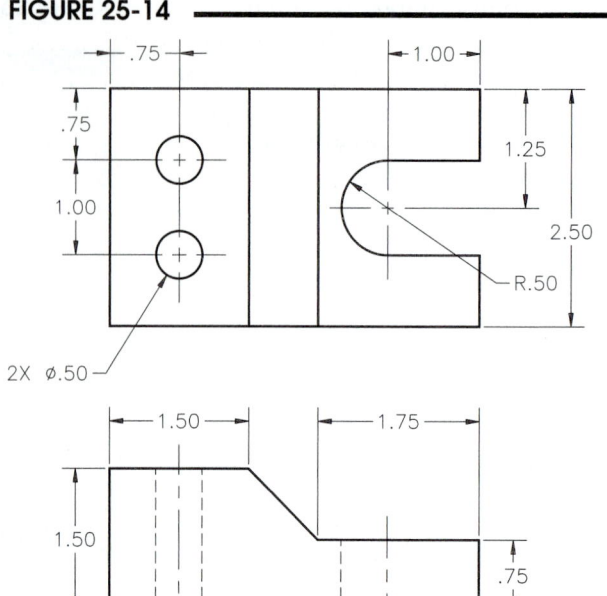

4. The next step involves creating an isometric framework of the desired object. In other words, draw an isometric box equal to the overall dimensions of the object. Using the dimensions given in Figure 25-14, create the encompassing isometric box with the *Line* command (Fig. 25-15).

    Use *ORTHO* to force isometric *Lines*. Watch the *Coords* display (in a relative polar format) to give the current lengths as you draw or use direct distance entry.

FIGURE 25-15

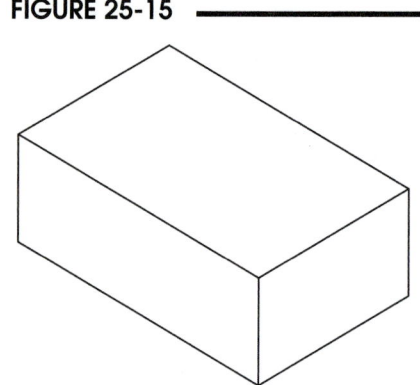

5. Add the lines defining the lower surface. Define the needed edge of the upper isometric surface as shown (Fig. 25-16).

FIGURE 25-16

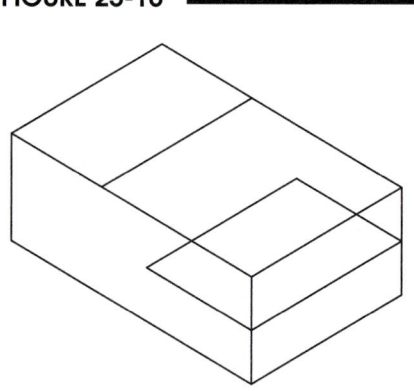

6. The <u>inclined</u> edges of the inclined surface can be drawn (with *Line*) only when *ORTHO* is *OFF*. <u>Inclined</u> lines in isometric cannot be drawn by transferring the lengths of the lines, but only by defining the <u>ends</u> of the inclined lines on <u>isometric</u> lines, then connecting the endpoints. Next, *Trim* or *Erase* the necessary *Lines* (Fig. 25-17).

FIGURE 25-17

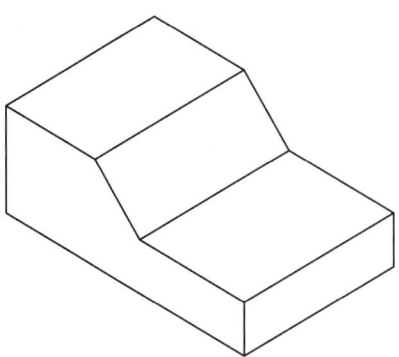

7. Draw the slot by constructing an *Ellipse* with the *Isocircle* option. Draw the two *Lines* connecting the circle to the right edge. *Trim* the unwanted part of the *Ellipse* (highlighted) using the *Lines* as cutting edges (Fig. 25-18).

FIGURE 25-18

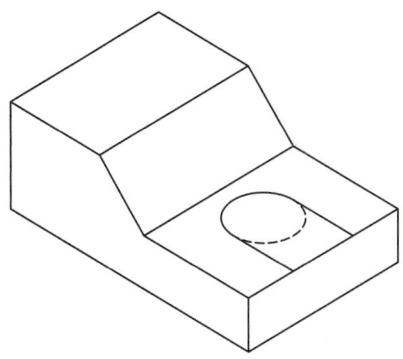

8. *Copy* the far *Line* and the *Ellipse* down to the bottom surface. Add two vertical *Lines* at the end of the slot (Fig. 25-19).

FIGURE 25-19

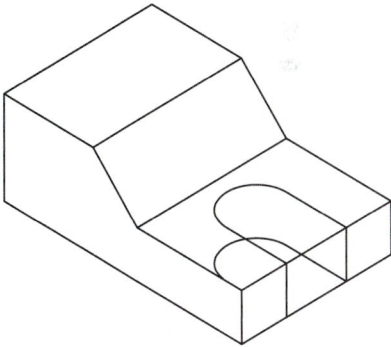

9. Use *Trim* to remove the part of the *Ellipse* that would normally be hidden from view. *Trim* the *Lines* along the right edge at the opening of the slot (Fig. 25-20).

FIGURE 25-20

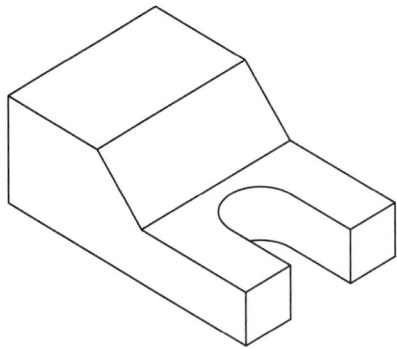

10. Add the two holes on the top with *Ellipse, Isocircle* option (Fig. 25-21). Use *ORTHO ON* when defining the radius. *Copy* can also be used to create the second *Ellipse* from the first.

Dimensioning Isometric Drawings in AutoCAD

Refer to Chapter 29, Dimension Styles and Variables, for details on how to dimension isometric drawings.

FIGURE 25-21

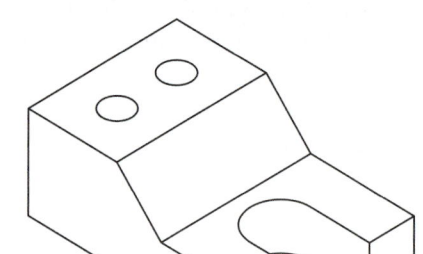

OBLIQUE DRAWING IN AutoCAD

Oblique drawings are characterized by having two axes at a 90 degree orientation. Typically, you should locate the <u>front face</u> of the object along these two axes. Since the object's characteristic shape is seen in the front view, an oblique drawing allows you to create all shapes parallel to the front face true size and shape as you would in a multi-view drawing. Circles on or parallel to the front face can be drawn as circles. The third axis, the receding axis, can be drawn at a choice of angles, 30, 45, or 60 degrees, depending on whether you want to show more of the top or the side of the object.

Figure 25-22 illustrates the axes orientation of an oblique drawing, including the choice of angles for the receding axis.

Another option allowed with oblique drawings is the measurement used for the receding axis. Using the full depth of the object along the receding axis is called <u>Cavalier</u> oblique drawing. This method depicts the object (a cube with a hole in this case) as having an elongated depth (Fig. 25-23).

FIGURE 25-22

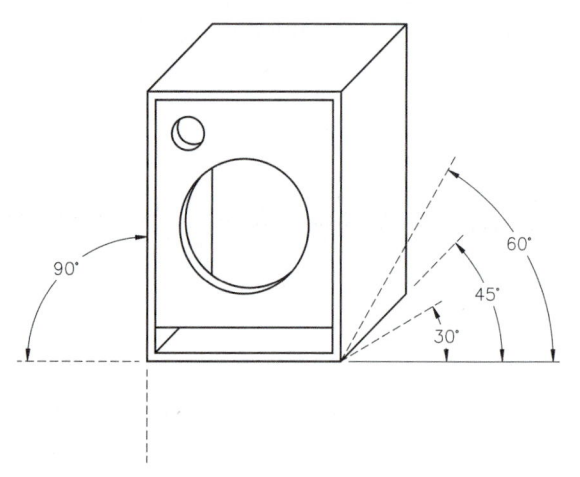

FIGURE 25-23

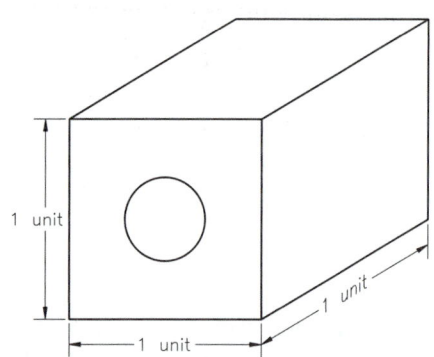

Using 1/2 of the true depth along the receding axis gives a more realistic pictorial representation of the object. This is called a <u>Cabinet</u> oblique (Fig. 25-24).

No functions or commands in AutoCAD are intended specifically for oblique drawing. However, *Polar Snap* and *Polar Tracking* can simplify the process of drawing lines on the front face of the object and along the receding axis. The steps for creating a typical oblique drawing are given next.

FIGURE 25-24

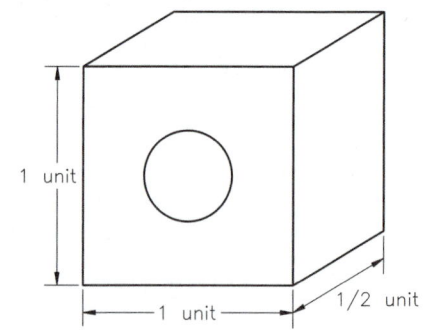

The object in Figure 25-25 is used for the example. From the dimensions given in the multiview, create a cabinet oblique with the receding axis at 45 degrees.

FIGURE 25-25

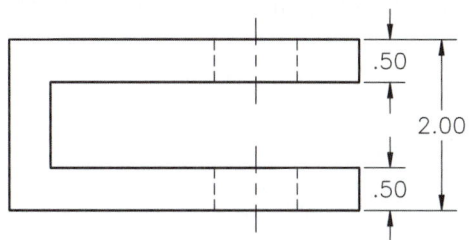

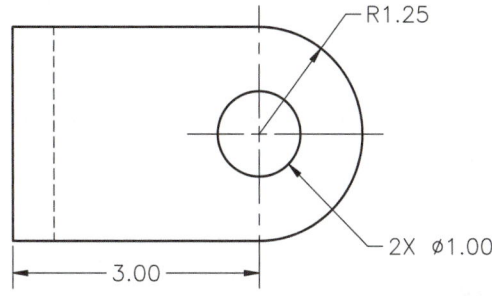

1. Create the characteristic shape of the front face of the object as shown in the front view (Fig. 25-26).

FIGURE 25-26

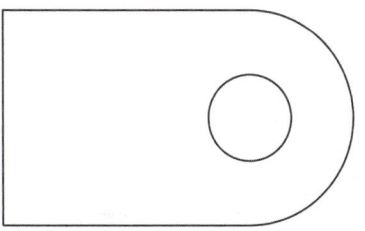

2. Use *Copy* with the multiple option to copy the front face back on the receding axis (Fig. 25-27). *Polar Snap* and *Polar Tracking* can be used to specify the "second point of displacement" as shown in Figure 25-27. Notice that a distance of .25 along the receding axis is used (1/2 of the actual depth) for this cabinet oblique.

FIGURE 25-27

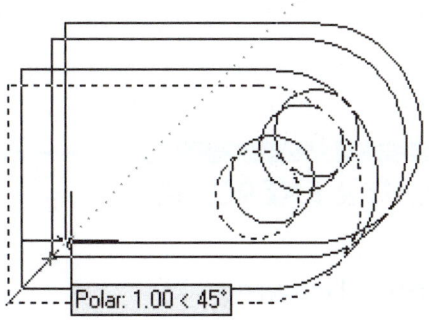

3. Draw the *Line* representing the edge on the upper-left of the object along the receding axis. Use *Endpoint OSNAP* to connect the *Lines*. Make a *Copy* of the *Line* or draw another *Line* .5 units to the right. Drop a vertical *Line* from the *Intersection* as shown (Fig. 25-28).

FIGURE 25-28

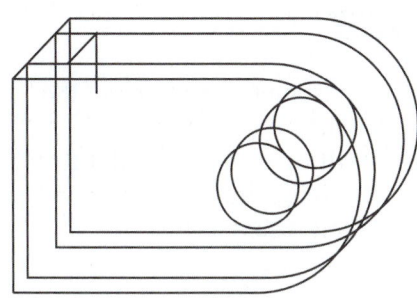

4. Use *Trim* and *Erase* to remove the unwanted parts of the *Lines* and *Circles* (those edges that are normally obscured) (Fig. 25-29).

FIGURE 25-29

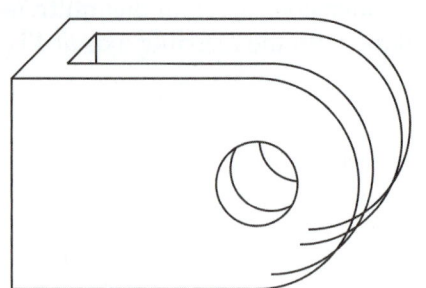

5. *Zoom* with a *window* to the lower-right corner of the drawing. Draw a *Line Tangent* to the edges of the arcs to define the limiting elements along the receding axis. *Trim* the unwanted segments of the arcs (Fig. 25-30).

FIGURE 25-30

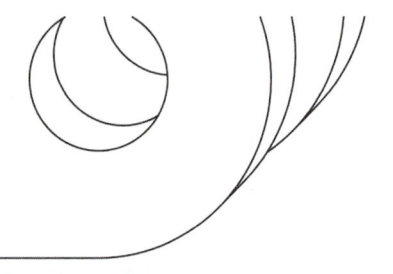

The resulting cabinet oblique drawing should appear like that in Figure 25-31.

FIGURE 25-31

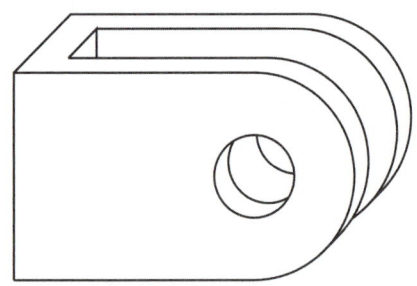

CHAPTER EXERCISES

Isometric Drawing

For exercises 1, 2, and 3 create isometric drawings as instructed. To begin, use an appropriate template drawing and draw a *Pline* border and insert the **TBLOCKAT.**

1. Create an isometric drawing of the cylinder shown in Figure 25-32. *Save* the drawing as **CYLINDER** and *Plot* so the drawing is *Scaled to Fit* on an A size sheet.

FIGURE 25-32

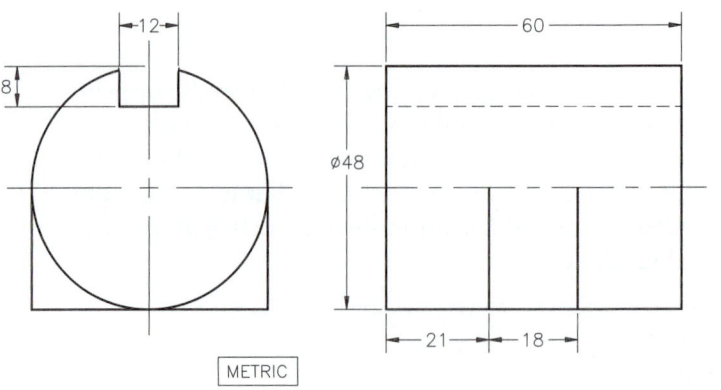

METRIC

2. Make an isometric drawing of the Corner Brace shown in Figure 25-33. *Save* the drawing as **CRNBRACE**. *Plot* at **1=1** scale on an A size sheet.

FIGURE 25-33

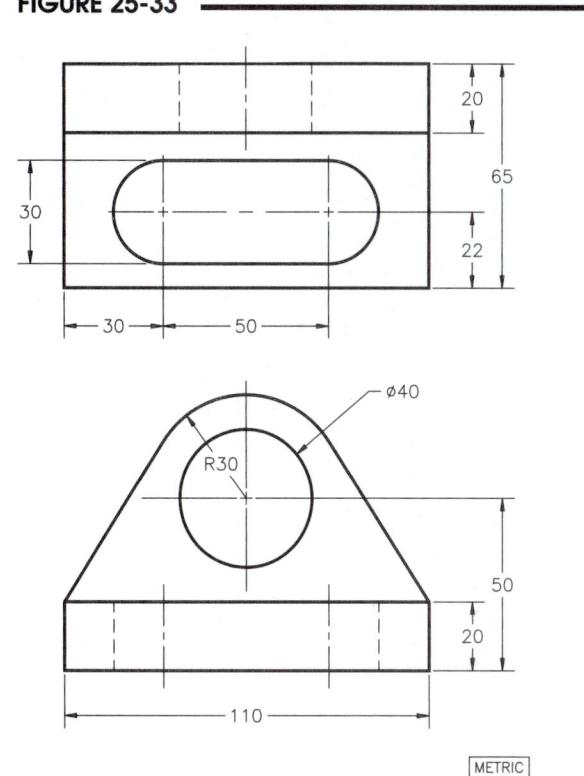

METRIC

3. Draw the Support Bracket (Fig. 25-34) in isometric. The drawing can be *Plotted* at **1=1** scale on an A size sheet. *Save* the drawing and assign the name **SBRACKET.**

FIGURE 25-34

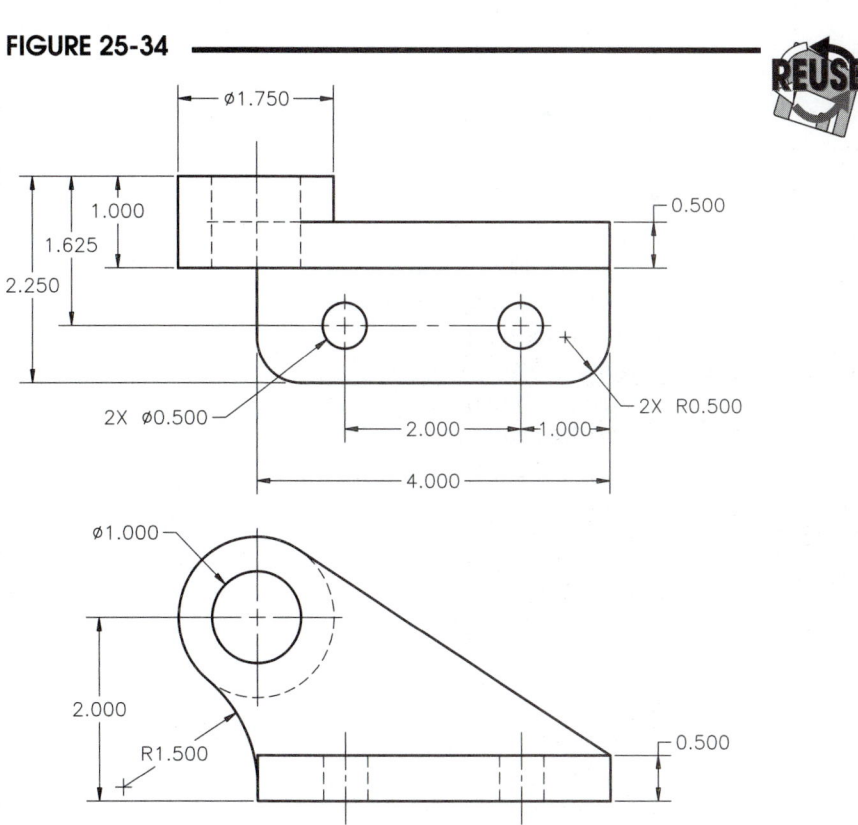

Oblique Drawing

For exercises 4 and 5, create oblique drawings as instructed. To begin, use an appropriate template drawing and draw a *Pline* border and *Insert* the **TBLOCKAT.**

4. Make an oblique cabinet projection of the Bearing shown in Figure 25-35. Construct all dimensions on the receding axis 1/2 of the actual length. Select the optimum angle for the receding axis to be able to view the 15 x 15 slot. *Plot* at **1=1** scale on an A size sheet. *Save* the drawing as **BEARING**.

FIGURE 25-35

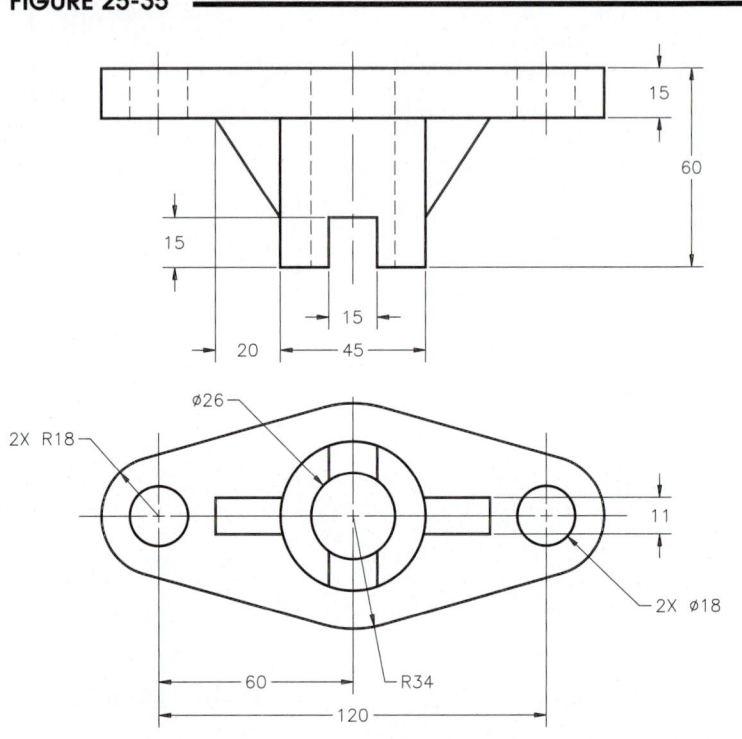

5. Construct a cavalier oblique drawing of the Pulley showing the circular view true size and shape. The illustration in Figure 25-36 gives only the side view. All vertical dimensions in the figure are diameters. *Save* the drawing as **PULLEY** and make a *Plot* on an A size sheet at **1=1**.

FIGURE 25-36

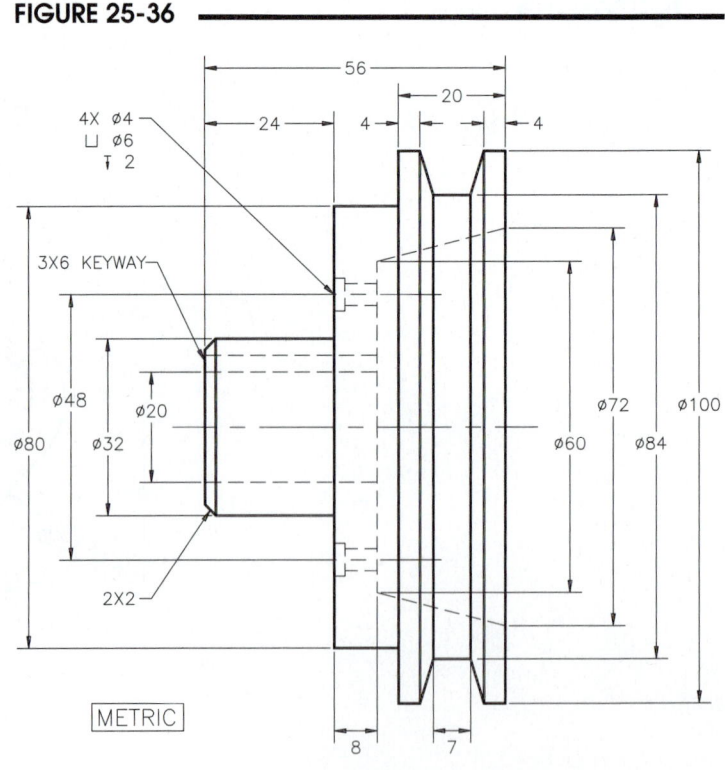

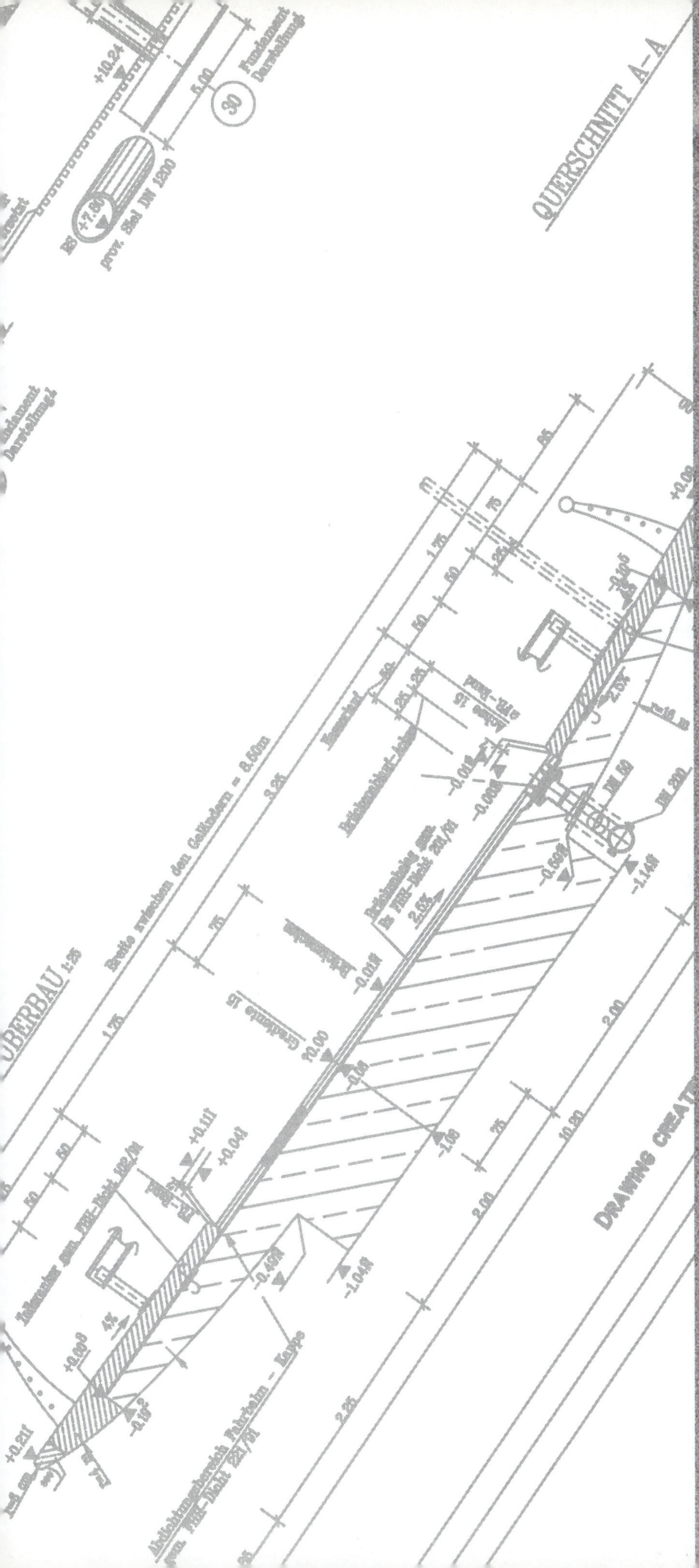

SECTION VIEWS

CHAPTER OBJECTIVES

After completing this chapter you should:

1. be able to use the *Hatch* command to select associative hatch patterns;

2. be able to specify a *Scale* and *Angle* for hatch lines;

3. know how to define a boundary for hatching using the *Pick Points* and *Select Objects* methods;

4. be able to use *-Hatch* to create hatch lines and discard the boundary;

5. be able to drag and drop hatch patterns from a Tool Palette and be able to create your own Tool Palettes;

6. know how to use *Hatchedit* to modify parameters of existing hatch patterns in the drawing;

7. be able to edit hatched areas using *Trim*, Grips, selection options, and *Draworder*;

8. know how to draw cutting plane lines for section views.

CONCEPTS

A section view is a view of the interior of an object after it has been imaginarily cut open to reveal the object's inner details. A section view is typically one of two or more views of a multiview drawing describing the object. For example, a multiview drawing of a machine part may contain three views, one of which is a section view.

Hatch lines (also known as section lines) are drawn in the section view to indicate the solid material that has been cut through. Each combination of section lines is called a hatch <u>pattern</u>, and each pattern is used to represent a specific material. In full and half section views, hidden lines are omitted since the inside of the object is visible.

Hatch lines are used also for many other applications. For example, architectural elevation drawings often include wall sections and use hatch lines to indicate walls that have been "cut through" to reveal the structural components. In addition, floor plans often contain a solid hatch pattern to indicate wall sections, as if the walls were cut through to reveal the "floor."

For mechanical drawings, a cutting plane line is drawn in an adjacent view to the section view to indicate the plane that imaginarily cuts through the object. Arrows on each end of the cutting plane line indicate the line of sight for the section view. A thick dashed or phantom line should be used for a cutting plane line.

This chapter discusses the AutoCAD methods used to draw hatch lines for section views and related cutting plane lines. The *Hatch* command allows you to select an enclosed area and select the hatch pattern and the parameters for the appearance of the hatch pattern; then AutoCAD automatically draws the hatch (section) lines. Existing hatch lines in the drawing can be modified using *Hatchedit*.

DEFINING HATCH PATTERNS AND HATCH BOUNDARIES

A hatch pattern is composed of many lines that have a particular linetype, spacing, and angle. Many standard hatch patterns are provided by AutoCAD for your selection. Rather than having to draw each section line individually, you are required only to specify the area to be hatched and AutoCAD fills the designated area with the selected hatch pattern. An AutoCAD hatch pattern is inserted as <u>one object</u>. For example, you can *Erase* the inserted hatch pattern by selecting only one line in the pattern, and the entire pattern in the area is *Erased*.

In a typical section view (Fig. 26-1), the hatch pattern completely fills the area representing the material that has been cut through. With the *Hatch* command you can define the boundary of an area to be hatched simply by pointing inside of an enclosed area.

FIGURE 26-1

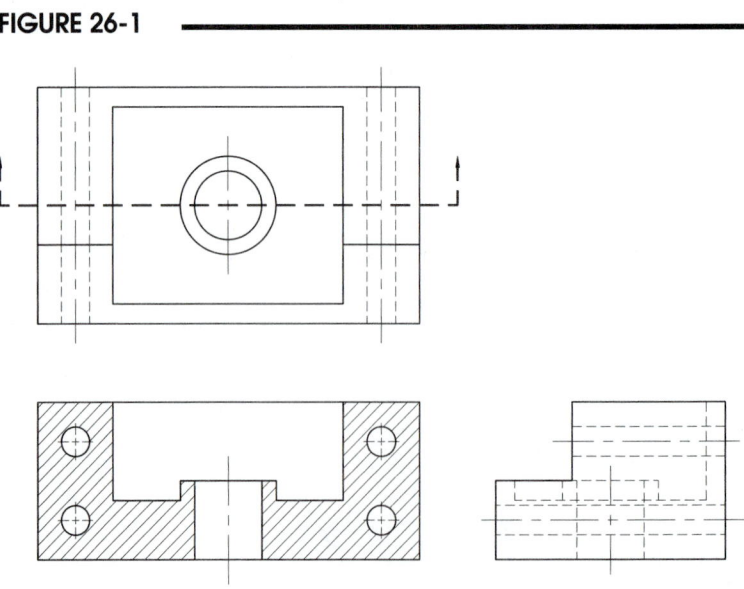

The *Hatch* command is used to create hatch lines in the area that you specify. *Hatch* operates in a dialog box mode. You can select the desired hatch pattern, then pick inside the area you want to fill, and *Hatch* automatically finds the boundary and fills it with the pattern. You can also drag and drop hatch patterns into a closed area using the Tool Palettes.

Hatch patterns created with *Hatch* are <u>associative</u>. Associative hatch patterns are associated to the boundary geometry such that when the shape of the boundary changes (by *Stretch, Scale, Rotate, Move, Properties, Grips,* etc.), the hatch pattern automatically reforms itself to conform to the new shape (Fig. 26-2). For example, if a design change required a larger diameter for a hole, *Properties* could be used to change the diameter of the hole, and the surrounding section lines would automatically adapt to the new diameter.

FIGURE 26-2

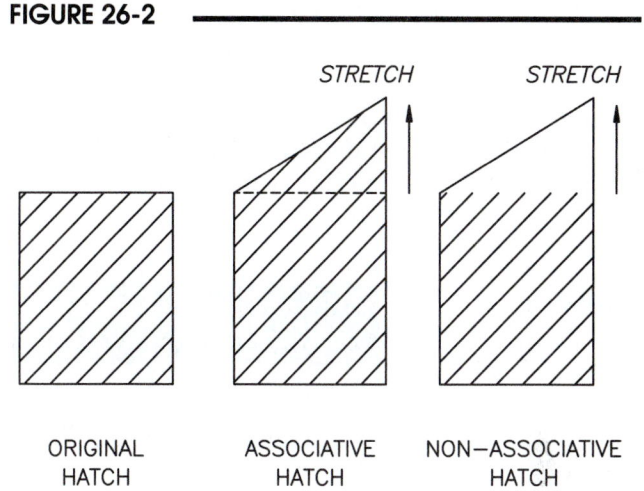

Once the hatch patterns have been drawn, any feature of the existing hatch pattern (created with *Hatch*) can be changed retroactively using *Hatchedit*. The *Hatchedit* dialog box gives access to the same options that were used to create the hatch (in the *Hatch and Gradient* dialog box). Changing the scale, angle, or pattern of any existing section view in the drawing is a simple process.

Steps for Creating a Section View Using the *Hatch* Command

1. Create the view that contains the area to be hatched using typical draw commands such as *Line, Arc, Circle,* or *Pline.* If you intend to have text or dimensions inside the area to be hatched, add them before hatching.

2. Invoke the *Hatch* command. The *Hatch and Gradient* dialog box appears (see Fig. 26-3).

3. Specify the *Type* to use. Select the desired pattern from the *Pattern* drop-down list or select the *Swatch* tile to allow you to select from the *Hatch Pattern Palette* image tiles.

4. Specify the *Scale* and *Angle* in the dialog box.

5. Define the area to be hatched by PICKing an internal point (*Pick Points* button) or by individually selecting the objects (*Select objects* button).

6. If needed, specify any other parameters, such as *Hatch origin, Gap tolerance, Island detection style,* etc.

7. *Preview* the hatch to make sure everything is as expected. Adjust hatching parameters as necessary and *Preview* again.

8. Apply the hatch by selecting *OK*. The hatch pattern is automatically drawn and becomes an associated object in the drawing.

9. If other areas are to be hatched, additional internal points or objects can be selected to define the new area for hatching. The parameters used previously appear again in the *Hatch and Gradient* dialog box by default. You can also *Inherit Properties* from a previously applied hatch.

10. For mechanical drawings, draw a cutting plane line in a view adjacent to the section view. A *Lineweight* is assigned or a *Pline* with a *Dashed* or *Phantom* linetype is used. Arrows at the ends of the cutting plane line indicate the line of sight for the section view.

11. If any aspect of the hatch lines needs to be edited at a later time, *Hatchedit* can be used to change those properties. If the hatch boundary is changed by *Stretch, Rotate, Scale, Move, Properties*, etc., the hatched area will conform to the new boundary.

Hatch

Pull-down Menu	Command (Type)	Alias (Type)	Short-cut	Screen (side) Menu	Tablet Menu
Draw *Hatch...*	*Hatch* or *-Hatch*	*BH* or *H*	...	*DRAW 2* *Hatch*	*P,9*

Hatch allows you to create hatch lines for a section view (or for other purposes) by simply PICKing inside a closed boundary. A closed boundary refers to an area completely enclosed by objects. *Hatch* locates the closed boundary automatically by creating a <u>temporary *Pline*</u> that follows the outline of the hatch area, fills the area with hatch lines, and then deletes the boundary (default option) after hatching is completed. *Hatch* ignores all objects or parts of objects that are not part of the boundary.

Any method of invoking *Hatch* yields the *Hatch and Gradient* dialog box (Fig. 26-3). Typically, the first step in this dialog box is the selection of a hatch pattern.

FIGURE 26-3

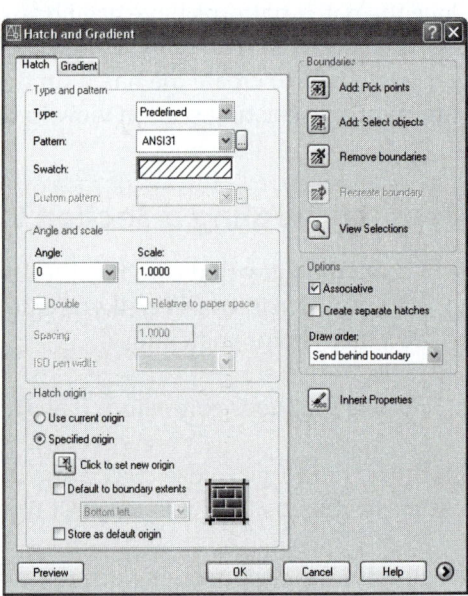

After you select the *Pattern* and other parameters for the hatch in the *Hatch and Gradient* dialog box, you select *Pick Points* or *Select Objects* to return to the drawing and indicate the area to fill with the pattern. While in the drawing, you can right-click to produce a shortcut menu, providing access to other options without having first to return to the dialog box (Fig. 26-4).

FIGURE 26-4

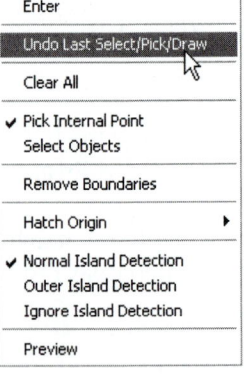

Hatch and Gradient Dialog Box—Hatch Tab

Type

This option allows you to specify the type of the hatch pattern: *Predefined*, *User-defined*, or *Custom*. Use *Predefined* for standard hatch pattern styles that AutoCAD provides (in the ACAD.PAT file).

Predefined

There are two ways to select from *Predefined* patterns: (1) select the *Swatch* tile to produce the *Hatch Pattern Palette* dialog box displaying hatch pattern names and image tiles (Fig. 26-5), or (2) select the *Pattern:* drop-down list to PICK the pattern name.

The *Hatch Pattern Palette* dialog box allows you to select a predefined pattern by its image tile or by its name. There are four tabs of image tiles: *ANSI, ISO, Other Predefined,* and *Custom*.

FIGURE 26-5

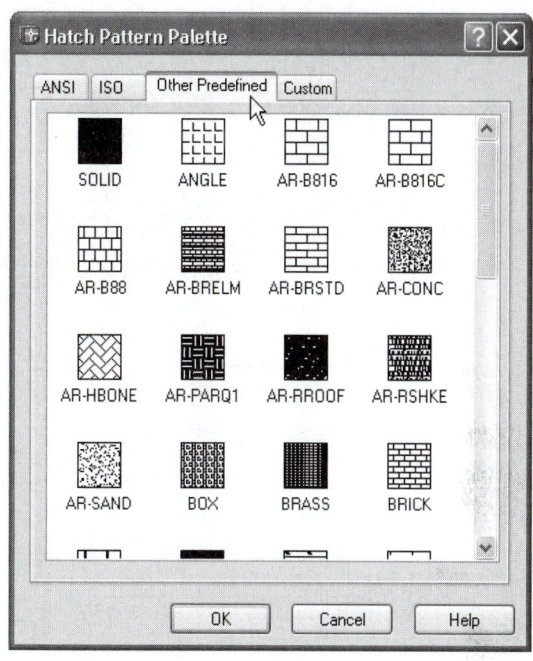

User-defined

To define a simple hatch pattern "on the fly," select the *User-defined* tile. This causes the *Pattern* and *Scale* options to be disabled and the *Angle, Spacing,* and *Double* options to be enabled. Creating a *User-defined* pattern is easy. Specify the *Angle* of the lines, the *Spacing* between lines, and optionally create *Double* (perpendicular) lines. All *User-defined* patterns have continuous lines.

Custom

Custom patterns are previously created user-defined patterns stored in other than the ACAD.PAT file. Custom patterns can contain continuous, dashed, and dotted line combinations. See the AutoCAD Customization Guide for information on creating and saving custom hatch patterns.

Pattern

Selecting the *Pattern* drop-down list (see Fig. 26-3) displays the name of each predefined pattern. Making a selection dictates the current pattern and causes the pattern to display in the *Swatch* window. The small button with ellipsis (…) just to the right of the *Pattern*: name drop-down list produces the *Hatch Pattern Palette*.

Swatch

Click in the *Swatch* tile to produce the *Hatch Pattern Palette* dialog box (see Fig. 26-5). You can select from *ANSI, ISO, Other Predefined,* and *Custom* (if available) hatch patterns.

Figure 26-6 displays each of the AutoCAD hatch patterns defined in the ACAD.PAT file. Note that the patterns are not shown to scale.

FIGURE 26-6

Hatch patterns are created using the current linetype. Therefore, the _Continuous_ linetype should be set current when hatching to ensure the selected area is filled with the pattern as it appears in the image tile. After selecting a pattern, specify the desired _Scale_ and _Angle_ of the pattern or ISO pen width.

Angle

The _Angle_ specification determines the angle (slant) of the hatch pattern. The default angle of 0 represents whatever angle is displayed in the pattern's image tile. Any value entered deflects the existing pattern (as it appears in the swatch) by the specified value (in degrees). The value entered in this box is held in the _HPANG_ system variable.

Scale

The value entered in this edit box is a scale factor that is applied to the existing selected pattern. Normally, this scale factor should be changed proportionally with changes in the drawing _Limits_. Like many other scale factors (_LTSCALE, DIMSCALE_), AutoCAD defaults are set to a value of 1, which is appropriate for the default _Limits_ of 12 x 9. If you have calculated the drawing scale factor, enter that value in the _Scale_ edit box (see Chapter 12 for information on the "Drawing Scale Factor"). The _Scale_ value is stored in the _HPSCALE_ system variable.

ISO hatch patterns are intended for use with metric drawings; therefore, the scale (spacing between hatch lines) is much greater than for inch drawings. Because _Limits_ values for metric sheet sizes are greater than for inch-based drawings (25.4 times greater than comparable inch drawings), the ISO hatch pattern scales are automatically compensated. If you want to use an ISO pattern with inch-based drawings, calculate a hatch pattern scale based on the drawing scale factor and multiply by .039 (1/25.4).

Double

Only for a _User-defined_ pattern, check this box to have a second set of lines drawn at 90 degrees to the original set.

Relative to Paper Space

This option changes the hatch pattern scale relative to paper space units. Checking this box automatically changes the _HPSCALE_ variable based on the proportion of the paper space units to model space units (viewport scale or _Zoom XP_ factor). Using this option, you can easily display hatch patterns at a scale that is appropriate for your layout. You can use this option only when you have a layout with a viewport and invoke _Hatch_ from the layout.

Spacing

This option is enabled if _User-defined_ pattern is specified. Enter a value for the distance between lines.

ISO Pen Width

You must select an ISO hatch pattern for this tile to be enabled. Selecting an _ISO Pen Width_ from the drop-down list automatically sets the scale and enters the value in the _Scale_ edit box. See "_Scale_."

Use Current Origin

By default, all hatch pattern origins correspond to the current coordinate system's 0,0,0 location. (0,0,0 is the default _HPORIGIN_ system variable setting.) This option is acceptable for most hatch patterns.

Specified Origin

Some hatch patterns, such as brick patterns, need to be aligned with a point on the hatch boundary. Figure 26-7 illustrates a brick pattern using the default (*Current origin*) and a *Bottom left* origin. Click this option to make the following options available so you control the starting location of hatch pattern generation.

Click to Set New Origin

Allows you to select a point in the drawing to specify the new hatch origin point.

Default to Boundary Extents

This option automatically calculates a new origin based on the rectangular extents of the hatch boundary. You can specify one of the four corners of the extents or the boundary's center.

Store as Default Origin

If you use a *Specified Origin*, check this box to store the value of the new hatch origin in the *HPORIGIN* system variable for future hatches in the drawing.

FIGURE 26-7

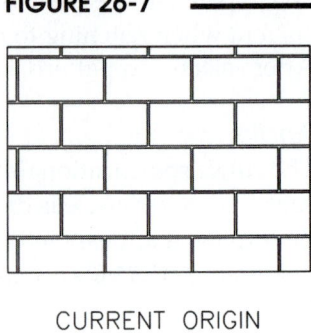

CURRENT ORIGIN

BOTTOM LEFT

Selecting the Hatch Area

Once the hatch pattern and options have been selected, you must indicate to AutoCAD what area(s) should be hatched. Either the *Add: Pick Points* method, *Add: Select Objects* method, or a combination of both can be used to accomplish this.

Add: Pick Points

This tile should be selected if you want AutoCAD to automatically determine the boundaries for hatching. You only need to select a point <u>inside</u> the area you want to hatch. The point selected must be inside a <u>closed shape</u>. When the *Add: Pick Points* tile is selected, AutoCAD gives the following prompts:

```
Pick internal point or [Select objects/remove Boundaries]: PICK
Selecting everything...
Selecting everything visible...
Analyzing the selected data...
Analyzing the internal islands...
Pick internal point or [Select objects/remove Boundaries]:
```

When an internal point is PICKed, AutoCAD traces and highlights the boundary (Fig. 26-8). The interior area is then analyzed for islands to be included in the hatch boundary. Multiple boundaries can be designated by selecting multiple internal points.

FIGURE 26-8

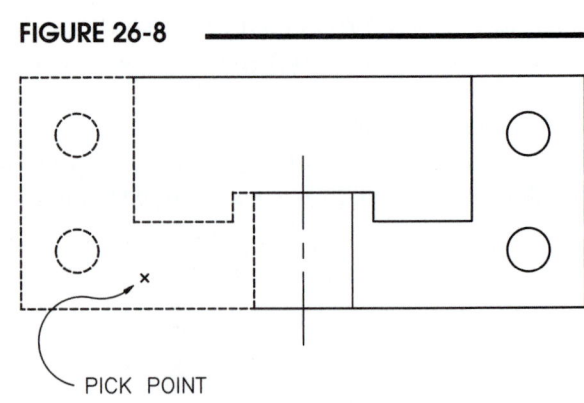

PICK POINT

The location of the point selected is usually not critical. However, the point must be PICKed inside the expected boundary. If there are any large gaps in the area, a complete boundary cannot be formed and a boundary error message appears (see *"Gap Tolerance"*).

Add: Select Objects

Alternately, you can designate the boundary with the *Select Objects* method. Using the *Select Objects* method, <u>you</u> specify the boundary objects rather than let AutoCAD locate a boundary. With the *Select Objects* method, no temporary *Pline* boundary is created as with the *Pick Points* method. Therefore, the selected objects must form a closed shape with no large gaps or overlaps. If gaps exist or if the objects extend past the desired hatch area (Fig. 26-9), AutoCAD cannot interpret the intended hatch area correctly, and problems will occur.

FIGURE 26-9

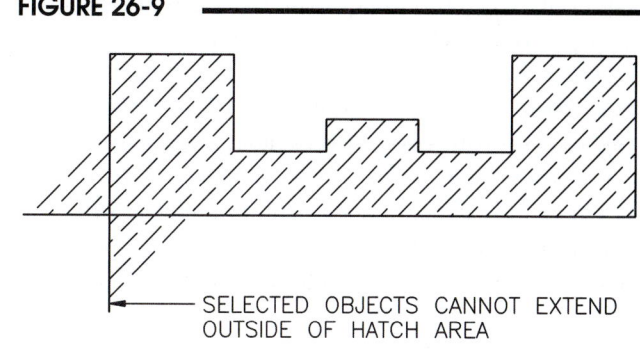

SELECTED OBJECTS CANNOT EXTEND OUTSIDE OF HATCH AREA

The *Select Objects* option can be used after the *Pick Points* method to select specific objects for *Hatch* to consider before drawing the hatch pattern lines. For example, if you had created text objects or dimensions within the boundary found by the *Pick Points* method, you may then have to use the *Select Objects* option to select the text and dimensions (if *Island Detection* was off). Using this procedure, the hatch lines are automatically "trimmed" around the text and dimensions.

Remove Boundaries

PICKing this tile allows you to remove previously selected boundaries or to select specific islands (internal objects) to remove from those AutoCAD has found within the outer boundary. If hatch lines have been drawn, be careful to *Zoom* in close enough to select the desired boundary and not the hatch pattern.

Recreate Boundary

This option is always disabled in the *Hatch and Gradient* dialog box, but is available for editing hatches. See *"Hatchedit."*

View Selections

Clicking the *View Selections* tile causes AutoCAD to highlight all selected boundaries. This can be used as a check to ensure the desired areas are selected.

Associative

This checkbox toggles on or off the associative property for hatch patterns. Associative hatch patterns automatically update by conforming to the new boundary shape if the boundary is changed (see Fig. 26-2). A non-associative hatch pattern is static even when the boundary changes.

Create Separate Hatches

By default, multiple hatches created with one use of the *Hatch* command are treated as one AutoCAD object, even when the hatches are within separate boundaries. In such a case, multiple hatch areas could be *Erased* by selecting only one of the hatched areas. Check *Create Separate Hatches* to ensure each boundary area forms a unique hatch object.

Draw Order

Although the draw order does not affect most hatching applications, this option is helpful when you assign solid or gradient hatch patterns and when hatch patterns (containing solid or gradient patterns) overlap each other or other objects (see *"Gradient* Tab"). For example, a gradient pattern could be created behind or in front of the boundary (Fig. 26-10). Your choice of *Do not assign, Send to back, Bring to front, Send behind boundary,* or *Bring in front of boundary* affects the current hatch pattern. You can also use the *Draworder* command to control the display order after creating the hatches (see *"Draworder"*).

FIGURE 26-10

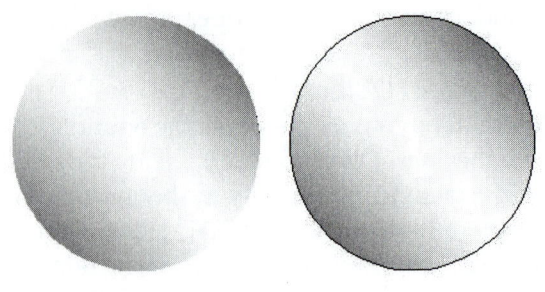

IN FRONT OF BOUNDARY   BEHIND BOUNDARY

Inherit Properties

This option allows you to select a hatch pattern from one existing in the drawing. This option operates similarly to *Matchprop* because you PICK the existing hatch pattern from the drawing and copy it to another area by selecting an internal point.

    Select hatch object: **PICK**
    Inherited Properties: Name <ANSI31>, Scale <1.0000>, Angle <0>
    Pick internal point or [Select objects/remove Boundaries]:

Preview

You should always use the *Preview* option after specifying the hatch parameters and selecting boundaries, but before you use *OK*. This option allows you to temporarily look at the hatch pattern in your drawing with the current settings applied and allows you to adjust the settings, if necessary, before using *OK*. After viewing the drawing, press the Esc key to redisplay the *Hatch and Gradient* dialog box, allowing you to make adjustments.

Expanded Dialog Box Options

Using the arrow button in the lower-right corner of the *Hatch and Gradient* dialog box you can expand the dialog box to reveal additional options as described below (Fig. 26-11).

FIGURE 26-11

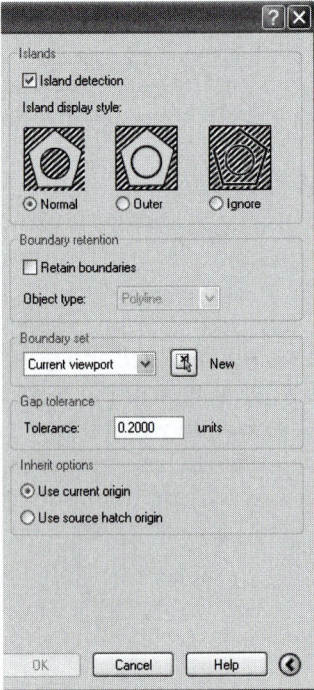

Island Detection

Check this box to cause AutoCAD to automatically detect any internal areas (islands) inside the selected boundary, such as holes (*Circles*), text, or other closed shapes. Island detection is performed according to the following options.

Normal

This should be used for most applications of *Hatch*. Text or closed shapes within the outer border are considered in such cases. Hatching will begin at the outer boundary and move inward, alternating between applying and not applying the pattern as interior shapes or text are encountered (Fig. 26-12).

Outer

This option causes AutoCAD to hatch only the outer closed shape. Hatching is turned off for all interior closed shapes (Fig. 26-12).

FIGURE 26-12

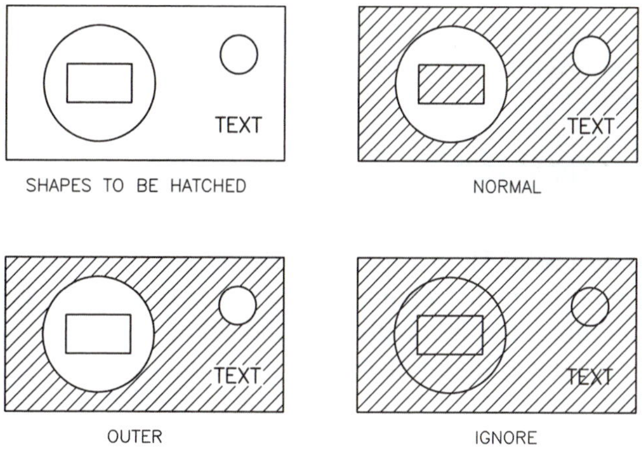

SHAPES TO BE HATCHED

NORMAL

OUTER

IGNORE

Ignore

Ignore draws the hatch pattern from the outer boundary inward, ignoring any interior shapes. The resulting hatch pattern is drawn through the interior shapes (Fig. 26-12).

Retain Boundaries

When AutoCAD uses the *Pick Points* method to locate a boundary for hatching, a temporary *Pline* or *Region* is created for hatching, then discarded after the hatching process. Checking this box forces AutoCAD to keep the boundary. When the box is checked, *Hatch* creates two objects—the hatch pattern and the boundary object. (You can specify whether you want to create a *Pline* or a *Region* boundary in the *Object Type* drop-down list.) Using this option and erasing the hatch pattern accomplishes the same results as using the *Boundary* command.

Object Type

Hatch creates *Polyline* or *Region* boundaries. This option is enabled only when the *Retain Boundaries* box is checked.

Boundary Set

By default, AutoCAD examines all objects in the viewport when determining the boundaries by the *Pick Points* method. (*Current Viewport* is selected by default when you begin the *Hatch* command.) For complex drawings, examining all objects can take some time. In that case, you may want to specify a smaller boundary set for AutoCAD to consider. Clicking the *New* tile clears the dialog boxes and permits you to select objects or select a window to define the new set.

Gap Tolerance

Tolerance is set to 0 by default. In this case, all boundaries selected to hatch with the *Hatch* command must be completely closed—that is, no hatch boundaries can contain a "gap." However, if gaps exist in the selected boundary, you can set a *Tolerance* value to compensate for any boundary gaps (Fig. 26-13). You can set the *Gap Tolerance* in the *Hatch and Gradient* dialog box or set the *HPGAPTOL* system variable. The *Gap Tolerance* must be greater than the size of the gap in drawing units.

FIGURE 26-13

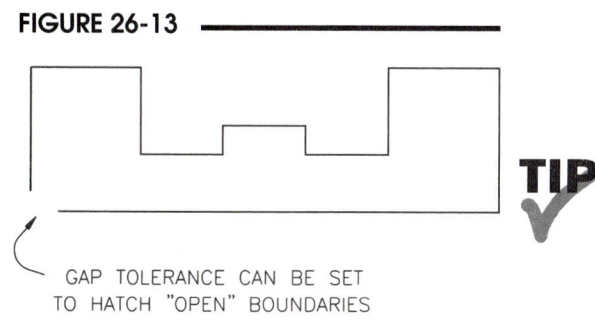

GAP TOLERANCE CAN BE SET
TO HATCH "OPEN" BOUNDARIES

TIP

If you attempt to hatch an open area when no *Gap Tolerance* is set or a gap exists that is greater than the setting for the *Gap Tolerance, a Boundary Definition Error* warning box appears. However, if a gap exists that is within the *Gap Tolerance*, the object will hatch correctly after answering *Yes* to the *Open Boundary Warning* box that appears.

Cutting plane lines should be drawn or plotted in a heavy lineweight. Two possible methods can be used to accomplish this: use a *Pline* with *Width* or assign a *Lineweight* to the line or layer. If you prefer drawing the cutting plane line using a *Pline*, use a *Width* of .02 or .03 times the drawing scale factor. Assigning a *Lineweight* to the line or layer is a simpler method. A *Lineweight* of .8mm or .031″ is appropriate for cutting plane lines.

For the example section view, a cutting plane line is created in the top view to indicate the plane on which the object is cut and the line of sight for the section view (Fig. 26-34). (The cutting plane could be drawn in the side view, but the top would be much clearer.) First, a *New* layer named CUT is created and the *Dashed linetype* is assigned to the layer. The layer is *Set* as the *Current* layer. Next, construct a *Pline* with the desired *Width* to represent the cutting plane line.

FIGURE 26-34

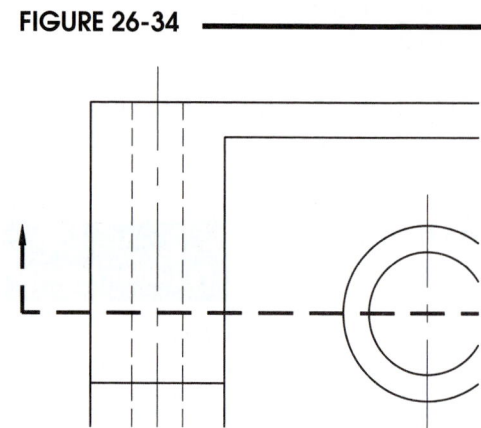

The resulting cutting plane line appears as shown in Figure 26-35 but without the arrow heads. The horizontal center line for the hole (top view) was *Erased* before creating the cutting plane line.

FIGURE 26-35

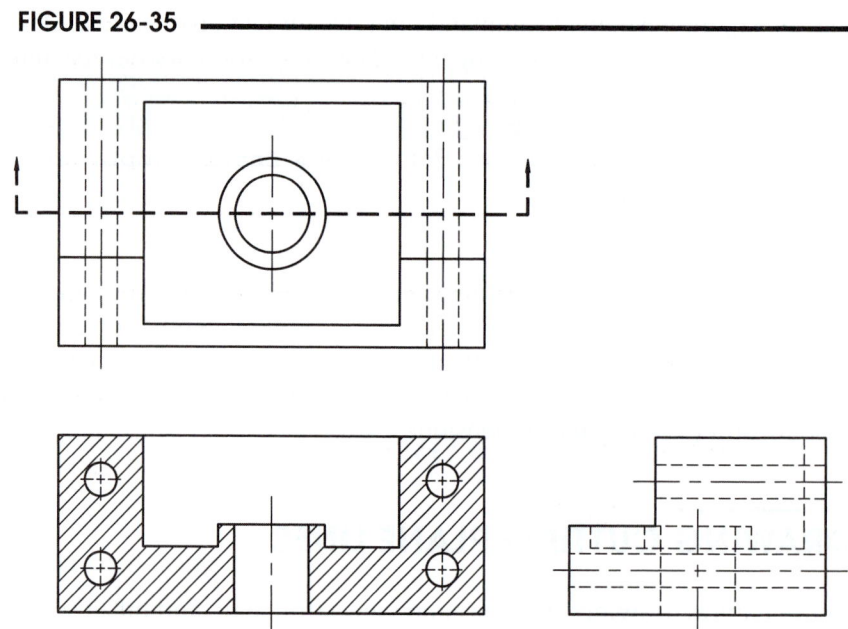

The last step is to add arrowheads to the ends of the cutting plane line. Arrowheads can be drawn by three methods: (1) use the *Solid* command to create a three-sided solid area; (2) use a tapered *Pline* with beginning width of 0 and ending width of .08, for example; or (3) use the *Leader* command to create the arrowhead (see Chapter 28, Dimensioning, for details on *Leader*). The arrowhead you create can be *Scaled*, *Rotated*, and *Copied* if needed to achieve the desired size, orientation, and locations.

CHAPTER EXERCISES

For the following exercises, create the section views as instructed. Use an appropriate template drawing for each unless instructed otherwise. Include a border and title block in the layout for each drawing.

1. *Open* the **SADDLE** drawing that you created in Chapter 24. Convert the front view to a full section. *Save* the drawing as **SADL-SEC.** *Plot* the drawing at **1=1** scale.

2. Make a multiview drawing of the Bearing shown in Figure 26-36. Convert the front view to a full section view. Add the necessary cutting plane line in the top view. *Save* the drawing as **BEAR-SEC.** Make a *Plot* at full size.

FIGURE 26-36

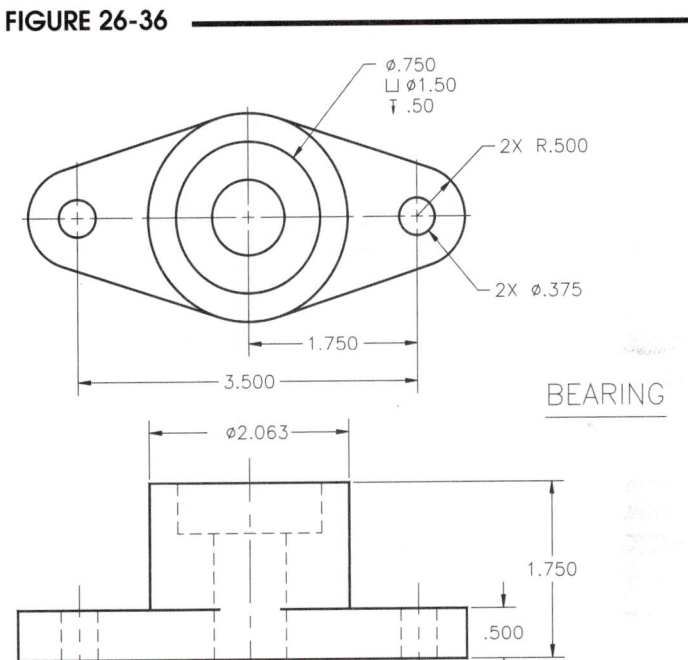

3. Create a multiview drawing of the Clip shown in Figure 26-37. Include a side view as a full section view. You can use the **CLIP** drawing you created in Chapter 24 and convert the side view to a section view. Add the necessary cutting plane line in the front view. *Plot* the finished drawing at **1=1** scale and *SaveAs* **CLIP-SEC**.

FIGURE 26-37

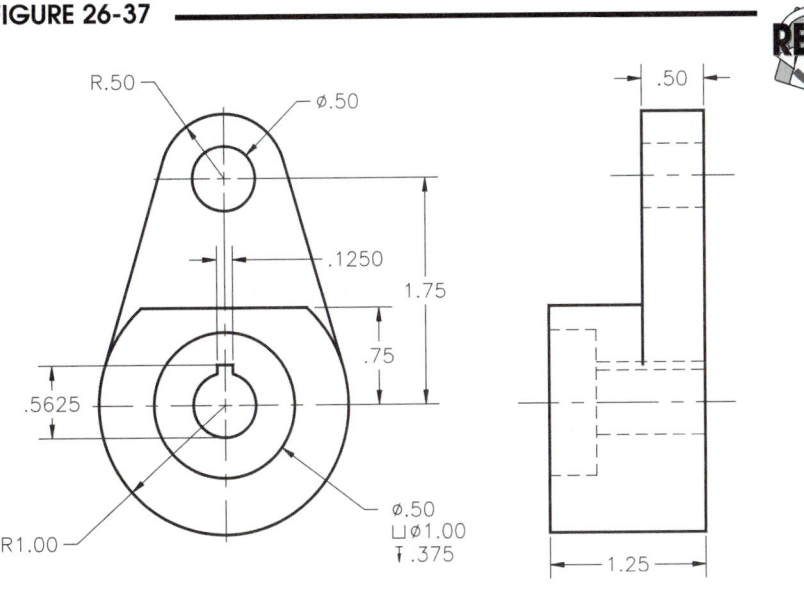

4. Create a multiview drawing, including two full sections of the Stop Block as shown in Figure 26-38. Section B–B' should replace the side view shown in the figure. *Save* the drawing as **SPBK-SEC**. *Plot* the drawing at **1=2** scale.

FIGURE 26-38 ━━━━━━━━

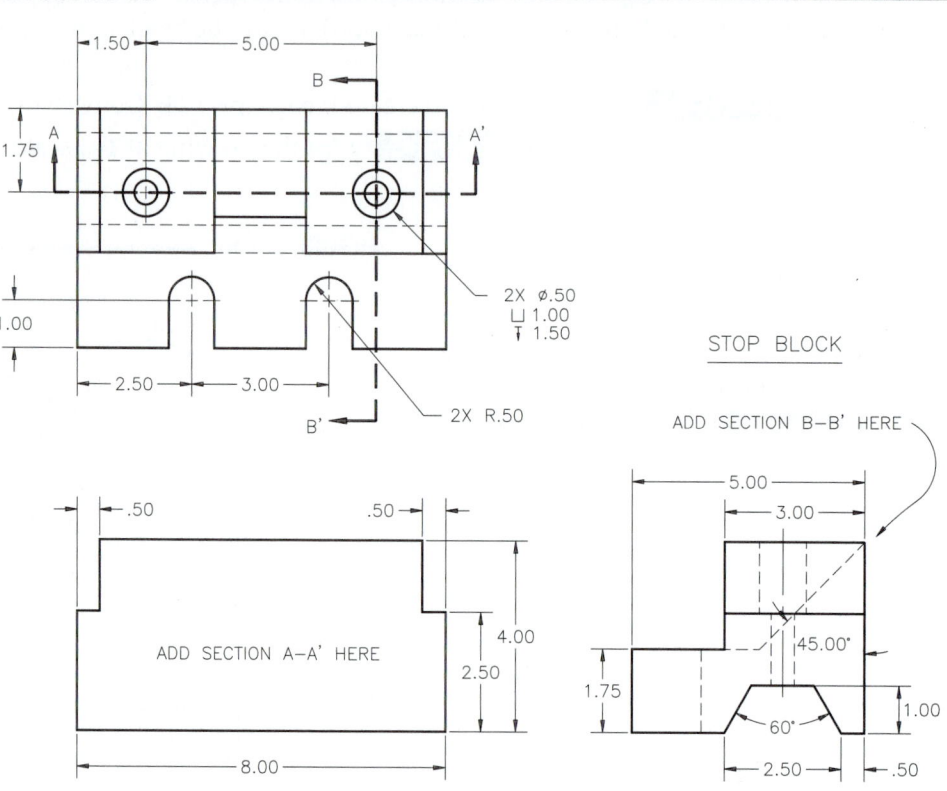

STOP BLOCK

ADD SECTION A–A' HERE

ADD SECTION B–B' HERE

5. Make a multiview drawing, including a half section of the Pulley (Fig. 26-39). All vertical dimensions are diameters. Two views (including the half section) are sufficient to describe the part. Add the necessary cutting plane line. *Save* the drawing as **PUL-SEC** and make a *Plot* at **1:1** scale.

FIGURE 26-39 ━━━━━━━━

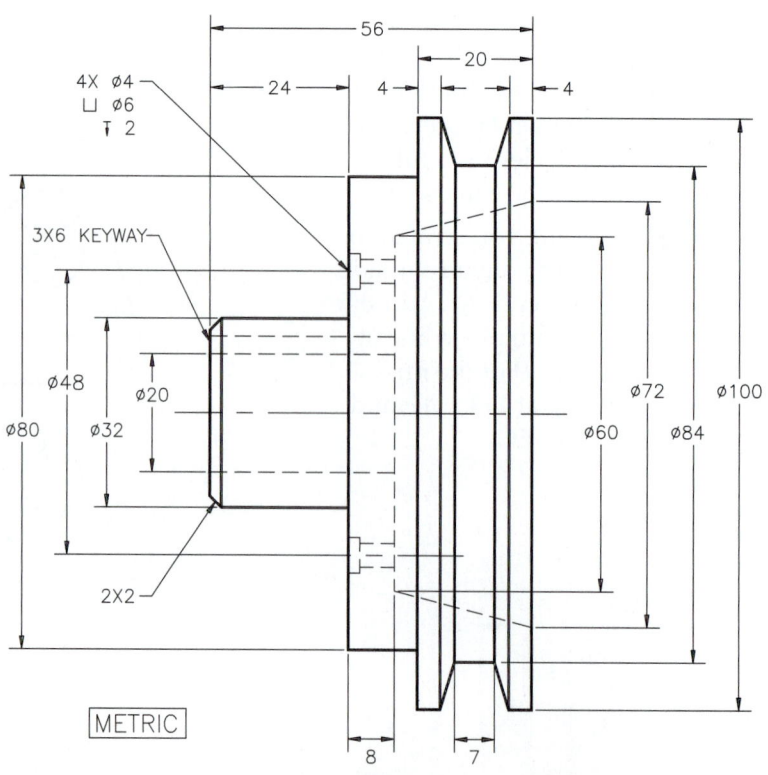

METRIC

6. Draw the Grade Beam foundation detail in Figure 26-40. Do not include the dimensions in your drawing. Use the *Hatch* command to hatch the concrete slab with **AR-CONC** hatch pattern. Use *Sketch* to draw the grade line and *Hatch* with the **EARTH** hatch pattern. *Save* the drawing as **GRADBEAM**.

FIGURE 26-40

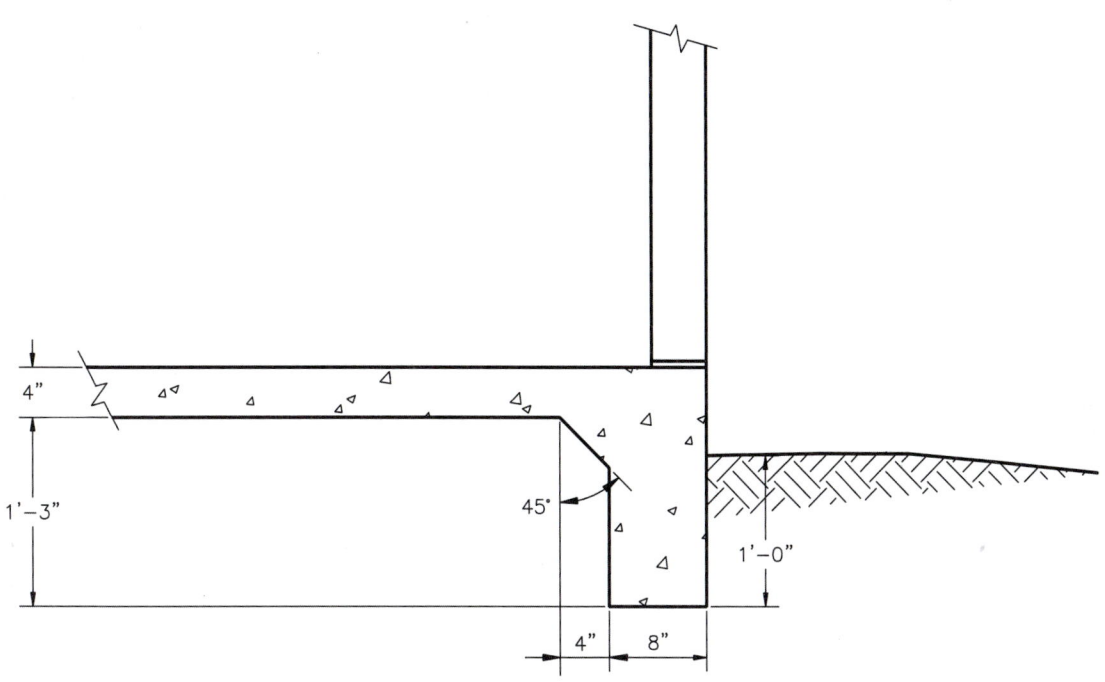

7. **Tool Palettes, Solid Fill**

A. *Open* the **OFFICE-REV** drawing that you last worked on in Chapter 21. Next, open the **Tool Palettes**.

B. Make a new *Layer* and name it **CARPET**. Accept the default color, linetype, etc. Make **CARPET** the *Current* layer.

C. Draw a *Line* across the doorway threshold of one of the two offices. Next, drag-and-drop a *Solid* pattern from the standard *Hatches* palette into the office. If problems occur, check to ensure that there are no gaps in the office walls.

D. When the solid pattern appears in the office, notice that the pattern may cover the outlines of the walls and furniture items. Use the *Draworder* command to move the new hatch pattern to the *Back*. Your drawing should look similar to that in Figure 26-41.

FIGURE 26-41

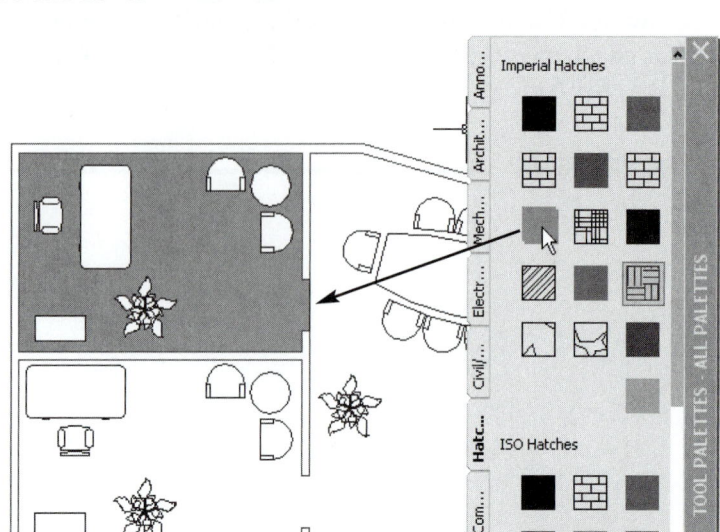

E. Using the same technique, fill the other two office spaces with carpet from the tool palette or use another wood floor pattern such as *AR-HBONE* from the *Hatch and Gradient* dialog box.

F. Use *SaveAs* to save and rename the drawing to **OFFICE-REV-2**.

8. **Solid Fill, *Fill* Command**

A. *Open* the **OFFICE-REV-2** drawing if not already open. Make a new *Layer* and assign the name **WALLS**. Accept the default color, linetype, etc. Make the **WALLS** layer *Current*.

B. Use the *Hatch and Gradient* dialog box to apply the *Solid* hatch pattern to the walls of the office.

C. Use the *Fill* command and turn *Off* the solid fill. Use the *Regen* command to apply the new *Fill* setting. Does *Fill* affect only the walls, or does it affect the other carpet solid fill patterns?

D. Turn *Fill On* and *Regen*. *Save* the drawing.

27

AUXILIARY VIEWS

CHAPTER OBJECTIVES

After completing this chapter you should:

1. be able to use the *Rotate* option of *Snap* to change the angle of the *SNAP, GRID*, and *ORTHO*;

2. be able to set the *Increment angle* and *Additional angles* for creating auxiliary views using *Polar Tracking*;

3. know how to use the *Offset* command to create parallel line copies;

4. be able to use *Xline* and *Ray* to create construction lines for auxiliary views.

CONCEPTS

AutoCAD provides no commands explicitly for the creation of auxiliary views in a mechanical drawing or for angular sections of architectural drawings. However, four particular features that have been discussed previously can assist you in the construction of these drawings. Those features are *SNAP* rotation, *Polar Tracking*, the *Offset* command, and the *Xline* and *Ray* commands.

In a mechanical drawing, an auxiliary view may be needed in addition to the typical views (top, front, side). An auxiliary view is one that is normal (the line-of-sight is perpendicular) to an inclined surface of the object. Therefore, the entire view is constructed at an angle by projecting in a 90-degree direction from the edge view of an inclined surface in order to show the true size and shape of the inclined surface. Architectural drawings often include a portion of a plan drawn at some angle, such as a ranch house with one wing at a 45-degree angle to the main body of the house. In either case and depending on the object, the view or rooms could be drawn at any angle, so lines are typically drawn parallel and perpendicular relative to that section of the drawing. Therefore, the *SNAP* rotation feature, *Polar Tracking*, the *Offset* command, and the *Xline* and *Ray* commands can provide assistance in this task.

FIGURE 27-1 ━━━━━━━━━━

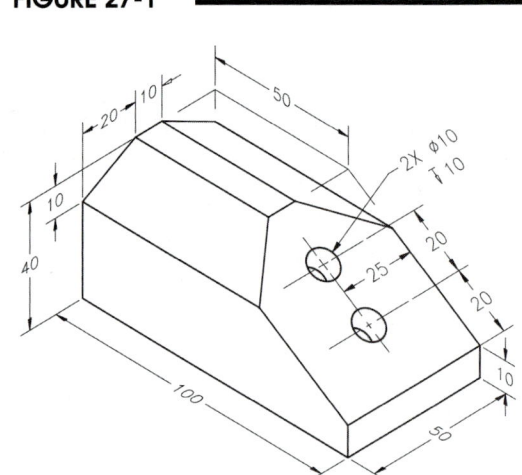

An example mechanical part used for the application of these AutoCAD features related to auxiliary view construction is shown in Figure 27-1. As you can see, there is an inclined surface that contains two drilled holes. To describe this object adequately, an auxiliary view should be created to show the true size and shape of the inclined surface. Although this chapter uses only this mechanical example, these same techniques would be used for architectural drawings or any other drawings that include views or entire sections of a drawing constructed at some angle.

CONSTRUCTING AN AUXILIARY VIEW

Setting Up the Principal Views

To begin this drawing, the typical steps are followed for drawing setup (Chapter 12). Because the dimensions are in millimeters, *Limits* should be set accordingly. For example, to provide enough space to draw the views full size and to plot full size on an A sheet, *Limits* of 279 x 216 are specified.

FIGURE 27-2 ━━━━━━━━━━

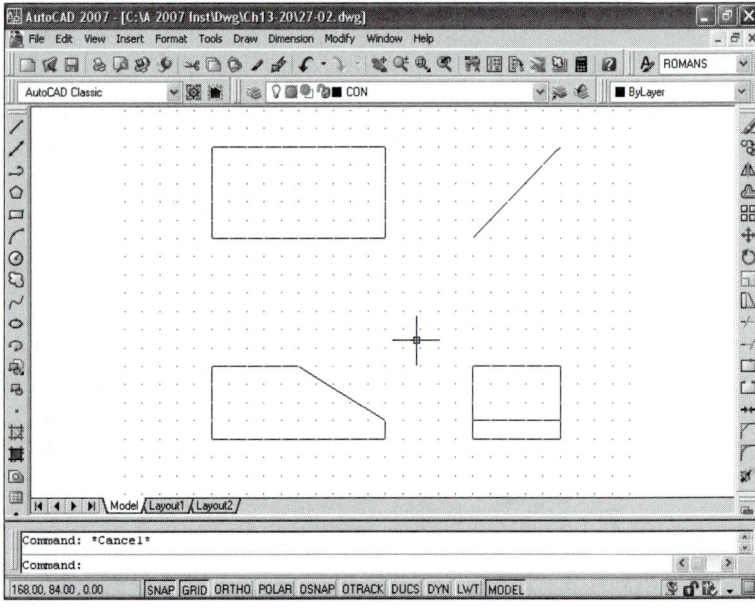

In preparation for the auxiliary view, the principal views are "blocked in," as shown in Figure 27-2. The purpose of this step is to ensure that the desired views fit and are optimally spaced within the allocated *Limits*. If there is too little or too much room, adjustments can be made to the *Limits*. Notice that space has been allotted between the views for a partial auxiliary view to be projected from the front view.

Before additional construction on the principal views is undertaken, initial steps in the construction of the partial auxiliary view should be performed. The projection of the auxiliary view requires drawing lines perpendicular and parallel to the inclined surface. One or more of the three alternatives (explained next) can be used.

Using *Snap Rotate* and *ORTHO*

One possibility to construct an auxiliary view is to use the *Snap* command with the *Rotate* option. This action permits you to rotate the *SNAP* to any angle about a specified base point. The *GRID* automatically follows the *SNAP*. Turning *ORTHO ON* forces *Lines* to be drawn orthogonally with respect to the rotated *SNAP* and *GRID*.

Figure 27-3 displays the *SNAP* and *GRID* after rotation. The command syntax is given below.

In this figure, the cursor size is changed from the default 5% of screen size to 100% of screen size to help illustrate the orientation of the *SNAP*, *GRID*, and cursor when *SNAP* is *Rotated*. You can change the cursor size in the *Display* tab of the *Options* dialog box.

NOTE: In AutoCAD 2007 the *Rotate* option of *Snap* does not appear as an option in the *Drafting Settings* dialog box or on the Command line. However, the *Rotate* option still functions by typing the letter *R* to activate it.

FIGURE 27-3

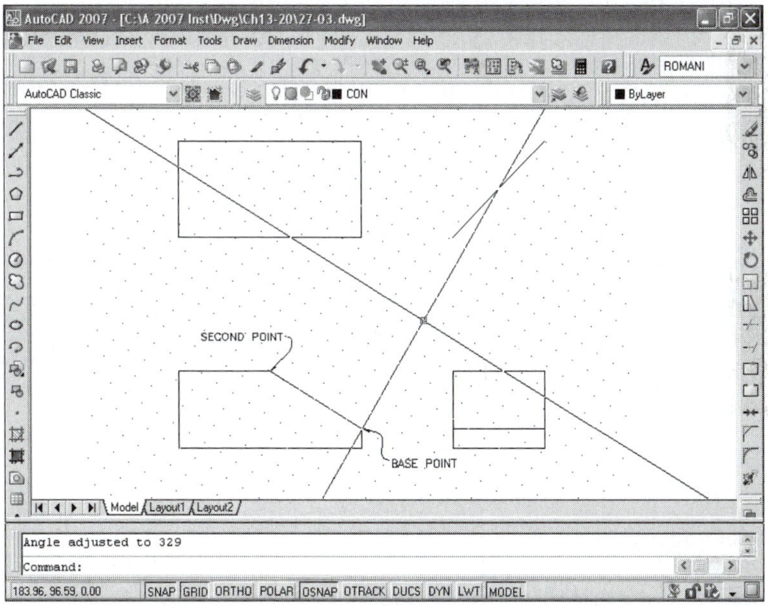

Command: **snap**
Specify snap spacing or [ON/OFF/Aspect/Style/Type] <0.5000>: **r**
Specify base point <0.0000,0.0000>: **PICK** or (**coordinates**) (PICK starts a rubberband line.)
Specify rotation angle <0>: **PICK** or (**value**) (PICK to specify second point to define the angle. See Fig. 27-3.)
Command:

PICK (or specify coordinates for) the endpoint of the *Line* representing the inclined surface as the "base point." At the "rotation angle:" prompt, a value can be entered or another point (the other end of the inclined *Line*) can be PICKed. Use *OSNAP* when PICKing the *Endpoints*. If you want to enter a value but don't know what angle to rotate to, use *List* to display the angle of the inclined *Line*. The *GRID*, *SNAP*, and crosshairs should align with the inclined plane as shown in Figure 27-3.

(An option for simplifying construction of the auxiliary view is to create a new *UCS* [User Coordinate System] with the origin at the new base point. Use the *3Point* option and turn on *ORTHO* to select the three points. See Chapter 36, User Coordinate Systems, for more information on this procedure.)

After rotating the *SNAP* and *GRID*, the partial auxiliary view can be "blocked in," as displayed in Figure 27-4. Begin by projecting *Lines* up from and perpendicular to the inclined surface. (Make sure *ORTHO* is *ON*.) Next, two *Lines* representing the depth of the view should be constructed parallel to the inclined surface and perpendicular to the previous two projection lines. The depth dimension of the object in the auxiliary view is equal to the depth dimension in the top or right view. *Trim* as necessary.

FIGURE 27-4

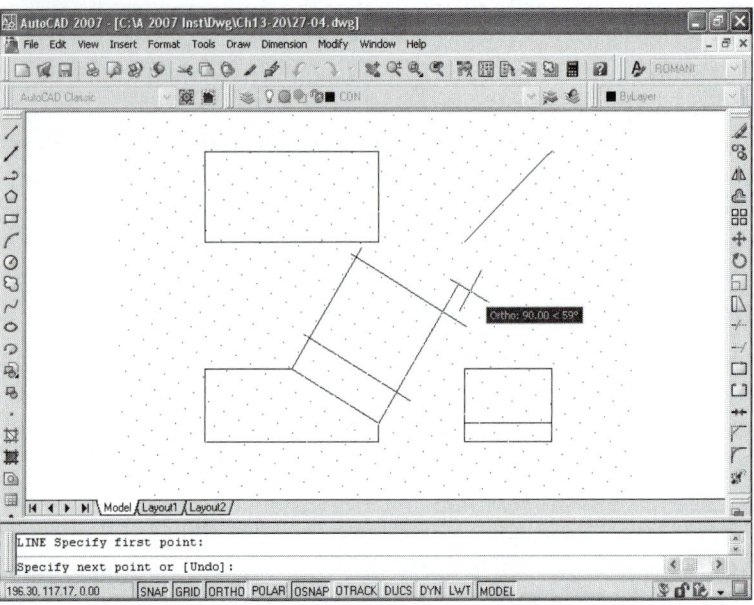

Locate the centers of the holes in the auxiliary view and construct two *Circles*. It is generally preferred to construct circular shapes in the view in which they appear as circles, then project to the other views. That is particularly true for this type of auxiliary since the other views contain ellipses. The centers can be located by projection from the front view or by *Offsetting Lines* from the view outline.

Next, project lines from the *Circles* and their centers back to the inclined surface (Fig. 27-5). Use of a hidden line layer can be helpful here. While the *SNAP* and *GRID* are rotated, construct the *Lines* representing the bottom of the holes in the front view. (Alternately, *Offset* could be used to copy the inclined edge down to the hole bottoms; then *Trim* the unwanted portions of the *Lines*.)

FIGURE 27-5

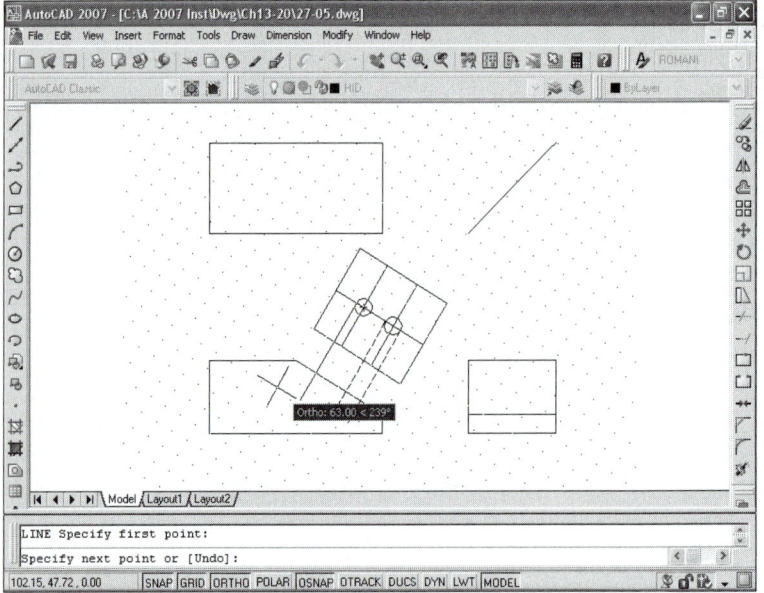

Rotating *SNAP* Back to the Original Position

Before details can be added to the other views, the *SNAP* and *GRID* should be rotated back to the original position. It is very important to rotate back using the <u>same base point</u>. Fortunately, AutoCAD remembers the original base point so you can accept the default for the prompt.

Next, enter a value of **0** when rotating back to the original position. (When using the *Snap Rotate* option, the value entered for the angle of rotation is absolute, not relative to the current position. For example, if the *Snap* was rotated to 45 degrees, rotate back to 0 degrees, not -45.)

Command: **snap**
Specify snap spacing or [ON/OFF/Aspect/Style/Type] <2.00>: **r**
Specify base point <150.00,40.00>: **Enter** (AutoCAD remembers the previous base point.)
Specify rotation angle <329>: **0**
Command:

Construction of multiview drawings with auxiliaries typically involves repeated rotation of the *SNAP* and *GRID* to the angle of the inclined surface and back again as needed.

With the *SNAP* and *GRID* in the original position, details can be added to the other views as shown in Figure 27-6. Since the two circles appear as ellipses in the top and right side views, project lines from the circles' centers and limiting elements on the inclined surface. Locate centers for the two *Ellipses* to be drawn in the top and right side views.

FIGURE 27-6

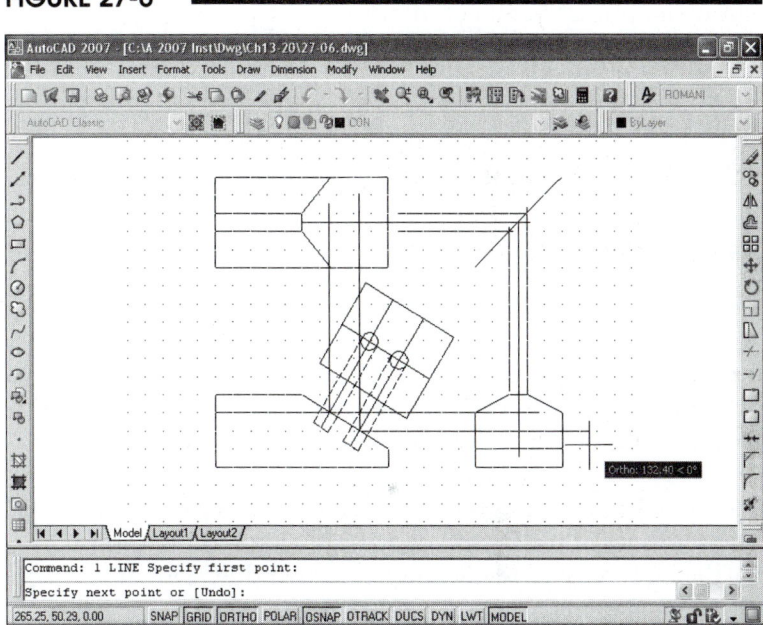

Use the *Ellipse* command to construct the ellipses in the top and right side views. Using the *Center* option of *Ellipse*, specify the center by PICKing with the *Intersection OSNAP* mode. *OSNAP* to the appropriate construction line *Intersection* for the first axis endpoint. For the second axis endpoint (Fig. 27-7), use the actual circle diameter, since that dimension is not foreshortened.

FIGURE 27-7

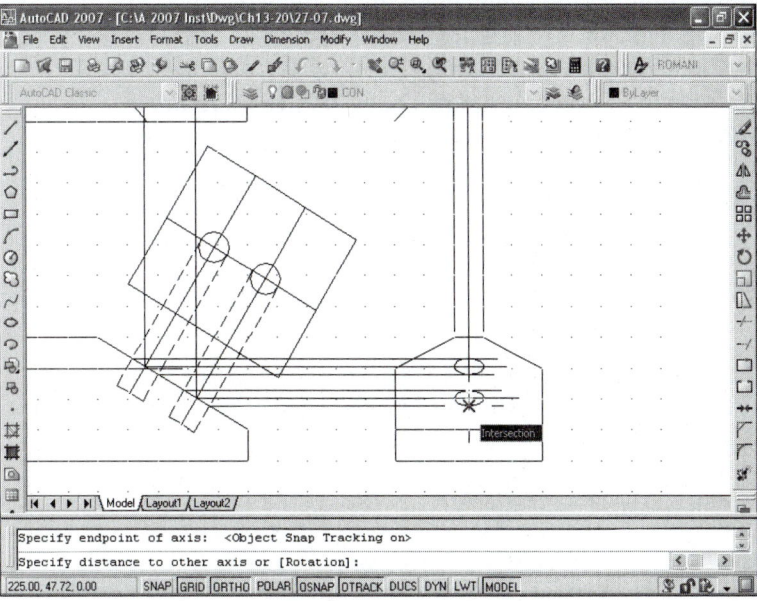

The remaining steps for completing the drawing involve finalizing the perimeter shape of the partial auxiliary view and *Copying* the *Ellipses* to the bottom of the hole positions. The *SNAP* and *GRID* should be rotated back to project the new edges found in the front view (Fig. 27-8).

At this point, the multiview drawing with auxiliary view is ready for centerlines, dimensioning, setting up a layout, and construction or insertion of a border and title block.

FIGURE 27-8

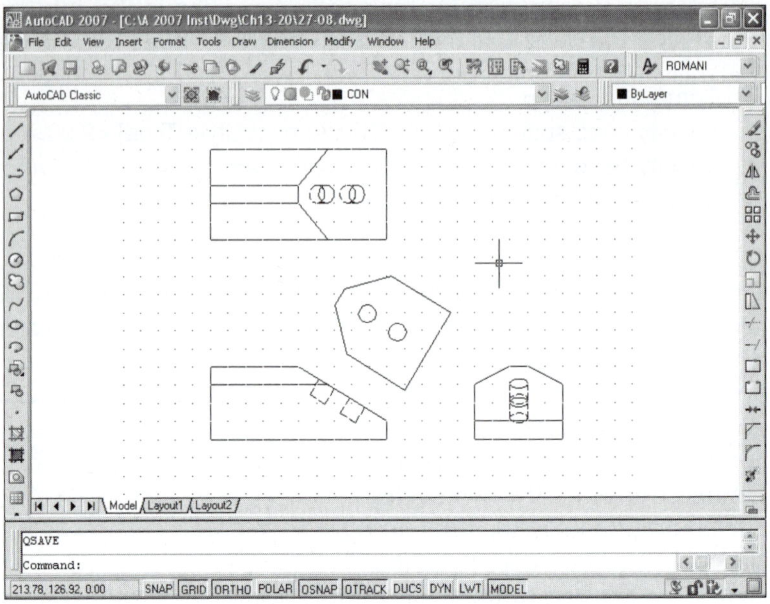

Using *Polar Tracking*

Polar Tracking can be used to create auxiliary views by facilitating construction of *Lines* at specific angles. To use *Polar Tracking* for auxiliary view construction, first use the *List* command to determine the angle of the inclined surface you want to project from. (Keep in mind that, by default, AutoCAD reports angles in whole numbers [no decimals or fractions], so use the *Units* command to increase the *Precision* of *Angular* units before using *List*.)

Once the desired angles are determined, specify the *Polar Angle Settings* in the *Drafting Settings* dialog box. Access the dialog box by right-clicking on the word *POLAR* at the Status Bar, typing *Dsettings*, or selecting *Drafting Settings* from the *Tools* pull-down menu. In the *Polar Tracking* tab of the dialog box, select the desired angles if appropriate from the *Increment angle* drop-down list. If the inclined plane is not at a regular angle offered from the drop-down list, specify the desired angle in the *Additional angles* edit box by selecting *New* and inputting the angles. Enter four angles in 90-degree increments (Fig. 27-9). Once the angles are set, ensure *POLAR* is on.

FIGURE 27-9

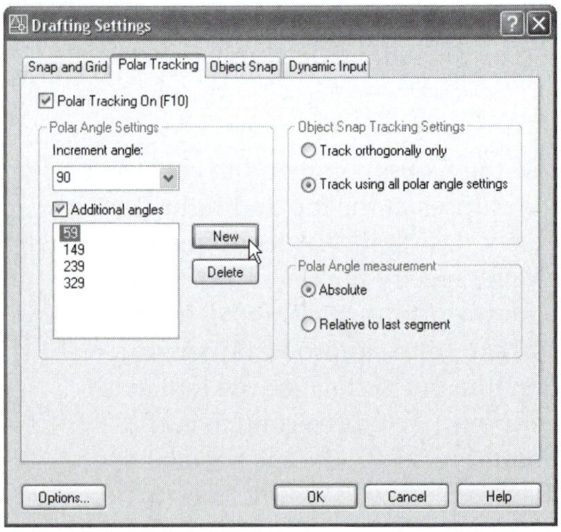

You may also want to set a *Polar Increment* in the *Snap and Grid* tab of the dialog box. This action makes the line lengths snap to regular intervals. (See Chapter 3 for more information on *Polar Snap* and *Polar Tracking*.)

Using *OSNAP*, begin constructing *Lines* perpendicular to the inclined plane. *Polar Tracking* should facilitate the line construction at the appropriate angles (Fig. 27-10).

FIGURE 27-10

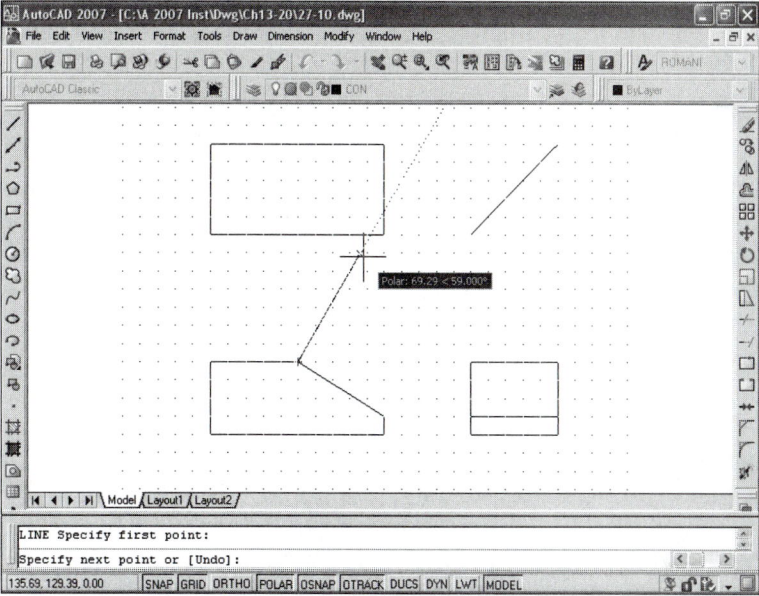

Perpendicular *Line* construction is also assisted by *Polar Tracking* (Fig. 27-11). Continue with this process, constructing necessary lines for the auxiliary view. The remainder of the auxiliary drawing, as described on previous pages (see Fig. 27-5 through Fig. 27-8), could be constructed using *Polar Tracking*. The advantage of this method is that drawing horizontal and vertical *Lines* is also possible while *Polar Tracking* is on.

Object Snap Tracking can also be employed to construct objects aligned (at the specified polar angles) with object snap locations (*Endpoint, Midpoint, Intersection, Extension*, etc.). See Chapter 7 for more information on *Object Snap Tracking*.

FIGURE 27-11

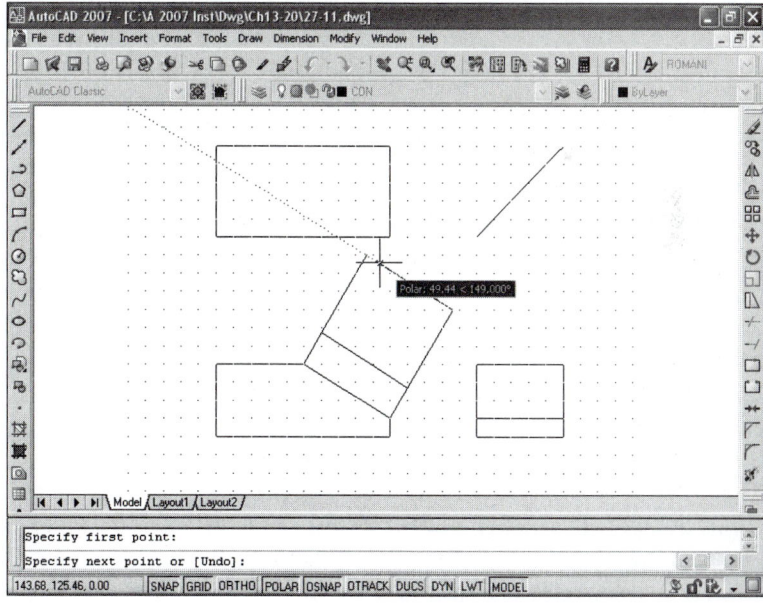

Using the *Offset* Command

Another possibility, and an alternative to the *SNAP* rotation, is to use *Offset* to make parallel *Lines*. This command can be particularly useful for construction of the "blocked in" partial auxiliary view because it is not necessary to rotate the *SNAP* and *GRID*.

Offset

	Pull-down Menu	Command (Type)	Alias (Type)	Short-cut	Screen (side) Menu	Tablet Menu
	Modify *Offset*	*Offset*	O	...	*MODIFY1* *Offset*	V,17

Invoke the *Offset* command and specify a distance. The first distance is arbitrary. Specify an appropriate value between the front view inclined plane and the nearest edge of the auxiliary view (20 for the example). *Offset* the new *Line* at a distance of 50 (for the example) or PICK two points (equal to the depth of the view).

 Note that the *Offset* lines have lengths equal to the original and therefore require no additional editing (Fig. 27-12).

FIGURE 27-12

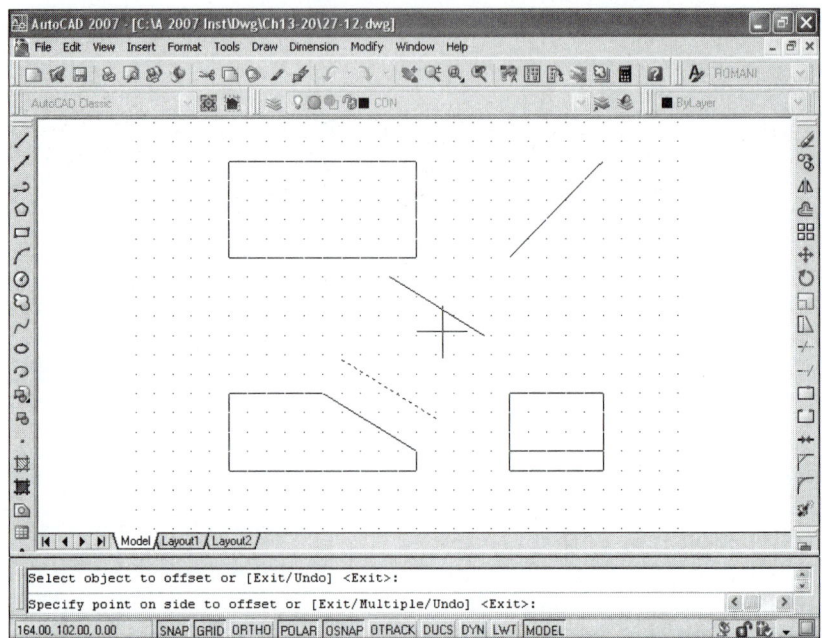

Next, two *Lines* would be drawn between *Endpoints* of the existing offset lines to complete the rectangle. *Offset* could be used again to construct additional lines to facilitate the construction of the two circles in the partial auxiliary view (Fig 27-13).

From this point forward, the construction process would be similar to the example given previously (Figs. 27-5 through 27-8). Even though *Offset* does not require that the *SNAP* be *Rotated*, the complete construction of the auxiliary view could be simplified by using the rotated *SNAP* and *GRID* in conjunction with *Offset*.

FIGURE 27-13

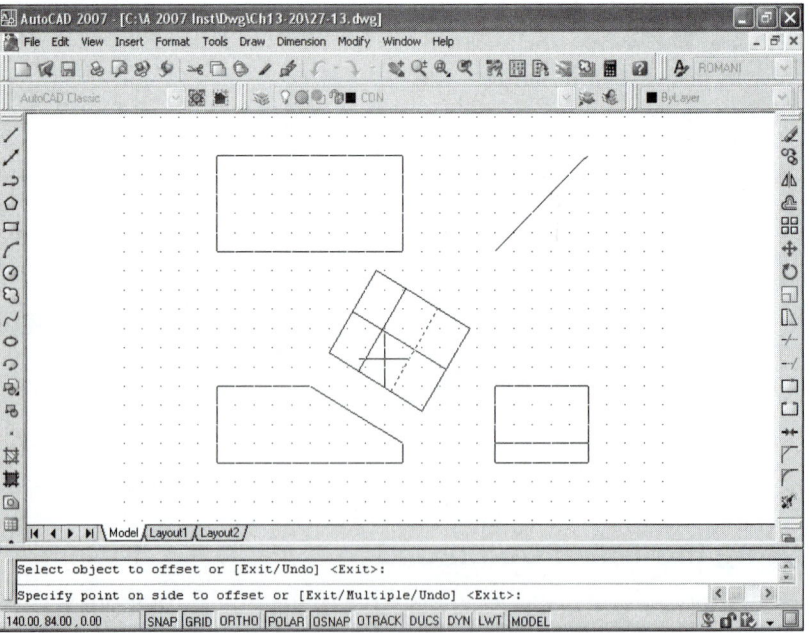

Using the *Xline* and *Ray* Commands

As a fourth alternative for construction of auxiliary views, the *Xline* and *Ray* commands could be used to create construction lines.

Xline

Pull-down Menu	Command (Type)	Alias (Type)	Short-cut	Screen (side) Menu	Tablet Menu
Draw *Construction Line*	*Xline*	*XL*	...	*DRAW 1* *Xline*	*L,10*

Ray

Pull-down Menu	Command (Type)	Alias (Type)	Short-cut	Screen (side) Menu	Tablet Menu
Draw *Ray*	*Ray*	...	...	*DRAW 1* *Ray*	*K,10*

The *Xline* command offers several options shown below:

Command: **xline**
Specify a point or [Hor/Ver/Ang/Bisect/Offset]:

The *Ang* option can be used to create a construction line at a specified angle. In this case, the angle specified would be that of the inclined plane or perpendicular to the inclined plane. The *Offset* option works well for drawing construction lines parallel to the inclined plane, especially in the case where the angle of the plane is not known.

Figure 27-14 illustrates the use of *Xline Offset* to create construction lines for the partial auxiliary view. The *Offset* option operates similarly to the *Offset* command described previously. Remember that an *Xline* extends to infinity but can be *Trimmed*, in which case it is converted to *Ray* (*Trim* once) or to a *Line* (*Trim* twice). See Chapter 15 for more information on the *Xline* command.

FIGURE 27-14

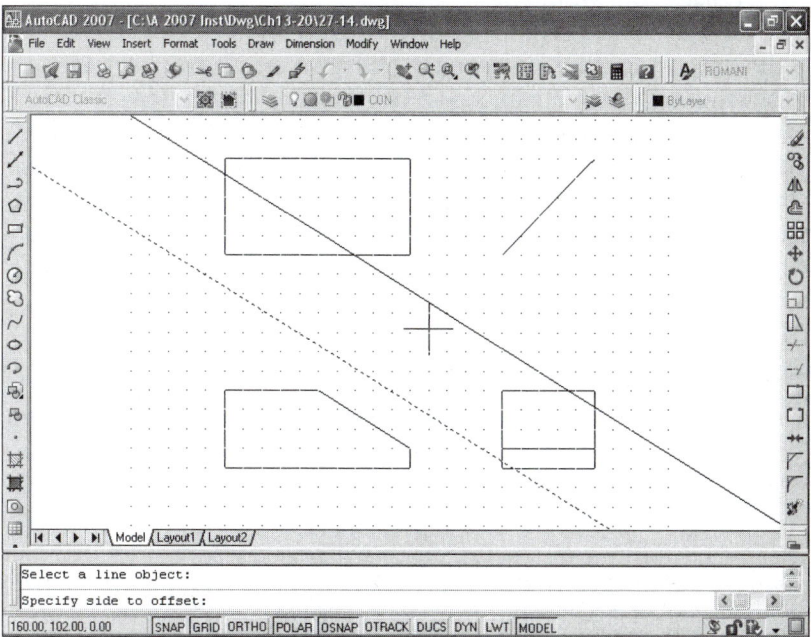

The *Ray* command also creates construction lines; however, the *Ray* has one anchored point and the other end extends to infinity.

> Command: **ray**
> Specify start point: **PICK** or (**coordinates**)
> Specify through point: **PICK** or (**coordinates**)

Rays are helpful for auxiliary view construction when you want to create projection lines perpendicular to the inclined plane. In Figure 27-15, two *Rays* are constructed from the *Endpoints* of the inclined plane and *Perpendicular* to the existing *Xlines*. Using *Xlines* and *Rays* in conjunction is an <u>excellent</u> <u>method</u> for "blocking in" the view.

There are two strategies for creating drawings using *Xlines* and *Rays*. First, these construction lines can be created on a separate layer and set up as a framework for the object lines. The object lines would then be drawn on top of the construction lines using *Osnaps*, but would be

FIGURE 27-15

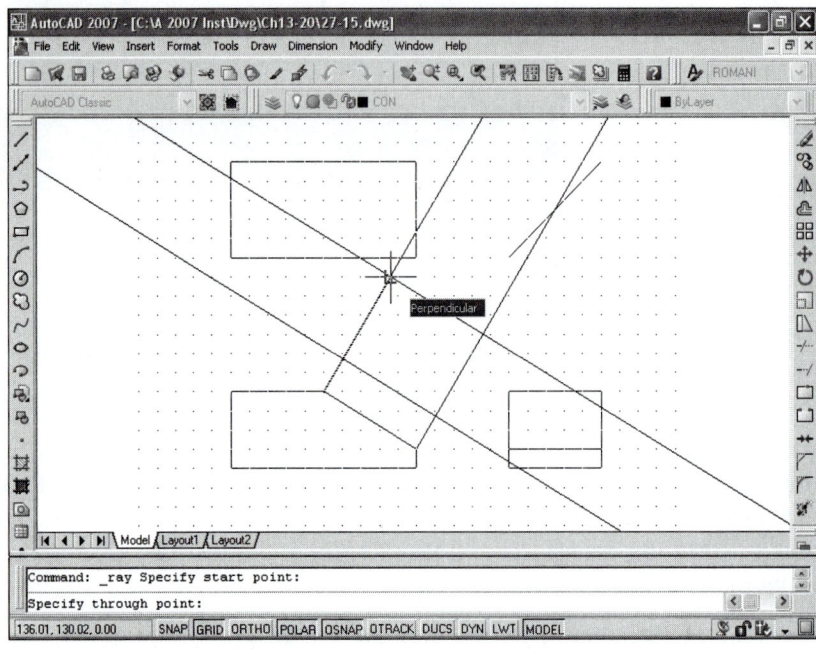

drawn on the object layer. The construction layer would be *Frozen* for plotting. The other strategy is to create the construction lines on the object layer. Through a series of *Trims* and other modifications, the *Xlines* and *Rays* are transformed to the finished object lines.

Now that you are aware of several methods for constructing auxiliary views, use any one method or a combination of methods for your drawings. No matter which methods are used, the final lines that are needed for the finished auxiliary view should be the same. It is up to you to use the methods that are the most appropriate for the particular application or are the easiest and quickest for you personally.

Constructing Full Auxiliary Views

The construction of a full auxiliary view begins with the partial view. After initial construction of the partial view, begin the construction of the full auxiliary by projecting the other edges and features of the object (other than the inclined plane) to the existing auxiliary view.

The procedure for constructing full auxiliary views in AutoCAD is essentially the same as that for partial auxiliary views. Use of the *Offset, Xline,* and *Ray* commands, *SNAP* and *GRID* rotation, and *Polar Tracking* should be used as illustrated for the partial auxiliary view example. Because a full auxiliary view is projected at the same angle as a partial, the same rotation angle and basepoint would be used for the *SNAP,* or the same *Increment angle* or *Additional angles* should be used for *Polar Tracking.*

CHAPTER EXERCISES

For the following exercises, create the multiview drawing, including the partial or full auxiliary view as indicated. Use the appropriate template drawing based on the given dimensions and indicated plot scale.

1. Make a multiview drawing with a partial auxiliary view of the example used in this chapter. Refer to Figure 27-1 for dimensions. *Save* the drawing as **CH27EX1**. *Plot* on an A size sheet at **1=1** scale.

2. Recreate the views given in Figure 27-16 and add a partial auxiliary view. *Save* the drawing as **CH27EX2** and *Plot* on an A size sheet at **2=1** scale.

FIGURE 27-16

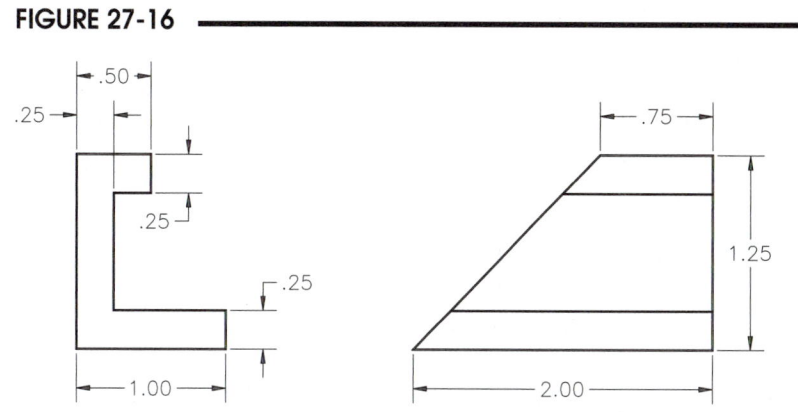

3. Recreate the views shown in Figure 27-17 and add a partial auxiliary view. *Save* the drawing as **CH27EX3.** Make a *Plot* on an A size sheet at **1=1** scale.

FIGURE 27-17

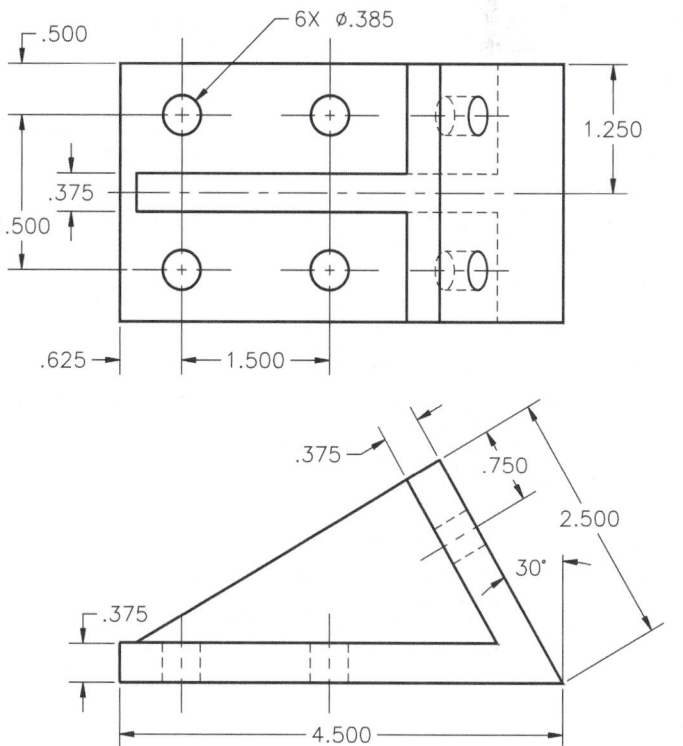

4. Make a multiview drawing of the given views in Figure 27-18. Add a full auxiliary view. *Save* the drawing as **CH27EX4**. *Plot* the drawing at an appropriate scale on an A size sheet.

FIGURE 27-18

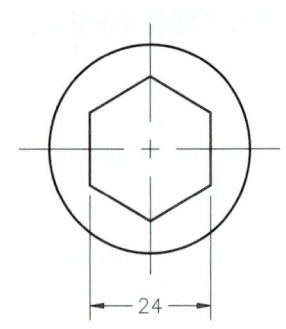

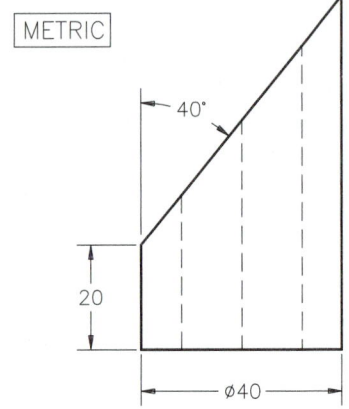

5. Draw the front, top, and a partial auxiliary view of the holder. During construction, note that the back corner indicated with hidden lines in Figure 27-19 is not a "clean" intersection, but is actually composed of two separate vertical edges. Make a *Plot* full size. Save the drawing as **HOLDER**.

FIGURE 27-19

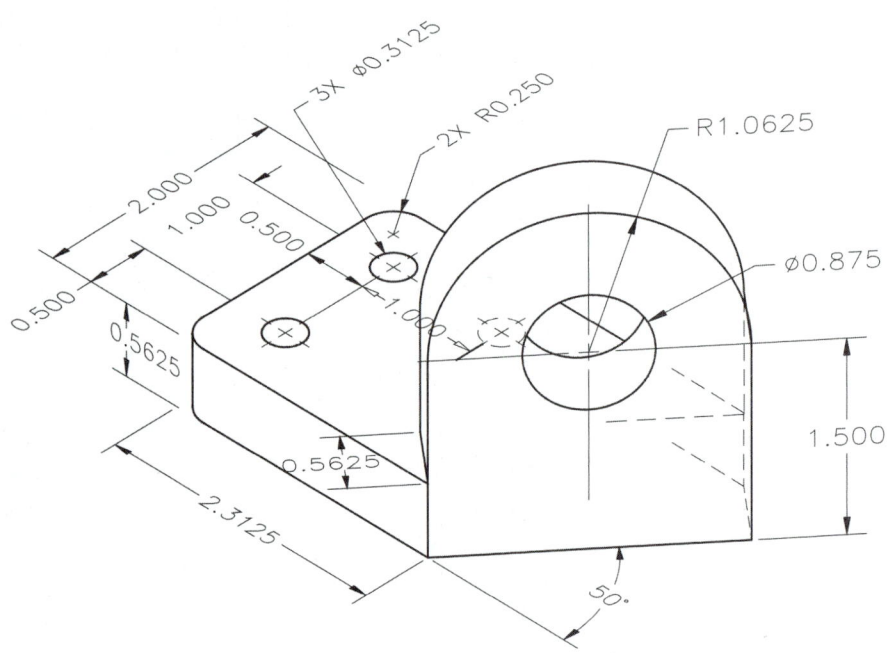

6. Draw three principal views and a full auxiliary view of the V-block shown in Figure 27-20. *Save* the drawing as **VBLOCK**. *Plot* to an accepted scale.

FIGURE 27-20

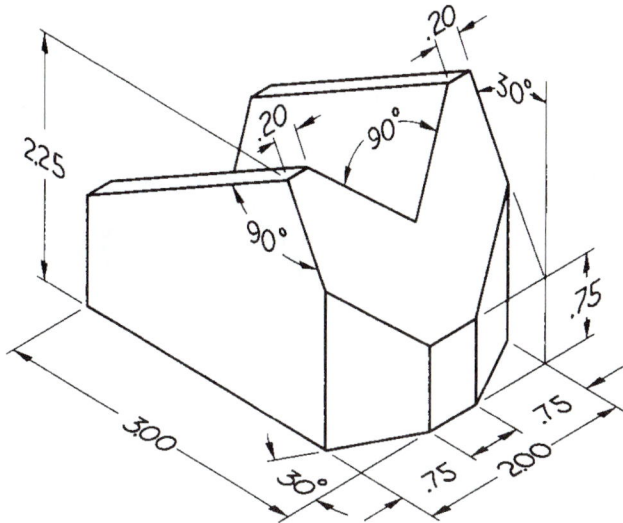

7. Draw two principal views and a partial auxiliary of the angle brace (Fig. 27-21). *Save* as **ANGLBRAC.** Make a plot to an accepted scale on an A or B size sheet. Convert the fractions to the current ANSI standard, decimal inches.

FIGURE 27-21

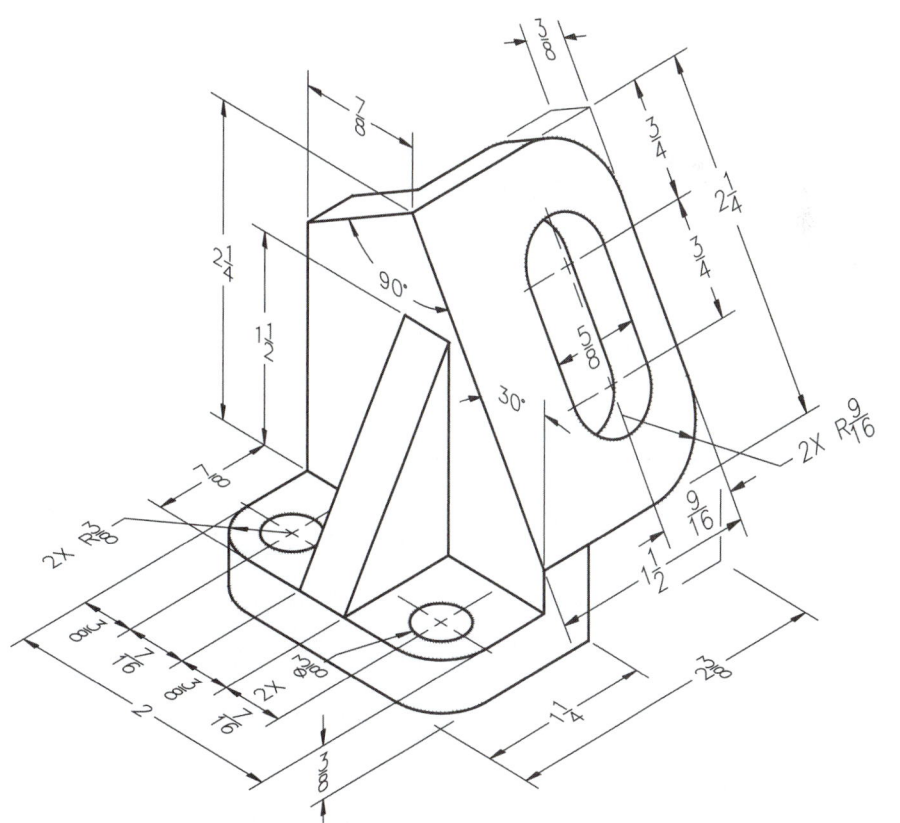

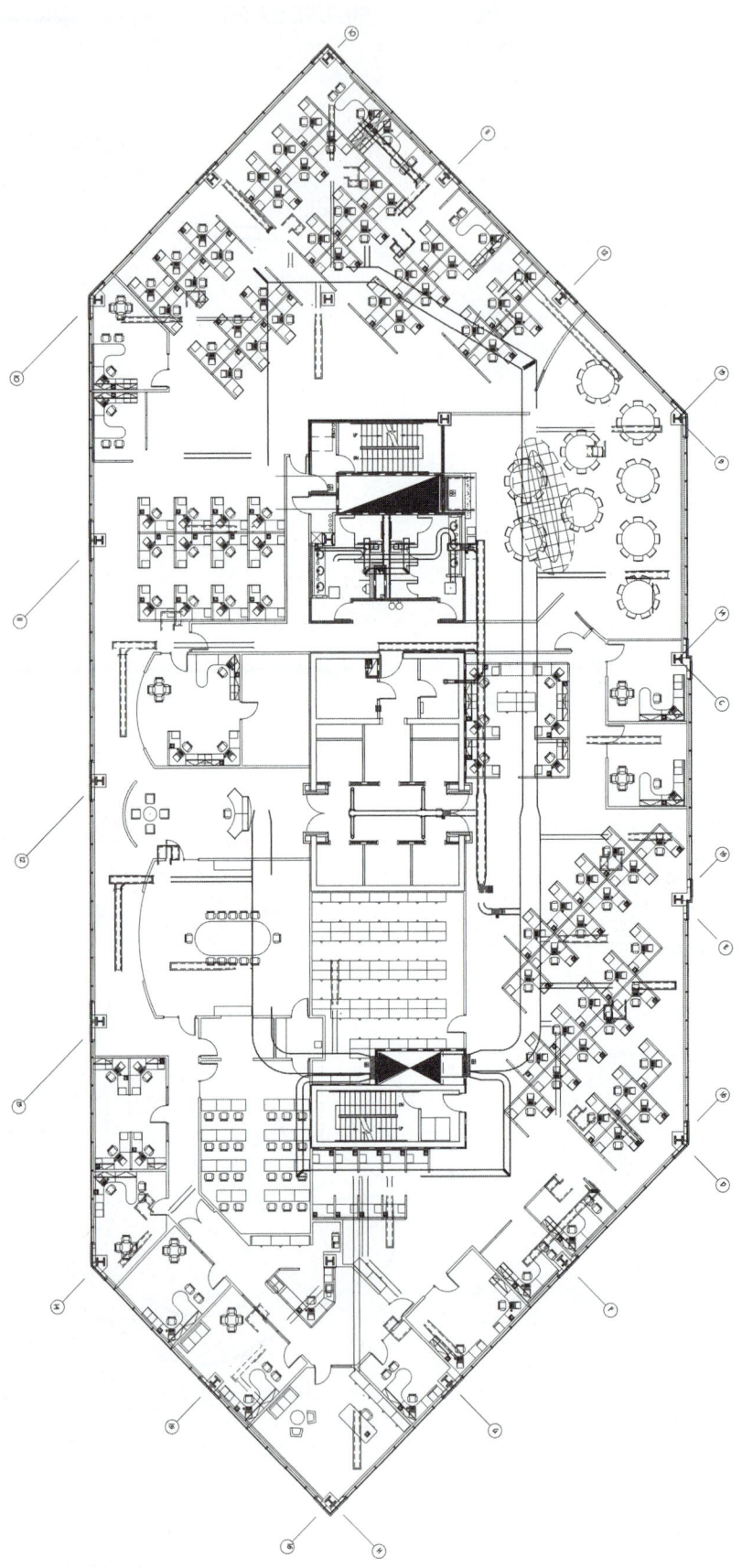

8TH FLOOR.DWG, Courtesy of Autodesk, Inc.

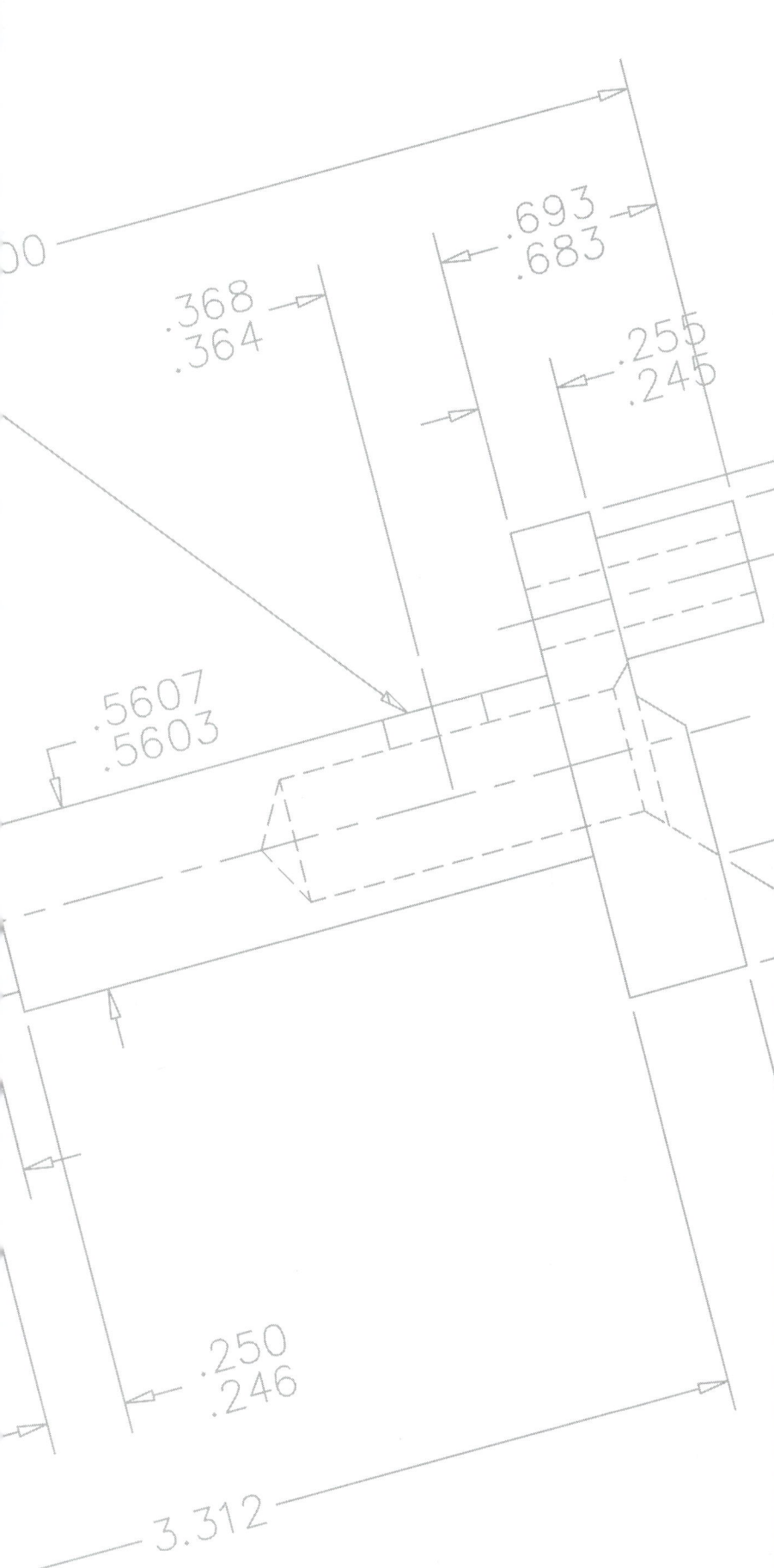

DIMENSIONING

CHAPTER OBJECTIVES

After completing this chapter you should:

1. be able to create linear dimensions with *Dimlinear*;

2. be able to append *Dimcontinue* and *Dimbaseline* dimensions to existing dimensions;

3. be able to create *Angular*, *Diameter*, and *Radius* dimensions;

4. know how to affix notes to drawings with *Leaders* and *Qleaders*;

5. know that *Dimordinate* can be used to specify Xdatum and Ydatum dimensions;

6. be able to use the new *Qdim* command to quickly create a variety of associative dimensions;

7. be able to create and apply geometric dimensioning and tolerancing symbols using the *Tolerance* command and dialog boxes;

8. know the possible methods for editing associative dimensions and dimensioning text.

CONCEPTS

As you know, drawings created with CAD systems should be constructed with the same size and units as the real-world objects they represent. In this way, the features of the object that you apply dimensions to (lengths, diameters, angles, etc.) are automatically measured by AutoCAD and the correct values are displayed in the dimension text. So if the object is drawn accurately, the dimension values will be created correctly and automatically. Generally, dimensions should be created on a separate layer named DIMENSIONS, or similar, and dimensions should be created in model space (see Chapter 29 for information on dimensioning in paper space).

The main components of a dimension are:

1. Dimension line
2. Extension lines
3. Dimension text (usually a numeric value)
4. Arrowheads or tick marks

FIGURE 28-1

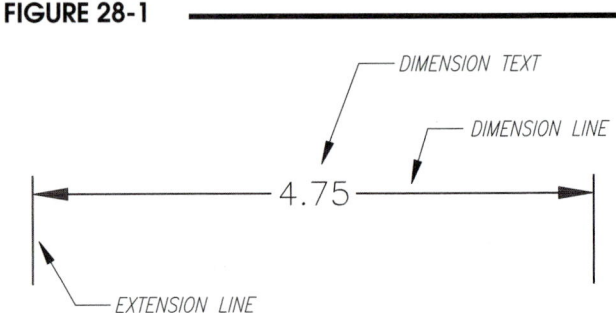

AutoCAD dimensioning is <u>semi-automatic</u>. When you invoke a command to create a linear dimension, AutoCAD only requires that you PICK an object or specify the extension line origins (where you want the extension lines to begin) and PICK the location of the dimension line (distance from the object). AutoCAD then measures the feature and draws the extension lines, dimension line, arrowheads, and dimension text (Fig. 28-1).

FIGURE 28-2

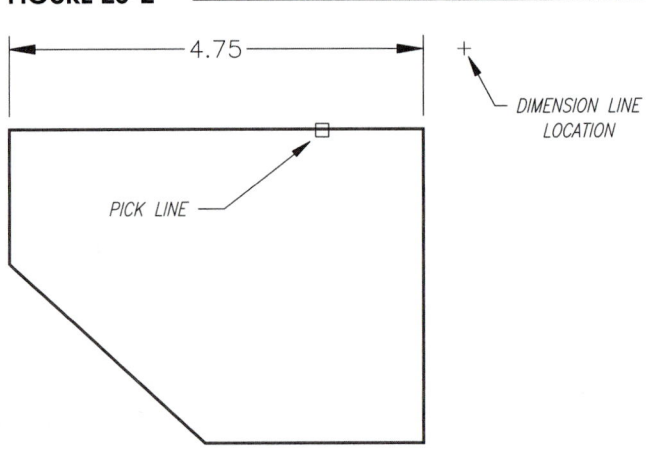

 For linear dimensioning commands, there are <u>two ways</u> to specify placement for a dimension in AutoCAD: you can PICK the <u>object</u> to be dimensioned or you can PICK the two <u>extension line origins</u>. The simplest method is to select the object because it requires only one PICK (Fig. 28-2):

> Command: **dimlinear**
> Specify first extension line origin or <select object>: **Enter**
> Select object to dimension: **PICK**

The other method is to PICK the extension line origins (Fig. 28-3). *Osnaps* should be used to PICK the object (endpoints in this case) so that the dimension is <u>associated</u> with the object.

> Command: **dimlinear**
> Specify first extension line origin or <select object>: **PICK**
> Specify second extension line origin: **PICK**

Once the dimension is attached to the object, you specify how far you want the dimension to be placed from the object (called the "dimension line location").

FIGURE 28-3

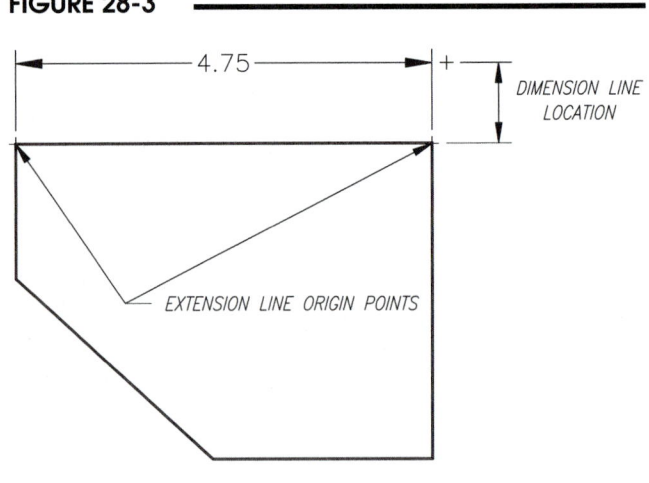

Dimensioning in AutoCAD is <u>associative</u> (by default). Because the extension line origins are "associated" with the geometry, the dimension text automatically updates if the geometry is *Moved, Stretched, Rotated, Scaled,* or edited using grips.

Because dimensioning is semi-automatic, <u>dimensioning variables</u> are used to control the way dimensions are created. Dimensioning variables can be used to control features such as text or arrow size, direction of the leader arrow for radial or diametrical dimensions, format of the text, and many other possible options. Groups of variable settings can be named and saved as <u>Dimension Styles</u>. Dimensioning variables and dimension styles are discussed in Chapter 29.

Dimensioning commands can be invoked by the typical methods. The *Dimension* pull-down menu contains the dimension creation and editing commands (Fig. 28-4). A *Dimension* toolbar (Fig. 28-5) can also be activated by using the *Toolbars* list.

FIGURE 28-4

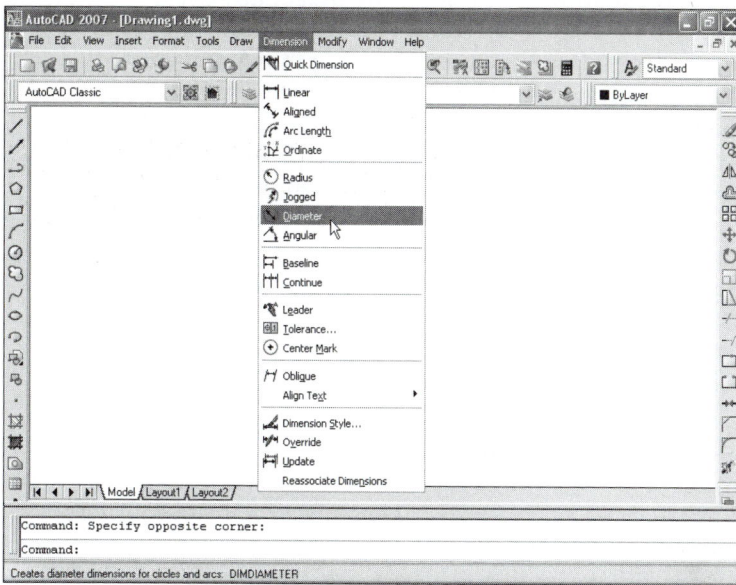

FIGURE 28-5

DIMENSION DRAWING COMMANDS

Dimlinear

Pull-down Menu	Command (Type)	Alias (Type)	Short-cut	Screen (side) Menu	Tablet Menu
Dimension Linear	*Dimlinear*	*DIMLIN or DLI*	...	*DIMNSION Linear*	*W,5*

Dimlinear creates a <u>horizontal, vertical, or rotated</u> dimension. If the object selected is a horizontal line (or the extension line origins are horizontally oriented), the resulting dimension is a horizontal dimension. This situation is displayed in the previous illustrations (see Fig. 28-2 and Fig. 28-3), or if the selected object or extension line origins are vertically oriented, the resulting dimension is vertical (Fig. 28-6).

FIGURE 28-6

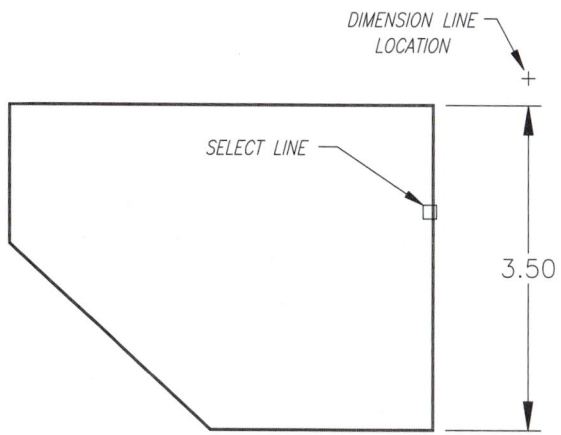

When you dimension an inclined object (or if the selected extension line origins are diagonally oriented), a vertical <u>or</u> horizontal dimension can be made, depending on where you drag the dimension line in relation to the object. If the dimension line location is more to the side, a vertical dimension is created (Fig. 28-7), or if you drag farther up or down, a horizontal dimension results (Fig. 28-8).

If you select the extension line origins, it is very important to PICK the <u>object's endpoints</u> if the dimensions are to be truly associative (associated with the geometry). *OSNAP* should be used to find the object's *Endpoint*, *Intersection*, etc.

FIGURE 28-7

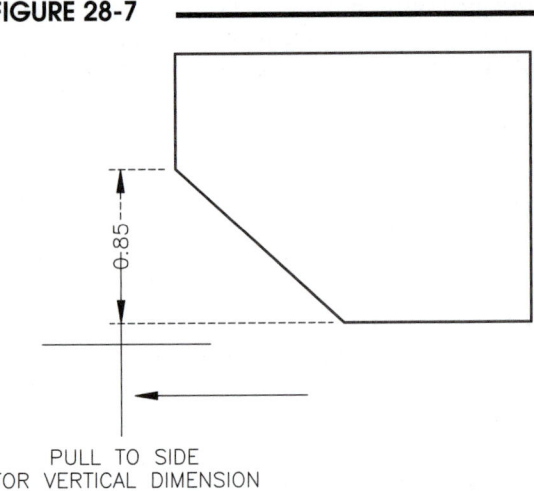

PULL TO SIDE
FOR VERTICAL DIMENSION

Command: ***dimlinear***
Specify first extension line origin or <select object>: **PICK** (use *Osnaps*)
Specify second extension line origin: **PICK** (use *Osnaps*)
Specify dimension line location or [Mtext/Text/Angle/Horizontal/Vertical/Rotated]: **PICK** (where you want the dimension line to be placed)
Dimension text = *n.nnnn*
Command:

When you pick the location for the dimension line, AutoCAD automatically measures the object and inserts the correct numerical value. The other options are explained next.

FIGURE 28-8

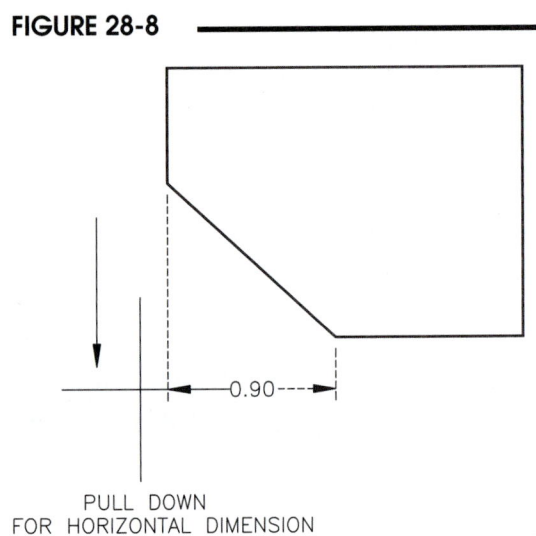

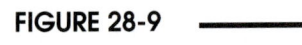

PULL DOWN
FOR HORIZONTAL DIMENSION

Rotated

If you want the dimension line to be drawn at an angle instead of vertical or horizontal, use this option. Selecting an inclined line, as in the previous two illustrations, would normally create a horizontal or vertical dimension. The *Rotated* option allows you to enter an <u>angular value</u> for the dimension line to be drawn. For example, selecting the diagonal line and specifying the appropriate angle would create the dimension shown in Figure 28-9. This object, however, could be more easily dimensioned with the *Dimaligned* command (see "*Dimaligned*").

FIGURE 28-9

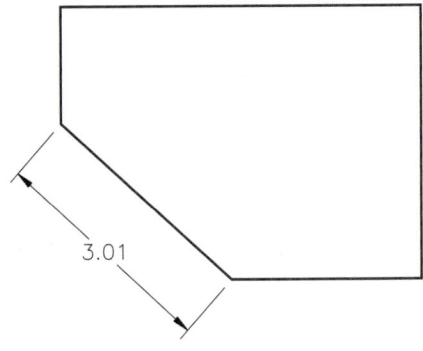

A *Rotated* dimension should be used when the geometry has "steps" or any time the desired dimension line angle is different than the dimensioned feature (when you need extension lines of different lengths). Figure 28-10 illustrates the result of using a *Rotated* dimension to give the correct dimension line angle and extension line origins for the given object. In this case, the extension line origins were explicitly PICKed. The feature of *Rotated* that makes it unique is that you specify the <u>angle</u> that the dimension line will be drawn.

FIGURE 28-10

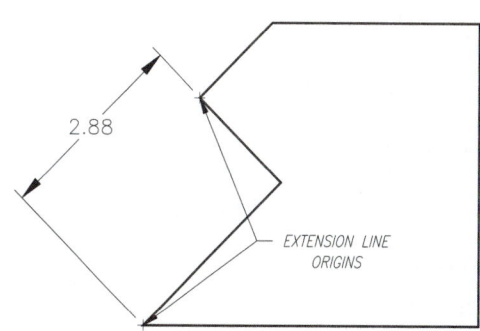

Text

Using the *Text* option allows you to enter any value or characters in place of the AutoCAD-measured text. The measured value is given as a reference at the command prompt.

> Command: ***dimlinear***
> Specify first extension line origin or <select object>: **PICK**
> Specify second extension line origin: **PICK**
> Specify dimension line location or
> [Mtext/Text/Angle/Horizontal/Vertical/Rotated]: ***T***
> Enter dimension text <*n.nnnn*>: (**text**) or (**value**)

FIGURE 28-11

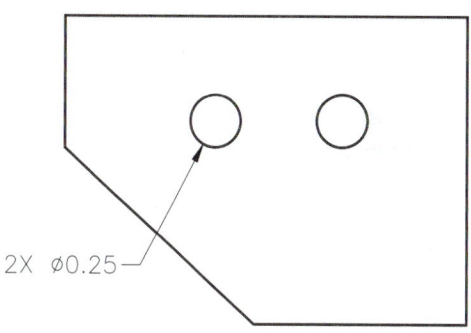

Entering a value or text at the prompt (above) causes AutoCAD to display that value or text instead of the AutoCAD-measured value. If you want to keep the AutoCAD-measured value but add a prefix or suffix, use the less-than, greater-than symbols (< >) to represent the actual (AutoCAD) value:

> Enter dimension text <0.25>: **2X <>**
> Specify dimension line location or [Mtext/Text/Angle/Horizontal/Vertical/Rotated]: **PICK**
> Dimension text = 0.25
> Command:

The "2X <>" response produces the dimension text shown in Figure 28-11.

NOTE: Changing the AutoCAD-measured value should be discouraged. If the geometry is drawn accurately, the dimensional value is correct. If you specify other dimension text, the text value is <u>not</u> updated in the event of *Stretching, Rotating,* or otherwise editing the associative dimension.

Mtext

This option allows you to enter text using the *Text Formatting* Editor. With the *Mtext* option, the actual AutoCAD-supplied text appears in the editor instead of being represented with less-than and greater-than symbols (<>). It is not recommended to change the AutoCAD-supplied text; however, you can enter any text before or after. Note that the AutoCAD-sup-

FIGURE 28-12

plied text appears in a dark background, whereas any text you enter appears with a lighter background (Fig. 28-12). Keep in mind that you have the power to change fonts, text height, bold, italic, underline, and stacked text or fractions using this editor.

Angle

This creates text drawn at the angle you specify. Use this for special cases when the text must be drawn to a specific angle other than horizontal. (It is also possible to make the text automatically align with the angle of the dimension line using the *Dimension Style Manager*. See Chapter 29.)

Horizontal

Use the *Horizontal* option when you want to force a horizontal dimension for an inclined line and the desired placement of the dimension line would otherwise cause a vertical dimension.

Vertical

This option forces a *Vertical* dimension for any case.

Dimaligned

Pull-down Menu	Command (Type)	Alias (Type)	Short-cut	Screen (side) Menu	Tablet Menu
Dimension Aligned	Dimaligned	DIMALI or DAL	...	DIMNSION Aligned	W,4

An *Aligned* dimension is aligned with (at the same angle as) the selected object or the extension line origins. For example, when *Aligned* is used to dimension the angled object shown in Figure 28-13, the resulting dimension aligns with the *Line*. This holds true for either option—PICKing the object or the extension line origins. If a *Circle* is PICKed, the dimension line is aligned with the selected point on the *Circle* and its center. The command syntax for the *Dimaligned* command accepting the defaults is:

FIGURE 28-13

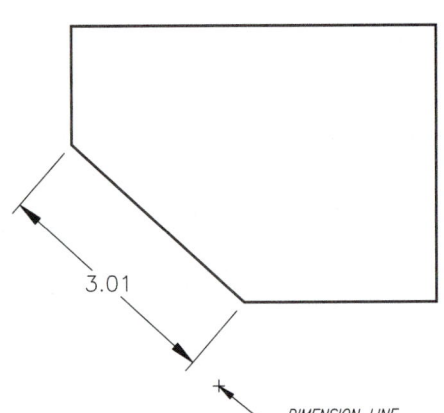

DIMENSION LINE
LOCATION

```
Command: dimaligned
Specify first extension line origin or <select object>: PICK
Specify second extension line origin: PICK
Specify dimension line location or [Mtext/Text/Angle]: PICK
Dimension text = n.nnnn
Command:
```

The three options (*Mtext/Text/Angle*) operate similar to those for *Dimlinear*.

Mtext

The *Mtext* option calls the *Text Formatting* Editor. You can alter the AutoCAD-supplied text value or modify other visible features of the dimension text such as fonts, text height, bold, italic, underline, stacked text or fractions, and text style (see "*Dimlinear*," *Mtext*).

Text

You can change the AutoCAD-supplied numerical value or add other annotation to the value in command line format (see "*Dimlinear*," *Text*).

Angle

Enter a value for the angle that the text will be drawn.

The typical application for *Dimaligned* is for dimensioning an angled but <u>straight</u> feature of an object, as shown in Figure 28-13. *Dimaligned* should not be used to dimension an object feature that contains "steps," as shown in Figure 28-10. *Dimaligned* always draws <u>extension lines of equal length</u>.

Dimbaseline

Pull-down Menu	Command (Type)	Alias (Type)	Short-cut	Screen (side) Menu	Tablet Menu
Dimension Baseline	Dimbaseline	DIMBASE or DBA	...	DIMNSION Baseline	...

Dimbaseline allows you to create a dimension that uses an extension line origin from a previously created dimension. Successive *Dimbaseline* dimensions can be used to create the style of dimensioning shown in Figure 28-14.

A baseline dimension must be connected to an existing dimension. If *Dimbaseline* is invoked immediately after another dimensioning command, you are required only to specify the second extension line origin since AutoCAD knows to use the previous dimension's first extension line origin:

    Command: **dimbaseline**
    Specify a second extension line origin or [Undo/Select]
    <Select>: **PICK**
    Dimension text = *n.nnnn*

FIGURE 28-14

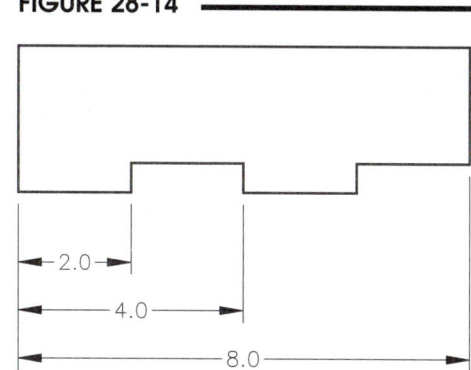

The previous dimension's first extension line is used also for the baseline dimension (Fig. 28-15). Therefore, you specify only the second extension line origin. Note that you are not required to specify the dimension line location. AutoCAD spaces the new dimension line automatically, based on the setting of the dimension line increment variable (Chapter 29).

FIGURE 28-15

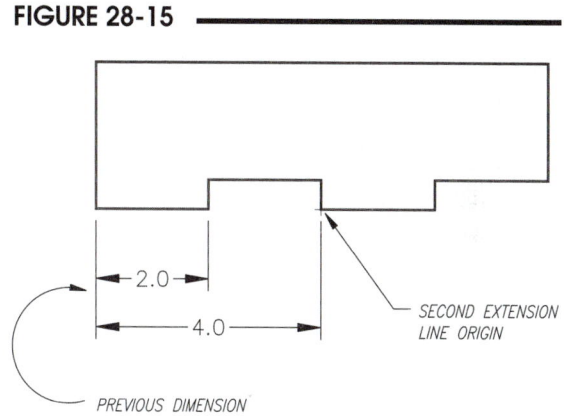

If you wish to create a *Dimbaseline* dimension using a dimension other than the one just created, use the "Select" option (Fig. 28-16):

    Command: **dimbaseline**
    Specify a second extension line origin or
    [Undo/Select] <Select>: **S** or **Enter**
    Select base dimension: **PICK**
    Specify a second extension line origin or
    [Undo/Select] <Select>: **Enter**
    Dimension text = *n.nnnn*

FIGURE 28-16

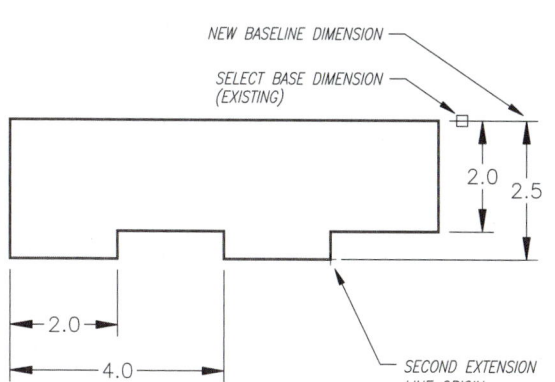

The extension line selected as the base dimension becomes the first extension line for the new *Dimbaseline* dimension.

The *Undo* option can be used to undo the last baseline dimension created in the current command sequence.

Dimbaseline can be used with rotated, aligned, angular, and ordinate dimensions.

Dimcontinue

Pull-down Menu	Command (Type)	Alias (Type)	Short-cut	Screen (side) Menu	Tablet Menu
Dimension Continue	Dimcontinue	DIMCONT or DCO	...	DIMNSION Continue	...

Dimcontinue dimensions continue in a line from a previously created dimension. *Dimcontinue* dimension lines are attached to, and drawn the same distance from, the object as an existing dimension.

Dimcontinue is similar to *Dimbaseline* except that an existing dimension's <u>second</u> extension line is used to begin the new dimension. In other words, the new dimension is connected to the <u>second</u> extension line, rather than to the <u>first</u>, as with a *Dimbaseline* dimension (Fig. 28-17).

The command syntax is as follows:

> Command: **dimcontinue**
> Specify a second extension line origin or [Undo/Select]
> <Select>: **PICK**
> Dimension text = *n.nnnn*

FIGURE 28-17

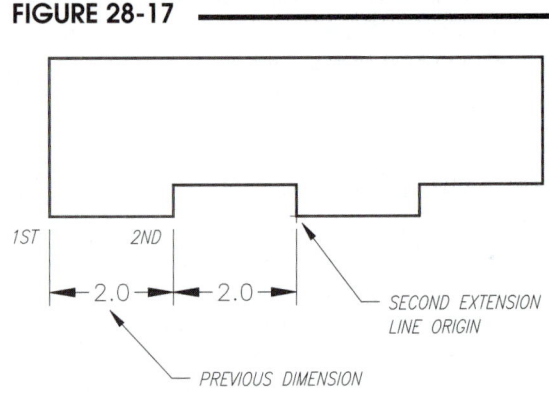

Assuming a dimension was just drawn, *Dimcontinue* could be used to place the next dimension, as shown in Figure 28-17.

If you want to create a continued dimension and attach it to an extension line <u>other</u> than the previous dimension's second extension line, you can use the *Select* option to pick an extension line of any other dimension. Then, use *Dimcontinue* to create a continued dimension from the selected extension line (Fig. 28-18).

The *Undo* option can be used to undo the last continued dimension created in the current command sequence. *Dimcontinue* can be used with rotated, aligned, angular, and ordinate dimensions.

FIGURE 28-18

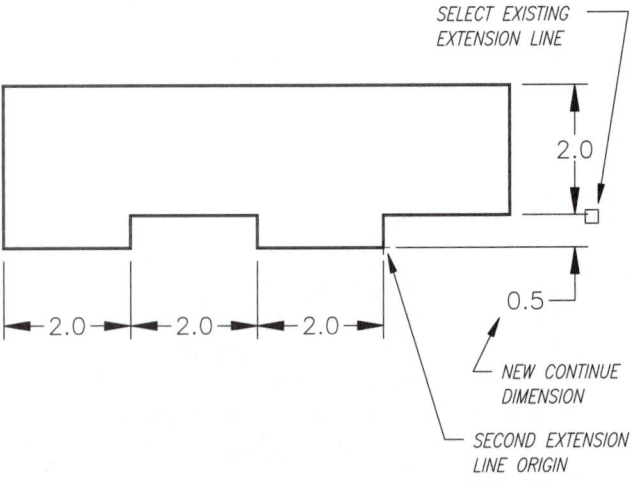

Dimdiameter

Pull-down Menu	Command (Type)	Alias (Type)	Short-cut	Screen (side) Menu	Tablet Menu
Dimension Diameter	Dimdiameter	DIMDIA or DDI	...	DIMNSION Diameter	X,4

The *Dimdiameter* command creates a diametrical dimension by selecting any *Circle*. Diametrical dimensions should be used for full 360 degree *Circles* and can be used for *Arcs* of more than 180 degrees.

FIGURE 28-19

```
Command: dimdiameter
Select arc or circle: PICK
Dimension text = n.nnnn
Specify dimension line location or [Mtext/Text/Angle]: PICK
Command:
```

You can PICK the circle at any location. AutoCAD allows you to adjust the position of the dimension line to any angle or length (Fig. 28-19). Dimension lines for diametrical or radial dimensions should be drawn to a regular angle, such as 30, 45, or 60 degrees, never vertical or horizontal.

A typical diametrical dimension appears as the example in Figure 28-19. According to ANSI standards, a diameter dimension line and arrow should point inward (toward the center) for holes and small circles where the dimension line and text do not fit within the circle. Use the default variable settings for *Dimdiameter* dimensions such as this. **TIP**

For dimensioning large circles, ANSI standards suggest an alternate method for diameter dimensions where sufficient room exists for text and arrows inside the circle (Fig. 28-20). To create this style of dimensioning, set the variables to *Text* and *Place text manually when dimensioning* in the *Fit* tab of the *Dimension Style Manager* (see Chapter 29).

FIGURE 28-20

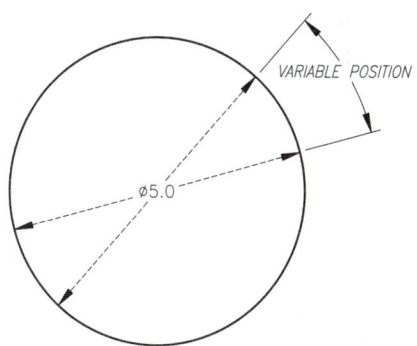

Notice that the *Diameter* command creates center marks at the *Circle's* center. Center marks can also be drawn by the *Dimcenter* command (discussed later). AutoCAD uses the center and the point selected on the *Circle* to maintain its associativity.

Mtext/Text

The *Mtext* or *Text* options can be used to modify or add annotation to the default value. The *Mtext* option summons the *Text Formatting* Editor and the *Text* option uses Command line format. Both options operate similar to the other dimensioning commands (see *Dimlinear*, *Mtext*, and *Text*).

FIGURE 28-21

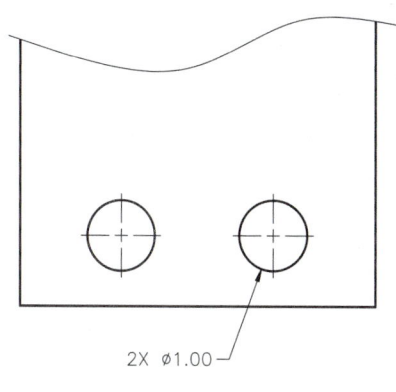

Notice that with diameter dimensions AutoCAD automatically creates the Ø (phi) symbol before the dimensional value. This is the latest ANSI standard for representing diameters. If you prefer to use a prefix before or suffix after the dimension value, it can be accomplished with the *Mtext* or *Text* option. Remember that the < > symbols represent the AutoCAD-measured value so the additional text should be inserted on <u>either side</u> of the symbols. Inserting a prefix by this method <u>does not override</u> the Ø (phi) symbol (Fig. 28-21). **TIP**

A prefix or suffix can alternately be added to the measured value by using the *Dimension Style Manager* and entering text or values in the *Prefix* or *Suffix* edit boxes. Using this method, however, <u>overrides</u> the Ø symbol (see Chapter 29).

Angle

With this option, you can specify an angle (other than the default) for the text to be drawn by entering a value.

Dimradius

	Pull-down Menu	Command (Type)	Alias (Type)	Short-cut	Screen (side) Menu	Tablet Menu
	Dimension Radius	*Dimradius*	*DIMRAD or DRA*	...	*DIMNSION Radius*	*X,5*

Dimradius is used to create a dimension for an arc of anything less than half of a circle. ANSI standards dictate that a *Radius* dimension line should point outward (from the arc's center), unless there is insufficient room, in which case the line can be drawn on the outside pointing inward, as with a leader. The text can be located inside an arc (if sufficient room exists) or is forced outside of small *Arcs* on a leader.

```
Command: dimradius
Select arc or circle: PICK
Dimension text = n.nnnn
Specify dimension line location or [Mtext/Text/Angle]: PICK
Command:
```

Assuming the defaults, a *Dimradius* dimension can appear on either side of an arc, as shown in Figure 28-22. Placement of the dimension line is variable. Dimension lines for arcs and circles should be positioned at a regular angle such as 30, 45, or 60 degrees, never vertical or horizontal.

When the radius dimension is dragged outside of the arc, a center mark is automatically created (Fig. 28-22). When the radius dimension is dragged inside the arc, no center mark is created. (Center marks can be created using the *Center* command discussed next.)

According to ANSI standards, the dimension line and arrow should point <u>outward</u> from the center for radial dimensions (Fig. 28-23). The text can be placed inside or outside of the *Arc*, depending on how much room exists. To create a *Dimradius* dimension to comply with this standard, dimension variables must be changed from the defaults. To achieve *Dimradius* dimensions as shown in Figure 28-23, set the variables to *Text* and *Place text manually when dimensioning* in the *Fit* tab of the *Dimension Style Manager* (see Chapter 29).

TIP

For very small radii, such as that shown in Figure 28-24, there is insufficient room for the text and arrow to fit inside the *Arc*. In this case, AutoCAD automatically forces the text outside with the leader pointing inward toward the center. <u>No changes</u> have to be made to the default settings for this to occur.

FIGURE 28-22

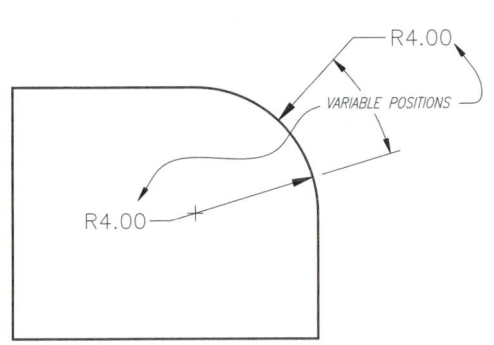

FIGURE 28-23

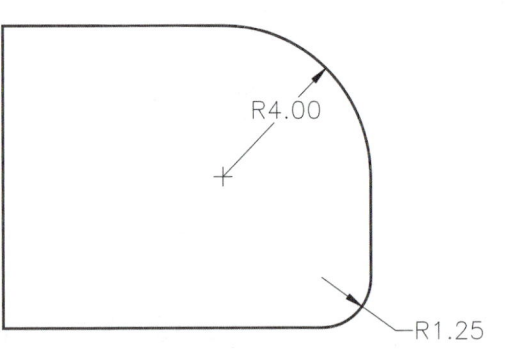

FIGURE 28-24

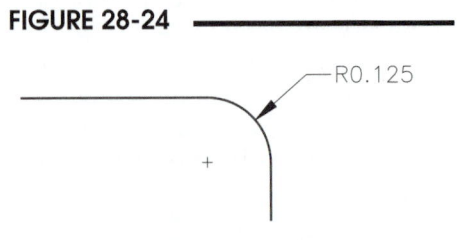

Mtext/Text

These options can be used to modify or add annotation to the default value. AutoCAD automatically inserts the letter "R" before the numerical text whenever a *Dimradius* dimension is created. This is the correct notation for radius dimensions. The *Mtext* option calls the *Text Formatting* Editor, and the *Text* option uses the Command line format for entering text. With the *Text* option AutoCAD uses the < > symbols to represent the AutoCAD-supplied value. Entering text inside the < > symbols overrides the measured value. Entering text before the symbols adds a prefix without overriding the "R" designation. Alternately, text can be added by using the *Prefix* and *Suffix* options of the *Dimension Style Manager* series; however, a *Prefix* entered in the edit box replaces the letter "R." (See Chapter 29.)

Angle

With this option, you can specify an angle (other than the default) for the text to be drawn by entering a value.

Dimjog

Pull-down Menu	Command (Type)	Alias (Type)	Short-cut	Screen (side) Menu	Tablet Menu
Dimension Jogged	Dimjogged	DJO	...	...	...

For an extremely large radius it may not be possible or practical to create a dimension line originating from the arc's actual center, as is traditionally done (see "*Dimradius*"). In such cases, a "jogged" radius dimension line is created. With this type of dimension, the dimension line is drawn from a hypothetical arc center point to the arc; however, the dimension line contains a "jog" to indicate that the origin of the dimension line is not the actual center.

FIGURE 28-25

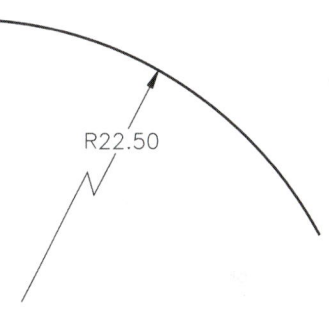

R22.50

The *Dimjogged* command creates a jogged radius dimension for extremely large arcs (Fig. 28-25). Since the actual center of the arc is not shown by the dimension, you must select the hypothetical center point ("center location override") to use for the origin of the dimension line. You can also specify where you want to place the jog along the dimension line. The *Mtext/Text/Angle* options operate similarly to the *Dimradius* command. See "*Dimradius*."

```
Command: dimjogged
Select arc or circle: PICK
Specify center location override: PICK (pick the desired hypothetical arc center)
Dimension text = nn
Specify dimension line location or [Mtext/Text/Angle]: PICK (place the location of the dimension text
and dimension line)
Specify jog location: PICK (pick the location for the "jog")
Command:
```

Dimarc

Pull-down Menu	Command (Type)	Alias (Type)	Short-cut	Screen (side) Menu	Tablet Menu
Dimension Arc Length	Dimarc	DAR	...	...	...

In some cases it may be necessary to dimension the length of an arc along its curve, similar to a partial circumference. The *Dimarc* command automatically measures this length and draws a dimension line concentric to the arc (Fig. 28-26). You select the arc to dimension and place the location of the dimension line.

FIGURE 28-26

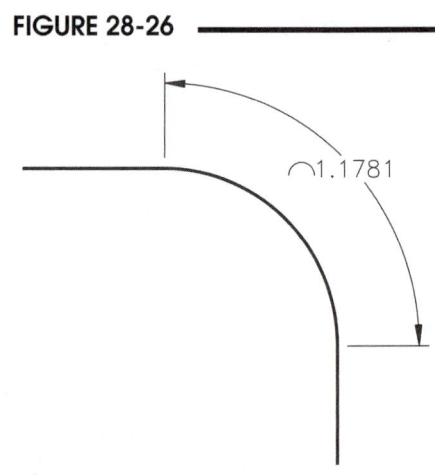

    Command: **dimarc**
    Select arc or polyline arc segment: **PICK**
    Specify arc length dimension location, or
    [Mtext/Text/Angle/Partial/Leader]: **PICK**
    Dimension text = *nn*
    Command:

The *Mtext/Text/Angle* options operate similarly to the *Dimradius* command (see "*Dimradius*"). Use the *Leader* option to draw an additional leader from the dimension text to the arc.

Although the *Dimarc* command by default measures the entire length of the selected arc, you can use the *Partial* option to dimension only part of an arc. You are prompted to select two points on the arc to measure. You can use *Osnaps* to select these two points. In this case, the extension lines for a *Partial* arc length are not perpendicular to the arc as they are with the default (full length) option.

    Specify arc length dimension location, or [Mtext/Text/Angle/Partial/Leader]: **p**
    Specify first point for arc length dimension: **PICK**
    Specify second point for arc length dimension: **PICK**
    Specify arc length dimension location, or [Mtext/Text/Angle/Partial]: **PICK**
    Dimension text = *nn*
    Command:

Dimcenter

Pull-down Menu	Command (Type)	Alias (Type)	Short-cut	Screen (side) Menu	Tablet Menu
Dimension Center Mark	*Dimcenter*	*DCE*	...	*DIMNSION Center*	*X,2*

The *Dimcenter* command draws a center mark on any selected *Arc* or *Circle*. As shown earlier, the *Dimdiameter* command and the *Dimradius* command sometimes create the center marks automatically.

The command requires you only to select the desired *Circle* or *Arc* to acquire the center marks:

    Command: **dimcenter**
    Select arc or circle: **PICK**
    Command:

No matter if the center mark is created by the *Center* command or by the *Diameter* or *Radius* commands, the center mark can be either a small cross or complete center lines extending past the *Circle* or *Arc* (Fig. 28-27). The type of center mark drawn is controlled by the *Mark* or *Line* setting in the *Lines and Arrows* tab of the *Dimension Style Manager*. It is suggested that a new *Dimension Style* be created with these settings just for drawing center marks. (See Chapter 29.)

FIGURE 28-27

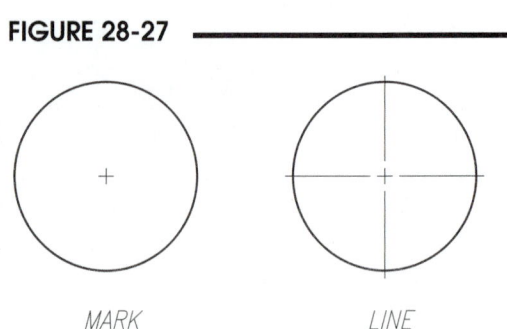

MARK   LINE

When you are dimensioning, short center marks should be used for *Arcs* of less than 180 degrees, and full center lines should be drawn for *Circles* and for *Arcs* of 180 degrees or more (Fig. 28-28).

FIGURE 28-28

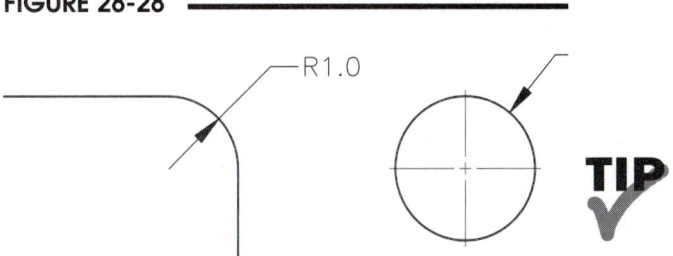

NOTE: Since the center marks created with the *Dimcenter* command are <u>not</u> associative, they may be *Trimmed*, *Erased*, or otherwise edited. The center marks created with the *Dimradius* or *Dimdiameter* commands <u>are</u> associative and cannot be edited.

Dimangular

Pull-down Menu	Command (Type)	Alias (Type)	Short-cut	Screen (side) Menu	Tablet Menu
Dimension Angular	Dimangular	DIMANG or DAN	...	DIMNSION Angular	X,3

The *Dimangular* command provides many possible methods of creating an angular dimension.

A typical angular dimension is created between two *Lines* that form an angle (of other than 90 degrees). The dimension line for an angular dimension is radiused with its center at the vertex of the angle (Fig. 28-29). A *Dimangular* dimension automatically adds the degree symbol (°) to the dimension text. The dimension text format is controlled by the current settings for *Units* in the *Dimension Style* dialog box.

FIGURE 28-29

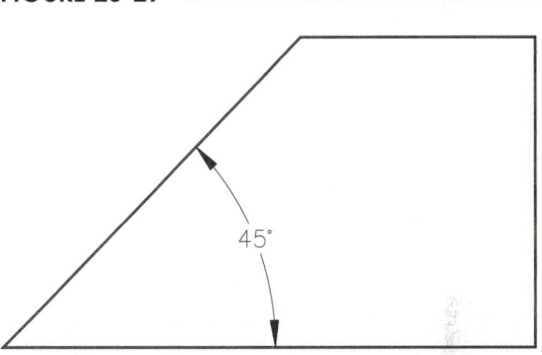

AutoCAD automates the process of creating this type of dimension by offering options within the command syntax. The default options create a dimension, as shown here:

    Command: **dimangular**
    Select arc, circle, line, or <specify vertex>: **PICK**
    Select second line: **PICK**
    Specify dimension arc line location or [Mtext/Text/Angle]: **PICK**
    Dimension text = *nn*
    Command:

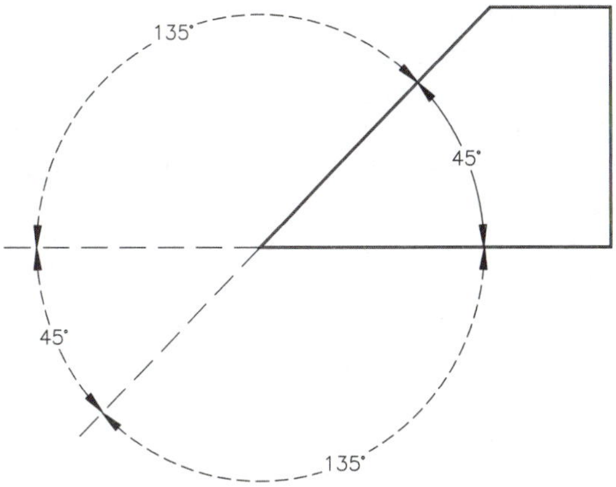

Dimangular dimensioning offers some very useful and easy-to-use options for placing the desired dimension line and text location.

At the "Specify dimension arc line location or [Mtext/Text/Angle]:" prompt, you can move the cursor around the vertex to dynamically display possible placements available for the dimension. The dimension can be placed in any of four positions as well as any distance from the vertex (Fig. 28-30). Extension lines are automatically created as needed.

FIGURE 28-30

The *Dimangular* command offers other options, including dimensioning angles for *Arcs*, *Circles*, or allowing selection of any three points.

If you select an *Arc* in response to the "Select arc, circle, line, or <specify vertex>:" prompt, AutoCAD uses the *Arc's* center as the vertex and the *Arc's* endpoints to generate the extension lines. You can select either angle of the *Arc* to dimension (Fig. 28-31).

FIGURE 28-31 ————————

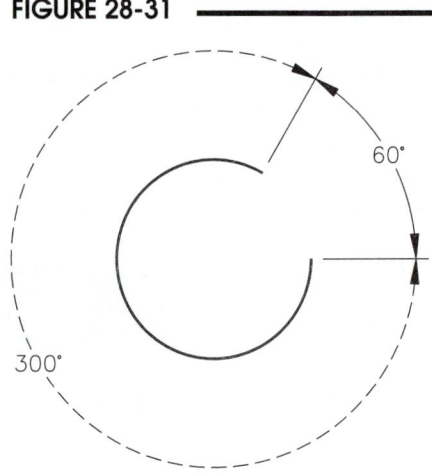

If you select a *Circle*, AutoCAD uses the PICK point as the first extension line origin. The second extension line origin does not have to be on the *Circle*, as shown in Figure 28-32.

If you press Enter in response to the "Select arc, circle, line, or <specify vertex>:" prompt, AutoCAD responds with the following:

> Specify angle vertex: **PICK**
> Specify first angle endpoint: **PICK**
> Specify second angle endpoint: **PICK**

This option allows you to apply an *Angular* dimension to a variety of shapes.

FIGURE 28-32 ————————

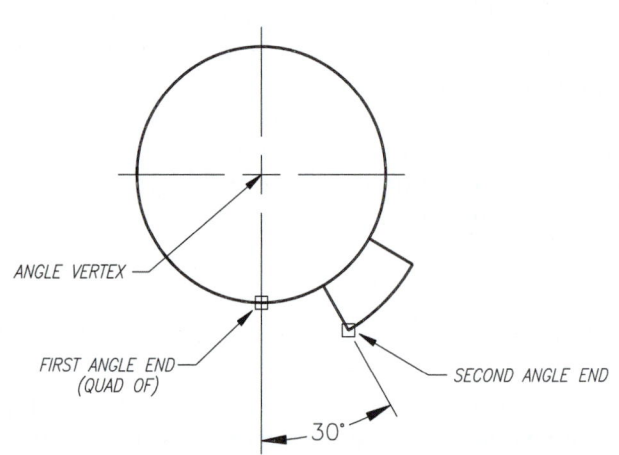

Leader

Pull-down Menu	Command (Type)	Alias (Type)	Short-cut	Screen (side) Menu	Tablet Menu
...	Leader	LEAD	...	...	...

The *Leader* command (not *Qleader*) allows you to create an associative leader similar to that created with the *Diameter* command. The *Leader* command is intended to give dimensional notes such as the manufacturing or construction specifications shown next.

FIGURE 28-33 ————————

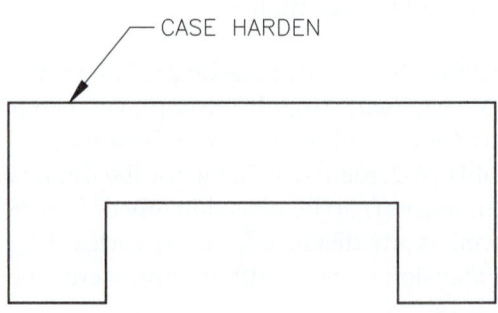

Command: *leader*
Specify leader start point: **PICK**
Specify next point: **PICK**
Specify next point or [Annotation/Format/Undo] <Annotation>: **Enter**
Enter first line of annotation text or <options>: **CASE HARDEN**
Enter next line of annotation text: **Enter**
Command:

At the "start point:" prompt, select the desired location for the arrow. You should use an *Osnap* option (such as *Nearest*, in this case) to ensure the arrow touches the desired object.

A short horizontal line segment called the "hook line" is automatically added to the last line segment drawn if the leader line is 15 degrees or more from horizontal. Note that command syntax for the *Leader* in Figure 28-33 indicates only one line segment was PICKed. A *Leader* can have as many segments as you desire.

If you do not enter text at the "Annotation:" prompt, another series of options are available:

Command: *leader*
Specify leader start point: **PICK**
Specify next point: **PICK**
Specify next point or [Annotation/Format/Undo] <Annotation>: **Enter**
Enter first line of annotation text or <options>: **Enter**
Enter an annotation option [Tolerance/Copy/Block/None/Mtext] <Mtext>:

Format

This option produces another list of choices:

Enter leader format option [Spline/STraight/Arrow/None] <Exit>:

Spline/STraight

You can draw either a *Spline* or straight version of the leader line with these options. The resulting *Spline* leader line has the characteristics of a normal *Spline*. An example of a *Splined* leader is shown in Figure 28-34.

FIGURE 28-34

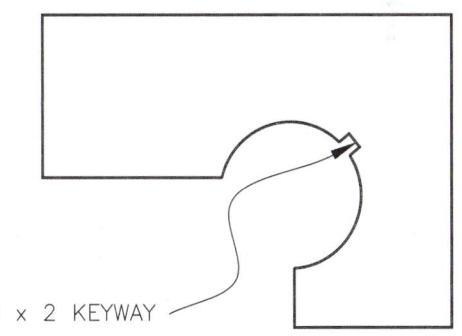

Arrow/None

This option draws the leader line with or without an arrowhead at the start point.

Annotation

This option prompts for text to insert at the end of the *Leader* line.

Mtext

The *Text Formatting* Editor appears with this option. Text can be entered into paragraph form and you can use the *Mtext* options (see "*Mtext*," Chapter 18).

Tolerance

This option produces a feature control frame using the *Geometric Tolerances* dialog boxes (see "*Tolerance*").

Copy

You can copy existing *Text* or *Mtext* objects from the drawing to be placed at the end of the *Leader* line. The copied object is associated with the *Leader* line.

Block

An existing *Block* of your selection can be placed at the end of the *Leader* line. The same prompts as the *Insert* command are used.

None

Using this option draws the *Leader* line with no annotation.

Undo

This option undoes the last vertex point of the *Leader* line.

A *Leader* is affected by the current dimension style settings. You can control the *Leader's* arrowhead type, scale, color, etc., with the related dimensioning variables (see Chapter 29).

Qleader

Pull-down Menu	Command (Type)	Alias (Type)	Short-cut	Screen (side) Menu	Tablet Menu
Dimension Leader	Qleader	LE	...	DIMNSION Leader	W,2

Qleader (quick leader) creates a leader with or without text similar in appearance to a leader created with *Leader*. However, *Qleader* allows you to specify <u>preset</u> parameters for the configuration of the text and the leader. These presets prevent you from having to specify the same parameters repeatedly when you create several leaders in a similar fashion for the current drawing. For example, you can specify what type of text object (*Mtext*, *Block*, *Tolerance*, etc.), how many line segments for the leader, what angle to draw the leader lines, what type of arrowheads, and other parameters for the leaders you create.

```
Command: qleader
Specify first leader point, or [Settings]: PICK (arrow location)
Specify next point: PICK (end of leader)
Specify next point: Enter
Specify text width <0.0000>: Enter
Enter first line of annotation text <Mtext>: Enter text or press Enter
Enter next line of annotation text: Enter text or press Enter
Command:
```

The command sequence above is used to create a one-segment leader with one string of text using the default *Qleader* settings. If you enter a value in response to the "Specify text width <0.0000>:" prompt, the value determines the width of the *Mtext* paragraph. As an alternative, you can press Enter to produce the *Text Formatting* Editor.

With *Qleader*, you specify the preset configurations by pressing Enter or entering the letter "s" at the first prompt: "Specify first leader point, or [Settings]<Settings>:." This action produces the *Leader Settings* dialog box. Settings you make in this dialog box control the appearance of subsequent leaders you create in the drawing until a setting is changed. The settings are <u>saved in the current drawing</u> and are not registered in your system. There are three tabs in the dialog box. The *Annotation* tab is described next.

Annotation Tab

This tab allows you to control presets for the appearance of the leader text as further explained below (Fig. 28-35).

FIGURE 28-35

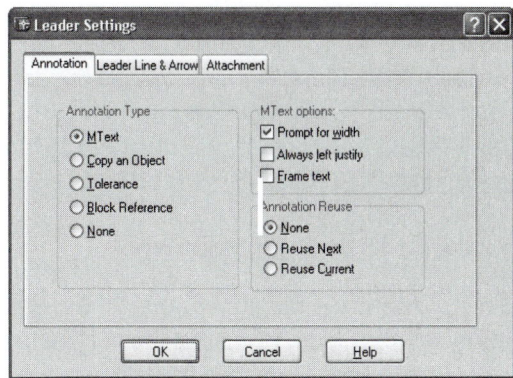

Annotation Type

This section specifies the type of text objects created (*Mtext*, *Copy an Object*, *Tolerance*, *Block Reference*, or *None*) and the related prompts when you use *Qleader*.

The *Mtext* option creates an *Mtext* object. Depending on your settings in the *Mtext Options* section, you can specify the width of each paragraph you create by responding to the "Specify text width <0.0000>:" prompt, use the *Text Formatting* Editor, or press Enter to input several individual lines of text.

Specifying *Copy an Object* causes *Qleader* to prompt you to select an existing text object to copy. The text object is then automatically attached to the leader end.

Use the *Tolerance* button to cause *Qleader* to produce the *Geometric Tolerance* dialog box. Here you specify text and symbol content for feature control frames that are attached to the leader end (see *"Tolerance"* for details on geometric tolerancing and this dialog box).

Use the *Block Reference* option if you want to attach a *Block* to the leader end. *Qleader* prompts you to specify an existing *Block* reference defined in the current drawing by producing the "Enter block name or [?]:" prompt. Enter a question mark (?) to list existing *Blocks* in the current drawing.

To create leaders without text attached, select *None* as the *Annotation Type*.

Mtext Options

This area is enabled only when *Mtext* is the specified *Annotation Type*. You can select none, one or two options in the section.

When *Prompt for width* is checked, this setting produces the "Specify text width <0.0000>:" prompt at the command line. Enter a value for the paragraph width or press Enter to produce the *Text Formatting* Editor. When this setting is not checked, the width prompt does not appear, but you can still produce the *Text Formatting* Editor by pressing Enter at the "Enter first line of annotation text:" prompt.

When a leader is drawn from right to left so it is attached to the right side of the text paragraph (Fig. 28-36), AutoCAD normally right-justifies the text. Checking the *Always Left Justify* option always draws the text left justified.

FIGURE 28-36

Always Left Justify is
checked

Always Left Justify is
not checked

The *Frame Text* option creates a frame around the text (Fig. 28-37).

FIGURE 28-37

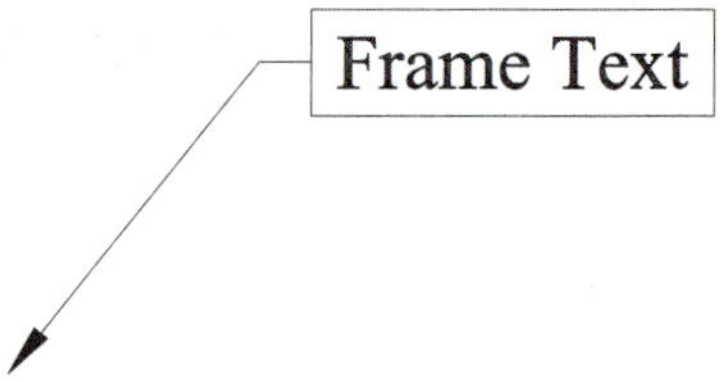

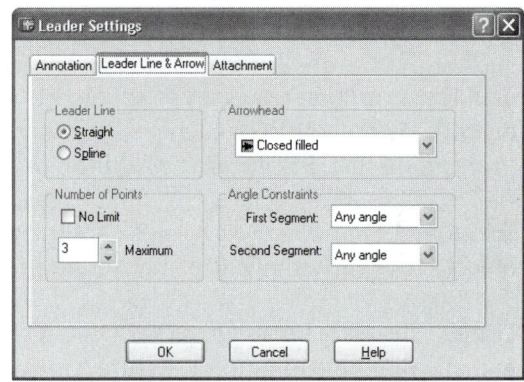

Annotation Reuse

When you want to add the same specification note or manufacturing note at several locations in a drawing, the *Annotation Reuse* option can save considerable time. To use the same annotation repeatedly, check *Reuse Next*. After setting this option, create one leader with the desired text. Subsequent leaders automatically appear with the same text. The *Reuse Current* button becomes active as a reminder.

Leader Line & Arrow Tab

This tab controls the appearance of the leader line and allows you to specify an arrowhead type to use (Fig. 28-38).

FIGURE 28-38

Leader Line

You can select from a *Straight* leader (*Line* segments) or a *Spline* (smooth curved) leader. Set the *Number of Points* to 2 (minimum setting) if you want one line segment (two end-points) or check *No Limit* if you are creating *Spline* leaders, since you need several points to control the shape of a curved leader.

Arrowhead

In the *Arrowhead* section, use the drop-down list to select from many different possibilities for the leaders.

Angle Constraints

This section allows you to specify (force) an angle for the *Qleader* segments to be drawn. For example, selecting 45 causes the *Qleader* to always be drawn at 45 degrees.

Attachment Tab

The *Attachment* tab (Fig. 28-39) is used for setting the attachment point when *Mtext* is the selected annotation type. This tab is disabled unless *Mtext* is specified in the *Annotation* tab.

FIGURE 28-39

The *Text Formatting Attachment* options can be specified for cases when the text is located on the right or left side of the leader (text is located on the right when a leader is drawn from left to right and vice versa). The default position for text attachment (Fig. 28-40) is *Middle of top line* for text on the right side and *Middle of bottom line* for text on the left side.

FIGURE 28-40

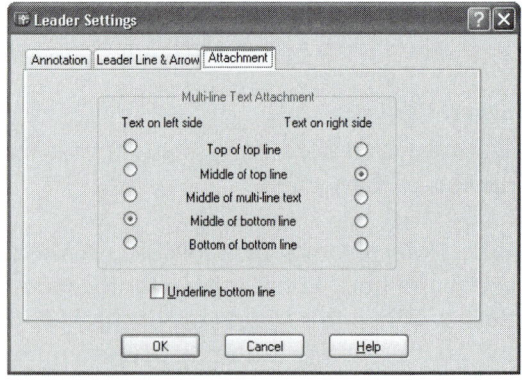

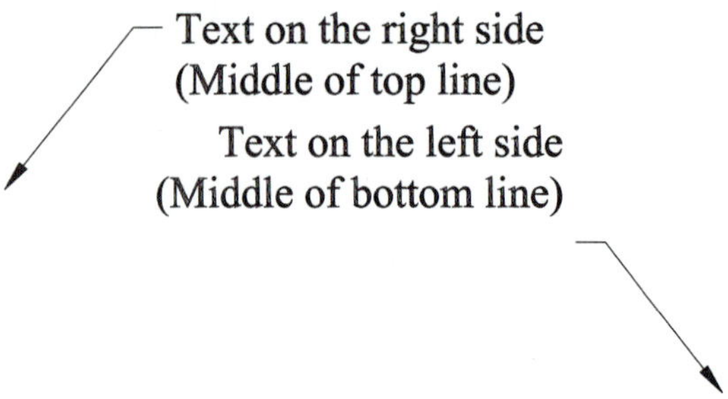

Text on the right side
(Middle of top line)

Text on the left side
(Middle of bottom line)

Dimordinate

Pull-down Menu	Command (Type)	Alias (Type)	Short-cut	Screen (side) Menu	Tablet Menu
Dimension Ordinate	Dimordinate	DIMORD or DOR	...	DIMNSION Ordinate	W,3

Ordinate dimensioning is a specialized method of dimensioning used in the manufacturing of flat components such as those in the sheet metal industry. Because the thickness (depth) of the parts is uniform, only the width and height dimensions are specified as Xdatum and Ydatum dimensions.

Dimordinate dimensions give an Xdatum or a Ydatum distance between object "features" and a reference point on the geometry treated as the origin, usually the lower-left corner of the part. This method of dimensioning is relatively simple to create and easy to understand. Each dimension is composed only of one leader line and the aligned numerical value (Fig. 28-41).

FIGURE 28-41

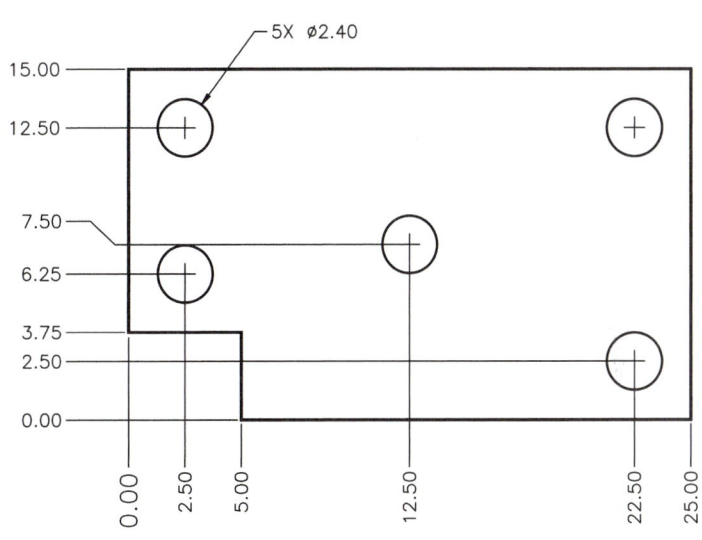

To create ordinate dimensions in AutoCAD, the *UCS* command should be used first to establish a new 0,0 point. *UCS*, which stands for user coordinate system, allows you to establish a new coordinate system with the origin and the orientation of the axes anywhere in 3D space (see Chapters 35 and 36 for complete details). In this case, we need to change only the location of the origin and leave the orientation of the axes as is. Type *UCS* and use the *Origin* option to PICK a new origin as shown (Fig. 28-42).

When you create a *Dimordinate* dimension, AutoCAD only requires you to (1) PICK the object "Feature" and then (2) specify the "Leader endpoint:". The dimension text is automatically aligned with the leader line.

FIGURE 28-42

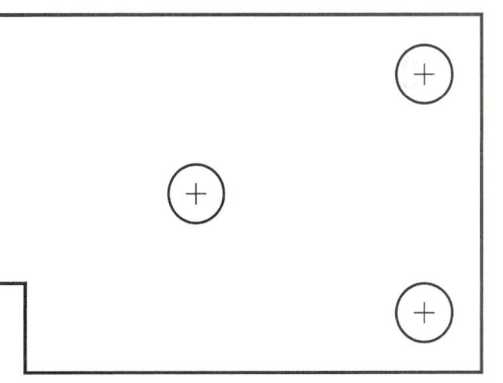

0,0 (UCS ORIGIN)

It is not necessary in most cases to indicate whether you are creating an Xdatum or a Ydatum. Using the default option of *Dimordinate*, AutoCAD makes the determination based on the direction of the leader you specify (step 2). If the leader is perpendicular (or almost perpendicular) to the X axis, an Xdatum is created. If the leader is (almost) perpendicular to the Y axis, a Ydatum is created.

Tolerance

Pull-down Menu	Command (Type)	Alias (Type)	Short-cut	Screen (side) Menu	Tablet Menu
Dimension Tolerance...	*Tolerance*	*TOL*	...	*DIMNSION Toleranc*	*X,1*

Geometric dimensioning and tolerancing (GDT) has become an essential component of detail drawings of manufactured parts. Standard symbols are used to define the geometric aspects of the parts and how the parts function in relation to other parts of an assembly. The *Tolerance* command in AutoCAD produces a series of dialog boxes for you to create the symbols, values, and feature control frames needed to dimension a drawing with GDT. Invoking the *Tolerance* command produces the *Symbol* dialog box and the *Geometric Tolerance* dialog box (see Fig. 28-57 and Fig. 28-58 on the next page). These dialog boxes can also be accessed by the leader commands. Unlike general dimensioning in AutoCAD, geometric dimensioning is <u>not associative</u>; however, you can use *Ddedit* to edit the components of an existing feature control frame.

Some common examples of geometric dimensioning may be a control of the flatness of a surface, the parallelism of one surface relative to another, or the positioning of a group of holes to the outside surfaces of the part. Most of the conditions are controls that cannot be stated with the other AutoCAD dimensioning commands.

This book does not provide instruction on how to use geometric dimensioning; rather, it explains how to use AutoCAD to apply the symbology. The ASME Y14.5M - 1994 Dimensioning and Tolerancing standard is the authority on the topic in the United States. The symbol application presented in AutoCAD is actually based on the 1982 release of the Y14.5 standard. Some of the symbols in the 1994 version are modified from the 1982 standard.

The dimensioning symbols are placed in a feature control frame (FCF). This frame is composed of a minimum of two sections and a maximum of three different sections (Fig. 28-55).

FIGURE 28-55

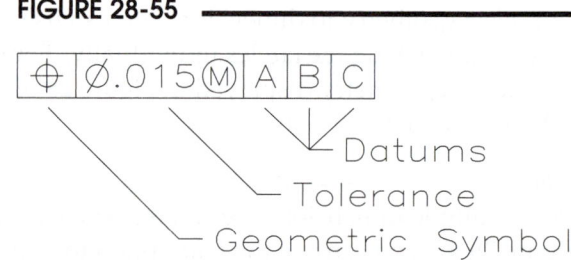

FEATURE CONTROL FRAME

The first section of the FCF houses one of 14 possible geometric characteristic symbols (Fig. 28-56).

The second section of the FCF contains the tolerance information. The third section includes any datum references.

FIGURE 28-56

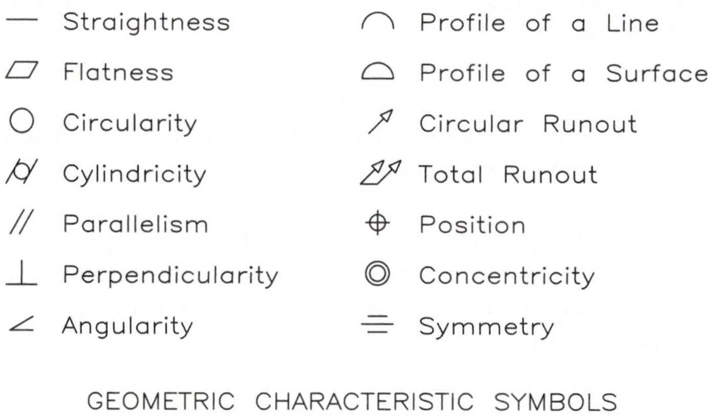

—	Straightness	⌒	Profile of a Line
▱	Flatness	⌓	Profile of a Surface
○	Circularity	⟋	Circular Runout
⌀	Cylindricity	⟋⟋	Total Runout
//	Parallelism	⊕	Position
⊥	Perpendicularity	◎	Concentricity
∠	Angularity	=	Symmetry

GEOMETRIC CHARACTERISTIC SYMBOLS

In AutoCAD, the *Tolerance* command allows you to specify the values and symbols needed for a feature control frame. The lines comprising the frame itself are automatically generated. Invoking *Tolerance* by any method produces the *Geometric Tolerance* dialog box (Fig. 28-57).

The *Geometric Tolerance* dialog box (Fig. 28-57) appears for you to specify the tolerance values and datum. The areas of the *Geometric Tolerance* dialog box are explained below.

FIGURE 28-57

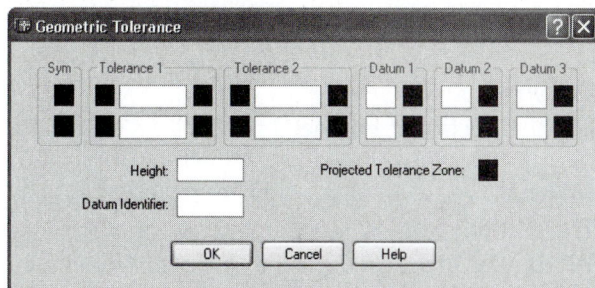

Sym
Click in this empty tile to produce the *Symbol* dialog box (Fig. 28-58). Here you select the geometric characteristic symbol to use for the new feature control frame. The selected symbol then appears in the *Sym* tile in the *Geometric Tolerance* dialog box.

FIGURE 28-58

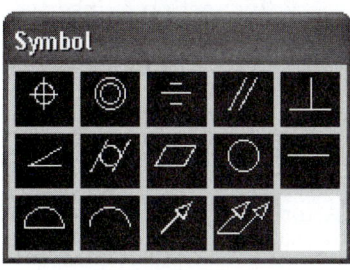

Tolerance 1
This area is used to specify the first tolerance value in the feature control frame. This value specifies the amount of allowable deviation for the geometric feature. The three sections in this cluster are:

Dia (first box)
If a cylindrical tolerance zone is specified, a diameter symbol can be placed before the value by clicking in this area.

Value (second box)
Enter the tolerance value in this edit box.

MC (third box)
A material condition symbol can be placed after the value by choosing this tile. The *Material Condition* dialog box appears for your selection (Fig. 28-59):

FIGURE 28-59

M	Maximum material condition
L	Least material condition
S	Regardless of feature size

Once you select a material condition or cancel, the *Geometric Tolerance* dialog box reappears.

Tolerance 2
Using this section creates a second tolerance area in the feature control frame. GDT standards, however, require only one tolerance area. The options in this section are identical to *Tolerance 1*.

Datum 1
This section allows you to specify the primary datum reference for the current feature control frame. A datum is the exact (theoretical) geometric feature that the current feature references to establish its tolerance zone.

Datum (first box)
Enter the primary datum letter (A, B, C, etc.) for the FCF in this edit box.

MC (second box)
Choosing the *MC* tile allows you to specify a material condition modifier for the datum using the *Material Condition* dialog box (Fig. 28-59).

Datum 2

This section is used if you need to create a second datum reference in the feature control frame. The options are identical to those for *Datum 1*.

Datum 3

Use this section if you need to specify a third datum reference for the current feature control frame.

Height

The edit box allows entry of a value for a projected tolerance zone in the feature control frame. A projected tolerance zone specifies a permissible cylindrical tolerance that is projected above the surface of the part. The axis of the control feature must remain within the stated tolerance.

Projected Tolerance Zone

Click this section to insert a projected tolerance zone symbol (a circled "P") after the value.

Datum Identifier

This option creates a datum feature symbol. The edit box allows entry of the value (letter) indicating the datum. A hyphen should be included on each side of the datum letter, such as "-A-".

After using the dialog boxes and specifying the necessary values and symbols, AutoCAD prompts for you to PICK a location on the drawing for the feature control symbol or datum identifier:

> Enter tolerance location: **PICK**

The feature control symbol or datum identifier is drawn as specified.

Editing Feature Control Frames

The AutoCAD-generated feature control frames are not associative and can be *Erased*, *Moved*, or otherwise edited without consequence to the related geometry objects. Feature control frames created with *Tolerance* are treated as one object; therefore, editing an individual component of a feature control frame is not possible unless *Ddedit* is used. If *Dblclkedit* is *On*, you can double-click on the FCF.

Ddedit is normally used to edit text, but if an existing feature control frame is selected in response to the "Select an annotation object" prompt, the *Geometric Tolerance* dialog box appears. The edit boxes in this dialog box contain the values and symbols from the selected FCF. Making the appropriate changes updates the selected FCF. (See Chapter 18 for detailed information on *Ddedit*.)

Basic Dimensions

Basic dimensions are often required in drawings using GDT. Basic dimensions in AutoCAD are created by using the *Dimension Style Manager* (or by entering a negative value in the *DIMGAP* variable). The procedure is explained briefly in the following example (application 5) and discussed further in Chapter 29, Dimension Styles and Variables.

GDT-Related Dimension Variables

Some aspects of how AutoCAD draws the GDT symbols can be controlled using dimension variables and dimension styles (discussed fully in Chapter 29). These variables are:

DIMCLRE	Controls the color of the FCF
DIMCLRT	Controls the color of the tolerance text
DIMGAP	Controls the gap between the FCF and the text, and controls the existence of a basic dimension box
DIMTXT	Controls the size of the tolerance text
DIMTSTSTY	Controls the style of the tolerance text

Geometric Dimensioning and Tolerancing Example

Six different geometric dimensioning and tolerancing examples are shown on the SPACER drawing in Figure 28-60. Each example is indicated on the drawing and in the following text by a number, 1 through 6. The purpose of the examples is to explain how to use the geometric dimensioning features of AutoCAD. In each example, any method shown in the *Tolerance* command table can be used to invoke the command and dialog boxes.

FIGURE 28-60

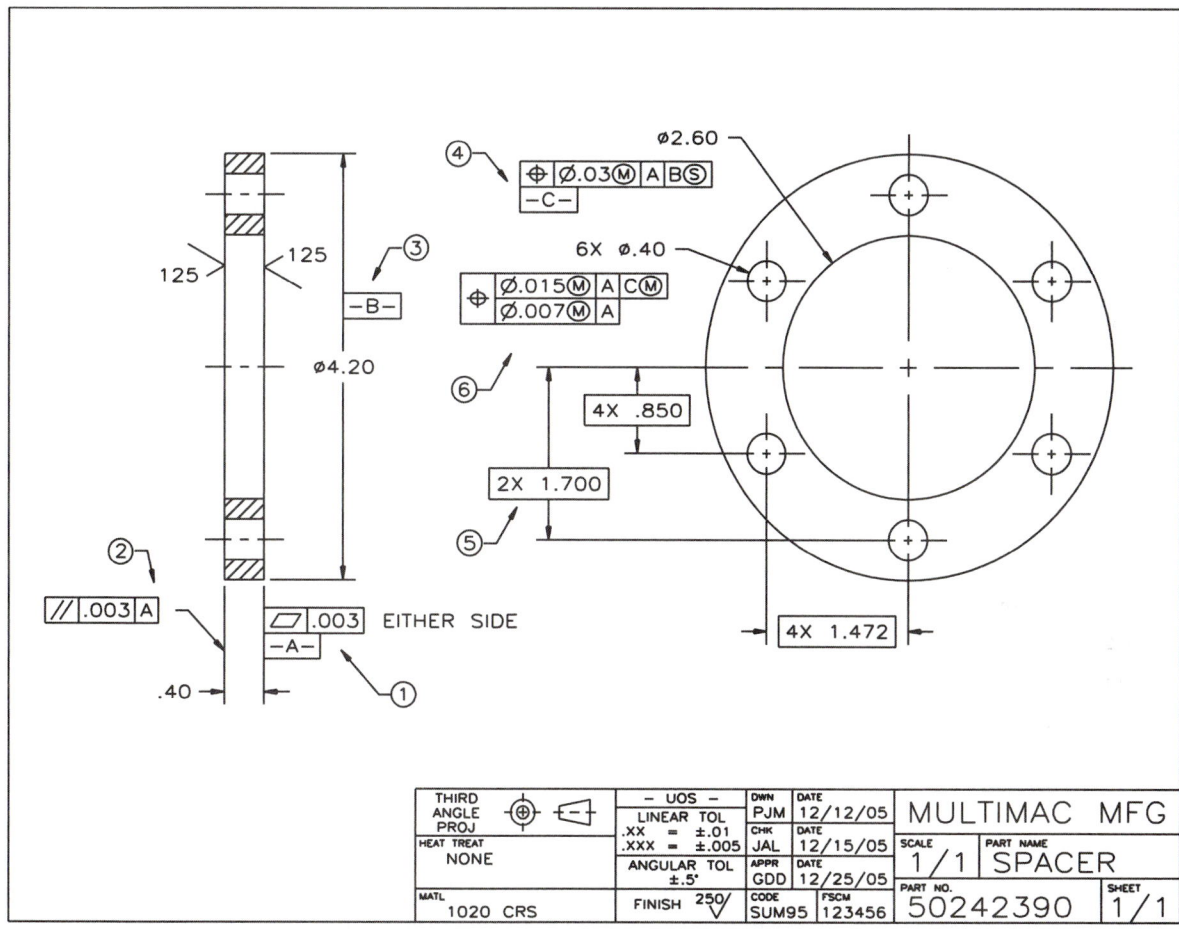

1. **Flatness Application**

   The first specification applied is flatness. In addition to one of the surfaces being controlled for flatness, it is also identified as a datum surface. Both conditions are applied in the same application (Fig. 28-61):

   FIGURE 28-61

   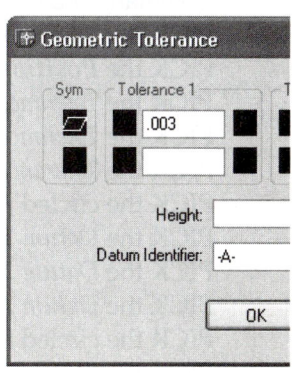

   Command: ***tolerance*** (The *Geometric Tolerance* dialog box appears.)
   PICK the ***Sym*** box. When the *Symbol* dialog box appears, select the ***flatness*** symbol. The *Geometric Tolerance* dialog box reappears.
   PICK the ***Tolerance 1*** edit box and type "**.003**" as the tolerance.
   PICK the ***Datum Identifier*** box and type "**-A-**".
   PICK the ***OK*** button. (The *Geometric Tolerance* dialog box disappears.)
   Enter tolerance location: **PICK** (PICK a point on the right extension line where you want the FCF to be attached, Fig. 28-62, on the next page.)
   Command:

Stretching a rotated dimension's extension line origin allows changing the length of the dimension as well as the length of the extension line. An aligned dimension's extension line origin grip also allows you to change the aligned <u>angle</u> of the dimension (Fig. 28-66).

FIGURE 28-66

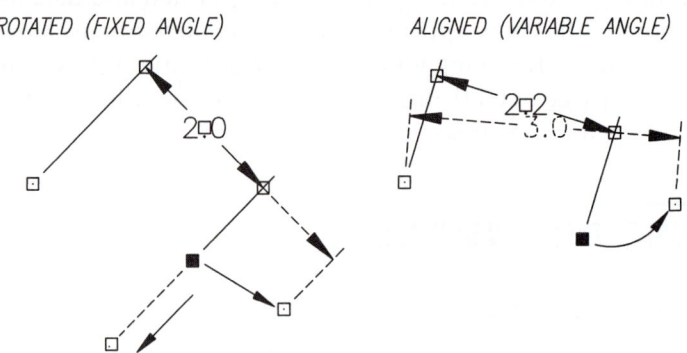

ROTATED (FIXED ANGLE)   ALIGNED (VARIABLE ANGLE)

Rotating the <u>center</u> grip of a radius dimension (with the ROTATE option) allows you to reposition the location of the dimension around the *Arc*. Note that the text remains in its original horizontal orientation (Fig. 28-67).

FIGURE 28-67

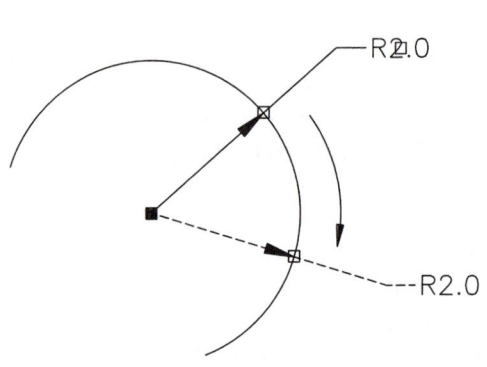

RADIUS

TIP
You can move the dimension <u>text</u> independent of the dimension line with grips for dimensions that have a *DIMTMOVE* variable setting of 2. Use an existing dimension, and change the *DIMTMOVE* setting to 2 (or change the *Fit* setting in the *Dimension Style Manager* to *Over the dimension line without a leader*). The dimension text can be moved to any location without losing its associativity. See "*Fit (DIMTMOVE),*" Chapter 29.

Many other possibilities exist for editing dimensions using grips. Experiment on your own to discover some of the possibilities that are not shown here.

Dimension Right-Click Menu

If you select any dimension, you can right-click to produce a shortcut menu (Fig. 28-68). This menu contains options that allow you to change the text position, the text precision, the *Dimension Style* of the dimension, and flip the arrows.

FIGURE 28-68

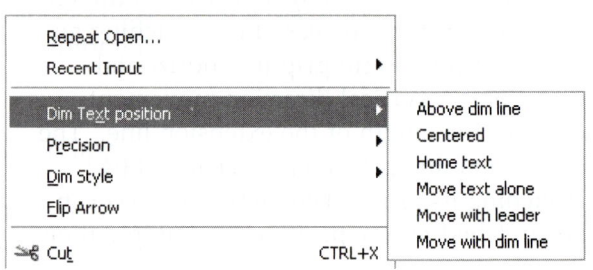

Dim Text Position

Several options are available on this cascading menu. The *Centered* and *Home Text* options are duplicates of the *Dimtedit* command options of the same names (see "*Dimtedit*"). The *Move text alone*, *Move with leader*, and *Move with dim line* options change the settings of the *DIMTMOVE* system variable for the dimension. The *Move* options operate best with grips on. For example, you can select the *Move text alone* option, make the text grip hot, then drag the text to any position in the drawing while still retaining the associative quality of the dimension (Fig. 28-69). (For more information on these options, see "*DIMTMOVE*" in Chapter 29.)

FIGURE 28-69

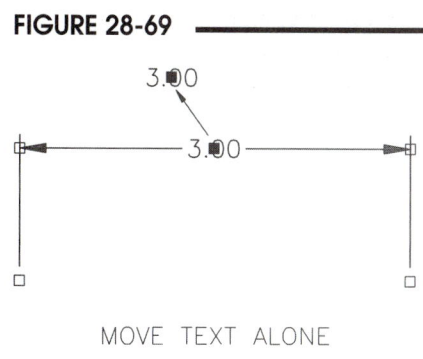

MOVE TEXT ALONE

Precision

Use this cascading menu to select from a list of decimal or fractional precision values. For example, you could change a dimension text value of "5.0000" to "5.00" with this menu. Your choice changes the *DIMDEC* system variable for the selected dimension. (For more information, see "*DIMDEC*" in Chapter 29.)

Dim Style

Use this option to change the selected dimension from one *Dimension Style* to another. Changing a dimension's style affects the appearance characteristics of the dimension. (For more information on *Dimension Styles*, see Chapter 29.)

Flip Arrow

Use this option to flip the arrow nearest to the point you right-click to produce the menu. In other words, if both arrows pointed from the center of the dimension out to the extension lines (default position), flipping both arrows would cause the arrows to be on the outside of the extension lines pointing inward (Fig. 28-70).

FIGURE 28-70

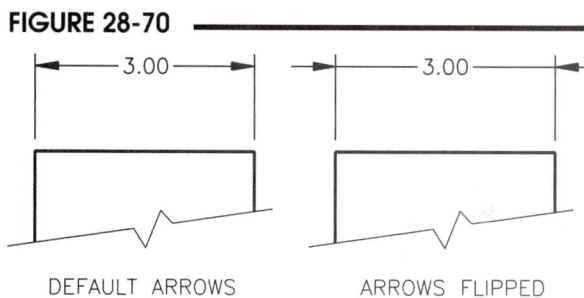

DEFAULT ARROWS   ARROWS FLIPPED

Although the *Centered* and *Home Text* options are duplicates of the *Dimtedit* command, the other options in this shortcut menu actually change system variables for the individual dimensions. These options do not change the variable settings for the *Dimension Style* of the selected dimension, but create dimension style <u>overrides</u> for the individual dimension. (For more information on *Dimension Styles* and dimension style overrides, see Chapter 29.)

Exploding Associative Dimensions

Associative dimensions are treated as one object. If *Erase* is used with an associative dimension, the entire dimension (dimension line, extension lines, arrows, and text) is selected and *Erased*.

Explode can be used to break an associative dimension into its component parts. The individual components can then be edited. For example, after *Exploding*, an extension line can be *Erased* or text can be *Moved*.

There are two main drawbacks to *Exploding* associative dimensions. First, the associative property is lost. Editing commands or *Grips* cannot be used to change the entire dimension but affect only the component objects. More important, Dimension Styles cannot be used with unassociative dimensions.

Dimension Editing Commands

Several commands are provided to facilitate easy editing of existing dimensions in a drawing. Most of these commands are intended to allow variations in the appearance of dimension <u>text</u>. These editing commands operate with associative and nonassociative dimensions, but not with exploded dimensions.

Dimtedit

Pull-down Menu	Command (Type)	Alias (Type)	Short-cut	Screen (side) Menu	Tablet Menu
Dimension Align Text >	Dimtedit	DIMTED	...	DIMNSION Dimtedit	Y,2

Dimtedit (text edit) allows you to change the position or orientation of the text for a single associative dimension. To move the position of text, this command syntax is used:

FIGURE 28-71

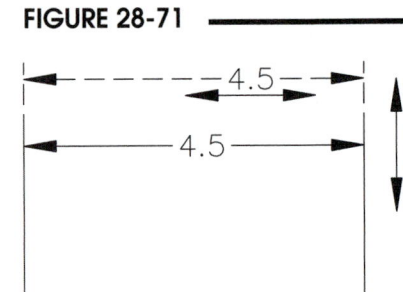

    Command: **Dimtedit**
    Select dimension: **PICK**
    Specify new location for dimension text or
    [Left/Right/Center/Home/Angle]: **PICK**

At the "Specify new location for dimension text" prompt, drag the text to the desired location. The selected text and dimension line can be changed to any position, while the text and dimension line retain their associativity (Fig. 28-71).

Angle

The *Angle* option works with any *Horizontal, Vertical, Aligned, Rotated, Radius,* or *Diameter* dimensions. You are prompted for the new <u>text</u> angle (Fig. 28-72):

FIGURE 28-72

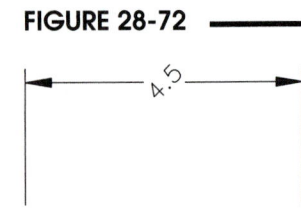

    Command: **dimtedit**
    Select dimension: **PICK**
    Specify new location for dimension text or [Left/Right/Center/Home/Angle]: **a**
    Specify angle for dimension text: **45**
    Command:

Home

The text can be restored to its original (default) rotation angle with the *Home* option. The text retains its right/left position.

Right/Left

The *Right* and *Left* options automatically justify the dimension text at the extreme right and left ends, respectively, of the dimension line. The arrow and a short section of the dimension line, however, remain between the text and closest extension line (Fig. 28-73).

FIGURE 28-73

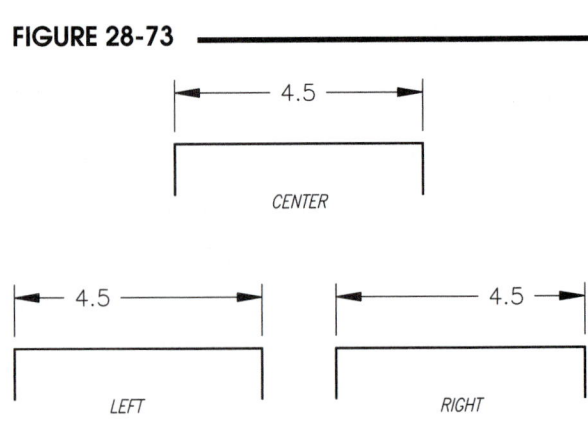

Center

The *Center* option brings the text to the center of the dimension line and sets the rotation angle to 0 (if previously assigned; Fig. 28-73).

This feature is important
feature occurs. For exam
Ø1.00" (see Fig. 28-21).

You can also enter specia
codes can be entered out<u>s</u>
dimensioning commands

Enter:	E
\U+2205	%
\U+00b0	%
	%
	%
\U+00b1	%
	%
\U+*nnnn*	

Associative Dimens

AutoCAD creates associa
that are associated to the
then <u>Osnap</u> to two points
and "Specify second exte
lines and are attached, or
ends of the extension line
extension line origins for
and the arrowhead tips a
mands are associative by

When you create the first
DEFPOINTS. The layer i
layer automatically, so dd

There are two advantage<u>s</u>
typical editing methods,
line components (dimens
automatically updates to
associative dimensions in
Chapter 29 for informatid

Examples of typical editir
Mirror, Move, Rotate, Scale
grip editing options.

Dimensions can have thre

Associative dimensions

Dimedit

Pull-down Menu	Command (Type)	Alias (Type)	Short-cut	Screen (side) Menu	Tablet Menu
Dimension Oblique	Dimedit	DIMED or DED	...	DIMNSION Dimedit	Y,1

The *Dimedit* command allows you to change the angle of the extension lines to an obliquing angle and provides several ways to edit dimension text. Two of the text editing options (*Home, Rotate*) duplicate those of the *Dimtedit* command. Another feature (*New*) allows you to change the text value and annotation.

> Command: **dimedit**
> Enter type of dimension editing [Home/New/Rotate/Oblique] <Home>:

Home

This option moves the dimension text back to its original angular orientation angle without changing the left/right position. *Home* is a duplicate of the *Dimtedit* option of the same name.

New

You can change the text value of any existing dimension with this option. When you invoke this option, the *Text Formatting* Editor appears with "0.0000" as the string of text in the dark background. Enter the desired text, press *OK*, then you are prompted to select a dimension. The text you entered is applied to the selected dimension.

The key feature of the New option is that the original <u>AutoCAD-measured value can be restored</u> in the case that it was "manually" changed for some reason. Simply invoke *Dimedit* and the *New* option, then <u>do not enter any new text</u> in the *Text Formatting* Editor, and press the *OK* button. As prompted next, select the desired dimension, and the original AutoCAD-measured value is automatically restored.

Rotate

AutoCAD prompts for an angle to rotate the text. Enter an absolute angle (relative to angle 0). This option is identical to the *Angle* option of *Dimtedit*.

Oblique

This option is unique to *Dimedit*. Entering an angle at the prompt affects the extension lines. Enter an absolute angle:

> Enter obliquing angle (press ENTER for none):

Normally, the extension lines are perpendicular to the dimension lines. In some cases, it is desirable to set an obliquing angle for the extension lines, such as when dimensions are crowded and hard to read (Fig. 28-74) or for dimensioning isometric drawings (see Chapter 29, "Dimensioning Isometric Drawings").

FIGURE 28-74

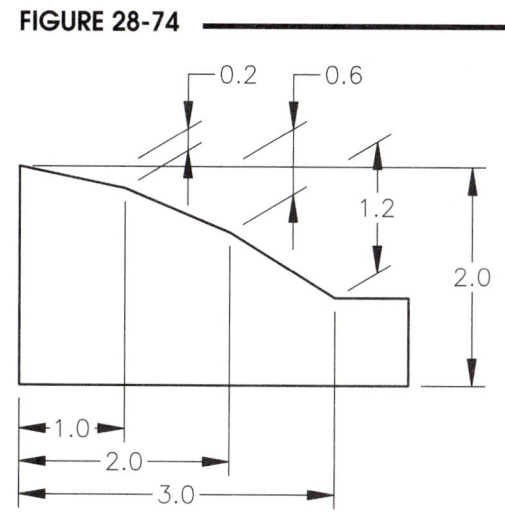

Properties

The *Properties* con
tively for a wide r
dimension produc
properties of the c

The dimension va
the *Lines & Arrow:*
sections. Each sec
(dimension variab
75). Each entry in
changing the entr

Each of the six cat
Alternate Units, an
Style Manager. Ty
Style Manager by s
draw the dimensi
settings for dimer
style appear diffe

Using the *Properti*
sions retroactively
viously created di
information on di
Dimension Styles

Customizing I

As discussed earl
plies as the defaul
In addition, the te
Dimedit.

The *Text* option ar
AutoCAD-measu
value with a dark
dark background
text is not associa
associative dimen

Text placed <u>outsic</u>

prefix/suffix to th
option, or the *Proj*
"2X <>") to retain
or a radius desigr
method <u>overrides</u>

Nonassociative dimensions

(*DIMASSOC* = 1) Nonassociative dimensions do not change automatically when the geometric objects they measure are modified. These dimensions have definition points, but the definition points must be included (selected) if you want the dimensions to change when the geometry they measure is modified. The existing dimensions in a drawing will update when the dimension style is modified. In AutoCAD 2000 and previous releases, these dimensions were called associative.

Exploded dimensions

(*DIMASSOC* = 0) This type of dimension is actually a group of separate objects (lines, arrows, text) rather than a single dimension object. These dimensions do not have definition points and do not change when the geometry is modified. Exploded dimensions do not update when the dimension style is changed. These dimensions can also be created using *Explode* on existing associative or nonassociative dimensions.

Characteristics of the three types of dimensions (associative, nonassociative, and exploded) are explained in the three illustrations here.

An associative dimension is fully associative; that is, if the geometry is changed (in this case by the *Scale* command) the related dimensions components (dimension lines, extension lines, and measured value) automatically update. The shape in Figure 28-76 is a *Pline*, so the *Scale* command affects the entire object. Note that only the single *Pline* object is selected to scale, <u>not the dimensions</u>.

FIGURE 28-76 ────────────

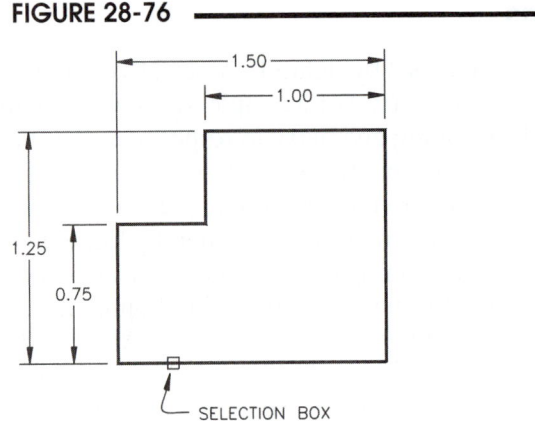

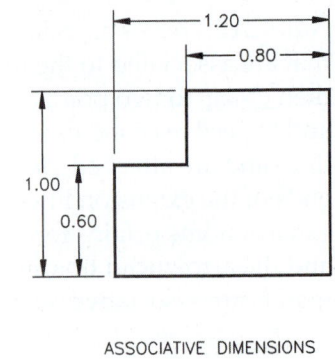

ASSOCIATIVE DIMENSIONS
AFTER *SCALE*

Figure 28-77 displays a similar action—using *Scale* to change an object—but with nonassociative dimensions. Nonassociative dimensions do not automatically change when the related objects are modified unless the dimension definition points are included in the selection set. Note in this case the *Pline* and only two vertical dimensions (1.25 and 0.75) are selected and are therefore scaled, while the horizontal dimensions (1.50 and 1.00) are not selected and therefore are not scaled.

FIGURE 28-77 ────────────

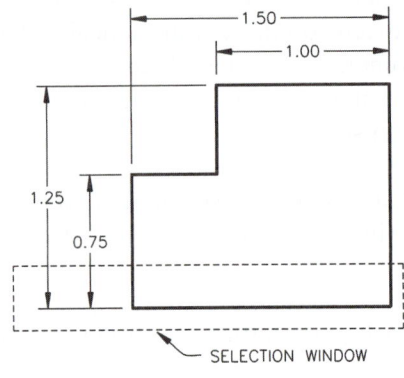

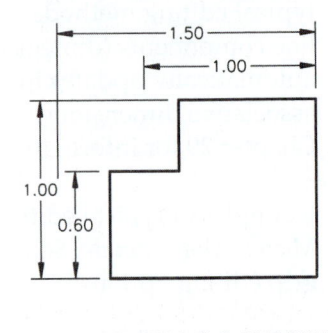

NONASSOCIATIVE DIMENSIONS
AFTER *SCALE*

The characteristics of exploded dimensions are explained in Figure 28-78. Exploded dimensions are composed of several individual and unrelated components (arrows, text, and lines). When *Scale* is used, only the selected individual dimension components are affected, so the resulting dimensions are unusable—essentially, the dimensions fall apart. In addition, exploded dimensions do not update if their original dimension style is modified.

FIGURE 28-78

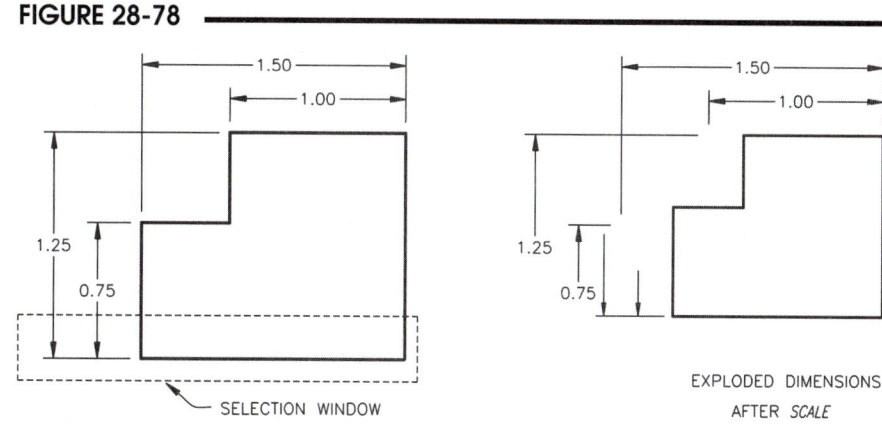

If you want to determine whether a dimension is associative or nonassociative, you can use the *Properties* palette or the *List* command. You can also use the *Quick Select* dialog box to filter the selection of associative or nonassociative dimensions. A dimension is considered associative even if only one end of the dimension is associated with a geometric object. Associativity is *not* maintained between a dimension and a block reference if the block is redefined. Associativity is not possible when dimensioning *Multiline* objects.

Remember that the associativity status for newly created dimensions (associative, nonassociative, or exploded) is controlled by the *DIMASSOC* system variable (2, 1, 0, respectively). You can set the *DIMASSOC* system variable at the Command line or use the *Options* dialog box (Fig. 28-79). In the *User Preferences* tab, a check in the *Make new dimensions associative* box (right center) sets *DIMASSOC* to 2, while no check in this box sets *DIMASSOC* to 1. The *DIMASSOC* setting does not affect existing dimensions, only subsequently created ones. The *DIMASSOC* variable is not saved with a dimension style.

FIGURE 28-79

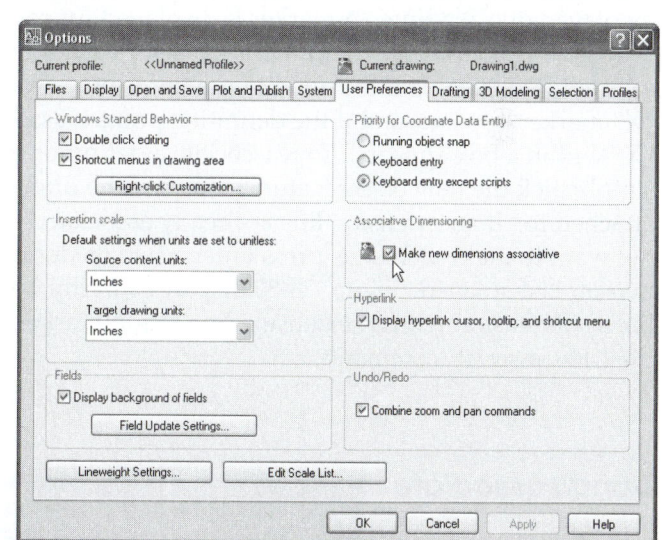

Dimreassociate

Pull-down Menu	Command (Type)	Alias (Type)	Short-cut	Screen (side) Menu	Tablet Menu
Dimension Reassociate Dimensions	Dimreassociate	...	...	...	...

With *Dimreassociate*, you can convert nonassociative dimensions to associative dimensions. In addition, you can change the feature of the drawing geometry (line endpoint, etc.) that an existing definition point is associated with. This command is especially helpful in several situations, such as:

TIP

if you are updating a drawing completed in AutoCAD 2000 or a previous release and want to convert the dimensions to fully associative AutoCAD 2007 dimensions;

if you created nonassociative dimensions in AutoCAD 2007 but want to convert them to associative dimensions;

if you altered the position of an associative dimension's definition points accidentally with grips or other method and want to reattach them;

if you want to change the attachment point for an existing associative dimension.

The *Dimreassociate* command prompts for the geometric features (line endpoints, etc.) that you want the dimension to be associated with. Depending on the type of dimension (linear, radial, angular, diametrical, etc.), you are prompted for the appropriate geometric feature as the attachment point. For example, if a linear dimension is selected to reassociate, AutoCAD prompts you to specify (select attachment points for) the extension line origins.

```
Command: dimreassociate
Select dimensions to reassociate ...
Select objects: PICK  (linear dimension selected)
Select objects: Enter
Specify first extension line origin or [Select object] <next>: PICK
Specify second extension line origin <next>: PICK
Command:
```

When you select the "dimensions to reassociate," the associative and nonassociative elements of the selected dimension(s) are displayed one at a time. A marker is displayed for each extension line origin (or other definition point) for each dimension selected (Fig. 28-80). If the definition point is not associated, an "X" marker appears, and if the definition point is associated, an "X" within a box appears. To reassociate a definition point, simply pick the new object feature you want the dimension to be attached to. If an extension line is already associated and you do not want to make a change, press Enter when its definition point marker is displayed. Figure 28-80 displays a partly associated linear dimension with its extension line origin markers during the *Dimreassociate* command.

FIGURE 28-80

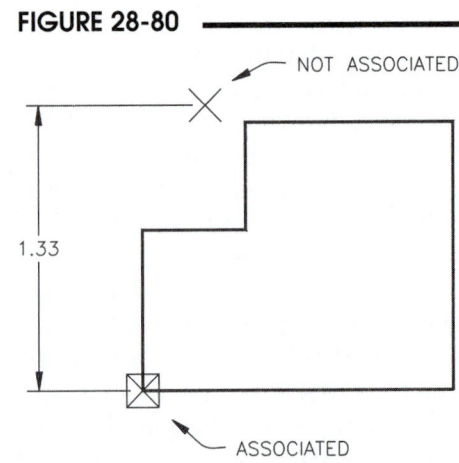

Dimdisassociate

Pull-down Menu	Command (Type)	Alias (Type)	Short-cut	Screen (side) Menu	Tablet Menu
...	Dimdisassociate	...	...	...	...

The *Dimdisassociate* command automatically converts selected associative dimensions to nonassociative. The command prompts for you to select only the dimensions you want to disassociate, then AutoCAD makes the conversion and reports the number of dimensions disassociated.

```
Command: dimdisassociate
Select dimensions to disassociate ...
Select objects: PICK
Select objects: Enter
nn disassociated.
Command:
```

You can use any selection method including *Qselect* to "Select objects." *Dimdisassociate* filters the selection set to include only associative dimensions that are in the current space (model space or paper space layout) and not on locked layers.

Dimregen

Pull-down Menu	Command (Type)	Alias (Type)	Short-cut	Screen (side) Menu	Tablet Menu
...	*Dimregen*	...	...	...	...

Dimregen updates the locations of all associative dimensions in the current drawing. *Dimregen* does not alter the associativity features of dimensions; it affects only the <u>display of associative dimensions</u>. *Dimregen* may be needed to regenerate the display of associative dimensions in three cases:

after panning or zooming with a wheel mouse within a viewport in a paper space layout when dimensions created in paper space are associated with drawing objects in model space (see Chapter 29, "Dimensioning in Paper Space Layouts");

after opening an AutoCAD 2002 to 2007 drawing that has been modified with a previous version of AutoCAD and the dimensioned objects have been modified;

after opening a drawing containing *Xrefs* if the external reference geometry is dimensioned in the current drawing and the *Xrefs* have been modified.

DIMENSIONING VARIABLES INTRODUCTION

Now that you know the commands used to create dimensions, you need to know how to control the appearance of dimensions. For example, after you create dimensions in a drawing, you may need to adjust the size of the dimension text and arrowheads, or change the text style, or control the number of decimal places appearing on the dimension text.

The appearance of dimensions is controlled by setting the dimensioning variables. You can ensure an entire group of dimensions appear the same by using one dimension style for the dimensions. A dimension style is the set of variables that all the dimension objects in the group reference.

Although you can change many settings that control the appearance of dimensions, there is one dimensioning variable that is the most universal, that is, the *Overall Scale* (*DIMSCALE*) of the dimensions. The *Overall Scale* controls all of the size-related features of dimension objects, such as the arrowhead size, the text size, the gap between extension lines and the object, and so on. Even though each of those features can be changed individually, changing the *Overall Scale* controls <u>all</u> of the size-related features of dimensions together.

To change the *Overall Scale*, open the *Dimension Style Manager* (by selecting it from the *Dimension* pull-down menu, *Dimension* toolbar, or by typing *D*). Select the *Modify* button, access the *Fit* tab, and enter the desired value in the *Use overall scale of* edit box (Fig. 28-81). This action changes the overall size for all dimensions created using that particular dimension style.

FIGURE 28-81

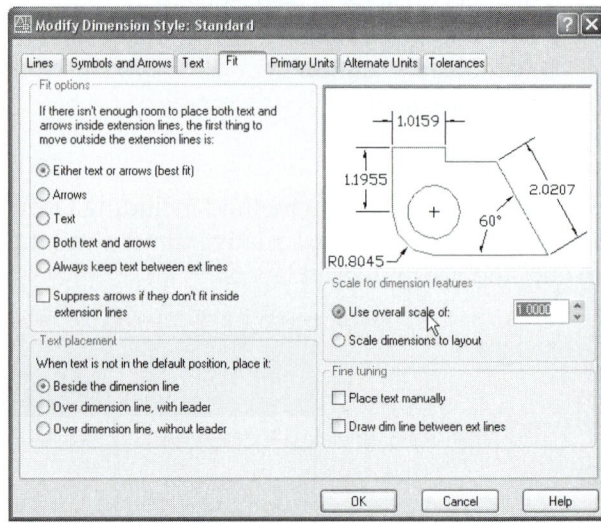

For example, assume you created a drawing and applied the first few dimensions, but noticed that the dimension text was so small that it was barely readable. Instead of changing the dimension variable that controls the size of just the dimension text, change the *Overall Scale*.

This practice ensures that the text, arrowheads, gaps, extension, and all size-related features of the dimension are changed together and are therefore correctly proportioned. Using the *Dimension Style Manager* to change this variable also ensures that all the dimensions created in that dimension style will have the same size.

Chapter 29 explains all the dimensioning variables, including *Overall Scale*, as well as dimension styles. While working on the exercises in this chapter, you will be able to create dimensions without having to change any dimension variables or dimension styles, although you may want to experiment with the *Overall Scale*. Knowledge and experience with <u>both</u> the dimensioning commands and dimensioning variables is essential for AutoCAD users in a real-world situation. Make sure you read Chapter 29 before attempting to dimension drawings other than those in these exercises.

CHAPTER EXERCISES

Only four exercises are offered in this chapter to give you a start with the dimensioning commands. Many other dimensioning exercises are given at the end of Chapter 29, Dimension Styles and Variables. Since most dimensioning practices in AutoCAD require the use of dimensioning variables and dimension styles, that information should be discussed before you can begin dimensioning effectively.

The units that AutoCAD uses for the dimensioning values are based on the *Unit format* and the *Precision* settings in the *Primary Units* tab of the *Dimension Style Manager*. You can dimension these drawings using the default settings, or you can change the *Units* and *Precision* settings for each drawing to match the dimensions shown in the figures. (Type *D*, select *Modify*, then the *Primary Units* tab.)

Other dimensioning features may appear different than those in the figures because of the variables set in the AutoCAD default STANDARD dimension style. For example, the default settings for diameter and radius dimensions may draw the text and dimension lines differently than you desire. After reading Chapter 29, those features in your exercises can be changed retroactively by changing the dimension style.

1. *Open* the **PLATES** drawing that you created in Chapter 9 Exercises. *Erase* the plate on the right. Use *Move* to move the remaining two plates apart, allowing 5 units between. Create a *New* layer called **DIM** and make it *Current*. Dimension the two plates, as shown in Figure 28-82. *Save* the drawing as **PLATES-D**.

FIGURE 28-82

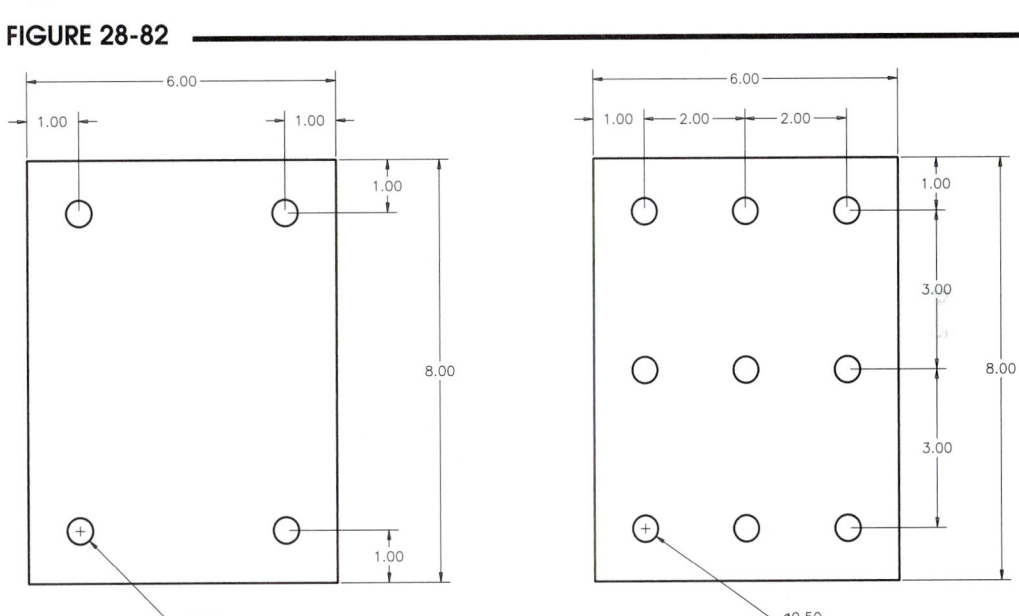

2. *Open* the **PEDIT1** drawing. Create a *New* layer called **DIM** and make it *Current*. Dimension the part as shown in Figure 28-83. *Save* the drawing as **PEDIT1-DIM**. Draw a *Pline* border of *.02 width* and *Insert* **TBLOCK** or **TBLOCKAT**. *Plot* on an A size sheet using *Scale to Fit*.

FIGURE 28-83

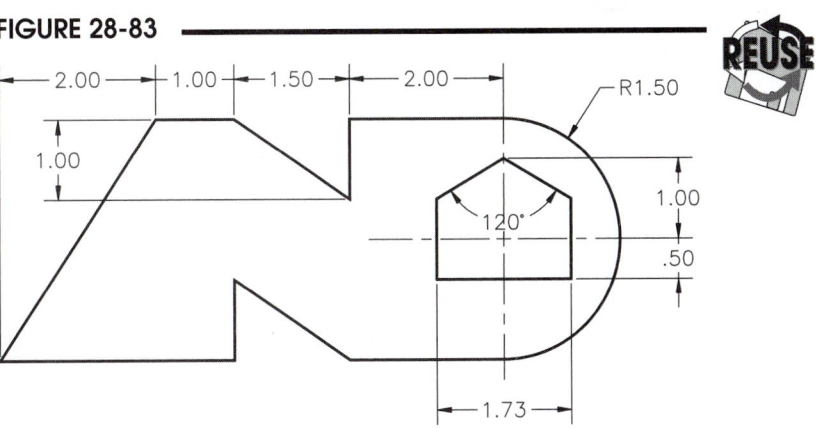

3. *Open* the **GASKETA** drawing that you created in Chapter 9 Exercises. Create a *New* layer called **DIM** and make it *Current*. Dimension the part as shown in Figure 28-84. *Save* the drawing as **GASKETD.**

Draw a *Pline* border with *.02 width* and *Insert* **TBLOCK** or **TBLOCKAT** with an **8/11** scale factor. *Plot* the drawing *Scaled to Fit* on an A size sheet.

FIGURE 28-84

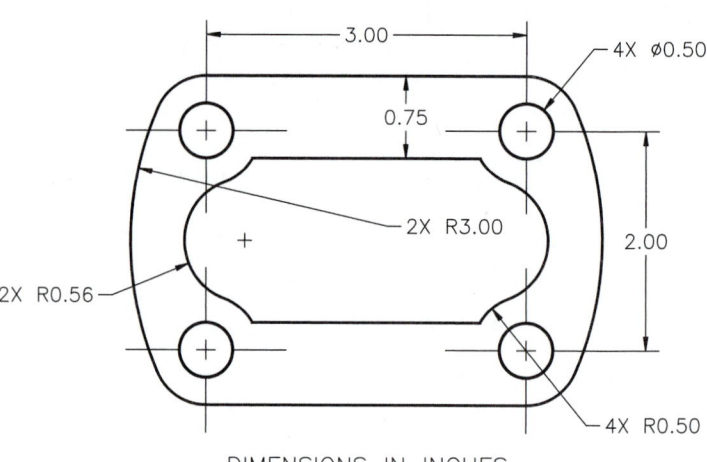

DIMENSIONS IN INCHES

4. *Open* the **BARGUIDE** multiview drawing that you created in Chapter 24 Exercises. Create the dimensions on the **DIM** layer. Keep in mind that you have more possibilities for placement of dimensions than are shown in Figure 28-85.

FIGURE 28-85

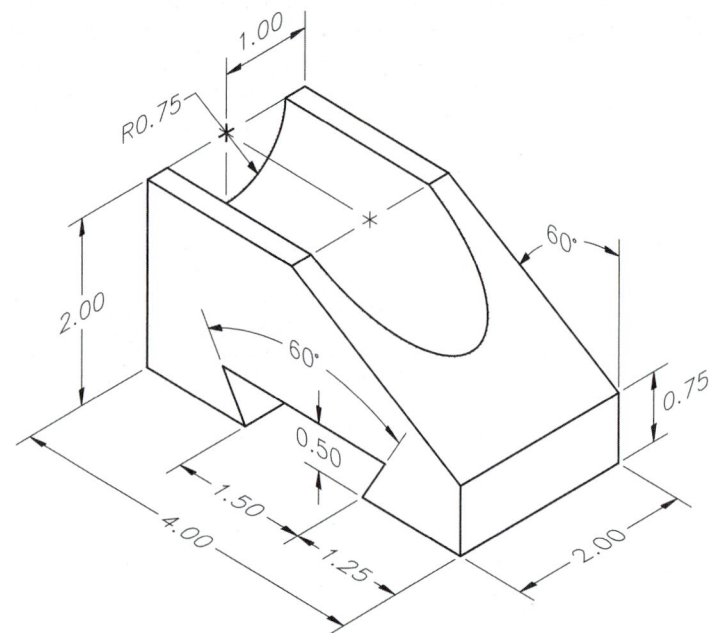

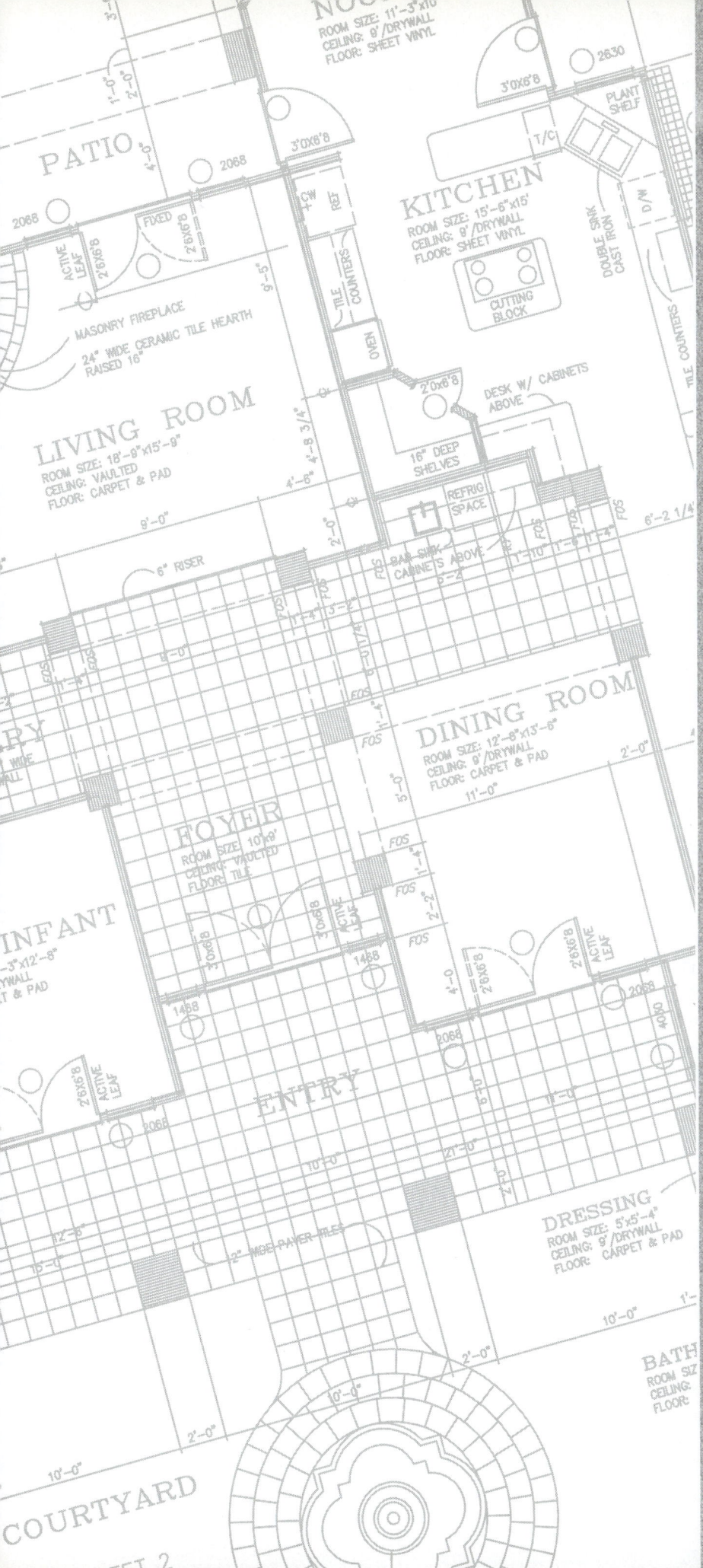

DIMENSION STYLES AND VARIABLES

CHAPTER OBJECTIVES

After completing this chapter you should:

1. be able to control dimension variable settings using the *Dimension Style Manager*;

2. be able to save and restore dimension styles with the *Dimension Style Manager*;

3. know how to create dimension style families and specify variables for each child;

4. know how to create and apply dimension style overrides;

5. be able to modify dimensions using *Update*, *Dimoverride*, *Properties*, and *Matchprop*;

6. know the guidelines for dimensioning;

7. be able to create associative dimensions in paper space;

8. know how to dimension isometric drawings.

CONCEPTS

Dimension Variables

Since a large part of AutoCAD's dimensioning capabilities are automatic, some method must be provided for you to control the way dimensions are drawn. A set of 79 dimension variables allows you to affect the way dimensions are drawn by controlling sizes, distances, appearance of extension and dimension lines, and dimension text formats.

An example of a dimension variable is *DIMSCALE* (if typed) or *Overall Scale* (if selected from a dialog box, Fig. 29-1). This variable controls the overall size of the dimension features (such as text, arrowheads, and gaps). Changing the value from 1.00 to 1.50, for example, makes all of the size-related features of the drawn dimension 1.5 times as large as the default size of 1.00 (Fig. 29-2). Other examples of features controlled by dimension variables are arrowhead type, orientation of the text, text style, units and precision, suppression of extension lines, fit of text and arrows (inside or outside of extension lines), and direction of leaders for radii and diameters (pointing in or out). The dimension variable changes that you make affect the appearance of the dimensions that you create.

FIGURE 29-1 ————

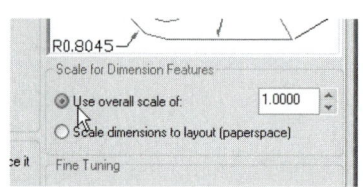

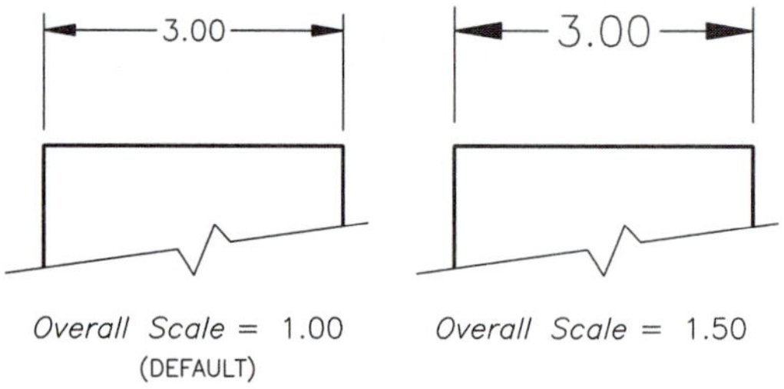

There are two basic ways to control dimensioning variables:

FIGURE 29-2 ————

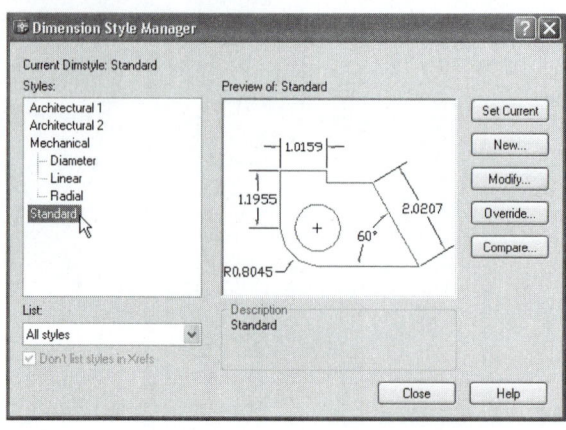

Overall Scale = 1.00   Overall Scale = 1.50
     (DEFAULT)

1.   Use the *Dimension Style Manager* (Fig. 29-3).

2.   Type the dimension variable name in Command line format.

The dialog boxes employ "user-friendly" terminology and selection, while the Command line format uses the formal dimension variable names but cannot be used for all settings.

Changes to dimension variables are usually made before you create the affected dimensions. Dimension variable changes are not always retroactive, in contrast to *LTSCALE*, for example, which can be continually modified to adjust the spacing of existing non-continuous lines. Changes to dimensioning variables affect existing dimensions only when those changes are *Saved* to an existing *Dimension Style* that was in effect when previous dimensions were created. Generally, dimensioning variables should be set before creating the desired dimensions although it is possible to modify dimensions and *Dimension Styles* retroactively.

Dimension Styles

Associative and nonassociative dimensions are part of a dimension style. The default (and the only supplied) dimension style is named STANDARD (Fig. 29-3). Logically, this dimension style has all of the default dimension variable settings for creating a dimension with the typical size and appearance. Similar to layers, you can create, name, and specify settings for any number of dimension styles. Each dimension style contains the dimension variable settings that you select. A dimension style is a group of dimension variable settings that has been saved under a name you assign. When you set a dimension *Current*, AutoCAD remembers and resets that particular combination of dimension variable settings.

FIGURE 29-3 ————

You can create a dimension style for each particular "style" of dimension (Fig. 29-4). For example, you may want to create one dimension style for most of your normal dimensions, another dimension style for limit dimensions (as shown in the diameter dimension), and another style for dimensions without extension lines (as shown in the interior slot). Each time you want to draw a particular style of dimension, select the style name from the list, make it the "current" style, then begin drawing dimensions.

FIGURE 29-4

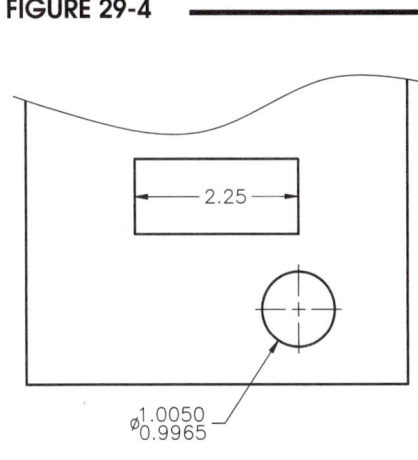

To create a new dimension style, select the *New* button in the *Dimension Style Manager* (see Fig. 29-3), then use the seven tabs that appear in the dialog box to specify the appearance of the dimensions. Selecting options in these tabs (see Fig. 29-14) actually sets values for related dimension variables that are saved in the drawing. When a dimension style is *Set Current*, the dimensions you create with that style appear with the dimension variable settings you specified. Creating dimensions with the current *Dimension Style* is similar to creating text with the current *Text Style*—that is, the objects created take on the current settings assigned to the style.

Another advantage of using dimension styles is that <u>existing</u> dimensions in the drawing can be globally modified by making a change and saving it to the dimension style(s). To do this, select *Modify* in the *Dimension Style Manager* or use the *Save* option of *-Dimstyle* after changing a dimension variable in Command line format. This action saves the changes for newly created dimensions as well as <u>automatically updating existing dimensions in the style</u>.

Dimension Style Families

In the previous chapter, you learned about the various <u>types</u> of dimensions, such as linear, angular, radial, diameter, ordinate, and leader. You may want the appearance of each of these types of dimensions to vary, for example, all radius dimensions to appear with the dimension line arrow pointing out and all diameter dimensions to appear with the dimension line arrow inside pointing in. It is logical to assume that a new dimension style would have to be created for each variation. However, a new dimension style name is not necessary for each type of dimension. AutoCAD provides <u>dimension style families</u> for the purpose of providing variations within each named dimension style.

The dimension style <u>names</u> that you create are assigned to a dimension style <u>family</u>. Each dimension style family can have six <u>children</u>. The children are <u>linear, angular, diameter, radial, ordinate, and leader</u>. Although the children take on the dimension variable settings assigned to the family, you can assign special settings for one or more children. For example, you can make a radius dimension appear slightly different from a linear dimension in that family. Do this by selecting the *New* button in the *Dimension Style Manager*, then selecting the type of dimension (*linear, radius, diameter*, etc.) you want to change from the *Create New Dimension Style* dialog box that appears (Fig. 29-5).

FIGURE 29-5

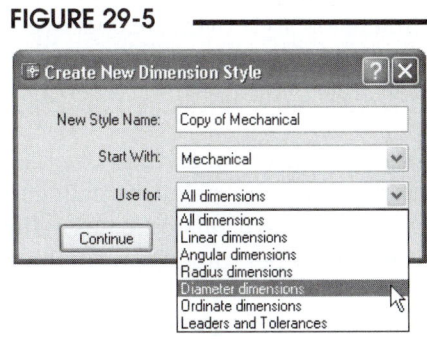

Select *Continue* to specify settings for the child. When a dimension is drawn, AutoCAD knows what type of dimension is created and applies the selected settings to that child.

Dimension styles that have children (special variable settings set for the *linear, radial, diameter*, etc. dimensions) appear in the list of styles in the *Dimension Style Manager* as sub-styles branching from the parent name. For example, notice the "Mechanical" style in Figure 29-3 has special settings for its *Diameter, Linear*, and *Radial* children.

For example, you may want to create the "Mechanical" dimension style to draw the arrows and dimension lines inside the arc but drag the dimension text outside of the arc for *Radial* dimensions (as shown in Fig. 29-6). To do this, create a new child for the "Mechanical" family as previously described, and set "Mechanical" as the current style. When a *Radius* dimension is drawn, the dimension line, text, and arrow should appear as shown in Figure 29-6.

FIGURE 29-6

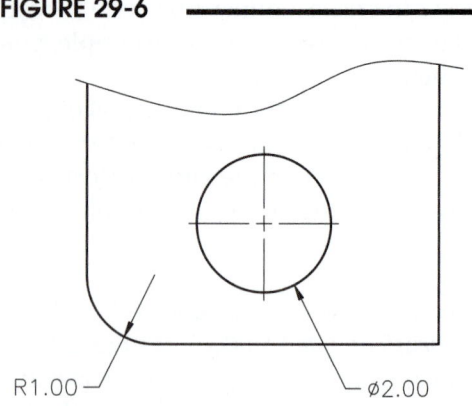

R1.00 —⏌   ⏌— ⌀2.00

In summary, a dimension style family is simply a set of dimension variables related by name. Variations within the family are allowed, in that each child (type of dimension) can possess its own subset of variables. Therefore, each child inherits all the variables of the family in addition to any others that may be assigned individually.

AutoCAD refers to these children as $0, $2, $3, $4, $6, and $7 as suffixes appended to the family name such as "Standard$0." AutoCAD automatically applies the appropriate suffix code to the dimension style name when a child is created. Although the codes do not appear in the *Dimension Style Manager*, you can display the code by using the *List* command and selecting an existing child dimension.

Child Type	Suffix Code
linear	$0
angular	$2
diameter	$3
radial	$4
ordinate	$6
leader	$7

Dimension Style Overrides

If you want only a few dimensions in a dimension style to have a special appearance, you can create a dimension style override. An override does not normally affect the parent style (unless you save the override setting to the style). Therefore, existing dimensions in the drawing are not affected by a dimension style override. You can create a dimension style override two ways.

The more practical method of creating a dimension style override is to select an existing dimension, then use the *Properties* palette to change the desired variable setting. This method (explained in "Modifying Existing Dimensions") creates a dimension style override <u>for the selected dimension only</u> and does not affect any other dimensions in the style.

Alternately, you can create a dimension style override in the *Dimension Style Manager*, then create the dimensions. First, make the desired style *Current*, then select the *Override* button. Proceed to specify the desired variable settings from the seven tabs that appear. A new branch under the family name appears in the *Dimension Style Manager* styles list named "<style overrides>" (see Fig. 29-8 under "Architectural 2"). Finally, create the new dimensions. To clear the overrides, simply make another style current. (Alternately, type any dimension variable name at the Command line and change the value. The override is applied only to the newly created dimensions in the current style and clears when another style is made current.)

DIMENSION STYLES

Dimstyle and Ddim

	Pull-down Menu	Command (Type)	Alias (Type)	Short-cut	Screen (side) Menu	Tablet Menu
	Dimension Style...	*Dimstyle* or *Ddim*	*DIMSTY, DST,* or *D*	...	*DIMNSION Dimstyle*	*Y,3*

The *Dimstyle* or *Ddim* command produces the *Dimension Style Manager* (Fig. 29-7). This is the primary interface for creating new dimension styles and making existing dimension styles current. This dialog box also gives access to seven tabs that allow you to change dimension variables. Features of the *Dimension Style Manager* that appear on the opening dialog box (shown in Fig. 29-7) are described in this section. The tabs that allow you to change dimension variables are discussed in detail later in this chapter in the section "Dimension Variables."

FIGURE 29-7

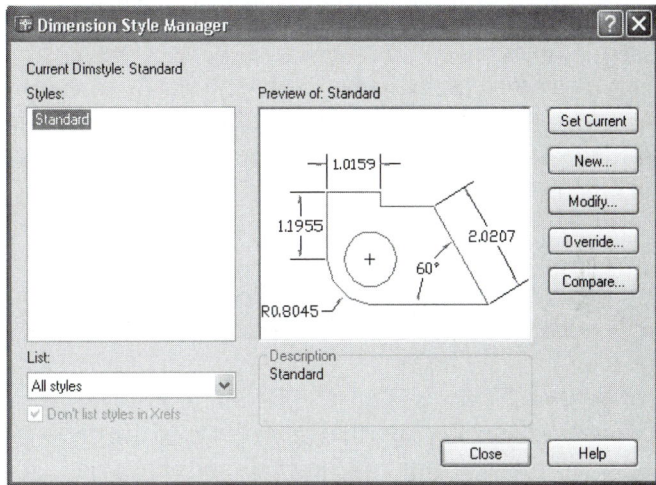

The *Dimension Style Manager* makes the process of creating and using dimension styles easier and more visual than using Command line format. Using the *Dimension Style Manager*, you can *Set Current*, create *New* dimension styles, *Modify* existing dimension styles, and create an *Override*. You also have the ability to view a list of existing styles including children and overrides; see a preview of the selected style, child, or override; examine a description of a dimension style as it relates to any other style; and make an in-depth comparison between styles. You can also control the entries in the list to include or not include Xreferenced dimension styles. The options are explained below.

Styles

This list displays all existing dimension styles in the drawing, including *Xrefs*, depending on your selection in the *List* section below. The list includes the parent dimension styles, the children that are shown on a branch below the parent style, and any overrides also branching from family styles.

The buttons on the right side of the *Dimension Style Manager* affect the highlighted style in the list. For example, to modify an existing style, select the style name from the list and PICK the *Modify* button. If you want to create a *New* style based on an existing style, select the style name you want to use as the "template," then select *New*.

Select (highlight) any dimension style from the *Styles* list to display the appearance of dimensions for that style in the *Preview* tile. Selecting a style from the list also makes the *Description* area list the dimension variables settings compared to the current dimension style.

TIP You can right-click to use a shortcut menu for the selected dimension style (Fig. 29-8). The shortcut menu allows you to *Set Current*, *Rename*, or *Delete* the selected dimension style. Only dimension styles that are unreferenced (no dimensions have been created using the style) can be deleted.

FIGURE 29-8

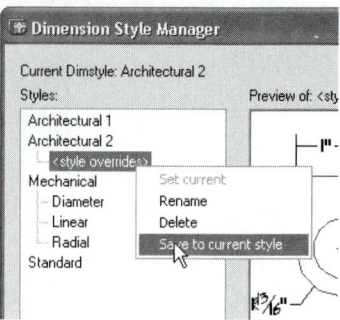

If you have created a style override (by pressing the *Override* button, then setting variables), you can save the overrides to the dimension style using the right-click shortcut menu. Normally, overrides are automatically discarded when another dimension style is made current. This shortcut menu is the only method available in the *Dimension Style Manager* to save overrides. (You can also use the *-Dimstyle* command to save overrides.)

List

Select *All Styles* to display the list of all the previously created or imported dimension styles in the drawing. Selecting *Styles In Use* displays only styles that have been used to create dimensions in the drawing—styles that are saved in the drawing but are not referenced (no dimensions created in the style) do not appear in the list.

Don't List Styles in Xrefs

If the current drawing references another drawing (has an *Xref* attached), the *Xref's* dimension styles can be listed or not listed based on your selection here.

Preview

The *Preview* tile displays the appearance of the high-lighted dimension style. When a dimension style name is selected, the *Preview* tile displays an example of all dimension types (linear, radial, etc.) and how the current variable settings affect the appearance of those dimensions. If a child is selected, the *Preview* tile displays only that dimension type and its related appearance. For example, Figure 29-9 displays the appearance of the selected child, *Mechanical: Radial*.

FIGURE 29-9

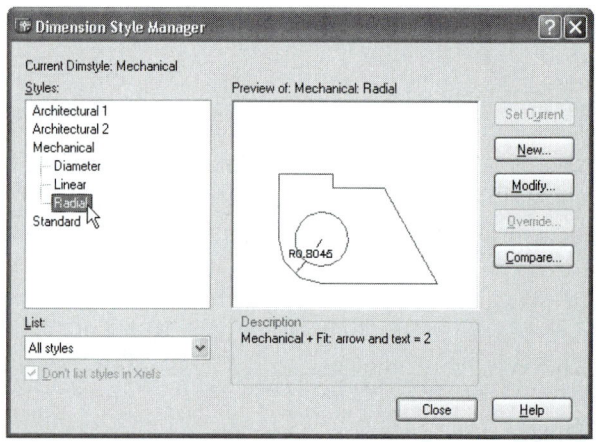

Description

The description area is a valuable tool for determining the variable settings for a dimension style. When a dimension style from the list is selected, the *Description* area lists the differences in variable settings from the current style. For example, assume the "Architectural 2" dimension style was created using "Standard" as the template. To display the dimension variable settings that have been changed (the differences between Standard and Architectural 2), make Standard the current style, then select Architectural 2 and examine the description area, as shown in Figure 29-10. You can use the same procedure to display the variable settings assigned to a child (the differences between the family style and the child), as shown in the *Description* area in Figure 29-9.

FIGURE 29-10

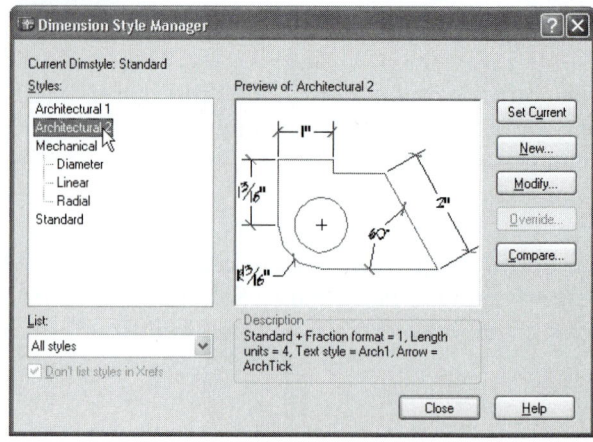

Set Current

Select the desired existing style from the list on the left, then select *Set Current*. The current style is listed at the top of the *Dimension Style Manager*. When you *Close* the dialog box, the current style is used for drawing new dimensions. If you want to set an *Override* to an existing style, you must first make it the current style.

New

Use the *New* button to create a new dimension style. The *New* button invokes the *Create New Dimension Style* dialog box (Fig. 29-11). The options are discussed next.

FIGURE 29-11

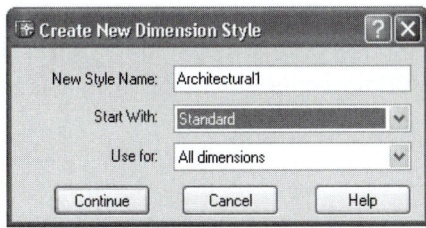

New Style Name

Enter the desired new name in the *New Style Name* edit box. Initially the name that appears is "Copy of (current style)."

Start With

After assigning a new name, select any existing dimension style to use as a "template" from the *Start With* drop-down list. Initially (until dimension variable changes are made for the new style), the new dimension style is actually a copy of the *Start With* style (has identical variable settings). By default, the current style appears in the *Start With* edit box.

Use For

If you want to create a dimension style family, select *All Dimensions*. In this way, all dimension types (*linear, radial, diameter*, etc.) take on the new variable settings. If you want to specify special dimension variable settings for a child, select the desired dimension type from the drop-down list (see Fig. 29-5).

Continue

When the new name and other options have been selected, press the *Continue* button to invoke the *New Dimension Style* dialog box. Here you select from the seven tabs to specify settings for any dimension variable to apply to the new style. See the "Dimension Variables" section for information on setting dimension variables.

NOTE: Generally you should not change the Standard dimension style. Rather than change the Standard style, create *New* dimension styles using Standard as the base style. In this way, it is easy to restore and compare to the default settings by making the Standard style current. If you make changes to the Standard style, it can be difficult to restore the original settings.

Modify

Use this button to change dimension variable settings for an existing dimension style. First, select the desired style name from the list, then press *Modify* to produce the *Modify Dimension Style* dialog box. Using *Modify* to change dimension variables automatically saves the changes to the style and <u>updates existing dimensions</u> in the style, as opposed to using *Override*, which does not change the style or update existing dimensions. See the "Dimension Variables" section for information on setting dimension variables.

Override

A dimension style override is a temporary variable setting that affects only new dimensions created with the override. An override does not affect existing dimensions previously created with the style. You can set an override only for the current style, and the overrides are automatically cleared when another dimension style is made current, unless you use the right-click shortcut menu and select *Save to current style* (see Fig. 29-8).

To create a dimension style override, you must first select an existing style from the list and make it the current style. Next, select *Override* to produce the *Override Current Style* dialog box. See the "Dimension Variables" section for information on setting dimension variables to act as overrides.

Compare

This feature is very useful for examining variable settings for any dimension style. Press the *Compare* button to produce the *Compare Dimension Styles* dialog box (Fig. 29-12). This dialog box lists only dimension variable settings; changes to variable settings cannot be made from this interface.

The *Compare Dimension Styles* dialog box lists the differences between the style in the *Compare* box and the style in the *With* box. For example, assume you created a new style named "Architectural 2" from the "Standard" style, but wanted to know what variable changes you had made to the new style. Select Architectural 2 in the *Compare* box and Standard in the *With* box. The variable changes made to Architectural 2 are displayed in the central area of the box. Also note that the settings for each variable are listed for each of the two styles.

FIGURE 29-12

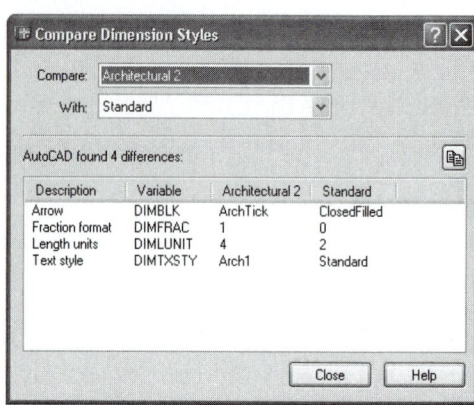

A useful feature of this dialog box is that the formal variable names are listed in the *Variable* column. Many experienced users of AutoCAD prefer to use the formal variable names since they do not change from release to release as the dialog box descriptions do.

You can also use the *Compare Dimension Styles* dialog box to give the entire list of variables and related settings for one dimension style. Do this by selecting the desired dimension style name from the *Compare* drop-down list and select *<none>* from the *With* list (Fig. 29-13).

FIGURE 29-13

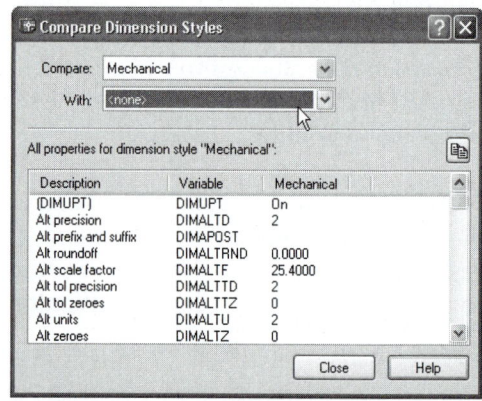

-Dimstyle

Pull-down Menu	Command (Type)	Alias (Type)	Short-cut	Screen (side) Menu	Tablet Menu
Dimension Update	*-Dimstyle*	...	...	*DIMNSION Dimstyle*	*Y,3*

The *-Dimstyle* command is the Command line equivalent to the *Dimension Style Manager*. Operations that are performed by the dialog box can also be accomplished with the *-Dimstyle* command. The *Apply* option offers one other important feature that is <u>not</u> available in the dialog box:

    Command: **-dimstyle**
    Current dimension style: Standard
    Enter a dimension style option [Save/Restore/STatus/Variables/Apply/?] <Restore>:

Save

Use this option to <u>create a new</u> dimension style. The new style assumes all of the current dimension variable settings comprised of those from the current style plus any other variables changed by Command line format (overrides). The new dimension style becomes the current style.

    Enter a dimension style option [Save/Restore/STatus/Variables/Apply/?] <Restore>: *s*
    Enter name for new dimension style or [?]:

Restore

This option prompts for an existing dimension style name to be restored. The restored style becomes the current style. You can use the "select dimension" option to PICK a dimension object that references the style you want to restore.

STatus

Status gives the <u>current settings</u> for all dimension variables. The displayed list comprises the settings of the current dimension style and any overrides. All dimension variables that are saved in dimension styles are listed:

    Enter a dimension style option [Save/Restore/STatus/Variables/Apply/?] <Restore>: *st*

    DIMASO   Off   Create dimension objects
    DIMSTYLE   Standard   Current dimension style (read-only)
    DIMADEC   0   Angular decimal places
    DIMALT   Off   Alternate units selected
    etc.

Variables

You can use this option to <u>list</u> the dimension variable settings for <u>any existing dimension style</u>. You cannot modify the settings.

    Enter a dimension style option [Save/Restore/STatus/Variables/Apply/?] <Restore>: *v*
    Enter a dimension style name, [?] or <select dimension>:

Enter the name of any style or use "select dimension" to PICK a dimension object that references the desired style. A list of all variables and settings appears.

~stylename

This variation can be entered in response to the *Variables* option. Entering a dimension style name preceded by the tilde symbol (~) displays the <u>differences between the current style and the (~) named style</u>.

For example, assume that you wanted to display the <u>differences</u> between STANDARD and the current dimension style (Mechanical, for example). Use *Variables* with the ~ symbol. (The ~ symbol is used as a wildcard to mean "all but.") This option is very useful for keeping track of your dimension styles and their variable settings.

    Command: *-dimstyle*
    Current dimension style: Mechanical
    Enter a dimension style option [Save/Restore/STatus/Variables/Apply/?] <Restore>: *v*
    Enter a dimension style name, [?] or <select dimension>: *~standard*
    Differences between STANDARD and current settings:

                    STANDARD   Current Setting
    DIMSCALE   1.0000   0.7500
    DIMUPT   Off   On
    DIMTXSTY   Standard   Mech1

Apply

Use *Apply* to <u>update dimension objects</u> that you PICK with the current variable settings. The current variable settings can contain those of the current dimension style plus any overrides. The selected object loses its reference to its original style and is "adopted" by the applied dimension style family.

> Enter a dimension style option [Save/Restore/STatus/Variables/Apply/?] <Restore>: **a**
> Select objects:

TIP This is a useful tool for <u>changing an existing dimension object from one style to another</u>. Simply make the desired dimension style current; then use *Apply* and select the dimension objects to change to that style.

Using the *Apply* option of *-Dimstyle* is identical to using the *Update* command. The *Update* command can be invoked by using the *Dimension Update* button, using *Update* from the *Dimension* pull-down menu, or typing *Dim*: (Enter), then *Update*.

DIMENSION VARIABLES

Now that you understand how to create and use dimension styles and dimension style families, let's explore the dimension variables. <u>Two primary methods</u> can be used to set dimension variables: the *Dimension Styles Manager* and Command line format. First, we will examine the dialog box (accessible through the *Dimension Styles Manager*) that allows you to set dimension variables. Dimension variables can alternately be set by typing the formal name of the variable (such as *DIMSCALE*) and changing the desired value (discussed after this section on dialog boxes). The dialog box offers a more "user-friendly" terminology, edit boxes, checkboxes, drop-down lists, and a preview image tile that automatically reflects the variables changes you make.

Changing Dimension Variables Using the Dialog Box Method

In this section, the dialog box that contains the seven tabs for changing variables is explained (see Fig. 29-14). The seven tabs indicate different groups of dimension variables. The tab names are:

Lines, Symbols and Arrows, Text, Fit, Primary Units, Alternate Units, Tolerances

Access to this dialog box from the *Dimension Style Manager* is accomplished by selecting *New, Modify,* or *Override*. Practically, there is only one dialog box that allows you to change dimension variables; however, you might say there are three dialog boxes since the title of the box changes based on your selection of *New, Modify,* or *Override*. Depending on your selection, you can invoke the *Create Dimension Style, Modify Dimension Style,* or *Override Current Style* dialog box. The options in the boxes are identical and only the titles are different; therefore, we will examine only one dialog box.

In the following pages, the heading for the paragraphs below include the dialog box option and the related formal dimension variable name in parentheses. The following figures typically display the AutoCAD default setting for a particular variable and an example of changing the setting to another value. These AutoCAD default settings (in the "Standard" dimension style) are for the ACAD.DWT template drawing and for *Start from Scratch, Imperial Default Settings*. If you use the *Metric Default Settings* or other template drawings, dimension variable settings may be different based on dimension styles that exist in the drawing other than Standard. See "Using *Setup Wizards* and Template Drawings" later in this chapter.

Lines Tab

The options in this tab change the appearance of dimension lines and extension lines (Fig. 29-14). The preview image automatically changes based on the settings you select in the *Dimension Lines* and *Extension Lines* sections.

FIGURE 29-14

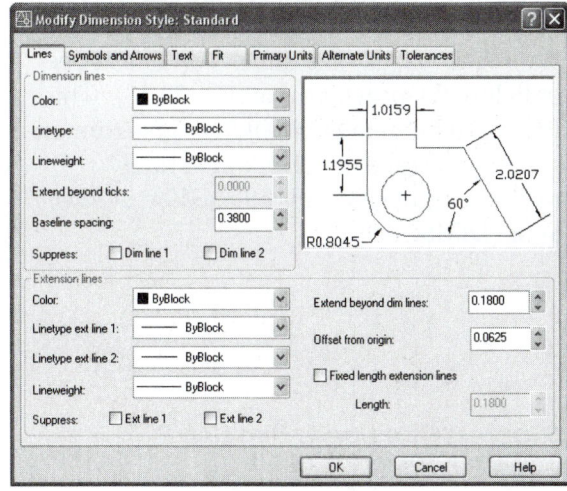

Dimension Lines Section (Lines Tab)

Color (DIMCLRD)

The *Color* drop-down list allows you to choose the color for the dimension line. Assigning a specific color to the dimension lines, extension lines, and dimension text gives you more control when color-dependent plot styles are used because you can print or plot these features with different line widths, colors, etc. This feature corresponds to the *DIMCLRD* variable (dim color dimension line).

Linetype (DIMLTYPE)

Generally, linetypes for dimensions are *Continuous*; however, for special applications you may want to change this setting (Fig. 29-15). Select from the drop-down list. This setting is stored in the *DIMLTYPE* variable.

FIGURE 29-15

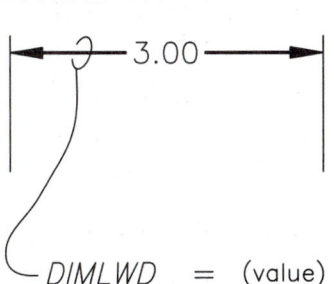

2006

Lineweight (DIMLWD)

Use this option to assign lineweight to dimension lines (Fig. 29-16). Select any lineweight from the drop-down list or enter values in the *DIMLWD* variable (dim lineweight dimension line). Values entered in the variable can be -1 (*ByLayer*), -2 (*ByBlock*), 25 (*Default*), or any integer representing 100th of mm, such as 9 for .09mm.

FIGURE 29-16

Extend beyond ticks (DIMDLE)

The *Extend beyond ticks* edit box is disabled unless the *Oblique, Integral,* or *Architectural Tick* arrowhead type is selected (in the *Arrowheads* section). The *Extend* value controls the length of dimension line that extends past the dimension line (Fig. 29-17). The value is stored in the *DIMDLE* variable (dimension line extension). Generally, this value does not require changing since it is automatically multiplied by the *Overall Scale* (*DIMSCALE*).

FIGURE 29-17

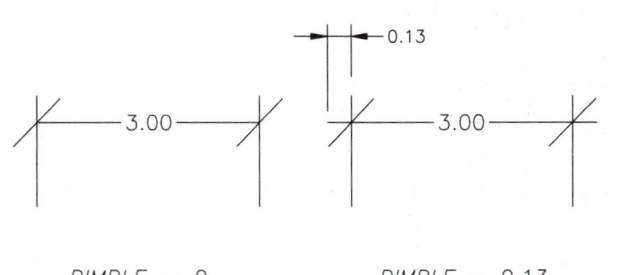

Baseline spacing (DIMDLI)

The *Baseline spacing* edit box reflects the value that AutoCAD uses in baseline dimensioning to "stack" the dimension line above or below the previous one (Fig. 29-18). This value is held in the *DIMDLI* variable (dimension line increment). This value rarely requires input since it is affected by *Overall Scale* (DIMSCALE).

FIGURE 29-18

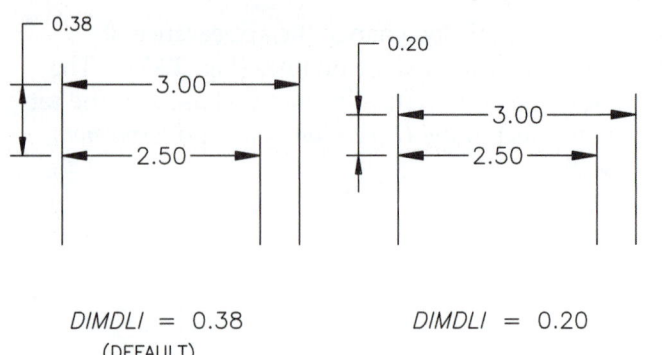

DIMDLI = 0.38
(DEFAULT)

DIMDLI = 0.20

Suppress Dim Line 1, Dim Line 2 (DIMSD1, DIMSD2)

This area allows you to suppress (not draw) the *1st* or *2nd* dimension line or both (Fig. 29-19). The <u>first</u> dimension line would be on the "First extension line origin" side or nearest the end of object PICKed in response to "Select object to dimension." These toggles change the *DIMSD1* and *DIMSD2* dimension variables (suppress dimension line 1, 2).

FIGURE 29-19

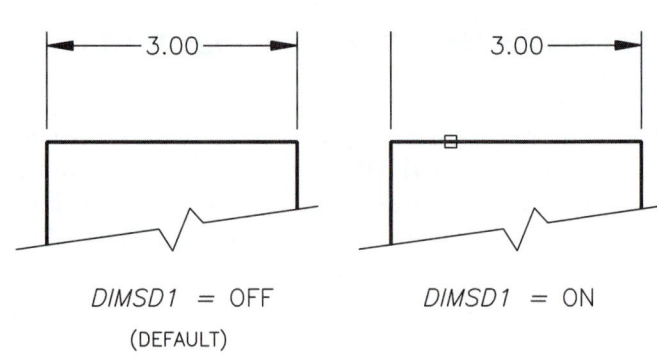

DIMSD1 = OFF
(DEFAULT)

DIMSD1 = ON

Extension Line Section (Lines Tab)

Color (DIMCLRE)

The *Color* drop-down list allows you to choose the color for the extension lines. You can also activate the standard *Select Color* dialog box. The color assignment for extension lines is stored in the *DIMCLRE* variable (dim color extension line).

Linetype Ext Line 1, Ext Line 2 (DIMLTEX1, DIMLTEX2)

These drop-down lists allow you to set the linetype for the first and second extension lines (Fig. 29-20). The linetype settings for extension line 1 and extension line 2 are stored in the *DIMLTEX1* and *DIMLTEX2* variables, respectively.

FIGURE 29-20

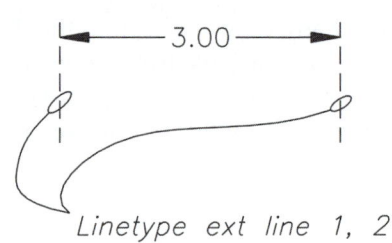

Linetype ext line 1, 2

Lineweight (DIMLWE)

This option is similar to that for dimension lines, only the lineweight is assigned to extension lines (Fig. 29-21). Select any lineweight from the drop-down list or enter values in the *DIMLWE* (dim lineweight extension line) variable (*ByLayer* = -1, *ByBlock* = -2, *Default* = 25, or any integer representing 100th of mm).

FIGURE 29-21

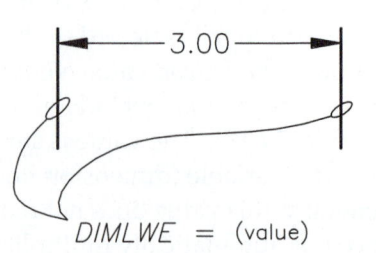

DIMLWE = (value)

2006

Suppress Ext Line 1, Ext Line 2 (DIMSE1, DIMSE2)

This area is similar to the *Dimension Line* area that controls the creation of dimension lines but is applied to extension lines. This area allows you to suppress the *1st* or *2nd* extension line or both (Fig. 29-22). These options correspond to the *DIMSE1* and *DIMSE2* dimension variables (suppress extension line 1, 2).

FIGURE 29-22

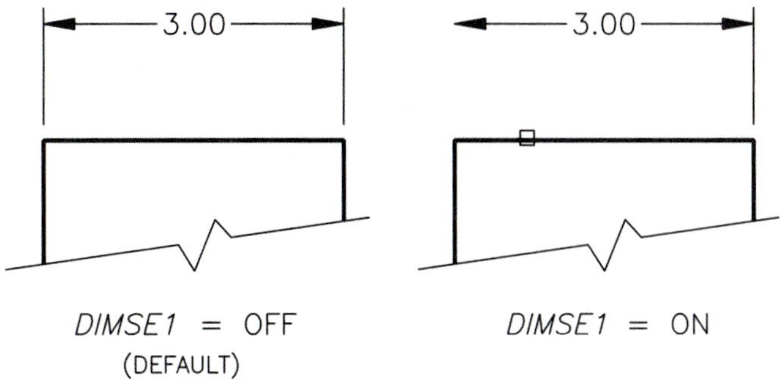

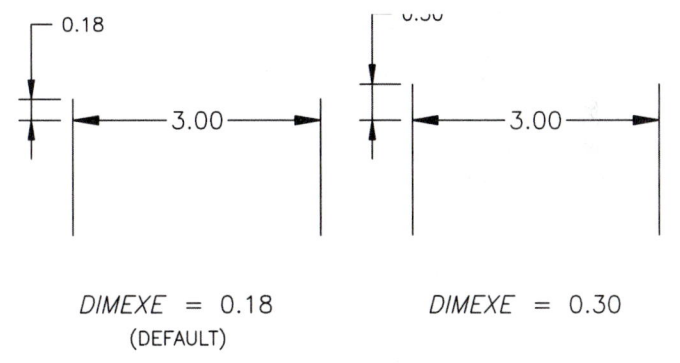

Extend beyond dim lines (DIMEXE)

The *Extend beyond dim lines* edit box reflects the value that AutoCAD uses to set the distance for the extension line to extend beyond the dimension line (Fig. 29-23). This value is held in the *DIMEXE* variable (extension line extension). Generally, this value does not require changing since it is automatically multiplied by the *Overall Scale* (DIMSCALE).

FIGURE 29-23

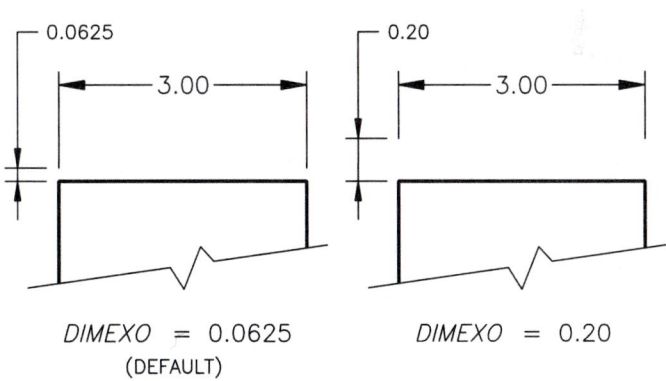

Offset from origin (DIMEXO)

The *Offset from origin* value specifies the distance between the origin points and the extension lines (Fig. 29-24). This offset distance allows you to PICK the object corners, yet the extension lines maintain the required gap from the object. This value rarely requires input since it is affected by *Overall Scale* (DIMSCALE). The value is stored in the *DIMEXO* variable (extension line offset).

FIGURE 29-24

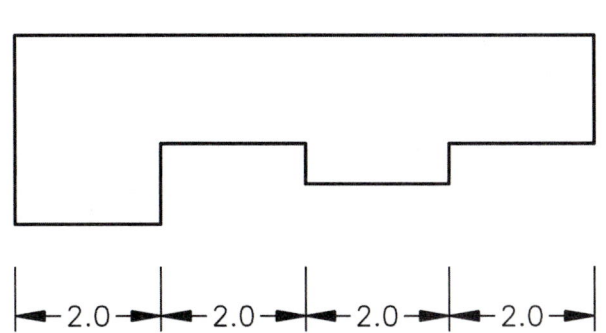

Fixed Length Extension Lines (DIMFXLON, DIMFXL)

This checkbox creates dimensions with extension lines all the same length (Fig. 29-25). The *Length* value determines the length of the extension lines measured from the dimension line toward the "extension line origin." The checkbox setting (on or off) is saved in the *DIMFXLON* variable, and the length value is stored in the *DIMFXL* variable.

FIGURE 29-25

Symbols and Arrows Tab

Use this tab to set your preferences for the appearance of arrowheads, center marks, and special arc and radius symbols (Fig. 29-26). The image tile updates to display your choices for the *Arrowheads* and *Center Marks* sections only.

FIGURE 29-26

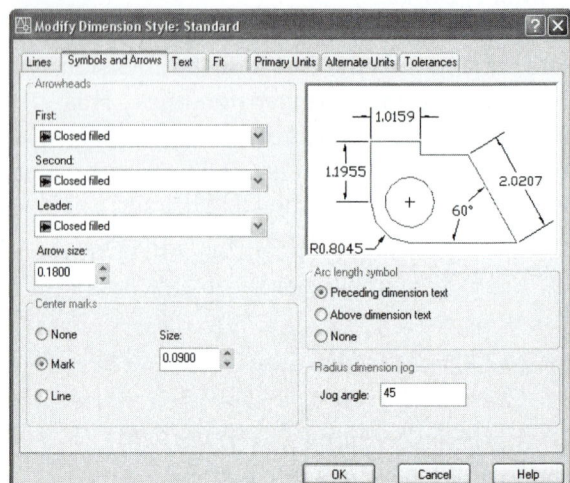

Arrowheads Section (Symbols and Arrows Tab)

First, Second (DIMBLK, DIMSAH, DIMBLK1, DIMBLK2)

This area contains two drop-down lists of various arrowhead types, including dots and ticks. Each list corresponds to the *First* or *Second* arrowhead created in the drawn dimension (Fig. 29-27). The image tiles display each arrowhead type selected. Click in the first image tile to change both arrowheads, or click in each to change them individually. The variables affected are *DIMBLK*, *DIMSAH*, *DIMBLK1*, and *DIMBLK2*.

FIGURE 29-27

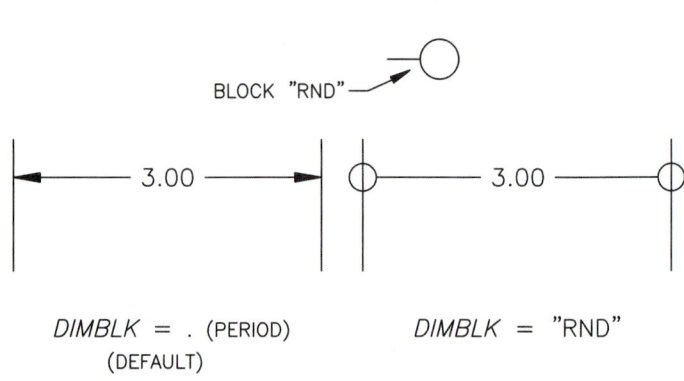

DIMBLK = . (PERIOD)
(DEFAULT)

DIMBLK = "RND"

DIMBLK specifies the *Block* to use for arrowheads if both are the same (Fig. 29-27). When *DIMSAH* is on (separate arrow heads), separate arrow heads are allowed for each end defined by *DIMBLK1* and *DIMBLK2*. You can enter any existing *Block* name in the *DIMBLK* variables. See "Dimension Variables Table" for details.

Leader (DIMLDRBLK)

This drop-down list specifies the arrow type for leaders (Fig. 29-28). This setting does not affect the arrowhead types for dimension lines. Select any option from the drop-down list or enter a value in the *DIMLDRBLK* variable (dim leader block). (For a list of arrowhead entries, see "Dimension Variables Table.")

FIGURE 29-28

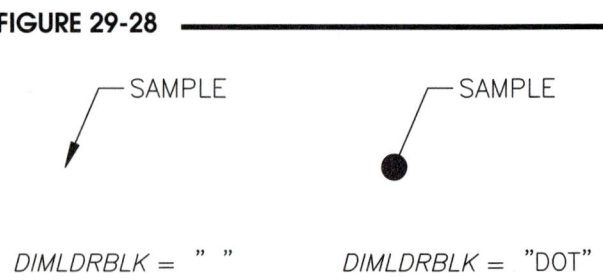

DIMLDRBLK = " "

DIMLDRBLK = "DOT"

Arrow size (DIMASZ)

The size of the arrow can be specified in the *Arrow size* edit box. The *DIMASZ* variable (dim arrow size) holds the value (Fig. 29-29). Remember that this value is multiplied by *Overall Scale* (*DIMSCALE*).

FIGURE 29-29

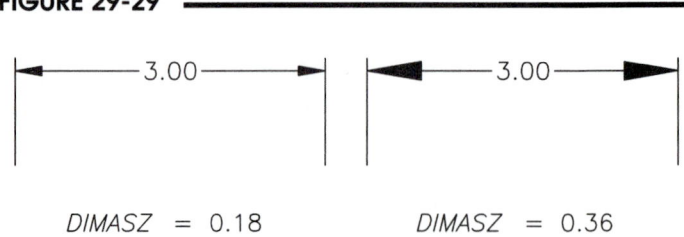

DIMASZ = 0.18
(DEFAULT)

DIMASZ = 0.36

Center Marks Section (Symbols and Arrows Tab)

Type (DIMCEN)

The list here determines how center marks are drawn when the dimension commands *Dimcenter*, *Dimdiameter*, or *Dimradius* are used. The image tile displays the *Mark*, *Line*, or *None* feature specified. This area actually controls the value of <u>one</u> dimension variable, *DIMCEN* by using a 0, positive, or negative value (Fig. 29-30). The *None* option enters a *DIMCEN* value of 0.

FIGURE 29-30

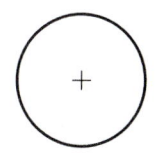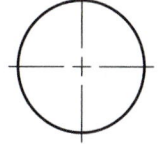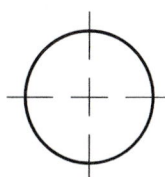

DIMCEN = 0.09   DIMCEN = −0.09   DIMCEN = −0.2
(DEFAULT)

Size (DIMCEN)

The *Size* edit box controls the size of the short dashes and extensions past the arc or circle. The value is stored in the *DIMCEN* variable (Fig. 29-30). Only positive values can be entered in this edit box.

Arc Length Symbol (DIMARCSYM)

Use this option to determine the presence and location of the symbol that appears for an *Arc Length* dimension. Specify the placement of the arc length symbol by selecting *Preceding Dimension Text* or *Above Dimension Text* (Fig. 29-31). Select *None* to suppress the display of the arc length symbol. The dialog box settings correspond to *DIMARCSYM* variable settings of 0 (preceding), 1 (above), and 2 (off).

FIGURE 29-31

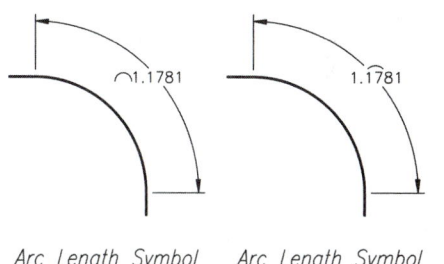

Arc Length Symbol   Arc Length Symbol
=Preceding   =Above

FIGURE 29-32

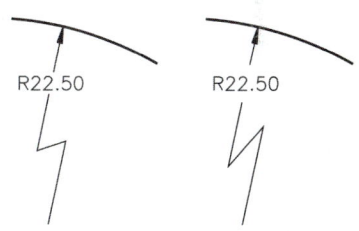

Jog Angle (DIMJOGANG)

This setting specifies the angle of the "jog" line for *Jogged* radius dimensions (Fig. 29-32). The *Jog Angle* value determines the angle between the dimension line and the jog line. The value is saved in the *DIMJOGANG* variable.

Jog Angle=60   Jog Angle=30

Text Tab

The options in this tab change the appearance, placement, and alignment of the dimension text (Fig. 29-33). The preview image automatically reflects changes you make in the *Text Appearance*, *Text Placement*, and *Text Alignment* sections.

FIGURE 29-33

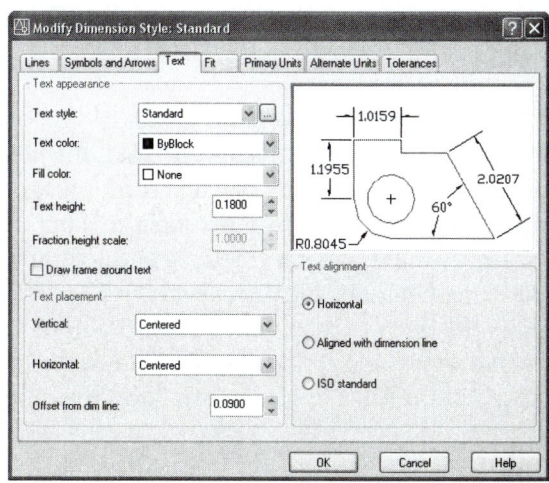

Text Appearance Section (*Text* Tab)

Text style (DIMTXSTY)
This feature of AutoCAD allows you to have different text styles for different dimension styles. The text styles are chosen from a drop-down list of <u>existing</u> styles in the drawing. You can also create a text style "on the fly" by picking the button just to the right of the *Text style* drop-down list, which produces the standard *Text Style* dialog box. The text style used for dimensions remains constant (as defined by the dimension style) and does not change when other text styles in the drawing are made current (as defined by *Style, Text,* or *Mtext*). The *DIMTXSTY* variable (dim text style) holds the text style name for the dimension style.

Text color (DIMCLRT)
Select this drop-down list to select a color or to activate the standard *Select Color* dialog box. The color choice is assigned to the dimension text only (Fig. 29-34). This is useful for controlling the text appearance for printing or plotting when color-dependent plot styles are used. The setting is stored in the *DIMCLRT* variable (dim color text).

FIGURE 29-34

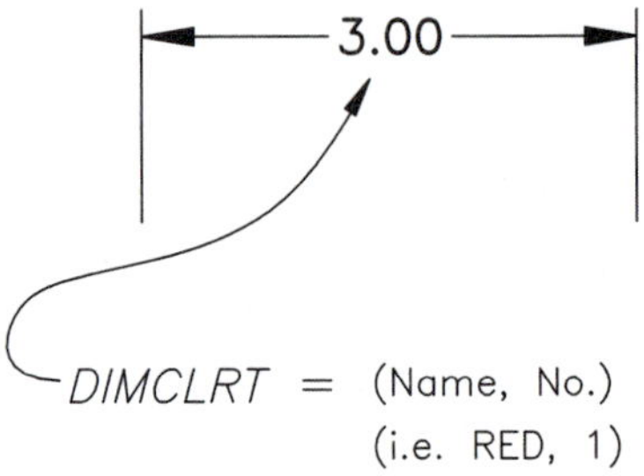

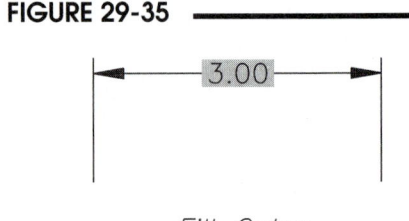

Fill Color (DIMTFILL, DIMTFILLCLR)
You can create a colored background for your dimensions using this option (Fig 29-35). Use the drop-down list to select a color or to activate the *Select Color* dialog box. You can also enter color name or number. The setting for type of background (none, drawing, or color) is saved in the *DIMTFILL* variable and the color value is saved in the *DIMTFILLCLR* variable.

FIGURE 29-35

Fill Color

Text height (DIMTXT)
This value specifies the primary text height; however, this value is multiplied by the *Overall Scale* (*DIMSCALE*) to determine the actual drawn text height. Change *Text height* <u>only</u> if you want to increase or decrease the text height in relation to the other dimension components (Fig. 29-36). Normally, change the *Overall Scale* (*DIMSCALE*) to change all size-related features (arrows, gaps, text, etc.) proportionally. The text height value is stored in *DIMTXT*.

FIGURE 29-36

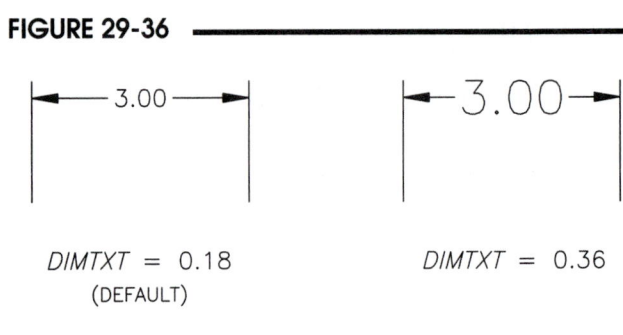

Fraction height scale (DIMTFAC)
When fractions or tolerances are used, the height of the fractional or tolerance text can be set to a proportion of the primary text height. For example, 1.0000 creates fractions and tolerances the same height as the primary text; .5000 represents fractions or tolerances at one-half the primary text height (Fig. 29-37). The value is stored in the *DIMTFAC* variable (dim tolerance factor).

FIGURE 29-37

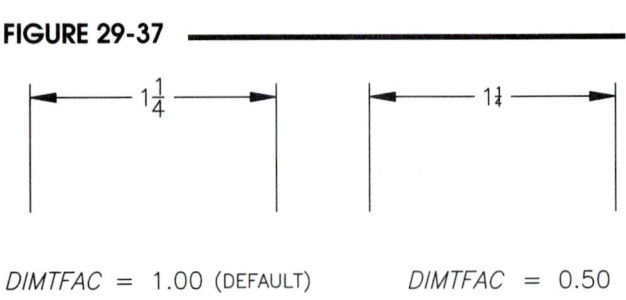

Draw frame around text (DIMGAP)

The "frame" or "gap" is actually an invisible box around the text that determines the offset from text to the dimension line. This option sets the *DIMGAP* variable to a negative value which makes the box visible (Fig. 29-38). This practice is standard for displaying a basic dimension. See *"Tolerance"* (geometric dimensioning and toler-ancing) in Chapter 28. See also *Offset from dimension line* in the *"Text Placement"* section.

FIGURE 29-38

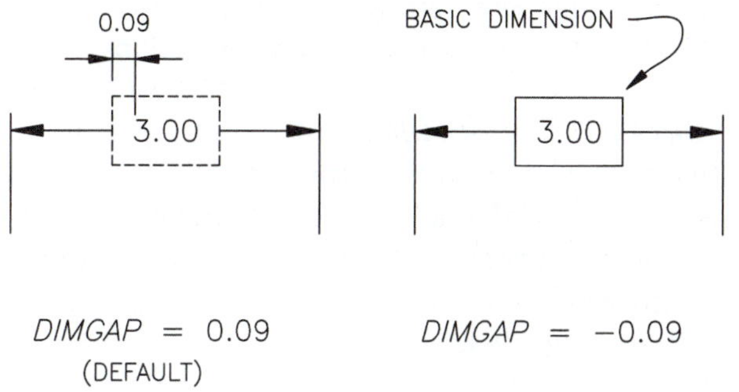

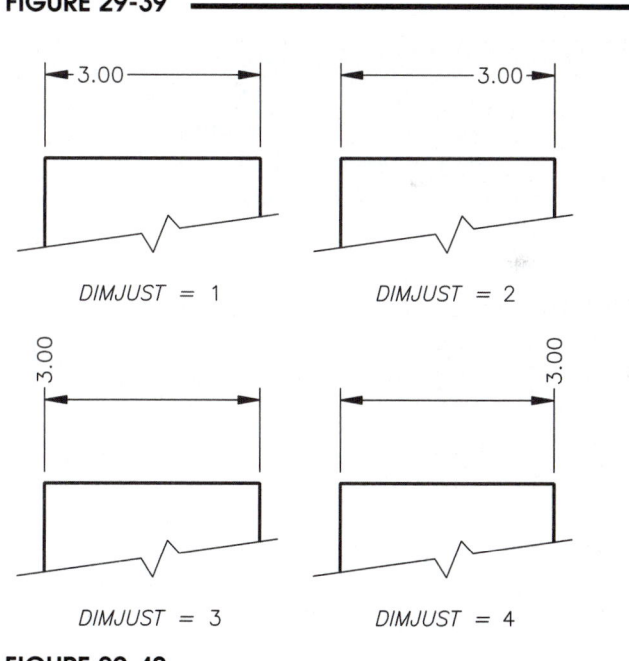

DIMGAP = 0.09
(DEFAULT)

DIMGAP = −0.09

Text Placement Section (Text Tab)

Vertical (DIMTAD)

The *Vertical* option determines the vertical location of the text with respect to the dimension line. There are four possible settings that affect the *DIMTAD* variable (dim text above dimension line). *Centered*, the default option (*DIMTAD* = 0), centers the dimension text between the extension lines. The *Above* option places the dimension text above the dimension line except when the dimension line is not horizontal (*DIMTAD* = 1). *Outside* places the dimension text on the side of the dimension line farthest away from the extension line origin points—away from the dimensioned object (*DIMTAD* = 2). The *JIS* option places the dimension text to conform to Japanese Industrial Standards (*DIMTAD* = 3).

Horizontal (DIMJUST)

The *Horizontal* section determines the horizontal location of the text with respect to the dimension line (dimension justification). The default option (*DIMJUST* = 0) centers the text between the extension lines. The other four choices (*DIMJUST* = 1-3) place the text at either end of the dimension line in parallel and perpendicular positions (Fig. 29-39). You can use the *Horizontal* and *Vertical* settings together to achieve additional text positions.

FIGURE 29-39

DIMJUST = 1

DIMJUST = 2

DIMJUST = 3

DIMJUST = 4

Offset from dim line (DIMGAP)

This value sets the distance between the dimension text and its dimension line. The offset is actually determined by an invisible box around the text (Fig. 29-40). Increasing or decreasing the value changes the size of the invisible box. The *Offset from dim line* value is stored in the *DIMGAP* variable. (Also see *"Draw frame around text."*)

FIGURE 29-40

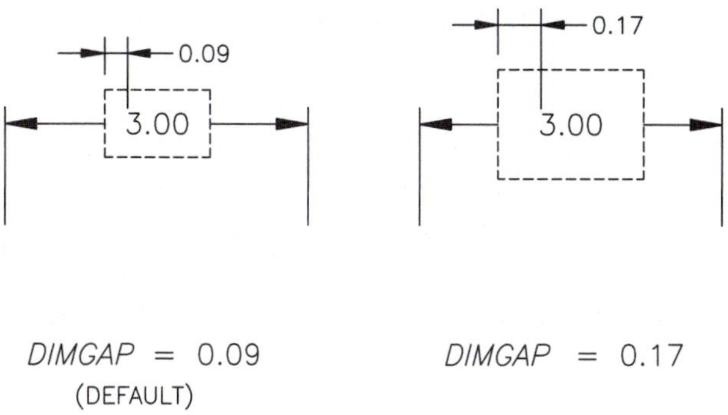

DIMGAP = 0.09
(DEFAULT)

DIMGAP = 0.17

Text Alignment Section (*Text* Tab)

The *Text Alignment* settings can be used in conjunction with the *Text Placement* settings to achieve a wide variety of dimension text placement options.

Horizontal (DIMTIH, DIMTOH)

This radio button turns on the *DIMTIH* (dim text inside horizontal) and *DIMTOH* (dim text outside horizontal) variables. The text remains horizontal even for vertical or angled dimension lines (see Fig. 29-41 and Fig. 29-42). This is the correct setting for mechanical drawings (other than ordinate dimensions) according to the ASME Y14.5M-1994 standard, section 1.7.5, Reading Direction.

FIGURE 29-41

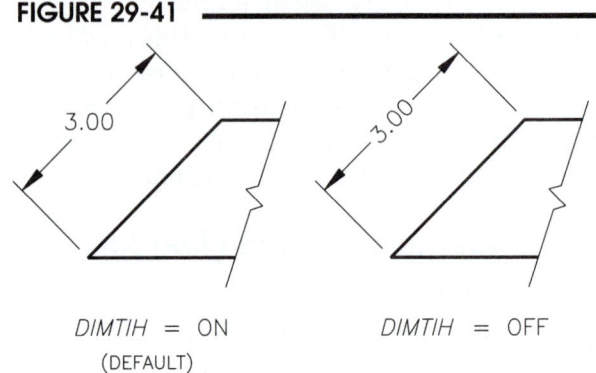

DIMTIH = ON
(DEFAULT)

DIMTIH = OFF

Aligned with dimension line (DIMTIH, DIMTOH)

Pressing this radio button turns off the *DIMTIH* and *DIMTOH* variables so the text aligns with the angle of the dimension line (see Fig. 29-41 and Fig. 29-42).

ISO Standard (DIMTIH, DIMTOH)

This option forces the text inside the dimension line to align with the angle of the dimension line (*DIMTIH* = off), but text outside the dimension line is horizontal (*DIMTOH* = on).

FIGURE 29-42

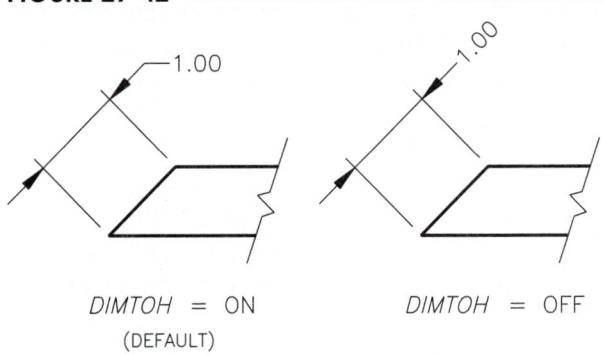

DIMTOH = ON
(DEFAULT)

DIMTOH = OFF

Fit Tab

The *Fit* tab allows you to determine the *Overall Scale* for dimensioning components, how the text, arrows, and dimension lines fit between extension lines, and how the text appears when it is moved (Fig. 29-43).

FIGURE 29-43

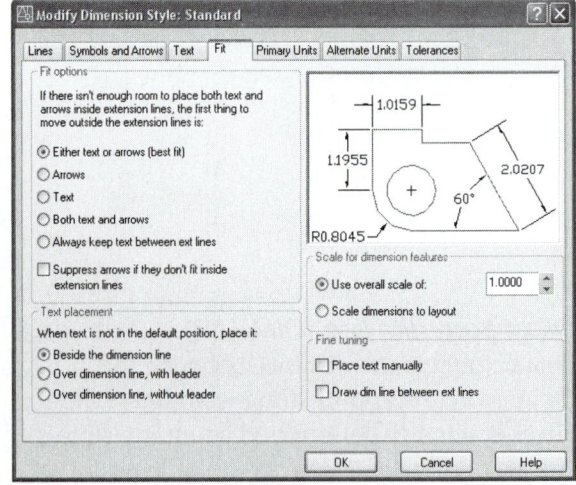

Scale for Dimension Features Section (*Fit* Tab)

Although this is not the first section in the dialog box, it is presented first because of its importance.

Overall Scale (DIMSCALE)

The *Overall Scale* value globally affects the scale of <u>all size-related features</u> of dimension objects, such as arrowheads, text height, extension line gaps (from the object), extensions (past dimension lines), etc. All other size-related (variable) values appearing in the dialog box series are <u>multiplied by the *Overall Scale*</u>. Notice how all dimensioning features (text, arrows, gaps, offsets) are all increased proportionally with the *Overall Scale* value (Fig. 29-44). Therefore, to keep all features proportional, <u>change this one setting rather than each of the others individually</u>.

FIGURE 29-44

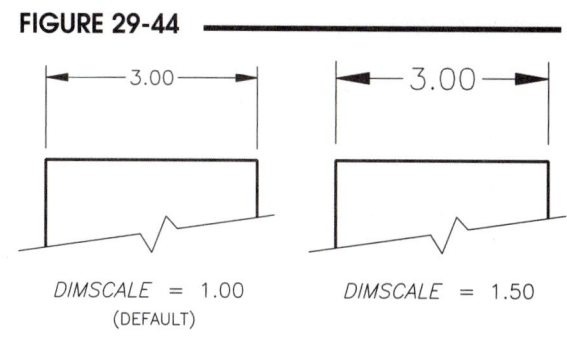

DIMSCALE = 1.00   DIMSCALE = 1.50
(DEFAULT)

Although this area is located on the right side of the box, it is probably the most important option in the entire series of tabs. Because the *Overall Scale* should be set as a family-wide variable, setting this value is typically the <u>first step</u> in creating a dimension style (Fig. 29-45).

Changes in this variable should be based on the *Limits* and plot scale. You can use the drawing scale factor to determine this value. (See "Drawing Scale Factor," Chapter 12.) The *Overall Scale* value is stored in the *DIMSCALE* variable.

FIGURE 29-45

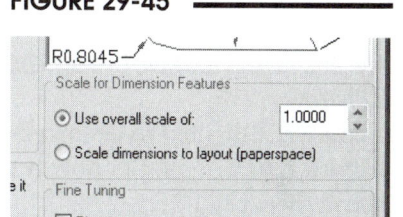

Scale dimensions to layout (paper space) (DIMSCALE)

Checking this box allows you to retroactively change the size of the dimensions in a viewport based on the viewport scale by selecting all the dimensions (in the viewport), then using *Update* from the *Dimensions* pull-down menu. The dimensions display in a new size based on the viewport scale. Checking this box sets *DIMSCALE* to 0. See Chapter 33 for applications for this setting.

Fit Options Section (Fit Tab)

This section determines which dimension components are <u>forced outside the extension lines only if there is insufficient room</u> for text, arrows, and dimension lines. In most cases where space permits all components to fit inside, the *Fit* settings have no effect on placement. *Linear, Aligned, Angular, Baseline, Continue, Radius,* and *Diameter* dimensions apply. See "*Radius* and *Diameter* Variable Settings."

Either the text or the arrows (DIMATFIT)

This is the default setting for the STANDARD dimension style. AutoCAD makes the determination of whether the text or the arrows are forced outside based on the size of arrows and the length of the text string (Fig. 29-46). This option often behaves similarly to *Arrows*, except that if the text cannot fit, the text is placed outside the extension lines and the arrows are placed inside. However, if the arrows cannot fit either, both text and arrows are placed outside the extension lines. The *DIMATFIT* (dimension arrows/text fit) setting is 3.

FIGURE 29-46

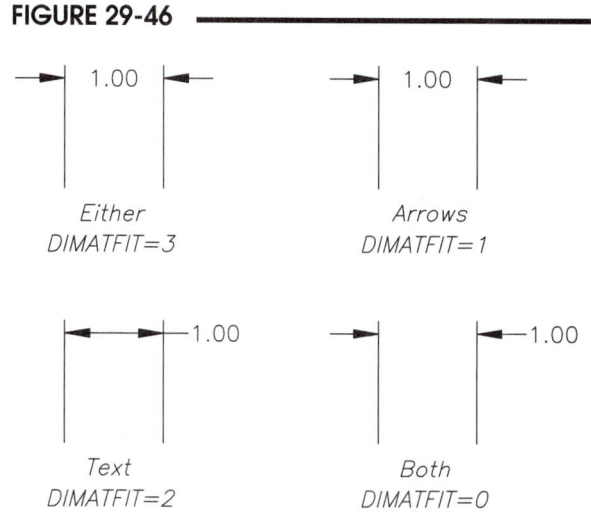

Arrows (DIMATFIT)

The *Arrows* option forces the arrows on the outside of the extension lines and keeps the text inside. If the text absolutely cannot fit, it is also placed outside the extension lines (see Fig. 29-46). This option sets *DIMATFIT* to 1.

Text (DIMATFIT)

The *Text* option places the text on the outside and keeps the arrows on the inside unless the arrows cannot fit, in which case they are placed on the outside as well (see Fig. 29-46). For this option, *DIMATFIT* = 2.

Both text and arrows (DIMATFIT)

The *Both text and arrows* option keeps the text and arrows together always. If space does not permit <u>both</u> features to fit between the extension lines, it places the text and arrows outside the extension lines (see Fig. 29-46). You can set *DIMATFIT* to 0 to achieve this placement.

Always keep text between ext lines (DIMTIX)

If you want the text to be forced between the extension line no matter how much room there is, use this option (Fig. 29-47). Pressing this radio button turns *DIMTIX* on (text inside extensions).

Suppress arrows if they don't fit (DIMSOXD)

When dimension components are forced outside the extension lines and there are many small dimensions aligned in a row (such as with *Continue* dimensions), the text, arrows, or dimension lines may overlap. In this case, you can prevent the arrows and the dimension lines from being drawn entirely with this option. This option suppresses the arrows and dimension lines <u>only</u> when they are forced outside (Fig. 29-48). The setting is stored as *DIMSOXD* = on (suppress dimension lines outside extensions).

FIGURE 29-47 ───────────────

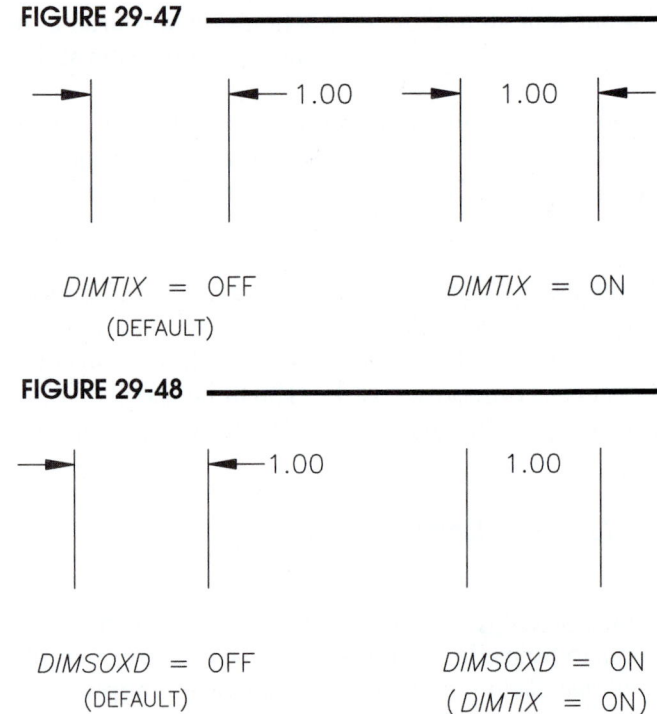

Text Placement Section (Fit Tab)

This section of the dialog box sets dimension text movement rules. When text is moved either by being automatically forced from between the dimension lines based on the *DIMATFIT* setting (the *Fit Options* above this section) or when you actually move the text with grips or by *Dimtedit*, these rules apply.

Beside the dimension line (DIMTMOVE)

This is the normal placement of the text— aligned with and beside the dimension line (Fig. 29-49). The text always moves when the dimension line is moved and vice versa. *DIMTMOVE* (dimension text move) = 0.

FIGURE 29-49 ───────────────

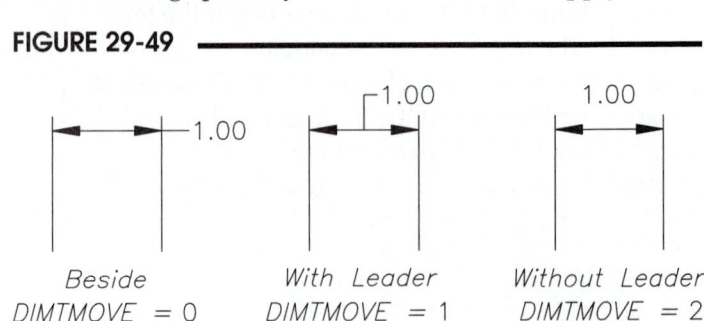

Over the dimension line, with a leader (DIMTMOVE)

This option creates a leader between the text and the center of the dimension line whenever the text cannot fit between the extension lines or is moved using grips (see Fig. 29-49). *DIMTMOVE = 1*.

Over the dimension line, without a leader (DIMTMOVE)

Use this setting to have the text appear above the dimension line, similar to *DIMTMOVE = 1*, but without a leader. This occurs only when there is insufficient room for the text between the extension lines or when you move the text with grips. *DIMTMOVE = 2*.

There is an important benefit to this setting (*DIMTMOVE = 2*). When there is sufficient room for the text and arrows between extension lines, this setting has no effect on the placement of the text. When there is insufficient room, the text moves above without a leader. In either case, if you prefer to move the text to another location with grips or using *Dimtedit*, the text moves as if were "detached" from the dimension line. The text can be moved independently to any location and the dimension retains its associatively. Figure 29-50 illustrates the use of grips to edit the dimension text.

FIGURE 29-50

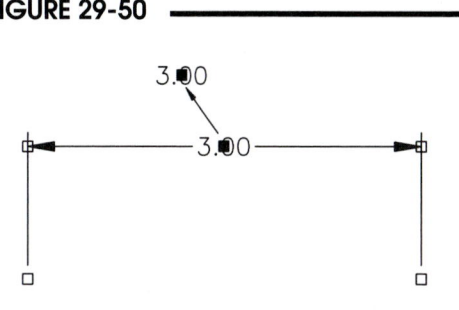

Without Leader (DIMTMOVE=2)

Fine Tuning Section (Fit Tab)

Place text manually when dimensioning (DIMUPT)

When you press this radio button, you can create dimensions and move the text independently in relation to the dimension and extension lines as you place the dimension line in response to the "specify dimension line location" prompt (Fig. 29-51). Using this option is similar to using *Dimtedit* after placing the dimension. *DIMUPT* (dimension user-positioned text) is on when this box is checked.

FIGURE 29-51

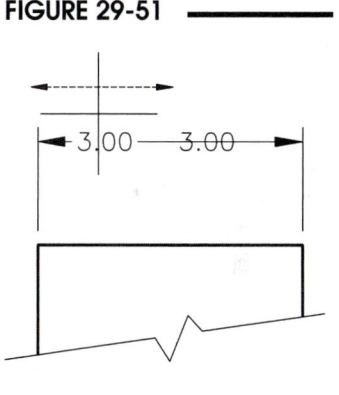

DIMUPT = ON

Always draw dim line between ext lines (DIMTOFL)

Occasionally, you may want the dimension line to be drawn inside the extension lines even when the text and arrows are forced outside. You can force a line inside with this option (Fig. 29-52). A check in this box turns *DIMTOFL* on (text outside, force line inside).

FIGURE 29-52

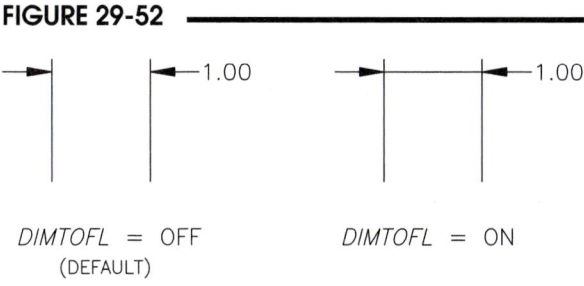

DIMTOFL = OFF
(DEFAULT)   *DIMTOFL = ON*

Radius and Diameter Variable Settings

For creating mechanical drawing dimensions according to ANSI standards, the default settings in AutoCAD are correct for creating *Diameter* dimensions but not for *Radius* dimensions. The following variable settings are recommended for creating *Radius* and *Diameter* dimensions for mechanical applications.

For *Diameter* dimensions, the default settings produce ANSI-compliant dimensions that suit most applications—that is, text and arrows are on the outside of the circle or arc pointing inward toward the center. For situations where large circles are dimensioned, *Fit Options* (*DIMATFIT*) and *Place text manually* (*DIMUPT*) can be changed to force the dimension inside the circle (Fig. 29-53).

For *Radius* dimensions the default settings produce incorrect dimensioning practices. Normally (when space permits) you want the dimension line and arrow to be inside the arc, while the text can be inside or outside. To produce ANSI-compliant *Radius* dimensions, set *Fit Options* (*DIMAT-FIT*) and *Place text manually* (*DIMUPT*) as shown in Figure 29-53. Radius dimensions can be outside the arc in cases where there is insufficient room inside.

FIGURE 29-53

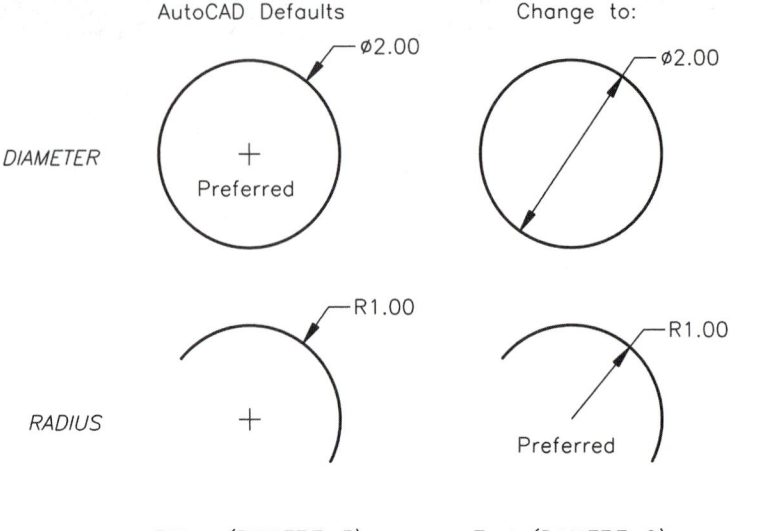

Primary Units Tab

The *Primary Units* tab controls the format of the AutoCAD-measured numerical value that appears with a dimension (Fig. 29-54). You can vary the numerical value in several ways such as specifying the units format, precision of decimal or fraction, prefix and/or suffix, zero suppression, and so on. These units are called primary units because you can also cause AutoCAD to draw additional or secondary units called *Alternate Units* (for inch <u>and</u> metric notation, for example).

FIGURE 29-54

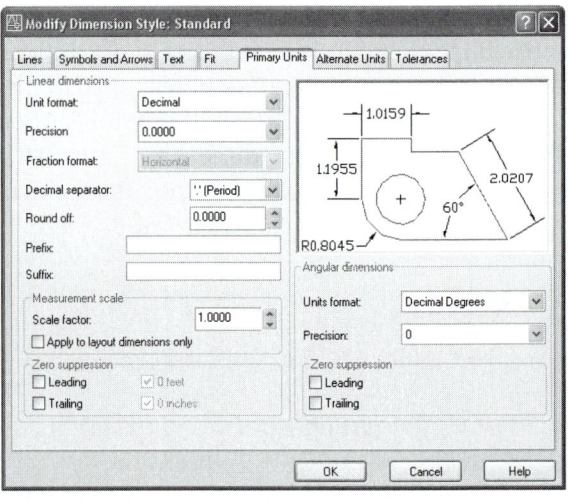

Linear Dimensions Section (*Primary Units* Tab)

This section controls the format of all dimension types except *Angular* dimensions.

Unit format (DIMLUNIT)

The *Units format* section drop-down list specifies the type of units used for dimensioning. These are the same unit types available with the *Units* dialog box (*Decimal, Scientific, Engineering, Architectural,* and *Fractional*) with the addition of *Windows Desktop*. The *Windows Desktop* option displays AutoCAD units based on the settings made for units display in Windows Control Panel (settings for decimal separator and number grouping symbols). Remember that your selection affects the <u>units drawn in dimension objects, not the global drawing units</u>. The choice for *Units format* is stored in the *DIMLUNIT* variable (dimension linear unit). This drop-down list is disabled for an *Angular* family member.

Precision (DIMDEC)

The *Precision* drop-down list in the *Dimension* section specifies the number of places for decimal dimensions or denominator for fractional dimensions. This setting does not alter the drawing units precision. This value is stored in the *DIMDEC* variable (dim decimal).

Fraction format (DIMFRAC)

Use this drop-down list to set the fractional format. The choices are displayed in Figure 29-55. This option is enabled only when *DIMLUNIT* (*Unit format*) is set to 4 (*Architectural*) or 5 (*Fractional*).

FIGURE 29-55

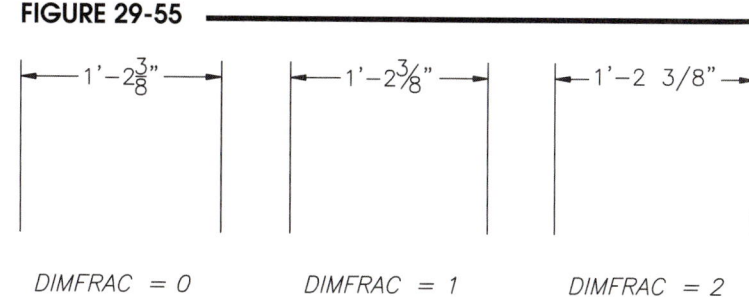

Decimal separator (DIMDSEP)

When you are creating dimensions whose unit format is decimal, you can specify a single-character decimal separator. Normally a decimal (period) is used; however, you can also use a comma (,) or a space. The character is stored in the *DIMDSEP* (dimension decimal separator) variable.

Round off (DIMRND)

Use this drop-down list to specify a precision for dimension values to be rounded. Normally, AutoCAD values are kept to 14 significant places but are rounded to the place dictated by the dimension *Precision* (*DIMDEC*). Use this feature to round up or down appropriately to the nearest specified decimal or fractional increment (Fig. 29-56).

FIGURE 29-56

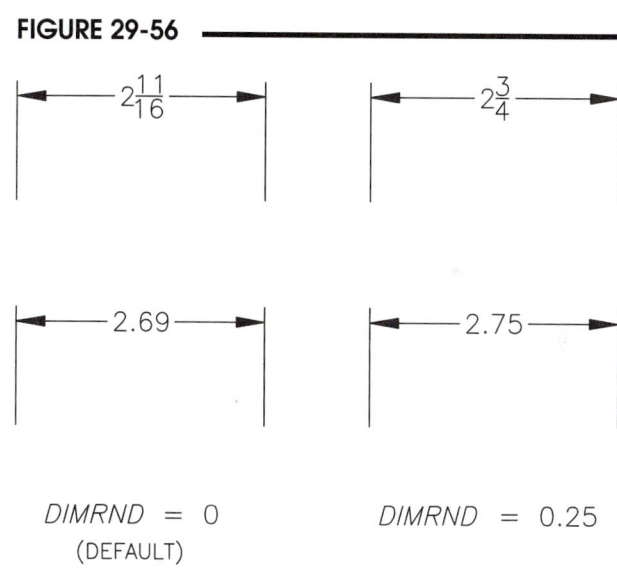

Prefix/Suffix (DIMPOST)

The *Prefix* and *Suffix* edit boxes hold any text that you want to add to the AutoCAD-supplied dimensional value. A text string entered in the *Prefix* edit box appears before the AutoCAD-measured numerical value and a text string entered in the *Suffix* edit box appears after the AutoCAD-measured numerical value. For example, entering the string " mm" or a " TYP." in the *Suffix* edit box would produce text as shown in Figure 29-57. (In such a case, don't forget the space between the numerical value and the suffix.) The string is stored in the *DIMPOST* variable.

FIGURE 29-57

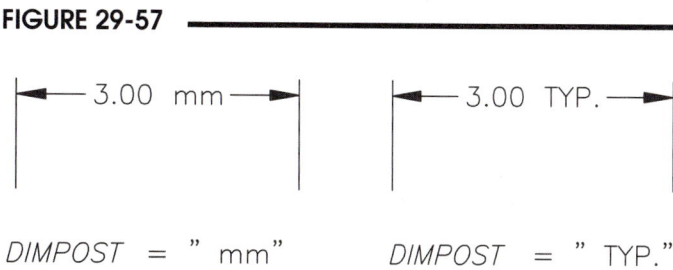

Alternate Unit Section (*Alternate Units* Tab)

Unit format (DIMALTU)
This drop-down list sets the format for alternate units. The setting is stored in the *DIMALTU* variable (dimension alternate units). This is the alternate units equivalent to *Units format* for *Primary Units*.

Precision (DIMALTD)
Use this drop-down list to set the decimal precision for alternate units. Note that the alternate units precision is controlled independently of the primary units precision (Fig. 29-62). The value is stored in the *DIMALTD* variable (dimension alternate decimals).

FIGURE 29-62

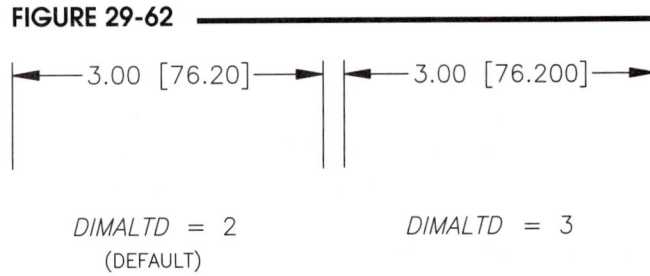

Multiplier for alt units (DIMALTF)
This is the alternate units equivalent for the primary units scale factor (*DIMLFAC*). In other words, the AutoCAD-measured primary units value is multiplied by this factor to determine the displayed value for alternate units (Fig. 29-63). You can also enter the desired multiplier in the *DIMALTF* (dimension alternate factor) variable.

FIGURE 29-63

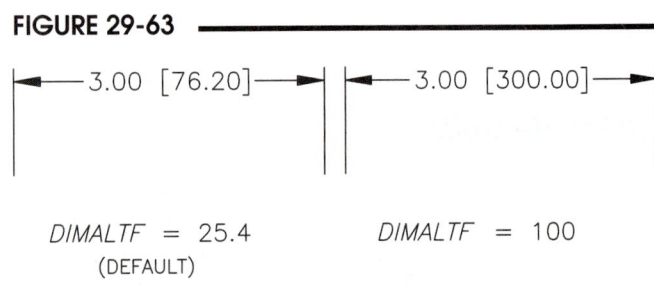

Round distances to (DIMALTRND)
If you do not want the alternate units value to be displayed as the actual measurement, but to be rounded to nearest regular increment, enter the desired increment in this edit box. For example, you may want alternate units to be displayed with two places of precision (to the right of the decimal) but to round to the nearest millimeter. Alternately, set the increment using the *DIMALTRND* variable (alternate rounding). This is the alternate units equivalent to the *Round off* option in the *Primary Units* tab (see Fig. 29-56).

Prefix/Suffix (DIMAPOST)
This section allows you to include a prefix and/or suffix with the alternate units measured value. For example, you may want to display "mm" after the alternate units value (Fig. 29-64). The prefix and/or suffix is stored in the *DIMAPOST* variable. (See also *Prefix/Suffix* in the *Primary Units* tab for more information.)

FIGURE 29-64

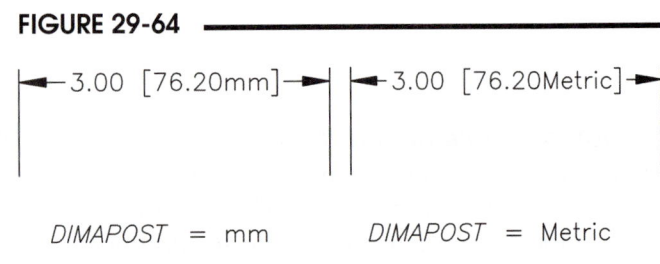

Zero Suppression Section (*Alternate Units* Tab)

The *Zero Suppression* section controls how zeros are drawn in alternate units values when they occur. A check in one of these boxes means that zeros are <u>not drawn for that case</u>. Your selection sets the value for *DIMALTZ*. See the description for *Zero Suppression, Primary Units* Tab for more information and illustration.

Placement Section (*Alternate Units* Tab)

This section only has two options; both are related to the *DIMPOST* variable. Normally, the alternate units values are placed *After primary value*. You can instead toggle *Below primary value* to yield a display as shown in Figure 29-65, right dimension. Selecting *Below primary value* sets the *DIMPOST* variable to "\x."

FIGURE 29-65 ——————

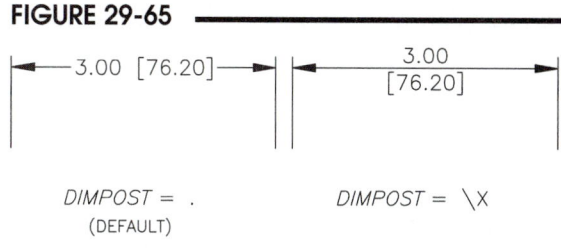

Tolerances Tab

The *Tolerances* tab allows you to create several formats of tolerance dimensions such as limits, two forms of plus/minus dimensions, and basic dimensions (Fig. 29-66). Most options in this tab are disabled until you select a *Method*.

FIGURE 29-66 ——————

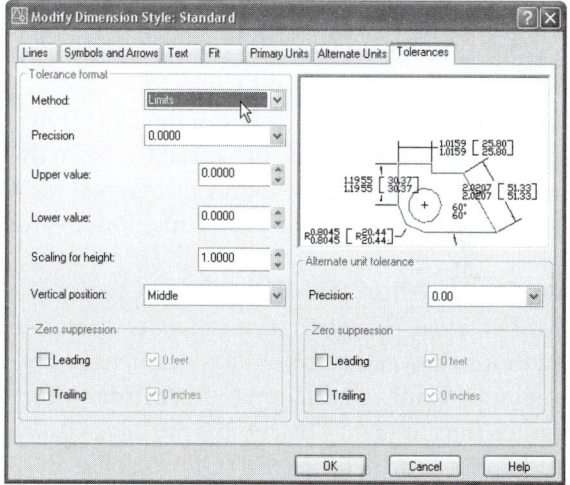

Tolerance Format Section (*Tolerances* Tab)

Method (DIMTOL, DIMLIM, DIMGAP)

The *Method* option displays a drop-down list with five types: *None, Symmetrical, Deviation, Limits,* and *Basic*. The four possibilities (other than *None*) are illustrated in Figure 29-67. The *Symmetrical* and *Deviation* methods create plus/minus dimensions and turn on the *DIMTOL* variable (dimension tolerance). The *Limits* method creates limit dimensions and turns the *DIMLIM* variable on (dimension limits). The *Basic* method creates a basic dimension by drawing a box around the dimensional value, which is accomplished by changing the *DIMGAP* to a negative value.

FIGURE 29-67 ——————

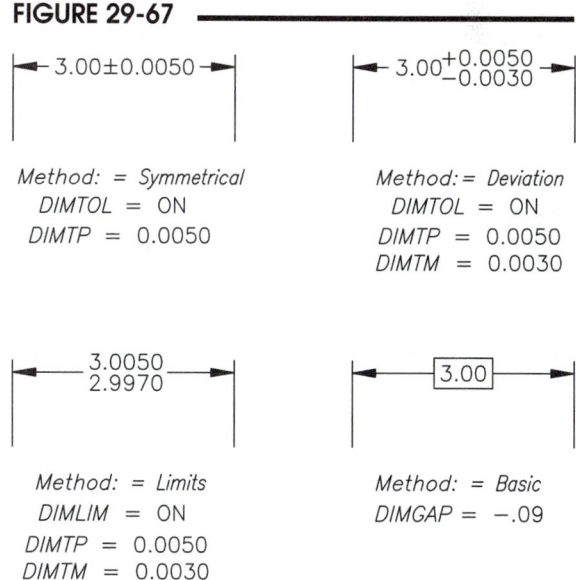

Precision (DIMTDEC)

Use the *Precision* drop-down list to set the precision (number of places to the right of the decimal place) for values when drawing *Symmetrical, Deviation,* and *Limits* dimensions. The precision is stored in the *DIMTDEC* variable (dimension tolerance decimals).

Upper value/Lower value (DIMTP, DIMTM)

An *Upper Value* and *Lower Value* can be entered in the edit boxes for the *Deviation* and *Limits* method types. In these cases, the *Upper Value* (*DIMTP*—dimension tolerance plus) is added to the measured dimension and the *Lower Value* (*DIMTP*—dimension tolerance minus) is subtracted (Fig. 29-67). An *Upper Value* only is needed for *Symmetrical* and is applied as both the plus and minus value.

Scaling for height (DIMTFAC)

The height of the tolerance text is controlled with the *Scaling for height* edit box value. The entered value is a <u>proportion</u> of the primary dimension value height. For example, a value of .50 would draw the tolerance text at 50% of the primary text height (Fig. 29-68). The setting affects the *Symmetrical*, *Deviation*, and *Limits* tolerance methods. The value is stored in the *DIMTFAC* (dimension tolerance factor) variable. Note that this is the same variable that controls the *Fraction height scale* for architectural and fractional values.

FIGURE 29-68

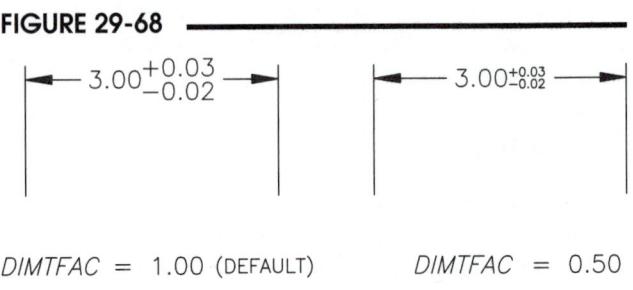

DIMTFAC = 1.00 (DEFAULT)   DIMTFAC = 0.50

Vertical position (DIMTOLJ)

For *Deviation*, *Symmetrical*, and *Limits* tolerance methods, you can control the placement of the tolerance values in relation to the primary units values (Fig. 29-69). The choices are *Top*, *Middle*, and *Bottom* and set the *DIMTOLJ* variable (dimension tolerance justification) to 2, 1, and 0, respectively. The *Top* option aligns the primary text and the top tolerance value, and the *Bottom* option aligns the primary text and the bottom tolerance value.

FIGURE 29-69

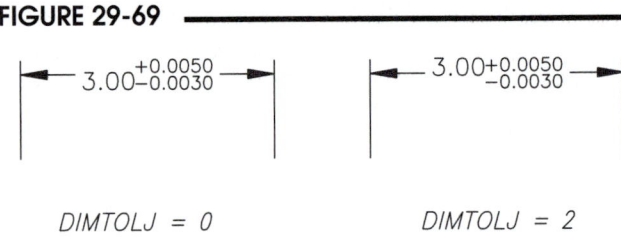

DIMTOLJ = 0   DIMTOLJ = 2

Zero Suppression (DIMTZIN) (Tolerance Format Section)

The *Zero Suppression* section controls how zeros are drawn in *Symmetrical*, *Deviation*, and *Limits* tolerance dimensions when they occur. A check in one of these boxes means that zeros are not drawn for that case. The *Zero Suppression* section here operates identically to the *Zero Suppression* section of the *Primary Units* tab, except that it is applied to tolerance dimensions (see *Zero Suppression*, *Primary Units* Tab for more information). The affected variable is *DIMTZIN* (dimension tolerance zero indicator).

Alternate Unit Tolerance Section (Tolerances Tab)

When you have specified that alternate units are to be drawn (in the *Alternate Units* tab) <u>and</u> you have turned on some form of tolerance *Method* (*Symmetrical*, *Limits*, etc.), the alternate units will automatically display as a tolerance along with the primary units. In other words, when alternate units are on, both primary and alternate units are drawn the same—with or without tolerances.

Precision (DIMALTTD)

This drop-down list sets the decimal precision for the alternate units tolerances (when alternate units and tolerances are on). The *DIMALTTD* variable (dimension alternate tolerance decimal) holds the setting.

Zero Suppression (DIMALTTZ)

When alternate units and tolerances are on, use these check boxes to determine how leading and trailing zeros are treated. The setting is stored in the *DIMALTTZ* variable (dimension alternate tolerance zeros). See *Zero Suppression* in the *Primary Units* Tab section for more information on zero suppression.

Changing Dimension Variables Using the Command Line Format

Dim...
(Variable
Name)

Pull-down Menu	Command (Type)	Alias (Type)	Short-cut	Screen (side) Menu	Tablet Menu
Tools Inquiry > Set Variable	(Variable Name)	...	...	TOOLS 1 Setvar	U,10

As an alternative to setting dimension variables through the *Dimension Style Manager*, you can type the dimensioning variable name at the Command: prompt. There is a noticeable difference in the two methods—the dialog boxes use different nomenclature than the formal dimensioning variable names used in Command line format; that is, the dialog boxes use descriptive terms that, if selected, make the appropriate change to the dimensioning variable. The formal dimensioning variable names accessed by Command line format, however, all begin with the letters *DIM* and are accessible only by typing.

Another important but subtle difference in the two methods is the act of saving dimension variable settings to a dimension style. Remember that all drawn dimensions are part of a dimension style, whether it is STANDARD or some user-created style. When you change dimension variables by the Command line format, the changes become overrides until you use the *Save* option of the *-Dimstyle* command or the *Dimension Style Manager*. When a variable change is made, it becomes an override that is applied to the current dimension style and affects only the newly drawn dimensions. Variable changes must be *Saved* to become a permanent part of the style and to retroactively affect all dimensions created with that style.

All other dimensioning controls are accessible using the dimension variables explained in the following Dimensioning Variables Table. All these dimension variables except *DIMASSOC* are included in every dimension style.

In order to access and change a variable's setting by name, simply type the variable name at the Command: prompt. For dimension variables that require distances, you can enter the distance (in any format accepted by the current *Units* settings) or you can designate by (PICKing) two points.

For example, to change the value of the *DIMSCALE* to **.5**, this command syntax is used:

```
Command: dimscale
Enter new value for dimscale <1.0000>: .5
```

Associative Dimensions

Since AutoCAD 2002, dimensions have full associativity. The *DIMASSOC* variable controls the associative feature. The *DIMASSOC* variable setting cannot be saved in a dimension style; therefore, one dimension style can contain associative and nonassociative dimensions, but not exploded dimensions (exploded dimensions can be created from a dimension style, but cannot be updated since they do not reference the style). The *Dimension Style Manager* does not provide access to the *DIMASSOC* variable. *DIMASSOC* must be changed by Command line format or by using the *Options* dialog box (see "Associative Dimensions" in Chapter 28).

Dimension Variables Table

This table is a summary of the 79 dimensioning variables in AutoCAD 2007. Each variable has a brief description, the type of variable it is, and its default setting (taken from the *AutoCAD Command Reference*). Remember that the dimension variables and the current settings can be listed using the *STatus* and *Variables* options of the *-Dimstyle* command.

Variable	Characteristics	Description
DIMADEC	Type: Integer Saved in: Drawing Initial value: 0	Controls the number of places of precision displayed for angular dimension text. -1 Angular dimensions display the number of decimal places specified by *DIMDEC* (linear dimension precision). 0-8 Angular dimension is drawn using the number of decimal places corresponding to the *DIMADEC* setting.
DIMALT	Type: Switch Saved in: Drawing Initial value: Off	When turned on, enables alternate units dimensioning. See also *DIMALTD, DIMALTF, DIMALTZ (DIMALTTZ, DIMALTTD)*, and *DIMAPOST*.
DIMALTD	Type: Integer Saved in: Drawing Initial value: 2	Controls alternate units decimal places. If *DIMALT* is enabled, *DIMALTD* governs the number of decimal places displayed in the alternate measurement.
DIMALTF	Type: Real Saved in: Drawing Initial value: 25.4000	Controls alternate units scale factor. If *DIMALT* is enabled, *DIMALTF* multiplies linear dimensions by a factor to produce a value in an alternate system of measurement.
DIMALTRND	Type: Real Saved in: Drawing Initial value: 0.00	Rounds off the alternate dimension units.
DIMALTTD	Type: Integer Saved in: Drawing Initial value: 2	Sets the number of decimal places for the tolerance values of an alternate units dimension.
DIMALTTZ	Type: Integer Saved in: Drawing Initial value: 0	Toggles suppression of zeros in tolerance values. 0 Suppresses zero feet and precisely zero inches 1 Includes zero feet and precisely zero inches 2 Includes zero feet and suppresses zero inches 3 Includes zero inches and suppresses zero feet To the preceding values, add: 4 Suppresses leading zeros 8 Suppresses trailing zeros
DIMALTU	Type: Integer Saved in: Drawing Initial value: 2	Sets the units format for alternate units of all dimension style family members except angular 1 Scientific 2 Decimal 3 Engineering 4 Architectural (stacked) 5 Fractional (stacked) 6 Architectural 7 Fractional 8 Windows® Desktop (decimal format using Control Panel settings for decimal separator and number grouping symbols)

Variable	Characteristics	Description
DIMALTZ	Type: Integer Saved in: Drawing Initial value: 0	Controls the suppression of zeros for alternate unit dimension values. *DIMALTZ* values 0–3 affect feet-and-inch dimensions only. 0 Suppresses zero feet and precisely zero inches 1 Includes zero feet and precisely zero inches 2 Includes zero feet and suppresses zero inches 3 Includes zero inches and suppresses zero feet 4 Suppresses leading zeros in decimal dimensions (for example, 0.5000 becomes .5000) 8 Suppresses trailing zeros in decimal dimensions (for example, 12.5000 becomes 12.5) 12 Suppresses both leading and trailing zeros (for example, 0.5000 becomes .5)
DIMAPOST	Type: String Saved in: Drawing Initial value: ""	Specifies a text prefix or suffix (or both) to the alternate dimension measurement for all types of dimensions except angular. For instance, if the current *Units* mode is Architectural, *DIMALT* is enabled, *DIMALTF* is 25.4, *DIMALTD* is 2, and *DIMAPOST* is set to "mm," a distance of 10 units would be edited as 10"[254.00mm]. To disable an established prefix or suffix (or both), set it to a period.
DIMARCSYM	Type: Integer Saved in: Drawing Initial value: 0	Controls the associativity of dimension objects. 0 Places arc length symbols before the dimension text 1 Places arc length symbols above the dimension text 3 Suppresses the display of arc length symbols
DIMASO	Type: Switch Saved in: Drawing Initial value: On	<u>Obsolete</u>. Retained in the product to preserve the integrity of scripts. See *DIMASSOC*.
DIMASSOC	Type: Integer Saved in: Drawing Initial value: 2	Controls the associativity of dimension objects. 0 Creates exploded dimensions. There is no association between the various elements of the dimension. The lines, arcs, arrowheads, and text of a dimension are drawn as separate objects. 1 Creates nonassociative dimension objects. The elements of the dimension are formed into a single object. If the definition point on the object moves, the dimension value is updated. 2 Creates associative dimension objects. The elements of the dimension are formed into a single object, and one or more definition points of the dimension are coupled with association points on geometric objects. If the association point on the geometric object moves, the dimension location, orientation, and value is updated. *DIMASSOC* is not stored in a dimension style. Drawings saved in a format previous to AutoCAD 2002 retain the setting of the *DIMASSOC* system variable. When the drawing is reopened in AutoCAD 2002 or later, the dimension associativity setting is restored. If a legacy drawing is opened in a current release of AutoCAD, the *continued next page...*

Variable	Characteristics	Description
DIMASSOC (continued)		*DIMASSOC* system variable takes on the value of the legacy drawing's *DIMASO* system variable.
DIMASZ	Type: Real Saved in: Drawing Initial value: 0.1800	Controls the size of dimension line and leader line arrowheads. Also controls the size of hook lines. Multiples of the arrowhead size determine whether dimension lines and text are to fit between the extension lines. Also used to scale arrowhead blocks if set to *DIMBLK*. *DIMASZ* has no effect when *DIMTSZ* is other than zero.
DIMATFIT	Type: Integer Saved in: Drawing Initial value: 3	Determines how dimension text and arrows are a arranged when space is not sufficient to place both within the extension lines. 0 Places both text and arrows outside extension lines 1 Moves arrows first, then text 2 Moves text first, then arrows 3 Moves either text or arrows, whichever fits best AutoCAD adds a leader to moved dimension text when *DIMTMOVE* is set to 1.
DIMAUNIT	Type: Integer Saved in: Drawing Initial value: 0	Sets the angle format for angular dimension: 0 Decimal degrees 1 Degrees/minutes/seconds 2 Gradians 3 Radians
DIMAZIN	Type: Integer Saved in: Drawing Initial value: 0	Suppresses zeros for angular dimensions. 0 Displays all leading and trailing zeros 1 Suppresses leading zeros in decimal dimensions (for example, 0.5000 becomes .5000) 2 Suppresses trailing zeros in decimal dimensions (for example, 12.5000 becomes 12.5) 3 Suppresses leading and trailing zeros (for example, 0.5000 becomes .5)
DIMBLK	Type: String Saved in: Drawing Initial value: ""	Sets the arrowhead block displayed at the ends of dimension lines or leader lines. To turn off arrowheads, enter a single period (.). Arrowhead block entries and *Override* Dimension Style dialog boxes are shown below. You can also enter the names of user-defined arrowhead blocks. "" closed filled "_DOT" dot "_DOTSMALL" dot small "_DOTBLANK" dot blank "_ORIGIN" origin indicator "_ORIGIN2" origin indicator 2 "_OPEN" open "_OPEN90" right angle "_OPEN30" open 30 "_CLOSED" closed

continued next page...

Variable	Characteristics	Description
DIMBLK (continued)		"_SMALL" dot small blank "_NONE" none "_OBLIQUE" oblique "_BOXFILLED" box filled "_BOXBLANK" box "_CLOSEDBLANK" closed blank "_DATUMFILLED" datum triangle filled "_DATUMBLANK" datum triangle "_INTEGRAL" integral "_ARCHTICK" architectural tick
DIMBLK1	Type: String Saved in: Drawing Initial value: ""	If *DIMSAH* is on, *DIMBLK1* specifies user-defined arrowhead blocks for the first end of the dimension line. This variable contains the name of a previously defined block. To disable an established block name, set it to a single period (.). For a list of arrowheads, see *DIMBLK*.
DIMBLK2	Type: String Saved in: Drawing Initial value: ""	If *DIMSAH* is on, *DIMBLK2* specifies user-defined arrowhead blocks for the second end of the dimension line. This variable contains the name of a previously defined block. To disable an established block name, set it to a single period (.). For a list of arrowheads, see *DIMBLK*.
DIMCEN	Type: Real Saved in: Drawing Initial value: 0.0900	Controls drawing of circle or arc center marks and centerlines by the *DIMCENTER, DIMDIAMETER,* and *DIMRADIUS* dimensioning commands. 0 No center marks or lines are drawn. <0 Centerlines are drawn. >0 Center marks are drawn. The absolute value specifies the size of the mark portion of the centerline. *DIMRADIUS* and *DIMDIAMETER* draw the center mark or line only if the dimension line is placed outside the circle or arc.
DIMCLRD	Type: Integer Saved in: Drawing Initial value: 0	Assigns colors to dimension lines, arrowheads, and dimension leader lines. Also controls the color of leader lines created with the *LEADER* command. The color can be any valid color number or special color label *BYBLOCK* or *BYLAYER*. Supply the color number. Integer equivalents for *BYBLOCK* and *BYLAYER* are 0 and 256, respectively.
DIMCLRE	Type: Integer Saved in: Drawing Initial value: 0	Assigns colors to dimension extension lines. The color can be any valid color number or the special color label *BYBLOCK* or *BYLAYER*. See *DIMCLRD*.
DIMCLRT	Type: Integer Saved in: Drawing Initial value: 0	Assigns colors to dimension text. The color can be any valid color number or the special color label *BYBLOCK* or *BYLAYER*. See *DIMCLRD*.
DIMDEC	Type: Integer Saved in: Drawing Initial value: 4	Sets the number of decimal places for the tolerance values of a primary units dimension. The precision is based on the units or angle format you have selected.

Variable	Characteristics	Description
DIMDLE	Type: Real Saved in: Drawing Initial value: 0.0000	Extends the dimension line beyond the extension line when oblique strokes are drawn instead of arrowheads.
DIMDLI	Type: Real Saved in: Drawing Initial value: 0.3800	Controls the dimension line spacing for baseline dimensions. Each baseline dimension is offset by this amount, if necessary, to avoid drawing over the previous dimension. Changes made with *DIMDLI* are not applied to existing dimensions.
DIMDSEP	Type: Single character Saved in: Drawing Initial value: Decimal point	Specifies a single-character decimal separator to use when creating dimensions whose unit format is decimal. When prompted, enter a single character at the Command line. If dimension units is set to *Decimal*, the *DIMDSEP* character is used instead of the default decimal point. If *DIMDSEP* is set to NULL (default value, reset by entering a period), AutoCAD uses the decimal point as the dimension separator.
DIMEXE	Type: Real Saved in: Drawing Initial value: 0.1800	Determines how far to extend the extension line beyond the dimension line.
DIMEXO	Type: Real Saved in: Drawing Initial value: 0.0625	Determines how far extension lines are offset from origin points. If you point directly at the corners of an object to be dimensioned, the extension lines stop just short of the object.
DIMFIT	Type: Integer Saved in: Drawing Initial value: 3	<u>Obsolete</u>. Has no effect in AutoCAD 2000 and later releases except to preserve the integrity of pre-AutoCAD 2000 scripts and AutoLISP routines. *DIMFIT* is replaced by *DIMATFIT* and *DIMTMOVE*.
DIMFRAC	Type: Integer Saved in: Drawing Initial value: 0	Sets the fraction format when *DIMLUNIT* is set to 4 (Architectural) or 5 (Fractional). 0 Horizontal 1 Diagonal 2 Not stacked (for example, 1/2)
DIMFXL	Type: Real Saved in: Drawing Initial value: 1.0000	Sets the total length of the extension lines starting from the dimension line toward the dimension origin. The length is set in drawing units.
DIMFXLON	Type: Switch Saved in: Drawing Initial value: Off	Controls whether extension lines are set to a fixed length. When *DIMFXLON* is on, extension lines are set to the length specified by *DIMFXL*.
DIMGAP	Type: Real Saved in: Drawing Initial value: 0.0900	Sets the distance around the dimension text when you break the dimension line to accommodate dimension text. Also sets the gap between annotation and a hook line created with the *LEADER* command. A negative *DIMGAP* value creates basic dimensioning—dimension text with a box around its full extents.

continued next page...

2007

Variable	Characteristics	Description
DIMGAP (continued)		AutoCAD also used *DIMGAP* as the minimum length for pieces of the dimension line. When calculating the default position for the dimension text, it positions the text inside the extension lines only if doing so breaks the dimension lines into two segments at least as long as *DIMGAP*. Text placed above or below the dimension line is moved inside if there is room for the arrowheads, dimension text, and a margin between them at least as large as *DIMGAP*: 2 * (*DIMASZ* + *DIMGAP*). *DIMGAP* also sets the gap between a tolerance symbol and its feature control frame.
DIMJOGANG	Type: Real Saved in: Drawing Initial value: 45 (90 for metric)	Determines the angle of the transverse segment of the dimension line in a jogged radius dimension. Jogged radius dimensions are often created when the center point is located off the page.
DIMJUST	Type: Integer Saved in: Drawing Initial value: 0	Controls horizontal dimension text position: 0 Center justifies the text between the extension lines. 1 Positions the text next to the first extension line. 2 Positions the text next to the second extension line. 3 Positions the text above and aligned with the first extension line. 4 Positions the text above and aligned with the second extension line.
DIMLDRBLK	Type: String Saved in: Drawing Initial value: ""	Specifies the arrow type for leaders. To turn off arrowhead display, enter a single period (.). For a list of arrowhead entries, see *DIMBLK*.
DIMLFAC	Type: Real Saved in: Drawing Initial value: 1.0000	Sets a global scale factor for linear dimensioning measurements. All linear distances measured by dimensioning (including radii, diameters, and coordinates) are multiplied by the *DIMLFAC* setting before being converted to dimension text. *DIMLFAC* has no effect on angular dimensions, and it is not applied to the values held in *DIMTM*, *DIMTP*, or *DIMRND*. If you create associative dimensions in paper space, the model space distance measured is multiplied by the absolute value of *DIMLFAC*. In model space, negative values for *DIMLFAC* are ignored, and the value 1.0 is used instead. For nonassociative dimensions, *DIMLFAC* must be set manually to accommodate viewport scaling.
DIMLIM	Type: Switch Saved in: Drawing Initial value: Off	When turned on, generates dimension limits as the default text. Setting *DIMLIM* on forces *DIMTOL* to be off.

2007

Variable	Characteristics	Description
DIMLTEX1	Type: String Saved in: Drawing Initial value: ""	Sets the linetype of the first extension line. The value is *BYLAYER, BYBLOCK,* or the name of a linetype.
DIMLTEX2	Type: String Saved in: Drawing Initial value: ""	Sets the linetype of the second extension line. The value is *BYLAYER, BYBLOCK,* or the name of a linetype.
DIMLTYPE	Type: String Saved in: Drawing Initial value: ""	Sets the linetype of the dimension line. The value is *BYLAYER, BYBLOCK,* or the name of a linetype.
DIMLUNIT	Type: Integer Saved in: Drawing Initial value: 2	Sets units for all dimension types except Angular. 1 Scientific 2 Decimal 3 Engineering 4 Architectural 5 Fractional 6 Windows desktop
DIMLWD	Type: Enum Saved in: Drawing Initial value: -2	Assigns lineweight to dimension lines. Values are standard lineweights. -3 Default (the *LWDEFAULT* value, integers represent 1/100th of a mm) -2 *BYBLOCK* -1 *BYLAYER*
DIMLWE	Type: Enum Saved in: Drawing Initial value: -2	Assigns lineweight to extension lines. Values are standard lineweights. -3 Default (the *LWDEFAULT* value, integers represent 1/100th of a mm) -2 *BYBLOCK* -1 *BYLAYER*
DIMPOST	Type: String Saved in: Drawing Initial value: ""	Specifies a text prefix or suffix (or both) to the dimension measurement. For example, to establish a suffix for millimeters, set *DIMPOST* to "mm"; a distance of 19.2 units would be displayed as "19.2mm." If tolerances are enabled, the suffix is applied to the tolerances as well as to the main dimension. Use <> to indicate placement of the text in relation to the dimension value. For example, enter <>mm to display a 5.0 millimeter radial dimension as "5.0mm." If you entered "mm <>", the dimension would be displayed as "mm 5.0." Use the <> mechanism for angular dimensions.
DIMRND	Type: Real Saved in: Drawing Initial value: 0.0000	Rounds all dimensioning distances to the specified value. For instance, if *DIMRND* is set to 0.25, all distances round to the nearest 0.25 unit. If you set *DIMRND* to 1.0, all distances round to the nearest integer. Note that the number of digits edited after the decimal point depends on the precision set by *DIMDEC*. *DIMRND* does not apply to angular dimensions.

Variable	Characteristics	Description
DIMSAH	Type: Switch Saved in: Drawing Initial value: Off	Controls use of user-defined arrowhead blocks at the ends of the dimension line: Off Normal arrowheads or user-defined arrowhead blocks set by *DIMBLK* are used. On User-defined arrowhead blocks are used. *DIMBLK1* and *DIMBLK2* specify different user-defined arrowhead blocks for each end of the dimension line.
DIMSCALE	Type: Real Saved in: Drawing Initial value: 1.0000	Sets the overall scale factor applied to dimensioning variables that specify sizes, distances, or offsets. It is not applied to tolerances or to measured lengths, coordinates, or angles. Also affects the scale of leader objects created with the *LEADER* command. 0.0 AutoCAD computes a reasonable default value based on the scaling between the current Model Space viewport and Paper Space. If you are in Paper Space, or in Model Space and not using the Paper Space feature, the scale factor is 1.0. >0 AutoCAD computes a scale factor that leads text sizes, arrowhead sizes, and other scaled distances to plot at their face values.
DIMSD1	Type: Switch Saved in: Drawing Initial value: Off	When turned on, suppresses drawing of the first dimension line and arrowhead.
DIMSD2	Type: Switch Saved in: Drawing Initial value: Off	When turned on, suppresses drawing of the second dimension line and arrowhead.
DIMSE1	Type: Switch Saved in: Drawing Initial value: Off	When turned on, suppresses drawing of the first extension line.
DIMSE2	Type: Switch Saved in: Drawing Initial value: Off	When turned on, suppresses drawing of the second extension line.
DIMSHO	Type: Switch Saved in: Drawing Initial value: On	<u>Obsolete</u>. Has no effect in AutoCAD 2000 and later releases except to preserve the integrity of pre-AutoCAD 2000 scripts and AutoLISP routines.
DIMSOXD	Type: Switch Saved in: Drawing Initial value: Off	When turned on, suppresses drawing of dimension lines outside the extension lines. If the dimension lines would be outside the extension lines and *DIMTIX* is on, setting *DIMSOXD* to on suppresses the dimension line. If *DIMTIX* is off, *DIMSOXD* has no effect.
DIMSTYLE	(Read-only) Type: String Saved in: Drawing	*DIMSTYLE* is both a command and a system variable. The *DIMSTYLE* system variable shows the current dimension style. To display the *DIMSTYLE* system variable, use the *Setvar* command. The *DIMSTYLE* system variable is read-only; you cannot change its value on the command line. To change the current dimension style, use the *DIMSTYLE* command.

Variable	Characteristics	Description
DIMTAD	Type: Integer Saved in: Drawing Initial value: 0	Controls vertical position of text in relation to the dimension line: 0 Centers the dimension text between the extension lines. 1 Places the dimension text above the dimension line except when the dimension line is not horizontal and text inside the extension lines is forced horizontal (*DIMTIH* = 1). The distance from the dimension line to the baseline of the lowest line of text is the current *DIMGAP* value. 2 Places the dimension text on the side of the dimension line farthest away from the defining points. 3 Places the dimension text to conform to a JIS representation.
DIMTDEC	Type: Integer Saved in: Drawing Initial value: 4	Sets the number of decimal places for the tolerance values for a primary units dimension. This system variable has no effect unless *DIMTOL* is set to *On*.
DIMTFAC	Type: Real Saved in: Drawing Initial value: 1.0000	Sets a scale factor used to calculate the height of text for dimension fractions and tolerances. AutoCAD multiplies *DIMTXT* by *DIMTFAC* to set the fractional or tolerance text height.
DIMTFILL	Type: Integer Saved in: Drawing Initial value: 0	Controls the background of dimension text. 0 No background 1 The background color of the drawing 2 The background specified by *DIMTFILLCLR*
DIMTFILLCLR	Type: Integer Saved in: Drawing Initial value: 0	Sets the color for the text background in dimensions. Color numbers are displayed in the *Select Color* dialog box. For *BYBLOCK*, enter 0. For *BYLAYER*, enter 256.
DIMTIH	Type: Switch Saved in: Drawing Initial value: On	Controls the position of dimension text inside the extension lines for all dimension types except ordinate dimensions: Off Aligns text with the dimension line. On Draws text horizontally.
DIMTIX	Type: Switch Saved in: Drawing Initial value: Off	Draw text between extension lines: Off The result varies with the type of dimension. For linear and angular dimensions, AutoCAD places text inside the extension lines if there is sufficient room. For radius and diameter dimensions, setting *DIMTIX* off forces the text outside the circle or arc. On Draws dimension text between the extension lines even if AutoCAD would ordinarily place it outside those lines.
DIMTM	Type: Real Saved in: Drawing Initial value: 0.0000	When *DIMTOL* or *DIMLIM* is on, sets the minimum (or lower) tolerance limit for dimension text. AutoCAD accepts signed values for *DIMTM*. If *DIMTOL* is on and *DIMTP* and *DIMTM* are set to the same value, AutoCAD

continued next page...

Variable	Characteristics	Description
DIMTM (continued)		draws a ± symbol followed by the tolerance value. If *DIMTM* and *DIMTP* values differ, the upper tolerance is drawn above the lower and a plus sign is added to the *DIMTP* value if it is positive. For *DIMTM*, AutoCAD uses the negative of the value you enter (adding a minus sign if you specify a positive number and a plus sign if you specify a negative number). No sign is added to a value of zero.
DIMTMOVE	Type: Integer Saved in: Drawing Initial value: 0	Sets dimension text movement rules. 0 Moves the dimension line with dimension text. 1 Adds a leader when dimension text is moved. 2 Allows text to be moved freely without a lead.
DIMTOFL	Type: Switch Saved in: Drawing Initial value: Off	When turned on, draws a dimension line between the extension lines, even when the text is placed outside the extension lines. For radius and diameter dimensions (while *DIMTIX* is off), draws a dimension line and arrowheads inside the circle or arc and places the text and leader outside.
DIMTOH	Type: Switch Saved in: Drawing Initial value: On	When turned on, controls the position of dimension text outside the extension lines: Off Aligns text with the dimension line. On Draw text horizontally.
DIMTOL	Type: Switch Saved in: Drawing Initial value: Off	When turned on, appends dimension tolerances to dimension text. Setting *DIMTOL* on forces *DIMLIM* off.
DIMTOLJ	Type: Integer Saved in: Drawing Initial value: 1	Sets the vertical justification for tolerance values relative to the nominal dimension text: 0 Bottom 1 Middle 2 Top This system variable has no effect unless *DIMTOL* is set to On.
DIMTP	Type: Real Saved in: Drawing Initial value: 0.0000	When *DIMTOL* or *DIMLIM* is on, sets the maximum (or upper) tolerance limit for dimension text. AutoCAD accepts signed values for *DIMTP*. If *DIMTOL* is on and *DIMTP* and *DIMTM* are set to the same value, AutoCAD draws a ± symbol followed by the tolerance value. If *DIMTM* and *DIMTP* values differ, the upper tolerance is drawn above the lower and a plus sign is added to the *DIMTP* value if it is positive.
DIMTSZ	Type: Real Saved in: Drawing Initial value: 0.0000	Specifies the size of oblique strokes drawn instead of arrowheads for linear, radius, and diameter dimension: 0 Draws arrows. >0 Draws oblique strokes instead of arrows. Size of oblique strokes is determined by this value multiplied by the *DIMSCALE* value.

Variable	Characteristics	Description
DIMTVP	Type: Real Saved in: Drawing Initial value: 0.0000	Adjusts the vertical position of dimension text above or below the dimension line. AutoCAD uses the DIMTVP value when DIMTAD is off. The magnitude of the vertical offset of text is the product of the text height and DIMTVP. Setting DIMTVP to 1.0 is equivalent to setting DIMTAD to on. AutoCAD splits the dimension line to accommodate the text only if the absolute value of DIMTVP is less than 0.7.
DIMTXSTY	Type: String Saved in: Drawing Initial value: "STANDARD"	Specifies the text style of the dimension.
DIMTXT	Type: Real Saved in: Drawing Initial value: 0.1800	Specifies the height of dimension text, unless the current text style has a fixed height.
DIMTZIN	Type: Integer Saved in: Drawing Initial value: 0	Controls the suppression of zeros in tolerance values. DIMTZIN values 0–3 affect feet-and-inch dimensions only. 0 Suppresses zero feet and precisely zero inches 1 Includes zero feet and precisely zero inches 2 Includes zero feet and suppresses zero inches 3 Includes zero inches and suppresses zero feet 4 Suppresses leading zeros in decimal dimensions (for example, 0.5000 becomes .5000) 8 Suppresses trailing zeros in decimal dimensions (for example, 12.5000 becomes 12.5) 12 Suppresses both leading and trailing zeros (for example, 0.5000 becomes 0.5)
DIMUNIT		<u>Obsolete</u>. Retained to preserve the integrity of scripts and AutoLISP routines. DIMUNIT is replaced by DIMLUNIT and DIMFRAC.
DIMUPT	Type: Switch Saved in: Drawing Initial value: Off	Controls cursor functionality for User Positioned Text: Off Cursor controls only the dimension line location. On Cursor controls the text position as well as the dimension line location.
DIMZIN	Type: Integer Saved in: Drawing Initial value: 0	Controls the suppression of zeros in the primary unit value. DIMZIN stores this value when you enter it on the Command line or set it under *Primary Units* in the *Dimension Style Manager*. DIMZIN values 0–3 affect feet-and-inch dimensions only. 0 Suppresses zero feet and precisely zero inches 1 Includes zero feet and precisely zero inches 2 Includes zero feet and suppresses zero inches 3 Includes zero inches and suppresses zero feet *continued next page...*

Variable	Characteristics	Description
DIMZIN (continued)		4 Suppresses leading zeros in decimal dimensions (for example, 0.5000 becomes .5000) 8 Suppresses trailing zeros in decimal dimensions (for example, 12.5000 becomes 12.5) 12 Suppresses both leading and trailing zeros (for example, 0.5000 becomes 0.5) *DIMZIN* also affects real-to-string conversions performed by the AutoLISP rtos and angtos functions. 8 Suppresses trailing zeros in decimal dimensions (for example, 12.5000 becomes 12.5) 12 Suppresses both leading and trailing zeros (for example, 0.5000 becomes 0.5) *DIMZIN* also affects real-to-string conversions performed by the AutoLISP rtos and angtos functions.

Using *Setup Wizards* and Template Drawings

ACAD.DWT Template Drawing

The previous dimension variable table gives default settings for the STANDARD dimension style in the ACAD.DWT template drawing. The ACAD.DWT drawing is the template or basis used when you begin AutoCAD with the *Startup* dialog box toggled off, choose *Start from Scratch, Imperial Default Settings* in the *Create New Drawing* dialog box, select the ACAD.DWT as a template drawing, or choose all the inch defaults in the *Setup Wizards*. In other cases, one or more dimension variables are automatically changed. In some template drawings, dimension styles other than STANDARD are available.

ANSI, ISO-25, DIN, and JIS Dimension Styles

The ANSI (American National Standards Institute) template drawings contain the STANDARD dimension style, which has the same variable settings as in the ACAD.DWT and as listed in the previous Dimension Variables table. If you select an ISO, DIN, or JIS template drawing, one or more dimension styles other than STANDARD may exist in the drawing. The following table lists the differences in the dimension variable settings listed by dimension style. All dimension variables not listed in the table have the same setting as the STANDARD style.

ISO-25 is the default ISO (International Standards Organization) template dimension style and the default style when you select *Metric* in *Start From Scratch*. DIN is the default dimension style for DIN (Deutsches Institut für Normung eV) templates. JIS is the default dimension style for JIS (Japanese Industrial Standard) templates.

Variable	Description	ANSI	ISO-25	DIN	JIS	Gb
DIMALTD	Alternate unit decimal places	2	4	2	2	4
DIMALTF	Alternate unit scale factor	25.4000	0.0394	0.0394	0.0394	0.0394
DIMALTTD	Alternate tolerance decimal places	2	3	2	2	4
DIMALTU	Alternate units	2	2	8	2	8
DIMASZ	Arrow size	0.1800	2.5000	2.5000	2.5000	5.0000

DIMCEN	Center mark size	0.0900	2.500	2.500	0.0000	5.0000
DIMDEC	Decimal places for dimensions	4	2	4	4	2
DIMDLI	Dimension line spacing	0.3800	3.7500	3.7500	7.0000	7.5000
DIMDSEP	Decimal separator	.	,	.	.	.
DIMEXE	Extension above dimension line	0.1800	1.2500	1.2500	1.0000	2.5000
DIMEXO	Extension line origin offset	0.0625	0.625	0.625	1.0000	1.0000
DIMGAP	Gap from dimension line to text	0.0900	0.6250	0.6250	0.0000	1.2500
DIMLUNIT	Linear unit format	2	2	6	2	2
DIMSOXD	Suppress outside dimension lines	Off	Off	Off	On	Off
DIMTAD	Place text above the dimension line	0	1	1	1	1
DIMTDEC	Tolerance decimal places	4	2	4	4	2
DIMTIH	Text inside extensions is horizontal	On	Off	On	Off	Off
DIMTIX	Place text inside extensions	Off	Off	Off	On	On
DIMTOFL	Force line inside extension lines	Off	On	On	On	On
DIMTOH	Text outside horizontal	On	Off	Off	Off	Off
DIMTOLJ	Tolerance vertical justification	1	0	1	1	0
DIMTXT	Text height	0.1800	2.5000	2.5000	2.5000	5.0000
DIMTZIN	Tolerance zero suppression	0	8	0	0	0
DIMZIN	Zero suppression	0	8	8	8	8

As you notice, most of the differences in variable settings are related to size. Sizes of individual dimension components are changed to show the arrows, gaps, offsets, increments, etc. appropriately for the English or metric system used. In this way, the *Overall Scale* (*DIMSCALE*) for each dimension style is set to 1 by default and can be easily changed to adjust for different plot scales and plot sheets. See "Table of AutoCAD-Supplied Template Drawing Settings," Chapter 12, for a more information on the template drawings.

Setup Wizards

When you use the *Advanced Setup Wizard*, no dimension styles are changed from the default settings; however, dimension style <u>overrides may be created</u>. For example, if you use the *Advanced Setup Wizard* to begin a drawing based on the ACAD.DWT template, then select *Architectural* units, an override to the STANDARD dimension style is created with *DIMLUNIT* set to 4 (*Architectural*). Other changes may occur depending on which template is used and which units are selected in the *Advanced Setup Wizard*. The *MEASUREINIT* system variable defines which template is used for the *Setup Wizards* (ACAD.DWT or ACADISO.DWT) and, therefore, which dimension style is used (STANDARD or ISO-25). No changes to dimensioning variables are made using the *Quick Setup Wizard*.

MODIFYING EXISTING DIMENSIONS

Even with the best planning, a full understanding of dimension variables, and the correct use of *Dimension Styles*, it is probable that changes will have to be made to existing dimensions in the drawing due to design changes, new plot scales, or new industry/company standards. There are several ways that changes can be made to existing dimensions while retaining associativity and membership to a dimension style. The possible methods are discussed in this section.

Modifying a Dimension Style

Existing dimensions in a drawing can be modified by making one or more variable changes to the dimension style family or child, then *Saving* those changes to the dimension style. This process can be accomplished using either the *Dimension Style Manager* or Command line format. When you use the

Modify option in the *Dimension Style Manager* or the *Save* option of the *-Dimstyle* command, the existing dimensions in the drawing <u>automatically update</u> to display the new variable settings.

Properties

Pull-down Menu	Command (Type)	Alias (Type)	Short-cut	Screen (side) Menu	Tablet Menu
Modify *Properties...*		*Properties*	PR or CH	*(Edit Mode)* *Properties* or Crtl+1	MODIFY1 *Property*	Y,12 to Y,13

Remember that *Properties* can be used to edit existing dimensions. Using *Properties* and selecting one dimension displays the *Properties* palette with all of the selected dimension's properties and dimension variables.

Using *Properties* you can modify any aspect of one or more dimensions, including text, and access is given to the *Lines and Arrows, Text, Fit, Primary Units, Alternate Units,* and *Tolerances* categories. Any changes to dimension variables through *Properties* result in <u>overrides to the dimension object only</u> but do not affect the dimension style. *Properties* has essentially the same result as *Dimoverride* (see "*Dimoverride*"). *Properties* can also be used to modify the dimension text value. *Properties* of dimensions are also discussed in Chapter 28.

Creating Dimension Style Overrides and Using *Update*

You can modify existing dimensions in a drawing without making permanent changes to the dimension style by creating a dimension style override, then using *Update* to apply the new setting to an existing dimension. This method is preferred if you wish <u>to modify one or two dimensions</u> without modifying all dimensions referencing (created in) the style.

To do this, use either the Command line format or *Dimension Style Manager* to set the new variable. In the *Dimension Style Manager*, select *Override* and make the desired variable settings in the *Override Current Style* dialog box. This creates an override to the current style. In Command line format, simply enter the formal dimension variable name and make the change to create an override to the current style, then use *Update* to apply the current style settings plus the override settings to existing dimensions that you PICK. The overrides remain in effect for the current style unless the variables are reset to the original values or until the overrides are cleared. You can <u>clear overrides</u> for a dimension style by making another dimension style current in the *Dimension Style Manager*.

Update

Pull-down Menu	Command (Type)	Alias (Type)	Short-cut	Screen (side) Menu	Tablet Menu
Dimension *Update*	*Dim* *Update*	*DIM* *UP*	...	*DIMNSION* *Update*	Y,3

Update can be used to update existing dimensions in the drawing to the current settings. The current settings are determined by the current dimension style and any dimension variable overrides that are in effect (see previous explanation, "Creating Dimension Style Overrides and Using *Update*"). This is an excellent method of modifying one or more existing dimensions without making permanent changes to the dimension style that the selected dimensions reference. *Update* has the same effect as using *-Dimstyle, Apply*.

Update is actually a Release 12 command. In Release 12, dimensioning commands could only be entered at the Dim: prompt. For example, to create a linear dimension, you had to type *Dim* and press Enter, then type *Linear*. In Release 13, all dimensioning commands were upgraded to top-level commands with the *Dim-* prefix added, so you can type *Dimlinear* at the command prompt, for example. *Update*, although very useful, was never upgraded. In a current release of AutoCAD, the command is given prominence by making available an *Update* button and an *Update* option in the *Dimension* pull-down menu. However, if you prefer to type, you must first type *Dim*, press Enter, and then enter *Update*. The command syntax is as follows:

```
Command: Dim
Dim: Update
Select objects: PICK (select a dimension object to update)
Select objects: Enter
Dim: press Esc or type Exit
Command:
```

For example, if you wanted to change the *DIMSCALE* of several existing dimensions to a value of 2, change the variable by typing *DIMSCALE*. Next use *Update* and select the desired dimension. That dimension is updated to the new setting. The command syntax is the following:

```
Command: Dimscale
New value for DIMSCALE <1.0000>: 2
Command: Dim
Dim: Update
Select objects: PICK  (select a dimension object to update)
Select objects: PICK  (select a dimension object to update)
Select objects: Enter
Dim: Exit
Command:
```

 TIP Beware, *Update* creates an override to the current dimension style. You should reset the variable to its original value unless you want to keep the override for creating other new dimensions. You can clear overrides for a dimension style by making another dimension style current in the *Dimension Style Manager*.

Dimoverride

	Pull-down Menu	Command (Type)	Alias (Type)	Short-cut	Screen (side) Menu	Tablet Menu
	Dimension Override	*Dimoverride*	*DIMOVER or DOV*	...	...	*Y,4*

Dimoverrride grants you a great deal of control to edit existing dimensions. The abilities enabled by *Dimoverride* are similar to the effect of using the *Properties* palette.

Dimoverride enables you to make variable changes to dimension objects that exist in your drawing without creating an override to the dimension style that the dimension references (was created under). For example, using *Dimoverride*, you can make a variable change and select existing dimension objects to apply the change. The existing dimension does not lose its reference to the parent dimension style nor is the dimension style changed in any way. In effect, you can override the dimension styles for selected dimension objects. There are two steps: set the desired variable and select dimension objects to alter.

```
Command: dimoverride
Enter dimension variable name to override or [Clear overrides]: (variable name)
```

Enter new value for dimension variable <current value>: (**value**)
Enter dimension variable name to override: **Enter**
Select objects: **PICK**
Select objects: **PICK** or **Enter**
Command:

The *Dimoverride* feature differs from creating dimension style overrides in two ways: (1) *Dimoverride* applies the changes to the selected dimension objects only, so the overrides are not appended to the parent dimension styles, only the objects; and (2) *Dimoverride* can be used once to change dimension objects referencing multiple dimension styles, whereas to make such changes to dimensions by creating dimension style overrides requires changing all the dimension styles, one at a time. The effect of using *Dimoverride* is essentially the same as using the *Properties* palette.

Dimoverride is useful as a "backdoor" approach to dimensioning. Once dimensions have been created, you may want to make a few modifications, but you do not want the changes to affect dimension styles (resulting in an update to all existing dimensions that reference the dimension styles). *Dimoverride* offers that capability. You can even make one variable change to affect all dimensions globally without having to change multiple dimension styles. For example, you may be required to make a test plot of the drawing in a different scale than originally intended, necessitating a new global *Dimscale*. Use *Dimoverride* to make the change for the plot:

    Command: *dimoverride*
    Enter dimension variable name to override or [Clear overrides]: *dimscale*
    Enter new value for dimension variable <1.0000>: **.5**
    Enter dimension variable name to override: **Enter**
    Select objects: (window entire drawing) Other corner: 128 found
    Select objects: **Enter**
    Command:

This action results in having all the existing dimensions reflect the new *DIMSCALE*. No other dimension variables or any dimension styles are affected. Only the selected objects contain the overrides. *Dimoverride* does not append changes (overrides) to the original dimension styles, so no action must be taken to clear the overrides from the styles. After making the plot, *Dimoverride* can be used with the *Clear* option to change the dimensions (by object selection) back to their original appearance. The *Clear* option is used to clear overrides from <u>dimension objects, not from dimension styles</u>.

Clear

The *Clear* option removes the overrides from the <u>selected dimension objects</u>. It does not remove overrides from the current dimension style:

    Command: *dimoverride*
    Enter dimension variable name to override or [Clear overrides]: *c*
    Select objects: **PICK**
    Select objects: **Enter**
    Command:

The dimension then displays the variable settings as specified by the dimension style it references without any overrides (as if the dimension were originally created without the overrides). Using *Clear* does not remove any overrides that are appended to the dimension style so that if another dimension is drawn, the dimension style overrides apply.

Matchprop

Pull-down Menu	Command (Type)	Alias (Type)	Short-cut	Screen (side) Menu	Tablet Menu
Modify *Match Properties*	*Matchprop*	*MA*	...	*MODIFY1* *Matchprp*	*Y,14 and Y,15*

Matchprop can be used to "convert" an existing dimension to the style (including overrides) of another dimension in the drawing. For example, if you have two linear dimensions, one has *Oblique* arrows and *Romans* text font (*Dimension Style* = "Oblique") and one is a typical linear dimension (*Dimension Style* = "Standard") as in Figure 29-70, "before."

You can convert the typical dimension to the "Oblique" style by using *Matchprop*, selecting the "Oblique" dimension as the "source object" (to match), then selecting the typical dimension as the "destination object" (to convert). The typical dimension then references the "Oblique" dimension style and changes appearance accordingly (Fig. 29-70, "after"). Note that *Matchprop* does not alter the dimension text value.

Using the *Settings* option of the *Matchprop* command, you can display the *Property Settings* dialog box (Fig. 29-71). The *Dimension* box under *Special Properties* must be checked to "convert" existing dimensions as illustrated above.

Using *Matchprop* is a fast and easy method for modifying dimensions from one style to another. However, this method is applicable only if you have existing dimensions in the drawing with the desired appearance that you want others to match.

FIGURE 29-70

BEFORE *MATCHPROP*

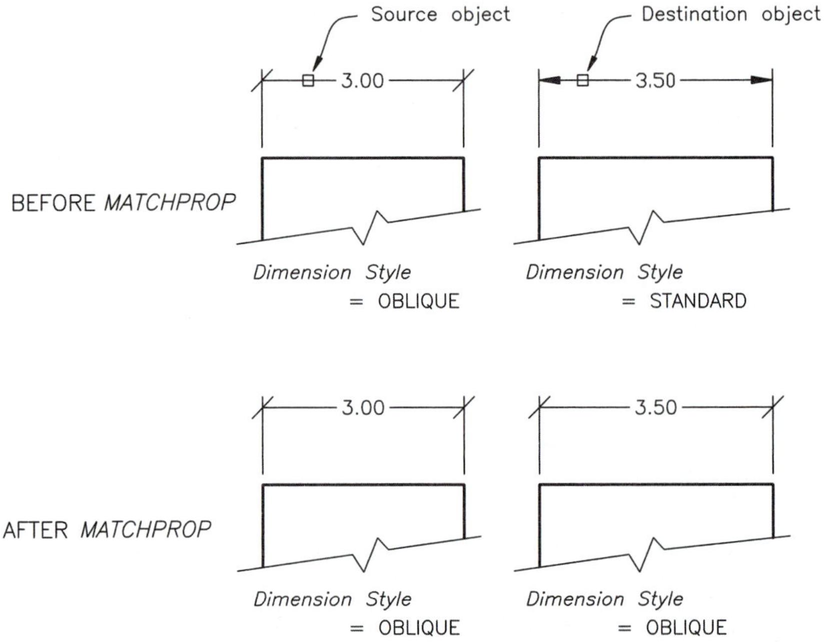

Source object — Destination object

3.00   3.50

Dimension Style = OBLIQUE   Dimension Style = STANDARD

AFTER *MATCHPROP*

3.00   3.50

Dimension Style = OBLIQUE   Dimension Style = OBLIQUE

FIGURE 29-71

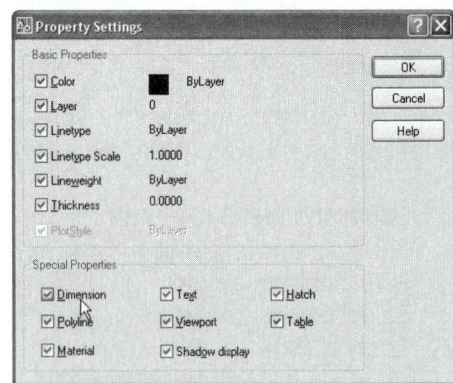

Dimension Right-Click Menu

If you select any dimension, you can right-click to produce a shortcut menu (Fig. 29-72). This menu contains options that allow you to change the text position, the text precision, and the *Dimension Style* of the dimension, and to flip the arrows. These options are also discussed in Chapter 28.

FIGURE 29-72

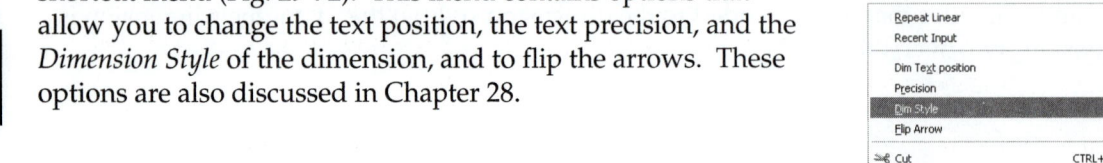

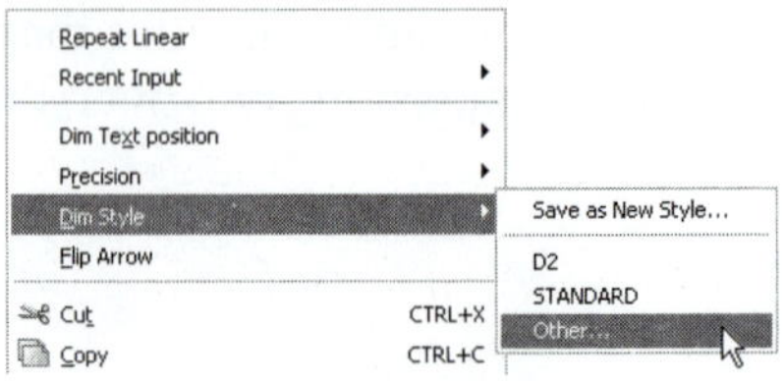

Dim Text Position

This cascading menu offers the *Centered* and *Home Text* options (duplicates of the *Dimtedit* command options), and the *Move text alone*, *Move with leader*, and *Move with dim line* options (*DIMTMOVE* variable settings).   Using a *Move* option creates a *DIMTMOVE* <u>dimension style override</u> for the individual dimension.

Precision

Use this cascading menu to select from a list of decimal or fractional precision values. For example, you could change a dimension text value of "5.0000" to "5.00" with this menu. Your choice creates a <u>*DIMDEC* dimension style override</u> for the selected dimension.

Dim Style

Use this option to change the selected dimension from one dimension style to another. You can easily create a new dimension style or change the dimension to another style, similar to using *Match Properties*. Select *Other* to produce a list of all the dimension styles contained in the drawing.

Flip Arrow

FIGURE 29-73

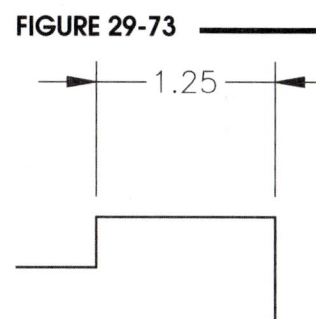

Use this option to flip the arrow nearest to the point you right-click to produce the menu. In other words, if both arrows pointed from the center of the dimension out to the extension lines (default position), flipping both arrows would cause the arrows to be on the outside of the extension lines pointing inward (Fig. 29-73). The *DIMATFIT* variable can also be used to affect the position of the arrows.

2006

GUIDELINES FOR DIMENSIONING IN AutoCAD

Listed in this section are some guidelines to use for dimensioning a drawing using dimensioning variables and dimension styles. Although there are other strategies for dimensioning, two strategies are offered here as a framework so you can develop an organized approach to dimensioning.

In almost every case, dimensioning is one of the last steps in creating a drawing, since the geometry must exist in order to dimension it. You may need to review the steps for drawing setup, including the concept of drawing scale factor (Chapter 12).

Strategy 1.   Dimensioning a Single Drawing

This method assumes that the fundamental steps have been taken to set up the drawing and create the geometry. Assume this has been accomplished:

> Drawing setup is completed: *Units, Limits, Snap, Grid, Ltscale, Layers,* border, and titleblock.
> The drawing geometry (objects comprising the subject of the drawing) has been created.

Now you are ready to dimension the drawing subject (of the multiview drawing, pictorial drawing, floor plan, or whatever type of drawing).

1.   Create a *Layer* (named DIM, or similar) for dimensioning if one has not already been created. Set *Continuous* linetype and appropriate color. Make it the *current* layer.

2.   Set the *Overall Scale (DIMSCALE)* based on drawing *Limits* and expected plotting size.

For 1/8" dimensions, for example, multiply the initial values of the size-related variables by .7; namely

Text Height	*(DIMTXT)*	.18 x .7 = .126
Arrow Size	*(DIMASZ)*	.18 x .7 = .126
Extension Line Extension	*(DIMEXE)*	.18 x .7 = .126
Dimension Line Spacing	*(DIMDLI)*	.38 x .7 = .266
Text Gap	*(DIMGAP)*	.09 x .7 = .063

Save these settings in your template drawing(s). When you are ready to dimension, simply multiply *Overall Scale* (1) times the drawing scale factor.

Although this method may seem complex, the drawing setup for individual drawings is simplified. For example, assume your template drawing contained preset *Limits* to the paper size, say 11 x 8.5. In addition, the previously mentioned dimension variables were set to produce 1/8" (or whatever) dimensions when plotted full size. Other variables such as *LTSCALE* could be appropriately set. Then, when you wish to plot a drawing to 1=1, everything is preset. If you wish to plot to 1=2, simply multiply all the size-related variables (*Limits, Overall Scale, Ltscale,* etc.) times 2!

DIMENSIONING IN PAPER SPACE LAYOUTS

Generally, dimensions are created in model space and are attached to model space geometry. In this way, you can make one or more layouts, each with one or more viewports that display the model space geometry, and the dimensions are visible by default in each of the layouts and viewports. Assuming the dimensions are created on a dimensioning layer or layers, you can control the display of the dimensions in each viewport using viewport-specific layer visibility controls in the *Layer Manager*. This strategy is used for almost all AutoCAD drawings previous to AutoCAD 2002, and will most likely be continued for most drawings in the future except for certain cases.

Since AutoCAD 2002, it is possible to create dimensions in a paper space layout <u>associated</u> with (attached to) model space geometry inside a viewport. To explain, a typical drawing includes geometry in model space (such as mechanical part views, a floor plan, or a diagram) and at least one paper space layout used to display and print the model geometry. You can create the dimensions in a layout (in paper space) and *Osnap* to objects in model space (inside a viewport). These new dimensions are fully associative and display the actual measurement value of the drawing objects <u>in model space units</u>.

This example drawing (Fig. 29-74) has some dimensions in model space (left viewport) and some dimensions in paper space (right viewport). Note that the dimensions for the small cutout have been omitted in model space (left viewport). All of the dimensions for the detail (right viewport) are created in paper space. Some of the dimensions for the detail are actually outside of the viewport border, but the two short horizontal dimensions (.5 and .25) appear to be in model space even though they were actually created in paper space. Note that all dimensions associated to model space geometry, whether created in model space or paper space, display the correct value of the model feature they are measuring.

FIGURE 29-74

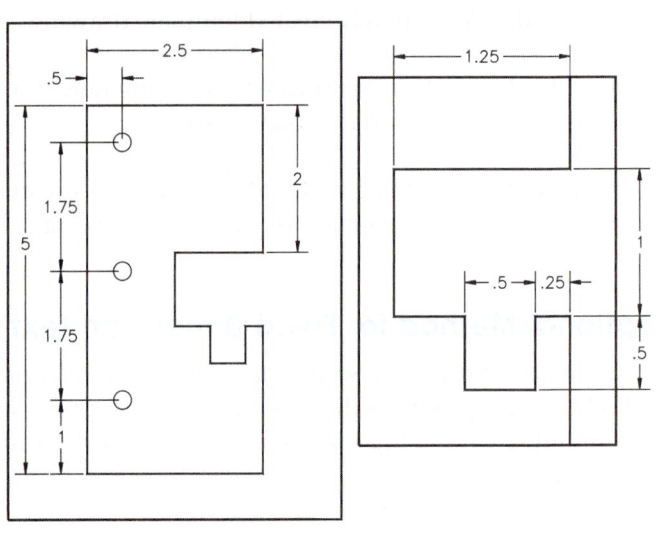

Figure 29-75 displays the completed drawing after adding some annotation and freezing the viewport border layer. Note that you cannot detect which dimensions are in model space and which are in paper space.

FIGURE 29-75

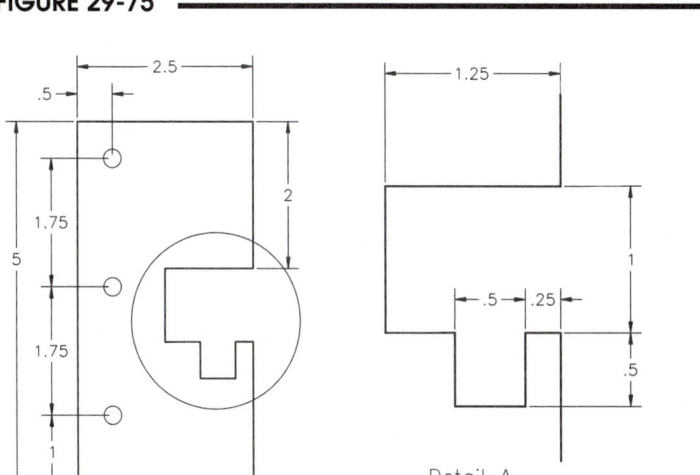

When to Dimension in Paper Space

Although it is still recommended to create dimensions in model space for most situations, there are some cases where creating dimensions in paper space could be used. Generally, whenever you want specific dimensions to appear for only one of several viewports, it may be useful to create those dimensions in paper space. Here are two specific examples.

If you have created a 3D solid model, then generated 2D views (a multiview layout) from the model with each view in a separate paper space viewport, consider creating the dimensions for each view in paper space.

If you want one or more detail views (small, enlarged sections of a larger drawing) each in a separate viewport or layout and at different scales than the full drawing display, consider creating dimensions for the detail views in paper space.

For the situations listed above, the advantages of creating dimensions in paper space attached to objects in model space (inside a viewport) are these:

Setting *DIMSCALE* or *Overall Dimension Scale* (the size of the dimension text, arrowheads, etc.) for paper space dimensions is simplified since you are concerned only with paper space units and not with the viewport scale (the scale of the geometry that appears in the viewports). This is especially important when you have several detail views of a drawing and each detail view is at a different scale than the full drawing view. Paper space dimensions appear in one size, whereas model space dimensions appear in different sizes when the drawing is scaled differently in each viewport.

Creating dimensions in paper space ensures that those dimensions appear only for that viewport but not for other displays of the model geometry appearing in other viewports and layouts. Therefore, you do not have to use different layers for different sets of dimensions and the *Layer Properties Manager* to set viewport-specific visibility (which dimension layers you want to appear in which viewports).

Dimensioning a 3D model requires that dimensions for each side or view of the model appear on different planes (created by using construction planes called User Coordinate Systems or UCSs in AutoCAD). For a multiview-type setup of a 3D model, dimensioning in paper space prevents having to use different UCSs as well as different layers for the dimensions that should appear in each view (on each plane of the 3D model). Also, you do not have to control viewport-specific layer visibility to ensure that dimensions for the top view do not appear in the front view, and so on.

TIP Although it appears that these advantages might outweigh the practice of dimensioning in model space, most drawings require that the dimensions appear in multiple viewports or layouts; therefore, dimensioning in model space is the more common practice. Dimensioning in paper space eliminates the possibility of showing the same dimensions in multiple viewports or layouts. See Chapters 33 and 42 for more information and applications of dimensioning in paper space.

Procedure for Dimensioning in Paper Space

Dimensioning in paper space differs from dimensioning in model space only in a few respects such as determining the *DIMSCALE* (*Overall Scale*) and the need to use *Dimregen* when zooming or panning in the viewport; therefore, the procedure for dimensioning in paper space is similar to that for dimensioning in model space. Here is a typical procedure to use when you need paper space dimensions.

1. Complete the model space geometry.

2. (Optional, depending on the situation and desired result.) Create a layer named DIM or DIMEN-SIONS and set it *Current*. Use the *Dimension Style Manager* to create a dimension style for the model space dimensions. Set the *Overall Scale* and other variables and save the style. Create the dimensions you need in model space. These dimensions are those that you want to appear in multiple viewports.

3. Set up the desired layouts and viewports and set the desired viewport scale for each viewport (so the model geometry appears in the desired scale in each viewport).

4. Create a layer named DIM-PS or DIMENSIONS-PAPERSPACE and set the layer *Current*.

5. Use the *Dimension Style Manager* to create a dimension style for the paper space dimensions by copying the style for model space dimensions and rename the style DIM-PS or similar. Set the *Overall Scale* so the dimensions print in paper space units. (Normally, an *Overall Scale* of 1 is used for paper space dimensions since the drawing is printed from the layout at 1:1.) The other dimension variables normally would not change (from the model space settings).

6. Activate the desired layout and ensure that paper space is active (the cursor appears in paper space, not in the viewport). With *Osnap* on, create the dimensions in paper space but *Osnap* the extension line origins, etc. to the objects in model space. The location of the dimension text can be placed outside or inside the viewport borders but the actual dimension objects are in paper space. Figure 29-76 displays two paper space dimensions, one has text inside the viewport border.

FIGURE 29-76

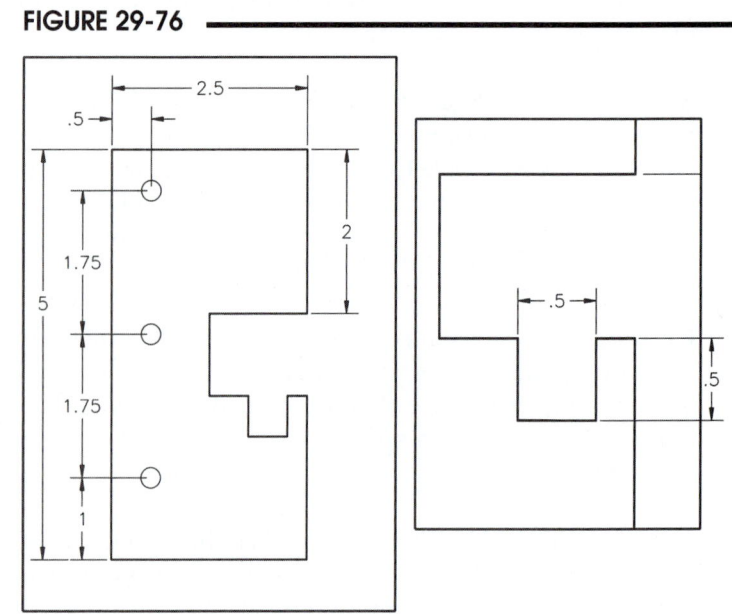

7. Complete the paper space dimensions. If you need to *Zoom* or *Pan* inside the viewport or change the viewport scale (using *Zoom XP* or the *Viewport Scale* drop-down edit box), you should use *Dimregen* to force AutoCAD to redisplay the location of the paper space dimension objects to align correctly with the new location of the model space geometry. Figure 29-77 illustrates the drawing after using *Pan* but <u>before</u> using *Dimregen*.

8. Complete any additional notation or objects in paper space. In the *Layer Properties Manager*, *Freeze* or disable plotting for the viewport border layer (see Fig. 29-75).

FIGURE 29-77

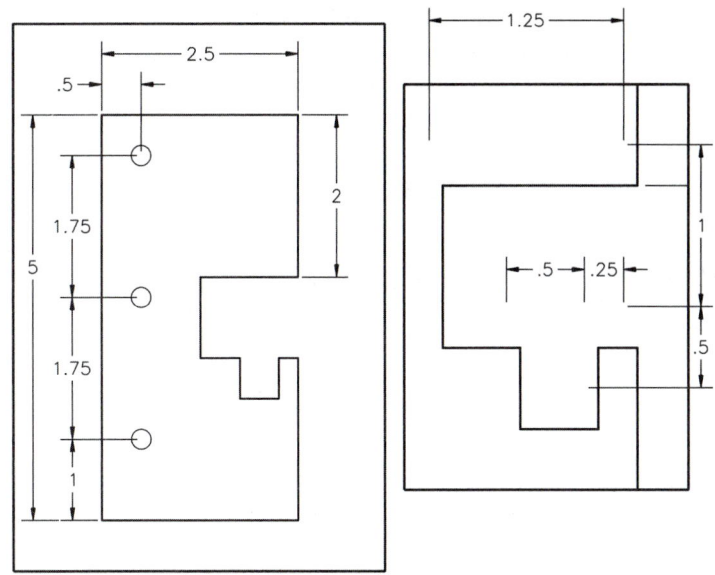

DIMENSIONING ISOMETRIC DRAWINGS

Dimensioning isometric drawings in AutoCAD is accomplished using *Dimaligned* dimensions and then adjusting the angle of the extension lines with the *Oblique* option of *Dimedit*. The technique follows two basic steps:

1. Use *Dimaligned* or the *Vertical* option of *Dimlinear* to place a dimension along one edge of the isometric face. Isometric dimensions should be drawn on the isometric axes lines (vertical or at a 30° rotation from horizontal; Fig. 29-78).

FIGURE 29-78

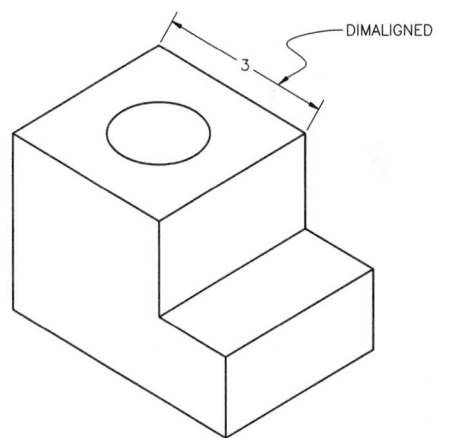

2. Use the *Oblique* dimensioning option of *Dimedit* and select the dimension just created. When prompted to "Enter obliquing angle," enter the desired value (**30** in this case) or PICK two points designating the desired angle (Fig. 29-79). The extension lines should change to the designated angle. In AutoCAD, the possible obliquing angles for isometric dimensions are **30, 150, 210,** or **330**.

FIGURE 29-79

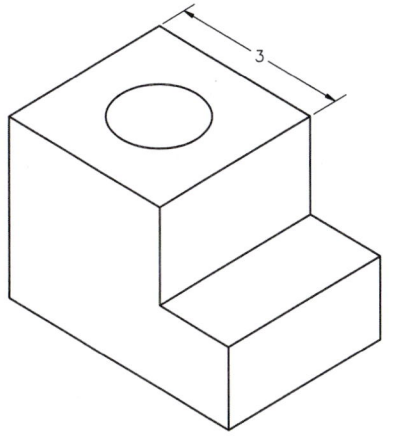

Place isometric dimensions so that they align with the face of the particular feature. Dimensioning the object in the previous illustrations would continue as follows.

Create a *Dimlinear, vertical* dimension along a vertical edge (Fig. 29-80).

FIGURE 29-80

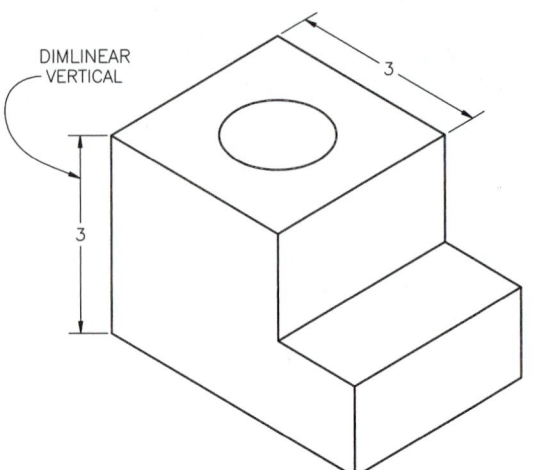

Use *Dimedit, Oblique* to force the dimension to an isometric axis orientation. Enter a value of **150** or PICK two points in response to the "Enter obliquing angle:" prompt (Fig 29-81).

FIGURE 29-81

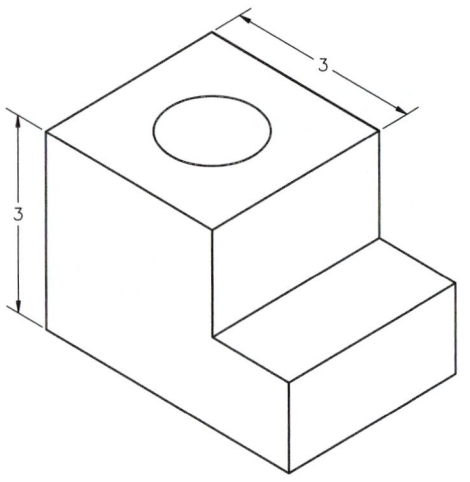

For isometric dimensioning, the extension line origin points <u>must</u> be aligned with the isometric axes. If not, the dimension is not properly oriented. This is important when dimensioning an isometric ellipse (such as this case) or an inclined or oblique edge. Construct centerlines for the ellipse on the isometric axes. Next, construct an *Aligned* dimension and *Osnap* the extension line origins to the centerlines. Finally, use *Dimedit Oblique* to reorient the angle of the extension lines (Fig 29-82).

FIGURE 29-82

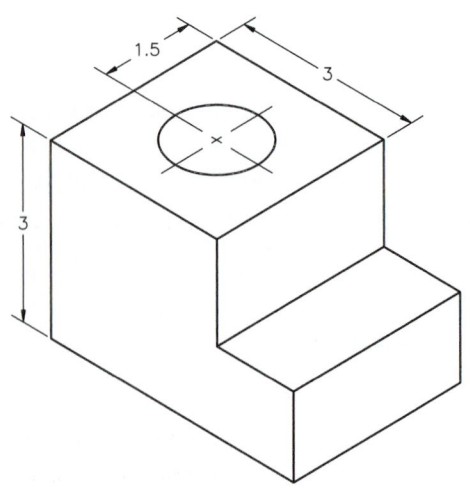

Using the same technique, other appropriate *Dimlinear vertical* or *Aligned* dimensions are placed and reoriented with *Oblique*. Use a *Leader* to dimension a diameter of an *Isocircle,* since *Dimdiameter* cannot be used for an ellipse (Fig. 29-83).

FIGURE 29-83

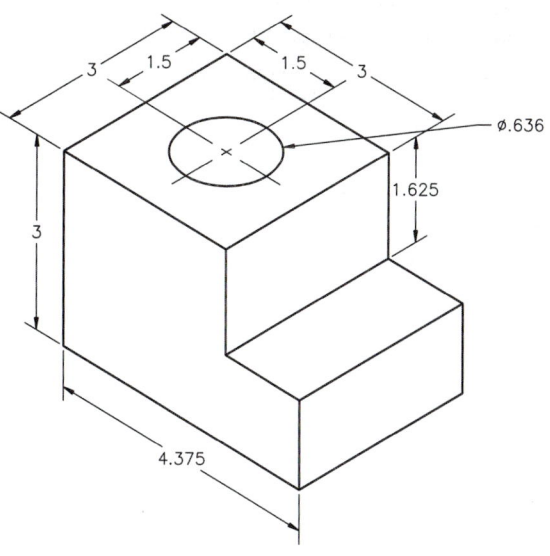

The dimensioning text can be treated two ways. (1) For quick and simple isometric dimensioning, <u>unidirectional</u> dimensioning (all values read from the bottom) is preferred since there is no automatic method for drawing the numerical values in an isometric plane. (2) As a better alternative, text *Styles* could be created with the correct obliquing angles (30 and -30 degrees). The text must also be rotated to the correct angle using the *Rotate* option of the dimension commands or by using *Dimedit Rotate* (Fig. 29-84). Optionally, *Dimension Styles* could be created with the correct variables set for text style and rotation angle for dimensioning on each iso-plane.

FIGURE 29-84

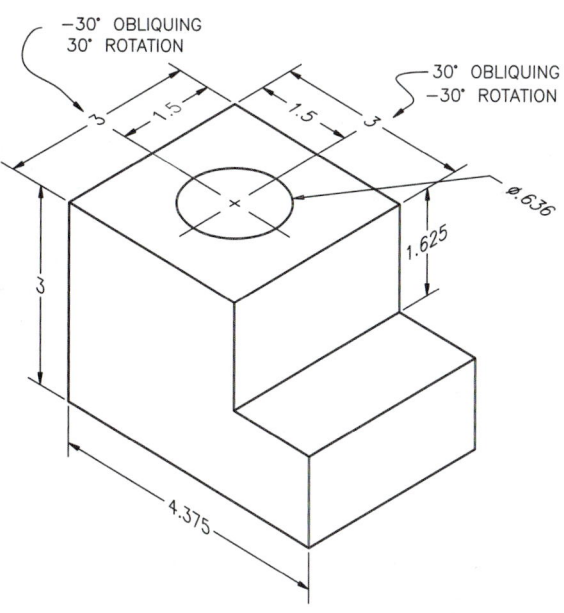

CHAPTER EXERCISES

For each of the following exercises, use the existing drawings, as instructed. Create dimensions on the DIM (or other appropriate) layer. Follow the "Guidelines for Dimensioning in AutoCAD" given in the chapter, including setting an appropriate *Overall Scale* based on the drawing scale factor. Use dimension variables and create and use dimension styles when needed.

1. **Dimension One View**

Open the **HAMMER** drawing you created in Chapter 13 Exercises and use the *Saveas* command to rename it **HAMMER-DIM.**   Add all necessary dimensions as shown in Chapter 13 Exercises, Figure 13-38. Create a *New* dimension style and set the variables to generate dimensions as they appear in the figure (set *Precision* to **0.00**, *Text Style* to *Romans* font, and *Overall Scale* to **1** or **1.3**). Create the dimensions in model space. Your completed drawing should look like that in Figure 29-85.

FIGURE 29-85

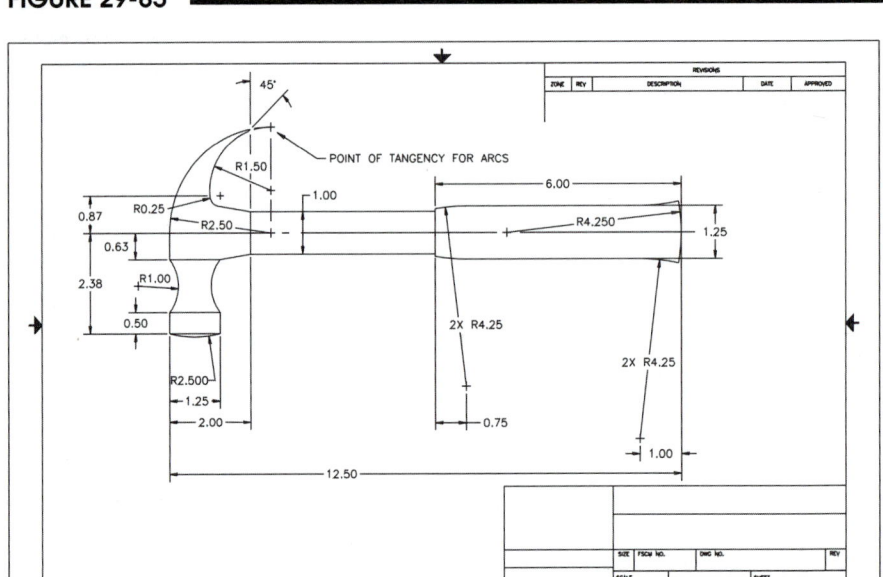

2. **Dimensioning a Multiview**

FIGURE 29-86

Open the **SADDLE** drawing that you created in Chapter 24 Exercises. Set the appropriate dimensional *Units* and *Precision*. Add the dimensions as shown. Because the illustration in Figure 29-86 is in isometric, placement of the dimensions can be improved for your multiview. Use optimum placement for the dimensions. *Save* the drawing as **SADDL-DM** and make a *Plot* to scale.

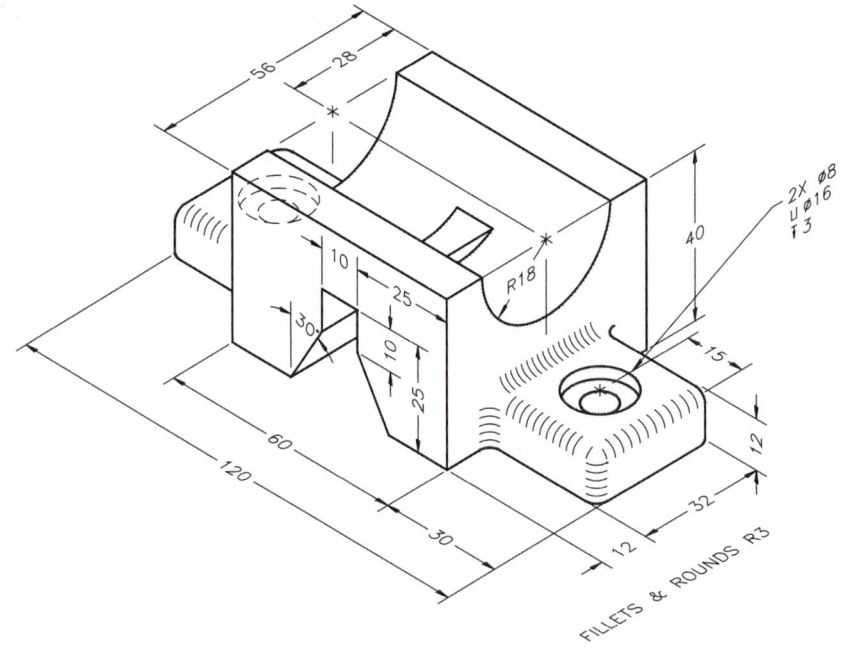

3. **Architectural Dimensioning**

Open the **OFFICE** drawing that you completed in Chapter 21. Dimension the floor plan as shown in Figure 21-87. Add *Text* to name the rooms. *Save* the drawing as **OFF-DIM** and make a *Plot* to an accepted scale and sheet size based on your plotter capabilities.

4. **Dimensioning an Auxiliary**

FIGURE 29-87

Open the **ANGLBRAC** drawing that you created in Chapter 27. Dimension as shown in Figure 29-87, but <u>convert the dimensions to</u> ***Decimal*** with **Precision** of **.000.** Use the "Guidelines for Dimensioning." Dimension the slot width as a *Limit* dimension—**.6248/.6255.** *Save* the drawing as **ANGL-DIM** and *Plot* to an accepted scale.

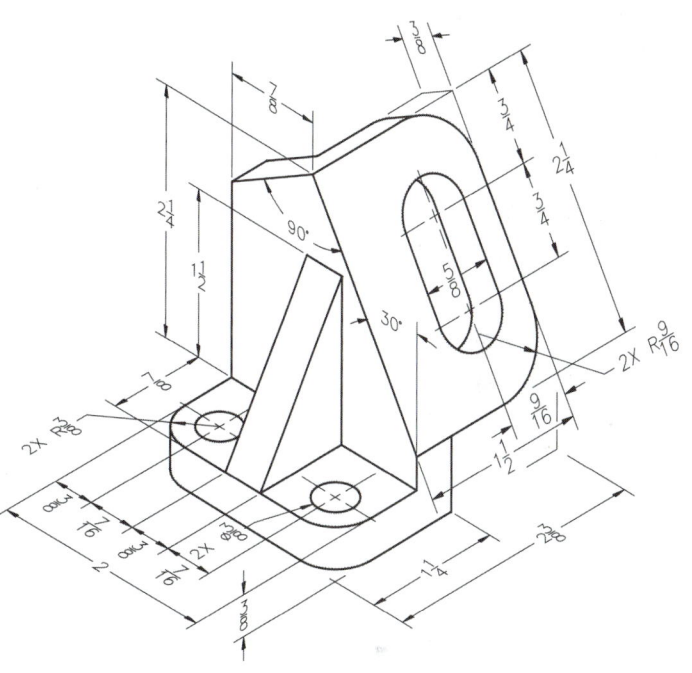

5. **Isometric Dimensioning**

FIGURE 29-88

Dimension the support bracket that you created as an isometric drawing named **SBRACKET** in the Chapter 25 Exercises. All of the dimensions shown in Figure 29-88 should appear on your drawing in the optimum placement. *Save* the drawing as **SBRCK-DM** and *Plot* to **1=1** on and A size sheet.

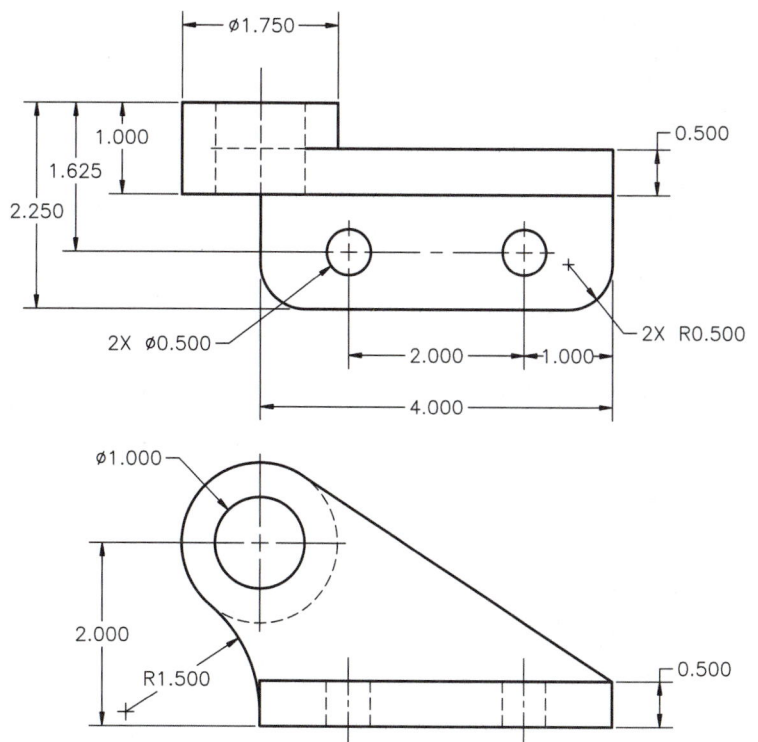

6. **Dimensioning a Multiview**

    Open the **ADJMOUNT** drawing that you completed in Chapter 24. Add the dimensions shown in Figure 29-89 but <u>convert the dimensions to *Decimal*</u> with **Precision** of **.000** and check **Trailing** under **Zero Suppression.** Set appropriate dimensional *Units* and *Precision.* Calculate and set an appropriate *Overall Scale.* Use the "Guidelines for Dimensioning" given in this chapter. Save the drawing as **ADJM-DIM** and make a *Plot* to an accepted scale.

    FIGURE 29-89

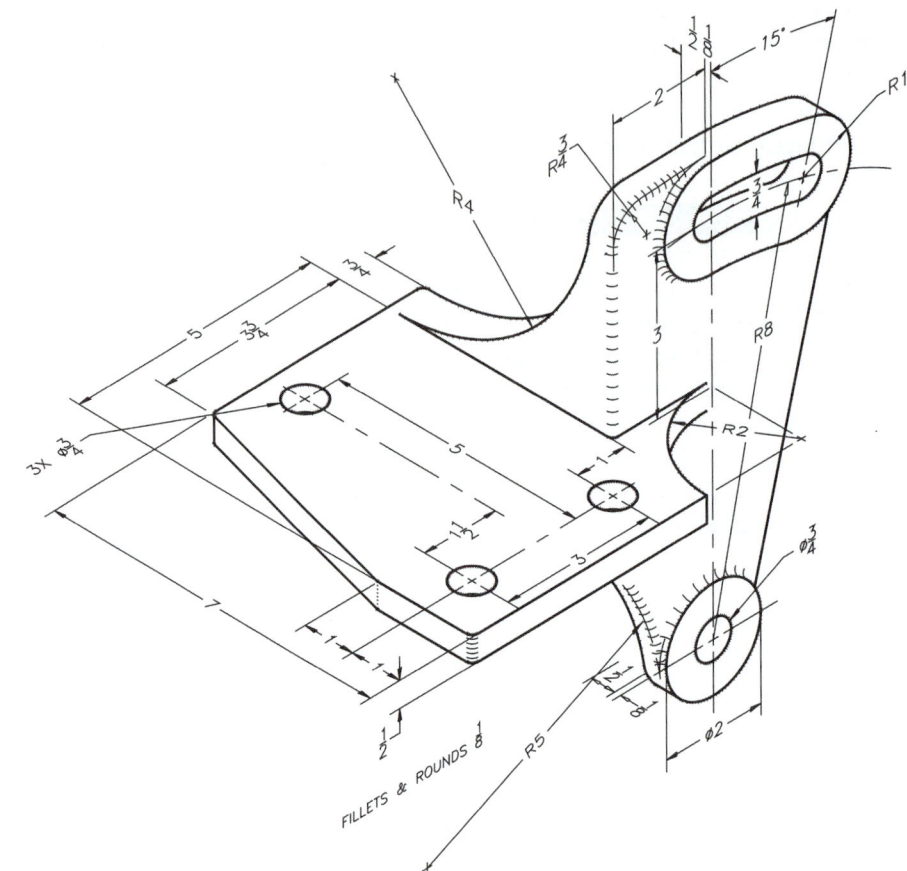

7. **Geometric Dimensioning and Tolerancing**

    This exercise involves Flatness, Profile of a Surface, and Position applications. *Open* the dimensioned **ANGL-DIM** drawing you worked on in Exercise 3 and add the following geometric dimensions:

    A. **Flatness** specification of **.002** to the bottom surface.

    B. **Profile of a Surface** specification of **.005** to the angled surface relative to the bottom surface and the right side surface of the 1/4 dimension. The 30° angle must be *Basic*.

    C. **Position** specification of **.01** for the two mounting holes relative to the bottom surface and two other surfaces that are perpendicular to the bottom. Remember, the location dimensions must be *Basic* dimensions.

    Use *SaveAs* to save and name the drawing **ANGL-TOL**.

8. **Geometric Dimensioning and Tolerancing**

This exercise involves a cylindricity and runout application. Draw the idler shown in Figure 29-90. Apply a **Cylindricity** specification of **.002** to the small diameter and a **Circular Runout** specification of **.007** to the large diameter relative to the small diameter as shown. The P730 neck has a **.07** radius and **30°** angle. *Save* the drawing as **IDLERDIM**.

FIGURE 29-90

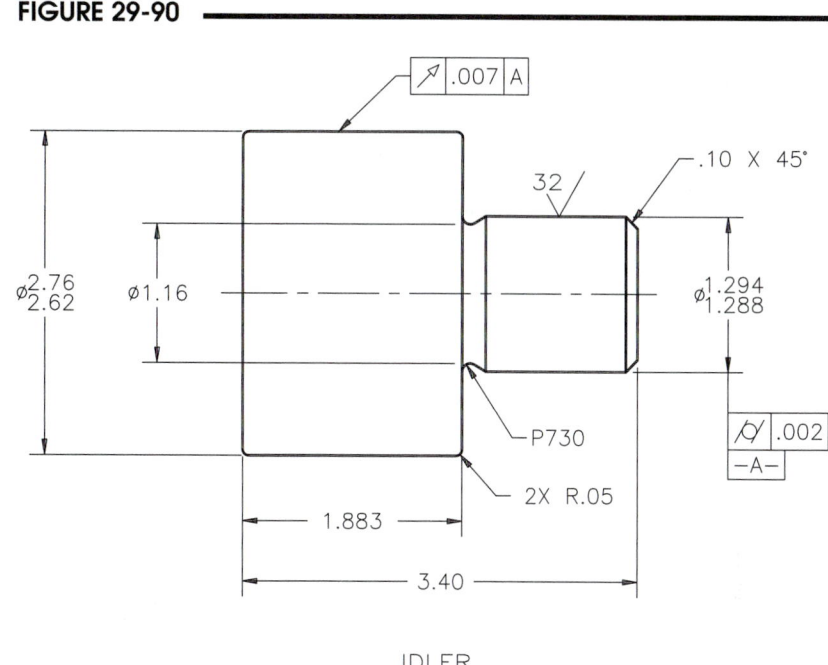

IDLER

9. **Dimension in Paper Space**

A. *Open* the **2BR-APT** drawing you worked with in Chapter 20 Exercises. Make a *New* dimension style and set the variables to generate dimensions as they appear in Chapter 15 Exercises, Figure 15-45. Create the dimensions in model space on a layer called **DIM**.

B. Create a *New Layout*. Use *Pagesetup* for the layout and select an appropriate *Plot Device* that can use a "**B**" size sheet, such as an HP 7475 plotter. Set the layout to a "**B**" size sheet. Next, use DesignCenter to locate and insert the **ANSI-B title block** from the Template folder. Make a new layer named **VPORTS** and create one viewport as shown in Figure 29-91. Set the viewport scale to display the apartment floor plan at **1/4"=1'** scale.

FIGURE 29-91

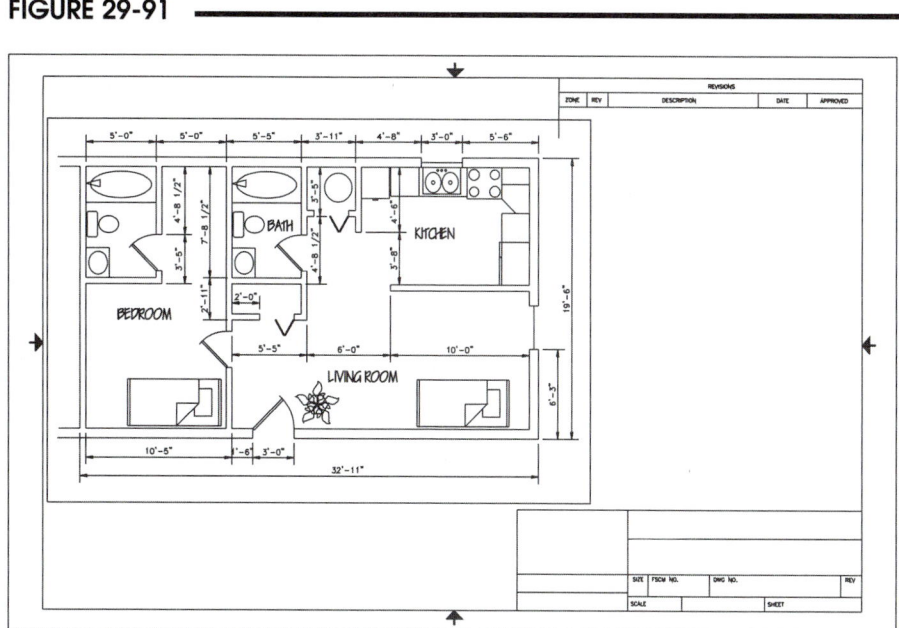

C. Create a smaller viewport on the right to display only one of the bathrooms at 1/2"=1' scale. Use the *Layer Manager* to turn off the display of the **DIM** layer for the new viewport. Create a new layer named **DIM-PS**. Create a *New* dimension style by copying the dimension style you created for model space, but change the *Overall Scale* to **1**. Create the dimensions for the plumbing wall (distance between centers of the fixtures) in paper space as shown in Figure 29-92. Label each viewport giving the scale. *Freeze* the **VPORTS** layer to achieve a drawing like that shown in Figure 29-92. Save the drawing as **APT-DIM**.

FIGURE 29-92

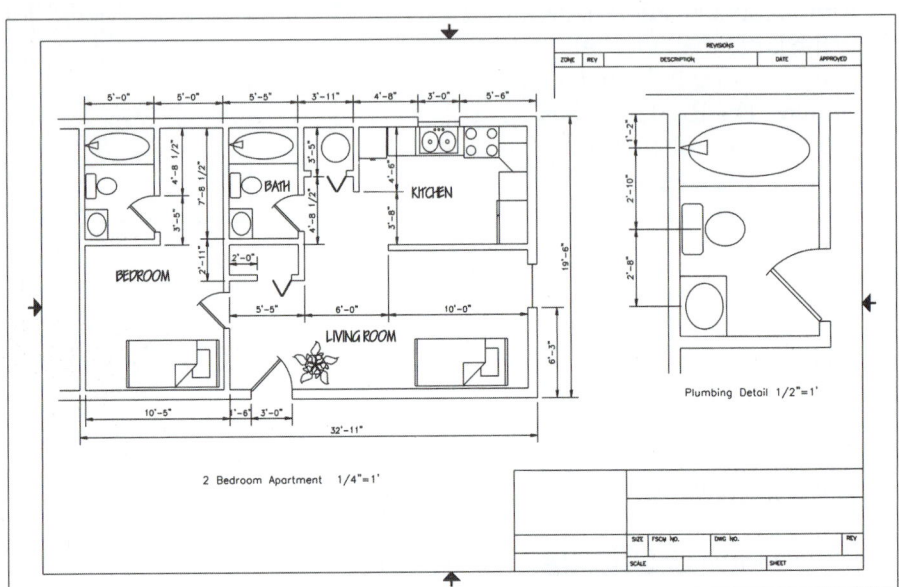

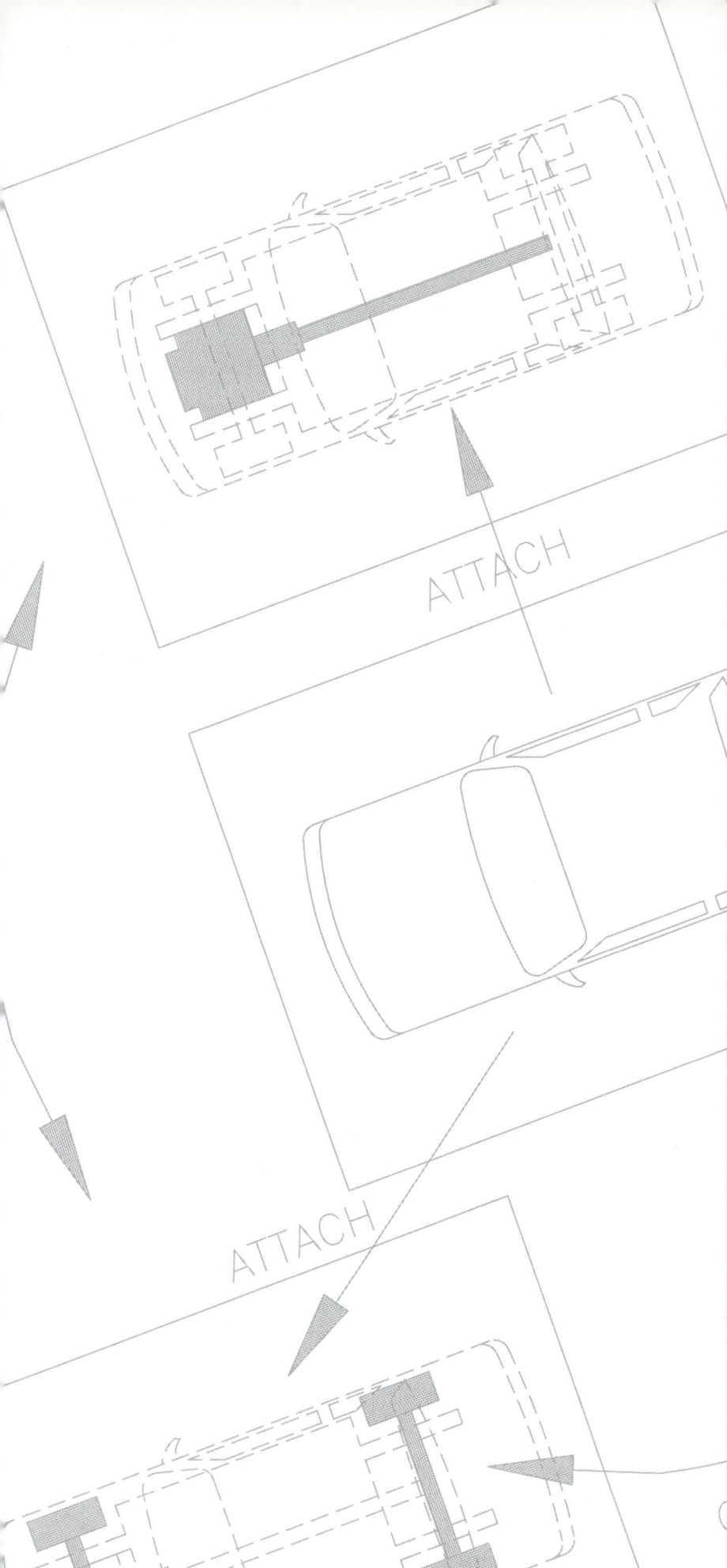

XREFERENCES

CHAPTER OBJECTIVES

After completing this chapter you should:

1. know the differences and similarities between an *Xrefed* drawing and an *Inserted* drawing;

2. be able to *Attach* an externally referenced drawing to the current drawing;

3. understand that an *Overlay* drawing cannot be nested;

4. be able to *Reload, Unload,* or *Detach* an externally referenced drawing;

5. be able to *Bind* an entire *Xrefed* drawing and *Xbind* individual named objects;

6. be able to *Xclip* an externally referenced drawing so only a portion of the *Xref* appears;

7. be able to control demand loading with *INDEXCTL* and *XLOADCTL*;

8. be able to use *Xopen* to open an *Xref* drawing for editing in a separate window;

9. know how to edit *Blocks* and *Xrefs* "in place" using *Refedit* and how to control in-place editing with *XEDIT*.

CONCEPTS

When you *Insert* a drawing as a *Block*, the inserted drawing becomes part of the current drawing—it is considered an "internal" reference. An *Inserted Block* (internal reference) is static—it does not change unless you *Explode*, edit, and redefine it. In contrast, an *Xref* (external reference) is a reference to another drawing. An *Xrefed* drawing is viewable from within the current drawing, but it is not *Inserted*. The *Xref* geometry and associated "named objects" do not become a part of the current drawing by default, although the *Xrefed* layer visibility can be controlled. When the original *Xref* drawing is changed, the changes are visible from within the current drawing.

INSERT	XREF
Any drawing can be *Inserted* as a *Block*.	Any drawing can be *Xrefed*.
The drawing "comes in" as a *Block*.	The drawing is visible as an *Xref*.
The *Block* drawing is a permanent part of the current drawing.	The *Xref* is not permanent, only "attached" or "overlayed."
The current drawing file size increases approximately equal to the *Block* drawing size.	The current drawing increases only by a small amount (enough to store information about loading the *Xref*).
The *Inserted Block* drawing is static. It does not change.	Each time the current drawing is opened, it loads the most current version of the *Xref* drawing.
If the original drawing that was used as the *Block* is changed, the *Inserted Block* does not change because it is not linked to the original drawing.	If the original *Xref* drawing is changed, the changes are automatically reflected when the current drawing is *Opened*.
Objects of the *Inserted* drawing cannot be changed in the current drawing unless *Refedit*, the Block Editor, or *Explode* is used.	The *Xref* can be changed in the current drawing if you use *Refedit* or use *Xopen* to edit the *Xref* in a separate window.
The current drawing can contain multiple *Block*s.	The current drawing can contain multiple *Xrefs*.
A *Block* drawing cannot be converted to an *Xref*.	An *Xref* can be converted to a *Block*. The *Bind* option makes it a *Block*—a permanent part of the current drawing.
A *Block* is *Inserted* on the current layer.	An *Xref* drawing's layers are also *Xrefed* and visibility of its layers can be controlled independently.
Any associated "named objects" of a *Block* can be used in the current drawing.	Any "named objects" of an *Xref* can be used in the current drawing only if *Xbind* is used.
Blocks can be nested.	An *Attached Xref* can be nested.
Any *Block* that is referenced by another *Block* is "nested."	An *Overlay Xref* cannot be nested.
A "circular" reference is not allowed with *Blocks* (X references Y and Y references X) because nested *Blocks* are also referenced.	Circular references are automatically detected (when the drawing you are *Xrefing* contains a nested *Xref* of the current drawing). Any nested *Xrefs* will load, but circular *Xrefs* are terminated.

INSERT	XREF
When a *Block* is inserted, you can control the properties of the block's layers.	When a drawing is *Attached* or *Overlayed*, you can control the properties of the *Xref's* layers.
You can *Xclip* the *Block* so that only portions of the entire *Block* drawing are visible.	You can *Xclip* the *Xref* so that only portions of the entire *Xref* drawing are visible.

As you can see, an *Xrefed* drawing has similarities and differences to an *Inserted* drawing. The following figures illustrate both the differences and similarities between an *Xrefed* drawing and an *Inserted* one.

The relationship between the current (parent) drawing and a drawing that has been *Inserted* and one that has been *Xreferenced* is illustrated here (Fig. 30-1). The *Inserted* drawing becomes a permanent part of the current drawing. No link exists between the parent drawing and the original *Inserted* drawing. In contrast, the *Xrefed* drawing is <u>not</u> a permanent part of the parent drawing but is a dynamic link between the two drawings. In this way, when the original *Xrefed* drawing is edited, the changes are reflected in the parent drawing (if a *Reload* is invoked or when the parent drawing is *Opened* next). The file size of the parent drawing for the *Inserted* case is the sum of both the drawings whereas the file size of the parent drawing for the *Xref* case is only the original size plus the link information.

FIGURE 30-1

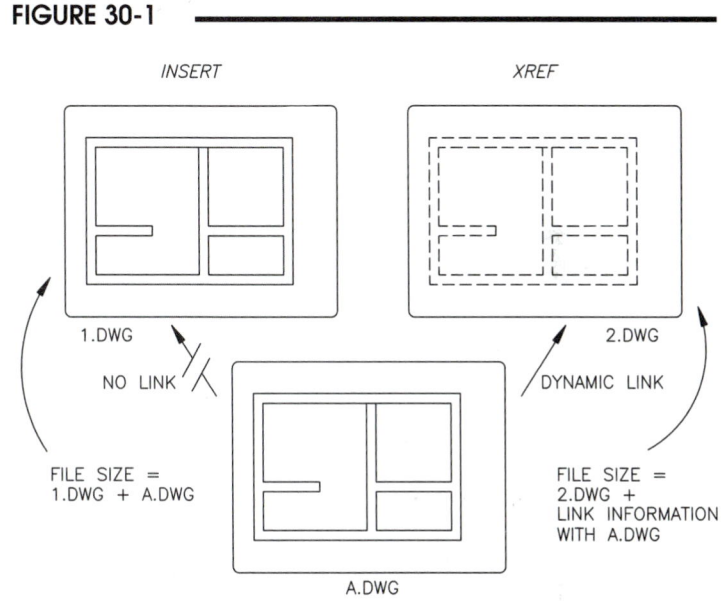

Some of the <u>similarities</u> between *Xreferenced* drawings and *Inserted* drawings are shown in the following figures.

Any number of drawings can be *Xrefed* or *Inserted* into the current drawing (Fig. 30-2). For example, component parts can be *Xrefed* to compile an assembly drawing.

FIGURE 30-2

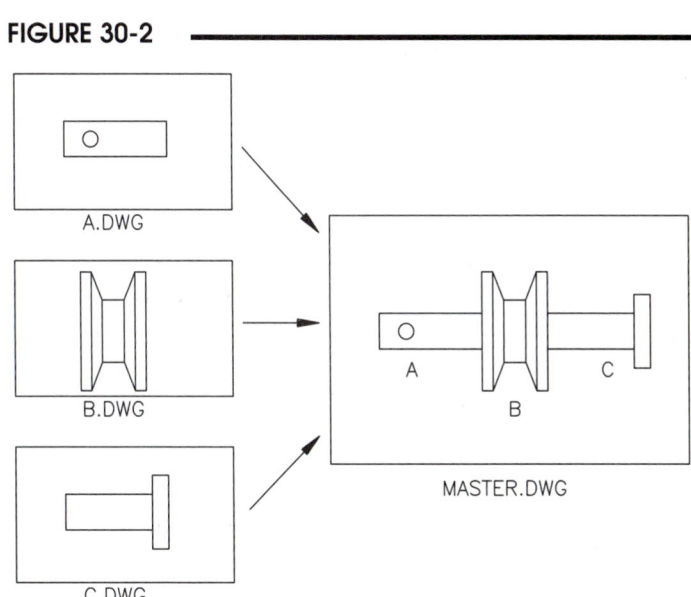

The current drawing can contain nested *Xref*s. This feature is also similar to *Block*s. For example, the OFFICE drawing can *Xref* a TABLE drawing and the TABLE drawing can *Xref* a CHAIR drawing. Therefore, the CHAIR drawing is considered a "nested" *Xref* in the OFFICE drawing (Fig. 30-3).

FIGURE 30-3

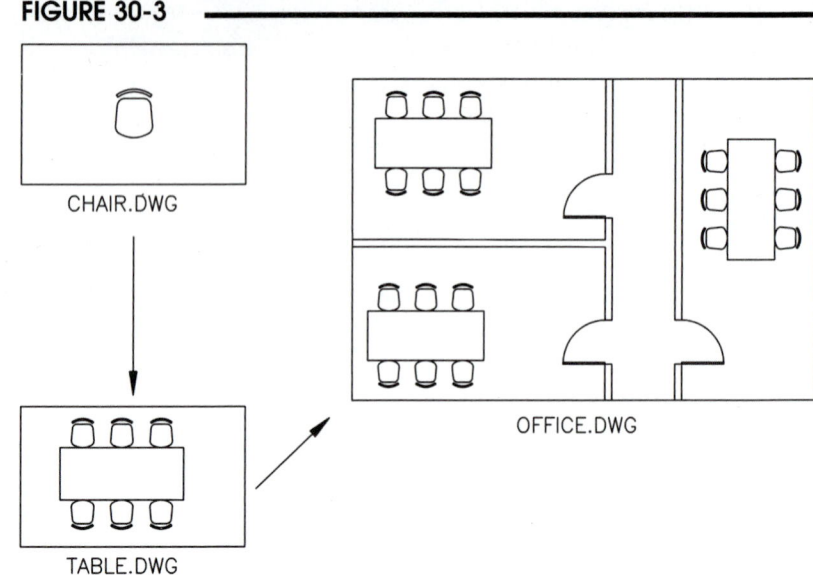

Xref drawings have many applications. *Xref*s are particularly useful in a networked office or laboratory environment. For example, several people may be working on one project, each person constructing individual components of a project. The components may be mechanical parts of an assembly; or electrical, plumbing, and HVAC layouts for a construction project; or several areas of a plant layout. In any case, each person can *Xref* another person's drawing as an external reference without fear of the original drawing being edited.

As an example of the usefulness of *Xref*s, a project coordinator could *Open* a new drawing, *Xref* all components of an assembly, analyze the relationships among components, and plot the compilation of the components (see Fig. 30-2). The master drawing may not even contain any objects other than *Xref*s, yet each time it is *Opened*, it would contain the most up-to-date component drawings.

In another application, an entire team can access (*Xref, Attach*) the same master layout, such as a floor plan, assembly drawing, or topographic map. Figure 30-4 represents a mechanical design team working on an automobile assembly and all accessing the master body drawing. Each team member "sees" the master drawing, but cannot change it. If any changes are made to the original master drawing, all team members see the updates whenever they *Reload* the *Xref* or *Open* their drawing.

FIGURE 30-4

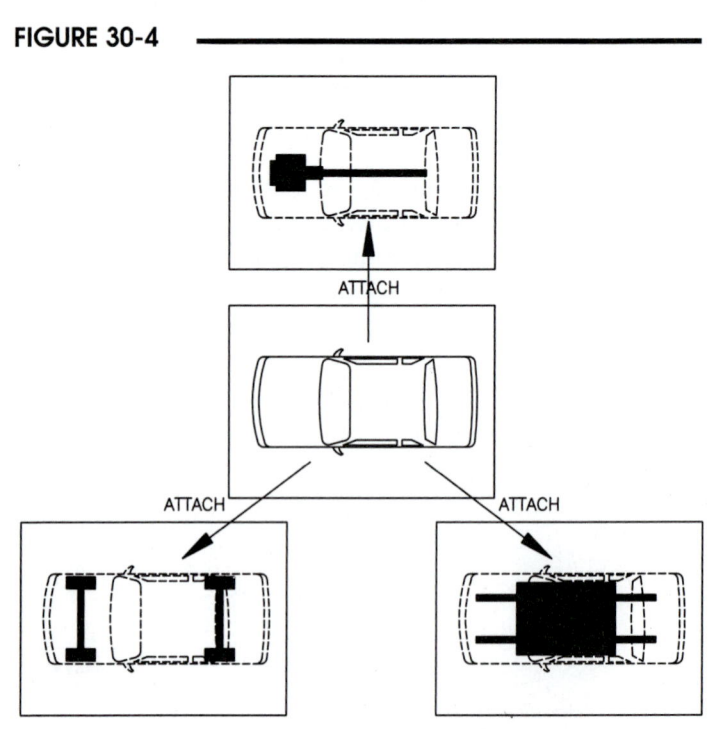

You also have the capability to clip portions of the *Xref* drawing that you do not want to see. Once you attach an *Xref*, you can use the *Xclip* command to draw a rectangular or polygonal (any shaped) boundary to determine what area of the *Xref* to view—all areas outside the boundary become invisible. For example, assume a site plane is *Xrefed* in order to calculate drainage around a house (Fig. 30-5).

FIGURE 30-5

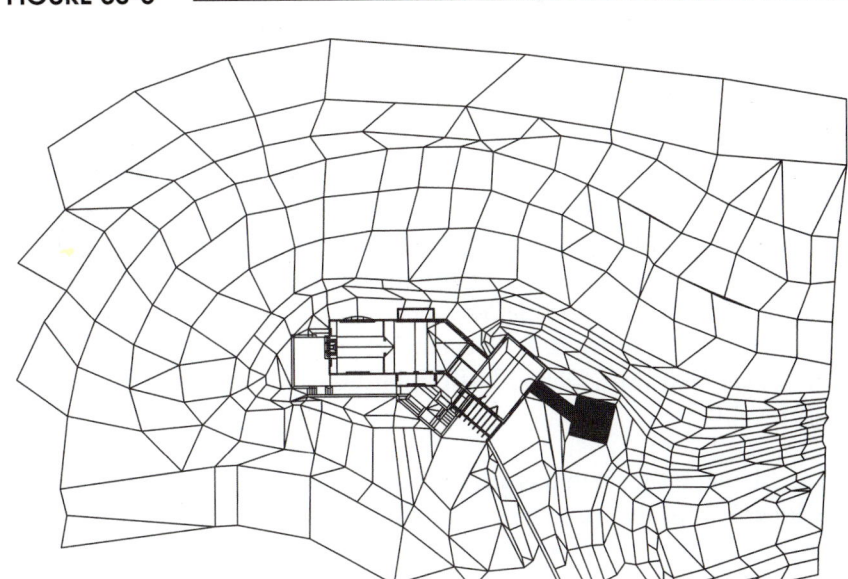

Xclip can be used to display only a portion of the Xref drawing needed for the calculations (Fig. 30-6).

FIGURE 30-6

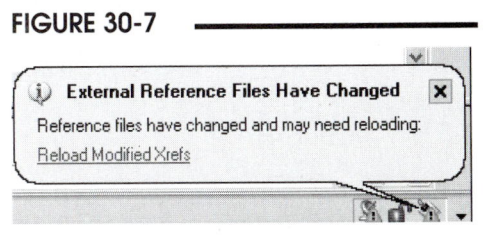

Internal features known as "demand loading," "spatial indexing," and "layer indexing" allow AutoCAD to <u>load only the portion of the drawing that appears in the clip boundary</u> into the current drawing. Demand loading and spatial and layer indexing can save a considerable amount of time and memory when Xreferencing large drawings such as maps or large floor plans, because only the small portion of the drawing that is needed (in the clip boundary) is actually loaded.

The named objects (*Layers*, *Text Styles*, *Blocks*, *Views*, etc.) that may be part of an *Xref* drawing become <u>dependent</u> objects when they are *Xrefed* to the parent drawing. These dependent objects cannot be renamed, changed, or used in the parent drawing. They can, however, be converted individually (permanently attached) to the parent drawing with the *Xbind* command.

Even though an *Xrefed* drawing's named objects (layers, linetypes, text styles, etc.) are attached, you cannot normally draw on or edit any of its layers. You do, however, have complete control over the <u>visibility</u> of the *Xref* drawing's layers. The *State* of the layers (*On*, *Off*, *Freeze*, and *Thaw*) of the *Xref* drawing can be controlled like any layers in the parent drawing.

If a drawing used as an *Xref* needs editing, three methods can be used. First, you can close the parent drawing, then *Open* the original external drawing, make the edits and *Save* the drawing. When the parent drawing is opened again, the new *Xref* is loaded. Second, <u>while in the parent drawing</u>, you can use the *Xopen* command to open the original (used as an *Xref*) to make the changes. In this case, when the *Xref* drawing is changed and saved, AutoCAD instructs you that a change has been made so you can reload the *Xref* in the parent drawing (Fig. 30-7). The parent drawing then reflects the new changes in the *Xref*. Third, you can use the *Refedit* command to edit the *Xref* from <u>within the parent drawing</u>, but only if the *Xedit* variable is set to 1 in the *Xrefed* drawing. Any changes made are "sent back" to the original *Xref* drawing.

FIGURE 30-7

External Reference Files Have Changed

Reference files have changed and may need reloading:
Reload Modified Xrefs

TIP

An *Xref* Example

Assume that you are an interior designer in an architectural firm. The project drawings are stored on a local network so all team members have access to all drawings related to a particular project. Your job is to design the interior layout comprised of chairs, tables, desks, and file cabinets for an office complex. You use a prototype drawing named INTERIOR.DWG for all of your office interiors. It is a blank drawing that contains only block definitions (not yet *Inserted*) of each chair, desk, etc., as shown in Figure 30-8. The *Block* definitions are CHAIR, CONCHAIR, DESK, TABLE, FILECAB, and CONTABLE.

FIGURE 30-8

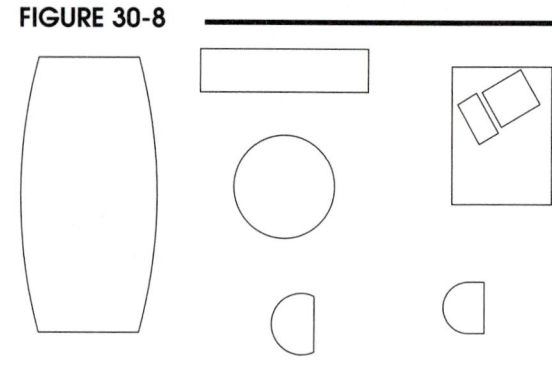

An architect on the team has almost completed the floorplan for the project. You use the *Attach* option of the *Xref* command to "see" the architect's floor plan drawing, OFFICEX.DWG. The visibility of the individual dependent layers can be controlled with the *Layer* command or *Layer Control* drop-down list. For example, layers showing the HVAC, electrical layout, and other details are *Frozen* to yield only the floor plan layer needed for the interior layout (Fig. 30-9).

FIGURE 30-9

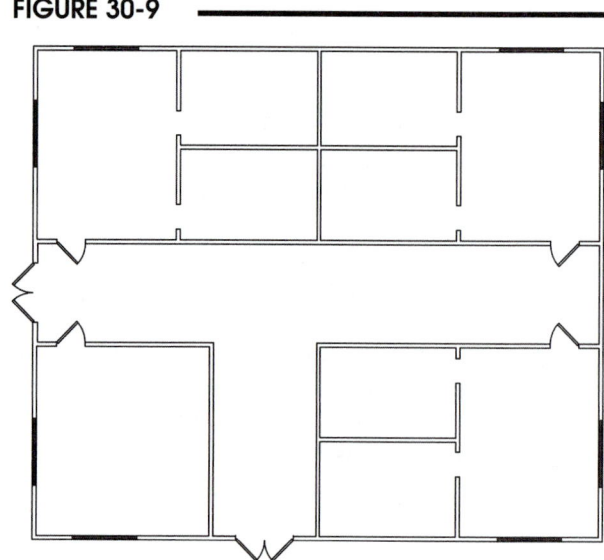

Next, you draw objects in the current drawing or, in this case, *Insert* the office furniture *Blocks* that are defined in the current drawing (INTERIOR.DWG). The resulting drawing would display a complete layout of the office floor plan plus the *Inserted* furniture—as if it were one drawing (Fig. 30-10). You *Save* your drawing and go home for the evening.

FIGURE 30-10

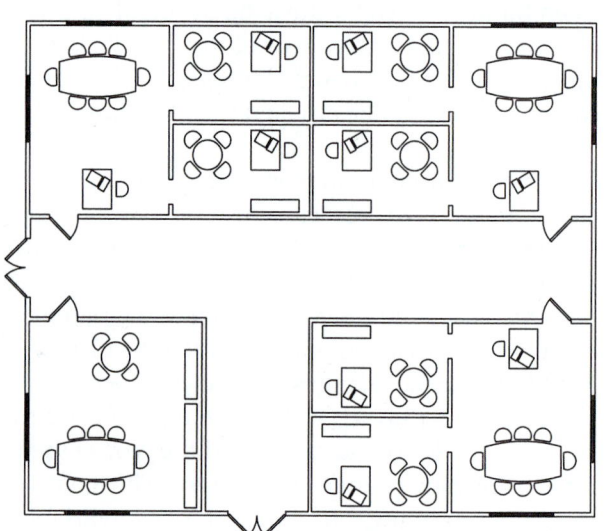

That evening, the architect makes a change to the floor plan. He works on the original OFFICEX drawing that you *Xrefed*. The individual office entry doors in the halls are moved to be more centrally located. The new layout appears in Figure 30-11.

FIGURE 30-11

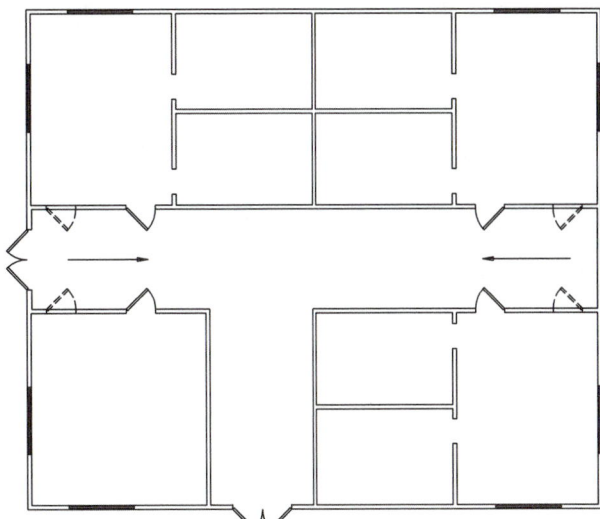

The next day, you *Open* the INTERIOR drawing. The automatic loading of the OFFICEX drawing displays the latest version of the office floor plan with the design changes. The appropriate changes to the interior drawing must be made, such as relocating the furniture by the hall doors (Fig. 30-12). When all parts of the office complex are completed, the INTERIOR (parent) drawing could be plotted to show the office floor plan and the interior layout. The resulting plot would include objects in the parent drawing and the visible *Xref* drawing layers.

FIGURE 30-12

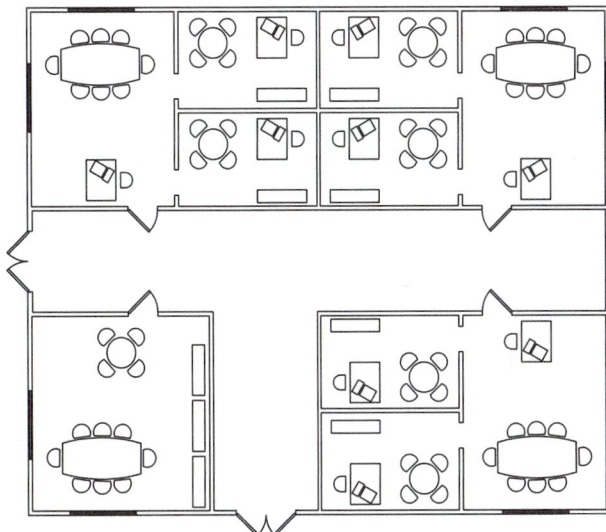

AutoCAD's ability to externally reference drawings makes team projects much more flexible and efficient. If the *Xref* capability did not exist, the OFFICEX drawing would have to be *Inserted* as a *Block*, and design changes made later to the OFFICEX drawing would not be apparent. The original OFFICEX *Block* would have to be *Erased*, *Purged*, and the new drawing *Inserted*.

The *Xclip* feature gives you the capability to isolate only a portion of an *Xrefed* drawing. For example, assume you wanted to discuss a particular room with the architect, so you open a *New* drawing and *Xref* the INTERIOR drawing. The OFFICEX drawing also is *Xrefed* because it is nested (previously *Xrefed* into the INTERIOR drawing) so the new drawing appears the same as the original INTERIOR drawing (see Fig. 30-12). To isolate only the room in discussion, *Xclip* is used to draw a rectangular boundary around the desired area (Fig. 30-13).

FIGURE 30-13

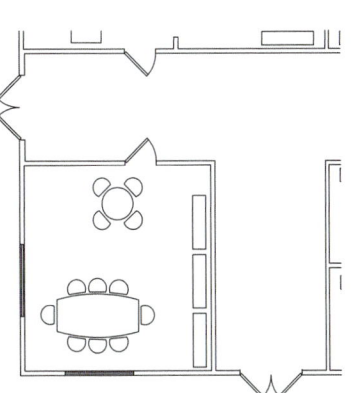

Often in the termination of a project, the final set of drawings is sent to the client. Because the *Xref* capability makes it feasible to work on a design as a set of individual component drawings, it is necessary to send the complete set of drawings to the customer or to combine the set into one drawing to submit. If you want to keep the set of drawings as *Xrefs* to the parent drawing, the *Xrefs* can be saved with a *Relative Path*. In this way the parent drawing can always locate and load the *Xrefs* as long as they were loaded onto another machine in the same directory structure relative to the location of the parent. Another strategy is to combine all the drawings into one drawing using *Bind*. *Bind* brings an *Xrefed* drawing into the current drawing as if it were *Inserted*, so it becomes a permanent part of the parent drawing (as a *Block*).

CREATING AND USING *XREFS*

The *Externalreferences* command and related commands can be invoked through the *Reference* toolbar (Fig. 30-14). If you prefer the pull-down menus, the *Insert* and *Modify* pull-downs contain *Xref* and related commands.

FIGURE 30-14

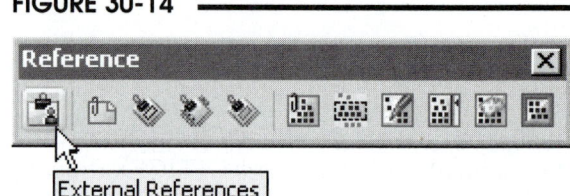

External-references or *Xref*						
	Pull-down Menu	**Command (Type)**	**Alias (Type)**	**Short-cut**	**Screen (side) Menu**	**Tablet Menu**
	Insert *External References*	*Xref* or *Externalreferences*	*XR* or *ER*	...	*INSERT* *Xref*	*T,4*

Both the *Externalreferences* command and the *Xref* command invoke the *External References* palette (Fig. 30-15). This palette allows you to create and manage external references (*Xrefs*). The central area of the palette lists drawings that are currently *Attached*. The options for managing the external references are available using a right-click shortcut menu, as shown in Figure 30-15.

FIGURE 30-15

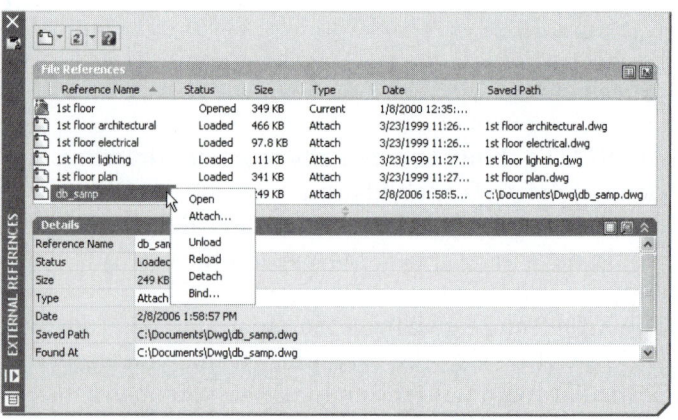

If you prefer to type, you can use the *-Xref* command (notice the hyphen prefix). Using *-Xref* displays the following prompt:

Command: ***-xref***
Enter an option [?/Bind/Detach/Path/Unload/Reload/Overlay/Attach] <Attach>:

Whether you use *Externalreferences*, *Xref*, or *-Xref*, the *Attach*, *Overlay*, *Detach*, *Reload*, *Unload*, *Bind*, and *Path* options are the same.

Attach

This (the default) option attaches one drawing to another by making the *Xref* drawing visible in the "parent" drawing. Even though the *Xref* drawing is visible, it cannot be edited from within the parent drawing unless *Refedit* is used. *Attach* creates a link between two drawings (Fig. 30-16).

FIGURE 30-16

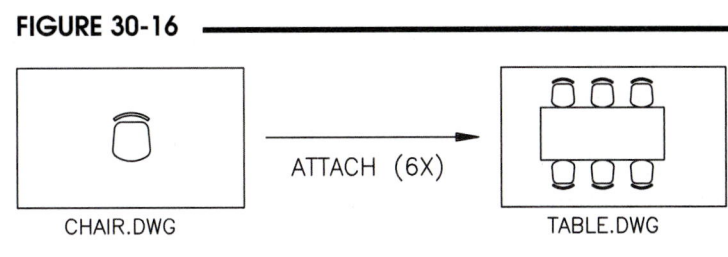

CHAIR.DWG   ATTACH (6X)   TABLE.DWG

Selecting the *Attach* option produces the standard *Select File* dialog box where you locate and select the .DWG file to attach (not shown). When you have selected the desired file to *Xref*, select *OK* to dismiss the *Select File* dialog box and produce the *External Reference* dialog box (see Fig. 30-17).

When the *External Reference* dialog box appears (Fig. 30-17), specify the *Reference Type—Attachment* or *Overlay*. Ensure the *Attachment* button is pressed. (See "*Overlay*" next.) (Typing the *-Xref* command and using the *Attach* option produces the *Select File* dialog box directly—the *External References* palette does not appear.) Once the desired file is specified, the *External Reference* dialog box appears again for further action.

FIGURE 30-17

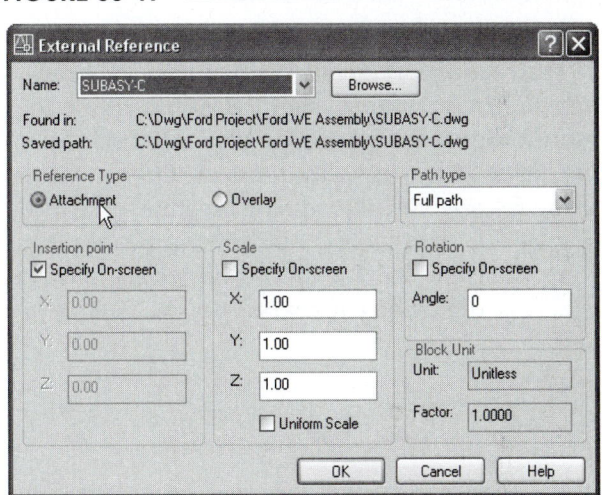

Notice the similarity of the edit boxes (Fig. 30-17) to the *Insert* command. The *Insertion Point, Scale, Rotation,* and *Block Unit* options operate identically to the *Insert* command. Similar to the action of *Insert,* the base point of the *Xref* drawing is its 0,0 point unless a different base point has been previously defined with the *Base* command.

You can also specify the *Path Type* to save with the *Xref* by selecting *Full Path, Relative Path* or *No Path* from the drop-down list. If you expect no changes to occur for the directory system used for the drawings and related *Xrefs, Full Path* is a suitable choice. However, if the drawing and related *Xrefs* might be sent to a client or another office or if the location on your system for the parent drawing and related *Xrefs* is expected to change, a better selection might be *Relative Path* or *No Path* (see "Managing Xref Drawings").

Attachment causes the specified drawing to appear in the current drawing similar to the way a drawing appears when it is *Inserted* as a *Block.* Each time the current drawing is *Opened,* the *Attached* drawing is loaded as an *Xref.* More than one reference drawing can be *Attached* to the current drawing. A single reference drawing can be *Attached* to any number of insertion points in the current drawing. The dependent objects in the reference drawing (*Layers, Blocks, Text styles,* etc.) are assigned new names, "external drawing | old name" (see "Dependent Objects and Names").

Overlay

An *Xref Overlay* is similar to an *Attached Xref* with one main difference—an *Overlay* cannot be nested. In other words, if you *Attach* a drawing that (itself) has an *Overlay*, the *Overlay* does not appear in your drawing. On the other hand, if the first drawing is *Attached*, it appears when the parent drawing is *Attached* to another drawing (Fig. 30-18).

FIGURE 30-18

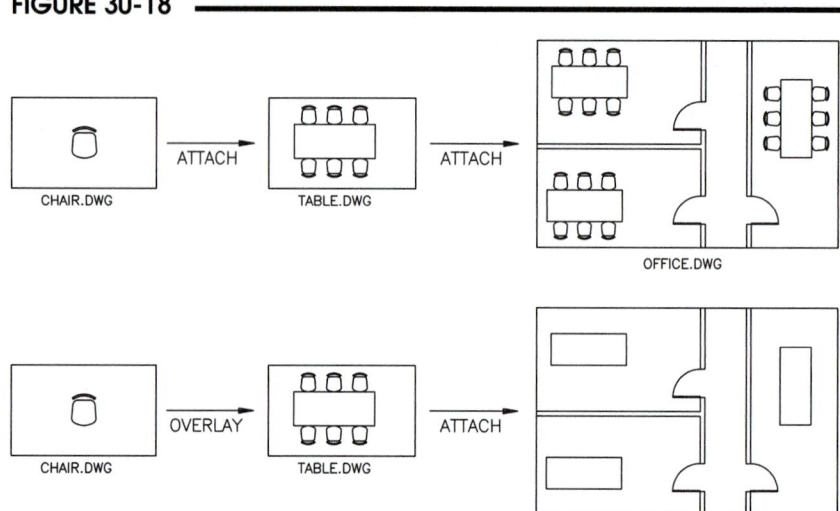

To produce an *Overlay* using the *External References* palette, you must first select the *Attach* option to produce the *Select File* dialog box. After selecting the desired file to overlay, press the *Overlay* button in the *External Reference* dialog box that appears (Fig. 30-19). If you type the *-Xref* command, simply use the *Overlay* option and the standard *Select File* dialog box appears.

FIGURE 30-19

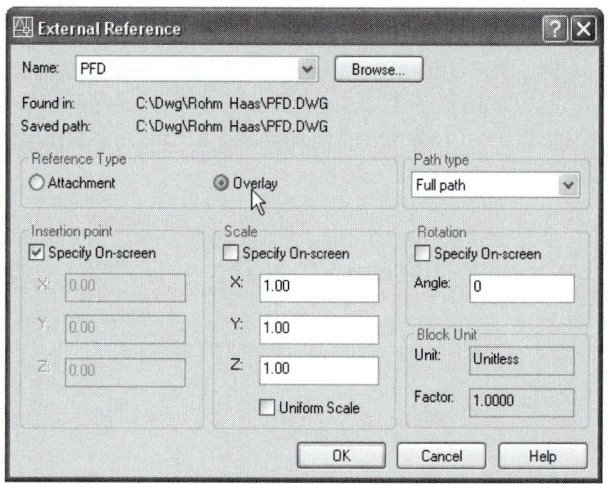

The *Overlay* option prevents "circular" *Xrefs* from appearing by preventing unwanted nested *Xrefs*. This is helpful in a networking environment where many drawings *Xref* other drawings. For example, assume drawing B has drawing A as an *Overlay*. As you work on drawing C, you *Xref* and view only drawing B without drawing B's overlays—namely drawing A. This occurs because drawing A is an *Overlay* to B, not *Attached*. If all drawings are *Overlays*, no nesting occurs (Fig. 30-20).

FIGURE 30-20

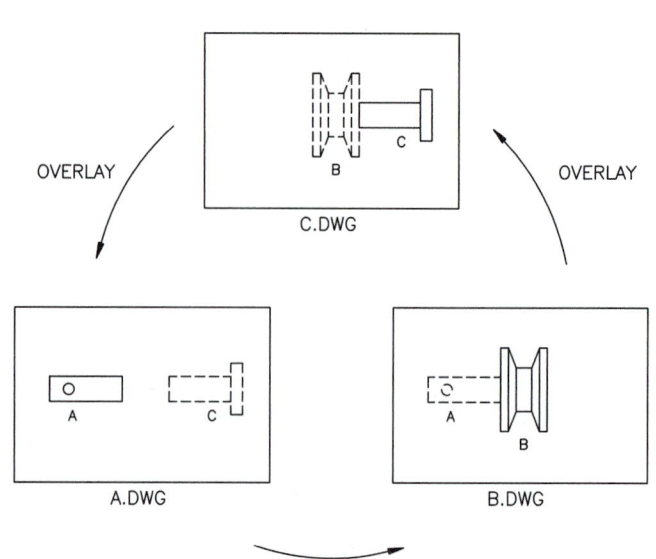

The overlay concept is particularly useful in a concurrent engineering project, where all team members must access each other's work simultaneously. In the case of a design team working on an automobile assembly, each team member can *Attach* the master body drawing and also *Overlay* the individual subassembly drawings (Fig. 30-21). *Overlay* enables each member to *Xref* all drawings without bringing in each drawing's nested Xreferences.

FIGURE 30-21

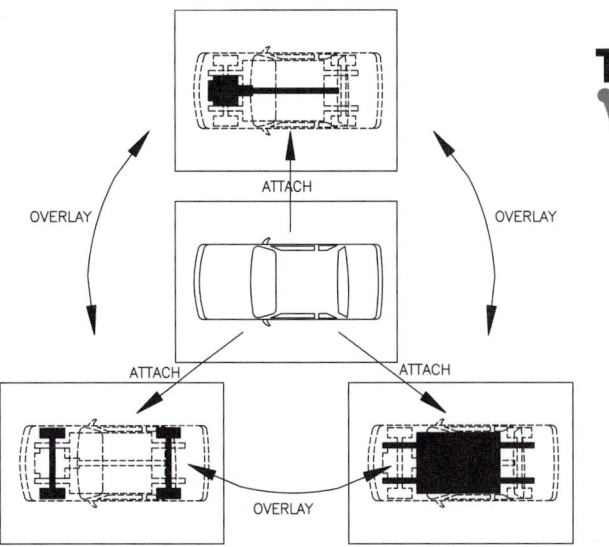

There are times when you cannot foresee the potential use of a drawing and therefore do not *Overlay* rather than *Attach*. For example, in a large design office it is quite possible to *Xref* a drawing on the network that includes your current drawing as an *Attached Xref*, thus causing a circular reference. If this occurs, an *AutoCAD Alert* dialog box appears giving notification of the circular reference and presents the option to continue or not (Fig. 30-22). If you respond with *Yes*, AutoCAD reads in the *Xref* and any nested *Xrefs* to the point where it detects the circularity, then terminates. Answering *No* cancels the *Attach*.

FIGURE 30-22

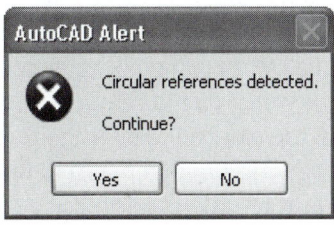

Detach

Detach breaks the link between the parent drawing and the *Xref*. Using this option causes the *Xrefed* drawing to "drop" from view immediately. When the parent drawing is *Opened*, the previously *Attached* drawing is <u>no longer</u> loaded as an *Xref*. Nested *Xrefs* cannot be *Detached* or selected to *Bind* since they are actually part of another drawing. You can *Detach* the drawing that contains the nested *Xref*.

Details/Preview

The lower section of the *External References* palette can display *Details* or a *Preview* depending on your selection (Fig. 30-23, right-center buttons). The *Details* option gives information about the highlighted *Xref* from the list above. The *Status*, *Size*, and *Date* fields are read-only; however, you can change the *Reference Name*, *Type* (*Attach* or *Overly*), and *Found At* (path) fields. For example, use the browse button at the right of the *Found At* field to locate a referenced drawing that has been moved to different folder or drive. In such a case, the *Status* of the *Xref* is listed above as *Not Found*, as shown in this figure for the "1st floor electrical" drawing.

FIGURE 30-23

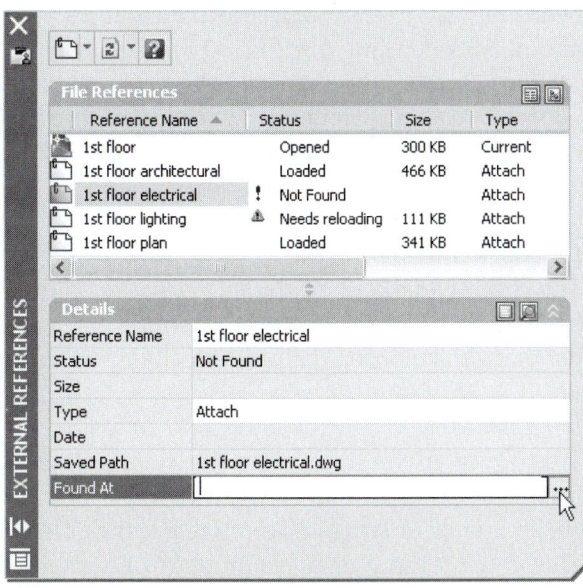

TIP When you archive (back up) drawings, remember also to back up the drives and directories where the *Xref* drawings are located (or use the *Bind* option when the project is finished). One possible file management technique is to specify a directory especially for *Xrefs* (see "Managing Xref Drawings"). In a networking environment, access rights to *Xref* directories must be granted.

Reload

This option forces a *Reload* of the external drawing at any time while the current drawing is in the Drawing Editor and ensures that the current drawing contains the <u>most recent version</u> of the external drawing. You are notified if a drawing has been changed and requires *Reloading* (see Fig. 30-23, "1st floor lighting" drawing). (See also "XREFNOTIFY.") In a networking environment, you may not be able to *Attach, Overlay, Bind,* or *Reload* a drawing based on whether the referenced drawing is currently being edited (open) by another person or by the setting of the *XLOADCTL* system variable in the referenced drawing (see "XLOADCTL").

Unload

The *Unload* option can be used to increase computing speed in a drawing. In cases when you need to view (load) drawings only on an as-need basis, *Unload* them. This causes the current (parent) drawing to open more quickly, speeds regenerations, and uses less memory. The unloaded *Xref* is not visible, but the link information remains in the current drawing. To view the *Xref* again, use *Reload*. This method is preferred over using *Detach* when you expect to use the *Xref* again.

Bind

The *Bind* option <u>converts</u> the *Xref* drawing to a *Block* in the current drawing, then terminates the external reference partnership. The original *Xrefed* drawing is not affected. The names of the dependent objects in the external drawing are changed to avoid possible conflicts in the case that the parent and *Xref* drawing have dependent objects with the same name (see "Dependent Objects and Names"). If you want to only bind selected named objects, use the *Xbind* command.

When the *Bind* button is selected (in the *External References* palette), the *Bind Xrefs* dialog box appears for you to select the *Bind Type* (see Fig. 30-28). This option determines how the <u>names of dependent objects</u> (such as layers and blocks) appear in the drawing after they become a permanent part of the parent drawing (see "Dependent Objects and Names").

Open

Using this option, you can open an externally referenced drawing in a separate window (of the same AutoCAD session), make any needed changes, then *Save* the drawing. When the opened drawing has been changed and saved, AutoCAD notifies you (in the system tray) that the drawing has been changed and needs reloading (see Fig. 30-7). (See also "XREFNOTIFY.") In addition, a message appears in the *Status* column of the *External References* palette reminding you to *Reload* (see Fig. 30-23). This method is generally easier than using *Refedit* for editing an *Xref*. *Opening* a drawing through the *External References* palette (*Xopen*) <u>overrides</u> the setting of the *XLOADCTL* and *XEDIT* variables (see "Demand Loading" and "Editing *Xrefs* and *Blocks*").

List View and Tree View

The central area of the *External References* palette is used to list and give the status of drawings that have been *Xrefed*. Use the two buttons in the upper-right corner of the palette to display the information in a list that can be ordered alphabetically by column (*List View*) or in a hierarchical tree structure (*Tree View*).

List View

This option (default) displays the *Xrefs* alphabetically by name. You can sort the list by column in forward or reverse order (*Reference Name, Size, Type, Date,* etc.) by selecting the column heading and clicking once or twice. You can resize the column width by dragging the column separator left or right. The *Status* column displays the state of each reference.

The following states are possible:

Status	Description
Loaded	Xref was found when drawing was opened or reloaded
Unload	Xref will be unloaded when dialog box is closed
Unloaded	Xref is currently unloaded
Needs Reloading	Xref has changed, use Reload to force reload when dialog box is closed
Reload	Xref will be reloaded when dialog box is closed
Loaded, recent changes	Xref was loaded but has been changed recently
Not found	Xref was not found in search paths when drawing was opened or reloaded
Unresolved	Xref was found but could not be read by AutoCAD
Unreferenced	Xref was attached but is erased
Orphaned	Nested Xref is attached to an Xref that is not loaded
Open	Xref will be opened for editing in another window when dialog is closed

Tree View

The *Tree View* shows the *Xrefs* in a hierarchical tree structure. This option is particularly useful if you have nested *Xrefs* (Fig. 30-24). Note that the icons also indicate the type and status of each referenced drawing (*Overlay, Attach, Reload, Unload, Not found*, etc.).

FIGURE 30-24

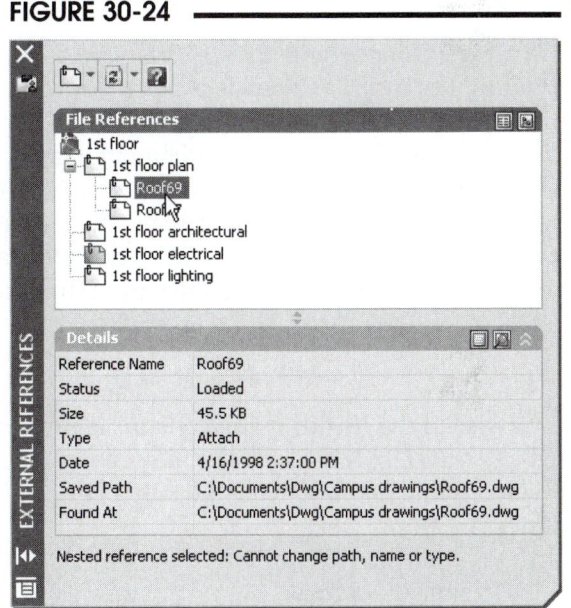

Dependent Objects and Names

The named objects that have been created as part of a drawing become dependent objects when that drawing is *Xrefed*. The named objects in an *Xref* drawing can be any of the following:

    Blocks
    Layers
    Linetypes
    Dimension Styles
    Text Styles

These dependent objects <u>cannot</u> be renamed or changed in the parent drawing. Dependent text or dimension styles cannot be used in the parent drawing. You cannot draw on dependent layers from the parent drawing; you can only control the visibility of the dependent layers. You <u>can</u>, however, *Bind* the entire *Xref* and the dependent named objects, or you can bind individual dependent objects with the *Xbind* command. *Xbind* converts an individual dependent object into a permanent part of the parent drawing. (See "*Xbind*.")

When an *Xref* contains named objects, the names of the dependent objects are changed (in the parent drawing). A new naming scheme is necessary because conflicts would occur if the *Xref* drawing and the parent drawing both contained *Layers* or *Blocks*, etc., with the same names, as would happen if both drawings used the same template drawing. When a drawing is *Attached*, its original object names are prefixed by the drawing name and separated by a pipe (|) symbol.

For example, if the DB_SAMP drawing is *Xrefed* and it contains a layer named PHONES, it is listed (in the parent drawing) as DB_SAMP | PHONES (Fig. 30-25).

FIGURE 30-25 ─────────────

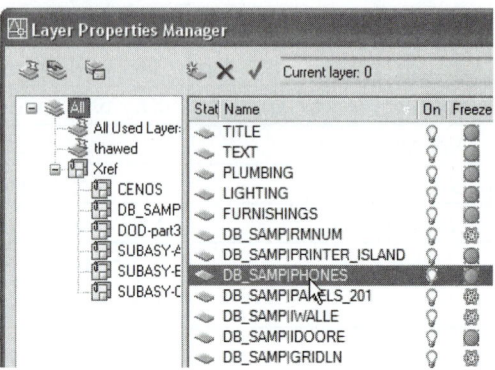

Because the dependent objects cannot be used in the parent drawing (you cannot insert a dependent block nor can you draw on a dependent layer), these objects cannot be renamed or deleted either. If you attempt to delete or rename a dependent layer, a warning appears (Fig. 30-26). (Keep in mind that dependent blocks, layers, etc. can be used if *Bind* or *Xbind* is used to make them a permanent part of the parent drawing. Additionally, DesignCenter can be used to drag-and-drop named objects into the parent drawing.)

FIGURE 30-26 ─────────────

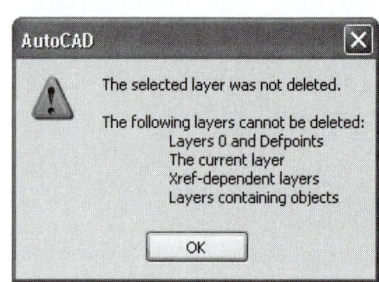

Likewise, if the *Xref* drawing contains a block named DESK2, it is renamed in the parent drawing to DB_SAMP | DESK2. Using the *-Block* command to reveal a listing of blocks in the current drawing yields the following display. Note the tally below of "User Blocks," "External References," and "Dependent Blocks":

```
Command: -block
Enter block name or [?]: ?
Enter block(s) to list <*>:

Defined blocks.
  "CENOS"                    Xref: resolved
  "DB_SAMP"                  Xref: resolved
  "DB_SAMP|CHAIR7"           Xdep: "DB_SAMP"
  "DB_SAMP|COMPUTER"         Xdep: "DB_SAMP"
  "DB_SAMP|DESK2"            Xdep: "DB_SAMP"
  "DB_SAMP|DESK3"            Xdep: "DB_SAMP"
  "DB_SAMP|DOOR"             Xdep: "DB_SAMP"
  "DB_SAMP|DR-36"            Xdep: "DB_SAMP"
  "DB_SAMP|DR-69P"           Xdep: "DB_SAMP"
  "DB_SAMP|DR-72P"           Xdep: "DB_SAMP"
  "DB_SAMP|KEYBOARD"         Xdep: "DB_SAMP"
  "DB_SAMP|RMNUM"            Xdep: "DB_SAMP"
  "DB_SAMP|SOFA2"            Xdep: "DB_SAMP"
```

"DOD-part306"   Xref: resolved
"SUBASY-A"   Xref: resolved
"SUBASY-B"   Xref: resolved
"SUBASY-C"   Xref: resolved

User Blocks	External References	Dependent Blocks	Unnamed Blocks
0	6	11	0

Command:

The same naming scheme operates with all dependent objects—*Dimension Styles*, *Text Styles*, *Views*, etc.

Dependent Object Names with *Bind* and *Xbind*

If you want to *Bind* the *Xref* drawing, the link is severed and the *Xref* drawing becomes a *Block* in the parent drawing. Or, if you want to bind only <u>individual named objects</u>, the *Xbind* command can be used to bring in specific *Blocks* or *Layer* names. The names are converted from the previous dependent naming scheme (using the *Xref* drawing name as a prefix with a | [pipe] separator) to another name depending on the command and option you use.

If you use *Xbind* or use *Bind* with the default *Bind* option, the names of dependent objects keep the *Xref* drawing name, but the | (pipe) symbol is changed to a $#$ separator, where # represents a number. For example, assume you use the *Bind* button (in the *External References* palette) with the DB_SAMP drawing from the previous example. The *Bind Xrefs* dialog box appears (Fig. 30-27). The default (*Bind*) option is the traditional naming scheme. That is, the *Xref* drawing name prefix stays with the object name.

FIGURE 30-27

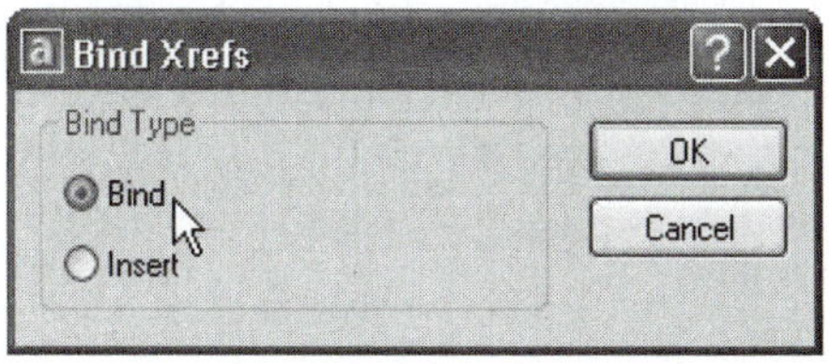

Using the *Bind* option, the resulting layer names would appear in the *Layer Properties Manager* listing (Fig. 30-28). Notice the dependent layer previously named DB_SAMP|CHAIRS is changed to DB_SAMP0CHAIRS. These layers <u>can be renamed</u> at this point because they are now a permanent part of the current drawing. Using the *Xbind* command to "import" selected named objects results in the same layering scheme illustrated here.

FIGURE 30-28

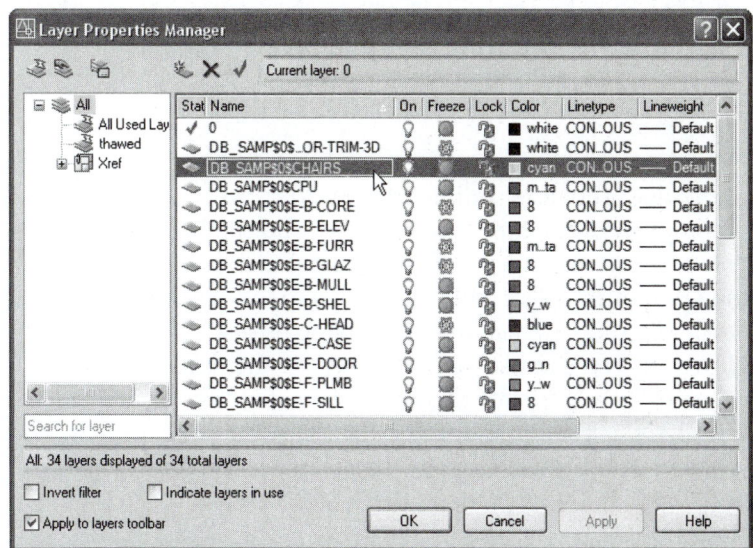

On the other hand, if you select the *Insert* option in the *Bind Type* dialog box, the *Xref* drawing name prefix is dropped from the object names when they are "imported." Therefore, the named objects assume their original name, assigned when they were created in the original drawing. Using the previous *Xref* example (DB_SAMP), the layer names would appear in the *Layer Properties Manager* listing as A-DOOR-TRIM-3D, CHAIRS, CPU, E-B-CORE, etc. (Fig. 30-29).

TIP

FIGURE 30-29

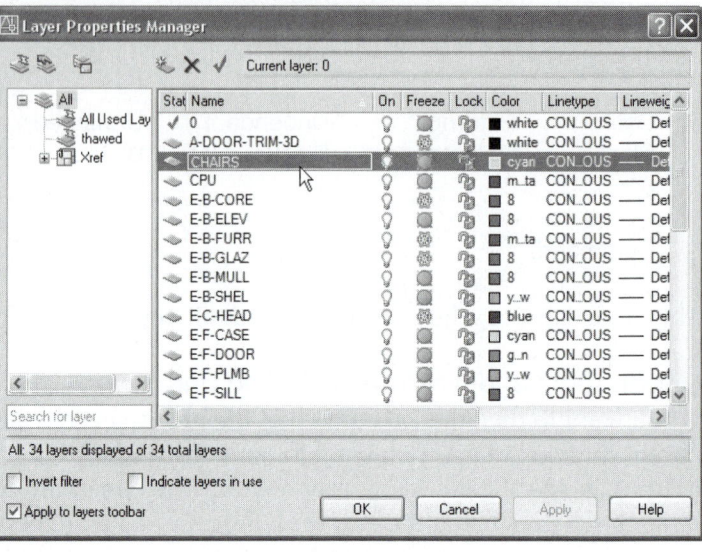

This method (*Bind, Insert*) is preferred if you want to *Bind* an *Xrefed* drawing that was created from the same template drawing as the current (parent) drawing. In this way, the same layers, dimension styles, text styles, etc. are not duplicated when you *Bind* the *Xref*. However, this method should not be used when two different *Blocks* exist with the same name or two layers exist with different linetype and color settings that should be maintained. When a dependent *Block* or layer having the same name as that in the parent drawing is "imported" with *Bind, Insert*, the *Block* definition and the layer properties of the parent drawing take precedence. Therefore, it is possible to accidentally redefine a *Block* or lose linetype and color properties of a layer with this scheme. There is no *Insert* option for *Xbind*.

Xattach

	Pull-down Menu	Command (Type)	Alias (Type)	Short-cut	Screen (side) Menu	Tablet Menu
	Insert *DWG Reference*	*Xattach*	*XA*	...	*INSERT* *Xref* *Attach...*	*T,4*

Xattach is a separate command from *Xref* but accomplishes the same action as *Xref, Attach*. If you use *Xref*, the *External References* palette appears, then you must select the *Attach...* tile to invoke the *Select File* and *External Reference* dialog boxes. However, you can use *Xattach* to invoke the *Select File* and *External Reference* dialog boxes directly (Fig. 30-30). Therefore, use the *Xattach* command when you want to *Attach or Overlay* an *Xref* but do not want to view or manage the list of current Xrefs.

FIGURE 30-30

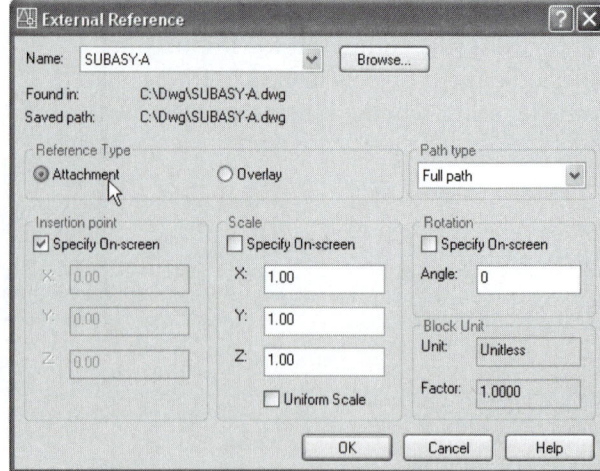

Xbind

	Pull-down Menu	Command (Type)	Alias (Type)	Short-cut	Screen (side) Menu	Tablet Menu
	Modify Object > External Reference >	*Xbind* or *-Xbind*	*XB* or *-XB*	...	*MODIFY1 Xbind*	*X,19*

The *Xbind* command is a separate command; it is not an option of the *Xref* command as are most other Xreference controls. *Xbind* is similar to *Bind*, except that *Xbind* binds an <u>individual</u> dependent object, whereas *Bind* converts the <u>entire</u> *Xref* drawing to a *Block*. Any of the listed dependent named objects (see "Dependent Objects and Names") that exist in a drawing can be converted to a permanent and usable part of the current drawing with *Xbind*. In effect, *Xbind* makes a copy of the named object since the original named object is not removed from the dependent drawing.

The *Xbind* command produces the *Xbind* dialog box (Fig. 30-31). All *Attached* and *Overlayed Xrefs* are listed. Expand any *Xref* to reveal the dependent object types (*Block, Dimstyle,* etc.) that can be imported. Expand each dependent object type to list the named objects of the type for that particular drawing. Select the desired object, then pick *Add ->* to add the object to the current (parent) drawing. The following object types are valid.

FIGURE 30-31

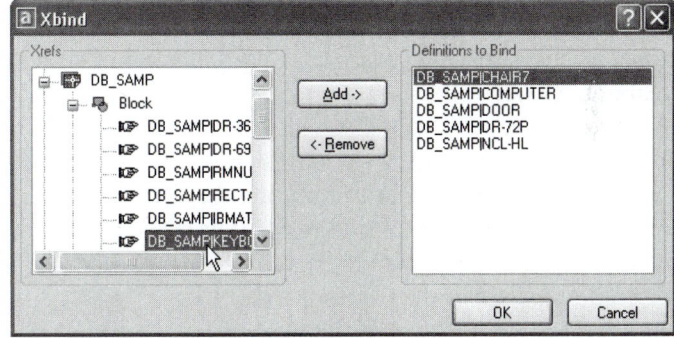

Block
Select this option to *Xbind* a *Block* from the *Xrefed* drawing. The *Block* can then be inserted into the current drawing and has no link to the original *Xref* drawing.

Dimstyle
This option allows you to make a dependent Dimension Style a permanent part of the current drawing. Dimensions can be created in the current drawing that reference the new Dimension Style.

Layer
Any dependent layer can be brought into the current drawing with this option. The geometry that resides on that layer does <u>not</u> become part of the current drawing, only the layer name and its properties. *Xbinding* a layer also *Xbinds* the layer's color and linetype.

Linetype
A linetype from an *Xrefed* drawing can be bound to the current drawing using this option. This action has the same effect as loading the linetype.

Textstyle
If a text style exists in an *Xrefed* drawing that you want to use in the current drawing, use this option to bring it into the current drawing. New text can be created or existing text can be modified to reference the new text style.

If you use (type) the *-Xbind* command, the Command line version of the command is invoked and individual dependent objects must be typed:

    Command: **-xbind**
    Enter symbol type to bind [Block/Dimstyle/LAyer/LType/Style]:

You must know and type the exact name of the dependent object name, including the dependent drawing name prefix and pipe character; therefore, this method is cumbersome compared to using the *Xbind* dialog box.

Adcenter

Pull-down Menu	Command (Type)	Alias (Type)	Short-cut	Screen (side) Menu	Tablet Menu
Tools DesignCenter	Adcenter	ADC	Ctrl+2	...	...

As an alternative to using *Xref*, *Xattach*, and *Xbind*, DesignCenter can be used to drag-and-drop drawings or named objects into the parent drawing. DesignCenter is accessed by any method shown in the table above.

Using DesignCenter to *Attach* or *Overlay* Xrefs

Normally when you drag-and-drop a .DWG file from the DesignCenter Content Area (window on the right) directly into the current drawing, it is *Inserted* as a *Block*. You can, however, use DesignCenter to attach *Xrefs*.

To *Attach* or *Overlay* an *Xref*, use the Tree View window (left window) to locate the folder that contains the drawing file you want to *Xref*, so all drawings contained in the folder appear in the Content Area (right window). Next, right-click on the desired drawing in the Content Area. Alternately right-drag (drag-and-drop, but holding down the right button) the desired drawing. This causes a shortcut menu to appear (Fig. 30-32). Select *Attach as Xref* from the menu. This causes the *External Reference* dialog box to appear (see Fig. 30-30). Note that you can specify *Attach* or *Overlay* in the *External Reference* dialog box. Proceed to specify the desired options in the dialog box and bring in the new *Xref*. The result is identical to using *Xattach* or the *External References* palette to *Attach* or *Overlay* the *Xref*.

FIGURE 30-32

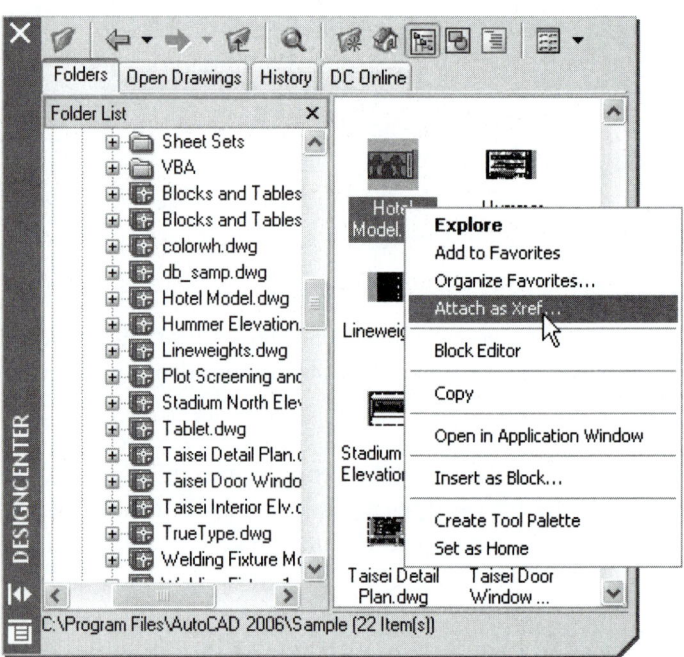

TIP Using DesignCenter as an Alternative to *Xbind*

DesignCenter can also be used to drag-and-drop named objects from any drawing into the current drawing. *Xbind* allows you to bring named objects into the current drawing but only from *Xrefed* drawings. With DesignCenter, you can drag-and-drop from any drawing. Named objects that can be imported with DesignCenter are *Blocks*, *Dimstyles*, *Layers*, *Layouts*, *Linetypes*, *Textstyles* and *Xrefs*.

To bring named objects into the current drawing, use the Tree View window (Fig. 30-33, left window) to locate the drawing file that contains the named objects you want to import, then expand the drawing (plus sign) so all named objects contained in the drawing appear in the Content Area. (right window). Next, drag-and-drop the desired named object from the Content Area. into the drawing area. This causes the selected named object to become part of the current drawing.

See Chapter 21 for more detailed information on DesignCenter.

FIGURE 30-33

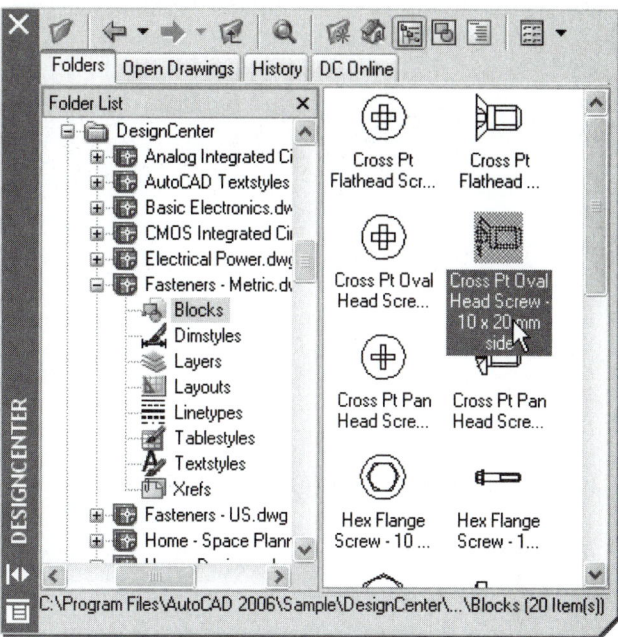

VISRETAIN

Pull-down Menu	Command (Type)	Alias (Type)	Short-cut	Screen (side) Menu	Tablet Menu
Tools Options... Open and Save Retain changes to Xref layers	VISRETAIN	...	...	TOOLS2 Option Open and Save Retain changes to Xref layers	...

This variable controls the dependent (*Xref* drawing) layer visibility, color, linetype, lineweight, and plot style settings in the parent drawing. Remember that you cannot draw or edit on a dependent layer without *Refedit*; however, you can change the layer's settings (*Freeze/Thaw, On/Off, Color, Linetype*, etc.) with the *-Layer* command, *Layer Properties Manager*, or *Layer Control* drop-down box. The setting of *VISRETAIN* determines if the settings that have been made are retained when the *Xref* is reloaded.

When *VISRETAIN* is set to 1 in the parent drawing, all the layer settings (*Freeze/Thaw, On/Off, Color, Linetype*, etc.) are retained for the Xreferenced dependent layers. When you *Save* the drawing, the settings for dependent layers are saved with the current (parent) drawing, regardless of whether the settings have changed in the *Xref* drawing itself. In other words, if *VISRETAIN*=1 in the parent drawing, the layer settings for dependent layers are saved. If *VISRETAIN*=0, then dependent layer properties settings are determined by the settings in the *Xref* drawing.

VISRETAIN also determines whether or not changes to the path of nested *Xrefs* are saved. You can change the path of an *Xref* or of a nested *Xref* (location AutoCAD searches for *Xrefs*) with the *Path* option of *-Xref* or the *Path type* drop-down list in the *External References* palette.

You can set *VISRETAIN* in the *Open and Save* tab of the *Options* dialog box. To set *VISRETAIN* to 1, select the checkbox next to *Retain changes to Xref layers* (Fig. 30-34, right center). No check appearing in the box means that *VISRETAIN* is set to 0.

FIGURE 30-34

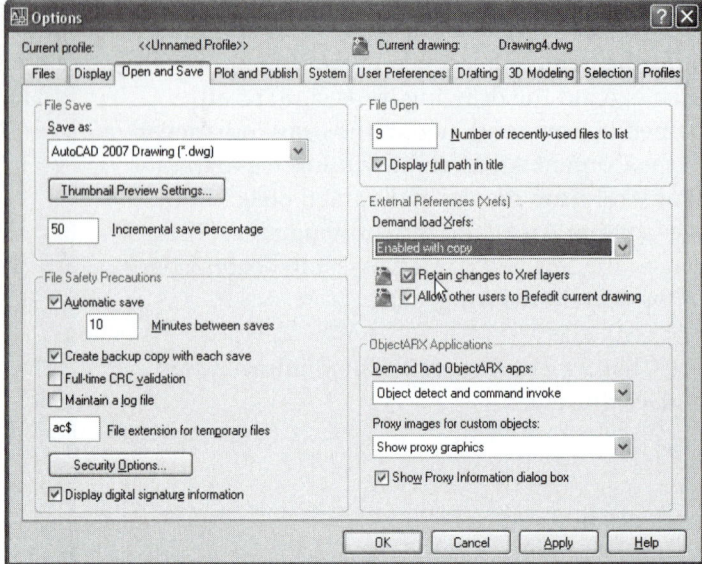

Xclip

Pull-down Menu	Command (Type)	Alias (Type)	Short-cut	Screen (side) Menu	Tablet Menu
Modify *Clip >* *Xref*	*Xclip*	*XC*	...	*MODIFY1* *Xclip*	*X,18*

When working with *Xrefs* you may want to display only a specific part of the attached drawing. For example, you may *Xref* an entire manufacturing plant layout (floor plan) in order to redesign an office space, or you may need to *Xref* a large map of a city just to design a new interchange at one location. In these cases, loading and displaying the entire *Xref* would occupy most of the system resources just for display, even though only a small area is needed.

Using the *Xclip* command, you can visually clip an <u>*Xref* drawing or a *Block*</u> to display only the portion inside the area defined by the clip boundary (Fig. 30-35). You can define the boundary as a rectangle or as a many-sided polygon of virtually any shape. All geometry inside the boundary is displayed normally. The portions of the *Xrefed* drawing (or *Block*) that fall outside the boundary are not only restricted from the display, but a feature called "demand loading" can be used to prevent the unwanted portions of the *Xref* from loading so system resources are not used unnecessarily (see "Demand Loading," "*INDEXCTL*" and "*XLOADCTL*").

FIGURE 30-35

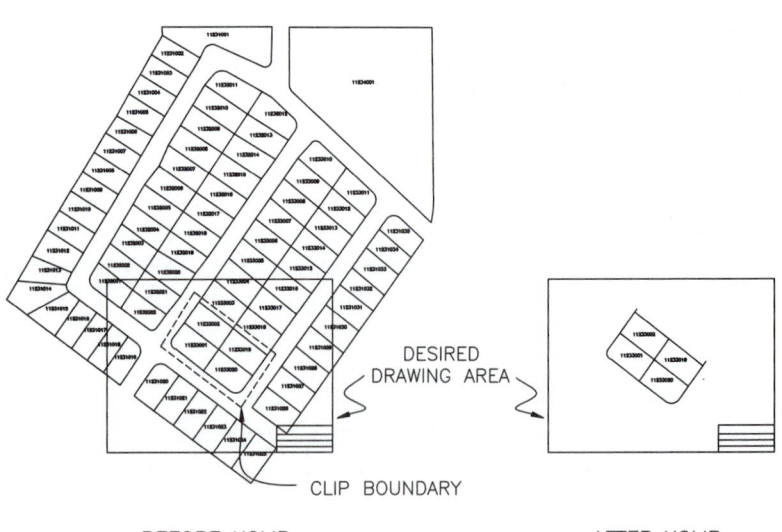

DESIRED DRAWING AREA

CLIP BOUNDARY

BEFORE *XCLIP*   AFTER *XCLIP*

Clipping does not edit or change the *Xref* in any way, it only prevents portions from being displayed:

```
Command: xclip
Select objects: PICK (the Xref Object)
Select objects: Enter
Enter clipping option [ON/OFF/Clipdepth/Delete/generate Polyline/New boundary] <New>: n
Specify clipping boundary: [Select polyline/Polygonal/Rectangular] <Rectangular>: r
Specify first corner: PICK
Specify opposite corner: PICK
Command:
```

The options of *Xclip* are as follows.

New Boundary

Rectangular
You define a rectangular boundary by PICKing diagonal corners. The clipping boundary is on the XY plane of the current UCS.

Polygonal
Select points as the vertices of a polygon. The polygon can have any shape and number of sides.

Select Polyline
Select an existing polyline (*Pline*) to use for the boundary. The existing polyline must consist of straight segments but can be open.

ON/OFF

ON displays only the *Xref* geometry inside the clipping boundary. *OFF* displays the entire *Xref*, ignoring the clipping boundary.

Clipdepth

Use this option to set front and back clipping planes on the *Xref* object. This option is useful for 3D geometry when you want to clip portions of the object behind a back clipping plane and in front of a front clipping plane. You must specify a clip distance from and parallel to the clipping boundary.

Delete

You can delete an existing boundary for the selected *Xref* or *Block* with this option. *Erase* cannot be used to delete a clipping boundary.

Generate Polyline

Automatically draws a *Pline* along an existing clipping boundary. The new *Pline* assumes the current layer, linetype, and color settings. This option is helpful if you want to change the clipping boundary using *Pedit* with the new *Pline*, then use the *Polyline* option to establish the new boundary.

XCLIPFRAME

	Pull-down Menu	Command (Type)	Alias (Type)	Short-cut	Screen (side) Menu	Tablet Menu
	Modify Object > External Reference > Frame	XCLIPFRAME	...	...	...	...

XCLIPFRAME is a system variable that can be used to control the display of the clipping boundary created by the *Xclip* command. For example, in some cases it may be desirable to display the boundary (Fig. 30-36), while in other cases no boundary may be preferred (Fig. 30-37). The options are 0 (turns clipping boundary off) and 1 (turns clipping boundary on).

FIGURE 30-36

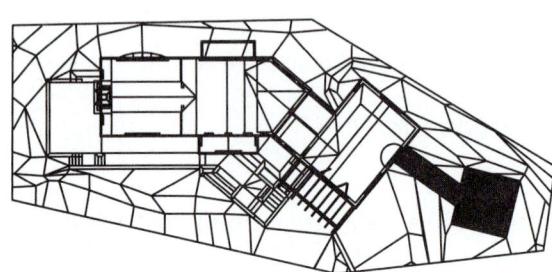

XCLIPFRAME = 1

FIGURE 30-37

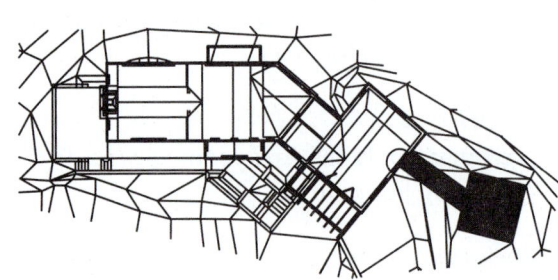

XCLIPFRAME = 0

Demand Loading

When *Xref* clipping is used to display only a selected area of an *Xrefed* drawing or block, it logically follows that the area outside the clipping boundary is of little or no interest in the current drawing. In order to conserve system resources and save time loading and regenerating these unwanted drawing areas, AutoCAD provides a feature called "demand loading."

The *Partialopen* and *Partiaload* commands allow you to open only part of a drawing based on *Layers, Views,* or by selecting a particular area of a drawing to load by specifying a boundary (see *"Partialopen"* and *"Partiaload"* in Chapter 2, Working with Files, for more information). These features are similar to the *Xclip* feature for *Xrefs* and can also take advantage of demand loading.

Demand loading allows AutoCAD to load only the information (about the *Xref* or partially opened drawing) that is needed to display the geometry in the clipping boundary. To take maximum advantage of demand loading, the demand loading feature must be enabled in the parent or partially opened drawing and the *Xref* drawings must be saved with layer and spatial indexes enabled. When a clipping boundary is formed, AutoCAD uses the layer index and spatial index in the *Xrefed* or partially opened drawing to categorize which objects are fully within, partially within, and not within the clipping boundary and which layers are visible and not visible.

Spatial Index   Updated in the *Xref* or partially opened drawing when a clipping boundary is formed to display a portion of the drawing's objects

Layer Index   Updated in the *Xref* or partially opened drawing when layer visibility is changed (*Freeze*, *Thaw*, *On*, *Off*)

Using the layer index and spatial index, AutoCAD determines which portions of the *Xrefed* or partially opened drawing file to load. This process provides performance advantages; however, certain demand loading controls must be used. The controls for demand loading are the *INDEXCTL* and *XLOADCTL* system variables.

INDEXCTL enables spatial indexing and layer indexing and should be set in the *Xref* drawings and in drawings intended to be used with *Partialopen* and *Partialload*. *XLOADCTL* is intended to be set in the parent drawing for an *Xref*, but has no usefulness for the *Partialopen* and *Partialload* features.

INDEXCTL

Pull-down Menu	Command (Type)	Alias (Type)	Short-cut	Screen (side) Menu	Tablet Menu
File *Saveas...* *Tools >* *Options...* *Index type*	*INDEXCTL*	...	...	*FILE* *Saveas* *Options...* *Index type*	...

When a drawing is used as an *Xref*, and particularly a clipped *Xref*, demand loading can increase performance in the parent drawing. Similarly, when *Partialopen* or *Partialload* is used to display a drawing, demand loading can also increase performance. In order for this to occur, spatial indexing and/or layer indexing must be enabled in the *Xref* or partially opened drawing(s). *INDEXCTL* controls how a drawing is loaded <u>when it is used as an *Xref*</u>. In other words, <u>*INDEXCTL* is set in the *Xref* drawing</u> (the drawing to be referenced), <u>not in the parent drawing</u> (the current drawing when using the *Xref* command).

Considering this in another way, assume you *Open* a drawing (parent drawing) and *Xref* two other drawings. Next, you create a clipping boundary around each *Xref*, so only a portion of each *Xref* is visible. Then you *Save* the parent drawing. In order to take advantage of demand loading and realize a performance increase when you *Open* the parent drawing again, layer and spatial indexing must first be enabled in <u>each of the two *Xrefed* drawings</u> for demand loading to occur. *Open* each (original) *Xref* drawing, and set *INDEXCTL* to 1, 2, or 3, then *Save*.

The *INDEXCTL* variable can be set by Command line format or can be set by using *SaveAs*, selecting *Tools*, then selecting the *Options...* button. The *Saveas Options* dialog box appears (Fig. 30-38) with four options for *Index Type*.

FIGURE 30-38

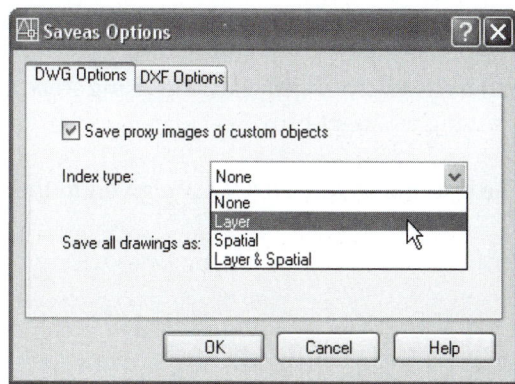

The *INDEXCTL* system variable is set to 0 by default, in which no spatial or layer indexes are created. The settings for *INDEXCTL* are as follows.

Setting	*Export Options* Setting	Effect
0	*None*	no indexes created
1	*Layer*	layer indexes created
2	*Spatial*	spatial indexes created
3	*Layer & Spatial*	both layer and spatial indexes created

To take full advantage of demand loading, set *INDEXCTL* to 3 in the drawing you anticipate using as an *Xrefed* or partially opened drawing.

XLOADCTL

Pull-down Menu	Command (Type)	Alias (Type)	Short-cut	Screen (side) Menu	Tablet Menu
Tools *Options...* *Open and Save* *Demand* *Load Xrefs:*	*XLOADCTL*	...	...	*TOOLS 2* *Options...* *Open and Save* *Demand* *Load Xrefs...*	...

The *XLOADCTL* variable set in the parent drawing (1) turns on or off demand loading for the current drawing and (2) controls what access other users have to *Xrefed* drawings. *XLOADCTL* is intended to be set in the parent drawing for an *Xref*, but has no usefulness for the *Partialopen* and *Partialload* features.

Demand loading is dynamic. If the clipping boundary is changed, the spatial index in the *Xrefed* drawing is automatically updated and if an *Xref* layer is frozen or thawed, the layer index is updated. This unique feature allows information from the *Xrefed* drawing to be loaded "on demand." However, there is one drawback to demand loading for *Xrefs*—that is, when demand loading is enabled, AutoCAD places a lock on all *Xrefed* drawings so that it can read and modify the information it needs (indexes) on demand. Therefore, other users can *Open* the *Xrefed* drawings as "read only" but cannot *Save* changes to them. Alternately, you can use the *Copy* option of demand loading, which creates a temporary copy of the *Xrefed* drawing(s) for use by the parent drawing and leaves the original *Xref* drawing available for use by others to edit and *Save*.

NOTE: *XLOADCTL* has no effect on <u>your</u> ability to use *Xopen* (or the *Open* option in the *External References* palette). In other words, if *XLOADCTL* is set to 1 (others cannot edit), you can still open and edit a currently *Xrefed* drawing using *Xopen*; however, other users can only open the original *Xref* drawing as "read-only."

The options of *XLOADCTL* are given in the following table.

Setting	*Options* Setting	Effect
0	*Disabled*	demand loading is disabled *Xref* drawing is loaded in its entirety others can access, edit, and save the *Xref* file
1	*Enabled*	demand loading is enabled locks original *Xref* drawing for use by parent drawing only
2	*Enabled with copy*	demand loading is enabled creates a temporary copy of the *Xref* for use by the parent drawing others can access, edit, and save original *Xref* file

The default setting for *XLOADCTL* is 2. If you are in a networking environment, a setting of 1 ensures others cannot alter the *Xref*, while a setting of 2 allows other users involved in a particular project (related to the *Xrefed* drawing) to continue working on the original *Xref* files.

XLOADCTL can also be set in the *Open and Save* tab of the *Options* dialog box (Fig. 30-39, upper right). The three options (*Disabled*, *Enabled*, and *Enabled with copy*) are explained in the previous table.

FIGURE 30-39

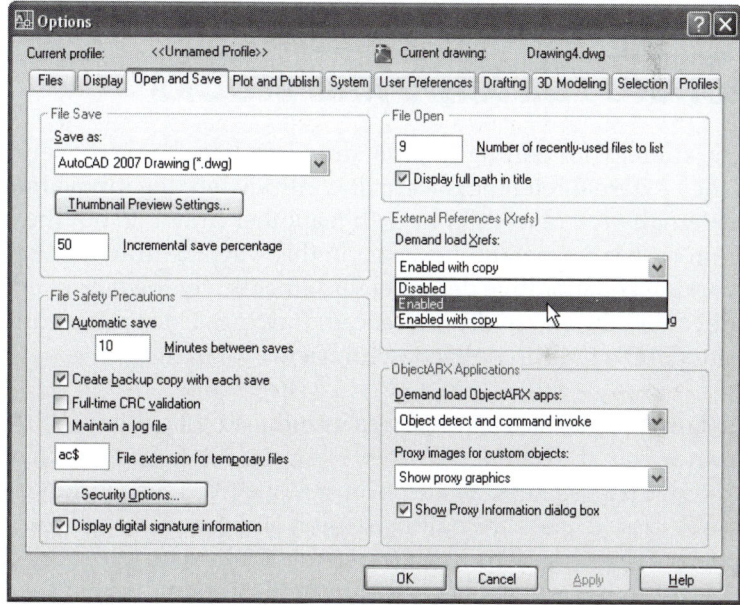

XLOADPATH

Pull-down Menu	Command (Type)	Alias (Type)	Short-cut	Screen (side) Menu	Tablet Menu
Tools *Options...* *Files* *Temporary External* *Reference File Location*	*XLOADPATH*	...	...	*TOOLS 2* *Options...* *Files* *Temporary* *External Refs..*	...

This system variable is used when *XLOADCTL* is set to 2 (*Enabled with copy*). Normally, when AutoCAD creates the copy of *Xref* drawings, the temporary copy is placed in the directory specified as *Temporary External Reference File Location* in the *Files* tab of the *Options* dialog box. You can use the *XLOADPATH* variable directly (in Command line format) or the *Options* dialog box to specify another location for temporary files. In the *Files* tab of the *Options* dialog box, expand *Temporary External Reference File Location*, highlight the previously specified location, then select the *Browse...* button to specify a new location for temporary files.

TIP

As a summary for setting demand loading and related controls, Figure 30-40 illustrates the appropriate drawing files (parent drawing and related *Xref* drawings) for setting the *XLOADCTL*, *XLOADPATH*, and *INDEXCTL* variables.

FIGURE 30-40

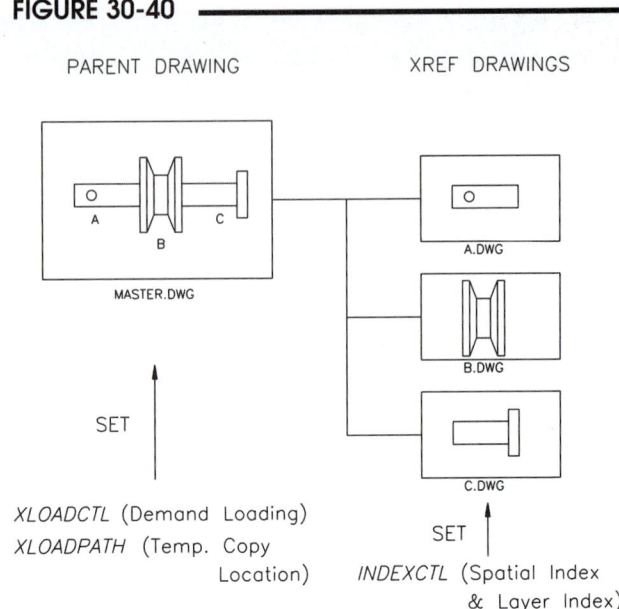

EDITING FOR *XREFS* AND *BLOCKS*

The strategy for editing external drawings has changed over the last several versions of AutoCAD. When external references were first introduced, the innovation was that any drawing could be viewed (externally referenced) from within another drawing, but the external drawing could not be changed from within the parent drawing. In this way, the drawings could be viewed by many users to facilitate working collaborative design environments, yet the lack of editing capabilities in the parent drawing prevented the externally referenced drawings from being edited by unauthorized users. This was AutoCAD's built-in protection for *Xrefs*.

In AutoCAD 2000, a concept was introduced that allowed *Xrefs* and *Blocks* to be edited "in place," or from within the parent drawing. First, use the *Refedit* command to select individual *Xrefs* or *Blocks* to be edited, then use *Refset* to add to or remove from the working set of objects, make the changes, and finally use *Refclose* to close the editing session and save the changes. The changes are "sent back" to the original external drawing. The *XEDIT* system variable (set in the original *Xref* drawing) controls whether or not *Refedit* can be used to make edits from the parent drawing, so security for the *Xrefs* can be maintained when desired.

Since AutoCAD 2004, a simpler and more direct method of editing *Xrefs* is available—the *Xopen* command (also see "*Xref*" and the *Open* option in the *External References* palette). Using *Xopen* <u>from within the parent drawing</u>, you can open any external reference drawing in a separate window, make changes, *Save*, and view the changes in the parent drawing immediately upon reloading the *Xref*. The *Xopen* command <u>overrides</u> the *XEDIT* variable, which in previous versions disabled editing capabilities from within the parent drawing. This method is clean and simple; however, there are no AutoCAD commands or system variables that can prevent unauthorized users from changing the original externally referenced drawings. Controls must be put in place by the CAD managers using network security measures to prevent unauthorized editing.

FIGURE 30-41

You can select any *Xref* object in the drawing and right-click to produce a shortcut menu (Fig. 30-41) with options for *Edit Xref in-place* (*Refedit* command), *Open Xref* (*Xopen* command), *Clip Xref* (*Xclip* command), and for producing the *External References* palette.

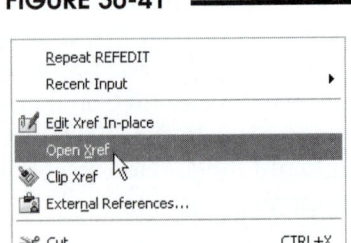

Xopen

Pull-down Menu	Command (Type)	Alias (Type)	Short-cut	Screen (side) Menu	Tablet Menu
Insert Xref Open	Xopen	...	...	...	...

The *Xopen* command allows a drawing currently being viewed as an *Xref* from within the parent drawing to be opened and edited in a separate window. In older versions of AutoCAD (2000 and 2002), only the *Refedit* command could be used to edit an *Xref* drawing from within the parent drawing.

If you use *Xopen* in Command line format, you are prompted to select the desired *Xref* you want to edit. After selection, you are presented with the new drawing window and the original externally referenced drawing.

```
Command: xopen
Select Xref: PICK
Command:
```

Alternately, use the *Open* option in the *External References* palette (Fig. 30-42). Selecting the shortcut menu option (*Open*) immediately opens the new window with the selected *Xref* drawing. You have full editing capabilities when you invoke *Xopen* by any method to open an *Xref* drawing.

When the changes are made and *Saved*, you are normally notified in the parent drawing that the *Xref* has changed. A notification icon or bubble appears at the system tray (depending on the setting of *XREFNOTIFY* system variable) indicating the name of the drawing that has changed and that it requires reloading (Fig. 30-43). To reload the drawing, use the *Reload* button in the *External References* palette (see Fig. 30-42). Also see "*XREFNOTIFY*."

NOTE: Opening a drawing using *Xopen* (by any method) <u>overrides</u> the setting of both the *XLOAD-CTL* and *XEDIT* variables. When *XEDIT* is set to 0 (in the original external drawing), the drawing currently being viewed as an *Xref* cannot be edited using *Refedit* in the parent drawing. When *XLOADCTL* is set to 1 (in the parent drawing), the drawing currently being viewed as an *Xref* cannot be edited by another user (technically, by another AutoCAD session, even if it is the same user as the person viewing the *Xref*). In other words, when viewing an *Xref* (in which its *XEDIT* variable is set to 0 and the parent's *XLOAD-CTL* variable is set to 1), you can use the *Xopen* command to edit the original drawing, but you cannot use the normal *Open* command or the *Refedit* command to edit the drawing.

FIGURE 30-42

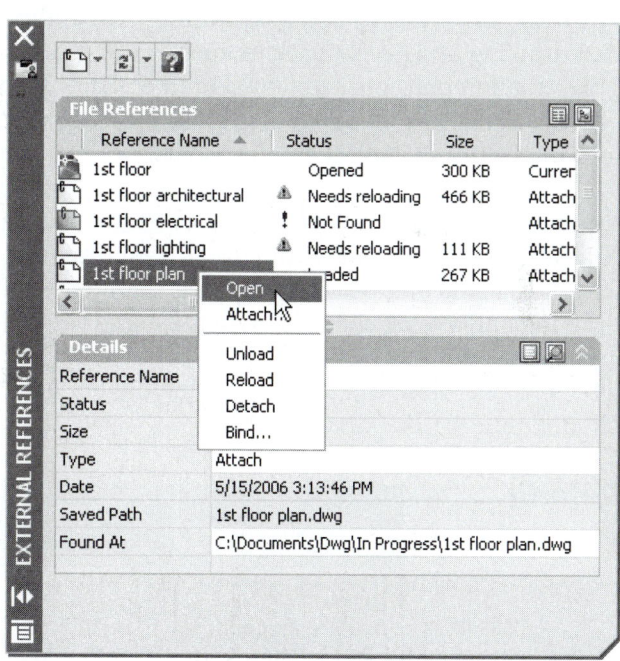

FIGURE 30-43

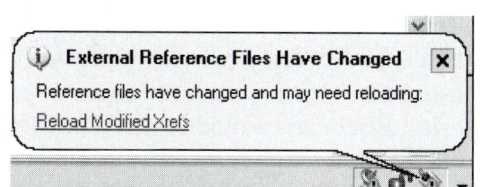

In-Place *Xref* and *Block* Editing

In AutoCAD 2000 and newer versions you can edit *Xrefs* and *Blocks* "in place," or from within the parent drawing using the *Refedit* command and a series of other commands and variables described next. Although these features can be used in AutoCAD 2007, the *Xopen* command is a simpler method for editing *Xrefs*; however, using *Refedit* to change *Blocks* from within the parent drawing is still a viable method since *Blocks* cannot be edited "in place" using *Xopen*. The commands and variables for editing *Blocks* or *Xrefs* "in place" are:

Refedit	(command)	Allows you to select individual objects in an *Xref* or *Block* and edit that geometry from within the current "parent" drawing
Refset	(command)	Allows you to add or remove objects to and from a working set while editing *Blocks* and *Xrefs* from the current drawing
Refclose	(command)	Closes the current reference editing session and saves or discards changes made to the *Xref* or *Block* geometry back to the original *Xref* drawing file or *Block* definition
XEDIT	(variable)	Controls whether a drawing file can be edited in place when referenced by another drawing
XFADECTL	(variable)	Controls the percentage of fading that occurs to objects not in the working set to edit

With the "in-place *Xref* and *Block* editing" capability, anyone who can *Xref* a drawing can also change that drawing and save it back to the original file <u>unless XEDIT is set to 0 in the original *Xref* drawing.</u> This situation presents a potential danger since the default setting for XEDIT is 1, which allows in-place editing. Therefore, it is suggested that the XEDIT system variable be changed to 0 for all template drawings in multi-user project environments.

Refedit is used to select *Xrefs*, *Blocks*, or <u>nested</u> *Xrefs* and *Blocks* to edit. When *Refedit* is used to change an *Xref* or nested *Xref* from within the current drawing and *Refclose* is used to save the changes, the original *Xref* drawing is updated with the new changes, which has the same effect as opening the original *Xref* drawing and saving changes. When *Blocks* defined in the current drawing or nested *Blocks* are edited and saved with this capability, the *Block* definition is saved in the original drawing containing the *Block*, which has the same effect as redefining the *Block* in its original drawing by the traditional method (see "Redefining *Blocks*," Chapter 21).

Refedit

Pull-down Menu	Command (Type)	Alias (Type)	Short-cut	Screen (side) MenuMenu	Tablet
Tools *In place Xref and* *Block Edit >* *Edit Reference*	*Refedit*	...	...	*MODIFY2* *Refedit*	...

The capability known as "in-place editing" allows you to change geometry in an *Xrefed* drawing or nested *Block* from within the current "parent" drawing. With in-place editing, changes to an *Xref* or nested *Block* can be made from within the parent drawing, then saved back to the original file.

<u>The *Refedit* command simply allows you to select which *Xref* or *Block* you want to edit and to pick the specific geometry you want to change.</u> Only one *Xref* or *Block* at a time (and its nested objects) can be selected for editing. The changes you make to the selected *Xref* or *Block* are accomplished using typical editing commands. Use the *Refclose* command to save the changes back to the original drawing file. *Xrefs* and nested *Blocks* can be edited in place only if their XEDIT system variable (in the original *Xrefed* drawing) is set to 1 to allow external editing.

Invoking *Refedit* by any method produces the "Select reference" prompt. Select any *Block* or *Xref* from the drawing area using the pickbox or window.

Command: **refedit**
Select reference: **PICK** (*Xref* or *Block*)
(*Reference Edit* dialog box appears)

After selecting an *Xref* or *Block*, the *Reference Edit* dialog box appears (Fig. 30-44). The name of the selected reference appears under *Reference name:* along with its preview image on the right.

FIGURE 30-44 ———————————————

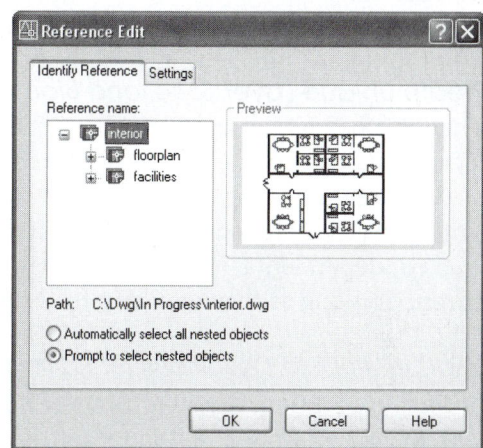

You can choose to *Automatically select all nested objects*, or you can choose that AutoCAD *Prompt to select nested objects*. If you want to edit only a few *Lines* or *Circles* in your *Xref* or *Block*, you should choose to be prompted. Select *OK* to return to the drawing editor.

The "Select nested objects" prompt requests that you select geometry <u>contained in</u> the selected reference object such as a *Line*, *Circle*, or *Block* that you want to edit. This becomes your working set. After selection, the working set becomes visually distinct because all other objects not in the working set become faded. For example, Figure 30-45 displays geometry comprising the table and chairs as the "nested objects," or working set, and the remainder of the drawing is faded. The amount of fading is controlled by the *XFADECTL* system variable (see "XFADECTL" later in this section).

FIGURE 30-45 ———————————————

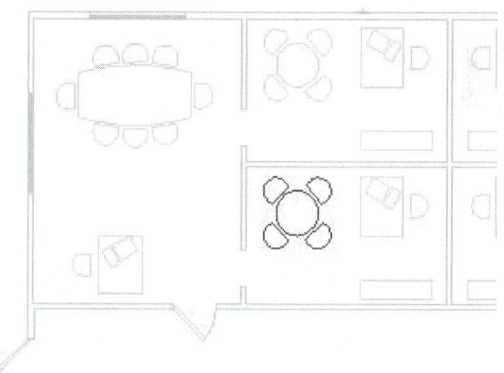

The working set can then be edited and changed in any way using normal editing commands and grips. To save changes made to the working set, use *Refclose* with the *Save* option. The changes are saved back to the original *Xref* drawing file or to the *Block* definition in the drawing that contains the *Block*. If a warning is issued when you select "nested objects" (Fig. 30-46), editing is not allowed because the *XEDIT* system variable is set to 0 in the original *Xrefed* drawing or original drawing containing the nested *Block*.

FIGURE 30-46 ———————————————

Refedit allows you to edit nested *Xrefs* and nested *Blocks*. When you select a reference object in response to the first *Refedit* prompt, "Select reference," the *Reference Edit* dialog appears and a hierarchical nesting structure is displayed for the reference objects (Fig. 30-47). Select the desired nesting level from the list to cause AutoCAD to highlight the selected reference in the drawing.

Settings Tab

The *Settings* tab (Fig. 30-48) provides the following options.

Create Unique Layer, Style, and Block Names

When this box is checked (default position), the layer and dependent object names of selected *Xrefs* are prefixed with the *Xref* drawing name and x, similar to the naming convention used when you *Bind* an *Xref*. If the box is not checked, dependent layer and object names appear in the current drawing as they would in the original *Xref* drawing.

Display Attribute Definitions for Editing

This feature lets you to edit <u>attribute definitions (tags)</u> in *Block* references from within the current drawing. When this box is checked, the attribute tags in *Blocks* (in the current drawing) or in nested *Blocks* are displayed instead of the attribute values themselves. You can select the attribute tags for editing along with the *Block* geometry. When changes are saved back to the *Block* reference, only <u>new</u> insertions of the *Block* are affected. Attributes of the previous *Block* insertions are not changed. See Chapter 22, Block Attributes, for more details on this subject.

FIGURE 30-47

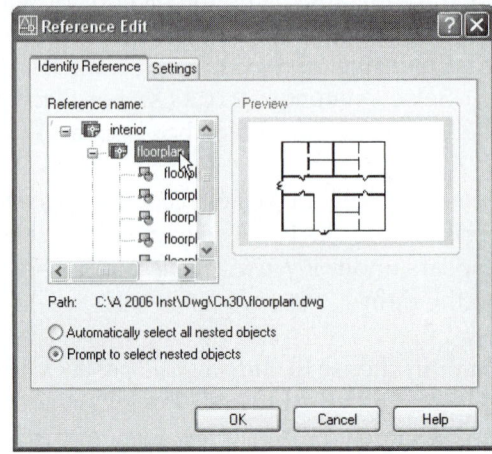

FIGURE 30-48

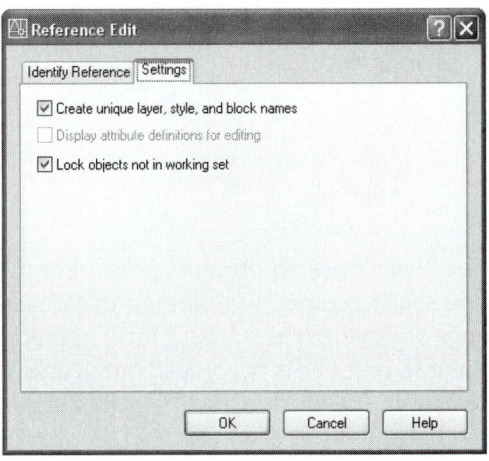

Lock Objects Not in Working Set

This option automatically locks all objects not in the working set including objects in the parent drawing (see *Refset*). Any selected locked objects behave as if they are on a locked layer. This feature prevents you from accidentally selecting and editing objects in the parent drawing.

If you select objects in an *Xref* as the working set, but they are on a locked layer, you must *Unlock* the layer to edit the geometry. When changes are saved back to the *Xref*, the layer state of the original *Xref* drawing is not changed—it remains locked.

The *Refedit* feature cannot be used with individual *Blocks* that have been inserted using *Minsert*.

Using *-Refedit* at the Command Line

Typing *-Refedit* produces the Command line version shown below.

```
Command: -refedit
Select reference: PICK
Select nesting level [Ok/Next] <Next>: O
Select nested objects: PICK
Select nested objects: Enter
Display attribute definitions [Yes/No] <No>: Enter
Use REFCLOSE or the Refedit toolbar to end reference editing session.
Command:
```

The Command line prompt to "Select nesting level [Ok/Next] <Next>:" serves the same function as the *Next* button or selecting from the hierarchical list in the *Reference Edit* dialog box. Note the last prompt, "Display attribute definitions [Yes/No] <No>:", provides the same level of access for editing attribute tags as provided in the dialog box version of the command.

Refset

Pull-down Menu	Command (Type)	Alias (Type)	Short-cut	Screen (side) Menu	Tablet Menu
Tools *In place Xref and* *Block Edit >* *Add to Workset* or *Remove from Workset*	*Refset*	...	...	...	...

You can *Add* or *Remove* objects from the working set using *Refset* while editing in place. You initially define a working set of objects to edit when you select objects at the "Select nested objects:" prompt during the *Refedit* command. The working set is visually distinct while the other objects in the drawing become faded. You can add or remove to or from this working set using *Refset*.

    Command: **refset**
    Transfer objects between the reference set and drawing...
    Enter an option [Add/Remove] <Add>: **PICK**
    1 Added to working set
    Command:

Add

Use this option or button from the *Refedit* toolbar to select objects to add to the working set. The selected objects change from faded to distinct and can then be edited.

Remove

This option or button from the *Refedit* toolbar allows you to select objects to remove from the working set. Objects removed from the working set become faded and cannot be edited.

Refclose

Pull-down Menu	Command (Type)	Alias (Type)	Short-cut	Screen (side) Menu	Tablet Menu
Tools *In-place Xref and* *Block Edit >* *Save Reference Edits*	*Refclose*	...	...	...	...

Use the *Refclose* command to *Save* back or *Discard* changes made to objects during an in-place editing session. In other words, use *Refedit* to select *Xrefs* and *Blocks* and objects within those references to edit, modify the geometry, then use *Refclose* to close the editing session and save or discard the modifications to the working set.

    Command: **refclose**
    Enter option [Save/Discard reference changes] <Save>: **Enter**
    Regenerating model.
    1 block instances updated
    (Block name) redefined.
    Command:

You can also *Save* or *Discard* edits by using the options on the *Tools* pull-down menu or the *Refedit* toolbar. The *Refedit* toolbar appears after a reference is selected for editing and disappears after changes to the reference are saved or discarded.

Save

 Saving changes saves the modifications made to the working set back to the original reference (*Xref* or *Block*) drawing file. The change is made without actually opening the reference drawing or recreating the block. If you decide to *Save* changes, you must confirm the dialog box that appears to complete the desired action (Fig. 30-49).

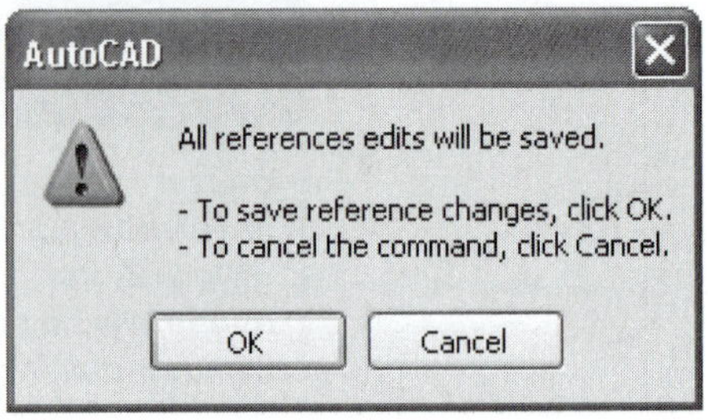

Using the *Save* option of *Refclose* does not save changes made to other objects (not in the in-place editing working set)—you must use the full *Save* command to save those changes. If you use *Refedit* and then *Refclose Save* to make edits to *Blocks* contained only in the current drawing, you must also *Save* the drawing.

FIGURE 30-49

Discard Reference Changes

 Use this option if you want to close the in-place editing session and discard all changes made to the working set. A dialog box appears similar to that shown in Figure 30-49. Any changes you have made to other objects in the current drawing (not in the working set) are not discarded.

Immediately after using *Refclose*, you can use the *Undo* command to return to the editing session.

XEDIT

Pull-down Menu	Command (Type)	Alias (Type)	Short-cut	Screen (side) Menu	Tablet Menu

Tools
Options...
Open and Save
Allow others to
  Refedit current
  drawing
	XEDIT	...	...	*TOOLS 2*
				Options...
				Open and Save
				Allow others
				  to Refedit
				  current drawing
					...

The value of this system variable determines whether or not a drawing can be edited in place when it is used as an *Xref*. XEDIT should be set in the original *Xref* drawing, not in the "parent" drawing. The possible settings are:

XEDIT = 0   The drawing cannot be edited while *Xrefed* by another drawing.
XEDIT = 1   The drawing can be edited in place when used as an *Xref* in another drawing.

You can change this system variable two ways: enter *XEDIT* at the Command line and change the value, or use the *Open and Save* tab of the *Options* dialog box (see Fig. 30-35). If you use the *Options* dialog box, placing a check in *Allow others to Refedit current drawing* sets the *XEDIT* variable to 1. The variable is saved in the drawing.

TIP The default setting of *XEDIT* in AutoCAD template drawings is 1, so in-place editing can occur without changing this variable. Keep in mind that in this case anyone who can *Xref* a drawing can also make an accidental or unauthorized change to that drawing and save it back to the original file. This situation presents a potential danger in multi-user project environments. Therefore, it is suggested that the *XEDIT* system variable be changed to 0 for all template drawings in multi-user project environments.

XFADECTL

Pull-down Menu	Command (Type)	Alias (Type)	Short-cut	Screen (side) Menu	Tablet Menu
...	XFADECTL	...	...	...	...

This system variable controls the percentage of fading during an in-place editing session. When *Refedit* is used and reference geometry is selected for editing, it becomes the working set and remains distinct (not faded). All other objects in the current drawing become faded (see Fig. 30-45). The default value is 50. The *XFADECTL* range is from 0 (0%, no fading) to 90 (90% fading). Fading ends when *Refclose* is used to close the in-place editing session.

You can set this variable by entering *XFADECTL* at the Command line or by accessing the *Display* tab of the *Options* dialog box (Fig. 30-50, lower right). Use the *Reference Edit fading intensity* slider to control the *XFADECTL* variable. *XFADECTL* is saved in the system registry, so once the variable is set, you do not have to reset it whenever new *Xrefs* are edited.

FIGURE 30-50

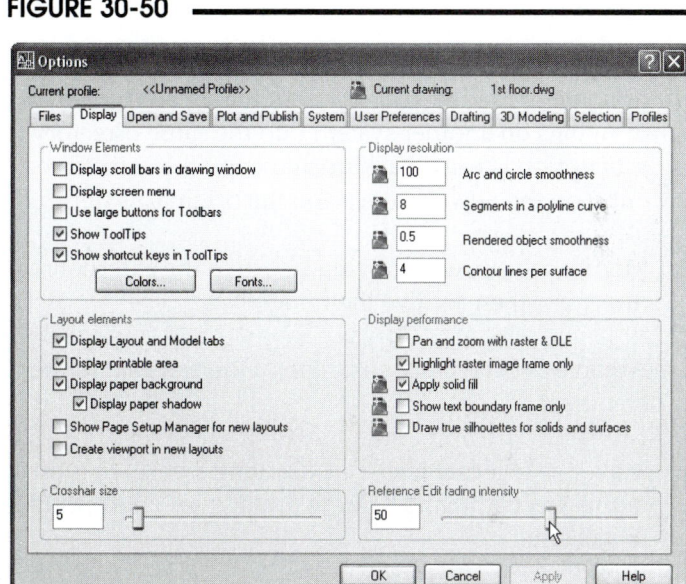

XREFNOTIFY

Pull-down Menu	Command (Type)	Alias (Type)	Short-cut	Screen (side) Menu	Tablet Menu
...	XREFNOTIFY	...	...	...	...

XREFNOTIFY is a system variable that controls the display of the notification icon and balloon message that appear at the lower-right corner of the Status bar when *Xrefs* have been updated or are not found. The settings are described below. This variable is saved in the system registry.

XREFNOTIFY = 0   *Xref* notification is disabled.

XREFNOTIFY = 1   *Xref* icon is enabled. This setting turns on the *Xref* icon in the Status bar when *Xrefs* are attached to the current drawing. If *Xrefs* are not found, the icon is displayed with a yellow alert symbol (!).

XREFNOTIFY = 2   *Xref* notification and balloon messages are enabled. This setting turns on the *Xref* icon, as in a setting of 1, but also displays balloon messages in the same area when *Xrefs* are modified.

MANAGING *XREF* DRAWINGS

TIP Because of the external reference feature, the contents of a drawing can be stored in multiple drawing files and directories. In other words, if a drawing has *Xrefs*, the drawing is actually composed of several drawings, each possibly located in a different directory. This means that special procedures to handle drawings linked in external reference partnerships should be considered when drawings are to be backed up or sent to clients. Several considerations are listed here:

1.   Create a project name for the drawing with the *PROJECTNAME* variable. The project name is stored in the drawing file. Notify the client so the client can create the same project name on his or her system (with the *Files* tab of the *Options* dialog box) to store the drawings and *Xrefs* (see "PROJECT-NAME").

2.   When *Attaching Xrefs* to a drawing intended to be used by clients, use the *No path* or *Relative path* options (in the *External Reference* dialog box). If the *No path* option is used, you can copy the *Xref* drawings on another system into the same directory (folder) as the parent drawing. If the *Relative path* option is used, back up and copy the same directory structure (from the parent drawing down only) on the new system as on the original system.

3.   Modify the current drawing's path to the external reference drawing so both are stored in the same directory; then archive them together.

4.   Archive the directory structure of the external reference drawing along with the drawing which references it.

5.   Make the external reference drawing a permanent part of the current drawing with the *Bind* option of the *Xref* command prior to archiving. This option is preferred when a finished drawing set is sent to a client.

6.   Use the *Reference Manager* to manage the paths of drawings and the related *Xrefs*. For example, if you copied the parent drawing and related *Xrefs* onto another system with a different directory structure than that saved in the *Full path* on the original system when the *Xrefs* were *Attached*, launch the *Reference Manager* on the new system. Using *Reference Manager*, you can change the path for a group of *Xrefs* globally. See "Reference Manager."

PROJECTNAME

Pull-down Menu	Command (Type)	Alias (Type)	Short-cut	Screen (side) Menu	Tablet Menu

Tools
Options...
Files
Project Files...
  Search Path
	PROJECTNAME	...	...	*TOOLS 2*
				Options
				Files
				Project Files..
				  Search Path
					...

Project names make it easier for AutoCAD users to exchange drawings that contain *Xrefs*. The *PRO-JECTNAME* variable allows you to specify a project name that AutoCAD uses to point to an alternate search path when trying to locate *Xrefs*.

If you load a drawing containing several *Xrefs* given to you by a colleague from another company, AutoCAD first attempts to locate the *Xrefs* by searching the hard-coded paths. (The path of an *Xref* is "hard coded" when the drawing is originally *Attached* or *Overlayed* or if the *Path type* option is used in the *External References* palette). The *Xrefs* are not found initially unless you have an identical directory structure on your system or network, and the *Xrefs* have been copied into the directories accordingly. If an *Xref* cannot be found because the drawing is not located in the hard-coded path, AutoCAD then uses the drawing's *PROJECTNAME* to search for the *Xref* in the location specified by your system registry.

There are two components to this scheme: (1) the *PROJECTNAME* variable is saved in the parent drawing file, and (2) the directory location for that project name is saved in your system registry. This scheme makes it possible for collaborating designers to exchange drawings containing *Xrefs* when the designers have different directory structures. The drawing that is exchanged contains the project name, but the location (path) for that project name can be different for each system (saved in the system registries). Each designer's system contains the same project name but has a different drive and/or directory mapping. In other words, if an *Xref* cannot be found in the hard-coded path, AutoCAD then searches for the *Xref* by using the *PROJECTNAME* variable in the drawing file, but the <u>location</u> for the project name is saved in your system registry.

The value for the drawing's *PROJECTNAME* variable (*Project1*, for example) can only be set using the *PROJECTNAME* variable in command line format. The *PROJECTNAME* variable is saved in the <u>parent drawing file</u>. However, you must use the *Files* tab of the *Options* dialog box (but <u>not</u> the *PROJECTNAME* variable in command line format) to specify the directory mapping (location), which is then stored in <u>your system registry</u>.

System Registry Project Name

To assign a search path to be saved in your system registry, you <u>must</u> use the *Files* tab of the *Options* dialog box (Fig. 30-51). You <u>cannot</u> use the *PROJECTNAME* variable or any other method of Command line entry to add, remove, or modify the directory location for a project name in the system registry. Using the dialog box, follow these steps.

1. Double-click *Project Files Search Path*, then select *Add*. A project folder named *Project1* appears (or *Projectx*, where x is the next available number).
2. Enter a new name or press Enter to accept *Projectx*.
3. Double-click to expand *Project1* (or other name you entered).
4. Select *Browse*. Use the *Browse for Folder* dialog box that appears (not shown) to locate the desired folder on your system where you want to store *Xref* drawings for the project.
5. Select *OK* in both dialog boxes to save the project name location to your system registry.

FIGURE 30-51

The action above assigns a project name to the <u>system registry</u>.

Saving the *PROJECTNAME* in the Drawing File

Use the *PROJECTNAME* variable at the command line to save the drawing's component of the project name. Do this <u>after</u> registering the project name in your system using the *Options* dialog box. The Command prompt appears as follows:

```
Command: projectname
Enter new value for PROJECTNAME, or . for none <" ">: project1
Command:
```

CHAPTER EXERCISES

1. *Xref Attach*

    Use the **SLOTPLATE2** drawing that you modified last (with *Stretch*) in Chapter 9 Exercises. The slot plate is to be manufactured by a stamping process. Your job is to nest as many pieces as possible within the largest stock sheet size of 30" x 20" that the press will handle.

    A. *Open* the **SLOTPLATE2** drawing and use the *Base* command to specify a basepoint at the plate's lower-left corner. *Save* and *Close* the drawing.

    B. Begin a *New* drawing using the
        C-D-SHEET prototype and assign the
        name **SLOTNEST**. Set *Limits* to **34** x
        22. Draw the boundary of the stock
        sheet (30" x 20"). *Xref Attach* the
        SLOTPLATE2 into the sheet drawing
        multiple times as shown in Figure 30-
        53 to determine the optimum nesting
        pattern for the slot plate and to mini-
        mize wasted material. The slot plate
        can be rotated to any angle but
        cannot be scaled. Can you fit 12
        pieces within the sheet stock? *Save*
        and *Close* the drawing.

    FIGURE 30-53

    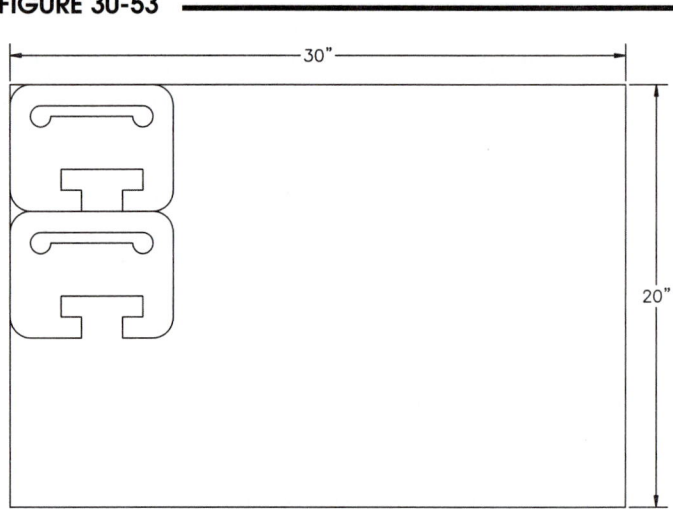

    C. In the final test of the piece, it was determined that a
        symmetrical orientation of the "T" slot in the bottom
        would allow for a simplified assembly. *Open* the
        SLOTPLATE2 drawing and use *Stretch* to center the
        "T" slot about the vertical axis as shown in Figure
        30-54. *Save* the new change and *Close* the drawing.

    FIGURE 30-54

    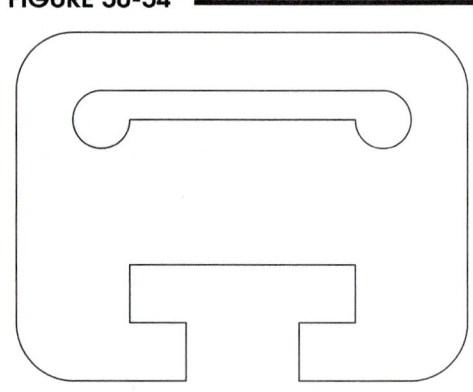

    D. *Open* the **SLOTNEST** drawing. Is the design change
        reflected in the nested pieces?

2. *Attach, Detach, Xbind, VISRETAIN*

    As the interior design consultant for an architectural firm, you are required to place furniture in an efficiency apartment. You can use some of the furniture drawings (*Blocks*) that you specified for a previous job, the office drawing, and *Insert* them into the apartment along with some new *Blocks* that you create.

    A. Use the **C-D-SHEET** as a *template* and set *Architectural Units* and set *Limits* of **48'** x **36'**.
        (Scale factor is 24 and plot scale is 1/2"=1' or 1=24). Assign the name **INTERIOR**. *Xref* the
        OFFICE drawing that you created in Chapter 21 Exercises. Examine the layers (using the
        Layer Control drop-down list) and list (**?**) the *Blocks*. *Freeze* the **TEXT** layer. Use *Xbind* to bring

the **CHAIR**, **DESK**, and **TABLE** *Blocks* and the **FURNITURE** *Layer* into the INTERIOR drawing. *Detach* the **OFFICE** drawing. Then *Rename* the new *Blocks* and the new *Layer* to the original names (without the **OFFICE0** prefix). In the INTERIOR drawing, create a new *Block* called **BED**. Use measurements for a queen size (60" x 80"). *Save* the INTERIOR drawing.

B. *Xref* the **EFF-APT2** drawing (from Chapter 18 Exercises) into the INTERIOR.DWG. Change the *Layer* visibility to turn *Off* the **EFF-APT2 | TEXT** layer. *Insert* the furniture *Blocks* into the apartment on *Layer* **FURNITURE**. Lay out the apartment as you choose; however, your design should include at least one insertion of each of the furniture *Blocks* as shown in Figure 30-55. When you are finished with the design, *Save* and *Close* the INTERIOR drawing.

FIGURE 30-55

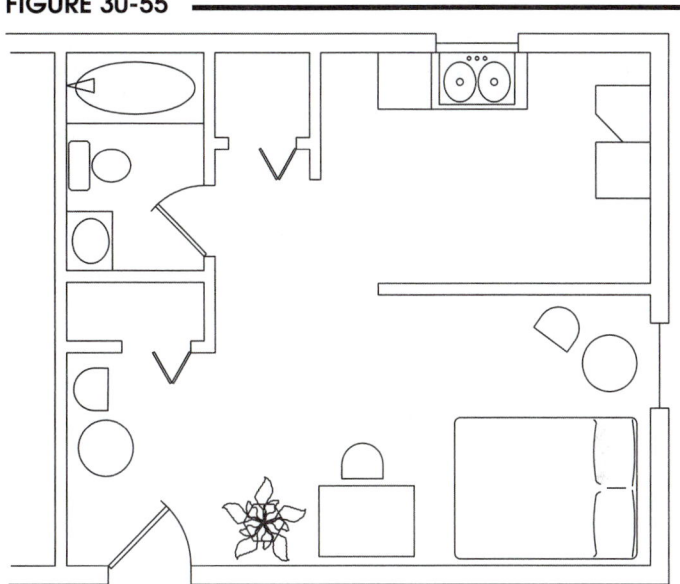

C. Assuming you are now the architect, a design change is requested by the client. In order to meet the fire code, the entry doorway must be moved farther from the end of the hall within the row of several apartments. *Open* the **EFF-APT2** drawing and use *Stretch* to relocate the entrance **8'** to the right as shown in Figure 30-56. *Save* and *Close* the EFF-APT2 drawing.

FIGURE 30-56

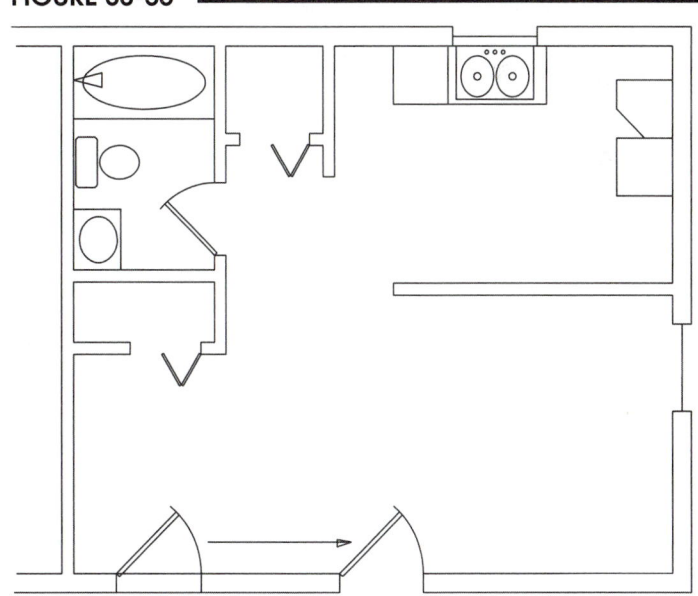

D. Next (as the interior design consultant), *Open* the **INTERIOR** drawing. Notice that the design change made by the architect is reflected in your INTERIOR drawing. Does the text appear in the drawing? *Freeze* the **TEXT** layer again, but this time change the *VISRETAIN* variable to **1**. Make the necessary alterations to the interior layout to accommodate the new entry location. *Save* the INTERIOR drawing.

3. **Xref Attach**

As project manager for a mechanical engineering team, you are responsible for coordinating an assembly composed of 5 parts. The parts are being designed by several people on the team. You are to create a new drawing and *Xref* each part, check for correct construction and assembly of the parts, and make the final plot.

A. The first step is to create two new parts (as <u>separate</u> drawings) named **SLEEVE** and **SHAFT**, as shown in Figure 30-57. Use appropriate drawing setup and layering techniques. In each case, draw the two views on <u>separate layers</u> (so that appearance of each view can be controlled by layer visibility). Use the *Base* command to specify an insertion point at the right end of the rectangular views at the centerline.

FIGURE 30-57 ━━━━━━━━━━━━━━━━━━━━━━━━━━━━━━━━

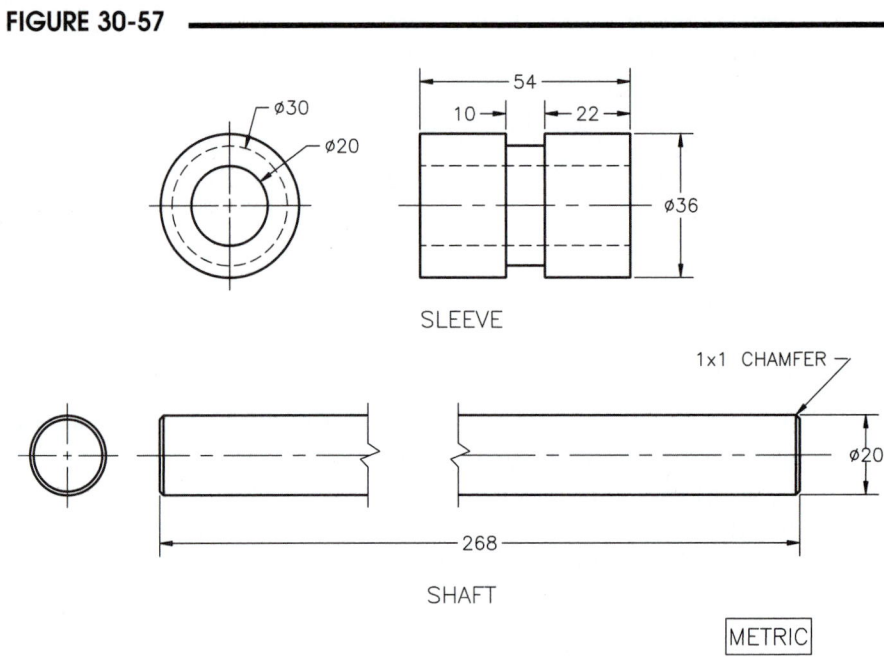

B. You will eventually *Xref* the drawings **SHAFT**, **SLEEVE**, **PUL-SEC** (from Chapter 26 Exercises), **SADDLE** (from Chapter 24 Exercises), and **BOLT** (from Chapter 9 Exercises) to achieve the assembly as shown in Figure 30-58. Before creating the assembly, *Open* each of the drawings and use *Properties* to move the desired view (with all related geometry, no dimensions) to a new layer. You may also want to set a *Base* point for each drawing.

FIGURE 30-58 ━━━━━━━━━━━━━━━━━━━━━━━━━━━━━━━━

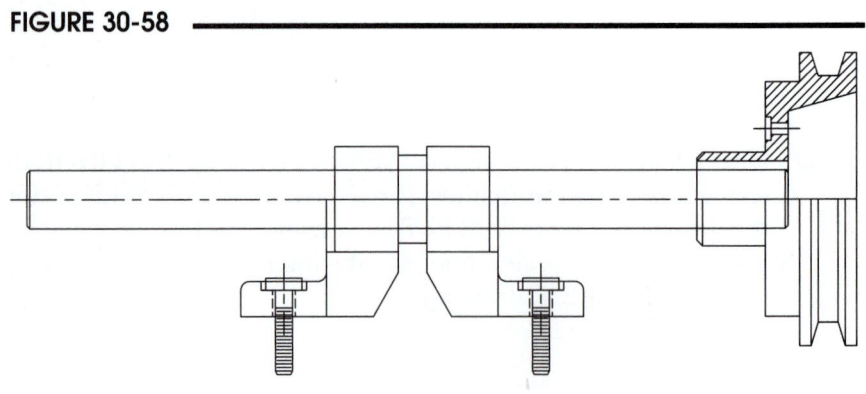

C. Finally, begin a *New* drawing (or use a *Template*) and assign the name **ASSY**. Set appropriate *Limits*. Make a *Layer* called **Xref** and set it *Current*. *Xref* the drawings to complete the assembly. The **BOLT** drawing must be scaled to acquire a diameter of **6mm**. Use the *Layer Control* drop-down list to achieve the desired visibility for the *Xrefs*. Set *VISRETAIN* to **1.** Make a *Plot* to scale. *Save* and *Close* the **ASSY** drawing.

4. *Xref Overlay*

A. Assume you are now the project coordinator/checker for the assembly. (If it is possible in your lab or office, exchange the SHAFT drawings [by diskette or by network] with another person so that you can

FIGURE 30-59

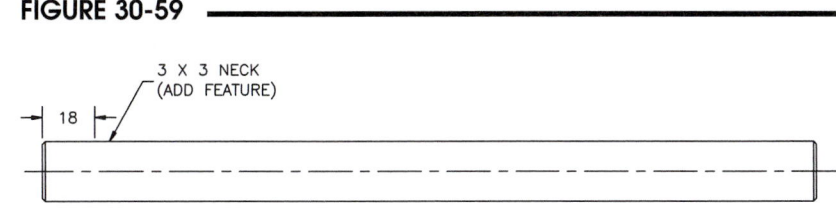

perform this step for each other.) Begin a *New* drawing and make a *Layer* called **CHECK**. Next, *Xref Overlay* the **SHAFT** drawing (of your partner). On the CHECK layer, insert some *text* by any method and a dimension similar to that shown in Figure 30-59. *Save* and *Close* the CHECK drawing. (When completed, exchange the CHECK and SHAFT drawings back.)

B. *Open* the **SHAFT** drawing. In order to see the notes from the checker, *Xref Overlay* the **CHECK** drawing. The instructions for the change in design should appear. (Notice that the *Overlay* option prevents a circular *Xref* since SHAFT references CHECK and CHECK references SHAFT.) Make the changes by adding the 3 x 3 NECK to the geometry (Fig. 30-60). *Save* and *Close* the **SHAFT** drawing (with the *Overlay*).

FIGURE 30-60

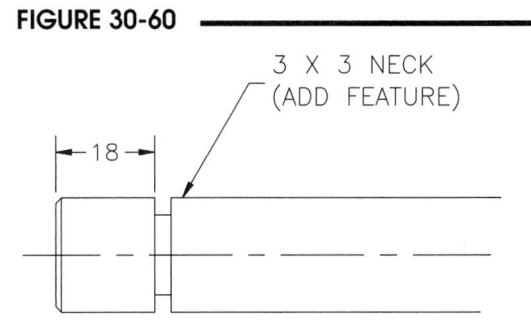

C. Now *Open* the **ASSY** drawing. The most recent *Xrefs* are automatically loaded, so the design change should appear (Fig. 30-61). Why don't the checker notes (from CHECK drawing) appear? Finally, *Save* the drawing and make another plot.

FIGURE 30-61

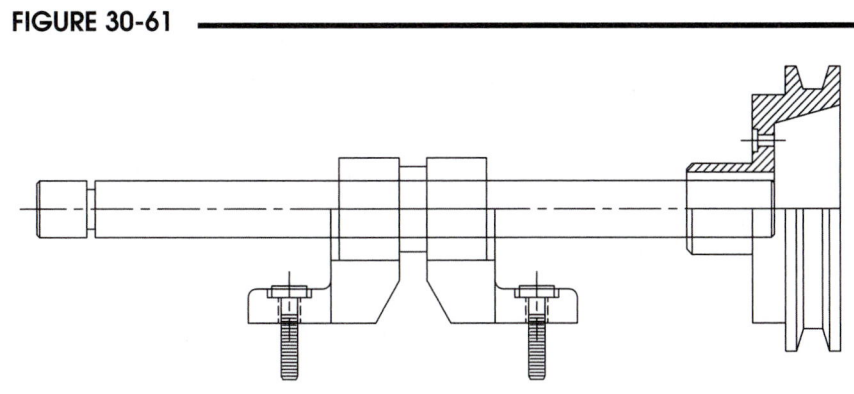

5. *Layouts, Vports*

This exercise serves as a review of the basic procedures and commands for using an *Xref* and creating layouts and viewports to display and print an existing drawing.

A. **Open** the **ASSY** drawing that you created in previous exercises (make sure all of the *Xrefed* drawings are also located in your working directory). Use *Xref* with the **Bind** option to bind all of the *Xrefed* drawings. Use *SaveAs* to save and name the drawing as **ASSY-PS**.

B. Activate a *Layout* tab. Use *Pagesetup* to select the paper size and to set the plotting options for the device you use to plot with. Make a *New Layer* named **TITLE** and set it as the *Current* layer. Draw a border and draw, *Insert*, or *Xref* a title block (in paper space).

C. Make a *New Layer* called **VIEWPORT** or **VPORT** and set it as *Current*. Next, use *Vports* to create one viewport. Make the viewport as large as possible while staying within the confines of the title block and border. Double-click inside the viewport to activate model space in the viewport, and use *Zoom XP* or the *Viewport scale* drop-down list (in the *Viewports* toolbar) to scale the model geometry times paper space (remember that the geometry is in millimeter units). Enter the scale in the title block (scale = XP factor or viewport scale).

D. Turn *Off* the **VIEWPORT** layer and *Save* the drawing (as ASSY-PS). *Plot* the drawing <u>from the layout</u> at **1=1**. Your drawing should appear similar to Figure 30-62.

FIGURE 30-62

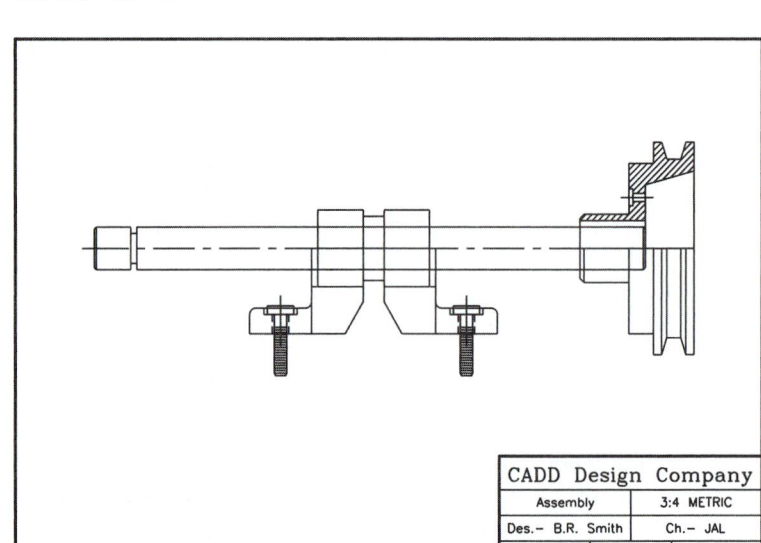

CADD Design Company		
Assembly	3:4 METRIC	
Des.– B.R. Smith	Ch.– JAL	
1/7/06	1/9/06	ASSY–PS

6. *DesignCenter, XCLIP, XLOADCTL, INDEXCTL*

A. **Open** the **DB_SAMP** drawing from the sample drawings provided with AutoCAD (if AutoCAD was installed with the "Full" installation and accepting all the other defaults, DB_SAMP.DWG is located in the C:\Program Files\AutoCAD 2007\Sample directory). Use *SaveAs* to change the name to **SAMP-OFFICE** and change the location of the drawing (using *SaveAs*) to your working directory.

B. Begin a *New* drawing. Use *Save* and assign the name **SAMP-CLIP**. Next, use *DesignCenter* to *Attach* as an *Xref* the **SAMP-OFFICE** drawing. Use an *Insertion point* of **0,0**, accept the defaults for scale factors and rotation angle, then *Zoom All*. The entire SAMP-OFFICE drawing should appear.

C. Create a clipping boundary using the *Xclip* command. Make a *Rectangular* boundary around room 6150 (lower right corner office). *Zoom* in to display only the displayed office, 6150. Now, use the *Layer* command or *Layer Control* drop-down list to *Freeze* the **SAMP-OFFICE | EMPLOYEE** layer. Your drawing should look like Figure 30-63.

FIGURE 30-63

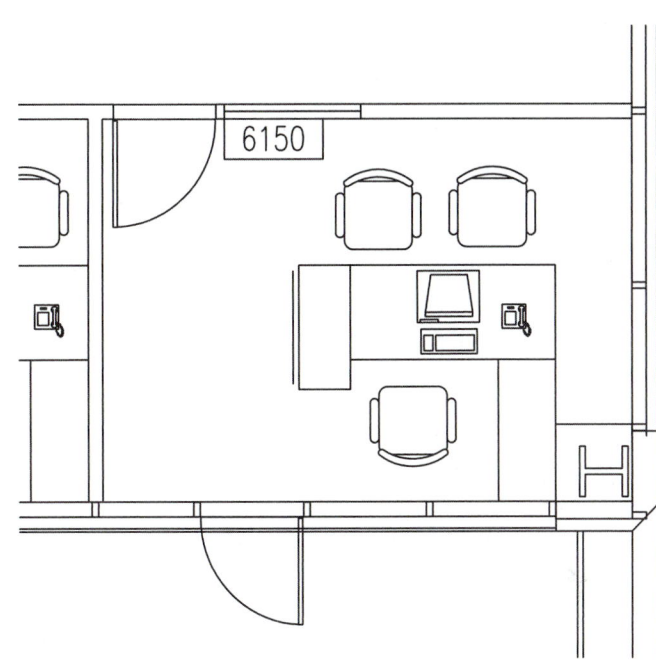

D. Use *XLOADCTL* or use the *Open and Save* tab of the *Options* dialog box to set the variable to **0** or *Disabled*. *Save* and *Close* the drawing (SAMP-CLIP).

E. Now *Open* **SAMP-CLIP** again and use your watch or count seconds to time how long it takes to load the drawing and Xref drawing. At this point, demand loading is disabled and layer and spatial indexing are not being utilized so the drawing and Xref loading time should be at its maximum for your system.

F. Next, *Close* the **SAMP-CLIP** drawing and *Open* the **SAMP-OFFICE** drawing. Use the *INDEX-CTL* variable or the *SaveAs* command with the *Options…* button to set indexing on for both layer and spatial indexes (a setting of 3). *Save* the SAMP-OFFICE drawing.

G. *Close* the **SAMP-OFFICE** drawing and *Open* the **SAMP-CLIP** drawing again. Use *XLOADCTL* or use the *Open and Save* tab of the *Options* dialog box to set the variable to **1** or *Enabled*. This action turns on demand loading, and you should now take advantage of the spatial and layer indexing. *Save* the drawing.

H. *Close*, then *Open* the **SAMP-CLIP** drawing again and time the loading process. You should experience a slight increase in the loading time speed now that demand loading is enabled. When using large (file size) Xrefs or multiple Xrefs, this feature can offer a big performance increase.

7. *Refedit*

A. *Open* the **SAMP-CLIP** drawing. Assuming you are planning to move into office 6150, your employer gives you the responsibility to arrange the furniture in your new office to your liking.

B. Use *Refedit* and select any wall at the "Select reference:" prompt. The *Reference Edit* dialog box should appear with the SAMP-OFFICE drawing listed as the *Reference name*. Select **OK** to dismiss the dialog box. At the "Select nested reference:" prompt, select two chairs. You should see all but the two chairs fade to a lighter color shade. Move the chairs to new locations in the office.

C   Use *Refset* to add other items, such as the desk, computer, and phone, to the working set as needed. Use *Move* and *Rotate* to position the other furniture items to new locations. Figure 30-64 shows one possible office arrangement. When you are finished, use *Refclose* to save your changes back to the SAMP-OFFICE drawing.

D. *Close* the **SAMP-CLIP** drawing. *Open* the **SAMP-OFFICE** drawing to verify that the changes you made to office 6150 were saved.

FIGURE 30-64

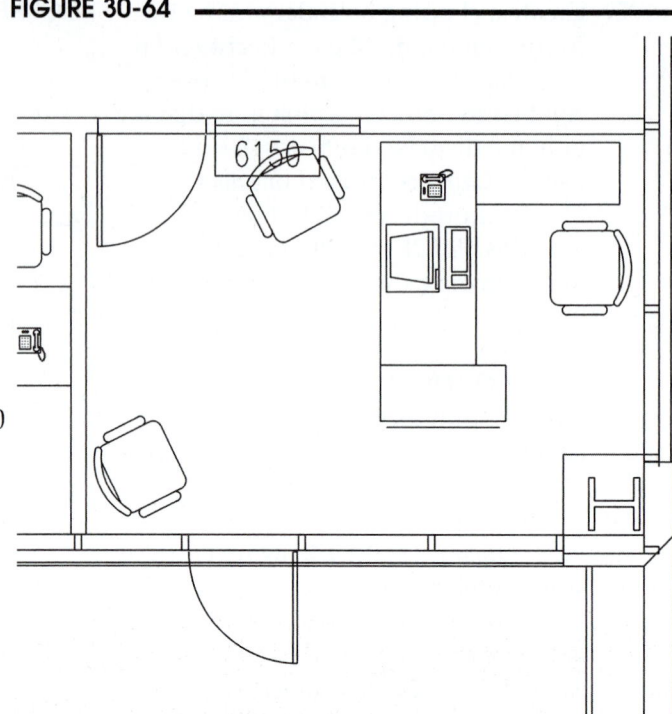

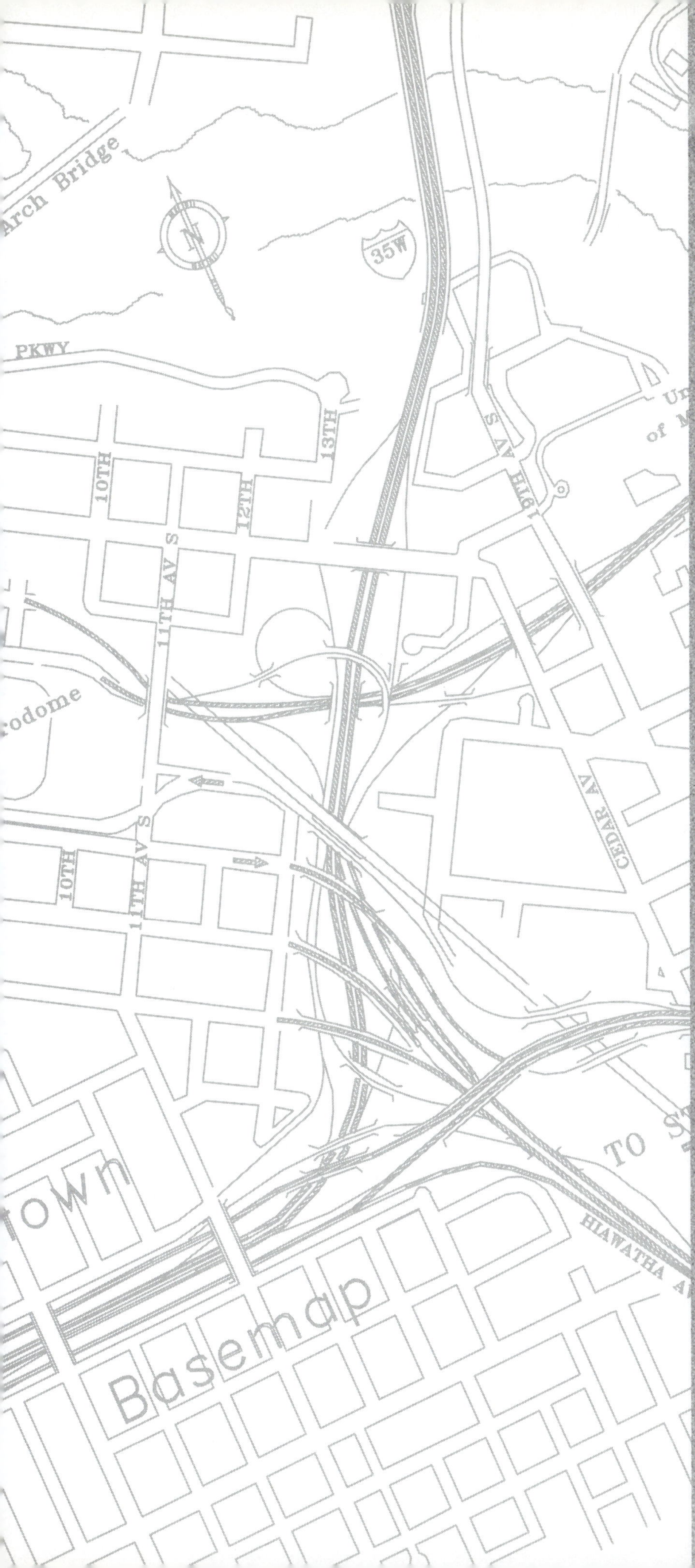

31

OBJECT LINKING AND EMBEDDING (OLE)

CHAPTER OBJECTIVES

After completing this chapter you should:

1. be able to copy and paste objects from one open AutoCAD drawing to another open AutoCAD drawing;

2. be able to use AutoCAD's Object Linking and Embedding (OLE) feature to copy information from a document (in a software program such as AutoCAD, Microsoft Word, or Microsoft Excel) and temporarily store it in the Clipboard;

3. be able to paste information from the Clipboard into a drawing;

4. be able to embed information from another software program into an AutoCAD drawing;

5. be able to link information from another software program into an AutoCAD drawing;

6. be able to manage OLE information in a drawing.

CONCEPTS

The term "graphics communication tool" is often used to describe AutoCAD's many advanced graphic capabilities. For many years people have been communicating design and conceptual information using AutoCAD's 2D and 3D graphic features. The Microsoft® Windows operating systems have now enabled AutoCAD users to extend their graphic communication abilities within, and beyond, the Drawing Editor.

 Microsoft Windows introduced the capability to "copy" and "paste" objects. As an AutoCAD user, there are two basic ways in which you can copy and paste objects. First, it is possible to copy and paste AutoCAD objects between multiple open drawings. Secondly, you can copy and paste between AutoCAD and other Windows software applications.

Cut and Paste Between AutoCAD Drawings

AutoCAD allows you to open multiple drawings at the same time in one AutoCAD session. One of the greatest advantages for doing this is the ability to copy objects from one drawing and paste them into another drawing. This feature can save hours of redrawing the same or similar geometry and increase your ability to collaborate with others on design ideas.

 When objects are copied from one AutoCAD drawing, the pasted objects in the receiving drawing are still normal AutoCAD objects. Once the objects have been pasted, there are no special functions or commands that you need to manipulate and edit them. There are also commands in AutoCAD specifically intended for you to handle the objects during the cut and paste process, such as *Copy with Base Point* and *Copy to Original Coordinates*. For example, Figure 31-1 displays the AutoCAD Drawing Editor with two drawings open. Objects from the drawing on the left are copied with the *Copyclip* command and placed into the drawing on the right with the *Pasteclip* command.

FIGURE 31-1

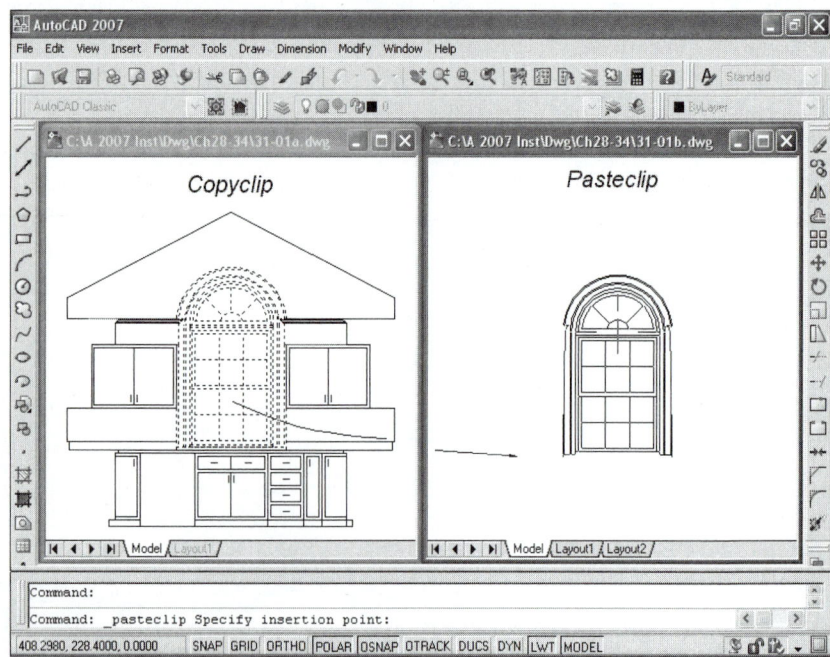

You can have multiple drawings open in one AutoCAD session, and copying and pasting between drawings is transparent—that is, even though the basic Windows Object Linking and Embedding (OLE) mechanism is being utilized, there are no special functions you need to use.

The following commands can be used to cut (erase), copy, and paste objects between AutoCAD drawings:

Menu Command	Formal Command	Result
Copy	*Copyclip*	copies selected objects to the Clipboard
Cut	*Cutclip*	copies object to the Clipboard and cuts (erases) them from the original drawing
Copy with Base Point	*Copybase*	copies objects to the Clipboard and asks for a base point
Copy Link	*Copylink*	copies all objects to the Clipboard
Paste	*Pasteclip*	pastes Clipboard objects into the drawing
Paste as Block	*Pasteblock*	pastes Clipboard objects into the drawing as a *Block*
Paste to Original Coordinates	*Pasteorig*	pastes Clipboard objects into the new drawing at the same location as the original drawing

Cut and Paste Between AutoCAD and Other Windows Applications (OLE)

You can also easily communicate 2D and 3D AutoCAD designs with other Windows applications and benefit from the Windows Object Linking and Embedding (OLE) feature. Unlike using only one software application (AutoCAD) to copy and paste, this concept combines objects created in two different software applications but displayed in only one. Microsoft developed the method called object linking and embedding, or OLE, to handle dissimilar objects. The standard OLE features of linking and embedding operate as smoothly and easily from within AutoCAD as they do in other Windows software such as Microsoft® Word, Excel, and others.

AutoCAD's support of the OLE features allows you to include data and information created in an application, such as Word, within your AutoCAD drawing. You can also take AutoCAD drawing information and use it within Word documents, Excel spreadsheets, and so on. This feature works best if all the applications using or sharing the common information support the OLE feature. But it is not absolutely necessary for a software application to support the OLE feature for that application to display AutoCAD drawing information by way of the copy and paste functions available within Windows.

AutoCAD's OLE features utilize the standard Windows *Copy, Cut,* and *Paste* commands common to all Windows programs. OLE, however, extends these Windows commands so information copied from one application (such as a Word document) and pasted into another application (like an AutoCAD drawing) can be *linked* back to the original source document, also known as the server. The source document is a separate document; it is only displayed within the client document (the AutoCAD drawing in this example).

OLE objects are treated by AutoCAD as any other object—that is, you can use the normal selection methods (pickbox, window, crossing window, etc.) to *Move, Erase, Copy, Scale,* or *Array* the OLE object. Select the OLE object by selecting the "border" around the object. The OLE object borders appear in the Drawing Editor, but do not appear in a plot or print.

Object Linking and Embedding enables you to create a bill of materials within, for example, Word, which is designed specifically for word processing, then paste it into an AutoCAD drawing. First, create and format the bill of materials within Word (Fig. 31-2). Next, *Copy* the formatted text onto the Clipboard and then *Paste* it into the AutoCAD drawing. In AutoCAD (the client), it will have the same appearance that it has in the Word (the source) document.

FIGURE 31-2

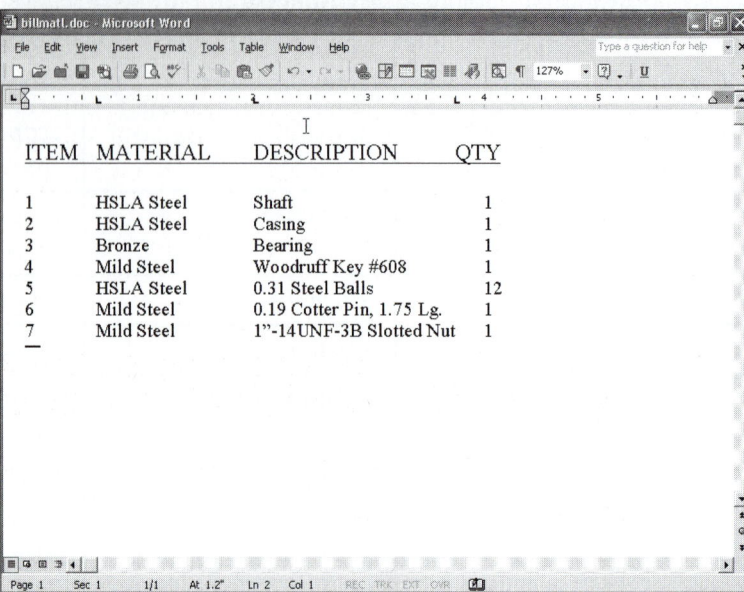

When a non-AutoCAD OLE object is pasted into AutoCAD (Fig. 31-03), you have the choice to paste it as an embedded or a linked object. A linked object is simply a picture of the original source document, whereas an embedded object is wholly contained in AutoCAD. Once the OLE object is pasted into AutoCAD as either an embedded object or linked object, you can double-click on the object in the drawing to cause the source application to open (in this case, Word) with the document loaded and ready for editing. Changing and saving the document causes the object in AutoCAD to change. If the object is embedded, only the AutoCAD object is changed—not the original source document. If the object is linked, the original source document is changed and is reflected in AutoCAD.

FIGURE 31-3

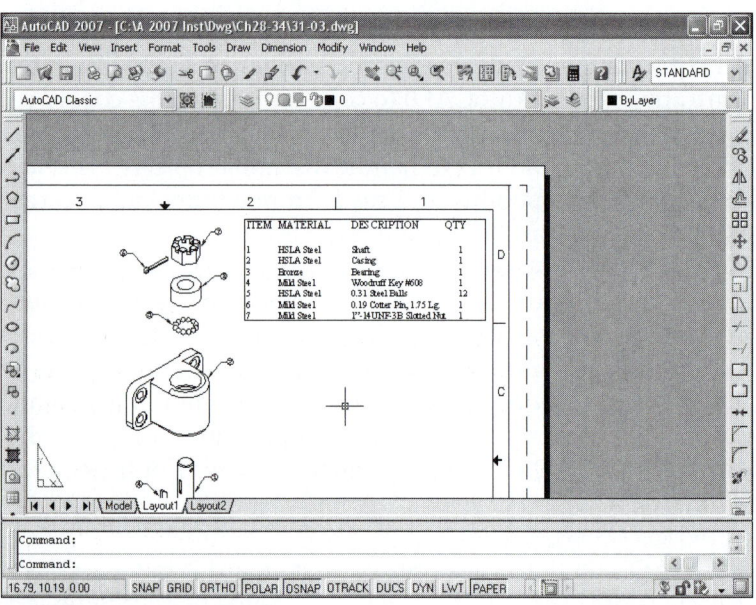

There are two strategies for pasting information from other applications into your AutoCAD drawing:

1. <u>Embed the object when pasting</u>
   Embedding is similar to using the AutoCAD *Insert* command. When embedded, the object becomes a permanent part of the drawing. If the source application that created the object supports OLE, you can double-click on the embedded object and the source application is started so you can edit the contents of the object for the insertion of the (embedded) objects in AutoCAD only. The original source document remains unchanged.

2. Link the object when pasting
    Linking is similar to the AutoCAD *Xref* command. When the object is linked, it is not part of the drawing; it is only displayed in the drawing. When you double-click on the linked object, the source application is started with the original document loaded and ready for editing. Any changes made to the original source document are reflected in the drawing automatically or upon a manual update, depending on your update setting in the OLE *Links* dialog box. Similar to an *Xref*, when the original document is edited, the updated version appears in the client application.

These AutoCAD commands are used to copy, cut, link, or embed objects between AutoCAD and other applications:

Menu Command	Formal Command	OLE Result
Copy	*Copyclip*	Embedded or linked
Cut	*Cutclip*	Embedded
Paste	*Pasteclip*	Embedded
Paste Special	*Pastespec*	Embedded or linked
Paste as Hyperlink	*Pasteashyperlink*	Hyperlink

The main difference between linked and embedded is how the information is stored—with the client (application that "displays" the document) or server (source application that created the document).

The Windows Clipboard

Windows manages the OLE features. The Windows Clipboard is a key mechanism used throughout the process of cutting and pasting within AutoCAD and linking and embedding other information in an AutoCAD drawing. The Clipboard is a Windows feature which stores information temporarily in memory so that it can be utilized (or pasted) into another application. This enables you to create the information only once and then share it with other documents where it is needed.

For example, assume the bill of materials mentioned earlier was created, spell checked, and formatted in Word, then selected and copied to the common memory area called the Clipboard. Then the information was pasted (similar to the AutoCAD *Insert* command) into an AutoCAD drawing with the same content and appearance it had in the original document. The same information could then be pasted (inserted) into a spreadsheet with the identical content and appearance it has in the AutoCAD drawing and in the original Word document that created it.

The Clipboard retains the most recent information that was cut or copied from an application currently running in the Windows environment. When another piece of information (e.g., graphical objects or text) is cut or copied, it replaces the last group of information stored within the Clipboard's memory area.

The options encountered when performing a paste vary depending upon the AutoCAD paste command being used (*Pasteclip, Pastespec*) and the type of object information being pasted from the Clipboard. The following table lists some common object types and options that are available when pasting objects into an AutoCAD drawing.

		Paste	Objects	As:		
Object Type:	AutoCAD Vectors	AutoCAD Block	AutoCAD Text	Bitmap	Picture	Image
AutoCAD Vector Objects (*Lines, Arcs, Circles, Blocks, Plines,* etc.)	✔	✔				
Other Vector Objects (primarily Windows metafile)		✔				
AutoCAD Text Objects (*Mtext, Text*)		✔	✔		✔	✔
Formatted Text Objects (such as Word text or Excel spreadsheet)		✔	✔	✔	✔	✔

To summarize, first create the document (to be copied) in its source application (Word, for example). While the document is open, use *Copy* or similar OLE command in that application to copy the document contents (called an OLE object) to the Clipboard. Next, open the client application (AutoCAD, for example) and use *Paste, Paste Special,* or a similar OLE command to paste the OLE object. The OLE object appears in the client application just as it did in its original format (in the source application). Later, you can double-click on the OLE object to open the original (source) application and modify the object. If the object is <u>linked</u>, the original source document is updated; if the object is <u>embedded</u>, only the OLE object is changed, but the source document (file) is not changed.

OLE RELATED COMMANDS

AutoCAD OLE related commands covered in this chapter are:

Cutclip, Copyclip, Copylink, Copybase, Pasteclip, Pasteblock, Pasteorig, Pastespec, Insertobj, Pasteashyperlink, Olelinks, Oleopen, and *Olescale*

Most OLE related commands are available from the *Edit* pull-down menu (Fig. 31-4). The *Insertobj* command is accessible from the *Insert* pull-down menu.

FIGURE 31-4

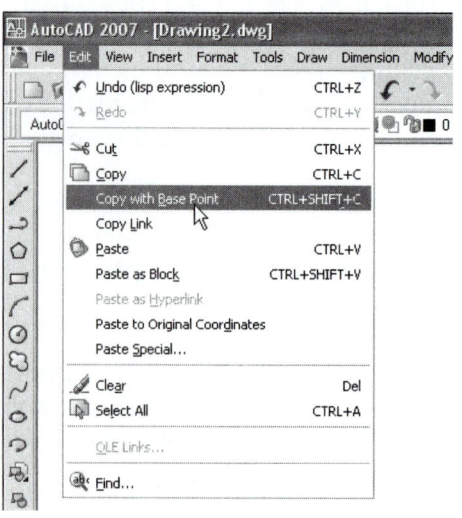

In addition, right-clicking in the drawing area when no commands are in use produces the default shortcut menu displaying the OLE commands (Fig. 31-5).

FIGURE 31-5

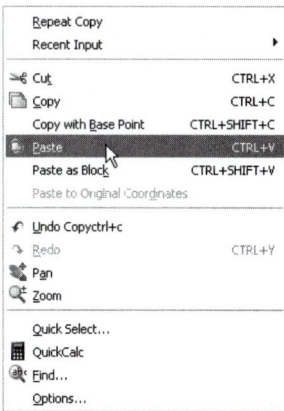

Copyclip

Pull-down Menu	Command (Type)	Alias (Type)	Short-cut	Screen (side) Menu	Tablet Menu
Edit *Copy*	*Copyclip*	...	*Ctrl+C* or (Default Menu) *Copy*	*EDIT* *Copyclip*	*T,14*

You can use the *Copyclip* command to select the objects from the current drawing to be placed on the Clipboard. *Copyclip* prompts you to "Select objects:". You can use any of the normal AutoCAD selection options (like *Window, Crossing, Fence, Group,* etc.). You can also pre-select the objects (known as Noun-Verb selection) you want placed on the Clipboard, and then issue the *Copyclip* command to avoid the "Select Objects:" prompt:

```
Command: copyclip
Select objects: PICK (Select objects to be copied from the drawing and placed on the Clipboard)
Select objects: Enter
Command:
```

AutoCAD objects that are selected with *Copyclip* remain in the drawing. Copies of the objects are temporarily stored on the Clipboard.

The *Copyclip* command can be used when you plan to paste the objects back into AutoCAD (using *Pasteclip, Pastespec,* or *Pasteblock*) or when you intend to link or embed them within other applications. When the objects are pasted into another application, you may have little control over where they are placed, depending on the host application.

When you paste the objects into the same or other AutoCAD drawing, you can drag the objects into the desired location. The *Copyclip* command does not prompt for a base point; therefore, pasting the objects involves using a base point at the lower-left corner of the set of objects (Fig. 31-6).

FIGURE 31-6

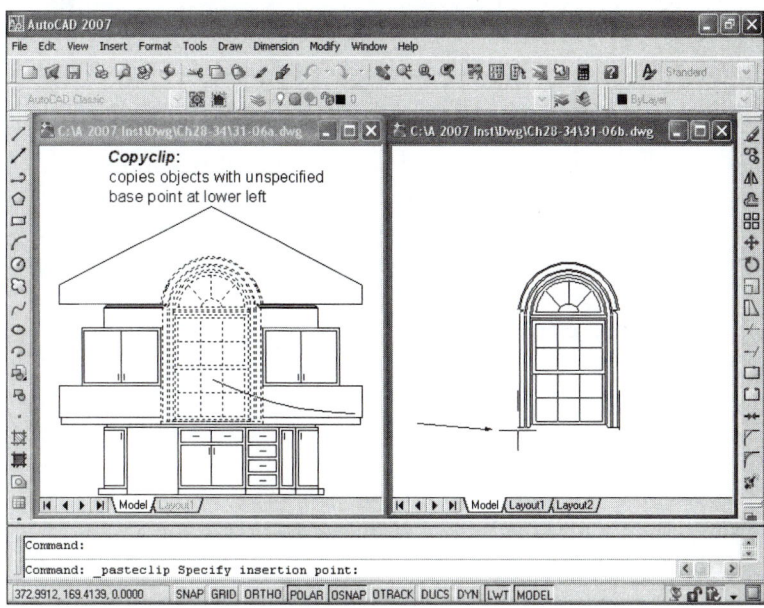

Copybase

Pull-down Menu	Command (Type)	Alias (Type)	Short-cut	Screen (side) Menu	Tablet Menu
Edit Copy with Base Point	Copybase	...	(Default Menu) Copy with Base Point	EDIT CopyBase	...

Copybase, similar to *Copyclip*, copies objects onto the Clipboard; however, *Copybase* allows you to specify a base point for the object(s). When you later paste the set of *Copybase* objects into a drawing using *Pasteclip, Pastespec*, or *PasteBlock*, you can locate the point of insertion using the previously specified base point (Fig. 31-7).

In contrast, *Copyclip* does not allow you to specify a base point but automatically determines a base point, usually in the lower left corner of the object set (see Fig. 31-6). Therefore, *Copybase* is generally preferred to *Copyclip* for typical operations within AutoCAD because you have more direct control of the insertion point.

FIGURE 31-7

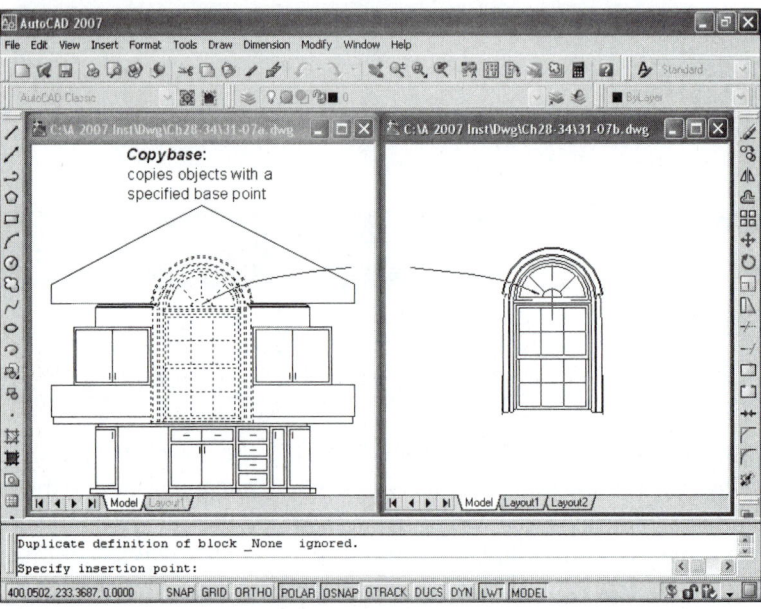

If you intend to use *Pasteorig* to paste the object set, it does not matter if you use *Copyclip* or *Copybase* since the object set is pasted to the original coordinates regardless of the base point. (See "*Pasteorig*," later in this chapter.)

Copylink

Pull-down Menu	Command (Type)	Alias (Type)	Short-cut	Screen (side) Menu	Tablet Menu
Edit Copy Link	Copylink	...	...	EDIT Copylink	...

The *Copylink* command is used to <u>copy the entire contents of the current drawing</u>, including objects on *Frozen* or *Locked* layers. You are not prompted to select individual objects. Whatever is contained in the drawing is copied to the Clipboard. This information can then be pasted into AutoCAD as normal AutoCAD objects or into another application as linked or embedded objects.

    Command: **copylink**
    Command:

No change is made to the objects in the current drawing. Copies of the objects are temporarily stored in the Clipboard.

Cutclip

Pull-down Menu	Command (Type)	Alias (Type)	Short-cut	Screen (side) Menu	Tablet Menu
Edit Cut	Cutclip	...	Ctrl+X	EDIT Cutclip	T,13

The *Cutclip* command places selected objects onto the Clipboard. However, unlike *Copyclip,* it <u>deletes</u> (*Erases*) the selected objects from the drawing. If desired, you can use the AutoCAD *Oops* command to "unerase" the objects deleted from the drawing by *Cutclip.* If you use *Oops,* you must use it before you delete any other objects from the drawing since *Oops* brings back only the last set of *Erased* objects.

```
Command: cutclip
Select objects: PICK (Select objects to be removed from the drawing and placed on the Clipboard)
Select objects: Enter
Command:
```

This command is typically used only when you do not need the selected items in the current drawing any longer but wish to use them in another drawing or another application. In this case, *Cutclip* allows you to clear (*Erase*) them out of the current drawing and then paste them into another application. For example, you may want to open a second AutoCAD drawing, then cut objects from the first drawing and paste them into the second drawing. This action helps to reduce the current drawing file size and also eliminates the need to draw the items again for another drawing. Since the objects temporarily remain on the Clipboard, the deleted items can be pasted into another drawing or document where they are needed.

Pasteclip

Pull-down Menu	Command (Type)	Alias (Type)	Short-cut	Screen (side) Menu	Tablet Menu
Edit Paste	Pasteclip	...	Ctrl+V	EDIT Pasteclp	U,13

You can copy information <u>from AutoCAD or another application</u> and paste it into your drawing with AutoCAD's *Pasteclip* command. If you copy AutoCAD objects and paste them into the same or another AutoCAD drawing, the objects are inserted as the original AutoCAD objects having the same properties. If you *Pasteclip* text from a word processing application or data from a spreadsheet application, the pasted object is embedded within the AutoCAD drawing.

Pasting AutoCAD Objects

First, you must use *Copyclip, Copybase, Copylink,* or *Cutclip* to copy AutoCAD objects to the Clipboard. Next, use *Pasteclip* to insert the objects into the same or another open AutoCAD drawing. AutoCAD prompts for the base point.

```
Command: pasteclip
Specify insertion point: PICK
Command:
```

The objects are inserted into the new location or new drawing. AutoCAD objects are pasted as the original objects and retain their original properties such as layer, color, linetype, text style, dimension style, and so on. Therefore, layers on which the pasted objects reside and other named objects associated with the pasted objects (linetypes, dimension styles, etc.) are added to the new drawing if they did not previously exist in the host drawing.

Pasting Non-AutoCAD Objects

If you have used Copy or similar OLE command from a word processing or spreadsheet program, for example, those objects are copied to the Clipboard. In AutoCAD, use *Pasteclip* to bring the objects into AutoCAD as <u>embedded</u> objects. No insertion base point is requested.

Command: **pasteclip**

When you copy objects that contain text from another application into an AutoCAD drawing using *Pasteclip*, the point size of the text in the source application is automatically converted to an approximate equivalent size in AutoCAD units. AutoCAD uses the appropriate units depending on whether the OLE object is inserted into a layout or into model space. If you *Paste* the object into a layout, the object is automatically scaled based on the layout's sheet size and units. If you *Paste* the OLE object into model space, the scale is based on the current page setup (if one has been set) and the *MSOLESCALE* system variable (see "*MSOLESCALE*").

FIGURE 31-8

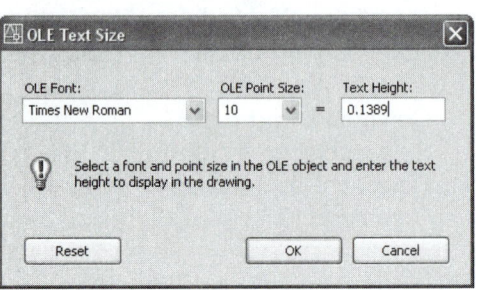

When *Pasteclip* is used with non-AutoCAD objects containing text, the *OLE Text Size* dialog box appears (Fig. 31-8). This dialog box gives you additional control of the scale of the OLE object by manipulating the *Text Height* for a particular *OLE Font* or the *OLE Point Size*. The *OLE Text Size* dialog box can be produced at a later time by highlighting an existing non-AutoCAD OLE object and using the *Olescale* command (see "*Olescale*").

The *OLE Scale* dialog box allows you to set the scale for the OLE object. Although it appears as if you can change the font and the point size, that is not the case—the *OLE Font* and *OLE Point Size* drop-down lists display only the fonts and sizes already contained in the OLE object. This tool is used to set the scale of the OLE object by specifying a *Text Height* value. Do this by selecting any combination of *OLE Font* and *OLE Point Size*, then entering the desired *Text Height* value. The *Text Height* value determines the scale for the entire OLE object.

In other words, assume the *Text Height* displayed a default value of 0.1389 for the 10 point Times New Roman font. Changing that value to 0.2778 would double the scale of the entire OLE object (0.2778 = 0.1389 x 2). Additionally, the *Text Height* for all other point sizes for all other fonts would automatically double. If you want to change the scale of the OLE object back to the original size, select *Reset*.

Remember that OLE objects in AutoCAD are treated as any other AutoCAD objects—that is, you can *Move, Erase, Copy,* or *Array* them. For example, instead of using the *OLE Scale* dialog box, you could use the *Scale* command to scale OLE objects.

Once the non-AutoCAD object is inserted, it becomes an embedded object. If the embedded object's source application supports OLE, the object is displayed in the original application's format. The embedded object can be edited using the original application that created it by double-clicking on the object.

Embedded objects are <u>copies</u> of the original information. They have no connection, or link, to the original source document in which they were created. Therefore, editing the original objects using the source application does not update the embedded (copied) object. Alternately, if you use the *Pastespec* command instead of *Pasteclip*, you have the choice of embedding or linking the objects.

Keep in mind that using *Pasteclip* to paste an AutoCAD object into another AutoCAD drawing <u>does not create an embedded object</u>. Pasted AutoCAD objects are inserted as exact copies of the original object set, complete with all associated properties and named objects, unless *Pasteblock* is used.

Pasteorig

Pull-down Menu	Command (Type)	Alias (Type)	Short-cut	Screen (side) Menu	Tablet Menu

Edit
Paste to
  Original
    Coordinates
	Pasteorig	...	(Default Menu)
			Paste to
			  Original
			    Coordinates
				EDIT
				PasteOri
					...

The *Pasteorig* command is designed specifically to be used for <u>copying AutoCAD objects into other</u> <u>AutoCAD drawings</u>. *Pasteorig* inserts the object(s) into the new drawing at the same coordinate position as the original drawing. Therefore, you are not prompted for an insertion point.

Command: **pasteorig**
Command:

To illustrate, open two AutoCAD drawings and use *Cutclip*, *Copyclip*, *Copylink*, or *Copybase* to copy objects from one drawing onto the Clipboard. Next, make the other AutoCAD drawing current by clicking in it and then use *Pasteorig*. The object(s) are pasted to the same coordinate location as in the original drawing (Fig. 31-9). This feature makes it simple to create or edit a second drawing based on geometry from a previously created drawing.

Note that there is no added control by using *Copybase* to specify a base point when you capture the original objects since *Pasteorig* automatically locates the new objects at their original coordinates regardless of a speci-

FIGURE 31-9

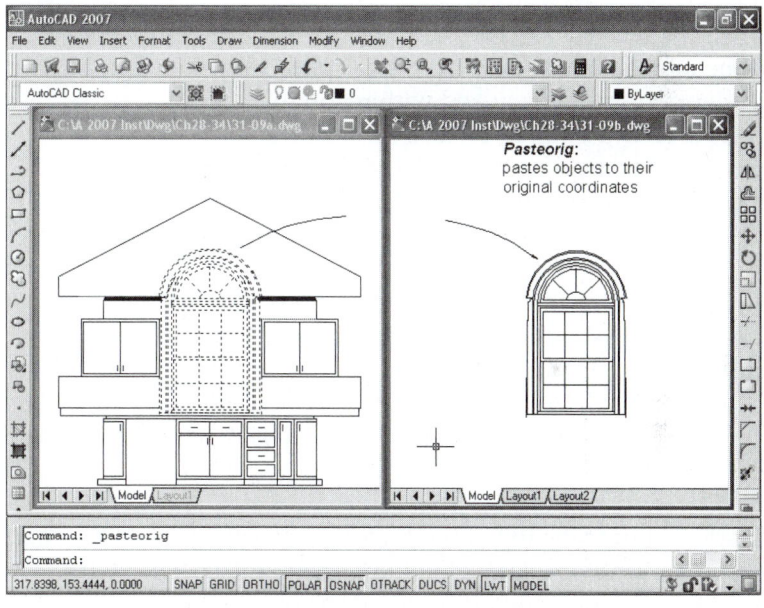

fied base point. *Pasteorig* does not operate for copying objects back to the original drawing. *Pasteorig* does not function for copying non-AutoCAD objects from other applications into AutoCAD.

Pasteblock

Pull-down Menu	Command (Type)	Alias (Type)	Short-cut	Screen (side) Menu	Tablet Menu
Edit *Paste as Block*	*Pasteblock*	...	(Default Menu) *Paste as Block*	*Edit* *PasteBlk*	...

Pasteblock is intended to be used for <u>copying AutoCAD objects into the same or other AutoCAD draw-</u> <u>ings</u>. *Pasteblock* operates essentially the same as *Pasteclip* except that *Pasteblock* inserts AutoCAD objects into AutoCAD as *Blocks*. This can be useful if you want to manipulate all objects from one paste operation as one object—a *Block*.

Pastespec

Pull-down Menu	Command (Type)	Alias (Type)	Short-cut	Screen (side) Menu	Tablet Menu
Edit Paste special...	Pastespec	...	...	EDIT Pastespe	...

Pastespec is intended to be used primarily to paste objects <u>from other applications</u> into AutoCAD.

The *Pastespec* command gives you the choice of embedding or linking objects that are pasted into the drawing. *Pastespec* produces the *Paste Special* dialog box (Fig. 31-10). Your selection of *Paste* or *Paste Link* determines if the objects from the Clipboard are <u>embedded</u> or <u>linked</u>, respectively.

If the information copied from an application that supports OLE is <u>linked</u> within your AutoCAD drawing, a "picture" of the original document is displayed that links (or references) the original source document that was used to create it.

FIGURE 31-10

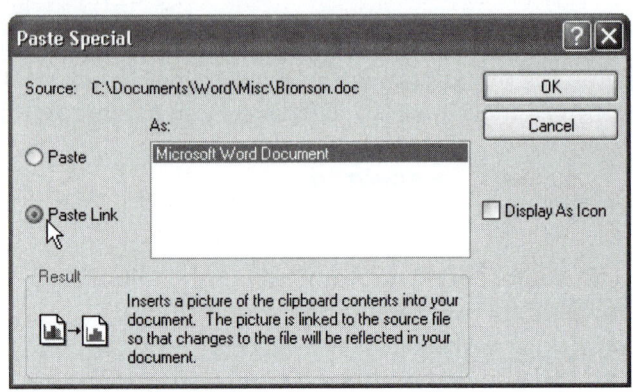

To use *Pastespec*, first you must use *Copy* or similar OLE command from another application (Word, Excel, AutoCAD, etc.) to bring objects onto the Clipboard. Next, use the *Pastespec* command in AutoCAD:

    Command: **pastespec** (The *Paste Special* dialog box appears)
    Command:

Displayed near the top of the *Paste Special* dialog box is the *Source:* application which created the object that is currently contained on the Clipboard (see Fig. 31-10). The two radio buttons on the left allow you to embed (*Paste*) or link (*Paste Link*) the objects.

The list of (Paste) *As:* options varies depending on the type of information contained on the Clipboard and the current *Paste/Paste Link* radio button you have selected (Fig. 31-11). If the current information on the Clipboard is from an AutoCAD drawing, the *Paste Link* option is <u>not</u> available (grayed out). All of the options in this dialog box are described in detail next.

FIGURE 31-11

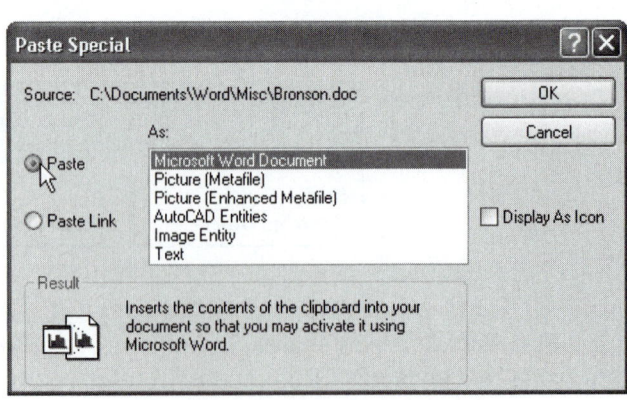

Paste

When the *Paste* radio button is selected, the object becomes <u>embedded</u> within the drawing. You can select the type of object to be inserted in the drawing by making a selection from the (Paste) *As:* list (see Fig. 31-11). The list of object types varies depending on the source of the objects on the Clipboard.

Paste Link

Selecting the *Paste Link* button causes AutoCAD to display a "picture" of the object in the drawing. The information displayed is not actually copied into the drawing; it is <u>linked</u> back to the source document from which it was originally copied. This option is available only when the information contained on the Clipboard was created with an application that supports OLE. When *Paste Link* is selected, only one choice appears in the (Paste) *As:* list—the source application (see Fig. 31-10).

Result

The *Result* box at the bottom of the dialog box describes the current object on the Clipboard. It also indicates how the pasted objects are managed in the drawing based on your selections within the dialog box (*Paste/Paste Link* and *Paste As:* list). This is a very helpful feature that dynamically explains your options as they are selected.

Display As Icon

You can display the source application's icon instead of displaying the actual information pasted from the Clipboard by choosing *Display As Icon* (Fig. 31-12). Since only the application's icon is visible in the drawing, you must double-click on it to view the linked or embedded information. This icon is a <u>shortcut</u> that points to the location of the information pasted from the Clipboard. This can be useful in a situation where the linked information is relevant to the project depicted in your drawing, but it is not necessary to display that information directly in the drawing.

FIGURE 31-12

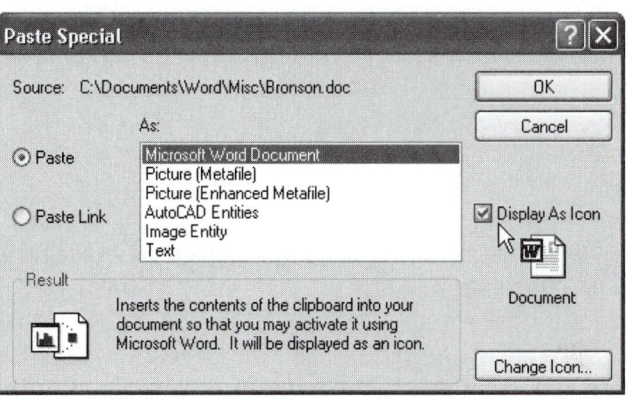

For example, you may be displaying a *Block* of a pump in the drawing. Specific text information such as the pump's specifications, manufacturer's model number, cost, and ordering information could be included within a Word document which is pasted into your drawing and displayed as an icon. The information is important for reference, but it is not necessary to display the information directly in your drawing.

If the information represented by the icon is <u>embedded</u>, double-clicking on the icon activates the application used to create it and displays the contents of the information that was pasted into the drawing in its original format. Changes made to the information as a result of this action are changed for the embedded information only and <u>are not reflected</u> in the source (original) document.

If the *Icon* information is linked, double-clicking on the icon activates the application used to create it and in turn displays <u>the original file</u> that contains the information. The information referenced by the pasted icon becomes highlighted in the source document. Once the source document is saved and closed, the updated version is linked to the icon displayed within your drawing. In this case, the source document itself has actually been changed.

Change Icon

Whenever you have selected the *Display As Icon* checkbox, a preview of the application's default icon is displayed directly beneath the checkbox area. A new button also appears labeled *Change Icon.* This allows you to change the icon if you so desire.

Insertobj

Pull-down Menu	Command (Type)	Alias (Type)	Short-cut	Screen (side) Menu	Tablet Menu
Insert OLE Object...	*Insertobj*	*IO*	...	*INSERT Insertob*	*T,1*

Insertobj allows you to paste specific information <u>from another application</u> into your drawing as <u>either a linked or embedded</u> object. It functions in much the same way as the *Pastespec* command. The key difference between *Insertobj* and *Pastespec* is that *Insertobj* allows you to paste a <u>file</u> without having the application or document currently open and a particular portion of it selected.

Because you are not copying objects from the Clipboard, you do not have to use *Copy* or similar OLE command before using *Insertobj*. *Insertobj* produces the *Insert Object* dialog box (Fig. 31-13).

FIGURE 31-13

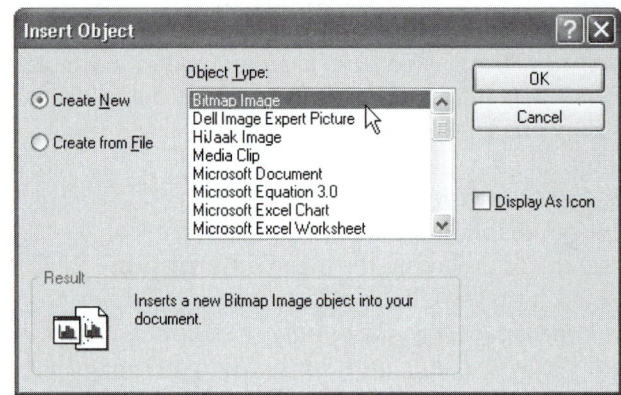

Command: **insertobj** (produces the *Insert Object* dialog box)
Command:

Create New
The options in this command afford you some flexibility. If the file doesn't exist, you can create a new file by selecting *Create New,* then the application you want to use to create it. The applications available to you (registered on your system) are listed in the *Object Type* listing. The selected application is opened when you select *OK* so that you can create the new document. When you are finished creating the new document and you close the application, the new document (object) is <u>embedded</u> into your drawing.

Create from File
When the *Create from File* radio button is selected (Fig. 31-14), you can select the *Browse* button to look for files available in your available drives and folders. After selecting a specific file, you can paste it into your drawing as an <u>embedded</u> or <u>linked</u> object. If the object is to be linked, the display of the linked object may be limited to an icon based on the file type and the OLE capabilities of the source application.

FIGURE 31-14

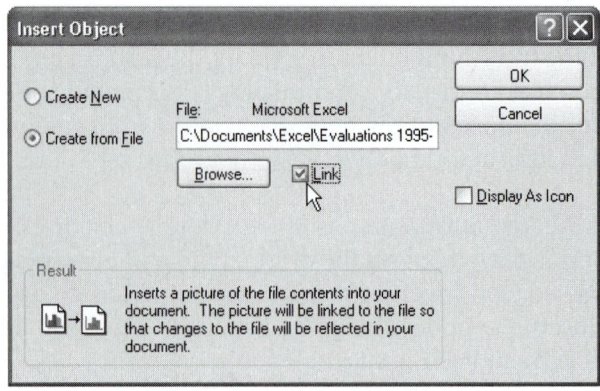

Pasteashyperlink

Pull-down Menu	Command (Type)	Alias (Type)	Short-cut	Screen (side) Menu	Tablet Menu
Edit **Paste as Hyperlink**	*Pasteashyperlink*	...	...	...	...

The *Hyperlink* command (not *Pasteashyperlink*) produces the *Insert Hyperlink* dialog box. That dialog box is used to create a link between an AutoCAD object and a Web page or another document. A hyperlink has a similar function to an embedded OLE object, in that opening the hyperlink opens the related document (usually your Web browser) and displays the Web page or other document. A hyperlink, however, is attached to an AutoCAD <u>object</u>, so passing the cursor over the object displays the hyperlink symbol and right-clicking on the object allows you to open the hyperlink. (See Chapter 23, Internet Tools and Collaboration, for more information on Hyperlinks.)

The *Pasteashyperlink* command or *Paste as Hyperlink* option from the *Edit* pull-down menu is another method of creating a hyperlink as an alternative to the *Insert Hyperlink* dialog box; however, you can use <u>*Pasteashyperlink* to link only to documents, not to Web addresses</u>. The resulting hyperlink is attached to an AutoCAD object. The main difference is that with the *Pasteashyperlink* command, you first open the document you want to link to, use *Copy*, then open AutoCAD and use *Paste as Hyperlink*. AutoCAD simply asks you to select an object to attach the link to.

```
Command: pasteashyperlink
Select objects: PICK
Select objects: Enter
Command:
```

The resulting hyperlink (to the remote document) is an <u>embedded</u> object in the drawing attached to a specific AutoCAD object. The procedure is as follows:

1.   open the source program, then open the desired document you want to link to;
2.   highlight some text, such as the title or first line of the document;
3.   select *Copy* from the program's *Edit* pull-down menu while leaving the document open;
4.   toggle to AutoCAD and select *Paste as Hyperlink* from the *Edit* menu;
5.   select the AutoCAD object you want to attach the document to.

Two conditions must exist for *Pasteashyperlink* to operate: 1) the document that you *Copy* from must be a previously saved document (not a Web page), and 2) the document that you copy from must be open at the time that you *Paste*.

The text that you select in step 2 appears at the cursor when you pass the cursor over the AutoCAD object. Opening the link produces the source program and displays the entire document that contains the text that you selected to *Copy*. Therefore, the selected text has significance only in that it appears in AutoCAD at the cursor, so you should select text that indicates the name or purpose of the linked document.

Pasteashyperlink is essentially another way to hyperlink a document to an object, but with this method you select the link from within the linked document rather than browsing for the document (file name) when establishing a *Hyperlink* from AutoCAD in the *Insert Hyperlink* dialog box.

MANAGING OLE OBJECTS

Olelinks

Pull-down Menu	Command (Type)	Alias (Type)	Short-cut	Screen (side) Menu	Tablet Menu
Edit OLE Links...	Olelinks	...	...	EDIT OLElinks	...

When you have *Linked* an object within an AutoCAD drawing, you can control if it is automatically or manually updated whenever changes are made to the original (source) document. The *Olelinks* command activates the *Links* dialog box with options for the maintenance of object links. If no objects are currently *Linked* within the drawing, the *OLE Links...* option in the *Edit* menu is grayed out and typing *Olelinks* invokes no response from AutoCAD.

> Command: **olelinks** (activates the *Links* dialog box if links exist in the current drawing)
> Command:

The following are features included in the *Links* dialog box (Fig. 31-15).

Links, Type, and Update Columns
The list box in the main section of the dialog provides a listing of all linked objects in the drawing. The *Links* column displays the source information (file name and path). The *Type* is the source application type. The *Update* column lists the current status of the update option.

FIGURE 31-15

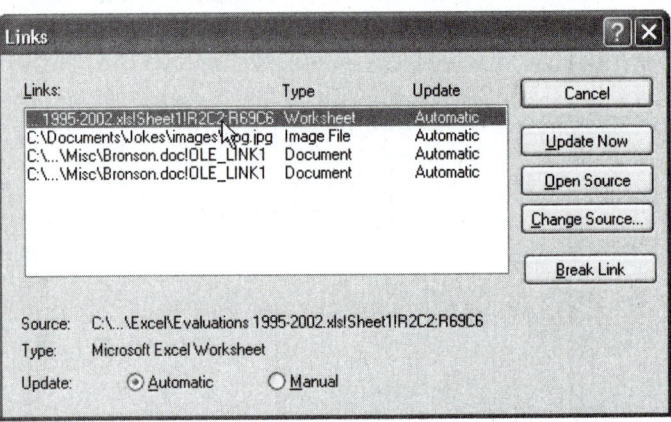

Source
Found in the lower-left corner of the dialog, this description specifies the drive, path, application, and additional information such as object type or selection area, depending on the source application.

Type
Also listed in the lower-left corner of the dialog is the *Type* of link for the highlighted item. This is the application-specific name (source application) of the currently selected linked object. The application-specific name is sometimes followed by an OLE link number assigned to the object by AutoCAD.

Update
Two *Update* radio buttons are located at the bottom of the dialog. The *Automatic* option causes updating of the linked object whenever its source document is modified while the drawing is open. Setting the *Update* function to *Manual* prevents the link from updating when revisions are made to the source. When *Manual* is selected, use *Update Now* to update the link.

Update Now
This button forces the links selected from the list to be updated. When you close the dialog box, listed data reflects the newly updated status of the source object.

Open Source
This option opens the currently selected link with the source application so it can be modified. The linked information is highlighted when the source application is opened.

Break Link

Use this option to sever the link of the object to its source file and automatically convert it to a "Static OLE object" or Windows metafile object. It can no longer be updated if the source is changed. The object can no longer be edited with the source application or with AutoCAD commands.

For example, if a spreadsheet displaying a bill of materials for the drawing is created in Excel and then linked inside the drawing, it can be modified at any time with Excel simply by double-clicking on the spreadsheet picture displayed on the drawing. If, however, the link is broken, the spreadsheet picture is converted to a metafile picture object and can no longer be edited using Excel. Static OLE objects can be edited or removed with normal AutoCAD commands.

Olescale

Pull-down Menu	Command (Type)	Alias (Type)	Short-cut	Screen (side) Menu	Tablet Menu
...	Olescale	...	...	...	...

The *Olescale* command produces the *OLE Scale* dialog box (see previous Fig. 31-8), which allows you to alter the scale of OLE objects in AutoCAD. This dialog box is the same dialog box that appears when you first paste a non-AutoCAD OLE object into AutoCAD using *Pasteclip* (see "*Pasteclip*").

Oleopen

Pull-down Menu	Command (Type)	Alias (Type)	Short-cut	Screen (side) Menu	Tablet Menu
...	Oleopen	...	...	...	...

Oleopen opens the OLE object's source application so you can edit the document. You must <u>first highlight the OLE object</u>, then use this command. You can also double-click on the OLE object to produce this command. The result of editing the OLE object differs based on whether the object is linked or embedded. If the OLE object is linked, the original source document is edited. If the OLE object is embedded, only the object in AutoCAD is changed, not the source document.

OLE Right-Click Shortcut Menu

You can highlight an OLE object, then right-click to produce the shortcut menu shown in Figure 31-16. The *OLE* cascading menu has several options.

Open

Use this option to open the OLE object's source application to edit the OLE object. This option invokes the *Oleopen* command.

FIGURE 31-16

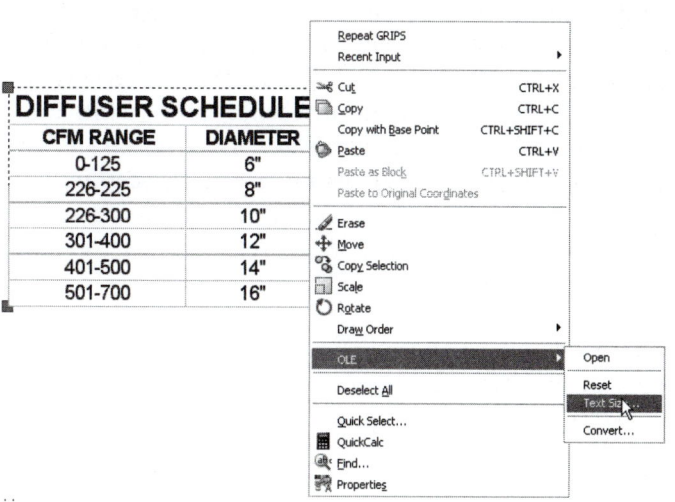

Reset

Use this option to reset the OLE scale to the original scale when the object was inserted.

Text Size

If the OLE object contains text, this option allows you to change the text size using the *OLE Scale* dialog box (see "*Pasteclip*"). When the text size is changed, the table containing the text (if existing) automatically adjusts accordingly.

Convert

This option produces the *Convert* dialog box (not shown) that operates only for embedded objects. The options available in the dialog box differ based on the selected OLE object type. For example, highlighting a Microsoft Excel Worksheet object would allow you to convert the object to a Microsoft Excel Chart.

Other AutoCAD Commands with OLE Objects

OLE objects in AutoCAD behave similarly to other native AutoCAD objects. For example, you can select an OLE object with the pickbox, window, crossing window, or other object selection method. You can add or remove the OLE objects to or from selection sets. Therefore, you can use standard Modify commands with OLE objects, such as *Erase, Move, Copy, Array,* and *Scale.*

OLE objects are sensitive to Object Snaps, such as *Endpoint, Midpoint,* etc. Therefore, you can draw *Lines,* for example, and *Osnap* to OLE objects. Or, you can *Scale* an OLE object that Osnaps to a predrawn rectangular area, or *Move* an OLE object that *Osnaps* to a title block or border.

You can also use grips to modify the size and location of the OLE object. Simply highlight the OLE object (a table in this case), and drag a grip to resize the table (Fig. 31-17). This technique is an alternate to using the *OLE Scale* dialog box.

FIGURE 31-17 ————

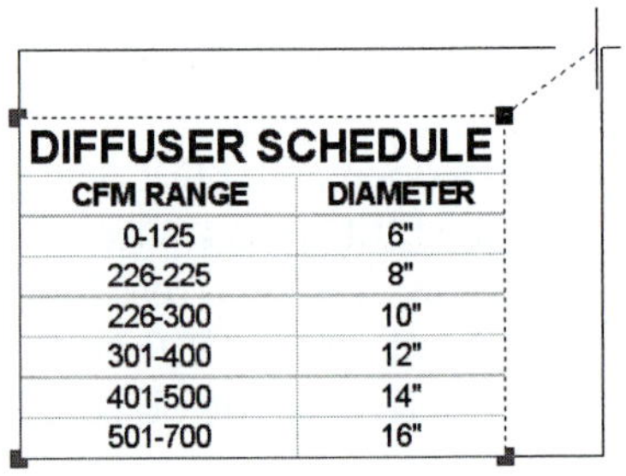

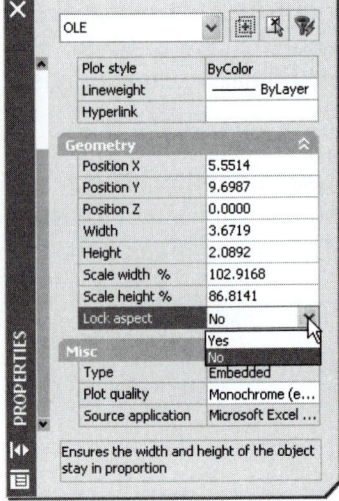

DIFFUSER SCHEDULE	
CFM RANGE	**DIAMETER**
0-125	6"
226-225	8"
226-300	10"
301-400	12"
401-500	14"
501-700	16"

Additionally, you can use the *Properties* palette to manipulate OLE features such as *Width, Height, Scale %,* and *Position* of the OLE object (Fig. 31-18). Normally, the *Lock Aspect* is set to *Yes* so scaling the OLE object using grips results in a uniform aspect ratio (width and height are scaled equally). Changing the *Lock Aspect* to *No* allows you to scale the OLE object non-uniformly using grips or *Properties.*

FIGURE 31-18 ————

OLE	
Plot style	ByColor
Lineweight	——— ByLayer
Hyperlink	
Geometry	
Position X	5.5514
Position Y	9.6987
Position Z	0.0000
Width	3.6719
Height	2.0892
Scale width %	102.9168
Scale height %	86.8141
Lock aspect	No
	Yes
	No
Misc	
Type	Embedded
Plot quality	Monochrome (e...
Source application	Microsoft Excel ...

Ensures the width and height of the object stay in proportion

OLE-RELATED SYSTEM VARIABLES

Four system variables affect the appearance of OLE objects in the Drawing Editor and in plots: *MSOLESCALE, OLEHIDE, OLEQUALITY*, and *OLESTARTUP*.

MSOLESCALE

When pasting an OLE object containing text into <u>model space</u>, the *MSOLESCALE* variable controls the initial size of the object. *MSOLESCALE* affects only current insertions, not previously pasted OLE objects. The initial value is 1.000. A value of 0.000 scales the OLE object by the *DIMSCALE* value. (If you *Paste* the object into a layout, the object is automatically scaled based on the layout's sheet size and units.)

OLEHIDE

OLEHIDE controls the visibility of OLE objects in AutoCAD. Your choice for this variable setting affects the AutoCAD Drawing Editor screen as well as printing and plotting. The possible settings are:

0   All OLE objects are visible (initial setting)
1   OLE objects are visible in paper space only
2   OLE objects are visible in model space only
3   No OLE objects are visible

The *OLEHIDE* system variable setting is saved in the system registry.

OLEQUALITY

OLEQUALITY sets the default quality level for plotting and printing OLE objects. This setting corresponds to the option selected in the *Plot and Publish* tab of the *Options* dialog box. When *OLEQUALITY* is set to 3 (default setting), the quality level is assigned automatically based on the type of object. For example, spreadsheets and tables are set to 0, color text and pie charts are set to 1, and photographs are set to 2. The four options are:

0   Monochrome
1   Low graphics
2   High graphics
3   Automatically Select

OLESTARTUP

OLESTARTUP determines whether the source application of an embedded OLE object loads when plotting. Loading the OLE source application may improve the plot quality, depending on the source application and type of OLE object. The possible settings are:

0   Does not load the OLE source application (initial setting)
1   Loads the OLE source application when plotting

The *OLESTARTUP* system variable setting is saved in the drawing file.

CHAPTER EXERCISES

1. **AutoCAD Objects:** *Copylink, Copyclip, Copybase, Pasteclip, Pasteorig*

    In this exercise you will gain experience copying and pasting between two open AutoCAD drawings. You will use existing drawings to copy most of the needed geometry to develop a half section view.

    A. *Open* the ▮**SADDL-DM**▮ drawing you created in the Chapter 29 exercises. Activate the *Model* tab, or if you created a titleblock in model space, *Freeze* its layer so you see only the geometry (including center lines and hidden lines) and dimensions. Also, *Freeze* layer **DIM**.

    B. With SADDL-DM open, begin a *New* drawing based on the *Start from Scratch, Imperial* settings. Use the *Window* pull-down menu to *Tile Vertically*. *Zoom Extents* in the SADDL-DM window.

    C. Use the *Copylink* command to automatically copy all objects from SADDL-DM. Use *Pasteorig* to bring the objects into the new drawing. This should create a display similar to that shown in Figure 31-19. Notice that all layers come into the new drawing when *Copylink* is used, including layers that are frozen. Since you do not want all these layers, *Exit* the new drawing <u>without</u> saving changes and begin a *New* drawing again using the *Imperial* default settings.

FIGURE 31-19

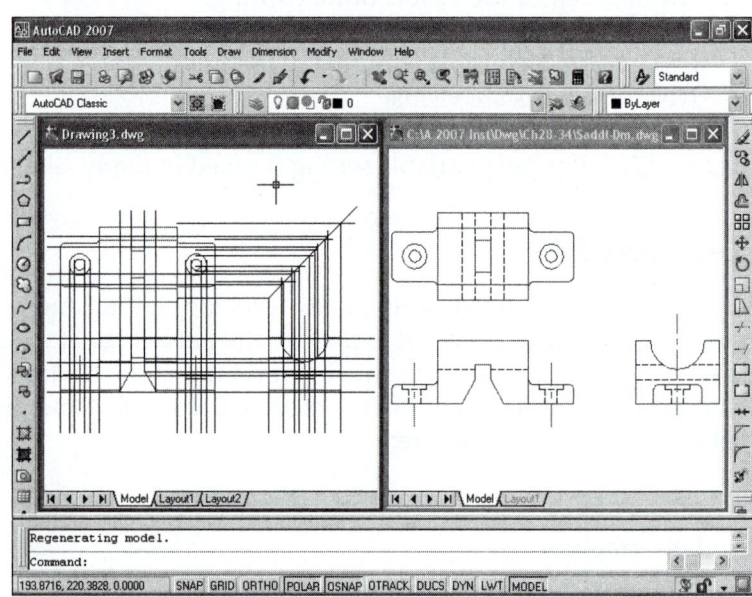

    D. Now use *Copyclip* and select only the visible geometry from the SADDL-DM drawing. Activate the new drawing and use *Pasteorig*. Perform a *Zoom Extents*. The two drawings should appear identical, except note that the new drawing's linetypes appear to be continuous. Change the *Ltscale* to **15**. *Save* the new drawing as **SADDLE-HALF SECTION**.

E. Next, *Close* SADDL-DM and <u>do not save changes</u>. *Open* the **SADL-SEC** drawing from Chapter 26 exercises, and *Tile Vertically*. In order to copy the cutting plane line from SADL-SEC, use *Copybase* and specify a base point at the *Center* of one of the two holes in the top view. Since the two drawings (SADDL-DM and SADL-SEC) may have different coordinate locations for the geometry, do not use *Pasteorig*, but use *Pasteclip* to place the cutting plane line in the SADDLE-HALF SECTION drawing at the *Center* of the appropriate hole. Your display should appear similar to that shown in Figure 31-20.

FIGURE 31-20

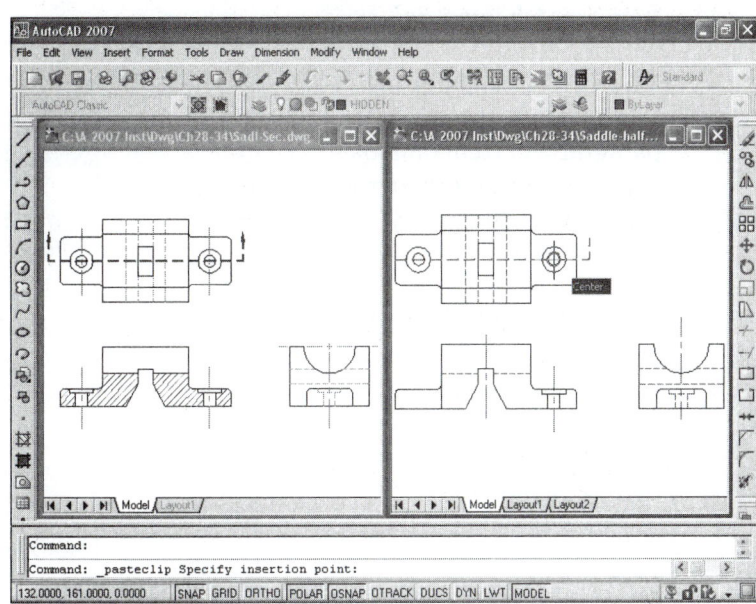

F. *Close* the SADL-SEC drawing and <u>do not save changes</u>. Maximize the new drawing and continue to generate a half section view. Your results should appear similar to that shown in Figure 31-21.

FIGURE 31-21

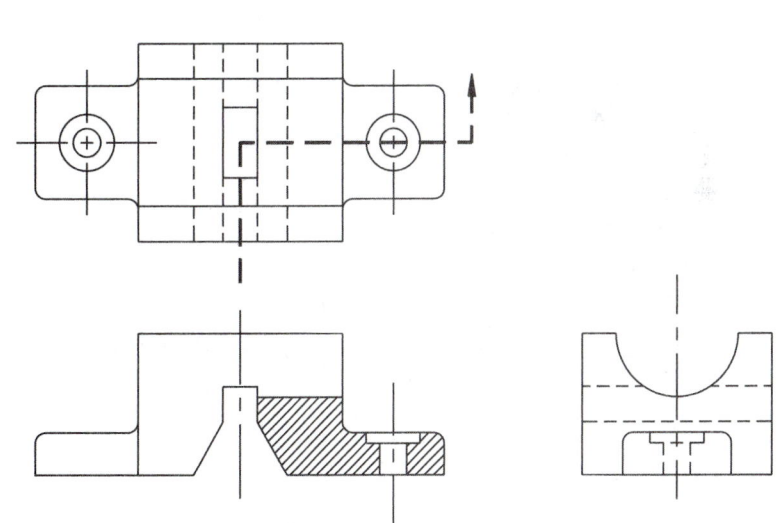

G. This technique (using copy and paste between AutoCAD drawings) is useful especially in cases such as this when you need geometry from <u>two or more drawings</u>. What is a better method to use when you need to create a drawing similar to only one other drawing?

2. *Copy, Pasteclip*

For this exercise, you will write a short paragraph of General Notes to be used in an AutoCAD drawing. Since the notes are text objects, it is easier to use a word processor to write and format the notes, *Copy* them to the Clipboard, and then use *Pasteclip* in AutoCAD to insert the notes into your drawing.

A. Begin a *New* drawing (*Start from Scratch, Imperial* default settings). Also, while AutoCAD is running, start **Word** and compose a short paragraph of General Notes similar to the note shown in Figure 31-22 (the word "copyrighted" is intentionally misspelled). After typing the notes, <u>highlight the entire block of text</u> and the select *Copy* from the *Edit* pulldown menu. This action copies the Word document onto the Clipboard.

FIGURE 31-22 ──────────────

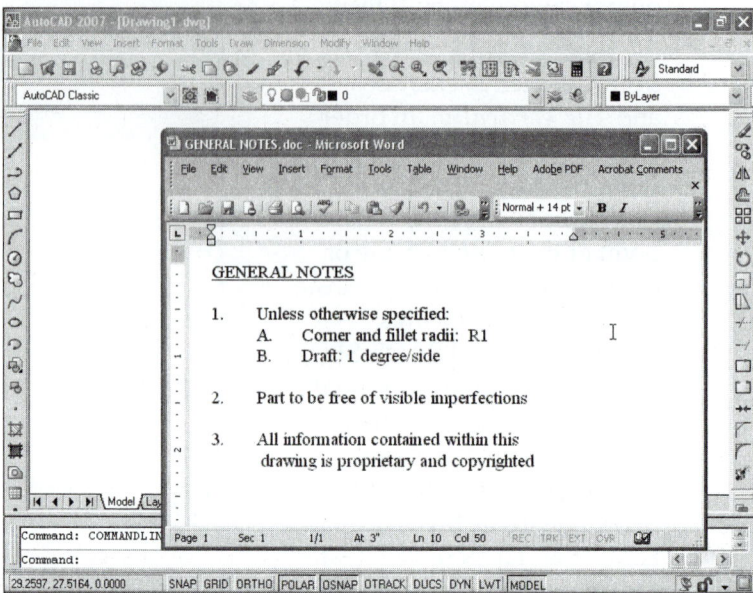

B. Now return to the new drawing you started in step A (you can use the **Alt+Tab** key sequence to switch to AutoCAD). In AutoCAD, use *Pasteclip*. When the OLE object "box" appears, select an appropriate insertion point. When the *OLE Scale* dialog box appears, select the *OLE Point Size* of *10* or *12*, and set the *Text Height* value to *.2. Zoom* if necessary to see the notes. Turn on Grips if not already on. Select the border around the object to display the grips at each corner. Select one of the corner grips and drag it in or out to resize the note. Notice the text within the pasted image changes proportionally as you change the box size.

C. The notes from the Word document are still stored within the current memory of the Clipboard. Select the *Mtext* command. When the *Text Formatting* Editor appears, do not enter in any text, but type a Ctrl+V (hold down the Control key and press the "v" key). Select the *OK* button in the *Text Formatting* Editor. What happened? Save this drawing as **PASTETXT**.

D. *Close* the PASTE TEXT drawing. *Open* the **SADDLE-HALF SECTION** drawing that you created in Exercise 1. Use *Pasteclip* to paste the note contained on the Clipboard. Use the grips to resize the note and move it to the open space in the upper-right corner. *Save* the drawing.

3. **Double-clicking on an Embedded Object**

   Assume that you printed the **SADDLE-HALF SECTION** drawing and noticed the misspelled word. Follow this procedure to change the spelling of the word "copywrighted" to "copyrighted."

   A. *Open* the **SADDLE-HALF SECTION** drawing if not already opened. Point to the pasted (embedded) note object and then double-click on it using the mouse. What happens?

   B. Now correct the misspelled word in **Word**. Then select *Exit* from the *File* pull-down menu in Word. Next, view the notes in AutoCAD. What happened to the notes? *Save* the drawing, but do not *Exit*.

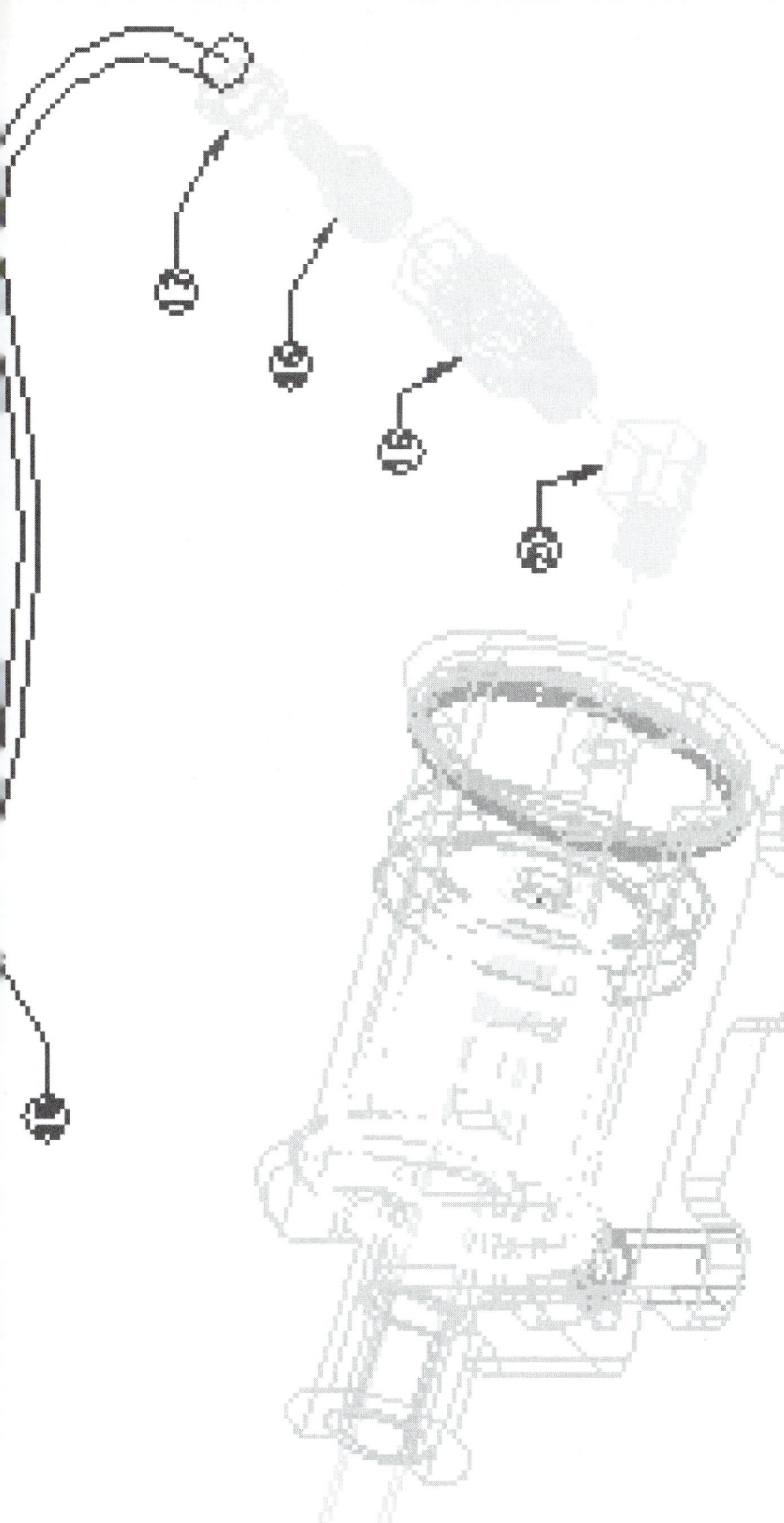

32

RASTER IMAGES AND VECTOR FILES

CHAPTER OBJECTIVES

After completing this chapter you should:

1. know the difference between raster files and vector files;

2. be able to *Imageattach* raster images within your drawing;

3. be able to use *Imageadjust* and *Imagequality* to control the appearance of an attached image;

4. be able to use *Imageclip* to create a clip boundary for an attached image;

5. know that certain images can be controlled for *Transparency*;

6. be able to *Import* vector file formats into your drawing and convert the contents into AutoCAD objects;

7. know the individual commands for importing and exporting specific file types (*Bmpout, 3dsin*, etc.);

8. be able to *Export* a variety of vector file formats from your drawing with this one command.

CONCEPTS

The previous chapter discussed how to utilize textual and graphic data from other applications within your AutoCAD drawings using AutoCAD's Object Linking and Embedding (OLE) feature. This chapter discusses how to utilize graphic information in the form of raster file and vector file formats within your drawings.

This chapter discusses two different but related topics: (1) attaching raster files and (2) importing and exporting vector file formats. When a raster image is attached to your drawing, the image that appears is actually linked to the original raster file similar to the way text or spreadsheet information in an AutoCAD drawing is linked to the original file when using OLE. When other types of graphic file formats are imported into your drawings, such as vector files, AutoCAD converts the contents of the file into AutoCAD objects and permanently adds them to the drawing.

Raster Files versus Vector Files

Raster images are "pictures" composed of many dots of various shades of gray to produce black and white images or many dots of various colors to produce color images. For example, when you look closely at a color or black and white picture in a newspaper, you can actually see the small dots that make up the image. The quality, or resolution, of the image depends upon the process used to create the image. Some software can produce images with many dots per inch (DPI) yielding a high quality (resolution) image that looks very distinct and clear. Images that have lower resolution have fewer dots per inch and produce a lower quality image. The number of dots used to create the image determines the resolution, or quality, of the picture. Figure 32-1 displays an enlarged view of a raster image to reveal the dots composing the lines and arcs.

FIGURE 32-1

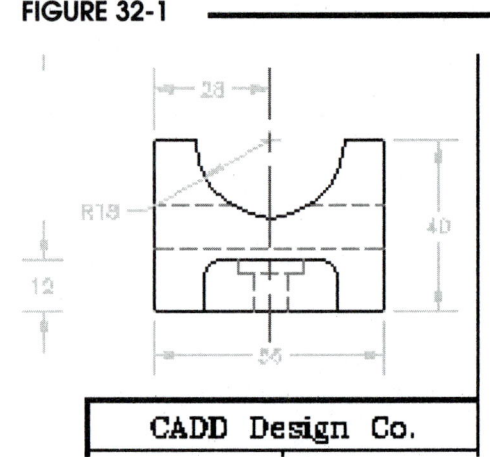

CADD Design Co.

A few of the applications that create raster images frequently attached within AutoCAD drawings are computer graphics or desktop publishing programs (used for creating company logos, for example), document management software, and mapping and geographic information systems (which produce aerial photographs or satellite photographs).

In contrast, CAD programs such as AutoCAD, which are the primary graphics creation and storage media for the engineering, architecture, design, and construction industries, generate vector data. A vector file defines a line, for example, as a vector between two endpoints. Therefore, lines in vector drawings do not enlarge when zoomed in like the dots in a raster image do (Fig. 32-2). The thickness of the displayed line on a computer monitor (for a vector drawing) is normally one pixel, the size of which is determined by the display device. Therefore, vector drawings are used when great detail is needed. Raster images are useful for insertion into a CAD (vector) drawing and can be helpful in the development of a CAD drawing but cannot be used for drawings that require detailed information because of the static configuration of the dots.

FIGURE 32-2

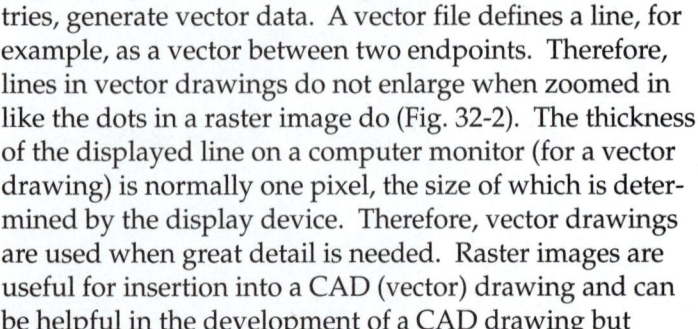
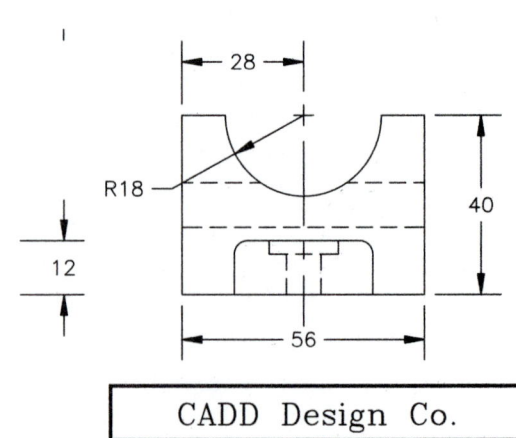

CADD Design Co.

AutoCAD stores its vector data in the .DWG format but can also create other forms of vector files such as .SAT or .WMF discussed in this chapter. (See Chapter 23, Internet Tools and Collaboration, for more information on vector and raster files.)

Attaching Raster Images and Importing Vector Files

Two techniques for utilizing raster file and vector file formats within AutoCAD drawings discussed in this chapter are:

1. attaching a raster image, which links the image to the original file, and
2. importing a vector file, which converts the file contents into AutoCAD objects.

Because AutoCAD can store graphic information only in a vector format, it treats an attached raster image differently because it is <u>not</u> vector information (made up of lines, arcs, and circles). An attached raster image in AutoCAD is actually a picture of the original raster file that was selected when attaching it to your drawing. Attached raster images are displayed within a <u>frame</u> in your drawing and are *linked* to (or reference) the original file. On the other hand, if an image is imported into AutoCAD, it must be converted to AutoCAD objects for AutoCAD to store the geometry as vector data.

Exporting Raster and Vector Files

Because AutoCAD's primary function is to create and manipulate vector data (AutoCAD objects), it specializes in handling vector file formats, especially the .DWG format. Although AutoCAD can create raster files such as .BMP, .TIF, .JPG, and .PNG, these formats have little usefulness for storing drawing information. Only images generated by AutoCAD's rendering capabilities are well suited for saving in a raster format (see Chapter 41, Rendering). Therefore, AutoCAD's file exportation capabilities are centered around vector formats such as .DWG, .DWF, .DXF, .PDF, .SAT, .STL, .WMF, and .EPS. These file formats are also discussed in this chapter.

Related Commands

This chapter discusses the following AutoCAD commands related to using raster images and vector file formats within your drawing:

Image	Inserts various types of raster images into your drawing
Imageattach	Attaches, or links a new raster image object and definition within your drawing
Imageframe	Controls whether the image frame is displayed on screen or hidden from view
Imageadjust	Allows you to control the brightness, contrast, and fade values of a raster image
Imageclip	Enables you to create new clipping boundaries for raster image objects
Imagequality	Allows you to control the display quality of raster images
Import	Imports a variety of vector file formats into AutoCAD
3Dsin, 3Dsout	Imports and exports 3D Studio files
Acisin, Acisout	Imports and exports ACIS solid models
Dxfin, Dxfout	Imports and exports AutoCAD drawings in an exchange format
Psout	Exports Encapsulated PostScript files
Bmpout	Exports bitmap (.BMP) raster files
Jpgout	Exports JPEG raster files
Tifout	Exports Tagged Image File Format (.TIF) files
Pngout	Exports Portable Network Graphic (.PNG) files
Export	Exports a variety of vector file formats from AutoCAD
DWG To PDF.pc3	Exports an AutoCAD drawing file to Adobe® PDF format

ATTACHING RASTER IMAGES

The procedure for attaching a raster image in AutoCAD is very similar to using the *Xattach* command (see Chapter 30, Xreferences). In fact, the tool for managing raster file images is the same tool you use for *Xrefs*—the *External References* palette. A raster image can be *Attached, Detached, Reloaded,* and the *Path* modified, just as you might do with an *Xref.* When you attach an image, it is displayed within a frame in the drawing. You can view the original image, but the image (and all the dots that define it) does not become a permanent part of the drawing. Like an *Xref,* you see only the picture of the original file within its frame. Since raster images normally have large file sizes, this method of "attaching" can save a tremendous amount of space within your drawing.

A few example applications of attaching a raster image to an AutoCAD drawing are listed here:

A. You want to display a digital (raster) picture of the true, or intended, object. For example, you can take a digital photograph of a house and attach it within your construction drawing of an addition to the house.

B. You want to use images of company or client logos within a titleblock.

C. You need to compare and verify site plan information, such as overlaying an aerial photograph to verify the contours shown on your AutoCAD drawing.

D. You need to compare or reconstruct an old "paper" drawing by scanning it (creating a raster image) and attaching it in an AutoCAD drawing.

You can edit the appearance of the attached image but not the individual parts (or dots) of the image within AutoCAD. You can adjust the quality, color, contrast, brightness, and transparency of an attached image with AutoCAD's various image editing commands and system variables. Attached raster images can also be edited with typical AutoCAD editing commands such as *Copy, Move, Rotate,* and *Scale.* Attached images can be edited with Grips as well. The special image attaching and editing commands are discussed next.

Activate the *Reference* toolbar to access buttons for attaching and manipulating raster images. The *Insert* and *Modify* pull-down menus also contain commands for attaching and manipulating raster images.

External-references

Pull-down Menu	Command (Type)	Alias (Type)	Short-cut	Screen (side) Menu	Tablet Menu
Insert *External References*	*Image* or *Externalreferences*	*IM* or *ER*	...	*INSERT* *Image*	*T,3*

Use the *External References* palette command to control the attachment of raster images within your drawing. There is no practical limit to the number or size of images you can attach within your drawing. You can have more than one image displayed within any viewport.

Although you can attach a variety of file types (BMP, TIF, JPG, GIF, TGA, etc.), AutoCAD determines how to handle the image (for display and editing) based upon the image file contents not the image file type. The following raster file types are accepted by the *External References* palette.

Image File Type	File Extensions	Description
BMP	*.BMP, *.RLE, *.DIB	Windows device-independent bitmap format
CALS-1	*.RST, *.GP4, *.MIL, *.CAL, *.CG4	Mil R-Raster-1
FLIC	*.FLC, *.FLI	Autodesk Animator FLIC
GEOSPOT	*.BIL	GeoSPOT with HDR and PAL files
GIF	*.GIF	Compuserve Graphic Interchange Format
IG4	*.IG4	Image Systems Group 4
IGS	*.IGS	Image Systems Grayscale
JPEG, JFIF	*.JPG	JPEG raster
PCX	*.PCX	PC Paintbrush Exchange
PICT	*.PCT	Macintosh PICT1, PICT2
PNG	*.PNG	Portable Network Graphics
RLC	*.RLC	Run-Length Compressed
TARGA	*.TGA	Truevision
TIFF	*.TIF	Tagged Image File Format

The *Externalreferences* command produces the *External References* palette (Fig. 32-3). With the palette, you can use the right-click shortcut menu to *Attach*, *Detach*, *Reload*, *Unload*, and *Open* raster images. The bottom half of the palette lists details about the images and allows you to rename images and to change the path of images. The functions of this palette for managing raster images are identical to those for managing *Xrefs* (see Chapter 30, Xreferences).

Tree View, List View

The two buttons located in the top-right corner of this dialog box allow you to control the appearance of the list of currently attached raster images in the center of the dialog box. The list can be displayed in a *List View* or *Tree View*.

FIGURE 32-3

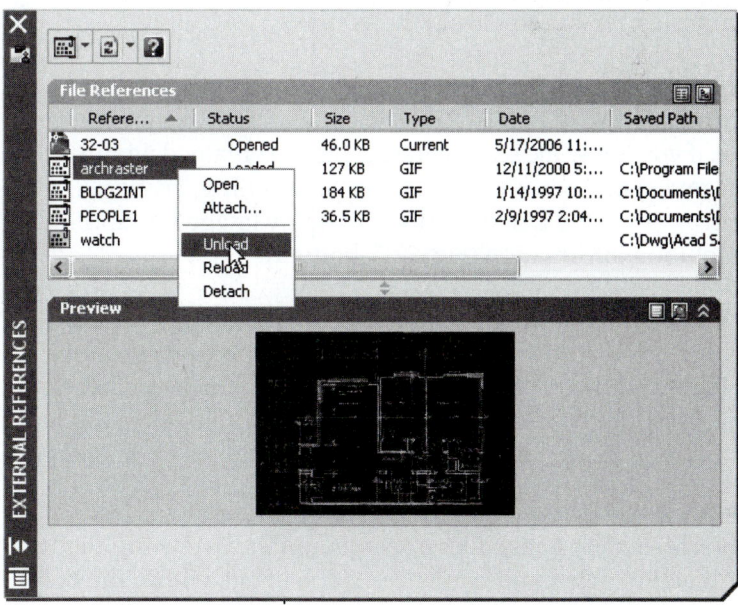

List View is the default setting for the list when you open the *External References* palette. When *List View* is active, column headings appear at the top of the list for *Image Name, Status, Size, Type, Date,* and *Saved Path* for each image currently attached to the drawing. When *Tree View* is selected, AutoCAD displays the list of attached images in a tree structure. This format allows you to view the list of images currently attached to the drawing in a hierarchical structure.

Attach

When you first activate the *External References* palette the only available option is *Attach*. Attach an image to the drawing in the same way you attach an *Xref*. Selecting the *Attach* option invokes the *Select Image File* dialog box (Fig. 32-4). Alternately, this dialog box can be opened directly using the *Imageattach* command.

From the *Select Image File* dialog box you can select from a variety of raster image file types (listed in the previous table). The features and functionality of the *Select Image File* dialog box are the same as those discussed in Chapter 30 for attaching *Xref* files. However, the *Select Image File* dialog box has the additional option of *Hide Preview / Show Preview* that allows you to

FIGURE 32-4

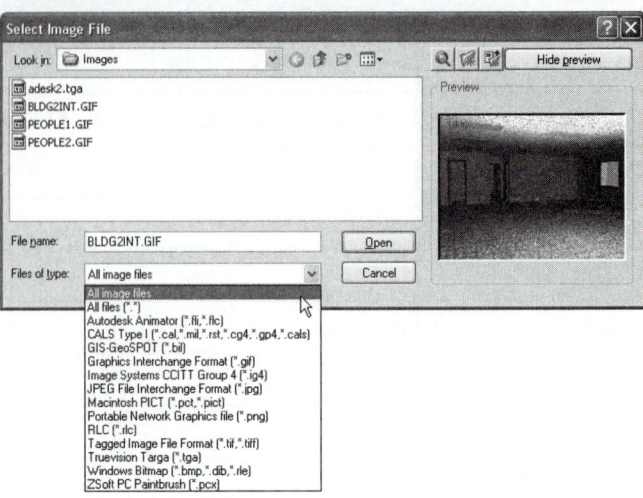

turn on and off the *Preview* display area. If the preview is turned off, the *Preview* display box does not display the thumbnail picture of the file currently selected from the file list. Use this feature to increase the speed at which you can select files from the dialog box when you know the image files by name. Turn the *Preview* display on again to view the thumbnail of the selected image file.

When an image file has been selected for attachment (select *Open* in the *Select Image File* dialog box), the *Image* dialog box appears for you to specify the attachment parameters (Fig. 32-5). Specifying the *Insertion Point* and *Rotation* angle is similar to inserting a *Block* or attaching an *Xref*. Use *Scale* factor to match the image geometry scale to the scale of the geometry in the drawing.

The *Scale* factor is based on the ratio of image units to the current AutoCAD units. You can enter a value in the edit box if you know the scale of the geometry contained within the image file, otherwise use the default setting, *Specify on screen*. Select the *Details* button to display image size and resolution information (Fig. 32-5, bottom).

FIGURE 32-5

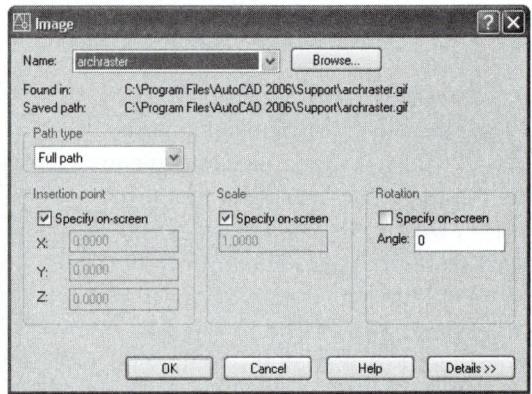

If you want to attach a second image in a drawing, selecting the *Attach* option may produce the *Image* dialog box directly. You must then select *Browse* to produce the *Select Image File* dialog box to select from folders and files. When the same image is attached more than once in a drawing, only one reference is displayed in the list in the *Select Image File* dialog box, but a different insertion point, scale, clipping boundary, etc. can be specified for each instance of the image in your drawing.

Detach

You can detach image files appearing in the list by highlighting the desired file(s), then selecting *Detach*. This action deletes all references to the image within the drawing (breaks the link to the image file), but does not delete the file from the storage device (local or network drive). This is essentially the same as detaching an *Xref* from a drawing.

Reload

You can *Unload* an image to remove its display while you work on the drawing, but the attachment information is not deleted from the drawing. You can then *Reload* the image picture by selecting the *Reload* button. *Reload* causes AutoCAD to reread the file referenced in the *Saved Path* and redisplay it within the image frame. This is also useful if the referenced image file has been changed and you want the changes to be displayed during the current drawing session.

Unload

This feature unloads the raster image picture from the display without deleting the attachment information from your drawing. Only the image's frame remains visible in the drawing. Unloading an image can dramatically decrease the amount of time AutoCAD needs to redisplay, or regenerate, the drawing since the image is not resident in current memory. Use this feature when you do not need to see the image in order to work on the drawing.

Unloading a drawing is not the same as turning off the display of the image. You can turn off the display of an image within the drawing using *Properties* (see "*Properties*," this chapter).

Details/Preview

The lower section of the *External References* palette can display *Details* or a *Preview* depending on your selection (Fig. 32-06, right-center buttons). The *Details* option gives information about the highlighted image from the list above. Most of the fields are read-only; however, you can change the *Reference Name* and *Found At* (path) fields.

FIGURE 32-6

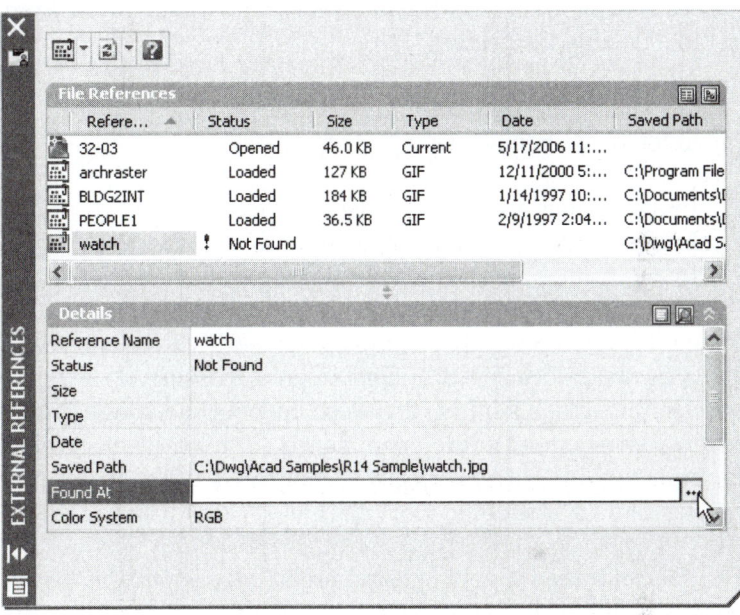

For example, use the browse button at the right of the *Found At* field to locate a referenced image that has been moved to different folder or drive. In such a case, the *Status* of the image is listed above as *Not Found*, as shown in this figure for the "watch" image.

You may want to do this, for example, if you have an older version of an aerial photograph currently attached and you want to temporarily display a newer photo of the same aerial view. You can *Browse*, find and select the newer raster image, then use the *External References* palette preview to verify that it is a more appropriate image to display in your drawing.

Imageattach

Pull-down Menu	Command (Type)	Alias (Type)	Short-cut	Screen (side) Menu	Tablet Menu
Insert *Raster Image* *Reference*	*Imageattach*	*IAT*	...	*INSERT* *Image* *Attach...*	...

Imageattach allows you to attach a raster image directly without having to use the *External References* palette first. When you use *Imageattach*, the *Select Image File* dialog box appears (see Fig. 32-4). The capabilities and operation of this dialog box were discussed previously regarding the *External References* palette.

Imageframe

Pull-down Menu	Command (Type)	Alias (Type)	Short-cut	Screen (side) Menu	Tablet Menu
Modify Object > Image > Frame	Imageframe	...	...	MODIFY1 Imagefrm	...

The *Imageframe* command allows you to control whether or not the image frames are displayed on the screen and in prints and plots. When you attach an image within your drawing it is displayed inside a "frame" that defines the image's outermost boundary. *Imageframe* controls only the frames. If image frames are not displayed, the images are still visible. The *Imageframe* setting is global; that is, all images in the drawing are affected.

        Command: **imageframe**
        Enter image frame setting [0, 1, 2] <1>:

0   Image frames are not displayed and not plotted.
1   Image frames are both displayed and plotted.
2   Image frames are displayed but not plotted.

Images can be modified by using the general editing commands (*Move, Copy, Scale, Rotate, Grips*, etc.) or by using special image editing commands (discussed next). The general editing commands require you to select the image's frame; therefore, turning the image frame off makes the image unselectable for editing with general editing commands. You must select the image's frame whenever AutoCAD prompts you to "Select objects:" to modify. The image becomes selectable again once you turn the display of its frame on.

The special image adjusting commands (*Imageadjust, Imageclip,* and *Imagequality*) allow you to select images whether the image frames are on or off.

Imageadjust

Pull-down Menu	Command (Type)	Alias (Type)	Short-cut	Screen (side) Menu	Tablet Menu
Modify Object > Image > Adjust	Imageadjust	IAD	...	MODIFY1 Imageadj	X,20

You can adjust an image's *Brightness, Contrast,* and *Fade* values by using *Imageadjust.* Use the *Imageadjust* command or double-click on an image's frame (if *Dblclkedit* is *On* and *Imageframe* is *On*) to produce the *Imageadjust* dialog box (Fig. 32-7). You can change the following settings.

Brightness
This slider and edit box control the brightness of the image. The edit box values range from 0 to 100. The higher the value, the brighter the image becomes. *Brightness* indirectly affects the range of *Contrast* for the image.

FIGURE 32-7

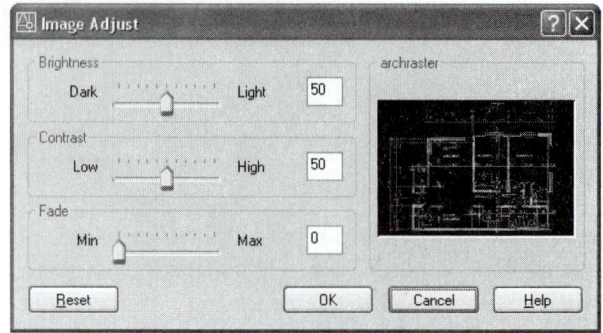

Contrast

Use this slider or edit box to control the contrast of the image. *Contrast* is the ratio of light to dark (for black and white images) or pure hue to no color (for color images). The higher the value, the more each pixel is forced to its primary or secondary color. *Contrast* indirectly affects *Fade*. The WATCH.JPG image is displayed in Figure 32-8 with *Contrast* set to 90 and *Brightness* set to 70.

FIGURE 32-8

Fade

The *Fade* effect of the image controls how much the image blends with the background color. The higher the value, the more the image is blended with the current background color. A value of 100 allows you to blend the image completely into the background. If you change the screen background color, the image will fade to the new color. For plotting, white is used for the background fade color.

Reset

Select this button to reset the values for *Brightness, Contrast,* and *Fade* back to the default settings of 50, 50, and 0, respectively.

Imageclip

Pull-down Menu	Command (Type)	Alias (Type)	Short-cut	Screen (side) Menu	Tablet Menu
Modify *Clip >* *Image*	*Imageclip*	ICL	...	*MODIFY1* *Imageclp*	X,22

You can clip an image with *Imageclip* in the same manner that you clip *Xrefs* with *Xclip*. When specifying the area to clip, you can use a rectangle or polygon to define the clip boundary. Everything outside the clip boundary is hidden from view. Figure 32-9 displays the WATCH.JPG image with a polygonal clip boundary.

FIGURE 32-9

Specifying the clip boundary rectangle or polygon is similar to using AutoCAD's *Rectangle* command or the *Window Polygon* option when selecting objects. *Rectangle* is the default boundary.

```
Command: imageclip
Select image to clip: PICK (select an image frame)
Enter image clipping option [ON/OFF/Delete/New boundary]
    <New>: Enter
Enter clipping type [Polygonal/Rectangular] <Rectangular>: Enter
Specify first corner point: PICK
Specify opposite corner point: PICK
Command:
```

Once a clipping boundary has been specified for an image you can turn the boundary *Off*. This action makes the entire image visible again (Fig. 32-10). You can also turn the boundary back *On*, which displays the image with its original clip boundary (see Fig. 32-9). Alternately, you can *Delete* the clip boundary so the entire image is visible.

If you use *Imageclip* and select an image that already has a boundary, the following prompt appears:

FIGURE 32-10

```
Command: imageclip
Select image to clip: PICK
Enter image clipping option [ON/OFF/Delete/New boundary]
    <New>: Enter
Delete old boundary? [No/Yes] <Yes>: Enter
```

Imagequality

Pull-down Menu	Command (Type)	Alias (Type)	Short-cut	Screen (side) Menu	Tablet Menu
Modify *Object >* *Image >* *Quality*	*Imagequality*	...	...	*MODIFY1* *Imagequa*	...

This command allows you to adjust the quality setting that affects display performance. High-quality images take longer to display. When you change the *Imagequality* setting, the display updates immediately without causing a *Regen*. The images attached to your drawing are always plotted using a high-quality display, regardless of the *Imagequality* setting. The setting you choose for *Imagequality* affects all images in the drawing globally.

```
Command: imagequality
Enter image quality setting [High/Draft] <High>: (option)
Command:
```

Transparency

Pull-down Menu	Command (Type)	Alias (Type)	Short-cut	Screen (side) Menu	Tablet Menu
Modify *Object >* *Image >* *Transparency*	*Transparency*	...	...	*MODIFY1* *Transpar*	...

Transparency allows you to control whether the background pixels in an image are transparent or opaque. Many image file formats allow images with transparent pixels. When setting image transparency to *On*, AutoCAD recognizes transparent pixels so that graphics on the screen (AutoCAD objects or another image) show through those pixels.

Transparency can be set for both bitonal and non-bitonal (Alpha RGB or gray-scale) images. When you attach an image to your drawing, the image's transparency setting is *Off* by default:

FIGURE 32-11

```
Command: transparency
Select image(s): PICK
Select image(s): Enter
Enter transparency mode [ON/OFF] <OFF>: (option)
Command:
```

This unique feature allows you to put one image on top of another, then set *Transparency* for the top image *On*. (Use *Draworder* to specify which image is "on top.") The transparency feature makes the "top" image appear in the same scene as the "background" image (Fig. 32-11, right image).

Properties

Pull-down Menu	Command (Type)	Alias (Type)	Short-cut	Screen (side) Menu	Tablet Menu
Modify Properties	*Properties*	*MO* or *PR*	(Edit Menu) *Properties...*	*MODIFY1 Property*	*Y,12 to Y,13*

Using *Properties* and selecting an image gives you the power to change many of the image's properties as you might with other image-related commands. The *Imageframe* must be *On* for you to select the image. The adjustable features of the image are listed in the *Image Adjust* section of the *Properties* palette (Fig. 32-12). You can change *Brightness*, *Contrast*, and *Fade*, similar to using the *Imageadjust* command. Select the ellipsis (...) just to the right of each of these edit boxes to produce the *Imageadjust* dialog box.

In the *Misc* section, you can set *Show clipped* and set *Transparency*—identical to the *On/Off* options of the *Imageclip* and *Transparency* commands. *Show image* allows you to turn on and off the display of the image.

FIGURE 32-12

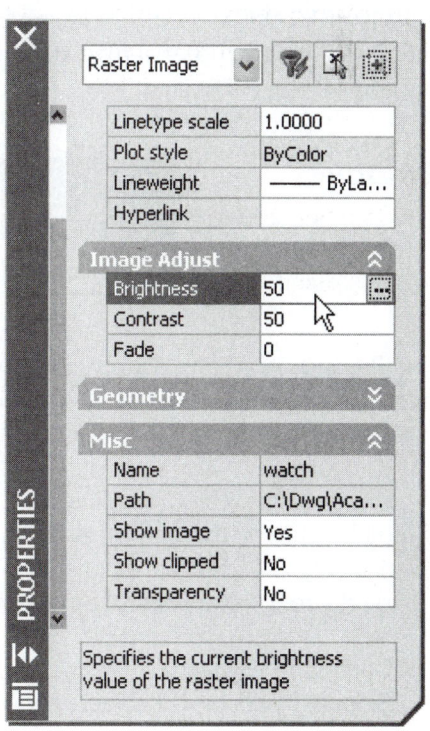

Draworder

Pull-down Menu	Command (Type)	Alias (Type)	Short-cut	Screen (side) Menu	Tablet Menu
Tools Draw Order>	Draworder	DR	...	TOOLS 1 Drawordr	T,9

The *Draworder* command is very useful when utilizing raster images in your drawing. Often you will attach the raster image so that it appears on top of another image or other AutoCAD objects. The *Draworder* command allows you to select an attached image and then specify the drawing order placement of the raster image as *Above, Under, Front,* or *Back*:

Command: **draworder**
Select objects: **PICK**
Select objects: **Enter**
Enter object ordering option [Above objects/Under objects/Front/Back] <Back>: (**option**)
Command:

For additional information about *Draworder,* see Chapter 26, Section Views.

Saveimg

Pull-down Menu	Command (Type)	Alias (Type)	Short-cut	Screen (side) Menu	Tablet Menu
Tools Display Image> Save...	Saveimg	...	...	TOOLS 1 Saveimg	...

The *Saveimg* (save image) command allows you to save the current drawing or rendered image as either a .BMP, .JPG, .PCX, .PNG, .TGA, or .TIF (raster) format. The saved image can later be viewed by using the *Replay* command or can be attached by using *Imageattach*. See Chapter 41, Rendering, for more information on this command.

Bmpout

Pull-down Menu	Command (Type)	Alias (Type)	Short-cut	Screen (side) Menu	Tablet Menu
...	Bmpout	...	...	...	...

The *Bmpout* command enables you to create a .BMP (bitmap) file of the drawing or of selected objects from the drawing. The resulting file can be read by many viewers such as Microsoft Internet Explorer or viewed as a raster image (using *Imageattach*) in an AutoCAD drawing.

Bmpout produces the *Create Raster File* dialog box (not shown), which is similar to the *Save Drawing As* dialog box but without options for saving as other file types. You can specify a location and a name for the new file. A .BMP file extension is automatically appended to the file name. When the dialog box clears, the following prompt appears at the Command line:

Command: **bmpout**
Select objects or <all objects and viewports>: **PICK** or **Enter**

You can select one or more objects or press Enter to write the entire drawing contents to the file. No other options exist for the command. See Chapter 41, Rendering, for other information on saving .BMP images generated by AutoCAD.

Jpgout

Pull-down Menu	Command (Type)	Alias (Type)	Short-cut	Screen (side) Menu	Tablet Menu
...	*Jpgout*	...	...	...	...

The *Jpgout* command allows you to generate a raster file in JPEG format. *Jpgout* produces the *Create Raster File* dialog box, identical to that appearing for the *Bmpout* command. When the dialog box clears, the prompt appears at the Command line, like the *Bmpout* command, allowing you to select specific objects or the entire drawing. (See "*Bmpout*.")

Pngout

Pull-down Menu	Command (Type)	Alias (Type)	Short-cut	Screen (side) Menu	Tablet Menu
...	*Pngout*	...	...	...	...

The *Pngout* command allows you to generate a .PNG (Portable Network Graphics) raster file of selected objects or of the entire drawing. The process is the same as that for using the *Bmpout* command or *Jpgout* command. (See "*Bmpout*.")

Tifout

Pull-down Menu	Command (Type)	Alias (Type)	Short-cut	Screen (side) Menu	Tablet Menu
...	*Tifout*	...	...	...	...

The *Tifout* command operates identically to the *Bmpout*, *Pngout*, and *Jpgout* commands but creates a raster file in Tagged Image File Format. See "*Bmpout*" for more information on this process.

IMPORTING AND EXPORTING VECTOR FILES

Importing vector file formats into your drawing converts the graphic information, or objects, within the selected file into AutoCAD objects and inserts them onto your drawing. The converted information becomes a permanent part of your drawing. Depending on the file being imported, this can dramatically increase the file size of your AutoCAD drawing.

Since importing vector files converts the contents into AutoCAD objects, you can edit them using typical AutoCAD editing methods. An imported object is initially defined in your drawing as a *Block* reference. The first step in editing imported vector files with AutoCAD is to *Explode* the block. You will find that AutoCAD often converts the information into *Lines*, *Arcs*, *Circles*, *Plines*, *Solids* (2D), or solids (3D).

Acisout

Pull-down Menu	Command (Type)	Alias (Type)	Short-cut	Screen (side) Menu	Tablet Menu
File Export... *.sat	Acisout	...	...	FILE Export *.sat	...

You can use the *Acisout* command to export AutoCAD solid or *Region* data from your drawing into an ACIS format. The ACIS file format allows AutoCAD to store solid model objects as an ASCII (.SAT) file type. The *Create ACIS File* dialog box that appears is the same as the *Select ACIS File* dialog box. *Acisout* ignores all selected objects that are not solids or *Regions*.

Stlout

Pull-down Menu	Command (Type)	Alias (Type)	Short-cut	Screen (side) Menu	Tablet Menu
File Export... *.stl	Stlout	...	...	FILE Export *.stl	...

You can create a file to be used with Stereo Lithography Apparatus for creating prototypes from AutoCAD solid models. The *Stlout* command creates an ASCII format STL file defining the 3D geometry. For more information, see Chapter 39, Solid Modeling Editing.

Dxfin

Pull-down Menu	Command (Type)	Alias (Type)	Short-cut	Screen (side) Menu	Tablet Menu
...	Dxfin	...	...	...	...

Autodesk initiated an interchangeable CAD drawing (vector file) standard in the early 1980s. This standard, called the Drawing Interchange File Format (.DXF), was intended to be utilized by AutoCAD and other CAD programs for exchanging vector drawing data in a simple ASCII format. Because the ASCII format can be used, DXF files can be very large. The DXF format has become one of the most widely used formats (along with IGES format) for exchanging drawing data.

 Importing vector data in .DXF format from other CAD applications is accomplished by using the *Dxfin* command. This command produces the *Select File* dialog box (not shown), similar to the other file import dialog boxes. <u>The *Dxfin* command must be used in a new drawing to function properly</u>.

Dxfout

Pull-down Menu	Command (Type)	Alias (Type)	Short-cut	Screen (side) Menu	Tablet Menu
Save As...	Dxfout	...	...	...	...

You can use the *Dxfout* command or the *SaveAs* command to export your drawing to a Drawing Interchange Format (*.DXF) file. *Dxfout* produces the *Save Drawing As* dialog box (not shown), the same dialog box that appears with the *SaveAs* command. To save the drawing as a .DXF, use the *Files of type* drop-down list and select one of the .DXF file type options. You can choose to export the drawing as an AutoCAD 2007, previous AutoCAD, or LT format for the .DXF.

When the *Save as type* box indicates the desired .DXF format you want to use, you can select *Options* from the *Tools* menu to produce the *Saveas Options* dialog box. Activate the *DXF Options* tab (Fig. 32-17).

FIGURE 32-17

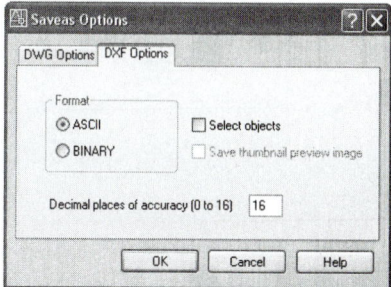

Here you can specify whether you want the interchange file to be created in ASCII format or binary format as well as the degree of accuracy used (in decimal places) for defining the geometry. Drawing Interchange Files, whether created as ASCII or binary format, are saved as a .DXF file type.

Because an ASCII file is composed of recognizable alphanumeric characters, the ASCII format .DXF files can be quite large, so the binary format is sometimes used. The advantage to the ASCII file is that it is readable in a text editor or word processor (see Fig. 1-4), whereas the binary format is not.

Dxbin

Pull-down Menu	Command (Type)	Alias (Type)	Short-cut	Screen (side) Menu	Tablet Menu
Insert *Drawing Exchange* *Binary...*	*Dxbin*	...	...	INSERT DXBin	...

The *Dxbin* command allows you to import Drawing Exchange Binary (.DXB) files into AutoCAD. These .DXB files are formatted with a specially coded binary format and are produced by programs such as AutoShade. *Dxbin* produces the *Select DXB File* dialog box (not shown).

Psout

Pull-down Menu	Command (Type)	Alias (Type)	Short-cut	Screen (side) Menu	Tablet Menu
...	*Psout*	...	...	...	...

You can export parts or all of your drawing as a PostScript (.EPS) file using *Psout*. PostScript files generated by *Psout* can contain a PostScript rendering of an AutoCAD model, for example. AutoCAD normally exports *Arcs, Circles, Plines,* and filled *Regions* as PostScript primitives instead of vectors. One exception is when the AutoCAD objects cannot be represented in PostScript, such as extruded objects, in which case they are output as vectors. In this case, the information is exported as wireframe images just as AutoCAD displays them in your drawing.

Psout produces the *Create PostScript File* dialog box (not shown). Selecting *Options* from the *Tools* menu allows you to set PostScript specific output options via the *Postscript Out Options* dialog box.

In the *Postscript Out Options* dialog box (Fig. 32-18, next page) you can specify what part of the drawing should be written to the .EPS file by selecting *Display, Extents, Limits, View,* or *Window* in the *What to plot* section. EPS files can contain a preview which speeds and simplifies the selection and display of a PostScript image in other software programs. The preview can be saved in EPSI or TIFF format, along with the size in pixels for the preview image. The *Prolog Section Name* specifies a name for the prolog section to be read from the ACAD.PSF file.

Because a PostScript file can contain all information necessary for printing to a specific size paper, that information can also be specified in this dialog box. For more information, see Using PostScript Files in the AutoCAD *User's Guide*.

FIGURE 32-18

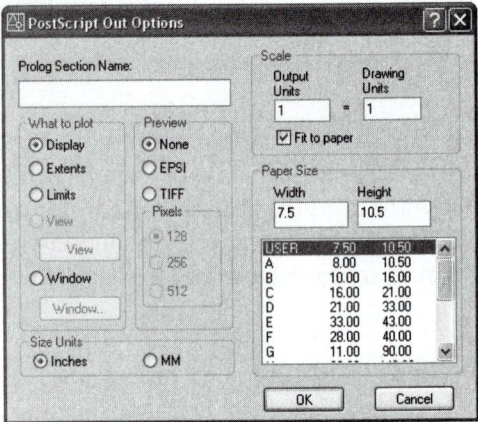

 As an alternative to *Psout*, if you configure a PostScript driver in the *Add-a-Plotter* wizard, you can output your drawings in PostScript format. To configure the PostScript driver, in the *Add-a-Plotter* wizard, select *Adobe* from the *Manufacturer* list, and select a *PostScript* level from the *Model* list.

Export

Pull-down Menu	Command (Type)	Alias (Type)	Short-cut	Screen (side) Menu	Tablet Menu
File *Export...*	*Export*	*EXP*	...	**FILE** *Export*	...

AutoCAD can *Export* your entire drawing or portions of it to many vector file formats other than .DWG. This can be accomplished by two methods: (1) use the *Export* command and select the desired file type to export, or (2) use the specific export command as discussed earlier, such as *Dxfout* or *Psout*. (.BMP raster files and .DXX data extract files can also be created using *Export*.)

Use the *Export* command to produce the *Export Data* dialog box (Fig. 32-19). Here you can choose the file format you want AutoCAD to create during the export process.

FIGURE 32-19

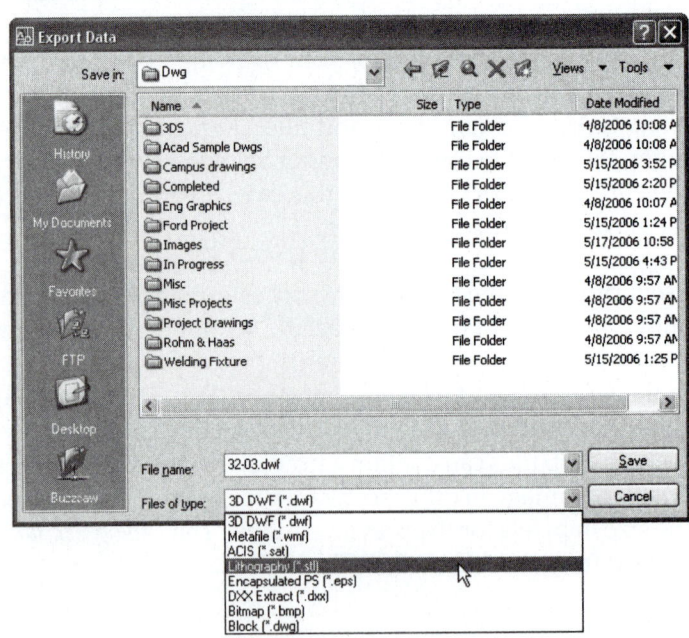

The available file formats that AutoCAD can export are listed in the following table along with the equivalent specific AutoCAD commands.

File Extensions	AutoCAD Equivalent Command	Description
3D DWF	PLOT	DWF file format
*.BMP	BMPOUT	Windows device-independent bitmap format
*.DWG	WBLOCK	AutoCAD drawing file format
*.DXX	ATTEXT	Attribute extract DXF format
*.EPS	PSOUT	Encapsulated PostScript format
*.SAT	ACISOUT	ACIS solid object format
*.STL	STLOUT	Solid object stereolithography format
*.WMF	WMFOUT	Windows Metafile format

Creating PDF Files

Using the *Plot* dialog box, you can create an Adobe PDF (Portable Document Format) file from an AutoCAD drawing. The process is similar to creating a .DWF file using the *DWF6 ePlot* driver in the *Plot* dialog box. Adobe PDF files can be read using Adobe Reader®, a viewer that can be downloaded free from www.adobe.com.

Adobe PDF files are extremely popular and widely used. They can be used for text documents, raster images, and vector images. Since Adobe Reader is free, reading an AutoCAD drawing in .PDF format is available for those without AutoCAD.

DWG To PDF.pc3 Driver

To create a .PDF file from an AutoCAD drawing, first open the desired drawing in AutoCAD. Next, use the *Plot* command to produce the *Plot* dialog box. In the *Printer/plotter* section, use the drop-down list to locate and select the *DWG ToPDF.pc3* printer driver (Fig. 32-20).

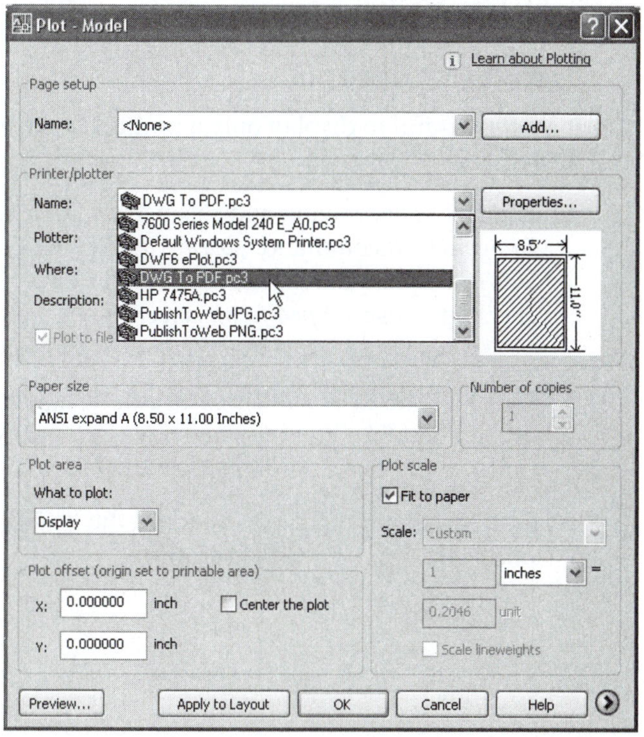

Creating the .PDF file is similar to creating a print or plot. That is, the options that affect a normal plot also affect the format for the .PDF file. For example, you have options for *Paper size*, *What to plot*, *Plot offset*, *Scale*, and so on.

FIGURE 32-20

CHAPTER EXERCISES

1. *Image Attach*

   You can use the *Image* command to enhance the appearance of your drawing by attaching a raster image such as a company logo, for example.

   A. *Open* the **SADDL-DM** drawing from Chapter 29 Exercises. *Zoom* in to the titleblock area in the lower-right corner of the drawing. You will attach a raster image in the titleblock as a company logo.

   B. Using the *External References* palette, select *Attach* from the right-click short-cut menu or select *Attach Image* from the options in the top-left corner of the palette. When the *Select Image File* dialog box appears, locate and select the **Finishes.Flooring.Tile.Square.Circular Mosaic.jpg** image from the **Textures** folder. (To find where your *Textures* folder is located, use *Options*, *Files* tab, then expand *Texture Maps Search Path*.) Select this file and *Open* it. In the *Image*

   FIGURE 32-21

   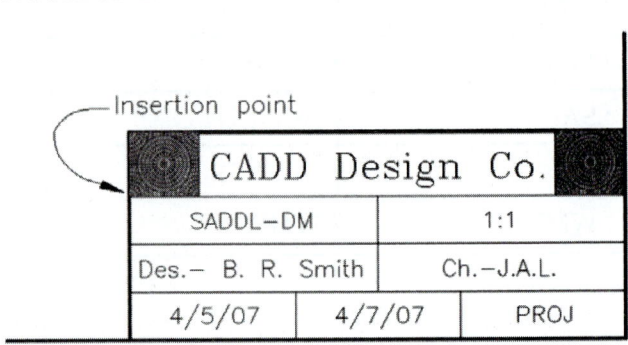

   dialog box choose **OK**. PICK the insertion point as shown in Figure 32-21 and scale the image by dragging it so that the image frame fits within the first text area of the title block.

   C. Select the attached image and *Copy* it to the other side of this title area. Compare your results to Figure 32-21.

   D. Use *Imageadjust* and select one of the image's frames. Set **Brightness** to **60** and **Contrast** to **80**. The logo should now appear as a pure black and white image. Do the same for the other image. *Zoom All* and save this drawing as **SADDL-DM2**.

2. *Image Attach, Imageclip*

   It is often useful to display only a portion of an attached image in your drawing. You can accomplish this with *Imageclip*.

   A. Begin a *New* drawing and select **Start from Scratch**, **Imperial** default settings. Use the **Imageattach** command. Locate the **WORLDMAP.TIF** file from the **Program Files/ AutoCAD 2007/Sample/VBA** folder, then select **Open**. In the *Image* dialog box, ensure **Full path** is checked. Ensure **Specify on screen** is checked in all boxes and attach the watch image at an *Insertion point* of **2,2**, a *Scale factor* of **5**, and *Rotation angle* of **0**.

   B. Now use the *External References* palette again to attach the map a second time to your drawing at the point **8,2**. Use the same *Scale factor* and *Rotation angle* as the first attachment, but this time specify these parameters in the *Image* dialog box.

   C. Notice that although you attached the WORLDMAP.TIF image twice it is referenced only once in the list of attached images.

D. Now use the *Imageclip* command to create the two clipping boundaries as shown in Figure 32-22.

FIGURE 32-22

First attachment of WORLDMAP.TIF with a rectangular clip boundary turned On.

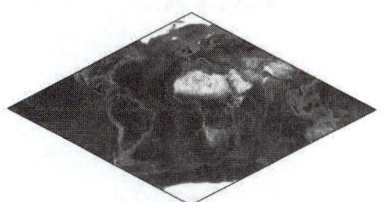

Second attachment of WORLDMAP.TIF with a polygonal clip boundary turned on.

E. After you have created the clip boundaries, use the *Imageclip* command again to turn the boundaries *OFF* and then back *ON* again. You will find that the clip boundary is saved even though you turn it *OFF* to display the entire image. Save this drawing as **CLIPPING**.

3. *Draworder*

Raster images are also useful for reference purposes during drawing construction. You can utilize aerial photographs, for example, in order to construct your drawing in relation to existing structures. When using images as reference information, you often need to reorder the display of objects on your drawing.

A. Begin a *New* drawing and select *Start from Scratch*, *English* default settings. Use the *Imageattach* command. Locate and *Open* the **ARCHRASTER.GIF** file from the **Program Files/AutoCAD 2007/Support** folder. Attach the image to your drawing at **1,1** with a *Scale factor* of **10**, and *Rotation angle* of **0**. *Zoom All*.

B. Use *Imageadjust*, select the image, and set the *Brightness* to **60** and *Contrast* to **70**.

C. Using the *Circle* command, draw a circle near the center of the image with a *Radius* of **1.5**. Use the *Hatch* command to hatch the object with a *Solid* pattern as shown in Figure 32-23.

FIGURE 32-23

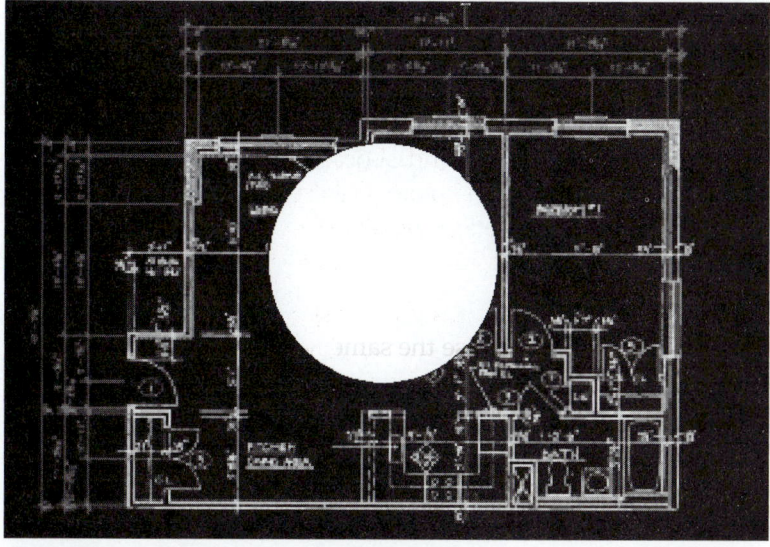

D. Notice that the solid hatch pattern is displayed on top of the image. Now use *Draworder*, select the solid hatch pattern and the *Circle*, and send them to the *Back*. They are no longer visible since the objects are now in back of the raster image. You can reverse the display of the objects by using *Draworder* again, selecting the raster image (by selecting its frame), and sending it to the *Back* of the display order. Save this drawing as **DWGORDER**.

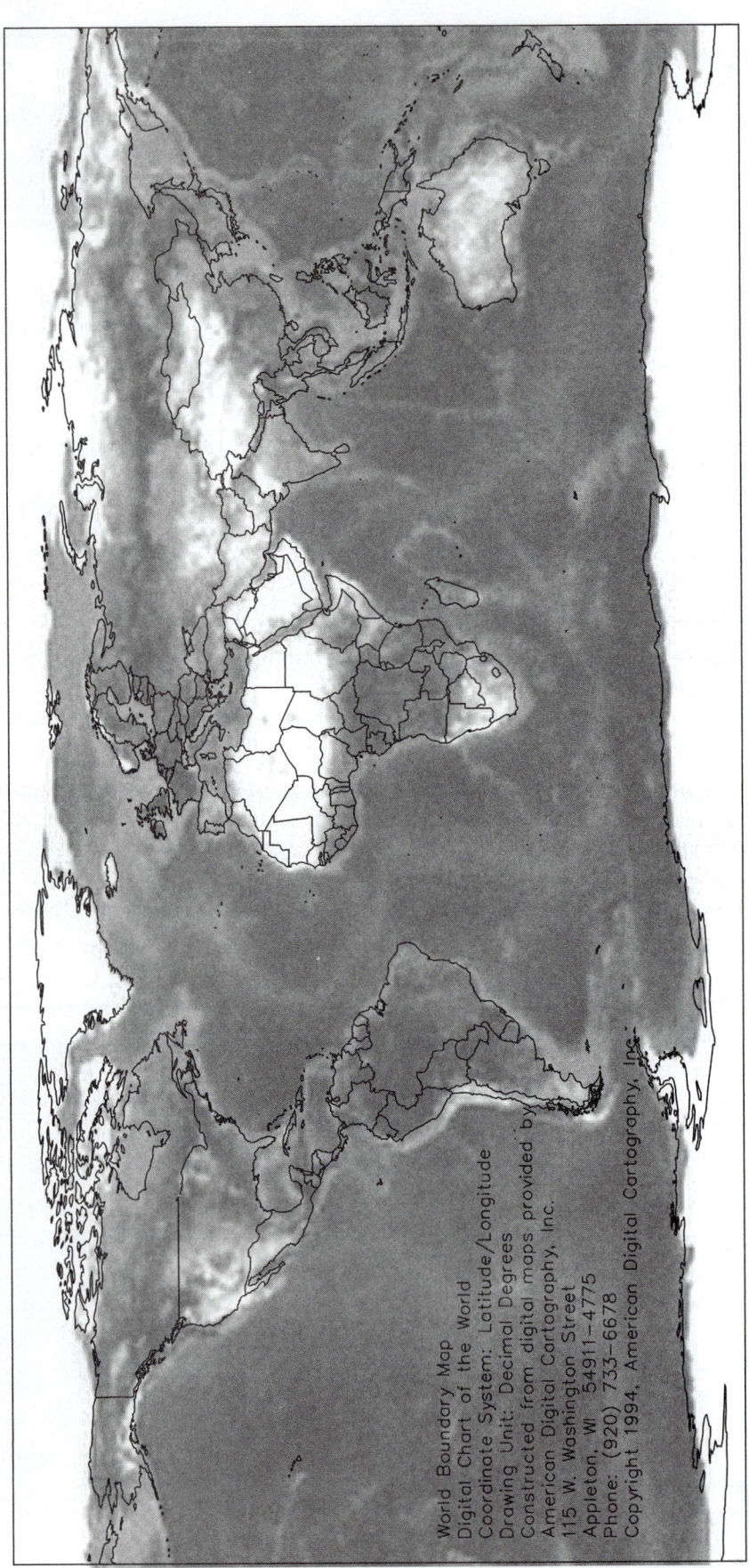

World Boundary Map
Digital Chart of the World
Coordinate System: Latitude/Longitude
Drawing Unit: Decimal Degrees
Constructed from digital maps provided by
American Digital Cartography, Inc.
115 W. Washington Street
Appleton, WI 54911-4775
Phone: (920) 733-6678
Copyright 1994, American Digital Cartography, Inc.

MAP2GLOBE.DWG, Courtesy of Autodesk, Inc.

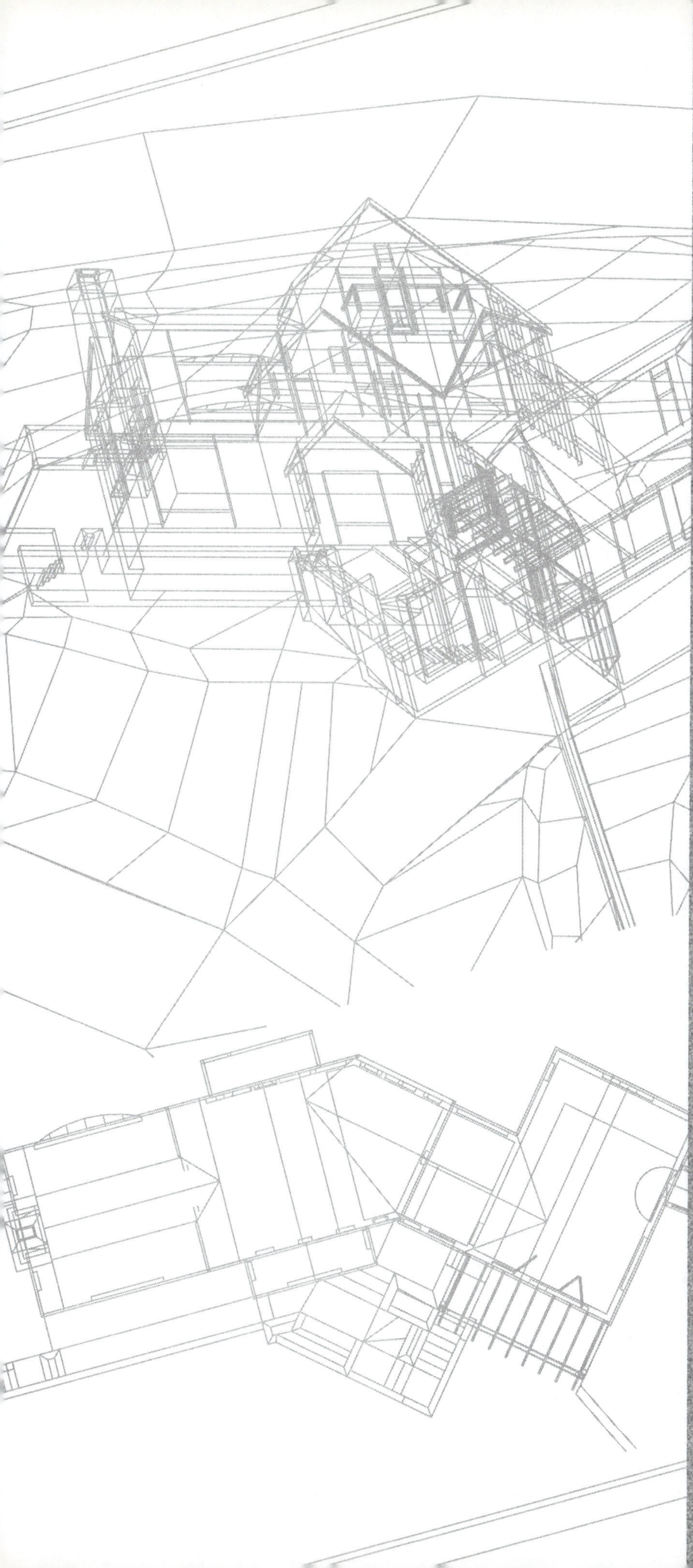

ADVANCED LAYOUTS AND PLOTTING

CHAPTER OBJECTIVES

After completing this chapter you should:

1. be able to use the *Layer Properties Manager* or the *Vplayer* command to control layer visibility in viewports;

2. be able to set *PSLTSCALE* for multiple viewports and layouts;

3. know how to dimension in paper space;

4. know the "Reverse Method" for determining the viewport drawing scale factor and how to use it to specify size-related settings;

5. know several applications for creating multiple viewports and layouts and considerations for doing so;

6. know the difference between color-dependent and named Plot Style Tables;

7. be able to create a new Plot Style Table using *Add-A-Plot Style Table Wizard*;

8. be able to edit a Plot Style Table using the Plot Style Table Editor;

9. be able to assign named plot styles to layers and objects;

10. know how to apply plot stamps to your prints and plots.

CONCEPTS

This chapter is concerned with advanced concepts of using layouts and plotting those layouts—specifically creating multiple viewports per layout, creating multiple layouts, attaching Plot Style Tables to layouts, and assigning plot styles to layers and objects. To isolate these complex ideas as an attempt to simplify their explanation, the chapter is divided into two sections: Advanced Layouts and Advanced Plotting.

Topics discussed in this chapter are arranged as follows:

Advanced Layouts

>   Layer Visibility for Viewports (*Layer, Current VP Freeze, New VP Freeze, Vplayer*)
>   Linetype Scale in Viewports (*Psltscale*)
>   Dimensioning in Paper Space
>   The "Reverse Method" for Calculating Drawing Scale Factor
>   Applications for Multiple Viewports
>   Considerations for Using Multiple Viewports and Layouts

Advanced Plotting

>   Plot Style Tables and Plot Styles
>   Plot Stamping

ADVANCED LAYOUTS

You are already familiar with creating and setting up drawings (Chapters 6 and 12), creating layouts and viewports (Chapter 13), and plotting layouts (Chapter 14). The ideas and examples used in those chapters are fundamental to your ability to set up and create drawings in AutoCAD; however, they were based on the premise of using only one large viewport per layout. The first half of this chapter deals with concepts related to setting up multiple viewports and multiple layouts.

To this point, you have learned the following. Objects that represent the subject of the drawing (model geometry) are normally drawn in model space. Dimensioning and related annotation (text) are traditionally performed in model space because they are directly associated to the model geometry. A layout simulates the sheet of paper you will print or plot on, so it is used to prepare the desired display of the model geometry for plotting. A viewport is created in a layout with the *Vports* command in order to "see" the model geometry. The display of the geometry in the viewport should be scaled to achieve an appropriate scale for the drawing. Since a layout represents the plotted sheet, you normally plot the layout at 1:1 scale. Therefore, the geometry displayed in the viewports is scaled to the desired plot scale by using the *Viewports* tool bar, the *Properties* palette for the viewport, or a *Zoom XP* factor. With the layout, you can use the *Page Setup* and *Plot* dialog boxes to specify and save the print or plot parameters such as plot device, paper size, and orientation, etc.

Now that you are experienced with those fundamentals, consider the following possibilities. There are many applications for creating multiple viewports per layout and/or creating multiple layouts in one drawing, each displaying different drawing geometry or the same geometry at different scales. For example, you may want to create one plot to show several views of one drawing, such as an overall view of an assembly and several "detail" views of individual components. Or you may want to show a floor plan and several views of individual rooms. You may have the need to show several separate drawings on one plot, which can be accomplished by *Xrefing* several drawings, creating several viewports, and controlling each viewport to display only the layers associated with one *Xref*. In another application, you may need to set up multiple layouts for creating multiple sets of plots, each layout displaying a different combination of layers. Using this scheme, for example, you can create and save a layout to plot the floor plan and the plumbing details for the plumbing contractor, a layout to plot the floor plan and the electrical layout for the electrical contractor, a layout for the floor plan and the HVAC layout, and so on.

When using multiple viewports and multiple layouts, the setup becomes more complex but offers more alternatives. Remember, you must scale the display of the geometry in each viewport and you must control the layer visibility in each viewport (which layers are visible in which viewports). When you have multiple viewports, you cannot effectively use the "drawing scale factor" to determine the viewport scale since that idea is based on preparing the drawing for only one large viewport. In addition, you cannot effectively control which layers are visible in which viewports using the global *Freeze/Thaw* or *On/Off* controls.

This chapter explains how you can control the viewport-specific layer visibility using the *Current VP Freeze* button and *New VP Freeze* button in the *Layer Properties Manager* or *Layer Control* drop-down list. In addition, another strategy for calculating the drawing scale factor for viewports is discussed. The "reverse method" is used to help determine appropriate values for *Ltscale*, *Dimscale*, text height, etc. based on the viewport scale.

This chapter also presents a totally different strategy you may want to consider when using multiple viewports and layouts, that is, creating dimensions in paper space, creating text in paper space, and setting *Psltscale* to paper space. This strategy prevents you from having to calculate and set the text height, *Dimscale*, and *Ltscale* for each viewport and having to control the layer visibility specific to each viewport.

Historically, the use of paper space, viewports, and layouts has changed from early AutoCAD releases having only a few capabilities (and therefore few applications) to current releases offering advanced capabilities and applications. In contrast to pre-AutoCAD 2000 releases, the common use of paper space, viewports, and layouts represents a major paradigm shift in the way drawings are prepared and plotted. Although this technology continues to advance and become more "user-friendly," these concepts are not simple to understand nor easy to carry out. Because there are currently different applications, strategies, and personal preferences for creating and setting up multiple viewports and layouts, this chapter presents the alternatives and discusses the advantages and disadvantages of each. In this way, you can make the appropriate decisions based on your specific applications, preferences, and level of technology.

Layer Visibility for Viewports

AutoCAD provides you with the capability to create multiple layouts. In each layout, you can create one or several viewports using *Vports* or *-Vports*. Even though you can have multiple viewports in a drawing, each viewport displays the same model space geometry, right? Well, yes and no. That is, there is only one model space, and by default the same model geometry is displayed in every viewport. You can, however, control what parts of the model space geometry are displayed in each viewport by (1) using *Zoom*, *Pan*, and other viewing commands to display different areas of model space and (2) using layer visibility to control what layers appear in which viewports. This section discusses layer visibility control for viewports.

Layer

Pull-down Menu	Command (Type)	Alias (Type)	Short-cut	Screen (side) Menu	Tablet Menu
Format Layer...	Layer or -Layer	LA or -LA	...	FORMAT Layer	U,5

The *Layer* command produces the *Layer Properties Manager* dialog box. This dialog box can be used to control which layers are visible in which viewports. Two small icon buttons in the *Layer Properties Manager* dialog box are used for viewport-specific layer visibility, *Current VP Freeze* and *New VP Freeze*.

Figure 33-1 displays the *Layer Properties Manager* showing the *Current VP Freeze* and *New VP Freeze* columns (last two columns). <u>These icons do not appear when the *Model* tab is current.</u> If a layout tab is selected and a viewport is active (the cursor appears in a viewport), use these icons to freeze or thaw layers <u>in the current viewport</u> or to freeze or thaw layers <u>in new viewports</u> (the selected layer will be frozen or thawed for any new viewports that are created).

FIGURE 33-1

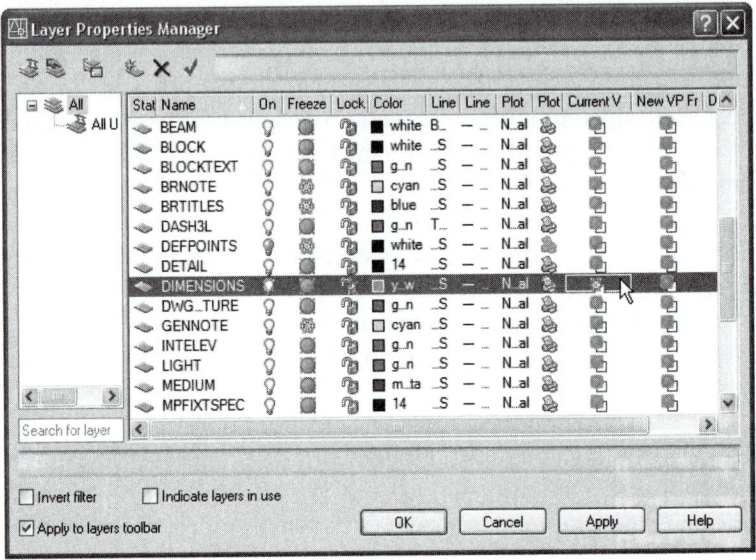

Current VP Freeze

For example, assume you have two viewports created in a layout and each viewport displays the same model space geometry (Fig. 33-2). Notice that in the left viewport all layers are visible compared to the right viewport, in which the dimension layer is not visible. To freeze the "Dimensions" layer in the right viewport only, first make the right viewport current, then use *Layer* to produce the *Layer Properties Manager* and select the *Current VP Freeze* icon to change this layer to a frozen state <u>for that viewport</u> (snowflake; see Fig. 33-1). If the left viewport were made active and then the *Layer Properties Manager* invoked, these icons for all layers would appear thawed (sunshine) for the current (left) viewport.

FIGURE 33-2

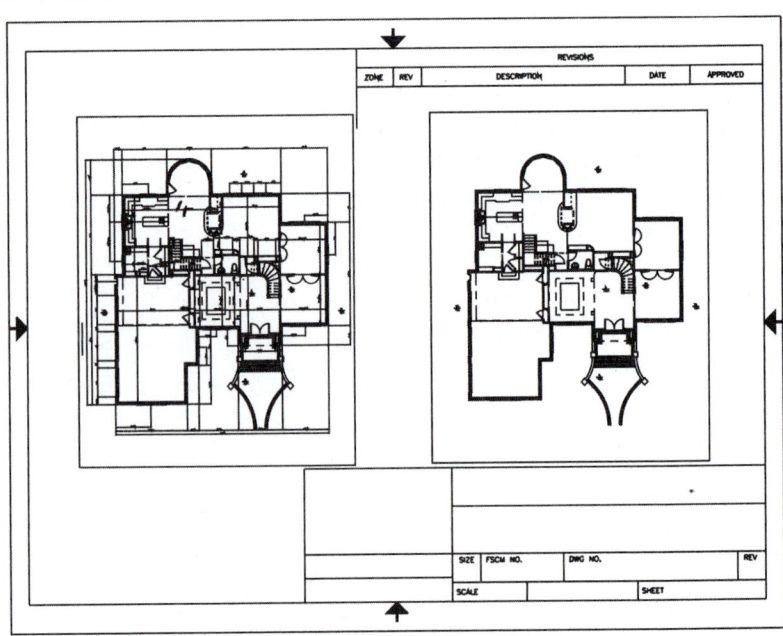

When a viewport is current, the *Current VP Freeze* icons that appear give the visibility state of layers <u>specific to the viewport that is current</u> when the list is displayed. In other words, the icons (*Current VP Freeze*) may display different information depending on which viewport is current when the list is viewed. Changing the *Freeze/Thaw* icon affects the layer's state globally (for all viewports in the drawing), while changing the *Current VP Freeze* icon affects the layer's state only <u>for the current viewport</u>.

A *Layout* tab itself is considered a paper space viewport when dealing with layer visibility. That is, if you are in paper space (*Pspace* is active, not in a viewport) and you PICK the *Current VP Freeze* icon, the highlighted layer becomes frozen for the entire layout (the current viewport), but not in the individual viewports or in the *Model* tab.

A *Freeze or Thaw in Current Viewport* icon appears in the *Layer Control* drop-down list (Fig. 33-3). This icon has the identical function as the *Current VP Freeze* icon in the *Layer Properties Manager*. The *New VP Freeze* button is not available in the drop-down list, only in the *Layer Properties Manager*.

FIGURE 33-3

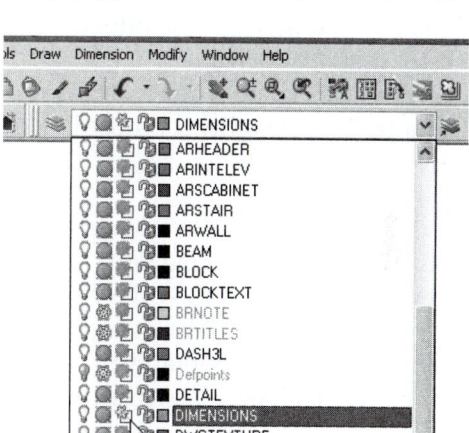

New VP Freeze

This icon prevents the selected layer from appearing in any viewports that are created from that time on. Any layers selected with the *New VP Freeze* attribute do not appear in subsequently created viewports.

This feature is helpful when you need several viewports, each to display a separate drawing. For example, assume you want to *Xref* several drawings, but only want one drawing to appear in each viewport. Assume you created one viewport and *Xrefed* a drawing into model space so it (all its layers) appeared in the viewport (Fig. 33-4). If you were to create a second viewport, the same drawing would normally appear in the second viewport just as all model space geometry normally appears in all viewports. If you were to *Xref* a second drawing into model space, it would also normally appear in <u>both</u> viewports. (See Chapter 30, Xreferences, for details on *Xref*.)

FIGURE 33-4

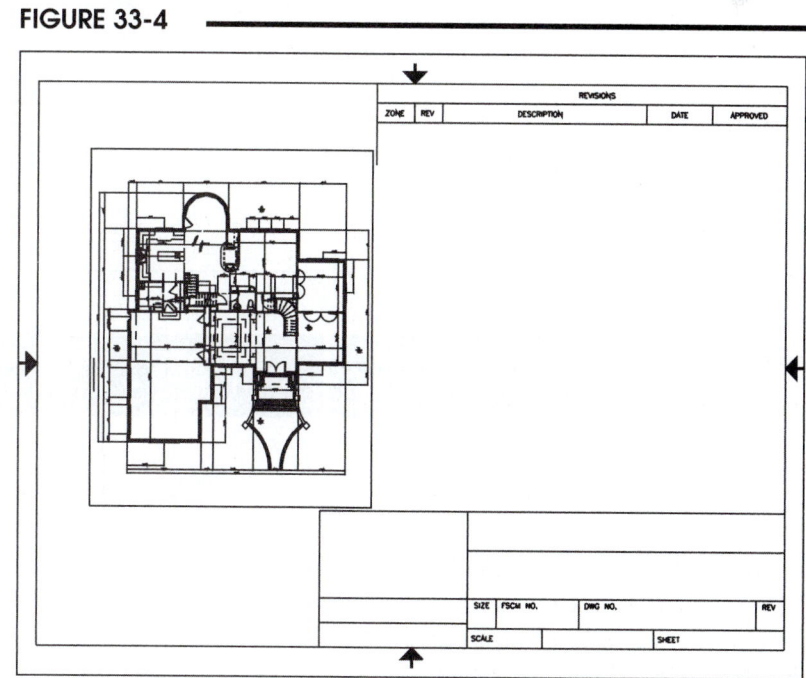

TIP

To avoid this problem, after you create one viewport and *Xref* the first drawing but before you create the second viewport, use *New VP Freeze* for all layers of the *Xref* drawing (Fig. 33-5, all "Xref1 | *" layers). (Unlike the *Current VP Freeze* icon, the *New VP Freeze* icon displays the same information for a specific layer no matter which viewport is current when the list is displayed.)

Note also that the *Layer Properties Manager* automatically includes a group filter for *Xref* layers (for any drawing containing *Xrefs*) to assist you in controlling visibility for those layers as a group.

Next, create the second viewport. The original *Xref* drawing does not appear in the second viewport (right) since all of its layers are frozen for that viewport and all other new viewports (Fig. 33-6).

FIGURE 33-5

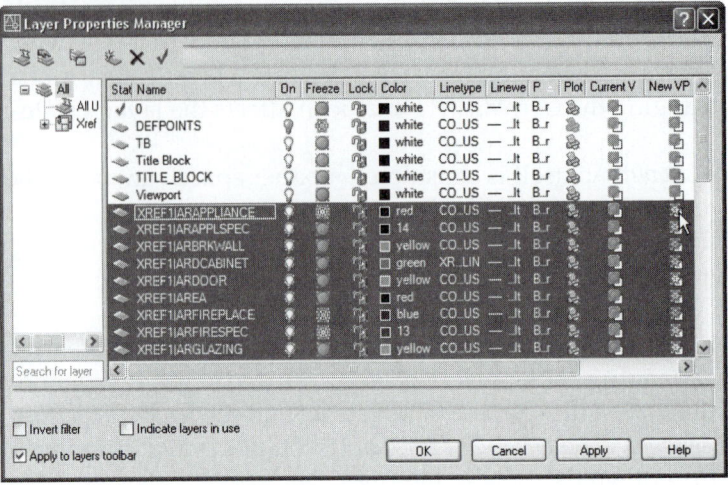

FIGURE 33-6

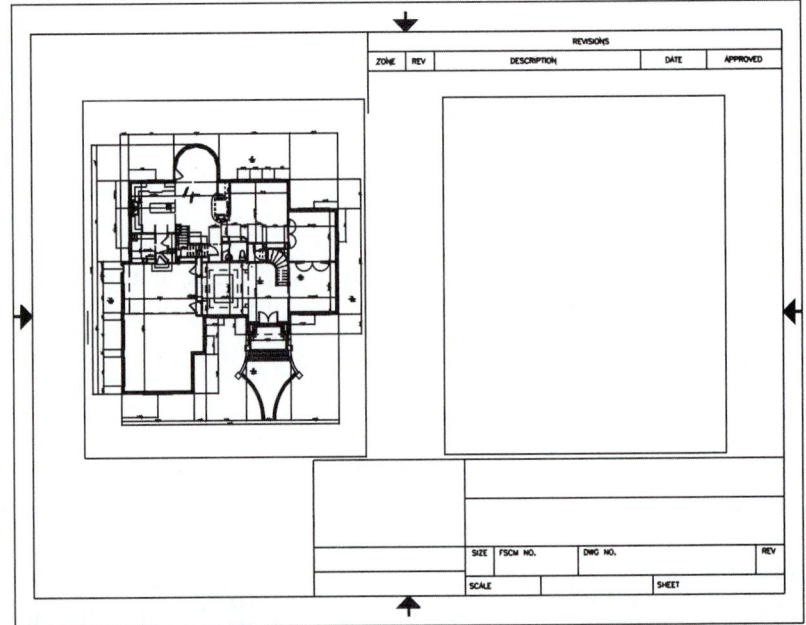

Vplayer

Pull-down Menu	Command (Type)	Alias (Type)	Short-cut	Screen (side) Menu	Tablet Menu
...	*Vplayer*	...	...	...	...

As an alternative to using the *Layer Properties Manager* or the *Layer Control* drop-down list, the *Vplayer* (Viewport Layer) command provides options for control of the layers you want to see in each viewport (viewport-specific layer visibility control). With *Vplayer* you can specify the state (*Freeze/Thaw* and *Off/On*) of the drawing layers in any existing or new viewport. Although the *Layer* command allows you to *Freeze/Thaw* layers <u>globally</u> and for <u>viewports</u>, *Vplayer* allows you to *Freeze/Thaw* only the <u>viewport-specific</u> layer setting.

A *Layout* tab must be set current for the *Vplayer* command to operate, and a layer's global (*Layer*) settings must be *On* and *Thawed* to be affected by the *Vplayer* settings. You can use *Vplayer* while you are in paper space or in any viewport. The simplest method, however, is to make the desired viewport current, then use *Vplayer*.

For example, you may have two viewports set up in a layout, such as in Figure 33-7. Assume you want to *Freeze* the DIM (dimensioning) layer for the right viewport, so you make that the current viewport (your cursor is in the right viewport). In order to control the visibility of the DIM layer for the current viewport, the command syntax shown below would be used.

FIGURE 33-7

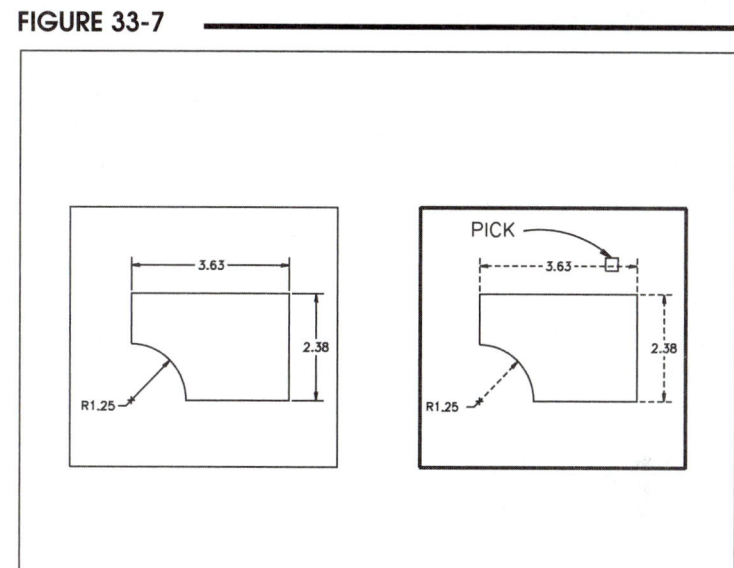

```
Command: vplayer
Enter an option [?/Freeze/Thaw/Reset/Newfrz/Vpvisdflt]: f
Enter layer name(s) to freeze or <select objects>: Enter
Select objects: (Select an object on the desired layer.  See Fig. 33-7.)
Select objects: Enter
Enter an option [All/Select/Current] <Current>: Enter
Enter an option [?/Freeze/Thaw/Reset/Newfrz/Vpvisdflt]: Enter
Command:
```

This action results in *Freezing* the DIM layer (containing the highlighted objects) in the right viewport.

You <u>do not</u> have to make the desired viewport current before using *Vplayer*—your cursor can be in paper space or in another viewport. Invoke *Vplayer* at any time. Assume that you are in the left viewport, but want to *Freeze* the DIM layer for the right viewport. In the command sequence on the next page, notice that AutoCAD allows you to switch to paper space to PICK the desired <u>viewport object</u> (viewport borders can be PICKed only from paper space). This sequence results in *Freezing* the DIM layer for the right viewport (Fig. 33-8).

FIGURE 33-8

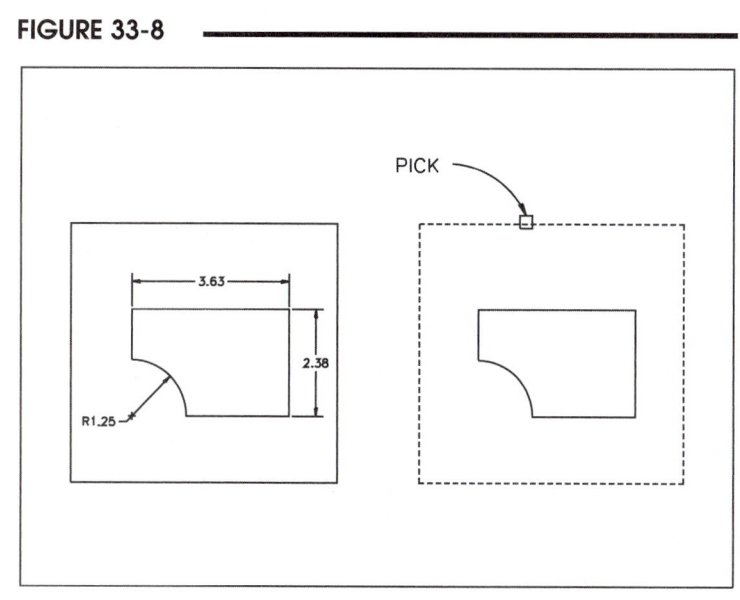

```
Command: vplayer
Enter an option [?/Freeze/Thaw/Reset/Newfrz/Vpvisdflt]: f
Enter layer name(s) to freeze or <select objects>: dim
Enter an option [All/Select/Current] <Current>: s
Switching to Paper space.
Select objects: PICK (Select the desired viewport object. See Fig. 33-8.)
Select objects: Enter
Switching to Model space.
Enter an option [?/Freeze/Thaw/Reset/Newfrz/Vpvisdflt]: Enter
Command:
```

Note that the *Select* option switches to paper space if necessary so that the viewport objects (borders) can be selected. The *Freeze* or *Thaw* does not take place until the command is completed and the drawing is automatically regenerated.

The *Vplayer* **Current** option automatically selects the current viewport. The **All** option automatically selects all viewports. The other options are explained next.

?

This option <u>lists the frozen layers</u> in the current viewport or a selected viewport. AutoCAD automatically switches to paper space if you are in model space to allow you to select the viewport object.

Freeze

This option controls the visibility of specified layers in the <u>current or selected viewports</u>. You are prompted for layer(s) to *Freeze*, then to select which viewports for the *Freeze* to affect.

Thaw

This option turns the visibility control back to the global settings of *On/Off/Freeze/Thaw* in the <u>Layer</u> command. The procedure for specification of layers and viewports is identical to the *Freeze* option procedure.

Reset

Reset returns the layer visibility status to the <u>default</u> if one was specified with *Vpvisdflt*, which is described next. You are prompted for "Layer(s) to Reset" and viewports to select by "All/Select/<Current>:".

Newfrz

This option <u>creates new layers</u>. The new layers are frozen in all viewports. This is a shortcut when you want to create a new layer that is visible only in the current viewport. Use *Newfrz* to create the new layer, then use *Vplayer Thaw* to make it visible only in that (or other selected) viewport(s). You are prompted for "New viewport frozen layer name(s):". You may enter one name or several separated by commas.

Vpvisdflt

This option allows you to set up <u>layer visibility</u> for <u>new</u> viewports. This is handy when you want to create new viewports but do not want any of the <u>existing</u> layers to be visible in the new viewports. This option is particularly helpful for *Xrefs*. For example, suppose you created a viewport and *Xrefed* a drawing "into" the viewport. Before creating new viewports and *Xrefing* other drawings "into" them, use *Vpvisdflt* to set the default layer visibility to *Off* for the <u>existing</u> *Xrefed* layers in new viewports. (See the example for *Layer, New VP Freeze*.) You are prompted for a list of layers and whether they should be *Thawed* or *Frozen*.

Linetype Scale in Viewports

Layouts give you the capability of creating several viewports that look into model space. These viewports can contain several views of one or more drawings at different scales. Because each viewport can display linetype scaling differently based on viewport scale, the *PSLTSCALE* variable is needed to control how multiple linetype scales appear in one layout.

PSLTSCALE

Pull-down Menu	Command (Type)	Alias (Type)	Short-cut	Screen (side) Menu	Tablet Menu
Format *Linetype...* *Details >>* *Use paper space units*	*Psltscale*	...	...	*FORMAT* *Linetype...* *Details >>* *Use p.s. units*	...

LTSCALE controls linetype scaling <u>globally</u> for the drawing, whereas the *PSLTSCALE* variable controls the linetype scaling for non-continuous lines for viewing in <u>paper space</u>. *LTSCALE* is a variable that accepts any value other than 0, whereas *PSLTSCALE* is a variable that accepts either a 0 (*Off*) or 1 (*On*). *LTSCALE* can still be used to change linetype scaling globally regardless of the *PSLTSCALE* setting.

If you want the linetype scale to always appear the same with respect to the drawing geometry, set *PSLTSCALE* to 0. This ensures that non-continuous (dashed) lines have a constant scale (based on the *LTSCALE*) with respect to the geometry regardless of how those lines are viewed. In other words, assume you created a hidden line between two points such that the line contained three short dashes. When *PSLTSCALE* is 0, you would always see three dashes whether they were viewed from the *Model* tab or whether the line appeared in two different viewports at two different scales. Even when you *Zoomed*, the line would have three dashes. With *PSLTSCALE* set to 0, the scale of non-continuous lines is controlled only by *LTSCALE*. A *PSLTSCALE* setting of 0 is suggested for drawings using only one viewport or when multiple viewports or layouts display the geometry at one scale.

If *PSLTSCALE* is set to 0, linetype spacing is controlled exclusively by the *LTSCALE* variable and is relative to the drawing units in model space (or in paper space when lines are created there). In other words, if *PSLTSCALE* is 0, the dashed lines would always have the same number of dashes for one *LTSCALE* setting. However, when the same lines are viewed in different viewports in different scales, the linetypes appear different lengths as in Figure 33-9.

FIGURE 33-9

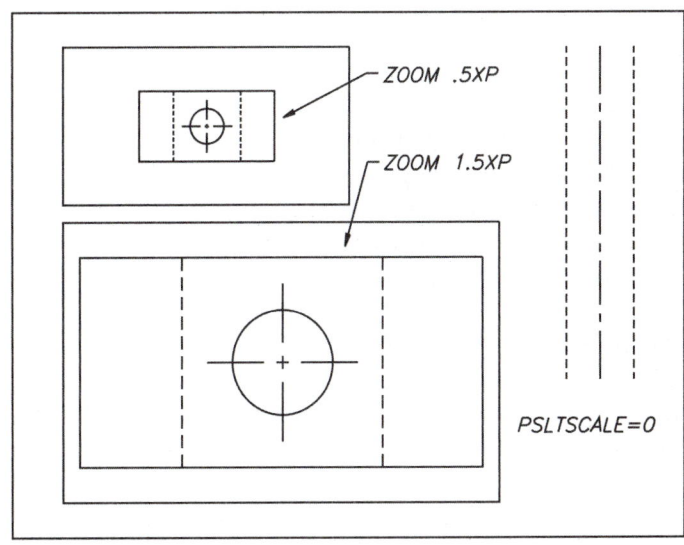

TIP

A *PSLTSCALE* setting of 1 (the default setting) is recommended when you are viewing the same geometry in multiple viewports at different scales. A *PSLTSCALE* setting of 1 makes all non-continuous lines appear to have the same linetype scaling, regardless of viewport scale (or *Zoom XP*) settings. Technically, *PSLTSCALE* set to 1 displays all non-continuous lines relative to paper space units so that all lines in paper space and in different viewports appear the same. Therefore, the model geometry in viewports can have different viewport scales (or *Zoom XP* factors), yet the linetypes display with the same scale (Fig. 33-10).

FIGURE 33-10

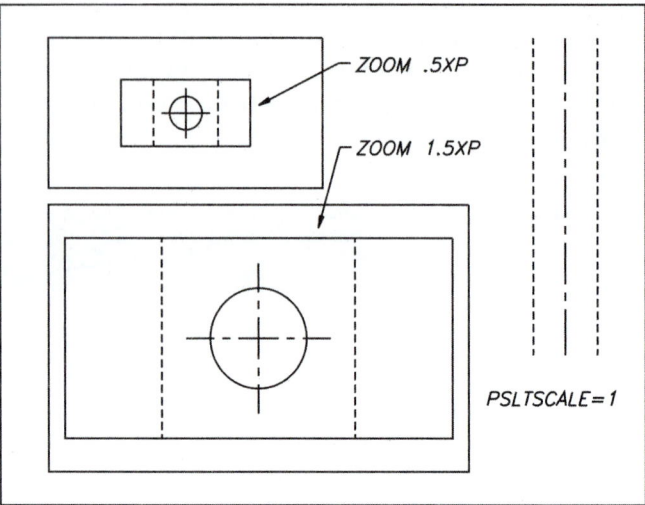

Set *PSLTSCALE* at the Command prompt by entering "Psltscale" and typing 1 or 0. *PSLTSCALE* can also be set using the *Linetype* command which produces the *Linetype Manager* (Fig. 33-11, lower left). You must use the *Show Details* button for this feature to appear. A check appearing in the *Use paper space units for scaling* checkbox sets *PSLTSCALE* to 1. The *Global scale factor* represents the *LTSCALE* value and the *Current object scale* represents the *CELTSCALE* value (see Chapter 12 for more information). If *PSLTSCALE* is changed, a *Regenall* must be used to display the effects of the new setting.

FIGURE 33-11

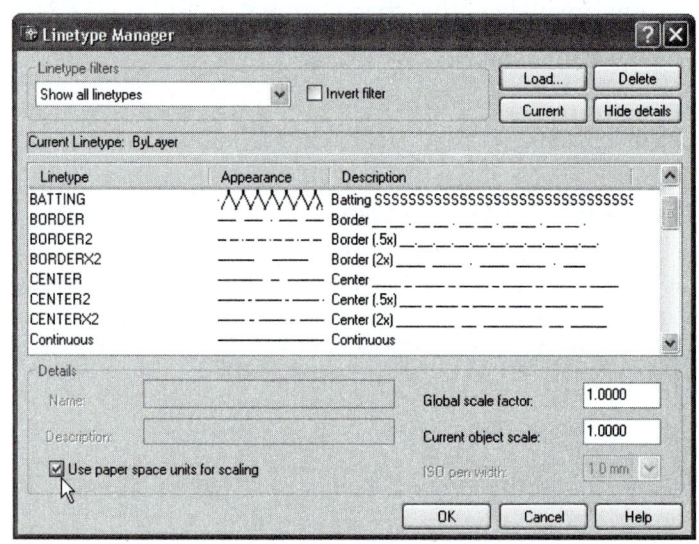

As an example, consider the crankshaft drawing (Fig. 33-12) that appears in its entirety in the top viewport at 1:1 and as a detail in the lower viewport at 2:1. Notice that the hidden line dashes are almost imperceptible in the top viewport, whereas the hidden lines in the lower viewport are much clearer. In this case, *PSLTSCALE* = 0, so the linetype scale is proportional to the model space geometry, no matter at what scale it is viewed.

FIGURE 33-12

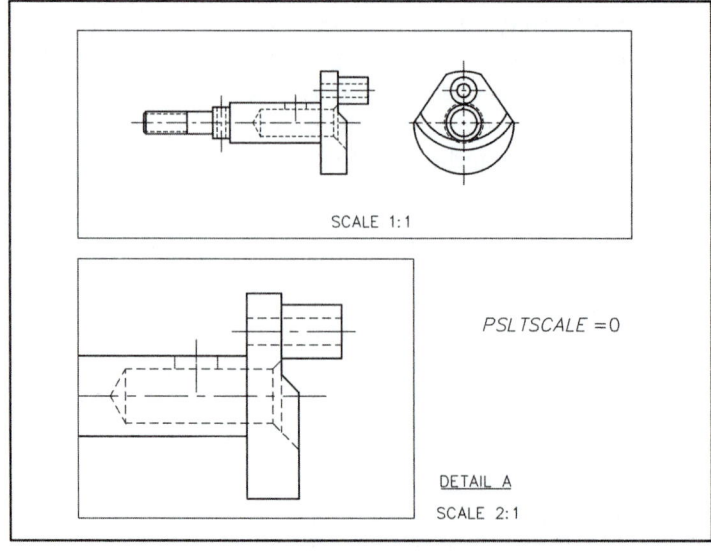

In contrast, when *PSLTSCALE* is set to 1 (Fig. 33-13), hidden line dashes in both viewports appear the same—they are automatically adjusted for the viewport scale. Another way to grasp this is that the linetype scale is proportional to paper space units, not model space units. In most cases where a layout contains multiple viewports, a *PSLTSCALE* setting of 1 is more practical.

FIGURE 33-13

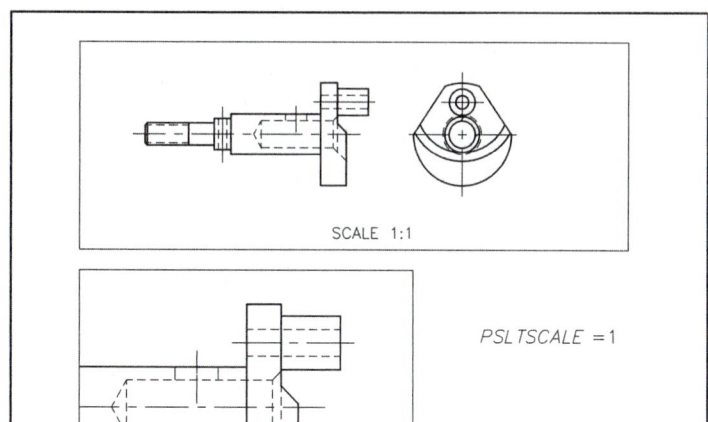

SCALE 1:1

PSLTSCALE =1

DETAIL A
SCALE 2:1

Dimensioning in Paper Space

When a drawing is presented in one viewport, or is presented in more than one viewport but at the same scale, dimensioning in <u>model space</u> is recommended. In these cases, dimensioning in model space is the simplest and most versatile option. However, in cases where a drawing is displayed in multiple viewports at different scales, dimensioning is more complex. Even though the geometry appears in different scales, dimensions should always appear at the same size (dimension text is typically .125" or 6mm).

There are two basic alternatives for creating dimensions displayed in multiple viewports at different scales: 1) put different sets of dimensions in <u>model space</u> on different layers and with different *Overall scales* (dimension styles), then control the viewport-specific visibility so the correct dimensions appear in the appropriate viewports, and 2) create the dimensions in <u>paper space</u> all in one size (for AutoCAD 2002 and newer). (See Chapter 29 for basic concepts on dimensioning in paper space.)

As an example for dimensioning in paper space, assume you created a mechanical drawing of a plate to be stamped (Fig. 33-14). You chose to show an overall view of the plate and a detail view giving the dimensions specific to the small notch. Therefore, a layout with two viewports is created, with the viewport scale for the full view set to 1:1 (left viewport) and viewport scale for the detail view set to 2:1.

FIGURE 33-14

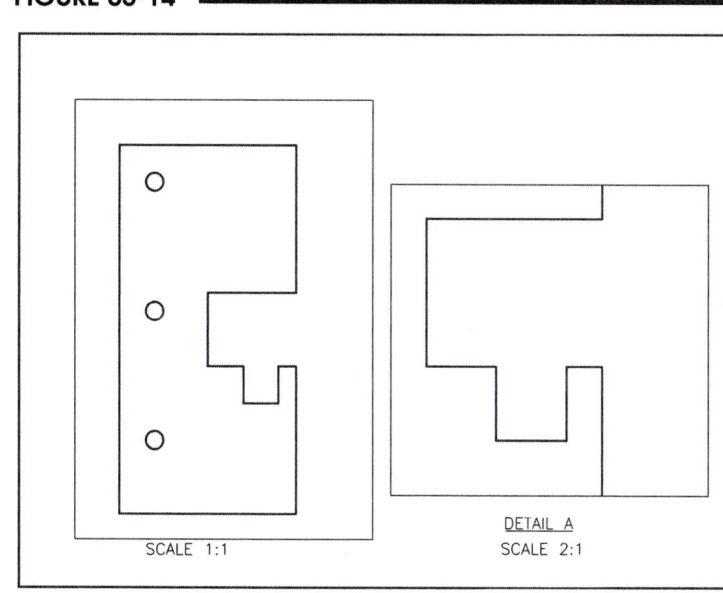

SCALE 1:1

DETAIL A
SCALE 2:1

Assume you then activate the *Model* tab and begin to dimension the object as you would normally. Checking the layout to see how the first few dimensions look, you notice that the dimensions appear in both viewports, and those in the detail view are also scaled 2:1 with the geometry (Fig. 33-15).

FIGURE 33-15

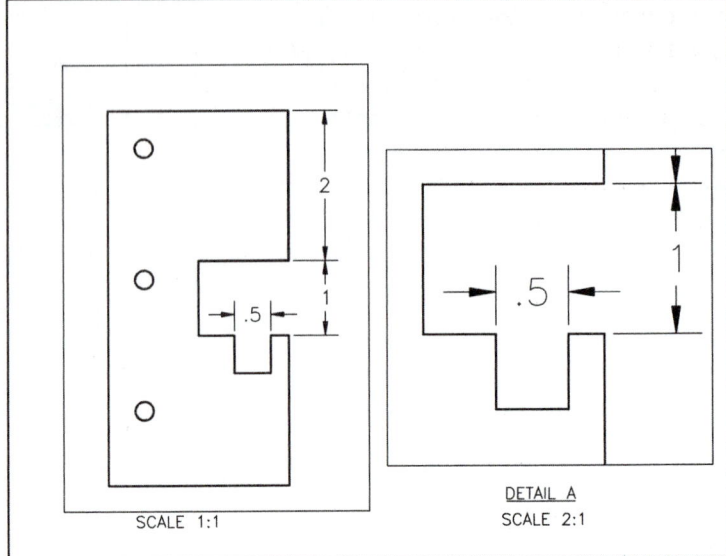

As one solution to the problem, you erase all the model space dimensions. Then, in the layout, you begin dimensioning <u>in paper space</u> with the dimensions associated with (attached to) the model space geometry (Fig. 33-16). You can use the same dimension style as before with an *Overall scale* of 1, since the layout will be printed 1:1. All dimensions have the same size, since they are all in the same space. Note that it <u>appears</u> that some dimensions are actually inside the viewport while others fall outside the viewport (when the viewport layers are frozen, this is not noticeable). In addition, dimensions attached to the geometry in the detail view reflect the actual model space measurement even though the geometry is scaled differently. Dimensions for the notch and for the overall part are placed in the appropriate viewport.

FIGURE 33-16

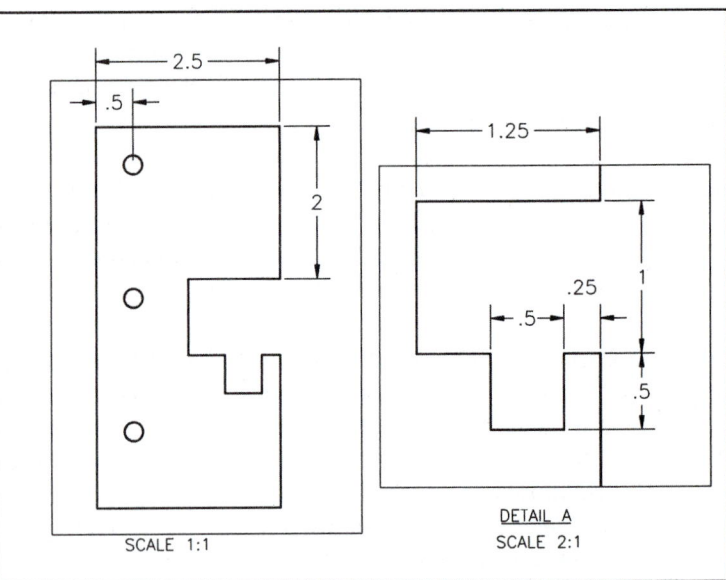

As the last step, the viewport borders layer is *Frozen* and detail area is indicated in the full view. Dimensioning in paper space allows you to dimension a drawing shown in different viewports at different scales (Fig. 33-17).

FIGURE 33-17

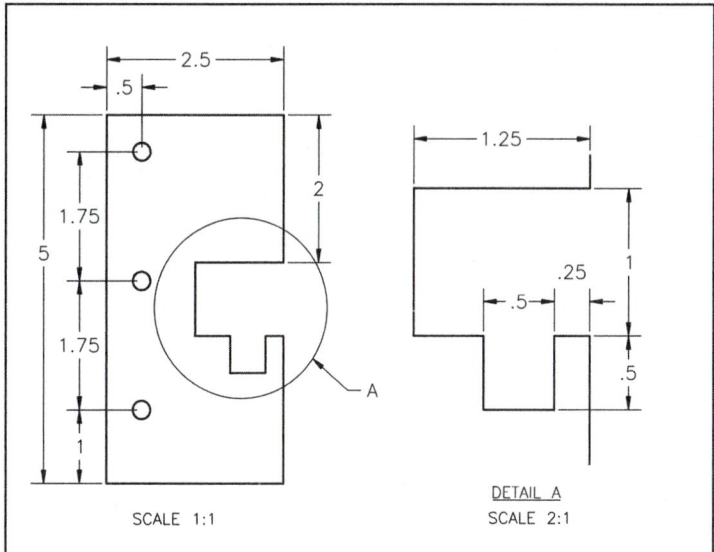

SCALE 1:1

DETAIL A
SCALE 2:1

Although dimensioning in paper space appears to be a simple solution to the complexities that arise when using multiple viewports, it is not always as simple as the previous example. Keep in mind that you may need more than two viewports or more than one layout. Some advantages and disadvantages of dimensioning in paper space are listed below.

Advantages to creating dimensions in a paper space layout include:
1.   you can use an *Overall scale* of 1, so calculation of the drawing scale factor (for dimensions, at least) is simpler;
2.   you can attach dimensions to objects in multiple viewports at different scales and all dimensions appear at one size, so you do not necessarily need more than one dimension style for one layout; and
3.   layer control is simplified, since all paper space dimensions appear in the layout and do not appear in other layouts.

Disadvantages to dimensioning in paper space include:
1.   each viewport and each layout requires its own set of dimensions, so you may have to draw all dimensions for every viewport and for every layout, even when you show the same geometry;
2.   you cannot *Copy* associative dimensions effectively from one viewport to another since copies are not associative;
3.   you cannot copy associative dimensions effectively from one layout to another layout unless you copy the entire layout with the *Layout* command;
4.   paper space dimensions are not useable if the drawing is used as an *Xref*, since only the model geometry of an *Xref* is displayed; and
5.   truly associative dimensioning in paper space is available only in AutoCAD 2002 and newer, so you cannot exchange drawings effectively using AutoCAD 2000 or older.

For more ideas and a more complete discussion of this topic, see "Applications for Multiple Viewports."

The ability to create paper space dimensions displaying model space measurements is available only since AutoCAD 2002. In previous releases of AutoCAD, dimensioning in paper space resulted in values measured in paper space units. Therefore, for example, the notch shown in the detail view in Figure 33-17 measuring .5 x .5 would be measured as 1 x 1 by a paper space dimension since the geometry is shown at 2:1 scale. In AutoCAD 2000 and previous releases, this condition could be corrected by changing the *DIMLFAC* variable to serve as a multiplier for the paper space units.

The "Reverse Method" for Calculating Drawing Scale Factor

Chapter 12 in this text gives the steps for drawing setup, including an explanation for calculating the "drawing scale factor" (DSF). Briefly, the DSF is determined when you set the *Limits*. Set the *Limits* to the sheet size you intend to print or plot on times some factor that will provide adequate space for you to draw the subject of the drawing full size. This factor, or multiplier, is the drawing scale factor. The reciprocal of the DSF is the plot scale. For example, if you plan to print on an 11 x 8.5 sheet and need enough space to draw an object 36" long, set the *Limits* to 11 x 8.5 times 4, or 44 x 34. So, the drawing scale factor is 4. The plot scale is then 1:4.

The drawing scale factor is important because there are many size-related variables and features (many that are set to a value of 1) that should change based on the size of the drawing and the intended plot scale. For example, the drawing scale factor gives you a guideline for the initial settings for *Ltscale*, dimension *Overall scale*, hatch pattern scale, text height, and so on.

This principle is useful for drawings that are to be set up in the model tab or in a layout with one viewport that occupies the sheet size (printable area of the layout). However, when you plan to have multiple viewports, this scheme does not work effectively. As a matter of fact, setting *Limits* is not a necessary step to begin drawing—setting *Limits* only prepares the drawing area with enough space to "see" those first several lines. Nevertheless, the method described in Chapter 12 does give you an understanding of the relationships and some guidelines to use for the initial settings.

Regardless of whether you plan to use only one viewport or multiple viewports, you need some method to help you determine how to set sizes of such things as text, dimension *Overall scale* (assuming you intend to create text and dimensions in model space), hatch pattern scale, etc. A different method for calculating drawing scale factor, called the "reverse method," is given here. The reverse method works well when:

- you plan to display the drawing in multiple viewports at different scales;
- you plan on using one viewport to fill the layout, but have not yet decided on a plot device or sheet size; or
- you are ready to begin drawing, but do not yet know some information such as what plot devices or sheet sizes you will use, what sizes you will create the viewports, or what scales you will display the geometry.

Steps for Calculating the Drawing Scale Factor by the "Reverse Method"

This method works by finding the viewport scale by trial-and-error using the *Viewport Scale* drop-down list. The viewport drawing scale factor is the reciprocal of the viewport scale.

1. Determine the approximate space you will need to create the subject of the drawing full size in the actual units. Set the *Limits* accordingly, but only to provide enough space. *Limits* do not have to be in any proportion or size. *Zoom All*.

   (As an alternative, you do <u>not</u> have to set the *Limits*. You can instead, *Zoom* out, estimating about how much area you will need to "see" the entire drawing.)

2. Begin creating the model geometry. Draw at least an outline of the area you'll need, "box in" the views, or draw the parts of the geometry that occupy the majority of the drawing area. You can continue to draw the geometry, until completion if desired, or at least until you are ready to dimension, hatch, add text, or for some reason require a drawing scale factor.

3. Activate a *Layout* tab. Use *Pagesetup* to assign a plot device and a sheet size. Activate the *Viewports* toolbar (including the *Viewport Scale* drop-down list. Use *Vports* to create the viewport or viewports that you need.

4. Activate the viewport for which you want to determine the drawing scale factor. *Zoom Extents*, or use real-time *Zoom*, to display the approximate area of the drawing you want in the viewport. Use one of the layer tools to set the desired viewport-specific layer visibility to ensure all objects that you intend to show appear in the viewport.

5. Use the *Viewport Scale* drop-down list to set (by trial and error) an appropriate scale for the viewport. You may need to *Pan* occasionally to "see" the desired area of the drawing. (If you are planning to create dimensions in model space or paper space, ensure you allow enough room either inside or outside the viewport.) Once the viewport scale is set, the drawing scale factor for that viewport is the reciprocal of the scale appearing in the *Viewport Scale* box. (Optional—once you have settled on the desired scale for the viewport, activate paper space, highlight the viewport border, and *Lock* the viewport using the *Properties* palette.)

6. To determine the <u>drawing scale factor for the viewport</u>, convert the viewport scale (calculate the reciprocal). Proportional scales (such as 1:2, 1:4, 1:8) can be converted directly by using the second number as the DSF (2, 4, 8, respectively). For feet and inch scales (for example, 1/4"=1'), multiply the denominator (bottom number of the fraction) by 12 (such as 4 x 12 = 48).

    As an alternate method, activate paper space if not already active. Highlight the viewport border and use the *List* command to give the "Scale relative to Paper space: *n.nnnn* xp" (where *n.nnnn* represents a value). This value is the viewport scale as a decimal value. To find the drawing scale factor for the viewport, calculate 1/*n.nnnn*. This is the drawing scale factor for that viewport.

7. Use the viewport drawing scale factor to set the size-related scales such as dimension *Overall scale* (if you are dimensioning in model space), *Ltscale* (if this is the main viewport), hatch pattern scale, text size (for model space text), and so on. For dimensions and text you create for the viewport in model space, use layers named DIM-1, TEXT-1, or similar, so visibility can be controlled for this and other viewports.

8. Repeat steps 4 through 7 for each viewport that will display the geometry at a different scale.

This method can be used for any case when you plot from a layout, including when you have only one viewport that occupies the entire layout.

Applications for Multiple Viewports

The capability of AutoCAD to create multiple viewports in one layout and multiple layouts in one drawing offers many possibilities for displaying a drawing. Several features of the same geometry can be shown using multiple viewports in concert with layer control for the viewports. In addition, one drawing can be printed and plotted in multiple ways using multiple layouts. It is even possible to print or plot several drawings in one layout using *Xrefs*.

A few examples of typical applications that make use of multiple viewports and/or layouts are:

*   architectural drawings that display several aspects of one drawing, such as an overall view of a floor plan and enlarged views of particular rooms or areas
*   mechanical drawings that display one view of the entire part or assembly and additional detail views of particular aspects of the part or individual components of the assembly

- civil engineering or construction drawings that show an overall plan and one or more detailed views of a specific aspect of the site or structure
- 3D mechanical drawings that show several views of one part or assembly, each view showing a different side or "view" of the 3D model, as in a multiview drawing (see Chapter 42)
- multiple architectural, engineering, construction, or design drawings displayed on one sheet, accomplished by *Xrefing* the drawings and displaying one drawing per viewport

Three examples of using multiple viewports or layouts are given next: 1) using two layouts to plot the same geometry at different scales, 2) using one layout to show the same geometry at different scales, and 3) using one layout to display *Xref* drawings in multiple viewports at different scales.

Using Two Layouts to Plot the Same Geometry at Different Scales

For the first problem, consider the drawing shown in Figure 33-18. This mechanical part was originally set up to plot only on a "C" size engineering sheet (22" x 17"). *Limits* were set to 22 x 17 so the geometry could be drawn full size, then the layout was set up with an ANSI C title block. Since the *Limits* size equaled the sheet size, viewport scale was set to 1:1 and the layout was plotted at 1:1. *Dimscale, Ltscale,* and other size-related variables were set to 1. Dimensions were created in model space. The resulting plot is a full size drawing of the part. The dimension text and other text as well as hidden line dashes are plotted at the ANSI standard .125" (1/8").

FIGURE 33-18

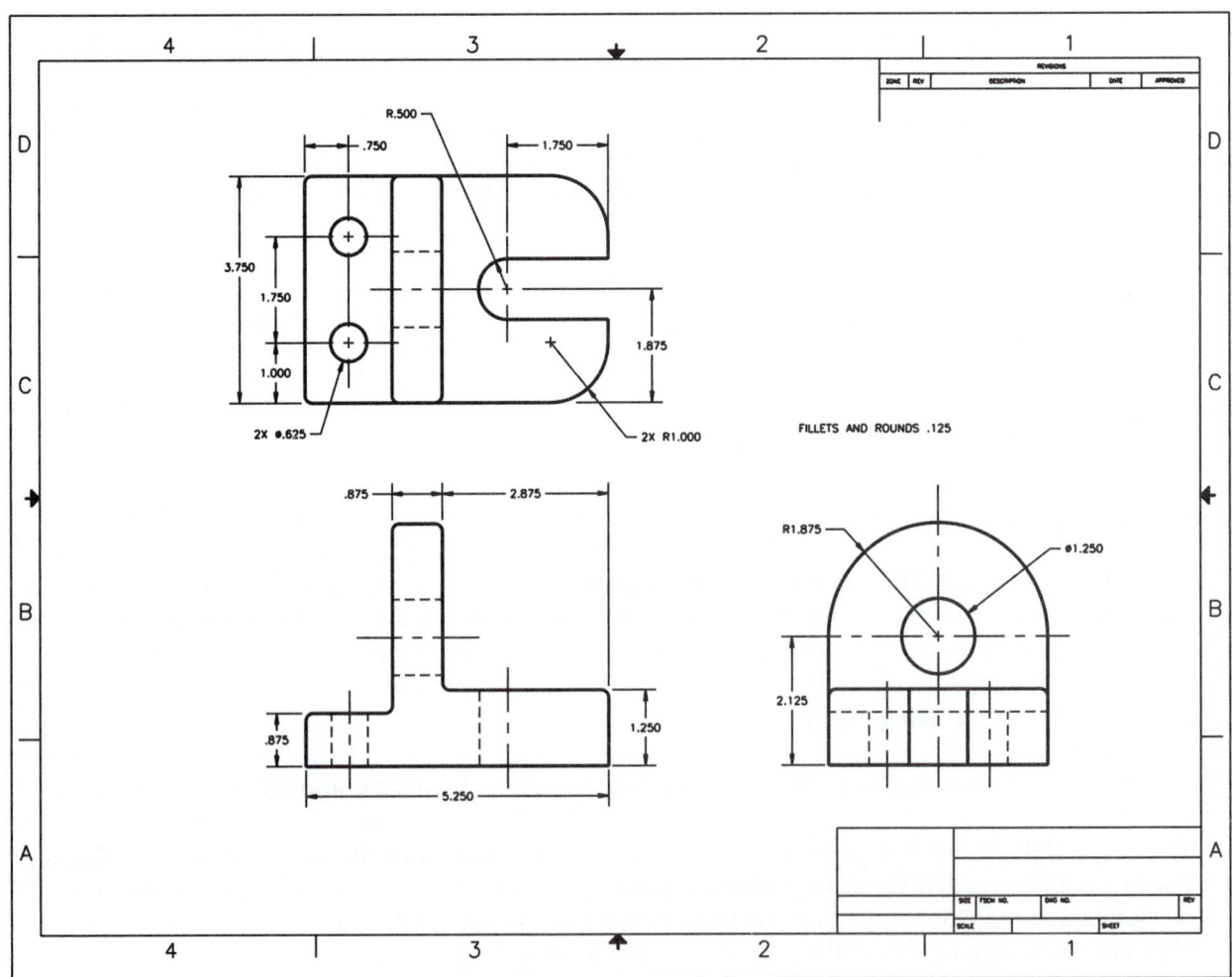

Assume the client calls and requests a fax of the drawing. You are now faced with printing the drawing on an ANSI "A" size 11 x 8.5 sheet in order to send the fax. A new layout is created with an ANSI A title-block and the geometry is scaled to fit the viewport at .375, or 3/8"=1" scale. However, making a print at this scale generates dimension text, other text, and hidden line dashes at about .047" (3/64")—much too small to be read easily. Here are several possible solutions.

Solution 1

Display the same model space dimensions in the new layout at a new *Overall scale*. In this solution, you have to make several changes such as setting the dimension *Overall scale* and *Ltscale* to 2.666 (the reciprocal of .375) in order to make the text readable and hidden lines recognizable. In addition, the placement of several dimensions must be adjusted to accommodate the (relatively) increased text size (Fig. 33-19). One problem with this solution now is that the drawing is prepared and saved only for A size prints since the *Ltscale* and dimension scale have changed.

FIGURE 33-19

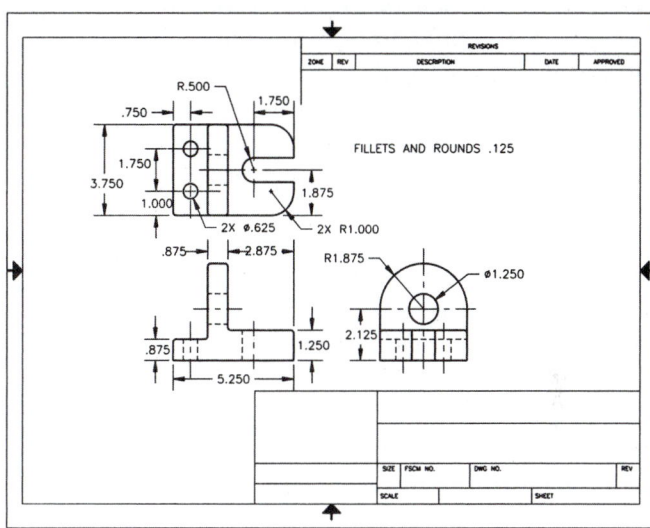

Solution 2

Create a duplicate file. Before making any changes (keeping the settings for the ANSI C layout) use *SaveAs*. Create the ANSI A layout and related settings in the new, separate drawing. Now you have a drawing prepared for A size prints and one for C size plots. The resulting problem is there are two drawings to update and maintain when needed.

Solution 3

Create a second layout. Create a <u>separate</u> set of the <u>model space</u> dimensions, and place them on a separate layer. Assign the new dimensions to a new dimension style and change the *Overall scale* appropriately for the new layout. Use viewport-specific layer visibility to display the correct dimension set in the appropriate viewports (layouts). Set *PSLTSCALE* to 1. The result is one drawing with all settings saved, ready to plot or print using either layout. The disadvantage is the amount of time to create, adjust, and manipulate visibility for the new dimensions.

Solution 4

Erase all model space dimensions. Set up the two layouts displaying the viewport geometry at different scales, and then create the dimensions in paper space. Two sets of dimensions are required, one for each layout. Set *PSLTSCALE* to 1. Again you have one drawing with all settings saved and ready to print or plot either layout. Once again, much time is needed to create the two sets of dimensions and the dimensions are unusable for new viewports, layouts, or *Xrefs*.

Solution 5

Assume that the dimensions were originally created in paper space. The existing ANSI C layout can be copied with the *Layout* command. This action gives you the needed second set of dimensions. However, these dimensions are <u>not associative</u>. The titleblock and viewport must be erased, the layout must be set up to the new device and paper size, and a new titleblock and viewport must be created. Then the existing (new) dimensions must be reassociated—almost as much trouble as creating all new dimensions from scratch. Once again, this solution has the same shortcomings as Solution 4.

Using Two Viewports to Display the Same Geometry at Different Scales

FIGURE 33-20

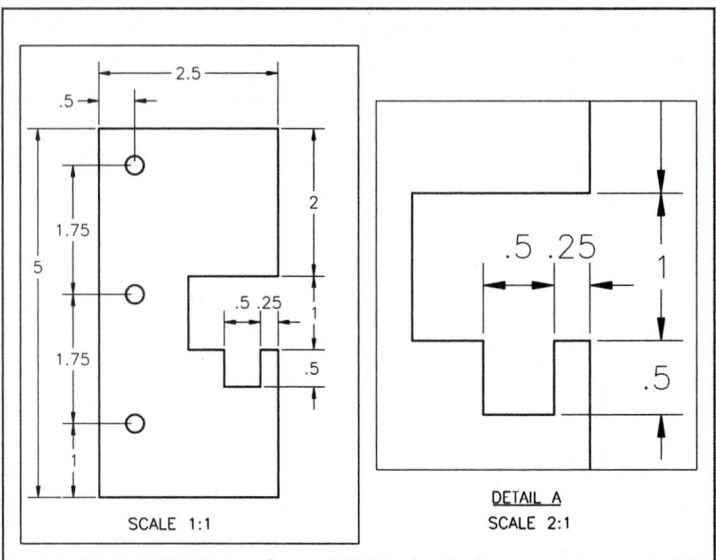

For the second application, consider the drawing of a plate (Fig. 33-20) showing a layout with two viewports, each viewport displaying some of the same geometry but at different scales. (This is the same example used to discuss creating dimensions in paper space.) Here, dimensions are <u>already created in model space</u> on a layer named DIM, or similar name. As you can see, the right viewport is intended as a detail view to show an enlarged view of the notched area of the plate at 2:1 scale. As a result, the dimensions are also shown at 2:1 scale.

Solution 1

As shown in Figures 33-15, 33-16, and 33-17, *Erase* all of the model space dimensions and create new dimensions in paper space. See the example in "Dimensioning in Paper Space."

Solution 2

Erase <u>only</u> the model space dimensions for the notch (appearing in the detail view), and then re-create them in <u>paper space</u> (Fig. 33-21). Note that the dimensions for the detail have been erased (see left viewport), and then new dimensions for the detail view were added, but <u>placed in paper space</u> (right viewport). Even though these dimensions appear to be in the viewport, they are actually in paper space—"on top of" the viewport. Create these dimensions on a separate layer, for example, DIM-2. Note that part of one model space dimension (large arrowhead) appears in the detail view. To correct for this, the DIM layer is frozen for the right viewport only (*Current VP Freeze*). The DIM-2 layer does not have to be frozen for the left viewport, since these dimensions appear only in the layout. One shortcoming is that you cannot display any view of the drawing with all dimensions.

FIGURE 33-21

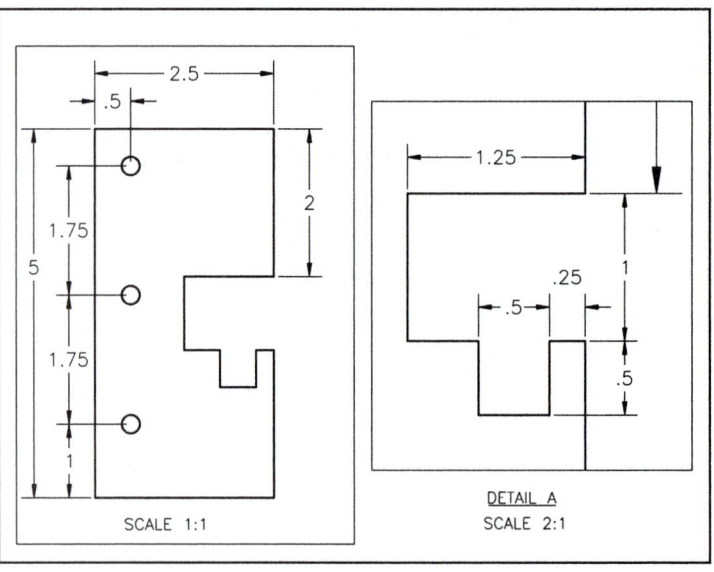

Solution 3

Similar to Solution 2, *Erase* only the model space dimensions for the notch. Re-create these dimensions in <u>model space</u>, but on a new layer and in a new dimension style with the *Overall scale* set to .5, since these dimensions will be displayed a 2:1. You can make the new dimension style by copying the existing one, then changing the *Overall scale*. Use viewport-specific layer visibility to display the appropriate dimensions in the appropriate viewports. In this case you can display all dimensions if needed, but at two different sizes. You must use *Matchprop* or a similar method to make all the dimensions the same scale if you need to show all in one view.

Solution 4

Use either Solution 2 or 3, but instead of erasing the dimensions for the notch, create a new layer named DIM-3 and move these dimensions to the new layer. This gives you the possibility of showing all dimensions in one view if needed. This is the only solution that provides a full set of model space dimensions.

Using One Layout to Plot *Xref* Drawings in Multiple Viewports at Different Scales

In this problem, one layout is created with three viewports to display and plot three *Xrefed* drawings at different scales. Although this sounds complex, it is a fairly straightforward procedure. The drawing typically contains no geometry (other than the *Xrefs*) and little or no text other than simple notes, title-block and border.

Begin a *New* drawing, immediately activate a *Layout* tab and use *Pagesetup*. Set up the layout for the desired plot device and sheet size. In most cases, a large sheet size is required since multiple drawings will be displayed and plotted. *Insert* the desired titleblock and border. Create the first viewport with the *Vports* command on layer VIEWPORTS. *Save* the drawing.

Before *Xrefing* the first drawing, it is important to set a common layer current, such as layer 0. (If layer VIEWPORTS is current when the drawings are *Xrefed* and that layer is later *Frozen*, the *Xrefs* will not appear.) Also, double-click inside the viewport since you want to *Xref* the drawing into model space, not paper space. Use *Xref* and *Attach* the desired drawing. In this example, the DOOR drawing is referenced (Fig. 33-22). Use the *Viewports* toolbar or other method to set the viewport scale.

FIGURE 33-22

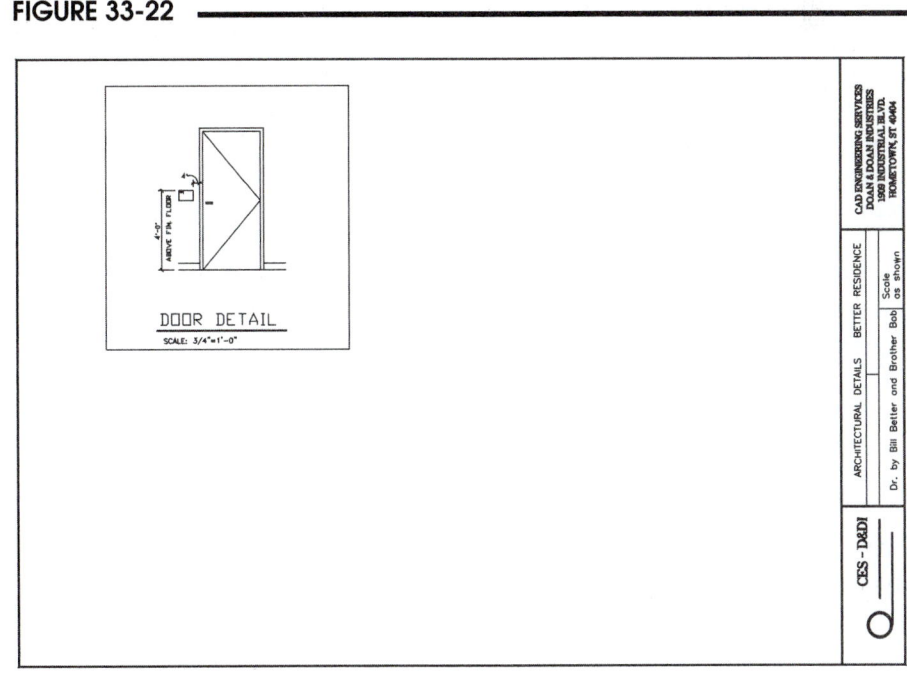

A second viewport is created on layer VIEW-PORTS. As soon as the new viewport is created, the DOOR drawing appears in the new viewport. The layer is changed to layer 0, and *Xref* is used to *Attach* the second drawing—WALL, in this case (Fig. 33-23). Note that without controlling for viewport-specific layer visibility, both drawings appear in both viewports. With the second viewport still active (your cursor is inside the viewport), use the *Layer Control* drop-down list or the *Layer*

FIGURE 33-23

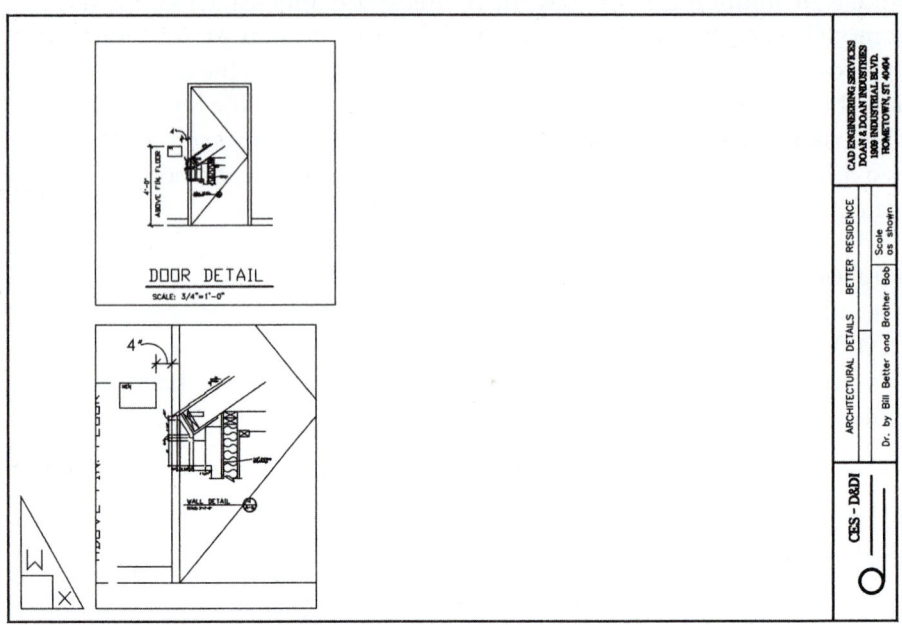

Properties Manager to freeze the DOOR|* layers (*Current VP Freeze*). Next, click in the DOOR viewport and freeze the WALL|* layers. <u>Additionally</u>, to prevent the *Xref* drawings from appearing in any <u>new</u> viewports you create, select *New VP Freeze* for all the DOOR|* and WALL|* layers. Use the *Viewports* toolbar or other method to set the viewport scale in the new viewport.

As in the previous step, create a new viewport on the appropriate layer, reset the current layer, then *Xref* the third drawing (SECTION) into model space. In this case, the previous two drawings do not appear in the new viewport; however, the third drawing appears on all three viewports (not shown). Use *Current VP Freeze* to freeze the SECTION drawing in the first two viewports. Set the viewport scale for the last viewport. *Freeze* the VIEWPORTS layer (globally) to com-

FIGURE 33-24

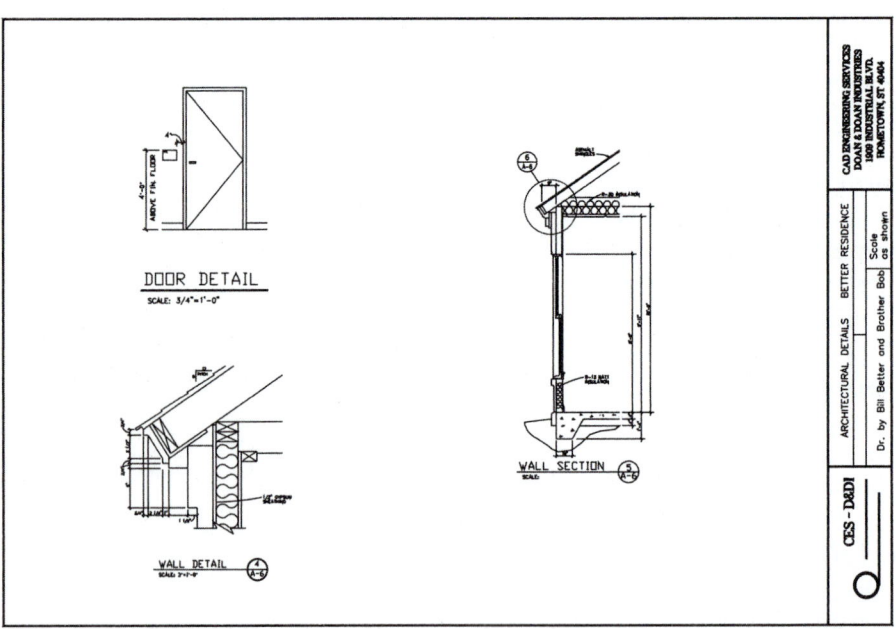

plete the drawing (Fig. 33-24). Remember to set the *VISRETAIN* variable to 1; otherwise, none of the viewports' layer visibility settings will be saved with the drawing.

Since, in this case, it was known for each of the drawings at what scales they were to be plotted, the scale for each could be set independently using the viewport scale. If dimensions were created for these drawings in model space and the recommended methods for setting DSF were used, the dimensions could be displayed effectively in this parent drawing and the dimension layers could be controlled for visibility. If the dimensions for the *Xrefed* drawings were created in paper space, they would be unusable in the parent drawing since you cannot display any layouts of an externally referenced drawing.

Considerations for Using Multiple Viewports and Layouts

It is difficult to provide guidelines you can use to set up drawings such as these because there are so many possible combinations of viewports and layouts <u>and</u> there are so many possible strategies to use for any one setup based on your preferences and the particular application. However, here are some general considerations:

1. If you need to show several aspects of one drawing on one sheet (print or plot), use multiple viewports in one layout.

2. If you need to show the same geometry but print or plot it in different scales or on different devices, create and save multiple layouts.

3. If you need to show several drawings on one sheet (print or plot), create multiple viewports on a layout, *Xref* the drawings, and control layer visibility to display one drawing per viewport.

4. Linetype scale can be shown either relative to model space geometry in all viewports and layouts (*PSLTSCALE* = 0), or can be scaled to appear the same for all viewports relative to paper space (*PSLTSCALE* = 1). *LTSCALE* and *PSLTSCALE*, however, are easily changed to suit a specific layout as needed.

5. Model space dimensions, text, and hatching can appear in multiple viewports.

6. Paper space dimensions and text appear only for the specific viewport or layout.

7. Model space dimensions, text, and hatching can appear in different viewports at the same scale without alteration.

8. Model space dimensions, text, and hatching intended to appear in multiple viewports of one layout in different scales must be duplicated at different scales, placed on separate layers, and controlled for viewport visibility.

9. Model space dimensions, text, and hatching intended to appear in multiple layouts in different scales can be reset (*Overall* scale) for each print, but may require adjustment and cannot be saved in one drawing file (unless duplicated on other layers).

10. Paper space dimensions and text can be easily created for each viewport or for each layout, but can be used effectively only for that particular viewport or layout.

PLOT STYLE TABLES AND PLOT STYLES

A plot style is an object property. A plot style controls how an object looks on a plotted drawing. By modifying an object's plot style, you can <u>override</u> that object's drawing color, linetype, and lineweight for the plot. You can also specify additional features such as line end styles, join styles, fill styles, dithering, gray scale, pen assignment, and screening. You can use plot styles if you need to plot the same drawing with different "looks."

When a drawing is plotted in AutoCAD 2000 or newer release, three main components are referenced to develop the finished plot:

1. layout settings you set up in the *Plot* dialog box and *Page Setup* dialog box (saved in the .DWG file);
2. plotter configuration information you specify for the plot device (saved in the device's .PC3 file); and
3. settings that are assigned in a Plot Style Table you attach (saved in .CTB and .STB files).

The third component, Plot Style Tables, are optional for plotting. This section discusses Plot Style Tables and plot styles.

Before explaining plot styles, it would help to consider how the appearance of objects is controlled in plotted drawings in pre-AutoCAD 2000 releases. In these releases, the <u>color</u> of an object determined how objects appeared in plotted drawings. For example, if you used a plotter with colored pens, it was normally configured so the object's color would plot with a pen of the same color. Alternately, you could configure the print or plot device so colors in the drawing would correspond to certain lineweights on the finished page. Normally, the objects' color is set to *BYLAYER*, so you can consider that the <u>layer's color</u> rather than the <u>object's color</u> determined the plotted appearance. The practice of associating colors with certain plotter pen numbers is known as making "pen assignments."

This strategy is still valid in current AutoCAD releases and will most likely be used for the majority of plots. As a matter of fact, several <u>color-dependent Plot Style Tables</u> are supplied with AutoCAD that utilize this same concept; that is, an object's color determines how the object appears in the plot. There is one limitation to this system, however. With a color-dependent Plot Style Table, all objects in the same color generally have the same appearance on the plotted sheet.

Current AutoCAD releases also offer <u>named Plot Style Tables</u>, in which the contained plot styles are not automatically assigned by color but can have names that you designate and can be assigned to <u>any</u> <u>object</u>. With named Plot Style Tables, objects that appear in one color in the drawing can be plotted with different appearances.

Plot styles, even the new color-dependent Plot Style Tables, offer many advantages. With plot styles you can assign line end styles, joint styles, fill patterns, gray scales, screen pattern percentages, and pen assignments. Therefore, a color-dependent Plot Style Table is simply a list of colors, and for each color in the drawing you can assign a specific lineweight, linetype, line end style, joint style, fill pattern, gray scale, screen percentage, color, and pen assignment for the plot.

Unlike some other object properties, plot style is <u>optional</u>. Objects will print or plot based on color even if no plot style is assigned to it or if no Plot Style Table is attached. If you assign a plot style to an object, and then detach or delete the Plot Style Table that defines that plot style, the plot style will have no effect on the object.

What is the Difference between a Plot Style and a Plot Style Table?

It is important to make the distinction between a plot style and a Plot Style Table because of how they are assigned or what they are attached to. This book uses lowercase letters for the terms "plot style" and upper- and lowercase for "Plot Style Table" to help make that distinction.

plot style   A property assigned to objects that determines the way the object appears in a plot. A plot style contains several appearance characteristics such as color, linetype, lineweight, line end style, joint style, fill pattern, gray scale, screen percentage, and pen assignment.

Plot Style Table   A Plot Style Table is a collection of many plot styles. A color-dependent Plot Style Table contains 255 plot styles, each plot style is a color. A Plot Style Table is "attached" to the *Model* tab or a *Layout* tab.

The Plot Style Table is attached to a <u>layout</u>. You can attach the same or a different Plot Style Table to each layout (the *Model* tab and all *Layout* tabs). You attach a Plot Style Table to a layout using the *Page Setup* dialog box. Since different layouts can have different Plot Style Tables attached, two similar layouts could be plotted with two different "looks." Alternately, one layout could be plotted with one Plot Style Table and later plotted with another Plot Style Table to achieve a different "look."

For example, Figure 33-25 displays the AutoCAD-supplied ACAD.CTB color-dependent Plot Style Table. The colors listed in the table (Color 1, Color 2, Color 3, etc.) are plot styles. Each color in the drawing takes on the appearance defined by its plot style, so everything drawn in Color 1 can have a particular plotted color, grayscale, pen number, screening, linetype, and so on, whereas everything drawn in Color 2 can take on a different appearance.

FIGURE 33-25

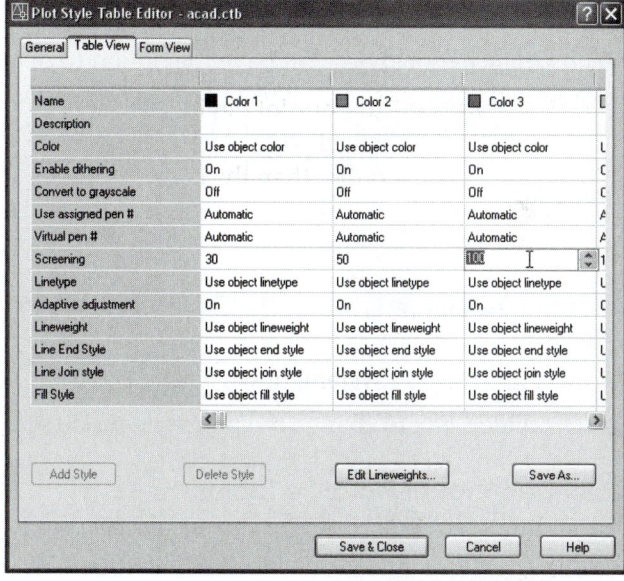

Color Plot Style Table Example

As an example, consider a construction drawing set up in one *Layout*, but the project has two phases. Using two Plot Style Tables, the layout could be plotted for the demolition phase showing one appearance, then later plotted during the construction phase with a different appearance.

The floor plan in Figure 33-26 illustrates the demolition phase. Walls to be demolished are most apparent, shown in 100% screen (dark), while unchanged walls are in 50% screen (medium), and new construction is barely noticeable in 30% screen (light). Note the Plot Style Table in Figure 33-25 is set up to plot Color 3 (demolition) in 100%, Color 2 (unchanged) in 50%, and Color 1 (new construction) in 30% screen.

FIGURE 33-26

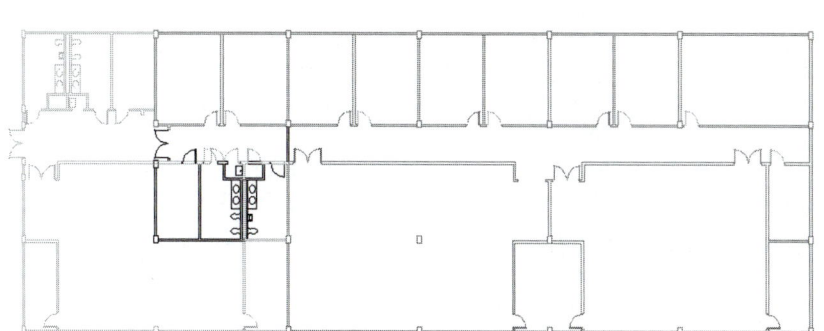

Figure 33-27 shows the same layout but for the construction phase. In this case, a second Plot Style Table is attached to the layout that applies the opposite screen percentages to Color 1 and Color 3, thereby emphasizing the walls needed for new construction (100% screen, dark) and fading the demolished walls (30% screen, light).

FIGURE 33-27

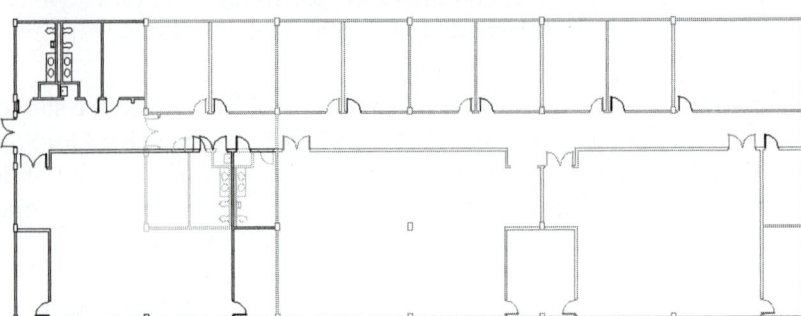

When you use plot styles you have the option to display the plot styles in the layout so you don't have to plot the drawing to see the results. You can choose to display plot styles by selecting *Display Plot Styles* on the *Plot Device* tab in the *Page Setup* dialog box. If you choose not to display plot styles in the drawing, you can view them using the *Full Preview* option in the *Plot* dialog box or by using the *Preview* command.

Color-Dependent and Named Plot Style Tables

There are two plot style modes: Color-Dependent and Named. Each drawing you open in AutoCAD is in either one mode or the other. You can change the mode for new drawings using the *Plotting* tab in the *Options* dialog box to change the plot style mode setting (see Fig. 33-28 and following explanation).

Color-Dependent Plot Style Tables
By default, AutoCAD continues to use object color to control output effects by creating color-dependent Plot Style Tables. For each color-dependent Plot Style Table there are 255 plot styles; each plot style is a color. You assign plot characteristics (lineweight, screening, dithering, end and joint types, fill patterns, etc.) to each plot style (see Fig. 33-25). When your drawing plots, the appearance of each line is based on its color in the drawing and the characteristics assigned to that color. In a color-dependent Plot Style Table, you cannot add, delete, or rename color-dependent plot styles. Color-dependent Plot Style Tables are stored in files with the extension .CTB. AutoCAD supplies several color-dependent Plot Style Tables (see "*Stylesmanager*").

Named Plot Style Tables
A named Plot Style Table is not based on a list of colors, but is based on a list of plot style names that you define. You can assign any plot style to any object regardless of that object's color. Each named Plot Style Table can contain many named plot styles. For each plot style name you define, you can assign specific line characteristics (lineweight, screening, dithering, end and joint types, fill patterns, etc.). A named Plot Style Table is similar to that shown in Figure 33-25, except the individual plot style names can be assigned instead of using color numbers (see Fig. 33-34, top row). Normally, you would then assign the named plot styles to layers in your drawing (although plot styles can be assigned to any object). Named Plot Style Tables are stored in files with the extension .STB. AutoCAD supplies several named Plot Style Tables.

Since AutoCAD can use only one of the two types of Plot Style Tables at a time (named or color-dependent), make the desired choice by using the *Plot and Publish* tab of the *Options* dialog box, then selecting the *Plot Style Table Settings* button on the bottom-right. This action produces the *Plot Style Table Settings* dialog box (Fig. 33-28, on the next page).

Default Plot Style Behavior for New Drawings

Use this section to control how plot styles affect drawings. This setting is saved with the drawing. Making a change here affects new drawings but not the current drawing. If you want to change the default plot style behavior for a drawing, select an option before opening or creating a drawing. Changing the default plot style affects only new drawings and Release 14 and earlier drawings when first opened in current AutoCAD releases.

FIGURE 33-28

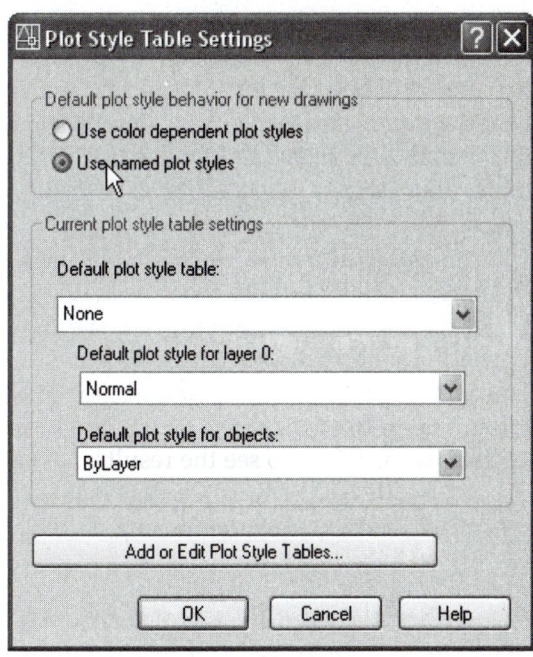

Use Color-Dependent Plot Styles
By default, AutoCAD uses color-dependent plot styles for new drawings and pre-AutoCAD 2000 drawings. You can also create and edit color-dependent plot styles by setting the *PSTYLEPOLICY* system variable to 1. If a drawing is saved with *Use Color-Dependent Plot Styles* as the default, you can change it to *Use Named Plot Styles* later. However, once a drawing is saved with *Use Named Plot Styles* as the default, you can change it to use color dependent plot styles with *Convertpstyles*.

Use Named Plot Styles
This check causes AutoCAD to use named plot styles as the default in new drawings. This action allows you to use the AutoCAD-supplied named Plot Style Tables or create new named Plot Style Tables. You can also set the *PSTYLEPOLICY* system variable to 0 to enable named Plot Style Tables. Once a drawing is saved with *Use Named Plot Styles* as the default, you can change it to use color dependent plot styles using *Convertpstyles*.

Default Plot Style Table
From this drop-down list, select the default Plot Style Table to attach to new drawings. If you are using color-dependent plot styles (*Use Color Dependent Plot Styles* is checked), this option lists all color-dependent Plot Style Tables (.CTB files) found in the search path as well as the value of *None*. If you are using named plot styles (*Use Named Plot Styles* is checked), this option lists all named Plot Styles Tables (.STB files).

Default Plot Style for Layer 0
Use Named Plot Styles must be checked to enable this drop-down list. Your choice sets the default plot style for Layer 0 for new drawings. The list displays the default value *Normal* and any plot styles defined in the currently loaded Plot Style Table. You can also control the default plot style for Layer 0 using the *DEFLPLSTYLE* system variable (saved in the drawing).

Default Plot Style for Objects
Use Named Plot Styles must be checked to enable this drop-down list. Here you set the default plot style that is assigned when you create new objects. The list displays a *BYLAYER*, *BYBLOCK*, and *Normal* style, and lists any plot styles defined in the currently loaded Plot Style Table. Alternately, you can set the *DEFPLSTYLE* system variable.

Add or Edit Plot Style Tables
Selecting this button invokes the *Plot Style Table Manager*, which is a separate system window. See *"Stylesmanager"* later in this chapter.

Attaching a Plot Style Table to a Layout

To attach a Plot Style Table to a *Layout* or *Model* tab, first select the desired *Model* tab or *Layout* tab. Invoke the *Page Setup Manager*, then select *New* or *Modify* to produce the *Page Setup* dialog box. In the upper-right corner of the *Page Setup* dialog box (Fig. 33-29), use the *Plot style table (pen assignments)* drop-down list to select the desired table (.CTB or .STB file).

Use the *Edit…* button to invoke the *Plot Style Table Editor*. (See the section titled "Editing Plot Styles Assigned to Objects.") Selecting the *New…* button invokes the *Add-A-Plot Style Table Wizard*. (See "Creating a Plot Style Table.") You can choose to *Display Plot Styles* if you want the layout to reflect the appearance of the plot styles in the layout (so the layout looks like a *Full Preview*).

FIGURE 33-29

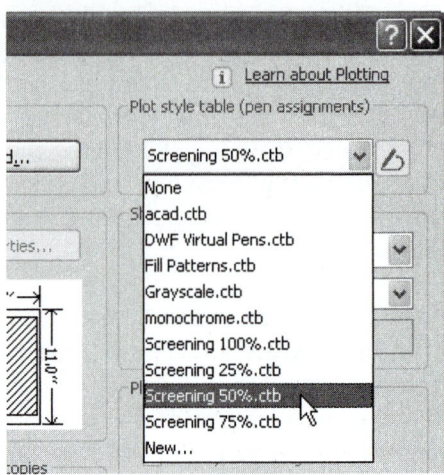

 If you want to change the name of a named Plot Style Table from Style 1, Style 2, etc. to something more descriptive (such as walls, electrical, plumbing), it should be done before opening a drawing that references the table. If the names are changed after the Plot Style Tables have been attached to a drawing, the plot style names in the drawing will not match the names of the styles in the attached Plot Style Table.

Stylesmanager

Pull-down Menu	Command (Type)	Alias (Type)	Short-cut	Screen (side) Menu	Tablet Menu
File *Plot Style* *Manager...*	*Stylesmanager*	...	...	...	...

The *Stylesmanager* command produces the *Plot Style Manager*. You create Plot Style Tables and edit Plot Style Tables with the *Plot Style Manager*. Selecting the *Plot Styles Manager* produces a window displaying the *Add-A-Plot Style Table Wizard* and the list of available Plot Style Tables (.CTB and .STB files). This separate application window runs independently of AutoCAD (Fig. 33-30).

Color-dependent Plot Style Tables are stored as .CTB files, and named Plot Style Tables are stored as .STB files. To find the location of the .CTB and .STB files on your system, use the drop-down list in the *Address* section at the top of the window, or use *Options*, *Files* tab, and expand the *Printer Support File Path*, then *Plot Style Table Search Path* sections.

FIGURE 33-30

To create a new Plot Style Table using the *Plot Style Manager*, select the *Add-A-Plot Style Table Wizard* (see Fig. 33-30). After creating new Plot Style Tables, they automatically appear in the window. See "Creating a Plot Style Table."

You can also use the *Plot Style Manager* to edit existing Plot Style Tables. Using the manager, select a Plot Style Table to edit by double-clicking its icon. Editing a table involves assigning the appearance characteristics (color, pen assignment, screening, linetype, lineweight, line end style, line join style, and fill style) for each plot style. See "Editing Plot Styles."

Creating a Plot Style Table

Many AutoCAD-supplied Plot Style Tables are available and have "generic" plot styles. These tables are intended for your use and can be edited for your needs using the *Plot Style Table Editor*. You can use the existing Plot Style Tables or use the *Add-A-Plot Style Table Wizard* to create other tables.

To create a new Plot Style Table, access the *Add-A-Plot Style Table Wizard* from the *Plot Style Manager* (see Fig. 33-30) or select the *Wizards* from the *Tools* pull-down menu.

In the *Add-A-Plot Style Table Wizard*, you can *Start from scratch*, *Use an existing plot style table*, *Use My R14 Plotter Configuration*, or *Use a PCP or PC2 file*. Selecting the first choice, *Start from Scratch*, you can create a color-dependent Plot Style Table or a named Plot Style Table (Fig. 33-31).

FIGURE 33-31

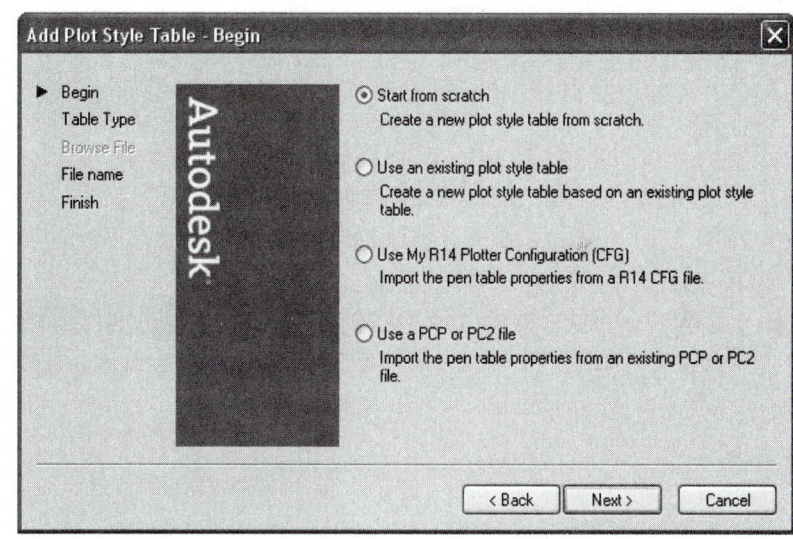

The *Add-A-Plot Style Table Wizard* guides you through the steps to easily create the new table. There are only a few steps involved with creating a new Plot Style Table. Once the table is created, your specific preferences on how objects appear in plots are achieved by editing the table and creating the individual plot styles using the *Plot Styles Table Editor*. The wizard contains explanatory text on each page, so no additional explanation is required here. If you feel that you do need help, look in the *AutoCAD User's Guide*. Once the table is created, edit it using the *Plot Style Table Editor* as explained in the next section.

Plot Style Table Conversion from Previous Releases

In releases of AutoCAD previous to 2000, the components needed for plotting were saved in the ACAD.CFG and the .PCP or .PC2 files. If you want, you can use the *Add-A-Plot Style Table Wizard* to convert your older Release 14 pen assignments to Plot Style Tables.

These new Plot Style Tables will then be available in the list in addition to the supplied Plot Style Tables. See Figure 33-31, last two options (*Use My R14 Plotter Configuration (CFG)* and *Use a PCP or PC2 File*). (See also *"Pcinwizard"* in Chapter 14.)

Editing Plot Styles

To edit plot styles, use the *Edit...* button in the *Page Setup* dialog box or double-click on a plot style file name (.CTB or .STB) from the *Plot Style Manager* or from Windows Explorer. This action invokes the *Plot Style Table Editor* (Fig. 33-32).

If the selected Plot Style Table is <u>color-dependent</u>, the assignment to objects is automatic; that is, each of the 255 plot styles in the table is already assigned to an object color. However, you can set the *Fill Style, Line Join Style, Line End Style, Lineweight, Line Type, Screening,* and so on for each color as shown. Therefore, each drawing color takes on the appearance of the selected options. You cannot change the plot style names (Color 1, Color 2, etc.) for a color-dependent Plot Style Table.

FIGURE 33-32

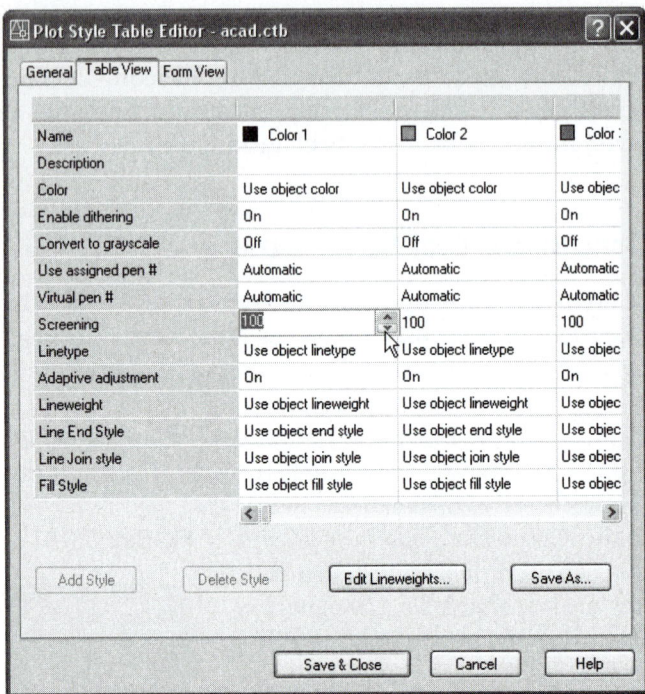

You can make the settings in either the *Table View* (see Fig. 33-32) or *Form View* (Fig. 33-33) tabs. These two tabs have identical options. Setting an option in one tab also makes the matching setting in the other tab.

FIGURE 33-33

If the selected Plot Style Table is a <u>named</u> Plot Style Table, the names are not automatically assigned as with the color-dependent tables. You assign the names for each plot style based on some descriptive feature or application. For example, you may want to name plot styles for an application, such as Plumbing, HVAC, Electrical, or possibly Phase 1 and Phase 2. You could also name the plot styles according to a feature you assign; for example, you may use the names Wide, Medium, and Thin to describe lineweights that are defined in the styles.

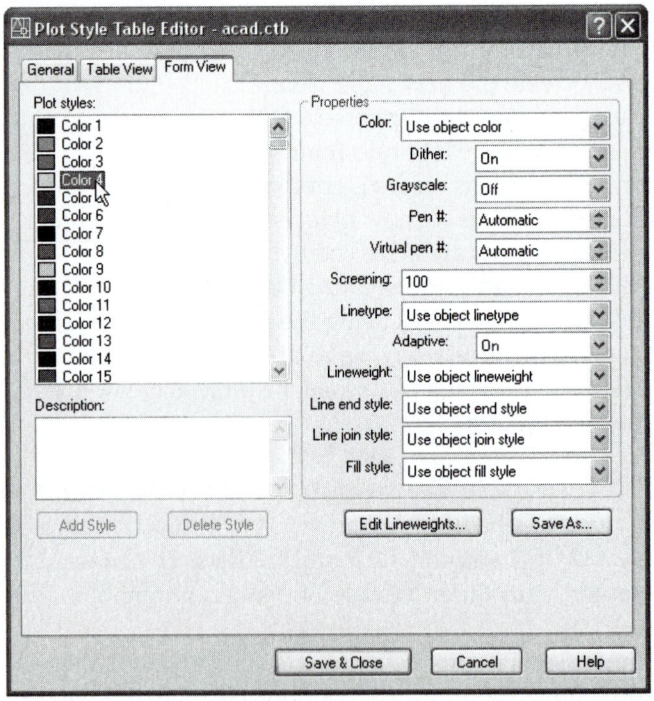

Figure 33-34 displays a named Plot Style Table in the *Plot Style Table Editor*. All of the same features (color, dither, grayscale, pen number, etc.) are available as in the color-dependent Plot Style Tables. The only additional feature is that <u>you assign the name</u> of each plot style, instead of the styles being automatically named for a color (Color 1, Color 2, etc.) as in a color-dependent table. By default, the plot styles are named Style 1, Style 2, etc. Create new plot styles by pressing the *Add Style* button. This action creates a new style named Style#, where # is the next number in the sequence. You can rename the styles by double-clicking on the name (top row).

You can make the settings in either the *Table View* (Fig. 33-34) or *Form View* (see Fig. 33-33) tabs. These two tabs have identical options. Setting an option in one tab also makes the matching setting in the other tab.

Named Plot Style Tables have some advantages over color-dependent tables. With named tables, you can assign specific characteristics to objects regardless of the objects' screen color, whereas with color-dependent tables all objects with the same screen color must appear the same in the plot (have the same plot style). Named plot styles are usually easier to assign to objects in complex drawings (when the drawing contains many colors and layers). See the discussion in the next section.

FIGURE 33-34

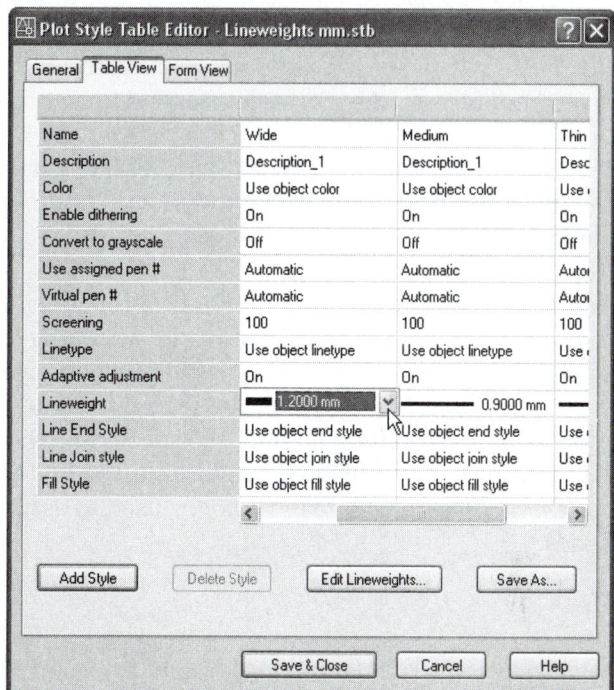

Assigning Named Plot Styles to Objects

Keep in mind that each drawing can have only one of the two types of Plot Style Tables attached—color-dependent or named Plot Style Tables. The desired table type should be set in the *Options* dialog box before creating a drawing (see "Color-Dependent Plot Style Tables" and "Named Plot Style Tables").

Color-Dependent Plot Style Tables

The process of assigning plot styles to objects is necessary only for named Plot Style Tables. This is based on the fact that <u>color-dependent plot styles are already assigned to colors</u>, not objects. In other words, each color-dependent plot style (named Color 1, Color 2, etc.) is automatically assigned to its matching color. Figure 33-35 explains this idea. Notice that in the figure displaying the *Layer Properties Manager*, the *Plot Styles* column is disabled since assignment is automatic. Drawings that have color-dependent Plot Style Tables attached display a <u>disabled</u> *Plot Style* drop-down list on the *Object Properties* toolbar.

FIGURE 33-35

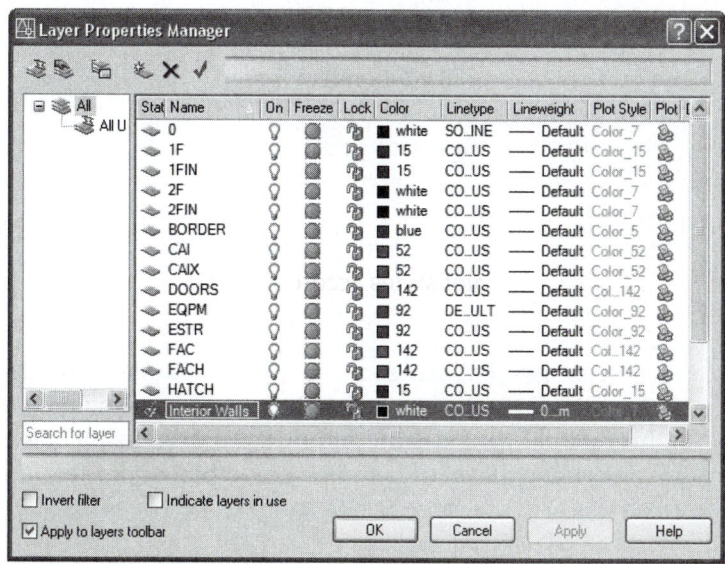

Named Plot Style Tables

Plot styles in a named Plot Style Table can be attached to any object. Individual plot styles within a named table are often assigned to layers. Assigning plot styles to individual objects offers much more flexibility, but can be a tedious job and create a fairly complex layout. If plot styles are assigned both to layers and objects, the plot style assigned to an object overrides the plot style assigned to the layer.

If you want to assign plot styles to individual objects (rather than to layers), three methods are available: the *Plot Style* drop-down list on the *Object Properties* toolbar, the *Properties* palette, or the *Plotstyle* command.

Figure 33-36 displays the *Plot Style* drop-down list on the *Object Properties* toolbar. Using this method, you would select the desired plot style from the list. The selected style becomes the <u>current</u> object-specific plot style— that is, any object created after the setting is made will have the selected plot style. This method is similar to specifying an object-specific color, linetype, or lineweight (see Chapter 11). The *Plot Style* drop-down list on the *Object Properties* toolbar cannot be used to assign color-dependent plot styles.

FIGURE 33-36

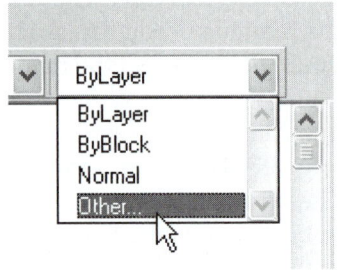

You can also use the *Plotstyle* command to assign object-specific plot styles (see "*Plotstyle*" next). Using *Plotstyle* has the same result as using the *Plot Style* drop-down list. The *Plotstyle* command can only be used to assign named plot styles.

Figure 33-37 displays the *Properties* palette. To use this method, select the desired object(s), then invoke the *Properties* palette. Select the desired plot style from the drop-down list. This method differs from the two previously described methods in that using the *Properties* palette is <u>retroactive for the selected objects only</u>, whereas the *Plotstyle* command and the *Object Properties* toolbar *Plot Style* drop-down list set plot styles for <u>newly created objects</u>.

FIGURE 33-37

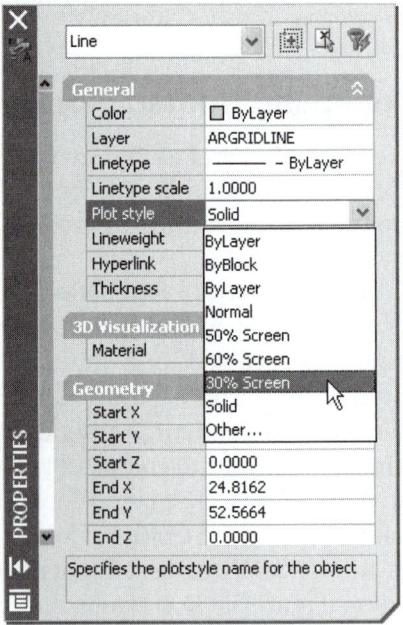

Making an assignment using any one of these three methods is similar to assigning colors, linetypes, and lineweights to individual objects rather than assigning the properties by the object's layer (*BYLAYER*). With this method, you lose the simplicity of organizing objects by layer and color; however, that is one of the underlying reasons for using named plot styles—so objects of different colors can be plotted in different ways.

Alternately, named plot styles can be assigned to layers. This is generally a simpler method, particularly for complex drawings with many objects. Figure 33-38 displays *the Layer Properties Manager* in which three plot styles, named Wide, Medium, and Thin, have been assigned to various layers. In this example, each of the three plot styles has a matching lineweight designated in the style.

FIGURE 33-38

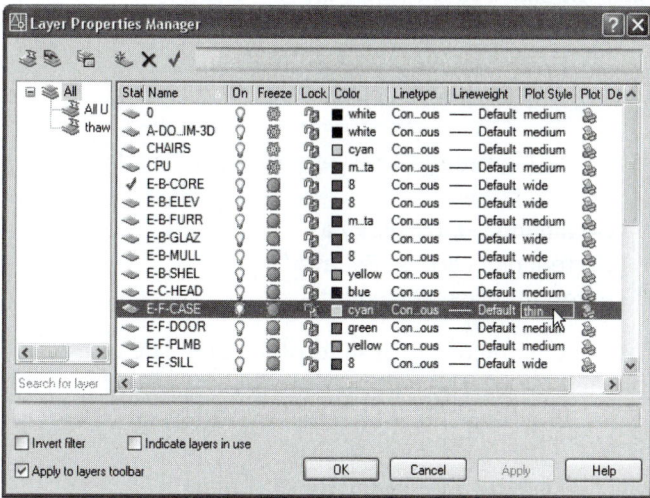

To assign plot styles from a named Plot Style Table to layers, click on the plot style name (usually *Normal*) in the *Plot Style* column for a specific layer. The *Select Plot Style* dialog box appears (Fig. 33-39) providing the list of plot styles to select from. All objects on the selected layer are then plotted with the parameters (screening, pen number, lineweight, line end joints, etc.) as previously designated in the plot style.

Note that the information given at the bottom of the dialog box indicates the *Active Plot Style Table* and the <u>layout</u> that it is *Attached to*. You can select a different Plot Style Table from this list to attach to the layout; however, doing so changes the setting previously made in the *Page Setup* dialog box. In other words, you can assign plot styles to objects and layers, but you cannot attach Plot Style Tables to individual objects or layers, only to layouts.

FIGURE 33-39

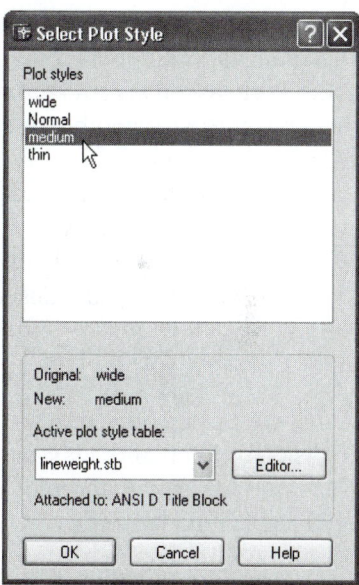

Plotstyle

Pull-down Menu	Command (Type)	Alias (Type)	Short-cut	Screen (side) Menu	Tablet Menu
Format *Plot Style...*	*Plotstyle* or *-Plotstyle*	...	...	...	...

As previously described briefly, the *Plotstyle* command is used to set the current named plot style for newly created objects. It has the same function as selecting a current plot style from the *Plot Style* drop-down list on the *Object Properties* toolbar. That is, the selected named plot style becomes the plot style assigned to all newly created objects.

The *Plotstyle* command produces the *Current Plot Style* dialog box (not shown). This dialog box is essentially the same as the *Select Plot Style* dialog box that appears when assigning plot styles to layers (see Fig. 33-39), except with *Plotstyle* the styles are assigned to objects. The *Plotstyle* command operates only for drawings that were created in named Plot Style Table mode. See the previous discussion, "Assigning Named Plot Styles to Objects."

To use the Command line version, type *-Plotstyle*. The command prompts are shown here.

    Command: **-plotstyle**
    Current plot style is "ByLayer"
    Enter an option [?/Current] :

Enter the name of the desired plot style. The question (?) option lists all plot styles in the currently attached table. The current drawing must be in named Plot Style Table mode and you must have a named Plot Style Table attached to the layout or viewport for this command to operate.

Named Plot Style Table Example

The following section is an example of using one <u>named</u> Plot Style Table to plot a drawing two times with different lineweights. Compare this example with the example given earlier in the chapter.

In the earlier example, two similar <u>color-dependent</u> Plot Style Tables were used to assign different screen percentages to different colors in the drawing. The two plots were created by attaching a different Plot Style Table to the layout before creating each plot. Using several color-dependent Plot Style Tables for one layout is a reasonable practice since the individual plot style names (color numbers) cannot change between Plot Style Tables. Therefore, all color-dependent Plot Style Tables have the same plot style names, although the specifications for each style (screen percentage in this case) can be different.

In the following example, only one named Plot Style Table is attached to a layout. To achieve different plots, the lineweight value is changed for individual plot styles to achieve the desired effects. In addition, objects to be plotted in different plot styles (lineweights) are drawn in the same color; therefore, color-dependent Plot Style Tables could not be used.

A drawing of an office building is completed and a layout is created to display and plot only the HVAC plan (Fig. 33-40). A plot displaying only the building core and the HVAC information is sent to the HVAC contractors for developing bids.

After a contractor is retained, it is discovered that one large duct unit must be specially manufactured. In order to make the unit more apparent for the manufacturer, a special plot must be made by applying a heavier lineweight to the unit.

FIGURE 33-40

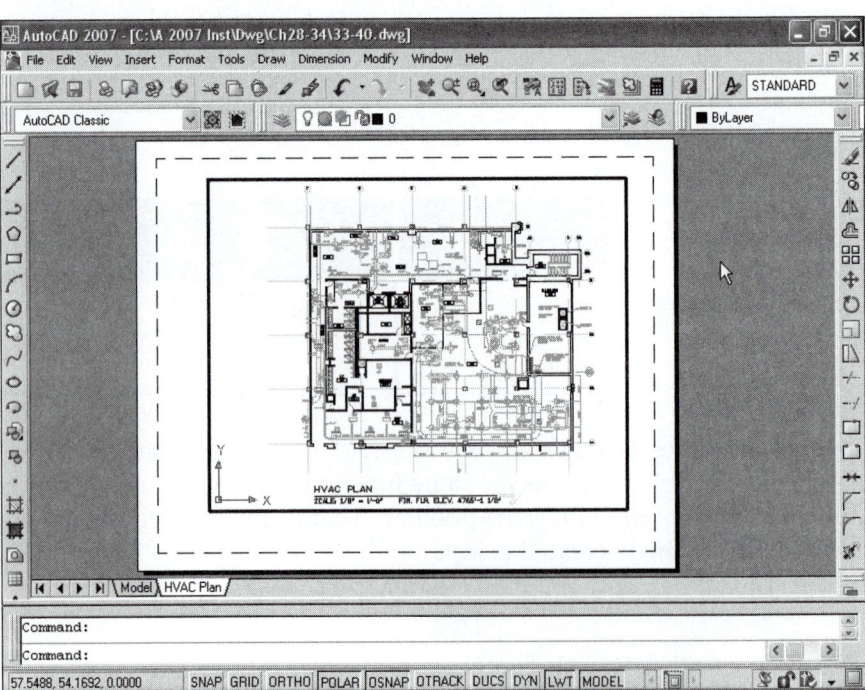

Since all of the HVAC objects are drawn in one color (see Fig. 33-40, light colored objects), a color-dependent Plot Style Table cannot be used. Instead, a named Plot Style Table is created called "LW mm" (Fig. 33-41). Several plot styles are created with different lineweight settings; each plot style is named according to weight (1.20mm, .90mm, .60mm, etc.). The Plot Style Table is then attached to the layout using the *Page Setup* dialog box.

FIGURE 33-41

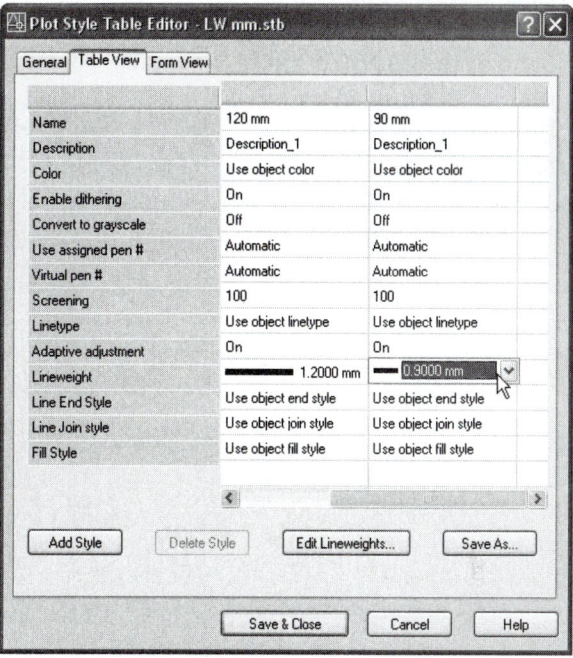

Next, objects representing the duct unit are selected and the *Properties* palette is invoked (Fig. 33-42). In the *Plot Style* row for all 24 objects [note the "All (24)" displayed at the top of the box], the desired plot style (named *90mm*) is selected and assigned to the objects.

FIGURE 33-42

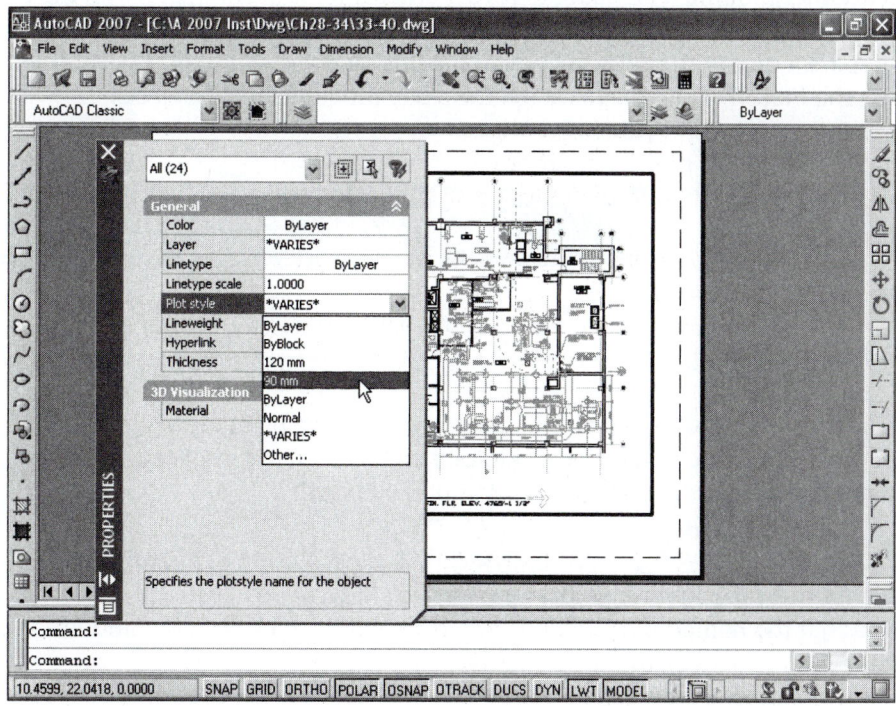

The finished plot reveals the special duct unit plotted in the wide lines (Fig. 33-43). Note that these objects are <u>plotted differently even though they share the layer and color of the other HVAC components.</u>

FIGURE 33-43

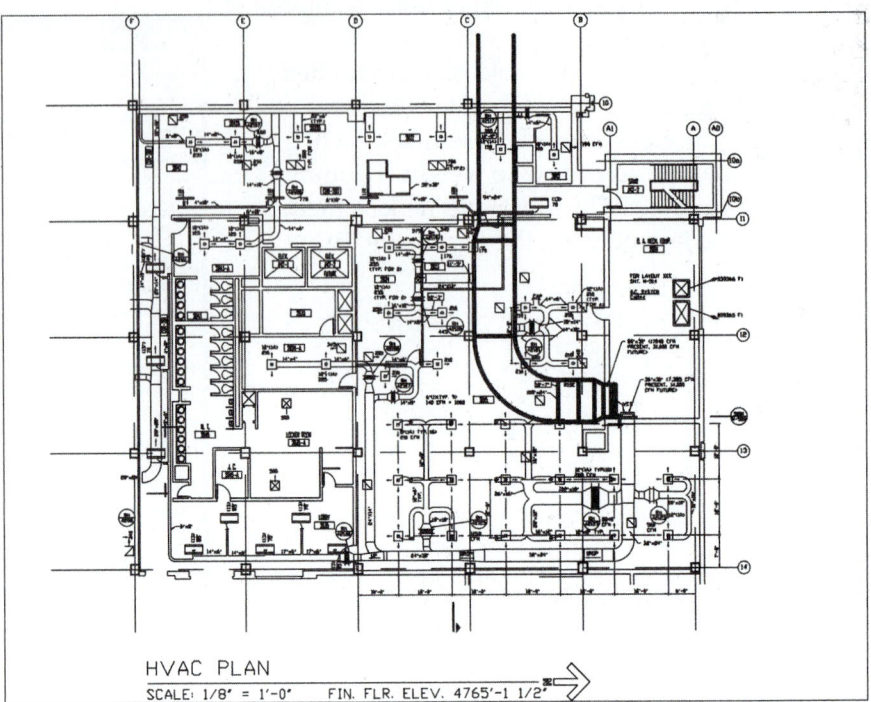

HVAC PLAN
SCALE: 1/8" = 1'-0"   FIN. FLR. ELEV. 4765'-1 1/2"

If another plot must be made to reveal other duct units in the layout, the process of applying the lineweight-based plot styles to particular objects within one layer can be used again. If you wanted to prepare a plot with uniform lineweights for all objects, it would be simple to select all objects and then use the *Properties* palette to assign the same plot style to all objects.

	Command (Type)	Alias (Type)	Short-cut	Screen (side) Menu	Tablet Menu
Pull-down Menu					
...	*Convertpstyles*	...	...	...	...

Convert-pstyles

Convertpstyles allows you to change <u>drawings</u> from using one type of Plot Style Table (named or color-dependent) to the other type of Plot Style Table. *Convertpstyles* is a Command line-driven utility. Type *Convertpstyles* in the drawing you want to convert.

Converting Drawings from Named Plot Style Tables to Color-Dependent Plot Style Tables

If a named Plot Style Table drawing is current when you use *Convertpstyles*, a small dialog box advises you that the named plot styles attached to objects, model space, and layouts will be detached.

    Command: **convertpstyles**
    Drawing converted from Named plot style mode to Color Dependent mode.
    Command:

After a drawing is converted to use color-dependent Plot Style Tables, you can assign a color-dependent Plot Style Table as you would normally. Plot styles are assigned by color.

Converting Drawings from Color-Dependent Plot Style Tables to Named Plot Style Tables

Before you use *Convertpstyles* in a color-dependent drawing, you should use *Convertctb* to convert the attached color-dependent tables to named tables. You can convert the color-dependent Plot Style Tables assigned to the drawing to named Plot Style Tables using *Convertctb* (see *Convertctb* next). After converting the tables, use *Convertpstyles* to reattach the newly converted named Plot Style Tables.

Assuming *Convertctb* has been used first in the drawing, type *Convertpstyles*. AutoCAD displays the standard file navigation dialog box for you to select the named Plot Style Table file to attach to the converted drawing. You must select a named Plot Style Table that was converted using *Convertctb* or created from a PC2 or PCP file. Normally you should select the named Plot Style Table that was converted from the color-dependent Plot Style Table that was assigned to the same drawing. The selected named Plot Style Table is assigned to model space and to all layouts. Drawing layers are each assigned a named plot style (from the converted Plot Style Table) that has the same plot properties that their color-dependent plot style had. After going through this conversion, you can change the named Plot Style Table assignment or assign other named Plot Style Tables to model space or layouts.

Covertctb

Pull-down Menu	Command (Type)	Alias (Type)	Short-cut	Screen (side) Menu	Tablet Menu
...	*Convertctb*	...	...	...	...

Convertctb is intended to be used in a drawing that you want to convert from using color-dependent Plot Style Tables to use named Plot Style Tables. First use *Convertctb* to convert the attached color-dependent <u>tables</u> to named tables, then use *Convertpstyles* in the drawing to convert the <u>drawing</u> (see *Convertpstyles*).

When you type *Convertctb* at the Command prompt in the drawing, the standard file navigation dialog box appears for you to select the desired color-dependent Plot Style Table, then reappears for you to specify a <u>name</u> (you should use the same name as the .CTB) and location for the new (now named) Plot Style Table. You should use *Convertctb* to convert all attached tables. The original color-dependent tables are not changed.

Convertctb creates one named plot style for each color that has unique plot properties, one named plot style for each group of colors that are assigned the same plot properties, and a default named plot style called NORMAL.

PLOT STAMPING

Plot stamping provides options for you to have an informational text "stamp" plotted on each drawing. For example, the drawing name, date, and time of plot can be added to each plot for tracking and verification purposes.

Plotstamp

Pull-down Menu	Command (Type)	Alias (Type)	Short-cut	Screen (side) Menu	Tablet Menu
...	*Plotstamp*	...	...	...	...

To enable plot stamping, you must use the expanded section of the *Plot* dialog box, then check *Plot Stamp On* (not shown).

To specify the text you want to appear on the plot, select the *Plot Stamp Settings* button that appears next to *Plot Stamp On* when you check it, or enter *Plotstamp* at the Command line. This action produces the *Plot Stamp* dialog box (Fig. 33-44).

FIGURE 33-44

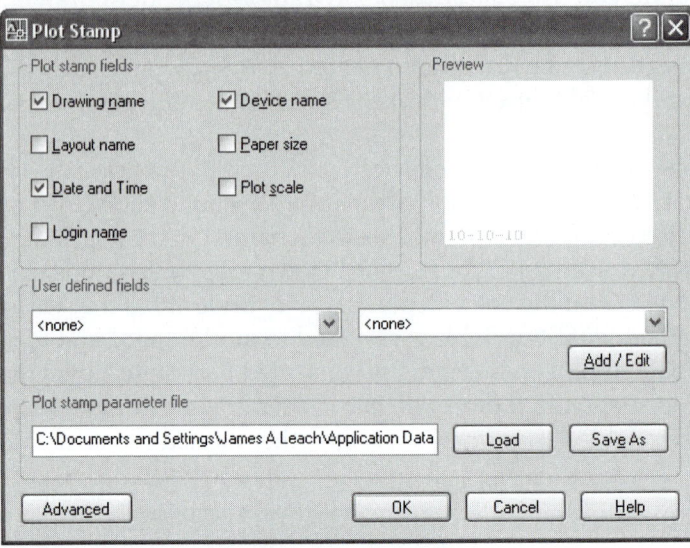

TIP Saving plot stamp parameters involves changing a plot stamp parameter (.PSS) file. Two .PSS files are supplied by AutoCAD, INCHES.PSS and MM.PSS. You can change either of these files or create new files. A simple method for creating your own .PSS files is to use either of the AutoCAD-supplied files as a template, make the desired changes in the *Plot Stamp* dialog box, then use the *Save As* button to save the changes to a new file name. (See *"Plot stamp parameter file."*)

The .PSS files are stored (by default for stand-alone installations) under your profile name in the Documents and Settings folder. To find the location of the .PSS files on your system, see the edit box in the *Plot stamp parameter file* section (lower-left corner), or use *Options*, *Files* tab, and expand the *Working Support File Search Path* section.

Plot stamp fields
In this area, check any items you want to appear in the plotted drawing. Each piece of text is added to the plot stamp separated by a comma. Content for *Drawing name*, *Layout name*, *Device name*, *Paper size*, and *Plot scale* is automatically obtained from the AutoCAD drawing. *Plot scale* may not be an appropriate item to select when you plot a layout since this text is actually the plot scale that appears in the *Plot* dialog box (normally set to 1:1), not the viewport scale (the actual plot scale of the geometry). Content for *Date and Time* and *Login name* are obtained from the computer's system data.

Preview
This image tile indicates only the location and orientation of the plot stamp, not the text content. You cannot see a preview of the actual plot stamp using the *Full Preview* button or the *Preview* command; you must make a plot.

User defined fields
You can specify additional text to include in the plot stamp by selecting the *Add/Edit* button. The *User Defined Fields* dialog box appears (not shown) for you to enter new text or edit existing user text fields. Once specified, this personalized text appears in the *User defined fields* drop-down lists (see Fig. 33-44). Two user defined fields can be added to a plot stamp.

Plot stamp parameter file
You can store the current plot stamp information in a file with a .PSS extension. For example, if you wanted to create a standard plot stamp for your company or school, first specify the plot stamp fields and advanced options you want, then save them to a .PSS file. Other users can then access the saved file and stamp their plots based on those standard settings.

To create a plot stamp parameter file, use either the AutoCAD-provided MM.PSS or INCHES.PSS file as a template, then save your changes to a new name using the *Save As* button. The *Load* button displays the standard file selection dialog box in which you can specify the location of the parameter file you want to use. The MM.PSS or INCHES.PSS files differ only in millimeter or inch units used for the default text height and offset.

Advanced

This button produces the *Advanced Options* dialog box (Fig. 33-45) where you can specify the *Location, Orientation,* and *Text properties* of the plot stamp. The *Location* and *Orientation* are reflected in the *Preview* tile of the *Plot Stamp* dialog box. The *X offset* and *Y offset* are the distances from the edge of the printable area or paper border you want the stamp to appear.

In the *Log file location* area you can indicate if you want to *Create a log file* and specify the name and location of the file. A log file is an ASCII text file that contains a record for each plot created, and each record lists the same items you specified for the plot stamp such as the date and time, drawing name, device name, and so on.

FIGURE 33-45

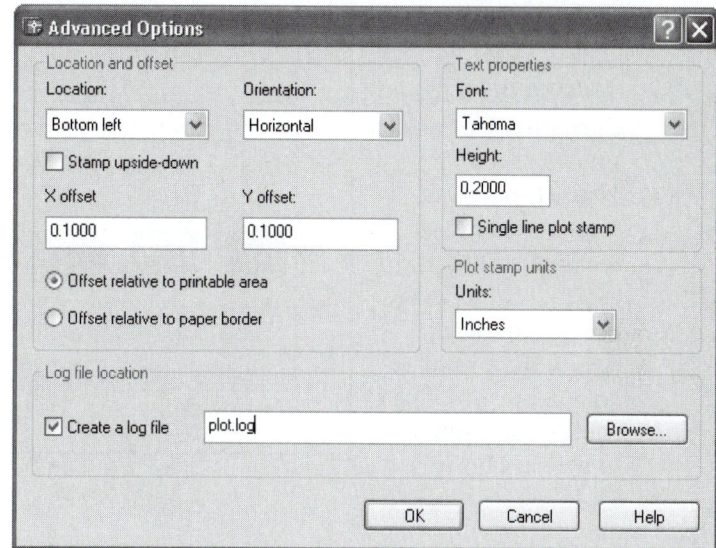

CHAPTER EXERCISES

1. **Multiple Viewports and Viewport-Specific Layer Visibility**

   A. *Open* the **Single Cavity Mold** drawing (download from the www.mhhe.com/leach Web site). Use *SaveAs* to rename the drawing **Ch33 SCM** and locate it in your working directory. Activate each of the layout tabs and the model tab to familiarize yourself with the drawing.

   B. Activate the **Plan of Ejector** tab. Right-click on the tab and select *Move or Copy* from the shortcut menu. In the *Move or Copy* dialog box, select *Plan of Ejector* and *Create a copy,* then *OK*. Now activate the new layout, right-click and *Rename* the tab **Ejector Details**.

   C. With the new layout activated, use *Pagesetup* to select an available printer as the *Printer/Plotter*. Select *Letter* as the *Paper Size*. Select *OK* to save the settings for the layout. Next, use the *Layer Control* drop-down list and select *Freeze or thaw in current viewport* to freeze layer **Bourder**. Make a test print of the layout. Note in the print that the dimensions are barely readable at that size, and the linetype scale is too large.

D. Using **grips**, select the existing viewport and **STRETCH** it to modify the viewport size to occupy only the right half of the printable area as shown in Figure 33-46. Next, make layer **LAYOUT-VIEWPORTS** the *Current* layer. Use *Vports* to make *Two: Horizontal* viewports and set the *Viewport Spacing* to **.25**. *Fit* the viewports into the left half of the printable area. Your drawing should look similar to that in Figure 33-46.

FIGURE 33-46

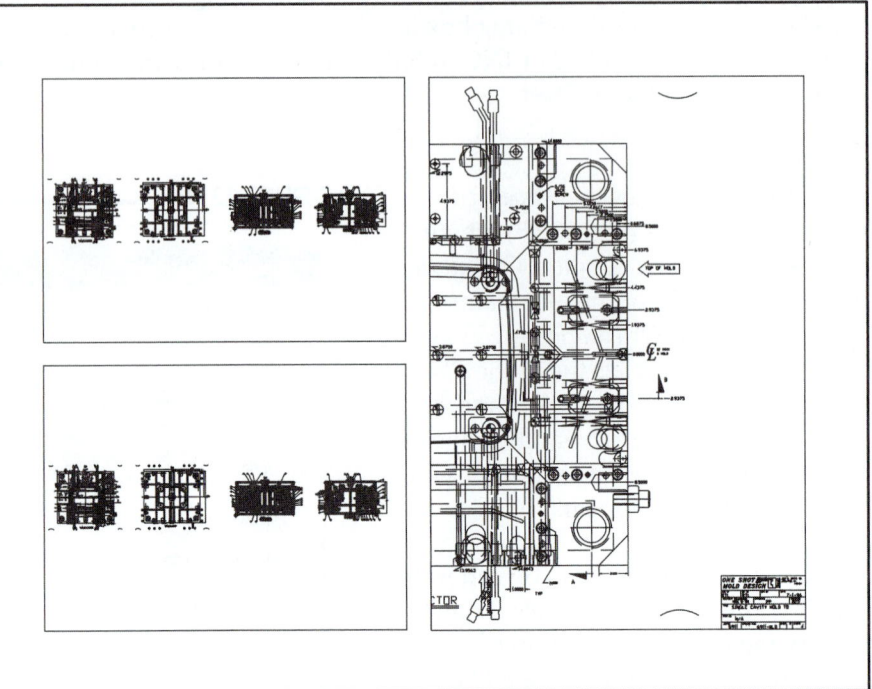

E. Use *Zoom* and *Pan* to display the upper-right corner of the mold in the small top viewport and the lower-right corner of the mold in the bottom viewport. Activate the *Viewports* toolbar and set the viewport scales as follows:

FIGURE 33-47

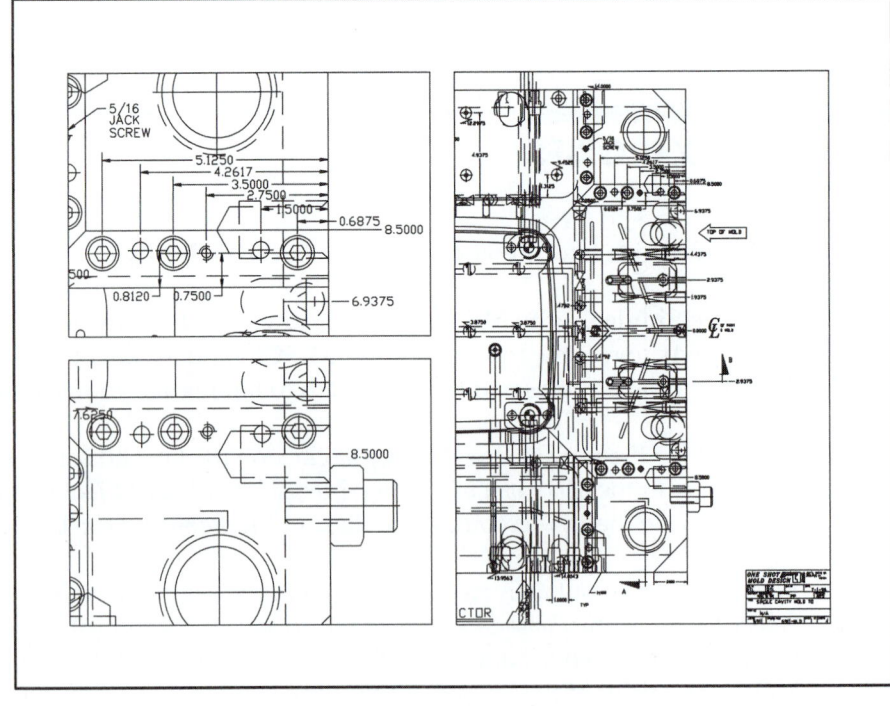

top left viewport	**1:2**
bottom left viewport	**1:2**
right viewport	**3:16** (.1875)

The resulting display should look like Figure 33-47.

F. Activate the right viewport and use the *Layer Properties Manager* and select *Current VP Freeze* to freeze these layers for the right viewport only:

A-H20
AIR
B-H20
DIMF
SLID1H20
SLID2H20

G. Activate paper space. On layer **NOTES**, add two circles and leaders to indicate Detail A and B. Make the text size **.12** and the text font *Romans*. Add the labels and scales for the two small viewports as shown in Figure 33-48. Set *LTSCALE* to **.5**. Finally, ensure *PSLTSCALE* is set to **1** so the linetype dashes appear equal in all viewports. *Save* the drawing and make a print of the layout.

FIGURE 33-48

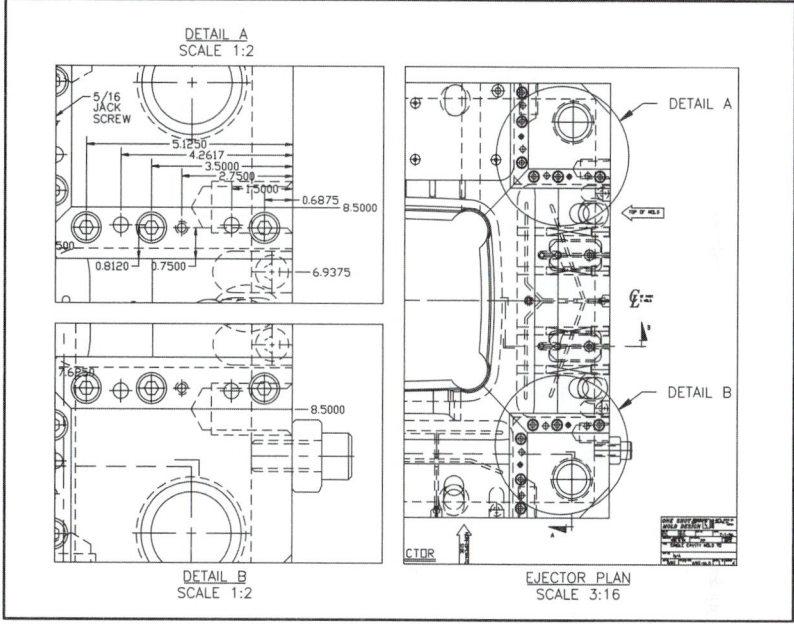

2. **Multiple Viewports and Dimensioning in Paper Space**

In this exercise, you will create a layout with two viewports to show an overall view of a drawing and a detail in a different scale. You will dimension the drawing both in model space and in paper space.

A. *Open* the **2BR-APT** you worked on last in Chapter 20 Exercises. Create a text *Style* using *Roman Simplex* font. Create a new dimension style named **ARCH-1** using the new font and with other settings to create architectural style dimensions as shown in Chapter 15, Figure 15-45. The drawing scale factor is **48**, so set the appropriate *Overall scale*. In model space, create dimensions for the apartment on the **DIM** layer. Make sure you add dimensions also to the new bedroom in your drawing. If you need assistance placing the dimensions, refer to Figure 33-49, on the next page. Use *SaveAs* and name the drawing **2BR-APT-DIM**.

B. Create a new layout using the *Template* option. Select the *ANSI B template* with an *ANSI B title block*. Next, use *Pagesetup* to specify a plot device and paper size for an *ANSI B* size sheet.

C. In the layout, create one viewport on the **VPORTS** layer that occupies about 2/3 of the layout area. *Zoom* and *Pan* as needed to show the entire floor plan. Set the viewport scale to **1/4"=1'**. If the entire floor plan does not appear in the viewport, make the necessary adjustments to the viewport border using **grips**.

D. Next, make a second smaller viewport on the right side of the layout. *Zoom* and *Pan* to display only the bathroom at the far left of the floor plan. Set the viewport scale to 1/2"=1'.

E. Make a new layer named **DIM-PS** and set the properties the same as layer DIM. Make a *New* dimension style called **ARCH-2** using ARCH-1 as a "template." Change the *Overall scale* to **1**. In paper space and on layer DIM-PS, create dimensions to indicate the distances between the wall and the three fixtures as shown in Figure 33-49. Label the viewports as shown, *Freeze* layer **VPORTS**, and *Save* the drawing. Make a *Plot*.

FIGURE 33-49 ━━━━━━━━━━━━━━

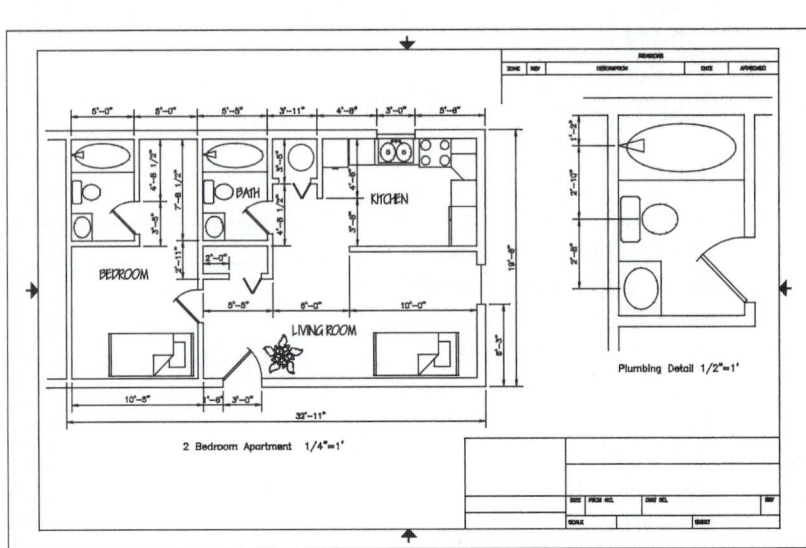

3. **Multiple Viewports, Dimensioning in Model Space, and Viewport-Specific Layer Control**

In this exercise, you will create a layout with three viewports to show an overall view of a drawing and two details in a different scale. You will dimension the drawing in model space and control the visibility for each viewport.

A. *Open* the **EFF-APT2** drawing that you worked on in Chapter 30 Exercises (with the door *Stretched* to the center of the front wall). *Erase* the plant. Create a text *Style* using *Roman Simplex* font. In model space, create dimensions for the interior of each room on the **DIM** layer, similar to those dimensions you created in Exercise 2. (You may consider also opening the **2BR-APT-DIM** drawing from the last exercise and using copy and paste to copy the needed dimensions to the new drawing. In this case, however, the dimensions will have to be reassociated.) *Zoom* in, and on another *New Layer* named **DIM2** give the dimensions for the wash basin in the bath and the counter for the kitchen sink.

B. To set up the layout, you have two choices: (1) use a *Layout Template,* or (2) create and set up the layout from scratch. In either case, use the template, plot device, and paper size of your choice.

C. Make a *New Layer* or use the existing layer named **VIEWPORT** and set it as the *Current* layer. Use *Vports* to make one viewport on the right side of the layout, occupying approximately 2/3 of the page. Use *Vports* again to make two smaller viewports in the remaining space on the left.

D. Activate the large viewport (in *Mspace*) and use *Zoom XP* or other viewport scale method to achieve the largest possible display of the apartment to an <u>accepted scale</u>. In the top-left viewport, use display controls to produce a <u>scaled</u> detail of the bath. Produce a <u>scaled</u> detail of the kitchen sink in the lower-left viewport, similar to that shown in Figure 33-50.

FIGURE 33-50

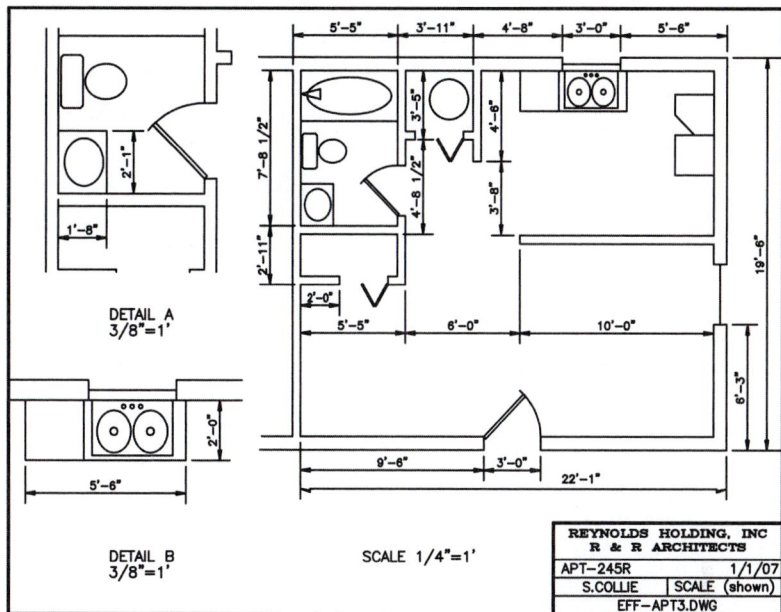

E. Use the *Layer Properties Manager* to *Current VP Freeze* layer **DIM2** in the large viewport and *Current VP Freeze* layer **DIM** in the two small viewports. Use *Properties* and select the dimensions that appear in the two detailed views and set a new *Dim scale overall* (in the *Fit* category) so the dimensions appear at the appropriate size for the viewport scale.

F. Return to paper space and *Freeze* layer **VIEWPORT**. Use *Text* (in paper space) to label the detail views and give the scale for each. *Save* the drawing as **EFF-APT3** and *Plot* from paper space at **1:1**. The final plot should look similar to Figure 33-50.

4. **Multiple Viewports, *Xref*, and Viewport-Specific Layer Visibility**

In this exercise, you will create three paper space viewports and display three *Xrefed* drawings, one in each viewport. Viewport-specific layer visibility control will be exercised to produce the desired display.

A. Begin a *New* drawing using a *Template* to correspond to the sheet size you intend to plot on. Activate a *Layout* tab. Use *Pagesetup* to select the paper size and to set the plotting options for the device you use to plot with. Create *New Layers* named **TITLE**, **XREF**, and **VIEWPORT**. On the **VIEWPORT** layer, use *Vports* to create a viewport occupying about 1/2 of the page on the right side.

B. Make the **XREF** layer *Current*. Double-click inside the viewport to activate model space in the viewport. In the viewport, *Xref* the **GASKETD** drawing from Chapter 28 Exercises. Use *Zoom* with an *XP* value or use the *Viewport scale* drop-down list (in the *Viewports* toolbar) to scale the gasket appropriately to paper space. Use the *Layer Properties Manager* to *Freeze* (globally) the **GASKETD | DIM** layer, then use *New VP Freeze* for all **GASKETD*** layers.

C. Option 1.   Change to *PS* (double-click in paper space) and create two more viewports on *Layer* **VIEWPORT** on the left side, each equal in size to about half of the viewport on the right. Set *Layer* **XREF** *Current* and *Xref* the **GASKETB** (Chapter 9) and the **GASKETC** (Chapter 16) drawings. Change to *MS* and use the *Layer Properties Manager* or *Vplayer* to constrain only one gasket to appear in each viewport.

Option 2.   Change to *PS* and create one more viewport on *Layer* **VIEWPORT** equal to one half the size of the viewport on the right. Set *Layer* **XREF** *Current* and *Xref* only the **GASKETB** drawing. Change to *MS* and use the *Layer Properties Manager* dialog box or *Vplayer* to set the **GASKETB** visibility for the existing and new viewports. Repeat these steps for **GASKETC.**

D. Use *Zoom XP* or the *Viewport Scale* drop-down list to produce a scaled view of each new gasket. Change to *PS* and *Freeze* layer **VIEWPORT**. Use *Text* to label the gaskets. *Plot* the drawing to scale. Your plot should look similar to Figure 33-51. *Save* the drawing as **GASKETS**.

FIGURE 33-51

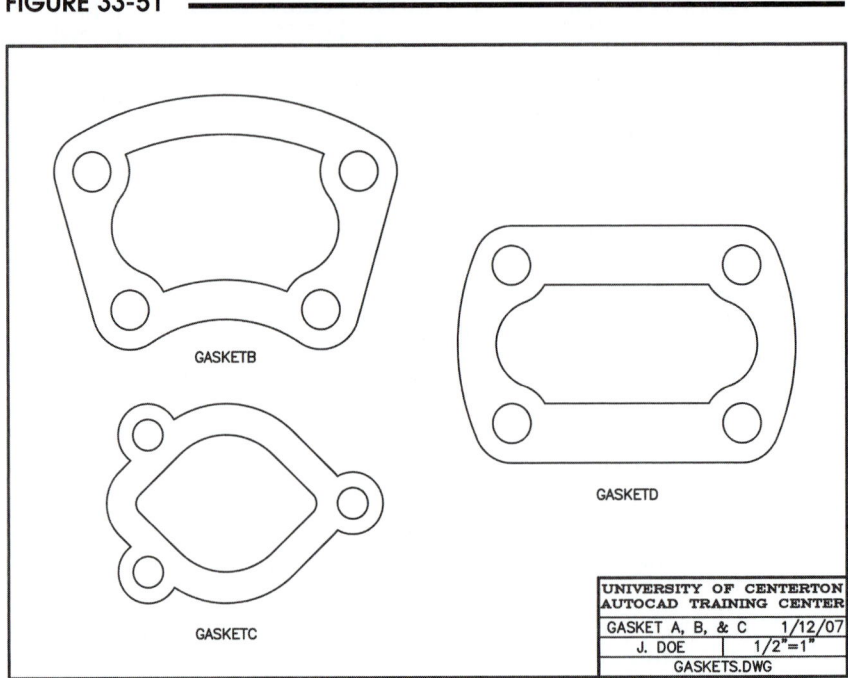

5. **Create a Color-Dependent Plot Style Table**

Start AutoCAD and begin a *New* drawing by any method. Open the *Plot Style Manager*. Use the *Add-A-Plot Style Table Wizard* to create a new Plot Style Table. In the *Begin* step, select *Start from Scratch*. In the next step, select *Color-Dependent Plot Style Table*, then proceed to assign the name **SCREEN PERCENTS**. When you are finished, check to make sure the new Plot Style Table appears in the *Plot Style Manager*.

6. **Edit a Color-Dependent Plot Style Table**

Open the *Plot Style Manager*. Double-click on the **SCREEN PERCENTS** Plot Style Table to cause the *Plot Style Table Editor* to appear. Use the *Table View* or *Form View* to assign *Screening* values of **100**, **75**, **50**, and **25** to plot styles **Color 1**, **Color 2**, **Color 3**, and **Color 4**, respectively. Select *Save & Close* to save the new settings. Reopen the new Plot Style Table to ensure your settings were saved.

7. **Create and Edit a Named Plot Style Table**

Open the *Plot Style Manager* if not yet open. Use the *Add-A-Plot Style Table Wizard* to create a new Plot Style Table. In the *Begin* step, select *Start from Scratch*. In the next step, select *Named Plot Style Table*, then proceed to assign the name **LINEWEIGHTS**. In the *Finish* step, select the *Plot Style Table Editor* button.

In the *Plot Style Table Editor*, create five plot styles named **1.20mm**, **.90mm**, **.60mm**, **.30mm**, and **.15mm**. Assign the respective settings from the *Lineweight* drop-down list. Select the *Save & Close* button. Finally, select the *Finish* button to close the wizard. In the *Plot Style Table Manager*, double-click on the new **LINEWEIGHTS.STB** file and check your settings.

8. **Set the Default Plot Style Behavior**

In AutoCAD, invoke the *Options* dialog box and select the *Plot and Publish* tab. Select the *Plot Style Table Settings* button in the lower-right corner. In the *Plot Style Table Settings* dialog box that appears, select *Use Color Dependent Plot Styles*. Under *Default Plot Style Table*, select *None*. Select the *OK* button to close the *Options* dialog box and save your changes. Now, new drawings that you create will use color-dependent Plot Style Tables by default, but can later be changed to use named Plot Style Tables.

9. **Assigning Color-Dependent Plot Styles**

A. *Open* the **HAMMER-DIM** drawing you created in Chapter 29 Exercises. Make a plot from the *ANSI B Title Block* layout as it exists from the previous exercise (before assigning plot styles). Note that the drawing plots with all lines in the same lineweight, making it difficult to discern between object lines and dimension lines (Fig. 33-52).

FIGURE 33-52

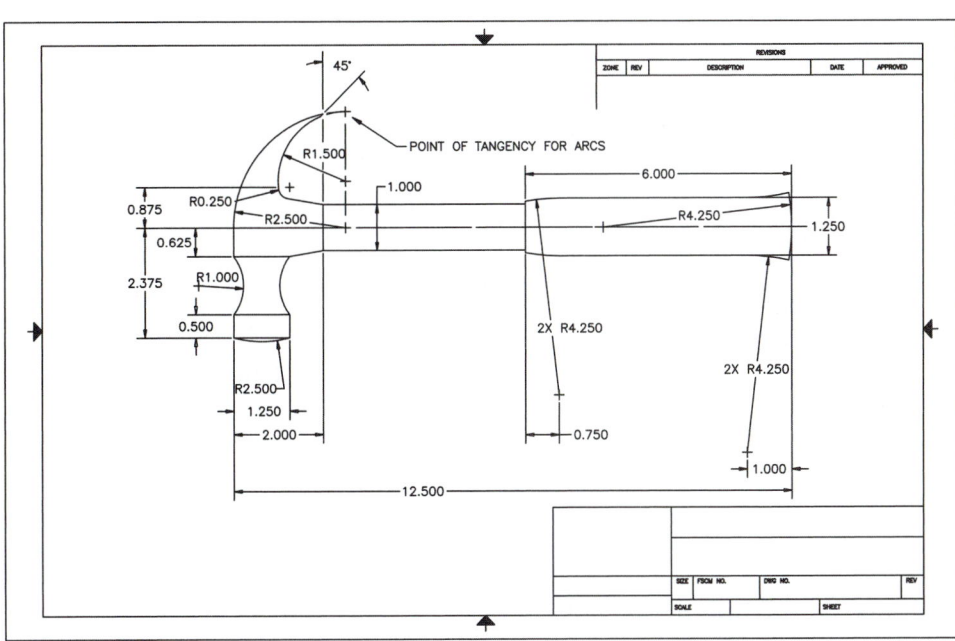

B. From the layout tab, invoke the *Page Setup* dialog box and attach the **SCREEN PERCENTS** Plot Style Table you created and edited in Exercises 5 and 6. Select the *Edit* button to view the plot styles settings (Color 1=100, Color 2=75, etc.). Use *SaveAs* and assign the name **HAMMER-PSTYLES**.

C. Use the *Layer Properties Manager* to see which plot styles are assigned to which layers. Since the plot styles are assigned by color, you will have to change the layer colors to achieve the desired screen percentages. Make the **GEOMETRY** layer color 1 (red) and the **DIMENSIONS** layer color 3 (green).

D. Use the *Page Setup* dialog box again and check *Display Plot Styles* in the expanded section. Check the layout to ensure the plot styles are assigned correctly. Your display should look similar to Figure 33-53. Finally, make a *Plot*. The plot should display the geometry at a 100 percent screen and the dimensions in a lighter screening. Save the drawing.

E. Experiment by changing the **DIMENSIONS** layer to color 3 or 4. Make plots to see the results.

FIGURE 33-53

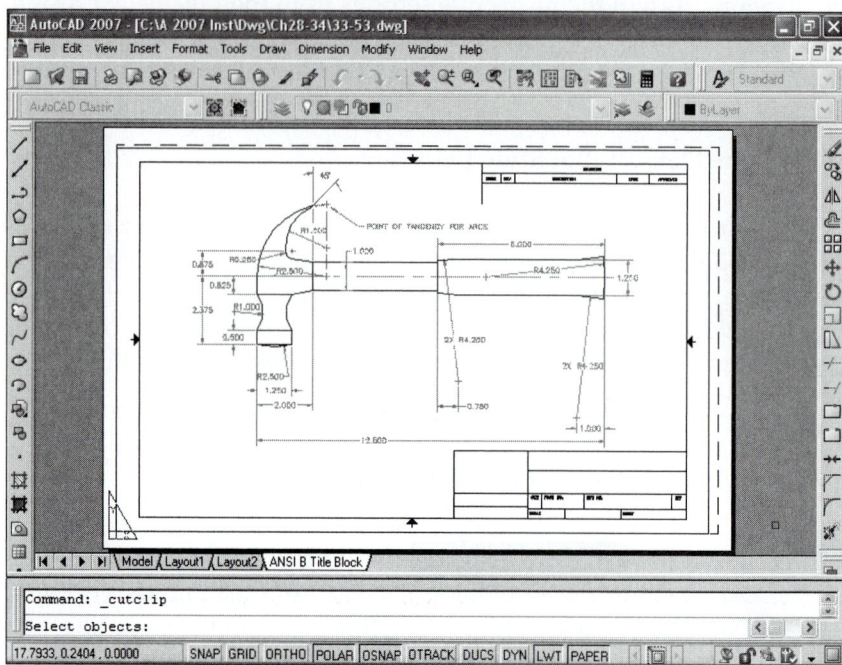

10. Assign Named Plot Styles

A. *Open* the **DB_SAMP** drawing from the Program Files/AutoCAD 2007/Samples folder. This drawing was created to use a named Plot Style Table. Assuming you are the general contractor, you are required to create a plot that displays the entire plan but highlights the building's concrete core.

B. From the layout tab, invoke the *Page Setup* dialog box and attach the **LINEWEIGHTS.STB** Plot Style Table you created and edited in Exercise 7. Select the *Edit* button to view the plot styles settings and style names. Also check *Display Plot Styles*. Use *SaveAs* and assign the name **VP_SAMP-PSTYLES** and save the file to your working directory.

FIGURE 33-54

C. Open the *Layer Properties Manager*. *Freeze* all the *Cyan* layers except **E-F-TERR**. Also *Freeze* layers **FILE_CABINETS** and **FURNITURE**.

D. Assign the **.15mm** plot style to all layers. Then assign the **.90mm** plot style to layer **E_B_CORE**. Your display should look like Figure 33-54 (this figure does not display the viewport). *Save* the drawing. Make a *Plot* and check your results.

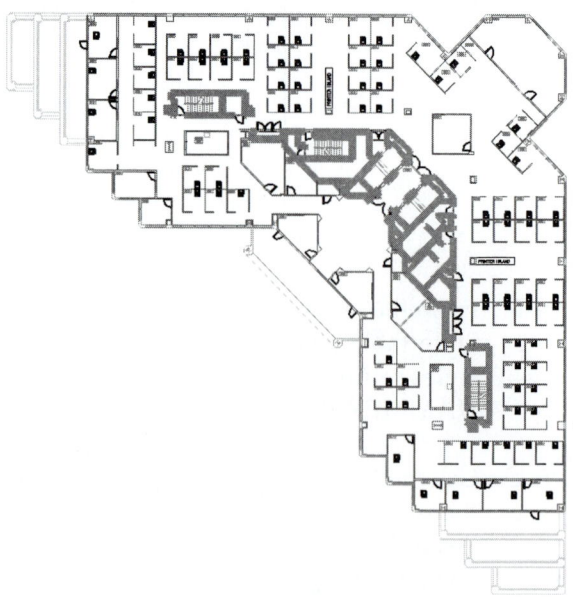

11. **Assign Named Plot Styles**

A. *Open* the **SADDL-DM** drawing from Chapter 29 Exercises. The model geometry and dimensions are drawn in the *Model* tab, and the titleblock and border are inserted into a layout tab.

B. Use Windows Explorer to search for the **ACAD.CTB**. Make a *Copy* of the file and *Rename* it **LINEWEIGHTS.CTB**. *Exit* Explorer.

C. In AutoCAD, select the layout tab and use the *Page Setup* dialog box to attach the new Plot Style Table that you created in the previous step. Also select *Display Plot Styles*. There should be no apparent change in the layout. Use *SaveAs* and assign the name **SADDL-DIM-PLOTSTYLES**.

D. Use any method to edit the new Plot Style Table so that the color of the geometry object lines have a heavy (thick) lineweight assigned and the dimension, hidden line, and center line colors have a light (thin) lineweight assigned. If the layout displays the geometry as you intend, make a *Plot* of the drawing. The resulting plot should look similar to Figure 33-55. *Save* the drawing.

FIGURE 33-55

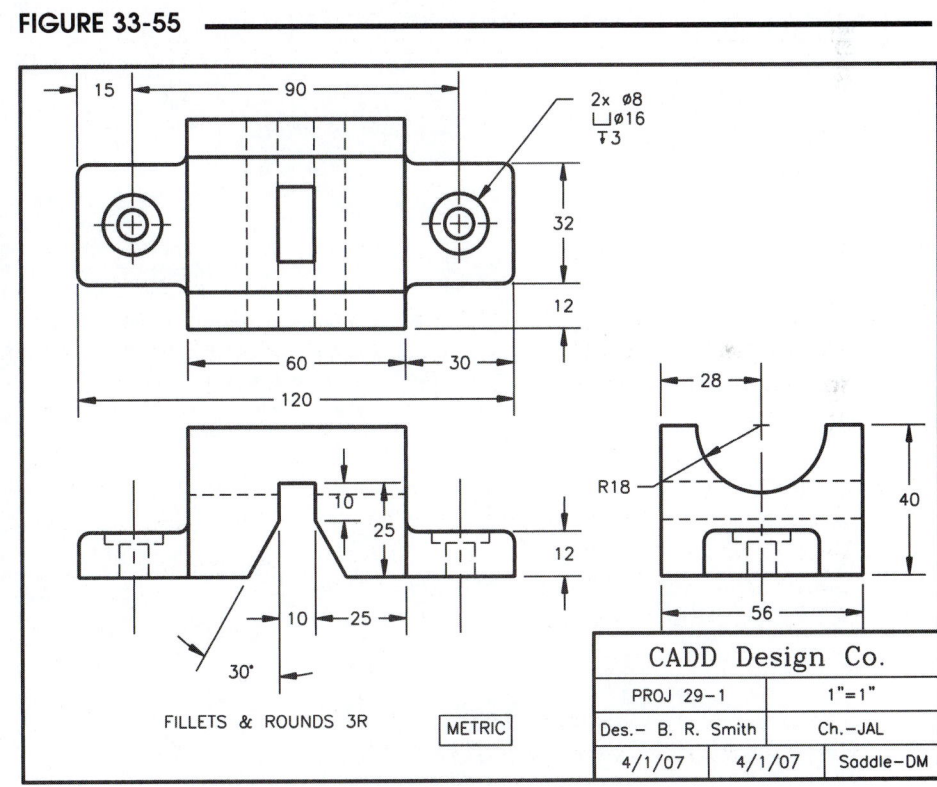

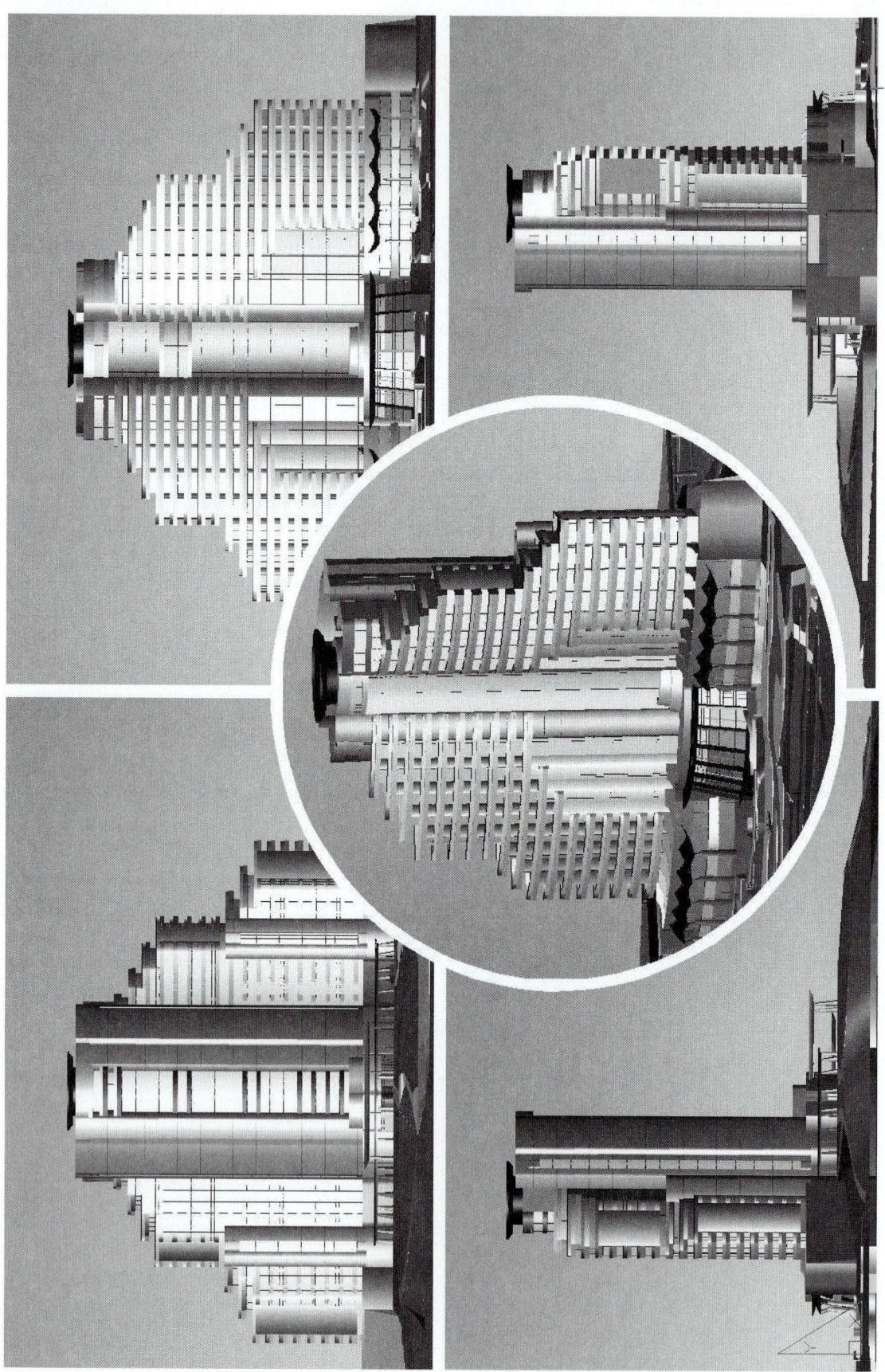

HOTEL MODEL.DWG, Courtesy of Autodesk, Inc.

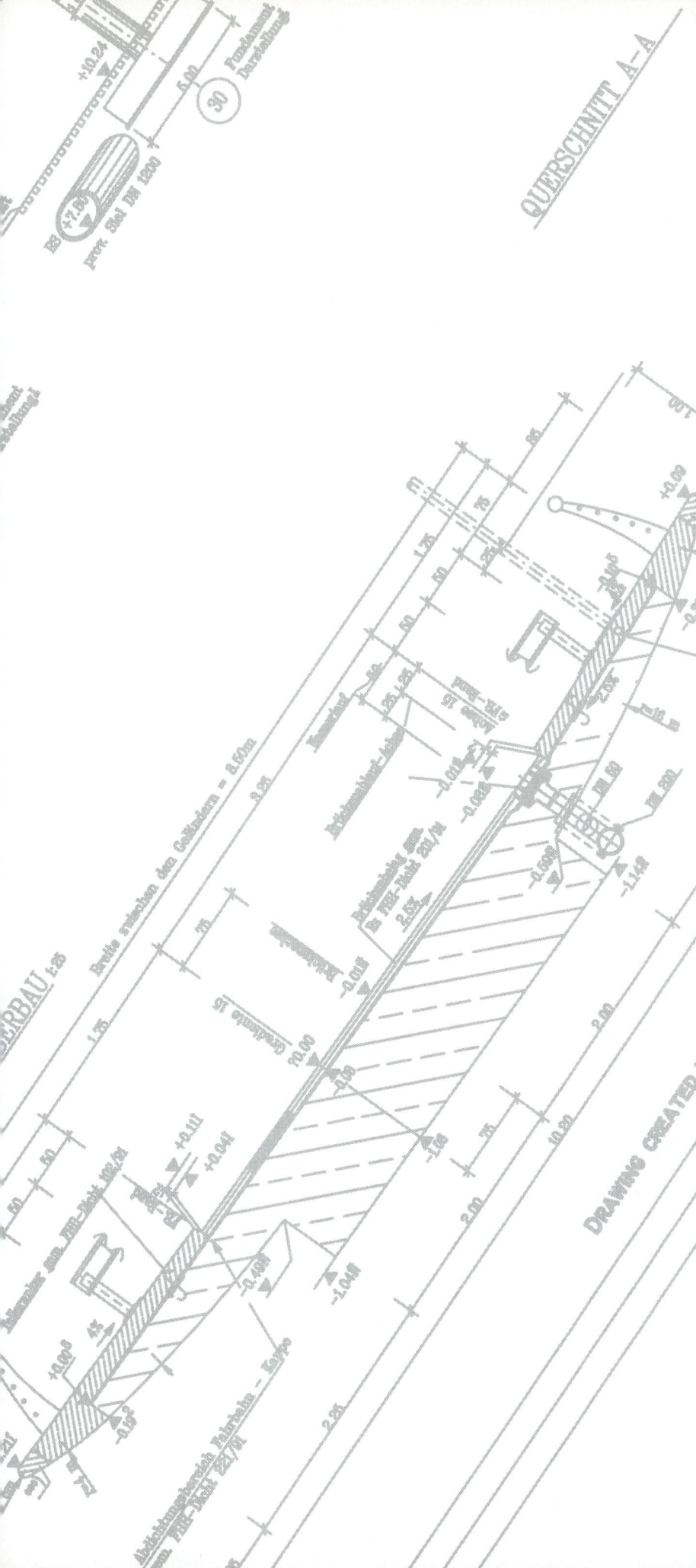

SHEET SETS

CHAPTER OBJECTIVES

After completing this chapter you should:

1. understand that a "sheet" is actually a layout in a drawing displaying a particular view;

2. know that a .DST file contains the information to manage a sheet set;

3. be able to use *Sheet Set Manager* to list, view, organize, label, print, and archive sheet sets;

4. be able to import layouts as sheets and to create new sheets from templates;

5. be able to rename and renumber sheets and create a sheet list table;

6. be able to use the *Create Sheet Set* wizard.

SHEET SETS CONCEPTS

Sheet sets have been used in industry for years—even before the advent of AutoCAD. Traditionally, the term "sheet set" is used to describe a set of bound drawing sheets all related to one project. For example, a set of house plans is a sheet set—a complete set of drawings including plans, elevations, and details needed to construct a house. Since engineers, architects, and designers prepare and deliver drawing sets to clients during and at the end of a design or construction project, a sheet set, therefore, can be a legally binding document that represents the product or result of the contracted job.

AutoCAD provides a tool called the *Sheet Set Manager* for you to manage multiple drawings as one set. The primary function of the *Sheet Set Manager* is to print, plot, or publish the set of drawings. In fulfilling that function, the *Sheet Set Manager* allows you to gather, list, organize, view, label, print, and archive a complete set of existing project drawings.

With older releases of AutoCAD, project coordinators usually organized projects by keeping all related drawings in one folder and assigning appropriate names. Another useful method of organizing "sheets" was to create one drawing with the plans, elevations, details, and sections in different areas of model space, then use multiple layouts to display and print the specific information. Although these are still good strategies, the *Sheet Set Manager* goes beyond those previous capabilities by allowing you to collect multiple drawings and multiple layouts into one sheet set.

Basic Principles

Learning to use AutoCAD sheet sets on your own can be difficult since much of the operation of the *Sheet Set Manager* is transparent to the user. In addition, the AutoCAD *Help* files and tutorials are not fully descriptive of what goes on "behind the scene," and they do not give the information in a logical sequence. However, once you understand a few simple concepts and the mechanics behind the *Sheet Set Manager*, you will be able to use sheet sets effectively.

A "Sheet" is a Layout

Each "sheet" in the *Sheet Set Manager* is an existing layout in a drawing. The *Sheet Set Manager* does not manage new objects called "sheets," it only manages existing layouts, but calls them sheets. Layouts are saved in drawing (.DWG) files; therefore, sheets are simply layouts that exist in drawing files, external to the *Sheet Set Manager*. One "sheet" references one layout. A layout cannot be used by more than one sheet.

The *Sheet Set Manager* Saves a .DST File

As stated earlier, the *Sheet Set Manager* allows you to locate, list, organize, view, label, print, and archive a group of drawings as a set. Generally speaking, the *Sheet Set Manager* does not physically collect, move, or contain .DWG files. Rather, the *Sheet Set Manager* simply saves a list of the sheets (layouts) and other information, such as the names of the drawings containing the layouts, the model geometry to view in the layouts, where the drawings are located, how the sheets are to be organized in the set, how the sheets are to be printed, and other information. The *Sheet Set Manager* saves its information in a "sheet set data file" or a .DST file. You do not have to *Save* a .DST file—the *Sheet Set Manager* automatically saves and updates the file as you work. No *Sheet Set Manager*-specific information is saved in a .DWG file.

The *Sheet Set Manager* Operates Independently of Drawing Files

The *Sheet Set Manager* is a feature that appears on the screen in palette format, similar to a tool palette. The *Sheet Set Manager* operates independently of related drawings in that it can be on the screen when no drawings are open (since no *Sheet Set Manager*-specific information is saved in a .DWG file). Using *Sheet Set Manager* to view the sheets listed in a .DST file actually opens the drawings containing the layouts. You can also open these and other drawings as you might normally without *Sheet Set Manager*. Closing a .DST file or the *Sheet Set Manager* does not close the related drawings, and closing the drawings does not close *Sheet Set Manager* or the .DST file.

Think of Sheet Sets as *Xrefed* Layouts

Although the *Sheet Set Manager* does not actually *Xref* layouts (it is not possible to *Xref* paper space in AutoCAD), you can use the *Xref* concept as an analogy. That is, sheet sets could be imagined as externally referenced paper space layouts. Each "sheet" is a layout contained in a drawing that is externally referenced into *Sheet Set Manager*. In other words, each sheet in the *Sheet Set Manager's* .DST file references a layout in a .DWG file. Those layouts, as any layouts, contain the drawing view, layer visibility, and scale information for the model space geometry. The model space geometry that is displayed in each layout can exist in the same drawing as the layout, or (as has been possible for many releases of AutoCAD) the model geometry displayed in a layout can exist within one or more externally referenced drawings (*Xrefed* to the layout drawing). To simplify, think of a "sheet" as a layout externally referenced into the *Sheet Set Manager*.

Another important component of this analogy is that the externally referenced layouts are not actually brought into a drawing as model space *Xrefs* might be. Instead, the list of sheets is stored in a .DST file, not in a .DWG file. A .DST file contains the list of externally referenced layouts, and the *Sheet Set Manager* saves and manages these references.

Two Sheet Creation Methods

You can create sheets directly in the *Sheet Set Manager* or by using the *Create Sheet Set* wizard (*NewSheetSet* command). Although there are other differences, both of these tools allow you to create sheets using existing layouts or create sheets from a layout template (see "Two Types of Sheets" next). The *Create Sheet Set* wizard is described near the end of this section because it is important to first have an understanding of the *Sheet Set Manager* before learning the uses and mechanics of the wizard.

Two Types of Sheets

In order to understand how to use the *Sheet Set Manager*, it is important to have an idea of the two types of sheets that can be created. First, a sheet can be created from an existing layout that already has specific model geometry displayed. This method is called the *Import Layout as Sheet* method (in *Sheet Set Manager*) or the *Existing drawings* method (in the wizard). Second, you can create new sheets from a layout template but without any specific model geometry displayed. This method automatically creates a new drawing file containing a new layout based on an existing layout in an existing drawing. Since no model geometry is displayed in the new layout, additional steps are needed to specify what drawing geometry you want to display in the layout.

Generally, if you have existing drawings with layouts already set up as you expect to plot or print them, it may be best to create the new sheet set with the *Import Layout as Sheet* or *Existing drawings* method. On the other hand, if you want to create a new sheet set so that all sheets in the set have the same layout settings (such as sheet size, title block, page layout, and so on), create sheets based on a layout template. In this way, the sheets can be managed, viewed, and plotted uniformly.

SHEET SET COMMANDS

There are only four commands that can be entered at the Command prompt related to sheet sets (described briefly below). All other sheet set-related operations are managed through the *Sheet Set Manager*.

OpenSheetSet	Using this standard file navigation dialog box, you can open a .DST file. Opening the .DST file automatically opens the *Sheet Set Manager*.
SheetSet	This command and the *SSM* alias open the *Sheet Set Manager*.
SheetSetHide	Use *SheetSetHide* at the Command prompt (or click the "X") to close *Sheet Set Manager*.
NewSheetSet	This command begins the *Create Sheet Set* wizard.

Opensheetset

Pull-down Menu	Command (Type)	Alias (Type)	Short-cut	Screen (side) Menu	Tablet Menu
File *Open Sheet Set...*	*Opensheetset*	...	...	...	...

The *OpenSheetSet* command simply produces a standard file navigation dialog box. Use it to locate and open any sheet set data (.DST) file. Opening the .DST file automatically invokes the *Sheet Set Manager* displaying the contents of the opened sheet set (see *"Sheet Set Manager"*).

NOTE: Three example sheet sets are provided in AutoCAD. Search in the Program Files/AutoCAD 2007/Sample/Sheet Sets folder for the Architectural, Civil, or Manufacturing folders, then locate and open the related .DST files within each.

When you open a .DST file, no drawing file is opened. You can open and close a related drawing using *Sheet Set Manager* or using traditional methods—it does not affect the .DST file. Closing a .DST does not close drawings that may be open. .DST files can be closed only using *Sheet Set Manager*.

The *Sheet Set Manager*

The *Sheet Set Manager* allows you to create and manage sheet sets. Each "sheet" in the *Sheet Set Manager* references a layout in a drawing. Within the *Sheet Set Manager* you can locate, list, organize, view, label, print, and archive a group of sheets as a complete project set. The information about the list of sheets and related drawings is saved in a sheet set data (.DST) file. The *Sheet Set Manager* automatically saves and manages the .DST files.

Sheetset

Pull-down Menu	Command (Type)	Alias (Type)	Short-cut	Screen (side) Menu	Tablet Menu
Tools *Palettes >* *Sheet Set Manager*	*Sheetset*	*SSM*	*Ctrl+4*	...	...

Use any of the options shown in the command table above (or use the *OpenSheetSet* command) to produce the *Sheet Set Manager*.

The *Sheet Set Manager* appears in a palette format (Fig. 34-1). Similar to other tool palettes, you can relocate the palette to any location on the screen, resize it for width or length, dock it to the right or left side of the Drawing Editor, or you can set the palette to *Auto-hide* to save screen space when not in use. To close the *Sheet Set Manager*, select the "X" at the top of the vertical title bar, select *Close* from the title bar shortcut menu, or enter the *SheetSetHide* command at the Command prompt. The *Sheet Set Manager* has three tabs on the side: *Sheet List*, *Sheet Views*, and *Model Views*.

FIGURE 34-1

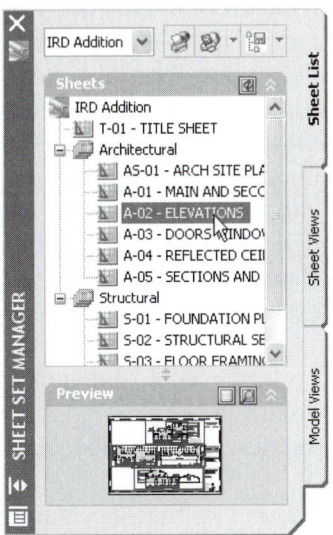

Opening an Existing Sheet Set (.DST) File in the *Sheet Set Manager*
To open an existing sheet set (.DST) file, use the drop-down list at the top of the *Sheet Set Manager* (Fig. 34-2) and select the *Open...* option to produce a standard file navigation dialog box to locate any .DST file. Alternately, select a file from the recently opened list. Once an existing sheet set file is open, the central area of the *Sheet Set Manager* on the *Sheet List* tab displays the list of sheets. Choose from currently open sheet sets by selecting from the list at the top of the drop-down list.

FIGURE 34-2

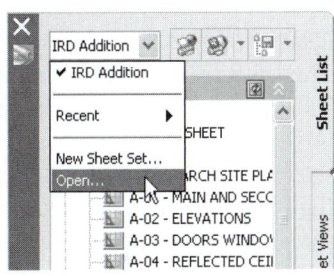

NOTE: Three example sheet sets are provided in AutoCAD for your practice. Search in the AutoCAD 2007/Sample/Sheet Sets folder for the Architectural, Civil, or Manufacturing folders, then locate and open a related .DST file within one of the folders.

Sheet List Tab
The *Sheet List* tab (Fig. 34-3) displays the sheets and their organizational structure in the sheet set. Also on this tab, options such as *Publish to DWF*, *Publish to Plotter*, and other plotting options are shown as icons at the top of the palette. In the *Sheets* section (central area), the top entry is the name of the sheet set. Under the sheet set name, both subsets and sheets can exist. Subsets are displayed with a multi-sheet icon, and sheets are displayed with a single-sheet icon. Subsets can contain nested subsets. The hierarchical structure can be easily changed by dragging and dropping sheets and subsets to any level or location in the list.

FIGURE 34-3

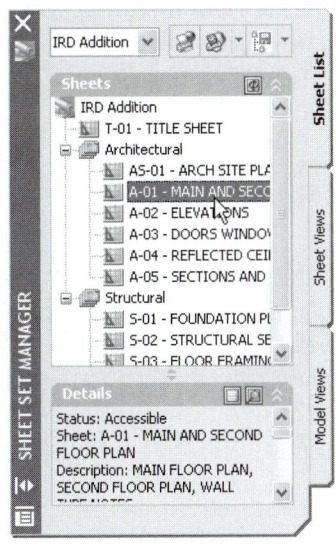

You can select the *Details* button to display details about the sheet such as availability status, description, named views in the drawing, drawing name and location, file size, last saved time and date, and sheet size. Alternately, display a preview (thumbnail bitmap) of any sheet by highlighting the sheet name in the list (in the *Sheets* section) and selecting the *Preview* button in the section below (see previous Fig. 34-1).

You can open a sheet (to view it in AutoCAD) by double-clicking on a sheet name from the list or by right-clicking and selecting *Open* from the shortcut menu (Fig. 34-4). Opening a sheet actually opens the .DWG file that contains the layout. The layout may display model geometry contained in the same drawing or model geometry from an Xrefed drawing (or, a newly created sheet may not show any geometry). Opening a drawing using the *Sheet Set Manager* is similar to using the *Open* command at the Command line (although a different part of the drawing may be displayed depending on the method used to open it). Once the sheet is open, you can edit the drawing if you want to make changes. <u>*Save* and *Close* the drawing using normal methods</u>—no *Close* option exists in the shortcut menu.

FIGURE 34-4

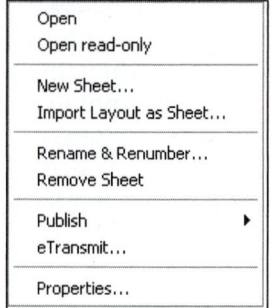

Multiple shortcut menus are available depending on whether you right-click on the sheet set name, a subset name, or a sheet name. Options in these menus are discussed in the following sections.

Remember that sheets can be created by importing existing layouts (with previously displayed model geometry) or making new sheets from a template, then specifying the model geometry you want displayed for the layouts (see previous discussion, "Two Types of Sheets"). The *Sheet List* tab displays all sheets; however, it is not readily apparent in the list of sheets which sheets are of which type.

Model Views Tab

The *Model Views* tab is the bottom tab on the *Sheet Set Manager* (Fig. 34-5). Use the *Model Views* tab to locate and list "resource" drawings whose geometry you might want to display on a sheet in the set. Simply put, <u>resource drawings are merely drawings listed in the path</u>. You can create a resource drawing by selecting the *Add New Location* button or option and then locating the folder containing the desired resource drawing.

FIGURE 34-5

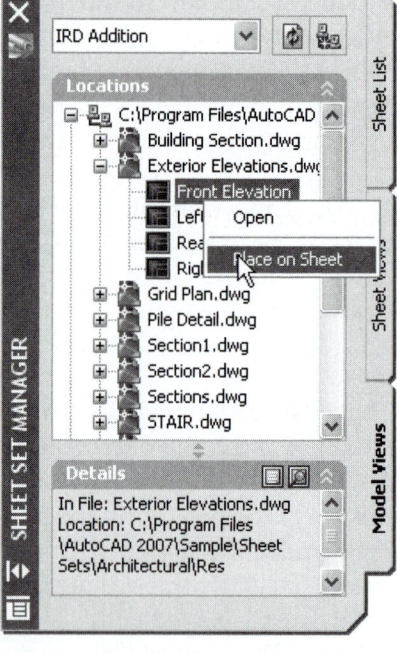

Once folders containing drawings have been located, the *Locations* section displays the list of resource drawings and the named model space views contained in the drawings. You can open drawings and views from the list by right-clicking and choosing *Open* from the shortcut menu. If you want to use a drawing as a resource drawing (to display in a sheet), <u>the drawing must contain at least one model space view</u>.

Using a drawing from the *Model Views* tab, you can automatically create a labeled, sized, and scaled viewport on a sheet. To place a drawing or named view on a sheet, first open the desired sheet from the *Sheet List* tab, then in the *Model Views* tab, right-click on the desired named view and select *Place on Sheet*. When you use a resource drawing to supply model geometry for a sheet, <u>the resource drawing is automatically *Xrefed* to the new drawing containing the layout for the sheet</u>. (See "To Place a View of a Drawing on a Sheet.")

Sheet Views Tab

The *Sheet Views* tab (the center tab) displays a list of named model space views (actually paper space viewports) that have been assigned to sheets in the set using the *Model Views* tab (Fig. 34-6). If no model space views have been assigned to sheets, no views are listed here. Remember that there are two ways sheets may have been created: by importing existing layouts with model space geometry for the layouts previously defined and by creating sheets using a layout template. <u>The *Sheet Views* tab displays only views assigned to sheets that were created from a template</u>.

Two possible displays in the *Sheet Views* tab are available by selecting the buttons near the top. Use the *View by category* list (Fig. 34-6) to create and display categories (subsets) for the views. Select *View by sheet* (not shown) to determine which sheets the views have been associated with. The purpose of the *Sheet Views* tab is to provide a means to organize the views by category, to rename and renumber the views, and to place view labels and callout blocks for the views on the sheets. A shortcut menu is available providing these options (see "To Organize Sheets, Subsets, and Views" and "To Place View Label and Callout Blocks").

FIGURE 34-6

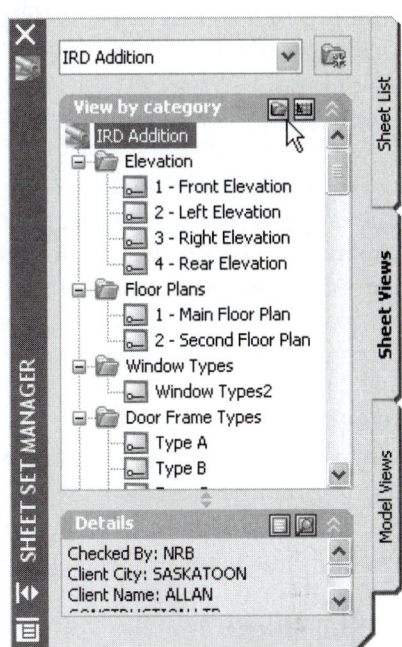

The Relationship of Layouts, Named Views, and Drawings to the *Sheet Set Manager*

The intention of Figure 34-7 is to explain the relationship of layouts, named views, and drawings to sheets in the *Sheet Set Manager*. Remember that sheets can be created by two basic methods: 1) by importing existing layouts (with previously displayed model geometry) or 2) by making new sheets from a layout template, then specifying the model geometry you want displayed for the layouts (see previous discussion, "Two Types of Sheets"). The *Sheet List* tab of the *Sheet Set Manager* displays and manages all sheets created by both methods. However, the *Sheet Views* tab and the *Model Views* tab are relevant only for sheets created from a layout template.

FIGURE 34-7

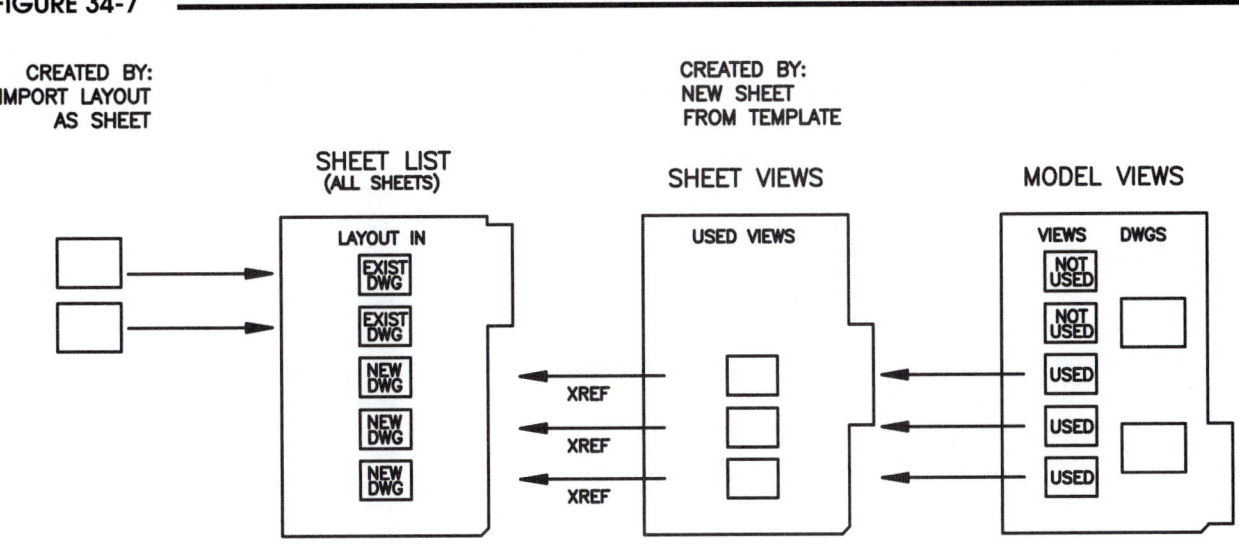

Note that on the left side of Figure 34-7, the formation of sheets created from existing layouts is relatively simple—that is, each sheet is simply a new name for an existing layout displaying previously defined model geometry. For each sheet, the referenced layout and the displayed model geometry can exist in the same .DWG file. (These layouts could instead display previously *Xref*ed geometry, but the *Xrefs* would not be listed in the *Sheet Views* or *Model Views* list).

 On the right side of the figure, the more complex formation of sheets created from a layout template is displayed. When a new sheet is created by this method, the *Sheet Set Manager* creates a new .DWG file to contain the layout. The model geometry selected to be displayed in one of these sheets is derived from a named model space view contained in another "resource drawing." Therefore, a sheet created from a layout template is always composed of one .DWG file for the layout and a second .DWG file that is automatically *Xref*ed containing the model geometry.

Sheet Creation Options

No matter which type of sheets exist or how sheets were created, all sheet names always appear in the *Sheet List* in the *Sheet Set Manager*. Remember from an earlier discussion that you can create two types of sheets and those sheets can be created either within the *Sheet Set Manager* or by using the *Create Sheet Set* wizard. The types of sheets, the options to select, and the creation methods to use are listed below.

Type of Sheet	Option to Select	Sheet Creation Method
importing existing layouts	*Import Layout As Sheet*	in the *Sheet Set Manager*
	Existing Drawings	in the *Create Sheet Set* wizard
new sheet from template	*New Sheet*	in the *Sheet Set Manager*
	An Example Sheet Set	in the *Create Sheet Set* wizard

Creating sheets directly in the *Sheet Set Manager* is described next. Creating sheets using the *Create Sheet Set* wizard is described near the end of this manuscript because it is important to first have an understanding of the *Sheet Set Manager* before learning the uses, mechanics, and shortcomings of the wizard.

To Create a Sheet, *Import Layout as Sheet*

FIGURE 34-8

To create a sheet using *Sheet Set Manager* by importing a layout from an existing drawing, right-click on any sheet or subset listed in the *Sheet List* and select *Import Layout as Sheet...* from the shortcut menu (Fig. 34-8). Use this method to create a sheet (as opposed to creating a sheet from a template) when you have an existing drawing with at least one layout previously set up to display some desired area of model geometry.

NOTE: A sheet set must be listed in the *Sheet Set Manager* to use this option (*Open* an existing .DST file from the Program Files/AutoCAD 2007/Sample/ Sheet Sets/ folder or use the wizard to create a new sheet set [see "*NewSheetSet*"]).

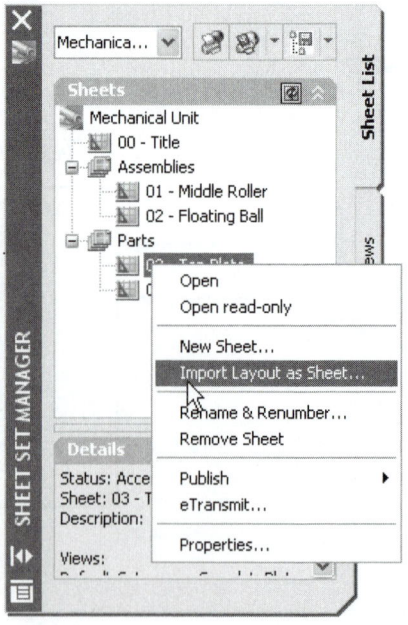

This action produces the *Import Layouts as Sheets* dialog box (Fig. 34-9). At the top of the box, select the *Browse for Drawings* button. A standard file navigation dialog box appears allowing you to select a drawing file. Once you select the desired drawing that contains the layout(s) you want to use as a sheet, all of the layouts contained in that drawing appear in the central list. Only one drawing at a time can be selected to populate the list—selecting a second drawing purges the previous list.

The *Import Drawings as Sheets* list shows all of the existing layouts contained in the selected drawing file. The columns give the *Drawing Name, Layout Name,* and *Status.* All rows that contain a check are ready for importation, and all rows are checked by default, so remove layouts for importation by removing the check. Any layout already used as a sheet is denoted in the *Status* column. (Remember that a layout can be used as a sheet only once, otherwise you must copy the layout if you want to use it for a second sheet.) Select the *Import Checked* button to import the checked layouts into the sheet set. After doing so, the selected layouts appear as new sheets in the *Sheet Set Manager Sheet List.*

FIGURE 34-9

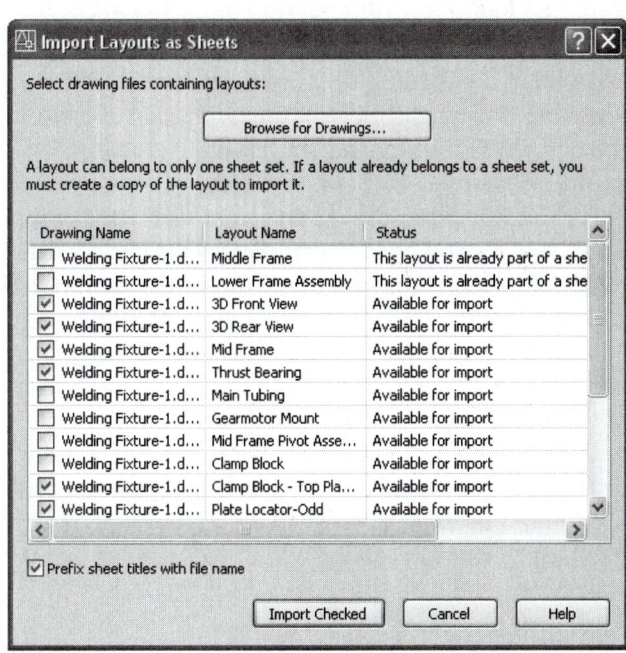

Prefix Sheet Titles with File Name
When this box is checked, the drawing file name is automatically added to the beginning of the sheet title. For example, if the layout name is "Mid Frame" and the drawing file is "Welding Fixture-1," the resulting sheet name is "Welding Fixture-1 – Mid Frame."

NOTE: For sheets created with the *Import Layout as Sheet* method, the *Sheet Views* or *Model Views* tabs of the *Sheet Set Manager* contain no information about the sheet, named views, or the drawing. In addition, no new drawings are created. The needed information about the layout is automatically added to the sheet data file (.DST).

To Create a Sheet, *New Sheet* (From Template)

Use this method of creating sheets if 1) you want to display a drawing that has no layouts previously set up or 2) if you have a drawing with layouts but want to create a sheet based on a layout template so multiple sheets in the set can have the same format. Generally, the template can contain information you want shared by many or all sheets created for the set, such as a title block, sheet size, page layout, and so on.

There are two steps to this process. First, you create the sheet and assign an appropriate name. AutoCAD automatically creates a new drawing file to contain the layout based on the specified template. Second, you must select a drawing and a named view in the drawing to place on the sheet (see "To Place a View of a Drawing on a Sheet" next for details on the second step).

NOTE: A sheet set must be listed in the *Sheet Set Manager* to use this option (*Open* an existing .DST file or use the wizard to create a new sheet set [see "*NewSheetSet*"]).

To begin the process of creating a new sheet from a layout template, right-click on any sheet or subset listed in the *Sheet List* and select *New Sheet…* from the shortcut menu (see previous Fig. 34-8).

This action produces the *Select Layout as Sheet Template* dialog box (Fig. 34-10). Enter the desired .DWT file in the edit box or use the browse button to select a .DWT file. All layouts contained in the template drawing appear in the layout list. Select the desired layout to use as a template layout.

Next, the *New Sheet* dialog box (Fig. 34-11) appears. Enter the desired *Number* and *Sheet Title* in the edit boxes. From this information, AutoCAD automatically creates the file name for the new drawing and enters the text in the *File Name* box. For example, entering "A-06" and "MISC DETAILS" in the two edit boxes causes AutoCAD to create a drawing file with the name of "A-06 MISC DETAILS." The new drawing that contains the new layout is created in the folder specified in *Folder Path*, and the template used for the sheet is specified in *Sheet Template*. To change the sheet template and folder path, see "To View and Edit Sheet Properties."

Once a new sheet has been created, you can double-click on the sheet name in the *Sheet List* of the *Sheet Set Manager* to display the new drawing

FIGURE 34-10

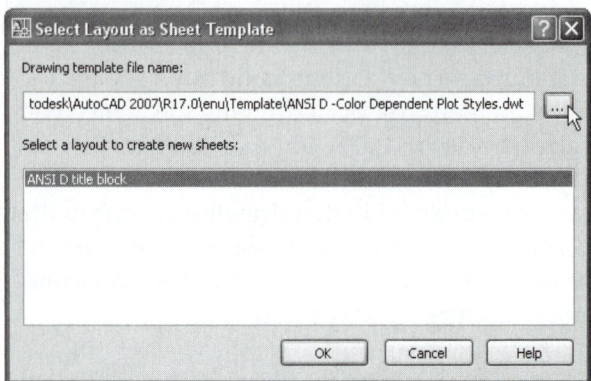

FIGURE 34-11

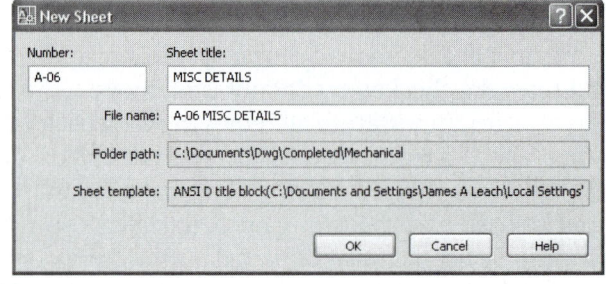

file containing the new layout. Note that activating the *Model* tab reveals that the new drawing contains no model space geometry. (See "To Place a View of a Drawing on a Sheet" next.)

When you create a new sheet using the *New Sheet* dialog box, you can decide whether or not to keep a relationship between the sheet number and the related drawing file that contains the layout. When you create a sheet from a template, the sheet number is automatically appended to the drawing file name as a prefix. The *Sheet Set Manager* allows you to keep this relationship even if you later decide to change the sheet number (see "To Rename and Renumber Sheets and Views"). If you do not want to keep a relationship between the sheet name and the related drawing file name, change the automatic entry in the *File Name* box by deleting the sheet number from the name.

To Place a View of a Drawing on a Sheet

Using the *New Sheet* option in the *Sheet Set Manager* (or using the *An Example Sheet Set* option in the wizard) creates a new drawing containing a layout, but the layout does not display any model geometry. As the next step, you must indicate what drawing and which named view in the drawing to display in the layout. There are three steps to this process: 1) locate the folder containing the drawing using the *Model Views* tab, 2) ensure the desired drawing contains at least one named model space view, and 3) use the *Sheet Views* tab to select a named view to place on the sheet. To place a view on a sheet, follow these steps:

1. The drawing containing the desired model space geometry for the sheet must be accessible from the *Model Views* tab of the *Sheet Set Manager*. Locate the folder containing the needed drawings by activating the *Model Views* tab, then selecting the *Add New Location…* option or icon button from the top of the tab. When the *Browse for Folder* dialog box appears, locate the folder containing the desired drawing, then select *OK*. The *Model Views* tab should display the list of drawings contained in the specified folder.

2. The drawing containing the desired model space geometry for the sheet must contain at least one model space view. In the *Model Views* tab, right-click on the desired drawing and select *See Model Space Views*. This action simply lists the existing views under the drawing name. If the desired view exists, you can click *Preview* to view a thumbnail image of the view or double-click on the view name to open the drawing and see the view. *Close* the drawing (using the *Close* command) and go to the next step.

    If the drawing contains no named model space views, double-click on the drawing name to open the drawing. (Alternately, you could use the *Open* command.) Use the *View* command to create the needed model space view, then *Save* and *Close* the drawing.

3. To place the view on the sheet, first activate the *Sheet List* tab and double-click on the desired sheet to open it. Next, return to the *Model Views* tab, right-click on the desired view, and select *Place on Sheet*. When the drawing view appears on the drawing sheet, move your cursor and pick the desired location, or if you need to change the scale, right-click and select the desired scale from the list, then place the view on the sheet.

Once the model space view is placed on the sheet, the view is listed as a view (paper space viewport) on the *Sheet Views* tab. You can use the *Sheet Views* tab to rename and renumber views as well as place callout blocks and labels on the sheets for the view. (See "To Rename and Renumber Sheets and Views" and "To Place View Labels and Callout Blocks.")

NOTE: Remember that creating a new sheet causes *Sheet Set Manager* to automatically create a new .DWG file that contains the new layout. When you place a named model space view on the sheet, *Sheet Set Manager* automatically <u>attaches the model geometry drawing to the layout drawing as an *Xref*</u>. Therefore, *Sheet Set Manager* automates the process of creating a layout, saving it in a new .DWG file, and using the *Xref* command to attach the drawing containing the model geometry. So, rather than creating the new layout in the drawing containing the geometry, *Sheet Set Manager* creates a new .DWG file just for the new layout, leaving the existing (*Xrefed*) .DWG file containing the model geometry unaltered.

To View and Edit Sheet Set Properties

You can view and edit properties of a sheet, sheet set, or subset. The properties can contain titles, descriptions, file paths, and custom properties that you define. The properties differ based on the "owner"—a sheet, a sheet set, or a subset.

In the *Sheet List* tab of the *Sheet Set Manager*, right-click on the name of the sheet set, subset, or sheet. In the shortcut menu that appears, select *Properties*. The *Sheet Set Properties* dialog box displays properties and values depending on what you selected. For example, Figure 34-12 displays properties for a sheet set. You can edit the values of any of these properties by clicking on a value. For entries that are text based, highlight the edit box and enter the desired text. For other properties, highlighting the property (left column) activates a browse button on the right (see Fig. 34-12).

FIGURE 34-12

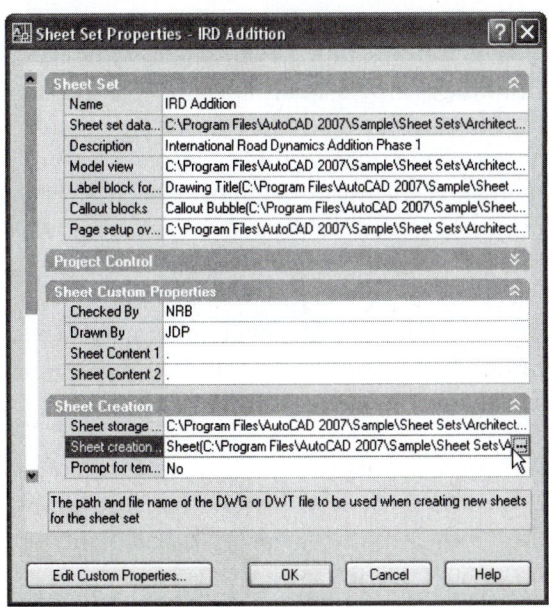

Particular properties that you may want to edit related to sheet creation are *Sheet creation template* or *Prompt for template*. Also, note that you can use this tool to locate and specify view label blocks, callout blocks and a page setup overrides file (see "*Publishing using Page Setup Override*").

TIP

Although you can change these properties through the *Sheet Set Manager*, some of the particular settings that appear initially are specified in the *Files* tab of the *Options* dialog box. There, under *Template Settings*, you can specify the location for the sheet set template files and the files used for the page setup overrides and the sheet creation layout template.

You can define custom properties for a sheet and a sheet set. Selecting the *Edit Custom Properties* button produces the *Custom Properties* dialog box (Fig. 34-13). Custom properties for the sheet set apply to the <u>entire set</u> and can include information specific to a project, such as the client contact and contract information.

FIGURE 34-13

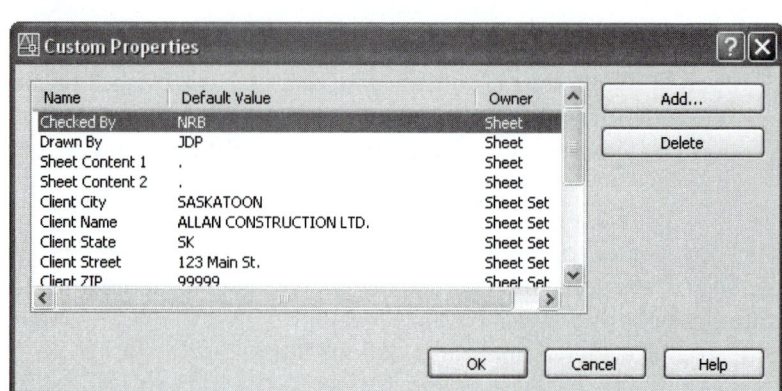

To Organize Sheets, Subsets, and Views

Organize Sheets

It is easy to create new subsets and organize sheets in the *Sheet List* tab of the *Sheet Set Manager*. You can organize sheets by dragging them to different subsets.

New Subsets

You can create a new subset by right-clicking on the sheet set name or on a subset name and selecting *New Subset* from the shortcut menu. This action produces the *Subset Properties* dialog box (Fig. 34-14). Enter the desired name in the *Subset* name edit box. If you want to specify a different folder or layout template from the one previously specified for the existing sheet set, remove the check from *Create folders relative to parent subset storage location*, then enter the information or select the browse button to locate the desired folder and .DWT file. When the *Prompt for Template* button is checked, AutoCAD prompts you to specify a drawing template file for creating new sheets rather than using the default drawing template file as specified.

FIGURE 34-14

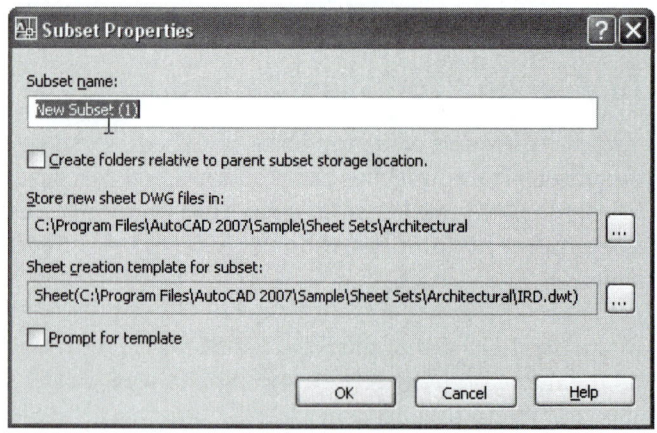

Remove Sheets

You can remove a sheet from the *Sheet List* by right-clicking on a sheet and selecting *Remove Sheet* from the shortcut menu. When a sheet is removed, the related drawing files containing the layout and geometry are not deleted even if the drawing file containing the layout was created by *Sheet Set Manager* or *Create Sheet Set* wizard.

Organize Views

In a large sheet set with many sheet views, all the views that have been placed on sheets are listed in the *Sheet Views* tab. (However, this tab does not list views that are contained in sheets created by the *Import Layout as Sheet or Existing drawings* methods.) These views are actually paper space viewports in layouts of named sheets in the *Sheet List* tab. Use the *View by sheet* display (button near the top) to display the sheet names and views for each sheet.

To organize the views by category, you must use the *View by category* display button (see Fig. 34-6). Make a new view category by right-clicking on the sheet set name at the top, then select *New View Category*. Once multiple categories exist, you can then drag and drop the existing views into the desired categories. Categories provide a means for you to organize views; however, the category names and related views do not have to correspond directly to the sheet names and related views. For example, you might have one category named "Plans" containing four views, but two sheets named "Plans 1" and "Plans 2," each displaying two views.

To Place View Label and Callout Blocks

As a common practice in industry, sheets in a sheet set are numbered in sequence such as pages in a book are numbered. Likewise, when several views appear on one sheet, they are numbered. View labels can be placed appropriately to designate the names and numbers of views. In this way, when a particular drawing view makes reference to some information from another sheet and view, a callout bubble gives that reference information. "View label block" is the term AutoCAD uses for the symbol (*Block*) that gives the view name and number. "Callout block" is the AutoCAD term for the symbol that is used to display references to other sheets and views. Other generic names used in industry for callout blocks are callout bubbles, reference tags, detail keys, detail makers, and building section keys.

Using the *Sheet Set Manager*, you can place view label blocks and callout blocks on your sheets. When this is done, it is important that names and numbers as they appear in the *Sheet Set Manager* correspond with the view labels and callouts. *Sheet Set Manager* controls these connections by automatically changing the information in the label and callout blocks on the sheets when you change the names and numbers as they appear in the *Sheet List* and *Sheet Views* (see "To Rename and Renumber Sheets and Views").

View Label Blocks

When working with a sheet set in which individual sheets contain multiple views, it is important to label the views with a name and number. You can easily accomplish this in *Sheet Set Manager* using the *Sheet Views* tab. First, the drawing sheet must be open to place a view label. Do this by double-clicking on the view name in the *Sheet Views* tab to open the corresponding sheet. (Alternately, you could open the sheet through the *Sheet List* tab.) Right-click on the appropriate view name and select *Place View Label Block* from the shortcut menu. Pick the desired location in the drawing (usually centered beneath the view) for the label to appear.

When placing view labels, you can change the scale for the block by entering "S" at the command line before picking the desired location. To change the text for the block, double-click on the block after insertion to produce the *Enhanced Attribute Editor*.

The information automatically entered for the view label is based on the view name and number (as specified in the *Sheet Views*). If you want to change the view name or number at a later time, you do not need to "manually" update information in the label and callout blocks. Instead use the *Rename & Renumber* option from the shortcut menu (see "To Rename and Renumber Sheets and Views"). When you regenerate the drawing, information in the label and callout blocks is automatically updated.

Callout Blocks

Callout blocks are placed in a drawing to specify a reference to another view, so the "bubble" lists the sheet number and view number to refer to. To place a callout block, there are two steps: first, using the *Sheet List* tab, open the sheet on which you want the callout to appear (double-click on the sheet name) and, second, in the *Sheet Views* tab, right-click on the view you want to refer to, and select *Place Callout Block* from the shortcut menu. In the cascading menu that appears, you may have options for bubbles with directional arrows, depending on the type of view (section, elevation, etc.). Pick the desired location in the drawing to place the block. You can change the scale for the block before insertion by entering "*S*" at the command prompt. The inserted block lists the sheet number and view number of the selected view (that you right-clicked). As with view label blocks, if you want to change the number of the view, use the *Rename & Renumber* option from the shortcut menu so information in the label and callout blocks is automatically updated when the drawing is regenerated.

TIP The blocks that are used by *Sheet Set Manager* when you place a view label block or a callout block are specified in the sheet set properties. You can change the block names or locations using the *Sheet Set Properties* dialog box (see "To View and Edit Sheet Properties" and Fig. 34-12).

To Rename and Renumber Sheets and Views

Sheets

To rename and renumber sheets, right-click on any sheet name in the *Sheet List* tab and select *Rename & Renumber* from the shortcut menu to produce the *Rename & Renumber Sheet* dialog box shown in Figure 34-15. You can change both the *Sheet Title* (sheet name) and the *Number*. You cannot change the folder path from this dialog box. Select *Next* or *Previous* to go to adjacent sheets in the subset.

FIGURE 34-15

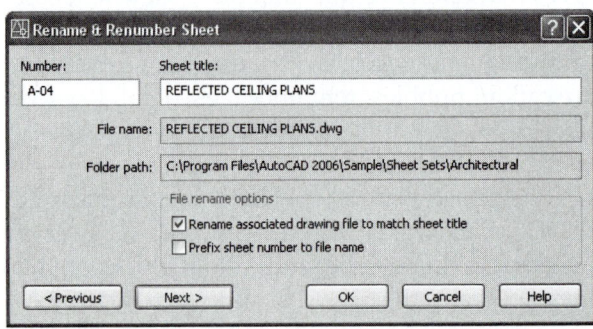

Generally sheets are numbered in sequence (as pages) for each subset. Using appropriate sheet numbers is important since callout blocks on one sheet may refer to another sheet. <u>Renumbering sheets using this tool automatically changes the related callout block numbers.</u> In such a case, when a change is made in the Sheet List, regenerate the drawing to see the new number in the drawing.

Rename Associated Drawing File to Match Sheet Title and *Prefix Sheet Number to File Name*
For each sheet in your sheet set there is a corresponding drawing file that contains the layout. If you prefer, you can maintain a relationship between the sheet name and the drawing file name that contains the sheet layout. This strategy usually involves using the sheet name or number as the title or as a prefix for the drawing file name. (For example, creating sheets with the *Create Sheet Set* wizard by default creates new drawing file names that include numbers corresponding to the sheet numbers (see previous discussion in "To Create a Sheet, *New Sheet* From Template"). The *Rename & Renumber Sheet* dialog box allows you manage this connection. For example, if you change the *Sheet Title*, you can maintain the naming scheme for the related drawing file by checking *Rename associated drawing file to match sheet*. If you change the sheet *Number*, check *Prefix sheet number to file name* to maintain the connection.

If you prefer, you could instead decide not to maintain a connection between sheets and related drawing file names. Because the sheet number is a sheet property saved only in the sheet data (.DST) file, you can use the *Sheet Set Manager* to reorder and renumber your sheets without affecting the corresponding drawing file names.

Views

Once the model space view is placed on a sheet, the view is listed as a paper space viewport in the *Sheet Views* tab. You can rename and renumber a paper space view by right-clicking on the desired view name, then select *Rename & Renumber* from the shortcut menu. In the

FIGURE 34-16

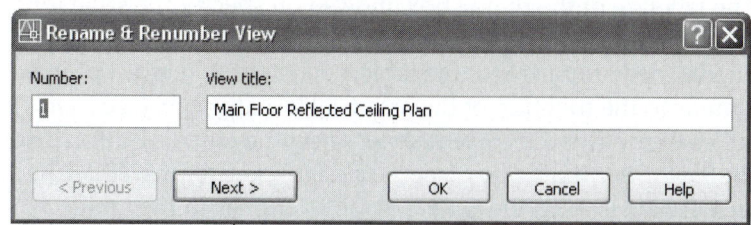

Rename & Renumber View dialog box that appears (Fig. 34-16), you can change the *Number* or the *View title*. Since the view numbers are included in callout blocks, changing a view name and number here automatically updates the callout blocks that may appear in the sheets. In this case, regenerate the drawing to see the new name and number.

To Create a Sheet List Table

When a complete sheet set is submitted to a client, a sheet list (sometimes called a sheet index) is usually included on the title sheet. The sheet list, similar to a table of contents, lists the sheets included in the set and can give information such as the sheet number, title, description, and other information. With the *Sheet Set Manager* you can easily create a sheet list table for a sheet set.

When you create a sheet list table, AutoCAD uses the *Table* feature, and each cell in the table can contain a *Field* (see Chapter 18). In this way, the table size and configuration are easily changeable, and the entries in the cells can be updated if you remove, add, or renumber sheets.

Insert Sheet List Table Dialog Box

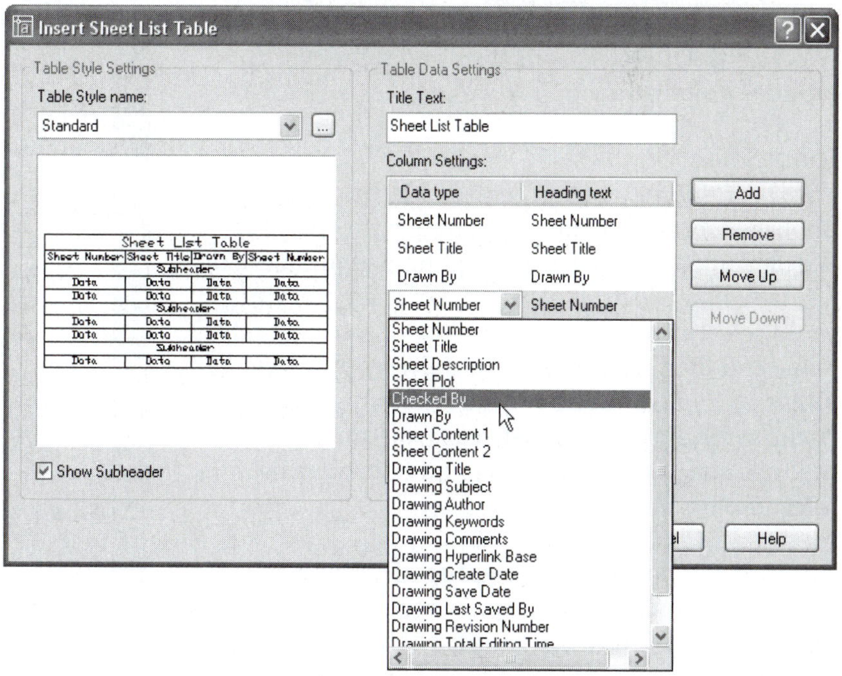

Assuming you want the sheet set table to appear on the title sheet, first you must open the title sheet. You can accomplish this in the *Sheet Set Manager* in the *Sheet List* tab by double-clicking on the title sheet name. With the title sheet open, right-click on the <u>sheet set name</u> (not the title sheet name), and in the shortcut menu, select *Insert Sheet List Table* to produce the *Insert Sheet List Table* dialog box (Fig. 34-17).

FIGURE 34-17

Table Style Settings
The left side of the dialog box allows you specify the style for the table. In this section you can select from existing table styles from the *Table Style Name* drop-down list. The table style defines the text, cell, and border properties of the table, such as text height, color, border lineweight, and so on. This feature is similar to the function of the *Insert Table* dialog box (see "*Table*" in Chapter 18). To select from existing styles or modify the current style, select the ellipsis button just to the right of the style name to produce the *Table Style* dialog box (see "*Tablestyle*" in Chapter 18). Check the *Show Subheader* box if you want the subset names from the *Sheet List* tab to appear in the table.

Table Data Settings
Specify the title you want to appear in the drawing for the table in the *Title Text* edit box. Under *Column Settings*, you can add columns to or remove columns from the sheet list table. You can specify the content of the columns by selecting options from the drop-down list. Information in the resulting sheet list table is automatically generated from data previously specified in the sheet list, sheet set properties, sheet set custom properties, sheet properties, sheet custom properties, other drawing properties, or system information. Some cells in the resulting table contain fields, so you can manually edit these cells in the sheet list table.

 Updating a Sheet List Table
When you add, remove, or renumber sheets, it is easy to update the sheet list table. After making changes to the sheet set, select any outside edge of the sheet list table, right-click, and select *Update Sheet List Table* from the shortcut menu. The sheet list table is automatically updated based on the new sheet set information; however, all previous manual edits are lost when you update the sheet list table.

Plotting and Publishing Sheet Sets

To access options for plotting and publishing, use the drop down list on the top of the *Sheet Set Manager* or select a sheet, sheet set, or subset and right-click to produce a similar menu (Fig. 34-18).

Sheet Selection
The sheet, subset, or sheet set that will be plotted, published, or used to prepare a transmittal package is determined by whatever is highlighted in the *Sheet List* tab of the *Sheet Set Manager*. For example, you could select a sheet name, subset name, or sheet set name and then select one of the *Publish* options or *eTransmit* to plot, publish, or create a transmittal package for those selected sheets. You can select multiple sheets using the Ctrl or Shift key during selection.

Additionally, you can save a group of sheets as a "sheet selection" and assign a name for the group. The sheet selection can be restored at a later time rather than having to select the individual members of the group repeatedly. Activating a sheet selection simply highlights the sheets in the group in the *Sheet List* for plotting, publishing, or using *eTransmit*.

FIGURE 34-18

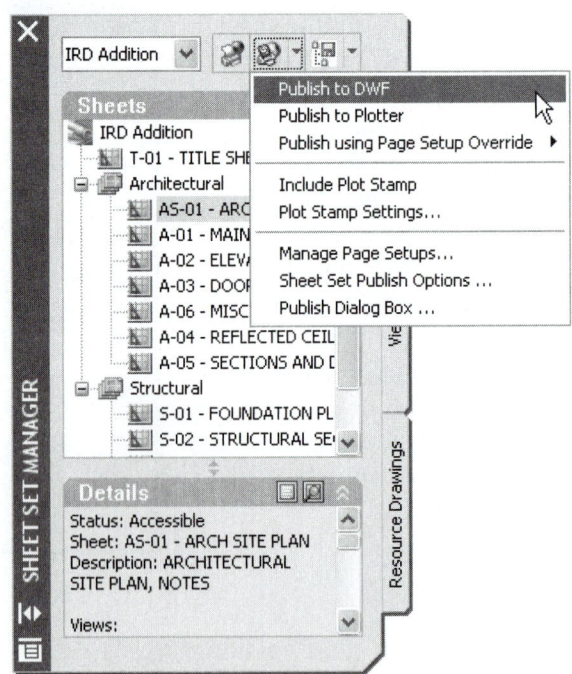

Create a sheet selection by highlighting several sheets or subsets, selecting the *Sheet Selections* button in the upper-right corner of the *Sheet List* tab, then selecting *Create...* from the menu. Enter the desired name for the group of sheets in the *New Sheet Selection* dialog box. Once the group is created, the name appears in the *Sheet Selections* list. Picking the group name simply highlights the individual members of the group.

Publish to DWF

Using this option causes AutoCAD to create a single-sheet .DWF or a multiple-sheet .DWF file, depending on what is highlighted in the *Sheet List*. The *Select DWF File* dialog box appears for you to specify the .DWF file name and location. You can view the .DWF file immediately after publication using the Autodesk DWF Viewer or by right-clicking on the *Plot/Publish Details* icon in the system tray and selecting *View DWF File...* from the shortcut menu.

Publish to Plotter

Selecting this option sends the sheet(s) to the plot or print device. The device used is based on what is specified in the page setup for the individual sheets.

When you create a sheet set using a layout template, you specify a drawing template (.DWT) file as the "sheet creation template." The template can contain the page setups that you want to use. In that way, those page setups are used when new sheets are created for the set.

Publish using Page Setup Override

Select this option if you want to publish the sheet(s) using a page setup override file rather than the page setup defined for the individual sheet(s). The page setup overrides file contains page setups that can be specified to override the page setups in each sheet. A page setup overrides file is a .DWT file. You can create your own file with the page setups you desire. The name and location of the page setup overrides file is specified in the *Sheet Set Properties* dialog box. Therefore, when you publish a sheet set, you can use the page setups defined in each drawing file or you can use the page setup overrides files to publish the sheets.

Include Plot Stamp

This option applies a plot stamp on each published sheet. Specify the settings using the *Plot Stamp* dialog box.

Plot Stamp Settings

The *Plot Stamp* dialog box appears when you select this option. Use the dialog box to specify the desired plot stamp to appear on the published sheets (see *"Plotstamp"* in Chapter 33).

Manage Page Setups

This option produces the *Page Setup Manager* (see Chapter 14). When the *Page Setup Manager* is opened from the *Sheet Set Manager*, only named page setups in the page setup overrides file that have *Plot Area* set to *Layout* or *Extents* are listed. Any of these page setups can be applied when you then *Publish* the sheet set. The *Current Page Setup* note indicates "Not applicable" if the *Page Setup Manager* is opened from the *Sheet Set Manager* since you cannot apply a new page setup to an entire sheet set after it has been created.

Sheet Set Publish Options

The *Sheet Set Publish Options* dialog box appears for you to specify .DWF file creation settings for the sheet set. This dialog is identical to the *Publish Options* dialog box described previously (see Chapter 23).

Publish Dialog Box

This option provides access to the *Publish* dialog box described previously (see Chapter 23).

To Transmit and Archive Sheet Sets

Because a sheet set is composed of many files, it could be difficult to gather all the related files and store them as one complete set. During project development and at the completion of a project it is likely that you may want to save the sheet set or transfer it for use at a different location. The *Sheet Set Manager* provides a tool for you to collect a sheet set and all related files for archival or for preparation of a transmittal package.

Creating a Transmittal Package

During design of a project or at the project completion you can publish your sheet set in a hardcopy format or .DWF format. However, rather than publishing the sheet set for a client or colleague, you may want to send or otherwise electronically transfer the actual .DWG files and all of the related files required to restore the sheet set at another AutoCAD station. The complete sheet set and reference files (sheet set data files, *Xrefs*, plot configuration files, font files, and so on) ready to transfer electronically is called a transmittal package. You can use the *Sheet Set Manager* to create a transmittal package of a sheet set using *eTransmit*. You can even include other types of files such a Word documents, Excel spreadsheets, or .PDF files in the transmittal package. You can also create a report of the transmittal package.

eTransmit (Create Transmittal Dialog Box)

The *eTransmit* option is available in the *Sheet Set Manager* in the *Sheet List* tab by right-clicking on a sheet, subset, sheet set name, or member of a *Sheet Selection*, then selecting *eTransmit* from the shortcut menu. Remember that the transmittal package will be created for the sheet or multiple sheets that are highlighted in the *Sheet List* (see "*Sheet Selections.*") The *Create Transmittal* dialog box appears (Fig. 34-19).

FIGURE 34-19

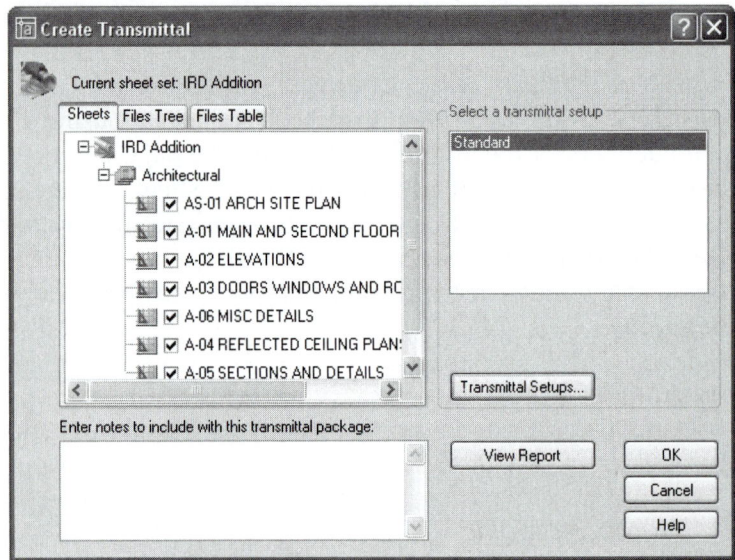

To create the transmittal, use the *Sheets*, *Files Tree*, and *Files Table* tabs to specify which sheets and related files to include in the transmittal package. The contents of these tabs and other options are explained below. Before you create the transmittal package, use the *Transmittal Setups* button to specify the file name, file type, and other parameters (see "*Transmittal Setup* Dialog Box" next). Selecting the *OK* button in the *Create Transmittal* dialog box creates the specified .ZIP file, .EXE file, or folders with related files in the location you specify.

Sheets
This tab lists the hierarchical structure of sheet sets, subsets, and sheets (by sheet name) to be included in the transmittal package. You can remove sheets from the list by removing the checks; however, sheets cannot be added.

Files Tree

This tab (not shown) lists all of the actual files by file name (not sheet names) in hierarchical tree format to be included in the transmittal package. All files associated with each drawing (such as related *Xrefs*, plot style tables, and fonts) are displayed. Possible file types include .DWG, .DWT, .DST, .FMP, .CTB, .STB, and .PC3. You can use the *Add File* button to include any type of file, including related non-AutoCAD files such as Word .DOC files or Excel .XLS files.

Files Table

This tab lists the same information as the *Files Tree* tab, but in a table format. In addition, the path of each file is given.

Enter Notes to Be Included with This Transmittal Package

You can use this edit box to enter any notes you want transmitted. These notes are included in the transmittal report. You can also specify a template of default notes (such as a waiver, copyright, or company information) to be included with all your transmittal packages by creating an ASCII text file called Etransmit.txt and saving it in the location specified in the *Support File Search Path* section in the *Files* tab of the *Options* dialog box.

View Report

AutoCAD automatically creates a report that is sent with each transmittal package. The report lists the time, date, sheet names, and all files included. Additional information is included related to the location of files needed for the transmittal package to work properly. If you have created a notes text file, these notes are also included.

Transmittal Setup Dialog Box

Selecting the *Transmittal Setups* button in the *Create Transmittal* dialog box provides a tool for you to create new setups or modify, rename, or delete old setups. Selecting *New* or *Modify* from the *Transmittal Setups* dialog box (not shown) produces the *New* or *Modify Transmittal Setup* dialog box (Fig. 34-20). Use this dialog box to specify many parameters of the transmittal package. Most of these options are self-explanatory; however, a few options are worth note. The *Transmittal package type* allows you to specify *Folder* (uncompressed files in a folder), *Self-Extracting Executable* (compressed, self-extracting executable file), or *Zip* (compressed .ZIP file requiring PKZIP or WinZip). The *Transmittal File Folder* specifies the location in which the transmittal package is created. Under *Transmittal Options*, it is recommended to check *Include Fonts* and *Include Sheet Set Data and Files*. (See also Chapter 23.)

FIGURE 34-20

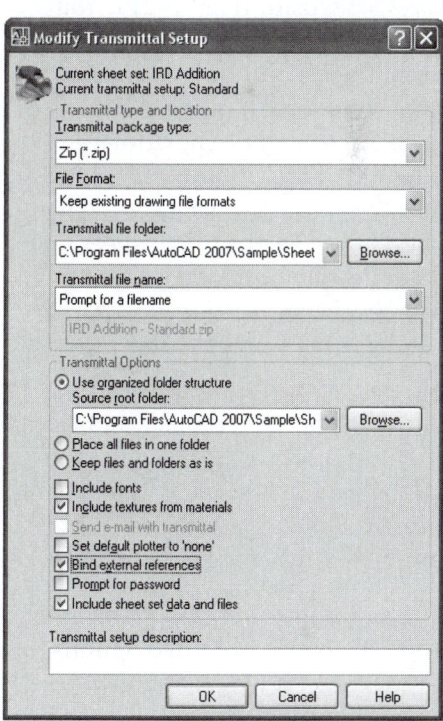

Archiving a Sheet Set

Archiving a sheet set is essentially the same as creating a transmittal package in that archiving a sheet set creates a .ZIP, .EXE, or folder and related files, depending on which option you select, in the location that you specify. The main difference is that you must right-click on the sheet set name to access the *Archive* option from the shortcut menu, whereas you can select a single sheet or combination of sheets to access the *eTransmit* option. Therefore, archiving is intended for saving a package containing the entire sheet set and related files.

To archive a sheet set, in *Sheet Set Manager* right-click the sheet set name (.DST file) and select *Archive* from the shortcut menu. Doing so produces the *Archive a Sheet Set* dialog box. This dialog box (not shown) is similar to the *Create Transmittal* dialog box (see previous Fig. 34-19) with the exception that you cannot save named archival setups (although you can save transmittal setups). The *Archive a Sheet Set* dialog box includes the *Sheets*, *Files Tree*, and *Files Table* tabs to specify which sheets and related files to include in the archival package. Although all files are initially included in the lists, you can uncheck sheets and files so they are not included in the archival set. Otherwise, all related files are saved. For information on these options, see previous discussion "*eTransmit* (*Create Transmittal* Dialog Box)" and Figure 34-19.

You can specify parameters for the archival package using the *Modify Archive Setup* button and the related dialog box (not shown). The options in this dialog box are the same as in the *Modify Transmittal Setup* dialog box. For information on these options, see previous discussion "*Transmittal Setup* Dialog Box" and Figure 34-20.

Newsheetset

Pull-down Menu	Command (Type)	Alias (Type)	Short-cut	Screen (side) Menu	Tablet Menu
File *New Sheet Set...*	*Newsheetset*	...	*New* *Sheet Set*	...	...

NewSheetSet is the formal command to produce the *Create Sheet Set* wizard (Fig. 34-21). You can also access *New Sheet Set...* from the *File* pull-down menu or from the shortcut menu in the *Sheet Set Manager*. <u>The primary function of the *Create Sheet Set* wizard is to create a sheet set data (.DST) file. The resulting sheet set data file may or may not actually contain sheets.</u>

FIGURE 34-21

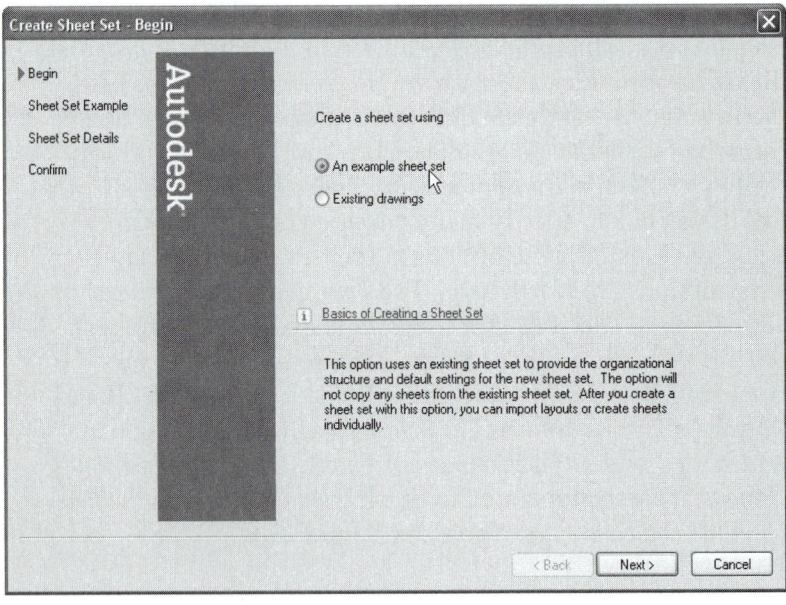

You can choose from two options for creating "sheet sets": *An example sheet set* or *Existing drawings*. Depending on which option you select, the next three steps vary as indicated on the left side of the wizard. Initially this wizard can be misleading since <u>only the *Existing drawings* option actually creates sheets—the *An Example Sheet Set* option creates subsets but no sheets</u>. Generally speaking, when the wizard is finished, the sheet set is most likely not complete, and you would therefore create new sheets within the *Sheet Set Manager* as discussed previously.

There are three reasons why you would use the *Create Sheet Set* wizard:

1. if you already have a large number of drawings containing model geometry, layouts, and page setups you want to use for the sheet set,
2. if you want to create a subset structure (with no sheets) based on AutoCAD-supplied hierarchy for an architectural, mechanical, or civil application, and
3. if you want to create sheets entirely within the *Sheet Set Manager*, but first need a "blank" .DST file (see NOTE below).

NOTE: As you know, the *Sheet Set Manager* provides all options for creating, manipulating, and managing sheets, and the *Sheet Set Manager* automatically manages the related .DST files. Oddly enough, you must open an existing .DST file to gain access to any options in the *Sheet Set Manager*, but you cannot create a .DST file from scratch using *Sheet Set Manager*—you must use the *Create Sheet Set* wizard to create a .DST file (see "To Create a Blank .DST File").

An Example Sheet Set

A sheet set example is actually a template of subsets. This option might better be titled "Create Subsets" since no sheets are created using this option. The resultant sheet set is actually not a set of sheets—only a structure of subsets based on a typical architectural, mechanical, or civil hierarchy. When you select this option, the three steps you will complete are *Sheet Set Example*, *Sheet Set Details*, and *Confirm*. You can also use this option to create a "blank" .DST file (see "To Create a Blank .DST File").

Sheet Set Example

In this step (not shown) you can select from AutoCAD-provided examples or browse to locate another example you have created and saved in the location specified in the *Files* tab of the *Options* dialog box (see "To View and Edit Sheet Set Properties"). The examples create a structure of architectural, manufacturing, or civil subsets, but no sheets. The AutoCAD-provided examples are *Architectural Imperial*, *Architectural Metric*, *Civil Imperial*, *Civil Metric*, *Manufacturing Imperial*, and *Manufacturing Metric*. Each example includes a complete set of sheet set properties including a layout template based on an imperial or metric sheet size.

Sheet Set Details

In this step (Fig. 34-22) you supply the desired name for the sheet set and a description (optional) as well as the location you want for the .DST file. This step provides access to the *Sheet Set Properties* dialog box (see previous Fig. 34-12) which defines the characteristics for the sheets that you will subsequently create. These settings can be changed if needed. For example, you could locate a folder to provide resource drawings, change the label block and/or callout block, specify a new page setup overrides file, and select a different .DWT file to use as the layout template file.

FIGURE 34-22

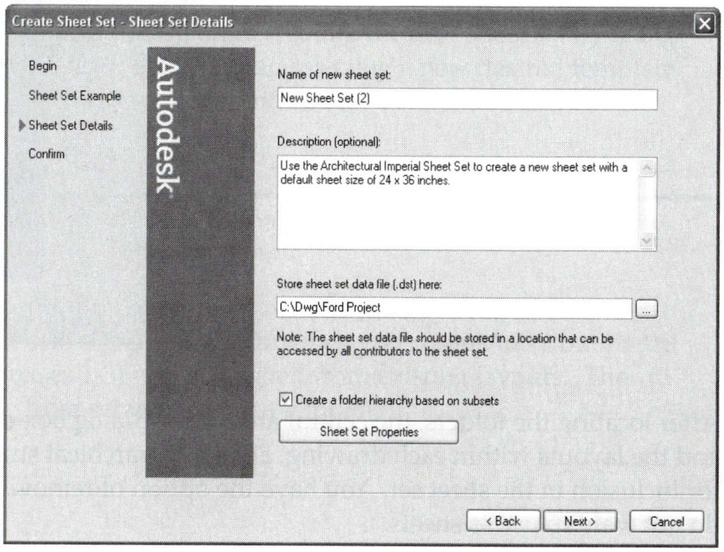

1. **Create a new sheet set using existing layouts.**

   A. Open the *Sheet Set Manager.* From the drop-down list at the top, select *New Sheet Set.* The *Create Sheet Set Wizard* appears. In the *Begin* step, select the *Existing Drawings* option, then select *Next.* In the *Sheet Set Details* step, assign a name for the new sheet set such as **Chapter 34 Exercise 1** or similar. Enter a description if desired. To designate the location for the .DST file, select the browse button just to the right of the edit box titled *Store sheet set data file (.dst) here* and select an appropriate location, such as the folder where you want to store your Chapter 34 Exercise drawings. (If you are working on a team project, select a network location that is accessible to other team members.) Select *Next.*

   B. In the *Choose Layouts* step, use the *Browse* button to locate the folder where you stored the drawings for the Chapter 29 Exercises. After selecting the folder, all the drawings contained in the folder should appear in the *Create Sheet Set* Wizard layout list. Since you do not want to include all the drawings, remove the checks from all drawings except **APT-DIM** and **HAMMER-DIM.** The list should appear similar to that in Figure 34-24. Note that more than one layout may exist for each drawing. Check all layouts for the two drawings. Select *Next.*

   FIGURE 34-24

   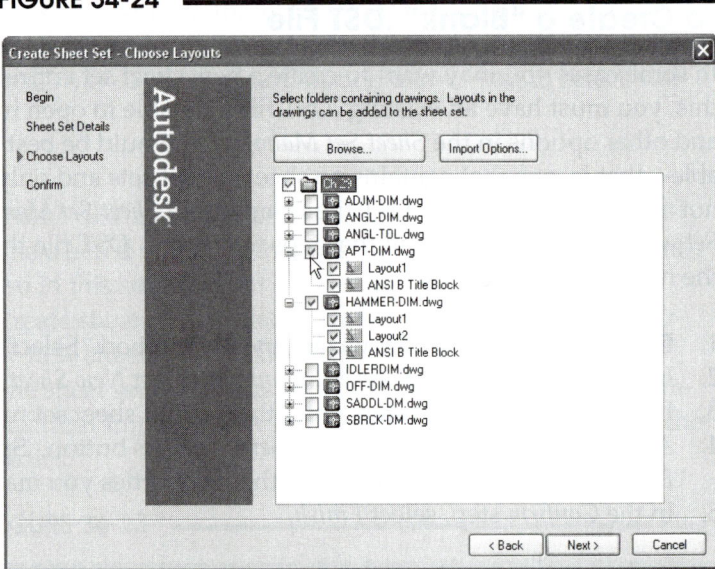

   C. In the *Confirm* step, confirm that the sheet set name, location, storage location, and data file location are correct. Select the *Finish* button to complete the process of creating the new sheet set. In the *Sheet List* of the *Sheet Set Manager*, a new sheet set named "Chapter 34 Exercise 1" (or similar) should appear (Fig. 34-25). Note that each layout from the two drawings (APT-DIM and HAMMER-DIM) is listed as a separate sheet in the new sheet set.

   FIGURE 34-25

   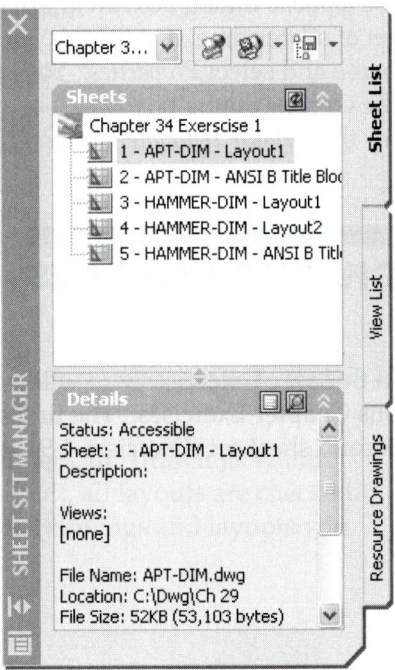

2. **Remove sheets and add new sheets.**

   A. Even though each layout is listed as a separate sheet in the *Sheet List*, some layouts may be duplicates or may not be appropriate for your sheet set. *Open* each layout from the list by double-clicking on the name or selecting *Open* from the shortcut menu. Determine which layouts display geometry in the layouts. You can delete sheets from the list by right-clicking on the name and selecting *Remove Sheet*. Select the *Sheet Views* tab and the *Model Views* tab and note that no entries appear for the sheet set since this sheet set was created from existing layouts.

   B. In the *Sheet List* tab of the *Sheet Set Manager,* right-click on the sheet set name (such as **Chapter 34 Exercise 1**) and select *Import Layout as Sheet* from the shortcut menu. In the dialog box that appears, use the *Browse for Drawings* button and locate the **SADL-SEC** drawing you created in the Chapter 26 Exercises. Select the *Import Checked* button to import the checked layout as a new sheet (Fig. 34-26). The drawing should appear in the *Sheet Set Manager's* sheet list as a new sheet in the set.

FIGURE 34-26

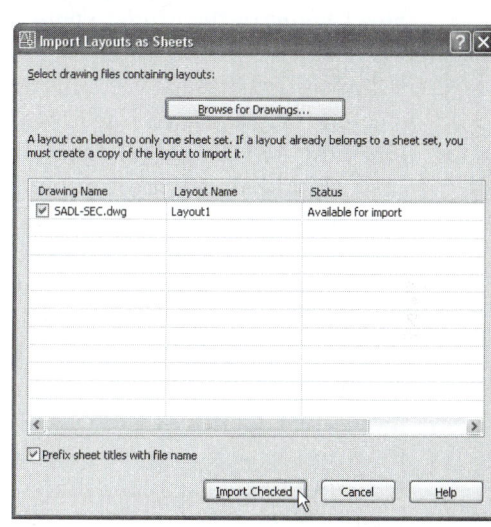

3. **Renumber sheets.**

   A. Since some sheets were removed and others added, the sheets are most likely not in the correct sequence and some sheets may not be numbered. Use drag-and-drop to reorder the sheets as you like. Renumber the sheets by right-clicking the desired sheet names and selecting *Rename & Renumber* from the shortcut menu.

   B. Repeat the procedure from step 2. B. to import any additional sheets you want from previous exercises. NOTE: Drawings that you select for importation as sheets must include at least one layout with a viewport displaying the model space geometry. Once the sheets have been added, reorder them and use *Rename & Renumber* as needed.

4. **Create a blank sheet set and specify a template.**

   A. First, create a new folder named **Chapter 34 Exercise 4**, or similar. Locate the following drawings you created from previous Chapter Exercises and copy them to the **Chapter 34 Exercise 4** folder:

   ASHEET.DWT (template file from Chapter 13)
   T-BLOCKAT (Chapter 22)
   T-PLATE (Chapter 9)
   GASKET B (Chapter 9)
   CLIP (Chapter 24)
   BARGUIDE (Chapter 24)
   BEAR-SEC (Chapter 26)
   CH27EX3 (Chapter 27)

B.   The first step in creating a new sheet set "from scratch" is to create a "blank" .DST file containing no sheets. Do this in the *Sheet Set Manager* by selecting *New Sheet Set* from the drop-down list at the top. The *Create Sheet Set* Wizard appears. In the *Begin* step, select the *An Example Sheet Set* option, then select *Next*. In the *Sheet Set Example* step, pick *Select a sheet set to use as an example*, and from the bottom of the list select *New Sheet Set*. Select the *Next* button. In the *Sheet Set Details* step, assign a name for the new sheet set such as **Chapter 34 Exercise 4** or similar. Enter a description if desired. Designate the location for the .DST file by selecting the browse button and locating the previously created folder (Chapter 34 Exercise 4). Select the *Next* button, then *Finish*. In the *Sheet Set Manager* the new sheet set name appears in the *Sheet List* tab; however, no sheets have yet been created for the set. Note that no entries appear in either the *Sheet Views* or the *Model Views* tab.

C.   Before adding sheets, you must set up a drawing to use as a template. In this way, all new sheets created will have the same layout specifications including page setups. To prepare the template, *Open* the **ASHEET.DWT** template drawing (already located in your working folder). This template should already be set up with a viewport in a *Layout* tab and have a page setup specified to plot on your printer or plotter at 1:1 scale. To be sure, use *Pagesetup* to check that the correct device and settings are specified. Also, activate a *Layout* tab, *Insert* the **TBLOCKAT.DWG** and create a border. *Save* the template.

D.   Activate the *Model Views* tab. Double-click on *Add New Location*. In the *Browse for Folder* dialog box that appears, locate the **Chapter 34 Exercise 4** folder containing all the drawings from the exercises. Select *Open*. The *Model Views* tab should now list all the drawings contained in the selected folder (Fig. 34-27).

FIGURE 34-27

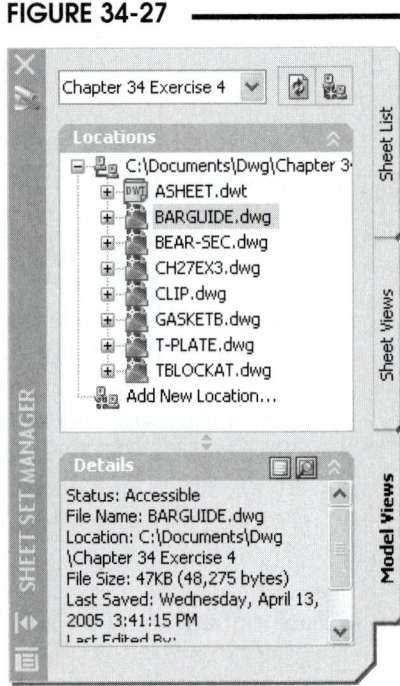

5.   **Create a new view and a new sheet.**

A.   Activate the *Sheet List* tab. Note that no sheets are displayed in the list. Activate the *Sheet Views* tab. Since no views have been assigned to sheets, no views are listed here. First you must create a model space view, then create a new sheet, and finally apply the view to the sheet.

B.   Activate the *Model Views* tab. *Open* the **BARGUIDE.DWG** by double-clicking or using the shortcut menu. In the *Model* tab, *Zoom All* to display the entire drawing limits. Using the *View* command, create a *New* view showing the *Current Display* and assign the name *Barguide Limits*. *Save* and *Close* the drawing. Note that in the *Model Views* tab, the new view name *Barguide Limits* is listed under the BARGUIDE.DWG drawing. Select the *Preview* button below to see a thumbnail preview.

C. Activate the *Sheet List* tab. Select the sheet set name, right-click, and select *New Sheet* from the menu. The *Select Layout as Sheet Template* dialog box appears (not shown) for you to select the desired template. Locate and select the **ASHEET.DWT** file, then select *OK*. When the *New Sheet* dialog box appears (Fig. 34-28), enter a *Number* and *Sheet Title* as shown. NOTE: It is helpful to use a sheet title somewhat <u>different</u> from the drawing name, such as **Barguide Multiview**. Once created, the new sheet should appear in the *Sheet List*.

FIGURE 34-28

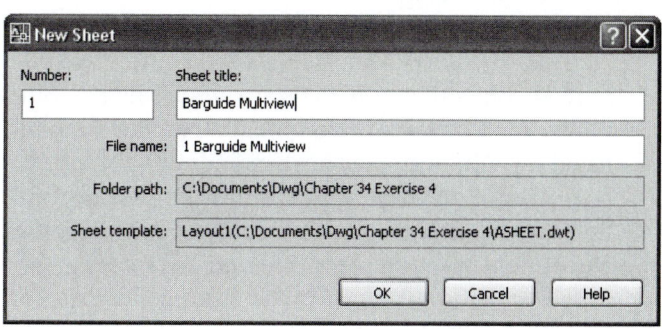

D. Finally, to specify the view to display in the new sheet, activate the *Sheet List* and double-click on the new sheet name to open it in the Drawing Editor. Next, activate the *Model Views* tab. Right-click on the *Barguide Limits* view listed under the BARGUIDE.DWG, and select *Place on Sheet*. Drag the display of the drawing toward the sheet. Before selecting the location, right-click to produce the scale list and select **1:1** from the list (Fig. 34-29). Locate the view appropriately on the sheet. (If you need to move the view, select the border around the view, right-click, and select *Move*.)

FIGURE 34-29

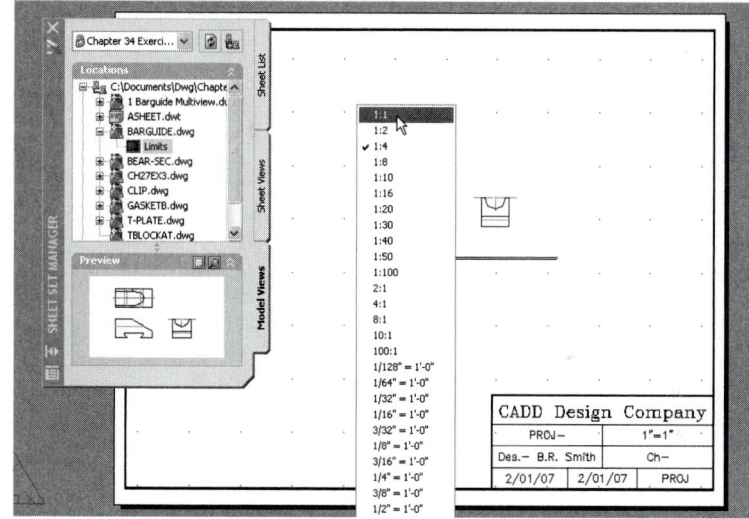

E. Activate the *Sheet Views* tab. Note that the new *View* name is displayed in the list. Activate the *Model Views* tab. Note that AutoCAD created a new drawing containing the new sheet (technically, the new drawing *Xrefs* the original drawing's *View*).

6. **Create additional views and sheets.**

Follow the procedure explained in steps 5. B, C, and D to create *Views* and *New Sheets* for each of the other drawings displayed in the *Model Views* tab (except ASHEET and TBLOCKAT). Note that the list of drawings you used for this sheet set contains two multiview drawings, one auxiliary drawing, one section drawing, and two stamping drawings. When you name the views and sheet titles, include the drawing name and the category of drawing. For example, for the BEAR-SEC drawing, name the view *Bearing Limits* and the sheet title *Bearing Section*. Follow that procedure for the other drawings.

7. **Create view categories.**

Activate the *Sheet Views*. Next select the *View by category button*. Note that only view names are listed. Since the view names include the drawing name the views can be easily identified even though the sheets names are not listed. Select the *New View Category* button located at the top-right corner of the *Sheet Views* tab. Create the following categories: **Multiview, Section, Auxiliary**, and **Stamping**. Next, drag and drop the named views into the appropriate categories (Fig. 34-30). Using view categories is especially helpful for architectural drawings that utilize *Callout Blocks* and *View Label Blocks*.

FIGURE 34-30

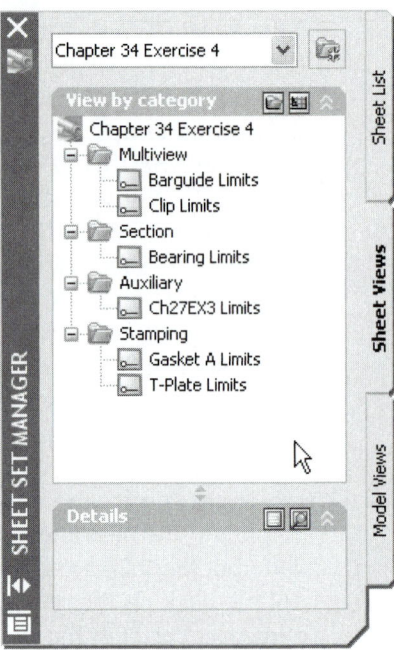

8. **Create a sheet list table.**

FIGURE 34-31

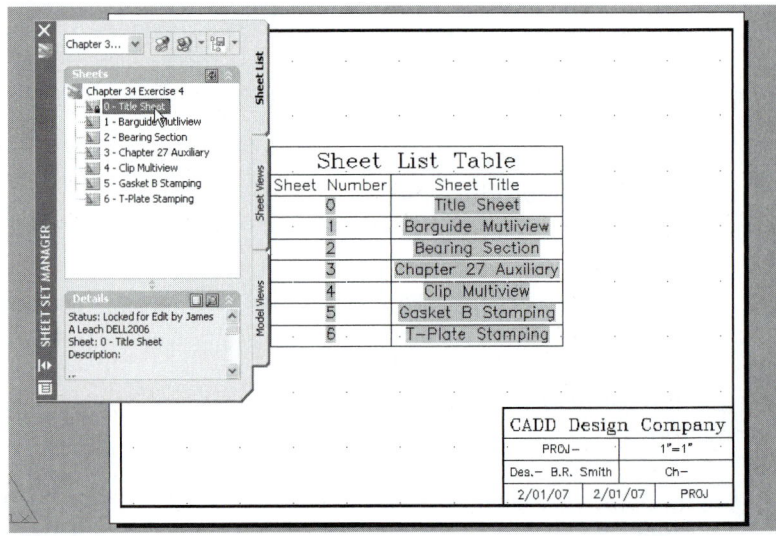

Create a new blank sheet following the procedure listed in Exercise 5. C. Enter the *Number* of the sheet as **0** and the *Sheet Title* as **Title Sheet**. When the new sheet is created, activate the *Sheet List* and double-click on the new sheet name to open it. Activate the drawing's *Layout* tab. With the new sheet open, in the *Sheet List* right-click on the sheet set name (**Chapter 34 Exercise 4**) and select *Insert Sheet List Table* from the shortcut menu. When the *Insert Sheet List Table* dialog box appears, customize the table style to your liking. Select the *OK* button and place the new table in the center of the drawing (Fig. 34-31). *Save* and *Close* the TITLE SHEET.DWG. Finally, in the *Sheet List*, drag and drop the **Title Sheet** so it is the first sheet listed in the set (see Fig. 34-31).

9. **Publish the sheet set.**

Because you used a template when you created each sheet, you can easily publish the set of drawings as one set. Select the *Publish* button from the top of the *Sheet Set Manager* and select *Publish Dialog Box*. When the dialog box appears, note that each sheet is listed with the default page setup. Select the *Plotter named in page setup* button. Next, select *Show Details*, then select any sheet to ensure the correct device is specified below. Select the *Publish* button to produce a complete plotted or printed sheet set.

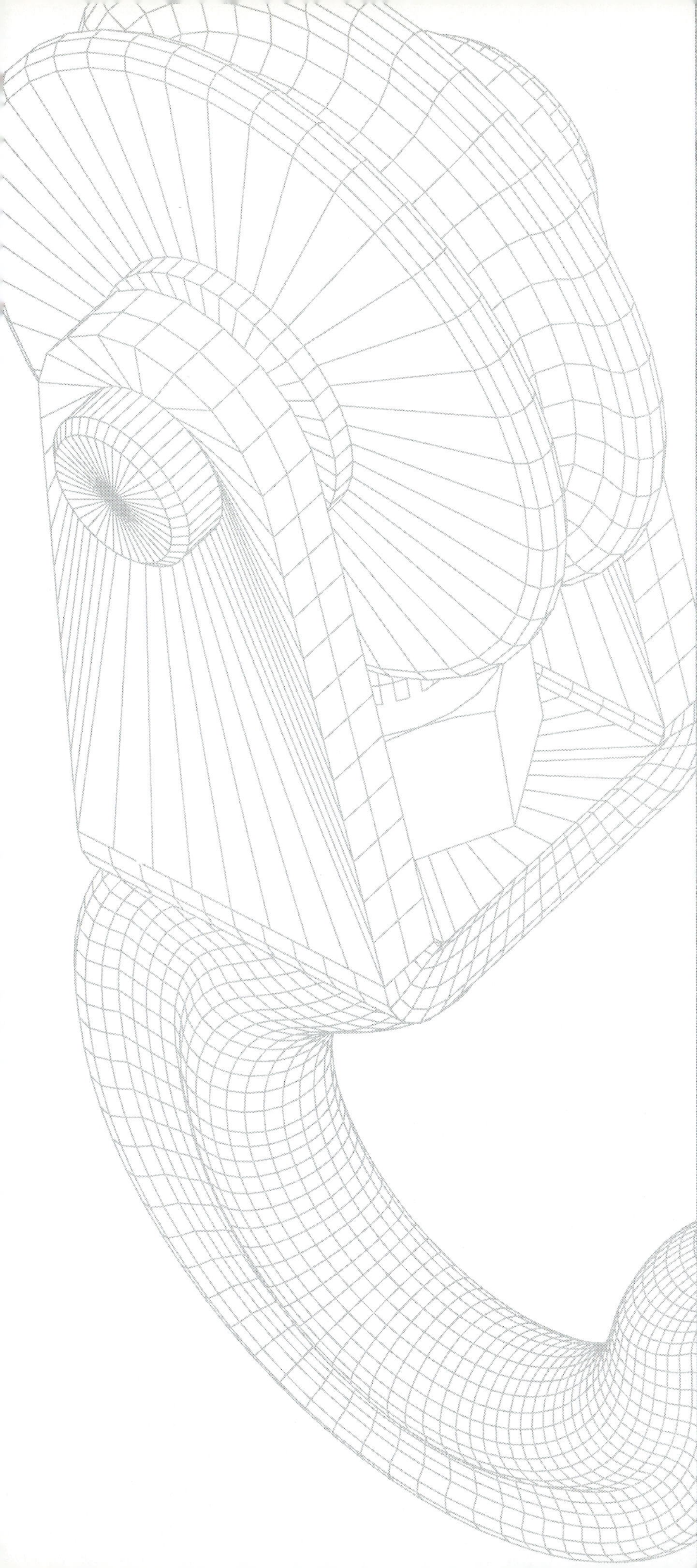

35

3D BASICS, NAVIGATION, AND VISUAL STYLES

CHAPTER OBJECTIVES

After completing this chapter you should:

1. understand the AutoCAD 3D interface;

2. know the characteristics of wireframe, surface, and solid models;

3. know the formats for 3D coordinate entry;

4. understand the orientation of the World Coordinate system, the purpose of User Coordinate Systems, and the use of the *Ucsicon*;

5. be able to use 3D navigation and view commands to generate different views of 3D models;

6. be able to use the *Visual Style control panel* and the *Visual Styles Manager* to specify variations of wireframe, hidden, and shaded displays of 3D models.

THE AutoCAD 3D INTERFACE

The 3D interface in AutoCAD 2007 changed dramatically from the previous version, AutoCAD 2006. In addition to the "look" of the interface, many of the functions for solid modeling construction, display, and rendering have been improved and are discussed in the following chapters.

The 3D interface consists of a large control panel on the right, called the *Dashboard*, and a drawing template displaying a grid in perspective mode intended especially for creating 3D objects (Fig. 35-1).

FIGURE 35-1

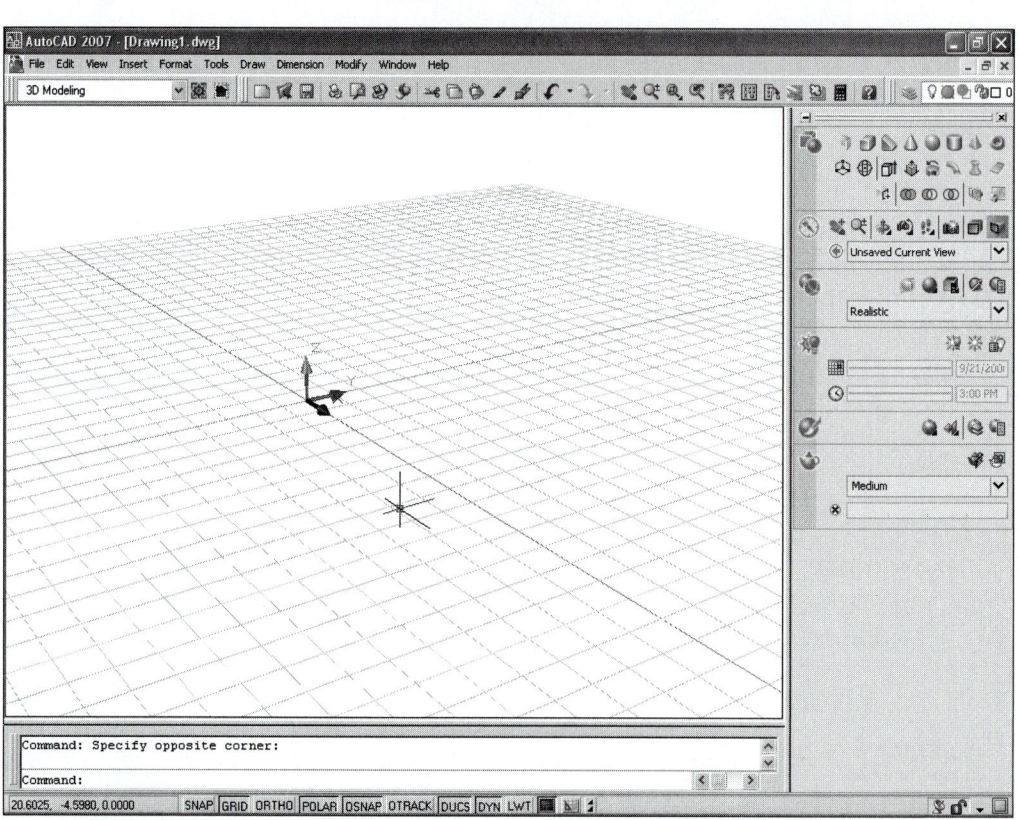

NOTE: When you open AutoCAD, you have the choice to begin working in the 2D (AutoCAD Classic) environment or in the new 3D environment (this occurs the first time AutoCAD is launched after installation, but can be suppressed for subsequent openings). It is likely that AutoCAD may have already been set up for you to begin working in 2D in your particular office or laboratory. In that case, follow the instructions below.

To initiate the 3D interface and begin a 3D drawing:

FIGURE 35-2

1. Produce the *Workspaces* toolbar by any method, such as right-clicking on any tool button and selecting *Workspaces* from the list. The *Workspaces* toolbar should appear on the screen. Select *3D Modeling* from the drop-down list (Fig. 35-2). This action should remove some of the 2D draw and edit toolbars and produce the *Dashboard*. (The *Dashboard* and *Dashboardclose* commands can also be typed to open and close the *Dashboard*.)

2. Use the *New* command to begin a new drawing. When the *Select Template* dialog box appears, select the ACAD3D.DWT or ACADISO3D.DWT template (Fig. 35-3). If you intend to create 3D drawings continually, you might want to use the *Options* dialog box and change the *Default Template File Name for QNEW* to one of these 3D templates.

FIGURE 35-3

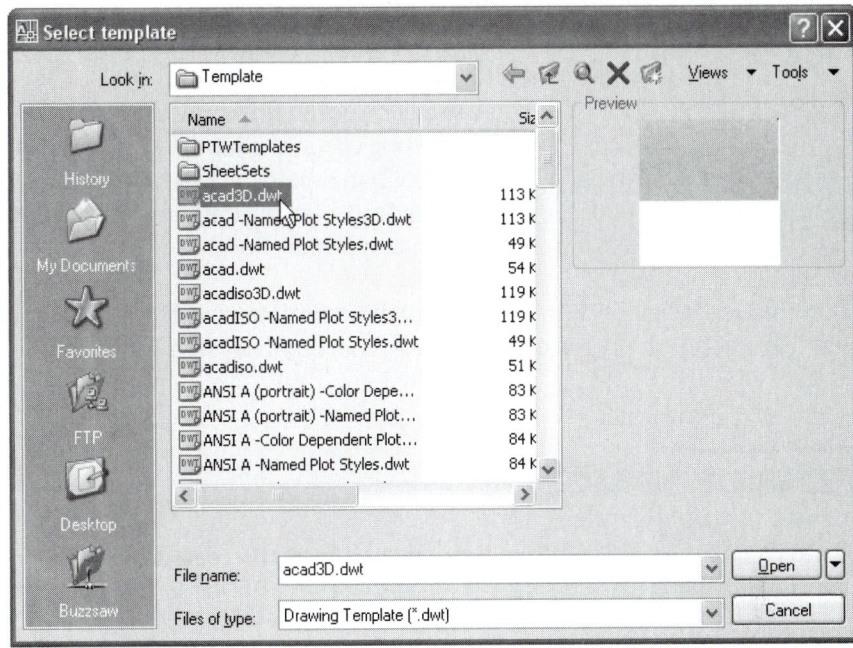

To initiate the AutoCAD classic (2D) interface and begin a 2D drawing:

1. Produce the *Workspaces* toolbar by any method (see Fig. 35-2). Select *AutoCAD Classic* from the drop-down list.

2. Use the *New* command to begin a new drawing. When the *Select Template* dialog box appears, select any non-3D template (not the ACAD3D.DWT or ACADISO3D.DWT template as shown in Fig. 35-3).

The *Dashboard*

The *Dashboard* consists of several individual control panels. Each control panel contains a set of related commands and/or system variable settings. The individual control panels are designated by horizontal separators as shown in Figure 35-4. Each control panel is displayed initially in its condensed form but can be expanded to display additional options by clicking on the down arrows (chevrons) that appear when you drag the pointer to the lower-left corner of the panel as shown in Figure 35-4, top panel. The six control panels that appear by default (from top to bottom) on the *Dashboard* are listed here.

FIGURE 35-4

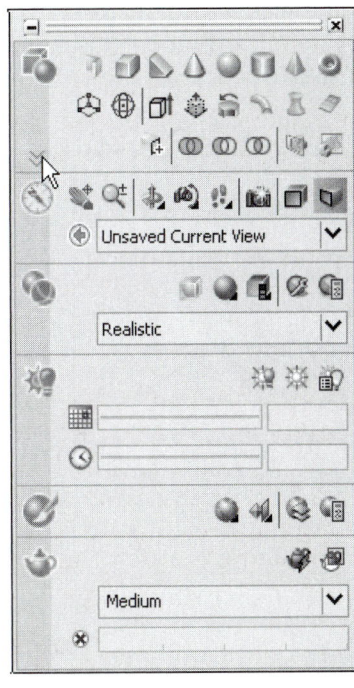

3D Make control panel   Use this control panel to create and edit solid model objects (see Chapters 38 and 39).

3D Navigate control panel   The *3D Navigate control panel* is used to change your view of the 3D model. For example, you can switch between perspective and parallel projection, select from a list of views, or change your orientation with *3Dorbit* and other commands. You can also create simple "walk through" animations (see Chapter 41, Rendering and Animation).

Visual Style control panel	Use the *Visual Style control panel* to change the display of your 3D objects to wireframe, hidden, and various "shaded" displays.
Light control panel	This control panel is used to create light sources for the model (see Chapter 41, Rendering and Animation).
Materials control panel	You can select, edit, and apply material finishes, textures, and images to individual 3D objects (see Chapter 41, Rendering and Animation).
Render control panel	This control panel helps you set the desired specifications for a rendering, such as shadow, background, and resolution settings (see Chapter 41, Rendering and Animation).

This chapter focuses on the *3D Navigation control panel* and the *Visual Style control panel*. But first, you will learn about the types of 3D models, coordinate systems, and forms of coordinate entry.

TYPES OF 3D MODELS

Three basic types of 3D (three-dimensional) models created by CAD systems are used to represent actual objects. They are:

1. Wireframe models
2. Surface models
3. Solid models

These three types of 3D models range from a simple description to a very complete description of an actual object. The different types of models require different construction techniques, although many concepts of 3D modeling are the same for creating any type of model on any type of CAD system.

Wireframe Models

"Wireframe" is a good descriptor of this type of modeling. A wireframe model of a cube is like a model constructed of 12 coat-hanger wires. Each wire represents an <u>edge</u> of the actual object. The <u>surfaces</u> of the object are <u>not</u> defined; only the boundaries of surfaces are represented by edges. No wires exist where edges do not exist. The model is see-through since it has no surfaces to obscure the back edges. A wireframe model has complete dimensional information but contains no volume. Examples of wireframe models are shown in Figures 35-5 and 35-6.

FIGURE 35-5 ━━━━━━━━━━━━━━━

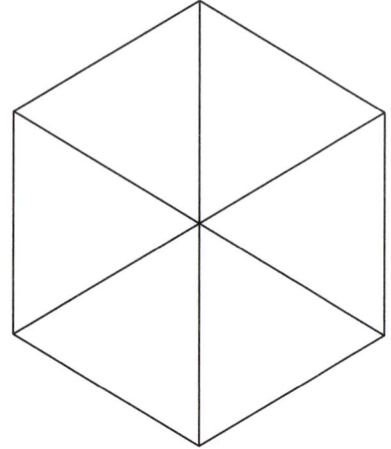

FIGURE 35-6 ━━━━━━━━━━━━━━━

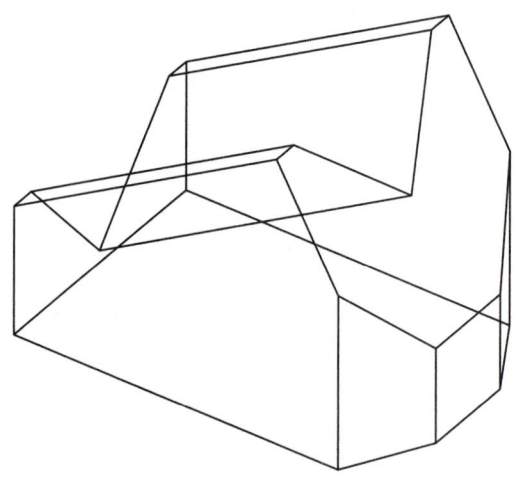

Wireframe models are relatively easy and quick to construct; however, they are not very useful for visualization purposes because of their "transparency." For example, does Figure 35-5 display the cube as if you are looking toward a top-front edge or looking toward a bottom-back edge? Wireframe models tend to have an optical illusion effect, allowing you to visualize the object from two opposite directions unless another visual clue, such as perspective, is given.

With AutoCAD a wireframe model is constructed by creating 2D objects in 3D space. The *Line, Circle, Arc,* and other 2D *Draw* and *Modify* commands are used to create the "wires," but 3D coordinates must be specified. The cube in Figure 35-5 was created with 12 *Line* segments. AutoCAD provides all the necessary tools to easily construct, edit, and view wireframe models.

A wireframe model offers many advantages over a 2D engineering drawing. Wireframe models are useful in industry for providing computerized replicas of actual objects. A wireframe model is dimensionally complete and accurate for all three dimensions. Visualization of a wireframe is generally better than a 2D drawing because the model can be viewed from any position or a perspective can be easily attained. The 3D database can be used to test and analyze the object <u>three dimensionally</u>. The sheet metal industry, for example, uses wireframe models to calculate flat patterns complete with bending allowances. A wireframe model can also be used as a foundation for construction of a surface model. Because a wireframe describes edges but not surfaces, wireframe modeling is appropriate to describe objects with planar or single-curved surfaces, but not compound curved surfaces.

Surface Models

Surface models provide a better description of an object than a wireframe, principally because the surfaces as well as the edges are defined. A surface model of a cube is like a cardboard box—all the surfaces and edges are defined, but there is nothing inside. Therefore, a surface model has volume but no mass. A surface model provides a better visual representation of an actual 3D object because the front surfaces obscure the back surfaces and edges from view. Figure 35-7 shows a surface model of a cube, and Figure 35-8 displays a surface model of a somewhat more complex shape. Notice that a surface model leaves no question as to which side of the object you are viewing.

FIGURE 35-7

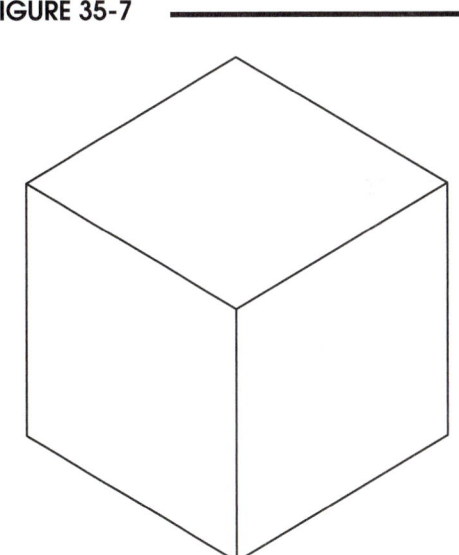

Surface models require a relatively tedious construction process. Each surface must be constructed individually. Each surface must be created in, or moved to, the correct orientation with respect to the other surfaces of the object. In AutoCAD, a surface can be constructed in a number of ways. Often, wireframe elements are used as a framework to build and attach surfaces. The complexity of the construction process of a surface is related to the number and shapes of its edges.

Because there are several generations of surfacing capabilities in AutoCAD, surface modeling in AutoCAD is somewhat ambiguous. Traditionally, surface modeling in AutoCAD was accomplished by creating polygon mesh surfaces such as a *Rulesurf*, *Tabsurf*, *Edgesurf*, or *Revsurf*. These older, "legacy," surfaces can be used to construct surface models as shown in Figure 35-8, but cannot be used in conjunction with solid models, and therefore these commands do not appear on the *Dashboard*. In addition, a region (see Chapter 15) can be used as a surface, but a region has different properties than a mesh and the *Region* command does not appear in any of the 3D modeling menus. Surface modeling using polygon meshes and regions is discussed in Chapter 40, Surface Modeling.

FIGURE 35-8

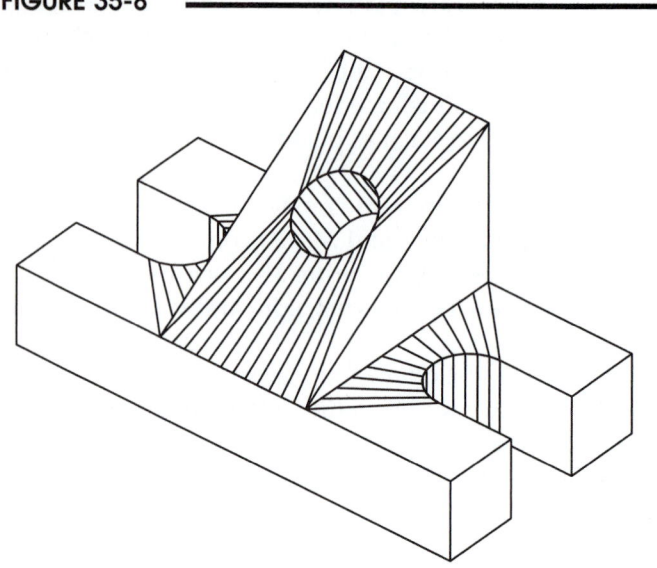

AutoCAD 2007 offers a number of newer surfaces much different than polygon meshes or regions. The *3D Make control panel* section of the *Dashboard* provides several commands that can be used to create surfaces as well as solid models. For example, *Extrude*, *Revolve*, *Sweep*, and *Loft* create a <u>surface</u> if applied to an "open" 2D object such as a *Line*, *Spline*, *Arc*, or *Pline*. However, *Extrude*, *Revolve*, *Sweep*, and *Loft* create a <u>solid</u> if applied to a "closed" 2D shape such as a *Circle*, *Ellipse*, *Region*, or closed *Pline*. In addition, only these newer surfaces (not regions or polygon meshes) can be used to "slice" solid models.

Surface models are capable of wireframe, hidden, or various shaded displays. A wireframe representation might be used during construction, but a hidden or shaded representation might be used to display the finished model. Wireframe, hidden, and various shaded displays are discussed later in this chapter (see "Visual Styles").

Solid Models

Solid modeling is the most complete and descriptive type of 3D modeling. A solid model is a complete computerized replica of the actual object. A solid model contains the complete surface and edge definition, as well as description of the interior features of the object. If a solid model is cut in half (sectioned), the interior features become visible. Since a solid model is "solid," it can be assigned material characteristics and is considered to have mass. Because solid models have volume and mass, most solid modeling systems include capabilities to automatically calculate volumetric and mass properties.

Techniques for constructing solid models are generally much simpler and more unified than those for constructing surface models. AutoCAD uses a process known as Constructive Solid Geometry (CSG) modeling. CSG is characterized by its simple and straightforward construction techniques of combining "primitive" shapes (boxes, cylinders, wedges, etc.) utilizing Boolean operations (*Union*, *Subtract*, and *Intersect*). CSG models store the "history" (record of primitives and Boolean operations) used to construct the models.

The CSG construction typically begins by specifying dimensions for simple primitive shapes such as boxes or cylinders, then combining the primitives using Boolean operations to create a "composite" solid. Other primitives and/or composite solids can be combined by the same process. Several repetitions of this process can be continued until the desired solid model is finally achieved. Figure 35-9 displays a solid model constructed of simple primitive shapes combined by Boolean operations. CSG construction and editing techniques are discussed in detail in Chapters 38 and 39.

Technically, AutoCAD is a <u>hybrid</u> modeler. That is, the solid modeling kernel (called ACIS) is used as a combination Boundary Representation (B-Rep) and CSG modeler. Boundary representation modeling defines a model in terms of its edges and surfaces (rather than by primitives and Booleans) and determines the solid model based on which side of the surfaces the model lies. The B-Rep capabilities (which are automatic and not apparent to the user) are used by AutoCAD for the display of the models (such as when a "mesh" representation is shown). However, the *Brep* command can be used to remove a composite solid's CSG history and reduce the editing capabilities of the model.

Solid models, like surface models, are capable of wireframe, hidden, or various shaded displays. For example, a wireframe display might be used during construction (see Fig. 35-9) and hidden or shaded representation used to display the finished model (Fig. 35-10). See "Visual Styles" in this chapter for more information.

FIGURE 35-9

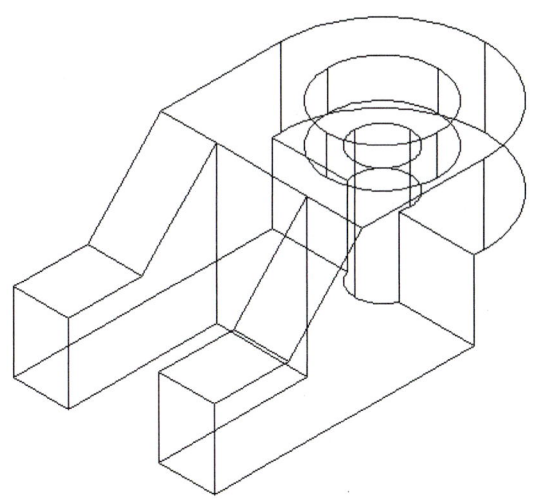

FIGURE 35-10

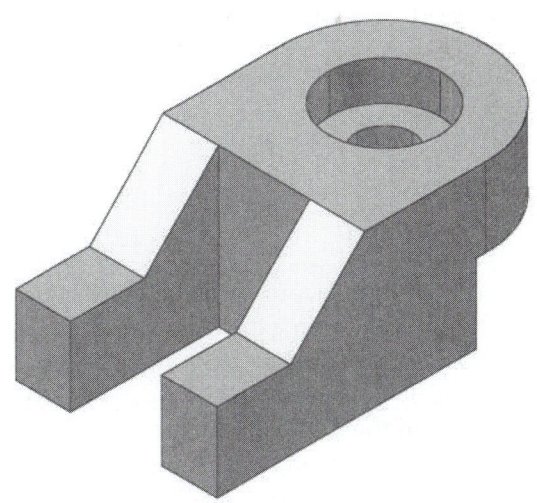

The new AutoCAD 2007 surfaces (those appearing on the *Dashboard*) can be used to *Slice* solids. For example, a *Lofted* surface can be used to *Slice* through a box solid to create a complex-curved face on a solid. To do this operation, only the new surface types (those created by *Extrude*, *Revolve*, *Sweep*, and *Loft* applied to an "open" 2D object such as a *Line*, *Spline*, *Arc*, or *Pline*) can be used. The older, pre-AutoCAD 2007 surfaces, such as polygon meshes and regions, cannot be used for this purpose. These new modeling methods are discussed in Chapter 39, Solid Model Editing.

2007

3D COORDINATE ENTRY

When creating a model in three-dimensional drawing space, the concept of the X and Y coordinate system, which is used for two-dimensional drawing, must be expanded to include the third dimension, Z, which is measured from the origin in a direction perpendicular to the plane defined by X and Y. Remember that two-dimensional CAD systems use X and Y coordinate values to define and store the location of drawing elements such as *Lines* and *Circles*. Likewise, a three-dimensional CAD system keeps a database of X, Y, and Z coordinate values to define locations and sizes of two- and three-dimensional elements. For example, a *Line* is a two-dimensional object, yet

FIGURE 35-11

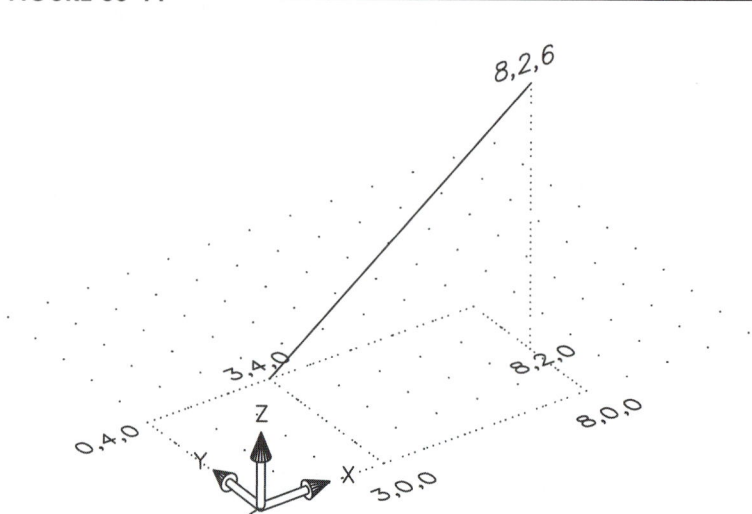

the location of its endpoints in three-dimensional space must be specified and stored in the database using X, Y, and Z coordinates (Fig. 35-11). The X, Y, and Z coordinates are always defined in that order, delineated by commas. The AutoCAD Coordinate Display (*Coords*) displays X, Y, and Z values.

The icon that appears (usually in the lower-left corner of the AutoCAD Drawing Editor) is the Coordinate System icon (sometimes called the UCS icon). When you draw in the 2D mode the icon indicates only the X and Y axis directions (Fig. 35-12).

FIGURE 35-12

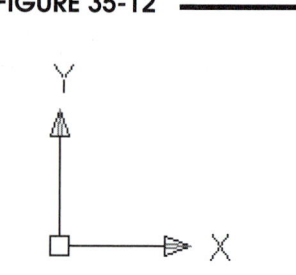

When you change your display of an AutoCAD drawing to one of the 3D *Visual Styles* or begin a drawing using the ACAD3D.DWT or ACADISO3D.DWT drawing template, the icon changes to a colored, 3-pole icon (Fig. 35-13). This UCS icon displays the directions for the X (red), Y (green), and the Z (cyan) axes and is located by default at the origin, 0,0. Technically, the setting for *Visual Style* controls which icon appears—the *2D wireframe* option displays the simple "wire" icon, whereas any 3D option displays the icon shown in Figure 35-13.

FIGURE 35-13

3D Coordinate Entry Formats

Because construction in three dimensions requires the definition of X, Y, <u>and</u> Z values, the methods of coordinate entry used for 2D construction must be expanded to include the Z value. The methods of command entry used for 2D construction are valid for 3D coordinates with the addition of a Z value specification. Relative polar coordinate specification (@dist<angle) is expanded to form two other coordinate entry methods available explicitly for 3D coordinate entry. The methods of coordinate entry for 3D construction follow:

1. **Mouse Input** | **PICK** — Use the cursor to select points on the screen. *OSNAP*, Object Snap Tracking, or point filters must be used to select a point in 3D space; otherwise, points selected are <u>on the XY plane</u>.

2. **Absolute Cartesian coordinates** | **X,Y,Z** — Enter explicit X, Y, and Z values relative to point 0,0.

3. **Relative Cartesian coordinates** | **@X,Y,Z** — Enter explicit X, Y, and Z values relative to the last point.

4. **Cylindrical coordinates (relative)** | **@dist<angle,Z** — Enter a distance value, an angle in the XY plane value, and a Z value, all relative to the last point.

5. **Spherical coordinates (relative)** | **@dist<angle<angle** — Enter a distance value, an angle <u>in</u> the XY plane value, and an angle <u>from</u> the XY plane value, all relative to the last point.

6. **Dynamic Input** | **dist,angle** or **X,Y,Z** — Enter a distance value in the first edit box, press Tab, enter an angular value in the second box. You can also enter X,Y,Z coordinates separated by commas to establish relative Cartesian coordinates.

7. **Direct distance entry** | **dist,direction** — Enter (type) a value, and move the cursor in the desired direction.

Cylindrical and spherical coordinates can be given <u>without</u> the @ symbol, in which case the location specified is relative to point 0,0,0 (the origin). This method is useful if you are creating geometry centered around the origin. Otherwise, the @ symbol is used to establish points in space relative to the last point.

Examples of each of the 3D coordinate entry methods are illustrated in the following section. In the illustrations, the orientation of the observer has been changed from the default plan view in order to enable the visibility of the three dimensions.

Mouse Input

Figure 35-14 illustrates using the <u>interactive</u> method to PICK a location in 3D space. *OSNAP* <u>must</u> be used to PICK in 3D space. Any point PICKed with the input device <u>without</u> *OSNAP* will result in a location <u>on the XY plane</u> (unless you are establishing a Dynamic UCS). In this example, the *Endpoint OSNAP* mode is used to establish the "Specify next point:" of a second *Line* by snapping to the end of an existing vertical *Line* at 8,2,6:

Command: *line*
Specify first point: **3,4,0**
Specify next point or [Undo]: *endpoint* of **PICK**

FIGURE 35-14

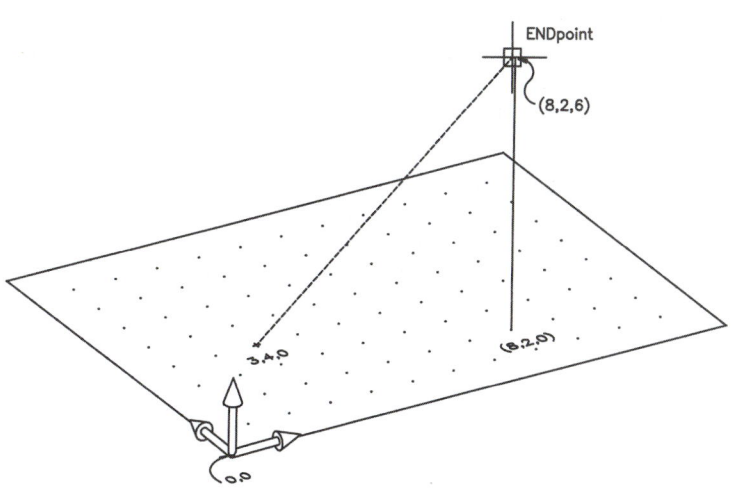

Absolute Cartesian Coordinates

Figure 35-15 illustrates the <u>absolute</u> coordinate entry to draw the *Line*. The endpoints of the *Line* are given as explicit X,Y,Z coordinates:

> Command: **line**
> Specify first point: **3,4,0**
> Specify next point or [Undo]:
> **8,2,6**

FIGURE 35-15

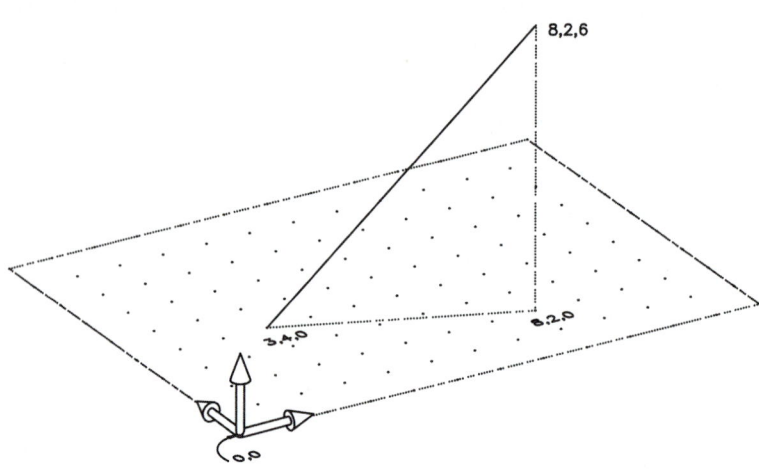

Relative Cartesian Coordinates

Relative Cartesian coordinate entry is displayed in Figure 35-16. The "Specify first point:" of the *Line* is given in absolute coordinates, and the "Specify next point:" end of the *Line* is given as X,Y,Z values <u>relative</u> to the last point:

> Command: **line**
> Specify first point: **3,4,0**
> Specify next point or [Undo]: **@5,
> -2,6**

FIGURE 35-16

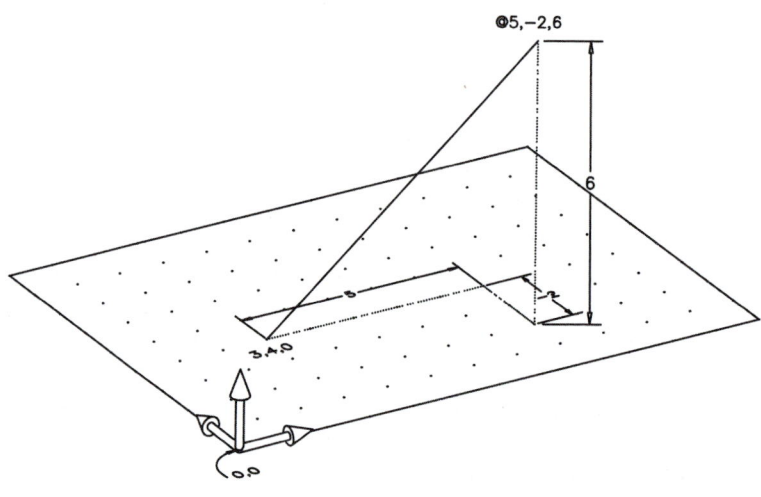

Cylindrical Coordinates (Relative)

Cylindrical and spherical coordinates are an extension of polar coordinates with a provision for the third dimension. Relative cylindrical coordinates give the distance <u>in</u> the XY plane, angle <u>in</u> the XY plane, and Z dimension and can be relative to the last point by prefixing the @ symbol. The *Line* in Figure 35-17 is drawn with absolute and relative cylindrical coordinates. (The *Line* established is approximately the same *Line* as in the previous figures.)

> Command: **line**
> Specify first point: **3,4,0**
> Specify next point or [Undo]:
> **@5<-22,6**

FIGURE 35-17

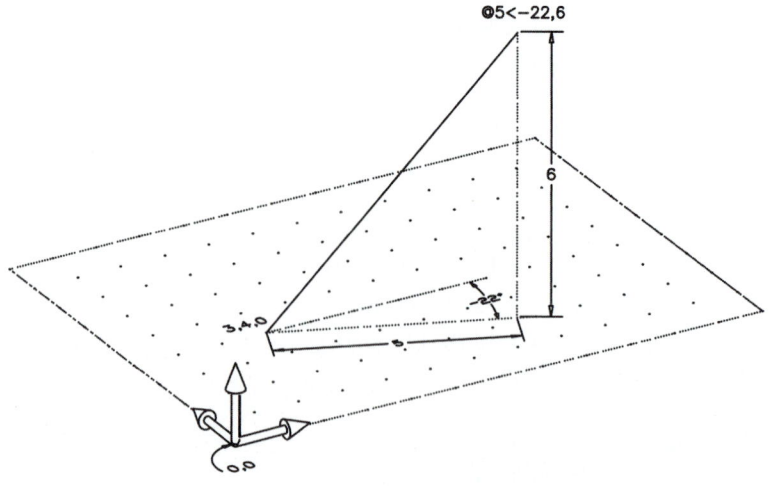

Spherical Coordinates (Relative)

Spherical coordinates are also an extension of polar coordinates with a provision for specifying the third dimension in angular format. Spherical coordinates specify a distance, an angle in the XY plane, and an angle from the XY plane and can be relative to the last point by prefixing the @ symbol. The distance specified is a 3D distance, not a distance in the XY plane. Figure 35-18 illustrates the creation of approximately the same line as in the previous figures using absolute and relative spherical coordinates.

FIGURE 35-18

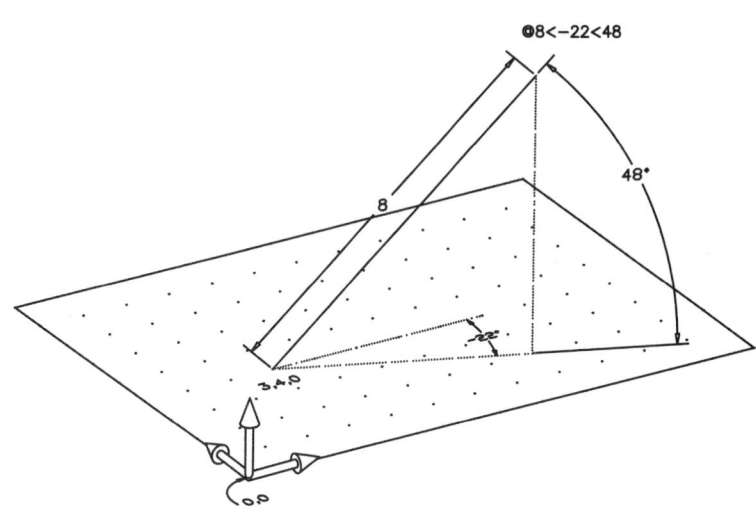

```
Command: line
Specify first point: 3,4,0
Specify next point or [Undo]:
@8<-22<48
```

Polar Tracking

Polar Tracking can be used when constructing and editing 3D models. When *POLAR* is turned on in the 3D drawing mode, the Polar Tracking vector appears in the current XY plane or in a plane parallel to the XY plane when designating points. If the "first point" or "base point" is located in 3D space, the resulting tracking vectors appear parallel to the coordinate system XY plane (in the XY plane of the point). By default, the tracking vectors appear for 90-degree increments, but can be set to any increment using the *Polar Tracking* tab of the *Drafting Settings* dialog box (see Chapter 3).

In AutoCAD 2007 the Polar Tracking vector capability has been extended to operate for the Z axis. For example, assume you used the *Move* command to move a small *Box* from atop a larger *Box* (Figure 35-19). With *POLAR* on, you could move the small box from the "base point" in a +Z or −Z direction as well as in the parallel XY plane polar tracking angles.

FIGURE 35-19

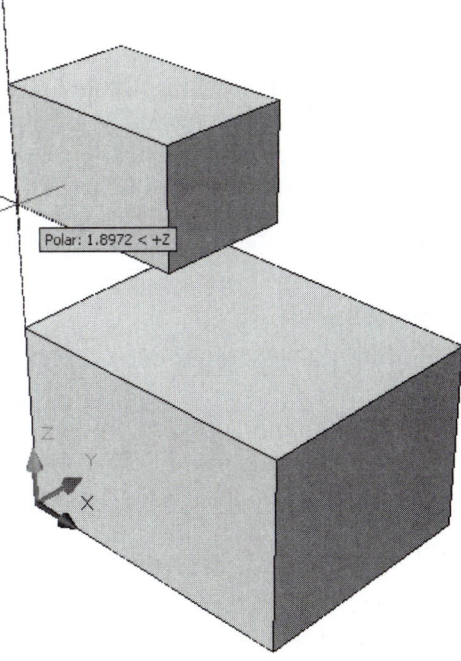

ORTHO

The *ORTHO* feature can also be used for drawing or editing in 3D space. When *ORTHO* is on, the "rubber band line" (from the last point to the current cursor position) is forced to an X, Y, or Z direction. This feature is similar to using *Polar Tracking* in 90-degree increments.

Direct Distance Entry

With Direct Distance Entry, entering a distance value and indicating a direction with the pointer normally results in a point on the current XY plane. However, by turning on *POLAR* or *ORTHO*, you can draw in an X, Y, or Z direction from the "last point." For example, to draw a square in 3D space with sides of 3 units, begin by using *Line* and selecting the *Endpoint* of the existing diagonal *Line* as the "first point:" as shown in Figure 35-19. To draw the first 3-unit segment, turn on *POLAR* or *ORTHO*, move the cursor in the desired X, Y, or Z direction, and enter "3". Next, move the cursor in the desired (90 degree) direction and enter "3," and so on.

FIGURE 35-20

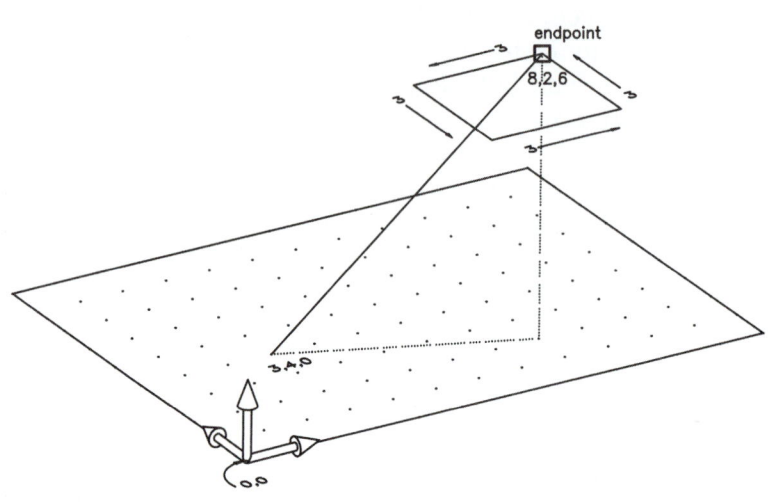

Object Snap Tracking

Object Snap Tracking functions in 3D similar to the way that *OSNAP* and Direct Distance entry operate—that is, when a point is "acquired" by one of the *OSNAP* modes, tracking occurs in the plane of the acquired point. To be more specific, when you acquire a point in 3D space (not on the XY plane), tracking vectors appear originating from the acquired object in a plane parallel to the current XY plane.

For example, assume a wedge was created with its base on the XY plane. You can track on a plane in 3D space parallel to the current XY plane that passes through the acquired point (Fig. 35-21). In other words, begin the "first point" of a line at a point on the XY plane (at the center of the base, in this case). For the "next point," acquire a point in 3D space with Object Snap Tracking. Note that tracking vectors are generated from that acquired point into space on an imaginary plane that is parallel with the current XY plane (and parallel with the X axis in this case).

FIGURE 35-21

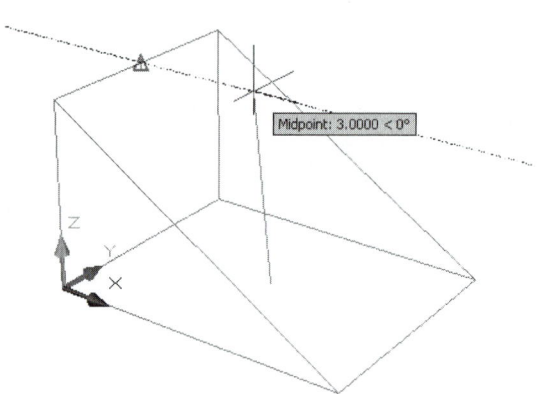

Point Filters

Point filters are used to filter X and/or Y and/or Z coordinate values from a location PICKed with the pointing device. Point filtering makes it possible to build an X,Y,Z coordinate specification from a combination of point(s) selected on the screen and point(s) entered at the keyboard. A .XY (read "point XY") filter would extract, or filter, the X and Y coordinate value from the location PICKed and then prompt you to enter a Z value. Valid point filters are listed below.

.X	Filters (finds) the X component of the location PICKed with the pointing device.
.Y	Filters the Y component of the location PICKed.
.Z	Filters the Z component of the location PICKed.
.XY	Filters the X and Y components of the location PICKed.
.XZ	Filters the X and Z components of the location PICKed.
.YZ	Filters the Y and Z components of the location PICKed.

The .XY filter is the most commonly used point filter for 3D construction and editing. Because 3D construction often begins on the XY plane of the current coordinate system, elements in Z space are easily constructed by selecting existing points on the XY plane using .XY filters and then entering the Z component of the desired 3D coordinate specification by keyboard. For example, in order to draw a line in Z space two units above an existing line on the XY plane, the .XY filter can be used in combination with *Endpoint OSNAP* to supply the XY component for the new line. See Figure 35-22 for an illustration of the following command sequence:

FIGURE 35-22

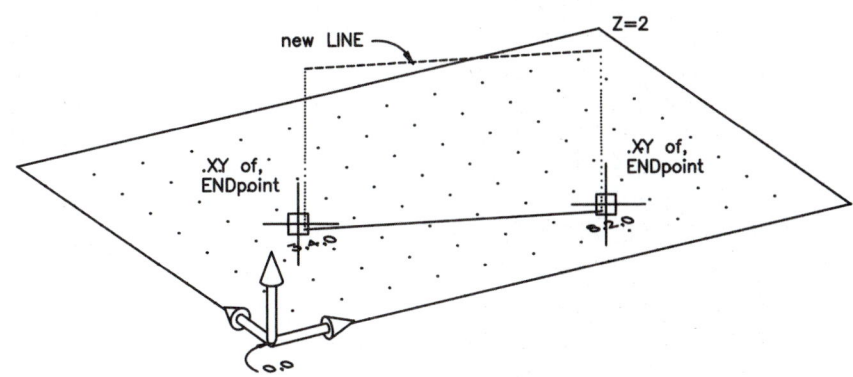

```
Command: line
Specify first point: .XY of
endpoint of PICK (Select one Endpoint of the existing line on the XY plane.)
(need Z) 2
Specify next point or [Undo]: .XY of
endpoint of PICK (Select the other Endpoint of the existing line.)
(need Z) 2
```

Typing **.XY** and pressing **Enter** cause AutoCAD to respond with "of" similar to the way "of" appears after typing an *OSNAP* mode. When a location is specified using an .XY filter, AutoCAD responds with "(need Z)" and, likewise, when other filters are used, AutoCAD prompts for the missing component(s).

Dynamic Input

In the normal Polar format (distance and angle edit boxes), Dynamic Input operates only in the current XY plane. You can, however, create User Coordinate Systems in 3D space and use Dynamic Input on the XY plane of these coordinate systems. See "Coordinate Systems" next. With the default settings, if you instead enter an X, Y, and Z coordinate set separated by commas, Dynamic Input accepts the coordinates as relative Cartesian coordinates (from the last point).

NOTE: Dynamic Input settings are made in the *Dynamic Input* tab of the *Drafting Settings* dialog box where you can change Dynamic Input to absolute coordinate mode for second points. You can also provide an additional edit box for a Z coordinate component if you change the *Show Z Field for Pointer Input* checkbox in the *Options* dialog box, *3D Modeling* tab.

COORDINATE SYSTEMS

In AutoCAD two kinds of coordinate systems can exist, the <u>World Coordinate System</u> (WCS) and one or more <u>User Coordinate Systems</u> (UCS). The World Coordinate System <u>always</u> exists in any drawing and cannot be deleted. The user can also create and save multiple User Coordinate Systems to make construction of a particular 3D geometry easier. <u>Only one coordinate system can be active</u> at any one time in any one view or viewport, either the WCS or one of the user-created UCSs. You can have more than one UCS active at any one time if you have several viewports active (model space or layout viewports).

The World Coordinate System (WCS) and WCS Icon

The World Coordinate System (WCS) is the default coordinate system in AutoCAD for defining the position of drawing objects in 2D or 3D space. The WCS is always available and cannot be erased or removed but is deactivated temporarily when utilizing another coordinate system created by the user (UCS). The icon that appears (by default) at the lower-left corner of the Drawing Editor (Fig. 35-9) indicates the orientation of the WCS. The icon, whose appearance (*ON, OFF*) is controlled by the *Ucsicon* command, appears for the WCS and for any UCS.

The orientation of the WCS with respect to Earth may be different among CAD systems. In AutoCAD the WCS has an architectural orientation such that the XY plane is a horizontal plane with respect to Earth, making Z the height dimension. A 2D drawing (X and Y coordinates only) is thought of as being viewed from above, sometimes called a <u>plan</u> view. This default 2D orientation is like viewing a floor <u>plan</u>—from above (Fig. 35-23). Therefore, in a 3D AutoCAD drawing, imagine this same orientation of objects in space; that is, X is the width dimension, Y is the depth dimension, and <u>Z is height</u>.

FIGURE 35-23

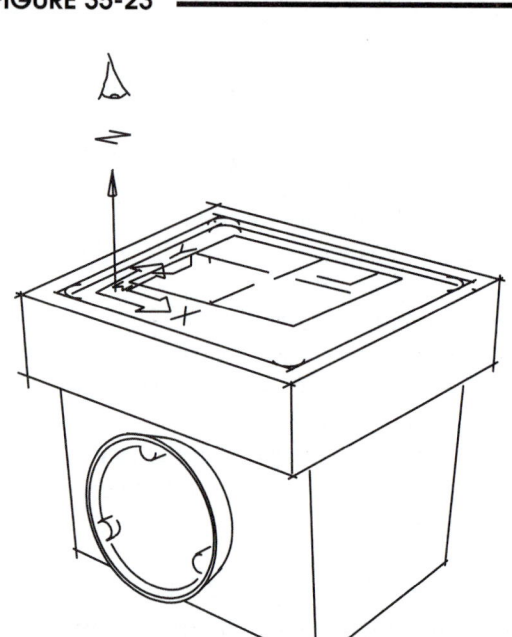

User Coordinate Systems (UCS) and Icons

There are no User Coordinate Systems that exist as part of the AutoCAD template drawings as they come "out of the box." UCSs are created to suit the 3D model when and where they are needed.

Creating geometry is relatively simple when having to deal only with X and Y coordinates, such as in creating a 2D drawing or when creating simple 3D geometry with uniform Z dimensions. However, 3D models containing complex shapes on planes not parallel with the XY plane are good candidates for UCSs. A User Coordinate System is thought of as a construction plane created to simplify creation of geometry on a specific plane or surface of the object. The user creates the UCS, aligning its XY plane with a surface of the object, such as along an inclined plane, with the UCS origin typically at a corner or center of the surface (Fig. 35-24).

FIGURE 35-24

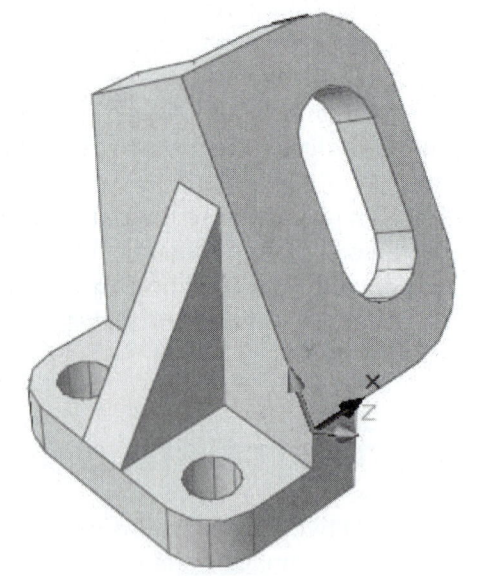

FIGURE 35-25

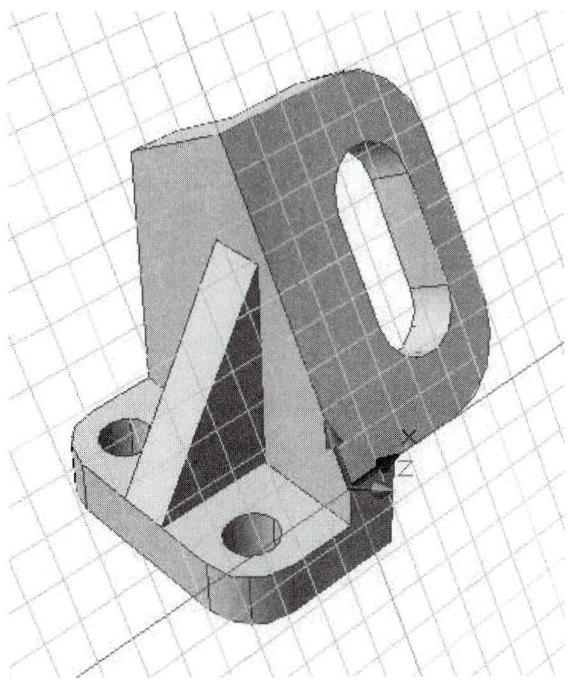

The user can then create geometry aligned with that plane by defining only X and Y coordinate values of the current UCS. The Coordinate Display (*Coords*) in the Status Bar always displays the X, Y and Z values of the <u>current</u> coordinate system, whether it is WCS or UCS. The *SNAP, GRID,* and *Polar Tracking* automatically align with the current coordinate system, providing *SNAP* points, tracking vectors, and enhanced visualization of the current XY "construction" plane (Fig. 35-25). Practically speaking, it is easier in most cases to specify only X and Y coordinates with respect to a specific plane on the object rather than calculating X, Y, and Z values with respect to the World Coordinate System. For example, a UCS would have been created on the inclined surface to construct the slotted hole shown in Figure 35-25.

User Coordinate Systems can be created by any of several options of the *UCS* command. Once a UCS has been created, it becomes the current coordinate system and remains current until another UCS is activated. Since it is likely you would alternate between different coordinate systems during construction and editing of the 3D model, it is suggested that UCSs be saved (by using the *UCS* Manager) for possible future geometry creation or editing.

Dynamic UCSs

In addition to static UCSs created with the *UCS* command, you can use temporary coordinate systems called Dynamic User Coordinate Systems; however, <u>Dynamic User Coordinate Systems operate only with solid models</u>. A Dynamic User Coordinate System, or DUCS, appears automatically only during the creation of a primitive such as a *Box, Wedge,* or *Cylinder*. Use the *DUCS* toggle on the Status line to activate the Dynamic UCS feature.

When the *DUCS* toggle is on, a Dynamic UCS appears on the surface you select to establish the base of a primitive. You can select which surface you want by hovering your cursor over any surface of an existing solid during a primitive command. For example, Figure 35-26 displays a Dynamic UCS appearing on the inclined surface while constructing a *Box* primitive. The <u>Dynamic UCS disappears after the primitive has been created</u>.

FIGURE 35-26

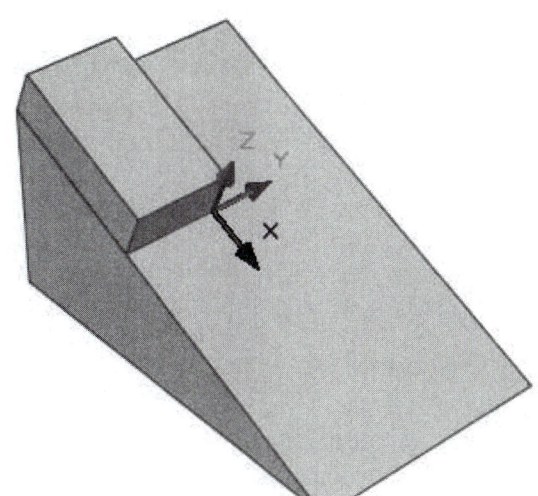

The Right-Hand Rule

AutoCAD complies with the right-hand rule for defining the orientation of the X, Y, and Z axes. The right-hand rule states that if your right hand is held partially open, the thumb, first, and middle fingers define positive X, Y, and Z directions, respectively.

More precisely, if the thumb and first two fingers are held out to be mutually perpendicular, the thumb points in the positive X direction, the first finger points in the positive Y direction, and the middle finger points in the positive Z direction (Fig. 35-27).

In this position, looking toward your hand from the tip of your middle finger is like the default AutoCAD viewing orientation in 2D mode—positive X is to the right, positive Y is up, and positive Z is toward you.

When you rotate geometry in 3D space or create a new User Coordinate System, it is possible that you want to rotate about the existing X, Y, or Z axis. In doing this you must indicate whether you want to rotate in a positive or negative direction. The right-hand rule is used to indicate the direction of positive rotation about any axis. Imagine holding your hand closed around any axis but with the thumb extended and pointing in the positive direction of the axis (Fig. 35-28). The direction your fingers naturally curl indicates the positive rotation direction about that axis. This principle applies to all three axes.

(Figures 35-27 and 35-28 are from *Technical Graphics Communication*, 2/E © 1997, Bertoline et al, McGraw-Hill. Reproduced with permission of The McGraw-Hill Companies.)

FIGURE 35-27

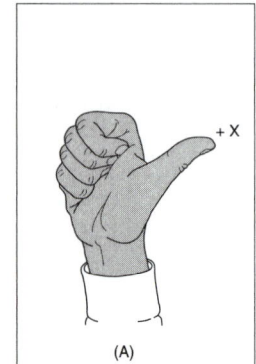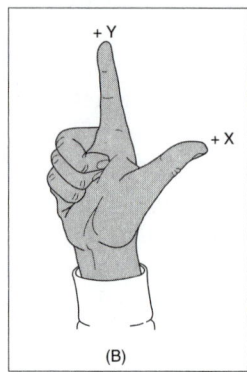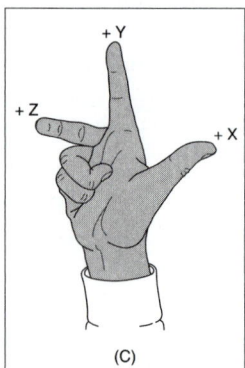

FIGURE 35-28

Ucsicon

Pull-down Menu	Command (Type)	Alias (Type)	Short-cut	Screen (side) Menu	Tablet Menu
View *Display >* *USC Icon*	*Ucsicon*	...	...	VIEW 2 *UCSicon*	*L,2*

The *Ucsicon* command controls the appearance and positioning of the Coordinate System icon. In order to aid your visualization of the current UCS or the WCS, it is <u>highly</u> recommended that the Coordinate System icon be turned *ON* and positioned at the *ORigin*.

    Command: **ucsicon**
    Enter an option [ON/OFF/All/Noorigin/ORigin/Properties] <ON>: **on**
    Command:

The *On* option causes the Coordinate System icon to appear (Fig. 35-29, lower left).

FIGURE 35-29

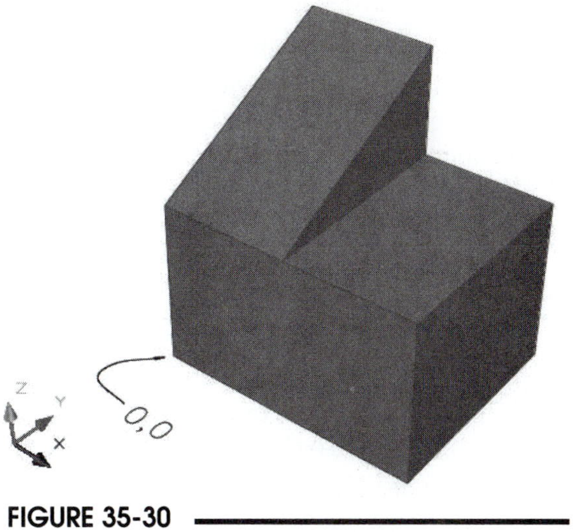

The *Origin* setting causes the icon to move to the origin of the current coordinate system (Fig. 35-30).

FIGURE 35-30

The icon may not always appear at the origin after using some viewing commands like *3Dorbit*. This is because the origin (0,0) may be located outside the border of the drawing screen (or viewport) area, which causes the icon to appear in the lower-left corner. In order to cause the icon to appear at the origin, try zooming out until the icon aligns itself with the origin.

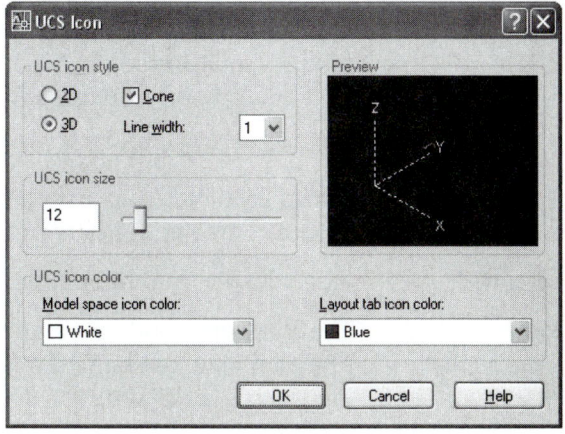

Options of the *Ucsicon* command are:

ON	Turns the Coordinate System icon on.
OFF	Turns the Coordinate System icon off.
All	Causes the *Ucsicon* settings to be effective for all viewports.
Noorigin	Causes the icon to appear always in the lower-left corner of the screen, not at the origin.
ORigin	Forces the placement and orientation of the icon to align with the origin of the current coordinate system.

Properties

The *Properties* option produces the *UCS Icon* dialog box (Fig. 35-31). Ironically, the *UCS Icon Style* and the *UCS Icon Size* have no effect for the 3-pole colored icon that appears in 3D mode. Also, it is actually the *Visual Style* option that determines which icon is displayed: the *2D wireframe* option displays the simple "wire" icon, whereas any 3D option displays the 3-pole colored icon. You can change the *UCS icon color* for model space and paper space. The default option for *Model space icon color* is either *Black* or *White*—the opposite of whatever you set for the *Model* tab background color in the *Options* dialog box.

FIGURE 35-31

3D NAVIGATION AND VIEW COMMANDS

A "view" is a particular display of a 2D drawing or a 3D model. The term "navigate" is descriptive of the process of achieving a view of a 3D model—that is, you as the observer move around, move closer to, or move farther from, the 3D model to arrive at a particular view. The commands listed below allow you to change the direction from which you view a 3D model or otherwise affect the view of the 3D model. Most of these commands appear on the *3D Navigation* control panel; however, many are accessed by other menus, toolbars, or Command line entry.

Pan	This command is the same as the real-time *Pan* used for 2D drawings. This command enables you to drag 3D objects about the view interactively.
Zoom	Identical to the real-time *Zoom* used for 2D drawings, you can enhance your view of 3D objects by interactively zooming in and out. For a view in perspective projection, *Zoom* also changes the amount of perspective as the objects appear nearer or farther.
3Dorbit	*3Dorbit* controls interactive viewing of objects in 3D. You can dynamically view the object from any point, even while the object retains its visual style. The constrained orbit (default) mode locks the vertical edges to a vertical position.
3Dforbit	The free orbit mode of *3Dorbit* allows you to dynamically view the object from any point; however, vertical edges do not remain vertical. You can "spin" the view around in this mode.
3Dcorbit	The *3Dcorbit* command enables you to set the 3D objects into continuous motion. This mode is a continuous free orbit mode since the vertical axis is not locked and objects appear to spin in any direction.
PERSPECTIVE	Actually a system variable, this setting toggles the current view between parallel projection and perspective projection.
3Ddistance	Use this command to make objects appear closer or farther away. This command performs the same function as *Zoom*, in that it changes the amount of perspective as if you moved closer to or farther from the 3D objects.
View Presets	You can access preset views that change the view of the 3D model, such as a top, front, or isometric displays.
View	The *View* command produces the *View Manager*. Here you can create new views and manage the specifications for the *Current View*, *Model Views*, and *Layout Views*. You can also select from preset orthographic views and isometric viewpoints. This dialog box provides access to view settings such as camera and target locations and lens length.
3Dclip	*3Dclip* allows you to establish front and/or back clipping planes in the *Adjust Clipping Planes* window.
Camera	You can create one or more cameras where each camera represents a view of the 3D model. You can specify a X,Y, and Z camera position, target position, and lens length for each camera. You can edit a camera's settings using grips or the *Properties* palette.
3Dswivel	This command simulates the effect of turning the current camera or view.
Vpoint	*Vpoint* is an old AutoCAD command that allows you to change your viewpoint of a 3D model. Three options are provided: *Vector*, *Tripod*, and *Rotate*.
Ddvpoint	Also a legacy command, this tool allows you to select a viewing angle in degrees.
Plan	The *Plan* command automatically gives the observer a plan (top) view of the object. The plan view can be with respect to the WCS (World Coordinate System) or an existing UCS (User Coordinate System).

2007

2007

Grid	The *Grid* command offers several settings that affect the appearance of the grid in 3D mode and control how the grid lines react to zooming.
Vports	You can use *Vports* effectively during construction of 3D objects to divide the screen into several sections, each displaying a different standard 3D view.

3D Navigation Control Panel

Many of the 3D navigation and view commands discussed in this section can be accessed through the *3D Navigation control panel* (Fig. 35-32). This control panel is the second panel from the top in the *Dashboard*, just below the *3D Make control panel*. In addition, a *3D Navigation* toolbar is available which contains several of the commands discussed in this section (Fig. 35-33). Other navigation and view commands can be accessed through the *View* pull-down menu.

FIGURE 35-32 ————————

FIGURE 35-33 ————————

When using the navigation commands such as *3Dpan*, *3Dzoom*, *3Dorbit*, *View*, *Vpoint*, *Plan*, etc., it is important to imagine that the <u>observer moves about the object</u> rather than imagining that the object rotates. The object and the coordinate system (WCS and icon) always remain <u>stationary</u> and always keep the same orientation with respect to Earth. Since the observer moves and not the geometry, the objects' coordinate values retain their integrity, whereas if the geometry rotated within the coordinate system, all coordinate values of the objects would change as the object rotated. The viewing commands change only the viewpoint of the <u>observer</u>.

Pan

Pull-down Menu	Command (Type)	Alias (Type)	Short-cut	Screen (side) Menu	Tablet Menu
View *Pan >* *Realtime*	*Pan*	*P*	*VIEW Pan*	*Pan*	*N,10-P,11*

The cursor changes to a hand cursor when this command is active. Identical to the real-time *Pan* command, you can click and drag the view of the 3D objects to any location in your drawing area. However, when any 3D visual style is displayed and *Perspective* is toggled on, *Pan* displays a true 3D effect. That is, panning slightly changes the direction from which the 3D objects are viewed. As an alternative to invoking this command, you can press and hold the mouse wheel to activate *Pan*. To end this command, press the Esc key or select *Exit* from the right-click shortcut menu.

Zoom

Pull-down Menu	Command (Type)	Alias (Type)	Short-cut	Screen (side) Menu	Tablet Menu
View *Zoom >* *Realtime*	*Zoom*	*Z*	*VIEW Zoom*	*Zoom*	*J,2-J,5*

This *Zoom* is identical to the real-time version of the *Zoom* command used in a 2D drawing. However, when viewing 3D objects in any 3D visual style and with *Perspective* toggled on, *Zoom* performs an actual 3D zoom; that is, zooming changes the distance the object appears from the observer. Technically, zooming changes the distance between the camera (observer) and the target (objects). Therefore, *Zoom* changes the size of the objects relative to screen size as well as the amount of perspective.

The *Zoom* cursor is a magnifying glass icon. Press and hold the left mouse button, then move up to zoom in (enlarge image) and down to zoom out (reduce image). As an alternative to invoking this command, you can turn the mouse wheel forward and backward to zoom in and out, respectively. To end this command, press the Esc key or select *Exit* from the right-click shortcut menu.

NOTE: In releases of AutoCAD previous to AutoCAD 2007, the *Zoom* command and the *3Dzoom* command changed only the size of the 3D objects relative to the screen but did not change the amount of perspective. In other words, the distance between the camera and target, and therefore the amount of perspective, did not change with *Zoom* or *3Dzoom*. The *3Ddistance* command was used to change the camera-target distance or the amount of perspective.

In AutoCAD 2007, *Zoom* and *3Ddistance* perform the same action—change the size of the objects on the screen and change the camera-target distance. Also see NOTE in "*3Ddistance.*"

3D Orbit Commands

AutoCAD offers a set of interactive 3D viewing commands, *3Dorbit*, *3Dforbit*, and *3Dcorbit*. When one of the three 3D orbit commands is activated, all commands and related options in the group are available through a right-click shortcut menu. These commands and options are discussed in the following sections. The 3D orbit commands all issue the same simple command prompt because they are interactive rather than Command-line driven.

```
Command: 3dorbit
Press ESC or ENTER to exit, or right-click to display shortcut-menu.
Regenerating model.
Command:
```

With these commands you can manipulate the view of 3D models by dragging your cursor. You can control the view of the entire 3D model or of selected 3D objects. If you want to see only a few objects during the interactive manipulation, select those objects before you invoke the command, otherwise the entire 3D model is displayed during orbit.

3Dorbit

Pull-down Menu	Command (Type)	Alias (Type)	Short-cut	Screen (side) Menu	Tablet Menu
View *Orbit >* *Constrained Orbit*	*3Dorbit*	*3DO*	*VIEW* *...*	*3Dorbit*	*R,5*

3Dorbit controls viewing of 3D objects interactively. After activating the command, simply hold down the left mouse button and drag the pointer around the screen to change the view of the selected 3D objects. Drag up and down to view the objects from above or below, and drag left and right to view the objects from around the sides. All objects in the drawing are displayed in the interactive display unless individual objects are selected before invoking the command. As stated in the Command prompt, press Esc, Enter, or select *Exit* from the right-click shortcut menu to end the command.

Using the *3Dorbit* command forces the interactive movement to a "constrained orbit." That is, the Z axis is constrained from changing so it always remains in a vertical orientation. This limitation keeps all vertical edges of the 3D objects in a vertical position, just as real objects are oriented (Fig. 35-34). Use *3Dforbit* (free orbit) to change the orientation of the vertical axis.

NOTE: Remember that you are changing the <u>view</u>. When moving the cursor about the screen, it appears that the 3D objects rotate; however, in theory the <u>target</u> remains stationary while the observer, or <u>camera</u>, moves around the objects.

FIGURE 35-34

2007

3Dorbit Shortcut Menu Options

During the *3Dorbit*, *3Dforbit*, or *3Dcorbit* commands, you can use the shortcut menu to access other 3D viewing commands and options (Fig. 35-35). Using commands from the shortcut menu does not exit the *3Dorbit* session (unless, of course, you select *Exit*).

Most of the options from this menu are discussed immediately below and in other sections of this chapter. The *Animation Settings*, *Walk*, and *Fly* options are discussed in Chapter 41, Rendering and Animation.

FIGURE 35-35

Exit		
Current Mode: Constrained Orbit		
Other Navigation Modes ▶	✔ Constrained Orbit	1
	Free Orbit	2
✔ Enable Orbit Auto Target	Continuous Orbit	3
Animation Settings …	Adjust Distance	4
Zoom Window	Swivel	5
Zoom Extents	Walk	6
Zoom Previous	Fly	7
✔ Parallel	Zoom	8
Perspective	Pan	9
Reset View		
Preset Views ▶		
Named Views		
Visual Styles ▶		
Visual Aids ▶		

Enable Orbit Auto Target
This option forces the center of the orbit (target) to the center of the objects. The resulting effect is that the observer is moving around the objects—the objects are the center of rotation. This setting is recommended for most applications. When this option is not checked, the center of the viewport becomes the target (center of rotation), disregarding the location of the objects.

Zoom Window
Select *Zoom Window* to zoom into a smaller area by defining a window, similar to the normal *Zoom* command with the *Window* option. With the *3Dorbit Zoom Window*, you must click and drag to define two diagonal corners of the window rather than click at each corner. The display remains only during the *3Dorbit* command and reverts to the previous display when you exit *3Dorbit*.

Zoom Extents
Select *More, Zoom Extents* from the shortcut menu to perform a typical *Zoom Extents*, similar to the normal *Zoom* command with the *Extents* option. The view is centered and sized so all selected objects are displayed in the 3D view.

Zoom Previous
This option displays the last view during the current *3Dorbit* session. Only one previous view is saved.

Reset View
You can use *Reset View* from the shortcut menu to undo all changes to the view during the *3Dorbit* session and reset to the view before using *3Dorbit* and still remain in the *3Dorbit* session.

Visual Aids, Compass
Selecting the *Compass* toggle from the shortcut menu produces a sphere composed of three circles in 3D space. The three circles represent rotation about each of the three axes, X, Y, and Z.

Visual Aids, Grid
This option toggles the grid lines that appear on the XY plane of the current UCS (see Fig. 35-1). By default, the number of grid lines changes as you zoom in or out. The number of grid lines and the adaptive behavior of the grid is controlled by the *Grid* command or the related system variables. See *Grid* later in this chapter. The *Grid* command cannot be activated during a *3Dorbit* session.

Visual Aids, UCS Icon
Use this option to toggle the UCS icon on and off. The UCS icon can enhance your orientation of the view because it changes as the view changes, even during perspective projection. Alternately, use the *Ucsicon* command to control the icon visibility.

Other options and commands available from this menu are discussed on the following pages.

3Dforbit

Pull-down Menu	Command (Type)	Alias (Type)	Short-cut	Screen (side) Menu	Tablet Menu
View *Orbit >* *Free Orbit*	*3Dforbit*	...	...	...	...

3Dforbit, or "3D free orbit," allows you to obtain a wide variety of views. The orbit movement is not constrained in any way as it is with *3Dorbit* where the Z axis always remains vertical. By holding down the left button on your pointing device and dragging your cursor, you can control the view of 3D objects by several methods depending on where you place your cursor with respect to the "arc ball."

When you invoke *3Dforbit*, the current viewport displays the selected objects and the "arc ball" (Fig. 35-36). The arc ball is a circle with each quadrant denoted by a smaller circle. Moving the cursor over different areas of the arc ball changes the direction in which the view rotates. The center of the arc ball is the target.

There are four possible methods of rotating your view depending on where you place the cursor and the direction you move it. The cursor icon changes to indicate one of these four methods. The four positions are as follows.

FIGURE 35-36

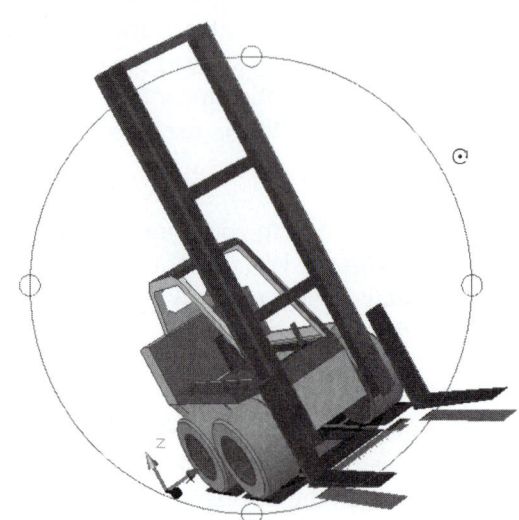

Inside the Arc Ball

 This icon is displayed when you move the cursor inside the arc ball. Imagine a sphere around the model. Click and drag the pointing device (hold down the left button) to rotate the sphere (and 3D model) around the target point (center of the sphere) in any direction. Note that the movement is not confined to the "constrained orbit" action (as with *3Dorbit*) and that the coordinate system Z axis (and therefore, vertical edges of objects) can be reoriented (see Fig. 35-36).

Outside the Arc Ball

 Move the cursor outside the arc ball, then click and drag to "roll" the display. This action rotates the view around the center of the arc ball. Imagine rotating the view about a Z axis protruding perpendicular from the screen.

Left and Right Quadrants

 This icon is displayed when you move the cursor over either of the small circles on the left or right quadrants of the arc ball, then click and drag to the left or right. This action allows you to rotate the view about a vertical axis that passes through the center of the arc ball or about an imaginary Y axis on the screen.

Top and Bottom Quadrants

 To rotate the view about a horizontal axis passing through the center of the arc ball, place the cursor over either of the small circles on the top or bottom quadrants of the arc ball, then click and drag up or down. This action represents rotation about an imaginary X screen axis.

Remember that the center of the arc ball is the target and remains stationary. The arc ball always appears in the center of your screen or current viewport. To center your objects in the arc ball, use the *Pan* (*3Dpan*) option or *Enable Orbit AutoTarget* option from the shortcut menu.

3Dcorbit

Pull-down Menu	Command (Type)	Alias (Type)	Short-cut	Screen (side) Menu	Tablet Menu
View *Orbit >* *Continuous Orbit*	*3Dcorbit*	...	...	...	...

The *3Dcorbit* command allows you to set the selected 3D objects into continuous motion. To set the 3D objects in motion, click in the drawing area and hold down the left button, drag the cursor in any direction, then release the button while the cursor is in motion. The objects continue to spin in the direction of your cursor motion. The speed of the cursor movement determines the speed of the spin. Change the direction and speed of the object movement by clicking and dragging the cursor again. You can stop the motion at any position and discontinue *3Dcorbit* by pressing Enter or the Esc key. Alternately, right-click and select *Exit* or other option.

PERSPECTIVE

Pull-down Menu	Command (Type)	Alias (Type)	Short-cut	Screen (side) Menu	Tablet Menu
Perspective	...	...	...	...	...

PERSPECTIVE is actually a system variable, not a command. The *PERSPECTIVE* variable can be changed by command line or by selecting the *Perspective Projection* and *Parallel Projection* buttons on the *3D Navigate control panel*. A setting of 1 turns on the perspective projection and a setting of 0 displays parallel projection.

Command: *perspective*
Enter new value for PERSPECTIVE <1>:

Parallel projection is used for 2D drawing and it
is automatically turned on when the *2D
Wireframe* visual style is set. However, parallel
projection displays 3D objects unrealistically
because parallel edges of an object remain parallel in the display. A good example of this theory
is shown in Figure 35-37 where the grid lines
remain parallel to each other, no matter how far
they extend away from the observer. Parallel
projection does not take into account the distance
the observer is from the object.

FIGURE 35-37

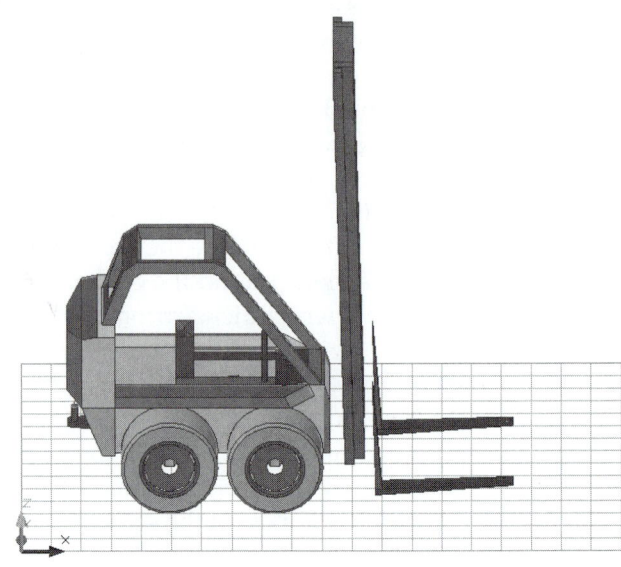

Perspective projection simulates the way we
actually see 3D objects because it takes into
account the distance the observer is from the
object. In AutoCAD, this distance is determined
by the distance from our eyes (camera) to the
objects (target). Therefore, since objects that are
farther away appear smaller, parallel edges on a
3D object appear to converge toward a point as
they increase in distance. Figure 35-38 shows the
same objects as the previous figure but in perspective projection. Perspective projection can be
displayed (*PERSPECTIVE* is set to 1) for any 3D
objects shown in any of the 3D visual styles.

NOTE: You can adjust the amount of perspective
using the *View Manager* (see "Adjusting the
Amount of Perspective").

FIGURE 35-38

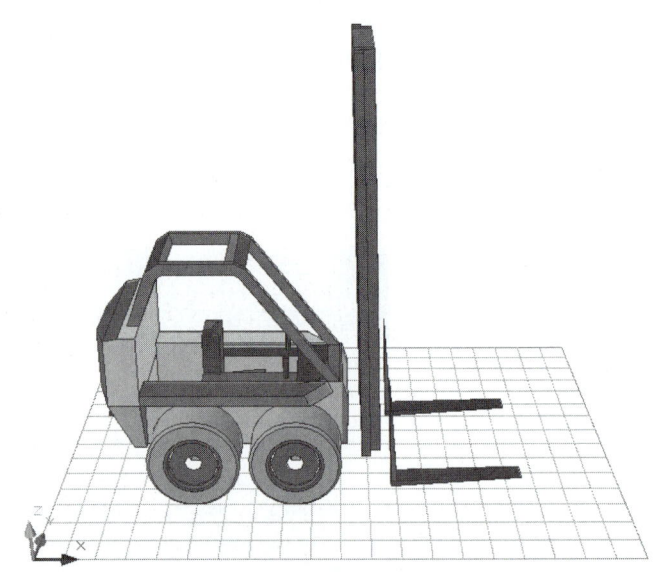

3Ddistance

Pull-down Menu	Command (Type)	Alias (Type)	Short-cut	Screen (side) Menu	Tablet Menu
View *Camera >* *Distance*	3Ddistance	...	...	...	...

The *3Ddistance* command changes the distance between the camera (observer) and the target (objects).
When perspective projection is on, the amount of perspective changes as a function of camera-target distance. When you invoke *3Ddistance*, the icon changes to a double-sided arrow shown in perspective.
Hold the left mouse button down and move the cursor upward to bring the objects closer (and to achieve
more perspective), or move the cursor down to move the objects farther away (and achieve less perspective).

In AutoCAD 2007, <u>the *Zoom* command serves the same function as *3Ddistance*</u>; therefore, you can use *Zoom* instead of *3Ddistance* to accomplish the same display. Normally, you think of using *Zoom* to change the size of objects on the screen; however, object size, camera-target distance, and amount of perspective are interrelated in AutoCAD 2007.

NOTE: Historically (in versions previous to AutoCAD 2007), neither *Zoom* nor *3Dzoom* changed the camera-target distance, but changed only the object size on screen. *3Ddistance* was the intended method for changing the camera-target distance. Therefore, you used *3Ddistance* to change the amount of perspective for the objects, then *3Dzoom* to adjust the size of the objects on the screen keeping the same perspective. The advantage of this older method was that you could adjust amount of perspective and object size <u>independently</u>. In AutoCAD 2007, you can adjust the amount of perspective independent of screen size by changing a camera's setting for *Lens length* or for *Field of View*. These settings can be changed in either the *3D Navigate control panel* or in the *View Manager* (see "Adjusting the Amount of Perspective.")

View Presets

The View Presets provide standard viewing directions for a 3D model, such as *Top*, *Front*, *Right*, and standard isometric viewing directions. The View Presets are not commands that can be entered at the Command line, but are selections you can make from any of several methods described here. The View Presets are available from the *View* toolbar (Fig. 35-39) and from the *3D Navigation control panel* drop-down list (Fig. 35-40). In addition, you can find the View Presets in the *3D Views* option from the *Views* pull-down menu (not shown) and in the *View Manager* (described in the next section).

Each of the 3D View options is described next. The FORKLIFT drawing is displayed in these figures in parallel projection for several View Presets. It is important to remember that for each view, imagine you are viewing the object from the indicated position in space. Imagine the observer changes position while the object remains stationary.

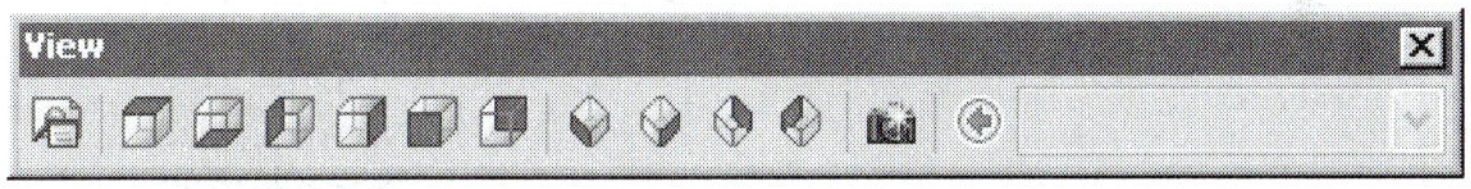

FIGURE 35-39

FIGURE 35-40

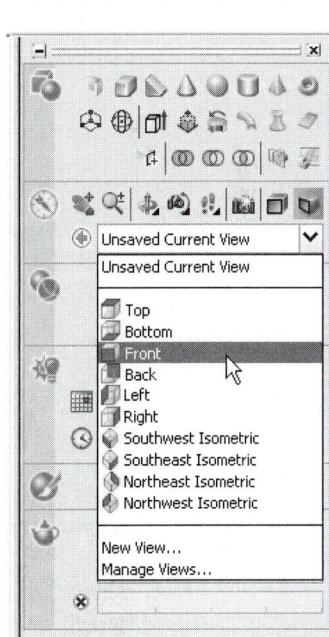

NOTE: View Presets display the object (or orients the observer) with respect to the World Coordinate System by default. For example, the *Top* option always shows the plan view of the WCS XY plane. Even if another coordinate system (UCS) is active, AutoCAD temporarily switches back to the WCS to attain the selected viewpoint.

NOTE: It is also important to note that if you use the *Top*, *Bottom*, *Front*, *Back*, *Right*, or *Left* option, and the *UCSORTHO* system variable is set to 1 (default setting), the UCS automatically changes to align with the view. Set *UCSORTHO* to 0 if you do not want to create a new UCS for each of these views (for more information on this setting, see Chapter 36, User Coordinate Systems).

Top

The object is viewed from the top (Fig. 35-41). Selecting this option shows the XY plane from above (the default orientation when you begin a *New* drawing). Notice the position of the WCS icon. This orientation should be used periodically during construction of a 3D model to check for proper alignment of parts.

Bottom

This option displays the object as if you are looking up at it from the bottom. The Coordinate System icon appears backward when viewed from the bottom.

Left

This option looks at the object from the left side.

Right

Imagine looking at the object from the right side (Fig. 35-42). This view is used often as one of the primary views for mechanical part drawing.

Front

Selecting this option displays the object from the front view. This common view is used often during the construction process. The front view is usually the "profile" view for mechanical parts (Fig. 35-43).

Back

Back displays the object as if the observer is behind the object. Remember that the object does not rotate; the observer moves.

SW Isometric

A *SW Isometric* is a view of the object as if the viewer were looking from a southwest position. Isometric views are used more than the "orthographic" views (front, top, right, etc.) for constructing a 3D model. Isometric viewpoints give more information about the model because you can see three dimensions instead of only two dimensions (as in the previously discussed views).

FIGURE 35-41

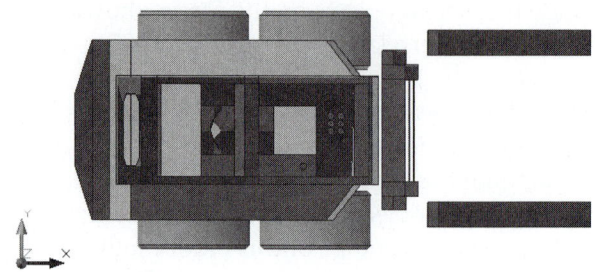

FIGURE 35-42

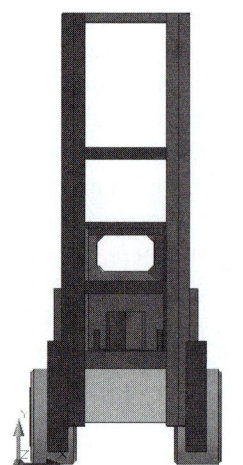

FIGURE 35-43

SE Isometric

The *Southeast Isometric* view is generally the first choice for displaying 3D geometry. This view shows the object as if the observer was located southeast in relation to the 3D objects. If the object is constructed with its base on the XY plane so that X is length, Y is depth, and Z represents height, this orientation shows the front, top, and right sides of the object. Try to use this viewpoint as your principal mode of viewing during construction. Note the orientation of the WCS icon (Fig. 35-44).

FIGURE 35-44

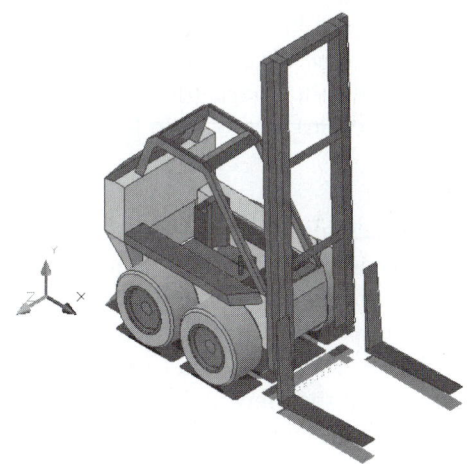

NE Isometric

The *Northeast Isometric* view shows the right side, top, and back (if the object is oriented in the manner described earlier).

NW Isometric

This viewpoint allows the observer to look at the left side, top, and back of the 3D object (Fig. 35-45).

FIGURE 35-45

View

Pull-down Menu	Command (Type)	Alias (Type)	Short-cut	Screen (side) Menu	Tablet Menu
View Named Views	View	V	...	VIEW 1 Ddview	...

The *View* command produces the *View Manager* which provides several functions useful for viewing 3D models. You can assign a name, save, and restore viewed areas of a drawing. You can access information and controls to the current view, any camera, and any saved views. You can also use this manager to adjust the amount of perspective displayed for selected views (see "Adjusting the Amount of Perspective").

For basic information on the *Set Current, New, Update Layers, Edit Boundaries,* and *Delete* functions of the *View Manger* see "*View*" in Chapter 10. The *View Manager* also provides access to the *View Presets* (described in the previous section) as shown in Figure 35-46.

FIGURE 35-46

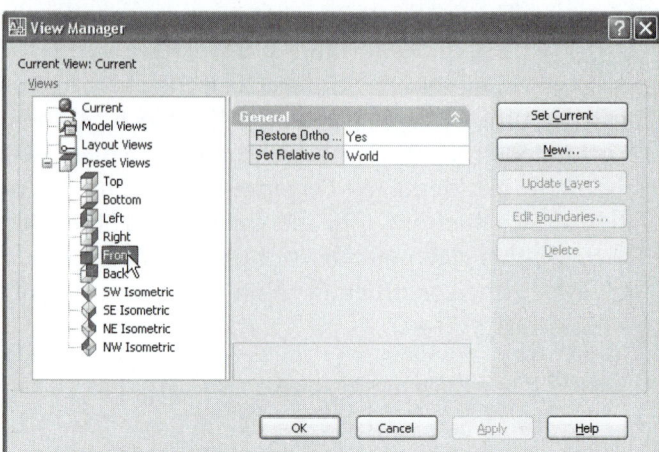

New

When you create a *New* view of a 3D model, AutoCAD actually <u>creates a camera</u> and inserts a camera icon into the drawing. The view settings and camera are saved under the view name. Therefore, you can access information about the view in the *View Manager* or by selecting the view's camera icon and using the *Properties* palette.

Preset Views

You can change your display to any of the *Preset Views* which are identical to those available from the *View* toolbar, the *3D Navigation control panel* drop-down list, or the *Views* pull-down menu, as explained earlier (see "View Presets"). To generate a 3D view using the *View Manager*, several steps are required. First, select the desired view preset from the list. Next, make the desired view current by selecting the *Set Current* button. Alternately, you can double-click on the desired view to make it current. Finally, select the *OK* button to produce the view.

It is strongly suggested that you ensure the *Set Relative to:* option is set to *World* (see Fig. 35-46). This setting causes the 3D view that is generated to be relative to the World Coordinate System. Using another option can make the resulting view difficult to predict. (See Chapter 36 for more information on this option and the *UCSBASE* system variable.)

Also keep in mind that if the *Restore Orthographic UCS* option is set to *Yes* when you use the *Top, Bottom, Front, Back, Right,* or *Left* option, the UCS automatically changes to align with the view (so the new XY plane is parallel with the screen). Change this setting to *No* if you do not want to create a new UCS with each orthographic view you select. (See Chapter 36 for information concerning the *Restore Orthographic UCS* option and the *UCSORTHO* system variable.)

Current and Model Views

The *View Manager* not only allows you to set views current, but provides access to information and controls for the current view and for any saved views. Note that both saved views and any cameras that have been created are listed under *Model Views*. You can control such settings as *Name, Category,* and *Visual Style,* as shown in Figure 35-47.

FIGURE 35-47

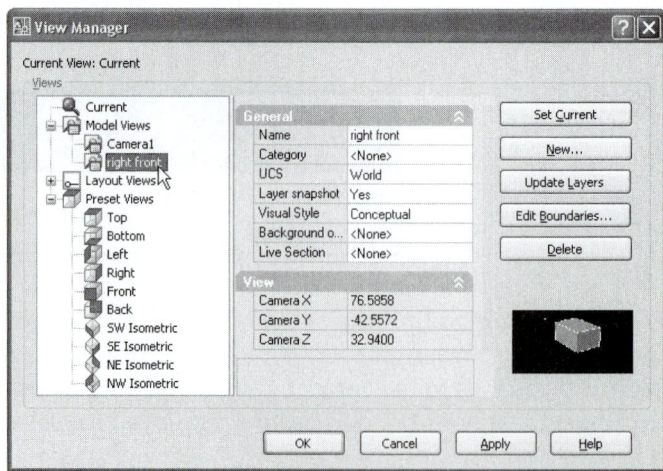

Note that the settings for *Camera* and *Target* are read-only in the *View Manager*. These values cannot be changed here since the *Camera* values are automatically generated using *3Dorbit* or similar navigation command, and the *Target* values can only be changed using the *3Dswivel* or *Camera* commands (and camera grips and *Properties* palette).

You can also change the settings for the *Lens length* and the *Field of View* for the *Current* view or any saved model view or camera. These settings affect the amount of perspective for the view (see "Adjusting the Amount of Perspective," next).

In addition, you can set up clipping planes for the *Current* view or any saved model view or camera. Clipping planes are imaginary planes perpendicular to the line of sight (parallel to the screen) for a particular view that mask the display of objects. You can set a front or back clipping plane. You can also accomplish this using the *3Dclip* command (see "*3Dclip*").

Adjusting the Amount of Perspective

Regardless of the size objects appear on the screen or plot, it is helpful to make some objects appear very large (such as a building) by increasing the amount of perspective, or to make some objects appear very small (such as a nut or bolt) by reducing the amount of perspective. To accomplish this, you must be able to adjust the amount of perspective that appears in a view independent of the size of the objects on the screen. However, in AutoCAD 2007, the *Zoom* and the *3Ddistance* commands change both the screen size and the appearance of perspective (camera-target distance) together. (See previous discussion in the "*3Ddistance*" section.)

In AutoCAD 2007 you can adjust the amount of perspective independent of the object's size on screen by changing a camera's setting for *Lens length* or the setting for *Field of View*. You can use either the 3D Navigate control panel or the *View Manager* to change these settings for the *Current* view (if no named views or cameras have been saved) and for named views and cameras. If you created a *Camera* or saved a *Model View*, you can also select the camera icon and use the *Properties* palette to change the settings.

Using the *3D Navigate Control Panel*

 In the expanded section of the *3D Navigate control panel*, just below the separator line (Fig. 35-48), the slider and two edit boxes are used to change the settings for *Lens length* or *Field of view*. You can use either the slider or one of the two edit boxes to change the values. These two settings are codependent, so changing one automatically changes the other. This action adjusts the amount of perspective for a given view without affecting the camera-target distance.

FIGURE 35-48

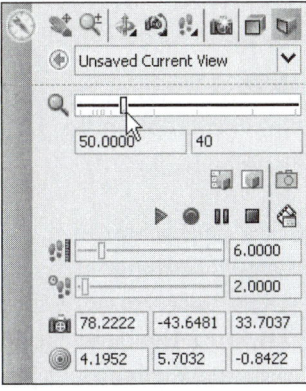

The default settings are 50 (mm) for *Lens Length* and 40 (degrees) for *Field of View* (the horizontal field of view). Remember that changing one value automatically changes the other. To increase the amount of perspective (so the object appears closer), decrease the *Lens Length* or increase the *Field of View*. To decrease the amount of perspective (so the object appears farther), increase the *Lens Length* or decrease the *Field of View*. Note that *Perspective* must be set to *On* to immediately view these changes.

Using the *View Manager*

To adjust the amount of perspective for a given view without affecting the camera-target distance, open the *View Manager* by any method. From the *View* list on the left side of the *View Manager*, select the *Current* view or any saved *Model Views* (saved cameras are listed under *Model Views*). In the *View* category, change the setting for *Lens length* or *Field of view*.

3Dclip

Pull-down Menu	Command (Type)	Alias (Type)	Short-cut	Screen (side) Menu	Tablet Menu
...	*3Dclip*	...	...	...	...

Clipping planes are invisible planes that can pass through the view. Anything in front of the front clipping plane or in back of the back clipping plane becomes invisible when the planes are toggled on. The clipping planes are always parallel with the screen and perpendicular to the viewing direction (the screen Z axis).

Using *3Dclip* produces the *Adjust Clip Planes* window (Fig. 35-49). The window displays the current view from above so the clip planes can be seen on edge (appear as horizontal lines). The white (or black) line on the bottom of the window represents the front clipping plane and the green line on the top (if visible) represents the back clipping plane. As you move the clip planes through the 3D objects, the results are displayed in the drawing area. To see the results of adjusting clipping planes, the clipping planes must be toggled on. The controls for the *Adjust Clip Planes* window are available if you select the icon buttons or right-click in the window to produce the shortcut menu.

FIGURE 35-49

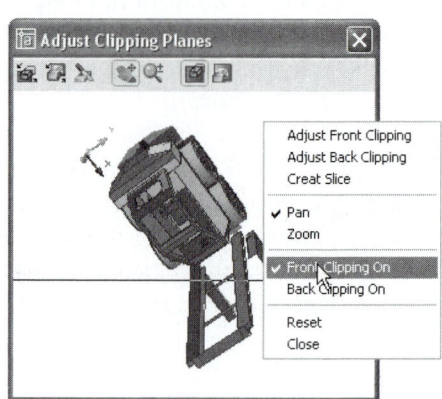

Front Clipping On

You must toggle the front clipping plane visibility on (depress the icon button or check the shortcut menu) if you want to see the results of adjusting the front clipping plane in the drawing area. Toggle this option on before using *Adjust Front Clipping*. Once the desired clipping is set, you can toggle this on or off for the display in the drawing area.

Back Clipping On

Toggle this option on before using *Adjust Back Clipping*. Like *Front Clipping On*, you must toggle the back clipping plane on to see the results of adjusting the back clipping plane in the drawing area. Once the desired clipping is set, you can toggle this on or off for the display in the drawing area.

FIGURE 35-50

Adjust Front Clipping

Move the horizontal black line near the bottom of the window upward to adjust the front clipping plane into the 3D objects. Any objects or parts of objects in front of the front clipping plane disappear. Notice how most of the fork and lift sections of the forklift are removed by the front clipping plane in Figure 35-50.

Adjust Back Clipping

Move the horizontal green line near the top of the window to adjust the back clipping plane. Generally, move the plane downward to pass through the 3D objects. Any objects or parts of objects in back of the back clipping plane disappear as shown in Figure 35-51.

FIGURE 35-51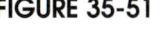

Slice

Make this selection if you want to move both front and back clipping planes together. First, adjust either clipping plane to determine the thickness of the slice, then toggle on *Slice*. You can achieve a view to display only a thin section of your 3D model.

When the *Adjust Clip Planes* window is dismissed, the clipping planes remain active (control the display) <u>during and after</u> the *3Dorbit* session. At any time, control the visibility of clipping planes using the *3Dorbit* shortcut menu by selecting *More*, then the *Front Clipping On* toggle or *Back Clipping On* toggle.

Camera

Pull-down Menu	Command (Type)	Alias (Type)	Short-cut	Screen (side) Menu	Tablet Menu
View Create Camera	Camera	CAM	...	...	...

For any 3D <u>view</u> in AutoCAD, the display is determined by the location of the observer, known as the camera, and the target, which is generally in the center of the 3D objects. For most views, such as those achieved through the use of *3Dorbit*, you determine the position of the camera by the particular observation point you generate. In this case, an <u>imaginary</u> camera is positioned at the eye of the observer. Also, AutoCAD automatically calculates the position of the target at the centroid (geometrical center) of the 3D objects. For a view created using *3Dorbit* or similar navigation command, AutoCAD creates only an <u>imaginary</u> camera.

You can instead create a physical camera in AutoCAD, including a camera icon in the 3D model, using the *Camera* command. You can alternately create a camera by saving a named *View* (see "*View*" discussed earlier). Using the *Camera* command, you can create a camera and position a target. In addition to the camera's location in 3D space, you can specify a camera name, a target location, the lens length, field of view, and clipping planes. Since a camera created with the *Camera* command appears in the view list in the *3D Navigation control panel* and in the *View Manager*, you can restore the camera view at any time. You can *Erase* a camera by selecting its icon.

2007

Camera allows you to set different <u>camera and target</u> locations by specifying 3D coordinate values or by picking points in 3D space. A camera icon appears when you select the camera location and a box representing the field of view appears when you select the target (Fig. 35-52). If you PICK points to define locations in 3D space, <u>use Osnap</u> modes. Alternately, use *List* to locate a desired geometry location before entering specific coordinate values.

FIGURE 35-52

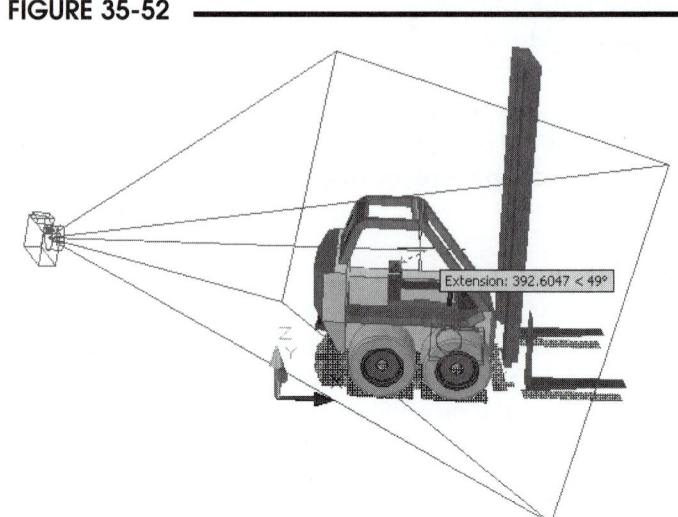

Command: **camera**
Current camera settings: Height=20.0000 Lens Length=50.0000 mm
Specify camera location: **PICK** or **coordinates**
Specify target location: **PICK** or **coordinates**
Enter an option [?/Name/LOcation/Height/Target/LEns/Clipping/View/eXit]<eXit>:

The options are briefly described below.

Name
Assign a name for the camera. Otherwise AutoCAD assigns a name such as "Camera1." The camera name appears in any list of views.

LOcation
Use this option to change the location of the camera. Enter coordinate values or PICK a location using *Osnaps*.

Height
This option keeps the camera's current X,Y location, but changes the *Camera Z* value for the current camera. This also resets the default *Height* value when creating subsequent cameras.

Target
Use this option to change the location of the target, which determines the direction the camera points. Enter coordinate values or PICK a location using *Osnaps*.

LEns
You can specify a particular *Lens length* in millimeters using this option. A camera's lens length affects the distance perception of the view, similar to changing the strength of a telescopic lens. Therefore, changing the *Lens length* effectively changes the amount of perspective for the view (see "Adjusting the Amount of Perspective"). Changing the *Lens length* value also automatically changes the camera's *Field of View* setting.

Clipping
You can establish *Front* and *Back* clipping planes for the camera. You are prompted for an "offset from target plane" in units. Clipping planes are invisible planes that can pass through the view. Any parts of the 3D model in front of the front clipping plane or in back of the back clipping plane becomes invisible when the planes are toggled on. (See "*3Dclip*" for more information.)

View

Answering *Yes* to this prompt automatically displays a view as seen by the camera and ends the *Camera* command.

A good application for the *Camera* command is an architectural 3D model. In this case, the model may represent a large object such as a house or building—an object that you might want to see from inside. In contrast, most mechanical applications are relatively small and only viewed from outside. For an architectural model, it would be appropriate to set the camera location inside a room and set different target locations at each end of the room.

Editing Camera Settings

Once a *Camera* (or named *View*) has been established, you can edit the settings such as lens length, field of view, camera position and target position. Two methods can be used to edit the settings: the camera's grips and the *Camera Preview*, or the camera's *Properties* palette.

Display Cameras

Before you can access the camera's grips or *Properties* palette, you must turn on the camera's icon so it appears in the drawing, if it is not already visible. Use the *Display Cameras* button in the expanded section of the *3D Navigate control panel* to turn on and off all the cameras in a drawing (see Fig. 35-48, right center). Alternately, select the *View* pull-down menu, then *Display, Camera*.

Camera Grips and *Camera Preview*

If you click on a camera icon in the 3D model, several grips appear and the *Camera Preview* window appears. You can select one of the following grips and drag it to establish a new location or setting: *Target location* grip, *Camera location* grip, *Camera & target location* grip and one of the *Lens length/FOV* grips. For example, Figure 35-53 displays dragging the *Lens length/FOV* grip. In addition, if you select an axis or plane of the Dynamic UCS that appears, you could change the *Target distance* or one of the previously mentioned grips along the selected axis or plane. (See Chapter 36, User Coordinate Systems, for more information on the Dynamic UCS.)

FIGURE 35-53

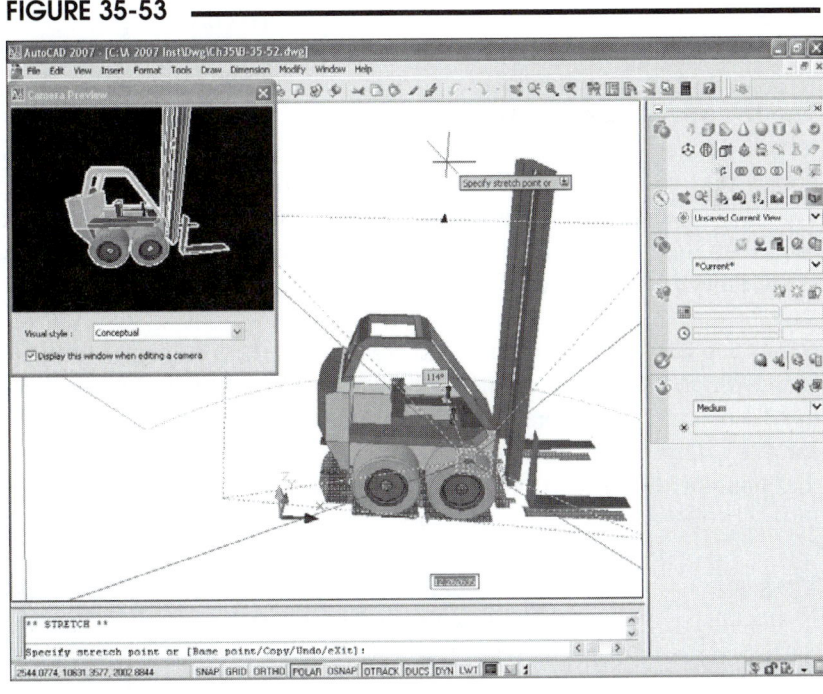

When you select a camera icon, the *Camera Preview* window also appears displaying the camera's current view (Fig. 35-53, upper-left corner). The view is dynamic so any changes to the view affected by dragging the grips are immediately displayed in the window. Note that you can specify a *Visual style* for the *Camera Preview*.

Camera *Properties* palette

As an alternative, you can edit the values for the camera settings using the *Properties* palette. Activate the *Properties* palette, then select the camera icon in the 3D model. Note that all the fields can be edited, including camera and target position edit boxes (Fig. 35-54).

FIGURE 35-54

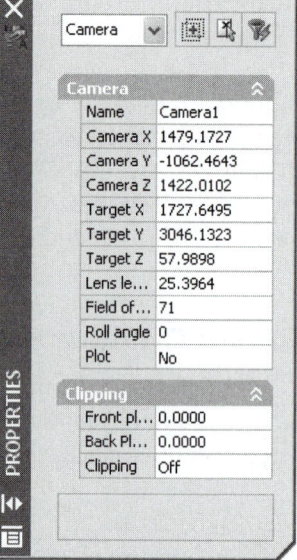

3Dswivel

Pull-down Menu	Command (Type)	Alias (Type)	Short-cut	Screen (side) Menu	Tablet Menu
View *Camera >* *Swivel*	*3Dswivel*	...	...	...	

The 3D view is determined by the location of the observer, known in AutoCAD as the camera, and the target, which is generally in the center of the 3D objects. The *3Dswivel* command changes the <u>target</u> for the current view. Using *3Dswivel* simulates the effect of turning a camera on a tripod to change the viewing area. For example, if you moved the cursor (arched arrow) to tilt the camera upward, the objects in the viewfinder would move down and vice versa. Likewise, if you turned the camera to the right, objects in the viewfinder would move to the left.

If you use the *Camera* command to create a camera, you specify the location of the *Target* in the process. However, if a camera is created by saving a *View*, the target is automatically calculated as the center of the 3D objects. In either case, you can later change a camera's target setting using *3Dswivel*, dragging the camera's *Target* grip, or using the camera's *Properties* palette. You cannot change the target for a view that has not been saved as a named view.

Vpoint

Pull-down Menu	Command (Type)	Alias (Type)	Short-cut	Screen (side) Menu	Tablet Menu
View *3D Views >* *Viewpoint*	*-Vpoint*	*-VP*	...	VIEW 1 *Vpoint*	*N,4*

Vpoint, like *3Dorbit*, allows you to achieve a specific view (or viewpoint) of a 3D object; however, *Vpoint* is an older command. Nevertheless, *Vpoint* can be used to specify an exact viewing angle or direction, whereas *3Dorbit* is strictly interactive.

```
Command: vpoint
Current view direction:  VIEWDIR=0.0000,0.0000,1.0000
Specify a view point or [Rotate] <display compass and tripod>:
```

View Direction

The default option of the *Vpoint* command is to enter X,Y,Z coordinate values. The coordinate values indicate the position in 3D space at which the observer is located. The coordinate values do not specify an absolute position but rather specify a <u>vector</u> passing through the coordinate position and the origin. In other words, the observer is located at <u>any point along the vector looking toward the origin</u>.

Even though the *Vpoint* command generates either a perspective or a parallel projection (whichever setting is current), *Vpoint* does not compute a true a distance. Therefore, the magnitude of the coordinate <u>values</u> is of no importance, only the relationship among the values. Values of 1,-1,1 would generate the same display as 2,-2,2.

Rotate

The name of this option is somewhat misleading because the object is not rotated. The *Rotate* option prompts for two angles in the WCS (by default), which specify a vector indicating the direction of viewing. The two angles are (1) the angle in the XY plane and (2) the angle from the XY plane. The observer is positioned along the vector looking toward the origin.

```
Command: vpoint
Current view direction:  VIEWDIR=0.0000,0.0000,1.0000
Specify a view point or [Rotate] <display compass and tripod>: r
Enter angle in XY plane from X axis <270>: 315
Enter angle from XY plane <90>: 35.3
Command:
```

The first angle is the angle <u>in</u> the XY plane at which the observer is positioned looking toward the origin. This angle is just like specifying an angle in 2D. The second angle is the angle that the observer is positioned <u>up or down from</u> the XY plane. The two angles are given with respect to the WCS.

Angles of **315** and **35** specified in response to the *Rotate* option display an almost <u>perfect isometric</u> viewing angle. An isometric drawing often displays some of the top, front, and right sides of the object. For some regularly proportioned objects, perfect isometric viewing angles can cause visualization difficulties, while a slightly different angle can display the object more clearly.

Display compass and tripod

Similar to *3Dorbit*, the *Display compass and tripod* method of *Vpoint* is a (somewhat) interactive method for adjusting your view; however, this option does not actually display the drawing geometry. Therefore, it is generally preferred to use *3Dorbit* instead of this option of *Vpoint*.

This option displays a three-pole axis system which rotates dynamically on the screen as the cursor is moved within a "globe." When you PICK the desired viewing position, the geometry is displayed in that orientation. Because the viewing direction is specified by PICKing a point without actually seeing the 3D objects, it is difficult to specify the desired viewpoint.

When the *Tripod* method is invoked, the current drawing temporarily disappears and a three-pole axis system appears at the center of the screen indicating the orientation of the X, Y, and Z axes for the new *Vpoint*. The axes are dynamically rotated by moving the cursor (a small cross) in a small "globe" at the upper-right of the screen. The center of the "globe" represents the North Pole, so moving the cursor to that location generates a plan, or top, view. The small circle of the globe represents the Equator, so moving the cursor to any location on the Equator generates an elevation view (front, side, back, etc.). The outside circle represents the South Pole, so locating the cursor there shows a bottom view. When you PICK, the axes disappear and the current drawing is displayed from the new viewpoint.

Ddvpoint

	Pull-down Menu	Command (Type)	Alias (Type)	Short-cut	Screen (side) Menu	Tablet Menu
	View *3D Views >* *Viewpoint Presets...*	*Ddvpoint*	*VP*	...	*VIEW 1* *Ddvpoint*	*N,5*

The *Ddvpoint* command is an older command used mostly before the introduction of *3Dorbit*. *Ddvpoint* produces the *Viewpoint Presets* dialog box (Fig. 35-55). This tool is an interface for attaining viewpoints that could otherwise be attained with the *3Dorbit* or the *Vpoint* command.

Ddvpoint serves the same function as the *Rotate* option of *Vpoint*. You can specify angles *From: X Axis* and *XY Plane*. Angular values can be entered in the edit boxes, or you can PICK anywhere in the image tiles to specify the angles. PICKing in the enclosed boxes results in a regular angle (e.g., 45, 90, or 10, 30) while PICKing near the sundial-like "hands" results in irregular angles (Fig. 35-55). The *Set to Plan View* tile produces a plan view.

The *Relative to UCS* radio button calculates the viewing angles with respect to the current UCS rather than the WCS. Normally, the viewing angles should be absolute to WCS, but certain situations may require this alternative. Viewing angles relative to the current UCS can produce some surprising viewpoints if you are not completely secure with the model and observer orientation in 3D space.

FIGURE 35-55

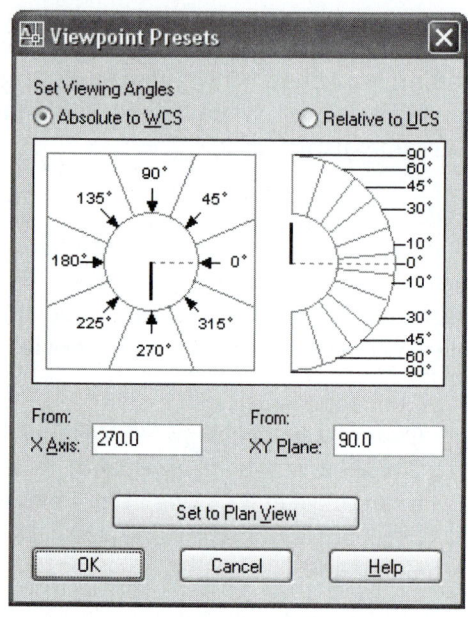

NOTE: When you use this tool, ensure that the *Absolute to WCS* radio button is checked unless you are sure the specified angles should be applied relative to the current UCS.

Plan

	Pull-down Menu	Command (Type)	Alias (Type)	Short-cut	Screen (side) Menu	Tablet Menu
	View *3D Views >* *Plan View >*	*Plan*	...	...	*VIEW 1* *Plan*	*N,3*

This command is useful for quickly displaying a plan view of any UCS:

```
Command: plan
Enter an option [Current ucs/Ucs/World] <Current>: (option) or Enter
Regenerating model.
Command:
```

Responding by pressing Enter causes AutoCAD to display the plan (top) view of the current UCS. Typing *W* causes the display to show the plan view of the World Coordinate System. The *World* option does <u>not</u> cause the WCS to become the active coordinate system but only displays its plan view. Invoking the *UCS* option displays the following prompt:

Enter name of UCS or [?]:

The *?* displays a list of existing named UCSs. Entering the name of an existing UCS displays a plan view of that UCS.

Grid

Pull-down Menu	Command (Type)	Alias (Type)	Short-cut	Screen (side) Menu	Tablet Menu
Tools *Draft Settings...* *Snap and Grid*	*Grid*	...	*F7 or Ctrl+G*	*Tools 2, Grid*	*W,10*

The *Grid* command has several options applicable to control the appearance of the grid that appears in any 3D visual style. Grid lines rather than grid dots are displayed when a *Visual Style* (*Vscurrent* command) is set to any option except *2D Wireframe* (see "Visual Styles"). The values that control the grid are stored in system variables that are saved with the drawing file. The prompt is as follows.

Command: **grid**
Specify grid spacing(X) or [ON/OFF/Snap/Major/aDaptive/Limits/Follow/Aspect] <0.5000>:

The following options affect the appearance of both the grid dots (*2D Wireframe*) and the grid lines (any 3D visual style); however, these options are especially applicable for 3D visual styles.

Major

The 3D grid contains major and minor grid lines. This value (5 is the default) specifies the frequency of major grid lines compared to minor grid lines. This setting is also controlled by the *GRIDMAJOR* system variable.

aDaptive

Use this option to control the density of grid lines when zoomed in or out.

Turn adaptive behavior on [Yes/No] <Yes>:

When *aDaptive* is turned on, the density of grid lines or dots is limited when you zoom out, which prevents the grid pattern from becoming too dense. This setting is also controlled by the *GRIDDISPLAY* system variable.

Allow subdivision below grid spacing [Yes/No] <No>:

If this subdivision option is turned on, additional, more closely spaced grid lines or dots are generated when you zoom in. The frequency of these grid lines is determined by the frequency of the major grid lines.

Limits

This option determines if the grid is displayed beyond the area specified by the *Limits* command.

Display grid beyond Limits [Yes/No] <Yes>:

Since *Limits* are generally not used for 3D drawings, the default option for the ACAD3D.DWT is *Yes*.

2007

Follow

This option controls if the grid plane changes to follow the XY plane of the Dynamic UCS. The default setting is *No*, in which case the grid plane remains with the current UCS XY plane. This setting is also controlled by the *GRIDDISPLAY* system variable.

Vports

Pull-down Menu	Command (Type)	Alias (Type)	Short-cut	Screen (side) Menu	Tablet Menu
View *Viewports >*	*Vports*	...	...	*VIEW 1* *Vports*	*M,3 and* *M,4*

The *Vports* command creates viewports. Either model space viewports or paper space (layout) viewports are created, depending on which space is current when you activate *Vports*. See Chapters 10 and 12 for basic information about *Vports* and using viewports for 2D drawings.

This section discusses the use of viewports to enhance the construction of 3D geometry. Options of *Vports* are available with AutoCAD specifically for this application.

When viewports are created in the *Model* tab (model space), the screen is divided into sections like tiles, unlike paper space viewports. Tiled viewports (viewports in model space) <u>affect only the screen display</u>. The viewport configuration <u>cannot be plotted</u>. If the *Plot* command is used, only the current viewport is plotted.

To use viewports for construction of 3D geometry, use the *Vports* command to produce the *Viewports* dialog box (Fig. 35-56). Make sure you invoke *Vports* in the *Model* tab (model space). To generate a typical ortho-graphic view arrangement in your drawing, first ensure that the current display is set to *Parallel projec-tion*. Next, invoke the *Viewports* dialog box and select three or four viewports from the list in the *Standard viewports* list, then select the *3D* option in the *Setup:* drop-down list (see Fig. 35-56, bottom center). This action produces a typical top, front, southeast isometric, and side view, depending on whether you selected three or four viewports.

Assuming the options shown in Figure 35-56 were used with the FORKLIFT drawing, the configuration shown in Figure 35-57 would be generated auto-matically. Note, however, that the objects are not sized proportionally in the viewports. Since a *Zoom Extents* is automatically performed, each view is sized in relation to its viewport, not in relation to the other views.

FIGURE 35-56

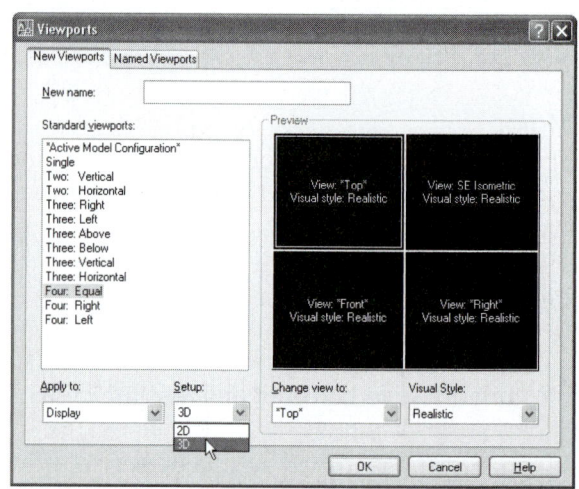

FIGURE 35-57

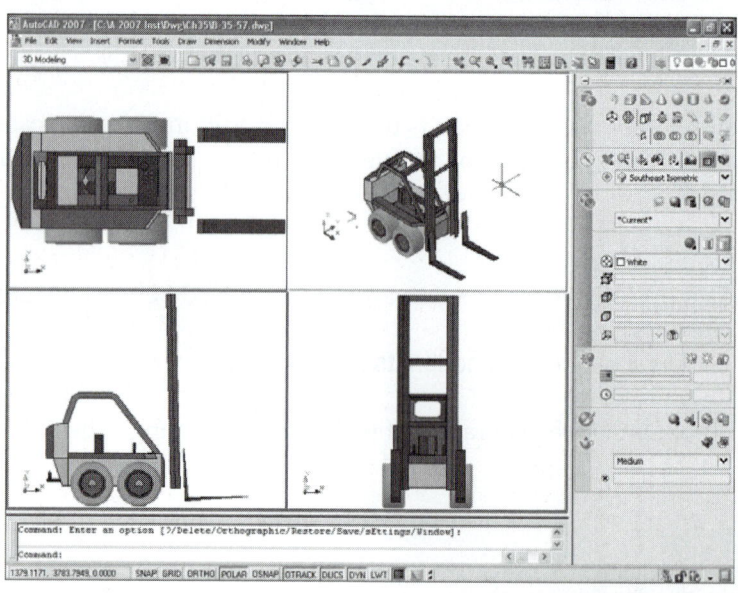

Next, if you desire all views to display the object proportionally, use *Zoom* with a constant scale factor (such as 1, not 1x) in each viewport to generate a display similar to that shown in Figure 35-58. You can then change other settings as desired for each viewport, such as *Visual style*, *Perspective*, etc.

NOTE: If *Perspective Projection* is turned on before using the *3D* option in the *Setup:* drop-down list, the objects are sized equally in the respective viewports.

NOTE: The setting of *UCSORTHO* has an effect on the resulting UCSs when the *3D Setup* option of the *Viewport* dialog box is used. If *UCSORTHO* is set to 0, no new UCSs are created when *Vports* is used. However, if *UCSORTHO* is set to 1, a new UCS is generated for <u>each new orthographic view created</u> so that the XY plane of each new UCS is parallel with the view. In other words, each orthographic view (not the isometric) now shows a UCS with the XY plane parallel to the screen. See Chapter 36 for more information on the *UCSORTHO* system variable.

FIGURE 35-58

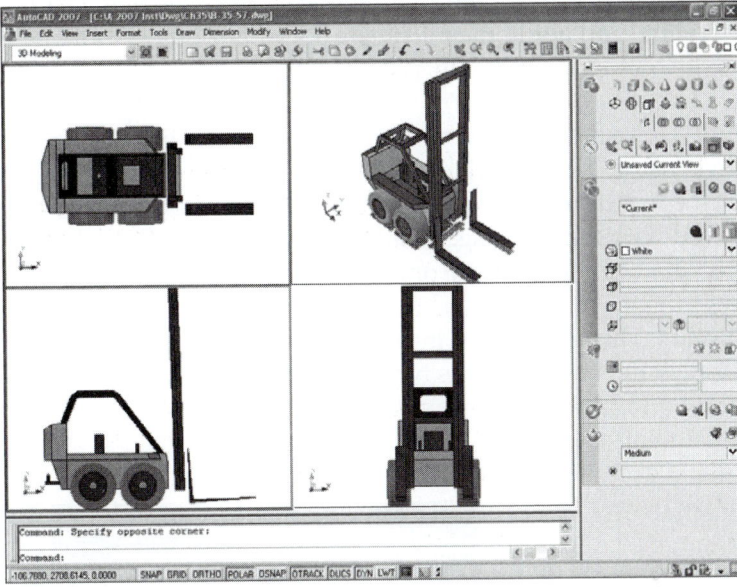

VISUAL STYLES

A "visual style" is a group of settings that specify the display of 3D objects. The display, or appearance, of 3D objects includes such characteristics as wireframe, hidden, or shaded displays, treatment of object edges, and application of shadows. Visual styles that display shaded surfaces on an object or that generate shadows generally increase the realistic appearance of 3D objects and can greatly enhance your visibility and understanding of the model, especially for complex geometry. Five preset "default" visual styles are available and can be applied to surface and solid models, whereas only wireframe displays are appropriate for wireframe models since those objects do not contain surfaces. You can specify a visual style for each viewport. When a visual style is applied, the model <u>retains</u> its appearance until you change it.

Applying visual styles replaces commands used in previous versions of AutoCAD such as *Shademode*, *Hide*, and *Hlsettings*. Instead of using many individual commands and typing system variables, you apply a visual style to the model and change the properties of the current visual style using the *Visual Style control panel* or the *Vscurrent* command. The *Visual Style control panel* is located near the center of the *Dashboard* (Fig. 35-59, shown in expanded mode). As soon as you apply a visual style or change its settings, you can see the effect in the viewport. Using the *Visual Style control panel* allows you to select from five default visual styles and modify the current style to your needs. Any changes are applied to a visual style called "*Current*."

FIGURE 35-59

2007

Changes you make to the *Current* style are saved with the drawing only if *Current* is the applied style when the drawing is saved. If you want to create new "named" visual styles, save them with the drawing, and apply them to any viewport or other drawing, use the *Visual Styles Manager* (see "*Visual Styles Manager*" later in this chapter).

NOTE: Visual styles are saved in the drawing file. Whether you make changes to the *Current* visual style using the *Visual Styles control panel* and save them with the drawing or whether you make changes to the five default styles or any new "named" styles using the *Visual Styles Manager*, visual styles are saved with the drawing. Visual styles can be applied across multiple drawings only if named styles are exported to a tool palette or are copied and pasted using the *Visual Styles Manager*.

NOTE: The illustrations in this section may appear differently than those shown in your computer display depending on the type and configuration of your display driver. See "Adaptive Degradation and Performance Tuning" near the end of this chapter.

Vscurrent

Pull-down Menu	Command (Type)	Alias (Type)	Short-cut	Screen (side) Menu	Tablet Menu
View *Visual Styles >*	*Vscurrent*	*VS*	...	...	...

The *Vscurrent* command is the Command-line equivalent to using the same options in the *Visual Styles control panel*. Note that the Command line prompt provides the same options as using the control panel drop-down list.

Command: **Vscurrent**
Enter an option
[2dwireframe/3dwireframe/3dHidden/Realistic/Conceptual/Other/cUrrent] <Current>:

The options are described in the following section "Default Visual Styles."

Default Visual Styles

The drop-down list in the *Visual Styles control panel* provides access to the five default visual styles: *2D wireframe*, *3D Wireframe*, *3D Hidden*, *Conceptual*, and *Realistic*. Although each of these styles can be modified to your liking and other new styles created, a description and illustration for each of these styles is given next.

FIGURE 35-60

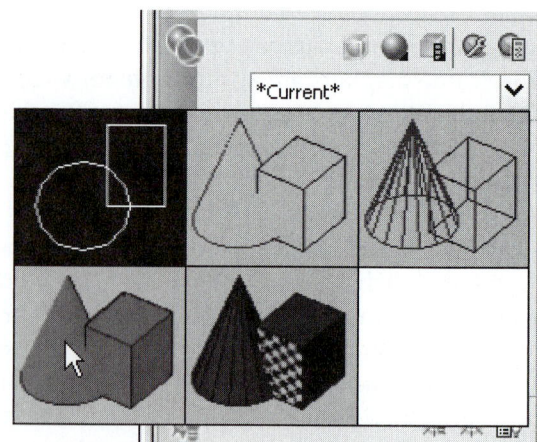

2D Wireframe

 This option displays only lines and curves. The 3D geometry is defined only by the edges representing the surface boundaries. The model is displayed in parallel projection—perspective projection mode is disabled. This display uses the black model space background typically used for 2D drawing. The UCS icon is displayed as a "wireframe" icon. Also, <u>linetypes and lineweights are visible</u>. Any raster and OLE objects are also visible in the display. Generally used for 2D geometry, this option can be used for 3D geometry when lineweights and linetypes are important (Fig. 35-61).

FIGURE 35-61

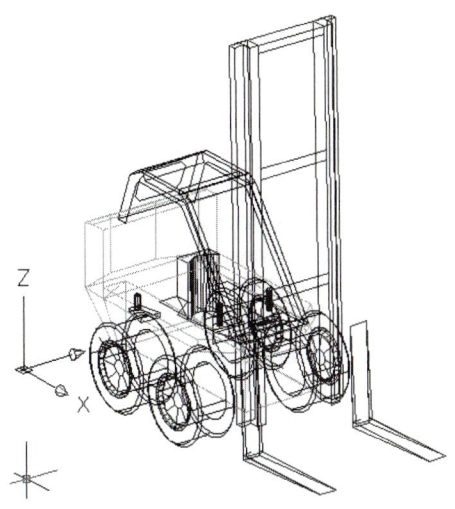

3D Wireframe

 Generally used to display 3D geometry when a wireframe representation is needed, this mode displays the objects using lines and curves to represent the surface edges (Fig. 35-62). Since all edges are displayed, this option is useful for 3D construction and editing. The shaded 3D UCS icon appears in this mode. Linetypes and lineweights are not displayed and raster and OLE objects are not visible. The *Compass* is visible if turned on. The lines are displayed in the material colors if you have applied *Materials*, otherwise they appear in the defined line or layer color. (See Chapter 41, Rendering and Animation, for information on using *Material*.)

FIGURE 35-62

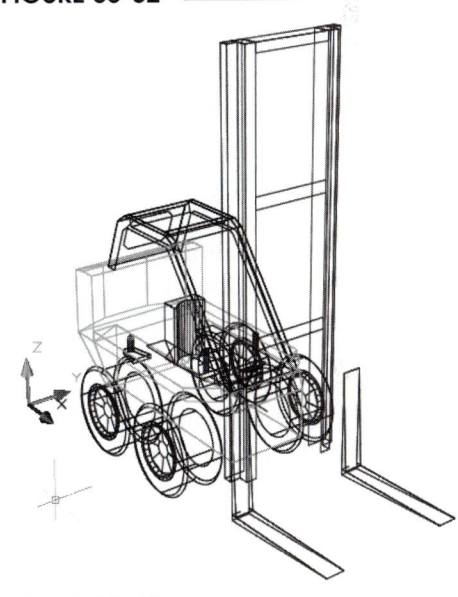

3D Hidden

 Use this mode to display 3D objects similar to the *3D Wireframe* representation but with hidden lines suppressed. In other words, edges that would normally be obscured from view by opaque surfaces become hidden (Fig. 35-63).

NOTE: The *3D Hidden* visual style replaces the older *Hide* command. *Hide* can still be used with the *2D Wireframe* style to remove obscured (hidden) lines from the display; however, *Hide* in this mode is not persistent—using *Regen* or *3Dorbit* cancels the display generated by *Hide*. You can use *Hide* to display meshed surfaces for curved edges. *Hide* is also affected by the setting of the *DISPSILH* variable; that is, if *DISPSILH* is set to 1, no mesh lines appear on curved surfaces.

FIGURE 35-63

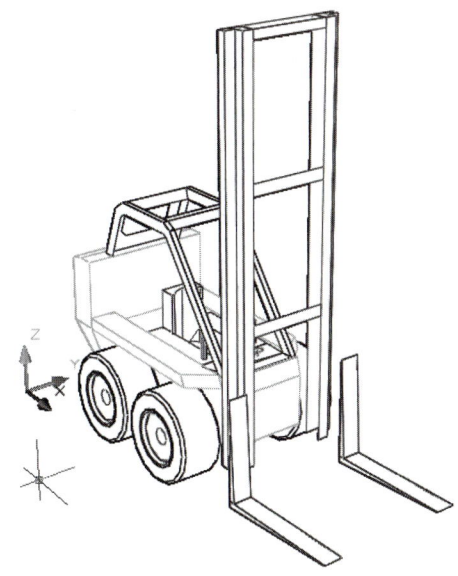

Isolines

In this case, the edges of the objects are displayed as well as the isolines that appear on curved surfaces. You can change the number of isolines using the *ISOLINES* system variable. The default *ISOLINES* setting is 4, so each of the holes in Figure 35-72 displays two visible lines (and two lines are not visible).

Facet Edges

This setting causes edge lines to appear at all facet edges as well as normal (non-curved) edges. The display is similar to that shown in Figure 35-72 (*Isolines* setting), except that additional lines would be displayed on all facet edges of curved surfaces (in the holes, for example). The *VSEDGESMOOTH* system variable (listed as the *Crease Angle* in the *Visual Styles Manager*) also affects the appearance of facet edges. *VSEDGESMOOTH* sets the angle at which facet edges within a face are <u>not</u> shown. If you turn on *Facet Edges* and facet edges do not appear in your model, change *VSEDGES-MOOTH* to a smaller angular value. *Facet Edges* must be on to display *Obscured Edges* and *Intersection Edges*.

Edge Color

If the edge mode is set to either *Isolines* or *Facet Edges*, you can use the drop-down list to set the color for the edges. The setting is stored in the *VSEDGECOLOR* system variable.

Edge Overhang

 Turn on this effect to make lines extend beyond their normal intersection (Fig. 35-73). This option gives a hand-drawn effect. When *Edge Overhang* is on, you can change the length of the overhang using the slider. The setting is stored in the *VSEDGEOVERHANG* system variable.

Edge Jitter

Edge Jitter gives a free-hand sketch effect as shown in Figure 35-74. The extent of the free-hand lines can be controlled using the slider—the higher the setting, the more "sketch" lines that appear. The settings are stored in the *VSEDGEJITTER* system variable. For a real hand-sketched effect, turn on both *Edge Jitter* and *Edge Overhang*.

FIGURE 35-72

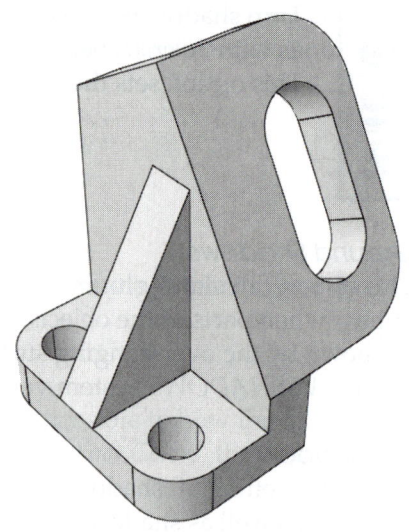

FIGURE 35-73

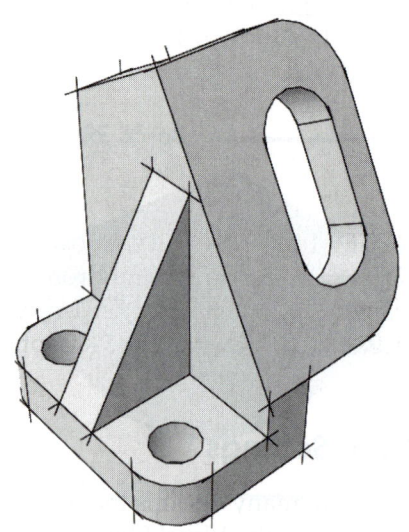

FIGURE 35-74

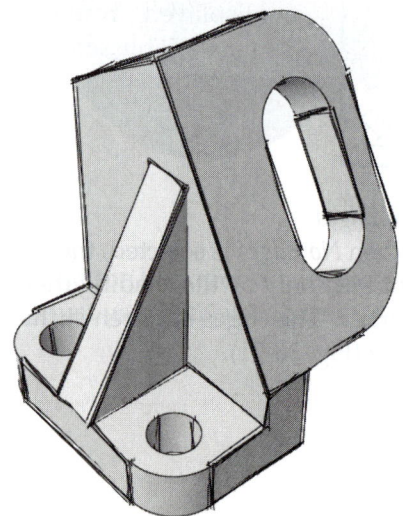

Silhouette Edges

If you want the outline edges only to appear around the object, turn *Silhouette Edges* on and set the edge mode to *No Edges*, as shown in Figure 35-75. You can, however, turn edges on (*Isolines* or *Facet Edges*), in which case the silhouette edges still appear heavier than the other edges. The on/off setting is stored in the *VSSILHEDGES* system variable. You can control the silhouette edge width using the slider (stored in the *VSSILHWIDTH* system variable). Silhouette edges are not displayed on wireframe or transparent objects.

FIGURE 35-75

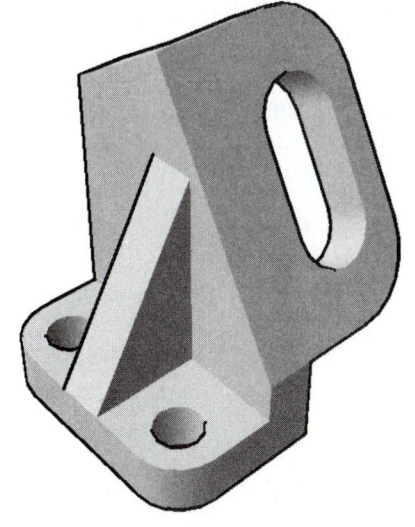

Obscured Edges

You can display the obscured edges (normally hidden edges) of your 3D model by turning *Obscured Edges* on (Fig. 35-76). To use this display, edge mode must be set to *Facet Edges* (not *Isolines* or *No Edges*). When *Obscured Edges* is on, you can set the desired color for those edges using the drop-down list. The *Color* option displays the *Select Color* dialog box, where you can set virtually any color for obscured edges. The setting is stored in the *VSOBSCUREDCOLOR* system variable.

FIGURE 35-76

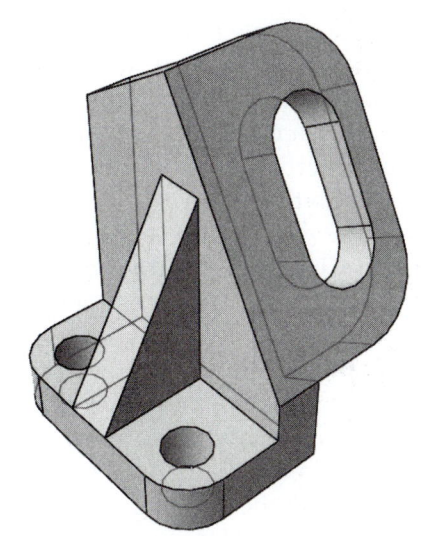

You can create a display that shows the obscured lines as <u>hidden</u> lines (or other non-continuous) similar to that in Figure 35-77. To do this, first set the edge mode to *Facet Edges* and set the face style to *No Face Style*. Next, set the *VSOB-SCUREDLTYPE* system variable to a value of 2. Other values result in other linetypes, such as dotted (3) or dashed (4).

FIGURE 35-77

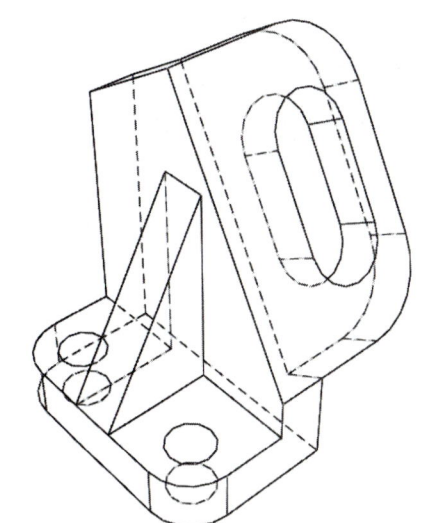

Intersection Edges

This display option is useful only when your 3D model contains multiple solids that intersect or overlap. In this case, turning on *Intersection Edges* displays lines of intersection wherever solids connect. The edge mode must be set to *Facet Edges* to use this option. The on/off setting is stored in the *VSINTERSECTIONEDGES* system variable. You can use the color drop-down list to specify the color for intersection edges (*VSINTERSECTIONCOLOR* system variable). If you prefer a non-continuous linetype for the edge lines, use the *VSINTERSECTIONLTYPE* system variable. To increase performance, turn off the display of intersection edges.

Visual Styles Manager

Pull-down Menu	Command (Type)	Alias (Type)	Short-cut	Screen (side) Menu	Tablet Menu
View *Visual Styles >* *Visual Styles* *Manager*	*Visualstyles*	*VSM*	...	...	...

The *Visual Styles Manager* (Fig. 35-78) gives you complete control of accessing, modifying, and creating visual styles. You can modify any style by selecting its image in the *Available Visual Styles in Drawing* section, then making changes to the face, environment, and edge properties below. Changes you make to a style are saved with the drawing.

NOTE: The *Visual Style control panel* (on the *Dashboard*) and the *Visual Styles Manager* operate differently in these respects:

1. Changes in the control panel made to the face, environment, and edge properties are immediately reflected in the drawing, whereas changes made to one of the visual styles in the *Visual Styles Manager* are not generally displayed in the drawing. However, changes made in the *Visual Styles Manager* can be displayed immediately in the drawing only if you first apply the selected visual style to the drawing, then make changes to that style in the *Visual Styles Manager*. You can instead display changes retroactively by first making the desired changes in the *Visual Style Manager*, then applying the style to the drawing using the *Apply the Selected Visual Style to Current Viewport* button.

2. Changes made to a visual style using the control panel are applied to a temporary style named *Current* but are not saved unless *Current* is the applied style when you save the drawing. However, using the *Visual Styles Manager*, changes to any style, including the five default styles, are automatically saved with that drawing only. Therefore changes made to any of the five default styles do not affect the default styles of other drawings. Named styles can be exported to a tool palette or copied and pasted for use in other drawings.

FIGURE 35-78

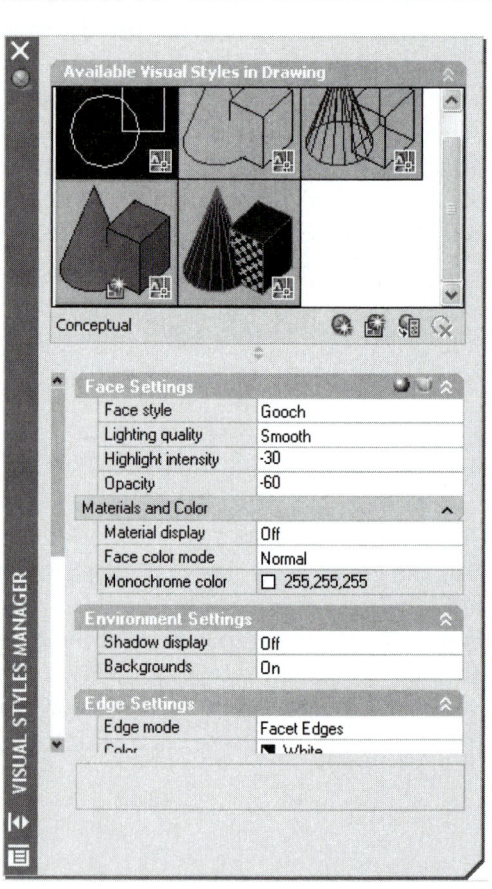

Available Visual Styles in Drawing

At the top of the palette-style interface, the *Available Visual Styles in Drawing* section displays image tiles of every visual style that is available in the drawing (the five default styles and any named styles). The selected visual style displays a yellow border and the name of the current style is displayed below the image tiles. The *Face Settings, Environment Settings,* and *Edge Settings* of the selected visual style are displayed in the panels below. The AutoCAD drawing icon may appear on an image tile. The five default visual styles that are shipped with the product display the product icon in the lower-right corner of the image tile. If one of the styles is applied to the drawing, an icon is displayed in that image tile (see Fig. 35-78, where the *Conceptual* style is applied to the viewport).

Shortcut Menu and Tool Strip Buttons

You can right-click in the image tile area to display the shortcut menu (Fig. 35-79). In addition, several buttons are available just below the image tiles.

FIGURE 35-79

Create New Visual Style

Use this option to create and save a "named" visual style. The *Create New Visual Style* dialog box appears (not shown) where you enter a name and an optional description. A new sample image is placed at the end of the panel and becomes the selected style. The initial settings for a new style are based on the *Realistic* style.

Apply Selected Visual Style to Current Viewport

This option applies the selected visual style to the current viewport. This button is useful if you want to immediately see how changes you make to the visual style affect the drawing.

Export the Selected Visual Style to the Tool Palette

Use this option if you want to use the selected visual style in other drawings. A tool (button) is created for the selected visual style and placed on the active tool palette. Therefore, open the desired tool palette before using this option. If the *Tool Palettes* window is closed, it is automatically opened and the tool is placed on the top palette.

Delete the Selected Visual Style

Use this option to remove the selected visual style from the drawing. A default visual style or one that is in use (currently applied) cannot be deleted.

Apply to All Viewports

This options applies the selected visual style to all viewports in the drawing.

Edit Name and Description

The *Edit Name and Description* dialog box is displayed so you can change the name, add a description, or change an existing description. The description is displayed in a tooltip when the cursor hovers over the sample image.

Copy

The current visual style and image is copied to the Windows Clipboard. You can then paste it into the *Available Visual Styles* panel or paste it into the *Tool Palettes* window to create a copy of the visual style.

Paste

After using *Copy*, you can paste a visual style into the *Available Visual Styles* panel or paste it into the *Tool Palettes* window. Since you can *Copy* and *Paste* between multiple open drawings in AutoCAD, this option provides another method for exporting and using a visual style in another drawing.

Printing 3D Models

You can print 3D models in any visual style. If you have a laserjet, inkjet, or electrostatic type printer configured as the plot device (not a pen plotter), you can select from *As Displayed*, *Wireframe*, *Hidden*, *3D Hidden*, *3D Wireframe*, *Conceptual*, or *Realistic* options from *Shadeplot* section of the (expanded) *Plot* dialog box. You could also select these options or the *Rendered* option from the *Properties* palette for the viewport.

Printing a Shaded Image From the *Model* Tab

If you want to print a shaded 3D drawing from the *Model* tab, use the *Shaded Viewport Options* section of the *Plot* dialog box, expanded section. Use the *Shade Plot* drop-down list to select from the *Wireframe*, *Hidden*, *3D Hidden*, *3D Wireframe*, *Conceptual*, or *Realistic* options (Fig. 35-83). Alternately, you can apply a visual style to the drawing display, then select *As Displayed* from this list to print the drawing as it appears on screen. You can also select from several resolution qualities for the print (see the "Shaded Viewport Options" section in Chapter 14).

FIGURE 35-83 ─────

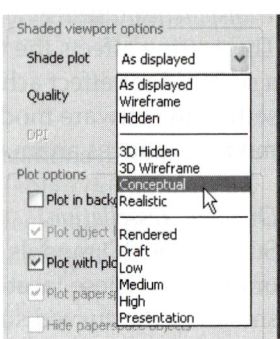

NOTE: The *Wireframe* option in the drop-down list is essentially the same as the *2D Wireframe* visual style. However, the *Hidden* option differs from the *3D Hidden* option based on your setting of the *DISP-SILH* system variable. If *DISPSILH* is set to 0 (the default), a *Hidden* display generates mesh lines on curved surfaces. The display of meshed lines is also affected by the *FACETRES* system variable setting.

Printing a Shaded Image From a *Layout* Tab

If you are printing from a layout and want to specify a shaded image for one or more viewports, use the *Shadeplot* option of the viewport's *Properties* palette. Here you can select from the same options (Fig. 35-84) as in the *Plot* dialog box.

FIGURE 35-84 ─────

Alternately, you can type the *-Vports* command to assign the *Shadeplot* option for one or more viewports. <u>You must be in a *Layout* for the *Shadeplot* options to appear in the Command prompt.</u>

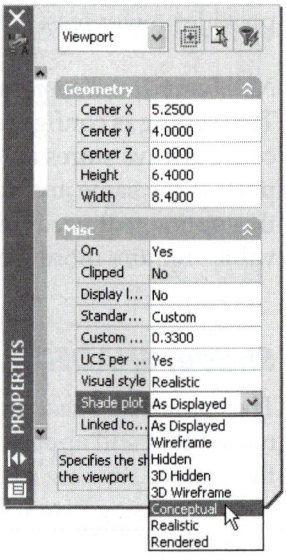

```
Command: -Vports
Specify corner of viewport or
[ON/OFF/Fit/Shadeplot/Lock/Object/Polygonal/Restore/2/3/4] <Fit>: s
Shade plot? [As displayed/Wireframe/Hidden/Visual Styles/Rendered]
<As displayed>: v
[3dwireframe/3dHidden/Realistic/Conceptual/Other] <Realistic>:
```

The *Rendered* option prints the drawing according to the settings made in the *Render* dialog box, including any specified lights, materials, and background (see Chapter 41 for more information on these features). If no rendering settings have been specified, the *Rendered* option produces a rendered image similar to the *Realistic* visual style.

CHAPTER EXERCISES

1. What are the three types of 3D models?

2. What characterizes each of the three types of 3D models?

3. What kind of modeling techniques does AutoCAD's solid modeling system use?

4. What are the seven formats for 3D coordinate specification?

5. Examine the 3D geometry shown in Figure 35-85. Specify the designated coordinates in the specified formats below.

FIGURE 35-85

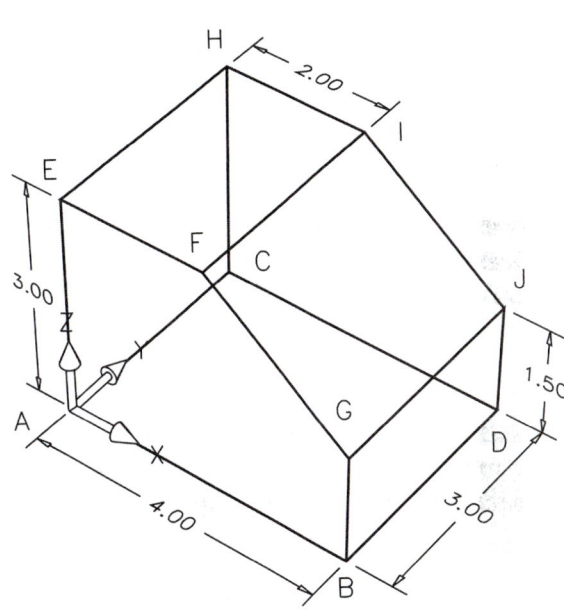

   A. Give the coordinate of corner D in absolute Cartesian format.

   B. Give the coordinate of corner F in absolute Cartesian format.

   C. Give the coordinate of corner H in absolute Cartesian format.

   D. Give the coordinate of corner J in absolute Cartesian format.

   E. What are the coordinates of the line that define edge F-I?

   F. What are the coordinates of the line that define edge J-I?

   G. Assume point E is the "last point." Give the coordinates of point I in relative Cartesian format.

   H. Point I is now the last point. Give the coordinates of point J in relative Cartesian format.

   I. Point J is now the last point. Give the coordinates of point A in relative Cartesian format.

   J. What are the coordinates of corner G in cylindrical format (from the origin)?

   K. If E is the last point, what are the coordinates of corner C in relative cylindrical format?

NOTE: For the following exercises you need to download the drawings from the McGraw-Hill/Leach Web site. Go to **www.mhhe.com/leach**, then find the page for **AutoCAD 2007 Instructor** and locate the exercise download drawings. Four of the drawings are used in these exercises: TRUCK MODEL, OPERA, WATCH, and R300-20. These drawings are previous release AutoCAD sample drawings and are provided courtesy of Autodesk.

6. In this exercise you will use a sample drawing and use the *Visual Style control panel* to practice with the five default visual styles.

   A. *Open* the **TRUCK MODEL** drawing. Select the *Model* tab if not already current.

B. In the *Visual Style control panel*, select the **3D Hidden** visual style. Your display should look similar to that in Figure 35-86. Note that perspective projection should be on. If perspective projection is not on, select the **Perspective Projection** button.

FIGURE 35-86

C. In the *Visual Style control panel*, select the **3D Wireframe** visual style. Do you see all the edges of the model?

D. Next, select the **2D Wireframe** visual style. Note that the display has been automatically changed to parallel projection mode for this visual style.

E. Type **Hide** at the command prompt. Do mesh lines occur on the truck?

F. Type **DISPSILH** and change the setting to **1**. Type **Hide** at the command prompt again. Did the mesh lines disappear? Change **DISPSILH** back to **0** and try the **Hide** command again.

G. Select the **3D Hidden** visual style again. Note that the truck is displayed in the parallel projection mode. Select the **Perspective Projection** button.

H. Next, select the **Conceptual** style. Finally, change the visual style to **Realistic**. Which of these two styles better display the contrast between surfaces?

I. **Exit** the Truck Model drawing and <u>do not save the changes</u>.

7. In this exercise, you open an AutoCAD sample drawing and use the *3D Navigate control panel* and the *View Manager* to practice with the preset views.

A. **Open** the **OPERA** drawing. Select the **Model** tab. **Freeze** layers **E-AGUAS**, **E-PREDI**, **E-TERRA**, **E-VIDRO**, and **O-PAREDE**. Use the *Visual Style control panel* to generate a **Conceptual** display.

B. Select a **Top** view from the *3D Navigate control panel*. **Zoom** in if necessary to fully display the objects on the screen. Examine the view and notice the orientation of the coordinate system icon.

NOTE: If *Zooming* is very slow, try changing the visual style to *3D Wireframe*.

C. Next use the **View** command to produce the *View Manager*. From the **Preset Views**, select a **SE Isometric** view, select **Set Current**, then **OK**. **Zoom** if necessary. Examine the icon and find north (Y axis). Consider how you (the observer) are positioned as if viewing from the southeast.

D. Use any method to produce a **Front** view. **Zoom** if needed. This is a view looking north. Notice the icon. The Y axis should be pointing away from you.

E. Generate a **Right** view. Produce a **Top** view again. **Zoom** whenever needed for a full-screen display.

F. Next, view the OPERA from a *SE Isometric* viewpoint once again. Generate a *3D Hidden* visual style. Your view should appear like that in Figure 35-87. Notice the orientation of the WCS icon on your screen.

FIGURE 35-87

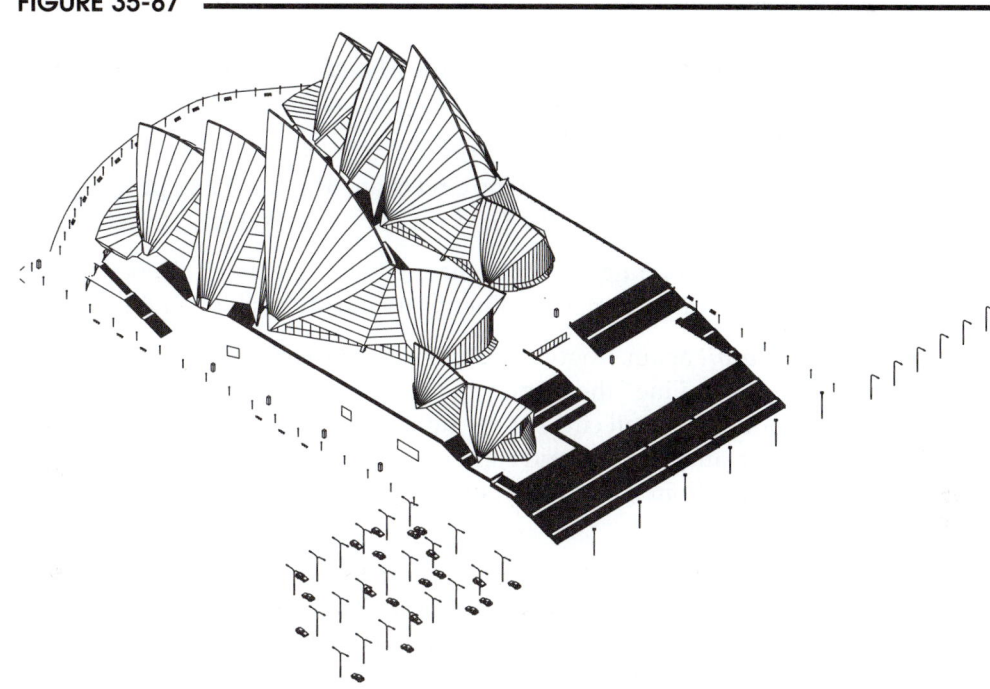

G. Finally, generate a *NW Isometric* and *NE Isometric*. In each case, examine the WCS icon to orient your viewing direction.

H. Experiment and practice more with *3D Navigate control panel* and the *View Manager*. <u>Do not save</u> the drawing.

8. In this exercise you will practice using the *3D Navigate control panel* and *Zoom*.

A. *Open* the **TRUCK MODEL** again that you used in Exercise 6. Do not change the current visual style throughout this exercise. Type "*UCS*," then use the *World* option. This action makes the World Coordinate System the current coordinate system. Use the *Ucsicon* command and ensure the icon is at the *Origin*. Notice the orientation of the truck with respect to the WCS. The X axis points outward from the right side of the truck, Y positive is to the rear of the truck, and Z positive is up.

B. Use the *3D Navigate control panel* to generate a *Top* view. Note the position of the X, Y, and Z axes. You should be looking down on the XY plane and Z points up toward you.

C. Generate a *SE Isometric* view. Note the position of the X, Y, and Z axes.

D. Use the *Parallel Projection* button to ensure perspective is turned off. Generate a *Right* view. Notice that AutoCAD automatically performs a *Zoom Extents* so all of the drawing appears on the screen. *Freeze* layer **BASE**. Now use *Zoom Extents* to display the truck at the maximum screen size.

E. Finally, use each of the following preset views in the order given and notice how the truck appears to change its position for each view. Remember that you should imagine that the viewer moves around the object for each view, rather than imagining that the object rotates.

SE Isometric, Top, SE Isometric, Bottom, SE Isometric, Right, SW Isometric, Left, SE Isometric, Front, SE Isometric

F. Close the drawing but <u>do not save</u> the changes.

9. Now you will get some practice with *3Dorbit* and the shortcut menu options.

A. *Open* the **WATCH** drawing. Activate the *Model* tab. Now generate a *SE Isometric* view to get a good look at the watch. Notice that the components are disassembled, similar to an exploded view assembly drawing.

B. Now invoke *3Dorbit*. Note that the *Grid* is on. Right-click to produce the shortcut menu, then under *Visual Aids* toggle off the *Grid*. In the shortcut menu, select *Visual Styles*, then *Realistic*.

C. Now drag your pointing device to change your viewpoint. Generate your choice for a view that shows the watch so all components are mostly visible. Right click and *Exit 3Dorbit*.

D. Activate *3Dorbit* again. Right click and select *Other Navigation Modes*, then *Free Orbit*. With this mode, try "rolling" the display by using your cursor outside the arc ball. Next, try placing your cursor on the small circle at the right or left quadrant of the arc ball. Hold down the left mouse button and drag left and right to spin the watch about an imaginary vertical axis. Try the same technique from the top and bottom quadrant to spin the watch about an imaginary horizontal axis.

E. Finally, switch back to *Constrained Orbit* and generate your choice of views that shows the watch so all components are visible. Right click and change to *Perspective* projection (you may then have to drag the cursor and change the display slightly to activate the perspective display). View the watch from several directions and examine the perspective effect. Arrive at a view that displays all of the components clearly.

F. Right click to produce the shortcut menu and select *Other Navigation Modes*, then *Continuous Orbit*. Hold down the left mouse button, drag the mouse, and release the button to put the watch in a continuous spin.

G. Finally, *Exit 3Dorbit*. *Close* the drawing and <u>do not save changes</u>.

10. In this exercise you will create four model space viewports with the *3D* option.

A. *Open* the **WATCH** drawing again. Activate the *Model* tab. Turn off **Grid**. Change the *Visual Style* to *Realistic*. Check the setting of the **UCSORTHO** variable by typing it at the command line and ensuring it is set to **0**. A setting of 0 means no new UCSs will be created when new views are generated.

B. Invoke the *Vports* command (make sure you are in model space). In the *New Viewports* tab of the *Viewports* dialog box, select **Four: equal**. At the bottom of the dialog box in the *Setup* drop-down list, select *3D*. Select the *OK* button.

C. Your display should show a top, front, right side, and southeast isometric view. Notice that the watch is not sized proportionally with respect to the other views. Activate the front viewport (bottom left). Type **Zoom** at the command line and enter a value of **1.5**. Do the same for the other two orthographic views. Your display should look like that shown in Figure 35-88 (on the next page).

D. Activate the isometric viewport. Change the *Visual Style* to *Conceptual*. This is a good viewport arrangement to use when you are constructing 3D models because you can see several views of the object on your screen at once. Remember this arrangement when you begin working with the exercises in the next several chapters.

E. *Close* the drawing but <u>do not save</u> the changes.

FIGURE 35-88

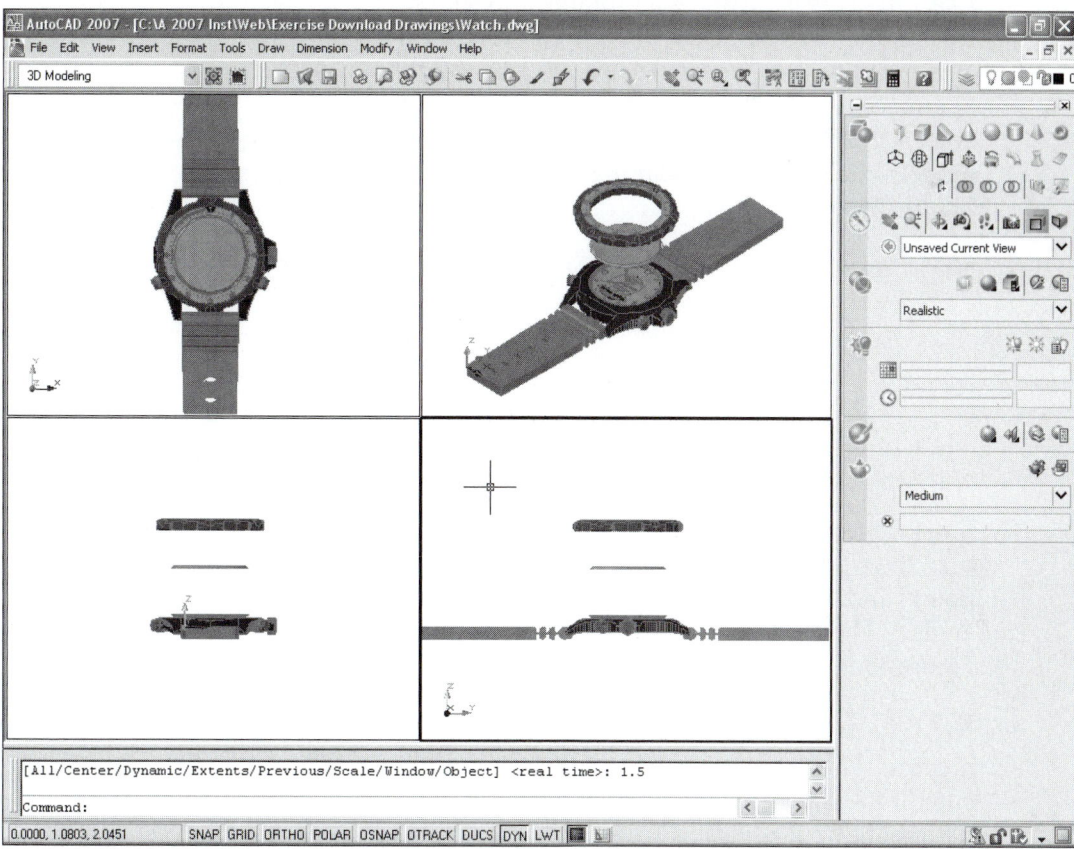

11. In this exercise, you will create a simple solid model to get an introduction to some of the solid modeling construction techniques discussed in Chapter 38. You will use this model again in these chapter exercises for practicing visual style edge and face settings. You will also use this model for creation of User Coordinate Systems in Chapter 36.

A. Start AutoCAD in the 3D environment (ensure you are using the *3D Modeling* workspace and that the *Dashboard* is on). Begin a *New* drawing using the ACAD3D.DWT template drawing. Use *Save* and name the drawing **SOLID1**. Also make sure the *Ucsicon* is *On* and set to appear at the *ORigin*. Use the *Layer Properties Manager* to create a new layer named **MODEL** and set the layer's color (in the *Index Color* tab) to **253**. Make **MODEL** the *Current* layer.

B. Type or select *Box* to create a solid box. When the "Specify first corner" prompt appears, enter **0,0,0**. Then move your cursor so the rectangle appears in positive X,Y space. At the "Specify other corner" prompt, enter **5,4**. Enter a value of **3** for the "height." Change your viewpoint to *SE Isometric*, then *Zoom* if necessary to view the box. Use *3Dorbit* to change the view <u>slightly</u>.

C. Change the visual style to *3D Wireframe*. Type or select the *Wedge*. When the prompts appear, use **0,0,0** as the "Corner of wedge." Then move your cursor again so the rectangle appears in positive X,Y space. At the "Specify other corner" prompt, enter **4, 2.5**. Specify **2** for the "height." Your solid model should appear similar to that shown in Figure 35-89.

FIGURE 35-89

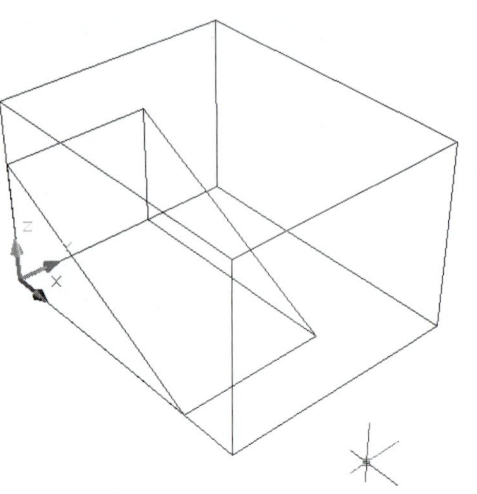

D. Use the *Rotate* command and select only the wedge. Use **0,0** as the "base point" and enter **-90** as the "rotation angle." The model should appear as Figure 35-90.

E. Now use *Move* and select the wedge for moving. Use **0,0** as the "base point" and enter **0,4,3** as the "second point of displacement."

FIGURE 35-90

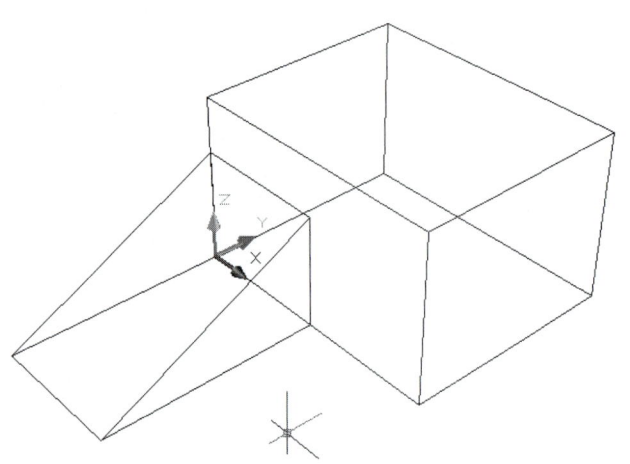

F. Finally, type or select *Union*. When prompted to "Select objects:", PICK both the wedge and the box. Change the visual style to *Conceptual*. The finished solid model should look like Figure 35-91. *Save* the drawing.

FIGURE 35-91

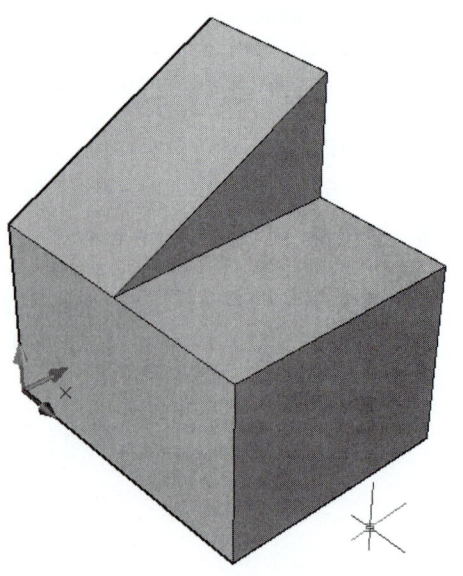

13. In this exercise you will use the SOLID1 drawing from the previous exercise and practice with the face and edge settings in the *Visual Style control panel* and the *Visual Styles Manager*.

A. If not already open, **Open** the **SOLID1** drawing. It should be set to the default *Conceptual* visual style. Use the **X-Ray Mode** toggle in the *Visual Styles control panel*. Your display should now appear as "see-through." Depending on your system's display driver and settings in the *Adaptive Degradation and Performance Tuning* dialog box, the display may have a "screen door" effect. Type **3Dconfig** to produce the dialog box, select **Manual Tune**, and examine the setting for **Enable hardware acceleration** at the top of the *Manual Performance Tuning* dialog box. If possible, change the setting, return to the drawing and toggle **X-Ray Mode** a couple of times. Do you notice a change in the object's appearance? Finally, produce the **Adaptive Degradation and Performance Tuning** dialog box again and return the setting to its previous state.

B. Make sure the default **Conceptual** visual style is set. Near the top center of the *Visual Style control panel*, find the flyout button for the four face color modes. Try each of the following face color settings and examine the resulting display: **Regular face colors**, **Monochrome mode**, **Tint mode**, and **Desaturate mode**. Next, use the **Layer Properties Manager** to change the MODEL layer color to **132**. Finally, experiment with the four face color settings again. Change the visual style to **Realistic**. Experiment with the four face color settings again and note the differences that occur while the **Realistic** style is current.

C. Expand the *Visual Style control panel* by hovering your pointer over the lower-left corner of the control panel and selecting the chevrons. Change the visual style several times between **Conceptual** and **Realistic**. Notice that the face style button automatically changes between *Realistic face style* and *Gooch face style*.

D. Set the visual style to **Conceptual**. Change the top button for edge settings (flyout) to **No edges**. Then change the setting to **Isolines** and to **Facet edges**. Finally, generate a free-hand sketch effect similar to that in Figure 35-92 using the following settings: **Isolines**, **Edge overhang** on, **Edge jitter** on, and **Silhouette edges** off. **Save** the drawing as **SOLID1 SKETCH**.

FIGURE 35-92

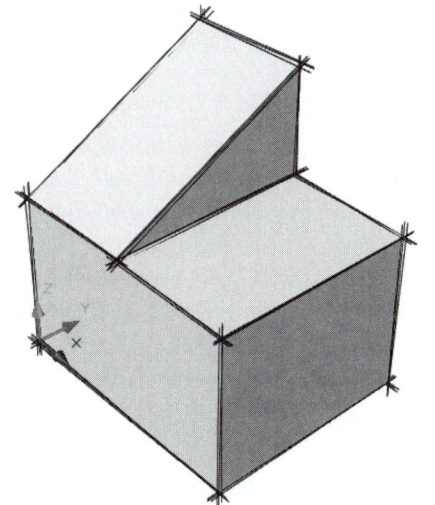

E. Produce the **Visual Styles Manager**. Arrange your drawing so you can see both the 3D object by using **Zoom** and/or **Pan** and by moving the *Visual Styles Manager* if needed. Select the default **3D Hidden** visual style from the manager. Next, use the **Apply Selected Visual Style to Current Viewport** button to display the visual style in the drawing.

F. Use **3Dorbit** and select a *SW Isometric* display. **Zoom** and/or **Pan** as necessary to see the object clearly again. Produce a display similar to that shown in Figure 35-93 by using the *Edge Modifiers* section of the *Visual Styles Manager* to set the *Halo Gap %* to **10**.

FIGURE 35-93

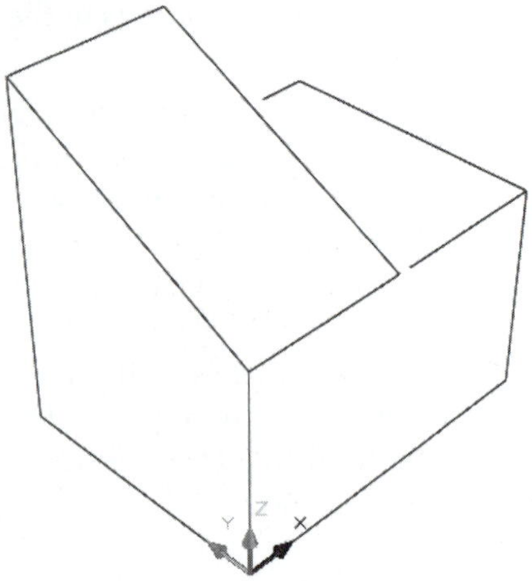

G. Finally, create a display similar to that in Figure 35-94 by changing the settings in the **Obscured Edges** section to *Visible* **Yes** and *Linetype* **Dashed**.

FIGURE 35-94

H. Use **Saveas** and change the name of the drawing to **SOLID1 HIDDEN**.

14. Use the R300-20 drawing to practice with the **3Dclip** command.

A. **Open** the **R300-20** drawing. Activate the **Model** tab. Use **Pan** and **Zoom** to enlarge the display to fill the screen. Set the visual style to **Conceptual**.

B. Use **3Dclip**. The *Adjust Clipping Planes* window should appear. Select the **Front Clipping On/Off** button (or you can right-click inside the window to select from a short-cut menu). Next, select **Adjust Front Clipping Plane**. Adjust the plane so it passes approxi-mately one-half of the way through the assembly in the lower-left corner of the drawing. Your display should appear similar to that shown in Figure 35-95. Notice how the interior features of the assembly are clearly shown. Finally, close the *Adjust Clipping Planes* window.

FIGURE 35-95

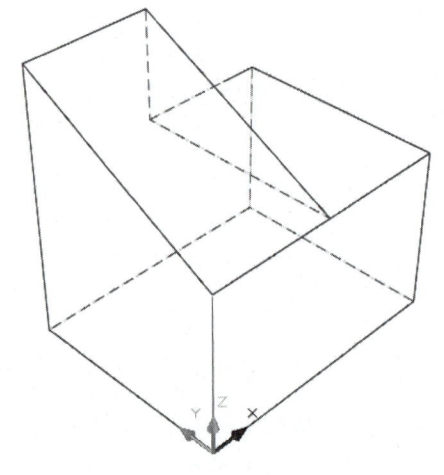

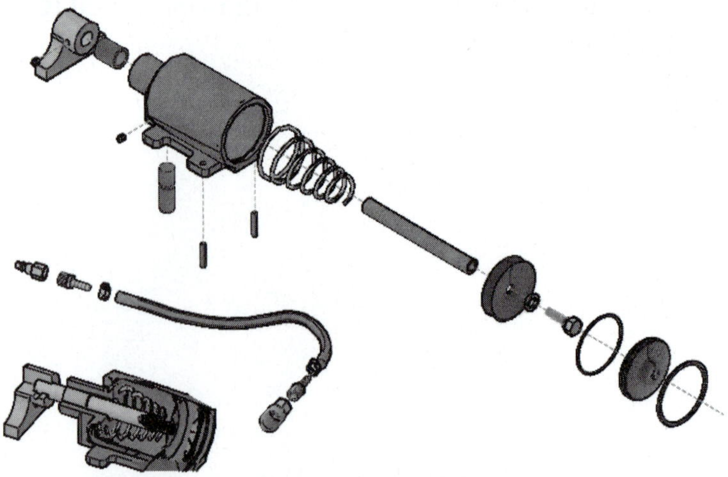

C. **Close** the drawing and <u>do not save</u> the changes.

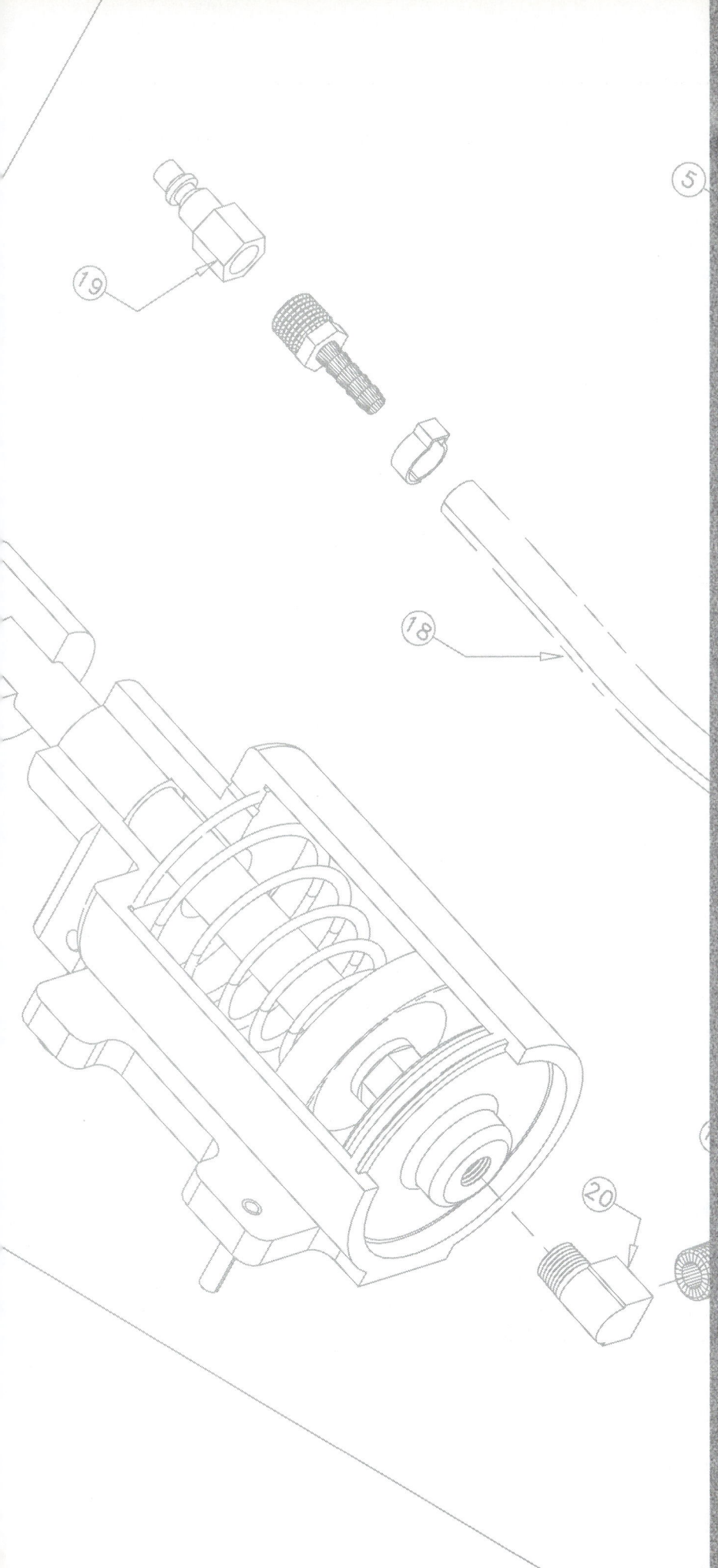

USER COORDINATE SYSTEMS

CHAPTER OBJECTIVES

After completing this chapter you should:

1. know how to create, *Save, Restore,* and *Delete* User Coordinate Systems using the *UCS* command;

2. know how to use the *UCS Manager* to create, *Save, Restore,* and *Delete* User Coordinate Systems;

3. be able to create UCSs by the *Origin, Zaxis, 3point, Object, Face, View, X, Y,* and *Z* methods;

4. be able to create *Orthographic* UCSs using the UCS Manager;

5. know the function of the *UCSVIEW, UCSORTHO, UCSFOLLOW,* and *UCSVP* system variables.

CONCEPTS

No User Coordinate Systems exist as part of any of the AutoCAD default template drawings as they come "out of the box." UCSs are created to simplify construction of the 3D model when and where they are needed. UCSs are applicable to all types of 3D models—wireframe models, surface models, and solid models.

When you create a multiview or other 2D drawing, creating geometry is relatively simple since you only have to deal with X and Y coordinates. However, when you create 3D models, you usually have to consider the Z coordinates, which makes the construction process more complex. Constructing some geometries in 3D can be very difficult, especially when the objects have shapes on planes not parallel with, or perpendicular to, the XY plane or not aligned with the WCS (World Coordinate System).

FIGURE 36-1

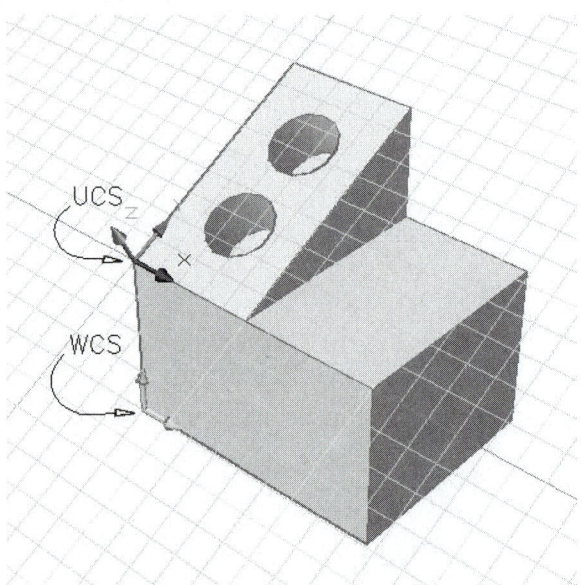

UCSs are created when needed to simplify the construction process of a 3D object. For example, imagine specifying the coordinates for the centers of the cylindrical shapes in Figure 36-1, using only world coordinates. Instead, if a UCS were created on the face of the object containing the cylinders (the inclined plane), the construction process would be much simpler. To draw on the new UCS, only the X and Y coordinates (of the UCS) would be needed to specify the centers, since anything drawn on that plane has a Z value of 0. The Coordinate Display (*Coords*) in the Status Bar always displays the X, Y, and Z values of the current coordinate system, whether it is the WCS or a UCS.

Generally, you would create a UCS aligning its XY plane with a surface of the object, such as along an inclined plane, with the UCS origin typically at a corner or center of the surface. Any coordinates that you specify (in any format) are assumed to be user coordinates (coordinates that lie in or align with the current UCS). Even when you PICK points interactively, they fall on the XY plane of the current UCS. The *SNAP, Polar Snap,* and *GRID* automatically align with the current coordinate system XY plane, enhancing your usefulness of the construction plane.

CREATING AND MANAGING USER COORDINATE SYSTEMS

You can create User Coordinate Systems several ways. The *Ucs* command is the traditional method for creating new coordinate systems and provides several options for orienting the new UCSs. The UCS Manager (*Ucsman* command) allows you to create orthographically positioned UCSs as well as name, save, and restore UCSs. You can also have UCSs created for you automatically when you create or restore an orthographic view.

You can create and save multiple UCSs in one drawing. Rather than recreating the same UCSs multiple times, you can assign a name and save the UCSs that you create so they can be restored at a later time. Although you can have multiple UCSs, only one UCS per viewport can be active at any one time.

Because you would normally want multiple UCSs for a 3D drawing, this chapter explains the features and assists you in creating and managing UCSs wisely. First the chapter discusses how to create UCSs, then naming, saving, and restoring UCSs are explained. At the end of the chapter, the system variables that control UCSs are discussed.

If you are constructing solid models, you can create Dynamic UCSs. Dynamic UCSs are temporary coordinate systems used for constructing a solid, but they cannot be saved or named. Dynamic UCSs are discussed in Chapter 38.

NOTE: When creating 3D models, it is recommended that you use the *Ucsicon* command with the *On* option to make the icon visible and the *Origin* option to force the icon to the origin of the current coordinate system.

Ucs

Pull-down Menu	Command (Type)	Alias (Type)	Short-cut	Screen (side) Menu	Tablet Menu
Tools *Move UCS* *New UCS >*	*Ucs*	...	...	*TOOLS 2* *UCS*	W,7

The *UCS* command allows you to create, save, restore, and delete User Coordinate Systems. There are many options available for creating UCSs. The *UCS* command always operates only in Command line mode.

FIGURE 36-2

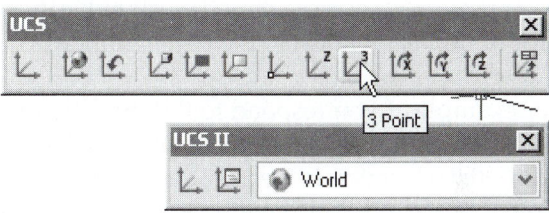

Command: **ucs**
Current ucs name: *WORLD*
Specify origin of UCS or [Face/NAmed/OBject/Previous/View/World/X/Y/Z/ZAxis] <World>:

Although the *UCS* command operates only in Command line mode, two *UCS* toolbars are available (Fig. 36-2). Choosing one of these icons activates the particular option of the *UCS* command.

You can also use the *Tools* pull-down menu to select an option of the *UCS* command. Suboptions are on a cascading menu (Fig. 36-3).

FIGURE 36-3

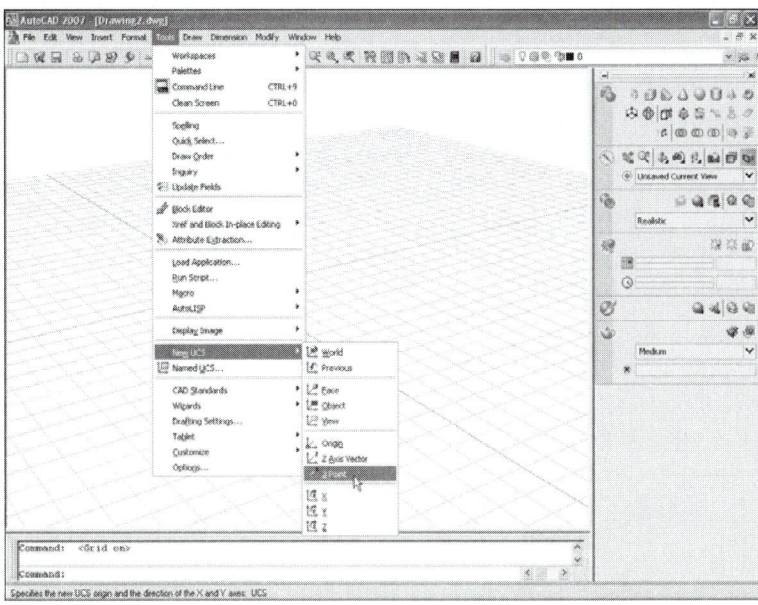

TIP When you create UCSs, AutoCAD prompts for points. These points can be entered as coordinate values at the keyboard or you can PICK points on existing 3D objects. *OSNAP* modes should be used to PICK points in 3D space (not on the current XY plane). In addition, understanding the right-hand rule is imperative when creating some UCSs (see Chapter 35).

NOTE: While learning these options, it is suggested that you turn off Dynamic UCS (*DUCS* on the Status bar). Dynamic UCS affects the orientation of the UCS as you select points in 3D space. Dynamic UCS is explained in Chapter 38.

The options of the UCS command are explained next.

  The default option of the *Ucs* command defines a new UCS using one, two, or three points.

    Command: **ucs**
    Current ucs name: *WORLD*
    Specify origin of UCS or [Face/NAmed/OBject/Previous/View/World/X/Y/Z/ZAxis] <World>:

If you specify a <u>single point</u>, the origin of the current UCS shifts to the point you specify. The orientation of the new UCS (direction of the X, Y, and Z axes) remains the same as its previous position, only the origin location changes. This option is identical to the *Origin* option.

For example, if you respond to the "Specify origin of UCS" prompt by PICKing a new location using *Osnap*, the new UCS's 0,0,0 location is the point you select (Fig. 36-4). Note that the X,Y,Z orientation is the same as the previous UCS position. At the following prompt, press Enter to accept the first point.

    Specify point on X-axis or <Accept>:

FIGURE 36-4

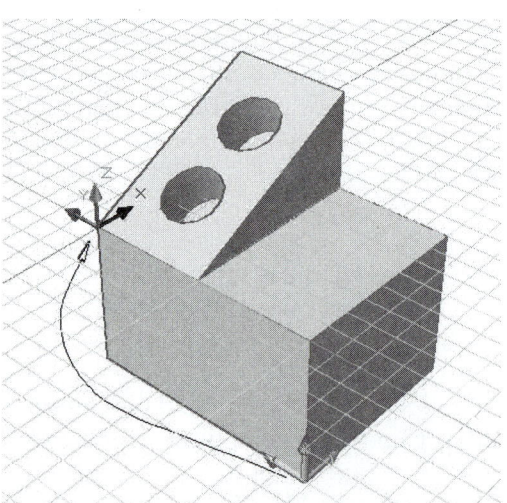

If you specify a <u>second point</u> (at the "Specify point on X-axis or <Accept>:" prompt), the UCS rotates around the previously specified origin point such that the positive X axis of the UCS passes through the new point you PICK.

For example, if you selected a second point as shown in Figure 36-5 as the "point on X axis," the coordinate system spins so the positive X direction aligns with the point. Press Enter to limit the input to two points.

    Specify point on XY plane or <Accept>:

FIGURE 36-5

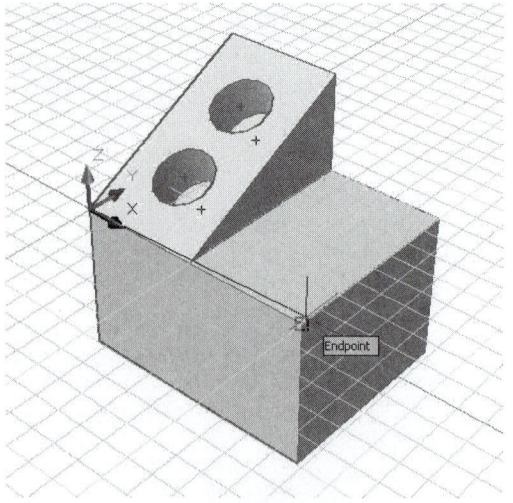

You can specify a <u>third point</u> at the prompt shown above. If you specify a third point, the UCS rotates around the X axis such that the positive Y half of the XY plane of the UCS contains the point. Remember that the first two points lock the origin and the direction for positive X, so the third point only specifies a location in the XY plane. This option (specifying three points at the default prompt) is identical to the separate *3Point* option.

For example, if the third point was specified on the inclined plane as shown in Figure 36-6, the XY plane of the resulting UCS would align itself with the inclined plane.

FIGURE 36-6

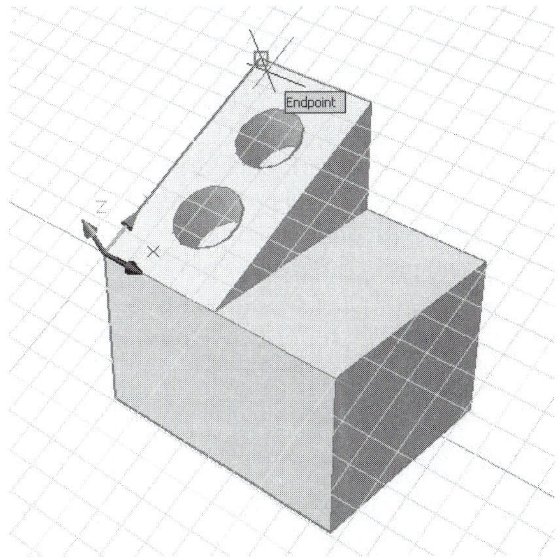

ORIGIN

The *Origin* option defines a new UCS by allowing you to specify a point for the new 0,0,0 location. The orientation of the X,Y, and Z axes remains the same as the previous UCS.

```
Command: ucs
Current ucs name:  *WORLD*
Specify origin of UCS or [Face/NAmed/OBject/Previous/View/World/X/Y/Z/ZAxis] <World>: o
Specify new origin point <0,0,0>:
```

This option is identical to using the default option to PICK one point, then pressing Enter to accept that point as the new origin.

ZAxis

You can define a new UCS by specifying an <u>origin</u> and a direction for the <u>Z axis</u>. Only the two points are needed. The X <u>or</u> Y axis generally remains parallel with the current UCS XY plane, depending on how the Z axis is tilted (Fig. 36-7). AutoCAD prompts:

```
Specify new origin point or [Object] <0,0,0>: PICK
Specify point on positive portion of Z-axis
<0.0000,0.0000,0.0000>: PICK
```

FIGURE 36-7

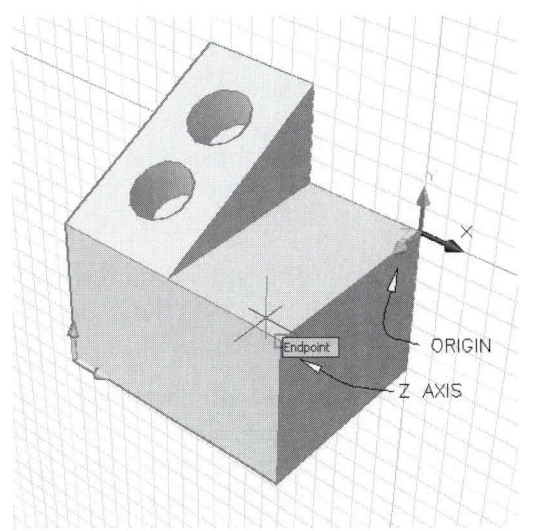

3point

This new UCS is defined by (1) the <u>origin</u>, (2) a point on the <u>X axis</u> (positive direction), and (3) a point on the <u>Y axis</u> (positive direction) or <u>XY plane</u> (positive Y). This is the most universal of all the UCS options (works for most cases). It helps if you have geometry established that can be PICKed with *OSNAP* to establish the points (Fig. 36-8). The prompts are:

FIGURE 36-8

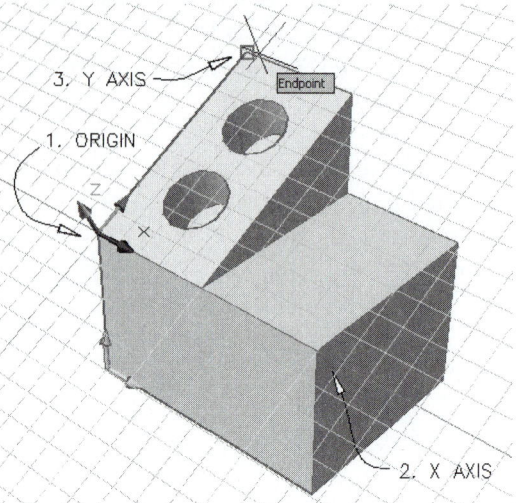

```
Command: ucs
Current ucs name: *WORLD*
Specify origin of UCS or
  [Face/NAmed/OBject/Previous/View/
  World/X/Y/Z/ZAxis] <World>: 3
Specify new origin point <0,0,0>: PICK
Specify point on positive portion of X-axis
  <0.000,0.000,0.000>: PICK
Specify point on positive-Y portion of the UCS XY plane
  <0.000,0.000,0.000>: PICK
```

Object

This option creates a new UCS aligned with the selected object. You need to PICK only one point to designate the object. The origin of the new UCS is nearest the endpoint of the edge (or center of circular shape) you select. The orientation of the new UCS is based on the type of object selected and the XY plane <u>that was current when the object was created</u>.

For example, assume you created a solid model using only the World Coordinate System. Then, using the *Object* option, you could quickly create new UCSs aligned with edges of the model; however, the XY plane of the new coordinate systems would remain parallel with the WCS XY plane (that was current when those edges were created). This option is extremely fast, but takes some experience and may require some trial-and-error experimentation to achieve the UCS orientation you intend.

AutoCAD prompts:

> Select object to align UCS:

The following list gives the orientation of the UCS using the *Object* option for each type of object (Fig. 36-9).

FIGURE 36-9

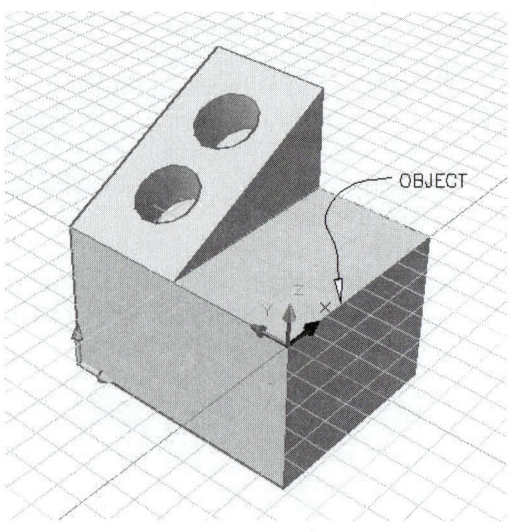

Object	Orientation of New UCS
Line	The end nearest the point PICKed becomes the new UCS origin. The new X axis aligns with the *Line*. The XY plane keeps the same orientation as the previous UCS.
Circle	The center becomes the new UCS origin with the X axis passing through the PICK point.
Arc	The center becomes the new UCS origin. The X axis passes through the endpoint of the *Arc* that is closest to the pick point.

Point   The new UCS origin is at the *Point*. The X axis is derived by an arbitrary but
                 consistent "arbitrary axis algorithm." (See the *AutoCAD Customization Guide*.)

Pline (2D)   The *Pline* start point is the new UCS origin with the X axis extending from the
                 start point to the next vertex.

Solid   The first point of the *Solid* determines the new UCS origin. The new X axis lies
                 along the line between the first two points.

Dimension   The new UCS origin is the middle point of the dimension text. The direction of
                 the new X axis is parallel to the X axis of the UCS that was current when the
                 dimension was drawn.

3Dface   The new UCS origin is the first point of the *3Dface*, the X axis aligns with the
                 first two points, and the Y positive side is on that of the first and fourth points.
                 (See Chapter 40, Surface Modeling, "*3Dface*.")

Text, Insertion, The new UCS origin is the insertion point of the object, while the new X axis is
or Attribute   defined by the rotation of the object around its extrusion direction. Thus, the object
                 you PICK to establish a new UCS will have a rotation angle of 0 in the new UCS.

Face

The *Face* option operates with <u>solids only</u>. A *Face* is a planar or curved surface on a solid object, **TIP**
as opposed to an edge. When selecting a *Face*, you can place the cursor <u>directly on the desired</u>
<u>face</u> when you PICK, or select an edge of the desired face.

Specify origin of new UCS or [Face/NAmed/OBject/Previous/View/World/X/Y/Z/ZAxis] <World>: *f*
Select face of solid object: **PICK**
(Select desired face of 3D solid to attach new UCS. Edges or faces can be selected.)
Enter an option [Next/Xflip/Yflip] <accept>:

Next
Since you cannot actually PICK a surface, or if you PICK an
edge, AutoCAD highlights a surface near where you picked or
adjacent to a selected edge (Fig. 36-10). Since there are always
two surfaces joined by one edge, the *Next* option causes
AutoCAD to highlight the other of the two possible faces. Press
Enter to accept the highlighted face.

FIGURE 36-10

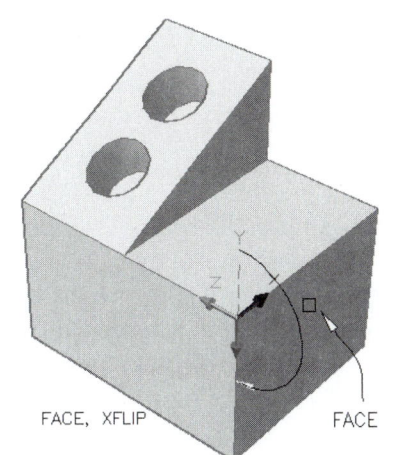

FACE, XFLIP   FACE

Xflip
Use this option to flip the UCS icon 180 degrees about the X axis.
This action causes the Y axis to point in the opposite direction
(see Fig. 36-10).

Yflip
This option causes the UCS icon to flip 180 degrees about the Y
axis. The X axis then points in the opposite direction.

View

   This *UCS* option creates a UCS parallel with the screen (perpendicular to the viewing angle) (Fig. 36-11). The UCS <u>origin</u> remains unchanged. This option is handy if you wish to use the current viewpoint and include a border, title, or other annotation. There are no options or prompts.

FIGURE 36-11

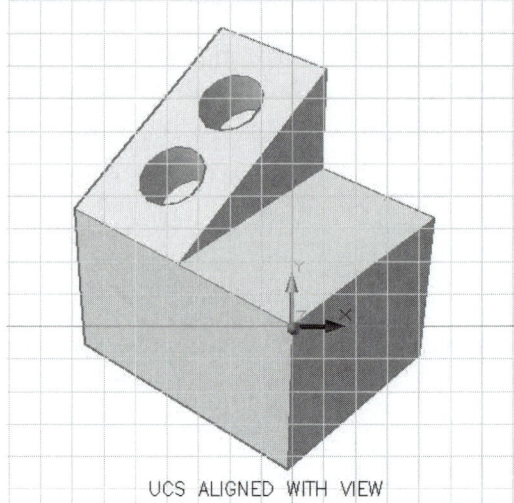

UCS ALIGNED WITH VIEW

X   Y   Z

     Each of these options rotates the UCS about the indicated axis according to the right-hand rule (Fig. 36-12).

The Command prompt is:

   Specify rotation angle about *n* axis <90>:

The angle can be entered by PICKing two points or by entering a value. This option can be repeated or combined with other options to achieve the desired location of the UCS. It is imperative that the right-hand rule is followed when rotating the UCS about an axis (see Chapter 35).

FIGURE 36-12

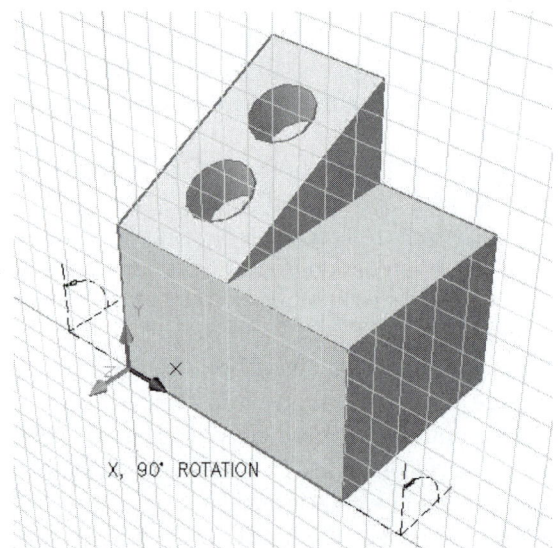

X, 90° ROTATION

Notice that in the prompt for the *X*, *Y*, or *Z* axis rotate options, the default rotation angle is 90 degrees ("<90>" as shown above). If you press Enter at the prompt, the axis rotates by the displayed amount (90 degrees, in this case). You can set the *UCSAXISANG* system variable to another value which will be displayed in brackets when you next use the *X*, *Y*, or *Z*, axis rotate options. Valid values are: 5, 10, 15, 18, 22.5, 30, 45, 90, 180. See "*UCSAXISANG*" in the System Variables section of this chapter.

Previous

   Use this option to restore the previous UCS. AutoCAD remembers the ten previous UCSs used. *Previous* can be used repeatedly to step back through the UCSs.

World

   Using this option makes the WCS (World Coordinate System) the current coordinate system.

Named

The *Named* option allows you to *Save* the current UCS, *Restore* previously named UCSs, and *Delete* named UCSs.

    Command: **ucs**
    Current ucs name: *WORLD*
    Specify origin of UCS or [Face/NAmed/OBject/Previous/View/World/X/Y/Z/ZAxis] <World>: **na**
    Enter an option [Restore/Save/Delete/?]:

Save

Invoking this option prompts for the name you want to assign for saving the current UCS. Up to 256 characters can be used in the name. Entering a name causes AutoCAD to save the <u>current</u> UCS. The *?* option and the UCS Manager list all previously saved UCSs.

Restore

Any *Save*d UCS can be restored with this option. The *Restored* UCS becomes the <u>current</u> UCS. The *?* option and the UCS Manager list the previously saved UCSs.

Delete

You can remove a *Save*d UCS with this option. Entering the name of an existing UCS causes AutoCAD to delete it.

?

This option lists the named UCSs (UCSs that were previously *Saved*). Also listed are the origin coordinates and directions for the X, Y, and Z axes.

NOTE: It is important to remember that changing a UCS or creating a new UCS does not change the display (unless the *UCSFOLLOW* system variable is set to 1). Only one coordinate system can be active in a view. If you are using viewports, each viewport can have its own UCS if desired.

Dducsp

Pull-down Menu	Command (Type)	Alias (Type)	Short-cut	Screen (side) Menu	Tablet Menu
...	Dducsp	...	...	TOOLS 2 Ucsp	W,9

The *Dducsp* command invokes the UCS Manager. It can be used to create new UCSs or to restore existing UCSs. The UCS Manager can also be used to set specific UCS-related system variables, described later.

The *Dducsp* command <u>activates specifically the</u> <u>Orthographic UCSs tab</u> of the UCS Manager (Fig. 36-13). Here you can create new UCSs. *Orthographic UCSs* <u>are</u> <u>not automatically saved</u> as named UCSs but are created as you select them; that is, you can create, save, and restore <u>additional but separate</u> UCSs with the names Front, Top, Right, etc.

FIGURE 36-13

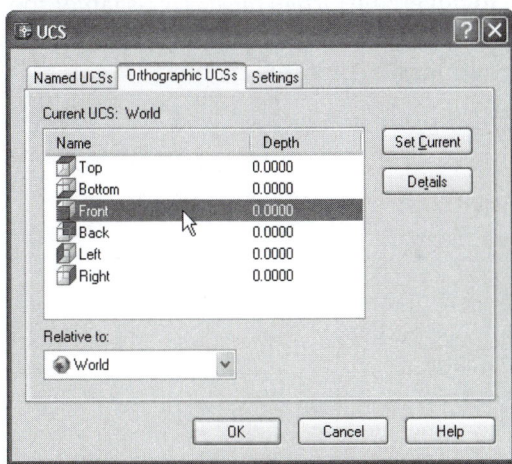

To create one of these *Orthographic UCSs*, you can <u>double-click</u> the desired UCS name, or highlight the desired UCS name (one click) and select the *Current* button. Press *OK* to return to the drawing to see the new UCS. Keep in mind that although the new UCSs are normal to the front, top, right, etc. orthographic views, they are <u>not necessarily attached to the front, top, or right faces of the objects</u>. The new UCSs typically have the same origin as the WCS, unless a new *Depth* is specified or a named UCS is selected in the *Relative to:* drop-down list.

Depth

Notice in the *Orthographic UCS* tab that each UCS can have a specified depth. *Depth* is a Z dimension or distance perpendicular to the XY plane to locate the new UCS origin. Select the *Depth* <u>value</u> of any UCS to produce the *Orthographic UCS Depth* dialog box (Fig. 36-14).

FIGURE 36-14

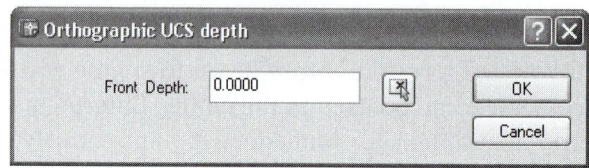

Details

Select the *Details* button to see the coordinates for the origin of the highlighted UCS. The direction vector coordinates for the X, Y, and Z axes are also listed. This has the same function as the *?* option of the *UCS* command.

Relative to:

The *Relative to:* drop-down list allows you to select any named UCS as the base UCS. When the *Front*, *Top*, *Right*, or other UCS option is selected, it is created in a orthographic orientation with respect to the origin and orientation of the selected named UCS. Typically, you want to <u>ensure *World* is selected as the base UCS</u> unless you have some other specific application. The setting in the *Relative to:* list is stored in the *UCSBASE* system variable.

Ucsman

Pull-down Menu	Command (Type)	Alias (Type)	Short-cut	Screen (side) Menu	Tablet Menu
Tools *Named UCS...*	*Ucsman*	*UC*	...	*TOOLS 2* *Ucsman*	*W,8*

The *Ucsman* command invokes the UCS Manager (Fig. 36-15). The UCS Manager has three tabs: *Named UCSs*, *Orthographic UCSs* and *Settings*. The UCS Manager can be accessed by several methods shown in the table above. However, the *Orthographic UCSs* tab of the UCS Manager is also accessed by the *Dducsp* command (see *Dducsp* command earlier this chapter).

FIGURE 36-15

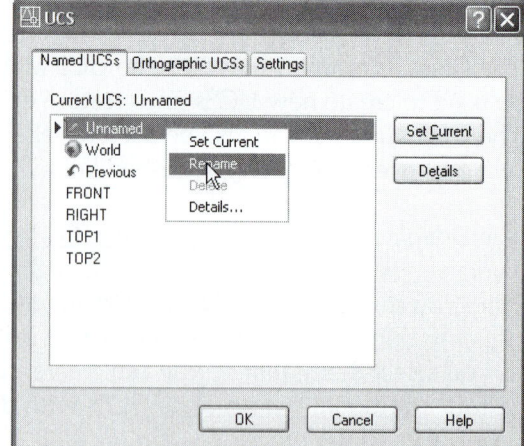

Named UCSs Tab

Ucsman produces the *Named UCS* tab of the UCS Manager (see Fig. 36-15). The list contains all named UCSs. Use this tab to make named UCSs *Current*, just as you would use the *Restore* option of the *Ucs* command. You can create a named UCS by first creating the UCS, then using the UCS Manager to *Rename* the *Unnamed* UCS as shown in Figure 36-15. You can also create a named UCS by using the *Save* option of the *UCS* command. In this tab you can <u>double-click</u> the desired UCS name or highlight (single-click) the desired name and then select *Set Current*. Either way, you must then select *OK* to restore the desired UCS. The *Details* button lists the origin coordinates and X, Y, and Z direction vectors of the highlighted UCS.

Settings Tab

The *Settings* tab can be accessed directly only by invoking the UCS Manager. However, you can use the *Dducsp* or *Dducs* commands or buttons explained earlier to produce the other tabs, then change tabs. The *Settings* tab (Fig. 36-16) allows you to control the UCS icon and two system variables.

FIGURE 36-16

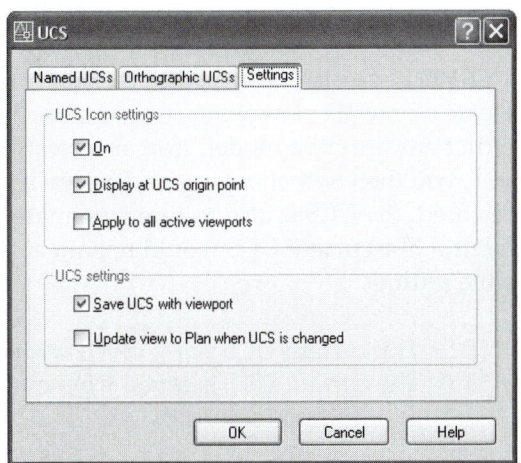

UCS Icon settings

Settings in this section are identical to controls of the *Ucsicon* command (see the *Ucsicon* command in Chapter 35). The *On* checkbox toggles the UCS icon on or off. *Display at UCS origin point* moves the icon to the coordinate system origin (otherwise it is located in the lower left corner of the screen) and has the identical functions as the *Origin* and *Noorigin* options of the *Ucsicon* command. *Apply to all viewports* is the same as the *All* option of the *Ucsicon* command; that is, the settings made in this section apply to the icons in all active viewports when multiple viewports are used.

UCS Settings

These two settings control two UCS-related system variables that affect only the <u>currently active viewport</u> when multiple viewports are used. *Save UCS with viewport* controls the *UCSVP* system variable and affects the <u>UCS of the current viewport</u>. When you are using multiple viewports, you have the option of locking the orientation of a UCS with a specific viewport so that when a new UCS is created or an existing UCS is restored, the locked UCS remains unchanged (that is, the UCS of the viewport that is active when the box is checked remains unchanged). *Update view to Plan when UCS is changed* controls the *UCSFOLLOW* system variable which affects the <u>View of the active viewport</u>. The viewport that is active when this box is checked will automatically change to a plan view of a new UCS whenever a new UCS is created or an existing UCS restored. When both boxes are checked, *Update view to Plan when UCS is changed* overrides *Save UCS with viewport*. (See "UCS System Variables" next in this chapter.)

UCS SYSTEM VARIABLES

AutoCAD offers a great amount of control and flexibility for UCSs, such as the ability to automatically create UCSs when creating views, saving UCSs with views, and having multiple UCSs current when using multiple viewports. Because of these options, it is easy to be overwhelmed if you are not aware of the system variables that control UCSs. This section is provided to help avoid confusion and to offer guidelines for UCS use. Not all UCS system variables are explained here, but only those that are typically accessed by a user.

UCSBASE

UCSBASE stores the name of the UCS that defines the origin and orientation when *Orthographic* UCSs are created using the *UCS, Orthographic* option or the *Orthographic UCSs* tab of the UCS Manager. Normally the World Coordinate System is used as the base UCS, so when a new orthographic UCS is created it is simply rotated about the WCS origin to be normal to the selected orthographic view (top, front, right, etc.). *UCSBASE* can also be set in the *Relative to:* drop-down list of the *Orthographic UCSs* tab of the UCS Manager (see Fig. 36-13).

 For beginning users, it is suggested that this variable be left at its initial value, *World*. This variable is saved in the current drawing.

UCSVIEW

UCSVIEW determines whether the current UCS is saved with a <u>named</u> view. For example, suppose you create a new 3D viewpoint, such as a southeast isometric, to enhance your visualization while you construct one area of a model. You also create a UCS on the desired construction plane. With *UCSVIEW* set to 1, you then *Save* that view and assign a name such as RIGHT. Whenever the RIGHT view is later restored, the UCS is also restored, no matter what UCS is current before restoring the view. Without this control, the current UCS would remain active when the RIGHT view is restored. *UCSVIEW* has two possible settings.

    0   The current UCS is not saved with a named view.
    1   The current UCS is saved whenever a named view is
         saved (initial setting).

 UCSVIEW is easily remembered because it saves the *UCS* with the *VIEW*. Remember that <u>the view must be named to save its UCS</u>.

UCSORTHO

This variable determines whether an orthographic UCS is automatically created when an orthographic view is created or restored. For example, if *UCSORTHO* is set to 1 and you select or create a front view, a UCS is automatically set up to be parallel with the view. In this way, the XY plane of the UCS is always parallel with the screen when an orthographic view is restored. The settings are as follows.

    0   Specifies that the UCS setting remains unchanged when an orthographic view is restored.
    1   Specifies that the related orthographic UCS is automatically created or restored when an ortho-
         graphic view is created or restored (initial setting).

 In order for *UCSORTHO* to operate as expected, you must select a preset orthographic view from the *View* pull-down menu, *Orthographic* option of the *View* command, or an orthographic view icon button. *UCSORTHO* does <u>not</u> operate with views created using *Vpoint*, *Plan*, the *3Dorbit* commands, or for any isometric preset views. *UCSORTHO* operates for views as well as viewports. *UCSORTHO* is saved in the current drawing.

A good exercise you can use to understand *UCSORTHO* is to use *Vports* to create four tiled viewports in model space and select the *3D Setup*. With *UCSORTHO* set to 1, the arrangement shown in Figure 36-17 would result. Here, each orthographic view has its own orthographically aligned UCS. Note the grid and UCS icon display in each viewpoint.

FIGURE 36-17

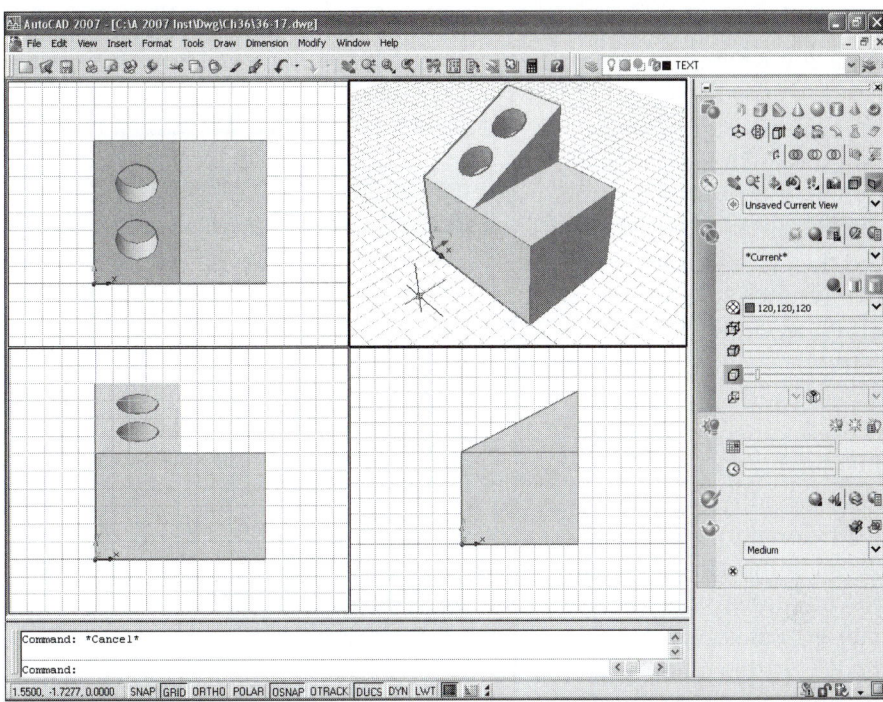

If *UCSORTHO* is set to 0 and the *Vports* command is used to create a *3D Setup*, the arrangement shown in Figure 36-18 would be created. Notice that the WCS is the only coordinate system in all viewports. Note the display of the grid and the UCS icons.

FIGURE 36-18

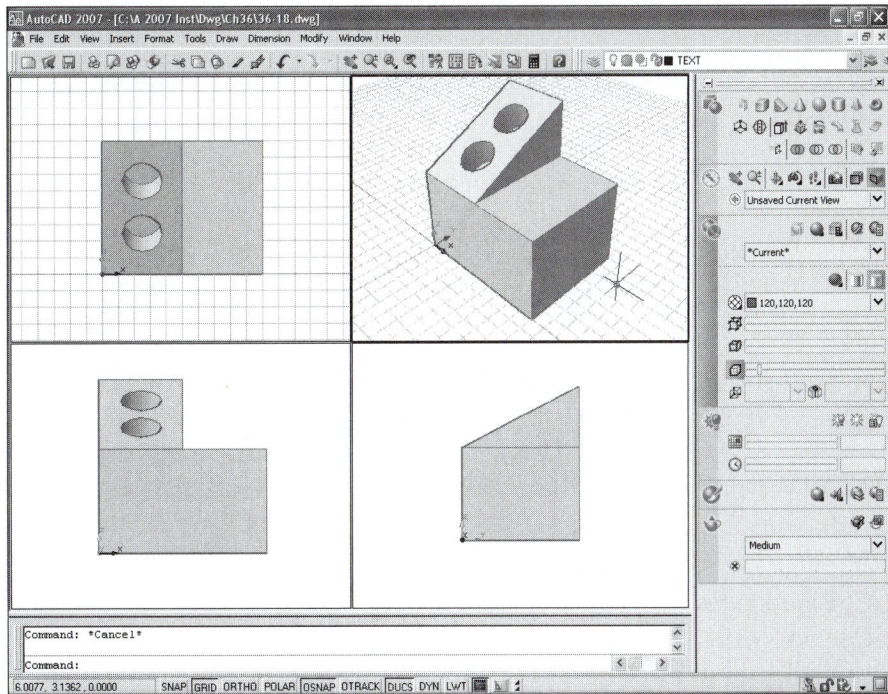

A point to help remember *UCSORTHO* is <u>when the view changes to an orthographic view, the UCS changes</u>.

UCSFOLLOW

The *UCSFOLLOW* system variable, if set to 1 (*On*), causes the <u>plan view</u> of a UCS to be displayed auto-matically <u>when a UCS is made current</u>. *UCSFOLLOW* can be set separately for each viewport.

For example, consider the case shown in Figure 36-19. Two tiled viewports (*Vports*) are being used, and the *UCSFOLLOW* variable is turned *On* for the <u>left</u> viewport only. Notice that a UCS created on the inclined surface is current. The left viewport shows the plan view of this UCS automatically, while the right viewport shows the model in the same orientation no matter what coordinate system is current. If the WCS were made active, the right viewport would keep the same viewpoint, while the left would automatically show a plan view of the new UCS when the change was made.

FIGURE 36-19

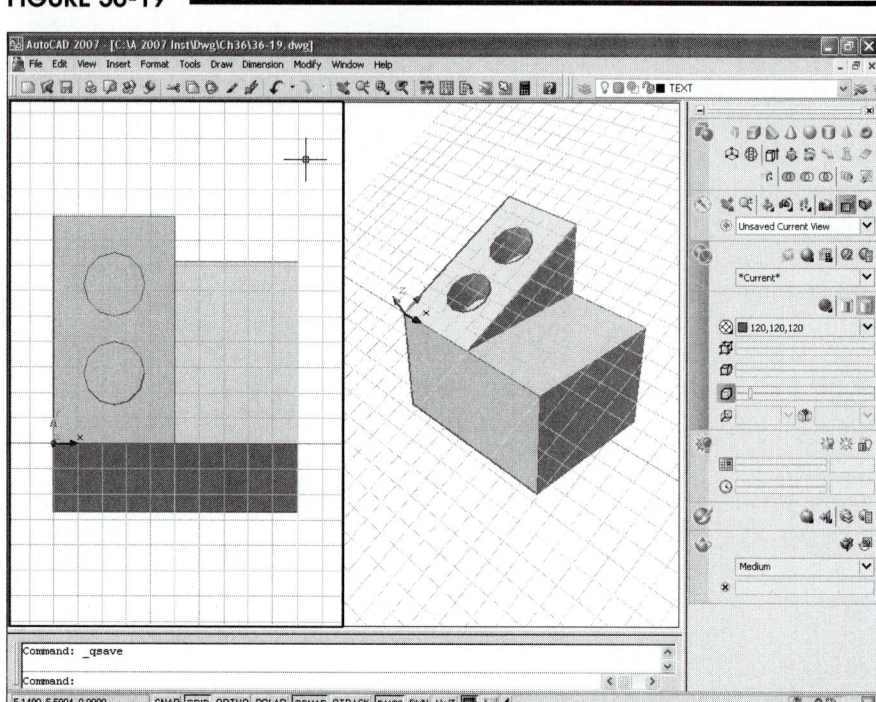

NOTE: When *UCSFOLLOW* is set to 1, the display does not change <u>until</u> a new UCS is created or restored.

The *UCSFOLLOW* settings are as follows.

> 0   Changing the UCS does not affect the view (initial setting).
> 1   Any UCS change causes a change to plan view of the new UCS in the viewport current when *UCSFOLLOW* is set.

The setting of *UCSFOLLOW* affects only model space. *UCSFOLLOW* is saved in the current drawing and is viewport specific.

A point to help remember *UCSFOLLOW* is <u>when the UCS changes, the view follows</u>.

UCSVP

UCSVP operates with model space and paper space viewports and is viewport specific (<u>can be set for each viewport</u>). It determines whether the UCS in the active viewport remains locked or changes when another UCS is made current in another viewport. *UCSVP* is similar to *UCSVIEW* where the UCS is saved with the *View*, only with *UCSVP* the UCS is saved with the viewport. The possible *UCSVP* settings are shown below.

> 0   The UCS setting is not locked to the viewport so the UCS reflected becomes that of any other viewport that is made active.
> 1   The viewport and UCS are locked. Therefore, the UCS is saved in that viewport and does not change when another viewport displaying another UCS becomes active (initial setting).

For example, suppose *UCSVP, UCSORTHO,* and *UCSVIEW* are all set to 0 (off) so there is no relationship established between views, viewports, and UCSs. In that case, only one UCS in the drawing could be active and would therefore appear in all viewports, similar to the arrangement shown in Figure 36-20. If a new UCS were created, it would appear in all viewports.

FIGURE 36-20

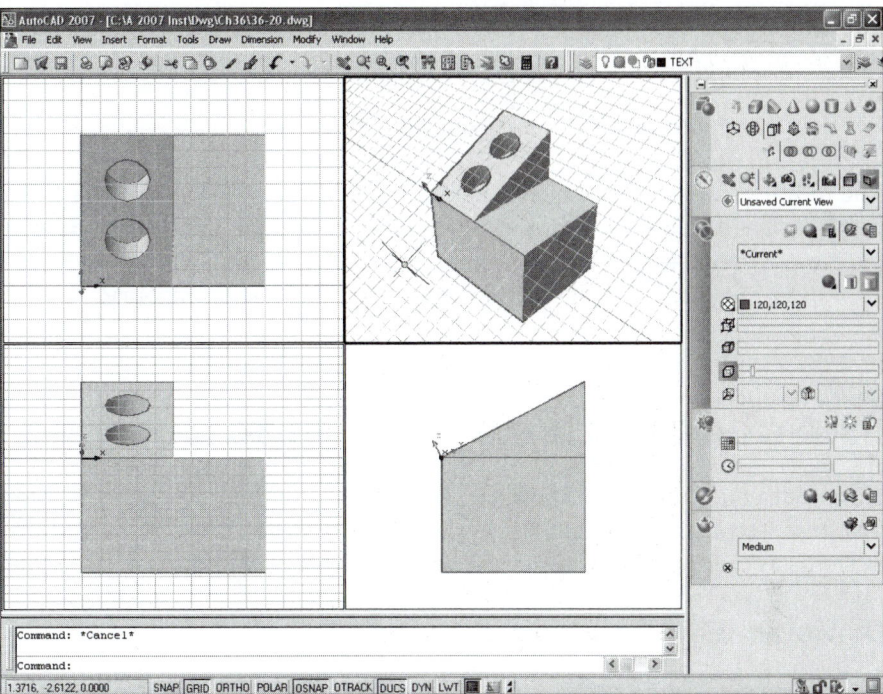

Assume *UCSVP* is then set to 1 in the upper-right (isometric) viewport. Suppose then another viewport were made active and the WCS restored. Now the WCS appears in all viewports <u>except</u> the isometric viewport where the UCS is locked (*UCSVP*=1) as shown in Figure 36-21.

FIGURE 36-21

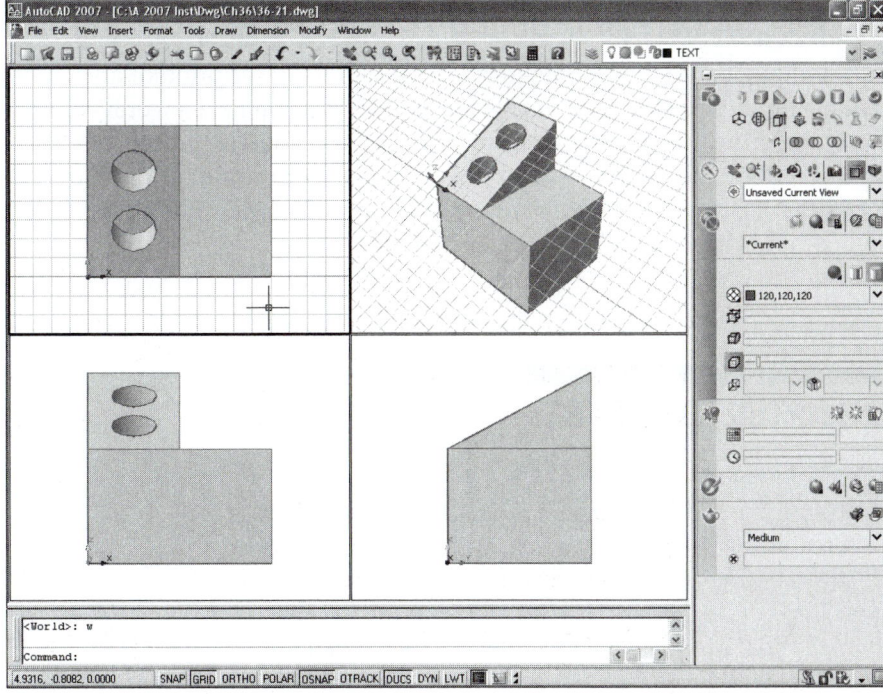

Remember that *UCSVP* saves a UCS to a <u>viewport</u>, whereas *UCSVIEW* saves a UCS to a named *View*.

UCSAXISANG

This variable stores the default angle when rotating the UCS around one of its axes using the *X*, *Y*, or *Z* options of the *UCS* command. When one of these options is used, the following prompt is displayed.

Specify rotation angle about *n* axis <90>:

The initial value is 90, so when you use one of these options to rotate the UCS, you can press Enter to rotate 90 degrees instead of entering a value. If you want another angle to appear as the default (in the < > brackets), enter the value in the *UCSAXISANG* variable. Valid values are: 5, 10, 15, 18, 22.5, 30, 45, 90, 180. The variable setting is saved in the system registry.

USER COORDINATE SYSTEMS AND 3D MODELING

Now that you understand the basics of 3D modeling discussed in Chapters 35 and 36, you are ready to progress to constructing and editing wireframe, surface, and solid models. The use of User Coordinate Systems as discussed in Chapter 36 are applicable to <u>all types</u> of 3D modeling.

Dynamic User Coordinate Systems

In AutoCAD 2007, you can create Dynamic UCSs. Dynamic UCSs, however, can be created only with solid models. Dynamic UCSs, are similar to other UCSs, but they are created "on the fly" during a primitive creation command such as *Box*, *Wedge*, or *Cylinder*. Therefore, dynamic UCSs are temporary coordinate systems that cannot be saved or named. Dynamic UCSs are discussed in Chapter 38.

The Progression for Learning 3D Modeling

In the sequence from an elementary to a complete description of an actual object, the three types of models range from wireframe to surface to solid. The next several chapters, however, present the modeling types in a different progression: wireframe models, solid models, then surface models. This is a logical order from simple to complex with respect to command functions and construction and editing complexity. That is to say, wireframe modeling is the simplest and surface modeling is the most tedious and complex type of 3D modeling.

CHAPTER EXERCISES

1. Using the model in Figure 36-22, assume a UCS was created by rotating about the X axis 90 degrees from the existing orientation (World Coordinate System).

    A. What are the absolute coordinates of corner F?

    B. What are the absolute coordinates of corner H?

    C. What are the absolute coordinates of corner J?

FIGURE 36-22

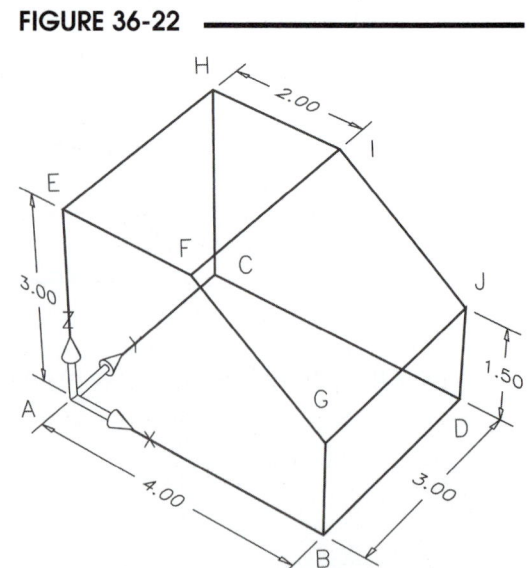

2. Using the model in Figure 36-22, assume a UCS was created by rotating about the X axis 90 degrees from the existing orientation (WCS) and then rotating 90 degrees about the (new) Y.

    A. What are the absolute coordinates of corner F?

    B. What are the absolute coordinates of corner H?

    C. What are the absolute coordinates of corner J?

3. *Open* the **SOLID1** drawing that you created in Chapter 35, Exercise 11. View the object from a *SE Isometric* view. Make sure that the *UCSicon* is *On* and set to the *ORigin*.

    FIGURE 36-23

    A. Create a *UCS* with a vertical XY plane on the front surface of the model as shown in Figure 36-23. *Save* the UCS as **FRONT**.

    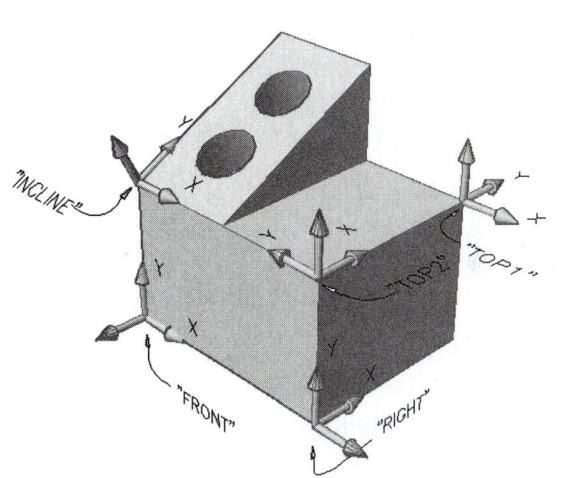

    B. Change the coordinate system back to the *World*. Now, create a *UCS* with the XY plane on the top horizontal surface and with the orientation as shown in Figure 36-23 as TOP1. *Save* the UCS as **TOP1**.

    C. Change the coordinate system back to the *World*. Next, create the *UCS* shown in the figure as TOP2. *Save* the UCS under the name **TOP2**.

    D. Activate the *WCS* again. Create and *Save* the **RIGHT** UCS.

    E. Use *SaveAs* and change the drawing name to **SOLIDUCS**.

4. In this exercise, you will create two viewport configurations, each with a different setting for *UCSORTHO*. In the first configuration, only one UCS exists, and in the second configuration, a new UCS is created for each viewport.

    A. *Open* the **SOLID1** drawing again. Use *SaveAs* and change the name to **VPUCS1**. Set the *UCSORTHO* variable to **0** so no new UCSs are created when you set up viewports. Ensure you are in model space and use the *Vports* command to create *Four Equal* viewports. Also in the *Viewports* dialog box, select the *3D Setup*. The resulting drawing should look like that in Figure 36-18. Notice that only one UCS appears in all viewports. Next, use any viewport and create a new UCS in the same orientation as the "Front" UCS shown in Figure 36-23. What happens in the other viewports? *Close* **VPUCS1** and *Save* the changes.

B. *Open* the **SOLID1** drawing again. Use *SaveAs* and change the name to **VPUCS2**. This time, set the *UCSORTHO* variable to **1** so when you set up viewports new orthographically oriented UCSs are created. Set up the same viewport configuration as in the previous step. Your new viewport configuration and the resulting new orthographic UCSs should look like that in Figure 36-17. In the isometric view create a new UCS on the inclined plane (like the "Incline" UCS in Fig. 36-23). What happens to the UCSs in the other viewports? *Close* **VPUCS2** and *Save* the changes.

5.   Using the same drawing, make the *WCS* the active coordinate system. In this exercise, you will create a model and screen display like that in Figure 36-19.

A. Set the *UCSORTHO* variable to **0** so no new UCSs are created when you set up viewports. Create **2** *Vertical* tiled viewports. Use *Zoom* in each viewport if needed to display the model at an appropriate size. Make sure the *UCSicon* appears at the origin. Activate the left viewport. Set *UCSFOLLOW On* in that viewport only. (The change in display will not occur until the next change in UCS setting.)

B. Use the right viewport to create the **INCLINE UCS,** as shown in Figure 36-23. *Regenall* to display the new viewpoint.

C. In either viewport create two cylinders as follows: Type *cylinder*; specify **1.25,1** as the *center*, **.5** as the *radius*, and **-1** (negative 1) as the *height*. The new cylinder should appear in both viewports.

D. Use *Copy* to create the second cylinder. Specify the existing center as the "Basepoint, of displacement." To specify the basepoint, you can enter coordinates of **1.25,1** or **PICK** the center with *Osnap*. Make the copy **1.5** units above the original (in a Y direction). Again, you can PICK interactively or use relative coordinates to specify the second point of displacement.

E. Use *SaveAs* and assign the name **SOLID2**.

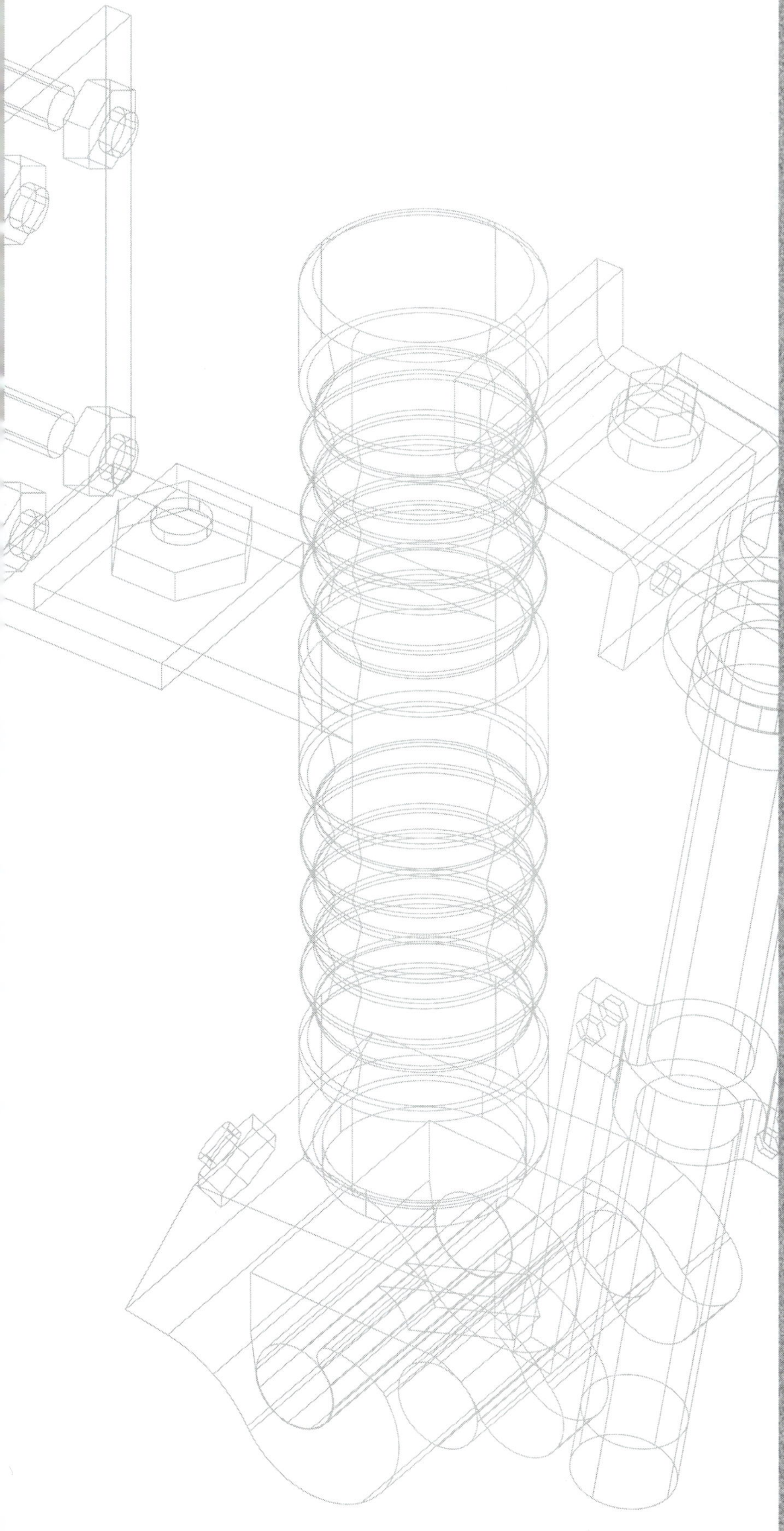

CHAPTER

37

WIREFRAME MODELING

CHAPTER OBJECTIVES

After completing this chapter you should:

1. understand how common 2D *Draw* and *Modify* commands are used to create 3D wireframe models;

2. gain experience in specification of 3D coordinates;

3. gain experience with 3D viewing commands;

4. be able to apply point filters to 3D model construction;

5. gain fundamental experience with creating, *Saving*, and *Restoring* User Coordinate Systems;

6. be able to create, *Save*, and *Restore* tiled viewport configurations;

7. be able to manipulate the Coordinate System icon.

CONCEPTS

Wireframe models are created in AutoCAD by using common draw commands that create 2D objects. When draw commands are used with X ,Y, and Z coordinate specification, geometry is created in 3D space. A true wireframe model is simply a combination of 2D elements in 3D space.

A *Line*, for example, can be drawn in 3D space by specifying 3D coordinates in response to the "Specify first point:" and "Specify next point or [Undo]:" prompts (Fig. 37-1):

FIGURE 37-1

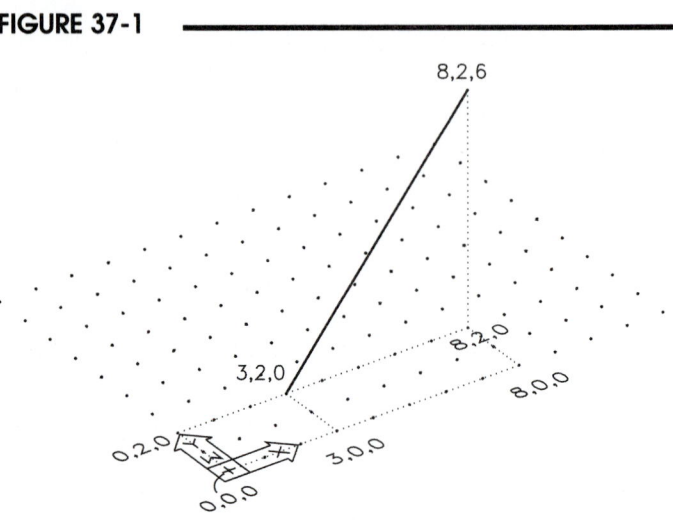

Command: **line**
Specify first point: **3,2,0**
Specify next point or [Undo]: **8,2,6**
Specify next point or [Undo]: **Enter**
Command:

The same procedure of entering 3D coordinate values can be applied to other draw commands such as *Arc, Circle,* etc. These 2D objects are combined in 3D space to create wireframe models. No draw commands are used <u>specifically</u> for creating 3D wireframe geometry (with the exception of *3Dpoly, Helix,* and *Xedges*). The use of 3D navigation, User Coordinate Systems (UCS), and 3D coordinate entry are the principal features of AutoCAD, along with the usual 2D object creation commands, and they allow you to create wireframe models. Specification of 3D coordinate values, 3D display and viewing (Chapter 35), and the use of UCSs (Chapter 36) aid in creating wireframe models as well as surface and solid models. The *3Dpoly* command is discussed at the end of this chapter.

WIREFRAME MODELING TUTORIAL

Because you are experienced in 2D geometry creation, it is efficient to apply your experience, combined with the concepts already given in the previous three chapters, to create a wireframe model in order to learn fundamental modeling techniques. Using the tutorial in this chapter, you create the wireframe model of the stop plate shown in Figure 37-2 in a step-by-step fashion. The Wireframe Modeling Tutorial employs the following fundamental modeling techniques:

FIGURE 37-2

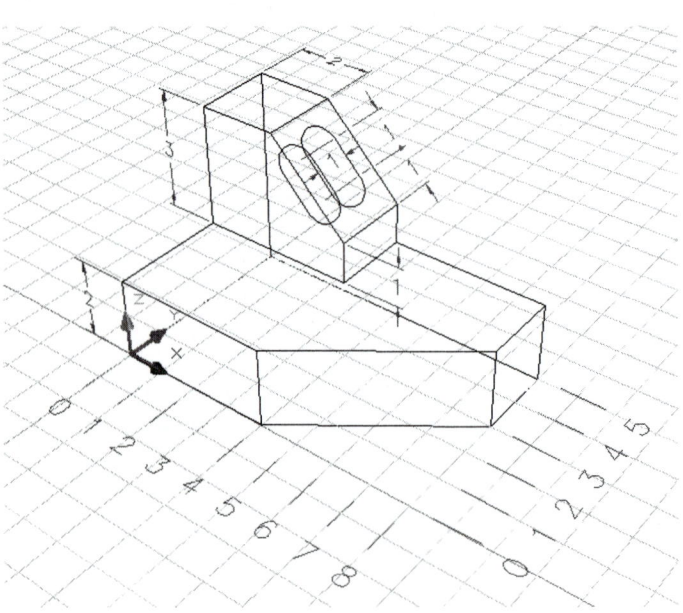

3D navigation commands
3D coordinate entry
Polar Snap
Polar Tracking
Vports
User Coordinate Systems

It is suggested that you follow along and work on your computer as you read through the pages of this tutorial. The Wireframe Modeling Tutorial has 22 steps. Additional hints, such as explicit coordinate values and menus that can be used to access the commands, are given on pages immediately following the tutorial (titled "Hints"). Refer to "Hints" only if you need assistance.

1. Set up your drawing.

   A. Start AutoCAD. Begin a new drawing in 3D mode. Use **Save** and name the file **WIREFRAME**.

   B. Examine the figure on the previous page showing the completed wireframe model. Notice that all of the dimensions are to the nearest unit. For this exercise, assume generic units.

   C. Keep the default **Decimal Units**. Set **Units Precision .00** (two places to the right of the decimal).

   D. Keep the default *Limits* (assuming they are already set to 0,0 and 12,9 or close to those values). It appears from Figure 37-2 that the drawing space needed is a minimum of 8 units by 5 units on the XY plane.

   E. Turn on **Grid Snap** or **Polar Snap** and **Polar Tracking**. Set the increments to 1.

   F. Turn on the *GRID*.

   G. Make a layer named **MODEL**. Set it as the **Current** layer. Set a **Color**, if desired.

2. Draw the base of the part on the XY plane.

   A. Use **Line**. Begin at the origin (0,0,0) and draw the shape of the base of the stop plate. (It is a good practice to begin construction of any 3D model at 0,0,0.) Use **Grid Snap** or **Polar Snap** as you draw. See Figure 37-3.

   B. Use the **3Dorbit** command to generate a view similar to that shown in Figure 37-3.

FIGURE 37-3 ──────────

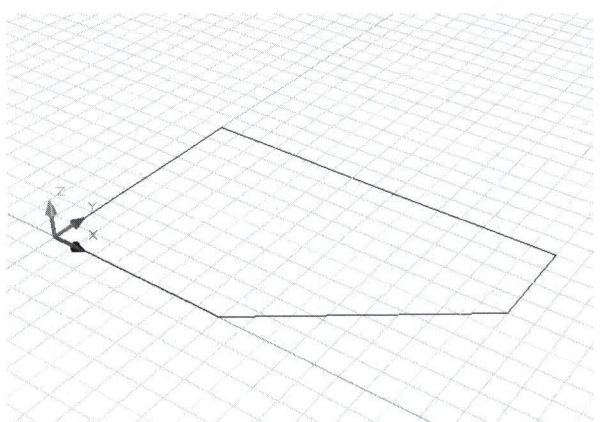

3. Begin drawing the lines bounding the top "surface" of the base by using absolute coordinates and *Polar Tracking*.

   A. Draw the first two *Lines*, as shown in Figure 37-4, by entering absolute Cartesian coordinates. You cannot PICK points interactively (without *OSNAP*) since it results in selections only on the XY plane of the current coordinate system. Do not exit the *Line* command when you finish.

FIGURE 37-4 ──────────

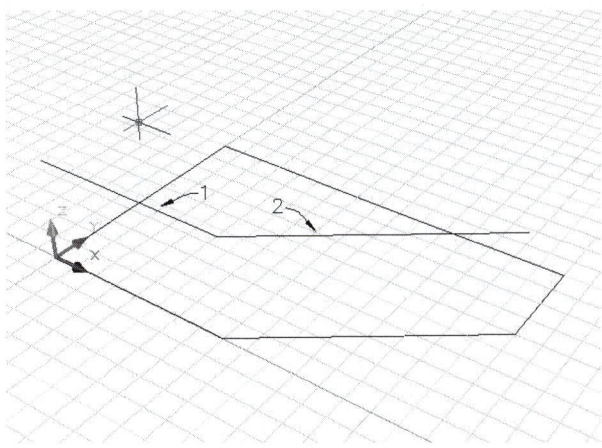

B. Draw *Lines* 3 and 4 using *Polar Snap* and *Polar Tracking* (Fig. 37-5). (If you have trouble using *Polar Snap* in *Perspective* mode, use **Direct Distance Entry**.) Do not exit the *Line* command.

C. Complete the shape defining the top surface (*Line* 5) by using the **Close** option. The completed shape should look like Figure 37-5.

FIGURE 37-5

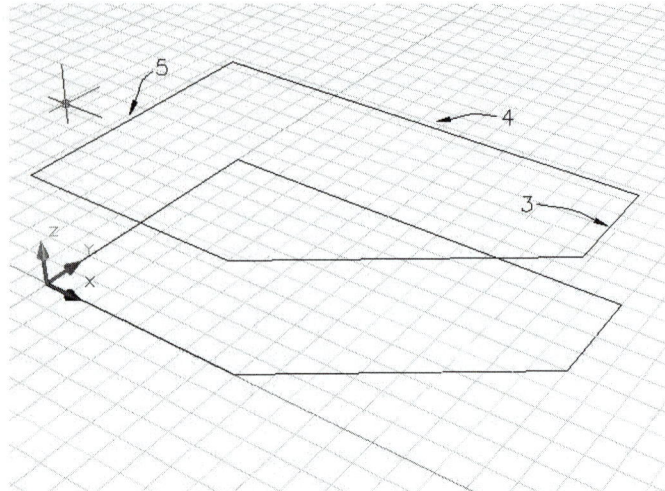

4. There is an easier way to construct the top "surface"; that is, use *Copy* to create the top "surface" from the bottom "surface."

A. *Erase* the 5 lines defining the top shape.

B. Use *Copy* and select the 5 lines defining the bottom shape. You can PICK **0,0,0** as the "base point." The "second point of displacement" can be specified by entering absolute coordinates or relative coordinates or by using *Polar Tracking* along the Z axis. Compare your results with Figure 37-5.

5. Check your work and *Save* the drawing.

A. Use *Plan* to see if all of the *Lines* defining the top shape appear to be aligned with those on the bottom. The correct results at this point should look like one shape since the top shape is directly above the bottom.

B. Use *Zoom Previous* to display the previous view.

C. If you have errors, correct them. You can use *Undo* or *Erase* the incorrect *Lines* and redraw them. See "Hints" for further assistance. You may have to use *OSNAP* or enter absolute coordinates to begin a new *Line* at an existing *Endpoint*.

D. *Save* your drawing.

6. Draw the vertical *Lines* between the base and top "surfaces" using the following methods.

A. Enter absolute coordinates (X, Y, and Z values) to draw the first two vertical *Lines* shown in Figure 37-6 (1, 2).

B. Use another method (*Osnap*) to draw the remaining three *Lines* (3, 4, 5). You have learned that if you PICK a point, the selected point is on the <u>XY plane</u> of the current coordinate system. *OSNAP*, however, allows you to PICK points in <u>3D space</u>. Begin each of the three *Lines* ("Specify first point:") by using *Endpoint* and picking the appropriate point on the line ends. The second point of each *Line* ("Specify next point or [Undo]:") can be specified also by using *OSNAP* (use *Endpoint*).

FIGURE 37-6

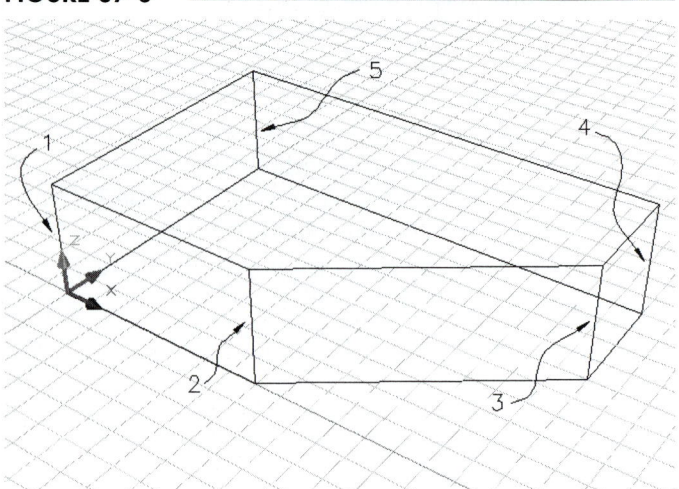

C. Once again, there is an easier method. *Erase* four lines and <u>leave one</u>. You can use *Copy* to copy
the one vertical *Line* around to the other positions. In response to the "base point or displacement:" prompt, select the base of the remaining *Line* (on the XY plane) with *Grid Snap* on. In
response to "second point of displacement:" make the multiple copies.

D. Use a *Plan* view to check your work. *Zoom Previous*. Correct your work, if necessary.

7. Draw two *Lines* defining a vertical plane. Often in the process of beginning a 3D model (when
geometry has not yet been created in some positions), it may be necessary to use absolute coordinates.

A. Use absolute Cartesian coordinates to
draw the two *Lines* shown in Figure
37-7. Refer to Figure 37-2 for the
dimensions.

B. Use *Pan* and *Zoom* if needed to
improve your display of the model to
show the new *Lines* clearly. Your
display should look similar to Figure
37-7.

FIGURE 37-7 ———————————————

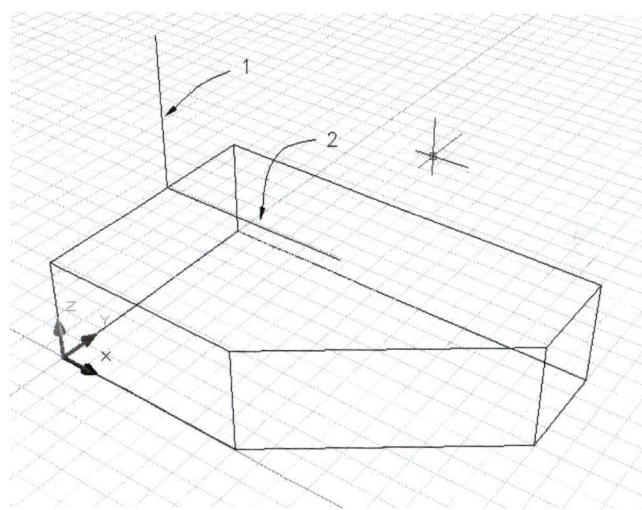

8. Rather than drawing the next objects by defining absolute coordinates, you can establish a vertical
construction plane and then PICK points with *Grid Snap* and *GRID* or *Polar Snap* and *Polar Tracking*
on the construction plane.

A. This can be done by creating a new
UCS (User Coordinate System).
Invoke the *UCS* command by typing
UCS or by selecting it from the menus.
Use the *3point* option and select the
line *Endpoint*s in the order indicated
(Fig. 37-8) to define the origin, point on
X axis, and point on Y axis.

FIGURE 37-8 ———————————————

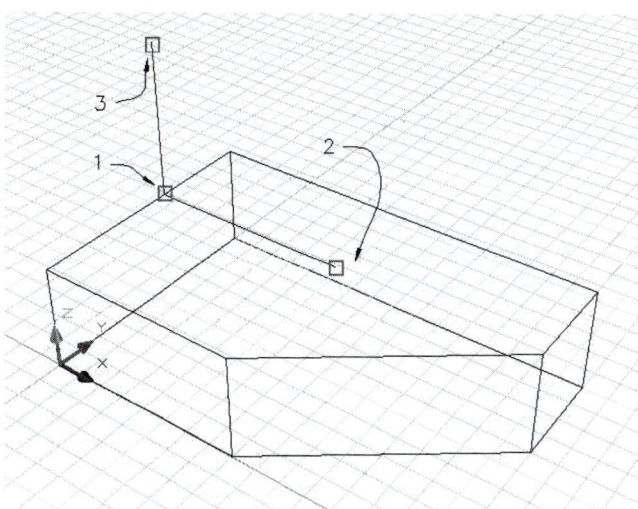

B. The Coordinate System icon should appear in the position shown in Figure 37-9. Notice that the *GRID* and *SNAP* have followed the new UCS. Move the cursor and note the orientation of the cursor. You can now draw on this vertical construction plane simply by PICKing points on the XY plane of the new UCS!

If your UCS does not appear like that in the figure, *UNDO* and try again.

FIGURE 37-9

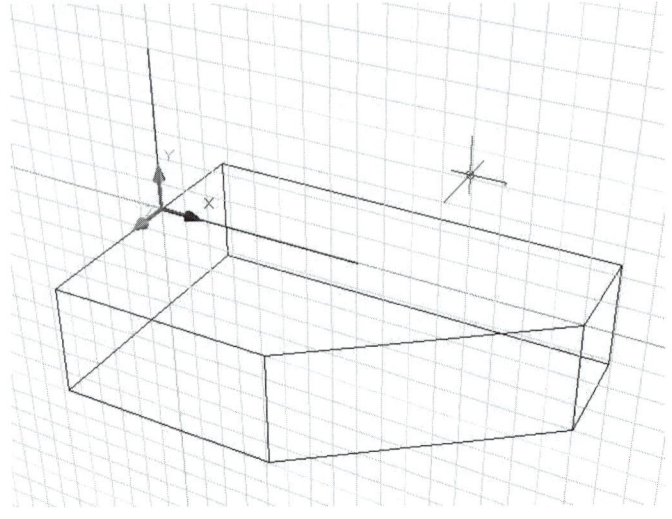

C. Draw the *Lines* to complete the geometry on the XY plane of the UCS. Refer to Figure 37-2 for the dimensions. The stop plate should look like that in Figure 37-10.

9. Check your work; save the drawing and the newly created UCS.

A. View your drawing from a plan view of the UCS. Invoke *Plan* and accept the default option "*<current>*." If your work appears to be correct, use *Zoom Previous*.

B. Save your new UCS. You can use either the *UCS Manager* or the *UCS* command to do this.

FIGURE 37-10

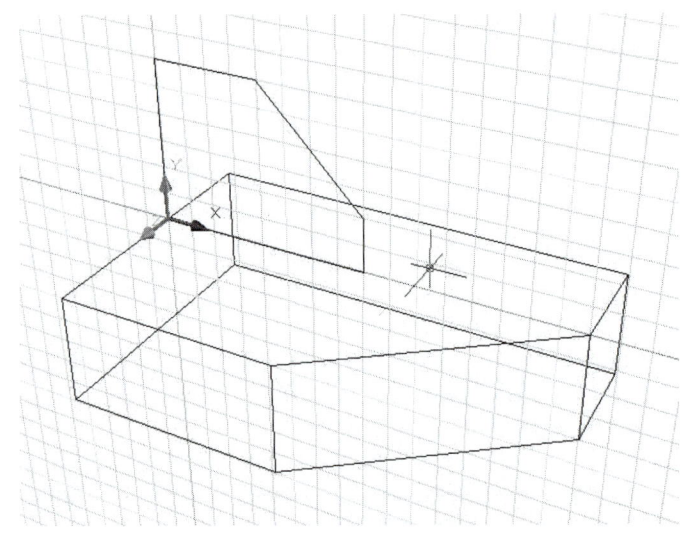

10. Create a second vertical surface behind the first with *Copy* (Fig. 37-11). You can give coordinates relative to the (new) current UCS.

A. Invoke *Copy*. Select the four lines (highlighted) comprising the vertical plane (on the current UCS). Do not select the bottom line defining the plane. Specify **0,0** as the "base point:" and give either absolute or relative coordinates (with respect to the current UCS) as the "second point of displacement." Note that the direction to *Copy* is negative Z.

FIGURE 37-11

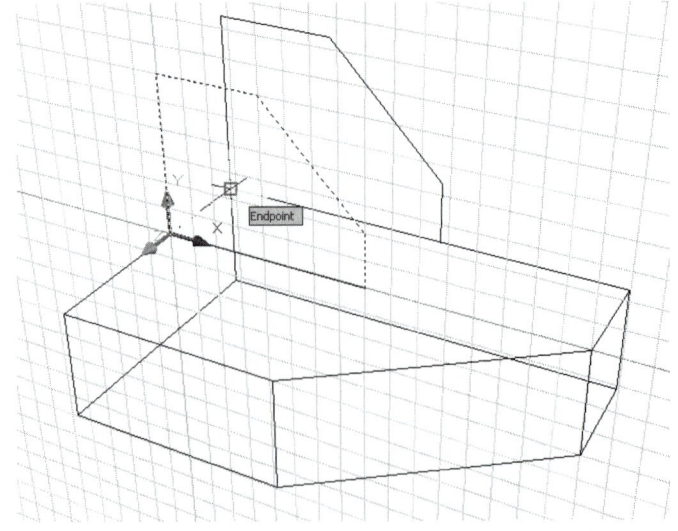

B. Alternately, using *Copy* with *OSNAP* would be valid here since you can PICK (with *OSNAP*) the back corner. Another alternative is to turn on *POLAR* and track along the negative Z axis, then enter **2** as the direct distance.

11. Draw the horizontal lines between the two vertical "surfaces" using one of several possible methods.

FIGURE 37-12

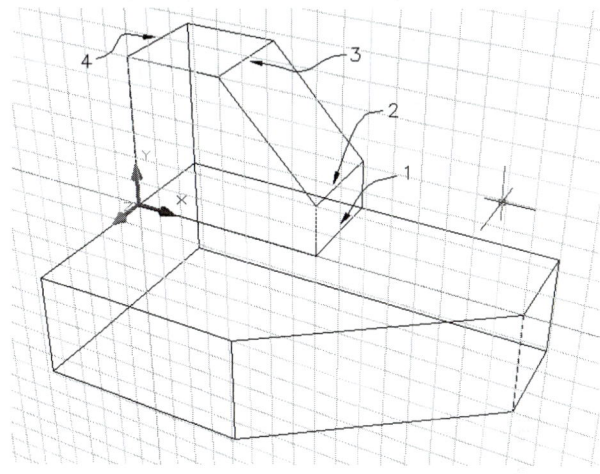

A. Use the *Line* command to connect the two vertical surfaces with four horizontal edges as shown in Figure 37-12. With *SNAP* on, you can **PICK** points on the XY plane of the UCS. Use *OSNAP* to **PICK** the *Endpoint*s of the lines on the back vertical plane.

B. As another possibility, you could draw only one *Line* by the previous method and *Copy* it to the other three locations. In this case, **PICK** the "base point" on the XY plane (with *SNAP ON*). Specifying the "second point(s) of displacement" is easily done (also at *SNAP* points).

12. A wireframe model uses 2D objects to define the bounding edges of "surfaces." There should be no objects that do not fulfill this purpose. If you examine Figure 37-13, you notice two lines that should be removed (highlighted). These lines do not define edges since they each exist between coplanar "surfaces." Remove these lines with *Trim*.

FIGURE 37-13

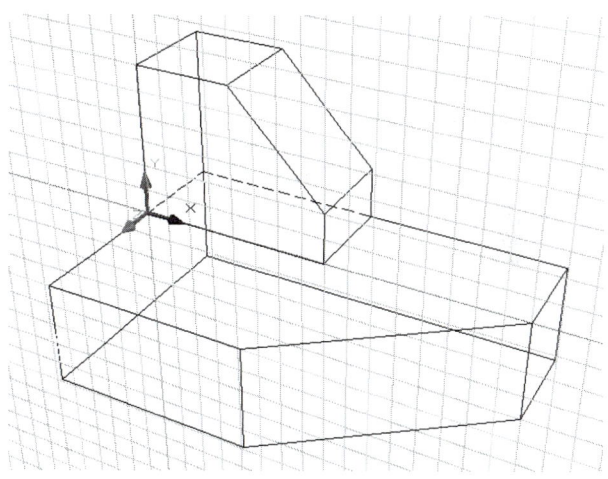

Trimming 2D objects in 3D space requires special attention. You have two choices for using *Trim* in this case: *Projection* = *None* or *View*. You can use the *View* option of *Projection* (to trim any objects that appear to intersect from the current view) or you must select the vertical *Line* as a cutting edge (because with the *None* option of *Projection*, you cannot select a cutting edge that is perpendicular to the XY plane of the current coordinate system). In other words, when *Projection=None*, you can select only the vertical *Line* as the cutting edge, but when *Projection=View*, either the vertical or horizontal *Line* can be used as the cutting edge (Fig. 37-14).

FIGURE 37-14

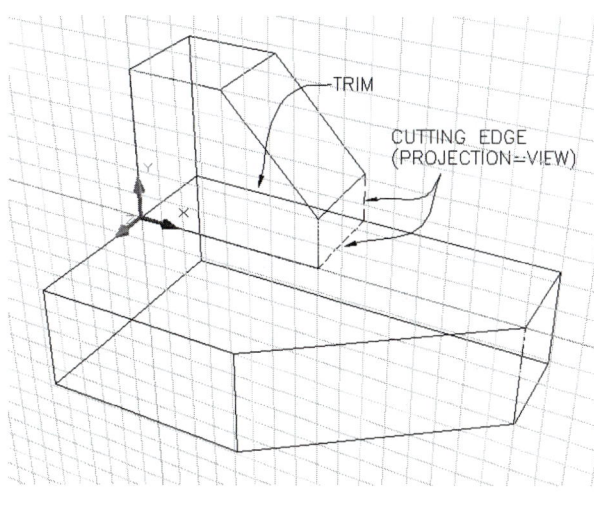

A. Use *Trim* and select either *Line* (highlighted, Fig. 37-14) as the *cutting edge*. Change the *Projection* to *View* and *Trim* the indicated *Line*s. Use *Trim* again to trim the other indicated horizontal *Line* (previous figure).

B. Change back to the WCS. Invoke *UCS*; make the *World* Coordinate System current.

13. Check your work and *Save* your
    drawing.

FIGURE 37-15

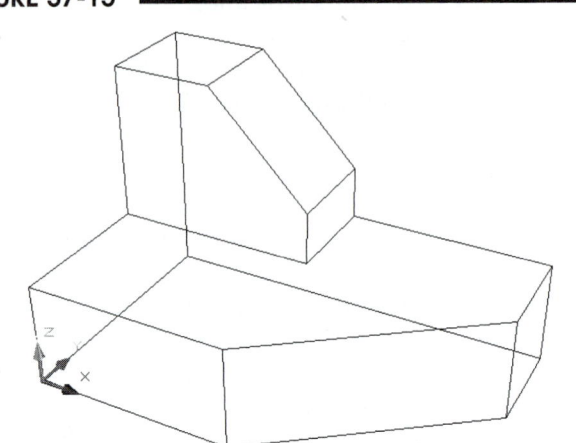

A. Make sure your wireframe model
    looks like that in Figure 37-15.
B. Turn *GRID* off, then use *Save* to
    secure your drawing.

14. It may be wise to examine your model from other viewpoints. As an additional check of your work
    and as a review of 3D viewing, follow these steps.

    A. Invoke the *View Manager*. Create a *New* view of the current display and give the name **VIEW 1**.
    B. Invoke *UCS* and *Restore* the UCS you saved called **FRONT**. Now use *Plan* with the *Current
        UCS* option. This view represents a front view and should allow you to visualize the alignment
        of the two vertical planes constructed in the last several steps.
    C. Next, use *UCS* and restore the *World* Coordinate System. Then use *Plan* again with the *Current
        UCS* option. Check your model from this viewing angle.
    D. Finally, use the *View Manager* to restore **VIEW 1**. This should return to your original viewpoint.

15. Instead of constantly changing *Vpoints*, it would be convenient to have several *Vpoints* of your model
    visible on the screen <u>at one time</u>. This can be done with the *Vports* (Viewports) command. In order
    to enhance visualization of your model for the following constructions, divide your screen into four
    sections or viewports with *Vports* and then use the *3D Setup* option.

    A. First, type **UCSORTHO** at the Command line and change the setting to **0**. This ensures no new
        UCSs will be created when new orthographic viewports are created. Next, use the *Vports*
        command to produce the *Viewports* dialog box. In the *New Viewports* tab, select *Four Equal*.
        Also select *3D* in the *Setup* drop-
        down list. When you select *OK*, your

FIGURE 37-16

screen should display the three
typical orthographic views (top,
front, and right side) and an isometric
view.

    B. Make the upper-right viewport active
        and use *3Dorbit* to adjust the display
        to show all the lines of the model
        clearly. Next, make each of the other
        viewports active (not the upper-right
        viewport) and set *Parallel Projection*
        on. In order to make the three views
        align and ensure the views are the
        same size, use *Zoom* in each view-
        port and enter a factor of **1**.

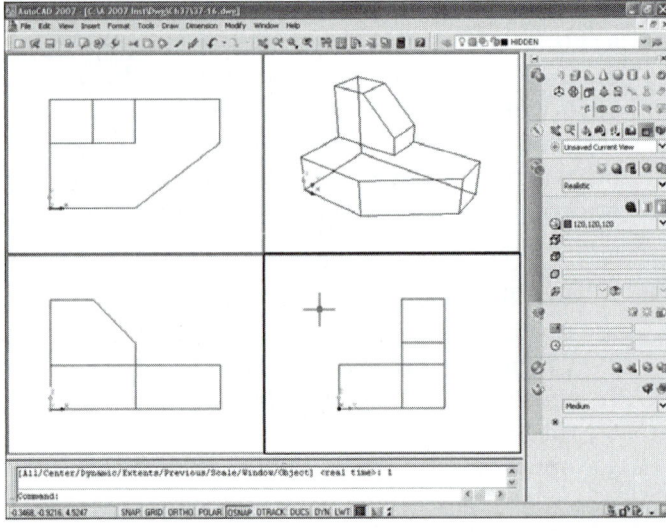

Note that by using viewports you can achieve multiple views of the model; however, since *UCSORTHO* is set to 0, only one UCS appears in all four viewports. The Coordinate System icon appears in all viewports, indicating the current coordinate system orientation. The *Ucsicon ORigin* and *All* settings force the icons to appear at the origin in all viewports.

16. In the next construction, add the milled slot in the inclined surface of the stop plate. Because the slot is represented by geometry parallel with the inclined surface, a UCS should be created on the surface to facilitate the construction.

A. Make the upper-right viewport current. Use the *UCS 3point* option. Select the three points indicated in Figure 37-17 for (1) the "origin," (2) "point on the positive X axis," and (3) "point on the positive Y axis portion on the XY plane." Make sure you *OSNAP* with the *Endpoint* option.

FIGURE 37-17 ───────────────

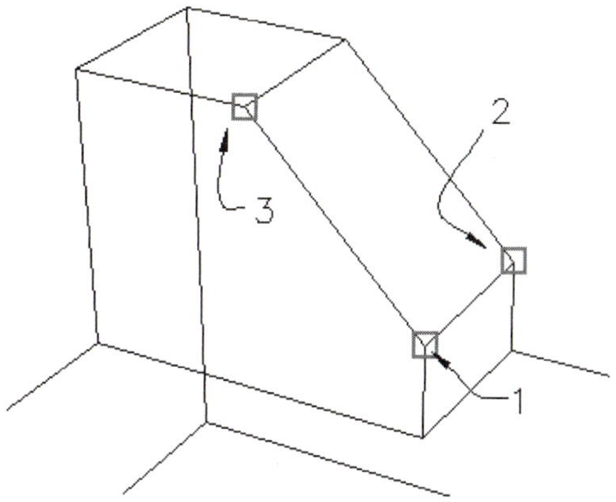

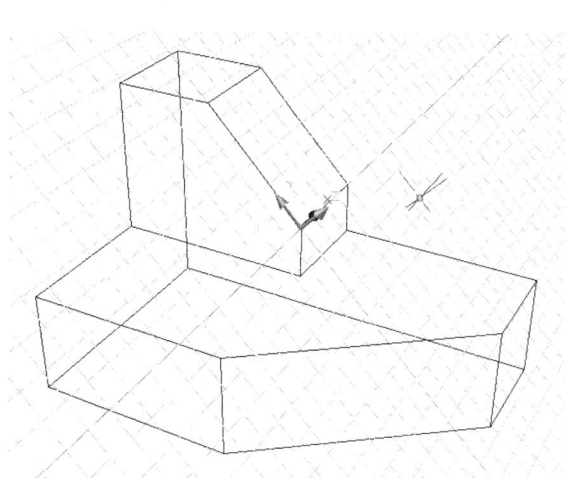

B. The icon should move to the new UCS origin. Turn on **GRID**. Your view should look like that in Figure 37-18. Notice that the new coordinate system most likely appears in all viewports (unless the *UCSVP* variable has been changed to 0). (If all your viewports do not display the new UCS, proceed anyway.)

FIGURE 37-18 ───────────────

C. Save this new UCS by using the **UCS** command with the *Save* option. Assign the name **AUX** (short for "auxiliary view").

(Remember to check the "Hints" if you have trouble.)

17. To enhance your visualization of the inclined surface, change the UCS and viewpoint for the lower-right viewport.

A. Activate any viewport other than the upper-right viewport. To make the new UCS appear in the viewport, use the **USC** command or the **UCS Manager** to set the AUX coordinate system current. This must be done in the other two viewports. All viewports should display the new "AUX" UCS.

C. Type **E** for *Erase*. *Erase* the last four vertical lines, leaving the first. Type **CP** for *Copy*. Select the remaining vertical line. When asked for the "base point:," select the bottom of the line (on the XY plane—make sure *SNAP* is *On*). At the "second point of displacement:" prompt, pick the other locations for the new lines.

D. *View* pull-down, **3D Views >**, **Plan View >**, **World UCS**. Examine your work. Type **Z** for *Zoom*. Type **P** for *Previous*.

H7.   A. Line 1   Specify first point: **0,3,2**
                Specify next point or [Undo]:   **0,3,5**
          Line 2   Specify first point: **0,3,2**
                Specify next point or [Undo]:   **4,3,2**

B. No help needed.

H8.   A. From the *Tools* pull-down menu, select **New UCS>**. Select the **3 Point** option. When prompted for the first point ("origin"), invoke the cursor (*OSNAP*) menu and select **Endpoint**. **PICK** point 1 (Fig. 37-8). When prompted for the second point ("point on positive portion of the X-axis"), again use **Endpoint** and **PICK** point 2. Repeat for the third point ("point on positive Y portion of the UCS XY plane").

B. If you have trouble establishing this UCS, you can type the letter **U** to undo one step. Repeat until the icon is returned to the origin of the WCS orientation. If that does not work for you, type **UCS** and use the **World** option. Repeat the steps given in the previous instructions. Make sure you use *OSNAP* correctly to **PICK** the indicated **Endpoint**s.

C. When constructing the *Lines* representing the vertical plane, ensure that *Grid Snap* and *GRID* (F7) are *ON*. Notice how you can PICK points only on the XY plane of the UCS with the cursor. This makes drawing objects on the plane very easy. Use the **Line** command to draw the remaining edges defining the vertical "surface."

H9.   A. *View* pull-down menu, **3D Views >**, **Plan View >**, **Current UCS**. Check your work. Then type **Z**, then **P**. Correct your work if necessary.

B. Type **UCS**. Type **NA** for *Named*, then **S** for the *Save* option. Enter the name **FRONT**.

H10.   A. Type **CP** or select **Copy** from the **Modify** pull-down menu. Select the indicated four *Lines* (see Fig. 37-11). For the base point, enter or **PICK 0,0**. For the second point of displacement, enter **0,0,-2**.

B. Alternately, perform the same sequence except, when prompted for the "second point of displacement:", select **Endpoint** from the *OSNAP* menu and **PICK** the farthest corner of the top plane.

H11.   A. Type **L** for *Line*. At the "Specify first point:" prompt **PICK** point (a) on the UCS XY plane (Fig. 37-27). You should be able to *SNAP* to the point. At the "Specify next point or [Undo]:" prompt, select the **Endpoint** *OSNAP* option and **PICK** point (b). Repeat the steps for the other 3 *Lines*.

FIGURE 37-27

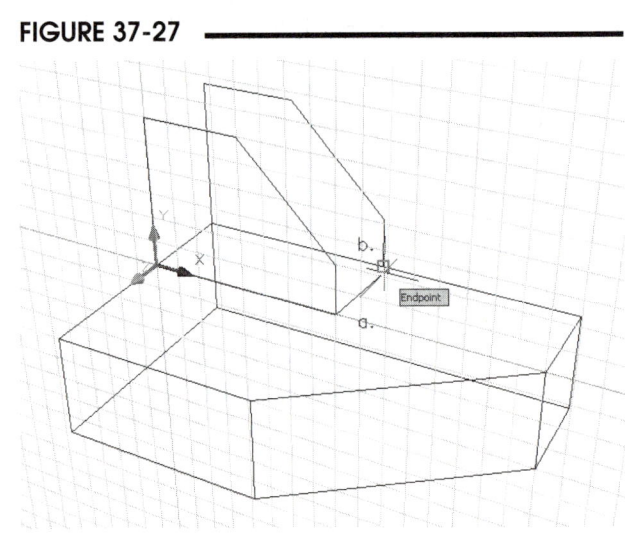

B. Alternately, draw only the first *Line* (line 1, Fig. 37-12). Type *CP* for *Copy*. Select the first *Line*. Type *M* for *Multiple*. At the "base point:" prompt, **PICK** point (a) (Fig. 37-27). Since this point is on the XY plane, PICK (with *Grid Snap ON*). For the second points of displacement, **PICK** the other corners of the vertical surface on the XY plane (with *SNAP ON*).

H12.   A. *Modify* pull-down, **Trim**. Select either the vertical or horizontal *Line* (as indicated in Fig. 37-14) as the cutting edge. Change *Projection* to *View*. Complete the trim operation. With *Projection* now set to *View*, *Trim* the other *Line* indicated (see Fig. 35-13).

      B. Type *UCS*, then *W* for *World*. The WCS icon should appear at the origin (see Fig. 37-15).

H13.   A. If your wireframe model is not correct, make the necessary changes before the next step.

      B. *File* pull-down, **Save**.

H14.   A. *View* pull-down menu, **Named Views** option. In the *View Manager*, select the *New* button. Enter the new name in the *New View* dialog box. Select *OK* to dismiss both dialog boxes.

      B. *Tools* pull-down, **Named UCS....** In the *Named UCS* tab, select **FRONT**. PICK the *Current* tile, then *OK*. Make sure you PICK <u>Current</u>. Then, from the *View* pull-down, select *3D Views >*, *Plan View >*, *Current UCS*. Examine your figure from this viewpoint.

      C. Invoke the *UCS* dialog box again and set *World* as *Current*. Next, use the *View* pull-down menu again to see a *Plan View* of the *World* coordinate system.

      D. Invoke the *View Manager* again. Select **VIEW 1** from the *Views* list, then press *Set Current*. Select *OK* to close the dialog box.

H15.   A. No help needed for creating viewports. To experiment with creating and erasing *Lines*, invoke the *Line* command and **PICK** points in the current viewport (wherever the cursor appears). Now, use *Line* again and **PICK** only the "Specify first point:." Move your cursor to another viewport and make it current (**PICK**). **PICK** the "Specify next point or [Undo]:" in the new viewport. Notice that you can change the current viewport <u>within</u> a command to facilitate using the same draw command (or edit command) in multiple viewports. *Erase* the experimental *Lines*.

      B. Click inside the upper-right viewport, then invoke *3Dorbit* to change the display. Next, make each of the other viewports active one at a time and select the *Parallel Projection* button in the *3D Navigate control panel*. Again use *Zoom* in each viewport and enter **1** at the first prompt.

H16.   A. **PICK** the upper-right viewport. Type *UCS*, then **3**. When prompted for the "origin," select *Endpoint* from the *OSNAP* menu and **PICK** point 1 (see Fig. 37-17). Use *Endpoint OSNAP* to select the other two points.

      B. No help needed.

      C. Select the *Tools* pull-down menu, **Named UCS....** When the *UCS* dialog appears, double-click the current UCS "**Unnamed**." In the edit box, double-click and type over the **Unnamed** with the name **AUX**. Select *OK*.

H17.   A. If you use the *UCS* command, enter *NA* for the *Named* option, then *R* for *Restore*. Using the *UCS Manager*, select **AUX** from the list, then press the *Set Current* button, then *OK*. Repeat this process for all viewports.

      B. **PICK** the lower-right viewport. Select *View* pull-down, *3D Views >*, *Plan View >*, *Current UCS*.

      C. Type *Z*. Make a Window around the entire inclined surface.

H18.   A. Check to see if *Grid Snap* and *GRID* are *ON*. If not, right-click on the word *SNAP* on the Status Bar and select *Grid Snap*. Make sure the word *SNAP* appears recessed. Turn on *GRID* (press **F7**).

B. Select the *Draw* pull-down menu, then *Circle >* with the *Center, Radius* option. For the center, **PICK** point **1,1** (watch the *Coords* display). Enter a value of **.5** for the radius. Repeat this for the second *Circle*, but the center is at **1,2**.

C. Turn *SNAP* off. Type *L* for *Line*. At the "Specify first point:" prompt, select the *Quadrant OSNAP* mode and **PICK** point (a) (Fig. 37-28). At the "Specify next point or [Undo]:" prompt, use *Quadrant* again and **PICK** point (b).

FIGURE 37-28

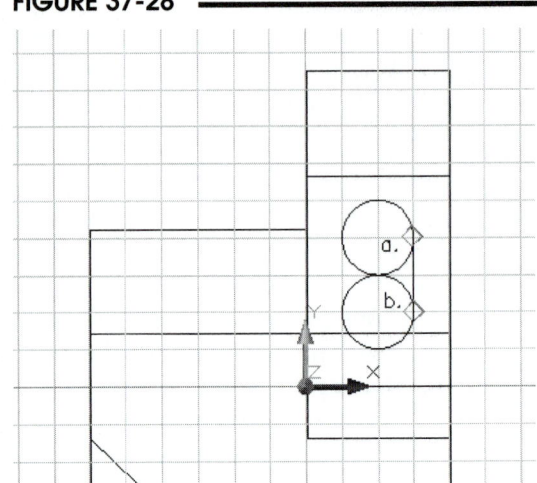

D. *Modify* pull-down, *Trim*. Select the two *Lines* as *Cutting edges* and *Trim* the inner halves of the circles, as shown in Figure 37-21.

H19.   A. Type *CP*. Select the two arcs and two lines comprising the slot. At the "base point or displacement:" prompt, enter **0,0,0**. At the "second point of displacement:" prompt, enter **0,0,-1**.
B. No help needed.

H20.   A. Type *Vports*. In the *Viewports* dialog box, enter **FTIA** in the *New name* edit box. Select *OK*.
B. **PICK** the upper right viewport. Type *Vports*. Choose the *Single* option. Activate *Vports* again. The name FTIA should appear in the *Named Viewports* tab. Select *FTIA*, then *OK*. Finally, type *U* to undo the last action.
C. Select the *Tools* pull-down menu and then *Named UCS...*. **PICK** *World* from the list and make it *Current*.

H21.   *File* pull-down, *Save*.

H22.   *File* pull-down, *Exit* AutoCAD.

COMMANDS

The draw and edit commands that operate for 2D drawings are used to create 3D elements by entering X,Y,Z values, or by using UCSs. *Pline* cannot be used with 3D coordinate entry. Instead, the *3Dpoly* command is used especially for creating 3D polylines.

3Dpoly

	Pull-down Menu	Command (Type)	Alias (Type)	Short-cut	Screen (side) Menu	Tablet Menu
	Draw *3D Polyline*	*3Dpoly*	*3P*	...	DRAW 1 *3Dpoly*	O,10

A line created by *3Dpoly* is a *Pline* with 3D coordinates. The *Pline* command allows only 2D coordinate entry, whereas the *3Dpoly* command allows you to create <u>straight</u> poly-line segments in 3D space by specifying 3D coordinates. A *3Dpoly* line has the "single object" characteristic of normal *Plines*; that is, several *3Dpoly* line segments created with one *3Dpoly* command are treated as one object by AutoCAD.

FIGURE 37-29

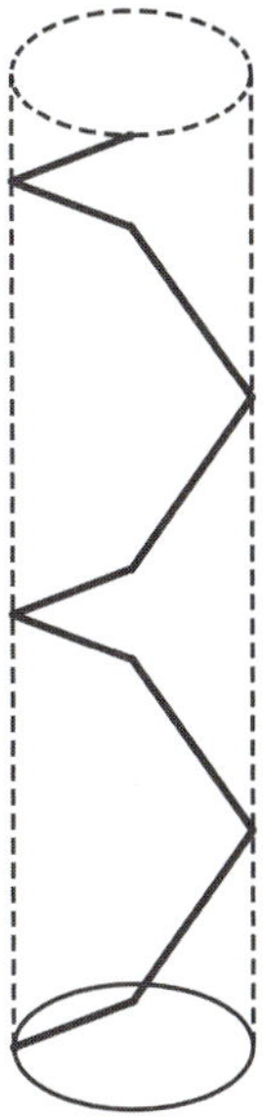

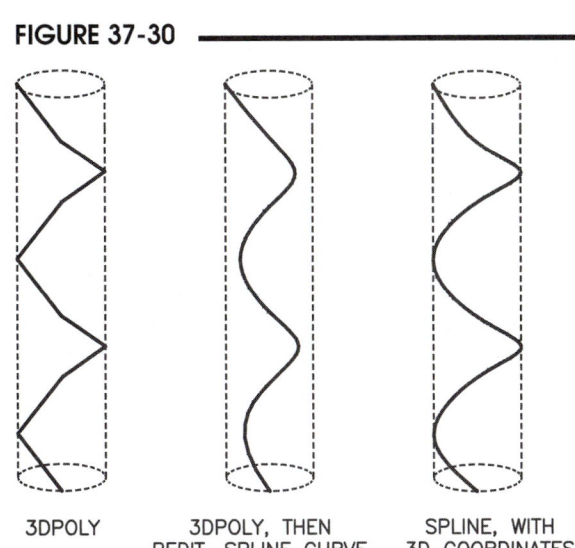

> Command: **3dpoly**
> Specify start point of polyline: **PICK** or **(coordinates)**
> Specify endpoint of line or [Undo]: **PICK** or **(coordinates)**
> Specify endpoint of line or [Undo]: **PICK** or **(coordinates)**
> Specify endpoint of line or [Close/Undo]: **Enter**
> Command:

If you PICK points or specify coordinate values, AutoCAD simply connects the points with straight polyline segments in 3D space. For example, *3Dpoly* could be used to draw line segments in a spiral fashion by specifying coordinates or PICKing points (Fig. 37-29).

Close

This option closes the last point to the first point entered in the command sequence.

Undo

This option deletes the last segment and allows you to specify another point for the new segment.

A *3Dpoly* does <u>not</u> have the other features or options of a *Pline* such as *Arc* segments and line *Width*.

Special options of *Pedit* can be used with *3Dpoly* lines to *Close*, *Edit Vertex*, *Spine curve*, or *Decurve* the *3Dpoly*. You can also use grips to edit the *3Dpoly*.

FIGURE 37-30

Although some applications may require a helix, *3Dpoly* and *Pedit* <u>cannot</u> be used in conjunction to do this. The *Spine curve* option of the *Pedit* command converts the *3Dpoly* to a polyline version of a B-Spline, not a true NURBS spline. A *Spline*-fit *Polyline* does not pass through the vertices that were specified when the *Pline* or *3DPoly* was created. The *Spline* command, however, <u>can</u> be used to specify points in 3D to create a true helix (Fig. 37-30).

NOTE: With AutoCAD 2007, you can use the *Helix* command to create a true helix. *Helix* is easier and more flexible to use than using *3Dpoly* for this purpose.

3DPOLY   3DPOLY, THEN   SPLINE, WITH
               PEDIT, SPLINE CURVE   3D COORDINATES

Lines and *Arcs* <u>cannot</u> be converted to *3Dpoly* lines using *Pedit*. You can, however, convert *Lines* and *Arcs* to *Plines* if they lie on the XY plane of the current UCS. Spline-fit *Plines* and *3Dpoly* lines can be converted to NURBS splines with the *Object* option of the *Spline* command, but the conversion does not realign the new *Spline* to pass through the original vertices.

Helix

Pull-down Menu	Command (Type)	Alias (Type)	Short-cut	Screen (side) Menu	Tablet Menu
Draw *Helix*	*Helix*	...	...	DRAW 1 *Helix*	...

The *Helix* command creates a true 3-dimensional helical shape. Technically, a *Helix* is a wireframe object, similar to a *3Dpline*. You can control several parameters that determine the size and shape of the *Helix*.

        Command: **helix**
        Number of turns = 3.0000   Twist=CCW
        Specify center point of base: **PICK** or (**coordinates**)
        Specify base radius or [Diameter] <1.0000>: (**value**)
        Specify top radius or [Diameter] <1.0000>: (**value**)
        Specify helix height or [Axis endpoint/Turns/turn Height/tWist] <1.0000>: (**value**) or **option**
        Command:

Axis Endpoint

Rather than specifying a height, you can specify the endpoint location for the *Helix* axis. The *Axis Endpoint* can be located anywhere in 3D space and defines both the length and orientation of the *Helix*.

Turns

"Turns" is the number of revolutions for the *Helix*. The default value is 3. Use this option to specify any number, with a maximum of 500. If no value is entered, *Turns* is dependent on *Turn Height*, if that option is used.

Turn Height

This value determines the height for one complete revolution within the *Helix*. When a value is specified for *Turn Height*, the number of *Turns* in the *Helix* is automatically adjusted accordingly. However, if the *Turns* value has been specified, you cannot enter a value for the *Turn Height*.

Twist

Normally, a *Helix* turns counter-clockwise (*CCW*) from base to top. Use this option to specify whether the *Helix* is drawn in the clockwise (*CW*) or the counter-clockwise (*CCW*) direction.

NOTE: Although a *Helix* has some wireframe applications, it is most useful as a "framework" for a 3D solid. Typically, *Helix* is used to create a "path," then *Sweep* is used to create a helical solid. *Sweep* allows you to sweep a shape, a *Circle* for example, along a path such as a *Line*, *Pline*, *Spline*, or *Helix*. See "*Sweep*" in Chapter 39 for more information on creating a solid helix.

2007

Xedges

Pull-down Menu	Command (Type)	Alias (Type)	Short-cut	Screen (side) Menu	Tablet Menu
Modify 3D Operations > Extract Edges	Xedges	...	...	...	...

The *Xedges* command allows you to quickly and easily create entire wireframe models from existing solid models. Use solid modeling techniques to create the desired model, then use *Xedges* to extract the edges, or wireframe elements. *Xedges* extracts edges from any solid, region, or surface. *Xedges* is simple to use since there are no options.

```
Command: xedges
Select objects: PICK
Select objects: Enter
Command:
```

You can also select individual edges and faces to extract. Do this by holding down the Ctrl key while selecting edges and faces (subobject selection). See Chapters 38 and 39 for information on creating and editing solid models and using subobject selection.

CHAPTER EXERCISES

1–4.   **Introduction to Wireframes**

Create a wireframe model of each of the objects in Figures 37-31 through 37-34. Use each dimension marker in the figure to equal one unit in AutoCAD. All holes are 1 unit in diameter. Begin with the lower-left corner of the model located at 0,0,0. Assign the names **WFEX1**, **WFEX2**, **WFEX3**, and **WFEX4.**

FIGURE 37-31 ——————————

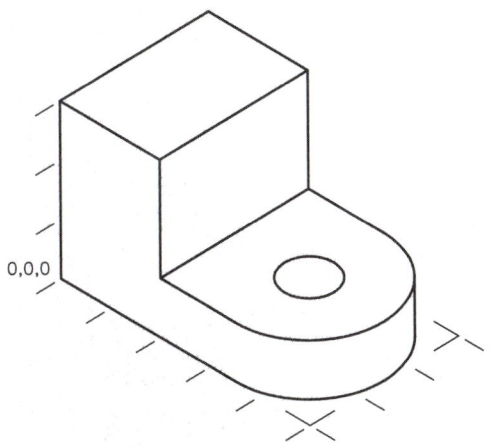

FIGURE 37-32 ——————————

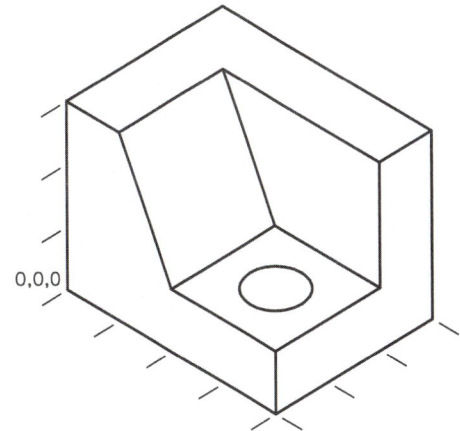

FIGURE 37-33 ━━━━━━━━━━━━━━━━━━

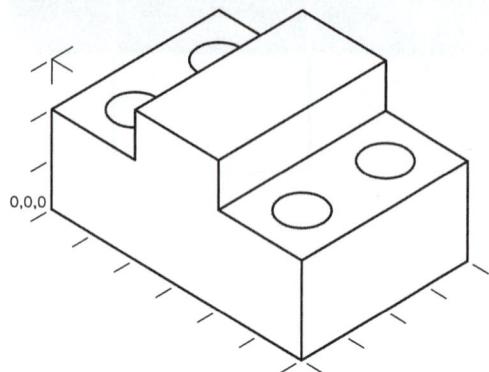

0,0,0

FIGURE 37-34 ━━━━━━━━━━━━━━━━━━

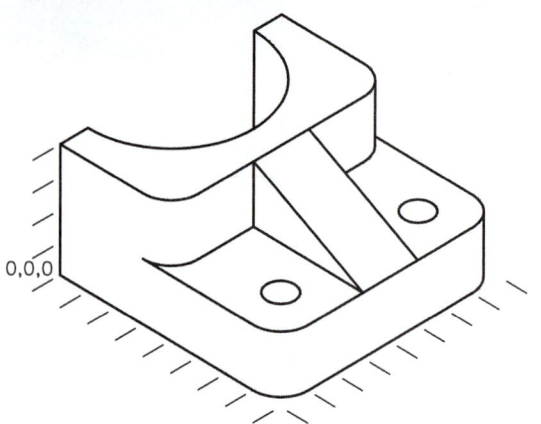

0,0,0

5. Create a wireframe model of the corner brace shown in Figure 37-35. *Save* the drawing as **CBRAC-WF**.

FIGURE 37-35 ━━━━━━━━━━━━━━━━━━

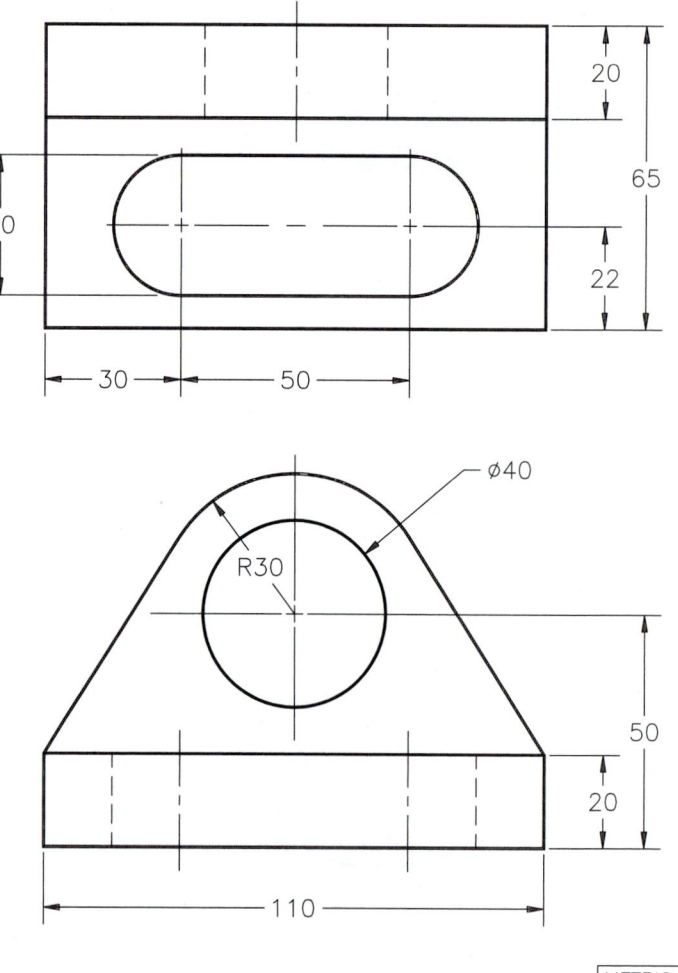

METRIC

6. Make a wireframe model of the V-block (Figure 37-36). *Save* as **VBLCK-WF**.

FIGURE 37-36 ━━━━━━━━━

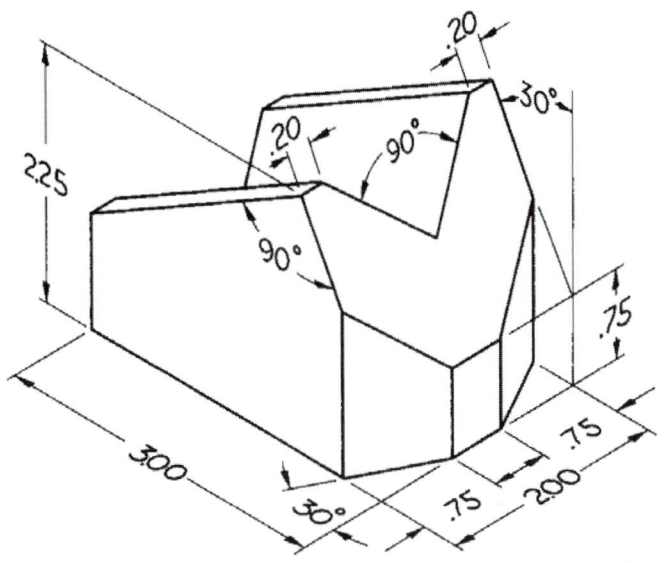

7. Make a wireframe model of the bar guide (Fig. 37-37). *Save* as **BGUID-WF.**

FIGURE 37-37 ━━━━━━━━━

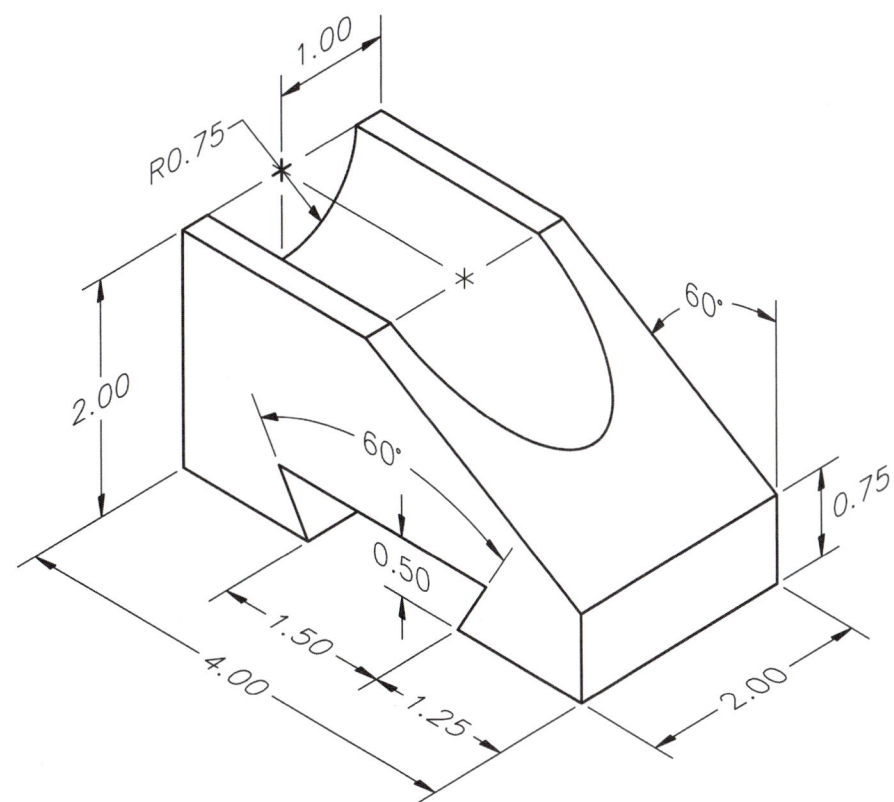

8.   Make a wireframe model of the angle brace (Fig. 37-38). *Save* as **ANGLB-WF**.

FIGURE 37-38 ━━━━━━━━━━━━━━━━━━━━━━━━

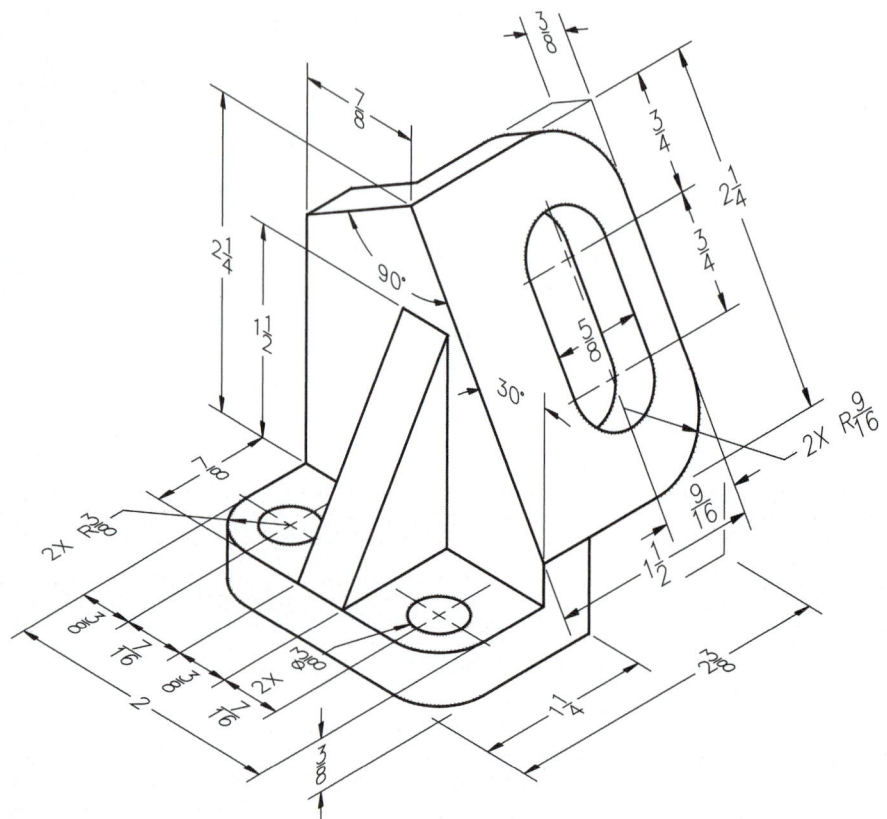

SOLID MODELING CONSTRUCTION

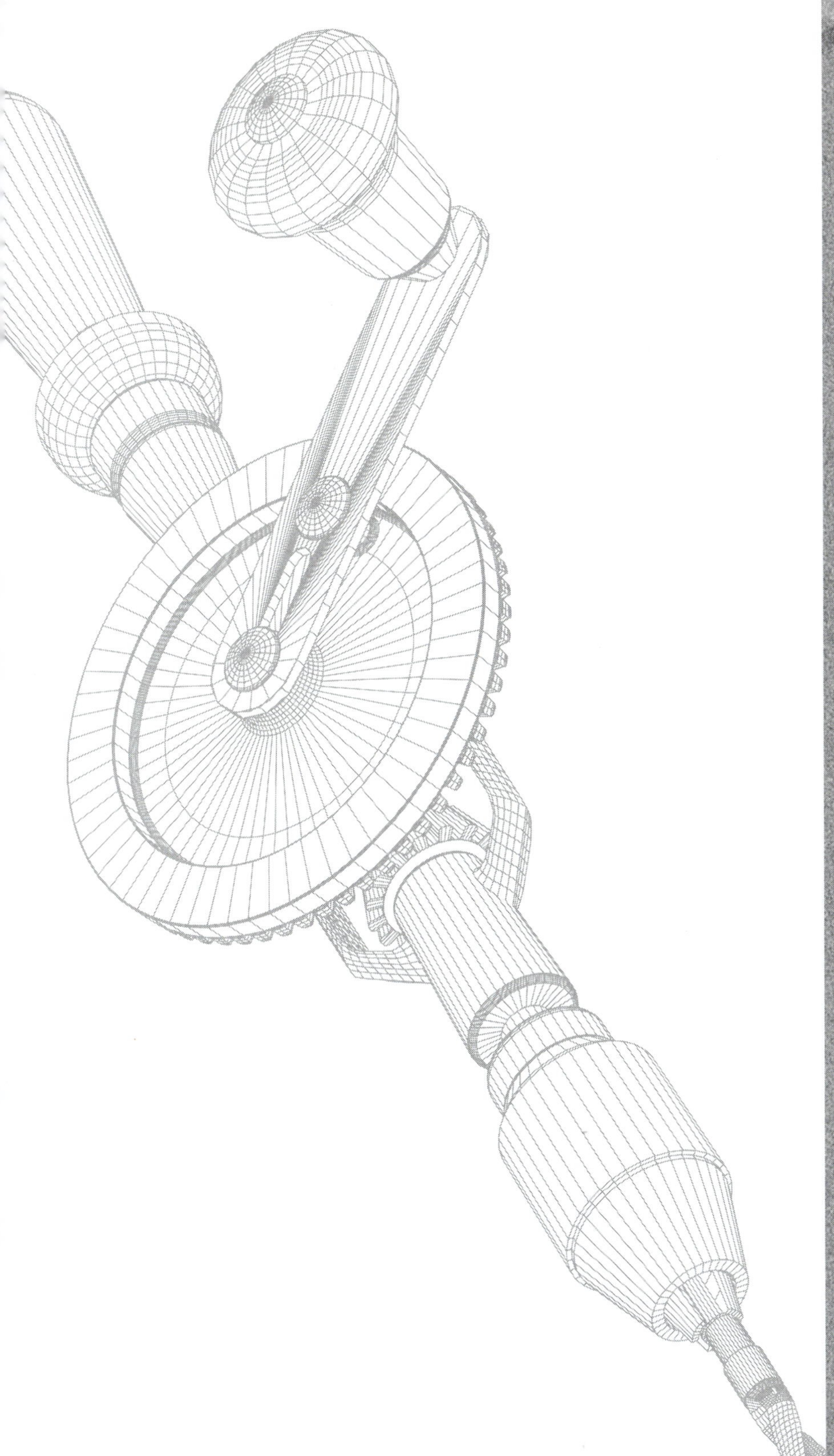

CHAPTER OBJECTIVES

After completing this chapter you should:

1. be able to create solid model primitives using the following commands: *Box*, *Wedge*, *Cone*, *Cylinder*, *Torus*, *Pyramid*, *Polysolid*, and *Sphere*;

2. be able to create swept solids from existing 2D shapes using the *Extrude*, *Loft*, *Revolve*, and *Sweep* commands;

3. be able to create Dynamic User Coordinate Systems;

4. be able to move solids in 3D space using *3Dmove*, *3Dalign*, and *3Drotate*;

5. be able to create new solids in specific relation to other solids using *Mirror3D* and *3DArray*;

6. be able to combine multiple primitives into one composite solid using *Union*, *Subtract*, and *Intersect*;

7. know how to create beveled edges and rounded corners using *Chamfer* and *Fillet*.

CONCEPTS

A solid model is a virtually complete representation of the 3D shape of a physical object. Solid modeling differs from wireframe or surface modeling in two fundamental ways: (1) the information is more complete in a solid model and (2) the method of construction of the model itself is relatively easy.

The basic solid modeling construction process is to use simple 3D solids, called "primitives," then combine them using "Boolean operations" to create a complex 3D model called a "composite solid." This modeling technique is known as Constructive Solid Geometry, or CSG, modeling. Although the process is relatively simple, you have tremendous flexibility during construction using tools such as UCSs and Dynamic UCSs, commands for moving and rotating, grip editing, and many other editing commands and techniques. AutoCAD uses a solid modeling engine, or "kernel," called ACIS.

Solid Model Construction Process

The process used to construct most composite solid models follow four general steps:

1. Construct simple 3D <u>primitive</u> solids such as *Box*, *Cylinder*, or *Wedge*, or create 2D shapes and convert them to 3D solids by *Extrude*, *Revolve*, *Sweep*, or *Loft*.

2. Create the primitives <u>in location</u> relative to the associated primitives using UCSs or <u>move, rotate, or mirror</u> the primitives into the desired location relative to the associated primitives.

3. Use <u>Boolean operations</u> (*Union*, *Subtract*, and *Intersect*) to combine the primitives to form a <u>composite solid</u>.

4. Make necessary design changes to features of a composite solid using a variety of editing tools such as grips, *Solidedit*, and using surfaces to *Slice* solids.

This chapter discusses the basic construction process organized into the following sections: 1) Solid Primitives Commands, 2) Dynamic User Coordinate Systems, 3) Commands to Move, Rotate, and Copy Solids, and 4) Boolean Operations. Each section is briefly described next. Chapter 39 discusses advanced editing techniques, such as grip editing, editing faces and edges, sectioning, and display variables.

Solid Primitives Commands

Solid primitives are the basic building blocks that make up more complex solid models. The commands that create 3D primitives "from scratch" are:

Box	Creates a solid box or cube
Cone	Creates a solid cone with a circular or elliptical base
Cylinder	Creates a solid cylinder with a circular or elliptical base
Polysolid	Creates a series of connected solid segments having a specified height and width
Pyramid	Creates a solid right pyramid with any number of sides
Sphere	Creates a solid sphere
Torus	Creates a solid torus
Wedge	Creates a solid wedge

Commands that create solid primitives from existing 2D shapes are:

Extrude	Converts a closed 2D object such as a *Pline, Circle, Region,* or *Planar Surface* to a solid by adding a Z dimension
Loft	Creates a solid using a series of 2D cross sections (curves or lines) that define the shape
Revolve	Converts a closed 2D object such as a *Pline, Circle,* or *Region* to a solid by revolving the shape about an axis
Sweep	Creates a solid by sweeping a closed planar profile along an open or closed 2D or 3D path

Dynamic User Coordinate Systems

As you learned in Chapter 36, the *UCS* command creates a User Coordinate System that is static—that is, it remains active until you change to another coordinate system. This chapter discusses how to create and control Dynamic Coordinate Systems.

A Dynamic User Coordinate System (DUCS) is a <u>temporary</u> UCS that appears automatically when you use a primitive command such as *Box, Cylinder,* or *Pyramid* and disappears when you complete the command. You can use the *DUCS* button on the Status bar to toggled this feature on or off. You cannot use the *Ucs* command to create a Dynamic User Coordinate System. A DUCS appears only with some primitive solid creation commands, but not with surface or wireframe models.

For example, suppose you wanted to create a *Box* on the surface of an existing solid. When you use the *Box* command, you are prompted for the "first corner." With *DUCS* on, you can create a temporary UCS on an existing solid to begin drawing the *Box*. You determine the location of the XY plane of the DUCS by "hovering" over any planar face of an existing solid. You can also specify the direction of the X and Y axes based on the location of the cursor in relation to the highlighted plane. Once the DUCS icon appears in the desired orientation, PICK to specify the "first corner" for the *Box*. The DUCS disappears once the new *Box* is complete.

Commands to Move, Rotate, and Mirror Solids

It is not always efficient or practical to create every primitive solid initially in the desired location and/or orientation. In may cases primitives are created in another position, then they are moved, rotated, or otherwise aligned into the desired position before combining them into composite solids. Also, some primitives can be created from others by mirroring or arraying existing solids. This section of the chapter discusses the following commands:

Move	The command used for moving objects in 2D can also be used in 3D space
3Darray	Makes a polar array about any axis or a rectangular array with rows, columns, and levels
3Dalign	Connects two solids together by matching three alignment points on each solid
3Dmove	A temporary grip tool allows you to restrict the movement to an axis or a plane
3Drotate	A temporary grip tool allows you to specify an axis about which to rotate
Mirror3D	Creates a mirror copy of solids by specifying a mirror plane

Boolean Operations

Primitives are combined to create complex solids by using Boolean operations. The Boolean operators are listed below. An illustration and detailed description are given for each of the commands.

Union	Unions (joins) selected solids.
Subtract	Subtracts one set of solids from another.
Intersect	Creates a solid of intersection (common volume) from the selected solids.

2007

Primitives are created at the desired location or are moved into the desired location before using a Boolean operator. In other words, two or more primitives can occupy the same space (or share some common space), yet are separate solids. When a Boolean operation is performed, the solids are combined or altered in some way to create one solid. AutoCAD takes care of deleting or adding the necessary geometry and displays the new composite solid complete with the correct configuration and lines of intersection.

Consider the two solids shown in Figure 38-1. When solids are created, they can occupy the same physical space. A Boolean operation is used to combine the solids into one composite solid and it interprets the resulting utilization of space.

Union
Union creates a union of the two solids into one composite solid (Fig. 38-2). The lines of intersection between the two shapes are calculated and displayed by AutoCAD.

Subtract
Subtract removes one or more solids from another solid. ACIS calculates the resulting composite solid. The term "difference" is sometimes used rather than "subtract." In Figure 38-3, the cylinder has been subtracted from the box.

Intersect
Intersect calculates the intersection between two or more solids. When *Intersect* is used with *Regions* (2D surfaces), it determines the shared <u>area</u>. Used with solids, as in Figure 38-4, *Intersect* creates a solid composed of the shared <u>volume</u> of the cylinder and the box. In other words, the result of *Intersect* is a solid that has only the volume which is part of both (or all) of the selected solids.

Chamfer and Fillet
The *Chamfer* command creates an angled edge between two surfaces, and a *Fillet* creates a rounded edge between two surfaces. Although the *Chamfer* command and the *Fillet* command are not Boolean operators, they are included in this section. With each of these commands, you specify dimensions to create a wedge primitive (with *Chamfer*) or a rounded primitive (with *Fillet*) and AutoCAD automatically performs the Boolean needed to add or subtract the primitive from the selected solids.

FIGURE 38-1

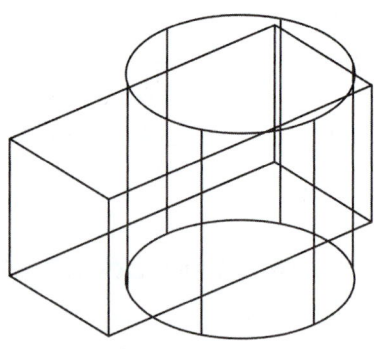

FIGURE 38-2

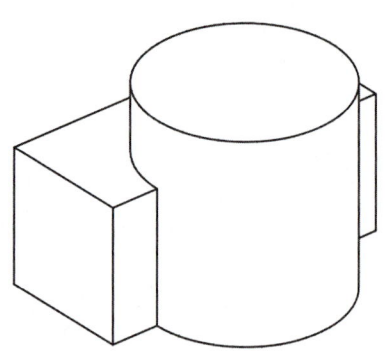

FIGURE 38-3

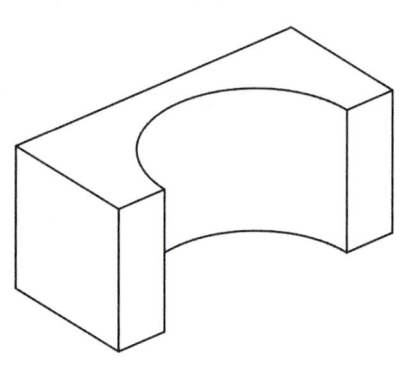

FIGURE 38-4

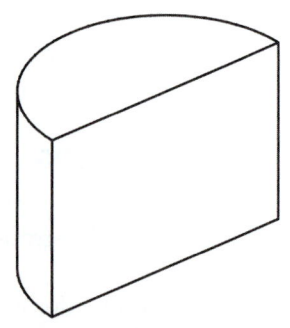

SOLID PRIMITIVES COMMANDS

This section explains the commands that allow you to create primitives used for construction of composite solid models. The commands allow you to specify the dimensions and the orientation of the solids. Once primitives are created, they are combined with other solids using Boolean operations to form composite solids.

NOTE: If you use a pointing device to PICK points, use *OSNAP* when possible to PICK points in 3D space. If you do not use *OSNAP*, the selected points are located on the current XY construction plane, so the true points may not be obvious. It is recommended that you use *OSNAP* or enter values.

The solid primitives commands can be accessed several ways. At the top of the *Dashboard* is the *3D Make control panel* that contains the solid primitives commands as well as other related editing commands. Figure 38-5 shows the control panel in its expanded configuration.

FIGURE 38-5

From the pull-down menus, select *Draw* then *Modeling* to produce the solid primitives commands on a cascading menu (Fig 38-6). Commands for moving solids and editing solids (including the Boolean commands) are located on menus branching from the *Modify* pull-down menu (not shown).

FIGURE 38-6

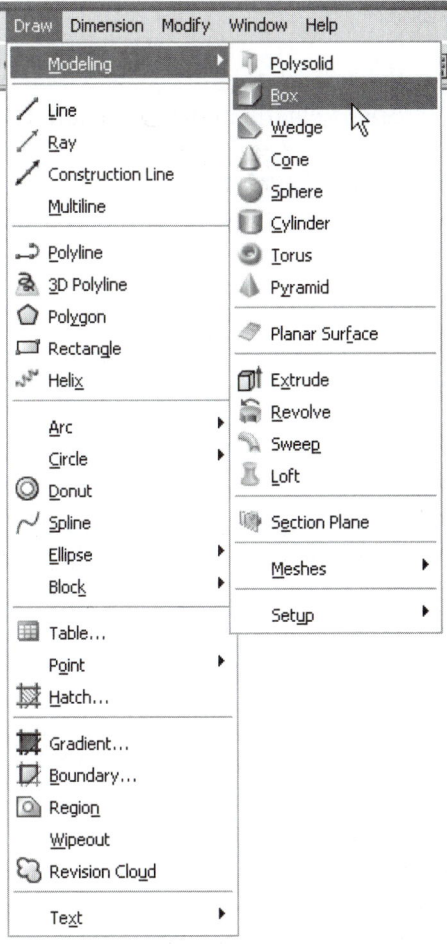

A toolbar is available that contains the solids primitives commands as well as the three Boolean commands. Note the name of the toolbar is *Modeling* (Fig. 38-7). A second toolbar is available (not shown) exclusively for editing solids.

FIGURE 38-7

In addition to these methods, each of the solid modeling commands are accessible by typing in the command name at the Command prompt, such as *Box* or *Wedge*.

Box

Pull-down Menu	Command (Type)	Alias (Type)	Short-cut	Screen (side) Menu	Tablet Menu
Draw *Modeling >* *Box*	*Box*	...	...	DRAW 2 SOLIDS *Box*	J,7

Box creates a solid box primitive to your dimensional specifications. The <u>base of the box is always oriented parallel to the current XY plane</u> (WCS, UCS, or Dynamic UCS). You can specify dimensions of the box by PICKing or by entering values. The box can be defined by (1) giving the corners of the base, then height, (2) by locating the center, then the corners and height, or (3) by giving each of the three dimensions.

>   Command: **box**
>   Specify first corner or [Center]: **PICK** or (**coordinates**) or **C**

Specify first corner

You can either PICK or supply coordinate values for the first corner. AutoCAD then responds with:

>   Specify other corner or [Cube/Length]:

If you PICK the other corner interactively, the length and width of the base are parallel with the X and Y axes. AutoCAD then requests the height.

>   Specify height or [2Point] <0.0000>:

Again, you can PICK or enter a value at this prompt. The *2Point* option allows you to PICK two points as the height distance.

FIGURE 38-8

Figure 38-8 shows a *Box* with the first corner at 0,0 and the other corner at 5,4. The height is 3. This box was created interactively by picking points. Note that the box is oriented with the edges aligned with the X,Y, and Z axes.

Length

After supplying the first corner, you can use the *Length* option to specify three dimensions for the box: the length, width, and height. Note that the first two values supplied (*Length* and *Width*) are the dimensions of the base, but either can be longer or shorter (*Length* in this case is not necessarily the longer of the two dimensions, as the term implies). Next, specify the height.

> Specify other corner or [Cube/Length]: **l**
> Specify length: **PICK** or (**value**)
> Specify width: **PICK** or (**value**)
> Specify height or [2Point] <0.0000>: **PICK** or (**value**)

NOTE: Using the *Length* option, the <u>direction</u> of the length and width is <u>determined by the current location of the crosshairs</u>; therefore, the edges of the box's base are not aligned with the X and Y axes unless *POLAR* or *ORTHO* is on. In addition, the box can be created in negative X,Y, and Z directions.

Cube

The *Cube* option requires only one dimension to create the *Box*. You can PICK a distance or enter a value. In either case, the <u>direction of the edges of the base is determined by the current location of the crosshairs</u>.

Center

With this option you first locate the center of the box, then supply the dimensions of the box by PICKing or entering values.

> Command: **box**
> Specify first corner or [Center]: **c**
> Specify center: **PICK** or (**coordinates**)
> Specify corner or [Cube/Length]:

FIGURE 38-9 ――――――

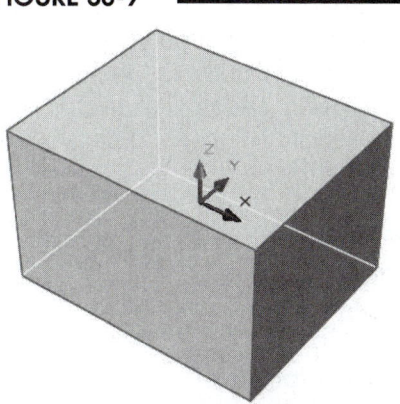

Figure 38-9 displays a *Box* created using the *Center* method. Note that the center is the <u>volumetric</u> center, not the center of the base.

Cone

Pull-down Menu	Command (Type)	Alias (Type)	Short-cut	Screen (side) Menu	Tablet Menu
Draw Modeling > Cone	Cone	...	...	DRAW 2 SOLIDS Cone	M,7

FIGURE 38-10 ――――――

Cone creates a right circular or elliptical solid cone ("right" means the axis forms a right angle with the base). By the default method, you specify the center location, radius (or diameter), and height. With this option, the orientation of the cylinder is determined by the current UCS so that <u>the base lies on the XY plane and height is perpendicular</u> (in a Z direction). Alternately, a different orientation can be defined by using the *Axis endpoint* or *3Ppoint* option.

Center Point

Using the defaults (PICK or supply values for the center, radius, and height), the cone is generated in the orientation shown in Figure 38-10. The default prompts are shown as follows:

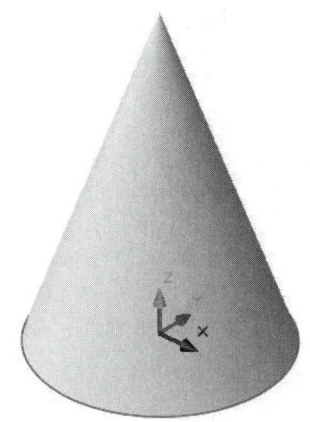

Command: *cone*
Specify center point of base or [3P/2P/Ttr/Elliptical]: **PICK** or (**coordinates**)
Specify base radius or [Diameter]: **PICK** or (**coordinates**)
Specify height or [2Point/Axis endpoint/Top radius]: **PICK** or (**value**)

3P/2P/Ttr

Note that you can define the location and diameter of the base using the *3Point*, *2Point*, or *Ttr* (tangent, tangent, radius) methods. These options work similar to creating a *Circle* by the same methods. Using the *2Point* and *Ttr* methods, the *Cone* base always lies in a plane parallel with the UCS, no matter which points in 3D space are selected. However, the *3Point* option allows a different orientation such that the base of the *Cone* lies in the plane defined by the three points. These options and principles operate similarly for creating a *Cylinder* or *Torus*.

Elliptical

FIGURE 38-11

This option draws a cone with an elliptical base. You specify two axis endpoints to define the elliptical base. The elliptical cone in Figure 38-11 was created using the following specifications :

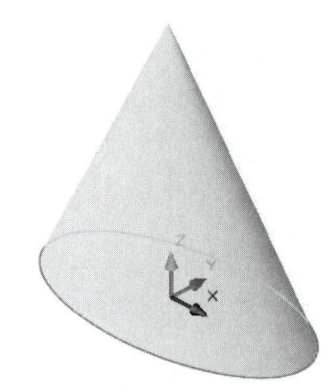

Command: **cone**
Specify center point of base or [3P/2P/Ttr/Elliptical]: **e**
Specify endpoint of first axis or [Center]: **c**
Specify center point: **0,0**
Specify distance to first axis: **2**
Specify endpoint of second axis: **1**
Specify height or [2Point/Axis endpoint/Top radius]: **3**
Command:

2Point

When specifying the height, you can use this option to interactively specify two points to determine the height distance.

Axis endpoint

FIGURE 38-12

Invoking the *Axis endpoint* option (after the base has been established) displays the following prompt:

Specify apex point: **PICK** or (**coordinates**)

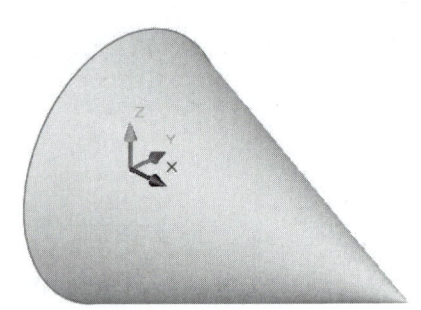

Locating a point for the axis endpoint defines the height and orientation of the cone. The axis of the cone is aligned with the line between the specified center point of the base and the apex, and the height is equal to the distance between the two points. Figure 38-12 shows a *Cone* with the base at 0,0 and the *Axis endpoint* located on the X axis.

FIGURE 38-13

Top radius

The *Top radius* option allows you to create a frustum (a truncated right cone with the top plane parallel to the base). The following prompt is displayed:

Specify height or [2Point/Axis endpoint/Top radius]: **t**
Specify top radius <0.0000>: **PICK** or (**value**)
Specify height or [2Point/Axis endpoint]: **PICK** or (**value**)

The top radius is the radius for the circle at the top of the frustum. Figure 38-13 shows a *Cone* with a base radius of 2 and a top radius of 1.

Cilinder

Pull-down Menu	Command (Type)	Alias (Type)	Short-cut	Screen (side) Menu	Tablet Menu
Draw Modeling > Cylinder	*Cylinder*	...	...	*DRAW 2 SOLIDS Cylinder*	L,7

Cylinder creates a cylinder with an elliptical or circular base with a center location, diameter, and height you specify. Default orientation of the cylinder is determined by the current UCS, such that the circular plane is coplanar with the XY plane and height is in a Z direction. However, the orientation can be defined otherwise by the *Axis endpoint* or *3Point* option.

FIGURE 38-14

Center Point
The default options create a cylinder in the orientation shown in Figure 38-14:

    Command: **cylinder**
    Specify center point of base or [3P/2P/Ttr/Elliptical]: **PICK** or (**coordinates**)
    Specify base radius or [Diameter] <0.0000>: **PICK** or (**value**)
    Specify height or [2Point/Axis endpoint] <0.0000>: **PICK** or (**value**)
    Command:

3P/2P/Ttr
You can define the location and diameter of the base using the *3Point*, *2Point*, or *Ttr* (tangent, tangent, radius) methods. These options work similar to creating a *Circle* by the same methods. If you use the *2Point* and *Ttr* methods, you can select any points in 3D space to define the base of the *Cylinder*; however, the base always lies in a plane parallel with the UCS. On the other hand, if you use the *3Point* option, the base lies in the same plane as the three specified points. These options and principles are the same for creating a *Cone* base diameter or *Torus* diameter.

Elliptical
This option draws a cylinder with an elliptical base. The elliptical cone in Figure 38-15 was created using the following specifications.

FIGURE 38-15

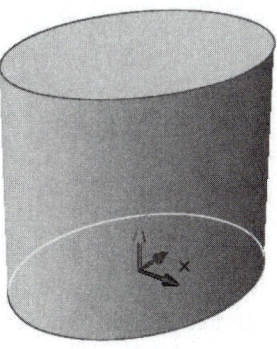

    Command: **cylinder**
    Specify center point of base or [3P/2P/Ttr/Elliptical]: **e**
    Specify endpoint of first axis or [Center]: **c**
    Specify center point: **0,0**
    Specify distance to first axis <0.0000>: **2**
    Specify endpoint of second axis: **1.5**
    Specify height or [2Point/Axis endpoint] <0.0000>: **3**

2Point
When specifying the height, you can use this option to interactively specify two points to determine the height distance.

Axis endpoint

Invoking the *Axis endpoint* option (after the base has been established) allows you to determine both the height and orientation for the cylinder.

FIGURE 38-16

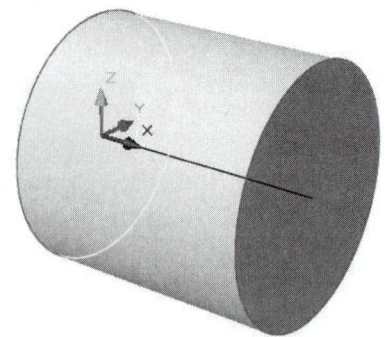

```
Command: cylinder
Specify center point of base or [3P/2P/Ttr/Elliptical]: PICK
(Endpoint)
Specify base radius or [Diameter] <0.0000>: 1.5
Specify height or [2Point/Axis endpoint] <0.0000>: a
Specify axis endpoint: PICK (Endpoint)
```

The cylinder in Figure 38-16 was created using the *Axis endpoint* method. Here, the *Line* along the X axis was previously established. The *Cylinder* was created by using *Endpoint* to snap to the two endpoints for the "center point of base" and "axis endpoint" (see the prompts listed above).

Wedge

Pull-down Menu	Command (Type)	Alias (Type)	Short-cut	Screen (side) Menu	Tablet Menu
Draw *Modeling >* *Wedge*	*Wedge*	*WE*	...	DRAW 2 SOLIDS *Wedge*	N,7

Creating a *Wedge* solid primitive is similar to creating a *Box* (see "*Box*"). Note that the prompts are the same, and therefore the methods for creating a *Wedge* are identical to those for creating a *Box*. The difference between a *Wedge* and a *Box* with the same dimensions is the *Wedge* contains half the volume of the *Box*—a *Wedge* is like a *Box* cut in half diagonally.

FIGURE 38-17

```
Command: wedge
Specify first corner or [Center]: PICK or (coordinates)
Specify other corner or [Cube/Length]: PICK or (coor-
dinates)
Specify height or [2Point] <0.0000>: PICK or (value)
Command:
```

Similar to a *Box*, a *Wedge's* base is always parallel to the XY plane of the current coordinate system (WCS, UCS, or Dynamic UCS). Additionally, using the default option ("Specify first corner"), the *Wedge* always slopes down toward second ("other") corner. Notice the orientation of the *Wedge* in Figure 38-17, which was created by PICKing the "first corner" at the origin and the "other corner" in an XY-positive direction.

Length

As an alternative to specifying two corners and a height, you can use the *Length* option to specify three dimensions for the *Wedge*: the length, width, and height. NOTE: Using the *Length* option, the direction of the length and width is determined by the current location of the crosshairs; therefore, the edges of the *Wedge's* base are not aligned with the X and Y axes unless *POLAR* or *ORTHO* are on. In addition, the *Wedge* always slopes down toward the direction indicated for the *Length*.

Cube

The *Cube* option requires only one dimension to create the *Wedge*. You can PICK a distance or enter a value. In either case, with this option the <u>Wedge always slopes down toward the direction indicated by the current location of the crosshairs.</u>

Center

With this option you first locate a "center" for the *Wedge*, then supply the dimensions by PICKing or entering values.

Command: **box**
Specify first corner or [Center]: **c**
Specify center: **0,0**
Specify corner or [Cube/Length]:

Figure 38-18 displays a *Wedge* created by specifying 0,0 as the *Center* method (see prompts above). Note that the <u>designated center is not the volumetric center, but the center of the sloping side of the *Wedge*.</u> If you imagine the *Wedge* as one half of a *Box*, the center would be the volumetric center for the imaginary *Box* (see "*Box*").

FIGURE 38-18 ⎯⎯⎯⎯⎯⎯⎯⎯

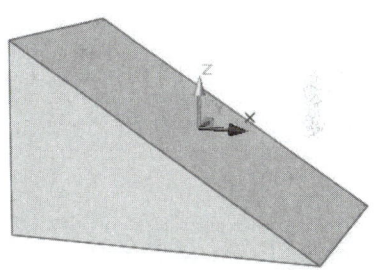

Sphere

Pull-down Menu	Command (Type)	Alias (Type)	Short-cut	Screen (side) Menu	Tablet Menu
Draw Modeling > Sphere	*Sphere*	...	...	*DRAW 2 SOLIDS Sphere*	*K,7*

Sphere allows you to create a sphere by defining its center point and radius or diameter.

FIGURE 38-19 ⎯⎯⎯⎯⎯⎯⎯⎯

Command: **sphere**
Specify center point or [3P/2P/Ttr]: **PICK** or (**coordinates**)
Specify radius or [Diameter] <0.0000>: **PICK** or (**value**)

Creating a *Sphere* with the default options and using 0,0 as the center would create the shape shown in Figure 38-19.

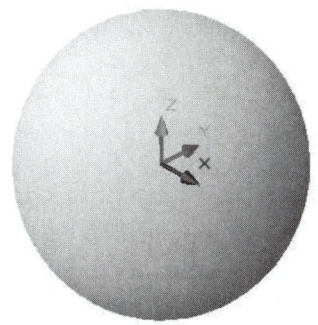

3P/2P/Ttr

As options, you can define the location and diameter of the *Sphere* using the *3Point*, *2Point*, or *Ttr* (tangent, tangent, radius) methods. For example, you could use the *3Point* option and *Osnap* to create a *Sphere* location and diameter based on three points on existing solids. These options are similar to the same options for creating a *Cone* base, *Cylinder* base, or *Torus*.

Torus

Pull-down Menu	Command (Type)	Alias (Type)	Short-cut	Screen (side) Menu	Tablet Menu
Draw Modeling > Torus	Torus	TOR	...	DRAW 2 SOLIDS Tours	O,7

Torus creates a torus (donut shaped) solid primitive. Three specifications are needed: (1) the center location, (2) the radius or diameter of the torus (from the center of the *Torus* to the centerline of the tube) and (3) the radius or diameter of the tube. AutoCAD prompts:

```
Command: torus
Specify center point or [3P/2P/Ttr]: PICK or (coordi-
nates)
Specify radius or [Diameter] <0.0000>: PICK or (value)
Specify tube radius or [2Point/Diameter] <0.0000>: PICK or
(value)
Command:
```

FIGURE 38-20

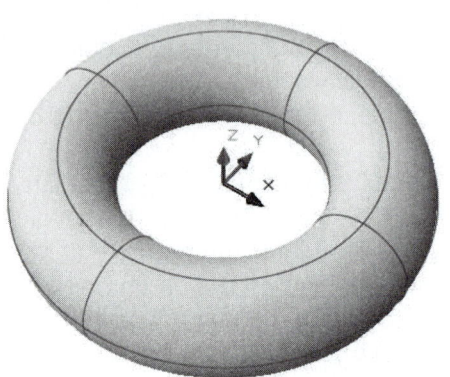

Using the default ("Specify center point") option, the *Torus* is oriented such that the *Torus* diameter (tube centerline) lies in the XY plane of the current UCS and the rotational axis of the tube is parallel with the Z axis of the UCS. Figure 38-20 shows a *Torus* created using the default orientation and the center of the torus at 0,0,0. The *Torus* has a torus radius of 3 and a tube radius of 1. (To improve the visibility of this figure, the Visual style edge setting is *Isolines*.)

A self-intersecting torus is allowed with the *Torus* command. A self-intersecting torus is created by specifying a tube radius equal to or greater than the torus radius. Figure 38-21 illustrates a torus with a torus radius of 3 and a tube radius of 3.

FIGURE 38-21

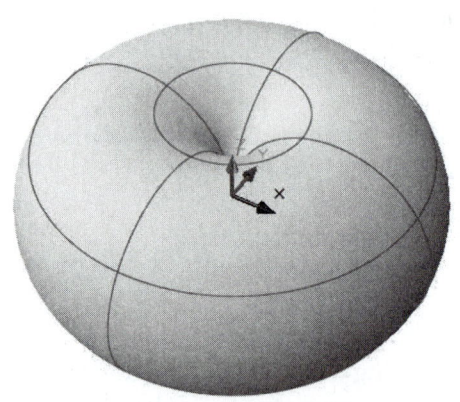

3P/2P/Ttr

Similar to the same options for drawing a *Circle*, you can define the location and diameter of the *Torus* using the *3Point*, *2Point*, or *Ttr* (tangent, tangent, radius) methods. For example, you could use the *3Point* option and *Osnap* to create a *Torus* location and diameter based on three points on existing solids. Using the *2Point* and *Ttr* methods, the *Torus* always lies in a plane parallel with the UCS, whereas the *3Point* option allows a different orientation. These options are similar to the same options for creating *Cone* base or *Cylinder* base.

Pyramid

Pull-down Menu	Command (Type)	Alias (Type)	Short-cut	Screen (side) Menu	Tablet Menu
Draw Modeling > Pyramid	Pyramid	PYR	...	DRAW 2 SOLIDS Pyramid	...

The *Pyramid* command creates a right pyramid with any number of sides.

With the default options, you specify the center of the base, the base radius, and the height. The resulting *Pyramid* has its base parallel to the XY plane of the current UCS, unless the *Axis endpoint* method is used.

FIGURE 38-22

```
Command: pyramid
4 sides  Circumscribed
Specify center point of base or [Edge/Sides]: PICK or (coor-
dinates)
Specify base radius or [Inscribed] <0.0000>: PICK or (value)
Specify height or [2Point/Axis endpoint/Top radius] <0.0000>:
PICK or (value)
Command:
```

Figure 38-22 shows a four-sided pyramid with the base located at 0,0, a base radius (inscribed) of 2, and a height of 4.

Edge

Use the *Edge* option to PICK two points in space to define the length and orientation of one edge. The *Pyramid* base is generated on the left side of the defined line going from the first to the second point.

```
Specify center point of base or [Edge/Sides]: e
Specify first endpoint of edge: PICK or (coordinates)
Specify second endpoint of edge: PICK or (coordinates)
```

Sides

Use this option to specify the number of sides (not including the base) for the *Pyramid*. Any practical number of sides can be specified with 3 as the minimum.

Inscribed/Circumscribed

This feature specifies how the base radius is drawn—either inscribed within or circumscribed about an imaginary circle. This method is similar to defining the size of a *Polygon*.

2Point

This option is the same as for a *Cone* or *Cylinder*. You can PICK two points to specify the height of the *Pyramid* such that the resulting height is equal to the indicated distance; however, the *Pyramid* does not change its orientation from that set by the base radius.

Axis endpoint

This option sets the height as well as the orientation for *Pyramid*. The height is equal to the distance between the center of the base and the axis endpoint selected. For example, Figure 38-23 shows a *Pyramid* similar to that in the previous figure, but with the *Axis endpoint* selected at 4,0. This option is the same as for a *Cone* or *Cylinder*.

FIGURE 38-23

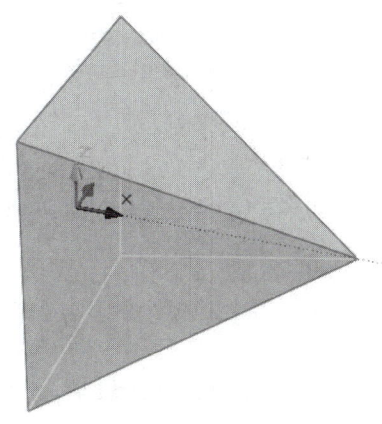

Top radius

Specifying a value for the *Top radius* creates a truncated pyramid such that the top surface is parallel with the base. Figure 38-24 displays a *Pyramid* with a base radius (inscribed) of 2, a *Top radius* of 1, and a height of 3.

FIGURE 38-24

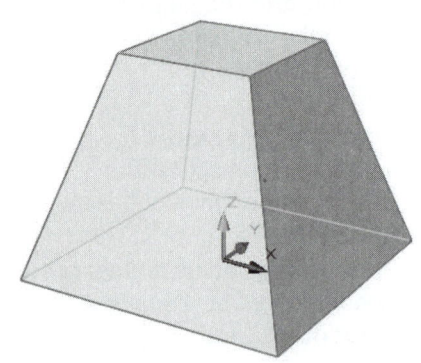

Polysolid

Pull-down Menu	Command (Type)	Alias (Type)	Short-cut	Screen (side) Menu	Tablet Menu
Draw *Modeling >* *Polysolid*	*Polysolid*	*PSOLID*	...	*DRAW 2* *SOLIDS* *Polysolid*	...

The *Polysolid* command can be used to <u>draw</u> a solid. *Polysolid* can also be used to <u>convert</u> a *Line*, 2D *Pline*, *Arc*, or *Circle* to a solid. Drawing a solid with *Polysolid* is similar to drawing a *Polyline*. However, a *Polysolid* is a rectangular shape that has a pre-scribed <u>height</u> as well as width. You interactively select points to determine the lengths of the sections as you draw. A *Polysolid* can have curved segments, but the profile (cross section) is always rectangular by default. For example, Figure 38-25 displays a *Polysolid* created by selecting five points for the segment endpoints and using the default *Height* and *Width* settings. See the following prompts.

FIGURE 38-25

```
Command: polysolid
Specify start point or [Object/Height/Width/Justify] <Object>: PICK
Specify next point or [Arc/Undo]: PICK
Specify next point or [Arc/Undo]: PICK
Specify next point or [Arc/Close/Undo]: PICK
Specify next point or [Arc/Close/Undo]: Enter
Command:
```

As you can see, *Polysolid* is an excellent command for <u>drawing walls</u>—having a prescribed height and width (rectangular cross section), but having different lengths that you PICK.

Height/Width

Use these options to set the height and width for subsequent *Polysolids* that you draw or convert.

2007

Command: **polysolid**
Specify start point or [Object/Height/Width/Justify] <Object>: **h**
Specify height <4.0000>: (**value**)

Alternately, use the *PSOLHEIGHT* and *PSOLWIDTH* system variables to set the default height and width for subsequent solids. The settings are not retroactive for previously drawn or converted segments.

Object

With the *Polysolid* command, you can use the *Object* option to convert an existing object into a *Polysolid*. An existing *Line*, 2D *Pline*, *Arc*, or *Circle* can be converted to a *Polysolid*. When the 2D object is converted, it assumes the currently specified height and width.

Specify start point or [Object/Height/Width/Justify] <Object>: **o**
Select object: **PICK**

Justify

Use this option to set the solid to be generated on the left, right, or center of the points you PICK when defining the segment endpoints. The justification is based on the direction from the first to the second point you PICK when drawing the segments.

Command: **polysolid**
Specify start point or [Object/Height/Width/Justify] <Object>: **j**
Enter justification [Left/Center/Right] <Center>:

Drawing Arcs

You can draw arc segments as you create a *Polysolid* using the *Arc* option when prompted for the "next point." The *Arc* option adds an arc segment to the solid such that the default starting direction of the arc is tangent to the last drawn segment as shown in Figure 38-26. The process of drawing arcs and the command line options are similar to drawing a *Pline*.

FIGURE 38-26

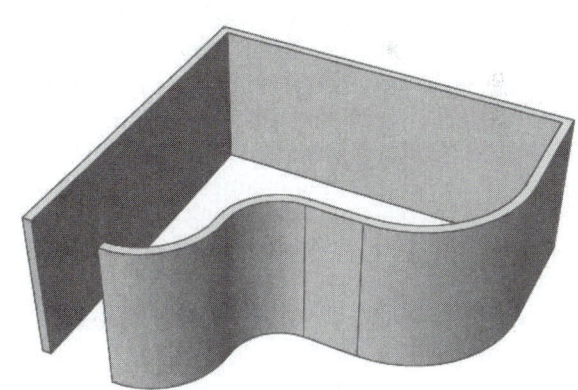

Specify the next point or [Arc/Close/Undo]: **a**
Specify endpoint of arc or [Close/Direction/Line/Second point/Undo]:

For a full explanation of these options, see *"Pline"* in Chapter 8.

Extrude

Pull-down Menu	Command (Type)	Alias (Type)	Short-cut	Screen (side) Menu	Tablet Menu
Draw *Modeling >* *Extrude*	*Extrude*	*EXT*	...	*DRAW 2* *SOLIDS* *Extrude*	*P,7*

Extrude means to add a third dimension (height) to an existing 2D shape. For example, if you *Extrude* a *Circle*, the resulting shape is a cylinder. If you *Extrude* a *Line*, the resulting shape is a planar surface. Therefore, *Extrude* can create a surface or a solid, depending on the 2D shape used. If you *Extrude* an open 2D shape (one having endpoints), such as a *Line*, *Arc*, or open *Pline*, a surface is created. (These ideas will be explained fully in Chapters 39 and 40.) However, if you *Extrude* an existing closed 2D shape such as a *Circle*, *Polygon*, *Rectangle*, *Ellipse*, *Pline*, *Spline*, *Region*, or planar surface, the resulting extrusion is a solid. For example, Figure 38-27 displays a closed *Pline* (before) to create a solid (after) using *Extrude*. Extruding solids is discussed in this section.

FIGURE 38-27

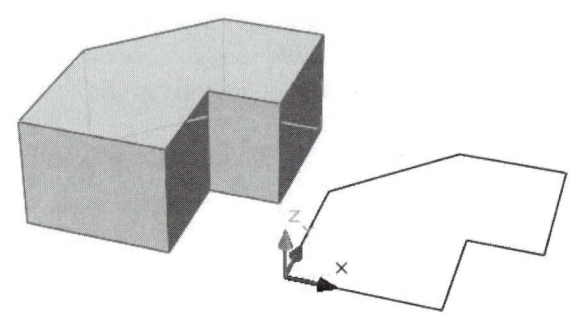

Extrude can simplify the creation of many solids that may otherwise take much more time and effort using typical primitives and Boolean operations. The versatility of this command lies in the fact that any closed shape that can be created by or converted to a *Pline*, *Spline*, *Region*, surface, etc., no matter how complex, can be transformed into a solid by *Extrude*. The closed 2D shape cannot be self-intersecting (crossing over itself). For example, the original shape used for the extrusion in Figure 38-28 (a closed *Pline*) contains straight and curved segments. Note that the original 2D shape was created on a vertical plane (the UCS XY plane), then extruded perpendicularly.

FIGURE 38-28

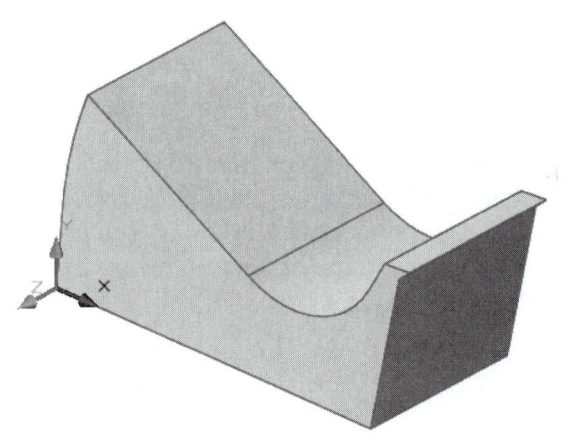

First, create the closed 2D shape, then invoke *Extrude*.

```
Command: extrude
Current wire frame density:  ISOLINES=4
Select objects to extrude: PICK
Select objects to extrude: Enter
Specify height of extrusion or [Direction/Path/Taper angle]: PICK or (value)
Command:
```

By default the 2D shape is consumed (deleted) by the *Extrude* command. The *DELOBJ* system variable controls if the 2D shape is deleted by *Extrude*.

Taper angle

A taper angle can be specified for the extrusion. Entering a <u>positive</u> taper angle causes the sides of the extrusion to slope <u>inward</u> (Fig. 38-29, right), while a negative taper angle causes the sides to slope outward. (Fig. 38-29, left). This is helpful, for example, for developing molded parts that require a slight draft angle to facilitate easy removal from the mold.

FIGURE 38-29

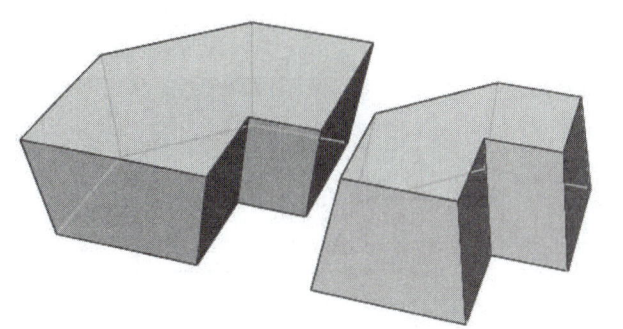

Direction

Normally, with the default method of *Extrude*, the selected 2D shape is extruded perpendicular to the plane of the shape regardless of the current UCS orientation, as shown in the previous three figures. However, you can use the *Direction* option to specify an <u>extrusion direction other than perpendicular</u>. Select two points to determine the direction.

```
Specify height of extrusion or
[Direction/Path/Taper angle]: d
Specify start point of direction: PICK or (coordinates)
Specify end point of direction: PICK or (coordinates)
```

FIGURE 38-30

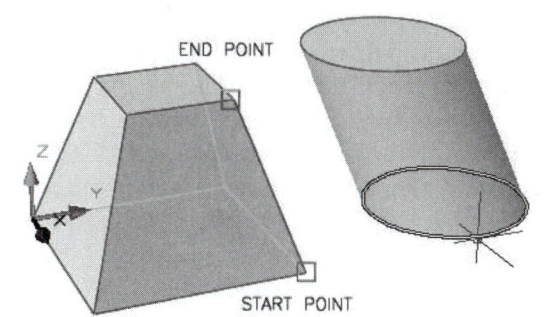

The two points of direction can be anywhere in 3D space and can be determined by selecting points on existing objects. The two points only indicate direction and do not change the orientation or position of the original 2D shape. For example, Figure 38-30 extrudes a *Circle* using the *Direction* option to *Osnap* to the *Endpoints* along the edge of a *Pyramid*.

Path

The *Path* option allows you to sweep the 2D shape called a "profile" along an existing line or curve called a "path." The path can be composed of a *Line, Arc, Ellipse, Pline, Spline*, or *Helix* (or can be different shapes converted to a *Pline*). The path does not have to be in a plane (2D) but can be in 3D space, such as in the form of a helix.

FIGURE 38-31

```
Specify height of extrusion or
[Direction/Path/Taper angle]: p
Select extrusion path or [Taper angle]: PICK
```

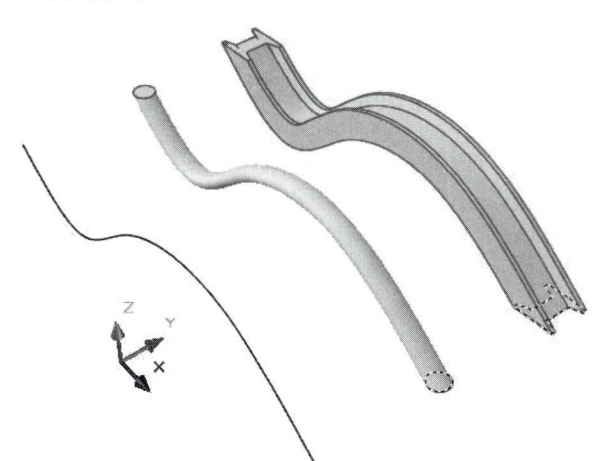

Figure 38-31 shows a *Circle* and a closed *Pline* (on the XY plane) extruded along a *Spline* path (that lies in the XZ plane). In this case, the same *Spline* path was selected for each extrusion. The path temporarily moves to the location of the profile for the extrusion. The 2D shapes that are extruded (shown highlighted) are consumed by default, but the path is not deleted. Notice how the original 2D shapes that were extruded change orientation as they extrude along the path.

The *Sweep* command can also be used to create solids similar to *Extrude* with the *Path* option. See "*Sweep*."

Revolve

	Pull-down Menu	Command (Type)	Alias (Type)	Short-cut	Screen (side) Menu	Tablet Menu
	Draw Modeling > Revolve	*Revolve*	*REV*	...	*DRAW 2 SOLIDS Revolve*	*Q,7*

Revolve allows you to revolve an open or closed 2D shape (profile) about a selected axis. Like *Extrude*, the resulting shape can be a surface or solid depending on the original shape used. *Revolve* creates a <u>solid</u> if the original profile used is "<u>closed</u>," such as a *Pline, Polygon, Circle, Ellipse, Spline*, or a *Region* object. (Using *Revolve* with "open" 2D shapes to create surfaces is discussed in Chapters 39 and 40.) By default, the profile is consumed (deleted) when the solid is created, but it can be retained by setting the *DELOBJ* system variable to 0. The command syntax for *Revolve* (accepting the defaults) is:

FIGURE 38-32

Command: **revolve**
Current wire frame density: ISOLINES=4
Select objects to revolve: **PICK**
Select objects to revolve: **Enter**
Specify axis start point or define axis by [Object/X/Y/Z] <Object>: **PICK**
Specify axis endpoint: **PICK**
Specify angle of revolution or [STart angle] <360>: **Enter** or (**value**)
Command:

Using the default option of *Revolve*, you must specify two points to define the axis to revolve about. You can select two points on the 2D object as the axis. The profile in Figure 38-32 (right) was used to create the 3D object (left) by selecting two points on the vertical edge.

To define the axis of revolution using the default option or the *Object* option, you can select points on other objects or specify coordinates. The <u>profile is always revolved about an axis in the same plane as the profile</u>. However, the two points that define the axis for revolving do not have to be coplanar with the 2D shape as long as one of the points lies in the same plane and the axis is not perpendicular to the 2D shape. In this case, *Revolve* uses an axis <u>in the plane</u> of the revolved shape aligned with the direction of the selected points or object. The other options for selecting an axis of revolution are described next.

Object
A *Line* or single segment *Pline* can be selected for an axis. The positive axis direction is from the closest endpoint PICKed to the other end.

X/Y/Z
Uses the positive X, Y, or Z axis of the current UCS as the axis of rotation. However, keep in mind that an axis perpendicular to the plane of the profile cannot be used.

Start angle

After defining the axis of revolution, *Revolve* requests the number of degrees for the object to be revolved. Any angle can be entered. For example, one possibility for revolving a profile is shown in Figure 38-33 where the profile is generated through 270 degrees.

By default, the object is revolved from its current plane in a counter-clockwise direction. If you revolve the object less than 360 degrees, you can use the *Start angle* option to specify an angular value from the current 2D shape position to begin the *Revolve*.

FIGURE 38-33

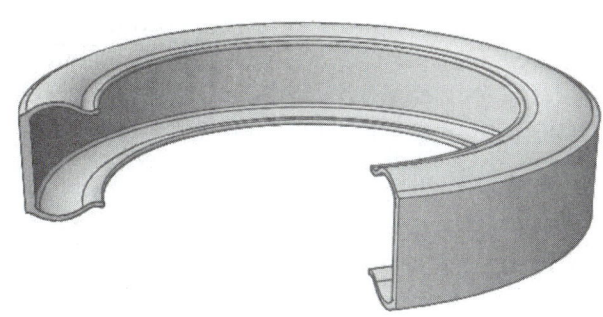

Sweep

Pull-down Menu	Command (Type)	Alias (Type)	Short-cut	Screen (side) Menu	Tablet Menu
Draw Modeling > Sweep	*Sweep*	...	...	DRAW 2 SOLIDS *Sweep*	...

The *Sweep* command allows you to create a new solid or surface by sweeping an open or closed planar curve (profile) along an open or closed 2D or 3D path. *Sweep* is similar to using *Extrude* with the *Path* option; however, *Sweep* has more versatility. Like *Extrude* and *Revolve*, you create a <u>solid</u> from a "closed" 2D shape, and you create a <u>surface</u> from an "open" 2D shape. (Creating surfaces with *Sweep* is discussed in Chapters 39 and 40.) To *Sweep*, simply select an existing profile and a path.

```
Command: sweep
Current wire frame density: ISOLINES=4
Select objects to sweep: PICK
Select objects to sweep: Enter
Select sweep path or [Alignment/Base point/Scale/Twist]: PICK
Command:
```

With *Sweep*, you can use a *Circle*, *Ellipse*, *Region*, closed *Pline*, or a surface as the profile to create a swept solid. The path can be any one of the following: *Line*, *Arc*, *Pline*, *Circle*, *Ellipse*, *Spline*, *Helix*, or edges of solids. The path can be closed and does not have to be planar. By default settings, the profile is deleted when swept, but the path is not deleted. The *DELOBJ* system variable setting determines if the profile, path, neither, or both are deleted.

The *Sweep* command automatically moves the profile into position at the end of the path. By default, the orientation of the profile is adjusted to sweep in a direction <u>perpendicular</u> to the path along its entire length. For example, figure 38-34 displays the before and after stages of sweeping a *Circle* profile along a *Helix* path. Note that the *Circle* was created in the same plane as the base of the *Helix*, but was automatically moved into position and realigned perpendicularly to the path as it was swept.

FIGURE 38-34

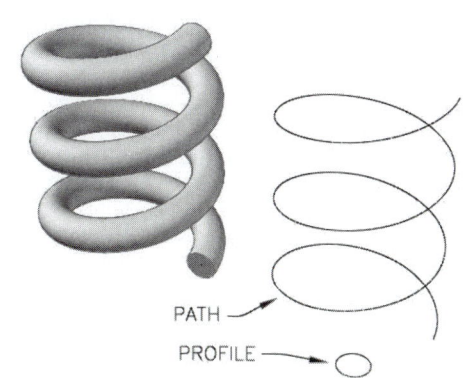

PATH

PROFILE

Alignment

Different from the *Extrude* with the *Path* option, *Sweep* provides an option for how the profile is swept along the path. *Alignment* specifies whether the profile is aligned to be perpendicular to the direction of the sweep path or left in its current orientation. By default, the profile is aligned.

> Align sweep object perpendicular to path before sweep [Yes/No] <Yes>:

If you answer *No* to this option, the profile is left in its original orientation with respect to the <u>first point</u> along the sweep path. From that point, the profile is realigned along the length of the path to keep the same orientation with respect to the path as the path turns. For example, Figure 38-35 displays a comparison between the two options for *Alignment*. Notice the "flat" appearance of the helix on the left (not aligned), yet the profile turns along the length of the path.

FIGURE 38-35

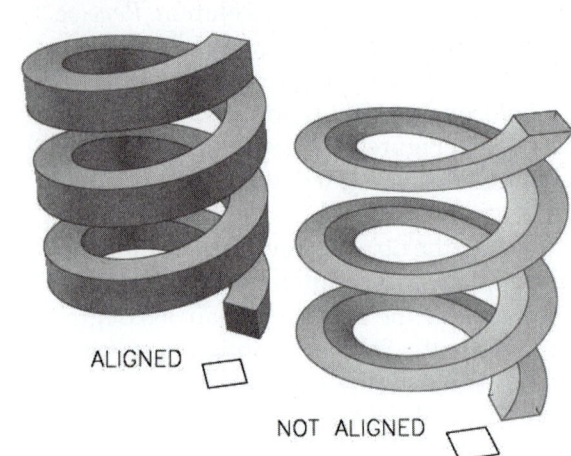

ALIGNED

NOT ALIGNED

Base Point

By default, the center of the profile is used as the base point when it is swept along the path. For example, in Figure 38-34, the center of the *Circle* becomes the base point. Use the *Base point* option to specify a new base point for the objects to be swept. If the specified point does not lie on the plane of the selected objects, it is projected onto the plane.

Scale

You can use this option to scale the <u>profile</u> for a sweep operation. Applying a scale factor causes the profile to be tapered uniformly as it is swept from the start to the end of the sweep path (Fig. 38-36, right). You can specify a scale factor or enter *R* for the reference option.

> Enter scale factor or [Reference] <1.0000>:

The *Reference* option of *Scale* allows you to specify the scale by picking two points or entering values.

FIGURE 38-36

TWIST: 180

SCALE: .5

Twist

Using this option, you can twist the profile as it is swept along the length of the path. You specify a value for the twist angle. The twist angle specifies the amount of rotation along the entire length of the sweep path (Fig. 38-36, left). Specify an angle value less than 360.

> Enter twist angle or allow banking for a non-planar sweep path [Bank] <0.0000>:

Bank

Bank causes an automatic, unspecified rotation (twist) of the profile being swept along a non-planar sweep path (such as a 3D *Pline*, 3D *Spline*, or *Helix* sweep path). For example, using the *Bank* option, a profile would twist 45 degrees for each turn of a *Helix*.

Loft

Pull-down Menu	Command (Type)	Alias (Type)	Short-cut	Screen (side) Menu	Tablet Menu
Draw Modeling > Loft	Loft	...	...	DRAW 2 SOLIDS Loft	...

With the *Loft* command, you can create a new solid or surface by specifying a series of cross sections. Similar to *Extrude*, *Revolve*, and *Sweep*, the resulting shape (surface or solid) depends on if the 2D shape(s) are "open" (having endpoints) or "closed" (one connected loop). *Loft* creates a solid when the cross-section shapes are <u>closed</u>.

The cross sections define the shape of the resulting solid. You must specify at least two cross sections when you use the *Loft* command. *Loft* then draws a solid in the space between (connecting) the closed cross sections. *Loft* will "morph" between the shapes to create a smooth transition. For example, Figure 38-37 displays a lofted solid between three cross sections: a square *Pline* at the base, a *Circle* in the middle, and a 6-sided *Polygon* at the top. The solid was created using the following prompts.

FIGURE 38-37

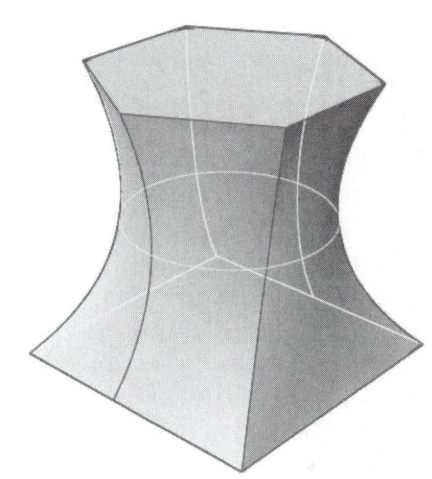

```
Command: loft
Select cross-sections in lofting order: PICK
Select cross-sections in lofting order: PICK
Select cross-sections in lofting order: PICK
Select cross-sections in lofting order: Enter
Enter an option [Guides/Path/Cross-sections only]
<Cross-sections only>: Enter
Command:
```

Closed cross sections that can be used with *Loft* to create solids are *Circle*, closed 2D *Pline*, closed 2D *Spline*, and *Ellipse*. (Keep in mind that *Plines* include *Rectangles* and *Polygons*.) The *DELOBJ* system variable setting determines whether the cross sections, guides, and path are automatically deleted when the solid is created.

Cross-sections only

When you use the *Cross-sections only* option, the *Loft Settings* dialog box is displayed (Fig. 38-38). Here you can select from several methods for controlling the contour of a lofted surface at its cross sections.

FIGURE 38-38

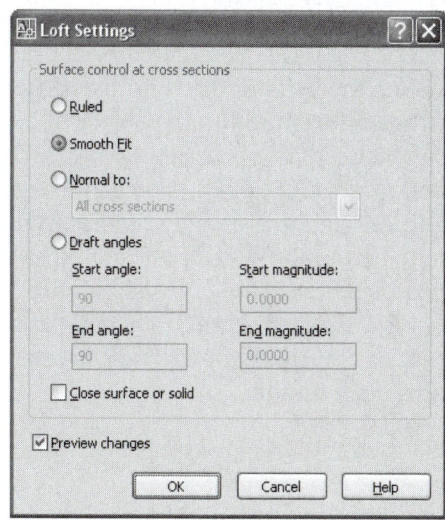

Ruled
This option specifies that the solid or surface is ruled (straight) between the cross sections and has sharp edges at the cross sections. See Figure 38-39, left.

Smooth Fit
Specifies that a smooth solid or surface is drawn between the cross sections and has sharp edges at the start and end cross sections. See Figure 38-39, center.

FIGURE 38-39

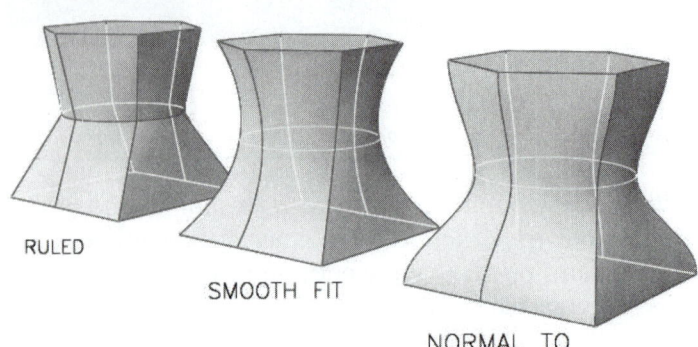

RULED

SMOOTH FIT

NORMAL TO

Normal to
Controls the surface normal (perpendicular) of the solid where it passes through the cross sections. You can determine that surfaces are normal to *All cross sections* or specific cross-sections, such as *Start Cross Section*, *End Cross Section*, and *Start and End Cross Sections*. See Figure 38-39, right.

Draft Angles
This section controls the draft angle and magnitude of the first and last cross sections of the lofted solid or surface. The draft angle is the beginning direction of the surface. The 90 (degree) setting is defined as 90 degrees from the cross section and 0 is defined as outward from the plane of the curve. The *Magnitude* controls the relative distance of the surface from the start cross section in the direction of the draft angle before the surface starts to bend toward the next cross section. Figure 38-40 displays two lofted solids, each created from the <u>same</u> two cross sections. As you can see, the

FIGURE 38-40

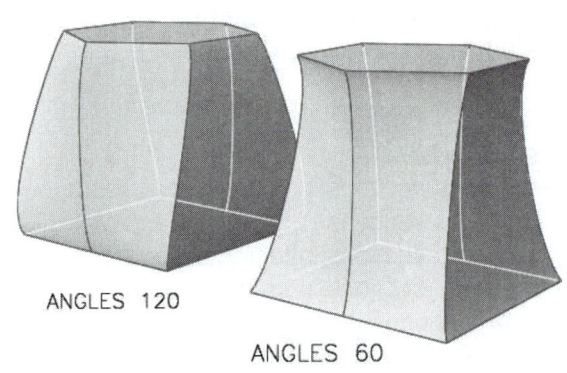

ANGLES 120

ANGLES 60

left solid's surfaces protrude outward applying the angle of greater than 90 for both the start and end profiles (120 in this case), whereas the solid on the right curves inward with angles of less than 90 (60 in this case).

Guides
With the *Guides* option, you can select multiple curves to define the contours of the solid or surface. Guide curves are lines or curves that define the form of the solid by adding wireframe information to the object. Objects that can be used as guides are *Splines* (2D or 3D) and *Plines* (2D or 3D). You can use guide curves to control exactly how points are matched up on corresponding cross sections to produce predictable results. Figure 38-41 displays a shape that might be created with guides that might be difficult to attain through other *Loft* options.

FIGURE 38-41

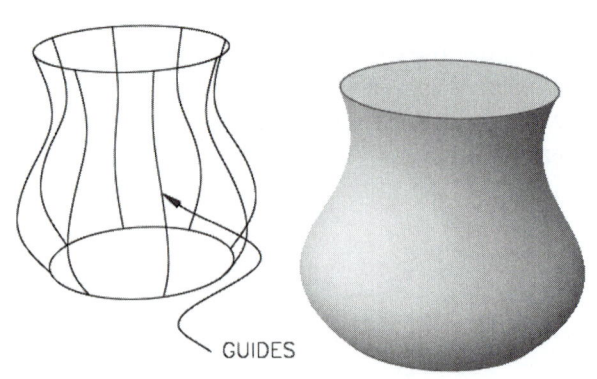

GUIDES

2007

In order for guides to work correctly, each guide curve must intersect each cross section, start on the first cross section, and end on the last cross section. You can select any number of guide curves for the lofted surface or solid.

Path

With the *Path* option, you can select a single path curve to define the shape of the solid. This method is similar to *Sweep*, except here you use multiple profiles (cross sections) to determine the shape of the solid along the path. The path curve must intersect the planes of all the profiles. Objects that can be used as a *Loft Path* include *Spline* (2D or 3D), *Helix*, *Circle*, *Ellipse*, *Pline* (2D or 3D).

For example, the solid in Figure 38-42 was created using the *Path* option of *Loft*. Note that the path curve is a *Spline*. Also note that although the profiles are similar, they are of varying sizes and orientations.

FIGURE 38-42

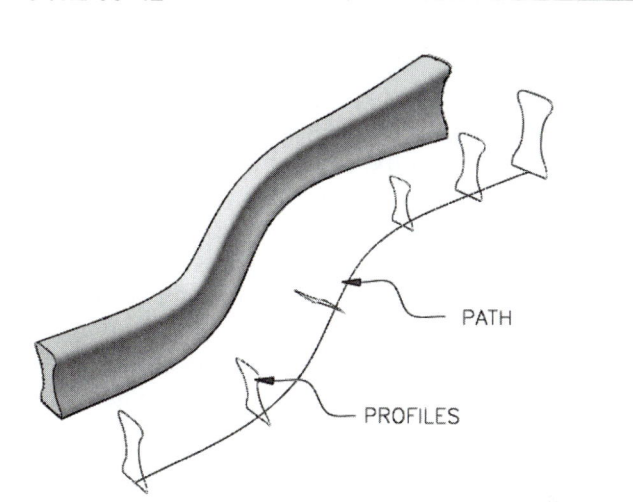

PATH

PROFILES

DYNAMIC USER COORDINATE SYSTEMS

As explained in the introduction of this chapter, the process of solid modeling construction involves either creating primitives in the desired location and orientation or moving, rotating, or copying the primitives to the desired location and orientation. Normally, the desired approach is to create the primitives directly in the desired location and orientation. To do this, User Coordinate Systems must be used.

Two kinds of User Coordinate Systems are available during the solid modeling construction process. Firstly, you can create a User Coordinate System using the *UCS* command, as you learned in Chapter 36. This UCS is static—that is, it remains active until you change to another coordinate system. Secondly, you can create a Dynamic User Coordinate System (DUCS). A DUCS is a <u>temporary</u> UCS that remains active only during a solid primitive creation command such as *Box* or *Wedge*, then it disappears. Dynamic User Coordinate Systems are most beneficial during solid model construction and editing. This section discusses how to create and control Dynamic Coordinate Systems.

Use the *DUCS* button on the Status bar to toggle this feature on or off (Fig. 38-43). If *DUCS* is toggled on, a Dynamic User Coordinate System appears <u>automatically</u> during primitives commands. If you right-click on the *DUCS* toggle, you can turn on the display of the crosshair labels (X, Y, Z).

FIGURE 38-43

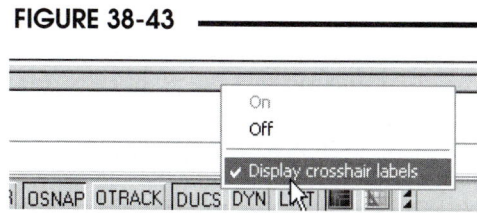

Using Dynamic User Coordinate Systems

Three conditions must exist for a Dynamic Coordinate System to appear:
1. The *DUCS* toggle must be on.
2. You must invoke one of these primitive commands: *Box*, *Wedge*, *Cone*, *Sphere*, *Cylinder*, *Pyramid*, *Torus*, and *Polysolid*.
3. You must hover your cursor over a planar surface of an existing solid during the first prompt of the command.

Dynamic Coordinate System Example

With the *DUCS* feature, you can temporarily align the XY plane of the UCS with a plane on a solid model. For example, assume you wanted to create a *Wedge* on the top surface of an existing *Box* and orient the *Wedge* so it sloped down toward the left vertical face. When you use the *Wedge* command, you are prompted for the "first corner." With *DUCS* on, you can create a temporary UCS on the top face of the *Box* to begin drawing the *Wedge*. You specify the desired XY plane of the DUCS by "hovering" over any planar face of the existing solid until it becomes highlighted and the crosshair labels appear indicating the direction for the DUCS XY plane and Z axes (Fig. 38-44, top surface). Note the location of the current UCS.

The orientation of the DUCS (direction of the X, Y, and Z axes) is based on the edge that is crossed when you approach the desired plane with the cursor. For example, compare Figures 38-44 and 38-45 and note the top surface's edge that is crossed and the resulting crosshair XYZ orientation. The crosshair XYZ orientation may be critical for creating some primitives such as a *Wedge*. Once the *DUCS* icon appears in the desired orientation, PICK to specify the "first corner" for the *Box*.

In Figure 38-46, the *Midpoint* of the right edge is selected as the "first corner." Note that once the first corner is established, the UCS icon appears in that location and orientation. Next, the *Endpoint* at the far left corner is selected as the "other corner" for the base of the *Wedge*.

Proceeding with the *Wedge* command, the "height" is specified next to complete the primitive. Figure 38-47 displays the wedge while the height is being PICKed. Note the UCS icon still located in the selected "first corner" location.

NOTE: Notice in Figure 38-47 that the *Wedge* slopes down toward the –X axis. Remember that a *Wedge* always slopes down in the direction of the second corner.

Once the primitive command is complete, the *DUCS* disappears. The normal (non-dynamic) UCS becomes active again and the UCS icon returns to its previous location and orientation (see Fig. 38-44).

FIGURE 38-44

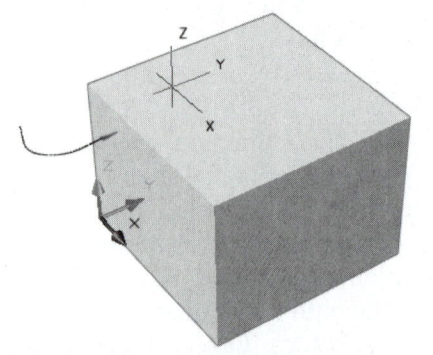

FIGURE 38-45

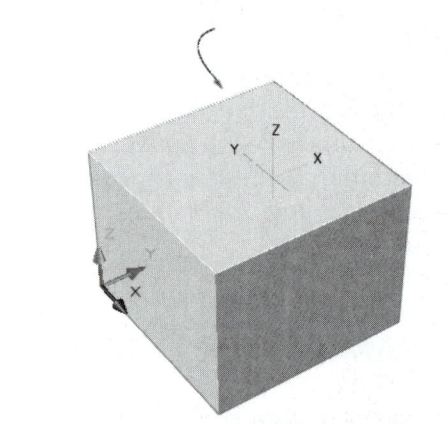

FIGURE 38-46

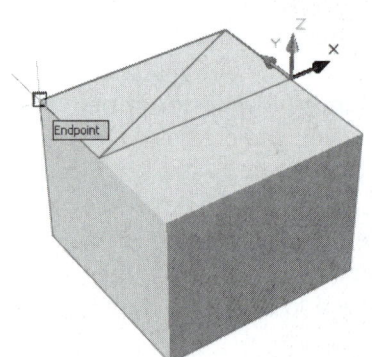

FIGURE 38-47

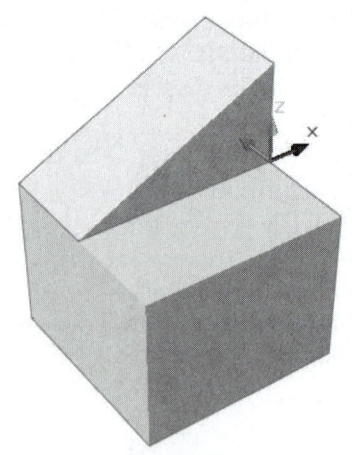

Dynamic UCS Principles

Several principles of Dynamic UCSs are followed:

A. The X axis of the Dynamic UCS is always aligned with an edge of the face and the positive direction of the X axis always points toward the right half of the screen.
B. Only the front faces of a solid are detected by the Dynamic UCS.
C. If *SNAP* mode is turned on, it aligns temporarily to the Dynamic UCS.
D. You can temporarily turn off the Dynamic UCS by pressing F6 or Shift+Z while moving the crosshairs over a face.

COMMANDS FOR MOVING SOLIDS

When you create the desired primitives, Boolean operations are used to construct the composite solids. However, the primitives must be in the correct position and orientation with respect to each other before Boolean operations can be performed. You can either create the primitives in the desired position during construction (by using UCSs) or move the primitives into position after their creation. Several methods that allow you to move solid primitives are explained in this section.

Move

Pull-down Menu	Command (Type)	Alias (Type)	Short-cut	Screen (side) Menu	Tablet Menu
Modify *Move*	*Move*	*M*	...	*MODIFY2* *Move*	*V,19*

The *Move* command that you use for moving 2D objects in 2D drawings can also be used to move 3D primitives. Generally, *Move* is used to change the position of an object in one plane (translation), which is typical of 2D drawings. *Move* can also be used to move an ACIS primitive in 3D space <u>if</u> *OSNAP*s or 3D coordinates are used.

Move operates in 3D just as you used it in 2D. Previously, you used *Move* only for repositioning objects in the XY plane, so it was only necessary to PICK or use X and Y coordinates. Using *Move* in 3D space requires entering X, Y, and Z values or using *OSNAP*s.

For example, to create the composite solid used in the figures in Chapter 36, a *Wedge* primitive was *Moved* into position on top of the *Box*. The *Wedge* was created at 0,0,0, then rotated. Figure 38-48 illustrates the movement using absolute coordinates described in the syntax below:

```
Command: move
Select objects: PICK
Select objects: Enter
Specify base point or displacement: 0,0
Specify second point of displacement or <use first point as displacement>: 0,4,3
Command:
```

FIGURE 38-48

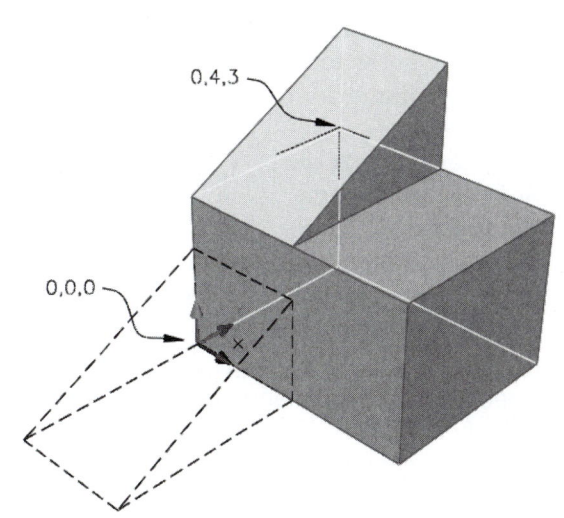

Alternately, you can use *OSNAPs* to select geometry in 3D space. Figure 38-49 illustrates the same *Move* operation using *Endpoint OSNAPs* instead of entering coordinates.

FIGURE 38-49

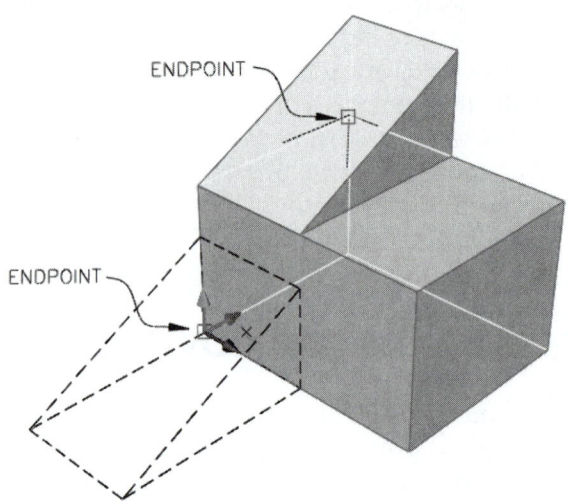

ENDPOINT

ENDPOINT

3Dalign

Pull-down Menu	Command (Type)	Alias (Type)	Short-cut	Screen (side) Menu	Tablet Menu
Modify *3D Operations >* *3D Align*	*3Dalign*	*3AL*	...	*MODIFY2* *3D Align*	...

3Dalign provides a means of aligning one 3D solid with another. *3Dalign* is very powerful because it automatically performs 3D <u>translation and rotation</u> if needed. All you have to do is select the points on two 3D objects that you want to align (connect).

Aligning one solid with another can be accomplished by connecting source points (on the solid to be moved) to destination points (on the stationary solid). Alternately, you can short cut this procedure by indicating a source plane or destination plane instead of selecting 3 points. You should always use *Osnap* modes to select the source and destination points to assure accurate alignment.

3Dalign performs a translation (like *Move*) and two rotations (like *Rotate*), each in separate planes to align the points as designated. The motion performed by *3Dalign* is actually done in three steps. For example, assume you wanted to move the *Wedge* on top of the *Box* and also reorient the *Wedge*. The command syntax for 3D alignment is as follows:

```
Command: 3dalign
Select objects: PICK
Select objects: Enter
Specify source plane and orientation ...
Specify base point or [Copy]: PICK  (S1)
Specify second point or [Continue] <C>: PICK (S2)
Specify third point or [Continue] <C>: PICK (S3)
Specify destination plane and orientation ...
Specify first destination point: PICK  (D1)
Specify second destination point or [eXit] <X>: PICK  (D2)
Specify third destination point or [eXit] <X>: PICK  (D3)
```

In this case, as shown in Figure 38-50, the "base point" (S1) is connected to the first destination point (D1). This first motion is a translation. <u>The base point and the first destination point always physically touch.</u>

FIGURE 38-50

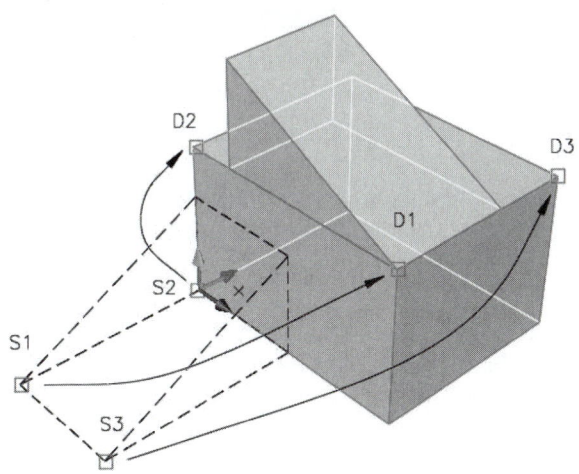

Next, the vector defined by the first and second source points (S1 and S2) is aligned with the vector defined by the first and second destination points (D1 and D2). The length of the segments between the first and second points on each object is not important because AutoCAD only considers the <u>vector direction</u>. This second motion is a rotation along one axis.

Finally, the third set of points are aligned similarly (S2-S3 with D2-D3). This third motion is a rotation along the other axis, completing the alignment.

In some cases, such as when cylindrical objects are aligned, only two sets of points have to be specified. For example, if aligning a shaft with a hole, the first set of points (source and destination) specify the attachment of the base of the shaft with the bottom of the hole. The second set of points specify the alignment of the axes of the two cylindrical shapes. A third set of points is not required because the radial alignment between the two objects is not important. When only two sets of points are specified, AutoCAD asks if you want to "continue" based on only two sets of alignment points.

You can also use the "source plane" method to align the *Wedge* with the *Box* as shown in Figure 38-51. This method uses the first source point (S1), or "base point," to indicate the XY plane of the source object. Once this base point is selected, press Enter to begin specifying the destination points, as shown in the following prompts.

FIGURE 38-51

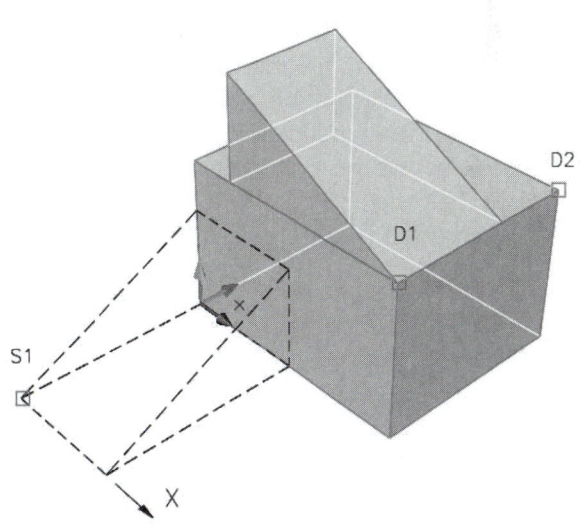

```
Command: 3dalign
Select objects: PICK
Select objects: Enter
Specify source plane and orientation ...
Specify base point or [Copy]: PICK (S1)
Specify second point or [Continue] <C>: Enter
Specify destination plane and orientation ...
Specify first destination point: PICK (D1)
Specify second destination point or [eXit] <X>:
PICK (D2)
Specify third destination point or [eXit] <X>: Enter
```

Only the first two destination points need to be specified. *3Dalign* uses the positive X direction of the source plane to make the alignment between the first two destination points (D1 and D2). Since the planar orientation of the source and destination planes is the same, no other rotation is needed, so pressing Enter completes the alignment.

3Dmove

	Pull-down Menu	Command (Type)	Alias (Type)	Short-cut	Screen (side) Menu	Tablet Menu
	Modify 3D Operations > 3D Move	3DMove	3M	...	...	...

3Dmove can be used like to the *Move* command to move objects in 3D space. Even the prompts for *Move* and *3Dmove* are identical. However, using *3Dmove*, you can <u>constrain movements only to a particular axis or plane</u>. For example, you can use *3Dmove* to move a solid along the X axis only.

Using *3Dmove*, once you select the object(s) to move, the <u>move grip tool</u> is displayed. Next, select the desired base point, just as you would with the *Move* command. The move grip tool locks to the selected "base point." For example, in Figure 38-52, the move grip tool appears during the "Specify base point" prompt. In this case, one *Endpoint* of the *Wedge* is at the base point.

FIGURE 38-52

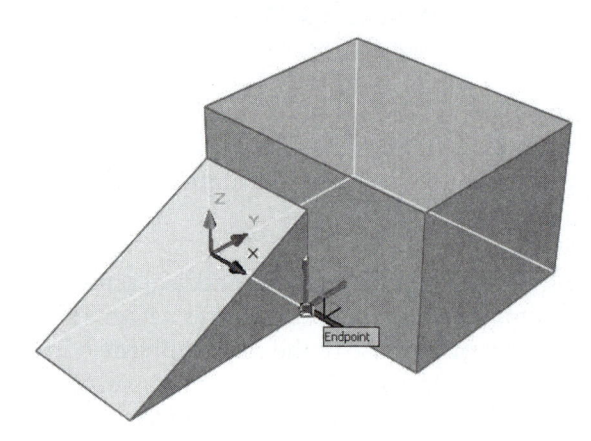

```
Command: 3dmove
Select objects: PICK
Select objects: Enter
Specify base point or [Displacement]
<Displacement>: PICK
Specify second point or <use first point as dis-
placement>: PICK
```

When the move grip tool appears, select any <u>axis</u> to constrain the movement of the selected objects to the indicated grip tool axis. Once selected, the axis changes to a yellow-gold color. For example, selecting the X axis of the grip tool displays a vector along the axis (Fig. 38-53, left). Likewise, selecting the Y or Z axis highlights the appropriate move vector (Fig. 38-53, center and right).

FIGURE 38-53

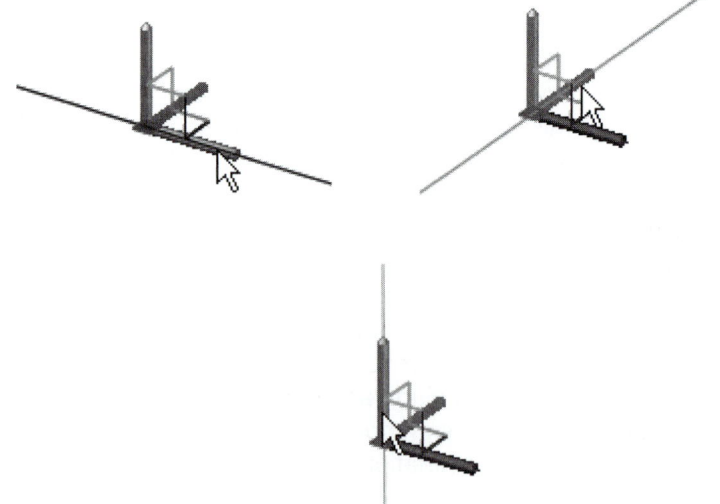

You can also constrain the movement to a <u>plane</u>. Do this by selecting one of the small squares on the move grip tool. For example, if you want to move the selected object within the XZ plane, select the small square connecting the X and Z axes (Fig. 38-54).

FIGURE 38-54

From the previous *Box* and *Wedge* example, selecting the Z axis on the move grip tool would constrain movement of the *Wedge* to a positive or negative Z direction, as shown in Figure 38-55. Next, move the *Wedge* to the desired location on the Z axis, then PICK or enter a value as the "second point."

FIGURE 38-55

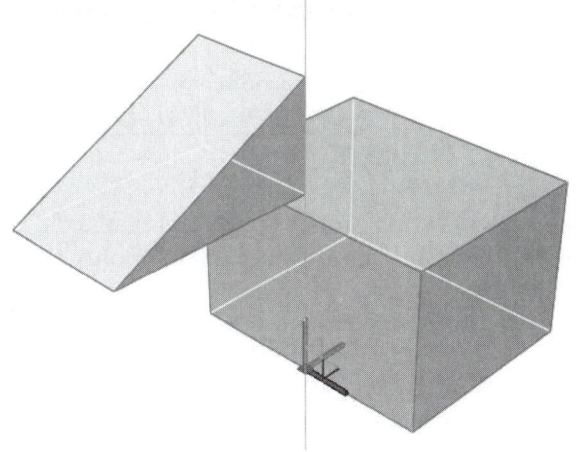

You can also use the Dynamic UCS feature with *3Dmove*. Either turn on the *DUCS* toggle on the Status bar or press Ctrl+D to turn on Dynamic UCS. This action causes the move grip tool to realign its orientation as you move the pointer over faces of solids (Fig. 38-56). The grip tool orients itself to the current plane depending on which edge of the face the pointer crosses. When the move grip tool displays the desired orientation, PICK to place the grip tool at the desired edge or corner (normally using *Osnap*).

FIGURE 38-56

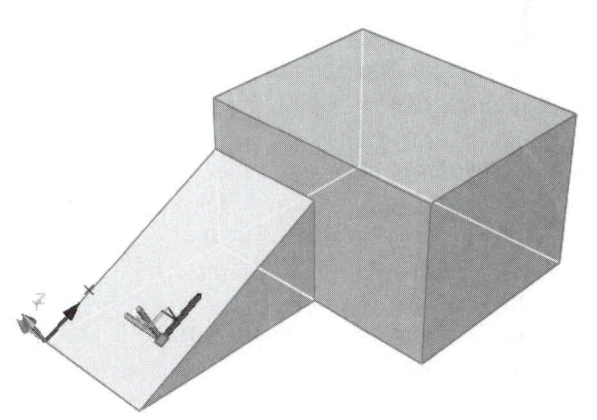

Once the grip tool is located on the desired plane, you can select any axis or plane indicator on the tool to constrain the direction of the move operation. For our example (Fig. 38-57), once the tool is oriented on the inclined plane of the *Wedge*, the movement of the *Wedge* is constrained to the grip tool's X axis. You can then PICK the desired "second point." If you enter coordinate values, they are relative to the selected plane.

FIGURE 38-57

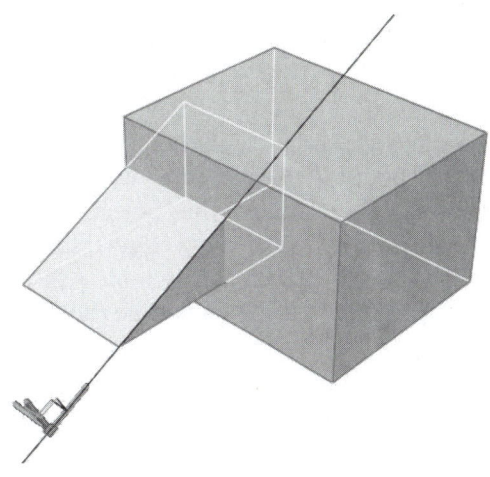

3Drotate

Pull-down Menu	Command (Type)	Alias (Type)	Short-cut	Screen (side) Menu	Tablet Menu
Modify *3D Operations >* *3D Rotate*	*3Drotate*	3R	...	...	W,22

3Drotate allows you to rotate solids about an axis. You specify the axis to rotate about by using the *3Drotate* grip tool. This grip tool displays only three possible rotational axes. However, since the grip tool can be located anywhere in 3D space and in any orientation using the *DUCS* feature, virtually any axis can be specified for the rotate action. The grip tool appears at the "Specify base point" prompt.

```
Command: 3drotate
Current positive angle in UCS: ANGDIR=counterclockwise  ANGBASE=0
Select objects: PICK
Select objects: Enter
Specify base point: PICK
Pick a rotation axis: PICK  (Select an axis handle to specify the rotation axis)
Specify angle start point: PICK
Specify angle end point: PICK
Command:
```

The rotate grip tool appears at the base point you specify. At the "Pick a rotation axis" prompt, select one of the circular "handles." Once a circular handle is selected, the related axis is displayed. For example, Figure 38-58 (left) displays selection of the X axis handle. Note the cursor location in the figure. When the handle is selected, the related axis vector remains on the screen until the rotate action is complete. Figure 38-58 (right) displays selection of the Z axis handle.

FIGURE 38-58

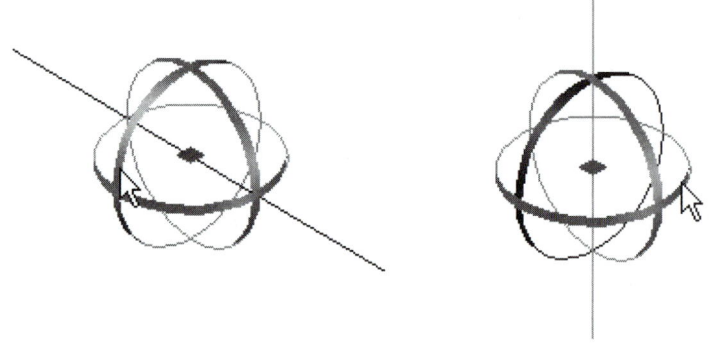

For example, suppose you wanted to rotate the wedge shown in Figure 38-59 about the Z axis at its nearest corner. Using *3Drotate*, select the wedge at the "Select objects" prompt, then select the nearest corner of the wedge at the "Specify base point" prompt to locate the *3Drotate* grip tool. At the "Pick a rotation axis" prompt, select the Z axis handle (note the cursor location).

FIGURE 38-59

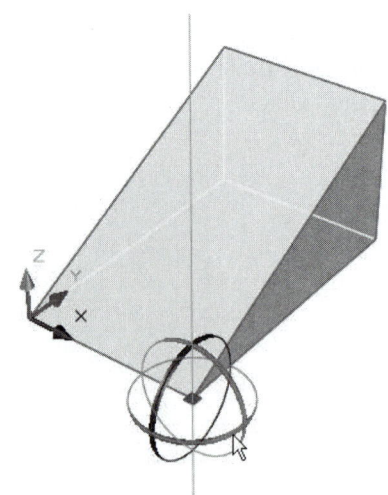

Next, you must specify an "angle start point." In our example, the *Endpoint* at the far right corner on the base is selected as the "angle start point." Finally, you can rotate the object about the selected point in real time at the "Specify angle endpoint" prompt, as shown in Figure 38-60.

FIGURE 38-60

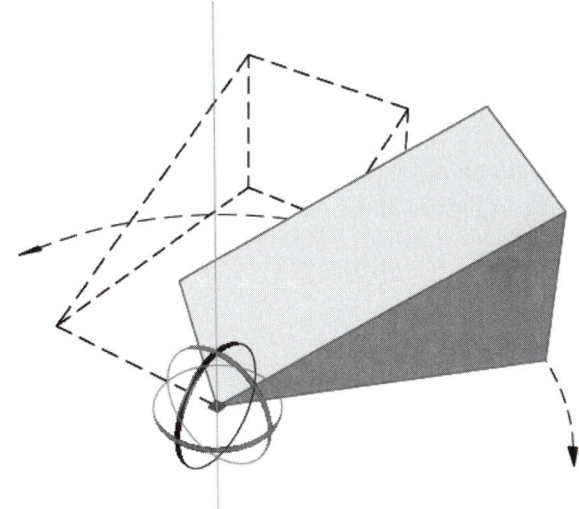

You can use the Dynamic UCS feature with *3Drotate* to orient the *3Drotate* grip tool on any surface of a solid. In other words, if *DUCS* is on, you can hover over any surface of a solid at the "Specify base point" prompt to align the grip tool with the high-lighted surface. Generally, the XY plane of the grip tool aligns with the selected face of the solid, so the Z axis is perpendicular to the face. For example, Figure 38-61 displays using the *3Drotate* command with *DUCS* on such that the orientation of the grip tool is aligned with the inclined surface of the wedge.

FIGURE 38-61

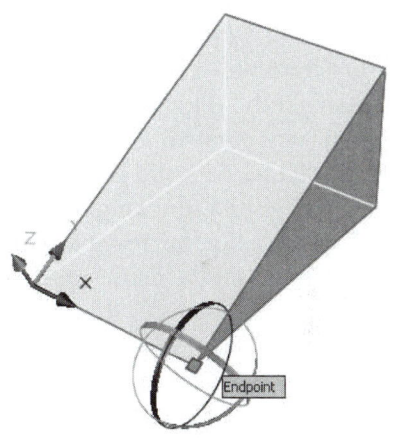

Endpoint

2007

Mirror3D

Pull-down Menu	Command (Type)	Alias (Type)	Short-cut	Screen (side) Menu	Tablet Menu
Modify *3D Operations >* *3D Mirror*	*Mirror3D*	...	...	...	*W,21*

Mirror3D operates similar to the 2D version of the command *Mirror* in that mirrored replicas of selected objects are created. With *Mirror* (2D) the selected objects are mirrored about an axis. The axis is defined by a vector lying in the XY plane. With *Mirror3D*, selected objects are mirrored about a <u>plane</u>. *Mirror3D* provides multiple options for specifying the plane to mirror about:

Command: **mirror3d**
Select objects: **PICK**
Select objects: **Enter**
Specify first point of mirror plane (3 points) or [Object/Last/Zaxis/View/XY/YZ/ZX/3points] <3points>:
PICK or (**option**)

3points

The *3points* option mirrors selected objects about the plane you specify by selecting three points to define the plane. You can PICK points (with or without *OSNAP*) or give coordinates. *Midpoint OSNAP* is used to define the 3 points in Figure 38-62 A to achieve the result in B.

FIGURE 38-62

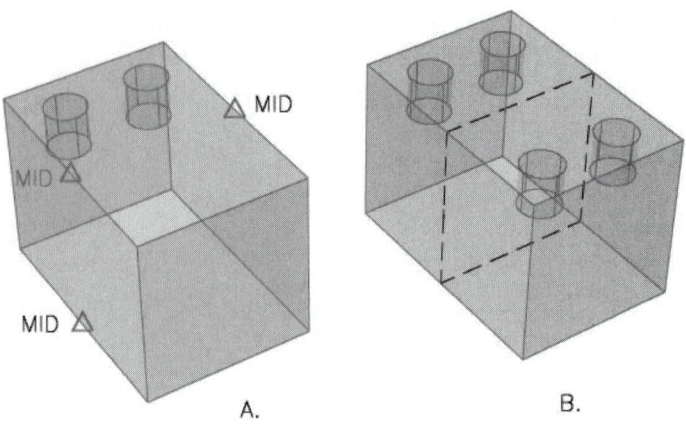

A. B.

Object

Using this option establishes a mirroring plane with the plane of a 2D object. Selecting an *Arc* or *Circle* automatically mirrors selected objects using the plane in which the *Arc* or *Circle* lies. The plane defined by a *Pline* segment is the XY plane of the *Pline* when the *Pline* segment was created. Using a *Line* object or edge of a solid is not allowed because neither defines a plane. Figure 38-63 shows a box mirrored about the plane defined by the *Circle* object. Using *Subtract* produces the result shown in B.

FIGURE 38-63

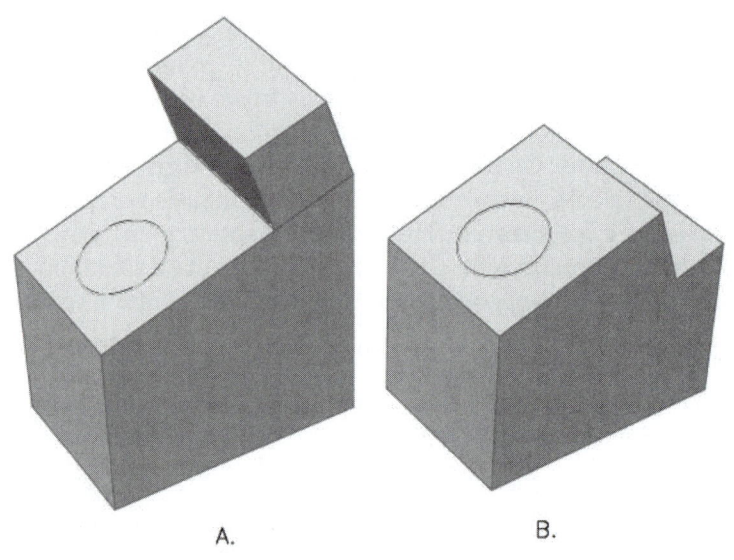

A. B.

Last

Selecting this option uses the plane that was last used for mirroring.

Zaxis

With this option, the mirror plane is the <u>XY</u> plane perpendicular to a Z vector you specify. The first point you specify on the Z axis establishes the location of the XY plane origin (a point through which the plane passes). The second point establishes the Z axis and the orientation of the XY plane (perpendicular to the Z axis). Figure 38-64 illustrates this concept. Note that this option requires only two PICK points.

FIGURE 38-64

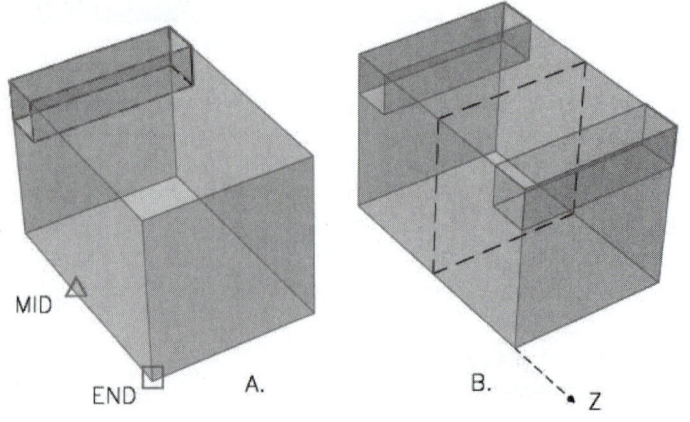

A. B.

View

The *View* option of *Rotate3D* uses a mirroring plane <u>parallel</u> with the screen and perpendicular to your line of sight based on your current viewpoint. You are required to select a point on the plane. Accepting the default (0,0,0) establishes the mirroring plane passing through the current origin. Any other point can be selected. You must change your viewpoint to "see" the mirrored objects (Fig. 38-65).

FIGURE 38-65

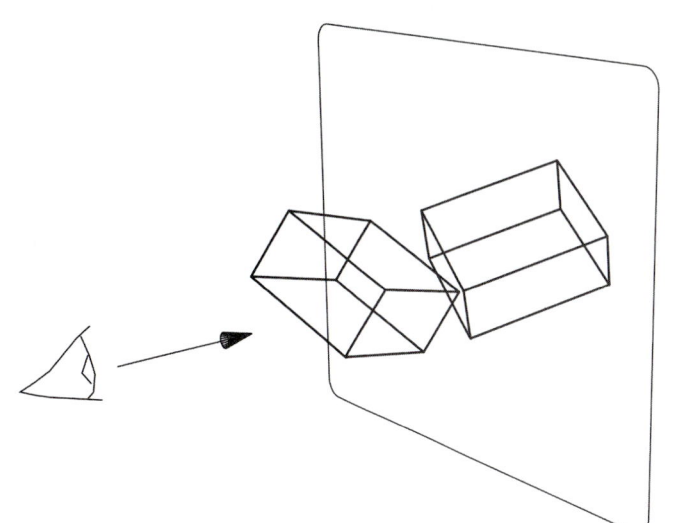

XY

This option situates a mirroring plane parallel with the current XY plane. You can specify a point through which the mirroring plane passes. Figure 38-66 represents a plane established by selecting the *Center* of an existing solid.

FIGURE 38-66

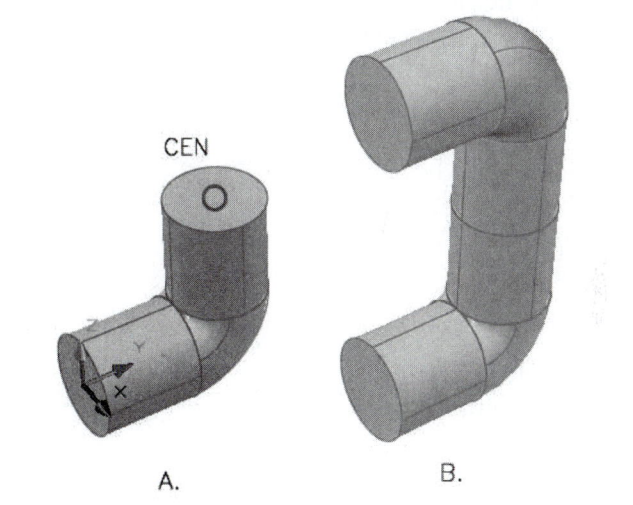

YZ

Using the *YZ* option constructs a plane to mirror about that is parallel with the current YZ plane. Any point can be selected through which the plane will pass (Fig. 38-67).

FIGURE 38-67

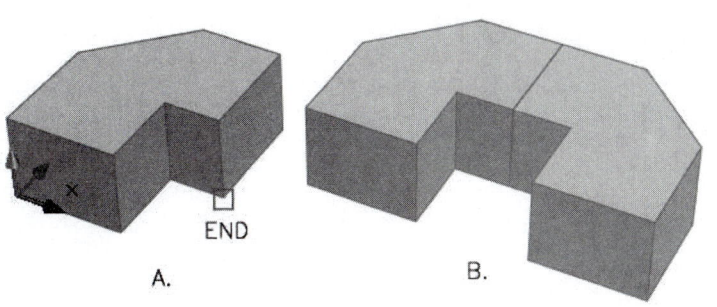

ZX

This option uses a plane parallel with the current ZX plane for mirroring.

3Darray

	Command (Type)	Alias (Type)	Short-cut	Screen (side) Menu	Tablet Menu
Pull-down Menu					
Modify *3D Operations >* *3D Array*	*3Darray*	*3A*	...	*MODIFY2* *3Darray*	*W,20*

Rectangular

With this option of *3Darray*, you create a 3D array specifying three dimensions—the number of and distance between rows (along the Y axis), the number/distance of columns (along the X axis), and the number/distance of levels (along the Z axis). Technically, the result is an array in a <u>prism</u> configuration <u>rather than a rectangle</u>.

```
Command: 3darray
Select objects: PICK
Select objects: Enter
Enter the type of array [Rectangular/Polar] <R>: r
Enter the number of rows (—-) <1>: (value)
Enter the number of columns (|||) <1>: (value)
Enter the number of levels (...) <1>: (value)
Specify the distance between rows (—-): PICK or (value)
Specify the distance between columns (|||): PICK or (value)
Specify the distance between levels (...): PICK or (value)
Command:
```

The selection set can be one or more objects. The entire set is treated as one object for arraying. All values entered must be positive.

Figures 38-68 and 38-69 illustrate creating a *Rectangular 3Darray* of a cylinder with 3 rows, 4 columns, and 2 levels.

FIGURE 38-68

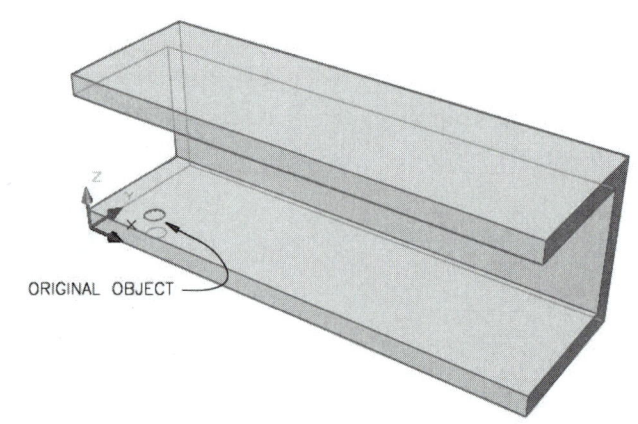

ORIGINAL OBJECT

The cylinders are *Subtracted* from the extrusion to form the finished part (Fig. 38-69).

FIGURE 38-69

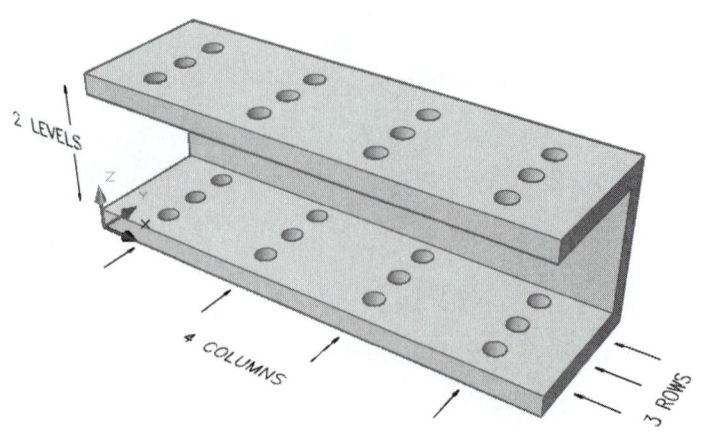

Polar

Similar to a *Polar Array* (2D), this option creates an array of selected objects in a <u>circular</u> fashion. The only difference in the 3D version is that an array is created about an <u>axis of rotation</u> (3D) rather than a point (2D). Specification of an axis of rotation requires two points in 3D space:

Command: **3darray**
Select objects: **PICK**
Select objects: **Enter**
Enter the type of array [Rectangular/Polar] <R>: **p**
Enter the number of items in the array: **(value)**
Specify the angle to fill (+=ccw, -=cw) <360>: **PICK** or **(value)**
Rotate arrayed objects? [Yes/No] <Y>: **y** or **n**
Specify center point of array: **PICK** or **(coordinates)**
Specify second point on axis of rotation: **PICK** or **(coordinates)**
Command:

In Figures 38-70 and 38-71, a *3Darray* is created to form a series of holes from a cylinder. The axis of rotation is the center axis of the large cylinder specified by PICKing the *Center* of the top and bottom circles.

FIGURE 38-70

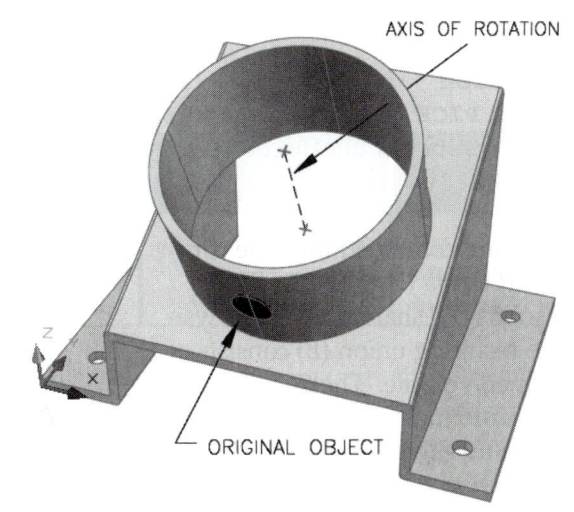

AXIS OF ROTATION

ORIGINAL OBJECT

After the eight items are arrayed, the small cylinders are subtracted from the large cylinder to create the holes.

FIGURE 38-71

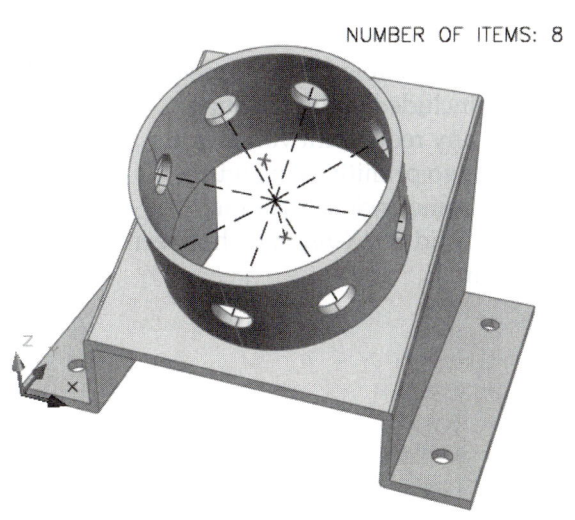

NUMBER OF ITEMS: 8

Chamfer

	Pull-down Menu	Command (Type)	Alias (Type)	Short-cut	Screen (side) Menu	Tablet Menu
	Modify Chamfer	Chamfer	CHA	...	MODIFY2 Chamfer	W,18

Chamfering is a machining operation that bevels a sharp corner. *Chamfer* chamfers selected edges of an AutoCAD solid as well as 2D objects. Technically, *Chamfer* (used with a solid) is a Boolean operation because it creates a wedge primitive and then adds to or subtracts from the selected solid.

When you select a solid, *Chamfer* recognizes the object as a solid and <u>switches to the solid version of prompts and options</u>. Therefore, all of the 2D options are not available for use with a solid, <u>only the "distances" method</u>. When using *Chamfer*, you must both select the "base surface" and indicate which edge(s) on that surface you wish to chamfer:

    Command: **chamfer**
    (TRIM mode) Current chamfer Dist1 = 0.5000, Dist2 = 0.5000
    Select first line or [Polyline/Distance/Angle/Trim/Method/mUltiple]: **PICK** (Select solid at desired edge)
    Base surface selection...
    Enter surface selection option [Next/OK (current)] <OK>: **N** or **Enter**
    Specify base surface chamfer distance <0.5000>: (**value**)
    Specify other surface chamfer distance <0.5000>: (**value**)
    Select an edge or [Loop]: **PICK** (Select edges to be chamfered)
    Select an edge or [Loop]: **Enter**
    Command:

When AutoCAD prompts to select the "base surface," only an edge can be selected since the solids are displayed in wireframe. When you select an edge, AutoCAD highlights one of the two surfaces connected to the selected edge. Therefore, you must use the "Next/<OK>:" option to indicate which of the two surfaces you want to chamfer (Fig. 38-80 A). The two distances are applied to the object, as shown in Figure 38-80 B.

FIGURE 38-80

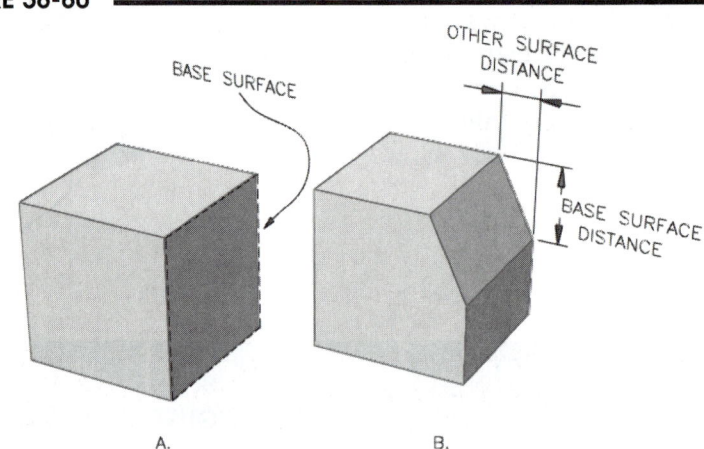

A.

B.

You can chamfer multiple edges of the selected "base surface" simply by PICKing them at the "Select an edge or [Loop]:" prompt (Fig. 38-81). If the base surface is adjacent to cylindrical edges, the bevel follows the curved shape.

FIGURE 38-81

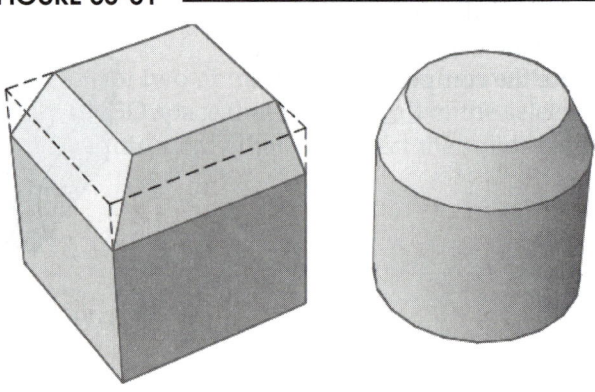

A.

B.

Loop

The *Loop* option chamfers the entire perimeter of the base surface. Simply PICK any edge on the base surface (Fig. 38-82).

FIGURE 38-82

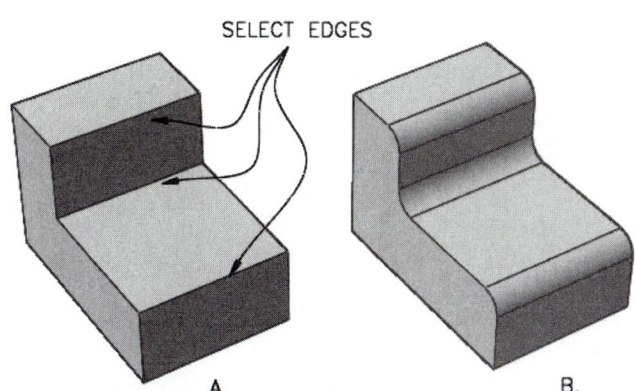

Select an edge or [Loop]: **1**
Select an edge loop or [Edge]: **PICK** (Select edges to form loop)
Select an edge or [Loop]: **Enter**
Command:

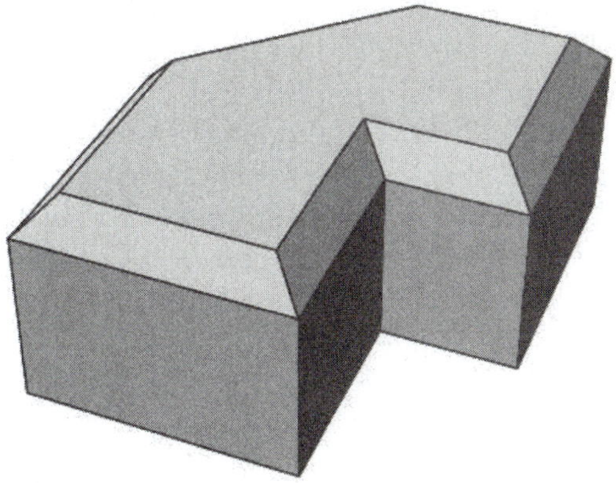

Edge

The *Edge* option switches back to the "Select edge" method.

Multiple

The *Multiple* option simply repeats the *Chamfer* command.

Fillet

	Pull-down Menu	Command (Type)	Alias (Type)	Short-cut	Screen (side) Menu	Tablet Menu
	Modify *Fillet*	*Fillet*	*F*	...	*MODIFY2* *Fillet*	*W,19*

Fillet creates fillets (concave corners) or rounds (convex corners) on selected solids, just as with 2D objects. Technically, *Fillet* creates a rounded primitive and automatically performs the Boolean needed to add or subtract it from the selected solids.

When using *Fillet* with a solid, the command <u>switches</u> to a special group of prompts and options for 3D filleting, and the <u>2D options become invalid</u>. After selecting the solid, you must specify the desired radius and then select the edges to fillet. When selecting edges to fillet, the edges must be PICKed individually. Figure 38-83 depicts concave and convex fillets created with *Fillet*. The selected edges are highlighted.

FIGURE 38-83

SELECT EDGES

A.   B.

Command: **fillet**
Current settings: Mode = TRIM, Radius = 0.5000
Select first object or [Undo/Polyline/Radius/Trim/Multiple]: **PICK** (Select desired edge to fillet)
Enter fillet radius <0.5000>: (**value**)
Select an edge or [Chain/Radius]: **Enter** or **PICK** (Select additional edges to fillet)
Command:

Curved surfaces can be treated with
Fillet, as shown in Figure 38-84. If
you want to fillet intersecting concave
or convex edges, *Fillet* handles your
request, provided you specify all
edges in <u>one</u> use of the command.
Figure 38-84 shows the selected edges
(highlighted) and the resulting solid.
Make sure you select <u>all</u> edges
together (in one *Fillet* command).

FIGURE 38-84

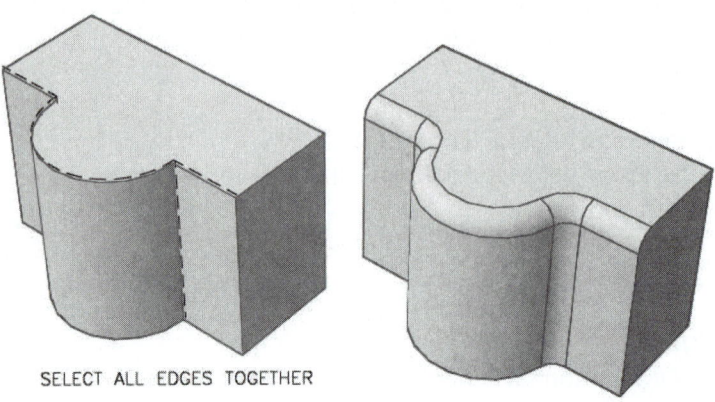

SELECT ALL EDGES TOGETHER

Chain

The *Chain* option allows you to fillet a
series of connecting edges. Select the
edges to form the chain (Fig. 38-85). If
the chain is obvious (only one direct
path), you can PICK only the ending
edges, and AutoCAD will find the
most direct path (series of connected
edges):

FIGURE 38-85

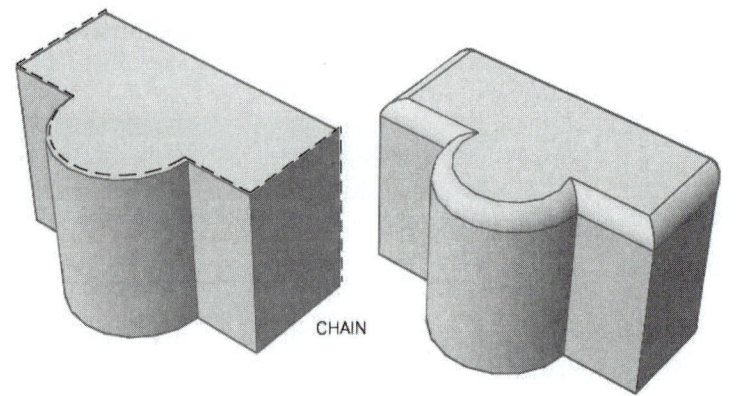

CHAIN

> Select an edge or [Chain/Radius]: *c*
> Select an edge chain or
> [Edge/Radius]: **PICK**

Edge

This option cycles back to the "<Select edge>:" prompt.

Radius

This method returns to the "Enter radius:" prompt.

Multiple

The *Multiple* option causes the *Fillet* command to repeat.

DESIGN EFFICIENCY

Now that you know the complete sequence for creating composite solid models, you can work toward
improving design efficiency. The typical construction sequence is (1) create primitives, (2) ensure the
primitives are in place by using UCSs or any of several move and rotate options, (3) combine the primitives into a composite solid using Boolean operations, and (4) make necessary design changes to composite solids using a variety of editing tools (see Chapter 39). The typical step-by-step, "building-block"
strategy, however, may not lead to the most efficient design. In order to minimize computation and construction time, you should <u>minimize the number of Boolean operations</u> and, if possible, the <u>number of
primitives</u> you use.

For any composite solid, usually several strategies could be used to construct the geometry. You should plan your designs ahead of time, striving to minimize primitives and Boolean operations.

For example, consider the procedure shown in Figure 38-74. As discussed, it is more efficient to accomplish all unions with one *Union*, rather than each union as a separate step. Even better, create a closed *Pline* shape of the profile; then use *Extrude*. Figure 38-76 is another example of design efficiency based on using an *n*-way Boolean. Multiple solids can be unioned automatically by selecting them at the select objects "to subtract from" prompt of *Subtract*. Also consider the strategy of reverse drafting, as shown in Figure 38-78. Using *Extrude* in concert with *Intersect* can minimize design complexity and time.

In order to create efficient designs and minimize Boolean operations and primitives, keep these strategies in mind:

- Execute as many subtractions, unions, or intersections as possible within one *Subtract*, *Union*, or *Intersect* command.

- Use *n*-way Booleans with *Subtract*. Combine solids (union) automatically by selecting <u>multiple</u> objects "to subtract from," and then select "objects to subtract."

- Make use of *Plines* or regions for complex profile geometry; then *Extrude* the profile shape. This is almost always more efficient for complex curved profile creation than using multiple primitives and Boolean operations.

- Make use of reverse drafting by extruding the "view" profiles (*Plines* or *Regions*) with *Extrude*, then finding the common volume with *Intersect*.

CHAPTER EXERCISES

1. What are the typical four steps for creating composite solids?

2. Consider the two solids in Figure 38-86. They are two extruded hexagons that overlap (occupy the same 3D space).

   A. Sketch the resulting composite solid if you performed a *Union* on the two solids.

   B. Sketch the resulting composite solid if you performed an *Intersect* on the two solids.

   C. Sketch the resulting composite solid if you performed a *Subtract* on the two solids.

FIGURE 38-86

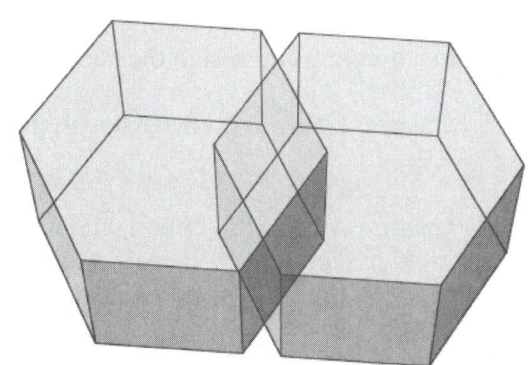

3.   Begin a drawing and assign the name CH38EX3.

FIGURE 38-87

A.   Create a *box* with the lower-left corner at **0,0,0**. The *Lengths* are **5, 4,** and **3**.

B.   Create a second *box* using a Dynamic UCS as shown in Figure 38-87 (use the *ORigin* option). The *box* dimensions are **2 x 4 x 2**.

C.   Create a *Cylinder*. Use a Dynamic UCS as in the previous step. The *cylinder Center* is at **3.5,2** (of the *UCS*), the *Diameter* is **1.5**, and the *Height* is **-2**.

D.   *Save* the drawing.

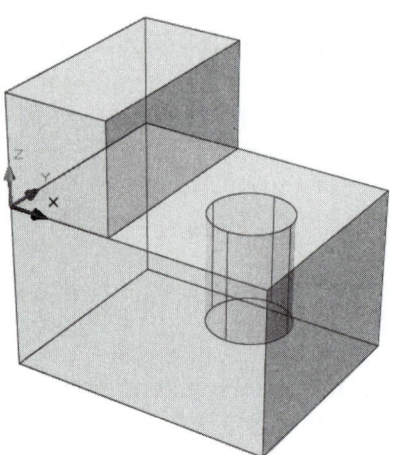

E.   Perform a *Union* to combine the two boxes. Next, use *Subtract* to subtract the cylinder to create a hole. The resulting composite solid should look like that in Figure 38-88. *Save* the drawing.

FIGURE 38-88

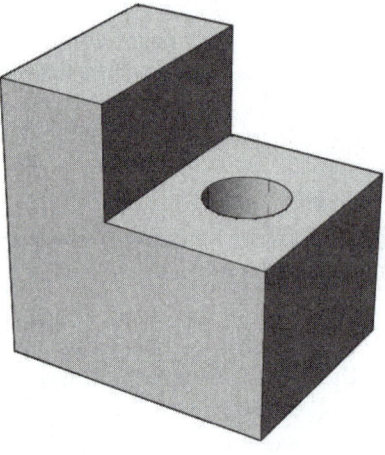

4.   Begin a drawing and assign the name CH38EX4.

FIGURE 38-89

A.   Create a *Wedge* at point **0,0,0** with the *Lengths* of **5, 4, 3**.

B.   Create a *3point UCS* option with an orientation indicated in Figure 38-89. Create a *Cone* with the *Center* at **2,3** (of the *UCS*) and a *Diameter* of **2** and a *Height* of **-4**.

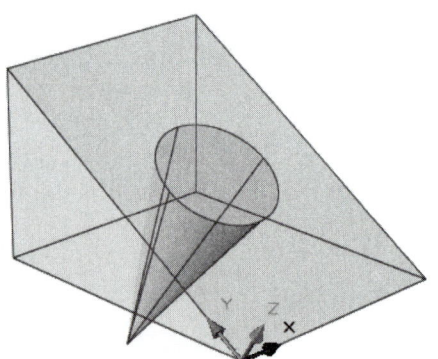

C. *Subtract* the cone from the wedge. The resulting composite solid should resemble Figure 38-90. *Save* the drawing.

FIGURE 38-90

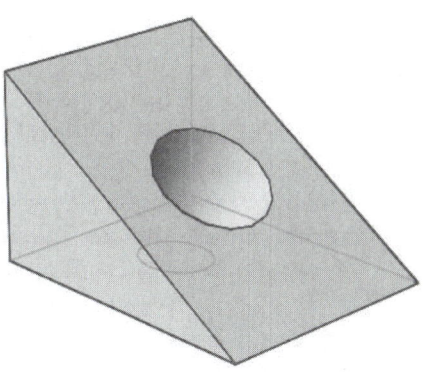

5. Begin a drawing and assign the name **CH38EX5.** Display a *Plan* view.

A. Create 2 *Circles* as shown in Figure 38-91, with dimensions and locations as specified. Use *Pline* to construct the rectangular shape. Combine the 3 shapes into a *Region* by using the *Region* and *Union* commands <u>or</u> converting the <u>outside</u> shape into a *Pline* using *Trim* and *Pedit*.

FIGURE 38-91

B. Change the display to an isometric-type view. *Extrude* the *Region* or *Pline* with a *Height* of **3** (no *taper angle*).

C. Create a *Box* with the lower-left corner at **0,0**. The *Lengths* of the box are **6, 3, 3**.

D. *Subtract* the extruded shape from the box. Your composite solid should look like that in Figure 38-92. *Save* the drawing.

FIGURE 38-92

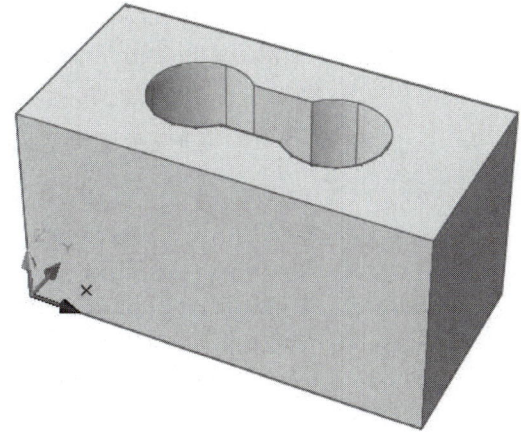

6. Begin a drawing and assign the name CH38EX6. Display a *Plan* view.

    A. Create a closed *Pline* shape symmetrical about the X axis with the locational and dimensional specifications given in Figure 38-93.

    B. Change to an isometric-type view. Use *Revolve* to generate a complete circular shape from the closed *Pline*. Revolve about the **Y** axis.

    C. Create a *Torus* with the *Center* at **0,0**. The *Radius of torus* is **3** and the *Radius of tube* is **.5**. The two shapes should intersect.

FIGURE 38-93

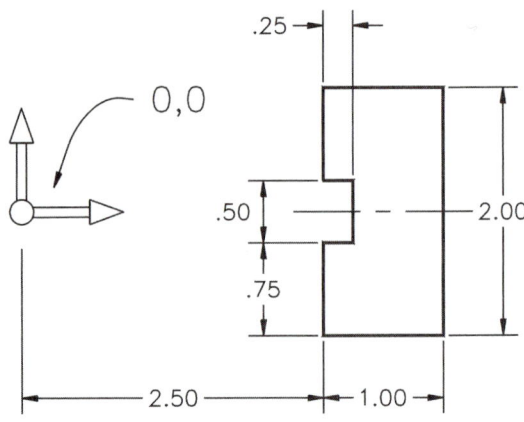

    D. Generate a display like Figure 38-94.

    E. Create a *Cylinder* with the *Center* at **0,0,0**, a *Radius* of **3**, and a *Height* of **8**.

    F. Use *3DRotate* to rotate the revolved *Pline* shape **90** degrees about the X axis (the *Pline* shape that was previously converted to a solid—not the torus). Next, move the shape up (positive Z) **6** units with *3DMove*.

    G. Move the torus up **4** units with *3DMove*.

FIGURE 38-94

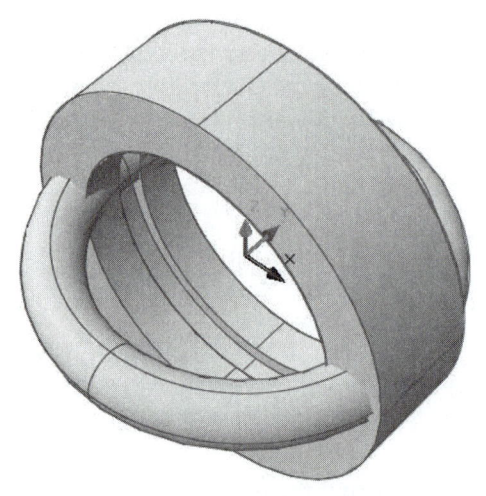

    H. The solid primitives should appear as those in Figure 38-95.

FIGURE 38-95

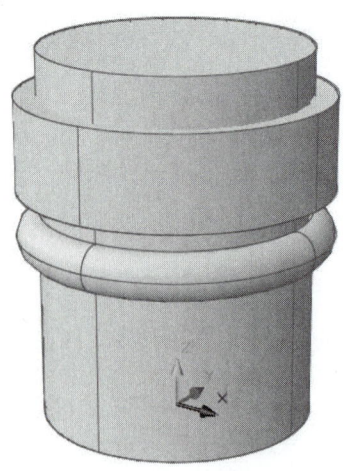

I. Use *Subtract* to subtract both revolved shapes from the cylinder. The solid should resemble that in Figure 38-96. *Save* the drawing.

FIGURE 38-96 ——————

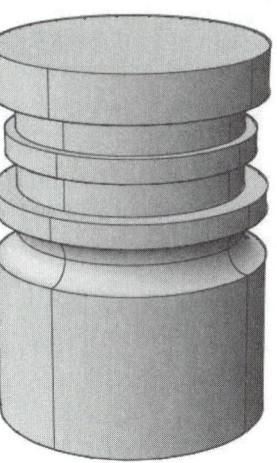

7. Begin a *New* drawing or use a *Template.* Assign the name FAUCET1.

A. Draw 3 closed *Pline* shapes, as shown in Figure 38-97. Assume symmetry about the longitudinal axis. Use the WCS and create 2 new *UCS*s for the geometry. Use *3point Arcs* for the "front" profile. *Save* the drawing as **FAUCET2**.

FIGURE 38-97 ——————

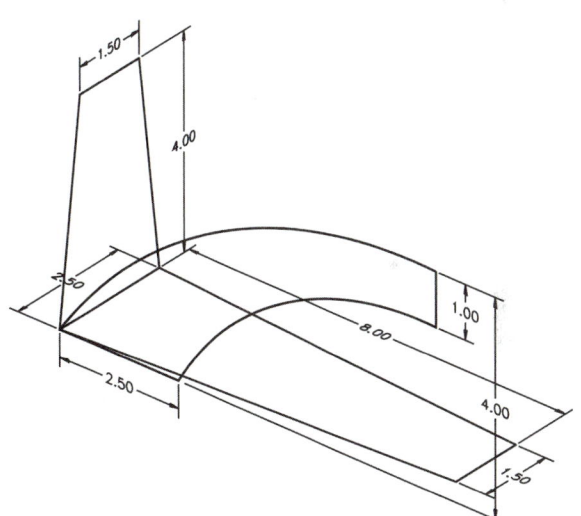

B. *Extrude* each of the 3 profiles into the same space. Make sure you specify the correct positive or negative *Height* value.

FIGURE 38-98 ——————

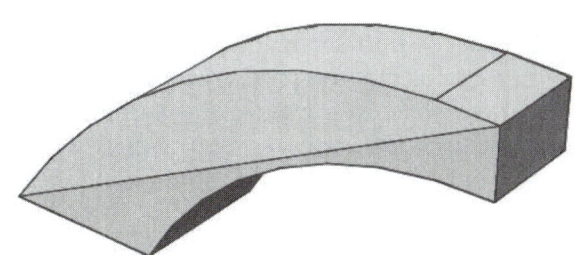

C. Finally, use *Intersect* to create the composite solid of the Faucet (Fig. 38-98). *Save* the drawing. Use *Saveas* and assign the name **FAUCET1**.

D. (Optional) Create a nozzle extending down from the small end. Then create a channel for the water to flow (through the inside) and subtract it from the faucet.

8. *Open* the **FAUCET2** drawing you saved in the last exercise. From this drawing you will create an improved faucet using the *Loft* command.

A. *Erase* the bottom profile (on the XY plane). Create a *Rectangle* on the XY plane as shown in Figure 38-99 using the *Endpoints* of the other profiles on the XY plane for the "first corner" and "other corner."

FIGURE 38-99 ──────────────

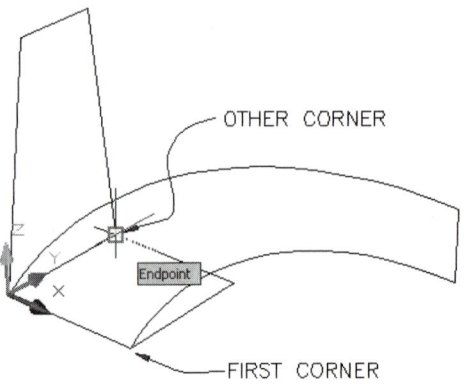

B. Create a new *UCS* at the spout end of the faucet as shown in Figure 38-100. Create a *Rectangle* as shown with *Dimensions* of **1.5** and **1.0**. Next, use *3Dmove* to move the new *Rectangle* **.5** units in a positive X direction relative to the current UCS.

FIGURE 38-100 ──────────────

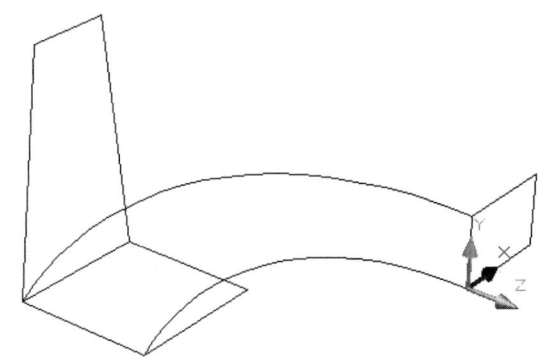

C. Change the *UCS* back to *World*. *Erase* the remaining two original profiles and keep only the two new *Rectangles*. Draw a diagonal *Line* between the opposite corners (*Endpoints*) for each *Rectangle*, as shown in Figure 38-101. Next, create a *Spline* by defining three points—the two ends are at the *Midpoints* of the diagonal *Lines* and the middle point is at 4.15, 1.25, 2.66. Finally, *Erase* the two diagonal *Lines*.

FIGURE 38-101 ──────────────

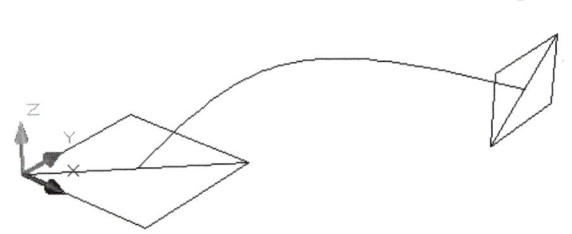

D. Use the *Loft* command. At the "Select cross-sections in lofting order:" prompt, select the two *Rectangles*. Use the *Path* option. At the "Select path curve:" prompt, select the *Spline*. Your completed faucet should look like that in Figure 38-102.

FIGURE 38-102 ──────────────

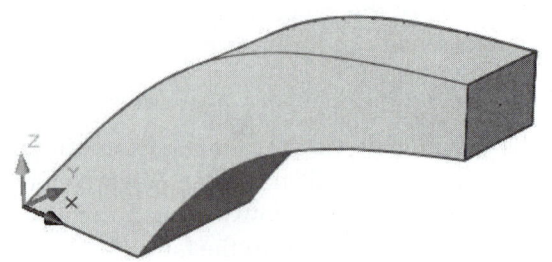

9. Construct a solid model of the bar guide in Figure 38-103. Strive for the most efficient design. It is possible to construct this object with one *Extrude* and one *Subtract*. Save the model as **BGUID-SL.**

FIGURE 38-103

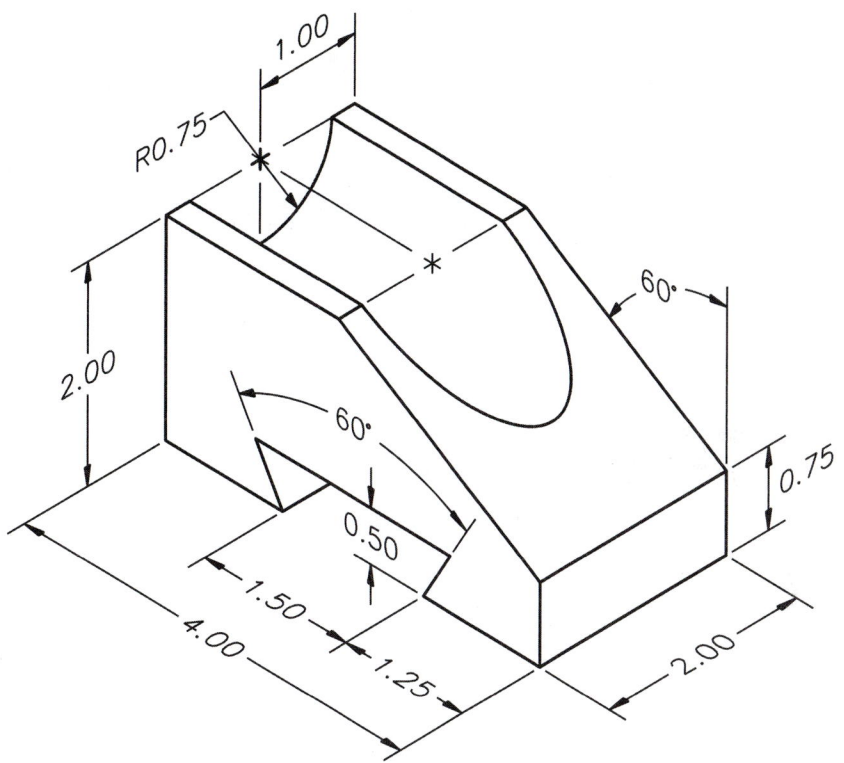

10. Make a solid model of the V-block shown in Figure 38-104. Several strategies could be used for construction of this object. Strive for the most efficient design. Plan your approach by sketching a few possibilities. Save the model as **VBLOK-SL.**

FIGURE 38-104

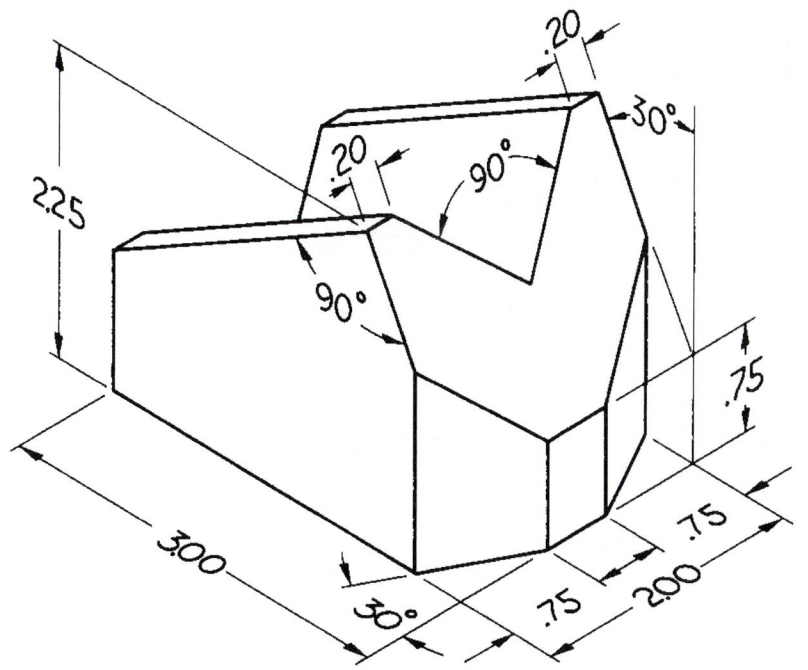

11. Construct a composite solid model of the support bracket (Fig. 38-105) using efficient techniques. Save the model as **SUPBK-SL**.

FIGURE 38-105

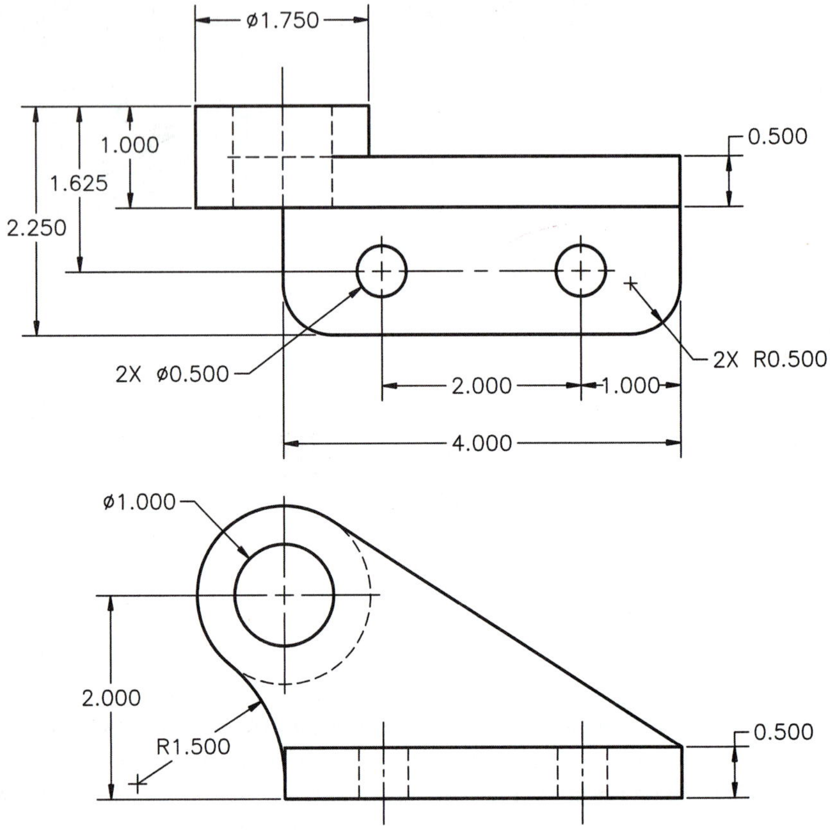

12. Construct the swivel shown in Figure 38-106. The center arm requires *Sweeping* the 1.00 x 0.50 rectangular shape along an arc path through 45 degrees. Save the drawing as **SWIVEL.**

FIGURE 38-106

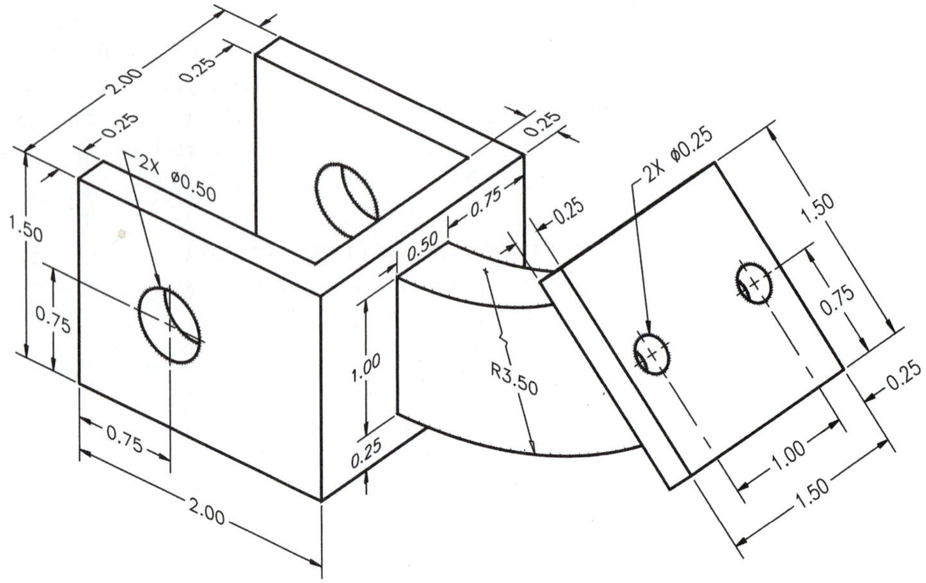

13. Construct a solid model of the angle brace shown in Figure 38-107. Use efficient design techniques. Save the drawing as **AGLBR-SL**.

FIGURE 38-107

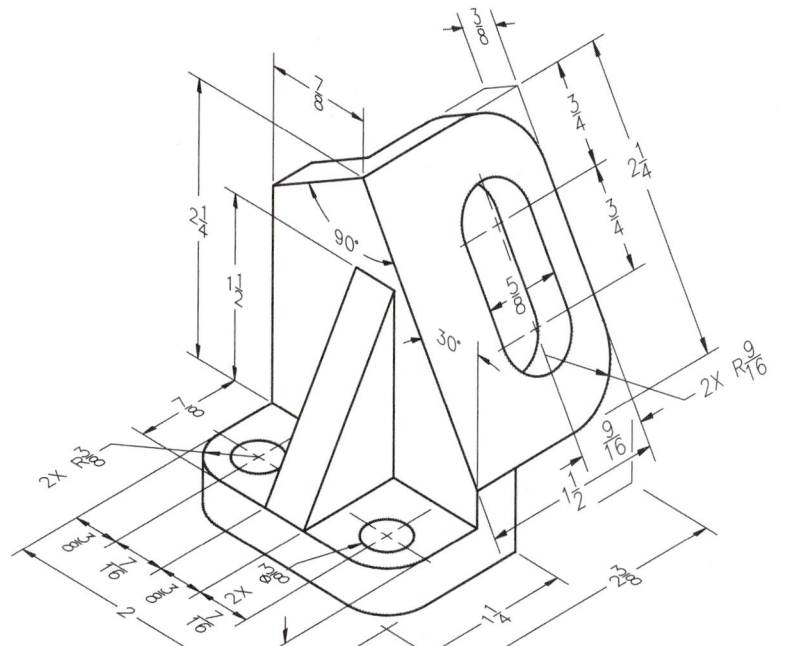

14. Construct a solid model of the saddle shown in Figure 38-108. An efficient design can be utilized by creating *Pline* profiles of the top "view" and the front "view," as shown in Figure 38-109. Use *Extrude* and *Intersect* to produce a composite solid. Additional Boolean operations are required to complete the part. The finished model should look similar to Figure 38-110. Save the drawing as **SADL-SL**.

FIGURE 38-108

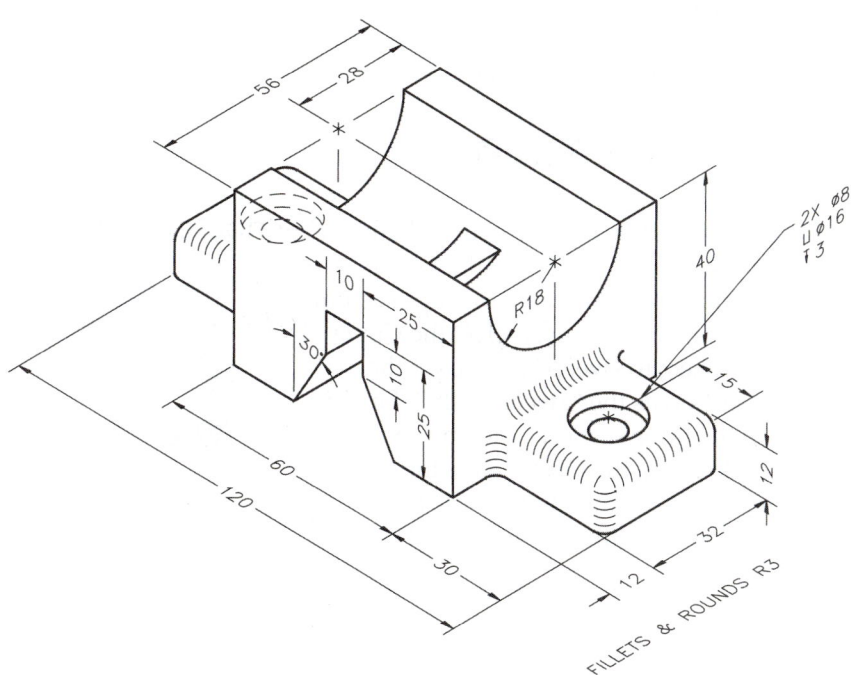

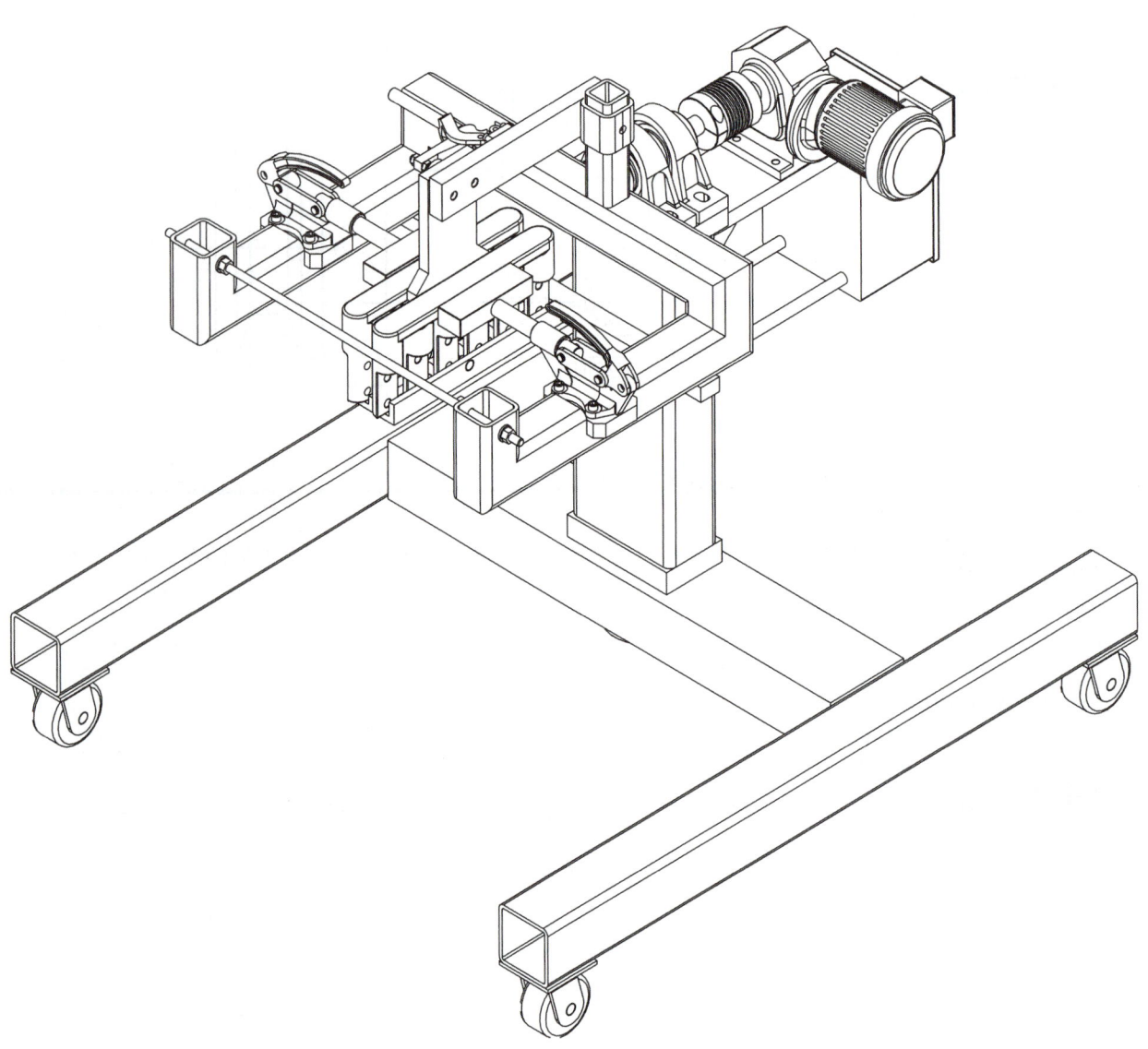

WELDING FIXTURE MODEL.DWG, Courtesy of Autodesk, Inc.

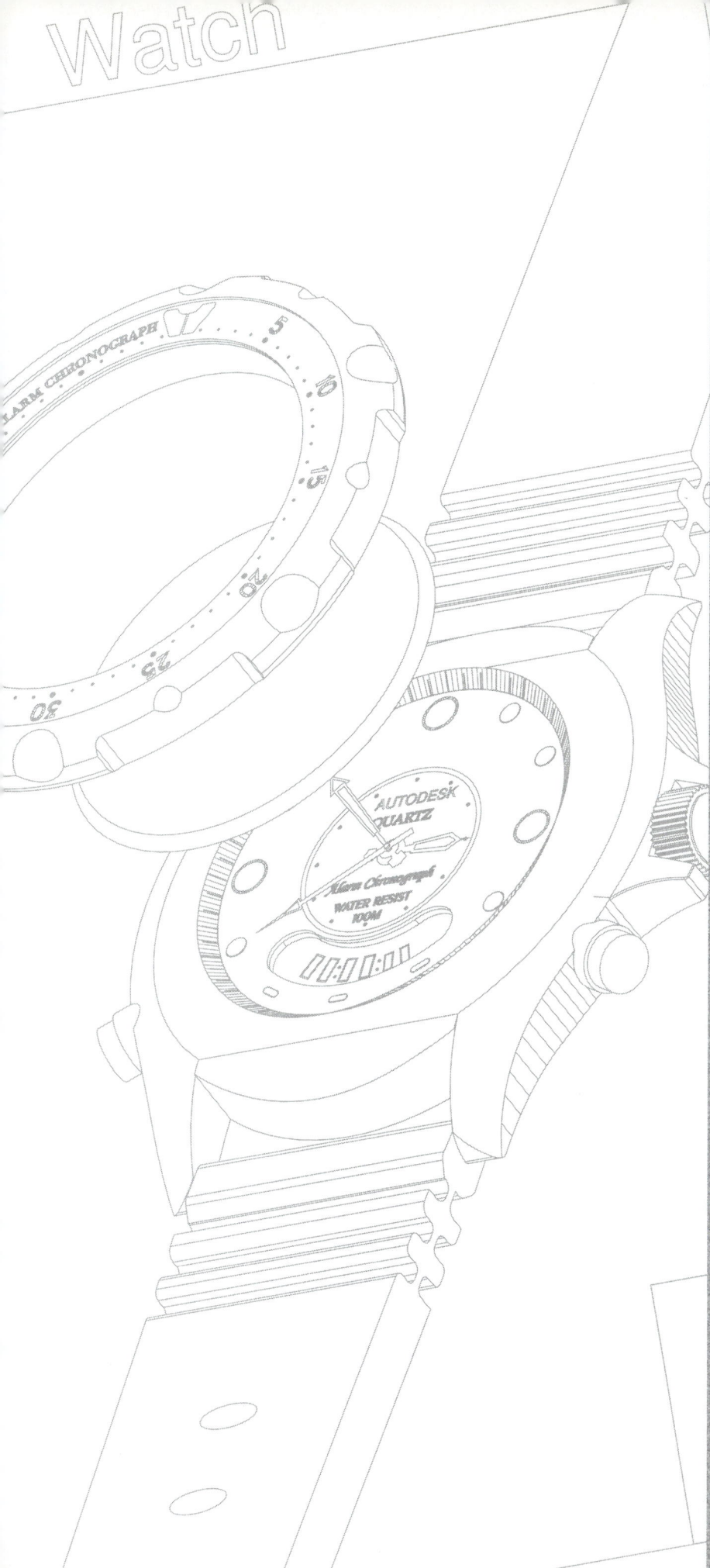

39

SOLID MODEL EDITING

CHAPTER OBJECTIVES

After completing this chapter you should:

1. be able to edit the properties of solid models using the *Properties* palette;

2. be able to edit primitive solids and composite solids using grips and the move and rotate grip tools;

3. be able to use subobject selection to edit faces, edges, and vertices of solids;

4. know how to use *Solidedit* to accomplish a variety of solid editing tasks;

5. be able to use the utility solid editing commands: *Presspull, Interfere,* and *Imprint;*

6. know how to pass a live slicing plane through solids using *Sectionplane;*

7. be able to create surfaces using *Planesurf, Extrude, Revolve, Sweep, Loft,* and use the surfaces to *Slice* solids;

8. be able to calculate and export solid geometry data.

CONCEPTS

As you learned in Chapter 38, the basic solid modeling construction process is to create simple 3D solids, called "primitives," then combine them using "Boolean operations" to create a complex 3D model called a "composite solid." A number of primitive types and construction techniques were discussed as well as strategies to augment the construction process, including using Dynamic User Coordinate Systems.

This chapter discusses several options you can use to edit composite solids, including editing individual primitives, faces, edges, and vertices. In addition, advanced solid editing techniques are discussed involving creating planar and complex-curved surfaces, then shaping solids by "slicing" them using these surfaces.

The topics in this chapter are arranged as follows:

> Editing Primitives using the Properties Palette
> Editing Solids Using Grips
> Solids Editing Commands: *Solidedit, Slice, Presspull, Sectionplane, Interfere,* and *Imprint*
> Using Surfaces with Solids
> Using Solids Data

Using the information from Chapter 38 and the solid editing techniques discussed in this chapter, you should be able to create practically any shape or configuration of solid model.

EDITING PRIMITIVES USING THE *PROPERTIES* PALETTE

You can use the *Properties* palette to edit basic dimensional characteristics of solid primitives. Even if a primitive has been combined into a composite solid using a Boolean operation, it can be edited by holding down the Ctrl key and selecting the desired primitive as a "subobject." Editing primitives using the *Properties* palette is fairly simple and straight-forward. This section explains the following procedures:

> Editing *Properties* of individual primitives
> Composite solid subobject selection
> Editing *Properties* of composite solid subobjects (primitives)

The illustrations in this section display grips on the selected primitives only to indicate which primitives are selected for editing. Grips are not discussed, however, until the next section of this chapter.

NOTE: Grips can be turned <u>on or off</u> when using the *Properties* palette to edit solids. Grips have no effect on editing capabilities using the *Properties* palette.

Editing *Properties* of Individual Primitives

The basic dimensional and positional properties of individual solids are available for editing using the *Properties* palette. The procedure is simple: with the *Properties* palette displayed on the screen, select the desired primitive, then change the desired values in the *Geometry* section of the *Properties* palette.

For example, assume you want to change the height of a *Box* primitive. Selecting the box displays the primitive's properties in the *Properties* palette (Fig. 39-1). (Notice that the grips appear on the primitive when selected, but the grips are not needed to access the *Properties* palette.) Find the *Height* edit box in the *Geometry* section of the palette and enter the desired value.

FIGURE 39-1

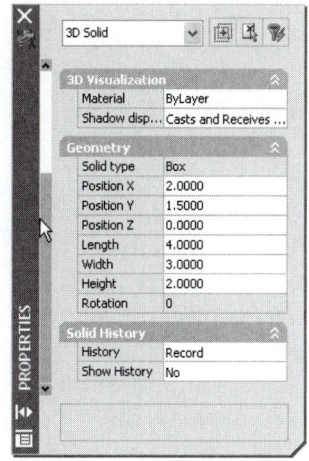
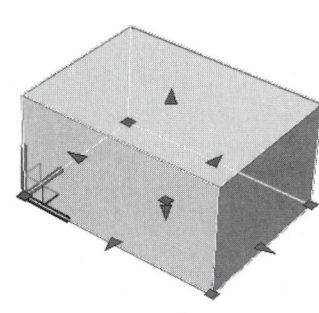

Notice that changing a dimension or positional value in the *Properties* palette immediately changes the primitive. In this case, the *Height* of the box was changed from 2.0000 to 4.0000.

FIGURE 39-2

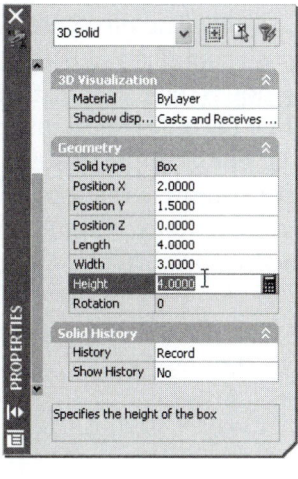
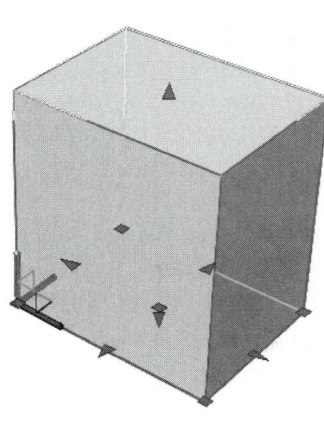

The positional values of primitives can be changed by the same method. Values that appear in the *Position X*, *Position Y*, and *Position Z* edit boxes represent the center of the base. This location is also indicated by a square grip at the base's center. Note that in our example the move grip tool appears at 0,0,0. In Figure 39-3 the *Position Y* value was changed from 1.5000 to 4.0000.

FIGURE 39-3

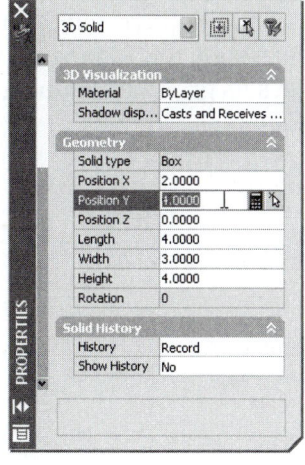
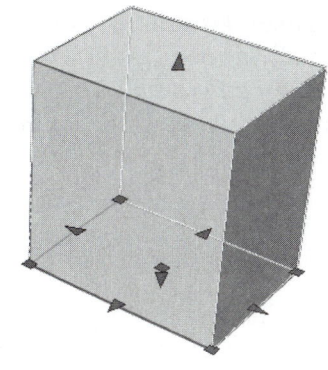

2007

Although all primitives have *Position X*, *Position Y*, and *Position Z* properties, each primitive type has distinct dimensional values. For example, a *Cylinder* has a field for *Radius*, *Height*, and *Elliptical* (Fig. 39-4). If *Elliptical* is set to *Yes*, additional fields appear for *Major radius* and *Minor radius*.

A *Cone* primitive has similar dimensional properties to a *Cylinder* with the addition of a dimensional value for *Top radius* (see "Grip Editing Primitives"). A *Pyramid*, as another example, has property fields for *Base radius*, *Top radius*, (number of) *Sides*, *Height*, and *Rotation*. Therefore, each primitive contains properties based on the dimensional features of the shape and the specifications given to create the solid.

FIGURE 39-4

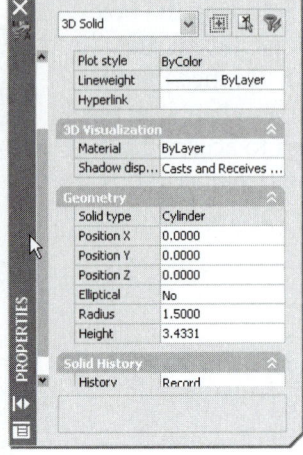
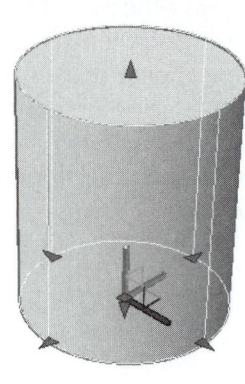

Composite Solid Subobject Selection

A <u>subobject</u> is a component of a solid. Therefore, an individual primitive that is part of a composite solid is a subobject. You can select any subobject by <u>holding down the Ctrl key</u>, then hovering the cursor over a subobject until it becomes highlighted, then using the left mouse button to PICK it. For example, assume a composite solid is constructed from two *Box* primitives as shown in Figure 39-5. Note that the two *Boxes* are combined by a *Union* since there are no edges displayed between the two primitives.

FIGURE 39-5

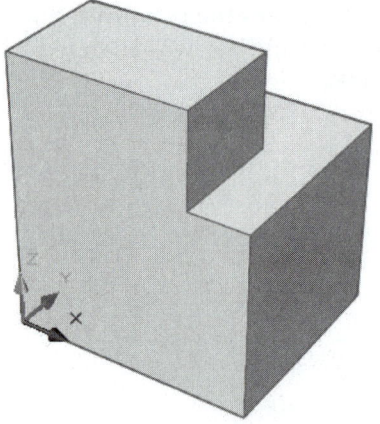

Holding down the Ctrl key forces the cursor to change from the crosshairs to a small pickbox. Hovering the small pickbox over the smaller *Box* primitive on top displays a "highlighted" effect for the small *Box* only, as shown in Figure 39-6. While it is highlighted, PICKing the small *Box* primitive displays its grips (assuming grips are turned on) as shown in Figure 39-7.

FIGURE 39-6

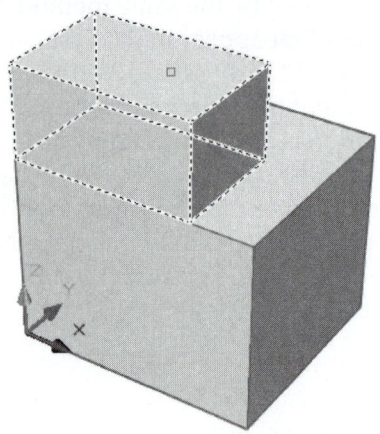

Editing *Properties* of Composite Solids

To change the properties of a composite solid subobject using the *Properties* palette, hold down the Ctrl key, hover over the desired primitive until it becomes highlighted, then PICK. The *Properties* palette should display the locational and dimensional properties of the primitive solid only, not the properties of the composite solid (Fig. 39-7). Although the primitive's grips are displayed when it is selected, grips do not have to be turned on to edit the primitive using *Properties*.

FIGURE 39-7

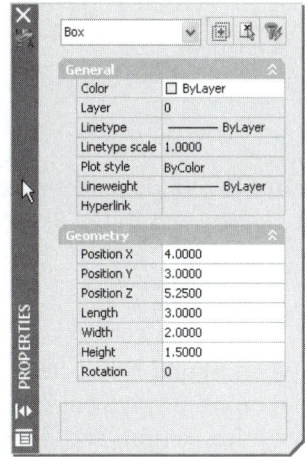
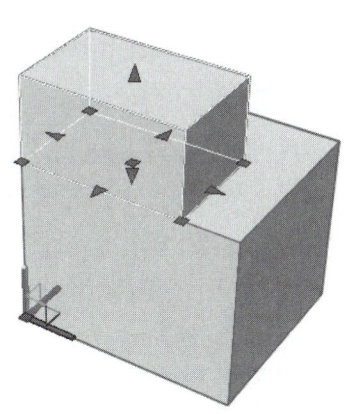

For example, assume you want to change both the location and dimensions of the subobject (*Box* primitive). Once the subobject is selected, change the desired values in the *Geometry* section of the *Properties* palette (Fig. 39-8). Note that in our example, the *Position Y* was changed from 3.0000 to 3.5000, so the *Box* moved in from the left edge of the base feature. In addition, the *Length* value was changed from 3.0000 to 2.0000. Note that the *Length*, *Width*, and *Height* are relative to the center of the base of the primitive; therefore, the length was reduced by 0.5000 on each end of the subobject.

FIGURE 39-8

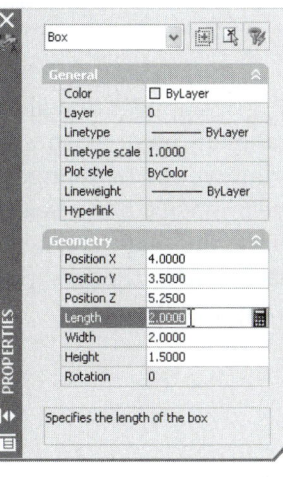
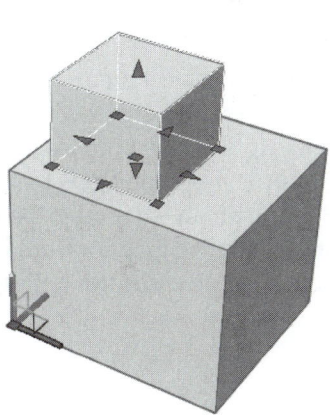

The important concept to note is that when you edit individual primitives (subobjects) of a composite solid using *Properties* or grips, the composite solid remains one unit. In other words, although the small box in our example changed dimensions and location, it is still *Unioned* to the large box.

EDITING SOLIDS USING GRIPS

The most intuitive method for editing solids is using the grips available on solids. When you select a solid, the grips appear. The grips allow you to change the solids in different ways, depending on the type of primitive and the type (shape) of grip.

You can also edit individual subobjects within solids. "Subobject" is the term AutoCAD uses to describe a component of a solid. You can select composite solids to edit, individual primitives contained in composite solids, and even individual faces, edges, and vertices.

There are numerous possibilities for editing solids using grips since this editing method is so dynamic and flexible. Every situation cannot be described in this section; therefore, only the basic principles and most common possibilities are discussed. This section will explain the following concepts:

Grip editing primitive solids
Move and rotate grip tools
Grip editing composite solids
Subobjects and selection
Grip editing faces, edges, and vertices
Selecting overlapping primitives
History settings

NOTE: To use grip editing features for solids, make sure grips are turned on. Do this by setting the *GRIPS* variable to 1 or checking *Enable Grips* in the *Selection* tab of the *Options* dialog box.

Grip Editing Primitive Solids

Assuming grips are turned on, when a solid is selected, one or more grips appear. Each type of solid primitive displays its own special types of grips based on the geometric characteristics of the primitive. For example, a *Box* is defined by width, height, and depth, so appropriate grips appear allowing you to change those characteristics. A *Cylinder*, however, is defined by a base diameter and height, so grips especially for those features appear. Several examples of solids and the related grips are described in this section.

Box

When you select a *Box* primitive, grips appear that allow you change the dimensional and locational characteristics. The shape of the grip informs you how the primitive can be changed. For example, note the grips that appear when a *Box* is selected (Fig. 39-9). The arrow-shaped grips indicate the shape can be changed by stretching the grip forward or backward in the specified direction. Square grips indicate motion in any direction.

FIGURE 39-9

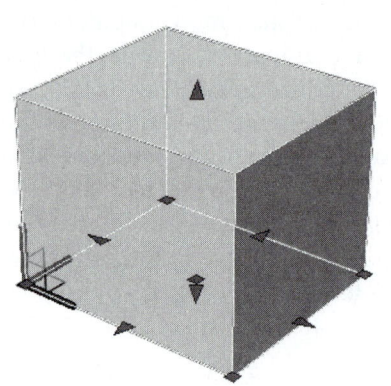

Change the dimensions of the *Box* using any of the arrow-shaped grips. For example, select a grip to make it hot, then stretch the face of the solid by moving the grip inward or outward as illustrated in Figure 39-10. The arrow grips stretch <u>only</u> in the indicated direction. Polar Tracking does not have to be turned on, but may help indicate the current length and can be used in conjunction with Polar Snap. You can also enter a value at the command prompt to specify a distance to stretch the solid.

Specify point location or [Base point/ Undo/eXit]: **PICK** or (**value**)

FIGURE 39-10

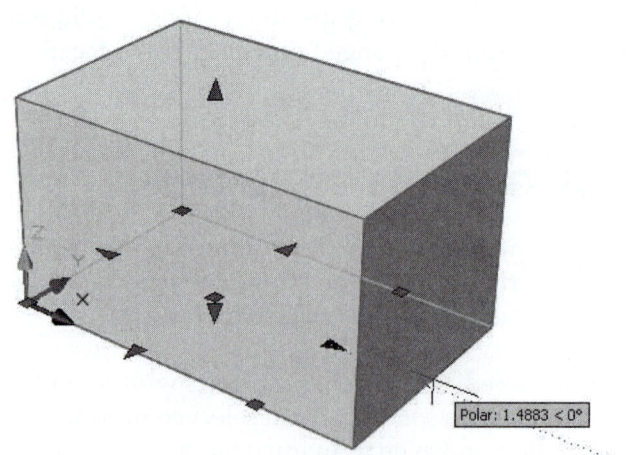

Stretch any other face inward or outward using the appropriate arrow grip. For example, change the height of the *Box* by stretching the arrow grip on top or bottom (Fig. 39-11).

FIGURE 39-11

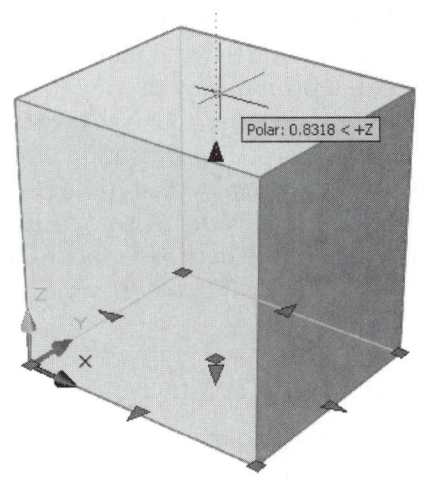

The square grips indicate that you can stretch the corner of the solid in any direction. For example, you can select a corner (square) grip to make it hot, then shorten one dimension while lengthening another (Fig. 39-12). The same prompt appears at the Command line where you can enter coordinates for the new location of the corner.

FIGURE 39-12

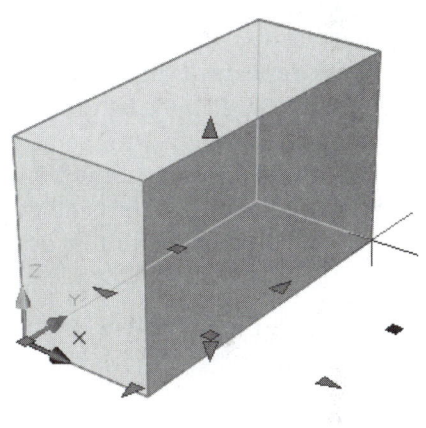

Specify point location or [Base point/Undo/eXit]: **PICK** or (**coordinates**)

If you select the square grip at the center of the base, the location of the entire primitive can be changed as shown in Figure 39-13. Polar Tracking can be used to ensure movement along a specified angle.

When the solid's "base" grip is selected (the square grip at the center of the base), the Command prompt changes to the standard grip options.

FIGURE 39-13

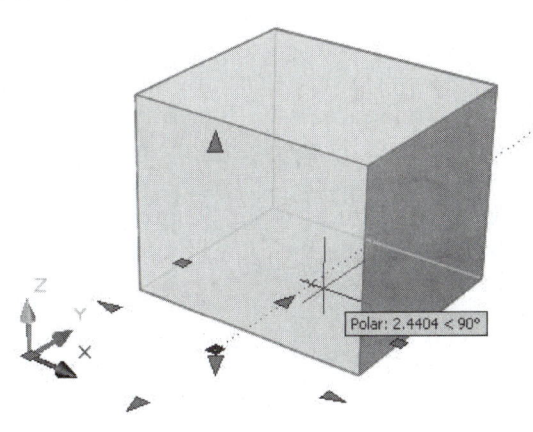

** STRETCH **
Specify stretch point or [Base point/
Copy/Undo/eXit]:

Cycle through the five options (STRETCH, MOVE, ROTATE, MIRROR, and SCALE) to change the entire primitive using the desired option. For example, you can rotate the entire primitive by making the base grip hot, then pressing Enter or spacebar to cycle to ROTATE, then dynamically rotating the solid in the current XY plane (Fig. 39-14). Keep in mind that STRETCH and MOVE accomplish the same results. Also, MIRROR has an effect on the solid only if the Copy suboption is used.

FIGURE 39-14

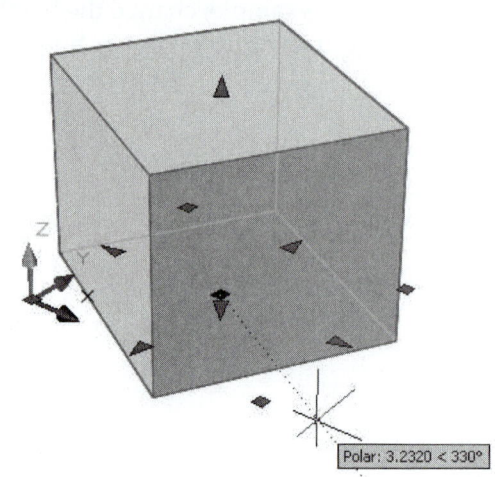

Wedge

Editing a *Wedge* primitive is similar to editing a *Box* in that the grips that appear on each are similar. The height grip on a *Wedge* appears at the highest point of the solid and can be used to stretch the height as shown in Figure 39-15. All other grip options are the same as for a *Box*, including the prompt that appears on the Command line. Remember that selecting the grip at the center of the base allows you to cycle through the five normal grip options (STRETCH, MOVE, ROTATE, MIRROR, and SCALE).

FIGURE 39-15

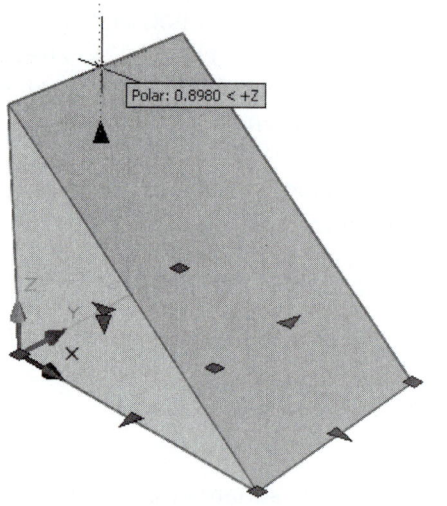

Cone

Grips that appear on a *Cone* primitive when selected are displayed in Figure 39-16. Remember that a *Cone* is dimensionally defined by specifying the base radius and height; therefore, grips appear that allow you to change those dimensional characteristics. For example, Figure 39-16 illustrates dragging the top grip downward to change the height of the solid.

FIGURE 39-16

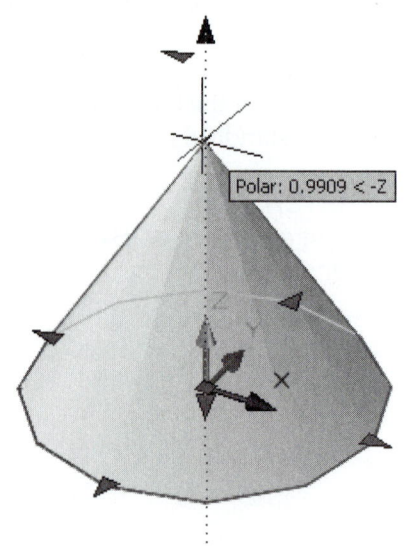

Also remember that a *Cone* can have a top radius. In such a case, the cone is a frustum (truncated cone). Even if a top radius was not defined initially and a normal cone was created, a top radius grip appears (see previous Fig. 39-16). Figure 39-17 displays the process of dragging the top radius grip to create a frustum.

FIGURE 39-17

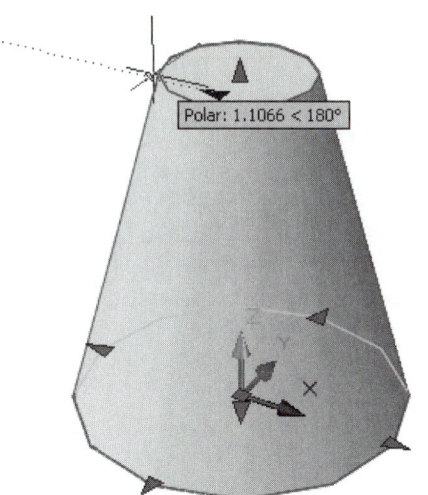

Sphere

In the case of a *Sphere*, four arrow-shaped grips appear (not shown). Therefore, you can change only the diameter dimension. In addition, a center grip appears for changing location.

Cylinder

A *Cylinder* is defined by the base radius and height, so the related grips appear when the primitive is selected as shown in Figure 39-18. You can change the height at the top or bottom grip or change the radius using one of the four arrow grips. Move the solid or access the five normal grip options by selecting the grip at the center of the base.

FIGURE 39-18

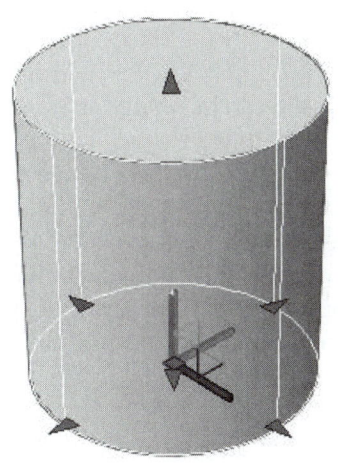

Pyramid

A *Pyramid* is also defined by the base radius and height. As you can see in Figure 39-19, selecting a *Pyramid* primitive reveals grips at each corner and along each side of the base; however, you can changes only the base diameter with these eight grips. Selecting any one of the corner grips or side grips changes all corners and sides uniformly. One side or corner cannot be changed individually because a *Pyramid* primitive is a right pyramid (the apex is always directly above the center of the base).

FIGURE 39-19

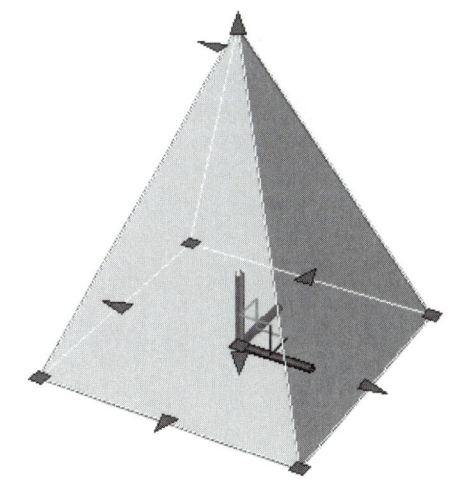

Note that a *Pyramid* can have a top radius, similar to a *Cone*. Therefore, a related top radius grip can be used to create a top radius for a *Pyramid* originally defined with no top radius, as shown in Figure 39-20. Likewise, you can change a previously defined top radius to a *Pyramid* with an apex (top radius of 0).

NOTE: Although a *Pyramid* can be defined initially with as many sides as you want, you cannot change the original number of sides using grips. You can, however, use the *Properties* palette to reset the original number of sides to any practical number.

FIGURE 39-20

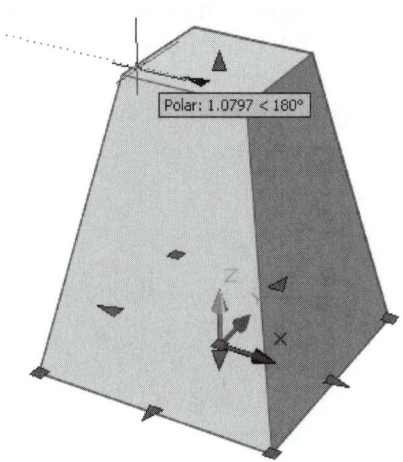

Torus

When a *Torus* primitive is selected for grip editing (not shown), four arrow grips appear allowing you to change the tube radius. In addition, one arrow grip allows you to change the torus radius. You can also use the grip at the center to change the location of the primitive or to access the five normal grip editing options (STRETCH, MOVE, ROTATE, MIRROR, and SCALE).

Polysolid

A *Polysolid* can be altered in a number of ways using grips. Remember that a *Polysolid* is a multi-segmented shape and has a predefined width (between "walls") and a predefined height. Like a *Pline*, a *Polysolid* is created by specifying points for each segment. When selected, a *Polysolid's* corner grips allow you to change the location of the segment vertices. For example, Figure 39-21 illustrates stretching a corner grip to change the location of one vertex.

FIGURE 39-21

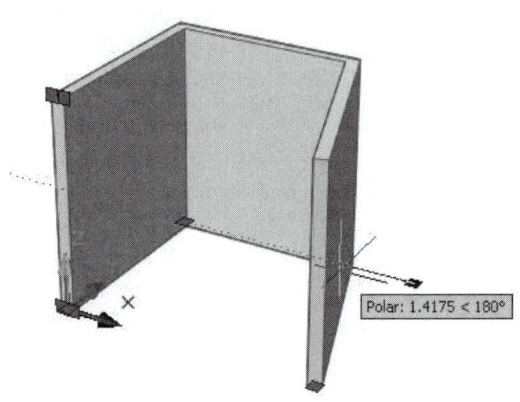

Grips that appear on a *Polysolid* allow you to change both the width and height for all segments. For example, you can change the width at the bottom of the *Polysolid* using the appropriate square grip. Note in Figure 39-22 that the bottom width is changed uniformly for all segments with one grip.

FIGURE 39-22

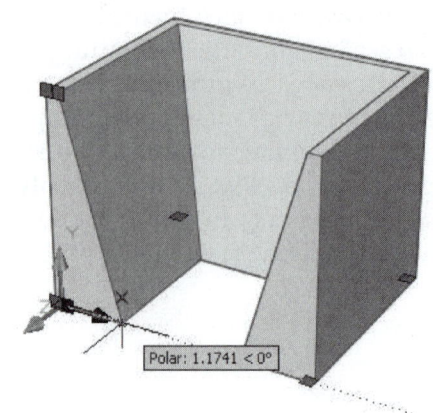

Similarly, dragging one of the top width grips changes the width for all segments uniformly. In this case as shown in Figure 39-23, the top width is increased uniformly as well as the height for all outside faces.

FIGURE 39-23

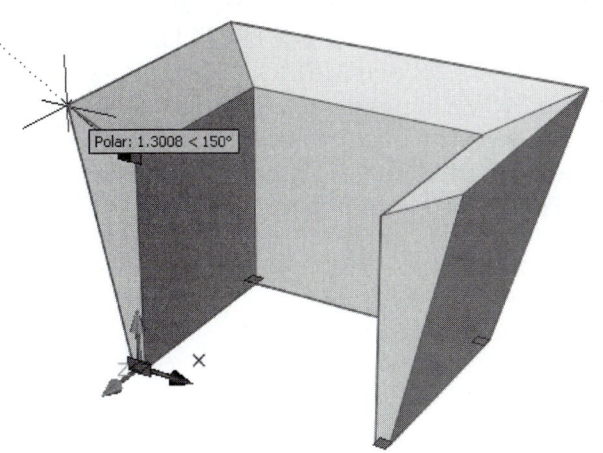

Polar: 1.3008 < 150°

Move and Rotate Grip Tools

Once the desired solids (primitives or composite solids) have been selected for editing, two things happen—the cold grip(s) appear and the move grip tool appears at the current UCS location. At this point, you have a choice to use the normal grip options (STRETCH, MOVE, ROTATE, SCALE, MIRROR) or use the move grip tool and rotate grip tool. *DUCS* has no effect on these options. Follow one of these procedures:

A. To use the normal grip options, simply select the base grip for the solid, face, edge, or vertex to make the grip hot. The STRETCH option appears at the command prompt. You can cycle through all <u>five</u> options by pressing Enter.

B. To use the move and rotate grip tools, hover the cursor over the solid's base grip (the square grip generally at the center of the base) until the move tool automatically relocates to the grip location. Select an axis or plane on the tool. The STRETCH option appears at the command prompt; however, only the STRETCH, MOVE, and ROTATE options can be used. If an axis of the grip tool is selected, the selected solid stretches, moves, or rotates along the axis. If a plane from the tool is selected (the small square area connecting two axes), the selected solid stretches, moves, or rotates within the plane.

Using the Normal Five Grip Options

To use the normal grip options, for example, assume a *Box* primitive solid is selected for editing. Once selected, the solid becomes highlighted and the grips appear as cold grips (Fig. 39-24). Note that the move grip tool appears at the UCS location. At this point, select the grip at the center of the base and make it hot so the STRETCH option appears at the Command prompt.

FIGURE 39-24

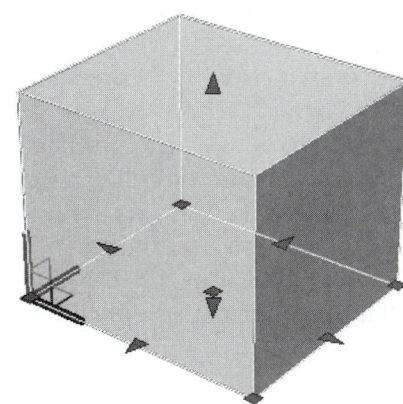

Once the base grip is hot, the move grip tool disappears and the UCS reappears. You can cycle through all five normal grip options: STRETCH, MOVE, ROTATE, SCALE, and MIRROR. Figure 39-25 displays a MOVE in progress. Using the normal grip options, when you cycle to the ROTATE grip option the selected solid rotates about the hot grip in the XY plane of the grip as shown previously in Figure 39-14.

FIGURE 39-25

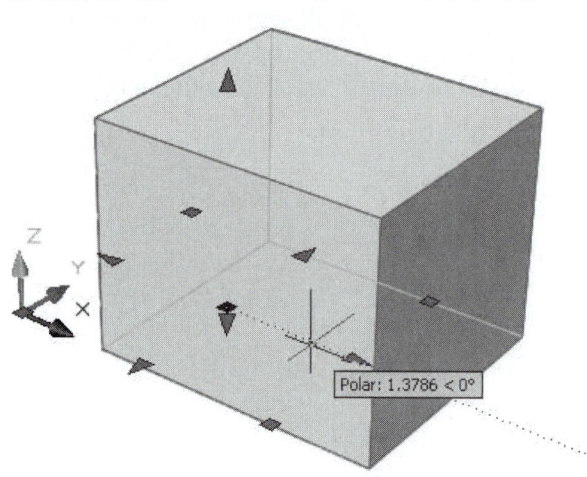

Using the Move and Rotate Grip Tools

On the other hand, to use the move and rotate grip tools with our example, assume the *Box* primitive is selected for editing as displayed previously in Figure 39-24. Rather than selecting the base grip to make it hot, <u>hover over the grip</u> until the move tool automatically relocates to the location of the base grip. At this point, select the desired axis or plane on the grip tool. Figure 39-26 displays this process during selection of the X axis of the grip tool. The grip is now considered hot and the STRETCH option appears at the Command prompt. Either the STRETCH or MOVE option could be used to move the solid along the grip tool's X axis. Pressing the Esc key allows you to select a different axis or plane on the move grip tool.

FIGURE 39-26

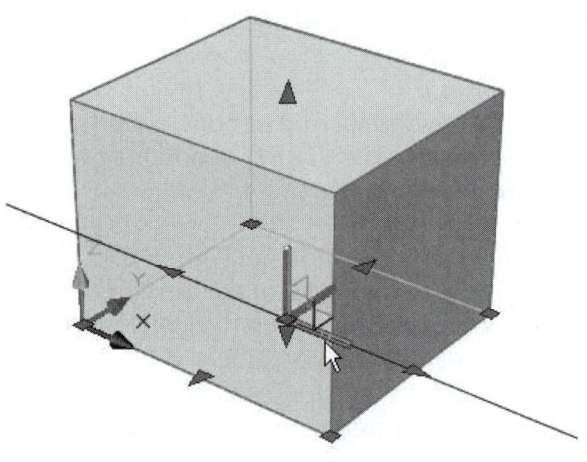

When the move or rotate grip tool is used, <u>only the STRETCH, MOVE, and ROTATE options appear</u> on the Command line when you press Enter to cycle through the options. If you cycle to ROTATE, the move grip tool is automatically replaced by the rotate grip tool. Figure 39-27 illustrates rotating the *Box* about the grip tool's X axis as previously selected. You can select a different axis to rotate about by first pressing the Esc key, then selecting a different axis on the rotate grip tool.

FIGURE 39-27

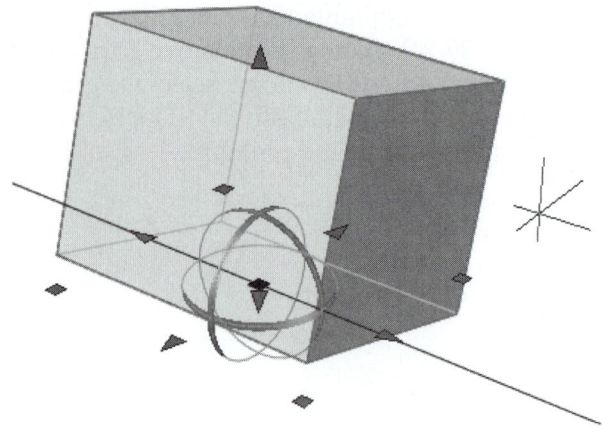

Grip Editing Composite Solids

When you select a composite solid for editing with grips, only the one grip appears at the center of the base as shown in Figure 39-28. The other grips that appear on individual primitives do not appear by default, but can be selected by holding down the Ctrl key and hovering the cursor over a primitive, then selecting it (see "Subobjects and Selection" next).

FIGURE 39-28

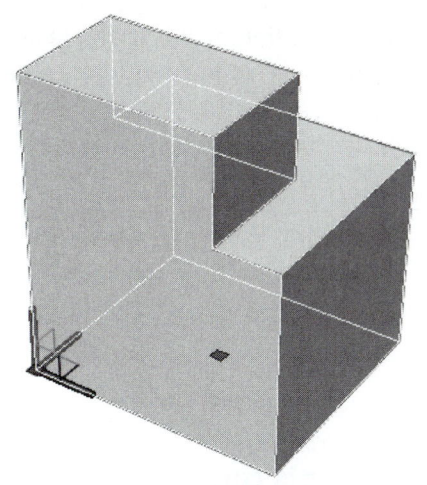

The single base grip that appears when a composite solid is selected, however, allows you to edit the composite solid in the same way as described in previous section "Move and Rotate Grip Tools." That is, if you select the base grip so it becomes hot, you can cycle through all five normal grip options: STRETCH, MOVE, ROTATE, SCALE, and MIRROR. For example, using the normal grip options, once the base grip is hot, you cycle to the ROTATE option to rotate the composite solid in the XY plane of the grip (Fig. 39-29.)

FIGURE 39-29

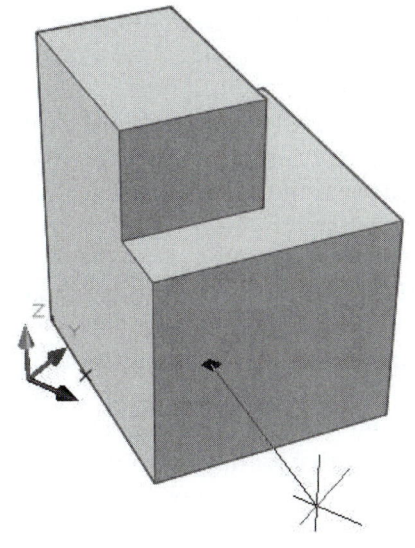

Similar to editing primitives with grips, to use the move and rotate grip tools, hover over the grip until the move tool automatically relocates to the location of the base grip. At this point, select the desired axis or plane on the grip tool (Fig. 39-30). You can press the Esc key to select a different axis or plane on the move grip tool. Remember that only the STRETCH, MOVE, and ROTATE options appear on the Command line when you use the move and rotate grip tools.

FIGURE 39-30

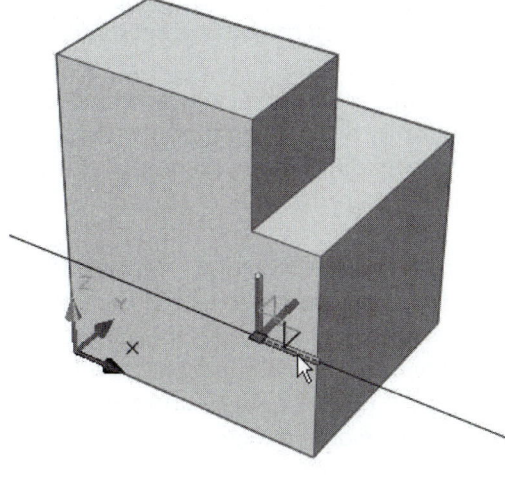

For example, if you cycle to ROTATE, the move grip tool is automatically replaced by the rotate grip tool. By default the composite solid rotates about the selected axis (Fig. 39-31). You can select a different axis to rotate about by first pressing the Esc key, then selecting a different axis on the rotate grip tool.

FIGURE 39-31

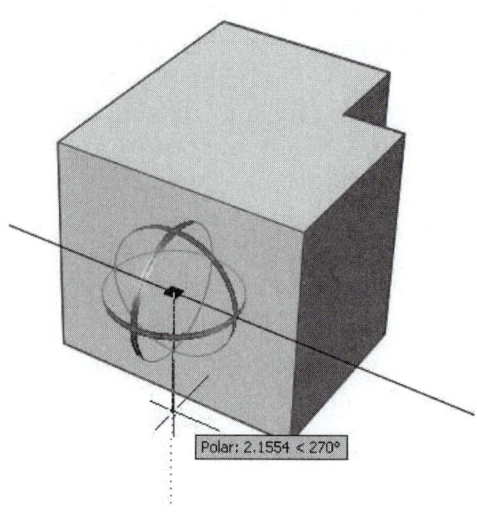

Polar: 2.1554 < 270°

Subobjects and Selection

A "subobject" is a component of a solid. A primitive within a composite solid is considered a subobject. Also, a face, edge, or vertex is a subobject of a primitive solid. The grips that appear on subobjects are exclusive to that type of subobject depending on the type of subobject: edge, vertex, face, or type of primitive.

As you already know, to edit a solid (not a subobject) using grips, you simply select the solid with the cursor for grips to appear. However, to select a subobject, you must hold down the Ctrl key, hover the cursor over the desired subobject until it becomes highlighted, then use the left mouse button to PICK the subobject.

With the default settings, there can be <u>two levels of subobjects</u>. Selecting the desired subobjects depends on the "level" of the subobject. For a first level subobject, hold down the Ctrl key, then PICK it. To select a second level subobject, repeat this process. The two levels are outlined here:

<u>First level subobject</u>, select once using Ctrl:
Primitive within a composite solid
Edge, face or vertex of a stand-alone primitive solid

<u>Second level subobject</u>, select twice using Ctrl:
Edge, face, or vertex of a primitive within a composite solid

NOTE: A second level subobject is available only when the solid's *History* property is set to *Record*. See *"History* Settings."

FIGURE 39-32

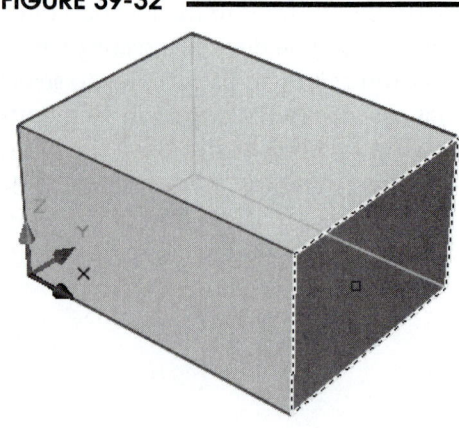

Selecting a first-level subobject requires a single selection. For example, to select a primitive within a composite solid, hold the Ctrl key to highlight, then select the primitive. This sequence is displayed previously in Figures 39-5, 39-6, and 39-7.

To select an edge, face, or vertex within a stand-alone primitive (a first level subobject), hold the Ctrl key to highlight, then select the desired edge, face or vertex. Figure 39-32 displays using Ctrl to highlight a face of a *Box* primitive. Once selected, the grip related to the feature appears. A face, edge, and vertex each have only one grip.

Selecting a second-level subobject (a face, edge, or vertex within a composite solid) requires a double selection. For example, assume a composite solid is constructed of two *Box* primitives and you want to edit one edge of the composite solid. First press and hold the Ctrl key, hover over and select the desired primitive as shown in Figure 39-33.

FIGURE 39-33

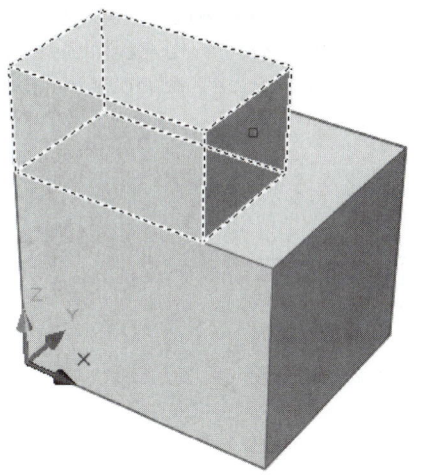

Once the grips appear on the primitive, select the desired edge, face, or vertex using the same method. That is, press and hold the Ctrl key to hover over and select the desired edge, face, or vertex. In our example, once grips appear on the previously selected primitive, press Ctrl to hover and select the desired edge of the highlighted primitive. The edge grip appears as shown in Fig. 39-34.

FIGURE 39-34

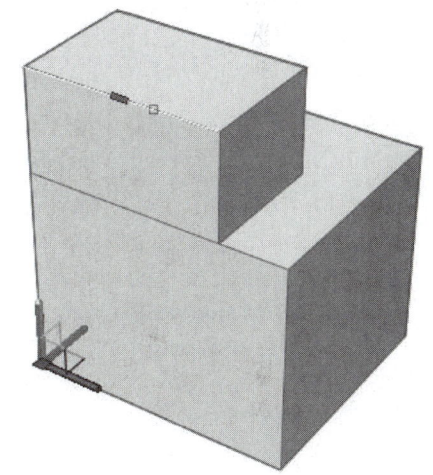

Grip Editing Faces, Edges, and Vertices

An individual face, edge, and vertex on a solid has a distinct grip shape. Each type of grip can be used to edit the object feature in predefined ways. Similar to using grips to edit primitive solids and composite solids, you can chose to edit with or without the move and rotate grip tools. The methods for activating the move and rotate tools for faces, edges, and vertices are the same as previously described for primitives and composite solids. When not using the move and rotate grip tools, all five normal grip options are available (STRETCH, MOVE, ROTATE, SCALE, and MIRROR). Using the move and rotate grip tools, only the STRETCH, MOVE, and ROTATE options can be used.

Edge Grips

FIGURE 39-35

Several examples for using edge grips are described and illustrated in this section; however, many more possibilities exist. Using our composite solid example, assume the top right edge grip was selected for editing. Without using the move and rotate grip tools, you can cycle to all five grip options. Using the STRETCH option would allow you to stretch the selected edge in any direction as shown in Figure 39-35. Note that other edges and faces are also affected; however, only the primitive containing the selected edge is changed.

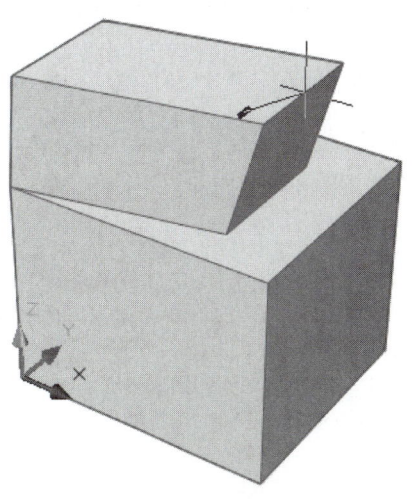

Cycling to the ROTATE option allows you to rotate the edge. Note in Figure 39-36 that the face containing the selected edge is rotated but the length of the edge itself does not change. Other primitives in a composite solid are not affected.

FIGURE 39-36

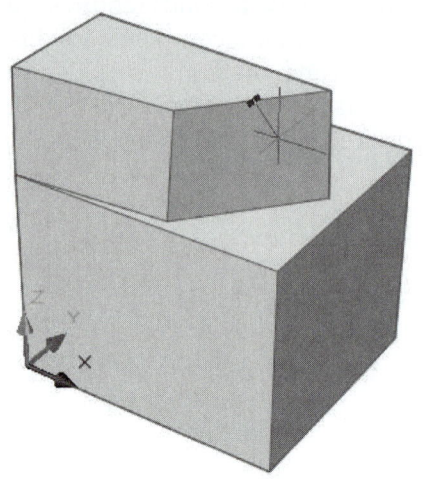

The SCALE option allows you to change the length of the selected edge (Fig. 39-37). Other edges and faces are affected but only those on the primitive containing the edge.

When the MOVE (with or without the move and rotate grip tools) and MIRROR grip options are used with an edge grip, the entire primitive is moved or mirrored without any other changes or deformations.

FIGURE 39-37

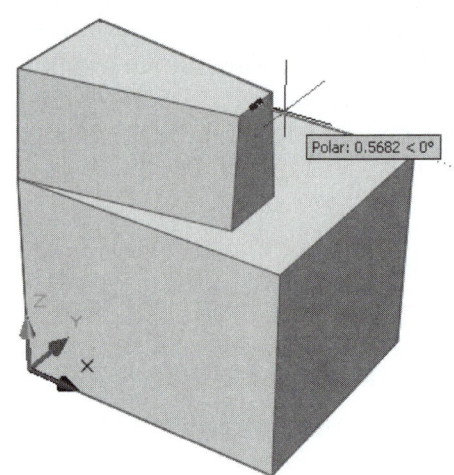

If you want to use the move and rotate grip tools, press Ctrl and hover over the grip until the move tool relocates to the grip position. Select the desired axis or plane on the grip tool. For example, you could select the grip tool's Z axis to ensure the selected edge grip moves only in the Z direction, as indicated in Figure 39-38. Press the Esc key to select another axis or plane.

FIGURE 39-38

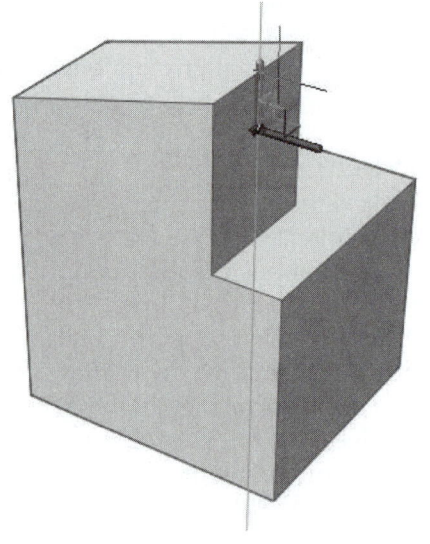

When you cycle to the ROTATE option, the rotate grip tool appears. By default, rotation occurs about the previously selected axis. Press Esc to select another axis on the rotate tool. In Figure 39-39, the rotate tool's X axis is used. Beware since the selected edge does not change length using any option other that SCALE; therefore, the primitive may become deformed depending on the axis of rotation as shown in Figure 39-39.

FIGURE 39-39

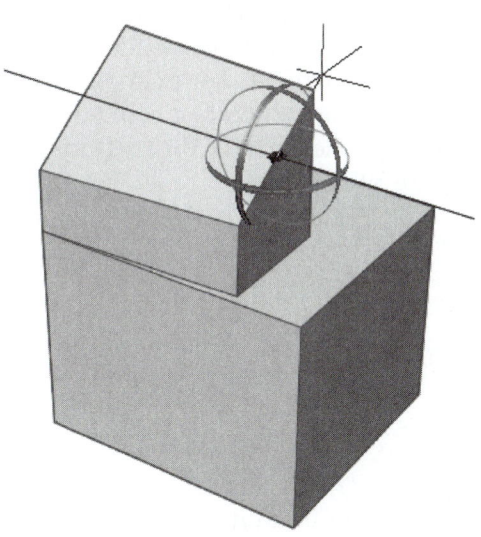

Another possibility for using an edge grip is shown in this example of a *Polysolid* primitive (Fig. 39-40). Note that one edge grip has been selected for editing. Using the STRETCH mode with the move grip tool allows a change to be made to the width of a single segment of the *Polysolid*. This is in contrast to changing the width of all segments using the primitive's grips normal grips (not the subobject edge grip) as shown previously in Figures 39-22 and 39-23. Because there are so many types of primitive and composite solid combinations, the possibilities for editing with grips are practically endless.

FIGURE 39-40

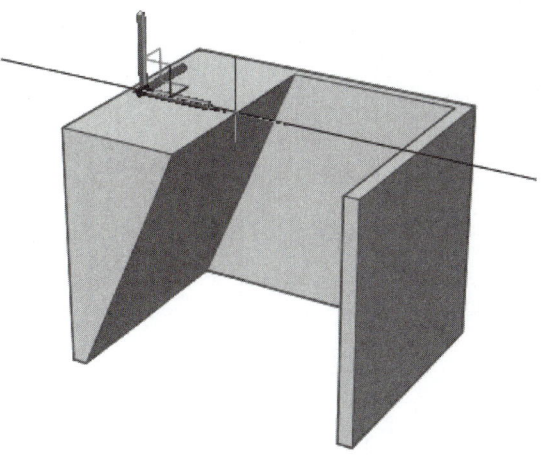

Face Grips

A face grip of a primitive solid or of a composite solid can be selected using the same procedure as explained earlier (using the Ctrl key). Faces of solids are usually planar but can include some curved faces. Grips that appear on faces of solids appear as a circle on the plane of the face as shown on the vertical front face of the *Wedge* primitive in Figure 39-41.

FIGURE 39-41

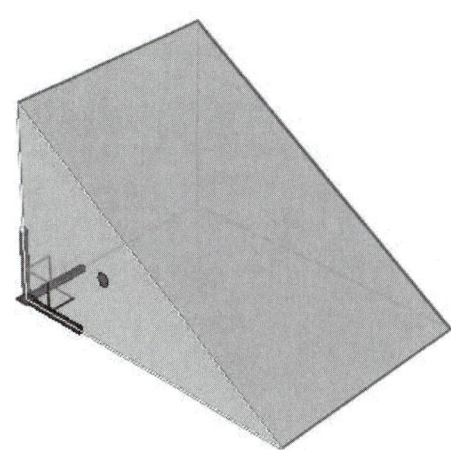

One of the most useful applications of using face grips is to STRETCH a face perpendicular to the plane of the face, similar to the way a shape would be *Extruded*. For example, using the move grip tool, the vertical face could be STRETCHed to change the depth of the *Wedge* as shown in Figure 39-42.

FIGURE 39-42

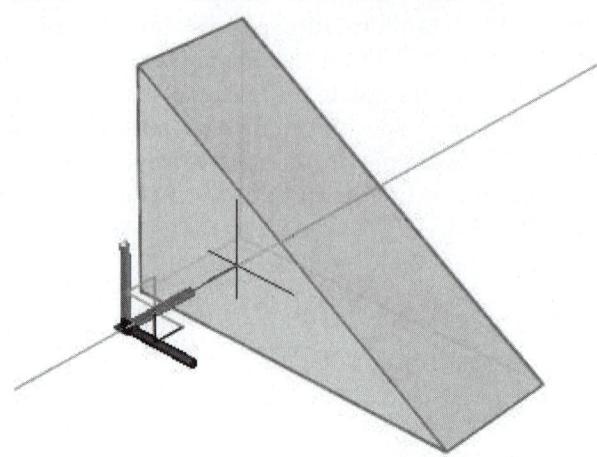

Another application is shown in Figure 39-43. In this case, the right face of a Pyramid (with a top radius) was selected. The illustration shows the face being STRETCHed using the move grip tool.

FIGURE 39-43

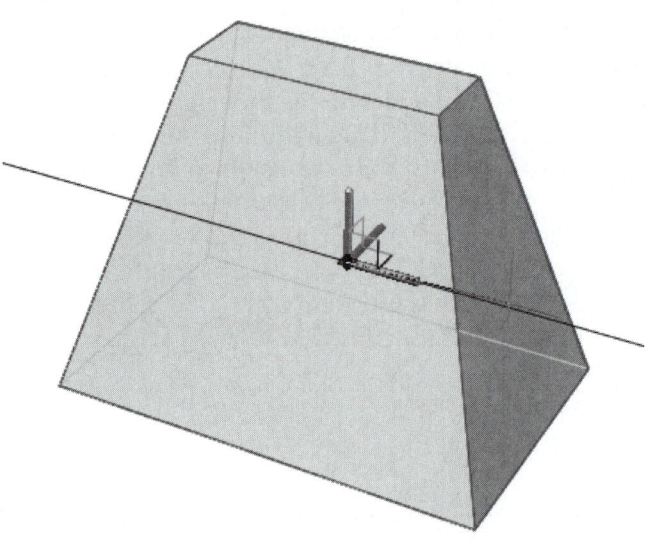

Vertex Grips

A vertex grip can be selected by using Ctrl to select any vertex of a solid. For example the upper-right corner vertex has been selected on the *Box* primitive shown in Figure 39-44.

FIGURE 39-44

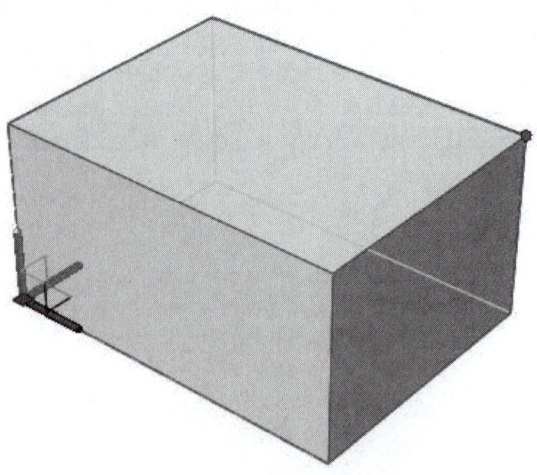

A vertex grip can be STRETCHed as shown in Figure 39-45. Only the selected vertex is affected along with any connected faces and edges. Interestingly, only the STRETCH can be used effectively with vertex grips. Since a vertex represents one point (with no length or dimension), it cannot be scaled or rotated. If the MOVE or MIRROR option is used, the entire solid is affected.

NOTE: Keep in mind that multiple grips can be selected. For example, you may want to ROTATE two faces together. Accomplish this by holding the Ctrl key down while selecting each face. Additionally, any combination of grip types can be selected. For example, you could select one face grip, one edge grip, and one vertex grip.

FIGURE 39-45

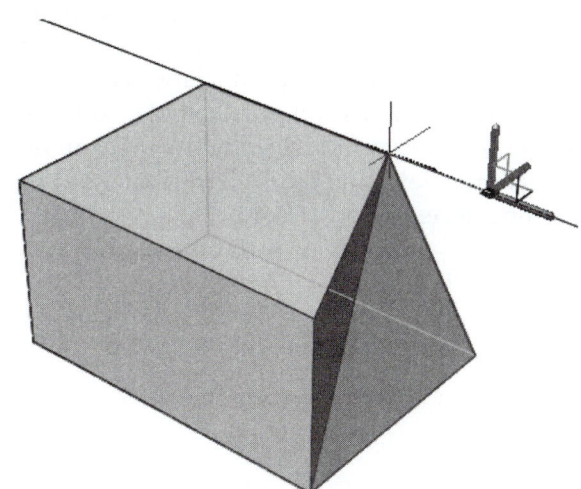

Selecting Overlapping Subobjects

In some cases the subobject you want to select may be behind, or overlapping, another subobject, and therefore, difficult to PICK. When you "hover" over composite solids while holding Ctrl, the solid in the foreground is detected first. For example, the location of the pickbox shown previously in Figure 39-33 would select the top *Box* primitive by default.

When subobjects are overlapping and selection preview is turned on (hovered objects become highlighted), you can cycle through the subobjects by rolling over the subobject on top to highlight it, and pressing and holding the Ctrl key and then pressing the spacebar continuously. When the required subobject is highlighted, left-click to select it. For example, note the location of the pickbox in Figure 39-46. Making this selection would result in selection of the top *Box*, as shown previously in Figure 39-33. However, holding down the Ctrl key and pressing the spacebar would cycle to select the large *Box* primitive under the small *Box* as shown in Figure 39-46. If selection preview is turned off, and more than one subobject is under the pickbox, you can cycle through the subobjects until the correct one is selected by pressing and holding Ctrl+spacebar and then left-clicking.

FIGURE 39-46

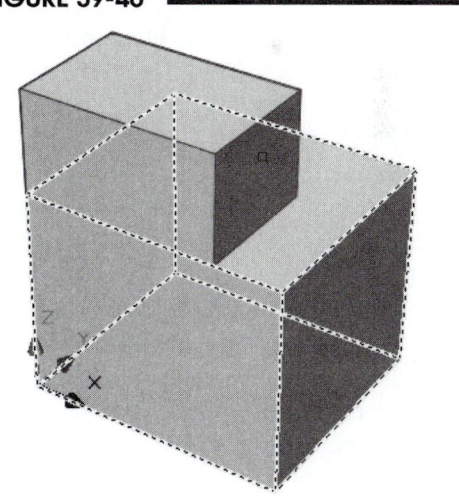

History Settings

Solid History

By default, a composite solid's *History* is set to *Record*. This default setting allows you to select individual primitives that make up the composite solid, as described in the previous sections. Using this scheme, selecting a second level subobject (a face, edge, or vertex within a primitive) requires a second Ctrl and PICK.

However, if you always want to select faces, edges, and vertices within composite solids without first selecting the individual primitive, you can change the solid's *History* setting to *None* using the *Properties* palette (Fig. 39-47). With this setting, you can select any face, edge, or vertex without having to select the primitive first. For example, you could directly select the top face on the large *Box* primitive without having to first select the *Box* itself. Remember that using the *Properties* palette allows you to change properties for the selected object(s) only.

FIGURE 39-47

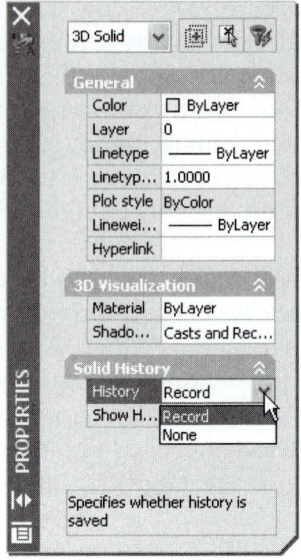

For example, assume the *History* setting for the composite solid shown in Figure 39-48 was changed to *None*. Selection of the L-shaped face would be possible only with this setting. Otherwise, with a *History* setting of *Record*, the entire faces of the *Box* primitives would be selected and highlighted.

NOTE: Once *History* is set to *None*, the history of the composite solid's construction is deleted and cannot be retrieved. Setting the *History* back to *Record* begins a new record.

FIGURE 39-48

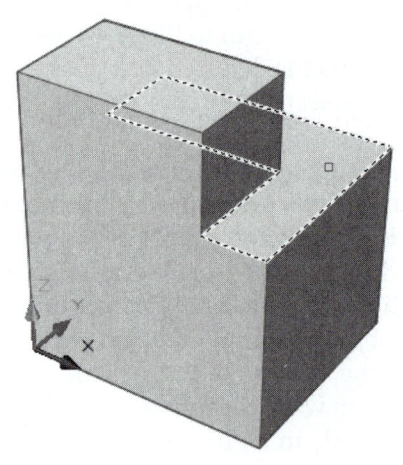

SOLIDHIST

The *SOLIDHIST* system variable sets the history for <u>all</u> solids, both new and existing. This setting overrides the *History* setting in an individual solid's *Properties* palette. When *SOLIDHIST* is set to the default of 1, composite solids retain the history of the original objects contained in the composite.

0   Sets the *History* property to *None* for all solids. No history is retained.
1   Sets the *History* property to *Record* for all solids. All solids retain a history of their original objects.

FIGURE 39-49

Show History

A composite solid is created by using multiple primitives combined by Boolean operations. For example, assume a composite solid was created by using *Subtract* to remove a small *Box* primitive from a large *Box* primitive as shown in Figure 39-49. Normally, all of the individual primitives are not visible unless you select them as subobjects. In our example, the small *Box* primitive is not visible.

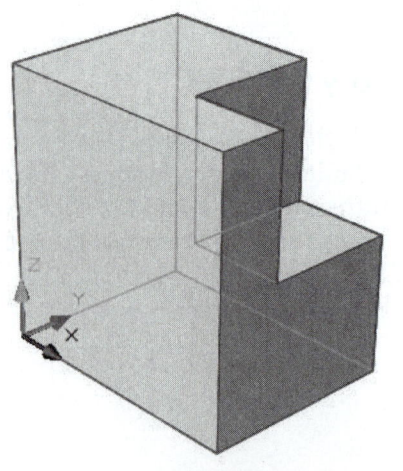

If a composite solid's *Show History* property is set to *Yes* in the *Properties* palette, the individual primitives that make up the composite solid can be viewed. To do this, select the composite solid and change its *Show History* setting to *Yes* as shown in Figure 39-50.

FIGURE 39-50

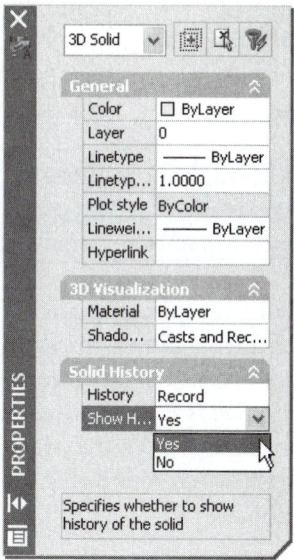

When a composite solid's *Show History* setting is set to *Yes*, the individual primitives that were used to create the solid are visible at all times as shown in Figure 39-51.

FIGURE 39-51

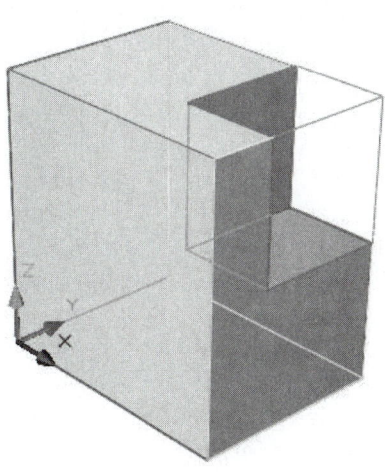

SHOWHIST

The *SHOWHIST* system variable controls the *Show History* property for <u>all</u> solids in a drawing. Depending on the setting, an individual solid's *Show History* setting in the *Properties* palette can be overridden.

0   Sets the *Show History* property to *No* (read-only) for all solids. Overrides the individual *Show History* property settings for solids. You cannot view the original objects that were used to create the solid.

1   Does not override the individual *Show History* property settings for solids.

2   Displays the history of all solids by overriding the individual *Show History* property settings for solids. You can view the original objects that were used to create the solid.

SOLIDS EDITING COMMANDS

In addition to using grips to edit composite solids, AutoCAD offers several commands that allow you to change the geometry.

Solidedit	This command has options for extruding, moving, rotating, offsetting, tapering, copying, coloring, separating, shelling, cleaning, checking, and deleting faces and edges of 3D solids.
Slice	Use this command to cut through a solid using a plane or a surface. You can keep both parts of the solid or delete part.
Presspull	This unique command combines creating an extrusion and performing a Boolean all in one command. The extrusion is created on a solid surface and then pushed into the solid's face (*Subtract*) or pulled out from the face (*Union*).

2007

2007

2007

Sectionplane	Use this command to create an imaginary variable cutting plane, similar to a clipping plane, to view the inside of a solid.
Interfere	*Interfere* checks two or more solids to see if they overlap in 3D space. You can create a new solid from an overlapping volume.
Imprint	Imprint creates an "edge" element on a solid that can be used for editing with grips.

Although most of the solid modeling commands are available using the *3D Make control panel* in the upper-right corner of the Drawing Editor, some of the editing commands are not and must be invoked through another menu or by typing.

For example, all options of the *Solidedit* command are accessible only from the *Solids Editing* cascading menu from the *Modify* pull-down menu (Fig. 39-52).

FIGURE 39-52

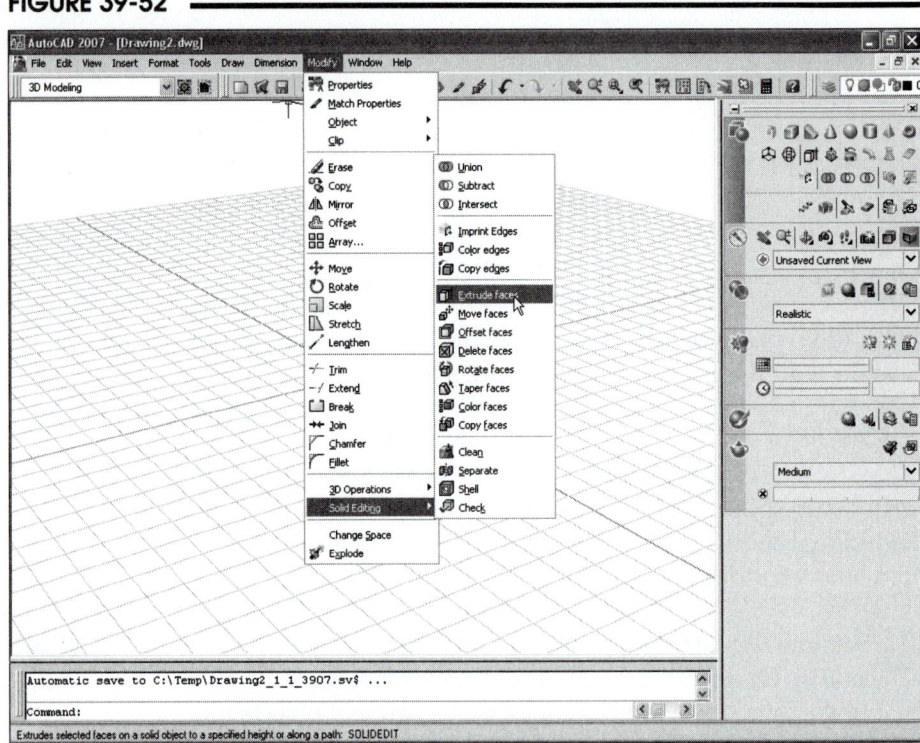

Solidedit

Pull-down Menu	Command (Type)	Alias (Type)	Short-cut	Screen (side) Menu	Tablet Menu
Modify *Solids Editing >*	*Solidedit*	...	...	...	...

The *Solidedit* command has several "levels" of options. The first level prompts you to specify the type of geometry you want to edit—*Face, Edge,* or *Body.* The second level of options depends on your first level response as shown below.

```
Command: solidedit
Solids editing automatic checking:  SOLIDCHECK=1
Enter a solids editing option [Face/Edge/Body/Undo/eXit] <eXit>: f
Enter a face editing option
[Extrude/Move/Rotate/Offset/Taper/Delete/Copy/coLor/Undo/eXit] <eXit>:
```

Solidedit operates on three types of geometry: *Edges, Faces* and *Bodies*. An *Edge* is defined as the common edge between two surfaces that has the appearance of a line, arc, circle, or spline (Fig. 39-53 A). *Faces* are planar or curved surfaces of a 3D object (Fig. 39-53 B). A *Body* is defined as an existing 3D solid or a non-solid shape created with a *Solidedit* option.

FIGURE 39-53

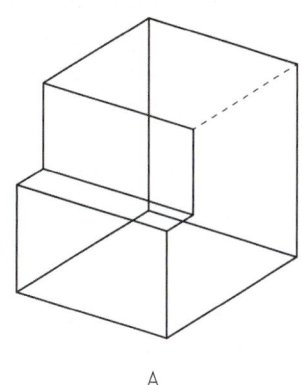 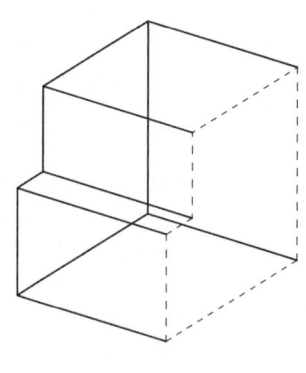

A.   B.

Object selection is an integral part of using *Solidedit*. For example, once you have specified the type of geometry and the particular editing option, the *Solidedit* command prompts to select *Edges, Faces,* or a *Body*.

> Select faces or [Undo/Remove]: Select a face or enter an option
> Select faces or [Undo/Remove/ALL]: Select a face or enter an option

When you are prompted to select *Faces* or *Edges,* select the desired face or edge directly with the pickbox. You will most likely go through a series of adding and removing geometry until the desired set of lines is highlighted. Remember that you can hold down the Shift key and select objects to *Remove* them from the selection set.

If you use this command often, it may help to change the composite solid's *History* setting to *None* using the *Properties* palette (see previous discussion, "History Settings"). Setting the *History* to *None* allows you to use the Ctrl key while hovering over solids to select individual faces and edges rather than the entire primitives that make up the solid.

Ensure that you highlight exactly the intended geometry before you proceed with editing. If an incorrect  set is selected, you can get unexpected results or an error message can appear such as that below.

> Modeling Operation Error:
>   No solution for an edge.

For example, many of the *Face* options allow selection of one or multiple faces. In Figure 39-54 both selection sets are valid faces or face combinations, but each will yield different results. Figure 38-54 A indicates selection of one face and B indicates selection of several faces defining an entire primitive. Add and remove faces or edges until you achieve the desired geometry.

FIGURE 39-54

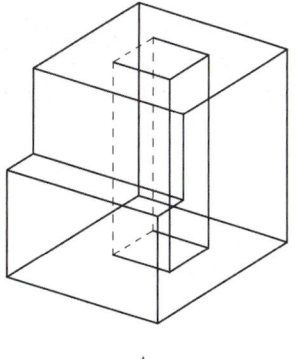 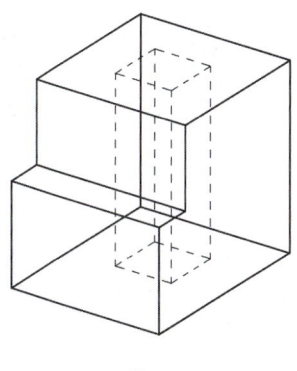

A.   B.

Face Options

The *Face* options edit existing surfaces and create new surfaces. A *Face* is a planar or curved surface. *Faces* that are part of existing 3D solids can be altered to change the configuration of the 3D composite solid. Individual *Faces* can be edited and multiple *Faces* comprising a primitive can be edited. New surfaces can be created from existing *Faces*, but entirely new independent solids cannot be created using these editing tools.

Extrude

The *Extrude* option of *Solidedit* allows you to extrude any *Face* of a 3D object in a similar manner to using the *Extrude* command to create a 3D solid from a *Pline*. This capability is extremely helpful if you need to make a surface on a 3D solid taller, shorter, or longer. You can select one or more faces to extrude at one time.

```
Command: solidedit
Solids editing automatic checking:  SOLIDCHECK=1
Enter a solids editing option [Face/Edge/Body/Undo/eXit] <eXit>: f
Enter a face editing option
[Extrude/Move/Rotate/Offset/Taper/Delete/Copy/coLor/Undo/eXit] <eXit>: e
Select faces or [Undo/Remove]: PICK
Select faces or [Undo/Remove/ALL]: PICK  or remove
Select faces or [Undo/Remove/ALL]: Enter
Specify height of extrusion or [Path]: (value)
Specify angle of taper for extrusion <0>: Enter or (value)
Solid validation started.
Solid validation completed.
```

For example, Figure 39-55 illustrates the extrusion of a face to make the 3D object taller, where A indicates the selected (highlighted) face and B shows the result. This extrusion has a 0 degree taper angle.

FIGURE 39-55 ────────────

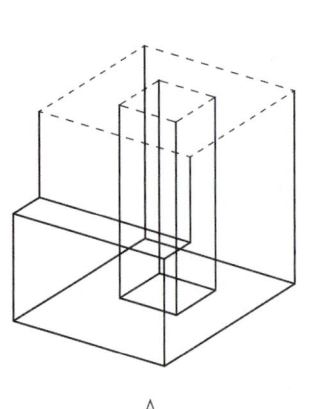 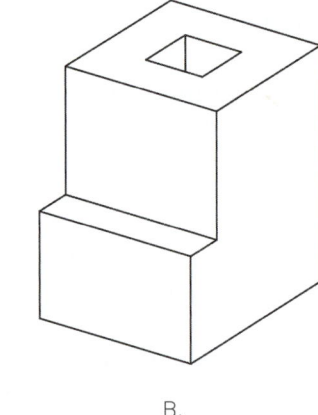

A.   B.

Specifying a positive angle tapers the face inward (Fig. 39-56 A) and specifying a negative angle tapers the face outward (B) *Height* is always perpendicular to the selected *Face*, not necessarily vertical.

FIGURE 39-56 ────────────

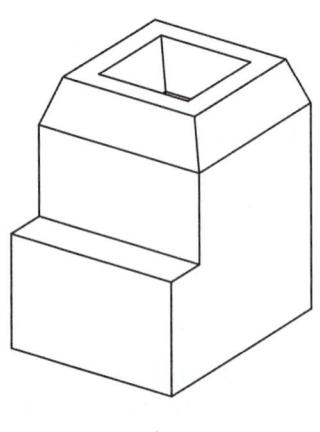 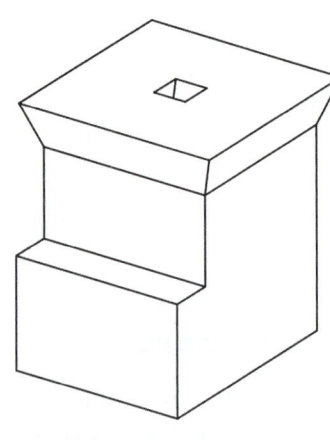

A.   B.

An internal face can be selected for extrusion. In Figure 39-57, a single face is selected (A) and the *Height* value specified is greater than the distance to the outer face of the solid, creating an open side on the object (B). *Extrude* allows an internal face to pass through an external face as shown in Fig. 39-57 B. Several other options, such as *Move, Rotate,* and *Offset* also allow the interior primitives to pass through the bounding box.

FIGURE 39-57

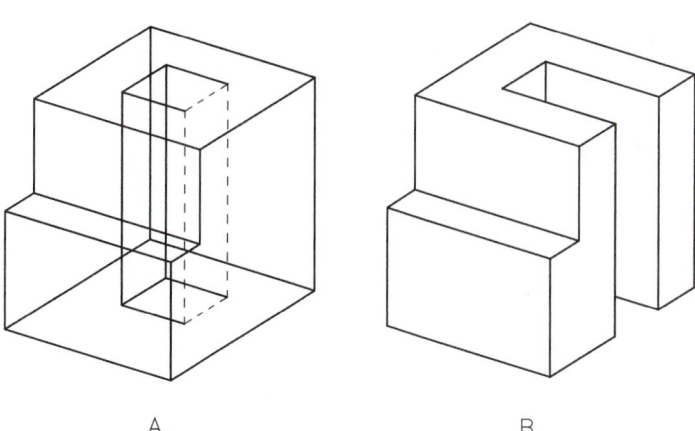

A.

B.

Faces can be extruded along a *Path*. The *Path* object can be a *Line, Circle, Arc, Ellipse, Pline,* or *Spline*. For example, Figure 39-58 illustrates extrusion of a face along a *Line* path (A) to yield the result shown on the right (B).

FIGURE 39-58

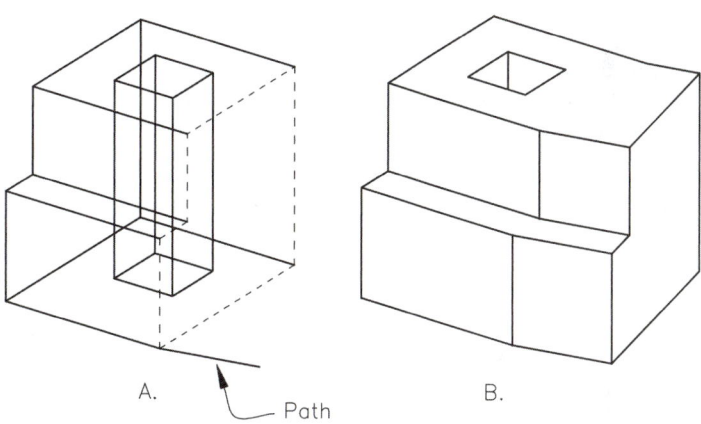

A.

Path

B.

NOTE: Entering a <u>positive</u> value at the "Specify height of extrusion or [Path]:" prompt <u>increases</u> the volume of the solid and entering a <u>negative</u> value <u>decreases</u> the volume of the solid. For example, entering a positive value for extruding the face in Figure 39-57 A would result in a smaller hole.

Move

With the *Move* option, you can move a single face of a 3D solid or move an entire primitive within a 3D solid. This capability is particularly helpful during geometry construction or when there is a design change and it is required to alter the location of a hole or other feature within the confines of the composite 3D model.

```
[Extrude/Move/Rotate/Offset/Taper/Delete/Copy/coLor/Undo/eXit] <eXit>: m
Select faces or [Undo/Remove]: PICK
Select faces or [Undo/Remove/ALL]: PICK  or remove
Select faces or [Undo/Remove/ALL]: Enter
Specify a base point or displacement: PICK
Specify a second point of displacement: PICK
```

For example, Figure 39-59 illustrates using the *Move* option to relocate a hole primitive (four faces) within a composite solid. In this case you must ensure all faces, and only the faces, comprising the primitive are selected.

With the *Move* option the internal features <u>can</u> be moved to extend into or past the "bounding box" of the composite solid, as they can with *Extrude* (see Fig. 39-57).

FIGURE 39-59

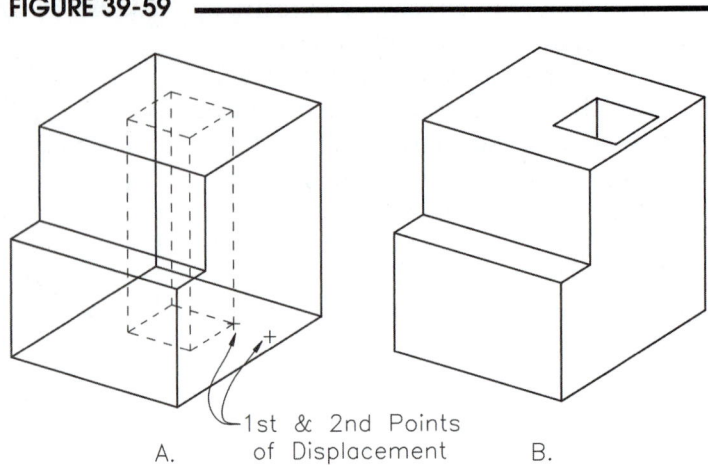

A.   1st & 2nd Points of Displacement   B.

Another possibility for *Move* is to move only one or selected faces to alter the configuration of an internal feature of a composite solid. An example is shown in Figure 39-60, where only one face is selected to move. Compare the results in Figures 39-59 B and 39-60 B.

FIGURE 39-60

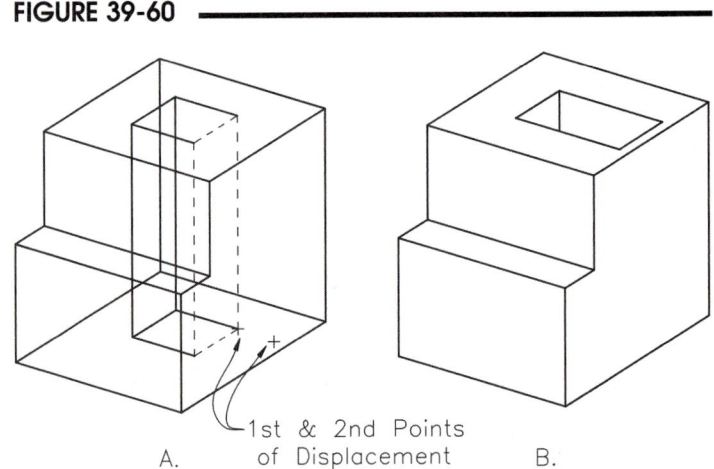

A.   1st & 2nd Points of Displacement   B.

Keep in mind that *Move* can be used to move a singular external face, achieving the same result as *Extrude* (see previous Fig. 39-55).

Offset

The 2D *Offset* command makes a parallel copy of a *Line, Arc, Circle, Pline*, etc. The *Offset* option of *Solidedit* makes a 3D offset. A simple application is making a hole larger or smaller.

```
[Extrude/Move/Rotate/Offset/Taper/Delete/Copy/coLor/Undo/eXit] <eXit>: O
Select faces or [Undo/Remove]: PICK
Select faces or [Undo/Remove/ALL]: PICK  or remove
Select faces or [Undo/Remove/ALL]: Enter
Specify the offset distance: PICK or (value)
```

You can offset through a point or specify an offset distance. Specify a positive value to increase the volume of the solid or a negative value to decrease the volume of the solid.

Figure 39-61 demonstrates how an internal feature, such as a hole, can be *Offset* to effectively change the volume of the hole. Since positive values increase the size of the solid, a negative value was used here. In other words, holes inside a solid become smaller when the solid is *Offset* larger.

FIGURE 39-61

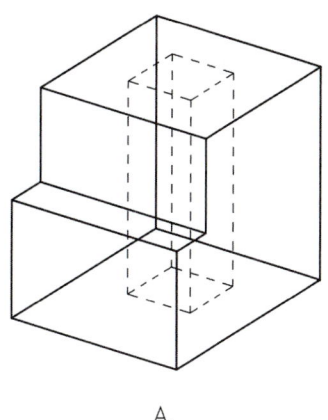
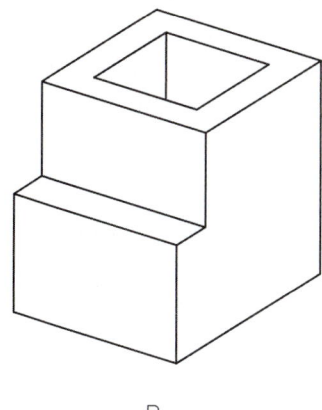

A.   B.

Delete

This option of *Solidedit* deletes faces from composite solids. Although you cannot delete one planar face from a solid, you can delete one curved face, such as a cylinder, that comprises an entire primitive. You can also delete multiple faces that comprise a primitive or entire feature.

[Extrude/Move/Rotate/Offset/Taper/Delete/Copy/coLor/Undo/eXit] <eXit>: *d*
Select faces or [Undo/Remove]: **PICK**
Select faces or [Undo/Remove/ALL]: **PICK** or remove
Select faces or [Undo/Remove/ALL]: **Enter**
Solid validation started.
Solid validation completed.

As an example, Figure 39-62 A displays two selected faces to be deleted from a composite solid. The result is displayed in B. In this case, <u>both faces must be highlighted</u> for the deletion to operate.

For the same composite solid, if all <u>four</u> internal faces comprising the hole are selected, as shown in previous Figure 39-61 A, the hole could be deleted.

FIGURE 36-62

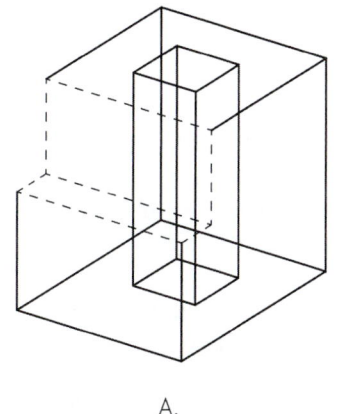
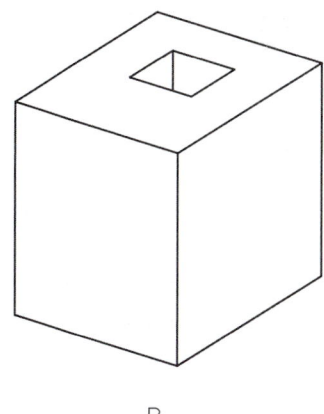

A.   B.

Rotate

Rotate is helpful during geometry construction and for design changes when components within a composite solid must be rotated in some way.

[Extrude/Move/Rotate/Offset/Taper/Delete/Copy/coLor/Undo/eXit] <eXit>: *r*
Select faces or [Undo/Remove]: **PICK**
Select faces or [Undo/Remove/ALL]: **PICK** or remove
Select faces or [Undo/Remove/ALL]: **Enter**
Specify an axis point or [Axis by object/View/Xaxis/Yaxis/Zaxis] <2points>: **PICK**
Specify the second point on the rotation axis: **PICK**
Specify a rotation angle or [Reference]: **PICK** or (**value**)

Several methods of rotation are possible based on the selected axis. These options are essentially the same as those available with the *3DRotate* command (see *"3DRotate"* discussed previously).

For example, Figure 39-63 illustrates the rotation of a primitive within a composite solid. Here, the primitive is rotated about two points defining one edge of the hole.

NOTE: Positive and negative rotation complies with the right-hand rule. Also keep in mind that when PICKing two points to define the axis of rotation, positive direction on the axis is determined from the first to the second point picked.

One face of a composite solid can be selected for rotation within a composite solid (Fig. 39-64 A). The selected face is rotated about two points defining one edge of the face.

FIGURE 39-63

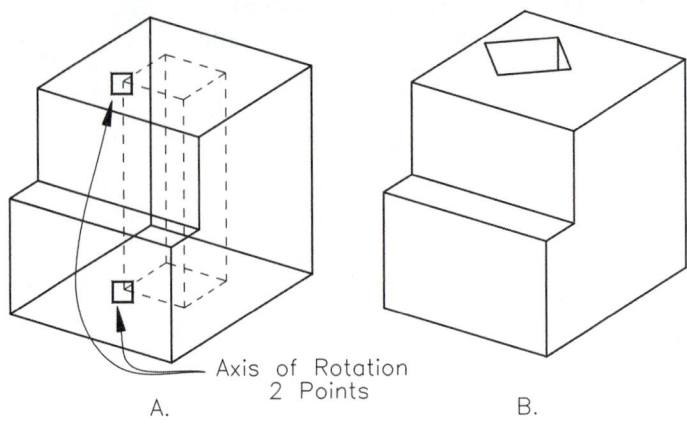

Axis of Rotation
2 Points
A.   B.

FIGURE 39-64

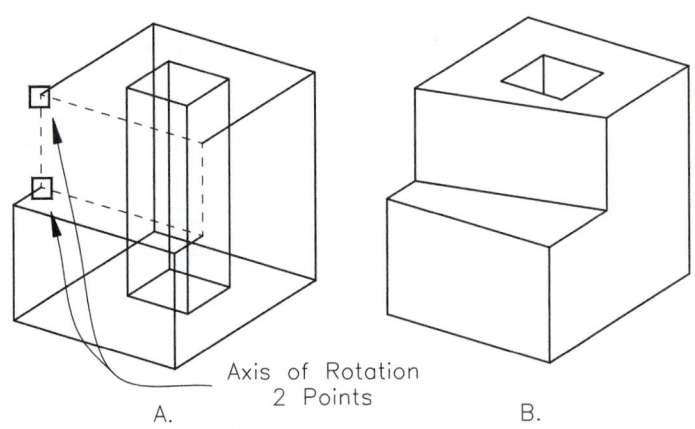

Axis of Rotation
2 Points
A.   B.

Taper

Taper angles a face. You can use *Taper* to change the angle of planar or curved faces (Fig. 39-65).

```
[Extrude/Move/Rotate/Offset/Taper/Delete/Copy/coLor/Undo/eXit] <eXit>: t
Select faces or [Undo/Remove]: PICK
Select faces or [Undo/Remove/ALL]: PICK  or remove
Select faces or [Undo/Remove/ALL]: Enter
Specify the base point: PICK
Specify another point along the axis of tapering: PICK
Specify the taper angle: Enter or (value)
```

TIP

The rotation of the taper angle is determined by the order of selection of the base point and second point. Although AutoCAD prompts for the "axis of tapering," the two points actually determine a reference line from which the taper angle is applied. The taper starts at the first base point and tapers away from the second base point. The rotational axis passes through the first base point and is <u>perpendicular</u> to the line between the base points. <u>The line between the base points does not specify the rotational axis, as implied.</u>

FIGURE 39-65

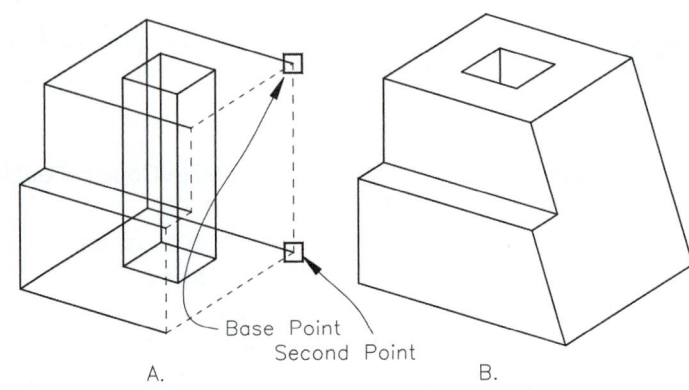

Base Point
Second Point
A.   B.

To explain this idea, Figure 39-65 A (previous page) illustrates the selection of a planar face, the base point, and the second point. An angle of –20 degrees (negative value) is entered to achieve the results shown in B. Positive angles taper the face inward (toward the solid) and negative values taper the face outward (away from the solid).

For some cases, either *Taper* or *Rotate* could be used to achieve the same results. For example, both Figure 39-65 B and previous Figure 39-64 B could be attained using *Taper* or *Rotate* (although different points must be selected depending on the option used).

Curved faces can also be selected for applying a *Taper*. Figure 39-66 A illustrates a cylindrical hole primitive. The entire cylinder primitive is selected as one *Face* object. Note the selection of the base point and second point. *Taper* converts the cylindrical hole into a conical hole (B).

FIGURE 39-66

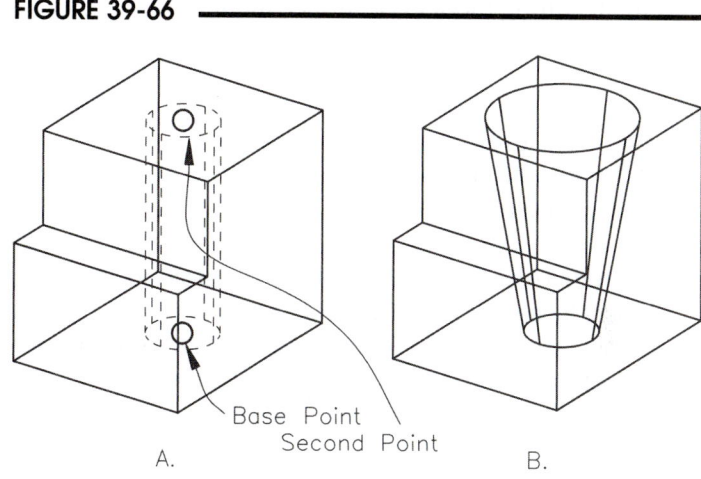

Base Point

Second Point

A.

B.

Copy

As you would expect, the *Copy* option copies *Faces*. However, the resulting objects are *Regions* or *Bodies*.

```
[Extrude/Move/Rotate/Offset/Taper/Delete/Copy/coLor/Undo/eXit] <eXit>: c
Select faces or [Undo/Remove]: PICK
Select faces or [Undo/Remove/ALL]: PICK  or remove
Select faces or [Undo/Remove/ALL]: Enter
Specify the base point or displacement: PICK  or (value)
Specify a second point of displacement: PICK
```

Although any *Face* can be selected (planar or curved), Figure 39-67 shows how *Copy* can be used to create a *Region* from a planar face. Typically, the *Copy* option would be used to create a new *Region* from existing 3D solid geometry. The *Extrude* command could then be used on the *Region* to form a new 3D solid from the original face.

If a curved *Face* is selected to *Copy*, the resulting object is a *Body*. This type of body is a curved surface, not a solid.

FIGURE 39-67

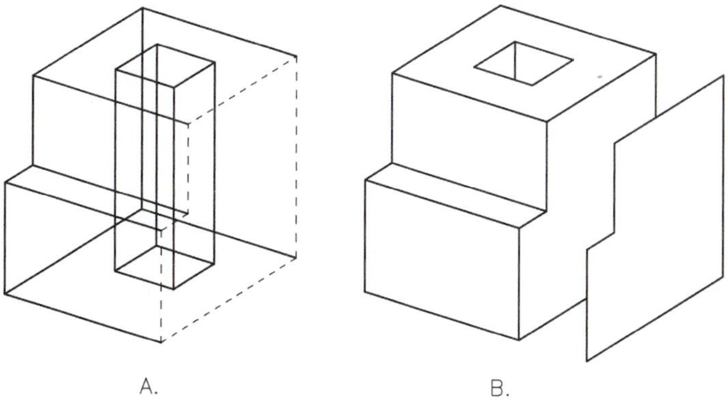

A.

B.

For example, Figure 39-70 A displays a 3D solid with a *Circle* that is on the same plane as the vertical right face of the solid. The *Imprint* option is used, the solid is selected, and the *Circle* is selected as the "object to imprint." Answering "Y" to "Delete the source object," the resulting *Imprint* is shown in B.

FIGURE 39-70

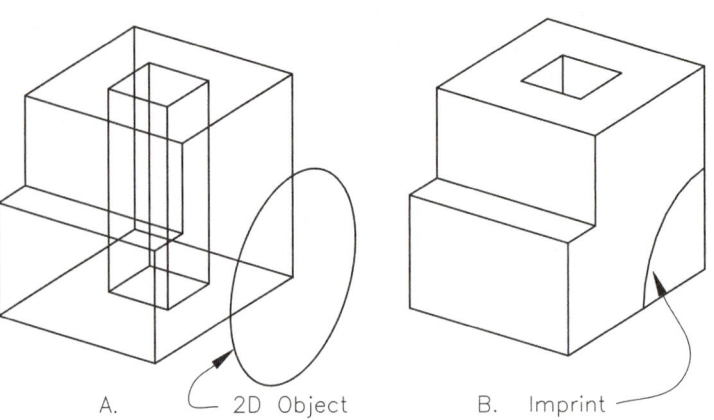

A.   2D Object   B.   Imprint

The *Imprint* creates a separate *Face* on the object—in this case, a total of two coplanar *Faces* are on the same vertical surface of the object. The new *Face* can be treated independently for use with other *Solidedit* options. For example, *Extrude* could be used to create a new solid feature on the composite solid (Fig. 39-71).

FIGURE 39-71

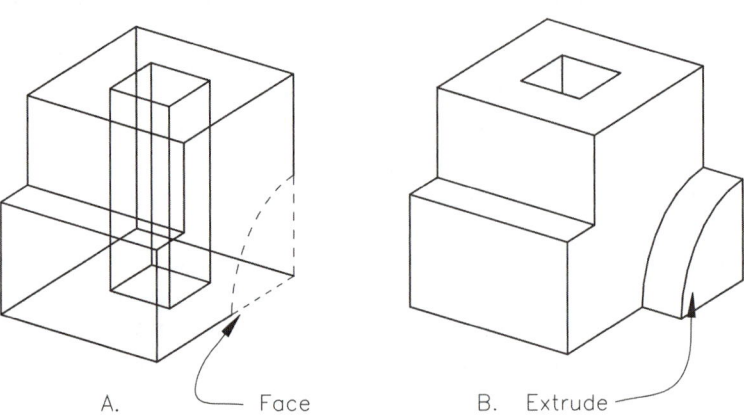

A.   Face   B.   Extrude

Clean

The *Clean* option deletes any 2D geometry on the solid.

For example, you may use *Imprint* to "attach" 2D geometry to a *Body* (solid) for the intention of using *Extrude*. *Imprinted* objects become permanently attached to the *Body* and cannot be removed by *Erase*. If you need to remove an *Imprint*, use the *Clean* option to do so. One use of *Clean* removes all *Imprints*.

> [Imprint/seParate solids/Shell/cLean/Check/Undo/eXit] <eXit>: **l**
> Select a 3D solid: **PICK**

Separate

It is possible in AutoCAD to *Union* two solids that do not occupy the same physical space and have two distinct volumes. In that case, the two shapes appear to be separate, but are treated by AutoCAD as one object (if you select one, both become highlighted).

TIP Occasionally you may intentionally or inadvertently use *Union* to combine two or more separate (not physically touching) objects. *Shell* and *Offset*, when used with solids containing holes, can also create two or more volumes from one solid. *Separate* can then be used to disconnect these into discrete independent solids. *Separate* <u>cannot</u> be used to disconnect or "break down" *Unioned*, *Subtracted*, or *Intersected* solids that form one volume.

> [Imprint/seParate solids/Shell/cLean/Check/Undo/eXit] <eXit>: **p**
> Select a 3D solid: **PICK**

Shell

Shell converts a solid into a thin-walled "shell." You first select a solid to shell, then specify faces to remove, and enter an offset distance. An example would be to convert a solid cube into a hollow box—the thickness of the walls equal the "shell offset distance." If no faces are removed from the selection set, the box would have no openings. Faces that are removed become open sides of the box.

```
[Imprint/seParate solids/Shell/cLean/Check/Undo/eXit] <eXit>: s
Select a 3D solid: PICK
Remove faces or [Undo/Add/ALL]: PICK
Remove faces or [Undo/Add/ALL]: PICK
Remove faces or [Undo/Add/ALL]: Enter
Enter the shell offset distance: (value)
Solid validation started.
Solid validation completed.
```

A positive value entered at the "shell offset distance" prompt creates the wall thickness outside of the original solid, whereas a negative value creates the wall thickness inside the existing boundary.

An example is shown in Figure 39-72. The left object (A) indicates the solid with all the selected faces highlighted and the faces removed from the selection set not highlighted. A negative offset distance is used to yield the shape shown on the right (B).

Don't expect to achieve the desired results the first time. As with most *Solidedit* options, the process of adding and removing faces is an inexact procedure since any edge you select can highlight either one of two faces.

FIGURE 39-72

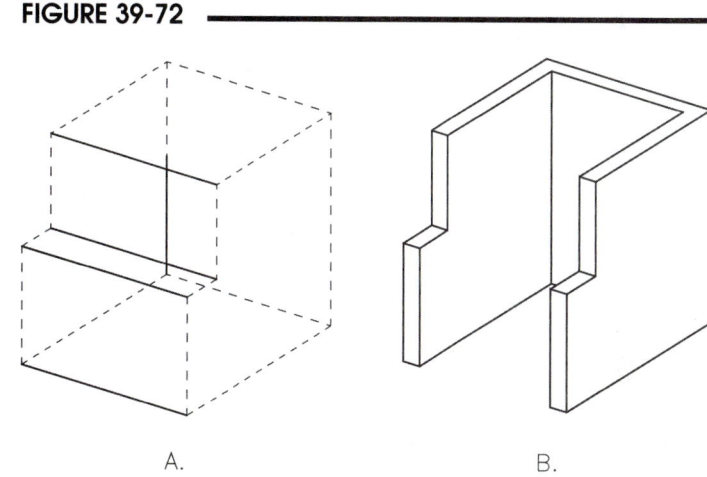

A.   B.

Check

Use this option to validate the 3D solid object as a valid ACIS solid, independent of the *SOLID-CHECK* system variable setting.

```
[Imprint/seParate solids/Shell/cLean/Check/Undo/eXit] <eXit>: c
Select a 3D solid: PICK
This object is a valid ACIS solid.
```

The *SOLIDCHECK* variable turns the automatic solid validation on and off for the current AutoCAD session. By default, *SOLIDCHECK* is set to 1, or on, to validate the solid. The *Solidedit* command displays the current status of the variable as shown in the command syntax below.

```
Command: solidedit
Solids editing automatic checking: SOLIDCHECK=0
Enter a solids editing option [Face/Edge/Body/Undo/eXit] <eXit>
```

Slice

	Pull-down Menu	Command (Type)	Alias (Type)	Short-cut	Screen (side) Menu	Tablet Menu
	Modify 3D Operations > Slice	Slice	SL	...	DRAW 2 SOLIDS Slice	...

The *Slice* command is a separate command (not part of the *Solidedit* options). However, *Slice* is a useful command for solids construction and editing for special cases and is therefore discussed in this chapter for those purposes.

A solid that contains oblique surfaces is a likely candidate for using *Slice*. For example, the simplest method of construction for the solid shown in Figure 39-73 is to create the rectilinear shapes (using *Box*), then use *Slice* to create the oblique surfaces.

With the *Slice* command you can cut a solid into two "halves" by defining a plane in any orientation. You can keep both sides of the resulting solid(s) or PICK one side to keep. (If in doubt as to which side to keep or how to select it, keep both sides, then use *Erase* to remove the unwanted side.)

    Command: **slice**
    Select objects to slice: **PICK**
    Select objects to slice: **Enter**
    Specify start point of slicing plane or [planar Object/Surface/Zaxis/View/XY/YZ/ZX/3points] <3points>:
    (option specific prompts)
    Specify a point on desired side or [keep Both sides] <Both>: **PICK** or **Enter**
    Command:

FIGURE 39-73

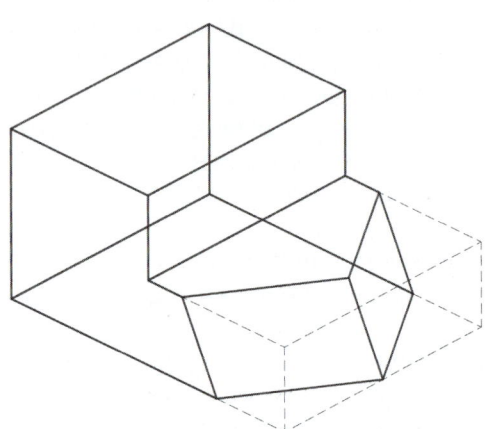

One oblique surface for the solid (Fig. 39-74) is created by first drawing a 2D *Line* on the solid, then defining a *3point* slicing plane using *OSNAPs* (*Endpoint*, *Endpoint* of *Line*, then *Midpoint* of solid edge). This figure shows the resulting solids after using *Slice* and keeping both sides. Duplicating this procedure for the other oblique surface creates the solid shown previously in Figure 39-73.

Slice can be used for many applications, such as an alternative to using *Wedge*, *Chamfer*, or *Solidedit*, *Taper*. Additionally, if you wanted to cut a section of a solid away, such as creating a notch or a half section (cutting away one quarter of an object), consider making multiple slicing planes with *Slice*, keeping all sides, *Erase* the unwanted section, then using *Union* to join the sections back into one solid.

FIGURE 39-74

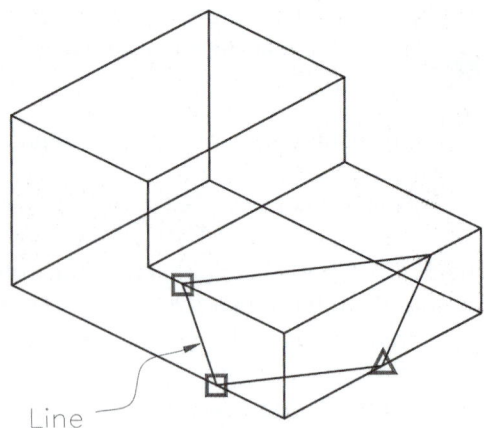

Line

Slice can also be used to create a true solid section. Slice cuts a solid or set of solids on a specified cutting plane. The original solid is converted to two solids. You have the option to keep both halves or only the half that you specify. Examine the object in Figure 39-75.

FIGURE 39-75

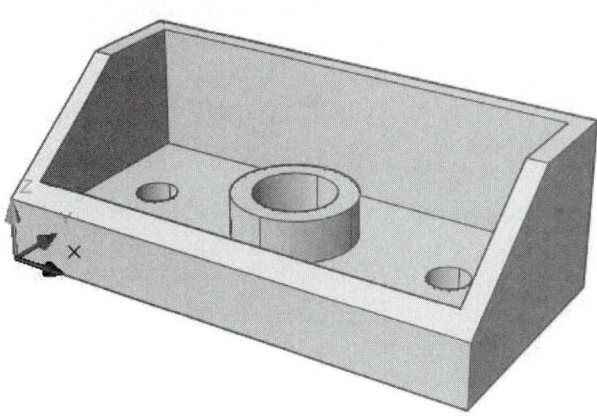

To create a true section of the example, use Slice. The ZX method is used to define the cutting plane midway through this solid. Next, the new solid to retain was specified by PICKing a point on that geometry. The resulting sectioned solid is shown in Figure 39-76. Note that the solid on the near side of the cutting plane was not retained.

Entering **B** at the "keep both sides" prompt retains the solids on both sides of the cutting plane. Otherwise, you can pick a point on either side of the plane to specify which half to <u>keep</u>.

FIGURE 39-76

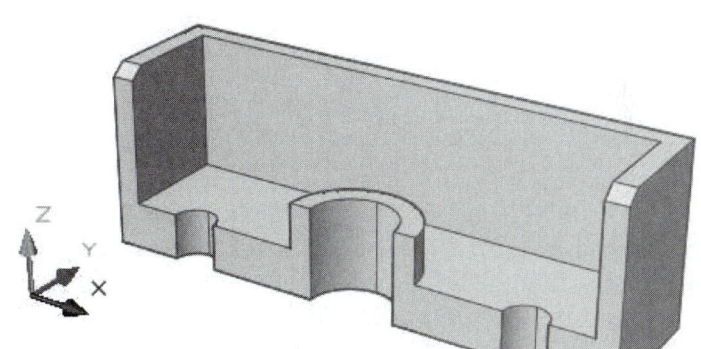

Surface option

Another very important and useful application for the Slice command is using surfaces to slice solids. That is, instead of slicing a solid using an imaginary plane (XY, YZ, ZX, or 3points options), you can select an existing surface as the slicing plane using the Surface option. The existing surface can be one created by commands such as Extrude, Sweep, Revolve, or Loft. Using this method to slice a solid allows you to create solids containing a variety of complex curved shapes. See "Slicing Solids with Surfaces" later in this chapter.

Presspull

Pull-down Menu	Command (Type)	Alias (Type)	Short-cut	Screen (side) Menu	Tablet Menu
...	Presspull	...	Ctrl+ Alt	...	...

Presspull is an unusual command because it accomplishes two tasks: 1) it converts a closed 2D shape such as a Circle, Pline, Polygon, etc. into an extruded solid, and 2) it performs a Boolean Union or Subtract based on whether you "pull" or "push" the extrusion. The closed 2D shape to extrude must lie in a plane. Presspull is simple to use since there are only two steps. The command prompts you only during the first step. Note that you can use Ctrl+Alt to activate this command.

2007

Command: **presspull**
Click inside bounded areas to press or pull. **PICK**
1 loop extracted.
1 Region created.
(At this point, "press" or "pull" the new extrusion, then specify a height.)
Command:

For example, assume that you create a closed 2D *Pline* on the top surface of a *Box* primitive as shown in Figure 39-77. Using the *Presspull* command, you select inside the closed *Pline* as shown in the figure. *Presspull* converts the closed *Pline* into a region.

FIGURE 39-77 ───────────────

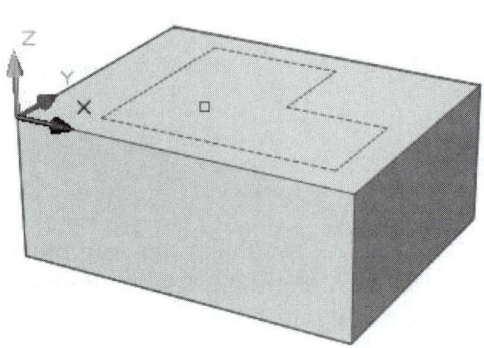

Next, you simply "pull" or "press" to create a new extruded solid from the 2D shape. In other words, *Presspull* converts the *Region* into an extruded solid when you drag the cursor out from or into the solid. In addition, *Presspull* automatically performs the appropriate Boolean operation (*Union* or *Subtract*). In the example shown in Figure 39-78, a new extruded solid is formed by pulling the new *Region* out from the surface of the *Box*. Once you PICK or enter a value to specify the height of the extrusion, *Presspull* automatically performs a *Union* to combine the new extrusion with the *Box*.

FIGURE 39-78 ───────────────

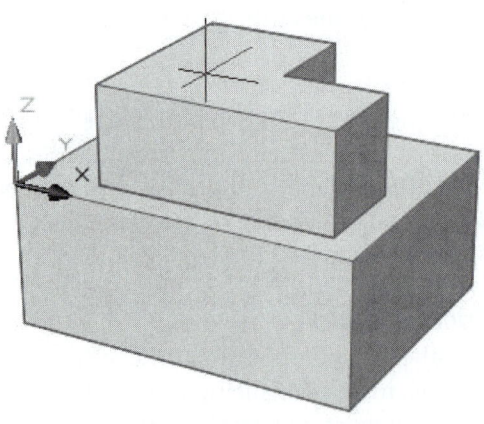

On the other hand, if you "press" the solid into the *Box*, *Presspull* performs a *Subtract*. In Figure 39-79, the new solid is formed by pulling the *Region* down into and through the *Box*. At this point the height has not been specified so no Boolean operation has been performed. Two solids exist at this point.

FIGURE 39-79 ───────────────

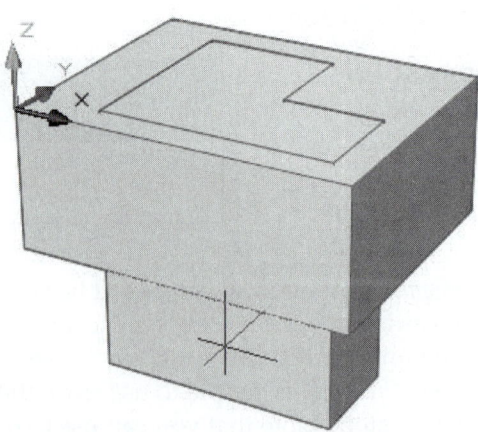

When you PICK or enter a value to define the height of the extrusion, the Boolean is automatically performed. In this case (Fig. 39-80), the new extruded solid is *Subtracted* from the *Box* to form a new composite solid.

FIGURE 39-80

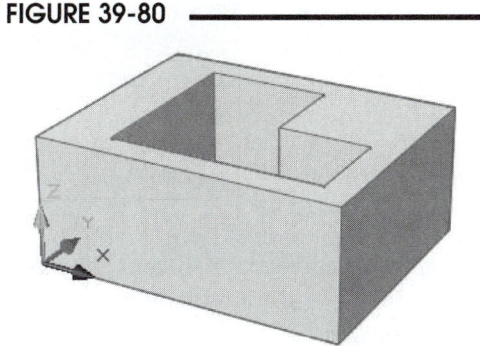

The 2D shapes that can be used with *Presspull* do not have to be single objects. *Presspull* works like the *Boundary* command so any areas bounded by other edges can be selected. You can press or pull any of the following types of bounded areas:

- Areas that could be selected using *Hatch* or *Boundary* by picking a point (with zero gap tolerance)
- Areas enclosed by crossing coplanar, linear geometry, including edges and geometry in blocks
- Closed *Plines*, *Regions*, 3D faces, and 2D solids that consist of coplanar vertices
- Areas created by geometry (including edges on faces) drawn coplanar to any face of a 3D solid

As another example, *Presspull* can be used to create an extruded solid not connected to another solid. In addition, the boundary does not have to be an object such as a closed *Pline* or *Region* since the needed *Region* is automatically created by *Presspull*.

For example, examine the 2D shapes (*Circle*, *Line*, and *Pline*) shown in Figure 39-81. Although two closed areas exist within these shapes, the shapes are not all closed and do not lie on the plane of an existing solid. Nevertheless, *Presspull* can be used to create a new extruded solid from one of the closed areas.

FIGURE 39-81

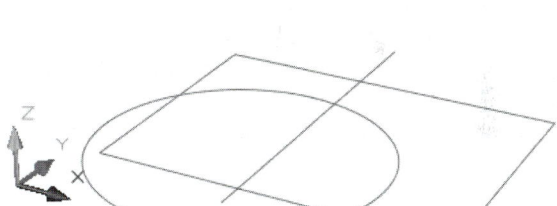

Invoking *Presspull*, then selecting inside one of the closed areas automatically creates a *Region* that can be pulled to form a new solid as shown in Figure 39-82. No Boolean operation is performed by *Presspull* since the new solid was not created from the plane of an existing solid.

FIGURE 39-82

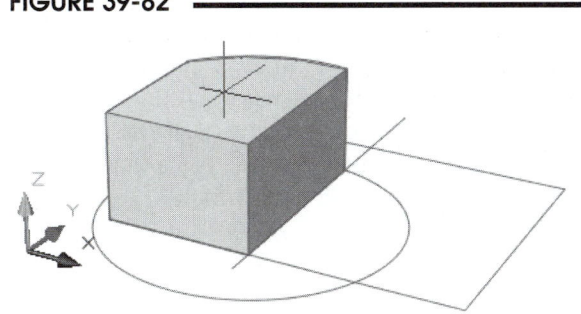

NOTE: Selection Preview must be on for *Presspull* to work. Use the *Selection* tab of the *Options* dialog box to control Selection Preview.

Sectionplane

Pull-down Menu	Command (Type)	Alias (Type)	Short-cut	Screen (side) Menu	Tablet Menu
Draw Modeling > Section Plane	*Sectionplane*	*SPANE*	...	...	...

Sectionplane creates a translucent clipping plane that can pass through a solid to reveal the inside features of the solid. The section plane is actually a section <u>object</u> that can be moved or revolved to dynamically display the solid's interior features. The section plane can contain "jogs" to create an offset cutting plane. The section object is manipulated using grips that appear on the section plane object.

> Command: **sectionplane**
> Select a face or any point to locate section line or [Draw section/Orthographic]: **PICK**
> Specify through point: **PICK**

You can select a face to create a section plane or specify two points to create the ends of the section plane. Selecting a face on a solid aligns the section object parallel to that face. For example, selecting the right face at the first prompt would create the section plane shown in Figure 39-83.

FIGURE 39-83 ────────────

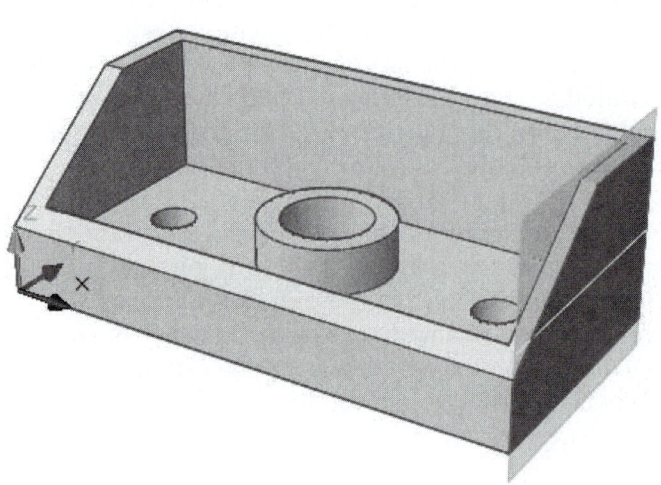

Selecting any point that is not on a face creates the first point of a section object independent of the solid. The second point creates the section object. For example, selecting two points just to the right and left of the composite solid shown in Figure 39-84 would produce a section plane crossing through the center of the solid. Note that the section plane is created perpendicular to the XY plane.

FIGURE 39-84 ────────────

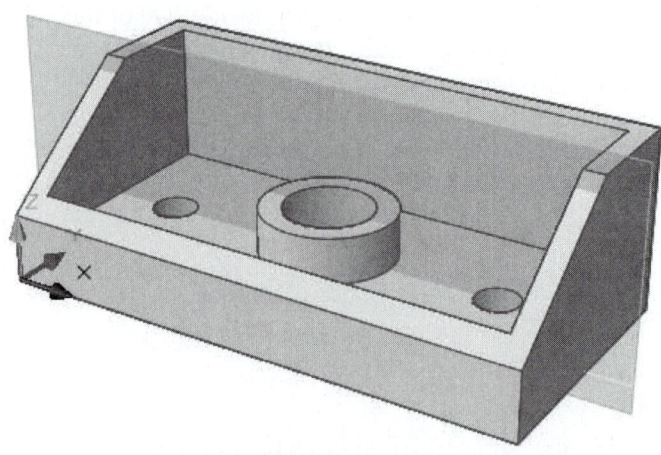

Section Plane Grips

Note that selecting the section plane object makes its grips appear (Fig. 39-85). Grips can be used to rotate the section plane to a different orientation, change the length of the section plane, flip the section plane, and drag the section plane into a new position. For example, Figure 39-85 displays the direction grip being dragged through the solid. PICKing the point at the center of the hole would move the section plane to that location, and the object would appear cut in half as shown in the figure.

FIGURE 39-85

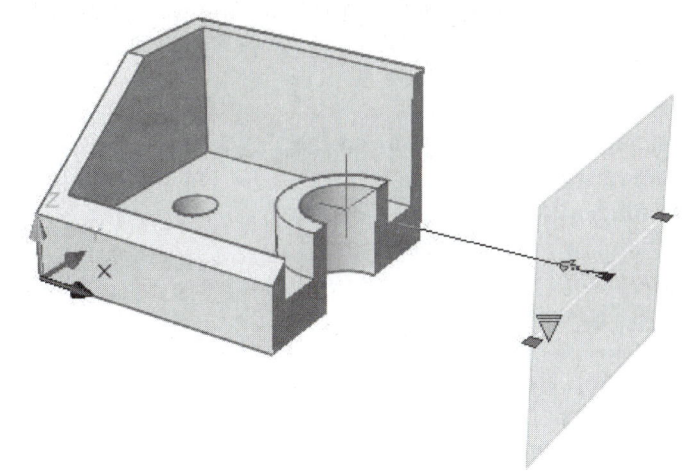

Once one of the square grips at the end of the section plane object is made hot, the normal five grip options appear (STRETCH, MOVE, ROTATE, SCALE, and MIRROR). For example, you can rotate the section plane by making a square grip hot, then cycling to the ROTATE option as shown in Figure 39-86.

FIGURE 39-86

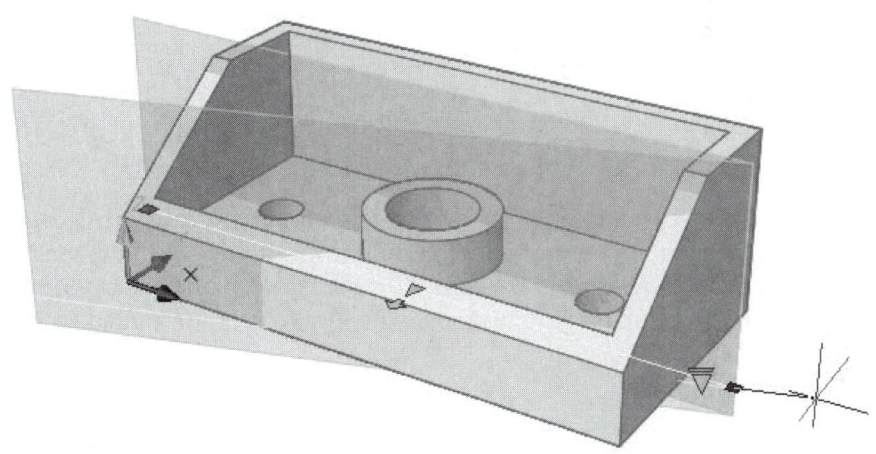

Live Sectioning

To make solid geometry on one side of the section plane appear invisible, the *Live Section* property for the section plane object must be set to *Yes*. When a section plane is created using the "face" option, *Live Section* property is set to *Yes* by default as shown previously in Figure 39-85. However, when you create a section plane by selecting points, *Live Section* is set to *No*. For example, note that in Figure 39-84 the section plane passes through the center of the solid, yet both sides of the plane are visible.

FIGURE 39-87

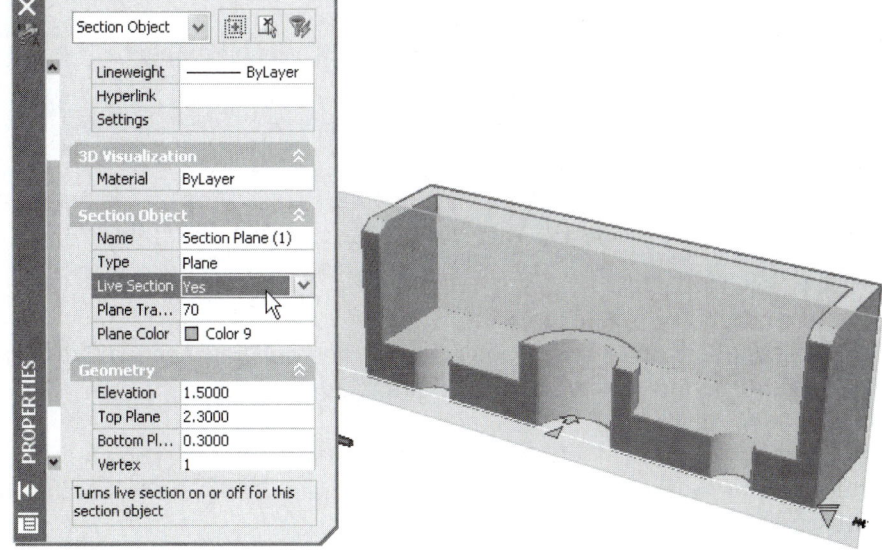

Using the *Properties* palette to set the *Live Section* to *Yes* makes the solid geometry on one side of the plane invisible as shown in Figure 39-86. The "flip" arrow grip determines which side of the section plane is invisible.

Draw Section

The *Draw section* option allows you to create an offset section plane. Offset section planes are used when "jogs" in the section line are needed to pass through multiple features of the solid as shown in Figure 39-88.

FIGURE 39-88

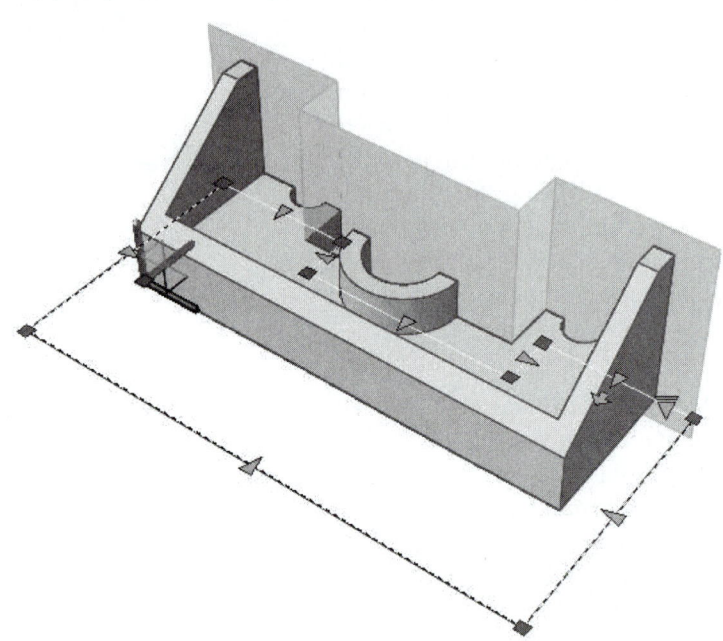

```
Command: sectionplane
Select face or any point to locate
section line or [Draw
section/Orthographic]: d
Specify start point: PICK
Specify next point: PICK
Specify next point or ENTER to com-
plete: Enter
Specify next point in direction of
section view: PICK
```

This option creates a section object with live sectioning turned off. Note that the section boundary is turned on by default (see "Section Object States"). If you select the section plane object so the grips appear, you can change several features of the section plane object. Stretch grips for each section of the section plane object allow you to change the location for the jogged sections (Fig. 39-88).

Orthographic

This option is very useful for creating typical sections. *Orthographic* aligns the section object to a specific orthographic orientation relative to the UCS. The resulting section contains all 3D objects in the model. If a single composite solid is located correctly with respect to the USC (lower-left corner at 0,0,0 and front side along the X axis), AutoCAD uses the center of volume of the solid (or multiple solids) to create the new section plane object.

FIGURE 39-89

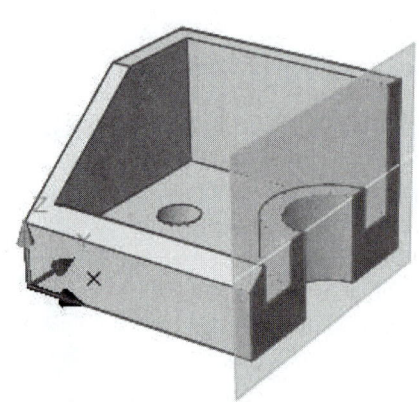

```
Command: sectionplane
Select face or any point to locate section line or [Draw section/Orthographic]: o
Align section to: [Front/Back/Top/Bottom/Left/Right]:
```

For example, using the *Right* option creates the section plane shown in Figure 39-89. Using the *Front* option produces a section plane as shown in Figure 39-87. Note that this option creates a section object with *Live Sectioning* turned on.

Section Object States

Section objects have three states:
Section Plane (a 2D plane), *Section Boundary* (a 2D box), and *Section Volume* (a 3D box). Grips that appear for each state allow you to make adjustments to the length, width, and height of the cutting area. Solids inside the cutting area are affected by the section plane object, whereas solids outside the area are not affected. Change the section state by selecting the state grip as shown in Figure 39-90.

FIGURE 39-90

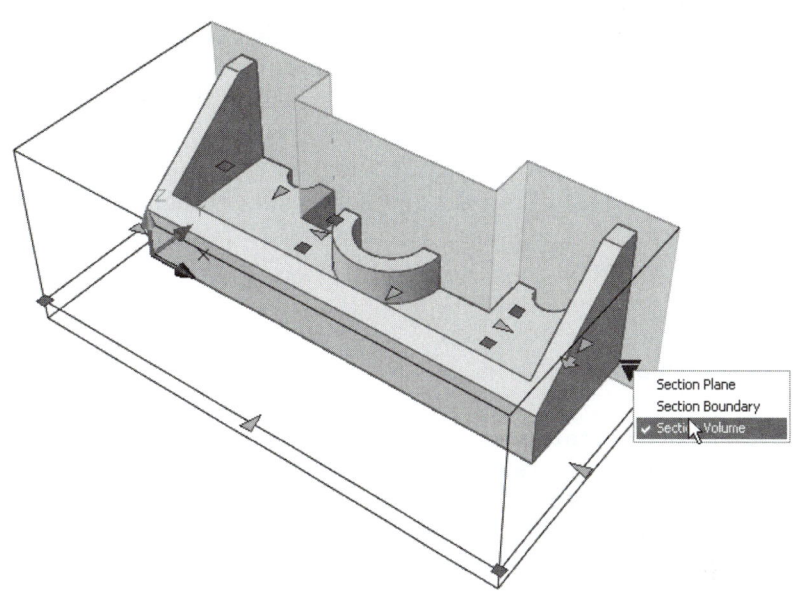

Section Plane	The cutting plane extends infinitely in all directions. Only the section line and transparent section plane indicator are displayed (Fig. 39-89).
Section Boundary	The cutting plane along the Z axis extends infinitely. A 2D box shows the XY extents of the cutting plane (Fig. 39-88).
Section Volume	A 3D box shows the extents of the cutting plane (Fig. 39-90).

Interfere

Pull-down Menu	Command (Type)	Alias (Type)	Short-cut	Screen (side) Menu	Tablet Menu
Modify *3D Operations >* *Interference Checking*	*Interfere*	*INF*	...	*DRAW 2* *SOLIDS* *Interfer*	...

In AutoCAD, unlike real life, it is possible to create two solids that occupy the same physical space. *Interfere* checks solids to determine whether or not they interfere (occupy the same space). If there is interference, *Interfere* reports the overlap and allows you to create a new solid from the interfering volume, if you desire. For example, consider the two solids shown in Figure 39-91. The two shapes fit together as an assembly. The locating pin on the part on the right should fit in the hole in the left part.

FIGURE 39-91

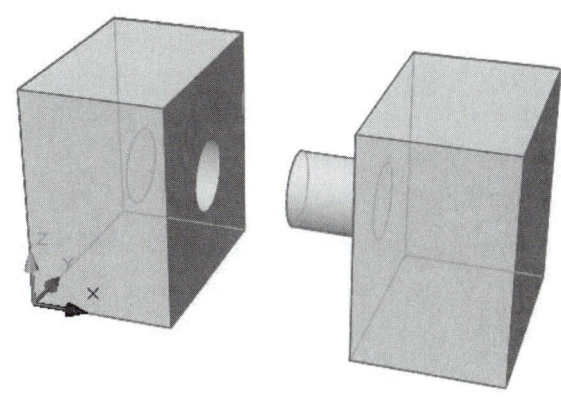

Sliding the parts together until the two vertical faces meet produces the assembly shown in Figure 39-92. There appears to be
some inconsistency in the assembly of the hole and the pin.
Either the pin extends beyond the hole (interference) or the hole
is deeper than necessary (no interference). Using *Interfere*, you
can check for an overlap. If no interference is found, AutoCAD
reports "Objects do not interfere." If an overlap is found,
Interfere automatically creates a new solid equal in shape and
size to the interfering volume. The original solids are not
changed in any way.

Normally, you specify two sets of solids for AutoCAD to check
against each other:

FIGURE 39-92

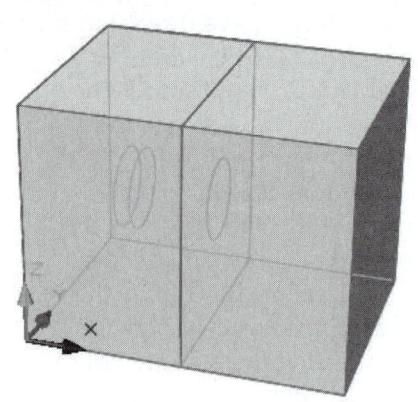

```
Command: interfere
Select first set of objects or [Nested selection/Settings]: PICK
Select first set of objects or [Nested selection/Settings]: Enter
Select second set of objects or [Nested selection/checK first set] <checK>: PICK
Select second set of objects or [Nested selection/checK first set] <checK>: Enter
```

When the possible interference has been checked, the
Interference Checking dialog box appears (Fig. 39-93). In
addition, the display in the Drawing Editor changes to
zoom in on the solid(s) of interference. If multiple inter-
ferences occur, you can use the *Previous* and *Next* options
to highlight each area. Remember that new solids are
created equal to the interfering volume. Use these new
solids to change and correct your solid model or delete
the solids by checking *Delete interference objects created on
close*.

FIGURE 39-93

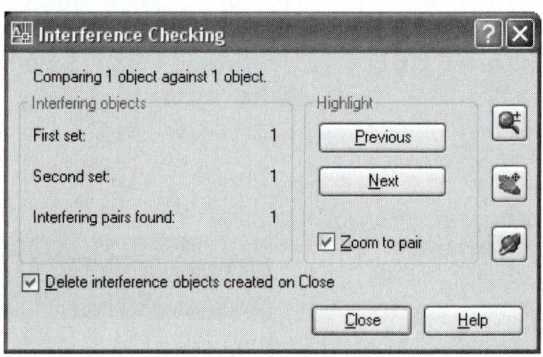

Nested Selection
Using this option during the prompts allows you to select
individual solid objects that are nested in Blocks and Xrefs.

Settings
If you use the *Settings* option during the first prompt, the
Interference Settings dialog box appears. Here you can change the
Visual Style for the interference objects and for the viewport.
Selecting Highlight interfering pair highlights the original solids,
whereas selecting *Highlight interference* highlights the solid(s) of
interference.

Check
Use this option to produce the *Interference Checking* dialog box.

FIGURE 39-94

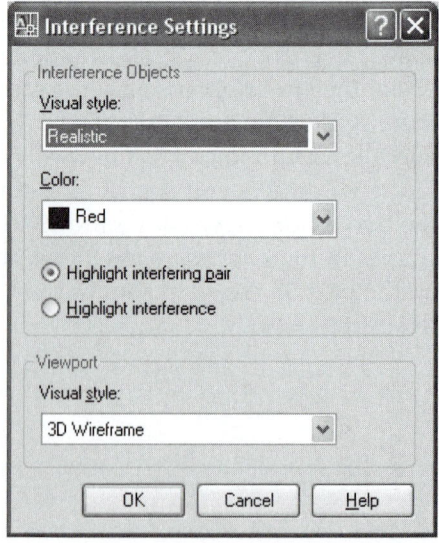

Imprint

	Pull-down Menu	Command (Type)	Alias (Type)	Short-cut	Screen (side) Menu	Tablet Menu
	Modify *Solid Editing >* *Imprint Edges*	*Imprint*	...	...	...	...

The *Imprint* command can be used to create new edges on the solids that can be edited with grips. As you remember, *Imprint* is an option of the *Solidedit* command; however, *Imprint* is also a stand-alone command and has direct access through the *Imprint* button on the *3D Make control panel*. The *Imprint* command operates exactly like the *Imprint* option of *Solidedit* (see previous coverage of *Solidedit*). That is, you can make an imprint from any *Arc, Circle, Line, Pline, Ellipse, Spline, Region,* or solid that lies in the plane of a solid.

The advantage of using *Imprint* to create new edges on a solid is that the new edges can be used to change the solid using grip editing methods. For example, assume you added *Line* to the top of an existing box, then used *Imprint* to "convert" the line to a new edge on the solid as shown in figure 39-95.

```
Command: imprint
Select a 3D solid: PICK
Select an object to imprint: PICK
Delete the source object [Yes/No] <N>: y
Select an object to imprint: Enter
Command:
```

FIGURE 39-95

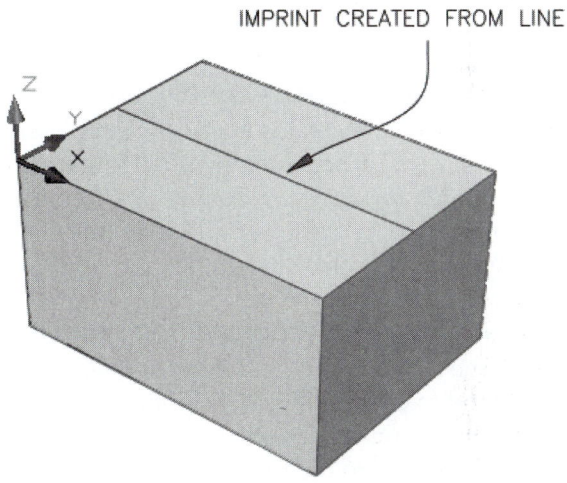

IMPRINT CREATED FROM LINE

Once the new edge is created, use the Ctrl key to hover the new edge, then PICK to select the edge only. Next, hover over the edge grip until the grip tool moves to the edge grip location, then select the desired axis on the grip tool. The solid shape can be changed using the grip tool to STRETCH the new edge. For example, the grip tool's Z axis is selected in Fig. 39-96 and the new edge is moved down to create the desired solid shape.

NOTE: For best results, use the move and rotate grip tool when editing a solid using an added edge created by *Imprint*.

FIGURE 39-96

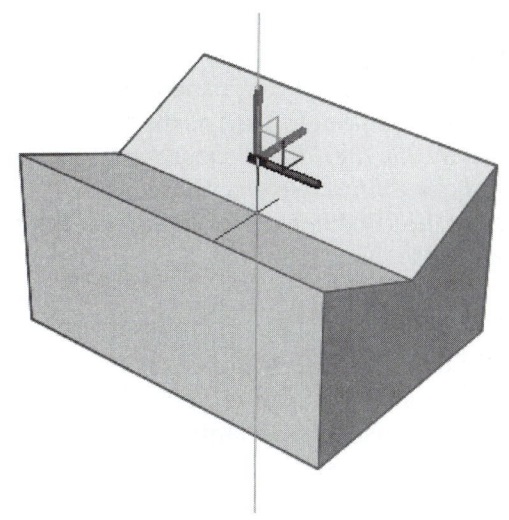

USING SURFACES WITH SOLIDS

Several commands were discussed in Chapter 38 that create solids from 2D shapes. These same commands can be used to create surfaces:

Extrude, Revolve, Sweep, Loft

When you use *Extrude, Revolve, Sweep,* or *Loft,* the determining factor in whether the resulting shape is a solid or a surface is if the original 2D shape is "open" or "closed." If you use a "closed" shape as the defining geometry, such as a *Circle, Ellipse, Polygon, Rectangle,* or other closed *Pline* or *Spline,* the resulting shape is a <u>solid</u>. However, if the shape to be extruded, revolved, swept, or lofted is an "open" shape (having two endpoints) such as a *Line, Arc,* open *Pline,* or open *Spline,* the resulting shape is a <u>surface</u>.

Extrude, Revolve, Sweep, or *Loft* generally are used to create curved surfaces. An additional command, *Planesurf,* can be used to create a planar (flat) surface. *Planesurf* uses a "closed" 2D area such as a closed *Pline* or points you PICK to create a surface.

Creating Surfaces Compatible with Solids

You must use the *Extrude, Revolve, Sweep, Loft,* or *Planesurf* command to create a surface usable with solids. The newly created surfaces can be used to cut through solids using *Slice* or can be converted to solids using the *Thicken* command. However, legacy AutoCAD surfaces, such as those created with *Rulesurf, Revsurf, Edgesurf, Tabsurf,* or other "polygon meshes," are not compatible with solids and cannot be used to create or edit solids (see Chapter 40, Surface Modeling, for more information on these legacy surfacing commands).

Extrude, Revolve, Sweep, and *Loft* Surface Examples

Using the *Extrude, Revolve, Sweep,* or *Loft* commands to create surfaces is similar to using the same command to create solids. The options within each command are practically the same whether you are intending to create surfaces or solids. The difference is the open or closed characteristic of the original geometry. Refer to the *Extrude, Revolve, Sweep,* and *Loft* commands in Chapter 38 for information on the options within each command.

A variety of surfaces can be created that are compatible with solids. Not only can you use the *Extrude, Revolve, Sweep,* or *Loft* commands, but a number of open 2D shapes can be used as the defining geometry such as a *Line, Arc, Pline, Spline,* partial *Ellipse* (about any 2D object with two end points). For example, consider the 2D shapes (*Arc, Pline,* and *Spline*) in Figure 39-97. Note that each of these shapes is "open."

FIGURE 39-97

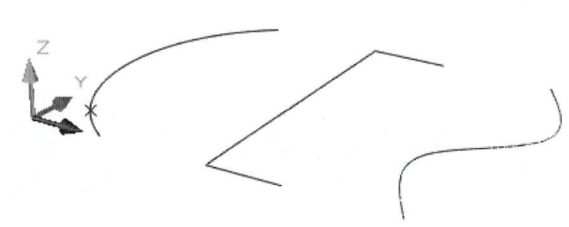

Using *Extrude* with each of these shapes would create surfaces as shown in Figure 39-98. These surfaces were created using the default option of specifying a height. Other possibilities exist using the *Direction* and *Path* options.

FIGURE 39-98

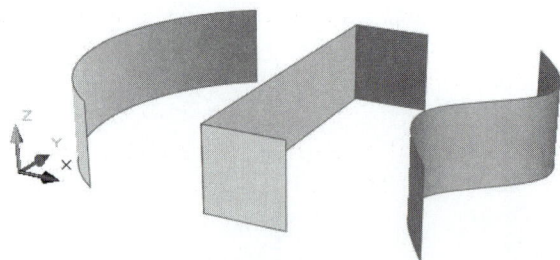

Using the original 2D shapes shown previously in Figure 39-97 and using the *Revolve* command for each of these shapes with an "angle of revolution" of 45 degrees would result in the surfaces shown in Figure 39-99.

FIGURE 39-99

The *Sweep* command could be used with the original 2D shapes and an additional *Arc* used as the path to generate the three new surfaces shown in Figure 39-100. Keep in mind other possibilities using the *Scale* and *Twist* options.

FIGURE 39-100

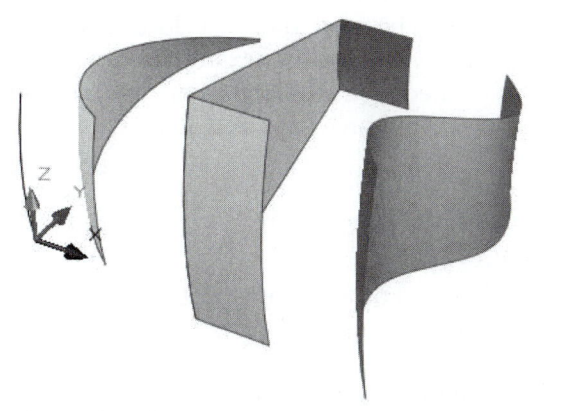

Remember that the *Loft* command requires several "cross sections" to loft through. In this case, three *Arcs* were used (Fig. 39-101, left) to create a lofted surface (right) using the *Cross sections* option. Other possibilities exist using the *Guides* and *Path* options.

FIGURE 39-101

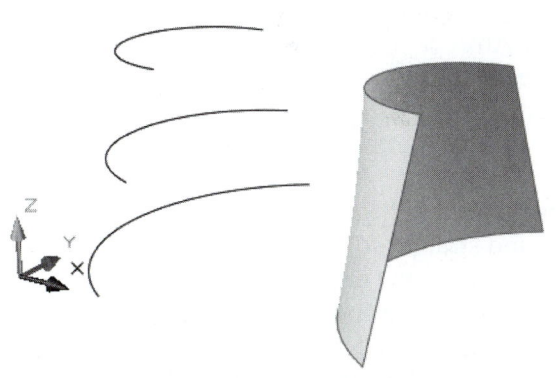

Planesurf

Pull-down Menu	Command (Type)	Alias (Type)	Short-cut	Screen (side) Menu	Tablet Menu
Draw *Modeling >* *Planar Surface*	*Planesurf*	...	...	*DRAW 2* *SURFACES* *Planar Surface*	...

The *Planesurf* command creates a planar (flat) surface. You can create the surface one of two ways: specify diagonal corners of a rectangle (default method) or select one or more objects that form an enclosed area (*Object* option). Using the default option, specify diagonal corners of a rectangle as shown in Figure 39-102. The prompts are as follows.

FIGURE 39-102

Command: **planesurf**
Specify first corner or [Object] <Object>: **PICK**
Specify other corner: **PICK**
Command:

Object

Use the *Object* option to create a planar surface by selecting one or more existing 2D objects that form a closed shape. The shape can be a <u>single</u> object such as a *Circle, Ellipse,* closed *Pline,* or closed *Spline.*

The *Object* option, however, can be used with a closed area and can be defined by <u>more than one object</u>. In this case, the objects must form a closed shape with no gaps or overlaps (similar to the criteria needed to form a *Joined Pline* or a *Region*). Objects that can be used to form a closed shape include *Line, Arc, Pline, Spline, Ellipse.* For example, Figure 39-103 displays a *Planesurf* created by the *Object* option. In this case, the closed area is defined by three *Lines* and an *Arc.*

FIGURE 39-103

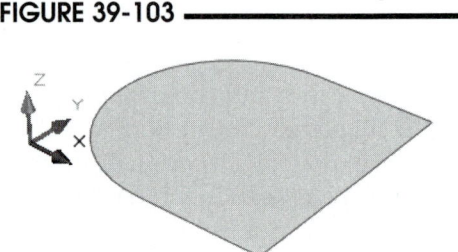

The *SURFU* and *SURFV* system variables control the number of lines displayed on the surface. The *DELOBJ* system variable controls whether the object(s) you select are automatically deleted when the surface is created or whether you are prompted to delete the object(s).

Slicing Solids with Surfaces

Combining surfaces with solids allows you to create solids with planar faces and <u>complex curve</u> faces. This is accomplished by using the *Slice* command with the *Surface* option to slice through solids using surfaces. Any shape of surface can be used, including complex curved surfaces such as a lofted surface. However, <u>only</u> surfaces created by *Extrude, Revolve, Sweep, Loft,* and *Planesurf* can be used with *Slice.* Legacy AutoCAD surfaces, such as those created with *Rulesurf, Revsurf, Edgesurf,* and *Tabsurf,* cannot be used for this purpose (see Chapter 40, Surface Modeling, for more information on these commands).

Slicing solids with surfaces is a fairly easy and straight-forward process. Simply ensure the desired surface to be used as the cutting plane is located correctly with respect to the solid, then use *Slice* with the *Surface* option and select the surface as the slicing plane.

FIGURE 39-104

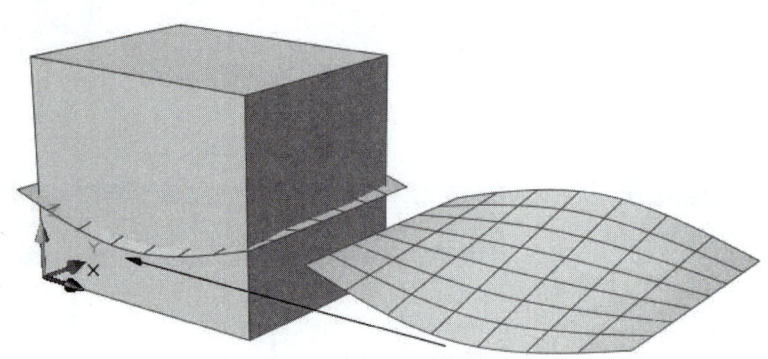

For example, assume you wanted to *Slice* a *Box* with a lofted surface. Figure 39-104 illustrates the lofted surface moved into position so the *Box* passes through the surface.

Command: **slice**
Select objects to slice: **PICK**
Select objects to slice: **Enter**
Specify start point of slicing plane or [planar Object/Surface/Zaxis/View/XY/YZ/ZX/3points] <3points>: **s**
Select a surface: **PICK**
Select solid to keep or [keep Both sides] <Both>: **PICK**
Command:

Following the command prompts shown above, the *Slice* command is invoked and the *Surface* option is used to specify the lofted surface as the slicing plane. Keeping only the bottom half of the *Box*, the resulting solid is shown in Figure 39-105.

FIGURE 39-105

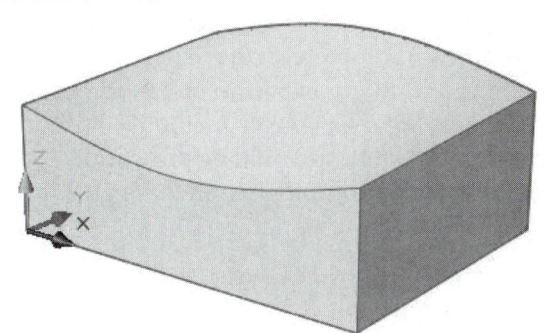

Slicing Solids with Surfaces Example

Using *Slice* with the *Surface* option is an extremely powerful technique for generating contoured shapes. For example, examine the series of illustrations shown next that depict the creation of a toy car body (courtesy of Autodesk and J. D. Mather).

The car body begins with a *Box* (Fig. 39-106). Two lofted surfaces are generated to represent the contour of the front and rear panels (for each surface two horizontal curves, top and bottom, are lofted using connecting guides connecting the curves). *Slice* is then used with the *Surface* option to shape the front and rear of the car.

FIGURE 39-106

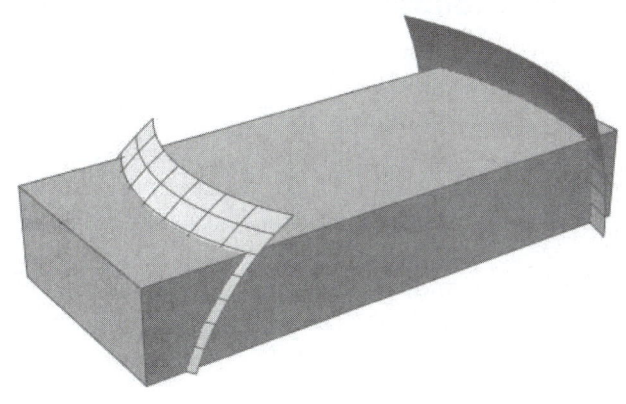

After the front and rear are sliced, two additional surfaces are created (Fig. 29-107). The surfaces are created using *Sweep* (by sweeping an almost vertical *Arc* along a slightly curved horizontal path). Next, *Slice* is used to create the contour for the left and right sides of the car.

FIGURE 39-107

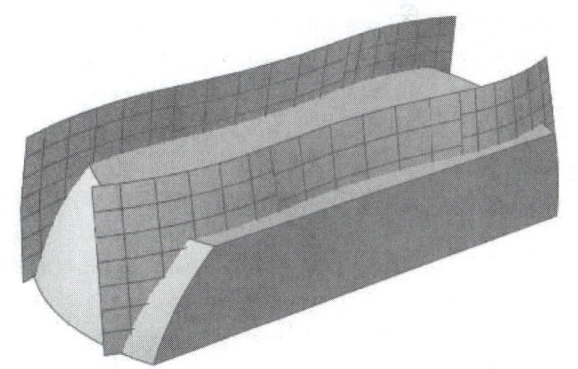

When the side contours are created, another lofted surface is created (Fig. 39-108). This new surface is used to slice away the top portion of the car body to create a smooth, sloping contour.

FIGURE 39-108

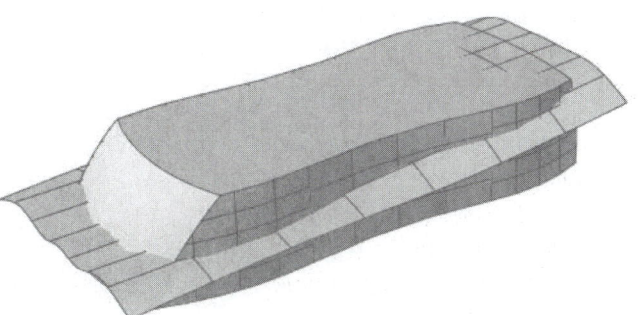

Finally, the *Radius* command is used to round all the corners (Fig. 39-109). The resulting shape of the car body is completely contoured—not something you might expect to be created using Constructive Solid Geometry methods. Without studying these steps it would be difficult to imagine how this complex geometry began as a *Box*.

FIGURE 39-109

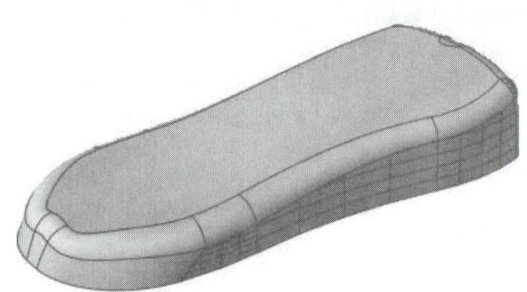

Creating Solids from Surfaces

You can convert a surface into a solid using one of several commands. For example, *Extrude* and *Presspull* can be used to convert a surface into a solid by adding an extrusion height. Using *Extrude* and *Presspull*, however, require the defining surfaces to be planar (flat) surfaces. In addition to using surfaces as defining geometry, remember that *Extrude* can use any single closed 2D object (such as a closed *Pline*) to create a solid. Also, *Presspull* requires only a closed area defined by any edges forming a boundary, similar to the criteria for a *Boundary* or *Hatch*. (See *Extrude* in Chapter 38 and *Presspull* in this chapter.)

Another command, *Thicken*, can convert surfaces into solids by adding a "thickness" value. The advantage of using *Thicken* is that any surface, planar or contoured, can be converted to a solid. *Thicken* is explained next.

Using the *Extrude, Presspull,* and *Thicken* commands, <u>only</u> surfaces created by *Extrude, Revolve, Sweep, Loft,* and *Planesurf* can be converted into solids.

NOTE: Most "legacy" AutoCAD surfaces, such as those created with *Rulesurf, Revsurf, Edgesurf, Tabsurf,* and other polygon meshes, cannot be converted to solids by any method. Only a few specific types of legacy 3D objects can be converted into solids using the *Convtosolid* command (see Chapter 40, Surface Modeling, for more information on these commands).

Thicken

Pull-down Menu	Command (Type)	Alias (Type)	Short-cut	Screen (side) Menu	Tablet Menu
Modify *3D Operations >* *Thicken*	*Thicken*	...	...	DRAW 2 SOLIDS *Thicken*	...

Thicken can add thickness to any surface created by *Extrude, Revolve, Sweep, Loft,* and *Planesurf*. The process of adding thickness converts the surface into a solid. Similar to adding height through extrusion, the thickness is applied uniformly to the specified surface. In the case of a contoured surface, thickness is added in a direction perpendicular to each point on the surface. The command is simple to use.

```
Command: thicken
Select surfaces to thicken: PICK
Select surfaces to thicken: Enter
Specify thickness <0.0000>: (value)
Command:
```

Note that *Thickness* can be applied to planar surfaces (not shown) as well as contoured surfaces. Figure 39-110 displays the conversion of three contoured surfaces into solids. The original surfaces were created using *Extrude* to create the surfaces from "open" 2D shapes (see previous Fig. 39-98 and related discussion). The resulting solids shown here are planar and single-curved solids.

Another example illustrates the effect of using *Thicken* to convert single- and double-curved surfaces into compound curved solids. The original surfaces were created using *Revolve* to create the surfaces from "open" 2D shapes (see previous Fig. 39-99 and related explanation).

The *DELOBJ* system variable controls whether the object(s) you select as the defining surfaces are automatically deleted when the solid is created or whether you are prompted to delete the object(s).

FIGURE 39-110

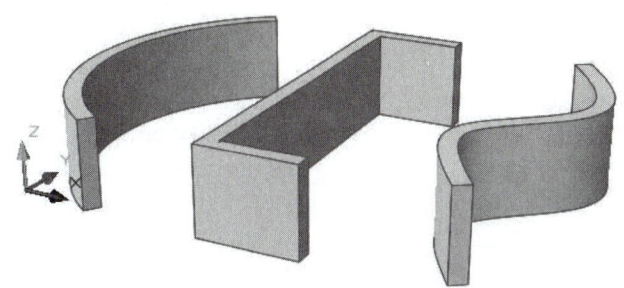

FIGURE 39-111

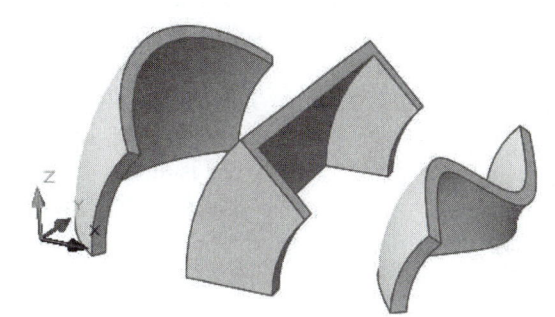

2007

USING SOLIDS DATA

AutoCAD offers a few commands that allow you to extract, import, and export data from solid model geometry. The *Massprop* command calculates a variety of properties for selected solids. Commands available for exporting and importing solid model geometric data for use with other programs are *Stlout*, *Acisin*, and *Acisout*.

Massprop

Pull-down Menu	Command (Type)	Alias (Type)	Short-cut	Screen (side) Menu	Tablet Menu
Tools Inquiry > Region, Mass Properties	Massprop	...	...	TOOLS 1 Massprop	U,7

Since solid models define a complete description of the geometry, they are ideal for mass properties analysis. The *Massprop* command automatically computes a variety of mass properties.

Mass properties are useful for a variety of applications. The data generated by the *Massprop* command can be saved to an .MPR file for future exportation in order to develop bills of material, stress analysis, kinematics studies, and dynamics analysis (Fig. 39-112).

FIGURE 39-112

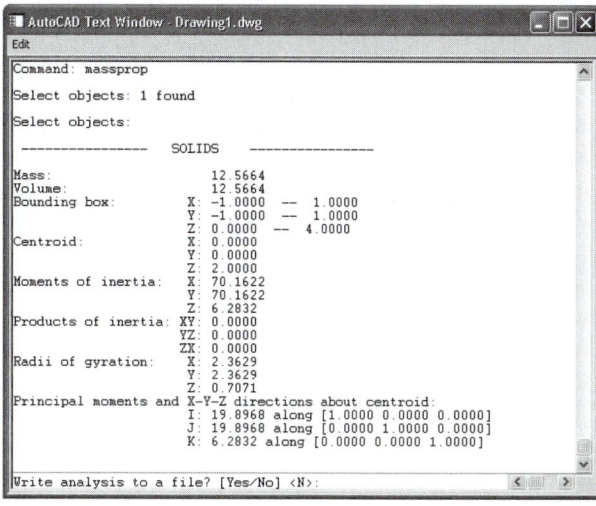

Applying the *Massprop* command to a solid model produces a text screen displaying the following list of calculations:

Mass

Mass is the quantity of matter that a solid contains. Mass is determined by density of the material and volume of the solid. Mass is not dependent on gravity and, therefore, different from but proportional to weight. Mass is also considered a measure of a solid's resistance to linear acceleration (overcoming inertia).

Volume

This value specifies the amount of space occupied by the solid.

Bounding Box

These lengths specify the extreme width, depth, and height of the selected solid.

Centroid

The centroid is the geometrical center of the solid. Assuming the solid is composed of material that is homogeneous (uniform density), the centroid is also considered the center of mass and center of gravity. Therefore, the solid can be balanced when supported only at this point.

Moments of Inertia

Moments convey how the mass is distributed around the X, Y, and Z axes of the current coordinate system. These values are a measure of a solid's resistance to <u>angular</u> acceleration (mass is a measure of a solid's resistance to <u>linear</u> acceleration). Moments of inertia are helpful for stress computations.

Products of Inertia

These values specify the solid's resistance to <u>angular</u> acceleration with respect to two axes at a time (XY, YZ, or ZX). Products of inertia are also useful for stress analysis.

Radii of Gyration

If the object were a concentrated solid mass without holes or other features, the radii of gyration represent these theoretical dimensions (radius about each axis) such that the same moments of inertia would be computed.

Principal Moments and X, Y, Z Directions

In structural mechanics, it is sometimes important to determine the orientation of the axes about which the moments of inertia are at a maximum. When the moments of inertia about centroidal axes reach a maximum, the products of inertia become zero. These particular axes are called the principal axes, and the corresponding moments of inertia with respect to these axes are the principal moments (about the centroid).

Notice the "Mass:" is reported as having the same value as "Volume." This is because AutoCAD solids cannot have material characteristics assigned to them, so AutoCAD assumes a density value of 1. To calculate the mass of a selected solid in AutoCAD, use a reference guide (such as a machinist's handbook) to find the material density and multiply the value times the reported volume (mass = volume x density).

Stlout

Pull-down Menu	Command (Type)	Alias (Type)	Short-cut	Screen (side) Menu	Tablet Menu
File Export... *.stl	Stlout	...	...	FILE Export *.stl	...

The *Stlout* command converts an AutoCAD (ACIS) solid model to a file suitable for use with rapid proto-typing apparatus. Use *Stlout* if you want to use an ACIS solid to create a prototype part using stereo lithography or sintering technology. This technology reads the CAD data and creates a part by using a laser to solidify microthin layers of plastic or wax polymer or powdered metal. A complex 3D prototype can be created from a CAD drawing in a matter of hours.

Stlout writes an ASCII or binary .STL file from an AutoCAD ACIS solid model. The model must reside entirely within the positive X,Y,Z octant of the WCS (all part geometry coordinates must be positive):

```
Command: stlout
Select a single solid for STL output:
elect objects: PICK
Create a binary STL file ? [Yes/No] <Y>: Enter N to create an ASCII rather than binary file.
Command:
```

AutoCAD displays the *Create .STL File* dialog box for you to designate a name and path for the file to create.

When you design parts with AutoCAD to be generated by stereolithography apparatus, it is a good idea to create the profile view (that which contains the most complex geometry) parallel to the XY plane. In this way, the aliasing (stair-step effect) on angled or curved surfaces, caused by incremental passes of the laser, is minimized.

Acisin

Pull-down Menu	Command (Type)	Alias (Type)	Short-cut	Screen (side) Menu	Tablet Menu
Insert ACIS File...	Acisin	...	...	INSERT ACISin	...

The *Acisin* command imports an ACIS solid model stored in a .SAT (ASCII) file format. This utility can be used to import .SAT files describing a solid model created by AutoCAD or other CAD systems that create ACIS solid models. The *Select ACIS File* dialog box appears, allowing you to select the desired file. AutoCAD reads the file and builds the model in the current drawing.

Acisout

Pull-down Menu	Command (Type)	Alias (Type)	Short-cut	Screen (side) Menu	Tablet Menu
File Export... *.sat	Acisout	...	...	FILE Export *.sat	...

This utility is used to export an ACIS solid model to a .SAT (ASCII) file format that can later be read by AutoCAD or other CAD systems utilizing the ACIS modeler. The *Create ACIS File* dialog box is used to define the desired name and path for the file to be created.

CHAPTER EXERCISES

1. **Editing Primitives using** *Properties*

    Open the CH38EX3 drawing that you created in Chapter 38 Exercises. Invoke the *Properties* palette. Hold down the **Ctrl** key while hovering and selecting the hole (*Cylinder* primitive) so properties for the primitive appear in the palette (Fig. 39-113). Edit the hole to have a new *Radius* of **1.0000** and a *Height* of **1.0000**. (Note that the "height" is the dimension in the negative Z direction in this case.) Use *SaveAs* and name the drawing **CH39EX1**.

FIGURE 39-113

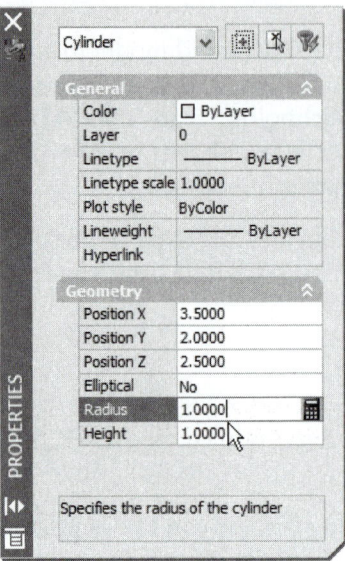

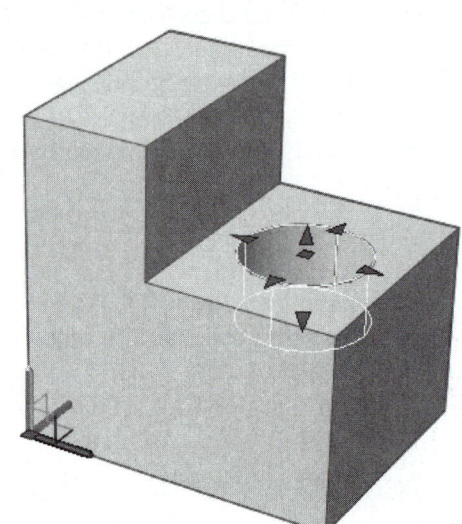

2. **Editing Primitives using** *Properties*

    If not already open, *Open* the **CH39EX1** drawing that you just created in the previous exercise. Invoke the *Properties* palette again. This time, hover and select the to *Box* primitive so its properties appear in the palette (Fig. 39-114). Edit the *Width* dimension of the *Box* to a value of **3.0000**. Use *SaveAs* and name the drawing **CH39EX2**.

FIGURE 39-114

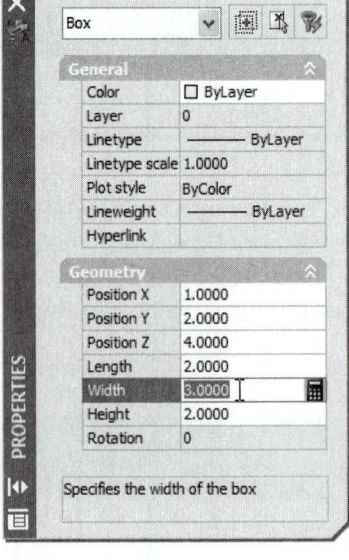

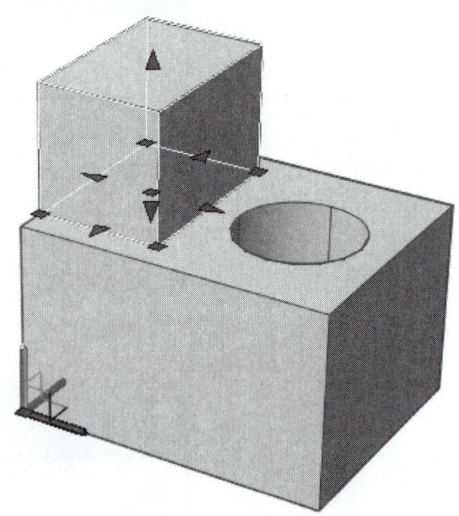

3. **Editing Primitives using Grips**

FIGURE 39-115

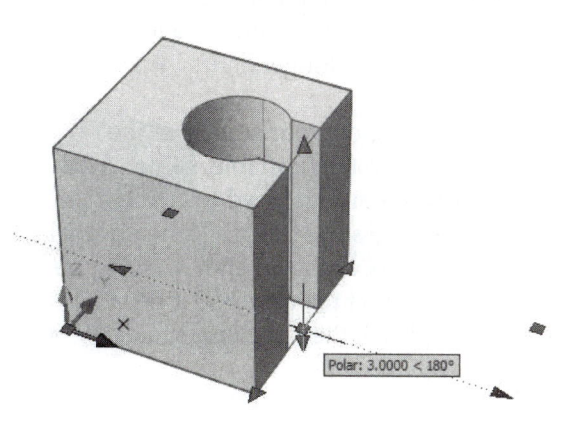

Open the CH38EX5 drawing that you created in Chapter 38 Exercises. Turn on *POLAR*. Hold down the **Ctrl** key while hovering and selecting the *Box* primitive. Select the grip in the center of the base on the right side to make it hot, then stretch the grip 3 units toward the center of the solid (Fig. 39-115). This action effectively "cuts" the solid in half. Use *SaveAs* and name the drawing **CH39EX3**.

4. **Editing Primitives using Grips**

FIGURE 39-116

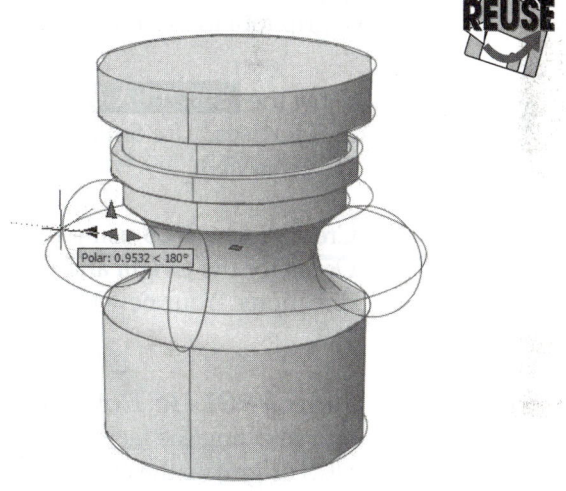

Open the CH38EX6 drawing that you created in Chapter 38 Exercises. Ensure *POLAR* is on. Hold down the **Ctrl** key while hovering and selecting the *Torus* primitive (Fig. 39-116). Select the appropriate grip to stretch the <u>tube</u> radius, not the torus radius. Enter a value of **.75** to increase the total tube radius (originally .50) to 1.25. Invoke the *Properties* palette to ensure the new *Tube Radius* is a value of **1.2500**. Use *SaveAs* and name the drawing **CH39EX4**.

5. **Editing Subobjects using the Move Grip Tool**

FIGURE 39-117

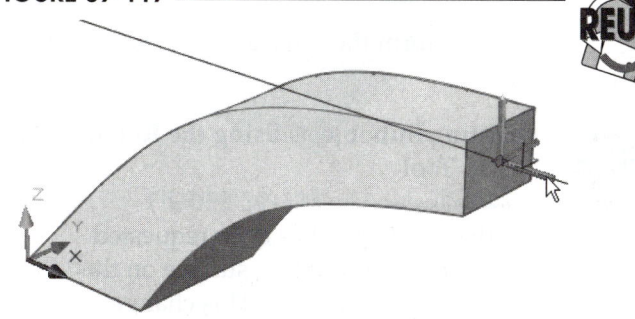

For this exercise, assume you are working as a designer of kitchen and bathroom fixtures. You have been contracted to design a kitchen faucet based on the style of a recently designed bathroom faucet. You can easily alter the existing bathroom design using the move grip tool to extend the faucet for use with a kitchen sink.

A. *Open* the **FAUCET1** drawing that you created in Chapter 38 Exercises. Use *SaveAs* and assign the name **FAUCETK**.

B. Turn on *POLAR*. Hold down the **Ctrl** key while hovering and selecting the vertical <u>face</u> at the end of the faucet (Fig. 39-117). A small circular grip should appear on the face. Hover the cursor over the grip until the move grip tool relocates to the grip (Fig. 39-117). Select the X axis of the move grip tool.

B. Invoke *Sectionplane*. Use the *Orthographic* option and align the section to the *Right* side. A new section plane should appear as shown in Figure 39-124.

FIGURE 39-124 ———

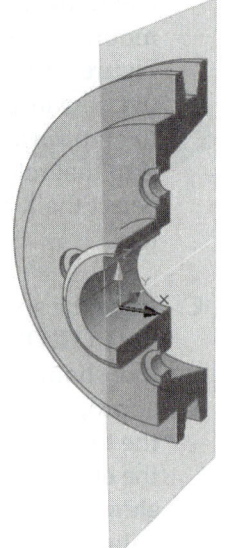

C. Experiment with the section plane by selecting it to activate the grips, then using the move grip tool to pass the section plane through the pulley to reveal its inner features.

D. Invoke the *Properties* palette, select the section plane object, and set the *Live Section* property to *No*. The section plane should appear in the drawing, but the pulley should not appear cut. *Save* the drawing.

10. **Using** *Sectionplane*

A. *Open* the **SADL-SL** drawing from the Chapter 38 Exercises. Use *SaveAs* and assign the name **SADL-SC**.

FIGURE 39-125 ————————————

B. Invoke *Sectionplane* to create a new section plane like that shown in Figure 39-125.

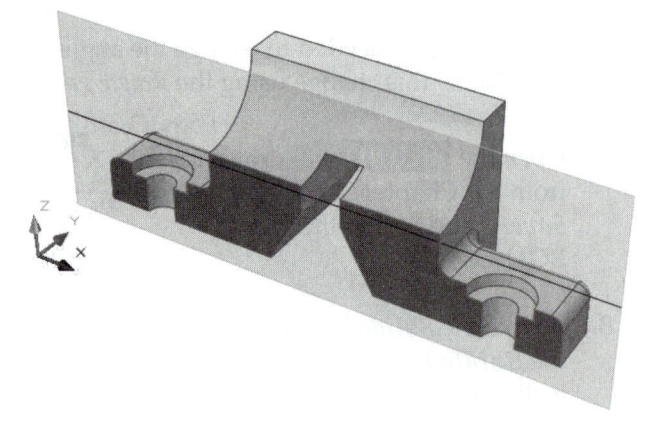

C. If you have problems, ensure your UCS is identical to that shown in Figure 39-125. Also check to make sure *Live Sectioning* is on.

D. Experiment with the section plane by selecting it to activate the grips, then using the move grip tool to pass the section plane through the pulley to reveal its inner features. *Save* the drawing.

11. **Using Surfaces to** *Slice* **Solids**

A. In this exercise you will create a soap dispenser. Begin a *New* drawing using the **ACAD3D .DWT**. Draw an *Ellipse* on the XY plane with the center at 0,0 and having a major axis of 3 (axis endpoint distance of 1.5 along the X axis) and a minor axis of 1.25 (axis endpoint distance of .625 along the Y axis). *Extrude* the *Ellipse* with a height of **3** units. *Save* the drawing as **SOAPDISP**.

FIGURE 39-126 ————————

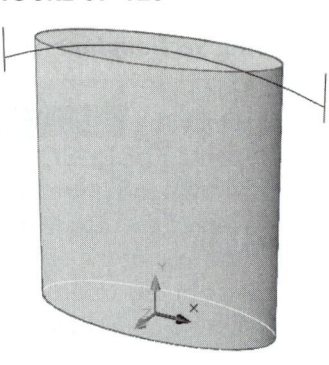

B. Make a new *UCS* by rotating about the *X* axis **90** degrees as shown in Figure 39-126. Turn *Grid Snap* on. *Zoom* out and *Pan* the elliptical solid to the top of the screen. Next, draw a *Circle* on the UCS XY (vertical) plane with the *Center* at **0,-3,0** and a *Radius* of **6** (so it passes through the top center of the solid). Finally draw a *Line* on each side of the solid and *Trim* the *Circle* using these lines. Your model should look like that in Figure 39-126. *Erase* the 2 *Lines* you used for trimming.

C. Now you will repeat the process from step B to create another arc but in a different plane. Make a new **UCS** by rotating about the *Y* axis **90** degrees as shown in Figure 39-127. Next, draw another *Circle* on the new UCS XY (vertical) plane again with the *Center* at **0,-3,0** and a *Radius* of **6** (so it also passes through the top center of the solid and the first *Arc*). Use the same method as before to **Trim** the *Circle* on each side of the solid. *Erase* the 2 *Lines* you used for trimming. Your model should look like that in Figure 39-127.

FIGURE 39-127

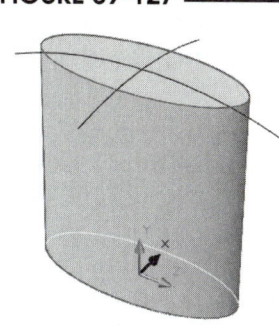

D. Set the **UCS** back to **World**. Create a surface using *Sweep*. Select the second *Arc* you created (along the short axis of the elliptical primitive) as the object to sweep. Set the **Alignment** option to **No**. Select the second *Arc* as the sweep path. Your new swept surface should look like that in Figure 39-128. You may have to **Erase** one or both *Arcs* used to define the sweep (depending on your *DELOBJ* setting). *Save* the drawing.

FIGURE 39-128

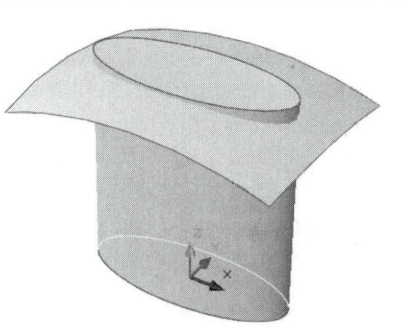

E. Use *Slice* to slice the top portion of the elliptical solid. Use the *Surface* option and select the swept surface as the slicing plane. Keep only the bottom of the solid. Create a new layer named **Surface**, change the swept surface to that layer, then turn the layer *Off*. Your solid should look like that in Figure 39-129 (*Isolines* are turned on in this figure to emphasize the curved top surface).

FIGURE 39-129

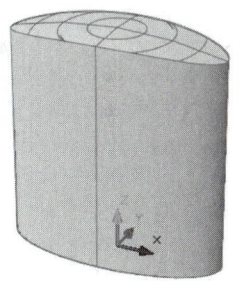

F. Draw a *Circle* with the *Center* at **0,0,0** (the center of the base of the solid) and a *Radius* of **.4**. *Extrude* the *Circle* to a *Height* of **3.6**. **Union** the two solids. Use *Fillet* with a *Radius* of **.1** to round the bottom elliptical edge and the top elliptical edge. *Fillet* the intersection along the top of the elliptical shape and the base of the cylindrical top with a *Radius* of **.05**. Finally, use the *Shell* option of *Solidedit* to "hollow out" the soap dispenser. At the "Remove faces" prompt, select only the top circular face (you must select the <u>edge</u> of the top circle, then hold down Shift to deselect the vertical cylindrical face). Enter a shell offset distance of **.1**. Your soap dispenser should look like that in Figure 39-130. *Save* the drawing.

FIGURE 39-130

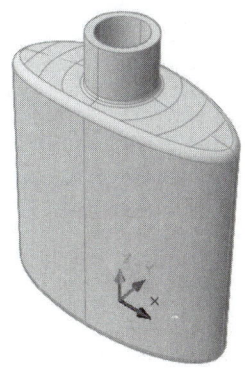

G. Optional: Create a pump for the top of the bottle to dispense the soap.

12. **Creating Surfaces and *Slicing* Solids**

A. In this exercise you will create a computer mouse. Begin a *New* drawing using the **ACAD3D .DWT**. Create a new *UCS* by rotating about the *X* axis **90** degrees. Generate a *Plan* view. Draw a *Spline* shaped like that in Figure 39-131. Note that the bottom-left end of the *Spline* is at 0,0,0. *Save* the drawing as **MOUSE**.

FIGURE 39-131

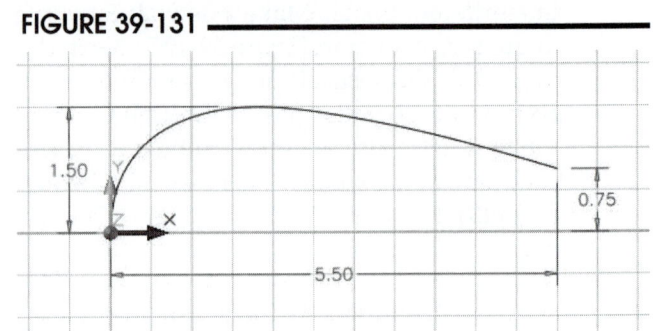

B. Set the *UCS* to *World*. Use *3Dorbit* to change back to a 3D view. Use *Copy* to make a copy of the *Spline*. Specify **0,0,0** as the base point and **0,-1.25,0** as the second point of displacement (*Polar Tracking* may help with this).

C. Use the *Scale* command to scale the newly copied *Spline*. Specify **2.5,-1.25,0** as the base point and **.75** as the scale factor. Next, *Copy* this *Spline* to the other side of the original *Spline* a total distance of **2.5** units. Your drawing should look similar to Figure 39-132.

FIGURE 39-132

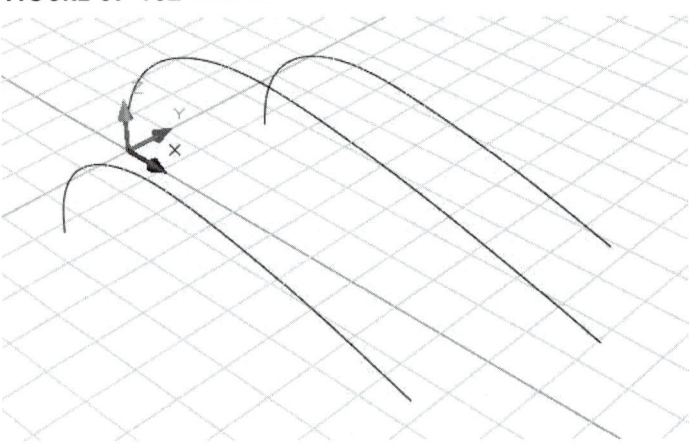

D. Use *Loft* to create a surface connecting the three splines using the *cross sections only* option. When *the Loft Settings* dialog box appears, use the *Smooth fit* option.

E. Create a *Box* with diagonal corners on the XY plane at **0,-1.25** and **4.75,1.25**. Specify **2.0** as the height. *Slice* the *Box* using the *Surface* option. Keep only the bottom of the *Box*. Create a *New Layer* named **Surfaces**, change the lofted surface to the new layer, then *Freeze* the layer. Next, use *3Dorbit* to examine your surface from different viewing angles. Viewed from the back, your new solid should look similar to that in Figure 39-133.

FIGURE 39-133

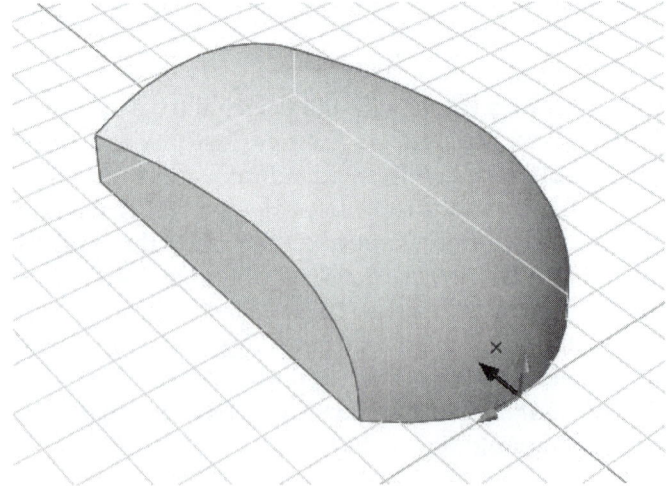

F. Now you need three surfaces to slice away the outer edges of the mouse. Begin by creating a long, sweeping *Arc* just inside the front edge of the base of the mouse such that the closest part of the arc is approximately .125 units from the front edge. Ensure the *Arc* is symmetrical and extends equally past each side of the mouse. Draw an almost vertical *Line* from the *Midpoint* of the *Arc* to the *Midpoint* of the top outside edge of the mouse. This *Line* should slope slightly outwards to the top of the mouse as shown in Figure 39-134. Using the *Arc*, *Sweep* a surface upward. Set the *Alignment* to *No*. Use the near vertical *Line* as the path.

FIGURE 39-134

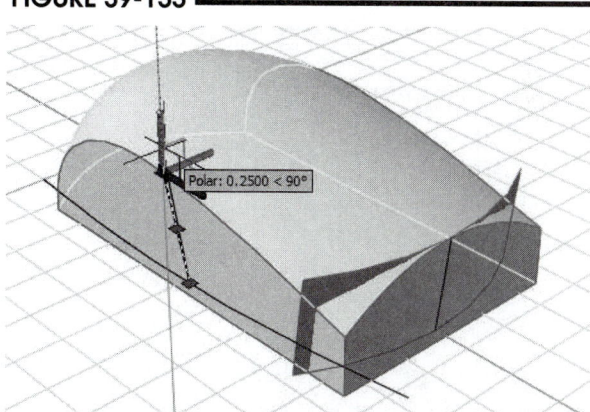

G. Next, create a similar *Arc* near the inside edge of the side of the mouse base, then connect its *Midpoint* with a *Line* to the *Midpoint* of the top side edge. To ensure this *Line* extends past the top surface of the mouse, use the move grip tool, select the vertical axis, and move the end of the *Line* up .25 units as shown in Figure 39-135.

FIGURE 39-135

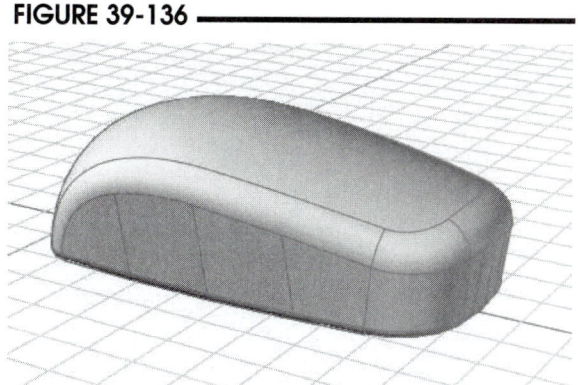

H. Next, create a swept surface using the same method you used to create the surface on the front of the mouse. *Mirror* this surface to the other side of the mouse using the X axis as mirror axis. Now that you have created the three surfaces, use *Slice* with each of the surfaces to cut away the outer edges of the mouse. Next, move the surfaces to the **Surfaces** layer. *Erase* any construction geometry used to define the surfaces.

I. Finally, use *Fillet* with a .75 radius to round the front (nearly) vertical edges. Use a *Fillet* of .3 to round the *Chain* along the top edge. Use a *Fillet* of .1 to round the *Chain* along the bottom edge. Your completed mouse solid should appear similar to that shown in Figure 39-136.

FIGURE 39-136

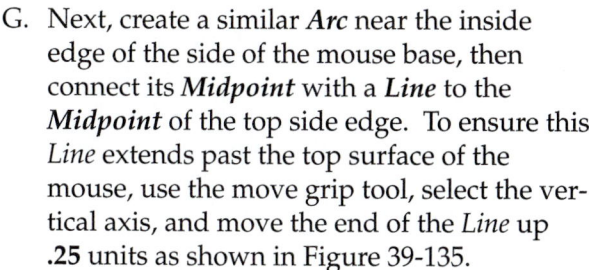

J. Optional: Create two mouse buttons using *Slice* and create a mouse wheel.

13. **Calculating *Mass Properties***

Open the **SADL-SL** drawing that you created in Chapter 38 Exercises. Calculate *Mass Properties* for the saddle. Write the report out to a file named **SADL-SL.MPR**. Use a text editor to examine the file.

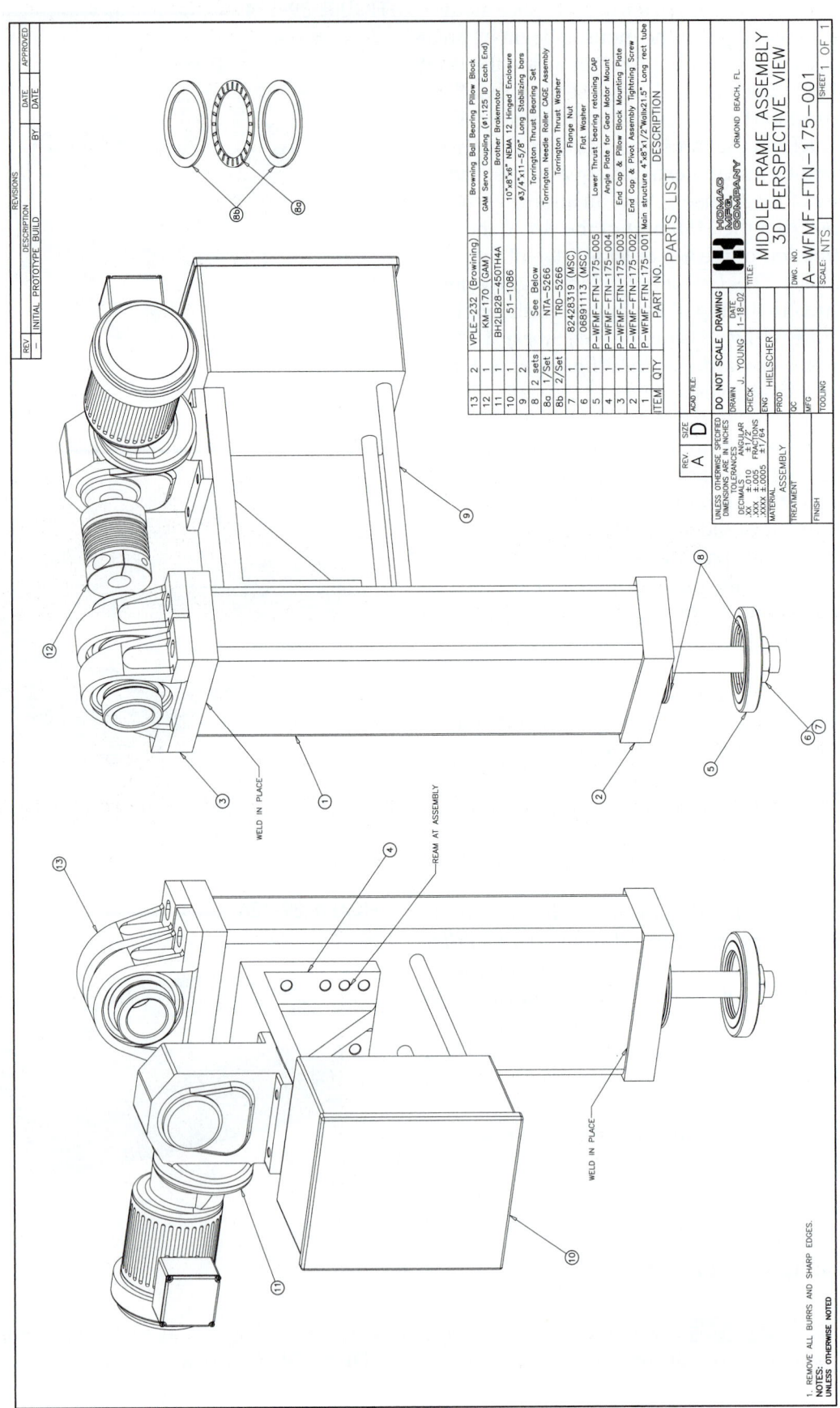

ITEM	QTY	PART NO.	DESCRIPTION
13	2	VPLE-232 (Browning)	Browning Ball Bearing Pillow Block
12	1	KM-170 (GAM)	GAM Servo Coupling (Ø1.125 ID Each End)
11	1	BH2LB28-450TH4A	Brother Brakemotor
10	1	51-1086	10"x8"x6" NEMA 12 Hinged Enclosure
9	2	See Below	#3/4"x11-5/8" Long Stabilizing bars
8	2 sets	See Below	Torrington Thrust Bearing Set
8a	1/Set	NTA-5266	Torrington Needle Roller CAGE Assembly
8b	2/Set	TRD-5266	Torrington Thrust Washer
7	1	82428319 (MSC)	Flange Nut
6	1	06891113 (MSC)	Flat Washer
5	1	P-WFMF-FTN-175-005	Lower Thrust bearing retaining CAP
4	1	P-WFMF-FTN-175-004	Angle Plate for Gear Motor Mount
3	1	P-WFMF-FTN-175-003	End Cap & Pillow Block Mounting Plate
2	1	P-WFMF-FTN-175-002	End Cap & Pivot Assembly Tightning Screw
1	1	P-WFMF-FTN-175-001	Main structure 4"x8"x1/2"Walkx21.5" Long rect tube

PARTS LIST

REVISIONS

REV	DESCRIPTION	BY	DATE	APPROVED
-	INITIAL PROTOTYPE BUILD			

UNLESS OTHERWISE SPECIFIED
DIMENSIONS ARE IN INCHES
TOLERANCES
DECIMALS   ANGULAR
.XX ±.010   ±1/2°
.XXX ±.005   FRACTIONS
.XXXX ±.0005   ±1/64

DO NOT SCALE DRAWING

DRAWN J. YOUNG DATE 1-18-02
CHECK
ENG HIELSCHER
PROD
QC
MFG
TOOLING

MATERIAL
TREATMENT
FINISH

MIDLAND MFG COMPANY   ORMOND BEACH, FL.

TITLE:
MIDDLE FRAME ASSEMBLY
3D PERSPECTIVE VIEW
ASSEMBLY

DWG. NO.
A-WFMF-FTN-175-001

REV A   SIZE D   SCALE: NTS   SHEET 1 OF 1

NOTES:
1. REMOVE ALL BURRS AND SHARP EDGES.
UNLESS OTHERWISE NOTED

WELD IN PLACE
REAM AT ASSEMBLY

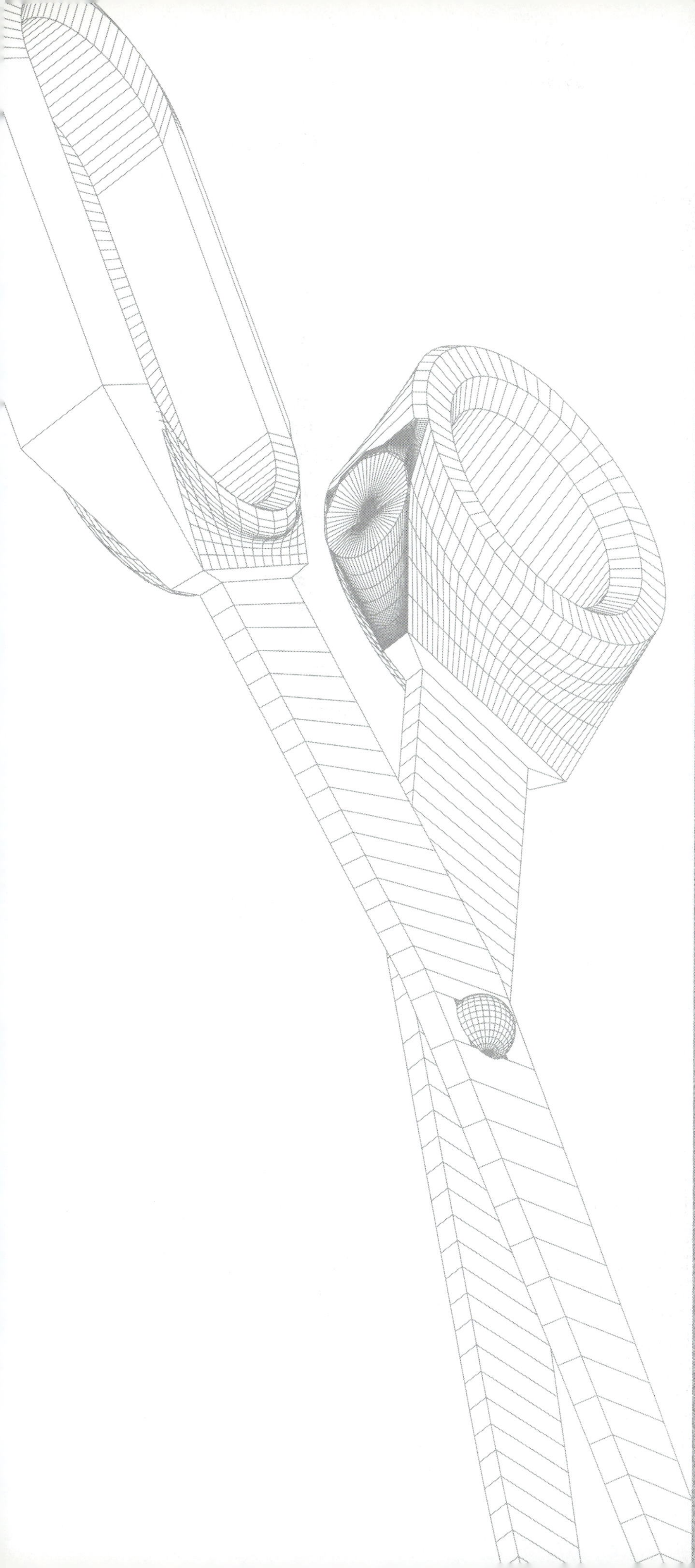

SURFACE MODELING

CHAPTER OBJECTIVES

After completing this chapter you should:

1. be able to create planar surfaces bounded by straight edges using *3Dface*;

2. be able to edit *3Dfaces* with grips and create invisible edges between *3Dfaces* using *Edge*;

3. be able to create meshed surfaces bounded by straight sides using *3Dmesh*;

4. be able to create geometrically defined meshed surfaces using *Rulesurf, Tabsurf, Revsurf,* and *Edgesurf*;

5. be able to edit polygon meshes (*3Dmesh, Rulesurf, Tabsurf, Revsurf,* and *Edgesurf*) using *Pedit, Explode,* and Grips;

6. be able to create *Regions* for use with surface models;

7. be able to use *Thickness* and *Elevation* to create surfaces from 2D draw commands.

CONCEPTS

Surface models do not describe physical objects as completely and as accurately as solid models. A surface model can be compared to a cardboard box—it has surfaces but is hollow inside. If a surface model is sectioned (cut in half), there is only air inside, whereas a solid model can be sectioned to reveal its interior features. Therefore, surface models have volume, whereas solid models have volume and mass.

Compared to a wireframe model, a surface model is more sophisticated because it contains the description of surfaces as well as the description of edges. Surface models can contain descriptions of complex curves, whereas "surfaces" on a wireframe are assumed to be planar or single curved. Surface models are superior to wireframe models in that they provide better visualization cues. Since surfaces are opaque, surface models have a "solid" look, even though they have no inside space defined.

Using AutoCAD, the construction and editing of surfaces is limited and tedious compared to construction and editing of either wireframe or solid models. Surface modeling techniques are especially inferior compared to solid modeling, primarily because each surface comprising a surface model must be constructed individually. Many of the older, "legacy," surfaces must be defined by first creating the edges that bound the surfaces.

Recent AutoCAD releases are rarely used to construct complete surface models. Not only are surface modeling techniques limited, but the completed models are not as realistic, robust, or useful as solid models for downstream applications such as rendering, animations, simulations, or manufacturing. However, a few applications for surfacing may exist, such as editing older drawings or working with clients who use early releases of AutoCAD.

New Surfaces and Legacy Surfaces

New Surfaces

There are several generations of surfacing capabilities in AutoCAD. You are already familiar with the most recent surfacing commands because these were discussed in Chapters 38 and 39. They are:

> *Planesurf, Revolve, Loft, Sweep,* and *Extrude*

These commands are unique in that you can use them to create surfaces, but you can also use them to create and edit solid models. For example, if you *Extrude* a closed 2D shape, such as a closed *Pline*, a solid results. However, if you *Extrude* an open 2D shape, such as a *Line*, a surface results. These newer surfaces can also be used with solid models by using the surfaces to *Slice* away parts of solids.

Legacy Surfaces

There are several older types of surfaces in AutoCAD called "legacy" surfaces. These surfaces are more limited and more difficult to create and work with than the newer surface commands. This chapter discusses legacy surfaces. The information is organized as follows:

Non-meshed surfaces defined by straight edges (usually planar)

> *3Dface*   A surface defined by 3 or 4 straight edges

Meshed surfaces (polygon meshes)

> *3Dmesh*   A planar, curved, or complex surface defined by a mesh

Geometrically defined meshed surfaces (polygon meshes)

Edgesurf   A surface defined by "patching" together 4 straight or curved surfaces

Rulesurf   A surface created between 2 straight or curved surfaces

Revsurf   A surface revolved from a 2D shape about an axis

Tabsurf   A surface created by sweeping a 2D shape along a path

Other objects that can be used as surfaces

Region   Region can be used to simplify construction of surface models. Boolean operations simplify creation of surfaces with holes or complex outlines.

Thickness   AutoCAD's first method of creating 3D shapes

FIGURE 40-1

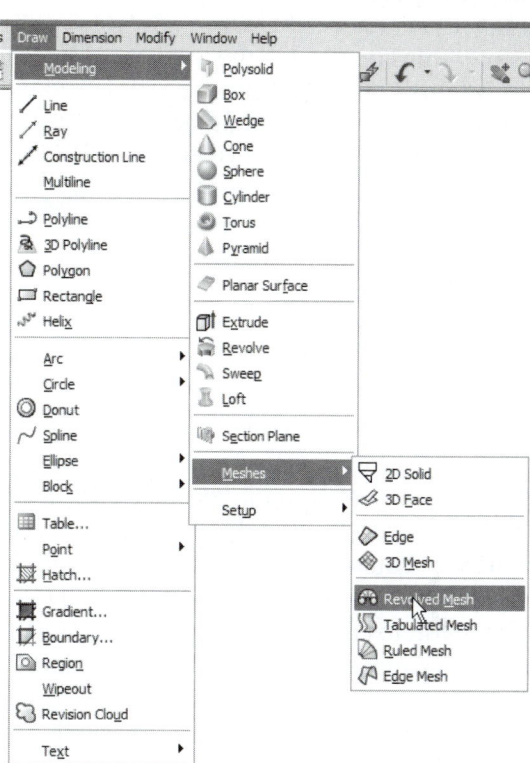

The legacy surface modeling commands can be accessed easily from the *Draw* pull-down menu (Fig. 40-1). No toolbars exist for legacy surfaces.

NON-MESHED SURFACES WITH STRAIGHT EDGES

3Dface

Pull-down Menu	Command (Type)	Alias (Type)	Short-cut	Screen (side) Menu	Tablet Menu
Draw Modeling > Meshes > 3D Face	*3Dface*	*3F*	...	*DRAW 2 SURFACES 3Dface*	*M,8*

3Dface creates a surface bounded by three or four straight edges. The three- or four-sided surface can be connected to other *3Dface*s within the same *3Dface* command sequence, similar to the way several *Line* segments created in one command sequence are connected. Beware, there is <u>no *Undo*</u> option for *3Dface*. The command sequence is:

Command: **3dface**
Specify first point or [Invisible]: **PICK** or (**coordinates**)
Specify second point or [Invisible]: **PICK** or (**coordinates**)
Specify third point or [Invisible] <exit>: **PICK** or (**coordinates**)
Specify fourth point or [Invisible] <create three-sided face>: **PICK** or (**coordinates**)
Specify third point or [Invisible] <exit>: **PICK** or (**coordinates**)
Specify fourth point or [Invisible] <create three-sided face>: **PICK** or (**coordinates**)
Specify third point or [Invisible] <exit>: **Enter**
Command:

After completing four points, AutoCAD connects the next point ("third point:") to the previous fourth point. The next fourth point is connected to the previous third point, and so on. Figure 40-2 shows the sequence for attaching several 3Dface segments in one command sequence.

FIGURE 40-2

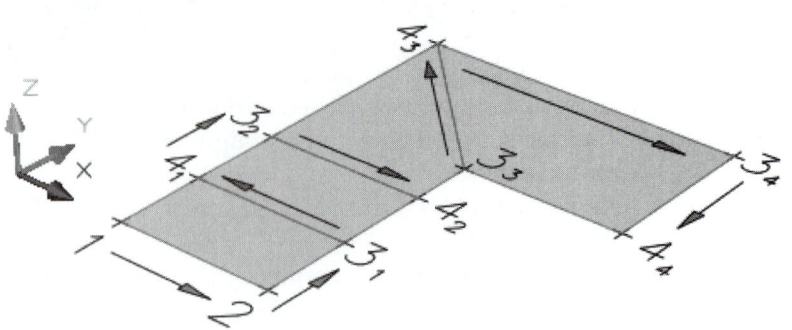

In the example, each edge of the 3Dface is coplanar, as is the entire sequence of 3Dfaces (the Z value of every point on each edge is 0). The edges must be straight but not necessarily coplanar. Entering specific coordinate values, using point filters, or using OSNAP in 3D space allows you to create geometry with 3Dface that is not on a plane. The following command sequence creates a 3Dface as a complex curve by entering a different Z value for the fourth point (Fig. 40-3):

FIGURE 40-3

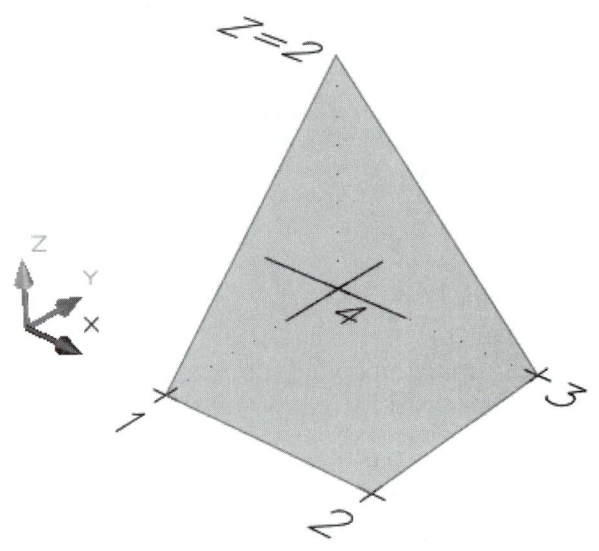

```
Command: 3dface
Specify first point or [Invisible]: PICK
Specify second point or [Invisible]: PICK
Specify third point or [Invisible] <exit>: PICK
Specify fourth point or [Invisible] <create three-sided face>: .XY
of PICK
(need Z): (value)
Specify third point or [Invisible] <exit>: Enter
Command:
```

Although it is possible to use 3Dface to create complex curved surfaces, the surface curve is not visible as it would be if a mesh were used. A 3Dmesh provides superior visibility and flexibility and is recommended for such a surface.

The lines between 3Dface segments in Figure 40-2 are visible even though the segments are coplanar. AutoCAD provides two methods for making invisible edges. The first method requires that you enter the letter "I" immediately before the first of the two points that define an invisible edge. For the shape shown in Figure 40-4, enter the letter "I" before points 3_1, 3_2, and 3_3. The second, and much easier, method for defining invisible edges of 3Dfaces is to use the Edge command.

Edge

Pull-down Menu	Command (Type)	Alias (Type)	Short-cut	Screen (side) Menu	Tablet Menu
Draw Modeling > Meshes > Edge	*Edge*	...	...	*DRAW 2 SURFACES Edge:*	...

The first method for creating invisible edges described above takes careful planning. The same action shown in Figure 40-4 can be accomplished much more easily and <u>retroactively</u> by this second method. AutoCAD provides the *Edge* command to create invisible edges after construction of the *3Dface*. The Command prompt is:

FIGURE 40-4

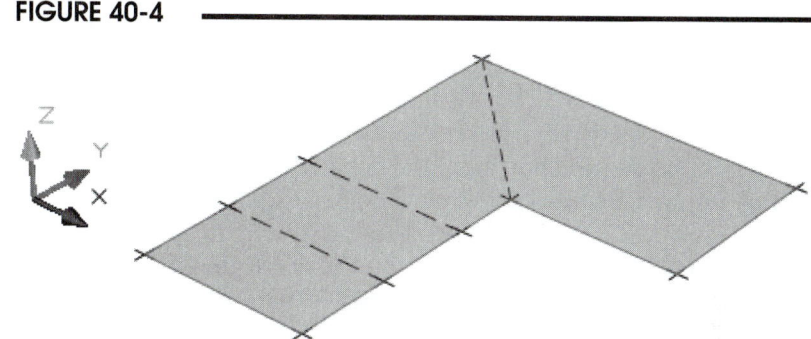

Command: **edge**
Specify edge of 3dface to toggle visibility or [Display]: **PICK**

Selecting edges converts visible edges to invisible edges. <u>Any</u> edges on the *3Dface* can be made invisible by this method. The *Midpoint OSNAP* option is automatically invoked to simplify selection of the edges. The *Display* option causes selected invisible edges to become visible again.

Applications of *3Dface*

3Dface is used to construct 3D surface models by creating individual (planar) faces or connected faces. For relatively simple models, the faces are placed in space as they are constructed. A construction strategy to use for more complex models is to first construct a wireframe model and then attach surfaces (*3Dfaces* or other surfaces) to the wireframe. The wireframe geometry can be constructed on a separate layer; then the layer can be turned off after the surfacing is complete.

FIGURE 40-5

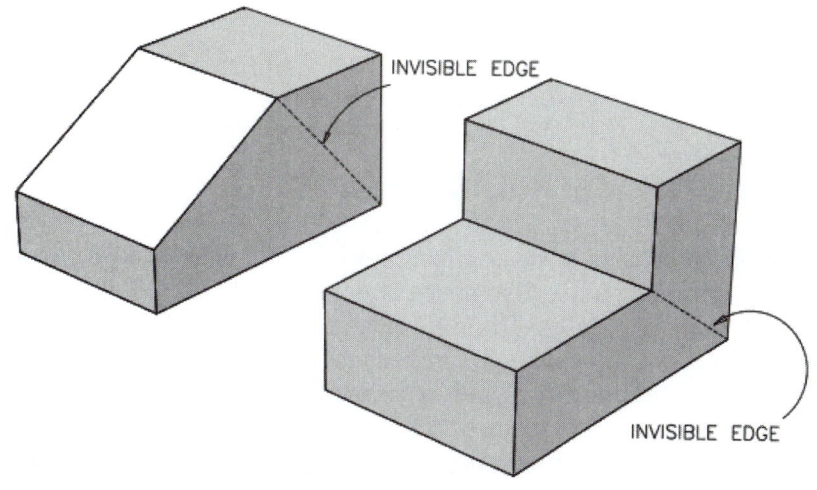

Figures 40-5 and 40-6 show some applications of *3Dface* for construction of surface models. Remember, *3Dface* can be used <u>only</u> for construction of surfaces with straight edges.

The edges made invisible by *Edge* are displayed in these figures in a hidden linetype. *Conceptual* visual style has also been used to enhance visualization. You may want to use visual styles periodically to enhance the visibility of surfaces.

FIGURE 40-6

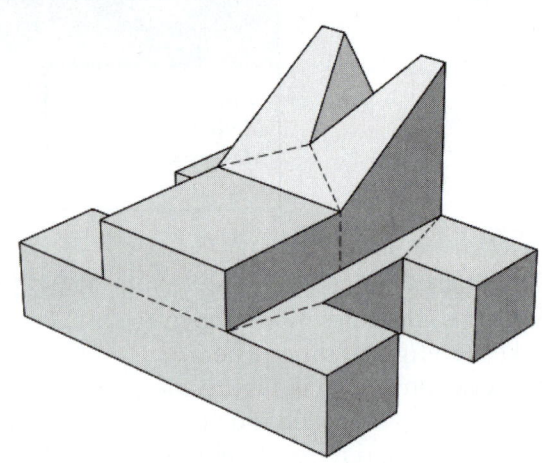

Editing *3Dfaces*

Grips can be used to edit *3Dfaces*. Each *3Dface* has three or four grips (one for each corner). Activating grips (PICKing the *3Dface* at the open Command prompt), makes the individual *3Dface* surface become high-lighted and display its grips. If you then make one of the three or four grips **hot**, all of the normal Grip editing options are available, that is, STRETCH, MOVE, ROTATE, SCALE, and MIRROR. The single activated *3Dface* surface can then be edited. MOVE, for example, could be used to move the single activated surface to another location.

The STRETCH option, however, is generally the most useful grip editing option, since all of the other options change the entire surface. The STRETCH option allows you to change one (or more) corner(s) of the *3Dface* rather than the entire surface, as with the other options. Grips always appear <u>parallel to the current UCS XY plane</u> and operate in their respective planes. For example, in Figure 40-7 the grips are aligned parallel to the current coordinate system. Therefore, STRETCHing the hot grip (shown solid) allows you to change the shape of the *3Dface*, but restricts the movement parallel to the current XY plane. Changing to a different UCS orientation allows you to STRETCH in another plane.

FIGURE 40-7

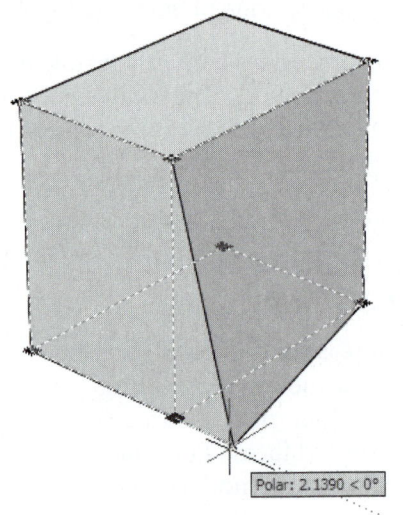

CREATING MESHED SURFACES

3Dmesh can be used to create a complex surface defining an irregular shape bounded by <u>four</u> sides (Fig. 40-8). The surface created with *3Dmesh* differs from a *3Dface* because the surface is defined by a "mesh." A mesh is a series of vertices (sometimes called "nodes") arranged in rows and columns con-nected by lines. *Pedit* or grips can be used to edit the resulting mesh (see "Editing Polygon Meshes," this chapter).

FIGURE 40-8

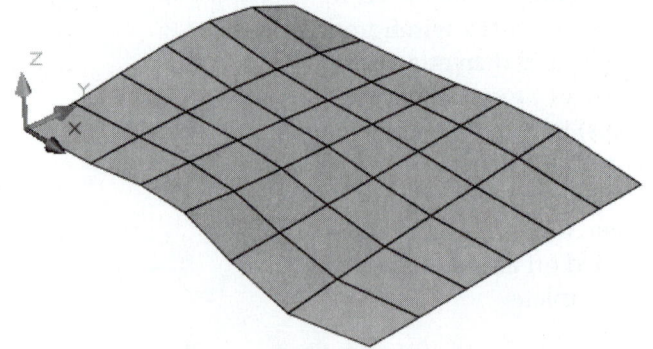

3Dmesh

	Pull-down Menu	Command (Type)	Alias (Type)	Short-cut	Screen (side) Menu	Tablet Menu
	Draw Modeling > Meshes > 3D Mesh	*3Dmesh*	...	...	*DRAW 2 SURFACES 3Dmesh:*	...

3Dmesh is a versatile tool for generating a complex surface. It is used to create a complex surface defining an irregular shape bounded by <u>four</u> sides (see Fig. 40-9). *Pedit* or grips can be used to edit the *3Dmesh* retroactively.

FIGURE 40-9

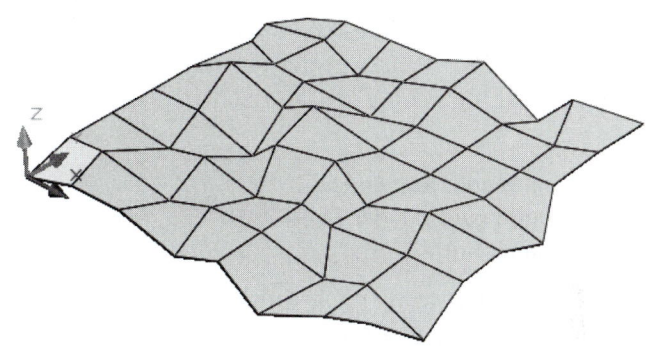

3Dmesh prompts for coordinate values for the placement of <u>each</u> vertex, allowing you to create a variety of shapes. A numbering scheme is used to label each vertex. The first vertex in the <u>first</u> row in the M direction is labeled *0,0*, the second in the first row is labeled *0,1*, and so on. The first vertex in the <u>second</u> row is labeled *1,0*, the second in that row is *1,1*, and so on.

Since each vertex can be given explicit coordinates, a surface can be created to define any shape. An example is shown in Figure 40-9. This command is useful for developing geographical topographies. The command syntax for this method of defining a *3Dmesh* is as follows:

```
Command: 3dmesh
Enter size of mesh in M direction: (value)
Enter size of mesh in N direction: (value)
Specify location for vertex (0, 0): PICK or
(coordinates)
Specify location for vertex (0, 1): PICK or (coordinates)
Specify location for vertex (0, 2): PICK or (coordinates)
(continues sequence for all vertices in the row)
Specify location for vertex (1, 0): PICK or (coordinates)
Specify location for vertex (1, 1): PICK or (coordinates)
Specify location for vertex (1, 2): PICK or (coordinates)
(continues sequence for all vertices in the row)
Specify location for vertex (2, 0): PICK or (coordinates)
Specify location for vertex (2, 1): PICK or (coordinates)
Specify location for vertex (2, 2): PICK or (coordinates)
(continues sequence for all vertices in the row)
(sequence continues for all rows)
Command:
```

Obviously, using *3Dmesh* is very tedious. For complex shapes such as topographical maps, an AutoLISP program can be written to read coordinate data from an external file to generate the mesh.

CREATING GEOMETRICALLY DEFINED MESHES

Geometrically defined meshes are meshed surfaces that are created by "attaching" a surface to existing geometry. In other words, geometrically defined meshed surfaces <u>require existing geometry</u> to define them. *Edgesurf, Rulesurf, Tabsurf,* and *Revsurf* are the commands that create these meshes. The geometry used may differ for the application but always consists of either two or four objects (*Lines, Circles, Arcs,* or *Plines*).

Controlling the Mesh Density

Geometrically defined mesh commands (*Edgesurf, Rulesurf, Revsurf,* and *Tabsurf*) do not prompt the user for the number of vertices in the M and N direction. Instead, the number of vertices is determined by the settings of the *SURFTAB1* and *SURFTAB2* variables. The number of vertices includes the endpoints of the edge:

> Command: **surftab1**
> Enter new value for SURFTAB1 <6>: (**value**)
> Command:

 SURFTAB1 and *SURFTAB2* must be set <u>before</u> using *Edgesurf, Rulesurf, Revsurf,* and *Tabsurf.* These variables are <u>not</u> retroactive for previously created surface meshes.

The individual surfaces (1 by 1 meshes between vertices) created by geometrically defined meshes are composed of a series of <u>straight</u> edges connecting the vertices. These edges do not curve but can only change direction between vertices to approximate curved edges. The higher the settings of *SURFTAB1* and *SURFTAB2*, the more closely these straight edges match the defining curved edges.

 Note that the defining edges of geometrically defined mesh <u>do not become part of the surface</u>; they remain separate objects. Therefore, you can construct surface models by utilizing existing wireframe model objects as the defining edges. If the wireframe model is on a separate layer, that layer can be *Frozen* after constructing the surfaces to reveal only the surfaces.

Edgesurf

Pull-down Menu	Command (Type)	Alias (Type)	Short-cut	Screen (side) Menu	Tablet Menu
Draw *Modeling >* *Meshes >* *Edge Mesh*	*Edgesurf*	...	...	DRAW 2 SURFACES *Edgsurf:*	R,8

An *Edgesurf* is a meshed surface generated between <u>four existing</u> edges. The four edges can be of any shape as long as they have connecting endpoints (no gaps, no overlaps). The four edges can be *Lines, Arcs,* or *Plines*. The four edges can be selected in any order. The surface that is generated between the edges is sometimes called a "Coon's surface patch." *Edgesurf* interpolates the edges and generates a smooth transitional mesh, or patch, between the four shapes.

The Command syntax is as follows:

> Command: **edgesurf**
> Current wire frame density: SURFTAB1=6 SURFTAB2=6
> Select object 1 for surface edge: **PICK**
> Select object 2 for surface edge: **PICK**
> Select object 3 for surface edge: **PICK**
> Select object 4 for surface edge: **PICK**
> Command:

Figure 40-10 displays an *Edgesurf* generated from four planar edges. Settings for *SURFTAB1* and *SURFTAB2* are 6 and 8.

The relation of *SURFTAB1* and *SURFTAB2* for any *Edgesurf* are given here:

SURFTAB1	first edge picked
SURFTAB2	second edge picked

FIGURE 40-10

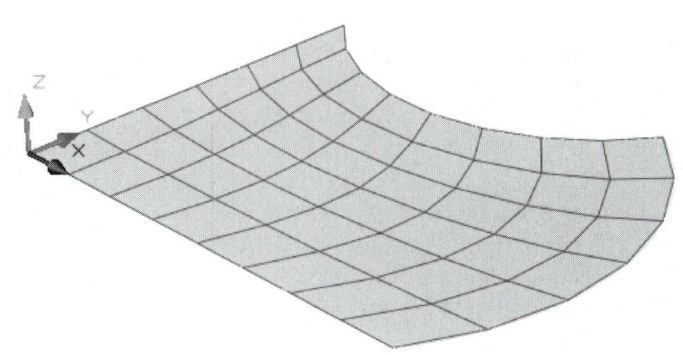

Edges used to define an *Edgesurf* may be <u>nonplanar</u>, resulting in a smooth, nonplanar surface, as displayed in Figure 40-11. Remember, the edges may be any *Pline* shape in 3D space.

FIGURE 40-11

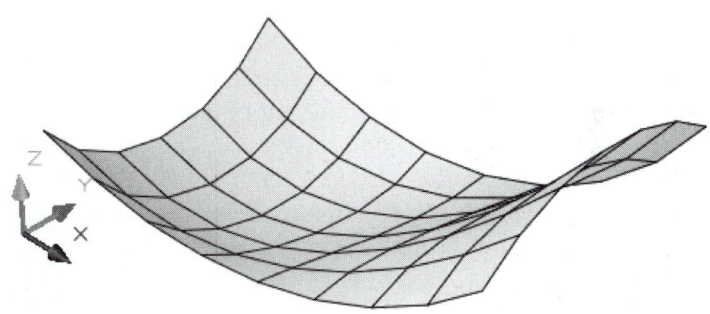

Rulesurf

Pull-down Menu	Command (Type)	Alias (Type)	Short-cut	Screen (side) Menu	Tablet Menu
Draw *Modeling >* *Meshes >* *Ruled Mesh*	*Rulesurf*	...	...	DRAW 2 SURFACES *Rulsurf:*	Q,8

A *Rulesurf* is a polygon meshed surface created between <u>two edges</u>. The edges can be *Lines*, *Arcs*, or *Plines*. The command syntax only asks for the two "defining curves." The two defining edges must <u>both</u> be open or closed.

Figure 40-12 displays two planar *Rulesurfs*.

FIGURE 40-12

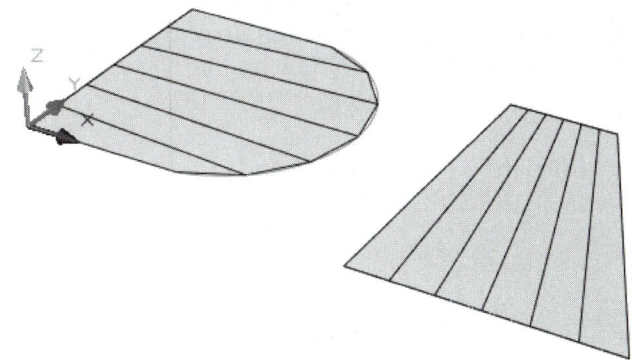

Figure 40-13 displays possibilities for creating *Rulesurfs* between two edges that are nonplanar.

The Command syntax for *Rulesurf* is as follows:

```
Command: rulesurf
Current wire frame density:  SURFTAB1=6
Select first defining curve: PICK
Select second defining curve: PICK
Command:
```

FIGURE 40-13

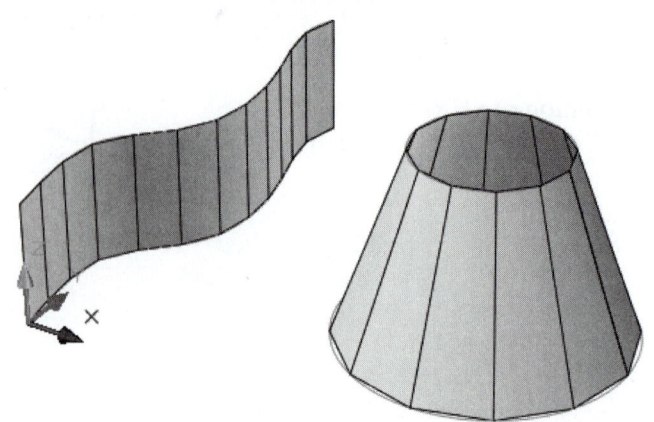

A cylinder can be created with *Rulesurf* by utilizing two *Circles* lying on different planes as the defining curves. A cone can be created by creating a *Rulesurf* between a *Circle* and a *Point* (Fig. 40-14), or a complex shape can be created using two identical closed *Pline* shapes (Fig. 40-14).

FIGURE 40-14

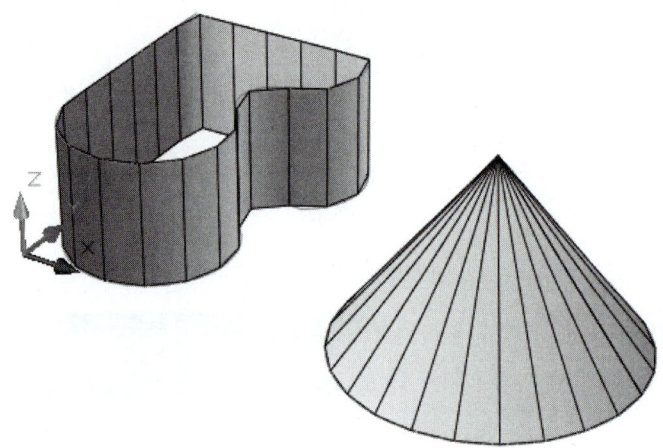

For open defining edges (not *Circles* or closed *Plines*), the *Rulesurf* is generated connecting the two <u>selected</u> ends of the defining edges. Figure 40-15 illustrates the two possibilities for generating a *Rulesurf* between the same two straight nonplanar *Lines* based on the endpoints selected.

Since the curve is stretched between only two edges, the number of vertices along the *defining curve* is determined only by the value previously specified for *SURFTAB1*.

SURFTAB1	*defining curve*
SURFTAB2	not used

FIGURE 40-15

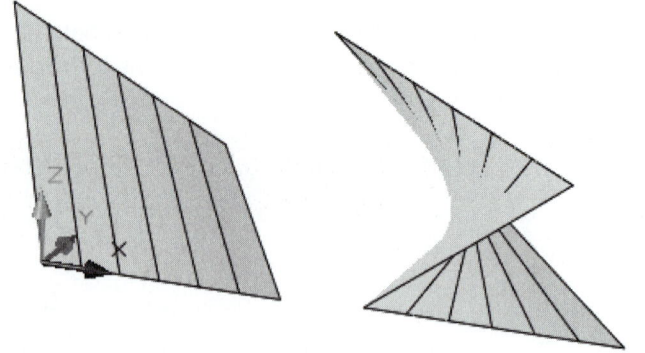

Tabsurf

	Command (Type)	Alias (Type)	Short-cut	Screen (side) Menu	Tablet Menu
Pull-down Menu					
Draw Modeling > Meshes > Tabulated Mesh	*Tabsurf*	...	...	*DRAW 2 SURFACES Tabsurf:*	*P,8*

A *Tabsurf* is generated by an existing *path curve* and a *direction vector*. The *curve path* (generatrix) is extruded in the direction of, and equal in length to, the *direction vector* (directrix) to create a swept surface. The path curve can be a *Line, Arc, Circle, 2D Polyline,* or *3D Polyline*. The direction vector can be a *Line* or an open 2D *Pline*. If a curved *Pline* is used, the surface is generated in a direction connecting the two end vertices of the *Pline*.

A *Circle* can be swept with *Tabsurf* to generate a cylinder (Fig. 40-16). The *Tabsurf* has only *N* direction; therefore, the current setting of *SURFTAB1* controls the number of vertices along the *path curve*.

The relation of *SURFTAB1* and *SURFTAB2* for *Tabsurf* are:

SURFTAB1	*Path curve*
SURFTAB2	not used

The *Tabsurf* is generated in the direction <u>away from</u> the end of the direction vector that is selected.

The Command sequence is:

Command: **tabsurf**
Select object for path curve: **PICK**
Select object for direction vector: **PICK**

FIGURE 40-16

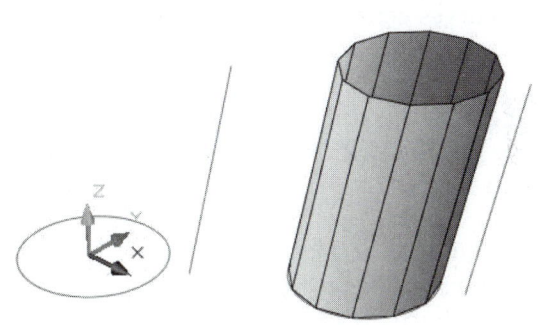

FIGURE 40-17

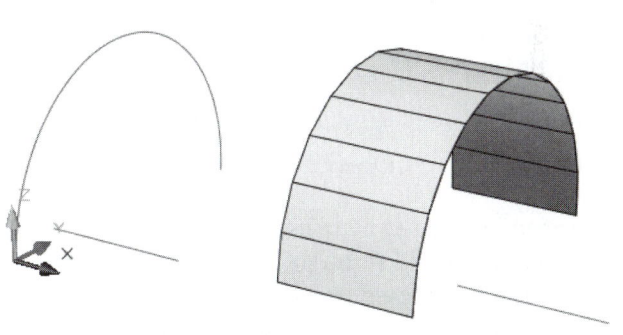

Revsurf

	Command (Type)	Alias (Type)	Short-cut	Screen (side) Menu	Tablet Menu
Pull-down Menu					
Draw Modeling > Meshes > Revolved Mesh	*Revsurf*	...	...	*DRAW 2 SURFACES Revsurf:*	*O,8*

This command creates a surface of revolution by revolving an existing path curve (*object to revolve*) around an *axis*. The path curve can be a *Line, Arc, Circle,* 2D or 3D *Pline*. The *axis* can be a *Line* or open *Pline*. If a curved *Pline* is used as the *axis*, only the endpoint vertices are considered for the axis of revolution.

```
Command: revsurf
Current wire frame density: SURFTAB1=6  SURFTAB2=6
Select object to revolve: PICK
Select object that defines the axis of revolution: PICK
Specify start angle <0>: Enter or (value)
Specify included angle (+=ccw, -=cw) <360>: Enter or (value)
Command:
```

The structure of the command allows many variations. The *object to revolve* can be <u>open or closed</u>. The *axis* can be in any plane. The *start angle* and *included angle* allow for complete (closed) or partial (open) revolutions of the surface in any orientation. Figures 40-18 and 40-19 show possible *Revsurfs* created with path curves generated through 360 degrees.

FIGURE 40-18

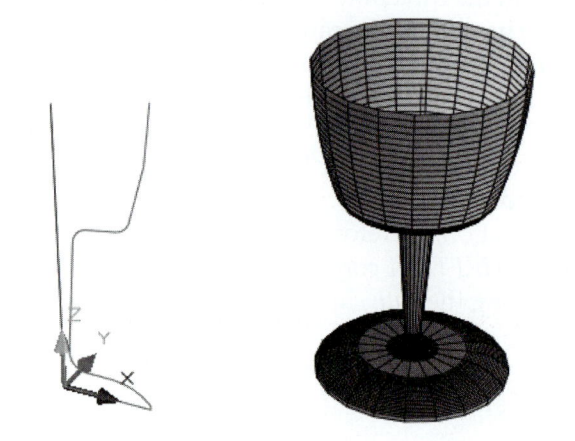

The wine glass was created with an open path curve generated through 360 degrees (Fig. 40-18).

The number of vertices in the direction of revolution is controlled by the setting of *SURFTAB2*. The *SURFTAB1* setting controls the number of vertices along the length of the revolved object.

FIGURE 40-19

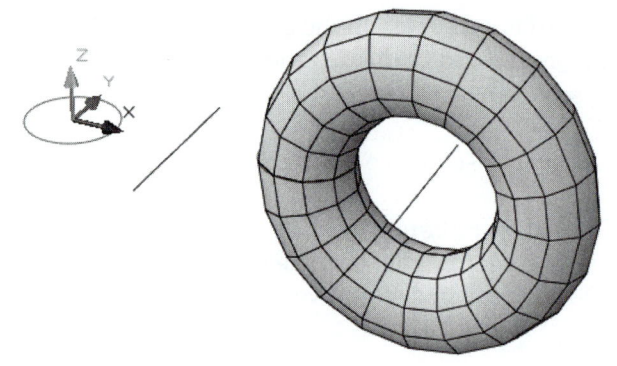

The relation of *SURFTAB1* and *SURFTAB2* for *Revsurf* is given here:

SURFTAB1   Axis of revolution
SURFTAB2   Object to revolve

A torus can be created by using a closed path curve generated through 360 degrees (Fig. 40-19). In this case, the surface begins at the path curve, is revolved all the way around the axis, and closes on itself.

When generating a *Revsurf* through <u>less than</u> 360 degrees, you must specify the included angle. Entering a positive angle specifies a counterclockwise direction of revolution, and a negative angle specification causes a clockwise revolution. As an alternative, you can pick different points on the *axis of revolution*. The end of the line PICKed represents the end nearest the origin for positive rotation using the right-hand rule.

EDITING POLYGON MESHES

There are several ways that existing AutoCAD polygon meshed surfaces (*3Dmesh, Rulesurf, Tabsurf, Revsurf,* and *Edgesurf*) can be changed. The use of *Pedit*, Grips, and *Explode* each provide special capabilities.

Pedit with Polygon Meshes

One of the options of *Pedit* is *Edit vertex*, which allows editing of individual vertices of a *Pline*. Vertex editing can also be accomplished with *3Dmeshes*, *Revsurfs*, *Edgesurfs*, *Tabsurfs*, and *Rulesurfs*. *Pedit* does <u>not</u> allow editing of *3Dfaces*. See Chapter 16 for information on using *Pedit* with 2D objects.

        Command: **pedit**
        Select polyline: **PICK** (select the surface)
        Enter an option [Edit vertex/Smooth surface/Desmooth/Mclose/Nclose/Undo]:

Each *Pedit* option for editing Polygon Meshes is explained and illustrated here.

Edit Vertex

Invoking this option causes the display of another set of options at the Command prompt, which allows you to locate the vertex that you wish to edit:

        Current vertex (0,0).
        Enter an option [Next/Previous/Left/Right/Up/Down/Move/REgen/eXit] <N>:

Pressing Enter locates the marker (X) at the next vertex. Selecting *Left/Right* or *Up/Down* controls the direction of the movement in the *N* and *M* directions (as specified by *SURFTAB1* and *SURFTAB2* when the surfaces were created; Fig. 40-20).

FIGURE 40-20

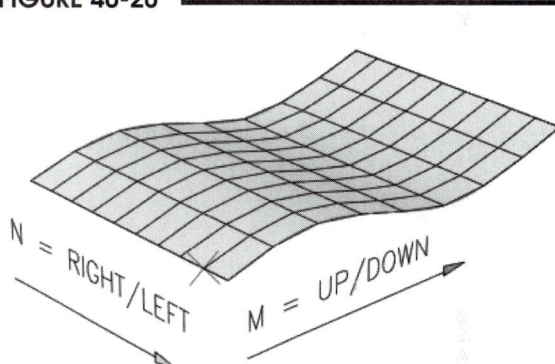

Once the vertex has been located, use the *Move* option to move the vertex to any location in 3D space (Fig. 40-21).

FIGURE 40-21

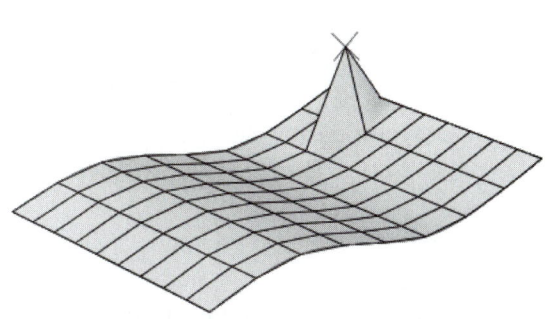

Smooth surface

Invoking this option causes the surface to be smoothed using the current setting of *SURFTYPE* (described later) (Fig. 40-22).

FIGURE 40-22

Desmooth

The *Desmooth* option reverses the effect of *Smooth*.

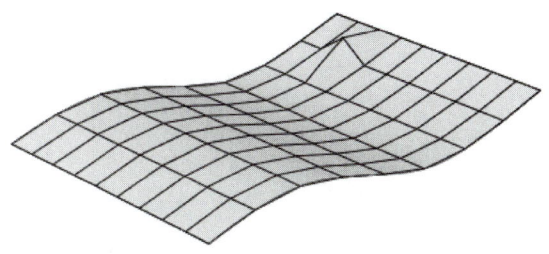

Mclose/Nclose

The *Mclose* and *Nclose* options cause the surface to close in either the *M* or *N* directions. The last and first set of vertices automatically connect. Figure 40-23 displays a half-cylindrical surface closed with *Nclose*.

FIGURE 40-23

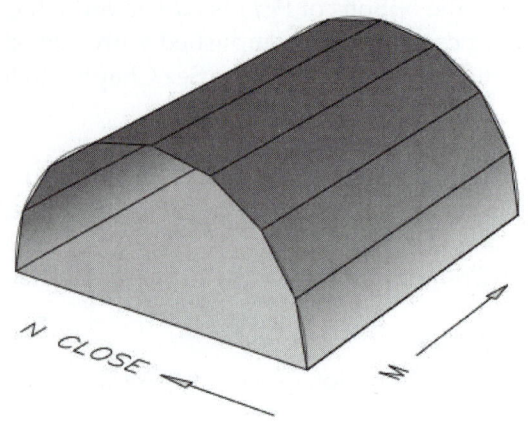

Undo

Undo reverses the last operation with *Pedit*.

eXit

This option is used to keep the editing changes, exit the *Pedit* command, and return to the Command prompt.

Variables Affecting Polygon Meshes

SURFTYPE

This variable affects the degree of smoothing of a surface when using the *Smooth* option of *Pedit*. The Command prompt syntax is shown here:

FIGURE 40-24

    Command: **surftype**
    Enter new value for Surftype <6>: (**value**)

The values allowed are **5**, **6**, and **8**. The options are illustrated in the following figures.

Original polygon mesh (Fig. 40-24)

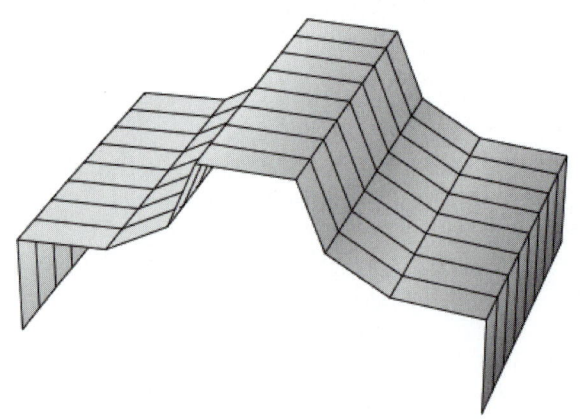

SURFTYPE=5
Quadratic B-spline surface (Fig. 40-25)

FIGURE 40-25

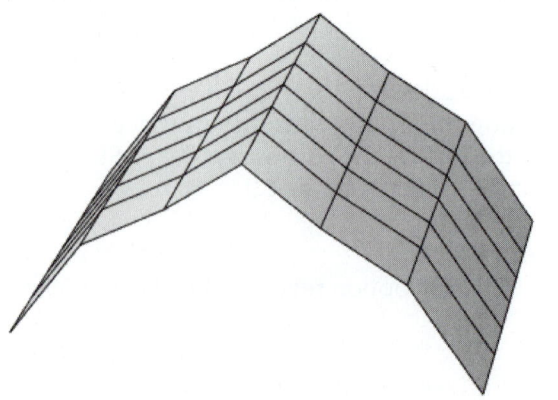

SURFTYPE=6
Cubic B-spline surface (Fig. 40-26)

FIGURE 40-26

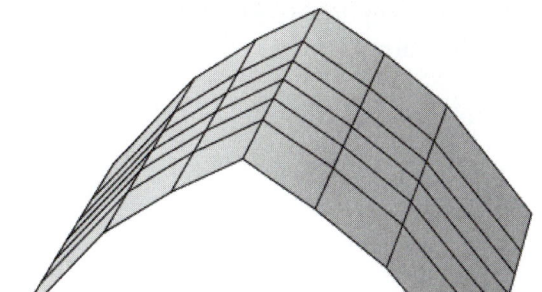

SURFTYPE=8
Bezier surface (Fig. 40-27)

FIGURE 40-27

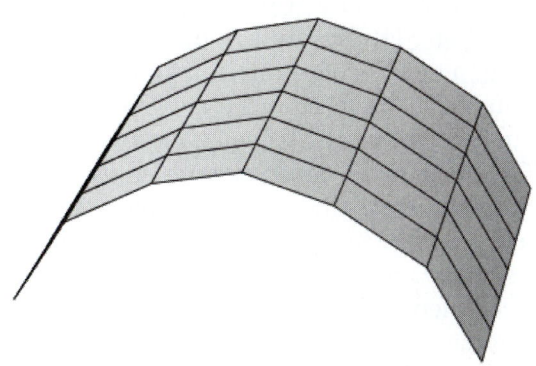

NOTE: Changing *SURFTYPE* by Command line format sets the smoothing only for <u>new</u> polygon meshes. If you want to change an <u>existing</u> polygon mesh, use *Properties*, then select *Quadratic*, *Cubic*, or *Bezier* from the *Fit/Smooth* drop-down list.

SPLFRAME

The *SPLFRAME* (spline frame) variable causes the original polygon mesh to be displayed. The values allowed are **0** for off and **1** for on. *SPLFRAME* can also be used with 2D polylines when using the *Spline* option.

Editing Polygon Meshes with Grips

Grips can be used to edit the individual vertices of a polygon mesh (*3Dmesh, Edgesurf, Rulesurf, Tabsurf,* and *Revsurf*). Normally, the STRETCH option of Grip editing would be used to stretch (move) the location of a single vertex grip. The other Grip editing options (MOVE, SCALE, ROTATE, MIRROR) affect the entire polygon mesh. Selecting the mesh at the open Command prompt activates all of the grips (one for each vertex). Selecting one of the individual grips again makes that one **hot** and able to be relocated with the STRETCH option.

Keep in mind that grips always appear <u>parallel to the current UCS XY plane</u> and operate in their respective planes. Therefore, you can control which plane the grips are STRETCHed by:

1.   STRETCH each grip alone in the grip's current plane;
2.   turn on *POLAR* to STRETCH a single grip in a Polar Z direction;
3.   use the move grip tool to specify the plane in which to STRETCH a grip.

For example, Figure 40-28 illustrates the use of the move grip tool to STRETCH in a positive Z direction.

FIGURE 40-28

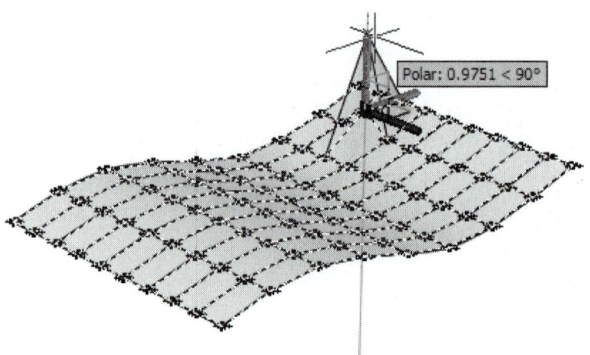

Remember that <u>any grip editing option other than STRETCH affects the entire polygon mesh</u>. For example, if you want to rotate the entire mesh in the grip's plane, you could select a grip, then cycle to the ROTATE option. However, you can use the rotate grip tool by hovering over the grip until the move tool appears, then PICK an axis on the move grip tool (such as in Fig. 40-28), then cycle to the ROTATE option so the rotate grip tool appears. Rotate about any one of the axes by pressing the Esc key and selecting the desired axis (Fig. 40-29).

FIGURE 40-29

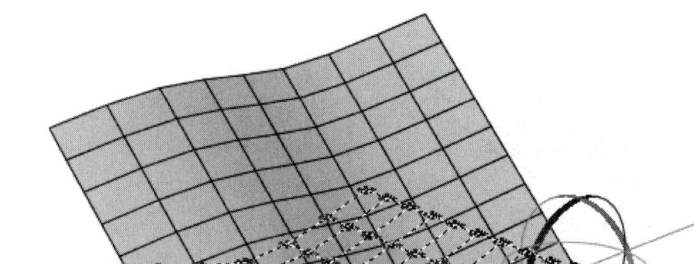

Editing Polygon Meshes with *Explode*

The *Explode* command can be used to allow editing *3Dmesh*es, geometrically defined meshes (*Edgesurf, Revsurf*, etc.), 3D primitives (*3D Objects...*), and *Regions*.

When used with other AutoCAD objects, *Explode* will "break down" the object group into a lower level of objects; for example, a *Pline* can be broken into its individual segments with *Explode*. Likewise, *Explode* will "break down" a 3D polygon mesh, for example, into individual 1 by 1 meshes (in effect, *3Dfaces*). *Explode* will affect any polygon meshes (*3Dmesh, Edgesurf, Revsurf, Tabsurf, Rulesurf*). *Explode* converts a polygon mesh into its individual 1 by 1 meshes (*3Dfaces*).

CONSTRUCTION TECHNIQUES FOR POLYGON MESHES AND *3DFACES*

Surfaces are created individually and combined to form a complete surface model. The surfaces can be created and moved into place in order to "assemble" the complete model, or surfaces can be created <u>in place</u> to form the finished model. Because polygon meshes require existing geometry to define the surfaces, the construction technique generally used is:

1. Construct a wireframe model (or partial wireframe) defining the edges of the surfaces on a separate layer.
2. Attach polygon mesh and other surfaces to the wireframe using another layer.

The creation of UCSs greatly assists in constructing 3D elements in 3D space during construction of the wireframe and surfaces. The sequence in the creation of a complete surface model is given here as a sample surface modeling strategy.

Surface Model Example

Consider the object in Figure 40-30. To create a surface model of this object, first create a partial wireframe.

FIGURE 40-30 ――――――――――――

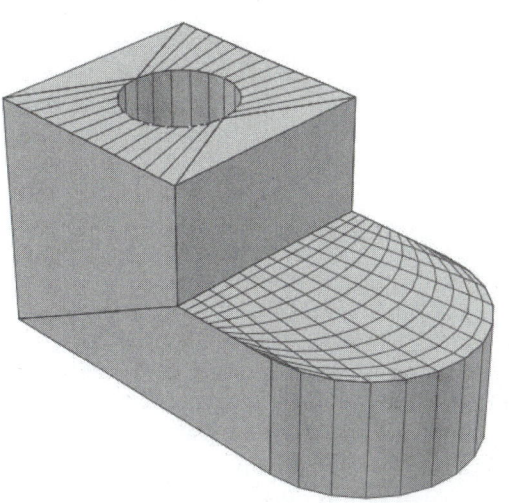

1. Create part of a wireframe model using *Line* and *Circle* elements. Create the wireframe on a separate layer. Begin the geometry at the origin. Use *3Dorbit* to enable visualization of three dimensions (Fig. 40-31).

FIGURE 40-31 ――――――――――――

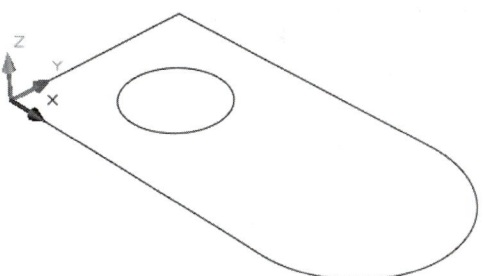

2. *Copy* the usable objects to the top plane by specifying a Z dimension at the "second point of displacement" prompt (Fig. 40-32).

FIGURE 40-32 ――――――――――――

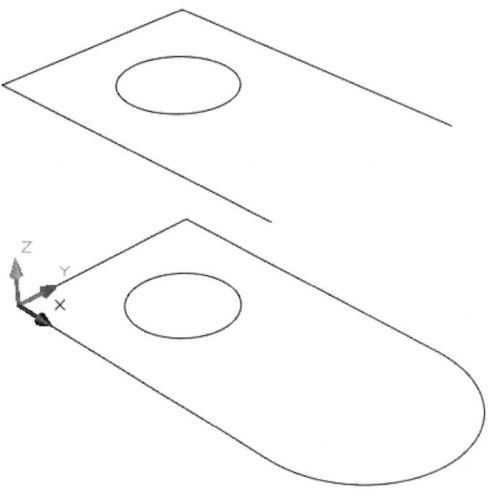

3. Create a UCS on this new plane using the *Origin* option. *OSNAP* the *Origin* to the existing geometry. Draw the connecting *Line* and *Trim* the extensions (Fig. 40-33).

FIGURE 40-33

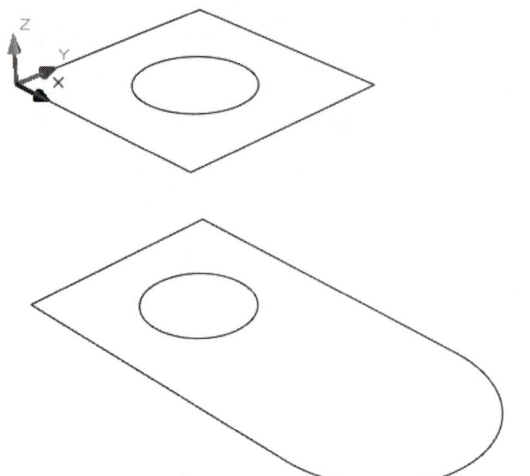

4. *Copy* the *Arc* and two attached *Lines* specifying a Z dimension for the "second point of displacement." *Copy* the *Line* drawn last (on the top plane) specifying a negative Z dimension. *Trim* the extending *Line* ends. A new *UCS* is not necessary on this plane (Fig. 40-34).

    This is not a complete wireframe, but it includes all the edges that are necessary to begin attaching surfaces.

FIGURE 40-34

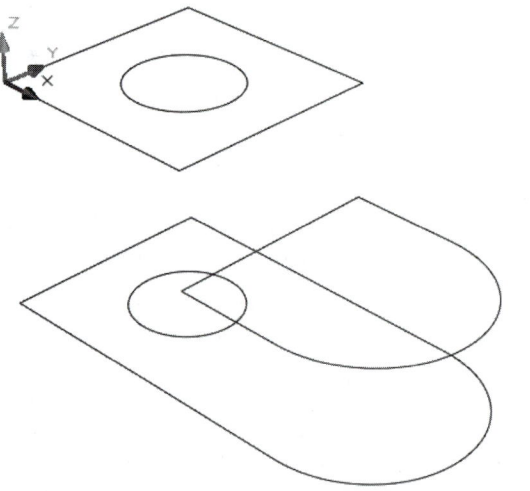

5. On a new layer, a 2-segment *3Dface* is created by PICKing the *Endpoints* of the wireframe objects in the order indicated. Remember that *3Dface* segments connect at the third and fourth points (Fig. 40-35).

FIGURE 40-35

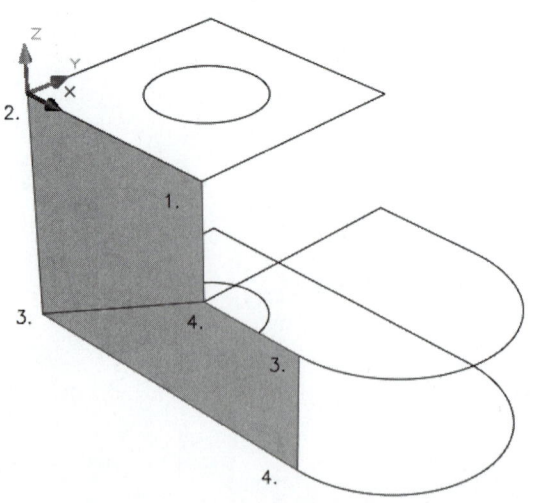

6. The *3Dface* is *Copied* to the opposite side of the model. A vertical *3Dface* is created between the top and the intermediate plane (Fig. 40-36). Another vertical *3Dface* is created on the back.

FIGURE 40-36

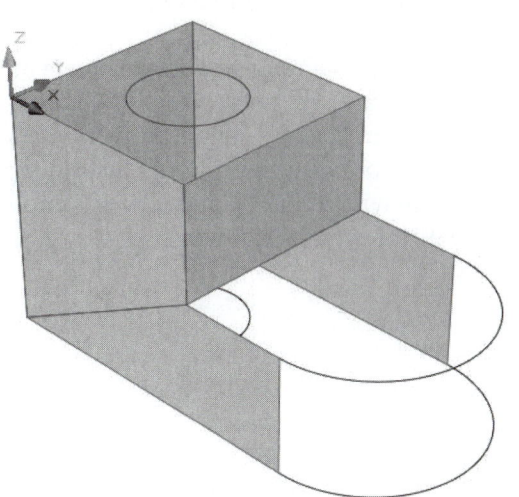

7. *Rulesurf* is used to create the rounded surface shown between the two *Arcs* (Fig. 40-37). The *SURFTAB1* and *SURFTAB2* variables were previously set to 12.

FIGURE 40-37

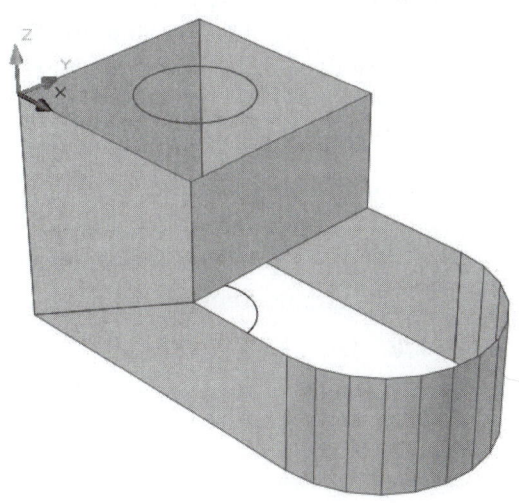

8. Next, a *Rulesurf* is used to create the cylindrical hole surface between the *Circles* on the top and bottom planes (Fig. 40-38).

FIGURE 40-38

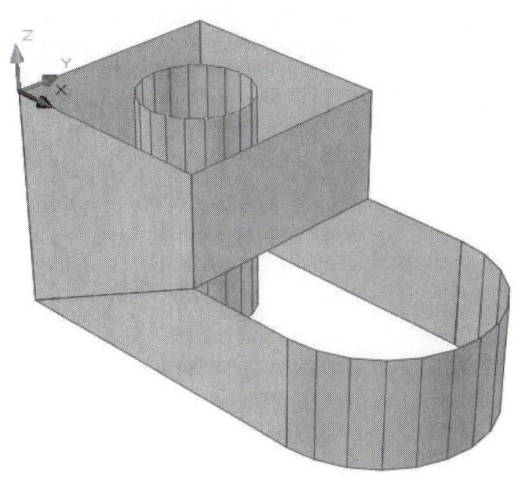

9. The surface connecting the *Arc* and three *Lines* is created by *Edgesurf* (Fig. 40-39).

FIGURE 40-39

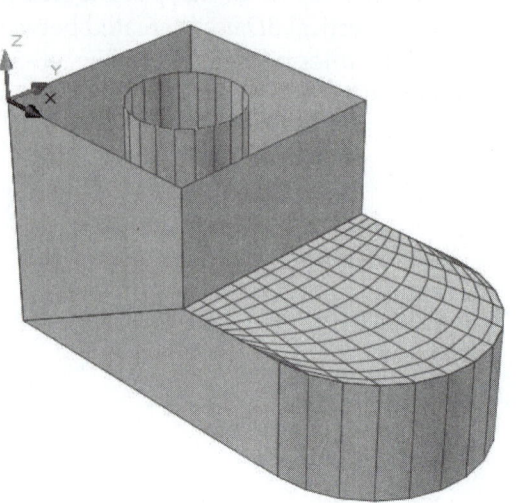

10. The only remaining surfaces to create are the top and bottom. <u>There is no</u> polygon mesh, *3Dface*, or *3Dmesh* that can be generated to create a simple surface configuration such as that on top—a surface with a hole! It must be accomplished by creating several adjoining surfaces.

Several methods are possible for construction of a surface with a hole. (The top surface has been isolated here for simplicity.) Probably the simplest method is shown here (Fig. 40-40). First, the *Circle* must be <u>replaced</u> by two 180 degree *Arcs*. Next a *Rulesurf* is created between one
Arc and one edge as shown.

FIGURE 40-40

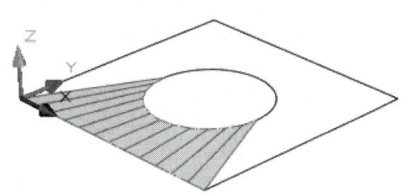

11. The top plane is completed by creating another *Rulesurf* and two triangular *3Dfaces* adjoining the *Rulesurfs* (Fig. 40-41). (Wireframe elements are <u>not</u> required for the *3Dface*.)

FIGURE 40-41

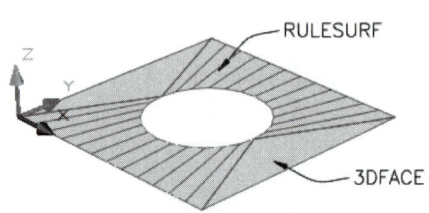

RULESURF

3DFACE

12. The surface model is shown here (Fig. 40-42). The bottom surface (not visible here) was created by *Copying* the top plane surfaces and the intermediate plane *Edgesurf*. The layer containing the wireframe elements has been *Frozen*.

FIGURE 40-42

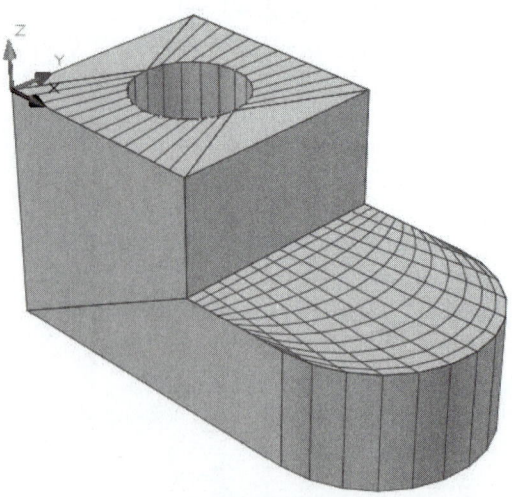

13. Using the *Visual Style control panel*, the final surface model is displayed with the *No edges* option (Fig. 40-43). The *No edges*, *Facet edges*, and *Isolines* options each depict the polygon mesh lines differently.

FIGURE 40-43

Limitations of Polygon Meshes and *3Dface*

From the previous example of the creation of a relatively simple 3D surface model, you can imagine how construction of some shapes with *3Dface* or polygon meshes can be somewhat involved. The particular characteristics of a surface that cause some difficulty in creation are the surfaces that embody many edges or that have holes, slots, or other "islands." Examine the surfaces shown in Figure 40-44. Although some of these shapes are fairly common, the construction techniques using *3Dfaces* and polygon meshes could be very involved.

3Dface and polygon meshes would <u>not</u> be an efficient construction technique to use in the creation of shapes such as these. Another AutoCAD feature, Region Modeling, can be used to easily create such shapes. Region models can be combined with *3Dfaces*, polygon meshes, and other surfacing techniques to create surface models.

FIGURE 40-44

TIP

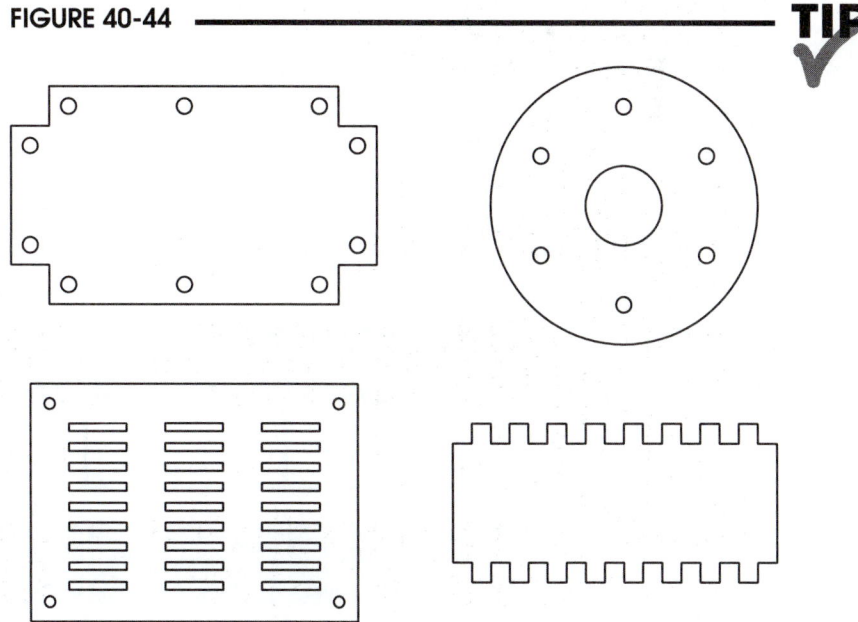

REGION MODELING APPLICATIONS

Region modeling is sometimes defined in the AutoCAD documentation as "2D solid modeling," <u>not</u> "surface modeling." This is because region modeling applies solid modeling techniques, namely, Boolean operations, to 2D closed objects. Practically, however, *Regions* can be considered surfaces.

A region is created by applying the *Region* command to <u>closed</u> 2D objects such as *Plines*, *Circles*, *Ellipses*, and *Polygons* (see Chapters 15 and 16). *Regions* are always <u>planar</u>. <u>Composite</u> regions can be constructed by combining multiple *Regions* using Boolean operations. AutoCAD allows you to utilize the Boolean operations *Union*, *Subtract*, and *Intersect* with *Regions* to create planar surfaces with holes and complex shapes. These surfaces can then be combined with other surfaces (regions, *3Dfaces*, or polygon meshes) on other planes to create complex surface models.

 Region modeling solves the problem of creating relatively simple surfaces that are not easily accomplished with the polygon mesh commands such as *3Dmesh*, *Rulesurf*, *Revsurf*, *Tabsurf*, and *Edgesurf*. Region modeling makes creation of surfaces with islands (holes) especially easy.

For example, consider the creation of a rectangular surface with a circular hole. This problem is addressed in the previous Surface Model Example (see Fig. 40-41). Using the polygon mesh commands, this relatively simple surface requires four *Lines*, two *Arcs* as a wireframe structure, then two *Rulesurfs* and two *3Dfaces* to complete the surface.

With the Region Modeler, the same surface can be created by using *Subtract* with a circular *Region* (create a *Circle*, then use *Region*) and a rectangular *Region* (draw a closed rectangle with *Line*, *Pline*, or *Rectangle*, then use *Region*; Fig. 40-45).

FIGURE 40-45

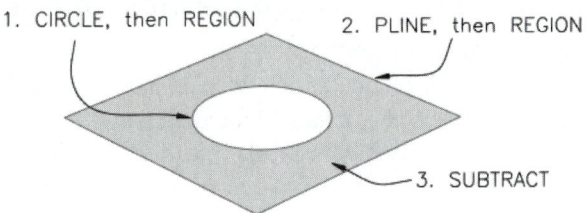

 Consider the surfaces shown in Figure 40-44. Imagine the amount of work involved in creating the surfaces using commands such as *3Dface*, *3Dmesh*, *Rulesurf*, *Revsurf*, *Tabsurf*, and *Edgesurf*. Next, consider the work involved to create these shapes as regions. Surfaces such as these are very common for mechanical applications for surface models.

USING *THICKNESS* AND *ELEVATION*

Thickness

Pull-down Menu	Command (Type)	Alias (Type)	Short-cut	Screen (side) Menu	Tablet Menu
Format *Thickness*	*Thickness*	TH	...	*FORMAT* *Thicknes*	V,3

Elevation

Pull-down Menu	Command (Type)	Alias (Type)	Short-cut	Screen (side) Menu	Tablet Menu
...	*Elevation*	...	...	...	...

Thickness and *Elevation* were introduced with Version 2.1 (1985) as AutoCAD's first capability for creating 3D objects. *Thickness* and *Elevation* are still useful for creating simple surfaces. Most 2D objects (*Line*, *Circle*, *Arc*, *Point*, *Pline*, *Rectangle*, *Polygon*, etc.) can be assigned *Thickness* or *Elevation* property. Only newer (since Release 13) objects such as NURBS curves (*Spline*, *Ellipse*), *Xline*, *Ray*, and *Mlines* cannot possess *Thickness* but <u>can</u> possess *Elevation*. *Thickness* and *Elevation* are usually assigned before using a 2D command; however, they can also be changed later using *Change* or *Properties*.

Thickness is a value representing a <u>Z dimension</u> for a 2D object. The new object is created using the typical draw command (*Line, Circle,* or *Arc*, etc.), but the resulting object has a uniform Z dimension as assigned. For example, a two unit high cylinder can be created by setting *Thickness* to 2 and then using the *Circle* command (Fig. 40-46).

FIGURE 40-46

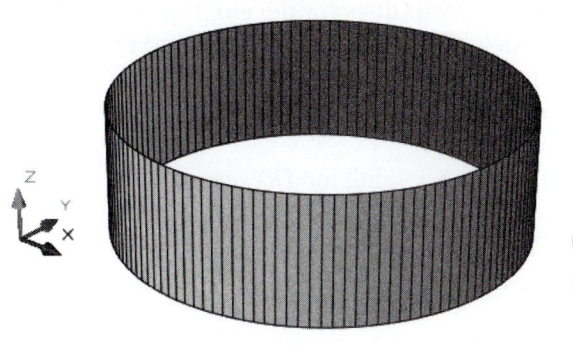

Thickness is <u>always perpendicular</u> to the XY plane of the object. Until later versions of AutoCAD, this posed a major limitation with the use of *Thickness*. This limited cylinders, for example, to having a vertical orientation <u>only</u>; no cylinders could be created with a horizontal orientation. The term "2 ½-D" was used to describe this capability of CAD packages during this phase of development.

Today, with the ability to create User Coordinate Systems (UCSs), *Thickness* can be used to create surfaces in any orientation. Therefore, the effectiveness of *Thickness* and *Elevation* commands is increased. Figure 40-47 displays a cylinder oriented horizontally, achieved by creating a UCS having a vertical XY plane and a horizontal Z dimension. UCSs can be created to achieve any orientation for using *Thickness*.

FIGURE 40-47

With a *Thickness* set to a value greater than **0**, commands normally used to create 2D objects can be used to create 3D surfaces having a uniform Z dimension perpendicular to the current XY plane. Figure 40-48 shows a *Line, Arc,* and *Point* possessing *Thickness* property. Note that a *Point* object with *Thickness* creates a line perpendicular to the current XY plane.

FIGURE 40-48

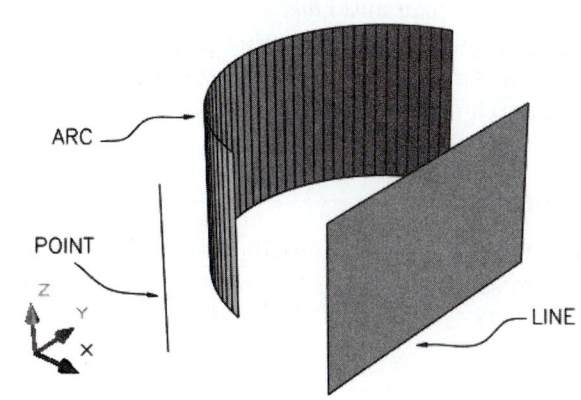

The ability to create *Plines* with *Thickness* offers many possibilities for developing simple surface models (Fig. 40-49). (Keep in mind that 2D shapes created with *Line* and *Arc* objects can be converted to *Plines* with *Pedit*.)

Assigning a value for *Thickness* causes all subsequently created objects to have the assigned *Thickness* property. *Thickness* must be set back to **0** to create normal 2D objects.

FIGURE 40-49

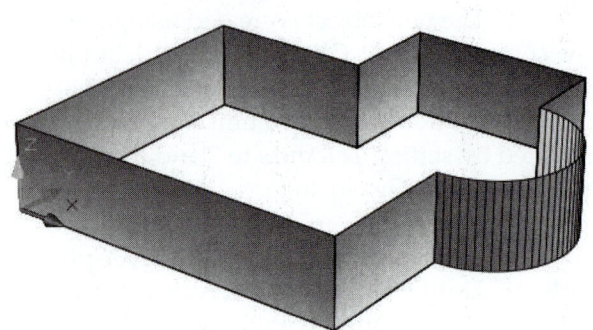

Elevation controls the Z dimension of the "base plane" for subsequently created objects. Normally, unless an explicit Z value is given, 2D objects (with or without *Thickness*) are created on the XY plane or at an *Elevation* of **0**. Changing the value of the *Elevation* changes the "base plane," or the elevation, of the construction plane for newly created objects.

Figure 40-50 displays the same objects as in the previous figure except the *Circle* has an *Elevation* of 2 (level with the top edges of the other objects).

Like *Thickness*, *Elevation* should be set back to 0 to create geometry normally on the XY plane.

FIGURE 40-50

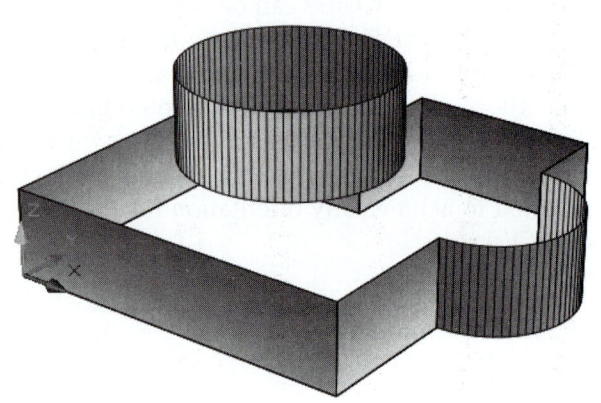

Changing *Thickness* and *Elevation*

Because *Thickness* and *Elevation* are properties that 2D objects may possess (like *Color* and *Linetype*), they can be changed or assigned <u>retroactively</u>. *Elevation* can only be changed retroactively using the *Change* command with the *Properties* option (see Chapter 16). *Thickness* can be changed using the *Properties* palette.

Properties

Using the *Properties* command (see Chapter 16) produces a dialog box providing the ability to change the *Thickness* of <u>existing</u> objects (Fig. 40-51).

 As an alternative to setting *Thickness* before creating surfaces, the 2D geometry can first be created and then changed to 3D surfaces by using this method to retroactively set *Thickness*.

FIGURE 40-51

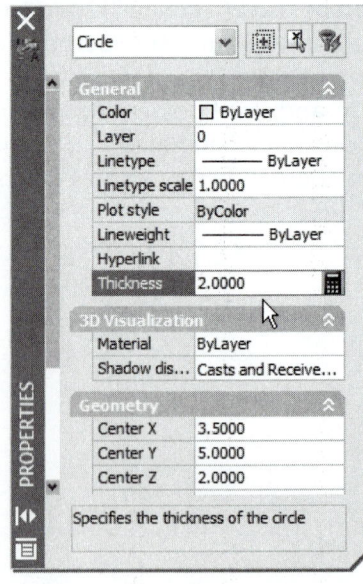

Figure 40-52 displays 2D *Pline* shapes. The *Properties* palette can be used to change the *Thickness* of the shape to create a 3D surface model.

FIGURE 40-52

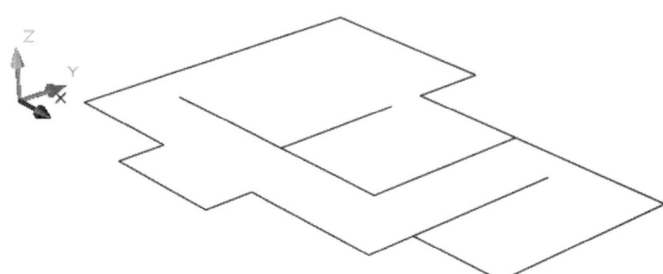

Figure 40-53 shows the same *Pline* shapes as in the previous figure after the *Thickness* property was changed to generate the 3D surface model.

FIGURE 40-53

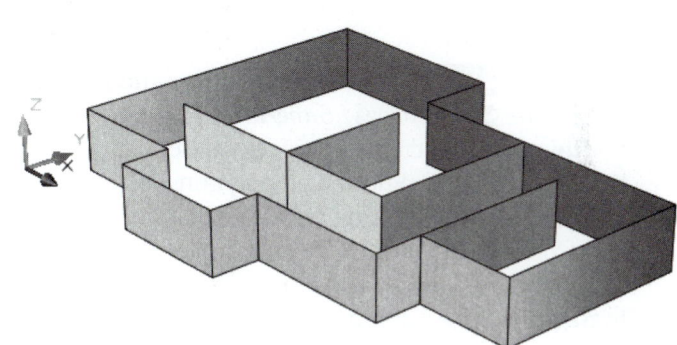

If you want to change an object's *Elevation*, the *Change* command can be used. The Command syntax is as follows:

```
Command: change
Select objects: PICK
Select objects: Enter
Specify change point or [Properties]: p
Enter property to change
[Color/Elev/LAyer/LType/ltScale/LWeight/Thickness/Material]: e
Specify new elevation <0.0000>: (value)
Enter property to change [Color/Elev/LAyer/LType/ltScale/LWeight/Thickness/Material]: Enter
Command:
```

CHAPTER EXERCISES

1–3. Create a surface model of each of the objects in Figures 40-54 through 40-56. These are the same objects you worked with in Chapter 37, Wireframe Modeling. You can use each of the Chapter 37 Exercises wireframes (**WFEX1**, **WFEX2**, and **WFEX3**) as a "skeleton" on which to attach surfaces. Make sure you create a *New Layer* for the surfaces. Use each dimension marker in the figure to equal one unit in AutoCAD. Locate the lower-left corner of the model at coordinate **0,0,0**. Assign the names **SURFEX1**, **SURFEX2**, and **SURFEX3**. Make a plot of each model in an appropriate visual style.

FIGURE 40-548

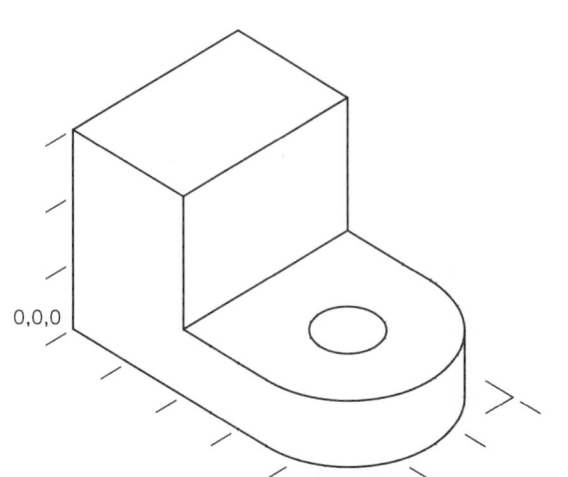

0,0,0

FIGURE 40-55

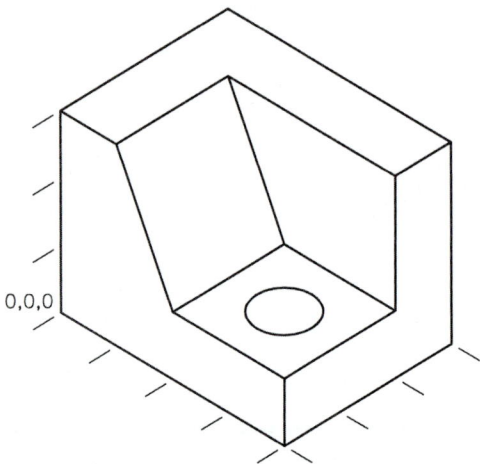

0,0,0

FIGURE 40-56

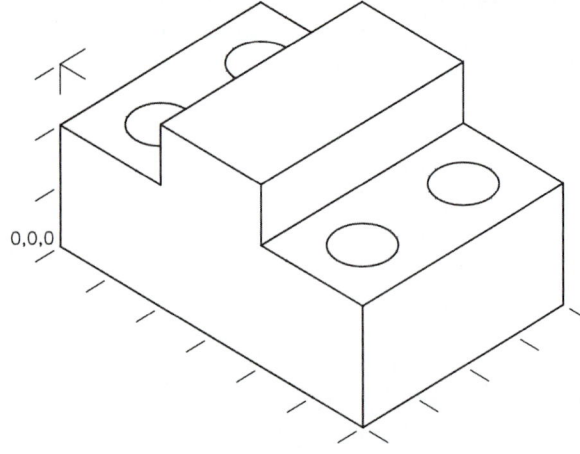

0,0,0

4. *Open* the WFEX4 drawing that you created in Chapter 37 Exercises. *SaveAs* **SURFEX4**. For this exercise, create a surface model using a combination of regions and polygon meshes. Use the existing wireframe geometry. Convert the *Lines*, *Arcs*, and *Circles* representing <u>planar</u> surfaces to closed *Plines*, then to *Regions*. Use polygon meshes for the curved surfaces. *Save* the drawing. Generate a descriptive viewpoint with *Vpoint* or *3Dorbit*. Activate a *Layout* and insert a titleblock and border. Create one *Vport* to display the model. *Plot* the drawing.

FIGURE 40-57

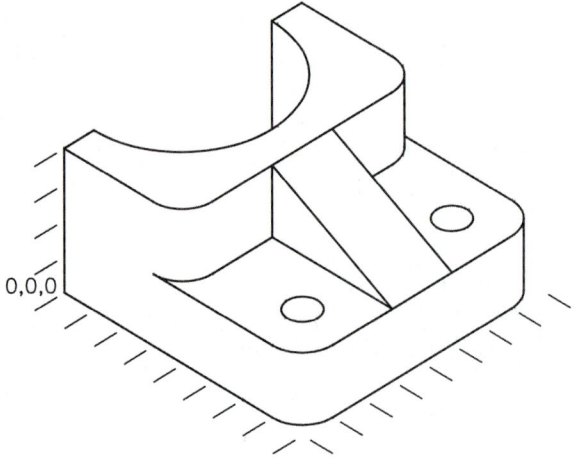

0,0,0

5. Create a surface model of the bar guide shown in Figure 40-58. You may want to use the BGUID-WF drawing that you created in Chapter 37 Exercises as a wireframe foundation. Make a *New Layer* for the surface model. Generate a descriptive viewpoint with *Vpoint* or *3Dorbit*. Activate a *Layout* and insert a titleblock and border. Create one *Vport* to display the model. *Plot* the model in an appropriate visual style. *Save* the surface model as **BGD-SURF**.

FIGURE 40-58

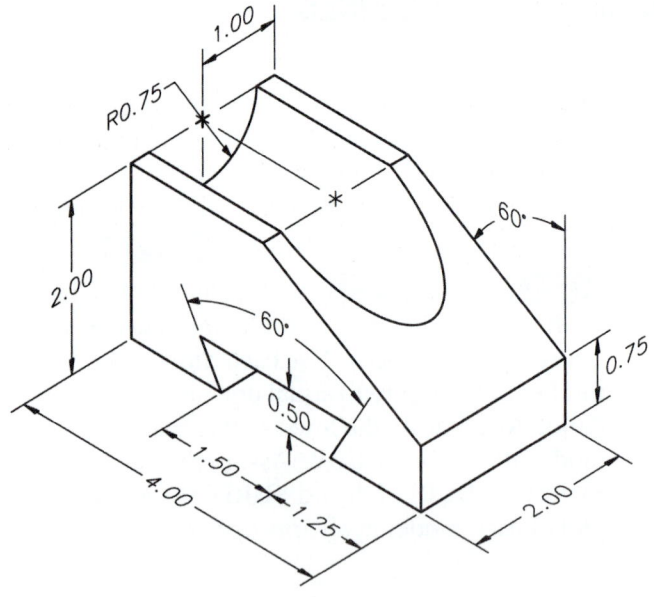

6. Create a surface model of the saddle in Figure 40-59. There is no previously created wireframe model, so you must begin a *New* drawing or use a *Template*. The fillets and rounds should be created using *Rulesurf* or *Tabsurf*. *Revsurf* must be used to create the complex fillets and rounds at the corners. Generate a descriptive viewpoint with *Vpoint* or *3Dorbit*. Activate a *Layout* and insert a titleblock and border. Create one *Vport* to display the model. Plot the drawing. *Save* the drawing as **SDL-SURF**.

FIGURE 40-59

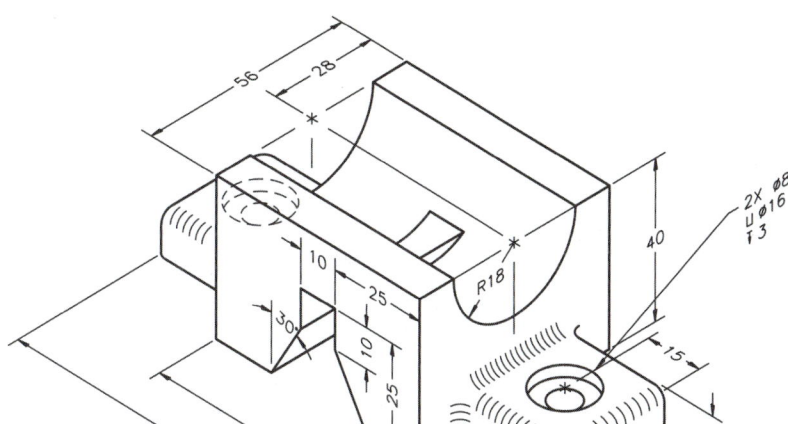

7. Begin a *New* drawing (or use a *Template*) for the pulley shown in Figure 40-60. This is the same object that you worked on in Chapter 38 Exercises. As a reminder, the vertical dimensions are diameters. Use *Revsurf* to create the curved (cylindrical) surfaces. Use *Regions* to create the vertical circular faces that would appear on the front and back (the surfaces containing the 4 counterbored holes or the keyway). Generate a descriptive viewpoint with *Vpoint* or *3Dorbit*. Activate a *Layout* and insert a titleblock and border. Create one *Vport* to display the model. *Save* the drawing as **PUL-SURF**.

FIGURE 40-60

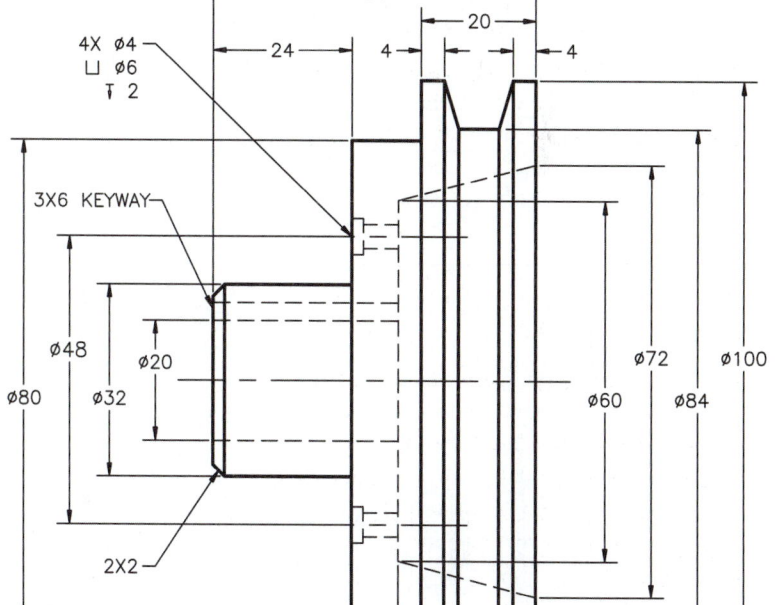

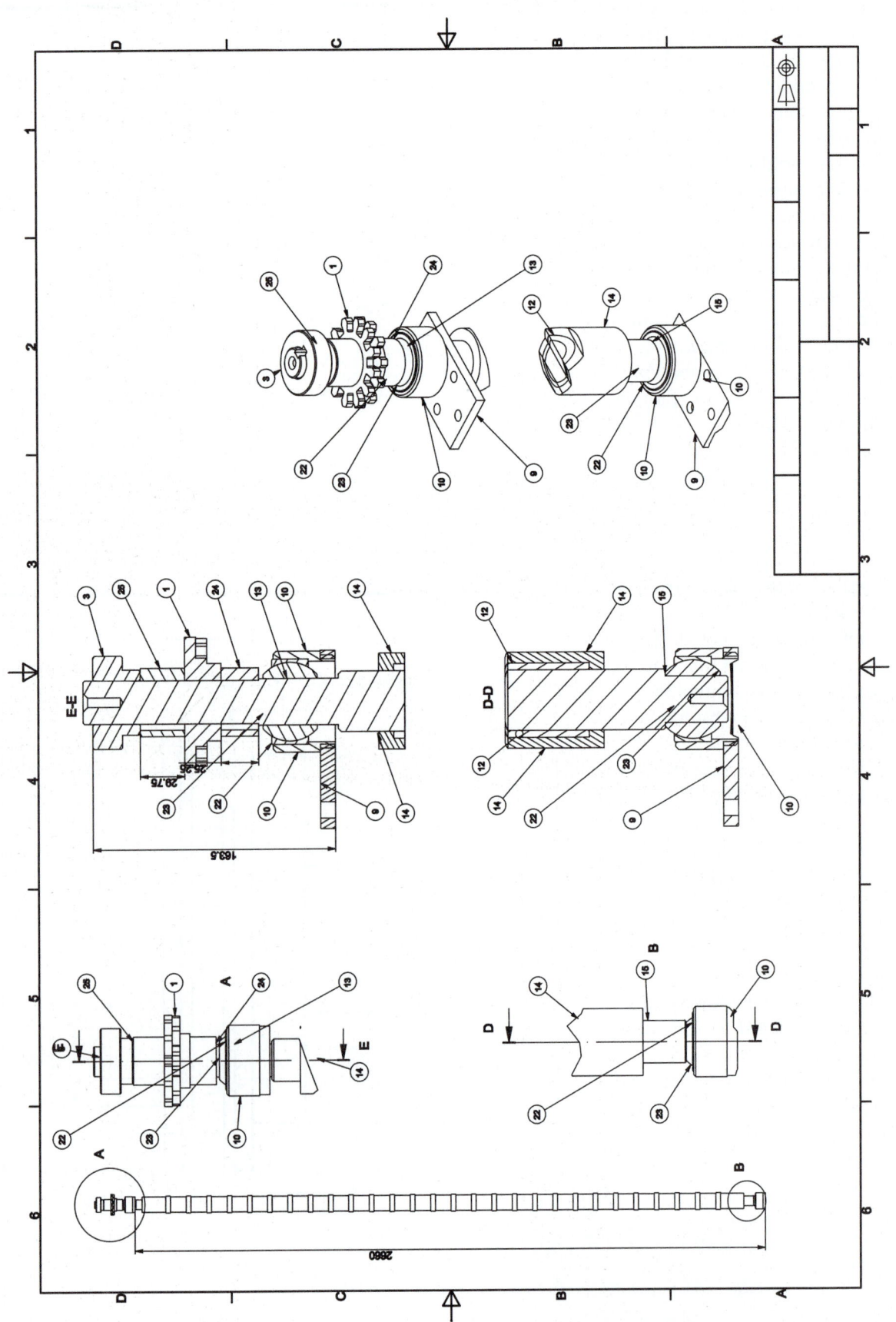

DRIVE ROLLER.DWG, Courtesy of Autodesk, Inc.

41

RENDERING AND ANIMATION

CHAPTER OBJECTIVES

After completing this chapter you should:

1. know the typical steps for rendering a surface or solid model using *Render*;

2. be able to create a variety of backgrounds for a rendering with *Fog* and *Background*;

3. understand and be able to create and set parameters for the three types of lighting: *Point Light*, *Distant Light*, and *Spotlight*;

4. be able to select *Materials* and *Attach* them to to object, faces, or layers;

5. be able to adjust material properties such as *Color*, *Shininess*, *Translucency*, *Self-illumination*, *Opacity*, and other properties;

6. be able to map raster images to objects and adjust the image orientation;

7. be able to specify parameters and create a rendering using the *Render* command;

8. be able to create animations using *3D Walk* and *3D Fly*.

CONCEPTS

As you know, AutoCAD provides the ability to enhance the visualization of a <u>surface</u> or a <u>solid</u> model using *Visual Styles*. *Visual Styles* allow you to add color and lighting to enhance the visibility of 3D objects. However, another AutoCAD feature called *Render* adds another level of realism to surface and solid models by giving you control of backgrounds, lights, colors, materials, and other effects.

Render involves the use of *Light* sources and the application of *Materials* for the 3D model. You control the type, intensity, color, shadows, and positioning of lights in 3D space. Render also provides a variety of materials for attachment to objects or layers and allows you to adjust the properties of the material. You can use digital images and map them onto surfaces of objects. With *Render* you can create a *Background* and use *Fog*.

Render is intended to be used as a visualization tool for 3D models. Visualization of a design before it has been constructed or manufactured can be of significant value. Presentation of a model with appropriate light sources and materials can reveal aspects of a design that might otherwise be possible only after construction or manufacturing. For example, a rendered architectural setting may give the designer and client previews of color, lighting, or spatial relationships. Or, a rendered model may allow the engineer to preview the functional relationship of mechanical parts in a manner otherwise impossible with wireframe representation. Thus, this "realistic" presentation, made possible by rendering a model, can shorten the design feedback cycle and improve communication among designers, engineers, architects, contractors, and clients.

With AutoCAD *Render* you can create "photo-realistic" displays. The *Render* command uses the *Lights* and *Materials* you assign and calculates the display based on the parameters that you specify. The ability to calculate shadows and reflection of color and light from one surface to another characterizes this "photo-realistic" quality.

You can render to a viewport, to a separate rendering window, or to a file. The rendered screen image created with the *Render* command can be saved as one of a number of file formats and replayed at a later time. The *Render* and *Advanced Render Settings* palettes provide control of format and quality of the rendered image.

The commands associated with rendering are available from the *Light control panel*, *Materials control panel*, and the *Render control panel* on the *Dashboard* (Fig. 41-1). The commands associated with rendering are also available from the *View* pull-down menu and *Render* toolbar.

FIGURE 41-1

Creating a Rendering with Default Settings

FIGURE 41-2

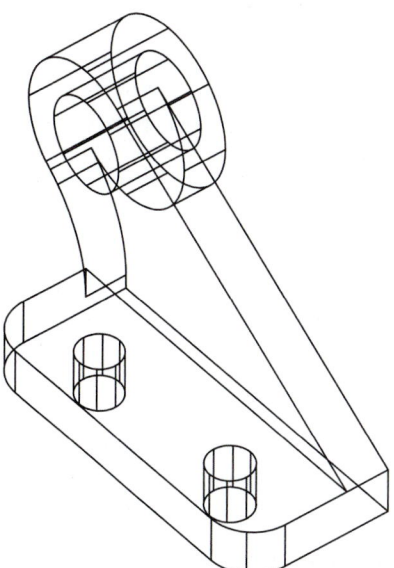

Creating a realistic rendering often takes more time that creating the 3D model. Although many parameters, adjustments, and steps can be used to create a realistic rendering (discussed later), you can use the *Render* command to create a rendering without specifying any background, lights, or materials.

To create a rendering, you must first create a 3D model, then use commands such as *Vpoint*, *3Dorbit*, *Camera*, *Pan*, and/or *Zoom* to generate a representative view of the function or feature of the model you want to display. For example, Figure 41-2 displays the support bracket from the desired viewpoint in wireframe representation.

Next use the *Render* command to produce a rendering of your 3D objects (Fig. 41-3). You can use the *Render* command to generate a rendering using default settings. AutoCAD creates the rendering using the current background color, the current object color, and default light sources. The default light sources are not capable of casting shadows and appear in fixed positions located behind and on each side of the current viewpoint, called "over-the-shoulder" lighting. No materials are used for a default rendering. Accepting the default "destination," the rendering is created in a separate window giving information about the image.

FIGURE 41-3

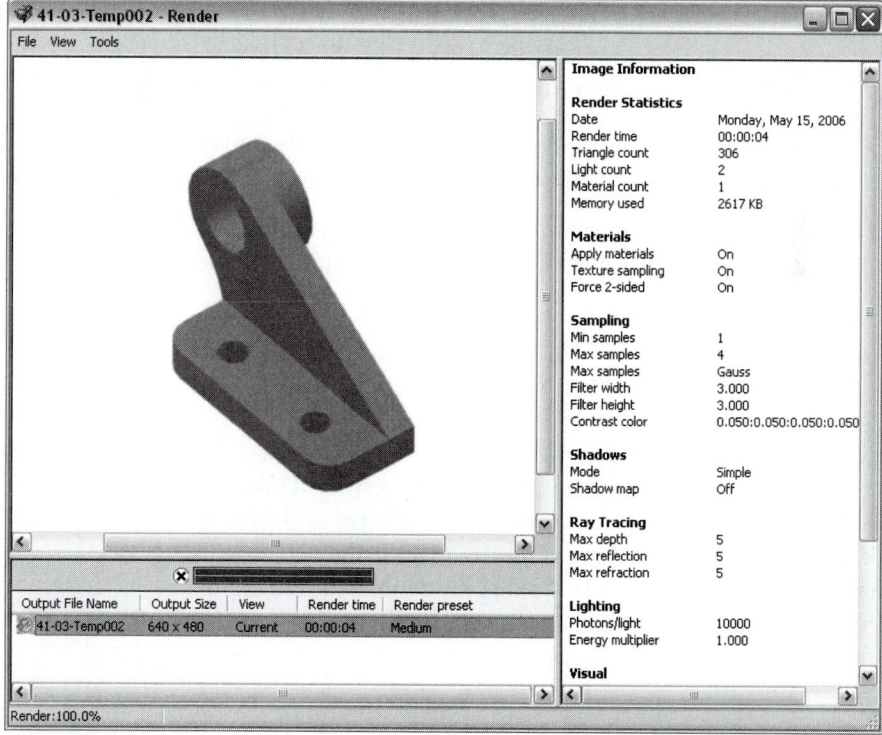

Typical Steps for Using Render

Although a rendering of a 3D surface or solid model can be created easily using default settings, the following steps are normally taken to create a rendering with the desired effects:

1. You must first create or open an existing 3D model. Both surface and solid models can be rendered.

2. Use any 3D navigation commands (*3Dorbit*, *Zoom*, *Pan*) needed to specify the desired viewing position. *Perspective mode* is allowed. If you want to create several renderings from different viewpoints, use *View* to save each display to a named view so it can be restored at a later time.

3. Set the background for the object to be rendered. You can use a solid color or gradient color as the background. You can specify an image (.BMP, .TIF, or other file format) as the background to merge with the objects to be rendered. You can alternately specify a *Render Environment*, which gives a *Fog* appearance.

4. Use the *Lights control panel* to create one or more lights. Parameters such as light type, intensity, color, and location (position in space) can be specified. A light source representing the sun can be specified by geographic location and time of day. You can specify if the lights cast shadows.

5. Use the *Materials control panel* to specify the desired material(s) for the object(s). The material properties such as color, reflective qualities, mapped images, and other attributes can be adjusted. You can preview the material as you adjust the parameters.

6. Specify the desired format and quality for the rendering using the *Advanced Rendering Settings* palette. Several options are available including the *Destination* for the rendering, such as rendered to a *Window*, *Viewport* or *File*.

7. Use the *Render* command to create the rendered image to the specified destination.

8. Because rendering normally involves a process of continual adjustment to light and material parameters, any or all of steps 2 through 7 can be repeated to make the desired adjustments. Don't expect to achieve exactly what you want the first time.

9. Once your test renderings approach the results you are looking for, apply the higher rendering qualities such as reflections and refractions to achieve the highest quality for the final rendering.

10. Use the *Render* command again to create the final rendered display or file. Usually, you should render to the screen until you have just the image you want; then save to a file if desired.

Color Systems

There are several methods for choosing and assigning colors in the rendering for such applications as background colors, material colors, and light colors. These systems appear throughout the many palettes and dialog boxes. These concepts are presented here. In each case, current color adjustments are displayed in a square color box or preview image tile appearing in the dialog box or palette.

Index Color (ACI)
The index colors (AutoCAD Color Index) is the original system used to select a color in AutoCAD. You can select from a color palette of 255 colors, ranging from an index value of 1 to 255. The first seven colors are "standard" colors that can be specified using a name or the index value. Figure 41-4 shows the *Index Color* tab of the *Select Color* dialog box. Select one of the color swatches to specify the color you want to use.

FIGURE 41-4

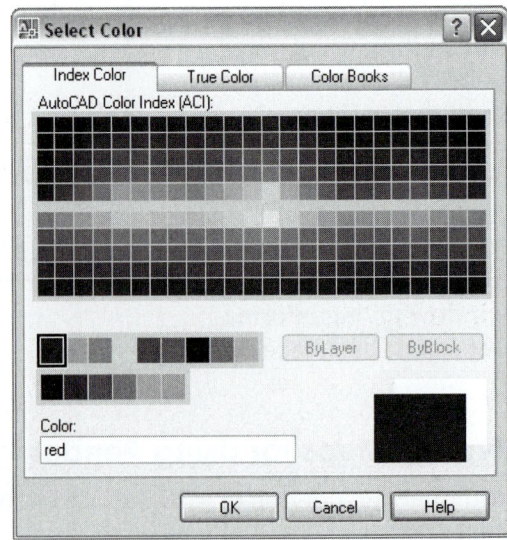

True Color, RGB Color Model

True colors can be selected by one of two methods: *RGB* and *HLS*. To use the *RGB* color model, select the *True Color* tab in the *Select Color* dialog box and select *RGB* from the *Color Model* drop-down list (Fig. 41-5). Adjust the sliders and values accordingly for each of the color components to specify the color you want to use. The *RGB* color model determines the resultant color by its red, green, and blue components. Each of the components can be specified with a value range of 0 to 255. A value close to 0 means that less of that particular color is used to create the final color, while a number closer to 255 means that more of the color is used. If a value of 0 is specified for all three components, a pure white color is created. If 255 is specified, a pure black results.

True Color, HLS Color Model

True colors can also be specified using *Hue*, *Saturation*, and *Luminance* to determine the color. The final mix of values creates an RGB color value that AutoCAD uses. To use the *HLS* color model, select the *True Color* tab in the *Select Color* dialog box and select *HLS* from the *Color Model* drop-down list (Fig. 41-6). Then adjust the sliders and values accordingly for each of the color settings to specify the desired color.

Hue

Hue is the primary color that is being selected based on the wavelength of light within the spectrum of visible light. The range is from 0 to 360.

Saturation

Saturation controls the purity of the color mix (*Hue* and *Luminance*). A value close to 100 for saturation means that the color specified is purest, and a value close to 0 mixes only a small amount of the hue to result in a "washed out" appearance. The range for saturation is 0 to 100.

Luminance

Luminance is used to specify the amount of white or black mixed with the hue. A value close to 0 for luminance means the color is dark or near black, and a value close to 100 means the color is light or near white. The range of luminance is 0 to 100.

Color Books

Color books are predefined colors used by the print, paint, and graphics industries. The popular industry standard systems are *Pantone*, *DIC*, and *RAL*. These are useful if you want to match specific color that has been specified by an interior designer, for example. To use a color from a color book, select the *Color Books* tab in the *Select Color* dialog box and then select a color book from the *Color Book* drop-down list (Fig. 41-7). Then use the slider and the color swatches to specify the color you want to use.

FIGURE 41-5

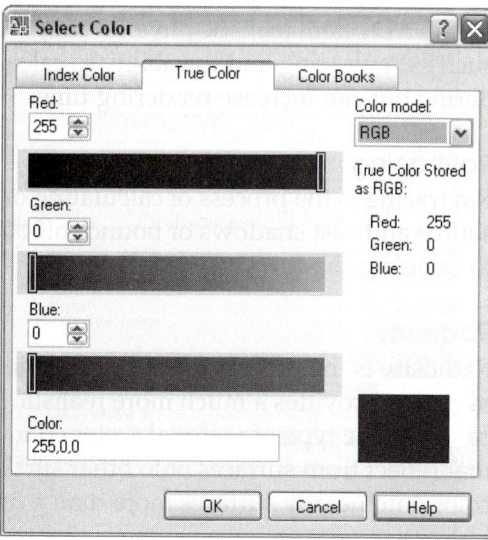

FIGURE 41-6

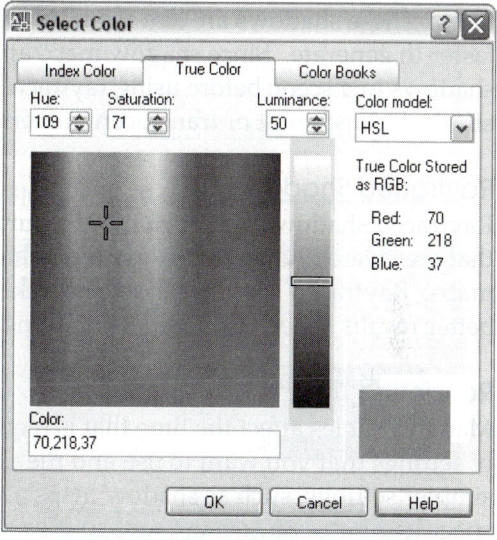

FIGURE 41-7

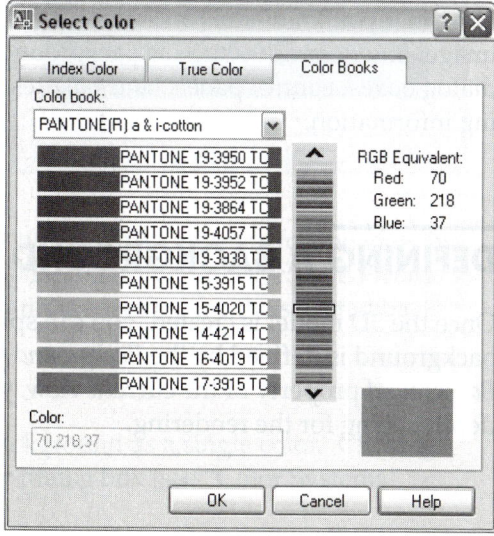

Adjust Background Image Dialog Box

Once an image has been selected in the *Background* dialog box, you can select the *Adjust Image* button to display the *Adjust Background Image* dialog box (Fig. 41-14).

Image Position

The *Image Position* drop-down list allows you can select *Center*, *Stretch* or *Tile* to modify the selected image. If *Center* or *Tile* is selected, the adjustment sliders enable you to adjust the image's *Offset* and *Scale*.

Sliders Adjust

Select either the *Offset* or *Scale* option and then adjust the *X* and *Y Scale* sliders to adjust the image. As you adjust the sliders, the preview displays the changes you make to the image.

Maintain Aspect Ratio When Scaling

The *Maintain Aspect Ratio When Scaling* checkbox forces the *X* and *Y Scale* sliders to move as one.

Reset

The *Reset* button is used to reset the *X* and *Y Scale* sliders for the current *Sliders Adjust* option.

FIGURE 41-14

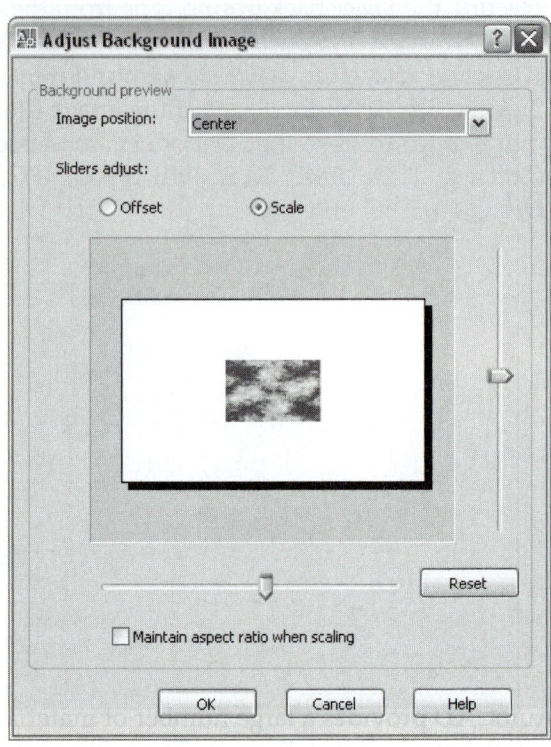

Renderenviron-ment

	Pull-down Menu	Command (Type)	Alias (Type)	Short-cut	Screen (side) Menu	Tablet Menu
	View > *Render >* *Render Environment*	*Render-environment*	*FOG*	...	*VIEW 2* *Render* *Environment*	*P,2*

The *Renderenvironment* command displays the *Render Environment* dialog box (Fig. 41-15). This tool is used to create the perception of depth in a rendering by using a "fog" effect. The current viewing direction and specified color are used to determine how the "fog and depth cue" are applied to the rendering. AutoCAD accomplishes this effect by assigning more of the color to the parts of the model that are farther from the observer (in the view) and assigning less color to geometry that is nearer.

FIGURE 41-15

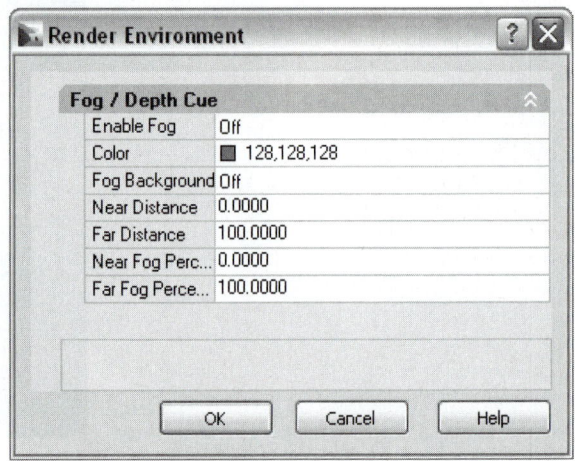

Enable Fog

This option controls if fog is applied or not when rendering. *Fog* must be *On* in order for the other settings in the dialog box to be used.

Color

Specify the color for the fog in this section. Select a color from the *Color* drop-down list or choose the *Select Color* option to display the *Select Color* dialog box (see "Color Systems" earlier this chapter).

Fog Background

Turn this *On* to apply the fog to both the background and the geometry.

Near Distance and Far Distance

The *Near Distance* and *Far Distance* values are used to specify nearest distance from the current view (camera) that the fog begins and the farthest distance from the current view the fog ends.

Near Fog Percentage and Far Fog Percentage

FIGURE 41-16

The *Near Fog Percentage* and *Far Fog Percentage* values are used to define the amount of opacity that is applied to the fog. The range of values is from 0 (no fog) to 100 (maximum fog). Setting the *Near Fog Percentage* to a low value and the *Far Fog Percentage* to a higher value makes the objects at the far distance less visible. For example, notice how in Figure 41-16 the color (white) is applied more heavily to the far end of the object to give the illusion of distance.

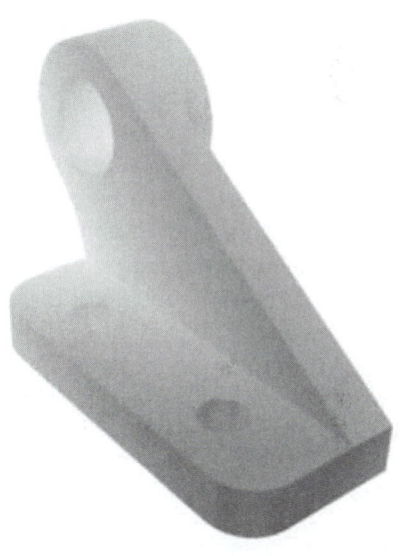

CREATING LIGHTS IN A MODEL

AutoCAD utilizes two different types of lighting systems: default lighting and user-defined lighting. Every drawing utilizes default lighting. User-defined lighting is enabled when you add at least one user-defined light to a model.

AutoCAD provides three different types of user-defined lights: *Point Light*, *Distant Light*, and *Spotlight*. A special type of *Distant Light* is *Sunlight*. *Point Lights* emit light in all directions from the light location, similar to a light bulb. On the other hand, *Distant Lights*, *Spotlights*, and *Sunlight* emit light in a specific, user-defined direction, such as when a spotlight or the sun shines from a particular direction. Similar to real life, the light generated from *Point Lights* and *Spotlights* "fall off," or decrease as a function of distance from the source. However, the light generated from *Distant Lights* and *Sunlight* does not decrease in intensity with distance.

By default, if any user-defined lights are on, default lighting automatically turns off. Only one type of lighting system, default lights or user-defined lights, can be enabled at one time. You can force the default lighting only to be on by setting the *DEFAULTLIGHTING* system variable to 1. In this case, the sun and other user-defined lights do not generate light, even if they are turned on. The setting of this system variable is viewport-specific.

Default Lighting

By default AutoCAD creates one or two "distant" lights that emit light from behind the upper-left and upper-right corners of the current viewpoint of the model—these lights follow the view direction when it changes. These lights are imagined as "over the shoulder" lights. AutoCAD provides only one control for the default lighting. You can switch between two default lights or a single default light using the *DEFAULTLIGHTINGTYPE* system variable (1 = two lights, 0 = one light). *DEFAULTLIGHTINGTYPE* is viewport-specific.

User-Defined Lighting

User-defined lights are lights that you specifically create and place in a model. The three types of lights you can create are *Point Lights*, *Spotlights* and *Distant Lights* (*Sunlight* is considered a special type of *Distant Light*).

Point Light

A *Point Light* is inserted at the location you specify, but no direction vector is needed since this light shines in all directions. Point light is typical of the light emitted from a light bulb. Point light falls off (grows dimmer) as the distance between the light source and surfaces increases, so surfaces close to the light are brighter than those at a distance. The objects in Figure 41-17 are illuminated by only one *Point Light* above the nearest corner and slightly to the right side. Notice how the light grows dimmer the farther the surfaces are from the light source.

FIGURE 41-17

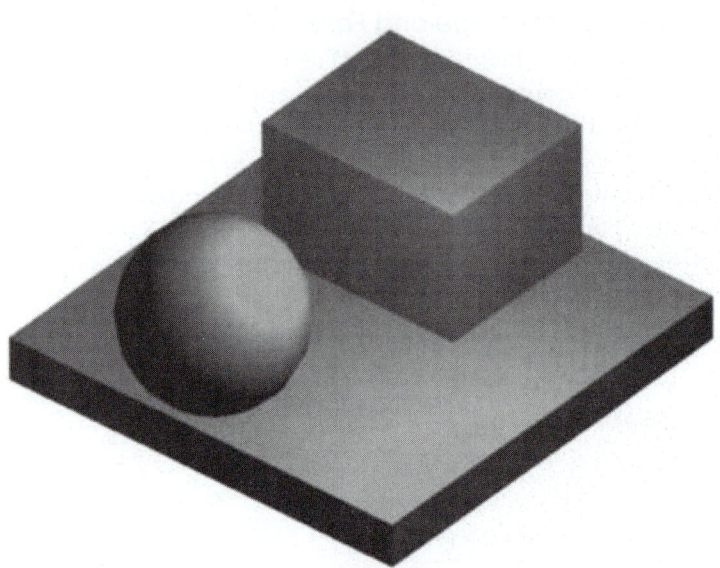

Spotlight

Spotlights have characteristics much like a spotlight used for stage lighting. You can specify the light's *Location* in space as well as the point that it shines toward (*Target*). Like point lights, the light falls off as the illuminated surfaces become farther from the light *Location*. Similar to stage spotlights, the beams of light from the source (*Location*) to the *Target* form a "cone" of light. Figure 41-18 shows the box and sphere illuminated by a *Spotlight* shining from directly above the objects. Notice the circular illuminated area formed by the cone of light. The angle of a *Spotlight's* cone of light can be adjusted for full illumination and for light fall-off.

FIGURE 41-18

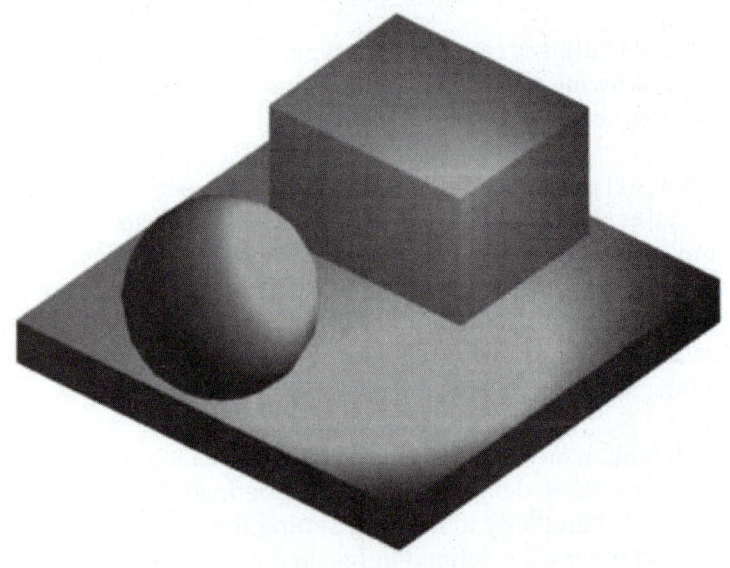

Distant Light

A *Distant Light* shines in one direction and has parallel rays. A *Distant Light* is defined only by a direction vector, not a light location. The vector determines the direction of the parallel beams of light which extend to infinity. Similar to sunlight, the light intensity does not fall off measurably (grow dimmer) for surfaces farther away. Illumination of a surface is determined only by the angle between the surface and the light beams—surfaces that are perpendicular to the beams are more brightly illuminated.

Figure 41-19 illustrates a box and sphere illuminated by <u>only</u> one *Distant Light* shining from above and to the right. Notice how each surface is illuminated differently because of the angle of the surface to the direction vector, yet each individual flat surface is illuminated evenly across the entire surface.

FIGURE 41-19

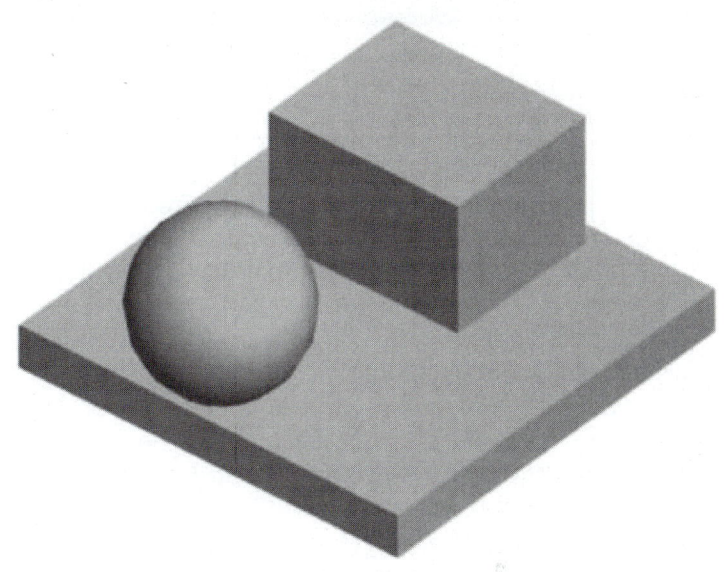

Lighting Parameters

Point Lights, *Spotlights* and *Distant Lights* affect the brightness or color of a surface in a model depending on the following parameters:

1. The *Intensity* of the light can be specified. The higher the *Intensity*, the brighter the rendered surfaces appear.

2. The *Color* of the light can be specified. The lighter the *Color*, the brighter the rendered surfaces appear.

3. The distance between the light and the surface affects brightness (*Point Lights* and *Spotlights* only).

4. The angle between the light source and the surface affects the brightness of the surface. A surface that shines the brightest is one in which the light is shining perpendicular to the surface. The more angled the surface is from perpendicular, the darker it appears.

5. When surfaces have an assigned material with a high *Reflective* attribute, the brightness of the reflected light is determined by the angle formed by the light source, object surface, and observer's viewpoint.

6. A surface's brightness can also be affected by a shadow or light reflected from another surface (for *Raytraced* displays).

7. If a surface has a material attached with a *Self-Illumination* attribute, the surface brightness is increased.

For dramatic effects, only one light might be used; however, for most renderings several lights or a combination of types of lights is used. Typically, a mix of lights is used. Now that you understand the types of lights and the effects of each type on a rendered object, examine the commands that let you insert and control the lights.

Light

	Pull-down Menu	Command (Type)	Alias (Type)	Short-cut	Screen (side) Menu	Tablet Menu
	View > *Render >* *Light*	*Light*	...	...	*VIEW 2* *Light*	*O,1*

When the *Light* command is started, you are prompted to select one of three options at the command line or the dynamic input tooltip: *Point*, *Spot*, or *Distant*. Once you specify the type of light you want to create, the command options for that particular type of light are displayed.

        Command: **light**
        Enter light type [Point/Spot/Distant] <Spot>:

As an alternative to using the *Light* command you can use the *Pointlight* command to create a *Point Light*, the *Spotlight* command to create a *Spotlight*, or the *Distantlight* command to create a *Distant Light*. Alternately, the *Light control panel* of the *Dashboard* palette allows you to create the different types of lights and access other commands and settings related to lights (Fig. 41-20).

FIGURE 41-20 ───────

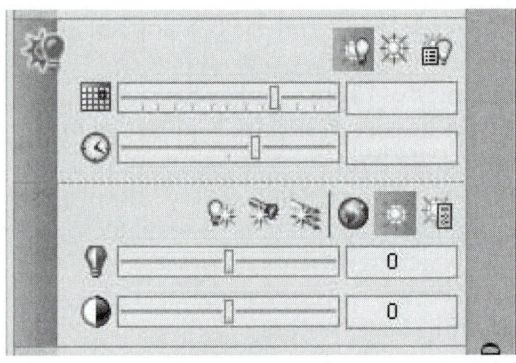

When you first start the *Light*, *Pointlight*, *Spotlight*, or *Distantlight* command, the *Viewport Lighting Mode* message box is displayed asking if you want to disable default lighting for the current viewport (Fig. 41-21). Clicking *Yes* turns off default lighting and sets the *DEFAULTLIGHTING* system variable to a value of 0. Clicking *No* keeps default lighting on and clicking *Cancel* ends the current command. Alternately, you can set this variable using the *Viewport lighting mode* toggle near the upper-right corner of the *Light control panel*.

FIGURE 41-21 ───────

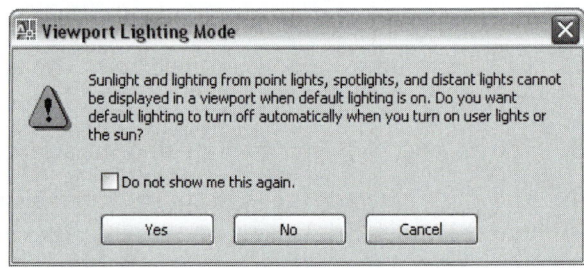

When you create lights for a model several properties of the lights should be adjusted to achieve the desired lighting effects for your model. The following is a list of the properties and options that are available for all three light types:

        Enter an option to change [Name/Intensity/Status/shadoW/Attenuation/Color/eXit] <eXit>:

Properties for lights can be specified two ways:

1. use the options that appear at the command prompt when the light is created; or
2. use the *Properties* palette for the light at any time after the light is added to the model.

Name

Assigning a new light a name is optional, but it is recommended to a assign a descriptive name for each light. If you do not assign a name, AutoCAD assigns a default name based on the type of light that is created. For example, for a *Point Light* AutoCAD assigns the name *Pointlight#*, where # represents a number based on the current number of point lights in the model.

Intensity

Intensity is a value that specifies the brightness of the light; the higher the value, the brighter the light. Keep in mind the resulting brightness of a <u>surface</u> illuminated by a *Point* light is based on several variables: intensity, color, distance to the surface, and so on.

Status

Indicates if a light is on or off when the model is rendered.

Shadow

This property indicates if a light casts shadows. Some lights may be used to only emit light, while others emit light and cast shadows. Along with toggling the shadows on or off, you also have the option of creating *Sharp* or *Soft* shadows. *Soft* shadows are more realistic, but take longer to calculate than *Sharp* shadows. If *Soft* is selected, you should specify the map size based on a list of memory sizes in the range of 64 to 4096 and softness of the shadows on a scale of 1 to 10.

Attenuation

Point Lights and *Spot Lights* possess a fall-off quality similar to actual lighting. That is, the farther the light beams travel from the light source, the dimmer the light becomes. *Attenuation* specifies the rate at which the light diminishes. Three options are available: *None*, *Inverse Linear*, and *Inverse Squared*. You can also specify a start and end limit for the attenuation. The start limit is used to determine at what distance the light starts to diminish, and the end limit is used to control the distance at which the light ends.

None

This option forces the light emitted from a point light to keep full intensity no matter how far the rendered surfaces are from the source. This is similar to a *Distant Light*.

Inverse Linear

This option sets an inverse linear relationship between the distance and the illumination. The brightness of a rendered surface is inversely proportional to the distance from the source. Put another way, there is a 1 to 1 relationship between distance and darkness. Generally, *Inverse Linear* attenuation is the easiest to work with and appears realistic for most AutoCAD renderings.

Inverse Squared

This is the greatest degree of light fall-off. The amount of light reaching a surface is inversely proportional to the square of the distance. In other words, if the rendered surfaces are 2, 3, and 4 units from the light source, the illumination for each will be $1/2^2$, $1/3^2$, and $1/4^2$ ($1/4$, $1/9$, and $1/16$) as strong. This option simulates the physical law explaining light fall-off.

Color

This option specifies the color of light that is emitted. Using the standard *Color* dialog box, you can select a *True Color* using *RGB* or *HSL* color models, an *AutoCAD Color Index* (*ACI*) color, or a color from a *Color Book*.

Hotspot and *Falloff*

These two options determine characteristics of the cone of light emitted from a *Spotlight*. The *Hotspot* is the area of the model that receives full illumination (Fig. 41-22). The size of this area is determined by the angular value. The light begins to fall off just outside the *Hotspot* until approaching no illumination at the *Falloff* cone. The area outside the *Falloff* cone receives no illumination. The *Falloff* angle must be equal to or greater than the *Hotspot* angle.

FIGURE 41-22

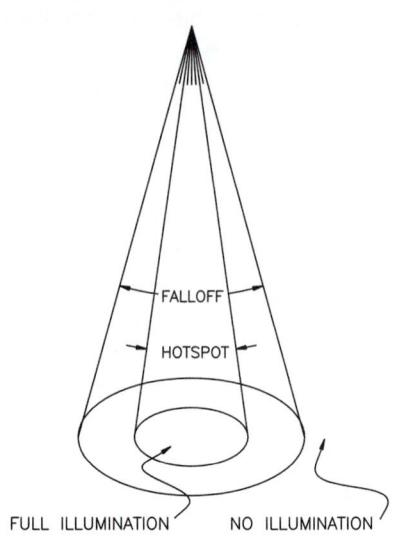

Creating a *Point Light*

To create a *Point Light* use either the *Pointlight* command or the *Light* command with the *Point* option. You are prompted for the source location (insertion point) for the light. Once the *Point Light* location is specified you can specify the properties of the light or press Enter to exit the command. You can specify properties for the light at a later time using the *Properties* palette.

> Command: **light**
> Enter light type [Point/Spot/Distant] <Distant>: **point**
> Specify source location <0,0,0>: **PICK**
> Enter an option to change [Name/Intensity/Status/shadoW/Attenuation/Color/eXit]
> <eXit>:

FIGURE 41-23

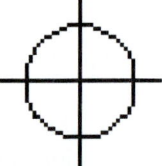

A symbol called a light glyph is displayed in the model to representing the location of the *Point Light* (Fig. 41-23). Light glyphs do not appear in a rendering or print.

Creating a *Spotlight*

To create a *Spotlight* use the *Spotlight* command or the *Light* command with the *Spot* option. You are prompted for the light source location (insertion point). Since a spotlight emits rays toward a specific location, you must also specify a *target* for the light.

> Command: **light**
> Enter light type [Point/Spot/Distant] <Distant>: **spot**
> Specify source location <0,0,0>: **PICK**
> Specify target location <0,0,-10>: **PICK**
> Enter an option to change [Name/Intensity/Status/Hotspot/Falloff/shadoW/Attenuation/Color/eXit]
> <eXit>:

FIGURE 41-24

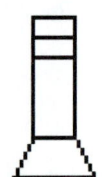

Once you indicate the source location and target, you can specify the properties of the light such as *Hotspot* and *Falloff*, or you can exit the command and specify the properties later using the *Properties* palette. A *Spotlight* glyph (flashlight symbol) is displayed in the model to represent the type of light, the source location, and the direction of the rays (Fig. 41-24). The glyph does not appear in renderings or prints.

Creating a *Distant Light*

You can use either the *Distantlight* command or the *Light* command with the *Distant* option to create a *Distant Light*. Unlike a *Point Light* or a *Spot Light*, a *Distant Light* has no source location. Instead, you must specify a direction vector since a *Distant Light* emits parallel rays of light. The vector between the *FROM* point and the *TO* point establish the direction of the light rays.

> Command: **light**
> Enter light type [Point/Spot/Distant] <Distant>: **distant**
> Specify light direction FROM <0,0,0> or [Vector]: **PICK**
> Specify light direction TO <1,1,1>: **PICK**
> Enter an option to change [Name/Intensity/Status/shadoW/Color/eXit] <eXit>:

You can then specify additional properties for the *Distant Light* or press Enter to exit the command and specify properties later using the light's *Properties* palette. *Distant Lights* do not have glyphs.

Using *Sunlight*

Sunlight is a special kind of *Distant Light*. Using the *Sun Properties* palette and a *Geographic Locator*, you are able to control the properties of the "sun" and geographical properties of the 3D model such as northern direction, the current time zone, and longitude and latitude values relative to the 3D model. This feature allows you to simulate life-like scenarios of lighting and shadows for your 3D model. For example, you can use these variables to calculate the effects of the path of the shadow from a skyscraper during specific periods of the day or year.

Sunproperties

Pull-down Menu	Command (Type)	Alias (Type)	Short-cut	Screen (side) Menu	Tablet Menu
View > *Render >* *Light >* *Sun Properties*	*Sunproperties*	...	...	...	...

The *Sunproperties* command displays the *Sun Properties* palette (Fig. 41-25). This palette provides complete control of the settings that affect the simulation of the sun. In general, the basic properties for the sun are similar to a *Distant Light*, in that the rays of light are parallel and do not "fall off" as they increase in distance from the source. However, the geographical properties provide you with the means to achieve life-like simulations.

General

These properties are common to all light types (*Point Lights*, *Spot Lights*, etc.).

Status

The *Status* indicates if sunlight is *On* or *Off* when the 3D model is rendered.

Intensity Factor

This value specifies the brightness of the light from the sun; the greater the factor, the brighter the light.

FIGURE 41-25

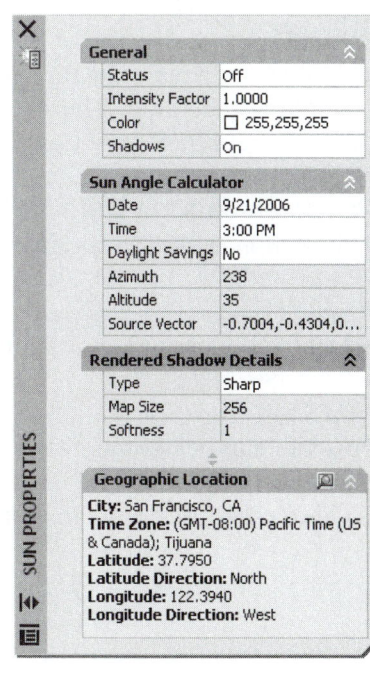

Color
Use this edit box to specify the color of light that is emitted from the sun. You can select from the standard options in the *Color* dialog box (*True Color*, *AutoCAD Color Index [ACI]*, or *Color Book*).

Shadows
This setting designates if the sun casts shadows (*On*) or not (*Off*). Generally, turn shadows *Off* for test renderings and *On* for final renderings.

Sun Angle Calculator
This section provides several options that change the lighting effects based on the angle of the sunlight. Most of these settings adjust the properties of the sun based on the geographic location of the model.

Date
Specify a calendar date to be used to calculate the angle for the sun with respect to the other geographic properties of the model. Calendar date can also be specified interactively using the *Date* slider on the *Light control panel* of the *Dashboard* palette.

Time
The time of the day is also used to calculate the angle of the sunlight. Time can also be specified interactively using the *Time* slider on the *Light control panel* of the *Dashboard* palette.

Daylight Savings
Yes or *No* options (and the *Date* setting) determine whether or not Daylight Savings Time affects the sun angle.

Azimuth
Azimuth is the bearing in degrees from north in a clockwise direction; for example, due east is 90 degrees. This property is read-only since it is based on the other input fields and the Geographic Location (see "*Geographiclocation*").

Altitude
This section displays the current altitude value of the sun, or the angle the sun appears in degrees vertically from the horizon. The maximum value is 90 degrees (directly overhead). This property is read-only since it is dependent on the other input fields and the *Geographic Location* (see "*Geographiclocation*").

Source Vector
This section displays the direction of the source vector used to calculate the direction from which the sunlight is generated. This property is also read-only and dependent on other input fields and the *Geographic Location* (see "*Geographiclocation*").

Rendered Shadow Details
This section is used to specify how AutoCAD calculates the shadows cast by the sun.

Type
Select from *Sharp* or *Soft* shadows created when the *Shadows* property is set to *On*.

Map Size
This value indicates the size of the shadow map used to calculate *Soft* shadows. Larger map sizes produce more detail but take longer to calculate. This property is read-only when the *Shadows* property is *Off* or when *Type* is set to *Sharp*.

Softness
Softness determines the level of blending (of shadow and light) along the edge of the *Soft* shadows. This property is read-only when the *Shadows* property is *Off* and when *Type* is set to *Sharp*.

Geographical Location

This section displays the current geographical location on Earth for the location of the model, which in turn influences the calculation for sunlight. This information is read-only in the *Sun Properties* palette since it is based on the setting in the *Geographic Location* dialog box. Use the *Launch Geographic Location* button in the top left of this section to display the *Geographic Location* dialog box. For more information on the *Geographic Location* dialog box see "*Geographiclocation*" next.

Geographic-location

Pull-down Menu	Command (Type)	Alias (Type)	Short-cut	Screen (side) Menu	Tablet Menu
View > *Render >* *Light >* *Geographic Location...*	*GEOGRAPHIC-LOCATION*	*GEO*	...	...	...

The *Geographiclocation* command displays the *Geographic Location* dialog box (Fig. 41-26) which allows you to specify the location on Earth for the model. The model's geographic location is used to determine the angle from which the "Sun" casts light and how shadows are defined in the model. In the dialog box, you can specify the location for your model manually by entering latitude and longitude values, or by selecting a location on a regional map. The *Geographic Location* dialog box also allows you to specify the northern direction of the drawing without having to rotate the entire model. Time zones are calculated based on the latitude and longitude values, although some locations do not have clearly defined time zones due to local time standards.

FIGURE 41-26

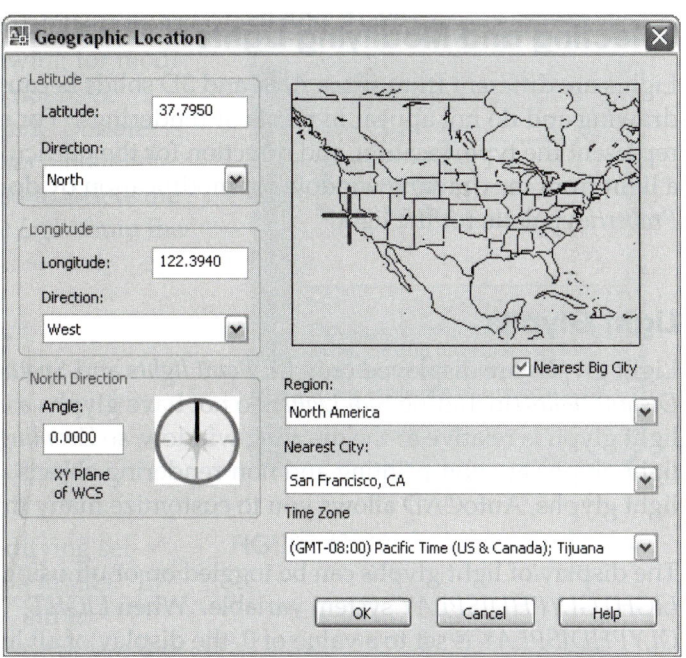

Latitude

The *Latitude* text box holds the current latitude value, and the *Direction* drop-down list allows you to specify which side of the equator the latitude value corresponds to.

Longitude

The *Longitude* text box holds the current longitude value, and the *Direction* drop-down list allows you to specify which side of the prime meridian the longitude value corresponds to.

North Direction

The *Angle* text box and *North Direction Preview* indicate the northern direction based on the current UCS of the drawing.

Material Editor

The *Material Editor* section provides you with the tools to define properties for a material. Available properties include *Diffuse* and *Ambient* color, *Shininess*, *Translucency*, and *Opacity*. The particular options available are based on the material template selected from the *Template* drop-down list.

Template

The *Template* drop-down list controls which material properties are available for modifying in the *Material Editor*. There are four general material templates: *Realistic*, *Realistic Metal*, *Advanced*, and *Advanced Metal*. In addition, AutoCAD provides 22 other templates as a basis for creating more specific materials. A few examples of these material-specific templates are glass, mirror, paint, stone, wood, and masonry.

For information on using the *Material Editor* to create and modify materials, see "Creating New and Modifying Existing Materials." This section describes the material property options available in the *Material Editor* as well as the available templates.

Material-attach

Pull-down Menu	Command (Type)	Alias (Type)	Short-cut	Screen (side) Menu	Tablet Menu
...	*Materialattach*	...	...	...	...

The *Materialattach* command displays the *Material Attachment Options* dialog box (Fig. 41-33) and allows you to assign a material to a layer. The *Material Attachment Options* dialog box displays the available materials and layers in the current drawing. To assign a material to a layer, drag a material from the *Materials List* on the left and drop it on a layer in the *Layers List* on the right. Use the *Detach* button to remove the material assignment from the layer. In this case, the "Global" material is assigned to the layer.

FIGURE 41-33

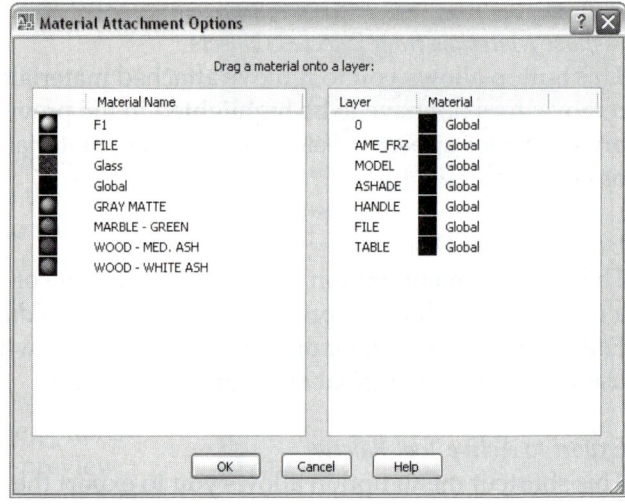

Materials List

The left side of the dialog box displays a list of the materials available in the drawing. The list includes all materials, both attached and not used.

Layers List

This list displays all the layers created in the drawing. If a material has not been attached to a layer, its *Material* is listed as "Global." If materials have been attached to objects only, the same "Global" assignment appears for the layers.

CREATING AND MODIFYING MATERIALS

When a 3D model is rendered, AutoCAD's rendering "engine" calculates the rendered display based on an array of interrelated physical principles and drawing properties. All properties specified for the non-geometrical objects in the model (background, lights, and materials) are taken into consideration and their effects on the 3D geometry are computed.

Materials comprise the most involved set of properties that can be designated for the 3D model. Materials can be defined by as many as 11 properties: *Diffuse, Ambient, Specular, Shininess, Refraction, Translucency, Self-illumination, Diffuse Map, Opacity, Reflection,* and *Bump Map.* All of the properties for a material can be adjusted in the *Material Editor* section of the *Materials* palette. The particular material template chosen from the *Template* drop-down list determines which properties can be modified.

Diffuse (Color)

Diffuse color is the primary color of a material. (Some materials also can have *Ambient* and *Specular* color.) Clicking on the *Diffuse* swatch displays the *Select Color* dialog box (see "Color Systems" earlier in this chapter). Selecting a color for the material from any of the color systems makes the surfaces appear in that color for the rendering. Selecting the *By Object* option disables the color swatch, in which case the material uses the color assigned to the object or object's layer.

Technically, *Diffuse* is a property of light reflection. *Diffuse* color reflects light in a diffuse manner—that is, light rays are scattered in all directions from the surface of the material producing a dull appearance, in contrast to a hard, highly reflective surface. For darker colors, light rays in the model emitted from *Point Lights, Distant Lights,* etc. are absorbed so the colored surface appears dark and dull (Fig. 41-34, left sphere). Selecting a lighter color for a material makes the surfaces brighter since it reflects more light (right sphere).

FIGURE 41-34

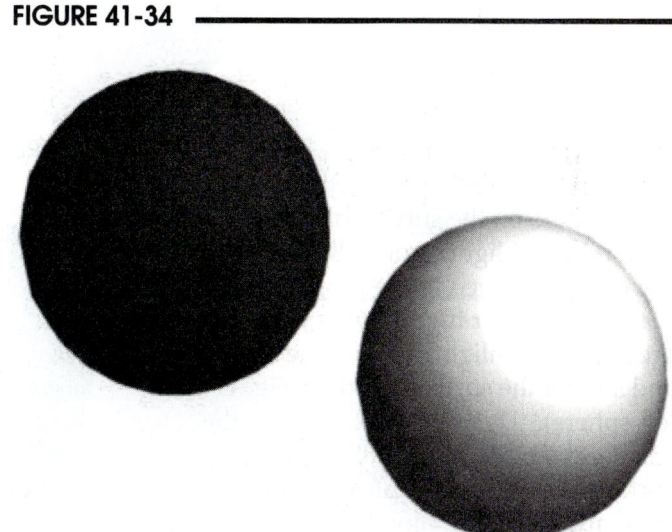

Ambient (Color)

The *Advanced Metal* and *Advanced* material templates allow you to specify a *Diffuse* color and an *Ambient* color. Clicking on the *Ambient* swatch displays the *Select Color* dialog box. Select the desired color from any of the color systems.

FIGURE 41-35

Ambient color is color that is illuminated only by ambient light. *Ambient* light is background light that illuminates all surfaces, such as when you increase the global brightness using the *Lights* control panel *Brightness* slider. Because most of the surfaces are illuminated by *Point Lights, Distant Lights,* and *Spotlights,* adjusting the *Ambient* color can be used to control the color and intensity of material "in the shadows." The *Ambient* color can be the same as the *Diffuse* color; however, assigning a dark *Ambient* color makes surfaces in the "shadows" appear darker and adds contrast (Fig. 41-35, left sphere), while a high *Ambient* color increases the overall light reflectivity of the surface and makes the "shadows" brighter (right sphere).

Grain Thickness
Use this value to specify the relative thickness of the two color bands that are used to generate the grains of the wood material. Although you have control of only this one value, AutoCAD renders the grain so that the *Color 1* grain is about twice the thickness of the *Color 2* grain.

Marble Dialog Box

FIGURE 41-46

The *Marble* dialog box (Fig. 41-46) allows you to define a material with the appearance of marble. To define a marble material, select *Marble* from the *Texture map* drop-down list in the *Diffuse Maps* section and then click the *Edit Map* button.

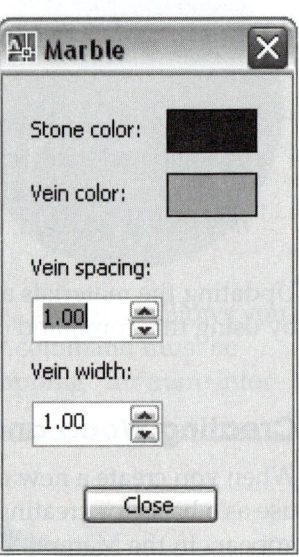

Stone Color and *Vein Color*
Marble has two color characteristics: the color of the body of the marble and the color of the long "veins" or "swirls" that run through the stone color. A white or black color is typical for the veins in marble. Clicking on the *Stone Color* or *Vein Color* swatches displays the *Select Color* dialog box (see "Color Systems"). Although in reality, marble contains a certain amount of "swirl" or randomness to the color and the vein pattern, there are no options to control these properties, unlike the *Wood* material property controls.

Vein Spacing
This value specifies the spacing <u>between</u> veins for the marble material.

Vein Width
Use this value to adjust the <u>width</u> of the veins for the marble material.

IMAGE MAPPING

Image mapping is the process of projecting a 2D image onto the surface of a 3D object. You can use a variety of image file formats, specifically .TGA, .BMP, .RLE, .DIB, .PNG, .JPG, .JPEG, .TIF, .GIF, and .PCX file formats.

FIGURE 41-47

Figure 41-47 displays a simple image file named CHECKER.TGA. Obviously the image is composed of only four squares of alternating black and white. When image files are mapped, they are automatically aligned and repeated across the surface. If you want more control of how the image is mapped, use the *Materialmap* command to adjust the placement and rotation of mapped images. *Materialmap* can also be used to control the placement and rotation of standard and custom materials. See "*Materialmap*" at the end of this section for more information.

Image mapping is accomplished using the *Materials* palette. There are a total of four possible types of maps that can be associated with a single material: texture maps (diffuse maps), opacity maps, reflection maps, and bump maps. Image maps increase the time AutoCAD takes to generate a rendering.

Texture Maps

A texture map is the projection of an image (such as a tile pattern) onto an object (such as a table or floor). Texture maps define colors for the surface of the object, as if the bitmap image were "painted" onto the object. For example, you could apply an image of a brick pattern to a horizontal flat surface to create the appearance of a brick floor. Texture maps conform to the attached object's surface geometry and therefore are affected by light and shadow. These capabilities can result in some very realistic renderings. Figure 41-48 shows the CHECKER.TGA image file mapped to the right sphere.

FIGURE 41-48

Reflection Maps

A reflection map (also referred to as an "environment map") simulates the effect of a scene reflected on the surface of an object. For example, an image used as a background (maybe a sky scene) can be reflected on an object with a highly reflective surface (such as chrome-plated metal). For reflection maps to render well, the object's material should have high *Reflection* and *Shininess*, and the image file used for the reflection bitmap should have a high resolution.

Figure 41-49 shows the CHECKER.TGA image file mapped onto the right sphere, and a reflection of the right sphere appearing on the left sphere. The reflection was accomplished by setting a high *Reflective* and *Shininess* value to the material for the left sphere. Although the checker pattern is not actually attached (mapped) to the left sphere, the material attached to the left sphere is considered an environmental map since it takes on the appearance of the environment.

FIGURE 41-49

Opacity Maps

Opacity maps create the illusion of opaque and transparent areas on the attached object. (Since most objects are naturally opaque, think of an opacity map as a "transparency" map.) AutoCAD uses the color value (lightness or darkness) of the mapped image to determine opacity or transparency. Images using white and black colors give the most effective results. Pure white areas in an opacity map are translated as completely opaque, while pure black areas are treated as transparent.

FIGURE 41-50

For example, if an image of a white square with a black circle were mapped onto a surface, the white area would be seen as solid (opaque) and the black area would appear as a hole in the surface. If an opacity map is a color image (not black and white), only the equivalent gray-scale values of the colors are used for the opacity calculation. Opacity maps do not necessarily change the color of the opaque areas of the mapped object.

Figure 41-50 shows the CHECKER.TGA image file attached as an opacity map to the right sphere. Notice how the black and white squares of the original map correspond to the "see through" and opaque areas of the sphere, respectively.

Bump Maps

Bump maps are similar to opacity maps in that a change in the value (gray scale) of the image is translated to a change in the surface appearance of attached object. However, in this case the brightness values of a bump map image are translated into apparent changes in the height of the surface of an object. Similar to opacity maps, using white and black colors gives the most effective results. For example, consider mapping an image of white text with a black background onto a surface. The resulting rendering would give the text (white image) the appearance of being raised (or embossed) against a flat (black) background. The surface has not actually changed, although the appearance of the surface (due to the map) has changed.

FIGURE 41-51

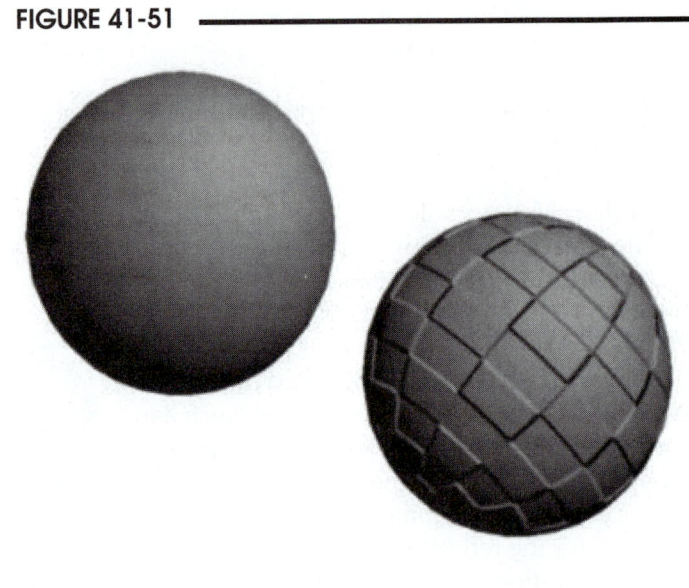

Figure 41-51 shows the CHECKER.TGA image file mapped on the right sphere. Note that the image's black and white "block" pattern appears as raised and lowered surfaces on the right sphere.

Combination Maps

You can apply maps in combination. For example, you can apply a wood grain bitmap as both a bump map and a texture map on a paneled wall to give the wall both the "feel" and the color of wood. In addition, an opacity map (a white surface with black dots) could be used to create the illusion of worm holes in the wood.

All maps have a blend slider that specifies how much they affect the rendering. For example, a texture map with a blend value less than the maximum (100) allows some of the material's surface colors and other properties to show through. Lower blend values reduce the effect of the mapped image.

Adjust Bitmap Dialog Box

The *Adjust Bitmap* dialog box (Fig. 41-52) is used to adjust how the image is mapped to the object. You can control the map scale, size, and rotation. The *Adjust Bitmap* dialog box is available only through the *Materials* palette in the *Material Editor*. If an image has been selected to be mapped (as a *Bump map*, *Texture map*, etc.), a button appears to the right of the *Select* button labeled *Adjust scaling/tiling, offset, and rotation values of bitmap*. Select this button to produce the *Adjust Bitmap* dialog box. The *Adjust Bitmap* dialog box is similar to the *Adjust Background Image* dialog box but offers more specific controls for the image.

Bitmap Scale

The *Bitmap Scale* section allows you to control the size of the image relative to the object to which it is mapped. You can also control if and how often the image is repeated to form a pattern.

FIGURE 41-52

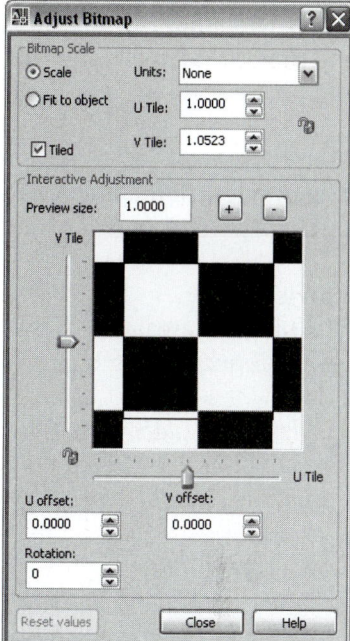

Scale

Checking this radio button produces the *Units* pull-down list where you can specify a real-world standard unit such as feet or inches. If you are using tiled bitmaps (repeated patterns), simulating bricks, stonework, tile, and wallpaper, select the *Tiled* checkbox to create a repeating pattern. Adjust the *Width* and *Height* values to stretch the single image shape.

Units

This list is enabled only if *Scale* is selected. Select the desired units to use for the size of the image relative to the object. When *None* is selected from the *Units* drop-down list, a fixed scale is used based on the *U Tile* and *V Tile* values. If one of the other values other than *None* is selected from the *Units* drop-down list, the image is scaled using the specified real-world units based on the *Width* and *Height* values.

Fit to Object

Use this option when applying a single image to be mapped, such as a billboard or a wall painting. Using this option and removing the *Tiled* check and setting the *U Tile* and *V Tile* values to 1.000 automatically adjusts the image to fit the surface or object size. You can stretch the image by adjusting the *U Tile* and *V Tile* values.

Tiled

If this box is checked the image repeats to create a pattern. The number of repeats is based on the object size and the scale and values in the edit boxes. In general, check this box to use a pattern in conjunction with the *Scale* option. Uncheck the box for the *Fit to Object* option.

Lock Aspect Ratio

Use this "lock" icon to maintain the aspect ratio (length to width ratio) of the image. The lock ties together the *U Tile* and *V Tile* values or the *Width* and *Height* values so if one value is changed the other value changes at the same amount or ratio.

U Tile and V Tile

Images use a set of UVW coordinates which are similar to XYZ coordinates, but UVW coordinates affect how the image is moved and rotated. The *U Tile* and *V Tile* values control the aspect ratio of the image which in turn can affect tiling. *U Tile* scales the image along the U axis, which is similar to the X axis except it applies only to moving and rotating the image. The *V Tile* represents the tiling value of the image in the V axis. As the *U Tile* and *V Tile* values are changed, the *U Tile* and *V Tile* sliders in the *Interactive Adjustment* section update accordingly.

Width and Height

The *Width* and *Height* values control the size of the image on an object or an individual face based on the units specified in the *Units* drop-down list.

Interactive Adjustment

The *Interactive Adjustment* section allows you to adjust the scale, rotation, and offset of the image interactively with sliders or by clicking and dragging in the preview area. Clicking and dragging the image in the preview allows you to adjust the *U Offset* and *V Offset* values, while right-clicking and dragging in the preview allows you to rotate the image and change the *Rotation* value.

Preview Size
The preview of the image can be adjusted by using the plus (+) and minus (-) buttons to zoom in and out.

Preview
The preview area allows you to see how the image is affected by the adjustments as they are made in the *Adjust Bitmap* dialog box.

U Tile and *V Tile* Sliders
The *U Tile* and *V Tile* sliders allow you to adjust the tiling of the image along the U and V axes. As you adjust the sliders, the *U Tile* and *V Tile* text boxes update accordingly.

Lock Aspect Ratio
Use this "lock" icon to maintain the aspect ratio (length/width proportion) of the image. This action locks the *U Tile* and *V Tile* sliders together so as one value is changed the other value changes at the same amount or ratio.

U Offset and V Offset
The *U Offset* and *V Offset* values are used to adjust the starting (insertion) point of the image along the U and V axes. The value can be adjusted using the text boxes or by clicking and dragging the image in the preview area.

Rotation
Changing this value rotates the image around the W axis (Z axis) of the UVW coordinate system. The rotation of the image has no effect when mapping the image to spherical and cylindrical shapes. For these cases, use the *Materialmap* command. See "*Materialmap*" later in this section for more information.

Reset Values
Use this button to restore the original values of the dialog box when the dialog box was opened.

Removing and Disabling an Image Map

To remove an image map from a material, use the *Material Editor* in the *Materials* palette. Click the *Delete map information from material* button to the right of the *Adjust scaling/tiling, offset, and rotation values of bitmap*. Once the button is clicked, the image map information is removed from the material.

For some cases it may be desirable to make a test rendering without displaying the image map. You can temporarily disable the image map from the material for a particular rendering without detaching the image. Do this by removing the check from the check box for the particular image map. For example, if you attached an image as a diffuse map, remove the check next to *Diffuse map* to temporarily disable the image.

Materialmap

Pull-down Menu	Command (Type)	Alias (Type)	Short-cut	Screen (side) Menu	Tablet Menu
View > *Render >* *Mapping*	*Materialmap*	*SETENV*	...	VIEW 2 *Mapping*	*R,1*

The *Materialmap* command allows you to reposition and readjust the orientation of a material that has been applied to an object or individual face. *Materialmap* is command-line oriented rather than dialog box oriented. You do, however, use grips interactively in the drawing area to reposition and readjust the image map with respect to the surface or object.

When you use *Materialmap*, you are first prompted for the type of material mapping that you want to apply. The *Materialmap* command supports four mapping shapes: *Box*, *Planar*, *Spherical*, and *Cylindrical*. After the mapping option is specified, you are prompted to select objects or faces that contain the map you want to adjust. Each of the mapping options should be matched to the geometry of the selected object. For example, use the *Cylindrical* option to adjust the position of an image map on a cylindrical solid.

```
Command: materialmap
Select an option [Box/Planar/Spherical/Cylindrical/copY mapping to/Reset mapping]<Box>: (option)
Select faces or objects: PICK
Select faces or objects: Enter
Accept the mapping or [Move/Rotate/reseT/sWitch mapping mode]:
```

During the "Select faces of objects:" prompt, you can select an individual face by pressing the ctrl key while hovering over the desired face. Once objects or faces are selected (at the last prompt shown above), grips appear on the object as well as the move or rotate grip, depending on the object or option. The *Move* and *Rotate* options produce the move grip tool and the rotate grip tool, respectively.

For example, assume an image was mapped to a *Box* and the image was repeated using the *Adjust Bitmap* dialog box. However, you wanted to adjust the scale and rotation of the pattern appearing on the top face of the box. Do this by using the *Materialmap* command with the *Planar* option. Use the ctrl key to select the top face of the box. Figure 41-53 displays the grips and rotate grip tool that appear and allow you to interactively adjust the image pattern. Selecting one of the four corner grips stretches the pattern. Using the rotate grip tool allows you to rotate the pattern orientation with respect to the top face.

FIGURE 41-53

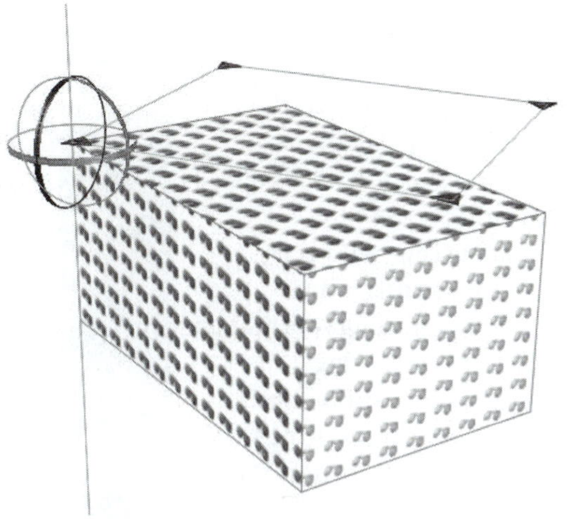

Remember that it is important to select the option that matches the geometry of the selected object. Every mapping option allows you to select from *Move*, *Rotate*, *Reset*, and *Switch mapping mode* options. Use the *Copy mapping to* option to copy the mapping adjustments (scale, proportion, rotation) to another similar object.

Box Mapping

Box mapping is used to adjust an image mapped on a box-shaped object. The adjustments affect all surfaces of the box unless an individual surface mapping is adjusted using the *Planar* mapping option. You can adjust the map by moving and rotating in three dimensions.

2007

Planar Mapping

Use *Planar mapping* to adjust an image mapped onto a 2D surface. Use this option to adjust one surface even if the material was originally mapped to the entire 3D shape. Rotation and movement are allowed only in the selected plane.

Spherical Mapping

Spherical mapping is used to map an image onto a spherical object. Rotation and movement of the image is allowed in three dimensions.

Cylindrical Mapping

Cylindrical mapping is used to adjust an image mapped onto a cylindrical object. The image can be moved and rotated in three dimensions.

RENDERING AND SPECIFYING PREFERENCES

Finally, after you have assigned a *Background* to a named view, inserted *Lights*, and created and attached *Materials*, you are ready to generate a rendering. To create a rendering, use the *Render* command. The rendering "engine" computes a display of the geometry based on assigned materials, light sources, and all associated parameters. Using the *Render* command is the simplest step for creating a rendering. However, you can specify any additional technical options affecting the output using the *Advanced Render Settings* palette. See *"Rpref"* for more information on the *Advanced Render Settings* palette.

Render

Pull-down Menu	Command (Type)	Alias (Type)	Short-cut	Screen (side) Menu	Tablet Menu
View > *Render >* *Render*	*Render*	*RR*	...	VIEW 2 *Render*	*M,1*

The *Render* command is used to collect all the assigned information concerning materials, lighting, background and other settings needed to calculate the final rendered output. During the rendering process the *Render window* (Fig. 41-54) is displayed by default which shows the current calculation process, recent rendering history, and information about the current rendering settings. See the *"Renderwin"* section for more information about the *Render window*.

FIGURE 41-54

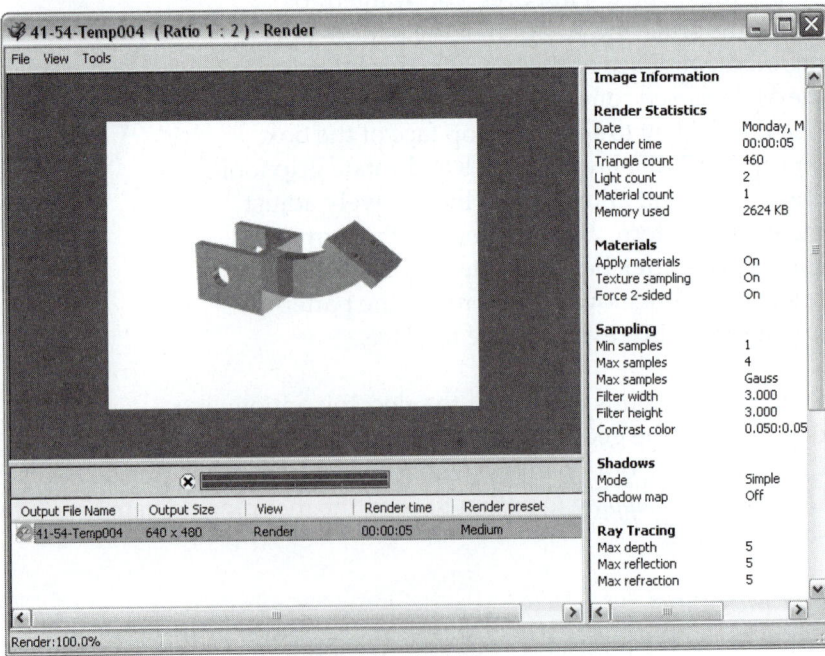

If no specific rendering preferences have been set in the *Advanced Render Settings* palette, the *Render* command uses the settings specified by the *Render Presets*. AutoCAD comes with five *Render Presets* that you can choose from; however, you can create your own settings and save them with the drawing.

The five *Render Presets* supplied with AutoCAD offer a range of speed and visual quality: *Draft, Low, Medium, High,* and *Presentation.* The *Render control panel* of the *Dashboard* palette (Fig. 41-55) offers you access to the *Render Presets* and many other controls and settings that you can use to adjust the rendering output. These controls are discussed in the following sections.

FIGURE 41-55

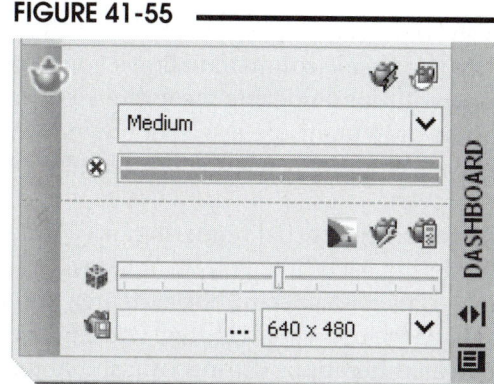

Renderwin

Pull-down Menu	Command (Type)	Alias (Type)	Short-cut	Screen (side) Menu	Tablet Menu
...	*Renderwin*	RW	...	...	...

The *Renderwin* command produces the *Render Window* (see previous Figure 41-54). The *Render Window* is divided into three main areas: *Image pane, Statistics pane,* and *History pane.* The *Render Window* is also displayed when the *Render* command is used and the *Destination* is set to *Window* or an *Output file.*

Image pane

The *Image pane* is typically the largest section of the *Render Window.* This area, located in the upper left, is where the rendered image is constructed. The image that is displayed is the actual size of the image file based on the specified *Output size* in pixels. You can use the *Zoom +* and *Zoom –* menu items from the *Tools* menu to zoom in to and out from the rendered image.

Statistics pane

The *Statistics pane* appears along the right side of the window. The settings that were used to create the rendered image are displayed in the pane. Select one of the images or temporary images in the *History pane* to display the rendering in the *Image pane* and the related statistics in the *Statistics pane.* The display of the *Statistics pane* can be toggled on or off by selecting *Statistics Pane* from the *View* menu.

History Pane

Along the bottom of the window is the *History pane.* Here you can access the different renderings that have been created for the current drawing session. By default, all renderings that are created during the session are assigned temporary file names. Right-clicking on one of the temporary file names listed in the *History pane* allows you to save the image to a file by selecting *Save* from the right-click menu. The right-click menu also has options that allow you to *Render Again, Remove from the list, Delete output file,* and *Make the rendering settings current.*

Progress Meter

The *Progress Meter* shows the time left to complete the current rendering. To the left of the *Progress Meter* is the *Cancel* button that can be selected to abort the current rendering in progress.

Rendercrop

Pull-down Menu	Command (Type)	Alias (Type)	Short-cut	Screen (side) Menu	Tablet Menu
...	*Rendercrop*	RC	...	...	...

The *Rendercrop* command allows you to render only a section of the drawing area without rendering the entire view. You are prompted to specify the area of the drawing using a selection window. For example, Figure 41-56 displays a cropped rendering created using *Rendercrop*.

Performing a partial rendering has a time advantage over computing a full rendering each time you want to test the effect of lighting, background, or materials for a specific object or area of the drawing. By selecting a smaller area of the drawing window, you can save yourself considerable time for repeated iterations during your initial rendering or drawing setup.

FIGURE 41-56

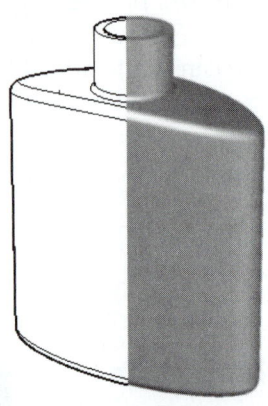

Rref

Pull-down Menu	Command (Type)	Alias (Type)	Short-cut	Screen (side) Menu	Tablet Menu
View > Render > Advanced Render Settings	*Rpref*	RPR	...	VIEW 2 *Rpref*	R,2

The *Rpref* command produces the *Advanced Render Settings* palette (Fig. 41-57). This palette lists the settings that the *Render* command uses during the rendering process. These settings control a variety of technical parameters that affect the rendering output. The parameters are described below.

Render Presets

The list at the top of the palette displays the available *Render Presets* and any custom presets saved with the drawing.

Render

This button initiates the *Render* command directly from the *Advanced Render Settings* palette.

Render Context

This top category of options involve output destination and size.

Save File
This small button on the right side of the *Render Context* title bar toggles the *Output File Name* edit box to an input field. If a file name is supplied, the rendered image is saved to an image file.

FIGURE 41-57

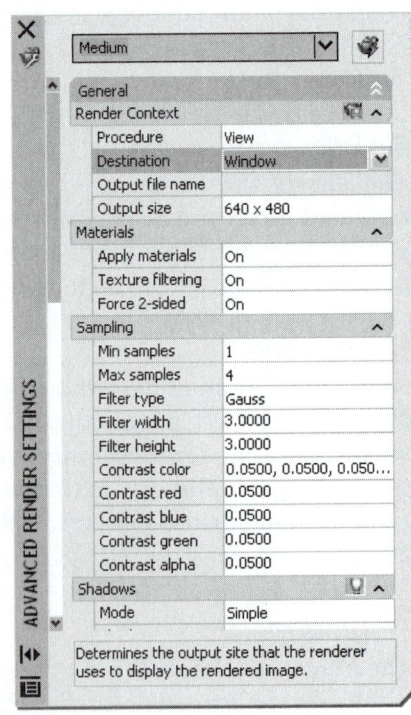

Procedure
The selected option here specifies what area of the drawing should be processed during rendering: *View, Crop,* or *Selected. View* renders the current view of the 3D model to the *Render Window. Crop* is available only when the destination of the rendering is specified as the viewport and allows the user to specify the region of the current view to be rendered. *Selected* allows you to render only selected objects in the 3D model.

Destination
The destination of the rendering process occurs in either the *Render Window* or the current *Viewport* (drawing area).

Output file name
The input field is available only when the *Save File* option (button) is enabled. Indicate the file name for the rendered image. You can select one of the following image formats: BMP, PCX, TGA, TIF, JPG, or PNG.

FIGURE 41-58

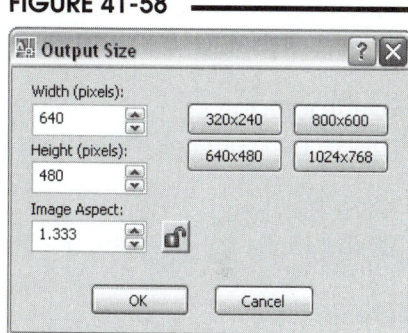

Output size
This drop-down list displays a choice of standard output sizes (320x240, 640x480, 800x600, and 1024x768), or you can choose *Specify Output Size* to produce the *Output Size* dialog box (Fig. 41-58). The dialog box allows you to pick standard sizes or specify a custom size. Enter the *Width* and *Height* values in pixels to define a custom image size for output.

Materials

This section allows you to turn on or off parameters that affect materials during the rendering process. Turning off materials speeds up the rendering time and may be useful for test renderings.

Apply materials
Here you can choose whether attached materials are used (*On*) or if materials are ignored (*Off*) during the rendering process. If *Off*, the 3D model is rendered using the Global material only.

Texture filtering
If set to *Off*, texture maps are disabled during rendering.

Force 2-sided
If *On*, faces are rendered on both the front and back side. Setting this to *Off* can save rendering time, but might also not render some faces visible in the view.

Sampling

The *Sampling* section contains the settings that determine how sampling should be handled during the rendering process. Sampling is the process of repeated calculations to determine the color of each pixel in the final image. Normally each pixel will have been calculated, or "sampled," many times. Sampling is used as an antialiasing technique that smoothes out the jagged edges. The rendering engine first samples the scene color at locations within the pixel or along the pixel's edge, then uses a filter to combine the samples into a single pixel color. If needed, see *Rpref* in AutoCAD *Help* for more information.

Shadows

You can use this section to determine how shadows are calculated during the rendering process.

Enable button
The button at the upper-right corner of this section toggles whether or not shadows are calculated during the rendering process.

Mode
There are three different shadow modes available: *Simple, Sort,* and *Segments. Simple* mode specifies that shadows are calculated in a random order. *Sort* mode specifies that shadows are calculated from the object back to the light. *Segments* mode specifies that shadows are calculated based on the volume of the shadow between the object and light.

Shadow map
You determine whether shadow-mapped (*On*) or raytraced (*Off*) shadows are calculated during the rendering process.

Ray Tracing
This section contains the settings that determine how a rendered image is shaded.

Enable button
The button at the upper-right corner toggles whether or not raytracing is used for the rendering.

Max depth
This value limits the combination of reflection and refraction. Tracing of a single ray stops when the total number of reflections and refractions reaches the maximum depth.

Max reflections
The setting determines the maximum number of times a ray of light can be reflected. A lower number decreases the accuracy of the rendering and the amount of time it takes to create the rendered image, a higher number increases both the accuracy of the rendering and the amount of time required.

Max refractions
Set the maximum number of times a ray of light can be refracted. A lower number decreases the accuracy and the time required, whereas a higher number improves the accuracy of the rendering but increases the time needed to create the rendered image.

Indirect Illumination
This group of settings determines how indirect lighting affects the model during the rendering process. The group includes *Global Illumination, Final Gather,* and *Light Properties.* The options in this group are technical and beyond the scope of most users. If needed, see *Rpref* in AutoCAD *Help* for more information.

Diagnostic
This group of options is used to troubleshoot the rendering process by providing diagnostic tools that illustrate how the rendering is calculated. You can project grids and enable area maps to visually define the rendering process. If needed, see *Rpref* in AutoCAD *Help* for more information.

Processing
To render the scene, the image is subdivided into tiles. This section contains the settings that control the size and order of rendered tiles. If needed, see *Rpref* in AutoCAD *Help* for more information.

Renderpresets

Pull-down Menu	Command (Type)	Alias (Type)	Short-cut	Screen (side) Menu	Tablet Menu
...	*Renderpresets*	*RP*	...	...	...

The *Renderpresets* command displays the *Render Presets Manager* (Fig. 41-59), which allows you to specify and save a custom rendering preset. Rendering presets define specifications of how the rendering process is performed. The *Render Presets Manager* contains many of the same properties that are found on the *Advanced Render Settings* palette, with a few exceptions. The *Render Presets Manager* allows you to create new custom presets and delete render presets from the drawing.

FIGURE 41-59

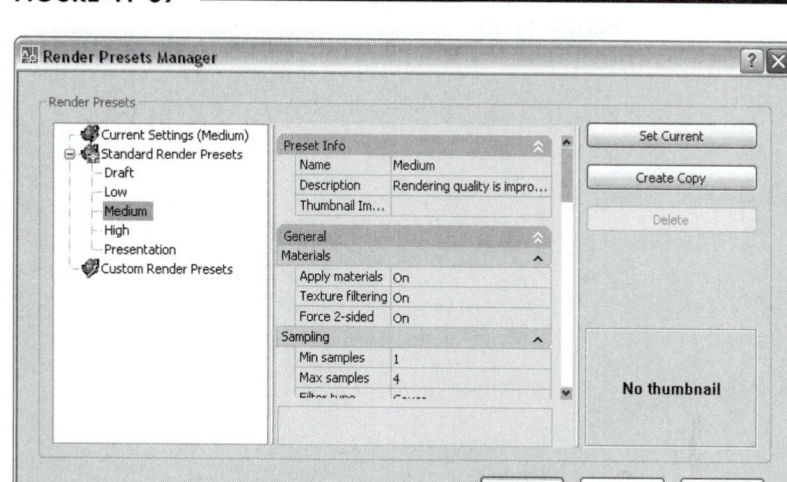

The list on the left displays the *Standard Render Presets* and a list of *Custom Render Presets* that you create. Create a custom render preset by making a copy of an existing preset, then modify the settings for your needs.

Property Panel

The central area contains the same properties that are displayed in the *Advanced Render Settings* palette. In addition, the *Preset Info* section lists the *Name*, *Description*, and a *Thumbnail Image* (optional) to represent the preset.

Set Current

Select any standard or custom preset from the list, then use this button to make it current. The subsequent renderings use the selected preset specifications.

Create Copy

The *Copy Render Preset* dialog box is produced (Fig. 41-60) that allows you to create a new render preset. The copy is based on the current preset from the *Render Presets* list. You must assign a name for the new render preset, and you can optionally provide a description. The new render preset is added under the *Custom Render Presets* node in the *Render Presets* list.

FIGURE 41-60

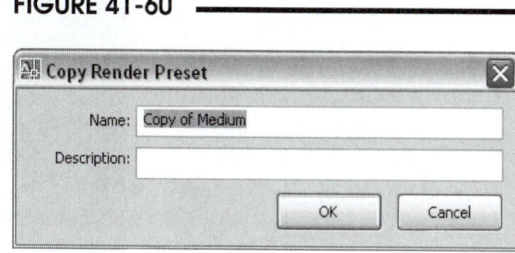

Delete

Select any custom render preset from the list and use this button to delete it from the drawing. You can delete only custom render presets.

Rendering Example

A solid model of an intake housing and a machinist's file are used here as an example for creating a *Background*, inserting *Lights*, creating *Materials*, using a *Bump Map*, setting rendering preferences, and creating a rendering (refer to "Typical Steps for Using Render").

2007

1. After the geometry is completed, *3Dorbit* is used to provide an appropriate viewing orientation for the rendering. If desired, the view can be saved using the *View* command. Figure 41-61 displays the solid models ready to begin the rendering process. Notice that the drawing includes three objects: the intake housing, a machinist's file, and a surface (only the back edge of the *Box* appears) that is used to simulate a table top, which will provide a surface for shadows and reflection for the intake housing and file.

FIGURE 41-61

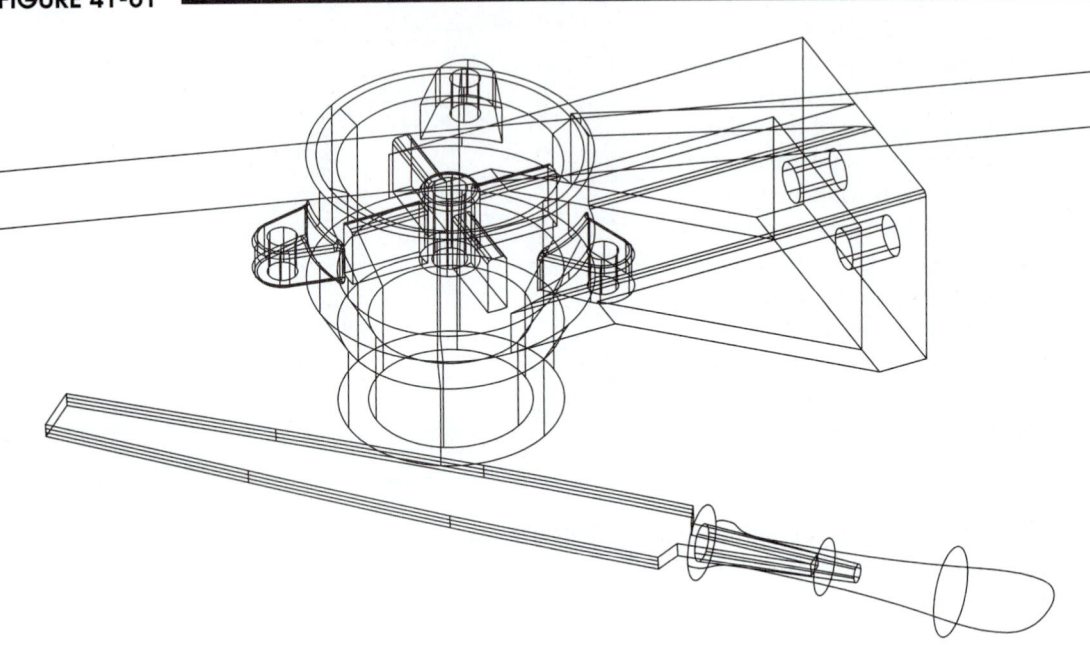

2. Begin the rendering setup process by setting up a *Background* (Fig. 41-62). Using the *Gradient* option, select the *Three color* option. Set the *Rotation* for the gradient to a value of 90 so the bands of the background run vertically instead of horizontally to provide some directional contrast from the solids in the model. A simple rendering is generated to check the background using the *Render* command with the *Medium* render preset. The rendering is generated with only the default materials and lights (using the Global material and the two "over-the-shoulder" lights). To speed up the rendering process, the *Selected*

FIGURE 41-62

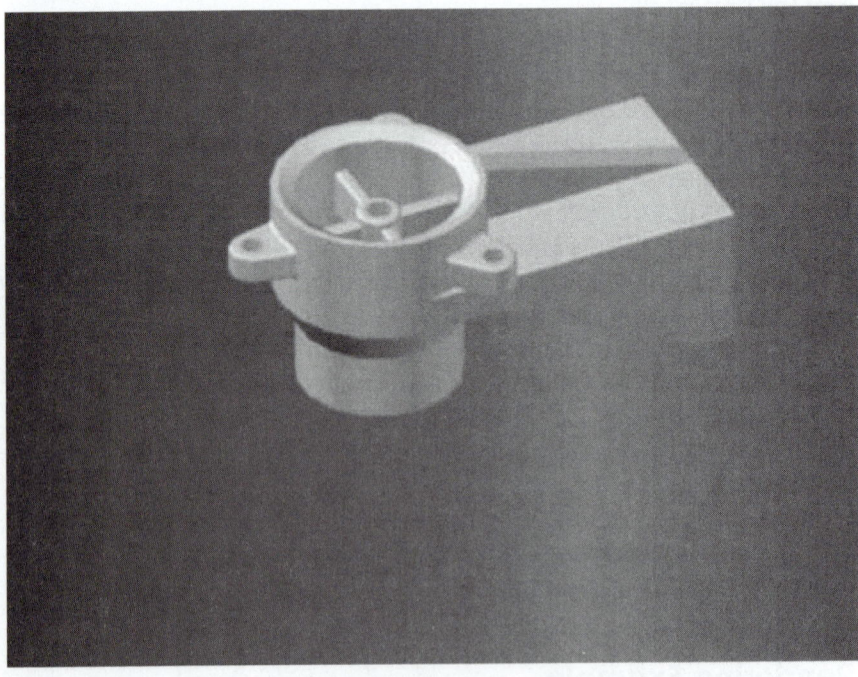

option is specified in the *Procedure* section of the *Advanced Render Settings* palette.

3. Next, a *Light* is inserted to produce both accent lighting and shadows (Fig. 41-63). For this example, a *Point Light* is positioned in the upper right. Using a cool color for the light (such as a light blue or purple color) can reinforce the appearance of metallic surfaces. The *Shadows* option is enabled for the light, although the *Medium* render preset has shadows disabled for this rendering. The light position illuminates the front, top, and right sides of the intake housing. Notice how the point light illumination falls off along the top

FIGURE 41-63

surface. Since only one light is used, the top and right side surfaces appear very bright, while the back of the cylinder has no light. Also notice the "floating" effect of the model due to the lack of shadows.

A material is selected and modified to represent a smooth, reflective metal for the intake housing (Fig. 41-63). In the *Materials* palette, the *Shininess* property is set to a high value. A hard, dark, and reflective material is selected for the table top. Only the intake housing and table top are selected for the rendering (use the *Selected* option under the *Procedure* property in the *Advanced Render Settings* palette). Generating a rendering displays the results of the light position and intensity as well as material characteristics (Fig. 41-63). You can make several adjustments to the *Light* and *Material* settings, then generate "rough" renderings until the desired effect is obtained.

4. Next, *Materials* are assigned to the file and handle (each created as a separate model). The handle is assigned a wood material, and the metal portion of the file has a standard metallic material attached (Fig. 41-64). *Shadows* are toggled off to speed rendering time when checking and adjusting material characteristics for the file and handle.

FIGURE 41-64

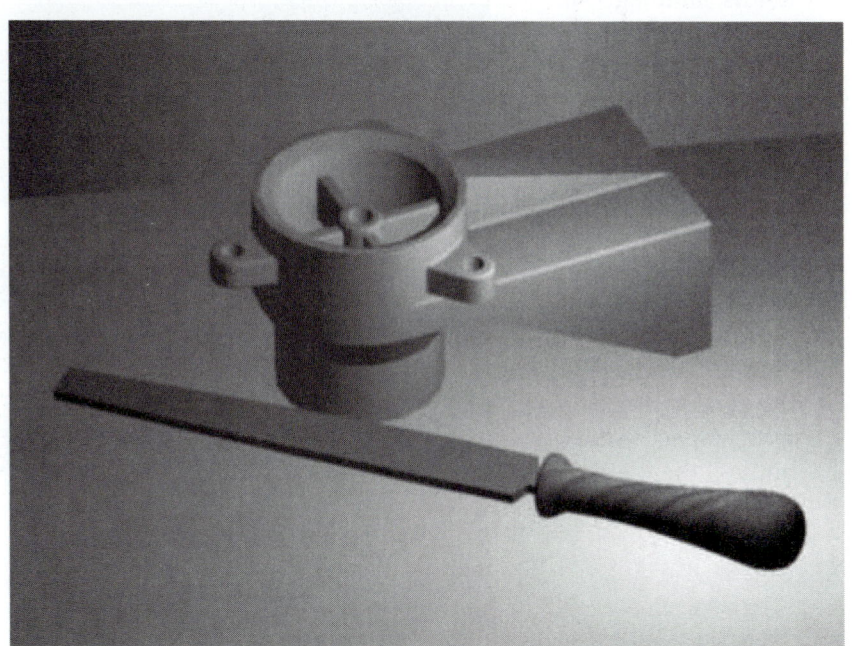

5. A rendering with only one point light can create drastic contrast from one side of the scene to the other—some portions of the model are well illuminated, while others are dark. Therefore, a second *Light* is inserted into the drawing to provide more even and realistic illumination over the surfaces of the model. This second light is a *Distant* light positioned on the left side and having a direction set to illuminate

FIGURE 41-65

the left side of the models (Fig. 41-65). The new light compensates for the previously dark left side of the models; however, with two or more new lights, a scene can have a bright, washed-out effect. *Intensity* for the lights should be adjusted to provide the desired illumination. Make several *Intensity, Location,* and *Color* adjustments, if necessary, to achieve the desired effects. A useful technique is to use a slightly different color for the two lights; for example, with the *Distant Light,* use a warm tone (light yellow).

Shadows have been enabled in the *Advanced Render Settings* palette to generate shadows (both lights in the 3D model are set to cast shadows). Shadows can make a dramatic difference in the realism of the rendered image. Note that the objects no longer "float," but now appear to be sitting on the table. Also note both sets of shadows are useable, one from each light source.

6. For the final adjustments, an image file (of a grid pattern) is used as a *Bump Map* to provide the serrated edges on the surface of the file model. The *Materialmap* command is then used to position (rotate) the bitmapped image at the desired angle with respect to the body of the file model.

FIGURE 41-66

For this final rendering (Fig. 41-66), raytracing was enabled and the *Presentation* render preset was used so reflections can be generated. Notice that the table surface now acts like a mirror due to its *Reflection* property evident by the reflection of the background colors.

Also note that the view is shown as a perspective view in this rendering. This view can be accomplished by changing the *Perspective* system variable value from 0 to 1. It is generally faster to work with perspective viewing off, and then turn it on during the final rendering.

The *Shadows* option for each of the lights has been adjusted using the *Properties* palette for the final rendering. For example, you may want to generate shadow maps, which are necessary to create "soft" shadows. Figure 41-66 displays *Raytraced Shadows*.

Don't expect to achieve excellent results when you create your first few renderings. Many options are available; therefore, a surprising amount of time can be spent making adjustments and generating "rough" renderings. During the initial phases you might want to use the *Low* render preset or even use the *Rendercrop* command to save time. Until the parameters are fully adjusted for creating the final rendered image, you should avoid the *High* and *Presentation* render presets since they require additional rendering time.

SAVING, PRINTING AND VIEWING RENDERINGS

Renderings can be saved to files in a number of ways. Images can be saved from the *Render Window* after the *Render* command is used, but if you render directly to the drawing area you can use the *Saveimg* command to save the rendered image to a standard image file format. After an image is created, you can use one of the many free image viewing programs to view the files. You can even upload images to a website so they can be viewed on a web page. Alternately, use *Shade Plot* with the *Rendered* or *As Displayed* option to print the rendered image.

Saveimg

Pull-down Menu	Command (Type)	Alias (Type)	Short-cut	Screen (side) Menu	Tablet Menu
Tools > Display Image > Save	*Saveimg*	...	...	*TOOLS 1 Saveimg*	...

The *Saveimg* command allows you to save a rendered image in a drawing window to an image file. Using *Saveimg* produces a standard file navigation dialog box titled *Render Output File*. You can save the image to a .BMP, .PCX, .TGA, .TIF, .JPG, or .PNG file format. You can view the file in an image editor or viewer that supports the file format, or you can insert the image into AutoCAD using the *External References* palette (see Chapter 32, Raster Images and Vector Files).

FIGURE 41-67

Once a file name, location, and file format have been specified in the *Render Output File dialog box*, click the *Save* button to open an options dialog box based on the file format that was selected. Figure 41-67 shows the options dialog box that is displayed when a .BMP file format is selected. Specify the available options for the file format as desired, and click *OK* to create the image file.

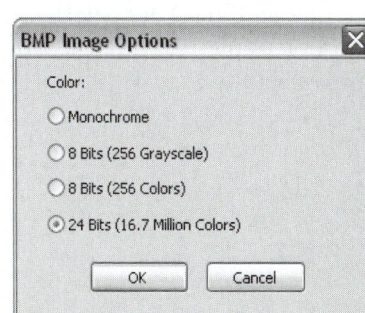

If you want to create an image file directly rather than first rendering in the drawing area, enable the *Save File* toggle on the *Advanced Render Settings* palette or under the *Render control panel* of the *Dashboard* palette (see "Saving a Rendered Image to a File" next).

Saving a Rendered Image to a File

You can render directly to an image file by selecting the *Save File* toggle in the *Advanced Render Settings* palette or using the *Save rendering to file* toggle in the *Render control panel* of the *Dashboard* palette. This method does not avoid having to render to the *Render Window*, but it can save you from accidentally forgetting to save the rendered image.

As an alternative, once the rendering has been processed in the *Render Window*, you can right-click on the image and select *Save* from the shortcut menu. Clicking *Save* from the right-click menu displays the *Render Output File* dialog box (see "*Saveimg*" for more information).

Printing a Rendered Image with *Shade Plot*

To make a print of an image already rendered on the screen, use the *Plot* command and select the *As Displayed* option of the *Shade Plot* section of the *Plot* dialog box (extended side). You can also make a rendered print (with or without previously generating a rendered display to the screen) using the *Rendered* option of *Shade Plot*. The *Rendered* option prints the drawing according to the settings made in the *Advanced Render Settings* palette, including any specified lights, materials, and background.

FIGURE 41-68

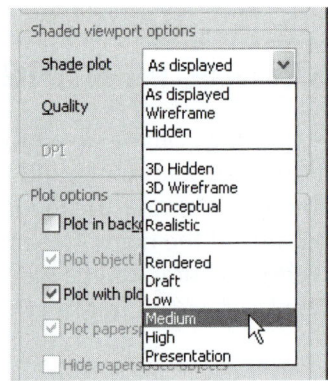

Printing a Rendered Image from the *Model* Tab

To create a rendered plot from the *Model* tab, use the *Shaded Viewport Options* section of the *Plot* or *Page Setup* dialog boxes (expanded section on the right side). Select *Rendered* from the *Shade Plot* drop-down list (Fig. 41-68). You can also select a specific visual style or render preset from the *Shade Plot* list. Alternatively, you could use the *Render* command prior to starting the *Plot* dialog box and then select *As Displayed* from the *Shade Plot* drop-down list. Assuming you have previously created and checked the rendering, using the *Rendered* option or a render preset option saves you time by not having to recreate the rendering before you plot. The *Quality* drop-down list allows you to specify resolution settings for the rendered plot. Choose from six resolution settings: *Draft*, *Preview*, *Normal*, *Presentation*, *Maximum*, and *Custom*.

FIGURE 41-69

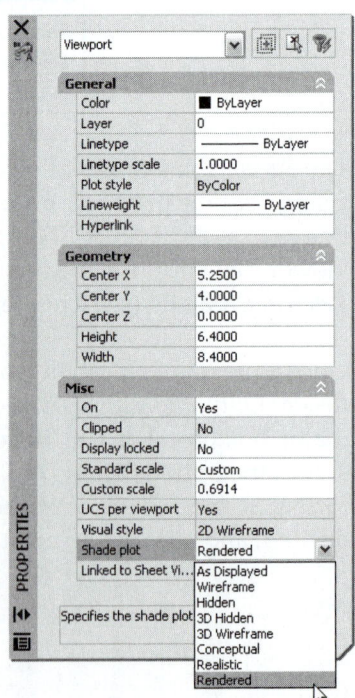

Printing a Rendered Image from a *Layout* Tab

If you are printing from a layout and want to specify a rendered image for one or more viewports, use the *Shadeplot* option of the viewport's *Properties* palette (Fig. 41-69). Select the viewport object (border), then invoke the *Properties* palette. The *Properties* palette lists the same options as the *Plot* or *Page Setup* dialog box except that no render presets are available.

Miscellaneous Rendering Tips

Setting Rendering Options for Visual Styles

The *Realistic* visual style is capable of displaying lights, shadows, materials, textures, and transparency without having to use *Render*. The display is not as life-like as creating a rendering, but it does help to save time when positioning lights and attaching materials in the 3D model.

Although the *Realistic* visual style allows you to see lights, shadows, materials, textures, and transparency, *Hardware Acceleration* must be enabled. This involves a few steps. To configure your display device for hardware acceleration, select the *System* tab in the *Options* dialog box. Next, select *Performance Settings* under the *3D Performance*. This selection produces the *Adaptive Degradation and Performance Tuning* dialog box (Fig. 41-70).

FIGURE 41-70

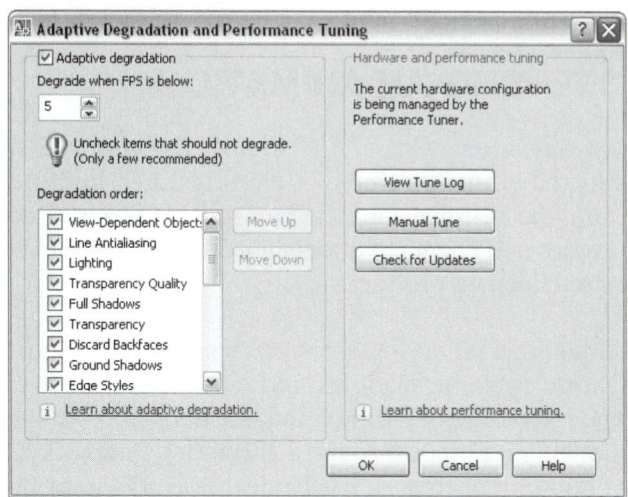

On the right side of the dialog box, select the *Manual Tune* button to display the *Manual Performance Tuning* dialog box (Fig. 41-71). The *Manual Performance Tuning* dialog box allows you to enable or disable hardware acceleration and specify which effects are managed by the graphics card. Check the *Enable hardware acceleration* box if it is not disabled (some graphics cards are not capable of properly supporting hardware acceleration for AutoCAD). Once enabled, turn on the desired *Effects* in the list. The *Effects* that can be enabled can vary based on the capabilities of your computer.

Effects that are supported but not recommended have a yellow warning icon next to them. In these cases AutoCAD has determined that your graphics card may be able to handle the feature, but may cause some problems. If the graphics card cannot handle the *Effect*, it is disabled and a red alert icon is displayed.

FIGURE 41-71

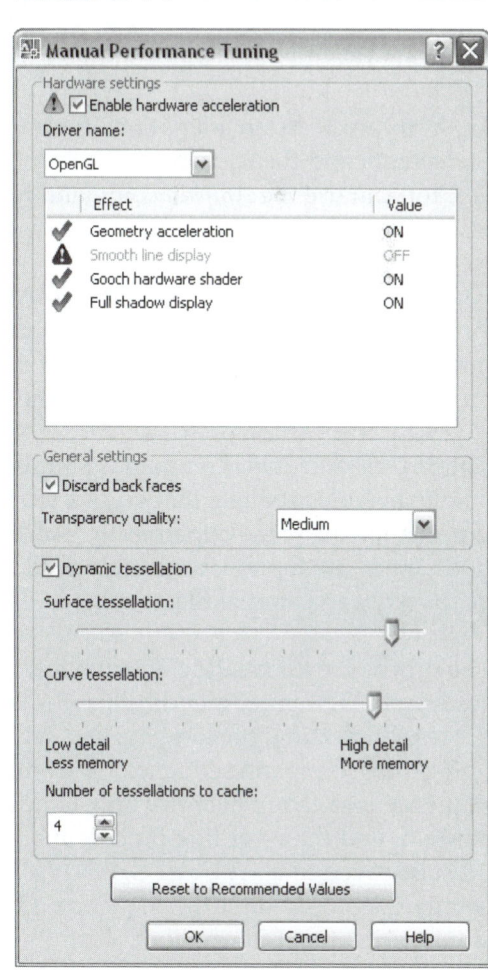

Viewres and FACETRES

The value you set with the *Viewres* command controls the accuracy of the display of circles, arcs, and ellipses. To ensure you get a good quality rendering, try increasing the value before rendering drawings that contain arcs or circles. The range is between 1 to 20,000, with 1000 being the default setting.

The *FACETRES* system variable controls the smoothness of meshed solids as well as the smoothness of shaded and rendered curved solids. It is linked to the value set by the *Viewres* command. When *FACETRES* is set to 1, there is a one-to-one correspondence between the viewing resolution of circles, arcs, and ellipses, and the tessellation of solid objects. When *FACETRES* is set to 2, the tessellation will be twice that set by *Viewres*, and so on.

When you raise and lower the value of *Viewres*, objects affected by both *Viewres* and *FACETRES* are affected. When you raise and lower the value of *FACETRES*, only solid objects are affected.

CREATING AN ANIMATION

Viewing a rendering of a 3D model is an effective way to visualize a design concept before it is constructed or manufactured. Colors, textures, lighting, shadows, and reflections help you better understand the 3-dimensional relationships of forms. Renderings shorten the design cycle by allowing colleagues and clients to understand and communicate particular aspects of the design before seeing the actual finished product.

Although a rendering helps you visualize a design, it is a static image. Animations also use all the color, texture, lighting, shadows, and reflective qualities of a rendering, but add motion. To get an even better understanding of 3-dimensional relationships, an animation can be created to simulate the effect of moving around or through a 3D model. Animations can simulate the effect of walking through a house or building or rotate a mechanical assembly to provide a better understanding of the function.

An animation in AutoCAD is created by turning on a camera to record motion. AutoCAD temporarily records the view the camera "sees" during an animation. You must then save the animation to a file and replay it using another video playback program such as Microsoft Windows Media Player.

There are two methods for creating an animation in AutoCAD.

1. You can create an animation "from scratch" using the *3D Navigation control panel* to control the camera and the motion.
2. You can use the *Anipath* command to create an animation based on assigning a path to control the motion of a camera.

Creating an Animation with the *3D Navigation control panel*

An animation is created by turning on the "camera" to record the view movement that you control using one or more 3D navigation commands such as *3Dcorbit*, *Pan*, *Zoom*, *3Dwalk*, or *3Dfly* to change the view around the model. The animation is the same view you see in the drawing window during the process.

The *3Dwalk* and *3Dfly* commands simulate walking or flying around or through a 3D model and display a *Position Locator* palette that gives a top view of the "camera" position. The camera records the view as you change the scene dynamically. Additional commands give you the tools to control how the animations are created, previewed, and saved to a file. The animation is not saved with the drawing; it must be saved as an external file.

The procedure for creating an animation using this method begins with starting a 3D navigation command. The process ends when the 3D navigation command sequence ends; therefore, the entire process of creating, previewing, and saving the animation to a file must be done before exiting the 3D navigation command sequence. The animation is not saved with the drawing file, but the animation sequence is saved in memory only temporarily while a 3D navigation command is running. You can pause or end the recording process and review the animation in a preview window, but the animation must be saved to a .AVI, .MPG, .MOV, or .WMV file format before exiting the last 3D navigation command you used for the sequence. Once the 3D navigation command is exited, the temporarily stored animation information is lost.

Creating an Animation using *Anipath*

This method for creating an animation is less complex and can result in a smoother animation. This method is based on controlling the view of the camera using a path. The path is an object such as an *Arc*, *Circle*, *Spline*, etc. A path can control either the motion of the camera or the motion of the target or both. The *Anipath* command produces the *Motion Path Animation* dialog box where you link the camera and target each to a point or a path. You can also select the specifications for saving the animation to a file when all the necessary adjustments have been made.

The *Anipath* method has another advantage over creating an animation from scratch using the *3D Navigation control panel*. That is, more information about the animation is saved in AutoCAD, such as the path attachment information. Therefore, it is easier to make adjustments for the animation to achieve the desired results before saving the animation to a file.

Because the process of controlling, creating, previewing, and saving animations involves a number of commands and a specific sequence, an overview of the basic steps for creating an animation using the 3D Navigation control panel is given first. The explanation of the particular commands and options follows.

Steps for Creating an Animation Using the *3D Navigation Control Panel*

1. Display the *Dashboard* palette. Expand the *3D Navigation control panel* so all the options are displayed. (To expand the panel, click the double-arrows located just below the "compass" icon.)

2 Click the *Animation Settings* button under the *3D Navigation control panel* and specify the desired animation settings for the preview (visual style, resolution, rate, file format).

3 If you intend to use the *3Dwalk* or *3Dfly* commands when creating the animation, ensure the *Step Size* and *Steps Per Second* values on the *3D Navigation control panel* are adjusted appropriately.

4. Set up the opening scene with the view you want displayed in the opening frame of the animation. Since AutoCAD does not offer any post-production tools, so you cannot edit the "video" in AutoCAD once it is recorded.

5. Ensure *Perspective Projection* is on, then start one of the 3D navigation commands, such as *Pan*, *Zoom*, *3Dorbit*, or *3Dwalk*.

6. Begin recording by selecting the *Start Recording Animation* button on the *3D Navigation control panel*. (When you begin a 3D navigation command, the button turns red indicating that you can begin recording.)

7. As you change the view using any of the 3D navigation commands, AutoCAD records the current movement in the view and saves it temporarily for creating the final animation. You can switch between 3D navigation commands during the animation; for example, you could switch from "walking" with *3Dwalk* to turning the view with *3Dorbit*. If you want to make any adjustments during recording, you can click the *Pause Recording Animation* button to temporarily pause recording.

8. You can review the temporarily saved animation in a preview window by clicking the *Play Animation* button. If needed, you can exit and return to the drawing window to add more sequences to the animation.

9. Once you are ready to generate and save the animation to a file, select the *Save Animation* button. The *SaveAs* dialog box is displayed and allows you to give the animation a name and location for the file.

10. Click the *Animation Settings* button to specify any changes in the final animation file that will be generated. Click *Save* to create the animation file. The *Creating Video* dialog box is displayed showing the progress of recreating the frames of the animation as they are being generated and saved to the file.

11. *Exit* the current 3D navigation command to end the process and purge the temporarily saved animation information from memory.

12. Generating the animation you want is an iterative process, similar to creating multiple renderings until you achieve the desired results. However, if you use this method to create an animation, nothing is saved in AutoCAD, so to make changes you must recreate the process.

Controlling Animation from the *3D Navigation Control Panel*

The *3D Navigate control panel* on the *Dashboard* palette contains controls that are helpful for changing the current camera view or managing and creating an animation (Fig. 41-72). The top part of the control panel gives you access to navigation commands such as *3Dcorbit*, *Pan*, *Zoom*, *3Dwalk*, or *3Dfly*. It is important to have quick access to these commands since recording an animation keeps the entire time sequence, including the time it takes to switch commands.

The expanded section of the control panel provides access to additional animation controls, especially the record, pause, and playback buttons. In addition, you can control the camera and target settings and step values as an alternative to the related dialog boxes.

FIGURE 41-72

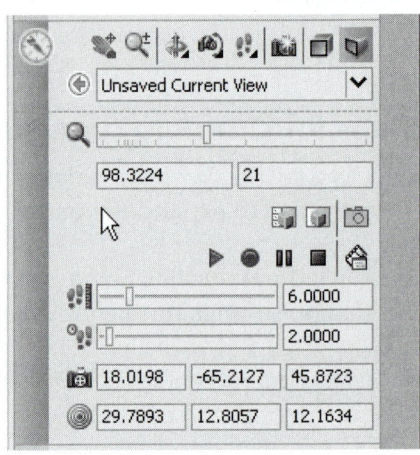

Once you activate a 3D navigation command to start the animation, turn on the round red *Start Recording Animation* button on the *Dashboard* (see previous Fig. 41-72). During the animation, you can *Pause Recording Animation* and resume again. At any time when the recording is paused you can see a preview of the animation using the triangular green *Play Animation* button. This button produces the *Animation Preview* where you can watch the animation the has been created to that point (Fig. 41-73). The drop-down list provides the standard visual styles for the preview only—these options do not affect the final saved version of the animation.

Keep in mind that this tool displays the current animation that is temporarily in memory. The actual animation should be saved using the *Save Animation* button, where you designate a file format, name and location for the animation file. If you want to save the current animation, <u>do not exit the current 3D navigation command until the animation is saved to a file</u> (see "Steps for Creating an Animation Using the *3D Navigation Control Panel*").

FIGURE 41-73

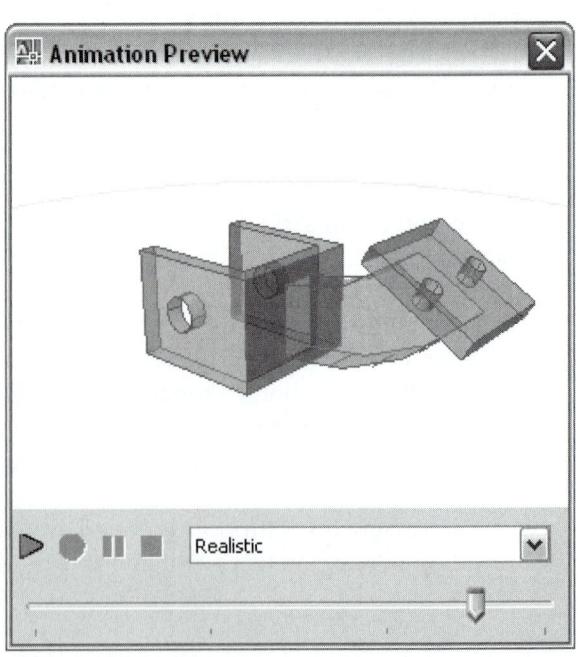

3Dwalk

Pull-down Menu	Command (Type)	Alias (Type)	Short-cut	Screen (side) Menu	Tablet Menu
View > *Walk and Fly >* *Walk*	*3Dwalk*	*3DW*	...	...	...

The *3Dwalk* command allows you to set up an animation as if you are walking through the model. In real life, as you walk along the floor through a house or building, your eyes maintain a constant distance from the floor; therefore, your movement is relatively 2-dimensional. The *3Dwalk* command restricts the camera movement in the animation to two dimensions—that is, theoretically the camera movement is restricted to a plane parallel to the XY plane of the model.

The *3Dwalk* command can be used only in perspective mode. If the drawing is not currently in perspective mode, a *Warning* message box (Fig. 41-74) is displayed prompting you to toggle on *Perspective projection*.

FIGURE 41-74

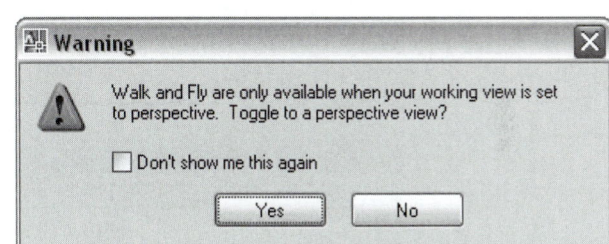

The *3Dwalk* command allows you to interactively adjust the camera view along the XY plane as if you are walking through the 3D model. The command uses a combination of keyboard input and mouse movement to create the movement of the camera through the current view. While the command is active, the large crosshairs in the drawing window indicate the target of the camera (Fig. 41-75).

FIGURE 41-75

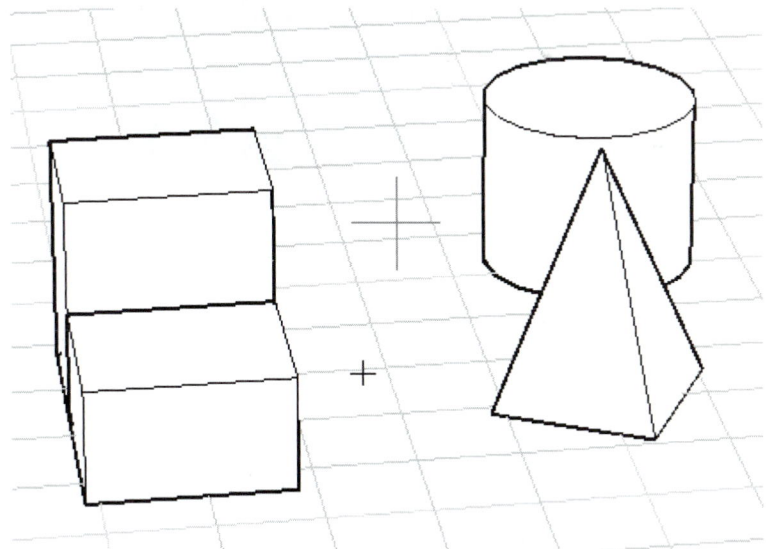

Once the command has been started and perspective mode is active, the *Walk and Fly Navigation Mappings* dialog box is displayed. This dialog box explains the key strokes and mouse input that the command accepts while it is running. Click *Close* to return to the drawing window. See "Walk and Fly Navigation Mappings" for more information on the keyboard input and the *Walk and Fly Navigation Mappings* dialog box.

By default after you have been returned to the drawing window, the *Position Locator window* is displayed that gives you an overhead display of the camera position and target (view direction) in the 3D model. See "*Position Locator*" for more information on the *Position Locator* palette. Both the *Walk and Fly Navigation Mappings* dialog box and *Position Locator window* can be toggled on or off using the right-click shortcut menu when the *3Dwalk* command is active.

NOTE: To set the height of the camera during *3Dwalk*, use the *Camera Position Z* field in the *3D Navigation control panel* expanded section.

3Dfly

Pull-down Menu	Command (Type)	Alias (Type)	Short-cut	Screen (side) Menu	Tablet Menu
View > Walk and Fly > Fly	3Dfly	...	...	...	...

The *3Dfly* command is virtually identical to the *3Dwalk* command; however, theoretically, *3Dfly* allows you to interactively adjust the current camera view in a Z (height) direction. You can accomplish this by holding down the mouse wheel and "panning" up or down to change the height position of the camera view in the 3D model. This action automatically resets the *Camera Position Z* field in the *3D Navigation control panel* expanded section. Ironically, this feature is available in the *3Dwalk* command as well. This is similar to the 3-dimensional freedom provided by *3Dorbit*.

Walk and Fly Navigation Mappings

FIGURE 41-76

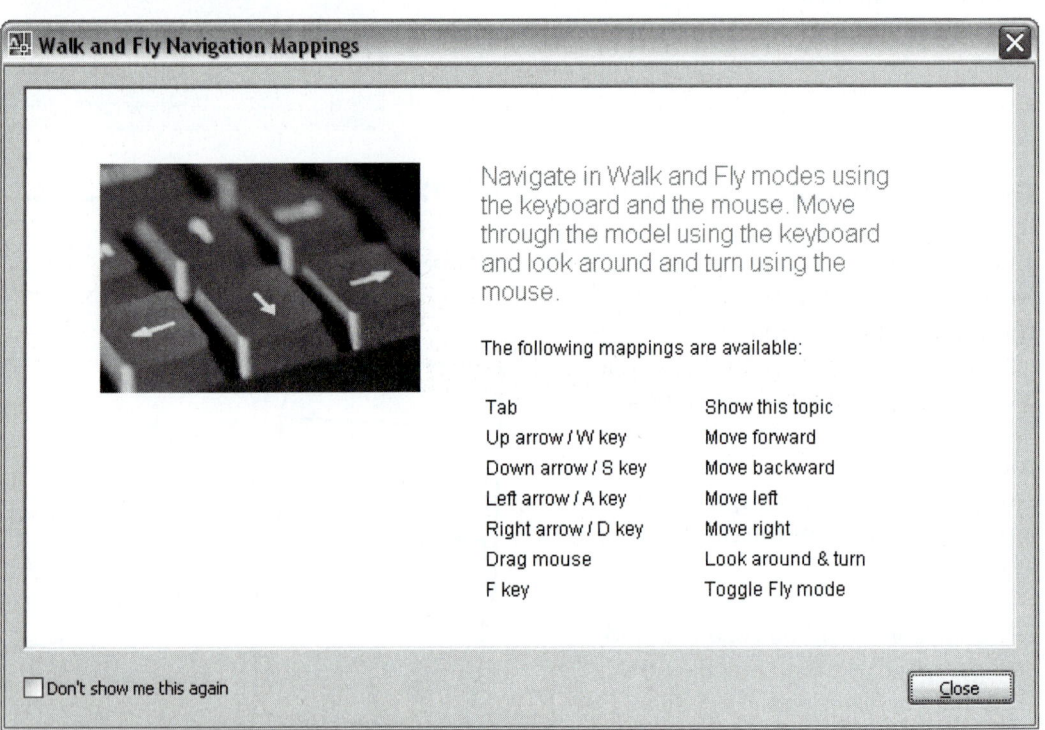

The *Walk and Fly Navigation Mappings* dialog box (Fig. 41-76) gives instructions for using the keyboard input and mouse movements that are used to change the view during the *3Dwalk* and *3Dfly* commands. The key stroke groups (arrow keys or W, S, A, D keys) are set up for use on either the left or right side of the keyboard, and are similar to keystrokes commonly used for computer games.

Although some information is given for mouse-driven movement, the description is not complete. You can move forward and backward in the model (similar to performing a *Zoom*) by rolling the mouse wheel. You can also hold down the left button and drag to perform a limited "turn" (similar to *Pan*).

Once these instructions are learned, check *Don't show me again*, since it is likely you will activate *3Dwalk* and *3Dfly* multiple times while switching between 3D navigation commands.

Position Locator

The *Position Locator window* (Fig. 41-77) is used to give you a dynamic overhead view in the 3D model as you walk or fly through it. The locator displays the position of the camera, target of the camera, and lens length/field of view ratio (camera angle). The *Position Locator* is available only for *3Dwalk* and *3Dfly*.

FIGURE 41-77

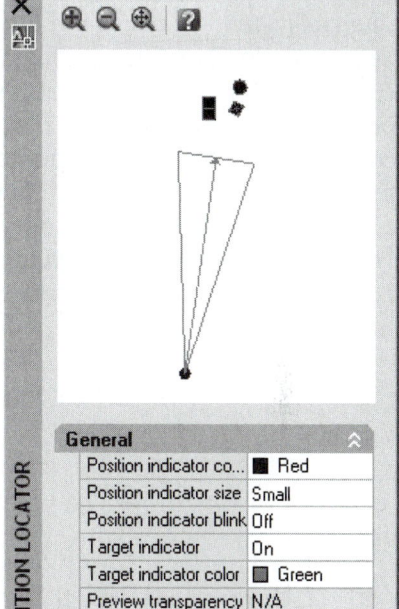

Preview area

You can dynamically change the camera by clicking and dragging in the preview area. You can drag the entire camera/target triangle or change the camera position or target position independently.

Zoom In/ Out/ Extents

These buttons enlarge and reduce the view of the preview. The drawing view is not changed.

Position indicator color

This field specifies the color of the dot used to indicate the camera position in the locator.

Position indicator size

Change the size of the dot representing the camera to *Small*, *Medium*, or *Large*.

Position indicator blink

On or *Off* options specify if the position indicator blinks in the preview area or not.

Target indicator

This option toggles the display of the triangular target indicator that shows the current viewing direction in the 3D model.

Target indicator color

Here you can choose a color for the target indicator.

Preview transparency

This option specifies the amount of transparency that objects have in the preview area. The transparency effect is only active when software acceleration is enabled, but you can see the transparency effect only when a visual style that displays obscured lines or shaded surfaces is active, such as *Conceptual* or *3D Hidden*.

Preview visual style

Specifies the visual style that should be used to display the objects from the drawing in the preview.

Walkflysettings

Pull-down Menu	Command (Type)	Alias (Type)	Short-cut	Screen (side) Menu	Tablet Menu
View > Walk and Fly > Walk and Fly Settings	*Walkflysettings*	...	...	...	...

The *Walkflysettings* command displays the *Walk and Fly Settings* dialog box (Fig. 41-78). Options here include controlling the distance that is covered per step (each press of an arrow key) while the *3Dwalk* or *3Dfly* command is active, and the number of steps that occur for each second one of the arrow keys is held down. This dialog box has settings that control the display for the *Walk and Fly Navigation Mappings* dialog box and *Position Locator window* when the *3Dwalk* or *3Dfly* command is first started.

FIGURE 41-78

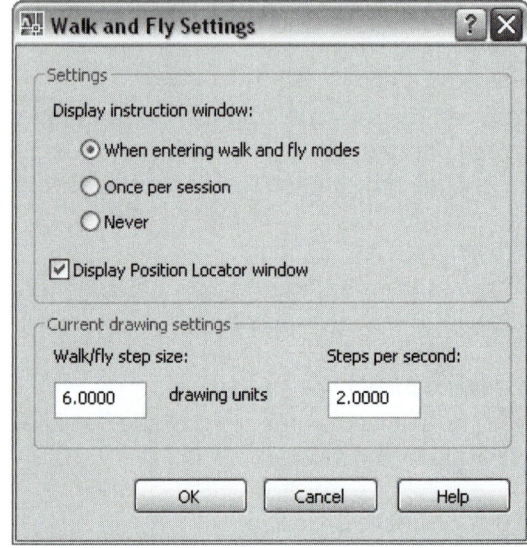

Display instruction window
Specifies when the *Walk and Fly Navigation Mappings* dialog box is displayed. The options include *When entering walk and fly modes*, *Once per session*, and *Never*. Select *Once per session* or *Never* once you become familiar with the keyboard and mouse movements.

Display Position Locator window
This checkbox toggles the display of the *Position Locator* window when the command *3Dwalk* or *3Dfly* is started.

Walk/fly step size
This value specifies the size of a step in drawing units when an arrow key is pressed during the use of the *3Dwalk* or *3Dfly* commands. The *STEPSIZE* system variable holds the current value.

Steps per second
This value specifies the number of steps per second that occur while one of the arrow keys is pressed and held down while the *3Dwalk* or *3Dfly* commands are in use. The *STEPSPERSEC* system variable holds the current value. This value has no effect on the frames per second used to generate the animation .AVI, .MPG, or .WMV file.

Creating an Animation Using *Anipath*

Anipath provides the key to creating a good animation. *Anipath* allows you to set an exact path for the animation, rather than repetitively trying to achieve a smooth movement using interactive 3D navigation commands. The *Anipath* command allows you to create an animation in which the camera movement is defined by a fixed path. Since you define the path using an AutoCAD object such as a *Spline* or *Circle*, a smooth and predictable motion results. In addition, it is possible to improve the animation through repeated adjustments to the path or other parameters, rather than having to repeat interactive 3D navigation commands to achieve the desired animation by trial and error.

The camera can be put in motion along a path and the target can be anchored to a fixed point or vice versa. Alternately, both the camera and target can be put in motion to follow independent paths. Before you use the *Anipath* command, create the desired path object(s) in the model. A path object can be a *Line*, *Arc*, *Elliptical arc*, *Circle*, *Polyline*, *3D Polyline* or *Spline*.

Anipath

Pull-down Menu	Command (Type)	Alias (Type)	Short-cut	Screen (side) Menu	Tablet Menu
View > *Motion Path* *Animations*	*Anipath*	...	...	...	...

The *Anipath* command displays the *Motion Path Animation* dialog box (Fig. 41-79) to specify the parameters needed to create and save the animation. The options are described next.

FIGURE 41-79

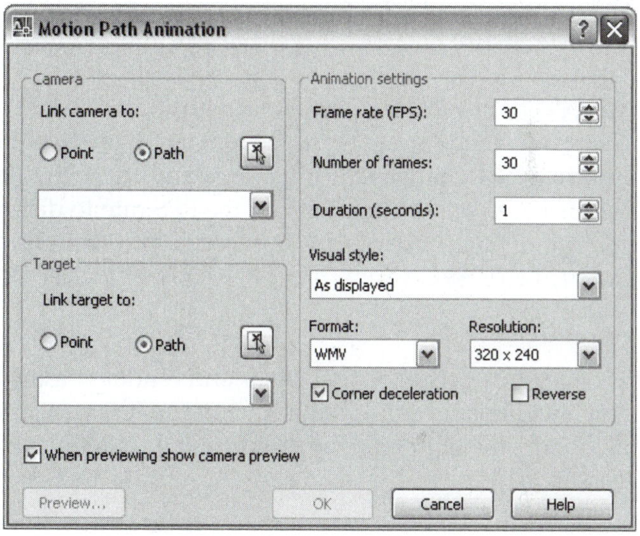

OK

When you have made all the desired settings and previewed the animation, use the *OK* button to write the animation to a file. Selecting *OK* produces the *Save As* dialog box, which allows you to assign a name and location for the file in .AVI, .MPG, .MOV, or .WMV format. Use the *Preview* button before you use *OK* to verify that all settings create an animation as you expect. You can access the *Animation Setting* dialog box from the *Save As* dialog box if needed (see "Saving an Animation to a File" next).

NOTE: Although the *Anipath* method for creating animations has many advantages, there is one caution. When you use the *Motion Path Animation* dialog box and specify the paths, points, and other settings, no settings are saved if you exit the dialog box unless you first save the animation to a file. If you exit the dialog box to make adjustments to the model without first saving a file, then use *Anipath* again, all settings in the dialog box must be specified again.

Link camera to

Link the camera to a fixed point or to a path. The *Point* option anchors the camera to a fixed point in the 3D model. Select the *Path* button to move the camera along a path in the 3D model.

Pick Point/Select Path

Click this button (on the right) to return to the drawing area to select a point in the drawing or to select the object to use as the camera's path of travel.

Point/Path List

If points or paths have previously been designated, this list displays the names of all previously defined paths and points in the 3D model that are available. Paths are assigned a name when you create them; points are stored with a name and as coordinates.

Link target to

This section allows you to link the target to a fixed point or to a path in the 3D model. If the camera is already linked to a fixed point, then the target must be linked to a path. The options here are identical to the *Camera* section above.

Animation settings

This section controls the specifications for the animation file that will be generated when you select *OK*.

Frame rate (FPS)
This value specifies the number of frames per second. A higher FPS rate creates a smoother animation; however, it also creates a larger file size and requires more time to generate.

Number of frames/ Duration (seconds)
These two values are interdependent and based on the *Frame rate (FPS)*. That is, changing one value automatically changes the other. For any one FPS rate, you can specify either the *Number of frames* or the *Duration (seconds)*. Generally, set the FPS rate to determine the "smoothness" of the animation, then determine the length of time in seconds you want the animation to run and enter that value in the *Duration* edit box.

Visual style
Choose from a list of visual styles and render presets for the animation file. Keep in mind that the preview displays only standard visual styles, whereas the final animation can be rendered using materials, shadows, mapped images, and so on, as long as an appropriate render preset is selected (*High* or *Presentation*).

Format
Select the type of animation file to create. The options are .AVI (Animation Video), .MOV (Apple QuickTime Movie), .MPG (MPEG Video), or .WMV (Windows Movie).

Resolution
Specify the overall width and height in pixels for the animation. Five resolution options are offered: 160x120, 320x240, 640x480, 800x600, and 1024x768.

Corner deceleration
If this box is checked, AutoCAD automatically slows down the camera as it moves through a corner to keep the animation smooth.

Reverse
Check this box to reverse the direction the camera or the target moves along its path.

Preview

This and the related option (below) are misleading. A normal preview appears in the drawing area, not in the *Animation Preview* (unless *When previewing show camera preview* is checked). A normal preview displays the motion of the camera in the drawing area regardless of whether the camera follows a path or follows the path of the target.

When previewing show camera preview
When you press the *Preview* button, this option (if checked) displays the *Animation Preview* (see Fig. 41-73) as well as the normal camera preview in the drawing area.

Saving an Animation to a File

The *Animation Setting* dialog box is the last interface you use before the temporarily saved animation is generated with the desired presets and saved to a file. This dialog box appears <u>only</u> through the *Save As* dialog box (not shown). Selecting the *Animation Settings* button in the *Save As* dialog box produces the *Animation Settings* dialog box (Fig. 41-80).

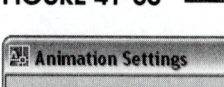

FIGURE 41-80

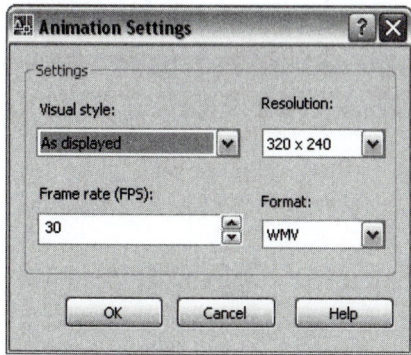

AutoCAD offers two methods for saving an animation to a file. The two methods follow the two methods you might use to create an animation. The first method is to create an animation using the *3D navigation control panel*, then use the *Save Animation* button to produce the *Save As* dialog box. The second method is to use the *Anipath* command—selecting the *OK* button in the *Motion Path Animation* dialog box (see Fig. 41-79) produces the *Save As* dialog box. However, in that case, you may have already specified the file format parameters and do not need to access the *Animation Settings* dialog box.

The *Animation Settings* dialog box offers the options needed to regenerate and save the animation to a file.

Visual Style
Choose from a list of visual styles and render presets that are used to generate the animation file. Note that the final animation can be rendered using materials, shadows, mapped images, and so on, as long as an appropriate render preset is selected (*High* or *Presentation*).

Frame rate (FPS)
Specify the desired number of frames per second. The greater the FPS rate, the smoother the animation; however, the cost is a larger animation file that takes more time to generate.

Resolution
Specifies the overall width and height of the animation in pixels. There are five resolution options: 160x120, 320x240, 640x480, 800x600, and 1024x768. Once again, the higher the resolution, the more realistic the animation. The compromises are a larger animation file and more time required to generate it.

Format
Four file formats are: .AVI (Animation Video), .MOV (Apple QuickTime Movie), .MPG (MPEG Video), or .WMV (Windows Movie).

Playing Back an Animation

AutoCAD cannot play back animations. Animation playback requires a specific video player program. However, these programs are not hard to find, are easy to install, and many programs are free and downloadable from the Internet. In fact, your computer most likely already has a program that will replay video files. If you are not sure if your computer already has a video player installed, you can try double-clicking on the rendered file to launch the default player for your system.

2007

The standard file formats and a few popular players that accept the files are listed here.

.WMV   Microsoft Windows Media Player 9 or higher
.AVI   QuickTime player or Windows Media Player
.MPG   QuickTime player or Windows Media Player
.MOV   Apple QuickTime player

CHAPTER EXERCISES

1. *Render, Lights, Materials*

    Open the **PULLY-SL** drawing that you created in Chapter 38 Exercises. Refer to the "Typical Steps for using Render" in this chapter to create a rendering. As an aid, follow these brief suggestions.

    Use *3Dorbit* to show the best view of the pulley. Create a *Distant Light* located to the left side and slightly above the pulley. Create a *Point Light* located to the right side and above. Define and attach a *Material* representing steel. The material should have a high *Reflection* value and a high *Shininess* value to make the material appear hard and reflective. Using the *Render* command with a *Render Preset* of *Low*, perform several renders and adjust the positions of the lights and their intensity values until you achieve the desired results. Experiment with some of the other properties for materials. When you have the best settings, perform a rendering using the *Presentation Render Preset* to achieve results similar to those shown in Figure 41-81. Finally, save the rendered image by right-clicking on the temporary file in the *History* pane in the render window and choosing *Save*. Give the image file the name **PULLY-REND.TIF** . Double-click on the **PULLY-REND.TIF** file to open the default image viewer on the computer. Use *SaveAs* and rename the drawing to **PULLY-REND**.

FIGURE 41-81

2. *Render, Lights, Materials*

    Open the **SADL-SL** drawing that you created in Chapter 38 Exercises. For this rendering, turn on *Perspective Projection* and use *3Dorbit* to generate a suitable viewpoint. Set up two lights: a cool-colored *Distant Light* with a vector of **0.4924,0.4132,-0.7660**, and a warm-colored *Point Light* at **-100,-300,100**. Create and attach a material called *Blue Metallic Metal*. The material should use the *Advanced Metal* material template and have a light blue-gray *Color*. Modify the material so it has a low *Reflection* value (**10**), and a mid-range *Shininess* value (**50**).

FIGURE 41-82
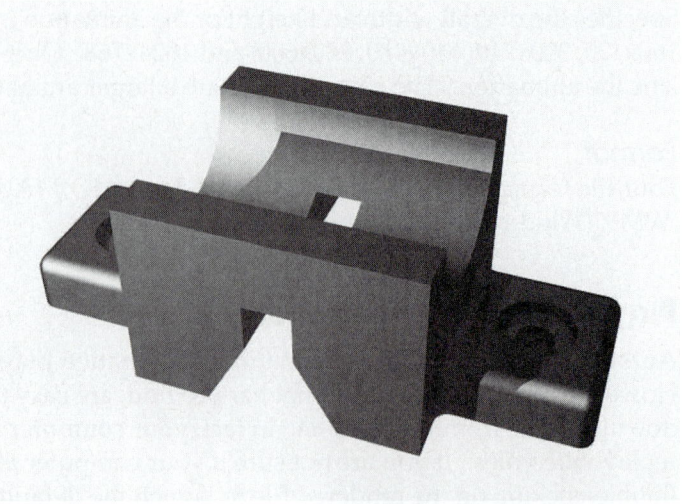

Create several low quality renders using the *Low Render Preset* to adjust the intensity values of the lights. The *Point Light* should provide slightly more light (at the near end of the model), and the *Distant Light* should provide slightly less (on the far side). The varying light intensities with the color effects give the proper impressions of distance. Also make necessary adjustments to the *Material* properties. If the entire rendering is too dark, you can try using the *Brightness multiplier* slider. When you have a good combination, compare your image to Figure 41-82 and create a rendering using the *Render Preset* set to *Presentation*. Save the rendering to the file name **SADL-REND.TIF**. Use *SaveAs* and rename the drawing to **SADL-REND**.

3. *Background, Render, Lights, Materials, Shadows, Render Preferences, Lights List*

Open the **PULLY-REND** drawing that you created in the first exercise. Save the existing viewpoint (from the previous exercise) to a named *View*. Create another view to show the back of the pulley. Try to show several of the through-holes in the view. Save the new viewpoint to a named *View*. Use the *Background override* option of the named *View* and create a *Gradient* background with *Three colors*. Use the *Rotation* option to bring the bands of color into a vertical orientation. Create a few quick renderings to check and adjust the background.

Create one more *Light* with a *Location* of your choosing. Launch the *Advanced Render Settings* palette, and select the "light bulb" icon next to *Shadows* and *Raytracing* to toggle those features off. In order to create the new rendering, launch the *Lights in Model* palette, and select one of the previous lights that were in the first exercise and turn it off with the *Properties* palette. Experiment to see which of the combination of two lights gives the best results. Adjust the *Intensity* and *Color* values of the lights as necessary. When you have the best settings, turn *Shadows* and *Raytracing* back on before creating the final rendering. Create the final rendering using the *Render Preset* set to *Presentation*. Save the rendering with the file name **PULLY-R-BK.TIF**. Use *SaveAs* and rename the drawing to **PULLY-R-BK**.

4. *Render, Spotlight, Material*

Open the **SADL-REND** drawing that you created in the second exercise. Change *Viewres* to **1000** and *FACETRES* to **1.0**. Make a solid *Box* using the *Center* option (locate the center at **0,-70,0**) with a *Length* of **600**, *Width* of **300**, and a *Height* of **-20**. Then use *Rotate* with the *Last* selection option and rotate the *Box* **20** degrees. The *Box* will serve as a base or "table top" for the saddle to rest on and provide a background for the *Spotlight* you will soon create. Create a new *Material* using the *Advanced* material template with a low *Reflection* value (**18**), a low *Shininess* value (**32**), and select a medium gray *Color*. *Attach* the new material to the "table top." Create a *Background* for the drawing using the *Gradient* option, or use the *Image* option and find a suitable image from the *Textures* or *Windows* folders. Create a few quick renderings to check and adjust the background and materials as necessary. Don't forget that disabling *Shadows* and *Raytracing* in the *Advanced Render Settings* palette can help speed rendering times.

Next, insert a *Spotlight* with a *Location* of **200,-200,300** and a *Target* of **60,0,0**. Assign a warm color to the *Spotlight*. Make sure that *Shadows* are *On* for the light and set the *Type* to *Soft* and a *Map Size* of **1024**. Lower the *Intensity* of the *Point Light* (that exists in the drawing from the previous exercise) to a value of **0** so the *Spotlight* provides the primary illumination from this viewing direction. Perform a few renderings using the *Medium Render Preset* with *Shadows Off* to test the angles of the *Hotspot* and *Falloff*. Make adjustments to those values so the saddle is fully illuminated, but the light falls off just beyond the saddle. Perform several additional renderings to test and adjust the lighting parameters. Any of the several parameters can be adjusted: *Intensity*, *Color*, *Shadow*, *Type* and *Map Size*, and *Location*.

Turn *Shadows On* for all lights in the drawing. Modify the material you created for the "table top" by adjusting the *Reflection* and *Shininess* values to give it a mirror-like finish. Create the final rendering using the *Render Preset* set to *Presentation*. Try to achieve a reflective effect on the table top similar to that shown in Figure 41-83. Save the rendering with the file name **SADL-REND2.TIF**. Use *SaveAs* and rename the drawing to **SADL-REND2**.

FIGURE 41-83

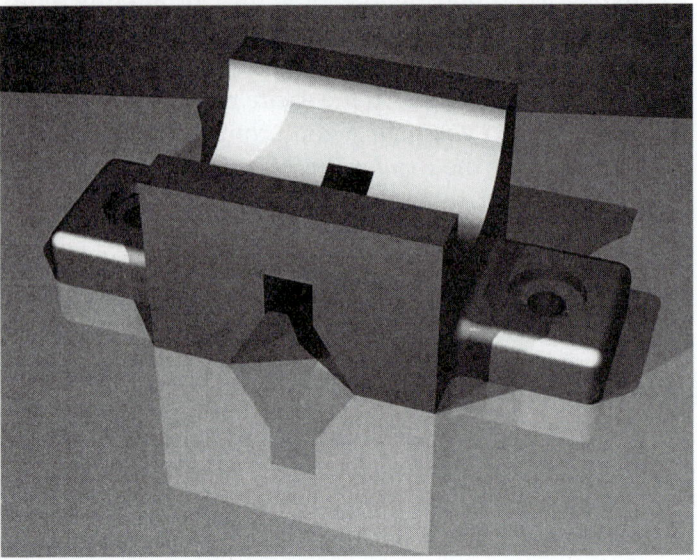

Optional (time permitting): Create several renderings, each with material and lighting adjustments, and save them from the *Render Window* using the names **SADL-REND3, SADL-REND4**, and so on. You can compare the different renderings using the *Render Window* or the default image viewer on your computer to determine the best renderings.

5. *Image Mapping, Materialmap*

NOTE: For this exercise you need to download the **FENCE.TGA** file available from the www.mhhe.com/leach Web site. Look for the Chapter Exercise Drawings in the Student Resources Center.

FIGURE 41-84

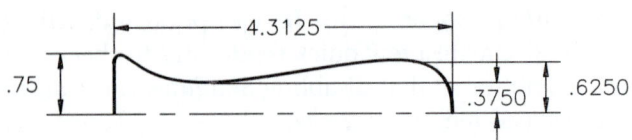

Create a machinist's file with a handle (similar to the one used as a rendering example earlier in the chapter). Use the dimensions shown at the top of Figure 41-84 to create half of the file using a *Pline* and *Spline*, then *Mirror* the shape along the long axis. Create a *Region* from the resulting shape, then use *Extrude* with a height of **3/16** to create a solid model of the file.

Create the handle solid model. First, draw the bottom shape shown in Figure 41-84 with a *Pline* and *Spline*, then use *Region* to create a single closed shape. Make the solid by using *Revolve*. Use *Move* to attach the handle to the file. Set *Viewres* to **1000** and *FACETRES* to **1.0**. Use *SaveAs* and rename the drawing to **FILE**.

Next, attach appropriate *Wood* and *Metallic* materials to the two solids. Create one *Point Light* and one *Distant Light*, then adjust the *Intensity* and *Color* for each light as desired. Make adjustments to the materials and lights as needed by creating a few quick renderings.

Modify the file materials by using the *Bump Map* property. Locate the **FENCE.TGA** image file you downloaded and set the *Bump Map Slider* to **-333**. Adjust the scale and tiling of the image map to: *U Tile* to **0.25**, *V Tile* to **0.75**, set *Bitmap Scale* to *Scale* and *Fit to Object*. Then, apply a *Spherical* map with *Materialmap* and rotate the mapping approximately **45** degrees. Set the *Render Preset* to *Medium* and create a rendering, make adjustments as necessary. Create the final rendering using the *Render Preset* set *Presentation*. Try to achieve a realistic file such as that shown in Figure 41-85. Save the rendering with the file name **FILE.TIF**. Use *Save* to save the changes to the drawing.

FIGURE 41-85

6. *Animation, Anipath*

   NOTE: For this exercise you need to download the CAMPUS.DWG file available from the www.mhhe.com/leach website. Look for the Chapter Exercise Drawings in the Student Resources Center.

   Open the **CAMPUS** drawing file. Use *SaveAs* to rename the file to **CAMPUS2** and save it to your working directory. Activate the *Model* tab. Switch to a *Plan* view so you can see the entire campus model. Draw a large *Circle* around the entire campus with the circle center at the center of campus. Draw the circle with Z coordinate of **0**. *Copy* the *Circle* up to a height of **55'-0"** (turn on *POLAR* to assist moving the circle in a Z direction only). Use *Anipath* and assign the new circle as the *Path* for the *Camera*. *Osnap* to the *Center* point of the original circle as the *Point* for the *Target*. Use a value of approximately **600** for the *Number of Frames* to create a short, but yet smooth animation. Set the *Visual Style*, *File Format*, and *Resolution* as desired. Keep in mind that the larger the resolution, the longer it will take to create the animation file. Assign the name **CAMPUS** for the animation file.

7. Open the **ADJMT-SL** or other solid or surface model that you would like to render. Achieving skill with Render requires time and experimentation. Experiment with different *Backgrounds*, *Lights*, *Materials*, *Images*, and *Animations*. Have fun.

8. *Modeling, Materials, Lights, Render*

Create a scene similar to that shown in Figure 41-86. Use *Spheres* for the balls; create the glass by *Revolving* a profile; create the rack by *Extruding* two shapes and *Subtracting* the inner shape; and create the pool cue by *Revolve, Cone,* or *Extrude* with a taper angle. Create appropriate *Lights* and *Attach* appropriate *Materials*. Save the drawing and assign it the name **POOL**.

FIGURE 41-86

42

CREATING 2D DRAWINGS FROM 3D MODELS

CHAPTER OBJECTIVES

After completing this chapter you should:

1. be able to use *Mvsetup* to create viewports showing the standard engineering "views" of an existing 3D model;

2. be able to use *Solview* to create viewports and layers for use with *Soldraw*;

3. be able to use *Soldraw* to project 3D solids onto 2D planes, complete with hidden and visible lines;

4. be able to use layers created with *Soldraw* to complete a dimensioned 2D multiview drawing;

5. be able to create a 2D or 3D profile, complete with hidden lines, from any view of a solid model using *Solprof*;

6. be able to use *Flatshot* to create a "flattened" 2D view of a solid model.

CONCEPTS

This chapter discusses the use of AutoCAD features that assist in the creation of a 2D drawing from a 3D model. Several features in AutoCAD provide a means for creating a 2D drawing from a 3D model: *Mvsetup, Solview, Soldraw, Solprof,* and *Flatshot. Mvsetup* can be used with any type of 3D model—wireframe, surface, or solid. *Solview, Soldraw, Solprof,* and *Flatshot* are used specifically with AutoCAD solid models, primarily for converting 3D geometry to 2D drawings.

Mvsetup is an automated routine that can be invoked by typing at the command prompt. It operates with any type of 3D model. This routine automatically sets up paper space viewports. It also has a series of options that allow alternate arrangements for the viewports. If you use the *Standard Engineering* option, *Mvsetup* creates the viewports and automatically places the 3D model correctly with a top, front, right side, and isometric viewpoints, similar to the *3D Setup* option of the *Vports* command. Although the viewports are set up and the 3D geometry is arranged for you, other operations are required to complete the 2D drawing with hidden lines and dimensions.

The *Solview* and *Soldraw* commands used together are far more automated and powerful than any other alternative for creating complete multiview (orthographic) drawings from AutoCAD solid models. *Solview* is similar to *Mvsetup* in that it automatically sets up the orthogonal viewpoints of the model in paper space viewports. It differs from *Mvsetup*, however, because you can select the desired orthogonal "views" and their placement, including section and auxiliary views. *Solview* also creates layers for use with *Soldraw*. The *Soldraw* command is used as the next step to project the solid model geometry onto a plane to create the 2D views on the newly created layering scheme. These new layers are complete with hidden lines and correct viewport-specific visibility settings. Dimensions can be drawn as an additional "manual" step on the layers provided by *Solview*.

Solprof (solid profile) creates a profile of an AutoCAD solid model. Rather than creating multiple orthogonal views, *Solprof* can use any view point of a solid, such as an isometric type viewpoint, to create a profile. *Solprof* provides a choice of using the profile to create a 3D wireframe model or to create a "flat" view projected onto a 2D plane. *Solprof* creates new layers for the profile geometry, complete with correct hidden lines if desired.

A new command, *Flatshot*, is similar to *Solprof*. It also projects solid geometry onto a 2D plane, therefore creating a "flat," 2D view. *Flatshot* is somewhat easier to use for creating a "flat" view because you can select the location, scale, and rotation angle for placing the view.

In summary, using *Solview* and *Soldraw* is the superior method for creating complete 2D multiview drawings from AutoCAD solid models. It is the most universal method and accomplishes the most for you automatically, including setting up layers for dimensioning. *Mvsetup*, however, can be used with wireframe or surface models to set up standard engineering views, but it cannot project the model onto a 2D plane with hidden and visible lines. *Flatshot* is an easy way to create individual "flat" views of a solid model from any orientation (viewpoint). Although *Solprof* can also create "flat" views, it is an excellent method to use if you want to create a wireframe model directly from a solid model.

Topics in this chapter are arranged as follows:

Using *Mvsetup* for Standard Engineering Drawings
Using *Solview* and *Soldraw* to Create multiview Drawings
Using *Solprof* and *Flatshot* with Solid Models

USING *MVSETUP* FOR STANDARD ENGINEERING DRAWINGS

Mvsetup is a program that assists you in setting up paper space viewports. The *Standard Engineering* option can be used to set up a 3D model in viewports, each viewport having a different standard view. This application of *Mvsetup* accomplishes similar results as the *3D Setup* option of *Vports*. See Chapters 13, 33, and 35 if you need more information about layouts, paper space, and *Vports*.

Mvsetup

Pull-down Menu	Command (Type)	Alias (Type)	Short-cut	Screen (side) Menu	Tablet Menu
...	*Mvsetup*	...	...	...	...

This section discusses the use of *Mvsetup* using the *Standard Engineering* option—the option for creating engineering drawings from 3D models. This option of *Mvsetup* operates with <u>any type</u> of 3D model. It automatically creates four viewports in a layout and displays the model from four *Vpoints* (front, top, side, and isometric views).

Mvsetup is a similar but older method of creating paper space viewports than the *Vports* command with the *3D* option; however, *Mvsetup* provides other options that allow you to insert a titleblock and scale and align the geometry. The sequence is given here.

Typical Steps for Using the *Standard Engineering* Option of *Mvsetup*

1. Create the 3D part geometry in model space.

2. Create a layer for viewports and a layer for the titleblock (named VPORTS and TITLE, for example). Set the viewports layer current.

3. Activate a *Layout* tab and use the *Page Setup* dialog box to set the *Plot device* and *Paper size*.

4. Invoke *Mvsetup* and use *Options* to set the *Mvsetup* preferences.

5. Use *Title block* to *Insert* or *Xref* one of many AutoCAD-supplied or user-supplied borders and titleblocks.

6. Use the *Create* option to make the paper space viewports. Select the *Standard Engineering* option from the list.

7. Use *Scale viewports* to set the viewport scale factor (*Zoom XP* factor) for the model geometry displayed in the viewports.

8. The model geometry that appears in one viewport can be aligned with the model in adjacent viewports if necessary using the *Align* option.

9. From this point, other AutoCAD commands must be used to create dimensions for the views or to convert some lines to "invisible" lines.

Mvsetup **Example**

To illustrate these steps, a 3D wireframe model (WFEX3 drawing from Chapter 37 Exercises) is used as an example (Fig. 42-1).

FIGURE 42-1

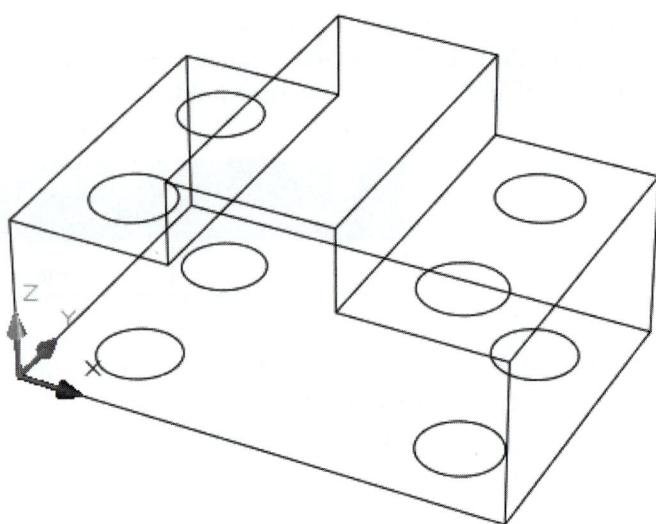

1. Create the 3D part geometry in model space. The wireframe is shown here as it exists in model space viewed from an isometric-type viewpoint.

2. Create a VPORTS layer for viewports and a TITLE layer for the titleblock. Set the VPORTS layer current.

3. Activate the *Layout1* tab. Erase any viewports. Use the *Page Setup* dialog box. Set the *Plot device* and the *Paper size* you intend to use.

4. Invoke *Mvsetup* and use *Options* to set the *Mvsetup* preferences. Specifically, the *Layer* option is used to specify the layer for the titleblock.

```
Command: mvsetup
Enter an option [Align/Create/Scale viewports/Options/Title block/Undo]: o
Enter an option [Layer/LImits/Units/Xref] <exit>: l
Enter layer name for title block or [. (for current layer)]: title
Enter an option [Layer/LImits/Units/Xref] <exit>: Enter
Enter an option [Align/Create/Scale viewports/Options/Title block/Undo]:
```

5. The *Title* block option is used to insert a titleblock and border in paper space.

```
Enter an option [Align/Create/Scale viewports/Options/Title block/Undo]: t
Enter title block option [Delete objects/Origin/Undo/Insert] <Insert>: i
Available title blocks:...
0:   None
1:   ISO A4 Size(mm)
2:   ISO A3 Size(mm)
3:   ISO A2 Size(mm)
```

(Thirteen titleblock options are displayed here.)

```
Enter number of title block to load or [Add/Delete/Redisplay]: 8
Create a drawing named ansi_b.dwg? <Y>: n
```

The previously selected options produce a titleblock and border appearing in paper space, as shown in Figure 42-2. The 3D model is not visible because viewports have not yet been created.

FIGURE 42-2

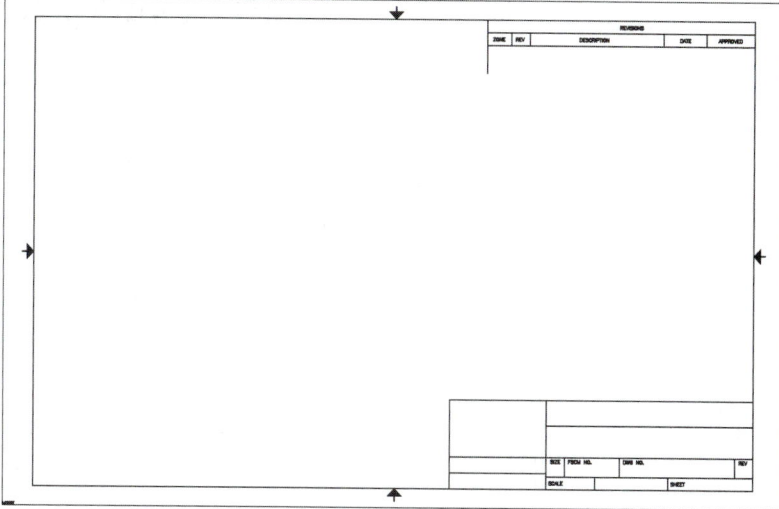

6. Use the *Create* option of *Mvsetup* to create the desired viewport configuration:

> Enter an option [Align/Create/Scale viewports/Options/Title block/Undo]: **c**
> Enter option [Delete objects/Create viewports/Undo] <Create>: **Enter**
> Available layout options: . . .
> 0:   None
> 1:   Single
> 2:   Std. Engineering
> 3:   Array of Viewports
> Enter layout number to load or [Redisplay]: **2**
> Bounding area for viewport(s). Default/<First point >: **PICK**
> Specify opposite corner: **PICK**
> Specify distance between viewports in X direction <0.0>: **Enter** or (**value**)
> Specify distance between viewports in Y direction <0.2>: **Enter** or (**value**)

Select the "*2: Std. Engineering*" option. PICK two corners to define the bounding area for the four new viewports. The action produces four new viewports and automatically defines a viewpoint for each viewport. The resulting drawing displays the standard engineering front, top, and right side views of the 3D model in addition to an isometric-type view, as shown in Figure 42-3.

FIGURE 42-3

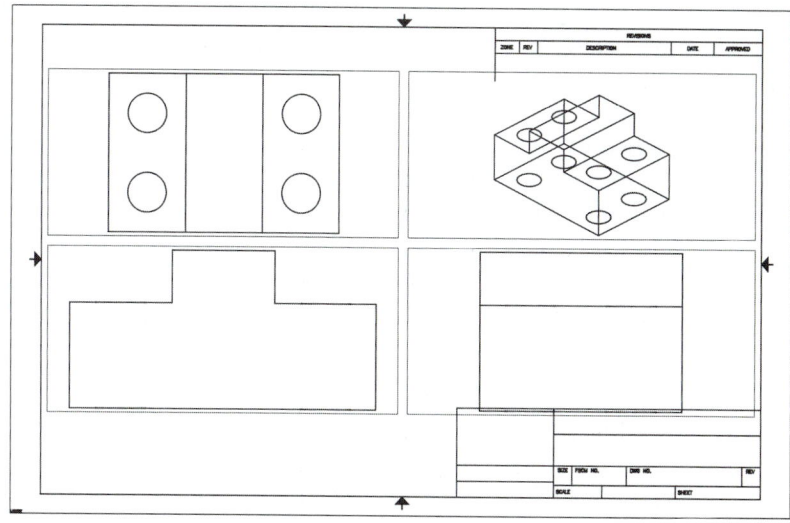

7. Notice in Figure 42-3 that the geometry in each viewport is displayed at its maximum size, as if a *Zoom Extents* were used. The *Scale viewports* option is used to set a scale for the geometry in each viewport:

    > Enter an option [Align/Create/Scale viewports/Options/Title block/Undo]: **s**
    > Select the viewports to scale...
    > Select objects: **PICK** (Select the viewport border objects)
    > Select objects: **Enter**
    > Set zoom scale factors for viewports. Interactively/<Uniform>: **u**
    > Set the ratio of paper space units to model space units...
    > Enter the number of paper space units <1.0>: **Enter** or (**value**)
    > Enter the number of model space units <1.0>: **Enter** or (**value**)
    > Enter an option [Align/Create/Scale viewports/Options/Title block/Undo]:

    Make sure you select the <u>viewport objects</u> (borders) at the "Select viewports to scale:" prompt. The *Uniform* option ensures that the 3D model will be scaled to the same proportion in each viewport. The ratio of paper space units to model space units is like a *Zoom XP* factor; for example, a ratio of 1 paper space unit to 2 model space units is equivalent to a *Zoom 1/2XP*.

The above action produces the scaling of the 3D model space units relative to paper space units, as shown in Figure 42-4.

FIGURE 42-4 ───────────────────────

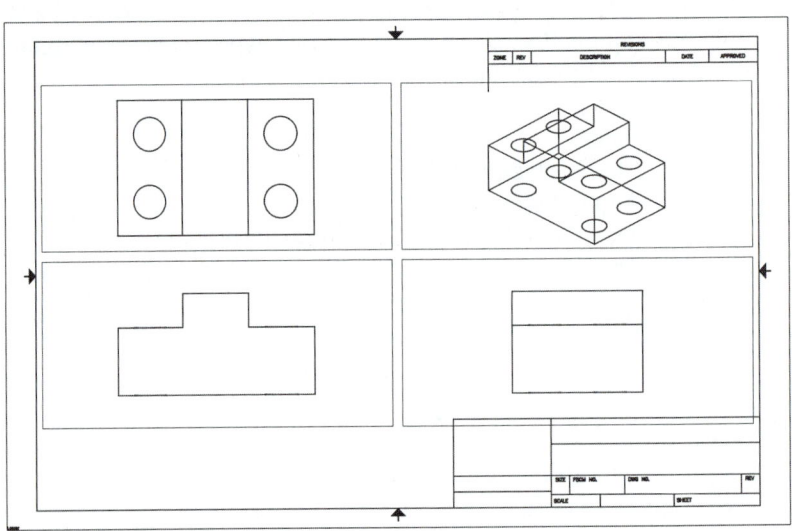

8. The *Align* option of *Mvsetup* can be used to orthogonally align the views if they are not automatically aligned by the *Scale viewports* option. The *Align* option allows you to *OSNAP* to an object in one view to accomplish a *Horizontal* or *Vertical* alignment with an object in another viewport. The *Horizontal* option can be used to align the model appearing in the front and side views and the *Vertical* option is used to select model geometry for alignment of the top and front views.

An additional step is required for this example in order to display the isometric view as desired within the viewport. *Zoom Extents, Zoom XP,* and *Pan* can be used to locate and size the model appropriately.

Finally, the VPORTS layer created for inserting the viewports can be turned *Off* or *Frozen* to display the views without the viewport borders, as shown in Figure 42-5.

FIGURE 42-5

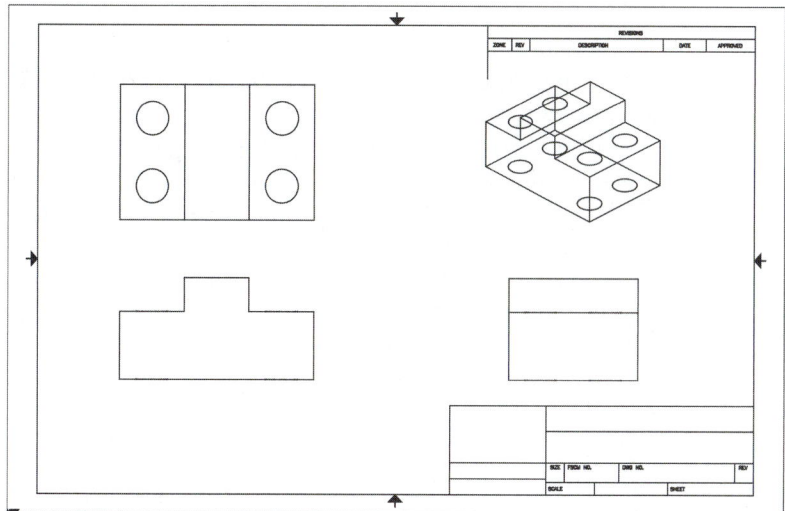

9.   This is as far as *Mvsetup* goes. As you notice, the result is not a complete conventional multiview drawing because there are no dimensions nor are there hidden lines representing the invisible edges of the object. Since the views are actually different *Vpoints* of a 3D model, hidden lines are not possible for solid models using *Mvsetup*, but hidden lines can be created for wireframe models.

Creating Hidden Lines for a 3D Wireframe Shown with *Mvsetup*

If a complete 2D drawing is to be constructed from the 3D wireframe model, some lines from the model must be converted to "invisible" lines, and/or other lines may have to be added. It is not possible to convert edges of a solid or surface model to a hidden linetype, but edges of a wireframe model can be converted. Because a wireframe model is composed of simple objects such as *Line, Arc,* and *Circle,* these objects can easily be changed to a hidden linetype or to a layer with a hidden linetype assigned. Probably the simplest way to create the "invisible" lines is to use *Properties,* select the desired objects, and change them to a layer having a hidden linetype assigned.

Better yet, it is desirable to create a separate hidden layer for each view (viewport). In this way, you have control over each view's hidden-layer visibility. For example, you may want to turn off just the isometric viewport's hidden lines or change that layer's linetype to continuous. Creating separate layers for each viewport's hidden linetypes may take more effort but affords you the most flexibility for displaying the drawing. Refer to Chapter 33, Advanced Layouts and Plotting, for more information on viewport-specific layer visibility.

Creating Dimensions for a 3D Model Shown with *Mvsetup*

If you use *Mvsetup* or any other method of creating 2D "views" of a 3D model in paper space viewports, dimensions can be created for each view. Two methods are possible.

1.   Create dimensions in paper space. This method is by far the simplest method. Dimensions can be drawn in paper space that are associated with (attached to) objects inside a viewport. Therefore, create the dimensions in paper space for each viewport to indicate measurements for each "view." In this way, separate layers, viewport-specific layer visibility, and multiple UCSs are <u>not</u> required to create and display each "view's" dimensions in the correct viewports and in the correct orientation. See Chapters 29 and 33 for information on creating dimensions in paper space and the advantages and disadvantages of doing so.

2. Create dimensions for each view on a separate layer and make each dimensioning layer visible only in the appropriate view (viewport).

    A. Create three new layers such as DIM-TOP, DIM-FRONT, and DIM-SIDE.

    B. If not existing, create and save three UCSs named TOP, FRONT, and SIDE, each with its XY plane parallel to the matching view.

    C. Starting with the front viewpoint, make the FRONT UCS active and make layer DIM-FRONT the current layer. Create the dimensions for the front view on this UCS and layer.

    D. Make layer DIM-FRONT frozen for all <u>other</u> viewports so the dimensions for the front view appear only in the front viewport.

    E. Moving to the TOP viewport, set its matching UCS and layer current. Create the dimensions for the view. Make layer DIM-TOP frozen for all other viewports.

    F. Do the same for the remaining side view. Finally, turn off the viewport layer and insert the title block.

USING *SOLVIEW* AND *SOLDRAW* TO CREATE MULTIVIEW DRAWINGS

The *Solview* command creates new layers and new views in paper space viewports for the existing 3D model space geometry. The *Soldraw* command creates new 2D objects on the new layers. *Soldraw* projects the model geometry onto a 2D plane with the appropriate continuous or hidden linetype. *Soldraw* can operate only with viewports that have been created with *Solview* (not with those created with *Mvsetup*). <u>*Solview* and *Soldraw* operate for solid models only</u>. The typical steps for using *Solview* and *Soldraw* to create a 2D drawing from a 3D model are described here.

Typical Steps for Using *Solview* and *Soldraw*

1. Create the part geometry in model space. Set up a UCS parallel to the desired profile (front) plane of the object. Also ensure that the HIDDEN linetype is loaded in the drawing.

2. Activate a *Layout* tab. Erase any viewports. Use the *Page Setup* dialog box to set the *Plot device* and the *Paper size* you intend to use.

3. Type *Solview* at the command prompt, select the *Setup View* icon button, or select *Modeling > Setup > View* from the *Draw* pull-down menu.

4. Use the *UCS* option to create the profile (front) view. You can select the location and scale for the view.

5. Use the *Ortho* option to create the other principal orthographic views. Usually a top and/or side view is needed.

6. If a section or auxiliary view is desired, use the *Section* or *Auxiliary* option to create the view in the desired location.

7. Invoke *Soldraw* by any method. *Soldraw* is used to project the "views" to a 2D plane, thus creating new 2D geometry on the appropriate layers (created by *Solview*).

8. *Freeze* the VPORTS layer. Create a new layer named TITLE and draw, *Insert,* or *Xref* a titleblock and border in paper space.

9. You can then create dimensions for the 2D drawing on the "*-DIM" layers prepared by *Solview*. Use the normal commands for creating and editing dimensions.

Solview

Pull-down Menu	Command (Type)	Alias (Type)	Short-cut	Screen (side) Menu	Tablet Menu
Draw *Modeling >* *Setup >* *View*	*Solview*	...	...	DRAW 2 SOLIDS *Solview*	...

Solview creates paper space viewports and automatically specifies the correct viewing angle (*Vpoint*) for each *viewport*. You select which views you want and the location for each. *Solview* also prepares new layers for subsequent use with the *Soldraw* command:

> Command: `solview`
> Enter an option [Ucs/Ortho/Auxiliary/Section]:

Ucs

The *Ucs* option creates a <u>view</u> normal (perpendicular) to the XY plane of a User Coordinate System. The *Ucs* option is generally the best way to create the first viewport from which other viewports can be created. All other *Solview* options require an existing viewport. You can select and readjust the view's location. You can also specify the size of the viewport.

Ortho

This option creates a principal orthographic view (top, side, bottom) from an existing viewport. New *Ortho* viewports are created by selecting an <u>edge</u> of an existing viewport object to project from. You can select and readjust the view's location and set the size of the viewport.

Auxiliary

This option is used to create an auxiliary view from an existing view. You are switched to model space to PICK two points to define the inclined plane used for the auxiliary projection. *OSNAPs* should be used for selection.

Section

This option creates a section view. *Solview* uses *Slice* to create the section at the cutting plane you define. The 3D geometry behind the cutting plane remains visible. Similar to the *Auxiliary* option, you PICK two points to define a cutting plane.

Solview creates the layers that *Soldraw* uses to place the visible lines, hidden lines, and section hatching for <u>each view</u>. *Solview* also creates layers for dimensioning that are set for visibility per viewport. Because of the possible complexity of the drawing, *Solview* places the visible lines, hidden lines, dimensions, and section hatching for each view on <u>separate layers</u>. This provides you with complete visibility control. The following naming convention is used for the layering scheme.

<u>Layer Name</u>	<u>Object Type</u>
view name–VIS	Visible lines
view name–HID	Hidden lines
view name–DIM	Dimensions
view name–HAT	Hatch patterns

(The *view name* is the original name that you specified for the view when it was created.)

Solview also creates a layer called VPORTS for the viewport objects. This layer should be reserved exclusively for the use of *Solview* and *Soldraw*. Do not draw or alter information on this layer.

Soldraw

Pull-down Menu	Command (Type)	Alias (Type)	Short-cut	Screen (side) Menu	Tablet Menu
Draw *Modeling >* *Setup >* *Drawing*	*Soldraw*	...	...	*DRAW 2* *SOLIDS* *Soldraw*	...

Soldraw is generally used immediately after *Solview*. The *Soldraw* command uses the 3D model in each viewport created by the *Solview* command and projects the profiles onto a 2D plane. (*Soldraw* actually uses the *Solprof* command.) New 2D objects are created as profiles and sections in the viewports. The new 2D objects are drawn in the appropriate *Continuous* or *Hidden* linetypes. If sectional views are included, hatch patterns are drawn using the current values of the *HPNAME*, *HPSCALE*, and *HPANG* variables. You can use *Soldraw* only with previously created *Solview* viewports.

Soldraw is very easy to use because all you have to do is select the viewports that you want to be affected by *Soldraw*. There are no options for *Soldraw*:

    Command: **soldraw**
    Select viewports to draw...
    Select objects: **PICK** viewport objects
    Command:

Solview and Soldraw Example

As the steps (given previously) for using *Solview* and *Soldraw* are explained here, an example AutoCAD solid model is used to illustrate the process (Fig. 42-6). The AutoCAD solid model is the angle brace shown in previous chapter exercises.

1. The first step to use *Solview* and *Soldraw* is to create a 3D model. When the geometry is complete, create a UCS parallel to the desired front view of your model, as shown in Figure 42-6. Your first view selected using *Solview* should be the profile (front) view.

2. Next, activate a *Layout* tab. Erase any viewports. Use the *Page Setup* dialog box to set the *Plot device* and the *Paper size* you intend to use. If desired, *Insert* or *Xref* a titleblock.

3. Invoke *Solview* at the Command prompt or select the *Setup View* icon button.

FIGURE 42-6

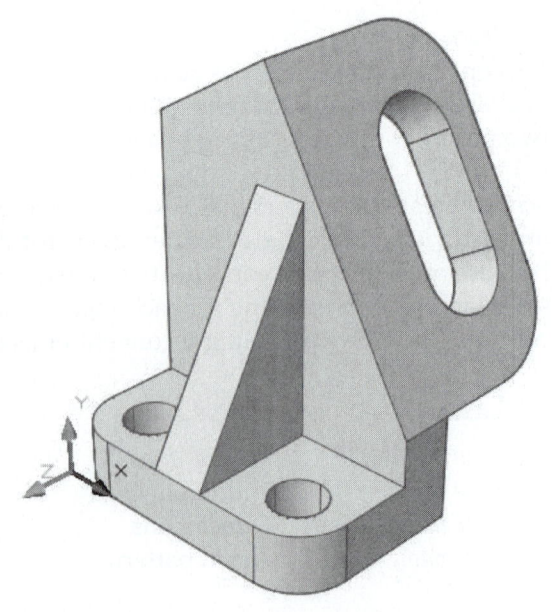

4. The *Current UCS* option is used to create a paper space viewport displaying the profile (front) view (see Fig. 42-7). The *UCS* option creates a view normal (perpendicular) to the current UCS XY plane.

> Command: **solview**
> Enter an option [Ucs/Ortho/Auxiliary/Section]: **u**
> Enter an option [Named/World/?/Current] <Current>: **Enter** (to use the current UCS)
> Enter view scale <1.0000>: **Enter** or (**value**)
> Specify view center: **PICK**
> Specify view center <specify viewport>: **PICK** (to readjust if necessary)
> Specify view center <specify viewport>: **Enter**
> Specify first corner of viewport: **PICK**
> Specify opposite corner of viewport: **PICK**
> Enter view name: **front**
> Enter an option [Ucs/Ortho/Auxiliary/Section]:

The "Enter view scale<1>:" value is the same value as a *Zoom XP* factor. That is, enter the proportion of model geometry units to paper space units.

PICK the desired "view center" to specify the desired location of the view. The model geometry appears after PICKing the view center. The "view center" can be readjusted if necessary.

At the "Specify first corner:" and "Specify opposite corner:" prompts, select the desired paper space viewport corners as indicated in Figure 42-7. Make sure you allow sufficient room for dimensions or other drawing objects you may want to include later. (It is OK for the viewports to overlap.) Finally, you must name the view. In this case, "FRONT" is used as the name of the new viewport.

FIGURE 42-7

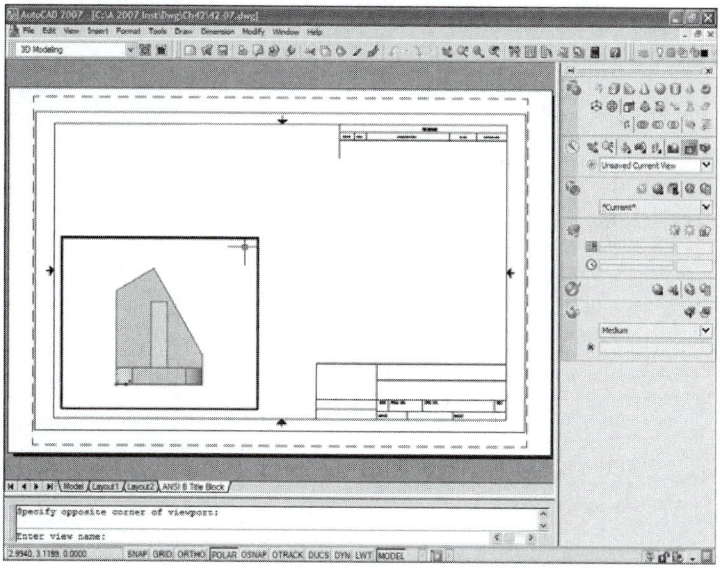

5. After the front view is created, you can create other views with the *Ortho* option. This option creates a new viewport orthographically aligned with the <u>edge</u> of the viewport object that you PICK:

> Enter an option [Ucs/Ortho/Auxiliary/Section]: **o**
> Specify side of viewport to project: **PICK** (the desired edge of viewport)
> Specify view center: **PICK**
> Specify view center <specify viewport>: **PICK** (to adjust if necessary)
> Specify view center <specify viewport>: **Enter**
> Specify first corner of viewport: **PICK**
> Specify opposite corner of viewport: **PICK**
> Enter view name: **top**
> Enter an option [Ucs/Ortho/Auxiliary/Section]:

As shown in Figure 42-8, PICK the edge of the viewport object representing the viewing direction for the new *Ortho* viewport. In other words, PICK the top edge of the front viewport to produce a top view. Notice the *Midpoint OSNAP* option is automatically invoked when you PICK the edge of the viewport. Next, PICK the view center for the top view (as before when the front viewport was established).

FIGURE 42-8

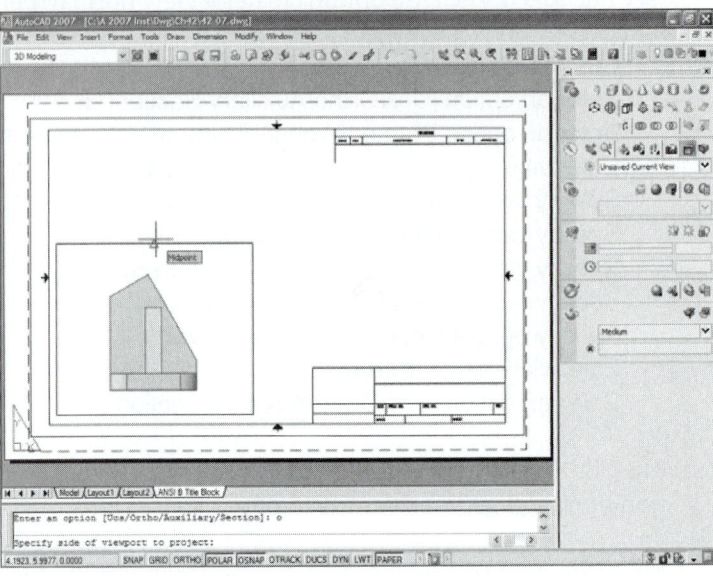

The new view then appears (Fig. 42-9). Specify the size for the viewport. Assign a name for this viewport such as "TOP."

FIGURE 42-9

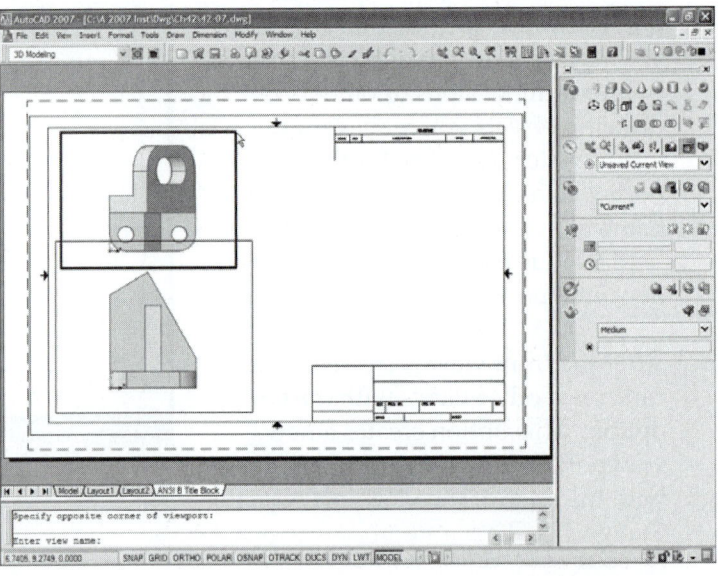

You can continue creating viewports with the *Ortho* option until the necessary principal views are established. For example, a right side view could be created by using the *Ortho* option, then PICKing the right edge of the front viewport object.

6. In this example, an auxiliary view is required to provide the necessary visual and dimensional information for the drawing. *Solview* is used with the *Auxiliary* option to create the viewport.

For this procedure you must select two points to define the auxiliary surface. You are automatically switched to model space for PICKing the surface's first and second points (*OSNAP*s should be used). For the example, the auxiliary surface is established as the upper right inclined plane as shown in Figure 42-10.

FIGURE 42-10

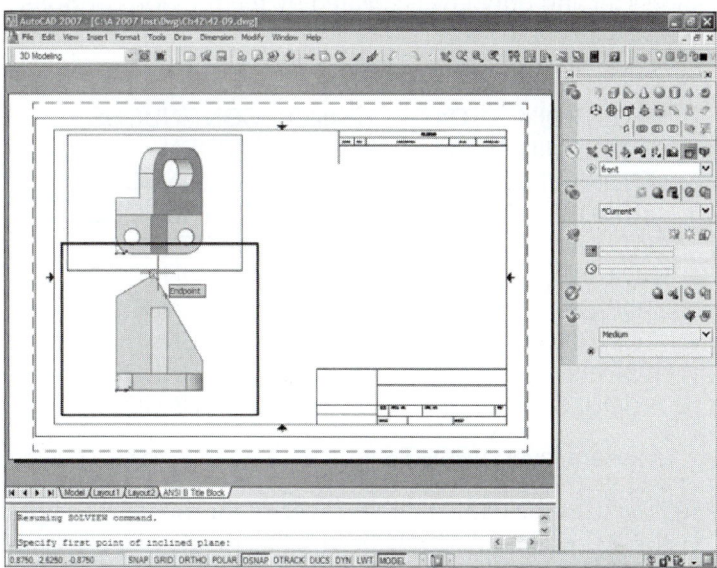

Next, PICK a point to define which side of the auxiliary surface to view the object from:

```
Enter an option [Ucs/Ortho/Auxiliary/Section]: a
Specify first point of inclined plane: PICK (use Osnaps)
Specify second point of inclined plane: PICK (use Osnaps)
Specify side to view from: PICK (on desired side of inclined plane)
Specify view center: PICK
Specify view center <specify viewport>: PICK (to adjust if necessary)
Specify view center <specify viewport>: Enter
Specify first corner of viewport: PICK
Specify opposite corner of viewport: PICK
Enter view name: aux
Enter an option [Ucs/Ortho/Auxiliary/Section]:
```

The resulting viewport with auxiliary view is shown in Figure 42-11.

The model geometry visible in the viewports is the existing 3D model. The model has not been changed in any way by *Solview*. However, as well as the viewports that are new, *Solview* creates new layers complete with viewport-specific layer visibility.

FIGURE 42-11

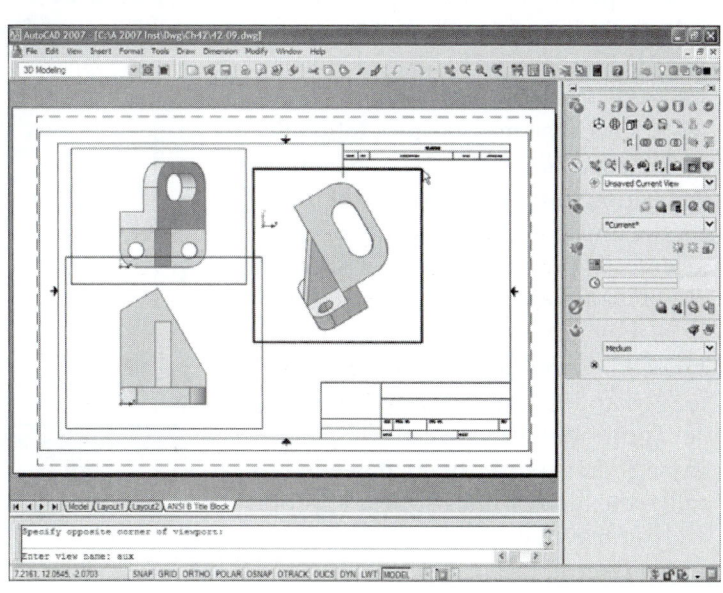

For our example, listing the layers reveals the work that was done by *Solview*. The following list shows <u>all</u> and <u>only</u> layers created by the previous options of *Solview*:

Layer Name	State	Color	Linetype
FRONT-DIM	On	7(white)	CONTINUOUS
FRONT-HID	On	7(white)	HIDDEN
FRONT-VIS	On	7(white)	CONTINUOUS
AUX-DIM	On	7(white)	CONTINUOUS
AUX-HID	On	7(white)	HIDDEN
AUX-VIS	On	7(white)	CONTINUOUS
TOP-DIM	On	7(white)	CONTINUOUS
TOP-HID	On	7(white)	HIDDEN
TOP-VIS	On	7(white)	CONTINUOUS
VPORTS	On	7(white)	CONTINUOUS

Current layer: VPORTS

Solview uses the name that you specify for the views as prefixes for the layer names. The layer visibility has automatically been set so that layers are visible only in the associated viewports (TOP-* layers are only visible in the top view, for example).

In addition, *Solview* creates a new UCS for each viewport with its XY plane parallel with the view (the *UCSVP* system variable is automatically set to 1 so each new UCS is saved with its viewport). In this way, when dimensions are created for each view, they are correctly aligned with the plane of the view.

7. Invoke the *Soldraw* command. Now that the views have been established by the *Solview* command, the model geometry is ready for projection onto 2D planes.

> Command: **soldraw**
> Select viewports to draw:
> Select objects: **PICK**
> Select objects: **Enter**

The viewport objects are PICKed in response to the "Select viewports to draw:" prompt. Normally you would <u>select all viewports</u>. PICK the viewport <u>objects</u> (borders) with the pickbox or Crossing Window, as shown in Figure 42-12. Because much computation is involved, *Soldraw* may take some time, depending on the complexity of the model and the speed of your computer system. When *Solview* has finished, the results may not be apparent—hidden lines may not display correctly if the HIDDEN linetype was not previously loaded and the *LTSCALE* variable is not set appropriately. You can complete these actions retroactively if needed to make the hidden lines appear appropriately in the views.

FIGURE 42-12 ──────────

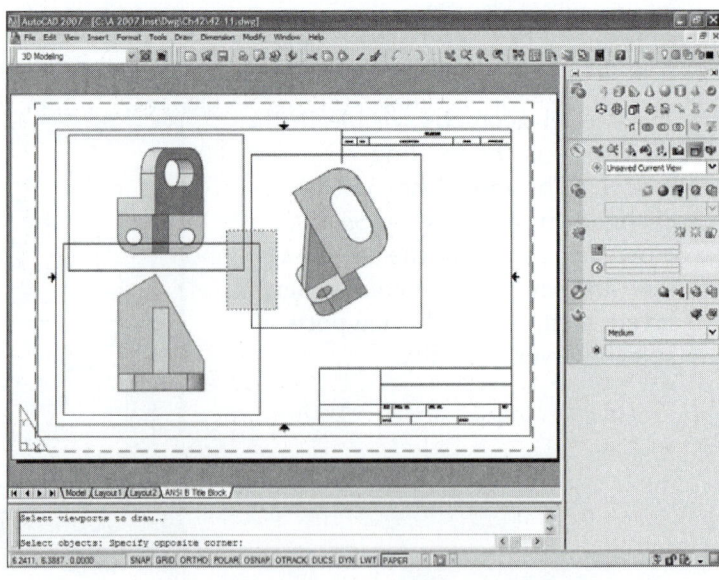

8. Use the *Layer Control* drop-down list or the *Layer Properties Manager* dialog box to control the visibility of the layers. *Freeze* the VPORTS layer to prevent the viewport borders from displaying, as shown in Figure 42-13.

9. If not done previously, create a new layer (called TITLE or BORDER, for example) and draw, *Insert,* or *Xref* a titleblock and border in paper space.

10. If you want to create dimensions for the new 2D views, use the normal dimensioning commands.

FIGURE 42-13

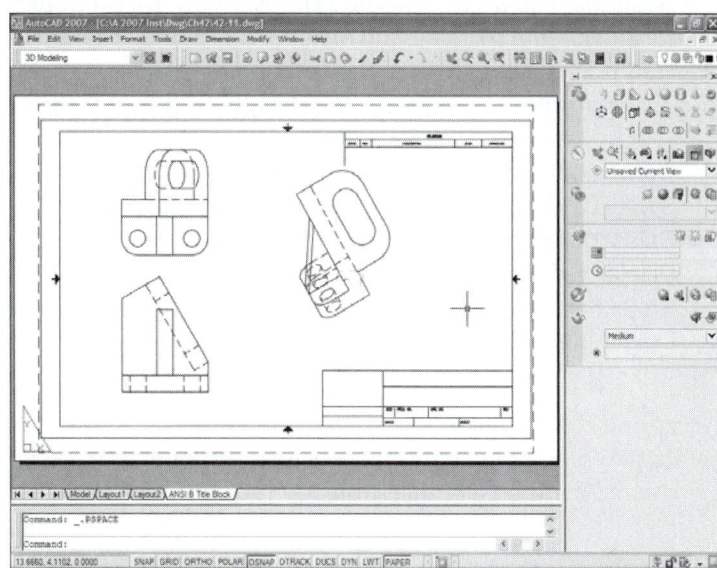

Dimensioning is accomplished in one of two ways:

A. Create dimensions in paper space. You can use this method to "attach" paper space dimensions directly to objects in model space (in the viewport). See Chapters 29 and 33 for more discussion on dimensioning in paper space.

B. Create dimensions inside each viewport. Since layers have already been created complete with the correct viewport-specific layer visibility and appropriate UCSs have been created with the correct orientation for each viewport, simply click in each viewport and create the needed dimensions for each view on the appropriate layer.

USING *SOLPROF* AND *FLATSHOT* WITH SOLID MODELS

Solprof

Pull-down Menu	Command (Type)	Alias (Type)	Short-cut	Screen (side) Menu	Tablet Menu
Draw *Modeling >* *Setup >* *Profile*	*Solprof*	...	...	*DRAW 2* *SOLIDS* *Solprof*	...

Solprof is a command that creates a profile of AutoCAD solids. *Solprof* generates a separate profile containing only the edges of straight and curved surfaces of a solid as seen from a particular viewpoint. The "profile" can be projected to a plane or can be generated as a wireframe model. The profile is generated on two new <u>layers that are automatically created</u>, one with hidden lines and one with visible lines. *Solprof* determines which profile edges should be projected as hidden or as visible lines and draws them on the appropriate layer. *Solprof* can be used in conjunction with *Mvsetup* to create standard engineering 2D drawings.

Solprof must operate <u>within a paper space viewport</u>. Activate a *Layout* tab and use the *Page Setup* dialog box to set the *Plot device* and the *Paper size* you intend to use. If desired, *Insert* or *Xref* a title block. Use *Vports* or *Mvsetup* to create the desired viewports. Then use the *Model* command or double-click inside the viewport to activate model space (inside the viewport). Then use *Solprof* and select the solids that you want profiled. Remember to <u>select the solids, not the viewport objects</u> as with *Soldraw*.

> Command: **solprof**
> Select objects: **PICK** (Select the <u>solid or solids</u> that you want profiled.)
> Select objects: **Enter**
> Display hidden profile lines on separate layer? [Yes/No] <Y>: **Enter** or **N**
> Project profile lines onto a plane? [Yes/No] <Y>: **Enter** or **N**
> Delete tangential edges? [Yes/No] <Y>: **Enter** or **N**

Accepting the *Yes* default for "Display hidden profiles on a separate layer?" prompt causes *Solprof* to create two *Blocks,* one with hidden lines and one with the linetype of the existing solid. Two new layers are created for the *Block* insertions:

> PV-*nx*   Visible profile layer
> PH-*nx*   Hidden profile layer

(Where PV designates the Profile Visible objects, PH designates the Profile Hidden objects, and *nx* represents a number and/or letter AutoCAD assigns to the viewports. This letter and number combination is called the "viewport handle." You can *List* the viewport object to display its handle.)

The *Hidden* linetype must be loaded into the drawing before *Solprof* can use it, or you can assign the HIDDEN linetype to layer PH-* after using *Solprof*. Answering *No* to the "Display hidden profiles on a separate layer?" prompt causes *Solprof* to create one *Block* drawn as visible lines with the linetype of the existing solid.

The next prompt, "Project profile lines onto a plane? <Y>:," allows you to create a 3D wireframe or a 2D profile. Answering *Yes* causes AutoCAD to project the profile onto a plane parallel to the screen, similar to the *UCS View* option. In other words, the solid as it appears from the current viewpoint is trans-formed to a 2D projection. This option is particularly useful for creating a 2D "view" from a solid model viewpoint. Answering *No* creates a 3D wireframe of the model.

The last prompt allows you to delete "tangential edges." These edges are lines between two tangent faces. Answering *No* to this prompt creates a line between a planar surface and a tangent curved surface. You should answer *Yes* to delete tangential edges for most applications.

 When *Solprof* has been used, the new objects may not be visible because *Solprof* does not change the layer visibility. To see the new *Solprof* layers, use the *Layer Properties Manager* dialog box or the *Layer Control* drop-down list to *Freeze* the original model layer(s). The new *Solprof* layers can then be viewed and edited if you choose.

NOTE: <u>When *Solprof* is activated in *Perspective Projection* mode, *Solprof* creates a 3D wireframe model</u>, even if you answer *Yes* to "Project profile lines onto a plane?" Therefore, you can create a "flat" view of the 3D object only when you activate *Solprof* in *Parallel Projection* mode and if you answer *Yes* to "Project profile lines onto a plane?"

Solprof 3D Wireframe Example

As an example, consider the solid model of the angle brace shown in Figure 42-14. In this case, the model has been completed and is displayed in model space. Also assume that the *Hidden* linetype has been loaded into the drawing.

FIGURE 42-14

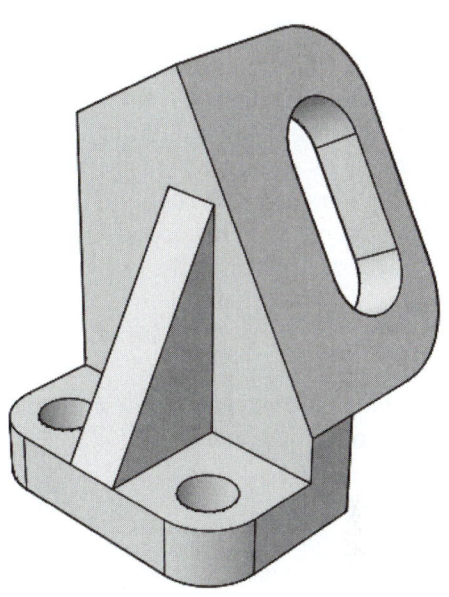

Assuming the model is displayed in the *Model* tab with the desired viewpoint, activate a *Layout* tab and use the *Page Setup* dialog box to set the *Plot device* and the *Paper size* you intend to use. If desired, *Insert* or *Xref* a titleblock. Use *Vports* to create the desired viewport(s). Then use the *Model* command or double-click inside the viewport to activate model space (inside the viewport).

Invoke *Solprof* while model space is active (the cursor is <u>in</u> the viewport) and select the solid.

The following responses are given to the prompts:

Display hidden profile lines on separate layer? [Yes/No] <Y>: **y**
Project profile lines onto a plane? [Yes/No] <Y>: **n**
Delete tangential edges? [Yes/No] <Y>: **y**

The resulting new layers are created as shown:

PH-1DD   *Hidden* linetype
PV-1DD   *Continuous* linetype

FIGURE 42-15

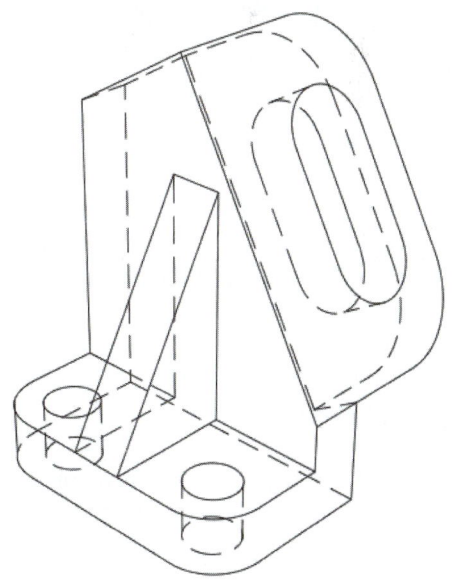

Layer visibility is <u>automatically</u> set to display the profile geometry only in the specific viewport (*Frozen* for new viewports).

Using the *Layer* command, the model layer must be *Frozen* to reveal the two profile layers. The resulting 3D wireframe geometry is displayed in its paper space viewport or in model space (Fig. 42-15).

This procedure can also be used to generate a 2D multiview drawing from an AutoCAD solid model. The process involves creating the viewports in a *Layout* using the *3D Setup* option of *Vports* or individually creating the viewports for the desired views—top, front, side, etc. Alternately, *Mvsetup* could be used to create the views. Next, use *Solprof* to create the views, complete with hidden and visible lines, for each viewport.

Keep in mind the power of using *Solprof* to automatically create a wireframe model from a solid model (answering *No* to, "Project profile lines onto a plane?"). In many cases when a wireframe is desired, it may be easier to create a solid model using the relatively simple CSG techniques, then use *Solprof* to automatically create the wireframe model from the solid. This method can be used when the model consists of multiple curved surfaces and complex lines of intersection (edges) between surfaces.

Flatshot

Pull-down Menu	Command (Type)	Alias (Type)	Short-cut	Screen (side) Menu	Tablet Menu
...	*Flatshot*	*FSHOT*	...	...	...

The result of using the *Flatshot* command is almost the same as using *Solprof*. According to AutoCAD *Help*, *Flatshot* creates a "flattened" view of all the 3D solids and regions in the current view. The term "flatshot" is somewhat misleading because the view that is created is a "flattened" 2D view only when used in *Parallel Projection* mode. *Flatshot* actually creates a 3D wireframe object when used in *Perspective Projection* mode. This occurrence is the same for the *Solprof* command. A comparison between *Flatshot* and *Solprof* is given at the end of this section.

The "view" created by *Flatshot* is a *Block*. The "view," like *Solprof*, is produced from the current viewing direction, similar to setting a new *UCS* with the *View* option and then projecting the view onto the new XY plane. However, *Flatshot* allows you to select any location for placing the resulting *Block* (view) on the <u>current</u> XY plane.

To create a <u>flattened</u> 2D view using *Flatshot*:

1. Set the projection mode to *Parallel Projection*. Next, set up the desired view of the 3D model using *3Dorbit* or *Vpoint*. Place objects you don't want captured on layers that are frozen since all 3D objects in the model space viewport are captured.
2. Invoke *Flatshot*. The *Flatshot* dialog box appears (see Fig. 42-18). Select the desired options in the dialog box and select the *Create* button. The "view" is then created but not yet placed in the drawing.
3. When prompted, specify an insertion point to place the view block. Adjust the basepoint, scale, and rotation if needed. The resulting block is projected onto the XY plane of the current UCS (see Fig. 42-16).
4. Use the *Plan* command with the *Current* option to see the new block correctly (see Fig. 42-17).

The command prompts for *Flatshot* are as follows:

```
Command: flatshot  (the Flatshot dialog box appears)
Units: Inches   Conversion:   1.0000
Specify insertion point or [Basepoint/Scale/X/Y/Z/Rotate]: PICK
Enter X scale factor, specify opposite corner, or [Corner/XYZ] <1>: Enter or (value)
Enter Y scale factor <use X scale factor>: Enter or (value)
Specify rotation angle <0>: Enter or (value)
Command:
```

FIGURE 42-16

Figure 42-16 displays the viewpoint used when *Flatshot* was invoked. At the "Specify insertion point" prompt, a point just to the left of the 3D model was selected. The resulting view is projected onto the current XY plane (step 3 above). Note the orientation of the *Ucsicon* in the figure.

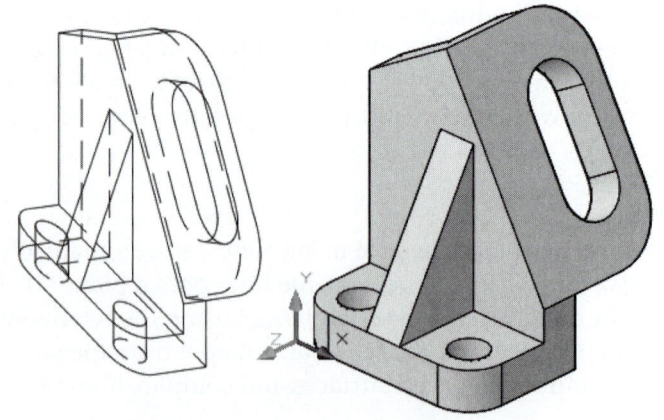

To see the resulting view correctly, you must change your viewpoint. The easiest method is to use the *Plan* command and the *Current* (default) option. This action displays the *Flatshot* view from a direction perpendicular to the current XY plane. Figure 42-17 displays the same drawing as the previous figure, but after using the *Plan* command. Note the orientation of the *Ucsicon* in this figure.

FIGURE 42-17

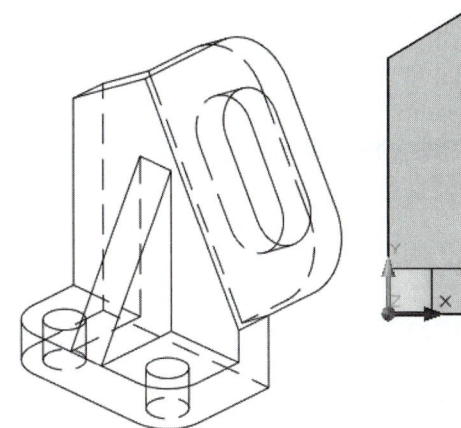

The *Flatshot* dialog box appears immediately when you start the command (Fig. 42-18).

FIGURE 42-18

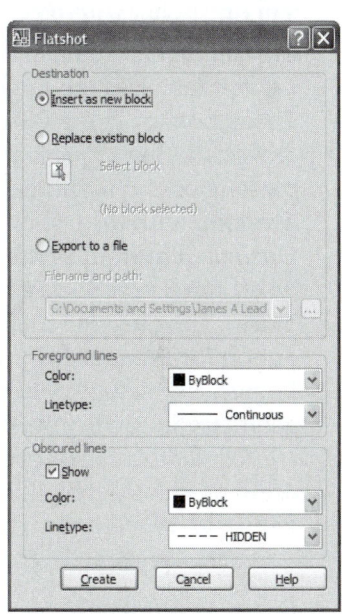

Destination
This section controls how the *Block* (view) is created. You can select *Insert As New Block*, *Replace Existing Block*, or *Export to File*.

Foreground Lines
This section allows you to specify the *Color* and *Linetype* of the visible lines that are created in the flattened view.

Obscured Lines
Using the *Show* checkbox, you control whether lines that are obscured in the drawing (hidden lines) are displayed in the flattened view. If not checked, obscured lines are not included in the view. If checked, you can set the *Color* and *Linetype* of the lines.

Flatshot versus *Solprof*

Flatshot compares to *Solprof* in these ways:

* *Flatshot* operates in model space, whereas *Solprof* operates only in a paper space layout.
* *Flatshot* automatically includes all solids and regions in the view, whereas *Solprof* requires selection of objects.
* For "flat" views, *Flatshot* places the view on the current XY plane, whereas *Solprof* projects onto a plane it creates perpendicular to the viewing direction (similar to the UCS View option).
* For 3D wireframe "views," *Flatshot* places the model orientation with respect to the current XY plane, whereas *Solprof* creates the wireframe model in the same location and orientation as the original solid.
* Both *Flatshot* and *Solprof* create "flat" views only from a *Parallel* projection and always create 3D wireframe objects from a *Perspective* mode; however, only *Solprof* can create a 3D wireframe from either *Parallel Projection* or *Perspective Projection* mode.
* Neither *Flatshot* or *Solprof* can project a view in *Perspective Projection* onto a "flat" 2D plane.

2007

In summary, use *Solprof* to create 3D wireframes and use *Flatshot* to create "flat" views. The decision is based on which planes the "views" are placed. Since *Flatshot* inserts the view on the current XY plane and allows you to select the location, it is more useable for 2D applications. Also, since *Solprof* creates a 3D wireframe in the same location and orientation as the original solid, it is easier to use for 3D applications.

CHAPTER EXERCISES

1. **Using *Mvsetup* with a Solid Model**

    Open the **BGUID-SL** drawing that you created in Chapter 38 Exercises. Use *SaveAs* to assign a new name, **BGUD-MVS**. Create two new *Layers* named **VPORTS** and **TITLE**. Make **VPORTS** the *Current* layer. Activate a *Layout* tab. Use the *Page Setup* dialog box to set the *Plot device* and the *Paper size* for plotting on a B-size sheet. Use *Mvsetup* with the following *Options*: *Layer* and *Title*. Insert the *B-size* sheet title-block. Then use the *Create* option for *Standard Engineering* setup. *Scale* the geometry in each viewport *Uniformly* to a factor of **1**. The drawing should look like Figure 42-19. *Save* the drawing.

    FIGURE 42-19

    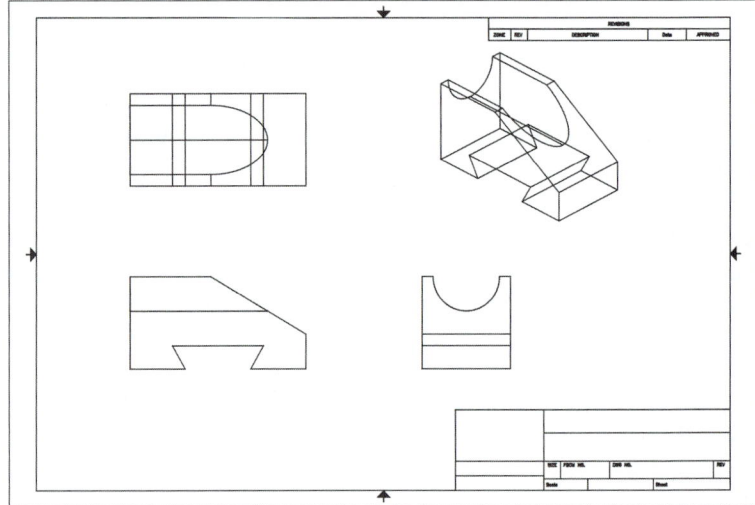

2. **Using *Mvsetup* with a Wireframe Model**

    FIGURE 42-20

    A. *Open* the **BGUID-WF** drawing that you created in Chapter 37 Exercises. Use *SaveAs* to assign a new name, **BGUD-MV2**. Make a *Layer* for the viewports and one for the titleblock. Set the viewports layer *Current*. Use *Mvsetup* with the same procedure as in the previous exercise.

    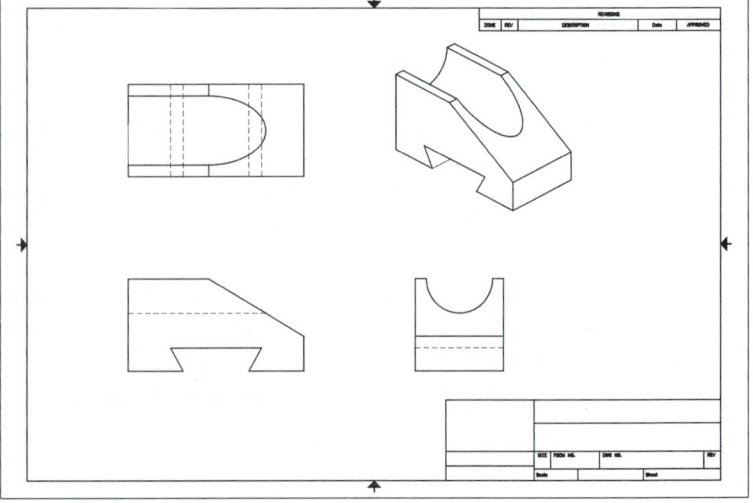

B. Convert the necessary lines to "invisible" lines by creating four new *Layer*s for hidden lines, one for each viewport. Assign the *Hidden* linetype to each new layer. Use the *Current VP Freeze* and *New VP Freeze* options to assign the correct viewport-specific layer visibility. Using *Properties*, change the appropriate lines to the matching layers.

C. *Freeze* the invisible edges of the wireframe model in the isometric viewport. *Save* the drawing. The completed drawing should look like Figure 42-20.

3. **Creating Dimensions in Paper Space**

Open the **BGUD-MV2** drawing if not already open. Use *Saveas* to save the drawing as **BGUD-MV3**. Create dimensions for the views <u>in paper space</u>. Create the dimensions on the DIM layer and draw the dimensions as shown in Figure 42-21. *Save* the drawing and *Plot* to scale.

FIGURE 42-21

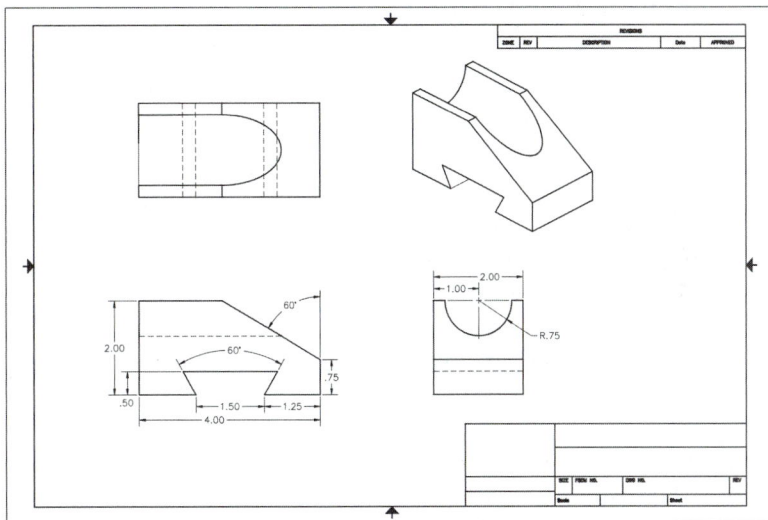

4. **Creating a 2D Drawing Using** *Solview* **and** *Soldraw*

A. *Open* the **SADL-SL** drawing from the Chapter 38 Exercises. Set up a *UCS* parallel with the front profile of the model (to create the front view as shown in Fig. 42-22). Activate a *Layout* tab. Use the *Page Setup* dialog box to set the *Plot device* and the *Paper size* for plotting on an **A**-size sheet. Change the *Printable area* units to *mm*. Next, make sure the *Hidden Linetype* is loaded and *LTSCALE* is set to **13**. Use *SaveAs* to give the name **SADDLE-SV**.

FIGURE 42-22

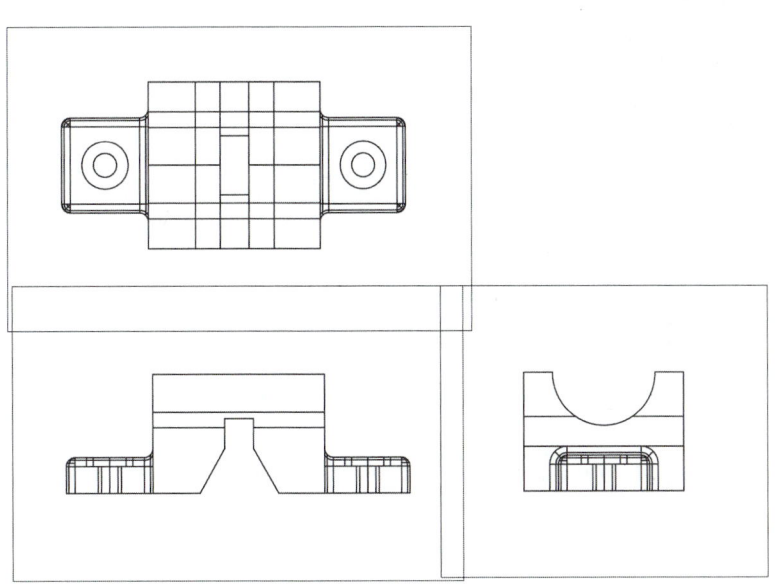

B. Use *Solview* with the *UCS* option to create the **FRONT** view. A *Scale* factor of **1** can be used. Use the *Ortho* option for the **TOP** and **RIGHT** views. If the views do not align orthogonally, use *Zoom Extents,* then *Zoom 1XP* in each viewport, or use *Mvsetup* with the *Align* option. Your results should appear like Figure 42-22. *Save* the drawing (as **SADDLE-SV**).

C. Next, use *Soldraw* to create the visible and hidden line geometry. When finished, *Freeze* the layer that the solid model is on. Insert or draw a title block and border. If the model does not fit within the border, use *Move* to move the viewport objects accordingly (make sure you *Move* two viewports together to ensure alignment). The drawing should appear like Figure 42-23. Use *SaveAs* and change the name to **SADDLE-SD**.

FIGURE 42-23

CADD Design Company		
Saddle	1:1	METRIC
Des.– B.R. Smith		JAL
2/1/06	2/3/06	SADDLE–SD

D. Next, dimension the drawing and add centerlines <u>in model space</u> (inside the viewports). Remember to use the layers that *Solview* created for the dimensions (one for each viewport). *Save* the drawing when you are finished (as **SADDLE-SD**). Make a *Plot* to scale.

5. Create a multiview drawing from the solid model you created of the angle brace (**AGLBR-SL** drawing from Chapter 38). Use *Solview* and *Soldraw* to create the multiviews. Use the *Auxiliary* option to create an auxiliary view. Create the dimensions on the **DIM** layers <u>in model space</u> (inside the viewports). *Save* the drawing as **AGLBR-SV**.

6. Use the solid model of the **SWIVEL** from Chapter 38 Exercises. Set up a layout for a B-size sheet and plan to print or plot the drawing at full size (1 = 1). Use *Solview* and *Soldraw* to create a front, top, and auxiliary view in a manner similar to that used in the previous exercises (4 and 5). Dimension the drawing (use Fig. 38-106 as a guide). Use *SaveAs* to assign the new name **SWIVEL-SD**.

7. **Using *Solprof* to Create a Wireframe Model**

A. *Open* the **PULLY-SL** drawing from Chapter 38 Exercises. Use *SaveAs* and assign the new name, **PUL-SOLPROF**.

B. In *Model* space, set the display mode to *Perspective Projection*. Activate a *Layout*. Use the *Page Setup Manager* to specify a print/plot device and paper size based on your system capabilities. *Insert* a title block and border if desired. Activate model space inside the viewport and use *3Dorbit, Zoom,* and *Pan* if needed to achieve a view similar to that shown in Figure 42-24. Set the display mode <u>in the viewport</u> to *Perspective Projection*.

FIGURE 42-24

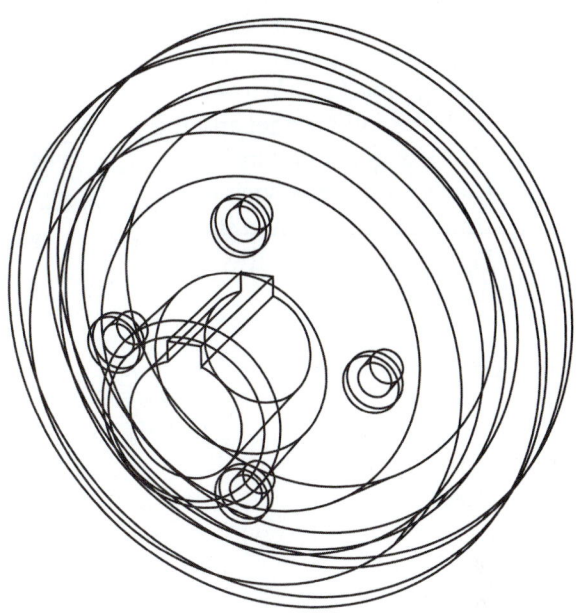

C. Activate *Solprof* and provide responses to the prompts as follows:

   Display hidden profile lines on separate layer? [Yes/No] <Y>: **n**
   Project profile lines onto a plane? [Yes/No] <Y>: **n**
   Delete tangential edges? [Yes/No] <Y>: **y**

D. Activate *Model* space. Make layer **PV-*nx*** the *Current* layer and *Freeze* the layer containing the solid model. Your wireframe model generated by *Solprof* should be visible as shown in Figure 42-24. Use *3Dorbit* to examine the wireframe model. *Save* the drawing. Note that *Solprof* was activated while in *Perspective Projection* mode. Also note that the resulting 3D wireframe can be viewed in *Perspective Projection* mode.

8. **Using *Flatshot* to Create a 2D View**

   A. *Open* the **PULLY-SL** drawing from Chapter 38 Exercises. Use *SaveAs* and assign the new name, **PUL-FLAT**.

   B. In *Model* space, set the display mode to *Parallel Projection*. Use *3Dorbit, Zoom,* and *Pan* if needed to achieve the desired view (something similar to that shown in Figure 42-24, but in *Parallel Projection* mode). *Load* the *Hidden* linetype if not already loaded. Create a *New Layer* and assign the name **2D-VIEW**. Assign an appropriate *Color* for the layer. Make the new layer the *Current* layer.

   C. Activate *Flatshot* and make the following selections in the *Flatshot* dialog box:

   Insert as new block
   *Foreground lines: **Bylayer** and **Continuous***
   *Obscured lines: **Show, Bylayer**, and **Hidden***

   When prompted for the insertion point, select a point on the current XY plane next to the solid model. Accept the defaults for X scale, Y scale, and rotation angle.

D. Use the *Plan* command and the *Current* option to generate a view point to correctly display the new, 2D view. Change the *Ltscale* to 24. Both the solid model and the 2D view generated by *Flatshot* should be visible similar to that shown in Figure 42-25. *Save* the drawing.

FIGURE 42-25

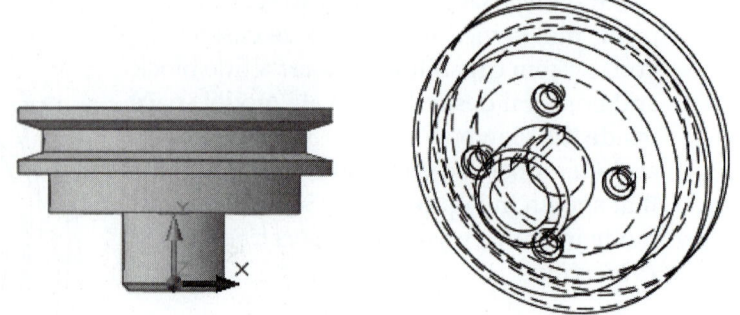

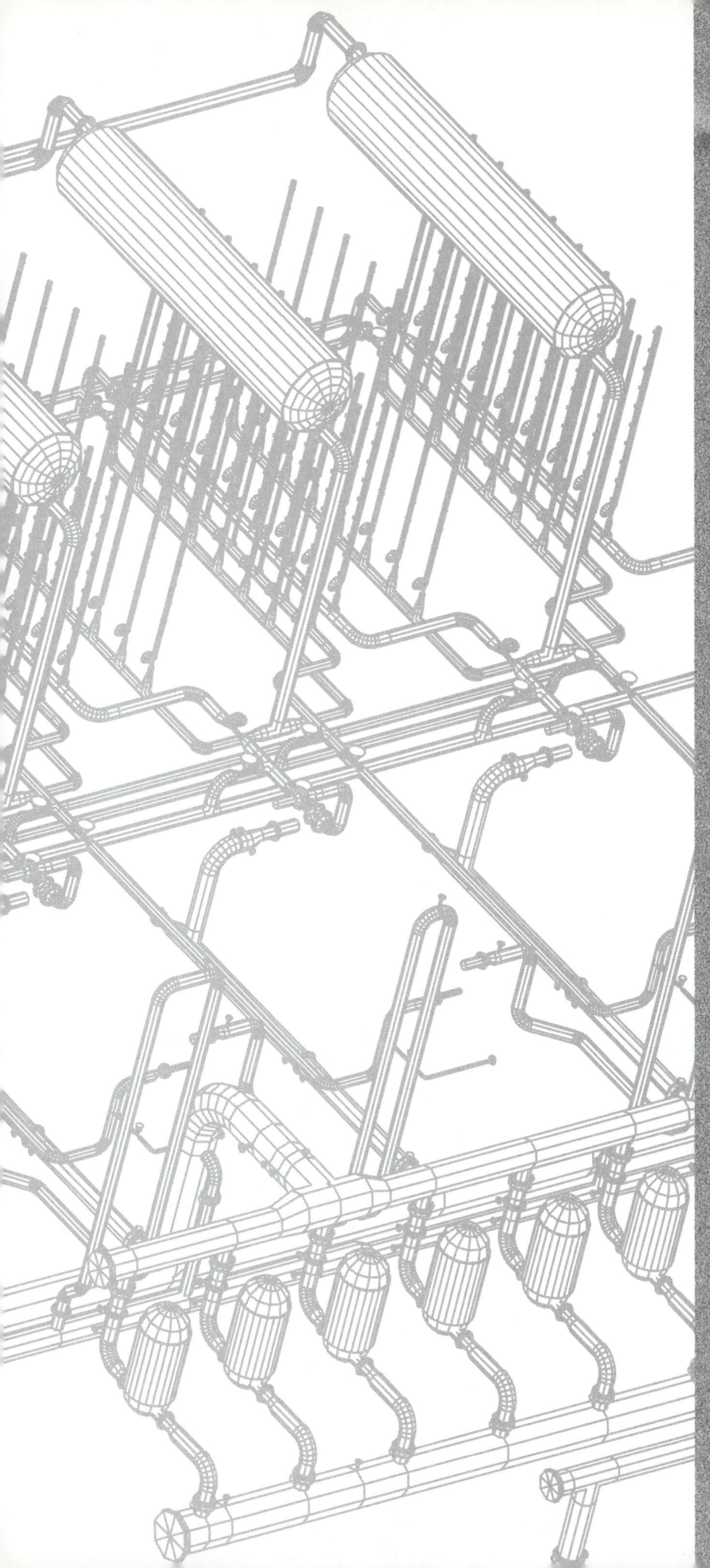

43

MISCELLANEOUS COMMANDS AND FEATURES

CHAPTER OBJECTIVES

After completing this chapter you should:

1. be able to manage named objects and files using wildcards;

2. be able to reinitialize I/O ports and the ACAD.PGP file using *Reinit* and be able to check the drawing for errors using *Audit;*

3. know how to use *Multiple* to automatically repeat commands;

4. be able to control dialog boxes with the *FILEDIA* system variable;

5. be able to create and view slides using *Mslide* and *Vslide;*

6. know how to create and run a *Script* for viewing slide shows;

7. know how to invoke the Quick Calculator for computing arithmetic expressions, locating points, and creating and editing geometry;

8. be able to use this chapter as a reference for the *Options* dialog box.

CONCEPTS

This chapter discusses several unrelated AutoCAD commands and features that are helpful for intermediate-level users. The following commands, variables, and features are covered in this chapter:

Miscellaneous Features
   Wildcards
   REGENAUTO
   Fill
   Multiple
   FILEDIA
   Reinit
   Audit

Using Slides and Scripts
   Mslide
   Vslide
   Creating and Using Scripts
   Creating a Slide Show
   Script

The Quick Calculator

The *Options* Dialog Box
   All tabs of the *Options* dialog box are explained and referenced to other chapters.

MISCELLANEOUS FEATURES

AutoCAD keeps objects in a drawing file other than those we know as graphical entities (*Line*, *Circle*, *Arc*, etc.). These are called "named objects." The named objects are:

Blocks	*Shapes*
Dimension Styles	*Table Styles*
Layers	*Text Styles*
Layouts	*User Coordinate Systems*
Linetypes	*Views*
Materials	*Viewport* configurations
Mline Styles	*Visual Styles*
Plot Styles	

Wildcard characters, the *Rename* command, and the *Purge* command can be used to access or alter named objects.

Using Wildcards

Whenever AutoCAD prompts for a list of names, such as file names, system variables, *Block* names, *Layer* names, or other named objects, any of the wildcards in the following wildcard list can be used to access those names. These wildcards help you specify a select group of named objects from a long list without having to repeatedly type or enter the complete spelling for each name in the list.

For example, the asterisk (*) is a common wildcard that is used to represent any alphanumeric string (group of letters or numbers). You may choose to specify a list of layers all beginning with the letters "DIM" and ending with any string. Using the *-Layer* command and entering "DIM*" in response to the ? option yields a list of only the layer names beginning with DIM.

Commands That Accept Wildcards

Several AutoCAD commands prompt you for a name or list of names or display a list of names to match your specification. Any command that has a ? option provides a list. Wildcards can be used with any of these commands and with many dialog boxes:

Attedit
-Block
-Dimstyle
-Insert
-Layer
-Linetype
-Lweight
-Rename
Setvar
-Style
UCS
-View
Visualstyles
Vplayer
-Xref

Valid Wildcards That Can Be Used in AutoCAD

The following list defines valid wildcard characters that can be used in AutoCAD:

Character	Definition
# (pound)	Matches any numeric digit
@ (at)	Matches any alpha character
. (period)	Matches any nonalphanumeric character
* (asterisk)	Matches any string, including the null string. It can be used anywhere in the search pattern—at the beginning, middle, or end of the string
? (question mark)	Matches any single character
~ (tilde)	Matches anything but the pattern
[...]	Matches any one of the characters enclosed
[~...]	Matches any character not enclosed
- (hyphen)	Used with brackets to specify a range for one character
' (reverse quote)	Escapes special characters (reads next character literally)

Below are listed one or more examples for each application of wildcard patterns:

Pattern	Will match or include . . .	But not . . .
ABC	Only ABC	
~ABC	Anything but ABC	
A?C	Any 3-character sequence beginning with A and ending with C	AC, ABCD, AXXC, or XABC
AB?	ABA, AB3, ABZ, etc.	AB, ABCE, or XAB
?BC	ABC, 3BC, XBC, etc.	AB, ABCD, BC, or XXBC
A*C	AAC, AC, ABC, AX3C, etc.	XA or ABCD
A*	Anything starting with A	XAAA
*AB	Anything ending with AB	ABX
AB	AB anywhere in string	AXB
~*AB*	All strings without AB	AB, ABX, XAB, or XABX
'*AB	*AB	AB, XAB, or *ABC
[AB]C	AC or BC	ABC or XAC
[~AB]C	XC or YC	AC, BC, or XXC
[A-J]C	AC, BC, JC, etc.	ABC, AJC, or MC
[~A-J]C	Any character not in the range A–J, followed by C	AC, BC, or JC

Wildcard Examples

For example, assume that you have a drawing of a two-story residential floor plan with the following *Layers* and related settings:

Layer name	State		Color	Linetype	Lineweight
"0"	on	-P	7 (white)	"CONTINUOUS"	Default
"1-ELEC-DIM"	on	-P	7 (white)	"CONTINUOUS"	Default
"1-ELEC-LAY"	on	-P	7 (white)	"CONTINUOUS"	Default
"1-ELEC-TXT"	on	-P	7 (white)	"CONTINUOUS"	Default
"1-FLPN-DIM"	on	-P	7 (white)	"CONTINUOUS"	Default
"1-FLPN-LAY"	on	-P	7 (white)	"CONTINUOUS"	Default
"1-FLPN-TXT"	on	-P	7 (white)	"CONTINUOUS"	Default
"1-HVAC-DIM"	on	-P	7 (white)	"CONTINUOUS"	Default
"1-HVAC-LAY"	on	-P	7 (white)	"CONTINUOUS"	Default
"1-HVAC-TXT"	on	-P	7 (white)	"CONTINUOUS"	Default
"2-ELEC-DIM"	on	-P	7 (white)	"CONTINUOUS"	Default
"2-ELEC-LAY"	on	-P	7 (white)	"CONTINUOUS"	Default
"2-ELEC-TXT"	on	-P	7 (white)	"CONTINUOUS"	Default
"2-FLPN-DIM"	on	-P	7 (white)	"CONTINUOUS"	Default
"2-FLPN-LAY"	on	-P	7 (white)	"CONTINUOUS"	Default
"2-FLPN-TXT"	on	-P	7 (white)	"CONTINUOUS"	Default
"2-HVAC-DIM"	on	-P	7 (white)	"CONTINUOUS"	Default
"2-HVAC-LAY"	on	-P	7 (white)	"CONTINUOUS"	Default
"2-HVAC-TXT"	on	-P	7 (white)	"CONTINUOUS"	Default

If you wanted to turn *Off* all the layers related to the second floor (names beginning with "2"), you can use the asterisk (*) wildcard as follows:

```
Command: -layer
Current layer:  "0"
Enter an option
?/Make/Set/New/ON/OFF/Color/Ltype/LWeight/Plot/PStyle/Freeze/Thaw/LOck/Unlock]: off
Enter name list of layer(s) to turn off: 2 *
Enter an option
[?/Make/Set/New/ON/OFF/Color/Ltype/LWeight/Plot/PStyle/Freeze/Thaw/LOck/Unlock]: ?
```

Layer name	State		Color	Linetype	Lineweight
"0"	on	-P	7 (white)	"CONTINUOUS"	Default
"1-ELEC-DIM"	on	-P	7 (white)	"CONTINUOUS"	Default
"1-ELEC-LAY"	on	-P	7 (white)	"CONTINUOUS"	Default
"1-ELEC-TXT"	on	-P	7 (white)	"CONTINUOUS"	Default
"1-FLPN-DIM"	on	-P	7 (white)	"CONTINUOUS"	Default
"1-FLPN-LAY"	on	-P	7 (white)	"CONTINUOUS"	Default
"1-FLPN-TXT"	on	-P	7 (white)	"CONTINUOUS"	Default
"1-HVAC-DIM"	on	-P	7 (white)	"CONTINUOUS"	Default
"1-HVAC-LAY"	on	-P	7 (white)	"CONTINUOUS"	Default
"1-HVAC-TXT"	on	-P	7 (white)	"CONTINUOUS"	Default
"2-ELEC-DIM"	off	-P	7 (white)	"CONTINUOUS"	Default
"2-ELEC-LAY"	off	-P	7 (white)	"CONTINUOUS"	Default
"2-ELEC-TXT"	off	-P	7 (white)	"CONTINUOUS"	Default
"2-FLPN-DIM"	off	-P	7 (white)	"CONTINUOUS"	Default
"2-FLPN-LAY"	off	-P	7 (white)	"CONTINUOUS"	Default
"2-FLPN-TXT"	off	-P	7 (white)	"CONTINUOUS"	Default
"2-HVAC-DIM"	off	-P	7 (white)	"CONTINUOUS"	Default
"2-HVAC-LAY"	off	-P	7 (white)	"CONTINUOUS"	Default
"2-HVAC-TXT"	off	-P	7 (white)	"CONTINUOUS"	Default

You may want to *Freeze* all layers (both floors) related to the electrical layout (having ELEC in the layer name). (Assume all the layers are *On* again.) You could use the question mark (?) to represent any floor number and an asterisk (*) for any string after ELEC as follows.

```
Command: -layer
Current layer:  "0"
Enter an option
?/Make/Set/New/ON/OFF/Color/Ltype/LWeight/Plot/PStyle/Freeze/Thaw/LOck/Unlock]: f
Enter name list of layer(s) to freeze: ?-elec*
Enter an option
[?/Make/Set/New/ON/OFF/Color/Ltype/LWeight/Plot/PStyle/Freeze/Thaw/LOck/Unlock]: ?
```

Layer name	State		Color	Linetype	Lineweight
"0"	on	-P	7 (white)	"CONTINUOUS"	Default
"1-ELEC-DIM"	Frozen	-P	7 (white)	"CONTINUOUS"	Default
"1-ELEC-LAY"	Frozen	-P	7 (white)	"CONTINUOUS"	Default
"1-ELEC-TXT"	Frozen	-P	7 (white)	"CONTINUOUS"	Default
"1-FLPN-DIM"	on	-P	7 (white)	"CONTINUOUS"	Default
"1-FLPN-LAY"	on	-P	7 (white)	"CONTINUOUS"	Default
"1-FLPN-TXT"	on	-P	7 (white)	"CONTINUOUS"	Default
"1-HVAC-DIM"	on	-P	7 (white)	"CONTINUOUS"	Default

"1-HVAC-LAY"	on	-P	7 (white)	"CONTINUOUS"	Default
"1-HVAC-TXT"	on	-P	7 (white)	"CONTINUOUS"	Default
"2-ELEC-DIM"	Frozen	-P	7 (white)	"CONTINUOUS"	Default
"2-ELEC-LAY"	Frozen	-P	7 (white)	"CONTINUOUS"	Default
"2-ELEC-TXT"	Frozen	-P	7 (white)	"CONTINUOUS"	Default
"2-FLPN-DIM"	on	-P	7 (white)	"CONTINUOUS"	Default
"2-FLPN-LAY"	on	-P	7 (white)	"CONTINUOUS"	Default
"2-FLPN-TXT"	on	-P	7 (white)	"CONTINUOUS"	Default
"2-HVAC-DIM"	on	-P	7 (white)	"CONTINUOUS"	Default
"2-HVAC-LAY"	on	-P	7 (white)	"CONTINUOUS"	Default
"2-HVAC-TXT"	on	-P	7 (white)	"CONTINUOUS"	Default

TIP

You may want to use only the layout layers (names ending with LAY) and *Freeze* all the other layers. (Assume all layers are *On* and *Thawed*.) The tilde (~) character can be used to match anything but the pattern given. The tilde (~) character is translated as "anything except."

```
Command: -layer
Current layer: "0"
Enter an option
?/Make/Set/New/ON/OFF/Color/Ltype/LWeight/Plot/PStyle/Freeze/Thaw/LOck/Unlock]: f
Enter name list of layer(s) to freeze: ~*lay
Enter an option
[?/Make/Set/New/ON/OFF/Color/Ltype/LWeight/Plot/PStyle/Freeze/Thaw/LOck/Unlock]: ?
```

Layer name	State		Color	Linetype	Lineweight
------------------	-----------		-------------	------------	------------
"0"	on	-P	7 (white)	"CONTINUOUS"	Default
"1-ELEC-DIM"	Frozen	-P	7 (white)	"CONTINUOUS"	Default
"1-ELEC-LAY"	on	-P	7 (white)	"CONTINUOUS"	Default
"1-ELEC-TXT"	Frozen	-P	7 (white)	"CONTINUOUS"	Default
"1-FLPN-DIM"	Frozen	-P	7 (white)	"CONTINUOUS"	Default
"1-FLPN-LAY"	on	-P	7 (white)	"CONTINUOUS"	Default
"1-FLPN-TXT"	Frozen	-P	7 (white)	"CONTINUOUS"	Default
"1-HVAC-DIM"	Frozen	-P	7 (white)	"CONTINUOUS"	Default
"1-HVAC-LAY"	on	-P	7 (white)	"CONTINUOUS"	Default
"1-HVAC-TXT"	Frozen	-P	7 (white)	"CONTINUOUS"	Default
"2-ELEC-DIM"	Frozen	-P	7 (white)	"CONTINUOUS"	Default
"2-ELEC-LAY"	on	-P	7 (white)	"CONTINUOUS"	Default
"2-ELEC-TXT"	Frozen	-P	7 (white)	"CONTINUOUS"	Default
"2-FLPN-DIM"	Frozen	-P	7 (white)	"CONTINUOUS"	Default
"2-FLPN-LAY"	on	-P	7 (white)	"CONTINUOUS"	Default
"2-FLPN-TXT"	Frozen	-P	7 (white)	"CONTINUOUS"	Default
"2-HVAC-DIM"	Frozen	-P	7 (white)	"CONTINUOUS"	Default
"2-HVAC-LAY"	on	-P	7 (white)	"CONTINUOUS"	Default
"2-HVAC-TXT"	Frozen	-P	7 (white)	"CONTINUOUS"	Default

Remember that wildcards can be used with many AutoCAD commands. For example, you may want to load a certain set of *Linetypes*, perhaps all the HIDDEN variations. The following sequence could be used:

```
Command: -linetype
 ?/Create/Load/Set: l
Linetype(s) to load: hid*
Linetype HIDDEN loaded.
Linetype HIDDEN2 loaded.
Linetype HIDDENX2 loaded.
```

Or you may want to load all of the linetypes except the "X2" variations. This syntax could be used:

```
Command: -linetype
?/Create/Load/Set: 1
Linetype(s) to load: ~*x2

Linetype BORDER loaded.
Linetype BORDER2 loaded.
Linetype CENTER loaded.
Linetype CENTER2 loaded.
Linetype DASHDOT loaded.
Linetype DASHDOT2 loaded.
etc.
```

Rename

The *Rename* and *-Rename* commands allow you to rename <u>any named object</u> that is part of the current drawing. See Chapter 21, Blocks, DesignCenter, and Tool Palettes.

Purge

Purge allows you to selectively delete any named object that is not referenced in the drawing. In other words, if the drawing has any named objects defined but not appearing in the drawing, they can be deleted with *Purge* or *-Purge*. See Chapter 21, Blocks, DesignCenter, and Tool Palettes.

Regenauto

Pull-down Menu	Command (Type)	Alias (Type)	Short-cut	Screen (side) Menu	Tablet Menu
...	*Regenauto*	...	...	...	...

When changes are made to a drawing that affect the appearance of the objects on the screen, the drawing is generally regenerated automatically. A regeneration can be somewhat time consuming if you are working on an extremely complex drawing or are using a relatively slow computer. In some cases, you may be performing several operations that cause regenerations between intermediate steps that you feel are unnecessary.

Regenauto allows you to control whether automatic regenerations are performed. *Regenauto* is *On* by default, but for special situations you may want to turn the automatic regenerations *Off* temporarily. The setting of *Regenauto* is stored in the *REGENMODE* variable (1=*On* and 0=*Off*). If *Regenauto* is *Off* and a regeneration is needed, AutoCAD prompts:

```
About to regen—proceed? <Y>
```

Fill

Pull-down Menu	Command (Type)	Alias (Type)	Short-cut	Screen (side) Menu	Tablet Menu
...	*Fill*	...	...	...	...

The *Fill* command controls the display of objects that are filled with solid color. Commands that create solid filled objects are *Donut, Solid,* and *Pline* (with *width*), *Hatch,* and text commands using TrueType fonts. *Fill* can be toggled *On* or *Off* to display or plot the objects with or without solid color. The command format produces this prompt:

Command: **fill**
ON/OFF <On>: (option)
Command:

Figure 43-1 displays several objects with *Fill On* and *Off*. The default position for the *Fill* command is *On*. If *Fill* is changed, the drawing must be regenerated to display the effects of the change. The setting for *Fill* is stored in the *FILLMODE* system variable.

Since solid filling a drawing with many wide *Plines, Donuts,* and *Solid* hatch patterns may be time consuming to regenerate, it may be useful to turn *Fill Off* temporarily. *Fill Off* would be espe-cially helpful for speeding up test plots.

FIGURE 43-1

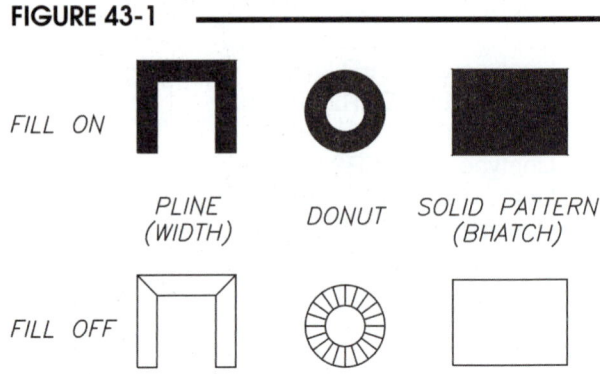

FILL ON

FILL OFF

PLINE (WIDTH)   DONUT   SOLID PATTERN (BHATCH)

See also Chapters 18 and 26 for using *Fill* with solid-filled TrueType fonts and solid hatch patterns.

Reinit

Pull-down Menu	Command (Type)	Alias (Type)	Short-cut	Screen (side) Menu	Tablet Menu
...	*Reinit*	...	...	...	...

The *Reinit* command reinitializes the input/output ports of the computer (to the digitizer) and reinitial-izes the AutoCAD Program Parameters (ACAD.PGP) file. Reinitialization may be necessary if you physically switch the port cable from the digitizer or if it loses power temporarily. If you edit the ACAD.PGP file, it must be reinitialized before you can use the new changes in AutoCAD (see Chap-ter 44 for information on the ACAD.PGP file).

Invoking the *Reinit* command causes the *Re-initialization* dialog box to appear (Fig. 43-2). You may check one of the boxes. PICKing the *OK* tile causes an immediate reinitialization.

If you have configured the digitizer as the "Current System Pointing Device," AutoCAD uses the Windows mouse driver; therefore, the "Digitizer" checkbox is disabled. If your pointing device is disabled and you need to use this dialog box, use TAB to move to the desired options, the space bar to check the boxes, and Enter to execute an *OK*.

FIGURE 43-2

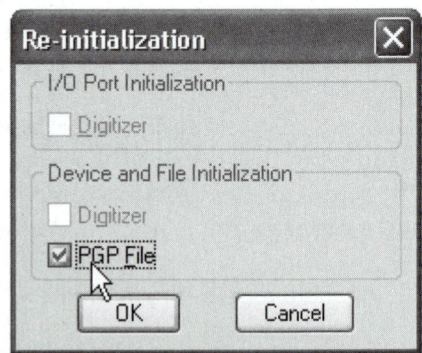

Multiple

Pull-down Menu	Command (Type)	Alias (Type)	Short-cut	Screen (side) Menu	Tablet Menu
...	*Multiple*	...	...	...	...

 Multiple is not a command, but rather a command modifier (or adjective) since it is used in conjunction with another command. Entering *"multiple"* at the Command prompt as a prefix to almost any command causes the command to automatically repeat until you stop the sequence with Escape.

For example, if you wanted to use the *-Insert* command repetitively, you enter the following:

Command: **multiple**
Enter command name to repeat: **-insert**
Enter block name or [?]:

The *Insert* command would repeat until you stop it with Escape. This is helpful for inserting multiple *Blocks*.

The multiple modifier repeats only the command, not the command options. For example, if you wanted to repeatedly use *Circle* with the *Tangent, Tangent, Radius* option, the *TTR* would have to be entered each time the *Circle* command was automatically repeated.

Audit

Pull-down Menu	Command (Type)	Alias (Type)	Short-cut	Screen (side) Menu	Tablet Menu
File *Drawing Utilities >* *Audit*	*Audit*	...	...	*FILE* *Audit*	*Y,24*

The *Audit* command is AutoCAD's diagnostic utility for checking drawing files and correcting errors. *Audit* will examine the current drawing. You can decide whether or not AutoCAD should fix any errors if they are found. The command may yield a display something like this:

Command: **audit**
Fix any errors detected? <N> **y**
 15   Blocks audited
Pass 1 100   objects audited
Pass 1 200   objects audited
Pass 1 282   objects audited
Pass 2 100   objects audited
Pass 2 200   objects audited
Pass 2 282   objects audited
Pass 3 100   objects audited
Pass 3 200   objects audited
Pass 3 282   objects audited
Total errors found 0 fixed 0
Command:

If errors are detected and you requested them to be fixed, the last line of the report indicates the number of errors found and the number fixed. If *Audit* cannot fix the errors, try the *Recover* command (see Chapter 2).

You can create an ASCII report file by changing the *AUDITCTL* system variable to 1. In this case, when the *Audit* command is used, AutoCAD automatically writes the report out to disk in the current directory using the current drawing file name and an .ADT file extension. The default setting for *AUDITCTL* is 0 (off).

FILEDIA

This system variable enables and disables FILE-related DIAlog boxes. File-related dialog boxes are those that are used for reading or writing files. For example, when you request to *Open* a file, the *Select File* dialog box appears (Fig. 43-3).

FIGURE 43-3

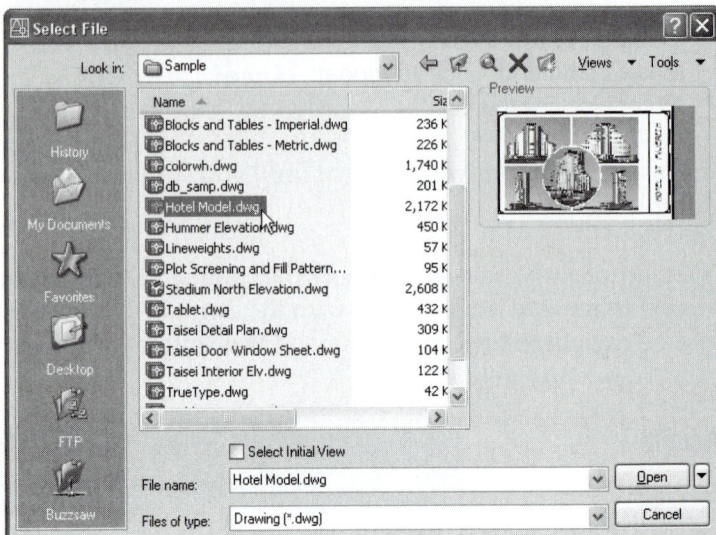

The settings for *FILEDIA* are as follows:

FILEDIA = 1   File dialog boxes appear when a file command is invoked (default setting).
FILEDIA = 0   File dialog boxes are disabled and command prompts are used.

The *FILEDIA* variable can be set to 0 to present the Command line prompt shown below instead of the *Select File* dialog box:

Command: **open**
Enter name of drawing to open:

If file dialog boxes are disabled, they can be invoked at the Command line by entering a tilde (~) symbol at the Command prompt. For example, when *FILEDIA* is set to 0, invoking the *Open* command by any method produces the command prompt; however, typing "~" (tilde) at the "Enter name of drawing to open:" prompt produces the *Select File* dialog box.

 Disabling the dialog boxes can be helpful for running a script file. Since dialog boxes require user input with a pointing device, some scripts require the Command line interface to be used instead of dialog boxes.

The *FILEDIA* variable setting is saved in the system registry.

ATTDIA

The *ATTDIA* variable controls the display of the *Enter Attributes* dialog box. The default setting is 0 (off). Ensure the setting is off if you are using a script to enter attributes. The variable is saved in the drawing file. (See Chapter 22, Block Attributes, for more information.)

USING SLIDES AND SCRIPTS

A slide is a "snapshot" of an AutoCAD drawing that can be made and saved to a file using the *Mslide* command. The resulting slide file contains only one image with no other drawing information. Slide files can later be viewed in AutoCAD using the *Vslide* command. A prepared series of slides can be presented at a later time, much like a slide show. This section also explains how to create a self-running slide show from your slide files by creating and running a *Script*.

Mslide

Pull-down Menu	Command (Type)	Alias (Type)	Short-cut	Screen (side) Menu	Tablet Menu
...	*Mslide*	...	...	...	...

Mslide is short for "make slide." A slide is a "screen capture" of the current AutoCAD display. The image that is saved is composed of the objects that appear in the drawing editor when the *Mslide* command is issued (excluding the menus, tool bars, grid, cursor, and Command line). If model or paper space viewports are active, the current viewport's image is captured. When a slide is made, the resulting image is saved as a file in the directory of your choice with an extension of .SLD.

Mslide is simple to use. Assume that you are working on a drawing—for example, Figure 43-4—and want to take a "snapshot" of the image using *Mslide*. Invoke the command by any method:

FIGURE 43-4

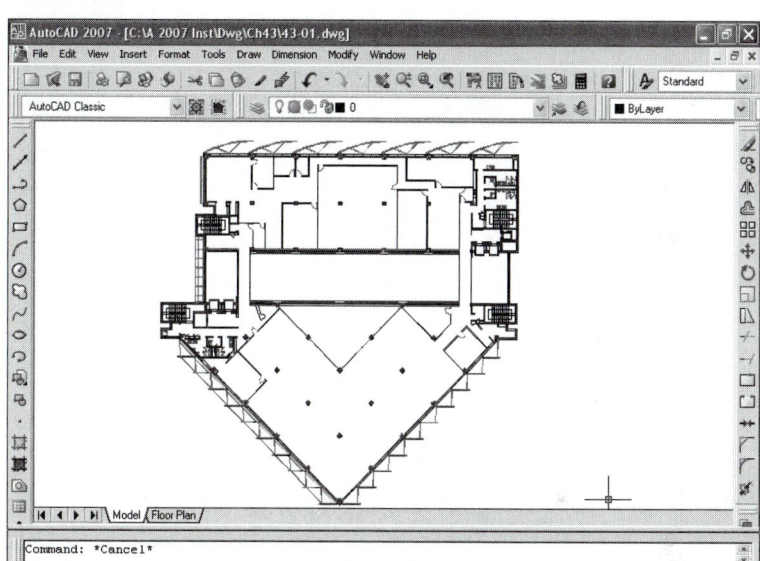

Command: **mslide** (The *Create Slide File* dialog box appears. Assign a name and directory.)
Command:

NOTE: *Vslide* and *Mslide* operate only in 2D mode (when the visual style is set to *2D Wireframe*).

When the *Create Slide File* dialog box appears (Fig. 43-5), assign a name for the slide. (If the *FILEDIA* variable is set to 0, the command line format is activated instead of the dialog box.) Only the file name is required because AutoCAD automatically assigns the .SLD file extension. The drawing image is saved without the grid, cursor, menus, toolbars, etc. Only the drawing image is saved in the .SLD; no other drawing information is saved. Therefore, an .SLD file is much smaller (file size in bytes) than the corresponding .DWG file. Use the *Vslide* command to view the slide again.

FIGURE 43-5

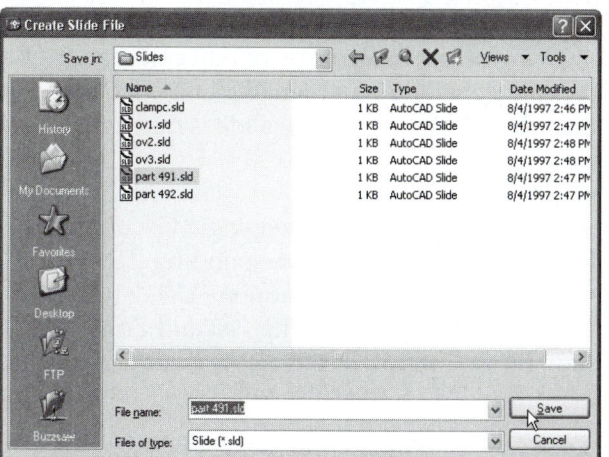

AutoCAD .SLD files are proprietary, meaning the file format was developed by AutoCAD to be used in AutoCAD. The .SLD file format is not widely used like a .TIF, .GIF, .JPG, etc. but can be viewed or converted by some software.

Vslide

Pull-down Menu	Command (Type)	Alias (Type)	Short-cut	Screen (side) Menu	Tablet Menu
...	Vslide	...	...	...	...

Vslide is short for "view slide." Use this command to view previously made AutoCAD slide (.SLD) files. Assuming previously created slide files are accessible, invoke *Vslide* by any method:

Command: **vslide** (The *Select Slide File* dialog box appears. Select the desired directory and slide name.)
Command:

When the *Select Slide File* dialog appears (Fig. 43-6), enter or select the desired directory and slide name. The selected slide then appears in the Drawing Editor (if viewports are active, it appears in the current viewport). Using the slide from the previous example (see *Mslide*), the slide image is "projected" in the Drawing Editor (Fig. 43-7).

FIGURE 43-6

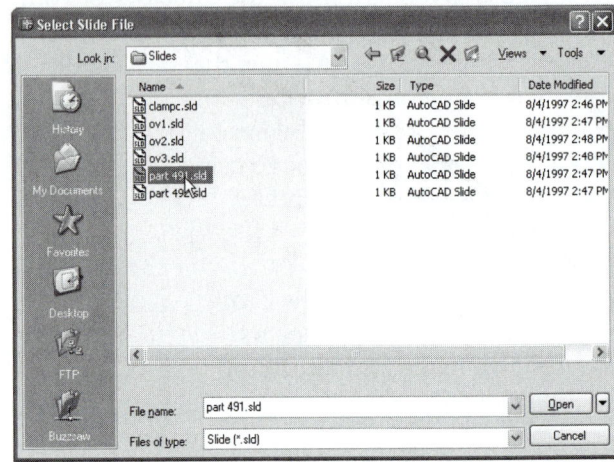

A slide file may be viewed during any AutoCAD session. If a drawing is in progress, the slide file is projected on top of the current drawing, much like a photographic slide is displayed on a projection screen or wall. Because a slide file contains only a description of the display, the slide file image cannot be edited. You cannot use *Pan* or *Zoom* to modify the display of the slide. In fact, any AutoCAD display command issued (*Redraw, Pan, Zoom*, etc.) will cause the slide to disappear and act upon the current drawing, not the slide. After a slide has been viewed, use *Redraw* to refresh the current drawing before using draw or edit commands. Because .SLD files are generally much smaller and contain less information than .DWG files, viewing a slide is much faster than loading a drawing.

FIGURE 43-7

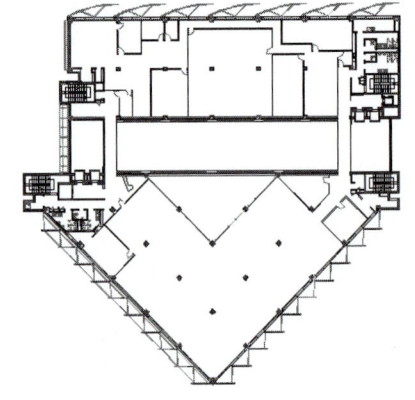

 Slide files are helpful for keeping a visual record of drawings, since the .SLD file is much smaller and faster loading than the corresponding .DWG file. The most common use for a series of slides is giving a presentation or a demonstration. Using prepared slides to present a complex drawing is much faster and more reliable than using *Pan* and *Zoom*. A script file (*.SCR) file can be created to "run" a series of slide files—a scripted slide show (see "Creating and Using Scripts").

Creating and Using Scripts

Scripts are text files that contain AutoCAD commands. A script file contains a user-specified listing of the desired AutoCAD commands. To run a script, use the *Script* command. AutoCAD reads and operates each command and option in the sequence it is listed in the script file. A script file contains essentially the same text that you would enter at the Command line during a drawing session. When a script is executed, each line of the script is echoed at the Command line.

Script files can be used for a number of purposes. If you want to give a presentation, you can create a self-running display of a series of slides by creating a script file. For this purpose, the script file repeats the *Vslide* command and the slide names. Since a script can contain any AutoCAD commands, you could make scripts to execute a series of frequently used commands—like a macro (this topic is discussed in Chapter 44, Basic Customization). You could even recreate an entire drawing by listing every command used to create the drawing in the script file.

There are three basic steps to creating and running a script:

1. In AutoCAD, determine the commands, options, and responses that you want to include in the script file. This can involve a "practice run" to determine each step. You may have to write down each command, option, and response. The *Logfileon* and *Logfileoff* commands can help with this process (see Chapter 44).

    If you are creating a slide show, the slides must be created first. A practice run of the show using *Vslide* can be helpful to ensure you know the correct slide sequence.

2. Use a text editor such as Windows Notepad or other program that can save an ASCII text file. Include all the desired commands. Save the file as an .SCR file extension.

3   In AutoCAD, use the *Script* command to locate, load, and run the script.

Creating a Slide Show

Assume that you have created the slides and know the sequence that you want to display them. You are ready to create the script file. Use any text editor or word processor that can create text files in ASCII format (without internal word processing codes). The script file should be saved using any descriptive file name but should have a .SCR file extension.

Begin the script file (in your text editor) by entering the *Vslide* command. When the *Vslide* command is issued in AutoCAD, the "Slide file:" prompt appears. Therefore, your next line of text (in the script file) should be the name of the desired slide file (SLIDE1, for example) without the .SLD extension. Upper- or lowercase letters can be used. For example, the script file at this point might look like this:

```
VSLIDE
SLIDE1
```

Each line or space in a script file is equivalent to one AutoCAD command, option, or response by the user. In other words, place a space or begin a new line for each time you would press Enter in AutoCAD. The file would be translated as:

Command: **VSLIDE** Enter
Slide file: **SLIDE1** Enter

You probably want to allow the slide to stay visible for a specific amount of time—for example, 5 seconds—then move on to the next slide. You may also want the slide show to be self-repeating automatically for use during presentations or an open house. A few script-controlling commands are given here that will enable you to more effectively present the slide show:

Delay   With the *Delay* command, you can enter the number of milliseconds you wish to have the current slide displayed. For example, an approximate 5 second display of the slide would be accomplished by entering "*Delay 5000.*" When entering the delay time, consider that the slide will remain displayed while the next *Vslide* command is issued, as well as during the time that it takes your system to retrieve and display the next slide.

NOTE: The *Delay* time may vary based on the speed of your computer. A delay time of 1000 may be less than 1 second on some computers.

Rscript   Typically entered as the last line of a script file, this command loops to the first line of the script and repeats the script indefinitely (until you interrupt with Backspace or Escape). Make sure you include an <u>extra space or line</u> after the *Rscript* line (to act as an Enter).

Resume   This command cannot be used in the script file, but must be typed at the keyboard during a script execution. You can interrupt a script, then type *Resume* to pick up with the script again. See *"Script"* next for details on using *Resume*.

An example script file is given here for displaying two slides (SLIDE1 and SLIDE2), then the drawing in the background, then repeating the show:

```
VSLIDE
SLIDE1
DELAY 2000
VSLIDE
SLIDE2
DELAY 500
REDRAW
DELAY 1000
RSCRIPT
```

In the script above, SLIDE1 displays for 2 seconds, SLIDE2 displays for .5 seconds, the *Redraw* causes the drawing (in the background) to appear and display for 1 second, and then the entire script repeats.

 TIP   Remember, for the script to operate correctly, no extra spaces can exist (at the end of lines, for example) and slide file names can have no spaces.

The drive and directory location of the slide files is significant. The slide files should be located in the path designated by the *Support File Search Path* section of the *Files* tab in the *Options* dialog box (Fig. 43-8). Otherwise, the path should be given in the script file before each slide name as shown below:

FIGURE 43-8

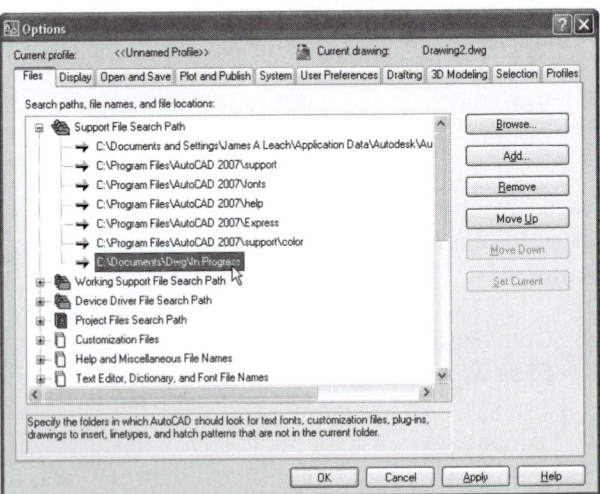

```
VSLIDE
C:\DWGS\IMAGES\SLIDE1
DELAY 2000
VSLIDE
C:\DWGS\IMAGES\SLIDE2
DELAY 500
REDRAW
DELAY 1000
RSCRIPT
```

Once the script is created and saved as an .SCR file, return to AutoCAD and test it by using the *Script* command.

Script

Pull-down Menu	Command (Type)	Alias (Type)	Short-cut	Screen (side) Menu	Tablet Menu
Tools Run Script...	Script	SCR	...	TOOLS 1 Script	V,9

The *Script* command allows you to locate, load, and run a script file. Invoking the *Script* command produces the *Select Script File* dialog box (Fig. 43-9):

Command: **script** (The *Select Script File* dialog box appears. Select the desired directory and script name.)
Command:

FIGURE 43-9

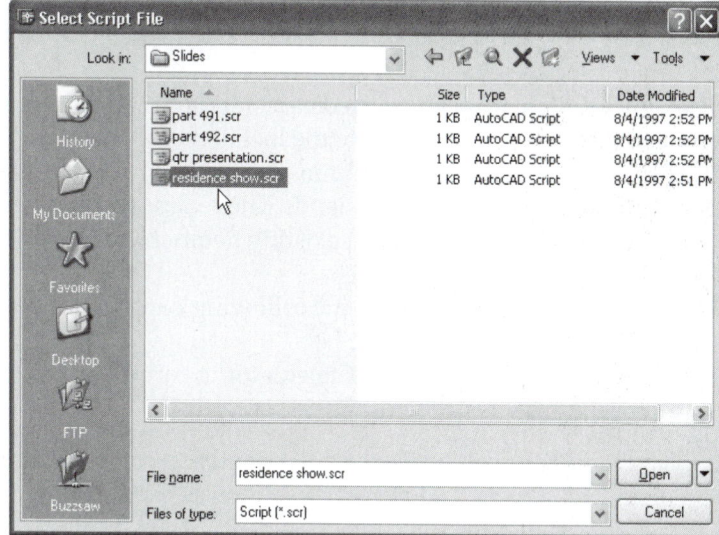

In the dialog box, the desired directory and file can be selected. Notice the default file extension of .SCR. (If the *FILEDIA* variable is set to 0, the dialog box does not appear and the "Script file:" prompt appears at the Command line instead.)

When the file is selected, the script executes. There is nothing more to do at this point other than watch the script (or begin your verbal presentation). If you are giving a verbal presentation, here are some other ideas:

1. Text slides can be created as an introduction, conclusion, and topic or section heading. Just use the *Mtext* or *Text* command to create the desired text screen in AutoCAD; then create a slide with *Mslide*.

2. To temporarily interrupt the script and view a particular slide, press the Backspace or Escape key. This enables you to discuss one slide indefinitely or stop to answer questions. Typing *Resume* causes the script to pick up again from the stopping point.

3. Remember that any AutoCAD drawing can be in the background while the slide show is projected over the drawing. The Backspace and *Resume* feature allows you to cut from the slides and return to the drawing for editing, *Pans*, or *Zooms*, then return to the slide show. You may want to have a drawing loaded in the background (an entire view of the drawing, for example) and show slides "on top" (close-ups, etc.).

NOTE: The *FILEDIA* variable does not have to be set to 0 before running a slide show script. Even though the *Vslide* command normally invokes a dialog box, the Command line format is automatically activated when using *Vslide* from a script. With scripts using other commands that activate dialog boxes, it may be helpful to set *FILEDIA* to 0.

Quickcalc

Pull-down Menu	Command (Type)	Alias (Type)	Short-cut	Screen (side) Menu	Tablet Menu
Tools QuickCalc	Quickcalc	QC	Ctrl+8	...	...

The *QuickCalc* command produces the *Quick Calculator* (Fig. 42-10). The Quick Calculator contains a *Number Pad* for typical numeric calculations, a Scientific pad providing typical scientific functions, a *Units Conversion* tool, and a section for creating and applying *Variables*.

The Quick Calculator can be used as a standard 10-key calculator and can also be used to calculate the location of points in a drawing. It can be used transparently (within a command) to calculate and pass values to the current command. The real power of the calculator is the ability to be used with existing geometry utilizing *OSNAPs*.

The Quick Calculator provides the following capabilities:

1. It can be used as a standard calculator to return values for arithmetic expressions.

2. The calculator can be used transparently to compute numbers, distances, angles, and vector directions as input for a command.

3. The calculator can be used transparently to pass coordinate data to the current command. This feature simplifies creating new geometry and existing geometry by enabling you to locate points that are not easily accessible by other means.

4. *Osnaps* can be used in the expressions to further enhance your ability to use existing geometry (coordinates) in calculations.

5. The calculator can be used to convert values from practically any unit type to another unit type.

6. Built-in functions are available as shortcuts or using *Osnaps*, creating geometry, and making mathematical calculations and conversions.

7. The calculator can be used as a programmable calculator because of its ability to store variables and interact with AutoLISP defined functions and expressions.

The Quick Calculator operates in two modes: the Basic Calculator Mode (see Fig. 43-11) and as a transparent command. The two modes are essentially the same. When you activate the calculator during another command (transparently), the calculations are passed to the current command when you use the *Apply* button (Fig. 43-11). During the Basic Calculator Mode, you can use the *Paste Value to the Command Line* button on the calculator's toolbar to accomplish the same task. The sections of the Quick Calculator are described next.

FIGURE 43-10

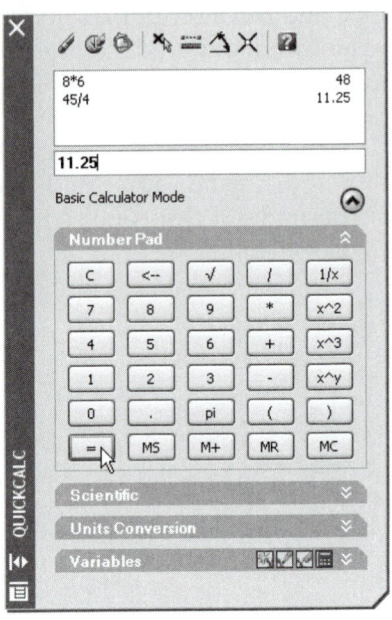

FIGURE 43-11

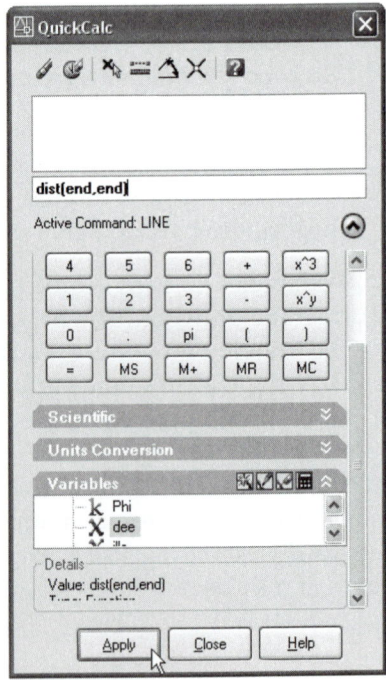

Toolbar

The toolbar at the top of the calculator has the following functions (from left to right):

Clear
Use this button to clear the input box.

Clear History
This button clears the history area (list of calculations just below the toolbar).

Paste Value to Command Line
This button is available only in Basic Calculator Mode. Use it to paste the value in the input box to the command line. When *QuickCalc* is used transparently during a command, this button is replaced by the *Apply* button at the bottom of the calculator.

Get Coordinates
Press this button, then pick a point in the drawing to paste the point's coordinates to the input box.

Distance Between Two Points
Use this button, then select two points in the drawing (using *Osnap* is recommended). The distance value is pasted to the input box.

Angle of Line Defined by Two Points
Using this option, you can return to the drawing to select two points. The angular value is pasted to the input box.

Two Lines Defined by Four Points
You are prompted for four points. Select two points on one line, then two points on a second line. The calculated intersection coordinate is pasted to the input box.

Number Pad

Use the number pad as you would any other calculator. Pressing the = (equals) button displays the calculation in the input box. Use the *Apply* button or *Paste Value to Command Line* button to pass the input box value to the current command or Command line. In addition to the typical numeric operators, the following functions are available.

sqt	(Square Root) Obtains the square root of a value.
1/X	(Inverse) Inverts any number or expression entered in the input box.
x^2	(X to the Power of 2) Squares a value.
x^3	(X to the Power of 3) Raises any number or expression entered in the input box to the power of 3.
x^y	(X to the Power of Y) Raises a number or expression entered in the input box to a specified power.
pi	Enters pi to 14 places in the input box.
MS	(Store in Memory) Stores the current value in the *QuickCalc* memory.
M+	(Add to Memory) Adds the current value to the value stored in the *QuickCalc* memory.
MR	(Restore Memory) If a value is currently stored in the *QuickCalc* memory, the value is restored to the input box.
MC	(Clear Memory) Clears the value currently stored in the *QuickCalc* memory.

2006

Scientific

The *Scientific* section of the Quick Calculator offers the following trigonometric, logarithmic, exponential, and other expressions (Fig. 43-12).

FIGURE 43-12

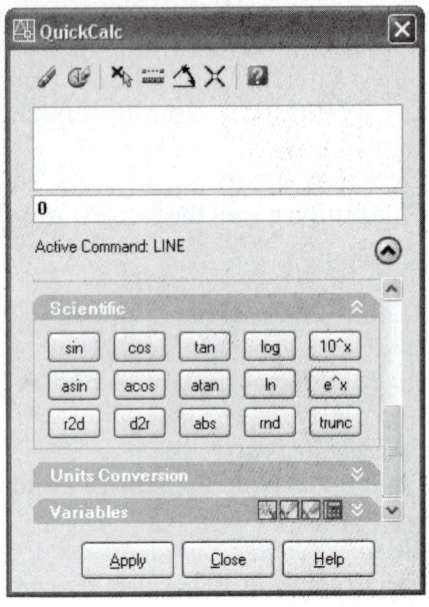

sin	Specifies the sine of the angle in the input box.
cos	Specifies the cosine of the angle in the input box.
tang	Specifies the tangent of the angle in the input box.
Log	Specifies the base-10 log of the value in the input box.
10^x	Specifies the base-10 exponent of the value in the input box.
asin	Specifies the arcsine of the number in the input box. The number must be between -1 and 1.
acos	Specifies the arccosine of the number in the input box. The number must be between -1 and 1.
atan	Specifies the arctangent of the number in the input box.
In	Specifies the natural log of the number in the input box.
e^x	Specifies the natural exponent of the number currently specified in the input box.
r2d	Converts angles in radians to degrees; for example, r2d (pi) converts the value of pi to 180 degrees.
d2r	Converts angles in degrees to radians; for example, d2r (180) converts 180 degrees to radians and returns the value of pi.
abs	Returns the absolute value of the number in the input box.
rnd	Rounds the number in the input box to the nearest integer.
trunc	Returns the integer portion (truncate) of the number in the input box.

Units Conversion

This section automatically converts units in a variety of forms. You can use only decimal values. Do not include unit designators (Fig. 43-13).

FIGURE 43-13

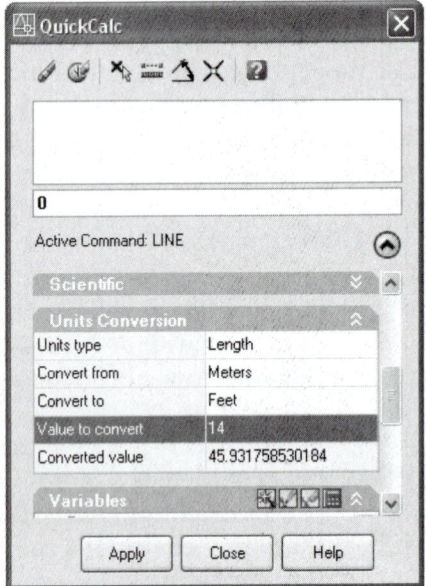

Units Type	Select from a list of length, area, volume, and angular values.
Convert From	Select the units of measurement to convert from.
Convert To	Select the units of measurement to convert to.
Value to Convert	Enter the desired value to convert from.
Converted Value	The calculated converted value appears in this box.

Variables

This powerful feature of the Quick Calculator provides access to predefined constants and functions. You can use the *Variables* area to define and store additional constants and functions.

Variables Tree

This tree lists the predefined shortcut functions. Any user-defined variables can be added to the tree. Shortcut functions are common expressions that combine a function with *Osnaps*. The power of the shortcut functions are evident when you invoke *QuickCalc* transparently. You can select any shortcut function to force it to appear in the input box, then use the *Apply* button to pass it to the current command. The following list describes the predefined shortcut functions in the tree.

dee	dist(end,end), or distance between two endpoints
ille	ill(end,end,end,end), or intersection of two lines defined by four endpoints
mee	(end+end)/2, or midpoint between two endpoints
nee	nor(end,end), or unit vector in the XY plane and normal to two endpoints
rad	rad, or radius of a selected circle, arc, or polyline arc
vee	vec(end,end), or vector from two endpoints
vee1	vec1(end,end), or unit vector from two endpoints

New Variable

Use this button to open the *Variable Definition* dialog box where you can define new variables. The dialog box is the same as that shown in Figure 43-15; however, when defining a new variable, all fields are empty.

Edit Variable

Select a variable from the tree, then use this button to produce the *Variable Definition* dialog box so you can make changes to the selected variable. Figure 43-15 shows the "mee" shortcut function.

Delete Variable

Select any variable and press this button to delete it from the tree.

Calculator Button

Select any variable, then press this button to return the variable to the input box.

FIGURE 43-14

FIGURE 43-15

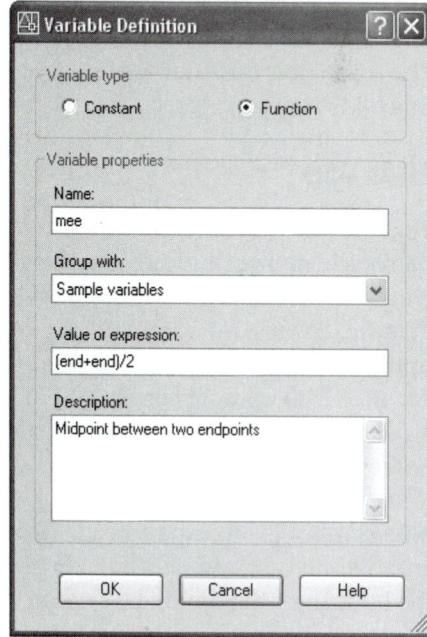

THE *OPTIONS* DIALOG BOX

Options

Pull-down Menu	Command (Type)	Alias (Type)	Shortcut Menu	Screen (side) Menu	Tablet
Tool *Options...*	*Options*	*OP*	(Default Menu) *Options...*	*TOOLS 2* *Options*	*Y, 10*

The *Options* command invokes the *Options* dialog box (see Fig. 43-16). The *Options* dialog box contains controls that allow you to specify how AutoCAD operates. Many of the options manage AutoCAD system variable settings that can also be changed by accessing the system variables directly at the Command line. Other options control the graphical user interface features that can only be changed through this dialog box.

Whether the *Options* dialog box settings change system variables or not, each setting is stored in one of two places:

1. the current drawing file
2. the system registry

Settings that are stored in the current drawing file are denoted in the dialog box with the blue AutoCAD icon. These settings can change when a new drawing is made current. Settings stored in the system registry (of the computer you are using) are generally user preferences and remain with that specific computer or user and therefore do not change when a new drawing is made current. These settings may be different on computers in a lab or office.

The following material explains the controls of the *Options* dialog box or refers you to other chapters of this text where the topics are discussed in more detail related to a specific feature of AutoCAD.

Files **Tab**

This tab allows you to specify the search paths, file names, and file locations of AutoCAD-related files (Fig. 43-16). The default settings are automatically created upon installation and are required for AutoCAD to operate correctly. For most cases these settings do not need changing; however, a few options may need customizing to suit your needs. Click on the plus symbol (+) to expand each heading and display suboptions. Help is given at the bottom of the dialog box when you select a folder.

FIGURE 43-16

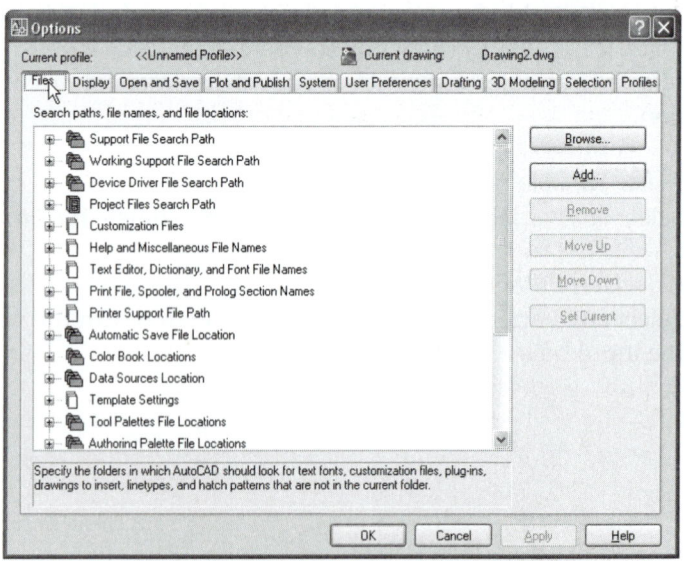

Support File Search Path

This section specifies where AutoCAD looks for text fonts, help files, menus, hatch patterns, plug-ins, linetypes, and drawings to insert when they are not found in the current drawing. For example, if you have a symbol library on a network drive, you can select the *Add* button to add the new folder. Do not *Remove* the default settings.

Working Support File Search Path

This set of directories is specific to your system and is the same as *Support File Search Path* for stand-alone configurations. For example, if you are running AutoCAD from a network, you can specify local directories to include in the search path.

Device Driver File Search Path

This option specifies where AutoCAD looks for device drivers for the video display, pointing devices, and printers and plotters. Do not change the defaults, but you can add other directories if additional driver directories are needed.

Project Files Search Path

The folder AutoCAD searches for *Xrefs* is listed here. See Chapter 30 for information on *XLOADPATH* and *PROJECTNAME*.

Customization Files

This section determines the names and locations of files related to menu customization. The *Main Customization File* is the ACAD.CUI file. Expand this section to find the default or change the location. The *Enterprise Customization File* specifies the location of an enterprise customization file. An enterprise customization file is an XML-based file that stores customization data. You can modify a customization file through the *Customize User Interface* dialog box. The .CUI files replace .MNU, .MNS, and .MNC files that were used to define menus in releases prior to AutoCAD 2006. See Chapter 45, Menu Customization.

Help and Miscellaneous File Names

This section specifies the location of the help and other files AutoCAD must access. You may want to change the initial URL found when the *Browser* command is used by setting the *Default Internet Location* (see Chapter 23, Internet Tools and Collaboration). The *Configuration File* path specifies the location of the configuration file used to store hardware device driver information.

Text Editor, Dictionary, and Font File Names

This section specifies the name and path for the text editor (*MTEXTED*), the main and custom dictionaries (*DCTMAIN*, *DCTCUST*), and the alternate font file and font mapping file (*FONTALT*, *FONTMAP*). See Chapter 18.

Print File, Spooler, and Prolog Section Names

The *Print File Name* allows you to specify a name for .PLT files other than the current drawing name. PLT files are created when you check *Plot to File* in the *Plot* dialog box. The name can also be specified in the *Plot* dialog box when the plot is made. The *Print Spool Executable* tells AutoCAD what application to use for print spooling. *PostScript Prolog Section Name* assigns a name for a customized prolog section in the ACAD.PSF file. The prolog section is used with the *Psout* command. (The *PSPROLOG* system variable is obsolete.)

Printer Support File Path

This section specifies search path settings for printer support files such as *Print Spooler File Location* (where AutoCAD writes print spool plots), *Printer Configuration Search Path* (the path for printer configuration files with a .PC3 extension), *Printer Description File Search Path* (the path for printer description files with a .PMP file extension), and *Plot Style Table Search Path* (the path for plot style table files with an .STB or .CTB file extension).

Automatic Save File Location

Here you can specify the path for the temporary automatic save file that is created if you select *Automatic Save* on the *Open and Save* tab. This value is also controlled by the *SAVEFILEPATH* system variable. See "*SAVETIME*" in Chapter 2, Working with Files.

Color Book Locations

AutoCAD supplies Pantone, DIC, and RAL color books that can be used when specifying colors in the *Select Color* dialog box. You can define other color book files in multiple folders for each path specified. This option is saved with the user profile.

Data Sources Location

This section specifies the path for database source files that can be attached to a drawing using *DBCONNECT*. Change the path if you want to connect to database files you supply that are in a different location.

Template Settings

The *Drawing Template File Location* specifies the path for AutoCAD template files (.DWT files). Since only one folder can be specified, add your own templates to this folder instead of changing this setting (see Chapter 12, Advanced Drawing Setup). The *Drawing Template File Name for QNEW* specifies what template AutoCAD uses when the *Qnew* command is used (see Chapters 2 and 6 for more information on *Qnew*). If you use sheet sets, you can use this section to specify the *Sheet Set Template File Location* and the *Default Template for Sheet Creation and Setup Overrides*. See Chapter 34, Sheet Sets, for more information.

Tool Palettes File Locations

This setting specifies the path for tool palette support files. An .ATC file (AutoCAD Tool Catalog) is automatically created and saved in this location when you customize your own tool palettes.

Authoring Palette File Locations

This section specifies the path for the block authoring palettes support files. The block authoring palettes are used in the Block Editor and provide tools for creating dynamic blocks. See Chapter 21 for more information.

Log File Location

If you create a log file (use *Logfileon* or select *Maintain a Log File* in the *Open and Save* tab), the file will be found in this location. This value is also controlled by the *LOGFILEPATH* system variable. See "*Logfileon*" in Chapter 44.

Plot and Publish Log File Location

Use this section to specify the path for the log file that is created if you select the *Automatically Save Plot and Publish Log* option in the *Plot and Publish* tab.

Temporary Drawing File Location

Here you can specify the location AutoCAD uses to store temporary files. Generally, a temporary folder created specially for temporary files such as C:\TEMP is suggested. The *TEMPPREFIX* (read-only) system variable also stores the current location of temporary drawing files. See "AutoCAD Backup Files" in Chapter 2.

Temporary External Reference File Location

Use this section to specify the location of your *Xref* files. See "*XLOADPATH*" in Chapter 30.

Texture Maps Search Path

The standard AutoCAD installation specifies the directories AutoCAD searches for rendering texture maps. Additional textures can be added to these directories. See Chapter 41, Rendering and Animation.

i-drop Associated File Location

Use this section to specify where files are placed when you drag and drop content from the Internet. The default setting is null; therefore, content is placed in the current directory (of the current drawing).

2006

Display Tab (Fig. 43-17)

Windows Elements
See Chapter 1, Getting Started, for information on these options.

Layout Elements
See Chapter 13, Layouts and Viewports, for more information.

Crosshair Size
Use this edit box or slider bar to change the size of the crosshair cursor. The value represents a percentage of the screen size, so entering 100 causes the crosshairs to extend completely across the drawing area.

Display Resolution
The following options control the related system variables.

FIGURE 43-17

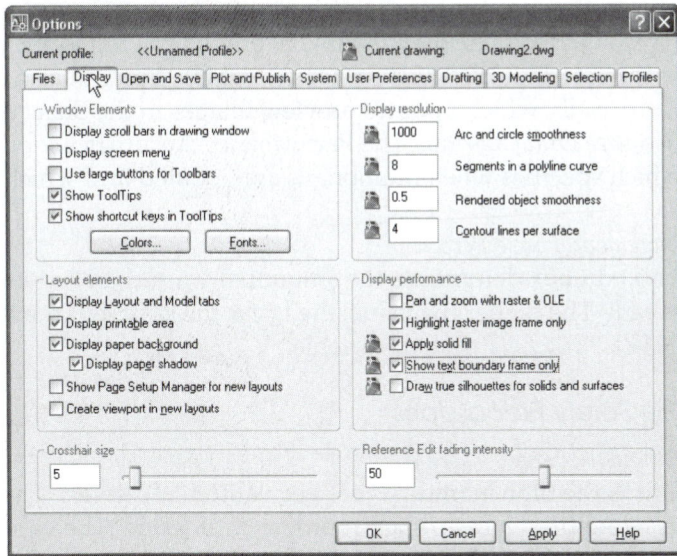

Arc and circle smoothness	*VIEWRES*	See Chapter 10
Segments in a polyline curve	*SPLINESEGS*	See Chapter 16
Rendered object smoothness	*FACETRES*	See Chapter 39
Contour lines per surface	*ISOLINES*	See Chapter 39

Display Performance
These options control the listed system variables.

Pan and zoom with raster & OLE	*RTDISPLAY*	Controls the display of raster and OLE images during real time *Zoom* or *Pan*
Highlight raster image frame only	*IMAGEHLT*	See Chapter 32
Apply solid fill	*FILLMODE*	See Chapters 15, 26, and 43
Show text boundary frame only	*QTEXT*	See Chapter 18
Show silhouettes in wireframe	*DISPSILH*	See Chapter 39

Reference Edit Fading Intensity
This slider controls the value (0-90) of the fading effect of *Xrefs* controlled by *XFADECTL*. See Chapter 30, Xreferences.

Open and Save Tab

(Fig. 43-18)

File Save

Save As:
Use this drop-down list to select the type of drawing format that is used when *Save* is invoked. You can also set this option in the *Save Drawing As* dialog box.

FIGURE 43-18

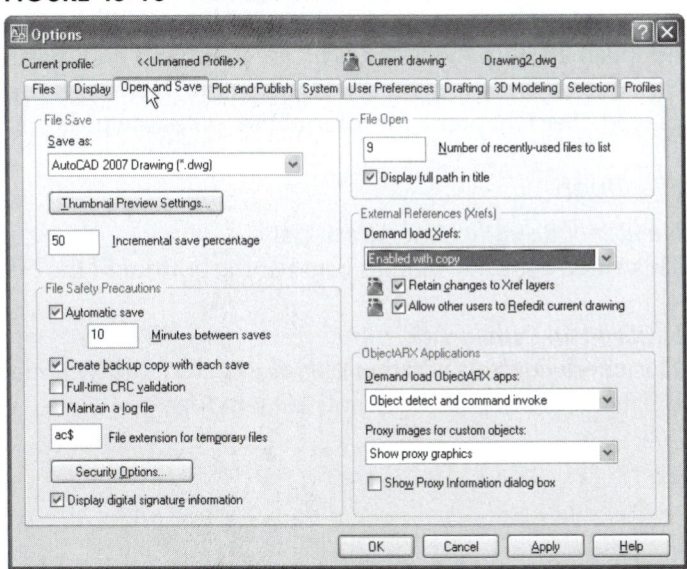

Thumbnail Preview Settings
This button produces a dialog box with three options. First, you can check *Save a Thumbnail Preview Image*, which if created displays an image of the drawing in the preview area of the *Select File* dialog box (stored in the *RASTERPREVIEW* system variable). You can also check *Generate Sheet, Sheet View, and Model View Thumbnails*, which updates preview images in the Sheet Set Manager's *Sheet List* tab, *View List* tab, and *Resource Drawings* tab. The Performance-Accuracy slider sets the *UPDATETHUMBNAIL* system variable, which specifies when thumbnails are updated and which thumbnails are updated.

Incremental Save Percent
This edit box determines the amount of wasted, or redundant, space that is tolerated when a drawing is saved. The higher the value, the faster the save, but the larger the file. The value is stored in the *ISAVEPERCENT* system variable.

File Safety Precautions

Automatic Save
This is the time in minutes between automatic saves. The automatic save feature creates temporary drawing files in case of an improper shutdown. The value is stored in the *SAVETIME* system variable. See Chapter 2, Working with Files.

Create Backup Copy with Each Save
When this box is checked AutoCAD creates a .BAK file from the previous .DWG file when a drawing is saved (*ISAVEBAK*). Speed up drawing saves by removing this check.

Full-Time CRC Validation
CRC (cyclic redundancy check) is an error-checking mechanism. If your drawings are being corrupted and you suspect a hardware problem or AutoCAD error, turn on this option. Checking this box causes a cyclic redundancy check to be performed each time an object is read into the drawing.

Maintain a Log File
This box determines if the command history is stored in a log file. See *"Logfileon"* and *"Logfileoff,"* Chapter 44.

File Extension for Temporary Files
Specify a file extension for files created when AutoCAD needs temporary file space.

Security Options
Selecting this button produces the *Security Options* dialog box where you can specify a password for the current drawing and attach a digital signature (ID certificate) to the current drawing. Passwords can also be specified with the *Securityoptions* command and in the *Save File As* dialog box. If *Display Digital Signature Information* is checked, when the drawing is opened, the *Digital Signature Contents* box is displayed. See Chapter 2 for information on passwords and Chapter 46 for information on digital signatures.

File Open

Number of Recently Used Files to List
This value controls the files listed at the bottom of the *Files* pull-down menu.

Display Full Path in Title
This check controls whether only the current drawing name or the name and full path (directory location) is listed at the top of the AutoCAD window.

External References (Xrefs)

Demand Load Xrefs:
The options are *Disabled*, *Enabled*, and *Enabled with copy*. The selected option determines if demand loading is enabled and if others can access a currently *Xrefed* drawing. The value is stored in the *XLOAD-CTL* variable. See Chapter 30, Xreferences.

Retain Changes to Xref Layers
Use this check to control the *VISRETAIN* system variable. If checked, layer visibility settings of *Xrefed* drawings made in the current drawing are saved with the current drawing (not the *Xref*). See Chapter 30, Xreferences.

Allow Other Users to Refedit Current Drawing
Check this box if you want to allow others to be able to use in-place editing (*Refedit*) with the current drawing. This setting controls the *XEDIT* system variable. See Chapter 30, Xreferences.

ObjectARX Applications

Demand Load ObjectARX Apps:
This drop-down box specifies if and when AutoCAD demand loads a third-party (add-on) application if a drawing contains custom objects created in that application. This setting controls the *DEMANDLOAD* system variable.

Proxy Images for Custom Objects
When third-party applications create new (not AutoCAD-native) objects, this setting determines how these proxy objects are displayed. Alternately, you can access the *PROXYSHOW* system variable.

Show Proxy Information Dialog Box
Check this box if you want a notice to appear if you open a drawing containing proxy objects and the application that created the objects is not available. This setting is stored in the *PROXYNOTICE* system variable.

Plot and Publish Tab

See Chapters 13, 14, and 33 for information regarding the options in this tab.

System Tab

3D Performance

This section provides only one button, *Performance Settings*. Selecting this button provides access to the *Adaptive Degradation and Performance Tuning* dialog box which controls settings that relate to configuration of the 3D graphics display system. See Chapters 35 and 41 for more information.

FIGURE 43-19

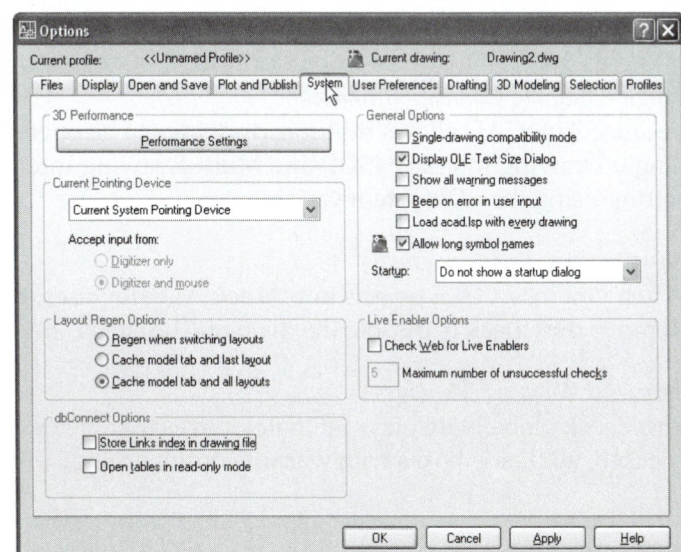

2007

Current Pointing Device
A list of the available pointing device drivers is accessible in this drop-down box. The options are as follows.

Current System Pointing Device: AutoCAD uses the Windows system pointing device.
Wintab Compatible Digitizer: This option is selected when a digitizer is used as the pointing device.

Accept Input From
If you are using a digitizer, you specify whether AutoCAD accepts input from both a mouse and a digitizer or ignores mouse input when a digitizer is set.

Layout Regen Options
These options specify how regenerations occur for the *Model* and *Layout* tabs. When you make changes to the drawing, the entire drawing is either regenerated when you switch tabs or the changes are saved to memory and displayed only for specific tabs that are activated. These options are also controlled by the *LAYOUTREGENCTL* system variable.

Regen When Switching Layouts
This setting regenerates the entire drawing each time you switch tabs.

Cache Model Tab and Last Layout
This option saves the changes to memory, displays the changes only for the two tabs, and suppresses a drawing regeneration. A full regeneration occurs when you switch to any other tab.

Cache Model Tab and All Layouts
Use this choice to regenerate the drawing only the first time you switch to each tab. For the remainder of the drawing session, the changes are displayed but saved to memory; regenerations are suppressed.

dbConnect Options

Store Links Index in Drawing File
When *Dbconnect* is used, this check stores the database index within the AutoCAD drawing file. This option can enhance performance during SQL queries. Speed up drawing opening and file size (for drawings with database information) by removing the check.

Open Tables in Read-Only Mode
When you connect to databases in AutoCAD, the database tables can appear within the AutoCAD drawing file. This choice specifies whether to open database tables in read-only mode.

General Options

Single-Drawing Compatibility Mode
Because AutoCAD allows multiple drawings to be open simultaneously, you can specify whether a Single-Drawing Interface (SDI) or a Multi-Drawing Interface (MDI) is enabled. You can also control this setting using the *SDI* system variable.

Display OLE Text Size Dialog
When you insert OLE objects into AutoCAD drawings, the *OLE Text Size* dialog box appears by default. Remove this check to disable the dialog box display. See Chapter 31.

Show All Warning Messages
This check globally displays all dialog boxes that include a *Don't Display This Warning Again* option. If checked, all dialog boxes with warning options are displayed regardless of the settings in each dialog box.

Beep on Error in User Input
Check this if you want AutoCAD to sound an alarm beep when it detects an invalid entry.

Load Acad.lsp with Every Drawing
If checked, AutoCAD loads the ACAD.LSP file when every drawing is opened. If this option is cleared, only the ACADDOC.LSP file is loaded into all drawing files. Clear this option if you do not want to run certain LISP routines (listed in the ACAD.LSP file) in specific drawing files. This setting is saved in the *ACADLSPASDOC* system variable.

Allow Long Symbol Names
AutoCAD named objects can include up to 255 characters. This check determines if long names are enabled for named objects such as layers, dimension styles, blocks, linetypes, text styles, layouts, UCS names, views, and viewport configurations. The value is stored in the *EXTNAMES* system variable.

Startup
Use this list to control if the *Startup* dialog box is displayed when you start AutoCAD. See Chapter 6, Basic Drawing Setup, and Chapter 12, Advanced Drawing Setup, for more information.

Live Enabler Options
If you use a drawing that contains non-AutoCAD-native objects created by third-party ObjectARX programs, you can use Object Enablers to allow you to display and use the custom objects in AutoCAD drawings even when the ObjectARX application that created them is unavailable. This section controls how AutoCAD checks for Object Enablers that are available on the Autodesk Web site. These settings are stored in the *PROXYWEBSEARCH* system.

Check Web for Live Enablers
Select this option to continually check for Object Enablers on the Autodesk Web site.

Maximum Number of Unsuccessful Checks
Specifies the number of times AutoCAD will continue to check for Object Enablers after unsuccessful attempts.

User Preferences Tab

(Fig. 43-20)

Window Standard Behavior
This section provides options for controlling keystroke and right-click behavior.

Double click editing
This checkbox toggles the double-click editing feature in the drawing area. (See Chapter 16, Modify Commands II, for more information on the *Dblclkedit* command.)

Shortcut Menus in Drawing Area
This option controls whether right-clicking in the drawing area displays a shortcut menu or issues Enter.

FIGURE 43-20

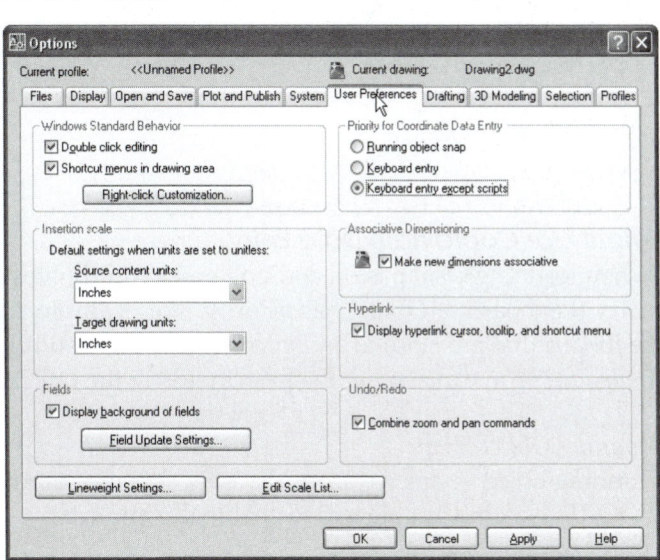

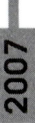

Right-Click Customization
This button displays the *Right-Click Customization* dialog box (Fig. 43-21). Here you can set preferences for shortcut menus that appear in the drawing area. Three basic types of shortcut menus are available: *Default mode*, *Edit mode*, and *Command mode* menus (see Chapter 1). The options here are self-explanatory. The settings are stored in the *SHORTCUTMENU* system variable.

FIGURE 43-21

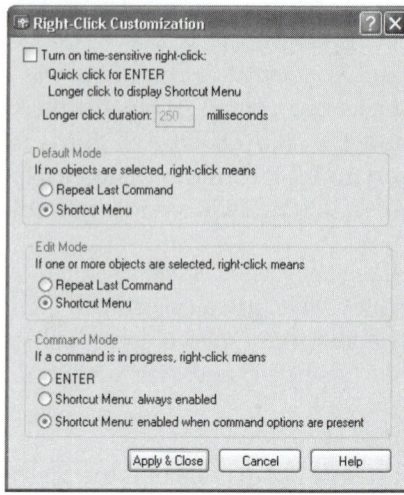

Insertion Scale

Source Content Units
When an object is inserted into the current drawing using drag-and-drop, automatic scaling occurs. This variable sets which units are used for the source object being inserted into current drawing when they are set to *Unitless* or are unspecified (in the *Block* dialog box). The value is stored in the *INSUNITSDEFSOURCE* system variable. See Chapter 21, Blocks, DesignCenter, and Tool Palettes, for more information.

Target Drawing Units
You can also specify which units to use for automatic scaling when *Blocks* or drawings are inserted into the current drawing when no insert units are specified (with the *Drawing Units* dialog box or *INSUNITS* system variable). The value is stored in the *INSUNITSDEFTARGET* system variable.

Fields

Display Background of Fields
When this box is checked a colored background appears around all text fields. See Chapter 18 for more information.

Field Update Settings
Use this button to produce the *Field Update Settings* dialog box (Fig. 43-22). The options here determine when fields are updated. You can check to have fields automatically updated when using the following commands: *Open, Save, Plot, eTransmit,* and *Regen*.

FIGURE 43-22

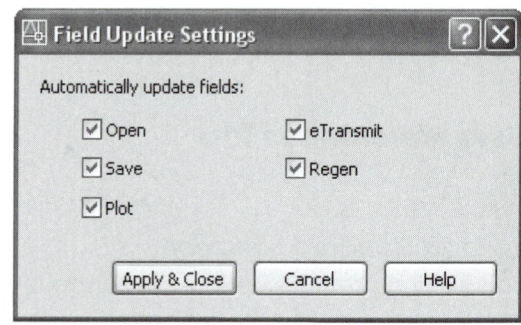

Priority for Coordinate Data Entry

When Running Osnap is on and you enter coordinates at the keyboard, this setting determines which entry (keyboard or *Osnap*) has priority. For example, if *Endpoint Osnap* is on and you enter values at the keyboard that are close to an *Endpoint*, will the resulting point be at the specified coordinates or at the *Endpoint*? You can also set the priority using the *OSNAPCOORD* system variable.

Running Object Snap
Running object snaps have priority over specific coordinates at all times. You can also set the *OSNAPCOORD* system variable to 0 to enable this priority.

Keyboard Entry
Coordinates that you enter at the keyboard always override running object snaps. You can also set the *OSNAPCOORD* system variable to 1 to prioritize keyboard entry.

Keyboard Entry Except Scripts
AutoCAD uses the specific coordinates that you enter at the keyboard rather than running object snaps, except in scripts. You can also set the *OSNAPCOORD* system variable to 2.

Associative Dimensioning

Check this box to *Make New Dimensions Associative*. Otherwise, dimensions are created as AutoCAD 2000 and earlier "nonassociative" dimensions. You can also set this option by using the *DIMASSOC* system variable. See Chapters 28 and 29.

Hyperlink

Display Hyperlink Cursor, Tooltip, and Shortcut Menu
If you drag the cursor across a hyperlinked object, the hyperlink cursor and tooltip appear alongside the crosshairs when this option is checked. When shortcut menus are enabled, a cascading hyperlink short-cut menu appears when you right-click. If this option is cleared, the hyperlink cursor and tooltip are never displayed and the hyperlink option on shortcut menus is not available.

Undo/Redo

When you use the *Undo* and *Redo* commands, each command is treated as one step of *Undo* or *Redo*. If you check *Combine Zoom and Pan Commands*, successive uses of *Zoom* and *Pan* are treated as one command for *Undo* and *Redo*.

Lineweight Settings

Use this button to specify options in the *Lineweight Settings* dialog box. See Chapter 11, Layers and Object Properties.

Edit Scale List

This button opens the *Edit Scale List* dialog box. See "*Scalelistedit*" in Chapter 13.

Drafting **Tab**

See Chapter 3 and Chapter 7 for information on this tab.

3D Modeling **Tab**

3D Crosshairs

This section provides options for the display of the "crosshairs" style cursor that appears in 3D mode.

Show Z Axis in Crosshairs
Checking this box forces the Z axis to be displayed in the crosshairs cursor.

Label Axes in Standard Crosshairs
When this box is checked the X, Y, and Z axes on the crosshairs cursor are labeled.

Show Labels for Dynamic UCS
Check this box to display axis labels on the crosshairs cursor for the Dynamic UCS even when the axis labels are turned off in the *Label Axes in Standard Crosshairs* box.

FIGURE 43-23

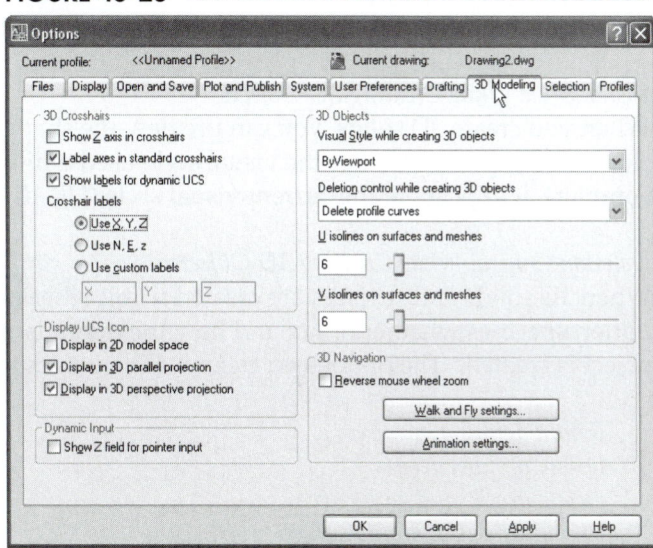

2007

Crosshair Labels

When axis labels are shown, you can choose from two standard schemes and one custom label scheme to display with the crosshairs cursor.

Use X, Y, Z
This option is the standard X, Y, and Z labels.

Use N, E, z
This option labels the axes with abbreviations for North, East, and Z elevation.

Use Custom Labels
Check this box to enable the edit boxes. Specify any labels for the axes with up to eight characters each.

Display UCS Icon

This section allows you to select in which modes you want to display the UCS icon.

Display in 2D Model Space
This check displays the UCS icon in model space for 2D drawing mode (technically, when the current visual style is set to *2D Wireframe*).

Display in 3D Parallel Projection
Check to display the UCS icon in model space for 3D drawing mode (when the current visual style is set to *3D Hidden*, *3D Wireframe*, *Conceptual*, or *Realistic*) when the projection style is set to *Parallel Projection*.

Display in 3D Perspective Projection
Check to display the UCS icon in model space for 3D drawing mode (when the current visual style is set to *3D Hidden*, *3D Wireframe*, *Conceptual*, or *Realistic*) when the projection style is set to *Perspective Projection*.

Dynamic Input

Show Z Field for Pointer Input
The single option in this section displays a field for the Z coordinate when using Dynamic Input (*DYN* is On).

3D Objects

Several options are available that control the display of 3D solids and surfaces.

Visual Style While Creating 3D Objects
When you create 3D solids, you can preview the shapes as you "drag" the cursor to specify the dimensions. This option specifies the visual style used while creating 3D solid primitives. When the solid is complete, it appears in the current visual style (*DRAGVS* system variable).

Deletion Control While Creating 3D Objects
When 3D solids and surfaces are created using other geometry (such as using cross-sections to *Loft*), this option specifies whether or not and how the defining geometry is automatically deleted once the 3D object is created. The options are below. (The values are stored in the *DELOBJ* system variable.)

Retain defining geometry
Delete profile curves
Delete profile and path curves
Prompt to delete profile curves
Prompt to delete profile and path curves

U Isolines on Surfaces and Meshes
This value sets the *Isolines* property in the U direction for surfaces and meshes (*SURFU* system variable).

V Isolines on Surfaces and Meshes
This value sets the *Isolines* property in the V direction for surfaces and meshes (*SURFV* system variable).

3D Navigation
This section sets walk, fly, and animation options for displaying 3D models. See Chapter 41, Rendering and Animation, for more information.

Selection Tab

See Chapter 20, Advanced Selection Sets, for information on this tab.

Profiles Tab

When you customize environment settings, such as color schemes and toolbar visibility and location, you can then save your settings as a profile using the *Profiles* tab in the *Options* dialog box (Fig. 43-24). Profiles are saved on the workstation, not in drawing files. If several people use the same workstation and also use the same login name, you can restore your particular environment by making your profile current. You could also create and save profiles to use with different projects.

FIGURE 43-24

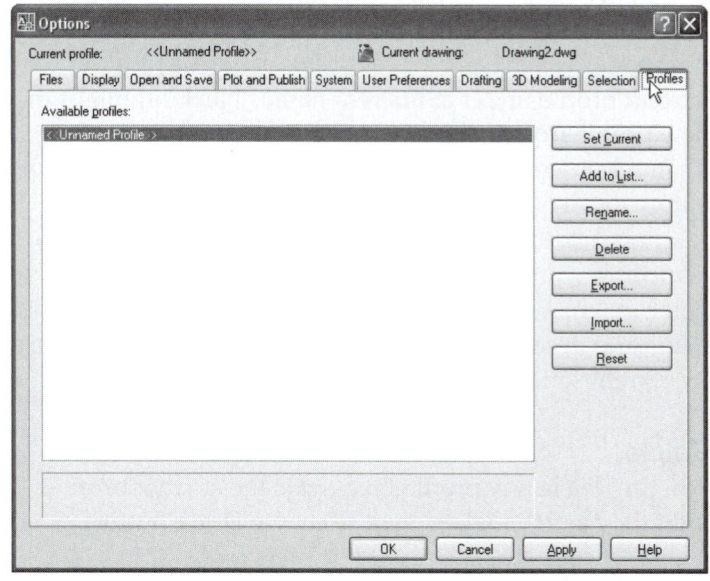

The profile information is stored in the system registry. Profiles can also be saved to a text file (an .ARG file), that can be imported to other workstations. Once you save a profile with the *Export* option, you can export or import the .ARG file to and from different computers. For example, you may want to transport your profile to different computers, or you may want to set up all stations in a lab or office with the same profile.

Any changes that are made to the environment are automatically updated to the current profile. By default, AutoCAD stores your current options in a profile named <<UNNAMED PROFILE>>. It is suggested, however, that you save your own profile under a different name.

If changes have been made to the environment and you want to restore the system defaults, highlight the <<UNNAMED PROFILE>>, then select *Reset* to make the <<UNNAMED PROFILE>> the system default profile. Next, make the <<UNNAMED PROFILE>> the *Current* profile.

To create your own profile, use the *Add to List* button to display the *Add Profile* dialog box. Enter the new name to first create the new profile. Then set up your desired environment and the settings are automatically added to your profile.

TIP

If you want to save those settings in an .ARG file, you must export the profile. For example, exporting a profile named "Myprofile" creates a MYPROFILE.ARG file. If changes are made to the environment and you then export the profile under the current profile name, AutoCAD updates the matching .ARG file with the new settings. If changes are unintentionally made to the environment, you can import the profile again into AutoCAD to restore the profile settings.

The individual buttons on the *Profiles* tab are explained next.

Available Profiles
This list displays a list of the available (previously created) profiles.

Set Current
This button makes the highlighted profile the current profile. To make a profile the current profile, select the desired profile from the list and choose *Set Current*.

Add to List
Use this button to create a new profile. First, select this button to display the *Add Profile* dialog box (Fig. 43-25). Enter the desired name in the *Profile Name* edit box, add a description, and *Apply & Close*. This action saves the current profile under a different name. Next, highlight the new profile from the list and set the new profile *Current*. Then set up your desired environment, and changes are automatically saved to the new current profile.

FIGURE 43-25

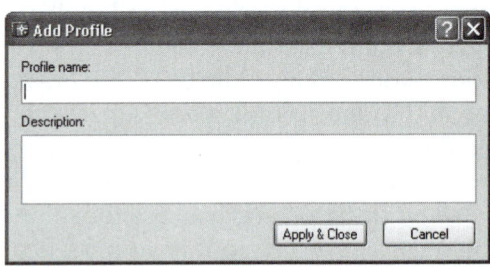

Rename
You can rename any profile. Selecting this button displays the *Change Profile* dialog box for changing the name and description of the selected profile. Use *Rename* when you want to rename a profile but keep its current settings.

Delete
You can delete any profile unless it is the current profile. First make any other profile current, then highlight the name to delete, then select the *Delete* button.

Export
This option allows you to export a profile, creating a file with an .ARG extension. Highlight the desired profile from the list and select *Export*. The *Export Profile* dialog box appears where you can enter a name for the .ARG file. Normally, you would want to assign the same file name as the name of the profile. The new profile can be used on other computers by *Importing* the profile. You can import the file on the same computer or a different computer.

Import
This option allows you to import a profile (an .ARG file) that has been saved with *Export*. You cannot import the current profile; therefore, first make any other profile *Current*. Select *Import* to produce the *Import Profile* dialog box (not shown). Locate and highlight the desired .ARG file and select *Open*. When the second (smaller) *Import Profile* dialog box appears, enter the name of the profile you want to create on your system (normally enter the same name as the assigned file name you are importing), then *Apply & Close*. To make the imported profile current, highlight that name in the *Available Profiles* list, and select *Current*.

Reset

Using this button *Resets* the values in the selected profile to the system default settings. Therefore, the profile that is highlighted in the *Available Profiles* list becomes reset to the defaults. Normally, you would want to *Reset* only the <<UNNAMED PROFILE>> to the system defaults.

First, highlight the desired profile to reset from the *Available Profiles* list. Then select *Reset*. Finally, you must make the reset profile *Current* to see the changes. If the highlighted name is also the current profile, a warning message appears (Fig. 43-26). Selecting *Yes* resets the current profile to the system defaults.

FIGURE 43-26 ──────────

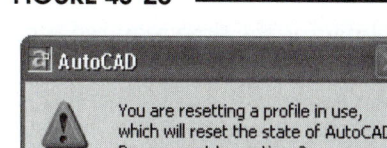

CHAPTER EXERCISES

1. **Wildcards**

   In this exercise, you will use wildcards to manage layers for a complex drawing. *Open* the AutoCAD **WILHOME** drawing that can be downloaded from the **www.mhhe.com/leach** Web site. Use the *Layer Properties Manager* dialog box to view the layers. Examine the layer names; then *Cancel*.

   A. Type the *-Layer* command. *Freeze* all of the layers whose names begin with the letters **AR**. (Hint: Enter **AR*** at the "...layer(s) to freeze:" prompt.) Do you notice the change in the drawing? Use the *Layer Control* drop-down box or list the layers using the *?* option to view the *State* of the layers.

   B. *Thaw* all layers (use the asterisk *). Now *Freeze* all of the layers <u>except</u> the architectural layers (those beginning with **AR**). (Hint: Enter **~AR*** at the "...layer(s) to freeze:"prompt.) Do only the architectural features appear in the drawing now? Examine the list of layers and their *State*. *Thaw* all layers again.

   C. Next, *Freeze* all of the layers except the wall layers (layers that have a "WALL" string). (Hint: Enter ~*WALL* at the ...layer(s) to freeze:" prompt.) Do only the grids appear in the drawing now? Examine the list of layers and their *State*. *(Layer 0 cannot be Frozen* because it is the Current layer.) *Thaw* all layers again and do <u>not</u> *Save* changes.

2. *Multiple* modifier

   Open the **OFF-ATT2** drawing and prepare to *Insert* several more furniture *Blocks* into the office. Type *Multiple* at the Command: prompt, then Enter, then *-Insert* and Enter. Proceed to *Insert* the *Blocks*. Notice that you do not have to repeatedly enter the *-Insert* command. Now try the *Multiple* modifier with the *Line* command. Remember this for use with other repetitive commands. (Do not *Save* the drawing.)

3. *Audit*

   With the OFF-ATT2 drawing open, use the *Audit* command to report any errors. If errors exist, use *Audit* again to fix the errors.

4. *Fill, Regen*

   Open the **SPCA Site Plan** sample drawing (from the www.mhhe.com/leach Web site). Then type *Regenall* and check the amount of time it takes for the drawing to regenerate (use your stopwatch or count the seconds). Now turn *Fill Off.* Also turn *Qtext On.* Make another *Regenall* as before to check the time. Is there any improvement?

   Imagine the time required for plotting this drawing using a pen plotter. *Fill* and *Qtext* can be set to save time for test plots. If you have a pen plotter, make a plot of the drawing with *Fill* and *Qtext* set in each position. How many minutes are saved with *Qtext On* and *Fill Off*? (Do <u>not</u> *Save* the changes to **SPCA Site Plan.**)

5. *Mslide, Vslide*

   A. Open your choice of the following sample drawings: **DB_SAMP, TRUETYPE,** or **TABLET.** (Assuming the default installation was used for AutoCAD, the drawings are located in the C:\Program Files\AutoCAD 2007\Sample directory.) Make eight to ten slides of the drawing, showing several detailed areas (using *Zoom* and *Pan*, etc.) and at least one slide showing the entire drawing. Experiment with *Freezing* layers and include several slides with some layers frozen. Assign descriptive names for the slides so you can determine the subject of each slide by its name. Strive to describe as much as possible about the drawing in the limited number of slides.

   B. When the slides of the drawing are complete, prepare at least two text slides to use as informative material such as an introduction, conclusion, or topic headings (make the text in one or more separate drawings, then use *Mslide*).

   C. Finally, use *Vslide* to view your slides and determine the optimum sequence to describe the drawing. Write down the order of the slides that you choose.

6. *Script*

   Use **Windows Notepad** or other text editor to create a script file to run your slide show. Include appropriate *delays* to allow viewing time or time for you to verbally describe the slides. Make the script self-repeating. Name the script **XXXSHOW.SCR** (where XXX represents your initials). In AutoCAD, use *Script* to run the show to "debug" it and optimize the delays.

7. **Quick Calculator**

   A. This exercise utilizes several *QuickCalc* functions. Begin by drawing the shape shown in Figure 43-27 using a **Pline**. Do not draw the dimensions. *Save* the drawing as **CAL-EX.**

FIGURE 43-27

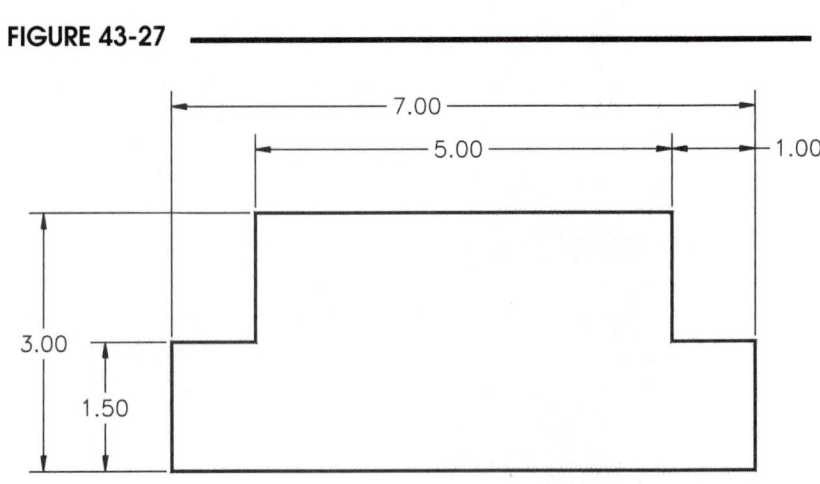

B. Draw the *Circle* in the center of the *Pline* shape with a **.5** unit *radius* (Fig. 43-28). The center of the *Circle* is located halfway between the *Endpoints* of the horizontal line segments. Use *QuickCalc* transparently to place the *Center* of the *Circle*:

Hint: Command: **Circle**
Invoke **QuickCalc** transparently. Double-click the *MEE* variable, then *Apply* it to the command.

FIGURE 43-28

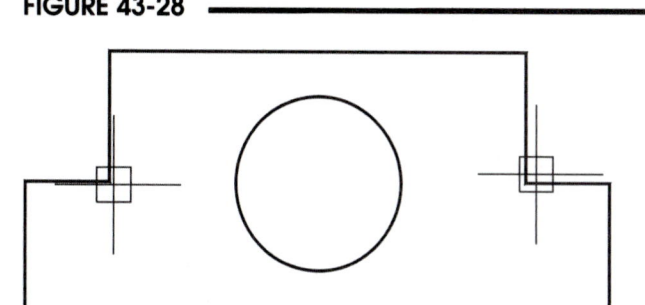

C. Draw the small *Circle* (Fig. 43-29) with a **.1 radius.** Its **Center** is located **.25** units to the right and **.25** units up from the lower-left corner of the *Pline* shape. Use *Quickcalc* transparently to place the *Circle*:

Hint: Command: **Circle**
Invoke **QuickCalc** transparently. Enter **end+[.25,.25]** into the input box, then *Apply* it to the command. Select the lower-left *Endpoint*.

FIGURE 43-29

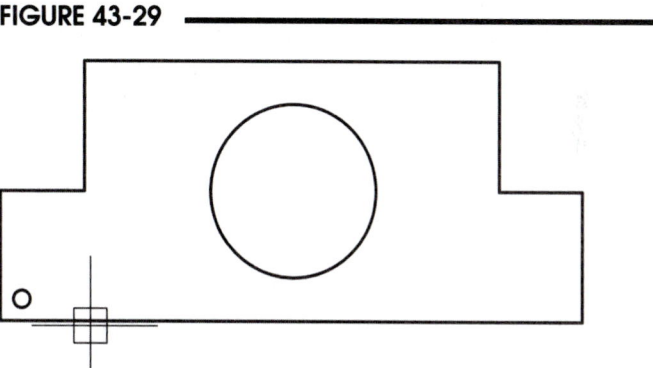

D. Create the *Circle* in the lower-right corner (Fig. 43-30). It has a **Radius** of **1.5** times the small circle.

Hint: Use *Circle*, then invoke *QuickCalc* transparently. Use the **MEE** variable to select the center of the *Circle* between P1 and P2. Invoke *QuickCalc* transparently again and use the **Rad** function and enter **rad*1.5** in the input box. Use *Apply* to select the first circle and paste the value to the command prompt. Press **Enter**.

FIGURE 43-30

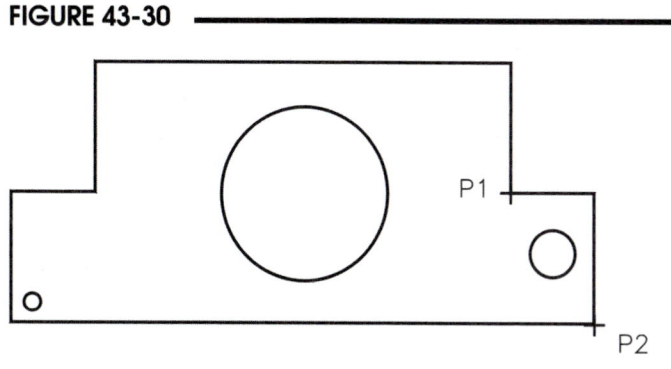

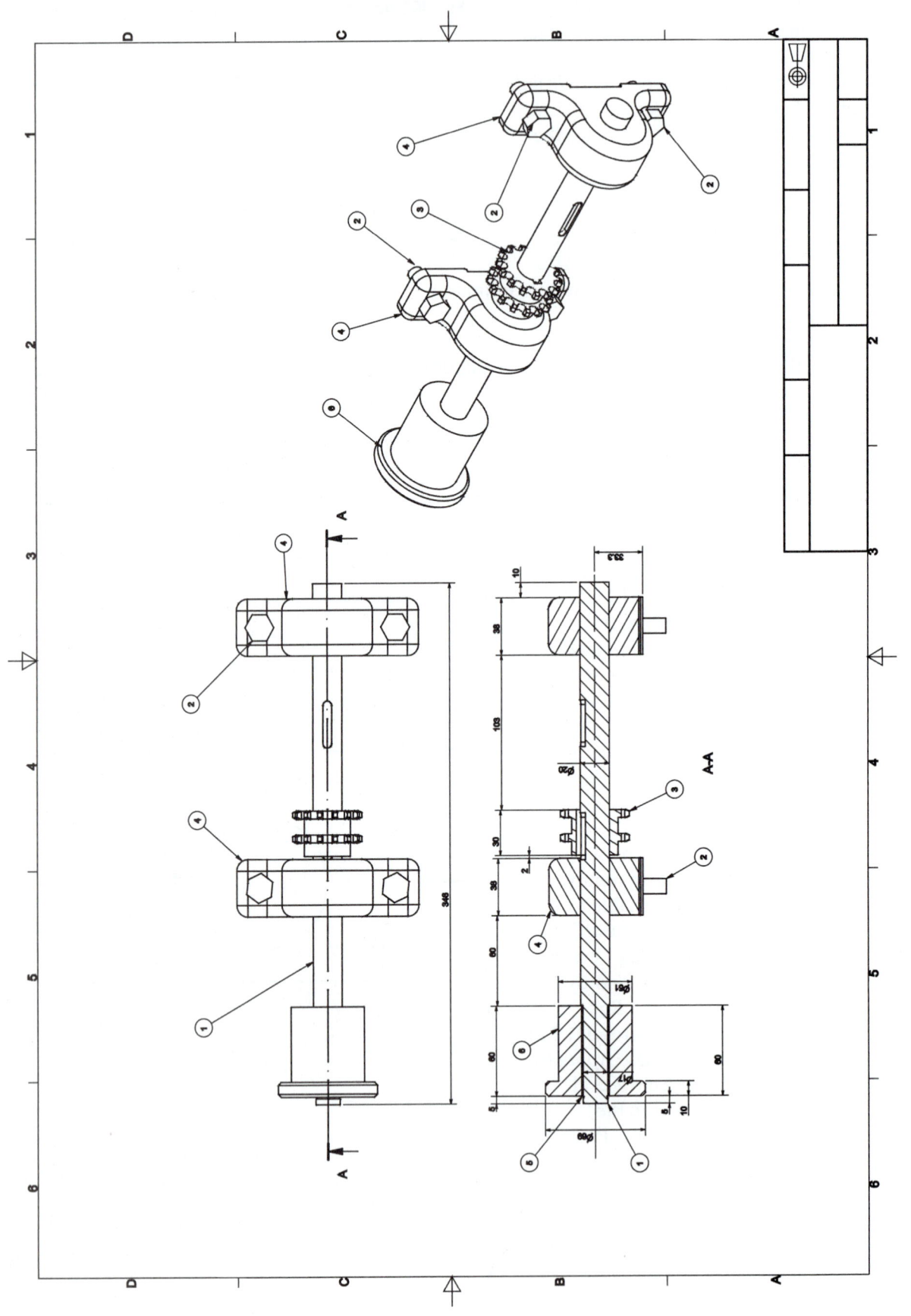

MIDDLE ROLLER.DWG, Courtesy of Autodesk, Inc.

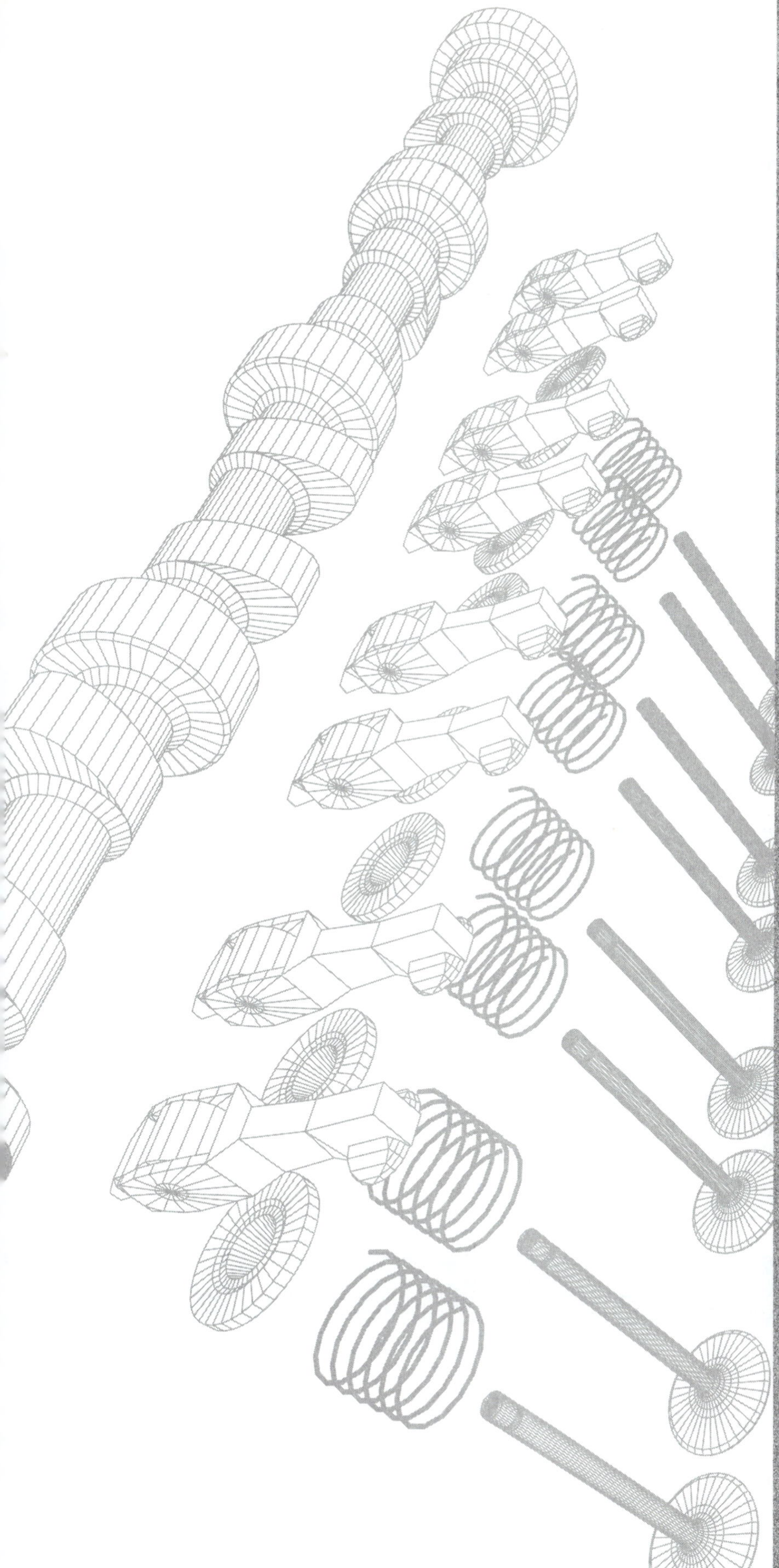

MISCELLANEOUS CUSTOMIZATION

CHAPTER OBJECTIVES

After completing this chapter you should:

1. have a basic understanding of the *Customize User Interface* dialog box and realize the customization possibilities within AutoCAD;

2. be able to customize an existing toolbar;

3. know how to accomplish other simple toolbar changes such as moving a command within a toolbar, moving commands to other toolbars, and adding separator bars in toolbars;

4. be able to create your own simple and complex linetypes;

5. be able to customize the ACAD.PGP file to create your own command aliases;

6. be able to create script files for accomplishing repetitive tasks automatically.

CONCEPTS

This chapter is offered as an introduction to customizing AutoCAD. Autodesk has written AutoCAD with an "open" architecture that allows you to customize the way the program operates to suit your particular needs. You have the ability to change any of the toolbars, tool buttons, menus, shortcut keys, mouse buttons, command aliases, and tool palettes as well as create new linetypes.

This chapter discusses the following miscellaneous customization topics and gives an introduction to the Customization User Interface.

    Introduction to the *Customize User Interface* dialog box
    Customizing an existing toolbar
    Creating your own simple and complex linetypes
    Customizing the ACAD.PGP file and creating command aliases
    Creating script files to use for automated activities

Chapter 45, Customization User Interface, follows through with the information given in this chapter by explaining the complete functions of the *Customization User Interface* dialog box. Chapter 45 is free and downloadable at www.mhhe.com/leach. In addition, you can download Chapter 46, CAD Management, and Chapter 47, Express Tools.

INTRODUCTION TO THE CUI

In releases of AutoCAD previous to 2006, two basic methods were used to customize the way commands were accessed. Inside AutoCAD, the *Toolbar* command invoked the *Customize* dialog box where you could customize toolbars, buttons, tool palettes, and keyboard shortcuts. However, if you wanted to customize pull-down menus, shortcut menus, screen menus, or digitizing tablet menus, it had to be done externally to AutoCAD using a text editor. Customizing these menus externally required a fair amount of time and expertise and was not attempted by most AutoCAD users.

Autodesk has redesigned the process of customizing toolbars and menus for AutoCAD 2006 and 2007. This new powerful feature called the *Customize User Interface*, or *CUI*, is a single dialog box that runs inside AutoCAD for customizing all aspects of AutoCAD command access, including toolbars, individual buttons, pull-down menus, shortcut (right-click) menus, keyboard shortcuts, mouse buttons, and legacy menus such as screen menus, image tiles, and digitizing tablet menus. This new *CUI* tool also allows you to create and save multiple "workspaces."

CUI or Toolbar

Pull-down Menu	Command (Type)	Alias (Type)	Short-cut	Screen (side) Menu	Tablet Menu
Tools *Customize >* *Interface...*	*CUI or* *Toolbar*	*TO*	...	*VIEW 2* *Toolbar*	*R,3*

Both the *CUI* and *Toolbar* commands produce the *Customize User Interface* (Fig. 44-1). This dialog box-type tool provides an easy and visual method to accomplish all aspects of command access customization. With this tool, any AutoCAD user can get involved with customization without the required tedium and expertise required in previous releases of AutoCAD.

In the top-left section (Fig. 44-1), note the list of *All Customization Files*, representing the customizable aspects of the AutoCAD interface: *Workspaces, Toolbars, Menus, Shortcut Menus, Keyboard Shortcuts, Mouse Buttons, LISP Files, Legacy* (menus), and *Partial CUI Files*. With this innovative strategy, not only are all aspects of customization controlled through this one interface, but all customization information can be held in one file, the ACAD.CUI file, although additional .CUI files are possible. One advantage to this strategy is that a single customization you make, such as creating a new macro describing the action of a new or changed command, can be applied to toolbars, pull-down menus, shortcut menus, and so on.

This chapter gives a only short introduction to the CUI tool by instructing you in the steps to customize an existing toolbar. Chapter 45, Customize User Interface (available for free download at www.mhhe.com/leach), dis-

FIGURE 44-1

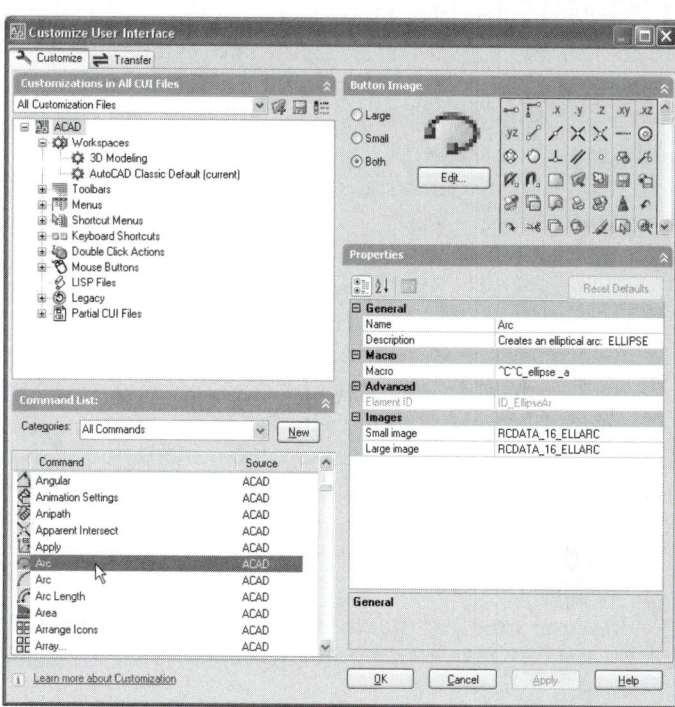

cusses the *Customize User Interface* tool exclusively and completely, including all aspects of customizing toolbars, buttons, pull-down menus, shortcut (right-click) menus, keyboard shortcuts, mouse buttons, legacy menus, and workspaces.

Customizing an Existing Toolbar

This short example is used as an introduction to the *Customize User Interface*. You will go through the simple process of adding an additional tool button to the *Modify* toolbar. Once you understand this process, you will be able to move on to learning the complete capabilities available for you to customize the AutoCAD user interface using this tool, as fully explained in Chapter 45.

To add a command button to an existing toolbar:

1. Use the *CUI* or *Toolbar* command to open the *Customize User Interface* window.
2. In the *All Customization Files* list (upper-left) expand the *Toolbars* section, then expand the desired toolbar you want to customize (*Modify*, for this example).
3. In the *Command List* below (lower-left), locate the desired command you want to add to the toolbar. Optional: Use the *Categories* drop-down list to narrow the search (for this example, select the *Modify* category.)
4. Select the command you want to add (*Ddedit* for this example), then drag and drop it into the desired toolbar. The location of the drop determines the command's location in the toolbar.
5. Select the *OK* button. If you want to continue with other customizations, press *Apply* to apply the action but keep the window open. Pressing *OK* or *Apply* causes AutoCAD to compile the new .CUI file and reload it for immediate use.

FIGURE 44-2

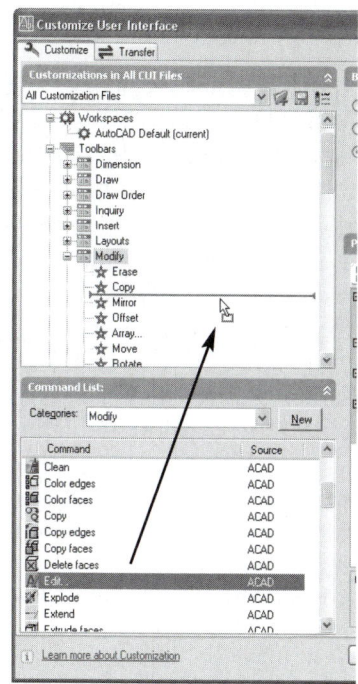

The previous action results in the customization of the *Modify* toolbar (Fig. 44-3). Note that the newly added command (*Ddedit*) is located between the *Copy* and *Mirror* commands. Compare this location with the "drop" location in Figure 44-2.

FIGURE 44-3

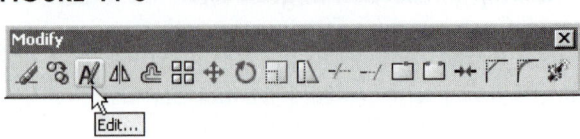

Other simple customizations that are possible for toolbars are listed below (Fig. 44-4).

FIGURE 44-4

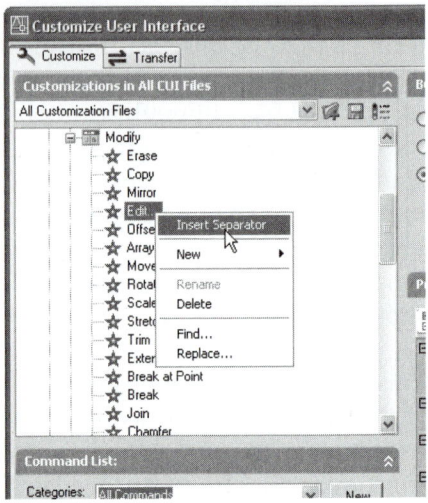

1. Relocate any existing command in a toolbar by dragging and dropping it to the new desired location.
2. Relocate any command to another toolbar by first opening the new toolbar in the list and dragging and dropping the desired command to the new toolbar.
3. Delete any command from a toolbar by right-clicking on the command, then selecting *Delete* from the shortcut menu.
4. Insert separator bars in toolbars by right-clicking at the desired location, then selecting *Insert Separator* from the shortcut menu.
5. Rename a toolbar (not a command) by right-clicking on the desired toolbar name, then selecting *Rename* from the shortcut menu.
6. Create a new toolbar or a new flyout by right-clicking, then selecting *New* from the shortcut menu (see Chapter 45 for more information).

To discover and learn about all of the possible ways to customize commands, buttons, toolbars, pull-down menus, shortcut menus, keyboard shortcuts, mouse buttons, and more, see Chapter 45. Chapter 45, Customize User Interface, Chapter 46, CAD Management, and Chapter 47, Express Tools, are available for free download at www.mhhe.com/leach.

CREATING CUSTOM LINETYPES

AutoCAD has a wide variety of linetype definitions from which you can choose (Chapter 11, Layers and Object Properties). These definitions are stored in an ASCII file called ACAD.LIN. The ACAD.LIN file describes both simple and complex linetypes. Simple linetypes can be composed of line segments, dots, and spaces only. Complex linetypes include text or graphical shapes within the definition of the linetype pattern. However, you are not limited to these linetype definitions. New linetypes can be easily created that contain the line spacing, text, and shapes that you need.

Linetypes are created by using a text editor or word processor to describe the components of the linetype. The Windows Notepad is a sufficient text editor that creates ASCII files (only alphanumeric characters without word processing codes) and can be used for this purpose. New linetype definitions are stored in ASCII files that you name and assign with an .LIN file extension. The file names should indicate the nature of the linetype definitions. For example, ELECT.LIN can include linetype definitions for electrical layouts, or CIVIL.LIN can contain linetype definitions for the civil engineering discipline. This scheme promotes better file maintenance than appending custom linetypes to the ACAD.LIN file.

Creating Simple Linetypes

A linetype definition is created by typing the necessary characters using a text editor and saving as an .LIN file. Examine the basic components of a simple linetype definition below. Each element of the linetype definition is separated by a comma (with no spaces). A linetype definition includes the following items:

1. The name of the linetype with a prefix of *

2. A description of the linetype (either text or symbols)

3. The alignment field

4. Numerical distances for dashes, dots, and spaces

5. The shape or text creation elements (contained in square brackets [ ])

A Simple Linetype Example

The following code is an example of creating a line that looks like line definition "EXAMPLE1"in the figure:

```
*example1,a sample linetype ____ . . _____ . __
A,1.0,-0.25,0,-0.1,0,-0.25,1.5,-0.25,0,-0.25,0.5,-0.5
```

The previous definition can be analyzed as follows:

1. The name of the linetype is "EXAMPLE1."

2. A description of the linetype is a "sample" of the linetype that you create with the keyboard characters: ____ . . _____ . __ (Note that this "graphical" description is created using the underscore and period characters.)

3. A   This is the alignment field.

4. 1.0   A positive value designates a line segment. This segment has a 1 unit length.

5. -0.25   A negative value denotes a gap or blank space. This gap has a length of 0.25.

6. 0   A value of 0 creates a dot.

7. repeat of the previous elements.

Figure 44-5 shows the "EXAMPLE1" linetype defined above.

FIGURE 44-5 ───────────────────

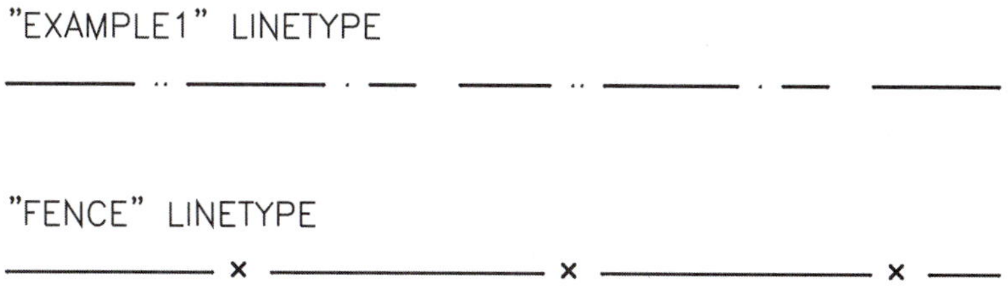

"EXAMPLE1" LINETYPE

───── .. ───── . ── ───── .. ───── . ── ─────

"FENCE" LINETYPE

───────── ✕ ───────── ✕ ───────── ✕ ──

Creating Complex Linetypes

In a complex linetype, two additional fields are possible in the definition and are placed in brackets [ ]. These fields are the Shape field and the String field. Each field has the following form:

    Shapes: [shape description, shape file, scale, rotation, Xoffset, Yoffset]
    Strings: ["text string," text style name, scale, rotation, Xoffset, Yoffset]

For linetypes containing shapes, the shape description and the compiled file name (.SHX) are provided in the first two places of the field. In comparison, a linetype containing a text string has the text string in quotes (" ") and the defined text style name in its first two places. If the text style does not exist, the current text style is used. The last four places of string and shape fields are the same. They are described as follows:

Scale	S=*value*	This is the scale factor by which the height of the text or shape definition is multiplied.
Rotation	R=*value* or A=*value*	A rotation angle for the text or shape is given here. The "A" designator is for an absolute rotational value, regardless of the the angle of the current line segment. The "R" designator is to specify a rotational value relative to the angle of the current line segment.
Xoffset	X= *value*	You can specify an X offset value from the end of the last element with this value.
Yoffset	Y=value	Specify a Y offset value from the end of the last element with this value.

Complex Linetype Examples

```
*PL,Property Line ----PL----PL----
A,0.5,-0.5,[PLSYM,SYMBOLS.SHX,S=1.5,R=0,X=-0.1,Y=-0.1],-0.5,0.5,1

*Fence,Fence line ----x----x----x----
A,1.5,-0.25,["x",STANDARD,S=0.2,R=0.0,X=0.05,Y=-0.1],-0.4,1.5
```

See Figure 44-5 for a sample of the "Fence" linetype defined above. For more examples of simple and complex linetypes, use the Windows Notepad to open the ACAD.LIN file. It is suggested that you make new files when you create your first linetypes rather than appending to the ACAD.LIN file.

CUSTOMIZING THE ACAD.PGP FILE

The ACAD.PGP (AutoCAD Program Parameter) file is an ASCII text file that provides two main features. It provides a link between AutoCAD and other programs by specifying what external programs can be accessed from within AutoCAD, and it defines command aliases—the one- or two-letter shortcuts for commands. In short, the ACAD.PGP file defines what commands can be typed at the AutoCAD Command prompt that are not native to AutoCAD.

Because the ACAD.PGP file is an ASCII file, you can edit the file to add your own command aliases and to define which external commands you want to use from within AutoCAD. Use any text editor, such as Windows Notepad or Wordpad, to view and modify the file. A portion of the AutoCAD ACAD.PGP file is shown in Figure 44-6.

FIGURE 44-6

Figure 44-6 displays only a small section of the ACAD.PGP file that defines the command aliases. The actual file is many pages long and contains primarily command aliases for just about every AutoCAD command. (See Appendixes B and C for the command aliases sorted by command and sorted by alias.) Autodesk's intention is for you to modify the file and to define the external commands and command aliases for personal productivity, although most of that work has been done for you.

All lines of this file that begin with a semicolon (;) are remarks only. These lines (more than a page) give useful information about modifying the file, the format for specifying external commands, and the format for specifying command aliases. Only lines without the semicolon (;) are read by the AutoCAD program. Defining external commands and command aliases are explained next.

Specifying External Commands

The External Commands section of the ACAD.PGP file determines what commands or other programs can be accessed from <u>within AutoCAD</u> by typing the command at the AutoCAD Command prompt. By examining this section of the file, you can see that the following DOS commands are already usable from within AutoCAD:

```
DEL,        DEL,        8,File to delete: ,
DIR,        DIR,        8,File specification: ,
SH,         ,           1,*OS Command: ,
SHELL,      ,           1,*OS Command: ,
START,      START,      1,*Application to start: ,
TYPE,       TYPE,       8,File to list: ,
```

The following Windows commands are also usable from within AutoCAD:

```
EXPLORER,   START EXPLORER, 1,,
NOTEPAD,    START NOTEPAD,  1,*File to edit: ,
PBRUSH,     START PBRUSH,   1,,
```

For example, if you prefer using Windows Explorer for file management, just type "Explorer" at the AutoCAD Command prompt to open it.

 You can use the *Tools* pull-down menu to edit the ACAD.PGP file in Notepad (Fig. 44-7). Select *Tools, Customize, Edit Program Parameters (acad.pgp)*. This feature makes customizing ACAD.PGP possible while AutoCAD is running.

FIGURE 44-7

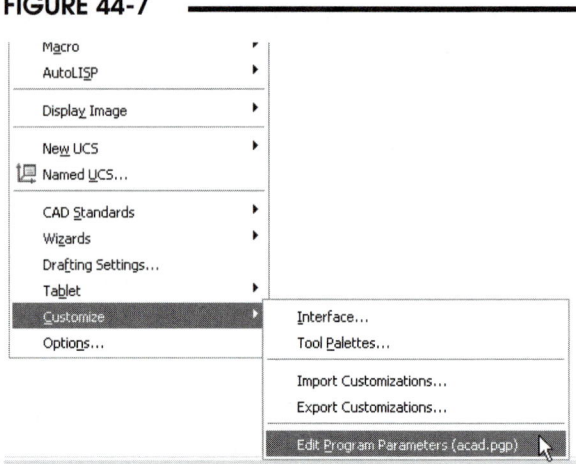

NOTE: If you edit ACAD.PGP while AutoCAD is running, use the *Reinit* command after editing to reinitialize ACAD.PGP to make the new changes usable (see Chapter 43 for information on *Reinit*).

If you prefer another text editor—Norton Editor, for example—you can change the ACAD.PGP file to allow "NE" to be typed at the command prompt. Do it by changing this:

```
NOTEPAD,   START NOTEPAD,   1,*File to edit: ,
```

to this:

```
NE,        START NE,        1,*File to edit: ,
```

Ensure you follow the format guidelines specified in the first page of the file:

```
<Command name>,[<Shell request>],<Bit flag>,[*]<Prompt>,
```

You should also use spaces (press the space bar) instead of tabs in the ACAD.PGP file.

Defining Command Aliases

Autodesk intends for you to define your own command aliases. After questioning users, Autodesk decided to define a command alias for almost all commands in AutoCAD. However, you can add more aliases, delete the defined aliases that you don't use, or alter the aliases that AutoCAD provides. The command alias section begins with these entries:

```
3A,            *3DARRAY
3DMIRROR,      *MIRROR3D
3DNavigate,    *3DWALK
3DO,           *3DORBIT
3DW,           *3DWALK
3F,            *3DFACE
3M,            *3DMOVE
3P,            *3DPOLY
and so on…
```

Because most commands already have an alias defined, it is probably more productive to learn the existing aliases rather than create new ones. There may be cases where it would be helpful to redefine existing aliases, however. For example, you may use *Trim* often but not *Mtext* (*TR* is the current alias for the *Trim* command, and *T* <u>and</u> *MT* for *Mtext*). In this case you could delete the *T* alias for *Mtext* (and just use *MT* for *Mtext*) and change *Trim* to *T*. To do this, follow the format specified at the beginning of the aliases section (give the alias and a comma, space over, and then give the AutoCAD command prefaced by an asterisk).

First, find and delete this:

        T,   *MTEXT

and change this:

        TR,   *TRIM

to this:

        T,   *TRIM

You may want to modify some of the other defined aliases. For example, you could change the alias C to invoke *Copy* rather than *Circle*. An alternative to *Circle* could be CI.

You can make comments anywhere in the ACAD.PGP by typing a semicolon (;) at the beginning of the line. This is helpful for making note of the changes that you have made.

One last suggestion: make a copy of the ACAD.PGP on diskette or in another directory before you begin experimenting, just in case.

CREATING SCRIPT FILES

A script file is an ASCII text file that contains a series of AutoCAD commands to be performed in succession. A script file is an external file that you create with a text editor and assign an .SCR extension. The script file is composed of commands in the sequence that you would normally use just as if you entered the commands at the AutoCAD command prompt. The *Script* command is used to tell AutoCAD to run the specified script file. When AutoCAD "runs" a script file, each command in the file is executed in the order that it is listed.

Script files have many uses, such as presenting a slide show, creating drawing geometry, or invoking a series of AutoCAD commands you use often—like a "macro." Script files are commonly used for repetitive tasks such as setting up a drawing. In fact, script "macros" may be incorporated directly into a menu file, thereby eliminating the need for an external .SCR file. However, in this chapter, only independent script files are addressed.

Chapter 43 discusses the *Script* command and the steps for creating scripts to present slide shows. A script for viewing slides uses the *Vslide* and *Delay* commands repetitively. If you are not familiar with the procedure for creating and running a script for this fundamental application, please review "Using Slides and Scripts" in Chapter 43.

Scripts for Drawing Setups

As you become familiar with AutoCAD, you realize several steps are repeated each time you set up a new drawing. For example, several commands are used and variables are set to prepare the drawing for geometry construction and for eventual plotting. Drawing setup is a good candidate for script application. Script files can be created to perform those repetitive drawing setup steps for you. Although an interactive AutoLISP routine would provide a more efficient and powerful method for this task, creating a script is relatively simple and does not require much knowledge or effort beyond that of the AutoCAD commands that would typically be used.

Keep in mind that template drawings that have many of the initial settings can be used as an alternative to creating a script for drawing setup. However, one advantage of a script (in general) is that it can be executed in any drawing and at any time during the drawing process.

For example, assume that you often set up drawings for plotting model space at a 1/4" = 1'-0" scale on a 24" x 36" sheet of paper. This involves setting *Limits, Grid, Snap, LTSCALE*, and *DIMSCALE*. Instead of executing each of these operations each time "manually," create a script to automate the process. Here are the steps:

1. Open your text editor from within AutoCAD or before starting AutoCAD. Creating scripts while AutoCAD is running is the most efficient method for writing and "debugging" scripts. In AutoCAD, use Notepad (or other text editor specified in the ACAD.PGP) to open your text editor.

2. Create the ASCII text file with the commands that you would normally use for setting up the drawing. Remember (from Chapter 43) that a space or a new line in the script is read as a Return (pressing the Enter key) while the script executes in AutoCAD. The script file may be written like this:

    ```
    LIMITS 0,0 1728,1152
    GRID 48
    SNAP 12
    LTSCALE 48
    DIMSCALE 48
    -DIMSTYLE SAVE STANDARD Y
    ```

3. Save the file with an .SCR extension. A good name for this file is SETUP48.SCR. Consider locating the file in a directory in the AutoCAD environment (as specified in the *Support Files Search Path* section of the *Files* tab in the *Options* dialog box) or in a location with other scripts.

4. Return to AutoCAD and use the *Script* command to test the script. Don't expect the script to operate the first time if this is your first experience writing scripts. It is common to have "bugs" in the script. For example, an extra space in the file can cause the script to get out of the intended sequence. An extra space in the script is hard to locate (since it is not visible). A good text editor allows the cursor to move only to places occupied by a character (including a space) and should be used for script writing.

Examining the commands in SETUP48.SCR, first, set the *Limits* using inch values. The "-DIMSTYLE SAVE STANDARD Y" line saves the *DIMSCALE* setting to the STANDARD dimension style. (If a dimension style other than STANDARD exists, the new *DIMSCALE* setting should be incorporated also into these dimension styles.) Keep in mind that scripts operate as if the commands were entered at the Command prompt, not as you might operate using dialog boxes.

Many other possibilities exist for drawing setup scripts. For example, you could set *-Units* (command) and the *DIMUNITS* variable. You may want to create a script for each plot scale and sheet size. It depends on your particular applications, hence the term "customization."

Possible applications for scripts are numerous. In addition to drawing setups, scripts can be used for geometry creation. It is possible to write scripts to create particular geometry "in real time" as the script executes. For example, repetitively drawn shapes could be created by scripts as an alternative to inserting *Blocks*. This idea is not generally used, but it may have advantages for certain applications. There are endless possibilities for script applications. Generally, anytime you find yourself performing the same tasks or series of commands repetitively, remember that a script can automate that process.

Special Script Commands

Creating scripts is a very straightforward procedure because the script file contains the AutoCAD commands that you would normally use at the command prompt for the particular operation. A few commands are intended for use specifically with scripts and other commands can be used as an aid to create or run script files. Those commands are listed next. Many of these commands are discussed in detail in "Using Slides and Scripts," Chapter 43.

Script	Use this command in AutoCAD to instruct AutoCAD to load and execute a script. The *Select Script File* dialog box appears, prompting you for the desired script file.
Delay	Include this command in a script file to force a delay (specified in milliseconds) until the next command in the script is executed. *Delay* is commonly used in script files that display slide shows.
Rscript	*Rscript* is short for "Repeat script." This command can be entered as the last line of a script file to make the script repeat until interrupted. *Rscript* is helpful for self-running slide shows.
Resume	If you want to interrupt a script with the Break or backspace key, type *Resume* to resume the script from the point at which it was interrupted. This command has no purpose in the script file itself but is intended to be typed at the keyboard.
Logfileon/ Logfileoff	These commands are used to turn on and off the log file creation utility in AutoCAD. When *Logfileon* is used, the conversation that appears at the command prompt (command history) is written to an external text file until *Logfileoff* is used (explained next in this chapter).

Logfileon/ Logfileoff

Pull-down Menu	Command (Type)	Alias (Type)	Short-cut	Screen (side) Menu	Tablet Menu
Tools *Options...* *Open and Save* *Maintain a log file*	*Logfileon* or *Logfileoff*	...	...	TOOLS 2 *Options...* *Open and Save* *Maintain a log file*	...

Logfileon and *Logfileoff* are used to turn on and off the creation of the AutoCAD log file. The log file is a text file that contains the command history—the text that appears at the Command line. When *Logfileon* is invoked, nothing appears to happen; however, any commands, prompts, or other text that appear at the command prompt from that time on are copied and written to the drawing's log file. Use *Logfileoff* to close the log file and discontinue recording the command history.

If the log file is on, AutoCAD continues to write to the log file, and each AutoCAD session in the log is separated by a dashed line. If you turn *Logfileoff* and *Logfileon* again, the new session is appended to the existing file. When a log file is created, it takes the current drawing's name by default. For example, if you are working on a drawing named PART35 and turn *Logfileon*, the log file is saved with a name similar to PART35_1_1_3281.LOG (where _1_1_3281 are automatically generated numbers). You can use the *Options* dialog box, *Files* tab, *Log File Location* section to locate the current location of the log files or to set a different location for log files (see Fig. 44-8).

The log file is an ASCII text file that can be edited. The file has no direct link to AutoCAD and can be deleted without consequence to AutoCAD operation. The log file cannot be accessed externally until *Logfileoff* is used or AutoCAD is exited.

A sample log file is shown here:

```
[ AutoCAD - Mon Dec 11 10:32:54 2006 ]-----------------------------------------

Command: line
Specify first point:
Specify next point or [Undo]:
Specify next point or [Undo]:
Specify next point or [Close/Undo]:
Specify next point or [Close/Undo]:
Command: _circle Specify center point for circle or [3P/2P/Ttr (tan tan radius)]:
Specify radius of circle or [Diameter]:
Command: _copy
Select objects: 1 found
Select objects:
Specify base point or [Displacement] <Displacement>:
Specify second point or <use first point as displacement>:
Specify second point or [Exit/Undo] <Exit>:
Specify second point or [Exit/Undo] <Exit>:
Command: _trim
Current settings: Projection=UCS, Edge=None
Select cutting edges ...
Select objects or <select all>: 1 found
Select objects: 1 found, 2 total
Select objects: 1 found, 3 total
Select objects:
Select object to trim or shift-select to extend or [Fence/Crossing/Project/Edge/eRase/Undo]:
Select object to trim or shift-select to extend or
[Fence/Crossing/Project/Edge/eRase/Undo]:
Select object to trim or shift-select to extend or
[Fence/Crossing/Project/Edge/eRase/Undo]:
Command: _qsave
```

The log file is especially helpful when you are writing scripts. When you create a script file, the commands and options listed in the file must be typed in the exact sequence that would occur if you were actually using AutoCAD. In order to write the script correctly, you have three choices: (1) remember every command and prompt exactly as it would occur in AutoCAD; (2) go through a "dry run" in AutoCAD and write down every step; (3) turn *Logfileon*; then execute a "dry run" to have all the steps and resulting text recorded.

In other words, turn *Logfileon* to create a rough script file. Then turn *Logfileoff*, open the log file, and edit out the AutoCAD prompts such as "Command:" and "Select objects:," etc. Finally, save the file to the desired name with an .SCR extension.

If you want to change the location of the log file that is created when you use *Logfileon*, use the *Files* tab of the *Options* dialog box (Fig. 44-8). In the dialog box, expand the *Log File Location* tree, select *Browse* to find the desired directory location, select *Apply*, then *OK*. This directory location is stored in, and can also be changed using, the *LOGFILEPATH* system variable.

Although more cumbersome for most applications than using the *Logfileon* and *Logfileoff* commands, you can also turn on and off the log file by toggling the *Maintain a log file* check box in the *Open and Save* tab of the *Options* dialog box.

FIGURE 44-8

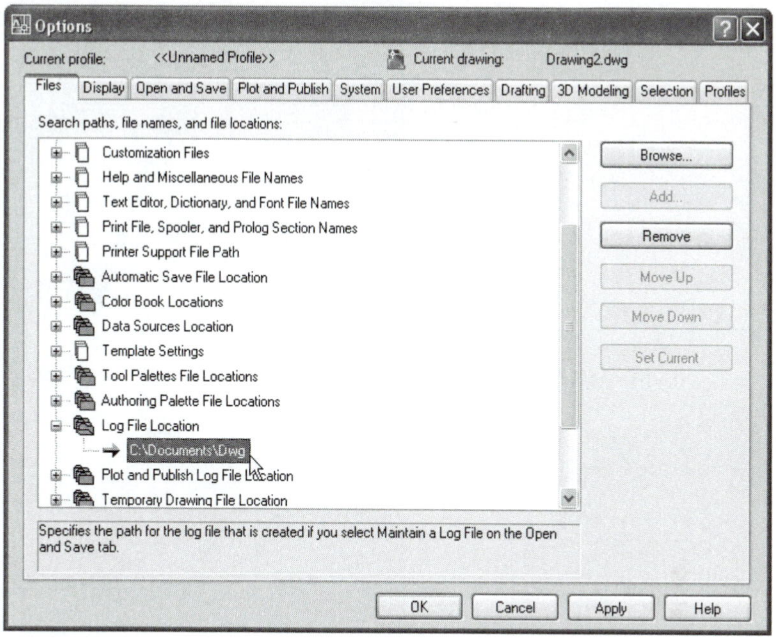

MORE CUSTOMIZATION

This chapter gives you instruction in some of the miscellaneous customizations available in AutoCAD, such as customizing linetypes, customizing command aliases in the ACAD.PGP file, and creating script files. In addition, this chapter serves as a short introduction to the customization capabilities available in the *Customize User Interface* dialog box. The complete capabilities for this unique tool are covered in Chapter 45, Customize User Interface. Chapter 45 is available for download at no charge at www.mhhe.com/leach.

ADDITIONAL CHAPTERS AND RESOURCES

Also at www.mhhe.com/leach, the following chapters and resources are available for download at no charge in the Student Resource Center:

Chapter 46, CAD Management
Chapter 47, Express Tools
Chapter Text/Review Questions in Word (.DOC) format, 25 questions per chapter
Chapter Exercises for Architectural Applications
Chapter Exercises for Mechanical Applications
Chapter Exercises for Civil/Electrical Applications

And for instructors only (in the Instructor Resource Center):

Solutions Manual for AutoCAD 2007 Instructor Chapter Exercises in .DWG and .DOC format
Answers for Chapter Text/Review Questions
Access to the images from the text (.TIF and .EPS files) for use in Power Point or other presentations

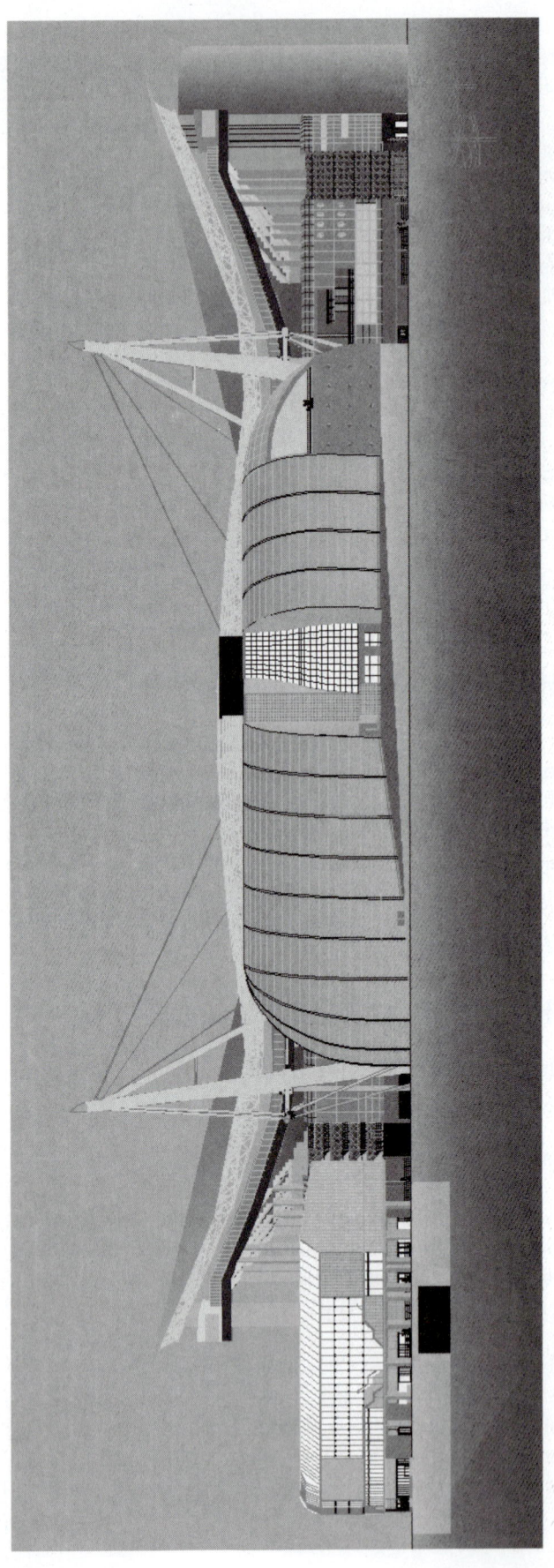

STADIUM NORTH ELEVATION.DWG, Courtesy of Autodesk, Inc.

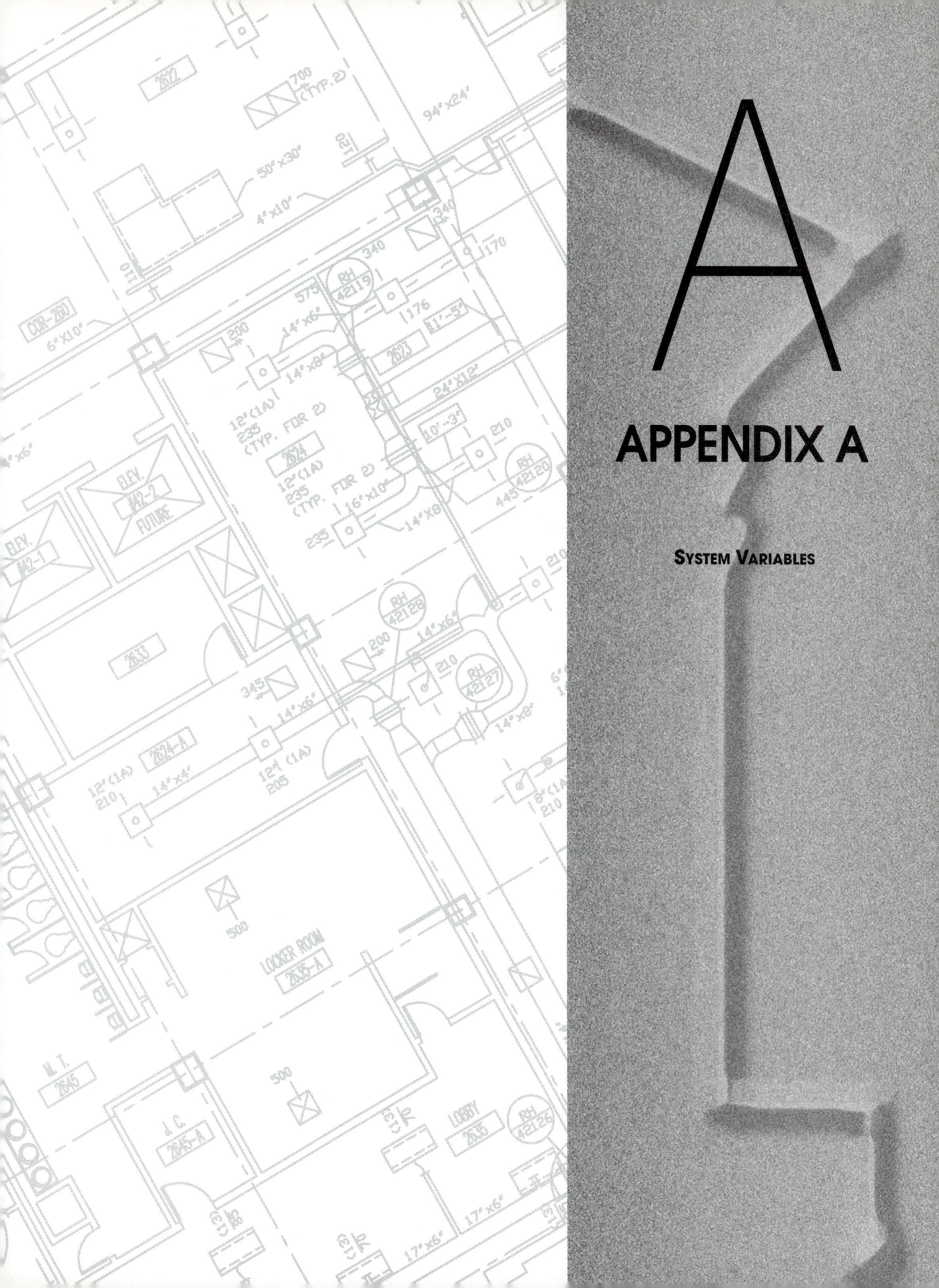

A

APPENDIX A

SYSTEM VARIABLES

APPENDIX A

System Variables

Variable	Characteristic	Description
3DDWFPREC	Type: Integer Saved in: Drawing Initial value: 2	Controls the precision of 3D DWF publishing. This system variable has a range from 1 to 6. Higher settings result in finer precision. *3DDWFPREC* Value Deviation Value 1 1 2 0.5 3 0.2 4 0.1 5 0.01 6 0.001 NOTE: Setting a *3DDWFPREC* value of 5 or 6 will create very large 3D DWF files or sheets in a multi-sheet DWF.
ACADLSPASDOC	Type: Integer Saved in: Registry Initial value: 0	Controls whether AutoCAD loads the acad.lsp file into every drawing or just the first drawing opened in an AutoCAD session. 0 Loads acad.lsp into just the first drawing opened in an AutoCAD session 1 Loads acad.lsp into every drawing opened
ACADPREFIX	(Read-only) Type: String Not saved	Stores the directory path, if any, specified by the ACAD environment variable, with path separators appended if necessary.
ACADVER	(Read-only) Type: String Not saved	Stores the AutoCAD version number. This variable differs from the DXF file $ACADVER header variable, which contains the drawing database level number.
ACISOUTVER	Type: Integer Not saved Initial value: 40	Controls the ACIS version of SAT files created using the *Acisout* command. Currently, *Acisout* supports only values of 15, 16, 17, 18, 20, 21, 30, 40, 50, 60, and 70.
ADCSTATE	(Read-only) Type: Integer Not saved Initial value: varies	Determines whether DesignCenter is active or not. For developers who need to determine status through AutoLISP. 0 DesignCenter is not active 1 DesignCenter is active
AFLAGS	Type: Integer Not saved Initial value: 0	Sets attribute flags for *Attdef* bit-code. It is the sum of the following: 0 No attribute mode selected 1 Invisible 2 Constant 4 Verify 8 Preset

2007

Variable	Characteristic	Description
ANGBASE	Type: Real Saved in: Drawing Initial value: 0.0000	Sets the base angle to 0 with respect to the current UCS.
ANGDIR	Type: Integer Saved in: Drawing Initial value: 0	Sets the positive angle direction from angle 0 with respect to the current UCS. 0 Counter-clockwise 1 Clockwise
APBOX	Type: Integer Saved in: Registry Initial value: 0	Turns the AutoSnap aperture box on or off. The aperture box is displayed in the center of the cursor when you snap to an object. This option is available only when Marker, Magnet, or Snaptip is selected. 0 Aperture box is not displayed 1 Aperture box is displayed
APERTURE	Type: Integer Saved in: Registry Initial value: 10	Sets *Aperture* command object snap target height, in pixels.
AREA	(Read-only) Type: Real Not saved	Stores the last area computed by *Area, List,* or *Dblist.*
ASSISTSTATE	Type: Integer Saved in: Not saved Initial value: Varies	Indicates whether the Info palette that displays *Quick Help* is active or not. 0 Not active 1 Active
ATTDIA	Type: Integer Saved in: Drawing Initial value: 0	Controls whether *Insert* uses a dialog box for attribute value entry. 0 Issues prompts on the command line 1 Uses a dialog box
ATTMODE	Type: Integer Saved in: Drawing Initial value: 1	Controls display of attributes. 0 Off makes all attributes invisible 1 Normal retains current visibility of each attribute: visible attributes are displayed; invisible attributes are not 2 On makes all attributes visible
ATTREQ	Type: Integer Saved in: Drawing Initial value: 1	Determines whether *Insert* uses default attribute settings during insertion of blocks. 0 Assumes the defaults for the values of all attributes 1 Turns on prompts or dialog box for attribute values, as specified at *Attdia*
AUDITCTL	Type: Integer Saved in: Registry Initial value: 0	Controls whether *Audit* creates an ADT file (audit report). 0 Prevents writing of ADT files 1 Writes ADT files

Variable	Characteristic	Description
AUNITS	Type: Integer Saved in: Drawing Initial value: 0	Sets units for angles. 0 Decimal degrees 1 Degrees/minutes/seconds 2 Gradians 3 Radians 4 Surveyor's units
AUPREC	Type: Integer Saved in: Drawing Initial value: 0	Sets number of decimal places for angular units displayed on the Status line or when listing an object.
AUTOSNAP	Type: Integer Saved in: Registry Initial value: 63	Controls AutoSnap marker, SnapTip, and magnet. Also turns on Polar and Object Snap Tracking, and controls the display of Polar and Object Snap Tracking ToolTips. The system variable value is the sum of the following bit values: 0 Turns off the AutoSnap marker, SnapTips, and magnet. Also turns off Polar Tracking, Object Snap Tracking, and ToolTips for Polar and Object Snap Tracking. 1 Turns on the AutoSnap marker 2 Turns on the AutoSnap SnapTips 4 Turns on the AutoSnap magnet 8 Turns on Polar Tracking 16 Turns on Object Snap Tracking 32 Turns on ToolTips for Polar Tracking and Object Snap Tracking
BACKGROUNDPLOT	Type: Integer Saved in: Registry Initial value: 2	Controls whether background plotting is turned on or off for plotting and publishing. By default, background plotting is turned off for plotting and on for publishing. _Value_ _Plot_ _Publish_ 0 Foreground Foreground 1 Background Foreground 2 Foreground Background 3 Background Background When -_Plot_, _Plot_, -_Publish_, and _Publish_ are used in a script (.SCR file), the BACKGROUNDPLOT system variable value is ignored, and -_Plot_, _Plot_, -_Publish_, and _Publish_ are processed in the foreground.
BACKZ	(Read-only) Type: Real Saved in: Drawing	Stores the back clipping plane offset from the target plane for the current viewport, in drawing units. Meaningful only if the back clipping bit in VIEWMODE is on. The distance of the back clipping plane from the camera point can be found by subtracting BACKZ from the camera-to-target distance.

Variable	Characteristic	Description
BACTIONCOLOR	Type: String Saved in: Registry Initial value: 7	Sets the text color of actions in the Block Editor. Valid values include *BYLAYER, BYBLOCK,* and an integer from 1 to 255. Valid values for True Colors are a string of integers each from 1 to 255 separated by commas and preceded by RGB. The True Color setting is entered as follows: RGB:000,000,000.
BDEPENDENCY-HIGHLIGHT	Type: Integer Saved in: Registry Initial Value: 1	Controls whether or not dependent objects are dependency highlighted when a parameter, action, or grip is selected in the Block Editor. Dependency highlighting displays objects with a halo effect. 0 Specifies that dependent objects are not highlighted 1 Specifies that dependent objects are highlighted
BGRIPOBJCOLOR	Type: Integer Saved in: Registry Initial Value: 141	Sets the color of grips in the Block Editor. Valid values include *BYLAYER, BYBLOCK,* and an integer from 1 to 255. Valid values for True Colors are a string of integers each from 1 to 255 separated by commas and preceded by RGB. The True Color setting is entered as follows: RGB:000,000,000.
BGRIPOBJSIZE	Type: Integer Saved in: Registry Initial value: 8	Sets the display size of custom grips in the Block Editor relative to the screen display. Valid values include an integer from 1 to 255. Use the *Regen* command to update the display size of custom grips in the Block Editor.
BINDTYPE	Type: Integer Not Saved Initial value: 0	Controls how xref names are handled when binding xrefs or editing xrefs in place. 0 Traditional binding behavior ("xref1 | one" becomes "xref0one") 1 Insert-like behavior ("xref1 | one" becomes "one")
BLIPMODE	Type: Integer Saved in: Drawing Initial value: 0	Controls whether marker blips are visible. 0 Turns off marker blips 1 Turns on marker blips
BLOCKEDITLOCK	Type: Integer Saved in: Registry Initial value: 0	Disallows opening of the Block Editor and editing of dynamic block definitions. When *BLOCKEDITLOCK* is set to 1, double-clicking a dynamic block in a drawing opens the *Reference Edit* dialog box. If the block contains attributes, double-clicking the block reference opens the *Enhanced Attribute Editor.* 0 Specifies that the Block Editor can be opened 1 Specifies that the Block Editor cannot be opened
BLOCKEDITOR	Type: Integer Saved in: Not saved Initial value: 0	Reflects whether or not the Block Editor is open. 0 Indicates that the Block Editor is not open 1 Indicates that the Block Editor is open

2006

2006

Variable	Characteristic	Description
BPARAMETERCOLOR	Type: String Saved in: Registry Initial value: 7	Sets the color of the parameters in the Block Editor. Valid values include *BYLAYER, BYBLOCK,* and an integer from 1 to 255. Valid values for True Colors are a string of integers each from 1 to 255 separated by commas and preceded by RGB. The True Color setting is entered as follows: RGB:000,000,000.
BPARAMETERFONT	Type: String Saved in: Registry Initial value: Simplex.shx	Sets the font used for parameters and actions in the Block Editor. You can specify either a TrueType font or a .SHX font (for example, Verdana or Verdana.ttf). You must add the .shx extension to specify an AutoCAD .SHX font. When specifying an Asian Big Font, use the following naming convention: an .SHX file followed by a comma (,), followed by the Big Font file name (for example, Simplex.shx,Bigfont.shx).
BPARAMETERSIZE	Type: Integer Saved in: Registry Initial value: 12	Sets the size of parameter text and features in the Block Editor relative to the screen display. Valid values include an integer from 1 to 255.
BTMARKDISPLAY	Type: Integer Saved in: Registry Initial value: 1	Controls whether or not value set markers are displayed for dynamic block references. 0 Specifies that value set markers are not displayed 1 Specifies that value set markers are displayed
BVMODE	Type: Integer Saved in: Not saved Initial Value: 0	Controls how objects that are made invisible for the current visibility state are displayed in the Block Editor. 0 Specifies that hidden objects are not visible 1 Specifies that hidden objects are visible but dimmed
CALCINPUT	Type: Integer Saved in: Registry Initial Value: 1	Controls whether mathematical expressions and global constants are evaluated in text and numeric entry boxes of windows and dialog boxes. 0 Expressions are not evaluated 1 Expressions are evaluated after you press the End key
CAMERADISPLAY	Type: Integer Saved in: Drawing Initial Value: 0	Turns the display of camera objects on or off. The value changes to 1 (to display cameras) when you use the *Camera* command. 0 Camera glyphs are not displayed 1 Camera glyphs are displayed
CAMERAHEIGHT	Type: Integer Saved in: Drawing Initial Value: 0	Specifies the default height for new camera objects. The height is expressed in current drawing units.
CDATE	(Read-only) Type: Real Not saved	Sets calendar date and time.

2006

2007

Variable	Characteristic	Description
CECOLOR	Type: String Saved in: Drawing Initial value: "BYLAYER"	Sets the color of new objects. Valid values include *Bylayer, Byblock,* and an integer from 1 to 255. Valid values for True Colors are a string of integers each from 1 to 255 separated by commas and preceded by RGB. The True Color setting is entered as follows: RGB:000,000,000.
CELTSCALE	Type: Real Saved in: Drawing Initial value: 1.0000	Sets the current object linetype scaling factor. This sets the linetype scaling for new objects relative to the *LTSCALE* setting. A line created with *CELTSCALE=2* in a drawing with *LTSCALE* set to 0.5 would appear the same as a line created with *CELTSCALE=1* in a drawing with *LTSCALE=1.*
CELTYPE	Type: String Saved in: Drawing Initial value: "BYLAYER"	Sets the linetype of new objects.
CELWEIGHT	Type: Integer Saved In: Drawing Initial value: "BYLAYER"	Sets the lineweight of new objects. -1 Sets the lineweight to "BYLAYER" -2 Sets the lineweight to "BYBLOCK" -3 Sets the lineweight to "DEFAULT" Other valid values that can be entered in millimeters include: 0, 5, 9, 13, 15, 18, 20, 25, 30, 35, 40, 50, 53, 60, 70, 80, 90, 100, 106, 120, 140, 158, 200, and 211. All values must be entered in millimeters. (Multiply a value by 2.54 to convert values from inches to millimeters.)
CENTERMT	Type: Integer Saved In: User settings Initial value: 0	Controls how grips stretch multiline text that is centered horizontally. *CENTERMT* does not apply to stretching multiline text by using the ruler in the In-Place Text Editor. 0 When you move a corner grip in centered multiline text, the center grip moves in the same direction, and the grip on the opposite side remains in place 1 When you move a corner grip in centered multiline text, the center grip stays in place, and both sets of side grips move in the direction of the stretch
CHAMFERA	Type: Real Saved in: Drawing Initial value: 0.0000	Sets the first chamfer distance.
CHAMFERB	Type: Real Saved in: Drawing Initial value: 0.0000	Sets the second chamfer distance.
CHAMFERC	Type: Real Saved in: Drawing Initial value: 0.0000	Sets the first chamfer length.

2006

Variable	Characteristic	Description
CHAMFERD	Type: Real Saved in: Drawing Initial value: 0.0000	Sets the first chamfer angle.
CHAMMODE	Type: Integer Not saved Initial value: 0	Sets the input method by which AutoCAD creates chamfers. 0 Requires two chamfer distances 1 Requires one chamfer distance and an angle
CIRCLERAD	Type: Real Not saved Initial value: 0.0000	Sets the default circle radius. A zero sets no default.
CLAYER	Type: String Saved in: Drawing Initial value: 0	Sets the current layer.
CLEANSCREENSTATE	(Read-only) Type: Integer Saved in: Not-saved Initial value: 0	Stores a value that indicates whether the clean screen state is on. 0 Off 1 On
CLISTATE	Type: Integer Saved in: Not saved Initial value: 1	Stores a value that indicates whether the command window is hidden or displayed. 0 Hidden 1 Displayed
CMDACTIVE	(Read-only) Type: Integer Not saved	Stores the bit code that indicates whether an ordinary command, transparent command, script, or dialog box is active. It is the sum of the following: 1 Ordinary command is active 2 Ordinary command and a transparent command are active 4 Script is active 8 Dialog box is active 16 DDE is active 32 AutoLISP is active (only visible to an ObjectARX-defined command) 64 ObjectARX command is active
CMDDIA	Type: Integer Saved in: Not saved Initial value: 1	Controls the display of dialog boxes for some commands. 0 Off 1 On
CMDECHO	Type: Integer Not saved Initial value: 1	Controls whether AutoCAD echoes prompts and input during the AutoLISP (command) function. 0 Turns off echoing 1 turns on echoing
CMDINPUTHISTORY-MAX	Type: Integer Saved in: Registry Initial value: 20	Sets the maximum number of previous input values that are stored for a prompt in a command. Display of the history of user input is controlled by the INPUTHISTORYMODE system variable.

2007

2006

2006

Variable	Characteristic	Description
CMDNAMES	(Read-only) Type: String Not saved	Displays the name of the currently active command and transparent command. For example, *LINE'ZOOM* indicates that the *Zoom* command is being used transparently during the *Line* command. This variable is designed for use with programming interfaces such as AutoLISP, DIESEL, and ActiveX Automation.
CMLJUST	Type: Integer Saved in: Drawing Initial value: 0	Specifies multiline justification. 0 Top 1 Middle 2 Bottom
CMLSCALE	Type: Real Saved in: Drawing Initial value: 1.0000 (English) 20.0000 (Metric)	Controls the overall width of a multiline. A scale factor of 2.0 produces a multiline that is twice as wide as the style definition. A zero scale factor collapses the multiline into a single line. A negative scale factor flips the order of the offset lines (that is, the smallest or most negative is placed on top when the multiline is drawn from left to right).
CMLSTYLE	Type: String Saved in: Drawing Initial value: "STANDARD"	Sets the multiline style that AutoCAD uses to draw the multiline.
CMMATERIAL	Type: String Saved in: Drawing Initial value: *Bylayer*	Sets the material of new objects. Valid values are *Bylayer* and the name of a material in the drawing.
COMPASS	Type: Integer Initial value: 0	Controls whether the 3D compass is on or off in the current viewport. 0 Turns the 3D compass off 1 Turns the 3D compass on
COORDS	Type: Integer Saved in: Drawing Initial value: 1	Controls when coordinates are updated on the Status line. 0 Coordinate display is updated as you specify points with the pointing device 1 Display of absolute coordinates is continuously updated 2 Display of absolute coordinates is continuously updated, and distance and angle from last point are displayed when a distance or angle is requested
CPLOTSTYLE	Type: String Saved in: Drawing	Controls the current plot style for new objects. If the current drawing is color-dependent (*PSTYLEPOLICY* is set to 1), *CPLOTSTYLE* is read-only and has a value of "BYCOLOR." If the (*PSTYLEPOLICY* is set to 0), *CPLOTSTYLE* can be set to the following values ("BYLAYER" is the default):

2007

Variable	Characteristic	Description
CPLOTSTYLE (continued)		"BYLAYER" "BYBLOCK" "NORMAL" "USER DEFINED"
CPROFILE	Type: (Read-only) Saved in: Registry Initial value: <<Unnamed Profile>>	Displays the name of the current profile.
CROSSINGAREA-COLOR	Type: Integer Saved in: Registry Initial value: 3	Controls the color of the selection area during crossing selection. The valid range is 1 to 255. The SELECTIONAREA system variable must be on.
CSHADOW	Type: Integer Saved in: Drawing Initial value: 0	Sets the shadow display property for a 3D object. To be visible, shadows must be turned on in the visual style that is applied to the viewport. 0 Casts and receives shadows 1 Casts shadows 2 Receives shadows 3 Ignores shadows NOTE: To display full shadows, hardware acceleration is required. When Geometry Acceleration is off, full shadows cannot be displayed. (To access these settings, enter 3Dconfig at the Command prompt. In the Adaptive Degradation and Performance Tuning dialog box, click Manual Tune.)
CTAB	(Read-only) Type: String Saved in: Drawing	Returns the name of the current (model or layout) tab in the drawing. Provides a means for the user to determine which tab is active.
CTABLESTYLE	Type: String Saved in: Drawing Initial value: Standard	Sets the name of the current table style.
CURSORSIZE	Type: Integer Saved in: Registry Initial value: 5	Determines the size of the cursor as a percentage of the screen size. Valid settings range from 1 to 100 percent. When set to 100, the cursor are full-screen and the ends of the cursor are never visible. When less than 100, the ends of the cursor may be visible when the cursor is moved to one edge of the screen.
CVPORT	Type: Integer Saved in: Drawing Initial value: 2	Sets the identification number of the current viewport. You can change this value, thereby changing the current viewport, if the following conditions are met: • The identification number you specify is that of an active viewport • A command in progress has not locked cursor movement to that viewport • Tablet mode is off

2006

2007

Variable	Characteristic	Description
DATE	(Read-only) Type: Real Not saved	Stores the current date and time represented as a Julian date and fraction in a real number: <Julian date>.<Fraction> For example, on January 29, 1993, at 2:29:35 in the afternoon, the DATE variable would contain 2446460.603877364. Your computer clock provides the date and time. The time is represented as a fraction of a day. To compute differences in time, subtract the times returned by *DATE*. To extract the seconds since midnight from the value returned by *DATE*, use AutoLISP expressions: **(setq s (getvar "DATE"))** **(setq seconds (* 86400.0 – s (fix s)))** The *DATE* system variable returns a true Julian date only if the system clock is set to UTC/Zulu (Greenwich Mean Time). *TDCREATE* and *TDUPDATE* have the same format as *DATE*, but their values represent the creation time and last update time of the current drawing.
DBCSTATE	(Read-only) Type: Integer Saved in: Drawing Initial value: 0	Stores the state of the *dbConnect Manager,* active or not active. 0 The *dbConnect Manager* is not displayed 1 The *dbConnect Manager* is displayed
DBLCLKEDIT	Type: Integer Saved in: Registry Initial value: 1	Controls the double click editing behavior in the drawing area. Double click actions can be customized using the *Customize User Interface* (CUI) editor. The system variable can accept the values of *On* and *Off* in place of 1 and 0. 0 Disabled 1 Enabled
DBMOD	(Read-only) Type: Integer Not saved	Indicates the drawing modification status using bit code. It is the sum of the following: 1 Object database modified 4 Database variable modified 8 Window modified 16 View modified
DCTCUST	Type: String Saved in: Registry Initial value: ""	Displays the path and file name of the current custom spelling dictionary.
DCTMAIN	Type: String Saved in: Registry Initial value: *varies by country*	Displays the file name of the current main spelling dictionary. The full path is not shown, because this file is expected to reside in the *support* directory. You can specify a default main spelling dictionary using *Setvar*. When prompted for a new value for *DCTMAIN*, you can enter one of the keywords below. *Keyword Language Name* enu American English

2007

Variable	Characteristic	Description
DCTMAIN (continued)		ena Australian English ens British English (ise) enz British English (ize) ca Catalan cs Czech da Danish nl Dutch (primary) nls Dutch (secondary) fi Finnish fr French (unaccented capitals) fra French (accented capitals) de German (Scharfes s) ded German (Dopple s) it Italian no Norwegian (Bokmal) non Norwegian (Nynorsk) pt Portuguese (Iberian) ptb Portuguese (Brazilian) ru Russian (infrequent io) rui Russian (frequent io) es Spanish (unaccented capitals) esa Spanish (accented capitals) sv Swedish
DEFAULTLIGHTING	Type: Integer Saved in: Drawing Initial value: 1	Controls default lighting in the current viewport. Default lighting is provided by a set of distant lights that follow the view direction. When default lighting is on, the sun and other lights do not cast light, even if they are turned on. The setting of this system variable is viewport-specific. 0 Auto: Default lighting is automatically turned off when point lights, spotlights, distant lights, or the sun are on 1 On
DEFAULTLIGHTING-TYPE	Type: Integer Saved in: Drawing Initial value: 1	Specifies the type of default lighting. The default lighting provided in AutoCAD 2006 and earlier releases used one distant light. The new default lighting uses two distant lights to illuminate more of the model and also adjusts ambient light. The setting of this system variable is viewport-specific. 0 Old type of default lighting 1 New type of default lighting
DEFLPLSTYLE	Type: String Saved in: Registry	Specifies the default plot style for new layers. If the current drawing is color-dependent (PSTYLEPOLICY is set to 1), DEFLPLSTYLE is read-only and has a value of "BYCOLOR." If the current drawing is named plot style mode (PSTYLEPOLICY is set to 0), DEFLPLSTYLE is writable and has a default value of "NORMAL."

2007

Variable	Characteristic	Description
DEFPLSTYLE	Type: String Saved in: Registry	Specifies the default plot style for new objects. If the current drawing is color-dependent (*PSTYLE-POLICY* is set to 1), *DEFPLSTYLE* is read-only and has a value of "BYCOLOR." If the current drawing is named plot style mode (*PSTYLEPOLICY* is set to 0), *DEFPLSTYLE* is writable and has a default value of "BYLAYER."
DELOBJ	Type: Integer Saved in: Drawing Initial value: 1	Controls whether objects used to create other objects are retained or deleted from the drawing database. 0 Objects are retained 1 Objects are deleted
DEMANDLOAD	Type: Integer Saved in: Registry Initial value: 3	Specifies if and when AutoCAD demand loads a third-party application if a drawing contains custom objects created in that application. 0 Turns off demand loading. 1 Demand loads the source application when you open a drawing that contains custom objects. This setting does not demand load the application when you invoke one of the application's commands. 2 Demand loads the source application when you invoke one of the application's commands. This setting does not demand load the application when you open a drawing that contains custom objects. 3 Demand loads the source application when you open a drawing that contains custom objects or when you invoke one of the application's commands.
DIASTAT	(Read-only) Type: Integer Not saved	Stores the exit method of the most recently used dialog box. 0 Cancel 1 OK
DIM...	See "Dimension Variables Table" in Chapter 29	
DISPSILH	Type: Integer Saved in: Drawing Initial value: 0	Controls display of silhouette curves of solid objects in wireframe mode. Also controls whether mesh is drawn or suppressed when a solid object is hidden. 0 Off 1 On
DISTANCE	(Read-only) Type: Real Not saved	Stores the distance computed by *Dist*.
DONUTID	Type: Real Not saved Initial value: 0.5000	Sets the default for the inside diameter of a donut.

Variable	Characteristic	Description
DONUTOD	Type: Real Not saved Initial value: 1.0000	Sets the default for the outside diameter of a donut. It must be nonzero. If *DONUTID* is larger than *DONUTOD*, the two values are swapped by the next command.
DRAGMODE	Type: Integer Saved in: Drawing Initial value: 2	Controls display of objects being dragged. 0 Does not display an outline of the object as you drag it 1 Displays the outline of the object as you drag it only if you enter **drag** on the command line after selecting the object to drag 2 Auto; always displays an outline of the object as you drag it
DRAGP1	Type: Integer Saved in: Registry Initial value: 10	Sets regen-drag input sampling rate.
DRAGP2	Type: Integer Saved in: Registry Initial value: 25	Sets fast-drag input sampling rate.
DRAGVS	Type: String Saved in: Drawing Initial value: Current visual style	Sets the visual style while creating 3D solid primitives and extruded solids and surfaces. You can enter a period (.) to specify the current visual style. *DRAGVS* can only be set to a visual style that is saved in the drawing. *DRAGVS* has no effect when the current viewport's visual style is set to *2D Wireframe*. The visual style specified for *DRAGVS* cannot be removed with the *Purge* command.
DRAWORDERCTL	Type: Integer Saved in: Drawing Initial value: 3	Controls the display order of overlapping objects. Use this setting to improve the speed of editing operations in large drawings. The commands that are affected by inheritance are *Break, Fillet, Hatch, Hatchedit, Explode, Trim, Join, Pedit,* and *Offset*. 0 Turns off the default draw order of overlapping objects: after objects are edited, regardless of their draw order, the objects are displayed on top until a drawing is regenerated (*Regen*) or reopened. This setting also turns off draw order inheritance: new objects that are created from another object using the commands listed above are not assigned the draw order of the original object. 1 Turns on the default draw order of objects: after objects are edited, they are automatically displayed according to the correct draw order. 2 Turns on draw order inheritance: new objects created from another object using the commands listed above are assigned the draw order of the original object. 3 Provides full draw order display. Turns on the correct draw order of objects, and turns on draw order inheritance.

2007

Variable	Characteristic	Description
DRSTATE	Type: Integer Saved in: Not saved Initial value: Varies	Determines whether the *Drawing Recovery* window is active or not. 0 The *Drawing Recovery* window is not active 1 The *Drawing Recovery* window is active
DTEXTED	Type: Integer Saved in: Registry Initial value: 0	Specifies the user interface for editing single-line text. 0 Displays the In-Place Text Editor 1 Displays the *Edit Text* dialog box
DWFFRAME	Type: Integer Saved in: Drawing Initial value: 2	Determines whether the DWF underlay frame is visible. 0 The DWF underlay frame is not visible and it is not plotted 1 Displays the DWF underlay frame and allows it to be plotted 2 Displays the DWF underlay frame but keeps it from being plotted
DWFOSNAP	Type: Integer Saved in: Registry Initial value: 1	Determines whether object snapping is active for geometry in DWF underlays that are attached to the drawing. 0 Object snapping is disabled for geometry in all DWF underlay attachments in the drawing 1 Object snapping is enabled for geometry in all DWF underlay attachments in the drawing
DWGCHECK	Type: Integer Saved in: Registry Initial Value: 0	Checks drawings for potential problems when opening them. 0 If a drawing that you try to open is found to have a potential problem, you are warned before the drawing is opened 1 If a drawing that you try to open is found to have a potential problem, or if it was saved by an application other than AutoCAD or AutoCAD LT, you are warned before the drawing is opened 2 If a drawing that you try to open is found to have a potential problem you are notified on the command line after the drawing is opened 3 If a drawing that you try to open is found to have a potential problem or if it was saved by an application other than AutoCAD or AutoCAD LT, you are notified on the command line after the drawing is opened
DWGCODEPAGE	(Read-only) Type: String Saved in: Drawing	Stores the same value as *SYSCODEPAGE* (for compatibility).
DWGNAME	(Read-only) Type: String Not saved	Stores the drawing name as entered by the user. If the drawing has not been named yet, *DWGNAME* defaults to "Drawing.dwg." If the user specified a drive/directory prefix, it is stored in the *DWGPREFIX* variable.

2006

2007

Variable	Characteristic	Description
DWGPREFIX	(Read-only) Type: String Not saved	Stores the drive/directory prefix for the drawing.
DWGTITLED	(Read-only) Type: Integer Not saved	Indicates whether the current drawing has been named. 0 Drawing has not been named 1 Drawing has been named
DYNDIGRIP	Type: Bit code Saved in: Registry Initial value: 31	Controls which dynamic dimensions are displayed during grip stretch editing. The *DYNDIVIS* system variable must be set to 2, which displays all dynamic dimensions. 0 None 1 Resulting dimension 2 Length change dimension 4 Absolute angle dimension 8 Angle change 16 Arc radius dimension
DYNDIVIS	Type: Integer Saved in: User settings Initial value: 1	Controls how many dynamic dimensions are displayed during grip stretch editing. *DINDIGRIP* controls which dynamic dimensions are displayed during grip stretch editing. 0 Only the first dynamic dimension in the cycle order 1 Only the first two dynamic dimensions in the cycle order 2 All dynamic dimensions, as controlled by the *DYNDIGRIP* system variable
DYNMODE	Type: Integer Saved in: User settings Initial value: 3	Turns Dynamic Input features on and off. When all features are on, the context governs what is displayed. When *DYNMODE* is set to a negative value, the Dynamic Input features are not visible, but the setting is stored. Press the *DYN* button in the status bar to set *DYNMODE* to the corresponding positive value. 0 All off 1 Only pointer input on 2 Only dimensional input on 3 All on When *DYNMODE* is set to anything but 0, you can turn off all features temporarily by holding down the temporary override key, F12.
DYNPICOORDS	Type: Switch Saved in: User settings Initial value: 0	Controls whether pointer input uses relative or absolute format for coordinates. 0 Relative 1 Absolute
DYNPIFORMAT	Type: Switch Saved in: User settings Initial value: 0	Controls whether pointer input uses polar or Cartesian format for coordinates. This setting applies only to a second or next point. 0 Polar 1 Cartesian

Variable	Characteristic	Description
DYNPIVIS	Type: Integer Saved in: User settings Initial value: 1	Controls when pointer input is displayed. 0 Only when you type at a prompt for a point 1 Automatically at a prompt for a point 2 Always
DYNPROMPT	Type: Integer Saved in: User settings Initial value: 1	Controls display of prompts in Dynamic Input tooltips. 0 Off 1 On
DYNTOOLTIPS	Type: Integer Saved in: User settings Initial value: 1	Controls which tooltips are affected by tooltip appearance settings. 0 Only Dynamic Input value fields 1 All drafting tooltips
EDGEMODE	Type: Integer Saved in: Registry Initial value: 0	Controls how *Trim* and *Extend* determine cutting and boundary edges. 0 Uses the selected edge without an extension 1 Extends or trims the selected object to an imaginary extension of the cutting or boundary edge Line, arc, elliptical arc, ray, and polyline are objects eligible for natural extension. The natural extension of a line or ray is an unbounded line (xline); an arc is a circle; and an elliptical arc is an ellipse. A polyline is broken down into its line and arc components, which are extended to their natural boundaries.
ELEVATION	Type: Real Saved in: Drawing Initial value: 0.0000	Stores the current elevation relative to the current UCS for the current space.
ENTERPRISEMENU	Type: String Saved in: Registry Initial value: "."	Stores the CUI file name (if defined), including the path for the file name.
ERRNO	(Read-only) Type: Integer Not saved Initial value: 0	Displays the number of the appropriate error code when an AutoLISP function call causes an error that AutoCAD detects. AutoLISP applications can inspect the current value of *ERRNO* with (getvar "errno"). The *ERRNO* system variable is not always cleared to zero. Unless it is inspected immediately after an AutoLISP function has reported an error, the error that its value indicates may be misleading. This variable is always cleared when starting or opening a drawing. See the *AutoLISP Developer's Guide* for more information.
EXPERT	Type: Integer Not saved Initial value: 0	Controls whether certain prompts are issued. 0 Issues all prompts normally 1 Suppresses "About to regen, proceed?" and "Really want to turn the current layer off?" 2 Suppresses the preceding prompts and "Block already defined. Redefine it?" (*Block*) and "A drawing with this name already exists. Overwrite it?" (*Save* or *Wblock*).

2006

2006

Variable	Characteristic	Description
EXPERT (continued)		3 Suppresses the preceding prompts and those issued by *Linetype* if you try to load a linetype that's already loaded or create a new linetype in a file that already defines it.
		4 Suppresses the preceding prompts and those issued by *Ucs Save* and *Vports Save* if the name you supply already exists.
		5 Suppresses the preceding prompts and those issued by the *Dimstyle Save* option and *Dimoverride* if the dimension style name you supply already exists (the entries are redefined).
		When a prompt is suppressed by *EXPERT*, the operation in question is performed as though you entered *y* at the prompt. The setting of *EXPERT* can affect scripts, menu macros, AutoLISP, and the command functions.
EXPLMODE	Type: Integer Not saved Initial value: 1	Controls whether *Explode* supports non-uniformly scaled (NUS) blocks. 0 Does not explode NUS blocks 1 Explodes NUS blocks
EXTMAX	(Read-only) Type: 3D Point Saved in: Drawing	Stores the upper-right point of the drawing extents. Expands outward as new objects are drawn, shrinks only with *Zoom All* or *Zoom Extents*. Reported in World coordinates for the current space.
EXTMIN	(Read-only) Type: 3D Point Saved in: Drawing	Stores the lower-left point of the drawing extents. Expands outward as new objects are drawn, shrinks only with *Zoom All* or *Zoom Extents*. Reported in World coordinates for the current space.
EXTNAMES	Type: Integer Saved in: Drawing Initial value: 1	Sets the parameters for non-graphical object names (such as linetypes and layers) stored in symbol tables. 0 Uses Release 14 parameters, which limit names to 31 characters in length. Names may include the letters A-Z, the numerals 0-9, and the special characters dollar sign ($), underscore (_), and hyphen (-). 1 Uses AutoCAD 2000 (and later releases) parameters. Names can be up to 255 characters in length, and may include the letters A-Z, the numerals 0-9, spaces, and any special characters not used by Microsoft Windows and AutoCAD for other purposes.
FACETRATIO	Type: Integer Not saved Initial Value: 0	Controls the aspect ratio of faceting for cylindrical and conic ACIS solids. A setting of 1 increases the density of the mesh to improve the quality of rendered and shaded models. 0 Creates an N by 1 mesh for cylindrical and conic ACIS solids

Variable	Characteristic	Description
FACETRATIO (continued)		1 Creates an N by M mesh for cylindrical and conic ACIS solids
FACETRES	Type: Real Saved in: Drawing Initial value: 0.5	Further adjusts the smoothness of shaded and rendered objects and objects with hidden lines removed. Valid values are from 0.01 to 10.0.
FIELDDISPLAY	Type: Integer Saved in: Registry Initial value: 1	Controls whether fields are displayed with a gray background. The background is not plotted. 0 Fields are displayed with no background 1 Fields are displayed with a gray background
FIELDEVAL	Type: Integer Saved in: Drawing Initial value: 31	Controls how fields are updated. The setting is stored as a bitcode using the sum of the following values: 0 Not updated 1 Updated on open 2 Updated on save 4 Updated on plot 8 Updated on use of *Etransmit* 16 Updated on regeneration
FILEDIA	Type: Integer Saved in: Registry Initial value: 1	Suppresses display of the file dialog boxes. 0 Dialog boxes are not displayed. You can still request a file dialog box to appear by entering a tilde (~) in response to the command's prompt. The same is true for AutoLISP and ADS functions. 1 File dialog boxes are displayed. However, if a script or AutoLISP/ObjectARX program is active, AutoCAD displays an ordinary prompt.
FILLETRAD	Type: Real Saved in: Drawing Initial value: 0.0000	Stores the current fillet radius.
FILLMODE	Type: Integer Saved in: Drawing Initial value: 1	Specifies whether multilines, traces, solids, solid-fill hatches, and wide polylines are filled in. 0 Objects are not filled in 1 Objects are filled
FONTALT	Type: String Saved in: Registry Initial value: "simple.shx"	Specifies the alternate font to be used when the specified font file cannot be located. If an alternate font is not specified, AutoCAD displays the Alternate Font dialog box for the following cases: 1. A Release 13 drawing is opened; *FONTALT* is not set or not found; and a TrueType, SHX, or PostScript font is not found for a defined text style. 2. A Release 14 drawing is opened, *FONTALT* is not set or not found, and a SHX or PostScript font is not found for a defined text style. For missing TrueType fonts in Release 14 drawings, AutoCAD automatically substitutes the closest TrueType font available.

Variable	Characteristic	Description
FONTALT (continued)		3. The *Browse* button is pressed in the *Options* dialog box when you specify an alternate font. AutoCAD validates the alternate font specified for *FONTALT*. If the font name or font file name is not found, the message "Font not found" is displayed. Enter either a TrueType font name (for example, Times New Roman Bold) or a TrueType file name (for example *timebd.ttf*). When a TrueType file name is entered for *FONTALT*, AutoCAD returns the font name in place of the file name) if the font is registered with the operating system.
FONTMAP	Type: String Saved in: Registry Initial value: "acad.fmp"	Specifies the font mapping file to be used. A font mapping file contains one font mapping per line; the original font used in the drawing and the font to be substituted for it are separated by a semicolon (;). For example, to substitute the Times TrueType font for the Romans font, the line in the mapping file would read as follows: **romanc.shx;times.ttf** If *FONTMAP* does not point to a font mapping file, if the FMP file is not found, or if the font file name specified in the FMP file is not found, AutoCAD uses the font defined in the style. If the font in the style is not found, AutoCAD substitutes the font according to substitution rules.
FRONTZ	(Read-only) Type: Real Saved in: Drawing	Stores the front clipping plane offset from the target plane for the current viewport, in drawing units. Meaningful only if the front clipping bit in *VIEWMODE* is on and the front-clip-not-at-eye bit is also on. The distance of the front clipping plane from the camera point is found by subtracting *FRONTZ* from the camera-to-target distance.
FULLOPEN	(Read-only) Type: Integer Not saved	Indicates whether the current drawing is partially open. 0 Indicates a partially open drawing 1 Indicates a fully open drawing
FULLPLOTPATH	Type: Integer Saved in: Registry Initial value: 1	Controls whether the full path of the drawing file is sent to the plot spooler. 0 Sends the drawing file name only 1 Sends the full path of the drawing file
GFANG	Type: Integer Not saved Initial value: 0	Specifies the angle of a gradient fill. Valid values are 0 through 360 degrees.
GFCLR1	Type: String Not saved Initial value: "RGB 000, 000, 255"	Specifies the color for a one-color gradient fill or the first color for a two-color gradient fill. Valid values are "RGB 000, 000, 000" through "RGB 255, 255, 255."

Variable	Characteristic	Description
GFCLR2	Type: String Not saved Initial value: "RGB 255, 255, 153"	Specifies the second color for a two-color gradient fill. Valid values are "RGB 000, 000, 000" through "RGB 255, 255, 255."
GFCLRLUM	Type: Real Not saved Initial value: 1.0000	Makes the color a tint (mixed with white) or a shade (mixed with black) in a one-color gradient fill. Valid values are 0.0 (darkest) to 1.0 (lightest).
GFCLRSTATE	Type: Integer Not saved Initial value: 1	Specifies whether a gradient fill uses one color or two colors. 0 Two-color gradient fill 1 One-color gradient fill
GFNAME	Type: Integer Not saved Initial value: 1	Specifies the pattern of a gradient fill. Valid values are 1 through 9. 1 Linear 2 Cylindrical 3 Inverted cylindrical 4 Spherical 5 Inverted spherical 6 Hemispherical 7 Inverted hemispherical 8 Curved 9 Inverted curved
GFSHIFT	Type: Integer Not saved Initial value: 0	Specifies whether the pattern in a gradient fill is centered or is shifted up and to the left. 0 Centered 1 Shifted up and to the left
GRIDDISPLAY	Type: Bitcode Saved in: Drawing Initial value: 3	Controls the display behavior and display limits of the grid. The setting is stored as a bitcode using the sum of the following values: 0 Restricts the grid to the area specified by the *Limits* command 1 Does not restrict the grid to the area specified by the *Limits* command 2 Turns on adaptive grid display, which limits the density of the grid when zoomed out 4 If the grid is set to adaptive display and when zoomed in, generates additional, more closely spaced grid lines in the same proportion as the intervals of the major grid lines 8 Changes the grid plane to follow the XY plane of the dynamic UCS. NOTE: Setting 4 is ignored unless setting 2 is specified.
GRIDMAJOR	Type: Integer Saved in: Drawing Initial value: 5	Controls the frequency of major grid lines compared to minor grid lines. Valid values range from 1 to 100. Grid lines are displayed in any visual style except *2D Wireframe*.

Variable	Characteristic	Description
GRIDMODE	Type: Integer Saved in: Drawing Initial value: 0	Specifies whether the grid is turned on or off. 0 Turns the grid off 1 Turns the grid on
GRIDUNIT	Type: 2D point Saved in: Drawing Initial value: 0.5000, 0.5000	Specifies the grid spacing (*X* and *Y*) for the current viewport.
GRIPBLOCK	Type: Integer Saved in: Registry Initial value: 0	Controls the assignment of grips in blocks. 0 Assigns a grip only to the insertion point of the block 1 Assigns grips to objects within the block
GRIPCOLOR	Type: Integer Saved in: Registry Initial value: 5	Controls the color of nonselected grips (drawn as box outlines). The valid range is 1-255.
GRIPDYNCOLOR	Type: Integer Saved in: Registry Initial value: 140	Controls the color of custom grips for dynamic blocks. The valid range is 1 to 255.
GRIPHOT	Type: Integer Saved in: Registry Initial value: 1	Controls the color of selected grips (drawn as filled boxes). The valid range is 1-255.
GRIPHOVER	Type: Integer Saved in: Registry Initial value: 3	Controls the fill color of a grip when the cursor pauses over the grip. The valid range is 1 to 255.
GRIPOBJLIMIT	Type: Integer Saved in: Registry Initial value: 100	Suppresses the display of grips when the initial selection set includes more than the specified number of objects. The valid range is 1 to 32,767. When set to 1, grips are suppressed when more than one object is selected.
GRIPS	Type: Integer Saved in: Registry Initial value: 1	Controls use of selection set grips for the Stretch, Move, Rotate, Scale, and Mirror grip modes. 0 Turns off grips 1 Turns on grips
GRIPSIZE	Type: Integer Saved in: Registry Initial value: 3	Sets the size of the box drawn to display the grip, in pixels. The valid range is 1-255.
GRIPTIPS	Type: Integer Saved in: Registry Initial value: 1	Controls the display of grip tips when the cursor hovers over grips on custom objects that support grip tips. 0 Turns off the display of grip tips 1 Turns on the display of grip tips
GTAUTO	Type: Integer Saved in: Registry Initial value: 1	Controls whether or not grip tools display automatically when selecting objects before starting a command in a viewport set to a 3D visual style. 0 Grip tools do not display automatically when selecting objects before starting a command. 1 Grip tools display automatically after creating a selection set before starting a command.

2006

2007

Variable	Characteristic	Description
GTDEFAULT	Type: Integer Saved in: Registry Initial value: 0	Controls whether or not the *3Dmove* and *3Drotate* commands start automatically when the *Move* and *Rotate* commands (respectively) are started in a 3D view. 0 *3Dmove* and *3Drotate* commands do not start automatically when the *Move* and *Rotate* commands (respectively) are started in a 3D view. 1 *3Dmove* and *3Drotate* commands start automatically when the *Move* and *Rotate* commands (respectively) are started in a 3D view.
GTLOCATION	Type: Integer Saved in: Registry Initial value: 0	Controls the initial location of grip tools when objects are selected prior to running the *3dmove* or *3Drotate* commands. 0 Places the grip tool at the same location as the UCS icon. Also aligns the grip tool with the UCS icon. 1 Places the grip tool on and aligned with the last selected object or subobject.
HALOGAP	Type: Integer Saved in: Drawing Initial value: 0	Specifies the distance to shorten a haloed line. The distance is specified as a percentage of one inch and is independent of *Zoom* level. A haloed line is shortened at the point where it will be hidden and is visible only when the *Hide* or *Hidden Shademode* is used.
HANDLES	(Read-only) Type: Integer Saved in: Drawing Initial value: On	Reports whether object handles can be accessed by applications. Because handles can no longer be turned off, has no effect except to preserve the integrity of scripts.
HIDEPRECISION	Type: Integer Not saved Initial Value: 0	Controls the accuracy of hides and shades. Hides can be calculated in double precision or single precision. Setting *HIDEPRECISION* to 1 produces more accurate hides because it uses double precision, but it also uses more memory and can affect performance, especially when hiding solids. 0 Single precision; uses less memory 1 Double precision; uses more memory
HIDETEXT	Type: Switch Saved in: Drawing Initial Value: 0	Specifies whether text objects created by the *Text*, *Dtext*, or *Mtext* commands are processed during a Hide. 0 Disables Hide processing of text objects. Text is not hidden and does not hide other objects unless the text has a thickness assigned to it. 1 Enables Hide processing of text objects. Text is hidden and text hides other objects. When dealing with text objects, legacy behavior is achieved by setting *HIDETEXT* to 0.
HIGHLIGHT	Type: Integer Not saved Initial value: 1	Controls object highlighting; does not affect objects selected with grips. 0 Turns off object selection highlighting 1 Turns on object selection highlighting

2007

Variable	Characteristic	Description
HPANG	Type: Real Not saved Initial value: 0	Specifies the hatch pattern angle.
HPASSOC	Type: Integer Saved in: Registry Initial value: 1	Controls whether hatch patterns and gradient fills are associative. 0 Hatch patterns and gradient fills are not associated with their boundaries 1 Hatch patterns and gradient fills are associated with their boundaries and are updated when the boundaries change
HPBOUND	Type: Integer Saved in: Registry Initial value: 1	Controls the object type created by *Bhatch* and *Boundary*. 0 Creates a region 1 Creates a polyline
HPDOUBLE	Type: Integer Not saved Initial value: 0	Specifies hatch pattern doubling for user-defined patterns. 0 Turns off hatch pattern doubling 1 Turns on hatch pattern doubling
HPDRAWORDER	Type: Integer Saved in: Not saved Initial value: 3	Controls the draw order of hatches and fills. Stores the *Draw Order* setting from the *Hatch* and *Fill* dialog box. 0 None. The hatch or fill is not assigned a draw order. 1 Send to back. The hatch or fill is sent to the back of all other objects. 2 Bring to front. The hatch or fill is brought to the front of all other objects. 3 Send behind boundary. The hatch or fill is sent behind the hatch boundary. 4 Bring in front of boundary. The hatch or fill is brought in front of the hatch boundary.
HPGAPTOL	Type: Real Saved in: Registry Initial value: 0	Treats a set of objects that almost enclose an area as a closed hatch boundary. The default value, 0, specifies that the objects enclose the area with no gaps. Enter a value, in drawing units from 0 to 5000, to set the maximum size of gaps that can be ignored when the objects serve as a hatch boundary.
HPINHERIT	Type: Integer Saved in: Drawing Initial value: 0	Controls the hatch origin of the resulting hatch when using *Inherit Properties* in *Hatch* and *Hatchedit*. 0 The hatch origin is taken from *HPORIGIN* 1 The hatch origin is taken from the source hatch object
HPNAME	Type: String Not saved Initial value: "ANSI31"	Sets a default hatch pattern name of up to 34 characters, no spaces allowed. Returns "" if there is no default. Enter a period (.) to set no default.

Variable	Characteristic	Description
HPOBJWARNING	Type: Integer Saved in: Registry Initial value: 10000	Sets the number of hatch boundary objects that can be selected before displaying a warning message.
HPORIGIN	Type: 2D point Saved in: Drawing Initial value: 0,0	Sets the hatch origin point for new hatch objects relative to the current user coordinate system.
HPORIGINMODE	Type: Integer Saved in: Registry Initial value: 0	Controls how *Hatch* determines the default hatch origin point. 0 Hatch origins are set using *HPORIGIN* 1 Hatch origins are set using the bottom-left corner of the rectangular extents of the hatch boundaries 2 Hatch origins are set using the bottom-right corner of the rectangular extents of the hatch boundaries 3 Hatch origins are set using the top-right corner of the rectangular extents of the hatch boundaries 4 Hatch origins are set using the top-left corner of the rectangular extents of the hatch boundaries 5 Hatch origins are set using the center of the rectangular extents of the hatch boundaries
HPSCALE	Type: Real Not saved Initial value: 1.0000	Specifies the hatch pattern scale factor; must be nonzero.
HPSEPARATE	Type: Integer Saved in: Registry Initial value: 0	Controls whether *Hatch* creates a single hatch object or separate hatch objects when operating on several closed boundaries. 0 A single hatch object is created 1 Separate hatch objects are created
HPSPACE	Type: Real Not saved Initial value: 1.0000	Specifies the hatch pattern line spacing for user-defined simple patterns; must be nonzero.
HYPERLINKBASE	Type: String Saved in: Drawing Initial value: " "	Specifies the path used for all relative hyperlinks in the drawing. If no value is specified, the drawing path is used for all relative hyperlinks.
IMAGEHLT	Type: Integer Saved in: Registry Initial value: 0	Controls whether the entire raster image or only the raster image frame is highlighted. 0 Highlights only the raster image frame 1 Highlights the entire raster image
IMPLIEDFACE	Type: Integer Saved in: Registry Initial value: 1	Controls the detection of implied faces. This variable must be set to 1 if you want to select and modify implied faces. 0 Implied faces cannot be detected. 1 Implied faces can be detected.

2006

2006

2007

Variable	Characteristic	Description
INDEXCTL	Type: Integer Saved in: Drawing Initial value: 0	Controls whether layer and spatial indexes are created and saved in drawing files. 0 No indexes are created 1 Layer index is created 2 Spatial index is created 3 Layer and spatial indexes are created
INETLOCATION	Type: Real Saved in: Registry Initial value: "http://www.autodesk.com"	Stores the Internet location used by *Browser*.
INPUTHISTORYMODE	Type: Bit code Saved in: Registry Initial value: 15	Controls the content and location of the display of a history of user input. The setting is stored as a bit code using the sum of the following values: 0 No history of recent input is displayed 1 History of recent input is displayed at the command line or in a dynamic prompt tooltip with access through Up Arrow and Down Arrow keys 2 History of recent input for the current command is displayed in the shortcut menu 4 History of recent input for all commands in the current session is displayed in the shortcut menu 8 Markers for recent input of point locations are displayed in the drawing
INSBASE	Type: 3D point Saved in: Drawing Initial value: 0.0000, 0.0000,0.0000	Stores insertion basepoint set by *Base*, expressed as a UCS coordinate for the current space.
INSNAME	Type: String Not saved Initial value: ""	Sets a default block name for *Insert* or *-Insert*. The name must conform to symbol naming conventions. Returns "" if no default is set. Enter a period (.) to set no default.
INSUNITS	Type: Integer Saved in: Drawing Initial value: 0	When you drag a block or image from Design-Center, specifies a drawing units value as follows: 0 Unspecified - Unitless 1 Inches 2 Feet 3 Miles 4 Millimeters 5 Centimeters 6 Meters 7 Kilometers 8 Microinches 9 Miles 10 Yards 11 Angstroms

2006

Variable	Characteristic	Description
INSUNITS (*continued*)		12 Nanometers 13 Microns 14 Decimeters 15 Decameters 16 Hectometers 17 Gigameters 18 Astronomical Units 19 Light Years 20 Parsecs
INSUNITSDEFSOURCE	Type: Integer Saved in: Registry Initial value: 1	Sets source content units value. Valid range is 0 to 20. (See *INSUNITS*.)
INSUNITSDEFTARGET	Type: Integer Saved in: Registry Initial value: 1	Sets target drawing units value. Valid range is 0 to 20. (See *INSUNITS*.)
INTELLIGENTUPDATE	Type: Integer Saved in: Registry Initial value: 20	Controls the graphics refresh rate. The default value is 20 frames per second. If you encounter problems related to graphics generation or timing, turn off the variable by setting it to 0. *INTELLIGENTUPDATE* works by suppressing the graphics update until the timer expires. Subsequent updates reset the timer. The performance improvement significantly affects updates for scripts and AutoLISP graphics. Those using regular AutoCAD commands will not see a noticeable difference in performance.
INTERFERECOLOR	Type: Integer Saved in: Drawing Initial value: 1	Sets the color for interference objects. Valid values include *Bylayer* and an integer from 1 to 255. Valid values for True Colors are a string of integers each from 1 to 255 separated by commas and preceded by RGB. The True Color setting is entered as follows: RGB:000,000,000. If you have a color book installed, you can specify any colors that are defined in the book.
INTERFEREOBJVS	Type: String Saved in: Drawing Initial value: Realistic	Sets the visual style for interference objects. *INTERFEREOBJVS* can only be set to a visual style that is saved in the drawing. The visual style specified for *INTERFEREOBJVS* cannot be removed with the *Purge* command.
INTERFEREVPVS	Type: String Saved in: Drawing Initial value: Wireframe	Specifies the visual style for the viewport during interference checking. *INTERFEREVPVS* can only be set to a visual style that is saved in the drawing. The visual style specified for *INTERFEREVPVS* cannot be removed with the *Purge* command.

2006

2007

Variable	Characteristic	Description
LAYOUTREGENCTL (*continued*)		2 The drawing is regenerated the first time you switch to each tab. For the remainder of the drawing session, the display list is saved to memory and regenerations are suppressed when you switch to those tabs. The performance gain achieved by changing the *LAYOUTREGENCTL* setting is dependent on several factors, including the drawing size and type, the objects contained in the drawing, the amount of available memory, and the effect of other open drawings or applications. When *LAYOUTRE-GENCTL* is set to 1 or 2, the amount of additional memory used is the size of the *Model* tab's display list multiplied by the number of viewports in each layout for which the display list is saved. If *LAYOUTREGENCTL* is set to 1 or 2 and performance seems slow in general or when you switch between tabs for which the display list is saved, consider changing to a setting of 0 or 1 to find the optimal balance for your work environment. NOTE: Regardless of the *LAYOUTREGENCTL* setting, if you redefine a block or undo a tab switch, the drawing is regenerated the first time you switch to any tab that contains saved viewports.
LEGACYCTRLPICK	Type: Integer Saved in: Registry Initial value: 0	Specifies the keys for selection cycling and the behavior for CTRL + left-click. 0 CTRL + left-click is used to select subobjects (faces, edges, and vertices) on 3D solids. 1 CTRL + left-click is used to cycle through overlapping objects. Disallows using CTRL + left-click to select subobjects on 3D solids.
LENSLENGTH	(Read-only) Type: Real Saved in: Drawing Initial value: 50.0000	Stores the length of the lens (in millimeters) used in perspective viewing for the current viewport.
LIGHTGLYPHDISPLAY	Type: Integer Saved in: Drawing Initial value: 1	Controls whether light glyphs are displayed. When this system variable is off, the glyphs that represent lights in the drawing are not displayed. 0 Off 1 On
LIMCHECK	Type: Integer Saved in: Drawing Initial value: 0	Controls creation of objects outside the drawing limits. 0 Objects can be created outside the limits 1 Objects cannot be created outside the limits
LIMMAX	Type: 2D point Saved in: Drawing Initial value: 12.0000,9.0000	Stores the upper-right drawing limits for the current space, expressed as World coordinate. *LIMMAX* is read-only when paper space is active and the paper background or paper margins are displayed.

2007

2007

Variable	Characteristic	Description
LIMMIN	Type: 2D point Saved in: Drawing Initial value: 0.0000, 0.0000	Stores the lower-left drawing limits for the current space, expressed as a World coordinate. *LIMMIN* is read-only when paper space is active and the paper background or paper margins are displayed.
LISPINIT	Type: Integer Saved in: Registry Initial value: 1	Specifies whether AutoLISP-defined functions and variables are preserved when you open a new drawing or whether they are valid in the current drawing session only. 0 AutoLISP functions and variables are preserved from drawing to drawing 1 AutoLISP functions and variables are valid in current drawing only
LOCALE	(Read-only) Type: String Not saved Initial value: "enu"	Displays the ISO language code of the current AutoCAD version you are running.
LOCALROOTPREFIX	(Read-only) Type: String Saved in: Registry Initial value: "pathname"	Stores the full path to the root folder where local customizable files were installed. These files are stored in the product folder under the Local Settings folder; for example, "C:\Documents and Settings\username\ Local Settings\Application Data\productname\version\language". The Template and Textures folders are in this location, and you can add any customizable files that you do not want to roam on the network. See *ROAMABLEROOTPREFIX* for the location of the roamable files.
LOCKUI	Type: Bit code Saved in: Registry Initial value: 0	Locks the position and size of toolbars and windows such as DesignCenter and *Properties* palette. Locked toolbars and windows can still be opened and closed and items can be added and deleted. To unlock them temporarily, hold down Ctrl. A lock icon in the status bar tray indicates whether toolbars and windows are locked. Right-click the icon to display locking options. The setting is stored as a bit code using the sum of the following values: 0 Toolbars and windows not locked 1 Docked toolbars locked 2 Docked windows locked 4 Floating toolbars locked 8 Floating windows locked
LOFTANG1	Type: Real Saved in: Drawing Initial value: 90	Sets the draft angle through the first cross section in a loft operation. The 0 direction is measured outward from the curve on the plane of the curve. The positive direction is measured toward the next cross section. Valid values include 0 to less than 360.

2006

2007

Variable	Characteristic	Description
LOFTANG2	Type: Real Saved in: Drawing Initial value: 90	Sets the draft angle through the last cross section in a loft operation. The 0 direction is measured outward from the curve on the plane of the curve. The positive direction is measured toward the previous cross section. Valid values include 0 to less than 360.
LOFTMAG1	Type: Integer Saved in: Drawing Initial value: 1	Sets the magnitude of the draft angle through the first cross section in a loft operation. Controls how soon the surface starts bending back toward the next cross section. Valid values include integers from 1 to 10.
LOGTMAG2	Type: Integer Saved in: Drawing Initial value: 1	Sets the magnitude of the draft angle through the last cross section in a loft operation. Controls how soon the surface starts bending back toward the next cross section. Valid values include integers from 1 to 10.
LOFTNORMALS	Type: Integer Saved in: Drawing Initial value: 1	Controls the normals of a lofted object where it passes through cross sections. This setting is ignored when specifying a path or guide curves. These settings can also be specified in the *Loft Settings* dialog box. 0 Ruled 1 Smooth 2 First normal 3 Last normal 4 Ends normal 5 All normal 6 Use draft angle and magnitude
LOFTPARAM	Type: Bitcode Saved in: Drawing Initial value: 7	Controls the shape of lofted solids and surfaces. 1 No twist (minimizes the twist between cross sections) 2 Align direction (aligns the start to end direction of each cross section curve) 4 Simplify (produces simple solids and surfaces, such as a cylinder or plane, instead of spline solids and surfaces) 8 Close (closes the surface or solid between the first and the last cross sections)
LOGFILEMODE	Type: Integer Saved in: Registry Initial value: 0	Specifies whether the contents of the text window are written to a log file. 0 Log file is not maintained 1 Log file is maintained
LOGFILENAME	(Read-only) Type: String Saved in: Drawing:	Specifies the path and name of the log file for the current drawing. The initial value varies depending on the name of the current drawing and where you installed AutoCAD.

2007

Variable	Characteristic	Description
LOGFILEPATH	Type: String Saved in: Registry	Specifies the path for the log files for all drawings in a session. You can also specify the path by using the *Options* command. The initial value varies depending on where you installed AutoCAD.
LOGINNAME	(Read-only) Type: String Not saved	Displays the user's name as configured or as input when AutoCAD is loaded. The maximum length for a login name is 30 characters.
LONGITUDE	Type: Real Saved in: Drawing Initial value: Varies	Specifies the longitude of the drawing model in decimal format. The default is the longitude of San Francisco, California. The valid range is -180 to +180. Positive values represent west longitudes. This value is not affected by the settings of the *AUNITS* and *AUPREC* system variables
LTSCALE	Type: Real Saved in: Drawing Initial value: 1.0000	Sets the global linetype scale factor (cannot equal zero).
LUNITS	Type: Integer Saved in: Drawing Initial value: 2	Sets linear units. 1 Scientific 2 Decimal 3 Engineering 4 Architectural 5 Fractional
LUPREC	Type: Integer Saved in: Drawing Initial value: 4	Sets the number of decimal places displayed for all read-only linear units and for all editable linear units whose precision is less than or equal to the current *LUPREC* value. For editable linear units whose precision is greater than the current *LUPREC* value, the true precision is displayed. *LUPREC* does not affect the display precision of dimension text.
LWDEFAULT	Type: Enum Saved in: Registry Initial value: 25	Sets the value for the default lineweight. The default lineweight can be set to any valid lineweight value in millimeters, including: 0, 5, 9, 13, 15, 18, 20, 25, 30, 35, 40, 50, 53, 60, 70, 80, 90, 100, 106, 120, 140, 158, 200, and 211. All values must be entered in millimeters. (Multiply a value by 2.54 to convert values from inches to millimeters.)
LWDISPLAY	Type: Integer Saved in: Drawing Initial value: 1	Controls whether the lineweight is displayed on the *Model* or *Layout* tab. The setting is saved with each tab in the drawing. 0 Lineweight is not displayed 1 Lineweight is displayed
LWUNITS	Type: Integer Saved In: Registry Initial value: 1	Controls whether lineweight units are displayed in inches or millimeters. 0 Inches 1 Millimeters

2007

Variable	Characteristic	Description
MAXACTVP	Type: Integer Saved in: Drawing Initial value: 64	Sets the maximum number of viewports that can be active at one time.
MAXSORT	Type: Integer Saved in: Registry Initial value: 1000	Sets the maximum number of symbol names or block names sorted by listing commands. If the total number of items exceeds this value, no items are sorted.
MBUTTONPAN	Type: Integer Saved in: Registry Initial Value: 1	Controls the behavior of the third button or wheel on pointing device. 0 The third button or wheel action on a pointing device supports the action defined in the AutoCAD menu (.mnu) file 1 The third button or wheel action on a pointing device supports panning by holding and dragging the button or wheel
MEASUREINIT	Type: Integer Saved in: Registry Initial value: varies by country	Sets the initial drawing units as English or metric. Specifically, *MEASUREINIT* controls which hatch pattern and linetype files an existing drawing uses when it is opened. Also controls which template is used. 0 English; uses the hatch pattern file and linetype file designated by the ANSIHatch and ANSILinetype registry settings 1 Metric; uses the hatch pattern file and linetype file designated by the ISOHatch and ISOLinetype registry settings
MEASUREMENT	Type: Integer Saved in: Drawing Initial value: 0	Sets drawing units as English or metric. Specifically, *MEASUREMENT* controls which hatch pattern and linetype file an existing drawing uses when it is opened. 0 English; AutoCAD uses the hatch pattern file and linetype file designated by the ANSIHatch and ANSILinetype registry settings 1 Metric; AutoCAD uses the hatch pattern file and linetype file designated by the ISOHatch and ISOLinetype registry settings The drawing units for new drawings are controlled by the *MEASUREINIT* registry variable (*MEASUREINIT* uses the same values as *MEASUREMENT*). The *MEASUREMENT* setting of a drawing always overrides the *MEASUREINIT* registry setting.
MENUCTL	Type: Integer Saved in: Registry Initial value: 1	Controls the page switching of the screen menu. 0 Screen menu does not switch pages in response to keyboard command entry 1 Screen menu does switch pages in response to keyboard command entry

Variable	Characteristic	Description
MENUECHO	Type: Integer Not saved Initial value: 0	Sets menu echo and prompt control bits. It is the sum of the following: 1 Suppresses echo of menu items (^P in a menu item toggles echoing) 2 Suppresses display of system prompts during menu 4 Disables ^P toggle of menu echoing 8 Displays input/output strings; debugging aid for DIESEL macros
MENUNAME	(Read-only) Type: String Saved in: Registry	Stores the menu file name, including the path for the file name.
MIRRTEXT	Type: Integer Saved in: Drawing Initial value: 0	Controls how *Mirror* reflects text. 0 Retains text direction 1 Mirrors the text
MODEMACRO	Type: String Not saved Initial value: ""	Displays a text string on the Status line, such as the name of the current drawing, time/date stamp, or special modes. Use *MODEMACRO* to display a string of text, or use special text strings written in the DIESEL macro language to have AutoCAD evaluate the macro from time to time and base the Status line on user-selected conditions.
MSOLECALE	Type: Real Saved in: Drawing Initial value: 1.0	Controls the size of an OLE object with text that is pasted into model space. *MSOLESCALE* controls only the initial size. If the scale factor value is changed, existing OLE objects in the drawing are not affected. Positive number scales by value. Zero (0) scales by *DIMSCALE* value.
MTEXTED	Type: String Saved in: Registry Initial value: "Internal"	Sets the name of the application to use for editing Multiline text objects. You can specify a different text editor for the *Mtext* and *Ddedit* commands. If you set *MTEXTED* to internal or to null, AutoCAD displays the internal *Multiline Text Editor*. You set *MTEXTED* to null by entering a period (.). If you specify a path and the name of the executable file for another text editor or word processor, AutoCAD displays that path and file name instead. If the Multiline text object is fewer than 80 characters, you can specify :lisped to use the LISP editor. Text editors other than the internal one show the formatting codes in paragraph text.
MTEXTFIXED	Type: Integer Saved in: Registry Initial value: 0	Controls the appearance of the Multiline Text Editor. 0 Displays both the Multiline Text Editor and the text within it at the size and position of the *Mtext* object in the drawing. Text too large or too small to be edited is displayed at a minimum or maximum size.

Variable	Characteristic	Description
MTEXTFIXED (continued)		1 Displays the Multiline Text Editor at a fixed position and size based on last use, and displays text in the editor at a fixed height.
MTJIGSTRING	Type: String Saved in: Registry Initial value: "abc"	Sets the content of the sample text displayed at the cursor location when the *Mtext* command is started. The text string is displayed in the current text size and font. You can enter any string of up to ten letters or numbers or enter "." (period) to display no sample text.
MYDOCUMENTSPREFIX	(Read-only) Type: String Saved in: Registry Initial value: "pathname"	Stores the full path of the My Documents folder for the user currently logged on. These files are stored in the product folder under the Local Settings folder; for example, "C:\Documents and Settings\username\My Documents".
NOMUTT	Type: Short Not Saved Initial Value: 0	Suppresses message display (muttering) when it would not normally be suppressed. Displaying messages is the normal mode of AutoCAD, but message display is suppressed during scripts, AutoLISP routines, and so on. 0 Resumes normal muttering behavior 1 Suppresses muttering indefinitely
NORTHDIRECTION	Type: Real Saved in: Drawing Initial value: Varies	Specifies the angle of the sun from north. This value is affected by the settings of the *AUNITS* and *AUPREC* system variables. NOTE: The angle is interpreted in the context of the world coordinate system (WCS). This value is completely separate from surveyor angular units, which are always set relative to the current UCS.
OBSCUREDCOLOR	Type: Integer Saved in: Drawing Initial value: 257	Specifies the color of obscured lines. Value 0 designates *Byblock*, value 256 designates *Bylayer*, and value 257 designates *Byentity*. Values 1-255 designate an AutoCAD color index (ACI). An obscured line is a hidden line made visible by changing its color and linetype and is visible only when the *Hide* or *Shademode* commands are used. The *OBSCUREDCOLOR* setting is visible only if the *OBSCUREDLTYPE* is turned *ON* by setting it to a value other than 0.
OBSCUREDLTYPE	Type: Integer Saved in: Drawing Initial value: 0	Specifies the linetype of obscured lines. An obscured line is a hidden line made visible by changing its color and linetype and is visible only when the *Hide* or *Shademode* commands are used. Obscured linetypes are independent of zoom level, unlike regular AutoCAD linetypes. Value 0 will turn off display of obscured lines and is the default. The linetype values are defined as follows: 0 Off 1 Solid 2 Dashed

2007

Variable	Characteristic	Description
OBSCUREDLTYPE (continued)		3 Dotted 4 Short Dash 5 Medium Dash 6 Long Dash 7 Double Short Dash 8 Double Medium Dash 9 Double Long Dash 10 Medium Long Dash 11 Sparse Dot
OFFSETDIST	Type: Real Not saved Initial value: 1.0000	Sets the default offset distance. <0 Offsets an object through a specified point >0 Sets the default offset distance
OFFSETGAPTYPE	Type: Integer Saved in: Registry Initial value: 0	Controls how to offset polylines when a gap is created as a result of offsetting the individual polyline segments. 0 Extends the segments to fill the gap 1 Fills the gaps with a filleted arc segment (the radius of the arc segment is equal to the offset distance) 2 Fills the gaps with a chamfered line segment
OLEFRAME	Type: Integer Saved in: Drawing Initial value: 2	Controls whether a frame is displayed and plotted on all OLE objects in the drawing. The frame on an OLE object must be displayed in order for grips to be visible. 0 Frame is not displayed and not plotted 1 Frame is displayed and is plotted 2 Frame is displayed but is not plotted
OLEHIDE	Type: Integer Saved in: Registry Initial value: 0	Controls the display of OLE objects in AutoCAD. 0 All OLE objects are visible 1 OLE objects are visible in paper space only 2 OLE objects are visible in model space only 3 No OLE objects are visible OLEHIDE affects both screen display and printing.
OLEQUALITY	Type: Integer Saved in: Registry Initial value: 1	Controls the default quality level for embedded OLE objects. 0 Line art quality, such as an embedded spreadsheet 1 Text quality, such as an embedded Word document 2 Graphics quality, such as an embedded pie chart 3 Photograph quality 4 High quality photograph
OLESTARTUP	Type: Integer Saved in: Drawing Initial value: 0	Controls whether the source application of an embedded OLE object loads when plotting. Loading the OLE source application may improve the plot quality. 0 Does not load the OLE source application 1 Loads the OLE source application when plotting

Variable	Characteristic	Description
ORTHOMODE	Type: Integer Saved in: Drawing Initial value: 0	Constrains cursor movement to the perpendicular. When *ORTHOMODE* is turned on, the cursor can move only horizontally or vertically relative to the UCS and the current grid rotation angle. 0 Turns off Ortho mode 1 Turns on Ortho mode
OSMODE	Type: Integer Saved in: Drawing Initial value: 0	Sets running object snap modes using the following bit-codes. 0 *None* 1 *Endpoint* 2 *Midpoint* 4 *Center* 8 *Node* 16 *Quadrant* 32 *Intersection* 64 *Insertion* 128 *Perpendicular* 256 *Tangent* 512 *Nearest* 1024 *Quick* 2048 *Apparent Intersection* 4096 *Extension* 8192 *Parallel* To specify more than one object snap, enter the sum of their values. For example, entering 3 specifies the *Endpoint* (bit code 1) and *Midpoint* (bit code 2) object snaps. Entering **16383** specifies all object snaps.
OSNAPCOORD	Type: Integer Saved in: Registry Initial value: 2	Controls whether coordinates entered on the command line override running object snaps. 0 Running object snap settings override keyboard coordinate entry 1 Keyboard entry overrides object snap settings 2 Keyboard entry overrides object snap settings except in scripts
OSNAPHATCH	Type: Integer Saved in: Registry Initial value: 0	Controls whether object snaps ignore hatch objects. The default setting, 0, improves performance. 0 Object snaps ignore hatch objects 1 Object snaps treat hatch objects the same as other objects
OSNAPZ	Type: Integer Saved in: Registry Initial value: 0	Controls whether object snaps are automatically projected onto a plane parallel to the XY plane of the current UCS at the current elevation. 0 *Osnap* uses the Z-value of the specified point 1 *Osnap* substitutes the Z-value of the specified point with the *Elevation* set for the current UCS

Variable	Characteristic	Description
OSOPTIONS	Type: Bitcode Saved in: Registry Initial value: 3	Automatically suppresses object snaps on hatch objects and geometry with negative Z values when using a dynamic UCS. The setting is stored as a bitcode using the sum of the following values: 0 Object snaps operate on hatch objects, and on geometry with negative Z values when using a dynamic UCS 1 Object snaps ignore hatch objects 2 Object snaps ignore geometry with negative Z values during use of a dynamic UCS
PALETTEOPAQUE	Type: Integer Saved in: Registry Initial Value: 0	Controls whether windows can be made transparent. When transparency is unavailable or turned off, all palettes are opaque. Transparency is unavailable when palettes or windows are docked, when transparency is not supported by the current operating system, and when hardware accelerators are in use. When transparency is available and turned on, you can use the *Transparency* option on the shortcut menu to set a different degree of transparency in individual palettes. 0 Transparency turned on by user 1 Transparency turned off by user 2 Transparency unavailable though turned on by user 3 Transparency unavailable and turned off by user
PAPERUPDATE	Type: Integer Saved in: Registry Initial Value: 0	Controls the display of a warning dialog when attempting to print a layout with a paper size different from the paper size specified by the default for the plotter configuration file. 0 Displays a warning dialog box if the paper size specified in the layout is not supported by the plotter 1 Sets paper size to the configured paper size of the plotter configuration file
PDMODE	Type: Integer Saved in: Drawing Initial value: 0	Controls how point objects are displayed. For information about values to enter, see *Point*.
PDSIZE	Type: Real Saved in: Drawing Initial value: 0.0000	Sets the display size for point objects. 0 Creates a point at 5% of the graphics area height >0 Specifies an absolute size <0 Specifies a percentage of the viewport size
PEDITACCEPT	Type: Integer Saved in: Registry Initial value: 0	Suppresses display of the "Object selected is not a polyline. Do you want to turn it into one <Y>?" prompt in *Pedit*. Entering with Y converts the selected object to a *Polyline*. When the prompt is suppressed, the selected object is automatically converted to a *Polyline*.

Variable	Characteristic	Description
PEDITACCEPT (*continued*)		0 The prompt is displayed 1 The prompt is suppressed
PELLIPSE	Type: Integer Saved in: Drawing Initial value: 0	Controls the ellipse type created with *Ellipse*. 0 Creates a true ellipse object 1 Creates a polyline representation of an ellipse
PERIMETER	(Read-only) Type: Real Not saved	Stores the last perimeter value computed by *Area, List,* or *Dblist.*
PERSPECTIVE	Type: Integer Saved in: Drawing Initial value: Varies	Specifies whether the current viewport displays a perspective working view. *PERSPECTIVE* is reset to 0 when the drawing file or DXF file is saved to a file format earlier than AutoCAD 2007. This system variable is valid only in model space. 0 Perspective view turned off 1 Perspective view turned on NOTE: Perspective view is not available in 2D Wireframe display. If you open a file from an earlier release that saved a perspective view in 2D Wireframe, it is displayed as a parallel projection.
PFACEVMAX	(Read-only) Type: Integer Not saved	Sets the maximum number of vertices per face.
PICKADD	Type: Integer Saved in: Registry Initial value: 1	Controls additive selection of objects. 0 Turns off *PICKADD*. The objects most recently selected, either individually or by windowing, become the selection set. Previously selected objects are removed from the selection set. Add more objects to the selection set by holding down Shift while selecting. 1 Turns on *PICKADD*. Each object selected, either individually or by windowing, is added to the current selection set. To remove objects from the set, hold down Shift while selecting.
PICKAUTO	Type: Integer Saved in: Registry Initial value: 1	Controls automatic windowing at the Select Objects prompt. 0 Turns off *PICKAUTO* 1 Draws a selection window (for either Window or Crossing selection) automatically at the Select Objects prompt
PICKBOX	Type: Integer Saved in: Registry Initial value: 3	Sets object selection target height, in pixels.
PICKDRAG	Type: Integer Saved in: Registry Initial value: 0	Controls the method of drawing a selection window. 0 Draws the selection window using two points. Click the pointing device at one corner and then at the other corner.

2007

Variable	Characteristic	Description
PICKDRAG (continued)		1 Draws the selection window using dragging. Click at one corner, hold down the pick button on the pointing device, drag, and release the button at the other corner.
PICKFIRST	Type: Integer Saved in: Registry Initial value: 1	Controls whether you select objects before (noun-verb selection) or after you issue a command. 0 Turns off *PICKFIRST* 1 Turns on *PICKFIRST*
PICKSTYLE	Type: Integer Saved in: Drawing Initial value: 1	Controls use of group selection and associative hatch selection. 0 No group selection or associative hatch selection 1 Group selection 2 Associative hatch selection 3 Group selection and associative hatch selection
PLATFORM	(Read-only) Type: String Not saved	Indicates which platform (operating system) of AutoCAD is in use.
PLINEGEN	Type: Integer Saved in: Drawing Initial value: 0	Sets how linetype patterns are generated around the vertices of a two-dimensional *Polyline*. Does not apply to *Polylines* with tapered segments. 0 *Polylines* are generated to start and end with a dash at each vertex 1 Generates the linetype in a continuous pattern around the vertices of the *Polyline*
PLINETYPE	Type: Integer Saved in: Registry Initial value: 2	Specifies whether AutoCAD uses optimized 2D polylines. *PLINETYPE* controls both the creation of new polylines with the Pline command and the conversion of existing polylines in drawings from previous releases. 0 Polylines in older drawings are not converted on open; *Pline* creates old-format polylines 1 Polylines in older drawings are not converted on open; *Pline* creates optimized polylines 2 Polylines in older drawings are converted on open; *Pline* creates optimized polylines *PLINETYPE* also controls the polyline type created with the following commands: *Boundary* (when object type is set to *Polyline*), *Donut, Ellipse* (when *PELLIPSE* is set to 1), *Pedit* (when selecting a line or arc), *Polygon*, and *Sketch* (when *SKPOLY* is set to 1).
PLINEWID	Type: Real Saved in: Drawing Initial value: 0.0000	Stores the default polyline width.
PLOTOFFSET	Type: Integer Saved in: Registry Initial value: 0	Controls whether the plot offset is relative to the printable area or to the edge of the paper. 0 Sets the plot offset relative to the printable area 1 Sets the plot offset relative to the edge of the paper

Variable	Characteristic	Description
PLOTROTMODE	Type: Integer Saved in: Registry Initial value: 1	Controls the orientation of plots. 0 Rotates the effective plotting area so that the corner with the Rotation icon aligns with the paper at the lower-left for 0, top-left for 90, top-right for 180, and lower-right for 270 1 Aligns the lower-left corner of the effective plotting area with the lower-left corner of the paper 2 Works the same as 0 value except that the X and Y origin offsets are calculated relative to the rotated origin position
PLQUIET	Type: Integer Saved in: Registry Initial Value: 0	Controls the display of optional dialog boxes and non-fatal errors for batch plotting and scripts. 0 Displays plot dialogs and non fatal errors 1 Logs non-fatal errors and does not display plot related dialog boxes
POLARADDANG	Type: String Saved in: Registry Initial value: null	Contains user-defined polar angles. You can add up to 10 angles. Each angle can be up to 25 characters, separated with semicolons (;). AutoCAD displays angles in the format set in the *AUNITS* system variable.
POLARANG	Type: Real Saved in: Registry Initial value: 90	Sets the polar angle increment. Values are 90, 45, 30, 22.5, 18, 15,10, and 5.
POLARDIST	Type: Real Saved in: Registry Initial value: 0.0000	Sets the snap increment when the *SNAPSTYL* system variable is set to 1 (polar snap).
POLARMODE	Type: Integer Saved in: Registry Initial value: 0	Controls settings for polar and object snap tracking. The value is the sum of four bitcodes: Polar angle measurements 0 Measure polar angles based on current UCS (absolute) 1 Measure polar angles from selected objects (relative) Object snap tracking 0 Track orthogonally only 2 Use polar tracking settings in object snap tracking Use additional polar tracking angles 0 No 4 Yes Acquire object snap tracking points 0 Acquire automatically 8 Press SHIFT to acquire
POLYSIDES	Type: Integer Not saved Initial value: 4	Sets the default number of sides for *Polygon*. The range is 3-1024.

Variable	Characteristic	Description
POPUPS	(Read-only) Type: Integer Not saved	Displays the status of the currently configured display driver. 0 Does not support dialog boxes, the menu bar, pull-down menus, and icon menus 1 Supports these features
PREVIEWEFFECT	Type: Integer Saved in: Registry Initial value: 2	Specifies the visual effect used for previewing selection of objects. 0 Dashed lines (the default display for selected objects) 1 Thickened lines 2 Dashed and thickened lines
PREVIEWFILTER	Type: Bit code Saved in: Registry Initial value: 7	Excludes specified object types from selection previewing. The setting is stored as a bit code using the sum of the following values: 0 Excludes nothing 1 Excludes objects on locked layers 2 Excludes objects in Xrefs 4 Excludes tables 8 Excludes multiline text objects 16 Excludes hatch objects 32 Excludes objects in groups]
PRODUCT	(Read-only) Type: String Not saved Initial value: "AutoCAD"	Returns the product name.
PROGRAM	(Read-only) Type: String Not saved Initial value: "acad"	Returns the program name.
PROJECTNAME	Type: String Saved in: Drawing Initial value: ""	Assigns a project name to the current drawing. The project name points to a section in the registry which can contain one or more search paths for each project name defined. Used when an xref or image is not found in its original hard-coded path. Project names and their search directories are created through the Files tab of the Preferences dialog box. Project names make it easier for users to manage xrefs and images when drawings are exchanged between customers, or if users have different drive mappings to the same location on a server. If the xref or image is not found at the hard-coded path, the project paths associated with the project name are searched. The search order is hard-coded path, then project name search path, then AutoCAD search path.

2006

Variable	Characteristic	Description
PROJMODE	Type: Integer Saved in: Registry Initial value: 1	Sets the current Projection mode for trimming or extending. 0 True 3D mode (no projection) 1 Project to the *XY* plane of the current UCS 2 Project to the current view plane
PROXYGRAPHICS	Type: Integer Saved in: Drawing Initial value: 1	Specifies whether images of proxy objects are saved in the drawing. 0 Image is not saved with the drawing; a bounding box is displayed instead 1 Image is saved with the drawing
PROXYNOTICE	Type: Integer Saved in: Registry Initial value: 1	Displays a notice when a proxy is created. A proxy is created when you open a drawing containing custom objects created by an application that is not present. A proxy is also created when you issue a command that unloads a custom object's parent application. 0 No proxy warning is displayed 1 Proxy warning is displayed
PROXYSHOW	Type: Integer Saved in: Registry Initial value: 1	Controls the display of proxy objects in a drawing. 0 Proxy objects are not displayed 1 Graphic images are displayed for all proxy objects 2 Only the bounding box is displayed for all proxy objects
PROXYWEBSEARCH	Type: Integer Saved in: Registry Initial value: 1	Specifies how AutoCAD checks for Object Enablers. Object Enablers allow you to display and use custom objects in AutoCAD drawings even when the ObjectARX application that created them is unavailable. *PROXYWEBSEARCH* is also controlled with the *Live Enabler options* on the *System* tab of the *Options* dialog box. 0 Prevents AutoCAD from checking for Object Enablers 1 AutoCAD checks for Object Enablers only if a live Internet connection is present
PSLTSCALE	Type: Integer Saved in: Drawing Initial value: 1	Controls paper space linetype scaling. 0 No special linetype scaling. Linetype dash lengths are based on the drawing units of the space (model or paper) in which the objects were created, scaled by the global *LTSCALE* factor. 1 Viewport scaling governs linetype scaling. If *TILEMODE* is set to 0, dash lengths are based on paper space drawing units, even for objects in model space. In this mode, viewports can have varying magnifications, yet display linetypes identically. For a specific linetype, the dash lengths of a line in a viewport are the same as the dash lengths of a line in paper space. You can still control the dash lengths with *LTSCALE*.

Variable	Characteristic	Description
PSOLHEIGHT	Type: Real Saved in: Drawing Initial value: 4 (imperial); 80 (metric)	Controls the default height for a swept solid object created with the *Polysolid* command. The value reflects the last entered height value when using the *Polysolid* command. You cannot enter 0 as the value.
PSOLWIDTH	Type: Real Saved in: Drawing Initial value: 0.25 (imperial); 5 (metric)	Controls the default width for a swept solid object created with the *Polysolid* command. The value reflects the last entered width value when using the *Polysolid* command. You cannot enter 0 as the value.
PSTYLEMODE	Read Only Saved in: Drawing Initial value: 1	Indicates whether the current drawing is in a Color-Dependent or Named Plot Style mode. 0 Uses named plot style tables in the current drawing 1 Uses color-dependent plot style tables in the current drawing
PSTYLEPOLICY	Type: Integer Saved in: Registry Initial value: 1	Controls whether an object's color property is associated with its plot style. The new value you assign affects only newly created drawings and pre-AutoCAD 2000 drawings. 0 No association is made between color and plot style. The plot style for new objects is set to the default defined in *DEFPLSTYLE*. The plot style for new layers is set to the default defined in *DEFLPLSTYLE*. 1 An object's plot style is associated with its color.
PSVPSCALE	Type: Real Not Saved Initial Value: 0	Sets the view scale factor for all newly created viewports. The view scale factor is defined by comparing the ratio of units in paper space to the units in newly created model space viewports. The view scale factor you set is used with the *Vports* command. A value of 0 means the scale factor is Scaled to Fit. A scale must be a positive real value.
PUBLISHALLSHEETS	Type: Integer Saved in: Registry Initial value: 1	Controls how the *Publish* dialog box list is populated when the *Publish* command is issued. 0 Only the current document's contents (layouts and/or model space) are automatically loaded in the publish list. 1 The contents (layouts and/or model space) of all open AutoCAD documents are automatically loaded in the publish list.
PUCSBASE	Type: String Saved in: Drawing Initial value: ""	Stores the name of the UCS that defines the origin and orientation of orthographic UCS settings in paper space only.
QCSTATE	Type: Integer Saved in: Not saved Initial value: Varies	Determines whether the *QuickCalc* calculator is active or not. 0 Not active 1 Active

Variable	Characteristic	Description
QTEXTMODE	Type: Integer Saved in: Drawing Initial value: 0	Controls how text is displayed. 0 Turns off Quick Text mode; displays characters 1 Turns on Quick Text mode; displays a box in place of text
RASTERDPI	Type: Integer Saved in: Registry Initial value: 300	Controls paper size and plot scaling when changing from dimensional to dimensionless output devices, or vice versa. Converts millimeters or inches to pixels, or vice versa. Accepts an integer between 100 and 32,767 as a valid value.
RASTERPREVIEW	Type: Integer Saved in: Registry Initial value: 1	Controls whether BMP preview images are saved with the drawing. 0 No preview image is created 1 Preview image created
RECOVERYMODE	Type: Integer Saved in: Registry Initial value: 2	Controls whether drawing recovery information is recorded after a system failure. 0 Recovery information is not recorded, the *Drawing Recovery* window does not display automatically after a system failure, and any recovery information in the system registry is removed 1 Recovery information is recorded, but the *Drawing Recovery* window does not display automatically after a system failure 2 Recovery information is recorded, and the *Drawing Recovery* window displays automatically in the next session after a system failure
REFEDITNAME	(Read-only) Type: String Not Saved Initial value: ""	Indicates whether a drawing is in a reference-editing state; also, stores the reference file name.
REGENMODE	Type: Integer Saved in: Drawing Initial value: 1	Controls automatic regeneration of the drawing. 0 Turns off *REGENAUTO* 1 Turns on *REGENAUTO*
RE-INIT	Type: Integer Not saved Initial value: 0	Reinitializes the digitizer, digitizer port, and *acad.pgp* file using the following bit-codes: 1 Digitizer I/O port reinitialization 4 Digitizer reinitialization 16 PGP file reinitialization (reload) To specify more than one reinitialization, enter the sum of their values, for example, 5 to specify both digitizer port (1) and digitizer reinitialization (4).
REMEMBERFOLDERS	Type: Integer Saved in: Registry	Controls the default path for the *Look In* or *Save In* option in standard file selection dialog boxes.

2006

Variable	Characteristic	Description
REMEMBERFOLDERS (continued)	Initial value: 1	0 This setting restores the legacy behavior of AutoCAD 2000 and previous releases. When you start AutoCAD by double-clicking an AutoCAD icon, if a *Start In* path is specified for the icon, that path is used as the default for all standard file selection dialog boxes. 1 The last used paths in each particular standard file selection dialog box are remembered across and within sessions. The *Start In* folder specified for the AutoCAD icon is not used.
REPORTERROR	Type: Integer Saved in: Registry Initial value: 1	Controls whether an error report can be sent to Autodesk if AutoCAD closes unexpectedly. Error reports help Autodesk diagnose problems with the software. 0 The Error Report message is not displayed, and no report can be sent to Autodesk 1 The Error Report message is displayed, and an error report can be sent to Autodesk
ROAMABLEROOT-PREFIX	(Read-only) Type: String Saved in: Registry Initial value: "pathname"	Stores the full path to the root folder where roamable customizable files were installed. If you are working on a network that supports roaming, when you customize files that are in your roaming profile they are available to you regardless of which machine you are currently using. These files are stored in the product folder under the Application Data folder; for example, "C:\Documents and Settings\username\Application Data\product-name\version\language".
RTDISPLAY	Type: Integer Saved in: Registry Initial value: 1	Controls the display of raster images during real-time zoom or pan. 0 Displays raster image content 1 Displays raster image outline only *RTDISPLAY* is saved in the current profile.
SAVEFILE	(Read-only) Type: String Saved in: Registry Initial value: ""	Stores current auto-save file name.
SAVEFILEPATH	Type: String Saved in: Registry Initial value: "C:\TEMP\"	Specifies the path to the directory for all automatic save files for the AutoCAD session. You can also change the path on the *Files* tab in the *Options* dialog box.
SAVENAME	(Read-only) Type: String Not saved Initial value: ""	Stores the file name and directory path of the current drawing once you save it.

Variable	Characteristic	Description
SAVETIME	Type: Integer Saved in: Registry Initial value: 10	Sets the automatic save interval, in minutes. 0 Turns off automatic saving >0 Automatically saves the drawing at intervals specified by the nonzero integer The *SAVETIME* timer starts as soon as you make a change to a drawing. It is reset and restarted by a manual *Save, Saveas,* or *Qsave.*
SCREENBOXES	(Read-only) Type: Integer Not saved	Stores the number of boxes in the screen menu area of the graphics area. If the screen menu is turned off, *SCREENBOXES* is zero. On platforms that permit the AutoCAD graphics area to be resized or the screen menu to be reconfigured during an editing session, the value of this variable might change during the editing session.
SCREENMODE	(Read-only) Type: Integer Not saved	Stores a bit-code indicating the graphics/text state of the AutoCAD display. It is the sum of the following bit values: 0 Text screen is displayed 1 Graphics area is displayed 2 Dual-screen display is configured
SCREENSIZE	(Read-only) Type: 2D point Not saved	Stores current viewport size in pixels (*X* and *Y*).
SDI	Type: Integer Saved in: Registry Initial value: 0	Controls whether AutoCAD runs in single- or multiple-document interface. Helps third-party developers update applications to work smoothly with the AutoCAD multiple-drawing mode. 0 Turns on multiple-drawing interface. 1 Turns off multiple-drawing interface. 2 (Read-only) Multiple-drawing interface is disabled because AutoCAD has loaded an application that does not support multiple drawings. SDI setting 2 is not saved. 3 (Read-only) Multiple-drawing interface is disabled because the user has set SDI to 1 and AutoCAD has loaded an application that does not support multiple drawings. (SDI was set to 1 before the application was loaded.) SDI setting 3 is not saved. If SDI is set to 3, AutoCAD switches it back to 1 when the application that doesn't support multiple drawings is unloaded.
SELECTIONAREA	Type: Integer Saved in: Registry Initial value: 1	Controls the display of effects for selection areas. Selection areas are created by the *Window, Crossing, WPolygon,* and *CPolygon* options of *Select.* 0 Off 1 On

Variable	Characteristic	Description
SELECTIONAREA-OPACITY	Type: Integer Saved in: Registry Initial value: 25	Controls the transparency of the selection area during window and crossing selection. The valid range is 0 to 100. The lower the setting, the more transparent the area. A value of 100 makes the area opaque. The *SELECTIONAREA* system variable must be on.
SELECTIONPREVIEW	Type: Integer Saved in: Registry Initial value: 3	Controls the display of selection previewing. Objects are highlighted when the pickbox cursor rolls over them. This selection previewing indicates that the object would be selected if you clicked. The setting is stored as a bit code using the sum of the following values: 0 Off 1 On when no commands are active 2 On when a command prompts for object selection
SHADEDGE	Type: Integer Saved in: Drawing Initial value: 3	Controls shading of edges in rendering. 0 Faces shaded, edges not highlighted 1 Faces shaded, edges drawn in background color 2 Faces not filled, edges in object color 3 Faces in object color, edges in background color
SHADEDIF	Type: Integer Saved in: Drawing Initial value: 70	Sets the ratio of diffuse reflective light to ambient light (in percentage of diffuse reflective light).
SHADOWPLANE-LOCATION	Type: Integer Saved in: Drawing Initial value: 0	Controls the location of an invisible ground plane used to display shadows. The value is a location on the current Z axis. The ground plane is invisible, but it casts and receives shadows. Objects that are located below the ground plane are shadowed by it. The ground plane is used when the *VSSHADOWS* system variable is set to display either full shadows or ground shadows.
SHORTCUTMENU	Type: Integer Saved in: Registry Initial value: 11	Controls whether Default, Edit, and Command mode shortcut menus are available in the drawing area. *SHORTCUTMENU* uses the following bitcodes. 0 Disables all Default, Edit, and Command mode shortcut menus, restoring R14 legacy behavior 1 Enables Default mode shortcut menus 2 Enables Edit mode shortcut menus 4 Enables Command mode shortcut menus. In this case, the Command mode shortcut menu is available whenever a command is active 8 Enables Command mode shortcut menus only when command options are currently available from the command line

2006

2007

Variable	Characteristic	Description
SHORTCUTMENU (*continued*)		To enable more than one type of shortcut menu at once, enter the sum of their values. For example, entering 3 enables both Default (1) and Edit (2) mode shortcut menus.
SHOWLAYERUSAGE	Type: Integer Saved in: Registry Initial value: 0	Displays icons in the *Layer Properties Manager* to indicate whether layers are in use. Setting this system variable to Off improves performance in the *Layer Properties Manager*. 0 Off 1 On
SHPNAME	Type: String Not saved Initial value: ""	Sets a default shape name; must conform to symbol naming conventions. If no default is set, it returns ""; enter a period (.) to set no default.
SHOWHIST	Type: Integer Saved in: Drawing Initial value: 1	Controls the *Show History* property for solids in a drawing. 0 Sets the *Show History* property to *No* (read-only) for all solids. Overrides the individual *Show History* property settings for solids. You cannot view the original objects that were used to create the solid. 1 Does not override the individual *Show History* property settings for solids. 2 Displays the history of all solids by overriding the individual *Show History* property settings for solids. You can view the original objects that were used to create the solid.
SIGWARN	Type: Integer Saved in: Registry Initial value: 1	Controls whether a warning is presented when a file with an attached digital signature is opened. If *SIGWARN* is on and you open a file with a valid signature, the digital signature status is displayed. If *SIGWARN* is off and you open a file, the digital signature status is displayed only if a signature is invalid. You can set the variable using the *Display Digital Signature Information* option on the *Open and Save* tab of the *Options* dialog box. 0 Warning is not presented if a file has a valid signature 1 Warning is presented
SKETCHINC	Type: Real Saved in: Drawing Initial value: 0.1000	Sets the record increment for *Sketch*.
SKPOLY	Type: Integer Saved in: Registry Initial value: 0	Determines whether *Sketch* generates lines or polylines. 0 Generates *Lines* 1 Generates *Plines*

2006

2007

Variable	Characteristic	Description
SNAPANG	Type: Real Saved in: Drawing Initial value: 0	Sets the snap and grid rotation angle for the current viewport. The angle you specify is relative to the current UCS. NOTE: Changes to this variable are not reflected in the grid until the display is refreshed. AutoCAD does *not* automatically redraw when variables are changed.
SNAPBASE	Type: 2Dpoint Saved in: Drawing Initial value: 0.0000, 0.0000	Sets the snap and grid origin point for the current viewport relative to the current UCS. NOTE: Changes to this variable are not reflected in the grid until the display is refreshed. AutoCAD does *not* automatically redraw when variables are changed.
SNAPISOPAIR	Type: Integer Saved in: Drawing Initial value: 0	Controls the isometric plane for the current viewport. 0 Left 1 Top 2 Right
SNAPMODE	Type: Integer Saved in: Drawing Initial value: 0	Turns Snap mode on and off. 0 Snap off 1 Snap on for the current viewport
SNAPSTYL	Type: Integer Saved in: Drawing Initial value: 0	Sets snap style for the current viewport. 0 Standard 1 Isometric
SNAPTYPE	Type: Integer Saved in: Registry Initial Value: 0	Sets the snap style for the current viewport. 0 Grid, or standard snap. 1 Polar snap. Snaps along polar angle increments. Used with polar tracking.
SNAPUNIT	Type: 2D Saved in: Drawing Initial value: 0.5000, 0.5000	Sets the snap spacing for the current viewport. If the *SNAPSTYL* system variable is set to 1, AutoCAD automatically adjusts the X value of *SNAPUNIT* to accommodate the isometric snap. **NOTE:** Changes to this system variable are not reflected in the grid until the display is refreshed. AutoCAD does *not* automatically redraw when system variables are changed.
SOLIDCHECK	Type: Integer Saved in: Not Saved Initial value: 1	Turns the solid validation on and off for the current AutoCAD session. 0 Turns off solid validation 1 Turns on solid validation
SOLIDHIST	Type: Integer Saved in: Drawing Initial value: 1	Controls the default *History* property setting for new and existing objects. When set to 1, composite solids retain a "history" of the original objects contained in the composite. 0 Sets the *History* property to *None* for all solids. No history is retained. 1 Sets the *History* property to *Record* for all solids. All solids retain a history of their original objects.

2007

Variable	Characteristic	Description
SORTENTS	Type: Integer Saved in: Drawing Initial value: 127	Controls *Options* object sort order operations. *SORTENTS* uses the following bit codes: 0 Disables *SORTENTS* 1 Sorts for object selection 2 Sorts for object snap 4 Clears all checkboxes 8 Sorts for *Mslide* slide creation 16 Sorts for *Regens* 32 Sorts for plotting 64 Clears all checkboxes To select more than one, enter the sum of their codes. For example, enter **3** to specify sorting for both object selection and object snap.
SPLFRAME	Type: Integer Saved in: Drawing Initial value: 0	Controls display of splines and spline-fit polylines. 0 Does not display the control polygon for splines and spline-fit polylines. Displays the fit surface of a polygon mesh, not the defining mesh. Does not display the invisible edges of 3D faces orpolyface meshes. 1 Displays the control polygon for splines and spline-fit polylines. Only the defining mesh of a surface-fit polygon mesh is displayed (not the fit surface). Invisible edges of 3D faces or polyface meshes are displayed.
SPLINESEGS	Type: Integer Saved in: Drawing Initial value: 8	Sets the number of line segments to be generated for each spline-fit polyline generated by *Pedit* Spline.
SPLINETYPE	Type: Integer Saved in: Drawing Initial value: 6	Sets the type of curve generated by *Pedit* Spline. 5 Quadratic B-spline 6 Cubic B-spline
SSFOUND	Type: String Saved in: Not saved Initial value: "."	Displays the sheet set path and file name if a search for a sheet set is successful.
SSLOCATE	Type: Integer Saved in: User settings Initial value: 1	Locates and opens the sheet set associated with a drawing when the drawing is opened. 0 Does not open a drawing's sheet set with the drawing 1 Opens a drawing's sheet set with the drawing *SSMAUTOOPEN* and *SSLOCATE* must both be set to 1 to open a sheet set automatically in the *Sheet Set Manager*.
SSMAUTOOPEN	Type: Integer Saved in: User settings Initial value: 1	Displays the *Sheet Set Manager* when a drawing associated with a sheet is opened. 0 Does not open the *Sheet Set Manager* automatically 1 Opens the *Sheet Set Manager* automatically *SSMAUTOOPEN* and *SSLOCATE* must both be set to 1 to open a sheet set automatically in the *Sheet Set Manager*.

Variable	Characteristic	Description
SSMPOLLTIME	Type: Integer Saved in: Registry Initial value: 60	Controls the time interval between automatic refreshes of the status data in a sheet set. The *SSM-POLLTIME* timer sets the time in seconds between automatic refreshes of the status data of sheets in a sheet set. Valid values are 20-600. The *SSMSHEET-STATUS* system variable must be set to 2 for the timer to operate.
SSMSHEETSTATUS	Type: Integer Saved in: Registry Initial value: 2	Controls how the status data in a sheet set is refreshed. The status data for sheets in the current sheet set includes whether a sheet is locked and whether a sheet is missing (or found in an unexpected location). This status data can be updated automatically for all sheets. To refresh the sheet set manually, use the *Refresh Sheet Status* button on the *Sheet List tab* of the *Sheet Set Manager.* 0 Do not automatically refresh the status data in a sheet set 1 Refresh the status data when the sheet set is loaded or updated 2 In addition, refresh the status data at a time interval set by *SSPOLLTIME*
SSMSTATE	Type: Integer Saved in: Not saved Initial value: Varies	Determines whether the *Sheet Set Manager* window is active or not. 0 Not active 1 Active
STANDARDS-VIOLATION	Type: Integer Saved in: Registry Initial value: 2	Specifies whether a user is notified of standards violations that exist in the current drawing when a non-standard object is created or modified. 0 Notification is turned off 1 An alert is displayed when a standards violation occurs in the drawing 2 Displays an icon in the status bar when you open a file associated with a standards file and when you create or modify non-standard objects
STARTUP	Type: Integer Saved in: Registry Initial Value: 0	Controls whether the *Create New Drawing* dialog box is displayed when starting a new drawing with the *New* and *Qnew* commands. Also controls whether the *Startup* dialog box is displayed when the application is started. If the *FILEDIA* system variable is set to 0, no dialog boxes are displayed. 0 Displays the *Select Template* dialog box, or uses a default drawing template file set in the *Options* dialog box on the *Files* tab 1 Displays the *Startup* and the *Create New Drawing* dialog boxes

Variable	Characteristic	Description
STEPSIZE	Type: Real Saved in: Drawing Initial value: 6.0000	Specifies the size of each step when in walk or fly mode, in drawing units. You can enter any real number from 1E-6 to 1E+6.
STEPSPERSEC	Type: Real Saved in: Drawing Initial value: 2	Specifies the number of steps taken per second when you are in walk or fly mode. You can enter any real number from 1 to 30.
SUNSTATUS	Type: Integer Saved in: Drawing Initial value: 1	Controls whether the sun casts light in the current viewport. 0 Off 1 On
SURFTAB1	Type: Integer Saved in: Drawing Initial value: 6	Sets the number of tabulations to be generated for *Rulesurf* and *Tabsurf*. Also sets the mesh density in the *M* direction for *Revsurf* and *Edgesurf*.
SURFTAB2	Type: Integer Saved in: Drawing Initial value: 6	Sets the mesh density in the *N* direction for *Revsurf* and *Edgesurf*.
SURFTYPE	Type: Integer Saved in: Drawing Initial value: 6	Controls the type of surface fitting to be performed by *Pedit* Smooth. 5 Quadratic B-spline surface 6 Cubic B-spline surface 8 Bezier surface
SURFU	Type: Integer Saved in: Drawing Initial value: 6	Sets the surface density in the *M* direction.
SURFV	Type: Integer Saved in: Drawing Initial value: 6	Sets the surface density in the *N* direction.
SYSCODEPAGE	(Read-only) Type: String Not saved	Indicates the system code page, which is determined by the operating system. To change the code page, see Help in your operating system.
TABLEINDICATOR	Type: Integer Saved in: User settings Initial value: 1	Controls the display of row numbers and column letters when the In-Place Text Editor is open for editing a table cell. 0 Off 1 On
TABMODE	Type: Integer Not saved Initial value: 0	Controls use of the tablet. 0 Turns off Tablet mode 1 Turns on Tablet mode
TARGET	(Read-only) Type: 3D point Saved in: Drawing	Stores a location (as a UCS coordinate) of the target point for the current viewport.
TBCUSTOMIZE	Type: Switch Saved in: Registry Initial value: 1	Controls whether toolbars can be customized. 0 Disables *Customize* and *Toolbar* and customization from the toolbar shortcut menu 1 Enables toolbar customization

2007

2006

2006

Variable	Characteristic	Description
TDCREATE	(Read-only) Type: Real Saved in: Drawing	Stores the time and date the drawing was created.
TDINDWG	(Read-only) Type: Real Saved in: Drawing	Stores the total editing time.
TDUCREATE	(Read-only) Type: Real Saved in: Drawing	Stores the universal time and date the drawing was created.
TDUPDATE	(Read-only) Type: Real Saved in: Drawing	Stores the time and date of the last update/save.
TDUSRTIMER	(Read-only) Type: Real Saved in: Drawing	Stores the user-elapsed timer.
TDUUPDATE	(Read-only) Type: Real Saved in: Drawing	Stores the universal time and date of the last update/save.
TEMPOVERRIDES	Type: Integer Saved in: Registry Initial value: 1	Turns temporary override keys on and off. A temporary override key is a key that you can hold down to temporarily turn on or turn off one of the drawing aids that are set in the *Drafting Settings* dialog box. 0 Off 1 On
TEMPPREFIX	(Read-only) Type: String Not saved	Contains the directory name (if any) configured for placement of temporary files, with a path separator appended.
TEXTEVAL	Type: Integer Not saved Initial value: 0	Controls the method of evaluation of text strings. 0 All responses to prompts for text strings and attribute values are taken literally 1 Text starting with an opening parenthesis [(] or an exclamation mark (!) is evaluated as an AutoLISP expression, as for nontextual input **NOTE**: *Text* takes all input literally regard less of the setting of *TEXTEVAL*.
TEXTFILL	Type: Integer Saved in: Registry Initial value: 1	Controls the filling of TrueType fonts while plotting, exporting with *Psout*, and rendering. 0 Outputs text as outlines 1 Outputs text as filled images
TEXTQLTY	Type: Integer Saved in: Drawing Initial value: 50	Sets the resolution of TrueType fonts while plotting, exporting with *Psout*, and rendering. Values represent dots per inch. Lower values decrease resolution and increase plotting speed. Higher values increase resolution and decrease plotting speed.

2006

Variable	Characteristic	Description
TEXTSIZE	Type: Real Saved in: Drawing Initial value: 0.2000	Sets the default height for new text objects drawn with the current text style (has no effect if the style has a fixed height).
TEXTSTYLE	Type: String Saved in: Drawing Initial value: "STANDARD"	Sets the name of the current text style.
THICKNESS	Type: Real Saved in: Drawing Initial value: 0.0000	Sets the current 3D solid thickness.
TILEMODE	Type: Integer Saved in: drawing Initial value: 1	Makes the *Model* tab or the last layout tab current. 0 Makes the last active layout tab (paper space) active 1 Makes the *Model* tab active
TIMEZONE	Type: Enum Saved in: Drawing Initial value: -8000	Sets the time zone for the sun in the drawing. The values in the table are expressed as hours and minutes away from Greenwich Mean Time. The geographic location you set also sets the time zone. If the time zone is not accurate, you can correct it in the *Geographic Location* dialog box or set the *TIME ZONE* system variable. -12000 International Date Line West -11000 Midway Island, Samoa -10000 Hawaii -9000 Alaska -8000 Pacific Time (US & Canada), Tijuana -7000 Arizona -7000 Chihuahua, La Paz, Mazatlan -7000 Mountain Time (US & Canada) -6000 Central America -6000 Central Time (US & Canada) -6000 Guadalajara, Mexico City, Monterrey -6000 Saskatchewan -5000 Bogota, Lima, Quito -5000 Eastern Time (US & Canada) -5000 Indiana (East) -4000 Atlantic Time (Canada) -4000 Caracas, La Paz -4000 Santiago -3000 Newfoundland -3000 Brasilia -3000 Buenos Aires, Georgetown -3000 Greenland -2000 Mid-Atlantic -1000 Azores -1000 Cape Verde Is. 0000 Casablanca, Monrovia -0000 Dublin, Edinburgh, Lisbon +1000 Amsterdam, Berlin, Bern, Rome, Stockholm

Variable	Characteristic	Description	
TIMEZONE *(continued)*		+1000	Belgrade, Bratislava, Budapest, Ljubljana
		+1000	Brussels, Madrid, Copenhagen, Paris
		+1000	Sarajevo, Skopje, Warsaw, Zagreb
		+1000	West Central Africa
		+2000	Athens, Beirut, Istanbul, Minsk
		+2000	Bucharest
		+2000	Cairo
		+2000	Harare, Pretoria
		+2000	Helsinki, Kyiv, Sofia, Talinn, Vilnius
		+2000	Jerusalem
		+3000	Baghdad
		+3000	Kuwait, Riyadh
		+3000	Moscow, St. Petersburg, Volgograd
		+3000	Nairobi
		+3300	Tehran
		+4000	Abu Dhabi, Muscat
		+4000	Baku, Tbilisi, Yerevan
		+4300	Kabul
		+5000	Ekaterinburg
		+5000	Islamabad, Karachi, Tashkent
		+5300	Chennai, Kolkata, Mumbai, New Delhi
		+5450	Kathmandu
		+6000	Almaty, Novosibirsk
		+6000	Astana, Dhaka
		+6000	Sri Jayawardenepura
		+6300	Rangoon
		+7000	Bangkok, Hanoi, Jakarta
		+7000	Krasnoyarsk
		+8000	Beijing, Chongqing, Hong Kong, Urumqi
		+8000	Irkutsk, Ulaan Bataar
		+8000	Kuala Lumpur, Singapore
		+8000	Perth
		+8000	Taipei
		+9000	Osaka, Sapporo, Tokyo
		+9000	Seoul
		+9000	Yakutsk
		+9300	Adelaide
		+9300	Darwin
		+10000	Brisbane
		+10000	Canberra, Melbourne, Sydney
		+10000	Guam, Port Moresby
		+10000	Hobart
		+10000	Vladivostok
		+11000	Magadan, Solomon Is., New Caledonia
		+12000	Auckland, Wellington
		+12000	Fiji, Kamchatka, Marshall Is.
		+13000	Nuku'alofa

Variable	Characteristic	Description
TOOLTIPMERGE	Type: Switch Saved in: User settings Initial value: 0	Combines drafting tooltips into a single tooltip. The appearance of the merged tooltip is controlled by the settings in the *Tooltip Appearance* dialog box. 0 Off 1 On
TOOLTIPS	Type: Integer Saved in: Registry Initial value: 1	Controls the display of tooltips. 0 Turns off display of tooltips 1 Turns on display of tooltips
TPSTATE	(Read-only) Type: Integer Not saved Initial value: varies	Determines whether the *Tool Palettes* window is active or not. 0 The *Tool Palettes* window is not active 1 The *Tool Palettes* window is active
TRACEWID	Type: Real Saved in: Drawing Initial value: 0.0500	Sets the default trace width.
TRACKPATH	Type: Integer Saved in: Registry Initial value: 0	Controls the display of polar and object snap tracking alignment paths. 0 Displays full screen object snap tracking path 1 Displays object snap tracking path only between the alignment point and From point to cursor location 2 Does not display polar tracking path 3 Does not display polar or object snap tracking paths
TRAYICONS	Type: Integer Saved in: Registry Initial value: 1	Controls whether a tray is displayed on the status bar. 0 Does not display a tray 1 Displays a tray
TRAYNOTIFY	Type: Integer Saved in: Registry Initial value: 1	Controls whether service notifications are displayed in the status bar tray. 0 Does not display notifications 1 Displays notifications
TRAYTIMEOUT	Type: Integer Saved in: Registry Initial value: 5	Controls the length of time (in seconds) that service notifications are displayed. Valid values are 0 to 10.
TREEDEPTH	Type: Integer Saved in: Drawing Initial value: 3020	Specifies the maximum depth, that is, the number of times the tree-structured spatial index may divide into branches. 0 Suppresses the spatial index entirely, eliminating the performance improvements it provides in working with large drawings. This setting assures that objects are always processed in database order, making it unnecessary ever to set the *SORTENTS* system variable.

Variable	Characteristic	Description
TREEDEPTH *(continued)*		>0 Turns on *TREEDEPTH*. An integer of up to four digits is valid. The first two digits refer to model space, and the second two digits refer to paper space.
		<0 Treats model space objects as two-dimensional (Z coordinates are ignored), as is always the case with paper space objects. Such a setting is appropriate for 2D drawings and makes more efficient use of memory without loss of performance.
		NOTE: You cannot use *TREEDEPTH* transparently.
TREEMAX	Type: Integer Saved in: Registry Initial value: 10000000	Limits memory consumption during drawing regeneration by limiting the number of nodes in the spatial index (oct-tree). By imposing a fixed limit with *TREEMAX*, you can load drawings created on systems with more memory than your system and with a larger *TREEDEPTH* than your system can handle. These drawings, if left unchecked, have an oct-tree large enough to eventually consume more memory than is available to your computer. *TREEMAX* also provides a safeguard against experimentation with inappropriately high *TREEDEPTH* values. The initial default for *TREEMAX* is 10000000 (10 million), a value high enough to effectively disable *TREEMAX* as a control for *TREEDEPTH*. The value to which you should set *TREEMAX* depends on your system's available RAM. You get about 15,000 oct-tree nodes per megabyte of RAM. If you want an oct-tree to use up to, but no more than, 2 megabytes of RAM, set *TREEMAX* to 30000 (2315,000). If AutoCAD runs out of memory allocating oct-tree nodes, restart AutoCAD, set *TREEMAX* to a smaller number, and try loading the drawing again. AutoCAD might occasionally run into the limit you set with *TREEMAX*. Follow the resulting prompt instructions. Your ability to increase *TREEMAX* depends on your computer's available memory.
TRIMMODE	Type: Integer Saved in: Registry Initial value: 1	Controls whether AutoCAD trims selected edges for chamfers and fillets. 0 Leaves selected edges intact 1 Trims selected edges to the endpoints of chamfer lines and fillet arcs
TSPACEFAC	Type: Real Not saved Initial value: 1	Controls multiline text line spacing distance measured as a factor of text height. Valid values are 0.25 to 4.0.

Variable	Characteristic	Description
TSPACETYPE	Type: Integer Not saved Initial value: 1	Controls the type of line spacing used in multiline text. At least adjusts line spacing based on tallest characters in a line; Exactly uses the specified line spacing regardless of individual character sizes. 1 At least 2 Exactly
TSTACKALIGN	Type: Integer Saved in: Drawing Initial value: 1	Controls the vertical alignment of stacked text. 0 Bottom aligned 1 Center aligned 2 Top aligned
TSTACKSIZE	Type: Integer Saved in: Drawing Initial value: 70	Controls the percentage of stacked text fraction height relative to selected text's current height. Valid values are from 25 to 125.
UCSAXISANG	Type: Integer Saved in: Registry Initial value: 90	Stores the default angle when rotating the UCS around one of its axes using the X, Y, or Z options of the UCS command. Its value must be entered as an angle in degrees (valid values are: 5, 10, 15, 18, 22.5, 30, 45, 90, 180).
UCSBASE	Type: String Saved in: Drawing Initial value: "World"	Stores the name of the UCS that defines the origin and orientation of orthographic UCS settings. Valid values include any named UCS.
UCSDETECT	Type: Integer Saved in: Drawing Initial value: 1	Controls whether dynamic UCS acquisition is active or not. 0 Not active 1 Active
UCSFOLLOW	Type: Integer Saved in: Drawing Initial value: 0	Generates a plan view whenever you change from one UCS to another. You can set *UCSFOLLOW* separately for each viewport. If *UCSFOLLOW* is on for a particular viewport, AutoCAD generates a plan view in that viewport whenever you change coordinate systems. Once the new UCS has been established, you can use *Vpoint*, *Dview*, *Plan*, or *View* to change the view of the drawing. It will change to a plan view again the next time you change coordinate systems. 0 UCS does *not* affect the view 1 Any UCS change causes a change to plan view of the new UCS in the current viewport The setting of *UCSFOLLOW* is maintained separately for paper space and model space and can be accessed in either, but the setting is ignored while in paper space (it is always treated as if set to 0). Although you can define a non-World UCS in paper space, the view remains in plan view to the World Coordinate System.

2007

Variable	Characteristic	Description
UCSICON	Type: Integer Saved in: Drawing Initial value: 3	Displays the user coordinate system icon for the current viewport using bit code. It is the sum of the following: 0 No icon displayed 1 On; icon is displayed 2 Origin; if icon is displayed, the icon floats to the UCS origin if possible 3 On and displayed at origin The *Ucsicon* command controls the visibility and placement of the UCS icon. Because entering *Ucsicon* at the Command prompt invokes the *Ucsicon* command, you must use the Setvar command to access the *UCSICON* system variable.
UCSNAME	(Read-only) Type: String Saved in: Drawing	Stores the name of the current coordinate system for the current space. Returns a null string if the current UCS is unnamed.
UCSORG	(Read-only) Type: 3D point Saved in: Drawing	Stores the origin point of the current coordinate system for the current space. This value is always stored as a World coordinate.
UCSORTHO	Type: Integer Saved in: Registry Initial value: 1	Determines whether the related orthographic UCS setting is automatically restored when an orthographic view is restored. 0 Specifies that the UCS setting remains unchanged when an orthographic view is restored 1 Specifies that the related orthographic UCS setting is automatically restored when an orthographic view is restored
UCSVIEW	Type: Integer Saved in: Registry Initial value: 1	Determines whether the current UCS is saved with a named view. 0 The current UCS is not saved with a named view 1 The current UCS is saved whenever a named view is created
UCSVP	Type: Integer Saved in: Drawing (viewport specific) Initial value: 1	Determines whether the UCS in active viewports remains fixed or changes to reflect the UCS of the currently active viewport. 0 Unlocked; UCS reflects the UCS of the current viewport 1 Locked; UCS stored in viewport and is independent of the UCS of the current viewport
UCSXDIR	(Read-only) Type: 3D point Saved in: Drawing	Stores the X direction of the current UCS for the current space.
UCSYDIR	(Read-only) Type: 3D point Saved in: Drawing	Stores the Y direction of the current UCS for the current space.

Variable	Characteristic	Description
UNDOCTL	(Read-only) Type: Integer Not saved	Stores a bit code indicating the state of the *Undo* feature. It is the sum of the following values: 0 *Undo* is turned off 1 *Undo* is turned on 2 Only one command can be undone 4 Turns on the Auto option 8 A group is currently active
UNDOMARKS	(Read-only) Type: Integer Not saved	Stores the number of marks that have been placed in the Undo control stream by the Mark option. The Mark and Back options are not available if a group is currently active.
UNITMODE	Type: Integer Saved in: Drawing Initial value: 0	Controls the display format for units. 0 Display fractional, feet and inches, and surveyor's angles as previously set 1 Displays fractional, feet and inches, and surveyor's angles in input format
UPDATETHUMBNAIL	Type: Bit code Saved in: Drawing Initial value: 15	Controls updating of the thumbnail previews in the *Sheet Set Manager*. The setting is stored as a bit code using the sum of the following values: 0 Does not update thumbnail previews for sheet views, model space views, or sheets 1 Updates model view thumbnail previews 2 Updates sheet view thumbnail previews 4 Updates sheet thumbnail previews 8 Updates thumbnail previews when sheets or views are created, modified, or restored 16 Updates thumbnail previews when the drawing is saved
USERI1-5	Type: Integer Saved in: Drawing Initial value: 0	*USERI1, USERI2, USERI3, USERI4,* and *USERI5* are used for storage and retrieval of integer values.
USERR1-5	Type: Real Saved in: Drawing Initial value: 0.0000	*USERR1, USERR2, USERR3, USERR4,* and *USERR5* are used for storage and retrieval of real numbers.
USERS1-5	Type: String Not saved Initial value: ""	*USERS1, USERS2, USERS3, USERS4, and USERS5* are used for storage and retrieval of text string data.
VIEWCTR	(Read-only) Type: 3D point Saved in: Drawing	Stores the center of view in the current viewport, expressed as a UCS coordinate.
VIEWDIR	(Read-only) Type: 3D vector Saved in: Drawing	Stores the viewing direction in the current viewport expressed in UCS coordinates. This describes the camera point as a 3D offset from the target point.

Variable	Characteristic	Description
VIEWMODE	(Read-only) Type: Integer Saved in: Drawing	Controls Viewing mode for the current viewport using bit code. The value is the sum of the following: 0 Turned off 1 Perspective view active 2 Front clipping on 4 Back clipping on 8 UCS Follow mode on 16 Front clip not at eye. If on, the front clip distance (*FRONTZ*) determines the front clipping plane. If off, *FRONTZ* is ignored, and the front clipping plane is set to pass through the camera point (vectors behind the camera are not displayed). This flag is ignored if the front clipping bit (2) is off.
VIEWSIZE	(Read-only) Type: Real Saved in: Drawing	Stores the height of the view in the current viewport, expressed in drawing units.
VIEWTWIST	(Read-only) Type: Real Saved in: Drawing	Stores the view twist angle for the current viewport.
VISRETAIN	Type: Integer Saved in: Drawing Initial value: 1	Controls the visibility, color, linetype, lineweight, and plot styles (if *PSTYLEPOLICY* is set to 0) of xref-dependent layers; specifies whether nested xref path changes are saved. 0 The layer table, as stored in the reference drawing (xref) takes precedence. Changes made to xref-dependent layers in the current drawing are valid in the current session only and are not saved with the drawing. When the current drawing is reopened, the layer table is reloaded from the reference drawing and the current drawing reflects those settings. The layer settings affected are *On, Off, Freeze, Thaw, Color, Ltype, Lweight,* and *Pstyle* (if *PSTYLEPOLICY* is set to 0). This setting also specifies that changes made to the paths of nested xrefs are for the current session only and are not saved with the drawing. 1 Xref-dependent layer changes made in the current drawing take precedence. Layer settings are saved with the current drawing's layer table and persist from session to session. Nested xref path changes are saved with the current drawing and persist from session to session.

Variable	Characteristic	Description
VPMAXIMIZEDSTATE	Type: Integer Saved in: Not saved Initial value: 0	Stores a value that indicates whether the viewport is maximized. The maximized viewport state is canceled if you start the *Plot* command. 0 Not maximized 1 Maximized
VSBACKGROUNDS	Type: Integer Saved in: Drawing Initial value: 1	Controls whether backgrounds are displayed in the visual style applied to the current viewport. 0 Off 1 On
VSEDGECOLOR	Type: String Saved in: Drawing Initial value: 7	Sets the color of edges in the visual style in the current viewport. The initial value is 7 (ACI Black).
VSEDGEJITTER	Type: String Saved in: Drawing Initial value: -2	Controls the degree to which lines are made to appear as though sketched with a pencil. Turn off the jitter effect by preceding the setting with a minus sign. 1 Low 2 Medium 3 High NOTE: Plot styles are not available for objects with the Jitter edge modifier applied.
VSEDGEOVERHANG	Type: Integer Saved in: Drawing Initial value: -6	Makes lines extend beyond their intersection, for a hand-drawn effect. The range is 1 to 100 pixels. Turn off the overhang effect by preceding the setting with a minus sign.
VSEDGES	Type: Integer Saved in: Drawing Initial value: 1	Controls the types of edges that are displayed in the viewport. 0 No edges are displayed 1 Isolines are displayed 2 Facet edges are displayed
VSEDGESMOOTH	Type: Integer Saved in: Drawing Initial value: 1	Specifies the angle at which crease edges are displayed. The range is 0 to 180.
VSFACECOLORMODE	Type: Integer Saved in: Drawing Initial value: 0	Controls how the color of faces is calculated. 0 *Normal*: Does not apply a face color modifier 1 *Monochrome*: Displays all faces in the color that is specified in the *VSMONOCOLOR* system variable. 2 *Tint*: Uses the color that is specified in the *VSMONOCOLOR* system variable to shade all faces by changing the hue and saturation values of the color. 3 *Desaturate*: Softens the color by reducing its saturation component by 30 percent
VSFACEHIGHLIGHT	Type: Real Saved in: Drawing Initial value: -30	Controls the display of specular highlights on faces without materials in the current viewport. The range is -100 to 100. The higher the number, the larger the highlight.

Variable	Characteristic	Description
VSFACEHIGHLIGHT (continued)		Objects with materials attached ignore the setting of VSFACEHIGHLIGHT when VSMATERIALMODE is on.
VSFACEOPACITY	Type: Real Saved in: Drawing Initial value: -60	Controls the transparency of faces in the current viewport. The range is -100 to 100. At 100, the face is completely opaque. At 0, the face is completely transparent. Negative values set the transparency level but turn off the effect in the drawing.
VSFACESTYLE	Type: Integer Saved in: Drawing Initial value: 1	Controls how faces are displayed in the current viewport. 0 No style applied 1 *Real*: As close as possible to how the face would appear in the real world 2 *Gooch*: Uses cool and warm colors instead of dark and light to enhance the display of faces that might be shadowed and difficult to see in a realistic display
VSHALOGAP	Type: Integer Saved in: Drawing Initial value: 0	Sets the halo gap in the visual style applied to the current viewport. The range is 0 to 100.
VSHIDEPRECISION	Type: Integer Saved in: Not-saved Initial value: 0	Controls the accuracy of hides and shades in the visual style applied to the current viewport.
VSINTERSECTION-COLOR	Type: Integer Saved in: Drawing Initial value: 7	Specifies the color of intersection polylines in the visual style applied to the current viewport. The initial value is 7, which is a special value that inverts the color (black or white) based on the background color. Value 0 designates *Byblock*, value 256 designates *Bylayer*, and value 257 designates *Byentity*. Values 1-255 designate an AutoCAD Color Index (ACI) color. True Colors and Color Book colors are also available. NOTE: *INTERSECTIONCOLOR* controls the color of intersection polylines when the visual style is set to *2D Wireframe*.
VSINTERSECTION-EDGES	Type: Switch Saved in: Drawing Initial value: 0	Controls the display of intersection edges in the visual style applied to the current viewport. NOTE: *INTERSECTIONDISPLAY* controls the color of intersection polylines when the visual style is set to *2D Wireframe*. 0 Off 1 On
VSINTERSECTION-LTYPE	Type: Integer Saved in: Drawing Initial value: 1	Sets the linetype for intersection lines in the visual style applied to the current viewport. The range is 1 to 11. 0 Off 1 Solid

2007

Variable	Characteristic	Description
VSINTERSECTIONLTYPE (continued)		2 Dashed 3 Dotted 4 Short Dash 5 Medium Dash 6 Long Dash 7 Double Short Dash 8 Double Medium Dash 9 Double Long Dash 10 Medium Long Dash 11 Sparse Dot
VSISOONTOP	Type: Integer Saved in: Drawing Initial value: 0	Displays isolines on top of shaded objects in the visual style applied to the current viewport. 0 Off 1 On
VSLIGHTINGQUALITY	Type: Integer Saved in: Drawing Initial value: 1	Sets the lighting quality in the current viewport. 0 Shows facets 1 Appears smooth
VSMATERIALMODE	Type: Integer Saved in: Drawing Initial value: 0	Controls the display of materials in the current viewport. 0 No materials are displayed 1 Materials are displayed, textures are not displayed 2 Materials and textures are displayed
VSMAX	(Read-only) Type: 3D point Saved in: Drawing	Stores the upper-right corner of the current viewport's virtual screen, expressed as a UCS coordinate.
VSMIN	(Read-only) Type: 3D point Saved in: Drawing	Stores the lower-left corner of the current viewport's virtual screen, expressed as a UCS coordinate.
VSMONOCOLOR	Type: String Saved in: Drawing Initial value: 255,255,255	Sets the color for monochrome and tint display of faces in the visual style applied to the current viewport. The initial value is white.
VSOBSCUREDCOLOR	Type: String Saved in: Drawing Initial value: Byentity	Specifies the color of obscured lines in the visual style applied to the current viewport. The initial value is Byentity.
VSOBSCUREDEDGES	Type: Integer Saved in: Drawing Initial value: 1	Controls whether obscured (hidden) edges are displayed. 0 Off 1 On
VSOBSCUREDLTYPE	Type: Integer Saved in: Drawing Initial value: 1	Specifies the linetype of obscured lines in the visual style applied to the current viewport. The range is 1 to 11. 0 Off 1 Solid 2 Dashed 3 Dotted 4 Short Dash 5 Medium Dash

2007

2007

Variable	Characteristic	Description
VSOBSCUREDLTYPE (continued)		6 Long Dash 7 Double Short Dash 8 Double Medium Dash 9 Double Long Dash 10 Medium Long Dash 11 Sparse Dot
VSSHADOWS	Type: Integer Saved in: Drawing Initial value: 0	Controls whether a visual style displays shadows. 0 No shadows are displayed 1 Ground shadows only are displayed 2 Full shadows are displayed NOTE: To display full shadows, hardware acceleration is required. When *Geometry Acceleration* is off, full shadows cannot be displayed. (To access these settings, enter *3Dconfig* at the Command prompt. In the *Adaptive Degradation and Performance Tuning* dialog box, click *Manual Tune*.)
VSSILHEDGES	Type: Integer Saved in: Drawing Initial value: 0	Controls display of silhouette edges of solid objects in the visual style applied to the current viewport. 0 Off 1 On
VSSILHWIDTH	Type: Integer Saved in: Drawing Initial value: 5	Specifies the width in pixels of silhouette edges in the current viewport. The range is 1 to 25.
VTDURATION	Type: Integer Saved in: Registry Initial value: 750	Sets the duration of a smooth view transition, in milliseconds. The valid range is 0 to 5000.
VTENABLE	Type: Integer Saved in: Registry Initial value: 3	Controls when smooth view transitions are used. Smooth view transitions can be on or off for panning and zooming, for changes of view angle, or for scripts. The valid range is 0 to 7. Below is listed the setting, for *Pan/Zoom*, for *Rotate*, for scripts. 0 Off, Off, Off 1 On, Off, Off 2 Off, On, Off 3 On, On, Off 4 Off, Off, On 5 On, Off, On 6 Off, On, On 7 On, On, On
VTFPS	Type: Real Saved in: Registry Initial value: 7.0	Sets the minimum speed of a smooth view transition, in frames per second. When a smooth view transition cannot maintain this speed, an instant transition is used. The valid range is 1.0 to 30.0.
WHIPARC	Type: Integer Saved in: Registry Initial value: 0	Controls whether the display of circles and arcs is smooth. 0 Circles and arcs are not smooth, but rather are displayed as a series of vectors.

2007

2006

Variable	Characteristic	Description
WHIPARC (*continued*)		1 Circles and arcs are smooth, displayed as true circles and arcs.
WHIPTHREAD	Type: Integer Saved in: Registry Initial value: 3	Controls whether to use an additional processor (known as multithreaded processing) to improve the speed of operations such as *Zoom* and *Pan* that redraw or regenerate the drawing. *WHIPTHREAD* has no effect on single processor machines. 0 No multithreaded processing; restricts regeneration and redraw processing to a single processor. This setting restores the legacy behavior of AutoCAD 2000 and previous releases. 1 Regeneration multithreaded processing only; regeneration processing is distributed across two processors on a multiprocessor machine. 2 Redraw multithreaded processing only; redraw processing is distributed across two processors on a multiprocessor machine. 3 Regeneration and redraw multithreaded processing; regeneration and redraw processing is distributed across two processors on a multiprocessor machine. NOTE: When multithreaded processing is used for redraw operations (value 2 or 3), the order of objects specified with the *Draworder* command is not guaranteed to be preserved for display but is preserved for plotting.
WINDOWAREACOLOR	Type: Integer Saved in: Registry Initial value: 5	Controls the color of the transparent selection area during *Window* selection. The valid range is 1 to 255. *SELECTIONAREA* must be on.
WMFBKGND	Type: Integer Not saved Initial value: 1	Controls whether the background display of AutoCAD objects is transparent in other applications when these objects are: • Output to a Windows metafile using the *Wmfout* command • Copied to the Clipboard in AutoCAD and pasted as a Windows metafile • Dragged and dropped from AutoCAD as a Windows metafile The AutoCAD defined values are: 0 The background is transparent 1 The background color is the same as the AutoCAD current background color
WMFFOREGND	Type: Integer Not saved Initial value: 0	Controls the assignment of the foreground color of AutoCAD objects in other applications when these objects are: • Output to a Windows metafile using the *Wmfout* command • Copied to the Clipboard in AutoCAD and pasted as a Windows metafile

2006

Variable	Characteristic	Description
WMFFOREGND *(continued)*		• Dragged and dropped from AutoCAD as a Windows metafile *WMFFOREGND* applies only when *WMFBKGND* is set to 0. The AutoCAD defined values are: 0 The foreground and background colors are swapped if necessary to ensure that the foreground color is darker than the background color 1 The foreground and background colors are swapped if necessary to ensure that the foreground color is lighter than the background color
WORLDUCS	(Read-only) Type: Integer Not saved	Indicates whether the UCS is the same as the World Coordinate System. 0 Current UCS is different from the World Coordinate System 1 Current UCS is the same as the World Coordinate System
WORLDVIEW	Type: Integer Saved in: Drawing Initial value: 1	Controls whether the UCS changes to the WCS during *3Dorbit, Dview,* or *Vpoint.* 0 Current UCS remains unchanged 1 Current UCS is changed to the WCS for the duration of *3Dorbit, Dview,* or *Vpoint* Command input is relative to the current UCS
WRITESTAT	Type: Read-only Not saved Initial value: 1	Indicates whether a drawing file is read-only or can be written to. For developers who need to determine write status through AutoLISP. 0 Can't write to the drawing 1 Can write to the drawing
WSCURRENT	Type: String Saved in: Not saved Initial value: AutoCAD default	Returns the current workspace name in the command line interface and sets a workspace to current.
XCLIPFRAME	Type: Integer Saved in: Drawing Initial value: 0	Controls visibility of xref clipping boundaries. 0 Clipping boundary is not visible 1 Clipping boundary is visible
XEDIT	Type: Integer Saved in: Drawing Initial value: 1	Controls whether the current drawing can be edited in place when being referenced by another drawing. 0 Can't use in-place reference editing 1 Can use in-place reference editing
XFADECTL	Type: Integer Saved in: Registry Initial value: 50	Controls the fading intensity for references being edited in place. 0 0 percent fading, minimum value 90 90 percent fading, maximum value
XLOADCTL	Type: Integer Saved in: Registry Initial value: 2	Turns xref demand loading on and off and controls whether it opens the original drawing or a copy. 0 Turns off demand loading; entire drawing is loaded

2006

Variable	Characteristic	Description
XLOADCTL (continued)		1 Turns on demand loading; reference file is kept open
		2 Turns on demand loading; a copy of the reference file is opened
		When *XLOADCTL* is set to 2, the reference copy is stored in the AutoCAD temporary files directory (defined by *Options*) or in a user-specified directory.
XLOADPATH	Type: String Saved in: Registry Initial value: ""	Creates a path for storing temporary copies of demand-loaded xref files. For more information, see *XLOADCTL*.
XREFCTL	Type: Integer Saved in: Registry Initial value: 0	Controls whether AutoCAD writes external reference log (XLG) files. 0 Xref log (XLG) files are not written 1 Xref log (XLG) files are written
XREFNOTIFY	Type: Integer Saved in: Registry Initial value: 2	Controls the notification for updated or missing Xrefs. 0 Disables *Xref* notification. 1 Enables *Xref* notification. Notifies you that *Xrefs* are attached to the current drawing by displaying the *Xref* icon in the lower-right corner of the application window (the notification area of the status bar tray). When you open a drawing, alerts you of missing *Xrefs* by displaying the *Xref* icon with a yellow alert symbol (!). 2 Enables *Xref* notification and balloon messages. Displays the *Xref* icon as in 1 above. Also displays balloon messages in the same area when *Xrefs* are modified. The number of minutes between checking for modified *Xrefs* is controlled by the system registry variable *XNOTIFYTIME*.
XREFTYPE	Type: Integer Saved in: Registry Initial value: 0	Controls the default reference type when attaching or overlaying an external reference. 0 *Attachment* is the default 1 *Overlay* is the default
ZOOMFACTOR	Type: Integer Saved in: Registry Initial value: 60	*ZOOMFACTOR* accepts an integer between 3-100 as valid values. The higher the number, the more incremental the change applied by each mouse-wheel forward/backward movement.
ZOOMWHEEL	Type: Integer Saved in: Registry Initial value: 0	Toggles the direction of transparent zoom operations when you scroll the middle mouse wheel. 0 Moves wheel forward zooms in; moving wheel backwards zooms out. 1 Moves wheel forward zooms out; moving wheel backwards zooms in.

Source: AutoCAD *Command Reference*

B

APPENDIX B

AUTOCAD 2006 COMMAND
ALIAS LIST SORTED BY
COMMAND

APPENDIX B

AutoCAD 2007 Command Alias List Sorted by Command

Command	Alias	Command	Alias
3DALIGN	3AL	DIMALIGNED	DAL
3DARRAY	3A	DIMANGULAR	DAN
3DFACE	3F	DIMARC	DAR
3DMOVE	3M	DIMBASELINE	DBA
3DORBIT	3DO	DIMCENTER	DCE
3DORBIT	ORBIT	DIMCONTINUE	DCO
3DPOLY	3P	DIMDIAMETER	DDI
3DROTATE	3R	DIMDISASSOCIATE	DDA
3DWALK	3DNAVIGATE	DIMEDIT	DED
3DWALK	3DW	DIMJOGGED	DJO
ADCENTER	ADC	DIMJOGGED	JOG
ADCENTER	DC	DIMLINEAR	DLI
ADCENTER	DCENTER	DIMORDINATE	DOR
ALIGN	AL	DIMOVERRIDE	DOV
APPLOAD	AP	DIMRADIUS	DRA
ARC	A	DIMREASSOCIATE	DRE
AREA	AA	DIMSTYLE	D
ARRAY	AR	DIMSTYLE	DST
-ARRAY	-AR	DIST	DI
ATTDEF	ATT	DIVIDE	DIV
-ATTDEF	-ATT	DONUT	DO
ATTEDIT	ATE	DRAWINGRECOVERY	DRM
-ATTEDIT	-ATE	DRAWORDER	DR
-ATTEDIT	ATTE	DSETTINGS	DS
BACTION	AC	DSETTINGS	SE
BCLOSE	BC	DVIEW	DV
BEDIT	BE	ELLIPSE	EL
BLOCK	B	ERASE	E
-BLOCK	-B	EXPLODE	X
BOUNDARY	BO	EXPORT	EXP
-BOUNDARY	-BO	-EXPORTTOAUTOCAD	AECTOACAD
BPARAMETER	PARAM	EXTEND	EX
BREAK	BR	EXTERNALREFERENCES	ER
BSAVE	BS	EXTRUDE	EXT
BVSTATE	BVS	FILLET	F
CAMERA	CAM	FILTER	FI
CHAMFER	CHA	FLATSHOT	FSHOT
CHANGE	-CH	GEOGRAPHICLOCATION	GEO
CHECKSTANDARDS	CHK	GEOGRAPHICLOCATION	NORTH
CIRCLE	C	GEOGRAPHICLOCATION	NORTHDIR
COLOR	COL	GRADIENT	GD
COLOR	COLOUR	GROUP	G
COMMANDLINE	CLI	-GROUP	-G
COPY	CO	HATCH	BH
COPY	CP	HATCH	H
CTABLESTYLE	CT	-HATCH	-H
CYLINDER	CYL	HATCHEDIT	HE
DBCONNECT	DBC	HIDE	HI
DDEDIT	ED	IMAGE	IM
DDGRIPS	GR	-IMAGE	-IM
DDVPOINT	VP	IMAGEADJUST	IAD

Command	Alias	Command	Alias
IMAGEATTACH	IAT	PSPACE	PS
IMAGECLIP	ICL	PUBLISHTOWEB	PTW
IMPORT	IMP	PURGE	PU
INSERT	I	-PURGE	-PU
-INSERT	-I	PYRAMID	PYR
INSERTOBJ	IO	QLEADER	LE
INTERFERE	INF	QUICKCALC	QC
INTERSECT	IN	QUIT	EXIT
JOIN	J	RECTANG	REC
LAYER	LA	REDRAW	R
-LAYER	-LA	REDRAWALL	RA
-LAYOUT	LO	REGEN	RE
LENGTHEN	LEN	REGENALL	REA
LINE	L	REGION	REG
LINETYPE	LT	RENAME	REN
-LINETYPE	-LT	-RENAME	-REN
LINETYPE	LTYPE	RENDER	RR
-LINETYPE	-LTYPE	RENDERCROP	RC
LIST	LI	RENDERPRESETS	RP
LIST	LS	RENDERWIN	RW
LTSCALE	LTS	REVOLVE	REV
LWEIGHT	LINEWEIGHT	ROTATE	RO
LWEIGHT	LW	RPREF	RPR
MARKUP	MSM	SCALE	SC
MATCHPROP	MA	SCRIPT	SCR
MATERIALS	MAT	SECTION	SEC
MEASURE	ME	SECTIONPLANE	SPLANE
MIRROR	MI	SETVAR	SET
MIRROR3D	3DMIRROR	SHADEMODE	SHA
MLINE	ML	SHEETSET	SSM
MOVE	M	SLICE	SL
MSPACE	MS	SNAP	SN
MTEXT	MT	SOLID	SO
MTEXT	T	SPELL	SP
-MTEXT	-T	SPLINE	SPL
MVIEW	MV	SPLINEDIT	SPE
OFFSET	O	STANDARDS	STA
OPTIONS	OP	STRETCH	S
OSNAP	OS	STYLE	ST
-OSNAP	-OS	SUBTRACT	SU
PAN	P	TABLE	TB
-PAN	-P	TABLESTYLE	TS
-PARTIALOPEN	PARTIALOPEN	TABLET	TA
PASTESPEC	PA	TEXT	DT
PEDIT	PE	THICKNESS	TH
PLINE	PL	TILEMODE	TI
PLOT	PRINT	TOLERANCE	TOL
POINT	PO	TOOLBAR	TO
POLYGON	POL	TOOLPALETTES	TP
POLYSOLID	PSOLID	TORUS	TOR
PREVIEW	PRE	TRIM	TR
PROPERTIES	CH	UCSMAN	UC
PROPERTIES	MO	UNION	UNI
PROPERTIES	PR	UNITS	UN
PROPERTIES	PROPS	-UNITS	-UN
PROPERTIESCLOSE	PRCLOSE	VIEW	V

Command	Alias	Command	Alias
-VIEW	-V	XATTACH	XA
VISUALSTYLES	VSM	XBIND	XB
-VISUALSTYLES	-VSM	-XBIND	-XB
VPOINT	-VP	XCLIP	XC
VSCURRENT	VS	XLINE	XL
WBLOCK	W	XREF	XR
-WBLOCK	-W	-XREF	-XR
WEDGE	WE	ZOOM	Z

AutoCAD 2007 Commands Called by Discontinued Commands or Aliases

2007 Command	Old Alias or Command	2007 Command	Old Alias or Command
ADCENTER	ADCENTER	INSERT	INSERTURL
ADCENTER	CONTENT	INSERT	DDINSERT
ATTDEF	DDATTDEF	LAYER	DDLMODES
ATTEDIT	DDATTE	LEADER	LEAD
ATTEXT	DDATTEXT	LINETYPE	DDLTYPE
BLOCK	ACADBLOCKDIALOG	LIST	SHOWMAT
BLOCK	BMAKE	MATCHPROP	PAINTER
BLOCK	BMOD	MATERIALMAP	SETUV
BOUNDARY	BPOLY	MATERIALS	FINISH
COLOR	DDCOLOR	MATERIALS	RMAT
COPY	CP	OPEN	OPENURL
DBCONNECT	AAD	OPEN	DXFIN
DBCONNECT	AEX	OPTIONS	PREFERENCES
DBCONNECT	ALI	OSNAP	DDOSNAP
DBCONNECT	ARO	PLOT	DWFOUT
DBCONNECT	ASE	PLOTSTAMP	DDPLOTSTAMP
DBCONNECT	ASQ	PROPERTIES	DDCHPROP
DIMALIGNED	DIMALI	PROPERTIES	DDMODIFY
DIMANGULAR	DIMANG	RECTANG	RECTANGLE
DIMBASELINE	DIMBASE	RENDERENVIRONMENT	FOG
DIMCONTINUE	DIMCONT	RENDERPRESETS	RFILEOPT
DIMDIAMETER	DIMDIA	RENDERWIN	RENDSCR
DIMEDIT	DIMED	SAVE	SAVEURL
DIMLINEAR	DIMLIN	SAVEAS	DXFOUT
DIMLINEAR	DIMHORIZONTAL	SHADEMODE	SHADE
DIMLINEAR	DIMROTATED	STYLE	DDSTYLE
DIMLINEAR	DIMVERTICAL	TEXT	DTEXT
DIMORDINATE	DIMORD	TILEMODE	TM
DIMOVERRIDE	DIMOVER	UCS	DDUCS
DIMRADIUS	DIMRAD	UCSMAN	DDUCS
DIMSTYLE	DIMSTY	UCSMAN	DDUCSP
DIMSTYLE	DDIM	UNITS	DDUNITS
DIMTEDIT	DIMTED	VIEW	DDVIEW
DONUT	DOUGHNUT	VPORTS	VIEWPORTS
DSETTINGS	DDRMODES	WBLOCK	ACADW-BLOCKDIALOG
DSVIEWER	AV		

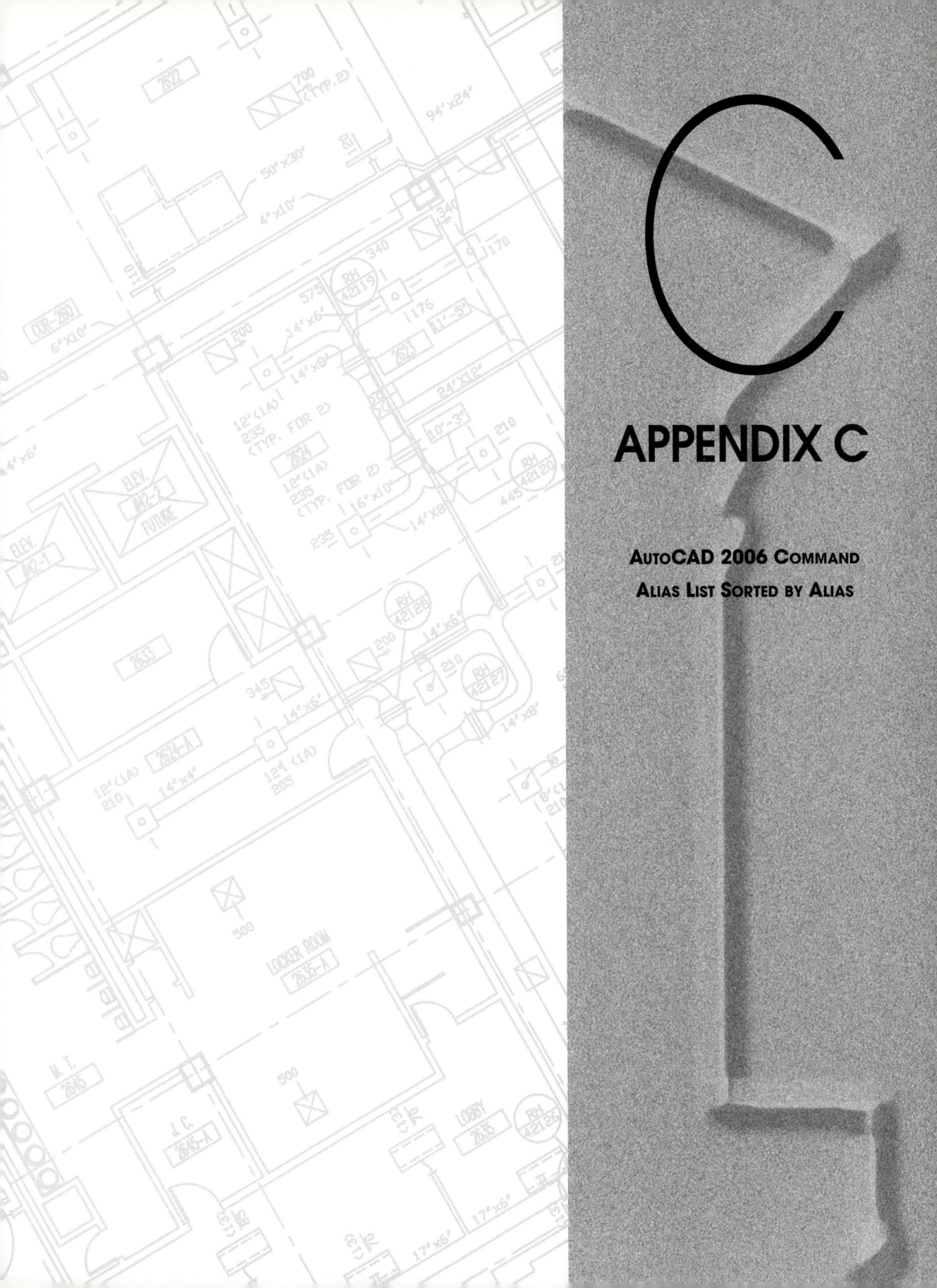

APPENDIX C

AutoCAD 2006 Command Alias List Sorted by Alias

AutoCAD 2007 Command Alias List Sorted by Alias

Alias	Command	Alias	Command
3A	3DARRAY	DC	ADCENTER
3DMIRROR	MIRROR3D	DCE	DIMCENTER
3DNAVIGATE	3DWALK	DCENTER	ADCENTER
3DO	3DORBIT	DCO	DIMCONTINUE
3DW	3DWALK	DDA	DIMDISASSOCIATE
3F	3DFACE	DDI	DIMDIAMETER
3M	3DMOVE	DED	DIMEDIT
3P	3DPOLY	DI	DIST
3R	3DROTATE	DIV	DIVIDE
A	ARC	DJO	DIMJOGGED
AC	BACTION	DLI	DIMLINEAR
ADC	ADCENTER	DO	DONUT
AECTOACAD	-EXPORTTOAUTOCAD	DOR	DIMORDINATE
AA	AREA	DOV	DIMOVERRIDE
AL	ALIGN	DR	DRAWORDER
3AL	3DALIGN	DRA	DIMRADIUS
AP	APPLOAD	DRE	DIMREASSOCIATE
AR	ARRAY	DRM	DRAWINGRECOVERY
-AR	-ARRAY	DS	DSETTINGS
ATT	ATTDEF	DST	DIMSTYLE
-ATT	-ATTDEF	DT	TEXT
ATE	ATTEDIT	DV	DVIEW
-ATE	-ATTEDIT	E	ERASE
ATTE	-ATTEDIT	ED	DDEDIT
B	BLOCK	EL	ELLIPSE
-B	-BLOCK	ER	EXTERNALREFERENCES
BC	BCLOSE	EX	EXTEND
BE	BEDIT	EXIT	QUIT
BH	HATCH	EXP	EXPORT
BO	BOUNDARY	EXT	EXTRUDE
-BO	-BOUNDARY	F	FILLET
BR	BREAK	FI	FILTER
BS	BSAVE	FSHOT	FLATSHOT
BVS	BVSTATE	G	GROUP
C	CIRCLE	-G	-GROUP
CAM	CAMERA	GD	GRADIENT
CH	PROPERTIES	GEO	GEOGRAPHICLOCATION
-CH	CHANGE	GR	DDGRIPS
CHA	CHAMFER	H	HATCH
CHK	CHECKSTANDARDS	-H	-HATCH
CLI	COMMANDLINE	HE	HATCHEDIT
COL	COLOR	HI	HIDE
COLOUR	COLOR	I	INSERT
CO	COPY	-I	-INSERT
CP	COPY	IAD	IMAGEADJUST
CT	CTABLESTYLE	IAT	IMAGEATTACH
CYL	CYLINDER	ICL	IMAGECLIP
D	DIMSTYLE	IM	IMAGE
DAL	DIMALIGNED	-IM	-IMAGE
DAN	DIMANGULAR	IMP	IMPORT
DAR	DIMARC	IN	INTERSECT
JOG	DIMJOGGED	INF	INTERFERE
DBA	DIMBASELINE	IO	INSERTOBJ
DBC	DBCONNECT	J	JOIN

Alias	Command	Alias	Command
L	LINE	RC	RENDERCROP
LA	LAYER	RE	REGEN
-LA	-LAYER	REA	REGENALL
LE	QLEADER	REC	RECTANG
LEN	LENGTHEN	REG	REGION
LI	LIST	REN	RENAME
LINEWEIGHT	LWEIGHT	-REN	-RENAME
LO	-LAYOUT	REV	REVOLVE
LS	LIST	RO	ROTATE
LT	LINETYPE	RP	RENDERPRESETS
-LT	-LINETYPE	RPR	RPREF
LTYPE	LINETYPE	RR	RENDER
-LTYPE	-LINETYPE	RW	RENDERWIN
LTS	LTSCALE	S	STRETCH
LW	LWEIGHT	SC	SCALE
M	MOVE	SCR	SCRIPT
MA	MATCHPROP	SE	DSETTINGS
MAT	MATERIALS	SEC	SECTION
ME	MEASURE	SET	SETVAR
MI	MIRROR	SHA	SHADEMODE
ML	MLINE	SL	SLICE
MO	PROPERTIES	SN	SNAP
MS	MSPACE	SO	SOLID
MSM	MARKUP	SP	SPELL
MT	MTEXT	SPL	SPLINE
MV	MVIEW	SPLANE	SECTIONPLANE
NORTH	GEOGRAPHICLOCATION	SPE	SPLINEDIT
NORTHDIR	GEOGRAPHICLOCATION	SSM	SHEETSET
O	OFFSET	ST	STYLE
OP	OPTIONS	STA	STANDARDS
ORBIT	3DORBIT	SU	SUBTRACT
OS	OSNAP	T	MTEXT
-OS	-OSNAP	-T	-MTEXT
P	PAN	TA	TABLET
-P	-PAN	TB	TABLE
PA	PASTESPEC	TH	THICKNESS
PARAM	BPARAMETER	TI	TILEMODE
PARTIALOPEN	-PARTIALOPEN	TO	TOOLBAR
PE	PEDIT	TOL	TOLERANCE
PL	PLINE	TOR	TORUS
PO	POINT	TP	TOOLPALETTES
POL	POLYGON	TR	TRIM
PR	PROPERTIES	TS	TABLESTYLE
PRCLOSE	PROPERTIESCLOSE	UC	UCSMAN
PROPS	PROPERTIES	UN	UNITS
PRE	PREVIEW	-UN	-UNITS
PRINT	PLOT	UNI	UNION
PS	PSPACE	V	VIEW
PSOLID	POLYSOLID	-V	-VIEW
PTW	PUBLISHTOWEB	VP	DDVPOINT
PU	PURGE	-VP	VPOINT
-PU	-PURGE	VS	VSCURRENT
PYR	PYRAMID	VSM	VISUALSTYLES
QC	QUICKCALC	-VSM	-VISUALSTYLES
R	REDRAW	W	WBLOCK
RA	REDRAWALL	-W	-WBLOCK

Alias	Command	Alias	Command
WE	*WEDGE*	*XC*	*XCLIP*
X	*EXPLODE*	*XL*	*XLINE*
XA	*XATTACH*	*XR*	*XREF*
XB	*XBIND*	*-XR*	*-XREF*
-XB	*-XBIND*	*Z*	*ZOOM*

Discontinued Commands or Aliases that Call AutoCAD 2007 Commands

Old Alias, Command	2007 Command	Old Alias, Command	2007 Command
AAD	*DBCONNECT*	*DIMBASE*	*DIMBASELINE*
ACADBLOCKDIALOG	*BLOCK*	*DIMCONT*	*DIMCONTINUE*
ACADWBLOCKDIALOG	*WBLOCK*	*DIMDIA*	*DIMDIAMETER*
ADCENTER	*ADCENTER*	*DIMED*	*DIMEDIT*
AEX	*DBCONNECT*	*DIMHORIZONTAL*	*DIMLINEAR*
ALI	*DBCONNECT*	*DIMLIN*	*DIMLINEAR*
ARO	*DBCONNECT*	*DIMORD*	*DIMORDINATE*
ASE	*DBCONNECT*	*DIMOVER*	*DIMOVERRIDE*
ASQ	*DBCONNECT*	*DIMRAD*	*DIMRADIUS*
AV	*DSVIEWER*	*DIMROTATED*	*DIMLINEAR*
BMAKE	*BLOCK*	*DIMSTY*	*DIMSTYLE*
BMOD	*BLOCK*	*DIMTED*	*DIMTEDIT*
BPOLY	*BOUNDARY*	*DIMVERTICAL*	*DIMLINEAR*
CONTENT	*ADCENTER*	*DOUGHNUT*	*DONUT*
CP	*COPY*	*DTEXT*	*TEXT*
DDATTDEF	*ATTDEF*	*DWFOUT*	*PLOT*
DDATTE	*ATTEDIT*	*DXFIN*	*OPEN*
DDATTEXT	*ATTEXT*	*DXFOUT*	*SAVEAS*
DDCHPROP	*PROPERTIES*	*FINISH*	*MATERIALS*
DDCOLOR	*COLOR*	*FOG*	*RENDERENVIRON-*
DDIM	*DIMSTYLE*		*MENT*
DDINSERT	*INSERT*	*INSERTURL*	*INSERT*
DDLMODES	*LAYER*	*LEAD*	*LEADER*
DDLTYPE	*LINETYPE*	*OPENURL*	*OPEN*
DDMODIFY	*PROPERTIES*	*PAINTER*	*MATCHPROP*
DDOSNAP	*OSNAP*	*PREFERENCES*	*OPTIONS*
DDPLOTSTAMP	*PLOTSTAMP*	*RECTANGLE*	*RECTANG*
DDRMODES	*DSETTINGS*	*RENDSCR*	*RENDERWIN*
DDSTYLE	*STYLE*	*RFILEOPT*	*RENDERPRESETS*
DDUCS	*UCS*	*RMAT*	*MATERIALS*
DDUCS	*UCSMAN*	*SAVEURL*	*SAVE*
DDUCSP	*UCSMAN*	*SETUV*	*MATERIALMAP*
DDUNITS	*UNITS*	*SHADE*	*SHADEMODE*
DDVIEW	*VIEW*	*SHOWMAT*	*LIST*
DIMALI	*DIMALIGNED*	*TM*	*TILEMODE*
DIMANG	*DIMANGULAR*	*VIEWPORTS*	*VPORTS*

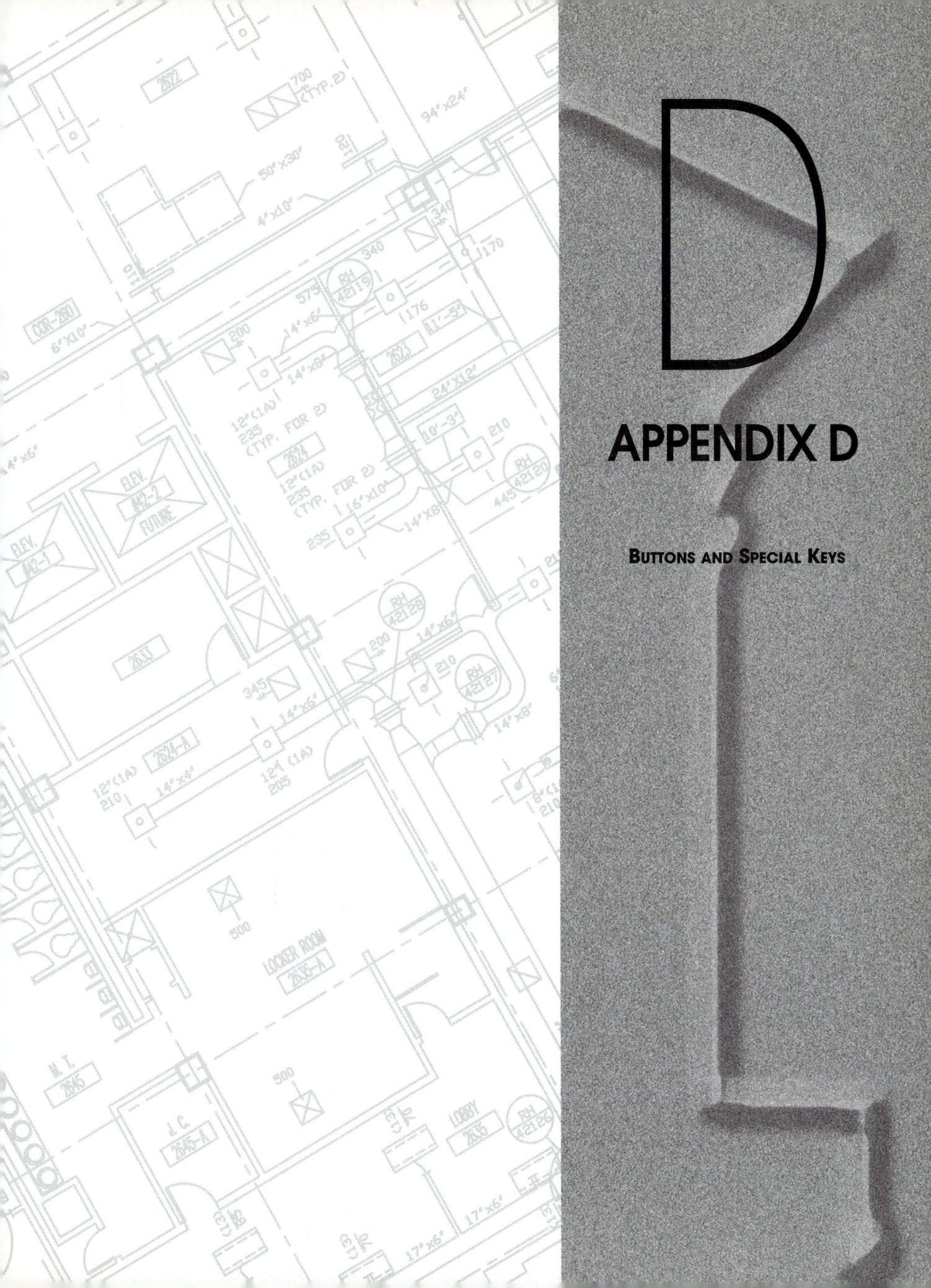

D
APPENDIX D

BUTTONS AND SPECIAL KEYS

APPENDIX D

BUTTONS AND SPECIAL KEYS

Mouse and Digitizing Puck Buttons

Depending on the type of mouse or digitizing puck used for cursor control, a different number of buttons are available. The *User Preferences* tab of the *Options* dialog box can be used to control the appearance of shortcut menus and to customize the actions of a right-click. In any case, the buttons have the following default settings.

#1 (left mouse)	**PICK**	Used to select commands or point to locations on screen.
#2 (right mouse)	**Shortcut menu or Enter**	Generally, activates a shortcut menu (see Chapter 1, "Shortcut Menus" and Chapter 2, "Windows Right-Click Shortcut Menus"). Otherwise, performs the same action as the Enter key on the keyboard.
press (center wheel)	*Pan*	Activates the realtime *Pan* command.
turn (center wheel)	*Zoom*	Activates the realtime *Zoom* command.

Function (F) Keys

Function keys in AutoCAD offer a quick method of turning on or off (toggling) drawing aids.

F1	*Help*	Opens a help window providing written explanations on commands and variables (see Chapter 5).
F2	*Flipscreen*	Activates a text window showing the previous command line activity (see Chapter 1).
F3	*Osnap Toggle*	If Running Osnaps are set, toggling this key temporarily turns the Running Osnaps off so that a point can be picked without using Osnaps. If no Running Osnaps are set, F3 produces the *Object Snap* tab of the *Drafting Settings* dialog box (see Chapter 7).
F4	*Tablet*	Turns the *TABMODE* variable on or off. If *TABMODE* is on, the digitizing tablet can be used to digitize an existing paper drawing into AutoCAD.
F5	*Isoplane*	When using an *Isometric* style *SNAP* and *GRID* setting, toggles the crosshairs (with *ORTHO* on) to draw on one of three isometric planes (see Chapter 25).
F6	*DUCS*	Toggles *Dynamic UCS* on and off (see Chapter 38).
F7	*GRID*	Turns the *GRID* on or off (see Chapter 1, "Drawing Aids").
F8	*ORTHO*	Turns *ORTHO* on or off (see Chapter 1, "Drawing Aids").
F9	*SNAP*	Turns *SNAP* (Grid Snap or Polar Snap) on or off (see Chapter 1, "Drawing Aids" and Chapter 3, "Polar Tracking and Polar Snap").
F10	*POLAR*	Turns Polar Tracking on or off (see Chapter 1, "Drawing Aids" and Chapter 3, "Polar Tracking and Polar Snap").
F11	*OTRACK*	Turns Object Snap Tracking on or off (see Chapter 7).
F12	*DYN*	Toggles Dynamic Input (see Chapter 1, "Drawing Aids" and Chapter 3, "Dynamic Input").

Control Key Sequences (Accelerator Keys)

Accelerator keys (holding down the Ctrl key and pressing another key simultaneously) invoke regular AutoCAD commands or produce special functions. Several have the same duties as F3 through F11.

Ctrl+A	*Select*	Selects all objects.
Crtl+B (F9)	*SNAP*	Turns *SNAP* on or off.
Ctrl+C	*Copyclip*	Copies the highlighted objects to the Windows clipboard.
Ctrl+D (F6)	*DUCS*	Toggles *Dynamic UCS* on and off (see Chapter 38).
Ctrl+E (F5)	*Isoplane*	When using an *Isometric* style *SNAP* and *GRID* setting, toggles the crosshairs (with *ORTHO* on) to draw on one of three isometric planes.
Ctrl+F (F3)	*Osnap Toggle*	If Running Osnaps are set, pressing Ctrl+F temporarily turns off the Running Osnaps so that a point can be picked without using Osnaps. If there are no Running Object Snaps set, Ctrl+F produces the *Osnap Settings* dialog box. This dialog box is used to turn on and off Running Object Snaps (discussed in Chapter 7).
Ctrl+G (F7)	*GRID*	Turns the *GRID* on or off.
Ctrl+H	*PICKSTYLE*	Toggles *PICKSTYLE* On (1) and Off (0) and toggles selectable *Groups* on or off.
Ctrl+J	*Enter*	Executes the last command.
Ctrl+K	*Hyperlink*	Activates the *Hyperlink* command.
Ctrl+L (F8)	*ORTHO*	Turns *ORTHO* on or off.
Ctrl+N	*New*	Invokes the *New* command to start a new drawing.
Ctrl+O	*Open*	Invokes the *Open* command to open an existing drawing.
Ctrl+P	*Plot*	Produces the *Plot* dialog box for creating and controlling prints and plots.
Ctrl+Q	*Exit*	Exits AutoCAD.
Ctrl+S	*Qsave*	Performs a quick save or produces the *Saveas* dialog box if the file is not yet named.
Ctrl+T (F4)	*Tablet*	Turns the *TABMODE* variable on or off. If *TABMODE* is on, the digitizing tablet can be used to digitize an existing paper drawing into AutoCAD.
Ctrl+U (F10)	*POLAR*	Turns *Polar Tracking* on or off.
Ctrl+V	*Pasteclip*	Pastes the clipboard contents into the current AutoCAD drawing.
Ctrl+W (F11)	*OTRACK*	Turns *Object Snap Tracking* on or off.
Ctrl+X	*Cutclip*	Cuts the highlighted objects from the drawing and copies them to the Windows clipboard.
Ctrl+Y	*Redo*	Invokes the *Redo* command.
Ctrl+Z	*Undo*	Undoes the last command.
Ctrl+0	*Clean Screen*	Toggles *Clean Screen* on and off.
Ctrl+1	*Properties*	Toggles the *Properties* palette.
Ctrl+2	*DesignCenter*	Toggles DesignCenter.
Ctrl+3	*Toolpalettes*	Toggles the *Tool Palettes* window.
Ctrl+4	*Sheet Set Manager*	Toggles the *Sheet Set Manager*.
Ctrl+5	*Info Palette*	Toggles the *Info Palette*.
Ctrl+6	*dbConnect*	Toggles the *DBConnect Manager*.
Ctrl+7	*Markup Set Manager*	Toggles the *Markup Set Manager*.
Ctrl+8	*QuickCalc*	Opens the *Sheet Set Manager*.
Ctrl+9	*Command Line*	Opens the *Command Line Window* dialog box.
Alt+F8	*Macros*	Opens the *Macros* dialog box.
Alt+F11	*Visual Basic Editor*	Opens the Visual Basic Editor.

Temporary Override Keys

Temporary override keys temporarily turn on or off one of the drawing aids that can otherwise be toggled on the Status Bar or in the *Drafting Settings* dialog box. Each function has a key combination for each (left and right) side of the keyboard.

Keyboard Left	Keyboard Right	Drawing Aid
Shift	**Shift**	Toggles *ORTHO*
Shift+A	**Shift+'**	Toggles *OSNAP*
Shift+X	**Shift+.**	Toggles *POLAR*
Shift+Q	**Shift+]**	Toggles *OTRACK*
Shift+E	**Shift+P**	*Endpoint* Osnap override
Shift+V	**Shift+M**	*Midpoint* Osnap override
Shift+C	**Shift+,**	*Center* Osnap override
Shift+D	**Shift+L**	Disable all snapping and tracking
Shift+S	**Shift+;**	Enables Osnap enforcement
	Shift+/	Toggles *DUCS*

Special Key Functions

Esc	The Escape key cancels a command, menu, or dialog box or interrupts processing of plotting or hatching.
Space bar	In AutoCAD, the space bar performs the same action as the Enter key or #2 button. Only when you are entering text into a drawing does the space bar create a space.
Enter	If Enter or Space bar is pressed when no command is in use (the open Command: prompt is visible), the last command used is invoked again.

E

APPENDIX E

COMMAND TABLE INDEX

Command Name (type)	Button	Pull-down Menu	Alias (type)	Short Cut	Screen (side) Menu	Tablet Menu	Chapter in this Text
3D		...	...	...	DRAW 2 Surface 3d Objec:	J,7-O,7	40
3DALIGN		Modify 3D Operations > 3D Align	3AL	...	MODIFY2 3D Align	...	38
3DARRAY		Modify 3D Operations > 3D Array	3A	...	MODIFY2 3Darray	W,20	38
3DCLIP		...	...	...	...	...	35
3DCONFIG		...	...	...	...	...	41
3DCORBIT		View Orbit > Continuous Orbit	...	...	...	...	35
3DDISTANCE		View Camera > Distance	...	...	...	...	35
3DFACE		Draw Modeling > Meshes > 3Dface	3F	...	DRAW 2 SURFACES 3Dface	M,8	40
3DFLY		View > Walk and Fly > Fly	...	...	...	...	41
3DFORBIT		View Orbit > Free Orbit	...	...	...	...	35
3DMESH		Draw Modeling > Meshes > 3Dmesh	...	...	DRAW 2 SURFACES 3Dmesh:	...	40
3DMOVE		Modify > 3D Operations > 3D Move	3M	...	...	...	38
3DORBIT		View Orbit > Constrained Orbit	3DO	...	VIEW 3dorbit	R,5	35
3DPAN		...	...	...	...	...	35
3DPOLY		Draw 3D Polyline	3P	...	DRAW 1 3Dpoly	O,10	37
3DROTATE		Modify 3D Operations > 3D Rotate	3R	...	MODIFY2 3DRotate	W,22	38
3DSIN		Insert 3D Studio...	...	...	INSERT 3DSin	...	32
3DSWIVEL		View Camera > Swivel	...	...	...	...	35

Command Name (type)	Button	Pull-down Menu	Alias (type)	Short Cut	Screen (side) Menu	Tablet Menu	Chapter in this Text
3DWALK		View > Walk and Fly > Walk	3DW	...	...	...	41
3DZOOM		...	...	...	...	...	35
ACISIN		Insert ACIS File...			INSERT ACISin		32, 39
ACISOUT		File Export... *.sat	... ...	... ...	FILE Export *.sat	... ...	32, 39
ADCENTER		Tools Palettes > DesignCenter	ADC	Ctrl+2	...	...	21, 30
ALIGN		Modify 3D Operations > Align	AL	...	MODIFY2 Align	X,14	16, 38
ANIPATH		View > Motion Path Animations	...	...	...	...	41
ARC		Draw Arc >	A	...	DRAW 1 Arc	R,10	8
AREA		Tools Inquiry > Area	AA	...	TOOLS 1 Area	T,7	17
ARRAY -ARRAY		Modify Array	AR, -AR	...	MODIFY1 Array	V,18	9
ASSIST		Help Info Palette	...	...	...	...	5
ATTDEF -ATTDEF		Draw Block > Define Attributes...	ATT, -ATT	...	DRAW 2 Attdef	...	22
ATTDISP		View Display > Attribute Display >	...	...	VIEW 2 Attdisp	L,1	22
ATTEDIT -ATTEDIT		Modify... Object > Attribute > Global	ATE, -ATE	...	MODIFY1 -Attedit	...	22
ATTEXT -ATTEXT		...	...	...	...	...	22
ATTREDEF		...	...	...	...	...	22
ATTSYNC		...	...	...	...	...	22
AUDIT		File Drawing Utilities > Audit	...	...	FILE Audit	...	43
BACTION		...	AC	...	...	...	21

Command Name (type)	Button	Pull-down Menu	Alias (type)	Short Cut	Screen (side) Menu	Tablet Menu	Chapter in this Text
BASE		*Draw* *Block >* *Base*	...	...	DRAW 2 *Base*	...	21
BATTMAN		*Modify* *Object >* *Attribute >* *Block Attribute* *Manager...*	...	...	...	...	22
BAUTHORPALETTE		...	...	...	...	...	21
BCLOSE		...	BC	...	...	...	21
BEDIT		*Tools* *Block Editor*	BE	...	...	...	21
BLOCK *-BLOCK*		*Draw* *Block >* *Make...*	B, -B	...	DRAW 2 *Bmake*	N,9	21, 22
BMPOUT		*File* *Export* *Bitmap (*.bmp)*	...	...	FILE *Export* *Bitmap (*.bmp)*	...	32
BOUNDARY *-BOUNDARY*		*Draw* *Boundary...*	BO, -BO	...	DRAW 2 *Boundary*	Q,9	15
BOX		*Draw* *Modeling >* *Box*	...	...	DRAW 2 *SOLIDS* *Box*	J,7	38
BPARAMETER		...	PARAM	...	...	...	21
BREAK		*Modify* *Break*	BR	...	MODIFY2 *Break*	W,17	9
BREP		...	...	...	...	...	38
BROWSER		...	...	...	...	...	23
BVHIDE		...	...	...	...	...	21
BVSAVE		...	BS	...	...	...	21
BVSAVEAS		...	...	...	...	...	21
BVSHOW		...	...	...	...	...	21
BVSTATE		...	VS	...	...	...	21

Command Name (type)	Button	Pull-down Menu	Alias (type)	Short Cut	Screen (side) Menu	Tablet Menu	Chapter in this Text
CAMERA		View, Create Camera	CAM	...	...	...	35
CHAMFER		Modify Chamfer	CHA	...	MODIFY2 Chamfer	W,18	9, 38
CHANGE		...	-CH	...	...	...	16
CHPROP		...	...	...	...	...	11, 16
CIRCLE		Draw Circle >	C	...	DRAW 1 Circle	J,9	3, 8
CLEANSCREENOFF		View Clean Screen	...	Ctrl+0	...	...	1
CLEANSCREENON		View Clean Screen	...	Ctrl+0		...	1
CLOSE		File Close	...	...	...	...	2
CLOSEALL		Window Close All	...	...	...	...	2
COLOR		Format Color...	COL	...	FORMAT Color	U,4	11
CONE		Draw Modeling > Cone		...	DRAW 2 SOLIDS Cone	M,7	38
CONVERTCTB		...	...	...	...	...	33
CONVERTTOLDLIGHTS		...	...	...	...	...	41
CONVERTTOLD-MATERIALS		...	...	...	...	...	41
CONVERTPSTYLES		...	...	...	...	...	33
COPY		Modify Copy	CO, CP	(Edit Mode) Copy Selection	MODIFY1 Copy	V,15	1, 4, 9
COPYBASE		Edit Copy with Base Point	...	(Default Menu) Copy with Base Point	EDIT CopyBase	...	31
COPYCLIP		Edit Copy	...	Ctrl+C or (Default Menu) Copy	EDIT Copyclip	T,14	31
COPYTOLAYER		Format, Layer Tools > Copy Objects to New Layer	...	...	...	...	11
COPYLINK		Edit Copy Link	...	...	EDIT Copylink	...	31
CUI		Tools Customize > Interface...	...	...	...	...	44, 45
CUTCLIP		Edit Cut	...	Ctrl+X	EDIT Cut	T,13	31

Command Name (type)	Button	Pull-down Menu	Alias (type)	Short Cut	Screen (side) Menu	Tablet Menu	Chapter in this Text
CYLINDER		Draw Modeling > Cylinder	...	...	DRAW 2 SOLIDS Cylinder	L,7	38
DASHBOARD		Tools Palettes > Dashboard	...	...	...	...	35
DBLCLKEDIT		...	...	...	...	...	16
DBLIST		...	...	...	...	...	17
DDEDIT		Modify Object > Text > Edit	ED	...	MODIFY1 Ddedit	Y,21	18
DDGRIPS		Tools Options... Selection	GR		TOOLS 2 Options... Selection	...	19
DDPTYPE		Format Point Style...	...	...	DRAW 2 Point Ddptype:	U,1	8
DDVPOINT		View 3D Views > Viewpoint Presets..	VP	...	VIEW 1 Ddvpoint	N,5	35
DIMALIGNED		Dimension Aligned	DAL	...	DIMNSION Aligned	W,4	28
DIMANGULAR		Dimension Angular	DAN	...	DIMNSION Angular	X,3	28
DIMARC		Dimension Arc Length	DAR	...	...	...	28
DIMBASELINE		Dimension Baseline	DBA	...	DIMNSION Baseline	...	28
DIMCENTER		Dimension Center Mark	DCE	...	DIMNSION Center	X,2	28
DIMCONTINUE		Dimension Continue	DCO	...	DIMNSION Continue	...	28
DIMDIAMETER		Dimension Diameter	DDI	...	DIMNSION Diameter	X,4	28
DIMDISASSOCIATE		...	DDA	...	...	...	28
DIMEDIT		Dimension Oblique	DED	...	DIMNSION Dimedit	Y,1	28, 29
DIMJOG		Dimension Jogged	DJO	...	...	...	28
DIMLINEAR		Dimension Linear	DLI	...	DIMNSION Linear	W,5	28
DIMORDINATE		Dimension Ordinate	DOR	...	DIMNSION Ordinate	W,3	28
DIMOVERRIDE		Dimension Override	DOV	...	...	Y,4	29
DIMRADIUS		Dimension Radius	DRA	...	DIMNSION Radius	X,5	28

Command Name (type)	Button	Pull-down Menu	Alias (type)	Short Cut	Screen (side) Menu	Tablet Menu	Chapter in this Text
DIMREASSOCIATE		Dimension Reassociate Dimensions	DRE	...	...	...	28
DIMREGEN		...	...	...	...	...	28
DIMSTYLE		Dimension Style...	D, DST	...	DIMNSION Ddim	Y,5	29
-DIMSTYLE		Dimension Update	...	...	DIMNSION Dimstyle	Y,3	29
DIMTEDIT		Dimension Align Text >	DIMTED	...	DIMNSION Dimtedit	Y,2	28, 29
DIST		Tools Inquiry > Distance	DI	...	TOOLS 1 Dist	T,8	17
DISTANTLIGHT		View > Render > Light > New Distant Light	...	...			41
DIVIDE		Draw Point > Divide	DIV	...	DRAW 2 Divide	V,13	15
DONUT		Draw Donut	DO	...	DRAW 1 Donut	K,9	15
DRAWINGRECOVERY		Drawing Utilities Recovery	DRM	...	...	...	2
DRAWORDER		Tools Display Order >	DR	...	TOOLS 1 Drawordr	T,9	26, 32
DSETTINGS		Tools Drafting Settings	DS, SE	Status Bar (Right Click) Settings...	TOOLS 2 Osnap..., or Grid or Polar	W,10	6
DSVIEWER		View Aerial View	AV	...	VIEW 1 Dsviewer	K,2	10
DWFATTACH		Insert, DWF Underlay	...	...	...	...	23
DWFCLIP		...	...	...	...	...	23
DWGPROPS		File Drawing Properties	...	...	...	...	2
DXBIN		Insert Drawing Exchange Binary...	...	...	INSERT DXBin	...	32
DXFIN		...	...	...	...	...	32
DXFOUT		File SaveAs DXF (*.dxf)	...	...	FILE SaveAs DXF (*.dxf)	...	32
EATTEDIT		Modify Object > Attribute > Single...	...	...	...	...	22
EATTEXT		Tools Attribute Extraction...	...	...	...	...	22

Command Name (type)	Button	Pull-down Menu	Alias (type)	Short Cut	Screen (side) Menu	Tablet Menu	Chapter in this Text
EDGE		Draw Modeling > Meshes > Edge	...	...	DRAW 2 SURFACES Edge:	...	40
EDGESURF		Draw Modeling > Meshes > Edge Mesh	...	...	DRAW 2 SURFACES Edgsurf:	R,8	40
ELEVATION		...	...	...	...	...	40
ELLIPSE		Draw Ellipse	EL	...	DRAW 1 Ellipse	M,9	15, 25
ERASE		Modify Erase	E	(Edit Mode) Erase	MODIFY1 Erase	V,14	4, 9
ETRANSMIT		File eTransmit...	...	...	...	...	23
EXPLODE		Modify Explode	X	...	MODIFY2 Explode	Y,22	16, 21
EXPORT		File Export...	EXP	...	FILE Export	...	32
EXTEND		Modify Extend	EX	...	MODIFY2 Extend	W,16	9
EXTERNAL-REFERENCES		Insert, External References	...	...	INSERT Xref	...	32
EXTRUDE		Draw Modeling > Extrude	EXT	...	DRAW 2 SOLIDS Extrude	P,7	38
FIELD		Insert Field...	...	...	...	...	18
FILL		...	...	...	...	...	43
FILLET		Modify Fillet	F	...	MODIFY2 Fillet	W,19	9, 38
FILTER		...	FI	...	ASSIST Filters	...	20
FIND		Edit Find...	...	(Default Menu) Find...	...	X,10	18, 22
FLATSHOT		...	FSHOT	...	...	...	42
GEOGRAPHIC-LOCATION		View Render > Light > Geographic Location...	GEO	...	...	...	41
GRID		Tools Drafting Settings... Snap and Grid	...	F7 or Ctrl+G	TOOLS 2 Grid Snap and Grid	W,10	6, 12, 35
GRIPS		Tools Options... Selection Enable Grips	...	...	TOOLS 2 Options Selection Enable Grips	...	19

Command Name (type)	Button	Pull-down Menu	Alias (type)	Short Cut	Screen (side) Menu	Tablet Menu	Chapter in this Text
GROUP -GROUP		...	G, -G	...	ASSIST Group	X,8	20
HATCH -HATCH		Draw Hatch...	H, BH, - H	...	DRAW 2 Bhatch	P,9	26
HATCHEDIT -HATCHEDIT		Modify Hatch...	HE, -HE	...	MODIFY1 Hatchedt	Y,16	26
HELP		Help Help	?	F1	HELP Help	Y,7	5
HIDE		View Hide	HI	...	VIEW 2 Hide	M,2	35
HYPERLINK		Insert Hyperlink...	...	Ctrl+K	...	...	23
ID		Tools Inquiry > ID Point	...	...	TOOLS 1 ID	U,9	17
IMAGE -IMAGE		Insert External References	IM, -IM	...	INSERT Image	T,3	32
IMAGEADJUST		Modify Object > Image > Adjust...	IAD	...	MODIFY1 Imageadj	X,20	32
IMAGEATTACH		Insert Raster Image Reference	IAT	...	INSERT Image	...	32
IMAGECLIP		Modify Clip > Image	ICL	...	MODIFY1 Imageclp	X,22	32
IMAGEFRAME		Modify Object > Image > Frame	...	...	MODIFY1 Imagefrm	...	32
IMAGEQUALITY		Modify Object > Image > Quality	...	...	MODIFY1 Imagequa	...	32
IMPORT		...	IMP	...	...	T,2	32
IMPRINT		Modify Solid Editing > Imprint Edges	...	...	...	...	39
INSERT -INSERT		Insert Block...	I, -I	...	INSERT Ddinsert	T,5	21
INSERTOBJ		Insert OLE Object...	IO	...	INSERT Insertob	T,1	31
INTERFERE		Modify 3D Operations > Interference Checking	INF	...	DRAW 2 SOLIDS Interfer	...	39

Command Name (type)	Button	Pull-down Menu	Alias (type)	Short Cut	Screen (side) Menu	Tablet Menu	Chapter in this Text
INTERSECT		Modify Solids Editing > Intersect	IN	...	MODIFY2 Intrsect	X,17	16, 38
JOIN		Modify Join	J	...	...	...	9
JPGOUT		...	...	...	...	...	32
JUSTIFYTEXT		Modify Objects > Text > Justify	...	...	...	...	18
LAYCUR		Format Layer Tools > Change to Current Layer	...	...	...	...	11
LAYDEL		Format Layer Tools > Layer Delete	...	...	...	...	11
LAYER -LAYER		Format Layer...	LA, -LA	...	FORMAT Layer	U,5	11, 12, 33
LAYERP		Format Layer Tools > Layer Previous	...	...	...	...	11
LAYFRZ		Format Layer Tools > Layer Freeze	...	...	...	...	11
LAYISO		Format Layer Tools > Layer Isolate	...	...	...	...	11
LAYLCK		Format Layer Tools > Layer Lock	...	...	...	...	11
LAYMCH		Format Layer Tools > Layer Match	...	...	...	...	11
LAYMCUR		Format Layer Tools > Make Object's Layer Current	...	...	...	...	11
LAYMRG		Format Layer Tools > Layer Merge	...	...	...	...	11
LAYOFF		Format Layer Tools > Layer Off	...	...	...	...	11
LAYON		Format Layer Tools > Turn All Layers On	...	...	...	...	11
LAYOUT -LAYOUT		Insert Layout >	LO	...	...	...	13
LAYOUTWIZARD		Insert Layout > Layout Wizard	...	...	...	...	13

Command Name (type)	Button	Pull-down Menu	Alias (type)	Short Cut	Screen (side) Menu	Tablet Menu	Chapter in this Text
LAYTHW		Format Layer Tools > Thaw All Layers	...	...	...	...	11
LAYULK		Format Layer Tools > Layer Unlock	...	...	...	...	11
LAYUNISO		Format Layer Tools > Layer Unisolate	...	...	...	...	11
LAYVPI		Format Layer Tools > Isolate Layer to Current Viewport	...	...	...	...	11
LAYWALK		Format Layer Tools > Layer Walk	...	...	...	...	11
LEADER		Dimension Leader	LEAD	...	DIMNSION Leader	R,7	28
LENGTHEN		Modify Lengthen	LEN	...	MODIFY2 Lengthen	W,14	9
LIGHT		View Render > Light...	...	...	VIEW 2 Light	O,1	41
LIGHTLIST		View Render > Light > Light List	...	...	...	...	41
LIMITS		Format Drawing Limits	...	...	FORMAT Limits	V,2	6, 11
LINE		Draw Line	L	...	DRAW 1 Line	J,10	3, 8
LINETYPE -LINETYPE		Format Linetype...	LT, -LT	...	FORMAT Linetype	U,3	11
LIST		Tools Inquiry > List	LS, LI	...	TOOLS 1 List	U,8	17
LOFT		Draw Modeling > Loft	...	...	DRAW 2 SOLIDS Loft	...	39
LOGFILEON / LOGFILEOFF		Tools Options... Open and Save Maintain log file	...	...	TOOLS 2 Options... Open and Save Maintain log file	...	44
LTSCALE		Format Linetype... Show details Global scale factor	LTS	...	FORMAT Linetype Show details Global scale factor	...	11, 12
LWEIGHT		Format Lineweight...	LW	...	...		11

Command Name (type)	Button	Pull-down Menu	Alias (type)	Short Cut	Screen (side) Menu	Tablet Menu	Chapter in this Text
MARKUP		Tools Palettes > Markup Set Manager	MSM	Ctrl+7	...	...	23
MASSPROP		Tools Inquiry > Mass Properties	...	...	TOOLS 1 Massprop	U,7	39
MATCHPROP		Modify Match Properties	MA	...	MODIFY1 Matchprp	Y,14-Y,15	11, 16, 29
MATERIALATTACH		...	...	...	...	...	41
MATERIALMAP		View > Render > Mapping	SETENV	...	VIEW 2 Mapping	R,1	41
MATERIALS		View Render > Materials...	RMAT	...	VIEW 2 Rmat	P,1	41
MEASURE		Draw Point > Measure	ME	...	DRAW 2 Measure	V,12	15
MINSERT		...	...	...	...	...	21
MIRROR		Modify Mirror	MI	...	MODIFY1 Mirror	V,16	9
MIRROR3D		Modify 3D Operations > Mirror 3D	...	...	MODIFY2 Mirror3D	W,21	38
MLEDIT -MLEDIT		Modify Object > Multiline...	...	...	MODIFY1 Mledit	Y,19	16
MLINE		Draw Multiline	ML	...	DRAW 1 Mline	M,10	15
MLSTYLE		Format Multiline Style...	...	...	DRAW 1 Mline Mlstyle:	V,5	15
MODEL		...	...	...	...	...	13
MOVE		Modify Move	M	(Edit Mode) Move	MODIFY2 Move	V,19	4, 9, 38
MREDO		...	...	...	...	5	
MSLIDE		...	...	...	...	...	43
MSPACE		...	MS	...	VIEW 1 Mspace	L,4	13
MTEDIT		...	...	...	...	...	18
MTEXT -MTEXT		Draw Text > Multiline Text...	T, MT, -T	...	DRAW 2 Mtext	J,8	18
MULTIPLE		...	...	...	...	...	43
MVIEW		...	MV	...	VIEW 1 Mview	M,4	33

Command Name (type)	Button	Pull-down Menu	Alias (type)	Short Cut	Screen (side) Menu	Tablet Menu	Chapter in this Text
MVSETUP		...	...	...	...	...	42
NEW	▢	File New...	...	Ctrl+N	FILE New	T,24	2
NEWSHEETSET		File New Sheet Set...	...	*(in Sheet Set Manager)* New Sheet Set	...	...	34
OFFSET	⬗	Modify Offset	O	...	MODIFY1 Offset	V,17	9, 27
OLELINKS		Edit OLE Links...	...	...	EDIT OLElinks	...	31
OLEOPEN		...	...	...	...	...	31
OLESCALE		...	...	...	...	...	31
OOPS		...	...	...	MODIFY1 Erase Oops:	...	5
OPEN	▤	File Open...	...	Ctrl+O	FILE Open	T,25	2
OPENDWFMARKUP	▦	File Load Markup Set...	...	...	...	...	23
OPENSHEETSET		File Open Sheet Set...	...	...	...	...	34
OPTIONS		Tools Options...	OP	*(Default Menu)* Options...	TOOLS 2 Options	Y,10	13, 20, 43
PAGESETUP	▥	File Page Setup Manager	...	...	...	V,25	13, 14
PAN -PAN	✋	View Pan	P, -P	*(Default Menu)* Pan	VIEW 1 Pan	N,11 - P,11	10, 35
PARTIALLOAD		File Partial Load	...	...	...	...	2
PARTIALOPEN		File Open Partial Open	...	...	FILE Open Partial Open	...	2
PASTEASHYPERLINK		Edit Paste as Hyperlink	...	...	...	...	23
PASTEBLOCK		Edit Paste as Block	...	*(Default Menu)* Paste as Block	EDIT PasteBlk	...	31
PASTECLIP	▣	Edit Paste	...	Ctrl+V	EDIT Pasteclp	U,13	31
PASTEORIG		Edit Paste to Original Coordinates	...	*(Default Menu)* Paste to Orig. Coordinates	EDIT PasteOri	...	31
PASTESPEC		Edit Paste Special...	PA	...	EDIT Pastespe	...	31

Command Name (type)	Button	Pull-down Menu	Alias (type)	Short Cut	Screen (side) Menu	Tablet Menu	Chapter in this Text
PCINWIZARD		Tools Wizards > Import Plot Settings...	...	...	...	...	14
PEDIT		Modify Object > Polyline...	PE	...	MODIFY1 Pedit	Y,17	16
PLAN		View 3D Views > Plan View >	...	...	VIEW 1 Plan	N,3	35
PLANESURF		Draw Modeling > Planar Surface	...	...	DRAW 2 SURFACES Planar Surface	...	39
PLINE		Draw Polyline	PL	...	DRAW 1 Pline	N,10	8
PLOT -PLOT		File Plot	PRINT	Ctrl+P	FILE Plot	W,25	14
PLOTSTAMP		File Plot Plot Device Plot Stamp Settings...	...	...	...	...	33
PLOTSTYLE -PLOTSTYLE		Format Plot Style...	...	...	...	...	33
PLOTTERMANAGER		File Plotter Manager...	...	...	...	...	14
PNGOUT		...	...	...	...	...	32
POINT		Draw Point >	PO	...	DRAW 2 Point	O,9	8
POINTLIGHT		View > Render > Light > New Point Light >	...	...	...	...	41
POLYGON		Draw Polygon	POL	...	DRAW 1 Polygon	P,10	15
POLYSOLID		Draw Modeling > Polysolid	PSOLID	...	DRAW 2 SOLIDS Polysolid	...	38
PRESSPULL		...	...	Ctrl+Alt	...	...	39
PREVIEW		File Plot Preview	PRE	...	...	X,24	14
PROPERTIES		Modify Properties	PR	(Edit Mode) Properties or Ctrl+1	MODIFY1 Property	Y,12-Y,13	11, 16, 18, 28, 29, 32
PSETUPIN -PSETUPIN		...	...	...	...	...	13
PSOUT		...	...	...	...	...	32
PSPACE		...	PS	...	VIEW 1 Pspace	L,5	13

Command Name (type)	Button	Pull-down Menu	Alias (type)	Short Cut	Screen (side) Menu	Tablet Menu	Chapter in this Text
PUBLISH		File Publish	...	...	...	...	23, 34
PUBLISHTOWEB		File Publish to Web...	PTW	...	...	...	23
PURGE		File Drawing Utilities > Purge >	PU	...	FILE Purge	X,25	21
PYRAMID		Draw Modeling > Pyramid	PYR	...	DRAW 2 SOLIDS Pyramid	...	38
QDIM		Dimension Quick Dimension	...	...	...	W,1	28
QLEADER		Dimension Leader	LE	...	DIMNSION Leader	W,2	28
QNEW		...	...	...	...	...	1, 2, 6
QSAVE		...	...	Ctrl+S	FILE Qsave	U,24-U,25	2
QSELECT		Tools Quick Select...	...	(Default Menu) Quick Select...	...	X,9	20
QTEXT		...	...	...	...	...	18
QUICKCALC		Tools QuickCalc	QC	Ctrl+8	...	...	43
QUIT		...	EXIT	...	FILE Quit	...	2
RAY		Draw Ray	...	...	DRAW 1 Ray	K,10	15, 27
RECOVER		File Drawing Utilities > Recover...	...	...	FILE Recover	...	2
RECTANG		Draw Rectangle	REC	...	DRAW 1 Rectang	Q,10	15
REDO		Edit Redo	...	Crtl+Y or (Default Menu) Redo	EDIT Redo	U,12	5
REDRAW		View Redraw	R	...	VIEW 1 Redraw	...	5
REDRAWALL		View Redraw	RA	...	VIEW 1 Redraw	...	5
REFCLOSE		Modify In-place Xref and Block Edit > Save Reference Edits or Discard Ref. Edits	...	...	...	...	21, 30

Command Name (type)	Button	Pull-down Menu	Alias (type)	Short Cut	Screen (side) Menu	Tablet Menu	Chapter in this Text
REFEDIT		Modify In-place Xref and Block Edit > Edit Reference	...	...	MODIFY2 Refedit	...	22, 30
REFSET		Modify In-place Xref and Block Edit > Add to Workset or Remove from Workset	...	...	...	...	30
REGEN		View Regen	RE	...	VIEW 1 Regen	J,1	5
REGENALL		View Regen All	REA	...	VIEW 1 Regenall	K,1	5
REGION		Draw Region	REG	...	DRAW 2 Region	R,9	15
REINIT		...	...	...	...	...	43
RENAME -RENAME		Format Rename...	REN - REN	...	FORMAT Rename	V,1	21
RENDER		View Render > Render...	RR	...	VIEW 2 Render	M,1	41
RENDERCROP		...	RC	...	...	...	41

RENDERENVIRON-MENT		View >
		Render >
		Render
		  Environment
			FOG	...	VIEW 2
					Render
					  Environment
						P,2	41

RENDERPRESETS		...	RP	...		...	41
RENDERWIN		...	RW	...	...	...	41
REVCLOUD		Draw Revision Cloud	...	...	...	...	15
REVOLVE		Draw Modeling > Revolve	REV	...	DRAW 2 SOLIDS Revolve	Q,7	38
REVSURF		Draw Modeling > Meshes > Revolved Mesh	...	...	DRAW 2 SURFACES Revsurf:	O,8	40
RMAT		View Render > Materials...	...	...	VIEW 2 Rmat	P,1	41
ROTATE		Modify Rotate	RO	(Edit Mode) Rotate	MODIFY2 Rotate	V,20	9
ROTATE3D		Modify 3D Operations > Rotate 3D	...	...	MODIFY2 Rotate3D	W,22	38

Command Name (type)	Button	Pull-down Menu	Alias (type)	Short Cut	Screen (side) Menu	Tablet Menu	Chapter in this Text
RPREF		View Render > Advanced Render Settings	RPR	...	VIEW 2 Rpref	R,2	41
RULESURF		Draw Modeling > Meshes > Ruled Mesh	...	...	DRAW 2 SURFACES Rulsurf:	Q,8	40
SAVE		File Save	...	Ctrl+S	...	U,24 - U,25	2
SAVEAS		File Save As...	...	...	FILE Saveas	V,24	2
SAVEIMG		Tools Display Image > Save...	...	...	TOOLS 1 Saveimg	...	32, 41
SCALE		Modify Scale	SC	(Edit Mode) Scale	MODIFY2 Scale	V,21	9
SCALELISTEDIT		Format Scale List...	...	...	...	...	13
SCALETEXT		Modify Object > Text > Scale	...	...	...	...	18
SCRIPT		Tools Run Script...	SCR	...	TOOLS 1 Script	V,9	43
SECTION		...	SEC	...	DRAW 2 SOLIDS Section	...	39
SECURITYOPTIONS		File Save As... Tools Security Options...	...	...	FILE Saveas Tools Security Options...	...	46
SECTIONPLANE		Draw Modeling > Section Plane	SPANE	...	...	...	39
SELECT		...	...	...	...	...	4
SETUV		View Render > Mapping...	...	...	VIEW 2 Mapping	R,1	41
SETVAR		Tools Inquiry > Set Variable	SET	...	TOOLS 1 Setvar	U,10	17
SHADE		...	SHA	...	...	N,2	35
SHEETSET		Tools Palettes > Sheet Set Manager	SSM	Ctrl+4	...	...	34
SKETCH		...	...	...	...	...	15
SLICE		Modify 3D Operations > Slice	SL	...	DRAW 2 SOLIDS Slice	...	38, 39

Command Name (type)	Button	Pull-down Menu	Alias (type)	Short Cut	Screen (side) Menu	Tablet Menu	Chapter in this Text
SNAP		Tools Drafting Settings... Snap and Grid	SN	F9 or Ctrl+B	TOOLS 2 Grid Snap and Grid	W,10	6, 12, 25, 27
SOLDRAW		Draw Modeling > Setup >	...	...	DRAW 2 SOLIDS Soldraw	...	42
SOLID		Draw Modeling > Meshes > 2D Solid	SO	...	DRAW 2 SURFACES Solid:	L,8	15
SOLIDEDIT		Modify Solids Editing >	...	...	...	...	38
SOLPROF		Draw Modeling > Setup > Profile	...	...	DRAW 2 SOLIDS Solprof	...	42
SOLVIEW		Draw Modeling > Setup > View	...	...	DRAW 2 SOLIDS Solview	...	42
SPACETRANS		...	...	...	...	...	18
SPELL		Tools Spelling	SP	...	TOOLS 1 Spell	T,10	18
SPHERE		Draw Modeling > Sphere	...	...	DRAW 2 SOLIDS Sphere	K,7	38
SPLINE		Draw Spline	SPL	...	DRAW 1 Spline	L,9	15
SPLINEDIT		Modify Splinedit	SPE	...	MODIFY1 Splinedt	Y,18	16
SPOTLIGHT		View Render > Light > New Spot Light	...	...	...	...	41
STATUS		Tools Inquiry > Status	...	...	TOOLS 1 Status	...	17
STLOUT		File Export... *.stl	...	...	FILE Export *.stl	...	32, 39
STRETCH		Modify Stretch	S	...	MODIFY2 Stretch	V,22	9
STYLE -STYLE		Format Text Style...	ST	...	FORMAT Style	U,2	18
STYLESMANAGER		File Plot Style Manager...	...	...	...	...	33
SUBTRACT		Modify Solids Editing > Subtract	SU	...	MODIFY2 Subtract	X,16	16, 38

Command Name (type)	Button	Pull-down Menu	ALIAS (type)	Short Cut	Screen (side) Menu	Tablet Menu	Chapter in this Text
SUNPROPERTIES		Draw Render > Light > Sun Properties	...	...	...	...	41
SWEEP		Draw Modeling > Sweep	...	...	DRAW 2 SOLIDS Sweep	...	39
SYSWINDOWS		Window	...	...	...	...	10
TABLE		Draw Table...	TB	...	...	...	18
TABLEDIT		...	...	...	...	...	18
TABLESTYLE		Format Tablestyle...	TS	...	...	...	18
TABSURF		Draw Modeling > Meshes > Tabuled Mesh	...	...	DRAW 2 SURFACES Tabsurf:	P,8	40
TEXT		Draw Text > Single Line Text...	DT	...	DRAW 2 Dtext	K,8	18
TEXTTOFRONT		Tools Bring Text and Dimensions to Front	...	...	...	...	26
THICKEN		Modify 3D Operations > Thicken	...	...	DRAW 2 SOLIDS Thicken	...	39
THICKNESS		Format Thickness	TH	...	FORMAT Thicknes	V,3	40
TIFOUT		...	...	...	...	...	32
TIME		Tools Inquiry > Time	...	...	TOOLS 1 Time	...	17
TOLERANCE		Dimension Tolerance...	TOL	...	DIMNSION Toleranc	X,1	28
TOOLBAR		View Toolbars...	TO	...	VIEW 2 Toolbar	R,3	44, 45
TOOLPALETTES		Tools Palettes > Tool Palettes Window	TP	Ctrl+3	...	...	16, 21, 26
TORUS		Draw Modeling > Torus	TOR	...	DRAW 2 SOLIDS Torus	O,7	38
TRANSPARENCY		Modify Object > Image > Transparency	...	...	MODIFY1 Transpar	...	32
TRAYSETTINGS		...	...	...	...	...	1

Command Name (type)	Button	Pull-down Menu	ALIAS (type)	Short Cut	Screen (side) Menu	Tablet Menu	Chapter in this Text
TRIM		Modify Trim	TR	...	MODIFY2 Trim	W,15	9
U		 Edit Undo	Crtl+Z or (Default Menu) Undo	 ...	 EDIT Undo	 T,12	 5
UCS		Tools Move UCS, or New UCS	...	...	TOOLS 2 UCS	W,7	36
UCSICON		View Display > UCS Icon >	...	...	VIEW 2 UCSicon	L,2	10, 35
UCSMAN		Tools Named UCS...	UC	...	TOOLS 2 Ucsman	W,8	36
UNDO		...	...	...	ASSIST Undo	...	5
UNION		Modify Solids Editing > Union	UNI	...	MODIFY2 Union	X,15	16, 38
UNITS -UNITS		Format Units...	UN -UN	...	FORMAT Units	V,4	6, 12, 25, 27
UPDATEFIELD		Tools Update Fields...	...	...	...	...	18
VIEW -VIEW		View Named Views...	V, -V	...	VIEW 1 Ddview	M,5	10, 35, 41
VIEWPLOTDETAILS		File View Plot and Publish Details	...	...	...	...	14

VIEWRES		Tools
		Options...
		Display
		Display
		    Resolution
			...	...	TOOLS 2
					Options
					Display
					Display
					    Resolution
						...	10

VISUALSTYLES		View Visual Styles > Visual Styles Manager	VSM	...	...	...	35
VPCLIP		Modify Clip > Viewport	...	...	...	...	13
VPLAYER		...	...	...	...	...	33
VPMAX		...	...	...	...	...	13
VPMIN		...	...	...	...	...	13
VPOINT		View 3D Views > Viewpoint	-VP	...	VIEW 1 Vpoint	N,4	35
VPORTS -VPORTS		View Viewports >	...	...	VIEW 1 Vports	M,3 and M,4	10, 13, 33, 35

Command Name (type)	Button	Pull-down Menu	ALIAS (type)	Short Cut	Screen (side) Menu	Tablet Menu	Chapter in this Text
VSCURRENT		View Visual Styles >	VS	...	...	...	35
VSLIDE		...	...	...	...	...	43
WALKFLYSETTINGS		View Walk and Fly > Walk and Fly Settings		...	...	...	41
WBLOCK -WBLOCK		File Export...	W, -W	...	FILE Export	W,24	21
WEDGE		Draw Modeling > Wedge	WE	...	DRAW 2 SOLIDS Wedge	N,7	38
WIPEOUT		Draw Wipeout	...	...	...	...	15
XATTACH		Insert Drawing Reference	XA	...	INSERT Xref Attach...	T,4	30
XBIND -XBIND		Modify Object > External Reference > Bind...	XB, -XB	...	MODIFY1 Xbind	X,19	30
XCLIP		Modify Clip > Xref	XC	...	MODIFY1 Xclip	X,18	30
XEDGES		Modify 3D Operations > Extract Edges	...	...	...	...	37
XLINE		Draw Construction Line	XL	...	DRAW 1 Xline	L,10	15, 27
XOPEN		Modify Xref and Block Editing Open Reference	...	...	...	...	2, 30
XPLODE		...	XP	...	...	...	21
XREF -XREF		Insert Xref Manager...	XR, -XR	...	INSERT Xref	T,4	30
ZOOM		View Zoom >	Z	(Default Menu) Zoom	VIEW 1 Zoom	J,2-J,5 or K,3-K,5	10, 35

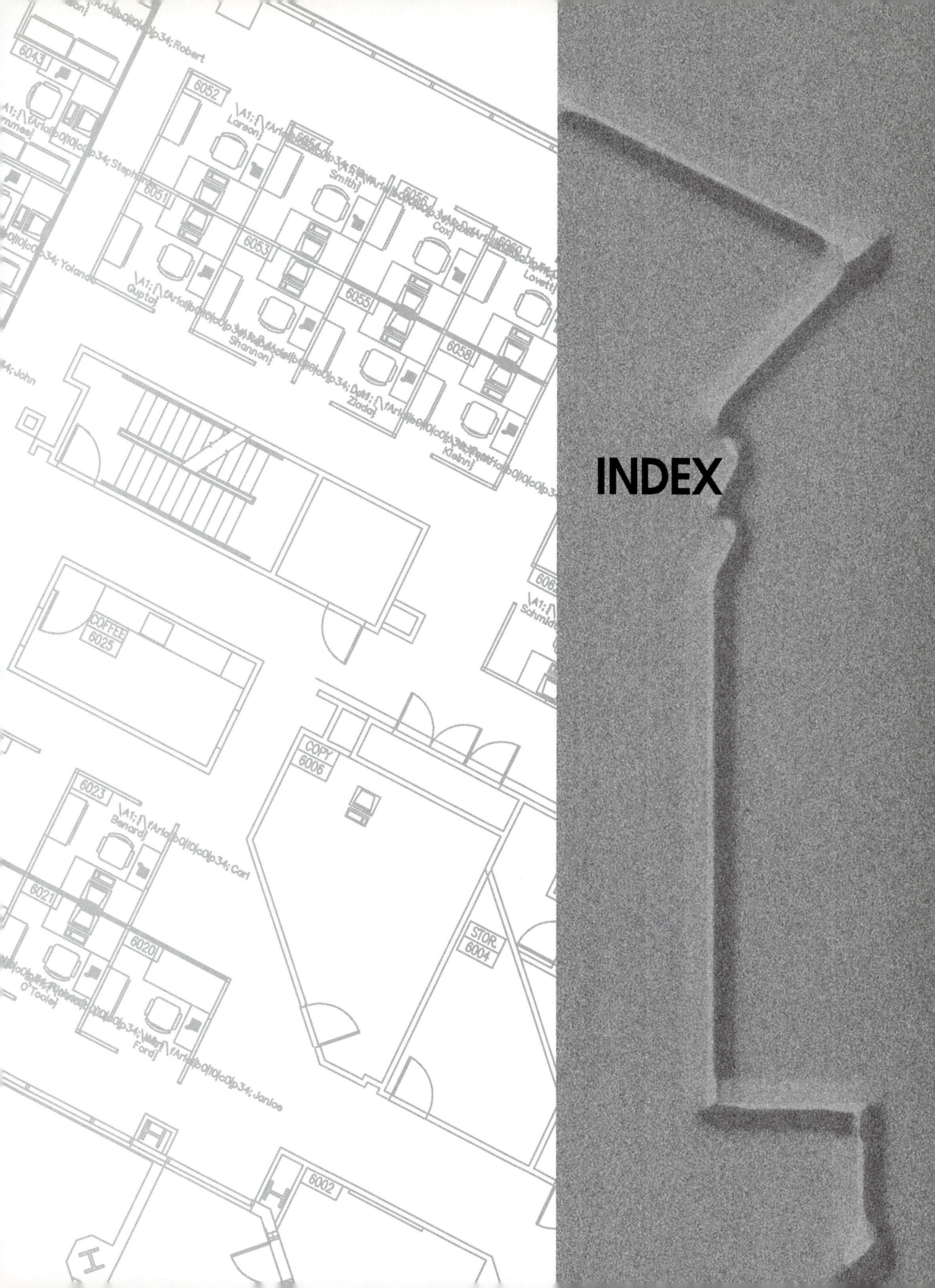

INDEX

@ (last point), 412
.3DS, 878, 879
.AVI, 1254, 1260, 1262, 1264
.BAK, 42
.BIL, 869
.BLK, 579
.BMP, 601, 869, 876, 883, 1213, 1236
.CDR, 601
.CG4, 869
.CGM, 601
.CLP, 601
.CMX, 601
.CSV, 579
.CTB, 332, 909
.DGN, 601
.DIB, 601, 869,1213, 1236
.DST, 934, 936, 955
.DWF, 582, 585, 588, 589, 851
.DWF and .DWG, Transferring over the Internet, 616
.DWF Files, Creating, 600
.DWF Files, Creating and Viewing, 600
.DWF Files, Viewing, 608
.DWG, 26, 33, 35, 42, 286, 503, 601, 883, 908, 934
.DWS, 33, 35
.DWT, 33, 35, 264, 266, 271, 1314
.DXB, 881
.DXF, 3, 33, 35, 867, 883
.DXX, 582, 883
.EPS, 867, 881, 883
.EXE, 594, 597, 951
.FLC, 869
.FLI, 869
.FMP, 447
.GEM, 601
.GIF, 601, 869, 1213, 1236
.GP4, 869
.HTM, 597
.ICO, 601
.IFF, 601
.IG4, 869
.IGS, 601, 869
.JPG, 588, 594, 595, 600, 601, 867, 869, 876, 1213, 1236
.MAC, 601
.MCS, 601
.MIL, 869
.MLN, 353
.MOV, 1254, 1262, 1264
.MPG, 1254, 1260, 1262, 1264
.MPR, 1165
.MSP, 601
.NIF, 601
.P10, 601
.PBM, 601
.PC2, 334, 335, 336
.PC3, 332, 333, 334, 335, 336
.PCL, 601
.PCP, 334, 335, 336
.PCT, 869
.PCX, 601, 869, 1213, 1236, 1250

.PGL, 601
.PIC, 601
.PLT, 314, 601
.PMP, 335
.PNG, 588, 594, 595, 867, 869, 876, 1213, 1236
.PRT, 601
.PSD, 601
.PSF, 881
.PSS, 922
.PTW, 597
.RAS, 601
.RLC, 869
.RLE, 869, 1213, 1236
.RST, 869
.RTF, 436
.SAT, 867, 878, 879, 883, 1167
.SGI, 601
.SHX, 419, 435, 564
.SHX Fonts, 435
.SLD, 1303
.STB, 332, 908
.STL, 880, 883, 1167
.SV$, 42
.TGA, 601, 869, 876, 1213, 1236
.TIF, 601, 869, 876, 1213, 1236
.TTF, 419, 564
.TXT, 436, 580, 598
.WMF, 883
.WMV, 1254, 1260, 1262, 1264
.WRL, 601
.XBM, 601
.XPM, 601
.XWD, 601
.ZIP, 594, 597, 951
2D Drawings, 622
2D Solid, 361
2D Wireframe, 1001
3D Align, 1088
3D Array, 1096
3D BASICS, NAVIGATION AND VISUAL STYLES, 961
3D Coordinate Entry, 968
3D Coordinate Entry Formats, 968
3D Environment, 962
3D Face, 1179, 1181
3D Free Orbit, 982
3D Hidden, 1001
3D Interface, 962
3D Make control panel, 963, 1067
3D Mesh, 1183
3D Models, Types of, 964
3D Move, 1090
3D Navigate control panel, 963, 989, 1256
3D Navigation and View Commands, 978
3D Navigation control panel, 979
3D Objects..., 1192
3D Orbit, 980
3D Orbit Commands, 980
3D Polyline, 1057
3D Rotate, 1092
3D Studio Max, 878

3D Views, 985
3D Viewing Commands, 980
3D Wireframe, 1001
3Dalign, 1065, 1088
3Darray, 1065, 1096
3Darray, Polar, 1097
3Darray, Rectangular, 1096
3Dclip, 978, 990
3Dclip, Adjust Back Clipping, 990
3Dclip, Adjust Front Clipping, 990
3Dclip, Back Clipping On, 990
3Dclip, Front Clipping On, 990
3Dclip, Slice, 990
3Dcorbit, 978, 983
3Ddistance, 978, 984
3Dface, 1179, 1181
3Dface, Applications of, 1181
3Dfaces, Editing, 1182
3Dfly, 1258
3Dforbit, 978, 982
3Dmesh, 1182, 1183
3Dmove, 1065, 1090
3Dorbit, 978, 980
3Dorbit, Enable Orbit Auto Target, 981
3Dorbit, Extents, 981
3Dorbit, Reset View, 981
3Dorbit, Visual Aids, Compass, 981
3Dorbit, Visual Aids, Grid, 981
3Dorbit, Visual Aids, UCS Icon, 981
3Dorbit, Zoom Previous, 981
3Dorbit, Zoom Window, 981
3Dorbit Shortcut Menu Options, 981
3Dpoly, 1057
3Dpoly, Close, 1057
3Dpoly, Undo, 1057
3Drotate, 1065, 1092
3Dsin, 878
3Dsin, 3Dsout, 867
3Dswivel, 978, 994
3Dwalk, 1257

A

Absolute Cartesian Coordinates, 3D, 969, 970
Absolute coordinates, 48
ACAD.CUI, 12, 1331
ACAD.LIN, 235, 1332
ACAD.DWT, 6, 29, 30, 31, 50, 93, 94, 95, 100, 227, 265, 267, 419, 422, 779
ACAD.LSP, 1319
ACAD.PAT, 659, 670
ACAD.PGP, 12, 1293, 1334
ACAD.PGP File, Customizing, 1334
ACAD.PSF, 881
ACAD3d.dwt, 267, 963
ACAD_ISOnW100 linetypes, 241
ACADISO.DWT, 6, 29, 30, 93, 94, 95, 227, 265, 267, 419, 422
ACADISO.LIN, 235
ACADISO3d.dwt, 267, 963
Accelerator Keys, 15
ACIS, 879, 967, 1064, 1087, 1149, 1167
Acisin, 879, 1167

Acisin, Acisout, 867
Acisout, 880, 1167
Acquire an object, 118
Acquisition Object Snap Modes, 118
Adaptive Degradation, 1011
Adaptive Degradation and
    Performance Tuning, 1011
*Adaptive Degradation and Performance
    Tuning* dialog box, 1253
Adcenter, 518, 816
Adcenter, Back/Forward, 523
Adcenter, Description, 523
Adcenter, Favorites, 521
Adcenter, Folders Tab, 519
Adcenter, History Tab, 520
Adcenter, Load, 521
Adcenter, Open Drawings Tab, 520
Adcenter, Preview, 522
Adcenter, Search, 521
Adcenter, Tree View Toggle, 520
Adcenter, Up, 522
Adcenter, Views, 523
Add, object selection, 72
Add-A-Plot Style Table Wizard, 910,
    912
Add-a-Plotter Wizard, 333
Adjust Amount of Perspective, 989
Adjust Bitmap, Bitmap scale, 1239
Adjust Bitmap, Interactive Adjustment,
    1240
Adjust Bitmap dialog box, 1238
Adjust Background Image dialog box,
    1214
Adjust Clip Planes, 990
ADVANCED DRAWING SETUP, 261
Advanced Layouts, 888
ADVANCED LAYOUTS AND
    PLOTTING, 887
ADVANCED SELECTION SETS, 479
Advanced Setup Wizard, 96
Advanced Render Settings, Diagnostic,
    1246
*Advanced Render Settings, Indirect
    Illumination*, 1246
Advanced Render Settings, Materials,
    1245
Advanced Render Settings, Processing,
    1246
*Advanced Render Settings, Ray
    Tracking*, 1246
Advanced Render Settings, Render,
    1244
*Advanced Render Settings, Render
    Context*, 1244
*Advanced Render Settings, Render
    Presets*, 1244
Advanced Render Settings, Sampling,
    1245
Advanced Render Settings, Shadows,
    1245
Advanced Render Settings palette,
    1244
ADVANCED SOLIDS FEATURES,
    1117
Aerial View, 210

Align, 382
Align, 3D, 1088
Alignment Parameter, 536
Alignment Point Acquisition, 126
Alignment vector, 118
All, object selection, 72
Ambient color, 1231
Angle, Advanced Setup Wizard, 97
Angle Direction, Advanced Setup
    Wizard, 97
Angle Measure, Advanced Setup
    Wizard, 97
Angle Override, 56
Angles in AutoCAD, 3
Animation, *3Dwalk*, 1254
Animation, *3Dfly*, 1254
Animation, Playing Back, 1263
Animation, Saving to a File, 1263
Animation Preview, 1256
Animation Setting, Visual Style, 1263
Animation Setting, Frame rate (FPS),
    1263
Animation Setting, Resolution, 1263
Animation Setting, Format, 1263
Animation Setting dialog box, 1263
Anipath, 1261
ANSI, 628, 629, 779
ANSI, (.DWT), 267
Aperture, 115
Apparent intersection, Object Snap,
    118
Apply to Layout, 318
Arc, 46, 146
Arc, 3Points, 147
Arc ball, *982*
Arc, Center, Start, Angle, 149
Arc, Center, Start, End, 149
Arc, Center, Start, Length, 150
Arc, Continue, 150
Arc, Start, Center, Angle, 147
Arc, Start, Center, End, 147
Arc, Start, Center, Length, 148
Arc, Start, End, Angle, 148
Arc, Start, End, Direction, 149
Arc, Start, End, Radius, 148
Arc ball, 982
Arcs or *Circles*, 150
Area, 410
Area, Add, Subtract, 411
Area, Advanced Setup Wizard, 97
Area, Object, 411
Area, Quick Setup Wizard, 96
Array, 180
Array, Polar, 181
Array, Rectangular, 181
Array Action, 543
Array dialog box, 180
ASCII, 436
ASME, 718
Assist, 85
Associative Dimensions, 731, 740,
    767
Associative Dimensions, Exploding,
    727
Associative Dimensioning, 1321

Associative Hatch, 657, 663
Attdef, 557, 573
Attdef, Attribute Prompt, 557
Attdef, Attribute Tag, 557
Attdef, Attribute Value, 558
Attdef, Constant, 559
Attdef, Invisible, 559
Attdef, Justify, 558
Attdef, Lock Position, 559
Attdef, Preset, 559
Attdef, Style, 558
Attdef, Verify, 559
Attdia Variable, 561
Attdisp, 562
Attdisp, Normal, 562
Attdisp, On/Off, 562
Attedit, 563
Attedit, Individual Editing, 568
Attedit, Multiple, 563
-*Attedit*, 566
-*Attedit*, Global Editing, 566
Attext, 580
Attext, CDF, 582
Attext, DXF, 582
Attext, SDF, 582
Attmode, 563
Attredef, 573
Attreq Variable, 562
Attribute Definition dialog box, 557, 560
Attribute Display, 562
Attribute Extraction wizard, 576
Attribute Extraction wizard, *Additional
    Settings*, 577
Attribute Extraction wizard, *Begin*, 577
Attribute Extraction wizard, *Export*, 582
Attribute Extraction wizard, *Finalize
    Output*, 578
Attribute Extraction wizard, *Finish*,
    579
Attribute Extraction wizard, *Save Template*,
    579
Attribute Extraction wizard, *Select
    Attributes*, 577
Attribute Extraction wizard, *Select
    Drawing*, 577
Attribute Extraction wizard, *Table Style*, 578
Attribute Extraction wizard, *Use Table*, 576,
    577
Attribute Extraction, 576
Attribute Location, Grip Editing, 575
Attributes, Creating, 556
Attributes, Displaying, 562
Attributes, Editing, 562
Attributes, Extracting, 576
Attributes, Redefining, 570
Attsync, 575
Audit, 1301
Auditctl, 1301
Authoring Palette File Locations, 1314
AUto, 70
AutoCAD 3D Interface, 962
AutoCAD Classic, 962, 963
AutoCAD DesignCenter, 517, 518, 816
AutoCAD DesignCenter, Inserting
    Layouts with, 287

AutoCAD Drawing Editor, 5
AutoCAD Drawing Files, 26
AutoCAD File Commands, 29
AutoCAD Objects, 46
AutoCAD Program Parameter
    (ACAD.PGP), 1300
AutoCAD Text Window, 18
Autodesk DWF Viewer, 588, 604
Automatic Save, Options, 1316
Automatic Save File Location, 1313
Autoscale, Scale Factors, 526
AutoSnap, 126
Autosnap aperture box, 126
Autosnap marker, 126
Autosnap tooltip, 126
AutoStack Properties dialog box, 431
AutoTracking tooltip, 126
Auxiliary View, Constructing, 690
Auxiliary View, Setting Up
    Principal Views, 682
AUXILIARY VIEWS, 681
Auxiliary Views, Full, Constructing,
    690
Available Visual Styles, 1009
Axonometric drawings, 642

B

Background, 1210
Background, Adjust Image, 1214
Background, Creating, 1210
Background, Environment, 1214
Background, Fog, 1215
Background, Gradient, 1212
Background, Image, 1213
Background, Solid, 1211
Background dialog box, *Solid* back-
    groud, 1211
Backup Files, 42
Baction, 533
Base, 503, 512
Base Point Parameter, 537
Basic Dimensions, GDT, 720
BASIC DRAWING SETUP, 91
Battman, 570
Battman, Apply, OK, 571
Battman, Edit, 571
Battman, Move up, Move Down, 571
Battman, Remove, 571
Battman, Settings, 571
Battman, Sync, 571
Bedit, 503, 513
Bedit, Action, 514
Bedit, Authoring Palettes, 514
Bedit, Close Block Editor, 514
Bedit, Define Attributes, 514
Bedit, Edit or Create Block Definition,
    514
Bedit, Parameter, 514
Bedit, Save Block As, 514
Bedit, Save Block Definition, 514
*Bedit, Update Parameter and Action
    Text Size*, 514

Beginning an AutoCAD Drawing, 26
Bezier surface, 1191
Big Fonts, 435
Block, 502, 503, 504
Block Attribute Manager, 570
BLOCK ATTRIBUTES, 556
Block Attributes, Creating and
    Inserting, 556
Block Color, Linetype, and *Lineweight*
    Settings, 505
Block command, Using to Create an
    Attributed Block, 560
Block Definition dialog box, 504
Block Editor, 574
Block Insertion with Drag-and-Drop,
    525
Block Insertion, Specified Coordinates,
    Scale, Rotation, 526
BLOCKS, DESIGNCENTER, AND
    TOOL PALETTES, 501
Blocks, Redefining, 512
Blocks and *Xrefs*, In-Place Editing, 824
Blocks Search Path, 515
Bmpout, 867, 876
Boolean Commands, *Regions*, 396
Boolean Operation Commands, 1065,
    1098
Boolean Operations, 1065
Border, 266
Boundary, 361
Boundary, Boundary Set, 362
Boundary, Island Detection, 362
Boundary, Object Type, 362
Boundary, Pick Points, 362
Boundary Representation, 967
Bounding Box, 1166
Box, 1064, 1068
Box, Center, 1069
Box, Cube 1069
Box, Length, 1069
Box, object selection, 72
Box, Specify first corner, 1068
Bparameter, 533
Bparameter Linear, 534
Break, 173
Break, at Point, 174
Break, Select, 2 Points, 174
Break, Select, Second, 174
B-Rep, 967
Brightness slider, 1226
Browse the Web, 589
Browser, 588, 589
Bullets and Lists, 430
Bump Maps, 1237
Button, 10
Buttons, Customizing, 1331
Byblock, 509, 551
Bylayer, 509
Bylayer, Color, 222
Bylayer, Linetype, 227
Bylayer drawing scheme, 223
Bylayer Linetypes, Lineweights and
    Colors, 630

C

Cabinet oblique, 650
CAD, 3
CAD Accuracy, 114
CAD Database, 3, 114
CAD/CAM, 114
Calculating Text Height for Scaled
    Drawings, 431
Calculations and Text Fields in Tables, 451
Calculator, 1308
CALS-1, 869
Camera, 978, 991
Camera, Clipping, 992
Camera, Height, 992
Camera, Lens, 992
Camera, Location, 992
Camera, Name, 992
Camera, View, 992
Camera Grips, 993
Camera Preview, 993
Camera Properties, 993
Camera Settings, Editing, 993
Cartesian coordinates, 2
Cavalier oblique, 550
CDF, 582
Cecolor, 238
Cell Shortcut Menu, Entering Formulas,
    457
Celtscale, 241
Celtscale, variables, 373
Center, Object Snap, 115
Centroid, 1166
Chamfer, 186
Chamfer, 3D, 1066, 1102
Chamfer, 3D, Edge, 1103
Chamfer, 3D, Loop, 1103
Chamfer, 3D, Multiple, 1103
Chamfer, Angle, 187
Chamfer, Distance, 187
Chamfer, Method, 186
Chamfer, Multiple, 188
Chamfer, Polyline, 188
Chamfer, Shift-select, 188
Chamfer, Trim/Notrim, 187
Chamfer, Undo, 188
Change, 380
Change, Point, 380
Change, Properties, 380
Change, Text, 381
Characters, Special Text, 438
Check Spelling dialog box, 440
Checkbox, 10
Choose Layouts, 954
Chprop, 380
Circle, 46, 50, 144
Circle, 2 Points, 145
Circle, 3 Points, 146
Circle, Center, Diameter, 145
Circle, Center, Radius, 145
Circle, Tangent, Tangent, Radius,
    146
Clean Screen, 22

Cleanscreenoff, 22
Cleanscreenon, 22
Close, 36
Closeall, 36
CNC (Computer Numerical Control), 114
Color, 238
Color, Bylayer, 238
Color, Linetype and *Lineweight*, 227
Color Books, 239, 1209
Color Books, *DIC*, 1209
Color Books, *Pantone*, 1209
Color Books, *RAL*, 1209
Color Control drop-down list, 238
Color Dependent Plot Style Tables, 910
Color dialog box, 227
Color property, changing, 373
Color Systems, 1208
Colors and Linetypes, Assigning, 222
Column and Row Settings, 451
COM, 335
Combination Maps, 1237
Command Aliases, Defining, 1336
Command Entry, 12
Command Entry Methods Practice, 19
Command Line, 7, 23
Command Line Input, 49, 54
Command Reference, Help, 83
Command Tables, 12
Command-Mode, Menu, 13
Commands, Methods for Entering, 12
Commands for Moving Solids, 1087
Commands to Move, Rotate, and Mirror Solids, 1065
Communication Center, 22, 588, 590
Communication Center, *Channels*, 591
Communication Center, *Settings*, 591
Communication Center dialog box, 591
Compare Dimension Styles dialog box, 746
Composite Solid Subobject Selection, 1120
Computer Numerical Control (CNC), 114
Computer-Aided Design, 3
Conceptual, 1002
Cone, 1064
Cone, 2Point, 1070
Cone, 3p/2p/Ttr, 1070
Cone, Axis Endpoint, 1070
Cone, Center Point, 1069
Cone, Elliptical, 1070
Cone, Top Radius, 1070
Configuring a Default Output Device, 109
Configuring Plotters and Printers, 332
Construction Layers, Using, 627
Construction Line, 342
Construction Techniques for Polygon Meshes and *3Dfaces*, 1192

Constructive Solid Geometry (CSG), 966, 1064
Constructive Solid Geometry Techniques, 1064
Content Area, *Adcenter*, 518
Continuous Orbit, 983
Contrast slider, 1226
Control Key Sequences, 15
Controlling Animation from the *3D Navigation Control Panel*, 1256
Controlling the Mesh Density, 1184
Convert dialog box, OLE, 860
Convertctb, 921
Converting Plot Style Tables, 920
Convertoldlights, 1226
Convertoldmaterials, 1235
Convertpstyles, 911, 920
Coordinate Data Entry, Priority, 1320
Coordinate Display, 6, 10
Coordinate Entry, 143
Coordinate Entry Methods, 47
Coordinate System Icon, 212, 968
Coordinate Systems, 2, 973
Coords, 10, 15
Copy, 12, 76, 175
Copy, Multiple, 177
Copy (OLE), 847
Copy Link, 850
Copy Render dialog box, 1247
Copy to Layer, 247
Copy with Base Point (OLE), 849
Copybase, 850
Copyclip, 849
Copylink, 850
Create a Sheet, 940
Create Camera, 991
Create New Dimension Style dialog box, 741
Create New Drawing dialog box, 30, 93
Create Sheet Set wizard, 952
Create Transmittal dialog box, 597, 950
CREATING 2D DRAWINGS FROM 3D MODELS, 1269
Creating a Distant Light, 1221
Creating a *Point Light*, 1220
Creating a Rendering with Default Settings, 1207
Creating a *Spotlight*, 1220
CREATING AN ANIMATION, 1254
Creating an Animation using *Anipath*, 1255, 1260
Creating an Animation with the *3D Navigation control panel*, 1254
Creating and Editing Tables, 450
Creating and Managing User Coordinate Sytems, 1024
Creating and Modifying Materials, 1230
Creating Lights in a Model, 1215
Creating Lights in a Model, *Defaultlighting*, 1215
Creating Lights in a Model, *Distant Light*, 1215

Creating Lights in a Model, *Point Light*, 1215
Creating Lights in a Model, *Spotlight*, 1215
Creating Lights in a Model, *Sunlight*, 1215
Creating Geometrically Defined Meshes, 1184
Creating Meshed Surfaces, 1182
Creating Surfaces Compatible with Solids, 1160
Creating the Table Text, 452
Creating Wood and Marble Materials, 1235
Crosshair, 3D, 1321
Crosshair, Labels, 1322
Crosshair, Size, 1315
Crosshairs, 6
Crossing Polygon, 71
Crossing Window, 70, 170
CSG (Constructive Solid Geometry), 966, 1064
Ctrl key, 1120
Ctrl+E, 645
Ctrl+O, 22
Cubic B-spline, 1191
CUI, 1330
Current and Model Views, 988
Current Entity Linetype Scale, 241
Current Pointing Device, Options, 1318
Cursor, 6
Cursor menu, 122
Cursor tracking, 10
Curve Tessellation, 1013
Custom Dictionary, 441
Custom Properties dialog box, 944
Custom tab, 41
Customer User Interface, 1330
Customization Files, 1313
Customize dialog box, 1330
Customize Menus, 1331
Customizing Buttons, 1331
Customizing Shortcut Keys, 1331
Customizing the AutoCAD Screen, 21
Customizing Toolbars, 1331
Cut (OLE), 845, 851
Cut and Paste, AutoCAD Drawings, 844
Cut and Paste, Other Windows Applications, 845
Cutclip, 851
Cylinder, 1064, 1071
Cylinder, 2Point, 1071
Cylinder, 3P/2P/Ttr, 1071
Cylinder, Axis Endpoint, 1072
Cylinder, Center Point, 1071
Cylinder, Elliptical, 1071
Cylindrical Coordinates (Relative), 969, 970

D

Dashboard, 962, 963, 999, 1067, 1206
Data Sources Locations, 1314
Datum, GDT, 721
Datum Identifier, 720
Datum Targets, GDT, 724

Dbconnect, 1314
DbConnect Options, 1318
Dblclkedit, 377
Dblist, 410
DC Online, 524
Dctcust, 441, 1313
Dctmain, 441, 1313
Ddedit, 419, 439
Ddgrips, 466
Ddim, 743
Ddptype, 151
Dducsp, 1032
Dducsp, Depth, 1032
Dducsp, Details, 1032
Ddvpoint, 978, 996
Default Lighting, 1216
Default Lighting, Defaultlightingtype, 1216
Default Menu, 13
Default Visual Styles, 1000
Default Visual Styles, 2D Wireframe, 1001
Default Visual Styles, 3D Hidden, 1001
Default Visual Styles, 3D Wireframe, 1001
Default Visual Styles, Conceptual, 1002
Default Visual Styles, Realistic, 1002
Defaultlighting, 1215, 1216, 1218
Define Attributes, 557
Defining a Background, 1210
Defining a Background, Background dialog box, 1211
Defining a Background, New View dialog box, 1211
Deflplstyle, 911
Degradation Order, 1011
Degrade When FPS Is Below, 1011
Delay, 1305, 1339
Delobj, 1165
Demand Load, Options, 1317
Demand Load Xrefs, 821, 1316
Demand Loading, 820
Demandload, 1317
Dependent Objects and Names, 813, 815
Design Efficiency, 1104
DesignCenter, 518
DesignCenter, Alternative to Xbind, 816
DesignCenter, Attach or Overlay Xrefs, 816
DesignCenter, Back/Forward, 523
DesignCenter, Creating Tool Palettes, 529
DesignCenter, DC Online, 524
DesignCenter, Description, 523
DesignCenter, Favorites, 521
DesignCenter, Folders Tab, 519
DesignCenter, History Tab, 520
DesignCenter, Inserting Layouts with, 287
DesignCenter, Load, 521
DesignCenter, Loading the Content Area, 524

DesignCenter, Open Drawings Tab, 520
DesignCenter, Preview, 522
DesignCenter, Resizing, Docking, and Hiding, 519
DesignCenter, Search, 521
DesignCenter, Tree View Toggle, 520
DesignCenter, Up, 522
DesignCenter, Views, 523
DesignCenter Blocks, Drawing Units, 97
DesignCenter Insert a Block, 525
DesignCenter Online, 524
DesignCenter Open a Drawing, 524
Device Driver File Search Path, 1313
Dialog Box Functions, 27
Dialog Boxes, 9
Dialog-Mode Menu, 14
Dictionary, 441
Digital Signatures, 35
Digitizing Tablet Menu, 10
Dimadec, 763, 768
Dimaligned, 700
Dimaligned, Angle, 700
Dimaligned, Mtext, 700
Dimaligned, Text, 700
Dimalt, 763, 768
Dimaltd, 764, 768, 779
Dimaltf, 764, 768, 779
Dimaltrnd, 764, 768
Dimalttd, 766, 768, 779
Dimalttz, 766, 768
Dimaltu, 764, 768, 779
Dimaltz, 764, 769
Dimangular, 707
Dimapost, 764, 769
Dimarc, 705
Dimaso, 769
Dimassoc, 752, 767, 769, 1321
Dimasz, 752, 770, 771, 779
Dimatfit, 757, 758, 770
Dimaunit, 763, 770
Dimazin, 763, 770
Dimbaseline, 701
Dimblk, 752, 771
Dimblk1, 752, 771
Dimblk2, 752, 771
Dimcen, 753, 771, 780
Dimcenter, 706
Dimclrd, 749, 771
Dimclre, 720, 750, 771
Dimclrt, 720, 750, 771
Dimcontinue, 702
Dimdec, 727, 761, 771, 780
Dimdiameter, 703
Dimdiameter, Angle, 703
Dimdiameter, Mtext/Text, 703
Dimdisassociate, 734
Dimdle, 733, 756
Dimdli, 750, 772, 780
Dimdsep, 761, 772, 780
Dimedit, 729
Dimedit, Home, 729
Dimedit, New, 729

Dimedit, Oblique, 729
Dimedit, Rotate, 729
Dimension, Align Text, 728
Dimension, Aligned, 700
Dimension, Angular, 707
Dimension, Arc Length, 706
Dimension, Baseline, 701
Dimension, Center Mark, 706
Dimension, Continue, 702
Dimension, Diameter, 703
Dimension, Jogged, 705
Dimension, Leader, 708
Dimension, Linear, 697
Dimension, Oblique, 729
Dimension, Ordinate, 713
Dimension, Overall Scale, 740
Dimension, Override, 781
Dimension, Qdim, 715
Dimension, Qleader, 710
Dimension, Radius, 704
Dimension, Tolerance, 718
Dimension, Update, 746
Dimension Drawing Commands, 697
Dimension Editing Commands, 728
Dimension Input, 56
Dimension line, 696
Dimension Overall Scale, 736
Dimension pull-down menu, 697
Dimension Right-Click Menu, 726, 784
Dimension Style, Modifying, 780
Dimension Style Children, 741
Dimension Style Families, 741
Dimension Style Manager, 741, 743, 748
Dimension Style Manager, Alternate Units tab, 763
Dimension Style Manager, Compare, 746
Dimension Style Manager, Description, 744
Dimension Style Manager, Fit tab, 736, 756
Dimension Style Manager, Lines tab, 749
Dimension Style Manager, List, 744
Dimension Style Manager, Modify, 745
Dimension Style Manager, New, 745
Dimension Style Manager, Override, 745
Dimension Style Manager, Preview, 744
Dimension Style Manager, Primary Units tab, 760
Dimension Style Manager, Set Current, 745
Dimension Style Manager, Styles, 743
Dimension Style Manager, Text tab, 753
Dimension Style Manager, Tolerances tab, 765
Dimension Style Overrides, 741, 781
Dimension Styles, 266, 740, 743
Dimension Styles, Template Drawings, 787
DIMENSION STYLES AND VARIABLES, 739
Dimension text, 696
Dimension Text Height, Fixed, 787
Dimension toolbar, 697
Dimension Variables, 740

Dimension Variables, Changing, 748, 767
Dimension Variables, GDT-Related, 720
Dimension Variables Introduction, 735
Dimension Variables Table, 767
DIMENSIONING, 695
Dimensioning, Guidelines, 785
Dimensioning a Single Drawing, 785
Dimensioning in Layouts, 788
Dimensioning in Paper Space, 778, 779
Dimensioning in Paper Space, Procedure, 790
Dimensioning Isometric Drawings, 791
Dimensioning Text, Customizing, 730
Dimensions, Associative, 731
Dimensions, Editing, 725
Dimensions, Grip Editing, 471, 725
Dimensions, Modifying Existing, 780
Dimetric drawings, 643
Dimexe, 751, 772, 780
Dimexo, 751, 772, 780
Dimfit, 772
Dimfrac, 761, 772
Dimgap, 720, 755, 765, 772, 780
Dimjog, 705
Dimjust, 755, 773
Dimldrblk, 752, 773
Dimlfac, 762, 773
Dimlim, 765, 773
Dimlinear, 697
Dimlinear, Angle, 700
Dimlinear, Horizontal, 700
Dimlinear, Mtext, 699
Dimlinear, Rotated, 698
Dimlinear, Text, 699
Dimlinear, Vertical, 700
Dimlunit, 760, 774, 780
Dimlwd, 749, 774
Dimlwe, 750, 774
Dimordinate, 713
Dimordinate, Mtext/Text, 715
Dimordinate, Xdatum/Ydatum, 713, 715
Dimoverride, 783
Dimpost, 761, 765
Dimradius, 704
Dimradius, Angle, 705
Dimradius, Mtext/Text, 705
Dimreassociate, 733
Dimregen, 735
Dimrnd, 761, 774
Dimsah, 752, 775
Dimscale, 94, 95, 263, 750, 757, 775
Dimsd1, 750, 775
Dimsd2, 750, 775
Dimse1, 751, 775
Dimse2, 751, 775
Dimsho, 775
Dimsoxd, 758, 775, 780

Dimstyle, 743, 775
-Dimstyle, 746
-Dimstyle, Apply, 748
-Dimstyle, Restore, 747
-Dimstyle, Save, 747
-Dimstyle, Status, 747
-Dimstyle, Variables, 747
Dimtad, 755, 776, 780
Dimtdec, 765, 776, 780
Dimtedit, 728
Dimtedit, Angle, 728
Dimtedit, Center, 728
Dimtedit, Home, 728
Dimtedit, Right/Left, 728
Dimtfac, 754, 766, 776
Dimtih, 756, 776, 780
Dimtix, 758, 776, 780
Dimtm, 766, 776
Dimtmove, 726, 758, 759
Dimtofl, 759, 777, 780
Dimtoh, 756, 777, 780
Dimtol, 765, 777
Dimtolj, 766, 777, 780
Dimtp, 766, 777
Dimtststy, 720
Dimtsz, 777
Dimtvp, 778
Dimtxsty, 754, 778
Dimtxt, 720, 754, 778, 780
Dimtzin, 766, 778, 780
Dimunit, 778
Dimupt, 759, 778
Dimzin, 762, 778, 780
DIN, 779
DIN, (.DWT), 268
Direct Distance Entry, 49, 56
Direct Distance Entry, 3D, 969, 972
Direct Distance Entry and Polar Tracking, 59
Direction Control dialog box, 99
Display Cameras, 993
Display compass and tripod, 995
Display Order, 664
Display Performance, 1315
Display Resolution, 211, 1315
Display Screen Menu, 8
Dispsilh, 1001, 1315
Dist, 412
Distance, 412, 984
Distant Light, 1217
Distant Light, Altitude, 1222
Distant Light, Azimuth, 1222
Distant Light, Geographic Location, 1223
Distant Light, Light Source Vector, 1223
Distant Light, Sun Angle Calculator, 1222
Divide, 351
Donut, 347
Double-click Edit, 377
Drafting Settings, 56, 104, 644
Drafting Settings dialog box, 56, 102, 124

Drafting Settings dialog box, Polar Tracking tab, 58, 130, 686
Drag and Drop, Using to Import Text Files, 436
Draw and Modify toolbars, 7
Draw Command Access, 142
Draw Commands, 142
Draw Commands, Locating, 46
DRAW COMMANDS I, 141
DRAW COMMANDS II, 341
DRAW COMMAND CONCEPTS, 45
Draw order, 876
Draw pull-down menu, 47, 142
Draw toolbar, 47, 142
Draw True Size, 3
Drawing Aids, 15
Drawing Cutting Plane Lines, 675
Drawing Editor, 5
Drawing Exchange Binary, 881
Drawing File Management, 42
Drawing Files, 26
Drawing Files, Naming, 26
Drawing Lines Using Polar Tracking and Polar Snap, 60
Drawing Lines Using the Three Coordinate Entry Methods, 50
Drawing Orientation, 317
Drawing Properties, 39
Drawing Properties dialog box, 39
Drawing Recovery, 42
Drawing Recovery Manager, 42
Drawing Scale Factor, 263
Drawing Scale Factor, Calculating, 323
Drawing Scale Factor, Reverse Method, 900
Drawing Setup, Steps, 92, 262
Drawing Setup Options, 93
Drawing Template Settings, 1314
Drawing Units dialog box, 97
Draworder, 674, 876
Drop-down list, 10
Dsettings, 58, 104
DSF, 323
Dsviewer, 210
Dsviewer, Auto Viewport, 211
Dsviewer, Dynamic Update, 211
Dsviewer, Global, 211
Dsviewer, Realtime Zoom, 210
Dsviewer, Zoom In, 211
Dsviewer, Zoom Out, 211
DUCS, 15, 1085
DWF Classic, R14 look, 602
DWF ePlot (Optimized for Plotting), 602
DWF ePlot, WHIP! 3.1 Compatible Version, 602
DWF eView (Optimized for Viewing), 602
DWF Properties, Gradient Resolution, 603
DWF Properties, Vector Resolution, 603
DWF Underlay, 610
DWF Underlay, To detach 610
DWF Viewer, 588, 600, 604, 608
DWF6 ePlot, 602
DWF6 ePlot Properties dialog box, 603
Dwfattach, 599, 610

Dwfclip, Delete, 612
Dwfattach, Insertion Point, 611
Dwfattach, Name, 611
Dwfattach, Path Type, 611
Dwfattach, Rotation, 611
Dwfattach, Scale, 611
Dwfattach, Select a Sheet, 611
Dwfattach, Shortcut Menu, 611
Dwfclip, 612
Dwfclip, New Boundary, 612
Dwfclip, On/Off, 612
DWG Reference, 814
DWG To PDF.pc3 Driver, 883
Dwgprops, 39
Dxbin, 881
Dxfin, 880
Dxfin, Dxfout, 867
Dxfout, 880
DYN, 15
Dynamic Blocks, 531
Dynamic Blocks, Assigning
    Parameters, 534
Dynamic Blocks, Creating, 533
Dynamic Coordinate System
    Example, 1086
Dynamic Input, 49, 969, 973
Dynamic Input, Polar Tracking, and
    Polar Snap, 63
Dynamic Input and Polar Tracking,
    62
Dynamic Tessellation, 1013
Dynamic UCS Principles, 1087
Dynamic USCs, 975, 1038
Dynamic User Coordinate Systems,
    975, 1038, 1065, 1085

E

Eattedit, 564
Eattedit, Apply, OK, Cancel, 565
Eattedit, Attribute, 564
Eattedit, Properties, 564
Eattedit, Text Options, 564
Eattext, 576
Edge, 170, 1181
Edge Jitter, 1006
Edge Mesh, 1184
Edge Overhang, 1006
Edge Settings, 1005
Edge Settings, Edge Color, 1005
Edge Settings, Facet Edges, 1005
Edge Settings, Isolines, 1005
Edge Settings, No Edges, 1005
Edgesurf, 1184
Edit Attributes dialog box, 561, 571
Edit box, 10
Edit Hyperlink dialog box, 592
Edit Reference, 826
Edit Scale List, 1321
Editing Camera Settings, 993
Editing Primitives Using the
    Properties Palette, 1118
Editing Properties of Composite
    Solids, 1121

Editing Properties of Individual
    Primitives, 1118
Editing Solids Using Grips, 1121
Editing Text, 438
Edit-Mode Menu, 13
Elevation, 1198
Ellipse, 349
Ellipse, Arc, 351
Ellipse, Axis End, 350
Ellipse, Center, 350
Ellipse, Isometric, 646
Ellipse, Parameter, 351
Ellipse, Rotation, 350
Enable Hardware Acceleration, 1012
Endpoint, Object Snap, 116
Enhanced Attribute Editor, 564, 929
Enter Attributes dialog box, 561
Entering Formulas Directly into Cells,
    457
*Entering Formulas Using Cell Shortcut
    Menu,* 458
*Entering Formulas Using the Field
    Dialog Box,* 457
Entering Other (Non-formula) Fields,
    458
Entity, 46
Environment Settings, 1004
ePlot, 588, 602
Erase, 74, 161
Etransmit, 588, 597, 950
Etransmit, Add File, 598
Etransmit, Files Table tab, 598
Etransmit, Files Tree tab, 598
Etransmit, Save As, 598
Etransmit, View Report tab, 598
Existing Drawings, 935, 954
Exit, 36
Explode, 381, 503, 507
Exploded Dimensions, 732
Export, 867, 882
Exporting Raster and Vector Files,
    867
Extend, 171
Extend, Crossing Window, 172
Extend, Edge, 172
Extend, Fence, 172
Extend, Projection, 172
Extend, Select all, 172
Extend, Shift-Select, 172
Extension, Object Snap, 119
Extension line, 696
External Commands, 1335
External Reference, Attach, 870
External Reference, Detach, 870
External Reference, Details/Preview,
    871
External Reference, Reload, 870
External Reference, Treeview, Listview,
    869
External Reference, Unload, 871
External Reference dialog box, 806
External References, 1317
External Text Editor, 437
External Text Format Codes, 437

Extnames, 1319
Extract Edges, 1059
Extract File, Creating, 581
Extrude, 1065, 1078, 1160
Extrude, Direction, 1079
Extrude, Path, 1079
*Extrude, Revolve, Sweep, and Loft
    Surface Examples,* 1160
Extrude, Taper Angle, 1079

F

F1, 14
F2, 14
F3, 14
F4, 14
F5, 15
F6, 15
F7, 15
F8, 15
F9, 15
F10, 15
F11, 15
F12, 15
Face Color Mode, 1003
Face Settings, 1003
Face Settings, Desaturate, 1003
Face Settings, Environment Settings, and
    Edge Settings, 1003
Face Settings, Monochrome, 1003
Face Settings, Regular Face Colors, 1003
Face Settings, Tint, 1003
Face Style, 1004
Face Style, Gooch, 1004
Face Style, Real, 1004
Face Style, Smooth, 1004
Faceres, 1253
Facet Edges, 1005
Facetres, 1315
Facets, 1004
FCF (Feature Control Frame), 718
Feature Control Frame (FCF), 718
Feature Control Frames, Editing, 720
Fence, 71
Field, 419, 449
Field Dialog Box, Entering Formulas, 457
*Field Dialog Box, Entering Other (Non-
    formula) Fields,* 458
Field of View, 989
Fields, 1320
File Commands, Accessing, 26
File Management, 42
File Navigation Dialog Box Functions,
    27
File Open, 1316
File Safety Precautions, 1316
File Save, 1315
Filedia, 1302
Files, Transferring over the Internet,
    616
Fill, 1299
Fillet, 183
Fillet, 3D, 1066, 1103
Fillet, 3D, Chain, 1104

Fillet, 3D, Edge, 1104
Fillet, 3D, Multiple, 1104
Fillet, 3D, Radius, 1104
Fillet, Multiple, 185
Fillet, Polyline, 184
Fillet, Radius, 184
Fillet, Shift-select, 186
Fillet, Trim/Notrim, 185
Fillet, Undo, 185
Fillets, Rounds and Runouts,
    Creating, 631
Fillmode, 1300, 1315
Fillmode for Solid Hatch Fills, 673
Filter, 487
Filter, Add Selected Object, 489
Filter, Add to List, 489
Filter, Apply, 489
Filter, Clear list, 489
Filter, Current, 489
Filter, Delete, 489
Filter, Delete Current Filter List, 489
Filter, Edit Item, 489
Filter, Layers, 228
Filter, Named Filters, 489
Filter, New Group, 229
Filter, Properties, 230
Filter, Save As, 489
Filter, Select, 488
Filter, Substitute, 489
Filter, X, 476
Filter, Y,Z, 477
Filter Definition, 230
Find, 419, 441, 565
Find, Find Text String, 442
Find, Options, 443
Find, Replace With, Replace, 442
Find, Select All, 442
Find, Select Objects, 442
Find, Zoom to, 442
Find and Replace dialog box, 441, 565
Find and Replace Options dialog box,
    443
Find dialog box, 33
Find/Replace, Text Formatting Editor,
    443
Flatness, GDT, 718
Flatshot, 1270, 1286
Flatshot, Destination, 1287
Flatshot, Foreground Lines, 1287
Flatshot, Obscured Lines, 1287
Flatshot dialog box, 1287
Flatshot versus *Solprof*, 1287
FLIC, 869
Flip Action, 543
Flip Parameter, 537
Flipscreen, 14
Floating Viewports, 278
Flyouts, 7
Fog, 1215
Fog, Color, 1215
Fog, Enable Fog, 1215
Fog, Fog Background, 1215
Fog, Near Distance and *Far Distance*,
    1215

Fog, Near Fog Percentage and *Far Fog
    Percentage*, 1215
Follow, 104
Font Handling, 603
Font Mapping, 447
Fontalt, 447, 1313
Fontmap, 447, 1313
Fonts, Alternate, 447
Fonts, Substituting, 447
Format Codes, External Text Editor,
    437
Free Orbit, 982
From, Object Snap, 121
FTP Protocol, 602
Full Shadows, 1005
Full-screen tracking vector, 126
Function Keys, 14

G

Gb, (.DWT), 268
GDT (Geometric Dimension and
    Tolerancing), 718
GDT-Related Dimension Variables,
    720
General Options, 1318
General tab, 40
Geographic Location, Lattitude, 1223
Geographic Location, Longitude, 1223
Geographic Location, Nearest Big City,
    1223
Geographic Location, Nearest City, 1223
Geographic Location, North Direction,
    1223
Geographic Location, Region, 1223
Geographic Location, Time Zone, 1223
Geographic Location dialog box, 1223
Geographicloation, 1223
Geometric Dimensioning and
    Tolerance Example, 721
Geometric Dimensioning and
    Tolerancing (GDT), 718
Geometric Tolerance dialog box, 719
Geometry Acceleration, 1013
GEOSPOT, 869
GETTING STARTED, 1
Global *Brightness* and *Contrast*, 1226
Gradient, 1212
Gradient fill, 666
Gradient tab, 666
Graphics Area, 6
Grid, 15, 16, 103, 265, 979, 997
Grid, Adaptive, 104, 997
Grid, Allow Subdvision, 104
Grid, Aspect, 104
Grid, Follow, 104
Grid, Limits, 104, 997
Grid, Major, 997
Grid, Major lines, 104
Grid, On/Off, 103
Grid, Snap, 103
Grid, Spacing, 103
Griddisplay, 997, 998
Gridmajor, 997

Grid Snap, 15, 50, 101
Grid Snap and Polar Tracking, 60
Gridunit, 94, 95
Grip Colors, 472
GRIP EDITING, 465
Grip Editing Composite Solids, 1129
Grip Editing Dimensions, 725
Grip Editing Faces, Edges, and Vertices,
    1131
Grip Editing Faces, Edges, and Vertices,
    Edge Grips, 1131
Grip Editing Faces, Edges, and Vertices,
    Face Grips, 1133
Grip Editing Faces, Edges, and Vertices,
    Vertex Grips, 1132
Grip Editing Options, 468
Grip Editing Primitive Solids, 1122
Grip Editing Primitive Solids, Box,
    1122
Grip Editing Primitive Solids, Cone, 1124
Grip Editing Primitive Solids, Cylinder,
    1125
Grip Editing Primitive Solids, Polysolid,
    1126
Grip Editing Primitive Solids, Pyramid,
    1125
Grip Editing Primitive Solids, Sphere, 1125
Grip Editing Primitive Solids, Torus, 1126
Grip Editing Primitive Solids, Wedge, 1124
Grip Size, 472
Gripblock, 473
Gripcolor, 472
Griphot, 472
Gripobjlimit, 473
Grips, 466
Grips, Activating, 467
Grips, Auxiliary Grid, 471
Grips, *Base*, 470
Grips, Cold and Hot, 467
Grips, *Copy*, 470
Grips, Dimensioning, 725
Grips, Editing Dimensions, 471
Grips, Enable, 473
Grips, Guidelines for Using, 472
Grips, Hover, 467
Grips, *Mirror*, 470
Grips, *Move*, 469
Grips, Ortho, 470
Grips, *Properties*, 470
Grips, *Reference*, 470
Grips, *Rotate*, 469
Grips, *Scale*, 469
Grips, *Stretch*, 469
Grips, *Undo*, 470
Grips, Using with Hatch Patterns, 672
Grips Features, 466
Grips Settings, 472
Grips Shortcut Menu, 468
Grips Tips, 473
Grips within Blocks, 473
Gripsize, 472
Griptips, 473
Ground Shadows, 1005
Group, 491

Group, Add, 494
Group, Changing Properties, 493
Group, Creating a Named, 491
Group, Description, 492, 494
Group, Explode, 494
Group, Find Name, 492
Group, Highlight, 492
Group, Include Unnamed, 492
Group, Name, 492
Group, New, 492
Group, Object selection, 72
Group, Rename, 494
Group, Re-order, 494
Group, Selectable, 492
Group, Unnamed, 493
Group Members, Reordering, 494
Group Selectable, 494
Groups Examples, 495

H

Halo Gap %, 1010
Hardware Acceleration, 1012, 1253
Hatch, 658
Hatch, Angle, 661
Hatch, Associative, 657, 663
Hatch, Boundary Set, 665
Hatch, Create Separate Hatches, 663
Hatch, Current Origin, 661
Hatch, Custom, 659
Hatch, Double, 661
Hatch, Draw Order, 664
Hatch, Gap Tolerance, 665
Hatch, Gradient, Angle, 667
Hatch, Gradient, Centered, 667
Hatch, Gradient, One Color, 666
Hatch, Gradient, Two Color, 666
Hatch, Inherit Options, 666
Hatch, Inherit Properties, 664
Hatch, Island Detection, 664
Hatch, ISO Pen Width, 661
Hatch, Pattern, 659
Hatch, Pick Points, 662
Hatch, Predefined, 659
Hatch, Preview, 664
Hatch, Recreate Boundary, 663
Hatch, Relative to Paper Space, 661
Hatch, Remove Boundaries, 663
Hatch, Retain Boundaries, 665
Hatch, Scale, 661
Hatch, Select Objects, 663
Hatch, Spacing, 661
Hatch, Specified Origin, 662
Hatch, Swatch, 659
Hatch, Type, 659
Hatch, User-defined, 659
Hatch, View Selections, 663
-Hatch, Advanced, 668
-Hatch, Command Line Option, 667
-Hatch, Draw Boundary, 668
-Hatch, Properties, 668
Hatch and Gradient dialog box, 658,
    667
Hatch and Gradient dialog box, *Hatch*
    tab, 659

Hatch Fills, Solid, 664
Hatch Pattern Palette dialog box,
    659
Hatch patterns, AutoCAD, 660
Hatch Patterns, Drag and Drop, 669
Hatch Patterns, Object Selection
    Features, 672
Hatch Patterns, Using Grips, 672
Hatch Patterns, Using Trim, 672
Hatch Patterns and Boundaries,
    Editing, 671
Hatch Patterns and Hatch
    Boundaries, 656
Hatchedit, 671
Hatchedit dialog box, 657, 671
Helix, 1057, 1058
Helix, Axis Endpoint, 1058
Helix, Turn Height, 1058
Helix, Turns, 1058
Helix, Twist, 1058
Help, 14, 82
Help, Contents, 83
Help, Index, 84
Help, Search, 84
Help and Miscellaneous File Names,
    1313
Help message, 7
HELPFUL COMMANDS, 81
Hidden and Center Lines, Drawing,
    629
Hide, 1001
Highlight Intensity, 1010
History Settings, 1135
History Settings, Solid History, 1135
HLS, 239
Horizontal Lines, Drawing, 50, 51
Hover Grip, 472
HP/GL Language, 314
Hpang, 661
Hpscale, 94, 95, 661
HTML, 594
Hyperlink, 40, 588, 592
Hyperlink, Email Address, 593
Hyperlink, Existing File or Web Page,
    593
Hyperlink, Options, 1321
Hyperlink, View of This Drawing, 593
Hyperlink Base, 40
Hyperlinkbase, 594, 593
Hyperlinks, .DWF Image, 593

I

ID, 412
i-Drop, 596, 617
i-Drop Associated Files Location, 1314
IGES, 880
Image, 1213
Image Adjust, 872
Image Adjust dialog box, 872
Image Clip, 873
Image Frame, 872
Image Mapping, 1236
Image Quality, 874
Image tile, 10

Image Transparency, 874
Imageadjust, 867, 872
Imageadjust, Brightness, 872
Imageadjust, Contrast, 873
Imageadjust, Fade, 873
Imageattach, 867, 871
Imageclip, 867, 873
Imageframe, 867, 872
Imagehlt, 1315
Imagequality, 867, 874
Import, 867, 878
Import Layout as Sheet, 940, 935
Importing External Text, 436
Importing Vector Files, 867
Importing with *Text Formatting Editor*,
    436
Imprint, 1138, 1159
Inclined Lines, Drawing, 51
Include Plot Stamp, 949
Index Color, 1208
Indexctl, 38, 821
Inetlocation, 589
In-Place Editing, *Xrefs* and *Blocks*, 826
In-place Xref and Block Edit, 574
INQUIRY COMMANDS, 407
Inquiry toolbar, 408
Insert, 503, 506
Insert, Object Snap, 116
Insert 3D Studio, 878
Insert a Block, DesignCenter, 525
Insert ACIS File, 879
Insert dialog box, 506
Insert Hyperlink dialog box, 592
Insert Layout dialog box, 296
Insert Object dialog box, 856
Insert OLE Object, 856
Insert Sheet List Table dialog box, 947
-Insert with *, 508
Inserting Attributed Blocks, 561
Inserting Drawings as Blocks, 506
Insertion Behavior, 451
Insertion Scale, 99, 1320
Insertobj, 856
Insunits, 1320
Insunitsdefsource, 1320
Insunitsdeftarget, 1320
Interfere, 1138, 1157
Interfere, Check, 1158
Interfere, Nested Selection, 1158
Interfere, Settings, 1158
Interference, 1158
Interference Checking, 1157
Interference Checking dialog box,
    1158
Interference Settings dialog box,
    1158
INTERNET TOOLS, 588
Intersect, 1065, 1066, 1100
Intersect, Regions, 398
Intersection, Object Snap, 116
Intersection Edges, 1008
Isavebak, 42, 1316
Isavepercent, 1316
ISO, (.DWT), 269
Isocircle, 646

Isolines, 1005, 1315
Isometric Drawing, Creating, 648
Isometric Drawing in AutoCAD, 644
Isometric Drawings, 642
Isometric Drawings, Dimensioning, 650, 791
Isometric Ellipses, 646
Isoplane, 15, 646, 647

J

JFIF, 869
JIS, 779
JIS, (.DWT), 269
Join, 174
JPEG, 869, 1213, 1236
Jpgout, 867, 877
Justifytext, 419, 445

K

Keyboard Entry Features, 12
Keyboard Input or *Units* Values, 99
Kinetix, 878

L

Last, object selection, 72
Last point, 412
Laydel, 255
Layer, 224, 890
Layer, Apply, 228
Layer, Color, 227
Layer, Current Viewport Freeze, New Viewport Freeze, 228
Layer, Current VP Freeze, 888, 890
Layer, Delete, 226
Layer, Description, 228
Layer, Freeze, Thaw, 226
Layer, Indicates Layers in Use, 226
Layer, Linetype, 227
Layer, Lineweight, 227
Layer, Lock, Unlock, 226
Layer, Make Object Current, 233
Layer, New, 225
Layer, New VP Freeze, 888, 890*Layer, OK*, 228
Layer, On, Off, 226
Layer, Plot Styles, 223, 228
Layer, Plot/Noplot, 228
Layer, Search for Layer, 228
Layer, Set, 226
Layer, Set Current, 226
Layer, Status, 226
Layer Control drop-down list, 224
Layer Control drop-down list, *Current Viewport*, 892
Layer Filters, 228
Layer Index, 821
Layer Previous, 234
Layer Properties Manager, 9, 224, 890
Layer Properties Manager, Current VP Freeze, New VP Freeze, 278
Layer Properties Manager, Layer List, 224
Layer property, changing, 373

Layer States, 231
Layer States, Blocks and *Xrefs*, 233
Layer States, Delete, 232
Layer States, Export, 232
Layer States, Import, 232
Layer States, New, 232
Layer States, Restore, 232
Layer States, Save, 231
Layer States Manager, 231
Layer Tools, 245
Layer Visibility, Viewports, 889
Layerp, 234
Layers, Using Construction Layers, 627
LAYERS AND OBJECT PROPERTIES, 221
Layerwalk dialog box, 248
Layfrz, 252
Layfrz, Block selection, 252
Layfrz, viewports, 252
Laycur, 246
Layiso, 246
Layiso, Off, 246
Layiso, Vpfreeze, 246
Laylck, 253
Laymch, 245
Laymcur, 233
Laymrg, 254
Layon, 254
Layout, 106, 285
Layout, ? (List Layouts), 286
Layout, Copy, 286
Layout, Delete, 286
Layout, New, 286
Layout, Rename, 287
Layout, Save, 287
Layout, Set, 287
Layout, Template, 286
Layout Elements, 1315
Layout Options and Plot Options, Setting, 108
Layout Regen Options, 1318
Layout Tab, Activate, 266
Layout Tabs, 18, 106
Layout Wizard, 283
Layout Wizard, Begin, 283
Layout Wizard, Define Viewports, 285
Layout Wizard, Orientation, 284
Layout Wizard, Paper Size, 284
Layout Wizard, Pick Location, 285
Layout Wizard, Printer, 284
Layout Wizard, Title Block, 284
Layoutregenctl, 1318
Layouts, 276
Layouts, Advanced, 888
Layouts, Considerations for Multiple, 907
Layouts, Creating, 283
Layouts, Setting up, 287
Layouts, Using Two, 902
Layouts and Printing, Introduction, 105
LAYOUTS AND VIEWPORTS, 275
Layouts and Viewports Example, 279

Layouts and Viewports, Guidelines for Using, 282
Layoutwizard, 283
Layoff, 252
Layoff, Block selection, 252
Layoff, Viewports, 252
Laythw, 252
Layulk, 253
Layuniso, 247
Layvpi, 253
Laywalk, 248
Laywalk, Filter, 249
Laywalk, Layer name edit box, 249
Laywalk, Purge, 249
Laywalk, Restore on Exit, 249
Laywalk, Select objects, 249
Laywalk, Shortcut menu, 250
Leader Settings dialog box, 710
Leader Settings dialog box, *Attachment* tab, 712
Leader Settings dialog box, *Leader Line and Arrow* tab, 712
Leader, 708
Leader, Annotation, 710
Leader, Arrow/None, 709
Leader, Block, 709
Leader, Copy, 709
Leader, Format, 709
Leader, Mtext, 709
Leader, None, 710
Leader, Spline/Straight, 709
Leader, Tolerance, 709
Leader, Undo, 710
Legacy Surfaces, 1178
Legacy Surfaces, 3Dface, 1178
Legacy Surfaces, 3Dmesh, 1178
Legacy Surfaces, Edgesurf, 1179
Legacy Surfaces, Region, 1179
Legacy Surfaces, Revsurf, 1179
Legacy Surfaces, Rulesurf, 1179
Legacy Surfaces, Tabsurf, 1179
Legacy Surfaces, Thickness, 1179
Lengthen, 167
Lengthen, Delta, 167
Lengthen, Dynamic, 168
Lengthen, Percent, 168
Lengthen, Total, 168
Lens Length, 989
Light, 1218
Light, Attenuation, 1219
Light, Attenuation, Inverse Linear, 1219
Light, Attenuation, Inverse Squared, 1219
Light, Color, 1219
Light, Falloff, 1220
Light, Hotspot, 1220
Light, Intensity, 1219
Light, Name, 1219
Light, Shadow, 1219
Light, Status, 1219
Light control panel, 964, 1206, 1218
Light Glyph, Edit Glyph Colors, 1225
Light Glyph, Glyph Preview, 1225
Light Glyph, Glyph Size, 1225
Light Glyph Appearance dialog box, 1224

Light Glyphs, 1224
Lightglyphdisplay, 1224
Light List, 1225
Lightlist, 1225
Lights in Model palette, 1225
Limits, 6, 99, 263
Limits, On/Off, 101
Limits Settings Table, Architectural, 327
Limits Settings Table, Civil, for ANSI Sheet Sizes, 330
Limits Settings Table, Mechanical, 326
Limits Settings Table, Metric, for ANSI Sheet Sizes, 329
Limits Settings Table, Metric, for Metric (ISO) Sheet Sizes, 328
Limits Settings Tables, 325
Limmax, 94, 95
Line, 46, 50, 143
Linear Parameter, 535
Linetype, 234
Linetype, Bylayer, 234
Linetype, Delete, 235
Linetype, Load, 235
Linetype Control drop-down list, 236
Linetype dialog box, 227
Linetype Manager, 234, 896
Linetype property, changing, 373
Linetype Scale, Viewports, 895
Linetype Scale property, changing, 373
Linetypes, Creating Complex, 1334
Linetypes, Creating Custom, 1332
Linetypes, Creating Simple, 1333
Linetypes, Lineweights, and Colors, Managing, 630
Linetypes, Lineweights, and Colors, Multiview Drawings, 630
Linetypes, Using, 628
Lineweight, 236
Lineweight Control drop-down list, 237
Lineweight dialog box, 227
Lineweight property, changing, 373
Lineweight Settings, 1321
Lineweight Settings dialog box, 236
Links dialog box, 858
List, 409
List box, 10
Live Enabler Options, 1319
Locking Viewport Geometry, 297
Loft, 1065, 1083, 1160
Loft, Cross-sections only, 1083
Loft, Guides, 1084
Loft, Path, 1085
Loft Settings dialog box, 1083
Log File Location, 1314
Logfileon/Logfileoff, 1339
Logfilepath, 1228, 1341
Lookup Actions and Lookup Tables, 546
Lookup Parameter, 537
Lookup Tables and Lookup Actions, 546

LPT, 335
Ltscale, 94, 95, 101, 240, 262, 263, 265, 740, 895
Lunits, 94, 95
Lwdefault, 237
Lwdisplay, 237
Lweight, 236
Lweight, Adjust Display Scale, 237
Lweight, Default, 237
Lweight, Units, 236
LWT, 237
Lwunits, 236

M

M2p, Object Snap, 121
Make Object's Layer Current, 233
Manage Page Setups, 949
Manual Performance dialog box, 1253
Manual Performance Tuning dialog box, 1012
Mapping, 1241
Mapping, Box, 1241
Mapping, Cylindrical, 1242
Mapping, Planar, 1242
Mapping, Spherical, 1242
Marble dialog box, 1236
Marker, 125
Markup, 588, 615
Markup Process, 614
Markup Set Manager, 588, 613
Markup Tools, 614
Mass Properties, 1165
Massprop, 414, 1165
Massprop, Bounding Box, 1166
Massprop, Centroid, 1166
Massprop, Mass, 1166
Massprop, Moments of Inertia, 1166
Massprop, Principle Moments, 1166
Massprop, Products of Inertia, 1166
Massprop, Radii of Gyration, 1166
Massprop, Volume, 1166
Massprop, X, Y, Z Directions, 1166
Match Properties, 244
Match Properties, Dimensioning, 784
Matchprop, 378
Matchprop, Dimensioning, 784
Material Attachment Options dialog box, 1230
Material Condition Symbol, 719
Material control panel, 1206
Material Display, 1010
Material Editor, 1230
Materials Tool Palette group, 1227
Materials Tool Palettes, 1227
Materials, 1228
Materials control panel, 964
Materials palette, 1228
Materialattach, 1230
Materialattach, Layers List, 1230
Materialattach, Materials List, 1230
Materialmap, 1241
Materials, *Ambient* (Color), 1231
Materials, *Bump Map*, 1234

Materials, *Diffuse* (Color), 1231
Materials, *Diffuse Map*, 1234
Materials, *Marble*, 1235
Materials, *Opacity*, 1233
Materials, *Reflection*, 1234
Materials, *Refraction Index*, 1233
Materials, *Self-illumination*, 1233
Materials, *Shininess*, 1232
Materials, *Specular*, 1232
Materials, *Translucency*, 1232
Materials, *Wood*, 1236
Maximize, 36
Mbuttonpan, 122, 202
MDI, 1318
Measure, 353
Measureinit, 5, 29, 31, 93, 94, 95
Mesh Density, 1184
Meshed Surfaces, Creating, 1182
Meshes, Geometrically Defined, 1184
Metafile, 878, 883
Microsoft Internet Explorer, 613
Mid Between 2 Points, Object Snap, 121
Midpoint, Object Snap, 117
Minimize, 36
Minsert, 503, 507
Mirror, 177
Mirror 3D, 1065, 1093
Mirror3d, 1065, 1093
Mirror3d, 3points, 1094
Mirror3d, Last, 1094
Mirror3d, Object, 1094
Mirror3d, View, 1094
Mirror3d, XY, 1094
Mirror3d, YZ, 1094
Mirror3d, Zaxis, 1094
Mirror3d, ZX, 1095
Mirrtext, 178
MISCELLANEOUS COMMANDS AND FEATURES, 1293
MISCELLANEOUS CUSTOMIZATION, 1329
Miscellaneous Rendering Tips, 1253
Mledit, 392
Mledit, Add Vertex, 394
Mledit, Closed Cross, 392
Mledit, Closed Tee, 393
Mledit, Corner Joint, 394
Mledit, Cut All, 395
Mledit, Cut Single, 395
Mledit, Delete Vertex, 395
Mledit, Merged Cross, 393
Mledit, Merged Tee, 394
Mledit, Open Cross, 393
Mledit, Open Tee, 394
Mledit, Weld, 395
Mline, 353
Mline, Justification, 354
Mline, Scale, 354
Mline, Style, 355
Mlstyle, 355
Mlstyle, Load, 356
Mlstyle, Modify, 356
Mlstyle, New, 356
Mlstyle, Rename and Delete, 356

Mlstyle, Save, 356
Mlstyle, Set Current, 356
Mlstyle, Styles List, 356
MM.PSS, 922
Model, 295
Model Space, 18
Model Space and Paper Space, 276
Model Tab, 18, 106
Model/Paper toggle, 294
Model View tabs, 938
Modify Command Concepts, 68
Modify II toolbar, 372
MODIFY COMMANDS I, 159
MODIFY COMMANDS II, 272
Modify Properties, 243
Modify pull-down menu, 160
Modify toolbar, 160
Modify Transmittal Setup, 599
Modifying a Light, 1225
Moments of Inertia, 1166
Motion Path Animation, Animation set-
    tings, 1262
Motion Path Animation, Link camera
    to, 1261
Motion Path Animation, Link target to,
    1262
Motion Path Animation, Preview, 1262
Motion Path Animation dialog box,
    1261
Mouse and Digitizing Puck Buttons,
    14
Mouse Input, 969
Mouse Wheel, Zoom and Pan, 201
Move, 75, 162
Move, 3D, 1065, 1087
Move Action, 539
Move and Rotate Grip Tools, 1127
Moving Solids, Commands, 1087
Mredo, 88
Mslide, 1303
Msolescale, 861
Mspace, 295, 282
Mtedit, 419, 439
Mtext, 423, 419
Mtext, At Least/Exactly, 425
Mtext, AutoCAPS, 428
Mtext, Background Mask, 428
Mtext, Bold, Italic, Underline, 428
Mtext, Bullets and Lists, 428
Mtext, Change Case, 428
Mtext, Character Set, 429
Mtext, Combine Paragraphs, 428
Mtext, Cut, Copy, Paste, 426
Mtext, Find and Replace, 428
Mtext, Font, 424
Mtext, Height, 424
Mtext, Import Text, 429
Mtext, Indents and Tabs, 427
Mtext, Insert Field, 429
Mtext, Justification, 426
Mtext, Line Spacing, 425
Mtext, Oblique Angle, 430
Mtext, Opaque Background, 426
Mtext, Overline, 428
Mtext, Remove Formatting, 428

Mtext, Rotation, 425
Mtext, Ruler, 425
Mtext, Select All, 428
Mtext, Stack Properties, 425
Mtext, Stack/Unstack, 429, 425
Mtext, Style, 424
Mtext, Symbol, 429
Mtext, Text Color, 424
Mtext, Undo, Redo, 425
Mtext, Width, 425
Mtext, Width Factor, 430
Mtext Shortcut Menu, 426
Mtexted, 425, 437
Multiline, 355
Multiline Edit Tools dialog box, 392
Multiline Style, 356
Multiline Style dialog box, 355, 356
Multiline Styles Example, 358
Multiline Text, 423
Multiline Text Format Codes, 437
Multilines, Editing, 358
Multiple, 1300
Multiple, object selection, 72
Multiple Attedit, 563
Multiple Redo, 88
Multiple Viewports, 901
MULTIVIEW DRAWING, 621
Multiview Drawing, Guidelines for
    Creating, 633
Mvsetup, 1270, 1271
Mvsetup, Creating Dimensions for a
    3D Model, 1275
Mvsetup, Creating Hidden Lines for
    a 3D Wireframe, 1275
Mvsetup, Typical Steps for Using the
    Standard Engineering Option, 1271
Mvsetup Example, 1272

N

N (No), 566
N size, 1103
Named Layer Filters, 224, 251
Named Plot Style Table Example,
    918
Named Plot Style Tables, 915
Named Views, 987
Navigation and View Commands,
    978
Navigation Control Panel, 979, 989
Nearest, Object Snap, 117
New, 29
New Group Filter, 229
New Layer, 225
New Layout, 285
New Page Setup dialog box, 321
New Sheet, 941
New Sheet Set, 952
New Subsets, 944
New Surfaces and Legacy Surfaces,
    1178
Newsheetset, 952
No (Global Editing), 566
No Edges, 1005
No Face Style, 1004

Node, Object Snap, 117
Nonassociative Dimensions, 732
Non-Uniform Rational Bezier Spline
    (NURBS), 348
Noun/Verb Command Syntax
    Practice, 77
Noun/Verb Selection, 481
Noun/Verb Syntax, 73
Number of Tessellations to Cache, 1013
NURBS, 348

O

Object, 46
Object ARX Applications, 1231
Object Embedding, 845
Object Group, 490
Object Grouping dialog box, 491
Object Groups, 490
Object Linetypes, Lineweights and
    Colors, 630
OBJECT LINKING AND EMBED-
    DING (OLE), 843
Object Linking, 745
Object Properties, 223
Object Properties, Changing, 242
Object Properties, Displaying, 233
Object Properties Toolbar, 7, 242, 379
Object Selection, Implied Windowing,
    481
Object Selection, Use Shift to Add, 481
Object Selection Filters, 484
Object Selection Filters dialog box, 487
Object Snap, Alignment Point
    Acquisition, 126
Object Snap, Aperture size, 126
Object Snap, Apparent intersection, 118
Object Snap, AutoSnap aperture box,
    126
Object Snap, AutoSnap marker, 126
Object Snap, AutoSnap tooltip, 126
Object Snap, AutoTracking tooltip, 126
Object Snap, Center, 115
Object Snap, Color, 126
Object Snap, Endpoint, 116
Object Snap, Extension, 119
Object Snap, From, 121
Object Snap, Full-screen tracking vector,
    126
Object Snap, Ignore Hatch Objects, 127
Object Snap, Ignore Negative Z object
    Snaps for Dynamic UCS, 127
Object Snap, Insert, 116
Object Snap, Intersection, 116
Object Snap, M2p, 121
Object Snap, Magnet, 125
Object Snap, Marker, 125
Object Snap, Mid Between 2 Points, 121
Object Snap, Midpoint, 117
Object Snap, Nearest, 117
Object Snap, Node, 117
Object Snap, None, 125
Object Snap, Parallel, 119
Object Snap, Perpendicular, 117
Object Snap, Polar tracking vector, 126

Object Snap, *Quadrant*, 118
Object Snap, *Replace Z value with current elevation*, 127
Object Snap, *Tangent*, 118
Object Snap, Temporary Override Keys, 123
Object Snap, *Temporary Tracking*, 120
OBJECT SNAP AND OBJECT SNAP TRACKING, 113
Object Snap Cycling, 125
Object Snap markers, 123
Object Snap Modes, 114, 115
Object Snap Modes, Acquisition, 118
Object Snap Modifiers, 120
Object Snap toggle, 124
Object Snap toolbar, 122
Object Snap Tracking, 127, 128
Object Snap Tracking, 3D, 972
Object Snap Tracking, Using to Draw Projection Lines, 624
Object Snap Tracking Settings, 130
Object Snap Tracking with Polar Tracking, 129
Object-specific drawing scheme, 223
Object-Specific Properties Controls, 234
Oblique Drawing in AutoCAD, 650
Oblique drawings, 643
Obscured Edges, 1007
Offset, 178
Offset, Auxiliary Views, 688
Offset, Distance, 179
Offset, Erase, 179
Offset, Layer, 179
Offset, Through, 179
Offset, Using for Construction of Views, 626
Offsetgaptype, 180
OLE, 845
OLE, *(Object Type) Object*, 860
OLE Links, 858
OLE Links dialog box, 858
OLE Objects, Managing, 858
OLE Related Commands, 858
OLE Right-Click Shortcut Menu, 859
Olehide, 861
Olelinks, 858
Olelinks, Break Link, 859
Olelinks, Open Source, 858
Olelinks, Source, 858
Olelinks, Type, 858
Olelinks, Update Now, 585
Olelinks, Update, 585
Oleopen, 589
Olequality, 861
OLE-Related System Variables, 861
Olescale, 859
Olestartup, 861
Oops, 86
Opacity, 1010
Opacity Maps, 1236
Open, 32
Open a Drawing from DesignCenter, 520

Open Sheet Set, 936
Open Xref, 824
Opendwfmarkup, 615
Opensheetset, 936
Options dialog box, 1312
Options dialog box, *Selection* tab, 466
Options, 1312
Options, Default Output Device, 288
Options, Display tab, 8, 23, 109, 277, 831, 1315
Options, Drafting tab, 125
Options, Files tab, 31, 437, 515, 531, 598, 1312
Options, General Plot Options, 288
Options, Open and Save tab, 41, 823, 830, 1315
Options, Plot and Publish tab, 288, 1317
Options, Plotting tab, 109
Options, Profiles tab, 1323
Options, Selection tab, 472, 481, 1323
Options, Support File Search Path, 515
Options, System tab, 1317
Options, User Preference tab, 1319
Order Group dialog box, 494
Organize Views, 944
Ortho, 15, 16
Ortho, 3D, 971
Ortho, Auxiliary Views, 683
Ortho, Isometric drawings, 645
Ortho and *Osnap*, Using for Projection Lines, 622
Orthographic & Isometric Views, 978
Orthographic UCS, 988
Osnap, 114
Osnap Applications, 130
Osnap Practice, 130
Osnap Running Mode, 123
Osnap Single Point Selection, 121
Osnap Toggle, 14
Osnap Tracking, 15
Osnapcoord, 1234
Other Table Editing Methods, 454
Otrack, 127

P

Page Setup, 319
Page Setup, Layout Name, 289
Page Setup dialog box, 289
Page Setup for Layouts, 289
Page Setups List, 320
Pagesetup, 289, 319
Pagesetup, Import, 321
Pagesetup, Modify, 321
Pagesetup, New, 321
Pagesetup, Set Current, 321
Palette, 7
Palette *Properties*, 528
Pan, 206, 978, 979
Pan, Down, 208
Pan, Left, 208
Pan, Mouse Wheel or Third Button, 202

Pan, Point, 207
Pan, Realtime, 207
Pan, Right, 208
Pan, Transparent, 208
Pan, Up, 208
Pan and *Zoom*, Mouse Wheel, 201
Pan and *Zoom* with *Undo* and *Redo*, 208
Pantone color, 239
Paper Sizes, Standard, 323
Paper Space, 18, 106
Paper Space and Model Space, 276
Paper Space Dimensioning, 757, 788, 897
Paper/Model toggle, 294
Paperupdate, 288
Parallel, Object Snap, 119
Parallelism, GDT, 718
Parameter, Custom Properties, 537
Parameters, Associating Actions, 538
Parameters and Actions, 532
Partial Load dialog box, 38
Partial Open dialog box, 37
Partialload, 38
Partialopen, 37
Passwords, 35
Paste, 847, 851
Paste as Block, 853
Paste as Hyperlink, 857
Paste Special, 854
Paste Special dialog box, 854
Paste to Original Coordinates, 850
Pasteashyperlink, 857
Pasteblock, 853
Pasteclip, 851, 852, 845
Pasteorig, 853
Pastespec, 854
Pastespec, Display As Icon, 855
Pastespec, Paste, 855
Pastespec, Paste Link, 855
Pastespec, Result, 855
Pasting AutoCAD Objects, 851
Pasting Non-AutoCAD Objects, 852
Pcinwizard, 336
PCL Language, 314
PDF Files, Creating, 883
Pdmode, 151
Pdsize, 151
Pedit, 383, 1189
Pedit, Close, 383
Pedit, Decurve, 385
Pedit, Desmooth, 1189
Pedit, Edit Vertex, 384, 386, 1189
Pedit, Exit, 386
Pedit, Fit, 385
Pedit, Join, 384
Pedit, Ltype gen, 386
Pedit, Mclose/Nclose, 1190
Pedit, Multiple, 383
Pedit, Open, 383
Pedit, Smooth surface, 1189
Pedit, Spline, 385
Pedit, Undo, 386
Pedit, Width, 384

Performance Tuner, 1011, 1012
Performance Tuner Log, 1011
Perpendicular, Object Snap, 117
Perspective, 978, 983
Pickadd, 377, 481, 482
Pickauto, 481, 483
Pickbox, 70
Pickdrag, 481, 484
Pickfirst, 13, 73, 377, 467, 481
Pickstyle, 493, 671, 673
PICT, 869
PICTORIAL DRAWINGS, 641
Pictorial Drawings, 2D Drawings, 643
Pictorial Drawings, Types, 642
Place View Label Block, 945
Plan, 978, 996
Plan View, 996
Planar Surface, 1161
Planesurf, 1161
Planesurf, Object, 1162
Pline, 152
Pline, Arc, 153
Pline, Close, 153
Pline, Halfwidth, 152
Pline, Length, 153
Pline, Undo, 153
Pline, Width, 152
Pline Arc Segments, 154
Plinegen, 386
Plines, Converting *Lines* and *Arcs*, 388
Plot, 108, 313
Plot, Add, 314
Plot, Apply to Layout, 318
Plot, Cancel, 318
Plot, Display, 315
Plot, Drawing Orientation, 313, 317
Plot, Extents, 315
Plot, Hide Paperspace Objects, 317
Plot, Layout Name, 313
Plot, Limits or *Layout*, 315
Plot, OK, 318
Plot, Page Setup dialog box, 313
Plot, Paper Size and Number of Copies, 314
Plot, Plot Area, 314
Plot, Plot Paper Space Last, 317
Plot, Plot Stamp On, 317
Plot, Plot Style Table (pen assignments), 316
Plot, Plot to File, 314
Plot, Plot with Plot Styles, 317
Plot, Plotter Configuration Editor, 314
Plot, Preview, 317
Plot, Printer/Plotter, 314
Plot, Save Changes to Layout, 317
Plot, Shade Plot, 316
Plot, Shaded Viewport Options, 316
Plot, View, 315
Plot, What to Plot, 314
Plot, Window, 315
Plot and Publish Log File Location, 1314

Plot and Publish tab, 1314
Plot Device, Set, 266
Plot dialog box, 108, 313
Plot dialog box, *Plot Settings* tab, 313
Plot dialog box or *Page Setup* dialog box, 321
Plot Drivers, 588
Plot in Backgroud, 316
Plot Object Lineweights, 317
Plot Offset, 313
Plot Options and Layout Options, Setting, 108
Plot Paperspace Last, 317
Plot Preview, 322, 324
Plot Scale, 264, 315
Plot Scale, Scale, 315
Plot Scale, Scale Lineweights, 316
Plot Stamp dialog box, 922, 949
Plot Stamp On, 317
Plot Stamp Settings, 949
Plot Stamping, 921
Plot Style, 908
Plot Style and Plot Style Table, Differences, 908
Plot Style Manager, 912
Plot Style Table, 907
Plot Style Table, Attach to Layout, 912
Plot Style Table, Attaching, 909
Plot Style Table, Color-Dependent, 910
Plot Style Table, Creating, 913
Plot Style Table, Named, 910
Plot Style Table Conversion, 913
Plot Style Table Editor, 912
Plot Style Table Example, Named, 918
Plot Style Tables, Color Dependent, 915
Plot Style Tables, Named, 910
Plot Style Tables and Plot Styles, 907
Plot Styles, 223
Plot Styles, Assigning to Objects, 908
Plot Styles, Editing, 914
Plot to Scale, 4
Plot with Plot Styles, 317
PlotScale, Custom Edit Boxes, 315
PlotScale, Fit to Paper, 315
Plotstamp, 921
Plotstamp, Advanced Options, 922
Plotstamp, Parameter File, 922
Plotstamp, Plot Stamp Fields, 922
Plotstamp, Preview, 922
Plotstamp, User Defined Fields, 922
Plotstyle, 917
Plotstyle property, changing, 373
Plotter Configuration Editor, 334
Plotter Manager, 332
Plottermanager, 332
Plotters and Printers, Configuring, 332
Plotting, 312
Plotting, Typical Steps, 312
Plotting and Publishing Sheet Sets, 948

Plotting Components, 332
Plotting *Layouts* to Scale, 324
Plotting *Model* Tab to Scale, 324
Plotting to Scale, 322
Plotting to Scale, Examples, 331
Plotting to Scale Guidelines, 324
Pngout, 867, 877
Point, 46, 151
Point Filters, 3D, 972
Point Light, 1216
Point Parameter, 535
Point Specification Precedence, 473
Point Style, 151
Point Style dialog box, 151
Pointer Input, 56
Polar, 15, 16
Polar Array, 181
Polar Format, 57
Polar Input Settings, 57
Polar Parameter, 535
Polar Snap, 16, 50, 57, 59, 101
Polar Snap, Polar Tracking, and Dynamic Input, 63
Polar Snap and Polar Tracking, 57
Polar Snap and *Polar Tracking*, Realignment of Views, 627
Polar Stretch Action, 541
Polar Tracking, 57
Polar Tracking, 3D, 971
Polar Tracking, Auxiliary Views, 686
Polar Tracking, Isometric Drawings, 645
Polar Tracking, Polar Snap, and Dynamic Input, 63
Polar Tracking, Using to Draw Projection Lines, 623
Polar Tracking and Direct Distance Entry, 59
Polar Tracking and Dynamic Input, 62
Polar Tracking and Grid Snap, 60
Polar Tracking and Polar Snap, 57
Polar Tracking Override, 59
Polar tracking vector, 126
Polar Tracking with Object Snap Tracking, 129
Polarang, 103
Polardist, 103
Polygon, 345
Polygon, Edge, 345
Polygon, Inscribe/Circumscribe, 345
Polygon Meshes, Editing, 1188
Polygon Meshes, Editing with *Explode*, 1192
Polygon Meshes, Editing with Grips, 1191
Polygon Meshes, *Pedit* with, 1189
Polygon Meshes, Variables Affecting, 1190
Polygon Meshes and *3Dface*, Limitations, 1197
Polyline, 175
Polysolid, 1064, 1076
Polysolid, Drawing Arcs, 1077

Polysolid, Height/Width, 1076
Polysolid, Justify, 1077
Polysolid, Object, 1077
Position, GDT, 718
Position Locator, 1259
Position Locator, Position indicator blink, 1259
Position Locator, Position indicator color, 1259
Position Locator, Position indicator size, 1259
Position Locator, Preview area, 1259
Position Locator, Preview Transparency, 1259
Position Locator, Preview Visual Style, 1259
Position Locator, Target indicator color, 1259
Position Locator, Zoom In/Out/Extents, 1259
Postscript, 881
PostScript Out Options dialog box, 881
Presspull, 1137, 1151
Preview, 322
Previous, object selection, 72
Primitives, Solids, 1064
Print, 313
Print File, Spooler, and Prolog Section Names, 1313
Printable Area, 323
Printer Support File Path, 1313
Printing, 313
Printing 3D Models, 1014
Printing a Rendered Image from a *Layout* Tab, 1252
Printing a Rendered Image from the *Model* Tab, 1252
Printing a Rendered Image with *Shade Plot*, 1252
Printing and Layouts, Introduction, 105
Printing and Plotting, 108
PRINTING AND PLOTTING, 311
Printing from a Layout Tab, 1014
Printing from the Model Tab, 1014
Printing the Drawing, 110
Profile, 1283
Profiles, Options, 1323
Project Files Search Path, 1313
Projected Tolerance Zone, GDT, 720, 724
Projection, 170
Projection and Alignment of Views, 622
Projectname, 832, 833, 1313
Properties, 243, 374, 439, 565, 730, 875, 977
Properties, Dimensioning, 781
Properties, Filter, 230
Properties, General group, 375
Properties, Geometry group, 376
Properties, Pickadd toggle, 377
Properties, Quick Select, 376
Properties, Select Objects, 376
Properties, Thickness, 1200

Properties of Objects, Changing, 373
Properties palette, 243, 295, 374, 439
Property Settings dialog box, 244, 784
Proxy Fonts, 736
Proxynotice, 1317
Proxyshow, 1317
Proxywebsearch, 1319
Psetupin, 290
Psltscale, 262, 265, 297, 895
Psout, 867, 881
Pspace, 294, 282
Psprolog, 1313
Pstylepolicy, 911
Publish, 588, 604
Publish, Add Sheets, 606
Publish, Load Sheet List, 606
Publish, Model Tab and Layout Tabs, 607
Publish, Move Sheet Up and Move Sheet Down, 606
Publish, Options dialog box, 607
Publish, Plot Stamp Settings, 606
Publish, Remove Sheets, 606
Publish, Save Sheet List, 606
Publish, Sheets to Publish, 605
Publish, Show Details, 607
Publish dialog box, 605, 949
Publish to DWF, 949
Publish to Plotter, 949
Publish to Web, 588, 594
Publish Using Page Setup Override, 949
Publishtoweb, 588, 594
Publishtoweb, Apply Theme, 596
Publishtoweb, Begin, 595
Publishtoweb, Create Web Page, 595
Publishtoweb, Enable i-Drop, 596
Publishtoweb, Generate Images, 597
Publishtoweb, Preview and Post, 597
Publishtoweb, Select Drawings, 596
Publishtoweb, Select Image Type, 595
Publishtoweb, Select Template, 596
Pull-down Menus, 8
Purge, 503, 515
Purge, Confirm each item, 516
Purge, Purge All, 516
Purge, Purge, 516
Purge, Purge nested items, 516
Purge, View Items, 516
Purge dialog box, 516
Pyramid, 1064, 1074
Pyramid, 2Point, 1075
Pyramid, Axis Endpoint, 1075
Pyramid, Edge, 1075
Pyramid, Inscribed/Circumscribed, 1075
Pyramid, Sides, 1075
Pyramid, Top Radius, 1076

Q

Qdim, 715
Qdim, Baseline, 716
Qdim, Continuous, 716
Qdim, DatumPoint, 717

Qdim, Diameter, 717
Qdim, Edit, 717
Qdim, Ordinate, 716
Qdim, Radius, 717
Qdim, Settings, 716
Qdim, Staggered, 716
Qleader, 710
Qleader, Annotation tab, 711
Qleader, Annotation type, 711
Qleader, Mtext Options, 711
Qleader Annotation Reuse, 712
Qnew, 6, 29, 31
Qsave, 34
Qselect, 485
Qselect, Append to Current Selection Set, 487
Qselect, Apply to, 486
Qselect, How to Apply, 487
Qselect, Object Type, 486
Qselect, OK, 487
Qselect, Operator, 486
Qselect, Properties, 486
Qselect, Select Objects, 486
Qselect, Value, 486
Qtext, 419, 446, 1315
Quadrant, Object Snap, 118
Quadratic B-spline, 1190
Quick Calculator, 1308
Quick Calculator, Number Pad, 1309
Quick Calculator, Scientific, 1310
Quick Calculator, Units Conversion, 1310
Quick Calculator, Variables, 1311
Quick Calculator toolbar, 1308
Quick Dimension, 715
Quick New, 31
Quick Select, 485
Quick Select dialog box, 485
Quick Setup wizard, 95
QuickCalc, 1308
Quit, 37

R

Radio button, 10
Radius and *Diameter* Variable Settings, 759
RAL color, 239
Raster Files, 601
Raster Files, Exporting, 867
Raster Files versus Vector Files, 866
Raster Image, 868
Raster Image Reference, 871
Raster Image Resolution, 603
Raster Images, Attaching, 867
Raster Images, External References, 868
RASTER IMAGES AND VECTOR FILES, 865
Ray, 344
Recover, 39
Rectang, 346
Rectangle, 346
Rectangular Array, 181
Redo, 88
Refclose, 570, 574, 829
Refclose, Discard References Changes, 830

Refclose, Save, 830
Refedit, 574, 826
-Refedit, 828
Reference Edit dialog box, 827
Reference Edit Fading Intensity, 1315
Reference Manager, 834
Reference toolbar, 806
Reflection Maps, 1236
Refset, 829
Refset, Add to Workset, 829
Refset, Remove from Workset, 829
Regen, 89
Regenauto, 1299
Region, 362
Region Modeling Applications, 1197
Reinit, 1300
Re-initialization dialog box, 1300
Relative Cartesian Coordinates, 3D, 969, 970
Relative coordinates, 48, 57
Relative polar coordinates, 48
Relative polar display, 10
Remove, object selection, 72
Remove Sheets, 944
Removing and Disabling an Image Map, 1240
Rename, 503 517
Rename and Renumber Sheet dialog box, 946
Rename and Renumber Sheets and Views, 946
Rename dialog box, 517
Render, 1013, 1206, 1208, 1242
Render, Color Systems, 1208
Render, Defining a Background, 1210
Render, Index Color (ACI), 1208
Render, Radiosity, 1210
Render, Raytraced Shadows, 1210
Render, Raytracing, 1210
Render, Rendering Time, 1210
Render, Shadow Maps, 1210
Render, Terms and Considerations, 1210
Render, True Color HLS, 1209
Render, True Color RGB, 1209
Render control panel, 964, 1206
Render Presets Manager, 1247
Render Presets Manager, Create Copy, 1247
Render Presets Manager, Delete, 1247
Render Presets Manager, Property Panel, 1247
Render Presets Manager, Set Current, 1247
Render Window, 1242
Render Window, History pane, 1243
Render Window, Image pane, 1243
Render Window, Progress Meter, 1243
Render Window, Statistics pane, 1243
Rendercrop, 1244
Renderenvironment, 1214
Renderenvironment, Color, 1215
Renderenvironment, Enable Fog, 1215
Renderenvironment, Far Distance, 1215

Renderenvironment, Far Fog Percentage, 1215
Renderenvironment, Fog Background, 1215
Renderenvironment, Near Distance, 1215
Renderenvironment, Near Fog Percentage, 1215
Renderenvironment dialog box, 1214
RENDERING AND ANIMATION, 1205
Rendering and Specifying Preferences, 1242
Rendering Example, 1247
Renderpresets, 1247
Renderwin, 1243
Rref, 1244
Resume, 1339
Retain changes to Xref layers, 818
Revcloud, 364
Reverse Method for Drawing Scale Factor, 900
Revision Cloud, 364
Revolve, 1065, 1080, 1160
Revolve, Object, 1080
Revolve, Start Angle, 1081
Revolve, X/Y/Z, 1080
Revolved Mesh, 1187
Revsurf, 1187
RGB, 239
Right-Hand Rule, 975
Rotate, 163, 995
Rotate, Reference, 164
Rotate Action, 542
Rotating Snap back to original position, 684
Rotation Parameter, 536
Rscript, 1306, 1339
Rtdisplay, 1315
Rtpan, 207
Rtzoom, 203
Ruled Mesh, 1185
Rulesurf, 1185
Run Script, 1307
Running Object Snap, 124
Running Object Snap toggle, 124

S

Save, 34
Save As, 34
Save Drawing As dialog box, 35, 259
Save Layer States, 250
Saveas, 34
Saveas Options dialog box, 881
Savefilepath, 41, 1313
Saveimg, 876, 1251
Saveimg, .BMP, 1251
Saveimg, .JPG, 1251
Saveimg, .PCX, 1251
Saveimg, .PNG, 1251
Saveimg, .TGA, 1251
Saveimg, .TIF, 1251
Savetime, 41, 413

SAVING, PRINTING AND VIEWING RENDERINGS, 1251
Saving a Rendered Image to a File, 1251
Saving an Animation to a File, 1263
Saving an AutoCAD Drawing, 26
Scale, 165
Scale, Reference, 165
Scale, Text, 443
Scale Action, 540
Scalelistedit, 299
Scalelistedit, Add, 300
Scalelistedit, Application of the Scale List, 300
Scalelistedit, Delete, 300
Scalelistedit, Edit, 300
Scalelistedit, Move Up and *Move Down*, 300
Scalelistedit, Reset, 300
Scaletext, 419, 443
Scaletext, Match Object, 445
Scaletext, Scale Factor, 444
Scaling Viewport Geometry, 295
Screen Menu, 8
Script, 1307, 1339
Script Commands, 1339
Script Files, Creating, 1337
Scripts for Drawing Setups, 1338
Scroll Bars, 208
SDF, 582
SDI, 1318
Section Boundary, 1157
Section Plane, 1154, 1156
Section View, Creating, 657
Section Volume, 1157
SECTION VIEWS, 655
Sectionplane, 1138, 1154
Sectionplane, Draw Section, 1156
Sectionplane, Live Sectioning, 1155
Sectionplane, Orthographic, 1156
Sectionplane, Section Object States, 1155
Sectionplane, Section Plane Grips, 1155
Security Options, 35, 1316
Select, 73
Select Color dialog box, 227
Select File dialog box, 32
Select Linetype dialog box, 227
Select Plot Style dialog box, 917
Select Template Dialog Box, 31
Selecting and Modifying Lights, 1224
Selecting Overlapping Subobjects, 1135
Selection Set Options, 69
Selection Set Variables, 480
SELECTION SETS, 67
Selection Sets, 68
Selection Sets Practice, 73
Selecturl, 594
Set Variable, 413
Setting Rendering Options for Visual Styles, 1253
Setup Commands, 97
Setup Wizards, Dimensioning, 779
Setvar, 413

Shade, 1013
Shade Plot, 316
Shade Plot, As Displayed, 316
Shade Plot, Hidden, 317
Shade Plot, Wireframe, 316
Shaded Viewport Options, 316
Shadedge, 1013
Shadedif, 1013
Shadow Maps, 1210
Shadow Options dialog box, 1219
Sheet List tab, 937
Sheet List Table, Create, 947
Sheet List Table, Updating, 948
Sheet Selection, 948
Sheet Set, Archiving, 951
Sheet Set, Example, 953
Sheet Set Commands, 936
Sheet Set Details, 953
Sheet Set Manager, 607, 934, 936
Sheet Set Properties, 943
Sheet Set Properties, View and Edit, 943
Sheet Set Publish Options, 949
SHEET SETS, 933
Sheet Sets, Transmit and Archive, 950
Sheet View tab, 939
Sheets, Organize, 944
Sheetset, 936
Shift + Left Button, 72
Shift-Select, 171
Shortcut Keys, Customizing, 1331
Shortcut Menu, *3Dorbit*, 981
Shortcut Menu, Command-Mode, 13
Shortcut Menu, Default, 13
Shortcut Menu, Dialog-Mode, 14
Shortcut Menu, Edit-Mode, 13
Shortcut Menu, *Mtext*, 431
Shortcut Menu, OLE, 849, 859
Shortcut Menus, 13
Shortcutmenu, 1320
Show History, 1136
Showhist, 1137
Silhouette Edges, 1007
Single, object selection, 72
Single Line Text, 420
Sketch, 359
Sketch Options, 359
Skpoly, 360
Slice, 1137, 1150
Slicing Solids with Surfaces, 1162
Slicing Solids with Surfaces Example, 1162
Slide Show, Creating, 1304
Slides and Scripts, Using, 1302
Snap, 15, 101, 264
Snap, Aspect, 102
Snap, Isometric, 644
Snap, On/Off, 102
Snap, Rotate, 102
Snap, Style, 102
Snap, Type, 102
Snap, Value, 102
Snap Marker, 115
Snap Rotate, Auxiliary Views, 683
Snap Tips, 115

Snaptype, 103
Snapunit, 94, 95, 103
Soldraw, 1270, 1278
Solid, 361
Solid Model Construction Process, 1064
SOLID MODEL EDITING, 1117
SOLID MODELING CONSTRUC-TION, 1063
Solid Models, 966
Solid Models, Analyzing, 1089
Solid Primitives Commands, 1067
Solidcheck, 1149
Solidedit, 1037, 1137
Solidedit, Body, Check, 1149
Solidedit, Body, Clean, 1148
Solidedit, Body, Imprint, 1147
Solidedit, Body, Separate, 1148
Solidedit, Body, Shell, 1149
Solidedit, Body Options, 1147
Solidedit, Edge, Color, 1147
Solidedit, Edge, Copy, 1146
Solidedit, Edge Options, 1146
Solidedit, Face, Color, 1146
Solidedit, Face, Copy, 1145
Solidedit, Face, Delete, 1143
Solidedit, Face, Extrude, 1140
Solidedit, Face, Move, 1141
Solidedit, Face, Offset, 1142
Solidedit, Face, Rotate, 1143
Solidedit, Face, Taper, 1144
Solidedit, Face Options, 1139
Solidhist, 1136
Solids, Commands for Moving, 1087
Solids, Creating from Surfaces, 1164
Solids Editing Commands, 1137
Solprof, 1270, 1283
Solprof 3D Wireframe Example, 1285
Solview, 1270, 1277
Solview, Auxiliary, 1277
Solview, Ortho, 1277
Solview, Section, 1277
Solview, Ucs, 1277
Solview and *Soldraw*, Typical Steps for Using *Solview* and *Soldraw*, 1276
Solview and *Soldraw* Example, 1278
Spacetrans, 419, 432
Spatial Index, 821
Special Key Functions, 15
Spell, 419, 440
Spelling, 440
Sphere, 1064, 1073
Sphere, 3D/2P/Ttr, 1073
Spherical Coordinates (Relative), 969, 971
Splframe, 386, 1191
Spline, 348
Spline, Close, 348
Spline, Fit Tolerance, 349
Spline, Grips, 391
Spline, Object, 385
Spline Data Points, Editing, 389
Splinedit, 389

Splinedit, Add, 390
Splinedit, Close/Open, 390, 391
Splinedit, Fit Data, 391
Splinedit, Move, 390
Splinedit, Move Vertex, 391
Splinedit, Purge, 390
Splinedit, Refine, 391
Splinedit, Reverse, 391
Splinedit, Tangents, 390
Splinedit, Tolerance, 390
Splinedit Control Points, Editing, 391
Splinesegs, 385, 1315
Splinetype, 385
Spotlight, 1216
Stack Properties dialog box, 429
Stacked Text, Creating, 431
Standard dimension style, 745, 757, 779
Standard toolbar, 7
Start from Scratch, 30, 93
Start from Scratch Imperial Settings, Table of, 94
Start from Scratch Metric Settings, Table of, 95
Starting AutoCAD, 5
Startup dialog box, 5, 29, 92
Startup Options, 92
Statistics tab, 41
Status, 408
Status Bar, 10
Status Bar Control, 17
Steps for Creating an Animation Using the *3D Navigation Control Panel*, 1255
Stepsize, 1260
Stepspersec, 1260
Stereolithography Apparatus, 114, 880, 1167
Stlout, 880, 1167
Stretch, 166
Stretch Action, 540
Style, 419, 433
Style, Backwards, 434
Style, Delete, 434
Style, Height, 434
Style, Obliquing angle, 434
Style, Rename, 433
Style, Upside-down, 434
Style, Vertical, 435
Style, Width factor, 434
Style Overrides, Dimensions, 742
Stylesmanager, 912
Subobjects and Selection, 1130
Subset Properties dialog box, 944
Subtract, 1065, 1066, 1099
Subtract, Regions, 397
Summary tab, 40
Sun Properties, 1221
Sun Properties, General, 1221
Sun Properties, Geographical Location, 1223
Sun Properties, Rendered Shadow Details, 1222
Sun Properties, Sun Angle Calculator, 1222

Sun Properties palette, 1221
Sunlight, Using, 1221
Sunproperties, 1221
Surface Model Example, 1193
SURFACE MODELING, 1177
Surface Models, 965
Surface Option, 1151
Surface Tessellation, 1013
Surftab1, 1184, 1185, 1186, 1187, 1188, 1190
Surftab2, 1184, 1185, 1186, 1187, 1188, 1190
Surftype, 1190
Surfu, 1162
Surfv, 1162
Sweep, 1065, 1081, 1160
Sweep, Alignment, 1082
Sweep, Base Point, 1082
Sweep, Scale, 1082
Sweep, Twist, 1082
Swivel, 994
System tab, 1317
Syswindows, 213
Syswindows, Arrange Icons, 214
Syswindows, Cascade, 213
Syswindows, Tile Horizontal, 213
Syswindows, Tile Vertical, 214

T

Table, 419, 451
Table Editing Methods, Other, 454
Table Style Settings, 451
Tabledit, 419, 453
Tables of Limits Settings, 326
Tablestyle, 419, 452
Tablestyle, Modify, 453
Tablestyle, New, 453
Tablestyle, Set Current, 452
Tablet, 14
Tabsurf, 1187
Tabulated Mesh, 1187
Tangent, Object Snap, 118
TARGA, 869
Target box, 115
Template, 94
Template, Use a, 30
Template Drawing Settings Table, 267
Template Drawings, 266
Template Drawings, Creating, 271, 266
Template Drawings, Dimensioning, 779, 787
Template Drawings, Using, 266
Template File, Creating, 580
Temporary Drawing File Location, 1314
Temporary External Reference File Location, 823, 1314
Temporary Files, Options, 1314
Temporary Override Keys for Object Snap, 123
Temporary Tracking, Object Snap, 120
Text, 419, 420

Text, Align, 421
Text, Center, 421
Text, Fit, 421
Text, Justify, 421
Text, Justify, 445
Text, Middle, 421
Text, ML, MC, MR, BL, BC, BR, 422
Text, Right, 421
Text, Scale, 443
Text, Start Point, 420
Text, Style, 422
Text, TL, TC, TR, 421
TEXT AND TABLES, 417
Text Attributes, 458
Text Creation Commands, 420
Text Editor, Dictionary, and Font File Names, 1313
Text Fields, 448
Text Fields and Calculations in Tables, 456
Text Fields and Tables, 448
Text Flow and *Justification*, 423
Text Formatting Editor, 423, 424
Text Formatting Editor, Importing Text, 436
Text Formatting toolbar, 424
Text Height, Calculating, 431
Text Justification, *TL, TC, TR, ML, MC, MR, BL, BC, BR*, 421, 423
Text Style, 433
Text Style dialog box, 433
Text Styles, 265
Text Styles, Fonts, External Text and External Editors, 433
Text Window, 18
Textfill, 419, 447
Textsize, 94, 95
Texttofront, 675
Texture Maps, 1236
Texture Maps Search Path, 1314
Thicken, 1164
Thickness, 1198
Thickness and *Elevation*, Changing, 1200
Thickness and *Elevation*, Using, 1198
Thickness property, changing, 373
Three-View Drawing, Guidelines for Creating, 633
TIFF, 869
Tifout, 867, 877
Tiled Viewports, 278
Tilemode, 276, 278
Time, 413
Time, Display, 413
Time, On/Off/Reset, 413
Titleblock, 266
Tolerance, 718
Tolerance, Datum Identifier, 720
Tolerance, Datum1, 719
Tolerance, Datum2, 720
Tolerance, Datum3, 720
Tolerance, Height, 720
Tolerance, Sym, 719
Tolerance, Tolerance1, 719

Tolerance, Tolerance2, 719
Tolerance, Projected Tolerance Zone, 720
Tool Palette *Properties*, 528
Tool Palettes, 527
Tool Palettes, Creating, 670
Tool Palettes, Creating and Managing, 529
Tool Palettes, Creating using DesignCenter, 529, 670
Tool Palettes, Creating using *Properties*, 530
Tool Palettes, *Customize, Delete, Rename*, 530
Tool Palettes, *Import* and *Export*, 531
Tool Palettes File Locations, 1314
Tool Palettes Window, 669
Tool *Properties*, 529
Tool Properties dialog box, 669
Tool Tip, 7
Toolbar, 1330
Toolbar, docked, 21
Toolbar, floating, 21
Toolbars, 7, 20
Toolbars, Customizing, 1331
Toolpalettes, 527, 669
Torus, 1064, 1074
Torus, 3P/2P/Ttr, 1074
Trackpath, 126
Transferring Files, Internet, 616
Transmittal Package, Creating, 950
Transmittal Setup, Modify, 598
Transmittal Setup dialog box, 598, 951
Transparency, 23
Transparency, image, 874
Transparency, Tool palettes, 528
Transparency Quality, 1013
Tray Settings, 22
Traysettings, 22
Trim, 169
Trim, Crossing Window, 170
Trim, Edge, 170
Trim, Erase, 171
Trim, Projection, 170
Trim, Shift-Select, 171, 172
Trim, Undo, 171
Trimetric drawings, 643
Trimmode, 185
True Color, 239, 1209
True Color, HLS Color Model, 1209
True Color, Hue, 1209
True Color, Luminance, 1209
True Color, RGB Color Model, 1209
True Color, Saturation, 1209
TrueType, 419
TrueType Fonts, 436
Two Sheet Creation Methods, 935
Types of 3D Models, 964
Typical Steps for Using Render, 1207

U

U, 86
UCS, 973, 1024, 1025
UCS, 3point, 1028

UCS, Dducsp, 1031
UCS, Face, 1029
UCS, Named, 1031
UCS, Object, 1027
UCS, Origin, 1027
UCS, Previous, 1030
UCS, View, 1030
UCS, World, 1030
UCS, X, Y, Z, 1030
UCS, Zaxis, 1027
UCS Icon, 212, 968, 976
UCS Icon Control, 976
UCS Icon dialog box, 212, 977
UCS Icons, 974
UCS Manager, 1031, 1032
UCS Manager, Named UCSs tab, 1033
UCS Manager, Settings tab, 1033
UCS System Variables, 1003
Ucsaxisang, 1038
Ucsbase, 988, 1032, 1034
Ucsfollow, 1031, 1036
Ucsicon, 212, 976
Ucsicon, All, 977
Ucsicon, Noorigin, 977
Ucsicon, On/Off, 977
Ucsicon, Origin, 977
Ucsicon, Properties, 977
Ucsman, 1032
Ucsortho, 985, 988, 999, 1034
Ucsview, 1034
Ucsvp, 1036
Undo, 72, 87, 171
Undo, Auto, 87
Undo, Back, 87
Undo, Begin, 87
Undo, Control, 87
Undo, End, 87
Undo, Mark, 87
Undo/Redo, 1321
Union, 1065, 1066, 1098
Union, Regions, 397
Unitmode, 99
Units, 97, 262
Units, Advanced Setup Wizard, 97
Units, Angles, 98
Units, DesignCenter Blocks, 99
Units, Insertion Scale, 99
Units, Precision, 98
Units, Quick Setup Wizard, 96
Units Values, Keyboard Input, 99
Update, 781
Updatefield, 450
URL, 589, 594, 613
User Coordinate System, 212
USER COORDINATE SYSTEMS, 1023
User Coordinate Systems, Creating and Managing, 1024
User Coordinate Systems (UCS) and 3D Modeling, 1038
User Coordinate Systems (UCS) and Icons, 974
User-Defined Lighting, 1216

User-Defined Lighting, *Distant Light*, 1217
User-Defined Lighting, *Lighting Parameters*, 1217
User-Defined Lighting, *Point Light*, 1216
User-Defined Lighting, *Spotlight*, 1216
Using *Materials* in the Model, 1226
Using *Mvsetup* for Standard Engineering Drawings, 1271
Using Solids Data, 1165
Using *Solprof* and *Flatshot* with Solid Models, 1283
Using *Solview* and *Soldraw* to Create Multiview Drawings, 1276
Using *Surfaces* with Solids, 1160
Using *the Move and Rotate Grip Tools*, 1128
Using *the Normal Five Grip Options*, 1127
Using *Thickness* and *Elevation*, 1198

V

Vector Files, 601
Vector Files, Exporting, 867
Vector Files, Importing, 867
Vector Files, Importing and Exporting, 877
Vector Files versus Raster Files, 866
Verb/Noun Syntax, 73
Vertical Lines, Drawing, 51
View, 208, 978, 987
View, Back, 986
View, Bottom, 986
View, Delete, 210
View, Edit Boundaries, 210
View, General, 210
View, Left, 986
View, New, 209, 988
View, NE Isometric, 987
View, NW Isometric, 987
View, Preset Views, 988
View, Presets, 978
View, Right, 986
View, Set Current, 209
View, SE Isometric, 986
View, SW Isometric, 986
View, Top, 986
View, Update Layers, 209
View, Window, 209
View Commands and 3D Navigation, 978
View Direction, 995
View Label Blocks, 945
View Manager, 208, 987, 988
View Options dialog box, Tool Palettes, 528
View Presets, 985
View Tune Log, 1012

VIEWING COMMANDS, 199
Viewplotdetails, 318
Viewpoint, 994
Viewpoint Presets, 996
Viewpoint Presets dialog box, 996
Viewport, 106
Viewport, Create, 266
Viewport Geometry, Locking, 297
Viewport Geometry, Scaling, 295
Viewport Scale, 281
Viewport Scale Control Drop-down list, 295
Viewports, 214, 277, 291, 998
Viewports, Considerations for Multiple, 907
Viewports, Creating Automatic, 109
Viewports, Floating, 278
Viewports, Multiple, 901
Viewports, Tiled, 278
Viewports, Using Two, 904
Viewports and Layouts, Guidelines for Using, 282
Viewports dialog box, 214, 291, 998
Viewports for *Xrefs*, 905
Viewports in Paper Space, Using, 291
Viewports Layer Visibility, 889
Viewports Linetype Scale, 895
Viewports toolbar, 267
Viewres, 211, 1253, 1315
Visibility Parameter, 537
Visibility States, 544
Visretain, 817, 1317
Visual Style control panel, 964, 999
Visual Style Properties, 1002
Visual Styles, 999
Visual Styles Manager, 1008
Visualstyles, Create New Visual Style, 1009
Visualstyles, Apply Selected Visual Style to Current Viewport, 1009
Visualstyles, Apply to All Viewports, 1009
Visualstyles, Copy, 1009
Visualstyles, Create the New Visual Style, 1009
Visualstyles, Delete the Selected Visual Style, 1009
Visualstyles, Edit Name and Description, 1009
Visualstyles, Export the Selected Visual Style to the Tool Palette, 1009
Visualstyles, Paste, 1009
Visualstyles, Reset to Default, 1009
Visualstyles, Size, 1009
Visualstyles, 1008
Vplayer, 892
Vplayer, ?, 894
Vplayer, Freeze, 894
Vplayer, Newfrz, 894
Vplayer, Reset, 894
Vplayer, Thaw, 894
Vplayer, Vpvisdflt, 894
Vpmax, 298

Vpmin, 299
Vpoint, 978, 994
Vports, 214, 216, 291, 979
Vports, ?, 216, 998
Vports, Delete, 216
Vports, Fit, 293
Vports, Join, 216
Vports, Lock, 293
Vports, Object, 293
Vports, On/Off, 292
Vports, Polygonal, 292
Vports, Restore, 216, 294
Vports, Save, 216
Vports, Shadeplot, 293
Vports, Single, 216
Vscurrent, 999
Vsedgecolor, 1006
Vsedgejitter, 1006
Vsedgeoverhang, 1006
Vsedges, 1005
Vsedgesmooth, 1006
Vsfacecolormode, 1003
Vsfacehighlight, 1010
Vsfaceopacity, 1010
Vslide, 1304
Vslightingquality, 1004
Vsfacestyle, 1004
Vshalogap, 1010
Vsintersectioncolor, 1008
Vsintersectionedges, 1008
Vsintersectionltype, 1008
Vsmaterialmode, 1010
Vsobscuredcolor, 1007
Vsobscuredltype, 1007
Vsshadows, 1005
Vssilhwidth, 1007

W

Walk and Fly Navigation Mappings, 1258
Walk and Fly Navigation Mappings dialog box, 1257
Walk and Fly Settings, 1260
Walk and Fly Settings, Display instruction window, 1260
Walk and Fly Settings, Display Position Locator window, 1260
Walk and Fly Settings, Steps per second, 1260
Walk and Fly Settings, Walk/fly step size, 1260
Walk and Fly Settings dialog box, 1260
Walkflysettings, 1260
Wblock, 503, 509
WCS, 973, 1024
Wedge, 1064, 1072
Wedge, Center, 1073
Wedge, Cube, 1073
Wedge, Length, 1072
Wheel Mouse, *Zoom* and *Pan*, 201
Why Set Up Layouts Before Drawing?, 107

Wildcards, 1294
Wildcards, Commands that Accept, 1295
Wildcards, Valid, 1295
Wildcard Examples, 1296
Window, 70
Window Polygon, 71
Windows Clipboard, 847
Windows Elements, 1315
Windows Right-Click Shortcut Menus, 28
Wipeout, 364
WIREFRAME MODELING, 1041
Wireframe Modeling Tutorial, 1042
Wireframe Models, 964
Wizard, 95
Wizard, Use a, 30
Wood dialog box, 1235
Working Support Files Search Path, 1313
WORKING WITH FILES, 25
Workspaces toolbar, 963
World Coordinate System (WCS), 974, 1024

X

X,Y,Z Coordinates, 2
X,Y,Z Directions, 1090
Xattach, 814
Xbind, 815
Xbind, Block, 815
Xbind, Dimstyle, 815
Xbind, Layer, 815
Xbind, Linetype, 815
Xbind, Textstyle, 815
XBind dialog box, 815
Xclip, 818
Xclip, Clipdepth, 819
Xclip, Delete, 819
Xclip, Generate Polyline, 819
Xclip, New Boundary, 819
Xclip, On/Off, 819
Xclipframe, 820
Xedges, 1059
Xedit, 830, 1317
Xfadectl, 831, 1315
Xline, 342
Xline, Ang, 343
Xline, Bisect, 344
Xline, Hor, 343
Xline, Offset, 344
Xline, Specify a point, 343
Xline, Ver, 343
Xline and *Ray*, Using for Construction Lines, 626
Xline and *Ray* Commands, Auxiliary Views, 689
Xloadctl, 822, 1317
Xloadpath, 823, 1313, 1314
Xopen, 825
Xplode, 508
Xplode, All, 509

Xplode, Color, 509
Xplode, Inherit from Parent Block, 509
Xplode, Layer, 509
Xplode, Linetype, 509
Xplode, Lineweight, 509
X-Ray Mode, 1003
Xref, 806
Xref, Attach, 807
Xref, Bind, 810
Xref, Clip, 818
Xref, Detach, 809
Xref, Details/Preview, 809
Xref, Open, 810
Xref, Overlay, 808
Xref, Path, 807
Xref, Reload, 810
Xref, Unload, 810
Xref, Xref found at, 809
Xref and *Blocks*, Editing, 824
Xref Drawings, Managing, 832
Xref Example, 804
Xref, List View, 810
Xref, Tree View, 811
Xref shortcut menu, 806
Xref versus *Insert*, 800
XREFCTL, 835
XREFERENCES, 799
XREFNOTIFY, 831
Xrefs, Creating and Using, 806
Xrefs and *Blocks*, In-Place Editing, 826
XY Parameter, 536

Y

Y (Yes), 567
Yes, (Individual Editing), 567

Z

Zoom, 202, 978
Zoom, 3D, 979
Zoom, All, 204
Zoom, Center, 205
Zoom, Dynamic, 206
Zoom, Extents, 204
Zoom, In, 205
Zoom, Mouse Wheel, 201
Zoom, Out, 205
Zoom, Previous, 206
Zoom, Realtime, 203
Zoom, Scale (X/XP), 205
Zoom, To Object, 205
Zoom, Transparent, 206
Zoom, Window, 204
Zoom and *Pan*, Mouse Wheel, 201
Zoom and Pan with Undo and Redo, 208
Zoom XP Factors, 296
Zoomfactor, 202

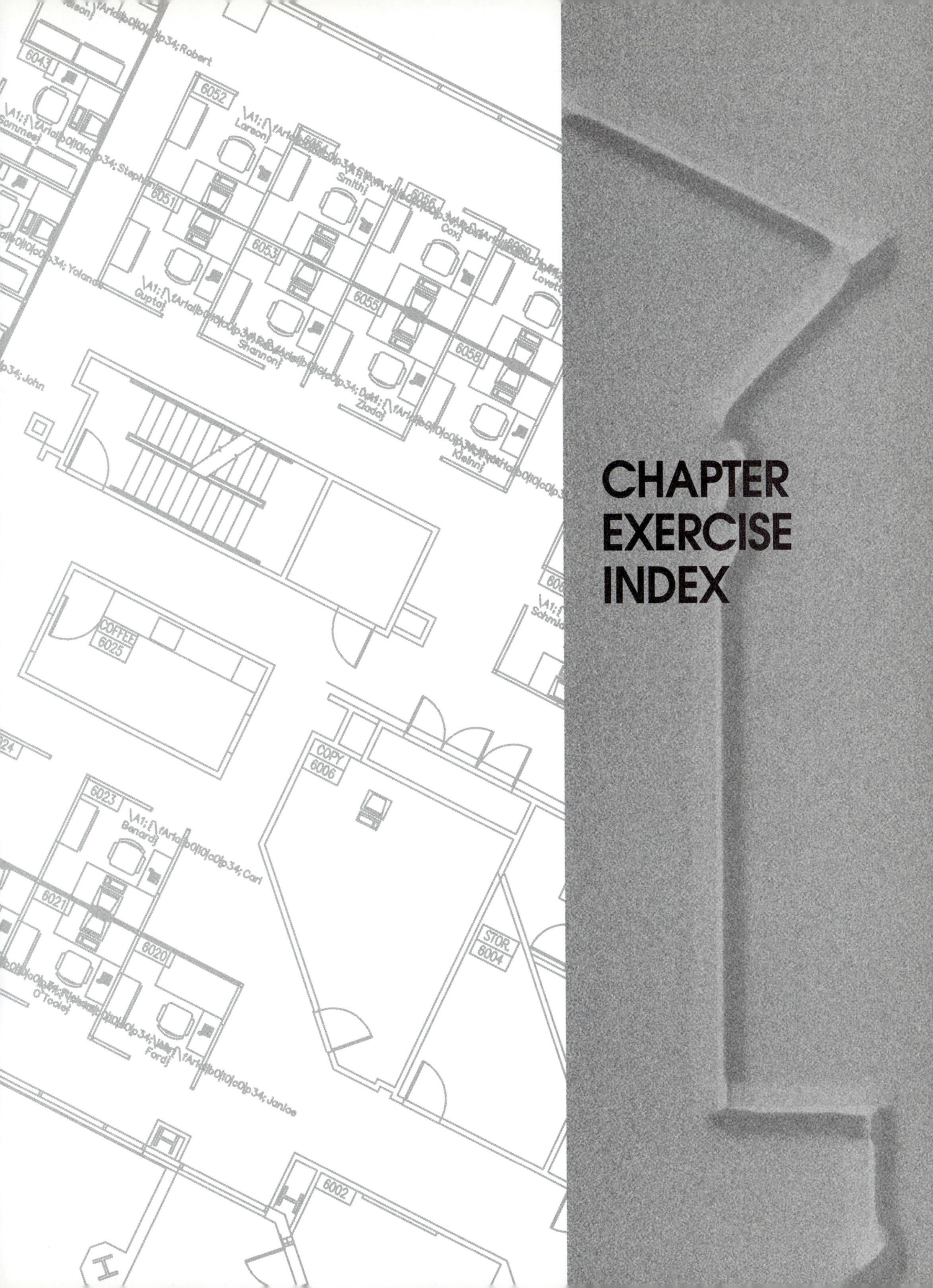

CHAPTER EXERCISE INDEX

CHAPTER EXERCISE INDEX

List of Drawing Exercise File Names (when files were created or changed)

2BR-APT.DWG, Ch. 20, 500
2BR-APT-DIM.DWG, Ch. 33, 925
ACAD3D.DWG, Ch. 39, 1172
ACWEBPUBLISH.HTM, Ch. 23, 618
ADJM-DIM.DWG, Ch. 29, 796
ADJMOUNT.DWG, Ch. 24, 640
ADJMT-SL.DWG, Ch. 38, 1115
AGLBR-SL.DWG, Ch. 38, 1113
AGLBR-SV.DWG, Ch. 42, 1290
A-METRIC.DWG, Ch. 6, 112
A-METRIC.DWT, Ch. 12, 272
A-METRIC.DWT, Ch. 13, 301
ANGLBRAC.DWG, Ch. 27, 693
ANGLB-WF.DWG, Ch. 37, 1062
ANGL-DIM.DWG, Ch. 29, 795
ANGL-TOL.DWG, Ch. 29, 796
APARTMENT.DWG, Ch. 12, 273
APT-DIM.DWG, Ch. 29, 798
APT-DIM.DWG, Ch. 34, 956
ARRAY.DWG, Ch. 9, 195
ARRAY1.DWG, Ch. 9, 195
ARRAY2.DWG, Ch. 9, 195
ASHEET.DWT, Ch. 12, 274
ASHEET.DWT, Ch. 13, 301
ASHEET.DWT, Ch. 18, 460
ASHEET.DWT, Ch. 34, 957
ASSY.DWG, Ch. 30, 839
ASSY-PS.DWG, Ch. 30, 840
BALL FIELD CH6.DWG, Ch. 6, 112
BALL FIELD CH7.DWG, Ch. 7, 139
BARGUIDE.DWG, Ch. 12, 273
BARGUIDE.DWG, Ch. 13, 301
BARGUIDE.DWG, Ch. 24, 639
BARGUIDE.DWG, Ch. 28, 739
BARGUIDE.DWG, Ch. 34, 957
BEARING.DWG, Ch. 25, 654
BEAR-SEC.DWG, Ch. 26, 677
BEAR-SEC.DWG, Ch. 34, 957
BGD-SURF.DWG, Ch. 40, 1202
BGUD-MV2.DWG, Ch. 42, 1288
BGUD-MV3.DWG, Ch. 42, 1289
BGUD-MVS.DWG, Ch. 42, 1288
BGUID-SL.DWG, Ch. 38, 1111
BGUID-SL-2.DWG, Ch. 39, 1171
BGUID-WF.DWG, Ch. 37, 1061
BILLMAT2.DWG, Ch. 18, 459
BILLMATL.DWG, Ch. 15, 368
BOLT.DWG, Ch. 9, 196
BOLT.DWG, Ch. 30, 838
BREAK1.DWG, Ch. 9, 190
BREAK2.DWG, Ch. 9, 190

BREAK3.DWG, Ch. 9, 190
BREAK-EX.DWG, Ch. 9, 190
BSHEET.DWT, Ch. 13, 302
BSHEET.DWT, Ch. 18, 460
CAL-EX.DWG, Ch. 43, 1326
CAMPUS2.DWG, Ch. 41, 1267
CAMPUS2.TIF, Ch. 41, 1267
CBRACKET.DWG, Ch. 9, 198
CBRACKET-PS.DWG, Ch. 13, 304
CBRACTXT.DWG, Ch. 18, 460
CBRAC-WF.DWG, Ch. 37, 1060
C-CIVIL-20.DWG, Ch. 14, 339
C-D-SHEET.DWT, Ch. 13, 302
C-D-SHEET.DWT, Ch. 18, 460
CH11EX2.DWG, Ch. 11, 257
CH11EX3.DWG, Ch. 11, 258
CH12EX4.DWG, Ch. 12, 274
CH14EX5.DWG, Ch. 14, 338
CH16EX2.DWG, Ch. 16, 399
CH19EX2.DWG, Ch. 19, 474
CH2 HORIZONTAL.DWG, Ch. 2, 43
CH2 INCLINED.DWG, Ch. 2, 44
CH2 VERT 2.DWG, Ch. 2, 44
CH2 VERTICAL.DWG, Ch. 2, 43
CH27EX1.DWG, Ch. 27, 691
CH27EX2.DWG, Ch. 27, 691
CH27EX3.DWG, Ch. 27, 691
CH27EX3.DWG, Ch. 34, 957
CH27EX4.DWG, Ch. 27, 692
CH33 SCM.DWG, Ch. 33, 923
CH38EX3.DWG, Ch. 38, 1106
CH38EX4.DWG, Ch. 38, 1106
CH38EX5.DWG, Ch. 38, 1107
CH38EX6.DWG, Ch. 38, 1108
CH39EX1.DWG, Ch. 39, 1168
CH39EX2.DWG, Ch. 39, 1168
CH39EX3.DWG, Ch. 39, 1169
CH39EX4.DWG, Ch. 39, 1169
CH3EX1.DWG, Ch. 3, 64
CH3EX2.DWG, Ch. 3, 65
CH6EX1A.DWG, Ch. 6, 110
CH6EX1B.DWG, Ch. 6, 111
CH6EX2.DWG, Ch. 6, 111
CH6EX3.DWG, Ch. 6, 111
CH7EX1.DWG, Ch. 7, 135
CH8EX.DWG, Ch. 8, 154
CH8EX3.DWG, Ch. 8, 155
CH8EX4.DWG, Ch. 8, 156
CH8EX5.DWG, Ch. 8, 156
CH8EX6.DWG, Ch. 8, 156
CLIP.DWG, Ch. 24, 638

CLIP.DWG, Ch. 34, 957

CLIPPING.DWG, Ch. 32, 885

CLIP-SEC.DWG, Ch. 26, 677

CONFTABL.DWG, Ch. 21, 550

CRNBRACE.DWG, Ch. 25, 653

CYLINDER.DWG, Ch. 25, 652

D-8-AR.DWG, Ch. 14, 339

DB_SAMPL2.DWG, Ch. 11, 255

DROPBAR.DWG, Ch. 38, 1114

DROPBAR-L.DWG, Ch. 39, 1170

DWGORDER.DWG, Ch. 32, 885

EFF-APT.DWG, Ch. 15, 366

EFF-APT2.DWG, Ch. 18, 459

EFF-APT2.DWG, Ch. 30, 837

EFF-APT3.DWG, Ch. 33, 927

EFF-APT-AREA.DWG, Ch. 18, 462

EFF-APT-PS.DWG, Ch. 15, 367

EFF-APT-PS.DWG, Ch. 18, 458

EFF-APT-REV.DWG, Ch. 15, 370

EXTEND1.DWG, Ch. 9, 190

EXTEND2.DWG, Ch. 9, 190

EXTEND3.DWG, Ch. 9, 190

FAUCET1.DWG, Ch. 38, 1109

FAUCET2.DWG, Ch. 38, 1109

FAUCETK.DWG, Ch. 39, 1169

FILE.DWG, Ch. 41, 1266

FILE.TIF, Ch. 41, 1267

FLANGE1.DWG, Ch. 20, 498

FLANGE2.DWG, Ch. 20, 498

FLOWDIAG.DWG, Ch. 9, 193

FRIENDS GRAPH.DWG, Ch. 9, 191

GASKETA.DWG, Ch. 9, 193

GASKETA.DWG, Ch. 16, 400

GASKETA2.DWG, Ch. 23, 619

GASKET-AREA.DWG, Ch. 15, 369

GASKETB.DWG, Ch. 9, 197

GASKETB.DWG, Ch. 34, 957

GASKETC.DWG, Ch. 16, 403

GASKETD.DWG, Ch. 28, 738

GASKETS.DWG, Ch. 33, 928

GEAR-REGION.DWG, Ch. 15, 369

GEAR-REGION 2.DWG, Ch. 16, 404

GRADBEAM.DWG, Ch. 26, 679

HAMMER.DWG, Ch. 13, 292, 306

HAMMER-DIM.DWG, Ch. 29, 794

HAMMER-DIM.DWG, Ch. 34, 956

HAMMER-PSTYLES.DWG, Ch. 33, 929

HANDLE.DWG, Ch. 15, 368

HANDLE2.DWG, Ch. 16, 403

HEADER.DWG, Ch. 7, 136

HOLDER.DWG, Ch. 27, 692

HOLDRCUP.DWG, Ch. 20, 498

IDLERDIM.DWG, Ch. 29, 797

INTERIOR.DWG, Ch. 30, 837

LIBRARY DESKS.DWG, Ch. 9, 196

LINEWEIGHTS.CTB, Ch. 33, 931

LINEWEIGHTS.STB, Ch. 33, 928

LINK.DWG, Ch. 8, 155

MANFCELL.DWG, Ch. 9, 194

MANFCELL.DWG, Ch. 14, 337

MATERIAL TABLE.DWG, Ch. 18, 462

MOVE1.DWG, Ch. 9, 188

OFF-ATT.DWG, Ch. 21, 548

OFF-ATT2.DWG, Ch. 22, 584

OFF-DIM.DWG, Ch. 29, 795

OFFICE.DWG, Ch. 21, 549

OFFICE-REN.DWG, Ch. 21, 552

OFFICE-REV.DWG, Ch. 21, 553

OFFICE-REV-2.DWG, Ch. 26, 680

PASTETXT.DWG, Ch. 31, 864

PEDIT1.DWG, Ch. 16, 401

PEDIT1-DIM.DWG, Ch. 28, 737

PERFPLAT.DWG, Ch. 9, 197

PFD.DWG, Ch. 21, 554

PIVOTARM CH7.DWG, Ch. 7, 135

PIVOTARM CH9.DWG, Ch. 9, 192

IVOTARM CH16.DWG, Ch. 16, 399

PIVOTARM CH24.DWG, Ch. 24, 638

PLANT.DWG, Ch. 21, 551

PLATE-AREA.DWG, Ch. 17, 415

PLATENEST.DWG, Ch. 16, 401

PLATES.DWG, Ch. 9, 194

PLATES2.DWG, Ch. 20, 497

PLATES-D.DWG, Ch. 28, 737

PLINE1.DWG, Ch. 8, 158

POLYGON1.DWG, Ch. 15, 365

POOL.DWG, Ch. 41, 1268

PULLEY.DWG, Ch. 25, 654

PULLY-R-BK.DWG, Ch. 41, 1265

PULLY-R-BK.TIF, Ch. 41, 1265

PULLY-REND.DWG, Ch. 41, 1264

PULLY-REND.TIF, Ch. 41, 1264

PULLY-SC.DWG, Ch. 39, 1171

PULLY-SL.DWG, Ch. 38, 1115

PUL-FLAT.DWG, Ch. 42, 1291

PUL-SEC.DWG, Ch. 26, 678

PUL-SEC.DWG, Ch. 30, 838

PUL-SOLPROF.DWG, Ch. 42, 1290

PUL-SURF.DWG, Ch. 40, 1203

RET-WALL.DWG, Ch. 8, 157

RET-WAL2.DWG, Ch. 15, 369

RET-WAL3.DWG, Ch. 16, 406

ROTATE1.DWG, Ch. 9, 189

ROTATE2.DWG, Ch. 9, 189

SADDL-DIM-PLOTSTYLES.DWG, Ch. 33, 931

SADDL-DM.DWG, Ch. 29, 794

SADDL-DM2.DWG, Ch. 32, 884

SADDLE.DWG, Ch. 24, 639

SADDLE.DWG, Ch. 30, 838

SADDLE-HALF SECTION.DWG, Ch. 31, 862

SADDLE-SD.DWG, Ch. 42, 1290

SADDLE-SV.DWG, Ch. 42, 1289

SADDL-REND.DWG, Ch. 41, 1265
SADDL-REND.TIF, Ch. 41, 1265
SADL-REND2.DWG, Ch. 41, 1266
SADL-REND2.TIF, Ch. 41, 1266
SADL-REND3.DWG, Ch. 41, 1266
SADL-REND4.DWG, Ch. 41, 1266
SADL-SEC.DWG, Ch. 26, 677
SADL-SEC.DWG, Ch. 34, 957
SADL-SL.DWG, Ch. 38, 1113
SADL-SL.MPR, Ch. 39, 1175
SAMP-CLIP.DWG, Ch. 30, 840
SAMP-OFFICE.DWG, Ch. 30, 840
SBRACKET.DWG, Ch. 25, 653
SBRCK-DM.DWG, Ch. 29, 795
SCALE1.DWG, Ch. 9, 189
SCALE2.DWG, Ch. 9, 189
SCHEM2.DWG, Ch. 22, 586
SCHEMATIC1.DWG, Ch. 9, 195
SCHEMATIC1B.DWG, Ch. 15, 367
SCREEN PERCENTS.CTB, Ch. 33, 928
SDL-SURF.DWG, Ch. 40, 1203
SHAFT.DWG, Ch. 30, 838
SLEEVE.DWG, Ch. 30, 838
SLOTNEST.DWG, Ch. 30, 836
SLOTPLATE CH8.DWG, Ch. 8, 155
SLOTPLATE2.DWG, Ch. 9, 189
SLOTPLATE2.DWG, Ch. 30, 836
SOAPDISH.DWG, Ch. 39, 1172
SOLID1.DWG, Ch. 35, 1019
SOLID1 HIDDEN.DWG, Ch. 36, 1022
SOLID1 SKETCH.DWG, Ch. 36, 1021
SOLID2.DWG, Ch. 36, 1040
SOLIDUCS.DWG, Ch. 36, 1039
SPACEPLT.DWG, Ch. 19, 463, 475
SPBK-SEC.DWG, Ch. 26, 678
STORE ROOM.DWG, Ch. 16, 404
STORE ROOM2.DWG, Ch. 18, 461
STORE ROOM3.DWG, Ch. 19, 477
SUPBK-SL.DWG, Ch. 38, 1112
SURFEX1.DWG, Ch. 40, 1201
SURFEX2.DWG, Ch. 40, 1201
SURFEX3.DWG, Ch. 40, 1201

SURFEX4.DWG, Ch. 40, 1202
SWIVEL.DWG, Ch. 38, 1290
SWIVEL2.DWG, Ch. 39, 1170
SWIVEL-SD.DWG, Ch. 42, 1206
TABLE-BASE.DWG, Ch. 7, 139
TABLE-BASE.DWG, Ch 9, 190
TABLET TEST.DWG, Ch. 10, 218
TBLOCK.DWG, Ch. 18, 460
TBLOCKAT.DWG, Ch. 22, 583
T-BLOCKAT.DWG, Ch. 34, 957
TEMP-A.DWG, Ch. 16, 401
TEMP-B.DWG, Ch. 16, 402
TEMP-C.DWG, Ch. 16, 402
TEMP-D.DWG, Ch. 16, 402
TEMP-E.DWG, Ch. 16, 402
TEMP-F.DWG, Ch. 19, 474
TEMPGRPH.DWG, Ch. 18, 459
T-PLATE.DWG, Ch. 9, 196
T-PLATE.DWG, Ch. 34, 957
T-PLATE2.DWG, Ch. 23, 619
TPLATEB.DWG, Ch. 19, 478
T-PLATE-MODEL.DWF, Ch. 23, 618
TRIM1.DWG, Ch. 9, 190
TRIM2.DWG, Ch 9, 190
TRIM3.DWG, Ch. 9, 190
TRIM-EX.DWG, Ch. 9, 190
VBLCK-WF.DWG, Ch. 37, 1061
VBLOCK.DWG, Ch. 27, 693
VBLOK-SL.DWG, Ch. 38, 1111
VP_SAMP.DWG, Ch. 13, 303
VP_SAMP-PSTYLES.DWG, Ch. 33, 930
VPUCS1.DWG, Ch. 36, 1039
VPUCS2.DWG, Ch. 36, 1040
WEDGEBK.DWG, Ch. 13, 308
WFEX1.DWG, Ch. 37, 1059
WFEX2.DWG, Ch. 37, 1059
WFEX3.DWG, Ch. 37, 1059
WFEX4.DWG, Ch. 37, 1059
WRENCH.DWG, Ch. 15, 367
WRENCH-REG.DWG, Ch. 16, 405
XXXSHOW.SCR, Ch. 43, 1326
ZOOM TEST.DWG, Ch. 10, 216